Ion Exchangers

Preface

This treatise on ion exchangers is, in a sense, a new edition of the book K. Dorfner, Ion Exchangers – Properties and Applications, which was the English translation of the third edition of K. Dorfner, Ionenaustauscher, Walter de Gruyter, Berlin, 1970. This fourth edition is being issued directly in English as there is no doubt that the international scientific and technological community can nowadays best communicate in this language. There will hardly be a country in which ion exchangers are not used and needed, depending on the extent of industrialization and scientific development, as a technical means or a scientific tool.

In order to achieve the utmost in expertise for every chapter and every detailed problem, the editor has chosen to form a team of authors. Thanks must be expressed that well-known and even great names in the field of ion exchangers and ion exchange have accepted the invitation to contribute. It was further the intention of the editor to maintain the character of a monograph to the greatest extent possible. We believe this has been achieved. The authors had full freedom in the arrangement of their chapters, with only some limitations on length. This freedom may, on the other hand, have resulted in differing opinions here and there, as well as in overlapping and repetition. The latter are considered to be of minor importance and are unavoidable in the preparation of such a volume within an acceptable time. Differing opinions, on the other hand, is more a positive aspect, making for fruitful discussion of the questions under consideration. Any pure mathematical approach to ion exchangers and ion exchange may be regarded by some as too extensive and too rigorous because the materials – whether they are natural, chemically modified, or even pure chemical products – can only be described by results from experimental work.

The relative independence of the various chapters should enable the reader to select the one or the other chapter of interest to him. The chapters on the applications of ion exchange resins, in particular, may be used in this way. It will be appreciated that for these chapters, experts were obtained who have summarized their experience of decades. Further, the necessity of an introduction into the field of ion exchangers and ion exchange has again in this expanded treatment been duly satisfied, in order to enable beginners to learn all that is necessary about this subject in a relatively short time. Further, the attempt was made to comprehensively place ion exchange as a phenomenon into the foreground, so that the results gained in one standardized product line might become more easily transferable to others where explanations and models are still needed to understand behavior and function. Synthetic ion exchange resins are defined and discussed for probably the first time with such consistency as reactive polymers, having the two main properties of storage capacity and selectivity. The result of all these endeavors is not only the revision and enlargement of the volume but a considerable improvement in its contents.

It is a long way from the first idea of revising a monograph to the point when the whole manuscript can be handed over to the publishers. It is also a matter of the resources available to enable one to finish such a work at all. If the resources are rather poor, only enthusiasm and tenacity can help. Invaluable and kind help as rendered in the beginning by Robert Auer von Brunkau of The Dow Chemical Company must therefore be gratefully acknowledged. Thanks are also due to Dr. George P. Herz, who lent much support during the preparation of the volume, especially when it was most needed. On the other hand the conviction that research begins at the desk provided continual motivation for the editing of such a treatise, despite unfortunate setbacks which were overcome by the belief that the value of ion exchangers and ion exchange goes far beyond water purification. The work in this field of basic and applied science must go on.

Mannheim, 1990 K. Dorfner

Contents in Brief

Contents

2.4 Treatment of Drinking Water with Ion Exchange Resins ... 835

Wolfgang Höll and Hans-Curt Flemming

2.5 Waste Water Treatment and Pollution Control by Ion Exchange .. 845

Friedrich Martinola

2.6 Ion Exchange Systems in Homes, Laboratories and Small Industries 859

Hans Träger

2.7 Ion Exchangers in Nuclear Technology ... 873

Günter Kühne

5 Ion Exchangers in Analytical Chemistry 1161

An Introduction to Analytical Applications of Ion Exchangers 1161

Janos Inczedy

5.1 Analytical Methods Based on Ion Exchange 1169

Günther Bonn and Ortwin Bobleter

1 Ion Exchangers

Konrad Dorfner
Mannheim, Federal Republic of Germany

The term ion exchangers is not very old. The first inorganic synthetic products suitable for the uptake of free alkalinity from solutions were called permutites, a term which was used for a fairly long time and may even now still be found employed synonymously with ion exchangers. It was the application of permutites to remove undesired bases from solutions that led to the designation base exchange, as the whole phenomenon was named when it was discovered. The first synthetic organic ion exchangers had already been developed; they were called base exchange resins, with the important difference that a distinction was made between anion and cation exchangers. With these products the term ionic exchange came into use, which was employed together with the much more common base exchange. The introduction of the term ion exchangers can most probably not be exactly dated beyond to say that, for the underlying process, the expression ionic or ion exchange had been used from early on.

Base exchange and together with it, base exchangers are expressions that continued in use for decades, regardless of the fact that the chemical products had changed dramatically. Base exchange was understood as the property of the trading of base-producing entities or cations by certain insoluble materials, either natural or synthetic, inorganic or organic. It was as well the early name for water softening, which was the first industrial ion exchange process. The term base exchange has, in the meantime, come to be more and more replaced by the chemically correct name water softening by sodium exchange and the base exchangers are called cation exchangers in the sodium form. It can only be hoped that the terms base exchange and base exchangers will disappear completely from the literature, since it is sometimes difficult enough to make clear that with respect to ion exchangers one always has to distinguish clearly between cation exchangers and anion exchangers.

Ion exchangers is the name given to insoluble electrolytes containing labile ions that easily exchange with other ions in the surrounding medium without any major physical change occuring in the electrolytes' own structure. The process taking place is usually called the ion exchange reaction and the labile ions in the ion exchangers exchange reversibly. The ion exchanging electrolyte is in general of a complex nature and is macromolecular. Since after dissociation, all electrolytes exist only as either cations or anions, the ionic sites present in a macromolecular matrix in the case of ion exchangers can also only be either cationic or anionic. As a result, the different matrices consisting of repeating units in the macromolecules and which include inorganic materials, highly complex organic materials of natural origin, and the comparatively simply-structured synthetic resins, carry an electrostatic charge in the form of a surplus charge or a fixed ion that is neutralized by the charge of the labile ion. These counterions are cations in a cation exchanger and anions in an anion exchanger. In other words – in the strictest sense with regard to the whole field – ion exchangers as such do not exist, but must first always be understood as either cation exchangers or anion exchangers. It is clear, therefore, that a cation exchanger consists of a macromolecular or polymeric anion with negative charges and labile cations and an anion exchanger of a macromolecular or polymeric cation with positive charges and labile anions. Collectively both may then be called ion exchangers.

Ion exchangers are thus insoluble polyelectrolytes having free ions that can reversibly interchange with ions in surrounding solutions. Generalized, a cation exchanger (for which the symbol] may be used) has negative anionic sites with cations A^+ electrostatically bound but free to undergo exchange with cations B^+, according to the following equation

$$]^-A^+ + B^+ \rightleftharpoons]^-B^+ + A^+;$$

correspondingly, an anion exchanger (for which the same general symbol may be used) has positive cationic sites with anions Y^- again electrostatically bound, but free to undergo exchange with anions Z^-, according to the following equation

$$]^+Y^- + Z^- \rightleftharpoons]^+Z^- + Y^-.$$

The double arrows indicate that an ion exchange process, if left to itself, reaches a state of equilibrium.

Ion exchangers are the tools for the established procedures and technologies of ion exchange. There are quite a number of different types of ion exchangers, but those based on solid organic polymers have become the most important. A binding nomenclature has, therefore, been found necessary, at least for the many analytical chemistry applications, and was issued by the International Union of Pure and Applied Chemistry (IUPAC), as Recommendations on Ion Exchange Nomenclature in 1971 [1]. Over 40 terms pertinent to ion exchangers and ion exchange were defined.

The solid organic polymeric ion exchangers usually called ion exchange resins predominate by far in technical applications. Their development and wide use in modern technology have made it necessary to elaborate at least a standardization of the test methods in order to have analytical methods to guarantee a qualitatively uniform functioning. Such test methods for ion exchangers are indispensable in research work, for the development of new or improved products, as well as in production to obtain reproducible product properties. The planning engineer cannot dimension an ion exchange unit without knowing the data of performance, and quality data are indispensable for operating the unit both in respect to examining the resins supplied and in following up on any drop in their efficiency, when they have been put into operation. Among others, the American Society for Testing and

Materials [2] and the West German Technical Standards Committee on Materials Testing (FNM) in DIN [3] have both elaborated standardized test methods as a quality control for ion exchange resins. It may be that this standardization of ion exchange testing methods may one day be extended to the production of standardized products which will be used by researchers everywhere instead of, as today, everyone's employing the resins from the next convenient source of supply.

The term ion exchange has often been closely related to or even used synonymously with adsorption and absorption. Exchange-adsorption was used instead of ion exchange during the transition in understanding from base exchange to modern cation exchange or anion exchange [4]. It is important first to distinguish between absorption and adsorption before one can identify ion exchange (Kunin).

Absorption refers to a uniform penetration of the component of a system into the, e. g., solid absorbing material following the old chemical rule of similia similibus solvuntur (like likes like), and with no change in concentration. Absorption is thus a process in which the solid engulfs another substance by distributing it very evenly throughout its own entire structure. The term adsorption on the other hand refers, e. g., to the interaction of a solute with a solid adsorbent through physical forces associated with the solid, resulting in a change in concentration of, e. g., a particular dissolved component of a system; it occurs mainly at the surface of the solid. Adsorption is thus characterized by an increase in concentration, which takes place at the interface of the liquid carrying the solute and the solid adsorbent.

It could be claimed, and it would make sense to do so, that ion exchange occuring at the surface only of a non-porous ion exchanger can safely be classified as adsorption. On the other hand, the exchange of a cation after it has wandered into the network of a porous cation exchange resin, could be regarded as absorption even though the mechanisms and energy factors for both cases are the same. But this would only be a phenomenological treatment: it disregards the fact that absorption, adsorption and ion exchange differ substantially, at least as far as synthetic polymers are concerned, in the mechanisms by which they retain solutes (Figure 1.1). All three are sorption processes. In absorption polar substances wander more easily into polar, e. g., plastic materials, just as non-polar substances wander more easily into non-polar plastics. In adsorption a solute is usually retained by the solid through physical (van der Waals) forces associated with the internal surface of the solid. In ion exchange a solute is retained by association or chemical reaction with ionic groups in the solid.

Still, so-called adsorbent resins have been described – and can be manufactured usually by condensation polymerisation – which are able to exchange ions only to a very slight extent, but can bind by adsorption organic coloring substances from solutions on their large surfaces. Most of these are amphoteric in character, which means that they contain both phenol and amine groups. Their main application has been similar to that of active carbon. Here, although the term adsorption for an ion exchange process may appear strange, phenomena associated with ion exchange have involved mechanisms other than the ionic exchange of ions. In the

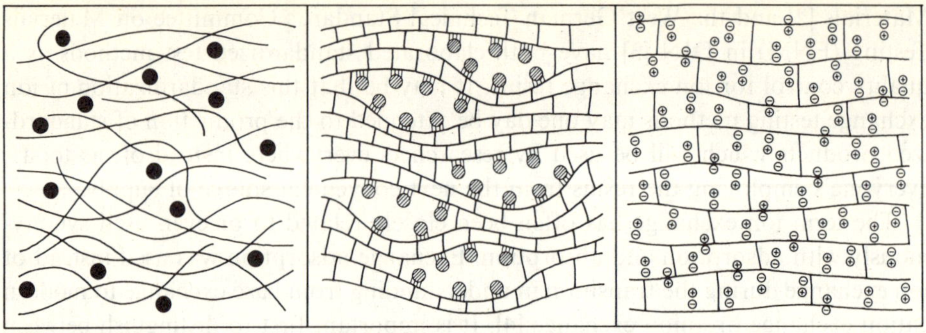

Figure 1.1 Sorption processes: examples of absorption, adsorption and ion exchange and their common way of schematic representation.

course of the increasing importance of addition polymerization materials, as opposed to condensation polymerisation products, a new class of porous polymeric sorbents has been developed, which are manufactured by suspension polymerization leading to polymer beads and which can be effectively used because of their porous structure as adsorbing media. These materials are truly non-ionic and are adsorbents because of their surface characteristics and van der Waals attractive forces. But over the years this group of polymeric adsorbents has been further developed, with polar groups – for instance hydroxyl, amine or halogens – being incorporated in a second reaction step into them. As a result, based on addition polymerization, products with a true porous structure are available.

The term ion exchangers denotes today insoluble polymeric or macromolecular substances with fixed ions. The reactive groups are dissociable and either present by nature in a naturally occuring exchange material or have later been incorporated during the manufacture of a synthetic ion exchanger. Ion exchangers are therefore reactive polymers or reactive macromolecules, but this term does not include macroions. In the case of ion exchangers as reactive polymers, the reactive groups are not functional, neither in the sense of the definition of functional groups in organic chemistry nor as defined in polymer chemistry. It is by reactions of polymers that ion exchangers can usually easily be obtained, and these are then called reactive polymers [5]. The fixed ions or reactive ionic groups of ion exchangers have the ability of undergoing ionic exchange reactions, but they are also able to store ionic species as counterions. As with all reactive polymers, the reactive ionic groups of ion exchangers are significant both because they are used to fulfil some sort of goal and are part of very important phenomena in nature.

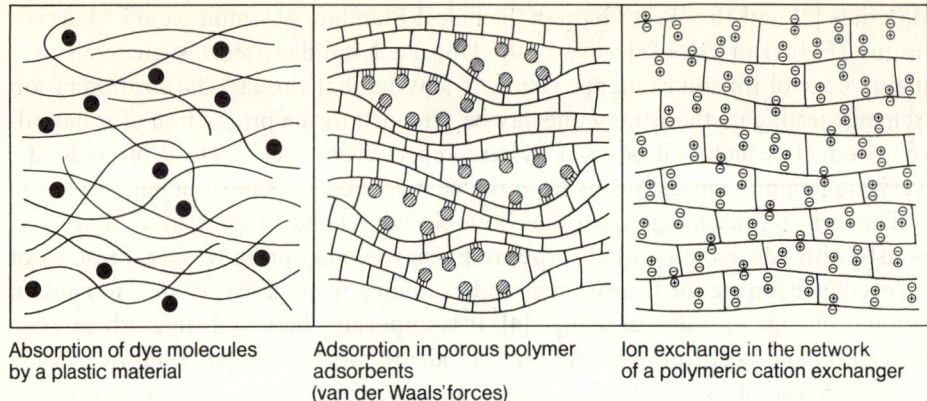

References

[1] Recommendations on ion exchange nomenclature. International Union of Pure and Applied Chemistry. Division of Analytical Chemistry. Commission on Analytical Nomenclature. Pure Appl. Chem. (1972), *29*, 619.

[2] ASTM Committee D-19 on Water, Pt. 31.

[3] Standardization of ion exchangers by DIN. (1984). Reactive polymers *3*, 80.

[4] R. Griessbach (1957). Austauschadsorption in Theorie und Praxis. Akademie-Verlag, Berlin (East).

[5] K. Dorfner (1986). Reactions of polymers, polymers with functional groups and reactive polymers. Synoptical review. Chemiker-Zeitung *110*, 109.

[6] K. H. Lieser (1975). Sorption mechanisms. NATO Adv. Study Inst. Ser., Ser. E, Vol. 13, 91 – 145.

1.1 Introduction to Ion Exchange and Ion Exchangers

Konrad Dorfner
Mannheim, Federal Republic of Germany

Introduction

Ion exchange is a phenomenon or process involving the reversible exchange of ions in solution with ions held by a solid ion-exchanging material, in which there is no directly perceptible permanent change in the structure of the solid. Best known are the ion exchange phenomena in soils and the ion exchange processes used for the treatment of water. Ion exchange is also used in many other chemical processes, including simple or chromatographic separations and catalysis, and is well-established in analytical chemistry. Ion exchange processes are employed as well in hydrometallurgy and metal recoveries, in the manufacture of sugar, in pharmacy and medicine as well as in biochemistry and biotechnology, and in agriculture and food processing. The ion exchange process usually involves either inorganic or organic ions in predominantly aqueous solutions and proceeds, as it can be assumed, in two steps, i. e., an in-change of the solute ions into the ion exchanger and then the ex-change of the previously present ions out of the ion exchanger. The dynamic state of equilibrium resulting from these two steps is an interchange of ions [1].

Due to the development of special synthetic ion exchange polymers, processes have been developed which, even if somewhat reluctantly, are also called ion exchange processes, by which separations primarily can be achieved. Among these are to be found the ion exclusion method, the ion retardation method, the exchange of non-electrolytes via a process known as ligand exchange, specific and chelate exchangers for chromatographic purposes, as well as for water and waste water treatment, as well as processes based on crown ether polymers. With regard to the so-called redox exchangers the position has been taken in this monograph that their analogy with ion exchangers is only formal, since ion exchange is, in principle, not a redox reaction. We shall, therefore, dispense with a description of it [2].

What can be achieved by an ion exchange process depends largely on the ion exchanger used. Not all ion exchanging materials, with respect to, e. g., temperature and stability, can answer the requirements of a given ion exchange process.

An ion exchanger can be any material that has the ability to uptake ionic species from a solution with the simultaneous release of a chemically equivalent number of ions into this solution. These materials include cellulose, dextran and agarose derivatives, clays and clay minerals, zeolites and other non-siliceous inorganic ion exchangers, materials based on coal and other organic natural products, synthetic liquid ion exchangers and, last but not least, ion exchange membranes. While all the types just mentioned are of greater or lesser utility, in special applications, in this text by “ion exchanger” is meant synthetic ion exchange resins such as are in daily use in the laboratory and in chemical technology, if not otherwise stated. These ion exchangers can be made by oneself, but they are also available as commercial ion exchangers in a wide range of products. Because of their importance, they will almost exclusively be discussed in this introduction. They provide the highest exchange capacity and are of greatest interest in the laboratory as well as in industrial applications.

1.1.1 History

The history of ion exchange and ion exchangers can be interpreted from different angles. One can, if one wants, trace the history of ion exchange in nature back to approximately 3 000 years B. C.; or one can begin with the curiosity-driven search for the explanation of phenomena observed in experimenting with soils; or one can begin with the preparation of ion exchangers by man and their use for advantageous and necessary ion exchange processes.

Ion exchange is widespread in nature and might even be seen as the basic process in nature. This is due to the fact that ionic compounds are by far more frequent in nature than neutral molecules. It could be said that for countless millions of years ion exchangers have been active in the solid lithosphere, in soils and in organisms. But the first historical evidence for the use of ion exchange is given in the interpretation of the miracle performed by Moses: “And he cried upon the

Lord; and the Lord shewed him a tree, which when he had cast into the waters, the waters were made sweet: there he made for them a statute and an ordinance, and there he proved them" (Moses, Exodus 15: 22–25). The bitter water at Marah was thus made potable, and it is now known that rotten cellulose is a good exchanger for magnesium ions. It was also already known to Aristotle, 384–322 B. C., that sea and impure drinking water can be purified by percolating them through layers of sand or rocks because the water loses part of its original salt content during this filtration. Later Sir Francis Bacon, in *Sylva Sylvarum*, described experiments in passing salt water through earth, saying that "it hath become fresh when drayened through twenty vessels". In the early nineteenth century Sir Humphry Davy, Lambuschini, and Huxtable studied extensively the ability of clays and soils to absorb components of manure liquor, as was similarily done by Gazzari in 1819, and Liebig found that clays had the ability to adsorb ammonia. Fuchs in 1833 reported that certain clays released potassium and sodium when treated with lime.

But credit for the true observation of the phenomenon of ion exchange is generally ascribed to Thompson and Way. In 1850, Thompson [3] and Way [4] reported on the observation that cultivated soil can exchange various bases, such as ammonium, calcium or magnesium. In 1845, Spence, a pharmacist from York working under contract to Thompson, had prepared a bed of sandy clay treated with ammonium sulfate in a glass column and allowed water to flow through it. Instead of ammonium sulfate, he obtained gypsum in the eluate. Thompson commented on this: "This was a complete surprise". But although Thompson published these results earlier than Way, he has been characterized as follows: "He made a dive into the ocean of science, and brought up one scientific pearl of great price, ... but he was never a man of science". From the description of the personality of, as well as the work done by, Spence it follows that he must be considered as the discoverer of ion exchange. Way, on the other hand, was a professor of chemistry and after the results of Thompson and Spence became known to him in 1848 he studied the base exchange properties of soils very intensively. He writes: "... the acid comes away in the liquid not in the free state, but combined with lime, which it has taken in exchange for its previous saturating base". The term base exchange was born, and the process was so thoroughly investigated by Way in the years 1850 to 1854 that for decades thereafter it was not followed by any other comparable work.

Around 1950 the discovery of ion exchange was celebrated in the chemical literature as being 100 years old. But these 100 years must be subdivided into important periods, in which not only the phenomenon as such was elucidated and appropriately designated as ionic or ion exchange but also ion exchange materials were synthesized for employment in specific applications.

Curiosity as to the nature of the reaction of manure with soils led to the discovery of base exchange in 1850, but application of the process could not be properly utilized until the concepts of reversibility and stoichiometry and of the role played by ions were clearly formulated or applied in the following decades.

The phenomenon of base exchange was interpreted as a chemical process by Henneberg and Stohmann [5], and by Eichhorn [6] in 1858, when they succeeded in confirming the reversibility of such exchange processes. Several other names could also be cited, but it was only in 1870 that Lemberg [7], in his classical work, widened the knowledge of the chemical character of base exchange by furnishing proof of the stoichiometry of base exchange processes. Based on experiments, he demonstrated that a number of natural minerals, particularly the zeolites, are capable of exchanging bases in equivalent quantities. In his most striking experiments he suceeded in converting leucite, $K[AlSi_2O_6]$, into analcime, $Na[AlSi_2O_6]$ using an NaCl-solution, and then analcime, using a KCl-solution, back into leucite. These experiments were the more striking because leucite contains little water while analcime contains approximately nine percent, and the experiments with these minerals showed, at all temperatures, an increase in water content when potassium was replaced by sodium. It should be noted that it was the opinion of the time that a reaction is chemical when its equivalency has been demonstrated. Why relatively little progress was made during the later second half of the 19th century is due to the fact that an authority like Liebig questioned Way's experimental results and considered base exchange a physical, not a chemical, process. The early physical school opposed the chemical interpretation of base exchange, did not agree with the stoichiometric views of the time, and even two conclusions arrived at by Way were found to be erroneous. In 1877 van Bemmelen published the first of a series of papers on absorption of salts by soils, further developing the point of view of Way and Lemberg [8]. The outstanding facts of van Bemmelen's are that in no case is there absorption of a neutral salt as a whole, but always an interchange of bases with the bases already present, and that the location of the exchange is in the silicates. It seems that mineralogists or, in modern terms, inorganic chemists were much earlier closer to an explanation of base exchange than were the agricultural chemists, but at the time the latter dominated the field.

After the work of Eichhorn and Lemberg base exchange had to be considered as reversible and stoichiometric, but the quantitative approach to relationships of the distribution of bases between the base-exchanging material and concentration in the solution went historically through an extensive evolutionary process during which already known or newly discovered physical-chemical phenomena or laws were applied. It is to be noted that in the book by Mulder [9] Berthelot's theorem of the state of equilibrium was applied for the first time to this field of investigation with a clarity which could not be much improved upon in modern times. This discovery was a guideline for several contemporary researchers for the understanding of the processes occurring in soils. But as the law of mass action was only discovered by Guldberg and Waage in 1867, the first attempts to solve mathematically the problem of the dependence of the quantity of the bases taken up in relation to the concentration were based on the assumption of a simple adsorption process. In both cases – the assumption of a simple adsorption process and the application of the law of mass action – contributions were made by various scientists over the following decades. Around 1859 both Boedeker and Wolff suggested empirical equations, with the difference being that in the former case the initial concentration and in the latter the equilibrium concentration was included as being proportional to the exchange. Kroeker suggested another empirical equation in 1892 [10]. Interest focussed more and more on the question of the effect of concentration on the final equilibrium. Influenced by investigations of Ostwald on the solubility of alkaline earth sulphates in mineral acids showing it to be an exponential function of the concentration of the acid, investigators expected the silicate reactions to be similar, but the concentration curves both with the pure silicates and with soils were not as a rule strictly exponential. To explain this, Hall and Ginningham [11] in 1907 then put forward a formula for ammonium salts, but this also held only for a comparatively restricted range of concentration. As the Freundlich adsorption isotherm was developed in 1905 and the Langmuir isotherm in 1916, it is understandable that Wiegner in 1912 [12] set up an equation closely related to the Freundlich type, and Vageler and Woltersdorf in 1930 [13] set up one closer to the Langmuir isotherm. A formal analogue with an adsorption mechanism based on the Langmuir equation was more or less finally set up by Boyd, Schubert and Adamson in 1947 [14].

It would seem to be worthwhile to try to find out how and when base exchange became ionic and then ion exchange. The very general comment that this was not the case until the theory of electrolytic dissociation was developed by Arrhenius in 1884 is certainly true. Up to this time and even thereafter the scientific disputes on base exchange were mainly concerned with the phenomenon as it occurred in soils and the various early attempts using synthetic materials to describe it chemically or physically and quantitatively. Very slowly a change from base exchange to cation exchange took place, first being clearly denoted so by Wiegner in 1912 [12]. But in these studies of base exchange it was noticed and emphasized that the so-called acid radicals were not adsorbed in contrast to what had been reported in particular for phosphoric acid by Rostworoski and Wiegner [15]. The retention of, for instance, phosphate in soils was explained quite speculatively and, together with investigations on the absorption of citrate, oxalate, carbonates, borates and nitrate, it was concluded that the absorption of acid radicals somehow falls into line with the absorption of bases. Being thoroughly convinced that a rational method for the analysis of soils could only be expected when the principles of the reactions in soils were clarified, further investigations on base exchange were taken up and it was proved by Ramann and Spengel that base exchange is an ionic reaction [16]. Base exchange had thus, through cation exchange, arrived at ionic or ion exchange [17] but the term ion exchange was identical with the current cation exchange. True inorganic anion exchangers in the early days of ion exchange were never made, but it took the ingenious spirit of Jenny to show, over investigations of permutites, that a corresponding anion exchange must as well be taken into consideration [18].

The law of mass action was somewhat reluctantly applied during the first 100 years of base exchange. It was Gans who first employed it, in 1908, but with little success, because of unsatisfactory conformity between experimental results and theory. Very detailed investigations on the base exchange equilibria by Rothmund and Kornfeld followed in 1918 [19] and Kerr was more successful in 1928 [20] with a true analysis of the applicability of the law of mass action. According to him the ionic exchange equilibrium could not be considered to be analogous to the simple equation of a double reaction with an insoluble precipitate, but he assumed that in base exchange the solid phases behaved as if they were dissolved in the solution, assuming further similar activity coefficients for the soluble species and insoluble species. Møller in 1937 still used the same approach, but in 1932 Vanselow had already presented a method based on the assumption that the two solids formed a series of continuous solid solutions and that the activity of each component was not unity, as Kerr had assumed, but is equal to its mole fraction in the solid [20]. Then Gapon in 1933 assumed for his formula that both the di- and monovalent ions behaved similarly on an equivalent basis on the absorbing surface and Kjelland in 1935 calculated from the equilibrium constants the free energies. Following investigations showed that the law of mass action is sometimes approximately obeyed within small concentration ranges. Thus Samuelson [21], Boyd, Schubert and Adamson [14], Marinsky [22], Kressman and Kitchener [23], and Duncan and Lister [24] finally succeeded, using a more rigorous application of the law of mass action to the phenomenon of ion exchange with more accurate data on "organic zeolites" and other newly available ion exchange resins.

All the initial fundamental studies of the mechanism, equilibria and rates of exchange reactions were difficult because of the chemical heterogeneity of the naturally occurring or poorly synthesized exchange materials available at the time; it was not known that for such physical-chemical investigations well-defined, uniform materials should be used. This is the reason why, on one hand, of the numerous experiments dealing with ion distribution between exchanger and solution only a small number is still of interest and, on the other, why Jenny for instance, at the beginning of the twentieth century emphatically welcomed such better defined synthetic products as the permutites.

It was Gans in 1905 who first synthesized suitable inorganic ion exchangers and called them permutites [25]. The fused sodium aluminosilicates, $Na_2O \cdot Al_2O_3 \cdot xSiO_2$, were granular cation exchangers stable in neutral media. Because of their solid glassy nature they had poor kinetics and were, after about ten years, discarded for modified natural mineral products, such as stabilized glauconites, bentonites and the synthetic amorphous sodium aluminosilicates [26]. The modified natural products had low operating capacities but the rate of regeneration was faster. On the other hand, the precipitated gels had a higher capacity to uptake ions but were very sensitive to low and high pH. The inorganic ion-exchanging materials were used for about forty years, then seemed to be completely abandoned because of limitations in their application; however, since 1950, many new developments have taken place in the field of inorganic ion exchangers.

An intermediate stage in the development of ion exchangers is represented by the period when ion exchange properties were recognized in various organic materials – for example, the sulfonated coals as cation exchangers stable in the hydrogen form were introduced in 1934 by Liebknecht [27]. In all cases these products contained $-SO_3^-$ groups but had carboxylic acid groups as well, resulting from oxidation of the coal structure, in the case when coal was sulfonated with sulfur trioxide instead of with concentrated fuming sulfuric acid. In the hydrogen form these exchangers were insoluble polymeric acids. They could be used both in the sodium and in the hydrogen cycle. The first anion exchanger developed and used at that time was a precipitated black oxidation product of aniline called emeraldine. But this product was not used for very long because it hydrolyzed to produce chlorine leakage.

The next milestone in ion exchange, after its discovery by Spence, Thompson and Way, was the discovery of Adams and Holmes [28] in 1935 that it is possible to make truly synthetic organic ion exchange resins. Their original materials were the formaldehyde condensation products of polyhydric phenols or aromatic diamines, which yielded the required insoluble cation or anion exchange resins. In combination these resins were, in principle, able to exchange hydrogen for other cations and hydroxyl for other anions in aqueous solutions, leaving only water. It was not long before Holmes produced a more strongly acidic cation exchanger, analogous to the sulfonated coals, from phenolsulfonic acid. Somewhat later, the anion exchangers were also improved, by using the more strongly basic guanidine. Recognizing the importance of the discoveries of Adams and Holmes, IG-Farbenindustrie in Germany purchased the patents on the subject and, since 1936, has continued developmental work in this field with the aim of systematic production of ion exchange resins to serve special purposes [29]. At the beginning of World War II the patents of Adams and Holmes were also given to the United States of America. Most of the cation exchangers developed there were based on phenol and formaldehyde as the resin matrix and $-SO_3H$ as the exchanging group. The weak base amines varied in composition and were made from aromatic amines and formaldehyde, phenol, formaldehyde and polyamines or acetone, formaldehyde and polyamines. The first resin to contain highly basic groups which removed silicic acid was a condensation product of epichlorohydrin and amines, developed in 1947. These resins were referred to as epoxy or medium basic resins.

The ion exchangers first obtained by polycondensation came increasingly to be replaced by polymerization products after 1945, when D'Alelio succeeded in incorporating sulfonic acid groups into a crosslinked polystyrene resin [30]. The use of these copolymers of styrene and divinylbenzene as matrices for attaching the exchanging groups brought a tremendous expansion of ion exchange processes worldwide, as a result of intense research on the part of the manufacturers of ion exchangers. In 1947 the invention of McBurney [31], based again on styrene divinylbenzene co-polymerization, led to the modern range of anion exchange resins. These unifunctional synthetic polymeric ion exchangers rapidly became the standard cation and anion exchangers. They were made in spherical bead form by the process of suspension polymerization, and were essentially monofunctional, in contrast to the granular polyfunctional resins of the Bakelite era. The new resins were also very stable over the total pH range and at higher temperatures.

Further developments led to improvements of ion exchange resins and the manufacture of resins with specific ion exchange properties. In 1948 Skogseid [32] developed by nitration, reduction and diazotization the first potassium specific exchanger; further, as new anion exchangers, pyridinium resins to be used in uranium recovery were synthesized. At this time investigators also began to attach a variety of exchanging groups to the resin matrix. The ease with which this could be done led researchers to investigate all conceivable types of properties, which they achieved simply by varying the reactive group on the matrix.

There is also the history of the applications of ion exchange and ion exchangers. The first suggestion to use ion exchange with a clay mineral was made by Harms in 1896; the aim was to exchange the calcium in the clay for the sodium and potassium in sugar solutions to reduce melassigenic formation, thus increasing the sugar yield. Gans then, in 1906, applied the ion exchange process successfully for water softening and described as further applications, in 1909, gold recovery from seawater, salt conversion, water purification, i. e. removal of manganese and iron, and manufacture of sugar [33]. Metzger in 1912 and Duggan in 1913 also attempted the recovery of gold and in 1915 a patent was taken out for the recovery of silver. In 1917 Folin and Bell employed for the first time ion exchange in analytical chemistry for the separation of ammonia from urine [34]. After the amorphous sodium aluminosilicates and greensands had become available – between 1914 and 1920 – water softening by ion exchange entered a new period of application lasting 30 years. Whitehorn showed first in 1923 that not only ammonia, but also other amines, including the amino acids histidine and lysine, were readily adsorbed by a synthetic aluminosilicate, either from aqueous solution or from solution in organic solvents [35]. In 1934 Austerweil and Jeanpost adsorbed copper from a solution with a greensand exchanger and then recovered the copper by elution with a salt, and Syrkin and Krynkina were issued a patent for the recovery of copper from a cuprammonium based liquor in 1937. The world's first commercial demineralising plant was set up in England in 1937 after – based on the work of Adams and Holmes – polycondensation cation as well as anion exchangers had become available. The recovery of vitamin B_1 from urine in an analytical procedure was first reported by Hennessy and Cerecedo in 1939, using a sodium aluminosilicate. The first pilot plant for the recovery of copper from cuprammonium waste liquors was built in Germany in 1939; the full-scale plant was built in 1941. By the use of improved synthetic resins both copper and water were recovered for reuse – the copper recovery was over 90% – and ammonia was also recovered for recycling. Griessbach and co-workers introduced in 1942 organic synthetic ion exchange resins in the hydrogen form into catalysis. The application of ion exchangers in chromatography was also suggested by Griessbach in 1939. Samuelson's systematic investigations from 1939 on showed the importance of ion exchangers in general analytical chemistry and in ion exchange chromatography. Condensation products were still used in 1943 to 1945 for the desalination of water for trucks in desert areas, desalting kits for fliers, and, as described by Applezweig, for the recovery of quinine from cinchona barks. Anion exchangers for concentrating metal ions were used in Germany prior to 1945, e. g. for the recovery of silver from photographic film rinses. When, in 1947, the work done at the Manhattan Project was published by the US Atomic Energy Commission it showed tremendous achievements using ion exchangers for the separation of the rare earths, the discovery and identification of element 61 and many radionuclide separations. The biochemical separation of nucleic acids in 1950 and of carbohydrates in 1952 further showed the great potential of ion exchangers. At the same time ion exchangers with styrene-divinylbenzene matrices in bead form with excellent physical and chemical properties were commercially developed, and innumerable applications became feasible. The main application of ion exchangers, water purification, became easier and widespread. D'Ans, Klement and Hein used ion exchange resins very successfully in the early fifties in preparative chemistry.

The conversion of calcium nitrate to sodium nitrate was carried out on an industrial scale in Norway in 1955. The first 100 years of ion exchange had opened the door wide to the countless applications now known and practised.

In historical reviews of ion exchange usually little attention is paid to technical developments beyond the first industrial applications of ion exchangers. Further, unfortunately, one can find in these reviews nothing about suggestions prior to that made by Adams and Holmes that with the new products deionized water could be obtained by passing even seawater through a combination of phenolic with basic condensation products, thus rendering it potable. Only some time later, when the anion exchangers had also been improved, did it become possible for the first time in practice to remove essentially all the ionic impurities from water, to obtain the equivalent of distilled water by simple percolation techniques. By this time, E. Leighton Holmes had become chief chemist of The Permutit Co. Ltd., and the world's first commercial demineralizing plant was set up by Friary, Holroyd and Healy Ltd., at Guilford in England. The fixed-bed ion exchange equipment consisted of two columnar tanks, its design and operation obviously having evolved from experience with sand filtration columns. "Ten years and one World War later, unifunctional addition polymer resins based on polystyrene had become the standard cation and anion exchangers" (manufactured under licence granted by General Electric to Dow Chemical Company, under the name Dowex 50, and to Rohm and Haas Company, under the name Amberlite IR-120; then the first commercial strongly basic anion exchange resin containing quaternary ammonium groups, capable of removing weakly ionized anions such as carbonic acid, silicic acid and organic acids, was manufactured by Rohm and Haas Company) [36]. Thus, for the first time, the removal from water of all inorganic constituents together with trace amounts of natural organic acids had been achieved. But very little was disclosed of the technically necessary modification of the equipment. For fixed-bed columns a basic design had evolved which has been in use since then. In 1949 the Rohm and Haas Company developed mixed bed deionizers, which they called Monobed, in which cation exchange resins and anion exchange resins were mixed in the same column permitting the production of high purity deionized water. This technology ushered in new age for the use of ultrapure water in various industries.Replacing the evaporator, the first large two-bed deionizer with a capacity of 190 000 m^3 per year seems to have been installed at a power plant in the USA in 1950; in 1951 in Japan the world's largest mixed bed deionizer was built, with a capacity of 600 m^3 per day. After that, many similar deionizers were installed throughout the world. The year 1957 may be considered as the beginning of ion exchange technology in nuclear technology. After several types of nuclear grade ion exchange resins had been made available by the Rohm and Haas Company, the world's first commercial nuclear power plant at Shipping Port Nuclear Power Station went into full operation, using mixed bed deionizers using Amberlite resins. From 1958 to 1961 the availability of macroporous cation and anion exchangers opened new areas in deionization systems, such as high flow rate deionization, external regeneration of condensate polishers, moving bed deionization and ultrapure water production. As early as 1962, in the Powdex Process, fine particle-size strongly acidic and strongly basic ion exchange resins were used as a new deionization system with, increasing the reaction velocity by a hundred-fold, much higher utilization than in ordinary processes, as well as removing suspended and colloidal matter. The following year, 1963, the Japan Organo Company and the Asahi Chemical Industry Company developed a new moving bed continuous ion exchange deionizer of, at first, a two-bed type, installing a commercial plant of 2 200 m^3 per day capacity, reducing the amount of ion exchange resins used, the regenerants consumed, and the land space required to half. After many improvements had been made, including a change from two-bed to a mixed bed, the technology was finally established in 1966 with the Continuous Mixed Bed Deionizer, which had a capacity of 2 400 m^3 per day. Another new technology for condensate water treatment – the mixed bed Ammonex Process, of the Cochrane Division of the Crane Company, USA – was developed in 1967, using the ammonium form of the cation exchange resin and the hydroxide form of the anion exchange resin. This process was first used at a power station; as a result, the removal of ammonium in the condensate was avoided, resulting in deionized water with less than 1 ppb of sodium, less than 5 ppb of silica, and with 90% of the iron sulfate removed. The operating time until sodium breakthrough was prolonged by a factor of about 10, and the operation cost was reduced by two-thirds. This resulted in a series of quick replacements of hydrogen/hydroxide form resins by ammonium/hydroxide resins in the mixed bed deionizers of existing plants. Reverting to developments in fixed bed technologies two techniques from 1968 should be mentioned: the stratified anion exchange system and the countercurrent regeneration (CCR) method. Kunin had pointed out that one of the main advantages of continuous ion exchange (CI) is the fact that it generally employs countercurrent regeneration, and to make a proper comparison of fixed bed and moving bed (CI) processes one has to take fixed bed systems using countercurrent regeneration. It was primarily this comparison which has resulted, in Europe, in the increased use of countercurrent fixed bed systems and a slight decline in the use of continuous ion exchange. In 1969 the first systematic investigation in successful practice of pyrogen removal from water by the combined use of reverse osmosis, ultrafiltration and ion exchange resins was made by Shimizu, Okkido et al. in Japan, opening a new way for the production of pyrogen-free water. Since then work on ion exchange technology had been mainly concerned with the development and improvement of existing processes and techniques rather than with the development of radical new ones. Much attention has been paid to one of the most interesting new processes – the Sirotherm Process – in which specially developed resins are regenerated thermally. This has proven particularly suitable for the partial desalination of brackish waters.

The theory and application of ion exchange is over 100 years old. There have been times of basic research and milestone inventions alternating with periods of quiet development. Table 1.1 gives a chronological survey of developments, including some but by no means all the names of those who have contributed to them.

Table 1.1 Chronological order of developments of ion exchange and ion exchangers

Year	Development	Names
1845–52	Discovery and description of base exchange in soils	Spence, Thompson; Way
1858	Base exchange explained as a reversible chemical process	Eichhorn
1876	Zeolites recognized as carriers of base exchange in soils; equivalence of exchange of bases proved	Lemberg
1901	Artificial zeolites used for removal of potassium from sugar juices	Harms, Rümpler
1901–02	Manufacture of sulphonated coals and suggestion for removal of potassium from sugar juices	Majert, Halse
1906–10	Industrial manufacture of sodium Permutite	Gans
1912–16	Siliceous gel permutites	de Brüm, Rüdorff, Engel
1914–1920	Greensands of New Jersey stabilized to Neo Permutite	Borrowman
1912–18	Base exchange explained as cation and then ionic exchange	Wiegner; Ramann, Spengel
1925	Colloidal systems as ion exchangers; ion swarm theory	Wiegner, Jenny
1934–39	Industrial manufacture of sulphonated coals for water softening filters	Liebknecht; Smit
1935	Invention of sulphonated and aminated condensation polymers	Adams, Holmes

Year	Development	Names
1936–45	Development and introduction of commercial condensation ion exchangers	Holmes; Griessbach
1942	Invention of sulphonated polystyrene polymerization cation exchangers	D'Alelio
1947	Invention of aminated polystyrene polymerization anion exchangers	McBurney
1948	Preparation of a sodium specific polystyrene cation exchanger chelating resin	Skogseid
1949–56	Development of carboxylic addition polymers as weak acid cation exchangers	Hale, Reichenberg; Topp, Thomas
1946–65	Development and investigations of ion exchange polymers	Bonhoeffer, Schloegl
1951–56	New zeolites as molecular sieves with ion exchange properties	Barrer; Breck et al.
1952–71	Invention and development of chelating polymers	Gregor; Pepper; Morris
1956	Development of cellulose ion exchangers	Peterson, Sober
1956–58	Preparation and studies of non-siliceous inorganic ion exchangers: insoluble salts, heteropolyacids	Buchwald, Thistlewhaite
1957–59	Development of dextran ion exchangers	Porath, Flodin
1957	Invention of macroporous ion exchange resins	Corte, Meyer; Kunin, Metzner, Bostnick
1957	Snake cage resins and ion retardation process	Hatch, Dillon, Smith
1958	Development of liquid ion exchangers for analytical and technical applications	Schindewolf; Brown et al.; Coleman; Moore; Allen
1959	Foundations laid for the new theoretical treatment of ion exchange	Helfferich
1960	Invention and development of isoporous ion exchange resins	Kressman, Millar
1961	Development of continuous ion exchange processes	Higgins; Asahi Chem. Ind.
1964	Thermally regenerable ion exchange resins and water desalination based on them	Weiss et al.
1967–70	Pellicular ion exchangers for chromatographic use	Horvath et al.; Kirkland
1975	Development of exchangers with crown ethers as reactive groups	Blasius et al.
1984	Industrial chromatographic resin systems Dowex^{R} Monosphere	Dow Chemical Company

This table is to be considered more as guideline than a complete representation of the history of ion exchange and ion exchangers. The complete history has been presented in a volume of its own [37]. The stormy development of ion exchange research after 1945 is evident from the graph of Figure 1.2, in which the number of scientific papers is plotted as a function of time (in years) (Kunin). The method

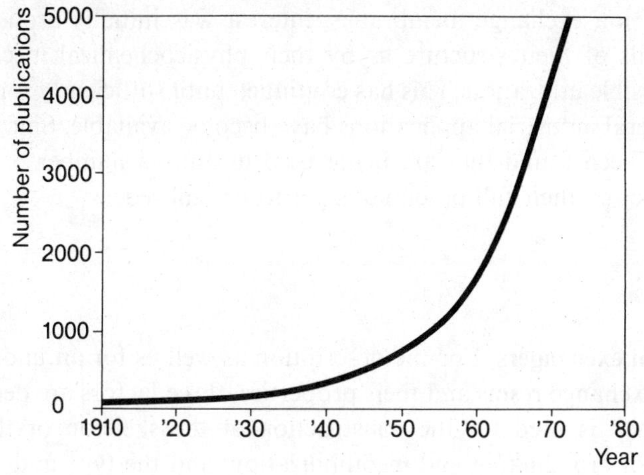

Figure 1.2 Number of publications on ion exchange as a function of time (Kunin).

of ion exchange has become an important modern tool widely accepted in the laboratory and in industry.

1.1.2 Types of ion exchangers

Ion exchange has become a general operating technique today. As indicated by the brief historical review at the beginning of this chapter, ion exchange was investigated first with inorganic materials; these found some industrial use, but were almost completely displaced by synthetic ion exchange resins. The demand for ion exchange materials that would satisfy the most diverse requirements led to the development of new products, which must be considered as separate ion exchanger types according to their special properties.

Among the available types, synthetic ion exchange resins have the greatest importance. But inorganic ion exchangers should not be neglected, since these have once again aroused interest because of their mechanical, thermal and chemical properties and because, in light of their mineralogical parallels, they offer information about numerous processes in soil science. The cellulose ion exchangers represent a separate type with their own characteristics related most closely to another group, i. e., the dextran ion exchangers. Both types have been rapidly accepted in ion exchange chromatography where they have become indispensable for analytical as well as preparative work. In addition, ion exchangers based on carbon have been produced and, together with a number of other materials with ion exchanging properties, have been investigated for their applicability. Liquid ion exchangers, as another type, probably are important primarily because of the technological possibilities they offer.

In the case of ion exchange membranes, interest was initially aroused not so much by questions of their structure as by their physicochemical mechanism of action and its possible utilization. This has continued until sufficiently sophisticated materials for general industrial applications have become available. Since then new membranes have been found that are being used in quite a number of industrial applications, although their full potential is yet to be realized.

1.1.2.1 Materials

Synthetic resin ion exchangers. For the description as well as for an understanding of synthetic ion exchange resins and their properties, three factors are decisive: The raw material which is used for the construction of the skeleton or the matrix, bridging agents for crosslinking and insolubilization, and the type and number of the ionogenic groups. Synthetic ion exchange resins chemically are insoluble polyelectrolytes with a high but not unlimited moisture regain capacity. With regard to their function one speaks of cation exchangers, anion exchangers, and amphoteric ion exchanger types.

Polymerization and polycondensation can in principle be used as the synthesis routes to form the matrix. At the present time, polymerization resins have become more important than polycondensation resins, as they have a higher chemical and thermal stability.

The most important starting material, which is used nearly exclusively in exchangers of the strongly acidic and strongly basic types produced on a large scale, is styrene. The styrene is polymerized with itself and with divinylbenzene into a polymeric molecule:

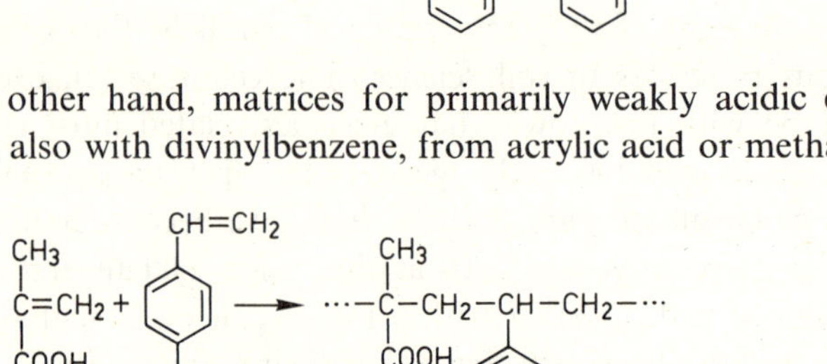

On the other hand, matrices for primarily weakly acidic cation exchangers are formed, also with divinylbenzene, from acrylic acid or methacrylic acid:

In polycondensation resins, the matrix is usually constructed of phenol and formaldehyde:

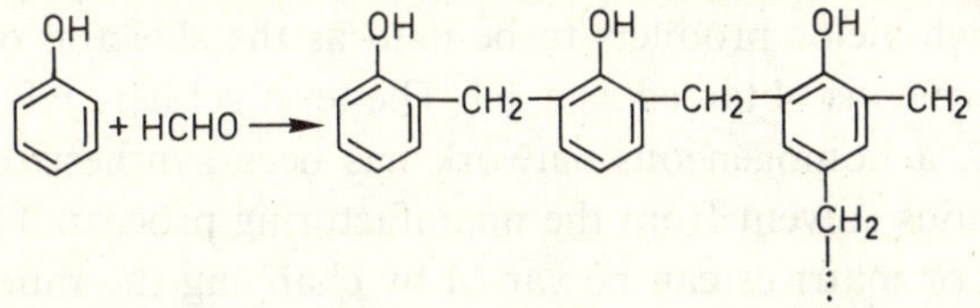

where the properties of the exchanger depend on the nature of the phenols, the quantity of starting materials used, and their side groups [38].

As far as the network of styrene-divinylbenzene copolymers is concerned, the reaction leading to their formation produces matrices which are not uniform in degree of crosslinking. Such ion exchange particles must accordingly be considered inhomogeneous substances in which relatively densly crosslinked regions, which are formed at the beginning of the polymerization process, are found connected with longer, more sparsely, crosslinked domains [39, 40]. From the standpoint of high polymer synthesis, it is probably impossible practically to obtain a uniform bridge distribution throughout the structure of a crosslinked copolymer. Naturally, the determination of significant thermodynamic values becomes difficult as a result and this must not be forgotten in all theoretical considerations. The degree of homogeneity of ion exchange structures depends on the purity, nature and properties of the starting materials used for their production, as well as on optimum conditions of polymer synthesis. In spite of the irregular structure of the matrix, however, it is possible to produce ion exchangers with a uniform distribution of the ionic groups [41]. Investigations of the ionogenic group distribution in common ion exchange resins established the heterogeneity of the gel type ion exchange resins contradicting the frequent assumption of a homogeneous gel that had been made in the study of the physicochemical properties of synthetic ion exchange resins. The ever-increasing experimental data of recent years suggests, however, that the heterogeneity of ion exchangers is of still a different nature, and at the molecular level of polymer chemistry it is basically associated with the structure of the initial copolymer [42].

Crosslinking naturally depends first of all on the quantity of divinylbenzene used as the crosslinking agent in production. Commercial ion exchangers of the gel type contain nominally between 2 and 12% divinylbenzene. The nature and degree of crosslinking have a decisive influence on the properties of ion exchange particles; additional details on this subject are described in the section on the properties of ion exchange resins.

With respect to the structure of the network of synthetic resin ion exchangers different types are now available with designations such as gel type, macroporous or macroreticular, and isoporous ion exchange resins. These terms refer to the polymeric organic part of the exchanger resin and will first be considered with regard to their basic properties as well as in comparison with each other.

Gel type ion exchange resins. During polymerization of styrene and divinylbenzene (or similar reactants), which yields products to be used as the skeleton of an ion exchanger, the network formed is obtained as a gel. The term gel here refers to the fact that, macroscopically, a homogeneous network has been synthesized that is elastic in nature and contains solvent from the manufacturing process. The properties of such copolymers or matrices can be varied by changing the ratios of the amounts of the individual monomers used during the synthesis. It is already obvious from this brief description that the term "gel type resins" refers to polymer matrices that do not contain pores. As a result it is important to pay close attention to the influence of the amount of crosslinking agent used during polymerization, since this factor influences markedly the very important property of swelling. A gel ion exchanger whose matrix contains a low proportion of divinylbenzene swells strongly in aqueous solution; this opens its structure widely, permitting large ions to diffuse easily into the exchanger, and at a rapid rate. Ion exchangers with matrices containing a higher proportion of crosslinking agent ($>10\%$ DVB) swell to a substantially lower degree in aqueous solutions. Porosity, which, strictly speaking, does not exist, may nonetheless be understood with respect to gel type resins as referring to the channels resulting from swelling that determine the size of the species, ion or molecule that may enter the structure, and the rate of diffusion and exchange. It is to be emphasized because of its importance, that gel type polymer structures have no appreciable porosity until they are swollen in a suitable medium; but such crosslinked polymers, as well as the ion exchangers derived from them, swell to a well-defined and reproducible degree in an appropriate solvent system, such as toluene for the basic copolymer, or water for the ion exchanger. This latent porosity may well be measured then in terms of the volume increase on exposure to the solvent. Other basic properties of gel type ion exchangers also depend on the amount of crosslinking. For example, the mechanical strength decreases with decreasing proportion of divinylbenzene, which means that the greater the crosslinking is the greater the mechanical strength is.

Macroporous (macroreticular) ion exchange resins. Macroporous ion exchangers are types in which a solvent is used during production from the monomers, so that a porous matrix structure is formed in the course of polymerization [43]. These macroporous structures, which have large internal surfaces, can be sulfonated very easily and completely and are much more resistant than gel type resins to osmotic shock. Furthermore, they are extremely uniform in external shape and, in contrast to the gel types, opaque. Pore sizes of several hundred nanometers in diameter and surface areas of up to 100 m^2/g and more can be obtained [44]. To prevent collapse of the structure, a larger portion of crosslinking agent needs to be used. However, in connection with the large internal porosity, this leads to a number of advantages such as a smaller swelling difference in polar and nonpolar solvents, a smaller loss of volume during drying, and a higher oxidation resistance. Because of their higher porosity, larger molecules can also penetrate the interior. The economy of macroporous ion exchangers is somewhat limited by their lower capacities and higher regeneration costs, but, their suitability for catalytic purposes is unique. Macroporous ion exchangers are fully developed products, which are in frequent use because of their advantages [45].

In Figure 1.3 the clear gel type standard ion exchange resin is shown for comparisons together with the opaque macroporous ion exchange resin. Figure 1.4 then shows electron micrographs of fragments of typical gel and macroporous resin beads and in Figure 1.5 a two-dimensional diagram is given showing the mesh size

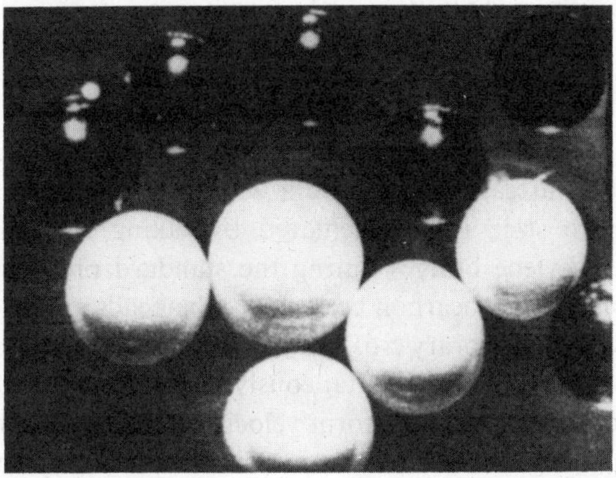

Figure 1.3 Types of ion exchangers. Gel type (above) and macroporous (below) ion exchange resin beads based on polystyrene-divinylbenzene.

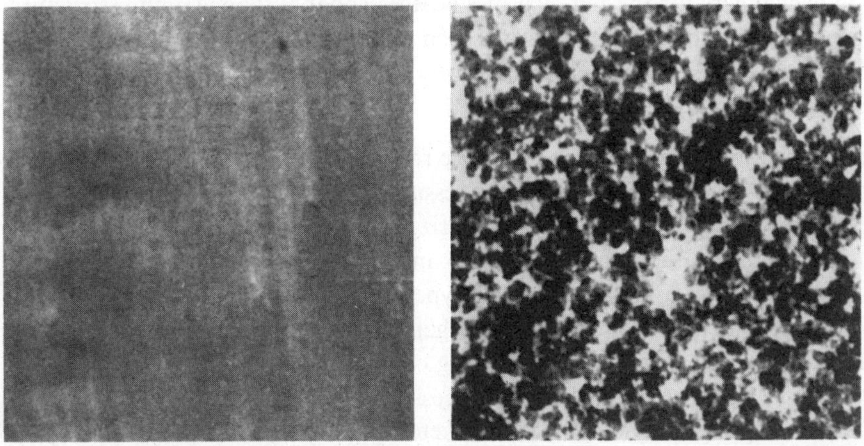

Figure 1.4 Electron micrographs of fragments of gel type (left) and macroreticular (right) ion exchange resins.

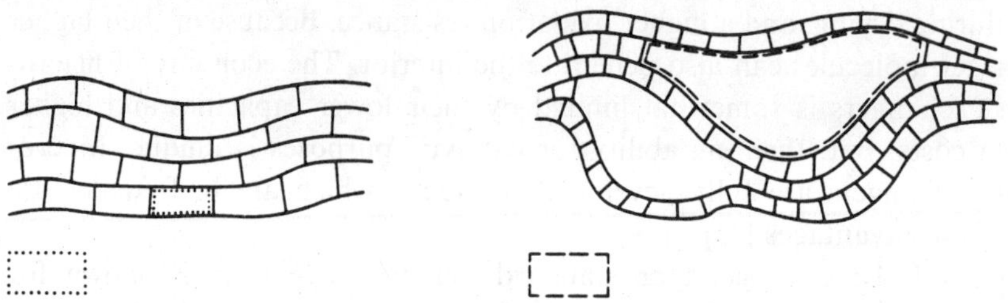

Figure 1.5 Two-dimensional diagram of ion exchange matrices. [dotted box] mesh size of gel type, [dashed box] pore size of macroporous exchange resins.

of a gel type matrix as compared with the pores of a macroporous resin. The latter should not lead to misconceptions about the three-dimensional structure of ion exchange matrices in general.

Isoporous ion exchange resins. The isoporous ion exchangers are a group in which crosslinking and pore structure are modified in a way to obtain polymers with a substantially uniform pore size. This is achieved by taking advantage of the possibility of forming methylene bridges during the standard process of chloromethylation of the preformed hydrocarbon beads [46]. The beads are manufactured using styrene, together with a temporary crosslinking agent which is unstable during chloromethylation, and is destroyed. Simultaneously, new crosslinks are formed, by a process which proceeds at a slow uniform velocity, and can be influenced to give an extremely even spacing of crosslinks. In practical application, the isoporous anion exchange resins show little sensitivity to organic fouling, have a higher capacity and regeneration efficiency, and a lower cost and price performance than the macroporous ion exchange resins. In some cases, they are particularly suited as anion exchangers for the removal of silicate. Probably the maximum structural homogeneity attainable up until now has been achieved in isoporous ion exchangers, since highly crosslinked regions are relatively rare and the crosslinking is more or less homogeneous [47].

The three structural models of ion exchange resins just described can be graphically depicted as shown in Figure 1.6, after Kressman. Following this representation a gel type ion exchange resin can be characterized by an alternation of expanded and contracted regions in the polymer network, in case the resin contains a solvent. In this swollen state ion exchangers of this type are characterized by a considerable specific permeability. Almost all their ionogenic groups are accessible to inorganic ions. If, however, the degree of crosslinking of such resins increases, the ionogenic groups may become screened off and inaccessible. The macroporous ion exchange resins are distinguished by a considerable heterogeneity, characterized by the presence of regions with higher and lower matrix densities. The resulting channels, which have a large internal area, exist also in the unswollen state, with the result

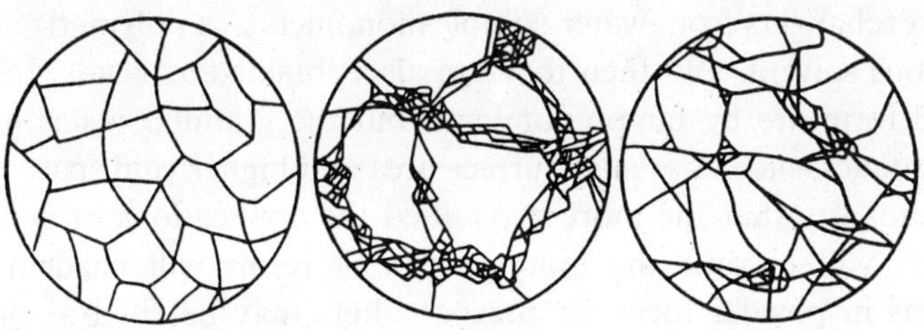

Figure 1.6 Structural models of gel type, macroporous, and isoporous (left to right) ion exchange resins, graphically depicted.

that when macroporous resins are hydrated, their degree of swelling does not change significantly. The regions with high matrix density are the reason for the considerable mechanical strength, and the porosity of macroporous ion exchange resins may be defined as an intrinsic porosity. In essence the structures of macroporous resins consist of agglomerates of quasi-spherical portions with interconnecting cavities between them. Kunin verified this by showing that the absorption of a solvent by macroporous resins takes place in two steps [48]. The isoporous ion exchange resins, then, are characterized by an enhanced permeability due to the consistent network they posess. This is certainly the case if, beyond methylene bridges, crosslinks with larger dimensions are used for their synthesis [49]. For all isoporous ion exchange resins the high permeability for organic ions is reflected in their capacity to sorb rapidly ions of higher molecular weight. But especially with respect to commercial isoporous ion exchange resins there are objections to a very strict classification of ion exchange matrices into gel type, macroreticular and isoporous, whereby a distinction of ion exchange materials into either gel type or macroporous is made [50].

For a complete understanding of practical classifications with respect to porosity it may be best to mention here that among anion exchangers a characterization is made depending upon the monomers used in building the polymer, with the porosity being expressed as the simple water retention capacity of the chloride form, and the ionogenic groups. Gel type quaternary ammonium resins are further divided into two broad classes, standard and porous. Standard types have less than fifty percent water retention in the chloride form, porous more than fifty percent. It is important to realize that here the term porous implies nothing about the nature of the polymer crosslinking mechanism (Fisher).

It must be added that there is still another way of classifying ion exchange resins, which is based upon the shape of the resin particles and their size. Polymerization ion exchangers are made by suspension polymerization in water and the resulting resins are beads. The modern range of ion exchangers consists nearly exclusively – at least in the basic products – of spherical types. The previous condensation resins were usually made in a bulk polymerization process and subsequently crushed to irregular particles of the desired size range. It is also possible to make polycondensation ion exchangers from water-soluble monomers in a polymerization process in a nonaqueous solvent, and then they are also obtained as beads. Ion exchange resins in bead form are by far predominant but the granular resins have certain inherent advantages, such as greater surface area and higher void space in columns. One can perhaps say that the more diversified the applications of synthetic resin ion exchangers will become the more classes of resins will be demanded. Ion exchange resins in powder form for precoat filters may be cited as one example here [51], and ultrafine ion exchange resins with a particle size of 0.5 – 1.5 μm are also available in the form of microspheres or agglomerates of microspheres. Because of their large surface, the exchange rate here is higher, due to the greater accessibility of ionic sites, although their application in columns is limited. Particles of this size, however, have their own field of application in the production of ion exchange papers and for incorporation into plastics, films, coatings, and fibers [52].

Pellicular ion exchangers. Synthetic ion exchange resins can in principle also be manufactured in other shapes than in beads or granules. The different shapes that can be chosen in forming the ion-exchanging materials are films, fibres, fabrics, tubes, foams or plates, etc. But in all these cases the whole material consists of the same polymer or copolymer, which has been converted into an ion exchanger. In addition to these, pellicular ion exchangers have been developed for special chromatographic purposes, with much higher chromatographic efficiency than conventional resins. Pellicular ion exchangers are composed of thin layers of polymeric exchange material bonded to glass beads. The latter enables the material to withstand the high pressure required and the thin layer of ion exchanging material yields fast kinetics and sharp separations at high speeds. The preparation of both anion and cation pellicular exchangers is quite easy. Glass beads of approximately 50 μm diameter are coated with a solution of styrene, divinylbenzene, and benzoylperoxide in ether and the solvent is evaporated. Polymerization and crosslinking is carried out at 90 °C in an aqueous suspension of the coated beads. The product is washed with water, acetone, benzene, and methanol, and then converted into the ion exchangers. The anion exchanger is prepared by chloromethylating the copolymer layer and reacting the product with dimethylbenzylamine. The pellicular ion exchanger is then sieved and washed with 2 mol/l NaOH and HCl solutions and dried. A pellicular cation exchanger is similarly prepared by sulfonating the coating of the glass beads with 98% sulfuric acid at 90 °C for 30 minutes. For making crosslinked poly(ethylene imine) pellicular exchangers, the water from commercial poly(ethylene imine) (Dow Chemical Co.) is removed by distillation with benzene. Glass beads are then coated with a solution of the dry poly(ethylene imine) in methylene chloride and the solvent is evaporated. The coated beads are exposed to methylene bromide vapor at 130 °C for 10 minutes in order to crosslink the polymer. The product is washed with acetone, water, diluted hydrochloric acid, then with water and dried (Horvath et al., [53]). With emphasis on controlled surface porosity in commercial pellicular ion exchangers (Zipax®; Trademark of DuPont de Nemours

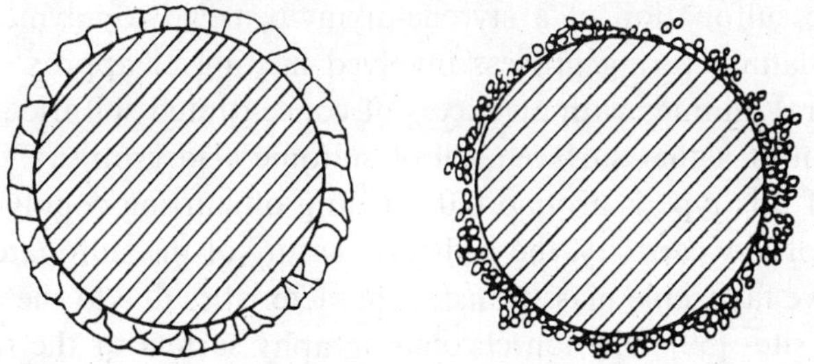

Figure 1.7 Pellicular ion exchangers with film-like (left) and porous (right) surface on solid inert core.

& Co., Wilmington, Del., USA) it was first shown that rapid separations of nucleotides and nucleic acid bases can be carried out due to the greatly improved mass transport effects [54]. The strong anion exchanger and the strong cation exchanger are still available for high pressure liquid chromatography since the hard, spherical nature of these exchangers permit their operation at high column input pressures. Figure 1.7 shows by way of comparison the structure of an ordinary pellicular ion exchanger and a superficially porous ion exchanger both built on a solid inert core. Further applications may be limited due to exchange capacities of only 0.162 mequ/ml as reported for similar ion exchangers [55].

Partial ionogenic ion exchange resins. It was as early as in 1952 that D. K. Hale stated during a Manchester ion exchange symposium that, when encountering difficulties in performing chromatographic separations with resins of low crosslinking, surface sulfonated polystyrene resins might be of help in making difficult separations [56]. A few researchers then investigated partially sulfonated ion exchange resin beads, but the results indicated that unless the beads had very low crosslinking and much care was taken, sulfonation occured to a different extent within single beads in the same reaction mixture. No further attention was paid to this knowledge during the development of ion exchange chromatography until 1969, when Skafi and Lieser prepared superficially sulfonated exchange resins that were highly crosslinked styrene-divinylbenzene copolymers. Because particle diffusion was thus cut down the exchange equilibrium was attained rapidly with these resins and aside from the fact that the capacity was several orders of magnitude smaller than that of common exchange resins some excellent practical separations were obtained [57]. Further investigation then described the preparation of low capacity, partially sulfonated macroporous cation exchange resins and the measurement of distribution data and selectivity coefficients of multivalent metal ions on these resins. Macroporous resins are chemically and mechanically stable: they swell and shrink very little, and the liquid content is essentially the same no matter what the liquid is. These properties made such resins ideal for modern forced-flow chromatography [58]. With the introduction of the so-called ion chromatography Small et al. described the preparation of a resin of very low cation exchange capacity by surface sulfonation of a styrene-divinylbenzene copolymer with 2% crosslinking. The manufacturing process involved heating at approx. 100 °C the copolymer for several minutes with an excess of concentrated sulfuric acid, which led to the formation of a thin surface shell of sulfonic acid groups. The capacity of a typical resin of this type is around 0.02 mequ/g of starting copolymer. Apart from being a resin of low capacity, the pellicular nature of the sulfonated material was expected to have favorable mass transfer characteristics due to the availability of all of the active sites [59]. The ion chromatography system of the same group of researchers uses a separation column containing a patented and commercially utilized anion exchange resin. For a simpler method called anion chromatography, which separates mixtures of anions without a suppressor column, an anion exchange resin with very low capacity was prepared by chloromethylation and amination with trimethylamine, starting from macroreticular highly crosslinked polystyrene beads XAD-1; this yielded a capacity of 0.007 mequ/g [60]. Systematic investigation of the preparation and ion chromatographic application of surface-aulfonated cation exchangers revealed, on the other hand, the effect of sulfonation temperature and time on the capacity of the materials from unswollen polystyrene-divinylbenzene copolymers, and it was found that ion-chromatographic separations can be advantageously influenced by optimization of the capacity and divinylbenzene content of the resins [61]. Finally it can be said that the efficiency of ion exchange chromatography is improved by decreasing the path length for the diffusion of ions to the place of sorption. This can be achieved by 1. granule size decrease, 2. using surface layer ion exchangers, and 3. using porous exchangers in which the diffusion is fast owing to the small dimensions of sorbent microgranules (which are immobilized by inert binder), and the presence of large transport pores [62].

New ion exchange chromatography gels based on vinyl polymers can fill the gap between the materials developed for use in HPLC and ion exchange chromatography with conventional ion exchangers. The matrix of these gels consists of hydrophilic vinyl polymers and, in contrast to other gels based on dextran, agarose or poly(acryl amide), contains pores between the intertwined polymer chains. This structure results in a mechanical stability hitherto unattainable with large-pore gels. Because of ether linkages and hydroxyl groups both the surface of the gel particles as the walls of the pores show strong hydrophilic properties. The basic gel is commercially available in five different pore sizes, covering in gel permeation chromatography separations in a molecular weight range from 100 to far beyond 10^7. By modification ion exchangers are obtained. The semirigid crosslinked gel has excellent permeability as a result of the highly stable gel matrix. The ion exchangers DEAE-650 and CM-650 are stable at pressures up to 7 bar. In DEAE-650 the diethylaminoethyl groups are bound to the hydroxyl groups of the matrix via ether linkages

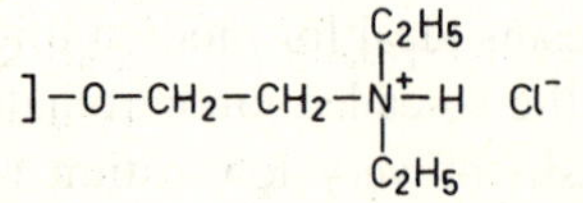

resulting in a weakly basic anion exchanger, and in CM-650 the carboxymethyl groups are also bound via ether linkages to the hydroxyl groups of the matrix

$$]-O-CH_2-COO^- \; Na^+$$

resulting in a weakly acidic cation exchanger. Both ion exchangers are available in two particle size ranges, i. e. Type S: particle size, moistened with water, 25 – 50 μm, and Type M: particle size, moistened with water, 45 – 90 μm.

The individual ionic groups of these ion exchangers have steep titration curves, as shown in Figure 1.8, and, as a result, do not have any buffer activity. The titration curves also show the ion exchange capacity and its dependence on pH.

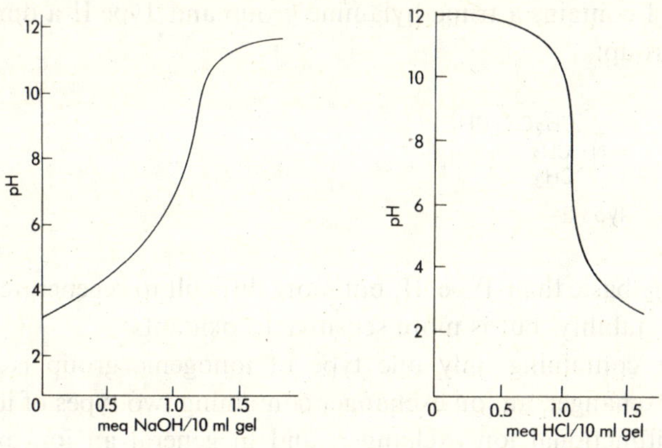

Figure 1.8 New ion exchanger gels. Titration curve of CM-650 (left) and DEAE-650 (right).

Further remarkable properties of these ion exchangers are: high pressure durability, negligible volume change with varying pH and ionic strength, high resolution at high flow rates, and the fact that the column can be packed to give uniform packing densities. The negligible change in the gel-bed volume means that the eluant can be changed or a gradient of pH or ionic strength can simply and rapidly be applied. Certain organic solvents can also be applied in the same manner. Packed in columns these ion exchangers can be regenerated in the presence of buffer solutions due to their high mechanical strength and chemical stability. Their exceptional mechanical strength permits these ion exchangers to be packed into stainless steel columns at packing pressures of 3 – 5 kg/cm², which permits the connection of an HPLC system [63].

Regardless of the role played by the matrix in determining the properties of an ion exchanger, the decisive factor is the ionogenic group. Up to now, the following groups have been incorporated into cation exchangers:

$-SO_3H$, $-COOH$, $-OH$,
$-PO_3H$, $-HPO_2H$, $-O-PO_3H_2$, $-AsO_3H_2$, etc.

and into anion exchangers:

$-NR_3OH$, $-NH(CH_3)_2OH$, $-NH_3OH$, $=NH_2OH$, $\equiv NOH$,

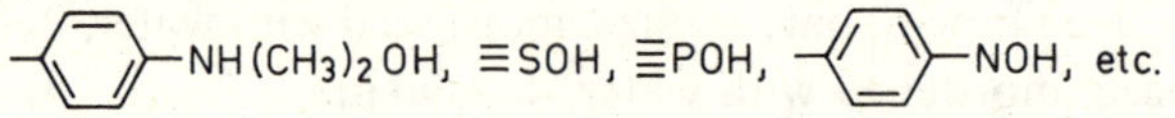

It is the ionogenic groups that confer the property of ion exchanger to the matrix. Depending on the acidity or basicity of the ionogenic group, one distinguishes between strong and weak acid and between strong and weak base ion exchangers. It has become accepted practice to distinguish between two types of strong base ion exchangers: Type I contains a trimethylamine group and Type II a dimethyl-β-hydroxyethylamine group:

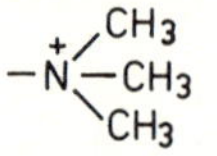

Type I

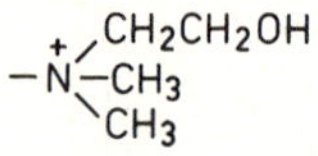

Type II

Type I is more strongly basic than Type II, but more difficult to regenerate. Type II has a higher thermal stability, but is more sensitive to oxidants.

An ion exchanger containing only one type of ionogenic group is called a monofunctional ion exchanger, an ion exchanger containing two types of ionogenic groups is called a bifunctional ion exchanger, and in general an ion exchanger containing more than one type of ionogenic group is called a polyfunctional ion exchanger (IUPAC). Cation exchangers containing two different ionogenic groups with the same charge, such as sulfonic acid and carboxylic acid groups, are in practice known as polyfunctional. For example, copolymers of acrylic acid or methacrylic acid with divinylbenzene have been sulfonated, and mixed sulfonic acid and phosphonic acid resins have also been produced. These would be called polyfunctional materials. In addition to cation and anion exchangers, amphoteric exchangers exist containing both acid and base groups. These would be called bifunctional materials. For example, condensation products of amines and phenols contain both very weakly acidic phenolic groups and basic amino groups. Bipolar exchange resins have also been prepared by the introduction of acidic and basic groups into the same matrix of a styrene-divinylbenzene copolymer.

A classification scheme based upon the type and number of ionic groups can therefore be propounded that includes – in a slight modification of an original proposal [64] – the important point that both anionic as well as cationic groups can be present as one (uni-) or as more than one (multi-) ionic site, this covering the fact that uniionic and multiionic materials of the same polarity must be distinguished:

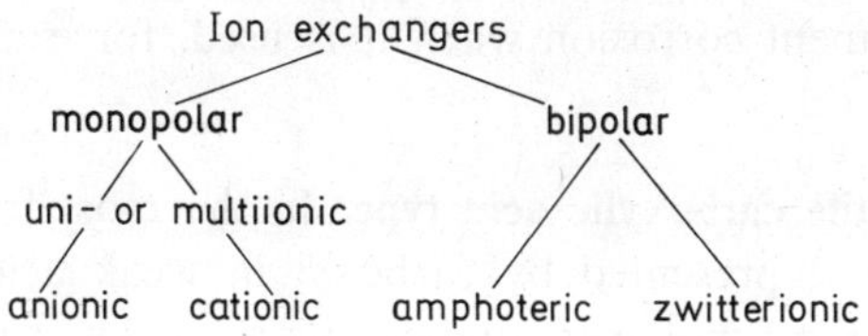

According to this classification the bipolar ion exchanging materials are easier to subdivide further into amphoteric and zwitterionic, without yet specifying whether in this group there is still more to be elucidated.

The conditions required for a matrix, especially insolubility, are satisfied by a number of polymers. Many of these polymers, however, do not contain chemical groups with an ion exchange capacity. Although such groups can be introduced into most polymers, the end products still cannot be used in ion exchange, since their water solubility increases together with the ionic and thus with the hydrophilic groups. Consequently, only the cited polymers and copolymers are suitable for the synthesis of ion exchangers. Exceptions from this general practice are to be found in work done on the synthesis of certain phosphorylated ion exchange resins based on low molecular weight polyethylene (PE-515, Dow Chemical Company), used for the adsorption of uranium. The commercial polyethylene is phosphorylated with PCl_3 and oxygen and then crosslinked with diamines and diols, yielding a product which is regular in size and insoluble in water. This resin shows a higher capacity than a commercial ion exchange resin of the styrene-divinylbenzene series [65].

Strong acid cation exchangers of the sulfonic acid type. The strong acid cation exchangers with ionogenic groups consisting of sulfonic acid have attained the greatest importance among ion exchangers produced from a matrix of styrene with divinylbenzene as the crosslinking agent, since they can be used industrially for water softening. They are produced by the sulfonation of suspension copolymer beads with sulfuric acid, sulfur trioxide, fuming sulfuric acid or chlorosulfonic acid. The SO_3-groups, which are the ionic groups yielding the cation exchange function, are probably primarily in para position and, for steric reasons, a double sulfonation is probably impossible. The ion exchange capacity of 5 mequ/g which is usually obtained confirms that only one aromatic ring on the average has been sulfonated, but experience has shown that at least one reactive site can be introduced for each aromatic ring in the copolymer, giving rise to the high exchange capacities.

The production of a strong acid cation exchanger of the sulfonic acid type involved a number of problems of detail which have only been solved with time. It is important to obtain undamaged crack-free beads. From the chemical standpoint, the sulfonation reaction leads to exchangers which contain hydrogen ions as the counter ions. By treatment with sodium hydroxide solution, the exchangers are then converted into the Na^+-form, in which form they are then used. Complete conversion into the Na-form is important because the hydrogen ions remaining in an exchanger may lead to equipment corrosion when it is used, for example, in water softening.

Weak acid cation exchangers of the carboxylic acid type. In the most frequently used forms, the ionogenic group represented by carboxyl in weak acid cation exchangers is provided by one of the copolymer components, i. e. primarily acrylic or methacrylic acid that has been crosslinked with divinylbenzene, so that another production step becomes unnecessary. With dissociation constants of between 10^{-5} and 10^{-7}, they are weak acids which can be very effectively converted from the salt into the acid form by means of strong acids. Because of their high selectivity for Ca^{2+} and Mg^{2+} ions, a regeneration with sodium chloride is not very effective and practically not feasible. The carboxyl cation exchangers are primarily suited for the removal of cations from basic solutions and for the splitting of weakly alkaline salts of polyvalent cations. However, structural modifications can also increase the acidity of carboxyl cation exchangers to such a degree that a splitting of sodium and potassium salts is possible [66]. In general, the acrylic acid resin is a slightly stronger acid than the methacrylic acid product and is therefore more useful in water treatment, especially for alkalinity reduction [67].

Phosphorus and arsenic-containing cation exchangers. Although these types of cation exchangers are not of great significance for large industrial applications, they are of interest as medium-strong acid exchangers because of their selectivity and high capacities. Once a phosphonic acid resin called Duolite ES-63 was available in limited quantities commercially. A number of routes can be used for making resins containing phosphonic, phosphonous, phosphinic, or phosphoric acid groups [68]. When phosphorous trichloride is reacted with styrene-divinylbenzene beads, the dichlorophosphine derivative is formed. This can then be converted to the monobasic phosphonous acid derivative by hydrolysis and then to the dibasic phosphonic acid derivative by oxidation. Alternatively, the dichlorophosphine can be oxidized and then hydrolyzed to give the dibasic acid [69]. The phosphonic acid resins are intermediate in acidity between the sulfonic and the carboxylic acid resins. The demand for this type of ion exchanger has been small, probably also because of its high price. Applications have been limited to special uses, for instance rare earth separations [70].

Arsenic-containing cation exchanger types have greater significance because of their selectivity, particularly with regard to their high affinity for uranium cations. On Soviet-made arsenic-containing ion exchangers the sorption of tantalum and niobium has been studied. A partition coefficient for tantalum more than 2–3 times that of the other exchangers compared [71] was obtained.

Polyfunctional cation exchangers. As mentioned before, polyfunctional or, preferably, multi-ionic cation exchangers are understood as containing two or more different ionic groups of the same charge, for instance sulfonic and carboxylic groups. The easiest approach to producing this type of bi-ionic exchanger is the sulfonation of copolymers of acrylic acid or methacrylic acid with divinylbenzene. Other multi-ionic resins contain the sulfonic and phenolic or the sulfonic and phosphonic groups. The polyfunctional cation exchangers have been evaluated in a broader sense by scientists in both academic and industrial institutions, while ion exchange materials for water treatment containing either the sulfonic and carboxylic groups for cation exchangers or various amino groups for anion exchangers dominated the interest of the manufacturers of ion exchange resins. Commercial bifunctional acid resins (and it should be noted that this term is also in use) are found among the ion exchangers produced in Russia under the names Kationit KBU-1 – a styrene-methacrylic polymerization resin with sulfonic and carboxylic ionic groups – and KU-1 – a polycondensation resin made from phenolsulfonic acid and formaldehyde.

Strong base quaternary ammonium anion exchangers. These anion exchangers are obtained by a relatively simple method from the chloromethylation products of styrene-divinylbenzene copolymers by their conversion with tertiary amines. The exchangers obtained are extremely stable and have a high exchange capacity. As strongly basic products they are also capable of exchanging silicate and carbonate. They are easily converted from the chloride form into the OH-form by treatment with NaOH. This regeneration is difficult, however, with Na_2CO_3, and almost impossible with NH_4OH. By conversion with trimethylamine one obtains the strongly basic anion exchanger known as Type I, which, in general, is the most common one. For reasons of regeneration feasibility, a Type II was finally developed with dimethylethanolamine which proved to be less stable, and did not become as important as expected, since the Type I resin is favored for the high temperature applications where chemical stability differences are most apparent [72]. Recently a new type of strongly basic anion exchanger based on an acrylamide matrix has been introduced, with test results showing an increase in the operating capacity of nearly a hundred percent as compared with conventional Type I resins [73].

Weak base anion exchangers of the amine type. This group of synthetic ion exchange resins comprises a rather complex range of products – there are more variations in composition and properties among weak base resins than in any other class of ion exchange material. The group includes materials with ionogenic groups of primary ($-NH_2$), secondary ($=NH$) and tertiary amine ($\equiv N$) functionality, individually or in mixtures. The latter would be themselves have to be classified as polyfunctional or multi-ionic weak base amine anion exchangers [74]. In building the matrix polycondensates and addition polymers serve for their production. Besides the previous aminated condensation products of phenol and formaldehyde, three principal types of weak base resins have been developed: Condensation products of epichlorohydrin with amines or ammonia, acrylic polymers, and amine derivatives of chloromethylated styrene-divinylbenzene copolymers. The ion exchangers thus obtained are suitable for sorbing strong acids with good capacity, but they are limited kinetically. The kinetics can be improved by the incorporation of about 10% of quaternary ammonium groups, thus creating another mixed-ionogenic or polyfunctional anion exchanger; it is then more a medium base resin with the good capacities of the weak base resin together with the superior kinetics required for acceptable rinse, especially in the practical application of water demineralization following the regeneration with caustic soda.

The classic weak base anion exchangers, because of their low basicity, sufficiently exchange only anions of strong acids, such as HCl or H_2SO_4; the anions of weaker acids, such as SiO_3^{2-} or HCO_3^-, are not extensively exchanged. For the same reason, however, these anion exchangers can also be converted into hydroxyl form by weak bases, such as sodium carbonate or ammonia. Their customary commercial chloride form is easily hydrolyzed. The weakly basic ionogenic groups cannot exchange neutral salt anions.

To improve the kinetic properties of weak base anion exchange resins the development has gone towards macroporous matrices with true porosity and high surface area. This structure has in commercial ranges almost displaced the gel structure used in the original styrene-divinylbenzene-based weak base resins.

Anion exchangers of the pyridine type. Gel type polymers with pyridine as the group with exchange activity are weakly basic anion exchangers. Little was published in the early years about their production and application. Materials of this type have always been expected to have good chemical, thermal, and radiation resistance as well as good kinetic characteristics [75]. It was then found that macroporous poly(vinylpyridine) anion exchangers possess a number of advantages, e. g., higher mechanical and osmotic stability, improved kinetic properties, over the above-mentioned gel type analogues [76]. Novel polyfunctional or multiionic anion exchangers on the basis of amino-substituted vinylpyridines have been described as being of interest as extractive sorbents for metals such as molybdenum, tungsten, gold, etc., with high capacity and chemical resistance, and as catalysts for certain chemical reactions [77]. Most of the research and development on anion exchangers of the pyridine type was as a result of the search for resins with specific exchange properties. This started with investigations into poly(1-hydroxy-4-vinylpyridine) ion exchangers [78] and led then mainly to the synthesis of amphoteric exchangers based on 4-vinylpyridine.

Depending on the counterion attached to the fixed group to maintain electroneutrality, the individual types can be classed under different forms. In the case of cation exchangers one speaks of an H-form or a Na-form and in anion exchangers of an OH-form or a Cl-form, etc. The exchangers can either be used in their delivered ionic form or they must be converted first into another form. The composition of the solution to be treated and the selectivity of an exchanger for certain ions generally determine the choice of an exchanger and its particular form. The choice of appropriate ion exchangers to solve a concrete chemical or other scientific problem is dictated by the application. The choice of a suitable resin for

Table 1.2 Conversion of ion exchange resins

Type of resin	Conversion	Reagent	Volume reagent per volume resin	Flow rate (cm/min)	Volume rinse water per volume resin	Method to test completeness
Strong acid Sulfonic acid cation exchanger	$Na^+ \rightarrow H^+$	2 N HCl	5	2	6	pH > 5 neutral
	$H^+ \rightarrow Na^+$	1 N NaCl,	5	2	6	Cl^- negative
	$Ca^{2+} \rightarrow Na^+$	1 N NaOH 2 N NaCl	5	2	6	Cl^- negative
Weak acid Carbocylic cation exchanger	$H^+ \rightarrow Na^+$	2 N NaOh	8	1	15	pH < 8
	$Ca^{2+} \rightarrow Na^+$	1 N NaCl	10	1	15	Cl^- negative
	$Ca^{2+} \rightarrow H^+$	0.5 N HCl	4	1	15	pH > 5
Strong base anion exchanger						
Type I	$Cl^- \rightarrow OH^-$	2 N NaOH	8	2	8	pH < 9
	$HCO_3^- \rightarrow OH^-$	2 N NaOH	4	2	8	pH < 9
	$OH^- \rightarrow$ formate$^-$	2 N formic acid	5	2	8	pH > 5
	$Cl^- \rightarrow$ formate$^-$	use $Cl^- \rightarrow OH^-$, then $OH^- \rightarrow$ formate$^-$				
	$Cl^- \rightarrow$ acetate$^-$	same as formate with HAc				
Type II	$Cl^- \rightarrow OH^-$	2 N NaOH	5	2	10	pH < 9
	$Cl^- \rightarrow NO_3^-$	1 N $NaNO_3$	4	2	6	Na flame test negative
Weak base tertiary amine anion exchanger	$OH^- \rightarrow Cl^-$	1 N HCl	3	1	4	pH > 6
	$Cl^- \rightarrow OH^-$	1 N NH_4OH	8	1	10	pH < 8

a given purpose may give rise to difficulties. It is to be hoped that the chapters of this volume dealing with the use of ion exchangers in industry and the various other fields will prove to be guidelines for the selection of an appropriate ion exchanger.

The conversion of an ion exchange resin from one ionic form to another is known as resin conversion and/or regeneration. The slight difference in meaning between these two terms lies in whether this process is carried out under economical aspects or not. In the laboratory the costs of the chemicals for converting a resin are usually negligible and it is more important to convert the resin sample quantitatively to a desired form using a considerable excess of the corresponding conversion reagent. Table 1.2 shows a survey of the common types of ion exchange resins and the conditions under which on a laboratory scale ion exchange resins are converted from one ionic form to another. Such conversions are normally performed in columns; it is then to be remembered that the resin may shrink, or it may swell by as much as hundred percent, depending on the conversion. Usually a test is carried out to determine the completeness of the conversion. As a rule, conversion is then complete when the first ion is no longer detected in the effluent. In addition, the easiest test method for completeness of conversion depends on the particular conversion: in many cases it can be evaluated by pH measurements or by simple qualitative tests. Rinsing the column using deionised water after completeness of conversion is the most common technique. In general, for any conversion for which the converting conditions are unknown, information can be gathered from the relative selectivities of the various counterions for the resin in use. If the conversion of a resin is to be achieved to an ionic form with a higher selectivity, the column is to be treated with two or three bed volumes of a 1M solution of the desired counterion. To convert to an ionic form with a lower relative selectivity for the resin in use, the volume of counterion solution needed will depend on the difference in selectivity, generally being one bed volume of 1M counterion solution for each unit difference in relative selectivity.

Certain special synthetic ion exchange resins require separate consideration, since they differ from customary resin types either because they have a special matrix or because of their particular reactive groups.

Specific and chelate ion exchangers. Specific ion exchangers are those types in which active groups have been introduced having the properties of a specific reagent. The specificity is based on the chemical structure of the ion exchanger itself and must not be confused with the phenomenon of selectivity. Because of its specificity, such an ion exchanger can sorb one ionic species to the exclusion of others under a broad range of conditions. Their mechanism of action will first be explained by a historical example.

Skogseid [79] subjected polystyrene to nitrogenation, reduction, conversion of the formed polyaminostyrene with picrylic chloride, and renewed nitrogenation, obtaining an exchanger containing the following building blocks (I):

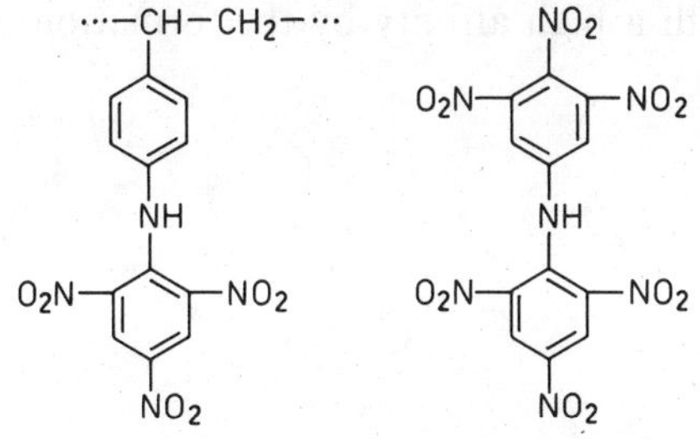

(I) (II)

Dipicrylamine (II) is a specific precipitating agent of potassium, and the above ion exchanger shows, as a result, an excellent specific uptake of potassium ions.

A next step in the development of specific ion exchangers was the report on a boron specific resin by Kunin and Preuss [80]. To introduce a respective group the principle was employed that boric acid reacts readily with polyhydric alcohols to form an acid considerably stronger than the original boric acid. The boron specific ion exchange resin was, therefore, synthesized by aminating a chloromethylated styrene-divinylbenzene copolymer with *N*-methylglucamine according to the following diagrammatic equation

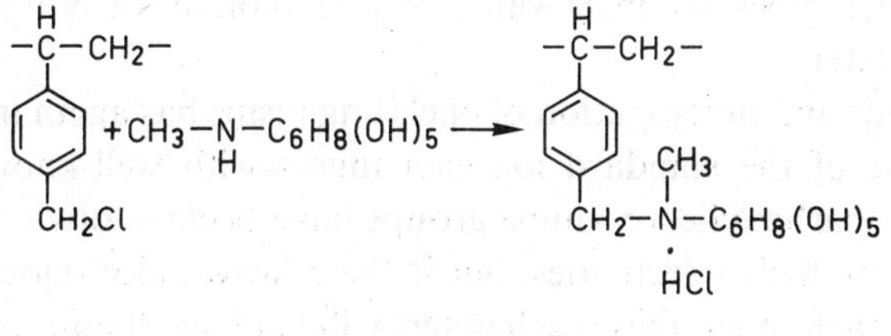

This resin shows a unique specificity on the grounds of the incorporated *N*-methylglucamine, which is known to form salts with acids and complexes with metals. Column tests indicated the utility of this ion exchange resin for removing boron from irrigation water and solutions containing unacceptable quantities of this trace element. The most important point is that the resin is readily regenerated with sulfuric acid.

Specific ion exchangers would, if seen in contrast to selective ion exchangers, include the entire group of chelate resins, whose specificity is based on the chelate or complex-forming reactive group. The best known commercial type is the iminodiacetate ion exchanger, Dowex A-1, in which the iminodiacetate groups are directly attached to the styrene matrix:

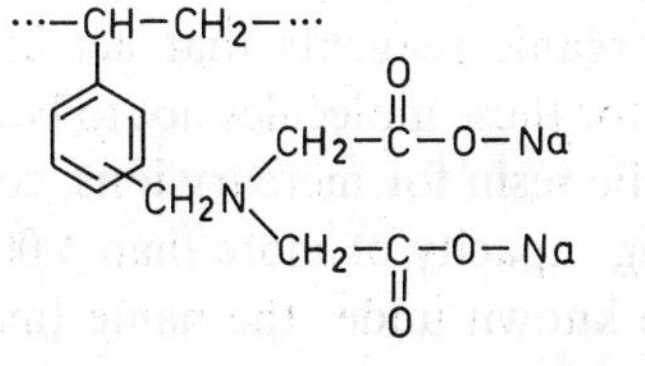

and which can fix polyvalent ions with a high affinity by the formation of heterocyclic metal chelate complexes:

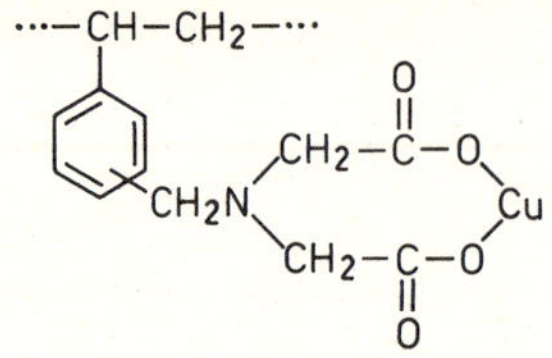

A series of specific and chelating ion exchangers has been produced, but their main application was in the field of analysis [81]. The above chelate-forming polymers have ion exchange properties by virtue of their carboxylic acid groups. It is therefore recommended to equilibrate them with the same buffer solution from which the cations are to be removed in order to avoid ion shifts during their application. This dependence on the pH is shown for Dowex A-1 in the fact that the uptake of ions is very low below pH 2, increases sharply from pH 2 to pH 4, and reaches a maximum above pH 4. Since a true ion exchange takes place, any metal removed from solution is replaced by an equivalent amount of the ions originally on the resin. For regeneration, e. g. in a column, the chelate ion exchanger is washed with two bed volumes 1 N HCl, 5 bed volumes water, two bed volumes 1 N NaOH and then five bed volumes water.

Interest in the synthesis and investigation of chelating resins has again increased, as limitations in the use of the standard ion exchangers with well-known polar groups, such as sulfonic, carboxylic, or amine groups have become obvious. These resins are characterized by their selectivities, but if the relative selectivities of ions in a mixture are close to each other, these exchangers will take up all ions, regardless of whether this is desired or not. As a result an increased amount of regenerant is needed than would be required if certain ions were removed specifically. Further, a regenerant effluent having a mixture of ions makes reuse of the specific ions more difficult. It may then be especially difficult to remove toxic from non-toxic metals in standard operations. The relative selectivity coefficients of the standard sulfonic cation exchangers are close to each other in magnitude and are often overestimated in practical application. With the requirements for emissions of many metal ions into the environment being in parts per billion, new types of exchangers are essential that are either highly selective or specific for a particular ion to the exclusion of other ions.

Several types of specific ion exchangers are already on the commercial market. The first one, based on dipicrylamine, and synthesized by Skogseid has not proven practical commercially. In principle, given organic reagents that act as specific precipitants of specific ions, there is no reason for these molecules not to be attached to a polymeric matrix. An example of a specific resin for mercury ions, containing thiol as an active group, with a high operating capacity of more than 1 000 mequ/l, and a leakage of about 2 μg/l, has become known under the name Imac TMR

[82]. In operation, when a solution containing 10 mg/l each of cadmium, lead, copper and mercury ions is percolated through a thiol exchanger bed, the mercury ions even replace the cadmium, lead and copper ions (in this order), and still continue to be removed from the solution until breakthrough occurs.

In contrast to the use of the term "chelate ion exchangers" for the resins mentioned above, the term should be reserved for resins carrying reactive groups that are able to form inner complexes, i. e. real chelates, with selected ions. These resins combine the processes of ion exchange and complex formation. Reactive groups resulting in chelate formation are usually introduced into a premanufactured macromolecular matrix of either gel or macroporous type. During this process the matrix structure must remain unchanged. Reactions involving the ring formation of chelates with metal ions are characteristic for chemical groups of only a few elements that can act as electron donors in exchanging groups, namely oxygen, nitrogen, sulfur, phosphorus or arsenic. Such chelate ion exchangers are then distinguished from ordinary type ion exchangers in that the affinity of a particular metal ion depends mainly on the nature of the chelating groups, with the size of the ion, its charge and other physical properties being of secondary importance. The strength of the binding forces in an ordinary exchanger is of the order of 8 to 12 kJ/mol; it is appreciably higher in a chelating ion exchanger being of the order of 60 to 100 kJ/mol. The exchange process in chelate ion exchangers is slower than that in ordinary type ion exchangers and seems to be controlled either by particle diffusion or by a second order chemical reaction. Chelate ion exchangers with macroporous matrix structure have faster exchange rates.

Among the synthetic ion exchange resins described above, those of the gel type still have the greatest importance, and continue to be used most frequently in industry. The production of an ion exchanger requires specialized knowledge, so that it would hardly be worthwhile to attempt homemade preparation. Ion exchangers also assume a sort of intermediate position between a chemical product, which is manufactured and further processed, and equipment serving for the manufacture of a particular product. They are highly sophisticated chemical materials, but they are used only as processing aids; they are therefore only the transfer agents of a chemical process without being completely changed or turned to another purposes. Because of the possibility of reversible conversion into different forms, they have diverse applications and it would seem that they should have an unlimited life. Naturally, like any material, they are subject to chemical and mechanical wear and have only a limited life.

The synthetic resin ion exchangers available today are commercial products with registered tradenames, made by various manufacturers. In the Russian literature, on the other hand, ion exchangers are usually referred to as cationites or anionites, etc. Since in this case it has become customary to designate individual types only by letters or numbers, such as SBS, EDE- 10 P, etc., this somewhat confusing dual nomenclature must be retained. Appendix 1 of this volume lists a selection of important commercial ion exchangers frequently cited in the literature. The number of available types is so large that it has become almost confusing. Data on their characteristics have been obtained from the technical literature of the manufacturers – the best source of information concerning the applicability of an available synthetic ion exchange resin.

The tables in the Appendix give the specification of an ion exchanger by covering the necessary data resulting from its properties. The acidity and basicity follow from the ionogenic group with its exchange activity and thus are immediately understandable. The production process determines the external form of an ion exchanger. While polymerization resins are generally delivered in the form of beads, polycondensation resins are available as milled granulates. The color of individual types ranges from white through yellow, brown, and dark brown to black. Important additional characteristics are their total exchange capacity, moisture content, particle size distribution, physical structure and stability, as well as data on elutable components. In practice, a knowledge of the intended application is necessary, which frequently needs to be determined specifically for special uses.

In conclusion, it may be said that the present significance of synthetic ion exchange resins results from their great mechanical strength and chemical resistance, high exchange rates, high capacity and the possibilities of varying their properties as a consequence of their synthetic nature. Their only disadvantage is a limited range of operating temperatures, although this range is sufficient for most purposes.

Snake-cage ion exchangers. The so-called snake-cage ion exchangers were developed as a separate group of ion exchange materials to permit the removal of salts from a single polyelectrolyte and regenerate the exhausted resin with water. In these snake-cage resins, a polymeric linear "snake" is formed by the polymerization of suitable monomers with a given charge in such a way that it is located in a polymeric crosslinked "cage" of opposite charge. Since the distance between charged groups determines their effect, considerable effort was needed to obtain a controlled structure of the effective interstices between negative and positive ionogenic groups. This was achieved by the polymerization of acrylic or methacrylic acid into different quaternary ammonium anion exchange resins [83]:

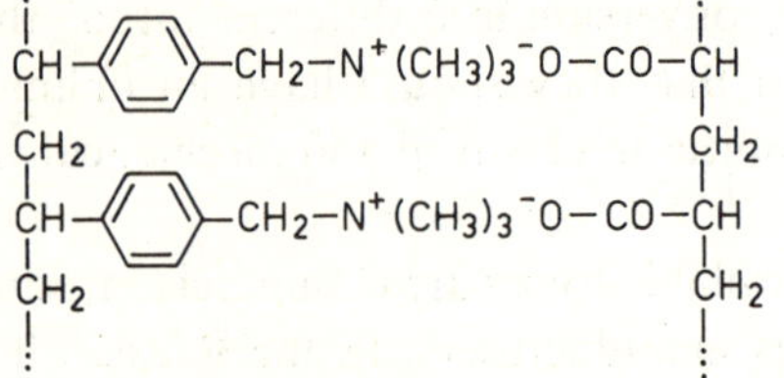

Such a snake-cage ion exchanger can exchange salts from aqueous solutions of organic compounds, such as glycerol, by simple sorption and can be regenerated subsequently by washing with water [84]. Since the snake-cage ion exchangers can also have a selective action, they permit the separation of electrolytes, such as NaCl and NaOH. The ion retardation process is carried out with snake-cage ion exchangers. With respect to classification of synthetic ion exchangers snake-cage resins belong to the amphoteric ion exchangers.

Inorganic ion exchangers. Historically, it was inorganic ion exchangers that were first studied for the practical applicability of ion exchange [85]. Synthetic inorganic ion exchangers were prepared by Gans on the basis of findings obtained on mineral exchangers. They consisted of fused sodium aluminosilicates and had, because of their glassy nature, very poor kinetics. As permutites or hydrothermal synthetic zeolites they were used for water softening, but to regenerate these exchangers took longer than the softening cycle. They were of great interest for scientific investigation of ion exchange but they lasted only about ten years on the industrial market. The inorganic ion exchangers that followed were then modified natural products such as greensands (glauconites), bentonites and clay type minerals, with low operating capacities but faster regeneration rates. The synthetic sodium aluminosilicates had a longer lifetime under practical application aspects. They were precipitated out of solutions consisting of alum, sodium aluminate and sodium silicate. These precipitated gels had higher capacities but were very sensitive to low and high pH and readily fouled by iron present in water.

The principle of ion exchange is the same for the synthetic ion exchangers [86] as for the modified mineral exchangers [87]. The skeleton carries an excess charge which is compensated for by mobile counterions, as they are called in ion exchange nomenclature. In the case of the zeolites, this was found to be a consequence of the fact that a part of the Si^{4+} building blocks is replaced by Al^{3+} in the silicate lattice. The lacking positive charge is replaced by alkali or alkaline earth ions, which are as counterions freely mobile in the inorganic skeleton. The natural and the older synthetic inorganic ion exchangers are no longer of industrial significance. Nevertheless, driven by the wish to obtain ion exchange materials that are stable at higher temperatures and radiation-resistant, intensive research and development work has resumed in this field in recent years. Based on the data available, this involves the production or modification, as well as the characterization and application, of clays and clay minerals, new zeolites and the combined group of non-siliceous inorganic ion exchangers. These groups of inorganic ion exchangers are, because of their increasing importance, treated in detail in separate chapters of this volume. For commercial inorganic ion exchangers see Appendix 1.

Cellulose, Dextran and Agarose ion exchangers. As a result of a small number of carboxyl groups in its structure natural cellulose has ion exchange properties. Numerous reaction products can be obtained by the oxidation of cellulose. These products include hydroxycellulose with 15% COOH-groups, and which, in powdered form, must theoretically be considered ion exchangers, although they find little practical use because of their solubility in water. Powdered cellulose ion exchangers, which are now used in columnar processes, have become very important as a result of the studies of Peterson and Sober on protein separation – as was already mentioned in the section on the history of ion exchange and ion exchangers. Cellulose ion exchangers have ion-exchanging properties just as do the synthetic ion exchange resins, but they differ from the latter in a few important points. They are treated therefore in detail in a separate chapter of this volume. For commercial cellulose ion exchangers see Appendix 1.

The polysaccharide dextran (poly-α-1,6-glucan), which is produced from sucrose by microbiological methods, consists of fibrous molecules that can be converted with epichlorohydrin and, as a result of this crosslinking, be transformed into polymers with a three-dimensional structure. According to the studies of Porath and Flodin, these can be used for the fractionation of water-soluble substances. These dextrans, which are obtained as beads from the suspension polymerization process, have proven to be useful as molecular sieves in gel filtration and in molecular sieve chromatography for chromatographic separations based on molecular weight. They can be obtained in standardized form under the tradename Sephadex. Since the number of hydroxyl groups is only insignificantly modified by crosslinking, it is possible to prepare esters and ethers by further reactions with suitable reactants and to introduce ionic groups into the dextran molecule, thus arriving at ion exchange materials with polysaccharide as the matrix. In an early study Porath and Lindner demonstrated that oxytocin and vasopressin can be separated on dextran diethylaminoethylether by the ion exclusion process. Dextran ion exchangers are treated in detail in a separate chapter of this volume.

Agar and agarose derivatives were first prepared for chromatography, electrophoresis and gel-bound enzymes by Porath and coworkers whose aim was to produce alkalistable and thermostable, insoluble spherical agar particles with a very low adsorption capacity. For this purpose the sulfate ester groups present in the original agar are removed by alkaline hydrolysis of crosslinked agar, in bead form, at an elevated temperature. The reaction is performed in the presence of sodium borohydride to prevent simultaneous oxidation. Further reduction of the adsorption capacity may be accomplished by treatment of the gel with lithium aluminum hydride in dioxane. It was found that desulfated and reduced agar and agarose gels can be packed in beds with excellent flow and molecular sieve properties [88].

Coal based ion exchangers and other materials with ion-exchanging properties. If wood, peat, firewood, soft coal, or anthracite is treated with oxidizing reagents or concentrated sulfuric acid, it is possible to introduce additional groups with an exchange capacity, beyond the existing OH- and COOH-groups with their exchange capacities. Also heating with caustic soda under anhydrous conditions permits the production of coals with ion exchange properties. Another group, the ammoniated coals, are also anion exchangers based on coal. The most important coal ion exchangers, which are also commercially available, are the sulfonated coals. These products contain a mixture of sulfonic, carboxylic and phenol groups of very different acid strengths. Depending on these different acid strengths, only phenol and carboxylic groups or, with a sufficient excess, also the sulfonic acid groups are transformed into the H^+-form during regeneration. If sulfonic acid groups are also present in the H-form, sulfonated coals are capable of neutral salt splitting according to the following equations:

$$\text{Coal-H} + NaCl \rightleftharpoons \text{Coal-Na} + HCl$$
$$2\ \text{Coal-H} + CaCl_2 \rightleftharpoons \text{Coal-Ca} + 2\ HCl.$$

If about 100% of theory is used for regeneration, only a conversion with the alkaline salts takes place according to:

$$2\ \text{Coal-H} + Ca(HCO_3)_2 \rightleftharpoons \text{Coal-Ca} + 2H_2O + CO_2$$

a reaction which is very economically used in the "starvation process" for partial deionization.

Among previously available commercial coal ion exchangers, the products Dusarit S and Imac C 19 of the former Imacti Industrielle Maatschappij Activit N. V., Amsterdam, should be mentioned. Dusarit S was a multiionic cation exchanger of high physical stability, in the form of granulate (0.3 – 1.2 mm) based on sulfonated coal and with the properties of activated carbon. It was delivered in the Na-form and served for the treatment of water and purification of glucose juices in the Na- and H-cycle. Imac C 19 was a multiionic strong and weak granular cation exchanger (0.3 – 1.2 mm) delivered in the H-form. It was used for the decarbonation of carbonate-rich water with a high regenerant effect. Table 1.3 lists the other properties of these coal ion exchangers, together with other commercial products. Occasionally, coal ion exchangers produced commercially or in the laboratory are also used for ion exchange chromatography.

Table 1.3 Coal ion exchangers (cation exchangers)

Designation	Capacity	Heat stability (°C)	Permissible range (pH)
Dusarit S	20 or 10 g CaO/l	80	0 – 14
Imac C 19	28 g CaO/l	80	4 – 14
Permutit S 53	10 – 15 g CaO/l	40	0 – 11
Permutit HI 53	1 – 28 g CaO/l	40	0 – 10
Zerolit Na	1.8 mequ/g		
Zeo-Carb HI		30	0 – 8

A number of other natural and industrial materials can also be converted into ion exchangers by chemical treatment. Tar, paper, cotton, pectin, tannin, and lignin are well-known examples. Under the influence of sulfuric acid, lignin is converted into lignic acid, with ion exchange properties which were recognized by Freudenberg at the time when the first synthetic ion exchange resins were produced. By treating lignin-containing substances, particularly sawdust, with sodium sulfite solutions, very inexpensive ion exchangers which can be used for the adsorption of radionuclides are obtained. Alginic acid has always been of some interest as a cation exchanger. Occasionally, it was also used for separations by ion exchange chromatography. The possible therapeutic applications of alginates are of particular interest because of their Sr-Ca-ion exchange reactions. Further details on carbonaceous as well as other ion exchange materials can be found in a separate chapter of this volume. Liquid ion exchangers and ion exchange membranes are also described in detail in separate chapters.

1.1.2.2 Structures

Structural chemistry is, in general, a part of physical chemistry that aims at the elucidation of the structures of molecules, in particular the arrangement and bonding of atoms. The term structure is used in many kinds of connexions in chemistry – in particular in the everyday language of organic chemistry by the structure of an organic compound the structural formula is actually meant. Structural formula are the words in the language of organic chemistry and they are basic to understanding and communicating everything else in the subject. Geometric structures are also important in inorganic chemistry, where crystalline structures especially are investigated in order to get an idea of the total molecular structure and the reactions that can be expected from the crystals.

The structure of organic polymers, in particular, is necessarily complicated because the possibilities of synthesizing such polymeric substances are numerous. In the simplest case one has what could be called a repetition of structural units along a chain. The next step in building polymeric structures is to form copolymers. Here a very wide variety of combinations is feasible, not only by the formation of branched macromolecules, but especially because, besides random copolymers, graft and block copolymers can be made up; further, there are many copolymers that do not fall under any distinct classification. It is also well known that polymer molecules can be linked together by including in the polymerization formulation a monomer having two or more functional groups. When polymer molecules are connected to one another by a sufficient number of interchain bonds, the network formed is, in effect, one giant molecule. The character of a network polymer depends primarily on the concentration of the crosslinking agent and the type and concentration of diluent present when the network is formed. The general principles of structural chemistry can hardly be applied to such network polymers.

It may sometimes be quite difficult to understand the title of a publication on ion exchangers that contains the word "structure", i. e. precisely what meaning this word has in the context of the paper. The synthetic organic ion exchangers are, in the majority, based on random polymeric networks consisting of styrene or an acrylic component crosslinked with divinylbenzene into which ionogenic groups have been incorporated. When the structures of such products are being discussed, and this also includes matrices that have been formed by the simpler but less practiced polycondensation reaction, it is important to note that in the literature several types of investigation of the structure of ion exchangers can be found. These comprise the following categories:

The structure of the complete ion exchanger, including an investigation into the distribution of the fixed ions.

The structure of the matrix copolymer or polycondensate network as it was formed during polymerization.

The structure relative to the fixation of the counterions.

The structure or structural changes due to interaction with solvents.

In an early attempt Grubhofer [89] suggested a model for the molecular structure of polystyrene ion exchangers crosslinked with divinylbenzene, which served to calculate the distances within the structure limiting ion diffusion through the ion exchanger. If x is the molar percentage of the crosslinking agent divinylbenzene, then the number of ethylene groups per crosslink between the points of attachment of the latter is $N_{Eth} = 50/x + 0.5$. The distance between two branching points in the network, when taking into account the thickness of the benzene ring at 4.5 Å, the length of two carbon-carbon bonds in the phenyl group at $2 \cdot 1.5$ Å, and the size of a single monomer unit at 2.2 Å is $a = 2.2N_{Eth} + 7.5$ Å. The angle between the main chain and the crosslink is assumed to be 90°. To obtain the dimensions of the cell accessible to the diffusion of ions, it is necessary to substract from the resulting cell diameter the thickness of the polystyrene chain, i. e. 15 Å. The results obtained for a high content of crosslinking agent in the resin agree well with independent estimates of the cell dimensions and with what is known from diffusion studies. For low content of the crosslinking agent they agree with data obtained by electron microscope studies of the same specimens in the dry state. This shows that the model describes correctly the fundamental structural characteristics of resins based on styrene and divinylbenzene. From the polymer chemistry point of view such a model is undoubtedly still a quite rough approximation, disregarding as it does a whole series of phenomena belonging to the formation of copolymer structures and especially to sulfonated ion exchange resin structures, e. g., the presence of amorphous regions, the entanglement of chains, non-uniform sulfonation or new reactions during sulfonation, etc.

Marcus and Kertes in their reflections on the structure of polystyrene-divinylbenzene-type ion exchangers based on the models shown in Figure 1.9 assume that the skeleton of such an exchanger, which consists of aliphatic $-CH_2CH-$ chains with attached benzene rings and the divinylbenzene bridges, is fairly elastic, and considerable changes of volume are, therefore, possible. Considering a fully swollen cation or anion exchange resin 10% crosslinked, the aromatic rings are connected to the skeleton in repeating units, about 8 Å apart, with the rings striving to be parallel to each other in order to minimize mutual interaction. The fixed groups

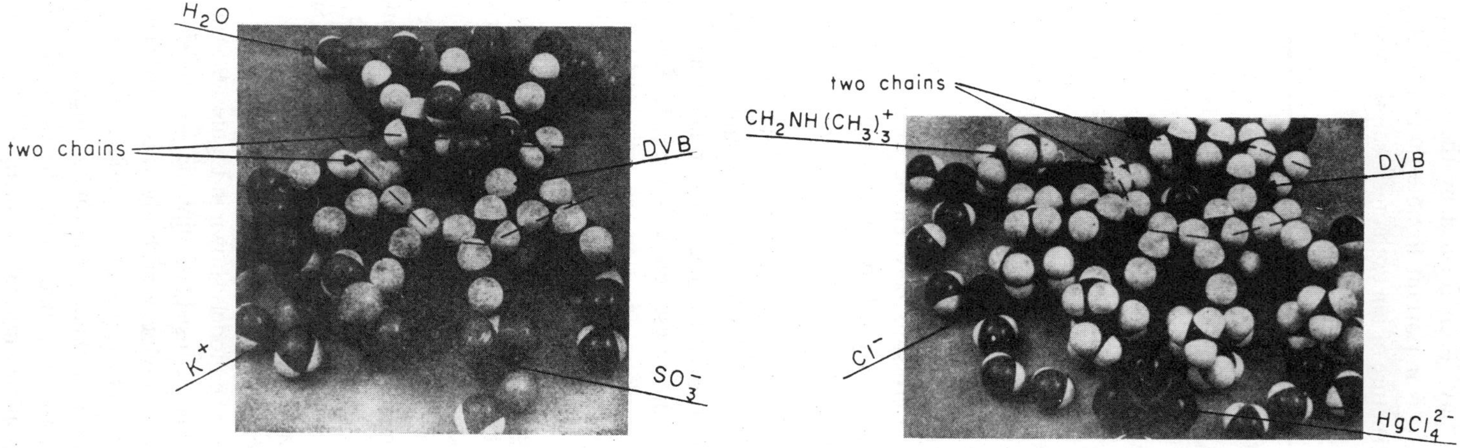

Figure 1.9 Three-dimensional models of polystyrene-divinylbenzene ion exchange resins with nominal 12% crosslinking showing the relative size of ionogenic groups, counterions, hydrocarbon skeleton and free space. Left: strong acid sulfonic cation exchanger K-form; right: strong base quaternary ammonium anion exchanger Cl-form. (From Ion Exchange and Solvent Extraction of Metal Complexes. Y. Marcus and A. S. Kertes. Wiley, London, 1969).

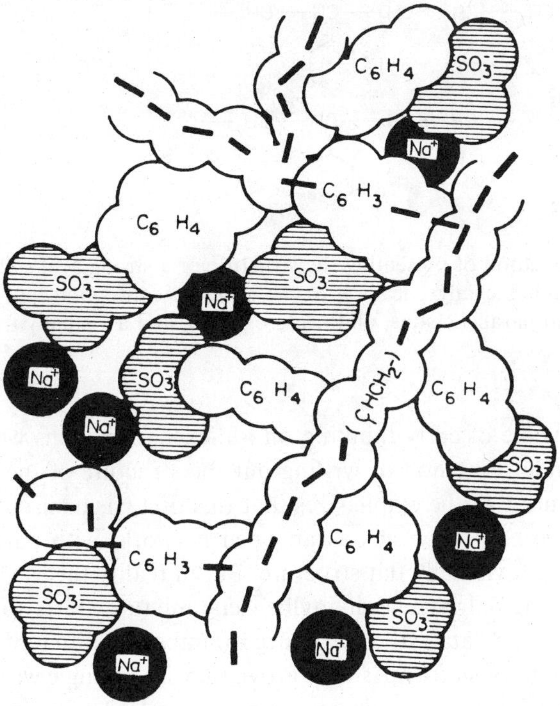

Figure 1.10 Two-dimensional projection of a strong acid sulfonic polystyrene-divinylbenzene cation exchanger with nominal 16% crosslinking. (From Molecular Science and Engineering, ed. A. R. von Hippel. M. I. T. Press, Cambridge, 1959).

are located at the ends of the benzene rings. Crosslinking has the effect of producing cavities 50 Å in average diameter which are lined by the exchange sites. These cavities are filled by water. With reference to the schematic representation of a fragment of a cation exchanger as shown in Figure 1.10 it should be pointed out that this two-dimensional projection of the model appears somewhat more crowded than the model actually is. But it must also be pointed out that an important feature of such ion exchange resins is the randomness of the crosslinks, resulting in a completely disordered structure of the exchanger. It is further assumed that there is a wide range of distances between sites, and that the environment of a site may vary from preponderately of organic material to freely available space occupied by mobile solvent. The flexibility of the chains, it is said, causes, of course, changes in the environment with time (and pressure?) [90].

But it would seem to be more appropriate to approach the structural questions of ion exchangers from a synthetic point of view in order to elucidate what the structure might be. In this light Arden has presented in a series of lectures in 1974 a philosophy of ion exchange resin structures [91]. Arden argues that it is probably

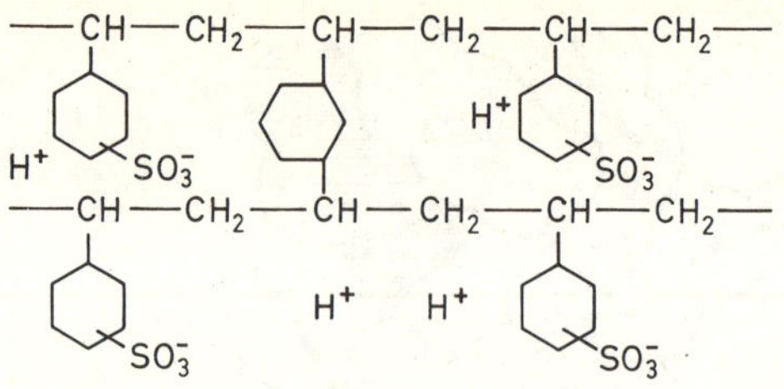

Figure 1.11 Standard method of presenting the structure of a strong acid sulfonic polystyrene-divinylbenzene cation exchanger leading to the basic misconception that the structure is an open network, with parallel chains, uniform crosslinks, and a virtually flat structure.

true to say that the whole of early thinking on ion exchange resins was incorrect, as a result of the standard method of writing out the structure of the polystyrene resin as shown in Figure 1.11; he emphasizes that this first basic misconception led investigators to conceive of the resins as an open network, with parallel chains, uniform crosslinks, and a virtually flat structure. But in reality this is very far from the truth. When the original styrene-divinylbenzene mixture starts to polymerize under the influence of the catalyst, a very large number of different crosslinked chains start to form simultaneously. As they grow, they intermingle with each other and grow through each other, as depicted in Figure 1.12. The diagrammatic molecule A on the left is 8% crosslinked, as is the one B in the middle. If these two molecules are separate from each other then they give a certain pore size which determines the measured characteristics of the resin. If the same two prototype molecules are formed so that they are entangled as shown in diagram C, then because the crosslinks are closed they cannot be separated from each other. They are still exactly 8% crosslinked, but the effective pore size is now only half of what it was before, and the apparent measured crosslinking is consequently considerably higher than 8%. In reality, while the original polymer is being manufactured the chains entangle and there are no resins which are not entangled, although the degree of entanglement may be modified by special procedures. The measured crosslinking of every resin in existence is in fact not the true crosslinking, but the

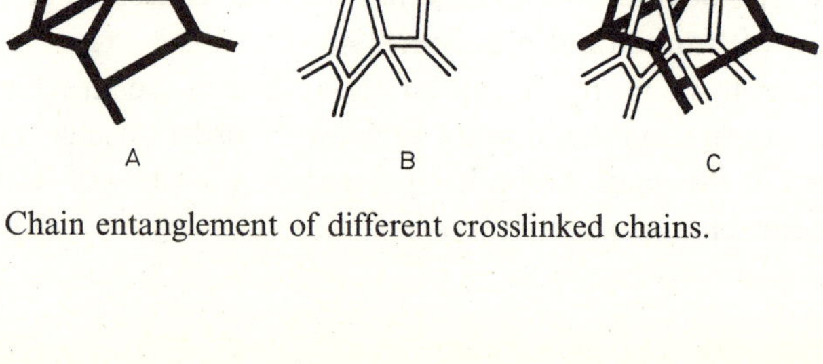

Figure 1.12 Chain entanglement of different crosslinked chains.

Figure 1.13 Progress of copolymerization from initiation to the final copolymer matrix.

sum of the two effects of divinylbenzene crosslinks together with the entanglement factor. It is in fact possible to deliberately introduce entanglement, and it has also been possible to synthesize so-called re-entangled structures. What is noticeable, however, is that as the degree of deliberate entanglement is increased, the resin becomes more brittle, and mechanically fragile. It follows that chain entanglement is a negative feature, which should be avoided as much as possible. The second basic misconception in at least the simplified picture of ion exchange macromolecules is that the crosslinking is uniform. This is not the case, since in the polymerization of styrene and divinylbenzene, there are three separate processes, not counting the ones due to the presence of ethylstyrene in the divinylbenzene: a) DVB-DVB polymerization, which is rapid, b) DVB-styrene polymerization, which is intermediate and c) styrene-styrene polymerization, which is slow. From this it results that, as polymerization starts at the catalyst centres, the first polymer to form has an excessively high DVB content, and is thus highly crosslinked. As the reaction proceeds, over a period of several hours, however, the later polymer is less and less crosslinked, until the final chains are completely linear. These process steps are shown in Figure 1.13. The final copolymer, and the exchangers made from it, are extremely heterogeneous, with highly crosslinked regions, surrounded by more loosly crosslinked areas. The disadvantages resulting from this heterogeneity will be discussed elsewhere. In fact, the development of newer resins was brought about by the problem of organic poisoning and the need to overcome this problem. Thus new types of resin structures were developed which had either been "disentangled" or which had "sintered", "spongelike" or "macroporous" structures. It is in fact possible, by the use of deliberate re-entanglement or disentanglement, to produce ranges of different resins, containing from 0% to over 30% DVB, that are very similar in their normal characteristics (Arden, Kressman and Millar). Figure 1.14 gives an overview of possible structures. If a resin material has a lower entanglement than a normal DVB polymer, it possesses a much larger pore size for the same DVB content. As a result, the water content of the drained resin – the factor used to measure apparent crosslinking – is very much higher for the same DVB content, and the volumetric capacity of the resin is correspondingly reduced. To compensate

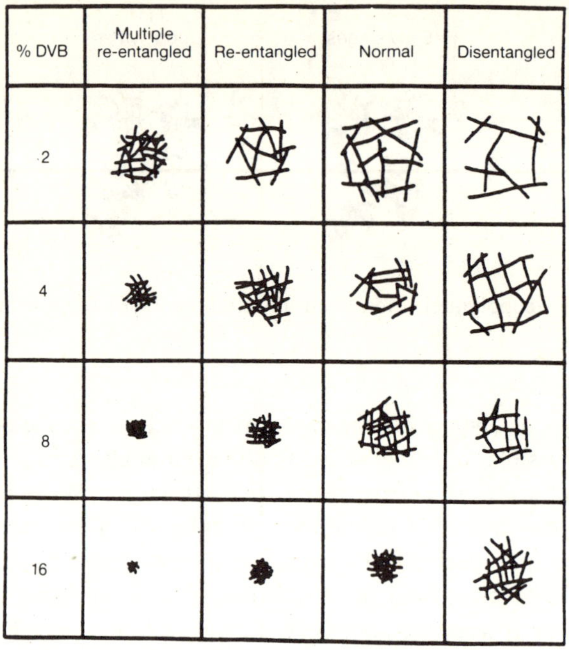

Figure 1.14 Possible structures obtained by the use or deliberate re-entanglement or disentanglement.

for this factor, the DVB content can be increased, and it is thus possible, by using increasing proportions of DVB under the special conditions of the polymerization, to produce very different resins, all having the same apparent crosslinking, but with increasing real crosslinking and decreasing entanglement. As the disentanglement increases, with increasing real crosslinking to compensate, the resultant resins become harder and tougher.

Whatever the method of polymerization used, it is possible to separate the polymer chains to a greater extent than the normal limits of intermolecular distance, so that the polymer beads contain holes (termed macropores) from the first stage of synthesis. The ion exchange resins made from these beads were already called "macroporous". In all these synthetic processes the polymers obtained have a structure that has also been called "sintered" or "sponge-like", as depicted in Figure 1.15. Decisive is – and this applies to several other methods of more recent years as well – that from the method of synthesis it follows that in the latter cases the structures must contain large pores.

A progressive summary similar to the one above was given by Millar of the discrepancies between the more or less ideal assumptions of a molecular model and a practical synthetic approach to the structure of ion exchange resins [92]. The complex of questions related to crosslinking and entanglement, as well as to pores

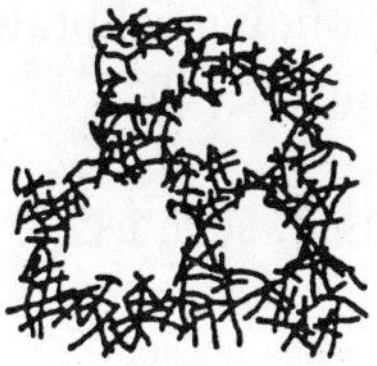

Figure 1.15 Sintered or sponge-like structure obtained by a polymerization technique separating the polymer chains.

and heterogeneity, are analyzed with an eye towards the validity of a number of assumptions made in simplifying kinetic expressions dependent on structure, with Millar emphasizing the non-ideality of real materials. With respect to crosslinking and entanglement from a synthetic point of view as a factor decisive for the structure of the resulting ion exchangers and their properties, the quality of the commercial DVB must be taken into consideration, when comparing, for instance, the early theoretical work of Pepper and his colleagues, or that of Gregor and others, with later results. Modifications in the structure of ion exchange polymers can be achieved by using solvents during their manufacture. In the production of commercial DVB, the divinylbenzene isomers are obtained mixed with their precursors in the original raw material, and are separated by distillation. Current commercial DVB contains about 60% DVB isomers, with the rest consisting almost entirely of ethylstyrene, and concentrates with up to 88% DVB isomers are available. In the late forties, however, the content of crosslinking component was only 25–35%, with the ethylstyrene content being about the same. The rest in those days was mainly the original diethylbenzene starting material. Thus – and this is the important point – the higher DVB content materials of those days were in fact solvent modified, with the resultant physicochemical differences described in Millar's paper, cited above [92]. Further reflection on pores and heterogeneity and basic considerations regarding the synthesis and formation mechanisms of crosslinked polymers have led to the conclusion that in synthetic organic polymers the occurrence of structural homogeneous networks seems to be the exception rather than the normal case [93].

A third approach to the elucidation of the structure of ion exchange resins is to run investigations based on chemical, physicochemical or physical methods. Chemical investigation of the network structure of crosslinked polymers would require the splitting of the polymer chains by unambiguous reactions. But up to now very little has been developed regarding chemical methods suitable for investigation of the structure of ion exchange resins [94]. It is obvious that results showing an almost linear correlation between the quantity of DVB in the polymerization starting material and in the pyrolysate of copolymers made from styrene and DVB to be converted to cation exchangers, can only be used as an indication for a characterization of the crosslinking conditions and do not allow any assertions as to the density of the network, etc.; they may, after a corresponding calibration, indicate the quantity of divinylbenzene incorporated into such a resin [95].

Physicochemical methods on the other hand, can be used, such as:

Measurement of the swelling in equilibrium

Application of the theory of capillary condensation

Measurement of the steric accessibility of ionic sites to organic ions with increase in their dimensions

Depression of the freezing point of the solvent in resins

Determination of the specific absorbing surface by the BET method, the porosity and the average pore diameter

Measurement of gas permeabilities

Microradiographic study of resin sections

Nuclear spin echo studies

Measurement of the hydrolysis kinetics of the ionogenic group

as well as physical methods, such as:

Determination of microhardness

Electron microscope methods

X-ray diffraction

Optical properties in transmitted and reflected polarised light beams

Absorption spectra in the visible region

Reflection spectra in the visible region

UV spectroscopy

IR spectroscopy

Nuclear magnetic resonance spectroscopy (NMR)

Electron spin resonance spectroscopy (ESR)

Mößbauer spectroscopy

Measurement of magnetic susceptibility

Temperature dependence of the dynamic mechanical moduli.

The results obtained are then used to propose more or less generalized structure models; but up to now it seems that none can be better used for the elucidation of the network structure of ion exchange resins than the investigation of the formation of the network polymer during its synthesis by measuring the consumption of the reactant and by investigating the changes in physicochemical or physical parameters during or after the incorporation of the ionogenic groups. Experience has shown that the structure of ion exchange resins depends as much on the method of preparation as on the chemical composition. As a result labelling of the exchangers with, e. g., radioactive uranium, and radiography were proven particularly effective in such studies, as they permit the elucidation of the heterogeneity of the internal structure of resin ion exchangers.

Contrary to resin ion exchangers the open-structured ion exchange materials, as for instance cellulose ion exchangers, consist of a loose network that permits the penetration of large molecules to the reactive sites from which the molecules are then readily eluted. Structural aspects can be derived from the native cellulose, either from wood or cotton fibers. In the native state, adjacent polysaccharide chains are extensively hydrogen-bonded, forming microcrystalline regions. These higher-oriented fibrillar chains are interspersed with amorphous regions of longer axial fibers with less hydrogen bonding. Hydrogen bonding between the neighboring cellulose chains, and especially in the fibril centers, provides dimensional stability for the cellulose matrix, and it is these forces that restrict the matrix to only moderate swelling, making cellulose insoluble in water. Limited acid hydrolysis results in preferential loss of the amorphous regions, yielding socalled microcrystalline cellulose. Within and in-between the cellulose chains are located "openings" or pores with a wide size range. When ionizable groups are introduced into such a matrix, the natural polymer cellulose becomes an ion exchange material with the open structure shown diagrammatically in Figure 1.16 a. By modification by mild acid hydrolysis the characteristic structure of cotton cellulose undergoes chain splitting and recrystallization within the interfibrillar regions, so that the crystallite fibrils are enhanced, causing much of the amorphous portion of the cellulose microstructure to be removed. During this regeneration microgranular cellulose is obtained, which is strengthened by crosslinking with epichlorohydrin, although the main structure-forming bonds are still hydrogen bonds. The impact of such modifications on the native structure of cellulose are shown in Figure 11.6 b. The ionogenic groups are attached by ether linkages directly to the glucose units of the

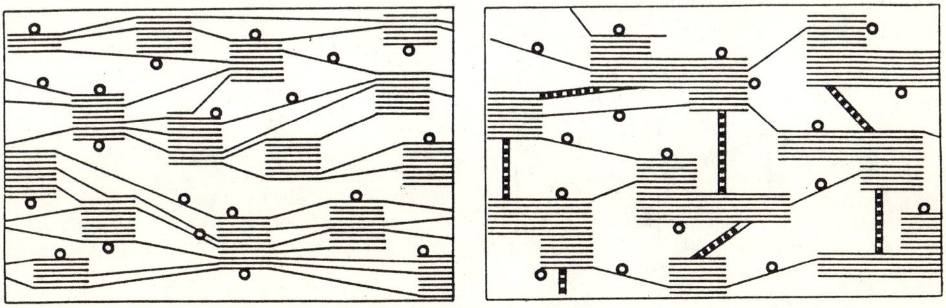

a b

Figure 1.16 Microstructures of ion exchange cellulose. Left: diagrammatic microstructure of a fibrous material, the solid lines representing the aggregates of carbohydrate chains and the dotted circles being the ion exchange sites; right: the same of a microgranular material with highly oriented regions, the wider dotted lines showing the crosslinks introduced and the dotted circles again being the ion exchange sites.

polysaccharide chains. The modification also results macrostructurally in the production of shorter particles that are dense and almost bead-shaped [96].

The best known exchanger structures are found among inorganic ion exchangers. It must even be emphasized that before the cation exchange behavior of inorganic ion exchangers can be fully understood a general appreciation of their structure is essential. However, since quite a number of inorganic ion exchangers are now available, structure must also be considered individually [97]. The principle of ion exchange is the same for the mineral exchangers as for the synthetic ion exchange resins [98]. The skeleton or crystal structure carries an excess charge which is compensated for by mobile counterions. In the case of the zeolites, this is known to be a consequence of the fact that a part of the Si^{4+} building blocks is replaced by Al^{3+} in the silicate lattice. The lacking positive charge is replaced by alkali or alkaline earth ions, which are present with free mobility in the mineral skeleton as counterions. As an illustration of these facts, the structure of chabazite is shown

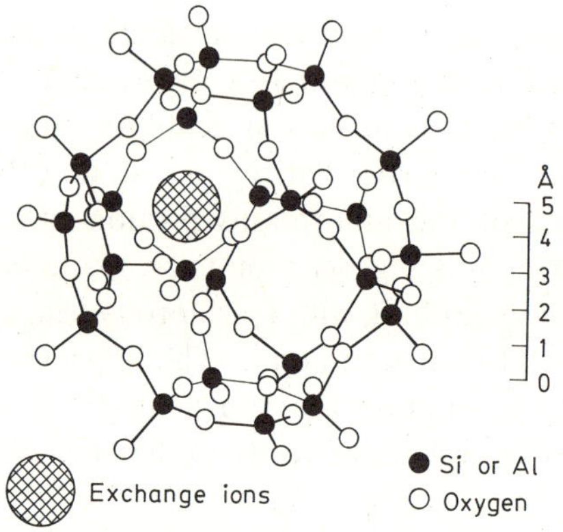

Figure 1.17 Stereostructure of chabazite.

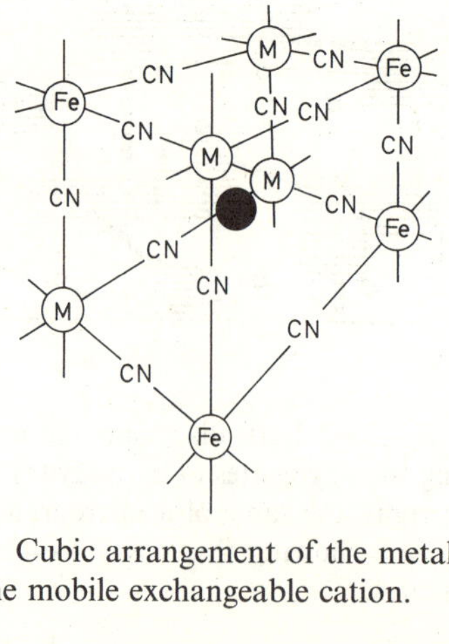

Figure 1.18 Cubic arrangement of the metal cyanide octaeders in the structure of ferrocyanide and the mobile exchangeable cation.

in Figure 11.7. As a further example, the complex salts based on ferrocyanide may be mentioned, which were developed in the search for new inorganic exchange materials. In relation to the general formula $M^{I}_{2}\,M^{II}Fe^{II}(CN)_6 \cdot n\,H_2O$ their structure is usually represented as shown in Figure 1.18. This figure shows the cubic arrangement of the metal cyanide octaeders with the mobile, exchangeable cation in the resulting cavity.

1.1.3 Fundamentals and definitions

Ion exchangers are solid or suitably insolubilized high molecular weight polyelectrolytes which can exchange their mobile ions for ions of the same charge from the surrounding medium. The resulting ion exchange is reversible and stoichiometric, with the displacement of one ionic species by another on the exchanger. Viewed in a different light, ion exchangers can be considered to be acids with a high molecular weight anion or bases with a high molecular weight cation, which can exchange their hydrogen or hydroxyl ions for equally charged ions, thus being converted into high molecular weight salts. If such a solid acid is neutralized by a base into a salt, however, the cations bound to the polyelectrolytes can again be displaced by other cations. The resulting process is known as cation exchange and the polyelectrolyte as the cation exchanger. In the second case, a solid base is obtained which is capable of hydroxyl exchange and which can be neutralized with an acid, and the anion from this acid which was bound first can again be displaced by another anion, a process then known as anion exchange. The polyelectrolyte on which this process takes place is called the anion exchanger.

The most widely used modern ion exchangers are organic materials based on synthetic polymers. The fundamental process can be described most easily for these and, on the other hand, all – even the general – phenomena of ion exchange can be more easily understood on the basis of these descriptions. Figure 1.19 shows some types of ion exchange resins.

The polymeric molecule of an ion exchanger is in the most general case – because of crosslinkages – a threedimensional network with a large number of attached ionogenic groups. Polymer molecules can be linked together by including a monomer with two or more functional groups in the polymerization. When polymer molecules are connected to one another by a sufficient number of interchain bonds, the network formed, is, in effect, one giant molecule. Crosslinks confer unique properties to such polymers. Networks that form in solution and remain intimately associated with the solvent are gels. Polymers whose crosslinks are permanent chemical bonds are insoluble in all nondegrading liquids, but can often swell enormously in liquids that would be solvents if the crosslinks were absent. The crosslinks in ion exchange resins occur randomly and the network structures can vary over very wide ranges. It is therefore quite difficult to depict ion exchangers in constitutional formula [99].

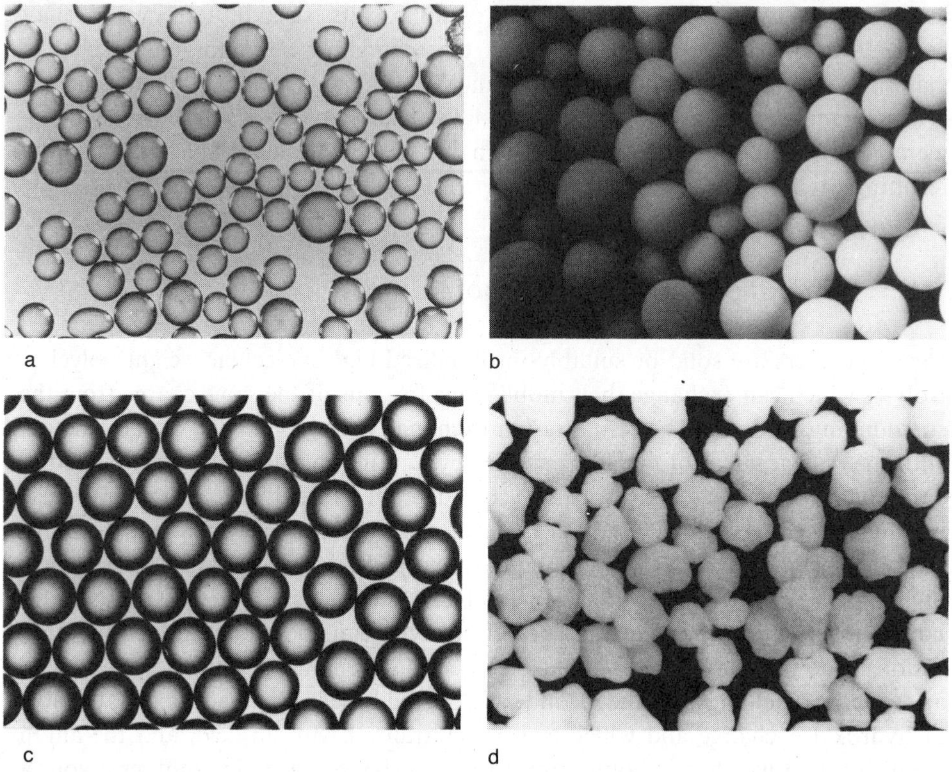

Figure 1.19 Some common types of ion exchangers: a) gel type strong acid cation exchanger; b) macroporous type strong base anion exchanger; c) Monosphere® cation exchanger with exceptional uniform bead size; d) granular form of a weak base anion exchanger. (®Trademark of the Dow Chemical Company).

The most important starting material used in synthetic resin ion exchangers is styrene. First, styrene is crosslinked with divinylbenzene into a polymeric network and then the reactive groups are introduced. One may therefore depict a cation exchanger containing sulfonic acid groups or an anion exchanger containing quaternary ammonium groups best by a quasi-constitutional formula that contains, irrespective of the structure, the three principal building entities, i. e. the basic polymer-forming constituent or chain, the crosslinking agent and the ionizable site. For instance, for a copolymeric styrene-divinylbenzene sulfonic acid cation exchanger the quasi-constitutional formula would be:

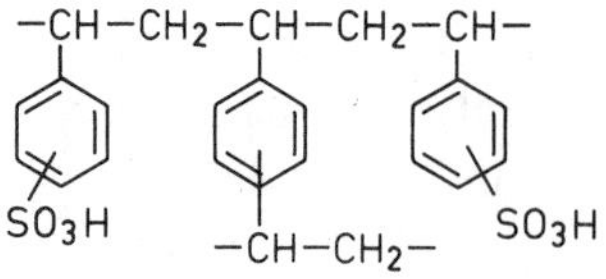

and for a styrene-divinylbenzene quaternary ammonium anion exchanger the quasi-constitutional formula would be:

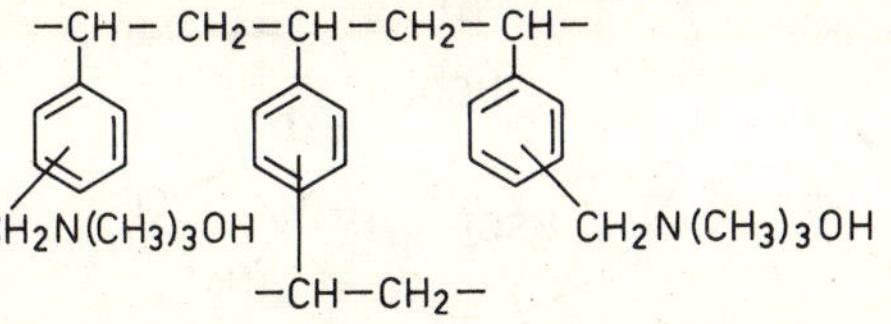

Matrices for synthetic resin ion exchangers are also formed from acrylic acid with divinylbenzene, and can be represented by the quasi-constitutional formula

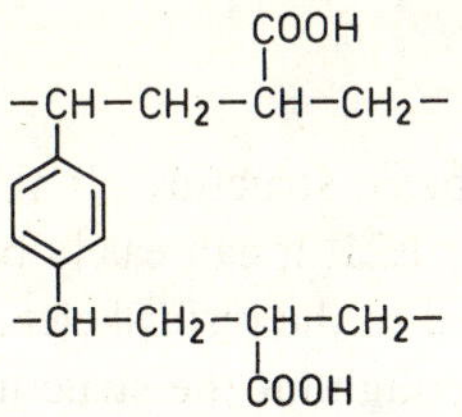

Such quasi-constitutional formula can be used advantageously for the representation of synthetic polymeric ion exchangers in order to depict their basic structural elements, and, above all, for writing down further reactions, as well as for more special ion exchange polymers.

A common depiction of the basic structure of synthetic ion exchange resins based on styrene and divinylbenzene with their anionic fixed groups in the case of cation exchangers or cationic fixed groups in the case of anion exchangers, is by figures showing schematically the charged matrix including the exchangeable ions in a circular outline to suggest the bead form of such ion exchangers, as in Figure 1.20. Such schematic drawings do not represent the true polymeric structure

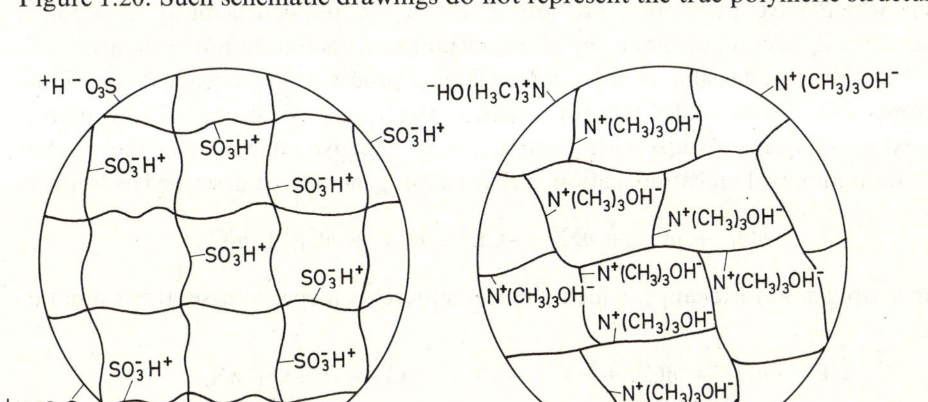

Figure 1.20 Schematic drawings of a strong acid cation exchanger (left) and a strong base anion exchanger (right) showing the charged matrix and the exchangeable ions.

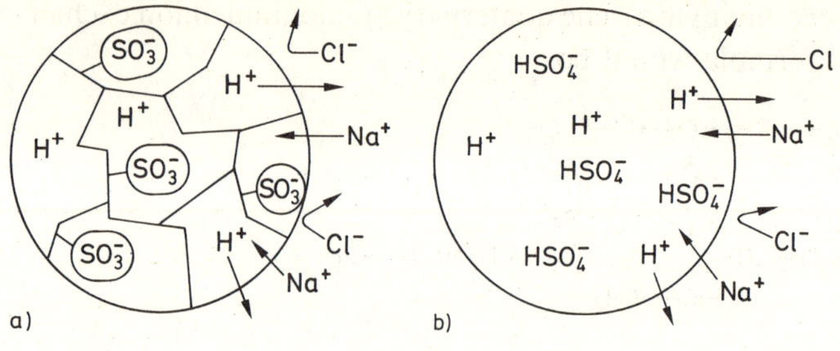

Figure 1.21 Comparison of cation exchange resin bead with sulfuric acid droplet. Left: schematic section through cation exchange resin bead; right: imaginary droplet of sulfuric acid surrounded by cation permeable membrane (Arden).

but are intended to give a visual impression of the basic structure. If the cation exchange resin is shown diagrammatically as in Figure 1.21, it can easily be understood that the sulfonic groups are immoveably attached to the resin skeleton, while the corresponding hydrogen ions are free to move throughout the structure. As a result, a sodium ion can enter freely, causing the ejection of a hydrogen ion, whereas a chloride ion approaching the surface of the bead is repelled by the fixed negative charges, and cannot enter. The effect is thus exactly as if the bead were a droplet of sulfuric acid solution, surrounded by a membrane through which cations, but not anions can pass (Arden).

Ion exchange processes require a mediating agent, generally water, in which the ions to be exchanged are dissolved. The ion exchanger in water contains water in its structure, so that when the exchange process is initiated, the solute ions penetrate the structure and immediately begin to exchange places, so to speak, with the original ions and, since electroneutrality must be maintained, join the co-existent ions which have penetrated previously. From the physicochemical standpoint, therefore, diffusion appears to be an important process during ion exchange.

It is therefore feasible to write ion exchange processes like chemical equations. If one uses the symbol CE^{n-} for a cation exchanger consisting of any matrix carrying ionogenic groups with a number n of negative charges, and the symbol C_1^+ for a monovalent cation, cation exchange can generally be described as follows:

$$CE^{n-} \cdot nC_1^+ + nC_2^+ + nX^- \rightarrow CE^{n-} \cdot nC_2^+ + nC_1^+ + nX^-;$$

for hydrogen ion exchange, which already represents a special case, this simplifies to:

$$CE^{n-} \cdot nH^+ + nC^+ + nX^- \rightarrow CE^{n-} \cdot nC^+ + nH^+ + nX^-,$$

becoming a process analogous to neutralization when the cation intended for the exchange is used in the hydroxide form, i. e., as a base:

$$CE^{n-} \cdot nH^+ + nC^+ + nOH^- \rightarrow CE^{n-} \cdot nC^+ + nH_2O.$$

Using the symbol AE^{m+} for an anion exchanger in the same manner, one obtains the general formula for anion exchange:

$$AE^{m+} \cdot mA_1^- + mA_2^- + mX^+ \rightarrow AE^{m+} \cdot mA_2^- + mA_1^- + mX^+;$$

for hydroxyl exchange, which also represents a special case:

$$AE^{m+} \cdot mOH^- + mA^- + mX^+ \rightarrow AE^{m+} \cdot mA^- + mOH^- + mX^+$$

and again, for the process analogous to neutralization:

$$AE^{m+} \cdot mOH^- + mA^- + mH^+ \rightarrow AE^{m+} \cdot mA^- + mH_2O.$$

As it will be seen, these fundamental formulas can be applied to all ion exchange processes and can be useful as a generalizing scheme when certain processes are to be interpreted as ion exchange events.

Nomenclature. The nomenclature on ion exchange and ion exchangers contains certain fundamental terms [100]. Those related to fundamentals are given below with their definitions, whereas other terms will be explained as the need arises. The following definitions are recommended (IUPAC):

Resin matrix:	The molecular network of an ion-exchange resin that carries the ionogenic groups.
Fixed ions:	In an ion exchanger, the non-exchangeable ions that have a charge opposite to that of the counter-ions.
Counter-ions:	In an ion exchanger, the mobile exchangeable ions.
Ionogenic groups:	In an ion exchanger, the fixed groupings that are either ionized or capable of dissociation into fixed ions and mobile counter-ions.
Co-ions:	In an ion exchanger, mobile ionic species with a charge of the same sign as the fixed ions.
Cation exchanger:	An ion exchanger with cations as counter-ions. The term cation-exchange resin may be used in the case of solid organic polymers.
Anion exchanger:	An ion exchanger with anions as counter-ions. The term anion-exchange resin may be used in the case of solid organic polymers.
Cation exchange:	Process of exchanging cations between a solution and a cation exchanger.
Anion exchange:	Process of exchanging anions between a solution and an anion exchanger.
Acid form of cation exchanger:	The ionic form of a cation exchanger in which the counter-ions are hydrogen ions (H-form) or the ionogenic groups have added a proton, forming an undissociated acid.
Base form of anion exchanger:	The ionic form of an anion exchanger in which the counter-ions are hydroxide ions (OH-form) or the ionogenic groups form an uncharged base, e. g. $-NH_2$.

Salt form of ion exchanger:	The ionic form of an ion exchanger in which the counter-ions are neither hydrogen nor hydroxide ions. When only one valence is possible for the counter-ion, or its exact form or charge is not known, the symbol or the name of the counter-ion without charge is used, e. g. sodium form (Na-form) tetra-methylammonium form, orthophosphate form. When one of two or more possible forms is exclusively present, the oxidation state may be indicated by Roman numerals, e. g. Fe(II)-form, Fe(III)-form.
Monofunctional ion exchanger:	An ion exchanger containing only one type of ionogenic group.
Bifunctional ion exchanger:	An ion exchanger containing two types of ionogenic group.
Polyfunctional ion exchanger:	An ion exchanger containing more than one type of ionogenic group.

In selecting terms used in the experimental and discussion sections of publications it is requested that one pays attention to these IUPAC recommendations on nomenclature (Inczedy).

Equilibria. If a system set up in the manner described above is left to itself an equilibrium forms, as in all chemical reactions; for the general case of an ion exchanger IE with the counterion C_1 and a solute ion C_2 for a binary exchange system, an equation can be written in the usual way for the reaction between the ion exchanger and the ionic solute:

$$IE \cdot C_1 + C_2 \rightleftharpoons IE \cdot C_2 + C_1.$$

For the further treatment of this equation a full generalization for all types of ion exchangers is unfortunately not possible, since the gel type ion exchange resins swell whereas the socalled macroreticular resins swell only little and zeolites practically not at all; clays and clay minerals swell only in one dimension and inorganic ion exchanger gels behave, depending on their structure, similarly to macroreticular or to common ion exchange polymers. All further statements are therefore related to the common gel type ion exchange resins and their equilibria with respect to swelling, ion exchange and sorption. In other words, when an ion exchange resin is immersed in a solution containing an electrolyte some or all of several processes may occur. Firstly, the resin swells by imbibing solvent from the external phase. Secondly, especially when an exchanger phase with one type of counterion and an external solution with a second but different counterion exist, an ion exchange process usually takes place. Thirdly, some electrolyte penetrates the ion exchanger gel to an extent which can no longer be explained by the underlying stoichiometry of ion exchange processes. All three processes can occur at the same time and obviously influence the total distribution of freely diffusible species in the system.

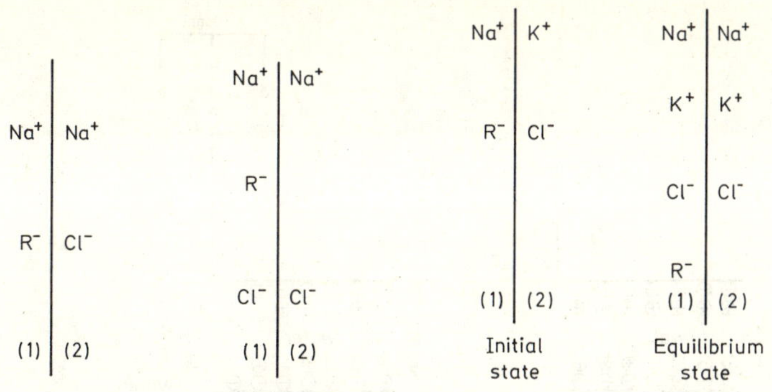

Figure 1.22 Donnan membrane equilibria.

Because the Donnan membrane equilibrium has often been applied to both molecular as well as theoretical models of ion exchange, this will be briefly described [101]. Donnan made use of the simple model shown in Figure 1.22, left, in which the vertical line indicates that the salt NaR is located on one side of the membrane which is impermeable to the anion R^-, while NaCl is on the other side. NaCl will then diffuse from (2) to (1), so that an equilibrium state is finally reached as shown in Figure 1.22, second left. This picture can also be extended to the case of an electrolyte with no common ion, in which only NaR is present on one side of the membrane and only KCl on the other (Figure 1.22, second right and right). KCl will then diffuse through the membrane in one direction, but NaCl will also permeate the membrane in the opposite direction. An equilibrium state results, which can be evaluated qualitatively as follows: the anion R^-, which cannot diffuse through the membrane, apparently attracts the cation of a second, quite different, electrolyte whose anion, however, apparently diffuses through the membrane to the same degree as R^- remains behind. This model can also be applied to ion exchange, if the interface between the exchange resin and the aqueous phase is seen as a membrane and the resin or the ion fixed on it is taken as the indiffusible component on one side. The counterion of the resin matrix is freely mobile. Therefore, if the exchanger is surrounded by an electrolyte solution containing a common or also not a common ion, an equilibrium will form according to the Donnan concept, which can therefore be treated according to the principles of membrane theory. For the case in which the membrane is permeable only to the solvent, the well-known theory of osmotic equilibrium, as well as the derivation of the osmotic pressure, result from the membrane theory. In ion exchange the capillary or swelling pressure plays the role of osmotic pressure, bringing about an equilibrium between the two phases with respect to the entire system.

The development of a Donnan potential at the interface of an exchanger resin and a solution can diagrammatically be illustrated as shown in Figure 1.23. Regarding a hypothetical pore of a cation exchanger before immersion on the left and after immersion in water on the right, the latter demonstrates that as the exchanger

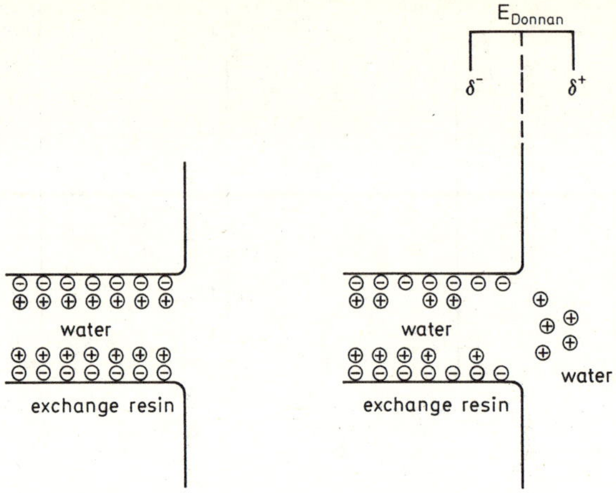

Figure 1.23 Development of a Donnan potential at an ion exchange resin and solution interface. Left: schematic pore of a cation exchanger in the dry state; right: the system after immersion in water.

comes to an equilibrium with a solution some diffusion of ions from the exchanger into the solution occurs, but only to a very small degree since only counterions are available for diffusion. Each counterion which does leave the exchanger phase leaves behind an uncompensated charge on the matrix of the exchanger. The ions which do diffuse out into the solvent remain near the surface of the resin in a diffuse double layer. A separation of charges takes place in this process and sets up an electrical potential across the exchanger and solution interface. This is the Donnan membrane potential. This situation makes it more difficult for the counterions to move because they would now have to do an increased amount of electrical work in wandering. But, in fact, very few ions are sufficient to set up quite substantial voltages. Moreover, the developed potential tends to return the counterions of the diffuse double layer to the fixed sites bound to the matrix. With respect to the capacity of the ion exchange resin the ions of this double layer never form a significant proportion so that it can still be considered that electroneutrality is maintained within the exchanger. But these few ions and the Donnan potential developed in their moving away are of the utmost importance for an understanding of all ion exchange processes. This is because what is called the ion exchange process has already begun when the counterions of the double layer are in dynamic equilibrium with those of the exchanger phase. Then foreign counterions introduced into the external solvent mix with and displace the original counterions of the diffuse layer and thus occupy sites within the exchanger. The presence of the Donnan potential at the resin and solution interface also explains naturally the fact that ion exchangers are permselective to their counterions, meaning that they virtually exclude all co-ions. This will be referred to in several contexts as the Donnan exclusion effect.

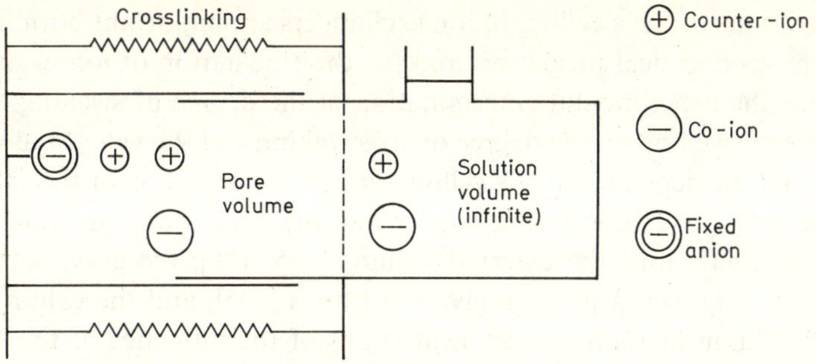

Figure 1.24 Gregor model for an ion exchange resin.

Further, for the phenomenological understanding of ion exchange, and because this approach has had a pronounced influence upon subsequent studies, the Gregor model of ion exchangers will be described [102]. As shown in Figure 1.24, in the matrix of the ion exchanger the crosslinking can be represented by elastic springs. The chemical components of this system are the matrix with the solvated fixed anion for the case of a cation exchanger, the solvated permeant ions, and the free solvent. The exchange resin is supposed to behave like a salt which becomes hydrated as the resin imbibes water and swells. During swelling the crosslinked network of the matrix exerts a pressure on the liquid inside the polymer structure, which is interpreted as an internal swelling pressure acting on the liquid within the resin. The counterions, on the other hand, dissociate, forming a concentrated electrolyte solution within the resin, thus causing by osmotic activity further amounts of water to enter the resin phase, which therefore continues to swell. The swelling of the resin is accompanied by a stretching of the crosslinked matrix or, following Gregor's model, an extension of the elastic springs. In Figure 1.24 it is further shown that co-ions can also penetrate the ion exchanger resins together with corresponding surplus counterions. On the basis of this model, whereby newly developed theoretical aspects have been disregarded, several phenomena of ion exchange and its equilibria can still be profitably discussed.

Swelling equilibria. The swelling of an ion exchanger gel after immersion into a solvent – in most cases water – is phenomenologically postulated as being, in principle, an osmotic process. In the literature this is referred to as the osmotic theory of the concept of osmotic pressure in the swelling of ion exchange resins. This theory postulates that the solvent retention is a reproducible, equilibrium quantity, which can be explained as the effort towards dilution made by the many existing fixed ions and counterions in the network or the tendency of these ionic groups to become hydrated. Water swelling must therefore increase with an increase in the number of ionogenic groups, up to the limits imposed by the crosslinking. The swelling equilibrium is reached when the tension of the network equals the osmotic pressure difference between the interior of the ion exchanger gel and the surrounding solution.

Investigations of the swelling of ion exchangers are important both for the study of their physicochemical properties and for the elucidation of ion exchange mechanisms. For the experimental determination of the degree of swelling, e. g. for the dependence of swelling on the degree of crosslinking and the valence of the counter-ion [103], and the dependence of swelling on the concentration of the solution [104], several methods have been found convenient and useful, among which the picnometric determination of the external volume [105, 106], the gravimetric measurements of the weight of a gel in a given salt form [105], and the calibrated tube or the sealed column method for determinations of the influence of the surrounding liquid on the bed volume of the gel [107] should be mentioned. But the most useful and quantitative characteristic of the interaction of an ion exchange resin with a solvent is given by the isopiestic determination of vapor sorption isotherms [108]. These so-called isopiestic lines are measured in the vapor pressure range from negligibly low values up to the vapor pressure of the pure solvent. In Figure 1.25 isotherms for different degrees of crosslinking are given for the H-form of sulfonated polystyrene resins, showing that the influence of crosslinkings is greater for $X > 0.6$. Isopiestic lines are in general of great importance for the evaluation of thermodynamic parameters that are usually based on the equilibrium pressure [109]. It can be said that swelling pressure and resin hydration studies have been a significant milestone in the theoretical treatment of ion exchange resins, for instance for the determination of the enthalpy of swelling [106, 110, 111]. It should also be mentioned that the application of the thermodynamic method to resins has been developed, at least to some extent, in analogy with the thermodynamics of the hydration of natural macromolecular substances [112, 113].

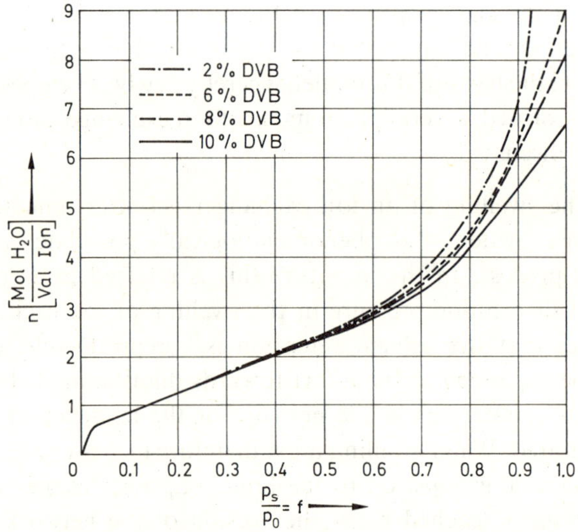

Figure 1.25 Water vapor sorption isotherms of ion exchange resins in the H-form (Dickel et al., 1959).

Mainly from the results of the studies of Gregor and Pepper and their coworkers, the influence of various factors on the swelling properties and equilibria have led to certain general conclusions. For polar solvents it can be concluded:

1. Swelling equilibria of gel type ion exchangers depend on the nature of the ionogenic site. Highly dissociated strong acid cation exchangers with the $-SO_3H$ group – as compared with the weak acid $-COOH$ group with the same degree of crosslinkage – swell more because of a higher number of osmotic active ions.
2. Swelling is greatly influenced by the nature of the counterions. Resin volumes change when the resins are converted to different ionic forms, with a sequence for monovalent counterions for the sulfonic acid type cation exchanger generally being $H^+ > Li^+ > Na^+ > K^+ > Cs^+ > Ag^+$; for polycarboxylic gels, one has, $Li^+ > Na^+ > K^+ > H^+$ (with the possibility of inversions).
3. The effect of counterions of higher valency on the swelling can even be negative, i. e. some ion exchangers can shrink when they are loaded with divalent ions. In general the swelling is lower with counterions of higher valence because of a lower number of osmotic active ions being present inside the exchanger. With polyvalent ions swelling is reduced usually according to the sequence $Na^+ > Ca^{2+} > Al^{3+}$.
4. Swelling equilibria of gel type resins depend largely on the degree of crosslinking, because higher crosslinking means a more rigid structure. Reasons for the decrease in swelling at higher degrees of crosslinking are: increase in the swelling pressure, changes in the number of bound water molecules, the difficulty of interpreting external volume and thermodynamic functions with the swelling pressure as variable.
5. Swelling is favored by an increase in the number of ionogenic groups (capacity) because of a higher ion concentration in the exchanger.
6. The external solution can determine the swelling with varying concentrations. Swelling is decreased by an increase in concentration of the external solution since this decreases the difference in the osmotic activity between the two phases.
7. Swelling of resins that are not immersed in water is dependent on the relative humidity, especially as 100% relative humidity is reached. Water sorption from solutions is always higher than swelling in water vapor.
8. The increase in resin swelling with increase in temperature is not great, even in the range of several tens of degrees, but is significant in certain operations, this being one of the reasons why thermodynamic treatments of swelling equilibria are difficult to make.

The mechanism of the uptake of water by ion exchangers of the gel type can be elucidated on the basis of swelling equilibria changes with regard to the exchange of different counterions. Interrelations between the degree of crosslinking and the hydration water of the mobile ions can be deduced from the Gregor model, leading to the conclusion that the higher the degree of crosslinking of the resin and, hence, the less the swollen volume, the closer will the mobile cations be forced towards the anionic groups fixed to the resin inducing a stronger interaction between them. By this interaction water molecules are displaced from the hydration shell of the counter-cation and its degree of hydration is reduced. The hydration number is not constant, but upper and lower limits do result [114]. For the salt forms of sulfonated resins the hydration of ion exchange resins is, on the other hand, regarded as a stepwise adsorption process. This is deduced from thermodynamic functions with the conclusion that the first water molecule is exclusively sorbed by an SO_3^--group fixed on the matrix, while the second and subsequent hydration shells are formed as a result of the hydration of the cation [115]. Analysis of the temperature variation of the water vapor adsorption isotherms has led also to the conclusion that, if there is one molecule per fixed group, then its interaction with the counterion and also the interaction of the counterion with the water molecule, is not a linear function of the ionic radius alone, i. e. is not purely coulombic, but factors such as polarisation may play an important role. The discussions along these lines have not been terminated yet [106, 111]. The mechanism of hydration has also been studied by physical methods. From the infrared spectra of thin films of sulfonated polystyrene it was earlier concluded that the hydration of the salt forms of sulfonated resins begins with the interaction of water with the counterion [116]. There are results of later years which would seem to contradict these findings but they are based on different substrates: they show that ion pairs and not individual ions are hydrated in sulfonated polymer films. If different explanations of the mechanism of primary hydration have to be applied it may well be that different mechanisms do exist for different ion exchanging polymers. NMR studies for the characterization of moisture absorption by hydrogen forms of Dowex 50 resins with different crosslinking show that water molecules in the resin phase are in a state of rapid if restricted motion. Water molecules in several different states in the resin would then have to be expected at low moisture content, with an exchange between the protons of the water within the resin and outside it. But the NMR method enables the heterogeneity of resins to be quantitatively determined.

From experimental results elucidating various relationships and from mechanisms obtained either from analysing differential thermodynamic functions or from interpreting the more rigorous investigations by a wide variety of physical methods, diagrammatic representations can be figured out, but this can easily lead to misconceptions due to oversimplification. It seems better therefore to refer only to the theoretical models which have been proposed and on which quantitative treatments of the swelling equilibria and other interrelated properties of ion exchangers have been based. The model proposed by Gregor was described above. This so-called osmotic theory was also applied by others [115], and was in part the subject of intense discussion over the following years [109, 117]. These may all be considered as mechanical models in contrast with molecular models as treated by Rice and Harris [118] and Katchalsky and coworkers [119]. The latter can be considered as extensions of the early polyelectrolyte theories where the resin gel is described as a

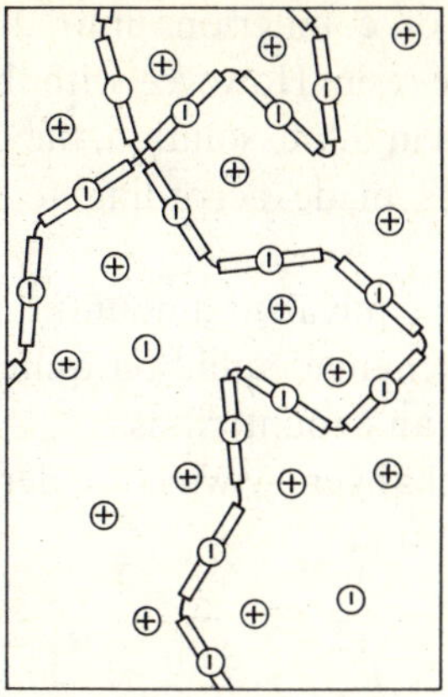

Figure 1.26 The Katchalsky molecular model.

network of rigid segments, each of them carrying only one ionic site, as shown in Figure 1.26. From thermodynamic considerations the necessary elasticity of the resin matrix for swelling is reflected by entropy changes, and the conformational entropy contribution is calculated in detail from the statistical theory of nonpolar gels (Flory). Due to their molecular nature electrostatic interactions are introduced in both models, in the Rice and Harris model for moderately and highly crosslinked resins as interactions between sites over the whole gel volume if no ion pairs have been formed, and in the Katchalsky model for slightly crosslinked polyelectrolyte gels as interactions between neighboring sites of the same chain. Both models require the introduction of empirical parameters, making it difficult to predict correctly and quantitatively theoretical values, which must be evaluated by experimental results (Rinaudo).

Ion exchange equilibria. Since ion exchange processes are stoichiometric and reversible, whenever an ion exchanger is brought into contact with an aqueous solution containing a counterion different from that initially bound to the resin, an exchange of ions occurs until equilibrium is reached, as shown in the previously given general equations. To describe this exchange process quantitatively well-known physical-chemical formulations can be applied, which fall essentially into two categories. One considers the exchange process as an adsorption phenomenon, the other classifies it as the interaction of coulombic forces in electrolytes, such as described either by the Donnan equilibrium theory or the law of mass action. Although it may be justifiable to treat an ion exchange process under operational aspects as an adsorption process using the Freundlich or Langmuir isotherms, this seems strange for ionic interactions concerning ion exchangers, and may be left to discussions in the literature (Cassidy [120] and Morris and Morris [121]). The Donnan membrane theory, with the Donnan exclusion effect, explains how co-ions are prevented from penetrating the interior of a high capacity ion exchanger and simultaneously force an equivalent quantity of counterions into the exchanger whenever the first counterions wander out of the resin. However, with the exceptions of ion exchange reactions of low capacity resins in dilute solution, the discrepancies between experimental results and calculations made according to a simplified Donnan membrane theory [122] are too great.

However, because in ion exchange reactions equivalent amounts of ions participate, the law of mass action describes ion exchange equilibria quite adequately. Given the ion exchange resin R with fixed ions and counterions A^+ – for instance in the case of a strong sulfonic acid cation exchanger – which undergo exchange with the counterions B^+, according to the equation

$$R \cdot A^+ + B^+ \rightleftharpoons R \cdot B^+ + A^+,$$

then, by application of the law of mass action, and taking into consideration the fact that the activities of ions in the resin phase cannot be evaluated accurately, an apparent exchange constant K_c is obtained

$$K_c = \frac{[R \cdot B^+][A^+]}{[R \cdot A^+][B^+]}$$

As this is not a true thermodynamic equilibrium constant it is called the equilibrium quotient. For practical purposes, and in order to measure the values analytically, the activity coefficients of the ions in both the resin and aqueous phase are usually ignored and the concentrations determined are used uncorrected. In the equation the brackets represent the concentrations of the reactants as expressed in some convenient units. Normality is often the unit chosen for the aqueous phase, while the concentration in the resin phase is usually expressed as equivalents of counterion per unit weight of exchanger

This empirical equilibrium quotient depends, however, on the properties of the ion exchanger and the nature and concentration of the electrolytes; as a result different empirical equilibrium quotients are obtained for different ions and this has led to the introduction of the concept of selectivity.

One quite often presented derivation of a thermodynamic ion exchange equilibrium constant and a selectivity relation is made in the following way. Based on the above equilibrium equation and considering that co-ions have no effect on this equilibrium – unless specifically complexing or, indirectly, on the activity coefficient in the solution – and therefore may be ignored, a thermodynamic equilibrium constant, K_{th}, may be written for the ion exchange

$$K_{th} = \frac{a_{RB} \cdot a_A}{a_{RA} \cdot a_B} = \frac{[RB] \cdot [A]}{[RA] \cdot [B]} \cdot \frac{\gamma_{RB} \cdot \gamma_A}{\gamma_{RA} \cdot \gamma_B};$$

this is generalized for the case where the exchanging ions have valencies z_A and z_B, to the thermodynamic equilibrium constant

$$K_{th} = \frac{(a_{RB})^{z_A} \cdot (a_A)^{z_B}}{(a_{RA})^{z_B} (a_B)^{z_A}},$$

where a represents the activity of the species involved and is, in turn, equal to the concentration of that species multiplied by an activity coefficient such that for a species i

$$a = [i] \cdot \gamma_i.$$

This extension gives the thermodynamic equilibrium constant as the usual product of two terms, the first being a concentration and the second an activity coefficient quotient. But in experimental practice use is mainly made of the concentration term only, as already mentioned above, which is then termed the selectivity coefficient because of the reasons outlined above as well. The selectivity coefficient is often given the symbol K_A^B, since it measures the tendency of the exchanger to prefer B over A. This means that

$$K_A^B = \frac{[RB] \cdot [A]}{[RA] \cdot [B]};$$

if K_A^B is greater than unity, the exchanger selects ion B, if it is less than unity, it selects ion A, and if it is equal to unity then the exchanger shows no preference for either ion.

Since the first investigations of ion exchange phenomena it has been shown that, in general, and in dilute solutions of electrolytes (< 0.1 N mol/l), an ion exchanger or an ion exchange resin will show a preference for one ion over another, that is, that the affinities of ions towards an ion exchanger are not equal. This phenomenon of preference is called selectivity of the ion exchanger for the counter-ion and is characterized by the selectivity coefficient, $k_{A/B}$. The IUPAC Recommendations on Ion Exchange Nomenclature define:

Selectivity coefficient, $k_{A/B}$: Equilibrium coefficient obtained by formal application of mass action law to ion exchange, characterizing quantitatively the relative ability of an ion exchanger to select one of two ions present in the same solution.

Exchange $[Mg^{2+} - Ca^{2+}]$:

$$k_{Mg/Ca} = \frac{[\overline{Mg}]\,[Ca]}{[Mg]\,[\overline{Ca}]}$$

Exchange $[SO_4^{2-} - Cl^-]$:

$$k_{SO_4/Cl} = \frac{[SO_4]_r\,[Cl]^2}{[SO_4]\,[Cl]_r^2}$$

Over-bars or subscript letters, "r", are used to designate concentrations in the ion exchanger. For exchanges involving counter-ions differing in their charges, the numerical value of

$k_{A/B}$ depends on the choice of the concentration scales in the ion exchanger and the solution (molal scale, molar scale, mole fraction scale, etc). Concentration units must be clearly stated in exchange of ions of differing charges.

Corrected selectivity coefficient, $k_{A/B}$: Concentrations of external solution in $k_{A/B}$ are replaced by activities.

For any given series of ions the selectivities can be arranged in a relative order and selectivity sequences can be established. For counterions of the same valence that do not form complexes, the affinities are ruled by electrostatic reciprocal effects with the fixed ion, by disturbance of the solvent structure in the exchanger and by changes in the swelling. For a general purpose cation exchanger and aqueous solutions usually the following selectivity sequences are valid for the most common cations:

$Tl^+ > Ag^+ \gg Cs^+ > Rb^+ > K^+ > NH_4^+ > Na^+ > H^+ > Li^+$
$Ba^{2+} > Pb^{2+} > Sr^{2+} Ca^{2+} > Ni^{2+} > Cd^{2+} > Cu^{2+} > Co^{2+} > Zn^{2+} > Mg^{2+} > UO_2^{2+}$;

likewise for anions with respect to a general purpose anion exchanger:

Citrate $> SO_4^{2-} > NO_3^- > I^- > B_r^- > Cl^- >$ formate $>$ acetate $> F^-$.

If measurements are to be made to investigate the ion exchange behavior of cations or anions it is common practice to determine the concentration distribution ratio D_c, the distribution coefficient D_g or the volume distribution coefficient D_v for which in the IUPAC Recommendations on Ion Exchange Nomenclature the following definitions are given:

Concentration distribution ratio D_c: The ratio of the total (analytical) concentration of a solute in the ion exchanger to its analytical concentration in the external solution. The concentrations are calculated per cm^3 of the swollen ion exchanger and cm^3 of the external solution.

Distribution coefficient D_g: The ratio of the total (analytical) amount of solute per gramme of dry ion exchanger to its concentration (total amount per cm^3) in the external solution.

Volume distribution coefficient D_v: The ratio of the total (analytical) amount of a solute in the ion exchanger calculated per cm^3 of column or bed volume to its concentration (total amount per cm^3) in the external solution. ($D_v = D_g \rho$, where ρ is the bed density, g of dry resin per cm^3 bed.) This quantity is most conveniently determined from column experiments and it is recommended to use the D_v values in describing the results from chromatographic separations.

The employment, especially of D_g,

$$D_g = \frac{\text{concentration of solute in resin}}{\text{concentration of solute in liquid}}$$

is, in practice, much easier than for instance trying to obtain the separability from selectivity sequences; from D_g it is, e. g., much more convenient to predict the column behavior of an ion, for any given set of conditions. The distribution coefficients of a large number of inorganic and organic ions have been measured over wide ranges of conditions and are available from the literature [123].

The ratio of the distribution coefficients for two different ions is then the separation factor, which is defined as

Separation factor $\alpha_{A/B}$: $\alpha_{A/B} = D_A/D_B$	Ratio between the disbribution coefficients of solutes A and B in a specified medium at a specified temperature. In exchange of counter-ions of equal charge the separation factor is equal to the selectivity coefficient provided that only one type of ion represents the analytical concentration (e. g., in exchanges of K^+ and Na^+, but not in systems where several individual species are included in the analytical concentrations);

this can also be written as

$$\alpha_{A/B} = \frac{\bar{C}_A \cdot C_B}{\bar{C}_B \cdot C_A}.$$

The separation factor, as the quotient of the distribution coefficients, is for practical purposes the most useful value. In general, if $\alpha_{A/B} > 1$ the ions A are more selectively taken up than the ions B by the ion exchanger, and the easier it will be to separate them on a column of ion exchange resin. In addition – and under the important restriction that the selectivity coefficient is different from the separation factor only in the case of the uptake of ions of different valence – from analogous considerations the various methods for the regeneration of ion exchangers can be dealt with [124]. In Table 1.4 average separation factors are given for a commercial polystyrene sulfonic acid ion exchange resin of various crosslinking in equilibrium with diluted chloride solutions.

Further fundamental facts are of interest with respect to ion exchange equilibria. In conclusion, it should be remarked that, as a rule, some properties of a counter-ion favor a particular uptake over others. These are, higher valence (especially in dilute solutions), smaller solvatized volume, greater polarizability, stronger complex formation with the fixed ion and weaker complex formation with co-ions. If a counterion has the tendency of being more easily associated with the fixed ion, then the exchange quilibrium is altered in its favor. Based on this principle (Le Chetalier) weak acid exchangers prefer, in contrast to strong acid exchangers, hydrogen ions as against almost all monovalent metal cations. This applies also to

Table 1.4 Mean separation factors of customary styrene sulfonic acid resins of different crosslinkages in equilibrium with dilute chloride solutions

Counterions	Crosslinking (% DVB)		
	4	8	16
Li^+/H^+	0.75	0.78	0.70
Na^+/H^+	1.2	1.5	1.6
NH_4^+/H^+	1.4	2.0	2.3
K^+/H^+	1.7	2.3	3.0
Cs^+/H^+	2.0	2.6	3.2
Ag^+/H^+	3.6	6.7	15.5
Zn^{2+}/Mg^{2+}	1.1	1.1	1.1
Co^{2+}/Mg^{2+}	1.1	1.1	1.1
Cu^{2+}/Mg^{2+}	1.1	1.2	1.3
Ni^{2+}/Mg^{2+}	1.2	1.3	1.3
Ca^{2+}/Mg^{2+}	1.4	1.6	2.1
Ba^{2+}/Mg^{2+}	2.5	3.5	5.9

chelating resins, as long as the reactive groups are weak acids; the metal ion can easily be exchanged against hydrogen ions, with the complex formation being very selective and very strong in neutral or alkaline solutions. On the other hand, the exchange quilibrium is altered unfavorably for the uptake of a counterion if it is removed either by association, complex formation or precipitation. Thus it is possible, due to the binding of $H^+ + OH^- \rightarrow H_2O$, to convert a weak acid exchanger with NaOH from the H^+-form into the Na^+-form, which is only partially feasible with NaCl. The addition of complex-forming agents to a solution with metal ions leads to complex-ion formation with selectivities not achieved otherwise; the most remarkable example for this is the separation of the rare earth and other fission products in spectroscopic purities, for which complex-ion formation and control of pH was the key to sharp differentiation among the rare earths by ion exchange processes [125]. The technique of changing the charge of a cation by complex-ion formation, thus modifying its attraction as an anion complex to an anion exchanger, e. g. $AuCl_4^-$, is of use in that this may happen to a varying degree for each species competing for the complexing reagent. Changes in the dielectric constant, the solvation and the structure of the solvent can deeply influence the selectivity on going over to nonaqueous or mixed-aqueous systems, which are used especially for analytical purposes; this is unpractical in technical applications because exchange velocities are too low. It must finally be mentioned that the thermodynamic equilibrium constant including both the activity coefficients and the correction factors is difficult to evaluate and is therefore treated in the chapter on the theory of ion exchange.

Apart from the application of mathematical relations for the quantitative description of ion exchange, a commonly used method of investigation of an estab-

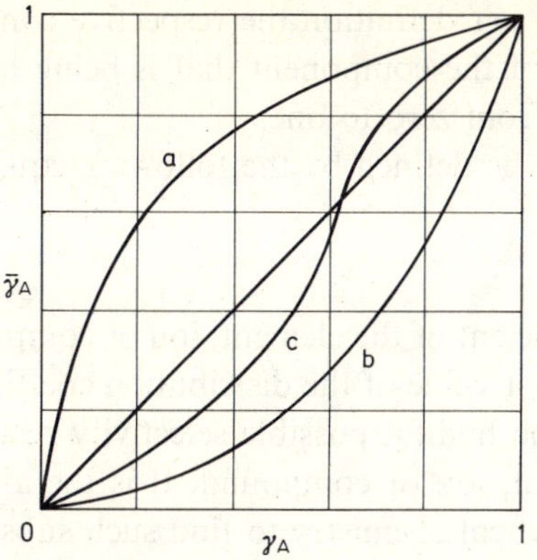

Figure 1.27 Ion exchange isotherms. $\bar{\gamma}_A$ and γ are the relative exchange concentrations in resin phase and solution, respectively; c shows a sigmoidal characteristic of the ion exchanger.

lished ion exchange equilibrium is to determine the concentrations of a solute in the solution and in the resin and to construct what is called an ion exchange isotherm. An ion exchange isotherm is defined as the concentration of a counter-ion in the ion exchanger expressed as a function of its concentration in the external solution under specified conditions and at constant temperature (IUPAC). For plotting it, the equivalent in part $\bar{\gamma}_A$ of the ion A in the exchanger is plotted as a function of the equivalent γ_A of the ion A in the solution, as shown in Figure 1.27. The numerical value of the equivalent is independent of whether the concentration is given molaly, molarly or in mole fractions. Figure 1.27 shows the three curves possible depending on various selectivities, i. e. a straight diagonal line in this so-called quadratic equilibrium representation if the state of equilibrium was reached without any selectivity, and non-linear curves for the possibilities of selectivity of one counterion over another. With respect to eventual connections between ion exchange and sorption, a sorption isotherm has been defined: it is the concentration of a sorbed species in the ion exchanger expressed as a function of its concentration in the external solution under specified conditions and at constant temperature (IUPAC). In accordance with the different forces that may become effective in relation to sorption processes, and which frequently appear side by side, the exchange plus superimposed sorption isotherms are (except for simple ions) often quite complicated, resembling complex equilibria. In practical studies much can be read from sorption isotherms, and sometimes even more from the subsequent desorption curves. Isotherms are also called either favorable, unfavorable or linear. Favorable isotherms, a in Figure 1.27, favor the uptake of the ionic species; unfavorable isotherms, b in Figure 1.27, do not favor uptake over the full range of concentration. The linear isotherm is obtained for $\bar{\gamma}_A = \gamma_A$. In addition there are also so-called sigmoidal isotherms, representing a selectivity reversal. Viewed practically, the sigmoidal isotherm is in general very undesirable. For an understanding

Figure 1.27 it must be mentioned that per definition the respective concentrations $\bar{y}_A$ and y_B are relative concentrations of the component that is being taken up by the exchanger, which therefore range from zero to one.

According to Lieser selectivity may be defined by the following equation

$$S = \log K_d\,(1) - \log K_d\,(\text{others}),$$

in which $K_d(1)$ is the distribution coefficient of the element, ion or compound under consideration and K_d(others) the highest value of the distribution coefficients of all other species that may be present. The highest possible selectivity results when a substance reacts with only one element, ion or compound. It is the aim of many scientists working in the field of analytical chemistry to find such substances. But substances of high selectivity are also very important in radiochemistry and nuclear or environmental technology where special radionuclides, long-lived fission products or trace amounts of elements or compounds are to be separated from other elements or compounds. High selectivity is not restricted to certain classes of compounds. Inorganic compounds, organic compounds or resins, and special groups show high selectivity for certain elements, ions or compounds. Lieser and coworkers have applied this equation to many separation problems.

Sorption equilibria. Sorption is defined with respect to ion exchangers as the uptake of electrolytes or non-electrolytes by ion exchangers through mechanisms other than pure ion exchange. In order to make the terminology clearer one has, in other words, to distinguish between non-exchange electrolyte adsorption and the adsorption of non-electrolytes. It is further important to realize that the terms electrolyte and co-ion uptake by an ion exchange resin are synonymous.

Discussing first the equilibrium between a resin containing the fixed ion with counterion A and an aqueous solution of an electrolyte containing the same ion A and co-ion X, the usual depiction shows that in both the above cases, either by diffusion or as a result of other special conditions, with the least quantity of counterions A migrating out of the resin or the uptake of co-ions X, the latter must, because of electroneutrality, be accompanied by an equivalent number of counterions A. Consequently, and because the fixed ion of the resin cannot migrate from the resin phase to the aqueous, the particular type of equilibrium called a Donnan equilibrium arises. The Donnan equilibrium that can be reached is governed by the Donnan potential established during the above-described migration processes, which tends to pull in ions A and to repel co-ions X. Qualitatively seen, the Donnan potential at the interface is negative for cation exchangers and positive for anion exchangers. Important is that the Donnan potential is greatest, and therefore co-ions almost entirely excluded, when the concentration difference between the resin phase and the external solution is large. A high exchange capacity, high crosslinking, and increasing dilution of the external solution are, therefore, factors favoring the exclusion of electrolyte from the exchanger. The potential required to counteract the initial migration of the counterion is least for those of higher valence and the effect of electrolyte (or Donnan) exclusion is consequently lower. Further, a given Donnan potential excludes multivalent co-ions to a greater extent than monovalent ions. An increase in the external electrolyte concentration is favorable for the penetration of electrolyte into the exchanger, but this is also accompanied by a shrinking of the resin due to a lowering of the water activity in the external solution relative to the exchanger phase. An equilibrium is reached, i. e. the Donnan equilibrium, at which the electrochemical potential of each ionic species is the same in both phases, with electroneutrality prevailing everywhere. A number of models have been proposed to describe the Donnan non-exchange electrolyte sorption equilibrium quantitatively. Ignoring the activity coefficients and the swelling pressure, it is generally accepted that from equating the electrochemical potentials of the co-ion in both phases and using the thermodynamic relationships concerned, the following equation is obtained as a general statement for the existing Donnan equilibrium for an electrolyte AX dissociating into v ions

$$\frac{\bar{C}_A^{v_A}}{C_A^{v_A}} = \frac{\bar{C}_X^{v_X}}{C_X^{v_X}},$$

This equation can be discussed for various conditions. At high concentrations of counterions in the resin and low concentrations of the same species in solution (C_A, $C_X \ll \bar{C}_A$) the uptake of co-ions (C_X), and implicitly of electrolyte, into the exchanger is small. This phenomenon is known as Donnan exclusion. It is larger for higher valence of the co-ion (v_X) and lower valence of the counterion (v_A). As the concentration of the solution is increased the Donnan exclusion becomes less effective and conditions for the situation known as electrolyte invasion are reached. In other words, the distribution coefficient

$$K_d = \frac{\bar{C}_X}{C_X}$$

of the co-ion or of the electrolyte increases with increasing concentration of the solution [126]. Some other consequences of the Donnan exclusion are also of interest to ion exchange fundamentals. Since it is caused by electrostatic effects and disappears at lower dissociation of the electrolyte or the exchangeable groups, cation exchangers, e. g., adsorb weak acids well but their dissociated salts only a little. In a similar way, weak acid exchangers adsorb strong acids well but their salts hardly at all. In all these cases the sorption depends greatly on the pH of the solution. As the internal structure of the ion exchange resins is highly decisive for all ion exchange phenomena, it must be concluded that the Donnan theory is valid only for homogeneous exchangers and constant electrical potentials within the resin, whereas under certain practical conditions it is not strictly applicable. Thus counterion association in the resin, specific interactions between the co-ions and the fixed ionic groups, and complex-ion formation in the external solution can greatly influence the degree of non-exchange electrolyte sorption.

The sorption of non-electrolytes is best brought into the picture by drawing attention to the fact that ion exchangers in the dry state can already sorb a great variety of substances besides water. Gas sorption by zeolites for drying and to remove polar impurities from non-polar gases is one example. For all sorptions of non-electrolytes either from the gas phase or from a liquid phase or solution it is a prerequisite for a large amount of sorption that there be a strong interaction with the constituents of the resin, the hydrocarbon network, the fixed and counter-ions, and any solvent already present. The interaction with the matrix occurs through van der Waals forces. Interaction with the fixed ions may occur through solvation, although this is not the only mechanism feasible, since solutes may be sorbed from aqueous solutions, where hydration is preferred over solvation by the solute. Hydrogen bonds formed between certain kinds of solutes, like alcohols, with the water of hydration, may then be the major factor. Because of the fixed ions the ion exchanger represents a strong polar medium which means that polar molecules are generally taken up and non-polar ones largely rejected. The properties of the ion exchange resin, i. e. degree of crosslinking, swelling and concentration of water-preferring ions, the kind of counterions, the solvent and additives in a solvent system all influence the sorption and allow room to vary the sorption conditions. For weak acid and weak base exchangers the same pH-dependent effects can be encountered as discussed for the sorption of electrolytes.

Besides investigations of the mechanisms of sorption of non-electrolytes by ion exchangers quite a number of investigations have been made in order to elucidate the sorbing behavior of individual chemical substances, as well as of homologous series. As typical for the sorption of non-electrolytes it must be mentioned that Langmuir adsorption isotherms are usually determined at 20 °C and in dependence on the pH. The Langmuir adsorption isotherms are, in most cases, convex and permit the calculation of the maximum sorption capacity. Included in the investigations of the sorption properties were many studies of the distribution of organic solvents between ion exchange resins and aqueous solutions. Among the individual substances studied phenol is an outstanding example, with a distribution coefficient higher than unity in strongly basic anion exchangers. For several aliphatic acids and alcohols in strong acid cation exchangers a general trend of a decreasing distribution coefficient with increasing external organic concentration and increasing crosslinking has been demonstrated. Because of the lack of a general quantitative theory for the sorption of non-electrolytes by ion exchange resins phenomenological solutions are presented [127]. Otherwise strong adsorption effects are to be expected for solutes that can form complexes with the counterions. For instance, anion exchangers in the Br^- or I^--form strongly absorb bromine and iodine, probably forming Br_7^- and I_7^- adducts in a stoichiometric process, reaching saturation of the resin beyond a certain concentration of solute, which is shown by a flattening out of the sorption isotherm.

Ligand exchange. This standard method developed by Helfferich is a special sorption procedure based on cation exchangers with transition metals as counterions. In this process the complex water molecules of the metallic ion bound to a cation exchanger of the carboxylic acid type are exchanged for ammonia. The method can be extended, with the ammonia being then exchanged for diamine. In contrast to ion exchange in general, this method – which is in accordance with complex chemistry – leads to an exchange of ligands on the counterion while the latter is retained in the resin phase. Carboxylic resins are used for this purpose, because of their higher selectivity compared with sulfonic acid exchangers. Ligand exchange is for the preparation of salts, and is suitable as well for the separation of amines, amino acids, etc. [128].

A very valuable approach to the fundamental question of ion exchange and the law of mass action was presented by Bobleter, yielding an equilibrium determination of cation exchangers through distribution and activity coefficients. It has often been denied that equilibrium constants and distribution coefficients are related to the law of mass action. It can, however, be shown that these coefficients and constants can be deduced directly from the law of mass action, provided ion activities and true molar concentration values are considered for the liquid part of the ion exchange phase [129]. The following is a gratefully included direct contribution from O. Bobleter, and the reader should note that slightly deviating notations have been used.

The exchange of the ion A for B on the monovalent exchange site R can be written

$$z_B\,A^{z_A} + z_A\,BR_{z_B} \rightleftharpoons z_B\,AR_{z_A} + z_A\,B^{z_B} \qquad (1)$$

whereby z_A and z_B represent the valency state of the ions A and B. The activities of the ions A and B must be equal in the solution and the ion exchanger phase. Therefore

$$a_{AR_{z_A}} = a_A zA \quad \text{and} \quad a_{BR_{z_B}} = a_B zB.$$

This leads directly to the thermodynamic equilibrium coefficient, ${}^0K_B^A$, which by definition is equal to (1):

$${}^0K_B^A \equiv \frac{a_{AR}^{z_B} \cdot a_B^{z_A}}{a_A^{z_B} \cdot a_{BR}^{z_A}} \equiv \frac{C_{AR}^{z_B} \cdot C_B^{z_A}}{C_A^{z_B} \cdot C_{BR}^{z_A}} \cdot \frac{\gamma_{AR}^{z_B} \cdot \gamma_B^{z_A}}{\gamma_A^{z_B} \cdot \gamma_{BR}^{z_A}} \equiv 1 \qquad (2)$$

(For reasons of simplicity the lower indices z_A and z_B have been omitted). In the second part of equation (2), the activities are replaced by the product of the corresponding concentrations C and the activity coefficients (γ). The molar equilibrium coefficient K_B^A is therefore

$$K_B^A = \frac{C_{AR}^{z_B} \cdot C_B^{z_A}}{C_A^{z_B} \cdot C_{BR}^{z_A}} = \frac{\gamma_A^{z_B} \cdot \gamma_{BR}^{z_A}}{\gamma_B^{z_A} \cdot \gamma_{AR}^{z_B}} \qquad (3)$$

A further simplification has been introduced: the ion A is present in traces only and B is the macroelectrolyte.

Activity coefficients in the outer solution. The activity coefficient of the macroelectrolyte ion is equal to that of the compound (BX)

$$\gamma_{B^{z_B}} = \gamma_{BX} \qquad (4)$$

Only γ_{BX} can be determined experimentally. The activity coefficients of many electrolytes in dependence on their concentration are listed in the literature.

The evaluation of the trace ion activity coefficient is more complicated, owing to the fact that its activity is influenced by the activity of the macroelectrolyte ion. By the use of the Harned rule this influence can be determined:

$$\log \gamma_{AX(0)} = \log \gamma_{AX(1)} + \alpha_{AX} \cdot C_{BX} \qquad (5)$$

In this equation, $\gamma_{AX(0)}$ indicates the activity coefficients of the neutral compound AX at the trace concentration, and $\gamma_{AX(1)}$ that of AX at the concentration at which the macroelectrolyte BX occurs in the solution. The α coefficients of several electrolytes at different macroelectrolyte concentrations are also given in the literature.

Activity coefficients in the ion exchanger phase. The exchanger loaded with the ion B has an activity coefficient of $_{BR}$. The determination of these values is relatively difficult, but at least for certain liquid exchanger systems the coefficients have been measured. In [129 a], for example, the activity coefficients for the Li-, Na- and K-salts of toluene sulfonic acid (Tsa) are given. It can be assumed that no essential differences exist between these liquid exchangers and an equivalent polystyrene DVB-crosslinked resin. The activity coefficients of the solid exchangers are therefore taken as equal to those of the liquid ones:

$$\gamma_{BR} = \gamma_{BTsa} \qquad (6)$$

Evaluation of the Kraus and Moore experiments (Chapter 5.1) has shown, however, that only the water-containing part of the exchanger phase is involved in exchange processes. This is comprehensible because both the polystyrene matrix and the toluene fraction of liquid exchangers can be classified as inert materials, which means that they don't contribute to the electrolyte activities.

In Figure 1.28 the activity coefficients for Li and Na toluene sulfonic acids are plotted. The original molal concentrations have been transformed into molar values after determination of the density of the solutions. The inert part of the exchangers is deducted by subtracting the volume of the toluene molecule for every fixed site exchanger group.

The water content in the exchanger phase can be calculated for strongly acidic polystyrene exchangers [129 b] by the equation

$$\frac{W^0}{\rho_0} = \overline{V^0} - 0.63 \qquad (7)$$

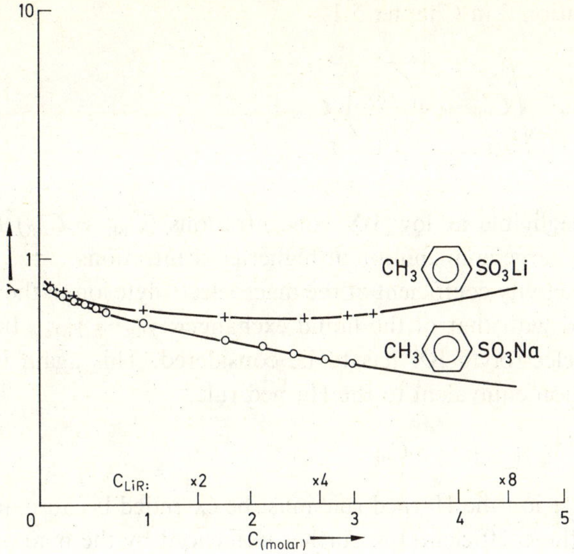

Figure 1.28 Activity coefficients of Na and Li toluene sulfonic acids in dependence on their concentration. The molal values were transformed into molar values and then the concentration was calculated for the inert free solution by deducting the volume of the toluene (conversion factor 1.12 per exchanger equivalent). For low outer electrolyte concentration the C_{LiR} values for three exchangers with different degrees of crosslinking (X2, X4, and X8) are indicated (Bobleter).

W^0 is the amount of water in grams of the specific exchanger volume V^0 (of 1 gram dry exchanger in the H^+ or Cl^--form) and ρ_0 the density of the water at the relevant temperature.

The concentration C_{BR} of the B-loaded exchanger is given by the relation of the capacity K_{sp} of the specific ion exchange volume to the water volume of the same amount of exchanger:

$$C_{BR} = \frac{K_{sp}}{W^0}\rho_0 = \frac{K_{sp}}{V^0 - 0{,}63} \qquad (8)$$

By this means the concentration is related to the amount of water in the exchanger phase.

A contraction (shrinking) of the exchanger phase usually occurs when the outer concentration of the electrolyte BX is increased. The specific volume of the exchanger must therefore be determined for every electrolyte concentration of interest. At low electrolyte concentrations Donnan exclusion is prevalent, but at higher concentrations Donnan invasion becomes increasingly important.

According to equation 3 in Chapter 5.1

$$\bar{C}_{BR} = \frac{C_{BR} \pm \sqrt{C_{BR}^2 + 4\left(\frac{\gamma_{BX}}{\bar{\gamma}_{BX}}\right)^2 C_{BX}^2}}{2} \tag{9}$$

– the invasion is negligible at low BX concentrations ($\bar{C}_{BR} \approx C_{BR}$) but it can constitute a relatively large contribution at higher concentrations.

In a first step the activity coefficient of the macroelectrolyte ion in the exchanger phase can be equated with that of the liquid exchanger: $\gamma_{BR} = \gamma_{BTsa}$. In addition, the invasion of the electrolyte BX has to be considered. This again is done by introducing an equation equivalent to the Harned rule.

$$\log \bar{\gamma}_{BR(1)} = \log \gamma_{BTsa(1)} + \alpha_1 \bar{C}_{BR} \tag{10}$$

In the case of the trace ion the Harned rule must be extended because, in addition to the BX invasion, the coefficients are further influenced by the macroelectrolyte concentration

$$\log \bar{\gamma}_{AR(0)} = \log \gamma_{ATsa(1)} + \alpha_1 \bar{C}_{BR} + \alpha_2 C_{BR} \tag{11}$$

The relation $\gamma_{BR(1)}/\gamma_{AR(0)}$ is of a special interest. From equation 10 and 11 it can be concluded that

$$\frac{\bar{\gamma}_{BR(1)}}{\bar{\gamma}_{AR(0)}} = \frac{\gamma_{BTsa(1)}}{\gamma_{ATsa(1)} \cdot 10^{\alpha_2 C_{BR}}} \tag{12}$$

Example of a Na^+/Li^+ exchange. According to equation 3, the equilibrium constant for the exchange of Na^+ trace ions for Li^+ is given by

$$K_{Li}^{Na} = \frac{C_{NaR}\, C_{Li^+}}{C_{Na^+}\, C_{LiR}} = k_d \frac{C_{Li^+}}{C_{LiR}} = \frac{\gamma_{NaCl(0)} \bar{\gamma}_{LiR(1)}}{\gamma_{LiCl(1)} \bar{\gamma}_{NaR(0)}} \tag{13}$$

The constant k_d represents the molar distribution coefficient. With increasing outer LiCl concentrations, the concentration C_{LiR} of the Li^+ ions adsorbed at the fixed exchanger sites increases, according to equation 8, due to the exchanger shrinking. As can be seen from Table 1.5 a the total Li concentration C_{LiR} is even higher owing to the LiCl invasion (shrinking and invasion according to equation 9).

The activity coefficients for the outer solution $\gamma_{NaCl(0)}/\gamma_{LiCl(1)}$ are calculated by equation 4 and 5. At low electrolyte concentrations they are close to 1, but they decrease considerably at higher concentrations of the solution (Table 1.5 b). The coefficient of equation 5 is between 0.034 and 0.037.

Equation 12 can be written

$$\frac{\bar{\gamma}_{LiR(1)}}{\bar{\gamma}_{NaR(0)}} = \frac{\gamma_{LiTsa(1)}}{\gamma_{NaTsa(1)} \cdot 10^{\alpha_2 C_{BR}}} \tag{14}$$

Table 1.5 a Concentrations of a Li^+ loaded cation exchanger (Dowex 50WX4) at different outer LiCl concentrations

Outer solution C_{LiCl} (mol/l)	Concentration in the exchanger phase (according to equations 8 and 9)	
	C_{LiR} (mol/l)	$\overline{C}_{LiR}$ (mol/l)
0.1	2.66	2.67
0.25	2.78	2.83
0.5	2.91	3.08
1.0	3.13	3.67
2.0	3.50	5.05
3.92	4.19	7.96

Table 1.5 b Activity coefficients and equilibrium constants of a Na^+/Li^+ exchange on Dowex 50WX4 (according to equation 14)

a	b	c	d
Outer solution C_{LiCl} (mol/l)	Activity coeff. in outer solution $\frac{\gamma\ NaCl\ (0)}{\gamma\ LiCl\ (1)}$	Activity coeff. in resin phase $\frac{\bar{\gamma}\ LiR\ (1)}{\bar{\gamma}\ NaR\ (0)}$	K_{Li}^{Na} (column b × c)
0.1	0.995	1.356	1.35
0.25	0.984	1.382	1.36
0.5	0.962	1.415	1.36
1.0	0.921	1.520	1.40
2.0	0.839	1.680	1.41
3.92	0.670	1.963	1.32
average			1.37

The γ values of Na and Li toluene sulfonic acid can be taken from Figure 1.28. The α_2 coefficient was estimated to be 0.01.

The relation $\bar{\gamma}_{LiR(1)}/\bar{\gamma}_{NaR(0)}$ thus obtained is also given in Table 1.5 b. The value of the equilibrium constant K_{Li}^{Na} resulting from these experiments was found to be 1.37 and lies very well within experimental errors.

According to equation 13, the molar distribution coefficient k_d is

$$k_d = K_{Li}^{Na} \frac{C_{LiR}}{C_{Li^+}} \tag{15}$$

In Figure 1.29 a the corresponding values are given for an increasing LiCl concentration in the solution. The logarithmic plot represents a straight line, which can be obtained only when correct thermodynamic parameters are used.

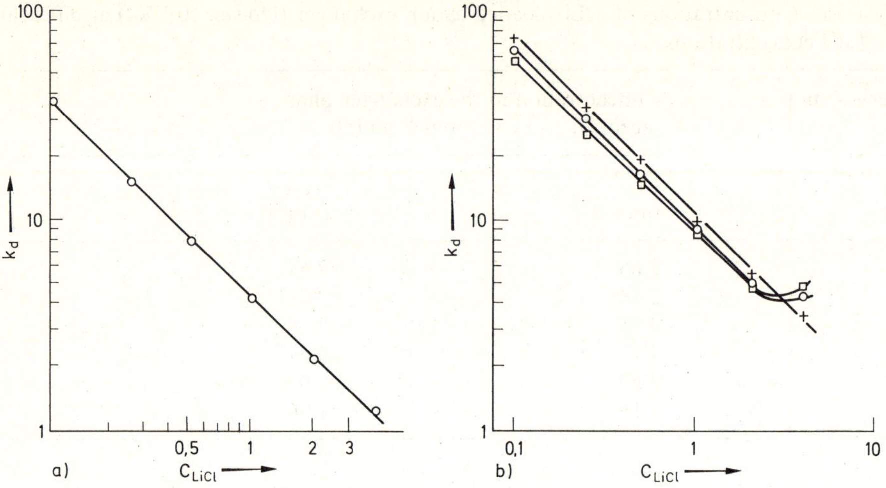

Figure 1.29 Distribution coefficients for the Na^+/Li^+ exchange. + Dowex 50WX8, ○ Dowex 50WX4 and □ Dowex 50WX2.
a) The molar distribution coefficient, k_d, is given in dependence on the LiCl concentration in the solution.
b) The weight distribution coefficient, k_D, is given in dependence on the LiCl concentration in the solution (Bobleter).

In Figure 1.29 b the weight distribution coefficients for the same experiments are given

$$k_D = {}^{wt}K_{Li}^{Na} \frac{M_{LiR}}{C_{Li^+}} \tag{16}$$

M_{LiR} is the specific capacity of the exchanger. Several characteristic features can be observed in this figure. The stronger crosslinked exchangers (Dowex 50W X 4 and 8) deviate greatly from linearity at higher LiCl concentrations. The parameter ${}^{wt}K_{Li}^{Na}$ is no longer a constant but changes for every measured point between 1.13 and 2.45.

To summarize, it can be said that the law of mass action is a good basis for the evaluation of thermodynamic ion-exchange parameters. The necessary condition, however, is that true molar concentration values are applied, which means that the shrinking or swelling, as well as Donnan invasion, are also taken into account. Under these premises equilibrium constants and distribution coefficients can be deduced in astonishingly good agreement with the experiments [129].

The fundamentals of ion exchange and ion exchangers require some further definitions related to or derived from the chemical or physical properties of ion exchangers. Given the general validity of the principles of electrical neutrality and stoichiometry, there must at first be a constant counterion content that is directly proportional to the fixed electrical charges of the matrix. In other words an ion exchanger has, depending on the number of fixed ions, a certain capacity for the uptake of counterions.

The capacity of an ion exchanger can, in general, be quantitatively expressed by the number of counterions relative to the number of exchange sites per unit weight of material. The capacity is usually expressed on a dry-weight or wet-volume basis and capacity ratings are assumed, unless otherwise stated, to relate to the hydrogen form of cation exchangers and the chloride form of anion exchangers. Depending on whether the capacity of an ion exchanger is regarded independent of or dependent on experimental conditions the following definitions are valid (IUPAC), with the term bed volume being used synonymously with column volume for the application of an ion exchanger in a packed column.

Theoretical specific capacity, Q_0:	Milliequivalents of ionogenic group per gramme of dry ion exchanger. If not otherwise stated the capacity should be reported per gramme of the H-form of a cation exchanger and Cl-form of an anion exchanger.
Volume capacity, Q_V:	Milliequivalents of ionogenic group per cm^3 (true volume) of swollen ion exchanger. (The ionic form of the ion exchanger and the medium should be stated.)
Bed volume capacity:	Milliequivalents of ionogenic group per cm^3 of bed volume determined under specified conditions (should always be given together with specification of conditions).
Practical specific capacity, Q_A:	Total amount of ions expressed in milliequivalents or millimoles taken up per gramme of dry ion exchanger under specified conditions (should always be given together with specification of conditions).
Break-through capacity of ion exchanger bed, Q_B:	The practical capacity of an ion exchanger bed obtained experimentally by passing a solution containing a particular ionic or molecular species through a column containing the ion exchanger, under specified conditions, and measuring the amount of species which has been taken up when the species is first detected in the effluent or when the concentration in the effluent reaches some arbitrarily defined value. The breakthrough capacity may be expressed in milliequivalents, millimoles or milligrammes taken up per gramme of dry ion exchanger or per cm^3 of bed volume.

The volume capacity Q_V and the bed volume capacity are related to the true volume and bulk volume respectively and depend on the type of counterions, the solvent and other conditions. The practical specific capacity, Q_A, covers, in contrast to the theoretical specific capacity, only the quantity of active fixed ions under specific conditions. The break-through capacity, Q_B, gives the uptake that can be achieved under practical conditions, whereby the requirements on the purity of the effluent have to be taken into consideration.

Porosity is, besides being an important fundamental property, one of the most difficult subjects to describe and explain in general terms for all ion exchanger types. No general definition is included in the IUPAC Recommendations on Ion Exchange Nomenclature except the term macroporous ion exchanger for which the definition is given as ion exchangers with pores which are large compared to atomic dimensions. The channels which enclose the ion exchanger skeleton are usually called pores, and represent the porosity. The term gel type resin refers to ion exchangers prepared with polymers that do not contain pores. The pore structure of these gels is quite difficult to describe since the distances between crosslinks and chains vary considerably. These resins are elastic three-dimensional polymers which cannot have a definite pore size. A steadily increasing resistance of the polymer network limits the uptake of ions and molecules of increasing size. The resins have no appreciable porosity until swollen in some suitable solvent such as water. Among the synthetic ion exchange materials due to a new polymerization technique, ion exchange resins with a crosslinked structure have been developed, which are entirely different from the conventional so-called homogeneous gels in that they have a rigid macroporous structure superimposed on the gel structure similar to those of conventional adsorbents. In reality, however, these synthetic ion exchangers are also gels. Ion exchange materials referred to as macroporous are very tightly crosslinked gels in which larger holes have been formed artificially. The term isoporous, which had been introduced, can obviously not be accepted without criticism, since all gel ion exchange resins are more or less isoporous. That ion exchange resins have a microporous structure is quite often claimed in the literature but this property has not yet been clearly defined. For application in high pressure liquid chromatography socalled porous types of ion exchangers have been developed – they are in fact superficially porous materials of a special kind, which are treated in detail elsewhere in this volume.

Much has been described already in the foregoing sections of the swelling of ion exchangers. Confining the following comments to the swelling behavior of synthetic gel type resins, the main fact with regard to their swelling is that it is in principle an osmotic process, whereby the osmotic activity of the internal ionogenic groups being present in high concentration causes (because of the inclination to become diluted) water to enter the resin structure, which therefore swells. This swelling of the resin is accompanied by a stretching of the crosslinked hydrocarbon matrix. The crosslinking of the exchanger network then limits the swelling, and the net restoring force of the crosslinks is interpreted as an internal swelling pressure acting on the pore liquid within the resin. Equilibrium is reached when the swelling pressure of the matrix counterbalances the osmotic pressure difference between the interior of the resin and the surrounding solution. This approach roughly explains the most important factors the swelling depends on. To express it quantitatively a weight swelling and a volume swelling have been defined (IUPAC):

Weight swelling in solvent, w_s: (e. g., w_{H_2O})	Gramme of solvent taken up by one gramme of dry exchanger.
Volume swelling ratio:	Ratio of the dry swollen volume to the true dry volume.

Swelling is necessary for the diffusion; if diffusion is hindered by too little swelling the exchange process becomes too slow.

Diffusion processes are essential for the whole field of ion exchange. In accordance with the above general equation for the ion exchange equilibrium the counterion A^+ moves during the exchange process from the ion exchanger into the solution and the ionic species B^+ moves from the solution into the exchanger. It was first recognized at the beginning of this century that basic ion exchange is essentially a redistribution of the exchanging counterions A^+ and B^+ by diffusion, a process therefore, that does not include an actual chemical reaction. In other words, basic ion exchange is in essence an equilibrium process that comes about by diffusion (Helfferich). In the following more or less only basic ion exchange and the role played in it by diffusion with respect to gel type ion exchange polymers can be regarded, neglecting for instance the diffusional sorption processes. In searching for mechanisms and models two have proven to be of basic importance, and have been used to develop the theory of the kinetics of ion exchange.

Based on extensive research into the rate factor in ion exchange Boyd and coworkers concluded that the ordinary exchange of ions is indeed not a chemical reaction of a kinetic order corresponding to its stoichiometric coefficients but a diffusion mechanism, in particular either diffusion in the adherent stagnant liquid layer to and from the exchanger particle or diffusion within the ion exchanger particle [130]. The discussions concerned were restricted to the exchange of counterions between ion exchanger particles and well-stirred electrolyte solutions, and did not include the much more complex behavior of columns. At that time it seemed evident that cases where a chemical exchange reaction is rate-controlling are extremely rare, if occurring at all. This basic diffusion concept is depicted in Figure 1.30. One comes to this figurative representation following Boyd's work by

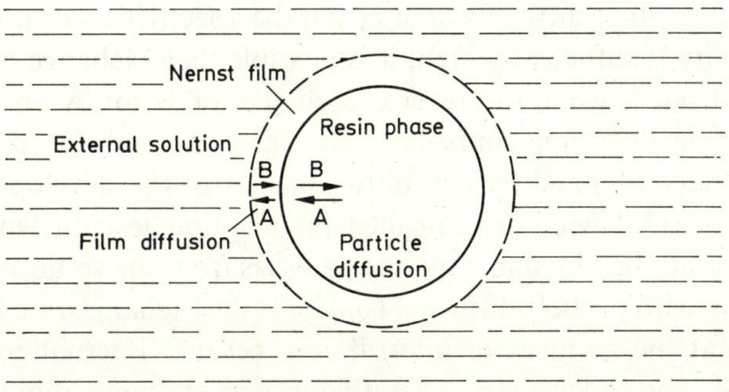

Figure 1.30 Film diffusion and particle diffusion in ion exchange kinetics.

assuming that at first, two different diffusion mechanisms exist during ion exchange. In the region of low external concentration (about 0.001 N), the rate is determined by the diffusion of ions through the liquid film enveloping the exchanger particle, while at high external concentrations (about 0.3 N), diffusion of ions through the exchanger particle itself represents the rate-controlling step. On the basis of this division, a distinction is made between film diffusion and particle or, better, gel diffusion. The outer boundary of the spherical exchanger particle represents the interface between the adhering Nernst film (the thickness of which was determined to be $3 \cdot 10^{-4}$ cm [130, Dickel]) which has film diffusion, and the exchanger particle, which has gel diffusion [131]. In an intermediate range of conditions both mechanisms may affect the rate factor of the exchange of ions. But if only film or gel diffusion are considered, especially under practical considerations, and it is to be determined which of them is rate-controlling, a quantitative criterion derived by Helfferich can be applied. These relations for the theoretical prediction of the rate controlling step are based on the effect of ion exchange capacity $\overline{C}$, solution concentration C, particle size r_0 (particle radius in cm), film thickness δ, diffusion coefficients D, and the separation factor α:

$$\text{Particle diffusion control} \quad \frac{\overline{C}\,\overline{D}\,\delta}{C\,D\,\mathrm{r}_0}(5 + 2\alpha_{\mathrm{A/B}}) \ll 1 \equiv N$$

$$\text{Film diffusion control} \quad \frac{\overline{C}\,\overline{D}\,\delta}{C\,D\,\mathrm{r}_0}(5 + 2\alpha_{\mathrm{A/B}}) \gg 1 \equiv N\,.$$

Using *the dimensionless number, N, which is to be called the Helfferich number*, the following criteria are obtained:

if $N \ll 1$ particle diffusion control

if $N \gg 1$ film diffusion control.

The thickness of the fictitious Nernst film is related to the engineer's Sherwood number by $\delta = 2r_0/Sh$. It is true that a low solution concentration favors film diffusion control, but a first difficulty arises from the selectivities of the exchanging ions. If the selectivity is sufficiently high, it is possible that exchange of A for B is film diffusion controlled, while the reverse exchange of B for A under identical conditions is particle diffusion controlled. An early method for distinguishing experimentally between film and particle diffusion control was developed by Kressman and Kitchener [132] with the so-called interruption test, in which the ion exchange process is stopped by removing the particles from the solution for a short period of time; only with particle diffusion control is the exchange rate higher upon reimmersion than at the moment of interruption, because internal concentration gradients have had time to level out. The difference in exchange rate that may be observed is schematically shown in Figure 1.31.

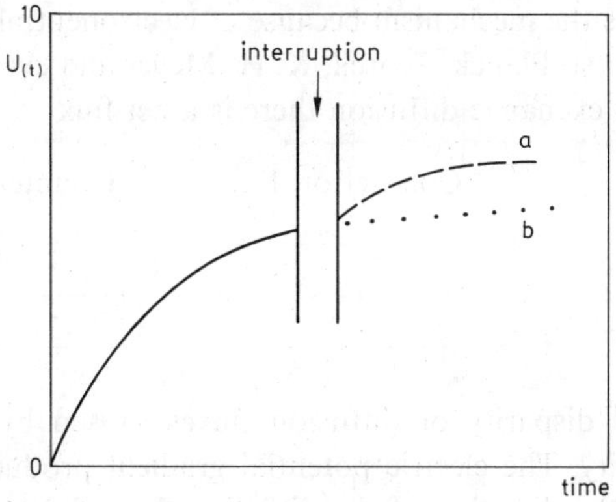

Figure 1.31 Diagrammatic representation of the interruption test and its effect on the observed exchange rate. After interruption the dashed line in continuation of the exchange indicates that the rate is controlled by concentration gradients in the particle or by mixed particle and film diffusion, and the dotted line that the unaffected film diffusion is rate determining.

Already with regard to this first model, and in particular with gel (particle) diffusion as the rate-controlling step in ion exchange materials, the problem of predicting diffusion coefficients in ion exchangers immediately arises. Diffusion in ion exchanger gels is slower than in solutions for several reasons. Part of the ion exchanger structure is not available for diffusion. The diffusion paths in the ion exchanger are more twisted and therefore longer. The size of ions influences their mobility in and through highly crosslinked and therefore denser portions of the gel, retarding larger ions more with increasing size of the diffusing species, preventing with very large species, the exchange equilibrium from being reached. These and several other conditions lead to a retardation in the diffusion in ion exchange resins and are also related to the question of whether a quasi-homogeneous model can be used for discussions of diffusion processes. The above model has been employed in various ways to relate the self-diffusion coefficients in ion exchanger gels to those in solution.

However, the diffusion of the counterions is – as was discovered some ten years later by Helfferich et al. – not as simple as it looks, and is moreover subject to the restriction of electroneutrality. This means that as ions of species A^+ initially present as counterions move into the solution, a stoichiometric amount of species B^+ must move into the exchanger to balance the electric charges of the fixed sites. The ions are carriers of electric charges and, therefore, their diffusion generates an electric potential – the diffusion potential – which influences the motion of the ions and which has to be taken into account in describing ion exchanging diffusion processes. Fluxes of counterions thus become coupled with one another, which is achieved by the corrective action of an automatically arising electric field, i. e. the diffusion potential. In ion exchange, diffusion processes are therefore the diffusion of ions with an electric field as the mechanism because of electroneutrality conservation [133]. Resorting to Nernst, Planck, Teorell, K. H. Meyer and even Warburg [134] it was shown that in ion exchange diffusion there is a net flux

	Counterion 1	Counterion 2
Diffusion	⟶	⟵
Transference	⟵	⟵
Net flux (equal in magnitude)	⟶	⟵

which results from an initial disparity of diffusion fluxes caused by a minute deviation from electroneutrality. The electric potential gradient produces electric transference of both ions, in the direction of the diffusion flux of the slower ion, which in effect retards the faster ion and accelerats the slower one. As the mechanisms of ion diffusion in the resin are a fundamental question for theoretical kinetic investigations, it cannot be ignored that, besides other refinements necessary, there are situations where ion exchange processes are accompanied by a chemical reaction – for instance the neutralization of a weak acid cation exchanger in free acid form by a strong base, $RCOOH + Na^+ + OH^- \rightarrow RCOONa + H_2O$. Taking this, although reportedly remote possibility into account – which will be treated below as ion exchange accompanied by reactions – and trying to sum up what the mechanism of ion diffusion in an ion exchange resin is it can for now be assumed that the ion diffusion inside the resin can be approached hypothetically by at least three major different mechanisms [135]. As shown in Figure 1.32 besides the gradient or Fickian diffusion mechanism based on Fick's theory, there are two other feasible mechanisms that may be expected to occur, namely the homogeneous or progressive conversion mechanism as discussed for noncatalytic heterogeneous solid fluid re-

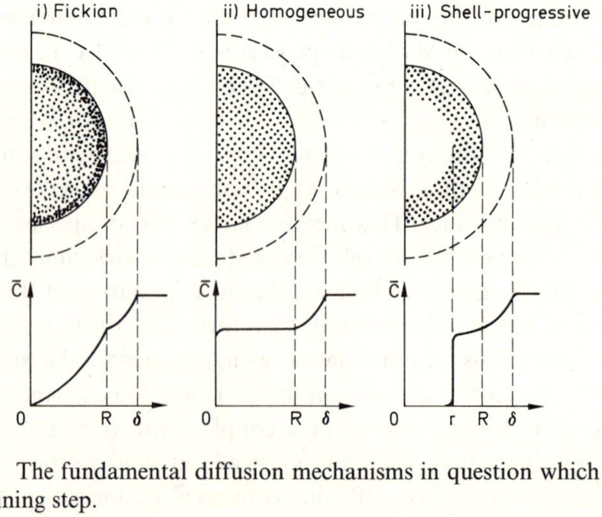

Figure 1.32 The fundamental diffusion mechanisms in question which may be basic as the rate determining step.

action models by Wen [136] and by Kunii and Levenspiel [137], or the shell progressive or unreacted core mechanism first developed by Yagi and Kunii [138]. From each of these hypothetical mechanisms theories of ion exchange kinetics can be derived, depending on whether they are based on Fick's laws [139], the Nernst-Planck equation, which contains in addition to the Fickian component a term for the electric field [133], or, in more recent times, the Stefan-Maxwell equation [140]. As far as values of diffusion coefficients are concerned, it has been shown that there is a relation between the effective diffusion coefficient $\overline{D}$ inside the resin and the diffusion coefficient D in aqueous solution, which as a quotient yield a retardation factor according to the following formula

$$\frac{\overline{D}}{D} = \frac{(2 - \varepsilon)^2}{\varepsilon^2},$$

with ε the fractional pore volume [141]. For this and similar relations a quasi-homogeneous model for the exchange polymer was employed in most cases. It has always been a question whether such a model can be used for the treatment of diffusion processes in resins and theories of kinetics based upon them. The various retarding effects have been taken into account in the quasi-homogeneous model by the use of the effective diffusion coefficients. For customary ion exchange resins this model is considered as entirely adequate, as has been shown by experimental results – very accurate tracer-diffusion measurements agree closely with theoretical rate equations. Nevertheless, for other non-uniform exchange materials it is necessary to determine the effective diffusion coefficients for a particular combination of ion and ion exchanger. The above equation is reasonably successful for univalent counterions, co-ions and small molecules in moderately crosslinked polymers, but predicts too little retardation in other cases in which other effects, as e. g. partial localization of the counterions by association, contribute significantly. In customary gel type ion exchange resins the diffusion coefficients increase with increasing particle size and decreasing crosslinking. If the degree of crosslinking decreases, swelling increases, and with an increasing water content of the exchanger phase the diffusion coefficients become increasingly similar to the values of ordinary dilute solutions. In most commercial ion exchange resins, the values of particle diffusion of univalent ions are approximately 1/3 to 1/10, those of bivalent ions 1/5 to 1/100 and those of trivalent ions 1/10 to 1/1000 of the diffusion values in aqueous solutions. Table 1.6 shows measured values of diffusion coefficients in various commercial ion exchangers. It can be seen that the values depend on the valence of the ion and the degree of crosslinking.

When it comes to the employment of diffusion coefficients in theoretical kinetic models, however, the very important question arises of whether they are constant or variable. This question must certainly be gone into in those discussions where diffusion is considered as the only rate control factor. The earlier models of ion exchange kinetics of particle diffusion controlled binary exchange employed con-

Table 1.6 Tracer particle diffusion coefficients (at different temperatures)

Ion exchanger	Counterion	Diffusion coefficient ($cm^2 \cdot s^{-1}$)	Activation energy (kcal/mol)	Source
Dowex 50X4	Na	$1.4 \cdot 10^{-6}$	4.93	G. E. Boyd et al., 1953
	Sr	$2.3 \cdot 10^{-7}$	5.64	
	Y	$7.5 \cdot 10^{-8}$	6.77	
	La	$6.9 \cdot 10^{-8}$	5.52	
Dowex 50W X8	Na	$1.6 \cdot 10^{-6}$		J. C. W. Kuo et al., 1963
Dowex 50X8	Zn	$6.3 \cdot 10^{-8}$	7.34	G. E. Boyd et al., 1953
	La	$9.2 \cdot 10^{-9}$		
Dowex 50X8.6	Na	$9.4 \cdot 19^{-7}$	6.50	
	K	$1.3 \cdot 10^{-6}$	5.22	
	Ag	$6.4 \cdot 10^{-7}$	5.90	
	Sr	$3.4 \cdot 10^{-8}$	8.28	
	Y	$3.2 \cdot 10^{-9}$	7.53	
Dowex 50X12	K	$4.3 \cdot 10^{-6}$		M. Tetenbaum et al., 1954
Dowex 50X16	Na	$2.4 \cdot 10^{-7}$	8.40	G. E. Boyd et al., 1953
	Ag	$1.1 \cdot 10^{-7}$	7.20	
	Sr	$3.0 \cdot 10^{-9}$	10.98	
	Zn	$1.2 \cdot 10^{-8}$	8.52	
KU-2X2	methylene blue	$2.78 \cdot 10^{-10}$	15.0	G. S. Libinson et al., 1963
KU-2	NH_4	$2.6 \cdot 10^{-6}$		L. N. Peletaev et al. 1986
PS-DVB X10	Na	$2.1 \cdot 10^{-6}$		G. Dickel et al., 1968
	H	$0.98 \cdot 10^{-6}$		
	K	$3.2 \cdot 10^{-6}$		
	H	$0.98 \cdot 10^{-6}$		
Dowex 50W X8	Na	$20.5 \cdot 10^{-7}$		M. G. Rao, 1974
	Mn	$2.22 \cdot 10^{-7}$		
	Sr	$1.95 \cdot 10^{-7}$		
PS-DVB X8	Na	$7.1 \cdot 10^{-7}$		P. Frölich, 1976
		$1.9 \cdot 10^{-7}$		
Dowex 1X8	Cl	$3.3 \cdot 10^{-6}$		J. Eliasek, 1968
	OH	$0.5 \cdot 10^{-6}$		
Kastel A 500	SO_4	$2.0 \cdot 10^{-6}$		L. Liberti et al., 1978
SBW	Cl	$2.0 \cdot 10^{-6}$		B. Mehlich et al., 1986
	OH	$0.3 \cdot 10^{-6}$		
Dowex 2X2	Br	$6.4 \cdot 10^{-7}$	5.0	B. A. Soldano et al., 1953
Dowex 2X6	Cl	$3.5 \cdot 10^{-7}$	6.8	
	Br	$3.9 \cdot 10^{-7}$	6.0	
	I	$1.3 \cdot 10^{-7}$	8.8	
	BrO_3	$4.6 \cdot 10^{-7}$	6.2	
	WO_4	$1.8 \cdot 10^{-7}$	7.2	
	PO_4	$5.7 \cdot 10^{-8}$	8.7	
Dowex 2X16	Br	$2.6 \cdot 10^{-7}$	9.3	
Dowex A-1	Na	$1.2 \cdot 10^{-7}$		A. Schwarz et al., 1964
	Zn	$6 \cdot 10^{-9}$		
	Co	$7 \cdot 10^{-9}$		

stant interdiffusion coefficients, for instance when Fick's first law was used as the flux equation with $\overline{D}$ = const. This approach is considered as correct for counterions with equal mobility $\overline{D}_A = \overline{D}_B = \overline{D}$. But on the basis of the Nernst-Planck model its validity for counterions with different mobilities, has been challenged, because here, as mentioned above, the effect of electrical-potential gradients is taken into account. In deriving the flux equations according to the Nernst-Planck model it has been found that the interdiffusion coefficient $\overline{D}$ is not constant, but depends, according to the relation

$$\overline{D} = \overline{D}_A \overline{D}_B \frac{(z_A^2 \overline{C}_A + z_B^2 \overline{C}_B)}{(z_A^2 \overline{C}_A \overline{D}_A + z_B^2 \overline{C}_B \overline{D}_B)},$$

on the relative concentrations of A and B (Helfferich). With vanishing concentration of either ion, the interdiffusion coefficient then approaches the diffusion of that ion. In other words, the interdiffusion rate is predominantly controlled by the ion which is in the minority. The corrective action of the electric field is mainly directed against the high-concentration species, so that the interdiffusion rate is predominantly controlled by Fickian diffusion of the low concentration species, as is shown on the left of Figure 1.32. Two limiting cases are still to be mentioned, namely, either where ions of equal mobility are involved or where the concentration of one counterion is always much smaller than that of the other, as in exchange of trace components, where the interdiffusion coefficient equals the diffusion coefficient of the trace ion: e. g., $\overline{C}_B \ll \overline{C}_A$, $\overline{D} = \overline{D}_0$. In film diffusion-controlled binary ion exchange the earlier models all used the Nernst film concept on the premise that film diffusion is a quasi-stationary process in a planar layer. In Boyd's theory, developed for isotopic exchange for the use of Fick's first law, constant diffusion coefficients were applied: $D_A = D_B = D$ = const. Adamson, as well as Dickel and Meyer and subsequent workers, later tried to include systems with counterions of different mobilities, with ion exchangers exhibiting selectivity. It was postulated that the fluxes of the ions in the film could obey Fick's law, but each with its own constant diffusion coefficient: D_A, D_B = const. The weaknesses of these approaches were discussed by Schlögl and Helfferich [142], who pointed out that a self-consistent treatment is achieved if one uses the three Nernst-Planck equations for A, B and the co-ion. Then again, for $D_A \neq D_B$ a concentration gradient of the co-ion appears, but now Fickian diffusion of the co-ion in one direction is compensated for by electric transference in the opposite direction. Schlögl's Nernst-Planck film model also clarifies the accumulation or depletion of electrolyte at the particle surface. The driving force here is the electrical-potential difference between bulk solution and particle surface, resulting from counterion interdiffusion. This force pushes co-ions into the film if the faster counterion is initially in the solution, and pulls co-ions out of the film in the opposite case (Helfferich). In a detailed study of the theory of film diffusion Bunzl and Dickel made measurements in small concentration intervals in which the equilibrium constant K can actually be considered constant [143]. The diffusion coefficients observed by Dunlop [144] in the ternary electrolyte for the system Li^+/K^+ were used as a basis for calculations and led to a quantitative agreement between calculated and measured values.

Transport phenomena in ion exchangers can most rigorously be treated by the thermodynamics of irreversible processes [145]. Diffusion processes in ion exchange are here considered as transport phenomena in a concentrated electrolyte, with the use of the chemical potential as the driving force for dealing with the simultaneous diffusion of several species. But since neither the transport coefficients nor the chemical potential gradients needed in the corresponding equations are generally known, these quantities are replaced by the diffusion coefficients, which are first determined empirically. It must again be emphasized that every diffusion coefficient refers to a given reference system [146]. Using the Nernst-Planck equation in thermodynamic transport theories obviously yields a differential equation in which, in addition to Helfferich's diffusion coefficient, there is a coefficient involving the heat of reaction [147].

Attempts to include selectivity into diffusion processes in ion exchange resins seem to show that the Nernst-Planck equations are only limiting laws valid for ideal systems. Helfferich corrected his relation for the diffusion coefficient with an additional term, which leads then to

$$\bar{D} = D_A D_B \frac{z_A^2 \bar{C}_A + z_B^2 \bar{C}_B + z_A \bar{C}_A z_B \bar{C}_B \left[\dfrac{\mathrm{dln}\, K_a}{\mathrm{d}\,(z_a \bar{C}_A)} \right]}{z_A^2 \bar{C}_A \bar{D}_A + z_B^2 \bar{C}_B \bar{D}_B},$$

where the actual existing gradients of activity coefficients have been expressed in terms of the molar selectivity coefficient $K_a = \dfrac{\bar{y}_B^{zA}}{\bar{y}_A^{zB}}$. But also according to Helfferich, the role of selectivity in particle-controlled diffusion processes is such that it does not affect the flux equation, since only the derivative $\mathrm{dln}K_a/\mathrm{d}(z_A\bar{C}_A)$ appears in the above equation and the absolute value of K_a is of no consequence. This relates to a constant boundary condition, so that the rate is independent of selectivity. But if the solution volume is not large compared with the amount of ion exchanger, so that C_A increases with conversion, the selectivity affects the boundary condition. Then, for any given solution concentration C_A, the boundary concentration $\bar{C}_A(r_0, t)$ is lower if the ion exchanger prefers B more strongly. In such systems selectivity for the ion B does indeed increase the rate and the conversion eventually attained. In film diffusion the influence of selectivity is rather different, as it has no effect on the flux equations, only on the boundary condition at the film/particle interface. In film diffusion the concentrations of both ions A and B at the interface are finite over the course of exchange, even if the concentration of A in the bulk solution remains negligible, and the boundary concentration $C_A(r_0,t)$ is thus affected by the selectivity even in this case. As a consequence, the rate of film-diffusion-controlled exchange increases with increasing selectivity for the ion B, even in cases in which particle-diffusion-controlled exchange is independent of such preference. This difference between the two mechanisms with respect to selectivity accounts for the appearance of the separation factor $\alpha_{A/B}$ in the Helfferich number N. It is important to note that the selectivity effect, which may be drastic, is taken into account differently in the different theories of ion exchange kinetics.

Very practically conceived investigations of the selective exchange of SO_4^{2-} against Cl^- on anion exchangers of different basicity by Liberti and coworkers under particle-diffusion-controlled conditions showed that, in general, because of the diffusive nature of the ion exchange mechanism, it must be concluded that selectivity may adversely influence the over-all kinetics of ion exchange processes. This must obviously be taken into account each time one is dealing with selective applications of ion exchange [148]. Two cases have been studied extensively: the chloride/sulfate exchange when particle diffusion is rate-determining because of the concentration conditions; and very dilute systems where resin selectivities for the divalent sulfate ion are drastically raised due to the electroselectivity effect. For the first case it was found that for this actual system also an apparantly variable interdiffusion coefficient can be determined experimentally, thus avoiding the problem of the knowledge of the individual ion diffusion coefficients for each exchange (which is the explanation for the lack of experimental evaluation of the Nernst-Planck model and the use, whenever possible, of Fick's simpler model); further, that from $\overline{D}$ vs. U plots the single ion diffusion coefficients can be extrapolated as limiting $\overline{D}$ values, and that the more rigorous rate laws for ion exchange can be applied to the investigated process. For the second case it was discovered that rate mechanisms do not seem to obey the well-known models based on diffusion only, but convincing evidence was obtained to show that the chemical reaction rate on fixed charges controls the overall process kinetics. In particular the activation energies – up to 16.83 kcal/equ, – are definitely beyond the range of usual diffusion controlled kinetics. Therefore, a spherically-symmetrical diffusion model with a billiard-ball-like mechanism of site-to-site jumping of ions inside the resin was suggested.

The determination of diffusion coefficients of ion exchangers is usually made by independent radioactive tracer or conductivity measurements. For the first case only measurements performed on spherical granules will be considered here. For such ion exchange materials the shallow bed technique is the most frequently used method. Initially introduced as the thin bed method for general kinetic investigations on cation exchange water softening it was adopted under the current term by Boyd and coworkers to study the self-diffusion of cations and anions, etc., in and through exchange polymers. This method makes possible the determination of rates of exchange down to very short times. Further, it makes it possible to obtain rate data under conditions of constant solution concentration of the exchanging ions. The first such experimental arrangement is shown in Figure 1.33. Later researchers have used the shallow bed method with more or less significant modifications, and with not only a small volume of exchanger but with as few as five resin beads.

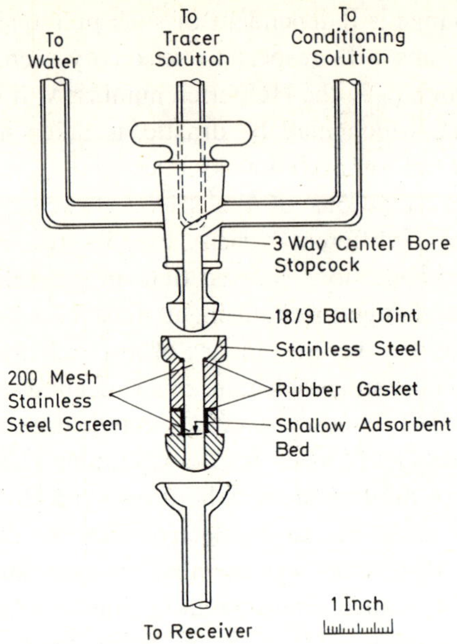

Figure 1.33 Equipment for the determination of exchange rates according to the shallow bed method.

Other experimental procedures can only be mentioned, for instance the disc method of Helfferich (for polycondensation ion exchange resins) or the shallow fluidized bed method of Streat. The diffusion coefficient is obtained from rate relations containing it, depending on the theoretical model used. Solving the many still unsolved problems and existing anomalies in describing diffusion in ion exchange and related ion exchange systems is a task remaining for ion exchange kinetics. Conductivity measurements for the determination of the mobilities in ion exchange materials were first carried out for investigations on membranes [149] but can as well be used for the determination of diffusion coefficients in ion exchanger beads [150]. In all these cases, evidently, quite special experimental equipment has to be designed, which may be the reason why for the determination of diffusion coefficients in ion exchangers the radioactive tracer method is much more frequently applied.

Among the ion exchange materials the diffusion processes occuring in macroporous ion exchange polymers deserve to be considered separately, because such resins are now widely used industrially. Macroporous (or porous) ion exchangers are heterogeneous and can be modelled as an agglomeration of microspheres within the bead macrosphere. In other words, a bead of a macroporous ion exchanger is formed of clusters of microspheres, which arrange in groups to give an open, porous structure containing comparatively large voids, called macropores. The microspheres also contain water, and it is essential to realize that ion exchange by diffusion proceeds within the macropores and the microspheres. In the macropores the self-diffusion coefficients of exchanging ions have values lower than, but comparable to, their values in the free aqueous solution. But in the microspheres, the diffusion coefficients are orders of magnitude lower than measured for even quite highly crosslinked gel type materials. It has been shown that diffusion coefficients in both phases are in agreement with the Nernst-Planck theory with regard to their dependence on composition. For a complete understanding, the fluxes in the macropores and the diffusion in the microspheres have to be brought together, resulting in parameters that affect the shape of the conversion vs. time curve, but which have a very simple physical explanation. If the parameter, β, is large, it means that macropore diffusion is rate limiting, while very small values indicate that diffusion into microspheres controls the overall exchange rate. α/β is the ratio of the relative uptake in the microspheres and macropores at equilibrium. When β is small, the uptake is a two-stage process. Diffusion in the macrostructure is very much faster than in the microstructure. The macropores are in equilibrium before any measurable uptake by the micropores occurs. If β is large, the composition of each small sphere is practically uniform, though changing over time in equilibrium with the composition of the macropore at that location. The other special case exists for $\alpha = 0$. In this situation there is no uptake by the microspheres and the model reduces to the simple homogeneous diffusion case (Patell and Turner, [150]). Thus it can be understood why the open, porous structure of the macroporous ion exchange resins does not lead to markedly improved exchange kinetics. The increased rate of diffusion into the centre of a bead through the macropores is counteracted by the much slower diffusion within the microspheres.

The possibilities of ion exchange accompanied by reactions mentioned only briefly above have been presented in a comprehensive analysis by Helfferich [151]. It has to be realized that the ion exchange processes described above as a redistribution of counterions is based, for theoretical treatments, on the assumption that the exchanging ions retain their identity. This is reasonable and justifiable for the simple and most extensively studied systems, such as the exchange of uni- or divalent cations with strongly acidic cation exchangers, a case that is very important in practice. This assumption is, however, questionable with counterions which are hydrolyzed, and is wrong when the ion exchange process is accompanied by chemical reactions such as, for instance, neutralization or complex formation. Helfferich has grouped the ion exchange processes accompanied by reactions into four general classes covering eleven typical cases. As presented in Table 1.7 processes like association, dissociation, precipitation and complex formation can accompany ion exchange, exercising – either by the binding or releasing of counterions – substantial influence. In class I the counterion released from the exchanger is consumed by a reaction. This is the case when H^+ or OH^- released by strong acid or strong base ion exchangers is removed by neutralization with solution of base or acid (reaction I a and I b), or when released H^+ or metal counterions form undissociated acid (reaction I c) or complexes (reaction I d) or are precipitated (reaction I e).

Table 1.7 Ion exchange with reaction. Classification and examples as grouped by Helfferich

Class	Reaction: Initial state		Reaction: Final state			Examples: Initial state		Examples: Final state	
I	$\boxed{-R + A}$	$+ B + Y \rightarrow$	$\boxed{-R + B}$	$+ AY$	a)	$\boxed{-SO_3^- + H^+}$	$+ Na^+ + OH^- \rightarrow$	$\boxed{-SO_3^- + Na^+}$	$+ H_2O$
					b)	$\boxed{-N^+R_3 + OH^-}$	$+ H^+ + Cl^- \rightarrow$	$\boxed{-N_3^+R + Cl^-}$	$+ H_2O$
					c)	$\boxed{-SO_3^- + H^+}$	$+ Na^+ + AcO^- \rightarrow$	$\boxed{-SO_3^- + Na^+}$	$+ AcOH$
					d)	$\boxed{2 - SO_3^- + Ni^{2+}}$	$+ 4\,Na^+ + EDTA^{4-} \rightarrow$	$\boxed{2 - SO_3^- + 2\,Na^+}$	$+ 2Na^+ + Ni(EDTA)^{2-}$
					e)	$\boxed{-N^+R_3 + Cl^-}$	$+ Ag^+ + NO_3^- \rightarrow$	$\boxed{-N^+R_3 + NO_3^-}$	$+ AgCl$
II	$\boxed{-R + A}$	$+ B + Y \rightarrow$	$\boxed{-RB}$	$+ A + Y$	a)	$\boxed{-COO^- + Na^+}$	$+ H^+ + Cl^- \rightarrow$	$\boxed{-COOH}$	$+ Na^+ + Cl^-$
					b)	$\boxed{-NH_3^+ + Cl^-}$	$+ Na^+ + OH^- \rightarrow$	$\boxed{-NH_2}$	$+ Na^+ + Cl^- + H_2O$
					c)	$\boxed{-N(CH_2COO^-)_2 + 2\,Ba^+}$	$+ Ni^{2+} + 2\,Cl^- \rightarrow$	$\boxed{-N(CH_2COO^-)_2Ni}$	$+ 2\,Na^+ + 2\,Cl^-$
III	$\boxed{-RA}$	$+ B + Y \rightarrow$	$\boxed{-R + B}$	$+ AY$	a)	$\boxed{-COOH}$	$+ Na^+ + OH^- \rightarrow$	$\boxed{-COO^- + Na^+}$	$+ H_2O$
					b)	$\boxed{-NH_2}$	$+ H^+ + Cl^- \rightarrow$	$\boxed{-NH_3^+ + Cl^-}$	
					c)	$\boxed{-N(CH_2COO)_2Ni}$	$+ 4\,Na^+ + EDTA^{4-} \rightarrow$	$\boxed{-N(CH_2COO^-)_2 + 2\,Na^+}$	$+ 2\,Na^+ + Ni(EDTA)^{2-}$
IV	$\boxed{-RA}$	$+ B + Y \rightarrow$	$\boxed{-RB}$	$+ A + Y$		$\boxed{-N(CH_2COO)_2Ni}$	$+ 2H^+ + 2\,Cl^- \rightarrow$	$\boxed{N(CH_2COOH)_2}$	$+ Ni^{2+} + 2\,Cl^-$

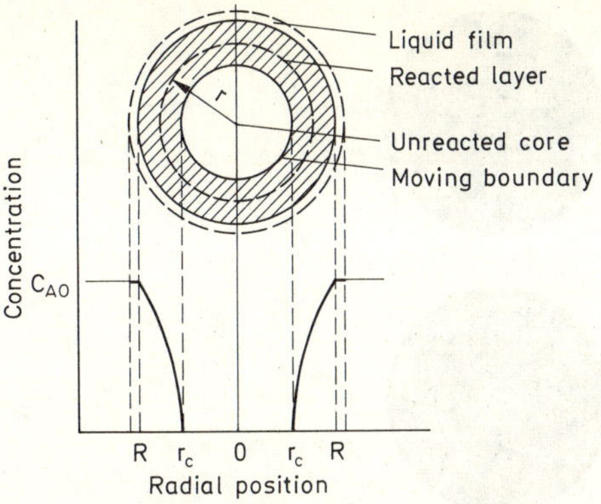

Figure 1.34 Schematic representation of the shell progressive mechanism in an ion exchanger bead according to Schmuckler and coworkers.

Helfferich predicted that the rate laws which could be derived for such ion exchange processes would appreciably differ from those for ordinary ion exchange in the absence of ionic reactions. This was later confirmed by the results of kinetic studies on the neutralization of strong acid and weak base ion exchange resins [152]. In class II and class IV the counterion to be taken up by the exchanger from the solution is characteristically consumed by a fast chemical reaction, and the interacting ions progress in shells into the resin matrix. Under the microscope a sharp moving boundary can be observed during the exchange process, when a completely reacted shell envelops an unreacted core. From a diffusion perspective, this is the so-called shell progressive mechanism (also known as shell-core, shrinking core or moving boundary mechanism) which is illustrated in Figure 1.34 [153], where the development of the concentration gradient from C_{Bo} to zero is traced. This model shows that the rate of these processes is controlled by the diffusion of the reactant through the reacted layer. The likelihood of a moving boundary in a spherical ion exchanger bead was first suggested by Helfferich for the elution of the sodium form of a weak acid carboxylic exchanger with hydrochloric acid (reaction II a). In this case a change of the resin from a dissociated to a non-dissociated species accompanies the exchange, but it was also realized that this was the exchange mechanism for, among others, the chloro-palladate-chloride interaction on a strong base anion exchanger and the chelation of copper on Dowex A-1. The description of the rate of ion interdiffusion in complex-forming ion exchangers (reaction II c) by using the isotherm yield dependences of the mass transfer rate on the exchange selectivity as the ratio between the dissociation constants for the complexes of the desorbed ion and the sorbed ion, was used as the parameter characterizing the exchange isotherm [154]. An optical verification of the sharp loading profiles inside the resin phase of weak electrolyte resins in their salt forms was presented by Hoell for weak acid cation exchangers or weak base anion exchangers (reaction II b), together with an isotherm close to rectangular in form. The corresponding fronts can be made visible

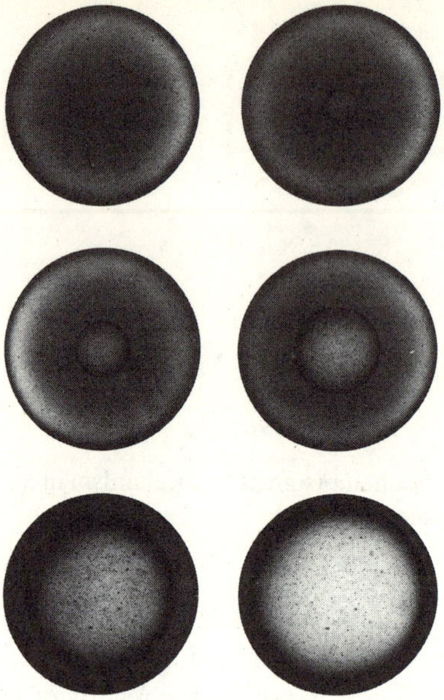

Figure 1.35 Succession of pictures of the exchange $Ca^{2+} \rightarrow H^+$ of a weak acid cation exchanger bead after 2, 4, 8, 11, 13, and 15 min with 1 N HCl (Höll).

in transparent resins, and observation is possible due to gradients of the refractive indices or to transitions in color [155]. Evaluation of photographs, as e. g. shown in Figure 1.35, provides exact evidence for the exchange processes in the resin types concerned. For reactions of class III two mechanisms are possible. One is dependent on electrolyte invasion caused by the counterion B and the co-ion Y – as a typical case (reaction III a) the OH^- ion as a co-ion is the minority ion in the shell, which means that ion exchange is controlled by diffusion of a co-ion. For more general considerations it is of interest to discuss what the rate control is for the ion exchange in systems with reaction (Helfferich). A chemical reaction may affect the rate of ion exchange in two ways. In the first case, the reaction rate is slow compared with diffusion. In the limiting condition, diffusion is fast enough to compensate for any concentration gradients within the ion exchanger bead. The reaction at the fixed site is then the rate-controlling factor and the rate of the ion exchange is entirely independent of particle size. In the second, the reaction may be faster than diffusion, and is binding ions on the diffusion of which the ion exchange depends. The binding of the ions then inhibits their diffusion and slows the exchange down. The rate of exchange is controlled by slow diffusion which, in turn, is affected by the equilibrium of the reaction. The rate, then, is diffusion-controlled and depends on particle size. This is also a situation which often, although not always, leads to a progressive shell mechanism. It is still possible, for all the above-described ion exchange and reaction processes to calculate fairly accurately all the necessary kinetic parameters, as well as to determine its extent of deviation from ordinary ion exchange, when one knows the nature of variation of the ion concentration through the diffusion zone [156].

Acidity and basicity. Ion exchangers have also been defined as high molecular weight solid acids or high molecular weight solid bases whereby these terms are equivalent to the more common expressions cation exchangers or anion exchangers. It is, therefore, of interest and considerable importance to examine in any exchange material the chemical properties from the point of view of acidity and basicity as well, and to determine the chemical nature of the ionogenic group. In principle, chemically, there are two fundamental types of ion exchanger, those in which under all normal conditions the ionogenic group is essentially completely dissociated and those where the extent of dissociation is markedly pH-dependent, corresponding to whether the ionogenic groups represent strong or weak acids, and strong or weak bases. Briefly summarizing, phenolic OH-groups or carboxylic acid groups remain partially or completely undissociated at low pH values, so that, for instance, the exchange capacity of exchangers with such ionogenic groups with respect to metal cations from acidic and neutral solutions is lower than from alkaline solutions. Anion exchangers with primary, secondary and tertiary amino groups exchange acid anions from alkaline and neutral solutions to a lesser degree than from acidic solutions. However, ion exchangers with strongly acidic or strongly basic groups – in the latter case quaternary ammonium groups – are capacity-independent over a broad pH range. Accordingly, they are also active over a wide pH range and can be used under acidic, neutral and alkaline conditions. Table 1.8 shows a summary of the effective pH ranges of customary types of ion exchangers. In other words, ion exchangers show a certain acidity or basicity as an important chemical property. To evaluate the acidity or basicity of any exchange material, inorganic or organic, so-called pH titration curves are determined, which involves the titration of the exchanger with either base or acid at various ionic strengths of the solution.

Table 1.8 Effective pH range of different customary types of ion exchangers

Type of ion exchanger	Approximate pH range
Strong acid cation exchanger	0–14
Weak acid cation exchanger	5–14
Weak base anion exchanger	0– 9
Strong base anion exchanger	0–14

These titration curves are an aid, moreover, in the determination of the total number of ionogenic groups and the number of types of ionogenic groups present. It may be added that optimum operating pH ranges have been obtained based on theoretical considerations from titration curves of cation (sulfonic and carboxylic) and

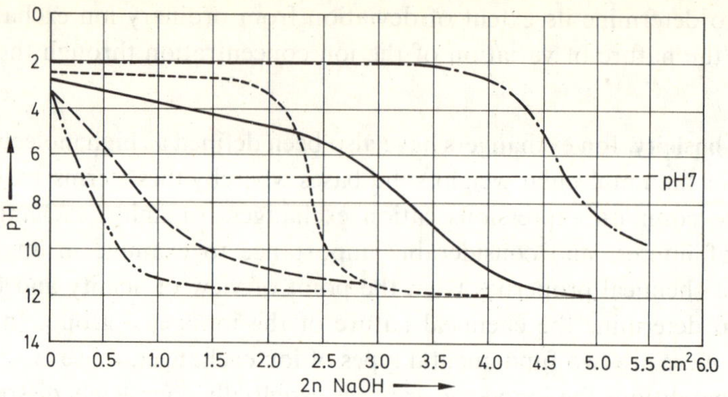

Key	Designation	Active Group
—·—·—	K-resin	-OH
---------	A-resin	- CH_2SO_3H
————	C-resin	- COOH
—··—··—	R-resin	- SO_3H
– – – –	Glauconite, dealkalized for comparison, test series for 20g	

Figure 1.36 pH-titration curves of some cation exchangers according to Grießbach.

anion (strong and medium base) ion exchange resins, and are recommended as follows: sulfonic pH = 4.7–9.3, carboxylic 7.3–9.3, medium base 4.7–8.0 and strong base 4.7–10.1 [157]. In the case of cation exchangers Griessbach has introduced the method of pH titration, again based on the fact that cation exchangers in the H-form can be titrated with alkalis like ordinary acids, with the advancing neutralization process being evaluated by measurement of the pH. To perform this direct titration, 5 g of the cation exchanger were first transformed into the H-form with hydrochloric acid, then washed with distilled water, and finally treated with 0.1 N NaCl solution as a neutral salt in a beaker. Small amounts of 2 N sodium hydroxide were added with vigorous stirring, and the pH adjustment was observed on the glass electrode. For an evaluation of the test results, the measured pH values were plotted as a function of the consumed quantity of alkali. Figure 1.36 shows the titration curves of some cation exchangers according to Griessbach [158]. The capacity of the exchanger can be read from these curves at the intersection of the curve with the line for pH 7. Moreover, these curves can be used to evaluate the exchangers for their acid strength, operating interval, and buffering power. Further, from pH titration curves information on the nature and concentration of the acidic or basic groups in both types of ion exchangers, cation as well as anion, can be obtained. A very important fact in this connexion is that, if the direct titration of a cation exchange resin in the H-form is attempted, there will be no pH change in the external solution until the endpoint of the titration is reached. The reaction for, e. g., a strong acid cation exchanger can be written $R-SO_3H + NaOH = R-SO_3Na + H_2O$, whereby the actual titration takes place inside the resin, and the pH changes cannot be measured directly. Griessbach has given an exact empirical expression describing the exchange reaction with $(Na^+)_R/(H^+)_R \cdot [(H^+)_S/(Na^+)_S]^n = K$, where the subscripts refer to the resin and external solution phases. Concentrations are used in place of activities, since the ratio of the activity coefficients of the univalent cations used is nearly equal to unity. But if a large excess of a neutral salt of sodium is added, the ratio $(Na^+)_R/(Na^+)_S$ becomes small, since n is 0.8–1.0 and K is 0.5–2.0 for most resins. The ratio $(H^+)_S/(H^+)_R$ must become very large which means that virtually all of the hydrogen ions are in the solution phase. Thus the addition of a neutral salt displaces an ion into solution upon dissociation, and yields results comparable to those which would be obtained if the resin were soluble. This is also the case for weak acid resins. For anion exchangers the same principle is valid. Therefore the direct pH titration of strong base anion exchangers is also carried out in the presence of a large excess of neutral salt. An experimental alternative but one definitely necessary, however, for the determination of pH titration curves of weak acid and weak base ion exchangers was adopted by Topp and Pepper in the method of weighed samples. Briefly, with cation exchangers, different amounts of base were added to weighed samples, 0.5 g, of the hydrogen form of the resin in separate flasks. After 48 hours of shaking, the pH of the solution was determined and aliquot samples were withdrawn for analysis. A parallel series of experiments was made in the presence of neutral electrolyte. The results obtained provide a basis for the interpretation of the titration curves of multiionic exchangers. Thus, resins containing both hydroxyl and sulfonic or carboxylic groups give titration curves which consist of segments characteristic for each ionisable group present. Anion exchangers available at that time were also studied with the conclusion, for instance, that a *m*-phenylenediamine-formaldehyde resin was ill-defined since the corresponding titration curve showed that it contained a variety of weakly basic groups, including primary aromatic amino and mixed aliphatic-aromatic secondary and tertiary amino groups [159].

Kunin and coworkers then studied in depth the acidity and basicity behavior of ion exchangers. For cation exchangers in the H-form titration curves for various ionogenic groups such as the methylene sulfonic $-CH_2SO_3H$, nuclear sulfonic $-SO_3H$, carboxylic $-COOH$, phosphonic $-PO_3H_2$, phosphonous $-PO_2H_2$, and phenolic $-OH$ groups were plotted, as shown in Figure 1.37 [160]. It can be seen that the titration curves, which were determined for different ionic strengths of the solution, are quite similar to the corresponding curves of the analogous soluble acids. The apparent ionization constants (pK_1) can be calculated from these curves from the relationship $pH = pK_1 + \log \text{salt/acid}$. It is important, however, to realize that upon changing the ionic strength or upon changing the base the titration curve of a cation exchange resin is a result not simply of the basic neutralization reaction $H^+ + OH^- \rightarrow H_2O$, but must be interpreted as a combination of two reactions, i.e.,

$$R{-}H + M^+ \rightarrow R{-}M + H^+$$
$$H^+ + OH^- \rightarrow H_2O.$$

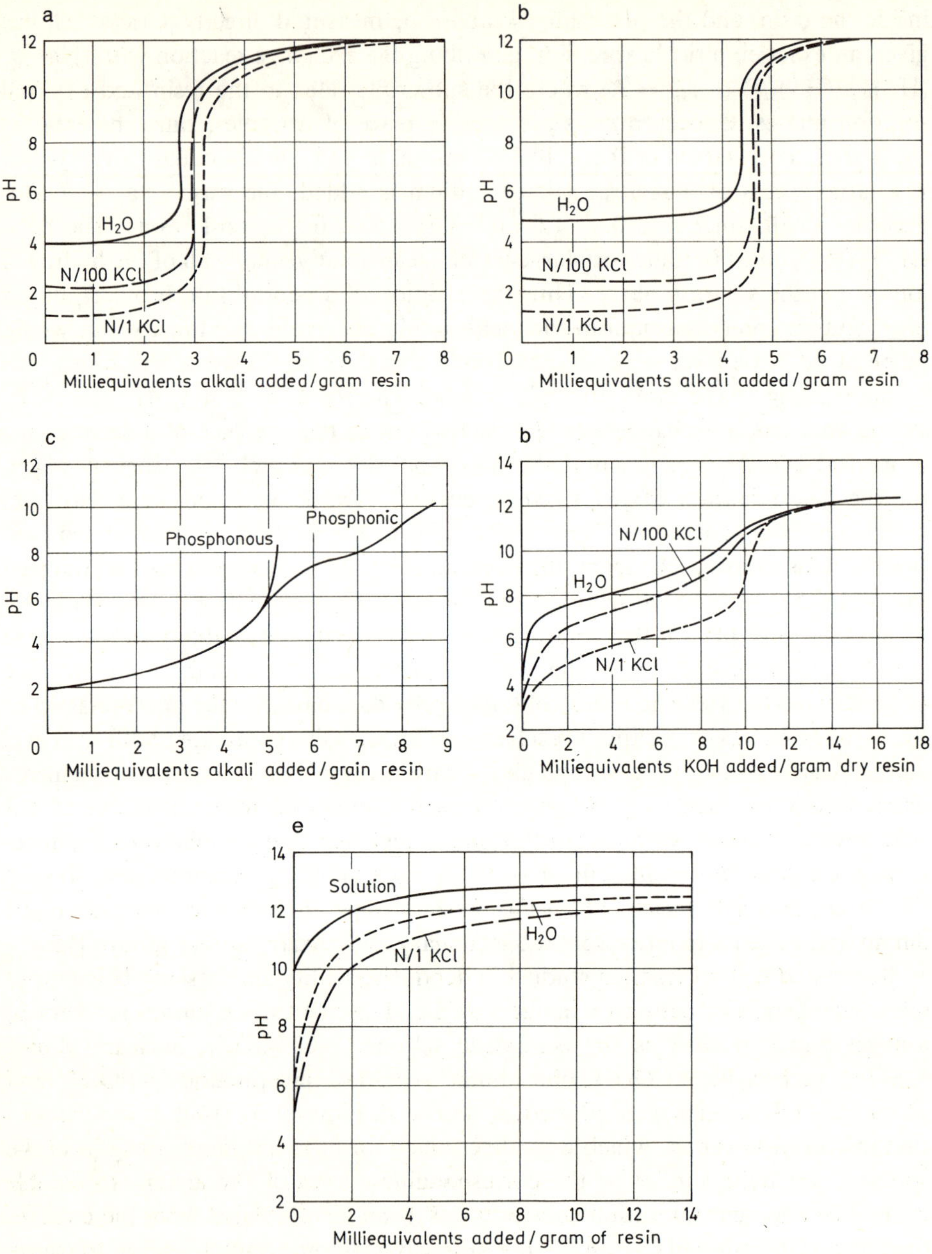

Figure 1.37 pH-titration curves of cation exchangers with various ionogenic groups: $-CH_2SO_3H$; $-SO_3H$, $pK_1 = 2$; $-PO_3H_2$, $pK_1 = 3$; $-PO_2H_2$, $pK_1 = 3$; $-COOH$, $pK_1 = 5-6$; and OH, $pK_1 = 10$ (Kunin).

It thus becomes dependent on the equilibrium, including the selectivity of the fixed site for the exchanging ion. However, since the equilibrium of the exchange is also dependent on the acid strength (pK value) of the acid group, the equilibrium constant becomes, at first sight, essentially a measure of the acid strength of the fixed group. From the curves in Figure 1.37 it can also be seen, as already mentioned, that different ionic strengths or, in other words, the presence of salt, have distinct effects upon the shape and position of the titration curve. The magnitudes of these effects are dependent on the acid strength of the ionogenic group. Gregor and Bregman [161] first determined the order of increasing acid strength for the three major resinous cation exchangers

$$SO_3H > PO_3H \cong PO_2H_2 > COOH > OH.$$

But at very high ionic strengths the titration curves of the various cation exchangers become quite independent of the ionic species.

Practically, the titration curves reveal that with cation resins of the sulfonic acid type the hydrogen ion behaves much like an alkali metal ion. However, the carboxylic group titrates with KOH between pH 7 and 11 in the absence of a salt, and is half-neutralized at about pH 9. In the presence of 0.1 N KCl, the titration curve extends from pH 4 to 10, half-neutralization being at about pH 7. The values are lowered by about 1 pH unit if $Ba(OH)_2$ is used in place of KOH [162]. The phenolic group in the phenol-formaldehyde resins is weaker and takes up base from about pH 10 upwards or pH 8 in 0.1 N KCl.

As described above Griessbach has already set up an empirical equation for the effect of the various cationic species upon the position and shape of the titration curve, based on the equilibrium constant; a somewhat similar one can be easily derived that qualitatively describes quite well the effect of concentration on the position of the titration curve. But such an approach does not take into consideration the physical chemistry of the polymer structures. Helpful for the field of ion exchangers in gaining knowledge of the whole electrochemical situation of the system, i. e. polymeric matrix versus ionized ionogenic groups, are similar endeavours in the field of polyelectrolytes, the latter being the soluble, polymeric (because they are not crosslinked) analogues of monomeric acids. The electrochemistry of these important reactive polymers has been studied intensively by Kern, Fuoss, Katchalsky, Mandel, Oosawa and several others.

It is state of the art for polyelectrolytes to use as a basis for potentiometric investigations the Henderson-Hasselbach equation derived from the dissociation equilibrium of monovalent electrolytes: the equation is $pH = pK_a + \log \alpha/(1-\alpha)$, with K_a the dissociation equilibrium constant according to the law of mass action and α the degree of neutralization calculated from the amount of base added. On investigating the specific effects of solutions of polyelectrolytes it was found necessary to modify this equation (Overbeck, Katchalsky and Spitnik, Mandel [163]); the form then used, in most cases, is the empirical Katchalsky and Spitnik version of the Henderson-Hasselbach equation: $pH = pK_a + n \log \alpha/(1-\alpha)$, with pK_a the apparent pK_a value of the polyacid and n another constant. Fisher and Kunin have attempted to bridge the gap between the polyelectrolytes and the polymeric ion exchangers by using this equation [164], and Kuhn et al. [165] as well as Helfferich, have, for the description of the neutralization curve of a carboxylic cation exchanger in the presence of a monovalent counterion, proposed the following equation

$$pH = pK_a + \log \frac{\alpha^2}{1-\alpha} + \log[\overline{X}] - \log[M^+],$$

with pH the pH measured in the equilibrium solution, K_a the apparent acidity constant, α the degree of dissociation, $[\overline{X}]$ the total concentration of ionogenic groups in the exchanger, and $[M^+]$ the concentration of counterions in the equilibrium solution. But in order to determine the nature of the binding between a carboxylic ion exchanger and a monovalent counterion including very weak dissociations, Auclair et al. used a system of equations for three different assumptions [166], relative to the different ways of binding.

For anion exchangers also titration curves show one of the important characteristic properties. The anion exchange resins that were originally and for quite some time available were of the weak base type, functioning only in acid media; it has become possible with the availability of strong base anion exchangers to extend the scope of anion exchange operations to neutral and alkaline media as well. Kunin and coworkers studied intensively the behavior of all types of anion exchange resins, and it became apparent that the range of basicities covers a range corresponding to strong, weak and very weak bases. Since it was realized that the ionic strength of the solution showed a marked effect on the titration curves, especially for the weakly basic exchangers, this was seen as an indication that the acid neutralization of these basic resinous exchangers might well proceed as a combination of two reactions, i. e.

$$RNH_3OH + Cl^- \rightleftharpoons RNH_3Cl + OH^-$$
$$OH^- + H^+ \rightarrow H_2O.$$

For anion exchangers in the OH-form the titration curves for various anionic groups are shown in Figure 1.38 [167]. The hydroxyl ion with the strongly basic resins behaves much like chloride. However, with weak base resins pH plays a special role, as a weak base resin passes into the undissociated base form at high pH values. Strong base anion exchangers as quaternary ammonium resins are ionized at all pH values and are fully neutralized with chloride from pH 8 downwards. The weakly basic groups $-NH_2$, $-NHR$ and $-NR'R''$ take up HCl between pH 7 and pH 1. With respect to the ionic strength or the presence of salts, the magnitude of the ionic strength effect increases as the basicity of the resin decreases. For example, the half-neutralization point of a resin can be at pH 4 with HCl, and at pH 5.5 with HCl plus 0.1 N KCl. However, acid strength and the anion valence also have pronounced effects on the nature of the titration curves. Increases in acid strength and the valence of the ion favor the replacement of the hydroxyl ion.

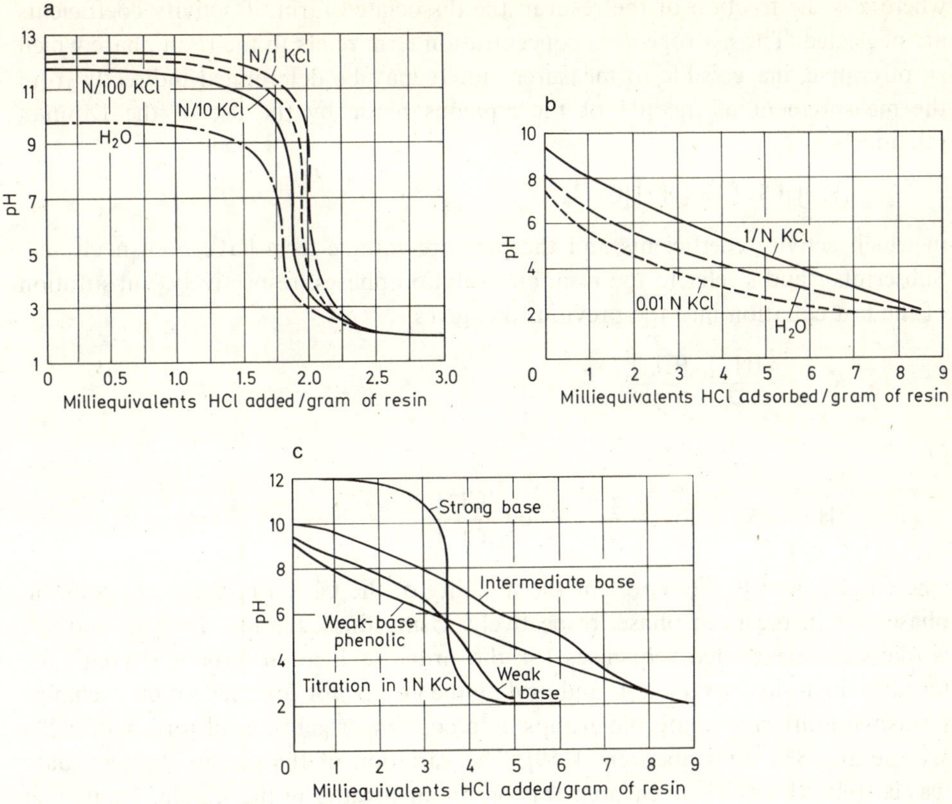

Figure 1.38 pH-titration curves of anion exchange resins: strong base, weak base and some resins of different basicity (Kunin).

Because, in spite of the increasing technological importance of weak base ion exchange resins, so little information concerning their basicities had been obtained, Kunin and coworkers examined the basicities of a series of weak base ion exchange resins [168]. In this work the precautions necessary and the techniques to be applied are critically reviewed. For the given calculations it would seem best to follow the descriptions exactly. The dissociation of the protonated form of a weakly basic resin may be described by the equilibrium

$$R\overset{+}{N}H \rightleftharpoons RN: + H^+$$

$$K_a = \frac{(RN:)\,(H^+)}{(R\overset{+}{N}H)},$$

where the parentheses indicate activities. This expression may be written as

$$K_a = \frac{\alpha\,[H^+]}{1 - \alpha},$$

where α is the fraction of the resin in the dissociated form, if activity coefficients are neglected. The hydrogen ion concentration term refers to the resin phase which is, of course, inaccessible to measurement. It may be determined indirectly from the measurement of the pH of the aqueous phase by the use of the Donnan relationship,

$$[H^+]_r[Cl^-]_r = [H^+]_s[Cl^-]_s,$$

in which activity coefficients and the pressure-volume term have been neglected. Subscripts r and s refer to the resin and solution phases, respectively. Substitution of the last equation into the previous one gives

$$K_a = \frac{\alpha[H^+]_s}{1-\alpha}\frac{[Cl^-]_s}{[Cl^-]_r}$$

or

$$pH = pK_a + \log\frac{\alpha}{1-\alpha} + \log\frac{[Cl^-]_s}{[Cl^-]_r}$$

where $[Cl^-]_s$ and $[Cl^-]_r$ represent the molality of the Cl ion in the outer solution phase and in the resin phase, respectively. Using this equation, Dragan and co-workers have reported sequences for the order of increasing base strength for tertiary alkyl-, hydroxyl alkyl- and aryl amine groups for fourteen anion exchange resins with differing ionogenic groups on a copolymer gel in bead form with 92% styrene and 8% divinylbenzene [169]. The variation of the derived basicity data was correlated with the measurements on retained water in the standard state and volume changes during transformation from Cl- to OH-form of the anion exchange resins. The results derived by the potentiometric titration curves in 1 M NaCl solutions, by the pH values of the water in equilibrium with the anion exchange resins in OH-form, by the studies of the ion exchange OH $\leftrightarrow$ Cl at equivalent concentrations, as well as by measurements of the volume changes from one ionic form to another, all lead to the same sequence of basicity for the fourteen anion exchange resins. In the series of strongly basic anion exchange resins synthesized with tertiary alkyl amines, the basicity decreased with the increase of the chain length of the alkyl substituent, inversely to the basicity of the free amine. The strongly basic anion exchange resins synthesized with hydroxy alkyl amines had a weaker basicity than those obtained with alkyl amines, and the replacement of an alkyl substituent at the quaternary nitrogen by a hydroxy alkyl substituent of a similar volume, decreased the basicity. The anion exchange resins synthesized with primary and secondary aromatic amines could not be titrated in a satisfactory way in an aqueous medium.

There have been repeated recent investigations based on titration curves. The two methods now established for titration curves are the method of direct titration or the method of individually weighed samples. The method of direct titration can be used for exchangers with easily dissociated ionogenic groups and with fast exchange kinetics. Exchangers containing weakly dissociated ionogenic groups and with slow exchange equilibration are treated by the method of individually weighed samples. The latter, however, can be used for any type of exchanger. Babkin and Kiseleva reviewed the testing of ion exchangers by stepwise potentiometric titration and Michaeli and Kinrot discussed the salt effects on pH titration curves of weak acid or weak base ion exchangers [170]. Of interest seems to be that the determination of ionization constants and exchange capacities of ion exchangers by potentiometric titration at 80 °C decreases the duration of the determination ten to twenty-fold, as compared with titrations performed at room temperature; the p*K* values of the cation exchangers examined were linearly dependent on the ionic strength of the solutions, and were virtually independent of temperature at 20 °C to 80 °C [171]. For the analysis of copolymers based on 2-methyl-5-vinylpyridine as well as of polyamine ion exchange resins from polyepichlorohydrin titration curves were used [172].

Ion exchange resin buffers. It has been mentioned that in the titration curves of weakly acidic ion exchangers the range of buffering activity can be clearly seen. This means that, in analogy to buffers in solution consisting of a weak acid and its salt, mixtures of an ion exchange resin in the H-form and a neutralized exchanger are also very good buffer systems. The same applies to a weak base anion exchanger mixed with a resin, e. g., in the SO_4-form. Comparing the buffering effects of ion exchange resins with those of weak acids and bases the advantages for the ion exchange technique are a wider buffering range, fewer ion effects on physiological experiments, low osmotic pressure effects and the ability to change the pH during

Table 1.9 The pH of Murashige and Skoog's culture medium containing various proportions of Amberlite IRC-50 charged with either hydrogen or calcium ions. The pH values for + cells are for five-day-old carrot cultures, the pH values are mean and standard error of three replicates. (From 174)

Grams Resin		Medium pH	
H^+	Ca^{2+}	− Cells	+ Cells
0.0	5.0	8.6 ± 0.08	8.0 ± 0.05
0.5	4.5	7.1 ± 0.08	6.6 ± 0.1
1.0	4.0	6.6 ± 0.02	6.2 ± 0.02
1.5	3.5	6.3 ± 0.02	5.9 ± 0.01
2.0	3.0	5.9 ± 0.01	5.7 ± 0.01
2.5	2.5	5.7 ± 0.03	5.4 ± 0.0
3.0	2.0	5.4 ± 0.04	5.2 ± 0.0
3.5	1.5	5.2 ± 0.03	4.9 ± 0.0
4.0	1.0	4.9 ± 0.03	4.7 ± 0.01
4.5	0.5	4.5 ± 0.06	4.4 ± 0.04
5.0	0.0	2.9 ± 0.03	2.9 ± 0.01
0.0	0.0	5.0 ± 0.12	4.6 ± 0.07

an experiment without changing the rest of the composition of the suspending media. The buffer capacity of ion exchangers has been determined by automated direct recording of potentiometric titration curves in a broad range of pH under conditions for maximum measurement reproducibility [173]. The practical applicability of ion exchange resin buffers was demonstrated in connexion with liquid plant cell cultures as a simple and effective method for reducing pH fluctuations in liquid cultures of carrot cells – it involves adding to the culture medium before autoclaving a cation exchanger system consisting of a mixture of the H- and Ca-forms of the carboxylic acid resin. The pH conditions achieved are shown in Table 1.9. The resin is nontoxic and nonabrasive to the carrot cells and can be stored in its treated form for an extended period of time [174].

Suspension effect. A phenomenon quite closely related to acidity and basicity of ion exchangers is the so-called suspension effect. Discovered by Jenny and interpreted first by Pallmann mainly on clay, the suspension effect was ascribed to a more or less strong adsorption of hydrogen ions, with its intensity decreasing with decreasing density of the adsorbed layer. Experimentally it can be shown that cation exchangers in the H-form give rise to a low pH when stirred in water around a glass electrode, while anion exchangers increase the pH because the latter supposedly contain adsorbed hydroxyl ions. When stirring is interrupted the resin particles sediment quickly and the supernatant liquid is neutral [175]. Chernoberezhskii and coworkers have intensively investigated the suspension effect of ion exchange resins in suspension and defined ΔpH = (pH of suspension – pH of equilibrium solution). Its magnitude of 2–3 pH units generally depends, among other things, on the concentrations and sizes of the solid phase particles, on the density of the electric charge, and on the concentration of electrolyte [176]. The two hypotheses persisting regarding the cause of the suspension effect have been critically re-examined, based on the data obtained by three independent methods [177]. The conclusion drawn is, that not the traditional hypothesis, which attributes the effect to a different activity of ions in the suspension than in the dialyzate, but the hypothesis that attributes the effect largely to a junction potential at the calomel-suspension boundary, is correct. The suspension effect for compacted ion exchanger resin beads was analyzed with very positive results in terms of an ionic, space charge generated phase boundary potential difference between the supernatant and slurry phases [177].

Exchange of monovalent ions for ions of higher valence. When an ion exchanger is described as a solid electrolyte having a "free" ion that can reversibly interchange with ions in solution, the example given is quite often that of a cation exchanger with fixed sulfonic acid groups on a polymeric matrix R in the sodium form, which is used for the important application of water softening, i. e. the exchange against calcium ions according to the equation

$$2\,RSO_3^-Na^+ + Ca^{2+} \rightleftharpoons (RSO_3^-)_2Ca^{2+} + 2\,Na^+.$$

Attention is not drawn, however, to the fact that this process might not be as easily explainable as the exchange of monovalent ions. In an attempt to give a more adequate explanation one may revert to the above-introduced practice of application of the law of mass action to the exchange process, in which monovalent and bivalent ions are involved as represented by the general equation

$$2\,R \cdot A^+ + C^{2+} \rightleftharpoons R_2 \cdot C^{2+} + 2\,A^+,$$

and then by the apparent exchange constant relationship

$$K_c = \frac{[R \cdot A^+]^2\,[C^{2+}]}{[R \cdot C^{2+}]\,[A^+]^2}$$

with $[RA^+]$ and $[RC^{2+}]$ the concentrations of the ions in the resin and $[A^+]^2$ and $[C^{2+}]$ their concentrations in solution. If the same units are used to express the concentration of the ions in the resin and in the solution, then K_c will be dimensionless, but its value will depend on the units employed. By introducing the equivalent fractions of the ion C this equation may also be written in the form

$$\frac{[\overline{X}_C]}{[1 - \overline{X}_C]} = K_c \frac{[\overline{C}^{2+}]}{[C^{2+}]} \frac{[X]}{[1 - X]^2}$$

where $[\overline{X}_C]$ and $[X_C]$ are the equivalent fractions of the ion C^{2+} in the resin and in solution and $[\overline{C}^{2+}]$ and $[C^{2+}]$ are the total concentrations of the exchanging ions in the resin and in the solution. Unlike the corresponding equation which can be derived for monovalent ions, this equation involves the term $[\overline{C}^{2+}]/[C^{2+}]$, so that the effective selectivity of the resin is determined by the value of $K_c([\overline{C}^{2+}]/[C^{2+}])$. Since $[\overline{C}^{2+}]$ is fixed by the exchange capacity it will be determined by the resin being used, and will not be affected by changes in the concentration of the external solution. The value of $[C^{2+}]$, however, will depend on the total concentration of the ions in the solution. It follows that the value of $[\overline{C}^{2+}]$ will depend not only on K_c and $[X]$ but also on $[C^{2+}]$, and the effective selectivity of the resin for the divalent ion will be inversely proportional to the concentration of the solution, which means that the relative uptake of the divalent ion will be much greater in dilute solution. The effect of solution concentration has been described in the literature either by non-linear curves, when the theoretical values of $[\overline{X}_C]$ are plotted against $[X_C]$ for different values of $K_c([\overline{C}^{2+}]/[C^{2+}])$, or as straight lines, if the appropriate mass action functions are plotted [178]. In exchange processes with strong acid cation exchangers involving monovalent and tervalent ions, the effect of the ionic strength of the solution is much greater. In this case the equivalent fraction of the tervalent ion in the resin will be inversely proportional to the square of the total solution concentration.

For weak acid carboxylic cation exchangers the exchange of monovalent for divalent counterions has also been intensively studied in order to determine the nature of the binding. For anionic polyelectrolytes and copper, as well as for alkaline earth cations, it had been postulated in prior investigations by one group of researchers that the binding is a complex formation, and by an other group that it is of electrostatic nature. For carboxymethylcelluloses and the divalent ions Mg^{2+}, Ca^{2+}, Sr^{2+}, and Ba^{2+} it was found that the degree of association depends little on the nature of these cations but is entirely determined by the charge density of the polyelectrolyte and should, therefore, be of purely electrostatic nature. In all the studies in question it was shown that the binding of a divalent cation and a weak acid polyelectrolyte takes place between two ionizable groups of the polyion, whereby these bound sites can be either on one polymer chain or on two different ones. Then with respect to weak and carboxylic cation exchangers and taking into consideration the different modes possible between compensating ions and dissociated ionogenic groups and the relative disposition of the carboxylic groups in the ion exchange resin, a series of equations was set up, representing the exchange reaction. From these it was concluded that carboxylic cation exchangers occur as diacids during the exchange of divalent ions, but that Ba^{2+} is only weakly bonded [179]. When one includes ions of higher valence than divalent in studies of metal ion adsorption on carboxylic cation exchangers (i. e. other than Cu^{2+}, Pr^{2+}, Ni^{2+} and Ca^{2+}), it seems evident that there is an additional charge formed after the exchange of H^+ for the metal ion. Further, it was found that the cations UO_2^{2+}, Bi^{3+}, ZrO^{2+} and Th^{4+} have high distribution coefficients even at pH 2.0 from aqueous acidic solutions with a weak acid cation exchanger which contradicts the idea that carboxylic exchangers have a high affinity for the hydrogen ion. The affinity for Cu^{2+} and Al^{3+} above pH 4.0 was comparable to the affinity for H^+ [180]. It might also be said that the more complicated nature of the binding of divalent cations on carboxylic ion exchangers is also shown by work done on the determination of the effective exchange constants of ions and the dissociation constants of complexes of these ions in the ion exchanger phase. The values of theoretical constants for the weakly acidic carboxylic exchanger KB-4P-3 for alkali metal ions were in good agreement with the experimental values whereas for alkaline earth ions the agreement was only satisfactory [181].

Anion exchange resins show basically similar effects towards divalent anions: in general polyvalent anions are preferentially taken up from a dilute solution. An investigation of the $Cl^- - X^{n-}$ exchange, where X^{n-} were NO_3^-, HSO_4^-, SO_4^{2-}, CO_3^{2-}, PO_4^{3-}, on a strong base anion exchanger, showed that the ion exchange selectivity of the multivalent anions is not controlled by electrostatic forces. It was concluded that the hydration effects are predominantly responsible for the selectivity, and a correlation was found between the ion hydration energy and selectivity constant values. These values decreased on this particular anion exchanger with increasing n value of X^{n-} [182]. In similar studies, Soldatov and coworkers have extended the investigation to the influence of crosslinking and the sorption of water with regard to the equilibrium of exchange of anions of different valences on a strong base anion exchanger. Using Dowex 1 X 1, 1 X 4 and 1 X 10 the ion exchange equilibria of the anions $SO_4^{2-} - Cl^-$, $SO_4^{2-} - OH^-$ and $PO_4^{3-} - OH^-$ and sorption of water of the Cl^--, OH^--, SO_4^{2-}- and PO_4^{3-}-forms were determined at 25 °C. With increasing crosslinking of the exchanger, the ion exchange constant of the $SO_4^{2-} - Cl^-$ system decreased and of the remaining systems changed in a complicated way [183]. The most conclusive investigation, however, of the exchange of monovalent for divalent anions as well as of the selectivity of anion exchangers is based on structural considerations of the resin phase, making use of the distance-of-charge-separation theory of Clifford and Weber. The results of this research demonstrate that it is the distance of fixed-charge separation in the resin which is the primary determinant of monovalent-divalent selectivity. The goal of the research was to determine the causes of sulfate and nitrate selectivity in anion resins and it was found that the resin matrix and the type of ionogenic group were the most important factors. Because the uptake of a divalent ion, e. g. sulfate, requires the distance of two closely-spaced positive charges, a resin is highly sulfate selective if the electrostatically-active nitrogen atoms (amine) are in the continuous polymer structure. It is hypothesized in this theory that this extreme divalent ion preference is due to the proximity (4.48 Å apart) of two active nitrogen atoms in the polymer backbone. This distance, 4.48 Å, is the nitrogen separation distance due to the single ethylene group in the amine monomers diethylenetriamine and triethylenetetraamine, commonly used to provide reactive sites and crosslinking in anion exchange resins:

$$|\longleftarrow 4.48\ \text{Å} \longrightarrow|\longleftarrow 4.48\ \text{Å} \longrightarrow|$$
$$NH_2 - CH_2 - CH_2 - NH - CH_2 - CH_2 - NH_2$$

Fixed pairs of properly-spaced positively charged ionic sites will tend to prefer single divalent anions to pairs of monovalent ions for both electrostatic and entropic reasons. In order to bring positive charges into close proximity within a resin one can: 1. Incorporate the amine ionic groups into the polymer chains, as opposed to having them pendant on the chains. 2. Minimize the size and number of organic groups attached to the N atom, i. e. minimize the size of the amine. 3. Minimize the resin flexibility, i. e. its ability to reorient to satisfy divalent counterions, by minimizing the degree of crosslinking [184]. This theory would also seem to provide as stimulus to further investigate both structure of the copolymer and distribution of the ionic groups as the actual chemical basis of ion exchange and ion exchange selectivity, since the distance-of-charge-separation theory is not restricted to divalent anion exchanges but also applies to cation exchanges and to polyvalent ions in general.

Ion exchange in multicomponent systems. Ternary and higher multicomponent systems in the field of ion exchange are understood as systems in which the exchange of one bound counterion is performed in a solution offering more than one ion for exchange. If one also includes ions of higher valence for both cation and anion exchangers, it is obvious that this field becomes quite wide, since several other chemical interactions and mutual ion effects must be expected and it would, therefore, be necessary to examine such systems with respect to equilibrium conditions, selectivity and diffusion conditions, etc. Although practically speaking, just such multicomponent systems are usually involved in ion exchange, they have unfortunately been described far less than binary systems, regardless of whether the approach has been theoretical, mechanistic or purely chemical.

One empirical effect found in ternary systems must be mentioned at the outset. This is that under certain conditions the concentration of a counterion in either phase can go beyond its eventual equilibrium value. This can occur if the respective counterion is more mobile than a competing one originating in the same phase. For the case that an ion exchanger in the A-form is in contact with a solution containing counterions B and C, of which B is the much more mobile one, one has that in the early stages of exchange, A and B exchange while C, because of its low mobility, barely begins to penetrate into the resin particle. C will subsequently exchange, although slowly, and will in part displace the B ions taken up earlier.

Several investigators have nonetheless attempted to describe and predict multicomponent ion exchange equilibria but because the number of experimental manipulations increases enormously with each additional ionic component, methods for the correlation of binary and ternary, as well as quaternary to multicomponent systems have been sought. Thus have J. Dranoff et al. (1957–8, 1961, 1963), V. S. Soldatov et al. (1968–9), S. E. Smith (1986), A. Jasz et al. (1961–3) and Gopala Rao et al. (1969, 1973–4) adopted a semi-empirical approach by ignoring the effect of the third component on the remaining pair of ions.

These and some other earlier investigators have claimed that the presence of a third ionic species does not interfere with the equilibrium relationship of the other two ions. This general assumption was disproved and some doubt arose about the validity of certain previous fundamental rules. V. S. Soldatov, V. A. Bichkova and coworkers (1970, 1973–5) and R. K. Bajpai et al. (1973) attempted a more rigorous method of prediction. In the former case, ternary data were predicted from binary results by taking into account the effect of the third component on the binary system. The method applied is similar to that used in the prediction of vapor-liquid equilibria and depends on the applicability of Harned's rule, which says that at constant ionic strength the ratio of the activity coefficients of electrolytes in a mixture is constant and equal to the ratio for the pure electrolyte solutions at the same ionic strength. The activity coefficients of the resin and salts in their ternary mixtures were calculated from the equilibrium of the three binary systems. In the latter case a method was proposed of predicting weighted activity coefficients in the resin phase in order to derive ternary data from binary equilibrium results. Furthermore, it was observed that ternary data could be predicted most successfully if the binary pairs containing the most favoured ion were used. Hence, it was found that for a selectivity sequence $A > B > C$, good correlation between ternary and derived binary data could be obtained for the ionic pairs $A-B$ and $A-C$. In an investigation on ion exchange equilibrium in polyionic systems by Danes and Danes [185] starting from the thermodynamic dependences between the activity coefficients in the ionic phase and the excess free mixing enthalpy of resinates correlations were deducted for the apparent ion exchange equilibrium constants in systems of more than two counterions. As is common practice, Gibbs' triangle with arrows for the graphical representation of the equilibrium in systems with three competing ions was used. The basis of the arrows represents the composition of a solution, which is in equilibrium with an ion exchange resin of given composition (arrow-head). But such a diagram is only valid for one total concentration of the solution and is by no means a closed description. Streat and coworkers presented some experimental ternary equilibrium data and showed the application of simple empirical graphical extrapolations to predict ternary results from the corresponding binary data. Two alternative graphical methods of prediction of ternary ion exchange equilibrium data were presented. In this paper ternary equilibrium data were obtained directly from actual experimental binary measurements for the three respective components. It was shown that methods assuming constant selectivity coefficients are unreliable in most cases. The method of prediction recommended in this work uses a triangular representation of the data. The accuracy and reliability of the graphical method of prediction were shown to be as good as a more rigorous calculation using derived resin-phase activity coefficients and it was suggested that these techniques are suitable for the prediction of equilibrium data in process design calculations for chromatographic separation processes [186]. Kolnenkov and coworkers have dedicated most of their work to the calculation of quaternary ion exchange systems using data for binary systems as well as ion exchange isotherms of binary systems [187]. The anion exchange equilibria and kinetics in the ternary system chloride-sulphate-phosphate was studied by Gregory because of its significance for the removal of phosphate from waters and effluents. In this work equilibrium and kinetic data were obtained from shallow bed experiments, using radioactively-labelled sulphate and phosphate and three strong base anion exchangers. The equilibrium uptake of phosphate, in the presence of sulphate and chloride, was always low and became still lower with decreasing pH. From the kinetic data it was found that the exchange of phosphate from a ternary system shows the above-mentioned empirical effect and can overshoot the equilibrium value. This effect was most apparent for resins with a low affinity for phosphate and at low flow rates [188]. A very interesting approach for the description and calculation of isolated and simultaneous ion exchange equilibria seems to be the one presented by Froelich et al. Based on the empirical equation $\bar{x}_B = [(1 + k)/(1 + kx_B)]\ x_B^n$ with k and n the respective characteristic constants under the restriction $-1 < k < n/(1-n)$, this equation can also be applied to systems with three counterions, if for the dependences of k and n on the total concentration c the equations

$$K = k_{00} + k_{10} \lg c + k_{01}\, y_C + k_{11}\, y_C \lg c$$

and

$$n = n_{00} + n_{10} \lg c + n_{01}\, y_C + n_{11}\, y_C \lg c,$$

are used, whereby y_C corresponds to the amount of $y_C = c_C/(x_A + x_C)$ of the counterion C in solution. Applied to the ternary system $H^+/Zn^{2+}/Cd^{2+}$ on a strong acid styrene-divinylbenzene cation exchanger, it was shown that the equation of isotherms used is suitable for the description of real systems [189]. Diffusion phenomena in ternary systems were also studied with the goal of verifying the Nernst-Planck equation for the description of ion exchange under the conditions of the gel-kinetics in a ternary ion exchange system [190]. To predict the selectivity in multicomponent systems with monovalent ions, equations have been derived by Soldatov and Bichkova, which have been proven to be applicable to non-ideal exchange systems with two or three exchangeable ions; they also made the suggestion for a method for predicting the ion exchange equilibria in multicomponent systems. Further, for the ternary systems $Zn^{2+}/Cd^{2+}/H^+$ and $Cu^{2+}/Ag^+/H^+$ on a strong acid cation exchanger, the possibility of predicting ternary data by use of the pair of binary exchange results containing the most preferred ion was examined by Sengupta and Paul using the graphical procedure of Streat. They found that the correlation between the experimental and the theoretically predicted results was reasonably good for both the above ternary systems [191]. The possibility of predicting multicomponent (n ion) exchange from ($n-1$) isotherms for the constituent pairs of ions was examined. This new approach treats ion exchange as a phase equilibrium by using standard procedures developed for solution thermodynamics. Surface effects were taken into account [191].

Ion exchange mechanisms. In organic chemistry the goal of investigating the mechanisms of a reaction is to predict what products will be formed, to understand how the rate of reaction depends on the structure of the reactants and how the spatial arrangement of groups in reactants is affected by the reaction. There are some general characteristics of mechanisms and it is sometimes defined that a reaction mechanism is a concise history of the detailed way in which reactant molecules are transformed into product molecules. Disregarding the formation of the polymer matrices of ion exchangers and the mechanisms involved there, it is assumed that in the ion exchange process itself important mechanisms must be involved. Within polymer chemistry certain mechanisms can also be determined relative to the reactions of polymers, but related only to chains or networks without reactive groups. Taking ion exchangers as one kind of reactive polymer, the determination of mechanisms aims at a direct observation and elucidation of the concrete steps that make up the overall ion exchange process [192]. It should be noted that "ion exchange mechanisms" and not "the mechanism of ion exchange reactions" was used. Ion exchange is a phenomenon in and of itself and ion exchange reactions may occur as part of the process of ion exchange depending on the chemical nature of the ion-exchanging material or the nature of the surrounding solution. A survey of the results of the determination of the mechanisms of ion exchange up to the present follows, there is still a long way to go until standard mechanisms for ion exchange will be obtained. Kinetic measurements and the elucidation of diffusion phenomena are only one field of study with regard to ion exchange mechanisms.

The most important methods applied for the investigation of ion exchange mechanisms are visible and UV light, IR spectroscopy, Raman spectroscopy, nuclear magnetic resonance (NMR), electron spin resonance (ESR), and magnetic susceptibility measurements [193], Mößbauer spectroscopy [194], proton magnetic resonance measurements [195], electron probe microanalysis [196] and extended X-ray absorption fine structure spectroscopy [197]. With regard to results based on the latter method it has been claimed that each of the other techiques has its limitations, giving only a very partial picture of the site occupied by the metal, whereas exafs spectroscopy is capable of probing bond distances and coordination numbers directly, whatever the state of the sample or its degree of loading. This would help to place ion exchange phenomena – obviously first with respect to inorganic materials – on a firm structural foundation. But there are still many questions regarding ion exchange and only the future will show which method can bring the best results under particular circumstances.

To begin with, since most of the ion exchange processes occur in aqueous media the state of water in swollen ion exchange resins is of general interest. The discussion of whether and to what extent water exists in swollen ion exchange resins as "bound" or "free" water was initially based on the enthalpy and entropy of swelling. Bound water is usually considered as the water that hydrates counterions. However, the enthalpy and entropy of swelling are the mean values of simultaneously occurring processes, such as simple sorption, dissociation of the ionogenic groups or ion pairs, dissolution of the "internal" solution formed and conformational changes in the chains of the resin matrix. It is, therefore, of interest to use other means to obtain information on the interaction of water with ion exchangers. Thus, in order to observe the relations directly, spectral methods, especially IR and NMR, have been applied, to study either the ion–water interaction or the hydration number of various ions. For cation exchangers of the sulfonic acid type it was found that the hydration numbers of, e. g., Na^+, K^+, Rb^+, Cs^+ and Mg^{2+} in the resin phase are lower than those found by the same method for cations in electrolyte solution. The hydration numbers in strong base anion exchangers are approximately one and are consistent with the hydration numbers of $N(CH_3)_4Cl$ and $N(CH_3)_4Br$ in solution. The hydration numbers increase with a decrease in crosslinking, i. e. as the resin phase becomes more dilute. In another work, where the hydration numbers of strong ion exchange resins were studied by the NMR method, it was shown that the hydration number of monovalent counter-cations in the cation exchanger varies around 3, and that the hydration number of anions increases in the series I^- to F^- [198]. For multivalent counter-cations the hydration numbers obtained are usually around 6. Concerning the state of water in swollen carboxylic resins without hydration of counterions but containing undissociated or ion-paired groups, the data obtained elsewhere were examined by using the Bradley isotherm and proton magnetic relaxation data. The latter were obtained with a pulse NMR spectrometer operating at 19.2 MHz for the longitudinal, T_1, relaxation times by the Carr-Purcell method and the transverse relaxation times, T_2, by the Carr-Purcell method as modified by Meiboom and Gill and by the Hahn method, indicating that water is bound to the matrix rather than to counterions. The data obtained seem to imply that water molecules in this resin (Zeo-Karb 226; 25% nominal DVB crosslinking) exist in a state of steeply decreasing restriction on rotational motion, perhaps multilayers, but this state can hardly be compared to that in liquid water; this seems at least to be true at that stage of swelling which corresponds to a water activity of 0.9 [199].

In describing the mechanism of the binding of counterions in ion exchangers the formation of ion-pairs has quite often been postulated. Originally it was claimed that a movable exchange ion forms an associated ion-pair with a fixed exchange group in an ion exchange resin, as represented by $A^- + R^+ = RA$, where A^- is a movable anion, R^+ the fixed exchange group, and RA is the ion-pair. Since R^+ and RA are both fixed to the resin matrix, and are at a finite distance of separation (7–10 Å), they do not possess translational degrees of freedom and should be regarded as separate, solid phases. The exchanger system has, then, four phases: the external solution, the internal solution and the two solid phases R^+ and RA [200]. The association of counterions into ion-pairs was also postulated at that time in order to explain the different electrochemical properties of two different ion exchange resins. However, besides providing one explanation for the selectivity of ion exchangers, the term ion-pair formation has remained quite vague.

There have been numerous studies of the mechanisms of the exchange of cations on strong acid cation exchangers. The aim of such investigations is to obtain additional insight into the ion exchange process from information concerning the structure of the sorbed ion. It might also be that such data could – by providing the correct formulation of chemical stages encountered in the exchange process – permit the quantitative mass-action expression of such equilibria. Furthermore, the Mößbauer spectroscopic study results confirm that the resin phase in strong acid cation exchangers with a degree of divinylbenzene crosslinking $\geqq 8\%$ must be regarded as a concentrated solution of a strong electrolyte. The counterions are hydrated in equilibrium with dilute aqueous solutions, and the solid electrolyte is completely dissociated. Spherical ion associations at higher crosslinkages occur also and are more pronounced for counterions with high charge densities. Depending on the moisture content, e. g. air-dried resin or dehydrated or hydrated, copper ions were found in different states in a KU 2X8 resin. Of interest is also the study of uranyl sulfate species or uranyl halide species sorbed on ion exchangers, as well as the cobalt coordination in cation exchange resins [201]. Metal coordination in general and the mechanisms of ion selectivity on ion exchangers would seem still be fruitful fields of research in ion exchange [202].

It is generally known that a characteristic feature of weak acid cation exchange resins is to show poor dissociation in the free acid form. Nevertheless, it is possible to exchange the proton of a weak acid carboxylic cation exchanger against metal cations. On the other hand, the exchange of protons and metal ions at macromolecular binding sites is a process of considerable importance in a variety of fields, not least in biochemistry. For this reason and because ion exchange resins offer a simple system for studies of the mechanism, Schowen and coworkers used this system for the characterization of certain features of the transition state for proton release from the resin, which state appears to be that for migration of a preformed hydronium ion through the matrix. The migration generates a hole which is rapidly occupied by the metal ion [203]. With respect then to the mechanism of the uptake of metal ions the state of hydrogen as compared with sodium can be examined by the paramagnetic probe method. The state of copper ions depends on the degree of hydration, which can be determined by ESR spectra. From the first investigations on visible and near-infrared spectra of divalent nickel, cobalt and copper ions sorbed on a cation exchanger with carboxylic groups, where it was assumed that there is at least a partial coordination of ions by the carboxylic groups [204], the intensive work of Chuveleva and coworkers on the mechanism of the sorption of metal ions has shown, e. g., for UO_2^{2+} that their interaction with carboxylic resins is the formation of complexes of the type

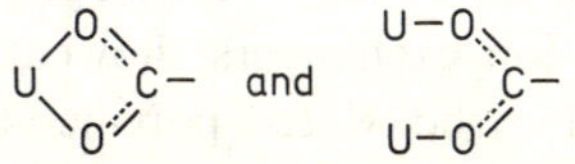

with more or less covalently bonded groups [205]. Numerous similar studies have followed, using the same methods [206] or one or the other of the above mentioned techniques [193, 202].

With regard to phosphoric acid type cation exchangers, study of the IR absorption of a resin in H- and Na-form saturated with U, Cu or Ni revealed that the sorption of U on the resin leads to the formation of a 4-membered coordination uranyl-phosphoryl ring compound, whereas the formation of similar 4-membered compounds in the resin containing Cu and Ni was not observed. The investigation of Cr(III) by EPR showed the formation of Cr(III) of distorted symmetry, revealing non-uniform distribution of sorbed Cr(III) ions in the exchanger matrix, and the study of iron(III) on phosphorus-containing ion exchangers by Mößbauer spectroscopy revealed that macromolecular complexes containing only O atoms in the 1st coordination sphere form on Fe(III) sorption on all types of these ion exchangers [207].

It could be expected that strong base anion exchangers would behave similarly to strong acid cation exchangers, since the states of sorbed species are simple, often being only the hydrated metal ion. Such conditions are not relevant for common anions, but one still has to distinguish between the uptake of customary anions and that of anionic metal complexes. For the first group NMR investigations of small molecules like CO_3^{2-}, HCO_3^-, formate, acetate, $C_2O_4^{2-}$ and benzoate showed that relaxation times are significantly lower than in bulk solution, suggesting that interaction with the resin limits molecular motion. An attempt was made to elucidate the equilibrium conditions of the system Zerolite FF$-H_3PO_4$ in water solutions by conductometric titration with a sodium base. It was found that the nature of the process occuring depended on the concentration of the phosphoric acid solution used [208]. The results on complex species sorbed on anion exchangers are numerous [193]. The behavior of molydenum was examined by IR spectroscopy; when sorbed from pH 7 or 9 solutions it polymerized in the exchanger on increasing concentration of the solutions; consequently the spectra reflect the presence of MoO_4^{2-} as well as of polymeric H_2MoO_4 with H bridges; the sorbed Mo composition is more complexed when sorbed from pH 5 solutions, consisting of paramolybdates and $Mo_8O_{26}^{4-}$ [209]. A new method of distribution analysis in anion exchange resin phases was developed – because of the peculiar nature of the internal solution in the resin – which involves equilibrium analyses on the simultaneous distribution among the three phases, i. e. the cation exchange resin phase, the anion exchange resin phase and the solution. It was applied to the visible absorption spectra and compositions of copper(II)-bromide and nickel(II)-thiocyanate sorbed in Dowex 1X4 [210]. The interpretation of mechanisms occurring in weak base anion exchangers seem to be much more difficult because these products are less well-defined reactive polymers. An infrared spectroscopic study of the sorption mechanism of vanadium(V) by porous vinylpyridine ion exchangers showed that the vanadium ions are sorbed as polyanions and a carboxylated polymer takes up vanadium as VO_2^+, forming an additional coordination bond between nitrogen and the sorbed cation.

In the following ion exchange mechanisms for chelating ion exchangers will be only briefly discussed for the two currently most important chelating polymers, iminodiacetic acid and amino phosphoric acid, with comments on some other resins that have been studied with interesting results. It has long been known that the kinetics of chelate resins are slower and that, since the selectivity for a given ion by a chelate is higher, the elution requires higher concentrations or higher dosages of the eluting agent. It is only now that spectral studies have been able to provide insight into the nature of the bonds formed and have been applied to examine the iminodiacetic and amino phosphoric resins. The iminodiacetic group dissociates differently depending on the pH

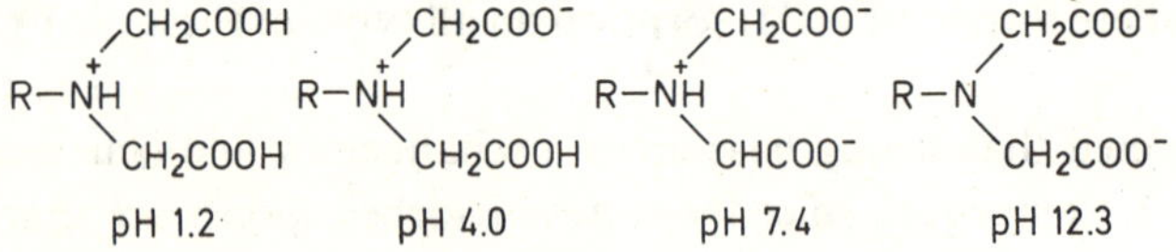

pH 1.2 pH 4.0 pH 7.4 pH 12.3

and accordingly acts differently. Visible light and ESR measurements have clearly shown that the complexes $CuCl_4^{2-}$ and $CoCl_4^{2-}$ are taken up from hydrochloric acid solutions, just as was found for the same species with Dowex 1. Heitner-Wirguin and coworkers have further found that the species formed from neutral solutions are chelated by the carboxylate groups, however no conclusive evidence was found for chelation through nitrogen as well. ESR measurements of copper ions on the chelating resin have shown that the bonding is partially ionic. For uranyl species more details are available: in neutral solutions the Dowex A-1 resin sorbs uranyl ions strongly, added salt leads to coordination of the neutral salt anion, and from the corresponding spectral shift the anions may be classified according to their coordinating power: $NO_3^- \approx SO_4^{2-} < Br^- < Cl^- < SCN^-$. The amino phosphoric ion exchanger is primarily an amphoteric material. Copper sorption and structure of the H-, Na- and Cu-forms were studied by IR spectroscopy. An equilibrium of the dissociated H-form was established. Sorption of copper involves complex formation. It must be realized that very little work has been done on the mechanisms by which metal ions are sorbed by chelating ion exchange resins. On two polycondensation resins containing 8-hydroxyquinoline or salicylaldehyde/salicylaldoxime and a macroporous polyhydroxamic acid copolymer the role of the co-ion was studied showing that it accompanies the ion sorbed but remains labile, being easily eluted by buffer solutions. Vernon, therefore, comes to the conclusion that, initially, the co-ion is intimately connected with the sorption of the metal and does not remain within the polymer lattice simply to effect charge neutralization. 1 : 1 metal : co-ion complexes are fairly commonplace but, with the statistical spatial distribution of chelating groups attached to a random polymer backbone, some 1 : 2 metal : resin sorption is inevitable. With structures involving regularity of chelating group attachment to the polymer backbone, the probability of a divalent co-ion spanning two sites and thereby neutralizing the charge on two sorbed metal cations is greatly increased. Finally it should clearly be seen that in chelating ion exchange the total metal sorption involves all species MR_n to $RM\text{-}X_{n-1}$, where X represents a monovalent co-ion. Each species will have its own metal-resinate stability constant and overlap of the different species of two sorbed metals results in a decrease in selectivity [211].

Effect of temperature on ion exchange. Temperature has an influence on ion exchange, but under the practical aspects of operating ion exchanger units an increase in temperature does not improve ion exchange rates sufficiently to justify the extra expense of installing special heating equipment. Certain advantages can be gained by operating at elevated temperatures if the solution is at this temperature initially or if the solution is viscous at normal room temperature. In more specific applications, a warm regenerate solution is beneficial. But since all ion exchange resins are subject to chemical degradation and structural changes at rather high temperatures, temperature limits should be determined from the technical literature on each resin.

With respect to equilibria, in particular the swelling equilibrium, the change in the solvent content of an ion exchange resin with temperature changes in the range of several tens of degrees is small. This leads to a low accuracy in the determination of the enthalpy of swelling from the temperature variation. Together with the effect of temperature on ion exchange equilibria, selectivity should be considered since this factor is influenced by the temperature. Bonner and coworkers and Kraus and coworkers have studied temperature dependence and have found that the selectivities of monovalent cations as against hydrogen in a Dowex 50 resin decreased with increasing temperature. In the same system divalent cations showed a slight, and trivalent ions a strong increase in selectivity [212]. The influence of temperature on the kinetics of ion exchange is also important, since an increase in temperature produces a substantial increase in diffusion. It was reported that the self-diffusion coefficient of sodium in a strong acid cation exchanger and of chloride in a strong base anion exchanger increases almost three-fold on increasing the temperature from 3 to 25 °C. The diffusion rate of complexed metal ions in an anion exchanger was reported to have been increased by a factor of approx. forty during a temperature increase from 20 to 180 °C. On the other hand the desorption of different cations from a strong acid cation exchanger was studied and it was found that increasing the desorption temperature from 35–40 °C to 50–55 °C reduces the desorption time from 9 to 2 hours. Most important is the influence of temperature on the dissociation of weakly dissociating ion exchange resins. Furthermore, the capacity in both resin types increases with increasing temperature. More important is that the dependence of the dissociation of weak acid cation exchangers and weak base anion exchangers on temperature may be used for a thermal regeneration of such resins, thus eliminating for the environment the problems caused by using regeneration chemicals.

Effect of pressure on ion exchange. External pressure exerted on an ion exchange resin has a certain influence on the ion exchange system. The exchange equilibria, which were the first properties examined under this aspect, are not influenced, because (from a thermodynamic point of view) there is no volume change of the exchanger phase with its surrounding solution. The hydrostatic pressure of a total ion exchange system has also been investigated with more or less the same result, except that some increase in the electrolyte dissociation, and change in the structure and properties of the solvent or the degree of solvation of ions were found. As far as ion exchanger powders are concerned the effect of pressure can become an operative factor because of the particle size. But what was theoretically derived and then experimentally proven is that if external pressure is applied to a styrene-divinylbenzene cation exchanger resin it influences swelling and the selectivity in the same way as an increase in the degree of crosslinking by divinylbenzene does. This could be shown for the ion exchanges hydrogen against copper and hydrogen against iron(III). As an application, regeneration can be accomplished without pressure, which results in better exchange kinetics and lower ion selectivities for this step. An external mechanical pressure simulates a further crosslinking of the matrix, also in the case of anion exchange resins, influencing in this way the selectivity of anion exchangers. But the effect was found to be not uniform, in contrast with cation exchange resins. The pressure applied causes here both an increase and a decrease in selectivity. Especially in the case of chromate anions high ion selectivities were observed [213]. Effects of pressure on ion exchange and ion pairing chromatography were also studied.

Ion exchange of ampholytes. In biochemistry, of all the applications of ion exchangers, ion exchange chromatography has been extensively used for the fractionation of complex mixtures of amphoteric molecules (also called ampholytes) such as amino acids, peptides, proteins, and nucleotides. In general, ampholytes can, with respect to their behaviour towards ion exchange materials, be treated either like organic bases or organic acids. For example, amino acids can be separated by displacement development on a strong acid ion exchange resin in the H-form with sodium hydroxide as displacing agent, or on a strong base resin with hydrochloric acid. With respect to ampholytes the early work with ion exchangers was pretty much empirical. Nevertheless attempts were made to find relations between ionization constants and sorption constants of amphoteric electrolytes, such as glycine, α-alanine, phenylalanine, and lysine. Further, a general method was suggested for determining the sorption equilibrium constant of an aqueous solution of an ampholyte-cation exchanger system. With proteins adsorption is confined to the surface of the particles and van der Waals interactions appear to be of considerable importance. The equilibrium principles of the sorption of nucleotides on anion and cation exchange resins was intensively studied. It was found that the mechanism of nucleotide sorption is not significantly different from the sorption of inorganic ions. Since the analysis of the sorption of partially deaminated resins has shown that, for a resin with a small sorption capacity the nucleotide sorption capacity approaches the full exchange capacity of the ion exchangers, one can expect that there will be a demand for special exchangers for other special molecules.

The fundamentals and definitions of ion exchange and ion exchangers are in principle all related to ion exchange in an aqueous solution. Nevertheless ion exchange between solids has also been investigated. A direct ion exchange has, for instance, been observed between α-Zr phosphate and several anhydrous metal halides. The exchange reaction occurs at 115 to 375 °C, and proceeds by continuous removal of the volatile acid. This type of exchange process between solids must be considered as a general property of ion exchangers in the H-form [225]. It might well be a matter of definition whether the exchange of ions between ion exchanger grains in fully deionized water can not also be considered an ion exchange between solids. Results of corresponding investigations are available showing mainly that exchange takes place on a limited surface area rather than over the whole surface of the grains [226].

Ion exchange with organic ions. That ion exchange is not confined to inorganic ions was first shown by Kressman and Kitschener in a study on the ion exchange equilibria of large organic cations on a phenolsulfonate resin. The affinity of some large organic cations, namely, tetramethylammonium, tetraethylammonium, trimethyl-n-amylammonium, phenyldimethylethylammonium, phenylbenzyldimethylammonium and the quininium ion increases with the size of the ion, which is in contrast to the simple inorganic cations. As an explanation of this it was suggested that van der Waals forces contribute largely to the affinity, with the electrostatic forces being less important. The rate of exchange, on the other hand, decreases with increasing ionic size. As the saturation capacity was found the same for the largest of the quaternary ions studied, namely phenylbenzyldimethylammonium, as for inorganic ions, this was taken as a clear indication that all the molecular pores within the resin are larger than the effective diameter of the ion. No physical adsorption was observed from solutions of the highly dissociated salts of large cations [214]. Although several studies of the binding of organic cations by cation exchange resins followed, virtually no investigations of the selectivities of anion exchangers for organic ions had been undertaken although it was becoming apparent that an understanding of the nature of such binding is of considerable importance: one of the chief problems associated with the use of anion exchange resins in water treatment is the virtually irreversible fouling of the resins by such naturally occuring materials as fulvic and humic acids. Further knowledge concerning the binding of organic ions by anion exchangers will aid in the proper development and selection of resins for use in situations in which organic fouling is a potential problem. In a pertinent study by Gustafson and Lirio, therefore, the influence of the chemical structure of the resin upon the equilibria and kinetics of binding of model organic species were evaluated.

The selectivities of poly (*N,N,N*-trimethylvinylbenzylammonium) chloride (I), the protonated form of poly (3-*N,N*-dimethylaminopropylacrylamide) (II) and poly-(3-*N,N,N*-trimethylammoniumpropylacrylamide) chloride (III) anion exchange

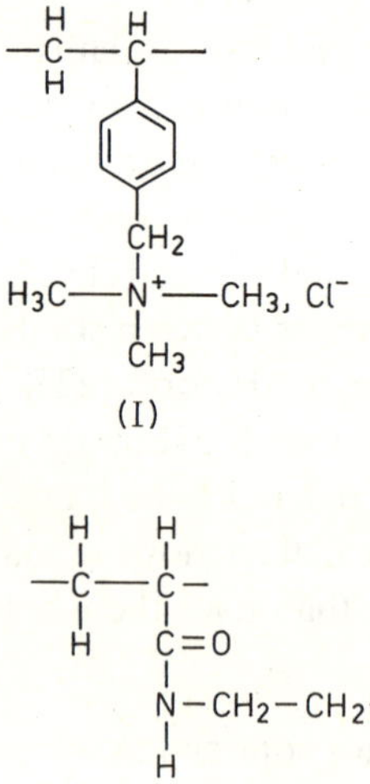

(I)

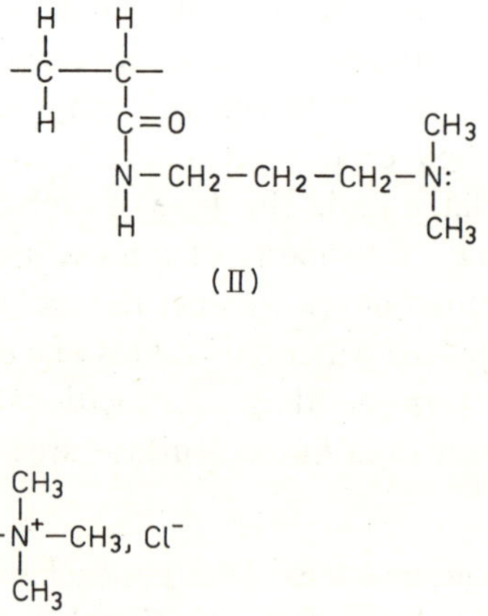

(II)

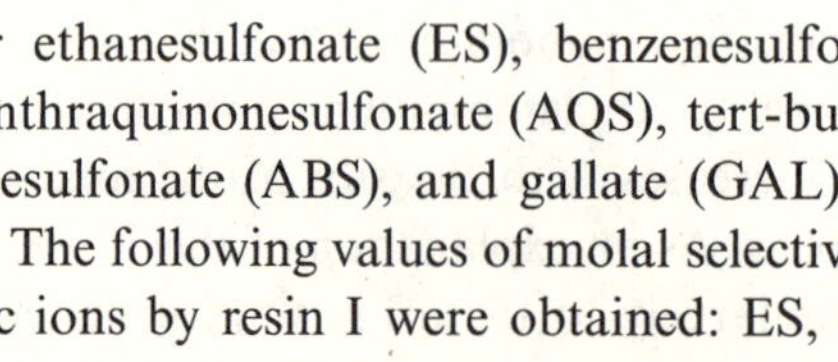

(III)

resins for ethanesulfonate (ES), benzenesulfonate (BS), 2-naphthalenesulfonate (NS), 2-anthraquinonesulfonate (AQS), tert-butylcatecholsulfonate (TBCS), dodecylbenzenesulfonate (ABS), and gallate (GAL), ions were measured at 25 °C and $\mu = 0.10$. The following values of molal selectivity coefficients, K^0_{Cl}, for the binding of organic ions by resin I were obtained: ES, 0.67; BS, 7.4, NS; 133, GAL; 150,

TBCS 400; AQS, 1290; ABS, 32,000. The selectivities of all the resins for organic species increase markedly as the number of aromatic rings in the organic ion increases. For species containing a single benzene ring the selectivity increases as the chain length of the aliphatic substituent increases. The selectivities of the above resins for a given organic species decrease in the order I > II ≈ III. The high selectivities observed are produced by a combination of electrostatic interaction and hydrophobic bonding. The influence of the latter effect is reduced considerably by the addition of nonaqueous solvents. The selectivity of resin III for naphthalenesulfonate decreases in various solvents in the order $H_2O \gg$ 50% $CH_3OH >$ 50% $C_2H_5OH >$ 50% n-$C_3H_7OH \approx$ 50% $(CH_3)_2CO$. With respect to kinetics, it is then of interest to note that Soldano and Boyd (1953) showed that, as the selectivity of polystyrenesulfonate resins for cations increases in the order $Na^+ < Zn^{2+} < Y3^+ < Th^{4+}$, the rate of diffusion within the resin phase decreases markedly. Similarly, the results of this investigation show that, as the selectivity increases in the order BS < NS < AQS, the rate of adsorption increases in the order AQS < NS < BS [215]. Ion exchange with organic compounds is furthermore always feasible, if the compounds themselves are ionisable, which means that an interaction with exchange materials can take place. Consequently, organic acids, bases, or salts may be taken up by the appropriate exchange material, thus, e. g., making possible their isolation and recovery from non-ionic organic compounds as well as from inorganic substances. The size and type of the organic molecule is a critical factor, so that cases are known where only part of the exchange sites are accessible to the ions to be adsorbed. An example is in the uptake of the triply-charged streptomycin ions on carboxylic acid resins: the exchange capacity of the exchangers was found to increase with swelling, but the total ion exchange capacity of the resins with respect to metal ions cannot be attained. This phenomenon is attributed to the fact that the sorption of the streptomycin ion is accompanied by the screening of part of the exchange sites of the cation exchanger because the dimensions of the organic ions exceed the average distance between adjacent ionogenic groups. A similar phenomenon was encountered in the uptake of nucleotides on strong base anion exchangers. The sorption capacity of Dowex 1 with respect to ATP, AMP and UMP depends little on the crosslinking and hardly changes over a range of 1 to 8% DVB. It is much lower than the total exchange capacity of the resins with respect to small ions, meaning that a large part of the active sites do not participate in the exchange processes. Partial deamination of the resins and the associated decrease in the total exchange capacity led to a slight increase in the relative exchange capacity with respect to the nucleotides. In these cases also part of the exchange sites are screened, which makes them inaccessible to interaction with the ions in solution. In evaluating the ion exchange behavior of a variety of resins for a range of organic ions it was found that the extent of exchange can be influenced by a great number of possible interactions, and that even small changes in the conditions under which exchange is conducted may have a significant influence on the performance of a system. No simple mechanistic model is successful in predicting even qualitatively the selectivity sequences or preferences for different systems [216]. It must also be mentioned that some organic compounds can undergo reactions leading to the formation of ionized addition compounds or complexes, and in this form enter ion exchange processes. Examples of this type of compound are given by the bisulfite addition complexes of aldehydes or ketones and the borate complexes of polyhydroxy substances. The feasibility of applying this ligand exchange technique for the sorption of organic compounds and its practical applicability has also been examined [217]. Further efforts to improve and diversify the handling of organic ions by ion exchangers were made with ion exchanger microdispersions, the employment of poly(4-vinyl pyridine) exchangers or by making use of the interaction between a resin containing dibenzo 18 crown 6 and organic molecules. As a whole ion exchange with organic ions lags behind the results available for inorganic ions. Not only for the ion exchange technology of biologically active naturally occurring compound separations, but also in other respects. This represents therefore a challenge to find ways of improving the efficiency of the ion exchange of organic ions. This general aspect applies also to the research on mechanisms of the binding of organic ions to ion exchange resins, although one example of progress in this direction is the analysis of proton magnetic resonance parameters, which permits measurement of the difference in the NMR spectrum of unfilled and filled ion exchangers and use of it as a criterion of the extent of the filling of ion exchangers with organic counterions [218].

Anomalous behavior of ion exchangers. In the literature on ion exchange quite a large number of papers deal with the anomalous behavior of ion exchangers. If, in the strictest sense, "regular behavior" is understood to refer to the behavior exhibited in the exchange of small ions on exchangers with a low degree of crosslinking and low capacity in very dilute solutions, and "anomaly" to mean any departure from this type of behavior, then quite a number of anomalies with respect to ion exchangers can be found. In fact all theoretical treatments of ion exchange processes seem to be restricted to such regular behavior and all deviations from corresponding results should be explainable by the one or the other anomaly. Following an early presentation certain anomalous reactions can be attributed to additional nearby forces, one of which, the polarization of the anionic carboxylic and phosphonic groups of cation exchangers, leads to the change in the affinity sequences of alkali ions. Ion pair formation has become the key expression at least for conditions under which the behavior can only be explained by assuming that part of the chemical bond between the anchor- and the counter-ion is of some homopolar nature. A frequently observed anomaly is the over-capacity of ion exchangers, especially that of synthetic ion exchange polymers, for organic acids and bases, which is referred to as molecular adsorption on ion exchange resins. Both the Weiss effect [219] and the Wheaton-Bauman effect [220] were regarded as the anomalous utilization of blocked cation exchangers, with the Wheaton-Bauman effect having become the basis of the ion exclusion process, which is used for the separation of ionized substances from nonionized or only weakly ionized substances when both are present in aqueous solution. Customary ion exchange resins are used – both cation and anion exchangers – and highly ionized exchangers generally produce better results than those of low ionization.

All sieving effects have as well been included among the anomalous behaviors of ion exchangers, with quite a number of potential applications for synthetic organic as well as inorganic ion exchange materials. Anomalies in ion exchange processes were found in the desalination of sea water as introduced by Calmon, based on a subsequent reaction of the exchanged ions $Ag-R + NaCl \rightarrow AgCl + Na-R$ and $Ba-R + MgSO_4 \rightarrow BaSO_4 + Mg-R$, or in the pre-loading of ion exchange materials as introduced by Samuelson and coworkers into analytical separations (which also included a subsequent reaction) or, in analogy with the latter, in the pre-loading of resins with boric acid to separate sugars or even in the use of the latter anomaly technically for the concentration of glycerin from dilute solutions. All ion exchange processes involving complex formation can be regarded as anomalous, including the very specific ion exchange materials, used e. g. one containing dimethylglyoxime. This Ni-specific resin takes up the metal but can hardly be regenerated. A review of the anomalous behavior of ion exchangers reveals not only that the related reactions are scientifically interesting, but that they can also lead to technically interesting potential applications [221]. The selectivity of ion exchangers can also be influenced anomalously, for instance in the case of weak base anion exchangers and the uptake of organic ions such as sulfacyl anions and sulfanilurea for chlorine due to the formation of H bonds between the tertiary amino groups of the polymers and the amides [222]. In a somewhat different sense it is to be considered an anomaly whenever the exchange material itself does not have the structure ideally assumed. Thus the pH-dependent distribution of carrier-free $^{144}Ce/^{144}Pr$, ^{91}Y, and ^{89}Sr between aqueous phases at constant ionic strength and styrene-divinylbenzene ion exchangers shows a secondary exchange function due to carboxyl groups on the polymer skeleton, the capacity of which is small in common resins, but which accounts for anomalous results in the literature [223]. Structural changes of ion exchangers in the process of ion exchange in relation to anomalous equilibria have been reviewed and the effect of ion exchanger inhomogeneity on the sorption isotherm of a nonexchange-electrolyte has led to criticism and analysis of the general properties of a model of a local-electroneutral ion exchanger. Pendent vinyl groups of styrene-divinylbenzene resins of high crosslink density undergo a side-reaction during chloromethylation resulting in aliphatic chlorides which differ strongly from aromatic chloromethyl groups in reactivity towards further modification (e. g., with dimethylamine), so that chlorine is still found in the resin by elemental analysis. It is still to be seen whether an anomalous behavior can be found, to which this new finding can be ascribed [224].

1.1.4 Procedures

The basic techniques used for the practical application of ion exchange are: 1. batch operation, 2. column processes, 3. continuous processes, and 4. fluidized bed operations. By these procedures a solution is brought into contact with an ion exchanger. For describing the details of each of these techniques, units and symbols are used which have been defined above in this chapter or will be defined as they appear in the text. To a great extent the ion exchange procedures described here are used for laboratory evaluations of ion exchange processes. It will, therefore, also be of interest to include mass transfer considerations in the descriptions, as they have been increasingly applied to the explanation of ion exchange processes. The fairly new technique of parametric pumping will be referred to to the degree that results have been accessible, and other original procedures will be briefly mentioned.

Quite often ion exchange is described as being directed up into two parts, the service run and the regenerating cycle. It will be seen that with each procedure various methods can be applied, especially in analytical chemistry and in preparative chemistry. Separations are probably the most widely used ion exchange methods in this respect. But nevertheless in very many industrial processes – not only in water treatment – it is usually the case that the procedure developed consists in principle of a service cycle and a regeneration cycle. During the service cycle either contaminating inorganic ions are separated and concentrated or valuable ionic species are separated and enriched by the ion exchange process, where the exchange is always for innocuous ions. In practice, then, one of the major problems associated with an ion exchange process is the cost of chemicals for regenerating the ion exchangers to their original ionic form; another problem inherent in regeneration is the safe handling and disposal of the resultant waste, which may contain highly toxic metals or concentrated salts. These problems justify devoting a section to the regeneration of ion exchangers.

1.1.4.1 Batch operation

Batch operation is the simplest ion exchange process. The ion exchanger is contacted with the electrolyte solution in any suitable vessel with stirring or shaking until an exchange equilibrium has been established between the counterions of the exchanger and the ions of the electrolyte:

$$\mathrm{IE \cdot C_1 + C_2X \rightleftharpoons IE \cdot C_2 + C_1X.}$$

The degree to which this process takes place depends on the equilibrium constant of the ion exchange system. After equilibrium has been attained, the ion exchanger is separated from the solution phase by filtration, centrifugation or settling. Both phases can then be analyzed for the content of the ions in question. But it is only in certain cases – which will be described below – that it is possible to approach quickly the full capacity of an exchanger in this so-called single batch process. If, however, additional ions are to be removed from the electrolyte solution or a quantitative exchange of ions present in the solution has to be obtained one can either use a large excess of the solid ion exchanger or add fresh ion exchanger in smaller portions successively to the solution. Separation must be performed again after each adjustment of the equilibrium. The latter method is called multi-stage batch process and it can be compared to discontinuous extraction and adsorption techniques. In practice the multi-stage batch process is laborious, time-consuming and may give rise to experimental errors. Its use is therefore recommended more for preparative purposes [227].

The single batch process is preferred even if it takes quite a long time to reach equilibrium. Its applicability is, however, limited to those exchange processes in which the equilibrium can be shifted strongly to the right of the above equation and is essentially brought to completion by a driving force such as the formation of weak electrolytes, insoluble products, or stable complexes. This is always true when a cation exchanger in the H-form is contacted with a metal hydroxide solution or an anion exchanger in the OH-form with an acid solution. In both cases, ion exchange is accompanied by the formation of water.

Despite certain advantages batch operation is rarely applied in normal ion exchange processes, but the single-stage batch process is quite suitable for the determination of physical constants and for basic investigations of diffusion phenomena. Distribution coefficients and selectivity coefficients are quite easily obtained from batch equilibrium data. For the determination of distribution coefficients resin samples, 0.1 to 1.0 g resin in the swollen state, are usually added to flasks containing different amounts of the solution, 10 to 100 ml, and are shaken for 10 to 15 hours using a mechanical shaker. The distribution of the ion under investigation is analyzed by suitable methods. Because several different samples can be processed simultaneously in a relatively short time, it is possible to follow the distribution of several solutes under many different sets of experimental conditions and to determine what effects the capacity and percent of crosslinking of the resin, the type of resin itself, the temperature, and the concentration and pH of the electrolyte in the solution have. The effect of the stirring rate or of the mechanical agitation in general has also been examined, with the finding that the distribution increased with the Reynold number up to a certain value and then remained essentially unchanged. In all cases it is advisable to ascertain, by repeated trials, the time to be allowed for equilibration. After taking suitable aliquots from the equilibrium mixture in which the concentration of the various ions in the aqueous phase have been determined, the resin concentrations are usually obtained by taking the difference. The apparatus applied in diffusion studies are variations of the batch operation. This is the case for the first limited batch method introduced by Kressman and Kitchener [228], as well as with batch experiments used to calculate the effective diffusion coefficients of the exchange of sodium with manganese on

A-type zeolite [229], and for a whole model which was derived from the ion exchange in batch and semi-batch reactors with internal diffusion under agitated conditions [230]. At least with regard to batch operations for smaller scale ion exchanger application it should be mentioned that the advantages of batch and fixed bed ion exchange techniques have been compared for different ion exchange processes. The fixed bed is preferable to the batch only where important ion exchanger deficit and moderate duration values are involved. The greater the output and solution normality, the more advantageous the batch processing becomes [231].

But in batch experiments under all conditions the degree of conversion depends on the selectivity (K_A^B) of the exchanger material for the ions A and B and their relative concentrations in the solution and resin batch. If, as an example, 1 ml of wet resin in the hydrogen form with a capacity of 2 mequ/ml is brought to equilibrium with 10 ml of 0.1 mol/l sodium chloride solution, the total exchangeable sodium in solution is 10 × 0.1/1000 equivalents or 1 milliequivalent. To calculate how much sodium (x) is exchanged with the resin at equilibrium the following equation can be used

$$K_H^{Na} = 2 = \frac{[\overline{Na}] \cdot [H]}{[\overline{H}] \cdot [Na]} = \frac{\left(\frac{x}{v_r}\right)\left(\frac{x}{v_s}\right)}{\left(\frac{(2-x)}{v_r}\right)\left(\frac{1-x}{v_s}\right)} = \frac{x_2}{(2-x)(1-x)},$$

with square brackets representing molar concentrations and v_r and v_s being the volume of resin and solution, respectively. Solving this equation for x gives $x = 0.763$, showing that 76% of the sodium ions are exchanged with hydrogen ions in this simple case of equilibration. Although the equilibrium and, inherently, the selectivity is positive for sodium, the full capacity of the exchanger sample has not been used yet, and several more equilibrations with fresh sodium chloride solution will be required for a complete conversion of the exchanger to the sodium form.

1.1.4.2 Column processes

The column process is the most important and most frequently used ion exchange technique. The ion exchanger is packed in a column, usually of glass, and all necessary operations are carried out in this bed. Basically, two techniques can be distinguished here, the descending and the ascending flow process. In the first case, the liquid moves down, in the second up, through the exchanger bed.

Given an exchanger column containing the ion exchanger with counterion C_1 one has the equipment arrangement shown schematically in Part I of Figure 1.39. The counterion C_1 of an exchanger is to be exchanged for the counterion C_2 of a solution in an overhead reservoir. As soon as solution with C_2 enters the exchanger (Figure 1.39, II), C_2 ions are exchanged by the exchanger. After a short time, the

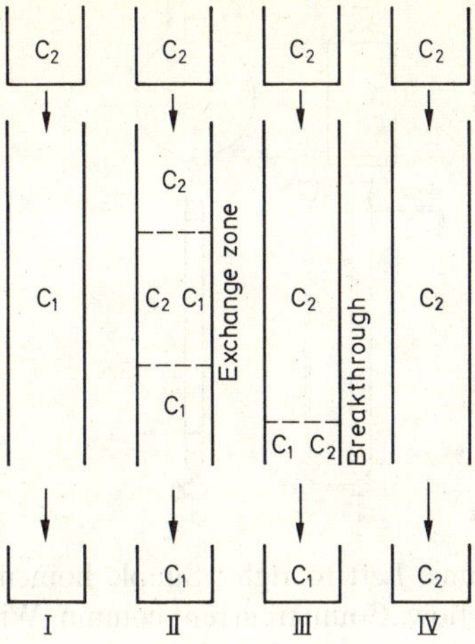

Figure 1.39 Ion exchange in the column process.

exchanger in the upper section of the column is completely loaded with C_2 as counterions. Additional C_2 ions flow unhindered through this part of the bed and reach the exchange zone farther down, where the C_1 counterions are exchanged with the still unexchanged C_2 ions. The liberated C_1 ions are eluted at the lower end of the column in a stoichiometric ratio. If this process is continued, the exchange zone in the column continues to migrate downwards until it reaches the lower end and the overall process has come to the point where C_1 and C_2 are simultaneously eluted from the column. Breakthrough takes place, at which the concentration of C_2 ions in the flow begins to increase until it finally reaches the same concentration as in the solution initially charged on the column (Figure 1.39, III). If C_2 ions continue to be charged (Figure 1.39, IV), no further ion exchange can take place, since the entire exchanger already has the C_2 form. As a result the C_2 ions flow through the column without hindrance.

During flow through the column, the ions which are to be exchanged continuously contact fresh ion exchanger, so that the equilibrium is increasingly shifted in the desired direction. Compared to a batch technique, ion exchange here becomes a complete and simple process.

Ion exchange columns. The ion exchange literature describes numerous types of columns which have been developed either because practical considerations led to new designs or special techniques made them necessary. In principle, the ratio of diameter to height in laboratory columns should be in the range from 1 : 10 to 1 : 20. It is essential, of course, that they be designed so that the liquids can flow through them easily.

A simple ion exchange column can be made with materials usually available in any laboratory. As shown in Figure 1.40, it consists of a simple glass tube provided with a bored stopper on both ends. The lower stopper is equipped with a glass

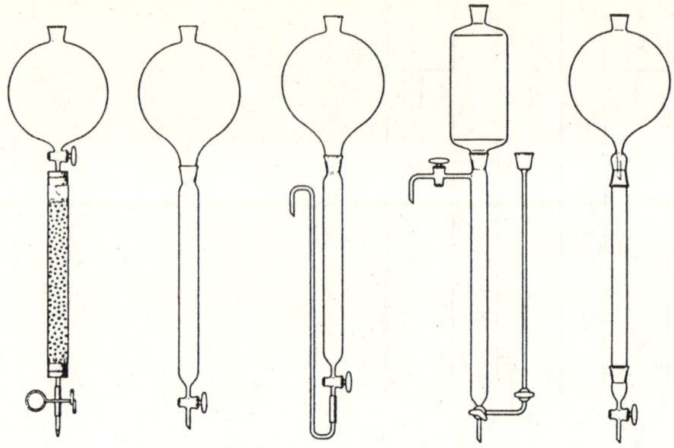

Figure 1.40 Laboratory ion exchange columns. Left to right: Simple homemade column. Column with ground joint. Column with overflow. Countercurrent column. Wickbold inversion column.

tube attached to the tip of a capillary as a dropping attachment via a flexible tube connection. The pinch-cock permits a regulation of the dropping rate. The upper end of the column is provided with an ordinary dropping funnel as a reservoir. Cotton balls inserted on both ends of the ion exchanger bed (cellulose, synthetic fiber, quartz, or glass fiber balls are used) prevent plugging of the discharge tube by ion exchanger particles, on the one hand, and turbulence of the ion exchanger bed on the other. It is also possible to have a glass blower produce a column with ground joints (Figure 1.40) on which reservoirs of different sizes can be attached. An overflow tube can be connected to the flexible tube provided under the discharge cock (Figure 1.40), thus preventing the drainage of the column. Several advantages are offered by a so-called countercurrent column (Figure 1.40), which can be used with an ascending flow technique as well as with the simple descending flow technique. This is obtained by a suitable adjustment of the three-way stopcock. The liquid flows through the column from the right attachment and is discharged through the left stopcock. The effect obtained is also known as reverse flow in adsorption and exchange processes (counterflow). The materials which have been exchanged near the top of the column do not need to cover the distance through the entire column bed and can thus be eluted more easily. The same end is achieved very simply by the Wickbold inversion column (Figure 1.40). A column version which is used very frequently in laboratories is the burette type. As shown in Figure 1.41, it consists of a simple burette with the necessary connection at the top for the loading of solutions. For ion exchange processes requiring high or low temperatures, the column and reservoir are equipped with a temperature-regulating jacket (Figure 1.42).

As a resin support in other types of columns described in the literature either a sintered glass plug or a sintered glass filter, which are more convenient than glass wool plugs for retaining the resin in the column, are used. To speed up an ion

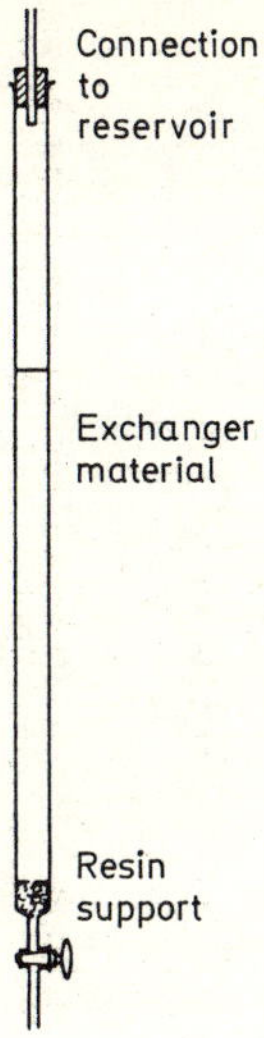

Figure 1.41 Ion exchange column of the burette type.

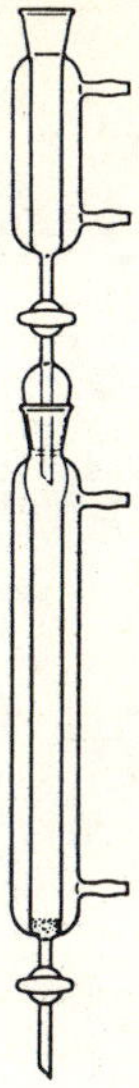

Figure 1.42 Heatable column.

exchange column procedure – especially useful for teaching – a simple apparatus was assembled in which, by means of a pipet filling bulb a slight suction is applied to the column so that an eight-fold reduction in time consumed, without a loss in precision, was attained [232]. If one prefers working with plastic material, for example in order to avoid contamination introduced from glass apparatus, the types of plastic column developed so far have been described [233]. For previously introduced exchange materials for chromatography a new column was developed for high speed working conditions [234]. This is a universal column with accessories, which enables the use and preparation of granular reactive polymers, their conditioning, modification, characterization and manipulation during the determination of fundamental characteristics, such as bed weights and volumes, weight and volume sorption capacities, swelling properties, and the changes of reactive polymers under the effect of the surrounding phase. The device permits work in an inert atmosphere or in a medium with a defined composition of gas and liquid phase [235].

Larger columns for testing ion exchange processes can be constructed in the same manner as described above using glass as the material. Dimensions of 50 mm diameter and 800 mm length have been recommended. Acid-washed sand or quartz is frequently used as the resin support. Experiments in columns of such dimensions usually already furnish results which can serve as the basis for the construction of production units.

Charging an ion exchanger into a column can be easily done with a bit of practice. The exchanger, present in any form, is first treated in a beaker with distilled water. The resulting swelling must always be carried out to prevent rupture

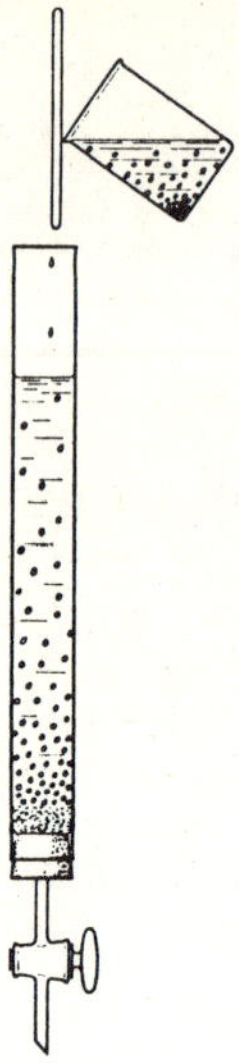

Figure 1.43 Charging of an ion exchange column.

of the column or too close a packing due to swelling in the column. Two hours are usually sufficient for swelling. The exchanger is then poured rapidly into the column (Figure 1.43), with care being taken that there is a uniform packing of the different-sized exchanger particles and that the exchanger is always covered with water to prevent the inclusion of air bubbles. Excess water is continuously suctioned from the column. If too much water has been removed accidentally from the column and air has come between the particles, it usually suffices to add water and to swirl the ion exchanger bed by tipping the column. Finally, a cotton ball is inserted on the top rim of the ion exchanger bed and the column is washed a few times with distilled water. Actually, several different techniques can be used to pack the preswollen ion exchange material into a column. If using the above process the liquid is drained down after only a portion is filled, it is called multi-stage bed packing. For the so-called single-stage bed packing a stirrer is placed in an attached funnel and kept in motion while adding the slurry of exchanger to the funnel; one lets the liquid drain from the column with a flow rate at the exit equal to or faster than the linear rate at which the exchanger bed settles in the column. In a process used mainly for chromatographic columns, called single-stage pump packing, a reservoir with exchanger slurry is placed above the column and while liquid from a pump is forced into the slurry reservoir the exchanger particles are forced into the chromatographic column at a velocity much greater than their settling rate. Pumping is continued until the column is fully packed. Which packing process is optimal depends on the type and the particle size of the exchanger material. A highly optimized process for packing ion exchange chromatographic columns has been developed, in which the resin particles are forced into the packed bed in a flowing fluid at a velocity much greater than the settling velocity of the particles. A column thus dynamically packed with anion exchange resin exhibited a more uniform particle size distribution throughout the column than one packed by gravitational settling methods [236]. The deformability of an ion exchange resin bed by compressive forces and the residual deformation of a resin bed have been attributed to a superpacking of the grains, and varies with the strength of the resin grains.

There are also commercial columns available mainly for chromatographic purposes, both analytical and preparative. These columns have advantages both for routine work and for a complete chromatographic assembly. All necessary accessories can also be obtained from relevant supply houses.

After having charged the column with the ion exchange material a conditioning of the resin is usually performed. Where the resin is in the required ionic form, the only conditioning necessary is a so-called backwash with deionized water. This step frees the resin bed of air pockets and removes debris and resin fines. One washes the resin with an upward flow of water, the rate of flow being adjusted so that the bed of resin expands to just less than twice its original volume. The resin particles should be free to move about in the flow of water, with any conglomerates being broken up by gently shaking the tube during the flow of water. When all fines and air bubbles have been removed one allows the resin to settle in the tube. Excess water can be drained off. When a column is taken into use after standing for some time the resin should be given a rinse with deionized water to remove any trace materials that may have leached out. If the column is in the required ionic form this is all that is normally required. Otherwise, a regeneration step can be included as part of the conditioning.

The particular processes of ion exchange which then take place in an exchange column during one cycle are:

1. Ion exchange or exhaustion
2. Washing of the exchanger bed
3. Regeneration or elution.

Ion exchange proper may take place in the most diverse forms, depending on the problem involved. Several examples will be given later in the section on laboratory and industrial applications. Washing of the exchanger bed is necessary between individual runs to remove excess reagent solution remaining in the bed. During regeneration, the exchanger is transformed back into its original form; if the ion which has been exchanged in the first run is to be recovered, it is eluted from the exchanger with a suitable liquid and collected in the eluate.

Three parameters serve to describe the dynamic and chemical processes taking place in exchange columns: Flow rate, pressure drop and breakthrough capacity. But before going into details regarding these terms the term *bed volume* (BV), one frequently encountered in ion exchange applications, must first be characterized. It represents the total volume of resin in a column, including the volume of both the beads of resin and the void space between the beads. Since the void space of closely packed spherical particles occupies a volume of about 35% of the total, a bed volume of solution passed through a column of ion exchange resin actually is equivalent to nearly three displacements of the void volume. The actual number of displacements of total solution in the bed is substantially less, however, since, of the remaining 65% of the bed occupied by resin beads, somewhere between 40% and 60% by weight is water (assuming that the system is a dilute aqueous solution) which is also subject to displacement. The term bed volume is a convenient one to use, but it is important that everyone using it have the same understanding of what it means. Unless otherwise specified, a bed volume refers to the volume occupied by a resin after it has been backwashed to achieve complete hydraulic classification, and then allowed to settle. Ion exchange resins typically are not uniform in particle size, but have a Gaussian distribution of particles ranging, e. g., between mesh sizes of 0.3 and 1.2 mm. Hydraulic classification, achieved through backwashing, results in the bed's having a maximum void volume, since each bead is in contact with beads of nearly identical size. This minimizes the possibility of small beads nesting in the voids between larger beads, which is the desired mode of operation for nearly all ion exchange beds, since it leads naturally to the lowest possible pressure drop. Therefore, bed volume is measured in this classified state, and all calculations involving bed volume are referred to this backwashed and settled condition. Since ion exchange resins swell and shrink when converted from one ionic form to another, the careful researcher or engineer will refer calculations to the backwashed and settled resin volume in the appropriate ionic form.

Another set of terms sometimes misunderstood relates to *flow rates*. Although the quantity of liquid flowing through a column can be expressed in drops per second and used repeatedly as a comparison criterion for one and the same column, a precise indication of the flow rate is given in $ml \cdot cm^{-2} \cdot min^{-1}$. If the flow rate is expressed only in $ml \cdot min^{-1}$, the diameter of the column must also be indicated. The linear flow rates in $cm \cdot min^{-1}$ or $cm \cdot s^{-1}$ are used less frequently. But other terms are also used to express flow rates. In the laboratory use of ion exchange columns it has been common practice to refer to volumes of liquid applied to the column in terms of bed volumes and to express flow rates in BV/minute. Here a simple way to measure flow rate is to run effluent from the column into a graduated cylinder and check the time required for a known volume to collect by using a stop-watch. If, e. g., 30 ml effluent is collected in 100 seconds, and the bed volume in the column is 120 ml, the flow rate is $(30 \times 60)/(100 \times 120) = 0.15$ BV/minute. Further, with respect to ion exchange column operations, the flow rate must be related to some characteristic of the resin bed. The two most important terms in this regard are: 1. Rate of flow per unit volume of resin, i. e., the space velocity; and 2. Rate of flow per unit of cross-sectional area of the column, i. e., the linear velocity, as already defined above. The former term is usually expressed not metrically but as gallons of flow per minute per cubic foot of resin in the bed (gpm/ft^3). It is a measure of the contact time of the flowing solution with the resin bed. Thus, a flow of 100 gpm through a column containing 25 ft^3 of resin is flowing at a rate of 4 gpm/ft^3, or $1/0.535 = 1.86$ minutes. It would be worthwhile to give some cross-references in this regard to other unit-systems. 1 gpm/ft^3 is equal to 8 bed volumes per hour (BV/h) or, in metric units, 8 liters per hour per liter of resin (l/h/l). For the laboratory, this may be more conveniently expressed as 0.133 milliliters of flow per minute per milliliter of resin (ml/min/ml). Continuing with velocity terms used, and staying with gallons the other velocity term important in ion exchange unit operations must also be briefly treated, i. e., the so-called *superficial linear velocity*. In the example given above of 100 gpm flowing through 25 ft^3 of resin it was calculated that the space velocity was 4 gpm/ft^3. But to calculate the pressure drop through the column of resin the superficial linear velocity must be determined. This is a function of the geometry of the column. Assuming that the 25 cubic feet of resin are contained in a column 3 feet in diameter and 6 feet high, with a resin depth of 3.5 feet, the cross-sectional area of the column is $(\pi/4)(3)^2$ or 7.07 ft^2 and the superficial velocity would typically be expressed as 100 gpm: 7.07 ft^2 = 14.1 gpm/ft^2. If this is then divided by 7.48 gal/ft^3, one obtains that the superficial linear velocity is 1.89 ft/min.

The resistance produced by friction leads to a *pressure drop* as a liquid flows through an ion exchange column, so that the flow rate and thus the volume flowing through per unit of time are reduced. The pressure drop depends on the particle size of the ion exchangers, as demonstrated in Figure 1.44 with the example of the cation exchanger Dowex 50X8 with particle sizes 20–50 mesh and 50–100 mesh. The pressure drop is a function of the apparent density, particle size, and shape. This is equally applicable to cation exchangers and anion exchangers. To determine the pressure drop, one would refer to the pressure drop curves for the particular ion exchange resin or polymeric adsorbent with which one is working. These are

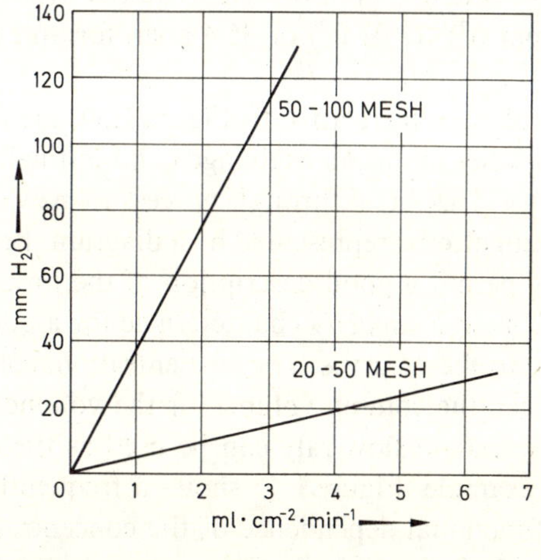

Figure 1.44 Pressure drop and flow rate of the same Dowex 50X8 ion exchanger of different particle sizes.

found in the technical literature for the product, and identify the pressure drop factor, in psi/ft, which relates to a superficial velocity of 14.1 gpm/ft^2 at the appropriate temperature. This value is then multiplied by 3.5 ft, the bed depth, to obtain the total pressure drop (exclusive of that resulting from valves and other fittings) which will occur at that flow rate and temperature in a bed of resin which has been hydraulically classified. Pressure drop curves in the technical literature nearly always refer to water, so that appropriate corrections for different viscosity and density must be made if the solution is not water. It is important to remember that changes in ionic form can result in changes in particle diameter and in bed height, so that a pressure drop curve relating to the proper ionic form should be used. These values are approximate, since the ionic form is rarely known with precision and since the particle size distribution of ion exchange resins varies slightly from one lot to another.

The superficial upflow velocity is also important, since it is the characteristic which defines the hydraulic expansion of an ion exchange resin or polymeric adsorbent. In the case mentioned above the freeboard of the unit is $(6-3.5)/3.5$ or 71% of the resin bed height. Supposing one wanted to backwash the 25 cubic feet of resin sufficiently to give a 40% hydraulic expansion. One has enough freeboard to permit this, so one would then refer to the hydraulic expansion curves in the technical literature for the particular resin one is working with, paying attention to temperature as a parameter. The backwash or hydraulic expansion curves normally relate to resin in the exhausted ionic form, since backwashing usually takes place after exhaustion and before regeneration. It is to be noted that if the solute being adsorbed is uncommonly dense, or if it causes greater than normal shrinking or swelling of the resin, the standard hydraulic expansion curves will not apply, and curves specific to the application must be derived experimentally. Assuming, however, that in the present case one has no such problem, one finds that at 60 °F one needs an upflow rate of 5 gpm/ft^2. The required overall flow rate, in this example, is therfore (5 gpm/ft^2) · (7.07 ft^2) or 35.4 gpm, for an expansion of 40% using water at 60 °F.

As indicated briefly in the explanations relative to Figure 1.39, the exchange of one species of counterion for another in an ion exchange column finally reaches a point where breakthrough occurs. This breakthrough process of new counterions which are no longer exchanged can also be represented by a diagram. Breakthrough curves in this diagram generally permit a good description of the processes in ion exchange columns since a *breakthrough capacity* characteristic for a column under given conditions can be assigned to these curves. The concentration, column length or time, or the equivalent fraction, the column volume, or the volume of solution flowing through the column at constant flow rate can be used as the coordinates of breakthrough curves. As an example, Figure 1.45 shows a frequently-used type of breakthrough diagram: the functional dependence of the concentration of ions retained in the loaded solution, i. e., ions which were not exchanged, upon the volume eluted from the column. Highly interesting and important are studies of

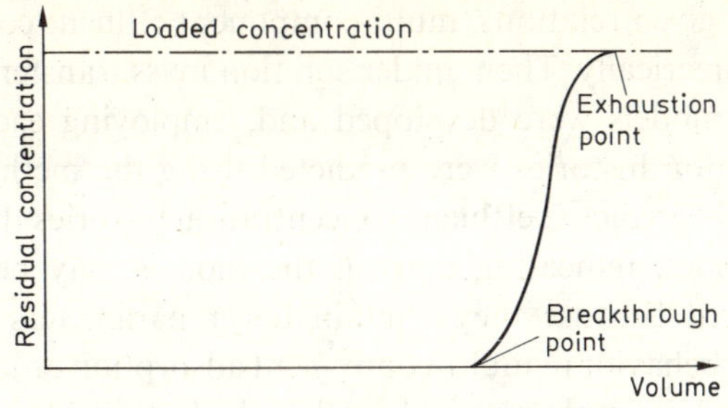

Figure 1.45 Breakthrough curve schematic.

the breakthrough behavior made by Bobleter. Cation exchangers of type Dowex 50X12, H-form, were loaded with solutions of approximately 10^{-2} M concentrations of strontium and cesium nitrate containing 0.01 to 0.5 M nitric acid. The concentration of the effluent solution was determined by measuring the activity of ^{90}Sr and ^{137}Cs. In the logarithmic scale chosen the logarithm of the concentration is proportional over a wide range to the volume of the effluent. Then a formula was derived for the linear region of these logarithmic breakthrough curves by using the given parameter of the exchanger column and the solution. Further, it was shown that the diffusion to the exchanger particle is the determining factor, when the distribution coefficients are high. The thickness of the diffusion layer is a function of time. In the case of lower distribution coefficients, an increase in the slope of the breakthrough curves occurs due to the back diffusion of the ions. As a further result an empirical formula was found that takes this influence into account. Thus it has been shown, that the breakthrough behavior can be described with the aid of parameters given by the type of exchanger, the dimensions of the column and the solution. There is, therefore, no further need to use the frequently-used height equivalent to a theoretical plate, which is itself dependent on many parameters [237]. Of substantial interest is that a mathematical description of breakthrough curves is feasible, and that diffusion coefficients can be calculated from them [238]. But even for binary systems ion exchange columns are unsteady (dynamic) systems that are rather difficult to describe accurately. The breakthrough behavior was therefore also examined by simulation, using the continuous system modeling program (CSMP). The simulation assumes that the column sometimes consists of a large number of mixed compartments, each of them containing a liquid and a resin phase, whereby exchange is governed by a diffusional mass transfer resistance. CSMP is designed for easy programming of sets of differential equations, and although good simulations were obtained in all cases, it created more difficulty than anticipated [239]. For ternary systems the column dynamics have been investigated and complete multi-component ion exchange diffusion equations have been derived based on irreversible thermodynamics and their approximations by the Nernst-Planck model or Fick's law model, both put in the same algebraic form.

From binary data and the given relations, multi-component effluent concentration histories were predicted numerically. Then, under solution mass transfer controlling conditions (0.05 N), four models were developed and, employing each of them, ternary effluent concentration histories were predicted using the method of characteristics. All of the models predicted effluent concentration histories that matched closely the experimental ones, indicating that all the models may be used. The simple models are preferred because they contain fewer parameters [240]. In a conceptual view of column behavior in multi-component adsorption or ion exchange systems Helfferich described a simple physical model which illustrates the concept of coherent waves, i. e., concentration variations, and their interference, as well as the application of convenient mathematical and graphical tools for prediction of column performance, especially in systems with arbitrary, variable entry conditions [241].

There are other types of ion exchange system than those making use of the simple descending flow technique (which is also called the cocurrent system). One which has already been briefly mentioned above is the countercurrent-flow ion exchange system, which is applied in preparative and large-scale processes. The principles here are simple and apparatus have often been described in the patent literature [242]. Another ion exchange system which must briefly be mentioned here is the mixed bed system. This term refers to a bed containing a cation as well as an anion exchanger which simultaneously removes cations and anions from the solution and replaces the salt by an equivalent quantity of water. Because of their importance in water treatment mixed bed filters will be treated in detail in other chapters of this volume.

The regeneration of ion exchangers has been a constant question and has most probably been discussed and treated in papers as thoroughly as the ion exchange itself. Improvements cost-wise in the operation of ion exchange units were first expected from the application of special regenerating and cleaning processes, and then it was realized that spent ion exchanger regenerants present a problem in pollution control. Water was considered as a regenerating agent for ion exchange processes in order to reduce environmental pollution; the use of water vapor for regeneration of ion exchangers was also presented. An ultrasonic process for regeneration of the exchange capacity of ion exchangers was studied, but it functioned only as a mechanical removal process for impurities. Thus, Kunin and Vassiliou investigated the regeneration of carboxylic cation exchangers with carbon dioxide, which depends on the acid strength of the resin, and which subsequently furnishes usable solutions of $NaHCO_3$ for the regeneration of weak base anion exchangers [243]. The regeneration of weak base anion exchangers with 4% aqueous ammonia solutions was favored, because this process causes neither unpleasant odors nor other disadvantages compared to NaOH regeneration. The process is recommended if the exchanger has already been damaged by humins, because an immediately effective remedy is provided by NH_3 [244]. In addition, regeneration with acid mixtures and by a mixture of solutions containing ions capable of complexing have been proposed, particularly to prevent precipitates on exchangers. The complexing was claimed to increase the regeneration effectiveness by a factor of more than three. Using nitric acid solutions on ion exchange resins can pose serious thermochemical hazards, especially when the volume of the resin is large. This danger is not widely known and applications involving the use of nitric acid as a regenerant or eluant for anion exchangers have been reported in the literature, although the original articles unfortunately contained no warning as to the dangers of using the acid, in spite of the fact that earlier reports, from the 1960s, described problems of vessel rupture, fire and explosion. Where nitric acid has to be used precautions to be observed should be explicitly described with even a precautionary checklist given to those responsible for or actually running an ion exchange unit where nitric acid is used [245]. The electrochemical regeneration of mixed-bed exchangers has been examined in quite a number of investigations. For instance, a pilot plant filter with electrochemical regeneration was developed. The anode was located inside the cation exchanger layer and the cathode inside the anion exchanger layer, thus effecting the regeneration by the formation of acids and bases on the electrodes. The electrochemical regeneration of ion exchange resins is indeed a fascinating idea. According to some patents, attempts have even been made to get along without a regenerating solution by shifting the adsorption and desorption equilibrium by varying the pressure or temperature. Sulfite and complexing agents have been used to clean exchange resins with iron oxide deposits and, when the general performance dropped, sodium hypochlorite solution has been used to improve the operation.

1.1.4.3 Continuous processes

In the continuous ion exchange processes – which are essentially reserved for industrial applications – the exchanger and liquid usually move in countercurrent columns. It is a characteristic and disadvantage of column processes that a large part of the upper ion exchange bed remains in the column without utilization during the run, while the lower part is undergoing ion exchange. The logical theoretical consequence of this situation is that the exhausted part of an ion exchange column should be continuously removed, with regeneration being performed immediately. This describes the characteristic feature of continuous or fluid-bed processes.

The technical difficulties which had to be overcome to realize such processes were great [246]. A number of apparently usable systems have been described, in which ion exchange proper and regeneration or elution were carried out simultaneously in different parts of the equipment. However, this process turned out not to be usable over an extended period of time, since the equipment was too costly and the ion exchange particles too unstable. If the problems of process engineering can be overcome, continuous ion exchange processes will find a broad field of application, and represent the future of ion exchange technology. This is already suggested by certain well-advanced techniques.

But it was felt that there was still a need for alternative types of contactor to the Higgins and Asahi types, particularly for treatment of ore slurries with ion exchange resins and for ion removal and recovery in the field of effluent treatment. There are several designs available with good economic prospects but this variety of continuous countercurrent liquid/solid contacting devices also presents a selection problem to the potential user. Equipment in use since the late 1960s is compared in the literature on the basis of chemical performance and throughput [247]. With respect to chemical engineering problems the prospect of increasing use of various types of continuous ion exchange equipment requires further development and testing of designs with the objective of formulating appropriate laboratory and pilot plant experiments, interpreting results fully and making design proposals for large scale plants for purposes of cost comparison. There are various limitations and deficiencies mainly involving the validity of certain assumptions, the lack of diffusion coefficient data for design purposes and the mathematical difficulties associated with multicomponent ion exchange [248]. It was shown in this paper how the theory of non-linear multicomponent chromatography can be used for defining and designing continuous separation. The influence of kinetic and dispersion factors was studied and preliminary experimental results were presented, comparing the performance of a moving and a fixed bed.

1.1.4.4 Fluidized beds

Initially what is now understood as a fluidized bed was nothing more than an agitated batch of ion exchange resin with a continuous flow of solution through the stirred bed, rather than a batchwise feeding as in the actual batch process. Solutions that contain relatively large amounts of solids cannot be processed in fixed column beds of ion exchange resin because the bed is rapidly plugged by the solids. Solutions have to be filtered to remove solids prior to passage through fixed beds of resin, but with some liquid-solids mixtures this is not practical. Special ion exchange contactors have been designed to process mixtures of liquid and solids, including a fluidized bed in which the resin is expanded by the flow of feed through the contactor. The fluidized bed permits the passage of solid through the bed [249].

Fluidized beds have found applications in continuous countercurrent processes as fluidized staged (compartmented) columns – as typified by the Himsley contactor. This contactor is designed to operate continuously by transferring the resin from stage to stage down the column, one stage at a time, while feed flow continues. Resin is removed on a time basis that is consistent with the rate of loading to equilibrium in each stage. The continuous feed flow is accomplished by recirculating the flow of feed solution around an individual stage by pumping the feed from below the stage to above the stage at such a rate that the net flow through the stage is downward. When completed, the process is repeated on the next higher stage. Regeneration is carried out countercurrently in a packed column. In a similar 5-stage laboratory model the highly rarefied fluidized cation exchanger beds were more than ten times as effective as packed resin beds. A multi-stage fluidized bed extraction column coupled with moving-bed regeneration and wash columns has also been built and tested [250].

In retrospect it is somewhat surprising that very little was published on batch fluidized columns or so-called deep fluidized beds. Results of experiments with a fluidized ion exchange column were however presented in which backmixing of the exchanger was prevented by horizontal perforated plates. Over a certain range of liquid velocities the resin distributes itself over a series of fluidized beds between which there is no transfer of resin. This reduces backmixing so that the column can be used for processes involving a fluidized batch of resin. It was found that the breakthrough of the column can be quite accurately simulated by a numerical model which describes the column as a series of well-mixed compartments with a mass transfer resistance between the phases and that breakthrough curves for systems with not too low selectivities are sufficiently exact for many process applications. A comparison with conventional packed columns is of interest. The volume of the fluidized ion exchange column is larger, the separation sharpness is less and the range of operating velocities is limited. The one important advantage that the process has is that liquids containing fine, suspended particles can be processed, if these do not tend to foul up the ion exchanger [251]. It is important to note that the breakthrough curves of deep fluidized and packed beds are similar. This applies at least to a plexiglass column 160 cm long, with a 3.5 cm inner diameter and containing a fluidized bed, used for the very simple exchange of chloride ions between a dilute aqueous solution of hydrochloric acid and a strong base anion exchanger, which was chosen in order to meet the basic assumptions of irreversible equilibrium and liquid side mass transfer control. After approximately 80% of the resin capacity is exhausted the internal resistance to mass transfer becomes important, and the overall rate of ion exchange is continuously slowed down [252]. Four highly interesting topics have been covered in one investigation by Lieser with respect to fluidized bed as an ion exchange procedure. The conical device containing the fluidized bed is schematically shown in Figure 1.46. The separation of the trace elements Fe, Ni, Cu, Zn and U from seawater in a fluidized bed of the chelating 2-hydroxyphenyl-(2)-azonaphthol (Hyphan®) on cellulose beads and on polystyrene beads was studied in parallel field experiments during a period of eight months. Samples from the incoming water and from the exchangers were taken from time to time for determining the trace element content. From the loading curves it follows that, at the beginning, the trace elements are taken up quantitatively on the cellulose beads whereas one half or less is fixed on the polystyrene beads. This is due to the fast exchange on cellulose beads compared to the slow exchange on the polystyrene ion exchanger. It was further shown that the loading curves of the individual elements can be calculated in the range where

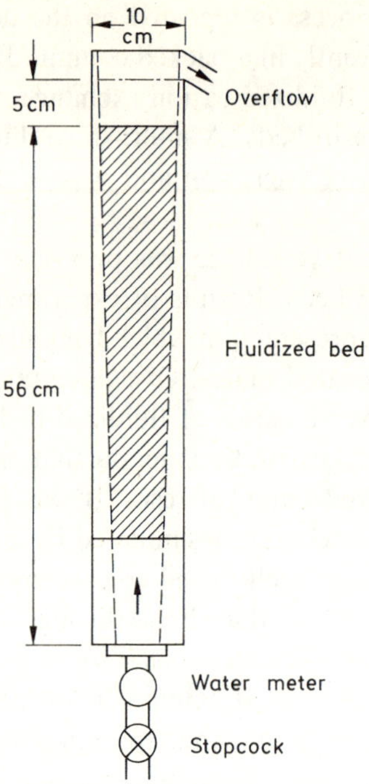

Figure 1.46 Fluidized bed (Lieser).

loading is proportional to concentration. The loading is restricted by the exchange equilibria. At the end of the experiments loading with the above-mentioned trace elements corresponded to about 40% of the capacity. Hyphan on cellulose beads is well suited for the separation of uranium from sea water, whereas the network of the polystyrene beads showed pronounced exclusion of the voluminous triscarbonato complex of uranium [253]. The preparative use of a fluidized bed and an anion exchanger as material was demonstrated for combinations of the broad spectrum, demand-type quaternary ammonium resin-polyiodide bactericide and virucide in the triiodide, pentaiodide and heptaiodide forms, using a fluidized bed method in a closed aqueous system at elevated temperatures. The triiodide disinfectant prepared by this method is equal to the product obtained by a room temperature batch contact process and can be prepared in a considerably shorter time. The apparatus used is relatively simple [254]. Several models have been presented for binary as well as ternary ion exchange systems and different design parameters have been studied. Last but not least a model has been proposed for ion exchange operations by means of solving the mass transfer equations in the liquid and solid phases numerically for a fixed bed and a single- or multi-stage fluidized bed. The experiments were carried out on a laboratory column in order to confirm the model calculations of the exchange of hydrogen for copper(II) ions on a strong acid cation exchanger. The possibilities and limitations of fixed and multi-stage fluidized bed have here been elaborated and compared. In some cases their performances are similar. Multi-stage fluidized beds can be used to reduce clogging problems caused in fixed beds by certain fluids [255].

1.1.4.5 Mass transfer

Mass transfer applied to ion exchange is a relatively new approach, aimed at describing and recording the transfer of solute from the solution to the ion exchanging material, i. e., in most cases, to the resin. Mainly from the perspective of chemical engineering ion exchange is preferably referred to as a mass transfer operation between a liquid and a solid phase, whereby an ionic species is transferred from the solution to the resin, provided a driving force exists. Two aspects should obviously be discussed with respect to mass transfer in ion exchange, i. e., what the maximum transfer is that can be achieved and how fast the mass transfer is. By including the fundamentals of ion exchange "reactions" into these considerations it is clear that the answers to these questions will be based on equilibria and diffusion.

Mass transfer analyses have now been applied to all the above mentioned modes of operation, i. e., batch operations, column processes, continuous processes, and fluidized beds, and even to areas of sole diffusion processes, for instance an intraparticle mass transfer in weak acid ion exchanger. Mass transfer is intended to eliminate the uncertainties of an empirical approach (with the use of safety factors, etc.) in the design of equipment for ion exchange processes and to help solve the problems of the modelling of ion exchange systems by providing a tool for the scale-up from the laboratory to large scale equipment. The concepts of mass transfer can be applied to ion exchange as it is, in essence, a diffusion process for reaching equilibrium. All diffusional mass transfer involves the migration of one substance through another under the influence of a "driving force". In ion exchange two immiscible phases have been brought into contact to enable mass transfer of the component from one phase to the other, i. e., it depends upon rate-process mass transfer. In ion exchange, the main mass transfer mechanism is diffusion in the fluid or solid phase, the diffusivity for a given ion being in the latter about an order of magnitude smaller than in the former. Mass transfer considerations of ion-exchange thus belong to ion exchange kinetics. Since, however, kinetics for dynamic performance investigations still pose unsolved problems, one wishes to probe beyond the simple, "ideal" case of binary exchange of strong electrolytes with strong acid or strong base resins. Progress, however, will depend on basing a model on the relation between the amount of mass transferred from one phase to another, and on the mathematical derivation of this relation. Usually the permeability of the medium and the mobility of the component in it have to be taken into account. The goal is to use the relation for calculating the necessary amount of the solid phase and the time required for the process.

A method of simultaneous evaluation of the interphase mass transfer coefficient, k_s, and the intraparticle diffusivity, $\overline{D}$, from batchwise sorption rate data has been proposed, first assuming that the isotherms or equilibrium relations are linear. An apparent mass transfer coefficient, k_{so}, which was determined by neglecting intraparticle diffuson resistances, has been used to evaluate the individual coefficients. Generally, however, the exchange isotherms and equilibrium relations of ion exchange are not linear. The above method had, therefore, to be extended to the case of non-linear isotherms. The values of the liquid-to-particle mass transfer coefficient $(k_A)_s$ and the ionic diffusivity $\overline{D}_A$ were evaluated from experimental results for the exchange of hydrogen and sodium ions with Dowex 50WX12 in a batch-wise stirred tank. The values obtained were found to be in good agreement with those estimated from published correlations [256]. Similar mass transfer studies of dissolution and sorption processes have been done, for optimization in reactors with a mixing device; previously there were also studies done on stirred suspensions, as a function of different factors. The mass transfer coefficient was found to be proportional to the Reynolds number, the Prandtl number, the diffusion coefficient and the diameter of the vessel. For fixed bed ion exchange processes it seems, first, to be necessary to determine an applicable model and then to study the mass transfer rates. The mass transfer of ions at a liquid film interface on a resin phase has thus been treated theoretically. It was found that ions having various diffusion coefficients necessitate the consideration of the gradient due to the electrical potentials, in addition to the conventional concentration gradient (driving force). The Nernst-Planck equation was modified to account for particular cases where the concentration of the external solution was either constant or variable. There was also simplified discussion on the case where the diffusion coefficient in the resin phase has to be taken into account [257]. The rate of uptake of solute at a particular level of a fixed ion exchange bed, however, is governed by the rate mechanism or combination of mechanisms applicable, by the relevant equilibrium isotherm, and by the values of the parameters entering the mathematical model that describes the mechanism. It is state of the art to use two simple models that permit manageable practical calculations, of which the first, called the f-mechanism, has the controlling resistance in the film, and diffusion inside the particle is relatively fast, so that the radial concentration gradient in the particle is virtually zero and the average concentration in the particle coincides with its concentration at the particle surface, where the fluid-phase concentration is in equilibrium with the concentration in the particle phase. The second, Glueckauf and Coates' linear driving force approximation, a relatively fast diffusion in the fluid and the major resistance in the film phase is assumed, is called the p-mechanism. Here, the concentration in the fluid phase is assumed to be uniform throughout, right up to the particle surface, where the concentration $Y(X)$ in the exchanger is in equilibrium with X. In the particle, the concentration then decreases linearly through an imaginary film to the average concentration in the particle, Y. Systems with both the f- and the p-mechanism have also been dealt with. In a third, relatively simple, model the fixed bed ion exchange process is considered as a second order reaction, with the corresponding kinetics. The attempt was made to solve the problem of mass transfer in terms of the *J* function [258]. Two other factors are important in modelling fixed bed ion exchange and the mass transfer related to them: pore diffusion and axial dispersion, and both a pore diffusion mass transfer mechanism and an axial dispersion mass transfer mechanism have been discussed [259]. Later the mass transfer in ion exchange in a fixed bed was studied in the combined diffusion kinetic region when the axial mixing is significant. A mathematical model was derived for the ion exchange process and a numerical method was proposed for its solution [260]. It was shown that agitated ion exchange beds can be described through the use of an overall transfer coefficient that is time-dependent. Basic mathematical relationships – although the actual mechanisms of the process are not known – were developed for the case in which the overall coefficient is a general function of time and they have been found to agree with experimental data on multi-stage units. These fundamental relationships offer a basis upon which small-scale results may be used in the design of large-scale agitated ion exchange units [261]. With the development of the more recent general theoretical analyses of ion exchange, resin phase mass transfer in a weak acid ion exchanger has been analyzed based on a shrinking core model. The Nernst-Planck equation was applied to the fluxes of ions diffusing through the shell and the analytical values for sorption and regeneration rates were derived by assuming a pseudo-steady-state diffusion. Those values are in agreement with the experimental results in R–H–NaOH systems for sorption and in R–Na–HCl, R–Cu–HCl, and R–Cu–H_2SO_4 systems for regeneration [262].

1.1.4.6 Ion exchange parametric pumping

Parametric pumping is a separation method based on a shift in the ion exchange equilibrium due to the action of a thermodynamic variable such as temperature or chemical potential. It is not a general procedure like those described above, but a special, newly applied separation method where a typical operation consists in passing a solution to be separated through a packed ion exchanger bed by making it flow alternately upwards when the column is hot and downwards when the column is cold. Thus parametric pumping is a cyclic process, which, as theory and experiment have shown, can produce substantial separations of liquid mixtures. This cyclic operation interacts constructively with the natural dependence of sorption equilibria on temperature to yield separations where each new cycle builds on the results of the previous one.

A summary follows of the principles of parametric pumping for thermal ion exchange – presented in detail elsewhere – as applied in batchwise operation, where heat is conveyed to the system using the heating and cooling jackets built around the ion exchange column and the storage tanks (see also Figure 1.47). A liquid contains the ions *i* and *j* which are to be separated. It is assumed that the

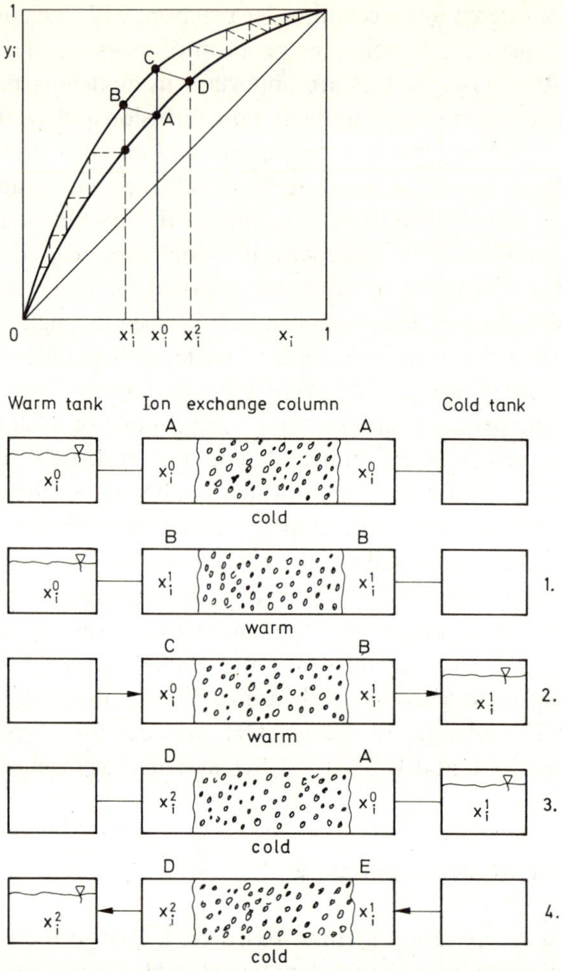

Figure 1.47 Schematic representation of ion exchange parametric pumping.

ion exchange equilibrium favors ion i, that the selectivity coefficient D_j^i increases with temperature and that adsorption and desorption phenomena of both the electrolyte and the solvent are negligible.

At the beginning of the cyclic operation, at the lower temperature level, the ion exchange column is equilibrated with the liquid to be separated, with a concentration of x_i^0. The same liquid, of the same composition, is also filled into the warm tank. The state of the ion exchange column on the equilibrium diagram is marked by point A. In the first subcycle both the ion exchange column and the warm tank are heated to the higher temperature level, without moving the liquid to be separated. At higher temperatures the ion exchange selectivity constant, T_j^i is higher, consequently j ions are released from the resin and i ions from the liquid are bound onto the resin. After the exchange of an equivalent amount of ions, the new composition of the liquid phase in the column becomes x_i^1. The state of the ion exchange column after the first stage is shown by point B on the equilibrium diagram.

In the second stage, the warm liquid from the hot tank is pumped through the hot ion exchange column into the cold tank. If the column dimensions, volume flowthrough, concentration and flow rate are appropriately selected, then the concentration at the inlet point of the column can be made x_i^0, equal to the concentration in the warm tank (point C in the equilibrium diagram), while at the exit point – and consequently in the cold tank – the concentration is x_i^1 (point B in the equilibrium diagram). This means that the solution in the cold tank is more concentrated with respect to ion *j* than the solution in the warm tank ($x_i^1 < x_i^0$).

In the third stage, the ion exchange column and the cold tank are cooled to the lower temperature level, without pumping the liquid phase. Selectivity at the feed point of the column becomes x_i^1 (point E in the equilibrium diagram). The concentration of the solution entering the warm tank is x_i^2, i. e., the solution is more concentrated with respect to ion i than the original solution.

Stages 1 through *n* make up the full cycle of the batchwise operated, thermal ion exchange parametric pumping separation method. This cycle is called a theoretical cycle. It is assumed in stages 1 and 3 of the theoretical cycle that new equilibrium concentrations can be achieved during thermostating (sections AB and CD), and in stages 2 and 4, during the flow of the liquid phase, that the new equilibrium composition of the liquid at the exit points is constant (x_i^1, B and x_i^2, D, respectively).

After the completion of a theoretical cycle, the concentration of ion *j* is higher in the cold tank and the concentration of ion *i* is higher in the warm tank, both relative to the original concentrations. When the theoretical cycles are repeated – dotted lines in the figure – both ion concentrations increase further. Theoretically, by successive repetition the separation factors, i. e., the ratio of the concentrations in the cold and warm tanks, could be increased to extreme values. In practice, however, after a certain number of cycles, a so-called quasi-steady-state condition is achieved. In this case, the separation factors will not increase any further and the concentrations in the respective tanks will not change. The extent of enrichment realized in batchwisely operated, thermal ion exchange parametric pumping depends on the operation parameters used, the column dimensions, the ion exchange resin and the system of ions to be separated [263].

Other laboratory apparatus described in the literature consist, e. g., of a jacketed glass column, 60 cm long × 1.1 cm inner diameter, connected at each end to a disposable, plastic 50 cm^3 syringe and a pump which displaces the reservoir syringes, forcing liquid through the bed; or a 52 cm long × 1.5 cm inner diameter column fitted with a water jacket to control temperature connected on top to a reservoir and at the bottom through a pump to another reservoir. In partial reflux operation, then, two columns connected in series can be used [264].

The advantages that parametric pumping offers over conventional ion exchange are that no regeneration solution is required – which means that in addition to an economy in chemicals the products are separated one from another without contamination by another chemical species – that low potential heat is used as the driving force for separation, which could be made available for 60 °C from warm waste process water and for 20 °C from river water, and that one has the possibility of sharp separation by adjusting the reflux ratio. As in distillation, parametric pumping can be carried out at either total or partial reflux, that is, with or without feed and product removal.

Parametric pumping was used to separate binary and ternary mixtures. Direct thermal parametric pumping was used to fractionate K^+-H^+ and $K^+-Na^+-H^+$ mixtures using Dowex 50XB as the ion exchanger. For separating proteins a semicontinuous pH-parametric pump was investigated, using the model system Hb and albumin on a Sephadex ion exchanger. Ion exchange parametric pumping has also been studied in order to separate metals in solution, such as silver and copper [265]. The concept of parametric pumping was introduced by R. H. Wilhelm and coworkers in 1966, while dealing with adsorption from a liquid phase which was modulated by temperature. Tondeur and Grevillot have presentated an extensive discussion of parametric ion exchange processes with respect to thermal ion exchange parametric pumping, i. e., the batch operating scheme for binary mixtures, modelling by the staged distribution approach, and modelling by the wave approach; as well as to experimental examples of the thermal ion exchange parapumps, i. e., effect of temperature on ion exchange equilibria, desalting experiments, and experimental separation of solutes; and to other forms of parametric pumping, i. e., cycling-zone and sorption, moving-bed parametric pumping, the use of other parameters than temperature (e. g., in pressure-swing adsorption) and ion exchange combined with other separation processes [266].

1.1.4.7 Various additional aspects of handling ion exchangers

The limitations and problems of ion exchange procedures (among others related to the technical setup of the apparatus) are to be found in bacterial growth and fouling of inorganic and organic ion exchangers. The degree of poisoning can be determined by spectrophotometry after contacting the swollen ion exchange material with an aqueous humic acid solution of pH 4.2–4.5, and eluting the latter by an alkaline solution. For the elimination of ion exchanger fouling by organic compounds, e. g., humic acids, ozone was found to be effective. Bacterial growth in large-size ion exchanger units (but also in filters for homes, laboratories and small industries) is a perpetual problem. For the desinfection of such units formaldehyde, chloramine T (*N*-chloro-*p*-toluenesulfonamido sodium) and peracetic acid have been recommended and used [267].

Another question is the waste partitioning of spent ion exchangers, especially of radioactively contaminated materials. Disregarding processes for dissolving radioactively contaminated organic ion exchange resins, acid digestion processes for treatment of organic exchange materials in hot (230–270 °C) conc. sulfuric acid containing nitric acid oxidant have been investigated to reduce the volume by converting it to a noncombustible residue. Residues were in fact obtained having less than four percent of their original volume and less than twenty-five percent of their original mass. The use of hydrogen peroxide at 260 °C, in the presence of sulfuric acid for acid digestion of ion exchange resins leads to almost complete oxidation. Compared to the solidification procedures mentioned briefly below incinerating a resin represents an extra processing step, but much smaller volumes result from it [268]. Solidification of ion exchange resin wastes has been investigated, making use of portland type and high alumina cements, a proprietary gypsum-based polymer-modified cement, a vinyl ester-styrene thermosetting plastic, bitumen, etc. In a new method for the treatment of radioactively contaminated, spent, ion exchange resins the radionuclides are transferred from the spent organic resins to zeolite/titanates. The latter are then sintered to a stable ceramic body of small volume, while the low-level eluted resins are either dewatered in transportable concrete containers or incinerated. It has seemed technically feasible to introduce this so-called PILO process on an industrial scale. Otherwise the waste disposal of spent ion exchange resins is a matter of general environmental control, depending on the history of the materials and the laws and regulations to be followed.

Special aspects of the use of ion exchangers have led to the development of special devices and apparatus. By a stirring cell, for instance, the exchange process can be traced in a secondary cycle, whose volume is negligibly small, by continuous measurement of the solution activity with corresponding radioactive labelling. A spiral ion exchanger for rapid radiochemical separations has been developed, with separation times as low as 10 ms where short-lived daughter products have to be separated from long-lived parent substances. An apparatus and method for separating and depositing specific materials from a flowing medium by using an endless belt with both ion exchange and filtering capability has been described in the patent literature. Likewise, an apparatus for deionization of a suspension has been described, with the ion exchanger material in a rotating basket so that the resin bed is not clogged by the suspended particles.

Additional special processes have also been developed tailored to solving special operating problems. A variant of the batch process, the resin in pulp (RIP) process, for the industrial recovery of uranium must be mentioned. In this case, the ion exchanger is charged into baskets and is moved up and down in the tanks which are filled with crude and regenerating solution. Although developed for uranium recovery, this process appears to be applicable also for other purposes [269]. In the United States the possibilities of the method with regard to its efficiency and cost have been investigated by the Argonne National Laboratory. A combined process of ion exchange resins and liquid ion exchangers exists in the form of the Eluex process [270], with a strong base anion exchanger and a tertiary amine in kerosene. This process is used very economically in several plants in combination with the RIP process for uranium ore treatment.

1.1.4.8 Regeneration of ion exchangers

After an ion exchange process has been developed and established the running of the process is done in a cycle of service and regeneration. There are cases in which an ion exchange resin, after having been used for a certain purpose, is discarded or incinerated but these are exemptions to the general rule.

As a practical example, the following equation for the softening of water illustrates the conventional ion exchange process

$$2\,R^-Na^+ + Ca^{++} \leftrightharpoons (R^-)_2\,Ca^{++} + 2\,Na^+,$$

where R^- represents the anionic fixed sites in the exchange resin. During the service cycle the reaction goes to the right, representing the exchange of the hardness-creating ions of the feed water for the innocuous sodium ions in the exchanger. After exhaustion, in order to use the exchanger again, the resin has to be regenerated by returning sodium to the resin. For this regeneration cycle a concentrated salt solution with at least an equivalent amount of sodium must be applied to replace the calcium on the resin. However, in order to have the reaction go back from right to left, a higher amount of salt is needed than indicated by the stoichiometry of the equation. Besides this problem, the calcium that has now been replaced may become a pollutant.

With regard to the different affinities of one and the same ion exchange resin towards different species in solutions to be treated and including the various types of ion exchangers, several methods for regeneration are employed to obtain optimum efficiency from the chemicals in use. Also the application of physical energy such as heat, pressure or electricity has been investigated, but is still in the early stages of development.

An industrial water-softening system requires, in fact, a salt equivalent of 1.5 times the theoretical or stoichiometric value, and only two-thirds of the salt dosage is converted to the calcium salts by actual regeneration of the exchanger, while one-third of the salt dosage is wasted and passes through the column into the regenerant waste stream unchanged. For example, a 100 ft^3 resin bed that is regenerated daily with a salt dosage of 6 lb/ft^3 will annually (300 days) use 180 000 lb of salt, of which 60 000 lb or approximately 28 tons are wasted unless recovered through multiple beds or other procedures. In summary, ion exchangers produce polluting wastes if the regenerant effluents cannot be utilized. This is one of the more serious limitations on ion exchange, especially in view of newer environmental protection laws as, for instance, the Federal Water Pollution Control Act in the USA, which limits the discharge of pollutants into receiving waters and into solid fill areas. Ion exchange produces a disposal problem of much smaller magnitude than when one has several original dilute solutions, but one that still must be taken into consideration.

Many techniques have been developed for utilizing the regenerant effectively, including countercurrent operation, recycling of unused regenerant, continuous operation, and the use of waste chemicals from other industrial processes, as e. g. waste hydrochloric acid, sulfuric acid and nitric acid. In a fixed bed system, where the resin does not move during the service cycle but remains as a compact bed after exhaustion, the regenerant is passed through the exhausted resin bed. Downflow regeneration is still quite often used, but upflow or countercurrent regeneration is certainly preferable. In practice the opinion predominates that complete regeneration of the resin bed is not advisable. When downflow regeneration is used, the very bottom of the bed is not fully regenerated. During the subsequent service cycle, water that has been purified in the top of the bed flows out through the bottom of the bed, releasing the least tightly-held ions onto the unregenerated resin. The degree of this leakage will depend on the previously used regenerant dosage and the concentration of the solution being treated. If countercurrent regeneratin is used, the partly regenerated resin is at the top of the resin bed and very much less leakage occurs during the subsequent service cycle. The regenerant flow rates are usually slow enough to ensure adequate time for exchange. The advantages of countercurrent regeneration are, therefore, lower leakage and improved effluent quality, lower regenerant, rinse and backwash consumption, and, finally, a reduction in the quantity of regenerant waste. The difficulty is that, on the one hand, leakage during a service cycle is in most cases not desired; but, on the other hand, to prevent leakage by regenerating a bed fully is too costly. As a technical solution of this problem dual beds or beds in series are frequently used.

To solve regeneration problems chemically one will try to use as little regenerant as possible, as well as to select chemicals which are inexpensive and environmentally safe. The so-called economical regeneration processes comprise techniques where the use of almost-equivalent amounts – the expression for this is close to 100% of the theory – is feasible. There are curves to be found in the ion exchanger literature giving the capacity of an exchange filter as a function of the quantity of the regenerant used. Such curves apply to the case in which the quantity of exchanger is calculated for a given performance, but used in a single filter. If, however, the same quantity of exchanger is distributed over two series-connected filters and the same quantity of acid as above used to regenerate the second filter first while the remaining excess of acid is used for the first filter, a considerably higher capacity is obtained. In the search for possibilities of saving regenerant, compound regeneration was long in use, in which considerable improvements were obtained for full desalination plants. Another step was then the series-connection of a weak and a strong electrolyte exchanger in separate tanks for sequential regeneration. This is based on the experience that weakly dissociated resins have a higher affinity for H-ions or OH-ions, respectively, so that the residual and unused hydrogen ions or hydroxyl ions of the strong electrolyte resins are employed for the regeneration of the weak electrolyte resins. Finally, with a logical extension of the advantages gained, the multilayer bed processes with filters were developed, in which weak acid and strong acid as well as weak base and strong base exchangers were arranged above each other in a single filter tank. Another development for optimal utilization of regenerants was the intensive fractionation process, with the principle of a repeated use of a regenerant in a cycle consisting of the exchange phases, i. e., loading and regeneration, separated by two displacement phases, i. e., concentration before and dilution after ion exchange. The intensive fractionation process has also been discussed in the chapter on preparative chemistry.

The recycling of regeneration chemicals was another alternative for reducing the amounts used. In the different cases investigated and applied it is necessary that the regenerant solutions used must be freed of ions or other substances they have taken up before being used again. Regenerant from water softening filters containing unused sodium chloride with calcium and magnesium was treated with caustic soda and lime to precipitate the calcium and magnesium, obtaining after filtration an alkaline salt solution, which is then neutralized with hydrochloric acid and reused for regeneration. Similarly, the neutral or alkaline sodium chloride solutions from the regeneration of adsorbent filters were recycled after they were freed from organic substances and the acidic regenerants of galvanic processes are freed by electrolytic separation of, e. g., copper and nickel and then used directly as acids for regeneration.

Special regenerating processes have led to improvements in the operation of ion exchange installations. Thus, Kunin and Vassiliou investigated the regeneration of carboxylic cation exchangers with carbon dioxide, which depends on the acid strength of the resin, but which also subsequently furnishes usable solutions of $NaHCO_3$ for the regeneration of weak base anion exchangers. A carbon dioxide regenerated ion exchanger process was developed by Hoell and coworkers for the removal of nitrate, sulfate and hardness from water (this is described in greater detail elsewhere in this volume). The regeneration of weak base anion exchangers with a 4% aqueous ammonia solution was suggested by Klump, as this would cause neither unpleasant odors nor other disadvantages compared with sodium hydroxide regeneration. The process was also recommended for when the exchanger has already been damaged by humins, because an immediately effective remedy is provided by NH_3 as has already been mentioned previously. Similar extensive investigations were carried out using ammonia as well as ammonium chloride or carbonate as volatile regenerants with respect to sea water and the demineralization of sugar solutions. Sulfite and complexing agents have been used to clean exchange resins with iron oxide deposits and, when the general performance has dropped, sodium hypochlorite solution has been used to improve the operation. With regard to the regeneration of cation exchange resins by complexing it might be mentioned that for strong acid cation exchangers solutions containing small additives of complexing ions such as citrate or tartrate have been used, with the complexing increasing the regeneration effectiveness by more than a factor of three. Ion exchange also offers many possibilities for utilizing water as an eluent through complex ion formation, novel ion exchangers, a change in valence, and charge potentials which are built up on the exchangers. In view of the emphasis of reducing the degree of pollution to the environment water as an eluent and therefore regenerant is an ideal solution.

Physical regeneration of ion exchange resins has, as already briefly mentioned above, been attempted by applying pressure, heat, adsorption heat, and electricity as energy sources. Because of their high volume in the loaded form it has proven possible to regenerate weak acid and weak base ion exchange resins by the application of pressure in a reaction of the squeezed-out counterions with water. Unfortunately the pressure needed is so high that the resin particles are pulverized. The regeneration of ion exchangers by the thermohydrolytic decomposition of the salt forms at reduced pressure has also been investigated. Thermal regeneration processes have repeatedly been studied, and in one case has led to the often-discussed Sirotherm process. The electrical regeneration of mixed-bed exchangers in bead form has quite often been investigated and described in the literature, for example in the work of V. P. Meleshko and coworkers and V. D. Grebenyuk and coworkers. An electrochemical regeneration of ion exchange resins is in principle possible, but the technical problems connected with it are great even when additional heat is applied. The degree of electrochemical regeneration was found to increase with increasing temperature for individual strong electrolyte as well as weak acid ion exchangers, with an opposite temperature effect for a mixed bed.

1.1.5 Applications of ion exchange

The applications of ion exchange and ion exchangers can be classified by methods or by field of application. In addition there are numerous other applications, actual as well as potential, which use ion exchange materials but do not actually involve an ion exchange procedure. Developments in the application of ion exchange depend mainly on the advances in fundamental understanding, advancements in the existing procedures or new devices, and improved ion exchange materials. As a result the range of application of ion exchange is vast, embracing both the laboratory and industry [271], as well as the whole field of ion exchangers as reactive polymers.

Among the general points in applying ion exchangers is that in the majority of the commercial uses, dilute solutions, with concentrations of less than 40 equ/m^3, are treated. When comparing laboratory and industrial applications of ion exchangers one further general point must be considered, relating in the different affinities for different ions that exchangers frequently exhibit. Theoretically, the stronger the affinity of a resin for a given ion, the more efficient it will be in the uptake of such an ion. This is not necessarily true practically, since total exploitation of an ion exchange process also involves regeneration or elution of the medium. Since the greater the affinity of an exchanger for a particular ion the more difficult and inefficient the reversal process becomes. In many commercial applications ion exchange processes become regeneration rather than sorption controlled, an aspect that applies particularly to water treatment.

The actual ion exchange methods can in a certain sense be seen in contrast to the classical chemical methods by which ionic species may be either removed or recovered from a solution by such processes as crystallization or precipitation and subsequent filtration. The efficiency and economics of these processes are frequently unfavorable where very dilute solutions are to be handled. In detail the most frequently used ion exchange methods would then be the following.

Ionic exchanger methods

Substitution
  of undesirable for desirable or innoccuous ions
Elimination
  of ionic impurities, e. g. nitrate from glyoxal, sodium from milk
Isolation
  of desirable ionic constituents from contaminated solutions
Purification
  of a main ionic constituent of accompanying impurities
Separation
  of ionic or of ionic and non-ionic species
Chromatographic separation
  of ionic species with similar exchange behavior
Recovery
  of, e. g., catalysts from process solutions or chromate from cooling-water circuits
Concentration
  of ionic species present in trace quantities
Extraction
  of valuable chemical substances and in hydrometallurgy; e. g., extraction of iodine
Removal
  of toxic substances and pollutants from effluents
Immobilization
  of drugs in a carrier function, enzymes etc.
Neutralization
Conversion
  of salts to acids or of one salt to an other
Slow release agent
  of ionically bound substances; e. g., long-term fertilizing of soil

Non-ionic exchanger methods

Ion exclusion
Ion retardation
Ligand exchange
Non-ionic adsorption
Ion sieving
Gas removal
Drying
Catalysis
Buffering

These methods and some newer ones such as the so-called ion chromatography will be dealt with in detail in various chapters of this volume.

The applications of ion exchangers can be characterized in different ways as either industrial in contrast to laboratory applications, with the latter either including analytical procedures or not; or with regard to substances or chemical products. The latter leads to a long list in which one can distinguish the most common applications of ion exchange, used in many existing units, the less common applications, to be found in a small number of units, and applications that are still under development. In the following list **fields of application** are grouped, most of which are dealt with at length in the various chapters of this volume. Such a list can, of course, never be complete.

Water treatment

Softening and deionization
Condensate treatment
Process water treatment
Drinking water treatment
Effluent treatment
Waste waters treatment
Brewing water treatment

Hydrometallurgy

Recovery and concentration of precious metals: silver; gold from gold-plating solutions [272]
Extraction and concentration of uranium and nickel from dilute leaching solutions [273]
Separation of rare earth metals and of heavy metals [274]
Purification of plutonium

Food industry

Sugar and sweetening agents
Wine-making
Spirits

Fruit juices
Citric acid
Sodium glutamate
Sorbitol deionization
Protein recovery [275]
Amino acid separation

Dairy industry

Milk whey demineralization, deacidification and decoloration
Purification of lactose
Removal of calcium resp. sodium from milk
Preparation of pure casein

Chemical process industry

Catalysis [276]
Purification of formaldehyde, glyoxal, phenol, glycerol, bisphenol A, etc.
Ethylene glycol demineralization
Boric acid manufacture

Pharmaceuticals

Alkaloids: Separation of opium derivatives, purification of morphine
Antibiotics: Separation and purification of streptomycin, bacitracin, neomycin [277]
Vitamins: Separation of vitamin B_{12}, recovery of ascorbic acid
Separation and purification of a number of other drugs
Carriers in the processing of pharmaceuticals with delayed action

Medicine

Reduction in serum cholesterol
Bipolar resins for blood preservation

Biochemistry and biotechnology

Immobilized enzymes [278]
Immobilized microorganisms
Virus and protein sorption [279]
Preparation of biologically active substances [280]

Preparative chemistry

SiO_2 sol manufacture
Supports for reagents in organic synthesis [281]

Analytical chemistry

General analytical chemistry
Ion exchange chromatography
Ion chromatography

Oil drilling

Lubrification of muds

Photography
- Silver from photographic baths [282]
- Recovery of color developers [283]
- Removal of hexacynoferrate

Corrosion inhibition
- Paints

Paints
- Fire-repellent additives
- Deionization for electrodeposition coating

Pickling baths
- Regeneration

Pollution control
- Detoxification
- Removal of toxic gases, vapors and aerosols from air
- Stabilization of metallic pollutant in industrial wastes
- Treatment of waste gases and waste waters [284]

Drying
- Drying agents for organic solvents [285]

Sea water
- Uranium from sea water [286]
- Metal extraction from sea water [287]

Separation of isotopes [288]

Sorption of gases [289]
- Removal of sulfur dioxide [290]
- Removal of mercaptans [291]
- Air deodorization [292]

Artificial soil
- Long-term fertilizing soil substituents
- Agents for regulating plant growth
- Improvement of the fertility of sandy soils
- Fibrous nutrient media for growing plants in space [293]

Builders for detergents

The development of the various applications of ion exchange from the traditional water treatment to special applications can also be demonstrated in some figures. Of the 60 000 m^3 of ion exchange resins produced throughout the world in 1969, 48 000 m^3 or 80%, were intended for use in conventional water treatment and 12 000 m^3 or 20% for special applications. By the year 2 000, world consumption is expected to reach 500 000 m^3, of which only 300 000 m^3 or 60% will be used in water treatment and 200 000 m^3 or 40% for special applications.

Some further remarks on the applications of ion exchange should be made with regard to utility in the sense that one has to distinguish between what is practical and impractical with respect to a particular problem. Basic knowledge regarding ion exchange shows that it is possible to use ion exchange to remove any ionic species from a solution. One must, however, usually be concerned with the economic utility of a proposed ion exchange process. What may be practical for analytical purposes, may be totally impractical for industrial usage, and what may be of considerable utility from a medical viewpoint, may be of little value where economics must be considered. An illustrative example concerns the adaptation of an analytical ion exchange chromatography technique to an industrial operation for the separation and recovery of a metal or biochemical compound. In such a case it cannot be ignored that huge quantities of ion exchange material and chemicals are required, because the ion exchange resin column merely serves as a fractionation device and only a small fraction of the exchange capacity can be initially used with the material to be fractionated, whereas adequate equipment is necessary for the large volume of resin. Feasibility reasoning without extensive experimentation is most probably a very important step in assessing the possible utility of a proposed ion exchange process. This is an exploratory step based upon a knowledge of the material to be treated and an elementary knowledge of ion exchange, where the interest is in ascertaining without prior experimentation whether or not one should seriously consider the use of ion exchange as a solution to a given problem. Firstly, the ionic charge of the species to be exchanged must be exactly known or the literature pertaining to the ionic system must be studied. Secondly, after having ascertained the nature of the ionic species to be exchanged or removed, one must consider what might be the appropriate ion exchanger or combination of ion exchangers required. Here the most difficult decision may be whether one should select a weak or strong electrolyte exchanger. The choice between both types involves both the acidity or basicity of the solution and the regeneration efficiency. Weak electrolyte ion exchange resins are much more effectively regenerated with acid or base than the corresponding strong electrolyte exchangers, and therefore for efficiency purposes, one always tries to employ a weak electrolyte ion exchange resin. The best known examples demonstrating that this is not always possible, are carbonic or silicic acid, which are too weakly acidic to be effectively taken up by a weak base anion resin, and which therefore require strong base anion exchange resins. Thirdly, the major factor, in the practical utility of any commercial ion exchange process is the exchange capacity that is realized, which must reach a certain level. All ion exchange materials are determined completely in their total ion exchange capacity, so that for a given resin the theoretical maximum capacity that one can realize in any ion exchange operation is automatically set. The maximum amount of solution, V_s, that can be processed between regeneration cycles by a given volume of exchanger, V_R, can then be calculated from the concentration (mequ/ml) of the ionic species to be removed, N_S, and the total exchange capacity, N_R (mequ/ml), from the relationships, $V_S \cdot N_S = V_R \cdot N_R$. But the utilization of the total exchange capacity of an ion exchange resin is rarely achieved except under highly uneconomical conditions: as a general rule only 50 to 75% is realized under practical conditions. There are, nevertheless, a number of economically successful ion exchange operations in which only 5 to 10% of the theoretical capacity is realized, as well as several cases in which almost all of the ion exchange capacity is realized.

1.1.6 Ion exchange models

Many materials have ion exchange properties and there has been considerable effort expended to find generalizing models of ion exchange. To review such efforts under kinetic, thermodynamic or other generalizing perspectives is useful for summarizing as well as simplifying the understanding of the action of ion exchanging materials. But in spite of the many developments in newer and better synthetic ion exchangers, sometimes tailor-made for a specific application, the existing technology of ion exchange seems far behind nature, which in three billion years of organism development has evolved the most complex molecular systems for the survival and reproduction of organisms and for the transport and storage of substances. The molecular species are involved in many physical and chemical processes, many of which also involve ion exchange processes. For instance, the wall of a plant cell can function as an ionic exchanger. If this is the case there must be a model applicable to explain the findings of investigations into this question. Ion exchange models should similarly be applicable to situations in the inorganic world involving structures through which ionic interchanges take place at low temperature, at high rates, and at high specificity, all in equilibrium with nature and without polluting the environment. For example, the accumulation of radium in ocean sediments can be modelled as a direct dependence between the value of cation exchange capacity of oceanic sediments and their ^{226}Ra content. Not only naturally occurring processes, however, can be explained by employing ion exchange models. This increases the interest in applying the concept of ion exchange also to well-known processes and to investigating how important properties can be influenced beneficially by the exchange of ions present. Ion exchange models should in the following, therefore, be understood as attempts to make clear cases in nature and in technology where ion exchange phenomena play an important or the decisive role.

Soil science, geochemistry and hydrogeology. The earliest references to ion exchange are in relation to soils and soil fertility. The role of cation exchange in soils and base exchange in geochemistry have been studied repeatedly [294]. But soils as such are not ion exchangers in the sense that they could be used in technical processes; therefore, their exchange behavior is explained by models and often explained by the exchange behavior of various soil constituents. Clay minerals, for instance, play an important role in arable soil. Soil fertility is favored whenever the exchangeable cations are predominantly calcium and magnesium ions – which are exchanged from the groundwater – because then the soil has the crumbly structure that allows the penetration of air and water. Sea water or the too extensive use of potash fertilizer causes the alkaline earth ions to be exchanged for sodium or potassium ions. The soil becomes too elutriated and, on drying, shrinks to tight, water and air-impermeable clods with poor fertility. The effect of pH on the contributions of organic matter and clay to soil cation exchange capacities or the rate processes in the desorption of phosphate from soil have also been explained by ion exchange models. Ion exchange has also been found as a factor in the geochemical differentiation of elements under hydrothermal conditions with chromatographic effects during solution filtration. Ion exchange reactions between clays and other natural components and water, ion exchange in sediments, the effects of clay mineral structure, and the equilibrium and selectivity of components of natural exchange systems can be modelled [295] as well as the transport, dispersion and cation exchange equilibria in ground water [296]. The cation exchange characteristics of the amorphous materials from soils have been examined as well, and a structural model for these amorphous materials was suggested that can explain the exchange properties [297]. To characterize the ion exchange in soils a model and corresponding equation based on an exchange constant, was presented. The exchange constant is a function of a number of physico-chemical characteristics of the soil and soil solution that may increase or decrease the volume of ion exchange [298]. Water participates strongly in geochemical processes part of which are ion exchange and adsorption processes both in the overlying stratum as well as in the groundwater. It was found that two conceptual models for diffusion-controlled and two-site kinetic adsorption are mathematically equivalent in describing ion exchange during transport through aggregated sorbing media. The models were used to simulate the observed asymmetry and tailing in the breakthough curves of $^{45}Ca^{2+}$, $^{36}Cl^{-}$, and T_2O in an oxisol soil [299]. Field observations have shown that ion exchange plays a significant role during infiltration of seawater into the shore, and this has been tested by laboratory investigations of the effect of infiltrating seawater on the chemical composition of groundwater. The ionic concentration of groundwater, during intrusion of salt water, changes according to the scheme $SO_4-Na \rightarrow Cl-Ca \rightarrow Cl-Mg$. A quantitative model was developed for inorganic ion exchange reactions in sea water and the resulting so-called φ rule was then used to study the relations between the element contents in stream water, seawater and the sediments on the ocean floor [300]. In a review of some chemical aspects in oceanography the conclusion was drawn that the processes of adsorption, complexformation, and ion exchange take place in the ocean, whereby most of the time these processes proceed simultaneously in any oceanographic chemical phenomenon, although one process might be more predominant than the others. A very interesting model has also been developed for ionic exchanges at the water-sediment interface in sea water.

Minerals. The ion exchange properties of minerals have often been described [301]. It is not the purpose of this paragraph to discuss minerals that are known as ion exchangers and have been used as such, or to explain here the ion exchange phenomenon on the basis of mineral ion exchangers, but to draw attention to processes relevant to mineralogy to which ion exchange models have been applied. (Mineral ion exchangers are treated in this volume under the subject inorganic ion exchangers.) During mineral formation ion exchange plays a major role in the geochemical differentiation of elements. The outer cations of feldspathoids are, for instance, only weakly bound to acid radicals and, therefore, can get involved in cation reaction even at room temperatures. This reaction occurs more readily than in spars, which exhibit cation exchange properties only at elevated temperatures. Ion exchange properties have also been observed in a series of nonsilicate minerals, such as apatite, amblygonite, and pyrochlore. The results obtained suggest that ion exchange reactions should be considered also in the mineral formation of phosphates, arsenates, vanadates, etc., not only in aluminosilicate formation. Results of investigations on mineral stability and the dissolution of minerals have also been interpreted from the perspective of ion exchange processes. It seems that the stability and exchange characteristics of minerals with ion exchange capacity remain poorly understood, because of the uncertainty of linkage between the equilibrium involving structural ions and the equilibrium involving the exchangeable ions. For mordenite it was found that the dissolution is caused directly by an ion exchange process, with zeolite acidity being the specific parameter. As a tabulation and evaluation of ion exchange data on smectites, certain zeolites and basalt has shown, classical models of ion exchange have been applied successfully to zeolite and smectite exchange reactions. The sorption behavior of basalts is more adequately explained by models of the interface, which take surface ionization and complexation into account [302]. In view of radioactive waste disposal in geological formations a method has been developed to predict sorption reactions of radionuclides on granitoid rock. A simple model was proposed to predict the sorption isotherm of a radionuclide in the presence of competing cations, such as potassium, sodium and manganese.

Glass and glassy materials. Ion exchange in glass was observed relatively early; it has since become used as a technical process for the strengthening of glass and glassy materials. Either at low temperature from solution or at elevated temperatures from molten salts the sodium ion of silicate glasses is exchanged for potassium or lithium. Based on the structure of silicate glasses, the rate of ionic transport in glass and electric potentials in glass have been investigated. The most characteristic features of ion exchange in glasses result from their network structure, which gives rise to cation exchange in a rigid anion lattice and low cation mobility. The rate of exchange is determined by interionic diffusion in glass, which is more like solid than liquid diffusion. Ion exchange can give rise to an electric potential in glass, which depends on ionic concentrations and mobilities in the glass and its selectivity for ions. The strengthening effect was modelled by studies of structural-chemical features of the surface layer of glass strengthened by ion exchange. By stuffing the surface with potassium ions compression is induced in the surface in tension, and it is assumed that the substitution of sodium ions for lithium ions provides a compressive stress surface layer so that, e. g., a glassy substance acquires a higher flexural strength. When bivalent ions are involved in ion exchange in glasses two processes have been described: one involves the behavior of the magnesium cation in magnesium-glasses, which is capable of diffusion exchange for two lithium ions from the melt, resulting in the crystallization of Li-aluminosilicates, which form compression stresses with a hardening effect of 3 – 3.5 times; the other involves a blocking effect by bivalent cations on glass surfaces of the K/Na exchange between glass layers near the surface and salt melts. The descriptions use a model yielding parameters which agree well with experimental data. It was found that an increase in the refractive index *n* of glasses after change in composition due to ion exchange, causing a change in thermal expansion and internal stresses, can easily be calculated theoretically. Owing to the proportionality between the change in *n* and internal stresses in the glass all stress could be measured with a refractometer and the diffusion process be followed [303]. Ion exchange in glasses can also be induced by electrolysis. On samples obtained by this process the ion diffusion, concentration distribution of potassium, resistivity change during electrolysis, stress distribution and breaking behavior were examined, showing that the fragments produced on breakage of glass in which potassium was introduced by electrolysis were – in comparison to those of the $Na^+ \rightarrow K^+$ glass where ion exchange resulted from diffusion – considerably smaller, with the edges of the fragments crumbling into blunt grains [304]. For the anomalous properties of the silver ion appearing in the ion-exchange strengthening of glass a model was also developed. It says that the increased selectivity of the silver ion in the ion-exchange strengthening of glass is caused by the formation of $AgOSiO_{3/2}$ dipoles facilitating the diffusion of Ag^+ into the glass. The increased abrasion resistance of glass strengthened by the $Na^+ \rightarrow Ag^+$ ion exchange is due to increased plasticity of the glass surface [305].

Inorganic chemistry. In general the investigation of inorganic hydrous oxides in inorganic chemistry from the ion exchange aspect offers additional information on the nature of these compounds, beyond their possible applications. In this connection, it is of interest to note transition phenomena between ion exchange and adsorption on oxides and similar materials, above all aluminum oxide. It has been demonstrated by Umland that these processes can be described by a formal exchange theory, so that the adsorption of dissociated salts may be considered as a simultaneous exchange reaction of hydrogen for metal cations and hydroxyl groups for acid anions. The presence of hydrogen and hydroxyl ions in Al_2O_3 may be deduced from various observations [306]. The adsorption-desorption behavior of Al_2O_3 of varying surface pH was also studied with aqueous solutions of anionic dyes, Orange G and Lissamine Green BN, and modelled as an ion exchange process. Quantitative desorption occurs with aqueous pyridine and desorption efficacy by inorganic anions was found in the order: $PO_4^{3-} > SO_4^{2-} > Cl^-$. There was a further discussion of the conditions under which Al hydroxide acts as a weak anion exchanger in the form $[Al(OH)_4]^-$. An investigation of selective sorption of inorganic ions by porous inorganic compounds and inorganic oxides and their derivatives in electrolyte solutions suggests that the difference in solubility is the driving force of sorption [307]. Iron oxides, especially magnetite, were studied with respect to their sorption properties, with the conclusion that they are of interest as ion exchangers, and an interpretation of the specific reactivity of manganese dioxide having a gamma or gamma-rho crystal structure suggests that the ion exchange capacity of MnO_2 affects the diffusion rate of protons into the cathode. With respect to sulfides, the ion exchange properties were manifested during flotation. A model was derived for the rate of ion exchange of the surface ions of an ionic solid and the electrical double layer in terms of the activities of the ions in solution and the short-range relative potential drop in the double layer adjacent to the solid surface. Data, obtained for $BaSO_4$ at 25 °C, confirmed this relationship. In a study of exchange phenomena on the surface of calcium phosphate precipitates in a sucrose solution of 15° Brix it was found that the surface of the prepared hydroxyapatite undergoes ion interchange yielding, depending on the pH, either anions or cations. The ions so produced cause a shift in the isoionic point. Studies of electrochemical reactions accompanied by ion exchange on electrodes have also been published. An ion exchange model of a mercury electrode covered with adsorbed organic ions has also been presented. The kinetic current of protonation in the presence of ionic surfactants depends on the process of ion exchange with the participation of the so-called ion exchanger layer formed during the adsorption of organic ions, which have long carbon chains. Even in such processes as the expansion of concrete in aboveground construction, bridge construction, and hydraulic structures a model can be developed based on physico-chemical considerations to explain the diffusion and exchange of alkali ions with silanol groups of the silica mineral. The elementary reactions and the kinetics of the so-called alkali/silica reaction had been relatively unknown and preventive measures to diminish the expansion of concrete were of empiric character only. The model has been tested in numerous experiments and it is in fair agreement with the mathematical calculations [308].

Organic chemistry. Ion exchange in organic chemistry has been treated extensively and there are numerous ways in which ion exchange can be applied in this field [309]. Briefly, it should be recalled that most organic compounds are nonionic and can, therefore, be purified of ionized impurities by two-stage deionization, just as water is purified. Organic compounds that are acids, bases, or salts on the other hand, can be taken up by ion exchangers, making possible various recovery and separation processes. All these are more or less simple applications of ion exchange and ion exchangers, and it is not the purpose of this paragraph to review them to any extent, but to draw attention to a few examples to demonstrate how ion exchange models have been and can be applied to the phenomena of organic chemistry. For example, in the field of pesticides, changes in the relative sizes of the carboxylic adsorption band in the IR spectra (near 1730 and 1610 cm^{-1}) of adsorbents before and after treatment with paraquat, reveal that ion exchange is the predominant mechanism for the uptake of the bipyridyl by H^+-ion saturated preparations of a humic acid from an organic soil. In this case the exchange occurs over two hydroquinone polymers, and by two carboxylic ion exchange resins. The reduced flow of water containing sodium alkylbenzene sulfonate through columns of coarse sand, silica gel, or soil aggregates, compared with water alone, has been attributed to partial blockage of the porous medium by a gel of Ca-alkylbenzenesulfonate, with the Ca being derived from the matrix by ion exchange. With continued flow, the exchangeable Ca was depleted, and the gel dissolved, increasing the flow rate except through there where ion exchange had caused irreversible swelling of the matrix. Pure C_{12} alkyl homologs of Na alkylbenzenesulfonate did not reduce the flow rate, i. e., precipitation of the sulfonates occurred only at higher calcium concentrations. As another example one has that the interaction of synthetic polyelectrolytes with surfactants follows an ion exchange mechanism, as was shown by potentiometric titrations of poly(acrylic acid), poly(methacrylic acid), and methacrylic acid-methylmethacrylate copolymer as weakly acidic polyelectrolytes with the cationic surfactant cetyltrimethylammonium bromide. The extent of this interaction is predominantly determined by temperature and by the content of hydrophobic groups in the polyelectrolyte. Reactions between interfacial substrates and ionic nucleophiles in oil-in-water microemulsions stabilized by ionic surfactants of opposite charge have also been dealt with by using an ion exchange model. This model, previously applied to similar reactions in normal micelles, assumes that the concentrations of free and bound counter-ion and co-ion (nucleophile) are related by an ion exchange equilibrium constant. When data for the reaction of a phosphate ester in a cetyltrimethylammonium bromide microemulsion are used, a value for the equilibrium constant of $0.1-0.2$ is obtained for OH^- and F^-, comparable to the values obtained in aqueous micelles. The model is also consistent with pK data in an anionic microemulsion system [310]. A whole framework has been developed for ion exchange in micellar solutions, for the quantitative analysis of the influence of charged micelles on reactions that involve exchangeable ionic species. In five cases model calculations based on the resulting mathematical expressions, which contain only experimentally accessible terms, were presented to exemplify the behavior predicted by the model. These calculations illustrated the role that a given set of experimental conditions plays in determining the concentration and kinetic behavior of exchangeable ionic reactants in the micellar pseudophase. Thus, a kinetic study of the alkaline hydrolysis by OH^- of a substrate, the *N*-methyl-4-cyanopyridinium ion – which resides exclusively in the intermicellar aqueous phase – provides direct evidence for the binding of OH^- to positive micelles, as well as selectivity coefficients for the binding of OH^- to hexadecyltrimethylammonium bromide micelles and for tetradecyltrimethylammonium chloride micelles. This was claimed to be the first quantitative description of the interaction of OH^- with positive micelles [311]. The pseudophase ion exchange model has also been used to explain the effects of cetyltrimethylammonium chloride micelles on the apparent basicity constant of benzimidazole and to explain the effects of added chloride ions on the apparent basicity constant for naphth[2,3]imidazole. The results have been used to estimate an intrinsic micellar basicity constant for benzimidazole, the value of which is similar to that in water. Similar measurements have been recorded with cetyltrimethylammonium nitrate and bromide.

Biology. Ion exchange models in this area shall in the following be subdivided into those relevant to plants and those to animal life. Because plant tissues contain acid groups able to take up and exchange cations, as well as basic groups to bind anions, many metabolic processes of plants involve, or are preceeded by, ion exchange. In essence it is due to the acid properties and the electronegative character of the root surface membrane that the young plant roots behave as acids exhibiting pronounced cation exchange and having a cation exchange capacity. The quantitative contribution of proteins to the cation and anion exchange capacity of plant roots and also the content of protein nitrogen were modelled with regard to growth conditions, the plant species, and the effects of various inhibitors such as chloramphenicol, NaF and $Pb(NO_3)_2$. A positive correlation was established between the exchange capacity of plant roots and the growth rate of the plant. The application of $Pb(NO_3)_2$ resulted in an increase in the anion exchange capacity by 19–27% and a decrease of the cation exchange capacity by 12–19%. In the roots of plants grown at low temperatures both the exchange capacity and the level of protein nitrogen increased. The treatment of roots by respiration inhibitors (NaF) and by inhibitors of protein synthesis (chloramphenicol) resulted in changes of cation and anion sorption on the root surface. These experimental results indicate a participation of protein substances in the anion and cation exchange capacity of plant roots. The velocity of protein metabolism in the roots does not affect the exchange capacity. Anion exchange capacity can be correlated with the intensity of respiration of the root [312]. Desorption of Mg^{2+} and Ca^{2+} by Ba^{2+} from intact roots of beans and yellow lupine was found to be strictly equivalent to the amount of H^+ formerly desorbed by Mg^{2+} and Ca^{2+} from the roots treated with 0.05 N hydrochloric acid, showing that the same ion exchange sites on the roots may be occupied by different cations, or in other words, that the ion exchange sites are not cation specific. As a consequence of these root properties, the uptake of mineral nutrients is intimately connected with an ion exchange between the root surface and the soil solution, or the soil particles in the case of contact exchange. A study of the nutrient absorption and its dependence upon various soil factors presents great difficulties because of the extremely complex composition of the root environment, but with the application of synthetic ion exchanger soil substitutes numerous studies of plant physiology focussing on the nutrient uptake have been performed, with highly interesting results. That the plant non-root tissues also possess cation exchange properties was established relatively early on air-dried, finely ground alfalfa. This means that materials prepared from plants also exhibit ion exchange properties. That they are influenced in their properties by the type of the counterion was demonstrated by the improvement of gelling properties of furcellaran by partial replacement of the calcium ions contained in it by potassium ions.

Cells and tissues, and their component parts, can behave as ion exchange systems. Ion exchange plays an important role, therefore, in the electrolyte metabolism of cells and in related functions involving the electrical properties of the cell membrane. Cells and tissues consist of solids, gels and membranes immersed in solutions containing relatively high concentrations of electrolytes. In the solid phases of the cells and tissues phosphate, sulfate, carboxyl and amino groups can be found, which are essentially some of the same known as ionogenic groups in ion exchange resins, which give, therefore, the solid phases the behavior of ion exchangers. In cells the membrane is the important part, but tissues are also highly compartmentalized by membranes, and, therefore, in physiology ion exchange is often referred to as the interchange of ions between two electrolyte solutions separated by a semipermeable membrane. In the simplest case, the two solutions are the extracellular and intracellular fluids and the membrane is the cell membrane. In some cases, however, the ion exchanges may occur across one or more layer of cells in a tissue, in which case there are a number of fluid compartments separated by membranes. It follows that ion exchange in cells and tissues has almost always been treated as ion exchange membrane processes, but living cells represent a far more complex system. Briefly, the predominant factor in the cell is its ability to transport certain ions against its activity gradients by an expenditure of energy derived from metabolic reactions. The mechanism by which such ion-pumps operate are still poorly understood, but they enable the cell to maintain an electrolyte distribution that is remarkably different from that of the surronding medium [313]. The cell can, however, be described as a biological ion exchange resin if the similarities between ion-accumulating cells and ion exchange resins are correlated. The data obtained indicate that ion exchange events extend beyond the surface membrane to the polyelectrolytes making up the fabric of the cell. The kinetic characteristics of biological ion exchange are amenable to analysis by a model commonly used for ion exchange resins. The theories of ion exchange equilibria currently used for ion exchange resins can be evaluated with respect to their suitability for adaptation to biological ion exchange. The concept of the cell as a biological ion exchange resin then provides a useful physical model [314]. Many other biological materials show ion exchange behavior. In the case of certain extracellular structures it is much easier to model and verify their ion exchange properties by standard ion exchange or chemical techniques. The data obtained with extracellular materials indicate, however, the extent and importance of ion exchange processes in the cells themselves, because extracellular and cellular components are composed of similar biochemical substances. Also mucus from the footsole of the freshwater snail *Lymneae stagnalis* behaves as a weak acid ion exchanger. The elucidation of the ion exchange properties of immobilized DNA would seem to be important, because from such examinations the interaction of metal ions with DNA in the form of reliable quantitative data on the stability constants of DNA-metal complexes can be gained. Such data make it possible to characterize the selectivity of DNA for ions and to compare the differences in the binding of cations to nucleic acids with their different influence on the structure and functions of the biopolymer. There have been investigations of the ion exchange properties of ion exchangers obtained by immobilization of DNA gels in polyacrylamide gel [315]. From the physiological point of view it is of interest that young and old erythrocytes show different rate constants of anion exchange as measured by $^{35}SO_4^{2-}$ efflux at 37 °C. The rate constant for $^{35}SO_4^{2-}$ efflux in exchange for Cl^- ions from old cells is approximately 20% greater than that from young, less dense cells. The cell water volume of older cells is also decreased. Based on these results and previously reported decreases of cell membrane area in older cells, it has been concluded that anion exchange is increased in older, denser human erythrocytes. The elucidation of mechanisms of bacterial attachment could be facilitated by investigations on bacterial adherence to anion exchange resins, providing a useful, rapid, in vitro screening assay for identifying putative antiadherence agents. Experiments with heparin and *Escherichia coli, Klebsiella ozaenae, Proteus mirabilis, Streptococcus fecalis* and *Pseudomonas aeruginosa* have provided additional existance that adherence to an anion exchange resin is similar to urinary bladder mucosa adherence. These studies indicate that bacterial adherence to an anion exchange resin responds to heparin and other chemical agents in a manner similar to response on the mucin-deficient rabbit urinary bladder [316]. Last but not least bone is an ion exchange system whose ionic exchange is quite different from ionic exchange in synthetic resins. Any model must, therefore, take into account that rather than an accumulation of ionizable groups linked covalently to a solid having no measurable solubility, the exchangeable ions of the bone crystal are linked principally by electrostatic forces to the other exchangeable ions. This "matrix" is capable of dissolution, the exchange capacity is not a fixed function of mass, and other variables pertain. In addition, the ion exchanging sites are embedded in an organic gel, the ground substance, which can influence the accessibility of the mineral crystals to ions in the circulating fluids. To develop a full ion exchange model for bone, the goal of which is an understanding of the ion exchange phenomena occurring within the skeleton, a knowledge of the chemistry and structure of bone is a prerequisite.

Paints and polymers. Paint is one of the important means of combating the corrosion of iron. In this regard models have been developed to describe dried paint film as an ion exchange system. Paint films are closest to ion exchange membranes, which makes it understandable that analogies between paint films and heterogeneous and homogeneous types of membrane have been discussed. Inclusion membranes do not correspond to any paint system, but structural analogies exist between a number of polymers used in paints and varnishes and ion exchange membranes. Since paint films have properties similar to those of ion exchange membranes, measuring methods applicable to the latter can be applied to the former to evaluate their protective efficiency. Thus, it was shown that the decrease in the resistivity value of an acrylic coating on a metallic support in contact with an electrolyte solution with increasing concentration indicated penetration of the electrolyte into the coating, which acted as an ion exchange membrane with poor permselectivity. The variation of resistance as a function of film thickness caused measuring problems. The water vapor permeability of TiO_2- or ZnO-pigmented coatings decreased with increasing pigment volume concentration (PVC), then became steady, before quickly increasing to almost the critical pigment volume concentration (CPVC). Similar important results mainly for water sorption and permeability have been presented for the ion exchange properties of polyurethane-based paint films [317]. With respect to the corrosion protection efficiency of alkyd-based paint media it was found that the ion exchange capacities of such coatings increased with increasing pH, electrolyte concentration, and temperature, and were dependent on the oil content of the binder. The ion exchange capacity can be related to corrosion resistance; it has a negative relation with the corrosion protection efficiency of the coatings and it can thus provide useful information about their ability to isolate the substrate from a corrosive environment. Results from investigations on the ion exchange capacities of three indigenous linseed oil-based alkyd resins indicated the formation of ionizable groups during the drying of the alkyd films. The ion exchange capacity can therefore be used to characterize the corrosion protection efficiency of alkyd based anticorrosive paints. A model based on ion exchange has also been discussed for the corrosion protection by paints in the automobile industry. The relation between pigment and binder in corrosion protecting coating systems as effected by the ion exchange capacity was determined for paints in which the binder was an alkyd, chlorinated rubber, epoxy, or polyurethane and the pigment was $Pb_3O_4-Fe_2O_3$, $BaSO_4-Fe_2O_3$, $Zn(PO_4)_2-BaSO_4$, $BaSO_4$, or TiO_2. For paints having the same binder, the absorption, permeability and ion exchange capacity depended on the pigmentation [318]. In a model set up for highest quality requirements it was suggested that it was better to describe the paint film as a solution than as a porous medium. The unpigmented polyurethane under investigation is a weak ion exchanger. Its capacity must be included in a model of mass transfer of molecules and ions through the film, which limits the corrosion rate of painted metals. A model for novel corrosion inhibitors phases the ion exchange activity occurring not in the paint film but onto the surface of porous inorganic oxides such as aluminas and silicas. The principle mechanism by which they then operate in organic coatings is again ion exchange, because aggressive ions permeating the paint film will be preferentially ion exchanged onto the inorganic support and the anticorrosive species released to protect the metal surface. It has not happened yet, but this model could give impetus to further research and development for the use of ion exchange in paints [319]. A characterization of latexes by ion exchange and conductometric titration has also been presented.

With respect to polymers it is not the purpose of this paragraph to return to a description of the reactions of polymers by which ionogenic groups can be introduced in order to convert them to ion exchange materials. This subject is treated extensively elsewhere in this volume. Nevertheless, ion exchange models have been applied to other treatments of plastic materials for the explanation of changes in properties. For galvanizing ABS copolymers a preceeding etching process results in the formation of groups with ion exchange activity. That ion exchange was occurring could be proved by means of $^{109}PdCl_2$ [320]. This may serve as an example for similar results. Ion exchange models have also been applied, e. g., to the method of end group determination using dyes for the analysis of fiber-forming polymers. One method was developed for a selective and direct determination of OH, COOH and SO_3H end groups in fiber-forming poly(ethylene terephthalate). The ion exchange method was found to be superior to hydrazinolysis and titrimetric methods since the former makes it possible to determine all the end groups of the polymer. An analytical procedure was also based on an ion exchange reaction for the determination of the uptake of colorless cationic retarders by acrylic fiber. Sodium ions in the fiber pass into the dye bath as a result of the exchange and are determined by atomic absorption analysis. Various theories of the tinctorial mechanisms occurring in wool and natural silk dyeing with acid dyes can also be discussed fruitfully taking into account ion exchange mechanisms. In general, the electrical double layer phenomenon at the liquor-substrate phase boundary affects the kinetics as well as the equilibrium in the dyeing of textiles. A simple model was developed to evaluate the effect of the double layer on the starting phase of dyeing and then tested using cotton and nylon 66 fibers and the ion exchanger Dowex SBR-P. The differences observed in the case of the substrate with a very high surface charge density, i. e., Dowex SBR-P, and a relatively low surface charge density, i. e., nylon 66 and cotton fibers, were discussed [321].

Ion exchange and the theory of the living state. The most fascinating aspect of ion exchange is its employment in the explanation of the living state. Up to now it seems that the association-induction model of G. Ning Ling is the most promising approach in the attempt to describe the living state as one complex system of materials sharing the necessary stringent qualifications. Ion exchange plays an important role here. For the description of the behavior of proteins in a dilute water solution of salt ions it was realized that one had to proceed considerably beyond the theory of Linderstrom-Lang in order not to ignore the very factors distinguishing the mechanisms of life from a physical model of simple dilute salt solutions. The relevancy of this is even more striking because it has long been known that living organisms are very sensitive to the differences between such chemically similar ions as K^+ and Na^+ [Ringer, 1833]. In the living state there also exists a protein-salt-water system and since protein, salt (ions), and water play major roles in living organisms, the latter must have other important attributes. Firstly, association between the behavior of proteins in dilute salt solutions as treated by the Linderstrom-Lang theory and the behavior of proteins, salt, and water in biological systems as described by the present model. Secondly, the nature of the amino acid components and their sequential order of arrangement and side chains give specificity to a protein so that the primary function of the highly polymerizable resonating chain is to provide a vehicle for the ready transmission of an inductive effect from one functional group to another. Treating this inductive effect quantitatively it becomes possible to calculate all the energies between a counter-ion and a fixed anion. Moreover, with a given inductive value of the anionic group, there will usually be a difference between the adsorption energies of one counter-ion, e. g., K^+, and another, e. g., Na^+. Thus, if association and induction are accepted as the two fundamental mechanisms of the present model one is led to the applicability of the concept of ionic association to living protoplasm in general. Living protoplasm is not, however, a simple dilute salt solution, but a three-dimensional lattice of protein, water and salt in which the charge-bearing protein molecules are immobilized into a fixed-charge system. In consequence, when charge fixation produces a sufficient degree of association, so-called short-range interactions dominate and selectivity of counterions follows. Then the living cell and its components can be discussed as a true fixed-charge system comparable to ion exchange resins with regard to the mechanisms involved in ionic specificity, behavior of proteins, selective accumulation of ions and nonelectrolytes, cellular electrical potentials, ionic permeability and diffusion, excitation and inhibition, contractive mechanism, enzyme action, drug and hormone action, antibody-antigen reaction, fertilization, chemical embryology, growth, differentiation, and cancer [322].

References

[1] G. Wiegner and H. Jenny (1927). Über Basenaustausch an Permutiten. (Kationenumtausch an Eugelen). Koll. Z. *42*, 268.

[2] K. Dorfner (1961). Redoxaustauscher. Chem.-Ztg. *85*, 80, 113.

[3] H. S. Thompson (1850). On the absorbent power of soils. J. Royal Agricult. Soc. Engl. *11*, 68.

[4] J. T. Way (1850). On the power of soils to absorb manure. J. Royal Agricult. Soc. Engl. *11*, 313. Id. (1852), ibid. *13*, 123. Id. (1855), ibid. *15*, 491.

[5] W. Henneberg and F. Stohmann (1858). Über das Verhalten der Ackerkrume gegen Ammoniak und Ammoniaksalze. Ann. Chem. Pharm. *107*, 152.

[6] H. Eichhorn (1858). Über die Einwirkung verdünnter Salzlösungen auf Silicate. Poggendorff Ann. Phys. Chem. *105*, 126.

[7] F. Lemberg (1870). Über einige Umwandlungen finnländischer Feldspate. Z. dtsch. geol. Ges. *22*, 355, 803. Id. (1876). Über Silikatumwandlungen. Ibid. *28*, 519.

[8] J. M. van Bemmelen (1878). Das Absorptionsvermögen der Ackererde. Landw. Vers.-Stat. *21*, 135. Id. (1879), ibid. *23*, 265.

[9] G. J. Mulder (1862). Die Chemie der Ackerkrume. Leipzig. Verlag J. J. Weber.

[10] D. B. Weisz (1932). Gegenüberstellung verschiedener Gleichungen über Austauschadsorption. Diss. Zürich Nr. 670.

[11] A. D. Hall and C. T. Gimingham (1907). The interaction of ammonium salts and the constituents of the soil. J. Chem. Soc. *91*, 677.

[12] G. Wiegner (1912). Zum Basenaustausch in der Ackererde. J. für Landw. *60*, 111, 197.
[13] P. Vageler and J. Woltersdorf (1930). Base exchange and acidity in soils. Z. Pflanzenernähr. Düng. Bodenk. *A15*, 329. Id. II. Preliminary experiments on permutites. Ibid. *A16*, 184.
[14] G. E. Boyd, J. Schubert and A. W. Adamson (1947). The exchange adsorption of ions from aqueous solutions by organic zeolites. I. Ion exchange equilibria. J. Amer. Chem. Soc. *69*, 2818.
[15] S. Graf Rostworowski and G. Wiegner (1912). Die Absorption der Phosphorsäure durch "Zeolithe" (Permutite). J. Landw. *60*, 223.
[16] E. Raman and A. Spengel (1916). Austausch der Alkalien und des Ammons von wasserhaltigen Tonerde-Alkalisilikaten (Permutiten). Z. anorg. Chem. *95*, 115. Id. (1918). Der Basenaustausch der Silikate II. Ibid. *105*, 82.
[17] G. Wiegner (1925). Dispersität und Basenaustausch (Ionenaustausch). Koll. Z. Erg. *36* (41), 341. Id. (1931). Some physicochemical properties of clays. I. Base exchange or ionic exchange. J. Soc. Chem. Ind. *50*, 65.
[18] H. Jenny (1927). Kationen- und Anionenumtausch an Permutitgrenzflächen. Koll. Beih. *23*, 428.
[19] V. Rothmund and G. Kornfeld (1918). Der Basenaustausch im Permutit. I. Z. anorg. allgem. Chem. *103*, 129. Id. (1919). II. Ibid. *108*, 215.
[20] H. W. Kerr (1928). The nature of base exchange and soil acidity. J. Amer. Soc. Agron. *20*, 309. Id. (1928). Identification and composition of the soil aluminosilicate active in base exchange and soil acidity. Soil Sci. *26*, 385. A. P. Vanselow (1932). The utilization of the base exchange reaction for the determination of activity coefficients in mixed electrolytes. J. Amer. Chem. Soc. *54*, 1307.
[21] O. Samuelson (1944). Diss., Tekniska Högskolan, Stockholm, pp. 56–67.
[22] J. A. Marinsky (1949). Diss. Report to the Office of Naval Research NR–026–001.
[23] T. R. E. Kressman and J. A. Kitchener (1949). Cation exchange with a synthetic phenolsulfonate resin. I. Equilibria with univalent cations. J. Chem. Soc. 1190.
[24] J. F. Duncan and B. A. Lister (1949). Ion exchange studies. I. The sodium-hydrogen system. J. Chem. Soc. 3285. Id. (1949). Ion exchange studies. II. The determination of thermodynamic equilibrium constants. Disc. Faraday Soc. No. 7, 104.
[25] R. Gans (1905). Zeolithe und ähnliche Verbindungen. Jahrb. preuß. geolog. Landesanstalt *26*, 179.
[26] L. F. Collins (1937). A study of contemporary zeolites. J. Amer. Water Works Assoc. *29*, 1472.
[27] O. Liebknecht (1934). DR Pat. 763 936.
[28] B. A. Adams and E. L. Holmes (1935). Absorptive properties of synthetic resins. J. Soc. Chem. Ind. *54*, 1. Id. (1938). Base-exchange resins. Chem. Age *38*, 117. Id. (1935). Improvements in and relating to the removal of components or constituents from liquids by adsorption or absorption. E. Pat. 450 308.
[29] R. Grießbach (1939). Über die Herstellung und Anwendung neuerer Austauschadsorbentien, insbesondere auf Harzbasis. Angew. Chem. *52*, 215.
[30] G. F. D'Alelio (1944). Production of synthetic polymeric compositions comprising sulphonated polymerizates of poly-vinyl aryl compounds and treatment of liquid media therewith. US Pat. 2 366 007.
[31] C. H. McBurney (1952). Resinous insoluble reaction products of tertiary amines with halo-alkylated vinyl aromatic hydrocarbon copolymers. US Pat. 2 591 573.
[32] A. Skogseid (1948). Noen derivater av polystyrol og deres andvendelse ved studium av ioneutvekslings-reaktioner. Diss. Oslo.
[33] R. Gans (1909). Über die technische Bedeutung der Permutite (der künstlichen zeolithartigen Verbindungen). Chem. Ind. *32*, 197.

[34] O. Folin and R. Bell (1917). Analysis of ammonia in urine. J. Biol. Chem. *29*, 329.
[35] J. C. Whitehorn (1923). Separation of some amino acids. J. Biol. Chem. *56*, 751.
[36] J. R. Millar (1973). Fundamentals of ion exchange. Chem. Ind. 5 May, 409.
[37] H. Shimizu (1980). Possible developments of ion exchange technology in 1980's. Yosui to Haisui *22*, 517 etc. (I – XIII).
[38] A. A. Vasilev (1968). Possible structural units of phenolsulfonic acid-formaldehyde ion exchangers. Zh. Prikl. Chim. (Leningrad) *41*, 1099.
[39] D. Braun and A. Y. Khym (1967). Über die Struktur vernetzter Polymerisate aus Styrol, p-Jodstyrol und Divinylbenzol. Kolloid-Z. *216/217*, 361.
[40] K. Dusek (1962). Ionenaustauschergerüste. III. Kopolymere des Styrols mit Divinylbenzol. Elastisches Verhalten der in Toluol gequollenen Kopolymeren. Coll. Czechoslov. Chem. Comm. *27*, 2841.
[41] K. M. Saldadze (1969). Structure and properties of ion exchangers. Paper presented at the Conference on Ion Exchange, London, July 1969.
[42] G. K. Saldaze and V. K. Varentsov (1974). Investigation of the ionogenic group distribution in ion exchange resins. J. Polymer Sci., Polym. Symp. *47*, 139.
[43] R. Kunin, E. Meitzner and N. Bortnick (1962). Macroreticular ion exchange resins. J. Amer. Chem. Soc. *84*, 305.
[44] K. A. Kun and R. Kunin (1964). Pore structure of some macroreticular ion exchange resins. Polym. Letters *2*, 587. J. Seidl et al. (1967). Makroporöse Styrol-Divinylbenzol-Copolymere und ihre Verwendung in der Chromatographie und zur Darstellung von Ionenaustauschern. Fortschr. Hochpolym. Forsch. *5*, 113.
[45] F. Martinola (1966). Macroporous ion-exchange resins for water conditioning. Effl. Water Treatm. J. 1966 (May/June), G. P. Herz (1965). Field experience with macroreticular resins. Effl. Water Treatm. J. 1965 (Sept.), 453. R. Kunin and J. Barrett (1979). Twenty years of macroreticular ion exchange resins. Proc. Int. Water Conf., Eng. Soc. West Pa. 40th, 183.
[46] T. R. E. Kressman (1966). Isoporous resins and the organic fouling problem. Effl. Water Treatm. J. 1966, 119.
[47] T. R. E. Kressman (1969). Properties of some modified polymer networks and derived ion exchangers. Paper presented at the Conference on Ion Exchange, London, July 1969.
[48] K. A. Kun and R. Kunin (1967). The pore structure of macroreticular ion exchange resins. J. Polymer Sci. *16C*, 1475.
[49] V. A. Dinaburg, G. V. Samsonov et al. (1968). Synthesis and study of the properties of macroreticular carboxyl ion exchange resins with N, N'-alkylenedimethacrylamides as crosslinking agents. Zh. Prikl. Khim. *41*, 891.
[50] S. Fisher (1984). Letter to the Editor. Reactive Polymers *2*, 239.
[51] F. Martinola (1973). Active filtration. Powdered ion exchangers and adsorbents as precoat media. Filtration and separation *10*, 420.
[52] B. J. Schultz and E. H. Crook (1968). Ultrafine ion exchange resins. Ind. Eng. Chem. Prod. Res. Dev. *7*, 120.
[53] C. G. Horvath, B. A. Preis and S. R. Lipsky (1967). Fast liquid chromatography: An investigation of operating parameters and the separation of nucleotides on pellicular ion exchangers. Anal. Chem. *39*, 1422. C. G. Horvath (1973). Pellicular ion exchange resins in chromatography. Ion Exch. Solvent Extr. *5*, 207 – 60. A review with 59 refs.
[54] J. J. Kirkland (1970). High speed separations of nucleotides and nucleic acid bases by column chromatography using controlled surface porosity ion exchangers. J. Chromatogr. Sci. *8*, 72.
[55] M. Seko et al. (1975). Manufacture of ion exchangers. Japan. Kokai 75 32 085; Asahi Chemical Industry Co., Ltd.

[56] D. K. Hale (1952). Symposium report: Principles and applications of ion exchange. Nature (London) *170*, 150.

[57] M. Skafi and K. H. Lieser (1970). Darstellung und Eigenschaften oberflächlich sulfonierter Harzaustauscher. Z. anal. Chem. *249*, 182.

[58] J. S. Fritz and J. N. Story (1974). Selectivity behavior of low capacity, partially sulfonated, macroporous resin beads. J. Chromatogr. *90*, 267.

[59] H. Small et al. (1975). Novel ion exchange chromatographic method using conductivity detection. Anal. Chem. *47*, 1801. T. S. Stevens and H. Small (1978). Surface sulfonated styrene-divinylbenzene: optimization of performance in ion chromatography. J. Liq. Chromatogr. *1*, 123.

[60] D. T. Gjerde et al. (1980). Anion chromatography with low-conductivity eluents. J. Chromatogr. *187*, 35.

[61] P. Hajos and J. Inczedy (1980). Preparation and ion chromatographic application of surface-sulphonated cation exchangers. J. Chromatogr. *201*, 253. Id. (1984). Preparations, examination and parameter optimisation of cation exchangers in ion-chromatography. In: Ion Exchange Technology. Ed. D. Naden and M. Streat. Ellis Horwood, Chichester UK, 1984, p. 450.

[62] G. V. Samsonov et al. (1984). Kinetic-dynamic analysis of the increase of effectiveness of preparative methods of ion exchange chromatography. Ionnyi Obmen Khromatogr. 1984, p. 94. Nauka, Leningr. Otd.: Leningrad, USSR.

[63] Brochure: Fractogel TSK® Ion Exchange Chromatography. E. Merck, Frankfurter Str. 250, 6100 Darmstadt 1, FR Germany.

[64] V. L. Bogatyrev (1980). Classification of ion exchangers. Deposited Doc. VINITI 3340-80. 4 pp. (Russ).

[65] J. Bartulin et al. (1982). Synthesis and adsorption properties of some specific phosphorylated resins for unranium. Hydrometallurgy *8*, 137.

[66] A. Chatterjee and J. A. Marinsky (1963). Dissociation of methacrylic acid resins. J. Phys. Chem. *67*, 41. Id. (1963). A thermodynamic interpretation of the osmotic properties of crosslinked polymethacrylic acid. Ibid. *67*, 47.

[67] S. M. Wheelwright (1982). Multivariant ion exchange applications of weak-electrolyte resins in water purification. Diss. Univ. California, Berkeley, CA, USA. 189 pp.

[68] British and United States Patents to various assignees: B. Pat. 726 918 (1955), 748 234 (1956), 757 398 (1956), 777 248 (1957); US Pat. 2 764 562 (1956), 2 837 488 (1958), 2 844 546 (1958), 2 911 378 (1959).

[69] A. A. Efendiev (1974). Phosphorus-containing selective ion exchangers. Issled. Obl. Kin. Model. Optim. Khim. Protsessov (USSR) *2*, 247. A review with 81 refs.

[70] M. Marhol (1976). Rare earth separation using selective ion-exchangers containing phospho-groups. Report INIS-mf-3321, 23 pp. (Eng).

[71] B. N. Laskorin et al. (1981). Sorption of tantalum and niobium by arsenic-containing ion exchangers. Khim. Tekhnol. Elementoorg. Soedin. Polim. 1981, 45.

[72] R. M. Wheaton and W. C. Bauman (1951). Properties of strongly basic anion exchange resins. Ind. Eng. Chem. *43*, 1088.

[73] F. Martinola (1981). A new type of strong base anion exchange resin. Vom Wasser *56*, 205 (Ger).

[74] D. C. Holliday and W. H. Lee (1976). Investigations into the behaviour of strong acid cation and mixed base anion resins. Proc. Ion Exchange Symp., No. 14, Cambridge, 1976.

[75] A. B. Pashkov et al. (1969). Synthesis and investigation of anion exchangers based on vinyl and alkylvinylpyridines. Conf. on Ion Exchange, London, 1969.

[76] T. K. Brutskus, K. M. Saldadze et al. (1974). Porous structure specific features of macroporous vinyl and alkylvinylpyridine based anion exchangers. Proc. 3rd Symposium on Ion-Exchange. 28 – 31 May, 1974. Balatonfured, Hung., p. 49.

[77] N. B. Galitskaya and A. B. Pashkov (1974). Amino-substituted vinylpyridine based anion exchangers. Proc. 3rd Symposium on Ion-Exchange. 28-31 May, 1974. Balatonfured, Hung., p. 55.
[78] A. Heller, Y. Marcus and I. Eliezer (1963). Poly-(1-hydroxy-4-vinylpyridinium) anion-exchangers. J. Chem. Soc. 1963, 1579.
[79] A. Skogseid (1946). Dissertation. Norwegian Technol. High School, Trondheim, Norway.
[80] R. Kunin and A. F. Preuss (1964). Characterization of a boron-specific ion exchange resin. Ind. Eng. Chem. Prod. Res. Dev. *3*, 304.
[81] G. Nickless and G. R. Marshall (1964). Polymeric coordination compounds. The synthesis and applications of selective ion-exchangers and polymeric chelate compounds. Chromatogr. Rev. *6*, 154. R. Hering (1965). Anwendung chelatbildender Ionenaustauscherharze in der analytischen Chemie. Z. Chem. *5*, 402. E. Blasius and B. Brozio (1969). Chelating ion exchange resins. In: Chelates in Analytical Chemistry. M. Dekker Publishing Co., New York.
[82] The Akzo Imac TMR Process for the removal of mercury from wastewater. Bull. Akzo Chemie, Amsterdam.
[83] M. J. Hatch et al. (1957). Preparation and use of snake-cage polyelectrolytes. Ind. Eng. Chem. *49*, 1812. US Pat. 3 041 292, 1962.
[84] F. Wolf and H. Mlytz (1968). Über den Einfluß der Struktur bipolarer Ionenaustauscherharze (Schlangenkäfigelektrolyte) auf die Trennung von Elektrolyt-Nichtelektrolytgemischen. J. Chromatogr. *34*, 59.
[85] R. Gans (1907). Reinigung des Trinkwassers von Mangan durch Aluminatsilicate. Chem.-Ztg. *31*, 355.
[86] S. B. Hendricks (1945). Base exchange of crystalline silicates. Ind. Eng. Chem. *37*, 625.
[87] U. Hofmann (1962). Die Besonderheiten des Ionenaustausches an Tonmineralien. In: K. Issleib (Hrsg.), Anomalien bei Ionenaustausch-Vorgängen. Berlin; Akademie-Verlag.
[88] J. Porath et al. (1971). Agar derivatives for chromatography, electrophoresis and gel-bound enzymes. I. Desulphated and reduced crosslinked agar and agarose in spherical form. J. Chromatogr. *60*, 167.
[89] N. Grubhofer (1959). Zur Struktur von mit Divinylbenzol vernetzten Polystyrol-Ionenaustauschern. Makromol. Chem. *30*, 96.
[90] Y. Marcus and A. S. Kertes (1969). Ion Exchange and Solvent Extraction of Metal Compounds. Wiley-Interscience, London, p. 243.
[91] T. V. Arden (1975). Structure of ion exchange resins. Inf. Chim. *143*, 205, 211 (Fr). English version made available as private communication.
[92] J. R. Millar (1983). On the structure of ion-exchange resins. In: Mass transfer and kinetics of ion exchange, ed. by L. Liberti and F. G. Helfferich. NATO ASI Ser., Ser. E. 1983, No. 71, p. 23.
[93] W. Funke (1968). Synthese und Bildungsmechanismus vernetzter Polymerer. Chimia *22*, 111.
[94] D. Braun (1979). Chemische Methoden zur Analyse von Netzwerken. Angew. Makromol. Chem. *76/77*, 351.
[95] E. Blasius et al. (1973). Charakterisierung von Ionenaustauschern auf Kunstharzbasis durch Pyrolyse-Gas-Chromatographie. Z. anal. Chem. *264*, 290. Id. (1975). Ermittlung des Vernetzungsgrades und Beladungszustandes von Ionenaustauschern auf Kunstharzbasis durch Pyrolyse-Gas-Chromatographie. Ibid. *277*, 9. Id. (1976). Charakterisierung von Ionenaustauschern auf Kunstharzbasis durch Pyrolyse-Massenspektrometrie. Talanta *23*, 301.
[96] C. S. Knight (1967). Some fundamentals of ion-exchange-cellulose design and usage in biochemistry. Advan. Chromatogr. *4*, 61.

[97] F. A. Belinskaya (1974). Structure of inorganic ion exchangers. Vestn. Leningrad. Univ., Fiz. Khim. (1), 94.
[98] S. B. Hendricks (1945). Base exchange of crystalline silicates. Ind. Eng. Chem. *37*, 625.
[99] H.-G. Elias (1968). Die Struktur vernetzter Polymerer. Chimia *22*, 101.
[100] Recommendations on ion exchange nomenclature. International Union of Pure and Applied Chemistry (IUPAC). (1972). Pure Appl. Chem. *29*, 617.
[101] F. G. Donnan (1911). Theorie der Membrangleichgewichte und Membranpotentiale bei Vorhandensein von nicht dialysierenden Elektrolyten. Z. Elektrochem. *17*, 572.
[102] H. P. Gregor (1951). Gibbs-Donnan equilibria in ion exchange resin systems. J. Amer. Chem. Soc. *73*, 642.
[103] C. Calmon (1952). Application of volume change characteristics of a sulfonated low crosslinked styrene resin. Anal. Chem. *24*, 1456.
[104] H. P. Gregor et al. (1951). Ion exchange resins. II. Volumes of various cation exchange resin particles. J. Colloid Sci. *6*, 245.
[105] H. P. Gregor et al. (1951). Determination of the external volume of ion exchange particles. Anal. Chem. *23*, 620.
[106] S. Lapanje and D. Dolar (1958). Thermodynamic functions of swelling of cross-linked polystyrenesulphonic acid resins. Z. physikal. Chem. NF *18*, 11.
[107] C. Braud and E. Selegny (1974). Interrelation of swelling and selectivity of ion-exchange resins. I. Determination of swelling of sulfonic resins in water. Separation Sci. *9*, 13. II. Relation between swelling and exchange insotherms of a sulfonic resin. Alkaline and alkaline-earth counterions. Ibid. *9*, 21.
[108] G. E. Boyd and B. A. Soldano (1953). Osmotic free energies of ion exchangers. Z. Elektrochem. *57*, 162.
[109] G. V. Samsonov and Y. A. Pasechnik (1969). Ion exchange and the swelling of ion-exchange resins. Russ. Chem. Rev. *38*, 547.
[110] H. P. Gregor et al. (1952). Ion exchange resins. V. Water-vapor sorption. J. Colloid Sci. *7*, 511.
[111] G. Dickel et al. (1959). Über das thermodynamische Verhalten von Kunstharz-Ionen-austauschern bei der Wasseradsorption. Z. physikal. Chem. NF *20*, 121. G. Dickel (1974). Water transport joined with ion exchange in resins. J. Polym. Sci., Polym. Symp. *47*, 179.
[112] H. B. Bull (1944). Adsorption of water vapor by proteins. J. Amer. Chem. Soc. *66*, 1499.
[113] A. D. McLaren and J. W. Rowen (1951). Sorption of water vapor by proteins and polymers: a review. J. Polym. Sci. *7*, 289.
[114] K. W. Pepper and D. Reichenberg (1953). The influence of the degree of crosslinking on certain fundamental properties of cation-exchange resins of the sulphonated crosslinked polystyrene type. Z. Elektrochem. *57*, 183.
[115] E. Glueckauf and G. P. Kitt (1955). A theoretical treatment of cation exchangers. III. The hydration of cations in polystyrene sulfonates. Proc. Roy. Soc. Ser. A *228*, 322. A. A. Tager (1960). A reason for the high sorption ability of ion-exchange resins. Vysokomol. Soed. *2*, 997. A. R. Gupta (1985). An interpretation of water sorption isotherms of ion exchange resins. Indian J. Chem., Sect. A *24*, 368.
[116] G. Zundel, H. Noller and G.-M. Schwab (1962). Folien aus Polystyrolsulfonsäuren und ihren Salzen. II. IR-Untersuchungen über Hydratation. Z. Elektrochem. *66*, 122. III. Zum Verständnis der Natur des Hydronium-Ions. Ibid. *66*, 129. VI. IR-Untersuchung der Be-, Al-, Ga-, In- und Tl-Form. Z. physikal. Chem. NF *35*, 199.
[117] J. E. Gordon (1962). Proton magnetic resonance and infrared spectra of some ion-exchange resin-solvent systems. J. phys. Chem. *66*, 1150. M. Rinaudo (1976). Swelling of polyelectrolyte gels and thermodynamic parameters of solvation. Charged Gels and

Membranes I, 91. A. Kellomaki et al. (1979). Proton NMR studies of ion exchange resins in light and heavy water. Acta Chem. Scand. Ser. A *33*, 671. M. Periyasami and W. T. Ford (1985). Rates of exchange of solvent in and out of crosslinked polystyrene beads. Reactive Polymers *3*, 351.

[118] S. A. Rice and F. E. Harris (1956). Polyelectrolyte gels and ion exchange reactions. Z. physikal. Chemie NF *8*, 207. F. E. Harris and S. A. Rice (1956). Model for ion exchange resins. J. Chem. Phys. *24*, 1258.

[119] A. Katchalski, S. Lifson and H. Eisenberg (1951). Equation of swelling for polyelectrolyte gels. J. Polymer Sci. *7*, 571. A. Katchalski and I. Michaeli (1955). Polyelectrolyte gels in salt solutions. J. Polymer Sci. *15*, 69.

[120] H. G. Cassidy (1957). Fundamentals of Chromatography. Interscience Publishers, Inc., New York, Chap. 9.

[121] C. J. O. R. Morris and P. Morris (1963). Separation Methods in Biochemistry. Interscience Publishers, Inc., New York, Chap. 8.

[122] O. Samuelson (1953). Ion Exchangers in Analytical Chemistry. John Wiley & Sons, Inc., New York, Chap. III.

[123] Y. Marcus and D. G. Howery (1976). Ion exchange equilibrium constants. Pergamon Press.

[124] F. Martinola (1977). Neue Aspekte zur Regenerierung von Ionenaustauschern. Z. Wasser- und Abwasser-Forsch. *10* (6).

[125] B. Tyburce et al. (1984). Representation of ion-exchange equilibria in the case of complexed solutes. Analusis *12*, 129 (Fr).

[125a] K. H. Lieser (1975). Selective separations of iodine, technetium, caesium and alkali-ions. Fresenius. Z. anal. Chem. *273*, 189.

[126] N. A. Polykhina et al. (1972). Nonexchange sorption of hydrochloric acid and its salts by strongly basic anion-exchange resins. Zh. Fiz. Khim. *46*, 928. Id. (1975). Determination of the nonexchange sorption of substances by ion exchange resins. V sb., Toriya i Praktika Sorbtion. Protessov. Voronezh. (10), 30.

[127] J. Hradil et al. (1982). Sorption of gases on macroporous 2,3-epoxypropyl methacrylate co-polymers modified with amino groups of various structure. Reactive Polymers *1*, 59. G. A. Chikin et al. (1984). Phenomenological model of the kinetics and dynamics of sorption of organic substances by ion exchangers. Zh. Fiz. Khim. *58*, 1734.

[128] F. Helfferich (1962). Ligand Exchange. I. Equilibria. J. Am. Chem. Soc. *84*, 3237. Id. (1962). II. Separation of ligands having different coordinative valences. Ibid. *84*, 3242.

[129] O. Bobleter et al. (1970). Gleichgewichtsbestimmung bei Kationenaustauschern über Verteilungs- und Aktivitätskoeffizienten. Ber. Bunsen-Ges. physikal. Chemie *74*, 1050. With further references.

[129a] R. A. Robinson and R. H. Stokes in J. Timmermann: The Physicochemical Constants of Binary Systems in Concentrated Solutions. Interscience Publishers, Inc., New York, p. 612.

[129b K. W. Pepper et al. J. Chem. Soc. (London) 1952, 3129.

[130] G. E. Boyd et al. (1947). The exchage adsorption of ions from aqueous solutions by organic zeolites. II. Kinetics. J. Am. Chem. Soc. *69*, 2836. Id. (1954). Self-diffusion of cations in and through sulfonated polystyrene cation-exchange polymers. Ibid. *75*, 6091. Id. (1954). Self-diffusion of anions in strong-base anion exchangers. Ibid. *75*, 6099. Id. (1954). Self-diffusion of water molecules and mobile anions in cation exchangers. Ibid. *75*, 6105. Id. Selfdiffusion of cations in hetero-ionic cation exchangers. Ibid. *75*, 6107. G. Dickel and L. von Nieciecki (1956). The kinetics of ion exchange in ion-exchange resins. II. Z. Elektrochem. *59*, 913.

[131] W. Nernst (1904). Theorie der Reaktionsgeschwindigkeit in heterogenen Systemen. Z. physikal. Chem. *47*, 52.

[132] T. R. E. Kressman and J. A. Kitchener (1947). Cation exchange with a synthetic phenolsulfonate resin. V. Kinetic. Disc. Farad. Soc. *7*, 90.

[133] F. Helfferich (1956). Kinetik des Ionenaustausches. Angew. Chem. *68*, 693. R. Schlögl and F. Helfferich (1957). Comment on the significance of diffusion potentials in ion exchange. J. Chem. Phys. *26*, 5. F. Helfferich and M. S. Plesset (1958). Ion exchange kinetics. A nonlinear diffusion problem. J. Chem. Phys. *28*, 418.

[134] W. Nernst (1889). Die elektrische Wirksamkeit der Ionen. Z. physik. Chem. *4*, 129. Id. (1888). Zur Kinetik der in Lösung befindlichen Körper. Ibid. *2*, 613. M. Planck (1890). Über die Erregung von Elektrizität und Wärme in Elektrolyten. Ann. Phys. und Chem. *39*, 161. T. Teorell (1935). An attempt to formulate a quantitative theory on membrane permeability. Proc. Soc. Exp. Biol. *33*, 282. K. H. Meyer and J. F. Sievers (1936). La permeabilite des membranes. I. Theorie de la permeabilite ionique. Helv. Chim. Acta *19*, 649. E. Warburg (1913). Über die Diffusion von Metallen in Glas. Ann. Physik. *40*, 327.

[135] L. Liberti (1983). Planning and interpreting kinetic investigations. In: Mass transfer and kinetics of ion exchange. NATO ASI Series, Ser. E. No. 71, ed. by L. Liberti and F. G. Helfferich, p. 188.

[136] C. Y. Wen (1968). Noncatalytic heterogeneous solid fluid reaction models. Ind. Eng. Chem. *60 (9)*, 34.

[137] D. Kunii and O. Levenspiel (1969). Fluidization Engineering. John Wiley, New York.

[138] S. Yagi and D. Kunii (1955). Chem. Eng. (Tokyo) *19*, 500.

[139] P. Pöpelt et al. (1977). Zur Ermittlung von Diffusionskoeffizienten in Festkörpern aus Sorptionsdaten. Mitt. Z. phys. Chem. (Leipzig) *258*, 97.

[140] E. E. Graham and J. S. Dranoff (1982). Application of the Stefan-Maxwell equation to diffusion in ion exchangers. 1. Theory. I & EC Fund. *21*, 360. Id. 2. Experimental results. Ibid. *21*, 365.

[141] J. S. Mackie and P. Mears (1955). The diffusion of electrolytes in a cation-exchange membrane. Proc. Roy. Soc. London *A232*, 498, 510.

[142] R. Schlögl und F. Helfferich (1957). The significance of diffusion potentials in ion exchange kinetics. J. Chem. Phys. *26*, 5.

[143] K. Bunzl und G. Dickel (1969). Zur Kinetik der Ionenaustauscher. II. Die Filmdiffusion bei differentiellen Umbeladungen. Z. Naturforsch. *24a*, 109. G. Dickel et al. (1983). Kinetics of isotopic and ion exchange in ternary mixtures of cation exchangers. Z. physikal. Chem. NF *135*, 185.

[144] P. J. Dunlop (1964). Recalculated values for the diffusion coefficients of several aqueous ternary systems at 25^0 C. J. Phys. Chem. *68*, 3062.

[145] K. Dorfner (1973). Ion Exchangers. Properties and Applications. Ann Arbor Science Publishers Inc., Sec. Printing 1973. pp. 249-255.

[146] G. S. Hartley and J. Crank (1949). Some fundamental definitions and concepts in diffusion processes. Trans-Farad. Soc. *45*, 801.

[147] G. Dickel (1983). The Nernst-Planck equation in thermodynamic transport theories. In: Mass transfer and kinetics of ion exchange. NATO ASI Series, Ser. E. No. 71, ed. by L. Liberti and F. G. Helfferich, p. 367.

[148] L. Liberti et al. (1974). Chloride-sulphate exchange on anion-exchange resins. Kinetic investigations. I. J. Chromatogr. *102*, 155. Id. (1978). Kinetic investigations. II. Particle diffusion rates. Desalination *25*, 123. Id. (1978). Chemical control in selective exchanges. Ibid. *26*, 181. Id. (1980). Ion Exchange kinetics in selective systems. J. Chromatogr. *201*, 43. F. G. Helfferich, L. Liberti et al. (1985). Anion exchange kinetics in resins of high selectivity. I. Analysis of theoretical models. Isr. J. Chem. *26*, 3.

[149] G. Manecke und K. F. Bonhoeffer (1951). Elektrische Leitfähigkeit von Anionenaustauschermembranen. Z. Elektrochem. *55*, 475. G. Manecke (1952). Elektrische Leitfähigkeit von Kationenaustauschermembranen. Z. physikal. Chem. *201*, 13.

[150] S. Patell and J. C. R. Turner (1980). The kinetics of ion-exchange using porous exchangers. J. Separ. Proc. Technol. *1*, 31.

[151] F. Helfferich (1965). Ion-exchange kinetics. V. Ion exchange accompanied by reactions. J. Phys. Chem. *69*, 1178.

[152] M. Gopala Rao and A. K. Gupta (1982). Ion exchange processes accompanied by ionic reactions. Chem. Eng. J. (Lausanne) *24*, 181.

[153] M. Nativ, S. Goldstein and G. Schmuckler (1975). Kinetics of ion-exchange processes accompanied by chemical reactions. J. inorg. nucl. Chem. *37*, 1951. G. Schmuckler (1984). Kinetics of moving-boundary ion-exchange processes. Reactive Polymers *2*, 103.

[154] A. I. Kalinichev et al. (1981). Investigation into the kinetic of ion exchange processes accompanied by complex formation. J. inorg. nucl. Chem. *43*, 787. Id. (1984). The rate of ion interdiffusion in complex-forming ionites for various convex and concave isotherms. In: Ion Exchange Technology (D. Naden and M. Streat eds.), p. 257. Ellis Horwood, Chichester, UK.

[155] W. Höll (1984). Optical verification of ion exchange mechanisms in weakly electrolyte resins. Reactive Polymers *2*, 93.

[156] Yu. S. Ilnitskii (1973). Effective coefficients of mutual diffusion of ions in a chemical reaction associated with ion exchange. Deposited. Publ. VINITI 6584–73, 16 pp.

[157] L. Stanisavlienvici (1972). Optimum pH range of ion exchangers. Energetica (Bucharest) *20*, 363.

[158] R. Grießbach (1939). Über die Herstellung und Anwendung neuerer Austauschadsorbentien, insbesondere auf Harzbasis. Angew. Chem. *52*, 215.

[159] N. E. Topp and K. W. Pepper (1949). Properties of ion-exchange resins in relation to their structure. Part I. Titration curves. J. Chem. Soc. 1949, 3299.

[160] R. Kunin (1949). Strong acid cation exchange resin. Anal. Chem. *21*, 87. R. Kunin and R. E. Barry (1949). Carboxylic, weak acid type, cation exchange resin. Ind. Eng. Chem. *41*, 1269. J. I. Bregman and Y. Murata (1952). Phosphonous and phosphonic cation exchange resins. J. Am. Chem. Soc. *74*, 1867.

[161] H. P. Gregor and J. I. Bregman (1947). Characterization of cation-exchange resins. 112th Meeting of the American Chemical Society Sept. 1947, Div. of Colloid Chemistry.

[162] P. G. Howe and J. A. Kitchener (1955). Fundamental properties of cross-linked poly(methacrylic acid) ion-exchange resins. J. Chem. Soc. 1955, 2143.

[163] J. Th. G. Overbeek (1976). Polyelectrolytes, past present and future. J. Pure & Appl. Chem. *46*, 91.

[164] S. Fisher and R. Kunin (1956). Effect of crosslinking on the properties of carboxylic polymers. I. Apparent dissociation constants of acrylic and methacrylic acid polymers. J. Phys. Chem. *60*, 1030.

[165] W. Kuhn et al. (1960). Über die quantitative Verwandlung von chemischer in mechanische Energie durch homogene kontraktile Systeme. I. Helv. Chim. Acta *43*, 502.

[166] B. Auclair et al. (1975). Determination de la nature de liaisons entre un echangeur d'ions carboxylique et des ions compensateurs monovalents. Bull. Soc. Chim. France (9–10), 1905.

[167] R. Kunin and F. X. McGarvey (1949). Equilibrium and column behavior of exchange resins. Strong base anion exchange resin. Ind. Eng. Chem. *41*, 1265. R. Kunin and R. J. Myers (1947). The anion exchange equilibria in an anion exchange resin. J. Am. Chem. Soc. *69*, 2874.

[168] R. L. Gustafson, H. F. Fillius and R. Kunin (1970). Basicities of weak base ion exchange resins. Ind. Eng. Chem. Fundam. *9*, 221.

[169] D. Dragan et al. (1974). The basicity of anion exchange resins with different functional groups. Ion Exch. Membr. *1*, 215.

[170] R. L. Babkin and L. V. Kiseleva (1974). Testing ion exchangers by stepwise potentiometric titration. Ref. Zh., Teploenerg. *1974*, Abstr. No. 12R14. I. Michaeli and A. Kinrot (1973). Isr. J. Chem. *11*, 271.

[171] N. G. Polyanski and G. V. Gorbunov (1976). Rapid potentiometric titration of carboxyl cation exchangers. Izv. Vyssh. Uchebn. Zaved., Khim. Khim. Tekhnol. *19*, 482.

[172] V. P. Barabanov et al. (1976). Potentiometric analysis of copolymers based on 2-methyl-5-vinylpyridine in a nonaqueous medium. Izv. Vyssh. Uchebn. Zaved., Khim. Khim. Tekhnol. *19*, 1143. B. A. Bolto and M. B. Jackson (1984). Polyamine ion-exchange resins from polyepichlorohydrin: resins with improved resistance to oxidation. Reactive Polymers *2*, 209.

[173] A. Murel et al. (1985). Measurement of the buffer capacity of ion exchangers. Eesti NSV Tead. Akad. Toim., Keem. *34*, 297.

[174] D. J. Styer and P. V. Ammirato (1984). Influence of an ion-exchange resin on cell suspension culture. Ann. N. Y. Acad. Sci. *435*, 344.

[175] H. Jenny (1946). Adsorbed nitrate ions in relation to plant growth. J. Colloid. Sci. *1*, 33.

[176] Yu. M. Chernoberezhskii et al. (1970). Suspension effect of ion-exchange resin suspensions. CA 74: 6775f. Id. (1973). Suspension effect IV. Temperature dependence of the suspension effect as a new method for determining the activation energy of the electrical conductivity of ion-exchange resins. Vestn. Leningrad. Univ., Fiz., Khim. (2), 11. Id. (1974). V. Relation of the magnitude of the suspension effect to the electrolyte (potassium chloride) concentration in a salt bridge. Ibid. (4), 105.

[177] R. A. Olsen and J. E. Robbins (1971). Cause of the suspension effect in resin-water systems. Soil Sci. Soc. Amer., Proc. *35*, 260. R. P. Buck and E. S. Grabbe (1986). Electrostatic and thermodynamic analysis of suspension effect potentiometry. Anal. Chem. *58*, 1938.

[178] T. R. E. Kressman and J. A. Kitchener (1949). Cation exchange with a synthetic phenolsulfonate resin. Part II. Equilibrium with multivalent cations. J. Chem. Soc. 1949, 1201.

[179] B. Auclair et al. (1975). Determination de la nature des liaisons entre un echangeur d'ion carboxylique et des ions compensateur divalents. Bull. Soc. Chim. France (9 – 10, Pt. 1), 1911.

[180] S. Ganapathy Iyer et al. (1973). Exchange of metal ions against hydrogen ion of a carboxylic acid exchanger. B. A. R. C. Report 720 (Bombay, India).

[181] R. N. Rubinshtein et al. (1974). Determination of true exchange constants and instability (dissociation) constants of the ionic complexes formed in the ion exchange phase from the dependence of apparent exchange constants on the degree of ion exchange. Izv. Akad. Nauk. SSSR, Ser. Khim. 1974, 1692.

[182] L. S. Yurkova et al. (1976). Sorption of differently charged ions by an AV-17 anion exchanger. Izv. Vyssh. Uchebn. Zaved., Khim. Khim. Tekhnol. *19*, 1799.

[183] Z. I. Sosinovich et al. (1977). Equilibrium of exchange of anions of different valences on a strongly basic anion exchanger. Vestsi. Akad. Navuk. BSSR, Ser. Khim. Navuk (5), 127.

[184] D. Clifford and W. J. Weber, Jr. (1983). The determinants of divalent/monovalent selectivity in anion exchangers. Reactive Polymers *1*, 77.

[185] S. and F. Danes (1972). Ion-exchange equilibrium in polyionic systems. Chim. Ind., Genie Chim. (15), 973 (Fr).

[186] W. J. Brignal et al. (1976). Representation and interpretation of multicomponent ion exchange equilibria. In: Theory and Practice of Ion Exchange; International Conference Cambridge July, 1976.

[187] V. P. Kolnenkov and N. S. Serzhinskaya (1976). Nature of ion exchanger-polyionic solution interface. Vestsi Akad. Navuk BSSR, Ser. Khim. Navuk (4), 126. Id. (1982). Study of the calcium (2+)ion-magnesium (2+)ion-sodium(1+)ion ion-exchange equilibrium on KU 2X8 at various temperatures. Ibid. (3), 45.

[188] J. Gregory (1976). Anion exchange equilibria and kinetics in the ternary system: chloride-sulphate-phosphate. In: Theory and Practice of Ion Exchange; International Conference Cambridge July, 1976.

[189] P. Frölich et al. (1982). Quantitative Beschreibung und Berechnung isolierter und simultaner Ionenaustauschgleichgewichte. Z. phys. Chemie (Leipzig) *263*, 979.

[190] J. Plicka et al. (1984). The kinetics of ion exchange sorption in ternary systems. In: Ion Exchange Technology (D. Naden and M. Streat, eds.) p. 331. Ellis Horwood, Chichester.

[191] V. S. Soldatov and V. A. Bichkova (1985). Binary ion exchange selectivity coefficients in multiionic systems. Reactive Polymers *3*, 199. Id. (1985). A method for predicting ion exchange equilibria in ternary ion exchange systems. Ibid. *3*, 207. M. Sengupta and T. B. Paul (1985). Multicomponent ion exchange equilibria. I. Zn^{2+}-Cd^{2+}-H^{+} and Cu^{2+}-Ag^{+}-H^{+} on Amberlite IR 120. Ibid. *3*, 217. A. L. Myers and S. Byington (1986). Thermodynamics of ion exchange: prediction of multicomponent equilibria from binary data. NATO ASI Ser., Ser. E 107; Ion Exchange: Science and Technology, p. 119.

[192] D. G. Katitskii and O. G. Ilyanaya (1975). Mechanism of ion exchange. Vsb., Fiz.-Khim. Izuch. Neorgan. Soedin. 1975, 91.

[193] C. Heitner-Wirguin (1977). Spectroscopic studies of ion exchangers. Ion Exch. Solvent Extr. *7*, 83.

[194] N. Greenwood and T. Gibb. (1971). Mössbauer Spectroscopy. Chapman and Hall, London, 1971.

[195] T. E. Gough, H. D. Sharma and N. Subramanian (1970). Proton magnetic resonance studies of ionic solvation in ion exchange resins. Can. J. Chem. *48*, 917 (1970). Ibid. *49*, 457 (1970). Ibid. *49*, 3948 (1971).

[196] S. J. B. Reed. Electron Probe Microanalysis. Cambridge University Press, Cambridge, 1975.

[197] L. Alagna et al. (1984). EXAFS Spectroscopy: a breakthrough in monitoring metal-support interactions in ion exchange materials? Ion Exchange Technology (D. Naden and M. Streat, eds.). Ellis Horwood, Chichester.

[198] V. V. Mank et al. (1972). Hydration numbers of ion-exchange resins studied by NMR method. Dokl. Akad. Nauk SSSR *203*, 1115.

[199] A. Narebska and K. Erdmann (1974). State of water in swollen ion-exchange resins according to Bradley isotherms and proton magnetic relaxation data. Proc. 3rd Symp. on Ion-Exchange, Balatonfüred, 28–31 May, 1974; p. 161.

[200] H. P. Gregor (1951). Ion-pair formation in ion-exchange systems. J. Am. Chem. Soc. *73*, 3537. Id. et al. (1955). Ibid. *77*, 2713.

[201] C. Heitner-Wirguin and M. Gantz (1973). Uranyl sulfate species sorbed on ion exchangers. J. inorg. nucl. Chem. *35*, 3341. Id. (1973). Uranyl halide species sorbed on ion exchangers. Isr. J. Chem. *12*, 723. C. Heitner-Wirguin (1977). Cobalt coordination in cation exchange resins. J. inorg. nucl. Chem. *39*, 2267.

[202] M. J. Hudson (1986). Coordination chemistry of selective-ion exchange resins. NATO ASI Ser., Ser. E 107 (Ion Exch.: Sci. Technol.), 35.

[203] J. F. Mata-Segrada et al. (1977). The transition state for exchange of protons and metal ions in a carboxylate ion-exchange resin. J. Am. Chem. Soc. *99*, 5916.

[204] T. Nortia and S. Laitinen (1968). Suomen Kem. *B41*, 136. S. Laitinen and T. Norita (1968). Ibid. *B41*, 253. Id. (1971). Ibid. *A44*, 79.

[205] E. A. Chuveleva et al. (1970). Mechanism of metal ion adsorption on complexing resins studied by infrared spectroscopy. II. Infrared spectra of carboxyl resins containing uranium. Zh. Fiz. Khim. *44*, 1990.

[206] J. A. Marinsky (1973). Equations for the evaluation of formation constants of complexed ion species in crosslinked and linear polyelectrolyte systems. Ion Exch. Solvent Extr. *4*, 227.

[207] E. A. Chuveleva et al. (1972). Zh. Fiz. Khim. *46*, 93. L. S. Molnikov et al. (1976). Dokl. Akad. Nauk SSSR 229, 387. A. L. Shvarts et al. (1980). Zh. Fiz. Khim. *54*, 1037.

[208] W. B. Smith et al. (1982). The carbon-13 NMR observation of small molecules. J. Magn. Reson. *46*, 172. J. Bartoszewicz et al. (1985). An attempt to elucidate the mechanism of ion exchange on the basis of conductometric titration curves of the system Zerolite FF-phosphoric acid. Pol. J. Soil Sei *16*, 31 (1983).

[209] L. V. Vasilenko and E. I. Kazantsev (1975). Composition of sorbed molybdenum ions on anion exchangers studied by an ir spectroscopic method. Izv. Vyssh. Uchebn. Zaved., Tsvetn. Metall. (1), 143.

[210] K. Yoshimura et al. (1977). Absorptiv spectra and composition of complexes sorbed in ion exchangers. II. Three-phase distribution analysis in copper (II)-bromide and nickel (II)-thiocyanate complex systems. J. inorg. nucl. Chem. *39*, 1697.

[211] F. Vernon (1982). The role of the co-ion in chelating ion exchange processes. Reactive Polymers *1*, 51.

[212] O. D. Bonner et al. (1957). Effect of temperature on ion-exchange equilibriums. I. The sodium-hydrogen and cupric-hydrogen exchanges. J. Phys. Chem. *61*, 1614. Id. (1959) II. The ammonium-hydrogen and thallous-hydrogen exchanges. Ibid. *63*, 1417. Id. (1959). III. Exchanges involving some bivalent ions. Ibid. 1420. K. A. Kraus et al. (1959). Temperature dependence of some cation-exchange equilibriums in the range O-200^0. Ibid. *63*, 1901. Id. (1960). Anion exchange studies. XXVI. A column method for measurement of ion-exchange equilibriums at high temperature. Temperature coefficient of the $Br^- - Cl^-$ exchange reaction. J. Chromatogr. *3*, 178.

[213] F. Wolf und K.-H. Mohr (1974). Zum Druckeinfluß auf das Quellungsgleichgewicht und die Ionenselektivität von Ionenaustauscherharzen. Vom Wasser *42*, 391. F. Wolf und P. Frölich (1974). Zur Druckabhängigkeit von Ionenaustauschreaktionen. Ibid. *43*, 443. Id. (1978). Einfluß des äußeren Druckes auf die Ionenaustauschselektivität eines starkbasischen Anionenaustauscherharzes. Ibid. *51*, 175.

[214] T. R. E. Kressmann and J. A. Kitchener (1949). Cation exchange with a synthetic phenolsulphonate resin. III. Epuilibrium with large organic cations. J. Chem. Soc. 1208.

[215] R. L. Gustafson and J. A. Lirio (1968). Adsorption of organic ions by anion exchange resins. Ind. Eng. Chem. Prod. Res. Dev. *7*, 116.

[216] M. J. Semmens (1975). A review of factors influencing the selectivity of ion exchange resins for organic ions. AIChE Symp. Ser. 71 (152), 214.

[217] M. Chanda et al. (1984). Ligand sorption of aromatic amines on resin-bound ferrous ion. Reactive Polymers *2*, 279.

[218] M. I. Lifshits and E. V. Komarov (1983). Effect of an organic sorbate on proton magnetic resonance parameters of a polymeric ion exchanger and water bound to it. Zh. Fiz. Khim. *57*, 2890.

[219] D. E. Weiss (1950). A novel method of using ion-exchange resins. Nature *166*, 66.

[220] R. M. Wheaton and W. C. Bauman (1953). Ion exclusion. A unit operation utilizing ion exchange materials. Ind. Eng. Chem. *45*, 228.

[221] R. Grießbach (1957). Anomale Reaktionen an Ionenaustauschern. Chimia *11*, 29.

[222] G. N. Altshuler and A. I. Fomchenkova (1972). Effect of the structure of an ion exchanger on the selectivity of exchange of sulfacyl anions and sulfanilurea for chlorine. Zh. Fiz. Khim. *46*, 2124.
[223] G. M. Armitage et al. (1976). Radioisotope studies of carboxyl sites on strong acid and strong base ion-exchange resins derived from polystyrene. Talanta *23*, 58.
[224] J. P. C. Bootsma et al.(1984). On the reaction of pendent vinyl groups during chloromethylation of styrene-divinylbenzene copolymers. Reactive Polymers *3*, 17.
[225] A. Clearfield and J. M. Troup (1970). Ion exchange between solids. J. Phys. Chem. *74*, 2578.
[226] A. V. Nikolaev et al. (1978). Contact exchange between grains of monopolar ion exchangers. Teoriya i Praktika Sorbtsion. Protsessov. (12), 3.
[227] U. A. Inogamov et al. (1970). Ion exchange preparation of potassium nitrate. CA 73: 72601b.
[228] T. R. E. Kressman and J. A. Kitchener (1949). Cation exchange with a synthetic phenolsulfonate resin. V. Kinetics. Disc. Faraday Soc. No. 7, 90.
[229] F. Danes and F. Wolf (1973). Zum Ionenaustausch von Na^+-Ionen gegen zweiwertige Kationen am zeolithischen Molekularsieb Typ A. III. Zur Kinetik des Ionenaustausches. Z. phys. Chemie (Leipzig) *252*, 15.
[230] V. A. Konstantinov et al. (1980). Model of the internal diffusion and mixed-diffusion process of ion exchange in batch and semibatch operated apparatus of ideal mixing. Deposited Doc. VINITI 3068.
[231] F. Danes (1971). The batch process application to ion exchange unit operation. I. Chem. Engng. Sci. *26*, 1277.
[232] B. Morelli and L. Lampugnani (1975). Ion-exchange resins. Simple apparatus. J. Chem. Educ. *52*, 572.
[233] J. D. Smith (1971). Plastic chromatography columns. Lab. Pract. *20*, 496.
[234] J. A. Schmit and R. A. Henry (1970). Applications of a new anion-exchange column for liquid chromatography. Chromatographia *3*, 497.
[235] S. Sevcik et al. (1979). Universal column for work with granular functional polymers. Chem. Listy *73*, 1285.
[236] C. D. Scott and N. E. Lee (1969). Dynamic packing of ion-exchange chromatographic columns. J. Chromatogr. *42*, 263.
[237] O. Bobleter (1965). Kinetische Untersuchungen über den Ionenaustauschvorgang unter Berücksichtigung der Uran-Spaltprodukt-Trennung. Ber. Bunsenges. *69*, 874. G. Dincler und O. Bobleter (1973). Das Durchbruchsverhalten von Ionenaustauschern. Z. physik. Chem. NF *86*,156.
[238] P. Frölich (1976). Berechnung von Diffusionskoeffizienten und Durchbruchskurven für den Fall der inneren Diffusion als geschwindigkeitsbestimmender Schritt (Gelkinetik) in Ionenaustauschersäulen. Acta hydrochim. hydrobiol. *4*, 495.
[239] A. P. van der Meer et al. (1982). CSMP simulation of the break-through of ion-exchange columns. Proc. Chem. Process Anal. Des. Using Comput. 1982, 7. 7.-7. 15, Delft, Neth.
[240] T. Vermeulen et al. (1980). Column dynamics of ternary ion exchange. I. Diffusional and mass transfer relations. Chem. Eng. J. (Lausanne) *19*, 229. Id. II. Solution mass transfer controlling. Ibid. *19*, 241.
[241] F. G. Helfferich (1984). Conceptual view of column behavior in multicomponent adsorption or ion-exchange systems. AIChE Symp. Ser. 80 (233), 1.
[242] V. I. Gorshkov (1983). Countercurrent-flow ion exchange. Zh. Vses. Khim. O-va. 28, 63.
[243] R. Kunin and B. Vassiliou (1963). Regeneration of carboxylic cation exchange resins with carbon dioxide. Ind. Eng. Chem. Prod. Res. Dev. *2*, 1.

[244] W. Klump (1967). Betrachtungen zu der Ammoniakregeneration schwachbasischer Anionenaustauscher. Energie *19*, 212.
[245] C. Calmon (1980). Explosion hazards of using nitric acid in ion-exchange eqipment. Chem. Eng. (New York) *87*, 271.
[246] N. Lengborn (1958). Continuous ion exchange. Svensk. Kem. Tidskr. *70*, 255.
[247] M. J. Slater (1969). A review of continuous counter-current contactors for liquids and particulate solids. Brit. Chem. Engng. *14 (1)*, 41. Id. et al. (1975). A comparison of continuous counter-current ion exchange equipment for regeneration of resins. Chemie-Ing. *47*, 588.
[248] M. J. Slater (1979). Continuous ion exchange plant design methods and problems. Hydrometallurgy *4*, 299. M. Bailly and D. Tondeur (1974). Multicomponent moving-bed ion exchange. Approach to the design of small-scale preparative columns. J. Chromatogr. *102*, 413.
[249] W. E. Prout and L.F.Fernandez (1961). Performance of anion resins in agitated beds. Ind. Eng. Chem. *53*, 449.
[250] M. J. Slater and P.Prudhomme (1972). Continuous ion exchange in fluidized beds. Can. Mines Br., Tech. Bull. 1972, 158.
[251] A. Buijs and J. A. Wesseligh (1980). Batch fluidized ion-exchange column for streams containing suspended particles. J. Chromatogr. 201, 319.
[252] T. Koloini and M. Zumer (1979). Ion exchange with irreversible reaction in deep fluidized bed. J. Chem. Eng. *57*, 183.
[253] K. H. Lieser and B. Gleitsmann (1982). Separation of heavy metals, in particular uranium, from sea water by use of anchor groups of high selectivity. II. Continuous flow in a fluidized bed. Fresenius Z. anal. Chem. *313*, 289.
[254] J. L. Labert et al. (1980). Preparation and properties of triodide-, pentaiodide-, and heptaiodide-quaternary ammonium strong base anion-exchange resins disinfectants. Ind. Eng. Chem. Prod. Res. Dev. *19*, 256.
[255] B. Biscans et al. (1985). Ion exchange resins in a fixed bed and in a multistage fluidized bed – comparison of processes. Chem. Eng. J. (Lausanne) *30*, 81.
[256] S. Goto et al. (1981). Simplified evaluations of mass transfer resistances from batch-wise adsorption and ion-exchange data. I. Linear isotherms. Ind. Eng. Chem. Fund. *20*, 368. Id. II. Nonlinear isotherms. Ibid. *20*, 371.
[257] T. Kataoka (1971). Liquid film mass transfer with or without resistances at liquid film and resin phase in ion exchange. Kogyo Yosui 1971 (155), 27 (Japan).
[258] H. Tan (1977). Calculation of J function by a pocket calculator. Chem. Eng. (New York) *84*, 158. L. P. van Brocklin and M. M. David (1975). Ionic migration effects during liquid phase controlled ion exchange. AIChE Symp. Ser. *71* (152), 191.
[259] A. E. Rodrigues (1983). Dynamics of ion exchange processes. In: Mass transfer and kinetics of ion exchange. Ed. by L. Liberti and F. G. Helfferich. NATO ASI Series, Ser. E No. 71, p. 259.
[260] M. M. Yusipov et al. (1984). Modeling and calculation of ion-exchange processes in a fixed bed. I. Region of combined diffusion kinetics. Uzb. Khim. Zh. 1984 (1), 58.
[261] J. Marchello and M. W. Davis (1963). Theoretical investigation of agitated ion exchange beds. Ind. Eng. Chem. Fund. *2*, 27.
[262] T. Kataoka and H. Yoshida (1981). Intraparticle mass transfer in weak acid ion-exchanger. Canad. J. Chem. Eng. *59*, 475.
[263] T. Szanya et al. (1985). Separation by ion-exchange and adsorption parametric pumping. I. Fundamentals of the parametric pumping method. Hung. J. Ind. Chem. *13*, 155.
[264] G. Grevillot and D. Tondeur (1977). Equilibrium staged parametric pumping. II. Multiple transfer steps per half-cycle reservoir staging. AIChE J. *23*, 840.

[265] T. J. Butts et al. (1973). Batch fractionation of ionic mixtures by parametric pumping. Ind. Eng. Chem. Fund. *12*, 467. H. T. Chen et al. (1977). A study of semicontinuous pH-parametric pumping in the model system hemoglobin-albumin on Sephadex ion exchangers. Pac. Chem. Eng. Congr., (Proc.) *2*, 54. G. Grevillot et al. (1984). Donnan partition parametric pumping. Reactive Polymers *2*, 71.

[266] D. Tondeur and G. Grevillot (1986). Parametric ion exchange processes (parametric pumping and allied techniques). NATO ASI Ser., Ser. E 107, Ion Exchange: Science and Technology, 369.

[267] Desinfektion von Kationenaustauschern mit Chloramin T. Bayer-Lewatit Information Bulletin. H. Schwab and H. Soldavini (1977). Desinfektion von Ionenaustauschern mit Peressigsäure Spezialqualität IA. Chemie-Technik *6*, 197.

[268] M. Matsuda et al. (1986). Decomposition of ion exchange resins by pyrolysis. Nucl. Technol. *75*, 187.

[269] R. F. Hollis and C. K. McArthur (1957). The resin-in-pulp method for recovery of uranium. Mining Eng. *9*, 442.

[270] J. W. Fisher and A. J. Vivyurka (1969). Combined ion exchange-solvent extraction process (Eluex) for ammonium diuranate production. Paper presented at Conference on Ion Exchange, London, 1969.

[271] L. H. Marcus (1980). Ion exchange resins. June, 1976-May, 1980 (citations from the Energy data base). See: Gov. Rep. Announce. Index (U.S.) 80 (26), 2625. Avail. NTIS.

[272] T. B. S. Giddey (1980). Ion exchange in the 1980s in South Africa. Proc. Natl. Meet.-S. Afr. Inst. Chem. Eng. 3rd 1980, 3F/1 – 3F/15. W. H. Waitz, jr. (1982). Ion exchange for recovery of precious metals. Plat. Surf. Finish. *69*, 56.

[273] D. J. King and P. M. Blythe (1979). The impact of the design of uranium extraction plants due to the use of a fluidized bed solid ion exchange loading system. CIM Bull. *72*, 135.

[274] F. Vernon (1979). Metal separation by chelating ion exchange. Acta Polym. *30*, 740. M. Knothe and S. Ziegenbalg (1982). Möglichkeiten und Probleme bei der technischen Anreicherung und Trennung von Metallen durch Ionenaustauscher. Z. Chem. *22*, 295.

[275] D. T. Jones (1975). Protein recovery by ion exchange. Food Process. Ind. *44*, 21, 23. A. D. A. Kanekanian and M. J. Lewis (1986). Protein isolation using ion exchangers. Dev. Food Proteins *4*, 135.

[276] Anon. (1985). Catalysis by functionalized porous organic polymers. Brochure Rohm and Haas Comp., Philadelphia, USA. F. J. Waller (1986). Catalysis with metal cation exchange resins. Catal. Rev.-Sci. Eng. *28*, 1. Id. (1986). Catalysis with a perfluorinated ion-exchange polymer. ACS Symp. Ser. *308*, 42. J. Klein et al. (1986). Der Einsatz von Ionenaustausch-Katalysatoren am Beispiel der Umsetzung von Stärkeprodukten. Chem.-Ing.-Techn. *58*, 436.

[277] P. A. Belter (1984). Ion exchange and adsorption in pharmaceutical manufacture. AIChE Symp. Ser. *80* (233), 110.

[278] P. A. Munro (1980). Potential applications of adsorbent support technology in New Zealand. Food Technol. N. Z. *15*, 7, 9, 11, 13.

[279] P. S. Porter and M. J. Semmens (1980). A review of virus and protein sorption by ion exchange resins. Environ. Int. *3*, 311.

[280] E. M. Savitskaya et al. (1982). Sorption of organic substances by ion exchangers of various nature. Pure Appl. Chem. *54*, 2169.

[281] G. Cainelli et al. (1980). Application of some polymer supported reagents to organic synthesis. Proc. IUPAC Symp. Org. Synth., 3rd 1980 (Pub. 1981), 19. Pergamon: Oxford, England.

[282] T. N. Hendrickson and G. A. Lorenzo (1981). Silver recovery by ion exchange. Symp. Recovery, Reclam. Refin. Precious Met., Proc. Paper 11. H. Meckl (1985). Experience

with ion exchange for silver recovery from photographic effluents. J. Imaging Technol. *11*, 51.
[283] J. W. Kleppe (1979). The application of an ion exchange method for color developer reuse. J. Appl. Photogr. Eng. *5*, 132.
[284] K. M. Saldadze (1975). Basic trends in the use of polymers for protecting the biosphere from pollution. Plast. Massy (5), 69.
[285] J. Bohorquez et al. (1982). Use of ion exchangers as drying agent for organic solvent. I. Adsorption equilibrium. Bull. Soc. Chim. Fr. (5–6. Pt. 1), I–193/I–196. Id. (1982). II. Kinetics of adsorption. Ibid. (5–6, Pt. 1), I–197/I–201.
[286] C. Bettinali and F. Pantanetti (1976). Uranium from sea-water: possibilities of recovery, exploiting slow coastal currents. Uranium Ore Proc. Advis. Group Meet. 1975, 213. IAEA: Vienna, Austria.
[287] J. Korkisch and I. Steffan (1976). Separation and concentration of seawater constituents on synthetic ion exchange resins. Strategies Mar. Pollut. Monit. 241. Wiley: New York. R. Lumbroso (1979). Ion exchange resins: recent progress. Surfaces *129*, 27.
[288] C. Sabau and A. Calusaru (1970). Literature survey on the separation of isotopes by ion exchange. Inst. Fiz. At. (Rom.) R. C.-6, 40 pp. (Eng). E. Gard and A. Calusaru (1977). Chemical and ion exchange in isotope separation with special reference to heavy elements. Isotopenpraxis *13*, 121. K. Heumann et al. (1979). Dependence of chlorine isotope separation on the degree of crosslinkage of a strongly basic anion exchange resin. Z. Naturforsch., B: Anorg. Chem., Org. Chem. *34B*, 406. R. Nakane (1980). Recent advances in isotope separation. Uranium enrichment and tritium separation by chemical exchange and plasma separation. Oyo Butsuri *49*, 754.
[289] J. Hradil et al. (1982). Reactive Polymers XLIV. Sorption of gases on macroporous 2,3-epoxypropyl methacrylate co-polymers modified with amino groups of various structure. Reactive Polymers *1*, 59.
[290] G. P. Buzanova et al. (1976). Removal of sulfur dioxide from gases by ion exchange sorption. Teoriya i Praktika Sorbtion. Protsessov. (11), 102. R. E. Anderson (1975). The triple alkali system for the removal of sulfur oxides from stack gases using ion exchange. Ion Exch. Membr. *2*, 99. L. V. Belyaevskaya et al. (1981). Study of the dynamics of sulfur dioxide sorption on ion exchange resins of different basicity and structure. Nauchn. Tr.-Mosk. Inst. Stali Splavov *131*, 76.
[291] J. D. Rutkowski et al. (1977). Study of selected methods for removal of mercaptans from the waste gases of kraft pulp plant. Pr. Nauk. Inst. Inz. Ochr. Srodowiska Politech. Wroclaw *43*, 73.
[292] T. Yoshino and M. Hamaguchi (1980). Ion exchange-type deodorization system. Kaukyo Gijutsu *9*, 563.
[293] V. S. Soldatov et al. (1985). Artifical substrates for plants based on fibrous ion exchange materials. Vestsi. Akad. Navuk BSSR, Ser. Khim. Navuk (6), 85.
[294] W. P. Kelley. Cation exchange in soils. Reinhold Publishing Company, New York, N. Y., 1949. R. Ranhama and Th. G. Sahama. Geochemistry. The University of Chicago Press, Chicago, Ill., 1950.
[295] M. M. Reddy (1977). Ion-exchange materials in natural water systems. Ion Exch. Solvent Extr. *7*, 165.
[296] H. D. Schulz (1981). Zweidimensionales Transport-Reaktions-Modell für Ionen im Grundwasser. Z. dt. geol. Ges. *132*, 585.
[297] G. S. R. Krishna Murti and T. V. Rao (1984). A model system for soil amorphous material. Agrochimica *28*, 257.
[298] N. I. Gamayunov (1985). Ion exchange in soils. Pochvovedenie (8), 38. D. L. Pisnki and L. T. Podgorina (1986). Isotherms of ion exchange sorption of calcium and lead by soils in model experiments. Agrokhimiya (3), 78.

[299] P. Nkedy-Kizza et al. (1984). On the equivalence of two conceptual models for describing ion exchange during transport through an aggregated Oxisol. Water Resour. Res. *20*, 1123.

[300] C. P. Chang et al. (1976). I. AΦ (Z/l,χ) rule of inorganic ion-exchange reactions in sea water. K'o Hsueh T'ung Pao *21*, 231 (Ch). Id. III. Application of the rule in marine geochemistry of elements. Ibid. *21*, 531.

[301] N. F. Chelishchev (1973). Ion-exchange properties of minerals. Nauka: Moscow, USSR. 203 pp.

[302] L. V. Benson (1980). Tabulation and evaluation of ion exchange data on smectites, certain zeolites and basalt. Report Lawrence Berkeley Laboratory-10 541. Avail. NTIS.

[303] W. G. French and A. D. Pearson (1970). Refractive index changes in glass by ion exchange. Amer. Ceram. Soc., Bull. *42*, 974.

[304] H. Ohta (1970). Ion exchange of glass by electrolysis. Hyomen *8*, 597 (Japan). Id. (1972). Strength and fracture behavior on ion-exchanged glass. Yogyo Kyokai Shi *80*, 159.

[305] A. M. Butaev (1981). Anomalous properties of silver ion appearing in the ion-exchange strengthening of glass. Fiz. Khim. Stekla *7*, 248.

[306] F. Umland (1956). Über die Wechselwirkung von Elektrolytlösung und γ-Al_2O_3. Entwicklung einer formalen Ionenaustauschtheorie für die Adsorption von Elektrolyten aus wäßriger Lösung. Z. Elektrochem. *60*, 711.

[307] M. P. Tiwari et al. (1979). Alumina as an ion exchanger and its applications. Part III. pH and adsorption of anionic dyes by alumina. J. Ind. Chem. Soc. *56*, 798. L. Stanisavlievici (1971). Retention of weak acids by anionic exchangers. Trib. CEBEDEAU *24* (331–332), 295. A. P. Lushina and V. B. Aleskovskii (1976). Ion exchange as the first stage in the transformation of solid substances in electrolyte solutions. Zh. Prikl. Khim. (Leningrad) *49*, 41.

[308] D. Hircke and G. Wolff (1974). Diffusion and ion exchange in alkali-silica reaction. Cem. Concr. Res. *4*, 609.

[309] C. Calmon and T. R. E. Kressman, editors. Ion exchange in organic and biochemistry (1957). Interscience Publishers, Inc., New York, N. Y., 1957.

[310] R. A. Mackay (1982). Reactions in microemulsions. The ion exchange model. J. Phys. Chem. *86*, 4756.

[311] F. H. Quina and H. Chaimovich (1979). Ion exchange in micellar solutions. 1. Conceptual framework for ion exchange in micellar solutions. J. Phys. Chem. *83*, 1844. Id. et al. (1979). 2. Binding of hydroxide ion to positive micelles. Ibid. *83*, 1851.

[312] M. B. Dontsov and A. N. Lyashenko (1975). Nature of the exchange capacity of plant roots. Biol. Nauki (Moscow) *18*, 142.

[313] E. g: J. E. Hesse et al. (1984). Proc. Natl. Acad. Sci. USA *81*, 4746. J. W. Schneider et al. (1985). Ibid. *82*, 6357. G. E. Shull et al. (1985). Nature *316*, 691. D. H. McLennan et al. (1985). Ibid. *316*, 696. K. Kawakami et al. (1985). Ibid. *316*, 733.

[314] R. Damadian (1971). Biological ion-exchanger resin. III. Molecular interpretation of cellular ion exchange. Biophys. J. *11*, 773.

[315] I. A. Kuznetsov et al. (1984). Ion-exchange properties of immobilized DNA. Reactive Polymers *3*, 37.

[316] M. R. Ruggieri et al. (1985). Further characterization of bacterial adherence to urinary bladder mucosa: comparison with adherence to anion-exchange resin. J. Urol. (Baltimore) *134*, 1019.

[317] H. Jullien (1969). Paint films and the ion exchange phenomenon. Double Liaison No. *172*, 669. Id. et al. (1974). Ion exchange properties of paint films. FATIPEC Congr. 1974, 12, 461. Id. et al. (1976). Ion exchange properties of polyurethane-based paint films. Dtsch. Farben-Z. *30*, 258.

[318] A. Koopmans (1980). New aspects in the relation between pigment and binder in corrosion protecting coating systems. Congr. FATIPEC 15th, III-204, 222.

[319] B. P. F. Goldie (1984). Novel corrosion inhibitors. Xth International Conference in Organic Coatings Science and Technology, Athens, Greece; Proc. p. 53.

[320] H. Orth and P. Kleinheins (1971). Untersuchungen über den Ionenaustausch an der Oberfläche eines gebeizten ABS-Copolymeren. Angew. Makromol. Chem. *19*, 99.

[321] B. Pacciarelli et al. (1986). Effect of electrical double-layer phenomenon on dying. Textilveredlung *21*, 51.

[322] G. Ning Ling (1962). A physical theory of the living state: the association-induction hypothesis. Blaisdell, New York, 1962.

1.2 Synthetic Ion Exchange Resins

Konrad Dorfner
Mannheim, Federal Republic of Germany

The history of synthetic ion exchange resins begins in 1935 with the work of Adams and Holmes on the manufacture of polycondensation products with cation and anion exchange properties. This work was accepted so quickly because it made possible water deionization by a new and attractive process other than distillation. It was then Griessbach who, in his classic paper [1], pointed out the potential uses of this new class of ion exchangers, which had such superior properties that not only water purification but also adsorption, complex formation, separations, other purifications, conversion, recovery and buffering became feasible. A new chemical technology was born and drew increasing attention after – following the ingenious idea of incorporating fixed ionic groups into a synthetic organic polymer – a wide variety of relatively stable and industrially useful products were made available. This enabled Samuelson in Sweden to do his systematic work on the application of ion exchange resins in analytical chemistry, showing the importance of ion exchangers as stationary phases in ion exchange chromatography. But the two biggest surprises came in 1947 (Calmon). Firstly, the work done on the Manhattan

Project by the U.S. Atomic Energy Commission was published showing unusual achievements with ion exchangers in the separation of the rare earths and other metal groups performed by elution with complexing agents. Then D'Alelio put forward another ingenious idea resulting in the commercial development of synthetic organic ion exchangers with styrene-divinylbenzene matrices as addition polymerization products in uniformly shaped bead form, with excellent physical and chemical properties. These resins had nearly twice the capacity of the phenolic cation exchange resins and about three times that of the inorganic zeolites in use at that time. In this period of rapid development of ion exchange resins, there also appeared the first polymerization strong base anion exchange resin invented by McBurney in 1952. Ion exchange technology was now able to meet the needs of the new power plants using boilers operating at critical pressures.

As will be seen, many more synthetic ion exchange resins have been developed than are used in large scale industrial applications. For the latter usually several years of testing are necessary. But the usefulness of synthetic ion exchange resins in other fields in science and technology has made them by far predominant in ion exchange technology.

1.2.1 Ion exchange resins as reactive polymers

In organic chemistry the term "functional groups" is used for those nonskeletal parts of chemical compounds representing the group of a molecule responsible for its characteristic properties and the reactivity of major interest. These groups are contrasted with the pure hydrocarbon skeletons. Because unsaturated compounds behave, on the one hand, differently from saturated compounds and are, on the other hand, of great importance for the manufacture of polymers, the C=C double bond is also considered as a functional group. The introduction of functional groups into parent substances and the use of such functional groups for further conversions is the interest of preparative organic chemistry [2]. The term functional groups has also been adopted in polymer chemistry. It is used to describe the chemical behavior or reactions of macromolecules and polymers, to elucidate structures of newly synthesized polymers and in the description of the synthesization of polymers by the conversion of existing functional groups into others. Initially, the functional groups of polymers were only occasionally called reactive entities or reactive groups but for some time this terminology has been used increasingly, so that the question may be raised of whether the threefold terminology reactions of polymers, polymers with functional groups and reactive polymers is justified, and whether reactive polymers should not have their own definition. Such a clear threefold terminology might well have far-reaching consequences.

Reactions of polymers were performed prior to the discovery of H. Staudinger's polymer analogous conversions, with the latter term having been introduced to focus on the fact that it is possible for macromolecules to undergo chemical reactions in such a way that both the size of the molecule and the degree of polymerization are not changed [3]. For decades, until around 1950, improvement of the newly developed plastics had first priority, when aspects of structure and their chemistry in general became more important. Plastics are generally understood to be highly stable and inert to all influences, i. e. free of chemical reaction. In contrast to this the reactions of polymers – with the exception of the work done for the preparation of synthetic organic ion exchange resins – were persued more out of sheer scientific interest [4]. In time, however, the modification of known and the synthesis of new macromolecular substances by reactions of polymers gained again in importance [5], and some reactions of polymers have required considerable practical importance, among which the production of poly(vinyl alcohol) by saponification of poly(vinyl acetate) is the most important [6].

If reactions of polymers are defined as chemical reactions in which at least one of the reacting substances is polymeric, then a general grouping of such reactions can be made. Here it is important to distinguish between polymers consisting of saturated carbon chains and those with heteroatomic chains. In detail the postulate of equal reactivities of groups in polymers as compared to the same groups in monomers can be considered as correct, but polymer reactions are so much slower that every single reaction of polymers must be regarded separately [7]. The division of reactions of polymers into reactions of polymers with monomers, reactions of polymers with polymers, and reactions of polymers without a chemical reaction partner covers all the reactions commonly known under special terms, as for instance polymer analogous reactions, etc., transfer reactions from polymer to polymer, etc., or thermal degradation, etc. [8, 9]. The domain of reactions of polymers has progressed but has still not yet come to an end. Without reactions of polymers the field of reactive polymers would only be partly accessible [10].

Chemically, polymers with functional groups can be considered as intermediates. This is because the functional groups of polymerizable monomers [11] can be dealt with phenomenologically such that the monomers used for a polymer reaction are essentially identical to the structural units present in the polymers. Monomers with several polymerizable bifunctional groups can be polymerized in such a way that one or more of these groups keep their polymerizable functionality within the polymer molecule. Polymers thus obtained can, therefore, with good reason be called polymers with functional groups.

It should be mentioned here that in the case of addition polymerizations the formation of either inert or functional polymers is easily overlooked because monomers with carbon-carbon double bonds prevail as starting materials. In the case of polycondensations the first-formed step-macromolecules still contain functional groups and continue to react, whereby the functional groups have to be understood as being the same as the polymerizable groups of the micromolecular reactants. But given the case of polycondensations starting from different monomers where one is bifunctional and the other multifunctional, then under controlled conditions functional groups in the polymer that are still polymerizable are left over. It seems that in both polycondensates and polyadducts one could, for a more precise understanding, speak of two kinds of functional groups: those, on the one hand, which in the sense of a definition of polymers with functional groups are present in order to enable these polymeric products to undergo a consecutive polymer reaction; and those that are not polymerizable but are, in the sense of functional groups of organic low molecular chemistry, suitable and applicable for further controlled reactions. In the first case the so-called prepolymers are a good example. Their definition from the Encyclopedia of Polymer Science and Technology is: A prepolymer is a partially polymerized substance, or one polymerized to a low degree of polymerization, for subsequent conversion to a high polymer. In other words, prepolymers are a group of polymers with lower molecular weight but with (it would be better to say "polymerizable") functional groups. A selection of inert polymers (that is plastic materials) and functional polymers (that is polymeric intermediates) or, in other words, plastics and resins, could be made, resulting in a comparison of the various polymeric chains, on the one hand, with precrosslinked macromolecules, on the other. But how confused the existing state of definitions and hence proper understanding of reactions of polymers, polymers with functional groups and reactive polymers is, may be seen from the following introduction to an interesting presentation: Considerable interest is being focused at present on synthetic polymers bearing reactive functional groups. By reactive functional groups I mean chemical functions enabling the polymer to chemically react with, or to exert a definite chemical action on, one or more components of surrounding medium, either in artificial or in biological systems. This definition is not absolute, since any chemical group may undergo chemical reactions. However, a polymer containing reactive chemical groups is usually prepared to be, and expected to act as a reactive substance and not merely as an inert material (P. Ferruti). Can this terminology of reactive functional groups in polymers be meaningfully introduced or has the change from functional polymers to reactive polymers occurred without an adequate definitional foundation?

Regarding polymers with functional groups, such as prepolymers, living polymers, resins, printing ink binders, coating materials, dispersions, liquid rubber products, metallized polystyrene, irradiation curable resins, photoresists, water soluble polymers, adhesives, sealants and thixotropic agents, it must be said that the term functional groups can and should have in polymer chemistry the same meaning as in general organic chemistry, which leads to the meaningful expression "polymers with functional groups". It is then easy to accept that polymers with functional groups are intermediates manufactured in huge quantities in chemistry, which are, therefore, to be contrasted with reactions of polymers, and which are further to be differentiated from reactive polymers.

The term reactive polymers started to appear almost faint-heartedly, in the chemical literature two decades ago. Then the generic term "conversions of polymeric substances" appeared, understood as reactions of polymers. The polymers with functional groups, which have become so important as chemical intermediates, are then quite often not mentioned at all. Usually it is stated that by conversions of polymers macromolecules or networks with special reactive groups are obtained and it is pointed out that, here, ion exchangers represent the oldest and best-known example. But the reactive polymers are a special group of polymeric substances containing entities of a special active significance. They are neither only plastic materials on which reactions have been performed, nor do they contain functional groups by means of which they can be converted into a plastic material. Their reactive groups are important because of their applicability.

The history of ion exchange resins shows how long it took until understanding made possible the naming and defining of reactive polymers. It was also difficult to understand the special features of the redox polymers from names like electron exchangers, redox exchangers or redoxites, i. e., their ability to perform redox reactions on a solid phase, that is, that it could clearly be seen that redox polymers represent a particular type of reactive polymer and have nothing in common with ion exchangers [12]. The same confusion results for chelating resins when these are called chelate ion exchangers – in accordance with their essential reaction, the formation of chelates. They are one independent type of reactive polymer definitively distinguished from ion exchangers and should be treated independently. Chelate polymers were arrived at by analogy (Skogseid). Quite consistently, they were called macromolecular complexing agents from the very beginning [13]. By analogy one arrived also at the other reactive polymers, for which it is a distinctive mark that for each, with the exception of perhaps one, a low molecular, organic or inorganic analogue can be found. It can be considered as coincidental that customary strongly acidic cation exchangers have catalytic properties. The types of reactive polymers that are included among the polymeric reagents comprise all those polymer-bound reagents that have not yet been elaborated enough to justify their classification as a particular type of reactive polymer (Table 1.10). The state of development of the individual types of reactive polymers varies not only with respect to their manufacture but also with respect to their practical importance.

Table 1.10 Reactive polymers and their monomer analogues

Reactive polymers	Monomer analogues
Ion exchangers	Precipitation reagents
Polyelectrolytes	Acids, bases, salts
Redox polymers	Oxidation and reduction reagents
Chelating polymers	Complexing agents
Polymeric catalysts	Catalysts
Polymer bound enzymes	Enzymes
Polymer drugs	Pharmaceutical active substances
Affinity chromatography supports	(as such not used)
Polymeric adsorbents	Sorbents
Polymer reagents	All kinds of reagents

Ion exchangers. These "resins" are with certainty the oldest synthetic reactive polymers. The manufacture and properties, applications in industry, preparative and analytical chemistry, and the theory of ion exchange processes have been described in numerous papers that are not easily accessible. Because of their reactive groups ion exchangers are not only able to exchange one ion against another but are also in a position to store the exchanged ion for as long as is desirable. This is to be found in natural processes also and is the reason why ion exchange must be considered a general phenomenon. Because of the reversibility of ion exchange processes ion exchange resins in technical exchange units are not simply a solitary reactant but an integral part of the whole system and process.

Polyelectrolytes. With regard to their main applications, that is flocculation, thickening, dispersion, hardness stabilization of water and inhibition of incrustations, these products are – by virtue of the lack both of crosslinkings and, thereby, of insolubility – not ion exchangers. Polyelectrolytes carry functionally active groups which make them soluble in water. As reactive polymers they act in a different way. Besides the phenomena of counterion condensation, self-swelling and others, the most important action of these reactive polymers is their boundary surface adsorption. The various reactions occurring can be summarized as adsorption onto the surface of colloidal suspensions. By similar reactions emulsified droplets are protected against flocculation. Only from theoretical models does it follow that polyelectrolytes perform ion exchange as a limiting state.

Redox polymers. The reactions of redox polymers are oxidations and reductions achievable on a solid phase. New developments in the past two decades have not overcome the problem of regeneration. As a result these reactive polymers have remained at the stage of possibly producible types and of laboratory application. For the so-called redox-ion exchangers which differ only in the kind of binding to the polymer matrix the same has to be said. Quite successful, however, has been water treatment with very cheap, one-way redox polymers based on natural raw materials [14]. It might well be that the recently introduced viologen (= dipyridyl derivative)-polymers are true electron exchangers since it seems feasible to use them to store electrons and to release them again [15].

Chelating polymers. With respect to ion exchange processes selectivity and specificity have been misunderstood over and over again. Low molecular chelating agents are defined as chemical compounds possessing the ability of binding metal ions quite specifically in cyclic compounds. If compounds able to form chelates are incorporated into a polymer carrier, then chelating polymers are obtained to which at least a very high specificity has to be ascribed. In this case the selectivity sequences of ion exchangers are replaced by the displacement sequences of the chelating polymers. Because of the kinetics of synthetic chelating polymers and naturally occurring chelate-forming polymers – alginic acids, chitin, and chitosan – these form an independent group of reactive polymers.

Polymeric catalysts. The efficacy of customary ion exchangers as catalysts is well-known. But further investigations have been made to prepare other polymeric catalysts in order to make use to an even greater extent of the advantages of catalysis with solid polymers. It might be of value to discuss whether catalysts can be called reactive at all since they, by definition, are never direct reactants. But perhaps the polymeric catalysts can be used to set up models that avoid reference to functional groups in relation to catalysts, referring instead to sites or groups of the catalyst as points of reaction. The activation of reactants and re-groupings brought about by the catalyst reveal that the catalyst is the reactive center and, in the case of polymeric catalysts, undoubtedly then a reactive polymer.

Polymer-bound enzymes. Why are enzymes macromolecules? [16]. The answer to this question is that as bio-catalysts the microenvironment for the active group in enzymes can only be provided by a large molecule if all functions are to be fulfilled. The binding of the enzymes to polymeric carriers in the well-known polymer-bound enzymes serves the enlargement of the enzyme particle and improves their handling in enzymatic reactors. Further research here in the field of reactive polymers will be devoted to the preparation of synthetic polymers with the properties of natural enzymes – a difficult but extraordinarily challenging task [17].

Polymer drugs. Ion exchangers in pharmacy are used, among other things, for manufacture of depot drugs. Modern polymer chemistry has surpassed this stage, since pharmacologically active polymers are synthesized and tested for their chemotherapeutic activity, as for instance in tumor research [18]. Pharmaceutically active substances are reactive in a perhaps complicated but special way. Polymer drugs are the analoguous reactive polymers.

Affinity chromatography supports. The affinity resins used in affinity chromatography are based on reactive polymers to which the necessary effectors have been covalently bound. The easily reversible complex formation is then usually executed with biological material that is to be purified. Natural polymers are often used as solid supports.

Polymeric sorbents. Porous polymeric sorbents or adsorption resins do not react by chemically well-defined groups in stoichiometric equivalents but by general adsorption of molecules brought into contact with them. The decisive factor for the final classification within reactive polymers could be that adsorption on synthetic polymeric sorbents is fully reversible and that, therefore, adsorption resins can be used in cycles. The adsorbents can usually be regained, for the most part unchanged, by mere regeneration. The question seems to be to what extent the sorption phenomenon can be explained as a chemical reaction and to what extent this term is extended also to include van der Waals forces.

Polymer reagents. It seems that quite often the designation polymeric reagent has been used for what above has been called reactive polymers. But those in this field at the early stages of the development of some of the above-described reactive polymers are aware that their origins are closely connected with ion exchangers. Most of them have now developed as an independent group of products. Considering the main fields of applications for reactive polymers such as biochemistry, organic synthesis, specific separations and analysis, it follows that the latter two are covered by the reactive polymer ion exchangers, chelating polymers, affinity chromatography supports and polymeric sorbents. With respect to the definition for polymeric reagents as given by Patchornik [19]: "A polymer reagent is a substance possessing the physical properties of a high polymer as well as the chemical properties of the attached reagent", then, the biochemical as well as the organic chemical syntheses would remain as the main fields for the so-called polymeric reagents. The great success of the Merrifield syntheses and impulses from them for further developments and applications would seem to confirm this [20].

The advantages of working with reactive polymers in contrast to their conventional low molecular analogues have often been described [21]. These studies often relate to insoluble high molecular matrices, which have been studied more than soluble reactive polymers [22]. Of the latter, perhaps with respect to polyelectrolytes much that is of interest can still be expected.

For a general characterization of reactive polymers one could try to set up a classification of the reactions of reactive polymers along the lines of work done by Satchell [23]. This is be, on the one hand, too early and on the other must be postponed because of its volume. It must at least be mentioned, firstly, that reactive polymers are capable of reactions surpassing the well-known reactions of low molecular chemistry and that of their reactive effects those on the microenvironment seem to be the most important. Immobilization by binding a species onto a polymer is not limited to enzymes but a general principle. It is also a fact that all reactive polymers are reciprocally inaccessible. The microenvironment is created by the crosslinking and structure of the polymers, with the latter influencing the reactivity of the reactive groups [24]. In the case of reactive polymers, in addition to general investigations of the molecular structure and the reactivity of chemical compounds or functional groups the structure of the matrix has to be taken into consideration as a characteristic property. One could, therefore, characterize the reactive polymers as polymeric products with which not only new application techniques can be created but also new reactivities (Table 1.11).

In order to arrive at a definition for reactive polymer, it should be noted that reactions of polymers, polymers with functional groups and reactive polymers are three different fields which eventually could be combined under the generic term conversions of polymers. The definition of reactive polymer is then, in any case, at first subjective but has to be formulated under the condition that the term polymeric reagent stands for only one group comprising numerous examples. Therefore the definition of reactive polymers should go beyond that given for polymeric reagents; further, the characteristics resulting from the possible secondary reactions should be included. With this in mind, reactive polymer could be defined in the following

Table 1.11 Classification of reactions of reactive polymers

Primary reactions	Secondary reactivities
Exchange reactions	easy reversibility; selective ion uptake; storage of the exchanged ions for optional elution
Polyanion and polycation formation	reactive binding onto coagulating polymer substances; non-ideal conditions for osmotic pressure, ion activity and electric transport; accumulation of (counter)ions
Redox reactions	differing redox potentials; reversibility
Complex bonds	specific bonds; storage of the bound ions
Catalyses	heterogeneous catalysis; three phase catalysis
Enzymatic reactions	high relative enzyme concentrations; higher stability of the enzyme group
Physiological reactions	e. g., antitumor activity
Adsorptions	reversible desorptions; selective elutions
Esterification with amino acids (Merrifield technique)	further solid phase reactions with amino acids for the synthesis of peptides
Polymeric transfer reactions	cascade reactions acc. to Patchornik
Apparant general polymer bound reagent reactions	e. g., infinite diluted reactions; intrapolymer reactions; improved Wittig reactions; synthesis of carbon chains

way: Reactive polymers are synthetic or semi-synthetic, insoluble or soluble polymeric substances capable of reactions surpassing the reactivity of their bound reactive groups, leading to a new, e. g., storing type of reaction, which can provide new reaction conditions to the surrounding system as well as new reaction sequences, so that they, in their total chemical behavior, approach general phenomena of the natural world.

1.2.2 Preparation and manufacture

A look at the preparation and manufacture of ion exchange resins should first devote some space to the basic requirements of such materials. From the definition given on various occasions in this volume it follows that for synthetic ion exchange resins the following factors are decisive and are, therefore, prerequisites to be met during synthesis. Synthetic ion exchange resins consist by definition of a high polymer bridged with suitable agents for crosslinking and a certain type and number of so-called ionogenic groups. The essential requirements are:

the resin must be highly polymeric and sufficiently crosslinked in order to be insoluble in water and other liquids and have good mechanic and thermic properties;

the resin, having either a gel or a porous structure, must be sufficiently hydrophilic to make it possible for ions to diffuse through the structure at a finite and acceptable rate;

the resin must contain an adequate number of accessible ionic exchange sites in order to yield a high exchange capacity;

the resin ion exchanger must be as stable as possible chemically, so as neither to undergo degradation during use nor to release parts or degraded parts of its structure;

the resin must have a particle size distribution tuned to the envisaged application;

the ion exchange resin in the swollen state should be denser than water.

These prerequisites are decisive for the physical and chemical properties of a synthetic ion exchange resin.

As developments over decades have shown there is somewhat more involved in the preparation of a suitable ion exchanger than synthesization of a highly crosslinked polymer containing ionogenic groups. Basically there seems to be very little new synthetic organic chemistry involved in the synthesis of ion exchange resins. But nevertheless the synthesis of practically and commercially suitable ion exchange resins may be described as partly science and partly art (Abrams). And despite the fact that the basic methods are conventional in organic synthesis and polymer chemistry, this does not imply that the synthesis of an ion exchange resin is an easy task. Developing processes for the commercial manufacture of high capacity and durable ion exchange resins under very narrow specifications necessary for their useful application is an exacting matter (Kunin). The principles involved in making the crosslinked polystyrene matrix are not too different from those used in the suspension polymerization of linear polystyrene. As another example, the sulfonation of the polymer is basically similar to the sulfonation of benzene. But since there exist only pure quantitative theories for the prediction of the necessary requirements for achieving a particular degree of crosslinking, a fixed particle size distribution, a predetermined rate of sulfonation or chloromethylation in a heterogenous system, etc., the synthesis of ion exchange resins has indeed remained an art rather than science. As a consequence each resin manufacturer employs modifications designed to achieve optimum conditions, which are closely guarded trade secrets. These special manufacturing conditions are used to enhance physical appearance, durability, porosity, favourable kinetics and other important properties. Developments in the syntheses of past decades may give an insight into what has so far been achieved.

Contrary to the historical development of synthetic ion exchange resins, which started with polycondensation products, to be followed by addition polymerisation materials, in the discussions that follow the reverse order has been chosen, out of recognition of the tremendous shift in importance of both groups. From a synthetic point of view there are two other general methods of synthesizing an ion exchange resin. The first way is to build the ionic groups into the resin during the polymerisation. The second way is first to form the polymer and then to incorporate the ionic groups into the hard polymer structure. The main advantage of the first way is that the resulting resin is a true homogeneous mass except for limitations due to the influence of the crosslinking. But this synthetic way has lost in importance to the second for many of the newer practically-employed ion exchange resins.

During the first thirty-five years of synthetic organic ion exchange resins their preparation and manufacture gave rise to certain goals [25]:

lowering the costs,

improving the regeneration efficiency,

increasing the exchange capacity of the exchangers,

decreasing the tendency for the resins to be poisoned,

ameliorating the osmotic and mechanical properties,

eliminating the health risks for the personnel working in manufacturing facilities.

Another author [26], reviewing the needs in ion exchange technology, listed the following: resins with very high exchange capacities for smaller units, improved throughput and longer cycles, formally stable resins for the treatment of hot solutions, e. g., condensates, foul-resistant resins for the treatment of organic loaded waters, special resins with adsorption capacity for organic contaminants for the pre-treatment of raw waters, the treatment of waste water and the purification of various products as, for instance, in food technology, radiation-resistant resins for nuclear technologies, resins with reduced regenerant demand for cost saving in chemicals, thermally and electrically regenerable resins for saving in regeneration chemicals and their wastes, resins with high metal specificities for the application of ion exchange in hydrometallurgy, manufacture of purest chemicals and waste water treatment, powdered resins for continuous processes, and resins loaded with indicators for easy use in small-sized units for homes and laboratories. There is a close connection between new developments and new applications of ion exchange resins [27]. The transition from polycondensation products to addition polymerization resins resulted in the possibility of adjusting the structure of the matrix to the requirements of the practical requirement at hand and led to experimental work aiming for a change in the pore structure of the resins. Work on the improvement of the matrix was also done to increase the resistance of the resins to the influence of temperature. Preparative chemists were able to introduce into the matrix (which is usually polystyrene-divinylbenzene) a variety of ionogenic groups, resulting in a broad range of both cation and anion exchange resins. Synthetic ion exchange resins in bead form had come to predominate but other special sizes, shapes and forms were developed, partly on demand and partly just in order to find out whether they might be of special use. The industry was able to offer high quality and durable materials, but more specialized technologies and newer applications continuously brought new demands for exchange resins with improved properties.

1.2.2.1 Polymerization ion exchangers

To introduce polymerization ion exchangers it seems appropriate to revert to the basic work of D'Alelio and his relevant patents. In the patent on the production of synthetic polymeric compositions comprising sulfonated polymerizates of poly-(i. e. multi-)vinyl aryl compounds it is stated that polymeric divinylbenzene and copolymers of divinylbenzene were already known. What is claimed as new is that these polymeric compounds can be sulfonated. In fact the polymerization of styrene was first observed by Simon in 1839, by Blyth and Hofmann in 1845 and by Berthelot in 1866; in 1935, when the science of high polymers began to make great strides, Staudinger and Husemann [28] discovered the styrene-divinylbenzene co-polymers. One may conclude in retrospect that there is at least conceptually a direct path leading from the sulfonation of coals by Liebknecht to the preparation of sulfonated bakelite-type ion exchangers by Adams and Holmes and then on to the sulfonation of copolymerizates of styrene and divinylbenzene by D'Alelio. Such a close connection was obviously not apparent to the patent examiner or the patent would not have been granted. In the patent on the preparation of copolymers from acrylic acid or an *a*-substituted derivative thereof and a polymerizable compound containing an unreacted double bond and at least one other polymerizable grouping, usually an acrylic compound, but including divinylbenzene it was clearly stated that the invention is, in particular, related to insoluble copolymers containing carboxyl groups [29]. With this invention two foundation stones were laid: first, for weak acid cation exchangers of the polymerization carboxylic acid group type; second, for a group of ion exchangers manufactured according to the principle of building the ionogenic groups into the resin during the polymerization. The main feature of these polymers was that they were crosslinked and therefore insoluble and infusible. Long before this invention acrylic acid and its derivatives had been polymerized, for the first time, in fact, by Otto Röhm in 1901. After this beginning the development of acrylic polymers concentrated on the manufacture of plastic materials and binders for lacquers and paints.

An invention is not yet a material that can be used. Bauman and coworkers developed the strong acid sulfonic-type ion exchangers for production on an industrial scale in the present forms at the Dow Chemical Company [30] and Kunin and coworkers at the Rohm and Haas Company investigated in detail the properties of the styrenic anion exchangers [31] and then the acrylic resins of the carboxylic, weak acid type [32] showing their suitability for several applications.

The availability of polymerization ion exchangers based on styrene and divinyl-benzene as well as acrylics and divinylbenzene as matrices brought a tremendous expansion of ion exchange techniques world-wide, mainly through intense research on the part of the manufacturers of ion exchangers. These resins were in bead form, could be readily manufactured in any mesh size desired, were easily reproducible in plants and were very stable over the total pH range. In their external appearance most synthetic ion exchangers are in bead form. Granular ion exchange resin materials are now seldom in use. The ion exchanger beads are strictly standardized and have in technical grades a particle size range usually between 0.3 and 1.2 mm in diameter. For special applications fine and ultrafine resins are of interest. Powdered resins are a special type. With respect to internal structure, the most important improvement was with the synthesis of the modified polymer matrices, the so-called macroporous resins. These resins have fixed pores built into the polymeric matrix before they are converted to ion exchangers. If the pores of ion exchange resins could be formed with identical shapes and diameters such resins could well be called isoporous. But the primary distinction made in synthetic polymerization ion exchangers, which is accepted world-wide, is that between gel type or macroporous resins as is shown in Figure 1.3 on page 23.

For preparation and manufacture there are in principle two factors to be considered, for both polymerization and polycondensation ion exchangers. These are:

> the construction of the matrix as a basic polymer and its crosslinking by suitable crosslinkers;

and

> the incorporation of ionogenic groups into the matrix for the exchange of ions.

For the latter, two pathways are feasible, both being used in practice: one can

> start with monomers containing the ionogenic groups in the monomer either per se or latently;

or one can

> incorporate the ionogenic group subsequently into the matrix by suitable reactions with the polymer.

The number of conceivable units for building ion exchange resins only by polymerization is quite large. Besides the actual building blocks of the resins there are quite a number of additional chemicals needed for synthesis. All these represent main aspects of developments in the synthesis of addition polymerization matrices.

Styrene-divinylbenzene copolymer matrices, gel type. Most ion exchange resins are prepared from styrene-divinylbenzene crosslinked copolymers; as a result, this type of matrix is still the most important. Copolymers of styrene and divinylbenzene have many advantages as matrices for ion exchange resins. Styrene monomer is relatively cheap and abundantly available. The copolymers have excellent physical strength and are not easily subject to degradation by oxidation, hydrolysis, or elevated temperatures. The aromatic ring in the polymer can be reacted with reagents producing ion exchangers as a result of the incorporation of ionogenic groups.

Styrene is further chemically well-described and is produced in huge quantities as feedstock for the ordinary plastic material polystyrene. Its history is quite interesting, as it was first discovered in the essential oil Storax liquidus from which its trivial name was derived. After some initial errors its main property, that is, the polymerization to a polymeric product, was clearly realized and, after 1920, it was consequently used for the production of polystyrene. But there were also considerable problems in manufacturing styrene in a large-scale synthesis at acceptable costs. This problem was finally solved by Mark and Wulff in a process using the catalytic dehydrogenation of ethylbenzene. On a commercial scale, direct dehydrogenation of ethylbenzene still accounts for the manufacture of approximately 90% of the world capacity of styrene. To retard polymerization of styrene, inhibitors are needed during distillation, as well as for storage and shipping. The most commonly used inhibitor is now 4-*tert*-butylcatechol which is employed in quantities of 10 to 50 ppm, depending on storage time and temperature. Styrene used for the manufacture of copolymers with divinylbenzene can also contain hydroquinone as an inhibitor, and this must be removed before polymerization. This is usually performed in an extraction-distillation process, but can also be done on ion exchangers. Styrene purified by this method polymerizes approximately three times faster [33]. It was discovered that removal of stabilizers in the monomers before polymerization is only absolutely necessary if the concentration is above 10 ppm. Below this concentration the styrene with inhibitor has no determinable undesirable influence on the structure and properties of the copolymer with divinylbenzene made from it.

On the other hand, divinylbenzene, often abbreviated DVB, has an important influence on the structure and thus on the properties of the copolymers. Commercial technical-grade DVB is produced only at a few places in the world from a technical mixture of diethylbenzene by catalytic dehydrogenation at elevated temperatures, followed by rectification of the crude reaction product to obtain the technical-grade divinylbenzene. According to the technical leaflet of the largest producer, The Dow Chemical Company, it is offered in two grades, DVB-22 and DVB-55, the -22 and -55 designating the approximate content of divinylbenzene; it consists in the 22-grade of 17.1 wt % of m-divinylbenzene and 8.2 wt % of p-divinylbenzene, and in the 55-grade of 36.4% m-isomer and 18.6% p-isomer. These products are stabilized with 1000 ppm 4-*tert*-butylcatechol [34]. For laboratory purposes divinylbenzene is available from several suppliers as a synthesis-grade chemical specified by gas chromatography with a content of DVB isomers of approximately 60% and of ethylvinylbenzene isomers of approximately 35%. These materials are stabilized with either 0.2% 4-*tert*-butylpyrocatechol or, more rarely, with hydroquinone. Either of the divinylbenzene isomers can be of interest for the preparation of a copolymer with styrene. Their preparative isolation has therefore been investigated. Wiley and coworkers have used for this purpose methods based on preparative gas chromatography, obtaining a 99.9% pure m-isomer from a commercial divinylbenzene mixture, but they were not able to separate readily a 99.9% pure p-isomer [35]. It may therefore be of special interest that Schwachula and co-workers isolated p-divinylbenzene by a chemical method from technical DVB by bromination, isolation and dehalogenation, described the preparative gas chromatographic isolation of m-ethylstyrene and m-divinylbenzene from technical grade DVB and isolated p- and m-divinylbenzene from technical DVB by separation with copper(I) chloride [36]. All this separation and isolation work was done out of the need to study the influence of either of the isomers on the structures and properties of the copolymers with styrene and the ion exchangers made from them.

For the actual manufacture of styrene-divinylbenzene copolymers as intermediates for the production of both cation and anion exchangers only a general description can be given because of the proprietary know-how as mentioned above. In the standard suspension polymerization performed in stainless steel or glass-lined kettles, styrene is mixed with divinylbenzene in relative quantities depending on the crosslinking desired in the final resin. In the case of the most widely used cation exchanger the monomer mixture contains 8% by weight divinylbenzene. Contrary to the production of thermoplastic materials, for the ion exchanger copolymers the optimum size and a maximum uniformity of the beads must already be achieved during the polymerization step. This depends on the size of the polymerization kettle and the stirrer, the velocity of stirring, the temperature conditions, the ratio of the aqueous phase to the monomer mixture, the type and quantity of the suspension stabilizer, of the initiator and of the monomers. Thus, the monomer mixture containing from 0.2 to 1.0% initiator, usually benzoyl peroxide, is agitated in water containing a small amount of suspending agent to disperse the monomer liquid into droplets of the proper size and to prevent agglomeration during the transition from liquid to solid phase. The size of the droplets is already determined during the first phase of the polymerization. It is approximately proportional to the interfacial tension, among other factors. By adding dispersing agents, the interfacial tension, the viscosity of the aqueous phase and the formation of a tenacious phase boundary layer can be positively influenced. Three classes of dispersing agents have been used: a) minimally soluble inorganic substances, e. g., calcium phosphate, magnesium silicate, barium sulfate, talc, or bentonite, b) hydrophilic synthetic organic polymers, e. g., poly(vinyl alcohol), or poly(carbonic acids), and c) natural high polymers, e. g., gelatin, hydroxyethylcellulose, methylcellulose, or starch. As radical-forming catalysts also azodiisobutyronitrile, other organic peroxides and other radical-forming substances soluble in styrene are used. Optimum control of the suspension polymerization based on optimum design of the agitation system and temperature control are essential for optimal yield of the particle size distribution and the general properties of the polymer intermediates. Particles larger than about 1 mm and smaller than about 0.25 mm are generally separated out and may be used for special purposes. The polymerization temperature lies between about 70 °C and the reflux temperature for periods ranging from about 3 – 12 h. Difficulties arise since pearl polymerization is accompanied by emulsion polymerization, which must be avoided as much as possible. The removal of the stabilizers of the monomers is only necessary if their concentration is above 10 ppm. After polymerization, the polymer beads are washed to remove any adhering dispersing agent, dewatered, and then dried. They are then ready for sulfonation to make cation exchangers or chloromethylation and amination to make anion exchangers.

In the practical manufacture of styrene-divinylbenzene beads by suspension polymerization it has been quite difficult to produce stable particles larger than 1.5 mm in commercial quantities and particles smaller than 0.5 mm (used mainly for scientific investigations). But it is of interest to have available beads of much smaller size for detailed studies of the sulfonation rates and the ion exchange characteristics of the sulfonated polymers and for use in various separation applications. Experiments have shown that poly(vinyl alcohol) with a low degree of sulfonation yields stable suspensions of fine droplets of styrene-divinyl-benzene mixtures in water. Thus transparent, spherical polystyrene beads crosslinked with 2, 4 and 8 moles% of commercial, pure m-, and pure p-divinylbenzene with a narrow particle size distribution were obtained by using poly(vinyl sulfate) suspending agent. The yields of 125 – 250 μm sieve cut were 70 – 90% while – and this is noteworthy as an example of the influence of the individual components of a polymerization formulation – the use of carboxymethyl cellulose dispersion agent under similar conditions yielded 500 – 800 μm size beads with a 40 – 50% yield. The difficulty encountered in the preparation of bead copolymers of the styrene-8 mole %-m-divinylbenzene system was eliminated by using poly(vinyl sulfate) suspending agent [37]. It has been claimed that in the presence of poly(vinyl acetate) or poly(acrylate ester) as a protective colloid a homo-disperse bead polymer with 12 – 24 μm particle size can be obtained, which could then be sulfonated or chloromethylated and aminated as usual [38]. As far as larger particles are concerned a method and apparatus for preparing large-grained adsorbents and coarse-grained ion exchange resins has been described in the patent literature. In one case it seems that the type of stirrer is decisive for particle size ranges of 1.0 – 2.5 mm and larger than 2.5 mm respectively with 22% and 77% yields; decisive in the other case is the presence of particles of weakly crosslinked copolymer of styrene and divinylbenzene to give resins with particle size 1.0 – 4.0 mm. Such particle sizes are suitable for the preparation of ion exchangers for use as gas adsorbents permitting rapid flow of gases, and as catalysts [39].

There seems to have been less investigation of styrene-divinylbenzene copolymers as a plastic material since such polymer materials are of minor importance. Investigations not on beads but on cylinders, 1 cm long and 1 cm in diameter, were nevertheless performed to elucidate the elastic behavior of such polymers after swelling in toluene [40]. Similar studies on styrene-divinylbenzene copolymers in bead form for their physical characteristics and or on their swelling behaviors in toluene led to a discussion of the structural features of such polymers [41]. And in a longer treatise based on an outline of the general principles for the topological description of a physicochemical system as the continuous medium of multiphase and multicomponent system and a mathematical model based on diagrams for the study of the swelling of divinylbenzene-styrene copolymer particles as used for the manufacture of ion exchangers in a low molecular weight solvent graphs showing the dependence of the relaxation of stresses in the surface layer of such polymers with 2, 5 and 8% DVB content on time [42].

The influence of divinylbenzene on the properties of the copolymer beads and the properties of the ion exchangers made from them has been studied intensively. The work of Wiley and co-workers [43] first showed that differences in the properties of pure meta- and pure para-divinylbenzene crosslinked polystyrene and their sulfonated products can be correlated with probable network structural differences. The p-DVB copolymerizes more slowly than does the meta isomer to give a crosslinked copolymer which swells less, sulfonates less rapidly, and, when sulfonated, gives an ion exchanger which has lower selectivity. On the basis of copolymerization data it was concluded that the network crosslinked with p-DVB is tighter and less uniform than that of the m-DVB crosslinked network. With respect to the use of commercial divinylbenzene, a structural implication was rather clearly established from the kinetic data. First, the para isomer polymerizes and copolymerizes more slowly than does the meta isomer. It is a monomer distinctly different from styrene. During copolymerization styrene and probably para DVB prefer p-DVB, so that the latter monomer will be rapidly exhausted from the polymerizing mixture. This would mean that, when the crosslinking stage is reached, the crosslinkages will be formed in bunches and probably some of the second vinyl groups are buried and thus unavailable for further crosslinkings. The result is predetermined heterogeneity. The meta isomer, however, presents a picture of more uniformly distributed divinyl units with, ultimately, more evenly spaced crosslinkages and a more homogeneous network. But a homogeneous network in styrene-divinylbenzene copolymers cannot, in general, be expected [43]. On the other hand, it must also be concluded that structural variations in the network can be produced by selected synthetic procedures, and these can be related to controlled variations in properties of the network (Wiley). Some time after Wiley Schwachula and coworkers [44] published their results in a large number of papers. Firstly, they also showed that there is a possibility for the structural determination of radically polymerized styrene and divinylbenzene copolymers by means of the system's kinetic data. The necessary kinetic data are gained from the copolymerization equation for binary systems of Mayo and Lewis, which contains the known r-values. This equation cannot be used, however, for the copolymerization of styrene and commercial divinylbenzene because the latter is too complex a mixture – as a result the polymerization reaction is a multicomponent copolymerization system. But for the copolymerization of styrene with m- or p-divinylbenzene separately this equation can be employed, resulting in a diagram as shown in Figure 1.48. This diagram shows that in the beginning of the polymerization more DVB is incorporated into the network

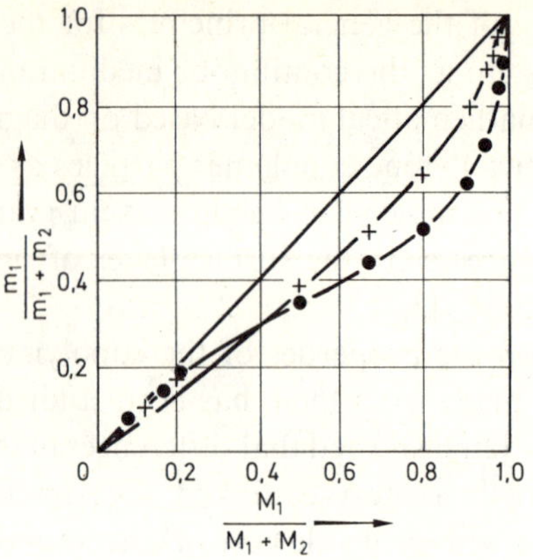

Figure 1.48 Copolymerization diagram of styrene and DVB isomers calculated with the *r*-values of Wiley. + Styrene (M_1) and *m*-DVB (M_2), $r_{12} = 0.65$, $r_{21} = 0.60$. ● Styrene (M_1) and *p*-DVB (M_2), $r_{12} = 0.14$, $r_{21} = 0.50$. M = monomers in mol, m = polymers in mol.

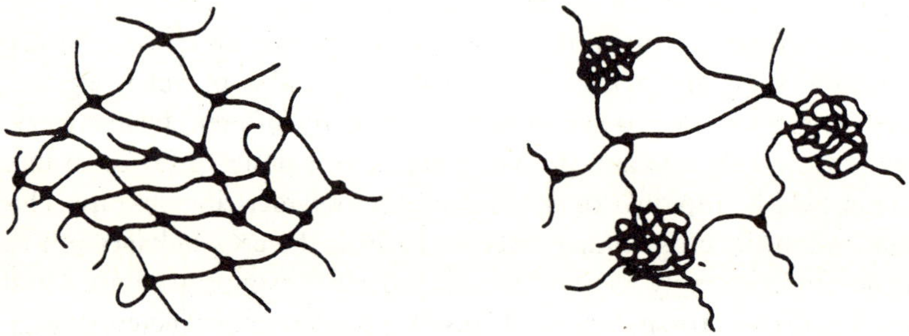

Figure 1.49 Schematic representation of networks of different structure. Left homogeneous and right inhomogeneous network.

than is in the initial monomer mixture, because both diagrams deviate considerably from the ideal straight line. Consequently a network is formed that has a microgel structure within its macrogel structure with inhomogeneities as shown in Figure 1.49, in comparison with an idealized relatively homogeneous strucutre. The properties of sulfonic cation exchangers containing copolymers of styrene and divinylbenzene isomers were comparatively studied. Swelling and granule degradation were found to be influenced by the divinylbenzene isomer used, i. e., m, p or (m + p)-DVB, in corresponding sulfonated styrene-divinylbenzene copolymers. The least swelling and granule degradation during sulfonation and during prolonged exposure to alkylphenols at 160 °C was observed in resins made from m-divinylbenzene. These samples also had the least decrease in static exchange capacity during the exposure to alkylphenols [45]. Taking all these results together it can be said that a great deal of knowledge has been accumulated relative to the influence of divinylbenzene [46] on the properties of the copolymer beads with styrene.

Styrene-divinylbenzene copolymer matrices, macroporous. The second stormy development in the field of styrene-divinylbenzene copolymers as matrices for ion exchangers began when researchers at at least three independent locations, accomplished, at about the same time, the polymerization in such a way as to yield a product with a (macro)porous structure. There was a need for such materials because it had not been possible to meet the requirements of the exchange of larger ions with the gel type products, nor were these materials resistant enough to fouling by organic matter present in natural surface water. In the first patent – using the terminology of today – on macroporous anion exchangers with high exchange rates and especially suitable for the exchange of higher molecular weight acids, materials were described which could be obtained by the polymerization of vinylaromatic compounds with a crosslinking agent such as divinylbenzene. What was new was that the suspension polymerization was carried out in the presence of at least 20% of a compound which dissolves the monomers but not the polymer. As such substances hydrocarbons, alcohols and nitrocompounds were mentioned. The polymers obtained were discribed – after chloromethylation and amination – as being opaque and porous anion exchangers with a spongy structure [47]. In the second patent, applied for only eight months after the first, it is described how a polymerized mass can be obtained consisting of an aggregation of microbeads defined by a network of microscopic channels extending through the mass. The process itself is performed, e. g., with styrene and divinylbenzene but in the presence of a liquid solvent or "precipitant" in which the monomer mixture is dissolved but which has so little solvating action on the crosslinked copolymer that under the conditions of the suspension polymerization phase separation of the product copolymer occurs. The novel feature of the process was described as being the selection of the solvent or "precipitant" that does not swell the resulting crosslinked polymer and does not greatly exceed the critical concentration. This precipitant should also be miscible in the aqueous medium, but its solubility should not exceed 20% in water. tert-Amylalcohol was used as a solvent in the example and the material obtained was described as white, opaque, spherical, porous particles. These polymerized pearls were converted both to cation and to anion exchangers. The cation exchangers were described as having low density, and as being macroreticular with approx. 17.5% voids. Both the ion exchange resins thus prepared have less volume change when converted, good exchange rates and efficiencies, and high resistance to oxidation and osmotic shock [48]. The third investigator had developed a so-called NONSOL process to produce macropores in copolymers and also used a solvent, e. g. heptane, which was miscible with the monomers but, did not swell the polymer, which thus precipitated out of solution as it was formed. The polymer thus had a "sintered" structure, rather similar to that of a fritted glass filter on the molecular scale; it was therefore macroporous, filled with air in the dry hydrocarbon polymer, and with water in the final ion exchange resin. For internal reasons this NONSOL technique was not put into practice and later the corresponding patent was abandoned, because the above-mentioned virtually identical first patent had been published eleven months before [49].

Although the terms "spongy", "sintered" and "macroreticular" can be found in the literature (especially in the early patents) as descriptive terms for macroporous styrene-divinylbenzene copolymer matrices (with "spongy" and "sintered" being more appropriate, since "macroreticular" meens "net-like", which is basically a two-dimensional concept) the term "macroporous" has become generally accepted. Styrene-divinylbenzene-based ion exchangers are classed throughout the world into gel type or macroporous.

Styrene-divinylbenzene copolymers used for the manufacture of ion exchangers which were prepared according to the so-called SOL process were called macroporous from the very beginning. The structures of these SOL resins had been disentangled and stretched by including in the monomer mixture a certain proportion of an organic solvent, such as toluene, which was completely miscible with the monomer, and which swelled the final polymer. Polymerization thus occurred in solution, so that as the chains formed, they were further apart than in a normal resin. The solvent was then removed by distillation from the solid produced. Because of lower chain-entanglement the resultant materials have much larger pore sizes for the same divinylbenzene content than normal suspension polymerized styrene-divinylbenzene polymers. Depending on the proportion of solvent used it is possible to obtain harder and tougher polymer beads which, after removal of solvent by distillation, contain air-filled holes of around 1 000 Å diameter. These were termed macropores. On a small scale such polystyrene beads, which are also highly crosslinked, with a large pore volume are produced from styrene and divinylbenzene with polyvinylpyrrolidone as a stabilizer, toluene as solvent, and AIBN as initiator, in particle sizes of 300 to 100 mesh. They are useful in gel permeation chromatography [50]. The first work on the preparation of macroporous styrene-divinylbenzene copolymers was followed by the development of quite a number of modified preparative methods in order to obtain suitable materials for chromatographic applications and for the preparation of ion exchangers [51]. For example, macroporous crosslinked copolymers convertible to ion exchange resins having enhanced mechanical and chemical properties and porosity were prepared by free radical polymerization of a mixture containing olefinic monomers and a precipitant immiscible with the monomers which was solubilized by the addition of a micelle-forming compound that was soluble with the monomer mixture [52]. Porous, crosslinked polymers having a high surface area, useful for the manufacture of ion exchangers, were prepared by copolymerization of a divinyl compound with a monovinyl compound in the presence of a homopolymer – for example a low molecular weight polystyrene – and a solvent that dissolved the homopolymer and swelled the copolymer (the homopolymer was subsequently extracted) [53]. Macroporous structures can also form when the divinylbenzene-styrene polymerization takes place in the presence of an inert inorganic filler such as pulverized calcium carbonate, which is later extracted from the copolymer beads. Polymers prepared in the presence of a solvent which solvates the polymer chains and participates in the copolymerization process by playing the role of a telogen (thereby shortening the mean length of the polymer chains) are classified as telogenated, macroporous matrices. The changes in such resins have been investigated, for instance for the case where the copolymerization was carried out in the presence of the telogenizing agent, CCl_4 [54]. But prior to all these investigations The Dow Chemical Company was granted a patent describing ion exchange resin particles of good porosity and strength obtained by dissolving the mixtures of styrene and divinylbenzene in, e. g., toluene; higher divinylbenzene content gives less volume change than known ion exchange resins and use of the inert solvent gives the desired high porosity [55]. Several other contributions to the study of the formation of macroporous copolymers emphasized that those inert solvents are attractive as precipitants in which the monomer mixture dissolves completely to yield a single phase, but which cannot dissolve or swell the resulting copolymer [56]. After in-depth studies of the preparation and properties of porous copolymers of styrene and divinylbenzene, the effects of comonomer ratio, and of the amounts of divinylbenzene and neutral solvent added to the modifier mixture on the porosity of styrene-divinylbenzene copolymers were examined with the goal of determining the parameters for preparing double porous styrene-divinylbenzene copolymers [57].

Twenty years after the invention of macroporous copolymers and the ion exchangers made from them it has been shown that these resins have pores of considerably larger size than those of the conventional gel type. In general, pore diameters go up to several hundred nanometers and the resin surface area may reach 500 m^2/g or higher. The resins are also in bead form and exhibit little volume change in solvents if prepared with a sufficiently high degree of crosslinkage (generally, considerably higher than with the gel type). The pores may be varied tremendously in both size and uniformity, with pore diameters up to 0.1 μm or larger. Because of the porosity much larger molecules may enter the bulk resin structure. Further, the apparent oxidation stability is improved because these materials are initially so highly crosslinked that the effects of a given amount of chain scisson are less apparent. Macroporous ion exchange polymers are now used routinely world-wide but poorer regeneration efficiencies, lower capacities, and higher regeneration costs are the price that has to be paid for the use of macroporous resins. Certain misconceptions, however, have to be clarified regarding their structure. It is only an idealization when macroporous ion exchangers in comparison with gel type resins are depicted as consisting of polymer chains with large pores. In reality, and from the conditions under which they are prepared, it must be concluded that in essence the macroporous ion exchange polymers have a gelular as well as a macroporous pore structure. Gelular structure refers then to the distances between the chains and crosslinks of the swollen gel structure, and macroporous structure to the pores that are not part of the actual chemical structure. The latter is the case because during manufacture, as the polymerization proceeds, phase separation occurs, leading ultimately to polymer beads containing micro-

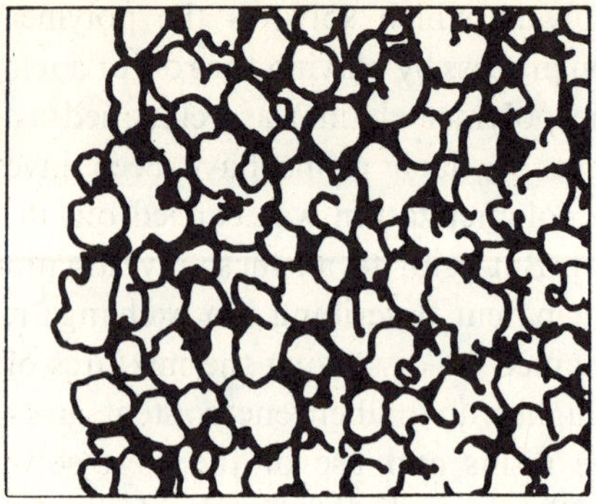

Figure 1.50 Diagrammatic structure of macroreticular ion exchange resin bead.

spheres cemented or joined together at points of contact which, in turn, form interconnecting pores. In such structures both the pores and the polymers are present as continuous phases. The average size of the pores is determined directly by the size of the microspheres and indirectly by the types and amounts of monomers and solvents used during the polymerization. Pore sizes ranging from about 50 Å to about 1 000 000 Å have been achieved. It must be mentioned that the term macroreticular was applied to these materials to distinguish them from gel type polymers made without polymer phase separations, whose molecular-size pores are referred to as gelular pores. The microspherical structure of the macroporous/macroreticular ion exchange resins is depicted in Figure 1.50; it has been verified by electronmicroscopic examinations. All of what has been described above with regard to marcoporous styrene-divinylbenzene copolymer matrices relates to ion exchange polymers obtained by a phase separation technique, which utilizes the addition of a solvent in which the monomers used for forming the polymer are soluble, but in which the polymer formed is insoluble. As it forms the polymer precipitates, building pores around the solvent. When the solvent is expelled, pores attaining the large size mentioned are obtained [58]. The second technique used for the preparation of porous ion exchangers consists in the incorporation of a linear polymer into the polymer bead during polymerization. Not disclosed in a first, early paper [59] but then, in a later one, on sulfonation or amination, it was determined that the linear polymer becomes a water soluble polyelectrolyte which is dissolved out, resulting in pores [60]. It should be mentioned that some call this type of resin second-generation macroporous.

Methylstyrene-divinylbenzene copolymer matrices. Anion exchange resins have also been prepared based on α-methylstyrene instead of styrene. The polyamines obtained are prepared by copolymerization of α-methylstyrene with divinylbenzene (catalyst azodiisobutyronitrile), chloromethylation (catalyst $ZnCl_2$) and corresponding ammonolysis. The developers feel that the raw material basis for the manufacture of ion exchangers has been broadened [61].

Fluorinated styrene-divinylbenzene and other fluorinated matrices. In the search to make ion exchange matrices both heat-resistant and oxidatively stable to the effects of peroxides and other active oxygen solutions poly(α,β,β-trifluoro styrene) has been found to be a suitable skeleton. Analogously to the styrene-divinylbenzene copolymer matrices, it was possible to prepare copolymers of α,β,β-trifluoro styrene with both styrene and with chlorotrifluoroethylene using an emulsion polymerization technique [62]. But the preparation of sulfonated ion exchangers from crosslinked poly(α,β,β-trifluoro styrene) was found to be much more difficult than the sulfonation of the corresponding styrene-divinylbenzene copolymers. This is because poly(α,β,β-trifluoro styrene) fails to undergo the analogous reaction in the temperature range 0 – 40 °C, and fails to react even with 30% oleum under any conditions. However, a very low degree of substituted product is obtained in the temperature range of 40 – 65 °C. However, with only an increase in the concentration of sulfonating reagent, $ClSO_3H$, polyelectrolytes of α,β,β-trifluoro styrene can be obtained whose final degree of sulfonation is very dependent on the temperature of sulfonation. From poly(α,β,β-trifluoro styrene) at 30 °C a crosslinked polysulfonic acid can be obtained with sulfone bridges as crosslinks and with sulfonic acid groups in meta substitution to the perfluorovinyl groups of the aromatic rings, as depicted in the following formular diagram

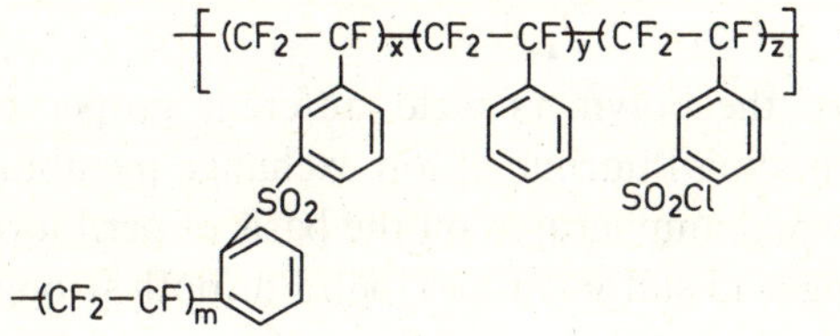

This finding gives a plausible explanation for the difficulty in achieving the ease of reaction observed with the sulfonation of polystyrene. Thus a new generation of high-capacity, oxidation-resistant ion exchange resins and membranes has been created. The oxidative stability of these perfluoroalkyl aromatic sulfonic acid polyelectrolytes was described with comparisons being drawn to their polystyrenesulfonic acid analogs and the difference in their oxidation-depolymerization stabilities in terms of benzylic carbon substituents [63]. In similar but later investigations α,β,β-trifluoro styrene-divinylbenzene copolymer was prepared by suspension polymerization of the monomers, with an increase in the yield of the polymer with an increase in the DVB concentration, reaching a maximum of 68% at approximately 4%. Analogously, an α,β,β-trifluoro styrene-styrene-divinylbenzene copolymer was prepared. Sulfonation of both copolymers gave cation exchange resins whose exchange capacity increased with the styrene-trifluoro styrene ratio [64]. Copolymerization of trifluoro styrene with other vinyl monomers, for instance *N*-vinyl-pyrrolidone, CHF = CFCN, CHF = $CFCOOCH_3$, has also been investigated. An increase in the number of fluorine atoms in the macromolecule was achieved by grafting poly(trifluoro styrene) onto fluoropolymer films containing copolymers of vinylidene fluoride and tetrafluoroethylene with perfluoropropylene. The sulfonation of polymers thus obtained results in membranes with improved mechanical and electrochemical properties.

Using acrylic acid or its derivatives fluorinated ion exchange materials containing carboxylic acid groups can also be obtained. For this purpose copolymerization or terpolymerization of the methyl or ethyl esters of trifluoroacrylic acid is carried out with ethylene or propylene. The preparation of fluorinated ion exchange resins from trifluoroacrylic acid-tetrafluoroethylene-ethylene terpolymer was also reported.

Perfluorinated cation exchanger polymers have become known under the tradename Nafion (E.I. du Pont de Nemours and Co., Wilmington, Del., USA). They are mainly produced as membranes comprised of copolymers of tetrafluoroethylene with a vinylsulfonylfluoride. The structure – following conversion of the sulfonylfluoride group into the corresponding sulfonic acid – is depicted in the following formular diagram

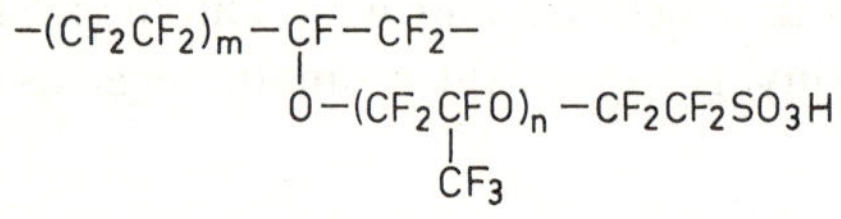

m = 5 to 12

Variations in the composition of the polymers yield different properties of the materials obtained [65]. Corresponding fluorinated ion exchange membranes containing carboxylic acid groups were manufactured on the basis of perfluorocarboxylic acid soon after the first strong acid sulfonic types [66]. For both strongly acidic as well as weakly acidic cation exchange polymers the most common monomers are sulfonic and carboxylic derivatives of vinyl ethers, with the general formula $F_2C{=}CFO[CF_2CF(CF_3)O]_m(CF_2)_n{-}X$ where X is the ion exchange group or its precursor. To manufacture membrane materials with sulfonic acid or carboxylic acid groups the following vinyl ethers are generally used: perfluoro(3,6-dioxa-4-methyl-7-octene)sulfonyl fluoride, perfluoro(3-oxa-4-pentene)sulfonyl fluoride, perfluoro(3-vinyloxypropane)-1-carboxylic acid or perfluoro(3-vinyloxyisopropoxypropane)-1-carboxylic acid. By copolymerizing these functional monomers with fluorinated olefins copolymers are formed constituting the basis for the preparation of fluorine-containing exchange materials having sulfonic or carboxylic acid groups or both.

Perfluorinated cation exchange membrane materials have been studied extensively [67]. Most of the work on fluorinated ion exchange materials concerns the manufacture mainly of cation exchange membranes. But further developments in the synthesis of ion exchange resins having a fluorinated matrix can be expected because such materials would result in new areas of application of ion exchange resins in fields such as high temperature catalysis, nuclear processes, and ion exchange in strongly acidic or oxidative media.

Brominated or so-called weighted styrene-divinylbenzene resins. The need for more suitable ion exchangers for continuous systems gave rise to a method for increasing the density by halogen substitution of ion exchange resins on the basis of crosslinked polystyrene [68]. The first continuous ion exchange systems were operated with conventional ion exchange resins. In view of the more exacting requirements to be met in continuous cycles, new ion exchangers with increased stability were developed. One development for new continuous systems was the synthesis of magnetic ion exchange resins. Materials with increased density are useful for continuous fluidized bed systems such as Fluicon, where the resin must find its path downwards against the upflowing liquids to be treated without the help of any additional external force. In a first synthesis bromostyrene was used. It was successful, yielding anion exchangers with a density of 1.20 g/ml as compared to 1.05 and 1.09 g/ml for conventional resins. A second way of synthesis is to brominate conventional matrices either before or preferably after the incorporation of the ion exchange groups or their precursors. A more or less optimum partial bromination of the aromatic ring for heavy resins was then investigated and commercialized so that a full range of high density products, strong base anion exchangers, weak base anion exchangers, and adsorbents is now available in industrial quantities [69]. This new generation of high-density ion exchangers and adsorbents results in significantly improved performance of actual processes such as fluidized and stratified beds. These new materials may also generate new applications.

Styrene-divinylbenzene and comonomer matrices. Experiments to obtain desired properties in ion exchange resins were based on using a comonomer in addition to the basic styrene and divinylbenzene as copolymerization constituents. In fact resin matrices based on styrene and technical grade divinylbenzene are actually multicopolymers because of the latter's being a complicated mixture. A Japanese patent describes porous styrene-divinylbenzene-ethylvinylbenzene copolymers as ion exchange resins containing sulfonic acid hydrozide groups. Styrene-divinylbenzene-maleic anhydride terpolymers containing variable amounts of DVB were used as matrices for biionic cation exchangers, which had a good ion exchange capacity [70]. Macroporous vinyl copolymers as ion exchanger matrices with improved flexibility can be manufactured out of styrene, divinylbenzene and (optionally) an acrylic comonomer, and hard crosslinked distinct copolymer globules are obtained from styrene, divinylbenzene and poly(dialkyldimethylammonium chloride) as thermosetting ion exchange resins after sulfonation [71]. Earlier systematic results of investigations mainly on acrylonitrile as the third component indicated that the stability of styrene-divinylbenzene copolymers can be increased substantially by the addition of 1% acrylonitrile, with a 5% addition increasing the stability even more. The equilibrium swelling of such copolymers was also studied, as an effect of the addition of acrylonitrile and ethylacrylate. It showed an increase, but this was dependent on the type of initiator, azobisisobutyronitrile or dilauroylperoxide, and its concentration. Azobisisobutyronitrile gave a linear swelling relation and dilauroylperoxide a curve with a minimum. Evaluation of the outer and inner structure swelling showed that the polymer structure was heterogeneous in the axial as well as radial direction, owing to uneven temperature during the polymerization reaction. Impurities in the divinylbenzene could also have had an effect. Therefore, the influence of p-divinylbenzene on the equilibrium swelling of styrene-divinylbenzene copolymers with acrylonitrile as third component, in contrast to technical divinylbenzene as the main crosslinker, was investigated. It is of interest to note that in this work the explanation of why crosslinking with p-divinylbenzene gave low equilibrium swelling values was that it is due to the presence of a binary system. It is to be noted that a ternary system is present when styrene is crosslinked with technical grade divinylbenzene, because of its content of ethylstyrene. The addition of acrylonitrile and the use of technical grade divinylbenzene resulted in improved homogeneity of the copolymer. This can be explained on the basis of the copolymerization parameters, which indicate a preferential growth with p-divinylbenzene resulting in heterogeneity. This was not the case with m-divinylbenzene. Dielectric measurements of styrene-divinylbenzene copolymers with polar additive were then carried out showing a linear dependence between the molar concentration of acrylonitrile as additive and the loss tangent. The loss tangent was also measured for methacrylate, 4-vinylpyridine, acrylate, and trichloroethylene. No dependence was found between the mechanical strength improvement and the dielectric magnitude [72]. To increase the particle strength the addition of a third component without an ion exchange active group (among others acrylonitrile especially) has also been claimed in a patent. On the other hand another patent specified that ion exchange resins with excellent mechanical strength are prepared by copolymerizing, for instance, styrene, technical grade divinylbenzene and ethylene glycol dimethacrylate, giving completely crack-free beads with an average compressive strength of 1358 g/bead vs. 63 g/bead in resins without the comonomer [72]. Such resins are not called terpolymers because of the above-mentioned composition of technical grade divinylbenzene.

Styrene-other type and size crosslinker copolymer matrices. Copolymerization and, as a result, crosslinking of styrene can be achieved with several multifunctional monomers other than divinylbenzene. This is another class of matrices which differ significantly from those with DVB. They are mainly long chains with the distance between the two vinyl groups exceeding appreciably the full length of the divinylbenzene molecule. In some cases the ion exchange resins with matrices manufactured in this way are called macronet polymers. Then they are highly permeable to large ions and also of theoretical interest.

The first step in the direction of matrices with stryene and crosslinking agents other than divinylbenzene, aiming for improvements in the production of ion exchange resins, was made by Kressman and Akeroyd as reported in a British patent specification in 1951. Divinyl aryl compounds, including the simplest of them, i. e., divinylbenzene, were very difficult and expensive to make and not available on the market. But certain aliphatic compounds that can be produced much more easily than divinylbenzene were envisaged to be used in place of it as crosslinking agents in the polymerization of styrene. The aliphatic compounds to be used were the diesters and vinyl esters of acrylic acid and alpha-substituted derivatives of acrylic acid, the divinyl esters of dibasic acids and of divinyl ether. Examples for these three classes of esters were, respectively, ethylene dimethacrylate, vinyl methacrylate and divinyl oxalate, with the preferred aliphatic compound being ethylene dimethacrylate. A characteristic feature of these esters is that they have two double bonds of approximately equal reactivity, which is further of the same order as that of divinylbenzene [73].

Multifunctional acrylate and methacrylate esters have become more readily available since such compounds are used as reactive diluents in UV-curable polymers for coatings and printing inks. Their use and effectiveness as crosslinkers in styrene copolymers for matrices to be employed for the preparation of ion exchangers has also been investigated. The purpose of these syntheses was to obtain a more uniform crosslinking and an improved elasticity. After a comparison of published figures for divinylbenzene and diisopropenylbenzene copolymers with up to then unreported figures for divinyl sulfone- and divinyl ketone-styrene copolymers, the following aliphatic esters were investigated:

Name	Code	Functionality
ethylene glycol dimethacrylate	EDMA	4
1,3-butylene glycol dimethacrylate	BGDMA	4
1,3-butylene glycol diacrylate	BGDA	4
triethylene glycol dimethacrylate	TEGDMA	4
neopentyl glycol dimethacrylate	NPGDMA	4
1,6-hexanediol dimethacrylate	HDDMA	4
1,6-hexanediol diacrylate	HDDA	4
trimethylolpropane trimethacrylate	TMPTMA	6
trimethylolpropane triacrylate	TMPTA	6
pentaerythritol tetramethacrylate	PETMA	8
pentaerythritol tetraacrylate	PETA	6.4
dipentaerythritol monohydroxy pentaacrylate	DPEHPA	8.9

Toluene swellings of these copolymers were measured and it was found that the most appropriate means of describing the crosslinking is the molar percent of excess vinyl groups in the monomer mixture. These groups are those not involved in the original incorporation into the growing chain. Based on this concept the results for crosslinkers of functionality $f = 4$ to $f = 8$ can be expressed in a single curve. If on the other hand the results are plotted as log swelling, i. e., toluene regain, vs. log of molar percent crosslinker, the most usual curve is concave to the intersection of the axes, as one has, e. g., for p-divinylbenzene, ethylene glycol dimethacrylate, trimethylol propane trimethacrylate, etc. Certain monomers, such as divinylsulfone, diisopropenylbenzene, divinylketone, show anomalous curves convex to the intersection of the axes, while the mixture of m- and p-divinylbenzene commercially available is an intermediate case [74]. Copolymers of styrene and dimethacrylic esters of some phenols, i. e., the dimethacrylates of 1,4-dihydroxybenzene, 2,2-bis(4-hydroxy-phenyl)propane and bis(4-hydroxyphenyl)diphenylmethane, were prepared and converted to strongly basic anion exchangers with chemical and thermal properties not inferior to those of styrene-divinylbenzene exchange resins [75]. The synthesis of styrene with *N,N'*-alkylenedimethacrylamides, *N,N'*-arylenedimethacrylamide and especially with hexamethylene-dimethacrylamide as copolymer (the latter for the sorption of uranyl ions) resulted in both gel type and macroporous resins for various ion exchangers [76].

Permeability or better enhanced permeability was the goal when the network of ion exchange matrices was altered in order to obtain ion exchangers with larger interchain spaces as a result of the use of crosslinking agents with different chain lengths. One group of researchers also used *N,N'*-alkylene dimethacrylamides to obtain sulfonic acid and carboxylic acid cation exchangers with one type having the following diagrammatic structure

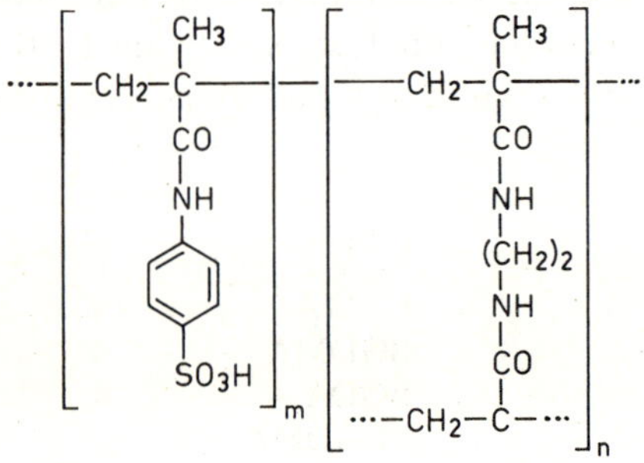

As can be seen, the permeability of such resins may be altered both by changing the amount of the crosslinking agent and by changing the length of its chain [77]. Enhanced homogeneity and, thus, better stability is another reason for changing from divinylbenzene to another crosslinking agent. A study of the preparation of strong acid cation exchangers based on copolymers of styrene and bis(4-vinyl-phenyl)methane satisfied this expectation, confirming the hypothesis that the main reason for influence on the stability of polymerization ion exchange resins resides in the ratio of the reactivity of the monomers, since exchangers based on a copolymer having an essentially homogeneous distribution of crosslinks posess the better mechanical stabilities [78]. New exchangers with improved osmotic stability, mechanical strength and sorption characteristics for high molecular weight organic ions were prepared from the divinyl compounds $H_2C{=}CCH_3RCCH_3{=}CH_2$ with R methylenedi-p-phenylene, ethylenedi-p-phenylene, or thiodi-p-phenylene [79]. The thermal degradation and kinetics of strong acid cation exchangers consisting of styrene and diisopropenyloxydiphenylene was also studied [80]. And ion exchangers with bis(1,3-butadienyl-1-yl)sulfide as crosslinking agent yielded because of its high purity, a better quality of exchangers than those prepared by crosslinking with divinylbenzene [81].

If it is possible to prepare ion exchangers with equally-sized pores, the term "isoporous ion exchanger" can be justly used as a third classification. This is a problem in the synthesis of copolymer matrices. It was shown as early as 1955 that styrene polymerizates can subsequently or during the sulfonation with chlorosulfonic acid be crosslinked by polymer aldehydes such as paraldehyde, metaldehyde, polyglyoxal or chloralhydrate. In retrospect it seems likely that by this synthetic pathway isoporous products could be obtained without the entanglement phenomenon of suspension polymerizates [82]. Similarly so-called isoporous macroreticular (which seems contradictory) ion exchangers were prepared from polystyrene by crosslinking with p-xylylene dichloride [83]. To emphasize that the properties of resins can be varied as desired by choosing a crosslinking agent with the proper chain length polystyrene was crosslinked with adipoyl chloride, succinoyl chloride, phosgene, biphenyl-4,4′-disulfonyl chloride, or dichloromaleoyl chloride to prepare resins which can be conventionally sulfonated, chloromethylated, or aminated to prepare ion exchange resins. It was claimed that the resins have a more homogeneous crosslinked structure, compared with prior-art resins crosslinked with divinylbenzene, and have a good ion exchange capacity and exchange rate. A greater porosity is claimed and may under special circumstances be obtained by using diisopropenyl benzene as crosslinker instead of divinylbenzene. Isopropenyl styrene was also used as the diolefin in the preparation of ion exchange resins but the aim here was to increase the mechanical strength of the resins.

Substituted styrene-divinylbenzene and other copolymer matrices. Of the substituted styrene derivatives vinyl benzyl chloride was most often investigated as the component to be copolymerized with divinylbenzene for the preparation of copolymers suitable for the manufacture of ion exchangers. In an earlier attempt to synthesize anion and chelating exchangers containing vinyl benzyl moieties, resins were prepared from poly[dimethyl(vinyl benzyl)sulfonium chloride], 1% by weight crosslinked with divinylbenzene, by heating with tris(hydroxymethyl)amino methane; some time later the same sulfonium chloride was reacted with a Schiff base prepared from paraformaldehyde and diethylenetriamine yielding a resin with a specific capacity for copper [84]. Macroreticular vinyl benzyl chloride polymers were obtained by copolymerizing vinyl benzyl chloride with divinylbenzene; they were further treated with either trimethylamine, to give a strongly basic anion exchanger or with triethylamine, pyridine or dimethylmonoethanolamine [85]. The problem of such types of ion exchangers seems to be the reduction of the chloride content and therefore the potential for corrosion. Anion exchangers based on copolymers of vinyl benzyl chloride with m-diisopropenylbenzene were described and their ion exchange properties were determined after amination with trimethylamine; similarly, the synthesis of vinyl benzyl chloride-divinylbenzene-acrylonitrile copolymers yielded – with pyridine or triethylamine – anionic and amphoteric resins having good physicomechanic and chemical properties [86]. The preparation of vinyl benzyl bromide-tertiary amine polymers and examination of their properties was also carried out as a spontaneous polymerization giving with 2-methyl-5-vinylpyridine or 2-vinylpyridine copolymers having a three-dimensional structure and containing a quaternary pyridine group. Such synthetic ion exchange resins had high mechanical strength and chemical resistance [87]. Polymers in bead form containing vinyl benzyl alcohol units prepared either from vinyl benzyl alcohol or by partial hydrolysis of vinyl benzyl chloride were thermally crosslinked to form insoluble matrices for strong base anion exchangers after having been heated with trimethylamine [88]. Further, cation exchangers were prepared by sulfonation of α-cyanostyrene-divinylbenzene copolymer or by hydrolysis of α-cyanostyrene-styrene-divinylbenzene copolymer [89].

Modified ion exchangers can also be prepared by polymer analogous reactions as sulfomethylated resins starting from macroporous styrene-divinylbenzene copolymers. In customary strong acid cation exchangers the SO_3H-group is fixed to the aromatic ring. It seemed to be of interest to investigate the extent an increased thermo-stability can be obtained when a methylene group is set in between the aromatic ring and the SO_3H-group. Sulfomethylation can be achieved by reaction of a chloromethylated resin with dimethyl sulfide and sodium sulfonate or, alternatively, by oxidation of polymer-bound thiol groups. Both methods give high conversions, as shown by infrared spectra and titration of the sulfonic acid groups [90].

Acenaphthene-divinylbenzene copolymer matrix. Searching for other polymer frameworks which might be suitable for ion exchange resins, acenaphthene was studied as a replacement for styrene and was polycondensed as well as polymerized with divinylbenzene. A polyacenaphthene-divinylbenzene sulfonic acid ion exchange resin with a capacity of 5,68 mequ/g was prepared. The resin was characterized by determination of solubility, swelling and selectivity coefficients with the cation pairs Li−H, Na−H, K−H, Mg−K, Ca−K, and UO_2−K. It was found that in spite of considerable water uptake the 10% crosslinked resin was more selective than a polystyrene-divinylbenzene sulfonic acid resin [91]. It was also found that ion exchangers prepared by sulfonating acenaphthene-divinylbenzene copolymers have higher radiation resistance and lower heat resistance [92], but this type of ion exchanger was then no longer investigated.

***N*-vinyl carbazole based matrices.** Poly(*N*-vinyl carbazole) is a high temperature-resistant thermoplastic material, as is its copolymer with styrene. Investigations showed that poly(*N*-vinyl carbazole) can be modified to give a cation exchange resin thermally stable at $\leq 250\,°C$ with a maximum capacity of 4.5 mequ/g, by sulfonation with 98.8% sulfuric acid at 30 °C [93]. A similar cation exchanger and also anion exchangers can be obtained from granular copolymerization products of vinylcarbazole, styrene and divinylbenzene [94]. Treating poly(*N*-vinyl carbazole) with PCl_3 at 76 °C or PBr_3 at 140 °C as phosphorylating agents and $AlCl_3$ as Friedel-Crafts catalyst gave a phosphonic acid cation exchanger with an ion exchange capacity of 3.2 mequ/g. The pH-metric titration curve showed two distinct dissociation stages and the differential thermal analysis curve showed no decomposition at ≤300 °C and a weight loss of 37.5% at 550 °C [95].

Pyridinium resins. Among other ion exchangers produced but of limited application were those based on the pyridinium resins. At first a new commercial-type ion exchange resin using pyridinium was introduced for uranium recovery, described only containing some pyridinium groups. This disclosure had been preceded however by a patent specification claiming the preparation of anion exchange resins from crosslinked copolymers of, e. g., 2-methyl-5-vinylpyridine with divinylbenzene by halogenation and subsequent amination [96]. Then Saldaze and coworkers investigated quite thoroughly vinylpyridine-based anion exchange resins as copolymers of divinylbenzene and 2-vinylpyridine, 4-vinylpyridine, 2-methyl-5-vinylpyridine, 4-methyl-2-vinylpyridine, 6-methyl-2-vinylpyridine, and 5-ethyl-2-vinylpyridine both in gel and in macroporous form. Both chemical stability and ability to sorb copper were studied. In general it was found that the above ion exchange resins are resistent to prolonged treatment with strongly acidic and alkaline solutions at 100 °C, with their properties not being changed after 720 hours in 10% H_2O_2 at 20 °C; mechanical destruction of the 0.1 – 0.3 mm fractions was negligible, and destruction of larger granules depended on the structure of the resin [97]. Of the various pyridine derivatives it seems that 4-vinylpyridine and 2-methyl-5-vinylpyridine may emerge as candidates of greater importance in copolymerizates with divinylbenzene, which would be represented by the following quasi-constitutional formulas:

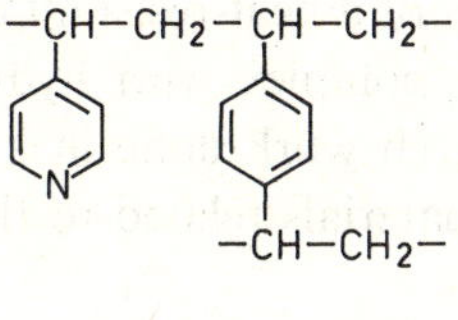

4-vinylpyridine-divinylbenzene copolymer

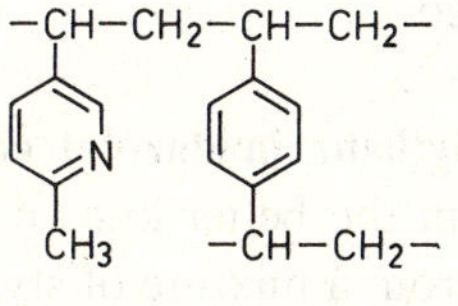

2-methyl-5-vinylpyridine-divinylbenzene copolymer

These two copolymers have been studied with respect to their rates of adsorption and diffusion rates with the ion concentration and resin porosity and a dependence on temperature and the ratio of quaternary ammonium resin form to its hydroxy form [98]. Another comprehensive work on the chromatographic and analytical properties of a gel type copolymer of 2-methyl-5-vinylpyridine with divinylbenzene crosslinking revealed that the pyridine moiety is basic, as indicated by its ability to complex ROH and RCOOH solutes. The composition, infrared absorption, swelling and gel permeation properties were also measured. Chromatographic affinity was directly related to the strength of the solute-gel complex. Solvents masked the pyridine gel in the order $CH_3CH_2OH > CHCl_3 > CCl_4$. The solute affinity varied inversely with the strength of the solvent-gel complex. The alcohol affinity correlated systematically with the proton donor strength. Gel capacity was high and several applications mainly to specific problems of purity analysis were described [99]. Further the adsorption of sulfur dioxide on poly(vinyl pyridine) resins with porous structures was investigated and a patent was granted for the removal of hexavalent chromium ions from an aqueous solution on ion exchange resins having pyridine groups and a porous, crosslinked structure. Crosslinked vinyl or alkylvinylpyridine copolymer hydrochlorides were treated with sulfonyl chloride in the presence of Friedel-Crafts catalysts to increase their sorption capacity for tin ions. Macroreticular anion exchangers based on copolymers of 2-methyl-5-vinylpyridine with certain dienes, such as dimethylacrylamides, with improved thermal and chemical stability were prepared, and subsequently chloroalkylated with benzyl chloride [100]. Other ion exchangers of the pyridinium type with improved selectivity for nickel and cobalt were prepared by chloromethylating crosslinked vinylpyridine polymers, treating the product with aqueous thiourea solutions, and then saponifying and optionally alkylating the product. Phosphonate derivatives of methylvinylpyridine copolymers have also been prepared, and the polymerization of vinyl benzyl bromide on a poly(2-methyl-5-vinylpyridine) matrix led to complete quaternization of the pyridine nitrogen and to the formation of a heat-resistant complex; replacement of the active bromium of the polymer complex with hydroxyl by saponification gave an ion exchange complex. Research work done in a broader sense on pyridinium resins has led to ion exchange materials related to the actual field of this type of ion exchangers.

Meanwhile, a crosslinked poly(4-vinyl pyridine) ion exchange resin has become available commercially in a normal as well as in an industrial grade, so that further investigation and possible applications can be expected.

Styrene-furfural based, poly(vinyl imidazole) and poly(benz imidazole) copolymer matrices. These resin matrices form a class apart from the better-known types or are very new and envisaged for special applications. From a mixture of styrene and furfural the furfural-based ion exchange resins were prepared as strong acid cation exchangers by sulfonation or weak acid phosphorous-containing materials, and as weakly basic anion exchanger by chloromethylation and amination [101]. With the poly(vinyl imidazole) resin an effective chelating resin is obtained: the imidazole group provides a useful nitrogen donor atom in resins intended for the extraction of copper, nickel and cobalt. The preparation and precursor product is analogous to that of polystyrene-based resins. Poly(benz imidazole) was initially developed as a flame-resistant fiber and is now produced in a variety of sample types for possible use in ion exchange, chromatographic and other applications [102].

Acrylic ion exchange resins. By choosing the acrylic molecule for the preparation of ion exchangers one introduces an aliphatic structure into the matrix which eliminates the interaction of an aromatic matrix (as in the case of styrene-based resins) with the aromatic moieties of, e. g., pollutants. On the other hand the acrylic molecule is unsaturated, which means that it can be polymerized. This is why weak acid cation exchangers based on acrylics have found broader interest. Divinylbenzene is also used mainly as the the crosslinking agent needed to build three-dimensional networks.

The acrylic components of the copolymers are either acrylic acid and methacrylic acid, acrylic acid esters, or acrylonitrile. Both acrylic acid and methacrylic acid have been used for the preparation of copolymers with divinylbenzene as cross-linking agent by a variety of patented techniques. One of the early acrylic ion exchange resins was made by suspension polymerization of methacrylic acid and divinylbenzene according to the following diagrammatic equation:

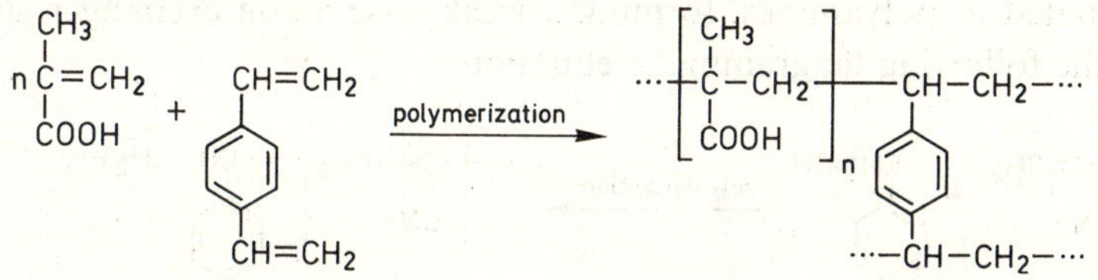

This initial polymerization product already contains pendent carboxylic acid groups, and can therefore serve as a weak acid cation exchanger without further reaction. If acrylic ester or acrylonitrile is used as the acrylic component, the cross-linked polymers must then be hydrolyzed to obtain the same ion exchanging material, according to the following equation:

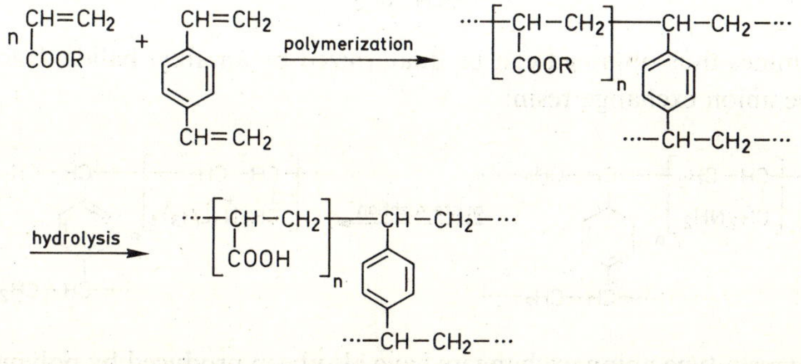

In the usual suspension polymerization of acrylic ion exchange resins (yielding gel type products) the copolymerization of acrylic acid derivatives – in contrast to resins based on methacrylic acid – with divinylbenzene is not as uniform and complete as with styrene, because the difference in the copolymerization parameters is much greater. As a consequence the polymeric material contains – after the hydrolysis of the intermediate material – detectable quantities of soluble non-crosslinked polyacrylic acid. The addition of secondary crosslinkers can reduce the release of solute polyacrylic acid to a non-detectable amount. Improvements are also obtainable by the application of post crosslinking. As far as acidity is concerned the acrylic acid resin is a stronger acid than the methacrylic acid product. As a consequence those resins deriving from methacrylic acid are appreciably less dissociated than those derived from acrylic acid. Beyond the dependence of the apparent pK_a value of a resin on the extent of crosslinking the thermodynamic pK_a values – corrected for internal concentration and ionic strength – are almost one pK unit apart, with the acrylic acid resins at about 4.8 ± 0.1, and the methacrylic resins at about 5.7 ± 0.2 [103].

If acrylonitrile is used, as it is done in practice by ion exchange resin manufacturers, as the precursor of acrylic acid resins, the products obtained are equivalent to those made from acrylic acid provided hydrolysis is essentially complete. The major differences between such materials lie in their internal structure, first in terms of gel crosslinking. The crosslinked acrylonitrile copolymers can also be reduced or hydrogenated to polyamines, forming a weak base anion exchanger, as can be seen from the following diagrammatic equation:

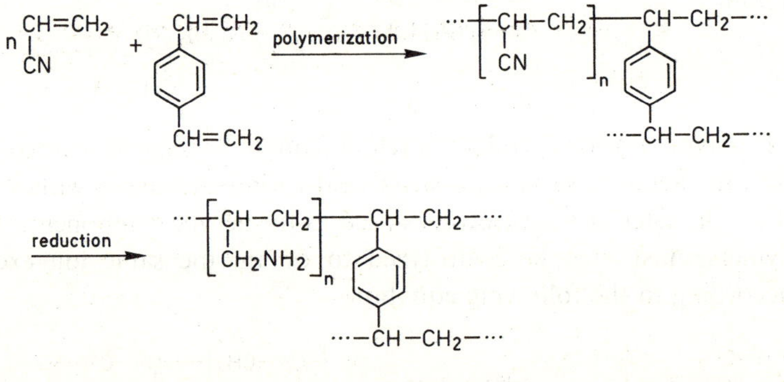

The polyamines thus obtained can be quaternized by an alkyl halide, leading to a strong base anion exchange resin:

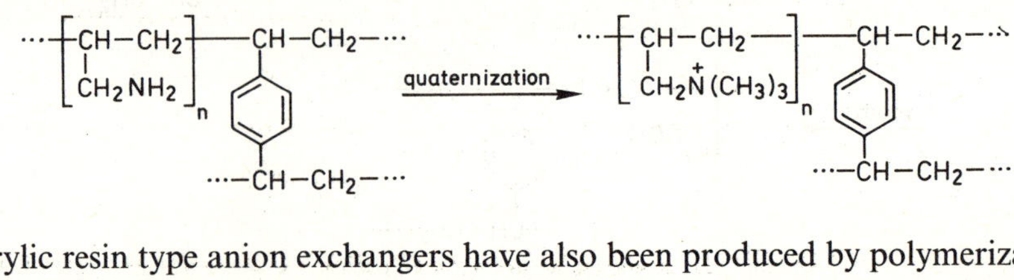

But acrylic resin type anion exchangers have also been produced by polymerization of acrylic and methacrylic acids in the presence of polyamines and by treatment of the crosslinked polymeric acrylic esters with polyamines. As not very much concrete information on these syntheses has been disclosed it can only be assumed that one synthesis starts with methyl acrylate and divinylbenzene reacting to form the cross-linked matrix containing perhaps the methyl ester group with dimethylaminopropylamine, yielding at first a weak base acrylic anion exchanger, as shown in the following formula:

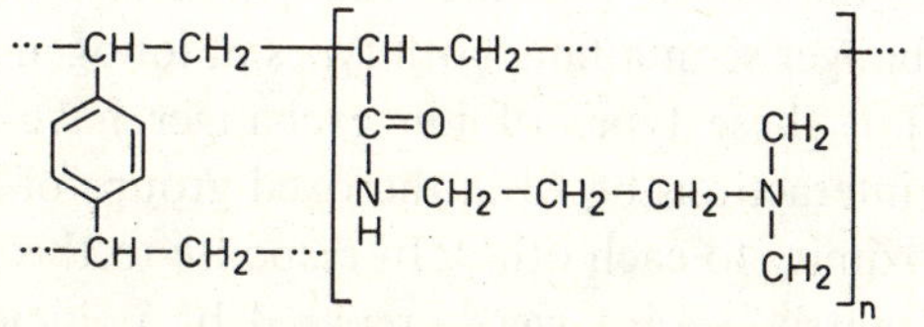

From this a strong base acrylic anion exchanger can then be derived by methylation, using either methyl chloride or methyl sulfate, which transforms it into the following formula:

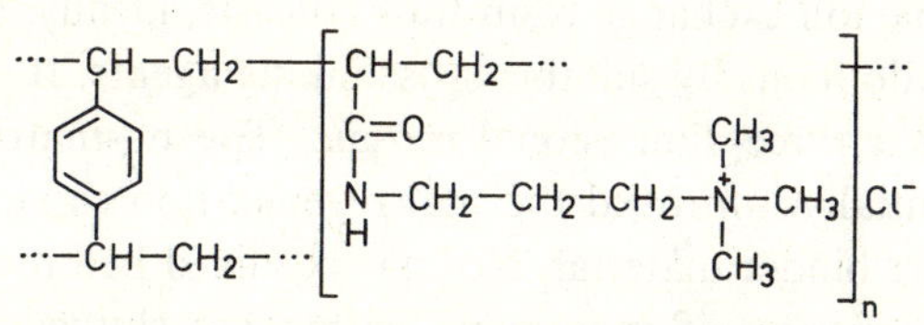

It must suffice to say that all major acrylic monomers, i. e., acids, esters, and nitriles, are used in the manufacture of the three acrylic ion exchange resin types: weakly acidic, weakly basic, and strongly basic.

It must be emphasized that with respect to structure and properties much less is known about the acrylic ion exchange resins. This may be due to the fact that they constitute only three percent of the cation exchange and anion exchange resins. The copolymer matrices or skeletal structures can be prepared both as gel type and as macroporous (macroreticular) types, with the latter slightly dominating the field. It has already been mentioned that there is a distinct difference in the acidity between the weakly acidic cation exchange resins based either on acrylic or methacrylic acid. But, in general, many of the desired properties of the weak acid cation exchangers are related to their acidity, as for example the ease of regenerating these resins with acid and their operational pH range. Weakly acidic cation exchangers exhibit a strong affinity for the hydrogen ion. Still, much remains to be studied concerning the fundamental nature of the acrylic ion exchange resins.

Other acrylic-based ion exchange resins. By scientific research other acrylic-based ion exchange resins have been synthesized and either merely examined for their ion exchange properties or applied on a laboratory scale for selective ion uptake, or in analytical chemistry for chromatographic separations. The results published are numerous and also very heterogeneous in their chemical nature.

Thermally regenerable ion exchangers. Under the trademark Sirotherm special ion exchange resins have been made available as thermally regenerable exchangers, together with an associated process. The resins are capable of salt adsorption because they contain both weakly basic and weakly acidic ionogenic groups within the same polymeric bead. The initial resins used for thermal regeneration were mixed beds of fine particle-sized weakly acidic and weakly basic exchangers consisting of microparticles of 10 to 20 μm in size; it was found, however, that there are severe mechanical problems in the handling of such fine particles. Amphoteric and snake-cage resins are ion exchanger scontaining both types of ionogenic groups within the same resin particle, but these types of ion exchanger have too low capacities for salt uptake due to interactions between the fixed groups of opposite charge which are in too close proximity to each other. In response to this problem composite ion exchange or "no-matrix" resins were prepared by polymerizing a mixture of basic and acidic monomers in such a way that the strong interaction between the oppositely charged monomers was avoided, and so that the ionogenic sites in the final polymer were not rendered inactive by internal salt formation. A no-matrix thermally regenerable ion exchange resin thus consists, ideally, of acidic domains separated from basic domains by suitable crosslinking agents. It is a resin particle with the character of a three-dimensional mosaic. The no-matrix resins were considered to have potentially improved thermal regeneration capacities because of the absence of an inert binder material. No-matrix mixed bed resins have been prepared by the polymerization of monomers within crosslinked polymer beads and by polymerization of mixtures of suitable monomers. To investigate the latter methyl acrylate-triallylamine systems, methacrylamide-triallylamine systems and acrylamide-triallylamine systems were prepared by various alternative routes. It became clear that no-matrix resins with good capacities can be obtained by the neutral precursor route. It was also shown that in order to achieve good capacities, a number of variables must be optimized, e. g., internal neutralization (which is a function of the solvent used in the polymerization), the acid to base ratio in the resin, the degree of crosslinking of the resin must be minimized, and the type of initiating system used carefully selected [104]. Further improved resins for ion exchange with thermal regeneration were then prepared with the goal of producing resins resistant to sulfite attack. Resins which are low in unsaturation and therefore resistant to sulfite attack can be prepared, for instance, from diallylamine and methyldiallylamine with bis(diallylamino) crosslinking agents. The use of polyether resins revealed that polyether amine resins (which are ten times more resistant to oxidation than the other materials used) show promise in this regard [105].

Magnetic ion exchange resins. To overcome the difficult problem of handling very small ion exchange resin beads D. E. Weiss devised the method of employing ion exchange resins with ferromagnetic fillers [106]. Historically magnetic resins must be considered a spin-off from the thermal regeneration research program. The main feature of magnetic ion exchange resins in this regard is that the magnetized forms of the small particle-sized resins flocculate very strongly to give agglomerates with sedimentation rates comparable to those of normal-sized beads. Secondly, it was shown that the fast reaction rates associated with the small size of the primary particles result from the flocs' readily dispersing upon agitation [107]. Magnetic resins have been prepared in a number of forms. Two groups of researchers have worked intensively on the manufacture of magnetic ion exchange resins, one in Australia and one in the USSR, obtaining homogeneous exchange materials as strong and weak cation as well as anion exchangers. Homogeneous magnetic ion exchange resin materials are those which contain magnetic material evenly distributed within the resin, which has been achieved by adding the magnetic material during bulk polymerization or suspension polymerization. The magnetic resins obtained by bulk polymerization are ground to granules. This leaves some magnetic particles exposed on the surface of the resin which are conveniently removed by appropriate washing with acid. The resins prepared by suspension polymerization are obtained as spherical beads. Optimization of the ferritization of sulfonated cation exchangers in the ferrous form as an interesting variation of the above mentioned subject is achieved by aftertreatment under alkaline oxidation conditions to make ferromagnetic domains [108]. Heterogeneous magnetic ion exchange resins, on the other hand, are obtained as shell-type resins by grafting the active polymeric chains onto a core of magnetic polymer pre-formed by embedding magnetic particles inside an inert crosslinked polymer [109]. The shapes of the various types of magnetic ion exchangers are shown in Figure 1.51. The particle size range of homogeneous

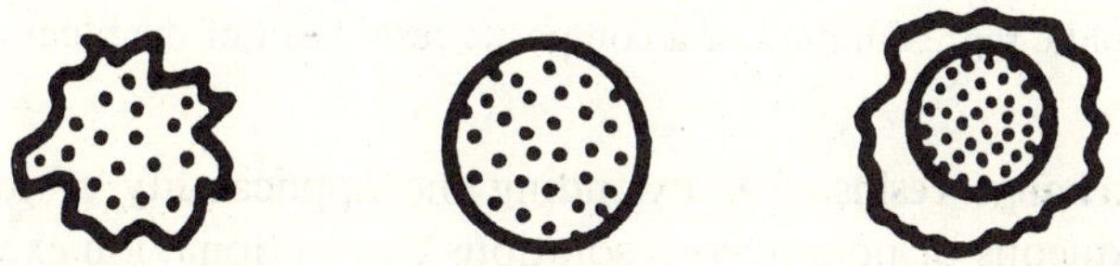

Figure 1.51 Types of magnetic ion exchangers. Left to right: homogeneous granules, homogeneous beads, and heterogeneous shell resins.

magnetic ion exchange resins is between 50 and 130 μm, and that of heterogeneous resins between 100 and 300 μm. A number of magnetic materials have been examined, with the priority being given to γ-Fe_2O_3 because of the small particle size of the commercially available material (about 0.1 μm) and its ready dispersion within the resin. Russian workers classify their magnetic resins in terms of magnetic susceptibility, and only when the magnetic susceptibility is above $10\,000 \cdot 10^{-6}$ cm^3/g are the resins used in ion exchange procedures. Magnetic versions of thermally regenerable ion exchangers have also been prepared and are still being tested in a novel continuous mode. If this is successful, then as one of the applications of magnetic ion exchange resins, the Sirotherm process is likely to be reborn [110]. Finally, magnetic ion exchangers can also be made from commercial ion exchangers, e. g., from Dowex 50WX8, 100–200 mesh, in H-form, by soaking it in an aqueous $FeSO_4 \cdot 7\,H_2O$ solution, then treating it with an aqueous alkali and heating it above room temperature with oxygen injection in order to achieve subsequent coating with ferrite [111].

"Plum pudding" ion exchange resins. Another spin-off of the efforts to find suitable ion exchangers for Sirotherm desalination are the so-called plum pudding resins, which were developed to meet the rate problems of this process. These ion exchangers are composite beads made by incorporating both the particle-sized acidic and basic exchangers in an inert "matrix" permeable to salt and water, as shown in Figure 1.52. The final particle is of conventional size and can be used in standard ion exchange equipment. Such particles have superior salt adsorption rates, which means that column operation at acceptable flow rates is quite feasible [112]. It was believed that these ion exchangers would make the Sirotherm process more successful.

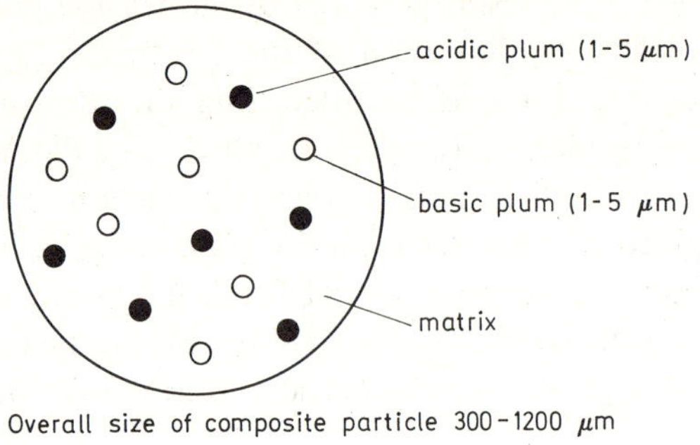

Figure 1.52 Schematic representation of a composite resin bead of the plum pudding type.

Oleophilic ion exchange resins. For extending the applicability of ion exchange resins to mixed aqueous or nonaqueous solutions conventional ion exchangers can be used (even in the presence of essentially anhydrous solvents) provided that the resins still contain their gel water. Macroporous (macroreticular) resins will exchange under completely anhydrous conditions provided that the solute is ionized. Another approach to ion exchange reactions in organic solvents held to be even more promising was to introduce into the exchanger matrix oleophilic groupings that cause the exchanger to swell in nonaqueous solvents. Such oleophilic ion exchange polymers were prepared by the copolymerization of methacrylic acid and dodecyl methacrylate, or of styrene and isobutylene followed by sulfonation, by the formation of a cage polymer of polystyrene about a linear, oil-soluble polymer of isobutylene or butyl rubber followed by crosslinking and sulfonation, by the acylation, e. g., with lauroyl groups, of polystyrene, followed by sulfonation, by the quaternization with long chain alkyl halides of polyvinylimidazole, or by the chloromethylation of polystyrene, followed by treatment with a long chain tertiary amine. Intensive studies of the resins suggested that these materials could open up entirely new fields of ion exchange application but so far they have been of no great technical importance [113].

Hybrid copolymers as ion exchanger matrices. These ion exchangers are obtained when acrylic acid monomers are polymerized and crosslinked in the pores of macroreticular styrene-divinylbenzene copolymers and the hybrid polymers obtained are chloromethylated and aminated to give regenerable ion exchangers useful for the desalination of water. The hybrid polymer has the same particle size as that of the starting pearls, and the final resins have anion exchange and cation exchange capacities [114]. For the removal of ions from a liquid containing a metal salt the hybrid ion exchange resins were regenerated by treatment with water at 90 to 95 °C.

Aliphatic ion exchange resins. By choosing the acrylic molecule for the preparation of ion exchangers one brings an aliphatic property into the matrix of an ion exchanger. Further interest in aliphatic ion exchange resins with sulfonic acid groups results primarily from the high capacities expected theoretically because of the large number of ionogenic groups per mass unit. Investigations of polymers made from vinylsulfonic acid and its derivatives led to vinylsulfonyl fluoride as the only compound suitable for suspension homopolymerization or copolymerization with commercial divinylbenzene. Polymer yield depended significantly on the type and concentration of the initiator. The best results were obtained with AIBN: 70 to 80% with 2.5 to 5.0 weight% of initiator in the monomer mixture. The overall capacity of the sulfonic acid units in alkaline saponified vinylsulfonyl fluoride-divinylbenzene copolymers was 3.8 – 4.5 mequ/g, corresponding to 62 – 68% vinylsulfonyl fluoride content in the copolymer. Ion exchangers with total capacity of approx. 5.5 mequ/g were obtained using pure isomers of divinylbenzene instead of the commercial product. Polymerization of vinylsulfonyl fluoride with triacryloyltriazine or bis(4-vinylphenyl)methane gave crosslinked products which, after saponification, had a total capactiy of 5.59 and 6.32 mequ/g, respectively. Polymerization of vinylsulfonyl fluoride with diallyl phthalate, divinyl sulfone, and p-diisopropenylbenzene gave only non-crosslinked soluble products [115]. The kinetic properties of weakly basic anion exchanger groups on an aliphatic matrix were examined by measuring the kinetic curves for hydrochloric acid sorption as a function of hydrochloric acid concentration – the results were comparable to those for ion exchangers with a polystyrene matrix [116].

Miscellaneous polymerization matrices as ion exchanger materials. Other polymerization products have been investigated with respect to their suitability as matrices for ion exchange materials. An extensive review would demonstrate the wide possibilities of basing ion exchangers on almost any polymerizate by the general procedure of preparing a matrix by polymerization of suitable monomers followed by incorporation of ionogenic groups. To get an overview of the many possibilities that have already become known one can divide the products up into, on the one hand, those prepared with the aim of obtaining, more or less, counterparts to customary ion exchangers based on a different polymer matrix and, on the other, those prepared to serve a special purpose. Those in the first group can be of economic interest those in the second are at least of interest as reactive polymers.

Ion exchangers were prepared by charring powdered polyethylene and heating it at 150 °C for three hours with 10 percent by weight 98% sulfuric acid to obtain a strong acid cation exchanger, as well as by sulfonation of styrene-grafted radiation polymerized porous polyethylene powder to obtain an exchange resin having an exchange rate ten times higher than that of a commercial chromatographic grade ion exchange resin; phosphorylated polyethylene ion exchangers were prepared for the recovery of uranium from copper ore solutions, using for the direct recovery of U_3O_8 the direct combustion of the charged resin; polypropylene as the core for ion exchangers in the form of a sulfonated divinylbenzene-propylene-styrene-graft copolymer served for the preparation of a cation exchanger. Sulfonated as well as aminated polybutadienes and similar elastomers have been synthesized and studied mainly with regard to their physicochemical properties as cation or anion exchangers, respectively. Poly(vinyl alcohol) can be obtained in bead form from poly(vinyl acetate) beads, and can be used as an ion exchanger; or it can be converted to a strong as well as a weak base anion exchanger; an anion exchange resin with improved sorption capacity for amylolytic enzymes was also prepared by treating PVA first with benzyl chloride and then with 1,2-epoxy-3-(diethylamino)propane; a photocurable ion exchange resin is obtained by treating PVA with an azide and sodium maleate or p-aminobenzaldehyde in the presence of sulfuric acid; porous acidic ion exchanger beads were manufactured by reaction of epichlorohydrin-crosslinked PVA beads with monochloroacetic acid, $POCl_3$, or sodium bromoethane sulfonate, and granular macroreticular PVA ion exchangers and carriers were prepared by using terephthalaldehyde as a crosslinking agent, which was then treated with p-$OHCC_6H_4N(CH_3)_2$, OHCCOOH, or succinic anhydride to add amino or carboxyl groups to the polymer. Chemical reactions of PVC can also be employed to prepare either (by sulfonation) sulfonic acid cation exchangers or (by reaction with 3-aminopyridine) the corresponding powdered ion exchange resins; wet dehydrochlorinated PVC yields carbonaceous products that can be sulfonated. Under the aspect of disposal and utilization of waste plastics, it can be used here to manufacture ion exchangers to be utilized as collectors of heavy metals. Polyhydroxymethylene was etherified with a carboxy-, sulfo-, amino-, or quaternary ammonium-containing compound in aqueous alkali to give materials useful as ion exchangers for biochemical, physiological, and medicinal use with better resistance than cellulose ethers to chemical and enzymatic attack. Poly(ethylene imine) has also been investigated as a matrix for the preparation of ion exchange resins; it was converted, either after modification with 2,5-divinyl-pyridine and crosslinking with methyl methacrylate and dibutyl itaconate or after crosslinking with epichlorohydrin, to anion exchange resins feasible for the recovery of uranium from copper minerals.

Poly(vinyl amine) resins crosslinked, for example, with esters of hexa-1,5-dien-3,4-diol, are anion exchangers with primary amino groups; they can be obtained by polymerization of isopropyl *N*-vinyl carbamate and saponification [117]. The synthesis of porous polyurethanes with ion-exchanging properties by polyaddition of toluene diisocyanates to ethylenediamine or triethanolamine has been reported. Synthesis of itaconic acid polymers and their application for the adsorption of gaseous substances especially amines by a macroreticular itaconic resin, showed that such a resin is superior to an itaconic acid-based gel type and to a commercial microporous resin in this respect. Ion exchangers made of vinyl heterocyclic monomers have been reviewed and an introduction to this group of exchanging materials is to be found in [118]. Macroporous styrene polymer particles which had been partially pyrolyzed to form nonreactive, carbonaceous adsorbents, were reacted to produce weak base, strong base, weak acid or strong acid, dense ion exchange particles, or their precursors, by halogenation, sulfonation, chloromethylation, chlorosulfonation and oxidation, alone or followed by aminolysis or other conventional reactions of polymers.

1.2.2.2 Incorporation of ionogenic groups

It has been mentioned several times already that the preparation and manufacture of synthetic ion exchange resins proceeds in most cases in two steps – the formation of the polymeric matrix and, subsequently, the incorporation of the ionogenic groups. The physical properties of the matrix substantially determine the characteristics of an ion exchanger, but decisive for its chemical mode of action are the ionogenic groups. They transform the hydrophobic matrix into a hydrophilic and swellable material. In principle, reactions of polymers take place during this operational step of the manufacture of ion exchangers, resulting, depending upon how the reaction is performed, in acidic cation exchangers or basic anion exchangers. The introduction of ionogenic groups on pre-formed crosslinked polystyrene resins is a reaction similar, in general, to those used in the reaction of linear polystyrene but it is often more difficult to carry out due to the insolubility of the resin, the presence of surface impurities left over from the various agents used in the emulsion polymerization, and the difficult characterization of the insoluble reactive product.

Disregarding low density popcorn polymers – which are of negligible importance for the preparation of ion exchange resins – the starting materials are either solvent swellable resin beads containing crosslinks (generally expressed as the initial percentage of crosslinker in the mixture of monomers) or rigid macroporous (macroreticular) beads. The solvent swellable gel type resin beads are characterized by a degree of swelling inversely proportional to their degree of crosslinking. This physical state of the copolymer who has a great influence on the incorporation of the ionogenic groups, as the swelling of the resin beads brings the copolymer to a state of complete solvation, thus allowing easy penetration of the three-dimensional network by molecules of the reagent. By controlling the swelling it is even possible to cause a reaction to occur only at a fraction of the available sites, yielding exchange materials in which the ionogenic groups are not evenly distributed throughout the bead but concentrated in the more accessible regions only. The large pores of the highly crosslinked, rigid macroporous (macroreticular) copolymer beads are an alternative to the swelling of the gel type copolymer beads. These copolymer beads have an effective surface area several hundred times larger than that of the gel type resins. Their handling for the incorporation of ionogenic groups is facilitated by a wider choice of solvents, since no swelling is required. Macroporous (macroreticular) resin beads having a much larger surface area are often less affected by surface impurities and react somewhat better than unwashed solvent swellable gel type resins.

Sulfonation. The sulfonation of crosslinked polystyrenes in both gel type and macroporous resins has been used extensively in the preparation and manufacture of strong acid cation exchangers containing the sulfonic acid group. Basic for laboratory preparations as well as for industrial-scale conventional batch sulfonation processes is the reaction of the copolymer beads with 95–98% sulfuric acid. In the laboratory a mixture of 0.2 g silver sulfate and 150 ml conc. sulfuric acid are heated to 80 to 90 °C; under stirring, 20 g of styrene-divinylbenzene copolymer are added and the mixture kept at 100 °C for three hours; it is then cooled to room temperature, left to stand for a few hours, diluted with 500 ml 50% sulfuric acid, cooled again, diluted with deionized water, and then washed with a surplus of water [119]. In large-scale manufacturing, the polymer beads and conc. sulfuric acid are charged into a jacketed, glass-lined kettle, the mixture is slowly heated to 80 to 100 °C at the start and to 130 to 150 °C to finish the reaction. The weight ratio of acid to polymer is important for obtaining a good mixing during the reaction – it is generally between 6 : 1 and 10 : 1. Because the reaction is exothermic, the temperatures must be carefully controlled to avoid cracking of the polymer beads. The end point of the sulfonation is checked by analysis and is reached when one sulfonic acid group per aromatic ring or an exchange capacity of 5.1 mequ/g of dry resin is obtained. The bulk of the sulfonated beads are then hydrated either by drainage of the concentrated acid and countercurrent displacement with successively lower concentrations of acid [120] or by replacement with a concentrated electrolyte solution which is gradually diluted with water [121]. The resin is then converted to its sodium form by neutralization with a slight excess of alkali. With respect to the problem of bead cracking the physical properties of the polymer beads are as important as the reaction conditions used in sulfonation, hydration and neutralization. Results here are largely empirical, but one among the factors which produces uncracked beads is the quality of the divinylbenzene used in manufacture.

As a result of the possibility of bead fracture other techniques have been developed, among which the application of swelling prior to sulfonation has been recommended for the laboratory and is used in large-scale manufacturing. In this case gel type polymer beads with 8% divinylbenzene crosslinking are swollen in perchloroethylene, trichloroethylene, methylene chloride or dichloroethane – the increase in volume can go as high as seventy percent. Beads swollen in perchloroethylene are more easily penetrated by sulfuric acid and those treated with methylene chloride react faster with chlorosulfonic acid. Thus, by pre-swelling the resin beads the reaction time can be reduced [122]. Experts can detect pre-swollen cation exchange resins and those sulfonated without solvent by microscopic examination. Products sulfonated with solvent appear smooth and homogenenous whereas those sulfonated without a solvent appear relatively rough and irregular. Pre-swollen beads seem to give about 0.1 mequ/ml lower capacity than the same beads not pre-swollen. A vast number of patents and publications describe the sulfonation of crosslinked polystyrene. Among other methods the following may be mentioned: sulfonation with gaseous chlorosulfonic acid or sulfur trioxide; a double sulfonation by sulfur trioxide in the dry state to give full penetration at normal temperature, followed by hot (100° C) sulfuric acid, which causes a temperature rise in the bead; homogeneous partial sulfonation with 100% sulfuric acid in nitrobenzene at room temperature; or continuous sulfonation.

The question of the rate of sulfonation is a separate one. In general one can sum that sulfonation rates decrease with increasing particle size and crosslinking. This implies that the sulfonic ion exchange groups are distributed throughout the entire three-dimensional resin matrix, and not just on the surface. It further indicates that the reaction rate is controlled by diffusion of the sulfonating agent into the volume of the resin, with sulfonation becoming more difficult as crosslinking increases. It follows further that even with complete sulfonation, the exchange capacities on a weight basis given in milliequivalents per gram dry resin decrease with increasing divinylbenzene content. But because, in this case, swelling decreases, the more highly crosslinked products have higher volume capacities in milliequivalent per milliliter. The rates of sulfonation of bead copolymers of styrene crosslinked with mixtures of m- and p-divinylbenzene, as 8 mole % mixed isomers, were found to be remarkably sensitive to the isomeric para/meta-composition of the mixture, with the rate being enhanced by a factor of four for compositions containing 20 to 40% para-isomer [123]. Higher sulfonation rates for macroporous as compared to gel type styrene-divinylbenzene copolymers were obtained in a sulfonation process at 20 °C for 1 to 1.5 hour with a 1 : 3 mole ratio of copolymer to chlorosulfonic acid and 8% DVB in the copolymers. After one hour of sulfochlorination under these conditions, macroporous resins had an ion exchange capacity of 5.4 mequ/g compared to 4.4 mequ/g for gel type under the same conditions. An increase in capacity is achieved when the sulfonation is performed with a mixture of concentrated sulfuric acid and fuming sulfuric acid in a ratio to the styrene-divinylbenzene copolymer of 1 : (3.2 – 3.5) : (3.6 – 4.0); this also results in a speed-up of the preparation. If ion exchange reactive groups are intended to be introduced solely onto the surface layers the preparation can be made by controlled sulfonation of the polymer in the absence of a swelling agent under the following conditions (as claimed in a patent): 15 g styrene-divinylbenzene macroreticular copolymer containing 10% DVB is stirred with 200 ml 95% sulfuric acid for three hours at 70 °C, washed and dried; it has an exchange capacity of 0.9 mequ/g.

Structural aspects in connection with the sulfonation reaction were investigated by Schmuckler and coworkers, who first studied the homogeneous sulfonation of styrene-divinylbenzene copolymers. In order to obtain homogeneously sulfonated cation exchangers the copolymers were sulfonated with oleum in a mixture of methylene chloride and nitromethane. Investigating the influences of the chemical interaction and of the diffusion process on the kinetic behavior of these systems, it was shown that raising the temperature of the sulfonation mixture markedly increased the chemical reaction rate, while the effect on the diffusion was small. Using a curve-fitting technique for the interpretation of the degree-of-sulfonation vs. time curves the results could serve as guidelines for the preparation of homogeneous, highly-sulfonated copolymers. However, it was shown in continuation of the studies of the sulfonation reactions of polystyrene and styrene-divinylbenzene copolymers with chlorosulfonic acid in comparison with oleum in organic solvents that sulfone bridges are formed under sulfonation with chlorosulfonic acid. The formation of sulfone bridges during the sulfonation of crosslinked polystyrene must be the result either of excess activity of this sulfonation reagent or of the proximity of the polymer chains to each other. It constitutes an undesirable side reaction. If the goal is to obtain a homogeneously sulfonated product, that is, one in which one sulfonate group is substituted on every aromatic ring, sulfonation with chlorosulfonic acid is, in consequence, unsuitable [124]. This definition of a homogeneously sulfonated product may not withstand close critical examination if the structure of gel type synthetic ion exchange resins is included. This is because investigation of the ionogenic group distribution in ion exchange resins has shown that there is a random alternation of high and low concentration of fixed groups. This is to be expected when it is born in mind that the structure of the initial copolymers has to be considered as being heterogeneous. Consequently, the sulfonation step in the preparation of ion exchangers does not alter the structure of the resin [125].

Carboxylation. The carboxylic group in a polymer matrix results in weak acid cation exchange resins. These are mainly made directly by copolymerization of acrylic or methacrylic acid with divinylbenzene or by hydrolysis of the crosslinked polymers of acrylic esters or acrylonitrile but there are also other ways of obtaining carboxylic ion exchange resins. Poly(vinyl benzyl) carbonic acid, for example, is based on styrene-divinylbenzene copolymers; it can be made either (as described in an early patent) from chloromethylated polystyrene by reaction with potassium cyanide followed by saponification, or by Friedel-Crafts acetylation of a 4% crosslinked styrene-divinylbenzene copolymer, reaction with morpholine and sulfur and hydrolysis according to the following diagrammatic equation:

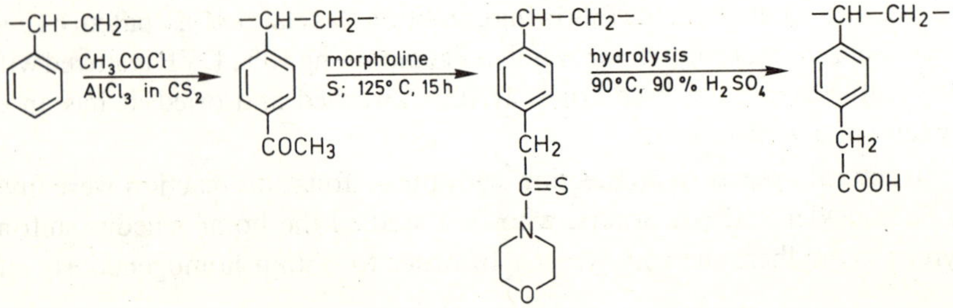

On the other hand, poly(vinyl benzoic acid) is obtained by oxidation of chloromethylated crosslinked polystyrene

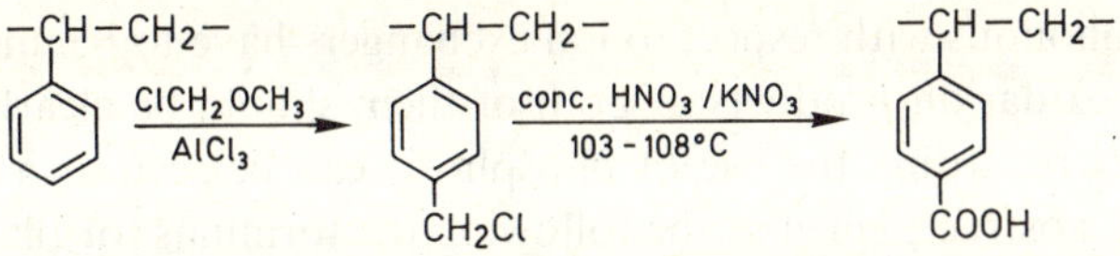

which, in order to increase the acid strength, can then be nitrated [126]. Crosslinked polystyrene was also carboxylated by treatment with 2-chlorobenzoyl chloride; the degree of substitution (24%) was determined by the increase in weight of the resin and chlorine analysis. The chlorobenzoylated resin was cleaved by *tert*-butanol and water in dimethoxyethane to give a quantitative yield of the polymeric acid. Carboxylic groups were also introduced by benzoylation, whereby the oxidation of the carbonyl group to the formyl group and subsequently carboxylic acid was improved by using acetic acid in methyl sulfochloride [127]. Via a Friedel--Crafts acetylation followed by oxidation with hypobromite, both a gel type and a macroreticular resin have been carboxylated, and carboxylated polystyrene can also be obtained by oxidation of a formylated resin using sodium or potassium dichromate in acetic acid. According to a Russian patent carboxylic cation exchangers are prepared by acetylation of styrene-divinylbenzene copolymers with acetic anhydride or acetyl chloride in the presence of a Friedel-Crafts catalyst with heating and subsequent oxidation with 25–30% nitric acid. Carboxyl group-containing ion exchangers were also prepared from chloromethylated p-styrene-divinylbenzene copolymer by hydrolysis and oxidation (KRK), reaction with sodiomalonate ester and saponification (KRDK), reaction with ethyl ammonoacetate and saponification (IRAU), or reaction with ethyl aminoenanthate and saponification (IRAE). Lithiation of crosslinked polystyrene resins either by following the reaction according to Braun of a halogenated polystyrene with an excess of *n*-butyllithium or by direct lithiation with a 1 : 1 mixture of *n*-butyllithium and *N,N,N′,N′*-tetramethylethylenediamine is the first stage for an analogous reaction of the lithiated resin with carbon dioxide to prepare polymers with caboxylic acid groups [128].

Phosphorylation. After cation exchange resins had been made available containing the sulfonic, carboxylic or phenolic acid group, Bregman and Murata in 1952 introduced phosphonous and phosphonic polymerization cation exchangers. They showed certain desirable chemical characteristics not found in the other exchangers [129]. Since then the preparation and manufacture of cation exchangers by phosphorylation of suitable matrices has been investigated repeatedly and it is sometimes said, especially in review articles, that resins containing phosphonous, phosphonic, phosphinic, or phosphoric acid groups have been made, but this is at first glance sometimes quite confusing. With respect to phosphonous and phosphinic the corresponding monomeric acids, phosphonous acid $HP(OH)_2$, and phosphinic acid $H_2P(O)(OH)$, are tautomeric, and it seems that the corresponding structures in an ion exchange resin have not been exhaustively examined. But it is further sometimes quite difficult to understand what is actually meant in descriptive articles since more misprints have happened here than in other cases. It would seem that phosphinic and phosphonous with respect to ion exchangers have to be understood as identical and that oxidation products of both of them should be clearly named phosphonic. Difficulty regarding the use of phosphoric can be seen as a misprint in, e. g., the title of an abstract, when in the following the formulas for phosphonic and phosphinic are given. Nevertheless, ion exchange resins containing phosphoric acid groups, $-O-P(O)(OH)_2$, have been reported. For the sake of completeness it should be metioned that phosphonium ion exchangers have been prepared as anion exchangers, and that the term phosphination is used for the introduction of phosphine groups on polymers.

Phosphinic and phosphonic resins can be prepared by the reaction of phosphorous trichloride with styrene-divinylbenzene copolymer beads, forming the dichlorophosphine derivative, which can then be converted to the phosphinic acid derivative by hydrolysis, and then to the phosphonic acid derivative by oxidation according to the following diagrammatic equation:

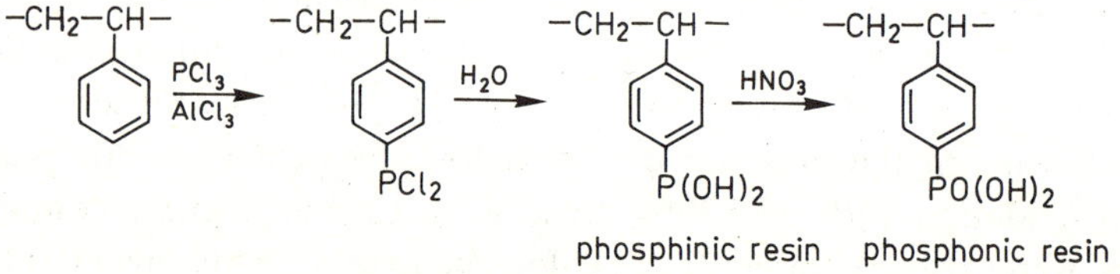

The oxidation can also be carried out before or during hydrolysis [130], before meaning that the copolymer is oxidized with chlorine before hydrolysis:

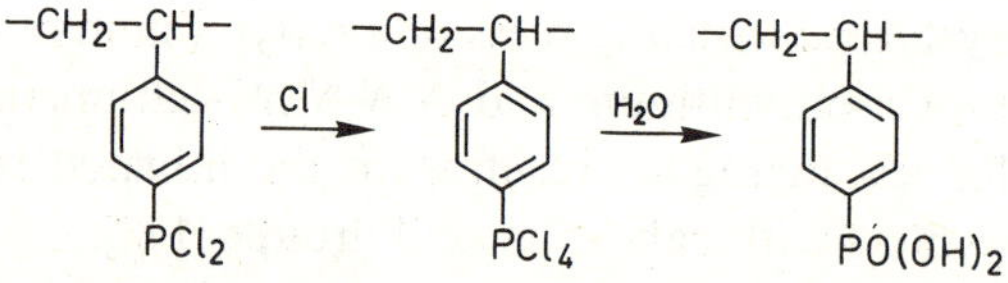

The phosphinic acid derivative is monobasic and the phosphonic acid dibasic; the latter is intermediate in acidity between the sulfonic and the carboxylic acid resins. A number of other routes were used to obtain the same materials, but resins were also prepared by phosphorylation of a chloromethylated resin, which yielded phosphoric acid ion exchangers with the active group bridged by a methylene group, i. e., $-CH_2PO(OH)_2$, to the aromatic nucleus [131]. When the same reaction, i. e., of divinylbenzene crosslinked chloromethylated polystyrene, was carried out with phosphorous trichloride in the presence of aluminium trichloride and subsequently hydrolysis and oxidation with nitric acid was performed, it resulted in the formation of a highly phosphorylated dibasic polystyrene phosphonate with a total sodium hydroxide exchange capacity of 8.0 mequ/g of resin. The probable course of the reaction was assumed to follow the following diagrammatic equation:

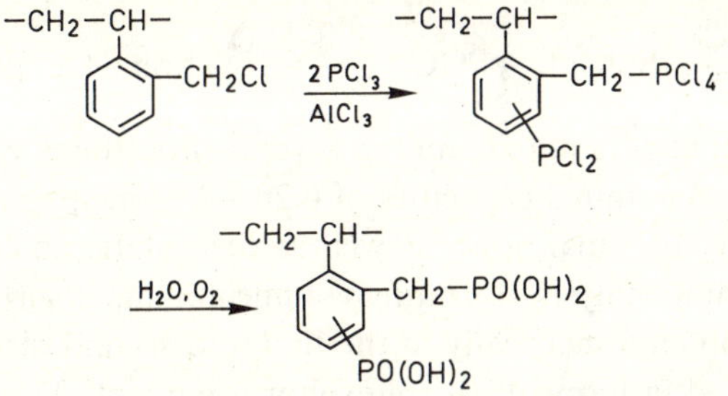

This is supported by the fact that a similar reaction occurs with benzyl chloride and phosphorous trichloride, that is, that phosphorylation occurs in the side chain as well as in the benzene nucleus. Under similar circumstances phosphorylation of polystyrene resulted in a polymer with an overall capacity of about 4.0 mequ/g of resin [131]. It was then demonstrated that resins containing phosphonic acid ionogenic groups show the same order of selectivity as do the corresponding monomers.

More detailed investigations of the properties of phosphonic as well as phosphinic ion exchangers resulted in selectivity sequences from equilibrium studies for monovalent, bivalent and trivalent ions (explained by the different polarizability of the phosphorus-containing ionogenic group) [132]; the extraction of Zn^{2+} was then used to compare phosphinic, sulfonic and carboxylic resins, showing a unique dependence of phosphinic acid resins on the extraction ratio due to ion pairing. It was also concluded that phosphinic acid resins function by both ion exchange and redox mechanisms, depending on the reduction potential of the metal [133]. After a critical review of the Friedel-Crafts reaction conditions otherwise used for the phosphorylation of copolymers to prepare P-containing ion exchangers, a new synthetic method was disclosed employing the copolymerization of phosphorus-containing monomers with divinylbenzene. Through a radical copolymerization of vinylphosphonous acid di-β,β'-chloroethylester, vinylacetate and technical DVB followed by the saponification of the ester groups these new phosphonic acid cation exchangers (KF series) were obtained [134]. A similar synthetic pathway – but yielding biionic, in this case phosphonic and carboxylic, cation exchangers – was reported for ion exchange resins based on α-phenylvinylphosphonic acid, which is copolymerized with acrylic acid [135]. With respect to true phosphoric acid cation exchangers the group $-O-PO(OH)_2$ should be reserved for this term. Such a cation exchange resin can be obtained by the reaction of phosphorous oxychloride with powdered polyvinylalcohol in dioxane medium. Another type of phosphoric acid cation exchanger was based on macroreticular poly(2–hydroxyethyl)-methacrylate (trade name Spheron, Lachema, Brno, Czechoslovakia; or Separon HEMA, Laboratory Instruments Works, Prague, Czechoslovakia), obtaining medium acidic phosphate cation exchangers by the reaction of phosphoryl chloride with Spheron or Separon HEMA (Sph–OH), dispersed in an organic solvent in the presence of a base, followed by hydrolysis:

$$\mathrm{Sph-OH + ClPOCl_2 \xrightarrow{-\,HCl} Sph-O-POCl_2 \xrightarrow{+\,H_2O} Sph-O-PO(OH)_2}.$$

This reaction was proved the most suitable among several procedures tried. The resulting cation exchangers had nominal capacities of 0.20–4.08 mequ/g and were characterized by their capacity for small ions, as well as their static and dynamic capacities for proteins, elemental analysis, operating volume and inner surface area. Some samples were titrated potentiometrically to the first and second dissociation points and the results discussed in terms of the phosphorus content. The problem that most samples exhibited a higher content of phosphorus than that corresponding to the capacity of small ions was explained by a partial reaction of phosphoryl chloride with two adjacent hydroxyl groups on the surface of Spheron according to the equation:

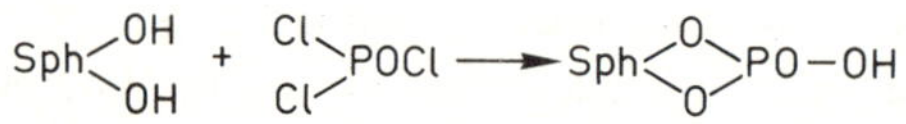

[136]. The synthesis of phosphorus-containing cation exchangers is a complex matter, as was again demonstrated in a study starting from commercial styrene-divinylbenzene, chloromethylated styrene-divinylbenzene, and acrylic acid-divinylbenzene copolymers. For the incorporation of the ionogenic groups into these copolymers $PCl_3/AlCl_3$ and dialkyl phosphites or trialkyl phosphites were used. This study was performed to elucidate the influence on selective ion exchange properties of the chemical and physical structure of the polymer matrix as well as of the structure of the phosphonic, phosphinic, and phosphoric ester ionogenic groups, especially when partly esterified, nuclear, or aliphatic-chain bound, etc., especially bound to the same matrix. To obtain comparison data, commercial sulfonic acid and carboxylic acid cation exchangers were tested under identical conditions. The difference between affinity and separation factors for the alkali metals, alkaline-earth metals, and transition and multivalent metals in acidic media was much greater for phosphorus-containing cation exchangers than for sulfonic exchangers. Carboxylic resins occupied a middle place due to the differing complexing ability of their ionogenic groups. For the same physical structure of the resin (e. g., microporous), the nuclei-bound phosphonic groups showed the best separation factor differences among various cations, as compared with the methylene group-bound phosphonic group. The aliphatic chain bound phosphonic group combined with a carboxylic group had even weaker distinguishing properties. Methylene group-bound phosphonic groups containing one active proton showed weaker separating abilities than diprotonic groups. The presence of an oxygen-bound phosphonic group diminished the separating ability of the resin. The positive effect of microporous structure over gel structure on affinity was much more

distinctive when an ester group was present [137]. Macroreticular cation exchange resins containing phosphoric acid groups were prepared by the reaction of glycidyl methacrylate-divinylbenzene copolymer beads with phosphoric acid or phosphorus oxychloride, and the adsorption behavior of metal ions was investigated. The phosphorylation of the polymer beads was effectively carried out by treatment with 85% phosphoric acid at 80 °C for three hours. The exchange resin obtained from beads with 2 mole % divinylbenzene showed high cation exchange capacity, salt splitting capacity, and adsorption capacity for copper, zinc, nickel, cadmium, and silver. On the other hand, an exchange resin obtained from poly(glycidyl methacrylate) beads had a high adsorption capacity for aluminium, iron, and UO_2^{2+} and had a higher selective adsorption for lithium than for sodium [137]. It was also reported that cation exchange resins made from *N*-vinylcarbazole-divinylbenzene copolymer phosphorylated in the presence of PCl_3 and a Friedel-Crafts catalyst are characterized by better hydrothermal stability and higher ion exchange capacity compared to their sulfonic acid analogues [137].

Chloromethylation. A key reaction in the preparation and manufacture of anion exchangers is the chloromethylation of crosslinked polystyrene which has, therefore, been investigated extensively. As procedures for the synthesis of chloromethylated crosslinked polystyrene two main routes have been elaborated and have been the subject of a number of papers and patents. In the conventional synthesis first employed by McBurney, the copolymer beads are chloromethylated with chloromethyl methyl ether in the presence of a catalyst such as tin chloride or aluminium chloride, according to the following diagrammatic equation:

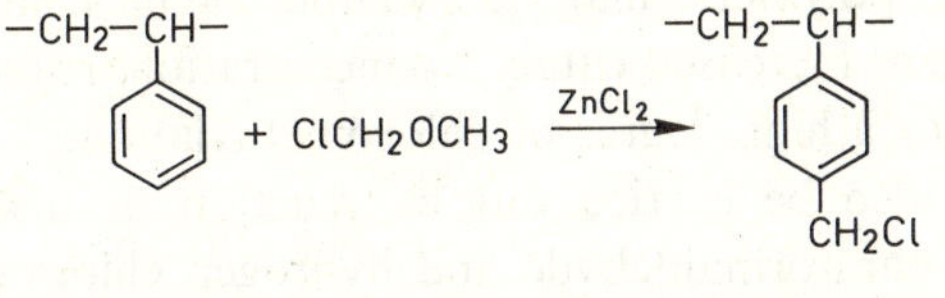

In the laboratory 20 g of styrene-divinylbenzene beads and a solution of 50 g chloromethyl methyl ether in 40 ml tetrachloroethylene are mixed and stirred for 30 minutes at room temperature; under continuous stirring at 40 – 60 °C, 10 g anhydrous zinc chloride is added over an hour; at this temperature the stirring is continued for two more hours and by continuously adding water the unreacted chloromethyl methyl ether is destroyed; the beads are then washed several times with water and vacuum dried at 50 °C. The chlorine content of the beads should be 15% by weight [119]. In large-scale manufacturing – according to what has been disclosed so far – the dry copolymer is placed in a reactor and chloromethyl methyl ether is added. This mixture is left undisturbed until all the beads are well-soaked and swollen with the ether-diethyl ether, which dissolves the aluminium trichloride catalyst, is charged to the reactor and stirred with the swollen beads. When external cooling has brought the temperature to about the freezing-point, aluminium trichloride is added and the mixture is allowed to react during stirring.

The chloromethylated beads are then filtered, washed and dried [120]. An extensive study of the chloromethylation of crosslinked polystyrene using the Friedel-Crafts alkylation with chloromethyl methyl ether in the presence of tannic chloride was also published very early [138]. The chloromethylation of styrene-divinylbenzene copolymers with all the three Friedel-Crafts catalysts mentioned above has been investigated further, because the reaction is accompanied by methylene bridging due to a secondary reaction which in the presence of a metal halide may take place between a chloromethyl group attached to an aromatic ring and an aromatic nucleus. Thus the degree of crosslinking increases, and the methylene bridging can account for up to 80% of the nominal crosslinking of numerous commercial anion exchange resins prepared from chloromethylated polymers. $ZnCl_2$ was found to be the most suitable of the catalysts; as an optimal condition swelling of the styrene-divinylbenzene copolymer in methylene chloride serves to keep the resin in the swollen state throughout the reaction; the methyl chloride also acts as a diluent reducing the probability of interaction between the chloromethyl groups and neighboring styrene chains. Nitromethane is added to the reaction in order to reduce the activity of the catalyst by complex formation. These reaction conditions were found to be optimal in that they resulted in a uniform product. That is one in which localized areas of high methylene bridging were avoided [139]. In general it seems that a very pure grade of chloromethyl methyl ether and of Friedel-Crafts catalyst should be used, if extensive methylene bridging is to be avoided. It was also mentioned in the literature that the chloromethylation reaction is easier to control when the reaction is performed in a mixture of chloroform or carbon tetrachloride and chloromethyl methyl ether. Maximum swelling allows rapid diffusion of the reactants into the polymeric matrix, and the chemical interaction becomes the rate-determining step. Increased catalyst concentration, reaction time and temperature are functions for a high degree of chloromethylation.

The chloromethylation can also be carried out by reaction of the styrene-divinylbenzene copolymer with paraformaldehyde and hydrogen chloride (as initially claimed by McBurney) which enables one to avoid handling of the toxic chloromethyl methyl ether. Similarly a macroporous copolymer was chloromethylated partially in the surface layer or to a maximum of 25% of the volume of the polymer phase after being pre-swollen in methanol and refluxed with 37% formaldehyde and concentrated hydrochloric acid, yielding a product with 2.24% chlorine content [140]. On the other hand, in a Japanese patent it was claimed that chloromethylation of styrene copolymer gives a higher degree of chloromethylene groups when the reaction is accomplished with chloromethyl methyl ether in the presence of paraformaldehyde and $AlCl_3$ as a Friedel-Crafts catalyst. Another chloromethylation procedure with the advantage that handling of the toxic chloromethyl methyl ether is omitted is to stir the copolymer with formaldehyde dimethyl acetal at 35 °C for one hour under a nitrogen atmosphere adding (after cooling to 0 °C) $SOCl_2$ and anhydrous $ZnCl_2$ dissolved in some formaldehyde dimethyl acetal, followed by raising the temperature to 35 °C for six hours. The chlorine content can be varied by changing the reaction time, temperature and amount of catalyst whereby longer time, higher temperature and more $ZnCl_2$ lead to higher conversions [141]. Earlier the reaction of styrene-divinylbenzene copolymers also with formaldehyde dimethyl acetal (and chlorosulfonic acid), but in addition with sulfuryl chloride (SO_2Cl_2) was presented as a new chloromethylation method. It has the disadvantage, however, that the increase in chlorine content is accompanied by darkening and even charring of the polymer beads [142]. A new process for chloromethylation was also described by Merrifield. This synthesis uses chloromethyl methyl ether with zinc chloride in tetrahydrofuran, yielding a copolymer with a very low degree of chloromethylation, which is particularly suited for use in the solid phase synthesis of certain peptides [143]. Another improved process was then developed enabling work with commercial chloromethyl methyl ether and affording good control of the degree of chloromethylation by the use of varying amounts of boron trifluoride etherate. This has the beneficial result that the reacted beads have an increased swelling in methylene chloride [144]. In a later investigation it was then shown that during the chloromethylation of styrene-divinylbenzene resins of high crosslinkage, a side reaction takes place with unreacted vinyl groups present in the resin. Addition of chloromethyl methyl ether to the vinyl groups results in aliphatic chlorides which differ strongly from aromatic chloromethyl groups in reactivity towards further reactions, with e. g. dimethylamine. This explains why infrared spectroscopy indicates complete conversion of benzylchloride residues, whereas chlorine is still found by elemental analysis [145].

The main object of chloromethylated styrene-divinylbenzene resins with respect to the preparation and manufacture of ion exchangers is their conversion to anion exchangers. For strongly basic anion exchange resins the conventional synthesis is the conversion of the chloromethylated copolymer by amination with trimethylamine or dimethylethanolamine, respectively, according to the following equations:

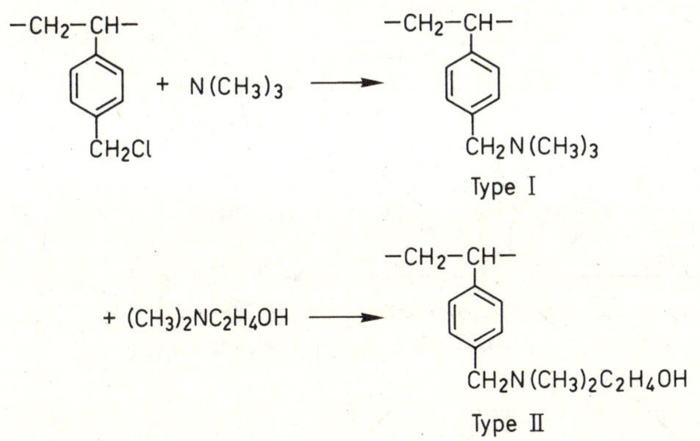

The main problem here was the above-mentioned methylene bridging of the chloromethylation. Because of this, the strong base resins often show anomalies not encountered with cation exchangers based on the same resin. Thus in this case the quantity of divinylbenzene used in making the copolymer does not reflect the actual crosslinking unless the degre of bridging can be controlled, and, indeed, strong base anion exchangers were made from linear polystyrene without the use of divinylbenzene as a crosslinking comonomer. Another interesting way of eliminating methylene bridging is based on the use of vinyltoluene instead of styrene for preparing the copolymer.

For weakly basic anion exchange resins the chloromethylated polystyrene can be reacted, for example, with a polyamine, according tho the following diagrammatic equation:

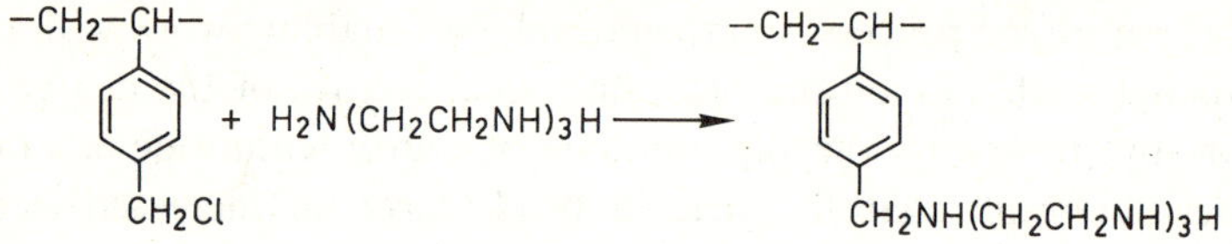

but with the unavoidable consequence of further crosslinking occuring through the polyamine bridge. As a result a very dense structure is obtained, which is relatively hydrophobic. Monoamines, such as dimethylamine, do not result in amine bridging, but this results in low exchange capacity and a resin that is still hydrophobic. A slightly basic anion exchanger was also prepared by treating chloromethylated styrene-divinylbenzene copolymer with monoethanolamine. It showed higher exchange capacity, chemical stability and alkaline resistance.

Numerous other reactions of chloromethylated styrene-divinylbenzene copolymer have been studied with the goal of synthesizing reactive polymers and in particular ion exchangers. Besides those mentioned briefly above already some of these may be further described. The chloromethylated

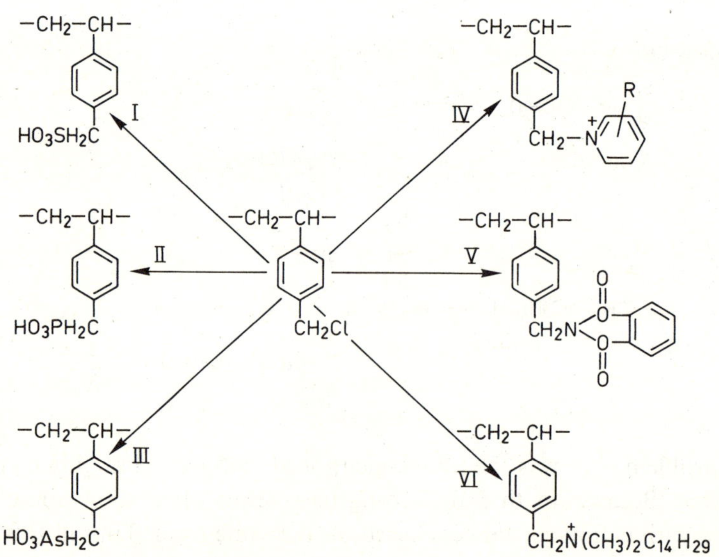

matrix is in essence hydrophobic; it is, therefore, difficult to introduce nucleophiles like sodium sulfite into it, unless a conversion to either a relatively unstable onium salt with dimethyl sulfide is used to facilitate the reaction or the synthesis is carried out with sodium sulfide in an organic solvent with a subsequent reaction with sodium sulfite to obtain a sulfonic acid cation exchanger (I). Phosphonic acid resins are often made by the Arbuzov reaction of, e. g., triethyl phosphite, the phosphonic ester thus formed being hydrolyzed to the free acid. The Arbuzov procedure is preferred because it leads to cleaner and better-defined phosphonic acid cation exchangers. Synthesis of arsenic-containing ion exchangers can also be performed with lithium dialkylarsenides and subsequent oxidation, yielding arsonic acid cation exchangers. On the other hand anion exchangers of the weakly basic pyridine-type were prepared from chloromethylated styrene-divinylbenzene copolymer in the presence of 5% aqueous sodium hydroxide, including pyridine-carboxylic acids and pyridine derivatives. Besides the reaction of chloromethylated styrene-divinylbenzene copolymer with 1-phenyl-3-methyl-5-pyrazolone and 1-phenyl-2,3-dimethyl-5-pyrazolone to obtain ion exchangers containing these groups, the interaction in the presence of potassium carbonate with phthalimide and its derivatives was used to obtain ion exchangers with a conversion of the CH_2Cl-groups of 93.4% for macroporous resin in DMF as solvent at 90 °C. Last but not least an amphophilic ion exchange resin can be prepared by treating chloromethylated styrene-2% divinylbenzene copolymer with $N(CH_3)_2C_{14}H_{29}$. It was studied for its molecular adsorption behavior (VI). Macroporous chloromethylated styrene-divinylbenzene polymer aminated successively with $C_{17}-C_{20}$ high molecular weight aliphatic amines and trimethylamine gives a highly basic hydrophobic anion exchanger.

That chloromethylation and subsequent conversion to anion exchangers by reaction with a variety of amines can also be applied to other matrices than styrene-divinylbenzene is demonstrated by the interaction of chloromethylated copolymers of styrene and *N,N'*-alkylenedimethacrylamides, yielding ion exchangers with $\leq$ 5.27 mequ/g capacity [146].

Ion exchangers by phase transfer catalysis. To overcome the problems in preparing reactive polymers by reaction of polymers – for instance from crosslinked poly(chloromethyl styrene) – the technique of phase transfer catalysis has been applied. Excellent degrees of conversion can be obtained with insoluble and hydrophobic polymers as well; further the reactions are generally quite clean, frequently giving better results than would be obtained under classical conditions. Phase transfer catalysis seem to contribute significantly to the prevention of side reactions in certain modifications of polymers. A number of modified crosslinked polystyrene resins have been prepared via phase transfer catalysis but no great interest has been shown so far in the preparation of ion exchangers using this technique, nor has a commercial resin been reported as having been made in this way [147].

Amination of acrylic matrices. In order to obtain acrylic anion exchange resins, thereby including the aliphatic nature of an acrylic-based, hydrophilic skeleton, copolymers of acrylamide and divinylbenzene can be aminated to yield a weakly basic anion exchange resin, as shown in the following diagrammatic structure:

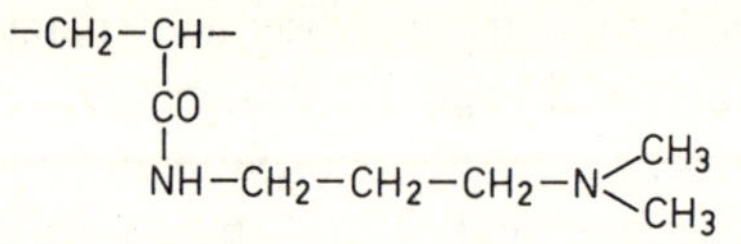

This is an unusal anion exchange resin in that it is the most basic of the weakly basic anion exchangers. From this anion exchange material a strongly basic anion exchange resin can be derived by methylation with either methyl sulfate or methyl chloride:

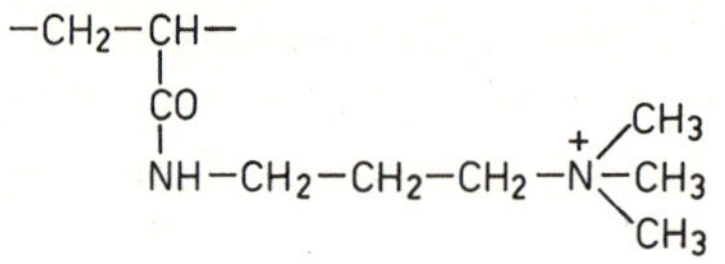

This is an exchange material which is as strongly basic as the styrenic analogue. These acrylic anion exchange resins foul to a much lower degree than do the styrenic resins when used in applications where high levels of organic compounds are present and may, because of their favourable regenerant consumption replace in time the styrenic strong base anion exchangers of Type II [148].

Nitration. Crosslinked polystyrene nitrated by mixtures of nitric and sulfuric acid is a precursor in the preparation of resins containing amino groups. The amination of the nitrated resin is accomplished by reduction with stannous chloride in acidic medium. The effect of pre-swelling was studied, showing that copolymer swelling was greatest in aromatic hydrocarbons and in halogenated organic compounds; the more swollen copolymer was easily nitrated, e. g., in the case with 2% DVB crosslinking 33.5% NO_3 and with 10% DVB 29.9% nitration. The amino resin is said to be unstable and should be used immediately to make derivatives since it decomposes readily in air and light. However, in reduction experiments of styrene-divinylbenzene copolymers only the polymers containing NO, NHOH and NH_2 groups were obtained, and the infrared specta showed that azoxy-, azo- and hydrazo-group-containing polymers were not formed. The increase in divinylbenzene group concentration from 2 to 8% in the copolymers decreased their ion exchange capacity from 6.8 to 5.0 mequ/g [149]. An ion exchanger was produced by treating crosslinked poly(amino styrene) copolymers with derivatives of monochloracetic acid in the presence of an HCl acceptor. To improve the sorption characteristics of the ion exchanger, to simplify the method of its production, and to increase the degree of transformation of amino groups, alkyl esters were used as the acid derivatives and calcium acetate was used for the HCl acceptor [150].

Mercuration. The synthesis of organomercury polymers as ion exchangers has been investigated. A copolymer of styrene with 5% DVB, dried at 100 °C and ground to 0.1 – 0.4 mm, was refluxed with 200 ml dioxane at 100 °C. The copolymer was filtered, 250 ml nitrobenzene and 10 ml acetic acid added, the mixture heated to 110 °C, treated in portions over five hours with 96 g $Hg(OOCCH_3)_2$ and kept then at this temperature for five hours. The resin grains were filtered, washed with hot nitrobenzene, acetic acid, 1N NaOH, 2N acetic acid, and water, and dried. The anion exchange properties of the resin obtained were also investigated, yielding very different distribution coefficients for different anions. In a second study a resin with a macroporous structure was mercurated giving an anion exchange resin with 2.12 mequ/g ion exchange capacity. The ion exchange rate in this resin was found to be greater than that of the previously prepared mercurated gel polymers [151]. It may be mentioned that resino-phenylmercuric chloride can also be used in solid phase organometallic synthesis.

Grafted ionogenic groups. A relatively new process for incorporating ionogenic groups into a styrene-divinylbenzene matrix is to graft functional monomers on the resins by means of either post-copolymerization with the residual double bonds of the copolymer or surface graft copolymerization. The latter has been described as the synthesis of ion exchange resins consisting of an inert nucleus with a thin shell of reactive material grafted onto the surface. The shell carrying the ion exchanging sites is substantially non-crosslinked. The exchange properties of such resins are characterized by fast equilibration, as was shown by rate curves for several types of ion exchanger structures [152]. If post-copolymerization of a functional monomer with the residual double bonds of a macroporous styrene-divinylbenzene resin is applied essentially two kinds of products can be prepared, highly grafted products but with loss of the macroporous character, or moderately (up to 20%) grafted but still macroporous resins. (The latter are in fact surface-grafted products.) As in both cases the grafted groups are not engaged in a highly crosslinked network they remain largely accessible. In conclusion it can be said that this method of incorporating ionogenic groups can be used provided the grafting monomer is easily copolymerized with styrene-like units. Further, the resin must be swellable by the monomer. It is also to be concluded that a part of the residual double bonds are buried inside highly crosslinked nodules, can only be acted upon very slowly over a long period and that upon dilution of the monomer or incomplete swelling only surface grafting can be obtained [153]. In order to obtain a non-swelling carboxylic ion exchange resin for the isolation of streptomycin, poly(methacrylic acid) was grafted on to styrene-divinylbenzene copolymers [154].

Multiionic ion exchange resins. The term multiionic is introduced here as a suggested replacement for the term polyfunctional, which has been used to describe ion exchange resins containing more than one type of ionogenic group. It should include the so-called bifunctional ion exchangers defined as containing two types of ionogenic groups (without its being stated, however, whether these ionogenic groups are of the same or of opposite charge). Instead of bifunctional the term biionic could be used, but in both cases, i. e., in multiionic as well as in biionic, it is to be understood that the ionogenic groups are of the same charge. This would seem to be a reasonable suggestion since, as is described below, there are also bipolar and amphoteric ion exchange resins which are defined as ion exchangers containing more than one ionogenic group but of opposite charge. Another factor decisive for this suggestion of a change in the nomenclature is that in this volume ionogenic groups are considered as reactive groups clearly distinguished from functional groups, as has been outlined in section 1.2.1 on ion exchangers as reactive polymers. An eminent authority in the field of ion exchangers, Kressman, wrote, before the nomenclature with the term polyfunctional was introduced, that the natural and modified natural materials are almost invariably multifunctional, since they carry more than one type of active group, although some are probably unifunctional. He further wrote that sulfonated coal and synthetic resins of the condensation type are multifunctional. Sulfonated coal cation exchangers usually contain $-SO_3H$, $-COOH$ and $-OH$ groups; synthetic resins of the condensation type usually contain $-OH$ groups together with either $-COOH$ or $-SO_3H$ groups. He also mentioned that at about that time unifunctional resins had been synthesized containing only one kind of active group. Further it was stated that, generally, the multifunctional resins are homofunctional, i. e. the several groups are all cationic or all anionic; further, that certain natural materials, notably leather and wool, contain both cationic and anionic groups and are thus heterofunctional. This implicitly-given nomenclature was certainly quite clear to someone new to ion exchangers.

Biionic ion exchangers of the polymerization type with known $-SO_3/-COOH$ group ratios have been prepared from butyl p-styrenesulfonate-methyl methacrylate-*p*-divinylbenzene copolymer. Investigation of selectivity indicated that the ionogenic groups had a mutual influence on each other, as was evidenced in the total exchange constant. The calculated exchange constants were somewhat higher than those from experimental data. The selectivity of the resins was investigated during complete dissociation – K^+-Li^+ exchange in basic medium – and during dissociation of only strongly acid groups – K^+-H^+ exchange [155]. The goal of more pronounced selectivity led to investigation of eventual synergistic effects in biionic ion exchangers. Thus, a sulfonic cation exchanger containing nitro groups in a polystyrene ring was prepared by nitration and it was found that the resin, which contained 1.6% nitrogen and 8.9% sulfur, showed an increased apparent selectivity toward the heavier alkali metals rubidium and cesium [156]. The dependence of sorptivity on the composition of the acrylic acid-diethyl vinylphosphonate copolymer even indicated that ion exchange resins having a narrow selectivity can be created by copolymerizing monomers containing different selective groups or containing only a part of the selective grouping, the rest of which is contained in a second monomer. The combination of such monomers in various ratios permits regulation of the selectivity, as, e. g., at pH 3.95 the copper sorptivity of such a resin decreases with increasing phosphorus concentration [157].

Bipolar ion exchange resins. These resins are synthetic insoluble crosslinked polymers carrying ionic groups both of positive and negative charge. This latter definition would certainly be comprehensible in the denomination heterobiionic. Mikes and Kovacs, investigating bipolar ion exchange resins very early on, defined them more precisely by emphasizing that in nonmonofunctional (heteroionic) ion exchange resins there are on the skeletion in certain cases, polar groups of opposite charge to those carrying out the main ion exchange reaction of the resin. These oppositely charged groups have a lower ionic strength than that necessary to accomplish practical ion exchange. Unlike these, in bipolar resins both the positive and negative reactive groups produce practical chemical ion reactions, in other words, their dissociation constants K_d are higher than $10^{-5}-10^{-6}$. This elaboration of properties certainly justifies calling such resins bipolar. Taking the latter as a precondition for bipolar ion exchange resins it was found that their single groups react independently and stoichiometrically although the environment may alter the conversion. There was further no evidence for the conclusion that only a part of the available active groups participate in the reactions, because the theoretical capacity, fixed by controlling the steps of the synthesis of the resin, was almost attained. With respect to selectivity it was found that the uptake of certain ions on bipolar resins differs from that achieved on monoionic ion exchangers of the same composition, and that a variation in the ratio of acidic to basic groups in the resin causes variation of the equilibrium data in certain ionic mixtures [158]. Bipolar ion exchange resins containing carboxy and quaternary ammonium groups were then manufactured by chloromethylation of an alkyl acrylate-styrene-divinylbenzene copolymer and amination with a tertiary amine yielding a material containing 3.3 mequ basic groups/g and 2.1 mequ acid groups/g [159]. Bipolar ion exchange resins with sulfonic acid groups and quaternary trimethylammonium groups were also synthesized based on both gel type and macroporous matrices. Their sorption capacity for humic acid was tested and was found to be higher, by a factor of at least two, than in the case of monoionic ion exchange resins. On the other hand, stronger interactions between the oppositely charged ionogenic groups of the bipolar resin matrix and the molecules of humic acid cause a reduction in sorption velocity [159].

Amphoteric polymerization resins. Amphoteric ion exchangers are, in a general sense, also bipolar or heteroionic, meaning that they contain both acidic and basic groups. But since the objective of most of them is to obtain complex formation within an immobile resin phase it is a prerequisite that they do not contain strongly basic ammonium groups, which are unsuitable for forming complexes. Further, the presence of strongly basic quaternary ammonium groups within the exchanging resin would reduce the capacity of complex formation of the acidic groups because of the formation of internal salt bonds. The decisive properties of amphoteric ion exchangers are increased selectivity for certain multivalent ions. Technological processes and methods in hydrometallurgy, nuclear energy, efficient purification and separation of physiologically active substances, etc., have led to numerous investigations for preparing and manufacturing amphoteric ion exchange resins.

The first amphoteric polymerization resins synthesized were amino carboxylic acid resins based on chloromethylated styrene-divinylbenzene copolymers, which, in the general terms of the patent specification concerned, are either reacted with ammonia or an amine followed by reaction with a halo-carboxylic acid or its salts or esters, or are treated with an amino carboxylic acid, its salts, or esters. In every case an exchanger with a high capacity for complex-forming metals is obtained [160]. The first specifically described amphoteric and selective ion exchanger was prepared apparently independently of each other at two locations from chloromethylated styrene-divinylbenzene with ethylenediamine, and reacted then with chloroacetic acid; it was characterized by the presence of both organic acid groups and basic nitrogen groups attached to the same resin matrix [161]. Diagrammatically the following equation can be postulated:

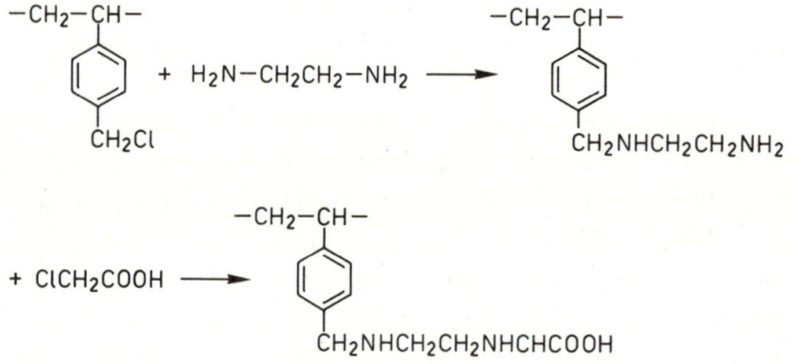

A probably fairly well-defined ion exchanger prepared by reaction of p-aminostyrene-DVB copolymer with 20% aqueous chloroacetic acid at 98 °C exhibits either complex formation or anion or cation exchange, depending on the pH of the medium. In weakly acid media, coordination compounds with d-elements are formed. In strongly acidic solutions, anion exchange occurs via proton addition to the nitrogen atom, and in strongly basic media cation exchange occurs at replaceable hydrogen atoms. If a weakly basic anion exchange resin containing tertiary amino groups is taken as parent material it can be quaternized with chloroacetate and the resulting amphoteric resin can be used for simultaneous removal of divalent cations and strong acid anions from water [162]. The homologous amphoteric ion exchangers prepared basically from *β*-aminopropionic acid were also investigated, showing interesting results. For isntance, for a resin prepared with *N*-methyl-*β*-aminopropionic acid chloromethylstyrene is treated with ethyl *N*-methyl-*β*-aminopropionate to form the stable, uniform amphoteric ion exchange resin, which hardly binds divalent cations at all and multivalent cations only as hydroxy compounds of uncertain composition. This was explained by the strong anion exchange properties of the active groups and the very instability of the metal complexes formed in the resin [163]. Other syntheses of amphoteric ion exchange resins containing aminopropionic residues were based on the reaction of polymers containing primary or secondary amino groups either with acrylonitrile at 75–85 °C for eight to ten hours followed by saponification of the product or on treatment with a mixture of malonic acid and formaldehyde, or again by treating chloromethylated styrene-divinylbenzene copolymer with alkylacrylonitrile and hydrolyzing the product with 60% sulfuric acid at 100 °C. This resulted in amphoteric ion exchangers with a high exchange capacity for uranyl anions and an insignificant sorption for iron (III), copper (II), nickel (II), and cobalt (II) cations [164]. Acrylonitrile is not a good swelling agent for the chloromethylstyrene and, therefore, the amination reaction may be incomplete. Of interest for distinguishing between amphoteric resins and chelating polymers is the study of the coordination properties of an ion exchange resin containing β,β'-iminodipropionic acid in comparison with that of the iminodiacetic acid and iminoaceticpropionic acid analogues. From capacity measurements it followed that the β,β'-iminodipropionic acid shows more amphoteric than chelating properties, in contrast to the two latter ones, and that only the iminodiacetic acid is to be classified as a chelating ion exchanger [165]. Polymerization resins with amphoteric anchor groups were also prepared by the direct reaction of chloromethylated styrene-divinylbenzene copolymers with amino acids. The prior stage towards these products was, for example, the treatment with esters of suitable amino acids, such as glycine or sarconic ethyl ester [166]. A study connected with similar attempts revealed that the reactivity, hence exchange capacity, of the halomethylated styrene-1.5% DVB copolymers increased in the following order: chloromethylated < bromomethylated < iodomethylated < chloromethylated in the presence of sodium iodide. Methyl esters of amino acids react readily with a chloromethylated macroreticular styrene copolymer, for instance the methyl ester of α-alanine, in the presence of sodium iodide at 50 °C in 6 : 1 dioxane-methanol mixture giving, after saponification, an ion exchange resin of 2.17 mequ/g ion exchange capacity. Subsequently the advantage of the use of tert-butyl esters of the amino acids was reported [167]. Further the reaction of L-cysteic acid with a chloromethylated copolymer of styrene yielded a sorbent for the chromatography of racemates, and asymmetric sorbents for the separation of racemic mixtures by ligand ion exchange chromatography were prepared with *N*-carboxymethyl-L-valine and with *N*-carboxymethyl-L-aspartic acid. Amino acid *N*-carboxyanhydrides with protected groups were grafted onto polymers with aminomethylated phenyl groups, followed by removal of the protecting groups, yielding graft copolymers useful as ion exchangers. The recovery of heavy metal ions from solutions was reported by passing the solutions through ion exchange resins containing lysine, cysteine, or aspartic acid as active groups followed by elution with mineral acids.

An entirely different approach to the synthesis of amphoteric and subsequently chelating ion exchangers begins with vinyl-benzylchloride and its derivatives. Thus, several vinyl-substituted amino acids were prepared as intermediates in the synthesis of amphoteric or complexing or chelating polymers. Most of the syntheses started with vinyl-benzylchloride and its derivatives. These products retain both the polymerizability of styrene as well as the ability to complex metal ions. The intention of this work is quite clear. In order to avoid the problems revealed in the early work on preparing amphoteric ion exchangers by reaction of copolymers with suitabe reactants, the synthesis of the requisite monomeric structures, followed by polymerization, was seen as the more attractive way. And since exchange resins based on polystyrene already had wide-spread applications, the synthesis of styrene derivatives containing corresponding groups was initiated. Also, since amino acids are known to be complexing agents for a number of metal ions, it seemed a logical choice to introduce them into a resin [168]. Vinyl-benzylchloride was then also used in a radical copolymerization with acrylonitrile to give a copolymer which was first converted to an anion exchanger by amination with pyridine. Hydrolysis of the aminated anion exchanger produced an amphoteric resin [169]. Starting from chloromethylated styrene, amphoteric ion exchangers were also prepared by reacting, in general terms, a polymer or monomer containing groups transformable into acid groups, a polymer or monomer capable of containing basic groups, and a non-conjugated vinyl monomer, followed by crosslinking and aminating the polymer, to resulting, in the case of ethyl acylate and dimethylamine, in a resin having both weak base and weak acid exchange capacities [170].

Corresponding to amino carboxylic acid resins, the first amino phosphonic acid polymerization resins were prepared either from esters of phosphoric acid and Schiff bases or from aminomethylpolystyrene crosslinked with 2% divinylbenzene by reaction with diethyl phosphonic acid to yield poly[1 – (diethyl-phosphono)ethyl aminomethylstyrene] which contained by weight 5.7% nitrogen and 9% phosphorus. After the sodium form of the amphoteric polymer was produced it showed an exchange capacity of 2.4 mequ Cu/g of sodium resin [171]. Similar investigations produced amphoteric ion exchangers with amino phosphonic groups, among which a selective ion exchanger with improved mechanical strength was prepared by treating crosslinked aminopolystyrene with a phosphorus methylol derivative. But better defined polymers with aminoethylphosphonic acid groups were obtained by reaction of chloromethylated styrene-divinylbenzene copolymer with diethyl *N,N*-dimethyl-*β*-aminoethylphosphonate or diethyl *N,N*-diethyl-*β*-aminoethylphosphonate and conversion of the diester group in polar solvents at elevated temperatures by an intramolecular reaction to the amphoteric polymer with monoester groups [172]. Amphoteric resins containing α-aminomethylphosphonic acid or α-aminoethylphosphinic acid groups can be prepared from aminated styrene-divinylbenzene copolymer with NaH_2PO_2 under different reaction conditions [173]. Investigating ion exchangers consisting of copolymers of α-phenylvinyl phosphonic acid with styrene and acrylonitrile an amphoteric ion exchanger was obtained from a vinylpyrrolidone copolymer. Reaction products of copolymers of methacrylate with divinylbenzene and diisopropenylbenzene with diamines gave, on phosphomethylation, amphoteric ion exchangers with a high selectivity for copper and uranyl ions from sulfate solutions. In another study an anion exchanger was first prepared by amination of chloromethylated styrene-divinylbenzene copolymer with *N*-methylfurfurylidenamine diethyl phosphonate. Treatment with dry hydrochloric acid led to cleavage of the ester group to the corresponding amphoteric ion exchanger [174]. A study of the sorption of copper on amphoteric ion exchangers containing phosphinates of various amino side groups by EPR, IR spectra and ion exchange data revealed that the sorption involves two mechanisms, which are ion exchange via hydroxyl groups and coordination with the phosphinato groups, or with amino groups when $NH-C_2H_4-NH_2$ is present [175]. Amphoteric amino phosphonic acid resins can also be prepared by phosphorylation of common anion exchangers with phosphorous acid in hydrochloric acid in the presence of formaldehyde. They can be made available as macroporous type resins with distinct sorption properties [176]. Amphoteric ion exchangers were also prepared by treatment of ethylene dimethacrylate-2-hydroxyethyl methacrylate copolymer with metaphosphoric acid, thus obtaining hydrophilic resins. Bromoacetalized polyvinylalcohol beads were treated with $POCl_3$ in chloroform and aminated to give amphoteric ion exchangers having cation and anion exchange capacity of 2.4 and 2.6 mequ/g, respectively.

Unique amphoteric and both amino phosphonic and amino carboxylic resins with rather high capacities were synthesized by copolymerization of ethylenimine-*N*-ethylphosphonic acid diethyl ester

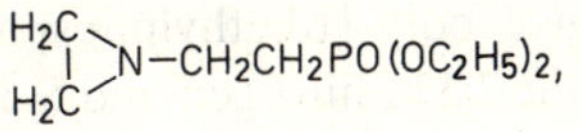

ethylenimine-*N*-acetic acid ethyl ester

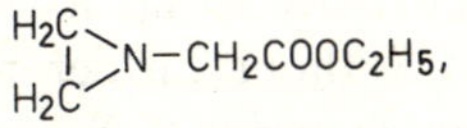

ethylenimine-*N*-propionic acid methyl ester

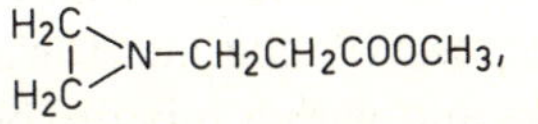

with *β,β'*-diethylenimine-1,4-diethylbenzene

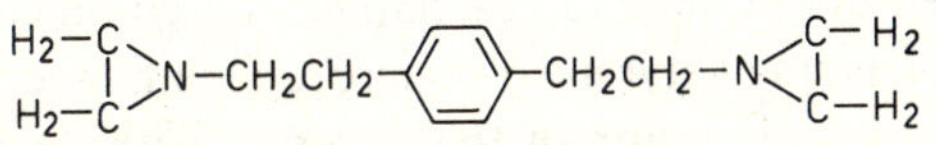

as crosslinking agent. By saponification of the ester groups of the copolymers amphoteric ion exchange resins of amino phosphonic acid and amino carbonic acid type with an especially uniform structure were obtained [177]. An elementary cell may be pictured

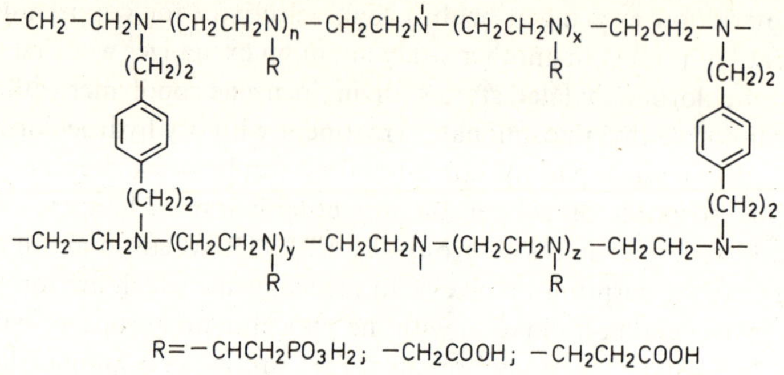

These resins (and similar ones obtained in a subsequent investigation) based on 2,4,6-tris(1-aziridinyl)-1,3,5-triazine and *β,β'*-di(1-aziridinyl)-1,4-diethylbenzene copolymerized with 2-(1-aziridinyl)butyric acid methylester and 1-aziridine succinic acid ester, respectively, as well as 2,4,6-tris(1-aziridinyl)-1,3,5-triazine with 1-aziridine acetic acid ethyl ester, 2-(1-aziridinyl)propionic acid methyl ester and -ethyl phosphonic acid diethyl ester were examined for the rate of uptake for magnesium, nickel, copper, and zinc ions.

Pyridine-containing amphoteric polymerization resins have been synthesized in various ways and with very different anionic groups. For the various syntheses the two methods of incorporating ionogenic groups into a resin have also been applied. Thus aminopolystyrene in xylene was refluxed with pyridine-2-carboxyaldehyde and reacted with diethyl phosphonic acid to give poly [(diethylphosphono) (2-pyridyl)-methylaminostyrene] containing by weight 7.5% nitrogen and 8.5% phosphorus and having after further treatment to a pH of 8–8.5 an exchange capacity of 2.5 mequ Cu/g of Na-resin [171]. Since none of the pyridine-containing amphoteric ion exchangers have proven to be of great importance so far a few of the many described in the literature will be mentioned without further reference. Treating a divinylbenzene-2-methyl-5-vinylpyridine copolymer with formaldehyde followed by oxidation with nitric acid an amphoteric ion exchanger can be obtained of which the capacity for calcium, copper and nickel ions increases with increasing carboxyl content. A carboxylic-type cation exchange resin (Amberlite H) was treated with thionyl chloride and pyridine for twenty hours and with a mixture of ethanol and pyridine again for twenty hours to give an esterified resin, which was treated with water containing L-lysine to give an amphoteric resin for optical modification of racemates. The synthesis of new amphoteric ion exchangers based on cyanopyridines and chloromethylated copolymers of styrene of macroreticular, macroporous, or gel structure was also reported by amination of the products with 3-cynaopyridine,4-cyanopyridine, or 2,5-dicyanopyridine and hydrolysis of the cyano groups to the carboxy groups. Investigating amphoteric ion exchangers with pyridine and phosphonic and phosphinic acid groups acid-base equilibrium parameter values were obtained which can be used for the calculation of concentration-distribution diagrams for various forms of anionic and cationic groups. Amphoteric ion exchangers having pyridinecarboxylic groups form strong complexes with copper(II) and consequently they are difficult to regenerate.

When one considers the demand in hydrometallurgy for sorbents exhibiting selective properties at increased mechanical strength, suitable for sorption of precious or semi-precious metals from slurries by means of ion exchangers, it can be understood why so many investigations have been carried out into amphoteric exchange resins made predominantly by polymer-analogous conversions but also starting from suitable monomers [178]. For the physico-chemical characterization of amphoteric ion exchangers their cation exchange capacity, anion exchange capacity, moisture content and density are usually determined, and pH titrations are carried out. Acid-base properties of amphoteric ion exchangers are the most important characteristics for the quantitative description of the processes of equilibrium adsorption of metal ions. Simultaneous fixation with respect to changes in concentration of hydrogen and sodium ions in the external solution allows the carrying out of the separate determination of the concentrations of acidic and basic groups of amphoteric ion exchangers and their dissociation constants. Spectroscopic methods applied to individual cases will be needed to elucidate the mechanisms of the sorption.

Specific and chelating polymerization ion exchangers. Standard ion exchangers prepared by the incorporation of ionogenic groups according to the description given in the preceding paragraphs have their limitations because if the relative selectivities of the ions in a mixture are close to each other, the resin will uptake all ions even if this is not desired. In technical applications this phenomenon results in such an increased regenerant demand that many theoretically feasible ion exchange processes have never been used. For this reason, exchangers that are either specific or highly selective for a particular ion to the exclusion of other ions are essential. Griessbach, in fact, predicted a few years after Adams and Holmes had shown that ionogenic groups can be attached to synthetic polymeric matrices, that resins would be prepared with sites more selective than the sulfonic and amino groups then existing. Further, in gravimetric quantitative analysis many ions are specifically determined through precipitation with specific organic molecules, such as nickel with dimethylglyoxime or potassium with dipicrylamine, and Skogseid got the idea while skiing in Norway to use the latter on styrene matrices to prepare a potassium specific resin which would act like a collector for this ion. In the following years two different perspectives on the development of specific or highly selective ion exchangers took form (Millar, [179]). On the one hand, Skogseid (and other workers later) attempted to prepare resins containing groups analogous to specific organic reagents of high selectivity, in the belief that the polymers would show a similar specificity or high selectivity in their ion exchange behavior [180]. On the other hand, Mellor [181], and later Gregor and Hale [182], advocated the preparation of resins containing groups notable for the range of ions they complex with whose selectivity is then obtained by making use of the pH-dependence of the stability of the various complexes formed. Hale, in fact, doubted at first the possibility of preparing completely specific resins, but then suggested that highly selective if not specific behavior might be expected from resins containing groups similar to those in monomeric chelating agents. The whole field of specific and chelating ion exchangers may still appear confusing and controversial if complexing phenomena are included, as these are already to be found in standard caboxylic, phosphorus-containing and amphoteric resins. Nevertheless the incorporation of truly specific or chelating sites onto a polymeric matrix in order to obtain a corresponding specific or chelating polymer is of great scientific and practical interest.

The potassium-specific resin of Skogseid has thus far not been practical for commercial use: a description of its preparation is given elsewhere in this volume. But many resins have been developed since then that are specific or at least very selective for specific ions. In principle there is no reason – given the existence of organic reagents for inorganic analysis that act as specific precipitants of specific ions – for these molecular groups not to be attached to polymeric matrices. Most remarkable seems to be a boron-specific ion exchanger developed by Kunin, based upon *N*-methylglucamine, which has a unique specificity for boron as boric acid [183]. This and other specific ion exchangers are listed below.

Specific group	Specific ion	Remarks
N-Methylglucamine	boron	See text above
8-Hydroxyquinoline	cobalt, copper, nickel	Several references in the ion exchange literature
8-Mercapto-quinoline	Pt elements	cf. CA 79 : 142 533
Rhodamine	Pt elements	Ibid.
β-Diketone	uranyl, copper, nickel, iron	Angew. Makromol. Chem. *116*, 195 (1983)
Thiostyrene	iron, silver	Previously by Akzo Chemical Co.
Thiolstyrene	mercury	Material for the TMR[R] process
Dimethylglyoxime	nickel	Several references in the literature
Alkylated amidines	nitrate	Dow Chemical Co., since 1971
Chromotropic acid	titanium	J. Polym. Sci *31*, 15 (1958)
Vinyl-substituted metal complexes	nickel, cobalt	Chemiker-Ztg. *108*, 255 (1984)

One can define a chelating ion exchanger as follows: it is an insoluble polymer to which a group known as a chelating agent in monomer chemistry has been attached.

This agent is able to form a ring incorporating metal cations by chemical bonding. The reaction of a chelating ion exchanger involves both ion exchange and chemical reactions. In other words only those chelating agents are suitable in which while metals are bound hydrogen is exchanged for the metal cations, since they can be regenerated by treatment with acid

$$\text{chelating agent-}H_2 + Me^{2+} \rightleftharpoons \text{chelating agent-Me} + 2\,H^+.$$

The active sites of almost all chelating resins are in fact weak acids, so that the metal ion, in spite of being strongly bound in the chelate formed, can easily be replaced again by hydrogen in neutral or alkaline solution. But in view of the fact that two steps – exchange and ring formation – are involved in chelate exchange of an ion, the kinetics are slower. Also since the selectivity for a given ion by a chelate is higher, the elution requires higher concentrations or higher dosages of eluting agent.

The majority of synthetic organic chelating ion exchangers are based on styrene-divinylbenzene copolymers. To incorporate the chelating ligands the pre-formed resin is chemically modified, a method preferred to other methods for synthesizing the large number of chelating ion exchangers prepared so far. Chloromethylation is the usual way of obtaining a suitable functional polymer, which is then treated with a chelating ligand to obtain a chelating ion exchanger. This reactions for the preparation of the frequently investigated iminodiacetic acid chelating ion exchanger are shown below: They include at one stage the conversion of chloromethylated polystyrene into its sulfonium salt – this being so-called sulfonium method – which is an important step, because the sulfonium salt is hydrophilic whereas the chloromethylated resin is not and would not react appreciably with the sodium salt of iminodiacetic acid:

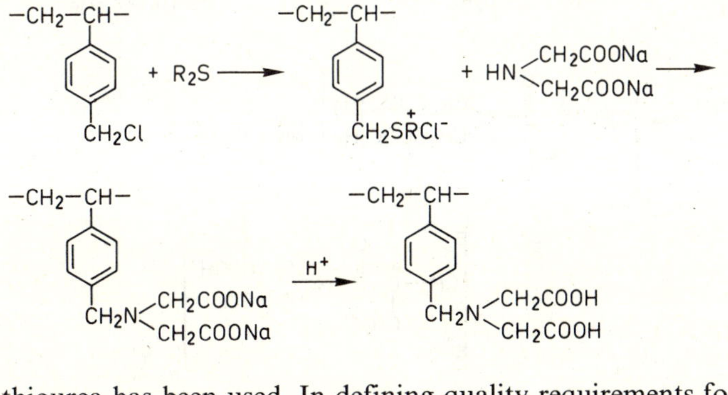

As R_2S thiourea has been used. In defining quality requirements for such chelate-forming resins it has been recommended that the resin should contain only built-in complexing groups with 1 : 1 stoichiometry with the metal ions to be separated and should be free of other groups. Thus, the above iminodiacetic resin should be prepared from chloromethyl polystyrene and esters of iminodiacetic acid, followed by saponification. Resins prepared by other methods, it has been claimed, contain other groups, which adversely affect the ion-specificity [184]. Before this method was applied at all the classical one invented by Skogseid was used, i. e. the crosslinked polystyrene is nitrated with nitric acid, reduced to poly(p-amino styrene) and reacted with chloroacetic acid:

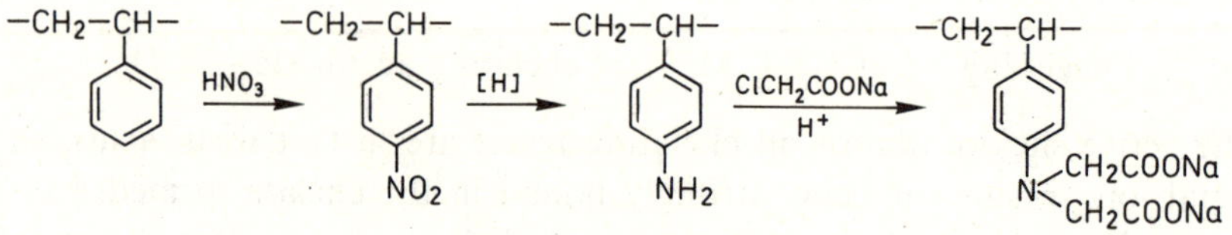

This procedure had the disadvantage that it involved a number of steps which increased the possibility of contamination of the resin with by- and intermediate-products, creating the problem of their removal. The final product is also different, having no methylene bridge between the aromatic nucleus and the iminodiacetate. Two other approaches have been employed starting from chloromethylated polystyrene: either reacting it with ammonia in dioxane followed by a reaction with chloroacetic acid sodium and bromoacetic acid or reacting it with iminodiacetonitrile followed by hydrolysis [185]. The ion exchanger containing iminodiacetic acid groups – in the literature also known under the term polycomplexon or komplexon ion exchange resins – has become the most important chelating exchanger. It is commercially available, e. g., as Dowex A-1, and has been repeatedly studied with respect to its complex-forming properties [186]. It is of interest to note that iminodiacetic acid has also been built into a polyacrylate resin according to the following diagrammatic equation:

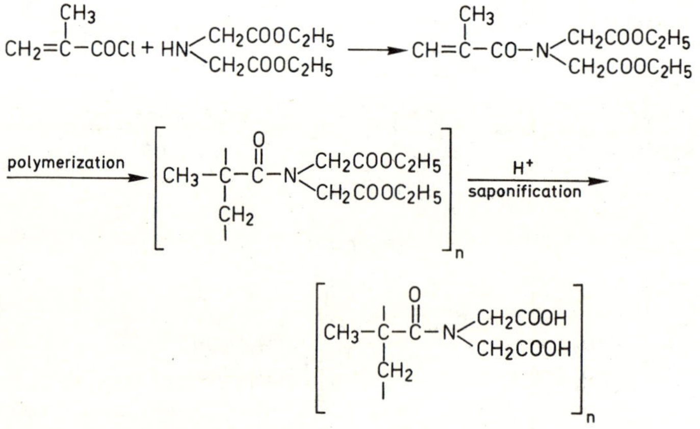

Three more of the more important chelating ion exchangers must be mentioned before a survey is given of the available chelating ion exchange resins. Amidoxime $-C(=NOH)NH_2$ has also been built into a polyacrylic matrix, resulting in a chelating ion exchange material in bead form of weak basicity. This chelating resin has apparently not yet found practical usage. A dithiocarbamate-containing chelating ion exchange resin $-NH-C\begin{smallmatrix}{=}S\\ {-}S-Na\end{smallmatrix}$ was synthesized for the selective complexation and determination of heavy trace metals. The resin can be prepared by the reaction of carbon disulfide with a polyaminepolyurea polyacrylic resin. A chelating resin with dithiocarbamate has also been made for the recovery of uranium from sea water. Starting again from a chloromethylated styrene-divinylbenzene copolymer and reacting it with thiourea an isothiouronium-containing resin $-CH_2-S-C\begin{smallmatrix}{=}NH\\ {-}NH_2\end{smallmatrix}$ can be prepared, which is considered to be the solution for the problem of metal mercury. This chelating ion exchanger forms extremely stable complexes with mercury and most noble metals. The resultant adduct cannot be broken up even with common regenerants so that the common procedure is to burn or wet-ash the resin to recover the exchanged metals.

The iminodiacetic acid chelating resin has become a well-known prototyp of aminopolyacetic acid chelating resins. Researchers have now also succeeded in preparing ion exchange resins containing ethylenediaminetetraacetic acid (EDTA) and even diethylenetriamine tetraacetic acid (DTTA). Resins with these two groups were prepared by copolymerization of the respective monomers with divinylbenzene as a crosslinker. These monomers are m-dibromoethylstyrene and *N,N*-bis(2-aminoethyl)-p-aminostyrene, respectively. The method of synthesis has been described in detail, from monomer synthesis to attachment of the chelating groups onto the copolymer matrix followed by after-treatment for each resin. The cation exchange capacity was found to be 3.5 mequ/g for EDTA-type and 5.1 mequ/g for DTTA-type resins, respectively [187]. Among monomeric chelating agents the aminopolycarboxylates are the main products used for chelating purposes, of which the predominant aminopolycarboxylates are the sodium salts of ethylene-diaminetatraacetic acid (EDTA), hydroxyethylenediaminetriacetic acid, diethylenetriaminepentaacetic acid and nitrilotriacetic acid (NTA). It would be of interest to have all these also as polymeric chelating agents available, which might be just a matter of time given the success of the new synthetic methods described above. An ion exchange resin with anchored nitrilotriacetic acid (NTA) was in fact prepared quite early on, but this product has not been studied extensively in its applications in any part of chemistry [188].

How enhanced the higher selectivity of a chelating ion exchange resin for one specific ion can be in comparison with others was demonstrated with two new materials from Dow Chemical Company. Water-insoluble chelate exchange resins with high selectivity for copper were prepared from chloromethylated derivatives of styrene-divinylbenzene copolymer and 2-picolylamines, thus containing the reactive group $-CH_2N-CH_2-$. Dowex XF-4195 was one such resin, manufactured with *N*-methyl-2-picolylamine in toluene to give a chelate exchange resin with a dry exchange capacity 3.1 mequ/g and tenfold greater selectivity for copper than for iron. But both new chelating ion exchange resins offer many interesting possibilities in the extractive metallurgy of copper, nickel, and cobalt. They are macroporous and their selectivity for copper and nickel over iron and most other cations in acid solution, plus their ability to be regenerated with sulfuric acid or aqueous ammonia, suggests that they should be useful in a number of hydrometallurgical applications [189]. In a comprehensive work Lieser and coworkers have shown that the above mentioned nitration and coupling procedure and the selection of reagents otherwise known for their chelating properties for inorganic trace analysis can be combined fruitfully for the synthesis of a number of chelating ion exchangers based on polystyrene. The following anchor groups were characterized by capacity, exchange rates and distribution coefficients: 4,5-dihydroxynaphthalenedisulphonic acid-(2,7) (chromotropic acid), alizarin, 1,2,5,8-tetrahydroxyanthraquinone (quinalizarin), morin, 1,5-diphenylthiocarbazone (dithizone), 1,5-diphenyl-carbazide, glyoxal-bis(2-hydroxyanil), 1-[thenoyl-(2′)]-3,3,3-trifluoroacetone (TTA), 5-(4-dimethylaminobenzylidene)-rhodanine (DMABR), ethylenediamine-N,N′-bis-(o-hydroxyphenylacetic acid) (Chel I), 1,2,3-trihydroxybenzene (pyrogallol), 4-methoxy-2-[thiazolyl-(2)-azo]-phenol (TAM), bis-sailicylidene)-1,2-diaminoethane (Salen), 8-hydroxyquinoline, pyrocatecholdisulphonic acid-(3,5) (Tiron).

A polystyrene exchanger with 2,4,6-triamino-1,3,5-triazine as an anchor group was then prepared and its properties investigated. Several of these chelating ion exchangers were used for separations, and for the separation and isolation of heavy metals [190].

Overviews of chelating ion exchangers usually classify the materials according to their donor atoms. In the following list this is done in a concentrated form for polymerization chelating ion exchangers.

1. Chelating polymerization ion exchangers containing nitrogen as the sole donor atom:

Chelating polymer	
N-Vinylimidazole	D. H. Gold and H. P. Gregor, 1959
Imidazole with styrene-divinylbenzene copolymer (S-DVB)	J. A. Wellemann et al., 1981
2-(2-Pyridyl)benzimidazole with S-DVB	Patents to Nat. Inst. Metall., Randburg, S. Afr.
N-(2-Pyridylmethyl)-bis-(2-pyridylmethyl)-2,2′-diaminophenol and *N*-(2-pyridylmethyl)-2,2′-diaminobiphenyl with S-DVB	L. R. Melby, 1975
2-Pyridyl-methylamine and *N*-2-(2-pyridyl) ethylenediamine in S-DVB	T. Yokoyama et al., 1983

6-Vinyl-2,2′-bipyridine and 4-vinyl-4′-methyl-2,2′-bipyridine	M. Furue et al., 1982
2,2′-Bipyridine in S-DVB	R. J. Card and D. C. Neckers, 1977
Polystyryl-1,5-naphthyridine and polystyrylpyrido(2,3-b)pyrazine	N. S. Valera and D. G. Hendricker, 1981
1,3-Bis(2′-pyridylimino)isoindolines in S-DVB	W. O. Siegl, 1981
8-Aminoquinoline in S-DVB	G. V. Myasoedova et al., 1977
Polystyryl-g-ethylenimine, DVB crosslinked and grafted	H. Egawa et al., 1971, 1973
Diethylenetriamine with S-DVB	D. E. Leyden et al., 1975, 1976
Formazans in PS, silica and cellulose matrices	A. Kettrup and M. Grote, 1976
Diphenylcarbazide in S-DVB	M. Dore et al., 1976
Diphenylcarbazone in S-DVB	W. Szczepaniak, 1965
Bis(3-aminopropylamine, bis(3-aminopropylphosphine, and diaminopropane	R. S. Drago et al., 1980

2. **Chelating polymerization ion exchangers containing nitrogen and oxygen as donor atoms:**

N-Methylaminoacetic acid in S-DVB	R. Hering, 1967
β,β′-Iminopropionic acid in S-DVB	L. Wolf and R. Hering, 1958
Sarcosine, methylaminopropionic acid, and iminodiacetic propionic acid in S-DVB	R. Hering et al., 1962, 1965, 1968
N-Acetoiminodicarboxylic acid and amino-acetone-*N,N*-diacetic acid in S-DVB	M. Marhol and K. L. Cheng, 1974
Diethylenetriaminetetraacetic acid in S-DVB	O. Szabadka et al., 1980
Poly(hydroxamic acid)	Numerous investigations; macroporous, e. g., F. Vernon and H. Eccles, 1975
Trihydroxamic acid grafted to several polymers	R. S. Ramirez and J. D. Andrade, 1974
N-Phenyl- and *N*-methylhydroxamic acid with macroporous S-DVB	R. J. Philips and J. S. Fritz, 1982
Poly(acrylamidoxime) crosslinked with DVB	M. B. Collela et al., 1980
Di-amidoxime in S-DVB	Y. Xia and J. Yan, 1982
o-Hydroxyoxime in phenyl acrylate	D. J. Walsh et al., 1983

Hydrazide in MMA-DVB	H. Egawa et al., 1977
Acrylic acid hydrazide	Japan. Pat., 1982
Tertiary amides in S-DVB	G. M. Orf and J. S. Fritz, 1978
Methyl 1-vinylimidazole-4,5-dicarboxylate and dimethyl-2-(p-vinylphenyl)imidazole-4,5-dicarboxylate with p-DVB	G. Manecke and R. Schlegel, 1976
Poly-1-(3,4-dicarboxy-pyrazol-2-yl)ethylene and 1-(4,5-dicarboxy-1,2,3-triazol-2-yl)ethylene with p-DVB	G. Manecke et al., 1971, 1979
Ethylenediamine in glycidylacrylate-ethylene-dimethacrylate and glycidylacrylate-methylene-bis-acrylamide	F. Svec et al., 1975, 1976, 1977
Schiff bases, e. g., 5,5′-methylene-disalicylaldehyde-1,2-propanediamine copolymer	F. A. Bottino et al., 1983

3. Chelating polymerization ion exchangers containing oxygen as the sole donor atom:

Sodium malonic acid monoester in S-DVB, hydrolyzed	S. B. Makarova et al., 1971
Poly(aldehyde carboxylic acids)	H. Haschke and E. Bader, 1971
Acetylacetone in S-DVB	S. Bhaduri et al., 1981, 1982
1-Theonyl-(2′)-3,3,3-trifluoroacetone in S-DVB	M. Griesbach and K. H. Lieser, 1980
β-Diketone groups in S-DVB macroreticular	A. Sugii et al., 1977
Poly(5,8-dihydroxy-3-vinyl-1H-benz/f/indazole-4,9-dione and its 1-methyl and 2-methyl derivatives	G. Manecke and E. Graudenz, 1973; F. K. Chow and E. Grushka, 1978
Poly(2-vinylpyridine-1-oxide), poly(4-vinylpyridine-1-oxide) and 2,2′-trimethylenedipyridine-1-oxide	K. A. Kadie and P. Holt, 1976

4. Chelating polymerization ion exchangers containing sulfur or nitrogen and sulfur as donor atoms:

Thiomethylstyrene crosslinked with DVB	J. R. Parrish, 1955; J. M. J. Frechet et al., 1979
Dithiocarbamates in various matrices	Z. Slovak, 1979; S. Siggia et al., 1974, 1977

Dithiocarbamates in S-DVB	M. Okawara and S. Sumitomo, 1958; K. Hirtani et al., 1981
Poly(iminoethylene)dithiocarbamate	P. C. H. Mitchell and M. G. Taylor, 1982
Aminoethylmercaptan and diethyldithiocarbamate in S-DVB	T. Saegusa et al., 1978
Disodium-4,4′-(4-di-azenediyl-5-mercapto-3-methyl-1,2-diazacyclopenta-2,4-dien-1-yl)dibenzene-sulfonate in Amberlite IRA-400	M. Chikuma et al., 1979, 1981
Thioglycoloyloxymethyl in S-DVB	R. J. Philips and J. S. Fritz, 1978
Isothiouronium in S-DVB	G. Koster and G. Schmuckler, 1967

Chelating polymerization ion exchangers with macrocyclic polyethers and related ligands. The chemistry of the macrocyclic crown ethers and cryptands as well as of the acyclic pendands, e. g., the glymes, has become a wide field since the pioneering work of Pedersen and Lehn in 1969. Hundreds of compounds have been synthesized since and examined for possible applications. Pedersen and Lehn found that the crown ethers can bind a variety of solutes including transition metal ions, lanthanoides, organic cations (such ammonium and diazonium) and even neutral molecules. Since the macrocyclic polyethers and related ligands are expensive, it became a reasonable choice to use them on polymeric supports so that they can be retrieved and recycled. For the exchange of ions the field of crown ethers was first studied by Kopolov, Hogen-Esch and Smid, who polymerized vinyl crown ether derivatives and vinylbenzoglymes. The following schematic formula are examples: poly(vinylbenzo-18-crown-6) (I) and poly(vinylbenzoglyme) (II) as well as a bicyclic crown ether, i. e., a cryptand (III):

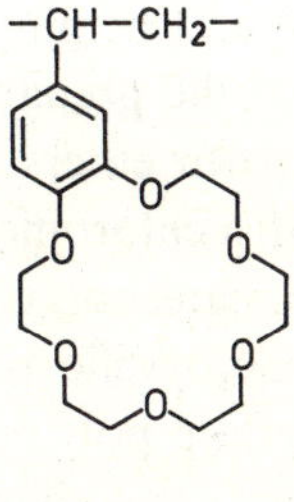

I

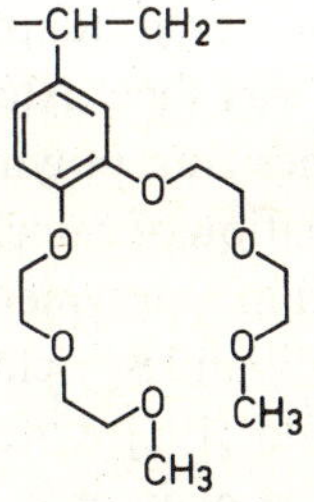

II

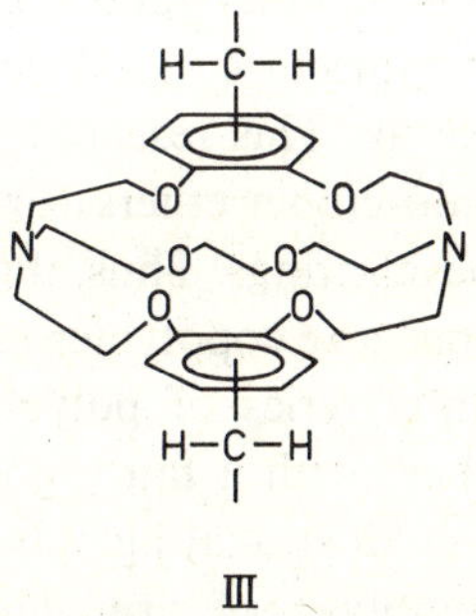

III

The products I and II are water-soluble and the binding of solutes to them is of greater interest under the aspect of polyelectrolytes and polysoaps [191]. Ion exchange resins can then be obtained by the crosslinking in a usual copolymerization reaction of monovinylbenzo crown compounds with divinyldibenzo crown compounds or divinyl benzene, leading to polystyrene-like matrices. When starting from chloromethylated polystyrene a substitution reaction can be performed, e. g., monobenzo crown compounds substituted by amino, hydroxymethyl or ω-bromoalkyl groups linked to the styrene-divinylbenzene matrix by $-C-NH-C-$, $-C-O-C$ or $-C-C-C-$ bonds. Besides the crown ethers with oxygen, as a heteroatom, there are also crown ethers with oxygen, nitrogen, and sulfur as heteroatoms, or crown ethers, cryptands and urea analogues with oxygen and nitrogen as heteroatoms. But all these reactive groups are neutral ligands which take up cations and anions simultaneously in order to maintain electroneutrality. All the products containing nitrogen in the reactive group show a complete protonation of the nitrogen at a pH below 2. Then there is no more complex formation with salts and due to the quaternary amino groups formed an anion exchanger results [192]. Warshawsky and coworkers synthesized a number of polymers, which they called polymeric pseudo-crown ethers, based on a one-step in-situ cyclisation reaction between a chloromethylated styrene-divinylbenzene copolymer and polyoxyalkylene under Williamson ether synthesis conditions, as shown in the following schematic equation:

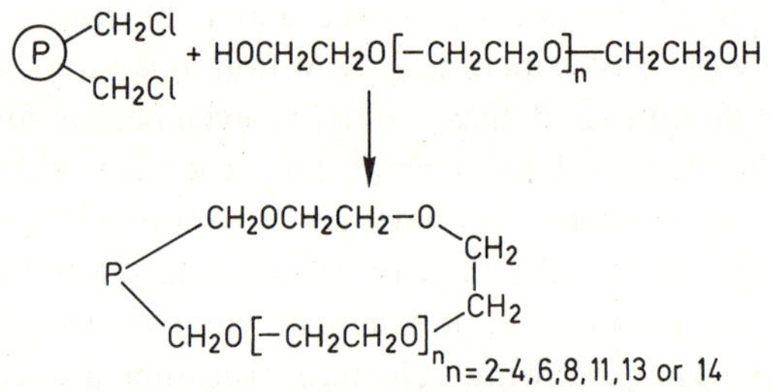

For this synthesis slightly crosslinked polymers with intermediate flexibility were best suited as they had as good accessibility due to high swelling and diffusion to large polyether molecules, good conformational adaptability and a minimum of further crosslinking. This research work was then extended to the preparation of polymeric pseudo-crown ethers for synthesizing polymeric crown ethers carrying pendant mycrocyclic rings. Thus, the alkylation of catechol with a chloromethylated styrene-divinylbenzene copolymer resulted in a polymer-bound catechol, which can be reacted with a series of polyglycol dihalides yielding macrocyclic polymeric benzocrown ethers with a fairly good yield [193]. Other chelating polymerization ion exchangers with special ligands may be of interest. A macrocyclic hexaketone for the strong uptake of uranyl cations has been bound to chloromethylated polystyrene in DMF using potassium carbonate as a condensing agent (see formula I below); this was used for extracting uranium directly from sea water in chloroform [194]. Also bound on chloromethylated polystyrene in DMF in the presence of

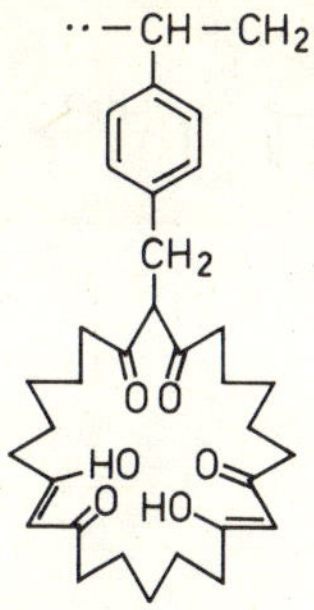

I

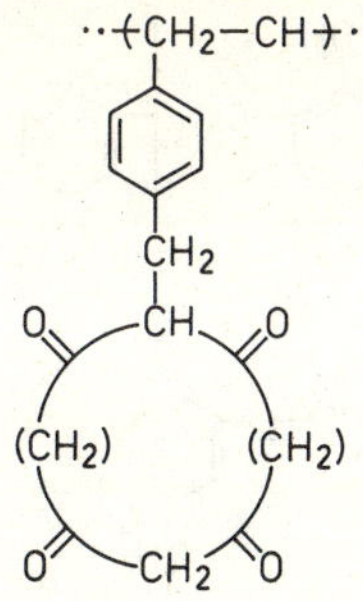

II

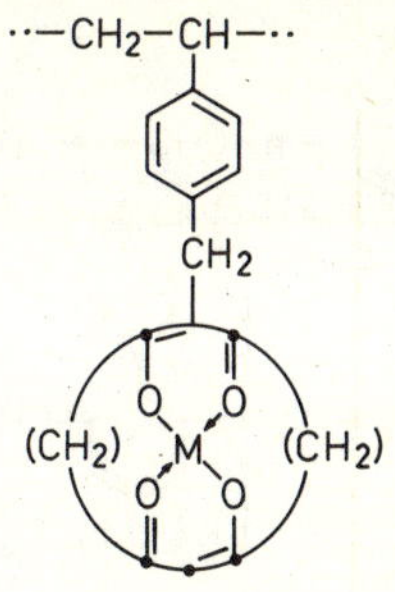

III

sodium hydride were 22- and 26-membered macrocyclic bis-1,3-diketones. This resulted in new chelating polymers of the structure shown in formula II above, which chelate metal ions such as copper, nickel and cobalt (as shown in formula III), with two ketonic groups undergoing enolization and subsequent deprotonation. Copper(II) has been eluted from such reactive polymers with 10% hydrochloric acid [195]. The commercially available cryptands, Kryptofix 221B and Kryptofix 222B (trade mark E. Merck, Darmstadt, FR Germany) bound to resins consisting of styrene and 2% divinylbenzene according to Merrifield are commercially available as chelating resins.

1.2.2.3 Polycondensation ion exchangers

As an introduction to polycondensation ion exchangers it would seem appropriate to review the pioneering work of Adams and Holmes of the Chemical Research Laboratory at Teddington, England. They first synthesized polycondensates of various polyhydric phenols with formaldehyde as the crosslinking agent, and demonstrated their capability for cation exchange on the very weakly dissociated phenolic OH-groups. This work was soon extended to the polycondensation of aromatic polyamines with formaldehyde to yield weakly basic materials which, therefore, could take up only strong mineral acids. Both these early materials had only partially dissociable ionogenic groups, and as a result their ion exchange capacity and their exchange rates were pH-dependent. What was needed, however, were strong acid and strong base ion exchangers. Thus from 1935 on, they prepared and manufactured quite a number of useful polycondensation ion exchangers. But by 1950 they had been overtaken in the market by polymerization ion exchangers based on polystyrene. Nevertheless, it would seem to be appropriate to describe briefly developments in the synthesis of polycondensation ion exchangers since there are cases where they are still in use [195, 196].

Pure phenolic polycondensation resins. The polymer formation of these simple weak acid cation exchangers may be diagrammatically represented as follows:

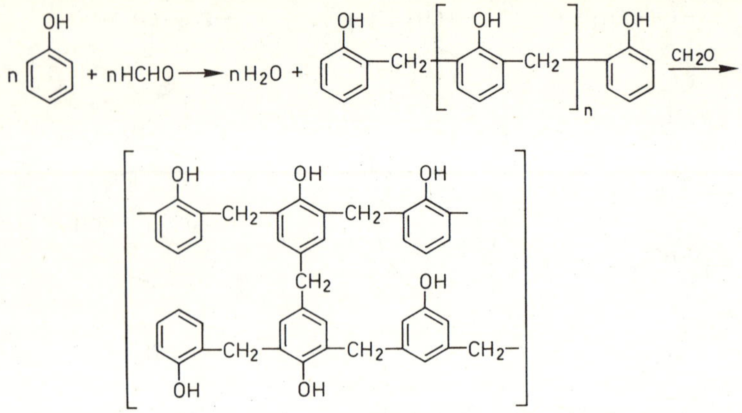

Water is formed along with the polymer, a general characteristic of polycondensation reactions. By adding a considerable excess of formaldehyde, the condensation reaction proceeds with the formation of long chains three-dimensionally bridged by methylene groups. As basic materials all kinds of phenols, especially phenol and resorcinol, (as well as tanning materials containing phenolic groups) can be used. During the elaboration of procedures for technical manufacture it became possible to increase the exchange capacity. In order to obtain thermostable cation exchangers, resins were prepared by heating phenol derivatives, such as ClC_6H_4OH and $Cl_2C_6H_3OH$, in the presence of metal chlorides. Phenol-formaldehyde resin ion exchangers can also be prepared in pearl form by suspension polymerization. Such ion exchanger pearls containing homogeneously distributed magnetic γ-iron oxide were prepared for the purpose of deionizing water. The ion exchange on phenolic ion exchangers containing only phenolic exchange groups was investigated with regard to the equilibrium of alkali metal exchange and it was found that these resins were less selective toward a Na-K mixture than were monoionic sulfonated ion exchangers, but the selectivity was significantly higher toward Rb and, especially, Cs [197]. The synthesis of porous cation exchangers based on phenol was also described. These materials are prepared by the reaction of phenol with paraformaldehyde in the presence of pore-forming agents, so that the structure of the resins consists of both discrete pores and gel structures having a high ion exchange rate in the initial stage due to an increased diffusion of ions [198]. The form obtainable by replacing the phenolic component by pyrocatechol was effective in the removal of antimony from solutions. Making use of the effect of structure of an ion exchange resin on its properties phenolic polycondensation resins were also prepared using m-cresol or p-cresol with formaldehyde, the latter mainly in view of a regular distribution of the ionizable groups [199].

Carboxylic polycondensation ion exchangers. A general synthetic means of obtaining carboxylic acid cation exchangers is to condense 1,3,5-resorcylic acid with phenol and an excess of formaldehyde having most probably the building block

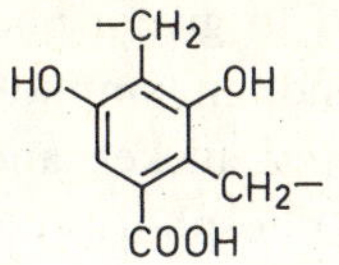

making the resin multiionic. The carboxylic acid group is bound to the aromatic ring directly. It may be mentioned that such resins had already been prepared in 1943, and were thus the first carboxylic ion exchangers. A carboxylic acid group bound directly to the aromatic ring can also be introduced into the finished polycondensate (as was developed very early by Farbenfabrik Wolfen), by treating it with carbon dioxide and sodium bicarbonate under pressure and at elevated temperatures [200]. A biionic cation exchanger also is produced from furfural aldehyde by polycondensation with hydrobenzoic acid having a ring-bound carboxylic group. The same applies to cation exchange resins resulting from reactions between furfural and salicylic acid. In order to obtain carboxylic polycondensation cation exchangers having the carboxy group in the side chain resorcinol-*O*-acetic acid was first prepared by etherification of the phenolic hydroxyl groups with chloroacetic acid and then polycondensed with formaldehyde. Such monoionic resins containing the reactive groups in side aliphatic chains bound with aromatic nuclei through oxygen, and not containing phenolic hydroxyl groups, can be synthesized by polycondensation of formaldehyde and phenol ethers, $PhOCH_2X$ or $PhOCH_2CH_2X$, with X = COOH, as well as SO_3H, PO_3H_2 or Me_3N^+, an idea that was taken up again and elaborated extensively [201]. The patent specification describes ion exchange resins with an increased degree of structuring and improved sorption properties which can also be prepared by polycondensation of benzyl bromide with furfural, followed by adding ionic groups to the polymer. Diphenyl oxide-formaldehyde polymers and ion exchangers based on them were reported to have been prepared by the polymerization of diphenyl oxide with formaldehyde or paraformaldehyde in the presence of sulfuric acid as a catalyst.

Phosphonic and phosphinic polycondensation ion exchange resins. The phosphorus-containing ion exchangers were also extended to types made by polycondensation. As already mentioned above phenol ethers bearing the PO_3H_2-group can be synthesized and converted by polycondensation with formaldehyde. Phenoxyalkylphosphonic acids $PhO(CH_2)_nPO(OH)_2$ (n = 1,2,3,6) were polycondensed with formaldehyde in the presence of 1% hydrochloric acid to give ion exchange resins containing $PO(OH)_2$-groups at the end of aliphatic side chains of various lengths, with the best kinetic ion exchange properties being observed for the resin from phenoxymethylphosphonic acid [202]. Bis(hydroxymethyl)phosphinic acid and its derivatives was used to obtain phosphorus-containing oligomers and polymers by polycondensation, one example being an Na bis(chloromethyl)phosphinate-1,4-dihydroxybenzene-formaldehyde copolymer ion exchanger [203]. Furfural and benzyl bromide were polymerized to yield a copolymer which was phosphorylated with

PCl_3 in the presence of $AlCl_3$ and oxidized with nitric acid to give a phosphonic acid-containing cation exchanger; an anion exchanger and an amphoteric ion exchanger were obtained as well, for the sorption of copper, nickel, and uranyl ions [204]. In continuation of the preceding investigations on phosphorus-containing ion exchangers based on aniline with formaldehyde and H_3PO_3, or by reacting formaldehyde, phenylaminoalkyl(aralkyl)phosphonic acid or its esters and phenol, ion exchangers with aminoalkylidenediphosphonate groups were prepared by condensation of *N*-arylaminoalkylidenediphosphonic acids in alkaline media with formaldehyde and resorcinol or phenol as a crosslinking agent [205].

Sulfonic acid polycondensation resins. The two principal ways feasible for obtaining strong acid sulfonic type polycondensation ion exchange materials are the sulfonation of phenolic polycondensates with sulfuric acid according to the diagrammatic equation:

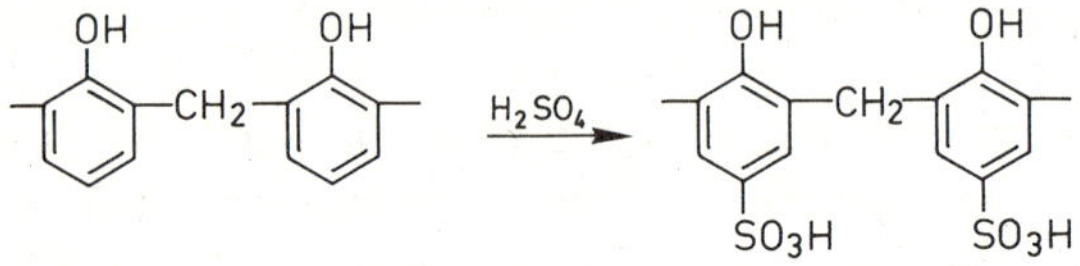

or the synthesization of phenolsulfonic acid followed by polycondensation with formaldehyde according to the schematic equation:

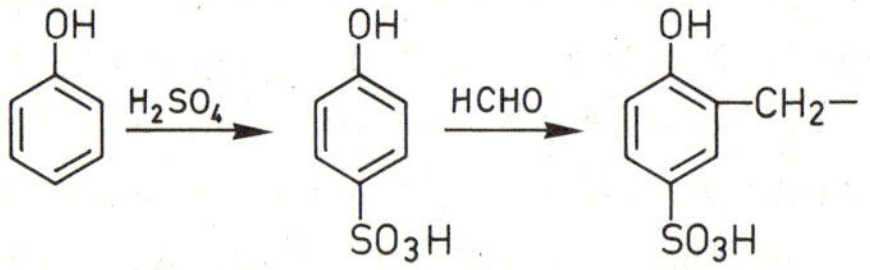

The latter is the more elegant and easier way. In both cases the materials obtained are multiionic, containing the weak acid phenolic and the strong acid sulfonic ionogenic group. The properties dependent on the reaction conditions of cation exchange resins formed by the acid condensation of phenolsulfonic acid with formaldehyde were examined thoroughly [206]. From these results optimum conditions for the manufacture of the resin can be derived, as well as indications for a structural representation of this type of ion exchanger. These aspects have been followed up mainly with respect to the exchange capacity of sulfonated ion exchangers based on phenol-formaldehyde condensates [207]. Another important synthetic way of obtaining strong acid sulfonic phenol resins consists in anchoring the sulfonic acid groups to the aromatic nucleus by means of formaldehyde as a bridging agent. This is achieved by simultaneous reaction of a phenol with formaldehyde and sodium sulfite, obtaining the sodium salt of an oxyphenylmethylene sulfonic acid, which is subsequently polycondensed with an excess of formaldehyde to a water-insoluble resin, as shown diagrammatically in the following equation:

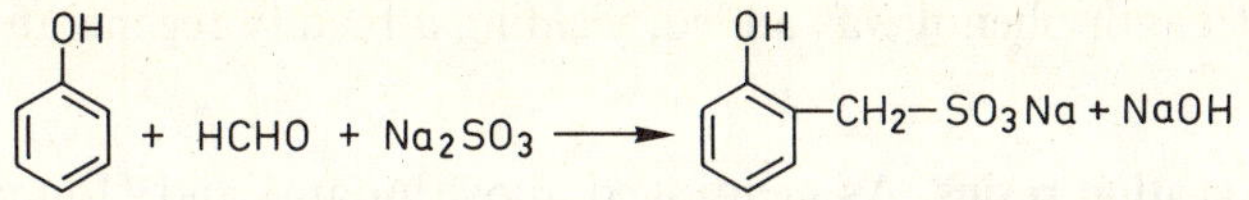

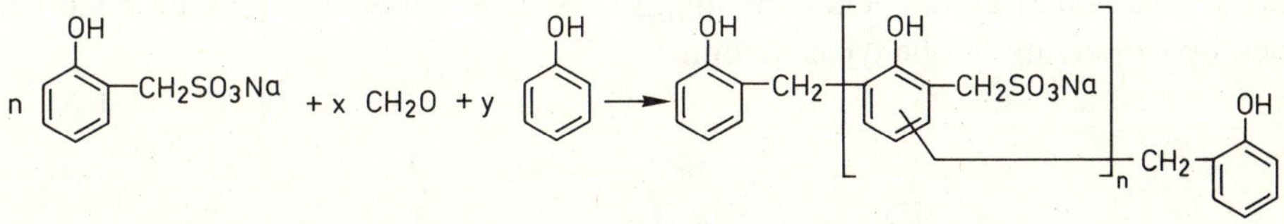

The reduced capacity encountered in applying sulfonating agents during the course of condensation can be avoided by additionally employing aldehydes bearing sulfonic acid groups, as, e. g., benzaldehyde-2,4-disulfonic acid, which, as an example, reacts according to the following diagrammatic equation with resorcinol:

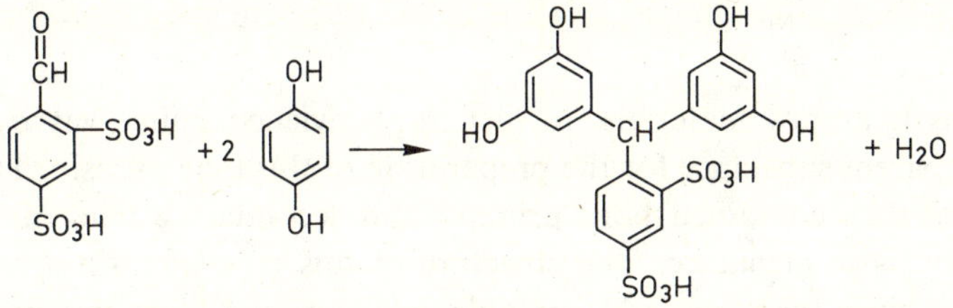

Making use of all these synthetic possibilities very powerful strong acid, biionic cation exchangers were developed and applied industrially. In a further process for manufacturing phenol sulfonic resins hydroxybenzylalcohol is sulfonated as the first step of the phenol-formaldehyde condensation:

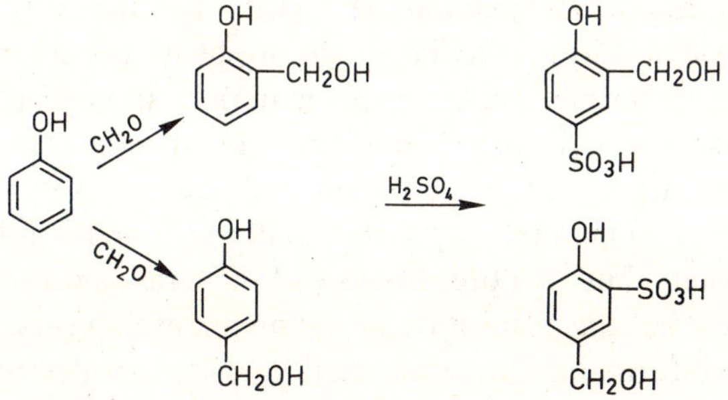

and subsequently polycondensed with an excess of formaldehyde, yielding a stable resin matrix. The above-described processes were all specified in European and American patents which have since expired. Sulfonic acid polycondensation resins were also prepared out of toluenesulfonic acid and, especially, naphthalenesulfonic acid with furfural, and a Russian patent specified acenaphthene as being sulfonated with the resulting sulfonic acid being condensed with formaldehyde; further, to improve the mechanical strength phenol was added, yielding a readily regenerated strongly acidic resin.

Aromatic amine polycondensation resins. As mentioned above Adams and Holmes had prepared aromatic polyamines crosslinked with formaldehyde, obtaining weakly basic anion exchangers consisting of the following diagrammatic building block by employing m-phenylenediamine

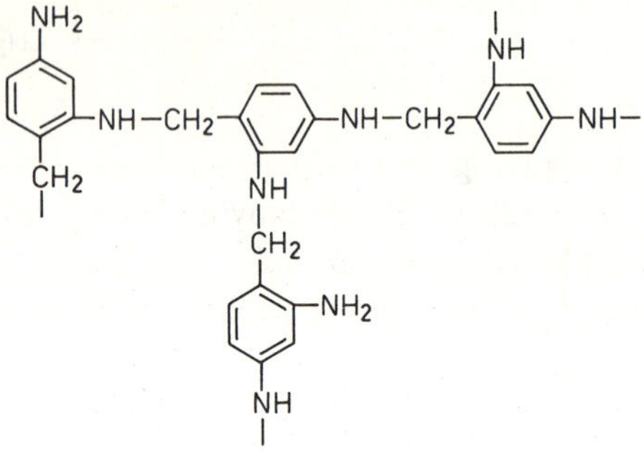

and formaldehyde as raw materials. In fact m-phenylenediamine was by far the most common parent substance for the preparation of this type of resin. They were multiionic since they contained both primary and secondary amino groups and were all weakly basic in nature. The structure of this type of resin seems to be rather complex, since reaction with formaldehyde can take place not only in the benzene ring but also through the amine groups. In order to increase the basicity of these anion exchangers, polyethylene polyamines, particularly tetraethylenepentamine, have been proposed, to give a higher concentration of amino groups and hence a higher capacity. Other manufacturing processes for the same resins included the condensation with guanidine derivatives or with components like dimethyl sulfate, hydrazine hydrate, etc., which increase the basicity. Further developments of aromatic amine polycondensation resins based on m-phenylenediamine led to spheroid or bead-shaped and porous ion exchange materials, used to decolorize sugar syrups, deodorize liquids and absorb ammonia. The condensation of m-phenylenediamine was also carried out in the presence of water-soluble inorganic salts such as lead, cobald or iron salts, to produce a porous condensation resin with high decolorizing capacity, but the pore structure of these resins was obviously unknown. In intensive studies of the adsorption of gas on ion exchangers, the pore structure of m-aminophenol-formaldehyde condensation resins was determined by condensation in the presence of a water-soluble inorganic salt, NaCl, $CaCl_2$ etc., with the pore structure being obtained by appropriate measurements [208]. This and similar types of polycondensation resins could perhaps be revived for special applications if their properties could be improved, for instance by a two-stage polycondensation process with or without pore-forming additives [209]. Another example could be the purification of waste waters of organic compounds by sorption on formaldehyde-m-phenylene sulfate-resorcinol resins. Anion exchangers were also prepared by Russian research workers by polycondensation of furfural with m-phenylenediamine, resulting in materials with satisfactory exchange capacities and excellent resistance to heat, chemicals and γ-irradiation.

Aliphatic amines and ammonia polycondensation resins. There is more variation in composition and properties among weak base anion exchange resins than among any other class of ion exchangers. For condensation resins monomeric amines, so-called polyethylene polyamines, or ammonia itself, can be seen as the parent materials, and if formaldehyde is used as the crosslinking agent the two principal possibilities are either the formation of a methylol group or of a Schiff base depending on the reaction conditions, as shown in the following equation:

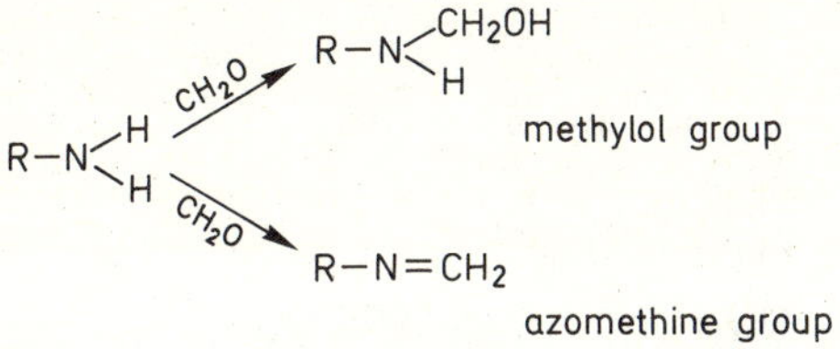

Both the methylol as well as the azomethine group can further react to form polycondensates, the ratio of primary, secondary and tertiary amino groups depending on the quantity of formaldehyde employed. Amino resins are in general based on melamine and urea as the amine components, as was also the synthesis of melamine-guanidine-formaldehyde anion exchangers [210]. Rather complex were the preparation of anion exchange condensation resins based on, e. g., ethylene dichloride as crosslinking agent and gaseous ammonia or polyethylene polyamines, which themselves are reaction products of ethylene dichloride and ammonia. Another variety of the aliphatic amine polycondensation resins was manufactured by condensation polymerization of polyacrylic acids with tetraethylenepentamine. Polycondensation of furfural with polyethylene polyamine gave a copolymer useful as an anion exchanger [211]. Inumerable variations are possible, and some of them have been exploited commercially, although most of these resins seem now to be obsolete.

Epichlorohydrin-based polycondensation anion exchangers. Of special importance for the preparation and manufacture of anion exchangers is the use of epichlorohydrin as a crosslinking agent. By condensing aliphatic amines and epichlorohydrin medium-strong and strong base anion exchangers can be produced. These resins are, therefore, also referred to as epoxy or medium basic resins. The first resins of this class to contain strong base groups were described in the United States in 1947 as the polycondensates of epichlorohydrin and tetraethylene pentamine, ethylene diamine, diethylene triamine, bis(3-aminopropyl)amine or a polyethylene polyamine [212]. In water treatment such resins removed silicic acid making possible a complete deionization. The development of these anion exchangers has been continued [213].

For the underlying chemical reaction the amine groups condense either with the chloromethyl group or with the epoxy group or even with both according to the equation:

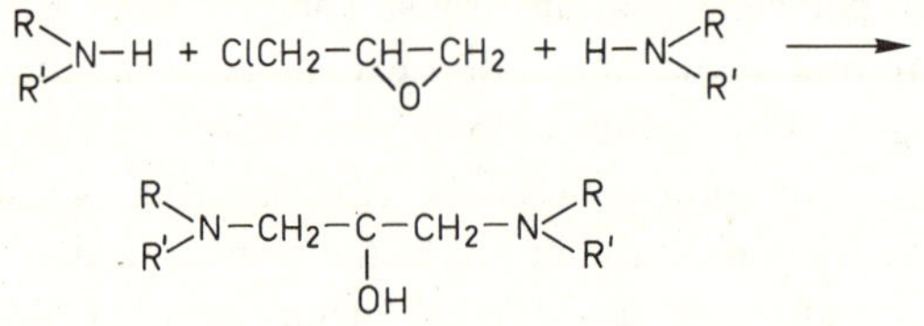

In the case of condensation with triethylenetetramine, for instance, the following diagrammatic network results:

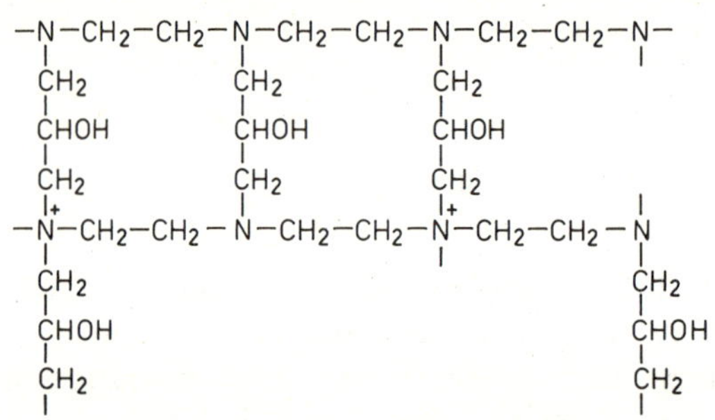

This biionic network contains both weak base (tertiary amine) and strong base (quaternary ammonium) groups. The presence of the latter and of the hydroxymethyl groups results in a highly hydrophilic gel type resin. By bulk polymerization granular materials are obtained, and beads are made by suspension polymerization in a solvent not miscible with water. One patent specification claimed the preparation of exchange resins with a porous structure by reacting polyethylenimines with epichlorohydrin in the presence of organic solvents. In this case the basicity represented by quaternary ammonium groups can be influenced by altering the ratio of the reactants. On the other hand, a weak base anion exchanger was specified as having been obtained in bead form by the reaction of epichlorohydrin with aqueous ammonia at elevated temperature, transferring the reaction product subsequently into an organic phase. By a similar but simplified method, The Dow Chemical Company has manufactured a granular material. New anion exchangers based on homo- and copolymers of epichlorohydrin can be prepared from polyepichlorohydrin and epichlorohydrin-styrene copolymer by crosslinking with aliphatic and aromatic di- or polyamines; it was shown that the purity of polyethylene polyamine affects the gelation time of epichlorohydrin-ethylenediamine copolymer ion exchangers [214]. Extending the preparation of weak base anion exchangers from epichlorohydrin onto general epoxy polymers, reaction products with polyalkylene polyamine can be obtained that are useful as anion exchangers [215].

But epichlorohydrin-based anion exchangers can also be obtained with vinylpyridines: as an example one has an epichlorohydrin-2-vinylpyridine copolymer with high quality parameters and high exchange capacity with the reaction involving both vinyl polymerization and opening of the epoxy group, leading to the formation of three-dimensional structures containing quaternized and free pyridine nuclei [216].

Aminated polycondensation products of phenol and formaldehyde. For the preparation and manufacture of weak base anion exchangers phenols have also been used (in spite of their weak acid properties), incorporated into an ion exchange material by their phenolic groups. This compensated for by employing amines of medium or high basicity. The aminated polycondensation resins of phenol and formaldehyde are also more durable than when condensation of amine, phenol and formaldehyde is conducted in a single run when the synthesis is carried out in a two-step reaction, as shown in the following diagrammatic equation:

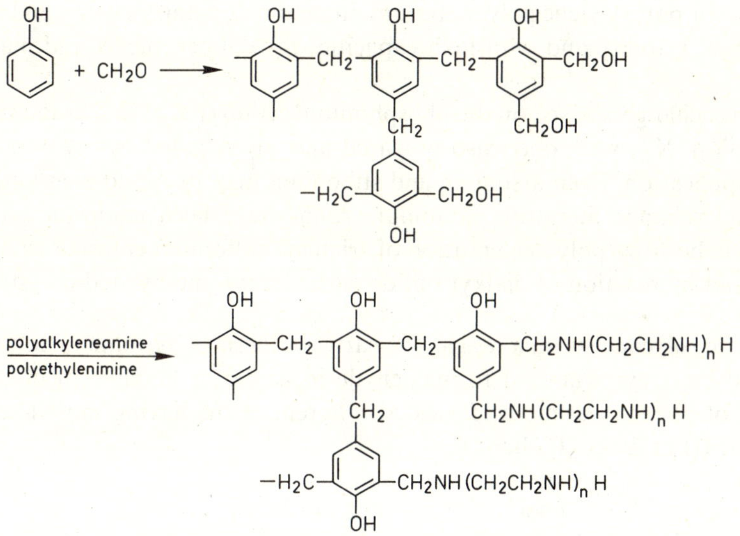

Chemical companies in America developed such anion exchangers early in the nineteen-forties, they were followed by others with modified processes. Thus the kinetic properties and mechanical strength of a product consisting of phenol, formaldehyde and polyethylene polyamine was improved by adding sodium chloride and sodium sulfate during its preparation by polycondensation. Finally very porous resins of this class of anion exchangers have gained commercial acceptance because of their increased reaction rates and good physical durability [217]. A porous phenol-formaldehyde-urotropine ion exchanger prepared by two-step polycondensation is used for the sorption of humic substances from water for electronic and semiconductor equipment [218].

Other polycondensation anion exchangers. New anion exchangers from diphenyl ethers or diphenyl oxide via chloromethylation and subsequent amination have been prepared. The polycondensation of 4,4′-bis(chloromethyl)diphenyl ether or 2,4′-bis(chloromethyl)diphenyl ether with polyethylene polyamine or polyxylylene polyamine gave ion exchange resins which had higher heat resistance, mechanical strength and chemical stability than ion exchangers based on epichlorohydrin and the same amines [219]. Multiionic anion exchangers with increased permeability and improved kinetic and sorption properties were prepared by condensation of a chloromethylated diphenyloxide formaldehyde oligomer with di- and polyamines [220]. Toluene-formaldehyde copolymers were used (under the name formolite) as the matrix for anion exchangers. They are obtained by chloromethylation and amination, and have a globular structure with uniformly distributed globules and a narrow distribution range of pores [221]. Silane-based ion exchangers can be found in the early ion exchange literature in the form of a benzyl-triolsilane p-sulfonic acid cation exchanger, but they were also investigated as anion exchangers by condensing phenyl-2,3-chloropropoxysilane with polyethylenimine, polyethylene polyamine or polyxylylene polyamine, obtaining ion exchangers with good chemical and thermal stability and with high capacities for copper, nickel and magnesium [222].

Anion exchangers based on the phosphonium group $(PY_4)^+X^-$, or the sulfonium group, $(SY_3)^+X^-$, were once also prepared and investigated but proved to be of limited application. Even arsonium and stibonium may be found mentioned in the older ion exchange literature. Sulfonium resins have been made on a relatively small scale both by polycondensation of trianisyl sulfonium chloride with formaldehyde and by reaction of dialkyl sulfide with a chloromethylated polystyrene.

Amphoteric polycondensation resins. The early synthesis of asymmetric ion exchangers based on L-tyrosine and formaldehyde in acidic or alkaline media with the addition of phenoxy acetic acid yielded different resins having most likely either methylene (I) or ether (II) bonds:

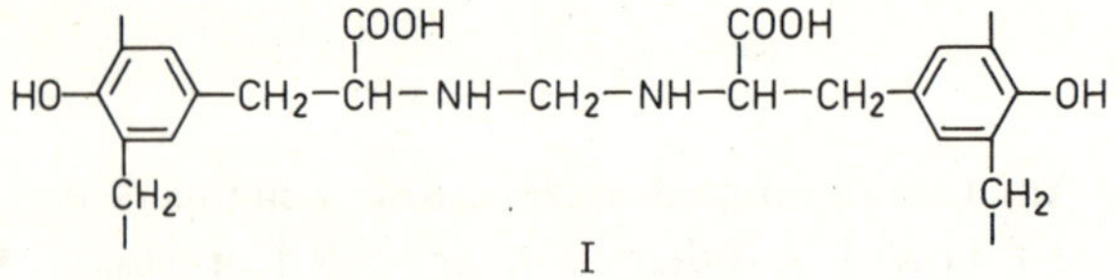

I

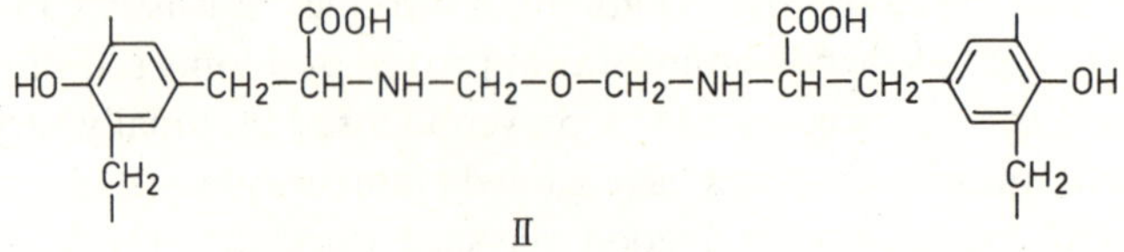

II

These resins have amphoteric properties because of their OH- and HOOC-groups, as well as their amino groups [223]. With the purpose then of synthesizing amphoteric ion exchange resins of aminophosphonic acid- and of aminocarbonic acid-type, ethylendiamine, triethylenetriamine and polyethylenimine were partially substituted by reaction with chloromethylphosphonic and/or chloroacetic acid, and then crosslinked by a condensation reaction with trimethylolphenol, phenol and formaldehyde or with epichlorohydrin (Manecke). Ion exchange resins with aminophosphonic acid groups were also synthesized by reaction of a crosslinked polyethylenimine with chloromethylphosphonic acid and chloromethylphosphonic esters. Starting with polyethylenimine and p-bromophenylarsonic acid a resin was produced which contained aminophenylarsonic acid groups. The composition of these amphoteric resins, their swelling and the uptake of Cu^{2+} ions by complex formation at pH 6 were investigated [224]. Although the field of amphoteric polycondensation resins was held to be of minor importance. There was continued synthesis of amphoteric resins based on polycondensation and polymerization ion exchangers, by reacting them with chloroacetic acid, the preparation of crosslinked amphoteric polyelectrolytes, the synthesis of anthranilic, salicylic, and gallic acid-melamine-formaldehyde amphoteric exchangers. Further the manufacture of amphoteric resins was achieved via the polycondensation of epichlorohydrin with maleic dihydrazide, curing the oligomer obtained with a mixture of hydrazine hydrate and nonaqueous polyethylene polyamine, and subsequently carboxymethylating with chloroacetic acid, or preparation from bisphenol A, formaldehyde and various melamine derivatives, with the latter being fibrous. New organophosphorus amphoteric polycondensation resins containing diaminoalkylphosphinic groups were prepared by means of a one-stage polycondensation of di- and polyamines with formaldehyde and H_3PO_2, and ion exchangers with aminoalkylidenediphosphonate groups were prepared by condensation of *N*-arylaminoalkylidenediphosphonic acids in alkaline media with formaldehyde and resorcinol or phenol as a crosslinking agent. Because most of the amphoteric resins have been prepared for an enhanced uptake of specific ions some of them will be discussed below under chelating ion exchangers.

Specific and chelating polycondensation ion exchangers. In a Japanese patent an ion exchange resin highly selective for potassium over sodium was claimed to have been synthesized by a mixture of p-nitrophenol and p-phenolsulfonic acid with formaldehyde: it was to show an uptake of potassium and sodium in the ratio of approximately 85 : 15, and was believed to act by a random structure like the dipicrylamine specific resin of Skogseid. This was published in 1957 and three years later in a review of the literature on the preparation of ion-specific polymers the comment was made that much of the work reported up to then had little significance, since the polymers were not well-defined and were poorly characterized [225]. It was then claimed in a US patent that boron values can be removed from aqueous media by contacting the media with a water-insoluble solid anionic ion exchange phenolic resin containing aromatic ortho-hydroxy carboxylic groupings, with the resin having been crosslinked with aldehydes in the presence of a cationic species.

Thiooxine group containing highly selective ion exchangers for heavy metal ions were prepared by polycondensation of 8-mercaptoquinoline with formaldehyde and various naphthols and phenols, which showed a decrease in uptake through the series mercury(II), copper(II), nickel(II), cobalt(II), zinc(II) and lead(II), according to the strength of intra-complex bonding of these metals [226]. Starting from formaldehyde and phenol sulfides or from phenol sulfide, methylol derivatives and polyethylenepolyamines the synthesis of highly selective sulfur-containing exchangers for the extraction and separation of molybdenum and tungsten were described, and inexpensive resin compositions were obtained by heating a urea-formaldehyde copolymer with carbon disulfide to be used for the removal of mercury from waste solutions. A polymer containing α-dioxime groups was obtained by condensing α-furildioxime with phenol and formaldehyde in the presence of sodium hydroxide. All cations for which α-furildioxime is a reagent were sorbed by this exchanger, i. e. lead(II) in 0.1 N hydrochloric acid, copper(II) in acetate buffer pH 5.2, and nickel(II), cobalt(II) and zinc(II) in an ammoniacal medium. A sulfonated phenol-formaldehyde resin was made highly selective for the uptake of titanium by the incorporation of chromotropic acid.

W. A. Klyachko synthesized the first polycondensation chelating ion exchange resin by polycondensation of phenol and formaldehyde, with ethylenedinitrilotetraacetic acid incorporated [227]. This well-known chelating agent was fixed as a solid solution within the phenol-formaldehyde condensate. This inferior method was independently altered by Gregor and coworkers and by von Lillin by selecting chelating agents which are capable of resin formation, thus becoming part of the polycondensed chelating resin matrix [228]. o-Aminophenol, anthranilic acid and/or m-phenylene diglycine, or anthranilic acid and/or 8-quinolinol were used as chelate monomers; they were followed by investigations on m-phenylenedinitrilo tetraacetic acid together with resorcinol, on the one hand [229], and o-aminophenol, resorcinol, 2,4-dihydroxybenzoic acid, 2,4-dihydroxyacetophenone and 8-quinolinol, on the other [230]. This gave chelating polycondensation resins with enhanced selectivity or even with specificity towards certain metal ions, which made them suitable therefore for chromatographic separations. The second important method for synthesizing chelating polycondensation resins – chemical conversion of polycondensation resins to chelating resins – was invented at about the same time; insoluble inter-linked condensation products of m-phenylenediamine and formaldehyde were treated with chloroacetic acid or its derivatives, forming ion exchangers selective in their action on various metals. The resins obtained had a copper adsorbance of 1.9 mmol per g of resin [231].

The use of polycondensation products as chelating resins has been widely attempted but the results obtained have often been disappointing. This may be due to a great extent to the imperfection of the resins obtained. The use of enclosure compounds yields resins of low capacity because the polycondensates give limited access to the exchanging species; in addition, during the elution and regeneration process loss of activity will occur because of the washing-out of complexing parts.

In the case of polycondensation resins incorporating a chelating ligand, for which in general aromatic amines and phenols are specially suited, the crosslinking phenols introduce additional reactive groups into the resin, which especially in alkaline media can cause further sorption, decreasing selectivity. The durability of such resins under physical and chemical stress conditions is also low. The production of polycondensation resins chemically converted to chelating resins by subsequent introduction of chelating groups is feasible only in a few cases.

It would be possible to present chelating polycondensation resins as reactive groups containing nitrogen, nitrogen and oxygen, oxygen, sulfur, nitrogen and sulfur, or phosphorus. Several examples of these resins, which very many publications have been devoted to, will be presented here. This literature contains a large number of polycondensation polymers that fall under the broader definition of polymers carrying pendent complexing ligands.

Chelating polycondensation ion exchangers with nitrogen as the sole donor atom are relatively few in number. Shepherol and Kitchener in 1957, and Nonogaki in 1958 prepared poly(aziridine) exchangers by the crosslinking of water-soluble poly(ethylene imine) with 1,2-dibromoethane, obtaining resins with a high affinity for copper(II). A chelating ion exchanger containing pyridylazoresorcyl groups was obtained by the polycondensation of 4-(2-pyridylazo)resorcinol with resorcinol and formaldehyde in 0.1 N sodium carbonate solution of which the potentiometric titration showed that it was a weak amphoteric resin. It forms chelates with metals and acts as well as a cation exchanger. The chelate formation is selective, with the most stable chelates being formed with palladium(II) and copper(II). The elution of these cations requires the use of conc. hydrochloric acid solutions. Other ions like mercury(II), calcium(II), zinc(II), cadmium(II), nickel(II), cobalt(II), aluminium(III), iron(III), zirconium(IV) and thorium(IV) formed somewhat weaker chelates [232]. Chelating ion exchangers selective for iron(III) ions were prepared from phenolic resin containing piperidine or piperazine groups. Thus, a solution of phenol, bisphenol A in ethanol and piperazine hexahydrate with formaldehyde was reacted under stirring; after 24 hours soaking in 0.1 M hydrochloric acid the polymer was converted to the free base with ammonia. When the resin was quaternized with sulfuric acid and contacted with a solution containing 100 ppm each of iron(III), copper(II), zinc(II), cobalt(II), nickel(II) and cadmium(II) ions, the removal of iron(III) was almost complete, while the concentration of the other ions remained the same [233].

Chelating polycondensation ion exchangers with nitrogen and oxygen as donor atoms were mentioned at the beginning of this paragraph. Iminodiacetic acid has always been one of the most interesting chelating agents for incorporation into polycondensation resins. In one case in the manufacture of phenol-formaldehyde copolymers containing iminodiacetic acid groups for chelating resins, *N*-(hydroxybenzyl)iminodiacetonitrile was formed and hydrolyzed to *N*-(hydroxybenzyl)iminodiacetic acid [234]. The alkaline synthesis of porous phenol-formaldehyde polymers containing iminodiacetic acid by mesans of the template technique was described – this technique induces porosity by the addition of a finely divided solid material insoluble under the reaction conditions, which is then removed by dissolution after the polymerization is complete [235]. Numerous investigations have been carried out to prepare amphoteric polycondensation resins with the objective of obtaining chelation within the immobile resin phase which does not contain as a distinct reactive group a monomeric chelating agent. In order to prove the equivalency of the mechanisms of the selective uptake of certain multivalent metals such as copper, nickel, cobalt, calcium and other nonferrous metals not enough is known yet about the structure of these complex-forming ion exchangers. With the goal of obtaining condensation polymers with complex chelating groups of higher structural perfection a new type of condensation resin was prepared – as mentioned above – from phenol, formaldehyde and piperazine. It may be a sufficiently cheap selective resin for copper and mercury and be practical for large-scale synthesis and application in industrial processes [236].

Chelating polycondensation resins with oxygen the sole donor atom can be divided into polyhydroxy compounds, polycarboxylic acid ethers, 1,3-diketones, fluorones, and flavones. Thus, pyrogallol-2,3,4-trihydroxybenzoic acid, 2,4-dihydroxybenzoic acid and 2,4-dihydroxybenzaldehyde, salicylic acid, resorcinol, and 2′,4′-dihydroxyacetophenone have been employed. With respect to salicylic acid, the condensation resin prepared exhibited selectivity for copper, aluminium, and uranyl ions in a manner predictable from the stability constants of the metal salicylates. The resin showed, as well, a higher selectivity for iron(III) than an equivalent p-hydroxybenzoic acid exchanger, and one of the salicylic acid condensation resins was used successfully in column operation to determine the iron and copper contents of a brine [237]. Polycondensation of formaldehyde with resorcinol in the presence of salicylic acid and boric acid, followed by leaching out of the boric acid, gave ion exchangers useful for the removal of arsenic and boron from aqueous solutions [238]. In a rather comprehensive work some new amphoteric ion exchangers have been synthesized by condensing salicylic acid, p-hydroxybenzoic acid, sulfosalicylic acid, gallic acid, 3-hydroxy-2-naphthoic acid, β-resorcylic acid, and resacetophenone with epichlorohydrin, employing ethylenediamine as a crosslinking agent. Some of the physiocochemical properties were investigated as well [239]. Using diethylenetriamine as a crosslinking agent, chelating amphoteric ion exchange resins were prepared by condensing salicylic acid, sulfosalicylic acid, catechol, 8-hydroxyquinoline, hydroquinone, 3-hydroxy-2-naphthoic acid, and anthranilic acid with epichlorohydrin; they were characterized then by their physicochemical properties. Further triethylenetetramine was investigated as a crosslinking agent [240]. A selective ion exchanger based on formaldehyde and 1,2,4-triacetoxybenzene with increased selectivity and sorption capacity for Group V and VI metals was prepared by precondensation of 1,2,4-triacetoxybenzene with formaldehyde at 1 : 2 – 4 molar ratios in aqueous ethanol of 33 – 50% concentration containing 0.7 – 1.8 mol/l sulfuric acid. The mixture was stirred at 50 – 80 °C for 1 – 4 hours and the crosslinking polymerization of the precondensate done under static conditions at 70–90 °C for 30–70 hours [241]. Furthermore, new chelating amphoteric ion exchangers were synthesized by condensing p-hydroxybenzoic acid, β-resorcylic acid, and anthranilic acid with epichlorohydrin, using tetraethylenepentamine as a crosslinker; the resins obtained were then characterized by their physicochemical properties [242].

Chelating polycondensation ion exchangers with sulfur as the sole donor atom would be represented by resins containing thiol groups. No further resins in this group have become known as having been made by polycondensation, beyond those mentioned above under specific polycondensation ion exchangers.

Chelating polycondensation ion exchangers containing nitrogen and sulfur as donor atoms can be synthesized from phenylthiourea with resorcinol and formaldehyde or 3-hydroxyphenylthiourea with phenol and formaldehyde; they can be used for the selective uptake of metals at different pH values, which can also be done with regard to various heavy metals by a mercaptobenzthiazole resin prepared by polycondensation with phenol and formaldehyde [243]. In the search for extracting resins for gold from sea water the condensation of an aminobenzenethiol with glyoxal gave a resin in which the hydroxyl groups of glyoxal bis(2-hydroxyanil) were replaced by thiol groups: it showed selectivity for gold(III), mercury(II), and silver(I) [244]. Other chelating resins containing nitrogen and sulfur as donor atoms were described about the same time, with 1,3-diphenyl-2-thiourea, 2-mercaptobenzimidazole, or p-dimethylaminobenzylidenrhodamine as reactive groups. Inexpensive resins have been sought for the treatment of waste waters and have been specified in patent applications as containing dithiocarbamate groups in m-aminophenol resins [245]. Similarly, for improving selective properties with respect to rare metal ions, condensates of aromatic amines with formaldehyde were treated with ammonium thiocyanate in an acid medium [246]. An ion exchanger with improved complexing properties was described as having been prepared by polycondensing epichlorohydrin with polyethylenepolyamine in the presence of 2-mercaptobenzothiazole [247].

Aminophosphonic acid polycondensation ion exchangers and their synthesis as metal-complexing polymers were described in early publications, yielding materials with low exchange capacities [248]. It was later reported, however, that the product of the polycondensation of phenol with polyethylenepolyamine was treated with phosphonic acid and formaldehyde to form ion exchangers with a high exchange capacity [249].

Chelating polycondensation ion exchangers with macrocyclic polyethers and related ligands. After the results of Kopolov, Hogen-Esch and Smid had been published, Blasius and coworkers reported on their work on the preparation and properties of ion exchangers based on polycondensed macrocyclic polyethers. Two exchangers prepared by polycondensation of dibenzo-18-crown-6 and dibenzo-24-crown-8with formaldehyde were first examined as complexing polymeric reagents. The easy synthesis yielded thermally stable materials having a capacity of 2.6 and 1.8 mequ/

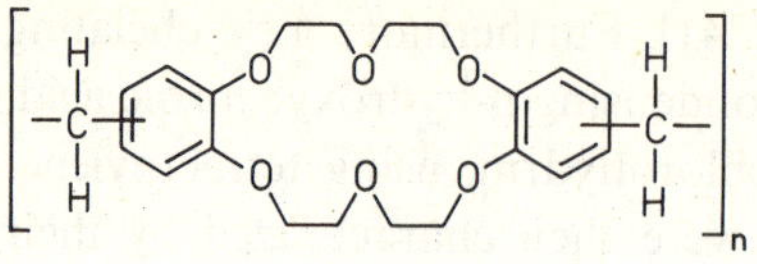

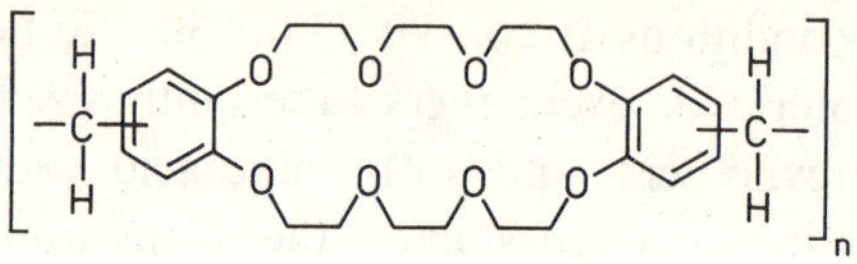

g air-dry resin, respectively. They showed a very different selectivity for the cations of alkali and alkaline earth elements and for variously substituted or structurally isomeric ammonium cations. Owing to the difference in partition coefficients, it was expected that separations would be possible in every individual group as well as among the groups [250]. In later extensive investigations a large number of polymeric macrocyclic crown ethers have been prepared by polycondensation either of a crown ether, a cryptand or a glyme with formaldehyde in formic acid, and additional crosslinking agents such as toluene, xylene, phenol and resorcinol have been used with all types of crown ethers containing oxygen, oxygen and nitrogen, or sulfur, and crpytands. Among all the polymers obtained the one with dibenzo-18-crown-6 as the reactive group is still the best studied. In creating epichlorohydrin-based chelating polycondensation ion exchangers Gramain and Frere have presented a number of polymeric crown ethers obtained from diazacrown ethers. The polymers are readily synthesized as well by reacting the cyclic diamines 1,4,10-trioxo-7,13-diazacyclopentadecane (diamine [21]H), 1,7,10,16-tetraoxa-4,13-diazacyclooctadecane (diamine [22]H) or a cryptand with epichlorohydrin, diepoxyoctane or the diglycidic ether of 2,2-bis(4-hydroxyphenyl)propane, yielding structures of repeating units as shown in the following diagrammatic formula:

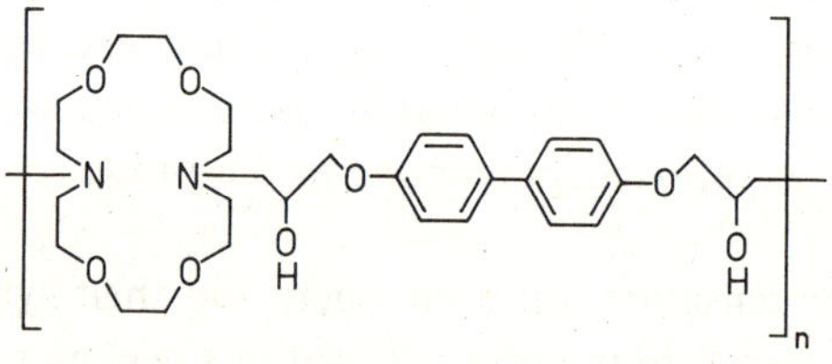

These extensive studies were then extended to the synthesis of polyesters and polyamides containing a backbone with diazacrown ethers. Depending on the starting diepoxide the polymers can be either water soluble or not. Their binding properties for alkali, alkaline earth, and some transition cations were studied and shown to be, generally, comparable to those of the analogous monomers. But in some cases new properties were also observed, due to various structural factors such as the nature of the neighboring groups and of the connecting chains [251]. Without trying to be complete in the presentation of these chelating polycondensation ion exchangers it should finally be mentioned that the condensing of Kryptofix to a linear polyacrylamide has been achieved by a Mannich reaction yielding the structure shown in the following diagrammatic formula:

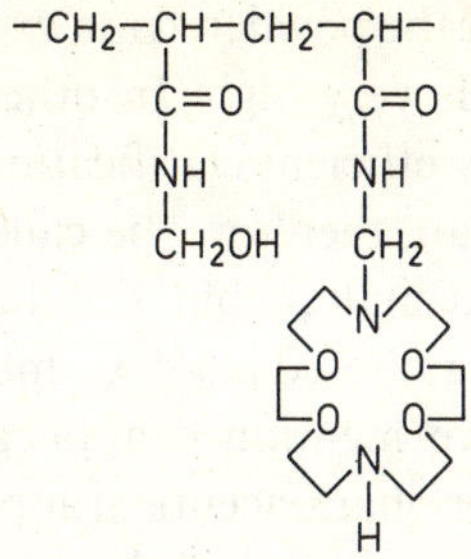

The interaction of the univalent cations lithium, sodium, potassium, and silver and of the bivalent cations barium and calcium with this reactive polymer was studied using ^{13}C NMR [252].

1.2.2.4 Special synthetic ion exchange polymers

The basic principle of ion exchange is always the same, but what can be altered is, firstly, particle size and the shape of ion exchanging materials. Even submicroscopic particles of anion exchange resin in spherical form have been prepared, which may be used with profit in any application where ground ion exchange resins are applicable. It can most probably be said that the main purpose for the development of special synthetic ion exchange polymers is to broaden the applicability of exchanging materials. If, for instance, the rates of exchange and regeneration are sufficiently high, it can bring extra benefits if the structure of the polymers is changed or if additives for special reactivities are incorporated, i. e., if special ion exchange polymers are created.

Powdered ion exchangers. One can easily obtain powders from normally-produced particle-sized beads by grinding or other mechanical procedures. It has been shown that particles of ground ion exchange resins with average diameters of 1 µm and smaller can be obtained. Such particles are not spherical and, as with all such materials, not uniform in size. The range of particle diameters may vary largely, depending on machine precision and the expenditure employed on the preparation of smaller particle size ranges. The essence of ground resin material is that even though large particles may constitute only a small fraction of the total number of the ground particles, they do represent a large fraction of the sample weight. Consequently such ground resins exhibit settling of a significant fraction by weight of the ion exchange material from aqueous suspension [253]. In their most important application – as precoat media for active filtration (Martinola) – powdered ion exchangers not only allow the removal of suspended matter during a filtration cycle but also act on the solution being filtered, thus purifying it chemically.

This important fact has to be taken into consideration in comparisons with inert but very effective materials such as diatomaceous earth, perlite or cellulose materials used as filtering aids. Because of the fine particle size of the powdered ion exchangers used as a precoat media, the chemical reaction rate is very high. In other words, the precoat filter unit can be used not only as a highly efficient clarification system but also as a chemical reaction unit. Such a process can, therefore, be called active filtration because it not only removes the suspended matter, but simultaneously changes the chemical composition of the solution being filtered [254]. Commercially available are powdered ion exchangers for the above mentioned application, as well as for the Powdex Process, and to a lesser extent for pharmaceutical applications such as drug carriers and tablet disintegrators (e. g. Lewasorb powder resins).

Several attempts have been made to prepare directly particles of ion exchanger beads smaller or much smaller than the 0.01 – 1.5 mm particle size range resulting from conventional emulsion polymerization. Recently remarkable progress seems to have been achieved in the preparation of small particle-size, spherical anion exchange resins. The process disclosed in a patent yields resins with narrow particle size ranges, having submicroscopic mean diameter values of 0.01 to 1.5 µm. The particles obtained are approximately spherical, have a relatively narrow particle size distribution, and they can be prepared in different particle sizes. At first strongly basic anion exchange emulsion resins were obtained but then particles of other anion exchange emulsion resins, and of cation exchange emulsion resins, with similar appearance and size distributions were obtained. The submicroscopic particles can finally be prepared as a dry powder ion exchange material. A wide variety of applications where up to now the obove-mentioned ground materials have been used and even where ground resins are currently unsuitable can be envisioned for this dry powder ion exchanger [255].

Fibrous ion exchangers. It has been argued that the applicability of ion exchangers could be broadened especially by fibrous ion exchange materials. Whether this will materialize or not is still to be seen, but much work has been done so far on the development of ion exchange fibers and fibrous ion exchange materials. Technically feasible ion exchange fibers must correspond in their chemical and physical properties to the customary ion exchange resins and if the exchange as well as the regeneration rate is substantially higher – as is claimed – or electrochemical regeneration becomes feasible, then the application of ion exchangers in a textile structure opens new possibilities. Chemical fibers comprise, on the one hand, different synthetic polymeric raw materials and, on the other, differently processed materials, such as monofilaments, fibers, non-wovens and fabrics. The same applies to fibrous ion exchangers so that, therefore, this term best covers the whole field. For example, in early, rather simple, work monofilaments containing an anion exchange resin were obtained by composite-spinning (either in a side-by-side or in a sheath-core relationship of the two components) from either spinning solutions or melts. One component contained the anion exchange resin dispersed in finely divided form with the other component being free of anion exchange resin and serving as reinforcement. As the reinforcing component viscose, cellulose ester and ether, vinyl resin, polyamide, polyester, polyolefins, etc., were tried; the resin-containing component was of the same material, containing monofilaments that made up over forty percent of its weight. With the exception of the strong base exchange capacity there was no loss in the capacity of the resin dispersed in viscous filaments. Later developments on fibrous ion exchangers were mainly made in the Soviet Union and in Japan; as a result it is difficult to make full use of the currently existing literature.

Fortunately Soldatov has reviewed the synthesis, structure and properties of new fibrous ion exchangers. As a result of extensive studies rather simple methods have been developed for the preparation of fibrous sulfonic acid cation exchangers, strong base anion exchangers, weak base and medium base anion exchangers, and weak acid cation exchangers with complexing properties. The materials discussed more specifically were the strong base, conventionally named SBF, the medium base, MBF, and the strong acid, SAF, fibers. The SAF and SBF exchangers are based on graft copolymers of polystyrene and polypropylene. The MBF exchangers have been synthesized by chemical modification of acrylic fibers. The main part of the work reported has been carried out on ion exchange fibers with a monofilament thickness from 20 to 40 mcm. All these materials have been used in various textile forms depending on the concrete problem to be solved. Filaments, fabrics, non-woven materials or staple fibers have been used in a variety of investigations. Table 1.12 gives the most important properties of the ion exchange fibers developed. The

Table 1.12 Properties of fibrous ion exchangers (Soldatov)

Ion exchanger	Ionogenic group	Exchange optimum (mequ/g)	Capacity maximum	Swelling (g/g opt.)	Tensile strength (kg/mm^2 opt.)	Elasticity modulus (kg/mm^2)	Elongation at rupture (%)
SAF	$-SO_3^-$	3.5	4.6	0.8	9	200	18
SBF	$-N(CH_3)_4^+$	3.2	4.6	0.8	16	200	45
MBF(I)	$-NH_3^+$	9	11	1.0	18	200	40
MBF (II)	$=NH_2^+$ $\equiv NH^+$	5–7	11	1.5	15	600	28

degree of sulfonation is usually about 1 sulfonic group per phenyl ring. At the same time more than one (up to 1.3) trimethyl ammonium groups per phenyl ring can be incorporated into the fibrous graft copolymers. Therefore, the exchange capacity of the strong base fibers is often higher than that of the commercial granulous resins of the same chemical type. The examination of the fibrous properties of these materials showed that their mechanical strength is due to the polypropylene skeleton used. The structure of the ion exchange fibers based on polypropylene strongly depends on the preparation procedure of the initial graft copolymer: it can be more or less dense, resulting in different mechanical and osmotic stabilities of the fibrous exchangers. All the exchangers presented are macroscopically homogeneous materials. The grafted polystyrene is regularly distributed throughout the polypropylene fibers, but, nevertheless, observation in an optical microscope with a magnification of 900 power shows a stripped structure with a period of about 5 – 6 mcm as is seen typically in the SBF-Cl sample with 173% grafted polystyrene. The X-ray diffraction analysis showed that the graft copolymers have a domain structure, the linear dimensions of the domains being in the range of 50 – 200 Å. The polypropylene matrix preserved its crystalline structure during manufacture. The strong acid and strong base ion exchange fibers have an inert base such as polypropylene; as a result, their chemical stability under common temperature conditions towards strong acids, bases and oxidants is similar to that of the common styrene-divinylbenzene resins. A careful investigation of the thermostability and the thermomechanical properties of fibers of this type revealed that they can be used in the same temperature range and under the same conditions as the granulous styrene-divinylbenezen resins. The selectivity of these fibrous exchangers towards inorganic ions appeared to be rather close to that of styrene-divinylbenzene resins that have the same swelling. It was mentioned, however, that there was not complete agreement on this [256].

That the then-current state and prospects for the development and use of ion exchange fibers was, to the author's knowledge, first reviewed in 1972 shows that corresponding work was begun only in the immediately preceding years. Polyvinyl alcohol is a fiber-forming material, used for the manufacture of vinal fibers. Sulfonated vinal fibers have been prepared and a sulfonated poly(vinyl alcohol) fiber support consisting of a bundle of superfine filaments, has been developed, which has a larger surface area and many more hydrophilic groups than customary ion exchange resins. It is far more suitable for the immobilization of enzymes [257]. Poly(vinyl alcohol) fibers containing carboxylic groups were found to have a higher static exchange capacity. The influence of plasticization in the drawing of PVAL fibers on the changes in ion exchange properties was studied. Modifications were also carried out on dehydrated fibers with diene structure by reacting them with maleic anhydride or Schiff bases to give ion exchangers [257] by graft copolymerization with acrylonitrile, acrylic acid or 2-methyl-5-vinylpyridine. The resulting ion exchangers were used for separating metals, dyes or antibiotics by (1) a radiochemical modification method for the graft copolymerization with methacrylic acid, (2) a simple polymer-analogous conversion of halogen derivatives in a treatment including phosphortrichloride, methylphosphonic acid dichloride and diiminoacetic acid, or (3) a treatment in a bath containing formaldehyde, sulfuric acid and sodium sulfate to prepare fibrous ion exchangers containing polyethylenimine, which gave a material with a good capacity for both acid and direct dyes and which is used in waste water purification processes. This latter fibrous ion exchanger is particularly interesting. With respect to the general properties of PVAL-based fibrous ion exchangers the proteolytic properties of a carboxylic material were studied, giving titration curves conforming to the Henderson-Hasselbach equation; the ion exchange fibers also had a higher heat of hydration, which increased with the increasing hydration capacity of the counterion and decreased with an increasing charge in the counter-ion. The static ion exchange capacity of chloride complexes of platinum metals decreased through the series palladium(II), platinum(II), ruthenium(IV), iridium(IV), and rhodium(III) ions; consequently, the fibers were recommended for the concentration of platinum metals from production solutions with a high salt content. The study of the kinetics of swelling led to a model that considered the fibrous ion exchange polymer as a viscoelastic body consisting of parallel elastic-viscous unit joints [258].

In similar early and rather simple work to that reported above polyacrylonitrile fibers were spun containing, as a solid mixture, finely dispersed ion exchangers of all types, 10 – 15% by weight. It was claimed that they could be woven into fabrics in order to be used in countercurrent or semi-selective processes. In a patent a true polyacrylonitrile ion exchange fiber was then described which was obtained by heating 91 mole % acrylonitrile, 7 mole % sodium sulfopropyl methacrylate and 2 mole % glycidyl acrylate at 50 °C for 24 hours in the presence of 0.005 mol/kg azobis (dimethylvaleronitrile). This yielded a copolymer which was spun and heat-treated to give a fiber insoluble in boiling water. The fiber adsorbed 32.9 mg/g iron(III) [259]. Work that followed then in the Soviet Union detailed the preparation and study of an ion exchange fiber and its modification with respect to water vapor sorption at 20° – 102 °C. The sorption isotherms for thirteen sorbents were given and the effect of crosslinking on sorption were discussed. Ion exchange fibers based on carboxy and hydroxyl groups containing acrylonitrile-2-methyl-5-vinylpyridine polymer were prepared with bisphenol A-epichlorohydrin crosslinkages or epoxy resins as crosslinkers forming a three-dimensional network; their properties due to that structure were examined [260]. Ion exchange fibers with a good sorption capacity based on polyacrylonitrile were also prepared from spinning solutions containing additional polyethylenimine or to obtain a weakly acidic cation exchanger, organosilicon compounds were used for saponification of the acrylonitrile-itaconic acid-methylmethacrylate copolymer. A sulfonate-group containing polyacrylonitrile (and polyethylene terephthalate) fiber was used for electrokinetic investigations on fibers, resulting in an adsorption model of ions to ion exchanging polymers [261]. Of interest for the manufacture of ion exchange fibers made of polyacrylonitrile is certainly that the hydrolysis rate of nitrile groups compared for sodium alkylsiliconates and sodium hydroxide depends on the diffusion of the saponification agent into the fiber as the limiting factor in the initial saponification stage. For the preparation of selectively acting ion exchange fibers of an aminophenol-type based on polyacrylonitrile the modification process was optimized, yielding treated fibers that retained their initial mechanical properties and which can also be used for fabric production. Amphoteric polyacrylonitrile ion exchange fibers with optimal properties were prepared by phosphorylation of amino-group containing materials in a batch containing phosphoric acid, formaldehyde and hydrochloric acid at 98 °C. The fiber had an ion exchange capacity 5.2 and 1.4 mmol/g for hydrochloric acid and sodium hydroxide, respectively, a tenacity of 9 cN/tex, and it withstood more than fifteen sorption-desorption cycles without deteriorating [262]. The manufacture of ion exchangers from acrylic fibers by treatment with quaternary ammonium compounds and hydrazine sulfate was also investigated and the use of new surface-modified ion exchange PAN fibers for the determination of transition metal traces in quartz optical glass and natural water samples by emission spectroscopy after preconcentration by sorption on them was reported. The modifying agents in this case were poly(epoxy amine) and poly (ethylenimine) and fibers with or without hydrolysis with sodium ethylsiliconate were tested [262].

The preparation of fibrous ion exchangers based on other polymers followed by the graft polymerization of methacrylic acid on polypropylene, polycaprolactam and poly(ethylene terephthalate). In the field of polyolefin fibers – which have an inert chemical nature – ion exchange fibers were prepared by dispersing a crosslinked styrene-divinylbenzene copolymer and polyethylene or polypropylene in a high-boiling solvent, e. g. tetralin, spinning the dispersion into a fiber from the melt, removing the residual solvent, and introducing ion exchanging groups by conventional methods. Similar processes of the combination of polyethylene and polystyrene either by grafting or blending led to sulfonic cation exchangers in fiber form, which were also cut into staple fibers and tested for their column behavior. They had higher breakthrough capacities than beads. Polypropylene fabrics containing carboxyl, sulfonic acid, phosphoric acid and trimethylaminomethyl groups were prepared by the radiation-initiated graft polymerization of acrylic acid or styrene in a gaseous phase with subsequent sulfonation, phosphorylation, chloromethylation and amination, yielding a high ion exchange capacity of up to 7 mequ/g, good kinetic characteristics, increased chemical and thermal stability and a uniform composition. Island-in-a-sea bicomponent fibers of polyolefin-polystyrene as well as colloidal ion exchangers in porous fibers were also reported. The production of modified polyamide fibers with ion exchange properties was most probably based on the same process of a conversion of the amide groups of the polyamide fibers as was described in more detail for the introduction of thioamide groups into the same fiber material, or for the preparation of ion exchanging polycaproamide fibers for which an aniline-formaldehyde modification was first presented. Fibrous ion exchangers were obtained from polyester fibers by impregnating them with a mixture containing styrene, divinylbenzene or o-divinylbenzyl chloride-p-divinylbenzyl chloride and a polymerization initiator and, after preprocessing, treating them with sulfuric acid. The chain structure of ion exchange polyester fibers grafted by acrylic acid was studied with regard to their ion exchanging properties. Rayon fibers have also been used as a basic material for the preparation of fibrous ion exchangers, e. g., either for a 2-methyl-5-vinylpyridine graft copolymer, or grafted with vinylsulfonic acid or methacrylic acid followed by alkylation with epichlorohydrin. To obtain ion exchanger fibers a 70 : 30 styrene-(chloro-α-methyl)styrene copolymer was spun to give a 75 den/25 filament yarn, cut and first treated with piperazine to give crosslinked fibers and then with chlorosulfonic acid to yield an ion exchanger with a capacity of 4.3 mequ/g, tenacity

2.95 g/den, and elongation 57%. A polystyrene-based ion exchange fiber which has a high ion exchange capacity and a high mechanical strength was prepared by using an island-in-a-sea type composite fiber, with the sea ingredient predominantly of polystyrene for ion exchange and the island ingredient of fiber-forming polypropylene for reinforcement, as the starting materials; the fundamental characteristics were then evaluated to explain the extremely high ion exchange rate for metal ions [263]. This fiber (Ionex) was then studied for its ability to adsorb and immobilize biologically active proteins. Carbonaceous fibrous ion exchangers obtained by oxidation of carbonaceous fibrous materials with crystal hydrate melts of aluminium and cobalt(II) nitrates have been found to have good sorption properties for ammonia, methylamine and dimethylamine; they can be used in many repeated cycles and be regenerated easily by simultaneous thermal treatment and evacuation [264]. For cationic dye removal from waste water, on the other hand, a fibrous silicon-containing ion exchanger was described. This material can be prepared by modification of polyacrylonitrile with sodium ethyl siliconate in the presence of hydrazine sulfate. Siloxane was formed on the fiber surface. The optimum ion exchange capacity of 5 mequ/g was obtained when the sodium ethyl siliconate concentration, the hydrazine sulfate concentration, and the process duration was 7%, 1.17–1.36%, and 125 minutes, respectively.

Amphoteric fibrous ion exchangers have also been prepared as already briefly mentioned in some cases above. For example fibrous ion exchangers were prepared by spinning an acrylonitrile-vinylpyridine copolymer. The chemical stability and capacity of the exchangers were increased by treatment with polyepoxy compounds during the spinning process. Subsequent saponification gave amphoteric exchange materials [265]. For another amphoteric ion exchange fibrous material PVC fibers were immersed in 80 parts trimethylenetetramine containing 20 parts water at 100 °C for 5 hours and in propane sultone at 50 °C for 2 hours. The synthesis of a rayon-based amphoteric fiber and examination of its possible use for the sorption of platinum metals was also reported. Polystyrene fibers were crosslinked with chlorosulfonic acid, chloromethylated or chlorosulfonated, aminated, and treated with thiocyanates to give fibers with complexing properties for metal ions; polyesters from 4-hydroxybenzoic acid and its derivatives or compositions containing polyesters as the main component were prepared, chloromethylated, treated with a thioether to form a sulfonium salt, and reacted with di-sodium iminodiacetate to form a fibrous chelating ion exchanger. A fibrous ion exchanger which is in fact amphoteric but can be used as a weak acid cation exchanger was obtained from commercial Vonnel V-17 acrylic fibers by crosslinking with hydrazine solution and then hydrolyzing residual nitrile groups with a sodium hydroxide solution to introduce carboxy groups. The cation exchange capacity is approximately 6–9 mequ/g, depending on the extent of the hydazine treatment [265]. Nitrogen and phosphorus containing amphoteric fibers have been made by the modification of PAN fibers with hydroxylamine and subsequent phosphomethylation. They have differing properties depending on the reaction conditions. A fibrous ion exchanger containing carboxylic and amine groups obtained from PAN fibers by treatment with sodium hydroxide and hydrazine and with a maximum ion exchange capacity of 6.2 mequ/g, can, after an amination lasting 120 minutes and saponification of the fibers be converted to a redox polymer by copper(II) sorption from 10% copper sulfate solution and reduction of sorbed Cu^{2+} to Cu^0 with sodium dithionite. The strongest fiber copper bond was attained at a NH_2-COOH ratio of 0.4 – 0.7. For the sorption of copper onto different types of amphoteric ion exchange fibrous materials from hydrometallurgical waste waters the types and conditions were reported [265].

With respect to non-woven materials it should be mentioned that fibrous ion exchangers with an ion exchange capacity of 4.5 – 5 mequ/g were prepared from nonwoven Nitron acrylic fibers by 60 minutes saponification of the nitrile groups at 96 °C with a solution containing sodium hydroxide and hydrazine sulfate in a 1 : 1 ratio. The modified non-woven acrylic fibers were used for the recovery of cationic dyes from the waste water of textile dyeing plants, in air purification, and for the sorption of transition metals [266]. Non-woven cloths with ion exchange properties have also been made of acrylic fibers containing in addition strong acid sulfonic groups; they have a thickness of 20 mm and a density of 0.06 g/cm^3. When such materials were packed in a column for removal of iron from a solution containing 50 ppm, the treated solution contained 0.7 ppm iron.

Ion exchanger films. The simpler method of combining a film forming material with an ion exchanging material was used to manufacture ion exchanger films. For example, polyethylene, phenolsulfonic acid-formaldehyde polymer and talcum were extruded at 200 °C to obtain a 0.2 mm thick porous film with a mean pore diameter of 0.1 µm and an exchange capacity of 2.43 mequ/100 cm^2. Such materials can be used for purifying waste water. But true ion exchanger films can be made by grafting poly(methacrylic acid) or poly(vinyl sulfoxy acid) onto poly(vinyl alcohol). Similarly, polyolefins were graft-copolymerized with hydroxystyrenes, and ion exchange groups were introduced onto the hydroxystyrene side chains to prepare ion exchange films. The synthesis of an ion exchange film or membrane by radiation-induced grafting of acrylic acid onto poly(tetrafluoroethylene) was reported, showing that the degree of grafting of acrylic acid onto the PTFE film increased with increasing radiation dose. The water absorption of the grafted film increased and the electric resistance decreased with an increase in the degree of grafting. A non-grafted layer remained in the center of the film at a low percentage of grafting and disappeared as the grafting proceeded. Interference microscopy showed that the grafting occured by gradual diffusion of monomer through successive layers of the film, which swells as the grafted front moves into the film [267]. Before this work, according to an American patent, perfluoro(2-fluorosulfonyl-3,6-dioxa-7-octene) was prepared from perfluoro(ethylene glycol)divinyl ether and sulfuryl fluoride, copolymerized with tetrafluoroethylene to give a clear film-forming copolymer that was heat-stable below 375 °C. The copolymer was treated with aqueous sodium hydroxide in methylsulfoxide to give a sodium salt ion exchanger which had specific resistances of 340 and 300 Ω · cm in 0.6 N potassium chloride and 25% sodium chloride, respectively. Very easy seems a process in which, according to a Japanese patent, polyethylene and polystyrene in a ratio of 2 : 1 are blended, extruded to form a film, drawn biaxially 200% in both directions, crosslinked with divinylbenzene, and sulfonated with sulfuric acid to prepare an ion exchange film. Thin films containing both anion and cation groups have been prepared by hydrolysis and quaternization of polymers plasma-polymerized from bis(dimethylamino)methylvinylsilane. Amide groups in these plasma polymers were hydrolyzed to carboxyl groups with an octylbromide/water mixture. Amine and amineoxide groups in the plasma polymers were quaternized by exposure to methylbromide vapor to give quaternary ammonium groups. The combination of hydrolysis and quaternization yielded thin films containing both anion and cation groups. It is expected that surfaces charged with anion and cation groups would offer an environment conducive to cell growth [268].

Ion exchanger foams. Ion exchangers in the form of a foam was first obtained by grafting a polyurethane foam with open cell structure with styrene and sulfonated. In comprehensive work on the production of homogeneous ion exchange foams for chromatography exchange groups were introduced onto phenol-formaldehyde polymers, styrene-grafted polyurethane and polyethylene foams by sulfonation, chloromethylation and amination. The maximum capacity of the sulfonated phenol-formaldehyde cation exchanger foams was 1.85 mequ/g, and that of styrene-polyurethane copolymer anion exchange foams was 2.2 mequ/g. Weak carboxylic ion exchange foams with a maximum capacity of 4.02 mequ/g were prepared by radiation grafting of polyethylene and polyurethane with methacrylic acid. Heterogeneous foams were prepared by foaming cation exchanger powder with the precursors of an open-cell polyether-type polyurethane foam, containing, e. g., 26% ion exchanger powder this having a capacity of 1.0 mequ/g [269]. In a patent it was claimed that a polyurethane foam impregnated in vacuo with 5 : 1 methacrylic acid-ethanol mixtures was irradiated with 1.3×10^5R γrays to give a grafted foam with ion exchange capability. A different way of obtaining a rigid ion exchanger foam was described in a paper on the application of a pyrolysis process to poly(*tert*-butyl *N*-vinylcarbamate) at 185 – 200 °C, giving a rigid foam having good thermal stability and ion exchanging properties. The foam contained cyclic urea units, primary amine units, and a small amount of urea crosslinks. The amounts of amine and residual carbamate units were determined as a function of pyrolysis by an ion exchange method and the results agreed with those expected from a random cyclization process. A different way of obtaining ion exchanger foams has been developed employing the physical impregnation of polyurethane foams by the liquid ion exchangers LIX 65 N and Kelex 100. The ion exchange properties of these foam-like resins were studied, and they appeared to be of potential interest in analysis and the determination of trace levels.

Several unique synthetic ion exchange polymers developed mainly for special applications will be described now. A molded ion exchange catalyst was prepared by extrusion of a powdered styrene-divinylbenzene copolymer, polypropylene and di-butyl phthalate at 170 – 190 °C into 10 mm high, 8 mm outer diameter, 6 mm inner diameter rings, which had, after chlorosulfonation, a static exchange capacity in terms of 0.1 N NaOH of 3.8 mequ/g, and which were used as dehydration catalysts for *tert*-butanol. Ion exchange resin moldings formed into a tube with an inner diameter of 60 mm and 5 mm thick were made from a mixture of strong acid powdered ion exchange resin, aqueous polyvinylalcohol and saponified monomethyl maleate-vinyl acetate copolymer. They showed after crosslinking by heat or electron beam and molding little change in exchange capacity. Further, in a US patent a method was described for preparing a cementitious ion exchange resin. Cation or anion exchange resin particles having an effective diameter of 0.1 – 1 mm were treated with a material selected from a group consisting of polyacrylate emulsions, poly(vinyl alcohol), and poly(vinyl acetate). The resulting material has a hydrophilic group. By drying the treated ion exchange resin into a cementitious mass, with an optional size reduction, to grains having an effective diameter of 2 to 50 mm were obtained.

The intensive search for thermally regenerable ion exchange resins led to another special type of synthetic ion exchange polymers called encapsulated ion exchangers. The first such thermally regenerable ion exchange materials were prepared by the encapsulation of crosslinked poly(*N*-alkyldiallylamine) microparticles by the copolymerization of ethyl acrylate and divinylbenzene. The thermally regenerable capacity of the resins, after hydrolysis of the ester groups, depended on the form of the polyamine microparticles during the encapsulation process. Thus, the thermally regenerable capacity of a resin prepared from the free base form of poly(propyldiallylamine) was 0.3 mequ/g compared with 1.2 mequ/g for a resin prepared from the hydrochloric form. The failure to achieve the theoretical maximum value of 2.1 mequ/g has been attributed to internal salt formation between the carboxylate and the protonated amino groups. In following work thermally regenerable resins were prepared by the encapsulation of polyacid microparticles in a polyamine matrix using several different preparative procedures; here the attempt was made to minimize the formation of the internal salt structures. The ion exchange resins containing crosslinked acrylic acid-divinylbenzene copolymer microparticles encapsulated within a crosslinked 1,6-bis(diallylamino)hexane-diallylamine copolymer matrix were obtained with thermally regenerable capacities of 1.2 to 1.3 mequ/g. The effect of using various salts of polyacrylic acid including cobalt(II), chromium(III), copper(II), iron(III), tin(II), lead(II), calcium(II) and allyalamine on the preparation and thermally regenerable capacity of the resins was studied. These variations were supposed to act as polymerization inhibitors preventing polymerization within the polyacid and making use of the counterion approach. In the latter approach it had been found that when a solution of acid and amine monomers is polymerized, the use of salts of the acid monomer rather than the free acid is effective in reducing the formation of internal salt structures. The best capacity of 1.8 mequ/g was obtained with the barium(II) salt, but the resin lacked sufficient physical strength. Only relatively minor improvements in strength were obtained by an increase in the amount of bis-diallylamino crosslinker from 5 to 15 mole % or by the use of the phosphate rather than the hydrochloric form of the diallylamine monomer. Post-crosslinking of the polydiallylamine matrix with a,a'-dichloroxylene did not significantly improve the physical strength of the resin but reduced the resin swelling, improving the thermally regenerable capacity on a volume basis. The hypothesis that high thermally regenerable capacities depend on the presence of low amounts of internal salt structures in the resin and therefore on low amounts of penetration of the amine monomers into the polyacrylic acid or poly(ethyl acrylate) microparticles was confirmed by the use of samples with different amounts of crosslinking. For example, the thermally regenerable capacities for resins prepared from polyacrylic acid with more than 15% divinylbenzene, 10–15% DVB and 10% DVB as crosslinkers were 0.65, 1.23 and 1.31 mequ/g, respectively. Similar, the thermally regenerable capacities for resins prepared from poly(ethyl acrylate) with more than 15% DVB, 10–15% DVB and 2.5% DVB as crosslinkers were 1.55, 1.79 and 2.10 mequ/g, respectively. The latter capacity of 2.1 mequ/g for a resin prepared from the ester indicated that all sites in the resin were used for the salt adsorption. The most important conclusion drawn was that it is still necessary for a compromise to be found between the thermally regenerable capacity on a weight basis and the amount of swelling of the resin. This because for these resins, in general, the greater the thermally regenerable capacity on a weight basis the greater the swelling and, therefore, the lower the capacity on a volume basis. In continuation of this work with emphasis not on encapsulation but on thermal regeneration an aqueous solution of diallyamine hydrochloride and 1,6-bis(*N,N*-diallylamino)hexane hydrochloride was absorbed into a macroporous polyacrylic resin bead and polymerized by γ-irradiation to give resin beads which were hard and strong, and which had a high density, low swelling and a thermally regenerable capacity of 0.81 mequ/g. On a volume basis the resin had a capacity of 0.22 mequ/ml which is equal to that of the previous best resin, but with the latter being physically weaker. The allylamine monomers, when absorbed into the macroporous polyacrylic acid, did not polymerize with a chemical initiator. In conclusion, this method of preparing thermally regenerable resins provided one of the best methods of making resins that are physically strong and show low swelling and, therefore, high capacity on a volume basis [270].

Ion exchanger capillaries. A new and very special type of ion exchanger in the form of capillary tubes was introduced by a Czechoslovakian group of researchers. These ion exchanging materials consist of an internal reactive layer and an external inert jacket. They were generally prepared from polyethylene tubes by grafting polystyrene onto the inner wall of the polyethylene capillary and introducing ion exchange groups by polymer-analogous transformations. At first a strong acid cation exchanger was prepared and studied, and it was shown that these materials can be used for the acidimetric determination of salts, for the concentration of ions as well as for chromatographic columns. Then the preparation and description of a strong base anion exchanger followed. With these anion exchanger capillaries the resistance of the reactive groups was tested and it was pointed out that the capacity of these materials can be continuously increased to much greater values. A new variety in the preparation of ion exchanger capillaries was the polycondensation process of aromatic hydrocarbons and formaldehyde onto the inner surface of polyvinylchloride tubes, which is not only simpler than the above method, but enables one to select more ion exchanger layers. For the characterization of the ion exchanger capillaries measurements of the conductivity were carried out on materials having different exchange capacities, ionic forms, and temperature. These results were used to describe the structure of the active layer and to attempt an electrophoretic separation of cations of various valences. Research and development on these special synthetic ion exchange polymers has come to a standstill; but it may well be that some day there will be a renascence of the work done in this project, perhaps in liquid ion exchange chromatography employing other ionogenic groups [271].

Special synthetic ion exchange polymers can in addition to their ion exchanging properties exhibit special reactivities. Optically active ion exchange resins can be prepared by the reaction of aniline, benzylamine, α-phenylethylamine, *N*-dimethylaniline or *N*-dimethylphenylethylamine with chloromethylated polystyrene and crosslinking with divinylbenzene [272]. A scintillating cation exchange resin has also been prepared, which adsorbs ions from aqueous solutions and fluoresces in response to radioactive ionic species that may be taken up. The emitted light is of a wavelength which activates a photomultiplier tube. By this means, radioactive isotopes can readily be counted by coupling the resin on which they are adsorbed to a photomultiplier tube and pulse counter. *N*-chloro-polyamides were prepared in a form favorable for the kinetics of heterogeneous redox reaction in aqueous system by polymerizing *N*-substituted acrylamides into ion exchange resins to obtain a snake-cage configuration. After *N*-chlorination these products were capable of oxidation and ion exchange [273]. For the separation of amines and acids by sterically hindered polymers, 2,6-di-*tert*-butyl-4-vinylphenol-styrene copolymer and 2,6-di-*tert*-butyl-4-vinylphenol polymer were used to separate sterically hindered amines from unhindered bases. 2,6-Di-*tert*-butyl-4-vinylpyridine-styrene copolymer effectively removed trace amounts of hydrochloric acid from acetylchloride and benzylchloride. Chemical disinfection by means of ion exchange resins seems now to be feasible because special anion exchange resins exhibiting strong bacterial properties have been developed. Similar to conventional ion exchange resins, these resins sorb microorganisms. However, as opposed to results obtained with conventional resins, the sorption is much stronger, indeed irreversible. It is followed by actual cell disruption/lysis. These new germicidal resins have become known as Insoluble Polymeric Contact Disinfectants, IPCDs [274].

1.2.2.5 Commercial ion exchangers

As mentioned elsewhere in this volume, of the synthetic ion exchange resins described above those of the gel type in bead form have the greatest importance since they continue to be used most frequently in industry. The production of an ion exchanger requires specialized knowledge, so that there is little purpose in attempting homemade preparation. Ion exchangers, as reactive polymers, occupy a kind of intermediate position between that of a chemical product which is manufactured, processed and then consumed, and equipment serving for the manufacture of a given product. They are highly sophisticated chemical materials but as they are used only as processing aids, they are the transfer agents of chemical processes without being significantly changed themselves or used for another purposes. Because of the possibility of reversible conversion into different forms, they have diverse applications and have in theory, an infinite life. Naturally, this is made impossible by the fact that like any material they are subject to chemical and mechanical wear. The possibility of secondary use of spent ion exchangers has repeatedly been considered. In one case it was reported that the sorption of copper, nickel and molybdenum by commercial ion exchangers which had already lost their exchange capacity in water purification was determined without any treatment of the exchangers. The sorption was determined in solutions containing sulfuric acid, under stationary conditions, at 20 °C and pH 4–5, by having 200 ml solutions in contact with the exchangers for twenty-four hours. But further results were unfortunately not given.

The ion exchangers available today are commercial products from various manufacturers. Some of the registered tradenames found in the literature, including some no longer used, are: Allasion, Amberlite, Chempro, De-Acidite, Diaion, Dowex, Duolite, Imac, Ionac, Kastel, Katex/Anex, Lewatit, Liquonex, Mykion, Permutit, Purolite, Varion, Vionit, Wofatit, Zeo-Karb, Zerolit. Except for some of the large international chemical companies active in this field there is a continuous change in the industry which makes it impossible to give a fully reliable listing of manufacturers and supply sources. In the Soviet Union and other countries of Eastern Europe, ion exchangers are usually referred to as cationites, anionites or polyampholites, etc. Since it has here become customary to designate industrial types only by often confusing letters or numbers or both, such as SBS, EDE-10P, etc., an unfortunately dual nomenclature has arisen [275]. It would seem, however, that at least for new products a change in this tradition is taking place, especially for special synthetic ion exchangers such as fibrous ion exchangers. Appendix 1 lists a selection of important commercial ion exchangers frequently cited in the literature. The number of available types is so large that it can become confusing, especially for someone new to the field. The data given in the Appendix on their characteristics have been compiled from the sales literature of the manufacturers – the best source of information concerning the applicability of a given synthetic ion exchange resin.

Designations used include certain specifications of an ion exchanger. Acidity and basicity follow from the ionogenic group with its exchange activity and thus are immediately understandable. The production process determines the external form of an ion exchanger. While polymerization resins are generally delivered in the form of beads, polycondensation resins are available as milled granulates. The color of individual types ranges from white through yellow, brown, and dark brown to black, and from transparent to translucent to opaque. Important additional characteristics are total exchange capacity, moisture holding capacity, particle size distribution, physical structure and stability, as well as data on elutable components. In practice, a knowledge of additional data is needed which depends on the intended application. In conclusion it may be said that the present importance of synthetic ion exchange resins is based on their great mechanical strength and chemical resistance, high exchange rates, high capacity and the possibility of varying their properties as a result of their synthetic nature. Their only disadvantage is their limited range of operating temperatures, although this range is also sufficient for most purposes.

A great amount of literature is available in which manufacturers of ion exchange resins present the preparation and properties of their materials either in technical information bulletins and leaflets or in papers presented by staff members on various occasions. As far as the production of ion exchange resins is concerned one manufacturing company has provided a more detailed look at the synthesis of ion exchange resins (see Figure 1.53). In the corresponding publication it is emphasized that, although manufacture of the strongly cationic and anionic resins shown there is based on such well-known unit processes as sulfonation, alkylation and amination, a great deal of ingenuity and empirical knowledge is needed for the synthesis of these polymers, particularly in controlling the rates and extent of the various reactions and in maintaining within a predetermined range the particle size distribution of the resin. "Understandably, resin producers are very secretive about process details" [276]. In a similar paper the development of ion exchange resin production at the state-controlled Nizhnii Tagil plant in the Soviet Union since 1954 has been described [277]. Whether in relation to this production or not cannot be said, but in a preceding paper it had been disclosed, that the amount of divinylbenzene has an effect on the properties of the cation exchanger KU-2-8. These resins were obtained by chlorosulfonation of styrene-divinyl benzene copolymers made from commercial divinylbenzene containing various amounts of diethylbenzene, as well as ethylstyrene and naphthalene. Increased concentration of diethylbenzene in the commercial DVB decreased monomer conversion into the styrene-divinylbenzene copolymers, facilitated their sulfonation, increased the swelling both of the copolymers and of the KU-2-8 resin, increased the exchange capacity of the exchanger, and accelerated its cation exchange ratio. With respect to the manufacture of ion exchangers in the Soviet Union of the polycondensation type the preparation of a modified EDE-10P resin with increased mechanical and chemical strength used in the purification of alcohols and sugar refinery syrups

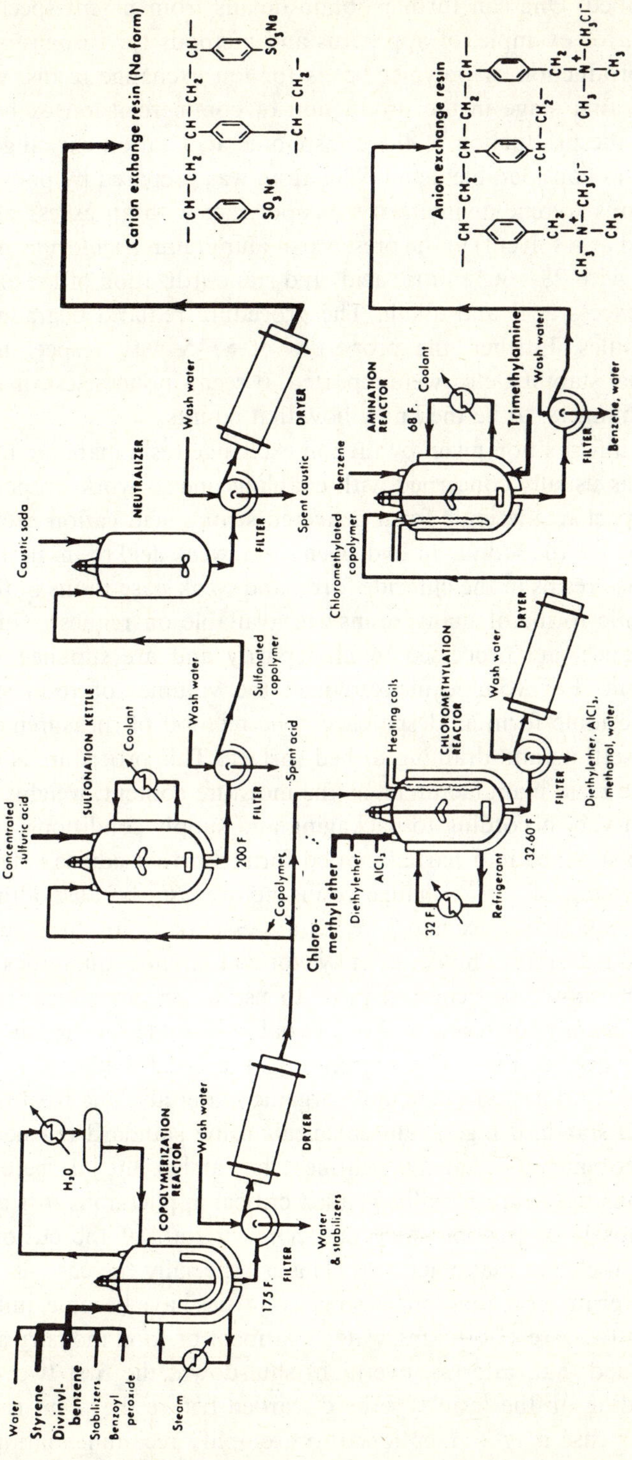

Figure 1.53 Commercial production of synthetic polymerization ion exchange resins using batchwise suspension polymerization, sulfonation, alkylation and amination. (From ref. 276).

was also described. One can further obtain details from patent specifications – the description, for example, of apparatus and methods for suspension polymerization in the production of polymer beads for ion exchange resins, which is the very important first stage in the production of commercial ion exchangers. The technology for the production of the phosphonic acid cation exchanger SF-5 has been published in considerable detail. This resin was prepared by phosphorylation of a macroporous styrene-divinylbenzene copolymer with an excess of phosphortrichloride (used as a solvent) in the presence of aluminium trichloride, hydrochloric acid, oxidation with 28–30% nitric acid, and standardization of the exchanger by washing with excess acid and alkali. The procedure reduced degradation of the exchanger granules. Further, the properties of SF-5 with respect to exchange capacity, thermal stability, etc., were reported. It seems nonetheless, in the production of ion exchangers, to be the know-how that counts.

The general information given by the ion exchange resin manufacturers in their sales literature is usually concerned with enabling one to work properly with the resin. With respect to the ionic form delivered strong acid cation exchangers are normally supplied in the sodium or hydrogen form, weak acid resins in the hydrogen form, strong base resins in the chloride form, and weak base resins in the free base form. Other ionic forms of many resins are available on request. Nuclear grade resins, for instance, are processed to high purity and are supplied in the fully regenerated form. For volume measurement the volumes of resin supplied are measured in the ionic form as despatched. The method of measurement requires back-washing, settling and draining to bed surface. This procedure is common to all ion exchange resin manufacturers, as the moisture content, weight and volume of the resin can vary according to packaging and supply conditions, inasmuch as some resins are supplied in a partially dried form. Further advice is usually given for the conditioning of ion exchange resins before use, because although most commercially available ion exchange resins are essentially insoluble and infusible polymers, standard resins, however, may contain minute quantities of soluble impurities which should be removed prior to use in certain critical applications. This is relevant mainly for resins of special quality suitable for drinking water and food processing applications. The objective of the conditioning is primarily the reduction of taste, odor, and extractable organic materials. The treatment may be carried out with standard regenerant chemicals using standard regeneration techniques. It is recommended that the treatment be carried out just before start-up. As the conditions may vary for all so-called critical applications it is advisable to cooperate as closely as possible with the technical staff of the customer service departments of the resin manufacturers. This is especially the case for food processing, since government regulations in some countries have become quite stringent. For instance, in the case of drinking water treatment or food industry applications it is recommended that, after an overnight shut-down, the first two to ten bed-volumes depending on the resin type be discarded before the plant is restored to service. Another case may be mentioned to exemplify recommendations given by ion exchange resin manufacturers for materials to be applied in the pharmaceutical industry. Cation exchangers for this industry are, for example, either strong acid styrene-divinylbenzene copolymers or weak acid poly(acrylic acid) in the sodium, potassium, ammonium, hydrogen or magnesium form with practically a food industry specification loaded to a minimum of 95% of their total capacity with the above-mentioned counterions, the rest being sodium or hydrogen. An additional purification is recommended, depending on the application, consisting of suspending the resins in water and drying them in air to a moisture content of 30% to achieve a very good trickling behavior. Depending on the requirements of an envisaged application the resins should be quality controlled with respect to moisture content, particle size distribution and degree of loading. A third example for special specifications for ion exchange resins involves their employment in nuclear chemical processes. Nuclear grade cation, anion, and mixed bed ion exchange resins are listed in the manufacturers' technical information literature as being available in various ionic forms for use in reactor coolant polishing loops. In addition to having closely controlled, (e. g., $-16 + 50$ US mesh), particle size distribution, these nuclear grade ion exchange resins are specified as having an extremely low heavy metal content. The fact that both gel type and macroporous type resins are available demands precise information on the part of the manufacturers. In evaluations of the fundamental and basic theories regarding the proper use of modern ion exchange resins macroreticular resins were compared to gel type resins in their resistance to oxidation and organic fouling, regeneration parameters, physical strength, colloidal silica entrapment, operating capacities, and leakage levels giving useful hints for an optimal usage of these competing types of resins [278]. For the storage and/or transportation of ion exchange resins at low temperatures recommendations have also been given: a water-miscible solvent with a melting point below 0 °C – e. g., methanol or acetone – should be added to a wet ion exchange resin to prevent a decrease in mechanical strength and ion exchange capacity during storage at low temperatures. There are numerous papers in which manufacturers of ion exchangers present newly developed resins with improved properties either for existing applications of ion exchange technology or for new applications for which ion exchange is able to provide a solution for existing problems.

Special details for the handling or improvements in operation of commercial ion exchangers can also quite often be found in publications supplied by the resin manufacturers. Thus, two improved methods for preparing pure ion exchange resins have been reported. For example, for preparing pure ion exchange resin AV-17-8ch from technical AV-17-8 two ways, an HCl-alkali method and an H_2SO_4-alkali method, have been reported. Each is a three-step process, requiring approximately thirteen hours. For the sulfuric acid method, the three stages consist of pretreatment with 5% sulfuric acid one, treatment with 5% sulfuric acid fifty and, finally, washing with water ten volumes per volume of resin. A study and comparison of the physicochemical properties of the ion exchangers KU-2X8 and KU-2X16 with respect to exchange capactiy, osmotic stability and IR spectra after contact with concentrated chromic acid solutions at room temperature indicated no change and satisfactory chemical stability. Ways have been described for the clean-up of fouled ion exchange resin beds with respect to different fouling agents. A resin has to be regarded as being organically fouled if the organic matter is irreversibly bound to it. The most common clean-up methods that can be practically applied to industrial ion exchange units involve treatment with alkaline brine solution or with alkaline hypochlorite. Organic matter not removed by normal sodium hydroxide regeneration can be removed by a salt solution; one of the most effective salts for this clean-up is sodium chloride. Its effectiveness is attributed to the high affinity of the chloride ion for the anion exchange resin site, so that the tightly-bound carboxyl group is displaced from the active site more completely than by the hydroxyl ion. The presence of sodium hydroxide in the brine solution improves the clean-up efficiency, an apparently synergistic effect. The brine is seen as effectively overcoming the attraction of the carboxylate ion for the amine site, while the high pH reduces the matrix adsorption of the organic molecule and increases its solubility. In certain cases of organic fouling, particularly where brine cleaning has not been carried out for some time or is not regularly used, the resin may become so heavily fouled that brine cleaning will be unsuccessful. In such cases, for certain types of anion resins, a sodium hypochlorite treatment may be employed. Care must be exercised in carrying out this treatment since the strong oxidizing effect of sodium hypochlorite solutions – it is recommended that they contain one percent free chlorine – can be detrimental. In general, this treatment is regarded as an extreme measure to be used only when other less aggressive methods have failed. However, if the resin is badly fouled with organic matter, the brunt of the oxidizing force of the sodium hypochlorite is taken up in oxidizing it, so that damage of the matrix by the hypochlorite is somewhat milder than one would imagine. It is important, however, that the cation component of mixed beds be completely exhausted, because acidic solutions liberate chlorine gas from sodium hypochlorite. The recommended contact time between the resin and the solution is about four hours, and a double or triple regeneration is required to ensure complete removal of the hypochlorite and to regenerate the resin. Iron clean-up treatment is also required in the operation of commercial resins, since it is almost invariably found that large quantities of iron are present with the organic matter adsorbed onto anion exchange resins. The iron clean-up uses either ten percent hydrochloric acid, or a sodium dithionite solution which reduces the ferric ion present to the more soluble ferrous form. A subsequent brine caustic wash gives a better result than that achieved by carrying out either of the two clean-ups separately [279]. It seems of interest to note that the study of the complexing of metal ions on the irradiated commercial amphoteric resin VPK showed that the complexing of strontium(II) on this exchanger pre-irradiated by $1 \cdot 10^{19}$ rad dose in water or nitric acid of pH 3.0 – 5.0, resulted in a stronger complex formation.

A great number of users' comments on properties and applications, and the interdependent influence of both, have also been published often resembling a kind of a counterpart to the supposed clarification provided by manufacturers' descriptions. With regard the buying of ion exchange resins the question has been raised whether they are more of a necessity or a nuisance. In an original paper critical reflections with illustrations of ion exchangers with too lax and too rigid specifications have been described. The following summarizing remarks are taken from the original paper. If a certain ion exchange material "brand X" has to be replaced it has to be born in mind that simply specifying "brand X" has two potential problems. First, the properties of one and the same ion exchanger type has changed over the years due for the most part to improvements in the methods of manufacturing resins. Following the "evolution" of a porous quarternary anion exchanger from 1958 to 1980 the variations appear at first glance to be small. They are primarily in the size of the resin beads, initially 0.35–0.42 mm and then 0.45–0.55 mm, which reflects the trend towards higher and higher flow rates. However, changes in bead size influence other hydraulic properties, in addition to pressure drop; and backwash expansion rates are dependent upon bead size (as well as upon temperature and linear flow). Thus the change in particle size of the same "brand X" may cause an increasing incidence of poor separation in make-up mixed beds if the rebedding with large size resins is done without adjusting the operating procedure to assure adequate bed expansion of the anion exchanger. Secondly, a multitude of sub-grades may be marketed with not only variations in bead size and level of fines, but also in ionic form, plant-washing techniques and even how the same brand is packed. For example, if at standard crosslinkage a cubic foot of resin in the H-form instead of the Na-form is supplied one has 7% less resin. Thus, if one elects to continue buying resin on a specification for "brand X" a qualifying description of at least the ionic form is in order. And unless bead size is totally unimportant – as it rarely is – information concerning it is essential as well.

With respect to batch-to-batch variations of a "brand X" of resin this is a function of the manufacturer, the particular product, but probably not the phase of the moon. Whoever has made large-scale batch polymerizations knows that at the end the material may well be off grade or that one can only hope to bring it within specification by blending. The practice of buying ion exchange resins has apparently taught the author that unless care is taken a shipment of resin may look like a warehouse clearance sale, with drums of resins from as many as six or seven batches in a single shipment. In this case it was found that variations in chemical properties were small but the size of the beads varied widely. Such a material in cation resin intended for a condensate polisher can, and did, cause serious separation problems.

An often raised question, especially when there are several bidders, is how to select replacement ion exchange resins, i. e. are there equivalents for "brand X". To this question the experienced authors of the report note that by far the most common specification is "brand X or equal". Following such a specification it becomes a question not only of defining the properties of "brand X" but, with increasing frequency, of deciding whether "brand Y" – being offered as a substitute at a lower price – is indeed equivalent. The examination of resin manufacturers' published charts of resin equivalency has led to the conclusion that some suppliers are obviously not fully aware of which products they are competing with. It can be said, however, that the general consensus is that cation exchange resins are more equivalent than anion exchange resins. A more detailed comparison of the properties of new sulfonic cation resins from different producers demonstrated that "brand Y" may not be equivalent to "brand X" in all cases. This is particularly evident in the values for highly crosslinked gel type resins. As far as anion exchangers are concerned the authors note that it would be a hopeless task to attempt to demonstrate the equivalency, or lack thereof, of anion resins in an elaborated table with a set of numbers similar to those given for cation resins. Essentially, for anion exchange resins, the question is "equivalency to do what?". This is because no two anion exchange resins are precisely equivalent in all operating situations. Two candidates may be equivalent on one water source and different on another. They may give a similar performance in a two-bed operation and different ones in a mixed bed. It can be shown by tabulating the data that there is more product scatter among anion resins than among cation resins; this is so even when such comparisons are limited to the trimethylamine quaternary resins based on styrene-divinylbenzene copolymers. The ion exchange industry tries to categorize quaternary resins as porous or standard grades, but this labelling does not represent a clear-cut distinction. However, a tabulation such as that mentioned above with regard to increasing order of water retention, shows that a porous resin can be expected to have a water retention of 53% or more. But as may also be seen, its other properties vary widely.

In summary, the buyer should write a specification following some rules for the producer's information. The material to be purchased has to be described by the manufacturers precisely and in a form readily understandable even if a trade name is used. The ionic form wanted by the buyer should specifically be given and it should be stated how the material specified is to be sampled. The parameters to be measured must be carefully selected to make sure that they describe the material wanted, whereby the number of properties to be measured should be kept to a minimum. It is also to be remembered that a number must have a known method by which it is obtained. The principal sin of a specification that is more a nuisance than helpful is that of omission. The single most descriptive characteristic of a sulfonated styrene-based cation exchanger is its water retention capacity in the specific ionic form. If this value is in the right range the capacity will be acceptable, the hydraulic properties will be normal and the performance at a given regeneration level can be given. A specification should, therefore, not be issued without it. In specifying anion exchange resins the designation Type I and Type II are well established for the two kinds of strong base resins. When a moisture content is given in a specification for a Type I resin the numerical value is almost always 60% maximum, but it seems advisable to analyse this figure in order to see if what one is buying meets the specification. During a corresponding test series it was found – this seems of interest to be noted – that the OH-form of a quaternary resin can be dried to yield a reproduceable value. Variations between triplicates were comfortably within the deviation expected for a Cl-form resin. Together with the total capacity expected the percent of conversion with the impurities carbonate and chloride is usually specified as well as the particle size distribution. A justification of the specification of carbonate and chloride in an anion exchange resin to be supplied in the OH-form is very questionable. Most of the comments on particle size distribution discussed above apply also to anion exchanger specifications. It is apparently common that specifications for anion exchangers describe a larger size anion resin. Further, the failure to mate the cation-anion properties also seems not to be uncommon and is a major reason for mixed bed separation problems.

The authors conclude that it is necessary to provide specifications if an ion exchange resin is purchased on the open market. Further, they are necessary if optimum performance of a demineralizer is to be achieved. They will certainly be given if buyers insist on their being met and are willing to reject material that doesn't meet specifications. On the question of whether generic specifications should be acceptable the authors conclude that they probably should be if they are limited to a narrow target such as the previously-used polishing resins for BWR plants. But condensate polishing in general is – according to ASTM D19.0801 – far too complex for a single set of specifications. This is because variations in unit design impose requirements that result in a specification either so narrow that it cannot be used everywhere or so broad that it guarantees optimum performance nowhere. But specifications can be taken from available materials which optimize the performance of a particular plant, operating on a specific water quality [280].

In another paper the same authors, Fisher and Otten, have turned a critical look to the measurement of extractables in new ion exchange resins. Relating in the main to nuclear grade ion exchange resins and materials used for food processing or for drinking water treatment the authors emphasize that over the years the feeling has developed that instead of testing the separate anion and cation components the testing should be done as much as possible on the material as it is actually going to be used. By using a hot extraction, a cold extraction and a column test, the components of a mixed bed were examined first separately and then as the final mixture. The mixtures were blended in the ratio of one equivalent of H-form cation to one equivalent of OH-form anion exchanger. The cation exchange resins were standard 8% gel type and the anion resins porous Type I quaternary gels. The components came from two different suppliers. As expected, the results obtained showed that a hot extraction removed a higher concentration of material than one carried out at ambient temperatures. Also, as expected, increasing the temperature has a greater effect on anion extractables than on cation extractables, due to decomposition of the ionogenic groups. The principal leachable material from either the components or the mixture was organic. The column tests proved to be primarily useful for following the organic parts. They showed differences similar to the cold leaching. The results of these investigations proved also that the leachables from the anion exchanger component are cationic, namely sodium from residual regenerant and amines, and the extractables from the cation exchanger component are anionic, namely low molecular weight sulfonates. Conditions in the mixed bed are such, however, that the anion leaching goes onto the cation resin and the cation leaching onto the anion resin, accounting for the fact that the leachables from the mixture are invariably lower than those of either component. It was further demonstrated that with the availability of greater analytical capabilities, the old tests which simply measure residue on evaporation have outlived their usefulness. Most of what comes off of nuclear grade ion exchange resins is organic in nature and much of it is volatile at 104 °C. The anion exchange resins are the major contributors of organic carbon in the form of low molecular weight amines and they are also the most likely source of sodium in the system. This is the case even when they are mixed with the H-form of a cation exchanger. It also became clear from these investigations that soluble leachables increase with storage time of the resins, particularly in the case of the OH-form of the strong base anion exchange resin [281]. For getting correlations between the extractable contents of the ion exchanger materials and in-plant performance it would seem necessary to obtain more information on the specific nature of the extractables rather than on their total nature. Some data on substances released into water by ion exchange resins has already been published. Investigations of the discharge of amines from a commercial weak base macroporous anion exchanger material consisting of a styrene-divinylbenzene matrix with tertiary amino groups in which the experimental results were compared with modelling computations for a particle collective with a probability distribution characteristic for the particle size showed that the discharge of these amines is a process controlled by diffusion, probably based on a distribution equilibrium in favor of the resin matrix. Having recognized these relations the cleaning of these resins is discussed and the conclusion drawn that it should be investigated again [281].

In accordance with what was said before about the intermediate status of an ion exchanger between that of a chemical product and a piece of equipment ion exchange resins have a limited lifetime. The limited lifetime of an ion exchange resin can be viewed from two perspectives: with regard to quality and to cost. The latter is of great importance for cost estimates of equipment and operating costs of ion exchangers in water treatment, depending mainly on the fouling of resins. The quality influences dependent on the lifetimes of ion exchange resins can lead to more difficult situations and even raise the question of whether their employment is feasible. Thus some consequences of the auto-degradation of polystyrene-based ion exchangers were studied and it was found that such commercial products of otherwise high quality undergo decomposition and form water soluble organic fragments. The vulnerability of such resins to degradation was a function of the degree of crosslinking. The infrared spectra of the fragments obtained had the essential features of the "pure" resin spectra. With a cation exchanger, sulfur analyses of the residues, in conjunction with cryoscopic measurements, allowed an estimate to be made of the residue fragment number-average molecular weight, and it was shown that the fragment size is appreciably influenced by the degree of crosslinking. Mass spectroscopic studies and determinations of the distributions of trace metal ions threw light onto the effect of this degradation on the remaining washed ion exchanger beads. The mass spectra and distribution results could be explained in terms of resin oxidation with the introduction of carboxyl groups on the resin matrix. The time scales suggested that the process of degradation was rapid. The mass spectrometric investigation provided a qualitative estimate for the age and the level of decomposition of stored samples of resins. It permits an estimate to be made of the efficacy of cleaning and conditioning of a resin and it also provides a potential method, by variation of the probe temperature, of examining aspects of the thermal stability of ion exchange resins [282]. The lifetime of synthetic resin ion exchangers was examined with regard to the radiation loading capacity and a model was described for calculating the lifetime of resins used for the purification of water containing radioactive impurities in the primary cooling circuit of nuclear reactors. An investigation of the alkali tolerance of Varion AP ion exchange resin in sodium hydroxide solutions showed that the strongly basic groups of this commercial anion exchanger decompose. This effect is decreased by keeping the alkali treatment short. In the presence of chloride or sulfate anions the decomposition rate is slower. This protecting effect of anions, which is enhanced by repeated treatment, depends on the anion distribution between exchanger and aqueous phase. Otherwise is it commonly known that the weakly basic anion resins of the gel type have a short operating life – in the range of three to five years – principally due to oxidation or resin deterioration rather than organic fouling. A short lifetime must be expected in applications in which the resins have to withstand severe chemical conditions. Thus, in the treatment of plating wastewater by an ion exchange method, both cation and anion exchange resins were easily oxidized with chromate, both in aqueous and in nonaqueous media such as formamide and formamide-ammonium formate. The rate of oxidation was greater in the nonaqueous than in the aqueous solutions and the rate increased with temperature. Oxidation resistance is in fact a general ion exchanger property and it will be treated in more detail below. Other general properties deviating from the general description of a commercial ion exchanger were reported; for instance, that the sulfonated cation exchanger SBS-1 is not mono- but multiionic. This was made evident by potentiometric titration over a seven day period with 0.5 N KOH-KCl, indicating a sulfonic exchange capacity of 2.58 mequ/g and a total capacity of 3.08 mequ/g. The increment of 0.50 mequ/g was attributed to carboxyl groups. Besides infrared spectroscopic investigations it was further found that the elution of VO_2^{2+} ion from SBS resin with 1N NH_4NO_3 was incomplete, which is characteristic of a carboxylic resin. For basicity testing of strong base anion exchangers a method was elaborated to determine the amounts of different strongly basic groups in currently used strong base anion exchangers. The method involves conversion under stirring into the base form of the weakly basic groups in the exchange resin – which is in the chloride form – by reaction with an ammonia solution containing sodium chloride. Because of the chloride added, the strong base groups are not affected by the ammonia. From chemical, microscopic, and bacteriological investigations on ion exchange resins it follows that mechanical wear and tear results in abrasion, fissuring and cracking in the granular structure as well as in surface damage. Hygienic problems include, then, increased bacterial numbers in the exchanger as well as taste problems resulting from solubilization of organic substances. A comparative investigation with different commercial ion exchangers for the extraction of iodide and bromide ions from aqueous solutions is also quite interesting. The anion exchangers Varion AD, Wofatit SBW, Wofatit L-150, Lewatit MP-500, and Amberlite IRA-402 after regeneration of the resins were examined. The extraction was done with solutions containing 0.724 mequ/l I^-, 16.4 mequ/l Br^- and mixtures of the two. The exchange capacity under static and dynamic conditions decreases in the sequence Varion AD → Amberlite IRA-402 → Lewatit MP-500 → Wofatit SBW → Wofatit L-150 (0.97, 0.87, 0.75, 0.54 and 0.22 mequ/cm^3 resin, resp.). The best regeneration efficiency was with Wofatit and the poorest with Varion AD. The stripping efficiency increases with the concentration of sodium hydroxide in the solution used to treat the resin. The joint extraction of Br^- and I^- ions was characterized by the preferential sorption of bromide. In a preliminary study of selected potential environmental contaminants ion exchange resins were included. A review of the literature published from 1953 through 1973 was conducted to prepare this preliminary report on environmental exposure factors with regard to the use and consumption of ion exchange resins; the health and environmental effects resulting from exposure to the substances, as well as applicable regulations and standards governing their use were evaluated [283].

Commercial ion exchangers must have a food regulation status if it is intended to use them in food processing or in the treatment of drinking water. Illustrative of this is work done on ion exchange and adsorbent resins for the removal of acids and bitter principles from citrus juices. Twenty commercially available ion exchange and adsorbent resins from two manufacturers were screened for their affinities for titratable acid and citrus bitter principles. Two anion exchange resins offered the prospect of simultaneous acid reduction and debittering, and two cation exchangers and fine adsorbent resins were moderate to powerful adsorbers of the bitter principles. Including materials from a preceding investigation it was possible to give a list of six weak base resins which were quite efficient for acid removal, of which all, except for ES 375 and IRA-35, comply with the United States FDA regulations for food contact use, being, therefore, potentially useful for commercial deacidification of citrus juices. Two anion exchange resins, IRA-401s and A 378, were found to combine the debittering and deacidifying functions and, as they comply with the United States FDA regulations, they offer commerical prospects of simultaneously deacidifying and debittering citrus juices [284]. For in-depth information on the use of ion exchange in the food industry of the United States the history of legal regulations regarding the use of ion exchangers and sorbents in the food industry has been discussed, and ion exchangers currently listed by the U. S. Food and Drug Administration (FDA) for use in food processing can be found tabulated there. The rationale for the regulation of ion exchange material use and over-regulations were discussed as well [285]. For the Federal Republic of Germany the corresponding legal regulations are contained in the Plastics Recommendation XXIV based on the Lebensmittel- und Bedarfsgegenständegesetz and controlled by the BGA (Bundesgesundheitsamt). In general the regulation says that there are no objections against the application of ion exchangers and sorbents for the treatment of drinking water and aqueous fluids which are used as food or for the manufacture and processing of foodstuffs as long as they are suitable for the envisaged purpose and meet the prerequisites outlined therein. An exhaustive description of the ion exchangers and related materials then permitted is given. Important is that if ion exchangers and sorbents are to be used for the treatment of water for drinking water purposes the advisory opinion of a testing laboratory has to be obtained [286]. Similar regulations may or may not exist in other countries, which must be checked in each individual case. For France and Italy some hints can be found in the older ion exchanger literature [287].

The purification of ion exchange resins when contaminated, i. e., the restoration of the properties of commercial ion exchangers, is another point of great interest. Industrial ion exchange columns sometimes become contaminated by various pollutants. Knowledge of these phenomena and appropriate techniques in forestalling any contamination insures long life and efficient use of the columns. Factors which have been discussed in this respect are the retention of solids suspended in the influents, precipitation on the resins due to either pH changes or supersaturation, adsorbed substances in such quantities that they exceed the extractibility of the regeneration step and organic contaminants. So-called resin cleaners have been offered commercially as ready-to-use liquid cleaners to remove troublesome deposits such as metallic oxides, silt and other suspended matter which collects on ion exchange beds. In a two-component recovery of ion exchangers, resins contaminated with colloidal or ionic compounds of iron, copper, cobalt and aluminium are treated with an aqueous solution containing 0.1 to 0.3% of a chelate-forming agent and a 1 to 11% aqueous solution of an acid that has been adjusted to pH 0.8 to 6.5 with ammonia. Combinations of sodium salts of nitrilotriacetic acid, sulfosalicylic acid and EDTA with hydrochloric acid, sulfuric acid, acetic acid, formic acid and citric acid have been tested and described in a patent application. For ion exchangers contaminated with sulfur compounds it was reported that for their purification Na_2S was added to a regenerating solution while the solution was heated to 50 to 60 °C. Under the patent heading "ultrasonic cleaning of resin" a method and apparatus are described for cleaning of contamination from ion exchange resins used for contamination removal from water condensed in steam turbines and from water in the recirculation system of a nuclear reactor. The resins are cleaned by establishing a counter-current flow between the resin and a carrier fluid and applying ultrasonic energy. A simple and inexpensive method using a hydrazine solution in Trilon B was described for the removal of deposited iron and KU-2-8 prior to a reactivation of the cation exchanger with sodium chloride. It is said that washing of iron-containing KU-2-8 with the complexon solution was used for the reduction of iron from ferric to ferrous form. Ion exchangers become poisoned by sulfur compounds in hydrometallurgy. According to a patent specification these poisons can be removed from the exchangers by 1 to 10 M nitric acid at 0 to 80 °C. Thus, anion exchangers that were used in the separation of the uranyltricarbonate anion from alkaline leachings and which gradually lost capacity owing to poisoning by sulfur compounds and organic substances were treated with 2 to 3 M nitric acid at 35 to 40 °C in a column stirred by air. Discharged nitric acid was used in the technological process to decompose the uranyltricarbonate complex and nitrose gases were adsorbed by an alkaline solution. With regard to fouling it should be mentioned that a method for assessing the rate of fouling of ion exchange resins has been outlined. The effect of the rate of fouling of the resin on plant output and the criteria for the most economical time of changing the resin have been given by means of graphs of typical changes of resin capacity with time. A detailed account of the measurement of the capacity of ion exchangers using water of fixed ionic content has been included in this presentation [288]. The influence of the presence of tannins in the influent on an ion exchange resin column has also been studied. In spite of the high resistance of the ion exchange resin in use to organic fouling agents, tannins caused clogging of the resin beds and permanently damaged the resin material. Although chlorine can oxidize tannins, the doses required are quite high, resulting in an undesirable high residual chlorine content in the treated water. Preliminary investigations showed that adsorption on high surface solids such as activated charcoal could substantially remove tannins from diluted solutions. By γ-irradiation a change in the sorption behavior of commerical ion exchange resins can be achieved. This was demonstrated for the sorption capacity of anion exchangers for gossypol and aliphatic fatty acids in methanol or benzene, and correlated with the γ-dosage given.

Bacteria growth on ion exchange resins has been thoroughly studied. Investigations with a strong acid cation exchanger focusing on the attachment and distribution of bacteria and operational possibilities of suppressing bacterial growth have shown that bacteria are not necessarily attached securely to the surfaces of ion exchange resins. They can, e. g., be easily washed off the strong acid cation exchanger Lewatit S 100. On further investigation it was concluded that they are distributed inhomogeneously in the resin bed in the form of nests. The number of colonies is normally higher in the upper region of the ion exchange column than in the lower. Possibilities for suppressing the growth of bacteria in an ion exchanger resin bed include continuous operation, as far as such is practicable, and keeping the resin in the regeneration agent during shut-off periods of operation. A 20% sodium chloride solution can suppress the growth of bacteria over a long period of time. Other measures, such as recirculated operation, backwashing and regeneration have proven themselves insufficient with respect to reduction or suppression of bacterial growth and cannot be applied instead of disinfection of the resin. It is of great importance that ion exchanger units be subjected to routine bacteriological monitoring in all cases where hygienic questions are involved with their use. The effectiveness of silver additives (oligodynamic effect) against bacterial growth during shut-off periods has also been investigated. The problem here is that considerable bacterial growth occurs in ion exchange resin beds during these periods. The bacteriostatic effect of very low silver concentrations was repeatable, but it was found that this effect decreases in intensity after a time. In the experiments reported this reduction occurred within several weeks. Evidence can be found for an increase in the tolerance of the bacterial flora towards silver ions – the adaption could be reproduced in vitro. Moreover was it found that the sensitivity of the bacteria population in the drinking water supply fluctuates quite strongly, i. e., between 5 and 50 μg/l Ag^+ as a toxic concentration limit, so that the development of silver-tolerant microorganisms in silver-treated ion exchange resin beds is further enhanced. According to the report it is, therefore, not possible to rely permanently on the germ-curbing effect of low silver concentrations. This is the second reason why it is of the utmost importance that ion exchanger units be subjected to routine bacteriological monitoring. The conclusion was that it is necessary to disinfect the resin beds periodically. The suitability of peracetic acid for the disinfection of ion exchangers was then investigated. Peracetic acid of 0.02% was found suitable for a satisfactory disinfection. Its use strongly reduced corrosive effects. Calcium ions seem to protect the bacteria, therefore the disinfection should be carried out with the sodium form. This disinfection, however, has no permanent effect and cannot prevent bacterial aftergrowth during new shut-off periods. A combination of silver and disinfectant can accomplish this until a new, silver-tolerant microflora has evolved. In this case the use of 0.02% peracetic acid is imperative, because higher concentrations dissolve the silver. The effectiveness of the disinfection procedure should also be monitored bacteriologically [289]. For the sterilization of ion exchangers with peracetic acid in a demineralization plant with filters loaded with the strong acid cation exchanger Wofatit KBS and the strong base anion exchanger Wofatit SBK it was earlier reported that the ion exchange capacities of both these resins were not affected adversely by sterilization with 0.2% acetic acid. Bacteria and their metabolites are often present in ion exchanger units for demineralizing water but flushing the Wofatit columns for 20 minutes with this acid concentration sterilized the resin adequately. An in-depth investigation into the washing-out of ion exchange resins after they have been disinfected with agents such as peracetic acid or formaldehyde revealed that this process proceeds in two phases. The first is the displacement of the disinfectant from the void spaces between the exchanger beads; the second is the removal of the disinfectant which had penetrated into the matrix of the exchanger during the disinfection, which is redelivered during the rinsing process according to the reversed concentration gradient. The diffusion process of the second phase is the limiting step for the whole operation. The washing-out efficiency is therefore primarily dependent on the rinsing time and not on the amount of rinsing water used. The results with strong acid cation exchangers show that it is much easier to free macroporous resins from peroxy acetic acid than gel type resins, corresponding to the different apparent diffusion coefficients of peroxy acetic acid in both these resin matrices [290].

1.2.3 Properties of ion exchange resins

Discussion of the properties of ion exchangers might well be limited to capacity, equilibrium and kinetic behavior, since it is these criteria that mainly determine the exchange processes. These properties are not sufficient, however, for characterizing synthetic polymeric products, which are prepared either in the laboratory or by complex batch processes in which a large number of characteristics are controlled by small variations in reactant composition, the degree of chemical completion, and overall processing techniques. It is precisely the possibility of predetermining the properties of synthetic exchanger materials during synthesis that has been the decisive factor in giving synthetic ion exchange resins their superior position as compared to natural ion exchange materials.

The prime objective of an ion exchanger is satisfied by a sufficiently high capacity, a rapid attainment of equilibrium and a satisfactory rate of exchange. But the question of selectivity, which appears already at this stage, suggests that a number of other factors that influence the properties of an ion exchanger must also be taken into consideration. For example, high quality in an ion exchange resin includes physical and chemical resistance and the characteristic of releasing none of its components to the surrounding media. The latter, in turn, can be given a positive rating only if oxidative, hydrolytic, thermal, or radioactive influences neither produce nor accelerate such a release.

The purpose here of presenting the properties, is to offer a basis for the evaluation of criteria important in research and practice. What is meant in the following by properties of ion exchange resins is not the resin structure and structural refinement and their determination mainly by spectroscopic methods, measurements that can nonetheless give answers to questions involving, for instance, hyperfine structures or which may be quite suitable for the study of kinetics and of the mechanisms of processes taking place within ion exchange materials. With respect to the properties of ion exchange resins it will be seen that the equipment needed for this experimental work is simple and inexpensive – a characteristic that applies generally to all ion exchange equipment. As for the kinetic behavior of ion exchange resins and the rate of exchange no standards exist; in the future, however, inclusion of this property – as has been already suggested above – in the technical specifications will be indispensible.

Quite often, a general characterization of a resin, in the sense of an identification, is needed. An unknown ion exchange resin can be characterized by treating a small sample of it with an excess of approximately 5% hydrochloric acid in a small column, followed by washing it briefly and drying it. From the dry sample an appropriate quantity is finely ground, incorporated into a potassium bromide disc and the infrared spectrum determined. By comparison with the spectra published by Whittington and Millar the general nature of the exchanger can usually be determined [291]. In this work the absorption spectra for fifty commercial ion exchange resins are given, including quaternary ammonium, tertiary and mixed amine, sulfonic, phosphonic, phosphinic, and carboxylic, as well as a number of experimental resins, including some of the solvent-modified types. A simple method was reported recently for the identification of ion exchange resins. The exchange materials are pretreated with sodium chloride and the change of color in $KMnO_4$ solution are observed. The resin is a cation exchanger when its color in $KMnO_4$ solution does not change and is an anion exchanger when the solution turns red [292]. Elemental analysis of ion exchange resins has also been carried out. The analytical determination of carbon, hydrogen, nitrogen, chlorine and oxygen in a number of commercial materials led to the proposing of stoichiometric formulas and tentative structures for the ion exchangers. For the perspective of practical application the determined amounts of nitrogen and chlorine were related to the ability of these exchangers to decolorize molasses. Where there is a need for the analysis of elemental impurities in an exchange resin several ways and methods have been employed. Firstly, an instrumental method based on neutron-activation was developed. The average deviation of the method was 8 to 12% and the sensitivity in the determination of elements was for zinc 3×10^{-7}, antimony 1×10^{-8}, iron 1×10^{-6}, cobalt 6×10^{-9}, scandium 1×10^{-10}, mercury 5×10^{-8}, chromium 1×10^{-7}, cesium 5×10^{-9}, sodium 1×10^{-7}, and bromine 5×10^{-8} g. There are cases when it is desirable to dissolve a resin – for example for remote removal of contaminated resin from nuclear processing facilities, for unplugging columns in nuclear processing facilities, for recovery of sorbed cations without elution from the resin, or for preparing counting samples of sorbed α and β emitters. That this is possible was discovered in tests used for evaluating the chemical stability of resin matrices where it was shown that crosslinked polystyrene resin matrices partially decompose by oxidative degradation. Besides this complete dissolution has been achieved by refluxing with 30 to 90% nitric acid, or by treatment at 60 °C with 50% hydrogen peroxide containing iron, or by digestion with 6% potassium permanganate, with preference being given to the dissolution of crosslinked polystyrenesulfonate cation exchange resins by dilute iron-hydrogen peroxide solution. The same laboratory described the iron-catalyzed dissolution of Dowex 50WX8 in two different papers. After the first intensive study, which reported on the mechanism as well, the dissolution of ion exchange resin by hydrogen peroxide was again described. Dowex 50WX8 slurry 40 vol% was effectively dissolved by a solution containing 0.001 M Fe^{2+} or Fe^{3+}, and 3 vol% H_2O_2 in 0.1 M HNO_3. Foaming pressurization is eliminated by maintaining the dissolution temperature $\leq$ 99 °C. Pre-mixing hydrogen peroxide with all reactants will not create a safety hazard, but operating with a continued feed of hydrogen peroxide is recommended to control the dissolution rate. It is further pointed out that a spent resin from chemical separation contains diethylenetriaminepentaacetic acid residue, and the resin must be washed with 0.1 M NH_4OH to remove excess of it before dissolution. Gamma irradiation of the resin up to 4 kWh/l did not change the dissolution rate significantly. Dissolution is complete in approximately one hour. With these investigations a dissolution process for routine operation is now available [293].

The measurement of properties even with standard test methods needs support procedures which must be clear in their efficacy and their accuracy in order to form a sound basis for whatever follows. Mass and weight versus volume should probably be discussed first in this regard. The ionic form of a resin affects both its mass and its volume. The measurement of mass or weight for ion exchange resins is more the object of critical consideration than that of volume, because the mass as measured depends on the moisture content of the resin particles. If, on the other hand, the resin particles are dried, too severe physical stress and possible chemical effects may render the dried samples useless for further testing. But since any specification for an ion exchange resin requires an accurate measurement of its quantity, the measurement of volume, rather than of mass, has therefore been chosen for the determination of various properties, except for the determination of the moisture content and relative density, where volume does not apply, and in the still frequently employed determination of the weight capacity.

It is appropriate, therefore, that the measurement of the volume of an ion exchange resin be described in this introduction. The volume of a quantity of resin can be measured by the use of measuring cylinders, the reading being taken either after the resin beads have settled or after they have settled and the cylinder been tapped to minimize the settled height. These two procedures are referred to as the free wet settled volume and the minimum tapped volume, respectively. For resin samples of constant particle size distribution two factors may influence with regard to the diameter of the volume-measuring vessel, the volume measured. Increasing the diameter – which reduces the bed height of the sample – reduces the packing of the bed and, therefore, increases the volume measured. Further, the wall effect decreases, reducing the void fraction in the bed and consequently reducing the volume measured. For the free wet settled volume, the two factors cancel each other out to some extent, so that there is little change in the measured volume as the diameter of the measuring vessel is changed. For the minimum tapped volume the bed is packed to the same extent regardless of vessel diameter and the wall effect predominates, causing an apparent decrease in volume as the diameter of the vessel increases. For the free wet settled volume it is, for the sake of consistency, important that resin samples of the same particle size distribution should be used, that the measuring cylinder should have a diameter greater than nine millimeters, i. e., ten times the diameter of the largest particle, and that a set of measuring cylinders should be kept exclusively for the measurement of resin volume. In practice the resin sample is poured into a measuring cylinder of approximately the desired capacity, which is then filled with deionized water and stoppered. The cylinder is turned upside down and then gently placed on a level surface. The resin beads are allowed to settle and a reading is taken after one minute. After this procedure has been repeated three times, the average value is recorded.

The second question that needs consideration before the properties of ion exchange resins are gone into is that of pre-conditioning freshly synthesized or fresh resins received from a manufacturer. A fresh resin can not be taken as representative of a resin that has aged through use in the application it was destined for. Firstly, a resin as received may be in a different form or not fully in the form as specified. Secondly, the exchange capacity and density of a resin are invariably affected by the first service cycles. A fresh resin, therefore, has to be subjected to several exchange cycles, both to stabilize the resin characteristics, i. e., size distribution, capacity and density, and to convert the resin to an ionic form, which serves as a reference for all the testwork. For example, for an anion exchange resin used in the sulfate form a method for pre-conditioning of the resin was designed as a compromise between the aging of a resin under plant conditions and the requirements of a fully reproducible procedure that can be implemented simply and rapidly. In the resulting batch procedure the resin sample is soaked in deionized water, contacted with 1 M sulfuric acid in a rolling bottle to remove chlorides, contacted with a fresh batch of 1 M sulfuric acid, and then contacted with 1 M sodium hydroxide. This process lasting 25 to 30 hours is repeated twice, with the second repetition stopping before the hydroxide contact. For the employment of ion exchangers in analytical chemistry a purification of the material may be necessary, which then, in most cases, includes pre-conditioning.

With regard to the testing of commercial ion exchangers, the development of these resins and their wide variety of applications in modern technology has made it necessary for set standards to be worked out. Standardization of test methods should be given first priority as a standardization of the requirements made on the resins; but this can only be considered after standardized methods for their testing have become available [294]. All these aspects are treated separately in a relevant chapter in this volume.

1.2.3.1 Moisture content and density

Synthetic ion exchange resins have a certain moisture content in the form of bound water resulting from the hygroscopic properties of the ionogenized resin. Beyond this, ion exchangers can take up free water or surface water, which, as it is unbound, can be removed by centrifuging and which has no influence on the exchanger properties. An early method for the determination of the moisture content was presented by Pepper and coworkers using a centrifuge as shown in Figure 1.61 and described there. This method has long been recommended for the determination of the moisture content (water regain). To distinguish between the free and the bound water drying procedures have as well been applied. For instance, the ion exchange resin Amberlite IRC-50 in the H-form – an acrylic acid-methacrylic acid-divinylbenzene copolymer – was dried by passing a stream of hydrogen through the resin. The resin lost weight for the first ninety-six hours at 0.33%/h with a total weight loss of 32.7%. For the next ninety-six hours the total loss was only 1.3%, and this slower loss was attributed to bound water. The surface water can otherwise be removed from the swollen ion exchanger sample by light centrifuging or even filter paper. With regard to the influence of the moisture content on the exchanger properties it was, e. g., found that the value of the dielectric coefficient is directly proportional to the water concentration in the resin.

The quantity of regained water depends on the nature, amount and form of the ionogenic groups, which have a positive influence on moisture uptake, and on the density of the network formed by crosslinking, which permits a moisture uptake only until the osmotic forces are compensated for. The resulting moisture content is expressed in percent of moisture per weight of wet resin, in percent of moisture per weight of dry resin, or in weight or mole number of water per equivalent of exchange capacity. With a higher degree of crosslinking (measured by the divinylbenzene content), less water is regained by the exchanger, as shown in Figure 1.54. The uptake of moisture from the atmosphere with different relative humidities depends on the ionic form of the exchanger, as shown in Figure 1.55.

A number of methods can be used to determine the regained water in ion exchange resins. Oven drying or azeotropic distillation are techniques generally

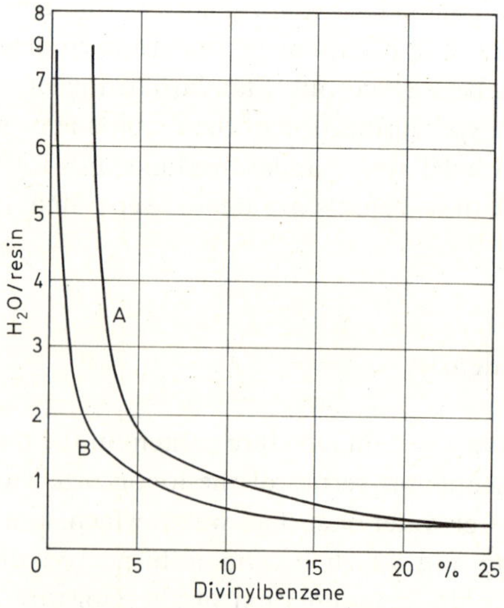

Figure 1.54 Moisture regain of cation exchangers of different degrees of crosslinking. A = strong acid sulfonic exchanger; B = weak acid carboxylic exchanger.

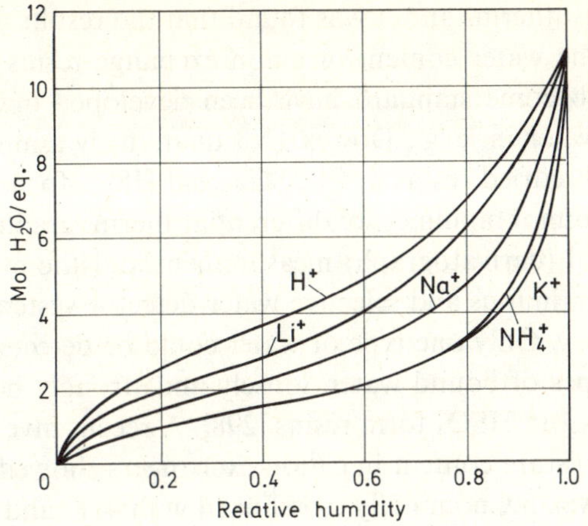

Figure 1.55 Moisture regain of a strong acid cation exchanger as a function of the relative humidity and ionic form.

used for high polymers. Their disadvantages can be circumvented by titration with Karl Fischer reagent, by which the water associated with ion exchange resins can be determined rapidly and satisfactorily in methanol or pyridine [295]. Thus, for instance, the moisture contents of Dowex AG50W resins in the H-, Li-, Na-, K-, Rb-, and Cs-forms were determined by using the Karl Fischer titration method, and the results were compared with the determinations by the NMR and drying methods. It was found that the moisture content decreases progressively in the order H > Li > Na > K > Rb > Cs, and that the direct titration with Karl Fischer reagent can give a rapid and accurate estimation of the water content for the sulfonated cation exchange resins, provided that the water content in the resin is not very low and errors up to 2% are acceptable [295]. However, all of these techniques fail when the residual water content is less than 100 ppm. Since a knowledge of such low moisture contents is important for, e. g., ion exchange in nonaqueous solvents, a method was developed using tritium-labeled water as the indicator. This method has a limit of detection of 10^{-10} mg H_2O [296]. A comparison with the Karl Fischer method showed that titration is more advantageous when the percent water is greater than 0.5 by weight.

For the sake of more completeness several other methods developed for the determination of the water uptake of ion exchangers will be summarized. Dickel and coworkers elaborated a method for spherical ion exchangers at complete saturation with water: the radius of the exchanger grains is determined by gravitational experiments and from the results, the weight of the sorbed water is calculated by the aid of the dry weight and the weight of the exchanger as measured in water. Measurements were conducted on different exchangers with different ion charges and the minimum and maximum water uptake were determined as a function of the ionic composition. The results were compared with those obtained with the aid of adsorption isotherms and it was found that the results agreed [297]. For the determination of the water content of anion exchange resins thermoanalytical methods that have become standard have been developed mainly for the investigation of adsorbed water in, e. g., Dowex 1X8 quaternary ammonium type anion exchange materials. Predried resins in Cl-, SO_4- and HSO_4-form were investigated by simultaneous thermographimetric, differential thermographimetric and derivative thermal analytical (derivatograph) measurement and the evolved gases were passed through the continuous and selective water detector system developed earlier in the same laboratory. Only one type of water could be detected in the Cl-form resin, whereas two types of bound water, loosely and strongly bound, could be distinguished in the SO_4- and HSO_4-form resins [298]. A recent investigation on the determination of the moisture content in cation exchangers showed that, when water in the H-form of Dowex 50, nominally crosslinked with 4, 8, and 12% DVB, was determined gravimetrically after centrifuging the resin samples at 500–4500 rpm, the results agreed well with those obtained by the Karl Fischer method, whereas the NMR method gave lower water contents.

The reverse process to moisture uptake is the drying of a resin. In a system for drying the ion exchangers KU-2 and AV-17 it was found that the optimal drying temperature for KU-2 cation exchanger in the H- or Na-form, and for the Cl-form of AV-17 anion exchanger, is 110 °C. The drying can be accelerated by drying agents like calcium chloride or phosphorpentoxide, and by being done in a vacuum. But because of the low heat resistance of the OH-form of the AV-17 as determined by its loss of selectivity and exchange capacity at higher temperatures, it is dried at $\leq$ 60 °C in the presence of drying agents in vacuo in a special drying apparatus [299]. The actual aim of many drying studies is to get an insight into the true amount of bound water in ion exchange resins. Thus, the process of the dehydration of the resin Dowex 50, H-form with different crosslinkings and bead sizes was studied by drying the resins over liquid nitrogen in the vacuum of a mercury diffusion pump for one or more than one year at 76, 100 and 123 °C. The beginning of pyrolysis was estimated gravimetrically and by gas analysis. The lowest content of residual water was attained with the smallest beads and was limited by the equilibrium of the thermal drying process and pyrolytic water formation. The last traces of water were determined radiometrically on drying the Na-form of the resin swollen in tritiated water.

The density of dry water-free ion exchange resin is about 1.2 g/ml for anion exchangers and about 1.4 g/ml for cation exchangers. Moisture uptake produces a characteristic density change in the various types, so that the density is about 1.3 g/ml in the strong acid polystyrene cation exchangers and about 1.1 g/ml in strong base anion exchangers. For practical purposes, the apparent or bulk density is decisive, amounting to 0.6–0.8 kg/l as a rule. Changes in the density of ion exchange resins are considered as irreversible reactions due to cleavage of the basic polymeric chains, the creation of new chains between structural elements, formation of microcracks in the polymeric skeleton, and formation of microcracks in and destruction of the ion exchange resin beads. For the separation of ion exchangers of differing densities various procedures and even apparatus can be employed.

With regard to practical applications, for the determination of the density and moisture content of resin beads, the mass of the hydrated beads is required. A satisfactory method is one in which the interstitial water is removed by vacuum-draining and the beads are blotted on absorbent paper to remove the surface water. The dry mass of resin beads is determined after the resin has been dried in a conventional oven at 110 °C for sixteen hours. Analytical methods for the determination of metals on resin are in some cases based on the mass of the dry resin.

Determination of the density of an ion exchange resin. A clean, dry pycnometer is weighed and tared. Approximately 25 ml of preconditioned resin are placed in a Büchner funnel containing filter paper. A vacuum is applied, the resin rinsed with deionized water and filtered for approximately 3 min. Then the resin is blotted on absorbent paper until it becomes free-rolling but not fully dried. The resin is immediately transferred to the tared pycnometer and reweighed. Then the pycnometer is carefully filled with deionized water and shaken occasionally to ensure that all the resin beads are wet. Care must be taken to avoid trapping air or losing resin beads when the stopper is inserted. The pycnometer is reweighed and the contents emptied, refilled with deionized water, and reweighed. The density is calculated by the formula:

$$d = \frac{m_2 - m_1}{(m_2 - m_4) - (m_1 - m_3)}$$

with m_1 = mass of pycnometer (g), m_2 = mass of pycnometer plus blotted resin (g), m_3 = mass of pycnometer plus blotted resin plus water (g), and m_4 = mass of pycnometer plus water (g).

Determination of the moisture content of an ion exchange resin. The blotted mass of the resin is determined as described above, using a tared beaker or weighing-bottle for weighing. Then this resin is dried in a thermostatically-controlled oven at 110 ± 5 °C for 16 hours. The resin is cooled in a desiccator and weighed. This yields the mass of the dried resin. The moisture content is then calculated by using the following expression:

$$\text{moisture content } [\%] = \frac{m_1\; m_2}{m_1} \cdot 100\,,$$

where m_1 = mass of blotted resin, and m_2 = mass of dried resin.

1.2.3.2 Particle size

Ion exchange beads and granulates. Ion exchange beads or granulates generally are marketed in particle sizes between 0.04 and 1 mm. Particle sizes of more than 1 mm tend to fragment during production. The listed measure of particle size is the diameter in mm or, as in the English-language literature, according to standard screen sizes in "mesh" values. The American standard screen size (US mesh) can be converted into millimeters by the following rule of thumb

$$\frac{16}{\text{mesh}} = \text{particle diameter in mm}$$

Table 1.13 offers a comparison of these two units of measurement. For British standard screens (BSS mesh) the following conversion formula applies:

$$\frac{12.2 - 15.5}{\text{BSS mesh}} = \text{particle diameter in mm}$$

Table 1.13 Particle size in US mesh and in mm

US mesh	Particle diameter in mm
16– 20	1.2 –0.85
20– 50	0.85–0.29
50–100	0.29–0.15
100–200	0.15–0.08
200–400	0.08–0.04

Table 1.14 Particle size in BSS mesh and in mm

BSS Mesh	Particle diameter in mm	
10	1.500	
20	0.735	
30	0.485	
40	0.360	
50	0.286	
60	0.235	Linearly
70	0.200	decreasing
80	0.173	factor
90	0.155	from
100	0.136	15.0 to
120	0.125	12.2
140	0.093	
160	0.079	
180	0.069	
200	0.061	

Table 1.14 compares BSS mesh values and particle diameters.

Ion exchange particles have a different volume in the dry and the wet states and therefore also have different particle sizes in the two states. This difference is due to the moisture-holding capacity of the exchange resins and depends on the nature of the ionogenic group and the degree of crosslinking. If the ionogenic group is known, swelling factors of a given exchanger can be stated as a function of the

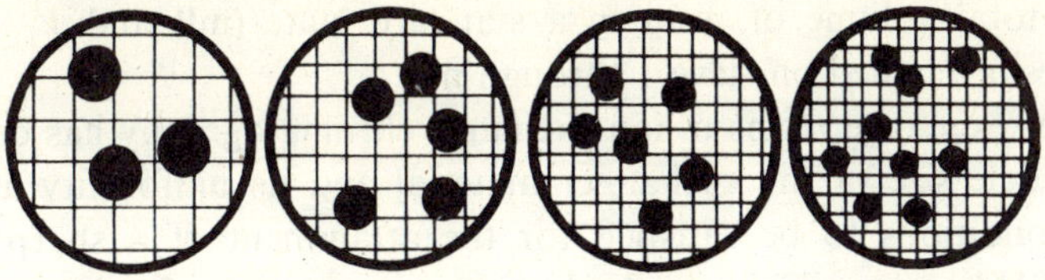

Figure 1.56 Different particle sizes as shown by retention on different screen sizes.

percentage content of crosslinking agent. A microscopic technique has been developed to investigate the particle volume [300]; it offers information on the volume in the swollen state, but as a function of the loading state.

The simplest way of determining the particle size of a dry exchanger is by a screen analysis using a standard set of screens. If customary screen sets are used, no further discussion is necessary. Figure 1.56 demonstrates how variations in particle size can be shown by retention on different screen sizes. A different method, which is more suitable for a comparison of particle sizes, makes use of the effective particle size and the uniformity coefficient. The effective size is a screen size which passes 10% of the total quantity while 90% is retained. The uniformity coefficient is obtained as the ratio of the mesh size in mm of the screen which passes 60% and the mesh size of the screen which passes 10%. This value offers information on the range of the particle size distribution, as the smaller the uniformity coefficient, the narrower the distribution.

In many cases, however, the particle size of the wet resin must be considered, since wet resins are used in most ion exchange applications. Wet screening is carried out with about 150 ml of an exchanger, after preswelling in distilled water for 2 h, being loaded onto the largest screen size on which it is to be classified. Particles of smaller particle size are now flushed through the screen with a flow of distilled water. This separates the particles of a size range above a certain upper limit. The fraction which has passed through the screen is then loaded onto the next screen and the previous treatment is repeated. The particles retained on this screen are flushed off with water; it is advisable to make use of a brush to remove particles adhering to the screen. After this procedure has been used with several screens, the result of the screen analysis is expressed in percent of wet exchanger retained by each screen. For greater detail a final fine sieving is made by manual operation of the individual fractions in a suitable vessel or a water-bath, e. g., 300 mm in diameter by 150 mm deep. Each sieve is alternately elevated and immersed by hand in the vessel for five minutes. Portions retained by the individual sieves are then transferred quantitatively into calibrated measuring cylinders for recording the free wet settled volume. For calculating the particle size distribution in percent retained on each individual sieve the following expression is used:

$$\underset{1 \rightarrow n}{\text{percent retained}} = \frac{V_i}{V_t} \cdot 100$$

where V_t = total volume of resin in a standard state (ml) and V_i = individual volume of resin retained on sieve 1 through n.

The use of exchange resins in ion exchange chromatography has demonstrated that the particle size of the exchanger material and its uniformity are the most important conditions to be satisfied for the attainment of a sharp separation. Hamilton [301] has described a method for the preparation of uniform particle size fractions which is carried out in the apparatus shown in Figure 1.57. The ion exchanger is charged into the separatory funnel and distilled water is passed through in an ascending flow. Since the rate of settling in water is proportional to the effective particle size, all fines are removed first, while the desired particles sizes and the oversize fractions remain in the separatory funnel. After changing the collecting beaker, the flow rate is increased and the desired particle size is flushed out. With careful manipulation, it is possible to obtain fractions in which the sizes of 60–80% of the particles do not deviate by more than ± 3 μm from the average.

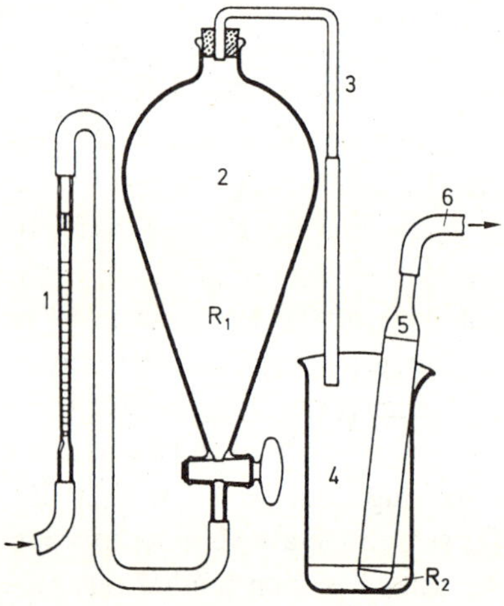

Figure 1.57 Apparatus of Hamilton for the recovery of uniform particle size fractions. 1 flow meter, 2 separating funnel, 3 overflow, 4 beaker, 5 ball filter, 6 pump connection; R_1 charged exchanger mixture, R_2 separated exchanger fraction.

About 40 liters of water are needed for the separation of narrow cuts from a 2 liter separatory funnel. A modification of this method was developed by Vassiliou and Kunin [302]. Two separatory funnels of different size are connected in series, and with the use of a circulating pump and a flowmeter a more rapid as well as simultaneous separation into different size fractions is obtained. Using a sedimentation method for the fractionation of ion exchange resins an apparatus was described in the Russian literature in which uniform fine-grain chromatographic exchange resins are prepared by sedimentation from aqueous suspensions. The sedimentation takes place in a transparent polyacrylate tube, 290 cm long and

80 mm in diameter, with a conical bottom to which a 45 mm wide glass tube is attached. The separated fraction of the exchange resin was transferred directly from the glass tube into a chromatographic column.

The required particle size depends on the operation to be performed and must be determined to optimize the ion exchange process. In the laboratory, a particle size of 0.3 – 0.5 mm is acceptable in most cases. This is the lowest limit for industrial installations. In investigations of the distribution coefficient of metal cations as a basis for chromatographic separations, particle sizes of 100 – 200 mesh are usually employed. Smaller particles are mechanically more stable than coarse fractions, a fact that needs to be kept in mind if the exchanger bed is moved or if it is subject to large volume changes. In any case, it is of advantage to maintain a uniform granulometry. For example, if ion exchange resins are to be milled for analytical purposes, a moisture content of about 40% is best. After swelling, the particles are milled in a meter mill at 13 000 – 15 000 rpm, then are screened or fractionated by the method of Hamilton.

The particle size may have a purely hydraulic or kinetic influence on the ion exchanger process. In the column process, the dependence of the flow rate on particle size is most apparent. Since the frictional resistance is higher with smaller particles sizes, the flow rate also decreases with decreasing particle size. To prevent a complete holdup of the liquid being filtered in the column, an overpressure must be applied on it by means of a suitable device. In an operation with a descending liquid front, a volume loss greater for small than for large particle sizes occurs at the head of the column. For an ascending liquid front, a volume expansion of the exchanger bed greater with small than with large particle sizes can be observed. The influence of particle size on the exchange rate can be seen in Figure 1.58 [303]. The exchange rate increases as the particle size decreases. The diffusion path lengths of the exchanging ions to and from the active sites are shorter with smaller particle sizes, so that exchange is more rapid. Hydraulic properties of ion exchange resins can be expressed by equations developed for packed beds. The equations are based on dimensional analysis, and the recognition of sensitive parameters can be illustrated by the equations, thus making their use important [304]. The Dow Chemical

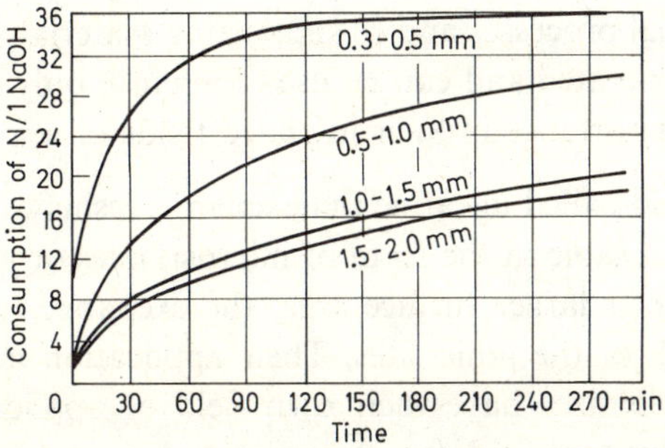

Figure 1.58 Influence of particle size on the exchange rate. Particle sizes given in mm.

Table 1.15 Screen Index, a statistical relationship, which is roughly proportional to the pressure drop across a resin bed (from Dow Chemical Company)

Screen mesh no.	Screen sizes (mm)	Percent retained	Pressure drop factor	% times factor
12	1.68	0	0.5	0
16	1.19	0.6	1	0.6
20	0.84	37.0	2	74.0
30	0.60	46.3	4	185.2
35	0.50	13.0	6	78.0
40	0.42	2.0	8	16.0
50	0.30	0.9	16	14.4
−50	−0.30	0.2	32	6.4
Sum = Screen Index =				373.6

Company, on the other hand, presents in its technical information a statistical relationship that has become known as Screen Index, and which is roughly proportional to the pressure drop across a resin bed. For the screen sizes shown in the first column of Table 1.15, the pressure drop essentially doubles for each succeedingly finer mesh. Arbitrarily setting 16 mesh equal to one establishes the pressure drop factors in the third column. As an example, a resin size distribution is shown in the second column. These second column percentages are multiplied by the third-column factors, and the results are recorded in the fourth column. The sum of the fourth-column numbers is the Screen Index. It can be shown that the column behavior of gel resins can be estimated from a knowledge of these parameters, but for macroporous resins other properties must be measured, for example the extent of electrolyte invasion.

Ion exchange powders. Grinding of ion exchanger beads or granulates leads to an ion exchanger powder with a very large surface area, an aspect which makes it a special type of ion exchanger. Powdered ion exchangers are finely divided resins of about 30 μm with an increased reaction rate and high adsorption power; they can be formed into compressed cakes or pellets. Powdered ion exchangers have found diverse applications in industrial processes and as supporting materials [305]. These ion exchangers cannot be regenerated and can be used therefore only once, but as a rule this is of subordinate importance in their intended fields of application.

Ultrafine ion exchange resins. So-called ultrafine ion exchange resins with a particle size of 0.5–1.5 μm are also available in the form of microspheres or agglomerates of microspheres. Because of their larger surface area, the exchange rate is higher due to the greater accessibility of the ionic sites. Their application in columns is limited. However, such particle sizes have their own field of application in the production of ion exchange papers and for incorporation into plastics, films, coatings, and fibers [253].

1.2.3.3 Crosslinking

The degree of spatial interlinkage of ion exchangers is determined by the production process. Commercial synthetic ion exchange resins of the gel type usually contain 2–12% divinylbenzene as a crosslinking agent. Since crosslinking has a controlling influence on several properties of an exchanger, this step offers an opportunity to produce during synthesis exchangers for special purposes.

Crosslinking influences not only the solubility but the mechanical stability, exchange capacity, water uptake and swelling behavior, volume change in different forms of loading, selectivity, and chemical as well as oxidation resistance of ion exchangers. Exchangers with a low degree of crosslinking are soft and mechanically unstable (in the swollen state), while highly crosslinked products are hard and brittle with an increased sensitivity to osmotic influences. The exchange capacity of a dry exchanger increases with a smaller degree of crosslinking and vice versa, while the exchange capacity per volume decreases with decreasing crosslinking because of more extensive swelling. With a higher degree of crosslinking, the moisture content and swelling behavior of exchangers decrease. The volume change of the loading form which changes during regeneration is greater with a low than with a high degree of crosslinking. The selectivity increases with higher degrees of crosslinking. It is also of interest that a higher degree of crosslinking improves chemical and oxidation resistance. In particular, however, attention is drawn to the exchange rate, which decreases with a higher divinylbenzene content, as demonstrated by Figure 1.59 for a strong acid cation exchanger based on polystyrene.

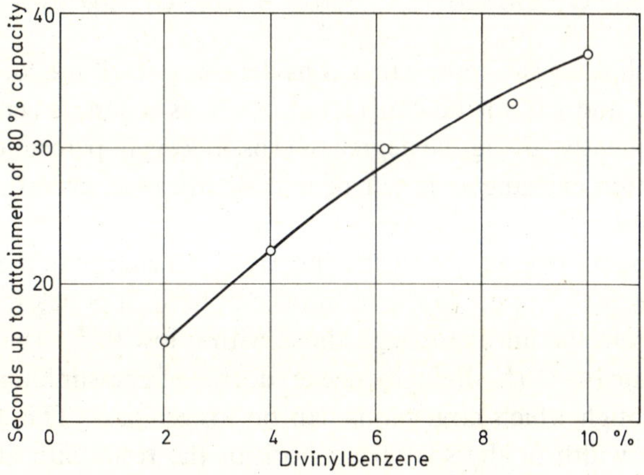

Figure 1.59 Exchange rate as a function of crosslinking.

In the literature, the degree of crosslinking usually is characterized by the value X% crosslinked. This expresses in a percentage the amount of crosslinking agent used to produce the ion exchanger. In the Dowex types, the degree of crosslinking can be recognized from the type identification. For example, Dowex 50X4 means that this exchanger consists of 96% styrene and other monovinyl monomers, and 4% divinylbenzene as crosslinking agent.

However, in an evaluation of crosslinking in cation and anion exchangers, it must be kept in mind that the crosslinking data do not have the same meaning for both types of ion exchangers. The degree of crosslinking of a polystyrene-sulfonic acid exchanger (cation exchanger) is primarily determined by the concentration of bridge-forming components in the monomers. Although additional crosslinkages can theoretically form during sulfonation by the formation of sulfur bridges, this effect is of less practical significance (although it cannot be overlooked) from the standpoint of the production of highly uniform products [306]. In the case of anion exchangers, however, additional crosslinking can be introduced by chloromethylation – which is the usual technique for the production of these exchangers – with the formation of methylene bridges between the benzene rings of parallel hydrocarbon chains. This new formation of crosslinkages takes place diagrammatically as follows, according to a Friedel-Crafts reaction:

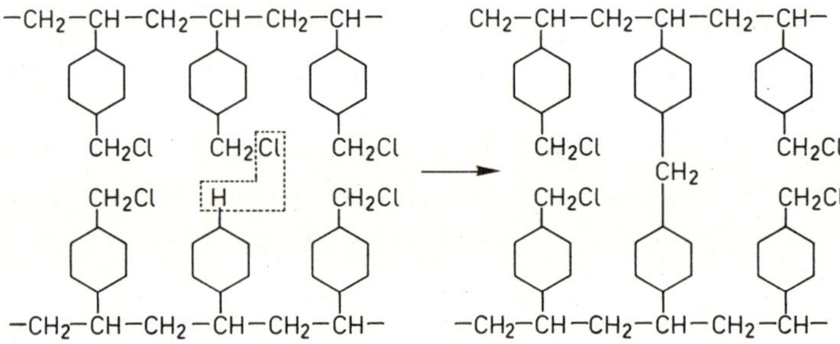

This additional crosslinking can represent a considerable part of the total crosslinking in the exchanger and cannot be considered solely as a supplementary contribution [307]. Consequently, giving the degree of crosslinking in percent of crosslinking agent used in anion exchangers is not of real significance, even though this is found in the literature.

Thus, by modifying the degree of crosslinking, i. e., by a change in the quantity of bridge-forming agent with respect to the monomer utilized, it is possible to obtain structures ranging from the linear through those with a low (0.5–1% divinylbenzene) and intermediate (4–10% divinylbenzene) degree of crosslinking to a closely packed network through which larger ions can no longer pass. This results from the decreasing mesh width of the spatial network of the resin with an increasing degree of crosslinking. The "pores" formed in the network of the exchanger thus have different sizes; for a normally crosslinked resin, it has been concluded that they are 10–20 Å. This property of ion exchangers – having a separation effect on ions as a result of pore size – is the basis for their use as ionic sieves [308].

If the unknown degree of crosslinking of a manufactured cation exchanger is to be determined, the method of Boyd and Soldano may be used. It has been found empirically that the following relation exists between the equivalent volume of the dry (V_t) and the equivalent volume of the swollen (V_q) ion exchange resin

$$\frac{V_q}{V_t} - 1 = \frac{k}{x}.$$

In the above, x is the degree of crosslinking and k is a constant for a given salt form. The following k-values can be used for polystyrenesulfonic acid exchangers

Cation	H^+	Li^+	NH_4^+	K^+	$(C_2H_5)_4N^+$
k	10.7	11.8	9.1	9.1	5.2

from which x is immediately obtained in percent divinylbenzene provided that the crosslinking falls into a range of 1–25%. The extent to which this or another relation applies to other cation exchanger types and anion exchangers has not yet been sufficiently investigated.

Attempts have repeatedly been made to determine crosslinking with physico-chemical methods of analysis. In the infrared spectrum, absorption band differences were found at 830 cm^{-1} and 797 cm^{-1} compared to pure m-DVB and p-DVB [309], indicating the degree of crosslinking with these isomers. Corresponding measurements should make it possible to determine most degrees of crosslinking. In one case the degree of crosslinking was determined by plotting the ratio of the absorption at 1605 and 2049 cm^{-1} vs. percent of (commercial) divinylbenzene. The polymer was dried in vacuo at 60 °C in the presence of phosphorpentoxide. The KBr pellets contained 0.1% KCNS as an internal standard.

In addition to this relationship between crosslinking and swelling and the above indicated effects on other properties of ion exchange resins, the neutral salt adsorption is also influenced and will be described more fully here. Neutral salt

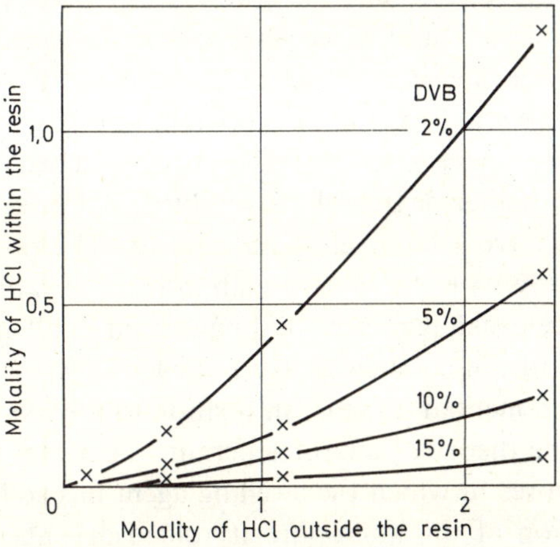

Figure 1.60 Neutral salt adsorption on a cation exchanger of different degrees of crosslinking.

adsorption is the phenomenon of adsorptive binding of neutral salts by ion exchangers, e. g., a cation exchanger in the presence of a high concentration of the external electrolyte can take up additional anions which entrain corresponding cations. This is shown in Figure 1.60 for the adsorptive fixation of hydrochloric acid on a polystyrenesulfonic acid resin as a function of the degree of crosslinking [310]. Numerous other influences of the crosslinking on the ion exchange behavior of solute inorganic and organic substances have been found. E. g., the sorption of alcohols and corresponding acrylates by various sulfonic cation exchangers was studied and the effect of the resin crosslinking on the ion exchange method of stability constant determination was evaluated by using distribution coefficient data for the resins Amberlite IR 120, 122 and 124 at 25 °C and ionic strength 0.1 M $NaClO_4$. The data for IR 120 and 122 – but not for IR 124 – were the same [311].

1.2.3.4 Porosity

As noted above, the porosity of ion exchange resins is directly related to the degree of crosslinking and the network formed as a result. The size of the capillary channels depend on the degree of crosslinking but is not uniform throughout an ion exchange particle.

Detailed studies of the "porosity" of ion exchangers have shown that some caution is indicated in the use of this term. As noted earlier, the network of exchanger resins contains voids which have a greater or lesser moisture regain capacity depending on the degree of crosslinking. In contrast, it has been possible to produce ion exchangers which represent true phase-dispersed systems. The two types differ in appearance. While the former are clear and transparent, the latter are opaque, and when combined with water have a cloudy opalescence. There are, however, problems in nomenclature here. Since no uniform terminology has become accepted for the above distinction, and since the custom of speaking of the "porosity" of the exchanger network with reference to the mesh width is widespread, in the former case reference here will be made to "apparent porosity", in the latter case to "real porosity", whenever necessary. The really porous ion exchangers show relatively little swelling with a higher degree of crosslinking; but because they are permeated by fine pores, they have a relatively high capacity. The pore sizes may be measured on the basis of results in the determination of pore radii in swollen gels, among other means. Since exchange resins that have been swollen in aqueous solution below 0 °C exhibit diffusion coefficients for radioactively-labeled counterions that show a more than exponential decrease in relation to the reciprocal value of the absolute temperature, the theory of mixtures permits the use of these results to determine the fraction of pores in which the swelling agent has solidified [312]. In a prior theoretical discussion of the microporosity of adsorbents the opinion was expressed that the concept of porosity or microporosity is applicable only to rigid solid adsorbents, and to the classical polymers and polymers that are capable of swelling, such as ion exchange resins, because these substances do not possess permanently existing pores. The isotherms of sorption of carbon tetrachloride and methanol vapors on ion exchange resins of the KU-2 type of different degree of crosslinking were also studied. The total quantity of pores in the investigated samples was calculated, and it was shown that the proper microporosity of ion exchange resins is extremely small. Comparison of experimentally obtained data indicated that the high sorption ability in the exchanger KU-2 cannot be explained by its microporosity but is associated with the ability of the ion exchange resin to swell in low molecular weight liquids [313]. Another term, "channel structure" was then introduced based on investigations of the porosity of ion exchange resins comprising sulfonated polystyrene crosslinked with divinylbenzene, determined as a function of the polymer structure. The term "channel structure" seemed more suitable than "gel structure" because of the larger pore diameter, smaller pore volume and smaller surface area [314].

Porosity also influences other ion exchanger properties, mainly the capacity and selectivity. The capacity would be much lower than it is if exchange resins had no pores and only the ionogenic groups at the surface were active for ion exchange. The high ion exchange capacities are obtained only because active ionogenic groups located in the interior of the pores also contribute to the capacity. A limiting factor here is the ratio of pore size to the size of the ions or molecules that are to be exchanged, since the resulting sieve effect leads to a selection of counterions on the basis of size. Consequently, larger ions are frequently found to have an effective capacity far below that obtained with small ions or which theoretical calculations lead one to expect [315].

Since the "porosity" has such far-reaching effects on the properties and thus on the application of ion exchangers, it is understandable that there is continuing discussion of its nature and degree in individual types of ion exchangers. To obtain a better understanding of the pore character, Mikes [316] observed the formation of the disperse system from the start of polymerization. In syntheses leading to true porous products, the structure is built up in individual steps: a molecular arrangement forms first, followed by intermolecular rearrangement and, finally, regrouping with the formation of visible pores. Kunin [317] pointed out the difficulties in nomenclature and the resulting confusion, which have been repeatedly mentioned in the present discussion. The distinction between "macroreticular" and "microreticular" for a description of the pore character appears very useful. "Macroreticular" pores or true porosity refers to structures in which the pores are larger than atomic distances and are not part of the gel structure. Their size and appearance depend little on the surrounding conditions. The expression "microreticular" refers to the apparent porosity of atomic dimensions which depends on the swelling behavior of the gel and, with regard to its size, on the surrounding conditions.

Measurements of pore sizes must be carried out to characterize pore structure. In addition to techniques for the determination of the total pore volume from the true density determined with a helium densitometer and the apparent density measured with a modified mercury porosimeter, a study of pore sizes requires a distinction between the specific surface O, pore volume V and pore radius r [318]. On the basis of model theory, the following relation exists between these parameters:

$$r = 2.7 \frac{V}{O}$$

The surface area is determined by the Brunauer-Emmet-Teller (BET) method or by the emanation method of Hahn. The pore volume is measured by the saturation values of carbon tetrachloride adsorption [319] or by a titration method [320], which lead to values that differ by less than 10%. The pore radii can then be calculated from the measured surfaces and pore volumes on the basis of the above equation. Using a method of van Bemmelen, Bachmann and Maier, the pore radii and their distribution can also be determined directly. In addition to other techniques for the investigation of porosity [321], Kun and Kunin [322] developed a modified mercury method in which the exchangers can be investigated in the hydrated state. Additional information on the pore structure is offered by electron microscopy [323], which furnishes results in good agreement with the indirect methods and gives data on the appearance, position and character of the pores.

Since both gel type and macroporous type ion exchangers are in widespread use investigations of their pores and pore structures have been made, usually in separate studies. An electron-microscope study of the porosity of a gel type anion exchanger was conducted, whereby the anion exchanger AV-17 was treated by a ferritization process to deposit magnetite in its pores. Electron microscopy of the magnetite-containing ferrite was then used to study the pore structure. It was suggested that this method would prove useful also in pore structure studies of other ion exchangers. The polymodal structure of ion exchangers with 10^{-7} to 10^{-8} cm pores was studied by using mercury porosity. The total pore volume and relative sub-µm pore volume of KU-23 increased from 0.47 to 1.20 ml/g and the bimodal structure changed to polymodal with increasing crosslinked density from 15 to 30%. The predominant radius of the micropores varied from 32 Å in KU-23-4-60 to 400 –900 Å in KM-2p [324].

An ion exchanger can be called macroporous or macroreticular (the two forms are used to indicate the particular way of preparing the matrix) on the basis of appearance, when the difference between the apparent and true density is a minimum of 0.05 g/cm^3, when the inner surface is 5 m^2/g and larger (BET), and when the pore radius is at last 50 Å [325] and it can be characterized by the relative micro- and macropore volumes and the macropore size distribution [326]. The latter intensive investigation further revealed that these density measurements and mercury penetration measurements can be carried out on a sulfonated macroreticular ion exchanger in the swollen state with different amounts of swelling agent. On taking up water the dry volume of this exchanger increases by swelling until a water content of about 33 wt.% is reached. Additional water then fills up the macropores to a final content of 58 wt.%. With the specific volume of the gel phase, the relative macropore volume of the fully swollen state was evaluated. The results of the mercury penetration method agree with those of the density measurements, taking into account the compressibility of the mercury itself and those of the swelling agent and the matrix. Corresponding correction factors could be obtained from measurements on "microreticular" ion exchange resins. Thus, the mercury penetration measurements provided additional information on the macropore size distribution of the swollen state. For the examined ion exchanger a logarithmic standard distribution with an arithmetically weighed mean pore radius of about 1100 Å was evaluated. The relative overall pore volume of 0.65 was not identical the value of 0.61 for the partition coefficient of ethylene glycol from dilute aqueous solutions, as only that part of water which is bound to the sulfonic group could be exchanged by ethylene glycol. Pure ethylene glycol substituted the bound water completely. The specific volume of the exchange resin thereby decreased in comparison to that of the water-swollen state.

1.2.3.5 Swelling

The volume of an ion exchanger depends on several facturs: (1) the surrounding medium (air, water, organic solvents), (2) the nature of the resin skeleton (type of matrix and crosslinking), (3) the charge density (nature and concentration of ionic groups), and (4) the type of counterions.

The volume change which takes place during transference from one medium to another, and which is influenced by the other factors, is known as swelling. This swelling is produced by osmotic pressure in the interior of the ion exchanger against the external, more dilute solution, so that the solvent uptake producing the swelling reduces the internal ion concentration. A distinction must be made between absolute swelling, in terms of the dried exchanger, and the difference in swelling volumes (breathing difference), i. e. the volume change of a swollen exchanger under different loads.

Absolute swelling takes places when an air-dried resin becomes wet; this must be kept in mind when an ion exchanger sample is prepared. This volume adaptation, which is also known as pre-swelling, can be measured in a graduated cylinder; in laboratory practice it plays a subordinate role since ion exchange resins are slurried into the columns only in the pre-swollen state. In industrial filter beds, pre-swelling must be performed carefully, possibly with the use of steam at elevated temperature, since undesirable malfunctions may otherwise occur, as a result of fragmentation of the exchanger beads or too close packing.

During absolute swelling a certain quantity of water is taken up by the exchanger. Pepper, Reichenberg and Hale [310] have described a centrifuge method for the determination of this quantity, the apparatus for which is shown in Figure 1.61. About 1 g air-dried resin is loaded into a glass filter attachment with a glass frit, and the assembly is immersed in distilled water for 1 h at 25 °C. The glass filter attachment is then inserted into a 15 ml centrifuge tube, closed with a rubber cap,

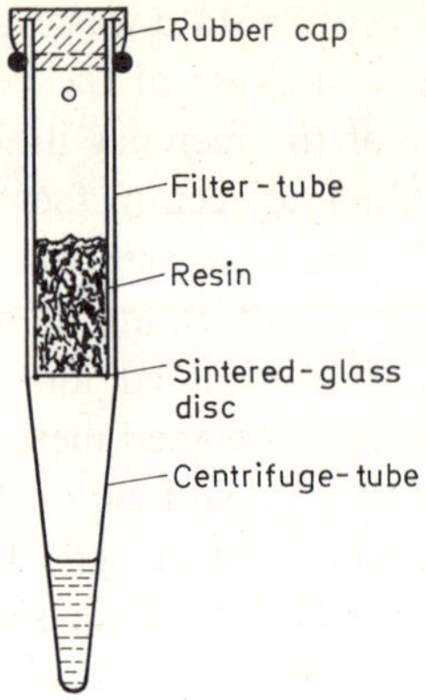

Figure 1.61 Assembly for the determination of the swelling water by the centrifuge method.

and centrifuged for 30 min at 2 000 rpm. The filter attachment is weighed, immersed once more in water, and again centrifuged. It is then dried overnight at 110 °C and brought to constant weight in vacuum at 1 mm Hg over phosphorus pentoxide at 10 °C. Subsequently – or better yet, before – the empty glass filter attachment is dried at 110 °C and weighed, and after centrifuging, it is weighed with water as a blank. The dry weight and wet weight of the sample can be easily calculated from these data, so that the quantity of water taken up in absolute swelling can be determined [327].

In this case, swelling depends on the quantity of regained water. This is influenced first of all by the matrix structure, i. e., the degree of crosslinking. While uncrosslinked polystyrene sulfonic acid undergoes unlimited swelling in water, the moisture regain capacity of the exchanger decreases with increasing crosslinking. If the water contains an electrolyte, swelling furthermore depends on the electrolyte concentration. When this concentration increases, the moisture uptake will increase, since the osmotic pressure difference between the external and internal solution is then smaller.

The swelling processes become more difficult to interpret when one changes from water to other solvents. Water with its strong dipole character interacts with the highly hydrophilic ionogenic groups. As the dipole character of a solvent becomes weaker, swelling of the exchanger will be less pronounced since its electrostatic solvation tendency decreases. The ion exchange skeleton of hydrocarbon chains, on the other hand, has a tendency to take up solvents that are less polar, so that weak acid cation exchangers in the H-form, for example, swell more in alcohol than in water. The weakly ionic character of the ionogenic groups and the resulting possibility of electrostatic solvation become subordinate to the affinity of alcohol and carboxyl groups [328].

The breathing difference is important in working with ion exchange columns. Its dependence on the species of counterion with a certain degree of crosslinking

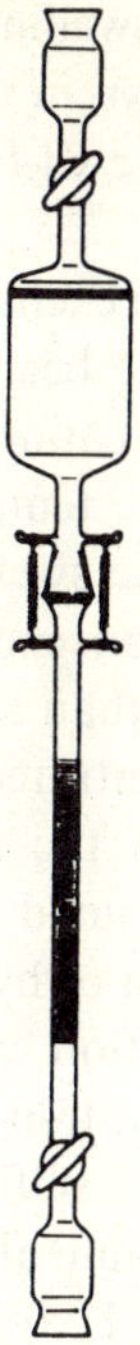

Figure 1.62 Apparatus for the determination of the breathing difference.

of the particular ion exchanger can be easily determined with an apparatus described by Blasius (Figure 1.62) [329]. For this purpose, a cation exchanger – for example Dowex 50X1 – can be used. A quantity of 4 g is charged into the 200 ml reservoir of the apparatus. The ground joints are then connected, and the ion exchanger is fed onto the glass frit with opened stopcocks and is charged to the desired state of loading. After complete loading, the assembly is inverted and the resin is flushed into the graduated tube with distilled water; the tube is also closed by a fused-in sintered glass disk. As soon as the resin volume has become constant under further washing, it is read off from the scale. For example, if K^+ has been used as the counterion in a first experiment and Fe^{3+} in a second, a volume decrease will be observed. This phenomenon follows the general rule that the moisture-holding capacity of this exchanger with a low degree of crosslinking decreases considerably and progressively when loaded with mono-, bi-, and trivalent ions. In a more highly crosslinked exchanger, the breathing difference in various loading states is smaller.

The conditions of the relation of swelling to the counterions is not always as simple as described in the above demonstration experiments. Although the rule has been accepted that the swollen volume of an exchanger is smaller the higher the valence of the counterion, the nature of the equivalent ions also plays a role, since equivalent ions may differ in size and solvation power [330]. If the counter-ions exhibit a higher degree of dissociation, swelling will be more pronounced, while ion pair formation and ion association will lead to a decrease in swelling. The weak acid cation exchangers have a smaller volume in the H-form than in the Na-form and the volume of weak basic anion exchangers is larger in the Cl-form than in the OH-form.

To obtain more information from swelling experiments on ion exchange and the behavior of ion exchangers a large number of investigations have been conducted. An examination of the dependence of the kinetics of strong acid cation exchangers showed that the swelling degree is practically independent of the temperature but that the swelling velocity as well as the activation energy of the swelling process quickly increased with the temperature. The H-form of the ion exchanger also swells to a higher degree when a solvent mixture is used rather than an individual solvent. The swelling of sulfonated cation exchangers in concentrated electrolyte solutions depended at first upon the activity of the water in the system. The observed marked divergence of the swelling isolines was attributed to the non-exchange sorption of the electrolytes. The non-exchange sorption of hydrogen ions contributed to the exchanger swelling in the form of hydration water in the exchanger, which amount increases with the electrolyte concentration in the outer solution. The partial molar swelling of concentrated solutions of sulfonated polystyrene salts was found to be approximately the same for the calcium and strontium salts but decreased in the order $Li > Na > K > Cs \gg Ag \sim Tl$ for these salts [331]. The swelling of a sulfonated styrene copolymer was also examined in solutions of linear sulfonated polystyrene. Water sorption and swelling of the solid ion exchanger in aqueous solutions of sulfonated polystyrene of different concentrations are a function of the degree of crosslinking of the copolymer. For crosslinking with $\leq 8\%$ DVB the sorption of the ion exchanger depended linearly on the concentration of the polyelectrolyte solution. Since the dependence of the ion exchanger swelling on the polyelectrolyte concentration was a function of the degree of crosslinking, a general equation for the dependence of the water sorption of the ion exchangers on the polyelectrolyte concentration for any given degree of crosslinking could be derived [332].

For the swelling of macroporous (macroreticular) and isoporous ion exchangers newer, interesting results have been presented. In a sorption-thermochemical study of the hydration of a macroporous sulfonic cation exchanger graphs were obtained for the adsorption isotherms of water vapor at 20 °C for the hydrogen, lithium, sodium, potassium, and calcium forms of the resin. Graphs were also given of the heats of wetting, energies of hydration, and differential entropies of wetting. The results are very similar to those obtained with a gel type sulfonic acid cation exchanger, the differences being due in general to the greater rigidity of the macroporous resin. Thus, heats of wetting are larger since energy is not expended in swelling the network. The initial hydration energies are quite high and equal. This is attributed to the hydration of the sulfonic acid groups common to all forms. The first part of the adsorption isotherms are more alike than for the gel type resin, the middle portions also are less separated, and the final portions do not turn upward so steeply [333]. Also swelling results for ion exchange resins with other and longer crosslinkers than divinylbenzene have been reported. Ion exchange resins based on *N*-(4-sulfophenyl)methacrylamide have macroreticular structures and swell more in water than other ion exchange resins with shorter crosslinks. Higher crosslinking in these resins also increases the enthalpy and entropy of swelling, which is caused by the decrease of the interchain attraction of these ion exchangers when the crosslink length is increased. The swelling capacity of polystyrene and styrene-divinylbenzene copolymer crosslinked with chlorodimethyl ether and 4,4′-bis(chloromethyl)biphenyl as ion exchangers increases with additional crosslinking in the swollen state, which is due to an effect of the crosslinking bridges on the energy of interaction between the polymer chains and consequently on the energy of the three-dimensional network. The swelling of isoporous polystyrene in poor solvents increased with increasing degree of crosslinking, contrary to the prediction by the known theories of swelling [334].

The kinetics of the swelling of carboxylic cation exchangers was studied by a microscopic method over the course of two reactions, namely the neutralization and exchange process, using 0.1 N HCl, LiOH, NaOH, and KOH solutions. The swelling of the weak acid cation exchanger in the neutralization process proceeds in at least two stages. The first extent of swelling, 0.5 – 0.8, is described by a first-order kinetic equation. The second stage can be described by a kinetic equation of a second order reaction. Compression of the cation exchanger in exchange reactions can also be described by a kinetic equation of a second order reaction.

The swelling behavior of a strong base anion exchanger, Dowex 1X8, was investigated by Inczedy. The water uptakes of completely swollen and air-dried anion exchange resins were investigated by traditional and thermoanalytical measurements. In examining the air-dried resin samples of the HSO_4-, H_2PO_4-, SO_4-,

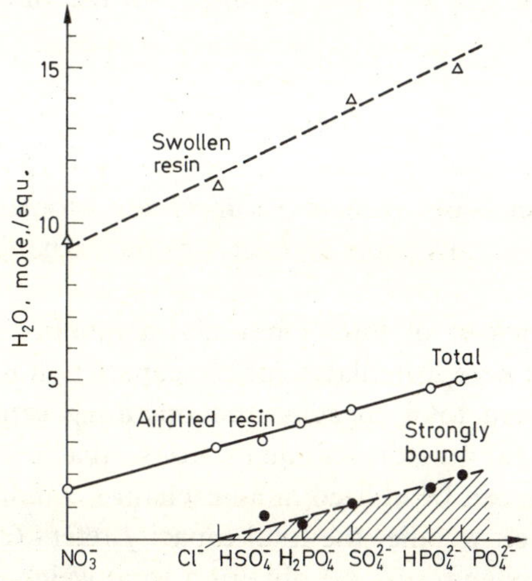

Figure 1.63 Water uptake of Dowex 1X8 in different forms at 22 °C (Inczedy).

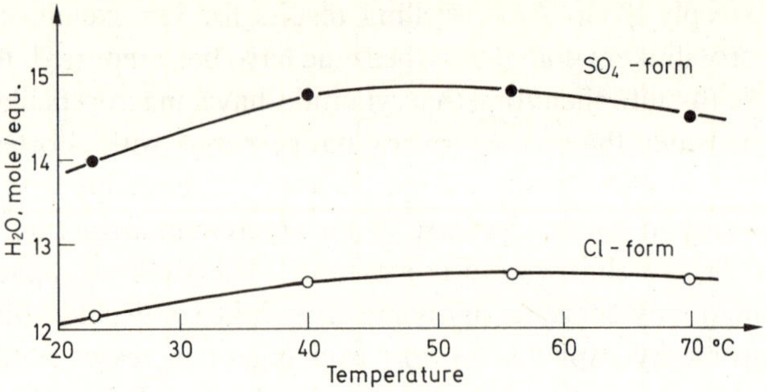

Figure 1.64 Water uptake of Dowex 1X8 in SO_4- and Cl-form at different temperatures.

HPO_4-, and PO_4-forms of the anion exchanger two forms of water, a strongly and a loosely bound, were found as depicted in Figure 1.63, showing also the water uptakes of the completely swollen NO_3-, Cl-, SO_4-, and PO_4-forms. It can be seen that the water uptake of the swollen resin and the strongly bound water increases in a very similar way. The water uptake – and hence the swelling of the swollen resin – changes a little with the temperature, as shown in Figure 1.64. This change (which has a maximum) is pronounced in the case of the SO_4-form of the anion exchange resin. The sequence of water uptake on the resins investigated was compared to the opposite order of the adsorption strengths of the anions and to the entropies of the hydrated anions. A conclusion was drawn regarding the role of the hydration of the anions in the selectivity sequence of anion exchange, i. e. that besides other effects such as polarization, H-bonding, etc., the free energy of hydration plays an important role. Competition between the electrostatic attraction and the hydration forces may be assumed. The radii of the hydrated ions have no significance whatsoever as regards the adsorption strengths of the anions [335].

1.2.3.6 Capacity

The capacity of an ion exchanger is one of its most important properties, since it permits a quantitative statement of how many counter-ions the exchanger can take up.

In the course of the development of ion exchangers, a number of different definitions and units of measure were formulated for the capacity. It is important, first of all, to distinguish between total capacity and operating capacity. Total capacity is obtained from the total quantity of counter-ions capable of exchange. Operating capacity is that which can be utilized in an exchange column under the selected conditions. Depending on whether the total capacity refers to the weight or the volume of dry or wet-swollen resin, one obtains a total weight capacity or a total wet volume capacity. Since the weight and volume of an exchanger can differ greatly in the dry and the swollen state, as is evident from the discussions on crosslinking and swelling, very different values are obtained for the capacity as a function of the parameter to which its determined value is referred. It is necessary, therefore, to indicate the specific units and conditions for which capacity values are being given.

The total capacity is an intrinsic property of an ion exchange resin. It is independent of the means of employing the resin and represents an ideal to be used largely for comparison. The total capacity, regardless of the units in which it is expressed, is always hundred percent of the available, or accessible, ion exchange capacity. It has, therefore, to be observed that, for many ion exchange resins, total capacity results from two or more ionogenic groups. Virtually all strong base anion exchangers are multiionic, containing at least a small portion (up to about 10%) of weak base amine groups. The total anion exchange capacity is thus made up of the weakly basic and the strongly basic (also called the salt splitting) capacity. When using total capacity measurements as a means of assessing the quality of a resin in use, it is important to know if the relative quantities of strongly basic and weakly basic capacities have changed. On the other hand, weakly basic anion exchangers sometimes contain slight portions of strongly basic anion groups, and these should be distinguished from one another in any precise determination of capacity. On the other hand some sulfonated phenolic ion exchange resins which were prepared at high sulfonation temperatures definitely have a higher capacity when titrated in the presence of calcium chloride than of sodium chloride, which must be ascribed to the presence of some weakly acidic carboxylic groups.

For laboratory and research purposes it has become increasingly popular to state the total capacity as a weight capacity in milliequivalents per gram of dry resin (mequ/g) or in milliequivalents per 100 g (mequ/100 g) of dry resin. First of alls, this is to be understood to refer to cation exchangers in the H-form and to anion exchangers in the Cl-form. A correct description of the ionic form is necessary as is a careful determination of the moisture content of the resin. The latter is important because while the dry weight may well be defined as the weight of the ion exchange resin in a state where all the removable water has been removed, the drying of a synthetic ionogenized polymer structure is not an easy task. Several drying methods have been elaborated for the different types of synthetic ion exchangers: strongly acidic, medium acidic, weakly acidic, strongly basic, weakly basic, and amphoteric resins. It must be mentioned once again that bringing ion exchange resin beads to the dry state means to dry them to a constant mass but not to treat them so that the ionically-bound water is also removed. Preference would be given to the method in which the quantity of resin required, which has already been converted into the required form, is washed and separated from surface water in a Buchner funnel, and dried to constant weight either at 110 °C in a thermostatically-controlled drying oven or better at 68 °C in a vacuum drying oven for 16 hours. This procedure may not be used with the hydroxyl form of anion exchangers because they decompose above 40 – 60 °C. These resins are converted to the chloride form before drying to avoid errors due to decomposition of the free-base form.

To determine the weight capacity in mequ/g exchanger of synthetic ion exchange resins, it is probably still most suitable in many cases to use the method described by Fisher and Kunin [336]. This procedure is particularly appropriate for routine tests, since it is relatively rapid and furnishes reliable and comparable values. Another method, the equilibrium titration curve, gives information which can be used to determine the effect of polymer structure on the chemical behavior of the polymer. But this method requires the setting up of a series of samples in contact with varying amounts of the ion to be exchanged and may require a period of one week to six months for a determination. Similar information may be obtained in certain cases by the direct titration method, e. g., when the ionogenic groups are highly ionized. To overcome the difficulties inherent in either titration method, a series of techniques have been developed involving the principle that one ion, present in large excess in a solution, is exchanged on the resin sites for another ion taken up during a preceding loading stage of the exchanger resin. The latter ion is chosen so that the resin can be easily and completely converted to the desired form, exchanged for the ion present in excess in solution, and analyzed readily by titration after the exchange reaction is complete. Methods of this type were first reported by Kunin and Myers and then modified to meet wider requirements.

The assembly shown in Figure 1.65, which was taken from the original literature, serves as the equipment. It consists of simple components available in every

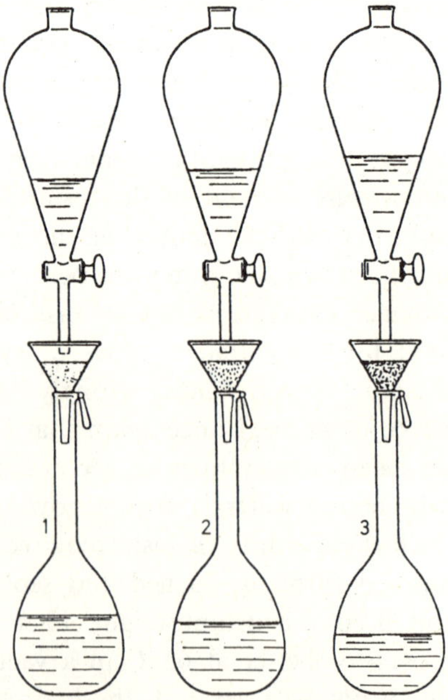

Figure 1.65 Equipment for the capacity determination according to Fisher and Kunin.

laboratory, in particular a short-stemmed 60° funnel which is placed in a 1 liter volumetric flask; filter paper of medium porosity is used. The procedure consists of the following steps.

Determination of the total weight capacity of a cation exchanger:

The cation exchanger (5 g) is transformed into the H-form by slow treatment with about 1 liter 1 N HNO_3 in the funnel shown in Figure 1.65. Subsequently, it is washed to neutrality with distilled water, and dried in air. Of this quantity, 1.000 ± 0.005 g is weighed into a dry 250 ml Erlenmeyer flask containing exactly 200 ml 0.1 N NaOH with 5% sodium chloride, and is allowed to stand overnight. An exchanger sample (1 g) of the same material is separately weighed into a weighing bottle, dried at 110 °C overnight, and weighed again to determine the percentage of solids. Of the supernatant liquid in the Erlenmeyer flask, 50 ml aliqots are back-titrated with 0.1 N H_2SO_4 against phenolphthalein. The capacity is then calculated by the formula

$$\text{capacity (mequ/g)} = \frac{(200 \cdot \text{normality}_{\text{NaOH}}) - 4\,(\text{ml}_{\text{acid}} \cdot \text{normality}_{\text{acid}})}{\text{sample weight} \cdot \dfrac{\%\ \text{solids}}{100}}.$$

It represents the total weight capacity of the exchanger in the dry H-form.

Nitric acid is used for the exchanger transformation into the H-form, since hydrochloric acid and sulfuric acid may lead to precipitation in a reaction with the heavy metal cations present in the exchanger. The resin must be completely in the H-form before weighing of the sample, since differences in equivalent weights of different ions would lead to errors. The standard sodium hydroxide solution is treated with 5% sodium chloride to obtain a complete exchange equilibrium by the excess of sodium ions. A reproducibility of ± 1% can consequently be obtained. This capacity determination measures the sum of both weakly and strongly dissociated groups.

Determining the capacity of strong acid groups present in an ion exchange resin either alone as concluded from the synthesis or together with largely undissociated acid groups leads in practice to the determination of the total strong acid cation exchange resin sulfonic acid capacity. An estimation of these highly dissociated groups, also called salt-splitting capacity, may be obtained by taking advantage of the well-known equilibrium reaction, which may be driven toward completion in the sodium form by a large excess of sodium ions if the resin is highly ionized. Under the same conditions the extent of reaction of weak acid groups – such as carboxylic groups – is zero to ten percent, depending on their acid strength.

As a procedure for determining the concentration of highly dissociated cationic groups, the resin is in the first (regeneration) step converted to the hydrogen form and rinsed as with the determination of the total capacity of a cation exchanger. A 5.000 ± 0.005 g sample is weighed into a funnel as in the assembly in Figure

1.65. At the same time a sample is weighed for the determination of the percentage of solids. The capacity sample in the funnel is leached with excatly 1 liter of 4% neutral sodium sulfate. Aliquots of 100 ml are taken for titration with standard 0.1 N sodium hydroxide, using phenolphthalein as the indicator. The capacity is then calculated by the formula

$$\text{capacity (mequ/g)} = \frac{\text{ml}_{\text{NaOH}} \cdot \text{normality}_{\text{NaOH}} \cdot 10}{\text{sample weight} \cdot \dfrac{\%\ \text{solids}}{100}}.$$

This method was found to have a reproducibility of 5 parts per 1000 by a single operator, with agreement of 1 part per 100 between operators.

Determination of the total weight capacity of an anion exchanger:

Approximately 10 g of an air-dried anion exchanger is transferred to the funnel of Figure 1.65 and converted to the Cl-form by slow treatment with 1000 ml 1 N HCl in the funnel. The chloride form thus obtained is consequently washed with alcohol until the filtrate is neutral to methyl orange, and is air-dried. Of this material, 5.000 ± 0.005 g is weighed into a fresh funnel and leached with exactly 1 liter of 1% ammonia solution. Then the resin is leached with exactly 1 liter of 4% sodium sulfate, and this effluent is collected in a separate flask. Using aliquots of 100 ml from each of the leachates, the chloride is titrated after neutralization with 0.1 N silver nitrate solution, using potassium chromate as an indicator. Separately, 1 g of the same initial exchanger sample is weighed into a weighing bottle, dried overnight at 100 °C, and weighed back to determine the percentage of dry solids. For both solutions the capacity is calculated by the formula

$$\text{capacity (mequ/g)} = \frac{\text{ml}_{\text{AgNO}_3} \cdot \text{normality}_{\text{AgNO}_3} \cdot 10}{\text{sample weight} \cdot \dfrac{\%\ \text{solids}}{100}}.$$

This represents the total weight capacity of the exchanger in the dry Cl-form.

The exchanger is converted into the Cl-form to prevent errors due to the different equivalent weights of different anion forms and to prevent the possibility of decomposition of the free base form during drying. The exchanger is washed with alcohol instead of water to avoid a possible hydrolysis of the salt form of weakly basic ion exchangers. The capacity calculated from the titration of the ammonia leachate approximates the weakly basic capacity of the resin; that calculated from the titration of the sodium sulfate leach represents the strongly basic capacity. The total anion exchange capacity is, as already mentioned, the sum of the two capacities. If only a total anion capacity is demanded, the leaching with ammonia solution may be omitted and the rinsed chloride form of the polymer may be leached directly with the sodium sulfate solution. The values obtained for anion exchange capacity in this case were found to be usually 2 to 3% lower than those obtained by the double determination. A reproducibility of 1 part per 100 was claimed as to be expected. Regeneration with chloride has been far more successful than regeneration with hydroxide. When the greatest accuracy is required increasing the concentration of the hydrochloric acid used as a regenerant is recommended.

The total capacity is stated in milliequivalents per milliliter (mequ/ml) of the wet settled resin (which means backwashed and then allowed to settle) referred to a specific counterion form to account for volume changes which occur as resins are converted from one form to another. This value is, of course, numerically identical to equivalents per liter of resin (equ/l). Other expressions are pound equivalents per cubic foot ($lb \cdot equ/ft^3$) of wet settled resin, again referred to backwashed resin in a stated ionic form or kilograins as calcium carbonate per cubic foot (kgr as $CaCO_3/ft^3$), a traditional term from the earlier days when ion exchange resins were used almost exclusively in water treatment, related also to backwashed resin in a stated form.

The capacities obtained by these methods are not in complete agreement with the so-called analytical capacity. Analytical capacity refers to one calculated by a sulfur determination for cation exchangers of the sulfonic acid type and by a nitrogen determination for anion exchangers. The deviations can be explained by the fact that active sites are no longer accessible to ion exchange due to pore plugging or constriction. In sulfonic acid exchangers, moreover, a part of the sulfur may be bound in the sulfone form. That steric hindrance generally influences the measured capacity is evident from the fact that highly crosslinked exchangers exhibit different capacities when loaded with ions of different sizes.

Other total capacity terms in weight and volume capacity measurements, particularly in specialized applications for ion exchange resins, have been developed by various industries. With respect to the use of ion exchange resins in the treatment of water, it may be mentioned once more that the methods have been standardized and adopted, for example by the American Society for Testing and Materials, ASTM, by DIN and others. The reader is urged to refer to these standard methods when questions of uniform testing arise. More details are given in Chapter 1.2.4. Another example of an industry having developed its own techniques for determining total capacity and expressing it in units which serve its own needs is the uranium producing industry, which typically expresses capacity quite simply in units of pounds of uranium as U_3O_8 per cubic foot of wet settled resin in the chloride form or as g U_3O_8 per liter of pre-conditioned free wet-settled resin in the sulfate form.

In contrast to the above discussed total capacity the operating capacity denotes the ion exchange capacity realized in a given application under specific conditions of use. For industrial purposes the operating volume capacity is most frequently determined; this is appropriately done in a semi-industrial or industrial filtering test. The reason for this is that in most industrial applications of ion exchange resins the resins are utilized as a bed in a well-designed column or operating tank through which the solution containing exchangeable ions is passed. In the practice of water softening, for example, the resin filter exchanges hardness ions to a degree somewhat lower than its total capacity. This can be observed at the point where there is a tolerable hardness leakage (breakthrough). Depending on how soon this point is reached this is called the operation capacity or column capacity of the filter, the capacity effectively used will be a portion of the total capacity, which can be substantial or very small. Kinetic phenomena, the "driving force", that is, the concentration in solution of the ion to be exchanged in relation to other ions of like charge, the selectivity of the resin for the species to be removed, the contact time, the temperature of the system, the presence or absence of competing ions, the completeness of regeneration in the preceding cycle, and the endpoint defined for the particular operation are all factors that make the operating capacity a complex value. Only when the flow is continued through a column of resin until there is no further uptake of ions and the ionic composition of the effluent from the filter is the same as that in the column feed has the so-called saturation capacity been reached. But this value, too, is specific to a given set of conditions. The saturation capacity is largely fixed by equilibrium relationships and is thus affected by the concentration of ions in the solution, both those which one wants to exchange and those which one doesn't want to be exchanged. The saturation capacity will be somewhat less than the calculated total capacity. The extent to which saturation capacity approaches total capacity depends on the equilibirum relationships prevailing in the system. In some cases the determination of the operating volume capacity in a semi-industrial filtering test is performed with calcium chloride solution, the capacity in g CaO/l being calculated directly from the quantity of calcium determined in the form of oxalate.

The capacity depends on various external conditions, among which the pH is the most important. Since the ionic groups represent strong or weak acids and strong or weak bases, they are naturally pH-dependent. Thus, carboxyl groups or phenolic OH-groups remain partially or completely undissociated at low pH values, so that the exchange capacity of exchangers with such groups with respect to metal cations from acid and neutral solutions is lower than from alkaline solutions. Anion exchangers with primary, secondary, and tertiary amino groups exchange acid anions from alkaline and neutral solutions to a lesser degree than from acid solutions. Ion exchangers with strong acid or strongly basic groups (in the latter case, quaternary ammonium groups), however, are capacity-independent over a

Table 1.16 Effective pH range of different types of ion exchangers

Type of ion exchanger	Approximate pH range
Strong acid cation exchanger	4–14
Weak acid cation exchanger	6–14
Weak base anion exchanger	0– 7
Strong base anion exchanger	1–12

broad pH range. Accordingly, they are also active over a wide pH range and can be used under acidic, neutral, and alkaline conditions. Table 1.16 shows for convenience once more a summary of the effective pH ranges of various types of ion exchangers.

Besides the exchange of ions, an additional ion adsorption may take place in ion exchangers, because the exchanger material adsorbs electrolytes beyond the capacity of the fixed ions (ionogenic groups). Although the adsorption capacity in most cases is substantially smaller than the exchange capacity, it should, strictly speaking, be determined and expressed separately from the total capacity. It is, however, difficult to distinguish between the exchange and the adsorption capacity.

In connection with the capacity and its dependence on the pH, it is important to note a method of pH titration of ion exchangers which permits a precise characterization of exchangers on the basis of the neutralization of their active groups. This experimental method, which was introduced by Griessbach, is based on the fact that cation exchangers in the H-form and anion exchangers in the OH-form can be titrated with alkalis and acids like ordinary acids and bases. The neutralization process is observed by measurement of the pH. The titration curves of monoionic strong acid or strong base exchangers resemble curves obtained by potentiometric, neutralization titration of strong mineral acids or sodium hydroxide.

To obtain the titration curve of an ion exchange resin, a quantity of the cation or anion exchanger is first transformed into the H- or OH-form with hydrochloric acid or sodium hydroxide, respectively, and then washed with distilled water. The actual titration curve can then be measured in two ways: either by the method of direct titration applicable to exchange resins with easily dissociated ionogenic groups and with fast exchange kinetics, or by the batch method of individually weighed samples, to be used for exchangers containing weakly dissociated ionogenic groups and with slow exchange equilibration. The latter method can be applied to any type of exchanger.

The method of direct titration employs 0.500 – 1.000 g of the swollen, vacuum-drained exchanger in a 150 ml beaker to which 50 ml of 0.5 – 1.0 N NaCl solution are added. Glass and reference electrodes are immersed into the suspension and while stirring with a magnetic stirrer the liberated acid or base is titrated. Equilibration should be permitted after each addition of the titrant. The stirrer is stopped and the corresponding pH value read. The titrant – which is usually 0.1 – 1.0 N in concentration – is added in 0.2 – 0.5 ml portions. The pH values measured are plotted on a graph with pH vs. ml of titrant added.

The method of individually weighed samples also requires 0.500 – 1.000 g of swollen and vacuum-drained or centrifuged resin which are weighed in dry glass or, better, polyethylene, tight-fitting bottles 100 ml in volume. 50 ml of 1 N NaCl and an increasing amount, i. e., 1 – 10 ml, of 1 N HCl or NaOH are added to each bottle. The content of each bottle is adjusted with deionized water to the same volume. The tightly closed bottles are equilibrated by shaking on a mechanical shaking apparatus. The shaking time depends on the type of exchanger. Two to

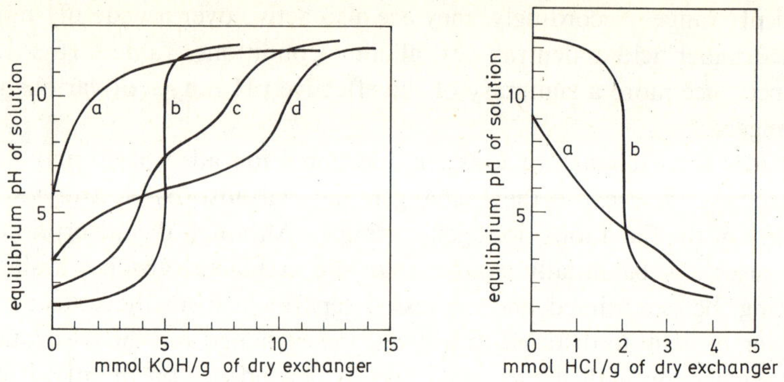

Figure 1.66 Titration curves. Left: cation exchangers: a) week acid OH-groups, b) strong acid SO_3H-groups, c) medium acid $PO(OH)_2$-groups and d) weak acid COOH-groups; right: anion exchangers: a) weak base anion exchanger, b) strong base anion exchanger.

three hours are sufficient for the equilibration of strong acid or strong base resins. Several days to some weeks may be necessary for other exchange resins. After equilibration has been reached, the pH is measured in the individual bottles and the values obtained are plotted analogously to the preceding procedure. During this time the dry solid of the resin is determined.

Figure 1.66 shows the titration curves of certain cation as well as anion exchangers as they can often be found in the ion exchange literature. The shape of the titration curve indicates the dependence of the exchange capacity on the pH of the external solution. The capacity of the exchanger can be read from these curves as the intersection of the curve with the line of pH 7. Moreover, these curves can be used to evaluate the exchangers for their acid strength, operating interval, and buffering power.

1.2.3.7 Selectivity

Selectivity is defined as the property of an ion exchanger to exhibit a preferential activity for certain ions. When an ionogenic surface such as in an ion exchange material is in equilibrium with a solution containing a mixture of counterions, the proportions of the different ions associated to the surface are not the same as their proportions in the bulk of solution. In the case of alkali ions, this means that these ions will be taken up with different readiness by a cation exchanger and that they can mutually displace each other. The exchange affinity of alkali ions towards a strong acid sulfonic resin increases according to the sequence

$$Li^+ < Na^+ < K^+ < Rb^+ < Cs^+.$$

Thus, lithium is taken up less easily than sodium, sodium less easily than potassium, etc.

Stoichiometric, reversible ion exchange exhibits all the characteristics of an equilibrium and follows, therefore, the law of mass action. Thermodynamic studies taking only into account dilute, e. g. 0.1 N solutions, and neglecting the penetration of co-ions (Donnan electrolyte) into the resin, have shown that ion exchange tends towards an equilibrium in which the quotient of the concentration ratios of the two exchangeable ions on the exchanger and in the surrounding solution becomes independent of concentration. This state in which the two ions assume a different distribution between the exchanger and the surrounding solution is characterized by a constant known as the selectivity coefficient.

The selectivity coefficient K_s is usually defined as the ratio of the ionic fractions of the exchanging ions A and B in the two phases:

$$K_S = \frac{Y_B}{Y_A} \cdot \frac{X_A}{X_B},$$

where Y is the concentration in the resin and X is the concentration in the solution. For the sake of clarity, it will be assumed that an exchanger is in the Na-form and is placed into KCl solution. The following reaction will then take place:

$$Na^+(\text{in resin}) + KCl(\text{solution}) \rightleftharpoons K^+(\text{in resin}) + NaCl(\text{solution}).$$

Its equilibrium is characterized by the contant K_s, which results from the above relation. Since the ion activity in the resin phase cannot be determined, K_s is not a thermodynamically defined equilibrium constant but only a coefficient defined according to practical requirements.

It must be kept in mind that the selectivity coefficient is not a true constant but is influenced by various factors, including: (1) the exchangeable ions (size and charge); (2) properties of the exchanger, i. e., particle size, degree of crosslinking, capacity, and type of ionogenic groups; (3) nature, i. e., total concentration as well as concentration ratio, of the existing ions both capable and incapable of exchange as well as the type and quantity of other substances in the solution, and (4) the reaction period. Various theories have been proposed in an attempt to obtain greater insight into the conditions which determine selectivity [337]. Since none of these theories furnishes quantitative information, only a few rules for practical use, which have been found empirically and by which the selectivity of an exchange process can be estimated, will be presented.

At low concentrations of aqueous solutions and at room temperature, all exchangers give preference to polyvalent ions over monovalent species in accordance with the selectivity sequence:

$$Na^+ < Ca^{2+} < La^{3+} < Th^{4+}.$$

The preferential uptake of polyvalent ions is concentration-dependent, however, and decreases with an increasing concentration of the solution.

For ions of identical valence, the selectivity at low concentrations and ordinary temperature increases with increasing atomic number according to the following sequences:

Li < Na < K < Rb < Cs
Mg < Ca < Sr < Ba
F < Cl < Br < I.

The selectivity for H^+- and OH^--ions depends on the strength of the acids or bases formed from the ionogenic group of the exchanger and the H^+- and OH^--ions. Weakly acidic and weakly basic ion exchangers are selective for H^+- and OH^--ions, since the solid acids and bases that form have undergone little dissociation and swelling.

The higher the degree of crosslinking, the higher will be the selectivity, i. e., the difference between the selectivity coefficients for different ions. It can be observed, however, that the selectivity decreases again at high crosslinking (starting with 15%). Some theorists ascribe so much significance to this selectivity reversal that they consider the phenomenon equivalent to that of selectivity itself [338]. In an advanced approach to the theoretical and quantitative description of ion exchange selectivity it is admitted that in the case of highly crosslinked resins (25% DVB), it must be taken into consideration that the presence of small amounts of impurities or some irregular ion exchange sites can lead to a strong deviation of the experimental data at the point of extrapolation to zero from the behavior predicted by the general equation derived [339]. Table 1.17 offers insight into the selectivity of a sulfonic acid exchanger having different but usual degrees of crosslinking with respect to different cations. The influence of crosslinking was also observed, for instance, in the sorption of different forms of arsenic(V) by an anion exchanger

Table 1.17 Relative selectivity coefficients of a sulfonic acid exchanger with different degrees of crosslinking

Cation	Crosslinking		
	4%*	8%*	10%*
Li	1.00	1.00	1.00
H	1.30	1.26	1.45
Na	1.49	1.88	2.23
NH_4	1.75	2.22	3.07
K	2.09	2.63	4.15
Rb	2.22	2.89	4.19
Cs	2.37	2.91	4.15
Ag	4.00	7.36	19.4
Te	5.20	9.66	22.2

*Percent of divinylbenzene as crosslinking agent

(Anionite AV-17). The selectivity of adsorption increased (and capacity decreased) when the concentration of divinylbenzene in the resin was changed from 2 to 8%; further, the selectivity of the exchange of alkaline earth metal ions on carboxyl ion exchangers depended, as a function of the mole fraction of metal ion, on the percentage of divinylbenzene in the resin. For this type of exchanger the highest selectivity was exhibited by $Ca^{2+}-H^+$ on all resins and for all pairs of ions on the material with 5% crosslinking [340].

With regard to ionic size and selectivity, the rule is accepted that larger ions, especially organic ions are taken up preferentially. Thus, in mono-, di-, and trisubstituted methyl-, ethyl-, and butylamines the selectivity coefficients increased with increasing length of the aliphatic part of the amine molecules, and in the anion exchange of chloro-substituted acetic acids the selectivity increased with the degree of chlorination. The maximum uptake of $[Co(NH_3)_6]^{3+}$ by a copolymer of methacrylic acid-divinylbenzene was studied. In the desorption from this loaded resin, the efficiency of cations increased in the order $Cs^+ < Rb^+ < K^+ < Na^+ < Li^+ < NH_4^+ < Mg^{2+} < Ca^{2+} < H^+$ and $Et_4N^+ < Me_4N^+ \ll$ cetyltrimethylammonium < cetylpyridinium. The logarithm of the selectivity coefficient increased linearly with increasing radius of the hydrated ions and decreasing Debye-Hückel parameter [341]. Finally, anionic metal complexes show a similar behavior, being exchanged with particular selectivity by anion exchangers.

To calculate the selectivity coefficient, the distribution of an ionic species and its concentration in the exchanger phase and the surrounding medium is determined after equilibration. The values found are plotted as molar fractions or percent equivalents in the exchanger and the solution. Figure 1.67 shows some representative examples.

Using this method of quadratic equilibrium representation, K_s can be easily determined. According to the above equation one has

$$K_S = \frac{Y_A}{Y_B} : \frac{X_A}{X_B} = \frac{Y_A}{Y_B} \cdot \frac{X_B}{X_A} = \frac{\text{area I}}{\text{area II}}$$

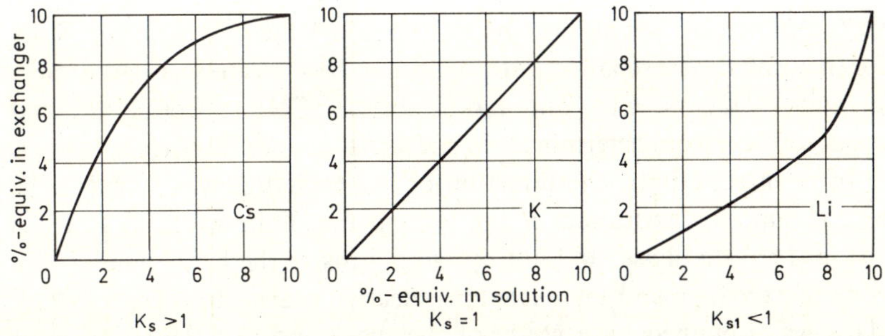

Figure 1.67 Determination of the selectivity coefficient by the quadratic equilibrium representation. NH_4^+ as reference ion, schematic.

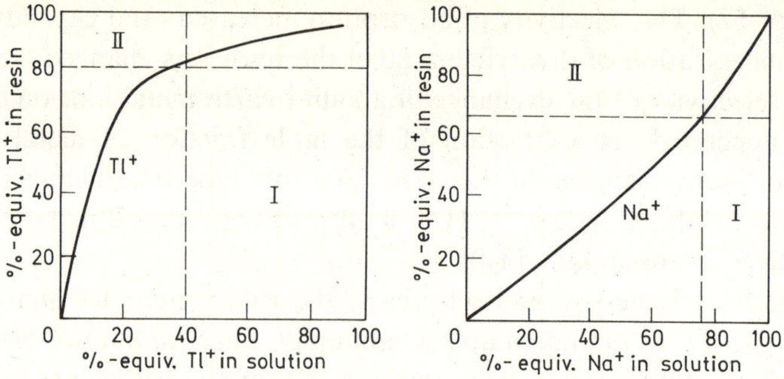

Figure 1.68 Calculation of the selectivity coefficient by the quadratic equilibrium representation.

in Figure 1.68, where the determination of K_s according to the method of Kressman and Kitchener [342] is shown for the example of thallium and sodium.

Two interests have dominated over the years in which ion exchange selectivity has been investigated: to explain and to improve selectivity. The infrared spectra of a strong base anion exchanger were determined to find the relation between ionogenic groups of Dowex 1X8 and the ions SO_4^{2-}, NO_3^-, CrO_4^{2-} and ClO_4^-, and to explain selectivity. Besides indicating a stronger interaction of CrO_4^2 and NO_3^- with the exchanger ions, the IR spectra gave information on the hydration of the resin. It was concluded that the selectivity of the exchanger is influenced by the hydration energy of the anions and by reaction with other substances in the system [343]. The selectivity coefficients as a function of temperature were determined for the exchange of n-alkylsulfates, $ROSO_3Na$, where $R = Me(CH_2)_n$ ($n = 0-7$), with the same strong base anion exchanger in the Cl-form. The selectivity coefficients were compared for the exchange processes involving the homologous series of $ROSO_3^-$ and RSO_3^- anions. The entropies of exchange increased with increasing chain length in a similar fashion for both families of ions. In conclusion the existence of a general hydrophobic selectivity effect active for all ionic solutes containing hydrocarbon chains was established [344]. To investigate the question of ion pairing versus ion hydration the values of the selectivity coefficients for heterovalent exchange with a strong base resin and radiotracer ReO_4^-, CrO_4^-, and WO_4^{2-} vs. Cl^-, and radiotracer $Cr(CN)_6^{3-}$, $Co(CN)_6^{3-}$, and $Fe(CN)_6^{4-}$ vs. macroCN^-, were determined. In such systems, contrary to early ideas on the nature of resin selectivity, the direction of the exchange was found to be determined by the superior hydration of the ions in the dilute external aqueous phase over that in the resin phase, and not by ion pairing in the latter phase [345]. The mechanisms of exchange and ion exchange selectivity seem, to a great extent, to coincide. This was obtained by a study of the mechanism of the sorption of transition metal ions by weak base anion exchangers from solutions of their complexes. In the non-protonic state a coordinated interaction of the ionogenic group of the ion exchanger with transition metal ions is the factor which determines the selectivity and in the protonic state the selectivity of the transition metal ion sorption is determined by the energy of electrovalent interaction [346]. For the study of the mutual effect of ions in ammonium-alkali cation-hydrogen ion systems on highly crosslinked sulfostyrene ion exchangers ternary plots were constructed to show the resin phase composition of $NH_4^+ - H^+ - M^+$ (M = Li, Na, K, Cs) at 25 °C and ionic strength of 0.1. These plots indicate the resin selectivity in ternary systems [347]. A so-called superselectivity in ion exchange membrane materials as an enhanced current efficiency beyond the Donnan limit was also discussed. In a critical examination, however, it was theorized that its origin is in the existence of a coarse-grained inhomogeneity in the distribution of the fixed ions. The results implied that superselectivity might be due to a inhomogeneity possibly resulting from dipole clustering of fixed ionic groups [348].

The problem of improving the selectivity of one ionic species over another is found, for example, in the practical application of ion exchangers for the removal of nitrates from water with anion exchangers in the chloride cycle. Strong base anion exchangers have a higher selectivity for the nitrate ion than for the chloride ion in dilute solutions. However, the selectivity, when sulfate is present, is $SO_4^{2-} > NO_3^- > Cl^-$, meaning that sulfate is preferred to nitrate. In investigations it was first shown that the selectivity for nitrate is improved in the presence of sulfate if the three methyl groups on the nitrogen are changed to three ethyl groups, i. e. from $R-CH_2-N(CH_3)_3^+$ to $R-CH_2-N(C_2H_5)_3^+$, as the distance between the nitrogen groups is increased. Then Clifford and Weber concluded selectivity of a resin depends on the degree of charge separation, which is the primary determinant of divalent-monovalent selectivity. Anion resins, particularly acrylics and epoxies, with closely-spaced N-containing ionogenic groups are inherently divalent-ion selective. This is because the uptake of a divalent anion, e. g. sulfate, requires the presence of two closely-spaced charges. Results from the extensive study of thirty commercial strong and weak base anion exchangers indicate that in order to bring positive charges into close proximity within a resin one can: (1) incorporate the amine groups into the polymer chaines as opposed to having them pendent on the chains, (2) minmize the size and number of organic groups attached to the N atom, i. e., minmize the size of the amine, and (3) maximize the resin flexibility, i. e., its ability to reorient to satisfy divalent counterions, by minimizing the degree of crosslinking. The distance-of-charge-separation theory is not restricted to divalent anion exchanges, but also applies to cation exchanges and to polyvalent ions in general [349]. Barron and Fritz, on the other hand, studied the effect of the structure of anion exchangers on the selectivity for various anions. They varied the pendent groups on the nitrogen and found that if this is a C_1 to C_6 group there is no change on the retention of several anions relative to chloride, but that retention of NO_3^- and I^- increases as the pendent groups on nitrogen increase in their number of carbons. For example, when the pendent group is butyl, the selectivity is the same for NO_3^- as for SO_4^{2-} [350].

1.2.3.8 Stability and attrition

In laboratory experiments, as well as in the industrial application of exchanger materials, the stability of the resins, which are usually used in cyclic processes, plays a decisive role: it influences not only the individual process but the total cost of an industrial process. Depending on the nature of the processes taking place in ion exchange, one must distinguish between physical and chemical stability; further, because of the use of ion exchangers in nuclear reactor engineering, their radiation resistance must also be considered. Beyond these, two other phenomena, irreversible adsorption and precipitation, can inactivate an ion exchanger, but these cases do not involve a question of stability. If the ionogenic groups of an exchange resin are poisoned by irreversible adsorption or if the further activity of the resin is suppressed by precipitates, the operating conditions need to be changed or, if the situation cannot be remedied in any way, other processes must be used. The question of stability, in contrast, concerns the physical and chemical properties of the resin.

The stability of an ion exchanger depends on the production method and is thus a question for ion exchanger manufacturers. The ion exchanger types available today have been refined to such an extent by many years of research and development that when the operating conditions recommended by the manufacturer are observed, many years of use can in most cases be expected.

Physical stability. The physical or mechanical stability of an ion exchange resin primarily resides in its strength to resist attrition. Good resins are characterized as being uniformly spherical, without internal cracks, resistant to mechanical compression, and with low brittleness. In practical application the ion exchanger beads must not fracture; otherwise, they will be flushed from the column or the exchanger tank. They must be resistent to osmotic influences, i. e., have a sufficient bursting and disintegration resistance under given loading and regeneration conditions. When this is not the case, too large a proportion of the monomer may be the reason for instability. A fundamental investigation on the structure of crosslinked copolymers showed that the penetration of solvent into the granules of styrene-p-divinylbenzene, stryrene-m-divinylbenzene or styrene-diisopropylenebenzene copolymer used as an ion exchanger matrix created tension at the swollen-nonswollen boundaries. This can be observed under a polarizing microscope and can be used for the evaluation and selection of materials having the least tension for use in stable ion exchange resins. The greater the microtension in the structure, the greater the relaxation time and the longer the existence of a polarized image. Relaxation time in the copolymers increased with an increased amount of crosslinking agent, and as a result the microtension was increased. This phenomenon was most clearly exhibited during the use of p-divinylbenzene as the crosslinking agent.

The thermal stability – at least of anion exchangers – depends on the degree of crosslinking; in weakly basic amine types of exchanger, thermal stability decreases with an increasing divinylbenzene concentration. Further studies of the thermal stability of ion exchange resins have demonstrated [351] that strong acid cation exchangers in all forms are completely stable in pure water up to 120 °C. The H-form becomes unstable at 150 °C, where the Li-form, however, is still stable for 30 days. At 180 °C the H-form is already highly sensitive and at 200 °C it exhibits pronounced degradation phenomena. Comparatively speaking, however, degradation still remains slight in the Li-form at 180 °C and even at 200 °C, which is also the case in the Na- and K-forms at the same temperatures. The thermal stability increases with decreasing divinylbenzene concentration. Weakly basic anion exchangers in the hydroxide and the salt form are stable for more than 30 days at 180 °C. In contrast, strongly basic exchangers in the OH-form are already unstable at 150 °C, while corresponding salt forms are still relatively stable. At higher temperatures, the resin is rapidly degraded. The strongly basic resins exhibit an increase in stability with decreased crosslinking. The influence of temperature in these cases falls under physical stability, since the test was performed in pure water. The ratings of stable or unstable, refer to a decrease in capacity and to degradation processes. The water-soluble degradation products detected consisted of sulfite and sulfate in the case of cation exchangers, and of trimethylamine and methanol in anion exchangers. As a test method for the determination of the thermal stability of ion exchangers a sample of a resin is heated with distilled water under reflux for 60 hours. As a coefficient of heat resistance the quotient of the minimum and maximum capacity, determined after and before the heating, respectively, or alternatively the quotient of the specific volumes of swollen resin determined before and after the heating is calculated. Instead of distilled water, solutions of bases or acids can be used to determine the heat resistance of ion exchangers in these media. Thus, in order to compare the temperature stability of different ion exchange materials it was recommended that materials of technical grade quality be used to $\leq$ 100 °C. Over 100 °C, materials of nuclear grade quality have to be used. H-form materials can be used to $\leq$ 110 °C, NH_4-forms up to 120 °C, and OH-forms up to 50 ° to 60 °C. HCO_3-form materials cannot be used in power stations [352]. In a separate investigation the effect of heat treatment in water on the physico-chemical properties of macroporous sulfonic cation exchangers was also studied. Here, too, it was found that the thermal stability of the cation exchangers in water decreased as the extent of their crosslinkage increased. The rate of splitting off the sulfonic group increased with the temperature of treatment. Generally, the thermal stability of a macroporous cation exchanger is higher than that of gels having the same extent of crosslinkage. The thermal treatment of the macroporous cation exchanger changed the structure of the spatial lattice and its porosity. This can be used for controlling the properties of these sorbents. Of interest is also the stability of phosphoric acid cation exchangers. Their thermal stability was determined either in open pyrex tubes or by sealing a dried exchanger with 2.5 ml liquid per milliequivalent of the exchanger (liquid = H_2O or a HNO_3 solution) in a pyrex tube. The concentration of phosphorus in the liquid phase, as H_3PO_4 and total phosphorus, and in the washwater of the solid residues (an oligomer), the exchange capacity, and the porosity of the residue were determined by conventional methods. The concentration of total phosphorus in the liquid phase is directly related to the loss of exchange capacity of the exchanger due to the thermal treatment [353].

A study of the low-temperature stability of ion exchangers including AV-17, KB-2 and KU-2 resins was also made by determining the amount of degraded granules after twelve freezing-thawing cycles at various moisture contents. After the twelve cycles the degree of degradation was 80, 75 and 62% for the ion exchanger types KB-4, KU-2 and AV-17, respectively. Most of the degradation occurred in the first five to seven cycles. The rate of degradation increased on increasing the moisture content to values higher than 35%. On freezing the exchangers at increasing moisture content, the loss in the static exchange capacity increased. At 15% moisture, the capacity decreased $\leq 10\%$, but on freezing an aqueous suspension, the capacity decreased by 25%. Other special ion exchange resins and a cation exchanger fiber were also tested [354].

If the general physical stability of exchanger particles is to be tested they can be subjected to alternate wetting and drying. Although such cyclic conditions do not

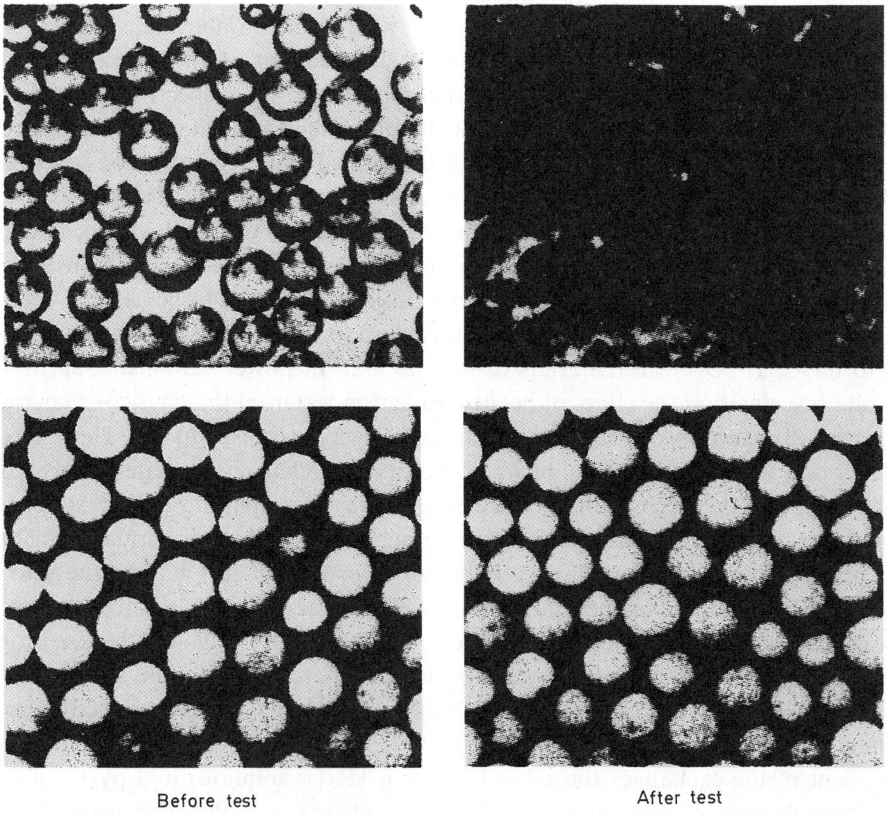

Figure 1.69 Drying test of two exchanger types of different physical stability.

occur in practice, they lead to valuable information. Figure 1.69 illustrates the behavior of a newly developed exchanger of very high quality and an older one of lower quality before and after a series of drying cycles [355]. In assessing mechanical resistance, accelerated testing methods should be applied at least to rate resins on a comparative basis regarding the manner and extent to which they withstand the continual cycling procedures of ion exchange processes. But although several general methods have been proposed, the effects of cycling a resin rapidly under the operation conditions for which it is likely to be used must be carefully interpreted, for the number of resin cycles is not the sole life-determining factor. Thus, if no attrition is apparent after 1000 resin cycles over an interval of one month, it cannot be assumed that there would be a resin attrition for 50 months at 20 cycles per month: some allowance must be made for normal ageing. At least two other methods have been used for assessing attrition resistance, as reported by Kunin on several occasions. The first involves grinding the resin in a ball mill or simply agitating it in a shaking device. Any results need careful interpretation because resins behave differently in the dry and the swollen states. For this reason the method can only rate resins according to their relative attrition susceptibilities. But a superior method is based on the swelling and shrinking cycles that follow repeated exhaustions and regenerations. In this method attrition of the resin beads may be expected to occur readily unless the resin has been adequately crosslinked during synthesis. (Still, beads can be shattered with relative ease even when there is a high degree of crosslinking.) The above-mentioned swelling and shrinking experiments can be executed by alternately wetting and drying resin samples placed on a turntable, the drying stage being effected under an infrared lamp. The resultant tendency to attrition can then be measured in terms of the increased area exposed by the resin samples as shown by sieve analysis performed before and after the test. After a series of such cycles, the samples are wet-screened and the extent of attrition

$$\left(\text{attrition number} = \frac{\text{average particle number}_{\text{final}}}{\text{average particle number}_{\text{initial}}} \times 100\right)$$

is ascertained by measuring the percentage of broken material passing through the smaller screen opening corresponding to the original screen cut. In a case specifically examined only 5 percent of the new resin Amberlyst 15 passed through US sieve size 16 compared with about 80 percent for the resin Amberlite IR 120 and about 95 percent for the resin Amberlite 122.

On the other hand it is reasonable to expect information on physical toughness – which is sometimes mistaken for mechanical strength – in a specification of ion exchange resins. One of the first measurements developed in an attempt to correlate toughness is the Chatillon Crush Test which determines the force required to fracture a resin bead of given size and ionic form between two parallel planes. The results obtained are usually expressed in terms of average strength per bead in

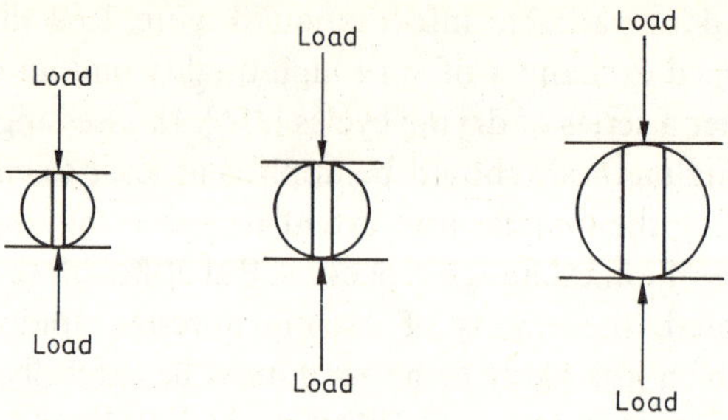

Figure 1.70 Schematic diagram of exchanger bead compression strength based on the cylinder concept.

grams per specific mesh size range and ionic form, with population sizes generally ranging from ten to fifty beads per test. The effect of particle size is significant. Other factors such as plasticity and flexibility, which are extremely difficult to determine, must be considered. The Chatillon test is still often used in conjunction with other specification values to insure proper toughness. The concept behind this test is that the sphere is assumed to be similar to a cylinder from a dynamic point of view and that therefore, the loading is distributed across the projected area of the bead between the Chatillon plates. Figure 1.70 is a schematic drawing of the loading situation with beads of various sizes. Despite the obviously high crush strength of the large diameter beads, large beads which contain cracks are fragile. There are contradictory opinions on the usefulness of the Chatillon test, but according to a study of Golden and Irving mechanical strength tests are indicative of the physical attrition and osmotic resistance of both gel and macroporous ion exchange resins [356]. The mechanical forces on a resin bead arise primarily from the pressure loss across the resin bed, but sudden changes of pressure such as those due to opening and shutting of valves must also be considered. To begin with, it can be said that mechanical forces are, on the whole, not strong enough to damage a perfect bead, gel or macroporous type. But particularly in a mixed bed system, where there is an interface during regeneration between the anion and cation resins, the resins at the interface are subject to an osmotic force brought about by a rapid swelling or de-swelling due to sudden contact with regenerant acid or alkali. If osmotic forces arise, they are sometimes considerably stronger than an ion exchange resin bead is capable of withstanding and cracked beads can result. Although only very fragile resins are likely to be completely broken by osmotic forces alone, a resin bead severely weakened by this means will then be much more liable to breakdown by mechanical forces, in particular under the higher forces resulting in high flow-rate conditions.

At this point field testing of physical resistance comes into the picture, because accelerated test methods are not fully reliable as a measure of results obtained under practical conditions, and a realistic assessment of resistance to mechanical attrition can only be made during actual process operations where large ion exchange resin beds are subjected to many types of flow rates through backwashing and regeneration stages. These cause a great deal of agitation and contribute greatly to mechanical disintegration. The field observations indicate that the toughness of a resin depends not only on its physical strength or osmotic shock resistance, but on the performance of the resin under stresses with regard to both mechanisms at once. These reportedly have a synergistic effect, which can be ten times or more as destructive to the resin bead as when either of the effects alone is measured. Therefore, resins which appear to be physically very strong and are able to undergo several cycles of osmotic shock with little change may break down quite rapidly in operating plants. Such factors as percent expansion, freeze-thaw stability and change in bead count quality and friability before and after various tests must be examined in order to get the whole picture. A satisfactory method for investigating field performance of ion exchange resins for water treatment is to study the influence of natural waters on the resins under conditions more closely related to field conditions. Although ion exchange resins have been used for a sufficiently long time for like assessment to be made under operation conditions, an accurate assessment of attrition is not available. It must be realizd that no single test is available from which resin selection can be automatically assured for rigorous conditions, although some equipment manufacturers have developed proprietary test procedures based on rapid cycling in analogous miniature system to gain accelerated life studies. The best estimate appears to be that commercial resins normally have an attrition rate of less than five percent per year. While resin operations are geared to the lowest agitation limits that will fluidize a resin, it is fortuitous that polymerization cation resins and the strong base anion resins are fairly resistant to breakdown. The weak base phenolic polyamine polycondensation resins are more susceptible to shock and during a particular cycling period one such resin was reduced in effective size from 0.4 to 0.1 cm^3, while a styrene-based weak base resin showed no signs of physical instability. Last but not least it must be mentioned that equipment involving hydraulic transfer also may subject an ion exchange resin to a pumping type of shear. In one pumping test a pump with a rubber impeller is used to circulate the resin continuously. The degree of plasticity may then play the decisive role, permitting the resin to be pushed through constrictions under hydraulic pressure without breaking.

Physical resin stability is certainly one of the many properties that contribute to the usefulness of ion exchange resins and has, therefore, been the subject of investigation for many years. This makes it understandable that many tests to determine resin stability have been developed. Resin stability can for example, be measured in terms of sphericity and whole uncracked beads, a test which is performed with the dried free-flowing ion exchanger beads on an inclined plane, by either weighing the beads that roll off as well as those that remain on the plane, and calculation the sphericity in percent, or examining microscopically the beads that roll off to determine which are cracked. Another stability test is the Dow

Chemical Company Attrition Test, in which the resin is pumped as a slurry with a pumping rate of approximately 900 ml/min through an apparatus designed to create attrition forces similar to those encountered in general commercial use. To measure sphericity after attrition in the Dow method the dried resin beads are placed on a tilted, vibrating plane surface. Whole spherical beads will roll off the surface, but broken or flattened beads will not. The percentage of whole beads is again defined as sphericity. When sphericity is measured after the attrition test this measurement is known as "sphericity after attrition". Resins with higher sphericity after attrition have proven better in their ability to resist fracture and fines formation in service. Most Dowex water treatment resins have a specification that states minimum percentages for sphericity, both before and after attrition.

One of the most important stability properties of ion exchange resins is their resistance to osmotic shock. Osmotic stability can be defined as the resistance of ion exchange resin materials under changes in osmotic pressure. The osmotic pressure changes that always occur in each loading and regeneration cycle act like a shock on the ion exchanger beads, a fact which led to the development of osmotic shock tests in order to enable one to predict the durability of a material envisaged for a special application or for checking a resin of replacement. The osmotic strength of a resin as shown in an osmotic shock test can also be considered as the regeneration stability. Several osmotic shock tests have been developed and described in the ion exchanger literature. The principle of each of them is to create cyclic conditions for the resin to be tested to obtain different osmotic pressures. Tests used for some time already are described in [357]. In the following a newly developed shock test for resistance to osmotic and hydromechanical stress will be described in more detail, as it is intended to become one of the standardized methods for testing ion exchange resins in the Federal Republic of Germany. Twenty-five ml of the ion exchanger resin to be tested are placed in a filter tube, which is a glass cylinder closed at the top and the bottom by sieves. Then acid and alkali flow alternately through the sample. The resin is loaded and regenerated over short intervals of time. A countercurrent washing process is interposed between these two steps. The change in the chemical state causes periodic volume changes which lead to both internal and external osmotic and hydromechanical stress in the resin, with the latter due to close packing. Depending on the stability of the resin, a smaller or larger amount is destroyed mechanically. The consequence is a reduction in the flow rate of the demineralized water, which is measured. All types of ion exchange resins can be tested by this method. The results of this shock test showed that the resins can be divided into four quality groups. Thus, an important advance indication is gained of both the quality of production batches of the same resin type and the suitability of a resin for the operating conditions in practice. Users of large amounts of resins should base their purchases on the mechanical resistance as shown in this shock test. As far as information on long-term behavior in practice for resins exists, it was found to correspond very well to the predictions based on the shock test results [358].

Chemical stability. The chemical stability of an ion exchanger is manifested in the chemical resistance of the ionogenic groups and in its general resistance to oxidation. While the sulfonic acid groups of strong acid cation exchangers of the styrene type hardly undergo a change under the influence of pH and temperature, this is no longer fully true for the same exchanger of the phenol type. One can also find a difference between the Na- and the H-forms: the latter is even more unstable. In the strongly basic anion exchangers the OH-form in particular tends towards irreversible degradation and a quality loss in the form of an ageing process. The quarternary ammonium group convert into tertiary amines and finally into groups without basic properties. Consequently, it is not advisable to allow such a resin in the OH-form to stand in a strong alkaline solution for an extended period of time. These influences on the chemical stability of ionogenic groups become evident in a loss of capacity and consequently in a quality loss of the treated solution. If an exchanger is degraded in its polymer skeleton because of insufficient oxidation resistance, this leads to a decrease in crosslinking and thus increased swelling. These oxidative attacks are enhanced by the presence of iron or copper. Exchangers of the styrene type with lower degrees of crosslinking are more susceptible to oxidation than those with higher crosslinking. The free chlorine present in ordinary city water also has an unfavorable influence on oxidation [359]. Data on the thermal stability of strong base anion exchangers in the nitrate form are useful and important for the production of nuclear fuels [360]. A critical temperature exists here above which the resins undergo spontaneous thermal decomposition, which can even lead to accidents.

Before going into greater detail on the resistance of synthetic ion exchange resins to oxidizing agents it should be noted that a simple laboratory experiment can serve as a test of oxidation resistance. The samples to be investigated are treated with 3% hydrogen peroxide solution at 45 °C for 72 hours. After this pre-treatment, the moisture regain of these samples is compared with that of untreated exchanger samples. Table 1.18 shows the results of such tests for three different exchangers. Exchanger A exhibits the lowest increase in water content and thus is most resistant to oxidation.

Table 1.18 Oxidation test of different exchangers

	Water content (%)		
	A	B	C
Untreated exchanger	47	45	45
H_2O_2 treated exchanger	54	81	75

Oxidizing agents are largely responsible for the chemical instability of ion exchange resins and many studies from the time of polycondensation resins to styrene-divinylbenzene copolymer or acrylate-based materials have been devoted to determining true resistance to various oxidizing agents. These chemicals can in principle attack a resinous ion exchanger either at the carbon chains or at the ionogenic groups, resulting in different reaction sequences with entirely different results. Styrene-divinylbenzene cation exchangers carrying sulfonic acid groups show a superior resistance in both respects to the condensation-type resins, which are sensitive even to less strongly oxidizing agents because of their labile phenolic hydroxyl groups. The carboxyl groups of weak acid cation exchangers as well are highly oxidation resistant, so that polymerization cation exchangers are attacked by oxidizing agents at the polymeric chains, a fact often observed in practical use of such exchangers. In the case of crosslinked polystyrene matrices the point of attack is often the tertiary methine group, which leads to a chain decomposition and a softening of the matrix resulting, finally, in its complete dissolution. The anion exchangers have a lower resistance, mainly due to the amine groups, but here again the styrenic products are the most resistant. For Type II resins a decrease in anion exchange capacity is observed or at least a conversion of the more labile ethanolamine groups into more stable tertiary amine groups. On the other hand a softening by chain breakages happens only in exceptional cases. From a synthetic point of view the oxidation resistance could only be increased for cation exchangers by increasing the strength of the chains, for instance by higher crosslinking; for anion exchangers this could be achieved by the incorporation of more stable groups. In applying ion exchange resins contact with oxidizing agents should be avoided whenever possible or their attacking potential be reduced to a minimum, this meaning that low temperatures be employed and that catalyzing agents be kept away. The following summarizes the most important influences of various oxidizing agents.

Since oxygen is a mild oxidizing agent it is fairly innocuous especially after the application of a trickling tower. Long-term tests have shown, however, that since it is present in any water to be treated prolonged life of an anion exchanger used in water deionization can only be expected if the water is first de-aerated. Styrenic as well as acrylic cation exchangers are de-crosslinked (condensate) at elevated temperatures, especially in the presence of iron as a catalyst. As a possible reaction leading to de-crosslinking the following mechanism for a sulfonated polystyrene cation exchanger was discussed [361]

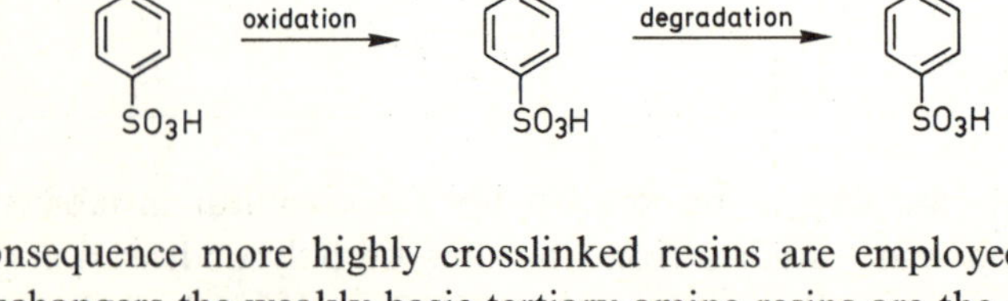

As a consequence more highly crosslinked resins are employed. Of the styrenic anion exchangers the weakly basic tertiary amine resins are the most durable.

Hydrogen peroxide can decompose the chains of cation exchangers in the same way as oxygen and can in the presence of iron even dissolve them. More highly crosslinked resins do not show a noticeable difference here. The interaction of hydrogen peroxide with anion exchangers can, on the other hand, be regarded under three aspects: the adsorption of hydrogen peroxide onto anion exchangers, the decomposition of the anion exchanger due to hydrogen peroxide treatment, and the catalytic decomposition of hydrogen peroxide in contact with the ion exchanger. Hydrogen peroxide is catalytically decomposed on the surface of a strong anion exchanger in the hydroxyl cycle. The surface activity of the transition complex $H_3O_4^-$ was found to be the limiting factor determining the rate of the decomposition process. Both strong anion exchangers in the Cl-form and weak base resins cause hydrogen peroxide decomposition. Hydrogen peroxide is also adsorbed onto the exchanger, expelling water from it in amounts that increase with the growing concentration of hydrogen peroxide in the solution. Strong anion exchangers in the hydroxyl cycle also adsorb hydrogen peroxide by means of ion exchange in the form of HO_2^-. The exchanging groups are decomposed either by hydrogen peroxide treatment, or by deamination; the rate of both processes increases with increasing hydrogen peroxide concentration.

Ozone, peroxodisulfate, chlorine, hypochlorite, chlorodioxide, sodium chlorate, bromine, chromic acid, dichromate, permanganate and nitric acid above 2.5 M are strongly oxidizing agents that greatly diminish the stabilty of even the polystyrene-based resins. Ozone and peroxodisulfate should be kept away from exchange resins. Chlorine is unfortunately present in almost all drinking water; strong acid cation exchangers are quite durable but anion exchangers are readily attacked by chlorine at the ionogenic groups. Reactions of sodium hypochlorite with anion exchangers based on styrene-divinylbenzene copolymers reduces the exchange capacity and the nitrogen content; investigation of the mechanism by IR spectroscopy pointed to an oxidation of the CH_2-group next to the amino group. Chromic acid present in waste waters and concentrates of the galvanic industry de-crosslinks cation exchangers unless they are very highly crosslinked; some medium base anion exchangers are quite resistant but have a markedly shorter life than in common water treatment. Nitric acid is a very efficient regenerant which is successfully applied in some ion exchange processes. Studies of the chemical stability of ion exchange resins in 1 N HNO_3 showed that strong acid sulfonic polymerization resins are resistant, the decrease in capacity being 1 – 3%. Strongly acidic resins of the polycendensation-type dissolve in 1 N HNO_3, as do the same resinous exchangers from acenaphthene and formaldehyde. Weakly acidic cation exchange resins containing COOH-groups are sufficiently resistant, as are also phosphonic polymerization-type cation exchangers. The chemical stability of cation exchange resins having a gel structure does not differ much in comparison with analogous macroporous resins. But at higher temperatures macroporous cation exchangers have a higher chemical stability in 10% nitric acid than those of gel structure. A remarkable change at room temperature takes place with strong acid cation exchangers when they are treated with 40% nitric acid. Important is the reaction of anion exchangers with nitric acid, since all kinds of substituted amino groups are more readily attacked resulting in substantially complete deamination. Anion exchangers in the NO_3-form should not be heated or dried and never be brought into contact with nitric acid of more than 1% in concentration.

A more extreme method than that described above of assessing oxidative stability is to warm the exchanger resins with 8 M nitric acid, i. e., 1 : 1 (v/v) concentrated acid: water, at 75 °C. The more stable resins take longer to dissolve. It has been claimed that one gel type resin took more than seventy days for complete dissolution compared with less than fourteen days for a more exposed macroreticular strong base resin.

Peracetic acid is used for the disinfection of ion exchangers; consequently, the stability of the materials concerned under contact with this disinfectant is of practical interest. The ion exchange capacities and water content of the following exchangers were studied: a gel type strong base anion exchanger Type I, a gel type strong base anion exchanger Type II, and a macroporous styrenic anion exchanger with tertiary amine groups, in both OH- and Cl-form; a gel type strong acid cation exchanger, a macroporous strong acid cation exchanger, in both Na- and H-form, as well as a weak acid acrylic macroporous cation exchanger, in H-form. They were treated with 0.2% peracetic acid for 4200 hours or 25 weeks, with a renewal after every 2 weeks. The anion exchangers did not decompose in peracetic acid, but this strong oxidizing agent attacked the matrix of cation exchangers, causing an increase in water content and a reduction in ion exchange capacity. The effects of the prolonged treatment by peracetic acid on the properties of the cation exchangers must be seen as an influence on the basic structure [362].

Reducing agents – unlike oxidizing agents – do not, in general, pose a threat to ion exchangers. Ion exchange resins are generally stable towards reducing agents such as hydrazine, sulfite, or formaldehyde; but reducing substances containing sulfur can destroy anion resins, and the free base forms are sensitive to aldehydes by causing their polymerization. Hydrogen in the presence of catalysts can reduce the sulfonic sites of strong acid cation exchangers but this happens either very seldomly in certain catalytic processes or is carried out deliberately as a reaction on the ion exchangers as reactive polymers.

Radiation stability. Because of the use of ion exchangers in nuclear reactors and for decontamination purposes, their radiation stability is of special interest. Basically, radiation damage in cation exchangers of the sulfonic acid type is caused mainly by the rupture of crosslinkages, and in anion exchangers by the degradation of ionogenic groups. According to Wiley and Devenuto [363], a degradation induced by Co^{60}-γ-irradiation in sulfonated styrene resins crosslinked with 4 and 8 mole % of m- or p-divinylbenzene isomers or their mixtures, including commercial divinylbenzene, leads to a capacity loss of 5–12.8% with radiation intensities of about 290 000 rad/h and total doses of $0.91 - 1.90 \cdot 10^8$ rad in the presence of water. This capacity loss is smaller (2–6%) in resins with up to 8% crosslinking and with irradiation in the dry state. Resins crosslinked with commercial divinylbenzene exhibit a somewhat greater capacity loss (11.2–11.9%) than others (10.2–10.7%) probably because of their alkylphenyl content. The moisture uptake also increases with irradiation, i. e., by 13–16% for 8% resins and by 2.7–3% for 4% resins exposed to $1.90 \cdot 10^8$ rad. The m-divinylbenzene crosslinked resin exhibits a somewhat smaller increase in moisture content (1.5% for 8% crosslinking and 0.8% for 4%), indicating a more stable network. These phenomena can be explained by desulfonation, hydroxylation and cleavage. According to corresponding studies by Hall and Streat [364] with strongly basic anion exchangers in the OH-form, a pronounced degradation is produced by Co-γ-irradiation in doses of up to 500 Mrad with the release of water-soluble aliphatic amines. In addition to tertiary, secondary, and primary amines, traces of formaldehyde were also detected. The quaternary ammonium groups form weakly basic amine groups, so that an increase in the weakly basic capacity occurs.

Generally, cation exchangers have a higher stability than anion exchangers and highly crosslinked exchangers are more stable than those with a low degree of crosslinking. Sulfonic acid cation exchangers are also more resistant to radiation than are carboxylic acid types. This is the result of an investigation of several carboxylic acid exchangers, that find application in the nuclear industry. The commercial resins were irradiated by γ-rays to a maximum dose of $1.2 \cdot 10^8$ rad in the H-form, as well in as the Na- and NH_4-forms, to determine the change in the exchange capacity. The salt forms of the materials are less resistant to γ-rays than is their H-form. The resins irradiated in an aqueous medium showed greater loss of capacity than those in an ethanol-water mixture. When the first commercial phosphonic acid cation exchanger Duolite C-63 was irradiated with ^{60}Co-γ-rays of a $7.05 \cdot 10^5$ rad/h dose the resin particles turned yellowish-brown and showed a linear decrease in capacity, the latter being determined by copper(II). With regard to radiation resistance, this exchange material was found to be between a strong acid cation exchanger like Dowex 50 and a strong base anion exchanger like Dowex 1. The loss in capacity was different depending on whether it was irradiated in air or in water with an absorbed dose of $8.7 \cdot 10^7$ rad. A strong base anion exchanger with pyridine as the ionogenic group on a vinylpyridine-divinylbenzene copolymer exhibited a higher degree of stability to ionizing radiation than any other strong base anion exchanger examined. The radiation stability of amphoteric ion exchangers was quite extensively studied. An amphoteric ion exchanger with an oxidized 2,5-divinylpyridine-divinylbenzene copolymer had a different radiation resistance in the NO_3-, Na- and Cu-forms. No rapid degradation of carboxylic groups and the polymer skeleton was observed on γ-irradiation with a dose of $2 \cdot 10^8$ to $1 \cdot 10^9$ rad of the air-dry NO_3- or Cu-forms or the water swollen NO_3-form. Irradiation of the Na-form under the same conditions decreased the exchange capacity and increased the swelling of the exchanger. A subsequent study of this exchange material by IR spectroscopy, however, showed that the number of oxo-groups was enhanced and that nitration of the benzene ring in the case of the NO_3-form in 7 N nitric acid occurs. The effect of ionizing radiation was also studied when the material was swollen in acetate buffer solutions containing various salts. The irradiation caused decarboxylation but the presence of sodium nitrate in the solutions during irradiation protected the macromolecular matrix from radiolysis [365].

With respect to the ion exchanger matrix in general it became apparent from comparisons of published results that strong acid cation exchange materials based on polycondensation structures are significantly more resistant than polymerization structures containing the same ionogenic group and in the same ionic form. With respect to polymerization structures, therefore, it was investigated whether the addition of ferrocene, cyclopentadienyl-mangan-tricarbonyl or vinylanthracene to the matrix could increase their radiation stability. It was found that this attempt yields better radiation resistance for strong acid cation exchangers but does not for anion exchangers. Furfural-based polycondensation ion exchangers were also studied for their radiation resistance. The weak acid cation exchangers from furfural-salicylic acid copolymer had higher radiation resistance than the strong acid products from furfural-p-toluenesulfonic acid, furfural-β-naphthalenesulfonic acid and furfural-hydrofuramide-polyethylene polyamine. The high radiation resistance of the weak acid salicylic material was attributed to the high energy of the C-C bonds, in comparison to the C-S bonds of the strong acid exchangers. The radiation resistance of the resins was again determined from changes in the exchanger capacity, weight, and swelling after γ-irradiation with a dose of $2 \cdot 10^8$ rad. In recent years numerous papers have been published on the radiation stability of synthetic ion exchange resins, often relating to a commercial ion exchanger brand. An extensive study of this literature reveals that the radiation decomposition of ion exchange materials has, unfortunately, the potential for a variety of undesirable consequences. Systematic efforts at identifying and resolving these problems and developing radiation-resistant ion exchangers seem extremely limited. There is, however, widely scattered information in the literature that is useful in designing and operating ion exchange-based process systems with a reasonable assurance of safety, and based on experimental data a better design and safer operation of synthetic organic ion exchange systems at nuclear material processing facilities can be expected [366].

These influences on the stability of the exchangers lead on the whole to their attrition during practical ion exchange processes. The resins are subject to physical and chemical deterioration. Their life can be extended if oxidative effects are reduced to a minimum. The life of an exchanger is measured either in m^3 solution/m^3 resin under given conditions or by the number of loading and regeneration cycles. The former method is recommended if chemical stability is to be determined, and the latter when physical influence is to be determined.

Martinola developed a method for the rapid routine control of anion exchangers during operation [367]. The various capacity values obtained offer an indication of the existing number of active sites and their base strength. Moreover, information is obtained concerning contamination of the resins by acid compounds and, with this method, measures to regenerate the exchangers in the laboratory can be tested. In strongly basic anion exchangers, a change in maximum capacity permits one to derive the change of effective volume capacity attainable under standard conditions. In weakly basic exchangers this is not possible, since other influences also play a role.

1.2.3.9 Electrochemical properties

Ion exchange resins as carriers of mobile ions, i. e., their counterions, in high concentration have good electrical conductivity based – as in the case of other concentrated electrolytes – on ionic conductivity. Numerous investigations of the conductivity of ion exchange resins and other effects related to it have been carried out because systems comprising a solid concentrated electrolyte in which one part is immobile, i. e., the fixed ionic sites, present uncommon properties, which have become most important in ion exchange resins in the form of membranes. As a consequence of the possible excess charge created by the counterions in the pore liquid of the exchanger resin the phenomena of electroosmosis, convection conductivity and electrokinetic potential – otherwise commonly-known effects of capillary systems – are found to a greater degree. These effects are caused by the mutual transport of solvent and excess charge during electrolysis or by the pressure drop. Ion exchange membranes are dealt with in depth in a separate chapter of this volume. The electrochemistry of granulated ion exchangers has been reviewed several times, as well as in an extensive monograph [368]. The following is a summary of the electrochemical properties of synthetic ion exchange resins.

One of the first conductivity tests of synthetic ion exchange resins was made in connection with investigations of the diffusion behavior of phenolsulfonic acid polycondensation resins in order to elucidate the mobility of counter-ions in such sulfonated materials. Six resins crosslinked with formaldehyde were prepared in cylindrical rods and their specific conductivities determined in the H-, Li-, Na-, K-, Rb-, Cs-, Sr- and Ba-forms. The corresponding equivalent conductivities of the cations in the resin phase were about one-third to one-eighth of those in aqueous solution of approximately 1 N concentration [369]. For polymerization ion exchangers the absolute values of the specific conductivity of common materials in the alkali- or Cl-form were determined to be 10^{-1} to $10^{-3}\,\Omega^{-1}\,cm^{-1}$. In principle it was found that the mobilities of ions in the exchanger are approximately proportional to the self-diffusion coefficients. This means, in other words, that for both values the same general rules are applicable. An insight into the factors determining specific conductivity can be obtained from results of an attempt to determine the self-diffusion coefficients of ions by measuring the electrical conductivity. On increasing the divinylbenzene content in the strong acid sulfonic cation exchanger from 4 to 16%, the self-diffusion coefficient decreased by a factor of four. The self-diffusion coefficient and therefore, the specific conductivity depended also on the charge and radius of the hydrated counterions: the H^+ ion had the greatest and the Mg^{2+} ion the smallest mobility. Under equilibrium conditions with increasing electrolyte concentration the specific conductivities increase and, correspondingly, the self-diffusion coefficients also increase. A relation between the electrical conductivity and the exchange capacity exists as well in that those exchangers have a high specific electric conductivity which have a high concentration of fixed ionic sites. The same rule is valid when the temperature is increased. A study of the effect of heat treatment on electrochemical properties was done in order to investigate the dependence of electrical conductivity on the duration of heat treatment (up to 72 hours). The ion exchangers tested were KU-2, AV-17, KU-1, and EDE-10P. When heated at 100 °C in distilled water over 72 hours the conductivity of the exchangers increased 8.35, 46, 32.4, and 92%, respectively. The determinations were carried out at the isoelectric point. Rinsing the same materials with acids and bases after treatment restored the original conductivity, except in case of the KU-1 cation exchange resin. Similar changes and the same anomalous behavior of the KU-1 resin were observed in studying swelling behavior. Formation within the resin matrix of the hydrophilic products of thermal degradation was assumed to explain a dilation of the matrix itself, and extraction from the matrix of such products by means of acid or basic rinsing accounted for the restoration of the original properties. The remarkable increase of the conductivity of the polycondensation anion exchanger EDE-10P was probably related to the increase in its swelling ability. Irreversible breaking of hydrogen bonds in the KU-1 cation exchanger due to the heat treatment appeared to explain the irreversible increase in electrical conductivity and in the swelling ability, in agreement with changes in the water retention ability previously observed. Gel-like particle formation, granule disintegration, and no practical alterations were observed on heating EDE-10P, AV-17, and KU-2 exchangers. With gel type ion exchange resins a lower degree of crosslinking leads to a higher swelling porosity and thus to the above-mentioned specific conductivity increase. The electric conductivity of so-called porous ion exchangers is not affected by the divinylbenzene content. The electric conductivity characteristics of true isoporous ion exchangers are intermediate between those of the macroporous and heteroporous resinous exchange materials.

One of the experimental method for the determination of the elctrical conductivity of ion exchange resins in the form of rods (membranes are here excluded as they are dealt with elsewhere in this volume) will be presented in the following as representative. According to Hills, Jacubovic and Kitchener a precision method was developed, which eliminates contact resistance and is accurate and convenient. It is, in principle, the same as was used by Shedlovsky for solutions. The ends of the rod or strip of the ion exchange resin are connected by two relatively large electrodes, a and d, to an a. c. Wheatsone network, Figure 1.71 left, while two small wire contacts, b and c, at intermediate positions, serve as potential probes.

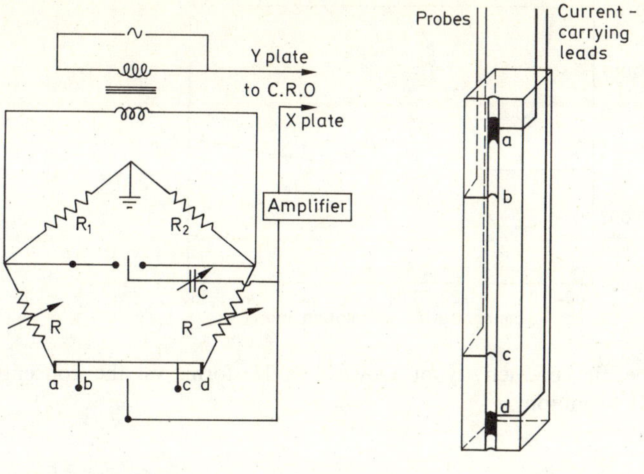

Figure 1.71 Equipment for the determination of the electric conductivity of ion exchanger rods. Left: bridge circuit with R_1 and R_2 equal ratio arms (R is a decade capacity box and C is a decade condenser box); right: the conductance cell.

The resistance between b and c is obtained by first making one and then the other of these points the corner of the bridge. The change in either of the decade resistors, R, required to re-balance the bridge is equal to twice R_{bc}. Any impedance at the probe contacts is immaterial. For ion exchangers in the form of rods, a Perspex cell having a grove of semicircular section has been used, Figure 1.71 right. Platinized platinum electrodes are cemented in the grove, the resin rods being held in place by elastic bands. The distance between the inner electrodes, b and c, is accurately known, as is the mean diameter of the rod, from independent measurements. The specific conductivity of several types of ion exchange resin were determined rapidly by this method, and with an accuracy of better than $\pm$ 0.2% [370].

For calculating the conductivity according to this method the following formula can be used, too:

$$\bar{\kappa} = \frac{\Delta l}{\Delta R \cdot q}$$

with Δl the difference in distance of the electrodes, ΔR the difference of the resistances measured, and q the cross-section of the rod. This method can also be applied to partially desolvated exchange resins.

Whenever ion exchange resins are in contact with an electrolyte solution it is from an electrochemical point of view of interest to examine how the specific conductivity depends on the concentration of the surrounding solution. It is further of interest to compare the conductivity of the resin with the conductivity of the solution under equilibrium conditions. In very dilute solutions the much higher ion concentration in the ion exchange resin takes effect and the conductivity of the

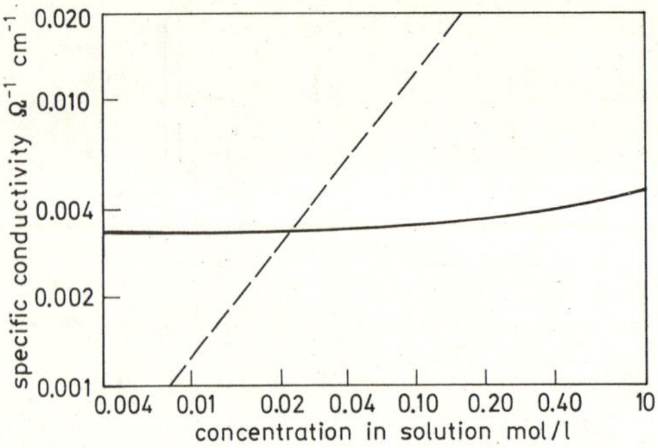

Figure 1.72 Specific conductivity of Dowex 50, Na-form, vs. the concentration of the surrounding NaCl-solution.

exchanger is greater than that of the solution. When the concentration of the solution is increased the concentration in the exchanger also increases, but not to the same degree: conductivity of the solution increases faster and finally outpaces that of the exchanger as shown in Figure 1.72. The difference in concentration between the two phases is now overcompensated for by the higher mobility of the ions in solution. At the equiconductivity point both the exchange resin and the solution have the same specific conductivity.

It is, further, of great phenomenological interest to study the transference numbers and the transport numbers of ion exchangers. Under the influence of an electrical field virtually only cations migrate through a cation exchanger and only anions through an anion exchanger. But in contrast to common electrolyte solutions a convective electric conductivity has to be taken into consideration in ion exchangers, which is caused by the fact that here the concentration of counterions predominates over that of co-ions almost totally. Figure 1.73 shows how the convection in an ion exchanger can be diagrammatically represented (Helfferich). The friction forces which the counterions exert on the molecules of the solvent in the pores of

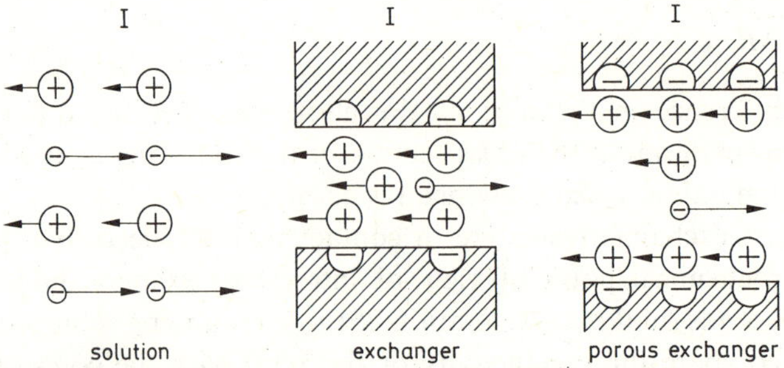

Figure 1.73 Schematic representation of the convection in an exchanger material as compared with a solution.

the exchanger are not compensated for by equal forces from the co-ions. Thus, the counterions force the pore liquid to move together with them. The extent of this convective motion depends on the concentration of the counterions, the strength of the electrical field and the structural motion resistance of the exchanger material. The convection of the pore liquid is greater than the mobility of the counterions relative to the pore liquid and the counterions migrate, therefore, faster relative to the matrix than would be the case due to their natural mobility. The conductivity of an exchanger resin is increased by the convection of the pore liquid which supports the migration of the counterions in the electric field. Phenomenologically, therefore, the mobility of the ions in an ion exchanger is in part due to electro-convective velocity and in part due to the so-called convective conductivity. The latter is high in exchangers of high capacity and low flow resistance; it is high as well if the system is in equilibrium with dilute solutions. In customary ion exchange resins the convective conductivity is estimated to contribute up to 40% of the total specific conductivity. Figure 1.73 also shows, in conclusion, a macroporous resin in which the surplus of cations is confined to a diffusive double layer on the walls of the pores; here only convection takes place, together with the surface conductivity.

The convective conductivity in ion exchangers cannot be easily calculated separately. Calculations including the specific conductivity, the convective conductivity and transference numbers have been carried out based on a model of an ion exchange resin as a charged sponge. It still seems difficult to correlate these values with the diffusion coefficient, at least in non-homogeneous synthetic ion exchange resins. Instead of making purely theoretical analyses it, therefore, seems appropriate that the mechanism of electric conductivity in ion exchangers has been re-examined [371].

The measurement of the specific electric conductivity of granular ion exchangers or ion exchange resin beads has been investigated, either because in many cases rods cannot be made out of the exchange material or because the electric conductivity of ion exchange columns was of interest. In an extensive work the nature of the conductance of columns of ion exchange resin spheres saturated with solutions of electrolytes of different concentration was examined, in order, in particular, to examine and if possible to confirm a model and mathematical equations based upon it. It was assumed that the equations would adequately describe the net specific conductivity of the ion exchange resin plug, together with the specific conductivity of the saturating solution. For a uniformly-sized ion exchange material the columnar cell shown in Figure 1.74 was used. This Plexiglas cell is filled with the resin particles between the perforated platinum electrodes, A and B. The system is then rinsed with distilled water by connecting side arms C and D to a circulating system consisting of two conductance cells leading to C and D, a solution reservoir, and a circulating pump. Rinsing is continued until the specific conductivities of the water entering and leaving the resin column are identical and less than $5 \cdot 10^{-6}\,\Omega^{-1} \cdot cm^{-1}$. For the purpose of these experiments, a specific conductivity of less than $5 \cdot 10^{-6}\,\Omega^{-1} \cdot cm^{-1}$ was considered to be negligible. The resistance of

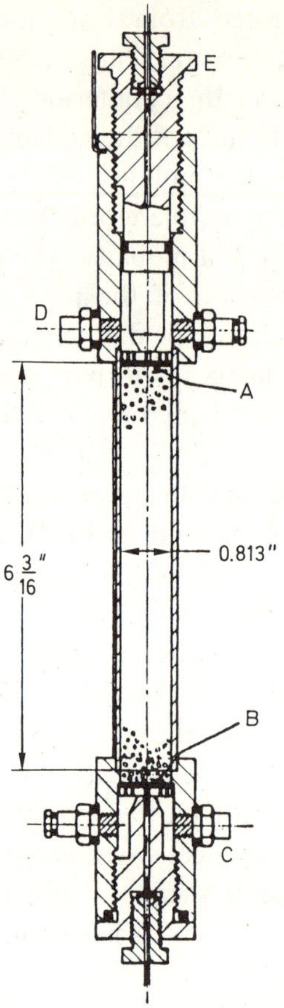

Figure 1.74 Conductivity cell for ion exchanger plugs. A, B perforated platinum electrodes; C, D side arms for circulating solutions; E micrometer screw.

the column is then measured, with the movable electrode adjusted to various settings before and after the point at which contact with the resin particle is made. For greater detail the reader is referred to the original paper [372]. To determine the value of κ_p, the specific conductivity of the resin plug saturated with the solution, the conductivities measured at different electrode distances (*d*) were plotted against *d*. For solutions of very low concentration, a definite break occurred in the conductivity-distance plot; the electrode distance at which the break occurred was taken as the contact point. Thus, the compaction of the resin particles was the result of gravity alone. In solutions of higher concentration no definite break was obtained. In these solutions the contact point was determined by observing the distance, *d*, at which the electrode appeared to make full contact with the resins. For sodium chloride solutions of widely varying conductivities representative curves of conductivity vs. distance between electrodes were measured. For solutions of high specific conductivity the conductivity changed only slightly with the electrode

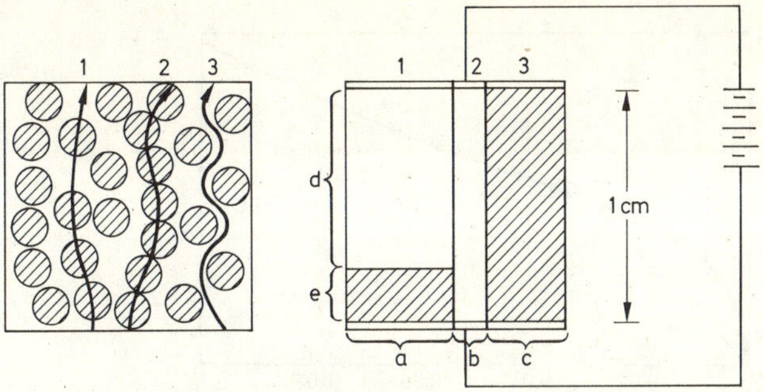

Figure 1.75 Electrochemical model of ion exchanger plug composed of conducting spheres and solution. Left: schematic representation of current path through plug: (1) represents current through solution and particles, (2) through beads in contact with each other, (3) current through solution; right: simplified model for the situation shown on the left.

distance. Therefore, a good value for κ_p could be obtained by visual observation of the contact point.

For understanding the specific conductivity of an aggregate of conductive particles saturated with a conducting electrolyte it was suggested that such a system is compatible with an equivalent resistance model comprising three elements in parallel, as shown in Figure 1.75. The first element, 1, comprises particles and solution in series with each other, the second element, 2, consists of particles in sufficiently close contact with one another to form continuously conduction paths, and the third element, 3, consists of fluid filling the interstices between the particles. The portions a, b and c of the three elements of the total cross-section of the plug as well as the portions d and e of exchanger and solution of the first element are constants which have to be determined empirically. But since $a + b + c = 1$ and $d + e = 1$, three equations for their determination are sufficient.

The specific conductivity, κ_p, of the plug containing conductive solids is the sum of the conductivities of the three elements

$$\kappa_p = \kappa_1 + \kappa_2 + \kappa_3 ,$$

with the three individual conductances being given by

$$\kappa_1 = \frac{a\,\bar{\kappa}\,\kappa}{d\,\kappa + e\,\bar{\kappa}}, \quad \kappa_2 = b\,\bar{\kappa} \quad \text{and} \quad \kappa_3 = c\,\kappa,$$

with $\bar{\kappa}$ the specific conductivity of the ion exchange resin and κ the specific conductivity of the solution. For strong acid cation exchanger beads with 8% crosslinking and aqueous solutions the following values were obtained over a wide range independent of particle size and the nature of the electolyte solution

$$a = 0.63,\ b = 0.01,\ c = 0.34,\ d = 0.95,\ e = 0.05.$$

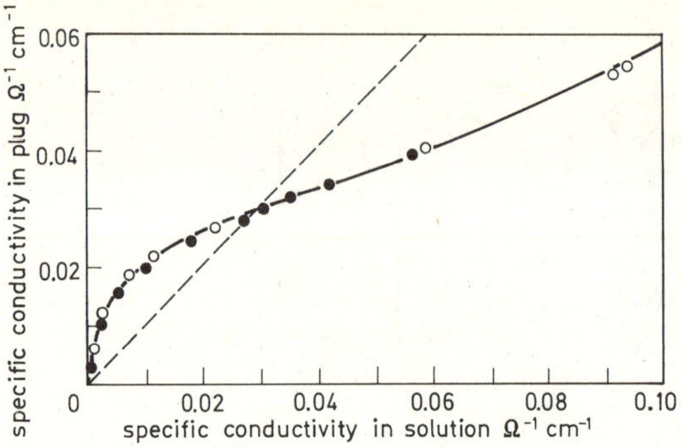

Figure 1.76 Specific conductivity of an exchanger plug as a function of the specific conductivity of the NaCl-solution being in equilibrium with the exchanger plug. ○ Dowex 50X8, ● Amberlite IR-120.

These values may as well be used for other spherical exchanger materials as an approximation. If the specific conductivity $\bar{\kappa}$ of the exchanger resin is known, the conductance of the plug can be calculated for each concentration of the solution by using these values.

The conductivity of the plug depends also on the solution conductivity, as shown in Figure 1.76. In case pure water fills the interstices of the plug ($\kappa \ll \bar{\kappa}$) the conductivity is due solely to the second element. But, as can be seen from the value of b, the conductivity is then low because of the small surface of contact of the spheres. By increasing the concentration of the solution the conductivities of the first and the third element increase rapidly. At the equiconductivity point, ($\kappa = \bar{\kappa}$), the conductivity of the plug is outpaced by that of the solution. At high concentrations of the solution ($\bar{\kappa} < \kappa$) the third element above all is responsible for the conductivity. The conductivity curve of the plug shows, above the equiconductivity point, an inflection point to higher conductivities which is explained by the fact that the conductivity $\bar{\kappa}$ of the exchange resin increases with the concentration of the solution. For the determination of the parameters of the conductivity of an ion exchanger column conductivity measurements such as those of Figure 1.76 are used. If the conductivity $\bar{\kappa}$ of the resin is not known its value can be obtained from the determination of the isoconductance point [373]. The measured value for $\kappa = 0$ gives b. From the differentiation of the above equation for κ_p with regard to κ it follows that the initial slope of the curve is $(a/e) + c$, and the slope of the curve at the equiconductivity point $cd + ae + ce^2$, when one neglects the concentration dependence of $\bar{\kappa}$. Thus, the three equations necessary are available. In further work on the same subject it was recommended that no electric contact be assumed between the channellar elements when the electrical conductivity is being modelled by the conductivity of these three elements. The effect of the geometric shape of ion exchanger grains on the electric conductivity of ion exchange columns was investigated as well. In evaluating the concept of three independent conductive circuits as described above, i. e., through resin, solution, and resin and solution, an expression was given for a real system including the specific electric conductivities through the column, the granular ion exchange resin, and the solution. Maxwell's equation for the limiting case of an exchanger's packing coefficient approaching zero was considered. The validity of the expression was tested experimentally by measuring the electric conductivity of ion exchange columns packed with a strong acid cation exchanger in the Na-, K-, or Mg-form or a strong base anion exchanger in the Cl-form in sodium chloride, potassium chloride or magnesium sulfate equilibrium solutions as a function of the solution concentration. Experimental and calculated results were compatible [374].

The electrochemical properties of ion exchange resin particles play a role as well in the elucidation of certain phenomena when the particles are used as model substances for the chemical investigation of colloids. The aggregation phenomenon, i. e., the interparticular interaction between cation and anion exchange resins when they are used for instance in a mixed bed, was observed in water purification apparatus. The aggregation, significant in extent, takes place when one mixes the H-form cation exchange particles with the OH-form anion exchange resin particles. In corresponding investigations it seemed that this aggregation was due to the electrostatic attraction force between the oppositely charged resin particles. This conclusion was supported by the fact that the aggregation was destroyed by the addition of salts and that the agglomerates are not formed in a non-polar medium in the complete absence of water. Moreover, it was demonstrated that the variation of the size ratio of particles influences markedly the aggregation behavior. In order to elucidate the mechanism of the aggregation phenomenon the electrokinetic potentials of ion exchange resins were measured by the method of streaming potential, and calculated according to the Helmholtz-Smoluchowsky equation. It was found that a cation exchange resin has higher zeta-potentials than an anion exchange resin and that the zeta-potentials are dependent on the ionic forms, especially on the valency of the counterions. The zeta-potentials are remarkably sensitive to the concentration of salt solutions. They decreased noticeably by the addition of 10^{-7} M sodium chloride and were scarcely detectable when the concentration of sodium chloride was increased to 10^{-3} M. It was concluded that the electrical double layer around the ion exchange resin particles is built up by the non-uniform ionic distribution inside and outside the particles and that the effective charge carried by the particles can be determined by a relation derived from Boltzmann's law. Values obtained for the resin Dowex 50WX8 in pure water were for the H-, Na-, Ca-, and Ba-form, 12.47, 21.59, 2.71, and 1.77, respectively, and for the resin Dowex 1X8 for the OH-, Cl-, and SO_4-form, 3.87, 1.24, and 0.92 mV, respectively. The conclusion was drawn that the difference in zeta-potentials for ions with the same valence might be related to the difference of the degree of dissociation [375]. Values of the effective charge carried by a particle of the resins examined were $1.58 \cdot 10^6$ for the univalent ions and $0.85 \cdot 10^6$ for divalent ions. These values seemed to be reasonable, taking into account the relation of the surface potential and the surface charge density. Other electrochemical properties investigated were the rheological behavior of suspensions of ion exchange resin particles with the electrophoretic mobility being measured by a microelectrophoretic technique in various potassium chloride solutions, and the Donnan potentials of single beads of ion exchanger which one tried to determine by puncturing single beads with microelectrodes [376]. It was reported that at $R_{el} \gg 1$, for the particles of an ion exchanger and for the non conducting particles with a thin double layer, the electrophoretic and the diffusion-phoretic mobility coincide and do not depend on the shape and the size of particles. The electrophoresis of Dowex 1 in the Cl-form dispersed in methanol and methanol-water mixtures was also studied and zeta-potentials were evaluated from the electrophoretic data obtained. The zeta-potentials were estimated for Zeokarb 226 in the H-form and Zeokarb 225 in the Na-, H- and Ba-forms in methanol-water mixtures from the corresponding data. These estimated potentials depended linearly on the applied electrical field during electrophoresis [377].

1.2.3.10 Behavior in nonaqueous solvents

In some cases nonaqueous solvents are treated with ion exchangers, particularly for purification and deionization purposes. Further, exchangers in nonaqueous solvents also serve as catalysts. These applications and the recovery of various compounds dissolved in solvents other than water have directed interest toward the behavior of ion resins in nonaqueous phases, leading to basic research in this area.

The early investigations of Bodamer and Kunin [378] showed that the behavior in nonaqueous solvents depends highly on the nature of the carbon chain of the exchange resin as well as on its porosity and crosslinking. In principle, ion exchange in nonaqueous solvents proceeds just as in water. However, since mass tranfer is slower, the flow rates in the columns must remain lower than in aqueous solutions. Ion exchange takes place more rapidly in polar solvents similar to water than in nonpolar ones. Provided that the pores of the resin are sufficiently large, so that exchange can take place at all, no measurable swelling of the resin needs to take place in nonaqueous solvents.

In tests with nonaqueous solvents, the exchanger is transformed into the acid and base or salt form by treatment with 10% hydrochloric acid or 4% sodium hydroxide solution and is washed to a neutral reaction. Cation exchangers in the Na-form are dried for 8 hours at 100 °C, while the other types are spread and air-dried for 48 hours at room temperature. This method does not, however, completely eliminate the water from the exchanger, and objections can be raised to the results obtained with it. It is, nevertheless, advisable to know the behavior of such pre-treated exchangers in nonaquenous solvents, since the large quantities of ion exchange resin needed in industrial plants cannot be subjected to prolonged and costly drying processes.

To evaluate the applicability of exchangers in nonaqueous media it is important firstly to investigate swelling, capacity, and selectivity. J. D. R. Thomas and other British researchers directed their attention to the solvent stability of ion exchange resins in the context of ion exchange in nonaqueous solvents. It was found that the observation in aqueous systems that salt forms of the resins are more able to withstand high temperatures than are free base and free acid forms also applies to basic resins in nonaqueous media like methanol and heptane. This general property extends to acidic resins in nonaqueous media as might also be the tendency for free base resins less resistant to capacity loss than the free acid types. Ion exchange resins are also degraded by normal erosive forces in nonaqueous solvents, but in organic media the hydrocarbon resin matrix may be expected to be especially sensitive to erosion. This could account for the tendency to particle attrition and emulsification at higher temperatures in methanol and heptane. The greater incidence of this feature in macroreticular resins lends support to the likelihood of this susceptibility of the hydrocarbon matrix. Nevertheless, stability studies at ambient temperature indicate that resins have adequate stability in nonaqueous media for a wide range of studies and applications [379].

The swelling of an ion exchange resin in a particular solvent depends on the dielectric properties of the solvent, and since the diffusion of various species in a solvent depends on the viscosity of the solvent, one should be able to predict the relative performance of ion exchange resins in a solvent from knowledge of the solvent's dielectric constant, the ionizing potential of the solvent and its viscosity. The swelling of ion exchange resins is ultimately associated with the ion exchange phenomenon, depending on factors such as capacity, crosslinking, nature of the counterions, nature and concentration of the solution in which the ion exchange resin is immersed, temperature and pressure.

For the determination of swelling in nonaqueous solvents, 30–40 ml of the dried ion exchanger is placed in a 100 ml graduated cylinder; subsequent shaking must be avoided. The volume is read with an accuracy of $\pm$ 0.5 ml. The exchanger is then covered with nonaqueous solvent up to the 100 ml mark. After 120 hours, the bed height of the exchanger is again read. The degree of swelling in percent is calculated from the final and the initial volumes. Table 1.19 and 1.20 show data on the swelling of some Amberlite resins in nonaqueous solvents. It can be seen that polar solvents produce more extensive swelling than nonpolar media. In hydrocarbons anion exchange resins swell more than cationic types.

From the numerical value of the dielectric constant, ion exchange resins should work no better in, e. g., liquid ammonia. However, the viscosity of this solvent is very low and provides evidence for using the ratio dielectric constant/viscosity as a measure for the performance of ion exchange resins in solvents other than water. The applicability of Dowex 50W, 1X8 and the cation exchanger Dowex ET-561 were tested in liquid ammonia at $-74\,^{\circ}$C and it was found that the swelling of the cation exchangers as well as of the anion exchanger is about the same as in water. This corresponds well to the D/η values of 8790 and 8590, respectively. For the

Table 1.19 Swelling of different Amberlite cation exchangers in nonaqueous solvents

Cation exchanger	% Swelling in						
	Water	Ethanol	Glycerol	Acetone	Glacial acetic acid	Benzene	Petroleum ether
IR-120 (H)	43	38	24	18	8	0	0
IR-120 Na	73	0	5	0	0	0	0
IR-112 Na	264	0		0		7	7
IR-105 (H)	107	100	120	73	55	0	
IR-105 Na	99	5	5	8	7	5	2
IRC-50 (H)	48	98		0	0	0	0
IRC-50 Na	202	0	6	1		3	1

Table 1.20 Swelling of different Amberlite anion exchangers in nonaqueous solvents

Anion exchanger	% Swelling in					
	Water	Ethanol	Acetone	Pyridine	Benzene	Petroleum ether
IRA-400 OH	37	63	25	20	18	5
IRA-400 Cl	45	63	20	28	11	5
XE-75 OH	220	140	40	20	7	5
XE-75 OH	200	130	23	32	3	3
IRA-401 OH	25	55	38		50	10
IR-4B OH	23	18	0	3	3	3
IR-4B Cl	73	5	0	3	3	0
IR-45 OH	31	52	40	50	35	15
IR-45 Cl	45	30	10	25	0	0
XE-76 OH	35	78	45		55	20

solvent dimethylsulfoxide it was found that swelling is slower but that the absolute increase in volume in water-free dimethylsulfoxide is higher than in aqueous media. Further, despite a great difference in dielectric constants, i. e., 19.9 for SO_2 and 78.3 for H_2O, the percent of swelling was the same in liquid sulfur dioxide as in water [380].

Tables 1.21 and 1.22 give information on changes in the capacity in nonaqueous solvents compared to aqueous media. Strontium petronate, the strontium salt of a sulfonated aliphatic hydrocarbon, was used to study purely cation exchange (Table 1.21). The values do not indicate maximum capacities since no excess of strontium petronate could be used because of its limited solubility; they represent the fraction of the added material that was exchanged. The acetic acid regain is of interest since acetic acid is bound to a greater extent than other substances, especially by benzene.

Table 1.21 Strontium-hydrogen exchange in nonaqueous solvents

Solvent	Added mval	Exchanged mequ/g resin		
		IR-105	IR-120	IRC-50
Ethanol	1.69	1.08	1.61	0.00
Acetone	0.59	0.27	0.53	0.00
Dioxane	1.85	0.00	1.26	0.00
Benzene	2.22	0.04	1.49	0.008

Table 1.22 Acetic acid sorption by different anion exchangers

Solvent	Added mval	Sorbed mequ/g resin				
		IR-4B	IR-45	XE-76	IRA-400	XE-75
Water	7.35	3.87	3.81	4.91	2.70	2.50
Ethanol	7.43	2.70	3.13	4.75	2.69	2.58
Acetone	7.50	0.33	3.08	4.10	3.01	2.52
Dioxane	7.71	0.49	4.06	4.06	2.12	1.94
Benzene	9.65	7.94	7.05	8.58	5.30	5.13
Total capacity		10.00	5.20	6.00	2.80	3.00

It has been postulated that several of the polar acetic acid molecules are arranged around a polar ionogenic group without forming a true ionic bond. The extraction of nitrogeneous bases from nonaqueous solutions by cation exchangers in the H-form was studied by using a solution of pyridine or o-phenanthroline in heptane and acetone and of quinoline in benzene. For the polycondensation resin KU-1 there was a clear dependence of the amount of uptake on the size of the granules. The polymerization resin KU-2 had a good capacity for the bases in all three solvents, the capacity being greater in the hydrocarbon solvents than in acetone. The sorption of acetic acid and its chloro derivatives was also studied extensively on a macroreticular cation exchange resin. The saturation capacity of acetic acid on the H-form was found to be about half the exchange capacity of the resin [381].

The selectivity of synthetic resin ion exchangers also differs in nonaqueous solvents and water. As an illustration, the reader is referred to the studies of Gable and Strobel [382], who investigated ion exchange equilibria in pure methanol. Higher selectivity coefficients were found than with water. It is assumed that ion solvation and ion pair formation are the causes. But the most comprehensive results on ion exchange equilibria on crosslinked Dowex 50W resin in anhydrous methanol are now available in a paper by D. Nandan and A. R. Gupta: the authors followed a suggestion of Marcus that important fundamental insight into ion exchange processes can be gained from studies of ion exchange equilibria in anhydrous nonaqueous systems, in particular the role of ion pairing in the resin can be clarified by a comparison of the exchange of ions in aqueous with that in anhydrous nonaqueous solutions. The selectivity coeffients for Li^+/H^+, Na^+/H^+, and K^+/H^+ cation exchange on Dowex 50WX4, Dowex 50WX8, and Dowex 50WX12 in anhydrous methanol at 25 °C were determined as a function of resin loading at constant ionic strength of 0.1 or 0.05 M. The thermodynamic equilibrium constants for Na^+ and K^+ are higher, and for Li^+ lower than in aqueous media. The free energies of transfer from water to methanol of the Dowex with different crosslinkings containing each of the four cations have been computed and compared with the free energy changes involved in the transfer of the chloride salts of these cations from water to methanol. The high selectivities observed for the potassium ion in the K^+/H^+ exchanges have been partly attributed to the ion pairing in the K^+-resinate in methanol [383]. But it is emphasizd that here "ion pair" should not be taken in the same sense as it is generally used in solution chemistry but should be interpreted the way Harris and Rice have used it in their statistical mechanical treatment of ion exchange. But the authors state that it might be generalized that ion pair formation is an important factor in these exchanges in nonaqueous media where one of the ionic forms of resinates involved in the exchange has a low free energy of transfer in going from aqueous to nonaqueous media. With respect to anion exchange in nonaqueous organic media the equilibrium coefficients and affinities of monovalent anions for the Wofatit SBW resin in methanol, ethylene glycol and dimethyl sulfoxide were determined, using specially derived equations. The affinities and equilibrium coefficients of the anions acetate, chloride, nitrite, perchlorate, bromide, chlorate, nitrate, iodide and rhodanide obtained on the OH-form of this resin of a particle size of 0.4 – 0.7 mm diameter at 20 °C with an exchange time of 30 min, in a medium of methanol ($K_B^A = 1.55-18.40$), ethylene glycol ($K_B^A = 1.35-6.75$) and dimethyl sulfoxide ($K_B^A = 0.02-0.04$) were investigated [384]. Using sulfonic acid styrene-divinylbenzene cation exchangers the ion exchange and isotope-effect of lithium in formamide, the affinity order for alkali metal/ammonium ions in methanol, ethanol, and acetone, the selectivity coefficients of alkali metals and alkaline earth ions including copper(II) and nickel(II) in liquid ammonia, which were found to be larger than in water, and the equilibrium constants in liquid sulfur dioxide have also been determined.

Since many organic compounds are not sufficiently soluble in water, a change to nonaqueous solvents is sometimes of advantage for analytical purposes. In addition to a number of other investigators, Inczédy and coworkers [385] have dealt with these questions. The authors studied the swelling behavior of strongly acidic cation resins and strongly basic anion exchangers in ethylene glycol, methanol, ethanol, dioxane, dimethylformamide, pyridine, and glacial acetic acid and applied these for the determination of several salts or organic acids and bases. Similar studies were also extended to macroporous ion exchangers and their applicability was demosntrated by the chromatographic separation of benzoic acid and anthranilic acid. Phipps [386] early used dimethylsulfoxide as the nonaqueous solvent for inorganic anions in a study which offers much additional information concerning ion exchange in nonaqueous solvents.

In order to elucidate the ion exchange-solvent system from a structural point of view nuclear magnetic resonance studies were carried out on a sulfonic acid styrene-divinylbenzene cation exchanger and a strong base anion exchanger containing the cations of hydrogen, lithium, sodium, potassium, rubidium, cesium, magnesium, calcium, strontium, and barium, and the anions fluoride, chloride, bromide, and iodide, respectively. The NMR spectra of cation exchangers containing 5–6 molecules of methanol per exchangeable ion were taken. The largest distribution of methyl and hydroxyl signals was observed for the H-form of the cation exchanger, providing evidence for the great mobility of alcohol particles in the ion exchanger phase. The smallest distribution was noted with the Li-form, due to a very stable solvation of these ions with alcohol. In the case of the remaining ions, no differences were noted in the position of methyl and hydroxyl bands. A similar behavior was observed for the anion exchanger except with fluoride ions, for which a displacement and splitting of the hydroxyl band took place towards weak fields, whereas the methyl band shifted towards strong fields. This is related to the different solvation character of the cations and anions [387]. The mechanism of methanol and propanol interaction with the H-forms of gel type (KU-2-4 and KU-2-20), and macroporous (KU-23) cation exchangers, and of methanol interaction with a weakly acid cation exchanger (KB-4) and a biionic polycondensation resin (KU-1) was studied by NMR at temperatures from room temperature down to $-105\,^{\circ}C$. The sorption study showed that at a maximum saturation 6 molecules of alcohol are sorbed onto each sulfonic acid group of the exchangers. The NMR spectra also reveal that only 2–3 alcohol molecules interact with one sulfonic acid group. On the side of the alcohols ROH_2^+ ions form on sorption. In the case of the macroporus exchanger, the sorption involves both protonated and non-protonated alcohol molecules. The presence of phenoxy radicals in the biionic polycondensation exchanger inhibits an interpretation of the NMR spectra of the sorption systems. More details on NMR studies of ion exchanger-solvent systems are to be found in the literature [388].

A few further examples of ion exchange behavior of inorganic and organic solute substances with ion exchange resins in nonaqueous solvents may be given. The distribution of magnesium, calcium, strontium, barium, gallium, indium, tin(II) and lead(II) was studied in a dimethyl sulfoxide media. The spectra of cobalt species sorbed onto Dowex 1X8, Cl-form, and Dowex 50WX8, Li-form, from acetone and ethanol solutions containing various concentrations of chloride ions were measured and interpreted. These spectra were compared to the spectra obtained by the sorption of cobalt ions from aqueous solutions of similar composition. From these studies an assignment of species sorbed from the nonaqueous media was possible [389]. During the extensive investigations of Marcus and coworkers on the anion exchange of metal complexes, the distribution of tracer concentrations of Re(VII), Cu(II), Cd(II) and Fe(III) between Dowex 1 and anhydrous methanol and ethanol solutions of hydrochloric acid and tetramethylammonium chloride, as well as the sorption of uranyl chloride from anhydrous methanolic, ethanolic and isopropanolic hydrogen chloride solutions was studied. The study of ion exchange in nonaqueous media was also extended to the anion exchange behavior of phenols on a De-Acidite FF resin in methanol as well as the sorption and separation of water-insoluble organic materials containing polar groups, for evaluating the role of macroreticular-macroporous resins for isolating polar constituents from petroleum fractions.

1.2.3.11 Behavior in mixed aqueous systems

The behavior of ion exchangers in mixed solvents seems in several respects to resemble that in nonaqueous media. As demonstrated by Bonner and Moorefield [390] in studies of silver-hydrogen exchange on Dowex 50 in ethanol-water and dioxane-water, the resin selectivity increases with the addition of an organic solvent to water, while the swelling volume of a resin in the H-form remains relatively constant in the mixed solvent. Kressman and Kitchener early published a note on equilibria between a nonaqueous solvent and a synthetic phenolsulfonate polycondensation resin, this being the first study under this aspect with a resinous exchanger. At that time the problem of equilibria was in the foreground. It was found that cation exchange occurs readily between the phenolsulfonate resin and cations in mixtures of water and nonaqueous solvents. The equilibria were found to obey the law of mass action, concentrations being used. But it was also found that more potassium ion passed into the solid phase in the $NH_4^+-K^+$ system in aqueous ethanol and aqueous acetone than in pure water. Gregor and coworkers then drew attention to the swelling behavior of polystyrenesulfonic acid resin in various forms in mixtures of water and methanol, ethanol, iso-propanol, and dioxane, finding that the more polar solvent, water, was sorbed preferentially, the degree of preference being greatest at low water content. The cationic state of the resin affected the selective sorption of the solvent to a marked degree. Swelling appeared to be determined largely by the dielectric constant of the sorbed solution, similar to that predicted by Bjerrum's theory of ion-pair formation. But Kressman and Kitchener, while speculating upon the state of ionization of the metal salts of the resin, had already pointed out that (according to the theory current then) incomplete dissociation of a salt in solution is not generally due to the existence of covalent linkages, but to the presence of ion pairs, triple ions, etc. It was assumed that a similar situation existed in the resin salts, which may, therefore, be regarded as completely ionized but almost entirely undissociated. For both macroreticular as well as gel type cation exchangers it was found later that water solvent partition in the two exchanger types in water-dioxane and water-iso-propanol was such that when the resins were initially saturated with water, the resin phase remained purely aqueous. In the presence of benzene or carbon tetrachloride, water remained fixed inside the resin. There have also been investigations of the chemical stability of ion exchange resins in mixed solvents, as well as on the effect of mixed solvents on ion exchange in mixed layers (bed) of ion exchangers.

More detailed data on the swelling of ion exchangers in mixed aqueous systems were presented for anion and cation exchangers, based on the observation that when 10% divinylbenzene crosslinked polystyrene resin in the hydrogen form was equilibrated for swelling with a 70% acetone in water mixture, the liquid taken up by the resin had the composition 72% water, 28% acetone, while the composition of the outer solution was 75% acetone and 25% water. With three sulfonated polystyrene resins, crosslinked with 2.25, 5.5 and 10% divinylbenzene, as well as for Zeo-Karb 225 and Dowex 2 the swelling data in water-acetone and water-dioxane mixtures were reported. The weights of water and acetone taken up by the dry hydrogen resins showed that for three cation exchange resins the water uptake fell continuously whereby for the 5.5% and 10% crosslinked resins the uptake rose fairly rapidly to a value that thereafter remained virtually constant. In the least strongly crosslinked resin, however, the acetone uptake showed a maximum when the molar fraction of acetone in the equilibrium solution had reached a value of approximately 0.5. For all resins it was concluded as well that about four moles of water per equivalent are firmly held by the resin, the remainder being readily displaced by acetone. In general it was found that the more highly crosslinked the resin and the greater its concentration of ionizing groups, the more strongly is acetone excluded from the resin phase. For the swelling of the anion exchangers in dioxane-water the same general feature was found, as the total swelling decreases only gradually until a molar fraction of 0.5 is reached in the equilibrium solution. Thereafter, decreasing amounts of both solvents are taken up, and the amount of swellling falls off rapidly. The plot of water uptake against the molar fraction in the resin phase was perceptibly curved. The rapid change in the slope of this line must occur somewhere below 0.15 g of water per gram of resin. The capacity of the resin was 3.8 mequ/g, which means that not more than two moles of water per equivalent are strongly held by the resin in its chloride form [391]. After several similar investigations as described above the swelling of a strong acid cation exchange resin in aqueous organic solutions of inorganic acids was studied. A resin fraction of 0.125 – 0.25 mm, hydrochloric acid, nitric acid and sulfuric acid of 0.2 N, and an organic solute, i. e., methanol, propanol, acetone, dioxane and dimethylformamide of 10 – 90%, were used for the experiments. The swelling of the resin increased with increasing dielectric permeability of the medium, while the nature of the acid did not influence the process. The swelling of dried weak acid carboxylic resin in the Li- and H-forms in water-methanol mixtures confirmed that also dry Li-forms of the cation exchanger swell only very slowly in methanol. In water-methanol of various concentrations it was found that the specific volume of the salt form is higher above a certain methanol concentration, and that of resins with divalent counterions in all methanol concentrations smaller than that of the H-form. More recent investigations on the selective swelling of ion exchangers in mixed solvents have shown that for a strong acid cation exchanger and a strong base anion exchanger in the H-, Cl-, NO_3-, SO_4-, and ClO_4-forms a preferential swelling by water occurs in mixed aqueous organic solvent mixtures. For the perchlorate form of the anion resin, the selective uptake of organic solvent was observed at $N_{methanol} < 0.2$ and at $N_{acetone} < 0.5$. Mineral acids decreased the selective uptake of water, especially in the case of resins in the sulfate form. This could be explained by the ability of the anion exchanger to take up mineral acids. The preferential sorption of water affected the sorption of nickel(II), cobalt(II), and copper(II). With respect to the same resins in various ionic forms an examination of the swelling in aqueous alcoholic mixtures with methanol, ethanol and propanol showed that maximum swelling occurs when the resin phase is enriched in alcohols, the reason for which lies in the maximum structural stability. For mixed media, changes in composition can change ion solvation almost completely along with the solution structure, i. e., from a water structure to an organic solvent structure. The invasion of lithium chloride into the anion exchanger Dowex 1X8, Cl-form, and the coincident swelling of the resin in aqueous methanol, ethanol, 1-propanol, and acetone was also measured. The data were compared with the invasion of hydrochloric acid into similar systems, and with the swelling in the absence of electrolyte, to elucidate the effect of the solvent composition on the invasion, and of the invasion on the total and selective swelling. The conclusion drawn was that the extent of ion-pairing of the invading electrolyte, and of the fixed ion at the exchange site, with the counterion, are the main factors determining the invasion and the swelling in the presence of the electrolyte [392].

A number of the studies of ion exchange performed in organic and mixed organic solvents have been done for the purpose of deducing principles of ion selectivity. However, mixed solvents have been preferred because, as a rule, equilibrium is reached very slowly, if at all, in strictly nonaqueous media. From investigations of ion exchange in mixed aqueous systems, it has become obvious that changes in the solvent system have a marked effect on relative ion selectivities. Alkali metal ion exchange on a strong acid sulfonic styrene-divinylbenzene cation resin in 0, 5, 10, 25, 50, 75, and 90% methanol-water, ethanol-water, and dioxane-water, and alkali metal barium ion exchange in acetic acid-water systems were investigated at 25 °C. In part, observed changes in the selectivity coefficient with solvent changes could be explained on the basis of dielectric constant effects on coulombic interactions, but the observed maxima and minima and changes in order of selectivity as the water content of the solvent was increased could be accounted for only by the electrostatic model, if it is assumed that ion solvation also changes in an appropriate manner. Some correlation between selectivity data and swelling data was noted. Radiotracer techniques had to be used to determine alkali metal exchange coefficients and gravimetric techniques for exchanges involving barium. From the solvent distribution data, there appeared to be no selective solvation of the resin phase [393]. The selectivity usually increases when one uses mixed aqueous organic solvents. To investigate this the importance of solvation in establishing the selectivity of a strong acid cation exchanger in the system Rb^+/H^+ and Mg^{2+}/H^+ in water-dimethylsulfoxide was investigated. The variations of the weight distribution coefficient, K_D, and the difference of the Walden products, $\delta(\Lambda^0\eta_0)$, of the exchanging ions as a function of the DMSO content were compared. The correlation between the K_D and the Walden product curves was excellent up to a DMSO concentration of 70%. It was concluded that solvation determines the selectivity in solvents that consist mainly of water, but not in a solvent that consists mainly of DMSO [394]. An extensive investigation on anion exchange in mixed solvents of univalent anions on the anion exchanger Wofatit SBW either with protonic solvents like alcohols, i. e., methanol, ethanol, propanol, and glycol, or dipolar aprotonic solvents, i. e., acetone, dimethylsulfoxide, and acetonitrile, by conductivity measurements at 20 °C for 30 minutes after addition of each of ten anions in an 80 : 20 and 20 : 80 mixture with water revealed interesting results for the relative affinities of the anions for the resin. The narrower range of values and the partial reversal of anion affinities for the resin was thought to be due to the composition of the ordered hydrate shell of the anions in solution with the counterions. The dielectric constant and dipole moment of the organic solvent influences the exchange reactivity of the exchanger. But no direct correlation between these parameters was found. Perturbed anions can be removed much faster from the resin [395]. The idea – developed earlier for an explanation of the special selectivity conditions in mixed dioxane-water system – that ion exchange resin selectivity occurs as the result of a competition of the exchanging ions for that phase providing the best solvation, (a competition won by the ion most in need of solvation, which is usually that ion with the highest charge density, the most acidic cation, or the most basic anion) has since been applied to the water-dioxane system. Usually, the diluted external aqueous phase provides the better solvating medium, so the smaller ion or, for ions of different structure, the more basic anion, goes to the external aqueous phase, relegating the other ion to the resin phase. But as dioxane replaces water, the mixed external solution becomes a poorer solvating agent, and since the resin phase preferentially takes up water, the latter becomes relatively a better solvating phase. Thus, the ion most in need of solvation tends to avoid the external solution; as a result separation factors should get smaller, which is what is observed experimentally. In fact, by properly choosing the nature of the macro-electrolyte and of the resin group, the normal aqueous-phase selectivity order of the halides can be reversed in dioxane-water mixtures [396].

There seems to be a remarkable behavior involving the kinetics of the uptake of ions by ion exchangers in that the internal diffusion coefficient increases with increasing particle size. This was presented for the uptake of copper(II) from a 0.1 N copper chloride solution in a 60 : 40 ethanol-water mixture by an anion exchange resin. The coefficient of internal diffusion, D_s, estimated by the Boyd method, increased from $1 \cdot 10^{-9}\ cm^2 \cdot s^{-1}$ for 0.2 mm grains to $4 \cdot 10^{-9}$ for 0.75 mm grains.

The behavior of many ions with strong acid and strong base exchangers differs considerably in mixed organic-aqueous systems compared with aqueous media. The large number of data available concerning the exchange behavior of cations and anions from aqueous and inorganic acid media have been supplemented by an extended series of studies on ion exchange in mixed organic-aqueous systems. The aqueous component can even consist primarily of organic acids of different normality which contain additions of various miscible organic solvents. The differences become apparent by the fact that the exchange capacity of a given metal ion increases with an increasing solvent concentration at constant acidity and in the presence of one and the same organic solvent. This is true for cation as well as anion exchangers and differs as a function of the inorganic acids and organic solvent present, although in many cases a dependence on the dielectric constant also exists. Pronounced differences in the exchange behavior of elements compared with pure aqueous media have been observed, particularly in mixtures with high concentrations of acetone or tetrahydrofuran and hydrochloric acid.

A large amount of data exists on the exchange behavior of ions in systems with methanol, ethanol, n-propanol, iso-propanol, n-butanol, iso-butanol, isomeric amyl alcohols, acetone, tetrahydrofuran, dioxane, ethyl glycol, methylethyl ketone, methyl-n-propyl ketone, diethyl ketone, ether, diisopropyl ether, acetic acid, propionic acid, mono-, di-, and tri-chloroacetic acid; much of this data was published by Korkisch and coworkers [397], and has been used mainly for chromatographic separations. However, the distribution coefficients determined on this basis probably can also be used for other purposes, e. g., in metallurgy or purifications [398].

If the concentration of organic solvent is high enough, one arrives at a range in which the advantages of ion exchange and solvent extraction become effective at the same time, as demonstrated by Korkisch [399]. This can be used as a basis for a new procedure combining the advantages of both techniques. Korkisch assumed as a theoretical explanation that the organic solvent, such as tetrahydrofuran, forms an oxonium salt with hydrochloric acid

$$\begin{matrix} CH_2-CH_2 \\ | \quad\quad\quad\ddot{O}| + H^+ Cl^- \\ CH_2-CH_2 \end{matrix} \rightleftharpoons \left[\begin{matrix} CH_2-CH_2 \\ | \quad\quad\quad\ddot{O}\longrightarrow H \\ CH_2-CH_2 \end{matrix} \right]^+ Cl^-$$

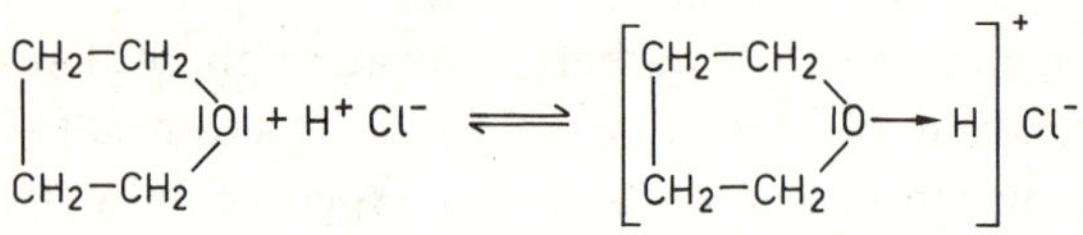

which becomes active as a liquid ion exchanger and undergoes an ion exchange reaction of the following type, for example, with the anionic Fe(III) chloride complex

$$\left[\begin{matrix} CH_2-CH_2 \\ | \quad\quad\quad\ddot{O}\longrightarrow H \\ CH_2-CH_2 \end{matrix} \right]^+ Cl^- + H^+ FeCl_4^-$$

$$\rightleftharpoons \left[\begin{matrix} CH_2-CH_2 \\ | \quad\quad\quad\ddot{O}\longrightarrow \\ CH_2-CH_2 \end{matrix} \right]^+ FeCl_4^- + H^+ Cl^-$$

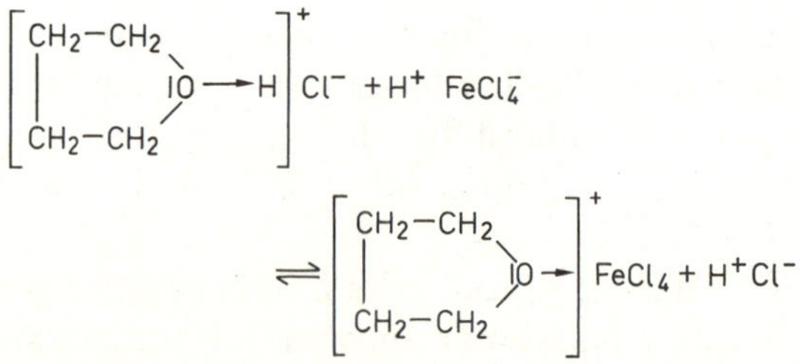

Here an ion association is formed with very little dissociation. The liquid anion exchanger competes with the solid cation or anion exchanger and, depending on the experimental conditions, either the liquid or the solid exchanger preferentially sorbs the metal ions. It can be assumed that this principle can be employed to solve many analytical and radiochemical problems by the use of other organic solvents or solvent mixtures and different inorganic or organic acids and bases, as well as of chelating complexants (Korkisch). Since then, however, divergent experimental results have been obtained. These imply that the mechanism of the combined system has been explained by Korkisch only for the behavior of Fe(III), Co(II), VO_2^{2+}, Th(IV) and Bi(III), which depends on solvent concentration and acidity. From experimental data with water-dioxane, water-ethanol, and water-acetone systems – which is presented in tables and diagrams as optical absorption curves – it is concluded that the proposed mechanism does not agree with the facts and that the principal factor to be considered is the action of the different organic solvents upon the ion exchange resins, which is especially noticeable in the modification of their swelling [400].

References

[1] R. Grießbach (1939). Über die Herstellung und Anwendung neuerer Austauschadsorbentien, insbesondere auf Harzbasis. Beihefte Z. Ver. Dt. Chem. Nr. 31, 1.

[2] The chemistry of functional groups (1964). (Patai, ed.) Wiley, New York.

[3] Das wissenschaftliche Werk von Hermann Staudinger (1969–1976). (M. Staudinger and H. Hopf, eds.). Hüthig and Wepf, Basel-Heidelberg.

[4] W. Kern and R. C. Schulz (1963). Methoden zur Umwandlung von natürlichen und synthetischen makromolekularen Stoffen. Houben-Weyl 14/2, 637.

[5] P. Schneider (1963). Umwandlung von Polymerisaten. Houben-Weyl 14/2, 661.

[6] D. Braun (1975). Reaktionen makromolekularer Stoffe. Kunststoff-Handbuch Bd. 1, 136. Hanser, München.

[7] Chemical reactions of polymers (1964). (E. M. Fettes, ed.) Interscience, New York.

[8] P. Rempp (1976). Recent results on chemical modification of polymers. Pure Appl. Chem. *46*, 9.

[9] N. A. Platé (1976). Problems of polymer modification and the reactivity of functional groups of macromolecules. Pure Appl. Chem. *46*, 49.

[10] G. G. Cameron (1980). Reactions of polymers: polymer modification. Macromolecular Chemistry, Vol. 1, SPR.

[11] H. Lüssi (1966). Beziehungen zwischen Struktur und Polymerisierbarkeit von Monomeren. Chimia *20*, 379.

[12] K. Dorfner (1961). Redoxaustauscher. Chemiker-Ztg. *85*, 80, 113.

[13] E. Bayer (1962). Struktur und Spezifität bei organischen Komplexbildnern. Synthese, Eigenschaften und Anwendung makromolekularer Komplexbildner. Anomalien bei Ionenaustauschvorgängen. (K. Issleib, ed.). Akademie-Verlag, Berlin.

[14] K. Dorfner (1977). Gegen Steinablagerungen und Korrosion. Chemie-anlagen + verfahren, 73.

[15] T. Endo et al. (1983). Syntheses of some novel polymers containing the viologen structure and their application to electrontransfer reactions in heterophases. Macromol. *16*, 881.

[16] L. Luisi (1979). Why are enzymes macromolecules? Naturwiss. *66*, 498.

[17] G. Wulff et al. (1977). Enzyme-analogue built polymers. Makromol. Chem. *178*, 2799. M. Mutter (1985). Die Konstruktion von neuen Proteinen und Enzymen – eine Zukunftsperspektive? Angew. Chem. *97*, 639.

[18] H. Ringsdorf et al. (1981). Pharmakologisch wirksame Polymere. Angew. Chem. *93*, 311.

[19] A. Patchornik (1982). Some novel developments in the use of polymeric reagents multiphase and multistep reactions. Nouveau J. Chim. *6*, 639.

[20] Polymer-supported reactions in organic synthesis (1980). (P. Hodge and D. C. Sherrington, eds.) Wiley, New York. N. K. Mathur, C. K. Narang and R. E. Williams (1980). Polymers as aids in organic synthesis. Acadamic Press, New York.

[21] G. Manecke (1977). Enzyme immobilization. Polym. Prepr. Am. Chem. Soc. Div. Polym. Chem. *18*, 535. L. Goldstein and G. Manecke (1976). The chemistry of enzyme immobilization. In Appl. Biochem. Bioeng. Vol. 1. Academic Press, New York.

[22] N.M.Weinschenker, G. A. Crosby (1975). Polymer reagents in organic chemistry. Ann. Rep. Med. Chem. *11*, 281. M. L. Hallensleben (1973). Polymere Reagentien. Angew. Chem. *85*, 457.

[23] D. P. N. Satchell (1977). The classification of chemical reactions. Naturwiss. *64*, 113.

[24] D. C. Sherrington (1982). The effect of polymer structure on the reactivity of bound functional groups. Nouveau J. Chim. *6*, 661.

[25] T. R. E. Kressman (1975). La synthese des resines echangeuses d'ions. Nouvelle tendances. Inf. Chim. *142*, 133.

[26] J. Diehlmann (1968). Dreißig Jahre Kunstharz-Ionenaustauscher. Plenary lecture of the symposium 30 Jahre Kunstharz-Ionenaustauscher, Leipzig, June 4–7, 1968. Publ. 1970. Akademie-Verlag, Berlin.

[27] G. Kühne (1967). Entwicklung und Anwendung von Ionenaustauschern. Chem. Ind. XIX/Juli 1967.

[28] H. Staudinger and E. Husemann (1935). Über hochpolymere Verbindungen. 116. Mitt.: Über das begrenzt quellbare Poly-styrol. Ber. deutsch. Chem. Ges. *68*, 1618.

[29] G. F. D'Alelio (1944). Process for removing cations from liquid media. US Pat. 2 340 111, Jan. 25, 1944.

[30] W. C. Bauman and J. Eichhorn (1947). Fundamental properties of a synthetic cation exchange resin. J. Am. Chem. Soc. *69*, 2830.

[31] R. Kunin and R. J. Myers (1947). The anion exchange equilibria in an anion exchange resin. J. Am. Chem. Soc. *69*, 2874.

[32] R. Kunin and R. E. Barry (1949). Carboxylic, weak acid type, cation exchange resin. Ind. Eng. Chem. *41*, 1269.

[33] B. Kolarz et al. (1974). Ion Exchangers. XVI. Removal of inhibitors from monomers on ion exchangers. Polimery (Warsaw) *19*, 417 (Pol).

[34] Divinylbenzene DVB-22 and DVB-55. Technical bulletin June 1954. The Dow Chemical Company, Midland, Mich., USA.

[35] R. H. Wiley and R. M. Dyer (1964). J. Polymer Sci. *A2*, 2503. R. H. Wiley, G. DeVenuto and T. K. Venkatachalam (1967). J. Gas Chromatogr. *5*, 590.

[36] G. Schwachula et al. (1967). Über die Darstellung von 1, 4-Divinylbenzol. Z. Chem. 7, 460. (1968). Die präparative gas-chromatographische Isolierung von m-Äthylstyrol und m-Divinylbenzol aus technischem Divinylbenzol. Chem. Techn. *20*, 622. (1978). Über die Isolierung von para- und meta-Divinylbenzol aus technischem Divinylbenzol. Chem. Techn. *30*, 144, 363.

[37] R. H. Wiley, K. S. Kim and S. P. Rao (1971). Microbead polymerization of styrene-divinylbenzene with sulfated poly(vinyl alcohol) suspending agent. J. Polymer Sci. *A1*, *9*, 805.

[38] J. Salat (1978). Ion exchange resins of homodispers divinylbenzene-styrene bead-polymer base. Hung. Teljes 15 875, 28 Dec 1978.
[39] G. Popov et al. (1983). Method and apparatus for preparing large-grained adsorbents. Ger (East) DD 158 906, 9 Feb 1983. Coarse grained ion exchange resins. Ger (East) DD 158 907, 9 Feb 1983.
[40] K. Dusek (1962). Ionenaustauschergerüste III. Kopolymere des Styrols mit Divinylbenzol. Elastisches Verhalten der in Toluol gequollenen Kopolymeren. Coll. Czechosl. Chem. Comm. *27*, 2841.
[41] K. P. Govindam and N. Krishnaswamy (1972). Porous polymers. Indian J. Technol. *10*, 63.
[42] V. V. Kafarov and I. N. Dhorokov (1977). Diagram principle for describing physicochemical systems. Gidrodin. Yavleniya Perenosa Dvukhfaznykh Dispersnyk Sist. 1977, 3. Irkutskii Politekh. Inst. Irkutsk, USSR.
[43] R. H. Wiley (1975). Crosslinked styrene/divinylbenzene network systems. Pure Appl. Chem. *43*, 57.
[44] G. Schwachula (1978). Der strukturbestimmende Einfluß des Divinylbenzols auf die Eigenschaften von Polymerisationsionenaustauscherharzen. Z. Chem. *18*, 242. G. Schwachula and G. Popov (1982). Influence of the structure of the matrix on the properties of ion exchangers. Pure Appl. Chem. *54*, 2103.
[45] P. S. Belov et al. (1973). Synthesis and comparative study of the properties of sulfonic cation exchangers containing copolymers of styrene and divinylbenzene isomers. Zh. Prikl. Khim. (Leningrad) *46*, 2031.
[46] R. N. Volkov et al. (1984). Optimization of the production of divinylbenzene. Prom-st. Sintek. Kauchuka, Shin i Rezinotekhn. Izdelii, Moskva 1984, 7.
[47] H. Corte and A. Meyer (1957). Herstellung von Anionenaustauschern mit Schwammstruktur. DAS 1 045 102, March 9, 1957. Farbenfabriken Bayer AG, Leverkusen.
[48] E. F. Meitzner and J. A. Oline (1958). Crosslinked copolymers for use in ion exchange resins. Brit. Pat. 932 125, July 24, 1963. US Appl. July 18, 1958. Brit. P. 932 126, divison of above. Fr. P. 102 478, 1962. Union of S. Africa Pat. Appl. 59/2393.
[49] T. R. E. Kressman and J. R. Millar in 1958 at the Permutit Company London. From T. V. Arden's lectures and private communication.
[50] D. C. Sherrington (1982). Small-scale production of macroreticular polystyrene beads. Macromol. Synth. *8*, 69.
[51] J. Seidl et al. (1967). Makroporöse Styrol-Divinylbenzol-Copolymere und ihre Verwendung in der Chromatographie und zur Darstellung von Ionenaustauschern Fortschr. Hochpolym.-Forsch. *5*, 113.
[52] L. D. Morse et al. (1971). US Pat. 3 627 708, Dec 14, 1971, Appl. 750 254, Aug. 05, 1968.
[53] Y. Fuchiwaki et al. (1971). Porous crosslinked copolymers useful as ion-exchange resins. Fr. Demande 2 044 630. Mitsubishi Chemical Industries Co. Ltd., 02 Apr. 1971.
[54] B. V. Moskvichev et al. (1974). Selectivity of trimethylammonium ion sorption on the telogenated sulfonic cation exchanger SVD-30-T. Zh. Fiz. Khim. *48*, 656.
[55] M. Mindick and J. Swarz (1965). Ion-exchange resin particles of good porosity and strength. US Pat. 3 549 562, 22 Dec 1970.
[56] Z. Vasicek (1974). A contribution to the study of formation of macroporous copolymers. Proc. 3rd Symp. on Ion-Exchange, Balatonfüred, 28-31 May 1974, p 393. Z. Vasicek et al. (1974). Macroporous copolymers. Czech. Pat. 169 290, 15 May 1977.
[57] B. Kolarz et al. (1977). Ion exchangers XVIII. Application of the factorial design for the determination of the parameters for preparing double porous styrene-divinylbenzene copolymers. Chem. Stosow. *21*, 183 (Pol).

[58] K. A. Kun and R. Kunin (1964). Pore structure of some macroreticular ion exchange resins. J. Polymer Sci. B2. Polymer Lett. *2*, 687. Id. (1967). The pore structure of macroreticular ion exchange resins. J. Polymer Sci. C2. Polymer Symp. *16*, 1457. Id. (1968). Macroreticular resins. III. Formation of macroreticular styrene-divinylbenzene copolymers. J. Polymer Sci. A1. Polym. Chem. *6*, 2689. F Martinola und A. Richter (1970). Macroporöse Ionenaustauscher und Adsorbentien zur Aufbereitung organisch belasteter Wässer. Jahrbuch vom Wasser, 37. Bd., 1.

[59] I. M. Abrams (1956). High porosity polystyrene cation exchange resins. Ind. Eng. Chem., *48*, 1469.

[60] F. X. McGarvey and M. C. Gottlieb (1979). Fundamentals and bacic theories regarding the proper use of modern day ion exchange resins. Proc. 40th Annual Int. Water Conf., Pittsburgh, p 175.

[61] J. I. Szanto (1963). Untersuchung der Synthese einiger Anionenaustauscherharze auf Basis *a*-Methylstyrol. Symp. Ionenaustauscher und ihre Anwendung. Balatonszeplak, Hung. 1963, 29.

[62] D. Livinston et al. (1956). Poly (α, *β*, *β*′-trifluorostyrene). J. Polymer Sci. *20*, 485.

[63] R. B. Hodgon, jr. (1968). Polyelectrolytes prepared from perfluoroalkylaryl macromolecules. J. Polymer Sci. A1. *6*, 171.

[64] A. V. Tevlina et al. (1972). Synthesis of α, *β*, *β*′-trifluorostyrenedivinylbenzene copolymers and ion exchangers based on them. Plast Massy 1972, 10 (Russ). E. I. Kraevskaya et al. (1983). Synthesis and study of polymer matrices based on α, *β*, *β*′-trifluorostyrene. Ionoobmen Materialy, M. 1983, 46 (Russ).

[65] W. Grot (1972). Perfluorierte Ionenaustauscher-Membrane von hoher chemischer und thermischer Stabilität. Chem.-Ing. Techn. *44*, 167. Id. (1975). Perfluorierte Kationenaustauscher-Polymere. Ibid. *47*, 61. Id. (1978). Use of perfluorosulfonic acid products as separators in electrolyte cells. Chem. Ing. Techn. *50*, 299.

[66] J. Surowiec (1984). New field of application for fluoropolymers manufacture of ion-exchange materials. Polimery (Warsaw) *29*, 302.

[67] A. Eisenberg and H. L. Yeager (1982). Perfluorinated ionomer membranes. ACS Symposium Series No. 180.

[68] E. Schmidt (1979). Ionenaustauscher für kontinuierliche Systeme. Vom Wasser, 53. Bd., 179.

[69] J. J. Wolff (1984). A new generation of anion exchangers and adsorbents: heavy resins. Reactive Polymers *2*, 13. M. N. Prajapati et al. (1986). Studies on high density anion exchangers. Reactive Polymers *4*, 205.

[70] M. A. Umarova and M. A. Askarov (1972). Ion exchange resins from products of the chemical reaction between styrene and maleic anhydride. Tr. Tashk. Politekh. Inst. *90*, 107.

[71] Th. J. Howell (1977). Hard crosslinked distinct copolymer globules. US Pat. Appl. 809 957, 27 Jun. 1977.

[72] G. Schwachula et al. (1967). Styrol-Divinylbenzol Kopolymere. I. Beständigkeit und Quellverhalten von Styrol-Divinylbenzol Kopolymerisaten. Plaste Kaut. *14*, 802. Id. (1967). II. Über den Einfluß verschiedener Zusätze auf das Quellverhalten von Styrol-Divinylbenzol-Kopolymerisaten. Ibid. *14*, 879. Id. (1968). III. Der Einfluß von p-Divinylbenzol auf das Quellverhalten in Styrol-Divinylbenzol-Kopolymerisaten Ibid. *15*, 33. F. Wolf et al. (1969). Dielektrische Messungen an Styrol-Divinylbenzol-Kopolymerisaten mit polaren Zusätzen. Ibid. *16*, 602. N. N. (1985). High-strength ion-exchange resins. Jpn. Kokai Tokkyo Koho JP 60 81 230 (85 81 230), Appl. 11 Oct 1983. Mitsubishi Chemical Industries Co. Ltd.

[73] T. R. E. Kressman and E. I. Akeroyd (1950). Improvements relating to the production of ion exchange resins. Brit. Pat. 694 778, July 29, 1953.

[74] J. A. Dale and J. R. Millar (1981). Cross-linker effectiveness in styrene copolymerization. Macromol. *14*, 1515.

[75] B. N. Truskin et al. (1968). Strongly basic anion-exchange resins from copolymers of styrene and dimethacrylic esters of some phenols. Tr. Mosk. Khim.-Tekhnol. Inst. No. 57, 124.

[76] E. E: Ergozhin et al. (1970). Macroreticular copolymers of styrene with N, N′-alkylenedimethacrylamides for the synthesis of ion exchangers. Izv. Akad. Nauk. Kaz. SSR, Ser. Khim. *20*, 44. Id. (1971). Anion exchangers. USSR Pat. 379 585, 20 Apr. 1973. Id. (1977). Study of the sorption of uranyl (2+) ion by ion exchangers of macroreticular structure. Ibid. *27*, 30.

[77] V. S. Yurchenko et al. (1968). Symp. Selective ion exchange sorption of antibiotics. Tr. Leningrad. Khim.-Farm. Inst. No. 25, 121. V. A. Dinaburg et al. (1968). Zhur. Prikl. Khim. *41*, 891. B. N. Truskin et al. (1968). Zhur. Prikl. Khim. *41*, 1293. K. B. Musabekov et al. (1970). Effect of bridge-forming components and counterions on the hydration of macroreticular sulfonic resins. Zhur. Fiz. Khim. *44*, 991.

[78] G. Schwachula and D. Lukas. (1974). Study of the preparation of strong acid cation exchangers based on copolymers of styrene and bis-(4-vinylphenyl) methane. Proc. 3rd Symp. on Ion-Exchange. 28-31 May, 1974. Balatonfüred, Hung.

[79] A. S. Tevlina et al. (1974). Ion-exchange resins by copolymerization of mono- and divinyl compounds and subsequent addition of ionic groups. USSR Pat. 521 288, 15 Jul. 1976.

[80] E. E. Ergozhin et al. (1976). Synthesis and study of new sulfonated cation exchangers based on styrene copolymers with some diisopropenylbenzene. Izv. Akad. Nauk. KazSSR, Ser. Khim. *26*, 35.

[81] A. K. Svetlov et al. (1982). Ion exchangers crosslinked by bis (1, 3-butadien-1-yl) sulfide. Zh. Prikl. Khim. (Leningrad) *55*, 2600.

[82] H. Stach (1955). Quellfeste Kationenaustauscher durch Vernetzung linearer Polysulfonsäuren. Angew. Chem. *67*, 786).

[83] R. V. Martsinkevich et al. (1975). Some sorption and selective properties of macroreticular isoporous ion exchangers based on polystyrene. Rasshir. Tezisy Dokl-Vses. Simp. Termodin. Ionnogo Obmena, 2nd 1975, 217.

[84] M. J. Hatch (1959). Anion- and chelate-exchange resins. US Pat. 3 013 994, Dec. 19, 1961. L. R. Morris (1959). Sulfonium-derived anion-exchange and chelating resins. US Pat 3 037 944, Jun. 5, 1962.

[85] J. H. Barrett (1973). Macroreticular vinyl benzyl chloride polymers. US Pat. 3 843 566, 22 Oct. 1974. M. A. Askarov et al. (1974). Synthesis and study of anion exchange resins consisting of copolymers of vinyl benzyl chloride with divinylbenzene. Uzb. Khim. Zh. *18*, 48.

[86] M. A. Askarov et al. (1976). Anion exchangers based on copolymers of vinyl benzyl chloride with m-diisopropenylbenzene. Vysokomol. Soedin., Ser. B *18*, 721. Id. (1972). Synthesis and study of vinyl benzyl chloride-acrylonitrile copolymers. Tr. Tashk. Politekh. Inst. *90*, 109.

[87] Kh. Rakhmatullaev et al. (1973). Polymerization and copolymerization of vinyl benzyl bromide. Tr. Tashk. Politekh. Inst. *107*, 135.

[88] S. F. Reed and D. L. Hundermark (1978). Polymer beads containing vinyl benzyl alcohol units and methods for thermally crosslinking them to form adsorbents and ion exchange resins. Eur. Pat. Appl. 7 791, Feb. 6, 1980.

[89] I. Gros (1970). Ion exchangers. Rom. Pat. 56 783, 30 Apr. 1974.

[90] F. Döscher et al. (1982). Sulfomethylierte Styrol-Divinylbenzol-Harze. Makromol. Chem. *183*, 93.

[91] D. Dolar et al. (1965). Ion exchange properties of a polyacenaphthylene-divinylbenzene sulphonic acid resin. Makromol. Chem. *84*, 108.

[92] A. I. Nikitina et al. (1971). Effect of matrix structure on the thermal stability of sulfonic-type cation exchangers in water. Tr. Tambovskogo Inst. Khim. Mashinostr. No. 7, 134.

[93] M. Biswas and K. J. John (1978). A cation exchange resin from poly(N-vinyl carbazole). Angew. Makromol. Chem. *72*, 57.

[94] V. P. Lopatinskii et al. (1971). Chemistry of carbazole derivatives. 49. Copolymerization of 9-vinylcarbazole with styrene and divinylbenzene. Izv. Tomsk. Politekh. Inst. *175*, 17.

[95] J. Pielichowski and E. Morawiec (1976). Chemical modification of poly(N-vinyl carbazole). J. Appl. Polym. Sci. *20*, 1803.

[96] A. H. Greer et al. (1958). New ion exchange resin for uranium recovery. Ind. Eng. Chem. *56*, 160. A. H. Greer (1957). Anion exchange resins from vinylalkyl, nitrogen-containing heterocyclic polymers. US Pat. 2 801 223, 30 Jul. 1957.

[97] K. M. Saldadze et al. (1969). Sorption of copper cations on vinylpyridine-based anion-exchange resins. Khim. Aktiv. Polim. Ikh. Primen. 1969, 87. Id. Chemical stability of vinylpyridine-based anion-exchange resins at high temperatures. Ibid. 193. Edited by K. M. Saldazze. Izd. Khimiya Leningrad. Otd: Leningrad, USSR.

[98] K. M. Saldadze et al (1970). Rate of cation adsorption by some anion-exchangers of the vinylpyridine series. Zh. Anal. Khim. *25*, 1462. Id. (1970). Effect of the ionic composition of vinylpyridine anion-exchangers on the sorption of cations of transition metals. Ibid. *25*, 1698. Id. (1974). Porous structure of copolymers of 2-methyl-5-vinylpyridine and divinylbenzene modified by a solvent. Kolloidn. Zh. *36*, 643. Id. (1975). Study of the sorption of metals by pyridine containing ion exchangers. Rasshir. Tezisy Dokl.-Vses. Simp. Termodin. Ionnogo Obmena, 2nd, p 74.

[99] D. H. Freeman et al. (1973). Interactive gel networks. Chromatographic and analytical properties with a pyridine funtional group. Anal. Chem. *45*, 768.

[100] L. N. Prodius et al. (1976). Macroreticular anion exchangers based on copolymers of 2-methyl-5-vinylpyridine with some dienes. Izv. Akad. Nauk Kaz. SSR, Ser. Khim. *26*, 70.

[101] N. Krishnaswamy et al. (1965). Furfural-based ion-exchange resins. I. Preparation and properties of a cation exchanger from furfural-styrene reaction products. J. Appl. Polymer Sci. *9*, 2655. Id. (1966). II. Cation-exchange resin from furfural-styrene reaction product. Indian J. Technol. *4*, 208. Id. (1972). Phosphorus-containing ion-exchange resins. I. Synthesis and characterization. Ibid. *10*, 185. Id. (1972). II. Equilibrium studies and some analytical applications. Ibid. *10*, 189. Id. (1975). Styrene-furfural-based weak-base anion exchanger. J. Appl. Polymer Sci. *19*, 2331.

[102] B. R. Green and E. Jaskulla (1984). Poly(vinylimidazole) – a versatile matrix for the preparation of chelating resins. In: Ion Exchange Technology. (D. Naden and M. Streat, eds). Horwood Ltd., Chichester, 1984, p 490. M. Chanda et al. (1985). Sorption of phenolics and carboxylic acid on polybenzimidazole. Reactive Polymers *4*, 39.

[103] R. Kunin and S. Fisher (1962). Effect of cross-linking on the properties of carboxylic polymers. II. Apparent dissociation constants as a function of the exchanging monovalent cation. J. Phys. Chem. *66*, 2275. A. Chatterjee and J. A. Marinsky (1963). Dissociation of methacrylic acid resins. J. Phys. Chem. *67*, 41.

[104] D. E. Weiss et al. (1966). Sirotherm. II. Properties of weakly basic resins. Aust. J. Chem. *19*, 561. III. Properties of weakly acidic ion-exchange resins. Ibid. *19*, 589. B. A. Bolto et al. (1976). Further rapidly reacting ion-exchange resins. J. Polymer Sci. Polym. Symp. *55*, 87. Id. (1976). New composite ion exchange resins made from neutral precursors. Ibid. *55*, 95.

[105] B. A. Bolto et al. (1984). Improved resins for ion exchange whith thermal regeneration. Reactive Polymers *2*, 5.
[106] D. E. Weiss et al. (1971). Ion exchange process. US Pat. 3 506 378.
[107] N. V. Blesing et al. (1970). Some ion exchange processes for partial demineralization. Ion-Exchange in the Process Industries. Soc. of Chem. Ind., London, p 371.
[108] Yu. N. Svyadoshch et al. (1971). Ferritization of strongly acidic polymerization sulfo cation exchangers. Tr. Nauch.-Issled. Proekt. Inst. Obogashch. Rud Tsvet. Metal. No. 5, 49.
[109] B. A. Bolto et al. (1976). Continuous ion exchange using magnetic shell resins. I. Dealkalization – laboratory scale. Ion Exch. Memb. *2*, 41.
[110] D. E. Weiss (1984). Private communication.
[111] K. Takeuchi et al. (1978). Magnetic ion-exchange resin. Jpn. Kokai Tokkyu Koho 78 146 986, 21 Dec.
[112] B. A. Bolto (1975). Sirotherm desalination. Ion exchange with a twist. Chem. Technol. *5*, 303.
[113] H. P. Gregor et al. (1965). Oleophilic ion exchange polymers. J. Amer. Chem. Soc. *87*, 5525, 5534, 5538. M. M. Emara and N.A. Farid (1985). Kinetic study of an oleophilic cation exchange resin in non-aqueous media. J. Indian Chem. Soc. 62., 40.
[114] J. H. Barrett and H. D. Clemens (1973). Hybrid copolymers and ion exchange resins made from them. Fr. Demande 2 255 334, 18 Jul 1975. US Pat. Appl. 426 769, 20 Dec 1973.
[115] L. Feistel et al. (1985). Polymere der Vinylsulfonsäure und ihre Derivate. I. Synthese, Homo- und Copolymerisation des Vinylsulfonsäurefluorids mit Divinylbenzen. Plaste Kaut. *32*, 51.
[116] G. N. Davydova et al. (1985). Kinetic properties of weakly basic anion exchangers on an aliphatic matrix type Am-6 and Am-7. Zh. Prikl. Khim. (Leningrad) *58*, 1387.
[117] W. Storck and G. Manecke (1967). Anionenaustauscher auf der Basis von Vinylamin. Makromol. Chem. *110*, 207.
[118] L. B. Zubakova et al. (1983). Ion-exchange materials made of vinyl heterocyclic monomers. Plast Massy 1983, 19.
[119] D. Braun, H. Cherdron and W. Kern. Praktikum der makromolekularen organischen Chemie. 3. Aufl. 1979. Hüthig Verlag, Heidelberg; p 317: Herstellung eines Kationenaustauschers durch Sulfonierung von vernetztem Polystyrol; ff.
[120] E. Guccione (1963). A look at the synthesis of ion-exchange resins. Chem. Eng. *70*, 138 (Apr. 15).
[121] W. C. Bauman (1949). Ion-exchange resins of stable granular form. US Pat. 2 466 675.
[122] R. F. Boyer (1950). Sulfonation of copolymers of monovinyl- and polyvinylaromatic compounds. US Pat. 2 500 149. R. M. Wheaton and D. F. Harrington (1952). Preparation of cation exchange resins of high physical stability. Ind. Eng. Chem. *44*, 1796.
[123] R. H. Wiley et al. (1968). Rates of sulfonation of bead copolymers of styrene crosslinked with mixtures of m- and p-divinylbenzene. J. Macromol. Sci., Chem. *2*, 407.
[124] D. H. Freeman et al. (1969). Homogeneous sulfonation of styrene-divinylbenzene copolymers with oleum in organic solvents. Isr. J. Chem. *7*, 741. S. Goldstein and G. Schmuckler (1972). Sulfone formation during the sulfonation of cross-linked polystyrene. Ion Exch. Mem. *1*, 63.
[125] G. K. Saldadze and V. K. Varentsov (1974). Investigation of the ionogenic group distribution in ion-exchange resins. J. Polymer Sci. Polym. Symp. *47*, 139.
[126] F. Wolf and H. Reuter (1968). Synthese und Prüfung neuartiger Carboxyl-Ionenaustauscherharze auf Polymerisationsbasis. Z. Chem. *8*, 26.

[127] C. R. Harrison et al. (1975). Introduction of carboxyl groups into crosslinked polystyrene. Makromol. Chem. *176*, 267.

[128] M. J. Farrall and J. M. J. Frechet (1976). Bromination and lithiation: two important steps in the functionalization of polystyrene resins. J. Org. Chem. *41*, 3877. T. M. Fyles and C. C. Leznoff (1976). The use of polymer support in organic synthesis. Can. J. Chem. *54*, 935.

[129] J. I. Bregman and Y. Murata (1952). Phosphonous and phosphonic cation exchange resins. J. Amer. Chem. Soc. *74*, 1867.

[130] T. R. E. Kressman and F. L. Tye (1955). Phosphonic acid cation-exchange resins. Brit. Pat. 726 918, 23 Mar 1955. I. M. Abrams (1958). Cation-exchange resins having phophorus acid functional groups. US Pat. 2 844 546, 22 Jul 1958.

[131] E. L. McMaster and W. K. Glesner (1956). Cation-exchange resins from vinyl aromatic copolymers. US Pat. 2 764 561, 25 Sept 1956. J. Kennedy and R. V. Davies (1956). The separation of uranium from heavy metals with phosphonic acid chelating resins. Chem. and Ind. (May 12), 378.

[132] V. S. Soldatov et al. (1968). Properties of phosphoric acid cation exchangers. CA 69:53985. T. V. Mekvabishvili et al. (1969). Thermal stability of phosphoric acid cation-exchange resins. CA 72:56218. G. S. Manecke et al. (1972). Neutralization of phosphoric acid cation-exchange resins. CA 77:66515. M. Marhol et al. (1974). Selective ion exchangers containing phosphorus in their functional group. I. Sorption and separation of some bivalent and trivalent ions. J. Radioanal. Chem. *21*, 177. 133.

[133] S. D. Alexandratos et al. (1985). Metal ion extraction capability of phosphinic acid resins: comparative study of phosphinic, sulfonic, and carboxylic resins using zinc ions. Macromol. *18*, 835.

[134] K. M. Saldadze et al. (1976). Synthesis of phosphoric acid cation exchangers and investigation of their physical and chemical properties. Proc. The Theory and Practice of Ion Exchange, 1976, Cambridge; ed. by M. Streat.

[135] M. Marhol et al. (1974). Selective properties and analytical use of the ion-exchange resin based on a-phenylvinylphosphonic acid. Proc. 3rd Symp. Ion-exchange, 1974, Balatonfüred, Hung.

[136] M. Marhol (1966). Ion-exchangers containing phosphorus in their functional group. J. Appl. Chem. *16*, 191. G. C. Daul et al. (1954). Ind. Eng. Chem. *46*, 1042. O. Mikes et al. (1983). Ion exchange derivatives of Spheron. IV. Phosphate derivatives. J. Chromatogr. *261*, 363.

[137] R. Bogoczek and J. Šurowiec (1981). Synthesis of phosphorus-containing Wofatit cation exchangers and their affinity toward selected cations. J. Appl. Polym. Sci. *26*, 4161. H. Egawa et al. (1985). Studies on selective adsorption resins. XXI. Preparation and properties of macroreticular chelating ion exchange resins containing phosphoric acid groups. J. Appl. Polym. Sci. *30*, 33239. M. Biswas and S. Bagchi (1985). A cation-exchange resin from phosphorylated N-vinylcarbazole-divinylbenzene copolymer. J. Polym. Mater. *2*, 123.

[138] K. W. Pepper et al. (1953). Properties of ion-exchange resins in relation to their structure. VI. Anion-exchange resins derived from styrene-divinylbenzene copolymers. J. Chem. Soc. 4097.

[139] S. Goldstein and G. Schmuckler (1973). Preparative aspects of the chloromethylated matrix of a uniform anion exchanger. Ion Exch. Mem. *1*, 135.

[140] M. Benes et al. (1977). Styrene-divinylbenzene copolymer with chloromethyl groups. Czech. Pat. 170 022, 15 Jun 1977.

[141] L. Galeazzi and A. Bursano-Busto (1975). Verfahren zur Chlormethylierung von Styrol-Divinylbenzol Copolymeren. Ger. Offen. 2 455 946.

[142] R. Hauptmann and G. Schwachula (1968). Versuche zur Chlormethylierung von polymeren aromatischen Kohlenwasserstoffen mittels Dimethylformal und Säurehalogeniden. Z. Chem. *8*, 227.

[143] R. S. Feinberg and R. B. Merrifield (1974). Zinc chloride-catalzed chloromethylation of resins for solid phase peptide synthesis. Tetrahedron *30*, 3209.

[144] J. T. Sparrow (1975). Improved procedure for the chloromethylation of polystyrene-divinylbenzene. Tetrahedron Lett. 1975, 4637.

[145] J. P. C. Bootsma et al. (1984). On the reaction of pendent vinyl groups during chloromethylation of styrene-divinylbenzene copolymers. Reactive Polymers *3*, 17.

[146] V. N. Prusova, E. E. Ergozhin et al. (1972). Interaction of chloromethylated copolymers of styrene and N, N′-alkylenedimethacrylamides with amines. Izv. Akad Nauk Kaz. SSR, Ser. Khim. *22*, 59.

[147] J. M. J. Frechet (1984). Chemical modification of polymers via phase transfer catalysis. In: Crown ethers and phase transfer catalysis in polymer science. Plenum Press 1984. Polymer Science and Technology, Vol. 24, pp 1-26. Id. et. al. (1979). Application of phase transfer catalysis to the chemical modification of cross-linked polystyrene resin. J. Org. Chem. *44*, 1774.

[148] R. Kunin (1983). Acrylic based ion exchange resins and adsorbents. Amber-hi-lites No. 173; Rohm and Haas Co., Philadephia, Pa., USA. F. Martinola (1981). Ein neuer Typ starkbasischer Anionenaustauscher. Vom Wasser 56. Bd., 205.

[149] E. E. Ergozhin and S. R. Rafikov (1970). Reduction of nitro derivatives of polystyrenes and their copolymers. Tr. Inst. Khim. Nauk. Akad. Nauk Kaz. SSR *28*, 103.

[150] V. G. Sinyavskii et al. (1974). Ion exchanger. USSR Pat. 522 193, 25 Jul 1976.

[151] W. Szczepaniak (1968). Organomercury polymers as ion exchangers. I. Anion-exchange properties of mercurated styrene-divinylbenzene copolymers. Chem. Anal. (Warsaw) *13*, 479. Id. (1971). II. Ion-exchange properties of a mercury containing divinylbenzene-styrene copolymer of macroporous structure. Poznan. Tow. Przyj. Nauk, Pr. Kom. Mat.-Przyr., Pr. Chem. *12*, 303.

[152] H. A. J. Battaerd and R. Siudak (1970). Synthesis and ion-exchange properties of surface grafts. J. Macromol. Sci., Chem. *4*, 1259.

[153] T. Brunelet et al. (1982). Functionalized resins. 2. Grafting of functionalized monomers on macroporous styrene-divinylbenzene resins. Angew. Makromol. Chem. *106*, 79.

[154] Chu Hsiu-Chang et al. (1978). Studies on the preparation method and properties of a nonswelling carboxylic ion exchange resin. Hua Hsueh Tung Pao 1978, 329.

[155] V. S. Soldatov et al. (1973). Selective properties of bifunctional ion exchangers containing sulfo- and carboxyl groups. Zh. Fiz. Khim. *47*, 976.

[156] R. Ratner et al. (1968). Selective cation exchange resin. J. Appl. Chem. *18*, 48.

[157] A. A. Efendiev et al. (1975). Reactivity of complexing polyelectrolytes. Vysokomol. Soedin., Ser. B. *17*, 6.

[158] J. A. Mikes and L. I. Kovacs (1962). Chemical properties of bipolar electrolyte exchange resins. J. Polym. Sci. *59*, 299.

[159] F. Wolf et al. (1981). Bipolar ion-exchange resins. Ger. (East) Pat. 147 671; Appl. 217 568, 12 Dec 1979. F. Wolf und E. Laqua (1977). Zur Moleküladsorption (Huminsäureadsorption) an bipolaren Ionenaustauscherharzen. Vom Wasser 48. Bd., 273.

[160] R. Böhm (1955). Ion exchangers which are selective for complex forming metal ions. Ger. (East) Pat. 10 801, 28 Nov 1955.

[161] D. K. Hale et al. (1957). Ion exchange resins. Brit. Pat. 767 821, 6 Feb 1957.

[162] N. N. (1984). Amphoteric ion exchange resin. Jpn. Kokai Tokkyo Koho JP 59 19 547, appl. 23 Jul 1982.

[163] G. Kühn and E. Hoyer (1967). Amphoteric ion exchange resins. I. Ion exchange resin with anchored *N*-methyl-β-aminopropionic acid. J. Prakt. Chem. *35*, 197.

[164] V. M. Balakin et al. (1977). Synthesis of ampholytes containing N-alkyl-β-aminopropionic acid groups. Deposited Doc. 1977, VINITI 1575 (Russ).

[165] G. Kühn and E. Hoyer (1967). Amphotere Ionenaustauscherharze. V. Eigenschaften eines Harzes mit Iminodipropionsäure als Ankergruppe. Makromol. Chem. *108*, 84.

[166] L. Wolf and R. Hering (1958). Über Ionenaustauscherharze mit komplexbildenden Ankergruppen. Chem. Techn. *10*, 661.

[167] V. A. Davankov et al. (1967). Reactivity of halomethylated copolymers of styrene and divinylbenzene. Izv. Akad. Nauk SSSR, Ser. Khim. 1967, 1612. Id. (1972). Synthesis and properties of ion-exchange resins based on neutral and hydroxyl-containing α-amino acids and chloromethylated macroreticular styrene copolymers. Vysokomol. Soedin., Ser. B. *14*, 276. Id. (1973). Use of tert-butyl esters of α-amino acids in the synthesis of unsymmetrical complexing ion exchangers. Ibid. *15*, 115.

[168] L. R. Morris et al. (1959). Synthesis of some amino acid derivatives of styrene. J. Amer. Chem. Soc. *81*, 377.

[169] M. A. Askarov et al. (1974). Study of copolymerization of vinyl benzyl chloride with acrylonitrile. Deposited Doc. 1974, VINITI 1161-74 (Russ).

[170] T. Itagaki et al. (1978). Amphoteric ion exchangers. Jpn. Kokai Tokkyo 78 99 294, 30 Aug 1978.

[171] J. Kennedy (1960). Ion exchange resin. Brit. Pat. 855 009, 23 Nov. J. Kennedy and G. E. Fricken (1961). Amphoteric ion-exchange resins. Brit. Pat. 859 834, 25 Jan. Id. J. Appl. Chem. *8*, 465 (1958).

[172] A. B. Davankov et al. (1973). Synthesis and study of polymeric complexons with aminoethylphosphonic acid groups. Vysokomol Soedin., Ser. A *15*, 1203 Id. (1974). Synthesis and study of ion exchangers with groups of β-aminoethylphosphonic acid and its derivatives. Ibid. *16*, 257.

[173] Zh. S. Amelina et al. (1974). Synthesis and study of ion exchangers with α-aminomethylphosphonic acid groups. Tr. Mosk. Khim.-Technol. Inst. *80*, 103. Yu. A. Leikin et al. (1976). Synthesis of an ion exchanger containing α-aminomethylphosphinic acid groups. Vysokomol. Soedin., Ser. A *18*, 364.

[174] Yu. P. Belov et al. (1977). Ion exchangers with groups of N-methyl-α-amino-2-furylmethylphosphonic acid and its diethyl ester. Izv. Akad. NaukSSR, Ser. Khim. 1977, 1861.

[175] I. Kolosova et al. (1982). Study of the complexing of copper ions with some phosphorus-containing polyampholytes. Koord. Khim. *8*, 1193.

[176] E. Dreipa et al. (1978). Sorption of uranium (VI) from sulfuric acid solutions by aminophosphoric acid ampholytes. Radiokhimiya *20*, 181. V. S. Soldatov et al. (1978). Acid-base and sorption properties of crosslinked macroporous polyampholytes containing amino groups and methylenephosphonic acid groups. Zh. Prikl. Khim. (Leningrad) *51*, 80.

[177] G. Manecke et al. (1962). Amphotere Ionenaustauscherharze. Teil I. Makromol. Chem. *55*, 51. Id. (1965). Teil III. Ibid. *82*, 146. Id. (1974). Teil V. Ibid. *175*, 1833.

[178] A. B. Pashkov et al. (1974). Production of high-molecular amphoteric compounds by means of polymer-analogous conversions. J. Polymer Sci., Symp. *47*, 147. M. K. Makarov et al. (1975). Some results and prospects for the synthesis, production and use of ANKB type amphoteric ion exchangers. Rasshir. Tezisy Dokl.-Vses. Simp. Temodin. Ionnogo Obmena, 2nd 1975, 78. B. Kapparov et al. (1983). Amphoteric ion exchangers. Tr. Inst. Khim. Nauk, Akad. Nauk Kaz. SSR *58*, 70.

[179] J. R. Millar (1975). Some aspects of chelating and complexing resins. Chem. and Ind. (London), 606.

[180] A. Skogseid (1946). New derivatives of polystyrene, including the dipicrylamine analogue exchanger. PhD Thesis, Norges Tekniske Hogskole, Trondheim, Norway.

[181] D. M. Mellor (1950). A possible method for the removal of trace elements from solutions. Australian J. Sci. *12*, 183.

[182] D. K. Hale (1956). Chelating resins. Research (London) *9*, 104.

[183] R. Kunin and A. F. Preuss (1964). Characterization of a boron-specific ion exchange resin. Ind. Eng. Chem. Prod. Res. Dev. *3*, 304.

[184] R. Hering (1968). Quality requirements and complex-chemical problems during application of chelate-forming ion exchangers. Symp. Kunstharz-Ionenaustauscher (pub. 1979), 503 (Ger). Akademie Verlag, Berlin, DDR.

[185] H. P. Gregor (1953). Chelate-forming ion exchange resins. XIII IUPAC Congr., Stockholm. K. W. Pepper and D. K. Hale (1954). Chelating resins. Chem. Eng. News *32*, 1897. S. L. S. Thomas (1954). Chelating resins. Ibid. *32*, 1897.

[186] G. Schmuckler (1977). Complexation in ion-exchange resins. Essays Anal. Chem. 1977, 371. N. N. Matorina et al. (1978). Effect of a chemical heterogeneity on the sorption properties of iminodiacetate ion exchangers. Zh. Fiz. Khim. *52*, 1735. Ö. Szabadka and J. Inczedy (1980). Equation for the evaluation of the protonation constants of complex forming resins. Abstr. 4th Symp. on Ion Exchange. 27-30 May 1980. Siofok, Hung. A. Nakashima et al. (1982). ESR and some spectral studies on coordination behaviors of paramagnetic metal ions for chelating resins. Bull. Chem. Soc. Jpn. *55*, 1811.

[187] K. Takeda et al. (1985). Synthesis of chelating resins containing aminopolyacetic acid moieties. Reactive Polymers *4*, 11.

[188] R. Hering (1961). Ein Ionenaustauscherharz mit verankerter Nitrilotriessigsäure. J. prakt. Chem. *14*, 285.

[189] K. C. Jones and R. R. Grinstead (1977). Properties and hydrometallurgical applications of two new chelating ion exchange resins. Chem. and Ind. (London) (15, Aug. 6), 637.

[190] M. Griesbach and K. H. Lieser (1980). Synthese von chelatbildenden Ionenaustauschern auf der Basis von Polystyrol. Angew. Makromol. Chem. *90*, 143. Id. (1980). Eigenschaften von chelatbildenden Ionenaustauschern auf der Basis von Polystyrol. Z. Anal. Chem. *302*, 109. Id. (1980). Abtrennung von Schwermetallionen mit Polystyrolaustauschern. Ibid. *302*, 181. Id. (1980). Quantitative Trennung von Schwermetallen mit Chelataustauschern auf der Basis von Polystyrol. Ibid. *302*, 184. K. H. Lieser and D. Thybusch (1981). Darstellung und Eigenschaften eines Polystyrolaustauschers mit 2.4.6-Triamino-1,3,5-triazin als Ankergruppe. Ibid. *306*, 100.

[191] J. Smid (1982). Binding of solutes to poly(vinylbenzocrown ether)s and poly(vinylbenzoglyme)s in aqueous media. Pure Appl. Chem. *54*, 2129.

[192] E. Blasius and K. P. Janzen (1982). Preparation and application of polymers with cyclic polyether anchor groups. Pure Appl. Chem. *54*, 2115.

[193] A. Warshawsky et al. (1979). Polymeric pseudocrown ethers. 1. Synthesis and complexation with transition metal anions. J. Amer. Chem. Soc. *101*, 4249. A. Warshawsky and N. Kahana (1982). Temperature-regulated release of alkali metal salts from novel polymeric crown ether complexes. J. Amer. Chem. Soc. *104*, 2663.

[194] I. Tabushi et al. (1979). Extraction of uranium from seawater by polymer-bound macrocyclic hexaketone. Nature *280*, 665. Y. Ito and T. Saegusa (1979). A new chelating polymer. J. Macromol. Sci., Chem. *A13*, 503.

[195] E. E. Ergozhin and S. M. Imakbekova (1979). Synthesis of ion exchangers by polycondensation. Tr. Inst. Khim. Nauk, Akad. Nauk KazSSR *49*, 57. A review with 162 refs.

[196] R. E. Anderson et al. (1982). Purification of Kraft pulp bleach plant effluents using condensate resins. Reactive Polymers *1*, 67.

[197] V. I. Gorshkov and I. Sh. Sverdlov (1975). Ion exchange on phenolic ion exchangers. I. Equilibrium of alkali metal ion exchange. Zh. Fiz. Khim. *49*, 2724.

[198] N. M. Khose and B. D. Dasare (1978). Synthesis and applications of porous condensate ion-exchange resins. Part I. Synthesis of porous cation exchangers based on phenol. Proc. Ion-Exch. Symp. 1978, 18-24. Ed. by G. T. Gadre, Bhavnagar, India.

[199] H. Jullien and J. Petit (1971). Effect of structure of ion-exchange resins on their properties. I. Synthesis of ion exchange macromolecules having regularly distributed ionizable groups. Bull. Soc. Chim. Fr. 1971, 848 (Fr).

[200] T. R. E. Kressman (1949). Improvements relating to cation-exchange resins. Brit. Pat. 618 251, Feb 18.

[201] N. N. Kuznetsova et al. (1969). Synthesis and properties of ion-exchange resins based on phenol ethers. Vysokomol. Soedin., Ser. A 9, 1751.

[202] K. P. Papukova et al. (1972). Synthesis of phenoxyalkylphosphonic acids and ion-exchange resins based on them. Zh. Prikl. Khim. (Leningrad) *45*, 1808.

[203] G. Borisov and I. Devedzhiev (1974). Use of bis (hydroxymethyl) phosphinic acid and its derivatives to obtain phosphorus-containing oligomers and polymers by polycondensation. Acta Chim. Acad. Sci. Hung. *81*, 169 (Russ).

[204] T. Tursunov and R. A. Nazirova (1977). Synthesis and study of polycondensation ion exchangers. Uzb. Khim. Zh. 1977, 32.

[205] K. A. Tetrov et al. (1983). Synthesis of ion-exchange resins with aminoalkylidenediphosphonate groups. Vysokomol. Soedin., Ser. B. *25*, 739.

[206] A. O. Jacubovic (1960). Cation exchange resins formed by the acid condensation of phenolsulfonic acid with formaldehyde. J. Chem. Soc. 1960, 4820.

[207] A. A. Vasilev (1968). Possible structural units of phenolsulfonic acid-formaldehyde ion exchangers. Zh. Prikl. Khim. (Leningrad) *41*, 1099. I. V. Samborskii et al. (1969). Exchange capacity of KU-1 cation exchanger. Tr. Voronezh. Gos. Univ. *72*, 133.

[208] I. Hashida and M. Nishimura (1972). Adsorption of gas on ion exchangers. III. Pore structure of m-aminophenol-formaldehyde condensation resins. J. Polym. Sci. Part A-1 *10*, 1975.

[209] V. A. Vakulenko et al. (1974). Chemical characteristics of polycondensation preparation of IA-1 type ion exchangers. Vysokomol. Soedin., Ser. B. *16*, 675.

[210] F. Polak et al. (1959). Synthesis of melamine-guanidine-formaldehyde anion exchanger. Przemysl Chem. *38*, 107. Id. (1960). II. Ibid. *39*, 307. Id. (1961). III. The influence of the molar ratio of melamine/guanidine on the properties of anion exchangers. Ibid. *40*, 153. Id. (1962). Structure of melamine-guanidine anion exchange resins. Ionenaustauscher in Einzeldarstell. *1*, 203. Id. (1964). Recovery of Ag from waste photographic solutions by using melamine-guanidine anion exchangers. Przemysl Chem. *43*, 164.

[211] R. A. Nazirova et al. (1975). Kinetics of polycondensation of furfural with polyethylene polyamine. Deposited Doc. VINITI 274-75.

[212] J. R. Dudley and L. A. Lundberg (1949). Anion-active resins. US Pat. 2 469 683, May 10, 1949. G. Manecke und K. F. Bonhöffer (1951). Elektrische Leitfähigkeit von Anionaustauschermembranen. Z. Elektrochem. *55*, 475.

[213] E. E. Ergozhin et al. (1979). Ion exchangers based on epoxy compounds. Tr. Inst. Khim. Nauk, Akad. Nauk Kaz.SSR *49*, 78. A review with 180 refs. Id. (1984). Aminoepoxy compounds and ion-exchange resins prepared from them. Ibid. *62*, 50. A review with 84 refs.

[214] E. E. Ergozhin et al. (1975). New anion exchangers based on homo and copolymers of epichlorohydrin. Rasshir. Tezisy Dokl.-Vses. Simp. Termodin. Ionnogo Obmena, 2nd 1975, 41. A. T. Chetverikova et al. (1979). Reactivity of polyethylene polyamines in the synthesis of anion exchangers of aminoepoxide type. Plast Massy 1979, 6.

[215] E. L. Ward (1980). Reaction product of an epoxide and a polyalkylene polyamide in bead form. US Pat. 4 189 539, 19 Feb 1980.

[216] M. A. Askarov et al. (1973). Anion exchangers from reaction products of epichlorohydrin with vinylpyridines. Plast Massy 1973, 10.
[217] D. H. Foster et al. (1974). Sorption of proteinaceous materials on weak base ion exchange resins. Am. Chem. Soc., Div. Org. Coat. Plast. Chem., Pap. *34*, 48.
[218] V. A. Vakulenko and E. P. Kuznetsova (1981). Synthesis and production technology of a sorbent of organic substances. CA 96:70050.
[219] E. E. Ergozhin et al. (1973). New anion exchangers from chloromethylated diphenyl ether and some polyamines. Izv. Akad. Nauk Kaz.SSR, Ser. Khim. *23*, 78. A. K. Kusainova and E. E. Ergozhin (1977). Polyfunctional anion exchangers based on oligomers of diphenyl ether and formaldehyde. Vysokomol. Soedin., Ser. A *19*, 2316.
[220] E. E. Ergozhin et al. (1976). Polyfunctional anion exchangers. USSR Pat. 536 198, 25 Nov 1976.
[221] Yu. V. Pokonova et al. (1977). Amination of chloromethylated formolites. Zh. Prikl. Khim. (Leningrad) *50*, 2328. Id. (1978). Study of the supramolecular structure of anion exchangers based on formolites. Ibid. *51*, 76.
[222] E. E. Ergozhin et al. (1980). Anion exchangers made of organosilicon compounds. Izv. Akad. Nauk Kaz.SSR, Ser. Khim. (3), 47.
[223] N. V. Chromov and C. P. Kzemikov (1961). Synthese asymmetrischer Ionenaustauscherharze auf der Basis von l-Tyrosin. Z. obsc. Chim. *31*, 2926.
[224] G. Manecke and H. Heller (1963). Amphotere Ionenaustauscherharze (Teil II). Makromol. Chem. *59*, 106.
[225] S. E. Bresler (1960). Development of synthesis and applications of ion-exchange and electron-exchange resins. Uspekhi Khim. *29*, 993. Id. (1960). Russ. Chem. Rev. *29*, No. 8, 469.
[226] E. E. Ergozhin et al. (1985). Polycondensation-type 8-mercaptoquinoline ion exchangers. Vysokomol. Soedin., Ser. B. *27*, 571.
[227] W. A. Klyachko (1951). Concerning selectivities of ion exchangers. Dokl. Akad. Nauk SSR *81*, 235.
[228] H. P. Gregor et al. (1952). Chelate ion exchange resins. Ind. Eng. Chem. *44*, 2834. H. von Lillin (1954. Über einen für Schwermetallionen selektiven Ionenaustauscher. Angew. Chem. *66*, 649.
[229] E. Blasius and G. Olbrich (1956). Komplexon-Austauscherharze, Herstellung und analytische Verwendung. Z. anal. Chem. *151*, 81.
[230] L. D. Pennington and M. B. Williams (1959). Chelating ion exchange resins. Ind. Eng. Chem. *51* (No. 6), 759.
[231] G. Naumann (1959). Ion exchangers with a chelate structure. Ger. Pat. (West) 1 051 498, Feb 26, 1959.
[232] W. Szczepaniak and J. Siepak (1969). Complex-forming ion exchanger containing pyridyl-azoresorcyl groups. Polymeri *14*, 538.
[233] J. H. Hodgkin (1979). Selective removal of ferric ions from an aqueous solution and selective metal-chelating resins for use therein. Eur. Pat. Appl. 30 106, 10 Jun 1981.
[234] M. Fuzimura and M. Kajigase (1975). Phenolic chelate resins. Japan Kokai 75 101 490, 12 Aug 1975.
[235] J. R. Kaczvinsky, J. S. Fritz et al. (1985). Synthesis and development of porous chelating polymers for the decontamination of nuclear waste. J. Radioanal. Nucl. Chem. *91*, 349.
[236] J. H. Hodgkin (1979). Synthesis of metal chelating polymers. Chem. and Ind. (5), 153.
[237] F. Vernon and H. Eccles (1974). Chelating ion exchangers containing salicylic acid. Anal. Chim. Acta *72*, 331.
[238] C. J. Wards et al. (1976). Selective ion exchangers. Brit. Pat. 1 430 578, 31 Mar 1976; Appl. 08 Jun 1972.

[239] R. N. Kapadia and A. K. Dalal (1982). Synthesis and physicochemical studies of some new amphoteric ion exchangers. J. Appl. Polym. Sci. *27*, 3793.

[240] R. N. Kapadia and M. V. Vyas (1983). Synthesis and characterization of some new chelating amphoteric ion-exchange resins. J. Appl. Polym. Sci. *28*, 983. M. V. Vyas and R. N. Kapadia (1984). Studies on some new chelating amphoteric ion exchange resins. Pap.-Semin. Chem. Behav. Appl. Weak Acid, Weak Base Chelating Ion Exch. 1984, 38; Dept. At. Energy, Bombay, India.

[241] N. Ya. Lyubman et al. (1978). Selective ion exchanger based on formaldehyde and 1,2,4-triacetobenzene. USSR Pat. 610 835, 15 Jun 1978.

[242] R. N. Kapadia and M. V. Vyas (1984). Chelating properties of amphoteric ion-exchangers. II. Exchangers derived from epichlorohydrin and p-hydroxybenzoic acid, β-resorcylic acid or anthranilic acid. Ind. J. Chem., Sect. A *23A*, 757.

[243] N. Hojo and H. Tsukioka (1954). Phenylthiourea resin. Research Repts. Fac. Textiles and Sericult., Shinshu Univ. *4*, 100. N. Hojo (1956). Selective chelate resin from 3-hydroxyphenylthiourea. Kogyo Kagaku Zasshi *59*, 631. Id. (1957). Selective chelate resin from mercaptobenzthiazole. Shinshu Daigaku Senigakubu Kenkyu Hokoku *7*, 166.

[244] E. Bayer (1964). Struktur und Spezifität organischer Komplexbildner. Angew. Chem. *76*, 76.

[245] N. Oda et al. (1972). Adsorbent for heavy metals or heavy metal compounds. Ger. Offen. 2 333 800, 17 Jan 1974.

[246] V. M. Balakin et al. (1974). Obtaining complexing ion exchanger by modification of polymers. USSR Pat. 449 075, 05 Nov 1974.

[247] T. V. Samborskii et al. (1975). Preparation of an ion exchanger based on mercapto-benzthiazole. Rashir. Tezisy. Dokl.-Vses. Simp. Termodin. Ionnogo Obmena, 2nd 1975, 70.

[248] J. Kennedy and G. E. Ficken (1958). Synthesis of metal-complexing polymers. II. Phosphonamide and α-aminophosphonate polymers. J. Appl. Chem. (London) *8*, 465.

[249] V. M. Balakin et al. (1973). Amphoteric ion exchanger. USSR Pat. 397 398, 22 Aug 1973.

[250] E. Blasius et al. (1974). Darstellung und Eigenschaften von Austauschern auf Basis von Kronenverbindungen. J. Chromatogr. *96*, 89. Id. Ibid. *201*, 147.

[251] P. Gramain and Y. Frere (1979). Synthesis and ion binding properties of epoxy polymers with diazacrown ethers. Macromolecules *12*, 1038. Id. (1981). Cation complexation properties of epoxy polymers and exchangers containing diazacrown ethers and a cryptate. Ind. Eng. Chem. Prod. Res. Dev. *20*, 524.

[252] A. Ricard et al. (1982). ^{13}C and ^{23}Na n.m.r. studies of the interactions between cations and kryptofix (2,2) bound to soluble or gel polyacrylamide. Polymer *23*, 907.

[253] B. J. Schultz and E. H. Crook (1968). Ultrafine ion exchange resins. Ind. Eng. Chem. Prod. Res. Dev. *7*, 120.

[254] F. Martinola (1973). Active filtration. Powdered ion exchangers and adsorbents as pre-coat media. Filtration sepn. *10*, 420.

[255] B. P. Chong (1984). Submicroscopic particles of anion exchange resin and anion exchange process using them. Europ. Pat. 0 009 397.

[256] V. S. Soldatov (1984). New fibrous ion exchangers for purification of liquids and gases. Stud. Environ. Sci. *23* (Chem. Prot. Environm.), 353 (Eng). V. M. Sedov et al. (1980). Ion-exchange properties and thermal stability of fibrous ion exchangers of VION type. Issled. po Khimii, Tekhnol. i Primeneniyu Radioaktiv. Veshchestv, L. 1980, 97 (Russ).

[257] H. Ichijo et al. (1983). Super fine filaments of sulfonated poly(vinyl alcohol) for enzyme immobilization. Seni Gakkaishi *39*, T532 (Eng).

[258] L. A. Volf et al. (1969). Thermally stable poly(vinyl alcohol) fibers. Khim. Volokna (3), 15. A. M. Maksimov et al. (1973). Modification of poly(vinyl alcohol) fibers by graft copolymerization. Zh. Prikl. Khim. (Leningrad) *46*, 2346. V. M. Mazovetskaya et al. (1977). Use of fibrous ion-exchangers containing polyethylenimine in wastewater purification processes. Issled. Obl. Khim. Polietilenimina Ego primen Promsti. 1977, 133. R. D. Chebotareva et al. (1980). Study of the kinetics of swelling of ion-exchange fibers. Kolloidn. Zh. *42*, 789.

[259] H. Tanaka et al. (1974). Acrylonitrile ion exchange resin. Japan Kokai 74 88 988, 26 Aug 1974.

[260] M. P. Zverev (1975). Work of the All-Union Scientific Research Institute of Synthetic Fibers in the area of ion-exchange fiber production. Khim. Volokna (5), 3. N. F. Kalyanova et al. (1978). Formation of a three-dimensional network in chemisorption fibers based on vinylpyridine copolymers. Vysokomol Soedin., Ser. B *20*, 461.

[261] W. Aichele et al. (1978). Electrokinetic investigations on fibers. 4. An adsorption model of ions to ion exchanging polymers. Makromol. Chem. *179*, 787.

[262] N. V. Bytsan et al. (1983). Preparation characteristics and some physicochemical properties of amino- and phosphorus-containing polyacrylonitrile ion-exchange fibers. Khim. Volokna (4), 40. T. M. Moroshkina et al. (1983). Use of new synthetic fibers for the concentration of small amounts of elements. Probl. Sovrem. Anal. Khim. *4*, 166.

[263] T. Yoshioka and M. Shimamura (1983). Studies of polystyrene-based ion exchange fiber. I. Bull. Chem. Soc. Jpn. *56*, 3726. Id. (1986). IV. A novel fiber-form material for adsorption and immobilization of biologically-active proteins. Ibid. *59*, 399.

[264] T. Ivanova et al. (1982). Study of the absorption of gases of basic nature by carbonaceous fibrous ion exchangers. Zh. Prikl. Khim. *55*, 544.

[265] M. P. Zverev et al. (1973). Ion exchange materials. USSR Pat. 407 921, 10 Dec 1973. T. Miyamatsu and N. Oguchi (1980). Study on fibrous ion-exchanger and its application. Part 4. Preparation of fibrous weak acidic cation exchanger. Seni Gakkaishi *36*, T308. N. V. Bytsan et al. (1982). Study of conditions for the synthesis of a phosphorus-containing polyampholyte. Deposited Doc., SPSTL 21, khp-D 81, 17-22 (Russ). N. A. Goncharova et al. (1982). Sorption of copper by different types of ion-exchange fibrous materials. Zh. Prikl. Khim. (Leningrad) *55*, 2095.

[266] G. G. Chernenko et al. (1979). Modification of nonwoven textile materials. Tekst. Prom-st (Moscow) (9), 51.

[267] I. Ishigaki et al. (1978). Synthesis of an ion exchange membrane by radiation induced grafting of acrylic acid onto poly(tetrafluoroethylene). Polym. J. *10*, 513.

[268] N. Inagaki and K. Ohishi (1985). Plasmapolymerized thin films containing anion and cation groups. Reactive Polymers *4*, 21.

[269] T. Braun et al. (1973). Ion-exchange foam chromatography. I. Preparation of rigid and flexible ion-exchange foams. Anal. Chim. Acta *64*, 45.

[270] B. A. Bolto et al. (1983). I. Thermally regenerable ion-exchange resins. The encapsulation of polyamine microparticles in a polyacid matrix. Reactive Polymers *1*, 119. Id. (1983). II. The encapsulation of polyacid microparticles in a polyamine matrix. Ibid. *1*, 129. Id. (1983). III. Filling a macroporous resin with crosslinked resin. Ibid. *1*, 139.

[271] J. Kupec and J. Stamberg (1967). Properties of strong acid capillaries. Chem. Listy *61*, 258. J. Kupec et al. (1968). Ionenaustauscher-Kapillaren. Paper Symp. 30 Jahre Kunstharz-Ionenaustauscher, publ. 1970, Akademie-Verlag, Berlin. J. Kupec et al. (1970). Ion exchange capillaries. VII. Electric conductivity of the cation-exchange layers and an attempt at electrophoretic separation. Coll. Czechoslov. Chem. Commun. *35*, 3659. J. Stamberg (1986). Private communication on the potential future applications.

[272] T. I. Rabek and J. Morawiec (1966). Reaction of some amines with chloromethylated and crosslinked polystyrene. Polim. Tworzywa *11*, 251. S. Inoue (1969). Optically active polymers. Kobunshi *18* (210), 596.
[273] B. U. Kaczmar (1973). Properties of N-chloropolyamides. Angew. Chem., Int. Ed. Engl. *12*, 430.
[274] M. B. Kril et al. (1984). Chemical disinfection by means of speciality ion exchange resins. In: Ion exchange Technology, ed. by D. Naden and M. Streat. Ellis Horwood Ltd.,Chichester; p 407.
[275] A. B. Paschkov and W. S. Titow (1958). Über die basischen Kennzahlen einiger sowjetischer Ionenaustauscher. Chimiceskaja Promyslen. *5*, 10. I. V. Samborskii and V. A. Vakulenko (1970). Domestic ion exchangers of the polycondensation type and their use in hydrometallurgy. Ekstr. Sorbtsiya Met. Nikelya, Kobalta Meti. 1970, 133. M. M. Senyavin (1980). Ionenaustauscher in der Technik und der Analyse anorganischer Materialien. Khimia, Moscow, 1980; p 54. A. B. Pashkov et al. (1973). Properties of ion exchange resins produced outside the USSR. NIITEkhim: Moscow; 69 pp.
[276] E. Guccione (1963). A look at the synthesis of ion exchange resins. Chem. Engng. *70*, No. 8, 138.
[277] V. A. Vakulenko and A. B. Paskov (1978). Development of ion exchange resin production. Proizvod. Pererab. Plastmass. Sint. Sinol. (7), 53.
[278] F. X. McGarvey and M. C. Gottlieb (1979). Fundamental and basic theories regarding the proper use of modern day ion exchange resins. Proc. Int. Water Conf., Eng. Soc. West. Pa.,4th, 175.
[279] G. M. Tilsley (1975). Clean-up of fouled ion exchange resin beds. Effluent Water Treatment J. *15*, 560.
[280] S. Fisher and G. Otten (1982). Specification for buying ion exchange resins: a necessity or a nuisance? Proc. Int. Water Conf., Eng. Soc. West Pa., 43rd, 393.
[281] S. Fisher and G. Otten (1984). Extractables in new resins: a critical look. Proc. Int. Water Conf., Pittsburgh, Pa, 1984. R. Wagner et al. (1984). Untersuchungen über die Amin-Abgabe von Anionenaustauscher-Harz. Z. Wasser. Abwasser-Forsch. *17*, 240.
[282] G. M. Armitage et al. (1972). Consequences of the autodegradation of polystyrene-based ion exchangers. Proc. Soc. Anal. Chem. *9*, 204. Id. (1973). Mass spectrometric study of the deterioration of polystyrene-based ion-exchangers. Talanta *20*, 315.
[283] F. D. Kover (1975). Preliminary study of selected potential environmental contaminants. Optical brighteners, methyl chloroform, trichloroethylene, tetrachloroethylene, and ion exchange resins. U. S. NTIS, PB Report 1975, PB-243910.
[284] R. L. Johnson and B. V. Chandler (1985). Ion exchange and adsorbent resins for removal of acids and bitter principles from citrus juices. J. Sci. Food Agric. *36*, 480.
[285] J. R. Millar (1983). Impact of food and drug legislation in the USA on the use of ion-exchange resins in food processing. Chem. Ind. (London) (21), 804.
[286] Kunststoffe-Empfehlung XXIV (1985). Ionenaustauscher und sorptiv wirkende Polymere für die Behandlung von Trinkwasser sowie von wäßrigen Flüssigkeiten, die als Lebensmittel oder bei der Herstellung und Verarbeitung von Lebensmitteln verwendet werden. Stand: 1.9.1984. 168. Mitt. Bundesgesundheitsbl. *28*, 24. Franck, Kunststoffe 34. Lfg. März 1985.
[287] T. Garlandi (1965). Norme sanitarie sulle resine scambiatrici di ioni. Mat. Plast. Elastom. *31*, 719.
[288] G. H. Mansfield (1975). Assessment and use of ion exchange resins. Effl. Water Treat. J. *15*, 123.
[289] H.-C. Flemming (1981). Bakterienwachstum auf Ionenaustauscher-Harz. Untersuchungen an einem stark sauren Kationenaustauscher. Teil I. Anhaftung und Verteilung der Bakterien; betriebliche Möglichkeiten der Wachstumsunterdrückung. Z. Wasser-Ab-

wasser Forsch. *14*, 132. Id. (1982). Teil II. Wirksamkeit von Silber-Zusätzen gegen die Nachverkeimung während Stillstandzeiten. Ibid. *15*, 259. Id. (1984). Teil III. Desinfektion mit Peressigsäure. Ibid. *17*, 229.

[290] R. Wagner and H.-C. Flemming (1984). Untersuchungen über das Auswaschverhalten von Ionenaustauscherbetten. Z. Wasser-Abwasser Forsch. *17*, 235.

[291] D. Wittington and J. R. Millar (1968). Infrared absorption spectra of ion-exchange resins. J. Appl. Chem. *18*, 122.

[292] D. Wang (1983). Identification of ion exchange resins. Huaxue Shijie *24*, 313.

[293] N. E. Bibler and E. G. Orebaugh (1976). Iron-catalyzed dissolution of polystyrenesulfonate cation-exchange resin in hydrogen peroxide. Ind. Eng. Chem. Prod. Res. Dev. *15*, 136. S. C. Lee (1981). Dissolution of ion exchange resin by hydrogen peroxide. Report DP-1594; avail. NTIS.

[294] G. Naumann (1976). Testing of ion exchangers. In: The theory and practice of ion exchange. Intern. Conference, Cambridge, 1976.

[295] F. X. Pollio (1963). Determination of moisture in ion exchange resins by Karl Fischer reagent. Anal. Chem. *35*, 2164. H. D. Sharma and N. Subramanian (1969). Ibid. *41*, 2063.

[296] E. Blasius and R. Schmitt (1968). Restwasserbestimmung in Ionenaustauschern mit Hilfe tritiummarkierten Wassers und Vergleich dieser Methode mit der Karl Fischer-Titration. Z. anal. Chem. *241*, 4.

[297] G. Dickel et al. (1969). Fallmethode zur Bestimmung der Wasseraufnahme bei Ionenaustauschern. Z. phys. Chem. NF *67*, 210.

[298] J. Kristof and J. Inczedy (1980). Determination of the water content of anion exchange resins using thermoanalytical methods. Abstracts 4th Symp. on Ion Exchange; 27-30 May, Siofok, Hung.

[299] I. F. Gleim et al. (1970). System for drying the ion exchangers KU-2 and AV-17. Ionnyi Obmen Ionity, p 58.

[300] D. H. Freeman and G. Scatchard (1965). Volumetric studies of ion exchange resin particles using microscopy. J. Phys. Chem. *69*, 70.

[301] P. B. Hamilton (1958). Ion-exchange chromatography of amino acids. Effect of resin particle size on column performance. Anal. Chem. *30*, 914.

[302] B. Vassiliou and R. Kunin (1963). Fine particle-sized ion exchange resins. Anal. Chem. *35*, 1328. D. A. Knyazev et al. (1969). Sedimentation method for the fractionation of ion exchange resins and filling of chromatographic columns. Zavod. Lab. *35*, 1446.

[303] K. Haagen (1953). Ionenaustausch an Carboxyl- und Sulfogruppen enthaltenden Kunstharzaustauschern. Z. Elektrochem. *57*, 178.

[304] F. X. McGarvey et al. (1984). Hydraulic properties of ion exchange resins. 45th Ann. Meeting Intern. Water Conference, Pittsburgh, Oct. 22-24.

[305] F. Martinola and G. Kühne (1969). Properties and application of powdered ion-exchange resins. Intern. Conference on Ion Exchange, London.

[306] O. D. Bonner and R. R. Pruett (1960). Variations in the structure of sulfonic acid type cation exchanger resins and the effect of these variations on their properties. Z. phys. Chem. NF *25*, 75.

[307] R. E. Anderson (1964). A contour map of anion exchange resin properties. Ind. Eng. Chem. Prod. Res. Dev. *3*, 85.

[308] E. Blasius et al. (1956). Ionensiebe. Kapillareigenschaften verschieden vernetzter Anionenaustauscher auf Kunstharzbasis. Angew. Chem. *68*, 671.

[309] D. H. Freeman et al. (1969). Recognition of crosslinking in the infrared spectra of poly(styrene divinylbenzene). Int. Conf. on Ion Exchange, London. H. Kveder (1970) Ir. Spectroscopic determination of the degree of crosslinking of ion-exchange resins based on styrene-divinylbenzene copolymers. Kem. Ind. *19*, 141 (Croat).

[310] K. W. Pepper et al. (1952). Swelling and shrinkage of sulphonated polystyrenes of different cross-linking. J. Chem. Soc. 3129.

[311] G. M. Armitage and H. S. Dunsmore (1973). Effect of resin crosslinking on the ion-exchange method of stability constant determination. J. Inorg. Chem. *35*, 1701.

[312] R. Schlögl and H. Schurig (1961). Eine experimentelle Methode zur Bestimmung der Porengröße in Ionenaustauschern. Z. Elektrochem. *65*, 863.

[313] A. A. Tager (1960). Porosity of ion-exchange resins. Vysokomolekulyarnye Soedineniya *2*, 994. Id. (1960). A reason for high sorption ability of ion-exchange resins. Ibid. *2*, 997.

[314] R. Böhm (1968). Untersuchungen der Porosität in stark sauren Kationenaustauschern mit einer Kanalstruktur. Kunstharz-Ionenaustauscher Symp., Plenar-Diskussionsvortrag (publ. 1970), 170.

[315] V. S. Yurchenko et al. (1968). Porosity and permeability of ion exchange resins for ions of organic substances. Tr. Leningrad. Khim.-Farm. Inst. No. 25, 121.

[316] J. A. Mikes (1969). Pore structure in ion-exchange materials. Conference on Ion Exchange, London.

[317] R. Kunin (1969). A critical examination of the pore structure of macroreticular ion exchange resins. Conference on Ion Exchange, London.

[318] W. Hein (1967). Untersuchungen der Kapillareigenschaften von Ionenaustauschern. Dissertation. Technische Universität Berlin.

[319] H. A. Benesi et al. (1955). Determination of pore volume of solid catalysts. Anal. Chem. *27*, 1963.

[320] A. Y. Mottlau and N. E. Fischer (1962). Measurement of pore volume by a titration technique. Anal. Chem. *34*, 714.

[321] R. Lumbroso and M. Reverbori (1964). Etude de la porosite d'un echangeur d'ion cationique du type copolymer styrene-divinylbenzene sulfone. Genie chim. *92*, 89.

[322] K. A. Kun and R. Kunin (1966). The pore structure of macroreticular ion exchange resins. J. Polym. Sci. *C16*, 1457.

[323] Z. Pelzbauer and V. Forst (1966). The electron-microscopic evaluation of the porosity of ion exchangers. Collect. Czech. Chem. Commun. *31*, 2338.

[324] B. N. Laskorin et al. (1983). The structure of porous ion exchangers. Plast. Massy (8), 17.

[325] F. Martinola and G. Siegers (1975). Innere Oberfläche und Adsorptionsverhalten makroporöser Ionenaustauscher. VGB Kraftwerkstechn. *55*, 1.

[326] G. Witte and J. Starmick (1980). Porosimetrie gequollener makroretikularer Ionenaustauscher. Angew. Makromol. Chem. *84*, 7.

[327] J. C. Parrish (1965). Measurement of water regain and macropore volume of ion-exchange resins. J. Appl. Chem. *15*, 280.

[328] J. Inczedy and E. Pasztler (1968). Swelling and salt splitting in non-aqueous media. Acta Chem. Acad. Sci. Hung. *56*, 9.

[329] E. Blasius and H. Pittack (1959). Ionensiebe. 2. Mitt. Capillar- und Ionensiebeigenschaften von Austauschern. Angew. Chem. 71, 445.

[330] E. Glueckauf and G. P. Kitt (1955). A theoretical treatment of cation exchangers. III. The hydration of cations in polystyrene sulphonates. Proc. Royal Soc. *228*, 322.

[331] E. O. Timmermann (1970). Concentrated polyelectrolyte solutions and ion exchangers. II. Swelling and partial molar volumes. Z. phys. Chem. (Frankfurt) *72*, 140.

[332] A. P. Platonov and V. S. Soldatov (1980). Swelling of sulfonated styrene copolymer ion exchangers in solutions of sulfonated polystyrene. Dokl. Akad. Nauk BSSR *24*, 341.

[333] E. F. Nekryach et al. (1972). Sorption-thermochemical study of the hydration of a macroporous KU-23 sulfonic cation exchanger. Ukr. Khim. Zh. (Russ. Ed.) *38*, 581.

[334] V. A. Davankov et al. (1976). On factor determining the swelling ability of crosslinked polymers, II. Angew. Makromol. Chem. *53*, 19.

[335] J. Inczedy (1978). Thermoanalytical investigation of ion exchange resins. The swelling water of anion exchange resins. J. Thermal. Anal. *13*, 257.

[336] S. Fisher and R. Kunin (1955). Routine exchange capacity determination of ion exchange resins. Anal. Chem. *27*, 1191.

[337] H. P. Gregor (1948). A general thermodynamic theory of ion exchange processes. J. Amer. Chem. Soc. *70*, 1293. Id. (1950). Gibbs-Donnan equilibria in ion exchange systems. Ibid. *73*, 642. F. E. Harris and S. A. Rice (1956). Model for ion exchange resins. J. Chem. Phys. *24*, 1258. S. A. Rice and F. E. Harris (1956). Polyelectrolyte gels and ion exchange reactions. Z. physik. Chemie (Frankfurt) *8*, 207. G. E. Boyd et al. (1961). A thermodynamic calculation of selectivity coefficients for strong base anion exchangers. J. Phys. Chem. *65*, 577. B. Chu et al. (1962). On anion exchange selectivity. J. Inorg. Nucl. Chem. *24*, 1405. J. S. Redinha and J. A. Kitchener (1963). Trans. Faraday Soc. *59*, 515. J. F. Millar et al. (1965). Theory of selective uptake of ions of different size by polyelectrolyte gels. Experimental results with potassium and quaternary ammonium ions and methacrylic acid resins. J. Chem. Physics *43*, 1783. J. A. Marinsky (1967). Prediction of ion exchange selectivity. J. Phys. Chem. *71*, 1572. M. Reddy and J. A. Marinsky (1971). Ion exchange selectivity coefficients. J. Macromol. Sci., Phys. *5*, 135. Id. et al. (1980). Hydration control of ion distribution in polystyrene sulfonate gels and resins. ACS Symp. Ser. 127, 387.

[338] D. Reichenberg and D. J. McCauley (1955). Properties of ion exchange resins in relation to their structure. VII. Cation exchange equilibria on sulphonated polystyrene resins of varying degree of crosslinking. J. Chem. Soc. 2731.

[339] V. S. Soldatov and V. A. Bichova (1983). Quantitative description of ion exchange selectivity in non-ideal systems. Reactive Polymers *1*, 251.

[340] V. S. Soldatov et al. (1972). Selectivity of the exchange of alkaline earth metal ions on carboxyl ion exchangers. Vestsi Akad. Navuk Belarus. SSR, Ser. Khim. Navuk (2), 25.

[341] J. L. Kanungo and S. K. Chakravarti (1972). Exchange characteristics of hexaamminecobalt (III) ion in Amberlite IRC-50. Kolloid-Z. Z. Polym. *250*, 891.

[342] T. R. E. Kressman and J. A. Kitchener (1949). Cation exchange with a synthetic phenolsulphonate resin. J. Chem. Soc. (London) 1190, 1201, 1208, 1211.

[343] R. Khristova (1972). Infrared spectra of strongly basic anion exchangers. God. Sofii. Univ., Khim. Fak. 1969–1970, 64, 179 (Ger).

[344] G. E. Janauer (1972). Anion-exchange selectivity in homologous series of hydrophobic ions. Monath. Chem. *103*, 605.

[345] J. Bucher et al. (1972). Selectivity in heterovalent anion exchange. Ion pairing vs. ion hydration. J. Phys. Chem. *76*, 2459.

[346] L. P. Karapetyan et al. (1975). Mechanism of the sorption of transition metal ions by weakly basic anion exchangers from solutions of their complexes. Rasshir. Tezisy Dokl.-Vses. Simp. Termodin. Ionnogo Obmena, 2nd, 18.

[347] V. A. Bichkova and V. S. Soldatov (1977). Mutual effect of ions in ammonium-alkali cation-hydrogen ion systems on highly crosslinked sulfostyrene ion exchangers. Vestsi Akad. Navuk BSSR, Ser. Khim. Navuk (2), 17.

[348] H. Reiss and I. C. Bassignana (1982). Critique of the mechanism of superselectivity in ion exchange membranes. J. Membr. Sci. *11*, 219.

[349] D. Clifford and W. J. Weber, jr. (1983). The determinants of divalent/monovalent selectivity in anion exchangers. Reactive Polymers *1*, 77.

[350] R. E. Barron and J. S. Fritz (1984). Effective functional group structure on the selectivity of low capacity anion exchange resins for monovalent ions. J. Chromatogr. *284*, 13. Id. (1984). For divalent ions. Ibid. *316*, 201.

[351] G. R. Hall et al. (1969). Thermal stability of ion exchange resins. Conference on Ion Exchange, London.

[352] O. Tjaelldin (1972). Temperature stability of different ion-exchange materials. Report 1972, SVF-2; Stiftelsen Varmetek. Forsk., Studsvik, Sweden.

[353] N. G. Polyanski et al. (1971). Effect of heat treatment in water on the physico-chemical properties of macroporous sulfonic cation exchangers. Zh. Fiz. Khim. *45*, 2858. Id. (1973). Thermal stability of phosphoric acid cation exchangers. Zh. Prikl. Khim. (Leningrad) *46*, 363.

[354] S. I. Laptev et al. (1976). Study of the low-temperature resistance of ion-exchange materials. Plast. Massy (1), 52.

[355] R. Kunin et al. (1962). Characterization of Amberlyst 15. Ind. Eng. Chem. Prod. Res. Dev. *1*, 140 (June).

[356] L. S. Golden and J. Irving (1972). Osmotic and mechanical strength in ion-exchange resins. Chem. Ind. (London) (21), 837.

[357] R. Rowe (1962). How to test cation resins for stability. Power Engng. *66*, 68. W. M. Alvino (1980). Stability of ion-exchange resins. 1. Resistance to osmotic shock. Ind. Eng. Chem. Prod. Res. Dev. *19*, 271. A. M. Prokhorova and L. V. Kostrova (1970). Osmotic strength of the Soviet strongly basic anion exchanger AV-17-8. Teploenergetika *17*, 10.

[358] R. Hochmüller (1984). Shock test for the determination of the resistance of ion exchange resins to osmotic and hydromechanical stress. In: Ion Exchange Technology. D. Naden and M. Streat, eds. Ellis Horwood Ltd., Chichester; 1984; p 472. Id. (1986). 47th Intern. Water Conf., Pittsburgh, PA.

[359] W. J. Blaedel and E. D. Olsen (1961). Application of chlorine in cation exchange separations. Anal. Chem. *33*, 531. L. F. Wirth et al. (1961). Ion exchanger stability under the influence of chlorine. Ind. Eng. Chem. *53*, 638.

[360] H. T. Fulham et al. (1969). Thermochemical instabilities in anion-exchange processing. Conference on Ion Exchange, London.

[361] G. P. Herz (1962). Der Einfluß organischer Substanzen auf Anionenaustauscher von Vollentsalzungsanlagen. Technische Überwachung *3*, 77.

[362] M. Falk et al. (1982). Stabilität von Wofatit-Ionenaustauschern gegenüber Peressigsäure. 2. Mitt. Über die Peressigsäure-Desinfektion von Ionenaustauschern. Pharmazie *37*, 387.

[363] R. H. Wiley and G. Devenuto (1965). Irradiation stability of sulfonated styrene resins crosslinked with various divinylbenzene isomers and mixtures thereof. J. appl. Polym. Sci. *9*, 2001.

[364] G. R. Hall and M. Streat (1963). Radiation-induced decomposition of ion-exchange resins. J. Chem. Soc. 1963, 5205. G. R. Hall et al. (1968). Ion exchange in nuclear chemical processes. Part 1. The effect of heat and ionizing radiation on resin performance. The Chem. Engr. No. 216. Trans. Inst. Chem. Engrs. *46*, 2.

[365] E. D. Kiseleva et al. (1978). Study of a change in the chemical structure of irradiated vinylpyridine carboxyl ampholyte by IR spectral methods. Dokl. Akad. Nauk SSR *238*, 877. Id. (1982). Change in ion exchange properties of ampholyte VPK during irradiation in intramolecular salt form. Zh. Fiz. Khim. *56*, 369.

[366] K. K. S. Pillay (1986). The effects of ionizing radiations on synthetic organic ion exchangers. J. Radioanal. Nucl. Chem. *97*, 135.

[367] F. Martinola (1965). Zur Überwachung von Anionenaustauschern im Betriebslabor. Berichtsheft VGB-Speisewassertagung 1965, 17.

[368] N. P. Gnusin and V. D. Grebenyuk (1972). Electrochemistry of granulated ion exchangers. Naukova Dumka: Kiev, Ukr SSR. 180 pp.

[369] A. O. Jakubovic et al. (1959). Ionic mobilities in ion-exchange resins. Trans. Faraday Soc. *55*, 1570.

[370] G. J. Hills et al. (1956). Determination of the electrical conductivity of crosslinked polyelectrolytes, resins, and gels. J. Polymer Sci. *19*, 382. Th. Shedlovsky (1930). A conductivity cell for eliminating electrode effects in measurements of electrolytic conductance. J. Am. Chem. Soc. *52*, 1806.

[371] V. A. Shaposhnik and I. V. Drobysheva (1977). Mechanism of the electric conductivity of ion exchangers. Nauch. Tr. Kuban. Un-t. *232*, 161.

[372] M. C. Sauer et al. (1955). Electrical conductance of porous plugs. Ion exchange resin-solution systems. Ind. Engng. Chem. *47*, 2187. K. S. Spiegler et al. (1956). Electrical potentials across porous plugs and membranes. Disc. Faraday Soc. *21*, 174.

[373] N. P. Gnusin et al. (1976). Determination of the isoconductance point of ion-exchange materials. Zavod. Lab. *42*, 709.

[374] N. P. Gnusin et al. (1973). Selection of the optimum variant of a three-line model. Ukr. Khim. Zh. (Russ. Ed.) *39*, 1072. Id. (1973). Effect of the geometric shape of ion exchange grains on the electric conductivity of ion exchange columns. Fiz.-Khim. Mekh. Liofilnost Dispersnykh Sist. No. 5, 24. Id. (1977). Electric conductivity of ion exchange columns. Elektrokhimiya *13*, 1712.

[375] M. Seno and T. Yamabe (1961). Aggregation of ion-exchange resin particles. Bull. Chem. Soc. Japan *34*, 1002. Id. (1964). Electro-kinetic potentials of ion-exchange particles. Nature *202*, 1110.

[376] M. Goldsmith et al. (1975). Biological ion exchanger resins. VI. Determination of the Donnan potentials of single ion-exchange beads with microelectrodes. J. Phys. Chem. *79*, 342; *80*, 2432 (1986); *80*, 2433 (1986).

[377] K. Singh and R. Kumar (1979). Electrophoresis of Dowex 1 (chloride form) dispersed in methanol-water mixtures and estimation of zeta-potentials. Ind. J. Chem., Sect. A *18*, 10.

[378] G. W. Bodamer and R. Kunin (1953). Behavior of ion exchange resins in solvents other than water. Ind. Eng. Chem. *45*, 2577.

[379] J. D. R. Thomas (1974). Ion-exchange in nonaqueous and mixed solvents. J. Chromatogr. 102, 209. G. R. Hall et al. (1970). In: Ion Exchange in the Process Industries. Soc. Chem. Ind., London, 1970. M. A. Minto (1971). M. Sc. Thesis, University of Wales, Uwist, Cardiff.

[380] A. T. Davydov et al. (1976). Sorption of alkali metal ions on the KU-2 cation exchanger in liquid sulfur dioxide. Zh. Fiz. Khim. *50*, 1030.

[381] S. Kamata (1970). Sorption of acetic acid and its chloro derivatives on macroreticular cation exchange resin. Kogyo Kagaku Zasshi *73*, 705.

[382] R. W. Gable and H. A. Strobel (1956). Nonaqueous ion exchange. I. Some cation equilibrium studies in methanol. J. Physic. Chem. 60, 513.

[383] D. Nandan and A. R. Gupta (1975). Lithium/hydrogen, sodium/hydrogen, and potassium/hydrogen ion exchange equilibriums on crosslinked Dowex 50W resins in anhydrous methanol. J. Phys. Chem. *79*, 180.

[384] O. Gürtler et al. (1972). Anion exchange on Wofatit SBW in organic media. Wiss. Z. Univ. Leipzig, Math.-Naturwiss. Reihe *21*, 81.

[385] J. Inczedy and E. Pasztler (1968). The analytical use of ion exchangers in organic solvents. I. Swelling and salt splitting in non-aqueous media. Acta Chim. Acad. Sci. Hung. *56*, 9. J. Inczedy and I. Högye (1968). II. Swelling and functioning properties of macroporous ion exchange resins. Ibid. *56*, 109.

[386] A. M. Phipps (1968). Anion exchange in dimethyl sulfoxide. Anal. Chem. *40*, 1769.

[387] A. A. Baran et al. (1971). NMR study of the reaction of methanol with ion exchange resins. Fiz.-Khim. Mekh. Liofilnost Dispersnykh Sist. No. 3, 31.

[388] V. V. Mank et al. (1977). NMR study of the interaction of alcohols with cation exchangers in H-form. Kolloidn. Zh. *39*, 379. G. S. Bystrov et al. (1976). Study of the ion exchanger-solvent system by a nuclear magnetic resonance method. Usp. Khim. *45*, 1621-45. Review with 91 refs.

[389] C. Heitner-Wirguin and N. Ben-Zwi (1970). Chloride species of cobalt (II) sorbed on ion exchangers. II. Nonaqueous media. Inorg. Chim. Acta *4*, 554.

[390] O. B. Bonner and J. C. Moorefield (1954). Ion exchange in mixed solvents. J. Physic. Chem. *58*, 555. T. R. E. Kressman and J. A. Kitchener (1949). Cation exchange with a synthetic phenolsulphonate resin. Part IV. A note on equilibria in presence of nonaqueous solvents. J. Chem. Soc. 1211. H. P. Gregor et al. (1955). Studies on ion-exchange resins. XII. Swelling in mixed solvents. J. Physic. Chem. *59*, 10. Y. Barbier and R. Rosset (1970). Properties of macroporous ion exchange resins. II. Water-solvent partition between the resin and solution phases. Bull. Soc. Chim. Fr. *11*, 4162.

[391] C. W. Davies and B. D. R. Owen (1956). Behavior of ion-exchange resins with mixed solvents. J. Chem. Soc. 1676.

[392] Y. Marcus and J. Naveh (1978). Swelling of anion exchangers in mixed solvents. 4. Swelling and the invasion of lithium chloride. J. Physic. Chem. *82*, 858.

[393] J. L. Pauley et al. (1969). Ion exchange processes in aqueos-organic media. Anal. Chem. *41*, 2047.

[394] R. Smits et al. (1973). Relationship between solvation as expressed by the Walden product and cation exchange selectivity in water-dimethylsulfoxide. Anal. Chem. *45*, 359.

[395] D. Fürtig et al. (1969). Anionenaustausch in gemischten Lösungsmitteln. I und II. Einwertige Anionen am Anionenaustauscher Wofatit SBW. Z. Chem. *9*, 192, 352.

[396] C. H. Jensen and R. M. Diamond (1971). Anion exchange in mixed organic-aqueous solutions. I. Dioxane-water. J. Phys. Chem. *75*, 79.

[397] J. Korkisch et al. (1966). Anion exchange properties of different metal ions in sulfuric acid solutions containing organic solvents. Z. anal. Chem. *215*, 86. Id. (1967). Selective cation exchange separation of cobalt in hydrochloric acid-acetone solution. Separ. Sci. *2*, 169. Id. (1968). Cation exchange behavior of several elements in hydrofluoric acid-organic solvent media. Talanta *15*, 19. Id. (1968). Cation exchange separation of thorium from rare earths and other elements in methanol-nitric acid medium containing trioctylphosphine oxide. Anal. Chem. *40*, 1952. Id. (1964). Ion exchange in non-aqueous solvents. Adsorption behavior of the uranium and other elements on strong-base anion-exchange resins from organic acid organic solvent media: method for the separation of uranium. Talanta *11*, 721.

[398] E. Blasius and R. Schmitt (1969). Sorption behavior of cation exchangers based on polystyrene in aqueous dioxane and dimethyl sulfoxide solutions. J. Chromatogr. *42*, 53. F. W. E. Strelow (1974). Partly non-aqueous media for accurate chemical analysis by ion exchange. Ion Exch. Membr. *2*, 37. Y. P. Denisov (1980). Possibilities for the use of ion exchange sorption from mixed solvents. Tr. Leningr. Politekhn. Int. (373), 65.

[399] J. Korkisch (1966). Combined ion exchange-solvent extraction: a new dimension in inorganic separation chemistry. Nature (London) *210*, 626. Id. (1966). Combined ion exchange-solvent extraction: a new separation principle in analytical chemistry. Österr. Chemiker-Ztg. *67*, 309.

[400] L. Vladescu and H. Hurwitz (1979). Combined ion exchange-solvent extraction. I. Critical study on experimental results. Rev. Roum. Chim. *24*, 1353.

1.2.4 Standardization of Test Methods for Ion Exchange Resins

Günter Kühne
Bayer AG, Leverkusen, Federal Republic of Germany

Introduction

The testing of ion exchange resins covers a wide range of laboratory activities [1]. This results firstly, because testing is based on chemical as well as on physical methods and secondly, because the applications of ion exchange resins are very varied. Naturally-occuring and industrially-produced products are fundamental sources. They exhibit essential features. The various ion exchange materials available are:

synthetic ion exchange resins
cellulose ion exchangers
dextran exchangers
inorganic ion exchangers
various other materials
liquid ion exchangers
ion exchange membranes.

The emphasis in this chapter will be on synthetic ion exchange resins.

But we also have to limit our interest in synthetic materials to synthetic organic resins in (solid) bead or granular form. The reason for this is because more than 90% of the ion exchangers in use are of this type, a fact that naturally is reflected in the scientific and practical work.

Of course, the purposes of testing, the test procedures and methods, as well as the measurements are similar to or the same as those for the other materials. However we find extensive and systematic standardization only for synthetic organic solid resins.

1.2.4.1 Definitions

Before we discuss the chemical, physical and engineering aspects of ion exchangers we must deal with the language. To recall the steps in the development of ion exchange we will return to the year 1935 as the starting point for synthetic organic resins. The descriptions and techniques employed at that time derived from existing activities, but they had to be modified to some extent in order to achieve correct understanding. Definitions had to be found which described the new science as exactly as possible.

Let us take capacity, as a case in point. Defined as "quantity taken up by a unit", people were used to employing any values relative to the material in use, e. g., active carbon, filter media, or inorganic ion exchangers, by weight. The new organic ion exchange resins, being swollen materials (water-soaked), were used in that state. Because they were sensitive to drying-out it made sense to relate the capacity value to the volume. The "operating volume capacity" became the main definition intended to remind people of the new measurement basis. But over the years people working in the ion exchange field became used to the fact that the values were volume-related and they came to omit the term "volume" from the expression, shifting then to new terms such as breakthrough capacity, operating capacity, or plant capacity. Other properties of the new generation of ion exchange resins were described using expressions from daily life. For example, the volume change of the resins was first called "breathing difference". Later on this was replaced by the technical expressions swelling and shrinking, still an explanation was still lacking of whether this increase or reduction in volume was caused by a change in water content, by osmotic influence, by change of solvent as swelling agent or by the exchange of counter-ions. If we follow this development we will see how useful a standardization of the language is. For those interested in the nomenclature development or, even more important, who have to read old articles, there is a table of old and new German expressions in the DIN standards [2].

1.2.4.2 Purposes of testing

All the tests carried out with ion exchange resins are means of answering scientific or technical questions arising from their nature and service. Classifying the purposes we can find the following:

development of new materials
production control and quality assurance
operating properties for product information sheets
individual operating conditions
references for new applications
product properties in plant service
acceptance control on receipt
control of plant service, plant conditions and operation safety.

1.2.4.3 Determination of contents and properties

Test methods enable the physical and chemical characteristics of a wide range of resins to be measured. They are applicable to various resins and resin manufacturers world-wide have made their performance of test methods available in their technical leaflets. Although the application of the methods has to be straightforward and within the capabilities of most laboratories concerned with or interested in the use of ion exchange resins, the interpretation of the results may require considerable experience in ion exchange technology, particularly when one attempts to predict the long-term performance of the resins. In the following list there are both ion exchange resin tests that yield direct measurements, and magnitudes obtained indirectly from such tests. For all the test results different units are used.

A. Physical properties – physical data

particle size distribution
- screen analysis
- resin grading
- wet screen analysis
- dry screen analysis
- cumulative analysis
- effective size
- uniformity coefficient

density
- bulk density
- true density
- apparent density
- specific gravity

volume determination
- vibrated resin volume
- minimum vibrated volume
- backwashed settled volume

moisture
- moisture retention capacity
- swelling water content
- water regain
- interstitial water

crosslinkage

porosity
- true porosity
- pore size
- pore diameter
- pore radii
- pore volume
- BET surface
- inner surface
- mercury porosity

volume change
- shrinking
- swelling
- compressibility
- elasticity

microscopic examination
- appearance
- physical form
- optical aspect
- colour

bead integrity
- perfect whole beads
- whole beads
- cracked beads
- broken beads

contamination

pressure drop
- head loss

stability tests by single or combined influences
- mechanical stability
- chemical stability
- osmotic stability
- thermal stability.

B. Chemical analytical measurements

elementary content

identification of resin matrices and chemical structure

identification of functional groups
degree of functionalization
chemical form – chemical state (degree of regenerated or exhausted form)
contamination during production
contamination during service.

C. Ion exchange reactions

total capacity
- active capacity
- salt splitting capacity
- strong acid capacity
- weak acid capacity
- strong base capacity
- weak base capacity

column performance test
- operating capacity
  - cocurrent regeneration
  - countercurrent regeneration
- leakage
- rinse water requirement.

D. Function of resins

ion exchange
adsorption
ion exclusion
ion retardation
chromatographic separation
release of ions (use as a carrier of ions and compounds).

E. Influence by contact with other substances

heterogeneous catalytic action by counterions
heterogeneous catalytic action as carrier of elements.

F. Filtration properties

pure mechanical filtration
filtration supported by chemical or physical surface properties.

G. Use as reaction reagent

irreversible or reversible chemical reaction of functional groups or counterions

1.2.4.4 Condition of materials being tested

Any test method has to define the conditions the resins have to be in before measurements are initiated. The actual state of the resin has to be included with the results. In cases where the tests are made with materials as they are found, this also has to be stated clearly together with the test results.

In practice we find the following resin conditions:

as delivered
after conditioning
after regeneration
after cleaning
after fractionization
after preparation.

1.2.4.5 Condition of ion exchange resins during or after the tests

With regard to the tests listed in section 1.2.4.3 the procedures of the test have greater or lesser influence on the condition of the material.

Condition by test	Test group 1.2.4.3X
reversible change	A, C, D, E, F, G
irreversible change	A, D, E, F, G
being destroyed	A, B

1.2.4.6 Timing of tests

Ion exchange resins are tested at various states of their useful life. Again this requires a careful selection of the tests:

for new resins from the manufacturer
for resins in operation
  regularly tested
  planned at regular intervals
  in case of operational changes
  in case of plant difficulties
  after use.

1.2.4.7 Plant operating results

Because an ion exchange plant very often has fixed operating conditions, service follows the requirements of standardized test methods. From plant performance records one gathers information on:

operating exchange capacity
pressure drop
regeneration efficiency
adsorption facilities
pressure drop
backwashed settled volume
volume change
compressibility
elasticity
leakage
rinse water requirement
filtration properties.

1.2.4.8 Standard test methods

Test methods published in the literature were first related mainly to the scientific work being done on the development of new resins in the laboratory. Further, information about factors influencing the performance of resins in operating plants were followed by explanatory comments and descriptions of practical test methods.

All the tests were based at first on existing chemical and physical methods, but very often needed alterations to adapt them to special cases in ion exchange practice. These chemical, physical and plant operational techniques for testing ion exchange resins interested different laboratories. Of course, resin manufacturers started testing services for their own products, but they also published the methods for the benefit of the resin market. Equipment manufacturers began testing to learn more for their plant design and calculations. Later the plant owners found it necessary to test resins and also instituted testing services for members of owners' associations and other interested parties. Gradually national institutes have taken up the task of collecting and adopting test methods. It is by their work that true standardization became operative. Finally, international interests have led to world-wide acknowledgement of test methods, which can be understood by most people because the international language of chemistry is used. Some examples of this work can be found in publications by

resin manufacturers:
Bayer
Dow

Duolite
Rohm & Haas
Zerolite, etc.

plant operators:
CEGB
VGB

independant engineers

national institutes:
ASTM [3]
DIN [4]
TGL [5]

international standards:
ISO [6].

1.2.4.9 Number of test methods

If we count the tests listed in section 1.2.4.3, 30 can be found. The principle methods known for these test procedures are about 120. But with individual alterations a number more like 250 is likely.

The total number of known standards and papers under revision for ion exchange resins is:

ASTM 15
DIN 12
TGL 25.

Given this substantial number it is understandable that standardization became necessary; nonetheless, standardization is still a very difficult, if important, aspect.

The industrial, national or international institutes must take into account for their decision the necessity for standardization of certain methods, the economic dimension of this problem and the possibility of finding acceptable methods practicable for most laboratories.

Theoretical aspects or just the impetus towards completeness should never be the deciding factor for more standards.

1.2.4.10 Test apparatus

For each test a certain type of laboratory equipment and a number of tools are necessary, but the aim in standardization should be to limit them as much as possible.

From the beginning, therefore, people should start with a catalogue of the intended kinds of tests, select from existing laboratory equipment a minimum of items which should also be standardized, and develop if necessary new test apparatus only when they can be used for many different determinations.

1.2.4.11 Description of the test methods

Institutes are accustomed, especially those working on standardization, to following a special form of description. The main headings here are:

scope and field of application of the method
concepts
units being used for the results
principles of the method
  in general
  in details
apparatus
reagents
sampling
sample preparation
settling volume
procedure and evaluation of the results
test report
references
previous editions
amendments
explanatory notes.

But before the sequence of tests is fixed in the final description of the standard a table should be set up showing the main operations, the quantities of chemicals being used, their concentration and volume, the time of action, the flow velocities, etc. It is also useful to include reference numbers in the sections and perhaps to finish with simple drawings of the equipment.

We can confirm from practical work in DIN [7] that this allows the test procedure to be carried out correctly, makes it easily understandable and free from unnecessary variations. It is further a practical guide in the laboratory if these are put up next to the test apparatus on water-repellant coated paper.

1.2.4.12 The importance of tests

Besides the technical or scientific reasons for testing there is another important aspect: business and commercial contracts. Before finalizing a contract we need to have data on the materials presented in a product information leaflet.

It is obvious that in large-scale production, either batchwise or continous, all the data have a certain range of values. Therefore tests should be made on samples from each production batch or in adequate intervals from continuous processes. Sampling itself is standardized already for any product and requires only adoption to special conditions of the ion exchange resins [8].

Table 1.23

Frequency: 1 = each sample, 2 = in intervals, 3 = only once

Group	Test	Frequency
Properties or characteristics:		
physical	physical form	1
	shipping weight	3
	screen analysis	1
	mechanical strength	2
	osmotic strength	2
	pressure loss	3
	bed expansion	3
	porosity	2
	volume change	3
	density	3
	moisture	2
chemical	ionic form	3
	operation temperature	3
	pH stability	3
	chemical resistance	3
	crosslinkage	3
	contamination	2
	total capacity	1
	functionalization	1
	adsorption	1
Operating or performance data/conditions/parameters	operating capacity	2
	rinse water requirement	2
	leakage	3
	bed depth	3
	rising space	3
	backwash expansion	3
	max. operating temp.	3
	flow rate	3
	regenerant quantity	3
	concentration	3
Product description	matrix	3
	functional group	3
	type	3
	colour	3
	shape	3

How often the tests have to be made depends on:

the type of data
technical data
characteristic figures
appearance
process control data
the variation of production
the necessary accuracy of the technical information, and
the actual reproducibility.

Finally, it is a legal question which variation of the settled values is permissible and what the warranties or guarantees are. Table 1.23 lists all the data of interest with an indication of frequency of testing required.

Depending on the resin property, experience and the laboratory situation, a test frequency at intervals can be carried out as follows:

on a selected sample when a certain number of samples were taken
on random samples
on a mixture of a certain number of samples.

The benefit is reduced laboratory work. Tests of mixtures allow one to learn of any variation that might be coming up. The test at intervals is possible when low production variations occur and if the limits for the data are wide enough.

Some of the resin properties need to be determined only once at the beginning, because with a fairly constant production – during which other tests are performed – they will not change very much, remaining within acceptable limits.

References

[1] G. Naumann (1976) in: The Theory and Practice of Ion Exchange, International Conference, University of Cambridge, 1976 (M. Streat, ed.), pp. 5.1–5.6, Society of Chemical Industry, London SW1.
[2] DIN 54 400, Definitions.
[3] American Society for Testing and Materials, Philadelphia, Pa. 19103, (United States of America), Water Treatment Materials, Particulate Ion-Exchange Materials, D 1782, D 2187 A–J, D 3087, D 3375.
[4] Deutsches Institut für Normung e. V., D-1000 Berlin 30, (Federal Republic of Germany), Normenausschuß Materialprüfung NMP 895, Series DIN 54 400 ff.
[5] VEB Chemiekombinat Bitterfeld, Bitterfeld (German Democratic Republic), Fachbereichstandard der Deutschen Demokratischen Republik, Group 145 470, TGL 22427/01 ff.
[6] International Organization for Standardization (Federation of National Standards Institutes).
[7] DIN 54 402, DIN 54 403, Determination of the Total Capacity.
[8] DIN 54 401, Sampling.

1.2.5 Laboratory Experiments and Education in Ion Exchange

Konrad Dorfner
Mannheim, Federal Republic of Germany

Introduction

Today's understanding of ion exchange and the advanced development of ion exchangers and ion exchange technology have resulted in a widespread use of ion exchange in laboratories and technical processes throughout the world. Ion exchange is also a general phenomenon in many areas of the natural sciences, but it is not, as we tried to make clear above, general inorganic or organic chemistry, but the chemistry of reactive polymers and macromolecules. Because of the wide dissemination and applicability of ion exchange processes it has become imperative to make their fundamentals available to students in chemistry and other relevant areas.

Teaching science is best done by demonstrating phenomena and facts through experiments. Education can mean communication of fundamentals and their consequences or training for the utilization of results. An ion exchange water treatment plant is one such result of the research and development of ion exchange. It seemed to be advisable to include in this volume a chapter presenting possible laboratory experiments and educational tools for the understanding and the propagation of the science of ion exchange. New ideas and developments may evolve from established experimental procedures. Ion exchangers are only one group in the range of reactive polymers and it can, therefore, be expected that experimental acquaintance with ion exchangers and the education in ion exchange may facilitate – based on fundamental knowledge gained, for instance, about the importance of diffusion processes in reactions between solids and liquids – the transition of knowledge onto the comprehension and use of other reactive polymers.

1.2.5.1 Laboratory experiments

The availability of ion exchange as a laboratory technique is neither a guarantee that all laboratory staff is familiar with it and its value nor that students, after having passed the first traditional courses, would know that many valuable ion exchange laboratory techniques have been developed and that a variety of well-characterized ion exchange materials are readily available. For students of chemistry and related sciences the situation has improved because at all levels of scientific education, starting already in secondary schools, courses are offered in which experiments with ion exchangers are demonstrated or practical work with ion exchangers can be done. In the following some laboratory experiments are described, which it might be useful to include in a special lecture or course for chemistry students on ion exchange and ion exchangers or be used to make oneself familiar with this field.

In the latter case Chapter 1.1 of this volume (introduction to ion exchange and ion exchangers) may be advantageously used. In the following no laboratory experiments will be included that have already been described elsewhere in this volume, either in Section 1.2.3 (properties of ion exchange resins) or in Chapter 1.1 (introduction to ion exchange and ion exchangers). Numerous laboratory experiments have, however, been developed and it will hopefully be understood that a selection can hardly meet all expectations and requirements. But the experiments chosen could form the nucleus of practical work to be associated with the study or teaching of ion exchange.

Ion exchange without columns. A newcomer to the field of ion exchange might get the impression from many textbooks and papers that the commonplace use of ion exchange resins is in some sort of column apparatus. A column setup is required for chromatographic-like separations and is the choice for most ion exchange processes after they have been elaborated and chosen for large scale application. But there is often little point in using columns for simple and general ion exchange. This should be kept in mind and can be easily demonstrated: an excess of the appropriate resin can be swirled in the test solution for several minutes, the resin filtered off and washed, with the wash being added to the eluate. Such a procedure is quicker and less laborious than a column technique. The resins can be regenerated and washed in bulk, no columns need be provided and packed, and cared for, and especially important, the volume of the eluate and washwater is minimized. Therefore, it has certainly been a reasonable recommendation that the concept of ion exchange be somewhat divorced from that of column apparatus in any introduction to ion exchangers. In this volume the benefits of ion exchange in columns have clearly been shown and in the following sections many ion exchange reactions without columns will be described.

Whatever procedure is preferred, an interesting pair of salts for demonstration is $NaBF_4$ and $NaBF_3OH$. The tetraflouroborate ion is stable in aqueous solution but the hydroxytrifluoroborate is readily hydrolyzed. One equivalent of either salt in solution yields one equivalent of base after reaction with the OH-form of the anion exchange resin Dowex 2X8. However, when one equivalent is reacted with the H-form of the cation exchange resin Dowex 50X8, the $NaBF_4$ yields one equivalent of HBF_4 but the $NaBF_3OH$ yields three equivalents of HF plus one of boric acid. The acid solutions so generated can be titrated with standard base solution to a phenolphthalein endpoint for the HBF_4 or HF, and similarly for the H_3BO_3, as if it were monoprotonic, if mannitol is added. The basic solutions can be titrated with standard acid to a methyl red endpoint [1].

Difference between sand filtration and ion exchange. A laboratory experiment can be arranged in two parallel columns in which the difference between sand filtration and sorption by an ion exchange resin is demonstrated. For this purpose extractions from artificially prepared polluted soils are filtered through the sand filter column and the exchanger column, the latter being filled with a strong base anion exchange resin in the OH-form. The sorbed herbicidal anions such as chloroacetate, chloropropionate, chlorobenzoate, or chlorophenoxyacetate are eluted with 5 – 10 ml 4 N sodium chloride solution. By analyzing the eluates and comparing them with the filtrates of the sand filter it can be demonstrated that with ion exchange as compared with the sand filter an increase in the concentration of herbicidal anions by three hundred to five hundred fold can be obtained. In employing the herbicides in the artificial soil samples in trace amounts it can further be demonstrated that the ion exchange process permits the determination of traces of sodium trichloroacetate (TCA), 2,2-dichloro propionic acid (Dalapon® or Dowpon®), 2,4-dichloro phenoxy acetic acid (2.4-D), TCB, dicamba, and other ionogenic herbicides in soil, water, and plant samples. A photometric analysis determines 2–3 mg and a spectrophotometric method determines 0.02 mg herbicide per kilogram of sample [2].

Ways of making ion exchangers. Ways of obtaining ion exchangers by polymerization or polycondensation synthesis are outlined in this volume in some detail in the chapter on synthetic ion exchange resins. It has been pointed out that a fairly wide range of commercial ion exchangers is available, comprising the customary types of cation as well as anion exchangers and even some specific and chelating resins. The sulfonation of a styrene-divinylbenzene matrix in spherical form can be carried out as a laboratory experiment as described above and the chloromethylation of the same polymerizate, which is a key reaction in the preparation of anion exchangers, has also been described. If the latter is chosen as a laboratory experiment care must be taken to avoid any accidents with the toxic choromethyl methyl ether; even better, one of the other chloromethylation reactions of the styrene-divinylbenzene copolymer with, e. g., paraformaldehyde and hydrogen chloride is to be preferred. The value of the chloromethylation reaction should doubtless be explained, in that numerous other reactions of chloromethylated styrene-divinylbenzene copolymers have been studied to synthesize reactive polymers as well as ion exchangers. The synthesis of strongly basic anion exchangers with ionogenic groups of Type I and II can be demonstrated in a laboratory experiment, as well as the transformation into the quaternary ammonium base.

Amination of chloromethylated crosslinked polystyrene beads. In a 100 ml three-necked flask with stirrer, reflux condenser, thermometer and gas inlet tube 20 g of chloromethylated resin are mixed with 50 ml benzene and boiled under reflux for 30 min. The mixture is then cooled to 30 to 35 °C and, while stirring, gaseous, dry trimethylamine is added while the temperature is slowly increased to 50 to 55 °C. After 4 h of gas injection the mixture is kept for 3 h at room temperature. The beads are filtered by suction, washed several times with benzene and dried in vacuo at 50 °C. After drying the beads are treated for 2 h with 10 ml 5% hydrochloric acid and thoroughly washed with water until acid free against methylene red. A determination of the anion exchange capacity can subsequently be demonstrated.

To obtain the quaternary ammonium base 5 g of the above material are transferred into a 150 ml Erlenmeyer flask, mixed with 25 ml 15% sodium hydroxide solution and stirred for 30 min. Then the beads are washed with water to remove the surplus alkali and washed to neutral on a Buechner funnel against phenolphthalein. This Type I strong base anion exchanger can be used for further experiments. For the preparation of a Type II strong base anion exchanger chloromethylated styrene-divinylbenzene copolymer is subjected to successive amination with *N*,*N*-dimethyl-2-hydroxyethylamine and trimethylamine [3].

Poly-*p*-aminostyrene is a convenient starting material to demonstrate the preparation of specific and related ion exchange materials. If it is not commercially obtainable it must be prepared from crosslinked polystyrene with 90% nitric acid and reduction of the poly-*p*-nitrostyrene with sodium sulfide according to the following equation using quasi-constitutional formulae

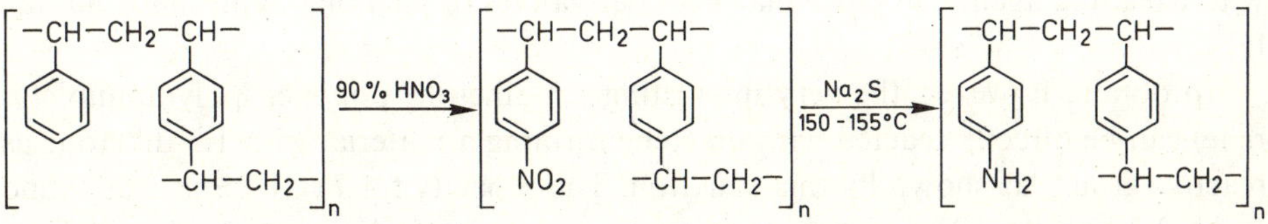

This is the method of Skogseid; the poly(*p*-amino styrene) once obtainable as a commercial product (Trolitul III) was, for instance, used to prepare redox resins by simple diazotation and coupling with suitable redox groups [4]. The synthesis of selective chelating exchangers on the basis of polystyrene was extensively investigated and described by Lieser and coworkers, who obtained such compounds by nitration of polystyrene, reduction, diazotation and coupling with chelating agents. The following processes were presented.

Nitration of polystyrene. 10 g of low crosslinked polystyrene or macroreticular polystyrene beads are added to 100 ml nitrating acid at 0 °C, consisting of 75% by weight 86% HNO_3 and 25% by weight 95–97% H_2SO_4. The reaction vessel is cooled with ice. The polystyrene is stirred 3 min with a magnetic stirrer and the macroporous resin shaken by hand. Then the mixture is poured onto 1 l small pieces of ice. The resin is filtered, washed with 0.1 M sodium hydroxide and distilled water and dried for 24 h in a drying pistol in vacuo over phosphor pentoxide. Yield in both cases 14 g. The color of the nitrated polystyrene is bright yellow. The degree of substitution can be expected to be one.

Reduction of poly(nitro styrene). In a 1 l three-necked flask containing a mixture of 130 g $SnCl_2 \cdot 2H_2O$, 132 ml fuming hydrochloric acid and 148 ml ethanol, 10 g of either poly(nitro styrene) are added and refluxed for 24 h. The reaction product is filtered on a glass frit, washed with distilled water, then with 1 l 2 M NaOH and again with distilled water, and dried for 24 h in a drying pistol in vacuo over phosphor pentoxide. Yield in both cases 8 g. The color of the poly(*p*-amino styrene) is bright brown. The degree of reduction according to elemenental analysis is 96% for the customary crosslinked polystyrene and 84% for the macroporous.

Diazotation of poly(amino styrene). To a mixture of 50 ml fuming hydrochloric acid and 50 ml distilled water cooled to 0 °C are added, firstly, while stirring with either a magnetic or a propeller stirrer, 10 g poly(amino styrene), and secondly, drop by drop a solution of 14 g $NaNO_2$ in 40 ml distilled water over 1 h. During the reaction the temperature is kept at 0 °C by an ice-salt bath. The diazonium compound thus formed is filtered and washed to neutrality with cold distilled water containing a few grams of sodium chloride. The degree of diazotation can be expected to be between 85 and 100%. The poly(*p*-diazonium chloride styrene) is always freshly prepared and immediately used for a further reaction.

By this means, a number of chelating polymers can be obtained by coupling with chelating agents, in particular with derivatives of phenol and aromatic amines [5].

To obtain, however, the very interesting 1,3-diketone polymer, poly(amino styrene) can be directly reacted with diketene forming a material with 1,3-diketone as reactive group, as shown by this equation. The capacity for Fe(III) ions was found to be 3.2 mmol/g. The distribution coefficients for UO_2^{2+}, Cu^{2+}, Ni^{2+} and Fe^{3+} ions have been presented as a function of pH [6].

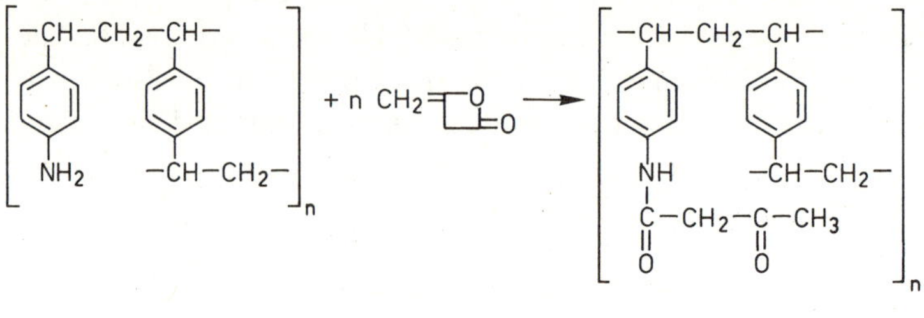

Reaction of poly(amino styrene) with diketene. 10 g poly(amino styrene) at 20 °C are suspended in a solution of 500 ml acetone and 0.5 ml pyridine. After 20 min 10 g diketene are dropped over the course of 10 min, and the mixture is stirred for 4 h at 20 °C. The reaction product is filtered, washed with methanol and dried in vacuo over phosphor pentoxide. 13.6 g of a light brown material can be obtained.

In the actual use of ion exchange resins column processes predominate. In Chapter 1.1 (introduction to ion exchange and ion exchangers) it is shown that simple ion exchange columns can be made in every laboratory from materials which are usually available. The use of a buret, in particular, as an ion exchange glass column is very easy. First, a plug of glass wool is pressed into a 100 ml buret. Then, an aqueous slurry of ion exchange resin is added to make a column approximately 20 cm long. Three 50 ml aliquots of the solution needed for the eventual conversion are then passed through the column and, finally, the column is rinsed with distilled water until it is free of ions. If a break in the experiment is necessary the column containing the resin should be kept under distilled water.

Comparison of batch and column techniques. With the example of the conversion of sodium chloride to hydrochloric acid it can be shown that when a batch technique is used, about 75% of the sodium chloride is converted to the acid, whereas with a column of resin there is 100% conversion to the acid. For this experiment a strong acid cation exchange resin in the hydrogen form is used. In the batch operation 25 ml of the wet resin plus 25 ml of deionized water are put into a 250 ml flask. 50 ml of 0.5 N sodium chloride is added and the mixture stirred for 20 min. The liquid is then filtered off from the resin into a 250 ml graduated flask, resin and filter paper carefully rinsed with deionized water and the volume made up to 250 ml. 20 ml aliquots are titrated with 0.1 N NaOH to determine free acid and in parallel with 0.1 N $AgNO_3$ to determine the chloride content, and the conversion is calculated. For the column process 25 ml of wet resin are put into the buret column, 40 ml of 0.5 N NaCl passed through over about 20 minutes, and collected in a 250 ml graduated flask. The column is rinsed with deionized water and the volume made up to 250 ml. 20 ml aliquots are titrated as before and the conversion is calculated.

Difference in acidity and basicity of weak and strong resin sites. This experiment demonstrates that strong sulfonic acid cation exchangers and strong quaternary amine base anion exchangers are always ionized and, therefore, may be used at any pH, while weak carboxylic acid cation exchangers are only ionized about pH 7 and can only be used for alkali treatment, and weak tertiary amine base anion exchangers absorb acids by a single addition and can, therefore, only operate at pH below 7. In small burets 5 ml of each of the wet ion exchangers are put and the cation resins are each regenerated with 6 bed volumes 1 N HCl and rinsed with 6 bed volumes deionized water at 1 bed volume per 3 min. 25 ml 0.1 N NaOH/0.1 N NaCl solution are pipetted on to each column and passed through it at 1 bed volume per 10 min, the eluate being collected in a 100 ml volumetric flask. The resins are rinsed with deionized water and the volumetric flasks filled up to the mark. 20 ml aliquots are titrated with 0.1 N NaOH using phenolphthalein indicator, and then the neutralized solution is titrated with 0.1 N $AgNO_3$ using potassium chromate indicator. It will be seen that both resins exchange sodium from sodium hydroxide for hydrogen to form water but only the strongly acidic resin gives an acidic eluate due to the exchange of hydrogen for sodium from the sodium chloride by splitting the salt.

The anion exchange resins are regenerated with six bed volumes 1 N NaOH and rinsed with 6 bed volumes deionized water at 1 bed volume per 3 min. 25 ml of 0.1 N HCl/0.1 N NaCl solution are pipetted onto each resin and passed through it at 1 bed volume per 10 minutes, the eluate being collected in a 100 ml volumetric flask. The resins are rinsed with deionized water and filled up to the mark. 20 ml aliquots are titrated with 0.1 N HCl and then, after neutralization, with 0.1 N $AgNO_3$. This experiment shows that both resins exchange chloride ions from hydrochloric acid for hydroxyl ions to form water, but only the strong base resin gives an alkaline eluate due to the exchange of hydroxyl ions for chloride from the sodium chloride.

Equilibrium studies in the presence of complexing agents. In the course of studies carried out on ion exchange equilibria it was shown that the chief factors determining the bond between an exchange resin and any ion are the charge on the ion and its hydrated radius. Thus, it was possible to compare the hydrated radii of ions by determining their relative equilibrium coefficients. Likewise, it was possible to compare ions of different charge and so to establish the well-known replacement series for all ions, similar to the lyotrophic series. For convenience such series are placed at the end of this paragraph. But for education in ion exchange it should also be demonstrated that the determination of the equilibrium constants and structures of complexes is an application of exchange resins which has proven to be very valuable. What is to be focussed on is that when a cation exchanger is allowed to compete with a complexing anion for a cation, it is possible to determine the dissociation constant of the complex by a few simple equilibrium experiments. For this purpose, it is preferable if the cation being investigated is present in very low concentration. Carrier-free radioisotopes or other isotopes present in concentrations below 10^{-3} or 10^{-4} M may be used conveniently for this purpose. By keeping the concentration low, a negligible amount of complexing agent and exchanger are combined with this substance, and they may thus be determined by making approximate calculations. In the first experiment for this laboratory application of ion exchange resins the dissociation constants for the citrate and tartrate complexes of strontium are determined. This element is first shaken with the exchanger in the ammonium form in a solution of several tenths molar ammonium chloride and a value for the exchange constant in the absence of citrate is obtained. Next, it is equilibrated with a solution of the same composition, which, however, contains citrate or tartrate of several tenths molar concentration, at a neutral pH. The distribution of the strontium ions between the exchanger and the solution in the absence and presence of the complexing agent is determined by counting the radioisotope of this element. From these data, the dissociation constants of the complexes are calculated. The values obtained agree closely with those reported in the literature [7].

Addendum: **Relative affinity orders** for various ions with various exchangers.
Strong acid cation exchanger, 8% crosslinked:

$Ba^{2+} > Pb^{2+} > Hg^{2+} > Cu^{2+} > Sr^{2+} > Ca^{2+} > Ni^{2+} > Cd^{2+} > Cu^{2+} > Co^{2+} > Zn^{2+} \simeq Cs^{+} > Rb^{+} > Fe^{2+} > Mg^{2+} \simeq K^{+} > Mn^{2+} > NH_4^{+} > Na^{+} > H^{+} > Li^{+}$.

Weak acid cation exchanger, carboxylic:

$H^{+} > Cu^{2+} > Co^{2+} > Ni^{2+} > Ca^{2+} > Mg^{2+} > Na^{+}$.

Strong base anion exchanger, Type I:

Benzene sulfonate > salicylate > citrate > I^- > phenate > HSO_4^- > ClO_3 > NO_3^- > Br^- > CN^- > HSO_3^- > BrO_3^- > NO_2^- > Cl^- > HCO_3^- > IO_3^- > formate > acetate > propionate > F^- > OH^-; Type II: nearly the same except iodide is less than phenate and hydroxide comes after bicarbonate.

Weak base anion exchanger:

$OH^- > SO_4^{2-} > CrO_4^{2-} > AsO_4^{2-} > PO_4^{3-} > MoO_4^{2-} > Cl^-$.

Chelate resin Dowex A-1:

$Pd^{2+} > Cu^{2+} > Fe^{2+} > Ni^{2+} > Pb^{2+} > Mn^{2+} > Ca^{2+} > Mg^{2+} > Na^{+}$.

Capacity for cations of different valencies. Selectivity should not be confused with capacity, which for, say, a cation exchange resin is equivalently the same for cations of different valencies. An easy practicable laboratory experiment can be used to demonstrate this. A strong acid cation exchanger is used to liberate acid from metal salts, and the total capacity of the resin is determined by titrating the liberated acid. About 1 g of dried resin is accurately weighed and placed in a flask together with 50 ml of deionized water. About 2 g of the metallic salt, for example sodium chloride in one case and magnesium chloride in the other, are added and the mixtures stirred for 10 min. The liberated acidity in the reaction flask is titrated using 0.1 N NaOH against methyl orange as indicator. The calculated resin capacity in mequ/g of dry resin will be found to be the same, within slight experimental errors. This may be repeated with other metal salts.

Determination of distribution coefficients. As it might be of importance to know how distribution coefficients are determined this should be demonstrated or done by the student himself in the laboratory. As outlined previously there are in fact three kinds of distribution coefficients which have been defined in the IUPAC Recommendations on Ion Exchange Nomenclature. The determination of the distribution coefficient then depends on whether a static (batch) or a dynamic (column) procedure is chosen. It is important that the distribution coefficient be presented for a given type of exchanger, if possible, with its history, the aqueous phase, e. g., 0.1 – 12 M HCl, the temperature, determination procedure of distribution coefficient and analyses of the elements. The most widely-used distribution coefficient is D_g, i. e., the ratio defined as the amount of an element per gram exchanger to the amount of the element per milliliter of the aqueous phase. This definition is chosen as the chemical species of an element in the resin phase is usually different from that in the aqueous phase.

Batch equilibration technique. Whatever the purpose of the knowledge of a distribution coefficient might be, its determination can be carried out analogously to the manner described in the following for calcium in hydrochloric acid on a commercial strong acid cation exchanger. Preparatory to its use the resin of selected particle size is purified by repeated washing with HCl, NaOH and water. The H-form of the cation exchanger is dried at 60 °C in an air bath. An exactly weighed amount of calcium carbonate is dissolved in excess hydrochloric acid, the solution evaporated to dryness and the residue taken up in hydrochloric acid to give a solution containing 10 mg of the element per milliliter. Analytical grade concentrated hydrochloric acid is used to prepare solutions of 0.5, 1.0, 2.0, 4.0, 6.0 and 8.0 M HCl. 1 g of resin and 1 ml of calcium solution are transferred into a 50 ml Erlenmeyer flask containing 24 ml of the HCl solution. The flask is stoppered and put into a thermostat at 21 ± 0.5 °C for a week, under occasional agitation. After filtering off the resin the amount of the element is determined by EDTA titration (NN indicator) and the weight distribution coefficient calculated. To obtain the volume distribution coefficient D_v, which is more suitable for chromatographic calculations, the formula

$$D_v = D_g \cdot \rho$$

is used, where ρ is the weight in grams of the dry resin in a resin bed volume, and is equivalent to the apparent density, which is about 0.45 for cation exchangers and 0.35 for anion exchangers. The D-values obtained can be plotted log D versus the concentration of the "eluant", i. e., in the above case, the molarities of HCl.

Column method. Determinations of the volumn distribution coefficient are also conveniently made by means of a column. This method is particularly suited when D_v is less than ca. 10. For the actual measurement small aliquots of the solution containing the ions of interest are added to columns of the resin pretreated with, e. g., the appropriate hydrochloric acid solutions. Then the columns are eluted with the same acid solutions. The values of D_v are computed from the number of geometric column volumes, i. e., height × cross sectional area, of effluent (eluant) at which the ion in question appears in maximum concentration. Since appropriate corrections are to be made for interstitial or void volume the whole method can be represented by the general formula

$$D_v = \frac{v}{A \cdot l} - \alpha,$$

where v is the volume in ml of eluant, which moves the maximum concentration of the band a distance l cm in the column of cross section A cm^2 and of void volume α. Because l in this equation cannot be measured the total length L of the column for which a volume v_{max} of eluant is necessary until the ion appears in maximum concentration in the effluent is used. This finally leads to the formula

$$D_v = \frac{v_{max}}{A \cdot L} - \alpha.$$

On the basis of this equation the volume distribution coefficient can be calculated from measurable data. The approximate value of v_{max} can be determined by adding the eluant and solution containing B ions in low concentration ($c_{B,O}$) continuously onto the column and establishing the volume of effluent belonging to the 50% breakthrough concentration $c_{B,O}/2$. The latter is equal to v_{max} in the ideal case.

Selectivity calculations. What is basically known about an ion exchange resin is its capacity. If a solution contains competing counterions one may or may not be preferentially exchanged after equilibrium has been reached. This is the phenomenon of ion exchange selectivity as described in more detail elsewhere in this volume. Selectivity relationships can be used in calculations to obtain an answer to the question of whether a proposed ion exchange process should be pursued. They can be useful even though only approximate selectivity data are available. Ion exchange calculations are based upon the mass action expression from which the term K_B^A as the selectivity coefficient immediately follows, for the reaction where A is the ion leaving the solution and going into the resin. Before coming to an example for laboratory experiments or demonstration purposes it must be emphasized that the mass action expression is more manageable when put in terms of equivalent fractions. For monovalent ion exchanges if the total ionic concentration (normality) of the solution is C, the equivalent fraction of ion A in solution is $X_A = \frac{[A]}{C}$, and the equivalent fraction of ion B in solution is $X_B = \frac{[B]}{C}$. Since only two exchangeable ions are present, $X_A + X_B = 1$. In the resin phase, $\overline{X}_A = \frac{[\overline{A}]}{\overline{C}}$, $\overline{X}_B = \frac{[\overline{B}]}{\overline{C}}$, and $\overline{X}_A + \overline{X}_B = 1$. If these values of $[\overline{A}]$, $[A]$, $[\overline{B}]$, and $[B]$ are substituted in the mass action expression, C and $\overline{C}$ drop out, and the expression can be rearranged to

$$\frac{\overline{X}_A}{1 - \overline{X}_A} = K_B^A \frac{X_A}{1 - X_A}.$$

This form of the relationship is the most useful for the discussion of exchange between monovalent cations or anions on fully ionized exchange resins. This relationship is also called the equivalent fraction form of the law of mass action expression for ion exchange; it allows the discussion of several important characteristics of the system:

1. $\overline{X}_A$ defines the limit to which the resin can be converted to the A-form when in contact with a solution of fractional composition X_A; the whole left side of the equation gives the state of the resin when the ion exchange process is run to the point where the effluent equals the influent.
2. $\overline{X}_A$ is alternately the maximum degree of regeneration that can be achieved with a regenerant of fractional composition X_A; if the regenerated resin has a composition $\overline{X}_A$, the purity of the effluent from the column in the next exhaustion can be estimated.
3. As the value of X_A approaches unity, the value of $\overline{X}_A$ approaches unity, regardless of the value of the selectivity coefficient K_B^A.
4. From the latter it follows that, if enough pure A solution is available, i. e., $X_A = 1.0$, the resin can be converted to the pure A-form, even though it is quite selective for the ions of B. This is theoretically true, but it becomes impractical if the selectivity coefficient is too unfavorable. In such cases the exchange front widens rather than sharpens as it moves down the column, with the effect that the exchange becomes very inefficient.

The terms C and $\overline{C}$ do not appear in the equivalent fraction form of the mass action expression for monovalent exchange, i. e., neither the total ionic solution concentration nor the total ion exchange capacity of the resin per unit volume affects the equilibrium distribution between monovalent ions. An example can best demonstrate the use of this mass action relationship.

Example: Removal of nitrate from a chloride well water. The permissible content of nitrate in drinking water is limited in most countries; e. g., in the USA to 0.9 mequ/l. A possible method for reducing the nitrate concentration in otherwise acceptable water is by ion exchange for chloride or sulfate ion. For demonstration the question can be put: What is the maximum volume of water of the following composition

cations	Ca^{2+}	Mg^{2+}	Na^+	total
mequ/l	1.0	1.0	2.5	4.5
anions	Cl^-	SO_4^{2-}	NO_3^-	total
mequ/l	3.0	0.0	1.5	4.5

that could be treated per liter of Type I strong base anion exchange resin? It is assumed that the exchange material has a total capacity of 1.3 equ/l and a selectivity for nitrate over chloride of 4. To calculate the fraction of the resin capacity that can be put into the nitrate form the mass action expression is used together with the composition of the solution. The equivalent fraction of nitrate in solution is equal to the nitrate concentration divided by the total anion concentration, or 1.5/4.5 = 0.33. Substituting this, and the value of $K_{Cl^-}^{NO_3^-}$, one gets:

$$\frac{\bar{X}_{NO_3^-}}{1-\bar{X}_{NO_3^-}} = K_{Cl^-}^{NO_3^-} \frac{X_{NO_3^-}}{1-X_{NO_3^-}} = 4\,\frac{0.33}{1-0.33} = 1.97 .$$

$$\bar{X}_{NO_3^-} = 0.66 .$$

This calculation indicates that a maximum of 66% of the resin sites can be loaded with nitrate ions from the given water. The maximum useful capacity for nitrate ion will be, therefore:

$$1.3 \text{ equ/l} \cdot 0.66 = 0.86 \text{ equ/l}.$$

The volume of water that can be treated per cycle will be

$$\frac{0.86 \text{ equ/l (resin)}}{1.5 \cdot 10^{-3} \text{ equ/l (water)}} = 5.7 \cdot 10^2 \text{ l/l}.$$

This is an attractive capacity but the regeneration of the nitrate form of the resin is inefficient, and cyclic operation may be marginal because of the large volume of regenerant required.

It might also be of interest to demonstrate calculations in monovalent-divalent systems with respect to selectivity. The equivalent fraction form of the mass action expression used above may also be used for exchanges between divalent ions such as magnesium and calcium. However, when the two exchanging ions differ in their charges the situation becomes more complex. The conventional mass action expression for such a reaction has already been given in this volume:

$$K_{A^+}^{B^{2+}} = \frac{[\overline{B}^{2+}] \cdot [A^+]^2}{[\overline{A}^+]^2 \cdot [B^{2+}]}.$$

When this equation is converted to the equivalent fraction form, the terms C and $\overline{C}$ do not drop out:

$$\frac{\overline{X}_{B^{2+}}}{(1 - \overline{X}_{B^{2+}})} = K_{A^+}^{B^{2+}} \frac{\overline{C}}{C} \frac{X_{B^{2+}}}{(1 - X_{B^{2+}})^2}.$$

This equation may be used as well for an exchange between a monovalent and a divalent ion on a fully ionized resin. Of importance is that in this case, the equilibrium resin composition is dependent on the selectivity coefficient, the ratio of ions in solution, and the total ionic concentrations of the solution and resin phases. The term $K_{A^+}^{B^{2+}} \frac{\overline{C}}{C}$ is called the apparent selectivity. For a given ion exchange resin, the total capacity of the resin, $\overline{C}$, is constant for strong acid and strong base resins that are ionized over the entire pH range. However, the total ionic concentration in solution may vary widely, and since C is in the denominator of the apparent selectivity term, as the solution concentration decreases, the left side of the equation becomes larger. This means that as the solution in contact with the resin becomes dilute, the resin converts more strongly to the divalent ion form. This monovalent-divalent effect not only allows the shifting of the equilibrium in the system to a more favorable value, it may also determine whether the approach to equilibrium is practical. If the above apparent selectivity is greater than unity, the exchange front is self-sharpening and the exchange is efficient. If this term is less than unity, the front is non-sharpening and attainment of equilibrium loading will be difficult. In many cases, this dependency on solution concentration is the decisive factor in whether an envisaged process is practical or not.

Example: Removal of nitrate from a sulfate well water. This example demonstrates how the ionic background in solution is often the decisive factor for whether a given ion can be removed by ion exchange on a practical basis. The question here can be: What is the maximum volume of water that can be treated per liter of strong base resin if the water has the following composition:

cations	Ca^{2+}	Mg^{2+}	Na^+	total
mequ/l	1.0	1.0	2.5	4.5
anions	SO_4^{2-}	Cl^-	NO_3^-	total
mequ/l	3.0	0.0	1.5	4.5

It is assumed that the total resin capacity is 1.3 equ/l and the selectivity coefficient for sulfate over nitrate is 0.04. The calculation is then as follows:

$$X_{SO_4^{2-}} = 3.0/4.5 = 0.67$$

$$C = 4.5 \cdot 10^{-3} \text{ equ/l}$$

$$\frac{\overline{X}_{SO_4^{2-}}}{1-\overline{X}_{SO_4^{2-}}} = K_{NO_3^-}^{SO_4^{2-}} \frac{\overline{C}}{C} \frac{X_{SO_4^{2-}}}{(1-X_{SO_4^{2-}})^2} = 0.04 \frac{1.3}{4.5 \cdot 10^{-3}} \frac{0.67}{(0.33)^2} = 213$$

$$\overline{X}_{SO_4^{2-}} = 0.93$$

$$\overline{X}_{NO_3^-} = 0.07\,.$$

Only 7% of the capacity of the resin will be available to the nitrate ion from this water, even though the resin is intrinsically selective for nitrate over sulfate. Naturally ocurring water seldom, if ever, has a pure chloride or sulfate background as in this and the previous example. However, it is apparent from them that the capacity of the resin for nitrate will vary widely depending on the ratio of chloride to sulfate and the total ionic concentration in the water being treated [8].

It is to be mentioned that the author of this guide for the selectivity calculations disclaims, because of the approximate nature of the mass action relationship used and the well-known nonideality of ion exchange resin behavior, the accuracy, precision, reliability, or universality of these calculations, but emphasizes that they are useful tools that can save work and embarrassment by suggesting the feasibility of a proposed process from a minimum of information. A relation that has been derived for the calculation of unknown selectivity coefficients from known values of tertiary reactions and which was confirmed by experimental investigation of the systems $H^+/Na^+/Ca^{2+}$, $H^+/Na^+/Co^{2+}$, and $H^+/Na^+/Al^{3+}$ with a strong acid cation exchanger appears to be quite useful and could also be presented and discussed [9].

Experiments with highly selective resins. Lewatit® TP 207 is a weakly acidic macroporous cation exchanger with chelate-forming imonodiacetate groups for the uptake of heavy metal cations from weakly acidic to weakly basic solutions. However, before carrying out laboratory experiments one should first become familiar with how to satisfy certain conditions imposed by the characteristics of this particular resin. The following provisions should be studied and possibly used as guidelines for working with highly selective ion exchange resins on a laboratory scale.

Instructions for Lewatit TP 207. This ion exchange resin is supplied in the di-Na-form, which means that the active groups are entirely loaded with sodium ions. If the exchanger in this form is subjected, for example, to a neutral effluent containing heavy metals, the treated water will show for many bed volumes a strong alkaline reaction with pH $>$ 10, which is undesirable in many instances. For this reason Lewatit TP 207 should be regenerated prior to its first application, followed by conditioning partly with sodium ions. For the selective adsorption of copper, vanadium and uranium in the form of their cations Cu^{2+}, VO^{2+} and UO_2^{2+}, Lewatit TP 207 can generally be applied in the H-form. Conditioning with sodium hydroxide solution is then unnecessary. The treated solution, i. e., the filtrate of the ion exchanger, in this case will have an acidic reaction. Special conditioning is not necessary in the presence of ammonia or with high concentrations of ammonia salts in an alkaline solution fed to the exchanger. To regenerate the new ion exchange resin as supplied, about 140 ml resin are filled into a column and regenerated with 250 ml 2 N HCl at a flow rate of maximum 500 ml/h. This is followed by rinsing at the same rate with 500 ml fully demineralized water. For conditioning the ion exchange resin which is now in the H-form, exactly 100 ml are placed into a measuring cylinder and tapped down under water and washed into a 500 ml beaker. The supernatant water is then carefully poured away, and the ion exchange resin stirred by a mixer for 2 h following the addition of 150 ml 1 N NaOH solution and approximately 2 g sodium chloride. This process will increase the volume of the ion exchange resin by about 25%. The ion exchanger, now in the mono-Na-form, is washed into the column and is ready for a subsequent loading cycle. The flow rate when loading the ion exchange resin should normally not exceed 1 l per 100 ml ion exchanger per h (service flow rate: 100 bed volumes per h). In special cases, i. e., with highly concentrated solutions, lower flow rates must be chosen. The regeneration of 100 ml Lewatit TP 207, i. e., the elution of the metals taken up, can be carried out, for example, with 150 – 200 ml 10% HCl solution, 150 – 200 ml 10 – 15% H_2SO_4 solution, or, with the necessary precautions with, 150 – 200 ml 10 – 12% HNO_3 solution. The rinse water requirements for 100 ml resin are 500 ml fully demineralized water with a flow rate the same as for the regeneration.

Experiment: Lewatit TP 208 is another weakly acidic macroporous cation exchanger with chelate-forming iminodiacetate groups. It is very suitable for the purification of brine. The only economical method of removing calcium and magnesium from brine, which is of decisive importance for the life expectancy of ion exchanger membranes and for the energy consumption in sodium chloride electrolysis, is to use highly selective ion exchange resins. The following experiment can be used to demonstrate that this resin with iminodiacetate groups satisfies the demands with regard to capacity that have to be made on an ion exchange resin for brine treatment. The resin shows a marked increase in the operating capacity with rising pH, and residual calcium content if ion exchanger membranes are used in sodium chloride electrolysis. The experiment described here is in fact the determination of the separation factor Ca/Na.

To convert the ion exchanger resin into the total sodium form exactly 50 ml ion exchange resin in the H-form in demineralized water is measured into a 50 ml graduated cylinder using a damping device, and then poured into a 1 l glass beaker. The supernatant water is suctioned off using a frit and 600 ml sodium hydroxide solution $c = 1$ mol/l is poured onto the ion exchanger specimen. Until the exchange reaction has ended, the ion exchange resin is stirred for about 5 h with an anchor mixer and then poured into a column of 22 mm diameter and 300 mm height. The excess caustic soda solution is displaced with 150 ml 1 molar sodium chloride solution. As a simulated brine for exhausting the ion exchange resin to equilibrium a solution is prepared containing sodium chloride $c(NaCl) = 5$ mol/l and calcium chloride $c(1/2\ CaCl_2) = 0.05$ mol/l. This solution is percolated with a throughput of 500 ml/h. Equilibrium is reached when the calcium content in the effluent is the same as in the influent. The excess brine is then displaced with 4 bed volumes of 0.1 molar sodium chloride solution followed by 1.5 to 2 bed volumes of demineralized water in order to wash the remaining sodium ions out of the resin bed. For desorbing the ion exchange resin and determining the calcium and sodium ions fixed to the resin, 100 ml 10% hydrochloric acid are percolated through the column with a throughput of 250 ml/h. The column is then rinsed with 250 ml demineralized water, and the desorbate and rinse water are collected together in a volumetric flask with a capacity of 1 l, which is topped up to the mark with demineralized water. The calcium and sodium contents are determined in aliqot parts. The molar ratio (1/2 Ca): Na, scaled up to 1 l ion exchange resin gives the separation factor = *c*(1/2 Ca) mol/l resin/*c*(Na) mol/l resin. A high separation factor indicates a high affinity of the resin for calcium under the experimental conditions. The latter can be varied with respect to pH and it will be found that as of pH 6 the separation factor is not dependent on the pH. It is therefore possible to determine a mean separation factor: for this resin it is 8.0. With the aid of this factor the maximum possible calcium adsorption can be calculated as 88% of the total capacity, or as 47.5 g Ca/l resin.

Diffusion experiments. To make clear that ion exchanges are not instantaneous, experiments can be performed to demonstrate dependence on the concentration and temperature of the solution used, and above all the diffusion of the exchanging ions concerned.

With regard to diffusion it can be imagined that ion exchange depends on the wandering of ions within the solution, across the Nernst or boundary film around the resin particle and within the resin bead to and from the exchange site. Under the conditions choosen in the following experiment these two effects are probably the controlling factors. In this experiment from a resin in the A-form the A^+ ions are released and are removed as they are produced. The rate of release of A^+ ions with respect to time is recorded. The results obtained by this method only demonstrate the principle in general terms and are not suitable for accurate determination of the reaction rates.

Experiment: About 1 g strong acid cation exchanger in the H-form is accurately weighed and transferred to a 250 ml beaker flask. 50 ml deionized water, 5 ml 0.1 N NaCl solution and methyl orange indicator are added. The stirrer and a stop watch are started immediately. Titrations of the released acidity are carried out by adding excess 0.1 N NaOH and noting the time at which the methyl orange end point is reached. The procedure is carried on until no further acidity is released from the resin. The values obtained are summarized in a graph of titre plotted versus time. The sodium chloride added at the beginning of the experiment acts as a displacement agent and may be varied so that a number of equilibrium curves will be successively obtained.

The process described in this experiment is usually understood as ordinary ion exchange where all of the counterions in any given resin bead are thought to diffuse simultaneously. But already in 1965 Helfferich described circumstances in ion exchange processes where a sharp moving boundary develops between the eluted and uneluted portion of the resin within an ion exchange resin bead. He anticipated that such processes would result when conversion of the resin involves consumption of the entering ion by a chemical reaction which was fast compared to the rate of diffusion. No experimental data were presented by Helfferich to confirm the hypothesis of a moving boundary, which was used by him for instance for dissociation, neutralization, hydrolysis, and complex formation. However, only a few years later Dana confirmed this hypothesis by observing the elution of the copper ammine complex from a sulfonated styrene-divinylbenzene copolymer cation exchange resin with sulfuric acid. It thus became possible to make a mathematical model of ion exchange with intraparticle rate control visible since the elution is marked by a change in the color of the resin from deep blue to light yellow. As the elution process takes place, a sharp moving color boundary develops between the eluted and the uneluted portion of the resin. The color change is due to the destruction of the copper ammine complex as the existing counterion by the hydrogen ions which penetrate the resin according to the following reaction equation (which takes place within the resin phase):

$$\text{R-Cu(NH}_3)_n^{2+} + n\text{H} \rightarrow \text{R-Cu}^{2+} + n\text{NH}_4^+.$$

Experiment: A commercial strong acid cation exchanger, e. g., Dowex 50WX8, 16–50 mesh, is preconditioned by first backwashing it in a column with water. This removes the fine particles and most of the broken particles. The ionic form of the resin is then changed several times by contacting it successively with 2.0 N sodium chloride, distilled water, 2.0 N sulfuric acid, distilled water, 2.0 N ammonium sulfate, distilled water, 0.25 M copper sulfate solution containing 2.5 mol/l ammonia, and again distilled water. In all cases twice the amount of solution required by the resin for complete stoichiometric exchange is used and the flow rate through the column is kept slow to insure a complete exchange of ions. The resin is removed from the column and stored in distilled water in a tightly stoppered bottle.

For the actual experiment a number of resin beads are selected that are not cracked, broken or non-sperical in shape. The selection is done under a microscope by putting a quantity of resin beads on a watch glass and covering them with distilled water. The light is reflected upwards from below the beads. Under these lighting conditions cracked and deformed beads can usually be detected even though the beads have a very deep blue color. The good beads are then removed from the watch glass with tweezers. A small amount of resin beads loaded with the copper tetrammine complex is put into a filtering crucible. The crucible is arranged for vacuum filtration using a water aspirator as the source of the vacuum. A small quantity of 5 N sulfuric acid is added to the resin in the crucible, which starts the elution of the copper tetrammine from the resin. However, before the resin is entirely eluted the acid solution is removed with vacuum and distilled water is immediately added to the crucible. The vacuum is left and water is added continuously until all of the acid has been removed. The resin beads are then removed and upon examination under the microscope they have a blue core with an outer transparent yellow shell. This is because the movement of the core inwards has been stopped. The resin bead containing the core can then be split in half using a sharp razor blade and the core diameter can be measured from the open face.

In extending this experiment and using a special cell constructed from the body of a number 3, oblique bore, ground-glass stopcock the acid elution of the dark blue cupric ammonia complex can be photographed at tuned intervals of several seconds and measured to provide rate data. Analysis of the data by means of a theoretical model will indicate that the overall elution process is controlled by a combination of internal and external mass transfer when acid concentrations less than 1.0 N are employed. At these acid concentrations the interdiffusion coefficient for the resin phase will be found to be $3 \cdot 10^{-6}$ cm^2/s, independent of acid concentration but somewhat dependent on particle size. Data collected at higher acid concentrations were, according to the original work, not amenable to analysis by the method employed. The model should be useful to demonstrate prediction of the time required to elute the cupric ammine complex or for designing equipment to carry out this process [10].

Analytical chemistry with ion exchangers. To demonstrate the application of ion exchangers in analytical chemistry numerous experiments could be selected from the literature. In the earlier days of ion exchange in the laboratory an experiment involving the column separation similar to the chromatographic separation of two ions suggested itself as suitable for ion exchange studies by students. An experiment of this type was completely different from those normally performed in undergraduate courses. Thus, for instance, the separation of copper and silver in a column conveniently made by packing an ordinary stopcock buret with resin was found to provide a suitable system for such studies. The silver ion was readily titrated with potassium thiocyanate using ferric alum as the indicator, while an iodometric titration of copper with sodium thiosulfate could easily be carried out. The copper and silver ions as silver and cupric nitrate in 0.1 N solution were absorbed at the top of the column of strong cation exchanger, washed with distilled water, and then eluted by passing a 1 M sodium nitrate solution through at the rather rapid rate of 3 to 4 ml/min.

A very simple experiment that freshmen students with even a poor laboratory background can perform to good profit is the determination of total salt concentration in water by the use of ion exchange resins. This experiment is performed in the usual way. It takes about two hours to regenerate the resin, rinse the column, percolate and elute the sample under test. But since teaching experience showed that the students spent too much time watching drops of solutions coming out the ion exchange column, a simple modified exchanger apparatus was assembled consisting of the resin column, a graduated pipet for collecting effluents, a pipet-filling bulb, and a Y junction for discarding effluents. This description of the apparatus demonstrates steps towards the automation of ion exchange in analytical chemistry. The experiment performed by the students using this apparatus then consisted of passing an unknown solution of potassium chloride through a strongly acidic cation exchange resin – e. g., Dowex 50W. This solution is converted into the corresponding mineral acid, and the effluent is titrated with a standard 0.1 N NaOH, using methyl orange as indicator. The results obtainable with this apparatus are, compared with the traditional procedure, about eight times faster, without a loss in precision [11].

Another simple ion exchange experiment having several features that make it attractive as an introduction to the use of ion exchange resins in analytical chemistry is the determination of calcium in dietary supplement tablets. Because of its simplicity, even students with relatively weak laboratory backgrounds can obtain satisfactory results, providing them with an opportunity to acquire experience in the analysis of a real world sample. During the experiment the students should repeatedly be reminded that low flow rates are necessary to ensure that efficient exchange occurs, because, as with most ion exchange experiments, rapid passage of the sample solution through the column is the most significant source of error.

Column preparation. A plug of glass wool is pressed into a 50 ml buret and then an aqueous slurry of strong acid cation exchanger, e. g., Dowex 50WX8, 20–50 mesh, capacity 1.8 mequ/ml, is added to make a column approx. 10 cm long. Three 25 ml aliquots of 3 M hydrochloric acid are then passed through the column to convert the resin to the H-form. Finally, the column is rinsed with distilled water until free of chloride.

Sample preparation and analysis. A weighted water-soluble calcium lactate tablet is ground to a fine powder using a mortar and pestle. An accurately weighed portion of this ground tablet (0.20–0.50 g) is then placed in a 150 ml breaker and dissolved in about 100 ml of hot (70–80 °C) distilled water. The resulting solution is then passed through 1 Whatman filter paper to remove traces of undissolved binders. The contents of the beaker and these rinsings are combined with the filtrate. The solution is allowed to cool to room temperature and then diluted to 250 ml with distilled water.

A 250 ml aliquot of this solution is passed through the exchange column at a rate of approx. 2–3 ml/min. This is followed by three 25.0 ml aliquots of distilled water eluted at the slightly faster rate of approx. 5 ml/min. The eluate is collected in a 250 ml Erlenmeyer flask. When approximately 100 ml have been collected, it is titrated with a standard (≈ 0.01 M) sodium hydroxide solution using phenolphthalein as an indicator. This procedure is then repeated for two additional 25.0 ml portions of the sample solution. From the average volume of sodium hydroxide solution required to reach the endpoint, the amount of calcium present in the tablet is calculated.

The author of this laboratory experiment notes that to enable students to complete the experiment in a three hour laboratory period, they were usually provided with columns containing the resin under distilled water, which have been previously prepared by student assistants. Reasonably good agreement among the values of EDTA titrations performed in parallel as a check on the students' results can be expected. These may be compared with the corresponding nominal values of calcium content provided by the manufacturers [12].

Softening or displacement and two-stage deionization. The demonstration of some common uses and applications of ion exchangers can include the most important ones, which are the softening and the deionization of water. It should be explained that the most important use of ion exchangers is more correctly termed displacement of ions present by the counterions of the resin, with the best known application being that of water softening. The two-stage deionization may be demonstrated by an experiment showing the preparation of deionized water in the common way with a strong acid cation exchanger column and a strong base anion exchanger column in series, where the hardness, the conductivity, and the pH before and after the water has been percolated through the two columns are measured. But, depending on the level, the experiment of deionization of a solution might as well be conducted by using colored ions enabling the audience to observe the formation of colored bands on the resins and the decolorization of the solution.

Water softening. For this experiment tap water is softened by a strong acid cation exchanger in the sodium form. 25 ml of the resin are put into a glass column of approximately 1.5 cm inner diameter. Tap water is passed through the column at one bed volume per five minutes. The resulting softened water is titrated with 0.1 M EDTA and eriochrome black T indicator or soap solution and compared with untreated tap water. The titration of the latter will indicate the presence of calcium and magnesium ions, the eriochrome black T indicator changing color from red to blue as these ions are complexed out of solution. In softened water, the eriocherome black T will be blue, since the hardness-causing calcium and magnesium ions are absent. If additionally or in parallel this experiment is carried out with the cation exchanger in the H-form the result can not only be followed by the measurement of the water hardness but also by the measurement of conductivity and pH. The latter will in this case show a change from 6 to 3.

Two-stage deionization of a solution. As mentioned above, in this experiment the deionization of a solution may be followed by using colored ions and observing the formation of colored bands of the resins and decolorization of the solution. A blue cation and a yellow anion are included in the solution, and the concentrations are chosen so that a green liquor is obtained for deionization. Extraction of the blue cation on the strong acid cation exchanger in the Na-form column is easily seen, and the resulting cation-free liquor is yellow. The yellow dichromate is then removed by a strong base anion exchanger in the Cl-form column, and the result is a colorless deionized water. 25 ml each of the resins are put into two glass columns of the same size as above and regenerated and washed. The bottom of the first strong acid cation column is connected up to the top of the strong base anion exchanger column, using rubber tubing, so that the effluent from the first is fed into the second column. The test solution containing 0.6 g/l $CuSO_4 \cdot 5H_2O$ and 0.08 g/l $K_2Cr_2O_7$ is passed through the resins at a flow rate of 50 ml per 15 minutes. The first column removes the potassium and copper, and a green band will form on it due to the copper. The eluate from this step containing sulfuric and chromic acids, is decolorized by the second column. An orange band is formed on this resin due to the chromate. Conductivity measurements of the effluent after treatment will verify that the solution has been deionized.

Preparing high purity substances. This experimental work can be used to demonstrate both the use of a mixed procedure as well as the preparation of a pure substance by ion exchange.

As an introduction to the use of a mixed bed of ion exchangers it should be mentioned that deionization with the conventional or reverse process can be accompanied by two undesirable effects: the formation of a temporary acidic or basic medium and the existence of a reversible exchange equilibrium during the first stage of deionization. A temporary acidic or basic condition can in many applications drastically alter the properties of a substance; and, because of the reversible nature of the exchange equilibria involved, it can be difficult to achieve complete deionization with a single pass through two columns of exchangers in series. Therefore, the third possibility of a mixed bed for deionization and purification is important: it involves contacting the solution batchwise or columnwise with a mixture of cation and anion exchangers in the hydrogen and hydroxide form, respectively. The mixed bed technique allows complete deionization by both maintaining a neutral medium at all times and giving the effectiveness of an overall irreversible equilibrium. Combinations involving strong acid cation exchange resins with strong base anion exchange resins are routinely used to produce water of extremely high quality, but mixed beds are not limited to the use of strong electrolyte resins. Such premixed resins are commercially available for the first combinations mentioned above but other combinations can be formulated by any user to achieve the necessary requirements. Examples of the latter are mixed beds of strong acid and weak base resins as applied to the separation of weak and strong acids, weak acid and strong base resins for the treatment of sugar, and weak acid and weak base resins employed for the deashing of pharmaceuticals. Mixed bed systems of the strong electrolyte type give the highest degree of deionization but the mixed bed of weak electrolyte resins can be quite useful for the gross deionization of a variety of solutions, e. g., for enzymes, sucrose, drugs and other chemicals; they also possess high regeneration efficiencies.

It is certainly of educational value to construct one's own mixed bed unit. In designing a larger laboratory ion exchange mixed bed that can be regnerated, several basic facts must be considered. First, a bed depth of at least 45 cm should be employed. Second, the unit should be at least 5 cm in diameter. Third, the design should be such as to permit a minimum flow rate of 7.5 to 30 l/min. For the construction a 120 cm section of a 10 cm glass pipe is used with lags with rubber tape to increase the diameter on both ends. On top a large tight-fitting rubber stopper is inserted, with central bore hole for a 1 cm hard rubber pipe as HCl inlet. This pipe should reach down 100 cm and have inserted at its end an 8 mm gas dispersion tube. Two more bore holes in the stopper are for a 0.6 cm air vent and a 0.6 cm NaOH inlet. On the bottom the column is closed with a 50 mesh stainless steel or plastic screen to hold the resin and it has a reducing piece ending at a 0.3 cm steel plate that holds a stopper with two bore holes, one for a 0.6 cm glass tube air inlet and one for a 1 cm glass tube backwash inlet in T-form as well the solution outlet (Figure 1.77). After the apparatus is constructed as described, 2 l of water and 1.6 l of strong acid cation exchanger are placed in the unit. The resin is then backwashed with water for 5 min at a flow rate sufficient to expand the resin by 75%. The resin is then permitted to settle and the internal distributor is adjusted so that it is just below the surface of the cation exchange resin. At this point, 3.2

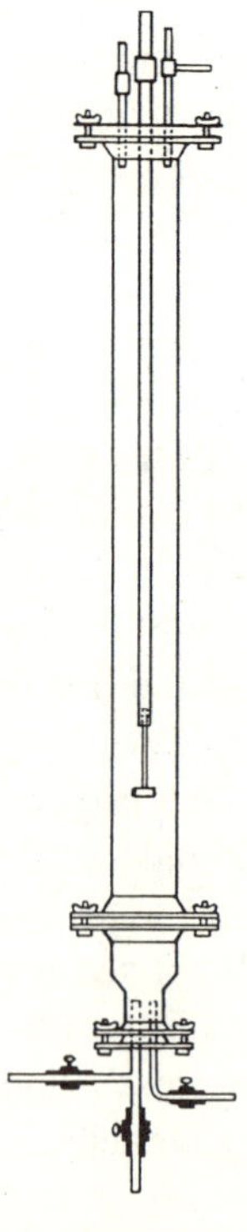

Figure 1.77 Mixed bed ion exchange column suitable also for the preparation of deionized water in the laboratory.

l of strong base anion exchanger and an amount of water sufficient to immerse the resin completely are added and the entire bed backwashed as before. The resin bed is permitted to settle and the water is drained to a point 2 cm above the bed. 6 l of 1 N NaOH are passed through the NaOH inlet at a flow rate of 400 ml/min. This is followed by 15 l of water. At this point, 6 l of 1 N HCl followed by 8 l of water are passed through the HCl inlet at a flow rate of 200 ml/min. While the acid is being fed into the unit, a slow flow of water, 20 ml/min, can be passed through the NaOH inlet in order to prevent back-diffusion of acid into the anion exchanger bed. Then, with the water level about 5 cm above the anion exchanger bed, filtered air is fed through the air inlet at a rate sufficient to mix the two resin components. Prior to using the mixed bed, several bed volumes of water should be passed through as a rinse. Conductivity of the rinse effluent can be checked to insure that the column is ready to use. After the mixed bed is exhausted, the unit must be backwashed at a rate sufficient to expand the bed 75 to 100% and for a period of time necessary to separate the two components. Sometimes difficulty is experienced in separating the components due to clumping, which occurs due to static attraction between cation and anion exchanger beads. If this occurs 1 l of a 10% NaCl solution can be passed through the bed and the backwashing continued. Regeneration should then be carried out as before. In this way a mixed bed column can be prepared, which can be used for the preparation of deionized water for the laboratory.

For purification – in this case deionization – of small quantities on occasion a commercial and regenerated mixed bed resin mixture can simply be placed in an ordinary laboratory ion exchange column and the solution to be purified passed through the resin and collected in a container. As an example, a urea solution, made by dissolving analytical grade material in deionized water to make a 7 M solution, can be passed over such a mixed bed ion exchanger column to remove isocyanate, which would otherwise react with amino and sulfhydryl groups in proteins, etc. This procedure is of importance in biochemistry where even stable enzymes devoid of sulfhydryl groups may be inactivated by long exposure even at moderate temperature to urea that contains an appreciable amount of cyanate. In such investigations special attention must be given to the use of urea as free as possible of cyanate or it can be purified easily in a mixed bed exchanger system. If, for example, an electrophoresis buffer containing 7 M urea is to be used, it is also recommended to freshly deionize the urea by a mixed bed column. The resin bed must, however, have certain minimum dimensions. A column, 40 cm long and 2.5 cm in diameter is recommended, to be used at a flow rate of 10 ml/min at room temperature for approximately 30 min, for this experimental demonstration of preparing high purity substances.

Catalysis. An experiment to demonstrate the catalytic application of an ion exchanger is the hydrolysis of ethyl acetate by means of a strong acid cation exchange resin. This is a particularly interesting example of preparative application for advanced students. It may be carried out at various temperatures to show the effect of temperature. The velocity constant of reaction, and energy of activation may also be determined. The reaction is followed by titrating the free acidity at timed intervals.

A stoppered 250 ml flask and reflux condenser for elevated temperature work is placed in a water bath at the required temperature, e. g., 25 °C, and 60 ml deionized water are added. Then 4 g of dried resin in the H-form are added and the stirrer started. This mixture is stirred for at least 15 min or until the water temperature equals the bath temperature. The stopper is removed and 2 ml of ethyl acetate are added by pipette. The stirrer is stopped and a 2 ml aliquot is removed from the flask by pipette. A stop watch is started when the addition is made, and the time and temperature taken when the first sample is removed. The stopper is replaced quickly and the stirrer restarted. 2 ml aliquots are removed at times of 1, 10, 20, 30, 40, 60 and 80 min. The experiment is then allowed to continue undisturbed for at least 24 h, when a final sample is taken. The progress of the experiment is followed by adding the aliquot to a flask containing about 50 ml of cool deionized water. The acidity is titrated at once using phenolphthalein as indicator and standard 0.1 N sodium hydroxide solution.

For the interpretation of the experiment a graph is plotted of titre against time in min, and temperature against time. If time $= t$, titre at time $t = T_t$, titre at 24 h $= T_{oo}$. A graph is plotted of log $(T_{oo} - T_t)$ against time; log $[CH_3COOC_2H_5] = -kt/2.303 + c$, slope $= -k/2.303$. Hence k, the velocity constant, can be found. From the Arrhenius relation $k = A \cdot e^{-E/RT}$ it can be shown that log $k_1/k_2 = E/R(1/T_2 - 1/T_1)$, where $E =$ energy of activation, $R =$ gas constant, and T_2 and $T_1 =$ temperature in K. Hence, since all these quantities are known, the energy of activation E can be calculated. Typical results from this experiment are $k_1 = 2.99 \cdot 10^{-3}$ min^{-1} at 23.8 °C, $k_2 = 10.22 \cdot 10^{-3}$ min^{-1} at 38.0 °C, and $E = 26.0$ kJ.

Sucrose hydrolysis by a strong sulfonic acid exchanger may be similarly investigated, with an informative account of the chemical and physical properties of ion exchange resins [13].

Laboratory experiments for modelling ion exchange processes. The modelling of ion exchange processes reflects the everlasting attempt to explain the facts and features of ion exchange. The first stages in the explanations of ion exchange kinetics and rate exchanges were non-mathematical models or diffusion. One major objective was to predict the column breakthrough of a predetermined concentration of a specific component in a multicomponent system. The most extensively used and discussed kinetic model is the progressive-shell shrinking-core model, which yields predictions for columnar mass transfer. Any model should allow the determination of concentration profiles of the exchanging ions for inner diffusion and autodiffusion under the effect of whatever driving force may be assumed. But is has also been found that the exchange rate may obey neither model equations for diffusion – whether film or particle – together with the Nernst-Planck diffusional model, nor chemically controlled mechanisms. In such cases it has been shown that in order to obtain independent information on the mechanism involved, an ion-diffusion study into the solid phase must be undertaken by means of autoradiography and X-ray microprobe analysis. These methods are the methods of choice if one wants to avoid the uncertainty always present when model equations are used and, moreover, to gain an insight into what is really occuring on the molecular level in the resin phase. The laboratory experiments in this connection are performable only under special circumstances. This is the case for X-ray microbe analysis, as an X-ray analyzer may not be available everywhere because of its high cost – in contrast with autoradiography which, at least for qualitative measurements, does not require expensive instruments.

Autoradiography. This technique, which requires the use of at least one radioactive species, makes use of thin sections of resin particle, loaded with proper amounts of the isotope, which are then contacted in the dark with special films sensitive to isotopic radiation. The magnified picture of the resin section at different values of exchange rate permits one to have a visible verification of the mechanism of isotope distribution inside the resin particle. A detailed description of the autoradiography of plutonium from a concentrated nitric acid solution onto a weak base anion exchanger resin (copolymerizate of a substituted vinylpyridine with approximately 10% divinylbenzene) is available from Streat and coworkers: it shows the concentration profile of plutonium within a single ion exchanger bead directly obtained by alpha-autoradiographs of sectioned samples [14]. For the investigation of the anion exchange of sulfate anions onto a weak base acrylic resin in the chloride form and the determination of the ion distribution in the resin phase the following procedure has been developed by Liberti and coworkers. After prefixed contact times, the reactor in use is suddenly separated from the reaction batch and kept rotating in the air for thirty seconds to remove the interstitial solution, it is then rapidly immersed into acetone. One resin bead is then embedded into a plastic material of comparable hardness to that of the ion exchange resin, and then the

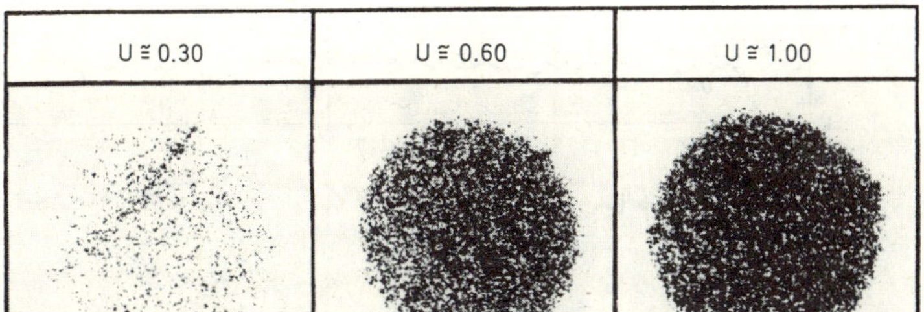

Figure 1.78 Typical autoradiograms for SO_4^{2-}-invasion into an anion exchange resin at different percentage of exchange (Liberti et al.).

whole mass is microtomized on a suitable microtome at 35 μm thick cuttings. The equatorial sections of each bead are laid down on a glass slide or a zinc support for autoradiography or X-ray microprobe analysis, respectively. For the autoradiographic determinations, the sections are contacted with a β-emitter sensitive film for 24 hours. After film processing by usual procedures, the pattern of SO_4^{2-} distribution is obtained. Figure 1.78 shows autoradiograms of sulfate distribution into equatorial sections of a single bead of the exchanger resin for direct ($Cl^- \rightarrow SO_4^{2-}$) exchange, which can be as well obtained for the reverse ($SO_4^{2-} \rightarrow Cl^-$) exchange, which gives direct information, i. e., modification of the solid phase, on the kinetics of ion exchange. Autoradiography is thus an interesting investigatory technique for obtaining detailed information that can be used to refine the theory of ion exchange kinetics in selective systems [15].

X-ray microprobe analysis. With X-ray microanalysis it is also possible to follow ion distribution within the resin phase. The advantage of this method is that it does not require the use of isotopes, as the energy of the electron beam of this technique may be adjusted so as to reveal selectively almost any element present in a section of a resin bead. The preparation of the sample is the same as in autoradiography, as briefly outlined above. More details regarding the carrying-out of the measurements may be taken from the application recommendations of the microanalyzer because low beam energy – for instance 10 kV and 30 nA – is required to avoid burning the sample. In an experiment where the bead cuttings were supported on a sulfure-free zinc slide to facilitate heat dissipation, results were obtained as shown in Figure 1.79. Because an X-ray microanalyzer is very expensive, its application to ion exchange investigations may be justified only under special circumstances, but where it is possible experiments carried out with it would be highly informative.

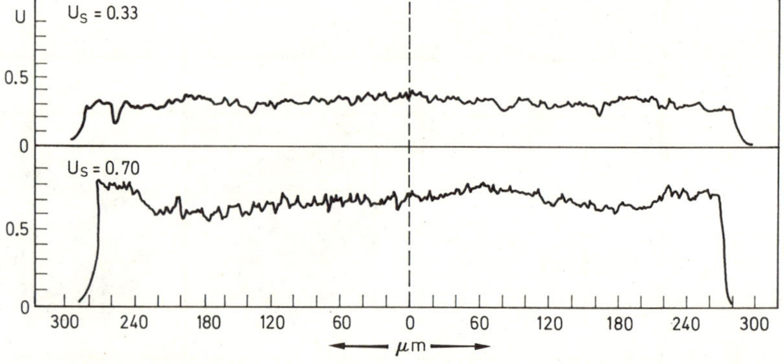

Figure 1.79 Distribution of sulphate at approximately 33% and 70% resin conversion measured by X-ray microprobe analysis (Liberti et al.).

Laboratory trials in preparation of large scale processes. Laboratory trials are costly and, therefore, before the laboratory study of an ion exchange process is decided upon one should investigate whether or not ion exchange is the method of choice. Since many different ion exchange processes are now established in industry the probability for a success of an already existing ion exchange application may be estimated already by analogy. The situation changes, when, for instance, a new ion exchange process for pollution control is discussed. In such a case ion exchange may have to be compared with alternative processes that could be used as well. An answer to the question whether an ion exchange process is more economic may, even before laboratory work is started, be expected.

An ion exchange process is most economical in the treatment of large volumes of solution containing low concentrations of ions to be removed with the goal of obtaining an effluent free of one specific ion, or type of ion. The nature of the resulting effluent may be a second important consideration, since it may prove to be an immediate problem or economical factor if the effluent stream cannot be reused, it must otherwise be taken care of giving rise to further costs instead of savings. From the very beginning these possibilities should be a part of the evaluation in the design of laboratory trials for a new ion exchange process, together with the question of the most suitable ion exchanger with respect to capacity and regeneration, the latter being evaluated under the aspect of the eventual waste created and its disposal, if the process can be run in cycles at all. A carefully designed basic program may permit an estimation of the operating costs if all of them can be studied on a laboratory scale, which would not then have to be re-examined in a pilot or larger scale operation.

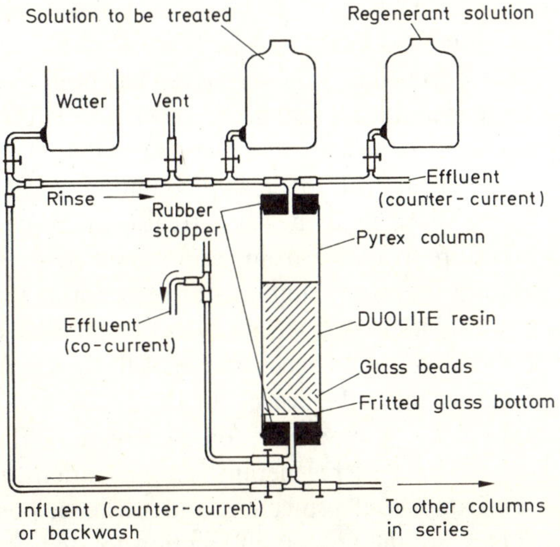

Figure 1.80 Typical laboratory setup allowing also for countercurrent operation.

The question of how small the column can be to produce data useful for the development of a large-scale process has been answered by a resin manufacturer: a resin volume of 50 to 100 ml is recommended for process evaluation on a laboratory scale. A column setup as shown in Figure 1.80 may be installed, which also allows for countercurrent (upflow) operation and in which several columns can be connected in series. The process flow rate will vary between 2 to 50 bed volumes per hour, according to the application. Smaller columns must be considered when very dilute solutions are concerned and especially where only limited amounts of solution are available. With careful work on the size of columns for resin quantities as mentioned above the determination of the throughput capacity and regeneration efficiency can be scaled up. Even small columns with 5 mm inner diameter have been used. Results from work at this level, however, should only be used in process feasibility decisions. Careful work implies the correct regulation of flow rates and collection of precise effluent volumes, and this becomes more difficult the smaller the columns.

The selection of the resin to be used is an important step in a process design study. This depends on the problem itself and can be influenced by many factors, such as elevated temperature, high flowrate applied, selectivity of the resin for one ion over an other, unusually concentrated solutions, and removal of large complex ions or recovery of organic compounds. After the resin type has been decided upon from knowledge of the ion exchange behavior of the ionic compounds in solution, i. e., either strong or weak anion or sulfonic or carboxylic cation, etc., the next decision will be to select for a given ionogenic type, based on a comparison of operational efficiency, either a gel or a macroporous resin. To compare a Type I and Type II strong base resin and a methacrylic and an acrylic acid carboxylic acid resin may also be advisable. Where weak base resins are to be employed a major screening program may be necessary before one can decide that a single resin, or perhaps two or three resins, should be included in detailed studies.

In initial experiments after a decision on the resin type has been made variables like flow rates, temperature, concentration, quantity and kind of regenerant, and coflow versus counterflow should be tested. In experiments that then follow the most promising combination of resins and operating conditions are then applied, which may be compared with those used in water treatment processes or with suggestions made in the literature available from resin manufacturers. The latter normally suggest types, concentrations and amounts of regenerants to be used for a specific resin. Changes of variables may further have to be studied, especially in the solution temperature and the type and concentration of the regenerant, as these are not always as predictable as the influence of an increased flow rate on the leakage and of the increased volume of regenerant on a decreased effluent concentration. Minimization of the volume of regenerant is today often an important question, with regard to environmental protection and waste disposal. For the documentation of test results, e. g., the volume of solution treated per volume of resin is plotted as a function of the amount of regenerant used or leakage values for a specific ion are plotted as a function of bed volumes treated. Monitoring studies of the process to be developed depend even more specifically on the problem at hand, for instance, whether deionization or removal of a single ion is to be achieved.

The final question and the most difficult part of laboratory experiments for large scale processes is the prediction of the resin life. Resin life evaluations involve repeated, usually short, cycles of loading and regeneration followed by a separate test program for the changes in properties of the resin under application. But with all imaginable precautions taken the prediction on the basis of laboratory experiments of resin life in absolute numbers of cycles to be expected is an uncertain and risky enterprise. It must nonetheless be said that ion exchange is a complicated process for which only experiment can be decisive.

1.2.5.2 Education in ion exchange

Ion exchange resins can be an exceptional educational tool. This was discovered in the late 1950's and early 1960's when the first elective courses in ion exchange were offered and when teachers began realizing that this topic can be of considerable use in their high school science programs. It was found that ion exchange experiments were most useful for teaching students not simply about ion exchange but also for illustrating the principles of electrolytes, the law of mass action, and of electroneutrality, diffusion, kinetics, polymers, and of introducing them to such important fields as water treatment, detergent chemistry, and ecology. The interest of the students was great, especially when it was brought to their attention how easily this topic can be connected with many phases of theoretical and applied chemistry. It can further be demonstrated that ion exchange is involved, both directly and indirectly, in many phenomena of everyday life. Starting with the topic of ion exchange a teacher can smoothly enter such diverse fields as water chemistry, life science, soils, atomic energy, sugar, milk products, etc. But as education in chemistry involves methods of teaching, an attempt will be made in the following to show how their requirements with respect to ion exchange and ion exchangers can be met in various cases.

For teaching larger audiences the use of an *overhead projector* has become common practice. It is, therefore, of interest that ion exchange can be demonstrated to large groups with the use of an overhead projector as well. The equipment needed consists of transparent plastic containers, stirring rods and containers for reagents, (dropper bottles and small plastic wash bottles are convenient). The reagents needed include a cation exchange resin in the H-form, bromthymol blue indicator (0.04% w/v), distilled water and dilute NaOH (0.001 M), HCl (0.001 M) and NaCl (0.1 M). The demonstration is then carried out in three stages:

First stage: Demonstration of the variations in color exhibited by the indicator with the base and the acid solutions: The bottom of the cell or container is covered with distilled water. Bromthymol blue indicator is added until the color of the projected solution is fairly intense. If the indicator is mid-range (greenish color) one drop of the NaOH is added. If the drop is added gently to the water plus indicator, the projection can even be used to demonstrate the reate of diffusion. Then the solution is stirred until the color is uniform and HCl is added dropwise with stirring until the acid color is developed.

Second stage: Demonstration of the ion exchange process: Dilute NaCl solution is poured into the second cell until the bottom is covered. Bromthymol blue indicator solution is added as in the first demonstration and one drop of NaOH is added to produce a blue color. A few beads of resin are scattered across the cell. The yellow color that then develops around each bead is evidence of acid, indicating that hydrogen ions have influenced the bromthymol blue. On stirring, the whole solution turns yellow. After the solution is stirred, a few drops of NaOH will momentarily turn the solution blue. The solution turns yellow again as more hydrogen ions are displaced from the resin. The exchange reaction can be written on the blackboard.

Third stage: Demonstration and verification of the role played by the sodium ions in the reaction: Distilled water and bromthymol blue are poured into the third container in the same manner as described in the first stage. A drop of NaOH is added and the solution is stirred before a few resin beads are added to the container. The rate of acidification in water is slow compared to the salt solution. The acidification in the cell is due to the hydrogen ions that have dissociated from the resin and that have been displaced by the sodium ions.

The exchange reaction can be written on the blackboard and further explanations can be given depending on the field in which the instructor is in. But by this excellent demonstration, which has been developed by Professor Himes of the Department of Agronomy of Ohio State University, ion exchange can be displayed in vivid color [16]. Another demonstration of ion exchange by means of an overhead projector employs strips of filter paper impregnated with an ion exchanger. A solution containing the ion to be exchanged is allowed to penetrate through the strips onto other strips of filter paper impregnated with appropriate indicators. The effect on the detecting strips of the solution that has passed through the ion exchanging paper is compared with that of the original solution. The respective differences in color are shown by the overhead projector [17].

In the *use of the magnetic board* to illustrate ion exchange it was found that a greater flexibility results from employing a mushroom-type overlap – this can be used to animate the concept of ion exchange. The practical demonstration is intended to illustrate that the process of ion exchange is an exchange of ions on a resin surface, as the cations are made of mushroom-type overlaps to slot into the resin surface. But this admittedly illustrative demonstration, which even includes a columnar animation, might well result in a misconception because it does not lead us to an understanding of the ion exchange material structure nor to the importance of diffusion in ion exchange processes [18].

The simulation of ion exchange structures and processes with computer graphics is a great future challenge for the understanding, demonstrating and teaching of ion exchange. Up to now simulation of the chemistry of zeolites by computer graphics has been developed. Zeolites have a unique structure and catalytic properties as well as excellent adsorption properties. With computer graphics, especially color graphics, the positions of cations and intercalated species, as well as all kinds of deformities, can be vividly demonstrated. By this means the selective character of these microporous solids can immediately be realized. Further, dynamic phenomena can interactively be investigated on the monitor [19]. In this area one should follow the ion exchange literature, as there will no doubt be similar computer graphics published for synthetic ion exchange resins and other ion exchange materials.

A very important aspect of education in ion exchange and ion exchange processes concerns persons who have to operate ion exchanger units. With the existing structure of the application of ion exchangers this applies above all to the operation of water treatment installations. Education has then, on the one hand, to be understood as the training of staff who may suddenly have been transferred to this new assignment or, on the other hand, of graduated students in general engineering who wish to enter the water treatment industry, whether as technologists or sales engineers, or as consulting engineers whose work will involve them in questions of water purification. The application of ion exchangers will always be predominantly in water treatment. There are myriads of papers on water management, the provision of drinking water and the upgrading of potable water for industry, medicine and research. With world-wide industrialization, purified water, i. e., water of a predetermined specification, has become the world's most essential raw material.

If ion exchange and also membrane technologies are applied for water purification training must combine the teaching of water knowledge, ion exchanger and membrane properties and general principles, as well as plant operation (covering plant design and maintenance of, e. g., different types of reverse osmosis modules). The field of saline water conversion, which is of particular interest to the countries lacking sweet water is another need to be served and then training on units for the upgrading and re-use of water. Beside the demand for various grades of purified water for hospitals, pharmaceuticals and cosmetic production, given the complexity of regulations for water quality, all those who are responsible for the supply of pure or ultra-pure water vital for the production of modern electronic components, will require education and training in water treatment technologies. In all these areas ion exchange plays a role. No attempt will be made here to analyse the education and training problems in these areas. Walter Lorch in England, who is responsible for much of the above information founded the School of Water Sciences, the first of its kind in the world. If the organization of similar institutions is envisaged in countries several chapters of this volume might well serve as a basis to elaborate corresponding lectures, and the whole volume can be recommended for further reading and an in-depth study of ion exchange and ion exchangers. In

Chapter 1.1 (introduction to ion exchange and ion exchangers) the fundamentals and principles of ion exchange as well as the types of ion exchangers are described in detail. Chapter 2.2 on raw water treatment with water softening, dealkalization and deionization together with Chapter 2.1 on general ion exchange technology and the detailed description of deionizing combination processes can serve as a valuable source for the demonstration of the many technical possibilities of the design and selection of water treatment plants in boiler feed water treatment or exploration of the operational needs of deionization equipment in industries where purified water is essential, for instance, textiles, paper making, air conditioning, microelectronics, pharmaceuticals, and hospitals. Laboratory experiments can be found in the literature for the explanation and use of mixed beds, as well as for the detailed presentation of design, the advantages of using ion exchangers and their employment in industrial water treatment. If condensate polishing is included, as it must be in any educational or training course in boiler feed water treatment, there is a separate chapter where probably more could be learned than in fact can be made use of even in courses for senior technologists who already have experience in the design and selection of water treatment plants for larger boilers. Depending on the location the partial desalination of brackish water may have to be discussed, too. With respect to the operation and maintenance of water treatment plants enough information has been gathered in this volume on the operation of ion exchange vessels insofar as they also operate as filters, as well as on the resin durability in water treatment resulting from solubility, chemical and thermal stability, mechanical strength, osmotic shock, and resin poisoning, together with clean-up procedures and sterilisation. The treatment of drinking water with ion exchange resins, waste water treatment and pollution control by ion exchange, the special requirement for ion exchangers in nuclear technology and, last but not least, ion exchange systems in homes, laboratories and small industries are outlined in such detail and presented with such expertise that an educational or training course on any level can be based on the respective chapters. With all the information that can be drawn from this volume for education in ion exchange for water treatment, it must also be mentioned that if ion exchange has to be taught with respect to other fields and applications – for instance for the sugar industry or biotechnology – there are other chapters covering exhaustively these topics as well.

References

[1] M. J. R. Clark (1976). Use of ion-exchange resins without columns. J. Chem. Educ. *53*, 770.

[2] I. K. Tsitovitch (1982). Use of ionites to control environmental pollution by herbicides. Khim. Shk. (2), 68.

[3] A. Carpov and D. Dragan (1974). Synthesis and characterization of strongly basic anion exchangers with functional groups of type I and II. Mater. Plast. (Bucharest) *11*, 336.

[4] K. Dorfner (1959). Untersuchungen über Redoxaustauscher. Dissertation Universität Marburg/Lahn.

[5] M. Griesbach and K. H. Lieser (1980). Synthese von chelatbildenden Ionenaustauschern auf der Basis von Polystyrol. Angew. Makromol. Chem. *90*, 143. Ibid. *113*, 129 (1983).

[6] M. G. Djanali and K. H. Lieser (1983). Darstellung und Eigenschaften eines Ionenaustauschers mit einem 1,3-Diketon als Ankergruppe auf der Basis von Polystyrol. Angew. Makromol. Chem. *116*, 195.

[7] J. Schubert and J. W. Richter (1948). The use of ion exchangers for the determination of physical-chemical properties of substances, particularly radiotracers, in solution. I. Theoretical. J. Phys. Coll. Chem. *53*, 340.

[8] R. E. Anderson (1975). Estimation of ion exchange process limits by selectivity calculations. AIChE Symp. Ser. 71, 152, 236.

[9] V. Ender and J. Bosholm (1985). Zur indirekten Ermittlung von Selektivitätskoeffizienten für Ionenaustauschreaktionen. Acta hydrochim. hydrobiol. *13*, 413.

[10] P. R. Dana (1969). Elution kinetics of the copper ammine-complex from a cation exchange resin. PhD Dissertation. Iowa State University, Ames, Iowa.

[11] B. Morelli and L. Lampugnani (1975). Ion exchange resins. A simple apparatus. J. Chem. Educ. *52*, 572.

[12] M. L. Dietz (1986). The determination of calcium in dietary supplement tablets by ion-exchange. J. Chem. Educ. *63*, 177.

[13] L. Nowakowski (1976). Cation exchanger as a catalyst for the hydrolysis of sugars. Chem. Szk. *22*, 45 (Pol).

[14] R. E. Bahu et al. (1976). A study of slow diffusing species in ion exchange resins. Proc. Int. Conf. on Theory and Practice of Ion exchange. Cambridge. Paper 19.1.

[15] L. Liberti et al. (1984). Diffusion versus chemically controlled kinetics in a selective system. The Cl^-/SO_4^{2-} exchange. Reactive Polymers *2*, 111.

[16] F. L. Himes (1976). Use of the overhead projector to illustrate ion exchange reactions. J. Agronom. Educ. *5*, 33.

[17] L. Nedzynski (1965). Ion exchangers in a chemistry lecture. Chem. SzK. No. *4*, 167 (Pol).

[18] T. Morris (1971). Use of magnetic board with mushroom type overlap models. Sch. Sci. Rev. *53*, 143.

[19] S. Ramdas et al. (1984). Simulation der Chemie von Zeolithen mit Computer-Graphik. Angew. Chem. *96*, 629.

1.3 Cellulose Ion Exchangers

Nikolaus Grubhofer
Heidelberg, Federal Republic of Germany

1.3.1 Development

Careful research has been devoted to the introduction of ion exchanging groups into cellulose for the purpose of improving dyeing characteristics and flame resistance. Credit, however, must be attributed to Herbert A. Sober and Elbert A. Peterson for the first systematic description of ion exchanging celluloses and their use as chromatographic adsorbents, which was presented in their classical paper (J. Am. Chem. Soc. 78, 751 (1956)). In cooperation with F. J. Gutter and Mary M. Wyckoff they were able to demonstrate how blood plasma proteins could be fractioned by chromatography on DEAE-cellulose with an efficiency far beyond that attained by conventional electrophoretic methods. Moreover, their methods are straightforward and easily adaptable to large scale production methods.

1.3.2 The cellulose matrix

Native cellulose consists of 2000 to 16000 glucose units linked by β-(1,4)-glycosidic bonds. The fibrous molecules run in parallel fashion and are linked by hydrogen bonds to crystalline regions. Only the hydroxyl groups in positions 2, 3 and 6 can be used for the introduction of functional groups. These hydroxyl groups, however, only become available for derivation after mercerisation (treatment with strong caustic), which results in a total of 25% of the groups becoming available.

Increasingly, regenerated cellulose which has been dissolved in xanthogenate or similar media and degraded to a degree of polymerization of 500–2000 glucose units is used as a matrix. Regenerated cellulose has a fibrous structure similar to native cellulose but the fibers are associated in a more random way ("microcrystalline"), and the particles are granular, more stable and therefore more suitable for technical applications [1, 2].

1.3.3 Cellulose particle structure

Cellulose Ion Exchangers contain a very loose hydrophilic molecular network, crosslinked by hydrogen bonds. Even very large molecules can penetrate freely and diffuse rapidly. The distance between the charged groups is approximately 5 nm, in contrast to conventional ion exchangers with about 1 nm distance between charged groups. Macromolecular poly-ions are attached either to only one or to a few sites of the exchanger. Selective desorption is possible under extremely mild conditions.

Stabilization of fibrous molecules is desirable to reduce softening or dissolution of the particle under adverse conditions such as low ionic strength, strong acid or alkali. From the many that have been tested, formaldehyde, epichlorohydrin and 1,3-dichloropropanol have proved to be useful crosslinking agents [2–4].

1.3.3.1 Basic cellulose ion exchangers

AE-cellulose (-O-C_2H_4-NH_2, aminoethyl cellulose). Its basicity is somewhat low compared to DEAE-cellulose. The cromatographic properties are slightly different [5, 6]. Its production, using aminoethyl sulfonic acid, yields ethylene imine as a by-product. For this reason manufacture has been dropped. Aminohexyl cellulose is being considered as a possible replacement.

BD- and BND-cellulose (benzoylated and benzoylated-naphthoylated DEAE-celluloses). These lipophilic anion exchangers were introduced by G. M. Tener and coworkers [7]. They are used for improved and highly specific separation of s-RNA [8–10]. M. Haber et al. [11] found that boiling of BD-cellulose prior to usage guarantees good quality, independent of storage.

DEAE-cellulose (-O-C_2H_4-N$(C_2H_5)_2$, diethylaminoethyl cellulose). It is widely used with best results for the chromatography of enzymes, serum components [12–19] and other proteins with neutral or acidic IP [20–25]. It is also used for the separation of polysaccharides [26–29], RNA [30–32], s-RNA [33, 34], t-RNA [35] and lipids [36–40], for the chromatography of amino acids [41], peptides [42–45], folic acid derivatives [46–48], phosphosaccharides [49, 50] or mononucleotides such as coenzyme A [51], FMN-FAD [52], AMP-ADP-ATP [53, 54]. The separation of poly- and oligonucleotides [55–57], and the purification of viruses [58, 59] and ribosomes [60, 61] are also known.

ECTEOLA-cellulose. This modified cellulose is a reaction product of epichlorophydrin, triethanolamine and cellulose. Its chemical structure is not yet fully known. It contains a variety of nitrogen functions of weak and medium basicity and possibly even some quaternary ammonium groups [62]. ECTEOLA-cellulose has been used successfully for the separation of RNA, DNA and higher nucleotides [7, 63–67]. It is occasionally used for the chromatography of proteins [68, 69], peptides [70], polysaccharides [71–74] and viruses [75].

ECTHAM-cellulose. This weakly basic adsorbent belongs to the group of rather unusual cellulose ion exchangers [76]. It is prepared by coupling tris(hydroxymethyl)aminomethane to cellulose with epichlorohydrin as a crosslinking agent [77].

GE-cellulose (-O-C_2H_4-NH-C=NH-NH_2-HCl, guanidoethyl cellulose). It is a strongly basic adsorbent useful for the fractionation of proteins of the globulin type [6, 78]. As it is based on AE-cellulose, whose production has been dropped, GE-cellulose is no longer available. Guanidohexyl cellulose could possibly become a replacement.

PAB-cellulose (-O-CH_2-C_6H_4-NH_2, para-aminobenzyl cellulose). PAB-cellulose contains primary aromatic amino groups and promises to be very useful for the chromatography of very high molecular nucleic acids. Its amino groups can be diazotized and coupled with proteins, enzymes, substrates, histones and antibodies to yield specific adsorbents [79–81].

PEI-cellulose. Polyethylenimine $(\text{-NH-}CH_2\text{-}CH_2\text{-})_n$ is adsorbed virtually irreversibly onto very weakly phosphorylated cellulose. The regular sequence of basic groups favours selective adsorption and desorption of polyanions such as oligonucleotides [82, 83]. PEI-cellulose is particularly popular for the TLC of nucleotides [84].

TEAE--cellulose. This cellulose is a reaction product of ethylbromide and DEAE-cellulose. Quaternarization is probably negligible [85]. The absence of carboxylic groups (masked as esters) may result in sharper separations [46, 86–88]. Fields of application are to be found in the chromatography of proteins and folic acid derivatives.

QAE-cellulose. QAE-cellulose is a rarely used strongly basic anion exchanger, in which the cellulose is substituted by 2-hydroxypropyl triethylammonium groups [89].

1.3.3.2 Acidic cellulose ion exchangers

CM-cellulose (-O-CH_2-CO_2H, carboxymethyl cellulose). This is the most important high capacity adsorbent for the fractionation of neutral and basic proteins [90]. The chromatography of peptides [91, 92], as well as the separation of lipids by thin layer chromatography [37], have been described.

Hydroxylapatite-cellulose. The precipitation of calcium phosphate on cellulose results in a modification of hydroxylapatite which is particularly suited to the adsorption of proteins [24, 93, 94], peptides [95], nucleotides [96, 97] and t-RNA [98]. The combination of hydroxylapatite with cellulose renders sufficiently high flow rates, which cannot be obtained using pure calcium phosphate.

P-cellulose. As each site is doubly charged, P-cellulose is more acidic than CM-cellulose and binds basic proteins very tightly. It can be processed to render particularly high exchange capacities without impairment of the original physical properties and flow characteristics [49]. P-cellulose is also suited to the chromatography of heavy metal ions [99], sugar phospates [100], peptides [101] and histones [102]. In addition, it can be used for enrichment and separation of inorganic ions, particularly of the transition elements [103].

SE-cellulose (-O-C_2H_4-SO_3H, sulfoethyl cellulose). Being strongly acidic, this cellulose derivative combines the usefulness of conventional polystyrene sulfonate resins with the virtues of cellulose ion exchangers [104]. Sulfopropyl cellulose (SP-cellulose) is more easily synthesized and can replace SE-cellulose.

1.3.3.3 Some highlights on the performance of cellulose ion exchangers

Proteins. The advent of cellulose ion exchangers had an impact on preparative and analytical protein chemistry, similar to that of the invention of paper chromatography on amino acid and nucleotide analysis. Literally hundreds of proteins and enzymes owe their final isolation and purification to cellulose ion exchangers. The sharp separations of protein subunits and cleavage products are very useful for the elucidation of the chemical protein structure. An outstanding example is the use of CM-cellulose for studies of antibody structure [105–108].

Nucleic acids. In this field, a similarly important part is played by cellulose ion exchangers. One only has to recall the revelation of the structure of t-RNA, for which R. W. Holley [34, 109] received the Nobel Prize. This work would have been infinitely more difficult without the frequent use of DEAE-cellulose. RNA and DNA are traditionally separated on ECTEOLA-cellulose columns since the work of A. Bendich et al. [110–112]. s-RNAs, however, are mostly prepared using DEAE-cellulose [35, 113–115], hydroxylapatite [97, 116] and special PAB-azo-histone-cellulose [102] have also been proposed. The lipophilic BD- and BND-celluloses [117] are promising tools for s-RNA chromatography.

Oligonucleotides. Complex RNA and DNA digests have been resolved into individual components by Staehelin [54, 118]. An important new technique was suggested by R. V. Tomlinson and G. M. Tener [56, 119] using concentrated urea and salt gradients on DEAE-cellulose for nucleic acids and nucleotides.

Polysaccharides. Substituted polysaccharides have been purified and separated on DEAE-cellulose [26, 28, 29, 120] and ECTEOLA-cellulose [71–74].

Lipids. Striking separations of different lipids have been obtained by G. Rouser [38] using DEAE-cellulose and gradient elution with organic solvents cf. C. E. Ballou et al. [121, 122]. This method seems to be very promising in the preparative as well as analytical chemistry of complicated lipid mixtures.

Low molecular weight substances. Cellulose ion exchangers also have an enormous potential for the field of low molecular weight substances. Very convincing examples of their use can be found in literature, e. g., sharp separations of AMP, ADP and ATP [118], flavine nucleotides [123], folic acid derivatives [124], oligonucleotides [54, 125], [126], nucleotides [123], amino acids [41], peptides [70], sugar phosphates [100] and inorganic cations [49].

Enzyme stabilization. Adsorption of labile proteins onto cellulose adsorbents and subsequent lyophilisation result in preparations of excellent stability [127]. Elution with appropiate buffers liberates the protein [128, 129].

1.3.3.4 Chelating celluloses

These cellulose derivatives are not necessarily ion exchanging celluloses but are usually described along with these materials. The introduction of an electrophilic group into the cellulose is the first step in transforming the cellulose into a corresponding derivative. Examples are cellulose azides from carboxymethyl cellulose by Curtius rearrangement [128].

Reagents with basic amino groups can be coupled to carboxyl groups of carboxylmethylcellulose in the presence of carbodiimide [130]. A method for the binding of 2-amino-4,6-dichloro-s-triazine to cellulose was developed by Kay and Lilly [131]. During the first stage the triazine derivative reacts with the hydroxyl group of cellulose. In a second stage the chlorine atom is replaced by an amino group. Proteins and other reagents containing an amino group can be bound to cellulose after acylation of the hydroxyl group with bromoacetylbromide [132].

In addition to the amino groups, diazonium groups are also useful for attaching chelating agents to cellulose. 1-(Chloromethyl)-4-nitrobenzene is bound to the hydroxyl group of cellulose. The reaction product is reduced, diazotized and coupled to a phenol derivative [80]. A second way of introducing a diazonium group is the reaction of carboxymethyl cellulose with 1-amino-4-nitro-benzene in the presence of *N,N'*-dicyclohexylcarbodiimide followed by reduction and diazotization [133]. Ethylsulfonyl(aniline)cellulose can also be used as an intermediate [134]. Diazonium cellulose can, for example, be coupled to dithizonate. The product is very useful for the recovery of copper, zinc and silver ions. Other products can be obtained by coupling chelating substances like oxine, cupferron, quinalizarin, 2-thenoyltrifluoracetone, phenylarsonic acid and *p*-dimethyl-amino-benzylidene rhodanine to diazonium cellulose [133].

Although widespread as a technique for ligand immobilization in affinity chromatography, the activation of hydroxyl groups in cellulose by cyanogen bromide is not yet a very common method for preparing chelating celluloses [135]. There is some experimental work left to do to avoid the undesirable formation of unreactive carbamate and crosslinked imidocarbamate.

Iminodiacetic acid with chelating properties for Cd, Co, Cu, Fe, Hg, Mn, Ni, Pb and Zn is coupled to cellulose using a aminoethyl cellulose [136] as a base. The resulting imino-diacetic acid-ethyl cellulose (IDE-cellulose) remains stable for years, can be stored in the hydrogen- or ammonium-form and also retains trace elements from soil extracts [137].

A special cellulose has been developed for the selective removal of uranium ions from simulated seawater containing 8-oxa-2,4,12,14-tetraoxopentadecane as the chelate-forming anchor group. Its synthesis and application are described by H. J. Fischer and K. H. Lieser [3]. Bis-salicylaldehyde ethylene diimine attached to cellulose can be used for the same purpose [138]. Heavy metals are separated from neutral, concentrated saline solutions by means of a cellulose containing 1-(2-hydroxyphenylazo)-2-naphthol as an anchor group (138). Traces of beryllium can be removed from a neutral solution using pyrogallol cellulose [138].

1.3.3.5 Affinity adsorbents

Para-aminobenzyl cellulose (PAB-cellulose) can be diazotized and coupled to proteins, substrates and other specifically adsorbing substances. Antibodies, for instance, can be adsorbed to PAB-azo-antigens, nucleotides to PAB-azo-histones and PAB-azo-guanine, tyrosinase to PAB-azo-tyrosine. PAB-azo-enzymes act on substrates without contaminating them.

For diazotization, 2 g of PAB-cellulose are added to 50 ml 2 N HCl and 10 ml of 14% $NaNO_2$ at 0 °C. The mixture is left to stand at this temperature for 1 h. The reaction product is filtered on a Buechner funnel and washed successively with 5% sodium acetate, 5% urea and finally water. For coupling with proteins, the aqueous solution is stirred gently for two hours at room temperature, or at 0 °C. The unreacted diazonium groups are blocked in the following way: 2 g of *β*-naphthol are dissolved in a small volume of 2 N NaOH and diluted with 2 l of water. 2 N H_3PO_4 is added until the pH is 8.0. After leaving to stand overnight at 0 °C, excess *β*-naphthol is removed. The cellulose material is added and stirred slowly at 0 °C for 6 h.

Oligodeoxynucleotide derivatives of cellulose are useful for the specific adsorption of nucleotides. Polynucleotides can be separated by virtue of the varying stabilities of the base-paired complexes formed between the components of the polynucleotide mixture and the cellulose-bound polynucleotide.

Poly(rA) containing m-RNA, for example, can be isolated and purified by affinity chromatography on dT-cellulose [139]. Other applications are the fractionation of oligoadenylic acids of different chain lengths [140] and the isolation of nucleoproteins [141, 142]. Oligodeoxycytidylic acid, oligodeoxyadenylic acid and oligodeoxyguanylic acid can be bound to cellulose as well [139].

1.3.3.6 Industrial applications

Ligno-cellulose wood-pulp exchangers, derived from the sulfite digestion of wood, are partially degraded sulfonated cellulose products which have industrial application in waste water treatment [143]. Cellulose ion exchangers can be applied to the large-scale recovery of protein from process effluents, such as serum albumin from abbatoir waste, lactalbumin from milk whey, and to the purification of eggwhite protein [144]. Batch adsorption in stir tanks was used to recover proteins for animal feed supplement [145, 146]. Humic and fulvic acids have been removed from surface waters [147].

1.3.3.7 Special cellulose ion exchange products for industrial application

Cellulose derivatives	Use	Reference
Amphoteric cellulose	Optimization of exchange capacity	148
Borate cellulose	Preparation of borate cellulose as a sorbent for blood stabilization	149
Blue Cotton	Cotton-bearing covalently linked copper-phthalocyanine, for removal of polycyclic mutagens, antibiotics and dyes	150 151 152
Carboxymethyl and DEAE-cellulose	Improvement of manufacture, reduced reagent consumption, improved flow properties	153
Copolymer of cellulose and acrylic acid	High capacity and stability, 1.25 meq/g, for treatment of waste water dyeing and electroplating	154

Cellulose derivates	Use	Reference
Copolymer of cellulose and methacrylic acid	Studies of pH and temperature dependence of ion exchange process for water treatment, 15% COOH groups	155
Copolymer of cellulose, polyacrylic acid and quaternary ammonium groups	Optimization of exchange capacity and affinity studies, for waste water treatment from dyeing	156 157 158
Crosslinked carboxymethyl cellulose	Crosslinking of CM-cellulose after preparing a thin film on a glass plate, 3.2 meq/g, for TLC	159
PAB-cellulose	For coupling of proteins and nucleic acids after diazotization	160
Phosphorylated cotton	Ion exchanging cotton fabric for removal of Ca-ions	161
Sulfate cellulose, cross-linked with epichlorohydrin	Optimization, high capacity of linked 1.9–2.2 meq/g	162
Sulfonated cellulose	Increased capacity	163

1.3.4 Chemical properties

1.3.4.1 Capacity determination

The adsorbent is dried overnight at 50 °C. 2 g of the dry adsorbent is then stirred for 10 min with 50 ml 0.5 N NaOH (basic exchangers) or 50 ml 0.5 N HCl (acidic exchangers) and washed with deionized water on a Buechner funnel until the effluent is neutral. The titration vessel is rinsed with a little water, and then 25 ml of 1.0 M NaCl is added, whereby partial salt splitting takes place. The adsorbent is either titrated to pH 3 with 0.1 N HCl or to pH 10 with 0.1 N NaOH under continuous stirring, using a pH-meter with glass electrodes.

1.3.4.2 Dissociation range

If the determination of capacity is carried out as described above with continuous recording of pH versus volume of titrant, the titration curves are obtained (Figures 1.81 and 1.82). Inflection points of the curves show approximate pK values indicating the pH at which 50% of the ionic groups are dissociated.

1.3.4.3 Particle size

The irregular shape of cellulose fragments makes the definition of particle size arbitrary and not too meaningful. Direct measurement of particle length with a microscope leads to the data given in the following table.

Typical properties of cellulose ion exchanger particles	Standard types
Length of 70% of the particles	50–200 μm
Diameter of 70% of the particles	10–20 μm
Moisture content	10–14%
Apparent density	0.1–0.2 (g/ml dry)

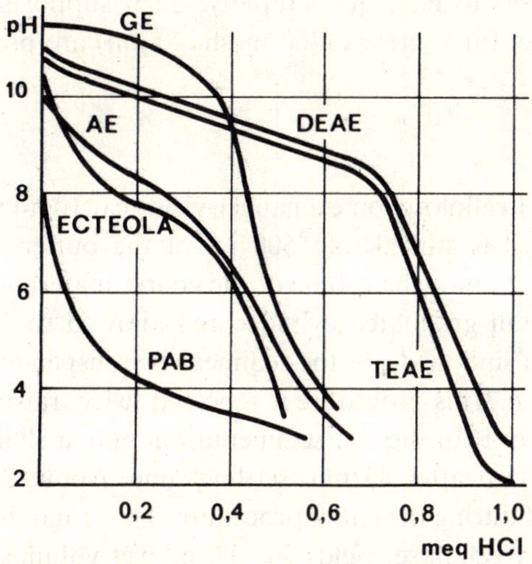

Figure 1.81 Typical titration curves with basic cellulose ion exchangers.

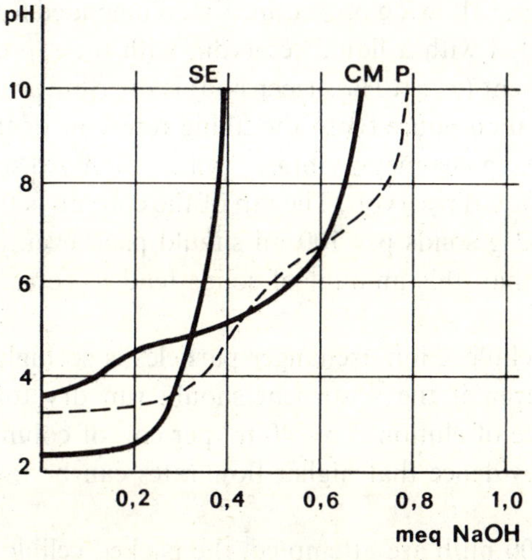

Figure 1.82 Typical titration curves with acidic cellulose ion exchangers.

1.3.5 Applications

1.3.5.1 Column characteristics

Columns with a height of 10 – 20 cm are generally used. Although longer and shorter columns [164] may be used, they do not offer any particular advantage. The ratio of length versus diameter should not exceed 10:1. Wide columns are faster and easier to operate and usually give higher concentrations in the effluent. Fine sinter-discs of glass, teflon or polyethylene are not suitable for cellulose ion exchangers since the pores tend to clog quite rapidly. Such supports should be protected by a filter-paper disc. Fine screens (400 mesh, 40 μm) are preferable.

1.3.5.2 Column techniques

Preparation of adsorbent. 1 g of cellulose ion exchanger yields ca. 10 ml wet column volume. 10 g of dry adsorbent is stirred into 500 ml of the buffer used at the beginning of elution. After 1 – 2 min, when some of the coarse material has settled, the slurry is decanted into a tall graduated cylinder and allowed to settle for 45 min. The turbid supernatant is siphoned off, the sediment is resuspended in 500 ml buffer and again left for 45 min. This procedure is repeated twice, raising the total number of sedimentations to four. In the last sedimentation only a slight turbidity should remain in the supernatant after 45 min settling time. About 20 – 30% of the cellulose will have been lost during the entire procedure. The remaining material, corresponding to 7 – 8 g of dry cellulose, yields 70 – 80 ml wet volume.

By contrast with native cellulose, the surface of the adsorbent cannot be flattened by knocking on the wall of the tube. Particle-size gradients are negligible during sedimentation of the adsorbent. Batchwise pouring of the adsorbent into the column is therefore of no advantage. The following procedure is recommended for preparing the column: The column is fitted with a liquid reservoir, with the tap closed. The required wet volume of adsorbent (as a 10% suspension) is distributed in the same volume of starting buffer and then poured into the filling reservoir of the column. A mechanical stirrer or, better, a magnetic vibrator which evenly distributes the cellulose is immersed into the liquid reservoir. The tap of the column is then opened and the slurry containing 1 – 2 g solids per 100 ml should pack evenly. Cellulose suspensions containing more than this amount of solids tend to form lumps and pack unevenly.

Capillary action between cellulose ion exchanger particles is so high that there is no danger of channeling even if the adsorbent should run dry for a certain amount of time. The usual rate of elution is 5 – 40 ml per cm^2 of column per h as induced by gravity. There is evidence that higher flow rates can be used without impairment of resolution.

If flow rates higher than 100 ml/h are attempted, the packed cellulose tends to collapse and builds up a fairly strong resistance against flow. Such a column must be repacked. It is a known fact that with pure water or with solutions of less than 0.01 molarity, swelling of the cellulose ion exchangers occurs with reduction of the flow rate. In this case, upward flow through the column can improve the situation.

The capacity of cellulose adsorbents for high molecular weight substances is very high. Loads of up to 100 mg of adsorbate per g of adsorbent may safely be recommended. So far there is no evidence that the application of adsorbate in trace quantities increases the sharpness of separation. It is therefore recommeded that higher loads be used when the concentration of the emerging substances becomes higher, analysis is simpler and more accurate, time of elution is shorter and the recovery of substances by means of eluent evaporation becomes easier.

Washing of column for reuse. Washing with 4 column volumes of 0.1 N NaOH and 4 volumes of water completely displaces proteins. Anion exchangers should receive a subsequent washing with 2 volumes of 0.1 N HCl in order to remove carbon dioxide, which may otherwise cause artifact peaks.

Choice of adsorbent. For proteins, DEAE-cellulose is by far the most popular adsorbent and is used in about 60% of all cases, followed by CM-cellulose (ca. 20%) and TEAE-cellulose (ca. 5%). DEAE-cellulose has also been applied in 75% of all cases outside the field of proteins. Based on those figures, a preliminary experiment with DEAE-cellulose is the logical approach, preferably applying the Boman [165] single-step elution technique at neutral pH. CM-cellulose is used particularly often for the chromatography of basic proteins. Recently, calcium phosphate has been recommended quite frequently as adsorbent for proteins and nucleic acids. Hydroxylapatite therefore deserves some consideration.

TEAE- and AE-cellulose may sometimes give results which differ from those obtained by using DEAE-cellulose but they have not yet been proved indispensable. ECTEOLA-cellulose is used almost exclusively for the separation of nucleic acids, nucleoproteins and nucleotides. Cellulose derivatives with strongly ionized groups, such as SE-, GE- and P-cellulose, are probably most useful for low molecular weight substances.

Choice of eluent. In order to avoid artifacts, it is advisable to examine whether elution with the most concentrated buffer could yield material which interferes with analysis. If this is the case, careful washing with this buffer is necessary.

Cellulose ion exchangers tend to bleed soluble polysaccharides into the aqueous solution (up to 30 ppm), especially those with a high degree of substitution and with highly ionized groups [166]. Impurities of this kind can be detected by mixing 1 ml of eluent with 2 ml of a 0.2% solution of anthrone in sulfuric acid without cooling. A green colour develops which can be diluted with glacial acetic acid or 50% sulfuric acid, for comparison with a known standard [167, 168]. Removal of polysaccharide impurities is possible by passing the eluent through a small column of Dowex 2X8, 200 – 400 mesh [166] (acidic adsorbents), or Dowex 50X8, 200 – 400 mesh (basic adsorbents). It should be noted that cellulase can degrade cellulosic adsorbents [169].

As a rule, adsorption equilibria on cellulose ion exchangers are much more sensitive to changes in the salt concentration than to changes in pH. A small shift in ionic strength can effect complete desorption. Therefore, the pH in many cases is kept constant during elution. Theoretical considerations [170] lead to the following recommendations for the proper choice of eluent:

1. Use cationic buffers (TRIS-HCl, piperazine-HCl) with anion exchangers, anionic buffers (phosphate, acetate) with cation exchangers.
2. Avoid working in a pH-range in the neighbourhood of the p*K* of the adsorbent used. With anion exchangers, use falling pH gradient, starting at about pH 8. With cation exchangers, use rising pH gradient, starting at about pH 5. The elution sequence with pH gradients discriminates between molecules of varying charge and is therefore more similar to an electrophoretic separation.
3. If possible, salt gradients should start with a molarity of at least 0.01. The concentration can be based on the buffering salts only, or they can also contain additional sodium chloride or similar salts. The upper limit is usually 0.5 M. The sequence of eluted molecules reflects adsorptive forces, molecular size and shape, rather than the charge of the molecules.

It is quite clear that satisfactory results may be achieved even under conditions which do not conform to those mentioned in the recommendations. Nevertheless, optimum results and absence of doubling peaks or artifacts can be expected by adhering to them. When planning to isolate the chromatographed fractions by evaporation of the solvent, the use of volatile buffers should be attempted, e. g. carbonic acid, carbonates, acetates and formates of the ammonium or triethylammonium ion [171, 182].

Choice of gradient. Relatively few cases have been published where elution was effected with one single solvent. Usually the composition of the eluent is changed during elution, either in a single step or in multiple steps, either discontinuously or continuously. Extensive discussions of gradient elution from the theoretical and practical points of view can be found in Alm, Williams and Tiselius [173]; Bock and Ling [174]; Clayton and Bushuk [175]; B. Drake [176]; E. Geiss [177]; Knedel [178]; Parr [179]; Peterson [180]; Popp et al. [181]; Rossett [182]; Schwab, Rieman and Vaughan [183]; Wolf and Sly [184].

If only one substance with established elution behaviour is to be purified routinely, mostly two-step procedures such as suggested by Boman [165] are acceptable, as they are fast and simple.

If, however, a complete analysis of a complex mixture of very different components, like serum or cell homogenates, is intended, gradients covering a wide range of buffer concentrations must be applied. Step-by-step discontinuous elution will render particular components in high concentrations at well-predictable positions in the chromatogram (see Figures 1.83 and 1.84).

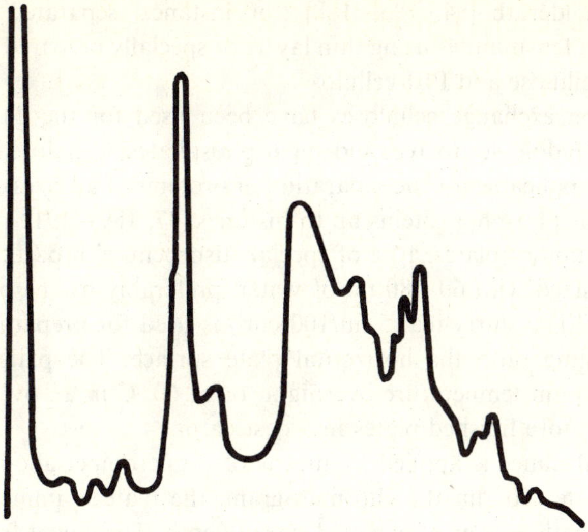

Figure 1.83 Continuous gradient.

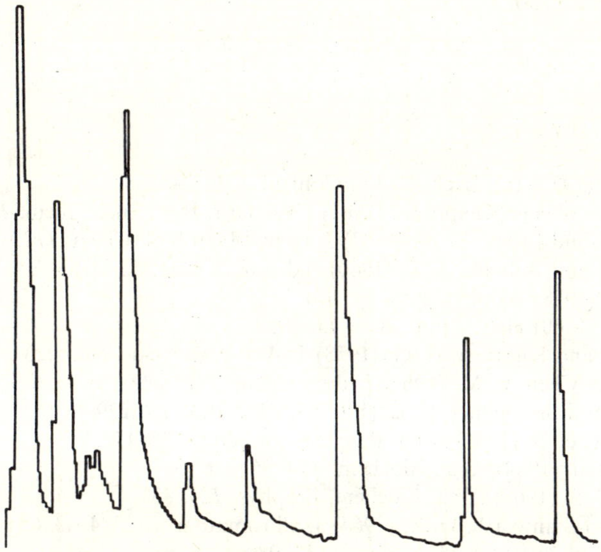

Figure 1.84 Step-by-step elution.

1.3.5.3 Thin layer chromatography

The advantages of this method, such as ultra-rapid separations, with a minimum of experimental effort and with less than one tenth of the quantity of material needed with paper chromatography, are also useful for work with cellulose ion exchangers. K. Randerath [84, 185–188], for instance, separated mixtures of nucleotides within a few minutes using thin layers of specially prepared ECTEOLA-cellulose, DEAE-cellulose and PEI-cellulose.

Furthermore, ion exchange celluloses have been used for the TLC of lipids, DNA, nucleotides, indole derivatives and sugar phosphates. Cellulose ion exchangers are universally applicable for the separation of proteins of all sizes, for peptides, including all iPs, for phosphoproteins up to histones [37, 189–191].

For the preparation of plates, 10 g of special adsorbent with particles below 30 μm are carefully mixed with 60–80 ml of water, preferably by the brief use of a high-speed blender. The slurry (ca. 5 ml/100 cm^2) is used for preparing the plates by subsequent casting onto the horizontal plate surface. The plates should be allowed to dry at room temperature overnight, or at 60 °C in an oven for 2 h. It is not necessary to store finished plates in a desiccator.

0.01 μmol of substance is applied by means of a micropipet about 3 cm from the edge of the plate. To run the chromatogram, the plate is immersed into an open beaker filled with a layer of eluent, 1 cm in depth. The eluent front migrates about 10 cm in 15 min. Nucleotides can be detected as dark spots at 253 nm. Proteins can be detected in the same manner after heating the wet chromatogram for a few minutes at 150 °C.

References

[1] Rowe, M. C. (1979) Ion Exch. Pollut. Control *2*, 203–11.
[2] Wegschneider, W. and Knapp, G. (1981) CRC Crit. Rev. Anal. Chem. *11*, 79–102.
[3] Fischer, H. J. and Lieser, K. H. (1983) Angew. Macromol. Chem. *112*, 1–14.
[4] Guthrie, J. D and Bullock, A. L. (1960) Ind. Eng. Chem. *52*, 935–7.
[5] Semenza, G. (1958) Biophys. Acta *24*, 401.
[6] Semenza, G. (1960) Helv. Chim. Acta *43*, 1057.
[7] Tener, G. M. and Khorana, H. G. (1958) J. Am. Chem. Soc. *80*, 6223.
[8] Fox, E. J. and Chen, C. M. (1967) J. Biol. Chem. *242*, 4490.
[9] Kelly, R. B. and Sinsheimer, R. L. (1967) J. Mol. Biol. *29*, 229.
[10] Ruelius, H. W. et al. (1968) Arch. Biochem. Biophys. *125*, 126.
[11] Haber, M. et al. (1984) Anal. Biochem. *139*, 363–6.
[12] Baumstark, J. S. (1968) Arch. Biochem. Biophys. *125*, 837.
[13] James, K. and Stanworth, D. R. (1964) J. Chromatogr. *15*, 324–336.
[14] Kessler, A. et al. (1960) J. Biol. Chem. *235*, 989.
[15] Moore, B. W. and McGregor, D. (1965) J. Biol. Chem. *240*, 1647.
[16] Peterson, E. A. and Kuff, E. L. (1969) Biochemistry *8*, 2916–23.
[17] Peterson, E. A. et al. (1961) Arch. Biochem. Biophys. *93*, 428.
[18] Sober, H. A. and Peterson, E. A. (1958) Fed. Proc. *17*, 1116.
[19] Sober, H. A. et al. (1956) J. Am. Chem. Soc. *78*, 756.
[20] Bomann, H. G. and Kaletta, U. (1958) Biochim. Biophys. Acta *24*, 619.
[21] Ellis, S. and Simpson, M. E. (1956) J. Biol. Chem. *220*, 939.
[22] Grossberg, A. L. et al. (1961) Arch. Biochem. Biophys. *93*, 267.
[23] Mandeles, S. (1960) J. Chromatogr. *3*, 256.

[24] Marchis-Moureu, G. et al. (1958) Bull. Soc. Chim. Biol. *40*, 2019.
[25] Moore, B. W. and Lee, R. H. (1960) J. Biol. Chem. *235*, 1359.
[26] Applegarth, D. A. and Dutton, G. G. S. (1964) J. Chromatogr. *15*, 246.
[27] Lipson, M. J. and Silbert, J. E. (1966) Biochem. Biophys. Acta *101*, 279.
[28] Mueller, M. et al. (1960) Z. Pflanzenernaehr. Duengung Bodenkunde *90*, 139.
[29] Ringertz, N. R. and Reichard, P. (1959) Acta Chem. Scand. *13*, 1467.
[30] Kersten, H. et al. (1964) Biochim. Biophys. Acta *80*, 521.
[31] McCordoquale, D. J. (1964) Biochim. Biophys. Acta *91*, 514.
[32] Portocala, R. P. and Popa, L. (1965) Biochim. Biophys. Acta *95*, 185.
[33] Dixit, S. N. et al. (1975) Biochemistry *14*, 1933–8.
[34] Holley, R. W. (1967) Methods Enzymol. *XIIA*, 596.
[35] Bergquist, P. L. et al. (1967) Methods Enzymol. *XIIA*, 660.
[36] Collins, F. D. (1963) Biochem. J. *88*, 319.
[37] Obreiter, J. B. and Stowe, B. B. (1964) J. Chromatogr. *16*, 226.
[38] Rouser, G. et al. (1961) J. Am. Oil Chem. Soc. *38*, 548.
[39] Svennerholm, L. and Thorin, H. (1962) J. Lipid. Res. *3*, 483.
[40] Vance, W. R. et al. (1966) Biochemistry *5*, 435.
[41] Buchanan, D. L. and Markow, R. T. (1960) Anal. Chemistry *32*, 1400.
[42] Cunningham, L. W. et al. (1957) Biochim. Biophys. Acta *26*, 660.
[43] Deeb, S. S. and Hager, L. P. (1964) J. Biol. Chem. *239*, 1024.
[44] Richard, A. J. and Kegeles, G. (1959) Arch. Biochem. Biophys. *80*, 125.
[45] Travis, J. and Liner, I. E. (1965) J. Biol. Chem. *240*, 1967.
[46] Kaufman, B. T. et al. (1963) J. Biol. Chem. *238*, 1498.
[47] Schertel, M. E. et al. (1964) J. Biol. Chem. *240*, 3154.
[48] Shiota, T. et al. (1964) J. Biol. Chem. *239*, 2259.
[49] Head, A. J. et al. (1958) J. Chem. Soc. (London) 3418–25.
[50] Pastore, E. J. and Friedkin, M. (1961) J. Biol. Chem. *236*, 2314.
[51] Cleland, W. W. (1964) Biochemistry *3*, 481.
[52] Massey, V. and Swoboda, B. E. P. (1963) Biochem. J. *338*, 474.
[53] Adam, A. and Moffat, J. G. (1968) Biochemistry *7*, 875.
[54] Staehelin, M. (1961) Arch. Biochem. Biophys. *49*, 20, 28.
[55] Bell, D. et al. (1964) Biochemistry *3*, 317.
[56] Tener, G. M. et al. (1967) Methods Enzymol. *XII A*, 398.
[57] Yamada, E. W. and Jakoby, W. B. (1959) J. Biol. Chem. *234*, 941.
[58] Hoyer, B. H. et al. (1958) Fed. Proc. *17*, 507.
[59] Levin, D. (1958) Arch. Biochem. Biophys. *78*, 33.
[60] Maeda, A. and Imahori, K. (1963) Biochim. Biophys. Acta *76*, 543.
[61] Selas, M. et al. (1967) J. Biol. Chem. *240*, 3988.
[62] Veder, H. A. and Pascha, C. H. (1961) Biochim. Biophys. Acta *47*, 408.
[63] Bosch, L. et al. (1960) Biochim. Biophys. Acta *41*, 454.
[64] Bradley, D. F. and Rich, A. (1958) J. Am. Chem. Soc. *78*, 5898.
[65] Olson, J. A. (1959) J. Biol. Chem. *234*, 5.
[66] Tanaka, K. (1958) Bull, Chem. Soc, Japan *31*, 393.
[67] Taussig, A. and Craeser, E. H. (1959) Arch. Biochem. Biophys. *83*, 436.
[68] Yoshida, A. and Freese, E. (1965) Biochim. Biophys. Acta *99*, 59.
[69] Zabin, I. (1963) J. Biol. Chem. *238*, 3300.
[70] Yanari, S. et al. (1960) Biochim. Biophys. Acta *45*, 595.
[71] Allalouf, D. et al. (1964) Biochim. Biophys. Acta *83*, 278.
[72] Bowness, J. M. (1961) Arch. Biochem. Biophys. *91*, 86.
[73] Lloyd, A. G. et al. (1963) Biochim. Biophys. Acta *69*, 496.
[74] Suzuki, S. et al. (1961) Biochim. Biophys. Acta *50*, 169.

[75] Craeser, E. H. and Taussig, A. (1957) Virology *4*, 200.
[76] Straetling, W. H. et al. (1976) Eur. J. Biochem. *66*, 423–33.
[77] Peterson, E. A. (1959) Anal. Chem. *31*, 857.
[78] Cabib, E. et al. (1965) J. Biol. Chem. *240F*, 2114.
[79] Brown, F. and Ward, D. N. (1959) Proc. Soc. Exp. Biol. Med. *100*, 701.
[80] Campbell, D. H. et al. (1951) Proc. Nat. Acad. Sci. U.S.A. *37*, 575.
[81] Lerman, L. S. (1953) Proc. Natl. Acad. Sci. U.S.A. *39*, 232.
[82] Randerath, K. (1963) Biochim. Biophys. Acta *76*, 622.
[83] Westley, J. and Green, J. R. (1959) J. Biol. Chem. *234*, 2325.
[84] Randerath, K. (1961) Angew. Chemie *73*, 674.
[85] Guthrie, J. D. et al. (1965) Anal. Chem. *13*, 1693.
[86] Boman, H. G. and Westlund, L. E. (1957) Arch. Biochem. Biophys. *70*, 572.
[87] Glomset, J. (1957) Acta Chim. Scand. *11*, 512.
[88] Wu, Y. V. and Scheraga, H. A. (1962) Biochemistry *1*, 698.
[89] Saraswathi, S. et al. (1978) J. Biol. Chem. *253*, 1024–9.
[90] Chung, H. and Mandeles, S. (1964) Biochim. Biophys. Acta *92*, 403.
[91] Deutsch, H. F. et al. (1961) J. Biol. Chem. *236*, 2216.
[92] Porter, R. R. (1959) Biochem. J. *73*, 119.
[93] Miyake, Y. et al. (1966) Biochim. Biophys. Acta *105*, 86.
[94] Tiselius, A. et al. (1956) Arch. Biochem. Biophys. *65*, 132.
[95] Bernardi, B. and Kuwasaki, T. (1968) Biochim. Biophys. Acta *160*, 301.
[96] Burness, A. T. H. and Vizoso, H. D. (1961) Biochim. Biophys. Acta *49*, 225.
[97] Hartmann, G. and Coy, U. (1961) Biochim. Biophys. Acta *51*, 205.
[98] Pearson, R. L. and Kelmers, A. D. (1966) J. Biol. Chem. *241*, 767.
[99] Folk, J. E. and Schirmer, E. W. (1963) J. Biol. Chem. *238*, 3884.
[100] Mendicino, J. and Vasarhely, F. (1963) J. Biol. Chem. *238*, 3528.
[101] Kress, L. F. et al. (1966) Biochim. Biophys. Acta *113*, 375.
[102] Brown, G. L. and Brown, A. V. (1958) Symp. Soc. Exp. Biol. *XII*, 6 Cambridge Uni. Press.
[103] Rode, G. C. and Campbell, M. C. (1962) Anal. Chim. Acta *27*, 422.
[104] Prager, M. D. and Speer, R. J. (1959) Proc. Soc. Exp. Biol. Med. *100*, 68.
[105] Gival, D. and Selq, M. (1964) Biochemistry *3*, 444–58.
[106] Goulian, M. and Lucas, Z. J. et al. (1968) J. Biol. Chem. *243*, 627–38.
[107] Utsumi, S. and Karush, F. (1965) Biochemistry *4*, 1766.
[108] Williamson, A. R. and Askonas, B. A. (1968) Arch. Biochem. Biophys. *125*, 401.
[109] Holley, R. W. et al. (1961) J. Biol. Chem. *236*, 197.
[110] Bendich, A. et al. (1955) J. Am. Chem. Soc. *77*, 3871.
[111] Bendich, A. et al. (1958) J. Am. Chem. Soc. *80*, 3949.
[112] Bendich, A. et al. (1958) Symp. Soc. Exp. Biol. *XII*, 31 Cambr. Uni. Press.
[113] Bock, R. M. and Cherayil, R. J. D. (1967) Methods Enzymol. *XIIA*, 638.
[114] Rafelson Jr., M. E. (1960) Arch. Biochem. Biophys. *90*, 68.
[115] Schapot, V. and Pitot, H. C. (1966) Biochim. Biophys. Acta *119*, 37.
[116] Ciferri, O. et al. (1964) Biochim. Biophys. Acta *87*, 508.
[117] Gillam, J. et al. (1967) Biochemistry *6*, 3043.
[118] Staehelin, M. (1961) Biochim. Biophys. Acta *49*, 11.
[119] Tomlinson, R. V. and Tener, G. M. (1963) Biochemistry *2*, 697.
[120] Neukomm, N. et al. (1960) Helv. Chim. Acta *34*, 64.
[121] Brennan, P. J. and Ballou, C. E. (1968) J. Biol. Chem. *243*, 2975.
[122] Brennan, P. J. and Ballou, C. E. (1967) J. Biol. Chem. *242*, 3046.
[123] Mitz, M. A. and Summaria, L. S. (1961) Nature *189*, 576.
[124] Usdin, E. and Porath, J. (1959) Ark. Kemi *11*, 41.

[125] Staehelin, M. et al. (1960) Arch. Biochem. Biophys. *85*, 292.
[126] Stevens, A. and Hilmoe, R. J. (1960) J. Biol. Chem. *235*, 3060.
[127] Mitz, M. A. and Yanary, S. S. (1956) J. Am. Chem. Soc. *78*, 2649–50.
[128] Hornby, W. E. et al. (1966) Biochem. J. *98*, 420.
[129] Wilson, R. J. A. et al. (1968) Biochem. J. *108*, 845.
[130] Weliky, N. et al. (1964) Immunochemistry *1*, 219.
[131] Kay, G. and Lilly, M. D. (1970) Biochim. Biophys. Acta *198*, 276.
[132] Jagendorf, A. T. et al. (1963) Biochim. Biophys. Acta *78*, 576.
[133] Bauman, A. J. et al. (1967) Anal. Chem. *39*, 932.
[134] Burba, P. and Lieser, K. H. (1976) Angew. Makromol. Chem. *50*, 151.
[135] Gstrein, H. et al. (1978) Paper *163*, Euroanalysis III.
[136] Horvath, Z. and Nagydiosi, G. (1975) J. Inorg. Nucl. Chem. *37*, 767.
[137] Horvath, Z. et al. (1977) At. Absorpt. Newsl. *16*, 152.
[138] Burba, P. (1982) Gewaesserschutz, Wasser, Abwasser *57*, 248–68.
[139] Gilham, P. T. (1971) Methods Enzymol. *21*, 191–7.
[140] Gilham, P. T. and Robinson, W. E. (1964) J. Am. Chem. Soc. *86*, 4985.
[141] Lindberg, U. and Sundquist, R. (1974) J. Mol. Biol. *86*, 451.
[142] Irwin, D. et al. (1975) Cell. *4*, 157.
[143] Wennerblom, B. A. and Joergensen, S. E. Ger. Pat. 2159863 (Dec. 2.1971).
[144] Jones, D. T. (1974) Process. Biochem. *9*, 17–9.
[145] Grant, R. A. (1974) Process. Biochem. *9*, 11–4.
[146] Grant, R. A. (1975) Effluent Water Treat. J. *15*, 616–19.
[147] Rowe, M. C. (1975) Effluent Water Treat. J.
[148] Ermolenko, J. N. et al. (1976) Vestsi Acad. Navuk BSSR, Ser. Khim. Navuk 5, 21.
[149] Ermolenko, J. N. and Luneva, N. K. (1977) Cellul. Chem. Technol. *11*, 647.
[150] Hayatsu, Y. et al. (1983) Gann *74*, 472–82.
[151] Hayatsu, Y. et al. (1983) Mut. Res. *119*, 233–8.
[152] Kobayashi, H. and Hayatsu, Y. et al. (1984) Gann *75*, 489–93.
[153] Zoebisch, B. et al. East Ger. Pat. 126869 (Aug. 17, 1977).
[154] Wattiez, D. et al. Ger. Pat. 2335213 (Juli 13, 1974); C.A. 81, P14405v.
[155] Tsibikov, V. B. (1973) Nov. Med. Priborostr. *1*; C.A. 83,32858r.
[156] Gangneux, A. et al. (1976) Eur. Polym. J. *12*, 535.
[157] Gangneux, A. et al. (1976) Eur. Polym. J. *12*, 543.
[158] Gangneux, A. et al. (1976) Eur. Polym. J. *12*, 551.
[159] Rinaudo, M. and Canova, P. (1974) C.R. Acad. Sci. Ser. *279*, 253.
[160] Schuetz, G. et al. (1977) Nucleic Acid Res. *4*, 71–84.
[161] Bauman, R. A. (Fr. Pat. 1508814 (May 1, 1968) C.A. 70, P21197c.
[162] Pastyr, J. and Kuniak, L. (1972) Cellul. Chem. Technol. *6*, 249.
[163] Tsypkina, M. N. and Boyarskaya, R. K. USSR Pat. 208928 (May 25, 1978).
[164] Taussig, A. and Craeser, E. H. (1957) Biochim. Biophys. Acta *24*, 448.
[165] Boman, H. G. and Westlund, L. E. (1956) Arch. Biochem. Biophys. *64*, 217.
[166] Doscher, M. S. and Wilcox, P. E. (1961) J. Biol. Chem. *236*, 1328.
[167] Dreywood, R. (1946) Ind. Eng. Chem. Anal. *18*, 499.
[168] Fischer, E. H. and Kohtes, L. (1951) Helv. Chim. Acta *34*, 1123.
[169] McClendon, J. H. (1961) Biochim. Biophys. Acta *48*, 398.
[170] Semenza, G. (1960) Chimia *14*, 325.
[171] Porath, J. (1955) Nature *175*, 478.
[172] Smith, M. et al. (1963) Methods Enzymol. *6*, 645–69.
[173] Alm, R. S. et al. (1952) Acta Chem. Scand. *6*, 826.
[174] Bock, R. M. and Ling, N. S. (1954) Anal. Chem. *26*, 1543.
[175] Clayton, J. W. and Bushuk, W. (1966) J. Chromatogr. *21*, 64.

[176] Drake, B. (1955) Ark. Kemi *8*, 1.
[177] Geiss, E. (1957) Z. Physiol. Chem. *308*, 74.
[178] Knedel, M. and Fateh-Moghadam (1965) GIT *9*, 675.
[179] Parr, C. W. (1954) Biochem. J. *56*, XXII.
[180] Peterson, E. A. and Sober, H. A. (1962) Methods Enzymol. *V*, 3.
[181] Popp, R. A. et al. (1966) Biochim. Biophys. Acta *115*, 113.
[182] Rossett, T. (1956) J. Chromatogr. *18*, 498.
[183] Schwab, H. et al. (1957) Anal. Chem. *29*, 1357.
[184] Wolf, C. R. et al. (1980) Biochim. Biophys. Acta *624*, 409–19.
[185] Randerath, E. and Randerath, K. R. (1964) J. Chromatogr. *16*, 111.
[186] Randerath, K. and Randerath, E. (1966) J. Chromatogr. *22*, 110.
[187] Randerath, K. and Weimann, G. (1963) Biochim. Biophys. Acta *76*, 129.
[188] Weimann, G. and Randerath, K. (1963) Experimentia *19*, 49.
[189] Bauer, R. D. and Martin, K. D. (1964) J. Chromatogr. *16*, 519.
[190] Beers, R. F. (1960) J. Biol. Chem. *235*, 2393.
[191] Dietrich, C. P. et al. (1964) J. Chromatogr. *15*, 277.

1.4 Dextran and Agarose Ion Exchangers

Gert-Joachim Strobel
Freiburg, Federal Republic of Germany

1.4.1 Sephadex ion exchangers

Sephadex ion exchangers are produced by introducing functional groups into Sephadex, a bead-formed gel prepared by crosslinking dextran with epichlorhydrin. The functional groups are attached to the glucose units of the dextran by stable ether linkages.

Sephadex is particularily suitable as a basis for an ion exchanger matrix. Since it is hydrophilic and shows very low non-specific adsorption, proteins, nucleic acids and other labile biological molecules are not adsorbed onto or denatured by the gel, so that high capacities are obtained. Good flow properties are obtained due to the bead form.

Sephadex ion exchangers are derived from either Sephadex G-25 or G-50. The G-types of Sephadex differ in their degree of crosslinking and, hence, in their degree of swelling. The degree of swelling is dependent on the presence of salts or detergents and will alter the performance of a packed ion exchange column.

Ion exchangers based on Sephadex G-25 are more tightly crosslinked than those based on Sephadex G-50 and therefore swell less and have greater rigidity. G-50 type ion exchangers are more porous and therefore have a higher capacity for molecules with molecular weights larger than 30000. The degree of swelling depends on the pH, the ionic strength of the buffers used and the nature of the counterion. It is thus not the same as that of the unsubstituted parent gel Sephadex G-50. Four different functional groups are used for substitution onto Sephadex, giving a total of eight different ion exchangers (Table 1.24).

Ion exchangers are termed "weak" or "strong" depending on the range of pH over which they carry a charge (Figure 1.85). The strong ion exchangers, QAE- and SP-Sephadex, are characterized by their constant capacities over a wide pH

Table 1.24 Ion exchanger types

Matrix porosity	Type of ion exchanger			
	Anion exchanger		Cation exchanger	
	weak	strong	weak	strong
G-25	DEAE-Sephadex A-25	QAE-Sephadex A-25	CM-Sephadex C-25	SP-Sephadex C-50
G-50	DEAE-Sephadex A-50	QAE Sephadex A-50	CM-Sephadex C-50	SP-Sephadex C-50

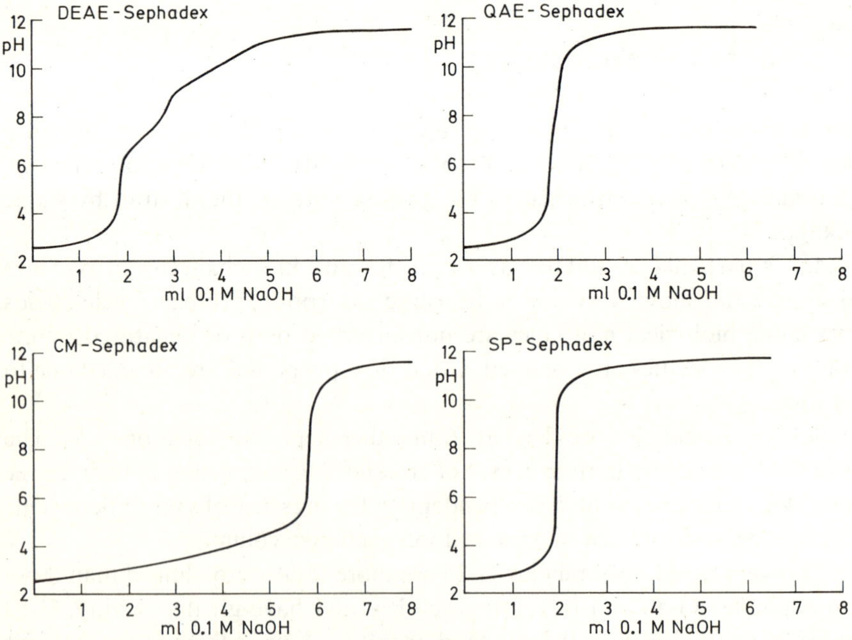

Figure 1.85 Titration of 0.1 gram ion exchanger in 50 ml 1 M KCl. (Work from Pharmacia, Uppsala, Sweden.)

range of 2–10. They have no buffering capacity in this pH range and therefore the sample loading capacity does not decrease at high or low pH values.

There exists a simple mechanism of interaction between the media and the solute due to the absence of intermediate forms of charge interactions.

Equilibration is quickly achieved since the ion exchanger cannot act as a buffer. Ion exchange experiments are therefore more controllable since the charge characteristics of the ion exchangers do not change with changes in pH. Data obtained from titration curves can easily be transferred to ion exchange experiments.

The "weak" ion exchangers start to lose their charge at pH values below 6 (CM-Sephadex) or above 9 (DEAE Sephadex); therefore these ion exchangers are recommended at around neutral pH, where they exhibit their highest capacity.

1.4.1.1 Chemical stability

As the Sephadex matrix consists of crosslinked sugar polymers, i. e. glucose, it is insoluble in all solvents such as water, salt solutions, organic solvents, alkaline and weakly acidic solutions. In strongly acidic solutions hydrolysis of the glycosidic linkages may occur and therefore pH values below 2 should be avoided, however for a short time lower pH values at low temperatures are tolerated. Sephadex ion exchangers can also be used in the presence of protein-denaturing solvents, which can be important when substances are to be separated on the basis of their electrostatic properties. Additional non-ionic interactions with the matrix often make a significant contribution to the separation obtained. Exposure to strong oxidizing agents or the dextran-degrading enzyme dextranase should be avoided. Long-term experiments with 0,1 M H_2SO_4 over a period of four weeks showed traces of free sugar in the supernatant (Pharmacia Quality Control Dept.).

During regeneration for a short time (several hours) Sephadex ion exchangers can be exposed to 0.1 M NaOH without detectible consequences in the performance. For long-term storage of the swollen ion exchanger antimicrobial agents are recommended (see Table 1.25).

Table 1.25 Recommended antimicrobial agents

Type of exchanger	Effective agents	Concentration (%)	Limitations
Anion	Phenyl mercuric salts	0.001	Use in weakly alkaline solutions only
	Chlorohexidine, e. g. Hibitane®	0.002	Generally applicable
Cation	Thimerosal, e. g. Merthiolate® ethyl mercurithiosalicyclate	0.005	Use in weakly acidic solutions only
Either	Chlorobutanol, e. g. Chloretone®	0.05	Use in weakly acidic solutions only

1.4.1.2 Physical stability

Due to the crosslinking, Sephadex ion exchangers can be autoclaved up to 30 min at 120 °C at neutral pH in the salt form. The chromatographic effect is not affected. During autoclaving small quantities of carbohydrates are released and have to be washed away with sterile buffer.

As Sephadex ion exchangers are soft gels, the pressure drop in a column should not exceed 1 bar when using G-50-derived gels and 2–3 bars when using G-25-based ion exchangers. A higher pressure drop could deform the beads into ellipsoids, causing a dramatic flow reduction and leading to cracks in the matrix beads, which then will generate fines clogging the column. If high flow rates are required during column operation bed heights under 30 cm are recommended.

1.4.1.3 Capacity

Sephadex ion exchangers exhibit high capacities for charged biomolecules, even at high ionic strength. This makes the gel ideal for proteins, which are stable only under such conditions. As the capacity depends on the number of charged groups on the matrix it may vary with pH.

The titration curve of DEAE Sephadex (Fig. 1.85) shows clearly that with increasing pH the functional groups available decrease rapidly. Ideally the optimal working range is in the steepest range of the curve. The total capacity of an ion exchanger, determined by titration with, e. g., NaOH not always is really available during protein purification. Due to the crosslinking of the matrix, pores of different diameters are generated which can exclude large molecules from entering the matrix. Charged groups of the ion exchanger coupled to the matrix inside the pores are then inaccessible for large globular molecules and the available capacity then is reduced for these large proteins. As a guideline for the available capacity see Table 1.26. If exact data are needed a determination of the available capacity for a specific protein is recommended.

The determination should be made in the range of the pH where the gel is to be used, as with changing pH the pore size of the matrix changes. Sephadex-based ion exchangers have a tendency to shrink with increasing ionic strength.

1.4.1.4 Choice of buffer pH

In order to bind the product to the ion exchanger under controlled conditions, a buffer with a given pH not too far away from the pK_a (± 0,5 pH units) is chosen. The ionic strength has to be low enough to enable a product to bind. Starting conditions should be sufficiently close to elution conditions so that elution is not time-consuming. The buffer pH should be chosen so that substances of interest

Table 1.26 Available capacity data

Ion exchanger	Protein M_r	Thyroglobulin 669×10^3	HSA 68×10^3	☒-lactalbumin 14.3×10^3	IgG 160×10^3	Bovine COHb 69×10^3	Ribonuclease 13.7×10^3
DEAE-Sephadex	A-25	1.0 mg/ml*	30 mg/ml*	140 mg/ml	N. D.	N. D.	N. D.
	A-50	2.0 mg/ml	110 mg/ml	50 mg/ml	N. D.	N. D.	N. D.
QAE-Sephadex	A-25	1.5 mg/ml	10 mg/ml*	110 mg/ml	N. D.	N. D.	N. D.
	A-50	1.2 mg/ml	80 mg/ml	30 mg/ml	N. D.	N. D.	N. D.
CM-Sephadex	C-25	N. D.	N. D.	N. D.	1.6 mg/ml*	70 mg/ml	190 mg/ml
	C-50	N. D.	N. D.	N. D.	7.0 mg/ml	140 mg/ml	120 mg/ml
SP-Sephadex	C-25	N. D.	N. D.	N. D.	1.1 mg/ml*	70 mg/ml	230 mg/ml
	C-50	N. D.	N. D.	N. D.	8.0 mg/ml	110 mg/ml	100 mg/ml

*Not recommended
N. D. = not determined

have a net charge opposite to that of the ion exchanger. If a pH change is necessary for elution, the starting conditions should be chosen in such a way that the new pH is still inside the buffering capacity. Ionic strengths of start buffers usually are recommended in the range of 10–50 mM depending on the pH and the buffer system.

1.4.1.5 Swelling properties

Sephadex A-25 and C-25 types, having a high matrix density, are characterized by a fairly rigid bead structure. Volume variation following deviation of pH or ionic strength is therefore limited and can be neglected in well-packed columns.

Sephadex A-50 and C-50 types, due to their reduced degree of crosslinking, are susceptible to significant volume variations. Swelling is greatest at low ionic strength, as repulsion between similarly charged groups on the matrix is then maximal (Figure 1.86). To minimize this effect a starting buffer of at least 0.1 M is recommended.

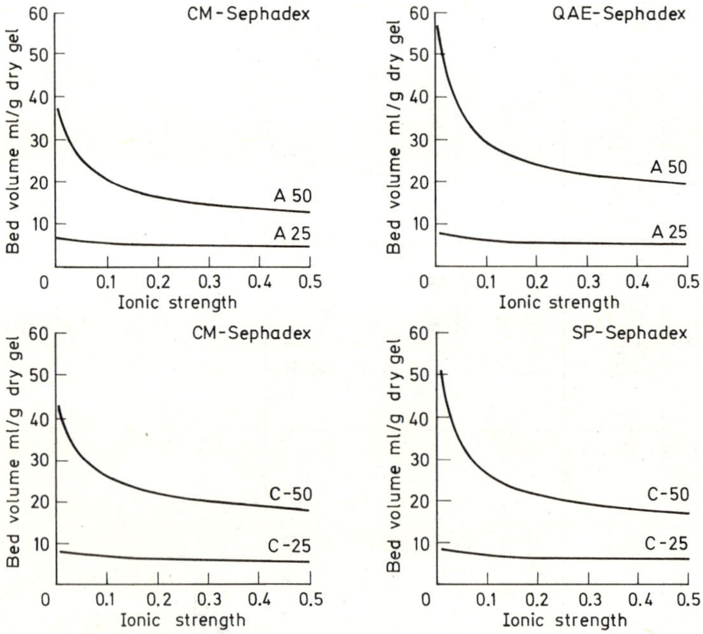

Figure 1.86 Bed volumes obtained from 1 gram dry ion exchanger as a function of ionic strength. Bed volumes were determined in a column 1.5 × 30 cm (K 15/30) at a constant pressure of 30 cm water.
DEAE- and QAE-Sephadex were measured in Tris-HCl buffer pH 7.6 with varying NaCl concentration. CM- and SP-Sephadex were measured in acetate buffer pH 4.3 with varying NaCl concentration. (Work from Pharmacia, Uppsala, Sweden.)

As the pH effects the charged groups on the matrix, repulsion is also pH-dependent. Swelling is therefore greatest at pH values where the ion exchange groups are fully dissociated; when pH values approach their p*K* swelling decreases. The strong ion exchangers (QAE and SP-Sephadex) do not swell at different pH values, as their functional groups remain charged over a wide pH range.

As a consequence of changing bed volumes ion exchange columns packed with A-50 and C-50 types gels cannot be automated in operation. Sepharose-based ion exchangers are the better choice in such a situation.

1.4.1.6 Operation

Sephadex ion exchangers are provided in dry form and have to be swollen in the starting buffer or at least in a 0.2 M salt solution (NaCl) before operation. As the ion exchangers are supplied in the salt form it is not necessary to pre-cycle (to treat with acid and alkali) the gel.

The amount of gel required for column packing is carefully stirred in an excess of starting buffer and heated in a boiling water bath. Do not use magnetic stirrers as they may damage the beads. Heating the gel deaerates the matrix and the swelling is completed in 1 – 2 h. If the gel is swollen at room temperature at least 24 h are necessary for complete swelling. The starting buffer must contain the same ion as that originally present in the ion exchanger. The supernatant should be replaced several times during the swelling period. Alternatively, the gel could be extensively washed on a funnel after swelling.

If another counter-ion is required suspend the ion exchanger in a solution of 0.5 M – 1 M of the salt of the new counter-ion. After sedimentation of the gel the procedure is repeated several times.

Packing a column. To obtain the best results from Sephadex ion exchangers the packing of the column has to be done with the greatest care.

For most of the ion exchange experiments gel bed heights of 20 – 30 cm are sufficient. Choose a column size where the required amount of gel packs in that height.

The swollen ion exchanger should be mixed with starting buffer so that it forms a slurry where air bubbles still can move up. Unless the gel has been swollen in a boiling water bath it should be degassed under vacuum. The temperature of the gel should be the same as the buffer used.

Before starting the packing procedure the column should be mounted vertically and the column outlet closed. Be sure that no air is left under the support net. Then pour the gel slurry into the column and open the outlet and either pump the buffer out or let it run until the surface is about 1 cm above the settled gel. Then close the outlet, mount the adapter of the column and carefully set it a few millimeters above the gel surface. Buffer then is pumped through the column under at least twice the flow-rate the column should be run at during the separation. The gel will then be packed to a final bed height after a few column volumes of buffer have passed through. The outlet is then closed and the adapter carefully set on the surface of the gel bed. The column is then ready for use after equilibration with start buffer, which requires at least 2 column volumes of buffer. To ensure that the column is equilibrated check pH and ionic strength of the eluent.

Regeneration of gel. Sephadex ion exchangers can be used repeatedly when regenerated carefully after the elution of bound protein. In order to remove all charged biomolecules from the gel, the column is washed with a high salt buffer (1 M NaCl) or/and the pH is changed. Low pH values favour the release of proteins from an anion exchanger, while a cation exchanger should be treated with high pH. 2–3 column volumes of buffer with extreme pH and ionic strength are in most cases sufficient, followed immediately washing with neutral buffer until the ion exchanger is equilibrated. The column then can be used for the next run. Treatment with extreme pH and high ionic strength will alter the volume of the gel. Before application of the above-mentioned buffers the adapters in a column have to be released so that the gel cannot destroy the column. With the equilibration buffer the gel will reach the same volume as before, but the packing of the gel bed must be checked with a test run before reuse. If cracks in the gel bed are observed the column must be repacked as described in the packing procedure.

Cleaning the gel and storage. With the above-described procedure most of the bound material can be removed from the gel. Sometimes proteins precipitate on the column and do not redissolve. In this case a treatment of the ion exchanger with NaOH is the only way to clean the gel. Sephadex ion exchangers can be treated with 0.1 M NaOH up to 40°C, which is very efficient for dissolving precipitated proteins. Also lipids, which have a tendency to stick to the gel matrix, are removed easily. The NaOH solution must not stay on the column for more than several hours in order to avoid chemical breakdown of the gel. A thorough wash with regeneration buffer will remove the NaOH and protect the gel. For long-term storage in the swollen state an antimicrobial agent should be included in the buffer (Table 1.25).

1.4.1.7 Large scale use

Sephadex ion exchangers have been used for many years for the production process of a wide range of pharmaceuticals. The biocompatibility of Sephadex ion exchangers and their good flow properties make them suitable for large-scale operations. A widely-used application is for purifying the prothrombin complex from blood plasma [1, 2], where DEAE Sephadex A-50 is used. Another batch operation is to purify human factor VII with DEAE- and SP-Sephadex [3].

Separations run on large columns (100 l and more) include the preparation of intravenous immunoglobulins from natural sera [4–6] or the purification of hyperimmune sera as anti-D, rabies, or varicella-zoster [7–9]. A rather new method describes the removal of hepatitis surface antigen from IgG preparations [10].

1.4.2 Sepharose® ion exchangers

DEAE- and CM-Sepharose Cl-6B Fast Flow and Q and S Sepharose are macroporous bead-form ion exchangers derived from crosslinked agarose gel Sepharose Cl-6B. A greater degree of crosslinking allows extremely high flow rates. There are several advantages in choosing Sepharose Cl-6B as the basis for an ion exchange matrix. The Sepharose matrix has much greater rigidity for equivalent porosity than a Sephadex gel would have. This is due to the difference in macrostructure of the two gel types. In Sephadex the dextran chains are arranged in a random fashion apart from where they are crosslinked, while in Sepharose the poly-saccharide chains lie in bundles. These bundles are further strengthened by intra-chain crosslinking. The resulting structure is macroporous and the total capacity of the gel is very good for molecules up to 4×10^6 in molecular weight. The gel has excellent flow properties and an extremely stable bed volume, which is relatively insensitive to changes in ionic strength and pH. The unspecific adsorption onto the matrix is very low. The gel characteristics are listed in Table 1.27.

1.4.2.1 Chemical stability

The crosslinked structure of Sepharose ion exchangers renders them chemically stable and practically insoluble in all solvents. They are stable in water, salt solutions and organic solvents in the range of pH 2–14. As with any DEAE ion exchanger, prolonged exposure to very alkaline conditions should be avoided due to the inherent stability of the DEAE-group as a free base. The ion exchangers can be used in non-ionic detergents such as 8 M urea and 8 M guanidine hydrochloride. Under oxidizing conditions, limited hydrolysis of the polysaccharide chains may occur. Sepharose also displays a notable resistance to microbial attack due to the presence of the unusual sugar 3,6-anhydro-L-galactose within the structure.

1.4.2.2 Capacity

Because of their macroporosity, Sepharose ion exchangers have high available capacity. Most proteins are of a lower molecular weight than the exclusion limit of the matrix (approx. 4 million), thereby making charged groups in the interior of the bead accessible for protein binding. Molecules with molecular weights above the exclusion limit of the matrix can still bind on to the surface of the matrix.

Table 1.27 Gel characteristics

	Q Sepharose Fast Flow	S Sepharose Fast Flow	DEAE-Sepharose Fast Flow	CM-Sepharose Fast Flow
Type of ion exchanger	anionic	cationic	anionic	cationic
Total capacity	0.18 – 0.25 mmol/ml	0.18 – 0.25 mmol/ml	110 – 160 μeq/ml gel	90 – 130 μeq/ml gel
Exclusion limit (globular proteins)	$\sim 4 \times 10^6$	$\sim 4 \times 10^6$	$\sim 4 \times 10^6$	$\sim 4 \times 10^6$
Bead form	Spherical, diameter 45 – 165 μm in wet form			
Bead structure	Crosslinked agarose, 6%			
Chemical Stability	Stable to all commonly used aqueous buffers – 1 M NaOH, 8 M urea, 8 M guanidine hydrochloride, 24% ethanol (tested at 40 °C for 7 days)			
pH Stability	pH 2 – 14			
Physical stability	Neglibible volume variation due to changes in pH or ionic strength			
Autoclavable	With Cl^- and Na^+, respectively, as counterions, at 121 °C, pH 7 for 30 min			

1.4.2.3 Regeneration of the gel and storage

The regeneration procedure of Sepharose ion exchangers is, in principle, the same as with Sephadex ion exchangers. Due to its high physical stability the gel can be regenerated in the column. Even a wash with 1 M NaOH to solubilize precipitated proteins or lipids can be performed while the gel remains in the column. The gel bed will not be affected by regeneration and columns packed with Sepharose ion exchangers are easily automated.

The gel can be stored in 20% ethanol without affecting the packed bed.

1.4.2.4 Large scale use

Sepharose Fast Flow ion exchangers are especially suitable for industrial use, where large volumes of raw material must be processed. Due to the high crosslinking of the matrix very high flow rates (400 cm/h, 15 cm bed height, 1 bar) can be obtained. The high substitution levels of the gels confer the high product binding capacities required in industrial processes.

References

[1] Suomela, H., Myllyä, G. and Raaska, E. (1977). Preparation and properties of a therapeutic factor IX concentrate. Vox Sang. *32*, 1–6.

[2] Chandra, S., Brummelhuis and H. G. J. (1981). Prothrombin complex concentrates for clinical use. Vox Sang. *41*, 257–273.

[3] Rao, L. V. M. and Bajaj, S. P. (1984). Purification of human factor VII utilizing O-(Diethylaminoethyl)-Sephadex and Sulfopropyl-Sephadex Chromatography. Anal. Biochem. *136*, 357–361.

[4] Suomela, H., Berglöf, J. H., Hämäläinen, E. et al. (1982). Immunoglobulin G for intravenous use, preparation and in vitro properties. Joint Meeting of the 19th Congress of the International Society of Haematology and the 17th Congress of the International Society of Blood Transfusion, Budapest, August 1–7. 1982.

[5] Condie, R. M. (1979). Preparation and intravenous use of undenatured human IgG. In: Immunoglobulins: characteristics and uses of intravenous preparations (Alving, B. M. and Finlayson, J. S., eds.) DHHS Publication No. (FDA) 80–9005. US Department of Health and Human Services, pp. 179–193.

[6] Suomela, H. (1980). An ion exchange method for immunoglobulin production. In: Methods of Plasma Protein Fractionation. (Curling, J. M., ed.). Academic Press, London, pp. 107–116.

[7] Hoppe, H. H., Mester, T., Henning, W. et al. (1973). Prevention of Rh-Immunization. Modified production of IgG anti-Rh for intravenous application by ion exchange) chromatography (IEC). Vox Sang. *25*, 308–316.

[8] Friesen, A. S., Bowman, J. M. and Price, H. W. (1981). Column ion-exchange preparation and characterization of Rh immune globulin (WinRho) for intravenous use. J. Appl. Biochem. *3*, 164–175.

[9] Friesen, A. S., Bowman, J. M. and Bees, W. C. H. (1982). Column ion-exchange chromatographic production of human immune globulins and albumin. Joint Meeting of the 19th Congress of the International Society of Haematology and the 17th Congress of the International Society of Blood Transfusion, Budapest, August 1–7, 1982.
[10] Zolton, R. P. and Padvelskis, J. V. (1984). Evaluation of anion exchange procedure for removal of hepatitis type B contamination from human gamma globulin products. Vox Sang. *47*, 114–121.

1.5 Zeolites

Michael Baacke and Akos Kiss
Degussa AG, Frankfurt am Main, Federal Republic of Germany

1.5.1 Definition of zeolites

Zeolites can be defined as crystalline hydrated alumosilicates of Group I and II metals of the Periodic Table (alkali metals and alkaline earth metals). Their structures are built from tetrahedral SiO_4 and AlO_4 units crosslinked by the sharing of oxygen atoms. The resulting three-dimensional networks are more or less open: there are channels and/or cavities accommodating water molecules and the cations necessary for charge balancing (the zeolite frameworks bear a negative charge created by partial substitution of Si^{IV} by Al^{III}). These alkali and alkaline earth metal cations have a certain mobility within the zeolite structure and can be exchanged for other cations.

The ion exchanging properties of zeolites were first reported in the middle of the 19th century by Thompson [1], Way [2] and Eichhorn [3]. Thus, Eichhorn investigated the exchange of Na^+ for Ca^{2+} by the chabazite/natrolite system. However, it was not until 50 years later that this aspect of zeolite chemistry was applied industrially. Water softening and removal of disturbing metal ions from boiler feed-water are essential in the generation of steam. This became more and more important at the time when industrialization was growing fast [4]. At first inorganic materials (crystalline as well as amorphous alumosilicates) were used for ion exchange. In time, complicated organic polymers became better available, and inorganic ion exchangers were replaced by polymers based on styrene-divinylbenzene copolymers. These materials can be regenerated more easily by strong acids. Several comprehensive reviews on the properties and applications of zeolites as ion exchangers can be found in the literature. The articles by Helfferich [5], Sherman [6–8] and Breck [10] are cited here without a claim of completeness in this list.

1.5.2 Properties of zeolites as ion exchangers

1.5.2.1 Stability

Zeolites have overall stable and solid frameworks. In particular, they are stable with regard to:

- high temperatures
- oxidizing/reducing agents
- ionizing radiation
- physical attack by swelling.

This framework stability makes the properties of zeolites in ion exchange processes rather predictable. Zeolites are relatively stable even at high pH levels. They are prepared at pH 12–13 and temperatures up to 300°C.

Just below these conditions they are usually rather stable.

1.5.2.2 Advantages of zeolites compared with organic ion exchangers

Compared with organic ion exchangers zeolites have the following advantages:

- they are more stable at elevated temperatures
- they are not attacked by ionizing radiation
- they are considerably more selective in certain separations.

Lack of knowledge of their properties as well as limited availability were responsible for the zeolites beginning to be widely used in industry only in the 1960s. Around this time simple procedures for the preparation of zeolites were developed, and large natural resources of several zeolite species were discovered. Since then commercial amounts of zeolites can be obtained inexpensively.

1.5.2.3 Selectivity

Zeolites are so well suited for ion exchange mainly because of their selectivity, which is a consequence of their "molecular sieve" property. For instance, big cations (Rb^+, Cs^+, organic cations) cannot diffuse into the channels and cavities of small-pore zeolites. Table 1.28 shows the pore openings of some hydrated zeolites.

Table 1.28 Pore openings of hydrated zeolites [7]

Zeolite	Pore opening (Å)
A	4.2
Analcime	2.6
Chabazite	3.7 × 4.2
Clinoptilolite	4.0 × 5.5
	4.1 × 4.7
Erionite	3.6 × 5.2
Linde L	7.1
Linde W	4.2 × 4.4
Mordenite	6.7 × 7.0
	2.9 × 5.7
Offretite	3.6 × 5.2
Phillipsite	4.2 × 4.4
	2.8 × 4.8
X	7.4
Y	7.4

If two cations – one larger, the other smaller than the pore entrance – interact with a zeolite, only the smaller one will be exchanged.

Ion exchange reactions are mostly done in aqueous solution. For this reason hydration must be taken into account when cation sizes are compared. Depending on the charge density of the cation the hydrate water molecules are bound more or less strongly. If the cations can pass the zeolite pores only after having lost the water molecules coordinated to them in solution, a distinct kinetic effect is observed, which varies with the ionic hydration energies.

Thus, the Ca^{2+}/Na-A zeolite exchange is often compared with the Mg^{2+}/Na-A zeolite exchange, although the Ca^{2+} ion is larger than the Mg^{2+} ion. Because of the higher charge density of Mg^{2+} the water ligands are bound more strongly to Mg^{2+} than to Ca^{2+}. The influence of hydration is especially obvious for La^{3+}. At room temperature no La^{3+}/Na-A exchange is observed. However, at elevated temperatures the hydration sphere is less tight, and exchange may occur [11]. In some zeolite frameworks certain sites are not accessible for big cations; partial replacement is observed then. For example, only 68 percent of Na^+ can be substituted by Rb^+ in Na-Y zeolite [12], since 16 Na^+ ions per unit cell are located within the sodalite cages, where Rb^+ cannot penetrate. The Na^+ ions undergoing exchange are those located in the supercages.

The molecular sieve effect can be used to evaluate the pore sizes of structurally uncharacterized zeolites. A series of cations differing in size are tested in ion exchange experiments for this purpose. As alkyl ammonium ions are available in a wide range of different diameters, they are usually the best choice. High selectivities were observed even in cases in which the pore is big enough to let pass both of two competing cations. For instance, the selectivity of zeolite A in the presence of

Ca^{2+} and Na^{+} can be as high as 100, if the exchange level is kept small (0–20 percent). At higher loadings (approx. 40), there is still a selectivity of 40, whereas the corresponding value of polystyrene sulfonic acids varies between 2,5 and 3.

It should have become clear by now, that there are features, besides the pore dimensions, which enhance the selectivity observed. When organic materials are used, the charge number of the cations is crucial. In general, this is also true for zeolites, but opposite selectivities (Ca < Na) are reported, too. The theory of the "anionic field strength of the exchange site", which was useful in similar considerations for glasses [13] proved to afford a reasonable approach for zeolites as well. According to this theory, the oxygen donor atoms of the zeolite networks compete with the hydrate water molecules for the cations. This is described by the equation:

$$G = (G^{A}_{zeo} - G^{B}_{zeo}) - (G^{A}_{sol} - G^{B}_{sol}) \qquad G\text{: free enthalpy.}$$

The first expression on the right side is attributable to the difference in G resulting from the interactions of two different cations with the zeolite framework, the second one represents the difference in hydration.

For zeolites with high local field strengths (aluminium-rich zeolites) the first expression is the more significant one. Small ions – even those having a high charge density and, consequently, high hydration energies – are favored by these zeolites. The second term becomes important for zeolites with weaker local fields (aluminium-deficient zeolites). Big hydrated ions are preferred in these cases. If the pore volume is small, only less voluminous hydrate spheres can exist within the zeolite framework. For this reason the affinity is highest for ions characterized by lower hydration energies.

Di- and trivalent cations are preferentially bound by zeolites having higher field strengths. In lower field types (low aluminium content) the anionic centers are too far distant from each other.

During the last few years much data about ion exchange equilibria have been obtained in industrial laboratories for more than 100 different zeolite types. Unfortunately, only few of the results obtained were published, covering mostly the investigations of the commercially important zeolites A, X, Y, chabazite, clinoptilolite and mordenite. In Table 1.29 some examples including references are listed.

Table 1.29 Investigation of ion exchange in different zeolites

Exchanging cations	Type of zeolite	References
H–Na	Mordenite	14
Na–Li	A	15, 16
Na–Li	Mordenite	14
Na–Li	X	17, 18
Na–Li	Y	18

Table 1.29 continued

Exchanging cations	Type of zeolite	References
Na – K	A	15, 16
Na – K	Analcime	19
Na – K	Chabazite	20
Na – K	Erionite	21
Na – K	Linde L	22
Na – K	Mordenite	14
Na – K	Offr./Erionite	23
Na – K	Phillipsite	24
Na – K	Stilbite	25
Na – K	X	17, 18
Na – K	Y	18
Na – Cs	A	15
Na – Cs	Chabazite	26
Na – Cs	Clinoptilolite	26
Na – Cs	Mordenite	26, 14
Na – Cs	Stilbite	25
Na – Cs	X	17, 18
Na – Cs	Y	27, 18
Na – Rb	A	15
Na – Rb	X	17, 18
Na – Rb	Y	18
Na – NH_4	Chabazite	28, 20
Na – NH_4	Clinoptilolite	29
Na – NH_4	Linde L	22
Na – NH_4	Linde W	30
Na – NH_4	Y	18
Na – Ag	A	31
Na – Ag	Analcime	19
Na – Ag	Mordenite	14
Na – Ag	X	32
Na – Tl	X	18
Na – Tl	Y	18
Na – Mg	A	34, 33, 35
Na – Mg	Mordenite	14
H – Na	Mordenite	14
Na – Ca	A	34, 33, 35
Na – Ca	Analcime	36
Na – Ca	Chabazite	26
Na – Ca	Clinoptilolite	26
Na – Ca	Erionite	21
Na – Ca	Mordenite	26, 14
Na – Ca	Offr./Erionite	23

Table 1.29 continued

Exchanging cations	Type of zeolite	References
Na – Ca	Phillipsite	24
Na – Ca	X	18
Na – Ca	Y	18
Na – Sr	Chabazite	20
Na – Sr	Clinoptilolite	26
Na – Sr	Mordenite	26, 14
Na – Sr	Y	27
Na – Ba	Mordenite	14
Na – Ba	X	18
Na – Mn	Mordenite	14
Na – Co	Mordenite	14
Na – Ni	Mordenite	14
Na – Cd	Y	37
Na – Y	X	38
Na – Ce	A	26
Na – Ce	X	27
Na – Ce	Y	27
Na – La	Mordenite	39
Na – La	X	18
Na – La	Y	18
K – Cs	Clinoptilolite	26
K – Cs	Linde L	22
K – Cs	Mordenite	26
K – Cs	Stilbite	25
K – NH_4	Clinoptilolite	29
K – NH_4	Offr./Erionite	23
K – NH_4	Y	40
K – Ba	Linde L	22
K – Co	Y	37
K – Ni	Y	37
H – Na	Mordenite	14
K – Zn	Y	37
K – Cd	Y	37
Rb – Cs	Clinoptilolite	26
Rb – Cs	Mordenite	26

Table 1.29 continued

Exchanging cations	Type of zeolite	References
$Rb-NH_4$	Y	40
$Cs-NH_4$	Clinoptilolite	26
$Cs-NH_4$	Mordenite	26
$Cs-NH_4$	Y	40
$Cs-Ca$	Clinoptilolite	26
$Cs-Ca$	Mordenite	26
$Cs-Sr$	Chabazite	26
$Cs-Sr$	Clinoptilolite	26
$Cs-Sr$	Mordenite	26
$Cs-Sr$	Stilbite	25
NH_4-Mg	Clinoptilolite	29
NH_4-Ca	Clinoptilolite	29
NH_4-Ca	Y	40
NH_4-Sr	Y	40
NH_4-Ba	Y	40
NH_4-Mn	Y	40
$Ca-Sr$	Clinoptilolite	26
$Ca-Sr$	Mordenite	26

The chemical composition of the zeolite used determines the maximal ion exchange capacity. The number of cations necessary for charge balancing corresponds to the zeolite framework charge caused by the substitution of silicon atoms for aluminium atoms. Consequently, the ion exchange capacity is inversely proportional to the silicon/aluminium ratio. This ratio is lowest for zeolite A, which has the highest ion exchange capacity known. It should be kept in mind that the capacities evaluated this way are theoretical quantities. The experimental capacity is usually lower for steric reasons. As we have discussed already, only a part of the sites can be occupied by the cations to be exchanged.

Recently Li and Rees have found a clear coherence between aluminium content and exchange capacity for a series of partially dealuminated zeolites [41]. This phenomenon was explained by an increase in the distance between the sites at lower aluminium contents. The influence of the site distance on the ion exchange properties is discussed in Section 1.5.2.7.

1.5.2.4 Influences on selectivity

In addition to the points discussed above, the selectivities and capacities observed in ion exchange processes are also influenced by

- the pH of the solution (H^+ is often a competing cation)
- the temperature of the solution (since for instance the hydrate water-cation bond strength is effected by temperature changes)
- the exact composition of the solution (relative concentrations).

The extent of the ion separation can be varied within wide ranges by the presence of other cations, by the choice of the solvent, by complexing agents, and by the kind and the concentration of the anions. Because of the high stability of the zeolite frameworks, these influences are fortunately less complicated and usually more predictable for zeolites than for organic ion exchange resins.

The addition of ammonia to a solution of Na^+ and Ag^+ ions provides an illustrative example. Without any additives, zeolite X has a strong preference for Ag^+. After the addition of ammonia, however, the selectivity is inverted ($Na^+ > Ag[(NH_3)_2]^+$). Selectivity changes like this are used for regenerations and separations which cannot be done otherwise. The highest exchange capacity of zeolites is often reached at elevated temperatures. Only under these conditions can large cations diffuse into the pores (after loss or loosening of the hydration sphere [42]).

At lower temperatures the exchange appears to be irreversible.

1.5.2.5 Multivalent cations

In ion exchange processes involving multivalent cations difficulties are encountered with respect to the pH value of the solution:

- on the one hand the pH value must be low enough to guarantee sufficient cation solubility
- on the other hand the pH level must not be too low: the zeolite framework will be disintegrated otherwise by hydrolysis and the expected exchange selectivities might be changed by high concentrations of competing protons
- multivalent cations may be precipitated as hydroxides. This is due to the hydrolysis reaction

$$M^+ \text{zeo}^- + H_2O = H^+ \text{zeo}^- + M^+ + OH^-.$$

These considerations apply primarily to trivalent cations, which require especially low pH levels. Most data known refer to exchange reactions of rare earth metal ions (Y^{3+}, Ce^{3+}, La^{3+}) and the more acid stable zeolites (zeolite X and Y, mordenite, clinoptilolite). Ion exchange reactions of multivalent cations with zeolite A are especially complicated due to the acid sensitivity of this zeolite. The following exchange pairs are found in the literature [33]:

Na^+/Ba^{2+}	Na^+/Mg^{2+}
Na^+/Cd^{2+}	Na^+/Zn^{2+}
Na^+/Co^{2+}	Na^+/Ni^{2+}

For zeolite A, attempts to replace Na^+ by Cu^{2+} or Fe^{3+} resulted in the disintegration of the zeolite network. Exchange of Cs^+, Na^+, and Sr^{2+} for Ce^{3+} was reported [26], but without X-ray investigations. It is not known whether the zeolite structure is retained in these cases.

Exchange of two divalent cations was possible [43]:

Ca^{2+}/Pb^{2+}
Ca^{2+}/Cd^{2+}
Ca^{2+}/Cu^{2+}.

Zeolite destruction was observed for:

Ca^{2+}/Hg^{2+}
Na^+/Al^{3+}
Na^+/Fe^{3+}.

Partial substitution of Na^+ by the trivalent cations mentioned was achieved when the reaction times were kept short [43].

In contrast to the results obtained for zeolite A, rare-earth metal ion exchange did not cause severe problems for faujasite-type zeolites (zeolites X and Y). These zeolites are sufficiently acid stable.

The exchange reactions

Na^+/Ce^{3+} [27]
Na^+/Y^{3+} [44]
Na^+/La^{3+} [11]

are good examples of the selectivity control by temperature variation. Only at around 80°C are all Na^+ cations replaceable by the trivalent rare-earth metal cations. This reaction is irreversible even at high Na^+ excess [45].

When Na^+ ions were exchanged for divalent cations like Zn^{2+}, Cd^{2+}, Mg^{2+} in zeolite Y, discontinuous isotherms were obtained, depending on the temperature chosen [46]. This observation suggests a two-step exchange of Na^+ in the supercages and in the sodalite cages. The concentration of the entering cations in the supercages must be high prior to the occupation of the sodalite cages through the smaller entrances. However, this effect requires further investigation.

1.5.2.6 Ion exchange in ternary systems

In the presence of more than two kinds of cations, ion exchange reactions become more complicated. Therefore, only relatively few data have been published on equilibria of this type. As an example we will discuss the investigation of the equilibria in the Na^+-NH_4^+-K^+-mordenite system [47]. In this context it should be pointed out that all exchange reactions at low pH values (especially reactions involving multivalent cations) are, strictly speaking, ternary exchange reactions.

The H_3O^+ concentrations are not negligible in these cases and must be included in the consideration of the exchange equilibria.

1.5.2.7 Thermodynamic aspects

Within the zeolite frameworks, there are several different sites, which are not equal in ion exchange. All these sites contribute to the termodynamic equilibrium characterized by the equilibrium constant K_a. From this constant, the parameters G^o (standard free enthalpy of the ion exchange reaction), H^o (standard enthalpy), and S^o (standard entropy) can be calculated.

The values obtained for G^o, H^o and S^o depend on many factors. For instance, a change in the Si/Al ratio may have a strong influence [34, 36]. Since the zeolites are usually present in the hydrated form in ion exchange reactions, the role of water must also be taken into consideration [48]. If the zeolites contain several cations, which are only partially exchangeable, the situation becomes even more complicated. In articles by Sherry [12], Golden and Jenkins [49], and Vasant and Uytterhoeven [50, 51], different interpretations for ion exchange reactions of this type are discussed.

The investigation of the thermodynamics of ion exchange is not a finished chapter of zeolite chemistry. The data obtained must also be seen in the context of kinetic measurements. A comprehensive review of the thermodynamic and kinetic aspects of zeolite ion exchange can be found in the plenary lecture by Barrer presented at the Fifth International Conference on Zeolites in Naples in 1980 [52].

1.5.2.8 Ion exchange isotherms

The results of selectivity measurements in ion exchange investigations are usually illustrated as isotherms: the mole fraction of one of the competing cations in solution is plotted against the mole fraction of the same cation in the zeolite. Thus, from ion exchange isotherms the relative loading of the zeolite by two different cations at different relative concentrations can be learned. The shape of the isotherms depends on the selectivity of the zeolite investigated. This is shown for some examples in Figure 1.87.

The differences between the isotherms can be interpreted in the following way:

Type A

Preference for the entering cation; the isotherm lies above the diagonal over the entire range of zeolite compositions.

Examples:

Na^+ – Ag^+ (zeolite A), [53]
Na^+ – Ca^{2+} (zeolite X), [17]
Na^+ – NH_4^+ (chabazite), [20]
Na^+ – NH_4^+ (clinoptilolite), [54].

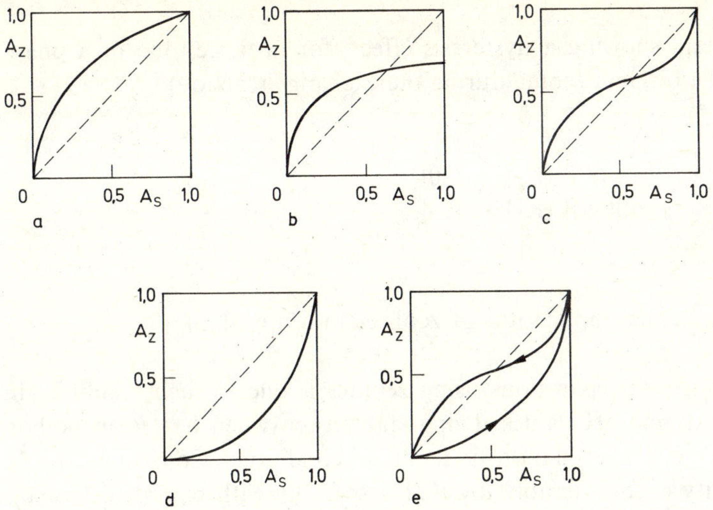

Figure 1.87 Different types of ion-exchange isotherms.

Type B
Initially, also preference for the entering cation; however, the exchange does not go to completion (because of sites inaccessible for the entering cation, for instance).

Examples:

$Na^+ - Cs^+$ (zeolite A), [15]
$Na^+ - Ba^{2+}$ (zeolite Y), [55]
$Na^+ - Cs^+$ (chabazite), [56].

Type C
Selectivity varying with the degree of exchange; the entering cation is preferred only initially.

Examples:

$Na^+ - K^+$ (zeolite A), [15]
$Na^+ - Li^+$ (chabazite), [20]
$Ca^{2+} - Sr^{2+}$ (clinoptilolite), [57].

Type D
Preference for the leaving cation over the entire range of zeolite compositions (reversal of A).

Examples:

$Na^+ - Li^+$ (zeolite A), [15]
$Na^+ - Ca^{2+}$ (phillipsite), [58]
$Na^+ - Li^+$ (zeolite Y), [59].

Type E
Ion exchange isotherm showing a hysteresis effect (for instance, due to a phase transformation of the original zeolite during the exchange reaction).

Examples:

$Na^+ - K^+$ (zeolite Z (K-F)), [60]
$Na^+ - Li^+$ (zeolite Z (K-F)), [60].

1.5.2.9 Limitations in the application of zeolites in ion exchange

The major problem in ion separations using zeolites is due to their solubility in aqueous media at extreme pH values. Being stable even when heated in boiling water, zeolites undergo dealumination under these conditions. Consequently, the ion exchange capacity is considerably lowered. Especially with regard to the stabilization of zeolites, the addition of powders as well as of stabilizing agents, is extremely important. However, these influences will not be discussed in this article.

Zeolites are almost insoluble between pH 5.5 and 10. Their solubility can be reduced even more by adding other soluble compounds in high concentrations. This is significant when large quantities of water are treated with small amounts of zeolites, as is the case when diluted solutions must be worked up. Moreover, zeolite ion exchangers should not be regenerated with strong mineral acids because of the solubility increase at low pH values. Depending on the time of treatment and the framework type, zeolites are dissolved rather rapidly under these conditions.

1.5.3 Special applications of zeolites

1.5.3.1 Removal of ammonium ions from waste water

One of the first commercial applications of zeolites to ion exchange was the removal of ammonium ions from the effluent of the second purification step of waste water. The possible use of NH_4^+-selective zeolites in technical dimensions was already demonstrated in the nineteen thirties by several studies [61, 62] and the running of a pilot plant. On the average, the proportion of N (from NH_4^+) was lowered to a residual value of 2–3 ppm. The applications of ion exchangers on the basis of organic resins for sewage water treatment was reported at a later time by Slechta et al. [63]. The results obtained with these materials, however, were not satisfying, as their selectivity for NH_4^+ is poor compared to the selectivity of zeolites.

Choosing the right zeolite. Having a high selectivity for NH_4^+, clinoptilolite was considered to be the optimal zeolite for the removal of ammonium ions. An additional reason for the choice of this naturally-occurring zeolite was its availability at low costs [28]. Recent studies compared different zeolites (clinoptilolite, erionite, mordenite, phillipsite) with regard to their selectivity and capacity for ammonium removal [30]. From these investigations it is evident that phillipsite as well as the synthetical zeolites F and W produced by Union Carbide Corporation are the types most suited for this purpose. Because of a higher selectivity for NH_4^+, these zeolites allow higher NH_4^+ loadings before breakthrough than clinoptilolite. For example, the capacity factor measured for phillipsite against clinoptilolite is 2.9.

Technical studies done with respect to the economic use of the new zeolite types showed an unexpected effect: the capacity for NH_4^+ exchange of zeolite F and clinoptilolite is much decreased after several application cycles. This phenomenon is due to the presence of Ca^{2+} and K^+. These cations are bound more tightly than NH_4^+; they are not completely removed in the standard regeneration procedures. For zeolite W, this effect is not observed. Thus, the capacity of zeolite W in dynamic experiments turned out to be three times as high as the capacity of the previously used clinoptilolite [30]. At the moment, further improvements are being investigated to achieve a semi-industrial scale.

Procedures. In the first pilot plant [29] around 95 percent removal of the ammonium ions was achieved. The exhausted zeolite was regenerated with alkaline solutions containing Ca^{2+} ions. These regenerating solutions could be reutilized after removal of ammonia by stripping with air. However, the NH_4^+-waste water problem had been transformed now into an ammonia exhaust air problem. Therefore, improvements became necessary:

- it was possible to oxidize the exchanged NH_4^+ ions electrolytically. However, this process turned out to be too expensive [64]
- washing of the exhaust air with sulfuric acid resulted in concentrated $(NH_4^+)_2SO_4$ solutions [65]
- the zeolite was treated with Na^+-containing solutions at pH 12, followed by steam stripping of the resulting exchange solutions [66].

It is not possible to exchange NH_4^+ in exhausted zeolites with sodium salts at pH 11, which is the most favorable with regard to the zeolite stability. Under these conditions, the volumes of the necessary exchange solutions would be too big. However, it became evident that even at pH 6 a decrease in ion exchange capacity and an increase in zeolite abrasion did not occur [67]. Besides relatively small plants in the United States of America and in Japan, the main American projects are

- the Upper Oceoquan Sewage Authority, Virginia (1979),
  and
- Tahoe-Truckee Sanitation Agency, California (1978).

Details about the procedures applied in these plants can be found in reports by Sherman [7, 8].

For the removal of toxic heavy metals (Cd, Co, Cr, Pb) Haberer and Stürzer proposed the use of naturally-occurring silicates related to zeolites [68]. The metals mentioned above could be removed by static adsorption. Even at very low starting concentrations, almost quantitative removal was achieved (the residual concentrations were in the ppb range).

1.5.3.2 Zeolites as builders in detergents

Many stagnant or slowly-flowing bodies of water are characterized by an excessive accumulation of organic matter. This phenomenon is termed eutrophication. It has a negative influence on the visual appearance, the functionability of the ecological system and the production of drinking water. Eutrophication is mostly caused by an excess of phosphates. Normally, the phosphates are enriched in rivers and lakes by soil erosion, washing away of fertilizers and municipal waste waters. In densely populated areas, a major part of the phosphates originates from detergents [69, 70]. For this reason, detergent manufacturers have undertaken great efforts to find a suitable builder as a substitute for phosphates. This component is intended to perform the main function of the phosphates, the reduction of the hardness of the water by complexation of the Ca^{2+} and Mg^{2+} ions. Such a substitute had to meet the following requirements:

- no decrease in detergent quality
- toxicological and ecological harmlessness
- economical application
- large-scale production from easily available raw materials.

Numerous substances which were tested for their suitability as phosphate substitutes did not meet all of the requirements specified. However, a breakthrough in this search was achieved when not only organic but also insoluble inorganic compounds were included in these studies. Here Na-A zeolite was found to be the material of choice [71, 72, 73]. This success was based in Germany on the joint efforts of Henkel (a detergent producer) and Degussa AG (a supplier of raw materials), and in the United States on the research activities of Procter and Gamble.

The first task of the zeolite is the removal of the Ca^{2+} and Mg^{2+} ions and heavy metal ions by ion exchange for "soft" Na^+ ions. Besides this main function, the zeolite also has some secondary effects on the washing process:

- generation of an alkaline pH
- adsorption of dirt on the particle surface
- coagulation of dirt pigments.

The Na-A zeolite proved to be especially suited for ion exchange in washing processes mainly because of its high affinity for Ca^{2+} ions even at low Ca^{2+} concentrations in the sewage water. Mg^{2+} ions are bound considerably more poorly. Because of the relevance for detergent chemistry, the ion exchange equilibria Na^+/Ca^{2+}, Mg^{2+} are extensively discussed in the literature [32, 74–77]. For this reason the subject will not be treated in this review. The Mg^{2+} uptake of Na-A zeolite is poor even at elevated temperatures [78]. In the United States, where the washing temperatures are lower than in Europe, the addition of Na-X zeolite was proposed [78, 79]. Such a mixture of Na-X and Na-A zeolites has a considerably better capability of binding Mg^{2+}, while the Ca^{2+} capacity is still satisfying. Since Na-A zeolite replaces detergent phosphates only in its main function – removal of the ions responsible for water hardness –, it was proposed to combine the zeolite with complexing agents [80]. By addition of various reagents of this type, the degree of Ca^{2+} exchange was increased considerably [81]. If phosphate is substituted only partially by Na-A zeolite, interactions between the two compounds are to be taken into account. Thus, there are the competing reactions of the formation of a Ca^{2+}/Mg^{2+} triphosphate complex and of the formation of the Ca^{2+}/Mg^{2+} exchanged zeolites. This situation results in a decrease in exchange capacity of the zeolite [82]. In addition to this, the rate at which exchange equilibrium is reached is slowed down extremely.

1.5.3.3 Separations of radioisotopes

For the treatment of solutions containing radioactive isotopes, organic ion exchange resins were used first. However, inorganic materials turned out to have clear advantages, as determined in several investigations [83, 84]. For instance, they have a higher stability to radioactive radiation, and they are more resistant to aqueous solutions at elevated temperatures. In 1913–1915, Ames reported the high selectivity and capacity of zeolites for the cations Ca^{2+} and Sr^{2+}, which are major components of radioactive waste waters [38, 85, 86]. Up to now, the following procedures have been applied:

- Lutze and Levi investigated the suitability of naturally-occurring volcanic tuff for selective adsorption of radionuclides, in particular ^{137}Cs [87]. The experiments were done in a pilot plant having a volume of 45 liters. It was established that the tuff can be used for this purpose. Thus, the activity of 1500 liters of water per liter of reactor volume was lowered to a residual value, which was 1 percent of the original activity. Further investigations dealt with the separations Ca/Sr, Ca/Ba and Ca/Ce.
- in Hanford, there was a yield of highly radioactive waste waters containing $^{137}Cs^+$ in concentrations of around 10^{-4} mol/l, besides high concentrations of sodium salts (approximately 4.5 mol/l). It was necessary to separate the Cs component before evaporation. Chabazite was chosen in this case because of its high selectivity for cesium [88]. The exchanger could be regenerated with warm solutions of ammonium salts.

  In the course of this regeneration step, the Na^+/Cs^+ ratio was determined to be smaller than 10, corresponding to an enrichment ratio higher than 4000. In further regeneration cycles, there was a decrease in exchange capacity, probably caused by the obstruction of the zeolite pores by precipitated aluminium salts. The solution resulting from ion exchange, which is enriched in Cs, is free of sodium, potassium and rubidium salts. Synthetical mordenite (zeolon) was used for this purpose [88, 89]. This second step is still in the stage of development (1 liter scale).

- for the purification of waste waters having only a low radioactivity, non-regenerative procedures are applied, too. Thus, natural clinoptilolite was used for the treatment of irradiated water from the Idaho National Engineering Laboratory [90]. After being exhausted, the zeolite was disposed of in a solid form. The volume of the radioactive waste was reduced by a factor of 600 in this way. The degree of purification was 99.5 percent.
- in the treatment of different sewage waters and distillates in Savannah River (Aiken) using ion exchangers on the basis of chabazite, 99 percent of the original amount of cesium could be separated. The capacity of the exchanger varied from 10^3 to 10^4 bed volumes, depending on the concentration [91].
- in Hanford, the process was yielding a condensated waste water containing ^{137}Cs, besides ^{90}Sr. The problem of sewage water purification was solved in this case by the use of a combination of a zeolite and an exchange resin. The zeolite (synthetical mordenite) bound more than 99.5 percent of the cesium, whereas the resin held back 80 percent of the strontium present [92]. This "combination exchanger" was regenerable in one step by a solution of a sodium salt.
- after the nuclear accident in Three Mile Island in 1979, zeolites were used for the cleanup of the contaminated water. Because of the strong public interest, reports dealing with this application of zeolites as well as with the different systems of purification and disposal were published in technical magazines and journals.

  Altogether, 3000 m^3 of radioactive water, containing ^{137}Cs in a concentration of around 1 ppm and ^{90}Sr in a concentration of 0.1 ppm, resulted from this accident [93]. Union Carbide Corporation was asked to take care of the clean-up problem. Many experiments were done in the laboratory before the decontamination could take place. With regard to selectivity, Linde IE-95 was chosen as the exchanger for ^{137}Cs, and Linde A-51 for the removal of ^{90}Sr. A mixture of 70 percent IE-95 and 30 percent A-51 turned out to have the highest exchange capacity [94]. 2500 m^3 of the contaminated water were treated with 1550 kilograms of the zeolite mixture. Up to now the zeolite has been stored at Three Mile Island; it will be molten down and finally disposed of after the nuclear reactions have faded.

References

[1] H. S. Thomson (1850), J. R. Agric. Soc. Engl. *11*, 68.
[2] J.T. Way (1850), J. R. Agric. Soc. Engl. *11*, 313.
[3] H. Eichhorn (1858), Poggendorf Ann. Phys. Chem. *105*, 126.
[4] R. Gans (1905), Jahrb. Koenig. Preuss. Geol. Landesanstalt (Berlin) *25*, 2.
[5] F. Helfferich (1962), Ion Exchange, Mc Graw-Hill, New York.
[6] J. D. Sherman (1979), Ion Exch. Pollut. Control *2*, 227.
[7] J. D. Sherman (1978), in J. D. Sherman, ed., Adsorption and Ion Exchange Separations, A. I. Ch. E. Symp. Series *74*, 98.
[8] J. D. Sherman (1984), in F. R. Ribeiro (ed.), Zeolites: Science and Technology, NATO ASI Series, Ser. E.: Appl. Sci. No. 80, Martinus Nijhoff, The Hague, p. 583.
[9] K. Dorfner (1970), Ionenaustauscher, p. 77. de Gruyter, Berlin.
[10] D.W. Breck (1974), Zeolite Molecular Sieves, p. 529, Wiley, New York.
[11] H. S. Sherry (1968), J. Coll. Interface Sci. *28*, 288.
[12] H. S. Sherry (1966), J. Phys. Chem. *70*, 1158.
[13] G. Eisenmann (1962), Biophys. J. *2*, 259.
[14] F. Wolf, H. Fürtig and K. Knoll (1971), Chem. Techn. *23*, 273.
[15] R. M. Barrer, L. V. C. Rees and D. J. Ward (1963), Proc. Roy. Soc. A273, 180.
[16] V. A. Federov, A. M. Tolmachev and G. M. Panchenkov (1964), Russ. J. Phys. Chem. *38*, 679.
[17] R. M. Barrer, L. V. C. Rees and M. Shamsuzzoha (1966), J. Inorg. Nucl. Chem. *28*, 629.
[18] H. S. Sherry (1969), in: Ion Exchange (Marinsky ed.), Marcel Dekker, New York, p. 89.
[19] R. M. Barrer and L. Hinds (1953), J. Chem. Soc. 1879.
[20] R. M. Barrer, J. A. Davies and L. V. C. Rees (1969), J. Inorg. Nucl. Chem. *31*, 219.
[21] H. S. Sherry (1979), Clays and Clay Minerals *27*, 231.
[22] P. A. Newell and L. V. C. Rees (1983), Zeolites *3*, 22.
[23] H. S. Sherry (1969), Ion Exch. Process Ind., Intl. Conf., Soc. Chem. Ind., p. 329.
[24] Y. Shibue (1981), Clays and Clay Minerals *29*, 397.
[25] L. L. Ames (1966), Can. Mineral. *8*, 582.
[26] B. W. Mercer and L. L. Ames (1963), Unclassified Hanford Laboratories Report HW-78461.
[27] L. L. Ames (1965), Can. Mineral. *8*, 325.
[28] L. L. Ames (1967), Proc. 13th Pacific NW. Indus. Waste Conf., 6.–7. 4. 1967, Wash. State. Univ., Publ. by Tech. Ext. Serv., Wash. State. Univ., Pullman, Wash., p. 135.
[29] B. W. Mercer, L. L. Ames and C. J. Touhill (1968), Am. Chem. Soc. Div. Water, Air, Waste Chem., Gen. Papers.
[30] J. D. Sherman and R. J. Ross (1980), in L. V. C. Rees (ed.), Proc. 5th Intern. Conf. Zeolites. Heyden, Philadelphia, p. 823.
[31] R. M. Barrer and W. M. Meier (1959), Trans. Faraday Soc. *55*, 130.
[32] D. W. Breck (1964), J. Chem. Ed. *41*, 678.
[33] D. W. Breck, W. G. Eversole, R. M. Milton, T. B. Reed and T. L. Thomas (1956), J. Am. Chem. Soc. *78*, 5963.
[34] S. A. I. Barri and L. V. C. Rees (1980), J. Chrom. *201*, 21.
[35] F. Danes and F. Wolf (1972), Z. Phys. Chem. (Leipzig), *251*, 329.
[36] L. L. Ames (1966), Amer. Mineral. *51*, 903.
[37] A. Maes, J. Verlinden and A. Cremers (1979), in: The Properties and Appliance of Zeolites (R. P. Townsend, ed.). The Chemical Society, Burlington House, London, p. 269.
[38] L. L. Ames (1965), J. Inorg. Nucl. Chem. *27*, 885.

[39] S. Hocevar and B. Drzaj (1980), in L. V. C. Rees (ed.), Proc. 5th Intern. Conf. Zeolites. Heyden, Philadelphia, p. 301.
[40] I. V. Baranova and A. M. Tolmachev (1977), Russ. J. Phys. Chem. *51*, 416.
[41] C.-Y. Li and L. V. C. Rees (1986), Zeolites *6*, 51.
[42] A. Maes and A. Cremers (1975), J. Chem. Soc., Far. Trans. I *71*, 265.
[43] B. H. Wiers, R. J. Grosse and W. A. Cilley (1982), Environ Sci. Technol. *16*, 617.
[44] E. N. Rosolovskaja, K. V. Topchieva and S. P. Dorozhko (1977), Russ. J. Phys. Chem. *51*, 861.
[45] H. S. Sherry (1976), in: Colloid and Interface Sci. Vol. 5 (M. Kerker ed.). Academic Press, New York, p. 321.
[46] R. W. Thomson and M. Tassopoulos (1986), Zeolites *6*, 9.
[47] R. P. Thownsend, P. Fletcher and M. Loisidou (1984), in: Proc. 6th Intern. Conf. Zeolites (D. Olson and A. Bisio, eds.). Butterworth, Guildford, p. 110.
[48] R. M. Barrer and R. P. Townsend (1985), Zeolites *5*, 287.
[49] T. C. Golden and R. G. Jenkins (1981), J. Chem. Eng. Data *26*, 366.
[50] E. F. Vansant and J. B. Uytterhoeven (1971), Trans. Far. Soc. *67*, 586.
[51] E. F. Vansant and J. B. Uytterhoeven (1971), Trans. Far. Soc. *67*, 2961.
[52] R. M. Barrer (1980), in: Proc. 5th Int. Conf. Zeolites (L. V. C. Rees, ed.), Heyden, Philadelphia, p. 273.
[53] H. S. Sherry and F. H. Walton (1967), J. Phys. Chem. *71*, 1457.
[54] R. M. Barrer, R. Papadopoulos and L. V. C. Rees (1967), J. Inorg. Nucl. Chem. *29*, 2047.
[55] H. S. Sherry (1968), J. Phys. Chem. *72*, 4086.
[56] L. L. Ames (1966), Can. Mineral. *8*, 572.
[57] L. L. Ames (1964), Amer. Mineral. *49*, 1099.
[58] R. M. Barrer and B. M. Munday (1971), J. Chem. Soc. A, 2904.
[59] B. K. G. Theng, E. Vansant and J. B. Uytterhoeven (1968), Trans. Faraday Soc. *64*, 3370.
[60] R. M. Barrer and B. M. Munday (1971), J. Chem. Soc. A, 2914.
[61] G. M. Gleason and A. C. Loonam (1933), Sewage Works J. *5*, 61.
[62] G. M. Gleason and A. C. Loonam (1934), Sewage Works J. *6*, 450.
[63] A. F. Slechta and G. L. Culp (1967), J. Water Pollu. Control Fed. *30*, 787.
[64] Battelle North West Laboratories, EPA Water Pollution Control Series Report No. 17010 ECZ 02/71 (1971).
[65] L. G. Kepple (1974), Water Sewage Works *42–43*, 62.
[66] W. A. Blain and R. W. Defore (1974), The Rosemount Advanced Wastewater Treatment Plant – An Approch to the Zero Discharge Goal, Central States Water Poll. Cont. Assoc., 47th Ann. Mtg., St. Paul, Mn.
[67] R. C. Polta and R. W. Defore (1976), Full Scale Physical Chemical Treatment Evaluation, Rosemount AWTP, 49th Ann. Conf., Water Pollu. Control Fed., Minneapolis, Mn.
[68] K. Haberer and U. Stürzer (1978), Naturwissenschaften *65*, 487.
[69] W. Nuemann (1971), Tenside *8*, 82.
[70] E. P. Frieser (1971), Allg. prakt. Chem. *22*, 219.
[71] DE-OS 2412837, Henkel KGaA (1974).
[72] US Pat. 3985669, Procter and Gamble (1974).
[73] M. Ettlinger and H. Ferch (1978), Manuf. Chem. Aerosol News (October), p. 51.
[74] L. V. C. Rees (1980), in R. P. Townsend (ed.), The Properties and Applications of Zeolites, Spec. Publ. No. 33, The Chemical Society, London, p. 218.
[75] K. Henning, J. Kandler and H. D. Nielsen (1977), Seifen, Oele, Fette, Wachse *103*, 571.
[76] F. Danes and F. Wolf (1972), Z. Phys. Chem. (Leipzig), *251*, 339.
[77] H. G. Smolka and M. J. Schwuger (1976), Coll. Polym. Sci. *254*, 1062.

[78] J. D. Sherman, A. F. Denny and A. J. Gioffre (1978), Soap Cosmet. Chem. Spec. *64*, 33–40.
[79] A. F. Denny, A. J. Gioffre and J. D. Sherman (1978), US Pat. 4094778.
[80] H. G. Smolka and M. J. Schwuger (1977), Tenside Detergents *14*, 222.
[81] M. S. Nieuwenhuizn, A. H. E. F. Ebaid, M. van Duin, A. P. G. Kieboon and H. van Bekkum (1984), Tenside Detergents *21*, 221.
[82] K. Pilchowski, K.-H. Bergk and F. Wolf (1984), Acta Hydrochim. et Hydrobiol. *12*, 203.
[83] R. Fullerton (1961), U.S. Atomic Energy Comm. Doc. No. HW-69256.
[84] J. W. Roddy (1981), Oak Ridge Natl. Lab. Rep. ORNL/TM-7782.
[85] L. L. Ames (1960), Amer. Mineral. *45*, 689.
[86] L. L. Ames (1967), J. Inorg. Nucl. Chem. *29*, 262.
[87] W. Lutze and H. W. Levi (1969), Ion Exch. Process Ind., Conf. Pap., 216–23.
[88] B. W. Mercer and L. L. Ames (1978), in: Natural Zeolites, Occurrence, Properties, Use (L. B. Sand and F. A. Mumpton, eds.). Pergamon Press, New York, p. 451.
[89] L. A. Bray and H. T. Fullam (1970), Molecular Sieve Zeolites I (2nd Int. Conf.) p. 450.
[90] E. Gallei, D. Eisenbach and A. Ahmed (1974), J. Catalysis *33*, 62.
[91] International Atomic Energy Agency, Wien, Technical Report Series No. *78*, 73 (1967) and No. *136*, 61 (1972).
[92] G. L. Hanson (1973), Proc. Natl. Conf. Complete Water Reuse, Washington, D. C., A. I. Ch. E.-E. P. A., p. 360.
[93] D. O. Campbell, E. D. Collins, L. J. King and J. B. Knauer (1980), Oak Ridge Natl. Lab. Rep. ORNL/TM-7448.
[94] E. D. Collins, D. O. Campbell, L. J. King and J. B. Knauer (1982), A. I. Ch. E. Symp. Series 213, 78, 9.

1.6 Clay Minerals as Ion Exchangers

Armin Weiss and Elfriede Sextl
Institut für Anorganische Chemie der Universität München, München, Federal Republic of Germany

Similar to zeolites clay minerals belong to the group of siliceous ion exchangers. In contrast to the zeolites, however, which exhibit porous, three-dimensional frameworks, clay minerals are layer silicates. They exist in different structural varieties and are primarily cation exchangers. To a lesser extent they can simultaneously exchange anions. This means that under certain conditions they should be able to desalinate water. Some clay minerals are also capable of exchanging electrons by redox reactions of surface ions; these minerals, therefore, can also act as redox exchangers.

1.6.1 Practical use of ion exchange properties of clay minerals

1.6.1.1 Cation exchange

Although the cation exchange capacity (CEC) of clay minerals is low compared to that of, e. g., industrial ion exchangers, it is of great significance in practical applications. The CEC and exchange properties of clay minerals are very important for arable soils. Ions supplied to soil by fertilizers are bound by ion exchange reactions, thus prolonging the residence time in the subsoil. The rate of leaching to the ground-water is highly decreased. Nevertheless the bonding strength allows exchange reactions with plant roots. Only in a very few cases are cations bound so strongly to the clay that the transfer to plants is reduced. In such cases the bonding can be regarded as more or less irreversible, for example in cesium ion fixation to illites and vermiculites [1, 2]. This aspect is especially interesting in the discussion of radioactive waste disposal [3].

The clay minerals exhibit different properties depending on the type of exchangeable cations attached to them. In the ceramic industry the type of cation has a critical effect on the green strength and dry bending strength. Polyvalent cations increase the dry bending strength considerably.

The following example illustrates the extent to which cation exchange may alter the properties of clay mixtures: Ceramic pastes with a plasticity suited for shaping with the potter's wheel can be liquefied by simple cation exchange reactions without any addition of water. The liquefied pastes can be used for the slip casting process [4].

The main component of bentonites, an industrially important group of clays, is montmorillonite, an intracrystalline, expandable clay mineral. If the cations originally present are exchanged for monovalent alkali metal ions, a wide range of applications can be realized. These alkali-bentonites are used in the foundry industry as binding agents for foundry sands or in the modern blast-furnace process for turning iron ores into pellets. Furthermore, the presence of monovalent cations increases the swelling ability of montmorillonite considerably. The thixotropy of suitable clay suspensions is increased and several applications become possible [5]. For instance, such montmorillonites (bentonites) may be used for thrilling muds in oil exploration and other deep thrillings.

In the past few years ion exchange reactions of clay minerals have gained additional significance in the oil industry. Appropriate exchange reactions are assumed to induce a shrinkage of swollen clays. Capillaries in oil-bearing rocks would then expand and oil mobility would increase.

Incorporation of suitable organic cations, especially of large quaternary ammonium ions, permits a high thixotropy in organic solvents as well. This had led to the use of organophilic clays as varnish thickener [6].

Exchange and adsorption properties of clay minerals change considerably when the exchangeable cations are replaced by hydrogen ions. In addition to cation exchange, hydrogen ions attack the crystal lattice. In this way bleaching earths are produced, which are used for cleaning vegetable oils.

Moreover, clay minerals may be used as desiccants because of the high solvation enthalpy of the exchangeable cations and its dependence upon the type of cations present.

Recently it has been shown that clays may play an important role in the self-cleaning processes of water. Cationic surfactants which get into the water in large quantities are bound in place of the exchangeable cations. They are thus removed from the water more or less permanently to equilibrium concentrations on the order of ppb.

1.6.1.2 Anion exchange

In addition to their ability to exchange cations, clay minerals can also exchange anions. The anion exchange capacity (AEC), however, scarcely reaches 15% of the CEC. Most important is the exchange for phosphates, especially in agriculture. Phosphates used as fertilizers remain in the soil for long periods of time because of exchange reactions on clay minerals [7, 8]. In industry, anion exchange reactions with polyphosphates and polyacrylates are used to stabilize clay suspensions towards "edge to face" flocculation. Moreover, the ability of clay minerals to exchange cations and anions simultaneously, i. e. their ion exchange amphotery, is important for understanding some of the colloidochemical applications [9].

1.6.1.3 Electron exchange

Related to ion exchange properties is the ability of some clays to exchange electrons, which is controlled either by the exchangeable cations or by metal ions at the crystal edges. For example, in iron-containing clays like nontronite, a reversible redox reaction is feasible at the crystal edges independent of the cation exchange reaction. Thus, this type of cationic exchanger also acts as a redox exchanger [10].

1.6.2 Differences in the exchange properties between clay minerals and polymeric organic exchangers

1.6.2.1 Organic polymers

Most of the synthetic organic ion exchangers consist in part of linear macromolecules with dissociable groups in regular or statistical sequence. They are connected more or less tightly to a three-dimensional framework by covalent bonds. The resulting network is so wide-meshed that it can take up surrounding solution: the organic ion exchanger swells three-dimensionally. At the same time conformational changes of the linear macromolecules occur, thereby influencing the distances between neighbouring charged groups. In this way these systems are able to adapt themselves to some extent to the counter-ions present.

Depending upon the size and valence of the ions being exchanged, appropriate conditions can be produced for each case (Figure 1.88a). This high cooperativity cannot, however, be generalized, but must be evaluated case by case.

1.6.2.2 Clay minerals

In contrast to organic polymers and to the three-dimensional zeolites, clay mineral ion exchangers have more or less rigid, two-dimensional structures. Most of the clay minerals have exchange sites on the outer surfaces of the crystallites only.

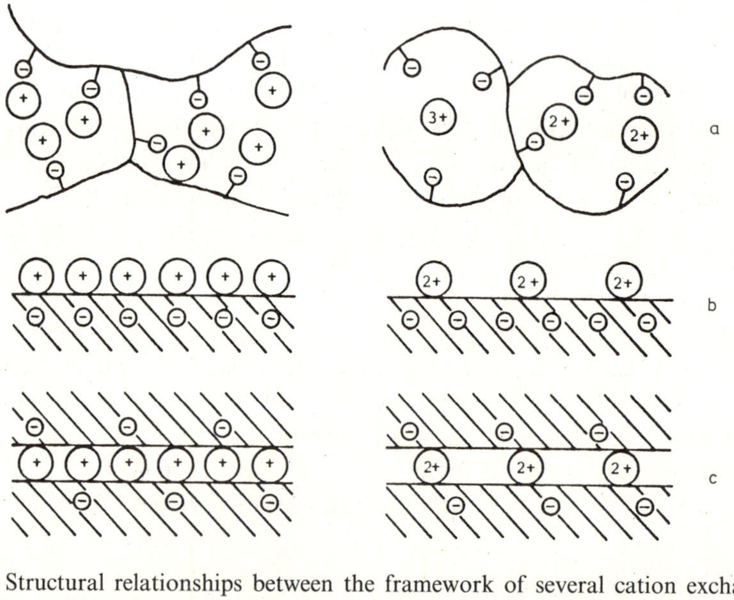

Figure 1.88 Structural relationships between the framework of several cation exchangers and their exchangeable cations a) polymeric organic cation exchanger b) surface of a clay mineral c) interlayer space of an expandable clay mineral.

There are random but definite distances between the exchange positions. Adaptation to valence or size of the counterions is not possible because there are no conformative degrees of freedom (Figure 1.88b). Only expandable clays, of which montmorillonite and vermiculite are important representatives, present additional accessible cations within the crystals. Changes in the solvation of these cations enable changes in the distances between neighbouring layers and consequently alterations in the distances between exchange sites on adjacent layers. The positions within one given layer cannot be altered (Figure 1.88c).

A special feature of clay minerals stems from their anisometric structure. The crystal surfaces differ significantly with regard to surface energy and chemical composition. Because of these differences, the surfaces can exhibit several kinds of dissociable groups. On the one hand there are completely dissociable positions arising from isomorphous substitutions of aluminum for silicon or of magnesium for aluminum or any other trivalent ion. Each of these isomorphous substitutions creates an excess negative charge, compensated by an exchangeable cation. These accompanying cations usually dissociate completely in water.

In addition, there are weakly acidic SiOH groups whose degree of dissociation depends upon the pH of the surrounding solution. Further dissociable sites originate when coordination polyhedra at the edges of the crystals are filled in, especially with aluminum and magnesium as central ions. As a further consequence of the anisometric structure, exchangeable cations and anions may be on the very same crystal, but distributed on crystallographically different faces. Clay minerals thus offer various types of dissociable groups and as a result, differ considerably in ion exchange properties from synthetic organic ion exchangers, which usually have only one type of dissociable group.

In accordance with their chemical nature, clay minerals are rather resistant to heat and radiation. But they are generally much more easily attacked by acids than are organic exchange resins [11].

1.6.3 Experimental methods

In most cases the same methods can be applied to ion exchange experiments with clay minerals as are usually employed for other ion exchangers [12]. Nevertheless four important points have to be considered in all experimental work:

1. The ion exchange capacities of most clays are considerably smaller than those of commercial organic exchangers. Thus more sensitive methods and much larger test samples are necessary.
2. The usually small crystal size of clay minerals causes difficulties when they are to be loaded onto columns. The flow rate through these columns is reduced to such an extent that the use of other methods or special precautions are advisable.
3. Special attention has to be paid to the high sensitivity of several clay minerals to acids. In most cases it is difficult to achieve a quantitative ion exchange with

hydrogen ions without considerable attack on the lattice. In general, Mg-rich silicates are more sensitive than Al-rich silicates.

4. Because of the considerable pH-dependence of the ion exchange capacity, it is reasonable to work in self-buffering systems. It is for this reason that ion exchange with ammonium ions is often preferred.

Kinetics of exchange reactions can vary significantly. If exchangeable cations are bound to the outer surfaces of the crystals only, equilibrium is attained in a short time (within minutes). But if exchange sites on surfaces within the crystals become accessible because of intracrystalline swelling, the rate is reduced significantly owing to the much slower cation diffusion in the interlayer spaces. This is especially true when ions are present which are capable of forming hydrogen bonds or whose dimensions induce important changes in the interlayer distances. In such situations it is best to carry out a few preliminary experiments in order to obtain criteria regarding expected reaction velocities. Table 1.30 shows several typical reaction times. In many cases, these times can be considerably shorter.

Table 1.30 Equilibration times for cation exchange in clay minerals [10, 12]

Mineral	Exchange reaction	Period of time for quantitative exchange	partial (80–95% of CEC) exchange
Kaolinite	Alkali metal ions in place of alkaline earth metal ions	ca. 5–10 d	ca. 8 h
Vermiculite	Li^+, Ca^{2+}, Ba^{2+} in place of Mg^{2+}	8–200 d (depending on particle size)	3 d (particle diameter ca. 50 μm) 90 d (particle diameter ca. 5 mm)
Mica	Ca^{2+} in place of Na^+	ca. 10 d (outer surface)	ca. 3 h (outer surface)
	Li^+ in place of Na^+	ca. 30 d (outer surface)	2–7 h (outer surface)
	Ba^{2+} in place of K^+	> 365 d (inner surface)	100–250 d (inner surface)

Expandable clay minerals reach ion exchange degrees of 70 to 90% very quickly. The reaction velocity then diminishes considerably and the exchange of the final 10 to 5% generally requires a long time. This must be taken into account in industrial processes, where only 70 to 80% of the CEC can be used on commercial time scales. Many experiments described in the literature refer to products which are not quantitatively exchanged, exhibiting exchange degrees of 85 to 95%. Swelling behaviour of clay minerals is considerably influenced by the types of exchangeable cation. Inversely, the selectivity of the ion exchange also depends on the actual state of swelling [13]. This subject is rather complicated, because it has to do with reactions of extremely high cooperativity. When large changes in interlayer distance occur in these systems, various selectivities may result in accordance with the number of bound cations.

In experimental work with clay minerals as ion exchangers, special attention has to be paid to their storage stability. They cannot be stored in the hydrogen form, which is the usual storage form of many other ion exchangers. The hydrogen ions would be replaced by magnesium, aluminum or iron ions from the structure, accompanied by partial decomposition of the crystals [14]. Similar reactions take place in natural clay deposits by weathering. While magnesium generally causes no problems, aluminum and iron can form polynuclear complex cations which are preferentially bound to the external surface. This often leads to results that are difficult to reproduce. In this case pretreatment of the samples with complexing agents for aluminum or iron is recommended (washing with EDTA, tartrate or citrate solutions).

1.6.4 Mica and mica-type clay minerals

1.6.4.1 Structure

For cation exchange reactions the group of mica-type clays is the most important one. These minerals are built of SiO_4 tetrahedra, condensed together to form two-dimensional sheets. Each tetrahedron is linked to three neighbouring tetrahedra by three of its oxygen atoms. These linking atoms lie approximately in the same plane. Out of this plane the Si-O bonds to the respective fourth oxygen atoms emerge almost vertically and always in the same direction. Two such tetrahedral layers are condensed together by magnesium or aluminum hydroxide layers. As the coordination number of these cations is six, this is referred to as an octahedral layer. Within these octahedra each Mg^{2+} or Al^{3+} is surrounded by four siloxane-oxygens and two OH^- ions in alternating two cis and one trans positions. The structure of mica and mica-type clay minerals can thus be described as a sequence of silicate layers, each of which consists of two tetrahedral layers, oriented 180° to one another, and condensed together by an octahedral layer of tri- and/or divalent cations. There are three octahedral sites per Si_4O_{10} unit. In order to obtain electroneutrality either two of these have to be occupied by trivalent or all three by divalent cations. Accordingly, one must distinguish between di- and trioctahedral compounds. Within the trioctahedral clay minerals, all of the octahedral sites are occupied; within the dioctahedral ones only two-thirds of the sites are filled (Figure 1.89).

The pure di- and trioctahedral forms are represented by pyrophyllite $Al_2(OH)_2Si_4O_{10}$ and talc $Mg_3(OH)_2Si_4O_{10}$ respectively. They possess electrically neutral layers and their ion exchange properties are therefore restricted to the so-

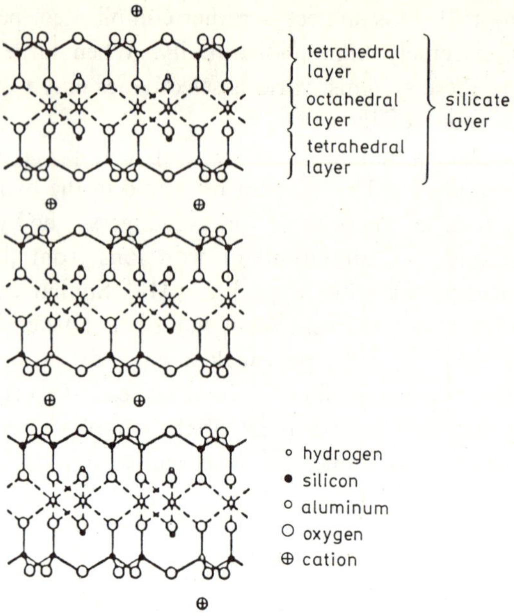

Figure 1.89 Schematic drawing of the structure of a dioctahedral mica-type clay mineral [10].

called "broken bonds" at the crystal edges. These "broken bonds" will be treated in detail later.

An essential property of this group of mica-type clay minerals is the possibility for isomorphous substitution. Within the tetrahedral layers, silicon may be replaced by aluminum or other trivalent cations which can be tetrahedrally coordinated by oxygen. Each substitution produces a permanent, localized excess negative charge, for which exchangeable cations generally compensate. Isomorphous substitution may also occur in the octahedral sheet. Mg^{2+} may, for example, replace Al^{3+}, or, in the group of trioctahedral minerals, Li^{+} may replace Mg^{2+}. Again localized excess negative charge will result. Substitutions may occur at the outer surface and within the crystals. In the latter case the cations necessary for charge compensation lie between the silicate layers. Their density is determined by the extent of isomorphous substitution. In Table 1.31 minerals belonging to this family are listed. They are described by the following general formula:

$$\underbrace{M^{m+}_{\frac{x+y}{m}}\,(H_2O)_n}_{\substack{\text{interlayer}\\ \text{space with}\\ \text{exchangeable}\\ \text{cations}}}\;\underbrace{\left\{\underbrace{(\overset{+3}{M},\ \overset{+2}{M})^{(6-x)+}_{2\ldots3}(OH)_2}_{\text{octahedral layer}}\;\underbrace{[(Si_{4-y}Al_y)O_{10}]}_{\text{tetrahedral layer}}\right\}^{(x+y)-}}_{\text{anionic silicate layer}}$$

Table 1.31 Compilation of ion exchange properties of mica-type clay minerals [15]

Charge per layer $x + y$ per (Si, Al)$_4$O$_{10}$-unit	Mineral	Exchangeable ions in the naturally occurring minerals	Cation exchange capacity (meq/100 g clay) total	on the outer surfaces,* pH-independent	on the inner surfaces, accessible by swelling	at the crystal edges,*** pH-dependent	Anion exchange capacity*** (meq/100 g clay)
0	Pyrophyllite, Talc	Mg^{2+}, Ca^{2+}; OH^-	< 2	< 2	0	< 1	<2
0.22–0.55 $x \gg y$ $x \ll y$	Smectites, Montmorillonites Beidellites	depending on deposit Ca^{2+}, Mg^{2+} or Na^+; OH^-	60 to 130	5 to 20	40 to 110	3 to 8	3 to 31
0.56–0.72 $x \ll y$	Vermiculites	Mg^{2+}, (Ca^{2+}, Na^+); OH^-	120 to 185	1 to 3	117 to 182	1 to 2	1 to 2
0.72–0.80	Illites	K^+, (Ca^{2+}, Na^+); OH^-	4 to 30	3 to ca. 25	0**	1 to 5	1 to 5
0.89–1.1 $x \sim 0$	Micas, (Muscovite, Biotite etc.)	K^+, (Ca^{2+}, Na^+, Mg^{2+}); OH^-	3 to 5	3 to 5	0**	versus 0	versus 0
1.8–2.0 $x \sim 0$	Brittle micas	Ca^{2+}, Mg^{2+}; OH^-	3 to 5	3 to 5	0**	versus 0	versus 0

* capacity inversely proportional to the crystal thickness
** no swelling possible; exceptions: exchange versus Ba^{2+} and RNH_3^+ in very slow reactions
*** capacity inversely proportional to the crystal diameter

1.6.4.2 Exchange capacity

The exchange capacity is mainly determined by the degree of isomorphous substitution. A rough survey is provided by the list in Table 1.31. The structure model in Figure 1.90 indicates schematically the sites of the exchangeable ions.

Pyrophyllite as dioctahedral and talc as trioctahedral minerals are the final members of this series. In this case $x + y$ is equal to 0.

Minerals of this group with $0{,}2 \leq x + y \leq 0{,}7$ are expandable by solvation of the interlayer cations. During the swelling process water layers are incorporated between the silicate sheets, in which the cations are able to move. Their low activation energy for interchange of sites makes them exchangeable with cations of the surrounding solution. In this case the exchange capacity is relatively high. It may reach maximum values of 185 meq/100 g clay (smectites, vermiculites). With increasing charge density, i. e., increasing exchange capacity, the electrical forces between adjacent layers increase in strength. There is a limiting value, varying with the type of interlayer cations, beyond which the hydration enthalpy of the cations is no longer sufficient to overcome the attractive electrostatic forces. The interlayer space remains inaccessible and ion exchange is restricted to the outer crystal surface. The thinner the crystals of the sample are, the higher the exchange capacity becomes in this case. Values of $x + y \leq 0{,}2$ involve weak electrostatic interaction between

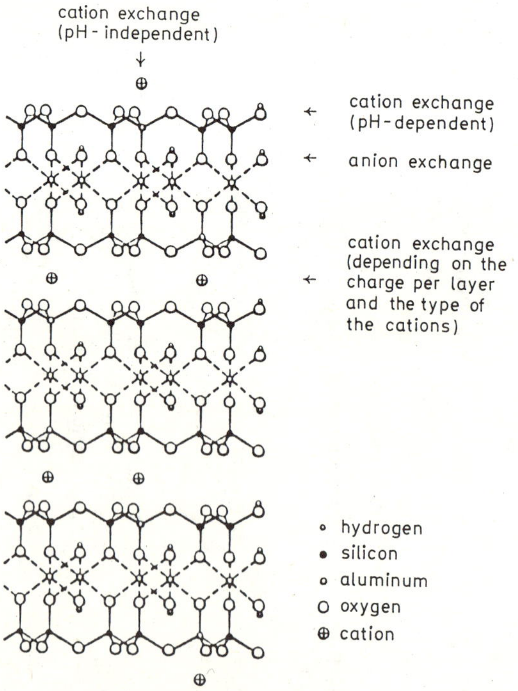

Figure 1.90 Ion exchange sites in a dioctahedral mica-type clay mineral [10].

adjacent layers with the van der Waals interaction predominating. Nevertheless the cation exchange capacity is small because the tendency to solvate does not suffice to take up water between the layers. Therefore the interlayer cations are immobile.

1.6.4.3 Exchange equilibria

The observed ion exchange equilibria depend greatly upon the charge density of the mica-type clays and upon the valence of the exchangeable ions. For monovalent ions the exchange equilibria are unfavourable, but for some clay minerals – namely the expandable ones – the interlayer expansion with monovalent cations can proceed to a greater extent, so that much larger diffusion cross-sections are possible. For divalent counterions the exchange equilibria are more selective, and the exchange isotherms are much steeper at low equilibrium concentrations. Although interlayer expansion remains restricted so that much smaller diffusion cross-sections are available for these cations between the layers, the mica-type clay minerals still exhibit an extremely high selectivity for divalent cations. For instance, in the case of clays contaminated with calcium carbonate, the relatively low solubility of calcium carbonate is sufficient to turn the sodium form of a montmorillonite into the calcium form. In addition to the valence of the exchangeable ions, the charge density of the mica-type clay minerals is of significant importance for the exchange equilibria. Clay minerals with high charge densities like illites, micas and brittle micas fix potassium, rubidium, cesium and ammonium ions more or less irreversibly. In a mixture with other ions, they are preferentially incorporated. At the same time their re-exchange is difficult. The total interlayer space of micas is generally blocked by their natural potassium content. Cation exchange is then restricted to the outer crystal surfaces. Ions blocking the interlayer space can be exchanged – though only in very slow reactions – for barium or n-alkylammonium ions [16]. Barium ions have radii of approximately the same size as those of potassium, ammonium and rubidium ions, but they carry twice as much charge. Therefore, on exchange only half of the interlayer cation sites is occupied. Consequently there is enough free space for interchange of sites and hence for the diffusion of water molecules into the interlayer spaces and the solvation of the barium ions. In the hydrated state with two water layers, the barium ions are then disposable again for exchange reactions [17].

Primary n-alkylammonium ions are also capable of removing the blockage of the interlayer space. In addition to the electrostatic interaction, each $R\text{-}NH_3^+$ ion may form three hydrogen bonds with the oxygen atoms of the silicate layers. The trigonal $\text{-}NH_3^+$ groups fit into the structure of the silicate layers especially well.

Mica-type clay minerals exhibiting medium charge densities, like minerals of the vermiculite group, incorporate at best two water layers. Therefore the difference between mono- and divalent cations is of little macroscopic effect. It should be emphasized that the quantitative exchange occurs very slowly, especially with large cations. Potassium, ammonium, rubidium and cesium ions are bound so strongly that water is squeezed out of the interlayer space, so that the conditions become similar to those for the micas described above.

In the group of the smectites (low-charged mica-type layer silicates) the charge of the exchangeable ions is of considerable importance. Divalent ions and ions of higher valence limit swelling properties. The interlayer distance increases at most to 1.2 nm. Monovalent cations, however, produce extensive diffuse electric double layers. In accordance with the concentration of electrolytes, the interlayer distance may increase by as much as several tens of nm. Then the properties of the molecules in the interlayer space approach properties of those in the bulk solution.

1.6.4.4 Sites of exchangeable ions

If the excess charge results from isomorphous substitutions within the tetrahedral and/or octahedral layers, the corresponding exchangeable cations enter sites at or near the basal planes. With expandable samples the sites of the inner basal planes are occupied as well. The number of exchangeable cations depends upon the charge of the layers. For non-expanding mica-type minerals only the external basal planes are accessible; therefore, the exchange capacity depends upon the thickness of the crystals.

At the margins of the tetrahedral layers, i. e. the prism surfaces, conditions are different. Formally "broken bonds", mentioned earlier, would be expected, because the Si-O-Si bonds do not continue at the crystal edges. Instead, the valences are saturated by SiOH units, exhibiting the characteristic low acidity of OH groups bound to silicon. The pK_s value generally lies between 8 and 10. This implies that the protons of the SiOH groups contribute to the cation exchange only at very high pH values, thus being responsible for a pH-dependent portion of the cation exchange capacity. Its order of magnitude depends on the diameter, not on the thickness of the crystals.

Like the pH-dependent portion of the cation exchange capacity, the number of the exchangeable anions is also related to the prism surfaces. Therefore their number increases with decreasing crystal diameter.

1.6.4.5 Special features in the experimental determination of exchange capacities

All exchange reactions involve changes in the solvation state of the clay minerals. This can result in alterations of the concentration of the co-ions in the equilibrium solution. With expandable mica-type clay minerals this alteration can be extraordinarily large and can lead to misleading results in the determination of exchange capacities. This becomes particularly important when the swelling state changes very much during the cation exchange reaction. With di- and multivalent counterions at any concentration (exceptions are observed in case the multivalent cations form complexes with strongly-bound anions), or with monovalent counter-ions at concentrations over 0.1 N, counter-ions (i. e., exchangeable ions) but no co-ions can enter the interlayer space. Increasing the interlayer distance therefore results in concentrating the co-ions in the equilibrium solution (salt-free water skin, negative adsorption [18, 19]). Inversely, if the interlayer distance decreases during the cation exchange reaction, water is pressed out of the solid phase, thereby diluting the equilibrium solution, and the concentration of the co-ions diminishes. Its careful determination is therefore necessary to examine exchange capacities accurately.

1.6.4.6 Applications

Very highly charged mica-type clay minerals do not play a significant role as direct ion exchangers in industrial applications. In nature, however, especially for soils, they are very important. All fertile arable soils contain these minerals and are thus capable of binding ions supplied by fertilizers for shorter or longer periods of time. This prevents immediate leaching of the ions to the ground-water. Such highly charged clay minerals can also be used for the elimination of radioactive cesium ions from the environment, because they are capable of fixing them irreversibly to a significant extent. Vermiculites, which are also frequently found in the soils, are less highly charged. In industry their specific (outer) surface can be extended considerably by an exfoliation process. These clay minerals are then used as ion exchangers in various ways. Their average exchange capacity is 150 meq/100 g clay. It must be taken into account that large monovalent cations like potassium, rubidium and cesium are bound almost irreversibly and that the exchange capacity can therefore be reduced to a fraction of its original value.

Certainly the most important group are the smectites, to which the montmorillonites belong. On the one hand, their ion exchange ability is used directly, on the other hand certain properties attained by these minerals after appropriate cation exchange are of special interest. Depending upon valence and type of the exchanged cations the properties of the exchanged forms vary greatly. Adequately prepared minerals can use the solvation tendency of exchangeable ions to act as desiccants. Moreover, they are employed as binding agents for foundry sands, as thixotropic auxiliary products for thrilling in oil exploration or as other thixotropic suspension agents.

Cation exchange with large organic cations permits new and various applications. The products obtained in this way are referred to as bentones or organophilic bentonites. The original cations are replaced by large bulky surfactant cations so that the total surface becomes organophilic. Thus, important thixotropic auxiliary products for anhydrous systems are obtained.

Minerals exhibiting very low isomorphous substitution (pyrophyllite and talc as extremes) show such a low exchange capacity that their exchange properties are not used in industry.

1.6.5 Chlorites

The structure of the chlorites is related to that of the mica-type clay minerals. The essential difference is that metal hydroxide layers are substituted for the interlayer cations (Figure 1.91).

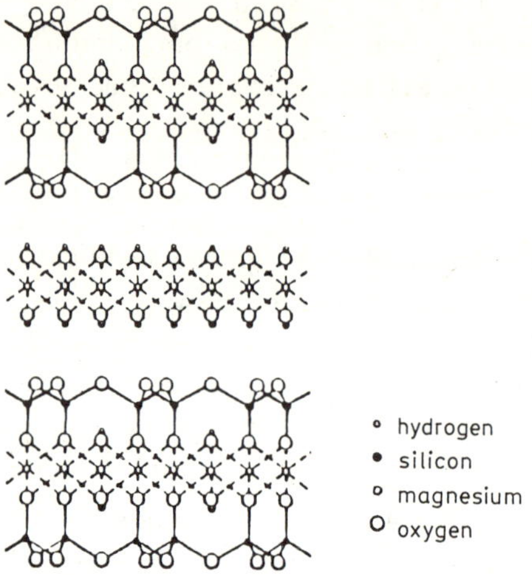

Figure 1.91 Schematic drawing of the chlorite structure.

If this hydroxide layer is electrically neutral, the total crystal is electrically neutral. Ion exchange reactions then occur only at the outer crystal surfaces and do not differ in any way from the exchange reactions occurring in the non-expandable clay minerals.

The metal hydroxide layers, however, are not always electrically neutral but may possess excess charge. Then, as in the group of the mica-type clay minerals, there are ions capable of being solvated. These cations become mobile in the solvating water layers and therefore exchangeable. This is the case with expandable chlorites like sudoite and corrensite, which are frequently found in certain soils.

In industry there are few applications for chlorites. But for soils their exchange capacity plays a significant role.

1.6.6 Transition mica-type minerals – chloritic minerals

Recently transitional links between mica-type clay minerals and chlorites have become a subject of great interest in industry. In the transition region there are no more coherent hydroxide layers between the mica-type silicate layers, but only islands of metal hydroxides or isolated groups of polynuclear cationic complexes which occupy the sites of exchangeable cations between the layers. The valence of these polynuclear cationic complexes is generally very high, so that they cannot be exchanged for ions of lower valence and consequently are bound more or less irreversibly. As large bulky groups, they stand between the silicate layers at rather large distances and thus look like pillars. For this reason, such products are referred to in the current literature as "pillared clays" [20].

Aluminum, iron or multivalent cationic zirconium and molybdenum complexes are suited very well as complex, multivalent cations. Any change of interlayer distance is restricted by the rigid, polynuclear complex cations. Thus channels are created, which exist even under extreme drying conditions. Moreover, within the pillars there are ions of unusual coordination, which permit special catalytic reactions with favourable kinetics. In this field it seems likely that important technical applications for catalytic purposes will be found.

1.6.7 Sepiolites and palygorskites

1.6.7.1 Structure

An important group of clay-related minerals comprises sepiolite and palygorskite. The idealized formulas are

$$Mg_8(Si_{12}O_{30})(OH)_4(H_2O)_4 \cdot (0-8)H_2O \text{ and}$$
$$Mg_5(Si_8O_{20})(OH)_2(H_2O)_4 \cdot (0-4)H_2O$$

respectively.

Palygorskite has been known for a long time in the literature and in industry as attapulgite or fuller's earth from Florida. The structures of sepiolite and palygorskite can be deduced from that of the mica-type clay minerals. As already described (p. (p. 499) the SiO_4 tetrahedra are linked together to form a two-dimensional layer, the so-called tetrahedral layer. Chemically palygorskite consists of twelve-membered Si-O rings, especially stable because of the favourable Si-O-Si valence angle (Figure 1.92a). In the group of mica-type clay minerals, the free vertices of all the tetrahedra within one sheet point in the same direction (Figure 1.92 b). Sepiolite and palygorskite differ in this one-sided orientation. In palygorskite the free vertices of one ribbon of twelve-membered rings point to one side, the vertices of the neighbouring ribbon to the opposite side. The ribbons correspond to the "double chains" of the amphiboles. These one-dimensional Si-O double chains are regularly condensed together by common Si-O-Si bonds (Figure 1.92 c, d). Sepiolite is principally constructed in the same way, the only difference being that there are ribbons of three instead of two condensed chains, within which the Si-O vertices are oriented in the same direction (Figure 1.92 e, f).

Palygorskite thus contains one-dimensional, ribbon-like units from the structure of the trioctahedral mica-type clay minerals. The structure of sepiolite is similar, but the ribbon-like units are wider by a factor of 1.5.

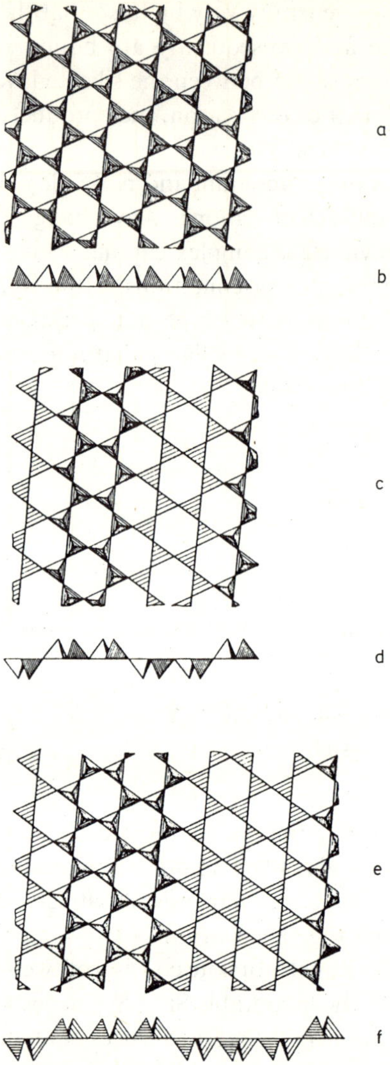

Figure 1.92 Comparison of the tetrahedral Si – O layers of clay minerals (a, b), palygorskite (c, d) and sepiolite (e, f) [21].

Condensation of these one-dimensional units results in three-dimensional structures with a preferred direction, characterized by parallel Si-O ribbons and channels. In palygorskite the width of these channels corresponds to two Si-O chains and in sepiolite to three. Their height is 0.64 nm, minus twice the van der Waals radius of oxygen (0.14 nm), i. e. ca. 0.36 nm (Figure 1.93).

Because of the one-dimensional character of the structure, the crystals generally form thin, extremely long needles or laths.

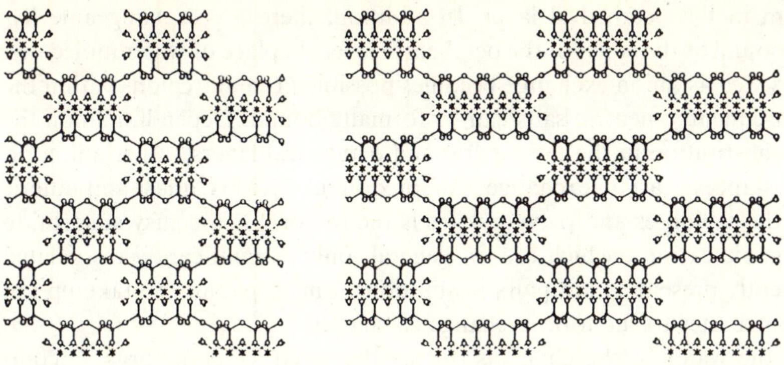

Figure 1.93 Schematic drawing of the structures of a) palygorskite and b) sepiolite (for an explanation of the symbols, see Figure 1.91).

1.6.7.2 Ion exchange

The structural similarities to the mica-type clay minerals imply analogies in their ion exchange behaviour. This is also connected with substitutions, which might take place in the tetrahedral or octahedral layers. In contrast to mica-type clay minerals, where isomorphous substitutions always produce negative charge, in palygorskites and sepiolites excess positive charge can arise as well. As trioctahedral minerals they contain magnesium in the octahedral layers, which can be replaced by trivalent aluminum [22]. Hence, several important aspects concerning their ion exchange properties result. On the one hand, at the outer surfaces – apart from lattice defects – mainly SiOH groups exist, which involve a pH-dependent cation exchange ability. On the other hand, the small structure-forming ribbons may show a pH-independent ion exchange. This occurs if silicon is replaced by aluminum at tetrahedral sites or if magnesium is replaced by aluminum or possibly by sodium at octahedral sites. But octahedral substitution rarely occurs and has not yet been incontestably proven for this group of clay minerals. If the substitution of aluminum for silicon extends to the total crystal volume, counter-ions must also be present in the channels. These channels are sufficiently high to allow the counter-ions to migrate within the water channels. Thus they become exchangeable [23].

Experimental results point to a wide variety of different sepiolites and palygorskites. Quantitative ion exchange via the narrow channels is slow, mainly caused by the rather large distance between the centres of excess charge created by substitutions.

An important feature of this group of clay minerals is the possibility that substitutions within tetrahedral and octahedral layers can compensate for one another. The replacement of silicon by aluminum in the tetrahedral layer may involve an internal charge compensation when aluminum simultaneously substitutes for magnesium in the octahedral layer. In addition, there are exchangeable hydroxide ions, bound at the edges of the octahedral layers in place of water molecules. Thus cation as well as anion exchange becomes possible in small regions within the channels. Consequently, neutral salts can be formally bound. Depending upon the extent of the substitution in the tetrahedral and octahedral layers, either anion or, as in most instances, cation exchange can predominate. This quasi-amphoteric behaviour of the sepiolites and palygorskites is the reason for the easy adsorption of salts of organic cations, which are, in general, only insignificantly dissociated and are frequently present as ion pairs in solutions. This capability to take up ion pairs often contributes to the total exchange capacity.

Moreover, the water in the channels can be displaced by polar organic compounds. But in most cases this is a pure adsorption process with the large accessible surface involving high exchange capacities.

However, it is not certain whether the scientific investigation of this subject can be regarded as completed.

1.6.8 Kaolinite and related minerals

1.6.8.1 Structure

All the clay minerals with ion exchange capabilities described in the previous chapters consist of silicate layers exhibiting the sequence tetrahedral-octahedral-tetrahedral. In most cases the average number of excess negative charges is the same in all tetrahedral layers, independent of regular or random distribution within each layer. There are a few exceptions to this ideal distribution. For example the rectorites (earlier known as allevardites) demonstrate variable extents of isomorphous substitution in the two tetrahedral layers belonging to one silicate sheet [24].

By contrast, silicate layers exhibit a highly asymmetric structure in kaolinite and its related minerals. The metal hydroxide layer is only condensed with one tetrahedral layer. The resulting silicate layers are then bordered by siloxane oxygens on one side and by hydroxide ions on the other (Figure 1.94). As a result, the corresponding surface properties and, therefore, the ion exchange properties, differ widely (Table 1.32).

1.6.8.2 Cation exchange

Under normal conditions these minerals are not expandable. Thus, the ion exchange is restricted to the outer crystal surface. The exchange sites, located on the basal planes, are independent of pH.

As in all the groups of clay minerals described here, there is also a pH-dependent cation exchange which is related to the "broken bonds", i. e., the SiOH groups at

Table 1.32 Compilation of ion exchange properties of kaolinite and structurally related clay minerals [25]

Mineral	Composition (idealized)	Cation exchange capacity (meq/100 g clay) pH-dependent*	pH-independent**	Anion exchange capacity (meq/100 g clay) on the prism surfaces*	on the basal planes**
Kaolinite (Nacrite, Dickite)	$Al_2(OH)_4Si_2O_5$	3 – 5	1 – 12	1 – 4	3 – 20
Halloysite	$Al_2(OH)_4Si_2O_5$		1 – 12		up to 80
Endellite	$Al_2(OH)_4Si_2O_5 \cdot 2\,H_2O$		1 – 12		up to 80
Chrysotile	$Mg_3(OH)_4Si_2O_5$		< 2		up to 30
Antigorite	$Mg_3(OH)_4Si_2O_5$		< 2		up to 30
Serpentine	$(Mg_{3-x}Al_x)\,(OH)_4(Si_{2-x}Al_x)O_5$; $x < 0.1$		< 2		up to 30
Amesite	$(Mg_2Al)\,(OH)_4(SiAl)O_5$	Practically, these minerals do not play any role. They are mostly contaminated by other minerals. Therefore reliable values do not exist.			
Garnierite	$Ni_3(OH)_4Si_2O_5$				
Cronstedtite	$(Fe_2^{2+}Fe^{3+})\,(OH)_4(SiFe^{3+})O_5$				
Chamosite	$(Fe^{2+}, Mg)_{2.3}(Fe^{3+}Al)_{0.7}(OH)_4(Si_{1.3}Al_{0.7})O_5$				

* capacity inversely proportional to the crystal diameter
** capacity inversely proportional to the crystal thickness

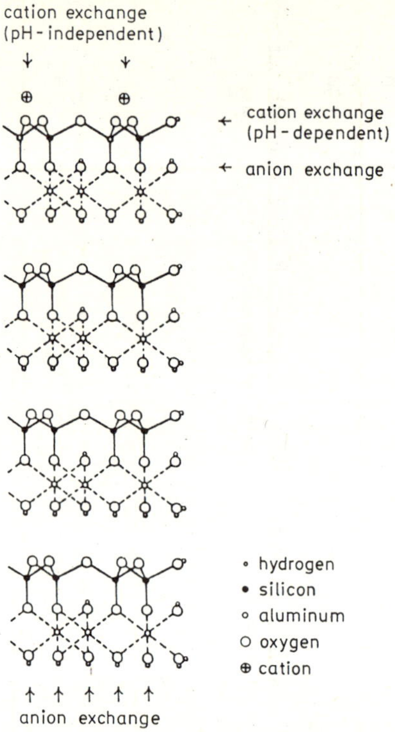

Figure 1.94 Schematic illustration of structure and ion exchange sites of a kaolinite crystal [25].

the crystal edges (Figure 1.94). The extent of the cation exchange may be determined in an anhydrous medium with sodium ethylate [25].

Only one of the two basal planes of the crystallites contributes to the pH-independent cation exchange. It was postulated earlier that there exists a specific adsorption of hydroxide ions thought to be responsible for the excess negative charge in these layers. In fact, however, the negative charge at the SiO-containing basal plane is created by substitutions of aluminum for silicon [26].

The trioctahedral minerals of this group, antigorite and chrysotile, should in principle also exhibit substitutions in the tetrahedral layers. However, negative charge in the octahedral layers would have to be due to the incorporation of lithium and sodium. As these minerals have been formed in nature under conditions of aluminum deficiency, substitutions of aluminum for silicon are improbable. Therefore the cation exchange capacity of the trioctahedral minerals of this group is usually much smaller than that of the dioctahedral ones, which have grown in the presence of aluminum. Isomorphous substitution of magnesium for aluminum at octahedral sites seems to occur rather infrequently.

Ion exchange experiments with colloidal cations have shown that these are usually bound to only one of the two basal planes of the crystals, the Si-O plane.

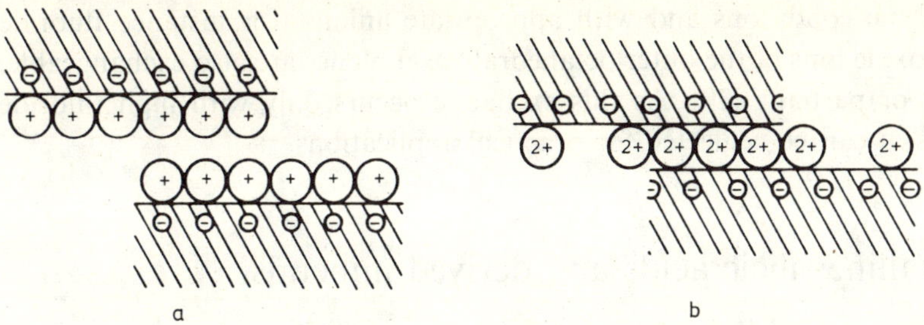

Figure 1.95 Kaolinite crystals with a) monovalent exchangeable cations: no face-to-face attraction between parallel crystals is possible. b) divalent exchangeable cations: face-to-face denting of parallel crystals occurs.

With a few crystals the colloidal cations may become fixed to both basal planes; with others such cations are not fixed at all on basal planes. Obviously this is connected with a twinning of the crystals. If (001) is the twinning plane, both external basal planes can be either Si-O or Al-OH planes. Consequently, either both planes exhibit cation exchange ability induced by substitutions, or neither of them do [27]. The portion of such twin-crystals is different for kaolinites from different deposits and varies between 0 and about 15%.

The valence of the exchangeable cations has a strong effect on rheological properties. From a practical point of view it is essential that kaolinite suspensions and dispersions of expandable mica-type minerals are affected in a completely different manner. This is due to the fact that in most cases isomorphous substitution on the outer Si-O basal planes of kaolinite crystals approaches the theoretically possible maximum value of about 0.25 nm^2 per excess charge [27]. Monovalent cations then produce a dense packing on the surface. This means that there is no electrostatic face-to-face attraction between parallel crystals (Figure 1.95a). Only every second site is occupied by divalent cations on the surface, thus permitting face-to-face denting (Figure 1.95b) [25]. The high degree of isomorphous substitution on the Si-O basal plane also implies incomplete cation exchange with very large organic cations for steric reasons.

1.6.8.3 Anion exchange

The anion exchange of minerals that belong to the kaolinite group is also more complicated than that of the mica-type clay minerals. The hydroxide ions at the edges of the octahedral layers are relatively easy to exchange and behave similarly to those on the surface of mica-type minerals [28]. The specific binding of phosphate and certain polymeric anions, especially polyphosphates and polyacrylates, is important for many applications.

Under extreme conditions and with appropriate anions (for instance fluoride ions) the hydroxide ions of the outer octahedral basal planes are also exchangeable, either entirely or partially [29]. As this exchange occurs only with high fluoride concentrations, it can be neglected for practical applications.

1.6.9 Crystalline silicic acids and derived minerals

In recent years a number of layer silicates has been discovered which belong to the expandable clay minerals. They are found mostly near soda lakes, such as Lake Magadi in Kenya. This group includes the minerals magadiite, kenyaite, makatite, kanemite, and synthetic products, e. g. ilerite, named after R. K. Iler [30].

Investigation of these minerals has shown that under appropriate conditions the cation exchange occurs quantitatively corresponding to their formulas. The ion exchange capacities are very high and reach values up to 800 meq/100 g clay. If large organic cations are used instead of small inorganic ones the exchange may be limited for steric reasons; even under favourable conditions only a fraction of the total exchange capacity is available. Table 1.33 lists the various synthetic and naturally occurring compounds and their cation exchange capacities.

Table 1.33 Compilation of characteristic properties of various naturally occurring minerals and synthetic compounds of the group of the crystalline silicic acids

Layer silicate	Composition	Exchange capacity (meq/100 g clay)		pK_s-value of the derived acid	
Kenyaite	$Na_2Si_{20}O_{41} \cdot x\ H_2O$	164 (H^+/Na^+; $x = 0$)*	[31]	−3 to 1.5	[32]
Magadiite	$Na_2Si_{14}O_{29} \cdot x\ H_2O$	216 (Na^+/H^+; $x = 6.8$)**	[33]	−5 to −3	[34]
		200 (H^+/Na^+; $x = 7.46$)**	[34]		
Ilerite (synthetic)	$Na_2Si_8O_{17} \cdot x\ H_2O$	314 (Na^+/H^+; $x = 8.9$)**	[33]	−3 to 1.5	[34]
Makatite	$Na_2Si_4O_8 \cdot x\ H_2O$	774 (H^+/Na^+; $x = 0$)*	[31]	−3 to 1.5	[34]
Kanemite	$NaHSi_2O_5 \cdot 3\ H_2O$	467 (Na^+/H^+; $x = 3$)*	[35]	2.3 to 3.3	[34]

* value calculated according to the cited literature
** experimentally determined value

Because silicates of this group can be synthesized rather easily, it is feasible that they will find use in a wide variety of industrial applications. Heavy metal ions could be fixed irreversibly by simple calcination. Like the smectites, these minerals become organophilic by ion exchange with organic cations [36]. Moreover, they can form intercalation complexes with a variety of polar organic molecules [31, 37]. Therefore, they could be used, for instance, as solid supports for pharmaceutically active compounds [38].

From a chemical perspective, the H-compounds are especially interesting. They represent silicic acids which formally contain Si-OH groups with variable acidities ($pK_s = -5.0$ to $pK_s = 3.3$) [34]. In all instances the pK_s is lower than that of orthosilicic acid ($pK_s = 9.75$).

The exploration of the chemistry of these pure layer silicates and crystalline silicic acids, however, has just begun.

1.6.10 Mixed-layer minerals

The foregoing discussion is based on the assumption that pure or nearly pure minerals are being considered. In nature, however, there are often rather large deposits of so-called mixed-layer minerals. These clay minerals are characterized by the fact that, within the very same crystal, layers of different clay minerals are found stacked together. Therefore, their properties generally correspond to the additive mixture of the various layer silicates involved.

In some cases with specific distribution of the excess charge energetically favourable, regular arrangements of the cations may occur.

Until now, however, these effects have not been taken advantage of technically. This is partly due to the fact that the synthesis of the mixed-layer minerals is still not understood in detail. The natural members of this group like rectorite, corrensite, sudoite etc. are mostly heterogeneous mixtures of different minerals and impurities.

1.6.11 Allophanes

In earlier discussions concerning clay minerals the allophanes played an important role. The term allophane covered amorphous, poorly characterized components of clays as defined in soil science. This definition is based on particle size: all particles smaller than 2 μm in diameter are referred to as clay. With increasing progress in X-ray crystal structure determination, it has been learned that most of these very fine soil particles are crystalline or paracrystalline. More recent investigations have shown that amorphous components of clays are actually found side by side with moderately crystalline minerals in volcanic soils. Due to vaguely defined structures, the ion exchange capacity of the compounds of this group is rather variable and differs from one deposit to the next. Therefore, it is difficult to make useful general statements. An important exception is the mineral imogolite which was discovered in Japan by Yoshinaga in the vicinity of active volcanoes [39, 40]. Since then it has been found in many soils of volcanic origin. Imogolite is a hydrated aluminosilicate which consists of bundles of long tubes of about 2.2 nm external and 1.0 nm internal diameter. It has already been successfully synthesized [41]. Depending upon conditions, its cation and anion exchange capacities can vary considerably. T. Henmi and K. Wada, for example, have obtained a CEC value of 30 meq/100 g for imogolite [42].

References

[1] A. Weiß and U. Hofmann (1951) Batavite. Z. Naturforsch. *6b*, 405–409.
[2] J. Barshad (1950) The effect of the interlayer cations on the expansion of the mica type of crystal lattice. Am. Mineral. *35*, 225–238.
[3] M. B. Hafez, A. A. Abdelrasoul and R. Algasmi (1982) Fixation of radioelements on illite and bentonite. Isotopenpraxis *18*, 321–322.
[4] M. A. Kessick (1980) Ion exchange and dewaterability of clay sludges. Int. J. Miner. Process *6*, 277–283.
[5] N. K. Mitra, P. K. Das Poddar, D. Naha and S. Ghatak (1979) Thixotropy study in bentonite suspension in relation to exchangeable cations. Indian. Ceram. *22*, 117–122.
[6] F. W. Berk and Co. Ltd. (1964) Pigmented coating compositions containing clays. Belg. Pat. 650,090.
[7] H. Bergseth (1967) Binding of phosphate to clay minerals and a study of its release by anion-exchange with ^{32}P. Kolloid-Z. Z. Polymere *215*, 52–56.
[8] J. A. Diez (1980) The effect of different types of clay on the dynamics of phosphorus in soil. Agrochimica *24*, 353–360.
[9] H. van Olphen (1977) An introduction to clay colloid chemistry, Wiley, New York.
[10] A. Weiß (1958) Cation-exchange ability of clay minerals, II. Cation-exchange in minerals of the mica, vermiculite, and montmorillonite groups. Z. anorg. allg. Chem. *297*, 257–286.
[11] A. Banin and S. Ravikovitch (1964) Kinetics of reactions in the conversion of sodium- or calcium-saturated clay to hydrogen-aluminum clay. Clays clay Miner. *14*, 193–204.
[12] A. Weiß (1958) Cation-exchange ability of clay minerals, I. Comparison of investigations. Z. anorg. allg. Chem. *297*, 232–256.
[13] H. Laudelout, R. Van Bladel and J. Robeyns (1972) Hydration of cations adsorbed on a clay surface from the effect of water activity on ion exchange selectivity. Soil Sci. Soc. Amer. Proc. *36*, 30–34.
[14] G. H. Cashen (1961) Electric charges on clays. Chem. Ind. (London) *28*, 1060.
[15] A. Weiß (1966) in: Ullmanns Encyklopädie der technischen Chemie (W. Foerst, ed.) pp. 583–597, Urban and Schwarzenberg, München, Berlin, Wien.
[16] H. Graf von Reichenbach (1973) in: Proc. Int. Clay Conf. 1972 (J. M. Serratosa, ed.) pp. 457–466, Div. Cienc. CSIC, Madrid.
[17] H. Graf von Reichenbach and C. I. Rich (1968) in: Trans. Int. Congr. Soil Sci., 9th (J. W. Holmes, ed.) pp. 709–719, Elsevier, New York.
[18] U. Hofmann and K. Giese (1939) Cation exchange by clay minerals. Kolloid-Z. *87*, 21–36.
[19] R. K. Schofield and H. R. Samson (1953) Deflocculation of kaolinite suspensions and the accompanying changeover from positive to negative chloride adsorption. Clay Min. Bull. *2*, 45–50.
[20] T. J. Pinnavia (1983) Intercalated Clay Catalysts. Science *220*, 365–371.
[21] A. Weiß (1964) in: Ullmanns Encyklopädie der technischen Chemie (W. Foerst, ed.) pp. 697–712, Urban and Schwarzenberg, München, Berlin.
[22] A. Corma, P. Perez Pariente, V. Fornes and A. Mifsud (1984) Surface acidity and catalytic activity of a modified sepiolite. Clay Miner. *19*, 673–676.

[23] V. E. Polyakov, Y. I. Tarasevich and F. D. Ovcharenko (1976) Ion-exchange equilibria in palygorskite. Sov. Prog. Chem. *42*, 43–46.
[24] A. Weiß, H. O. Becker and G. Lagaly (1970) in: Proc. Intern. Clay Conf. 1969, Vol. 2, pp. 67–73, Israel Univ. Press.
[25] A. Weiß (1959) Cation exchange ability of clay minerals, III. Cation-exchange in kaolinite. Z. anorg. allg. Chem. *299*, 91–120.
[26] R. H. S. Robertson, G. W. Brindley and R. C. Mackenzie (1954) Mineralogy of kaolin clays from Pugu Tanganyika. Am. Mineral. *39*, 118–138.
[27] A. Weiß and J. Russow (1963) in: International Clay Conference (I. Th. Rosenquist and P. Graff-Petersen, eds.) pp. 203–213, Pergamon Press, Oxford, London, New York, Paris.
[28] E. Halevy (1964) Exchangeability of hydroxyl groups in kaolinite. Geochim. Cosmochim. Acta *28*, 1139–1145.
[29] A. Weiß, A. Mehler, G. Koch and U. Hofmann (1956) Anion-exchange in clays. Z. anorg. allg. Chem. *284*, 247–271.
[30] R. K. Iler (1964) Ion-exchange properties of crystalline hydrated silica. J. Colloid. Sci. *19*, 648–657.
[31] G. Lagaly, H.-M. Riekert and H. H. Kruse (1986) in: Chemical Reactions in Organic and Inorganic Constrained Systems (R. Setton, ed.) pp. 361–379, Reidel, Dordrecht, Boston.
[32] K. Beneke and G. Lagaly (1983) Kenyaite – synthesis and properties. Am. Mineral. *68*, 818–826.
[33] F. Wolf and W. Schwieger (1979) Ion-exchange of monovalent cations in synthetic sodium polysilicates with layer structure. Z. anorg. allg. Chem. *457*, 224–228.
[34] H.-J. Werner, K. Beneke and G. Lagaly (1980) Acidity of crystalline silicic acids. Z. anorg. allg. Chem. *470*, 118–130.
[35] K. Beneke and G. Lagaly (1977) Kanemite – intracrystalline reactivity and relations to other sodium silicates. Am. Mineral. *62*, 763–771.
[36] G. Lagaly, K. Beneke and A. Weiß (1975) Magadiite and H-magadiite: I. Sodium magadiite and some of its derivatives. Am. Mineral. *60*, 642–649.
[37] K. Beneke, H.-H. Kruse and G. Lagaly (1984). Crystalline silicic acid with distinct intracrystalline reactivity. Z. anorg. allg. Chem. *518*, 65–76.
[38] G. Lagaly (1977) Crystalline silicic acid, its potassium salt, synthesis and uses. Ger. Pat. 2742912.
[39] N. Yoshinaga and S. Aomine (1962) Allophane in some Ando soils. Soil Sci. Plant Nutr. (Tokyo) *8*, 6–13.
[40] K. Wada and N. Yoshinaga (1969) Structure of imogolite. Am. Mineral. *54*, 50–71.
[41] V. C. Farmer, A. R. Fraser and J. M. Tait (1977) Synthesis of imogolite: a tubular aluminium silicate polymer. J. Chem. Soc. Chem. Commun. *13*, 462–463.
[42] K. Wada and M. E. Harward (1974) Amorphous clay constituents of soils. Advanc. Agron. *26*, 211–260.

1.7 Non-siliceous Inorganic Ion Exchangers

Karl Heinrich Lieser
Technische Hochschule Darmstadt, Fachbereich Anorganische Chemie und Kernchemie, Darmstadt, Federal Republic of Germany

1.7.1 General aspects of synthetic inorganic ion exchangers

1.7.1.1 Survey

The first monograph on inorganic ion exchangers was published in 1964 by Amphlett [1]. It already dealt with the most important groups of inorganic ion exchangers, i. e. clay minerals, zeolites, heteropoly acids, hydrous oxides and insoluble phosphates. Further investigations rounded off the knowledge of inorganic compounds with ion exchange properties and a comprehensive survey of these compounds may now be given. As the natural and synthetic silicates represent a very extensive group of compounds with ion exchange properties, they are dealt with in separate chapters.

All inorganic ion exchangers are sparingly soluble solid compounds in which the exchangeable species are bound by chemical bonds of more or less ionic character. The non-siliceous inorganic ion exchangers may be divided into the following groups:

- *Hydrous oxides* [1–5]. This group comprises the great number of sparingly soluble metal and semi-metal hydrous oxides. The groups responsible for ion exchange are the hydroxide groups on which various reactions with cations or anions may take place.
- *Acid salts* [1–5]. In this group (e. g. phosphates, arsenates, antimonates, molybdates, tungstates, hexacyanoferrates) acidic hydrogen atoms are exchanged for cations. A great variety of compounds belonging to this group is known.
- *Salts of heteropoly acids* [1, 3, 4]. Only a few heteropoly acids are sparingly soluble, for instance the ammonium salts of molybdo-phosphate and tungstophosphate. They exchange cations.
- *Other ionic compounds* [6]. All sparingly soluble salts (e. g., sulfates, halides, sulfides) exhibit ion exchange properties. Cations and anions may be exchanged. The extent of exchange increases with decreasing solubility of the compound formed by exchange. In favourable cases quantitative transformation into the less soluble compound by ion exchange is observed.

1.7.1.2 Exchange mechanisms

Hydrous oxides of many elements, M, are amphoteric. The reactive groups are hydroxide groups which may undergo various reactions, depending on the nature of the hydrous oxide and the pH (Ⓜ denotes an atom of the element M bound to one or more other M atoms via oxygen):

a) Protonation at low pH

$$Ⓜ\!-\!OH + H^+ \rightleftharpoons Ⓜ\!-\!OH_2^+ \qquad (1)$$

and simultaneous bonding of an anion X^-

$$Ⓜ\text{—}OH_2^+ + X^- \rightleftharpoons Ⓜ\text{—}OH_2^+X^- \tag{2}$$

The resulting species has anion exchange properties:

$$Ⓜ\text{—}OH_2^+\,X^- + Y^- \rightleftharpoons Ⓜ\text{—}OH_2^+\,Y^- + X^- \tag{3}$$

b) If $Ⓜ^+$ is a soft acid and the anion X^- a soft base, X^- may substitute the OH-group and enter the inner coordination sphere of M

$$Ⓜ\text{—}OH + X^- \rightleftharpoons Ⓜ\text{—}X + OH^- . \tag{4}$$

Again, this is an anion exchange.

c) Deprotonation at high pH

$$Ⓜ\text{—}OH + OH^- \rightleftharpoons Ⓜ\text{—}O^- + H_2O \tag{5}$$

and simultaneous bonding of a cation M_1^+

$$Ⓜ\text{—}O^- + M_1^+ \rightleftharpoons Ⓜ\text{—}O^-M_1^+ . \tag{6}$$

The resulting species has cation exchange properties:

$$Ⓜ\text{—}O^-M_1^+ + M_2^{n+} \rightleftharpoons Ⓜ\text{—}O^-M_2^{n+} + M_1^+ . \tag{7}$$

d) The hydrogen atom may also be substituted directly and predominantly covalent bonds may be formed

$$Ⓜ\text{—}OH + M_2^{n+} \rightleftharpoons Ⓜ\text{—}O\text{—}M_2^{(n-1)+} + H^+ . \tag{8}$$

Again, this is a cation exchange.

e) The same product as in eq. (8) is formed by a condensation reaction with the monohydroxo complex of M_2^{n+}

$$Ⓜ\text{—}OH + HOM_2^{(n-1)+} \rightleftharpoons Ⓜ\text{—}O\text{—}M_2^{(n-1)+} + H_2O . \tag{9}$$

This is a chemisorption process. Taking into account the hydrolysis reaction

$$M_2^{n+} + H_2O \rightleftharpoons HOM_2^{(n-1)+} + H^+ \tag{10}$$

and summing up eqs. (9) and (10), it is found that this sum (hydrolysis plus chemisorption) is identical with eq. (8) (ion exchange). For many di-, tri- and tetravalent ions M_2^{n+} it has been found that sorption proceeds parallel to the formation of hydroxo complexes [7–12], which leads to the conclusion that in these cases the concept of hydrolysis and chemisorption corresponds better to reality than that of ion exchange. The reactions described by eqs. (9) and (10) are called hydrolytic adsorption [7]. Interesting reactions of this kind have been described by Kautsky [8]: sorption of Th^{4+} on silica gel at pH 3 followed by sorption

of phosphoric acid on the thorium. Formally, the first reaction can be described as cation exchange and the second as anion exchange. But these reactions are better understood as the formation of thorium silicate on the surface of silica gel and the additional formation of thorium phosphate. Both may be considered as surface compounds. Keeping this concept of formation of surface compounds in mind, the "exchange" properties of hydrous oxides can be predicted on the basis of the low solubility of the surface compounds formed.
The ion exchange on acid salts is described by the equation

$$\overline{H} + M^+ \rightleftharpoons \overline{M} + H^+ \tag{11}$$

$\overline{H}$ is the exchangeable hydrogen in the acid salt. It was shown that an equivalent number of protons is set free in the exchange process. The structure of the acid salts remains unchanged.

In the case of heteropoly acids, mainly the ammonium salts are used because of their low solubility. The ion exchange reaction is

$$\overline{NH_4} + M^+ \rightleftharpoons \overline{M} + NH_4^+ . \tag{12}$$

As in the case of the acid salts, the structure of the compounds is not affected by the ion exchange.

The reactions taking place in other ionic compounds can be more complicated. All exchange reactions begin at the surface, where cations or anions from the solution may be exchanged for those in the surface layer of the solid, if the properties of the solid compound and the compound formed by exchange are similar, in particular with respect to solubility. Thus we may have cation exchange

$$\overline{M}_1 + M_2^{n+} \rightleftharpoons \overline{M}_2 + M_1^{n+} \tag{13}$$

or anion exchange

$$\overline{X}_1 + X_2^{n-} \rightleftharpoons \overline{X}_2 + X_1^{n-} . \tag{14}$$

The equilibrium constants K for the exchange at the surface according to eqs. (13) and (14) increase with decreasing solubilities of the compounds formed by exchange [13–17]. They are $K \approx 1$ for isotopic exchange. If mixed crystals are formed in which M_1^{n+} or X_1^{n-} are substituted by M_2^{n+} or X_2^{n-}, respectively, the exchange process proceeds into the interior of the crystals by recrystallization or by diffusion, until equilibrium is established between solution and solid. If the solubility of the compound formed by exchange is much smaller than that of the original solid, quantitative reaction may occur, even if mixed crystals are not formed. An example is the reaction

$$(AgCl)_s + I^- \rightarrow (AgI)_s + Cl^- \tag{15}$$

which leads to quantitative transformation of solid AgCl into solid AgI [18, 19].

1.7.1.3 Selectivity

The selectivity S may be defined by

$$S = \log D_1 - \log D_2 = \log T \quad (16)$$

where D_1 and D_2 are the distribution coefficients of the ion under consideration and other ions, respectively, and T is the separation factor [20]. In general, the selectivity of inorganic ion exchangers is higher than that of ion exchange resins. This is due to the fact that rather specific reactions may take place, as explained in Section 1.7.1.2.

An important criterion for the selectivity of hydrous oxides and of various ionic compounds is the possibility of formation of compounds or surface compounds of very low solubility under the given conditions. Ions which fulfill this condition will be bound with high selectivity.

In the case of acid salts and salts of heteropoly acids other criteria are decisive: ionic radius, charge and interaction with the surrounding medium. Thus, most compounds of these groups exhibit a high selectivity for monovalent cations, in particular Cs^+.

1.7.1.4 Kinetics

With respect to the rate of exchange two cases must be considered:

a) The exchange reaction is restricted to the surface. If an appreciable energy barrier is not involved in the exchange reaction, diffusion through the adherent solution layer (film diffusion) is the rate-determining step and the exchange proceeds rather fast. In this case, a typical half-time of the exchange reaction is 10 s (thickness of the adherent layer about 0.2 mm). This value increases with the square of the thickness of the adherent layer. It depends on the intensity of stirring or the flow rate of solution in a separation column. Typical examples are exchange reactions on the surface of compact hydrous Al_2O_3.
However, if the exchange reaction at the surface is hindered by an appreciable energy barrier, the exchange reaction itself becomes the rate-determining step. For heterogeneous exchange reactions at the surface of ionic crystals activation energies of the order of 20 to 80 kJ/mol have been found [13, 16, 21, 22], which are due to the stripping of at least parts of the solvation shell when the ions enter the surface. In this case a pronounced temperature dependence is observed, and at room temperature typical half-times for exchange are of the order of several minutes. Examples are exchange reactions of alkaline earth ions on the surface of alkaline earth carbonates or sulfates.

b) The exchange reaction proceeds into the interior of the particles. If the ratio $d^2/D_{eff} > 100$ s, where d is the diameter of the particles and D_{eff} the effective diffusion coefficient, the transport within the particles (gel diffusion) will be the rate-determining step. This transport depends on the porous structure of the particles and therefore to a great extent on the method of preparation. Many efforts have been made to prepare various hydrous oxides and acid salts of high porosity and nevertheless sufficient mechanical stability. Typical effective diffusion coefficients of the order of $D_{eff} \approx 10^{-8}\,cm^2s^{-1}$ are calculated for gel diffusion in amorphous and porous inorganic ion exchange materials using literature data. This corresponds to half-times of exchange of the order of 20 min for particles of 0.1 mm diameter. With increasing crystallinity D_{eff} decreases drastically to values of the order of $10^{-12}\,cm^2s^{-1}$ [5], which leads to half-times of exchange of the order of 20 min even for particles of 1 μm grain size. The fact that effective diffusion coefficients in inorganic exchangers are, in general, one order of magnitude or more lower than those found for typical cation and anion exchange resins is a great disadvantage of inorganic ion exchange materials.

In crystalline non-porous ionic compounds ion exchange proceeds into the interior of the particles by recrystallization or solid diffusion. In general, recrystallization brings about an appreciably faster transport. It depends very much on the size of the crystals and the degree of disorder and therefore on the method of preparation and pretreatment. The half-time of recrystallization may vary between one hour and many days [23].

1.7.1.5 Other properties

Some inorganic ion exchange materials are compact and react mainly at the surface (Al_2O_3, $BaSO_4$). Many have layered structures and the exchange takes place in the space between the layers (most of the acid salts). Others have three-dimensional structures with pores (most of the hydrous oxides).

In contrast to organic ion exchange resins, inorganic ion exchangers have relatively rigid structures. Layer structures may exhibit crystalline swelling by taking up ions of increasing diameter or organic molecules, but three-dimensional structures undergo little swelling or shrinking, a fact that leads to more pronounced exclusion by ion-sieve effects. The mechanical stability of inorganic ion exchangers is, in general, low, compared with organic ion exchange resins.

The main advantages of inorganic ion exchange materials are: High selectivity and high radiation stability. Furthermore, most compounds are resistant to oxidizing agents and organic solvents, and many of them can be prepared more cheaply than organic resins. Disadvantages are: Relatively low stability towards strong acid and/or base solutions, low mechanical stability, low abrasion resistance, slow exchange rates into the interior, medium or low capacities and measurable solubility of the compounds.

1.7.2 Hydrous oxides

1.7.2.1 General properties

The ion exchange properties of hydrous oxides were first studied in detail by Kraus and coworkers [24–26]. Several reviews on this subject have been published [1–5], as well as on ion exchange chromatography on hydrous oxides [27]. Figure 1.96 gives a survey of the elements, M, the hydrous oxides of which exhibit ion exchange properties. As can be seen from this figure all sparingly soluble oxides or hydrous oxides may be used as inorganic ion exchangers in the form of amorphous or crystalline material.

The term hydrous oxides refers to the solids existing in oxide-water systems. Hydrous oxides of stoichiometric composition are found only in a few cases, such as FeOOH or AlOOH. In most systems, however, there is no evidence of definite hydrates. MOH groups are identified by IR spectroscopy. Water molecules are rather strongly bound onto the MOH groups via hydrogen bridges and are given off only at elevated temperatures (>100 to 300 °C). The MOH groups are decomposed very slowly into oxides and water at still higher temperatures. Even after heating to 800 to 1000 °C small amounts of hydroxide groups may be found by measuring the exchange capacity [28]. It is evident that the ion exchange properties, in particular the ion exchange capacity and the kinetic behaviour, depend strongly on the method of preparation and the treatment before use.

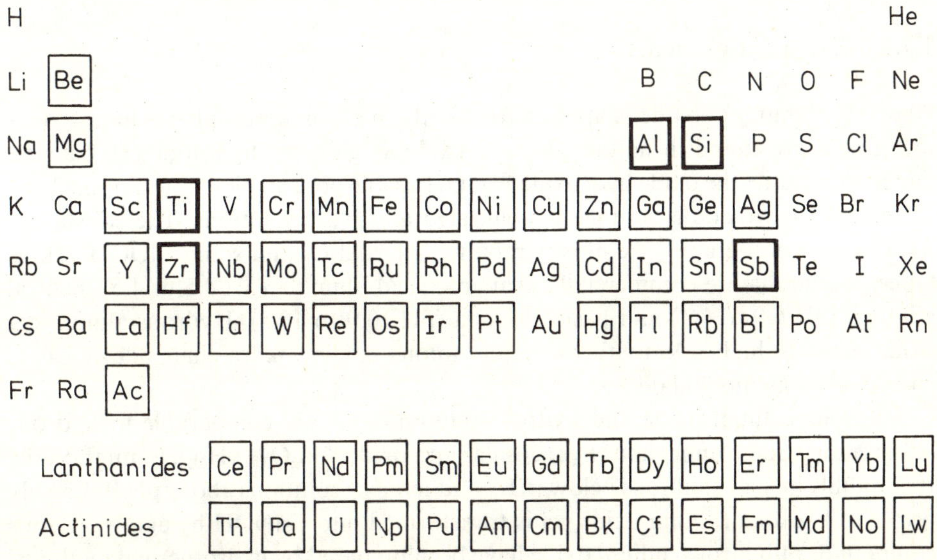

Figure 1.96 Elements whose hydrous oxides have ion exchange properties. (Those of greater practical importance are specially marked).

Hydrous oxides may be characterized by their specific outer and inner surface area, exchange capacity, density, pore-size distribution, X-ray diffraction spectra, IR-spectra and Mößbauer spectroscopy. Some hydrous oxides, such as silica gel and alumina gel, are activated by heating. This means that more free hydroxide groups with the desired exchange properties are available after treatment at higher temperatures.

Furthermore, all hydrous oxides may be characterized by a certain pH at which the overall charge on the surface is zero (zero point of charge or isoelectric point) [29]. This value depends on the nature of the cations and anions in solution and on the ionic strength. At lower pH protonation and anion exchange according to eqs. (1) and (3) are observed, at higher pH deprotonation and cation exchange according to eqs. (5) and (7).

1.7.2.2 Divalent elements

Hydrous beryllium oxide can be precipitated as an amorphous gel, whereas BeO exists in two crystalline forms, a stable orthorhombic and a metastable tetragonal form. The amorphous product exhibits amphoteric ion exchange behaviour. The applicability is restricted by dissolution below pH 6 and above pH 10.

$Mg(OH)_2$ is found as brucite in nature and has a layered structure of the CdI_2-type. It has anion exchange properties only and it dissolves below pH 10 and above pH 13.

1.7.2.3 Trivalent elements

Hydrous aluminium oxide is applied frequently in chromatographic columns. Various forms are found in nature, the hydroxide as gibbsite (hydrargillite), bayerite and norstrandite, the oxide hydroxide AlOOH as diaspore and boemite. Amorphous aluminium hydroxide is soluble in weak acid and basic solutions. The substances used in chromatography are mostly mixtures of γ-Al_2O_3 and some AlOOH. Three types of alumina are commercially available: acid alumina (pH 3.5 to 4.5), neutral alumina (pH 6.9 to 7.1) and basic alumina (pH 10.0 to 10.5). Hydrous aluminium oxide exhibits high selectivity for some anions carring more than one negative charge, such as molybdate.

Hydrous gallium oxide and hydrous indium oxide are comparable to hydrous aluminium oxide. Amorphous hydrous gallium oxide, $Ga_2O_3 \cdot xH_2O$, is amphoteric. It is soluble in acid solutions below pH 3 and in basic solutions above pH 9. Freshly prepared hydrous gallium oxide gel is transformed into GaOOH by aging. Various compounds of hydrous indium oxide have been prepared by hydrothermal methods: Cubic $In(OH)_3$, InOOH and $In_2O_3 \cdot xH_2O$. The latter is also amorphous and amphoteric and dissolves below pH 3 and above pH 10.

In the system iron(III) oxide/water various forms are known: amorphous hydrous iron(III) oxide $Fe_2O_3 \cdot xH_2O$, α-FeOOH (goethite), β-FeOOH and γ-FeOOH. The amorphous iron(III) hydroxide is amphoteric, whereas β-FeOOH shows only anion exchange [30]. Sorption of phosphate is strongly preferred. The hydrous oxides are stable down to pH 1.6 and are transformed slowly into α-Fe_2O_3 in strong alkaline solution. Pseudomorphous hydrous iron(III) oxide obtained by topochemical reaction from $Fe_2(SO_4)_3$ is also of interest as ion exchange material because of its high specific surface area and its pseudocrystallinity [31].

Hydrous manganese(III) oxide $Mn_2O_3 \cdot xH_2O$ resembles the iron(III) compounds. It is amphoteric and dissolves below pH 5. γ-MnOOH (manganite) is found in nature and has only cation exchange properties [30].

Lanthanum hydroxide may be obtained by precipitation. It shows anion exchange properties [30], but dissolves below pH 8 and is therefore of very limited use as ion exchanger. The same holds for the other hydroxides of the lanthanides.

Hydrous bismuth(III) oxide, $Bi_2O_3 \cdot xH_2O$ shows the behaviour of a strong basic anion exchanger [30]. Chloride ions are bound with rather high selectivity, which is due to the formation of BiOCl [30].

1.7.2.4 Tetravalent elements

Hydrous silica, silica gel, $SiO_2 \cdot xH_2O$, is widely used as an ion exchange material. It is stable in acid solution, but begins to dissolve above pH 10. The ion exchange behaviour of silica gel can be described as that of a weakly acidic cation exchanger [32]. A more realistic description of the behaviour involved is hydrolytic adsorption (see Section 1.7.1.2). The isoelectric point varies between about 1.5 and 3 [29]. Some experimental results indicate anion exchange behaviour below pH 3 [33]. Selective separations of cations on silica gel are possible in pH ranges where the ions to be separated form hydroxo complexes [8–10]. Polynuclear hydroxo complexes are not bound in appreciable amounts. For ions of the same charge, the following sequences of sorption have been reported:

$Li^+ > Na^+ > K^+ > Rb^+$; $Be^{2+} > Mg^{2+} > Sr^{2+} \approx Ba^{2+}$; $Al^{3+} > Ga^{3+} > In^{3+}$; $Sc^{3+} > Lu^{3+} > Yb^{3+} > Tm^{3+} > Er^{3+} > Y^{3+} > Tb^{3+} > Gd^{3+} > Nd^{3+} > Ce^{3+} > La^{3+}$ [2, 34–37].

Sorption of $Zn(NH_3)_4^{2+}$ and $Zn(en)_3^{2+}$ was found to be independent of pH in the range 6.0 to 9.2 and was used for determination of the specific surface area [38]. The sorption of Pa, Zr, Nb, Th and U on silica gel was also investigated [2, 39–41]. Separation of Pa from nitric acid solutions on $SiO_2 \cdot xH_2O$ or Vicor glass is of special interest for the reprocessing of thorium breeder fuels. The exchange capacity of hydrous silica depends on the number of available silanol groups and goes up to several mmol/g.

For special purposes, the surface of hydrous silica may be modified by treatment with various reagents [42, 43]. In this way, the surface is coated with new species, either by chemical bonding or by adsorption. Chelating groups bound on silica gel have proved to be useful for selective separations [44–46].

Hydrous titanium dioxide has found great interest in recent years for selective sorption of uranium from sea water [47–49]. Besides the amorphous $TiO_2 \cdot xH_2O$, 3 crystalline forms of TiO_2 are known, rutile, anatase and brookite. Hydrous titanium dioxide is amphoteric and stable in the pH range between 1 and 12. The isoelectric point varies between 3.5 and 6.5 [29]. For alkali ions the affinity decreases in the sequence $Cs^+ > Rb^+ > K^+ > Na^+$. In alkaline solutions the sorption of Sr^{2+} leads to the formation of a crystalline compound of composition Sr : Ti = 1 : 1 [2], corresponding to $SrTiO_3$. Several multivalent anions, in particular phosphate, are bound very strongly. Sorption capacities up to several mmol/g have been measured, depending on the methods of preparation and on the conditions. On the basis of kinetic studies [50, 51] effective diffusion coefficients for gel diffusion of about $4 \cdot 10^{-8}\,cm^2s^{-1}$ for Cu^{2+} and about $1 \cdot 10^{-8}\,cm^2s^{-1}$ for UO_2^+ have been calculated.

Hydrous zirconium dioxide behaves also as an amphoteric ion exchanger. Amorphous $ZrO_2 \cdot xH_2O$ and two crystalline forms are known. The isoelectric point of hydrous zirconium dioxide shows a rather broad variation between about 4 and 11 [52, 53]. Similar to hydrous titanium dioxide, hydrous zirconium dioxide is stable in the pH range between 1 and 12. The cation exchange behaviour is similar to that of $TiO_2 \cdot xH_2O$, but the distribution coefficient for uranyl ions is appreciably smaller. With respect to binding of anions, the selectivity for phosphate ions is rather high [1]. Furthermore, quantitative separation of I^-, Br^- and Cl^- ions has been reported [54].

Hydrous manganese dioxide is also often used as an ion exchange material. Various compounds obtained in the system MnO_2/water have structures close to α-, β-, γ- or δ-MnO_2. β-$MnO_2 \cdot xH_2O$ shows only cation exchange behaviour and is stable in the pH range between 2 and 12. In applying hydrous manganese dioxide it has to be taken into account that MnO_2 is a strongly oxidizing agent which oxidizes concentrated HCl and may also oxidize other species. Various separations on $MnO_2 \cdot xH_2O$ have been reported in the literature [2].

With respect to their use as ion exchangers hydrous cerium dioxide and hydrous thorium dioxide are very similar to the analogous compounds of Ti and Zr. Both exhibit cation and anion exchange properties [5]. They are stable in acid solutions down to pH 1 to 2 and in alkaline solutions. $CeO_2 \cdot xH_2O$, however, may act as an oxidizing agent. The selectivity for cations is similar to that found for $ZrO_2 \cdot xH_2O$ (e. g. $Cs^+ > Rb^+ > K^+ > Na^+$) and the selectivity for phosphate ions is high. From hydrous ThO_2 and CeO_2 membranes have been made for the demineralization of saline water [55]. Other hydrous oxides of tetravalent elements have also been proposed for the same purpose [2].

Hydrous tin dioxide exists as α-$SnO_2 \cdot xH_2O$ and β-$SnO_2 \cdot xH_2O$. Freshly prepared α-$SnO_2 \cdot xH_2O$ is soluble in acid solutions, whereas β-$SnO_2 \cdot xH_2O$ is relatively stable in concentrated nitric and sulfuric acid solutions. The compounds are amphoteric. They dissolve in alkaline solutions above pH 10. Hydrous tin dioxide can behave either like a cation or an anion exchanger. The selectivity for divalent transition metal ions and uranyl ions is relatively high and the selectivity sequence $Cu^{2+} > Zn^{2+} > Co^{2+} \geqslant Fe^{2+} > Ni^{2+} > Mn^{2+}$ resembles the order of the hydrolysis constants for the formation of the monohydroxo complexes [56]. For anions the following selectivity series has been reported: $PO_4^{3-} > C_2O_4^{2-} > SO_4^{2-} > Cr_2O_7^{2-} > [Fe(CN)_6]^{4-} > [Fe(CN)_6]^{3-} > Cl^- > MnO_4^- > Br^- > I^-$ [57].

1.7.2.5 Pentavalent elements

Hydrous antimony(V) oxide is the most important representative of this group. It is obtained by precipitation as an amorphous or glassy material which is gradually transformed into a crystalline solid by prolonged aging. Hydrous antimony(V) oxide is rather acidic and correspondingly the isoelectric point is very low (≈ 0) [29]. It is a cation exchanger and exhibits a high selectivity for Na^+. This property is easily understood on the basis of the conception presented in Section 1.7.1.2: sodium ions form a sparingly soluble antimonate(V).

The properties of $Sb_2O_5 \cdot xH_2O$ (antimonic acid) have been described in detail by Clearfield [5]. In the literature hydrous antimony(V) oxide is often called hydrated antimony pentoxide (HAP). The mole ratio $H_2O : Sb_2O_5$ in the various samples varies between about 3 and 6, depending on the conditions of preparation. For the crystalline antimonic acid a structure has been derived consisting of $Sb(OH)_6^-$ octehedra which are linked together. The solubility decreases with aging and degree of crystallization. The following selectivity sequences are reported in acid media: for alkaline ions $Na^+ > K^+ > NH_4^+ > Rb^+ > Cs^+ > Li^+$ and for alkaline earth ions $Ba^{2+} > Sr^{2+} > Ca^{2+} > Mg^{2+}$ [58]. The properties of HAP and sometimes also the selectivity depend on the preparation and on the treatment before use and the solubility decreases with the time and temperature of aging. In strongly acid solution transformation into crystalline material is observed. HAP is of special interest for the separation of Na^+ and Ag^+, even from concentrated mineral acid solutions [59] and from salt solutions, like sea water.

Hydrous bismuth(V) oxide is of little practical importance as an ion exchanger because of the low stability of the oxidation state +5 of bismuth.

Hydrous niobium(V) oxide and hydrous tantalum(V) oxide are obtained by precipitation as gel-like products. They are stable in acid and in alkaline solutions up to about pH 12. Like hydrous antimony(V) oxide, both show only cation exchange properties. The acidity decreases in the series $Sb_2O_5 \cdot xH_2O > Ta_2O_5 \cdot xH_2O > Nb_2O_5 \cdot xH_2O$.

1.7.2.6 Hexavalent elements

Hydrous molybdenum(VI) oxide and hydrous tungsten(VI) oxide are found as monohydrates $MoO_3 \cdot H_2O$, $WO_3 \cdot H_2O$ (yellow) and as dihydrates $MoO_3 \cdot 2H_2O$, $WO_3 \cdot 2H_2O$. The compounds exhibit cation exchange properties. Hydrous molybdenum(VI) oxide is of very limited use as an ion exchange material because it starts to dissolve at pH > 4. Hydrous tungsten(VI) oxide dissolves in neutral and alkaline aqueous solutions. The yellow compound $WO_3 \cdot H_2O$ has a crystalline structure, whereas for the compound with higher water content, $WO_3 \cdot xH_2O$ ($x \geqslant 2$), amorphous and crystalline structures are reported. Divalent ions are sorbed onto hydrous tungsten(VI) oxide in the sequence Mn^{2+}, $Ni^{2+} > Ca^{2+} > Sr^{2+} > Cu^{2+}$, $Zn^{2+} > Co^{2+}$, $Ba^{2+} > Mg^{2+}$ [60].

More information on hydrous oxides, in particular on their preparation, may be found in the literature [2, 4, 5].

1.7.3 Acid salts

1.7.3.1 General properties

A survey on acid salts exhibiting ion exchange properties is given in Table 1.34. All compounds have low solubility. They are obtained by precipitation in the form of gel-like materials which may be transformed by aging and heating into more or less crystalline solids. The crystal structures of many substances have been determined, and in most cases layered structures have been found. Some have fibrous structures. In general, the grain size of the crystallites is small (0.1 to 1 μm), but in favourable cases single crystals of about 1 mm diameter could be prepared.

The composition of most substances is non-stoichiometric. The ratio cationic constituents : anionic groups : water varies within a relatively broad range and the exchange properties depend on the composition. Stoichiometric or nearly stoichiometric compounds can only be obtained by refined procedures.

Table 1.34 Acid salts as inorganic ion exchangers (compounds of greater importance are printed in heavy type)

Anionic groups carrying the exchangeable hydrogen	Cationic constituents
Silicates	Zr (130, 131)
Phosphates	**Ti** (5), **Zr** (5), Hf (5), Ge (5), Sn^{IV} (5), Pb (5), Lanthanides (77), Cr (78–81), Ce^{IV} (5), Th (5), UO_2^{2+} (5)
Arsenates	Ti (5), Zr (5), Sn (5), Ce^{IV} (5), Th (5)
Antimonates	Ti (89–93), Zr (86–88), Sn^{IV} (94–96), Ce^{IV} (97, 98), Fe^{III} (99), Cr^{III} (100, 101), Ta (102)
Molybdates	Ti (112, 113), Zr (103–110), Sn (122, 123), Ce^{IV} (125)
Tungstates	Ti (112, 113), Zr (103–110), Sn (122, 123)
Hexacyanoferrates	Al, **Ti** (152–154), Fe (155), Co (140–148), Ni (140–148), Cu (140–148), Zn (140–148), Sb (161), **Mo** (149–151), Ag, Cd, Zr (156), Sn^{II}, Sn^{IV} (157–160), W, Pb, Bi

The cation exchange properties of all acid salts listed in Table 1.34 are due to the presence of hydrogen atoms bound to the anionic groups, as in HPO_4^{2-}, $HAsO_4^{2-}$ or $HFe^{II}(CN)_6^{3-}$. The acidity of these hydrogen atoms is very low. They are readily exchanged for other cations, preferably monovalent, but also divalent ions. Compounds in which the hydrogen in the acid salt is substituted by alkali ions may exchange these alkali ions for other cations.

The selectivity depends on the size of the hydrated cations in relation to the size of the cavities in the structure of the acid salts and the hydration energy of the cations, because in most cases these have to strip off at least a part of their hydration shell in order to fit into the cavities.

1.7.3.2 Phosphates

The properties and the exchange properties of zirconium phosphate (ZrP) have been investigated in great detail, and there are several overviews of this subject in the literature [1–5]. Several forms of ZrP have been investigated, amorphous ZrP of variable composition and the crystalline modifications α-ZrP, γ-ZrP, ζ-ZrP, η-ZrP and θ-ZrP. The amorphous products can be described by the formula $Zr(OH)_x(HPO_4)_{2-x/2} \cdot yH_2O$. The crystalline phases have a layered structure. The formula is $Zr(HPO_4)_2 \cdot nH_2O$ with $n = 1$ for α-ZrP, $n = 2$ for γ-ZrP, $n = 0$ for ζ- and η-ZrP and $n \approx 6$ for θ-ZrP. The idealized structure of α-ZrP is a combination of octahedra and tetrahydra. Zr is linked by six oxygen atoms to P and P is bound by three oxygen atoms to Zr, while the fourth oxygen atom carries the exchangeable hydrogen atom at the surface of the layer.

The exchange properties depend strongly on the method of preparation and the degree of crystallinity. At low loading, the sequence of distribution coefficients (sequence of selectivity) is $Cs^+ > Rb^+ > K^+ > Na^+ > Li^+$. The distribution coefficients for Cs^+ and Na^+ differ by about two orders of magnitude at maximum. On a strongly acidic cation exchange resin they differ by about one order of magnitude. Thus, the selectivity of the inorganic ion exchanger ZrP for the separation of alkali ions is appreciably higher. The sequence changes with the degree of loading. By quantitative loading of α-ZrP with alkali ions solid phases are obtained which contain higher amounts of water. They can be characterized by the spacing of the layers. As the effective diffusion coefficients in α-ZrP are very small ($D_{eff} \approx 10^{-12}$ to $10^{-13}\ cm^2s^{-1}$) [5], the practical use of α-ZrP is restricted to small particles of the order of 1 μm or less.

It should be mentioned that intercalation compounds are also formed by the layered structures of ZrP like those described for layered silicates.

Titanium phosphate (TiP) is similar in its properties to ZrP. α-TiP has the same structure and the same spacing of the layers as α-ZrP, but due to the smaller distance Ti-O the dimensions of the windows connecting the cavities are somewhat smaller in α-TiP and therefore the diffusion of larger cations is hindered. Na^+ ions are taken up more readily than Cs^+ ions, giving rise to the formation of $TiHNa(PO_4)_2 \cdot nH_2O$ and $Ti(NaPO_4)_2 \cdot nH_2O$ [61]. n may vary between $n = 1$ and 4, and the interlayer distance increases accordingly. In contrast to ZrP, TiP is not stable in alkaline solutions. The properties of γ-TiP ($Ti(HPO_4)_2 \cdot 2H_2O$) are analogous to those of γ-ZrP [62, 63]. The selectivity with respect to alkali ions is the same as that of γ-ZrP. The large layer spacing of γ-TiP also permits the exchange of Ca^{2+} and Sr^{2+}. Fibrous TiP was also prepared [5].

The properties of hafnium phosphate (HfP) are practically identical with those of ZrP (α-HfP $\simeq$ α-ZrP). The exchange behaviour of tin(IV) phosphate (SnP) has also been studied and α-SnP ($Sn(HPO_4)_2$) has been prepared [64]. As SnP readily decomposes in alkaline medium, the use as ion exchanger is very restricted. The same holds for $Si(HPO_4)_2$, $Ge(HPO_4)_2$ and $Pb(HPO_4)_2$ which are even more easily hydrolyzed in water.

Cerium(IV) phosphate and thorium phosphate have fibrous structures [5] or they are amorphous. The fibrous acid salts are of practical interest, because they may be used to prepare ion exchange papers, thin layers [65] or membranes [66, 67]. The structure of the fibers is not yet known. The composition corresponds to the formulas $Ce(HPO_4)_2 \cdot nH_2O$ and $Th(HPO_4)_2 \cdot nH_2O$ ($n \approx 1$ to 2). At low loading the selectivity sequences are $Cs^+ > K^+ > Na^+ > Li^+$ and $Ba^{2+} > Sr^{2+} > Ca^{2+} > Mg^{2+}$ [5]. Changes in these sequences are observed at higher loading. A rather high selectivity for other cations like Pb^{2+}, Cu^{2+}, Ag^+ and Tl^+ is found [68]. Microcrystalline cerium phosphates with mole ratios Ce : P = 1.15 to 1.55 and crystalline cerium(IV) phosphate sulfates have also been prepared [69–72]. The latter exhibit sieve properties with the selectivity sequence $Na^+ > Ag^+ > Sr^{2+} > Ba^{2+} > Cs^+ > Ca^{2+}$ [71, 73].

Uranyl phosphate is found in nature as autunite and can be prepared easily. Three types of phosphates are known, $UO_2(H_2PO_4)_2 \cdot nH_2O$, $UO_2(HPO_4) \cdot nH_2O$ and $(UO_2)_3(PO_4)_2 \cdot nH_2O$, but only $UO_2(HPO_4) \cdot nH_2O$ (HUP) has substantial ion exchange properties [5]. It is a crystalline compound with a layered structure in which the acidic H^+ ions are easily accessible and may be exchanged for other cations. HUP is soluble in phosphoric acid. It has favourable properties for use in columns. The selectivity sequence for alkali ions is $NH_4^+ > Rb^+ > Cs^+ > K^+ \gg Na^+$ [74]. NH_3 is taken up readily from the atmosphere [75]. Separation of the heavier alkali ions from Na^+ is easily achieved [74]. The capacity is in agreement with the theoretical value calculated from the formula $UO_2(HPO_4) \cdot nH_2O$ [76]. From experimental data [74] half-times of exchange of the order of several minutes for particle sizes of about 0.1 mm are calculated, which correspond to effective diffusion coefficients for interdiffusion of the order of $10^{-7}\ cm^2\ s^{-1}$.

The exchange properties of aluminium phosphate, rare earth phosphates [77], chromium phosphates [78–80] and several polyphosphates [81, 82] have also been studied, but the compounds are not well defined. Chromium tripolyphosphate is highly selective towards alkali ions in the order $Cs^+ > Rb^+ > K^+ > Na^+$ [81].

1.7.3.3. Arsenates

In general, the properties of the arsenates are similar to those of the phosphates, and the exchange properties of the arsenates have often been investigated parallel to those of the corresponding phosphates.

α-Zirconium arsenate, $Zr(HAsO_4)_2 \cdot H_2O$, has the same structure as α-ZrP and the ion exchange properties are nearly identical with the exception that Ba^{2+} ions are exchanged by the arsenate but not by the phosphate [83]. This is explainable by small changes in the lattice distances.

α-Titanium arsenate, $Ti(HAsO_4)_2 \cdot H_2O$, is less stable than α-TiP and undergoes hydrolysis, even in neutral solutions. Therefore, ion exchange is restricted to acid media ($pH \leqslant 5$). Fibrous titanium arsenate has also been prepared [5].

Tin(IV) arsenate, $Sn(HAsO_4)_2 \cdot H_2O$, like tin phosphate, is hydrolyzed in alkaline solution and therefore also of restricted use as an ion exchanger. Thorium arsenate, $Th(HAsO_4)_2$, has very pronounced ion sieve properties and is highly selective for Tin(IV) Li^+; other ions are not taken up [84]. In the system cerium(IV)/arsenic acid several crystalline products have been prepared with As : Ce ratios from 1 : 1 to 4 : 1 [85], like in the system cerium(IV)/phosphoric acid. Most of these substances are rather unstable in alkaline, and even in neutral solutions. The yellow compound with formula $Ce(HAsO_4)_2 \cdot 2H_2O$, however, is stable up to pH 10 [85] and may be used as a cation exchanger.

1.7.3.4 Antimonates, molybdates, tungstates and others

Ion exchange properties have been found for a great number of amorphous or microcrystalline antimonates, molybdates, tungstates and other compounds. In principle, the exchange properties are comparable to the properties of the phosphates.

Zirconium antimonate gels with mole ratios Sb : Zr ≈ 2 show high selectivity for Cs^+ and Sr^{2+} [86, 87]. In other work, the selectivity sequence $Na^+ > K^+ > NH_4^+ > Rb^+ > Cs^+ > Li^+$ has been reported for crystalline and amorphous zirconium antimonate [88]. This may be due to the presence of larger amounts of hydrous antimony oxide.

Titanium antimonate gels and semicrystalline compounds of various composition (Sb : Ti ≈ 1 to 3.6) were investigated with respect to capacities, distribution coefficients and application for separation of cations [89–93]. The exchange behaviour of tin(IV), cerium(IV), iron(III), chromium(III) and tantalum antimonate was also studied [94–102].

Zirconium molybdate and zirconium tungstate gels with mole ratios Mo or W : Zr ≈ 2 were investigated in great detail [103–110]. The crystalline compound of formula $Zr\ Mo_2O_7(OH)_2 \cdot 2H_2O$, however, is decomposed in alkaline media into molybdate and hydrous zirconium dioxide [111]. The structure of the crystals consists of a three-dimensional network [111].

Titanium molybdate and tungstate gels with mole ratios Mo or W : Ti $\approx$ 1 to 2 were also prepared and high selectivity of titanium tungstate for Pb^{2+} was reported [112, 113]. Both compounds were used for separation of various cations, in particular by paper chromatography [90, 114–121]. High selectivity for Pb^{2+} was also found in the case of tin(IV) molybdate and low exchange capacity for tin(IV) tungstate [122, 123]. Tin(IV) molybdate was applied for chromatographic separations [124]. Cerium(IV) molybdate is also reported to bind Pb^{2+} strongly [125]. The molybdates and tungstates of thorium and of chromium(III) were used for various separations of cations [126–129].

Zirconium silicate is relatively stable and has high ion exchange capacity [130, 131]. Mixed zirconium phosphate – silicate with mole ratio Zr : Si : P $\approx$ 1 : 4 : 0.8 was found to be selective for Pu separation [132]. Zirconium selenite and tellurate, $Zr(H_4TeO_6)_2 \cdot 4H_2O$, have also cation exchange properties. In the latter, one proton is exchangeable [133–135]. Amorphous titanium selenite was used for selective separations of Cd^{2+} [90, 119, 136], tin(IV) selenide for separation of various cations. Titanium vanadate was reported to be selective for Sr^{2+} [139].

1.7.3.5 Hexacyanoferrates

Selective sorption of Cs^+ was first reported for the hexacyanoferrate(II) of nickel [140], and the exchange properties of the hexacyanoferrates of a great number of elements (Al, Ti, Fe, Co, Ni, Cu, Zn, Sb, Mo, Ag, Cd, W, Pb and Bi) have been studied. Stabilities and exchange capacities vary according to the nature of the compounds and the method of preparation. But, in principle, the exchange properties of this group of compounds are very similar. In most cases very pronounced differences in the distribution coefficients for alkali ions have been found: $Cs^+ > Rb^+ > K^+ > Na^+$. At low loading, the selectivity as defined by eq. (16) is $S(Cs^+/Na^+) \geqslant 3$, whereas it is only $\leqslant 2$ for ZrP and appreciably smaller for hydrous oxides. Therefore, the main interest of these compounds is the separation or enrichment of heavy alkali ions. The distribution coefficients for Ag^+ and for alkaline earth ions are also rather high. The selectivity sequence of the latter is $Ba^{2+} > Sr^{2+} > Ca^{2+} > Mg^{2+} > Be^{2+}$. Exchange capacities are in the range of 1 to 4 meq/g, depending mainly on the method of preparation. The rate of exchange into the inner parts of the particles is relatively slow, as in the case of other acid salts.

The chemical stability of hexacyanoferrates(II) is limited. Whereas all compounds are relatively stable in acid media, hydrolysis is observed in neutral and alkaline solutions. It begins at lower pH for aluminum hexacyanoferrate(II) and at higher pH for the hexacyanoferrates(II) of titanium, zirconium and heavy metals. Due to hydrolysis and/or dissolution increasing amounts of hexacyanoferrate(II) ions are found in solution. In alkaline media all hexacyanoferrates (II) are decomposed into soluble hexacyanoferrates and hydrous oxides. Furthermore, hexacyanoferrate(II), preferably the dissolved species, is sensitive to oxidation into hexacyanoferrate(III).

The exchange properties of the hexacyanoferrates(II) of heavy metals like Zn, Cu, Ni, Co [140–148], of Mo (MoFe) [149–151] and of Ti (TiFe) [152, 154] were studied most thoroughly. It was shown that the properties vary with composition. For MoFe a structure was proposed [150]. The compound with the mole ratio Mo : Fe $\approx$ 2.4 was found to have most favourable properties [151]. The composition of gel-like precipitates of TiFe corresponded to the formula $(TiO)_2OH[HFe(CN)_6] \cdot 4H_2O$ and it was shown that the acidic hydrogen is exchanged for other cations [152]. All these compounds exhibited the above-mentioned selectivity for the heavy alkali ions in the sequence $Cs^+ > Rb^+ > K^+ > Na^+$, with a high preference for Cs^+ ions. It is assumed that unhydrated Cs^+ ions fit best into the cavities of the network [150]. Several hexacyanoferrates also showed high selectivity for Ag^+ ions [147, 148].

High selectivity for Cs^+ ions was also found for thin surface layers of iron hexacyanoferrate which were used in a $^{137}Cs/^{137m}Ba$ radionuclide generator [155]. Whereas Cs^+ ions are bound very strongly, Ba^{2+} ions can be eluted very fast and easily. Zirconium hexacyanoferrate(II) was used for the separation of Zn^{2+} from other divalent ions, for separation of Ca^{2+} and Ba^{2+} and for the separation of UO_2^{2+} from Th^{4+} and Zr^{4+} [156]. Tin(IV) and tin(II) hexacyanoferrates(II) were applied for various separations and the following selectivity sequences are reported for the tin(IV) compound: $Li^+ > K^+ > Na^+ > NH_4^+$ and $Ba^{2+} > Ca^{2+} > Mg^{2+}$ [157–160]. The hexacyanoferrate of antimony showed high selectivity for Sr^{2+} [161]. Comparison of the distribution coefficients of Cs^+ and Sr^{2+} ions for various hexacyanoferrates and phosphates in water (pH 7), 1 M HNO_3 and 1 M salt solutions [162] leads to the conclusion that the distribution coefficients of the hexacyanoferrates are 1 or 2 orders of magnitude higher than those of the phosphates.

1.7.4 Salts of heteropoly acids

1.7.4.1 General properties

The number of compounds with ion exchange properties in this class is much smaller than in the groups of the hydrous oxides and acid salts. Low solubility is a prerequisite for use as an ion exchanger. It is found only for some salts of heteropoly acids, whereas the free acids are very soluble. The salts formed with larger monovalent or divalent cations, such as K^+, NH_4, NR_nH_{4-n+}, Ag^+, Tl^+, Sr^{2+}, Ba^{2+}, Hg^{2+}, Pb^{2+} are sparingly soluble. In most cases, the ammonium compounds are used.

Table 1.35 gives a survey of the salts of heteropoly acids exhibiting ion exchange properties. Ammonium molybdophosphate (AMP) has been most thoroughly studied. It is a sparingly soluble crystalline yellow compound with the composition $(NH_4)_3[P(Mo_3O_{10})_4] \cdot nH_2O$ in which the phosphor atom is surrounded tetrahedri-

Table 1.35 Salts of heteropoly acids as ion exchangers

Heteropoly acid	Cations	Application
Molybdo-phosphate	(a) NH_4^+ (3, 163–169), 8-hydroxyquinolinium (3, 166) (b) K^+, Rb^+, Cs^+ (3, 168), $CH_3NH_3^+$, $(CH_3)_2NH_2^+$, $(CH_3)_3NH^+$, $(CH_3)_4N^+$ ($(Ch_3)_4N^+$) (3, 168)	(a) Separation Na/K/Rb/Cs (b) Separation Cs/Sr/Y
Tungsto-phosphate	NH_4^+ (3, 170–172), Tl^+ (3)	Separation Na/K/Rb/Cs
Molybdo-arsenate	NH_4^+ (3, 166, 173)	Separation Na/K/Rb/Cs
Tungsto-arsenate	(a) NH_4^+ (3, 166, 173) (b) Sn^{IV} (Sn : W : As = 12 : 5 : 2) (4)	(a) Separation Na/K/Rb/Cs (b) Separation Cu/Ni, Mg; Ba/Mg
Molybdo-silicate	NH_4^+ (3, 166, 173), 8-hydroxyquinolinium (3, 174), pyridinium (3, 174)	Separation Na/K/Rb/Cs
Tungsto-silicate	NH_4^+ (3, 166, 173)	Separation Na/K/Rb/Cs
Molybdo-germanate	NH_4^+ (3, 174), 8-hydroxyquinolinium (3, 174), pyridinium (3, 174)	Separation Na/K/Rb/Cs; ^{137}Cs/other fission products

cally by four groups of three MoO_6-octohydra, which are in turn connected via one or two oxygen atoms. The ammonium ions and water molecules are located between these $[P(Mo_3O_{10})_4]^{3-}$ complexes, and the ammonium ions may be exchanged for other, preferably monovalent, cations. Ammonium tungsto-phosphate (ATP) is isomorphous with AMP. All other compounds of heteropoly acids have similar structures. Variation with respect to the composition of the heteropoly acids, including mixed heteropoly acids, and variation of the cations, are possible. The compounds may be used for batch separation, in chromatographic columns, as thin layers or may be impregnated onto paper. The effective diffusion coefficients within the structures of the salts of heteropoly acids are rather low. On the other hand, the particle sizes of the crystals are, in general, very small, so that equilibration is achieved within reasonable times.

All salts of heteropoly acids are decomposed in alkaline solutions, AMP for instance into phosphate and molybdate. In strongly acid solutions the cations are exchanged for hydronium ions. If the water content is high, the compounds begin to melt in the temperature range of 40 to 100 °C and water is given off.

1.7.4.2 Ammonium molybdophosphate

Ammonium molybdophosphate (AMP) is obtained in the form of a yellow microcrystalline powder of composition $(NH_4)_3[P(Mo_3O_{10})_4] \cdot 5H_2O$. It has an appreciable oxidation potential [163], comparable to that of chromium(VI). Therefore, reducing species lead to the formation of lower oxidation states of Mo with their characteristic blue colour. For monovalent cations the following selectivity sequences were found: $Cs^+ > Rb^+ > K^+ > Na^+$ and $Tl^+ > Ag^+$ [164–166]. The selectivity as defined by eq. (16) for the separation of the alkali ions is appreciably higher than that found for phosphates (Section 1.7.3.2) and comparable to that for hexacyanoferrates(II) (Section 1.7.3.5). It was also observed that the ion exchange between different cations is governed by the law of mass action [167].

The sorption of di- and trivalent ions is more complex. However, mixtures of mono-, di- and trivalent cations (for instance Cs^+, Sr^{2+}, Y^{3+}) can be separated in a chromatographic column by application of various eluents [168]. Furthermore, separation of macro quantities of monovalent cations (Na^+, K^+, Rb^+, Cs^+) [167], divalent cations (Sr^{2+}, Ba^{2+}, Cd^{2+}, VO^{2+}) [169] and trivalent cations (Y^{3+}, In^{3+}) [169] by use of AMP was studied.

1.7.4.3 Salts of other heteropoly acids

The ion exchange properties of ammonium tungstophosphate (ATP) are very similar to those of AMP. Both have the same structure, but ATP is white and less sensitive towards reducing agents. Furthermore, ATP is reported to be more stable in strong acids and to have somewhat higher capacity for Cs^+ [170]. On the other hand, the distribution coefficient of Cs^+ is apparently somewhat higher for AMP than for ATP [166, 171, 172].

The properties of potassium, rubidium, cesium and the alkyl ammonium molybdophosphates were found to be approximately the same as those of AMP [168, 170]. Molybdoarsenates and tungstoarsenates [166, 173] have no advantages with respect to the corresponding phosphates. This also holds for the germanates [174]. Ammonium tungstoarsenate, ammonium molybdosilicate and ammonium tungstosilicate are rather unstable in aqueous solutions and easily form colloidal sols [166]. These compounds are therefore of very limited use as ion exchangers.

1.7.5 Other ionic compounds

1.7.5.1 General properties

Ion exchange is not only shown by the salts of heteropoly acids which have been discussed in the previous section, but by many other ionic compounds of low solubility, mainly those of much simpler composition, such as sparingly soluble sulfates, sulfides, halides and others. An overview is given in Table 1.36.

All exchange reactions on ionic compounds can be described by eqs. (13) (exchange of cations) or (14) (exchange of anions). Three kinds of exchange may be distinguished:

- exchange in the surface layer
- exchange throughout the crystals and formation of mixed crystals
- exchange with formation of a new phase.

Table 1.36 Overview of other sparingly soluble inorganic ionic compounds with ion exchange properties

Class of compounds	Examples of application
Sulfates	Separation of Ra, separation of radioactive Sr (13, 14, 17, 21); separation of chromate
Halides	(a) chlorides or bromides: separation of I (18–20) (b) fluorides: separation of Ca or Sr
Sulfides	Separation of traces of heavy metals (Ag, Cu, Hg) (182, 183), separation of Tc (184)
Carbonates	Separation of alkaline earths (16)
Oxalates	Separation of alkaline earths
Chromates	–
Phosphates*	–
Perchlorates	Separation of Fr^+, Cs^+ or Rb^+ (15, 17)
Tetraphenylborates	Separation of Fr^+, Cs^+ or Rb^+ (15, 17)

* Neutral phosphates of mono- and divalent elements

Exchange in the surface layer is observed if cations or anions have similar properties [13−17, 21, 22]. It is rather fast [22, 175, 176]. The exchange equilibrium depends on the relative solubilities of the compounds formed [13−16, 21]. The exchange capacity is proportional to the specific surface area and therefore inversely proportional to the size of the particles. The exchange capacity of the surface is sufficient for separation of microamounts of ions, in particular trace elements or radionuclides, but not sufficient for separation of macroamounts of ions.

Exchange throughout the crystals takes place if mixed crystals are formed [13−15, 21, 22, 177]. The equilibrium distribution of the ions between solid and solution depends on the relative solubilities. In this connection it has to be taken into account that the partial solubilities of the components of mixed crystals change continuously with the composition. In general, the exchange proceeds by recrystallization of the solid in the solution, not by solid diffusion [177, 178]. Therefore the exchange rate is found to be identical with the rate of recrystallization. As the latter depends on the grain size and the degree of disorder of the solid particles, the exchange rate is very small for large and well-crystallized particles, and equilibration between solid and solution may take many months.

Exchange with formation of a new phase is observed if the solubility of the compound formed by the exchange reaction is much smaller than that of the original solid and the structures of the compounds are different. In this case the thorough exchange proceeds by dissolution and reprecipitation, similar to recrystallization. This kind of exchange depends primarily on the size of the original crystals and the difference in solubilities and may proceed rather quickly (within several minutes).

Because in these systems ion exchange depends strongly on the solubilities and on the ability to form mixed crystals, very high selectivities are observed [20].

1.7.5.2 Sulfates

Heterogeneous ion exchange on sulfates has been thoroughly studied [13, 14, 17, 21]. The solubilities decrease in the sequence $CaSO_4 \cdot 2H_2O > SrSO_4 > PbSO_4 > BaSO_4 > RaSO_4$. Correspondingly, the equilibrium constants of the ion exchange equilibria

$$M_1SO_4(s) + M_2^{2+}(aq) \rightleftharpoons M_2SO_4(s) + M_1^{2+}(aq)$$

are $K > 1$ if the solubility of M_2SO_4 is smaller than that of M_1SO_4. ^{90}Sr may be separated on solid $BaSO_4$ or $SrSO_4$ by ion exchange [13, 14]. For $BaSO_4$ the equilibrium constant is $K < 1$, but the partial solubility of $SrSO_4$ for $BaSO_4/SrSO_4$ mixed crystals is smaller than the solubility of solid $SrSO_4$. Ra^{2+} is enriched by ion exchange on $BaSO_4$ ($K > 1$), because the solubility of $RaSO_4$ is smaller than that of $BaSO_4$. Thus, interesting separations of trace amounts of M^{2+} ions forming sparingly soluble sulfates are possible by isotopic or non-isotopic ion exchange.

Ca^{2+} ions are not exchanged at the surface of $SrSO_4$ or $BaSO_4$ in appreciable amounts, because of the relatively high solubility of $CaSO_4 \cdot 2H_2O$, and they are not incorporated, because $CaSO_4 \cdot 2H_2O$ and $SrSO_4$ or $BaSO_4$ do not form mixed crystals [13, 14]. Therefore, the ion exchange reactions on sparingly soluble sulfates are very selective [20]. The selectivity for separation of Ba^{2+} and Sr^{2+} is $S \approx 2$, the selectivity for separation of Ba^{2+} from monovalent or trivalent cations is $S > 5$.

By anion exchange, chromate ions are exchanged for sulfate ions on solid sulfates, because chromates and sulfates are isomorphous and form mixed crystals. The equilibrium constants of the ion exchange reaction

$$MSO_4(s) + CrO_4^{2-}(aq) \rightleftharpoons MCrO_4(s) + SO_4^{2-}(aq)$$

depend on the relative solubilities of $MCrO_4$ and MSO_4. Furthermore, traces of phosphate or arsenate can be enriched on the surface of $BaSO_4$ or $SrSO_4$ [179, 180].

1.7.5.3 Halides

Silver halides exhibit anion exchange. The solubilities decrease in the sequence AgCl > AgBr > AgI. Therefore, Br^- ions are enriched by AgCl. As AgCl and AgBr are isomorphous, mixed crystals are formed. AgI is not isomorphous with AgCl and AgBr, but it has a markedly lower solubility. Therefore, quantitative transformation is observed [18, 19]:

$$AgCl(s) + I^-(aq) \rightarrow AgI(s) + Cl^-(aq)$$

This reaction proceeds within minutes and allows effective separation of I^- from aqueous solutions with very high selectivity ($S > 5$) [20, 181]. All other anions, except chalkogenide ions which form sparingly soluble Ag_2S, Ag_2Se and Ag_2Te, do not interfere.

The ion exchange properties of lead halides are similar to those of silver halides.

The solubility of fluorides is very different from that of other halides. Thus, di-, tri- and tetravalent cations form sparingly soluble fluorides. In the case of divalent ions, the solubilities decrease in the order $BaF_2 > PbF_2 > MgF_2 > SrF_2 > CaF_2$. Therefore, in contrast to the sulfates, Ca^{2+}, Sr^{2+} and even Mg^{2+} ions are enriched on BaF_2.

1.7.5.4 Sulfides

The solubility of the sulfides of divalent metals decrease in the order MnS > FeS > NiS > CoS > ZnS > CdS > SnS > PbS > CuS > HgS. Accordingly, metals forming sulfides of lower solubility are enriched on sulfides of higher solubility by ion exchange. Separation can be enhanced by use of freshly prepared sulfides of high specific surface area. If the sulfides are precipitated in the presence of an excess of sulfide ions, surplus sulfide ions may be adsorbed on the surface and react with metal ions. In this way, very effective separations of traces of metals from solutions are possible, either by shaking the solutions with freshly precipitated sulfides or by filtering them through filter layers of sulfides. The separation may be due to ion exchange and/or chemical transformation. Thus, traces of Ag^+, Cu^{2+} or Hg^{2+} may be separated by use of filter layers of ZnS, MnS, CuS or PbS [182, 183] and Ag^+ or TcO_4^- [184] may be separated on CdS or ZnS because the sparingly soluble compounds Ag_2S or Tc_2S_7, respectively, are formed by exchange on the surface or by chemical reaction. The lower the solubility, the more effective is the separation.

1.7.5.5 Other compounds

The examples of ion exchange on ionic compounds are quite numerous. Carbonates [16] are not of great interest for separations in the laboratory, because they are dissolved in acid. Furthermore, they react with CO_2 to form soluble hydrogen carbonates and their solubility in water is not very much differentiated. The ion exchange properties of sparingly soluble oxalates and chromates have not been investigated in detail. Acid phosphates of tetra-, tri- and pentavalent elements exhibiting exchange of the acidic hydrogen have been discussed in Section 1.7.3. Neutral sparingly soluble phosphates of mono- and divalent elements such as $LiPO_3$, $Li_3PO_4 \cdot 1/2H_2O$ or $Ca_3(PO_4)_2$ and others are able to exchange cations, but they have not yet found practical application. Apatite shows cation and anion exchange.

High selectivity for the heavy alkali ions is observed in the case of sparingly soluble perchlorates and tetraphenylborates [15]. The solubilities decrease in the order $NaClO_4 \gg KClO_4 > RbClO_4 > CsClO_4$ and $NaBPh_4 \gg KBPh_4 > RbBPh_4 > CsBPh_4$. As the perchlorates as well as the tetraphenylborates do not form sparingly soluble compounds with other cations, the selectivity for separation of Fr^+, Cs^+ or Rb^+ from aqueous solutions is very pronounced [15, 17].

The sparingly soluble compounds formed with organic complexing agents are mentioned only to complete the picture. An example are the oxinates [46, 185–189], which exchange metal ions and, in agreement with the general concept, enrich the ions in the inverse order of the solubilities of the oxinates. This behaviour is representative for the great number of organic compounds which are able to exchange cations. They may be used in form of powders, as thin filters or supported on an organic or inorganic matrix.

References

[1] C. B. Amphlett (1964) Inorganic Ion Exchangers, Elsevier, Amsterdam.
[2] V. Veselý and V. Pekárek (1972) Synthetic Inorganic Ion Exchangers, I. Hydrous Oxides and Acidic Salts of Multivalent Metals, Talanta *19*, 219.

[3] A. Clearfield, G. H. Nancollas and B. H. Blessing (1973) New Inorganic Ion Exchangers, in: Ion Exchange and Solvent Extraction (J. A. Marinsky and Y. Marcus, eds.), Vol. 5, Marcel Dekker, New York.
[4] A. K. De, A. K. Sen (1978) Synthetic Inorganic Ion Exchangers, Sep. Science and Technol. *13*, 517.
[5] A. Clearfield (1982) Inorganic Ion Exchange Materials, CRS Press, Boca Raton, Florida.
[6] K. H. Lieser (1977) Nicht-silikatische Anorganische Ionenaustauscher, in: Ullmanns Encyklopädie der Technischen Chemie, 4. Aufl., Bd. 13, p. 294, Verlag Chemie, Weinheim.
[7] R. Fricke, H. Schmäh (1948) Z. Anorg. Chemie *255*, 253.
[8] H. Kautsky, H. Weßlau (1954) Z. Naturforschg. *9* b, 569.
[9] H. W. Kohlschütter, H. Getrost and S. Miedtank (1961) Z. Anorg. Allg. Chem. *308*, 190.
[10] H. W. Kohlschütter, S. Miedtank and H. Getrost (1963) Fresenius Z. Anal. Chem. *192*, 381.
[11] K. H. Lieser, I. Loc and S. Quandt (1976) Radiochim. Acta *23*, 133.
[12] K. H. Lieser, S. Quandt and B. Gleitsmann (1979) Fresenius Z. Anal. Chem. *298*, 378.
[13] K. H. Lieser and W. Hild (1962) Radioisotopes in the Physical Sciences and Industry, Proceedings IAEA Vienna p. 338.
[14] W. Hild and K. H. Lieser (1964) Z. Anorg. Allg. Chem. *331*, 133.
[15] K. H. Lieser and A. B. H. Hecker (1964) Radiochim. Acta *3*, 96.
[16] K. H. Lieser (1964) Radiochim. Acta *3*, 93.
[17] K. H. Lieser, Ph. Gütlich, W. Hild, A. Hecker and I. Rosenbaum (1965) Exchange Reactions, Proceedings IAEA Vienna p. 385.
[18] W. Hild and K. H. Lieser (1964) Naturwiss. *1*, 12.
[19] K. H. Lieser and W. Hild (1967) Z. Anorg. Allg. Chem. *350*, 327.
[20] K. H. Lieser (1975) Fresenius Z. Anal. Chem. *273*, 189.
[21] K. H. Lieser and W. Hild (1963) Z. Anorg. Allg. Chem. *320*, 117.
[22] K. H. Lieser, Ph. Gütlich and I. Rosenbaum (1965) Exchange Reactions, Proceedings IAEA Vienna p. 375.
[23] K. H. Lieser (1969) Angew. Chem. *81*, 206; Internat. Ed. *8*, 188.
[24] T. A. Carlson, K. A. Kraus, J. S. Johnson Jr. and H. O. Phillips (1955) Adsorption of rare earths on inorganic materials, ORNL 1583, 203.
[25] K. A. Kraus and H. O. Phillips (1956) J. Am. Chem. Soc. *78*, 249.
[26] K. A. Kraus, H. O. Phillips, T. A. Carlson and J. S. Johnson Jr. (1958) Ion Exchange Properties of Hydrous Oxides, Proc. 2nd Intl. Conf. Peaceful Uses of Atomic Energy, Vol. 28, Paper No. 15/p/1832, United Nations, Geneva, 3.
[27] M. J. Fuller (1971) Chromatogr. Rev. *14*, 45.
[28] H.-D. Greiling and K. H. Lieser (1984) Radiochim. Acta *35*, 79.
[29] G. A. Parks (1965) Chem. Rev. *65*, 177.
[30] M. Abe and T. Ito (1965) Nippon Kagaku Zasshi *86*, 817.
[31] H. H. Stamm, H. W. Kohlschütter (1965) J. Inorg. Nucl. Chem. *27*, 2103.
[32] J. Doleźal, J. Horáćék, J. Šramek, and Z. Šulecek (1966) Microchim. Acta 38.
[33] G. A. Parks (1967) Adv. Chem. Ser. *67*, 121.
[34] R. W. Dalton, J. L. McClanahan and R. W. Maatman (1962) J. Coll. Sci. *17*, 207.
[35] H. Ti Tien (1965) J. Phys. Chem. *69*, 350.
[36] R. W. Maatman (1965) J. Phys. Chem. *69*, 3196.
[37] L. A. Allen and E. Matijevic (1970) J. Colloid Interface Sci. *33*, 421.
[38] K. Unger and F. Vydra (1968) J. Inorg. Nucl. Chem. *30*, 1075.
[39] J. Doležal, J. Horaček, J. Šrámek and Z. Šulcek (1966) Microchim. Acta 38.

[40] M. Sakanoue and M. Abe (1967) Radioisotopes (Tokyo) *16*, 645.
[41] M. R. Zaki, I. Abd-El-Moheim. Z. Anorg. Allg. Chem. *360*, 208 (1968); *365*, 325 (1969).
[42] K. Unger and K. Berg (1969) Z. Naturforsch. B *24*, 454.
[43] K. Unger, K. Berg, D. Nyamah and Th. Lothe (1974) Colloid Polym. Sci. *252*, 317.
[44] D. E. Leyden and G. H. Luttrell (1975) Anal. Chem. *47*, 1612.
[45] J. R. Jezorek and H. Freiser (1979) Anal. Chem. *51*, 366.
[46] R. E. Sturgeon, S. S. Bermann, S. N. Willie and J. A. Desaulniers (1981) Anal. Chem. *53*, 2337.
[47] R. V. Davies, J. Kennedy, R. W. McLroy, R. Spence and K. M. Hill (1964) Nature *203*, 1110.
[48] N. Ogata (1971) J. Atom. Energy Soc. Jpn. *13*, 121.
[49] M. Kanno (1977) Bull. Soc. Sea Water Sci. Jpn. *31*, 115.
[50] C. Heitner-Wirguin and A. Albu-Yaron (1965) J. Appl. Chem. (London) *15*, 445.
[51] F. Ambe, P. Burba and K. H. Lieser (1979) Fresenius Z. Anal. Chem. *295*, 13.
[52] G. A. Parks (1967) Advan. Chem. Soc. *67*, 121.
[53] C. B. Amphlett, L. A. McDonald and M. J. Redman (1958) J. Inorg. Nucl. Chem. *6*, 236.
[54] S. Tustanowski (1967) J. Chromatogr. *31*, 268.
[55] K. S. Rajan and A. J. Casolo (1969) U.S. Patent 3,479,266.
[56] J. D. Donaldson and M. J. Fuller (1968) J. Inorg. Nucl. Chem. *30*, 1083.
[57] J. D. Donaldson and M. J. Fuller (1970) J. Inorg. Nucl. Chem. *32*, 1703.
[58] J. Lefebre and F. Gaymard (1965) C. R. Acad. Sci., Paris, *260*, 6911.
[59] F. Girardi and E. Sabbioni (1968) J. Radioanal. Chem. *1*, 169.
[60] A. K. De and K. Chowdhury (1978) Chromatographia *11*, 586.
[61] G. Alberti, U. Costantino and M. L. Luciani Giovagnotti (1980) Gazz. Chim. It. *110*, 61.
[62] G. Alberti, U. Costantino and M. L. Luciano Giovagnotti (1979) J. Inorg. Nucl. Chem. *41*, 643.
[63] E. Kobayashi (1979) Bull. Chem. Soc. Jpn. *52*(5), 1351.
[64] U. Costantino and A. Gasperoni (1970) J. Chromatogr. *51* 289.
[65] G. Alberti, M. A. Massucci and E. Torracca (1967) J. Chromatog. *30*, 579.
[66] G. Alberti and U. Costantino (1968) Italian Applic., 36739 A/67; U.S. Patent 5,728,775.
[67] G. Alberti and U. Costantino (1976) Italian Applic., 52585 A/70; U.S. Patent 3,985,611.
[68] G. Alberti, M. Casciola, U. Costantino and M. L. Luciani Giovagnotti (1976) J. Chromatog. *128*, 289.
[69] K. H. König and E. Meyn (1967) J. Inorg. Nucl. Chem. *29*, 1153.
[70] K. H. König and E. Meyn (1967) J. Inorg. Nucl. Chem. *29*, 1519.
[71] K. H. König and G. Eckstein (1969) J. Inorg. Nucl. Chem. *31*, 1179.
[72] G. Alberti, U. Costantino, F. Di Gregorio, P. Galli and E. Torracca (1968) J. Inorg. Nucl. Chem. *30*, 295.
[73] K. H. König and G. Eckstein (1972) J. Inorg. Nucl. Chem. *34*, 3771.
[74] V. Pekarek and M. Benesova (1964) J. Inorg. Nucl. Chem. *26*, 1743.
[75] M. G. Shilton and A. T. Howe (1980) J. Solid State Chem. *34*, 137.
[76] V. Pekárek and V. Veselý (1965) J. Inorg. Nucl. Chem. *27*, 1151.
[77] Sw. Pajakoff (1968) Monatsh. Chem. *99*, 1400.
[78] L. Zsinka and L. Szirtes (1969) Radiochem. Radioanal. Lett. *2*, 257; L. Zsinka, L. Szirtes, and V. Stenger (1970) ibid. *4*, 257.
[79] A. Holroyd and J. E. Salmon (1956) J. Chem. Soc. 1956, 269.
[80] J. P. Redfern and J. E. Salmon (1961) J. Chem. Soc. 1961, 291.
[81] D. Betteridge and G. N. Stradling. J. Inorg. Nucl. Chem. *29*, 2652 (1967); *31*, 1507 (1969).

[82] E. Kobayashi and T. Goto (1970) Kogyo Kagaku Zasshi *73*, 692.
[83] E. Torracca, U. Costantino and M. A. Massucci (1967) J. Chromatogr. *30*, 584.
[84] G. Alberti and M. A. Massucci (1970) J. Inorg. Nucl. Chem. *32*, 1719.
[85] G. Alberti, U. Costantino, F. Di Gregorio and E. Torracca (1969) J. Inorg. Nucl. Chem. *31*, 3195.
[86] H. C. Phillips and K. A. Kraus (1962) J. Am. Chem. Soc. *84*, 2267.
[87] J. R. Feuga and T. Kikindai (1967) R. C. Acad. Sci., Paris, Ser. C *264*, 8.
[88] S. N. Tandon and J. Mathew (1975) J. Radioanal. Chem. *27*, 315.
[89] M. Qureshi and V. Kumar (1970) J. Chem. Soc. A 1970, 1488.
[90] M. Qureshi, N. Zehra, S. A. Nabi and V. Kumar (1973) Talanta *20*, 609.
[91] M. Qureshi and V. Kumar (1971) J. Chromatogr. *62*, 431.
[92] L. Kosta, V. Ravnik and M. Levskek (1970) Radiochim. Acta *14*, 143.
[93] J. S. Gill and S. N. Tandon (1974) J. Radioanal. Chem. *20*, 5.
[94] M. Qureshi, V. Kumar and N. Zehra (1972) J. Chromatogr. *57*, 351.
[95] M. Qureshi, K. G. Varshney and R. P. S. Rajput (1975) Anal. Chem. *47*, 1520.
[96] M. Qureshi, N. Zehra and S. A. Nabi (1976) Z. Anal. Chem. *282*, 136.
[97] J. S. Gill and S. N. Tandon (1972) Talanta *19*, 1355.
[98] J. S. Gill and S. N. Tandon (1973) Talanta *20*, 585.
[99] J. P. Rawat and D. K. Singh (1976) Anal. Chim. Acta *87*, 157.
[100] M. Qureshi, R. Kumar and H. S. Rathore (1972) Talanta *19*, 1377.
[101] J. Mathew and S. N. Tandon (1976) Chromatographia *9*, 235.
[102] M. Qureshi, J. P. Gupta and V. Sharma (1973) Anal. Chem. *45*, 1901.
[103] K. A. Kraus, H. O. Phillips, T. A. Carlson and J. S. Johnson (1958) Proceedings of the Second International Conference on Peaceful Uses of Atomic Energy, Geneva, 1958, Paper No. 15/P/1832 United Nations Vol. 28, p. 3.
[104] K. A. Kraus, T. A. Carlson and J. S. Johnson (1956) Nature *177*, 128.
[105] H. J. Riedel. Radiochim. Acta *1*, 32 (1962); Nukleonik. *5*, 48 (1963); Ber. Kernforsch. Jülich, 32 (1962).
[106] M. H. Campbell (1965) Anal. Chem. *37*, 252.
[107] W. J. Maeck, M. E. Kussy and J. E. Rein (1963) Anal. Chem. *35*, 2086.
[108] J. M. P. Cabral (1960) J. Chromat. *4*, 86.
[109] S. Ahrland, J. Albertsson, L. Johansson, B. Nihlgard and L. Nilsson (1964) Acta Chem. Scand. *18*, 707.
[110] S. Ahrland, J. Albertsson, L. Johansson, B. Nihlgard, and L. Nilsson (1964) Acta Chem. Scand. *18*, 1357.
[111] A. Clearfield and R. H. Blessing (1972) J. Inorg. Nucl. Chem. *34*, 2643.
[112] M. Qureshi and H. S. Rathmore (1969) J. Chem. Soc. A, 2515.
[113] M. Qureshi and J. P. Gupta (1969) J. Chem. Soc. A, 1755; Chech, 2620.
[114] J. P. Rawat and S. Mujitaba (1975) Sep. Sci. *10*, 151.
[115] M. Qureshi and J. P. Gupta (1970) J. Chem. Soc. A, 2620.
[116] M. Qureshi and J. P. Gupta (1971) J. Chromat. *62*, 439.
[117] M. Qureshi, W. Hussain and F. Khan (1971) Experimentia *27*, 607.
[118] M. Qureshi, K. G. Varshney and F. Khan (1972) Sep. Sci. *6*, 559.
[119] M. Qureshi, K. G. Varshney and S. K. Kabiruddin (1972) Can. J. Chem. *50*, 2071.
[120] W. Hussain and M. Gulabi (1971) Sep. Sci. *6*, 737.
[121] S. W. Hussain (1972) Analusis *1*, 314.
[122] M. Qureshi and J. P. Rawat (1968) J. Inorg. Nucl. Chem. *30*, 305.
[123] M. Qureshi and K. G. Varshney (1968) J. Inorg. Nucl. Chem. *30*, 3081.
[124] J. P. Rawat and P. Singh (1974) Ann. Chim. (Rome) *64*, 873.
[125] A. K. De and S. K. Das (1976) Sep. Sci. *11*, 183.
[126] M. Qureshi and W. Hussain (1970) J. Chem. Soc. A, 1970, 1204.

[127] M. Qureshi and S. A. Nabi (1971) J. Chem. Soc. A, 1971, 139.
[128] A. K. De and K. Chowdhury (1976) Talanta *23*, 137.
[129] M. Qureshi, R. Kumar and H. S. Rathore (1972) Talanta *19*, 1377.
[130] T. P. Tang, P. Sun and K. Chen. Hua Hsueh 1965, 33; C. A. *65*, 14474g (1966).
[131] T. Tam, P. Sun and K. Chen. Hua Hsueh 1967, 9; C. A. *69*, 61782c (1968).
[132] D. Naumann (1961) Z. Chemie *1*, 247.
[133] M. J. Nunes da Costa and M. A. S. Jeronimo (1961) J. Chromatog. *5*, 546.
[134] A. E. Taylor and C. A. Jensen (1958) J. Am. Chem. Soc. *80*, 5918.
[135] L. Zsinka and L. Zsirtes (1969) Proc. 2nd Hungarian Conf. on Ion-Exchange, Balatonczcplak, Vol. II., Sept. 10–14, p. 377.
[136] A. P. Rao and S. P. Dubey (1972) Anal. Chem. *44*, 686.
[137] M. Qureshi and S. A. Nabi (1972) Talanta *19*, 1033.
[138] M. Qureshi, R. Kumar and V. Sharma (1974) Anal. Chem. *46*, 1855.
[139] M. Qureshi, R. Kumar and H. S. Rathore (1972) Anal. Chem. *44*, 1081.
[140] S. Z. Roginskij, M. I. Janovskij, O. V. Altshuler, A. E. Morokhorets and E. T. Malinina (1960) Radiochimija *2*, 431, 438.
[141] V. Kouřim, J. Rais and B. Million (1964) J. Inorg. Nucl. Chem. *26*, 1111.
[142] V. Kouřim, J. Rais and J. Stejskal (1964) J. Inorg. Nucl. Chem. *26*, 1761.
[143] S. Kawamura, H. Kuraku and K. Kurotaki (1970) Anal. Chim. Acta *49*, 317.
[144] M. T. G. Valentini, S. Màloni and V. Maxia (1972) J. Inorg. Nucl. Chem. *34*, 1427.
[145] C. Konecny and R. Caletka. J. Radioanal. Chem. *14*, 255 (1973); Anal. Abstr. *26*, 2997 (1974).
[146] L. C. Neskovic and M. Fedoroff (1976) J. Radioanal. Chem. *30*, 533.
[147] M. Wald, W. Soyka and B. Kaysser (1973) Talanta *20*, 405.
[148] J. Nagy. Izotoptechnika *16*, 674 (1973); Anal. Chem. *48*, (1976).
[149] D. Huys and L. H. Baestlé (1964) J. Inorg. Nucl. Chem. *26*, 1329.
[150] L. H. Baestlé, D. Van Deyck and D. Huys (1965) J. Inorg. Nucl. Chem. *27*, 683.
[151] L. H. Baestlé, D. Huys and D. Van Deyck (1966) J. Inorg. Nucl. Chem. *28*, 2385.
[152] K. H. Lieser, J. Bastian, A. B. H. Hecker and W. Hild (1967) J. Inorg. Nucl. Chem. *29*, 815.
[153] J. Bastian and K. H. Lieser (1967) J. Inorg. Nucl. Chem. *29*, 827.
[154] K. H. Lieser, J. Bastian and A. B. H. Hecker (1967) Fresenius Z. Anal. Chem. *228*, 98.
[155] H. Bernhard and K. H. Lieser (1969) Radiochim. Acta *11*, 153.
[156] J. S. Gill and S. N. Tandon (1973) J. Radioanal. Chem. *13*, 391.
[157] M. Qureshi, K. G. Varshney and A. H. Israili (1971) J. Chromatogr. *59*, 141.
[158] J. S. Gill and S. N. Tandon. J. Inorg. Nucl. Chem. *34*, 3885 (1972); Radiochem. Radioanal. Lett. *4*, 379 (1973).
[159] M. Qureshi, K. G. Varshney and F. Khan (1972) J. Chromatogr. *65*, 547.
[160] M. Qureshi, K. G. Varshney and F. Khan (1973) Sep. Sci. *8*, 279.
[161] R. Boeckl and K. H. Lieser (1973) Radiochim. Acta *20*, 51.
[162] I. Sipos-Galiba and K. H. Lieser (1980) Radiochem. Radioanal. Lett. *42*, 329.
[163] J. F. Keggin. Nature *131*, 908 (1933); ibid. *132*, 351 (1933).
[164] J. van R. Smit (1958) Nature *181*, 1530.
[165] J. van R. Smit, W. Robb and J. J. Jacobs (1959) Nucleonics *17*, 116.
[166] J. van R. Smit, J. J. Jacobs and W. Robb (1959) J. Inorg. Nucl. Chem. *12*, 95.
[167] J. van R. Smit, W. Robb and J. J. Jacobs (1959) J. Inorg. Nucl. Chem. *12*, 104.
[168] R. W. C. Broadbank, S. Dhabanandana and R. D. Harding (1961) J. Inorg. Nucl. Chem. *23*, 311.
[169] S. J. Harvie and G. H. Nancollas (1970) J. Inorg. Nucl. Chem. *32*, 3923.
[170] J. Krtil and V. Kouřim (1959) J. Inorg. Nucl. Chem. *12*, 367.

[171] J. Krtil (1961) J. Inorg. Nucl. Chem. *19*, 298.
[172] J. Krtil (1962) J. Inorg. Nucl. Chem. *24*, 1139.
[173] M. Lesigang (1964) Mikrochim. Acta 1964, 34.
[174] M. Lesigang and F. Hecht (1964) Mikrochim. Acta 1964, 508.
[175] K. H. Lieser, Ph. Gütlich and I. Rosenbaum (1965) Radiochim. Acta *4*, 216.
[176] K. H. Lieser, Ph. Gütlich and I. Rosenbaum (1966) Radiochim. Acta *5*, 38.
[177] K. H. Lieser, H. Braun and H. Mager (1976) J. Radioanal. Chem. *32*, 367.
[178] K. H. Lieser, H. Mager and G. Pallikaris (1977) Z. physik. Chemie, Neue Folge *105*, 35.
[179] K. R. Kar and G. Singh (1968) Mikrochim. Acta 1968, 560.
[180] K. R. Kar, M. M. Bhutani and G. Singh (1968) Mikrochim. Acta 1968, 1198.
[181] K. H. Lieser and W. Hild (1968) Radiochimica Acta *7*, 74.
[182] A. Disam, P. Tschöpel and G. Tölg (1979) Fresenius Z. Anal. Chem. *295*, 97.
[183] Z. Gregorowicz, H. Stec and J. Ciba (1980) Fresenius Z. Anal. Chem. *303*, 381.
[184] G. Zuber and K. H. Lieser (1974) Radiochim. Acta *21*, 60.
[185] K. L. Cheng and H. Y. Guh (1978) Mikrochim. Acta 1978, 55.
[186] D. E. Leyden and G. H. Luttrell (1976) Anal. Chim. Acta *84*, 97.
[187] M. M. Guedes de Mota, F. G. Römer and B. Griepink (1977) Fresenius Z. Anal. Chem. *287*, 19.
[188] M. M. Guedes de Mota, M. A. Jonker and B. Griepink (1979) Fresenius Z. Anal. Chem. *296*, 345.
[189] J. R. Jezorek and H. Freiser (1979) Anal. Chem. *51*, 366.

1.8 Non-synthetic Ion Exchange Materials

Konrad Dorfner
Mannheim, Federal Republic of Germany

Introduction

The evaluation of other ion exchange materials than those treated comprehensively in the preceding chapters involves a considerable number of natural organic substances, which prior to the manufacture of synthetic ion exchange resins had been examined for their ion exchange properties. In the following pages, coal, lignin and wood, peat and humic acid, alginic acid and alginates, tannins, pulp and paper, cotton, starch and pectin, and chitin and keratin will be treated separately and in some detail. This belongs in essence to the field of reactions of polymers, since modification of natural polymers for commercial uses through introduction of attached groups began with the nitration of cellulose in 1838. By reaction of cellulose with succinic acid to form cellulose succinate, as well as from casein and formaldehyde, cation exchangers were prepared. Olive-kernels as well as coffee-grounds have also been used as a basic material for the preparation of ion exchangers. Under the name Biosorbens M a new type of selective sorbent was reported to have been prepared by crosslinking a residual mycelium from penicillin cultures with urea-formaldehyde polymer or urea and formaldehyde. It was intended to be used to decontaminate radioactive wastewater from the preparation of nuclear material, e. g., uranium-mining waste water. It is assumed that the retention mechanisms for metals and anions by this sorbent involve ion exchange, chelate formation and sorption-coprecipitation processes. As examples for newer proteinaceous ion exchange materials, hydrolytically-stable gel ion exchanging sorbents were produced by treating granulated proteins with a solution of alkaline earth metal and then with a solution of a crosslinking difunctional reagent in an organic solvent or in water; or, in a two-stage process, crosslinkable thermosetting plastics were manufactured by the chemical and physical modification of plant proteins, such as those from wheat, soy, or corn, optionally mixed with rennet casein; they were first degraded to reduce their molecular weight by about fifty percent, then ionogenized, and finally crosslinked with formaldehyde. Edible ion exchange resins it was believed might be useful as carriers for antibiotics and to reduce stomach acidity were prepared by reacting a nitrogen-containing compound such as ammonia or taurine with aqueous dextrose solution, yielding a gummy product with significant acidity. Pullulan was crosslinked with epichlorohydrin and etherified with aminoalkyl halides, epoxy amines, bromomethanesulfonate or chloroacetate in the presence of sodium hydroxide to give products with ion exchange properties.

Of greater interest seems to be the utilization of waste materials as the basic raw materials for the production of ion exchangers. The already-mentioned coffee-grounds from soluble coffee production were thus recommended as the base for an ion exchange water softening material. The properties of powdered ion exchangers based on petroleum asphaltites containing $-N^+Me_3$, $-N^+Et_3$, $-N^+C_2H_5$, $-PO(OH)_2$, $-PO(H)OH$ and $-PO(ONa)_2$ groups were thoroughly examined. Asphaltites were also sulfonated or chloromethylated for the preparation of ion exchangers, and even the kinetics of the amination of chloromethylated reaction products of asphaltites with formaldehyde, used for the preparation of ion exchangers, was examined and a diffusion mechanism suggested from the low diffusion coefficient and activation energy of diffusion of trimethylamine in the resin. Cation exchangers were manufactured by sulfonating petroleum acid tars which had been preheat-treated at 160–80 °C for 15–30 min and tarry ion exchangers were obtained by heating at 350–800 °C the distillation residue from the manufacture of diisocyanates. Under the name Sindex C 26 a cation exchange resin has been prepared from acid sludge with active sulfonic groups. It has an affinity for monovalent and divalent cations that increased with increasing temperature and decreasing resin particle size. Because of the inexpensive starting material a process was elaborated by Dunlop to convert used tyres to ion exchangers for wastewater purification. The exchange capacities obtained were comparable with customary ion exchange resins, with a selectivity for copper and zinc ions.

1.8.1 Coal-based ion exchangers

The most significant modification of a natural product was undoubtedly the sulfonation of coal by Smit in Holland and Liebknecht in Germany to form cation exchangers that were used as sulfonated coals or carbonaceous zeolites for forty years for water softening, dealkalization and partial deionization. They were found to be stable at low pH and therefore suitable for these applications. Now, however, with the greater attention paid to the development of decontamination technologies derived from naturally-occurring materials that are inexpensive due to their ease of availability, sulfonated coals have again appeared as alternative materials to commercial ion exchange resins, when high performance is not essential and costs are the main consideration. Sulfonated coals, it is said, are still widely used in Eastern Europe in water softening for steam generation and in desalination units.

Numerous papers have been published on the preparation of carbonaceous ion exchange materials and very diverse materials have been used as starting materials. Anthracite, low-rank coals, coalites, bituminous coal, various types of coke and semi-coke, brown coal or lignites, charcoals from lignin and different woods or wood-cutting wastes, activated carbon, carbon blacks and graphite have to be distinguished in this regard.

Investigating the existing hydroxyl and carboxyl groups in low-rank coals for their acidic reactivity by aqueous ion exchange and non-aqueous titration the total acidity found by titration was higher than that obtained by ion exchange at pH 13. The strongest acidity determined by titration was also higher than the carboxyl-group contents obtained by ion exchange at pH 8.25. The weaker acidity determined by titration was greater than the equivalent of the phenolic group content deduced from the difference in ion exchange at pH 13 and pH 8.25 for two coals out of five, but was in reasonable agreement for the remaining three. These discrepancies are believed to arise because of the swelling of the coals and enhancement of acidic strength of phenolic groups in ethylenediamine used during the titration procedure. These effects may cause some stronger phenolic groups to titrate with the carboxylic groups and also allow the titration procedure to determine weaker and less accessible phenolic groups than the aqueous exchange method [1]. If coals are treated with oxidizing reagents it is possible to introduce additional groups with exchange capacities besides the existing hydroxyl and carboxyl groups. Cation exchangers were thus prepared by oxidizing the surface of Missouri bituminous coal with 6 M nitric acid through which air was bubbled. This treatment introduces oxygen-containing active groups onto the coal surface, producing an ion exchanger with acid-base properties and the capability of chelating even metal ions. The Na-form of the ion exchanger removed sulfuric acid from water, raising the pH from 1.5 to >8. Iron (II) was reduced from a concentration of 25 mg/l in pH 1.5 sulfuric acid (thus simulating mine water) to <0.2 mg/l by the Na-form of the ion exchanger. When passed over the acid form of the coal ion exchanger, copper and cadmium levels in neutral solution were lowered from 10 mg/l to 0.01 and 0.00 mg/l, respectively, and from 50 mg/l to 0.02 and 0.01 mg/l, respectively. However, the low ion exchange capacity of surface-treated coal limits its usefulness in treating water. Studies of ion exchangers made from coals, including the determination of pore-size distribution from the adsorption isotherm of methanol, as well as the distribution of pore-size and internal surface area, led to a discussion of the relations between the pore-size distribution and the degree of carbonization or the internal surface area. Similar shapes of the distribution curves for anthracites, coalites and lignites are obtained. Almost all pore diameters in the coals were less than 40 Å, and most were in the range of 10 – 20 Å. Heating with caustic soda also permits the production of coal with ion exchange properties. Thus coal with a low degree of carbonization was crushed to 0.3 – 5 mm particles, treated with a dilute aqueous solution of caustic alkali at room temperature or under heated conditions, and treated with an acid to yield an ion exchange material. For demonstrating its usefulness for the removal of heavy metal ions from industrial waste waters an aqueous solution containing 10 ppm cadmium(II) was passed through a column packed with the alkali-treated coal at 50 ml/min. After operating for one hour, the cadmium content of the treated solution was zero. Another group, the ammoniated coals, are also anion exchangers based on coal. Thus, lignite was heated with aqueous ammonia at pH 9.5 and 60 °C for one hour and treated with calcium chloride to prepare an ion exchange material. The chloromethylation and amination of coal, carried out consecutively, was also investigated and reported to give ion exchangers which can be compared more directly with synthetic ion exchange resins because of the similarity in preparation [2].

Amphoteric ion exchangers have also been prepared from coal. Ground coal, <200 mesh, was first chlorinated in water to yield a weak acid cation exchanger. After rinsing and drying in vacuo, the powder was then aminated with 30% aqueous trimethylamine or 45% aqueous diethylenetriamine under atmospheric or nitrogen pressure of 20 kg/cm^2, to obtain weak base anion exchange characteristics. The exchange capacity of the products was determined volumetrically, and was 1.2, 0.3 and 0.9 milliequivalents as total, anion and cation exchange capacities, respectively, for the triethylamine-treated exchanger, and 1.0, 0.4 and 0.6 milliequivalents as total, anion and cation exchange capacities, respectively, for the diethylenetriamine-treated exchanger. It was found that both –COOH and –OH act as cation exchangeable groups. Potentiometric titrations indicated that the products obtained are similar to the common aromatic amines. For a potential application the trimethylamine-treated exchanger showed a high chromium(VI) removal from aqueous solutions. Chlorinated coal showed low removal, as expected. But regeneration of the exchanger with HCl did not restore the exchangeability appreciably [3].

But, as already mentioned above, the most important coal ion exchangers, which may again become commercially available, are the sulfonated coals. They were in fact the first cation exchange materials found to be stable at low pH, and they have been used in the past as cation exchangers until they were replaced by synthetic resins. The manufacture of sulfonated coals was re-elaborated, for instance, for the sulfonation of thermally plasticized anthracite or saprotelic coal by heating at 100 – 110 °C in oleum (15% SO_3), giving ion exchangers that can be used for water softening. A batch pilot plant for the sulfonation reaction has been described, which was used in scaling up the process, consisting of a jacketed, stirred reactor, tanks for reagents, sulfur trioxide absorber, piping, feed pumps, and gravity collecting system. The plasticized coal containing organic additives (gas oil obtained by catalytic cracking, anthracene, etc.) is crushed and sieved to obtain a fraction containing granules 2 – 3 mm in diameter. These granules are suspended in 3.5 times their weight of oleum and sulfonated. The material is converted to hydrogen or sodium forms. The ion exchange capacity of these forms was stated as being up to 410 and up to 385 gequiv/m^3, respectively. The exchanger was regenerated with 1% hydrochloric acid or 10% sodium chloride solution. Similarly, the sulfonation of lignite was optimized by a modified Simplex method after a thorough study of the preparation of cation exchange materials by sulfuric acid reaction. As three significant variables, the influence of temperature, reaction time and H_2SO_4/coal ratio (v/w) on the exchange capacity of the sulfonated lignites was investigated by a modified Simplex method. As a statistical method this has the advantage of great simplicity, leading straight to the optimum point of the process. The exchange capacity of the parent lignite was increased to 3 mequ/g. The sulfolignites obtained have high specific surface areas and suitable stabilities, i. e., low leaching and swelling, for use in water treatment processes. The ionogenic groups were analyzed by discontinuous titrations and by Fourier transform infrared spectroscopy. The development of the specific surface area was studied by the *p*-nitrophenol method [4]. From practical application it was reported that the dynamic exchange capacity of sulfonated coals used as cation exchangers in water softening depends on both the nature of the ionogenic groups and the method of softening. Thus, the exchange capacity of the coal in magnetic softening was higher than that in conventional softening. Moreover, the exchange capacity increase caused by the magnetic activation was higher for the coal-containing sulfonic acid groups than for the coal-containing carboxylic acid groups. Calculations of the dependence of the capacity of regenerated KU-2-8 ion exchange resin and sulfonated coal and of filtrate quality on regeneration conditions agreed with the experimental data, verifying a model for calculating the characteristics of ion exchangers for the planning of water purification. Further, the ion exchange capacity of sulfonated coal used in the removal of sodium from water was found to be decreased by ten percent during regeneration with a 50% solution of ammonium nitrate vs. regeneration with a 2% aqueous solution of sulfuric acid. The amount of water required for washing NH_4NO_3-regenerated sulfonated coal is half that used in sulfuric acid regeneration. The waste water does not require neutralization prior to discharge. Coke was also used as a parent material for sulfonated coal ion exchangers as, for instance, for one produced by the sulfonation of comminuted shale tar coke with four times its weight of oleum, with the temperature raised to 280 °C; and for one from Egyptian petroleum coke, for which the best conditions for sulfonation were found to be a 1 : 4 coke/oleum ratio at 30 °C for twenty-six hours, yielding an ion exchange capacity of 1.77 mequ/g. With regard to lignite-based coal sorbents it should be mentioned that the kinetics of heterovalent cations comparatively have been studied, comparing natural coal and coal treated with nitric acid, sulfuric acid, or oleum. The ion exchange properties of the sorbents are incompletely used owing to the size of the pores and the degree of hydration of the ions. The rate of exchange is similar to that of synthetic carbons. The exchange rate of ions on coal is affected by the dissociation of ionogenic groups, the hydration of ions and the temperature.

"Carbonaceous cation exchanger" was the first designation used for charcoal ion exchangers; the charcoal itself may be obtained from various woods and similar natural products. Quebracho sawdust, cane sugar, coconut husks, lignin, peat and woodcutting wastes of Siberian larch have been carbonized, either by treatment with acid or by elevated temperatures of 300 °C to 600 °C, with conversion then mainly to cation exchangers by sulfonation with conc. sulfuric acid, 25% oleum or sodium bisulfite. The products obtained were quite often also called resins, perhaps because of the similarity in the cation exchange behavior to polycondensation ion exchange resins. The exchange capacities even of the earliest products were only a little less than that of polycondensation products and it was concluded from titration curves that they contain sulfonic and carboxylic acid groups. The mechanism of the sulfurization of charcoal from hydrolyzed lignin was studied and it was found that the sulfonation proceeds via electrophilic substitution in the aromatic rings of the charcoal macromolecules. Bone charcoal became quite useful as a starting material but also untreated bone charcoal showed sorptive properties for heavy metals. With only limited success, a commercial carbon ion exchanger DOU-1 was developed from wood charcoal, having a statistical ion exchange capacity of 2.5 – 6.0 mequ/g, thermostability up to 300 °C, density 1.7 – 1.9 g/cm^3, porosity 78 – 82%, ash content 2.0 – 2.5%, and electrical resistance $10^8 - 10^{11}\ \Omega \cdot cm^{-1}$. This material can be used in rare-metal industry, nutritive products, and water treatment. The oxidation of charcoal with nitric acid for a few hours at elevated temperatures introduces carboxyl, carbonyl and phenolic groups with ion exchange capacities of approx. 3 mequ/g. The determination of the exchange properties of vapor-activation oxidized peat charcoal showed that the selectivity of adsorption of organic compounds onto this material was not significantly affected by vapor-activation. Both species of charcoal cation exchangers can be converted to the sodium form by contacting simultaneously with 0.5 M sodium chloride and 0.05 M sodium hydroxide. The study of the ion exchange dynamics revealed quite complicated relationships [5]. Oxidized charcoal DOU-3S was found useful as a carbonaceous ion exchanger for the removal of alkaline earth metal impurities from inorganic reagents. Thus, the removal of calcium and magnesium impurities from pure grade sodium carbonate, pure grade potassium carbonate, and analytical grade sodium hydroxide solutions with DOU-3S charcoal gave reagents which surpassed the existing standards with respect to calcium and magnesium content.

Activated carbons may be regarded a unique class of ion exchange adsorbents [6]. This is even more the case when activated carbons – especially commercially available products – are oxidized by nitric acid, hypochlorite solution, hydrogen peroxide or air. Prior to oxidation the commercially activated carbon materials can be treated with hydrofluoric and hydrochloric acids for ash removal. While aiming to increase the sodium ion exchange capacity by oxidation at 90 °C for 7 hours with 8 N HNO_3 it was found that a chemically unstable material results, which easily peptizes in alkaline media; heat treatment in vacuo for three to four hours at 400 °C, however, increases the chemical stability of oxidized carbons. It was also reported that significant amounts of fulvic acids are released by nitric acid oxidation, but not by the other oxidizing agents, and that these fulvic acids play no significant part in the ion exchange process. Wet oxidation of porous activated carbon with nitric acid and hydrogen peroxide changes the pore structure. Investigations carried out in order to elucidate the ion exchange process of stable active carbons oxidized with nitric acid under standardized conditions, with respect to the relation between the capacity and NaOH solution pH, showed that free hydrogen groups in the carbon matrix interacted with sodium ions at pH $\sim$ 9.0, while carboxylic groups reacted at pH 9 – 11.3, and phenolic groups at $>$11.3. An increase of oxidation caused a decrease in the pH value at which carboxylic and phenolic groups became active. For the same pH and concentration values, the ion exchange capacity of the carbons for alkali and alkaline earth metals depended on the charge and the size of the adsorbed ion [7]. When active carbons are oxidized under identical conditions, the distribution of the acidic sites depends on the nature of the original material. This also follows from an investigation of the modification of the cation exchange properties of activated carbon by treatment with nitric acid in which it was found that the uptake of inorganic cations by high-surface-area activated carbon can be increased by an order of magnitude by controlled exposure to high concentrations of nitric acid at elevated temperatures. Distribution coefficients of cations are also increased, when compared to untreated activated carbon, which has also ion exchange properties. The acid strength of the active groups from the nitric acid treatment is greater than those of the starting material. Surface area measurements and small-angle neutron scattering indicate that the increase in effective ion exchange capacity is in the case of this material not accompanied by gross changes in the structure of the product. The ^{13}C NMR of solid samples suggested that the concentration of carboxyl and phenolic ionogenic groups in the activated carbon is increased by the treatment [8]. The properties of an activated carbon designated BAU treated with phosphoric acid were studied to determine if it can be used as a selective ion exchanger. The carbon treated at 450 and 700 °C with 40% H_3PO_4 contained 1.75 and 6.5% P_2O_5 and had exchange capacity 2.1 and 2.23 mequ/g, respectively. Washing with water decreased the P_2O_5 content from 1.75 to 0.55% without decreasing the sorptive properties. The exchange capacity increased continuously with increasing pH of the solutions. The phosphatized carbon was found to behave like a phosphorous-containing synthetic ion exchange resin [9]. A similar study was made on the properties of activated carbons prepared from furfuryl alcohol and modified with phosphortrichloride at 400 – 800 °C by introducing a phosphorus content of 0.7 – 3.14 wt.%. Measurements of surface area and phosphorus concentration on the surface were made together with a determination of ionic capacity, redox activity, and thermal analysis. Increase of modification temperature increases the amount of bound phosphorus and lowers the surface area. The surface phosphorus-carbon compounds decompose above 850 °C. The increased phosphorus surface concentration increases the ionic capacity and lowers the reduction capacity with respect to iron(III).

The ion exchange of carbon blacks had been known for many years, but it took quite some time before attempts were made to study the phenomena by the techniques that had proven profitable in the field of synthetic organic ion exchange resins. A simple technique was then described by which the ion exchange-pH curves of carbon blacks can be obtained. These ion exchange-pH curves were compared for acetylene blacks, channel blacks, furnace blacks, and a thermal black in 1.0 N KCl at pH 1 – 13. More information about the surface properties of carbon blacks is obtained by this technique than by other methods in more general use. A qualitative explanation of the shapes of the exchange curves was offered and the significance of the different shapes of the curves for different carbon blacks was discussed [10]. Some of the works on activated carbons also discuss the oxidation of carbon blacks. One of these works will be discussed. In an extensive study on the cation exchange on the surface of carbon blacks a potentiometric method in solutions of NaOH + NaCl (0.025 equ Na^+) was used to obtain plotting curves of cation exchange on customary carbon black types in the original state, oxidized by nitric acid and sulfuric acid, and also in one case after gasification and chemical treatments. The original carbon black at pH 7.0 had an insignificant cation exchange capacity of 0.005 – 0.008 mequ/g. At pH 10.5 it was equal to 0.045 mequ/g and practically the same for all carbon blacks. Oxidation and especially sulfation increased the cation exchange capacity, which agrees with the increase in specific surface area. After combustion loss of 11% and oxidation, the cation exchange capacities of the gasified sample at pH 7.0 and pH 10.5 were 0.155 and 0.245 mequ/g, respectively. After combustion loss of 23% and sulfation it had capacities of 0.57 and 0.82 mequ/g, respectively. Studies on graphite as a potential starting material for ion exchangers have been carried out in two ways. Several compounds that have particular ion exchange or chelation properties were adsorbed on granular graphite. The resulting materials exhibit ion exchange and chelation properties similar to typical commercial exchange resins used for separations involving metal ions; however, the exchange capacities are lower (approx. a thousand times) than those of commercial materials. Particular attention was devoted to the immobilization of several metals on Eriochrome BlackT-modified graphite columns. Further, the properties of chemically and physically modified graphite were investigated [11].

1.8.2 Lignin and wood

Technical lignin and dry hydrolysis lignin with about 20% moisture content are active sorbents, with the absorption capacity comparable to the capacity of industrial-grade activated carbon. The different acid groups in lignin can be determined with high-frequency titrations in water-acetone media. Such lignin materials have been recommended as sorbents for phenol in order to purify wastewaters, suggesting that the absorption of phenol by lignin occurs by physical adsorption. Bobleter and coworkers then investigated the manufacture of cation as well as anion exchangers based on lignin and wood as a possible means of producing inexpensive ion exchangers. By partial sulphitation of lignin and sawdust, cation exchangers were obtained that have total exchange capacities of 0.40 and 0.66 mequ/g, respectively, and therefore come close to the capacities of synthetic ion exchange resins. In order to obtain high yields and economic results, the reaction must be carried out under alkaline conditions (Na_2SO_3 + NaOH). The best temperature for reaction is approximately 150 °C. The exchange capacity of these products remains constant even after repeated charging. The lignin and wood exchangers are readily combustible and it is therefore possible to concentrate radioactive substances very effectively from large amounts of liquid. The decontamination of radioactively contaminated surfaces can effectively be carried out by using these exchangers. Estimates of costs based on laboratory experiments indicated that these methods of producing wood-based exchangers are considerably less expensive than the production of synthetic resins. The production of anion exchangers by the reaction of wood materials with amines was also systematically investigated. It was found that β-diethylamino-ethylchloride-hydrochloride is especially suitable; it was brought to reaction with sawdust at 100 °C. The wood substances were pretreated with 17–20% NaOH. Using this method very good exchange capacities of up to 1.2 mequ/g were obtained. Anion exchange materials made from finely-grained sawdust can be used successfully for the decontamination of different types of surfaces. Since it is usually not worthwhile to regenerate the ion exchangers which have been charged after liquid or surface decontamination, different possibilities for their further treatment were examined. The combustibility of the wood exchangers proved to be advantageous, since it results in a great reduction of the original volume, which would considerably reduce waste storage costs. [12]

Prior to this work experiments had been carried out to prepare cation exchangers from sulfite liquors and sulfuric acid or hydrochloric acid with formaldehyde, yielding an exchange capacity of 0.75 mequ Ca^{2+}/g and showing some selectivity towards divalent metal ions. Then nitrogen-containing lignin derivatives, for instance chlorolignin, reacted with hexamethylenediamine or polyethylenepolyamine and crosslinked with formaldehyde gave ion exchange resins which are selective sorbents for germanium. To increase the mechanical strength and volume capacity of lignin-based anion exchangers a concentrated sulfite liquor was treated in the presence of oxygen with aqueous ammonium hydroxide at 120–180 °C and 3–15 atm and pH 1–4. For elucidating sulfonation mechanisms the oxidative sulfonation of lignin and some guaiacyl monomers and dimers similar to its decomposition products was studied, showing that both the side chains and the aromatic rings were sulfonated simultaneously, the degree of sulfonation being limited by steric hindrance. The oxidative sulfonation of sprucewood dioxane lignin in a sodium bisulfite solution with nitric acid proceeds, most likely, by free radical substitution in the aromatic ring. The production of ion exchangers from industrial hydrolysis lignin was achieved by reacting chlorinated products with α-aminopyridine and butylamine, resulting in ion exchangers in high yield with 2.23 and

1.96 mequ/g statistical exchange capacity, using 0.1 N HCl and 20.4 and 19.8 mg/g copper ion adsorption, respectively. Group analysis and IR spectroscopy revealed that the amines react with carbonyl and chlorine groups of lignin [13]. The manufacture of ion exchange granulates based on lignin sludge has also been reported. And ion exchangers for the purification of industrial wastewaters were prepared by oxidation of hydrolytic lignin. The sorption capacities of cation exchangers obtained by oxidation with nitric acid, hydrogen peroxide, and ammonium peroxodisulfate for magnesium and calcium were 0.1853, 0.1785, and 0.1345 mequ/l, respectively.

Wood-based ion exchangers were originally of interest because of the seeming advantages of a granulated form with high capacity as compared to the, at that time, available cellulose ion exchangers. Thus, for instance, granulated balsa was reacted with sodium hydroxide and subsequently with diethyl-2-chloroethylamine to give ion exchangers for pharmaceutical applications. For acid removal beechwood hydrolysates as ion exchangers were claimed in a patent and the effect of surfactants on the ion exchange by alkali-impregnated birchwood chips was examined. To open new aspects in the cationization of lignocellulose materials 2-hydroxy-3-(trimethylammonio)propyl hemicellulose chloride was mainly used; it was isolated in a yield of 15.7% by fractional extraction combined with delignification of beech saw-dust. The latter previously alkylated with (3-chloro-2-hydroxypropyl)trimethylammonium chloride. After the reaction this component, which was substituted to the highest possible degree and had the highest possible exchange capacity, was able to be extracted from the beechwood with water [14]. The phosphorylation of spruce, pine, oak and beech sawdust was studied, and the products obtained were thoroughly tested, especially the phosphorylated spruce sawdust, which had the best chemical stability, as well as a theoretical capacity of 3.65 – 5.33 and a practical capacity of 1.24 – 4.59 mequ/g, for K^+ ions. The best properties for uranium sorption from wastewater were found for the mixed $H^+ - NH_4^+$ sorbent form, when 95% of the uranium was sorbed under static conditions. However, the sorption kinetics is very slow taking 7 to 20 days. The uranium elution was most effective with 1% solution of ammonium carbonate. The mechanical properties of this sorbent were found insufficient for column application. The mechanical and chemical properties deteriorated after twenty cycles [15]. Cedar sawdust was treated with (2-chloroethyl)diethylamine solution in hydrochloric acid to give a product containing diethylaminoethyl groups of 0.6 mmol/g and in a broader and more general study sawdust, which was obtained by treatment with Na_2SO_3 and $NaHSO_3$, was used as a cation exchanger to remove magnesium, cadmium and calcium from wastewater or to remove basicity from potable and groundwater, suggesting after comparison with Dowex 50WX8 and a formaldehyde-phenol-salicylic acid copolymer that sawdust is an economical adsorbent [16].

1.8.3 Peat and humic acid

Peat is a multiionic ion exchange material which can be involved in ion exchange and complex-formation processes. For determining the degree of participation of carboxyl groups in ion exchange in peat, the infrared spectroscopic method was used. The studies were carried out on cotton grass peat and cane peat and the IR absorbance at 1580, for Ca^{2+}, Ba^{2+} and Ni^{2+}, and 1600 cm^{-1}, for Cu^{2+} and Co^{2+} increased linearly with increasing adsorption. The adsorption of cations by peat depends on the pH of the interacting solution, with corresponding data for the system natural peat and Ca^{2+} being 62, 118, 173 and 235 mequ/100 g at pH 2.6, 4.0, 5.0 and 8.1, respectively. The cation exchange capacities of peat materials have been determined according to a modified Puustjärvi method and expressed in mequ/100 g air dried material [17]. In order to evaluate further the ion exchange properties of peat, electrophorosis was used to determine the dependence of its electrokinetic potential on its adsorption capacity. The sign of the electrokinetic potential depended on the nature of the adsorbed cation and its numerical value decreased with increasing charge of the cation. This specific character of the uptake of cations by peat results from a comparison of the sorption properties of ^{239}Pu, ^{95}Zr and ^{91}Y, showing that the sorption of ionic forms of Pu(IV) from acidic solutions ($0.03-0.5$ N HNO_3) leads to the formation of strong Pu-sorbent bonds. This means that the process differs from simple ion exchange by its irreversibility and by an esoteric form of the stoichiometry of the exchange. But results from investigations on the sorptive and ion exchange activity of peats with different degrees of mineralization and the ion exchange properties of a nonhydrolyzable peat residue may justify a real ion exchange concept in peat materials.

Mentioning humic acid in connection with ion exchange can easily be misunderstood, because this term is usually related to organic matter, generally referring to that part of organic material from soil which is soluble in aqueous alkaline solution but insoluble in acid solution or ethyl alcohol and which has been a continual problem in water treatment. But humic acid can also be prepared from various sources as a polymeric substance, often in the form of a chocolate-brown, dust-like powder which is slightly soluble in water, usually with much swelling, and which is further soluble in alkali hydroxides and carbonates, and in hot concentrated nitric acid (yielding a dark-red color). Oxidation of humic acid produces a mixture of phenolic aldehydes and acids such as vanillin, p-hydroxybenzaldehyde, syringaldehyde, *p*-hydroxybenzoic acid, vanillic acid, 3,5-dihydroxybenzoic acid and *m*-hydroxybenzoic acid. Humic acid itself results from the decomposition of organic matter, particularly dead plants, usually from turf. High-purity native humic acid for which a large structure has been proposed can be obtained from peat and lignites. In addition to illustrating the complex nature of the compound, the large structure also shows the presence of complexed heavy metals, such as iron (Figure 1.97). Significant structural differences found in some purified native humic acid samples prepared from lignites have been ascribed to different conditions of

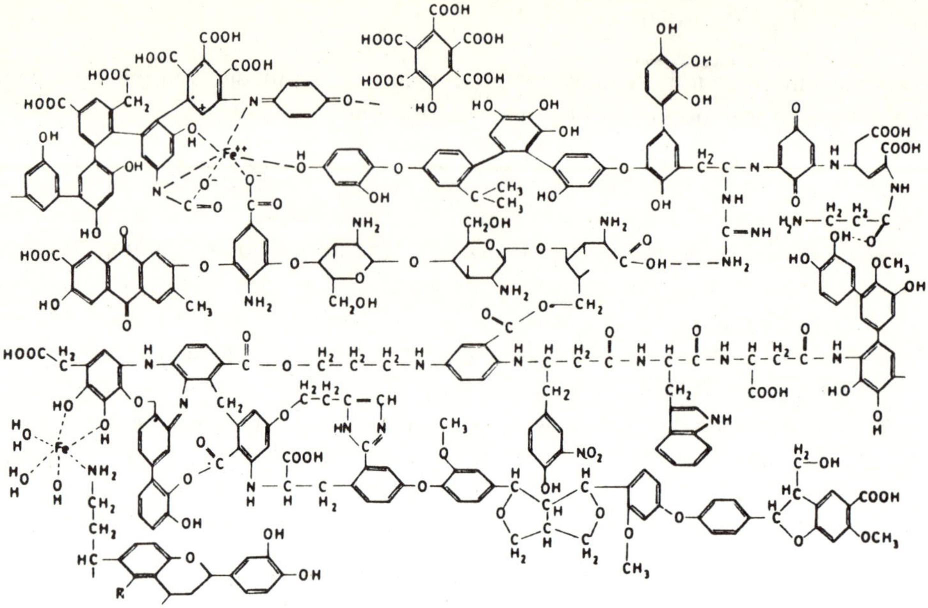

Figure 1.97 Proposed structure for humic acid illustrating a multiionogenic nature and showing the presence of complexing heavy metals, such as iron.

carbonization and marsh formation. Humic substances have ion exchange and also complexing properties and ion exchangers have been manufactured by granulation of humic acids. The effect of sorption of Na^+, Ca^{2+} and Fe^{3+} by humic acid from peat on the electrokinetic potential was studied by the electrophoretic method. In the case of Ca^{2+} and Fe^{3+} the potential first decreased to zero with increasing amounts of sorbed ions, then became positive and increased up to a constant value close to the zeta-potential value of humic acid in H-form. In the case of Na^+ the potential decreased with the increasing sorption of ions, but it did not become positive. It was concluded that the concentration dependence of the potential is due to neutralization by cations of negative charges formed in humic acid as a result of decarboxylations taking place during the decomposition of peat [18]. Using the radioactive tracer technique and assuming the presence of anionic sites in humic acid, equilibrium constants have been tabulated for a number of cations and the hydrogen ion. For the equilibrium reaction with magnesium the enthalpy, the entropy and the free energy of the exchange reaction were measured. In the presence of an increasing concentration of H^+ the sorption capacity for Ca^{2+} decreased.

To produce an ion exchanger from humic acid it can be sulfonated. This was shown in one case by treating peat humic acid with $H_2SO_4-BF_3$ complex of various mole ratios under experimental conditions between 10 to 360 minutes at 20 to 150 °C. The best reaction conditions were with a mole ratio $BF_3-H_2SO_4$ of 0.134, reaction temperature 100 °C and reaction time 90 minutes. Under these conditions an ion exchanger was obtained with an exchange capacity of 1.35 mequ/g. Cation exchangers having improved physicochemical properties and homogeneity were prepared by treating humic acids with furfural, by polycondensation of peat humic acid with furfural in an acid medium in presence of sawdust followed by drying at 280 – 310 °C in vacuo, and by reaction of humic acid with formaldehyde and phenol formaldehyde. So-called nitrohumic acid can be prepared by oxidation of sub-coal and the most various other starting materials such as asphalt, scrap tyres and even charred poly(vinyl chloride). In the first case on oxidation with nitric acid a product was obtained that adsorbs heavy metal ions like humic acid. Cadmium is adsorbed at low pH. Alkali salts of nitrohumic acid show a rapid velocity of ion exchange, because the intermolecular hydrogen bond decreases, polymolecular particles of nitrohumic acid become porous, and hydrogen ion in water decreases during ion exchange [19]. To improve the mechanical strength of nitrohumic acid processing into a nitrohumic acid-system seems to be necessary. Amphoteric ion exchangers were prepared by pre-treating humic acids with amines, e. g., phenylenediamine, and polycondensing the product with an aldehyde.

1.8.4 Alginic acid and alginates

Alginic acid is a natural polymer with a molecular weight of about 240 000. Making up a large portion of the cell walls of seaweeds, it can be obtained in the form of mixed salts of calcium, magnesium, and other cations, being very slightly soluble in water, tasteless, and capable of absorbing 200 – 300 times its weight of water, as well as salts to the extent of 60% of its weight. It resists hydrolysis but is soluble in alkaline solutions. As a 3% suspension in water it shows a pH between 2.0 and 3.4. The commercial product sodium alginate is the best-known alginic acid salt; it can be extracted from seaweed or kelp and dried to a cream-colored powder. It is soluble in water, forming a viscous, colloidal solution. In aqueous acid solutions it becomes insoluble when the pH is below 3.

The ion exchange properties of calcium alginate prepared from seaweed were investigated in acid as well as alkali salt solutions. When the ion exchange between calcium alginate and hydrogen ions was examined by use of various acids of different concentrations it was found that the power of strong acids to exchange calcium is generally high and changes very little, but that of weak acids varies according to the series of the dissociation constants. The exchange is almost complete in more than 0.2 N hydrochloric acid, but with the same acid in concentrations of less than 0.1 N an equilibrium is attained according to the law of mass action. Quantitatively calcium in calcium alginate is removed by use of 100 ml 0.2 N hydrochloric acid or 500 ml of 0.1 N hydrochloric acid. If the same alginate is treated overnight with 0.1 N aqueous solutions of various alkali salts, such as KCl, NaCl, Na_2SO_4, $NaNO_3$, sodium acetate, Na_2CO_3, $NaClO_3$, and NaOH at room temperature and the extent of cation exchange is determined, then the effect of the anions on the cation exchange followed the lyotropic series. Sodium carbonate and sodium hydroxide were different, however, in that alginic acid tends to dissolve into the solution. Neutral salts like sodium chloride also have a certain tendency to dissolve alginic acid, especially of lower molecular weights [20]. In searching for a possible increase in the sorption of radioactive strontium two samples of alginic acid were prepared from commercial sodium alginate, differing in their monomeric composition, molecular weight, and sorption capacity for strontium (90) and calcium (45). The value of the molecular weight in the 100 to 37 000 range did not have a noticeable effect on the sorption capacity. Of real practical interest can only be ion exchangers from alginates that have been made truly insoluble. To obtain such products sodium or calcium alginate, (the latter prepared from sodium alginate) was crosslinked with epichlorohydrin. Crosslinking of sodium alginate powder was used to prepare a microgranular resin. This gave quantitative results in the separation of $Ni^{2+}-Cu^{2+}$, $Co^{2+}-Cu^{2+}$, and $Ca^{2+}-Be^{2+}$ pairs by ion exchange chromatography. Its excessive swelling in basic media made it unsuitable for use in ammonia solution. When preformed calcium alginate beads were crosslinked, exchangers suitable for column operation in basic media were obtained, and exchange tests for copper(II) and nickel(II) ammonia complexes were performed. In continuing this work for obtaining useful alginic acid ion exchange materials calcium alginate beads were prepared with a dropping device and crosslinked with epichlorohydrin in order to obtain selective ion exchangers from natural polymers. The alginate was first activated with alcoholic soda to convert hydroxyl groups into alcoholates. This treatment also leads to considerable dehydration, and the consequent shrinkage in particle size was employed to evaluate the action of sodium hydroxide. The shrinking kinetics was studied as a function of temperature between 2° and 38 °C, initial bead diameter, and reagent concentration. A small number of particles were photographed during reaction in stagnant fluid. Their diameters were then measured and suitably classified. The shrinkage is dependent on mass transfer, its rate is directly proportional to the water concentration gradient on the surface of the particles, and is inversely proportional to their initial diameter at ≤ 3 mm [21].

The more general properties of alginates had already been studied prior to the above described work. Strontium-calcium selectivity was examined because there had been considerable medical interest in the ion exchange properties of alginates. The potential therapeutic application of alginates suggested the study of their calcium vs. strontium ion exchange reaction. The ion exchange properties were determined by dialyzing 2 ml of 1% sodium alginate against 50 ml of a solution containing a mixture of calcium chloride and strontium nitrate. The ionic strengths of the solutions were 0.66, and the equivalent ratio of the salts in the solution was chosen to give approximately equal amounts of the two cations bound to the alginate. The results were calculated as a selectivity coefficient. The correlation between the selectivity coefficients and the composition of the uronic acid of the alginates was very clear and indicated that the strontium selectivity of the alginates resulted from their content of L-guluronic acid residues. It was suggested that a polyguluronic acid might be the alginate preparation with the most selectivity for strontium compared to calcium, but up to then no polyguluronic acid had been isolated from brown algae or from other sources. But the ion exchange reaction involving calcium and magnesium ions was investigated for an alginate fragment containing 90% L-guluronic acid residues. Comparison of results of fractionation of the Ca-Mg-polyguluronates with a theoretical model for ion binding, involving near-neighbour auto-cooperative effects, suggested that calcium ions were selectively bound in long sequences between polyguluronate chains. An observed lack of reversibility in the binding reaction was explained by assuming that the interchain bridges containing calcium ion are kinetically very stable [22]. These researchers concluded in addition that alginate is both an ion exchanger and a gel-forming substance. Investigating the exchange mechanism which determines the retention capacity of alginic acid by chromatographic, pH, and viscosity measurements, performed with several metal ions, other researchers have been able to show that ion exchange is not the only mechanism, but that the influence of the two vicinal hydroxyl groups on the retention capacity of alginic acid is also important [23]. The interaction of fully swollen fibrous alginates with simple electrolytes confirmed various stoichiometric relationships, so that it could be shown that these gels are cation exchange materials, and equilibrium concentration coefficients with regard to the interchange of sodium and calcium be determined. It was further established that some of the exchange reactions give rise to marked alterations in the macroscopic shape, optical behavior, and elasticity of these gels. An attempt was made to show that ionic metatheses can be made responsible for a folding of initially straight segments of the chain-like alginate molecules, and that this, in turn, brings about the overall alteration of the gel properties [24].

Applications of alginic acid ion exchangers were first discussed for the quantitative separation of inorganic ions. Heavy metals were removed from a solution and the sulfate anion could be determined by titration of the hydrogen ions liberated from the alginic acid. Binary separations of copper and magnesium, copper and nickel, copper and zinc and ternary separations of iron, nickel and copper, and iron, copper and magnesium have been achieved with good results. When investigating the exchange of calcium and strontium ions on alginic acid from dilute solutions it was found that at 20 °C an equilibrium distribution is established within 10 min. The capacity for calcium, strontium and barium ions was 1.1, 1.3 and 1.4 mequ/g, respectively. The distribution coefficients were also determined for divalent cations under conditions for static adsorption as a function of pH and temperature so that separations were derived from these results [25].

1.8.5 Tannins

Sulfited tannins are employed internally in water treatment in, for instance, medium pressure boilers because of their solubility in water. An ion exchange model was applied and a mechanism discussed including redox reactions, but it is not the purpose of the current section of this volume to present these results. Condensed quebracho tannins are the main raw material in this field [26]. For the competing mimosa extracts sulfonation with sodium bisulfite or sodium sulfite increases the solubility, which is also influenced by the introduction of sulfite groups in the flavonoid units. The mechanism of sulfitation can be depicted in the following schematic reaction:

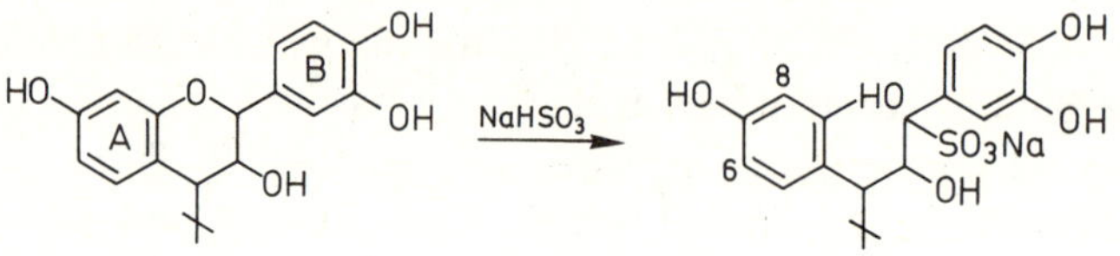

The introduction of a sulfonic acid group at the 2-position results in the opening of the heterocyclic ring, through hydrolysis of a benzyl ether link. The introduction of a polar $-SO_3Na$ group, and a phenolic $-OH$ group on the A-ring, confers increased solubility. The reactivity of the molecule is significantly modified by the conversion of the A-ring to a substituted resorcinol with highly reactive sites at the 8 and 6 positions.

Similar structures can be assumed in ion exchangers based in one way or another on tannins. It is known that Adams and Holmes prepared polycondensation ion exchange resins with ammonia as a catalyst and sulfited quebracho extracts and formaldehyde, which condensed at room temperature readily forming a gel which dried to a very hard black mass highly suitable for use as filter medium for the removal of calcium and magnesium in the base exchange water softening process. As an ion exchanger from bark an improved, high-capacity, cation exchange material was prepared by treating bark known to contain tannin with formaldehyde and a sulfite salt. Thus, a material having cation exchange properties was prepared from hemlock bark containing 12 to 14% tannin by mixing it with 20% aqueous formaldehyde, heating rapidly to 95 °C, and then adding hydrochloric acid to a pH of 1 while the temperature was kept at 95 °C for 20 min. The capacity was increased by treating the resulting product with 5% aqueous sodium bisulfite for 30 min at 130 °C in a pressure vessel, or by compressing the reacted bark at 140 °C and 4 000 lb./sq. in. for 20 min before the sulfiting step. Controlled rapid heating after the addition of formalin improves the distribution of tannin in the bark and minimizes extractable losses. Another example is the aqueous extract of the heartwood of *Rhus parviflora* containing a brown polymer soluble in ethanol and acetone, which on condensation with alkaline formaldehyde gave a resin with redox properties and which on sulfonation gave an ion exchanger. A gelatin-tannate cation exchanger can be prepared by mixing solutions of tannin and gelatin (2–10% and 0.02–1%, resp.), precipitating with various components, filtering, washing with water, drying in air, and grinding. By using this tannin bonded to gelatin, germanium can be separated selectively from large amounts of other elements over a wide pH range. This is of interest because the separation of Ge from diluted solutions by using ion exchange resins is not selective and the resin capacity is low. In the gelatin tannate an increase in the tannin content from 44% to 82% results in an increase in the sorption capacity from ≈ 20 to ≈ 55 mg Ge/g; pH has only a very slight influence on the capacity. The germanium-loaded gelatin tannate can be dissolved by boiling hydrogen peroxide in an ammonia medium, and germanium can then be determined photometrically [27]. Resins based on tannin and suitable for the selective separation of multivalent metals from aqueous solutions (described in a patent) consist of solid condensates prepared from plant tannins and aldehydes in the presence of catalyst. Thus, hydrolyzable plant tannin concentrate 90 kg, condensed plant tannin concentrate 10 kg, and 30% aqueous formaldehyde 40 kg were mixed and treated with enough concentrated sulfuric acid to cause the resin condensate to separate out in solid clumps. The resin is isolated, dried, and hardened 6 hours at 300 – 400 °C before granulation.

1.8.6 Pulp and paper

Aspects of ion exchange adsorption onto wood pulp of the most various origin have repeatedly been investigated. One aspect studied in recent years was the general effect of soluble electrolytes and even certain polyelectrolytes on fiber swelling, beating rate, fiber flocculation, drainage, and strength properties [28]. On the other hand the ion exchange equilibria of dispersed wood pulp and the ion exchange properties of pulp having carboxylic groups were investigated in aqueous solutions of sodium chloride, potassium chloride, calcium chloride, barium chloride, magnesium chloride, and aluminium chloride, using the Donnan membrane equilibrium theory. Observed were an ion exchange reaction at low concentration and an additional ion adsorption by the salt adsorption process at a higher concentration [29]. Pulp fibers were also phosphorylated (for instance, in a patented process) in order to increase their capillary suction pressure. Thus, a sheet of unbeaten, bleached softwood kraft pulp was immersed 30 minutes in aqueous phosphoric acid containing urea and dried at 160 °C for 10 min to phosphorylate the fiber. The sheet was washed and slurried in hot hydrochloric acid and the slurry treated with aqueous sodium carbonate to convert the ionogenic groups to the salt form. By this treatment the product exerted 60 cm water capillary suction pressure at 5 g water/g fiber, compared with 12 cm for an untreated, unrefined pulp. There is a change in the sorption properties of pulps during beating which can be related to their ion exchange properties. In a quite complex finding it was reported that beating breaks hydrogen bonds between wood pulp fibers and increases the sorption capacity of pulp towards the negatively-charged colloidal particles. The increase in sorption capacity is not exactly proportional to the decrease in hydrogen bond concentration because of the change in the surface area of the fibers and the alteration of their electrostatic properties. During beating hemicellulose fractions are first removed from the fiber surface and then readsorbed. These fractions contain most of the carboxylic groups of pulp. Beating changes the ion exchange properties of pulp towards electrolytes, such as KCl. The ion exchange capacity changes of pulp during beating depend on the initial concentration of hemicellulose and the sorptional properties of pulp fibers with respect to hemicellulose. To estimate quantitatively the sorption of sodium, magnesium, and calcium on kraft pulp its total ion exchange capacity was determined. For semichemical kraft cellulose pulp the capacity for magnesium ion as magnesium acetate and sodium ion as sodium hydroxide was 136 mequ/kg pulp when the equilibrium concentration of sodium ion was 17.5 mmol/l, and the Mg – Na distribution coefficient was 9 at a sodium concentration of ≤45 mmol/l and increased to 90 mmol/l. The Ca – Mg distribution constant was 2.9 when the sorption of calcium, magnesium and sodium ions by pulps was determined as a function of magnesium concentration at constant calcium and sodium concentrations. It was thus possible to describe the equilibria between pulp and surrounding cations by an ion exchange theory [30]. To obtain ion exchange materials from pulp comparable to resins attention was turned to the utilization of pulp wastes. Thus, chemical pulping waste consisting of solids and lignosulfonic acid is neutralized, heated one hour at 90 °C with 60% nitric acid, adjusted to pH 7 with sodium hydroxide, mixed with sodium alginate and rubber latex, adjusted to pH 2 with hydrochloric acid, and the decanted precipitation steam-pressed at 120 °C to give an ion exchange material with 3.18 mequ/g calcium(II) ion exchange capacity. The preparation of pulp-based ion exchangers in the form of fibers, threads, films and membranes was described in another patent: acrylate was grafted, e. g., onto spruce sulfite α-cellulose in concentrated sodium hydroxide with a persulfate-thiosulfate initiator to give an α-cellulose-poly(Na acrylate) graft copolymer fiber. After further treatment to a film material the IR spectra showed the presence of a large number of carboxylic acid groups.

Finished paper has ion exchange properties which were examined to elucidate their role in inorganic paper chromatography. The protons liberated by the exchange can be detected directly on the paper by various methods. The ion exchange effect of the paper can explain the form of the chromatograms in some cases. In addition to the pH-dependent dissociation of the acid groups of the cellulose, lactone formation also influences the exchange capacity [31]. Ion exchangers made from waste newsprint paper by carboxylation with citric acid seemed to be of interest for heavy metal removal from wastewater. Thus, as disclosed in a patent, the reaction product of waste newsprint, citric acid, and aluminium sulfate was dried, pulverized, and rinsed with 1 M sodium bicarbonate solution to prepare an ion exchanger having a copper(II) exchange capacity of 1.65 mequ/g. The exchanger was rinsed with a 10% calcium hydroxide solution and then a solution containing Zn^{2+} 0.5 ppm, Cu^{2+} 0.5 ppm, and Cd^{2+} 0.3 ppm. The effluent contained no metal. Starting from analytical grade material, paper was modified by carboxylation and crosslinking to enhance its ion exchange properties. To determine the properties obtained due to the carboxylic groups introduced and by the crosslinking, the coefficient of separation $K_{Cu/Ni}$ was determined as the quotient of the ion mobilities after 4 hours electrophoresis at 200 volts and pH 5.5. $K_{Cu/Ni}$ was 1.0 for untreated paper, 1.5 for carboxylated, and 3.5 for paper treated with formaldehyde in the presence of hydrochloric acid. Paper containing 3.5% and 8% carboxyl and treated with formaldehyde in the presence of aluminium trichloride as a catalyst had $K_{Cu/Ni}$ 4.8 and 8, respectively. The improvement of K by crosslinking does not result in destruction of the paper structure or affect adversely its chemical or mechanical properties. The treatment with formaldehyde can be achieved before or after the oxidation [32]. As special paper ion exchangers can be considered the so-called ion exchanger papers, which are papers loaded with ion exchange resins. Because of the almost exclusive use of these materials in analytical chemistry they will not be further discussed here.

1.8.7 Cotton and cotton products

Raw cotton subjected to γ-irradiation shows cation exchange properties, with a number of groups on the cottons exhibiting these properties and increasing with increased radiation dosage. The cation exchange properties of γ-irradiated cotton, determined by using radioactive tracer techniques, offer on the other hand a sensitive means of determining changes in the cottons. The irradiation was carried out under oxygen with cobalt-60 and the exchange tested on the carboxylic groups thus formed by the exchange of hydrogen for calcium. The latter can then be exchanged for other ions such as sodium, magnesium, or cobalt. Cation exchange properties of cotton receiving γ-dosages ranging from 0 to 100 megaroentgens have been reported on in the literature [33]. It is easily imaginable that similar cotton ion exchangers can be obtained by simple oxidation processes, etc., such as is familiar for other cellulose-containing materials.

More attention has been paid to cotton products than to raw cotton as a starting material to obtain ion exchange materials. Thus, the grafting of methacrylic acid onto cotton textiles gave an ion exchanging material which was further modified to give it bactericidal or blood-coagulating properties. Two grafting methods can be used: direct grafting during the irradiation with a fast electron beam, or post-irradiation grafting. Direct grafting gave $\leq 21\%$ weight increase of cotton under the optimum conditions described. Post-irradiation grafting gave only $\leq 1.6\%$ cotton weight increase [34]. By reacting non-woven cotton cellulose textile with hexamethylenetetramine, sodium hypophosphite, and melamine or urea in the presence of ammonium chloride, ion exchangers can be obtained which are non-flammable, partly crosslinked copolymers. That ion exchange must sometimes be counteracted to obtain other desired properties may be shown by one example. Cotton textiles were phosphorylated with various compounds in an attempt to obtain a resin-free, wash-resistant flameproof finish that would not change the feel of the treated material. Diammonium phosphate-urea, vinylphosphonic acid, phosphorous acid and its methyl esters all imparted a wash-resistant flameproof finish.

However, when the treated fabric was washed in hard water its flame resistance decreased owing to ion exchange with calcium and magnesium. This effect was reduced by after-treatment with heavy metal ions and also by esterification of the hydroxyl groups of the cellulose phosphate. Similarly to ion exchanger papers, cotton fabrics can also be loaded with ion exchange resins. Thus, a continuous-belt ion exchanger was invented for the removal of chromate from cooling tower blowdown waters. The chromates are removed continuously using ion exchange resin impregnated on cotton fabric belting. Cotton toweling cloth was treated with 5% sodium hydroxide solution, washed, dried, impregnated with a monomer solution prepared from 2 M dimethylaminoethyl methacrylate and less than 1 M 1.4-dichloro-2-butene, also containing potassium peroxodisulfate [35].

1.8.8 Starch and pectins

Starch has thoroughly been investigated as an ion exchanger. Naturally occurring starch, especially potato starch, contains phosphate groups which are, on the one hand, esterically bound to the starch matrix, and on the other have two exchangeable hydrogen ions that can undergo ion exchange according to the schematic equations:

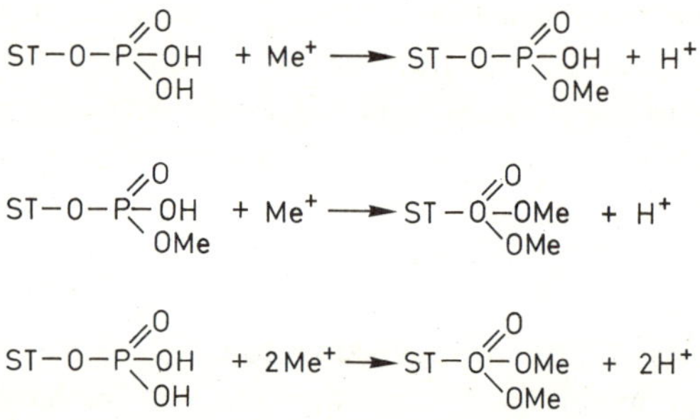

under the assumption that the starting material was completely in the H-form. This is not the case for the so-called native or commercial starches, which normally contain several cations such as potassium, sodium, magnesium, and calcium and which can, therefore, only be depicted by the following general formula:

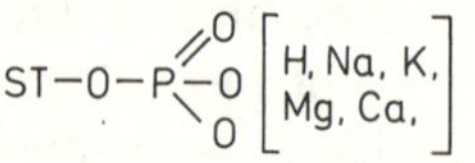

It is possible to exchange all the cations naturally present as counterions by hydrogen, leading to the so-called H-starch which is then used as the starting material for ion exchange investigations of starch. The starch granules have to some extent all the properties prerequisite for ion exchange processes. They have a porous sponge-like structure with pore widths of up to 150 Å, permitting access to larger and hydrated ions. The various counterions bound to the starch molecule via the phosphate group have a significant influence on the properties – for instance the viscosity – of starch. Quantitative ion exchange at the starch-bound phosphoric acid groups provide a convenient means for rapid routine determination of the phosphoric acid content of potato starch. Starting from H-starches having two replaceable hydrogen atoms, first one, then both hydrogen ions may be replaced by a univalent (alkali) metal to form primary and secondary alkali metal phosphates, respectively. With bivalent metals, such as calcium, magnesium or zinc, secondary phosphates are formed. In the alkalimetric method the starch is first converted to the H-starch with diluted hydrochloric acid and titrated with 0.02 N alkali to form first the primary and then the secondary metal phosphate. The alkali is then backtitrated. In a complexometric method H-starch is converted to the secondary phosphate complex with excess calcium ions, and is then decomposed with acid; the liberated calcium is titrated complexometrically. H-starch is formed at pH 2.9–3.2, the primary phosphate at pH 5–6, the secondary alkali metal phosphate at pH 8.5–9.0 and the secondary metal phosphate complex at a pH above 8.0 [36]. For the sake of completeness it may be mentioned that a natural starch is known, the so-called Florideae starch which contains the sulfate group as an anchor group.

According to a US patent, in order to obtain ion exchange starches which retain their original granular form, the native materials were first crosslinked with formaldehyde in a ratio of 50 parts starch to 3 parts paraformaldehyde by refluxing in 200 parts acetone and acidifying with hydrochloric acid to pH 2, for five hours with constant stirring. The crosslinked starch was filtered, washed with hot water and dried. For the modification to an ion exchanger, then, ten parts of the dried product are suspended in a solution of 17.5 parts monochloroacetic acid in 10 parts water. The suspension is then poured slowly with stirring into 100 parts 40% sodium hydroxide. After cooling, the product is filtered, washed with water, then acetone, and dried, giving a material having an ion exchange capacity of 1.33 mequ/g which can be used for ion exchange chromatographic separations [37]. Hydrolyzed starch-polyacrylonitrile graft copolymers were prepared as potassium salt in the form of films, which when placed in aqueous acid contracted to rubberlike films. The hydrated films remain intact except under high shear conditions. The films showed ion exchanging properties by releasing potassium and taking up magnesium ions. When starch is oxidized carboxylic groups are formed. For such materials chemical tests and infrared spectroscopy revealed the ion exchange nature of antibiotic (streptomycin, monomycin, kanamycin, oleandomycin, lincomycin and anethesin) binding through intermolecular hydrogen bonds between the hydroxyl groups of the antibiotics and the carboxyl groups of the starch. Increasing the number of carboxyl groups in the starch increased the sorption of the therapeutic agent. Semi-synthetic starch anion exchangers were also prepared starting from branched polyethylenimine diluted in water, mixed and treated with dry starch.

The mixture is homogenized and treated with sodium hydroxide, then, after mixing, epichlorohydrin is added and thus crosslinking takes place. The precipitated polymer in water is washed and dried to give a polyethylenimine starch anion exchanger with exchange capacity of 2.5 mequ/g and swelling volume of 6.0 – 7.0 ml/g.

Pectinic acid as an ion exchanger was first investigated under the general aspect of selective properties of ion exchange resins. Crosslinked pectins of varying ion exchange capacity were prepared by varying the fraction of methyl ester groups hydrolyzed in the polygalacturonic acid, followed by reaction with formaldehyde. These carboxylic ion exchangers showed a selectivity which decreased with increasing exchange capacity [38]. Since pectin is, similar to the alginates, a natural ion exchanger of outstanding properties the exchange of calcium(II), strontium(II), and barium(II) on pectate was intensively examined, but mainly from the point of view of application of sodium or calcium pectate as a therapeutic agent for decreasing the level of radioactive strontium in an organism. The interaction of calcium, strontium, and barium with the carboxyl groups of pectate was estimated from the selectivity coefficients for the exchange of calcium for strontium and calcium for barium in pectate and from the stability constants of calcium, strontium, and barium pectates in potassium chloride solutions of a constant ionic strength. These characteristic quantities were determined by using tetramethylmurexide as an auxiliary ligand. The equilibrium state of the cation exchange is affected by the type of the starting pectate; the exchange equilibrium is shifted toward the cation bound in the starting pectate. The affinity of cations for the carboxyl groups of pectate rises very gently in the order $Ca < Sr < Ba$. The pectic acid used as material was prepared from commercial citrus pectin by alkaline deesterification [39]. According to an other study, when the $K^+ - NH_4^+$ ion exchange selectivity of sunflower pectic acid was compared with that of the cation exchangers Dowex 50X10 and Dowex 50X8, the pectic acid ion exchanger exhibited a higher selectivity for potassium than the Dowex exchangers. According to a patent, crosslinked porous amylopectin spherical granule ion exchangers can be obtained by dissolving amylopectin in aqueous alkalis, suspending in nonsolvents or poor solvents, and crosslinking with epichlorohydrin, yielding gels capable of excluding different molecular weights depending on the molar quantity of epichlorohydrin employed.

1.8.9 Chitin and chitosan, kerogen, and keratin

Chitin and chitosan are the only basic natural polymers, because they contain free amino groups. Chitin is a polysaccharide consisting predominantly of unbranched chains of *N*-acetyl-D-glucosamine residues. It can be isolated as an amorphous solid, which is practically insoluble in water, dilute acids, dilute and concentrated alkalis, alcohol and other organic solvents but soluble in concentrated hydrochloric acid, sulfuric acid, phosphoric acid and anhydrous formic acid. Chitosan is partially acetylated chitin and is, therefore, the soluble derivative of chitin. Investigations on the acid capacity of chitin indicated that with the uptake of hydrochloric and sulfuric acid all amino and acetamide groups are involved. Thus, at pH 2.5 approximately 4.8 mequ acid radical/g chitin are bound. Below pH 2 an increase in acid capacity can be observed. Organic acids, on the other hand, are only slightly taken up, with less than 0.5 mequ/g. But the most specific property of chitin and chitosan is their power to form metal complexes with the amino group as electron donor; the free electron pairs of the chain oxygen atom in combination with the partial crystalline structure facilitate the fast chelate formation with transition metals and their radioactive isotopes. The chitin-chitosan polymer has already proven its applicability as a selective complexing sorbent even with ion concentrations in the ppb range. A comparative study of the chelating ability of chitosan, *p*-aminobenzylcellulose, and diethylaminoethylcellulose for a number of ions has been discussed. There is a strong interaction between molybdenum and these polymers in thiocyanate solutions and in seawater. By combining the sensitivity of graphite-furnace atomic absorption spectrometry with the efficiency of the selective collection of molybdenum on chitosan at pH 2.5, it was possible to determine the molybdenum concentration in as little as 50 ml of seawater [40]. Very general is the use of chitosan described in a Japanese patent entitled "Shaped Material Made of Chitin Derivative, Spherically Shaped Materials and Fibers Thereof, and Their Use". According to this patent the acylation of chitosan with acid anhydrides in organic solvent containing dispersants followed by precipitation gave gel-like spherical particles of diameter of 50 – 150 μm for use as ion exchangers, adsorbents for chromatography columns, or base materials for immobilization of enzymes, antibodies and antigens. The polymeric network of chitosan is used in Germany as the matrix for *E. coli* cells to obtain cell-biocatalysts. In a Russian patent the preparation of amphoteric ion exchangers is described: chitosan is treated with salicylaldehyde, with subsequent treatment of the product with a salt of either a halo-substituted alkanethio acid or a halo-substituted aliphatic carboxylic acid in the presence of alkali in a diluent medium.

Kerogen is the term used for insoluble, polymeric, organic substances of fossil sediments of virtually unknown structure and composition. The sulfonation of kerogen from Baltic shales with 95% sulfuric acid is correspondingly a complex process involving formation of SO_3H, COOH and phenolic OH groups and also degradation. The concentrations of sulfur and SO_3H groups in the sulfonated kerogen reach maximum values when the reaction is conducted at about 120 °C. Above 130 °C there is extensive degradation and loss of sulfur. The ion exchange capacity of the sulfonated kerogen is high and increases to about 6 mequ/g as the sulfonation temperature is increased to 160° to 170 °C [41].

Wool keratin has been treated with sodium sulfite and sodium dithionite, but the reaction products were not examined for their possible ion exchanging properties; nonetheless structures have been postulated here which could, because of the presence of sulfur adjacent to the inorganic group, be of special interest [42].

References

[1] T. P. Maher and H. N. S. Schafer (1976). Determination of acidic functional groups in low-rank coals: comparison of ion-exchange and non-aqueous titration methods. Fuel *55*, 138.

[2] B. I. Losev and V. V. Davydov (1976). Chloromethylation of coals with subsequent amination. Metallurgiya i Koksokhimiya. Resp. Mezhved. Nauch.-tekhn. SB. (51), 79.

[3] A. Äkama and S. Koike (1972). Amphoteric ion exchangers prepared from coal. I. Preparation and properties of amphoteric ion exchangers. Hokkaido-Ritsu Kogyo Shikenjo Hokoku (195), 1. Id. (1972). II. Ion-exchange behavior of chromium(VI). Ibid. (195), 5.

[4] G. D. Lyakhevich et al. (1973). Sulfonation of the thermally-plasticized coal in a pilot plant. Vestsi. Akad. Navuk Belarus. SSR, Ser. Khim. Navuk (1), 84. J. V. Ibarra and J. J. Lazaro (1985). Lignite sulfonation optimized by a modified Simplex method. Ind. Eng. Prod. Res. Dev. *24*, 604.

[5] V. M. Zverev et al. (1972). Dynamics of ion-exchange sorption on sulfonated and oxidized lignin coals. Zh. Prikl. Khim. (Leningrad) *45*, 765.

[6] V. A. Garten and D. E. Weiss (1957). The ion- and electron exchange properties of activated carbon in relation to its behaviour as a catalyst and adsorbent. Rev. pure appl. Chem. *7*, 69.

[7] B. N. Strashko and I. A. Kuzin (1968). Preparation of oxidized carbon, and a study of its ion exchange properties. Sin. Svoistva Ionoobmen. Mater. 1968, 303. Izd. Nauka: Moscow.

[8] J. S. Johnson et al. (1986). Modification of cation-exchange properties of activated carbon by treatment with nitric acid. J. Chromatogr. *354*, 231.

[9] Y. V. Karyakin and A. N. Amelin (1971). Properties of phosphatized carbon. Tr. Voronezh. Tekhnol. Inst. *19*, 97. J. Siedlewski et al. (1975). Study on the properties of activated carbons modified with phosphorus trichloride. Koks, Smola, Gaz *20*, 300.

[10] J. Caudle et al. (1971). Ion exchange of carbon blacks. Conf. Ind. Carbons Graphite, Pap., 3rd 1970, 168.

[11] T. M. Matviya (1982). Ion-exchange properties of chemically and physically modified graphite. Diss. Abstr. Int. B. 1982, 43, 714.

[12] O. Bobleter and K. Buchtela (1961). Ein Kationenaustauscher auf Holz- bzw. Ligninbasis für Entaktivierungszwecke. Atomkernenergie *6*, 476. Id. (1963). Anionenaustauscher auf Holz- bzw. Cellulosebasis zur Dekontamination von Oberflächen und Flüssigkeiten. Ibid. *8*, 415.

[13] E. Tzolova and Z. Christov (1978). Production of ion exchangers from industrial lignin. Holzforschung *32*, 142 (Fr.).

[14] M. Antal et al. (1984). New aspects in cationization of lignocellulose materials. II. Distribution of functional groups in lignin, hemicellulose, and cellulose components. J. Appl. Polym. Sci. *29*, 643.

[15] M. Marhol et al. (1969). Uranium sorption from natural water on phosphorylated sawdust. Ustav Jad. Vyzk. No. 2218-Ch, 41 pp. (Czech).

[16] Ki Suk Maeng and I Sik Kang (1979). Development and uses of competitive ion exchange resins. Hwahak Kyoyuk *6*, 97 (Korean).

[17] V. A. Thorpe (1973). Collaborative study of the cation exchange capacity of peat materials. J. Assoc. Offic. Anal. Chemists *56*, 154.

[18] N. I. Gamayunov and O. A. Gerashenko (1972). Electrophoretic studies of the humin substances in peat. Kolloid Zh. *34*, 802.

[19] M. Nakagawa (1973). Nitrohumic acid ion exchangers for removal of heavy metals from wastewater. Kinzoku *43*, 66 (Japan).

[20] S. Miyake (1958). The cation exchange reaction of calcium alginate in alkali salts solution. Kogyo Kagaku Zasshi *61*, 212. Id. (1958). Cation-exchange reaction of calcium alginate in acid solution. Ibid. *61*, 444.

[21] F. Ferrero et al. (1982). Selective ion exchangers from crosslinked alginate. Ann. Chim. (Rome) *72*, 73. Id. (1982). Shrinking kinetics of calcium alginate beads in alcoholic soda. Chem. Eng. Commun. *15*, 197.

[22] A. Haug and O. Smidsrod (1967). Strontium-calcium selectivity of alginates. Nature *215*, 757. Id. (1972). Dependence upon the gel-sol state of the ion-exchange properties of alginates. Acta Chem. Scand. *26*, 2063.

[23] D. Cozzi et al. (1969). The mechanism of ion exchange with alginic acid. J. Chromatogr. *40*, 130.

[24] I. L. Mongar and A. Wassermann (1952). Absorption of electrolyte by alginate gels without and with cation exchange. J. Chem. Soc. 492. Id. (1952). Influence of ion exchange on optical properties, shape, and elasticity of fully-swollen alginate fibres. Ibi.d. 500.

[25] H. Specker et al. (1954). Die quantitative Trennung anorganischer Ionen durch Ionenaustausch an Alginsäure. Z. anal. Chem. *141*, 33. M. Y. Dalmatova et al. (1971). Ion exchange equilibrium of radioisotopes of alkaline-earth elements in alginic acid. Radiokhimiya *13*, 131. Id. (1972). Ion exchange interaction of some divalent cations with alginic acid. Ibid. *14*, 741.

[26] K. Dorfner and E. Stahl (1977). Unpublished results based on investigations with thermofractography.

[27] A. M. Andrianov and V. P. Koryukova (1973). Separation of germanium from solutions using a gelatin tannate. Zh. Prikl. Khim. (Leningrad) *46*, 425.

[28] L. G. Vinogradova et al. (1977). Ion exchange adsorption on wood pulps of different origin. Khimiya i Tekhnol. Tsellyulozy (4), 10. J. W. Swanson (1965). Effects of soluble nonfibrous materials on formation and consolidation of paper webs. Consolidation Paper Web, Trans. Symp., Cambridge, Engl. *2*, 741.

[29] F. Onabe and J. Nakano (1970). Ion exchange equilibrium of wood pulp dispersed in an electrolyte solution. Kami Pa Gikyoshi *24*, 461.

[30] A. Ohlsson and S. Rydin (1975). Washing of pulps. 2. Sorption of sodium, magnesium, and calcium on kraft pulp. Sven. Papperstidn. *78*, 549.

[31] G. Ackermann and H. P. Frey (1968). The role of paper as ion exchanger in inorganic paper chromatography. Fresenius' Z. anal. Chem. *233*, 321.

[32] I. N. Ermelenko and V. I. Sokolova (1967). Ion exchange analytical paper from modified cellulose with crosslinking. Vestsi. Akad. Nauk Belarus. SSR. Ser. Khim. Navuk (3), 55.

[33] R. J. Demint and J. C. Arthur, jr. (1959). The effects of γ-radiation on cotton. III. Base-exchange properties of irradiated cotton. Textile Research J. *29*, 276.

[34] B. V. Drachec et al. (1972). Radiation grafting of a monomer on cotton using an electron accelerator. Radiats. Tekh. *8*, 16.

[35] D. A. Brown et al. (1976). Continuous belt ion exchanger: chromate removal from cooling tower blowdown waters. Appl. Polym. Symp. 1976, 29 (New Spec. Fibers), 189.

[36] S. Winkler (1960). Die ionenaustauschenden Eigenschaften der Kartoffelstärke. II. Die Bestimmung des Phosphorsäuregehaltes der Kartoffelstärke auf komplexometrischem und alkalimetrischem Wege. Stärke *12*, 35. Id. Die Stärke als Ionenaustauscher. Parey Verlag Berlin; 1971. In: Handbuch der Stärke in Einzeldarstellungen, Bd. VI-3.

[37] A. L. Bullock and J. D. Guthrie (1961). Ion-exchange starches which retain their original granular form. U.S. Patent 2 992 215, July 11.

[38] H. Deuel et al. (1953). Selective properties of ion-exchange resins. Z. Elektrochem. *57*, 172.

[39] R. Kohn and V. Tibensky (1971). Exchange of calcium, strontium, and barium ions on pectin. Collect. Czech. Chem. Commun. *36*, 92.
[40] R. A. Muzzarelli and R. Rocchetti (1973). Determination of molybdenum in sea water by hot graphite atomic absorption spectrometry after concentration on p-aminobenzylcellulose or chitosan. Anal. Chim. Acta *64*, 371.
[41] I. D. Cheshko et al. (1979). Sulfonation of kukersite shale kerogen. Khim. Tverd. Topl. (Moscow) (4), 74.
[42] G. Valk (1968). Über die Reaktion von Chlor, Natriumsulfit und Natriumdithionit mit Wollkeratin. Forschungsbericht NRW 1935. Westdeutscher Verlag Köln und Opladen.

1.9 Liquid Ion Exchangers

Erik Högfeldt
Department of Inorganic Chemistry, The Royal Institute of Technology, Stockholm, Sweden

Introduction

Ion exchanging groups can either be attached to a polymeric network (solid resin) or to a fragment polymer, either to the uncrosslinked polymer (polyelectrolyte) or to an organic radical hydrophobic enough to make the compound practically insoluble in the aqueous phase, but soluble in an organic solvent (liquid ion exchanger). In the following the discussion will be limited to liquid extractants with ion exchange properties. The emphasis will be on similarities between solid and liquid ion exchangers.

1.9.1 The extractants

The types of extractants treated in this section can be divided into two groups: strong or relatively strong acids and strong bases.

Acids. To this group belong various sulfonates where the hydrocarbon chains are long enough to make the extractant practically insoluble in the aqueous phase. An example is given by naphthalene sulfonates, such as

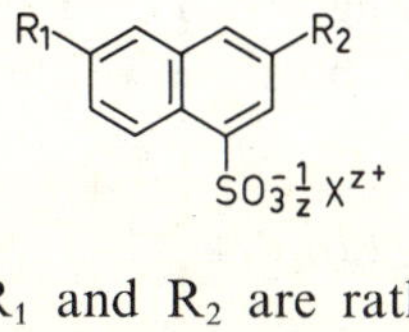

where R_1 and R_2 are rather long hydrocarbon chains, e. g., C_9H_{19}, $C_{12}H_{25}$ etc. In practice an isomer mixture containing also branched chains is obtained.

The commercially available dinonylnaphthalene sulfonic acid belongs to this group. Somewhat weaker acids are obtained by the dialkylphosphoric acids, $\begin{matrix} R_1O \\ R_2O \end{matrix}\!\!>\!POOH$ where again the hydrocarbon chains have to be rather long to achieve low solubility in the aqueous phase. A practically important compound is di-(2-ethylhexyl)phosphoric acid.

Bases. To this group belong long-chain tertiary amines and quaternary ammonium salts, R_3N and $R_4N^+X^-$. Commercially available are tri-*n*-octylamine, tri-*n*-dodecylamine (TLA) while salts of TLMA (trilaurylmethylammonium) can be synthesized from TLA. Commercial products are often a mixture of various amines.

The sulfonic acids are extensively aggregated to micelles. The dialkylphosphoric acids are dimerized. The degree of dimerization depends upon the compound as well as upon the diluent used. The amines are monomeric, whereas their salts, as well as the quaternary ammonium salts, are extensively aggregated, the latter to micelles, the former to dimers, trimers etc., in aromatic diluents with more extensive aggregation in aliphatic diluents (see below).

For sulfonates and dialkylphosphates the selectivity is mainly governed by electrostatic considerations, with highly-charged ions being preferred over more lowly charged ions. With amines, where metals often are extracted as anionic complexes, steric factors may be important and amines with a high selectivity for a certain metal have been found.

In the following the physical chemistry of the compounds mentioned above will be reviewed with emphasis on effects that can be treated quantitatively.

1.9.2 Basic physical chemistry

1.9.2.1 Ion exchange or extraction

Extractants with strong acid groups (sulfonic acids, etc.) or strong basic groups (quaternary ammonium groups) extract through ion exchange. For weak acids and bases ion-pairing or compound formation are important.

Sometimes the same system can be formulated either as extraction or as an ion exchange process. This is illustrated by the following example.

Long-chain ammonium salts act as liquid extractants – they are the liquid analogs to strong base anion exchangers.

Consider a tertiary ammonium salt (R_3NHCl) extracting zinc. Depending upon the predominant species in the aqueous phase the reaction can be written in two different ways:

$$2R_3NHCl(org) + Zn^{2+} + 2Cl^- \rightleftarrows (R_3NH)_2ZnCl_4(org) \quad (1)$$

$$2R_3NHCl(org) + ZnCl_4^{2-} \rightleftarrows (R_3NH)_2ZnCl_4(org) + 2Cl^- \quad (2)$$

At low chloride activities in the aqueous phase Zn^{2+} predominates and reaction (1) is assumed to occur. At high chloride activities reaction (2) is assumed to predominate. From the equilibrium point of view there is no difference between the two reactions, as it is always possible to obtain (2) from (1) and vice versa by adding or subtracting the reaction

$$Zn^{2+} + 4Cl_4^- \rightleftarrows ZnCl_4^{2-} . \quad (3)$$

Kinetically, there is a difference, as recently shown by Aparicio [1]. As a general rule, it can be stated that in the organic phase the most symmetrical, highly coordinated complex predominates over the whole range of ligand activities experimentally available in agreement with reactions (1) and (2) [2]. In the following the discussion will be limited to extraction systems that can be treated as ion exchange processes.

1.9.2.2 Water uptake by pure ionic forms

General. Both solid and liquid ion exchangers extract water from an aqueous phase. This is illustrated in Figure 1.98, where the number of water molecules per sulfonic acid group, W, is plotted against the water activity $\{H_2O\} = a$ for the system $HD-C_7H_{16}-H_2O$ [3] (HD = dinonylnaphthalene-sulfonic acid). The data were acquired isopiestically.

The same system was studied calorimetrically. In Figure 1.99 the integral heat of adsorption Q is plotted against W.

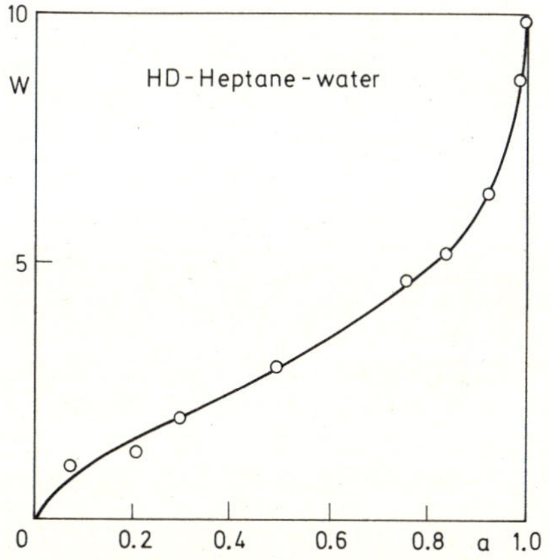

Figure 1.98 W plotted against the water activity, a, for the system $HD-C_7H_{16}-H_2O$. The curve has been computed using the constants in Table 1.37. Reproduced with permission from Pergamon Press.

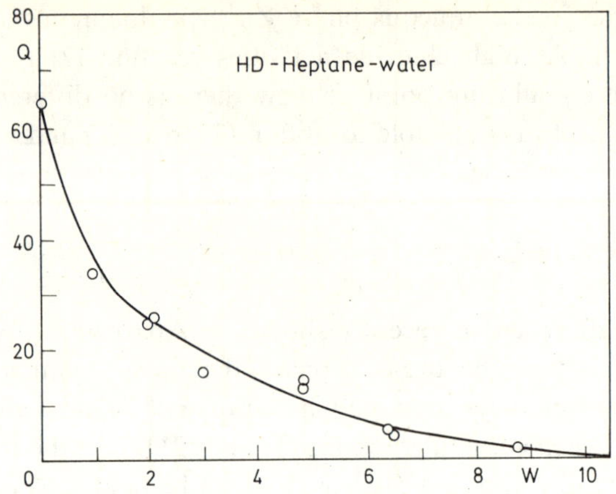

Figure 1.99 The integral heat of sorption (Q) in kJ/mol plotted against W for the system $HD-C_7H_{16}-H_2O$. The curve has been calculated from the q-values given in Table 1.37.

$W(a)$-data can be used to calculate integral free energies from

$$\frac{\Delta G}{RT} = \int_0^a \ln a dW = W\ln a - \int_o^a W d\ln a \qquad (4)$$

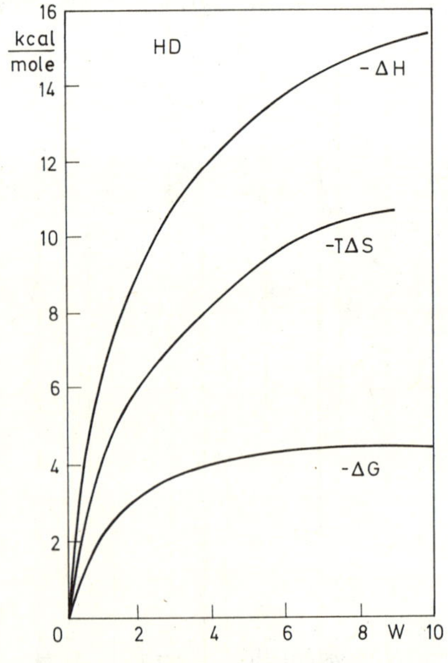

Figure 1.100 $-\Delta G$, $-\Delta H$ and $-T\Delta S$ plotted against W for the system $HD-C_7H_{16}-H_2O$. Reproduced with permission from Pergamon Press.

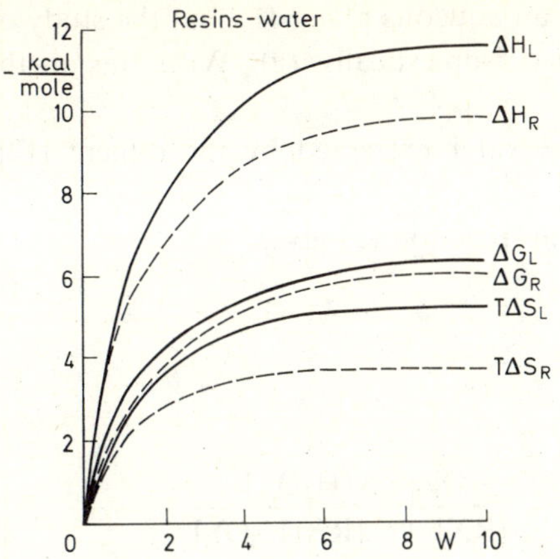

Figure 1.101 $-\Delta G$, $-\Delta H$ and $-T\Delta S$ plotted against W for the hydrogen form of Dowex 50X8. L refers to ref. [4] and R refers to ref [5]. Reproduced with permission from Pergamon Press.

which can be obtained from a plot of W against ln a or log a using graphical or numerical integration. The resulting curves for the data in Figures 1.98 and 1.99 are shown in Figure 1.100. Very similar curves are obtained for solid resins [3], as seen in Figure 1.101 for the hydrogen form of Dowex 50X8.

It is also possible to translate data like those in Figure 1.98 into activities and activity coefficients [6].

The BET equation has been applied by Kellomäki [7, 8] to water sorption by resins. However, at high water activities this isotherm does not fit the experimental data. It seems rather unlikely that the BET isotherm would fit a system where electrostatic as well as interactions through hydrogen bonding are important.

Water sorption as a stepwise hydration process. It has been shown [9, 10] that by assuming hydration to be a stepwise process most of the variation in excess free energy of ions is accounted for.

Application of this model to the liquid cation exchanger dinonylnaphthalenesulfonic acid (HD) in heptane showed ideal behavior at water activities so low that formation of the monohydrate was the only species to be taken into account [11]. The extension to higher water activities showed that the water uptake can be divided into two regions, those at low water activities, where primary hydrates are formed, and the region where n, the number of water molecules attached to the ion exchange site, is about 10 and larger. In this region the model can only be regarded as a practical way of summarizing experimental data, with no significance given to the n-values used.

The hydration reaction can be written

$$RX(org) + nH_2O(g \text{ or } l) \rightleftarrows RX(H_2O)_n(org)\,. \tag{5}$$

In practice water is extracted from an aqueous phase (1), but the study of reaction (5) is most conveniently performed isopiestically (g). With this method phase separation becomes no problem.

When studying reaction (5) the water extracted by the diluent [12] must be corrected for.

The law of mass action applied to reaction (5) gives

$$K_n = \frac{\{RX(H_2O)_n\}}{\{RX\}\,a^n}. \tag{6}$$

The expression for W becomes

$$\begin{aligned} W &= \frac{[H_2O]^{tot}_{org} - [H_2O]^{dil}_{org}}{[HD]_{tot}} = \frac{\Sigma n[RX(H_2O)_n]}{[RX] + \Sigma[RX(H_2O)_n]} \\ &= \frac{\{RX\}\,(\Sigma n\,(K_n/f_n)a^n}{\{RX\}\,(1/f_0 + \Sigma(K_n/f_n)a^n} = \frac{\Sigma n k_n a^n}{1 + \Sigma k_n a^n}, \end{aligned} \tag{7}$$

with f_0 = activity coefficient of RX and f_n that of $RX(H_2O)_n$; $a = \{H_2O\}$.

It is assumed that f_n is practically constant in the range where $RX(H_2O)_n$ contributes appreciably to W and that $f_0 = 1$ in the range where RX cannot be neglected.

For $n = 1$ eq. (7) gives

$$\frac{W}{1 - W} = k_1 a\,.$$

In Figure 1.102 $W/(1 - W)$ is plotted against $a \cdot 10^4$ for the system HD$-$$-C_7H_{16}-H_2O$ at very low water activities [11]. A straight line can be fitted to the data supporting the assumptions made about f_0 and f_1.

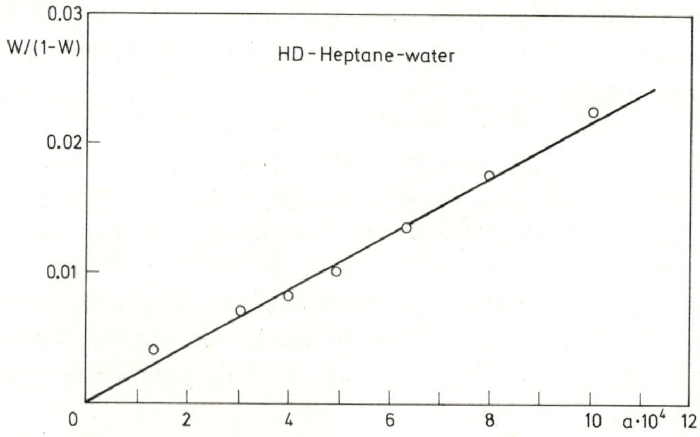

Figure 1.102 $W/(1-W)$ plotted against $a \cdot 10^4$ for the system $HD-C_7H_{16}-H_2O$ at very low water activities.

Table 1.37 Species and constants for the system HD-C_7H_{16}-H_2O. T = 298 K [HD] = 0.085 M. From ref. [3]

Species	k_n	q_n	ΔG_n (kJ/mol)	ΔH_n (kJ/mol)	$T\Delta S_n$ (kJ/mol)	ΔS_n (J/(K · mol))
HD	–	64.0	–	–	–	–
HD(H_2O)	21.4	25.1	– 7.60	–38.9	–31.3	–105
HD$(H_2O)_2$	200	25.1	–13.14	–38.9	–25.8	– 86
HD$(H_2O)_4$	200	22.6	–13.14	–41.4	–28.3	– 95
HD$(H_2O)_6$	1300	1.26	–17.78	–62.8	–45.0	–151
HD$(H_2O)_{14}$	300	–12.6	–14.14	–76.6	–62.4	–209
HD$(H_2O)_{26}$	420	–12.6	–14.98	–76.6	–61.6	–207

The results of fitting the data in Figures 1.98 and 1.99 to the model are given in Table 1.37.

It is possible that all hydrates up to $n \approx 6$ are formed. In using the model the minimum number of parameters is sought, giving preference to major species.

The full-drawn curve in Figure 1.98 has been calculated from eq. (7) with the constants given in Table 1.37.

The data in Figure 1.99 were fitted by the equation

$$Q = \Sigma q_n \alpha_n \tag{8}$$

where

$$\alpha_0 = 1/(1 + \Sigma k_n a^n), \; \alpha_n = k_n a^n/(1 + \Sigma k_n a^n), \tag{9 a,b}$$

and q_n a constant.

In Figure 1.99 Q approaches 64, as $W \rightarrow 0$. The values for ΔH_n given in Table 1.37 were calculated from

$$\Delta H_n = q_n - q_0 = q_n - 64.0 \text{ (kJ/equiv)}. \tag{10}$$

The large negative entropy values indicate that water is rather tightly bound to HD. Similar results are obtained for other cations [6].

In Figure 1.103 the fraction of HD present in each species, α_n, is plotted against W. The constants given in Table 1.37 have been used in the calculations, together with equations (9 a,b).

Infrared data indicate formation of H_3O^+ at low water activities [3] suggesting formation of

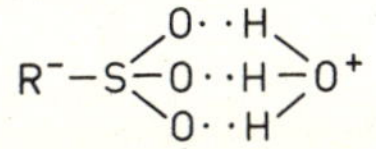

Zundel [13] studied infrared spectra of thin films of ion exchange resins and found no evidence of H_3O^+.

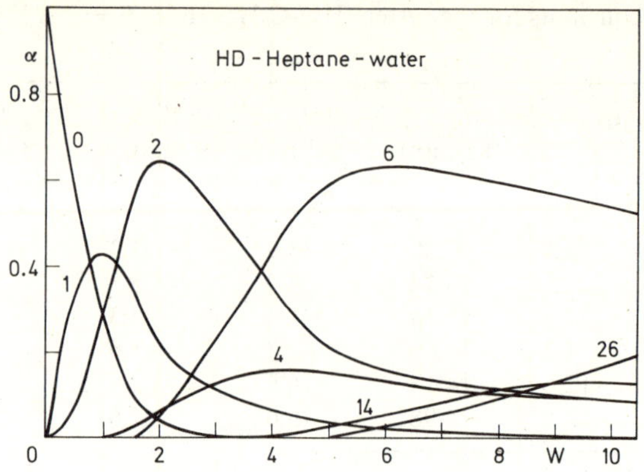

Figure 1.103 α_n plotted against W for the system $HD-C_7H_{16}-H_2O$. The curve marked 0 refers to unhydrated HD, 1 to $HD \cdot H_2O$ etc.

It deserves to be mentioned that infrared data belonging to unhydrated HD correlate nicely with the curve marked $0(n = 0)$ in Figure 1.103.

The model outlined above applies to solid resins as well. In Figure 1.104 W is plotted against a for the system Dowex $1X10-Cl^{-}-H_2O$ and in Figure 1.105 Q is plotted against W for calorimetric data on the same system [14]. The curves in Figures 1.104 and 1.105 have been calculated with the constants given in Table 1.38.

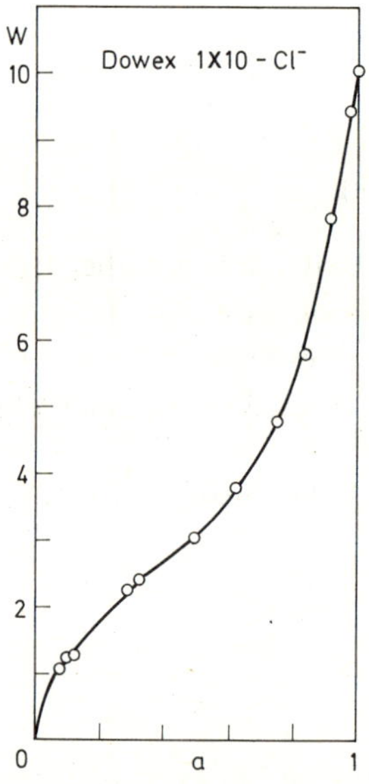

Figure 1.104 W plotted against a for the system Dowex $1X10-Cl^{-}$. Data from ref. [14].

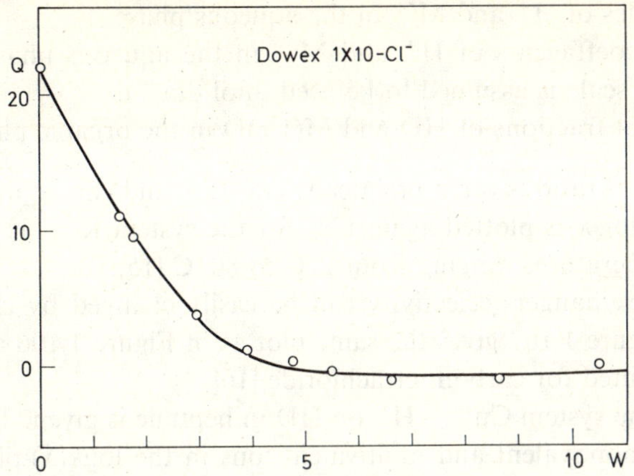

Figure 1.105 Q in kJ/mol plotted against W for the system Dowex 1X10$-Cl^-$$-H_2O$. Data from ref. [14].

Table 1.38 Species and constants for the system Dowex 1 × 10-Cl^--H_2O. T = 298 K. From ref. [14]

Species	k_n	q_n	ΔG_n (kJ/mol)	ΔH_n (kJ/mol)	$T\Delta S_n$ (kJ/mol)	ΔS_n (J/(K · mol))
RCl	–	22.0	–	–	–	–
$RCl(H_2O)$	48	12.55	− 9.60	− 9.46	+0.14	+ 0.5
$RCl(H_2O)_2$	250	12.47	−13.69	− 9.54	+0.15	+13.9
$RCl(H_2O)_4$	1840	−4.39	−18.64	−26.40	−7.76	−26.0
$RCl(H_2O)_{10}$	3000	0	−19.85	−22.01	−2.16	− 7.2
$RCl(H_2O)_{18}$	1700	0	−18.44	−22.01	−3.57	−12.0

1.9.2.3 Ion exchange

Cation exchange. Although organophosphorous esters like di(-2-ethylhexylphosphoric)acid also extract metals by ion exchange, no simple stoichiometry results and the number of extractant molecules involved in the process is not easy to predict. For this reason the discussion will be limited to HD as compared to strong acid resins.

Consider the ion exchange reaction

$$(1/z)M^{z+} + HD(org) \rightleftarrows M(1/z)D(org) + H^+ . \qquad (11)$$

It is advantageous to work with a dimensionless equilibrium "constant", κ, defined by

$$\kappa_{M,H} = \frac{\bar{x}_M^{1/z}\, N_H\, y_H}{\bar{x}_H \cdot N_M^{1/z}\, y_M^{1/z}} \qquad (12)$$

N_H, N_M = normalities of H^+ and M^{z+} in the aqueous phase.
y_H, y_M = activity coefficients of H^+ and M^{z+} in the aqueous phase. Here the molarity scale is assumed to be used (mol dm^{-3}).
$\bar{x}_H, \bar{x}_M$ = equivalent fractions of HD and $M(1/z)D$ in the organic phase.

The activity coefficient ratio is often practically constant and can be included in κ.

In Figures 1.106 log κ is plotted against $\bar{x}_K$ for the system $K^+ - H^+$ on HD in heptane at four temperatures ranging from 1 °C to 80 °C [15].

With liquid ion exchangers selectivity can be easily changed by changing the diluent (solvent). Figure 1.107 gives the same plot as in Figure 1.106 at 298K but with heptane substituted for carbon tetrachloride [16].

In Figure 1.108 the system $Cu^{2+} - H^+$ on HD in heptane is given. There is little difference between monovalent and multivalent ions in the log$\kappa(\bar{x})$-plots. Similar curves are also obtained for solid resins. In Figure 1.109 the system $Cs^+ - Na^+$ on Dowex 50 of various crosslinkings studied by Soldano et al. [17] is plotted. The meaning of the curves and crosses [18] will be discussed below.

Muraviev [19] studied the ion exchange between Na^+ and H^+ on HD dissolved in toluene and then adsorbed on a Amberlite XAD−2 resin. Such a system can be

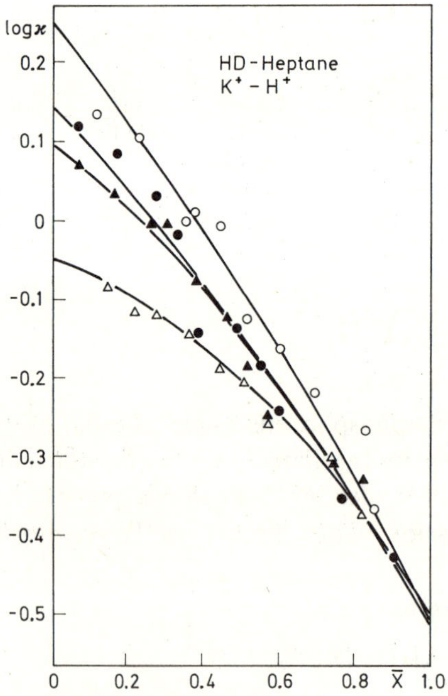

Figure 1.106 Logκ plotted against $\bar{x}_K$ for the system $K^+ - H^+$ on HD in heptane at various temperatures. After ref [15].
○ 1°C; ● 11°C; ▲ 25°C; △ 80°C.
The curves have been calculated from the constants given in Table 1.39.

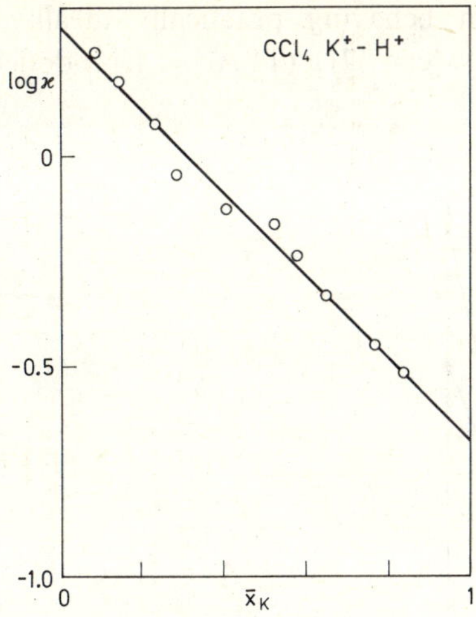

Figure 1.107 Logκ plotted against $\overline{x}_K$ for the system $K^+ - H^+$ on HD in carbon tetrachloride. The curve has been calculated from the constants in Table 1.39. Reproduced with permission from Pergamon Press.

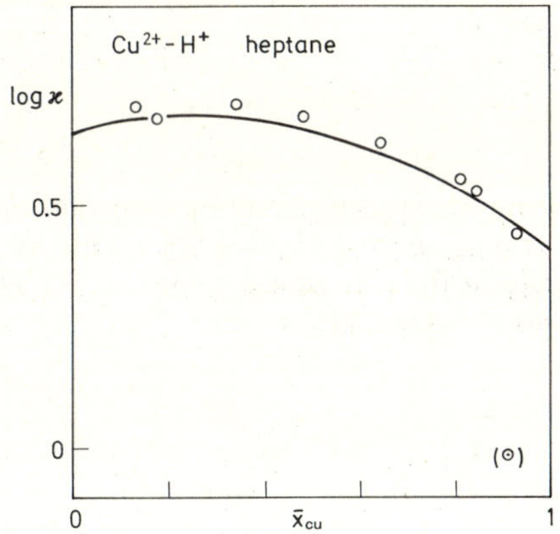

Figure 1.108 Logκ plotted against $\overline{x}_{Cu}$ for the system $Cu^{2+} - H^+$ in heptane. The curve has been calculated from the constants in Table 1.39. Reproduced with permission from Pergamon Press.

regarded as solid resin with easily changed capacity. In Figure 1.110 log κ is plotted against $\overline{x}_{Na}$ for this system and samples with capacity decreasing from bottom to top. Again curves similar to those in Figures 1.106–1.108 are obtained.

Anion exchange. Liquid anion exchange data are mostly treated as an extraction of anionic metal complexes. The protonated form of long-chain amines are known to aggregate in organic solvents. Sometimes data can be treated as if no aggregation occured [20, 21]. In other cases it must be taken into account [22, 23]. As example of a liquid anion exchange system behaving practically ideally, the system $TcO_4^- - NO_3^-$ on $TLAHNO_3$ in *o*-xylene [21] (TLA = tri-*n*-dodecylamine = trilaurylamine) can be mentioned.

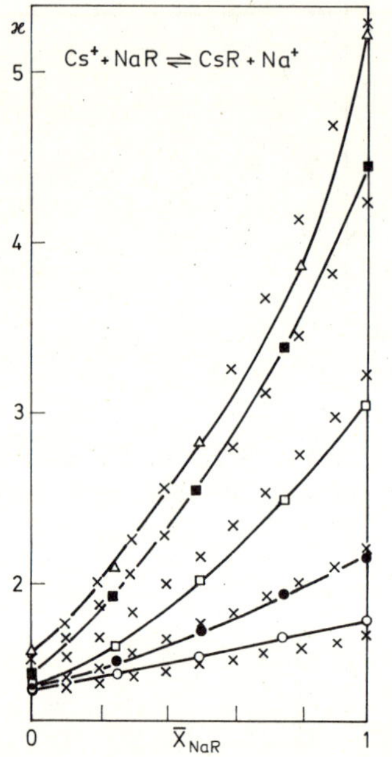

Figure 1.109 The system $Cs^+ - Na^+$ on Dowex 50 of various crosslinkings studied by Soldano et al. [17]. Reproduced with permission from Elsevier Science Publishers.
The full-drawn curves have been calculated from the constants in Table 1.39. The crosses have been calculated from the equations in Table 1.40.

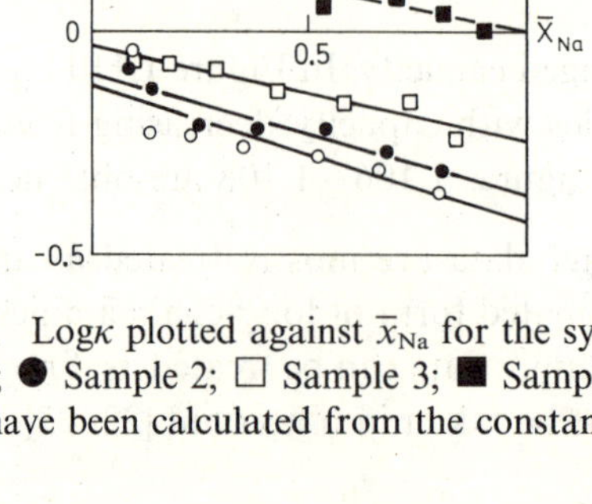

Figure 1.110 Logκ plotted against $\bar{x}_{Na}$ for the system $Na^+ - H^+$ on HD−XAD-2.
○ Sample 1; ● Sample 2; □ Sample 3; ■ Sample 4; △ Sample 5; ▲ Sample 6.
The curves have been calculated from the constants in Table 1.39. From ref. [19].

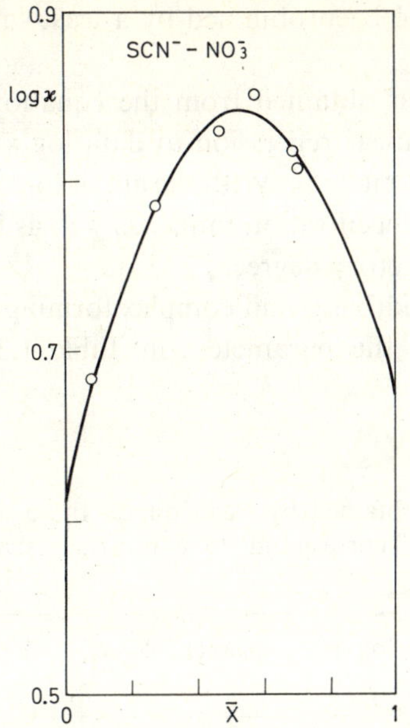

Figure 1.111 Logκ plotted against $\bar{x}_{NO3}$ for the system $SCN^- - NO_3^-$ on a pyridine resin. From ref. [24]. The curve has been calculated from the constants in Table 1.39.

Solid resins may also behave nonideally, as shown in Figure 1.111 for the system $SCN^- - NO_3^-$ on a pyridine resin [24].

All curves in Figures 1.106 – 1.111 can be fitted with one and the same model.

A model for ion exchange. A model to account for nonideal behavior in ion exchangers has been recently devised [25]. For any extensive thermodynamic property transformed into the corresponding molar property, Y, the following equation was obtained from Guggenheim's zeroth approximation:

$$Y = y_1\bar{x}_1^2 + y_2\bar{x}_2^2 + 2y_m\bar{x}_1\bar{x}_2 \tag{13}$$

y_1, y_2 = Y for the pure ionic forms ($\bar{x}_1 = 1$, $\bar{x}_2 = 1$), y_m refers to the mixed form.
$\bar{x}_1, \bar{x}_2$ = equivalent fractions of components 1 and 2, cf. eq. (12).

Most ion exchange data are plotted against $\bar{x}_1$ or $\bar{x}_2$, as in Figures 1.106 – 1.111.

Here the following equation can be used:

$$Y = y_1\bar{x}_1 + y_2\bar{x}_2 + B\bar{x}_1\bar{x}_2 . \tag{14}$$

From (13) and (14)

$$y_m = \frac{1}{2}[y_1 + y_2 + B], \tag{15}$$

B is a constant for taking account of deviations from linearity in the plot of $Y(\bar{x})$. The curves in Figures 1.106 – 1.111 have been calculated from eq. (14) with Y = log

κ. The constants given in Table 1.39 have been obtained by a least-squares fit to data log $\kappa(\bar{x})$.

The crosses in Figure 1.109 have been obtained from the equations given in Table 1.40 [18]. They were obtained by linear regression to data log $\kappa(0)$ and log $\kappa(1)$ as functions of the nominal DVB content in %. With the aid of four parameters the system can be fitted over the whole concentration range, as well as for a DVB-content from 2 to 16% to a rather satisfactory degree.

The extension of the model to weak acid(base) and complex forming extractants (resins) is obvious. It remains to relate the parameters in Table 1.39 to more fundamental properties.

Table 1.39 Constants in eqs. (14) and (15) obtained by least-squares fit to data log$\kappa(\bar{x})$. The ion given first under the heading "System" corresponds to $\bar{x}$. For the system K^+-H^+, $\bar{x} = \bar{x}_K$ and $\bar{x}_K = 0$ corresponds to $\kappa(0)$.

System	Exchanger	t (°C)	Sample number	log$\kappa(0)$	log$\kappa(1)$	logκ_m	log$K$	Ref.
K^+-H^+	HD$-C_7H_{16}$	1	–	0.25	−0.50	−0.05	−0.10	15
K^+-H^+	HD$-C_7H_{16}$	11	–	0.14	−0.51	−0.11	−0.20	15
K^+-H^+	HD$-C_7H_{16}$	25	–	0.10	−0.52	−0.08	−0.17	15
K^+-H^+	HD$-C_7H_{16}$	80	–	−0.05	−0.50	−0.13	−0.23	15
K^+-H^+	HD$-CCl_4$	25	–	0.31	−0.98	−0.34	−0.34	16
$Cu^{2+}-H^+$	HD$-C_7H_{16}$	25	–	0.65	0.40	0.78	0.61	16
Cs^+-Na^+	Dowex 50	25	–	–	–	–	–	17
Cs^+-Na^+	2% DVB	25	–	0.25	0.14	0.20	0.20	17
Cs^+-Na^+	4% DVB	25	–	0.33	0.14	0.24	0.24	17
Cs^+-Na^+	8% DVB	25	–	0.48	0.12	0.30	0.30	17
Cs^+-Na^+	12% DVB	25	–	0.65	0.16	0.41	0.41	17
Cs^+-Na^+	16% DVB	25	–	0.71	0.20	0.46	0.46	17
Na^+-H^+	HD$-$XAD2	25	1	−0.12	−0.42	−0.27	−0.27	19
Na^+-H^+	HD$-$XAD2	25	2	−0.10	−0.36	−0.23	−0.23	19
Na^+-H^+	HD$-$XAD2	25	3	−0.03	−0.25	−0.14	−0.14	19
Na^+-H^+	HD$-$XAD2	25	4	0.19	−0.02	0.09	0.09	19
Na^+-H^+	HD$-$XAD2	25	5	0.18	−0.05	0.90	0.34	19
Na^+-H^+	HD$-$XAD2	25	6	0.48	0.17	0.87	0.51	19
$SCN^--NO_3^-$	Pyridine resin	25	–	0.67	0.61	1.04	0.77	24

Table 1.40 log$\kappa(0)$ and log$\kappa(1)$ for the system Cs^+-Na^+ expressed as functions of the divinylbenzene content (%) in Dowex 50 resins. T = 298 K. From ref. [18]

Constant	Equation
$\kappa(0)$	$\log\kappa(0) = 0.1992 + 3.412 \cdot 10^{-2}$ (%)
$\kappa(1)$	$\log\kappa(1) = 0.1207 + 3.840 \cdot 10^{-3}$ (%)

Thermodynamics. A thermodynamic equilibrium constant for reaction (11) can be calculated from the integral free energy by [26, 27]:

$$\log K = \int_0^1 \log\kappa(\bar{x})\mathrm{d}\bar{x}. \quad (16)$$

From eqs. (14)–(16)

$$\log K = \frac{1}{3}[\log\kappa(0) + \log\kappa(1) + \log\kappa_{\mathrm{m}}]. \quad (17)$$

In Table 1.39 the log*K*-values obtained from eq. (17) are also given.

For the excess free energies of the two components third degree polynominals in $\bar{x}$ are obtained. For the case of linear log $\kappa(\bar{x})$, i.e. $B = 0$ the activity coefficients take the form

$$\log f_1 = \frac{1}{2} C\bar{x}_2^2$$
$$\log f_2 = \frac{1}{2} C\bar{x}_1^2 \quad (18\,\mathrm{a-c})$$
$$C = \log\kappa(1) - \log\kappa(0).$$

The system thus behaves like a regular solution.

Table 1.41 Water uptake during exchange between $R^+ - H^+$ on HD in three diluents as well as the solid resin KRS-8. $T = 298$ K. Ionic strength $I = 0.1$ (R,H)Cl. From ref. [29]

Diluent	R^+	w_1	w_2	w_m
Heptane	$C_2H_5NH_3^+$	2.6	10.3	2.3
	$(C_2H_5)_2NH_2^+$	3.7	10.3	1.7
	$(C_2H_5)_3NH^+$	8.9	10.3	1.5
	$(C_2H_5)_4N^+$	(~ 26)	10.3	–
Nitrobenzene	$C_2H_5NH_3^+$	1.8	6.1	2.8
	$(C_2H_5)_2NH_2^+$	0.9	6.1	1.6
	$(C_2H_5)_3NH^+$	0.8	6.1	1.6
	$(C_2H_5)_4N^+$	1.8	6.1	1.4
2-Ethyl-hexanol	$C_2H_5NH_3^+$	1.8	2.7	2.2
	$(C_2H_5)_2NH_2^+$	1.1	2.7	1.9
	$(C_2H_5)_3NH^+$	1.6	2.7	3.0
	$(C_2H_5)_4N^+$	2.0	2.7	2.4
KRS-8	$C_2H_5NH_3^+$	7.2	12.8	9.2
	$(C_2H_5)_2NH_2^+$	5.9	12.8	7.2
	$(C_2H_5)_3NH^+$	5.2	12.8	5.5
	$(C_2H_5)_4N^+$	5.2	12.8	3.9

w_1 refers to pure ammonium salt, w_2 to hydrogen form

According to the model the system is formally treated as a binary one. This is possible because both the water and diluent (solvent) activities are practically constant during ion exchange.

Soldatov and Bichkova recently extended the model to three ions [28].

Water extraction during ion exchange. Since different ionic forms extract different amounts of water, the water extraction must change during the ion exchange process. For solid resins the water often varies linearly with composition, which is the case for $B = 0$ in eq. (14). This process has been studied extensively for HD and to some extent for solid resins. Typical results are shown in Figure 1.112, where W is plotted against $\bar{x}$ for ethylammonium ions versus H^+ on HD in different solvents: a) heptane, b) 2-ethylhexanol, d) nitrobenzene. In c) the solid resin KRS-8 is shown. Very similar curves are obtained for HD and the solid resin. The data in Figure 1.112 have been fitted with the model, with the exception of the system

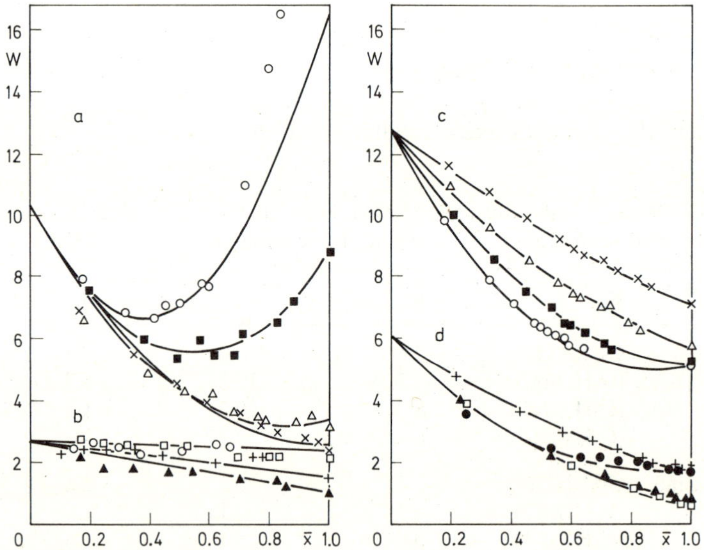

Figure 1.112 W plotted against $\bar{x}_{RD}$ for systems $R^+ - H^+$ on HD in:

a) Heptane; × $R^+ = C_2H_5NH_3^+$; △ $R^+ = (C_2H_5)_2NH_2^+$; ■ $R^+ = (C_2H_5)_3NH^+$; ○ $R^+ = (C_2H_5)_4N^+$

b) 2-Ethylhexanol; □ $R^+ = C_2H_5NH_3^+$; ▲ $R^+ = (C_2H_5)_2NH_2^+$; + $R^+ = (C_2H_5)_3NH^+$; ○ $R^+ = (C_2H_5)_4N^+$

d) Nitrobenzene; + $R^+ = C_2H_5NH_3^+$; ▲ $R^+ = (C_2H_5)_2NH_2^+$; □ $R^+ = (C_2H_5)_3NH^+$; ● $R^+ = (C_2H_5)_4N^+$

c) The system $K^+ - H^+$ on KRS-8.
× $R^+ = C_2H_5NH_3^+$; △ $R^+ = (C_2H_5)_2NH_2^+$ ■ $R^+ = (C_2H_5)_3NH^+$; ○ $R^+ = (C_2H_5)_4N^+$

The curves have been calculated from the constants in Table 1.41.

$(C_2H_5)_4N^+ - H^+$ on HD in heptane where $W(1) \approx 26$, which might imply large structural changes not possible to account for by the simple model. The constants found to fit the data are given in Table 1.41 [29].

1.9.2.4 Aggregation

Liquid cation exchangers. Compounds like HD aggregate in organic solvents. The common opinion is that the systems are monodisperse, supported by the ultracentrifugation study of BaD_2 in benzene by Kaufman et al. [30]. They found that the salt formed octamers. The loading method gave values from 7–15 for HD in different diluents [31]. For HD in heptane Morris [32] suggested heptamers.

Low angle scattering data by Soldatov and coworkers suggest aggregation numbers of the same order as above, but polydispersity of the system.

Aggregation changes with composition as shown in Figures 1.113, where the average aggregation number, $\bar{n}$, as determined by vapor phase osmometry, is plotted against $\bar{x}_{RD}$ for the system $(C_2H_5)_4N^+$ on HD in heptane [33]. The upper curve refers to a fully hydrated sample and the lower one to a dry sample. Presence of water enhances aggregation, indicating that water participates in forming the micelles.

Equ. (14) applies also here, but the numbers arrived at have only formal meaning.

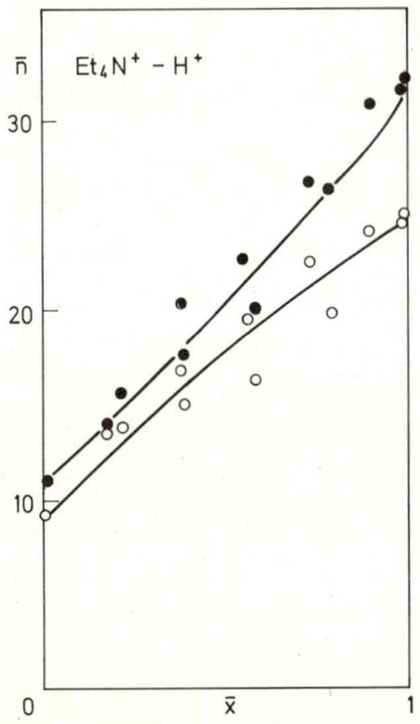

Figure 1.113 The average degree of aggregation, $\bar{n}$, plotted against $\bar{x}_{RD}$ for the system $(C_2H_5)_4N^+ - H^+$ in heptane.
● Wet sample; ○ dry sample.

Liquid anion exchangers. Quaternary ammonium salts are known to form large aggregates [34]. Tertiary ammonium salts might form large aggregates in aliphatic diluents [35]. In aromatics the aggregation is much less extensive. This is illustrated in Figure 1.114, where the average $\bar{p}$ is plotted against $\bar{q}$ for the reaction

$$p\text{TLA(org)} + q\text{H}^+ + q\text{NO}_3^- \rightleftarrows (\text{TLA})_p(\text{HNO}_3)_q\text{(org)}$$

in the diluent tertiary butylbenzene [36].

According to Figure 1.114 $p = q$ and the aggregates are $(TLAHNO_3)_n$. This is typical of ammonium salts. The data can be explained by formation of $(TLAHNO_3)_2$ and $(TLAHNO_3)_3$ with predominance of the trimer. The liquid anion exchangers are thus polydisperse and aggregated to a much lesser extent than sulfonic acids.

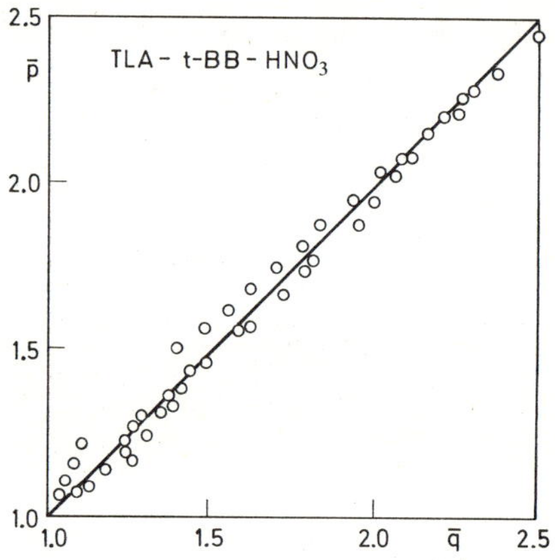

Figure 1.114 $\bar{p}$ plotted against $\bar{q}$ for the system TLA-*t*-BB-HNO_3 at 298 K; the ionic strength $I = 5$ M $(Na,H)NO_3$.

Concluding remarks. It has been shown above that cationic micelles with aggregation numbers in the range 10—20 can be described by the same model as solid resins. A possible explanation is that the ion exchange process is a site effect, with the micelle behaving as a separate phase like a solid resin.

For anion exchange this may not be possible in aromatic diluents, where the aggregation is small, cf. [22, 23]. On the other hand, more work is needed to understand why sometimes such systems can be treated as a separate phase, i. e., independent of aggregation.

1.9.3 Applications

Liquid ion exchangers can be used in practically every instance where solid resins can be used.

In analytical chemistry liquid ion exchangers have found extensive use, as illustrated by the review of Green [37]. An often used cation exchanger is di-2-ethylhexylphosphoric acid while tri-*n*-octylamine can be regarded as a typical anion exchanger in analytical work.

Many different ion-selective electrodes (ISE) have been constructed and employed in analytical chemistry, as well as in equilibrium analysis, i. e., the field of solution chemistry concerned with determination of stability constants of complexes. Although liquid membrane electrodes are less stable than solid-state electrodes, they are widely used. They require frequent rejuvenation (loss of extractant, etc.). With careful calibration they can be used even in the exacting field of equilibrium analysis. This is illustrated in a recent compilation of stability constants [38].

Solvent-impregnated resins were recently reviewed by Warshawsky [39], who treated preparation, analytical and technological applications. Each technological process has a range of applicability, where it is competitive with other methods. This is shown in Figure 1.115, where the logarithm of the metal ion concentration is plotted against flow rate. The new technique of employing liquid membranes deserves some comments. The liquid surfactant membrane (LSM) consists of droplets containing a thin organic film enclosing an aqueous phase and dispersed in another. The film consists of an extractant dissolved in a suitable diluent and a surfactant to ensure emulsion formation. The transport from one aqueous phase to the other is achieved via coupled extraction/reextraction (stripping) processes taking place simultaneously on both sides of the liquid membrane. Thanks to these coupled processes metal ions can be extracted against their concentration gradient. One problem is to achieve stable-enough emulsions. An alternative is the solid supported liquid membranes (SSLM), where the extractant is immobilized on

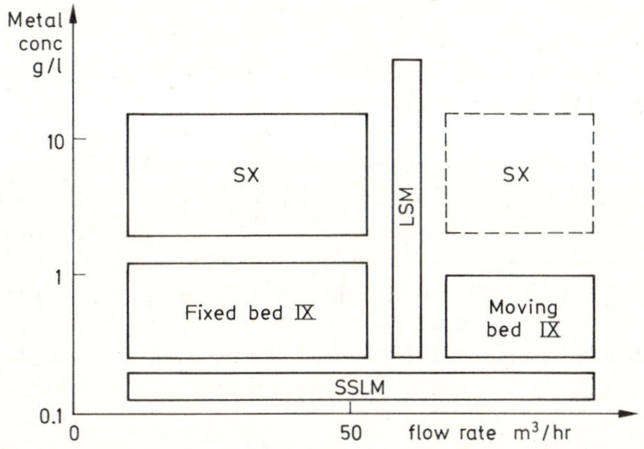

Figure 1.115 A semilogarithmic plot of metal concentration in g/l against flow-rate (m^3/h). The range of applicability shown for different ion exchange and solvent extraction techniques. IX = ion exchange; SX = solvent extraction; LSM = surfactant liquid membrane; SSLM = solid supported liquid membrane [41].

microporous inert supports interposed between the two aqueous solutions. As seen in Figure 1.115 LSM and SSLM are useful in different regions of metal concentration and flow rate. Both processes occur under nonequilibrium conditions, so kinetic parameters are of interest. Typical extractants used besides selective coordinating reagents are di(-2-ethylhexylphosphoric) acid and long-chain tertiary amines. The advantages of the LSM and SSLM processes are summarized in Table 1.42. These two techniques are presently being applied to a number of technical processes together with mechanistic studies in order to reveal fundamental parameters. A good idea of the problems presently under study can be obtained from recent conference proceedings such as reference [40].

Table 1.42 Advantages of liquid membrane processes

1. High selectivities
2. Ions can be pumped uphill
3. Fluxes are higher than with solid membranes
4. Low capital and operating costs
5. Expensive extractants can be used
6. High separation factors can be achieved in a single stage
7. High feed/strip volume ratios

References

[1] J. L. Aparicio (1986) Studies on the Kinetics of Solvent Extraction Processes, Thesis for licentiate degree, Department of Inorganic Chemistry, The Royal Institute of Technology, Stockholm, Sweden.

[2] E. Högfeldt (1966) Liquid Ion Exchangers, in: Ion Exchange, 1 (J. A. Marinsky, ed.) pp. 139–171, Dekker, New York.

[3] V. S. Soldatov, L. V. Yurevich and E. Högfeldt (1977) On the properties of solid and liquid ion exchangers-IV Thermodynamic and ir-investigation of the hydration of dinonylnaphthalene sulfonic acid in heptane. J. Inorg. Nucl. Chem. *39*, 2069–2073.

[4] S. Lapanje and D. Dolar (1958) Thermodynamic functions of swelling of crosslinked polystyrenesulphonic acid resins I. Resins in the hydrogen state. Z. phys. Chem., N. F., *18*, 11–25.

[5] J. S. Redinha and J. A. Kitchener (1963) Thermodynamics of ion-exchange processes. System H, Na and Ag with polystyrene sulphonate resins Trans. Faraday Soc. *59*, 515–529.

[6] E. Högfeldt, V. S. Soldatov and Z. I. Kuvaeva (1976) On the properties of solid and liquid ion exchangers III. The hydration equilibria of some salts of dinonylnaphthalene sulfonic acid. Chemica Scripta *10*, 210–214.

[7] A. Kellomäki (1978) Adsorption of protium and deuterium oxides on cation exchange resins. Acta Chem. Scand. *A32*, 747–751.

[8] A. Kellomäki (1980) Preferential adsorption of protium and deuterium oxides from solutions of diethyl ether and ethyl acetate on a sodium ion exchanger. Acta Chem. Scand. *A34*, 43–46.

[9] E. Högfeldt (1960) On the thermodynamics of hydrated protons. Activities, activity coefficients and the average degree of hydration. Acta Chem. Scand. *14*, 1597–1611.

[10] E. Högfeldt and L. Leifer (1975) On hydration and ion pair formation in simple electrolytes I. The systems $KCl-H_2O$, $KBr-H_2O$, $KI-H_2O$ and NH_4Cl-H_2O. Chemica Scripta *8*, 57–62.

[11] E. Högfeldt (1966) Hydration in liquid and solid cation exchangers. Nature *210*, 941–942.

[12] E. Högfeldt and K. Rasmusson (1966) The extraction of water and hydrochloric acid by aromatic hydrocarbons. Sv. Kem. Tidskr. *78*, 490–499.

[13] G. Zundel (1969) Hydration and Intermolecular Interaction, Academic Press, New York · London.

[14] Z. I. Sosinovich, L. V. Novitskaya, V. S. Soldatov and E. Högfeldt (1985) Thermodynamics of water sorption on Dowex 1 of different crosslinking and ionic form, in: Ion Exchange and Solvent Extraction, *9* (J. A. Marinsky and Y. Marcus, eds.) pp. 303–338, Dekker, New York.

[15] E. Högfeldt, Z. I. Kuvaeva and V. S. Soldatov (1978) On the properties of solid and liquid ion exchangers-V. The potassium-hydrogen exchange on dinonylnaphthalene sulfonic acid in heptane in the temperature range 1–80 °C. J. Inorg. Nucl. Chem. *40*, 1405–1407.

[16] Z. I. Kuvaeva, V. S. Soldatov, F. Fredlund and E. Högfeldt (1978) On the properties of solid and liquid ion exchangers-VI. Application of a simple model to some ion exchange reactions on dinonylnaphthalene sulfonic acid in different solvents and for different cations with heptane as solvent. J. Inorg. Nucl. Chem. *40*, 103–108.

[17] B. Soldano, Q. V. Larson and G. E. Myers (1955) The osmotic approach to ion exchange equilibria. II. Cation exchangers. J. Am. Chem. Soc. *77*, 1339–1344.

[18] E. Högfeldt (1984) A useful method for summarizing data in ion exchange. I. Some illustrative examples. Reactive Polymers *2*, 19–30.

[19] D. N. Muraviev and E. Högfeldt (1988) Stability and ion exchanger properties of Amberlite XAD-2 impregnated with dinonylnaphthalene sulfonic acid. Reactive Polymers *8*, 97–102.

[20] L. Kuca and E. Högfeldt (1968) On extraction with long-chain tertiary amines X. The mechanism of the extraction of trivalent iron by trilaurylammonium chloride. Acta Chem. Scand. *22*, 183–192.

[21] A. Beck, D. Dyrssen and S. Ekberg (1964) Extraction of Tc(VII) by trilaurylammonium nitrate in xylene. Acta Chem. Scand. *18*, 1695–1702.

[22] M. Aguilar and M. Muhammed (1976) On extraction with long-chain amines-XXVI. The extraction of $ZnCl_2$ by tri-*n*-dodecylammonium chloride dissolved in benzene. J. Inorg. Nucl. Chem. *38*, 1193–1197.

[23] M. Valiente and M. Muhammed (1984) On extraction with long-chain amines XXX. The extraction of copper(II) from chloride solutions by tri-*n*-laurylammonium chloride dissolved in toluene. Chemica Scripta *23*, 64–72.

[24] K. Gärtner (1963) Zur Thermodynamik der Austauschadsorption 1. Experimentelle Bestimmung der thermodynamischen Standardfunktionen für den Austauschvorgang an stark basischen Anionenaustauschern. Z. phys. Chem. (Leipzig) *223*, 132–140.

[25] E. Högfeldt and V. S. Soldatov (1979) On the properties of solid and liquid ion exchangers-VII. A simple model for the formation of mixed micelles applied to salts of dinonylnaphthalene sulfonic acid. J. Inorg. Nucl. Chem. *41*, 575–577.

[26] E. Ekedahl, E. Högfeldt and L. G. Sillén (1950) Activities of the components in ion exchangers. Acta Chem. Scand. *4*, 556–558.

[27] E. Högfeldt (1952) On ion exchange equilibria. II. Activities of the components in ion exchangers. Arkiv Kemi *5*, 147–171.

[28] V. A. Bichkova and V. S. Soldatov (1985) A method for predicting ion exchange equilibria in ternary ion exchange systems. Reactive Polymers *3*, 207–215.

[29] A. V. Mikulich, V. S. Soldatov and E. Högfeldt (1985) Ion exchange of liquid and solid cation exchangers with ethylammonium ions. Acta Chem. Scand. *A39*, 583–586.

[30] T. F. Ford, S. Kaufman and O. D. Nichols (1966) Ultracentrifugal studies of barium dinonylnaphthalenesulfonate-benzene systems. I. Sedimentation velocity. J. phys. Chem. *70*, 3726–3732.

[31] E. Högfeldt, R. Chiarizia, P. R. Danesi and V. S. Soldatov (1981) Structure and ion exchange properties of dinonylnaphthalene sulfonic acid and its salts. Chemica Scripta *18*, 13–43.

[32] D. F. C. Morris (1981) in: Proceedings of symposium on solvent extraction of metals, Hamamatsu, Japan, pp. 43–50.

[33] E. Högfeldt, A. V. Mikulich and V. S. Soldatov (1988) On the aggregation of tetramethyl and tetraethylammonium salts of dinonylnaphthalene sulfonic acid in heptane. Acta Chem. Scand. A42 301–303.

[34] Y. Marcus and A. S. Kertes (1969) Ion Exchange and Solvent extraction of Metal Complexes, Chapter 10, Wiley-Interscience, London, New York.

[35] E. Högfeldt, F. Fredlund and K. Rasmusson (1964) On extraction with long-chain tertiary amines VI. Aggregation of the trilaurylamine-nitric acid complex in *n*-octane and *n*-dodecane studied by two-phase emf-titrations. Trans. Royal Inst. Technol. *Nr 229*, pp. 1–16.

[36] E. Högfeldt and F. Fredlund, to be published.

[37] H. Green (1973) Talanta Review. Use of liquid ion-exchangers in inorganic analysis. Talanta *20*, 139–161.

[38] E. Högfeldt (1982) Stability Constants of Metal-Ion Complexes, Part A, Inorganic Ligands, IUPAC Chemical Data Series, No *21*, Pergamon Press, Oxford.

[39] A. Warshawsky (1981) Extraction with Solvent-Impregnated Resins, in: Ion Exchange and Solvent Extraction, *8* (J. A. Marinsky and Y. Marcus, eds.) pp. 229–310, Dekker, New York.

[40] Proceedings of International Solvent Extraction Conference (ISEC), Denver 1983.

[41] M. Muhammed (1986) Private communication.

1.10 Ion Exchange Membranes

Hideo Kawate, Kazuo Tsuzura and Hiroshi Shimizu
Asahi Chemical Industry Co., Ltd., Kawasaki and Japan Organo, Tokyo, Japan

1.10.1 Historical review

Functionally, the ion exchange membrane is related to both cell membrane biochemistry and electrochemistry. It is an ion exchange body in membrane form, thus providing special functional characteristics which cannot be obtained with ion exchange resins.

In 1925 Michaelis [1 – 4] performed studies on collodion membranes, and attempted an explanation of their ionic permselectivity as a sieving effect of their capillaries under the influence of the electric charge in the membrane. This research on collodion membranes was the starting point for the history of synthetic ion exchange membranes. Quantitative descriptions of the permselectivity based on the Donnan equilibrium theory were formulated by Meyer and Sievers [5] and by Teorell [6] in 1935 and 1936. In 1949, Sollner [7] attempted a quantification of bi-ionic potential, as a measure of membrane permselectivity between various ions of the same charge. Although this question is important for practical applications and has been studied by many others since then, quantification has not yet been achieved.

In 1940, Meyer [8] reported the achievement of electrodialysis in an alternating arrangement of cation-selective collodion membranes and anion-selective membranes of cellophane coated with proteins. At that time, it was not yet possible to obtain membranes applicable in industrial applications. With the report by W. Juda [9] of his fabrication of membranes from ion exchange bodies of synthetic resin in 1950, a period of rapid progress began, with research on both ion exchange membranes and electrodialysis based on these membranes.

The research efforts for development and utilization of ion exchange membranes composed of hydrocarbon polymers were directed mainly toward water desalination in the United States and Europe, while in Japan they were directed toward the production of salt from sea water, by companies such as Asahi Chemical, Asahi Glass, Tokuyama Soda, and Japan Organo.

The first plants utilizing ion exchange membranes for salt production began operation in 1960. Stable operation proved difficult, however, due to scaling which occurred when Ca^{2+} and SO_4^{2-} ions present in the sea water were concentrated by the ion exchange membrane together with the Na^+, Cl^-, and other ions. The scale tended to accumulate in the equipment and impede flow, and in extreme cases formed within the membrane and caused membrane rupture. A fundamental solution to the scaling problem was achieved in 1965, with the development and application of a membrane which effectively suppressed the concentration of sea water components leading to scale, and thus allowed attainment of the first truly commercial process. By 1973 the ion exchange membrane process had completely replaced the traditional process of salt production in solar fields in Japan, which had been labor-intensive and constantly subject to adverse weather, with no prospect of significant improvement.

The use of ion exchange membranes for water desalination is limited in Japan to small islands and other relatively isolated areas, due to the generally abundant rainfall. In other parts of the world, however, early research on the production and utilization of ion exchange membranes for this purpose was begun by organizations such as Ionics Corporation, the South Africa National Chemical Research Laboratory, and T. N. O., the Dutch Organization for Applied Scientific Research. The Kennedy Administration, in its space programs, lent its support to water conversion programs, and vigorous research and development projects proceeded with financial assistance from the Office of Saline Water of the U.S. government.

Since electrodialysis inherently involves concentration, the problem of scale formation also impeded early progress in water desalination applications. This problem was eliminated by the EDR technology developed by Ionics, which has led to the widespread use of the ion exchange membrane for water desalination.

As described in the following pages, current applications of ion exchange hydrocarbon membranes include electrodialysis, electrolysis, dialysis and fuel-cell functions. Nevertheless the largest use of these membranes remains the production of salt from sea water in Japan, and water conversion in the U.S., the Middle East, and other areas.

Even as the development of hydrocarbon membranes progressed, it became increasingly desirable to find other membrane materials free from their poor resistance to heat and chemicals. This goal of many years standing was achieved by the advent of Du Pont's Nafion perfluoro ion exchange membrane in 1971, which possessed the heat and chemical resistance necessary for use in chlor-alkali electrolysis. This membrane, with functional groups of sulfonic acid, provided low electric resistance but also exhibited low current efficiency. The invention of the perfluoro carboxylic acid ion exchange membrane by Seko [10] in 1974 brought a substantial increase in current efficiency, and led to the development of composite membranes possessing both the low electric resistance of the sulfonic acid membrane and the high current efficiency of the carboxylic acid membrane. The current efficiency of a membrane is governed largely by the surface-layer facing the cathode, and chlor-alkali electrolysis now generally employs composite membranes which have an ultra-thin carboxylic acid layer and a comparatively thick sulfonic acid layer.

1.10.2 Membrane types and methods of preparation

1.10.2.1 General classification

The three basic categories of ion exchange membranes are anion exchange, cation exchange, and combination anion-cation exchange membranes, in which the fixed-ion groups are respectively cations, anions, and a combination of the two. They can be further classified as shown below according to degree of chemical homogeneity, substrate polymer, and ion exchange group. The chemical structures of the three major types of ion exchange membranes currently in common use are shown in Figure 1.116.

Degree of chemical homogeneity

Non-homogeneous membranes

Fabricated [11 – 14] from a mixture of ion exchange resin particles and a binder.

Semi-homogeneous membranes

a) Inter-polymer membranes, fabricated [15 – 16] from a mixture of a polymer containing ion exchange groups and a binder in a solvent common to the two.
b) Graft and snake-cage membranes, produced by impregnation [17 – 18] of a thin polymeric membrane with a monomer that is then polymerized by exposure to radiation. The ion exchange [19] groups may be either present in the monomer or introduced after its polymerization.

Homogeneous membranes

Most practical membranes are of this type, and are composed of either hydrocarbon or fluorocarbon polymers.

a) Hydrocarbon membranes are prepared in the five basic steps of (1) polymerization, (2) crosslinking (three-dimensional), (3) membrane fabrication, (4) emplacement of reinforcing material, and (5) introduction of ion exchange groups.

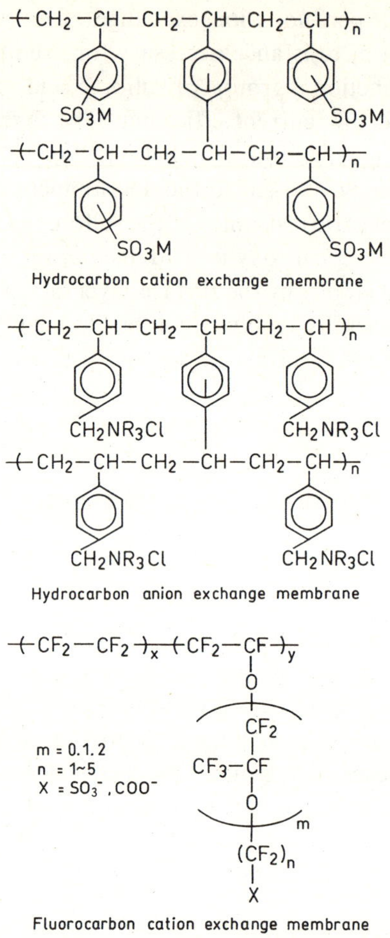

Figure 1.116 Typical chemical structures of ion exchange membranes.

The order of these steps varies according to the type of process and base materials, as follows.

Bulk polymerization process [20]:

Steps (1), (2), and optionally (4) concurrently, then (3), (5).

Paste process [21]:

Steps (3) and (4) concurrently, then (1), (2), (5).

Latex process [22–24]:

Step (1), followed by (3) and (4) concurrently, then (2) and (5).

b) Fluorocarbon membranes have come into increasing use in recent years. They are prepared by copolymerization of tetrafluoro ethylene and perfluorovinyl ether, followed by extrusion into sheets and emplacement of reinforcing material. The membrane form [25] is maintained by hydrophobic bonding and crystallization in the fluorocarbon framework, with no crosslinking.

Substrate polymers

Membrane substrates are generally either hydrocarbon or fluorocarbon polymers. Most of the practical hydrocarbon membranes in current use are obtained by polymerization of styrene monomers, while others are formed by condensation of phenol monomers. Fluorocarbon membranes are generally obtained by polymerization of tetrafluoro ethylene and perfluorovinyl ether.

Ion exchange groups

Cation exchange groups are generally either strongly acidic (sulfonic acid) groups or weakly acidic (carboxylic acid) groups. Anion exchange groups are generally either strongly basic (tertiary ammonium) groups or weakly basic (primary, secondary, or tertiary amine) groups.

1.10.2.2 Special-function ion exchange membranes

The three most important types of special-function ion exchange membranes are the sea water concentration membranes with univalent permselectivity, bipolar membranes for production of acids and bases, and mosaic membranes for piezodialysis.

Membranes with univalent permselectivity can be prepared by the formation on the membrane surface of either a thin, closely crosslinked layer [26–28] or a thin layer with a charge opposite [29–31] to that of the rest of the membrane. The latter form is generally preferred, because of its high permselectivity and low electric resistance.

Bipolar membranes consist of a cation exchange and an anion exchange layer lying back to back. They may be produced by heat bonding [32] a pair of membranes containing a polyethylene matrix binder, and also by forming cation exchange groups in one side of a styrenized polyethylene film by sulphonation and anion exchange groups in the other side by amination. [33]

Table 1.43 Methods of mosaic membrane preparation

Domain size (cm)	Method of preparation
≈ 1	Bonding
$\approx 10^{-1}$	Casting
$10^{-1}-10^{-3}$	Polymer blending Lamination Suspension
$10^{-4}-10^{-5}$	Latex-polyelectrolysis
$\approx 10^{-6}$	Block copolymerization

The mosaic membrane is composed of an array of anion and cation exchange domains with intervening electrically neutral regions. The main methods of fabrication are casting [34], deposition [35–36], light or radiation grafting [37–40], and block copolymerization [41–43]. The domain sizes which can be obtained with these and other methods of preparation are shown in Table 1.43.

1.10.3 Physico-chemical properties of ion exchange membranes

1.10.3.1 Properties and means of determination

Water content, exchange capacity, and Donnan equilibrium

The membrane is an electrolytic three-dimensional matrix of crosslinked or otherwise bonded polymers. When immersed in a solution, it absorbs water and effectively assumes a gel structure.

The water content of the membrane is expressed as the amount of water per gram of swollen membrane at equilibrium in pure water (g H_2O/g wet membrane). The exchange capacity is normally expressed as milliequivalents of exchange group per gram of dry membrane (mequ/g dry membrane). The fixed-ion concentration is expressed as the ion exchange capacity divided by the water content per gram of dry membrane. The relation between these three characteristics is shown in Figure 1.117.

For a membrane immersed in an aqueous electrolytic solution, the relation between the ionic activity in the membrane ($\bar{a}$) and that in the solution (a) can be expressed by the following Donnan equilibrium, with the electrolyte being separated into ν_A moles of cation A and ν_X moles of anion X

$$a_A^{\nu_A} \cdot a_X^{\nu_X} = \bar{a}_A^{\nu_A} \cdot \bar{a}_X^{\nu_X} . \tag{1}$$

With NaCl as the electrolyte and assuming the activity coefficient to equal 1, then

$$C_{Na} \cdot C_{Cl} = \bar{C}_{Na} \cdot \bar{C}_{Cl} . \tag{2}$$

It then follows that the ion concentration of a cation exchange membrane in a NaCl solution can be approximated as

$$\bar{C}_{Cl} = \frac{\sqrt{\bar{C}_r^2 + 4\,C^2} - \bar{C}_r}{2} \tag{3}$$

and

$$\bar{C}_{Na} = \bar{C}_r + \bar{C}_{Cl} \tag{4}$$

$\bar{C}_{Cl}$ = Cl^- ion concentration in membrane (co-ion concentration)
$\bar{C}_{Na}$ = Na^+ ion concentration in membrane (counter-ion concentration)
$\bar{C}_r$ = fixed-ion concentration of membrane
C = concentration of solution in equilibrium with membrane.

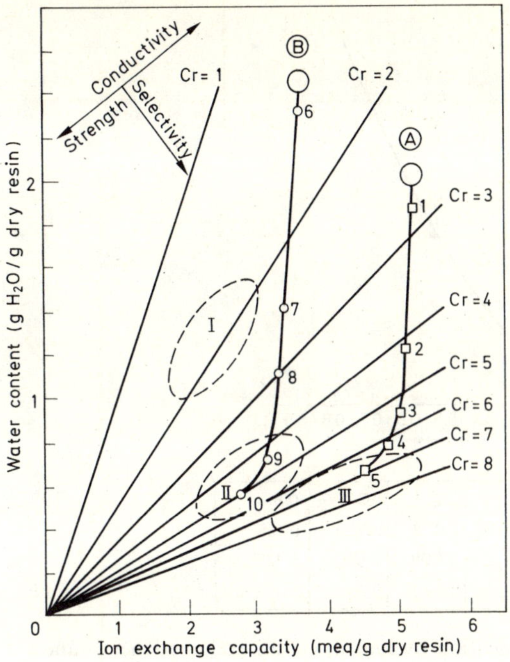

Figure 1.117 Fixed-ion concentration of ion exchange membranes.
Cr: Fixed-ion concentration (mequ/g H_2O)

Ⓐ Dowex-50	Ⓑ Dowex-1
1 DVB 5%	6 DVB 2%
2 DVB 8%	7 DVB 4%
3 DVB 12%	8 DVB 5%
4 DVB 17%	9 DVB 8%
5 DVB 33%	10 DVB 12%

I Nepton CR-51
II Amberplex C-1, A-1; Permaplex C-10, A-10
III Nepton CR-6; various Japanese membranes.

The co-ion concentration (Donnan salt concentration) can be found by removing the membrane from the solution after equilibrium, completely removing adsorbed liquid from the membrane, and then immersing the membrane in pure water. With a cation exchange membrane having a fixed-ion concentration $\overline{C}_r$ of 5 mequ/ml, increasing the concentration C of the solution results in an increase in the ratio $\overline{C}_{Cl}/\overline{C}_{Na}$ in the membrane as shown in Figure 1.118. Approximate determination from eq. (3) and (4) for a 0.5 M NaCl solution yields the values shown in Figure 1.119. As this illustrates, the intra-membrane anion (co-ion) concentration decreases with increasing fixed-ion concentration in the cation exchange membrane, while the intra-membrane cation (counter-ion) concentration increases.

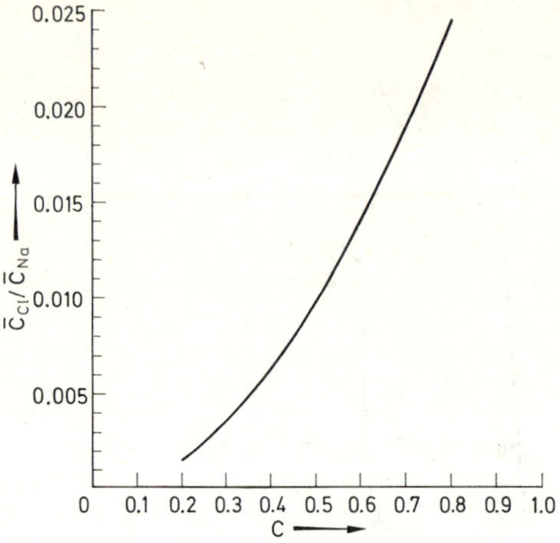

C (MNaCl)	0.2	0.5	0.8
$\bar{C}_{Cl}$ (meq/g H_2O)	0.008	0.050	0.125
$\bar{C}_{Na}$ (meq/g H_2O)	5.008	5.050	5.125
$\bar{C}_{Cl}/\bar{C}_{Na}$	1.6×10^{-3}	9.9×10^{-3}	2.44×10^{-2}

Figure 1.118 Donnan permeation by ions in cation exchange membrane.

C_r (meq/g H_2O)	2.0	3.0	5.0	7.0
$\bar{C}_{Cl}$ (meq/g H_2O)	0.118	0.081	0.050	0.036
$\bar{C}_{Na}$ (meq/g H_2O)	2.118	3.081	5.050	7.036
$\bar{C}_{Cl}/\bar{C}_{Na}$	5.6×10^{-2}	2.6×10^{-2}	9.9×10^{-3}	5.1×10^{-3}

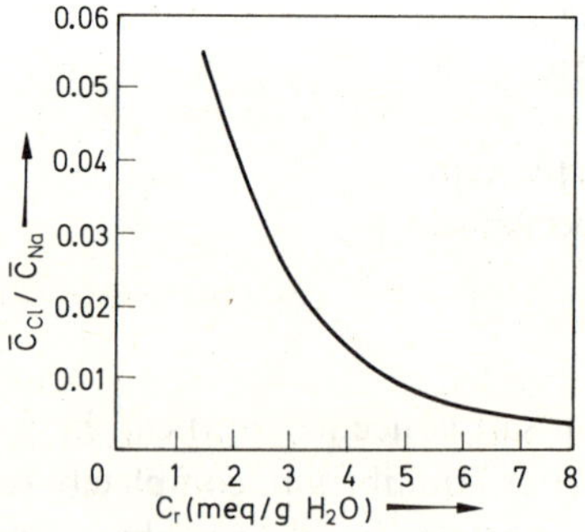

Figure 1.119 Cation exchange membrane fixed-ion concentration and Donnan permeation by ions (0.5N NaCl).

As indicated in Figure 1.117 [44] by the plots for the ion exchange resins, the fixed-ion concentration varies with the degree of crosslinking, and an increase in fixed-ion concentration results in higher membrane performance. Region I in this figure corresponds to the ion exchange resins obtained in the early stages of progress related to ion exchange membranes, as typified by phenol formaldehyde polycondensate membranes. Region II corresponds to the membranes which were used in early practical applications. Most of the styrene-based polymeric membranes in current use are represented by Region III. For homogeneous membranes as well, the values generally follow an isometric curve of fixed-ion concentrations dependent on the particular ion exchange resin and reinforcing material, with no change in selectivity but, as described below, with variation in electric resistance.

Conductivity

The electric conductivity of the membrane is the major determinant of electrodialytic energy consumption. It varies with the composition, concentration, and temperature of the electrolytic solution. Various methods have been proposed for its measurement, utilizing either direct or alternating current.

The most common method utilizes alternating current with the membrane in an electrolytic solution, in which the total resistance of the membrane and the solution in equilibrium is first measured, and the resistance of the solution alone is then measured after removal of the membrane. The difference between the two measured values is taken as the membrane resistance r, and the specific conductivity κ is then calculated from the membrane surface area S, thickness d, and area resistance R, as follows

$$R = r\,S, \text{ in } \Omega\text{ cm}^2 \tag{5}$$

$$\kappa = d/R, \text{ in S cm}^{-1}. \tag{6}$$

Measurement of the membrane resistance is performed with a cell of the type shown in Figure 1.120. The electrode separation is about 0.5 to 4 mm, partly because greater separation tends to result in larger measurement error related to

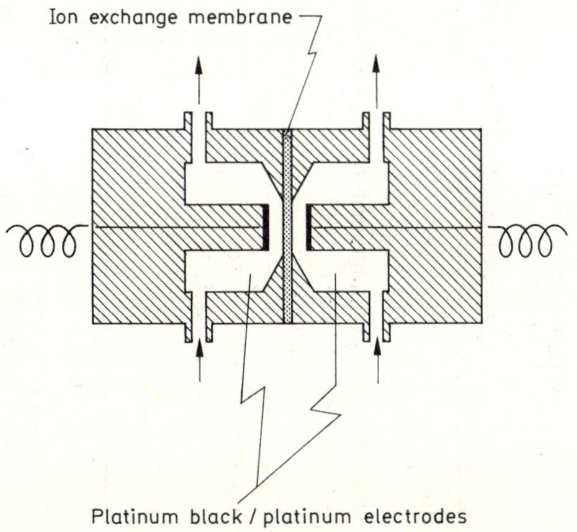

Figure 1.120 Cell for measurement of membrane resistance.

Table 1.44 Cation exchange membrane conductivity and electrolyte composition

	Electrolyte	HCl	KCl	NaCl	LiCl	$HgCl_2$	$CaCl_2$	Sea water
Membrane*	Specific electric conductivity ($\times 10^3\ \Omega^{-1}\ cm^{-1}$)	100	24	16	9.0	2.7	2.5	6.5
	( ): As ratio to that of Na	(6.25)	(1.50)	(1.000)	(0.56)	(0.17)	(0.16)	(0.41)
	Ion	H^+	K^+	Na^+	Li^+	Mg^{2+}	Ca^{2+}	–
Solution	Limiting equiv. conductivity of ion ($\times 10^3\ \Omega^{-1}\ cm^{-1}$)	349.8	73.5	50.1	38.6	53.0	59.5	–
	( ): As ratio to that of Na	(6.98)	(1.47)	(1.000)	(0.77)	(1.06)	(1.19)	–

* Aciplex CA-3

Table 1.45 Cation exchange membrane conductivity and electrolyte composition

	Electrolyte	NaCl	NaBr	NaI	$NaNO_3$	Na_2SO_4	Sea water
Membrane*	Specific electric conductivity ($\times 10^3\ \Omega^{-1}\ cm^{-1}$)	11.0	4.5	0.35	4.0	2.7	10.0
	( ): As ratio to that of Cl	(1.000)	(0.41)	(0.031)	(0.36)	(0.25)	(0.91)
	Ion	Cl^-	Br^-	I^-	NO_3^-	SO_4^-	–
Solution	Limiting equiv. conductivity of ion ($\times 10^3\ \Omega^{-1}\ cm^{-1}$)	75.35	78.14	76.8	71.46	80.0	–
	( ): As ratio to that of Cl	(1.000)	(1.02)	(1.01)	(0.94)	(1.05)	–

* Aciplex CA-3

the electric resistance of the solution, particularly with solutions of low concentration. Platinum black/platinum plate electrodes are normally employed. The electrolytic solution is usually aqueous sodium chloride of 0.5 M concentration. The alternating current is supplied by an alternating bridge, normally at a cyclic frequency of about 1000 Hz.

The influence of the electrolyte composition and concentration on membrane conductivity is especially marked for cation and anion exchange membranes in sea water or brine containing multivalent ions, as indicated by Tables 1.44 and 1.45.

The influence of temperature on the membrane conductivity is also quite marked, as shown in Figure 1.121 by the ratios of the electric resistance of the Aciplex CK-1 and CA-1 membranes at various temperatures.

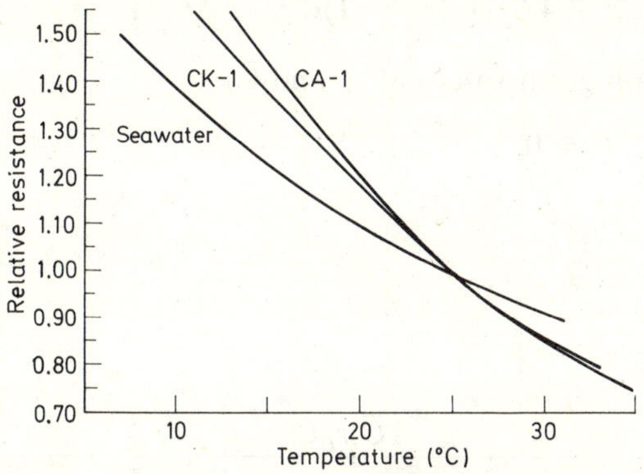

Figure 1.121 Relation of temperature and resistance for Aciplex CK-1 and CA-1.

Membrane transport number

The transport number of a membrane indicates its capability for the separation of counter-ions and thus is a major indicator of its performance in electrodialysis.

An expression for the transport number can be derived from the Donnan equilibrium as expressed in eq. (3) and (4) above, as follows. With the subscripts g and n referring, respectively, to the counter-ion and the co-ion, the ratio of their concentrations in the membrane r_C may be expressed as

$$r_C = \frac{\bar{C}_g}{\bar{C}_n} = \frac{\sqrt{\bar{C}_r^2 + 4C^2} + \bar{C}_r}{\sqrt{\bar{C}_r^2 + 4C^2} - \bar{C}_r}. \quad (7)$$

With $\bar{U}$ as the degree of mobility of the ions within the membrane and $\bar{r}_U$ as the ratio between counter-ion and co-ion mobilities,

$$\bar{r}_U = \bar{U}_g/\bar{U}_n, \quad (8)$$

and with the terms r defined as

$$r = r_C r_U, \quad (9)$$

the transport number for the counter-ion and the co-ion may be expressed, respectively, as

$$\bar{t}_g = \frac{\bar{U}_g \bar{C}_g}{\bar{U}_g \bar{C}_g + \bar{U}_n \bar{C}_n} = \frac{r_U r_c}{r_U r_c + 1} = \frac{r}{r + 1} \tag{10}$$

$$\bar{t}_n = \frac{\bar{U}_n \bar{C}_n}{\bar{U}_g \bar{C}_g + \bar{U}_n \bar{C}_n} = \frac{1}{r + 1}. \tag{11}$$

It is $\bar{t}_g$ which is referred to as the membrane transport number. In terms of its relation to the external solution, it may be expressed as

$$\bar{t}_g = \frac{r_U(\sqrt{\bar{C}_r^2 + 4C^2} + \bar{C}_r)}{(r_U + 1)\sqrt{\bar{C}_r^2 + 4C^2} + (r_U - 1)\bar{C}_r}. \tag{12}$$

Thus, for a membrane with given values of r_U and $\bar{C}_R$,

$$\bar{t}_g = 1 \qquad \text{if} \quad C = 0, \tag{13}$$

and

$$\bar{t}_g = \frac{r_U}{r_U + 1} \quad \text{if} \quad C = \infty. \tag{14}$$

Furthermore, since

$$\frac{d\bar{t}_g}{dC} = -\frac{8\bar{C}_r r_U C}{\{(r_U + 1)\sqrt{\bar{C}_r^2 + 4C^2} + (r_U - 1)\bar{C}_r\}^2 \sqrt{\bar{C}_r^2 + 4C^2}} \tag{15}$$

it follows that

$$\bar{t}_g' = 0 \quad \text{if} \quad C = 0, \tag{16}$$

and

$$\bar{t}_g' > 0 \quad \text{if}\ C > 0. \tag{17}$$

A plot of $\bar{t}_g$ against C will thus be of the form shown in Figure 1.122, which shows $\bar{t}_g$ to decrease with increasing external solution concentration, and to approach $r_U/(r_U + 1)$ at external solution concentrations approaching infinity.

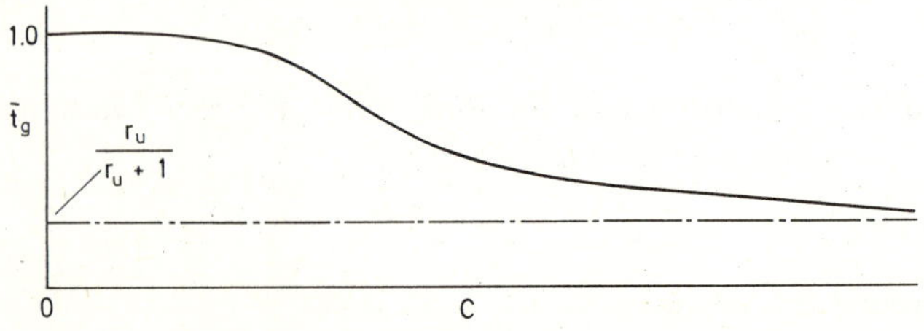

Figure 1.122 Membrane transport number and external solution concentration.

The transport number of a given membrane may be either the "static transport number", which is calculated from the membrane potential, or the "dynamic transport number", which is determined from measurement of the actual ion migration. In most cases, it is the static transport number which is given as a membrane characteristic.

Membrane potential and determination of transport number

When an ion exchange membrane is placed between two electrolytic solutions of different concentrations, a difference in electric potential arises between the two membrane surfaces. This is referred to as the membrane potential. In the simplest system, when the electrolytic solution is aqueous sodium chloride, the membrane potential may be expressed by

$$E_m = -(\bar{t}_+ + \bar{t}_-)\frac{RT}{F}\ln\frac{a_{\pm 2}}{a_{\pm 1}}, \tag{18}$$

where $\bar{t}_+$ and $\bar{t}_-$ are the transport numbers for the cation and the anion, respectively, as a ratio to the total equivalent volume of ions passing through the membrane; $a_{\pm 1}$ and $a_{\pm 2}$ represent the activity of the electrolyte on each side of the membrane at equilibrium; and R, T, and F are the gas constant, absolute temperature, and Faraday's constant, respectively.

The membrane potential is normally measured in the type of cell shown in Figure 1.123. The ratio of electrolyte concentrations is generally in the range of 2 to 10. It is necessary to maintain a stream of fresh electrolytic solution near each membrane surface, as the electrolyte tends to pass through the membrane with a resulting variation in electrolyte concentration on both sides of the membrane.

Calomel or silver-silver chloride electrodes are usually employed, and the difference in electric potentials is measured with a differential potentiometer. The measured membrane potential and eq. (7) determine the membrane transport number.

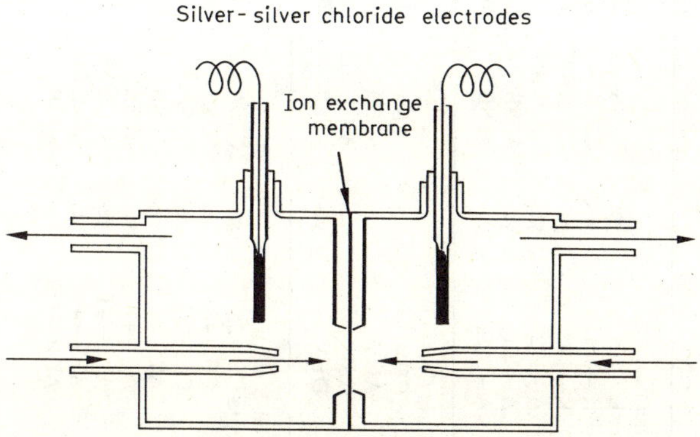

Figure 1.123 Cell for measurement of membrane electropotential.

Table 1.46 Characteristics of typical commercial membranes

Maker	Reinforcing	Type*	Thickness (mm)	Exchange capacity (mequ · g^{-1})	Water content (%)	Electric resistance (Ω · cm)	Transport number	Strength	Use
Asahi Chemical Co.				a)−1	b)−2	c)−1	d)−1	e)−1	
Aciplex K−101	yes	C, SA	0.21−0.23	1.8−2.0	27−28	1.8− 2.3	> 0.99	2.5−3.4	Desalination
K−301	yes	C, SA	1.5 −1.8	0.6−0.7	23−24	27 −33	> 0.93	2.8−5.3	Electrolytic reduction
K−172	yes	C−1, SA	0.11−0.13	1.5−1.6	20−30	1.9− 2.2	> 0.99	2.6−3.3	Concentration
A−172	yes	A−1, SB	0.13−0.15	1.8−1.9	24−25	1.7− 2.1	> 0.99	2.2−3.0	Concentration
A−201	yes	A, SB	0.22−0.24	1.4−1.6	26−27	3.6− 4.2	> 0.99	2.6−3.8	Desalination
A−211	yes	A, SB	0.43−0.48	0.9−1.1	32−34	6.7− 7.6	> 0.92	1.8−2.4	Desalination
Asahi Glass Co.				a)−1	b)−1	c)−1	d)−2	e)−2	
Selemion CHV	yes	C, SA	0.11−0.15	1.5−1.8	15−16	2.0− 3.5	> 0.91	3−5	Desalination, concentration
AHV	yes	A, SB	0.11−0.15	2.0−2.3	15−16	2.0− 3.5	> 0.93	3−5	Desalination
ASV	yes	A−1, SB	0.11−0.15	2.0−2.3	15−16	3.0− 4.5	> 0.95	3−5	Concentration
DHV	yes	A, SB	0.13−0.17					3−5	Diffusion dialysis
Tokuyama Soda Co.				a)−1	b)−1	c)−1	d)−3	e)−2	
Neosepta CL−25	yes	C, SA	0.15−0.17	1.5−1.8	25−35	2.2− 3.0	> 0.98	3−5	Desalination, concentration
CII−45	yes	C, SA	0.15−0.17	1.8−2.3	25−35	1.8− 2.5	> 0.98	3−5	Desalination, concentration
C66−5T	yes	C, SA	0.13−0.18	2.2−2.6	35−45	1.1− 1.7	> 0.98	2−4	Desalination, concentration
AV−4T	yes	A, SB	0.14−0.16	1.5−2.0	20−30	2.7− 3.5	> 0.98	6−7	Concentration
AF−4T	yes	A, SB	0.15−0.20	1.8−2.5	25−35	1.8− 2.5	> 0.98	6−7	Desalination, concentration
AVS−4T	yes	A−1, SB	0.15−0.17	1.5−2.0	25−30	3.7− 4.7	> 0.98	4−6	Desalination, concentration
AFS−4T	yes	A−1, SB	0.15−0.20	1.8−2.5	30−40	2.5− 3.2	> 0.98	3−5	Concentration
ACII−45T	yes	A, SB	0.14−0.20	1.3−2.0	20−35	2.0− 2.7	> 0.98	4−6	Desalination, concentration (Operable in high pH solution)
AHF				a)−2	b)−2	c)−2	d)−4	e)−2	
AHFion C−60	No	C, SA	0.30	1.5	0.35	5.0	0.80	3.1	
C−300	No	C, SA	0.15	0.6	0.17	5.0	0.85	3.9	
A−60	No	A, SB	0.30	1.6	0.28	6.0	0.82	3.1	
A−300	No	A, SB	0.30	0.6	0.17	4.6	0.86	7.7	

Maker and	Reinforcing	Type*	Thickness (mm)	Exchange capacity ($mequ \cdot g^{-1}$)	Water content (%)	Electric resistance ($\Omega \cdot cm$)	Transport number	Strength	Use
Ionac Chemical Co.						c)–3	d)–2	e)–2	
Ionac HC–3142	Yes	C, SA	0.15	1.06	–	9.1	0.94	13.0	
HC–3470	Yes	C, SA	0.30	1.05	–	9.6	0.96	13.4	
HA–3148	Yes	A, SB	0.18	0.96	–	10.1	0.90	13.4	
HA–3475	Yes	A, SB	0.30	0.74	–	18.0	0.94	14.4	
Ionics, Incorp.				a)–1	b)–1	c)–3	d)–5	e)–2	
Nepton 61 CZL 386	Yes	C, SA	0.60	2.7	0.40	11.0	0.94	8	Desalination
61 ATL 386	Yes	C, SA	0.60	2.7	0.46	11.0	0.92	8	Desalination
103 PZL 386	Yes	A, SB	0.60	1.8	0.46	12.0	0.95	10	Desalination
103 QZL 386	Yes	A, SB	0.63	2.1	0.36	9.0	0.97	10.8	Desalination
Du Pont				a)–1				e)–1	
Nafion 117	No	C, SA	0.18	0.91	–	–	–	36	
324	Yes	C, SA	–	0.67/o.91	–	–	–	50	NaCl electrolysis
423	Yes	C, SA	–	0.83	–	–	–	–	
901	Yes		–	–	–	–	–	–	NaCl electrolysis

C, SA : Strongly acidic cation permeable membrane
C–1, SA : Strongly acidic univalent cation permeable membrane
A, SB : Strongly basic anion permeable membrane
A–1, SB : Strongly basic univalent anion permeable membrane

a)–1 Exchange capacity : mequ/g (dry membrane)
a)–2 Exchange capacity : mequ/g (wet membrane)
b)–1 Water content : χ (gH_2O/g dry membrane)
b)–2 Water content : χ (gH_2O/g wet membrane)
c)–1 Electric resistance : equilibrated with 0.5N NaCl solution at 25 °C
c)–2 Electric resistance : Ωcm (equilibrated with 0.6N-KCl solution)
c)–3 Electric resistance : Ωml (equilibrated with 0.1N-NaCl solution)
d)–1 Transport number : Calculated from the membrane potential when placed between sodium chloride solutions of 0.1 mol/l and 0.2 mol/l
d)–2 Transport number : Calculated from the membrane potential when placed between sodium chloride solutions of 0.5 mol/l and 1.0 mol/l
d)–3 Transport number : Measured by electrophoresis with water; current density, 2 A/dm at 25 °C
d)–4 Transport number : Calculated from the membrane potential when placed between potassium chloride solutions of 0.5 mol/l
d)–5 Transport number : Measured by electrophoresis with sodium chloride solution of 0.5 mol/l
e)–1 Tensile strength : kg/mm^2
e)–2 Bursting strength : kg/cm

If a pair of silver and silver chloride electrodes are used, their potential difference is included in the measured value and must therefore be subtracted to obtain the membrane potential.

1.10.3.2 Properties of typical commercial membranes

The properties of representative membranes as given by their manufacturers are shown in Table 1.46.

1.10.4 Ion exchange membrane applications and principles

1.10.4.1 Principles of major applications

Electrodialysis

Basic system of electrodialysis

The principles of electrodialysis are illustrated in Figure 1.124. Salts generally dissolve in an aqueous solution to form anions and cations, as shown for sodium chloride in Figure 1.124 a. In the presence of an electric field between two electrodes, as in Figure 1.124 b, the cations move toward the cathode and the anions toward the anode (a process referred to as electrophoresis) thus constituting an electric current flow. Insertion of a cation exchange membrane (Figure 1.124 c) or an anion exchange membrane (Figure 1.124 d) between the two electrodes will then result in the selective transport of cations or anions through the membrane.

In electrodialysis, an alternating series of cation and anion exchange membranes are placed between the two electrodes, to achieve separation and concentration of the salt ions. As illustrated in Figure 1.124 e for sodium chloride, the salt is removed

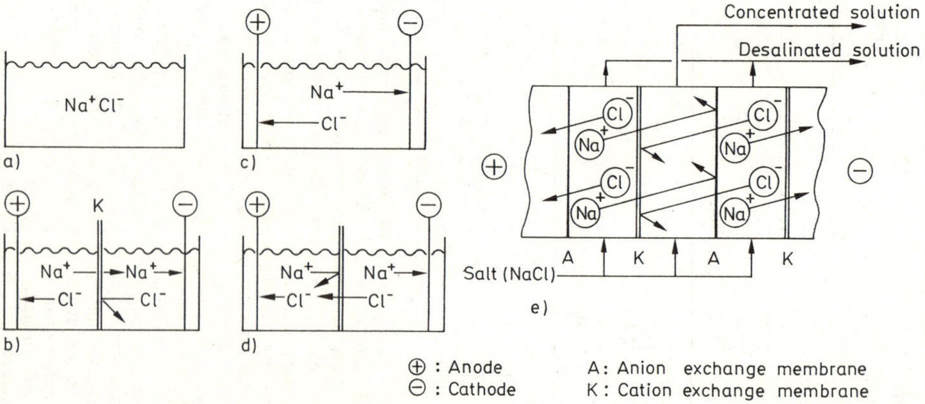

Figure 1.124 Basic principles of electrodialysis system.

from the desalting compartments and supplied to the adjacent concentration compartments. The electrodialyzer may thus be used to achieve either desalination or salt concentration.

Electrodialyzer

Electrodialyzers are generally of the sheet-flow type or the tortuous-path type, in reference to the pattern of solution flow between the membranes. In both types, the membranes may be arrayed either vertically or horizontally. The form of gasket required for each type is shown in Figure 1.125. The basic configuration of the two types is, in other respects, essentially the same. It is described below with reference to the sheet-flow vertical-membrane electrodialyzer shown schematically in Figure 1.126 and 1.127.

The main components of the electrodialyzer are as follows.

a) *Ion exchange membranes*. These contain portholes for supply of the solution to and from the desalting and concentration compartments.

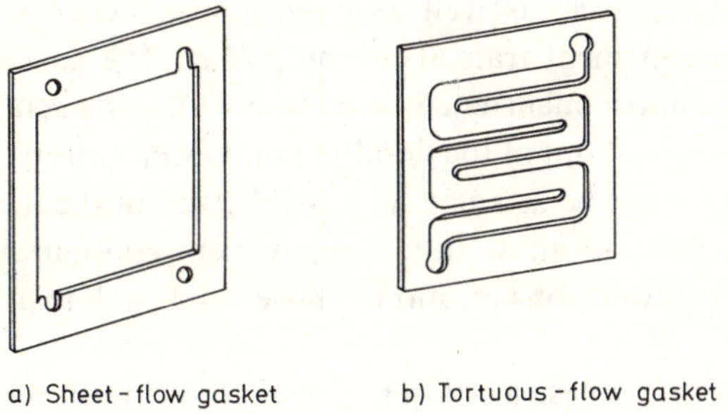

Figure 1.125 Electrodialyzer gasket forms.

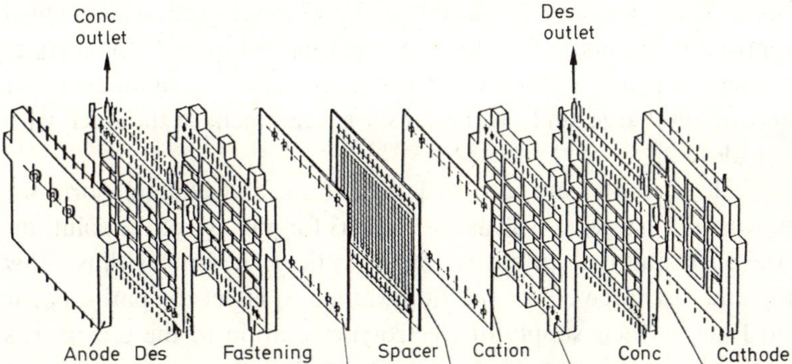

Figure 1.126 Schematic view of electrodialysis stack.

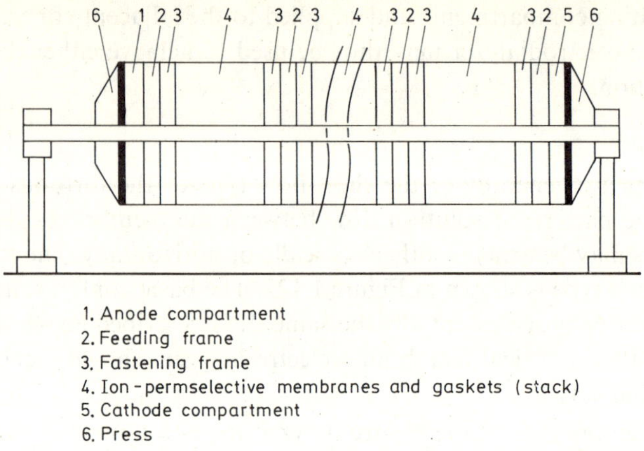

Figure 1.127 Large capacity electrodialyzer components.

b) *Gaskets.* Each gasket is in the form of a frame which contains portholes for solution supply and discharge, as well as passageways between some of the portholes and the compartment framed by the gasket. The gaskets thus serve to form part of the compartment enclosure, secure the adjacent membranes, and route the solution in and out of the desalting or concentration compartment. The gasket thickness should be as small as possible, within the limits necessary for uniform solution flow, to allow thin compartment configurations and the resulting reduction in power consumption. Those used in Japan are generally 0.5 – 0.75 mm thick.

c) *Separators.* The separator is basically a screen which fits into the area enframed by the gasket. It serves to maintain separation between the membranes in their electrically conducting areas, and also to create flow turbulence and thus effect increased diffusion in the solution.

d) *Fastening frames.* These frames are placed at the ends of each stack, which contains an appropriate number of membranes, gaskets and spacers, to facilitate handling and assembly. They are in the form of a lattice, and contain ports for solution supply and discharge, and protrusions for securement of the stack with bolts or turnbuckles.

e) *Feeding frames.* These frames are also in lattice form, and contain ports for solution supply and discharge, as well as short ducts for passage of the solutions to and from the ports and the open areas of the lattice. The frames thus allow external supply and discharge of the solutions, and also effect the pathway for electric current flow by their supply of electrolytic solution to the open areas within the lattice.

f) *Electrodes.* One electrode is installed at each end of the dialyzer, within an electrode frame containing electrical connectors and short ducts for independent passage of anolyte or catholyte to and from the space within the frame.

A large-capacity electrodialyzer of the type used for table-salt production is shown schematically in Figure 1.127. This configuration includes three stacks between the anode and the cathode. Each stack is composed of a series of membranes, gaskets and spacers between two fastening frames. The entire assembly is secured in the dialyzer by a hydraulic filter press.

Piezodialysis

As illustrated in Figure 1.128 a and b, the application of pressure on one side of an ion exchange membrane will give rise to an electromotive force ($\Delta\psi$) in the membrane proportional to the pressure difference (ΔP) between the membrane surfaces. The relationship is expressed as

$$\Delta\psi = -\beta\Delta P. \tag{19}$$

The value of β, which is termed the electric osmosis coefficient, is negative for anion exchange membranes and positive for cation exchange membranes.

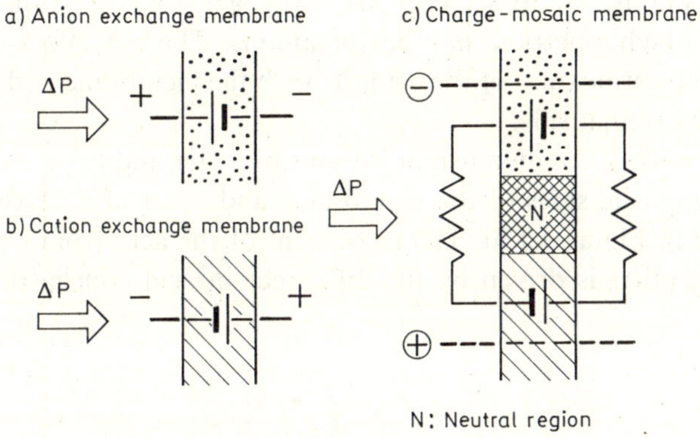

Figure 1.128 Basic principle of piezodialysis.

If a mosaic of anion and cation membranes is formed and placed in an electrolytic solution, a loop circuit will arise between the two with the solution acting as a resistor, as shown in Figure 1.128 c. As part of the resulting current, anions will move through the anion exchange membrane and cations through the cation exchange membrane into the area of lower pressure, thus resulting in desalting of the solution at the higher pressure and concentration of the solution at the lower pressure.

For effective piezodialysis, it is necessary to have a strong electric current in the loop circuit. For this purpose, the anion exchange, cation exchange and neutral regions must be small and densely arrayed, and the electric conductivities of the anion exchange and cation exchange regions must be similar. Various attempts have been made to achieve these characteristics, and a block copolymerization process has recently been reported [45] for attainment of a domain size of 10^{-6} cm^2.

Although piezodialysis has still not progressed beyond the research and development stage, it holds strong promise for salt concentration, desalination, separation of salt from organic materials, and other applications similar to those of electrodialysis, without the need for an external source of electric power.

Dialysis

Basic System

Diffusion dialysis is a form of membrane separation that has long been used in applications ranging from blood dialysis in artificial kidneys to recovery of alkalis from rayon production effluents. The mechanism of separation is based on the differing rates of diffusion of various substances through the membrane. It is thus a simple, energy-efficient process, with no need for external sources of electric energy or pressure.

With an anion exchange membrane, the diffusion dialysis process can be employed for the recovery of acids from a solution containing both an acid in dissociated form as hydrogen ions and anions, and a salt in dissociated form as metal ions and anions. Of these, only the hydrogen ions are thus left behind, together with a stoichiometrical number of anions. The net effect, as shown in Figure 1.129, is the permeation of the acid through the membrane and the retention of the salt on the original side.

In the actual process, countercurrent streams of water and the acid-salt solution are supplied to opposite sides of the membrane, and the acid is recovered continuously, as shown in Figure 1.130. The movement of the acid from one side of the membrane to the other is driven by the difference in acid concentration between the two sides.

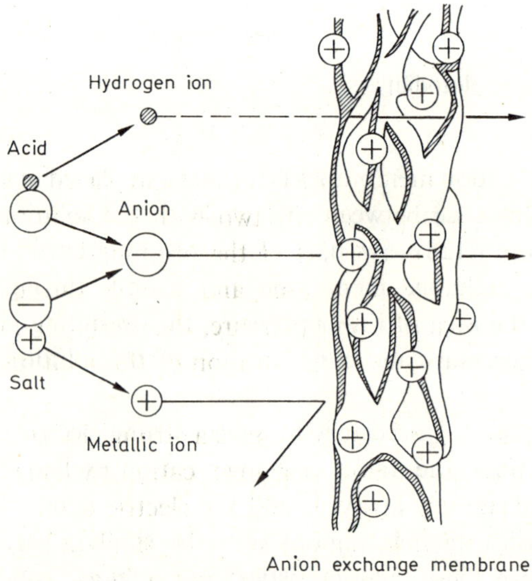

Figure 1.129 Permeation of acid through anion exchange membrane.

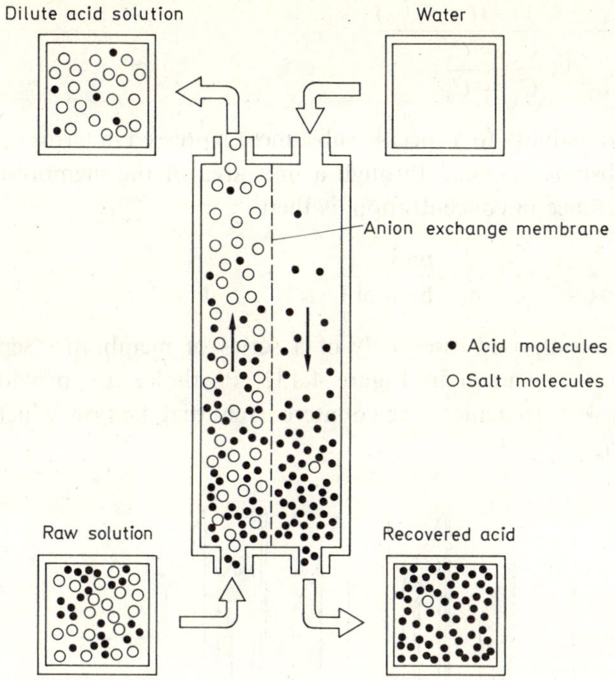

Figure 1.130 Basic principle of diffusion dialysis.

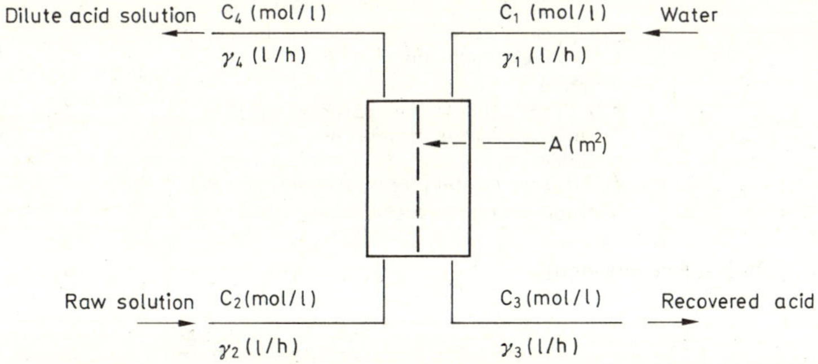

Figure 1.131 Computation of membrane permeability U.

Dialytic coefficient

The basic indicators of a membrane's performance in diffusion dialysis are its permeability to acid, referred to as the dialytic coefficient U_H, and its permselectivity, which is expressed as the ratio of its permeabilities to hydrogen and metallic ions, U_H/U_B. The membrane permeability U is generally determined from the concentrations (C, in mol/l), flow rates (r, in l/h), and membrane area (A, in m^2) indicated in Figure 1.131. The change in concentration across the membrane is first expressed as

$$\Delta C = \frac{(C_4 - C_1) - (C_2 - C_3)}{\ln \dfrac{C_4 - C_1}{C_2 - C_3}}. \tag{20}$$

The membrane permeability to a specific substance, expressed in terms of the molar amount of the substance passing through a unit area of the membrane per unit time and unit difference in concentration, is then

$$U = \frac{C_3 \cdot r_3}{A \cdot \Delta C}, \quad \text{in} \quad \frac{\text{mol}}{\text{m}^2 \cdot \text{h} \cdot \text{mol/l}}. \tag{21}$$

The dialytic cell is composed essentially of a series of membranes separated by gaskets and spacers as shown in Figure 1.132. Portholes are provided in the membranes and gaskets to achieve the countercurrent dialytic flow which is shown schematically in Figure 1.133.

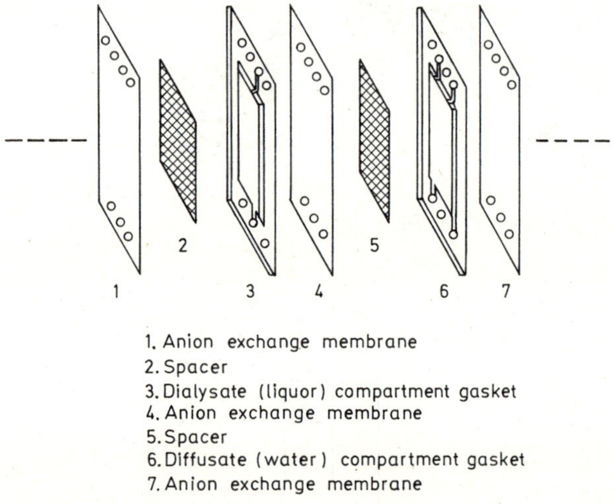

Figure 1.132 Dialyzer components.

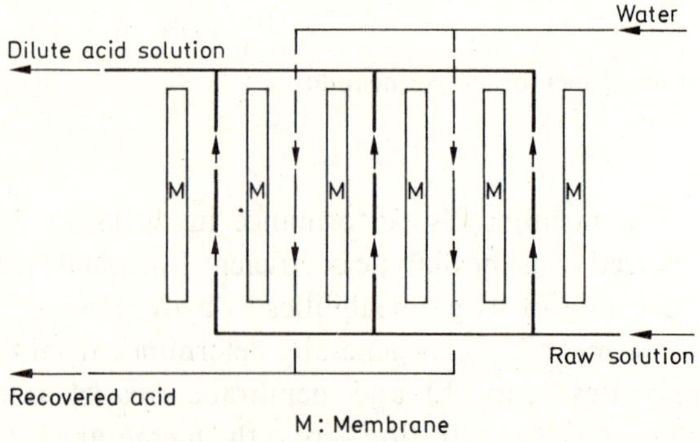

Figure 1.133 Dialyzer flow system.

Membrane electrolysis

The basic principles of membrane electrolysis may be illustrated by the electrolysis of aqueous NaCl. As shown in Figure 1.134, the application of voltage between the two electrodes results in the formation of chlorine gas at the anode, and both hydrogen gas and sodium hydroxide at the cathode. The cation exchange membrane allows permeation of Na^+ ions into the catholyte compartment, while blocking the movement of OH^- ions toward the anode. The reaction in the catholyte compartment is

$$2Na^+ + 2H_2O + 2e \rightarrow 2NaOH + H_2. \quad (22)$$

This process is now rapidly replacing other processes for the production of caustic soda, with fluorocarbon membranes and electrolyzers of either monopolar or bipolar configuration, as shown in Figure 1.135 a and b.

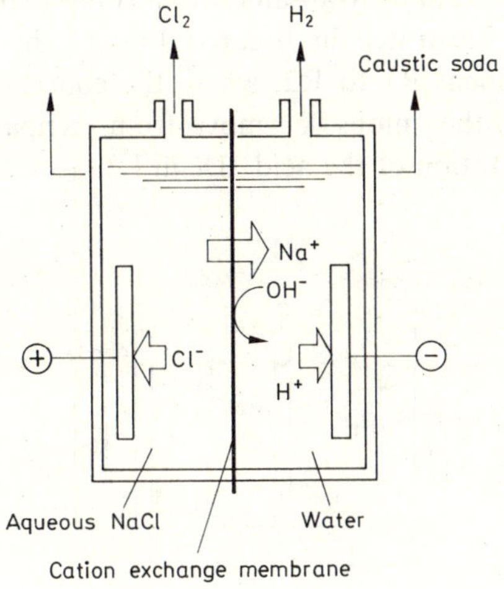

Figure 1.134 Basic principle of NaCl electrolysis.

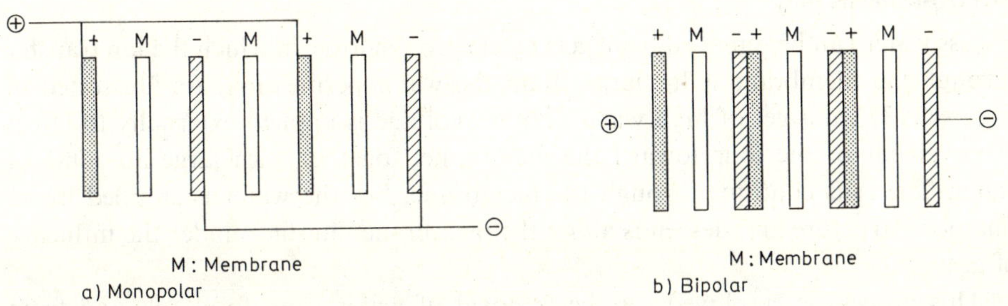

Figure 1.135 Electrolyzer configurations.

The same principles may be applied to the electrolytic reduction of uranium [46], in which the reaction in the catholyte compartment is

$$UO_2Cl_2 + 2HCl + 2H^+ + 2e \rightarrow UCl_4 + 2H_2O\,. \tag{23}$$

Similarly, for the electrolytic reduction and dimerization of acrylonitrile to obtain adiponitrile [47], one has

$$2CH_2 = CH-CN + 2H^+ + 2e \rightarrow NC-(CH_2)_4-CN\,. \tag{24}$$

The membrane itself may serve as an electrolyte, if it is coated with an anodic catalyst on one side and a cathodic catalyst on the other. Water electrolysis can be performed, with the evolution of oxygen at the anode and hydrogen at the cathode, by applying a voltage across these electrodes.

Electrolytic synthesis

An acid and base can be produced from neutral salts with a combination of bipolar and non-bipolar membranes, as illustrated in Figure 1.136. In this process, the cations M^+ move from compartment R1 to R2, where the concentration of the base MOH increases. Conversely, the anions X^- move from compartment L1 to L2, resulting in increased concentration of the acid HX in L2.

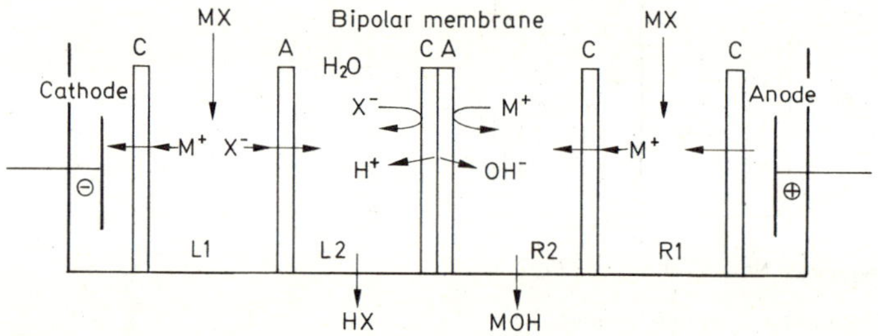

Figure 1.136 Electrolytic synthesis of acid and base with bipolar membrane.

Electro-osmosis [48]

Excess water can be removed from gels by electro-osmosis, in which the ion transfer through the membrane is far larger than the water permeation. As illustrated in Figure 1.137, in a gel of high water content voltage is applied externally to effect a movement of the ions toward the membrane, together with large amounts of water. The ions continue through the membrane, but the water is stopped by its fine-pore structure and descends along the membrane interface under the influence of gravity.

This process is employed for the removal of water from food gels and from cellulose pulps.

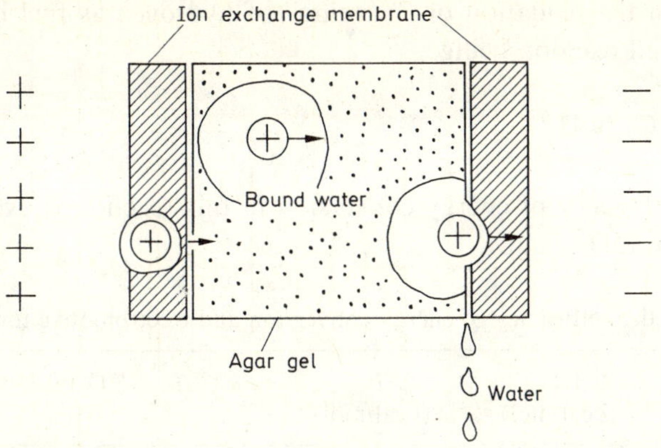

Figure 1.137 Basic principle of electro-osmosis.

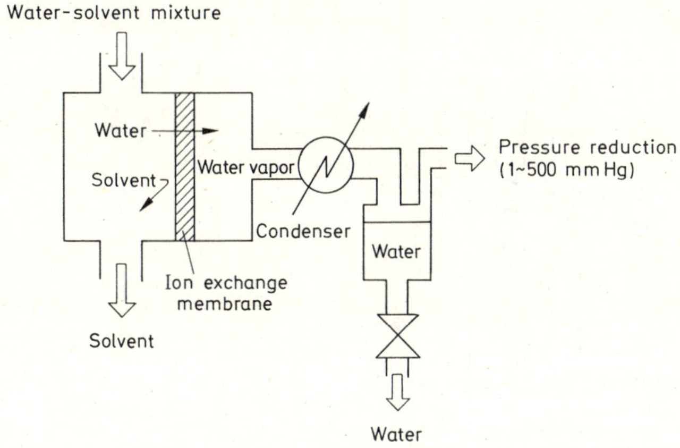

Figure 1.138 Water separation by pervaporation.

Pervaporation [49]

Water can be removed from organic solvents or sea water by pervaporation, in which a pressure differential is applied across an ion exchange membrane. As shown in Figure 1.138, the feed solution is supplied on one side of the membrane, and the pressure is lowered on the opposite side.

Water then selectively permeates the membrane at a high rate and collects as vapor under the lower pressure, which can then be condensed by cooling and removed from the system.

Pervaporation is used to separate water from mixtures with alcohols or 1,1-trichloroethane, and can also be used to concentrate sea water.

Fuel cells

Fuel cells may be functionally defined as batteries which directly convert a chemical energy input to an electrical output. The most widely studied fuel cells utilize the energy released in the oxidation or "burning" of hydrogen as fuel by oxygen or air, with the overall reaction being

$$H_2 + \frac{1}{2} O_2 \rightarrow H_2O . \tag{25}$$

The theoretical efficiency of energy conversion in this reaction is extremely high, as shown in Table 1.47.

Table 1.47 Theoretical efficiency of energy conversion and electromotive force

T(K)	Reaction	$-\Delta H$ (kcal/mol)	$-\Delta F$ (kcal/mol)	μ (%)	E (V)
298	g + g → g	57.798	54.636	94.53	1.18
298	g + g → l	68.317	56.690	82.98	1.23

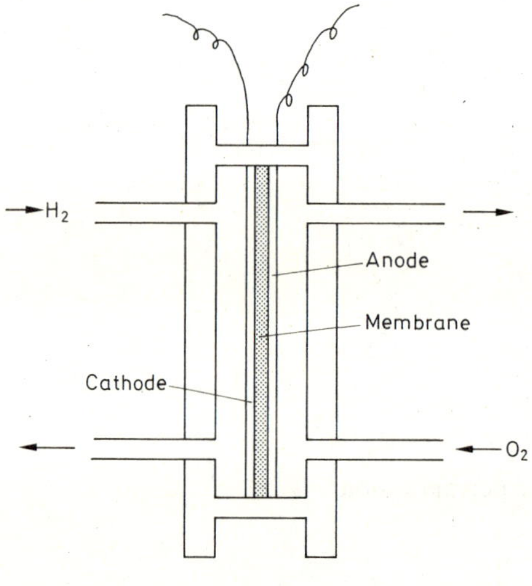

Figure 1.139 Fuel cell configuration.

As shown in Figure 1.139, the five main components of the fuel cell are the hydrogen gas compartment, cathode, cation or anion exchange membrane, anode, and oxygen gas compartment.

With a cation exchange membrane, the hydrogen ions formed at the cathode pass through the membrane and react with oxygen at the anode to form water, as follows.

At the cathode: $H_2 \rightarrow 2H^+ + 2e$ (26)

Through the membrane: $2H^+$

At the anode: $1/2\, O_2 + 2H^+ + 2e \rightarrow H_2O.$ (27)

If an anion exchange membrane is used, hydroxyl ions from the anode pass through the membrane and react with hydrogen at the cathode to form water, as follows.

At the cathode: $H_2 + 2OH^- \rightarrow 2H_2O + 2e$ (28)

At the anode: $O_2 + H_2O + 2e \rightarrow HO_2^- + OH^-$ (29)

At the anode: $HO_2^- \rightarrow OH^- + 1/2\, O_2$. (30)

In either case, the overall reaction is thus:

$$H_2 + \frac{1}{2} O_2 \rightarrow H_2O\,. \quad (31)$$

An important characteristic of the fuel cell is its reversibility; it can be used without modification for the electrolysis of water. This allows the replenishment of hydrogen and oxygen by the application of electric energy from sources such as solar cells, and thus, in effect, allows the repeated changing and discharging of the fuel cell. This makes it of particular interest in electric power systems for spacecraft. The first use of this type of system was in the Gemini program. It supplied about 100 KWh of electric power in eight days on Gemini 5, and about 200 KWh on Gemini 7. Numerous improvements in fuel cells have been made since that time, for increased performance and economy. Continuing improvements, together with freedom from the pollutants and noise associated with combustion-powered electrical generation, are expected to lead to a widening range of applications.

In addition to the hydrogen-oxygen fuel cells, others utilizing ion exchange membranes which are currently under development are methanol fuel cells and redox-flow fuel cells.

1.10.4.2 Physico-chemical considerations for electrodialysis

Material transfer

When an electric current is applied to an alternating series of cation and anion membranes, liquid is transferred through the membrane together with the ions, and alternating compartments of concentrated and desalinated solutions are formed. The study of this material transfer has long been based on the Nernst-Planck equation. In this treatment, the concentration of the brine moving into the concentration compartment is expressed as the ratio between the ionic flow and the volume flow, by the expression

$$C_b = \frac{|j_1^K| - |j_2^A|}{|J^K| + |J^A|}\,. \quad (32)$$

j = Ionic flow

J = Volume flow

K : Cation exchange membrane

A : Anion exchange membrane

1 : Cations

2 : Anions.

Furthermore, the volume flow J may be expressed as

$$J = J_{\mathrm{elosm}} + J_{\mathrm{osm}}, \tag{33}$$

elosm: Electro-osmosis
osm: Differential osmosis.

Oda [50] has proposed the following equations, in which he relates the ionic flow (ion transfer) to electrophoresis and electrolyte diffusion, and the volume flow (liquid transfer) to electro-osmosis and concentration differential osmosis

$$m = (\bar{t}_{\mathrm{K}} + \bar{t}_{\mathrm{A}} - 1)\frac{i}{F} - (D_{\mathrm{SK}} + D_{\mathrm{SA}})\,\Delta C \tag{34}$$

$$q = (\beta_{\mathrm{K}} + \beta_{\mathrm{A}})\,i + (D_{\mathrm{WK}} + D_{\mathrm{WA}})\,\Delta C \tag{35}$$

m = Ion transfer quantity
i = Current density
C = Difference between concentrations on two sides of membrane
t = Membrane transfer number
q = Liquid transfer quantity
D_{S} = Coefficient of electrolyte diffusion
β = Coefficient of electro-osmosis
D_{W} = Coefficient of concentration differential osmosis.

Eqn. (34) and (35) refer respectively to the phenomena expressed in the numerator and the denominator of eq. 32.

A similar treatment has been developed by Gorsa [51]. A description of the relation to the fixed-ion concentration has been developed by Smagin [52].

These relations have also been investigated empirically by Tanaka [53–54], with data obtained in experimental electrodialysis. As indicated in Figure 1.140 and

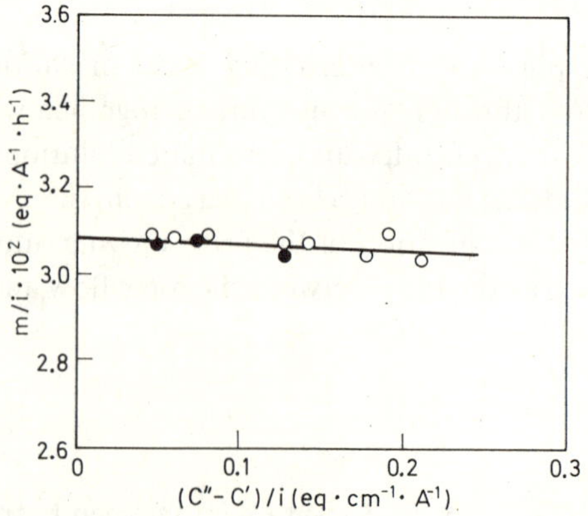

Figure 1.140 Plot of m/i vs. $(C'' - C')/i$, at different flow rates in desalting compartments. ○: 4 cm/s; ●: 2 cm/s; Temp.: 25 °C; Memb.: Aciplex CK-2 and CA-3.

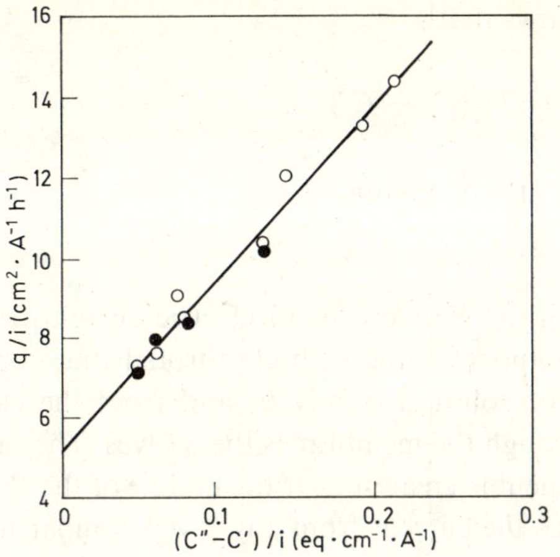

Figure 1.141 Plot of g/i vs. $(C'' - C')/i$, at different flow rates in desalting chambers. ○: 4 cm/s; ●: 2 cm/s; Temp.: 25 °C; Memb.: Aciplex CK-2 and CA-3.

1.141, his results show that a direct linear relation exists between m/i and $(C'' - C')/i$, and also between q/i and $(C'' - C')/i$, expressed by

$$\frac{m}{i} = \frac{\lambda - u(C'' - C')}{i} \tag{36}$$

$$\frac{q}{i} = \frac{\varphi + \varrho(C'' - C')}{i}, \tag{37}$$

in which λ, u, φ, and ϱ are the membrane characteristics defined by

$$\lambda = \frac{\bar{t}_K + \bar{t}_A - 1}{F(1 + \Delta i/i)} - \frac{\delta' + \delta''}{FD(1 + \Delta i/i)} \cdot [D_{SK}(\bar{t}_K - t_+) + D_{SA}(\bar{t}_A - t_-) \tag{38}$$

$$u = D_{SK} + D_{SA} \tag{39}$$

$$\varphi = \frac{\beta_K + \beta_A}{1 + \Delta i/i} + \frac{\delta' + \delta''}{FD(1 + \Delta i/i)} \cdot [D_{WK}(\bar{t}_K - t_+) + D_{WA}(\bar{t}_A - t_-)] \tag{40}$$

$$\varrho = W_{WK} + D_{WA}. \tag{41}$$

These characteristics have been found to be practically unaffected by electrodialysis conditions other than temperature.

Current efficiency and electric power consumption

Current efficiency and leak current

The theoretical current efficiency in the electrodialysis process can be expressed as

$$\eta = \frac{m}{i} F. \tag{42}$$

From eq. (11) and (12), it follows that

$$\eta = (\bar{t}_K + \bar{t}_A - 1) - \frac{F}{i} \cdot u(C'' - C') \tag{43}$$

which can be simplified to the approximation

$$\eta = \bar{t}_K + \bar{t}_A - 1 . \tag{44}$$

In the electrodialyzer, however, a relatively small electric current, referred to as leak current, travels from the anode to the cathode through the ports which are provided in the membranes for solution supply to and from the electrodialyzer compartments, rather than through the membranes themselves. The rate of current leakage from a given compartment is greatest near the middle of the electrodialyzer, as shown in Figure 1.142, since the current from any given compartment tends to travel a considerable distance before entering a later compartment and proceeding to the cathode through the succeeding membranes.

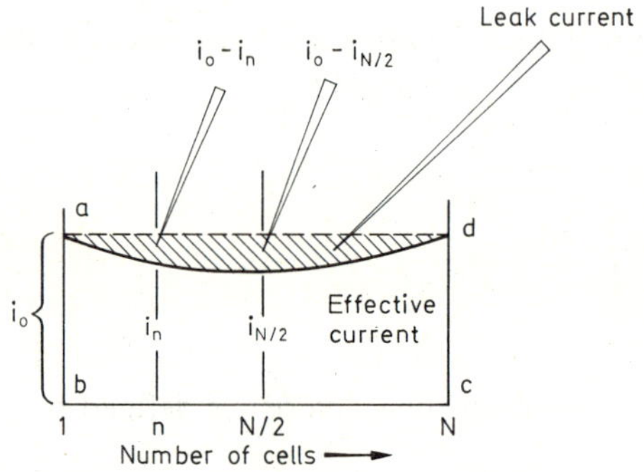

$$\frac{i_1}{i_0} = \frac{\sum_{n=1}^{n=N} (i_0 - i_n)}{N \cdot i_0}$$

Figure 1.142 Current distribution in electrodialysis stack.

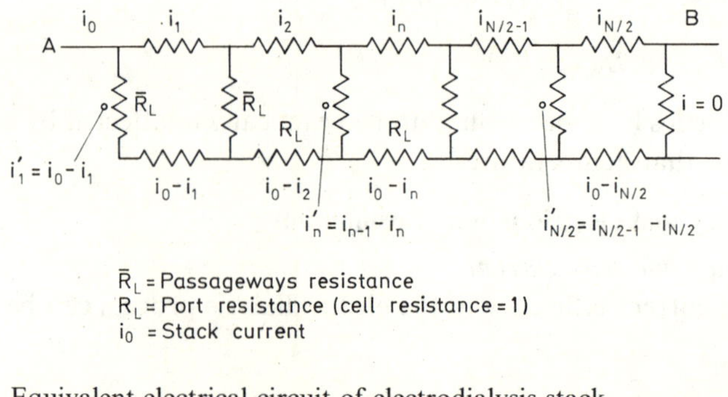

Figure 1.143 Equivalent electrical circuit of electrodialysis stack.

The path followed by the current in the electrodialyzer may be represented by the equivalent circuit shown in Figure 1.143. For this circuit the relation between the leak current i_L and the total current i_0 may be expressed [55] as

$$i_L = \frac{2(n+2)(n+1)}{24\gamma + 3n(1+\bar{\gamma})(n+2)} i_0 . \tag{45}$$

where $\gamma \equiv R_L/R$, $\bar{\gamma} \equiv \bar{R}_L/R$, and n is the number of membrane pairs, with R being the electric resistance of each pair of membranes and compartments, R_L that of the ports, and $\bar{R}_L$ that of the intervening passageways.

The actual current efficiency of the electrodialyzer, η' is therefore expressed as

$$\eta' = (\bar{t}_K + \bar{t}_A - 1)\left(1 - \frac{i_L}{i_0}\right) - \frac{F}{i} \cdot u \cdot (C'' - C') \tag{46}$$

or, in a simplified approximation, as

$$\eta' = (\bar{t}_K + \bar{t}_A - 1)\left(1 - \frac{i_L}{i_0}\right). \tag{47}$$

Unit power consumption

In electrodialytic salt concentration, the electric power consumption per ton of product electrolyte is given as $(E/n)(F/\eta M) \cdot 10^3$ kWh. In electrodialytic desalination, the consumption is $(E/n)(F/\eta \cdot \eta_{H_2O})(\Delta C)$ kWh per cubic meter of product water. In both cases, E represents the electrodialyzer voltage, and n the number of cells (or membrane pairs) in the electrodialyzer. As these expressions imply, the unit power consumption is proportional to the voltage drop across each cell (E/n). The term η_{H_2O} represents the ratio of product water to feed water, and which varies inversely with the amount of water flowing together with the ions from the desalting compartments to the concentration compartments. The term ΔC represents the difference in concentration between feed and product water.

The voltage drop across each cell can be expressed essentially in terms of the membrane and compartment resistances and the membrane potential, as

$$E/n = I \cdot (R_K + R_A + R_D + R_C) + E_m + E_{other}/n \tag{48}$$

I = Current (A)
R_K = Resistance of cation exchange membrane (Ω)
R_A = Resistance of anion exchange membrane (Ω)
R_D = Resistance of desalting compartment (Ω)
R_C = Resistance of concentration compartment (Ω)
E_m = Membrane potential (V)
E_{other} = Potential of frame, electrode compartments, etc. (V)
n = Number of cells.

Limiting current density

Under the applied voltage difference between the electrodialyzer electrodes, the movement of cations through the electrolytic solution is slower than their rate of transfer through the cation exchange membrane, thus tending to cause a cation deficiency near the membrane surface. As shown in Figure 1.144 this deficiency is alleviated by ion diffusion in the solution. The occurrence of the resulting gradient of ion concentration between the bulk solution and the membrane surface is referred to as concentration polarization.

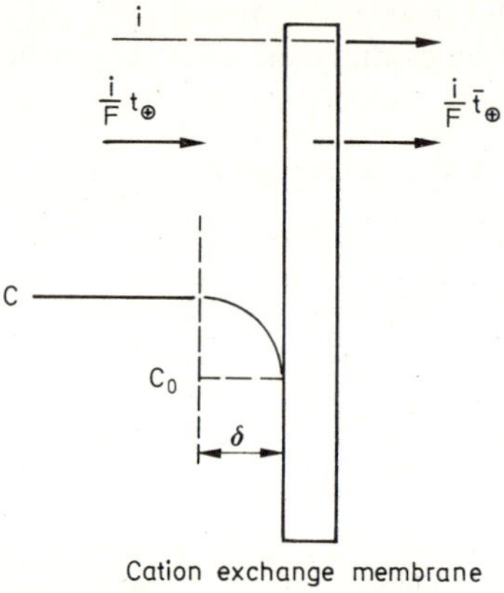

Figure 1.144 Ion transfer and concentration gradient at membrane interface with desalting compartment.

In the extreme case of concentration polarization, the ion concentration reaches zero at the membrane surface. The limiting current density $i_{\lim}$ at which this occurs may be derived theoretically [56] from the Nernst-Planck equation as applied to the ion balance in the region of the membrane, which is

$$J_i = -D_i\left(\frac{\mathrm{d}C_i}{\mathrm{d}x} + \frac{Z_i F}{RT} C_i \frac{\mathrm{d}\psi}{\mathrm{d}x}\right), \tag{49}$$

J_i = Flux of ion species i
D_i = Diffusion coefficient of ion species i
C_i = Concentration of ion species i (equ/unit volume)
x = Perpendicular distance from membrane surface
Z_i = Charge of ion species i
F = Faraday constant
ψ = Electric potential.

Furthermore, for an electrolyte of univalent ions ($C_+ = C_- = C$ in solution),

$$J_+ = -D_+\left(\frac{\mathrm{d}C}{\mathrm{d}x} + \frac{FC}{RT}\cdot\frac{\mathrm{d}\psi}{\mathrm{d}x}\right) \tag{50}$$

$$J_- = -D_-\left(\frac{dC}{dx} - \frac{FC}{RT}\cdot\frac{d\psi}{dx}\right). \tag{51}$$

In the membrane, the flux ratio J_+/J_- is equal to the transport number ratio ($\bar{t}_+/\bar{t}_-$), so that

$$\frac{J_+}{J_-} = \frac{-\bar{t}_+}{\bar{t}_-} = \frac{-\bar{t}_+}{(1-\bar{t}_+)}. \tag{52}$$

Substitution of eq. (52) into eq. (50) and (51) results in

$$J_+ = \frac{-2D_+}{1-(D_+\bar{t}_-/D_-\bar{t}_+)}\frac{dc}{dx} \tag{53}$$

$$J_- = \frac{-2D_-}{1-(D_-\bar{t}_+/D_+\bar{t}_-)}\frac{dc}{dx}. \tag{54}$$

The ratio of ionic diffusion in the solution may be expressed in terms of the transport numbers as

$$\frac{D_+}{D_-} = \frac{t_+}{t_-} = \frac{t_+}{(1-t_+)}, \tag{55}$$

and the diffusion coefficient D for a univalent electrolyte may be expressed as

$$D = \frac{2\cdot D_+\cdot D_-}{D_+ + D_-} = 2\left(\frac{1}{D_+} + \frac{1}{D_-}\right)^{-1} = 2D_+(1-t_+). \tag{56}$$

Substitution of eq. (55) and (56) into eq. (53) and (54), respectively, then yields

$$J_+ = -\frac{2D_+}{1-(t_+\bar{t}_-/t_-\bar{t}_+)}\frac{dc}{dx} = -\frac{2D_+(1-t_+)\bar{t}_+}{(\bar{t}_+ - t_+)}\frac{dc}{dx} = -\frac{D\bar{t}_+}{(\bar{t}_+ - t_+)}\frac{dc}{dx} \tag{57}$$

$$J_- = -\frac{D\bar{t}_-}{(\bar{t}_- - t_-)}\frac{dc}{dx} = \frac{D\bar{t}_-}{(\bar{t}_+ - t)}\frac{dc}{dx} \tag{58}$$

From Faraday's law, on the other hand, the current density i may be expressed as

$$i = F(J_+ - J_-). \tag{59}$$

It therefore follows that

$$i = -\frac{FD}{(\bar{t}_+ - t_+)}\cdot\frac{dc}{dx} = \frac{FD}{(\bar{t}_- - t_-)}\frac{dc}{dx}. \tag{60}$$

Given the width δ of the depletion layer, across which the concentration gradient occurs in the region of the membrane interface with the desalting compartment, the bulk solution concentration C_0, and the limiting current density i_{lim} at which the concentration at the membrane surface reaches zero, we thus find

$$i_{lim} = \frac{FD}{(\bar{t}_+ - t_+)} \frac{C_0}{\delta}. \tag{61}$$

At current densities higher than i_{lim} the ion diffusion becomes insufficient to carry the current through the membrane, and it is then carried instead by H^+ and OH^- ions formed by the dissociation of the water in the solution. The transport of these ions leads to reduced current efficiency, increased operating voltage, and deposit of insoluble hydroxides on the membrane.

It is therefore essential to operate the electrodialyzer at less than the limiting current density. In general this may be related to the total electropotential across the cell during current flow, by

$$\Delta E = -(2R_s + R_m)i + (\Delta E_b' + \Delta E_b'') + \Delta E_m + \Delta E_j \tag{62}$$

R_s = Electric resistance of the bulk solutions
R_m = Electric resistance of the membrane
$\Delta E_b'$ = Electropotential across depletion layer on one side of the membrane
$\Delta E_b''$ = Electropotential across concentration layer on other side of the membrane
ΔE_m = Membrane potential
ΔE_j = Potential difference between solutions.

This may then be rewritten as

$$\Delta E = -(2R_s + R_m)i + \frac{-FD/(\bar{t}_+ - t_+)}{\lambda} + C - C \ln \frac{1 + (i/i_{lim})}{1 - (i/i_{lim})}, \tag{63}$$

where

$C \equiv (RT/F)(t_- - t_+)$.
λ = Equivalent conductance. (64)

The total electropotential at various current densities may thus be plotted as the sum of the two terms on the right side of eq. (63), as shown in Figure 1.145. Although this equation implies that ΔE will become infinite in operation at the limiting current density, the increase in ΔE is more gradual in actual electrodialyzer operation, due to the occurrence of water splitting. For a given electrodialyzer, the value [57] of i_{lim} can be determined from the point of discontinuity in the plot of its operating voltage and current as shown in Figure 1.146 a, or more clearly by the plot shown in Figure 1.146 b.

In relation to eq. (61), experience in electrodialyzer operation has shown that $\delta = k/V^n$, where V is the rate of flow in the desalination compartment and k and n are constants; it has further been shown that with NaCl concentrations of 0.05 mol/l or less $i_{lim} \propto C$. Substitution of these relations into eq. (61) results in the more convenient expression

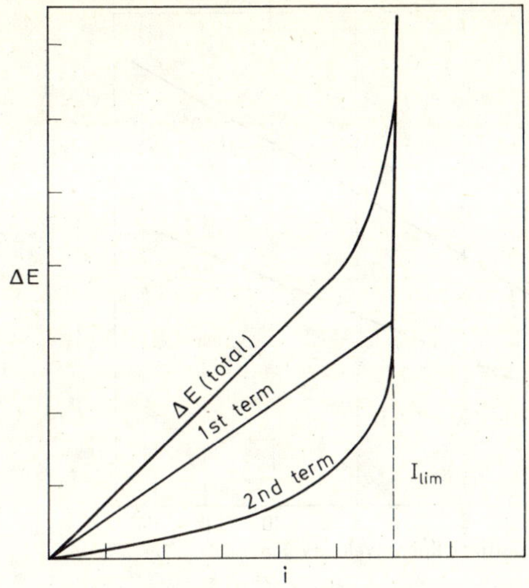

Figure 1.145 Current density and total electropotential.

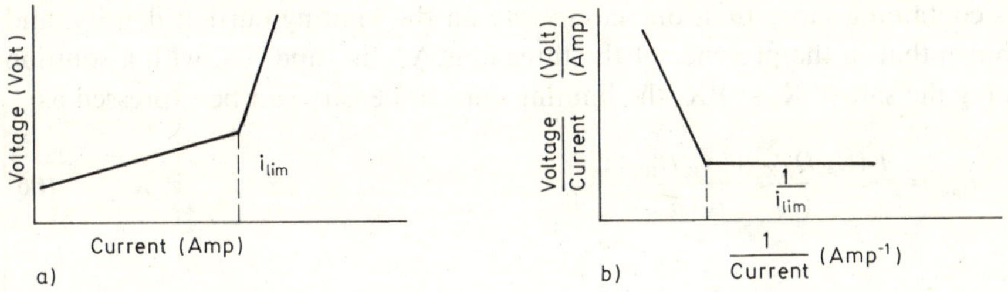

Figure 1.146 Limiting current density plots.

$$\left(\frac{i}{C}\right)_{\text{lim}} = \frac{FD}{k(\bar{t}_+ - t_+)} V^n . \tag{65}$$

Figure 1.147 [58] shows the relationship found empirically between $(i/C)_{\text{lim}}$ and the flow rate in the desalting compartment, with feed water of 1000 ppm NaCl and electrodialysis temperatures of 20 °C and 40 °C. As indicated by these plots, the limiting current density increases with increasing flow-rate and electrodialysis temperature, with a corresponding reduction in the tendency for occurrence of concentration polarization.

The use of appropriate spacers within the electrodialysis compartments has also been found to reduce the depletion layer width (δ) and thus to raise the limiting current density. In industrial applications, the use of these spacers results in values [59–63] of 0.3 to 1.0 for the exponent n. The use of ion-conducting spacers has also been reported [64–65], for a further increase in the limiting current density.

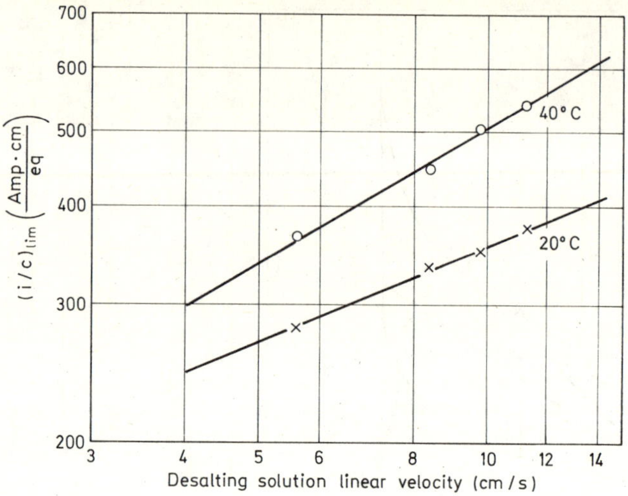

Figure 1.147 Relation between flow rate and $(i/C)_{lim}$.

Theoretical studies have also been conducted [66] to determine the effect of systems containing more than one electrolyte on the limiting current density, and have shown that in the presence of the three ions A^+, B^+ and X^+, with a solution containing the salts AX + BX, the limiting current density can be expressed as

$$i_{lim} = \frac{F(x_A D_{AX} + x_B D_{BX}) C}{(t_X - \bar{t}_X)\delta}, \tag{66}$$

where

$$C = Z_A C_A + Z_B C_B$$

$$\frac{Z_A C_A}{Z_B C_B} = x_A + x_B$$

$$x_A + x_B = 1.$$

This implies that the limiting current density varies linearly with the ratio of electrolyte concentrations, a relationship which has been confirmed experimentally for an NaCl + KCl system, as shown in Figure 1.148.

Scale deposition and prevention

Electrolytic salt concentration at membrane interface

In electrodialysis, scale deposition tends to occur at the interface between the membrane and the concentration compartment due to the high electrolytic salt concentrations in this region. A description of the concentration of scale components at the interface may be derived from the mass balance in this region.

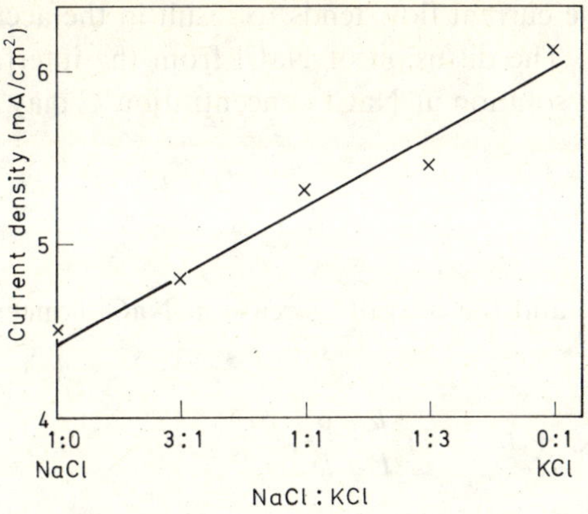

Figure 1.148 Relation between ratio of electrolyte concentrations and current density.

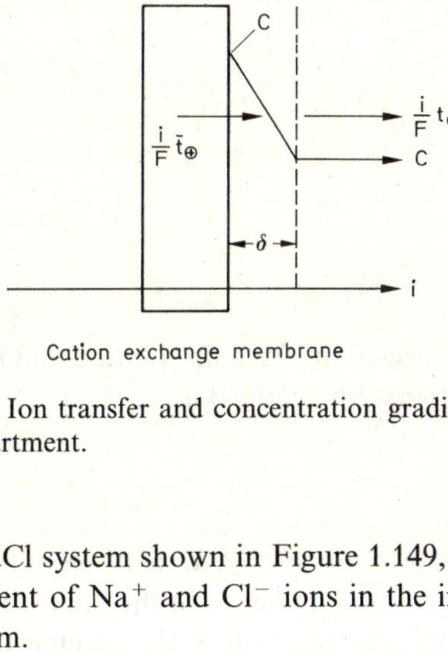

Figure 1.149 Ion transfer and concentration gradient at membrane interface with concentration compartment.

In the NaCl system shown in Figure 1.149, the mass balance of the electrophoretic movement of Na^+ and Cl^- ions in the interface region may be given as the following sum.

	Na^+ mass balance	Cl^- mass balance
Intramembrane transfer	$+\ \bar{t}_{Na}\, I/F$	$-\ (1 - \bar{t}_{Na})I/F$
Intrasolution electrophoresis	$-\ t_{Na}\, I/F$	$+\ t_{Cl} I/F$
Total mass balance	$(\bar{t}_{Na} - t_{Na})I/F$	$(\bar{t}_{Na} + t_{Cl} - 1)I/F.$

Since $t_{Na} + t_{Cl} = 1$, the mass balance for the Na^+ ions and for the Cl^- ions is then

$$\mathrm{Bal}_{El} = \frac{(\bar{t}_{Na} + t_{Cl} - 1)I}{F}. \tag{67}$$

As $t_{Na} \risingdotseq 1$ and $t_{Cl} \risingdotseq 1/2$, the current flow tends to result in the accumulation of NaCl in the interface region. The diffusion of NaCl from the interface, at NaCl concentration C', to the bulk solution at NaCl concentration C may be expressed as

$$\text{Dif.} = \frac{D(C' - C)}{\delta}. \tag{68}$$

At equilibrium, $\text{Bal}_{El} = \text{Dif.}$, and the overall increase in NaCl concentration may therefore be expressed as

$$\Delta C \equiv C' - C = (\bar{t}_{Na} + t_{Cl} - 1)\frac{I}{F} \cdot \frac{\delta}{D}. \tag{69}$$

Since $\bar{t}_{Na} \risingdotseq 1$, eq. (45) may, as an approximation, be generalized to eq. (46), to express the increase in concentration (ΔC_i) of any given cation species in the region of interface between the cation exchange membrane and the concentration compartment, or anion species in the corresponding interface region of an anion exchange membrane:

$$\Delta C_i = C_i' - C_i = t_i \frac{I}{F}\frac{\delta}{D_i}. \tag{70}$$

The term t_i may be defined as

$$t_i = \frac{X_i l_i}{\Sigma X_i l_i}, \tag{71}$$

where x_n is the molar fraction of ion species i in the bulk solution and l_i is its degree of mobility. Furthermore, the term D_i may be defined as

$$D_i = \frac{\Sigma D_{ki} x_k}{\Sigma X_k}, \tag{72}$$

in which D_{ki} is the diffusion coefficient of ion species k, as the counter-ion, to ion species i, and x_k is the molar fraction of ion species k, in the bulk solution.

It then follows that the increase in concentration of the counter-ion species in the region of interface, for both cation and anion exchange membranes, may be expressed as

$$\Delta C_k = C_k' - C_k = (\bar{t}_k - t_k)\frac{I}{F}\frac{\delta}{D_k}. \tag{73}$$

Limiting deposition concentration

Scale deposition will occur if the concentration of scale-forming ions at the membrane interface with the concentration compartment exceeds saturation.

The need for adjustment of the bulk solution may be determined by comparison of the ion concentration at the membrane surface, which is obtained as the sum of the bulk solution concentration and the concentration increase in the membrane

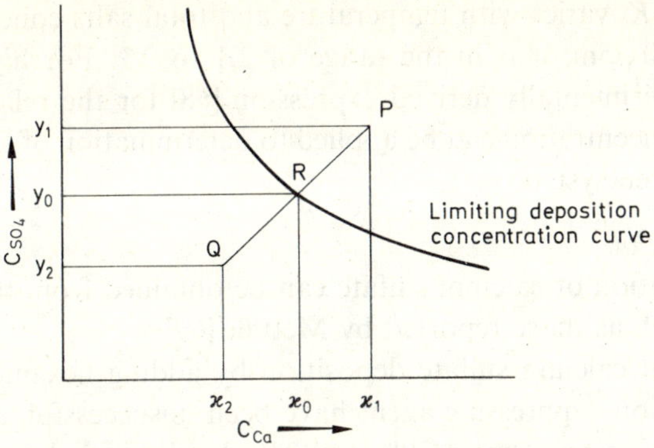

Figure 1.150 Diagram of limiting deposition concentration curve.

interface region, with the limiting deposition concentration curve, which is obtained by plotting the ion concentrations at saturation.

Calculation of the necessary adjustment is illustrated in Figure 1.150 for calcium sulfate. Point P is the intersection of the membrane surface concentrations x_1 and y_1, where x_1 is the sum of the Ca^{2+} bulk solution concentration and C_{Ca} from eq. (70) or (73), and y_1 the analogous sum for SO_4^{2-}. Point Q represents the desired membrane surface concentrations, which are normally 20% below those of the limiting deposition concentration curve; the amount of diluting solution (q) necessary to achieve these concentrations may, therefore, be expressed as

$$q_s = \frac{x_1 - x_0}{x_0 - x_2} q_d = \frac{y_1 - y_0}{y_0 - y_2} q_d , \tag{74}$$

where q_d is the volume of the bulk solution.

Calcium carbonate deposition

Calcium carbonate scale, also referred to as alkali scale, tends to occur at the membrane interface with the concentration compartment when the electrodialyzer feed water contains high concentrations of the alkaline ion HCO_3^-. It can be prevented by adding acid to either the concentration stream or the desalination stream, to lower the alkalinity. The deposition concentrations are given in the Langelier index [67]. Deposition is prevented by maintaining a pH at the membrane interface which is lower than pH_s, the pH at which $CaCO_3$ saturation occurs under the conditions obtaining at the interface. The term pH_s may be expressed as

$$pH_s = pCa + pAlk + (pK_2 - pK_s) , \tag{75}$$

pCa = Negative log of Ca concentration in $CaCO_3$ (ppm)
pAlk = Negative log of alkali M concentration in $CaCO_3$ (ppm)
pK_2 = Negative log of 2nd dissociation constant of carboxylic acid
pK_s = Negative log of $CaCO_3$ activity.

The value of $pK_2 - pK_s$ varies with temperature and total salts concentration. At 25 °C and 200 to 5000 ppm, it is in the range of 2.1 to 2.2. For high total salts concentration, an experimentally derived expression [68] for the relation between $pK_2 - pK_s$ and ion concentration can be applied to determination of the deposition characteristics of a given system.

Calcium sulfate deposition

The limiting concentration of calcium sulfate can be obtained from the saturation values by methods such as those reported by Mettler [69].

Attempts to prevent calcium sulfate deposition by adding hexametaphosphate or some other deposition-suppressing agent have been unsuccessful, as they lower the limiting concentration by at most 20% to 30%. In electrodialytic desalination it is therefore necessary to remove calcium from the raw feed solution by water softening. In the electrodialytic concentration of sea water for salt production, calcium sulfate deposition is prevented by utilization of cation and anion exchange membranes which are permselective to univalent ions and thus reduce the entry of Ca^{2+} and SO_4^{2-} ions into the concentration compartments.

Electrodialysis reversal

Electrodialysis reversal (EDR) may be employed to avoid the adverse effects of scale deposition and prevent fouling of the membrane by impurities from the raw feed solution. It employs periodic reversal of the electrodes, electrolyte streams, and electrodialyzer compartments, as illustrated in Figure 1.151, for removal of scale from the membrane surface facing the concentration compartment and impurities from the surface facing the desalting compartment.

EDR had been proposed as early as 1958 [70] but it was not until the early 1970s that the first practical EDR process, employing current reversal every 15 to 20 minutes, was developed [71] by Ionics Inc., based on advances in electrode materials and electrodialyzer control systems.

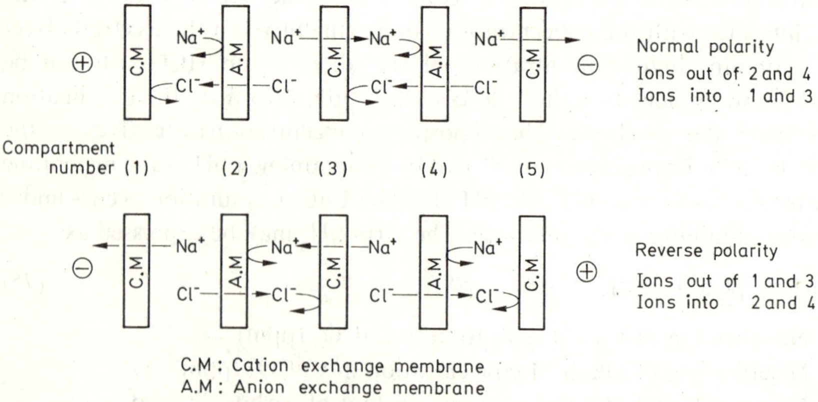

Figure 1.151 Basic principle of EDR.

The effectiveness and utilization of EDR systems have grown rapidly since then. Recent test operation has reportedly [72] resulted in 93% water recovery in desalination of a solution containing 800 ppm of calcium sulfate both with and without the addition of hexametaphosphate to the concentration stream, with calcium supersaturation reaching 440% in the concentration stream. This is in marked contrast to conventional electrodialysis, in which supersaturation of 150% has been found impractical even with hexametaphosphate addition. Recent operating capabilities with the EDR process are reportedly [73] in the range of:

Langelier index limit: +2, and
Calcium sulfate saturation limits: 220–250%.

Distribution of flow among cells

In electrodialysis, a uniform distribution of solution to all compartments is essential, to prevent the occurrence of water splitting. A theoretical description of the complex flow system of the electrodialyzer, involving the distribution of solutions to and from some 100 to 300 compartments through the conduits formed by the membrane and gasket ports, has been developed by Azechi [74, 75] based on the model shown in Figure 1.152.

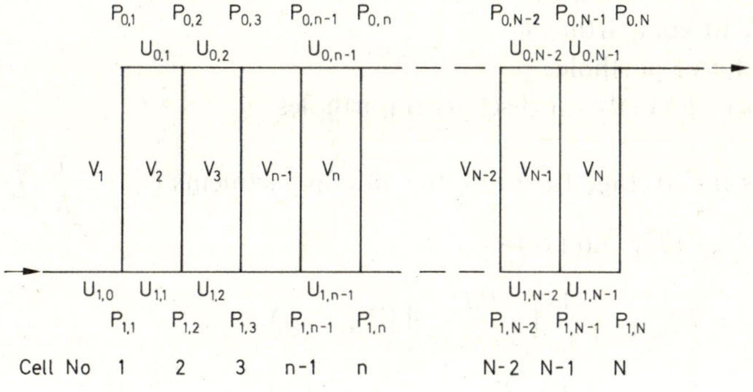

N : Total number of compartments
$P_{1,n}$: Pressure at inlet port of compartment n
$P_{0,n}$: Pressure at outlet port of compartment n
V_n : Flow rate in compartment n
$U_{1,n}$: Flow rate in inlet conduit between compartments n and n+1
$U_{0,n}$: Flow rate in outlet conduit between compartments n and n+1

Figure 1.152 Supply and discharge flows in electrodialysis stack.

In this model the rate of flow v through compartment n is related to the difference between the static pressures $P_{1,n}$ and $P_{0,n}$ at the upper and lower compartment ports, by the equation

$$P_{1,n} - P_{0,n} = K v_n^{\gamma}, \qquad (76)$$

where K and γ are constants. Furthermore, between the bifurcation points n and $n-1$ at the inlet ports of compartments n and $n-1$,

$$P_{1,(n-1)} + \frac{1}{2g} U_{1,(n-2)}^2 = P_{1,n} + \frac{1}{2g} U_{1,(n-1)}^2 + \frac{\varphi_a}{2g} U_{1,(n-1)}^2 \tag{77}$$

and for the outlet ports of the same compartments,

$$P_{0,(n-1)} + \frac{1}{2g} U_{0,n-2} = P_{0,n} + \frac{1}{2g} U_{0,n-1}^2 + \frac{\varphi_b}{2g} U_{0,(n-1)}^2, \tag{78}$$

where φ_a and φ_b are the coefficients of loss due to friction between the bifurcation points.

At the same time, the relation of the rate of flow v through compartment n to its supply and discharge flow rates v may be expressed as

$$\frac{1}{4} \Pi \Phi^2 m U_{1,n} = ld \left(\sum_{k=1}^{N} v_k - \sum_{k=1}^{n} v_k \right) \tag{79}$$

$$\frac{1}{4} \Pi \Phi^2 m U_{0,n} = ld \sum_{k=1}^{n} v_k, \tag{80}$$

l = Width of compartment
d = Length of compartment
φ = Diameter of porthole
m = Number of supply (or discharge) portholes.

If v_0 denotes the average flow rate for all compartments ($v_0 = \frac{1}{N} \sum_{k=1}^{N} v_k$), it then follows from eq. (77) and (79) that

$$\begin{aligned} P_{1,n} = P_{1,n-1} &+ \frac{1}{2g} \left(\frac{4ld}{\pi m \varphi^2} \right)^2 \left[\left(N v_0 - \sum_{k=1}^{n-2} v_k \right)^2 \right. \\ &\left. - \left(N v_0 - \sum_{k=1}^{n-1} v_k \right)^2 \right] - \frac{\varphi a}{2g} \left(\frac{4ld}{\pi m \varphi^2} \right)^2 \left(N v_0 - \sum_{k=1}^{n-1} v_k \right)^2 \end{aligned} \tag{81}$$

and it follows from eq. [78] and [80] that

$$\begin{aligned} P_{0,n} = P_{0,n-1} &+ \frac{1}{2g} \left(\frac{4ld}{\pi m \varphi^2} \right)^2 \left[\left(\sum_{k=1}^{n-2} v_k \right)^2 - \left(\sum_{k=1}^{n-1} v_k \right)^2 \right] \\ &- \frac{\varphi b}{2g} \left(\frac{4ld}{\pi m \varphi^2} \right)^2 \left(\sum_{k=1}^{n-1} v_k \right)^2. \end{aligned} \tag{82}$$

The flow rates at any compartment of the electrodialyzer can thus be determined by measurement of the pressure at appropriate locations and application of eq. (76), (81), and (82). The results of this determination for the Asahi Chemical SS-O electrodialyzer, which is used in salt production, are shown in Figure 1.153.

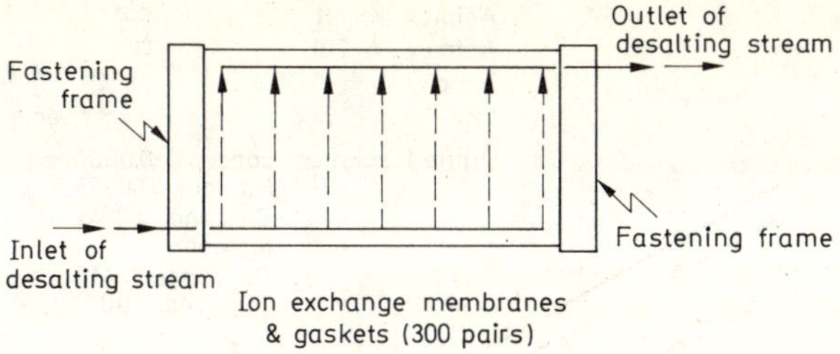

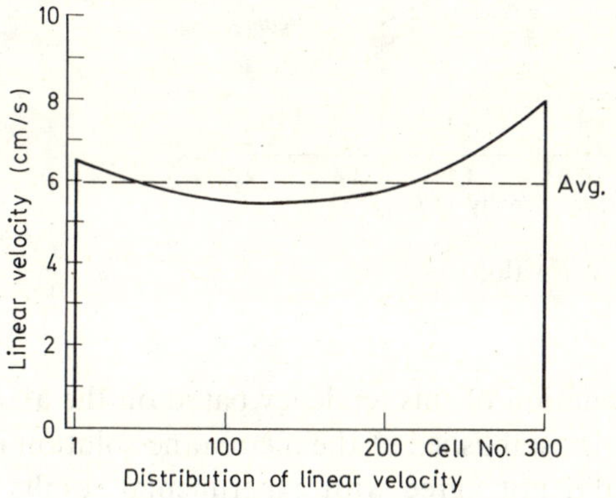

Figure 1.153 Distribution of linear velocity among cells (Stack of SS-O Model).

1.10.4.3 Special permselectivities

Permselectivity between ions of like charge

Determination for univalent and divalent ions

The permselectivity between ions of like charge is usually expressed by the filtering effect index, which is defined as

$$F_{B/A} = \left(\frac{\bar{t}_B}{\bar{t}_A}\right) \Big/ \left(\frac{C_B}{C_A}\right), \tag{83}$$

where $\bar{t}_i$ is the membrane transport number of ion species i, and C_i is its concentration in the diluent. The permselectivity between univalent and divalent ions is of particular importance in practical applications, because of its relation to scale deposition.

The value of $F_{B/A}$ is affected by current density, diluant flow rate, diluant concentration, temperature, and other electrodialysis conditions, as well as by the properties of the membrane itself.

Figure 1.154 shows the variation in $F_{B/A}$ with changes in current density, and illustrates its tendency to approach the ratio of intrasolution ion mobilities as the operating current density i approaches the limiting current density i_{lim}. Oren [76]

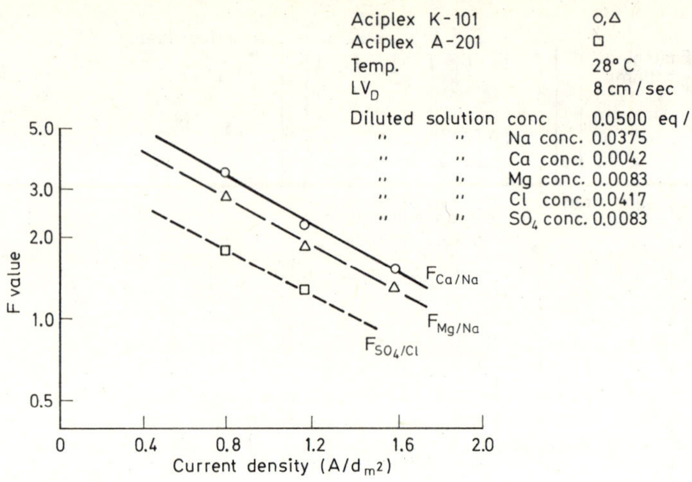

Figure 1.154 Current density vs. F value.

attempted a theoretical explanation of this tendency based on the assumption that an ion exchange equilibrium is established at the membrane-solution interface, but found that the predictions did not agree with experimental results obtained by radioisotope labelling of the ions involved. His study thus indicates that a simple equilibrium at the interface does not occur, and that the electric potential of the membrane surface must also be considered.

The filtering effect index $F_{B/A}$ provides a direct and convenient means for calculation of the concentration of the product solution from that of the desalting compartment supply solution or the raw feed solution. Since it is affected by the degree of desalination, however, calculation based on the permselectivity coefficient K, the differential of $F_{B/A}$, is more practical in cases where correction of $F_{B/A}$ for degree of desalination would be necessary.

The value of K has been shown experimentally to remain unchanged over a broad desalination range. For the system shown by Figure 1.155 K is defined by the expression

$$K_{\mathrm{Na}}^{\mathrm{Ca}} = \frac{\mathrm{d}\,E_{\mathrm{Ca}}/C_{\mathrm{Ca}}}{\mathrm{d}\,E_{\mathrm{Na}}/C_{\mathrm{Na}}}. \tag{84}$$

This may be rewritten as

$$K_{\mathrm{Na}}^{\mathrm{Ca}} \cdot \frac{-\mathrm{d}\,E_{\mathrm{Na}}}{C_{\mathrm{Na}}} = \frac{\mathrm{d}\,E_{\mathrm{Ca}}}{C_{\mathrm{Ca}}}. \tag{85}$$

Moreover, since $C_{\mathrm{Na}} = E_{\mathrm{Na}}/Q$ and $C_{\mathrm{Ca}} = E_{\mathrm{Ca}}/Q$, it follows that

$$K_{\mathrm{Na}}^{\mathrm{Ca}} \cdot \frac{-\mathrm{d}\,E_{\mathrm{Na}}}{E_{\mathrm{Na}}} = \frac{\mathrm{d}\,E_{\mathrm{Ca}}}{E_{\mathrm{Ca}}}. \tag{86}$$

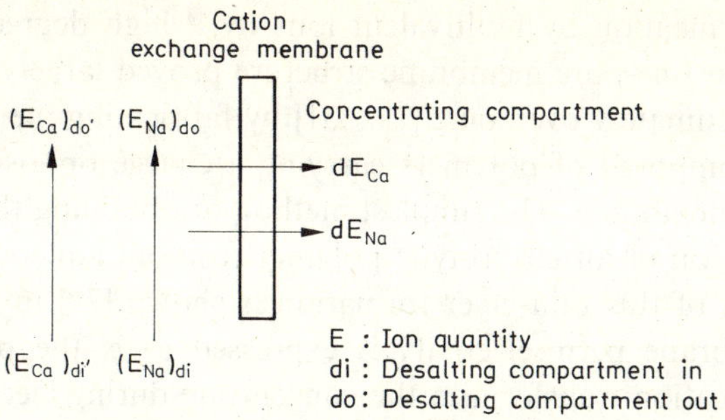

Figure 1.155 Ion quantities in desalting compartment.

Integration of both sides then yields

$$K_{Na}^{Ca}[-\ln E_{Na}]_{(E_{Na})_{di}}^{(E_{Na})_{do}} = [-\ln E_{Ca}]_{(E_{Ca})_{di}}^{(E_{Ca})_{do}}. \tag{87}$$

From this we may obtain

$$K_{Na}^{Ca} = \frac{\log\{(E_{Ca})_{di}/(E_{Ca})_{do}\}}{\log\{(E_{Na})_{di}/(E_{Na})_{do}\}} = \frac{\mathrm{p}E_{Ca}}{\mathrm{p}E_{Na}}$$

$$(E_{Ca})_{di} = (C_{Ca})_{di}Q_{di} \quad (E_{Ca})_{do} = (C_{Ca})_{do}Q_{do} \tag{88}$$

$$(E_{Na})_{di} = (C_{Na})_{di}Q_{di} \quad (E_{Na})_{do} = (C_{Na})_{do}Q_{do}$$

and the value of K_{Na}^{Ca} may thus be obtained from eq. (88) by measurement of these values.

From the obtained value of K_{Na}^{Ca}, for a desired degree of sodium ion removal, it is thus possible to calculate the calcium ion concentration at the desalting compartment outlet, using the equation

$$(E_{Ca})_{do} = (E_{Ca})_{di}\,10^{-K_{Na}^{Ca}\cdot \mathrm{p}E_{Na}}. \tag{89}$$

It must also be noted [77] that the value of K, like that of $F_{B/A}$, is influenced by the electrodialysis conditions as well as by the membrane characteristics.

Means of obtaining high permselectivity for univalent ions

Membranes with a high permselectivity for univalent ions are essential for effective electrodialysis in the production of salt from sea water, which involves a high degree of concentration from solutions containing multivalent ions such as Ca^{2+} and SO_4^{2-}. Their development has, therefore, been most rapid in Japan, where most table salt is obtained from sea water.

Early in the development of the salt production process, it was found that the passage of multivalent ions through the membrane lowered the current efficiency for NaCl to unacceptable levels, and quickly led to scale deposits in the concentration compartment which seriously impeded operation. Early attempts [78] at preventing membrane permeation by multivalent ions by a high degree of polymer crosslinking to obtain a fine-pore membrane structure proved largely ineffective.

The problem was ultimately overcome [29–31] by fabricating membranes with a thin surface layer composed of polymers carrying a charge opposite to that of the main body of the membrane. The simplest method of obtaining this layer is by electrophoretic deposition of an electrolytic polymer from an aqueous suspension or solution. The effect of this thin-layer formation is shown [79] in Figure 1.156. The increase in membrane permselectivity is expressed as ε, the proportion of univalent ion to total cation transfer into the concentrate during electrodialysis of a feed solution containing a cationic polymer formed by formalin condensation of dicyandiamide. As shown in the figure, the rise in ε is accelerated by increasing the concentration of the polycation $C^{\circ}x$ in the feed solution.

The configuration of the thin surface layer carrying a charge opposite to that of the main body of the membrane is shown [79, 80] schematically in Figure 1.157. Its barrier effect is apparently attributable to the requirement of higher energy levels for passage of multivalent ions than for univalent ions across the area of high electric potential shown [79] in Figure 1.158. Various methods have been reported [29–31], [81–84] for the formation of thin surface layers which are stable in long-term operation and cause no significant rise in electrodialytic voltage. One

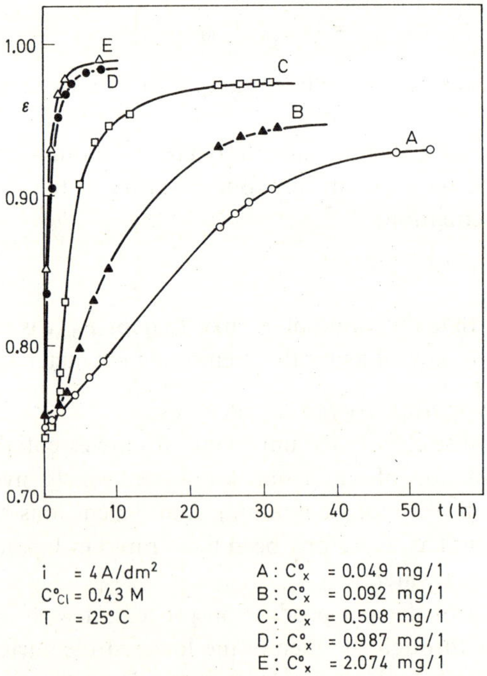

Figure 1.156 Change of ε with time.

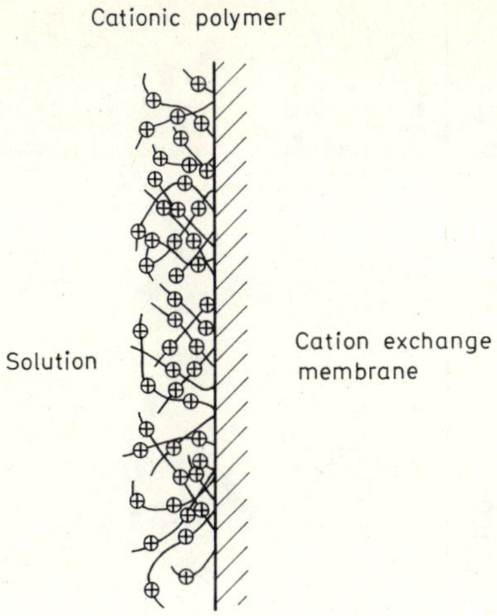

Figure 1.157 Cationic polymer layer on membrane surface.

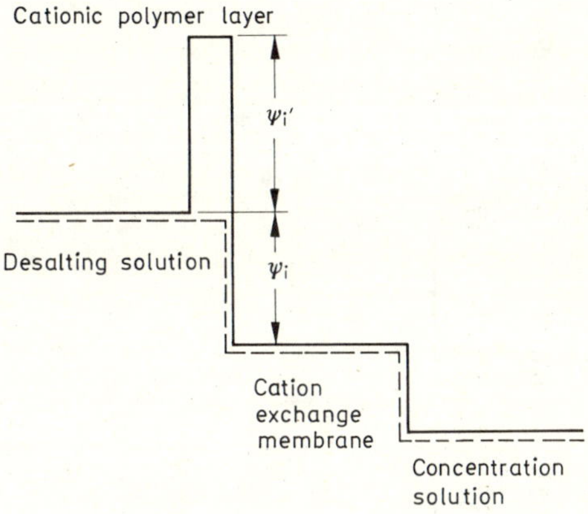

Figure 1.158 Potential profile across cation exchange membrane. Dotted line: profile in the absence of the cationic polymer layer.

of the most advanced is employed by Asahi Chemical in its fabrication of Aciplex K-172 and A-172 membranes, which are commercially used in combination for electrodialytic production of concentrated brine from sea water. The filtering effect indexes and product composition obtained with these membranes are shown in Table 1.48.

Permselectivities for inorganic ions and organic molecules

Desalination of organic substances

Effective, practical separation of inorganic salts from organic substances is achieved by electrodialysis with ion exchange membranes permselective to the salts.

Table 1.48 Typical analysis of concentrated brine and seawater

Product		Analysis							Filtering effect		
		Cl^-	SO_4^{2-}	Ca^{2+}	Mg^{2+}	K^+	Na^+	Total	F_{SO_4}	F_{Ca}	F_{Mg}
Concentrated brine	Concentration (N)	3.679	0.004	0.055	0.102	0.104	3.422	3.683	0.01	0.36	0.13
	Ratio (%)	99.9	0.1	1.5	2.8	2.8	92.9	100.0			
Seawater	Concentration (N)	0.500	0.051	0.019	0.098	0.009	0.425	0.551			
	Ratio (%)	90.7	9.3	3.5	17.9	1.7	6.9	100.0			

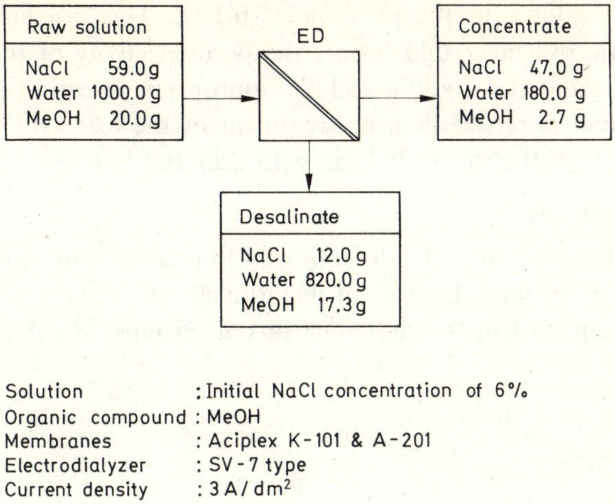

Figure 1.159 Mass balance for electrodialysis of mixed solution of organic compound and inorganic salt.

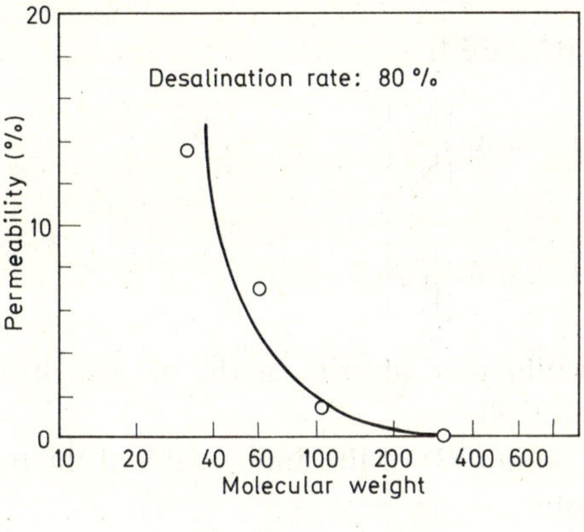

Figure 1.160 Relation between molecular weight and permeability of organic compounds.

Figure 1.159 shows the mass balance for 80% removal of NaCl from a 6% solution containing methanol as the organic substance, in electrodialysis with standard Aciplex K-101 and A-201 membranes. As indicated in Figure 1.160, the rate of organic material recovery increases with increasing molecular weight, and salt separation is possible with little or no membrane permeation by organic substances having molecular weights of 200 or more.

Special measures are necessary, however, for electrolytic organic materials carrying an electric charge and thus undergoing electrophoretic movement through the solution. Those with a molecular weight of 100 or less tend to pass through the membranes. The membrane impedes those of higher molecular weight, but some of these may adhere to the membrane surface. This organic fouling [85] increases the electric resistance and lowers the permselectivity of the membrane. Measures which prevent this fouling include appropriate selection of membrane polymer [86–88] and pore size [89], pretreatment of the feed [90] solution, and periodic washing [91] of the electrodialyzer with acid and base.

Amino acid electrodialysis

In the electrodialytic separation of salt from solutions containing amino acids, the pH is of particular importance because of its influence on amino acid dissociation, which involves the constituent amino and carboxylic groups. The dissociation may be expressed as

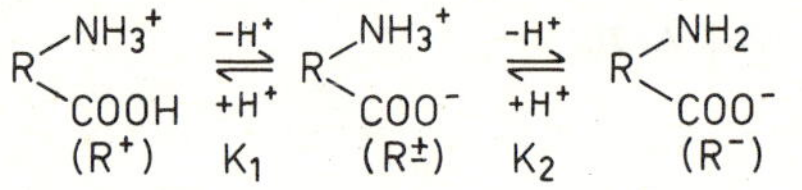

The equilibrium constants of this dissociation may be expressed as

$$K_1 = \frac{[R^\pm][H^+]}{[R^+]}; \; K_2 = \frac{[R^-][H^+]}{[R^\pm]}. \tag{91}$$

Since $-\log K = pK$, pK_1 and pK_2 are thus

$$pK_1 \equiv -\log K_1 = pH - \log\frac{[R^\pm]}{[R^+]} \tag{92}$$

$$pK_2 \equiv -\log K_2 = pH - \log\frac{[R^-]}{[R^\pm]}. \tag{93}$$

It is therefore possible to determine pK_1 and pK_2 as the pH which results in $[R^\pm] = [R^+]$ and $[R^-] = [R^\pm]$, respectively.

Furthermore, the isoelectric point (PI) of the amino acid, at the pH giving $[R^+] = [R^-]$, may be determined from

$$PI = \frac{pK_1 + pK_2}{2}. \tag{94}$$

For each amino acid, the pK and PI values can therefore be obtained from its acidic and basic titration curves. These values are shown in Table 1.49 for the main amino acids.

Each amino acid may thus be anionic, cationic, or nonionic, depending on the pH of the solution in which it is contained, and its dissociation and membrane permeation will vary with the pH. This variation is illustrated [92] in Figure 1.161, which shows the permeation of two membrane systems by glycine, one system being comprised of membranes containing strong acidic and basic groups and the other of membranes containing weak acidic and basic groups. As shown, the points of maximum permeation occur at pH 2.3 (pK_1) and pH 9.6 (pK_2).

Table 1.49 pK and PI values of amino acids

	pK_1	pK_2	pK_3	PI
Alanine	2.34	9.69		6.00
Arginine	2.17	9.04	12.48	10.76
Asparagine	2.02	8.80		5.41
Aspartic acid	1.88	3.65	9.60	2.77
Cysteine	1.96	8.18	10.28	5.07
Cystine	<1.00	1.70	pK_3 7.48 pK_4 9.02	4.60
Glutamic acid	2.19	4.25	9.67	3.22
Glutamine	2.17	9.13		5.65
Glycine	2.34	9.60		5.97
Histidine	1.82	6.00	9.17	7.59
Hydroxy proline	1.92	9.73		5.83
Isoleucine	2.36	9.68		6.02
Leucine	2.36	9.60		5.98
Lysine	2.18	8.95	10.53	9.74
Methionine	2.28	9.21		5.74
Phenylalanine	1.83	9.13		5.48
Proline	1.99	10.96		5.48
Serine	2.21	9.15		5.68
Threonine	2.71	9.62		6.16
Tryptophan	2.38	9.39		5.89
Tyrosine	2.20	9.11	10.07	5.66
Valine	2.32	9.62		5.96

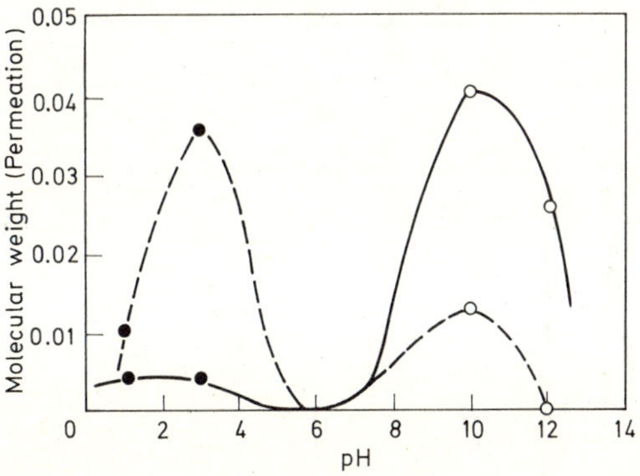

Figure 1.161 Membrane permeation by glycine.

One important application of electrodialysis is the desalination of soy sauce, which contains a large number of amino acids. The mass balance of a process employing Aciplex K-101 and A-201 membranes to reduce the NaCl content by 70% is shown in Figure 1.162. Table 1.50 shows the amino acid content before and after the desalination, as determined by the high performance liquid chromato-

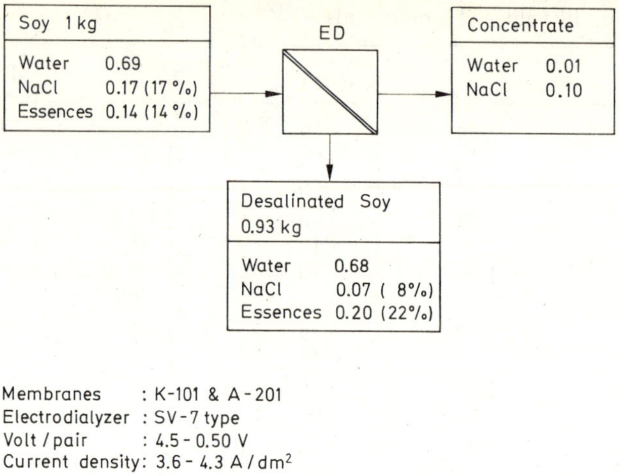

Figure 1.162 Mass balance for electrodialytic desalination of soy (70% desalination).

Table 1.50 Analysis of raw and desalinated soy

	s. g.	pH	NaCl (N)	Essences (wt%)	Total nitrogen (g/l)	Nitrogen in amino acids (g/l)	Reduced sugar (g/l)
Raw soy	1.180	4.55	2.91	13.8	1.60	9.0	3.55
Desalinated soy	1.123	4.76	0.86	21.7	1.76	9.6	4.56

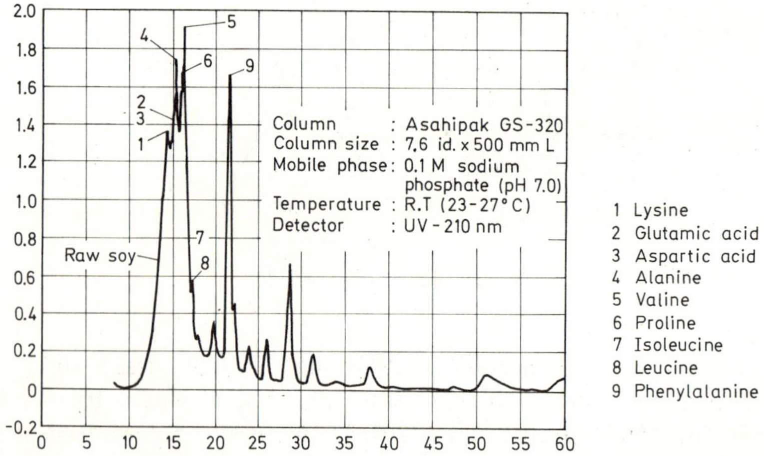

Figure 1.163 HPLC analysis of raw soy.

grams of Figure 1.163 and 1.164, obtained with an Asahipak GS 320 column. As indicated, the amino acids remain almost entirely in the desalinate, with almost no permeation to the concentrated NaCl solution.

The relation between the molecular weight of each of the amino acids with $PI = 5.5-6.5$ and the degree of membrane permeation is indicated in Figure 1.165. As shown, the permeation tends to decrease with increasing molecular weight. It should be noted that the permeation amounts are three times as large as those in Figure 1.160, since the amount of NaCl separated is three times as large. In this

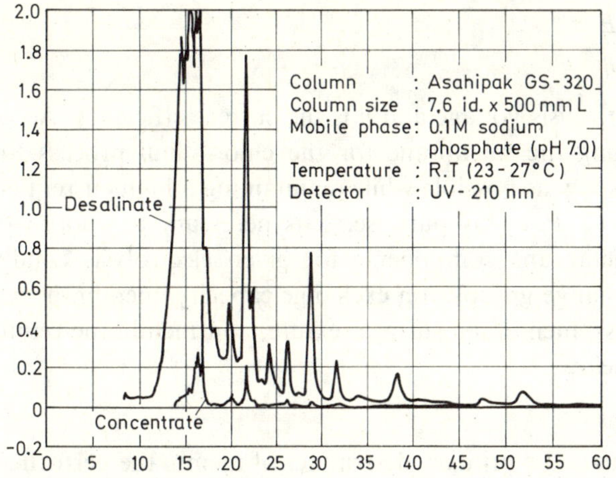

Figure 1.164 HPLC analysis of desalinated soy.

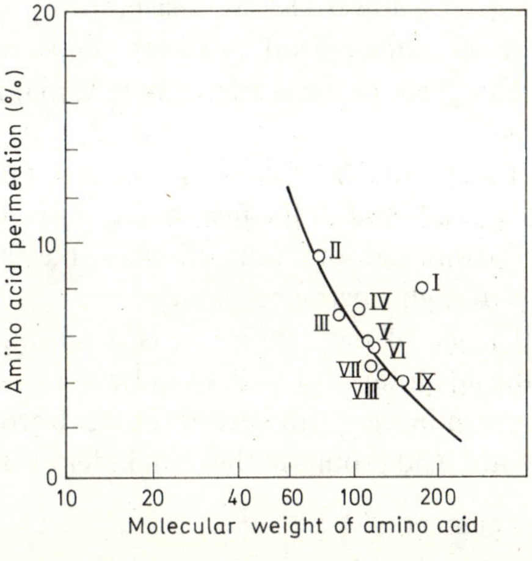

Figure 1.165 Relation between molecular weight and permeability of amino acids (I Arginine, II Glycine, III Alanine, IV Serine, V Proline, VI Threonine, VII Valine, VIII Isoleucine, IX Methionine).

plot, the location of the permeation ratio for arginine away from the curve of the other permeation ratios is due to its high *PI*, at a pH considerably different from that of soy sauce, and its resulting dissociation and electrophoresis.

1.10.4.4 Perfluoro ion exchange membranes for chlor-alkali process

The basic indicator of the economics of the chlor-alkali process is the amount of electric power necessary to produce one ton of caustic soda. This unit power consumption P may be expressed as

$$P = 670 \frac{E}{\eta}, \tag{68}$$

where E is the electrolysis voltage and η is the current efficiency. Accordingly, it is essential in designing the membrane for the chlor-alkali process to achieve an electric potential as low as possible while maintaining a high current efficiency and high product quality. For this purpose, it is necessary to choose the optimum combination of membrane components for given electrolysis conditions. These include the ion exchange groups, ion exchange capacity, membrane water content, membrane thickness, membrane surface texture, and membrane reinforcing material, as described below.

Ion exchange groups

The exchange group is the primary determiner of membrane performance. For the chlor-alkali process, composite membranes containing both carboxylic acid and sulfonic acid exchange groups are generally recognized as providing the closest approach to ideal performance. Because of their strong acidity and high hydrophilicity, sulfonic acid groups impart a lower electric resistance to the membrane than carboxylic acid groups. For the same reason, however, the fixed-ion concentration of sulfonic acid groups in the membrane is relatively low, making it difficult to obtain high current efficiency.

With carboxylic acid groups, the situation is effectively reversed. They are weakly hydrophilic, resulting in a low water content in the membrane. This imparts a high fixed-ion concentration to the membrane and thus allows achievement of high current efficiency, but also leads to high electric resistance.

Since the overall current efficiency of the membrane is determined largely by the surface facing the cathode, the advantages of both of these ion exchange groups can be obtained in a composite membrane, with carboxylic acid groups in a thin layer facing the cathode and sulfonic acid groups in the remainder of the membrane.

Ion exchange capacity and water content

As indicated by eq. (7), (10), and (11), a high current efficiency requires a high concentration of fixed ions in the membrane. This concentration is directly related to the ion exchange capacity (IEC) and the water content of the membrane. This

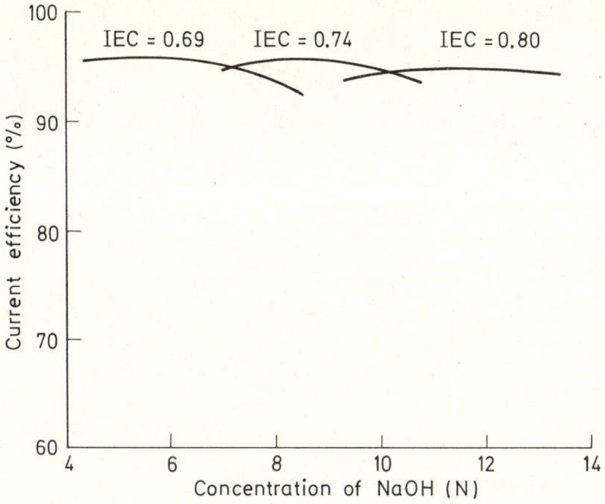

Figure 1.166 Relation between current efficiency, NaOH concentration, and ion exchange capacity (IEC meq/g dry resin). Electrolysis conditions: 4.0 kA/m^2, 90°C.

water content, however, is strongly influenced by the catholyte concentration. The optimum IEC of the membrane layer facing the cathode thus varies with the catholyte concentration, as shown in Figure 1.166 for the IEC of the thin carboxylic acid layers.

Membrane thickness

The optimum membrane thickness is determined by the desired membrane strength, electric resistance and purity of the product caustic soda. Increasing the thickness allows a higher product purity, but also results in a higher electric resistance.

Membrane surface texture

A fine-grain texturing of both membrane surfaces is desirable, to prevent the adhesion of hydrogen and chlorine gas bubbles which evolve in the catholyte and anolyte compartments during electrolysis, respectively. Adhesion of these bubbles to the membrane surface will locally prevent current flow, resulting in increased overall cell voltage and non-uniform current density in the membrane.

The advantages of membrane-surface texturing are illustrated in Figure 1.167. With a smooth, non-textured membrane, the cell voltage is relatively high, and increases markedly with close proximity between membrane and electrodes. With the textured membrane, the cell voltage tends to decrease toward a minimum with decreasing electrode gap and remain at the minimum despite further gap narrowing, due to the absence of gas bubble adhesion.

Reinforcing material

A woven, mesh-form material is generally implanted in the membrane to provide dimensional stability, mechanical strength, and operational safety. The reinforce-

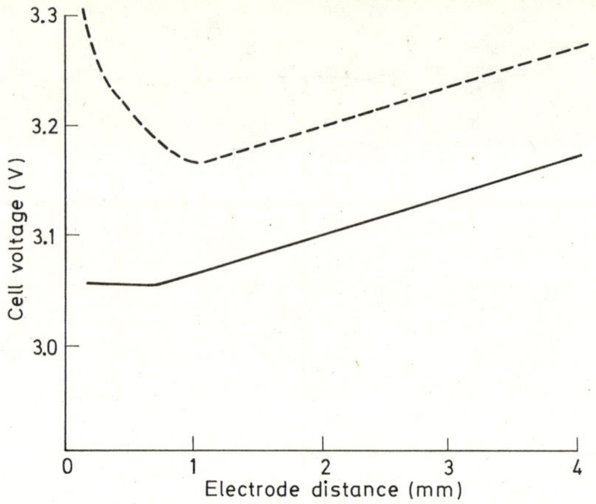

Figure 1.167 Electrode distance vs. cell voltage (--- Conventional membrane, – Rough surface membrane, 1.2 m × 2.4 m cell, current density: 4.0 kA/m², temperature: 90°C).

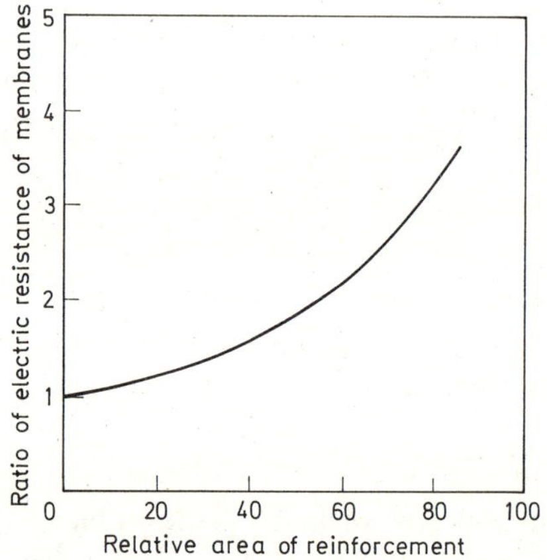

Figure 1.168 Relationship between electric resistance of membranes and relative area of reinforcement.

ment is usually composed of polytetrafluoro ethylene or some other acid-resistant polymer.

Although the reinforcing material tends to increase the overall electric resistance of the membrane, as indicated in Figure 1.168, its use is nevertheless required in order to obtain membranes of sufficient strength. The relation between the tensile strength of the membrane and the relative area of the reinforcement is illustrated in Figure 1.169.

Membrane reinforcement is particularly important for operational safety in industrial applications. Without reinforcement, any pinholes which may form in

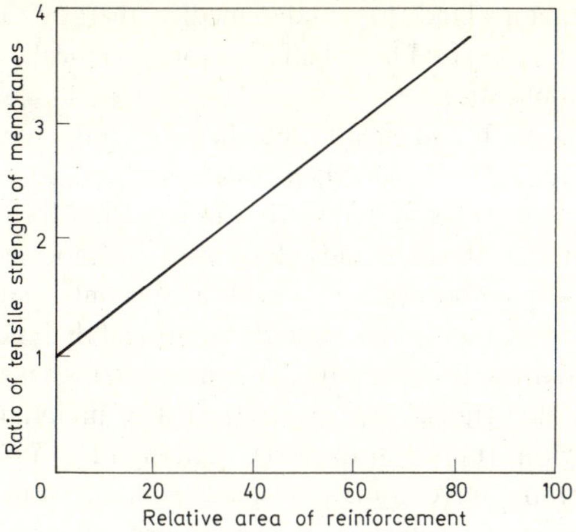

Figure 1.169 Relationship between tensile strength of membrane and relative area of reinforcement.

the membrane are apt to lead to unimpeded tears in the membrane, which will then permit a dangerous intermixing of hydrogen and oxygen gases in the cell compartments. The reinforcing material reduces this danger, since any tear which may originate from a pinhole will be confined to the area within the interstices of the reinforcing mesh.

1.10.5 Future prospects for functional membranes

Progress in synthetic functional membrane technology has been broad and rapid in the years since the first reported fabrication of a synthetic ion exchange membrane in 1950. In addition to ion exchange membranes, dialysis, reverse osmosis, gas separation, pervaporation, microfiltration, and ultrafiltration membranes are now employed in a broad range of applications. This progress is expected to continue.

Membrane separation processes consume little energy, as they generally require no material phase changes. Their use is therefore expected to grow as fuel and energy costs increase. Pervaporation is particularly promising in applications which would otherwise require distillation at high levels of energy consumption, such as the separation of substances with an azeotropic point, for which distillation is normally impossible. In the same way, it is also appropriate for concentrating dilute aqueous alcohols obtained through fermentation, which may become an important alternative energy source as fossil fuels become depleted in the future.

Gas separation membranes, such as those for oxygen enrichment and hydrogen removal, hold interesting implications for industrial use. The market for oxygen-enriched air is expected to surpass five million dollars with applications in efficiency enhancement of boilers and automobile engines, industrial fermentation processes, oxidation processes in the chemical industry, and in medical therapy. Gas separation membranes are also being investigated as substitutes for biological membranes, in artificial lungs and other applications.

Among dialytic membranes, liquid membranes have recently become a major focus of research and development, due to their functional similarity to biological membranes. These liquid membranes may be either in emulsion form or held in a supporting framework. Some of those already developed are reportedly similar in basic functions to biological membranes, which include not only material transfer by diffusion in accordance with Fick's law, as in conventional dialytic membranes, but also special permselectivities, transfer rates several times higher than those of Fick's law diffusion (facilitated transport), and transport of materials from lower to higher concentration regions (up-hill transport). Although the functional mechanisms of biological membranes have in many respects remained a mystery, it now appears that it will be possible to develop liquid membranes with similar functional effects.

With the growth of new industries based on biotechnology and C1 chemistry, the role of liquid and other functional membranes is expected to grow rapidly, in applications such as the concentration of methane from biogas and the separation of hydrogen from carbon monoxide.

In addition to their capability for material transfer and separation, functional membranes may also be developed for various types of energy conversion. Important progress is expected in the development of sensor, thermoelectric, piezoelectric, and photocell membranes, which respectively convert chemical energy, heat, mechanical pressure, and light to electricity, of artificial muscle and power generation membranes which convert chemical energy to mechanical energy, of photosynthesis and photolysis membranes which convert light to chemical energy, and of photocontrol membranes which convert light to mechanical energy.

Functional membranes will thus continue to grow rapidly in capability and importance, in a widening range of applications involving material transfer and energy conversion.

References

[1] L. Michaelis and A. Fujita (1925) Untersuchungen über elektrische Erscheinungen und Ionendurchlässigkeit von Membranen. II. Die Permeabilität der Apfelschale. Biochemische Zeitschrift, *158*, 28–37.
L. Michaelis and A. Fujita (1925) Untersuchungen über elektrische Erscheinungen und Ionendurchlässigkeit von Membranen. IV. Potentialdifferenzen und Permeabilität von Kollodiummembranen. Biochemische Zeitschrift, *161*, 47–60
L. Michaelis and A. Fujita (1925) Untersuchungen über elektrische Erscheinungen und Ionendurchlässigkeit von Membranen. VII. Die Permeabilität der Kollodiummembran für mehrwertige Kationen. Biochemische Zeitschrift, *164*, 23–30.

[2] L. Michaelis and Sh. Dokan (1925) Untersuchungen über elektrische Erscheinungen und Ionendurchlässigkeit von Membranen. Biochem. Zeitschrift, *162*, 258–265.

[3] L. Michaelis and K. Hayashi (1926) Untersuchungen über elektrische Erscheinungen und Ionendurchlässigkeit von Membranen. Biochem. Zeitschrift, *173*, 411–425.

[4] L. Michaelis (1926) Die Permeabilität von Membranen. Die Naturwissenschaften, *14*, 33–42.

[5] K. H. Meyer and J.-F. Sievers (1936) La permeabilite des membranes. I. Theorie de la permeabilite ionique. Helv. Chim. Acta., *19*, 649–664
K. H. Meyer and J.-F. Sievers (1936) La permeabilite des membranes. II. Essais avec des membranes selectives artificielles. Helv. Chim. Acta., *19*, 665–677.

[6] T. Teorell (1935) An attempt to formulate a quantitative theory of membrane permeability. Proc. Soc. Exptl. Biol. Med. *33*, 282–285.

[7] K. Sollner (1949) The origin of bi-ionic potentials across porous membranes of high ionic selectivity. J. Phys. and Colloid Chem., *53*, 1211–1226.

[8] K. H. Meyer and W. Strauss (1940) La permeabilite des membranes, VI. Sur le passage du courant electrique a travers des membranes selectives. Helv. Chim. Acta., *23*, 795–800.

[9] W. Juda and W. A. McRae (1950) Coherent ion exchange gels and membranes. J. Am. Chem. Soc., *72*, 1044.

[10] M. Seko (1980) Electrolysis of cation exchange membrane and use thereof in the electrolysis of sodium chloride. Japan. 1351 ('80)
M. Seko (1980) Electrolysis of cation exchange membrane and use thereof in the electrolysis of sodium chloride. Japan. 14148 ('80).

[11] J. Juda and W. A. McRae (1958) Ion-exchange materials. US Pat. 2636851.
G. W. Bodamer (1954) Permselective films of anion-exchange resins. US Pat. 2681319.
R. J. Malcolm (1956) Method of making low-resistance ion-exchange membranes. US Pat 2774108.

[12] N. Tamura (1976) Effects of post-treatment on the properties of heterogeneous cation exchange membranes synthesized from ground powder of cation exchange resins and polypropylene. Bull. Chem. Soc. Japan, *No. 7*, 1118–1124.

[13] W. Juda and W. A. McRae (1956) Ion-exchange materials. Japan. Pat. 8127 ('56).

[14] S. Sekino and K. Takano (1957) Process for producing of ion-exchange resin membranes. Japan. Pat. 4593 ('57).

[15] H. P. Gregor and D. M. Westone (1956) Specific transport across sulfonic and carboxylic interpolymer cation-selective membranes. Discussions Faraday Soc., *No. 21*, 162–173.

[16] H. P. Gregor, H. Jacobson, R. C. Shair and D. M. Wetstone (1957) Interpolymer ion-selective membranes. J. Phys. Chem. *61*, 141–147.

[17] T. Uchino and H. Hani (1960) Preparation of ion-exchange resin membranes from polyethylene-styrene graft copolymers obtained by γ-ray irradiation. Asahi Garasu Kenkyo Hokuku *10*, 17–28.

[18] T. Yanagida, K. Shinohara, H. Kawabe and T. Takamatsu (1963) Process for producing of anion-exchange membranes. Japan. Pat. 10993 ('63)
M. Yanagita, K. Shinohara, H. Kawabe and T. Takamatsu (1963) Process for producing of cation-exchange membranes. Japan. Pat. 11275 ('63).

[19] H. Hani and T. Uchino (1961) Process for producing of ion permselective membrane. Japan. Pat. 17637 ('61).

[20] Y. Tunoda, M. Seko, M. Watanabe, A. Ehara and T. Misumi (1957) Cation exchange resins. Japan. Pat. 4144 ('57).
Y. Tunoda, M. Seko, M. Watanabe, A. Ehara and T. Misumi (1957) Anion-exchange resins prepared in the presence of plasticizer and polymer. Japan. Pat. 4145 ('57).
Y. Tunoda, M. Seko, M. Watanabe, A. Ehara and T. Misumi (1957) New base polymers for ion-exchange resins, especially having large dimensions, and granular base polymers which are not fractured during their resins thereof. Japan. Pat. 6387 ('57).

Y. Tunoda, M. Seko, M. Watanabe, A. Ehara, T. Misumi and Y. Yamakoshi (1957) Cation exchange resins, especially having large dimensions and granular cation-exchange resins which are not fractured during their resins thereof. Japan. Pat. 10696 ('57).
Y. Tunoda, M. Seko, M. Watanabe, A. Ehara, T. Misumi and Y. Yamakoshi (1957) Anion exchange resins, especially having large dimensions and granular anion-exchange resins which are not fractured during their resins thereof. Japan. Pat. 10697 ('57).
Y. Tunoda, M. Seko, M. Watanabe, T. Misumi and Y. Yamakoshi (1960) Cation exchange resins, especially having large dimensions of horyzontality and granular cation-exchange resins which are not fractured during their resins thereof. Japan. Pat. 7290 ('60).
[21] Y. Mizutani, R. Yamabe and K. Motomura (1964) Preparation of ion-exchange membranes. Japan. Pat. 19542 ('64).
Y. Mizutani, W. Tejima, S. Akiyama and H. Ihara (1965) Preparation of ion-exchange membranes. Japan. Pat. 28951 ('65).
[22] T. Kuwata and S. Yoshikawa (1957) Preparation of anion-exchange resin membranes. Japan. Pat. 4591 ('57).
[23] S. Sekino, A. Nishihara and Y. Mineki (1958) Preparation of anion-exchange resin membranes. Japan. Pat. 1145 ('58).
[24] H. Hani and T. Hiraga (1966) Preparation of ion-exchange membranes. Japan. Pat. 5271 ('66).
H. Hani, T. Hiraga and A. Nishihara (1960) Preparation of ion permselective membranes. Japan. Pat. 13009 ('60).
H. Hani, A. Nishihara and Y. Oda (1961) Anioin-exchange resin membrane with permselectivity between anions. Japan. Pat. 15258 ('61).
[25] M. Seko (1980) Electrolysis of cation exchange membrane and use thereof in the electrolysis of sodium chloride. Japan. Pat. 1351 ('80).
M. Seko (1980) Electrolysis of cation exchange membrane and use thereof in the electrolysis of sodium chloride. Japan. Pat. 14148 ('80).
M. Seko, Y. Yamakoshi, H. Miyauchi, M. Fukumoto, K. Kimoto, K. Watanabe, T. Hane and S. Tushima (1982) Cation exchange membrane preparation and use thereof. Japan. Paat. 7165 ('82).
M. Seko, Y. Yamakoshi, H. Miyauchi, M. Fukumoto, K. Kimoto, K. Watanabe, T. Hane and S. Tushima (1982) Improved cation exchange membrane. Japan. Pat. 7166 ('82).
M. Seko, Y. Yamakoshi, H. Miyauchi, and K. Kimoto (1982) Improved cation exchange membrane. Japan. Pat. 7166 ('82).
[26] Y. Onoue, Y. Mizutani, W. Tejima, R. Yamane and S. Akiyama (1961) Improvement of cation exchange membrane for the permselectivity. J. Electro. Chem. Soc. Japan, *29*, 468–470.
[27] Y. Oda and A. Nishihara (1961) Cation exchange resin membrane with permselectivity between cations. Japan. Pat. 4210 ('61).
[28] R. Yamane, Y. Mizutani, K. Motomura and T. Ideo (1964) Preparation of SO_4^{2-} non permselective anion exchange membranes. J. Electro. Chem. Soc. Japan, *32*, 277–280.
[29] Y. Mizutani, R. Yamane and T. Sata (1965) A electrodialysis which can pass to select cations with low valence. Japan. Pat. 23607 ('71).
Y. Mizutani, R. Yamane, T. Sata and T. Ideo (1965) A electrodialysis which can pass to select anions with low valence. Japan. Pat. 30165 ('71).
[30] K. Mihara, T. Misumi, H. Miyauchi and K. Ishida (1970) Improved anion-exchange resin membranes and process for producing the same. Japan. Pat. 19980 ('70).
K. Mihara, T. Misumi, H. Miyauchi and K. Ishida (1970) Improved anion-exchange resin membranes and process for producing the same. Japan. Pat. 30693 ('70).
K. Mihara, T. Misumi, H. Miyauchi and K. Ishidi (1972) Improved cation-exchange resin membranes and process for producing the same. Japan. Pat. 3081 ('72).

[31] S. Tushima, T. Misumi and M. Murakoshi (1978) A Method for improvement in specific selective permeability of cation exchange membrane. Japan. Pat. 44155 ('78).
[32] F. B. Leitz (1971) Cationic-anionic ion-exchange membrane. US Pat. 3562139.
[33] L. T. C. Lee, G. J. Dege and Kang-Jen Liu (1977) High performance, quality controlled bipolar membrane. US Pat. 4057481 (1977).
[34] J. Shorr and F. B. Leitz (1974) Development of membranes and resins for piezodialysis. Desalination, 14 (1), 11 – 20.
[35] F. B. Leitz (1974) Piezodialysis, in: Membrane Separation Processes (P. Meares, ed.), Amsterdam, 1974.
[36] F. B. Leitz and J. Shorr (1972) Research on piezodialysis – third report. Res. Develop. Progr. Rept. No. 775, Office of Saline Water, U.S. Department of the Interior.
[37] A. Schidler (1972) Analysis and summary report of operation, Guantanamo naval base desalination facility. Res. Develop. Progr. Rept. No. 769, Office of Saline Water, U.S. Department of the Interior.
[38] A. Chapiro, G. Bex, A. M. Jendrychowsak-Bonamour and T. O'Neill (1969) Preparation of permselective membranes by radiation grafting of hydrophilic monomers onto poly(tetrafluoroethylene)films. Advan. Chem. Ser., *91*, 560 – 573.
[39] A. Chapiro, G. Bex, A. M. Jendrychowsak-Bonamour and S. Mizrahi (1976) Preparation of mosaic membranes by radiochemical grafting in polytetrafluoroethylene films. Eur. Polym. J., *12 (11)*, 773 – 780.
[40] A. Chapiro and A. M. Jendrychowsak-Bonamour (1980) Synthesis of permselective membranes by radiation induced grafting of hydrophilic monomers into poly(tetrafluoroethylene) films. Polym. Eng. Sci., *20 (3)*, 202 – 205.
[41] A. Schindler and H. Yasuda (1971) Mosaic membranes for desalination. U.S. Office Saline Water, Res. Develop. Progr. Rep., No. 689, 105.
[42] M. Kamachi, M. Kurihara and J. K. Stille (1972) Synthesis of block polymers for desalination membranes. Preparation of block copolymers of 2-vinyl pyridine and methacrylic acid or acrylic acid. Macromolecules, *5 (2)*, 161 – 167.
[43] T. Yamabe, K. Umezawa, Sh. Yoshida and N. Takai (1973) Piezodialysis, in: Proc. Int. Symp. Fresh Water from the Sea, 4th, Amaroussion, Greece (A. Delyannis, ed.) Vol. 4, pp. 475 – 483.
[44] S. Itoi (1981) Properties evaluating method of the ion-exchange membrane. Membrane, *6 (3)*, 185 – 196.
[45] A. Akimoto (1985) Development and future of pentablock copolymer membrane. Chem. Econ. Eng. Rev., *17 (1 – 2)*, 42 – 47.
[46] T. Segawa and I. Ito (1975) A Method of electrolytic reduction of uranyl chloride solution. Japan. Pat. 35029 ('75).
[47] M. Seko (1972) A new process electro-hydrodimerization of acrylonitrile, in: First pacific chemical engineering congress, Kyoto, Japan, Oct., 10 – 14, 1972 (The society of chemical engineerings, Japan and American institute of chemical engineerings) pp. 251 – 255.
[48] T. Inoue and T. Tanaka (1985) Method and apparatus for dehydration of water-containing substance by electro-osmosis, US Pat. 4549947.
[49] P. Aptel, J. Cuny, J. Jozcfonviez, G. Morel and J. Neel (1974) Liquid transport through membranes prepared by grafting of polar monomers onto poly(tetrafluoroethylene) films. II. Factors determining pervaporation rate and selectivity. J. Applied Polymer Sci., *18 (2)*, 351 – 364.
[50] Y. Oda and T. Yawataya (1955) On the electro-osmotic water transport through cation-exchange resin membranes. Bull. Chem. Soc. Japan, *28 (4)*, 263 – 269.
[51] G. Garsa and O. Kedem (1976) Electro-osmotic pumping in unit cells, in: 5th Inter. Symp. on Fresh Water from the Sea, *3*, 79 – 87.
[52] V. N. Smagin, V. A. Chukhim and V. A. Kharchuk (1983) Technological account of electrodialysis apparatus for concentration. Desalination, *46*, 283 – 290.

[53] Y. Tanaka and N. Kanai (1980) Effects of temperature on ion exchange membrane electrodialysis. Bulletin of the Society of Sea Water Science, Japan, *34*, 31–36.

[54] Y. Tanaka (1977) Material transport in an ion exchange membrane electrodialyzer for the concentration of sea water. Denki Kagaku Oyobi Kogyo Butsuri Kagaku, *45 (10)*, 630–635.

[55] J. R. Wilson (1960) Demineralization by electrodialysis. Butterworths Scientific Publication, p. 266–274.

[56] K. S. Spiegler (1971) Polarization at ion exchange membrane-solution interfaces. Desalination, *9 (4)*, 367–385.

[57] D. A. Cowan and J. H. Brown (1959) Effect of turbulence on limiting current in electrodialysis cells. Ind. Eng. Chem., *51*, 1445–1448.

[58] K. Mihara and M. Kato (1965) Design of desalting electrodialyzer using ion exchange membrane. Kagaku Kogaku, Japan, *29*, 432–438.

[59] X. W. Zhong, W. R. Zhang, Z. Y. Hu and H. C. Li (1983) Effect of characterizations of spacer in electrodialysis cells on mass transfer. Desalination, *46*, 243–252.

[60] A. Kitamoto and Y. Takashima (1968) Electro-osmosis, maximum attainable concentration, limiting current density, and energy efficiency in electrodialysis using ion-exchange membranes. Bull. Kagaku Kogaku, Japan, *32 (1)*, 74–82.

[61] Y. Tsunoda and M. Kato (1967) Compact apparatus for sea water desalination by electrodialysis using ion exchange membranes. Desalination, *3*, 66–81.

[62] R. Yamane, T. Sata, Y. Mizutani and Y. Onoue (1969) Ion-exchange membranes. XXVIII. Concentration polarization phenomena in ion-exchange membrane electrodialysis. 2. Effect of the condition of the diffusion boundary layer on the limiting current density and on the relative transport numbers of ions. Bull. Chem. Soc. Japan, *42 (10)*, 2741–2748.

[63] D. A. Cowan and J. H. Brown (1959) Effect of turbulence on limiting current in electrodialysis cells. Ind. Eng. Chem., *51*, 1445–1448.

[64] Ora Kedem (1975) Reduction of polarization in electrodialysis by ion-conducting spacers. Desalination, *16 (1)*, 105–118.

[65] O. Kedem, J. Cohen, A. Warshawsky and N. Kahana (1983) EDS-sealed cell electrodialysis. Desalination, *46*, 291–299.

[66] M. Seno, K. Yamagata, J. Shinoda and T. Yamabe (1966) Concentration polarization effect in the electrodialysis using ion-exchange membranes. II. Limiting current densities in various electrolyte and membrane systems. Denki Kagaku, Japan, *34 (10)*, 820–823.

[67] W. F. Langelier (1936) The analytical control of anti-corrosion water treatment. J. Am. Wat. Work. Assoc., *28*, 1500–1521.

[68] H. A. Stiff, Jr., and L. E. Davis (1952) A method for predicting the tendency of oil-field waters to deposit calcium carbonate. Trans. Am. Inst. Mining Met. Engrs., *195*, 213–216.

[69] A. V. Mettler and A. G. Ostroff (1967) Proximate calculation of the solubility of gypsum in natural brines from 28 degree to 70 degree. Environmental Science and Technology, *1 (10)*, 815–819.

[70] W. Juda and W. A. McRae (1958) Method of electrodialyzing aqueous solution. US Pat. 2863813.

[71] D. R. Brown (1981) Treating Bahrain Zone C groundwater using the EDR process. Desalination, *38 (1–2–3)*, 537–547.

[72] Melvin E. Mattson and Melvin Lew (1982) Recent advances in reverse osmosis and electrodialysis membrane desalting technology. Desalination, *41 (1)*, 1–24.

[73] William E. Katz (1983) State-of-the-art of the electrodialysis reversal (EDR) process. Int. Water Eng., *21 (1)*, 12–20.

[74] S. Azechi and Y. Fujimoto (1970) Flow characteristics of the practical-scale electrodialytic apparatus of filter-press type. Bull. Soc. Sea Water Sci., *23 (4)*, 134–147.

[75] S. Azechi (1970) A numerical analysis of permselectivity between two ions having the same electric charge on the basis of mass transfer model across ion exchange membrane. Bulletin of the Society of Sea Water Science, Japan, *24 (1)*, 25–36; *24 (2)*, 55–67.

[76] Y. Oren and A. Litan (1974) State of the solution-membrane interface during ion transport across an ion-exchange membrane. J. Phys. Chem., *78 (18)*, 1805–1811.

[77] Y. Tsunoda (1965) Electrodialysis for producing brine concentrates from sea water, in: 1st Inter. Symp. on Water Desalination, Washington, D.C., Oct. 3–9, 1965 (U.S. Department of the Interior) pp. 325–335.

[78] Y. Onoue, Y. Mizutani, R. Yamane and Y. Takasaki (1961) Selectivity of cation exchange membranes for $NaCl-CaCl_2$ system. J. Electro. Chem. Soc. Japan, *29*, 187–191.
Y. Onoue, Y. Mizutani, R. Yamane and Y. Takasaki (1961) Permselectivity of improved cation exchange membranes for $NaCl-CaCl_2$ system. J. Electro. Chem. Soc. Japan, *29*, 544–546.

[79] Y. Tanaka and M. Seno (1981) Treatment of ion exchange membranes to decrease divalent ion permeability. J. Memb. Sci., *8 (2)*, 115–127.

[80] T. Sata (1973) Modification of properties of ion exchange membranes. II. Transport properties of cation exchange membranes in the presence of water-soluble polymers. J. Colloid Interface Sci., *44 (3)*, 393–406.

[81] K. Mihara, T. Misumi, H. Miyauchi and K. Ishida (1970) Anion-exchange resin membranes. Japan. 30693 ('70).

[82] Y. Mizutani, R. Yamane, T. Sata and T. Ideo (1972) Cation-exchange resin membranes. Japan. 3801 ('72)
Y. Mizutani, R. Yamane, T. Sata and T. Ideo (1972) Cation-exchange resin membranes. Japan. 3802 ('72).

[83] K. Mihara, T. Misumi and H. Miyauchi (1973) Ion exchange membrane having an excellent relative-selective permeability between ions of same sign. Japan. 34676 ('73).

[84] S. Tsushima and T. Misumi (1978) Process for preparing improved cation-exchange membranes. US Pat. 4042496
S. Tsushima and T. Misumi (1978) Process for preparing improved cation-exchange membranes. Japan. Pat. 44155 ('78).

[85] E. Korngold, F. de Korosy, R. Rahav and M. F. Taboch (1970) Fouling of anionselective membranes in electrodialysis. Desalination, *8 (2)*, 195–220.

[86] A. Chiolle, L. Credali and P. Parrini (1973) Properties of new aliphatic anionic membranes for electrodialysis, in: Proc. Int. Symp. Fresh Water from the Sea, 4th, Amaroussion, Greece (A. Delyannis, ed.) Vol. 3, pp. 81–90.

[87] R. Ehara, Y. Takatori, N. Okonogi and M. Tomita (1974) Applications of ion-selective membranes to food processing. Shokuhin Kogyo, *17 (14)*, 58–69.

[88] E. Kusumoto (1979) Organic fouling ion-exchange membrane. Bulletin of the Society of Sea Water Science, Japan, *33*, 143–153.

[89] E. Korngold (1971) Prevention of colloidal fouling in electrodialysis by chlorination. Desalination, *9 (3)*, 213–216.

[90] K. Kusumoto, H. Ihara and Y. Mizutani (1976) Preparation of macroreticular anion exchange membrane and its behavior relating to organic fouling. J. Appl. Polym. Sci., Japan, *20 (12)*, 3207–3213.

[91] K. Okada, M. Tomita and Y. Tamura (1977) Electrodialysis in the treatment of dairy products. Part 2. Development of electrodialysis plant. "Milchwissenschaft" Tech. Inform., Morinaga, Milk, Ind. Co., *31 (1)*, 1–8.

[92] T. Yamabe (1973) Ion exchange. Membrane separation process with ion-exchange membranes. Bunseki Kagaku, *22 (4)*, 465–471.

1.11 Polymeric Adsorbents

Robert Kunin
Yardley, USA

1.11.1 Introduction

Although adsorption technology has been practiced commercially by the chemical industry for the past century, its importance has increased in recent years. This is the result, at least in part, of our increasing concern over pollution and the actions of various regulatory agencies in implementing legislation aimed at controlling the contamination of our air and water. The development of manufacturing techniques which involve the isolation of product from crude streams, illustrated by the adsorption of biologically active agents from fermentation broths, as well as the ever-increasing concern for purity of various industrial and fine chemicals, have also led to increased adoption of adsorption phenomena for achievement of separations in the processing industries.

1.11.2 General considerations

The term adsorption refers to the change of concentration of any particular component of a system which occurs at the surface of a solid or liquid. Adsorption phenomena can occur in any of the following heterogeneous systems: solid/liquid; solid/gas; liquid/gas. Adsorption normally refers to an increase in concentration which takes place at the interface in any of the above-noted systems. However, theoretically, one can conceive of a decrease in concentration which occurs at the interface, leading to the term negative adsorption. This would be the result of a greater adsorption of the solvent than the solute dissolved in the solvent. The substance being adsorbed or concentrated at the surface or interface is referred to as the adsorbate and the solid upon whose surface adsorption occurs is referred to as the adsorbent.

One must distinguish between the terms adsorption and absorption. The latter refers to a uniform penetration of some component of a system into the bulk of the liquid or solid absorbent with no change in concentration. It is the overall effect which is the determining factor that distinguishes between the two processes. For example, the phenomena of ion exchange have, over the years, been termed both adsorption and absorption by various investigators. If we consider adsorption strictly as a process confined to changes in concentration occurring at the interface of a solid-liquid system and absorption as a process involving solids engulfing substances throughout their entire structure, ion exchange processes may very well fall into both categories and hence the term sorption might better be applied to avoid controversy.

Ion exchange phenomena which occur at the surfaces of dense and non-porous crystals (kaolinite, a clay mineral, for example) may safely be classified as adsorption. On the other hand, the exchange of sodium for hydrogen ions in the porous, sulfonic acid, gel-phase, cation exchange resins could be regarded as absorption, even though the mechanisms and energy factors for the two cases are identical.

Although the use of the term adsorption for ion exchange processes may appear strange to those who traditionally think of adsorptive processes as involving adsorbents such as carbon and silica gel, and adsorbates such as iodine and water, over the years many phenomena associated with ion exchange have involved mechanisms and reactions other than the exchange of ions. For example, reactions involving the removal of phenol from water, hydroquinone from monomers and color bodies from sugar syrups, employ the chloride form of a strongly basic anion exchange resin. These reactions do not necessarily involve an ion exchange mechanism only and are truly classical examples of adsorption.

The increased use of selected ion exchange resins for non-ion exchange adsorptive processes, coupled with the successful commercial development of macroreticular ion exchange resins, led to a dramatic spin-off which is represented by the macroreticular porous polymers and the synthetic carbonaceous adsorbents.

In essence, macroreticular copolymers are prepared by suspension polymerization of a monomer and a crosslinking agent, typically styrene and divinylbenzene, in the presence of a solvent which is capable of dissolving the monomers but is a poor swelling agent for the crosslinked polymer being formed. As the polymerization proceeds, phase separation occurs, leading ultimately to polymer beads containing microscopic gel structures cemented or joined together at points of contact which, in turn, form interconnecting pores. In structures such as these, both the pores and the polymers are present as continuous phases. The average size of the pores and the surface area of the macroreticular crosslinked polymer are determined directly by the nature of the internal gel structures and indirectly by the types and amounts of monomers and solvents used in the polymerization. Macroreticular copolymers can be prepared from styrene, acrylic esters, pyridine, and other monomers by similar vinyl-type polymerization techniques. These can be converted to ion exchange resins by introducing suitable ionically functional sites through sulfonation or chloromethylation and aminolysis.

Macroreticular copolymers can also be used effecively as adsorbing media, in which case no ionically active groups are added to the structure. These porous polymers, or polymeric adsorbents, constitute a new and unique class of adsorbents owing to the wide range of pore structures that one can develop within the framework of a particular chemical system. For example, in the styrene-divinylbenzene class of polymeric adsorbents having surface areas ranging from 7 to 700 m^2/g and average pore diameters ranging from 50 to 1,000,000 Å, pore volumes can range from 10% to 90%. The surface characteristics are also well defined. The classical adsorbents such as the silicas, aluminas, and the various activated carbons do not offer this flexibility, nor do they offer surfaces which are chemically as well-defined.

It is of considerable interest to place this new class of adsorbents into perspective by briefly comparing them with the classical and/or commonly recognized adsorbents. In general, adsorbents are solids which possess high specific surface areas, usually well above 5 m^2 of exposed surface per g of solid and which fall into two major physical classes: porous and non-porous. The porous adsorbents consist of particles that are usually large (greater than 50 mesh), their high surface being a result of pores of varying diameters which "permeate" the particle. The diameters of these pores are larger than molecular distances. Non-porous adsorbents are usually finely divided solids (less than 10 μm) and the high surface area of such materials results from the fine state of subdivision that is achieved by various techniques such as grinding, precipitation, etc. The effect of the subdivision of the specific surface of a theoretical adsorbent is given in Table 1.51. Included are the theoretical surface areas for solids of several densities covering the range of most anion and cation exchange resins. These calculations are based upon particles of spherical shape using the relationship

$$A = \frac{3 \cdot 10^{-3}}{r\varrho}$$

A = specific surface area (m^2/g)
r = radius of particle (mm)
ϱ = density (g/cm^3)

The specific surface areas of several commercial porous and non-porous adsorbents are given in Table 1.52.

Table 1.51 Specific surfaces of spheres of varying density and particle size

Screen size (mesh)	Particle diameter (mm)	Specific surface (m^2/g) $\varrho = 1.1$	$\varrho = 1.3$	$\varrho = 1.5$
20	0.84	0.0065	0.0055	0.0048
30	.59	.0091	.0078	.0068
40	.42	.013	.011	.0095
50	.30	.018	.015	.013
70	.21	.027	.022	.019
100	.15	.036	.031	.027
200	.074	.074	.062	.054
325	.44	.124	.104	.091
Sub-sieve sizes (μm)				
40	.040	.14	.12	.10
20	.020	.27	.23	.20
10	.010	.55	.46	.40
1	.001	5.5	4.6	4.0

Table 1.52 Specific surfaces of typical porous and non-porous adsorbents

Porous adsorbent	Specific surface (m^2/g)	Non-porous adsorbent	Specific surface (m^2/g)
Granular carbons	500–2000	Carbon black	100
Silica gel	600	TiO_2 pigment	70–80
Bone char	60–80	ZnO pigment	1–10
Soils	10–100		
Asbestos	17		
Polymeric adsorbents	7–700		

It is obvious that the dense, non-porous adsorbents do not have sufficient surface area to be useful as adsorbents unless they are pulverized or prepared in very finely divided states. When the particle size is small enough to result in a substantial surface area, the usefulness of such materials is limited, since they cannot be used in typical columnar operations. Pressure drops are excessive and upflow operation is ruled out because the materials are too buoyant. The porous adsorbents are generally more useful because they can be used in columns, and in such applications the hydraulic considerations of pressure drop and expansion are manageable.

The polymeric adsorbents which combine high surface area with relatively large particle size are particularly well-suited for many applications under a wide variety of conditions owing to their unique adsorptive properties and particle size.

1.11.3 Physical properties of macroreticular polymeric adsorbents

The physical properties of the macroreticular polymeric adsorbents are best described in terms of their pore structure and surface area. Some of these properties are given in more detail in Table 1.53 and Figure 1.170.

Table 1.53 Pore structure characteristics of Amberlite polymeric adsorbents

Adsorbent	Chemical structure	Surface area (m^2/g)	Average pore diameter (Å)	Range of pore diameter[a] (Å)	Total porosity[b] (ml pores / ml resin)	(ml pores / g resin)
Non-polar						
Amberlite XAD-1	Styrene/DVB*	100	200	100–200	0.37	0.47
Amberlite XAD-2	Styrene/DVB	300	100	30–100	0.42	0.68
Amberlite XAD-4	DVB	725	48	30–100	0.45	0.96
Amberlite CAD-16	DVB	860	66	–	–	1.82
Polar						
Amberlite XAD-7	Acrylic	450	85	30–1800	0.55	0.97
Amberlite XAD-8	Acrylic	160	150	30–900	0.52	0.82

a) Pore size ranges between 5% and 95% of the total porosity at the maximum pore volume.
b) Calculated from apparent and skeletal densities.
*) Divinylbenzene

These data describe not only the surface area of porosities but also the pore size distribution of five commercially available macroreticular polymeric adsorbents.

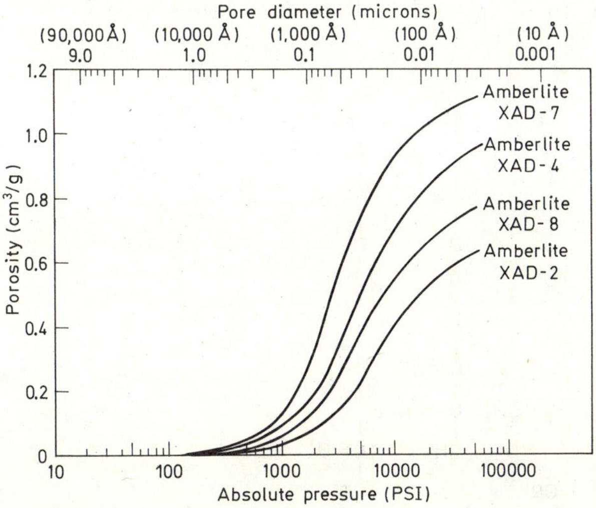

Figure 1.170 Pore distributions of some Amberlite adsorbents as determined by mercury intrusion.

1.11.4 Chemical structure of the macroreticular polymeric adsorbents

In contrast to the macroreticular ion exchange resins, the polymeric adsorbents are truly non-ionic and their adsorptive properties are totally dependent upon their surface characteristics. Figure 1.171, 1.172 and 1.173 describe the chemical structures of two of the more interesting classes of porous polymers used as polymeric adsorbents. One is based upon a styrene structure crosslinked with divinylbenzene and is extremely hydrophobic (Figure 1.171). The other is a crosslinked polymethacrylate structure which is more hydrophilic (Figures 1.172 and 1.173). These macroreticular polymeric adsorbents are dimensionally and chemically stable and chemically inert in virtually all environments. They are also highly insoluble, particularly after an initial conditioning step that removes trace levels of organic impurities.

The polarity of the polymeric structures can be related to the dipole moments of the monomers from which the polymers are derived. For example, Amberlite XAD-2 and Amberlite XAD-4, both of which are based upon a styrenic structure,

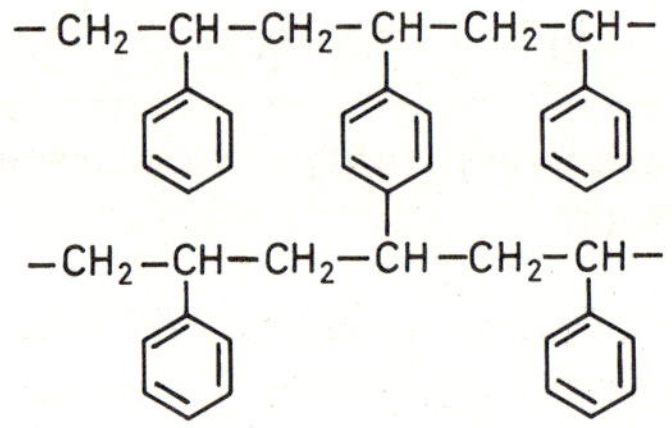

Figure 1.171 Structure of Amberlite XAD-2 and Amberlite XAD-4.

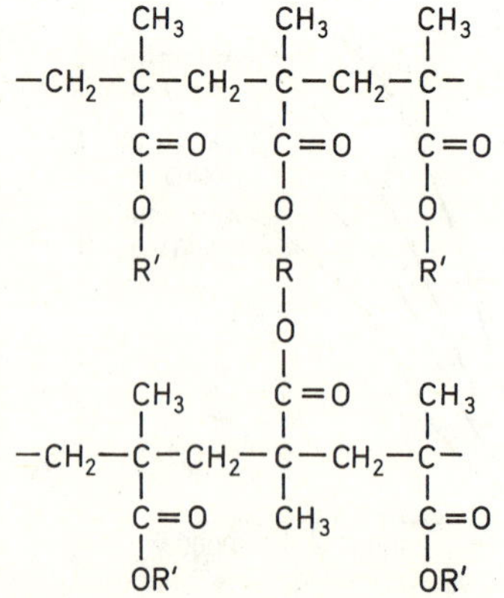

Figure 1.172 Structure of Amberlite XAD-7.

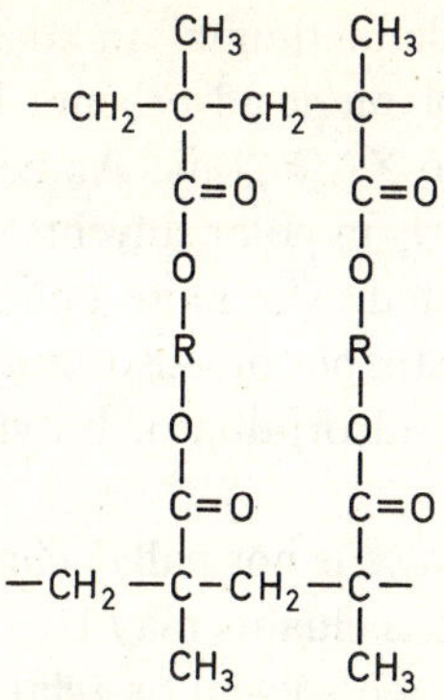

Figure 1.173 Structure of Amberlite XAD-8.

will exhibit polarity associated with a dipole moment of 0.3 debyes. The acrylate materials, Amberlite XAD-7 and Amberlite XAD-8, are derived from monomers whose dipole moments are approximately 1.8 debyes.

1.11.5 Adsorption properties of macroreticular polymeric adsorbents

In essence, the adsorptive properties of the polymeric adsorbents may be predicted from their chemical structures, the theoretical solubility parameters of the adsorbent, as well as the structure and solubility of the adsorbate in a particular solvent.

The solubility parameter is defined as the square root of the cohesive energy density, or

$$\gamma = \left(\frac{\Delta H_v - RT}{M/\varrho}\right)^{1/2}$$

γ = solubilitiy parameter
ΔH_v = heat of vaporization
R = gas constant
T = thermodynamic temperature
M = molar mass
ϱ = density

The solubility parameter is a measure of the energy required to dissociate molecules of a given compound. Since this is analagous to the energy required to dissolve a solute in a solvent the solubility parameter serves as a first approximation of solubility. Hence, materials with similar solubility parameters tend to be soluble in one another. As the values diverge, two products tend to be less soluble in one another.

From a structural viewpoint, the classical "rule of thumb" in adsorption technology – "like likes like" – holds true for the polymeric adsorbents. For example, the aromatic, or non-polar adsorbents, Amberlite XAD-2 and Amberlite XAD-4, are highly selective for aromatic solutes, particularly in polar solvents such as water where the solubility of these solutes is quite limited. The more polar adsorbents, Amberlite XAD-7 and Amberlite XAD-8, are capable of adsorbing more polar solutes from non-polar media, illustrated by the adsorption of butyric acid from hexane.

Elution of the adsorbate from the polymer surface is normally achieved through the use of organic solvents (although some aqueous eluants may be used). Elution may be predicted by the solubility parameter of the solvent as related to that of the solute. In general, lower solubility parameters are indicative of more effective eluting agents. The solubility parameters of several organic solvents are presented in Table 1.54.

Table 1.54 Solubility parameters of organic solvents

Solvent	Solubility parameter ($cal^{0.5}/cm^{1.5}$)
2-Butanone	9.3
2-Propanone	10.0
1-Butanol	11.4
1-Propanol	11.9
Ethanol	12.7
Methanol	14.5
Water	23.2

Methanol is almost always a good eluting agent, or regenerant, for the classes of organic solutes which are adsorbed effectively onto polymeric adsorbents. It will usually remove the adsorbate from the adsorbent in a minimum volume, and is readily available and relatively inexpensive.

The solubility of the adsorbate in the regenerant or elution agent is quite important. Not only must van der Waals' attractive forces, which bind the adsorbate to the adsorbent, be overcome, but the solubility of the adsorbate in the eluting solvent must be sufficiently high to permit rapid dissolution after the solvent diffuses to the adsorption site.

1.11.6 Applications of macroreticular polymeric adsorbents

Whereas the conventional carbonaceous and inorganic adsorbents are used widely throughout the chemical process industries, the macroreticular polymeric adsorbents have carved out a unique position for themselves in view of their unusual selectivities and physical properties. The adsorbents are currently being used for a host of

Table 1.55 Performance of Amberlite XAD-2 as an adsorbent for a series of pharmaceuticals

Adsorbate	Molecular weight	Concentration (ppm)	Capacity (mg/ml)	Peak (ppm)	Elution Elution req.[a] (Bed vol.)
Vitamin B_{12}	1355	15	5.2	7,200	2
Tetracycline	444	1000	44	30,000	2
Tetracycline	444	100	7.4	16,000	2
Oxytetracycline	460	100	7.6	17,600	2
Oleandomycin	688	100	40	20,000	2

a) Bed volumes of methanol required for complete elution of adsorbed pharmaceutical.
Polymeric adsorbents are also being used for the newer antibiotics classified as the cephalosporins.

applications in the pharmaceutical industry (recovery and purification of antibiotics, vitamins, and other biologicals), for the treatment of industrial wastes, for the in vivo treatment of blood (hemoperfusion), and for many problems of an analytical nature.

1.11.6.1 Pharmaceutical applications

Many of the newer and important pharmaceutical products, particularly those produced by fermentation processes, are currently being recovered and purified by using macroreticular polymeric adsorbent.

Production of vitamin B_{12}. Vitamin B_{12} ($C_{63}H_{88}CoN_{14}O_{14}P$, *M*r 1355) is currently being produced by a microbial fermentation process quite similar to that for the production of streptomycin and neomycin. It is also being recovered from the fermentation broth by means of Amberlite IRC-50, again in a manner similar to that for the antibiotics. Since vitamin B_{12} (cyanocobalamin) is essentially a nonionic compound, it is adsorbed and eluted by mechanisms other than ion exchange. For example, vitamin B_{12} is not adsorbed onto the sodium salt of Amberlite IRC-50; however, it is adsorbed effectively onto the acid form of the resin and from an acid medium (ca. pH 3). Impurities that are also adsorbed can be eluted from a carboxylic-exchanger with very dilute acid (1 N HCl) prior to the subsequent elution of the vitamin B_{12} with an acid/acetone/water solution. Mechanically, the overall operation is almost identical with that described previously for the production of streptomycin.

More recently, interest has developed in the use of the polymeric adsorbents for the recovery and purification of vitamin B_{12} in view of its solubility characteristics and the manner in which it behaves with Amberlite IRC-50. Table 1.56 compares the performances of Amberlite IRC-50 and the polymeric adsorbent, Amberlite XAD-2, with respect to the adsorption and elution of vitamin B_{12}.

Table 1.56 Comparison of Amberlite IRC-50 and Amberlite XAD-2 for processing vitamin B_{12}

Sorbent	Vitamin B_{12} Capacity (Column)[1] (mg/ml)		Elution	
	Break-through	Saturation	Peak conc. (ppm)	Elution volume (Bed vol.)
Amberlite IRC-50	0.03	0.14	150	5
Amberlite XAD-2	3.50	5.20	7200	2

[1] One gpm/cu. ft. flow rate

The superior behavior of the polymeric adsorbent is self-evident and it is understandable that new plants for producing vitamin B_{12} are now employing the polymeric adsorbents.

Processing of other antibiotics. The solubility characteristics of such antibiotics as oleandomycin and the tetracyclines suggest the use of the polymeric adsorbents for their recovery and purification. Table 1.57 summarizes some data describing the adsorption and elution performance of Amberlite XAD-2 for these antibiotics.

Table 1.57 Adsorption and elution of antibiotics with Amberlite XAD-2

Antibiotic	Capacity (Column)[1] (mg/ml)	Elution (Methanol)	
		Peak conc. (ppm)	Elution requirement (Bed vol.)
Oxytetracycline	7.6	17,000	2
Oleandomycin	40	20,000	2

[1] One gpm/cu. ft.; conc. 100 – 1000 ppm

Several of the newer, synthetic antibiotics derived from fermentation broth are purified by means of the polymeric adsorbents.

1.11.6.2 Phenolic wastes

The treatment of wastes containing phenol and substituted phenols is a universal problem owing to the widespread use of these chemicals. Many industrial wastes contain phenol, which not only creates pollution problems but also constitutes loss of a valuable raw material. The ideal waste treatment process is one that removes phenol from the waste and recovers the phenol in a usable form. This has now been accomplished on a commercial basis using the polymeric adsorbent, Amberlite

Table 1.58 Adsorption of phenol and substituted chlorophenols on Amberlite XAD-4 (temperature 25 °C; flow rate 0.5 gpm/ft^3)

Solute	Solubility in water (ppm)	Solute in influent (ppm)	Solute in influent (mmol/l)	Solute adsorbed (sbs/ft^3) Zero leakage	Solute adsorbed (sbs/ft^3) 19 ppm leakage
Phenol	82,000	250	2.7	0.78	0.83
m-Chlorophenol	26,000	350	2.7	2.40	2.53
2,4-Dichlorophenol	4,500	430	2.7	5.09	5.49
2,4,6-Trichlorophenol	900	510	2.6	12.02	13.81

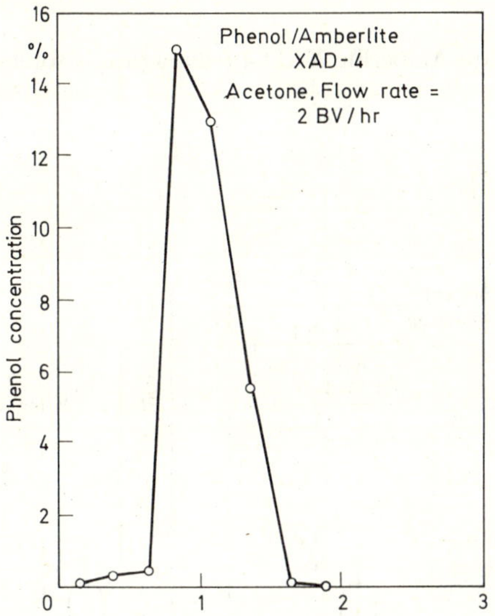

Figure 1.174 Acetone regeneration of Amberlite XAD-4 exhausted with phenol.

XAD-4. Table 1.58 and Figure 1.174 summarize the performance of Amberlite XAD-4 for adsorbing phenol and chlorophenols under varying conditions.

It is obvious that the performance of the polymeric adsorbent improves as the solubility of the phenolic compound decreases. Elution of the phenol is readily accomplished with various solvents such as methanol, ethanol or acetone (Figures 1.175 and 1.176).

Careful selection of the regenerating solvent often allows recycling back into the process a regenerant stream laden with the adsorbed organic compound. Thus, what would have been a non-productive pollution control step is transformed into a closed-loop materials recovery process.

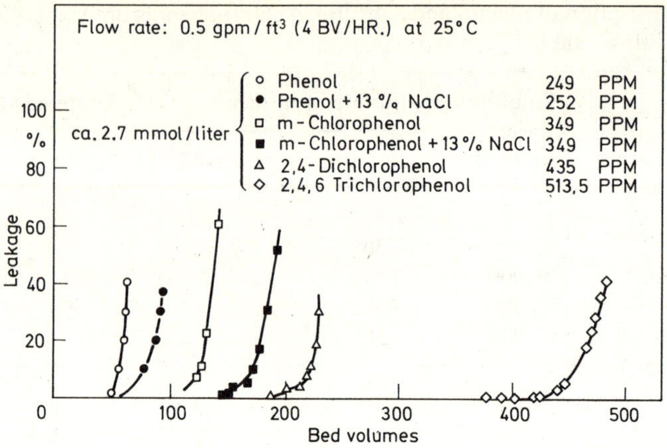

Figure 1.175 Column adsorption on Amberlite XAD-4 from aqueous solutions.

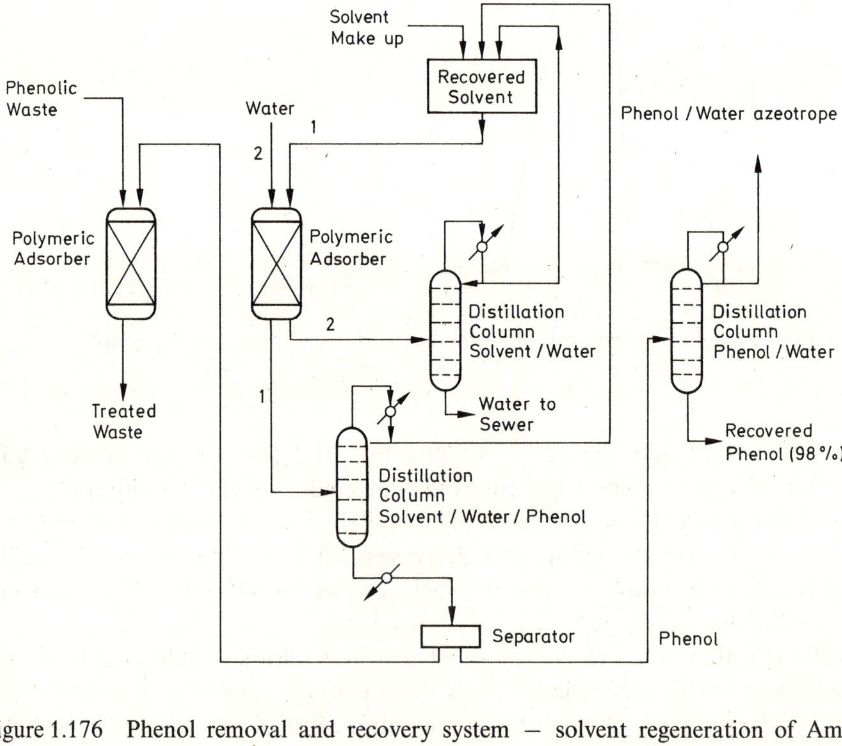

Figure 1.176 Phenol removal and recovery system – solvent regeneration of Amberlite adsorbent.

1.11.6.3 Trinitrotoluene (TNT) and chlorinated pesticide wastes

Figures 1.177–1.180 describe the performance of Amberlite XAD-4 for the treatment of the waste effluents from a munition plant and a chlorinated pesticide manufacturing plant. In both cases, the data support the excellent performance of the polymeric adsorbent, Amberlite XAD-4. In the case of the pesticide waste, the

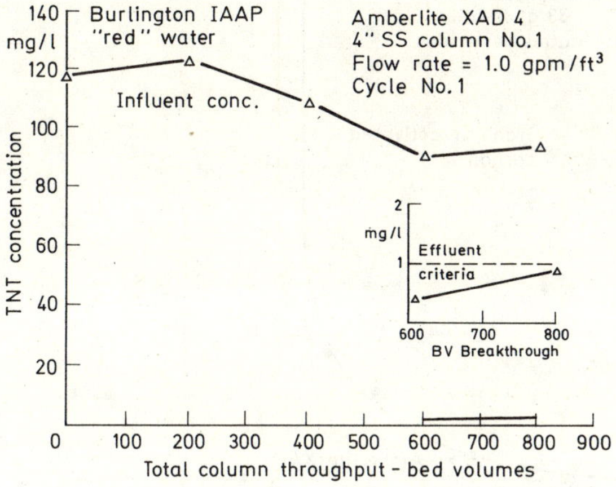

Figure 1.177 Adsorption of TNT by Amberlite XAD-4.

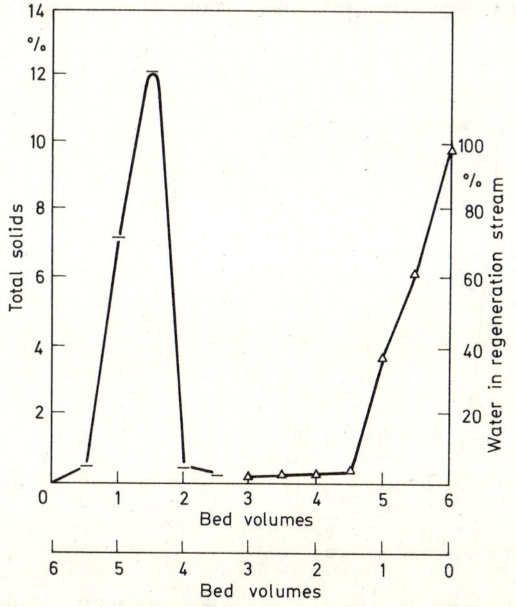

Figure 1.178 Acetone regeneration of Amberlite XAD-4, Cycle No. 3.

polymeric adsorbent is clearly superior to the conventional carbon adsorbents. In the case of the TNT waste, the use of carbon is clearly not indicated since desorption can only be achieved thermally and such practices are clearly unsafe for a carbon saturated with an explosive such as TNT.

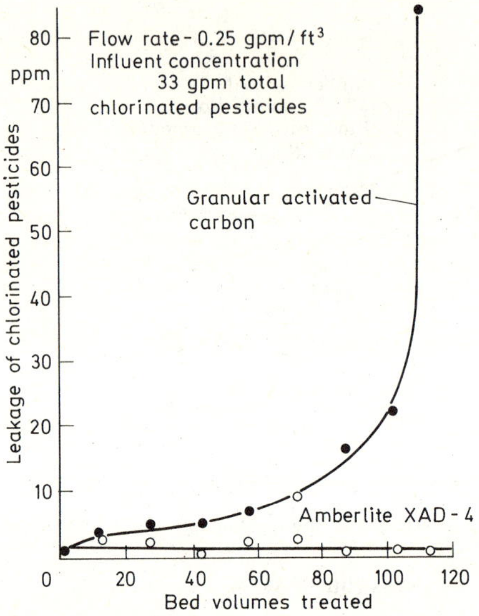

Figure 1.179 Treatment of waste effluent from manufacture of chlorinated pesticides.

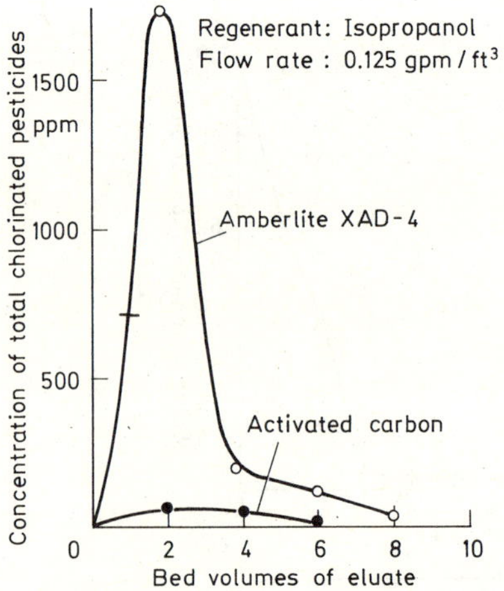

Figure 1.180 Elution of chlorinated pesticides from Amberlite XAD-4 and granular activated carbon with isopropanol.

1.11.6.4 Removal of noxious compounds from water

Regardless of the efficacy of industrial waste control systems, traces (parts per billion, ppb) of many noxious compounds are entering into the drinking water supplies. There has been much concern that has been expressed over this problem and several studies have been initiated on the use of adsorbents to remove these noxious compounds even though there has been no concrete evidence concerning their harmful nature at these trace levels. The macroreticular polymeric adsorbents have shown considerable effectiveness in removing trace levels of such compounds from water supplies.

A recent publication by Fritz and his associates at the Department of Chemistry of Iowa State University, Ames, Iowa summarizes the results of a comprehensive study on the use of Amberlite XAD-2 and Amberlite XAD-7 for identifying and removing a host of typical organic pollutants from water. Using both Amberlite XAD-2 and Amberlite XAD-7, Fritz and his associates have developed a method for extracting trace organic contaminants from potable water. They have demonstrated that Amberlite XAD-2 and Amberlite XAD-7 are capable of adsorbing weak organic acids and bases and neutral organic compounds quantitatively from water containing ppb to ppm concentrations of the compounds listed in Table 1.59.

Table 1.59 Compounds removed from water with Amberlite XAD-2 and Amberlite XAD-7

Methyl isobutyl ketone	Acenaphthylene
n-Hexanol	1-Methylnaphthalene
Ethyl butyrate	Methylindenes
Benzene	Indene
Naphthalene	Acenaphthene
Benzoic acid	2,2-Benzothiophene
Phenylenediamine	Isopropylbenzene
Phenol	Ethylbenzene
2,4-Dimethylphenol	Naphthalene
p-Nitrophenol	2,3-Dihydroindene
2-Methylphenol	Alkyl-2,3-dihydroindene
Aniline	Alkyl benzothiophenes
o-Cresol	Alkyl naphthalenes

The acidic components were desorbed with alkali and the basic components with acid. The neutral compounds were eluted with ether.

Although the work of Fritz and his associates was analytically oriented, it does point to the potential use of the Amberlite macroreticular, polymeric adsorbents for treating potable water contaminated with various noxious, organic compounds. These studies are being continued by many other investigators in the United States.

1.11.6.5 The removal of toxins from blood

Although not generally considered a waste treatment process, the removal of toxins from blood is the most important waste treatment process known to man. When our industrial pollution control and abatement systems do not perform efficiently or have not been introduced practically, man must depend upon his own pollution control system, the kidney to prevent these toxins from entering the blood system. The kidney, however, sometimes does not have the capacity or ability to treat the blood and cleanse it effectively of certain toxins that have been ingested, so that some device such as an artificial kidney must be employed.

Many drugs commonly implicated in intentional or accidental overdoses are removed by hemodialysis (artificial kidneys). This technique is cumbersome, slow, and requires large volumes of solutions and highly trained personnel. The various studies that have been discussed previously encouraged several investigators to experiment with ion exchange resins and adsorbents for removing these toxic materials by directly treating the blood of a patient and thereby avoiding the need for the hemodialysis procedure. If one considers the low rate of diffusion across membranes and the large surface area of membranes required for hemodialysis to be practical, and compares these factors with the rapid rate of adsorption and high surface area of ion exchange resins and polymeric adsorbents, it is obvious that treating blood directly with ion exchange resins and polymeric adsorbents (hemoperfusion) has many advantages over the process of dialysing blood through membranes (hemodialysis). The past 25 years of research have resulted in techniques for preparing resins and adsorbents that are sterile and free of pyrogen reactions. Further, one may now choose from a host of products to select an optimum product for the hemoperfusion of particular toxins that may occur in the bloodstream of intoxicated patients.

Although much of the above-described effort on hemoperfusion has been devoted to the use of ion exchange resins, the availability of macroreticular polymeric adsorbents has aroused much interest because of their (1) high surface areas, (2) inertness, and (3) ability to adsorb a spectrum of toxic drugs from the bloodstream without altering the ionic composition or pH of the blood. Rosenbaum [3, 4] of the Albert Einstein Medical Center (Philadelphia, PA) has been quite successful in removing toxins from blood by hemoperfusion through columns of the macroreticular Amberlite polymeric adsorbents. He has demonstrated through an exhaustive study on animals and on a number of humans that one can safely and readily remove toxins such as barbiturates and glutethimide from the bloodstream of comatose patients by the hemoperfusion of the blood over Amberlite XAD-2 and Amberlite XAD-4 polymeric adsorbents. He also compared the procedure with hemodialysis (artificial kidney) and found the hemoperfusion technique using the polymeric adsorbent to be less complicated and faster. Although carbons have also been used for hemoperfusion, their use has been found to be inherently troublesome due to the instability of the carbon particles.

The studies Rosenbaum performed with Amberlite polymeric adsorbents were conducted with resins that had been carefully treated to remove any potentially harmful impurities, micro-organisms, and pyrogens.

When the hemoperfusion tests were performed on several patients intoxicated with various barbiturates and glutethimide, comparative tests were made using hemodialysis. The patients responded well and the toxins were cleared from the patients much faster using the hemoperfusion techniques with the Amberlite XAD-2 resin adsorbent. Rosenbaum found further improvements when substituting Amberlite XAD-4 resin for the Amberlite XAD-2 resin.

The overall promise of hemoperfusion with ion exchange resins and polymeric adsorbents for the treatment of drug intoxication can best be summarized with the following quotation from Rosenbaum:

"The Amberlite XAD-2 resin hemoperfusion systems appears to be clinically superior to hemodialysis in the treatment of drug intoxication ... It results in higher clearance rates of intoxicants and is mechanically simpler and less expensive. In patients with overwhelming, life-threatening intoxication, hemoperfusion therapy may be of value in reducing coma time and the occurrence of residual complications, particularly pneumonitis. Moreover, the potential range for effective adsorption of toxins by the resin column has not been fully explored and, with the use of combinations of lipophilic, hydrophilic, anion, and cation exchange resins, may be broader than for hemodialysis."

Although the work of Rosenbaum represents the beginning of a new era in life-saving hemoperfusion techniques, it also culminates a quarter of a century of studies by others in the medical profession. It must be noted, however, that hemoperfusion using resins does not and cannot replace hemodialysis. Whereas it can take over the kidney and liver functions for a temporary period as in the case of drug intoxication, it cannot replace the hemodialysis (artificial kidney machine) procedure for those who have lost their kidney function permanently. In the future, however, hemoperfusion using resins may be a useful adjunct to hemodialysis in such cases. It is the author's opinion that this application represents the ultimate use of polymeric adsorbents for the treatment of wastes.

1.11.7 Summary

Macroreticular porous polymers are a comparatively recent development; their use as matrices for ion exchange resins dates from about 1960. Their use as polymeric adsorbents is even more recent, dating from the late 1960s. As a consequence, the commercial use of these materials as adsorbents is in its infancy. New uses, and new techniques, which take advantage of the unique adsorptive and elution properties of polymeric adsorbents are being developed at an increasingly rapid rate. Part of the impetus in this direction results from increasing emphasis upon the removal of trace organics from industrial waste waters and from certain aqueous process streams. One can be certain that these uses will continue to expand, just as has been the case with every new separative technique following its introduction.

It should be noted that, although the use of polymeric adsorbents is similar in many respects to the use of ion exchange resins, there are some important differences. To dwell briefly upon the similarities, both classes of materials function best when dealing with dilute systems, typically those in which the concentration of solute is below about 1000 ppm (0.1%). Both ion exchange resins and polymeric adsorbents are utilized in columns in which the media are contained as a packed bed and through which the solution to be treated is allowed to flow. Both systems require that the bed be removed from "loading" service and subjected to a separate regeneration or stripping step. Both will operate at space velocities of about 0.25 to 2.0 gpm/ft^3, and are usually operated to a predetermined endpoint prior to regeneration.

The fundamental difference between ion exchange resins and polymeric adsorbents lies in capacity determination. Whereas in the use of ion exchange resins one can always "look up" the total ion exchange capacity of a resin (even though the extent to which that capacity can be utilized must frequently be determined experimentally), this cannot normally be done in the case of the polymeric adsorbents. There is no readily titratible "base case" capacity for polymeric adsorbents. The surface area and pore diameter of any particular porous polymer are known, but the interdependence between these characteristics and capacity for a given solute is not well understood or known. It is therefore necessary in all cases to carry out experimental study programs to determine the operating capacity for a particular adsorbate under the proposed conditions of use. This, if done properly, will provide adequate data from which a successful commercial design can follow.

The next few years will see an increase in the use of porous polymers. We will learn more and more about the behavior of these adsorbents and the literature on the subject will increase. Perhaps in time we will have amassed a sufficient body of information so that the need for experimental data will have at least been somewhat reduced.

References

[1] J. S. Fritz (1972) Anal. Chem. *44*, 139.
[2] K. A. Kun & R. Kunin (1967) J. Poly. Sci. C. (16), 1457.
[3] R. Kunin (1980) Amber-Hi-Lites No. 163, Rohm and Haas Company, Philadelphia, Pa.
[4] J. Rosenbaum (1970) Trans. Amer. Soc. Artificial Organs *16*, 134.
[5] J. Rosenbaum (1972) Clin. Toxicology *5*, 331.

2 Ion Exchangers in Industry

An Overview of Industrial Applications

Robert Kunin
Yardley, USA

Introduction

Ion exchange technology has been used commercially on a worldwide basis for almost a century, and practically all industries and homes are dependent upon this technology either directly or indirectly. Progress made in any technology is usually measured in tonnages produced or money expended per year. Although ion exchange in its broad sense has shown much progress over the years in terms of turnover and tonnages, it is most significant to consider the progress in ion exchange technology in terms of the betterment of human lives. In other words, has humanity been bettered by progress in ion exchange technology? In many respects, the answer to this question is yes. Ion exchange technology has improved medical delivery systems in many areas, improved the purity of many drugs, decreased the cost of drugs, lowered the cost of electricity, and decreased the cost of various chemical and pharmaceutical processing operations. Of course, it must be recognized that there is a negative feature to ion exchange: when used on a regenerable basis, the waste regenerants add to the pollution load. As a result, the use of electrical and thermal energy for regeneration of ion exchangers still demands attention.

The ion exchange market

Some trends in the marketplace for ion exchange technology can be noted in the Figures 2.1 to 2.3, which present developments in ion exchange resin prices and production. The data contained in Figure 2.3 reported on a worldwide basis agree with the United States data assuming that the United States of America produces

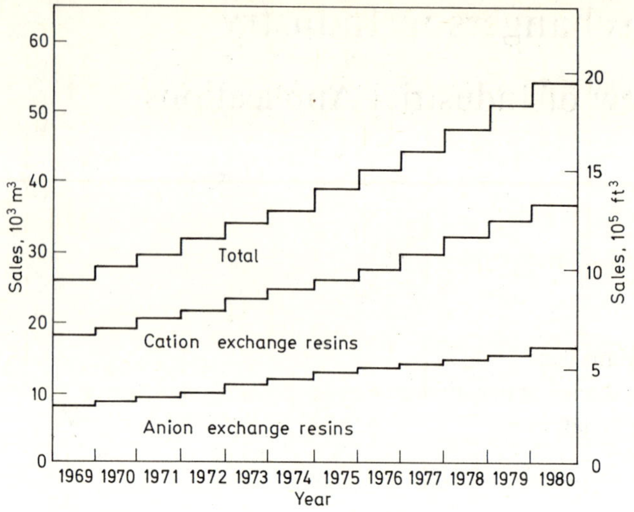

Figure 2.1 United States ion exchange resin sales trend – bulk basis by type [1].

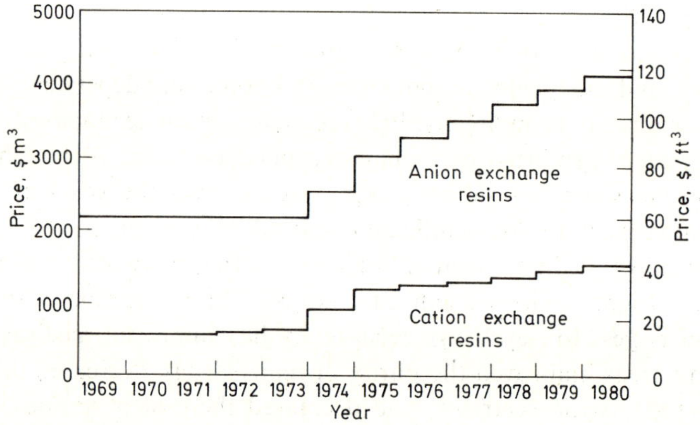

Figure 2.2 United States ion exchange resin price trends – bulk prices by type [1].

one third of the world's production. These data do not include any figures for the zeolites or molecular sieve zeolites. Data for these products indicate that production is currently well over a billion pounds per year, much of which is produced synthetically with a selling price of 25–30 ¢ per pound for powders and \$1–\$2 per pound for the pellets.

One may estimate the overall U.S. market for ion exchange systems, equipment plus ion exchangers, by assuming that the cost of ion exchangers is approximately 10% of the total value of the overall system. However, one should assume that half of the annual sales of ion exchangers are for replacing "spent" ion exchangers.

Area	Year						
	1967	1970	1974	1977	1981	1984	2000
N. America	35	40	78	50	61	90	250
W. Europe	16	19	35	30	32	45	100
E. Europe & U.S.S.R.	4	5	9	16	20	30	60
Japan & rest of Asia	3	4	10	7	17	35	50
Other	2	2	6	3	10	15	20
Total	60	70	138	125	140	205	500

Note: These figures are taken from a number of sources and offer only rough estimates of the quantities involved, and their geographical distribution, see, e. g., Reports on progress of applied chemistry, 55, 66–76, (1970), Reports on progress of applied chemistry, 59, 235–41, (1974), Chemicals, 8–11, 12–18, (1981).

Figure 2.3 Estimated world ion exchange production (1967–2000) [3] (Figures are given in thousands of cubic meters).

This calculation indicates that the field of ion exchange technology in the United States of America is easily a billion dollar business, without including the cost of the chemical regenerants such as caustic, soda ash, sulfuric and hydrochloric acids. On a worldwide basis, one may safely assume that the industry represents a multi-billion dollar industry. This must be compared with the estimate that at least 6000 lives are saved annually by the hemoperfusion process used for drug overdosages, a process based upon a spin-off of ion exchange technology.

Growth of ion exchange applications

It is of interest to examine that growth of ion exchange technology in terms of the major areas of applications. Those applications that are being used commercially are summarized in the following tables. These lists are most impressive and serve as a testament to the late Edgar Kirsopp, who looked upon ion exchange resins as "reactive polymers" and encouraged the research chemists to look far beyond water treatment for applications of ion exchange.

Water treatment. The list of water treatment applications alone is indeed impressive. (Table 2.1). Progress made in deionization has been quite unusual in that ion exchange has progressed from a technique primarily designed for the purpose of removing the dissolved inorganic impurities from water to a technique which also removes much of the organic matter as well as a good portion of the inorganic and organic colloids. In fact, by means of ion exchange one may now produce

Table 2.1 Examples of major ion exchange application: Water treatment

Water softening	Iron and manganese removal
Dealkalisation	Nitrate removal
Deionisation	Organic matter removal
Fluoride removal	Radioactivity removal
Color removal	Colloid removal
Oxygen removal	

water having the conductance of the pure water substance itself. This water quality can be achieved on a very large scale. Further, it has also been found that certain ion exchange resins are capable of removing bacteria and other biological activities. Interestingly, it is possible to produce sterile and pyrogen-free water; however, this method cannot be used for producing such water for injection into humans and animals since the technique is not reliable, as yet, and not approved by the U.S. Pharmacopeia and the FDA for this purpose.

Sugars and polyhydric alcohols. Ion exchange technology has made slow but steady progress in the sugar-refining industry around the world. In many respects the progress made in sugar refining has rivalled that made in water treatment. Currently, sugar juices and syrups are being softened (calcium removal), decolorized and deionized by means of ion exchange (Table 2.2). As bone char units "wear-out", they are being replaced by ion exchange resin units. The most dramatic change that has occurred in the sugar industry revolves around the use of ion exchange in the corn sugar industry for the purification of glucose and fructose, resulting in the production of the high fructose corn syrups (HFCS) currently replacing sucrose in the beverage industry. In the production of HFCS from corn, ion exchange technology is employed on a large scale, as shown in Table 2.3. The chromatographic step is one of the largest chromatographic operations now used commercially.

Another new innovation in the sugar industry involves the use of powdered ion exchange precoat formulations for the decolorization and decalcification of sugar syrups.

Table 2.2 Examples of major ion exchange applications: Sugars and polyhydric alcohols

Purification of cane, corn and beet sugar
Glycerine purification
Sorbitol

Table 2.3 Ion exchange process steps for production of HFCS

Deionization and decolorization of crude dextrose
Deionization and decolorization of crude enzymatically isomerized dextrose
Chromatographic separation of dextrose and fructose
Polishing of HFCS

Purification of pharmaceuticals. The growth of ion exchange technology in the pharmaceutical industry continues unabated. Most of the progress involves the use of ion exchange resins for the processing of new antibiotics and the recovery of the essential amino acids from fermentation liquors. Further, various ion exchange and adsorption processes are now employed for the production of vaccines and immobilized enzymes. Some progress has been made for the use of related technology directed towards the synthesis of insoluble disinfectants [4].

Hydrometallurgy. Because of decreased activity in the copper and uranium industries, progress in the use of ion exchange technology in hydrometallurgy has been limited to the recovery of gold and silver and to the treatment of wastes from the metal-finishing industries. Unfortunately, as our knowledge of the use of ion exchange technology for uranium and copper hydrometallurgy has been upgraded, the market for uranium and copper has deteriorated with but little to encourage its recovery before the end of the century (Table 2.4).

Table 2.4 Examples of major ion exchange applications: Hydrometallurgy (recovery and purification)

Uranium	Zinc
Thorium	Copper
Rare earths	Rhenium
Transition metals	Molybdenum
Transuranic elements	Nickel
Gold, silver, platinum	Cobalt
Chromium	

Solvent purification. One of the newer uses of ion exchange technology involves the use of ion exchange technology for the treatment of solvents and for purifying the ethanol distillates being used as fuel in automobiles. The macroreticular ion exchange resins and related polymeric adsorbents continue to show promise for the treatment of other organic process streams [2].

Catalysis. Although ion exchangers are generally recognized throughout the chemical and petrochemical industries as excellent heterogeneous catalysts, considerable progress has been made in the commercial use of sulfonic acid cation exchange resins for the production of isopropyl and secondary butyl alcohol, as well as for the production of methyl tertiary butyl ether, a gasoline additive. For example, whereas the inorganic zeolites have been used commercially as catalysts for many years, it is now recognized that ion exchange resins, particularly the sulfonic acid cation exchange resins, are excellent acid catalysts. We now find them being used commercially for alkylation, esterification and hydration reactions (Table 2.5). This mode of catalysis is now preferred for the production of methyl tertiary butyl ether and isopropyl alcohol [4].

Table 2.5 Examples of major ion exchange applications: Catalysis

Sucrose inversion	Epoxidation
Esterification	Hydration
Acylation	Hydrolysis
Condensation	

Agriculture. It is seldom realized that the farmers of the world are, by far, the largest users of ion exchange systems for the production of crops. The soil itself is a huge ion exchange system with a capacity for calcium, magnesium, potassium, ammonia, nitrates and phosphates. In fact, the soil system of the farmer is a huge regenerable system with the applied fertilizers serving as the regenerants [2]. Potential uses of ion exchange resins in agriculture are listed in Table 2.6.

More recently, "synthetic" soils based upon ion exchange resins have been used horticulturally for the purpose of supplying the soil nutrients and trace elements as well as for removing various toxic metals such as mercury, lead, etc.

Table 2.6 Examples of major ion exchange applications: Agriculture

Nutrient media
Sustained release of nutrients

Medicine. Ion exchange resins and the related polymeric adsorbents, as noted in Table 2.7, are used medically for a host of applications. Of significance is the fact that several products are now being produced as sterile and pyrogen-free media for use as hemoperfusion devices for treating cases of drug poisoning [2].

Table 2.7 Examples of major ion exchange applications: Medicine

Antacids	pH control
Sodium reduction	Potassium removal
Taste masking	Skin treatment
Sustained release	Toxin removal
Diagnostic	Hemodialysis
Tablet disintegration	Hemoperfusion

Waste treatment. Progress continues with respect to the use of ion exchange and polymeric adsorption technologies for treating waste effluents and contaminated well waters [4].

Analysis. The users of ion exchange for analytical purposes are legion; however, the widespread use of the "ion chromatograph" technique for the analysis of hundreds of organic and inorganic ionic species has been most dramatic. This technique, a combination of ion exchange chromatography with a novel ion exchange/conductivity detection system permits one to analyze parts per billion quantities of many ionic species.

New ion exchange materials

Much progress has been made in recent years in the direction of improving the uniformity and physical stability of ion exchange resins, as well as in the production of materials that are very selective for the alkaline earth and transition metals as well as for boric acid and borates. It is now possible to remove traces of these moieties from saturated solutions of sodium chloride, ammonium sulfate, etc. [2].

Conclusion

The continued growth of ion exchange technology industrially is assured and will be supported, it appears, by the continued development of new and improved ion exchange resins as well of inorganic zeolites.

References

[1] Encyclopedia of Chemical Technology, 3rd Edition, Vol. 13 (1978) (Kirk-Othmer, eds.) Wiley, New York.
[2] R. Kunin (1985) Ion Exchange Resins, Updated Edition, Krieger, Melbourne, Fl.
[3] Mass Transfer and Kinetics of Ion Exchange (1983) (L. Liberti and F. G. Helfferich, eds.), Martinus Nijhoff, The Hague, Netherlands.
[4] Ion Exchange Technology (1984) (D. Naden and M. Streat, eds.) Ellis Horwood, London, England.

2.1 General Ion Exchange Technology

Michael Streat
University of Technology, Department of Chemical Engineering, Loughborough, Leicestershire, Great Britain

Symbols

A	cross sectional area of column
a	interfacial area for mass transfer
C	liquid phase concentration
$\bar{C}$	resin phase concentration
D	liquid phase diffusion coefficient
$\bar{D}$	resin phase diffusion coefficient
d	resin bead diameter
$f_j(\bar{y}, t)$	resin conversion distribution function for stage j
h	bed height
h_E	height of exchange zone
k	resin phase pseudo mass transfer rate parameter
k_1	liquid phase pseudo mass transfer rate parameter
k_2	liquid phase pseudo mass transfer rate parameter
k_L	liquid phase mass transfer coefficient
K_c	selectivity coefficient
N_{OL}	number of overall liquid phase transfer units
N_{OR}	number of overall resin phase transfer units

$p_j(y)$	probability of resin at conversion $\bar{y}$ in stage j
r_o	resin bead radius
$r(\bar{y}, x_j)$	rate of sorption of resin
R	resin phase volumetric flow rate
S	liquid phase volumetric flow rate
t	time
u_L	superficial velocity of liquid
u_R	superficial velocity of resin
V	volume hold-up of resin
V_E	exchange zone volume
V_R	resin volume hold-up per stage
V_S	solution volume hold-up per stage
X	resin conversion
x	equivalent or molar ionic fraction in solution
x^*	equilibrium composition in solution
x_b	breakthrough solution composition
x_e	exhaustion solution composition
y	equivalent or molar ionic fraction in resin
y^*	equilibrium composition in resin
$\bar{y}$	resin conversion level
z	valency of ionic species
Z	bed height
α	separation factor
δ	liquid film thickness
ε	porosity of resin bed
τ	residence time in stage

Subscripts

A	ionic species A
B	ionic species B
j	stage j

2.1.1 Ion exchange equipment

2.1.1.1 Introduction

Ion exchange ranks as one of the foremost separation techniques for the treatment of relatively dilute aqueous solutions and for the treatment of certain specialised non-aqueous systems. In this sense, ion exchange can be compared with liquid-liquid extraction, ultrafiltration and reverse osmosis. Ion exchange usually operates in ionic solutions by counter-exchange of charged species and is governed by stoichiometry and by overall electrical neutrality. In this respect, it differs from adsorption which is governed by physical or chemical interaction in the pores of the solid sorbent. In all other respects, ion exchange and adsorption are similar, except that the former is driven by a chemical potential gradient in both the exhaustion and regeneration stages. Nevertheless, there is a close analogy between, for example, the sorption of uranium by a strong base resin and the sorption of gold from aqueous cyanide solutions onto active carbon particles. Both these processes have been developed in parallel in recent years and advances in equipment design for both ion exchange and sorption overlap to a certain extent.

Ion exchange is not confined to dilute solutions and it is possible to treat concentrated solutions provided the process technology is suitable. Continuous ion exchange renders it possible to treat concentrated solutions especially in hydrometallurgy; a good example is the 'Metsep' process [1] which was operated for a short time in South Africa. The potential to treat concentrated solutions exists but has not been widely exploited.

The design requirements for ion exchange process plant and equipment are given below.

1) The apparatus must ensure good contact between the resin phase and process solution at all times.
2) In most applications the residence time of the resin must be much larger than that of the process solution. This requires that the resin be retained in the reactor much longer than the process solutions. This is normally achieved in a packed column or a continuous countercurrent contactor.
3) To ensure low capital cost and compact plant design, the solution residence times should be short. Optimum selection of resin inventory and solution flow is of paramount importance.
4) In continuous ion exchange, the control of resin inventory and resin transfer within the contactor is extremely important.
5) Plant design should ensure efficient sorption, washing and regeneration of the resin to avoid unnecessary reagent loss and solution contamination.
6) Polymeric ion exchange materials are subject to abrasion and fracture and therefore mechanical stress must be avoided. Operational design of plants must recognise this problem quite apart from the associated complication of osmotic shock.

The selection of ion exchange equipment is totally process-dependent, although there are some simple ground rules. If a trace impurity is to be removed from a low-cost feed solution for the purposes of purification then a fixed bed or cascaded fixed bed plant is likely to be the optimum choice, e. g., boiler feed water treatment. If however, the flow-rate is large, then the capital cost of plant will rise and continuous countercurrent ion exchange may be warranted. However, continuous ion exchange is found to be advantageous only if the ion exchange selectivity is very favourable. Therefore, the separation of uranium or gold at trace concentrations by continuous ion exchange is very attractive, since the value of the recovered product is large when compared with the capital cost of plant and the inventory of resin. Nevertheless, it is probably true to say that more than 90% of all installations involve packed columns and the remainder are speciality applications involving a variety of moving bed and continuous multistage contactors.

2.1.1.2 Fixed beds

Fixed beds are most commonly used for ion exchange operations in a wide range of industrial applications. The great advantage of fixed beds is that they are simple to design and operate and the columns are readily fabricated. The ion exchange material remains located in the column and is not subjected to arduous hydraulic conditions. As a result the expected resin lifetime is long. In fact, there are some uranium plants in existence today with resin in service for over 20 years.

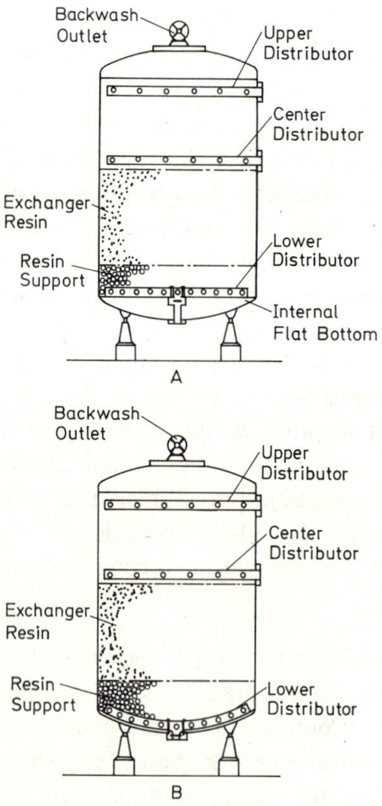

Figure 2.4 Details of a typical fixed bed column design. A Flat bottom type. B Dished bottom type. (Reproduced by permission from Academic Press; "Ion Exchange Technology", 1956; F. C. Nachod and J. Schubert [Editors].)

The simplest type of fixed bed is shown in Figure 2.4. The column is usually a parallel-sided pressure vessel containing dished ends. Materials of construction for the vessel are usually mild steel with a lining of rubber or plastic. In exceptional circumstances, stainless steel may be used and glass reinforced plastic may be acceptable for small-diameter columns. Single columns are rarely larger than 2 – 3 m in diameter and the bed depth is usually about 1 – 2 m. It is customary to provide freeboard of about 1 – 2 m above the upper surface of the bed to allow space for expansion of the bed during backwashing to clear trash and rubbish trapped in the upper void space.

The usual direction of flow is downwards so that the resin acts as a packed bed without any relative movement of the particles. The bed is irrigated at the top and solution leaves by a network of distributor pipes located in the dished base. The pipes are buried in sand or gravel which acts as a support medium for the ion exchange particles. This can cause problems due to the volume of process solution trapped in the bed of sand resulting in cross-contamination between cycles of operation. This can be overcome by embedding the distributor manifold in a false base, i. e., using concrete or cement, and providing distributor cups of plastic evenly spaced over the base of the resin column. Some installations use an elaborate manifold of distributor pipes in the dished base which is simply buried in ion exchange resin.

Regeneration of the resin can be performed in either downflow or upflow. Countercurrent regeneration is now common practice in water treatment and involves upflow of the regenerant without fluidising the resin particles. Many techniques exist for avoiding fluidisation of the particles, e. g., an outlet manifold at the upper surface of the bed, or expanding rubber bellows to retain the resin in a static condition.

Packed beds are periodically fluidised to backwash fine particles trapped in the upper void space. Clear water is fed in at the base and allowed to leave at the top of the column. In mixed beds, used for demineralisation, upflow is used to separate cation and anion resin to facilitate regeneration in the same column. Subsequently, the resin is remixed by air agitation.

Fixed columns of resin are cyclic in operation and must therefore be cascaded to maintain continuous flow of process solution. This requires extensive use of valves and pipework and the ancilliary instrumentation and control equipment necessary to maintain continuous operation. Although the column itself is simple to design and operate, the arrangement of valves, controls, etc., results in a relatively complex process plant. Also, scale-up is not always easy since the results obtained in a laboratory do not necessarily indicate the likely behaviour in a 2 – 3 m diameter column. Pressure drop in a column is important and most operations are restricted to flow-rates of about $60\ m^3\ m^{-2}\ h^{-1}$ in downflow. Larger flows incur enhanced pressure drop and the likelihood of severe channeling.

2.1.1.3 Cascaded fixed beds

The sequence of operations of a cascaded fixed bed plant for uranium purification is shown in Figure 2.5. Three columns are used in series for extraction, backwashing and split-flow elution. There are eight separate stages required to maintain pregnant solution flow and to operate a complete cycle of split-flow elution [2]. The sequence is as follows:

Resin loading:

stages (2) to (8) – pregnant feed flow through columns (a) and (b)

Resin washing and elution:

stage (1)	displacement of pregnant feed from (c) to (a)
	displacement of feed solution from (a) to pregnant feed tank
stage (2)	water wash of column (c)
stage (3)	first eluant to (c) to displace water
stage (4)	first eluant to (c) to eluant make-up
stage (5)	second eluant to (c) peak uranium concentration to precipitation
stage (6)	second eluant to (c) to become first eluant in next cycle
stage (7)	as for stage (6) to completely elute (c)
stage (8)	column (c) on stand-by to become column (b) in next cycle.

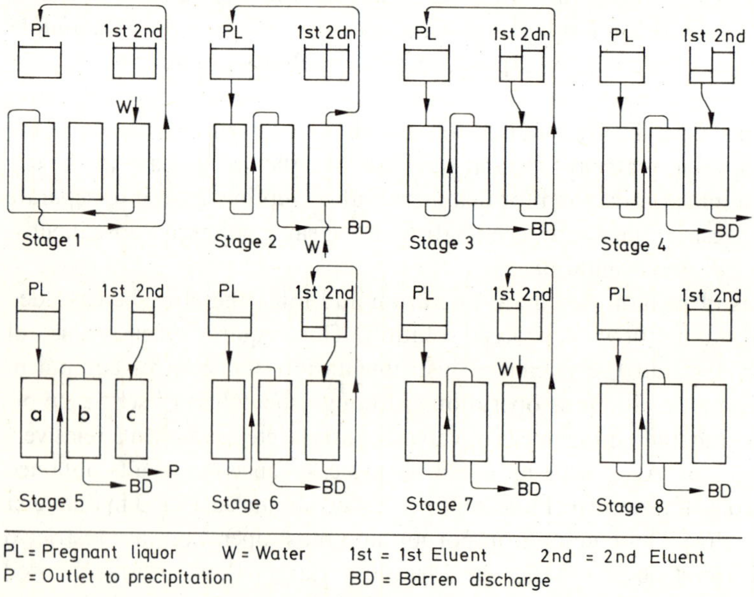

Figure 2.5 Cyclic operation of a 3-bed uranium extraction-elution process involving backwash and split elution. (Reproduced by permission from "Ion Exchange Sorption Processes in Hydrometallurgy", published for SCI by John Wiley, 1987 [© SCI].)

By analogy with other chemical engineering separation processes, fixed bed ion exchange is a multiple batch process. The major disadvantage of fixed bed operation is that the process is cyclic in operation and at any instant only a relatively small part of the resin is doing useful work. This leads to a high resin inventory and to a high capital cost for plant and equipment. The cyclic nature of the process could be overcome in continuous countercurrent operation. Several sophisticated ion exchange plants have been developed and operated in a range of process industries during the last 30 years.

2.1.1.4 Moving fixed beds

The earliest attempts to develop a continuous ion exchange contactor were based on the concept of a moving fixed bed reactor. In most embodiments, the fixed column of resin was moved intermittently in either downflow or upflow against a countercurrent flow of process solution. The technique was favoured for water treatment, effluent treatment, pollution control and the removal of trace metals or toxic materials from trade waste.

The foremost development is the 'Higgins' or 'Chem-Seps' contactor which comprises a closed loop containing fixed beds of resin in separate compartments capable of moving countercurrent to the process solution under the action of a hydraulic pulse [3]. Figure 2.6 is a schematic drawing of the contactor showing the

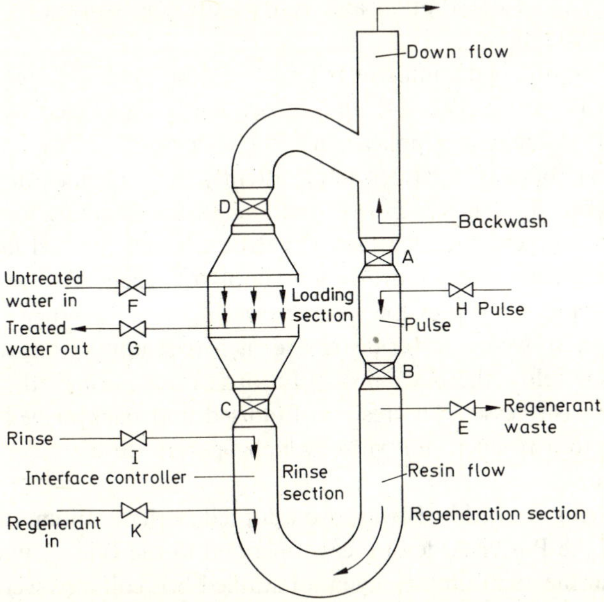

Figure 2.6 Schematic diagram of the Higgins Moving Bed Contactor. (Reproduced by permission from "Ion Exchange in the Process Industries" – Fig. 1, p. 122 by I. R. Higgins and R. C. Chopra – SCI 1970).

extraction, washing and regeneration sections. The liquid streams pass in and out of the contactor via screened collectors and these are sequenced by flow controllers to enable resin movement in countercurrent flow. Several sophisticated valves are needed to isolate sections of the loop during solution flow and resin transfer. These valves are required to close in a resin bed and to give little or no resin fracture or abrasion. Full-bore butterfly valves or sliding plate valves are used to restrict resin damage to about 1% of inventory per month. The extraction or loading section of the column is usually about 1 m in diameter though some larger plants have been built with the extraction chamber 2–3 m in diameter. The flow-rate can be high and depends only on the total pressure drop over the packed extraction section. Since comparatively shallow beds are used, typical flow-rates lie in the range 40–100 $m^3\,m^{-2}\,h^{-1}$. The valve positions during the operating cycles are as follows (see Figure 2.6):

1) Run cycle:
   Valves A, E, F, G, I, K open
   Valves B, C, D, H, closed
2) Pulse cycle:
   Valves B, C, D, H open
   Valves A, E, F, G, I, K closed

Despite the inclusion of a backwash cycle in the operating sequence, it is still necessary to clarify the feed solution and restrict the undissolved solids content below about 100–1000 ppm.

An alternative moving bed contactor is known as the 'Asahi' design which was first suggested by Porter in 1956 [4] and subsequently developed by the Asahi Chemical Company in Japan for application in the water and waste water treatment field [5]. The contactor also operates intermittently with normal flow upwards through a packed bed of resin which is pressed against a retaining top screen. At the end of the service cycle, the feed flow is valved off and the bed falls into the bottom part of the reaction vessel, drawing a supply of regenerated resin through a non-return valve into the top of the reactor (see Figure 2.7). When service flow is resumed, the resin above the lower distributor is forced upwards against the top screen and the resin below the bottom distributor is forced out of the reactor by pipeline to a backwash vessel. The resin is fluidised and backwashed to remove fines and then fed to a regeneration vessel which operates in a similar manner to the extraction column.

The idea of moving packed beds was also adopted in the uranium industry. An early example was the Porter-Arden process operated in the Blind River region of Ontario, Canada in the 1950's [6]. Conventional fixed bed columns were used, but exhausted resin was transported to an empty conventional column by hydraulic conveying. The resin was backwashed and then transported to a bank of elution columns, whilst a charge of eluted resin was transported to the vacant extraction

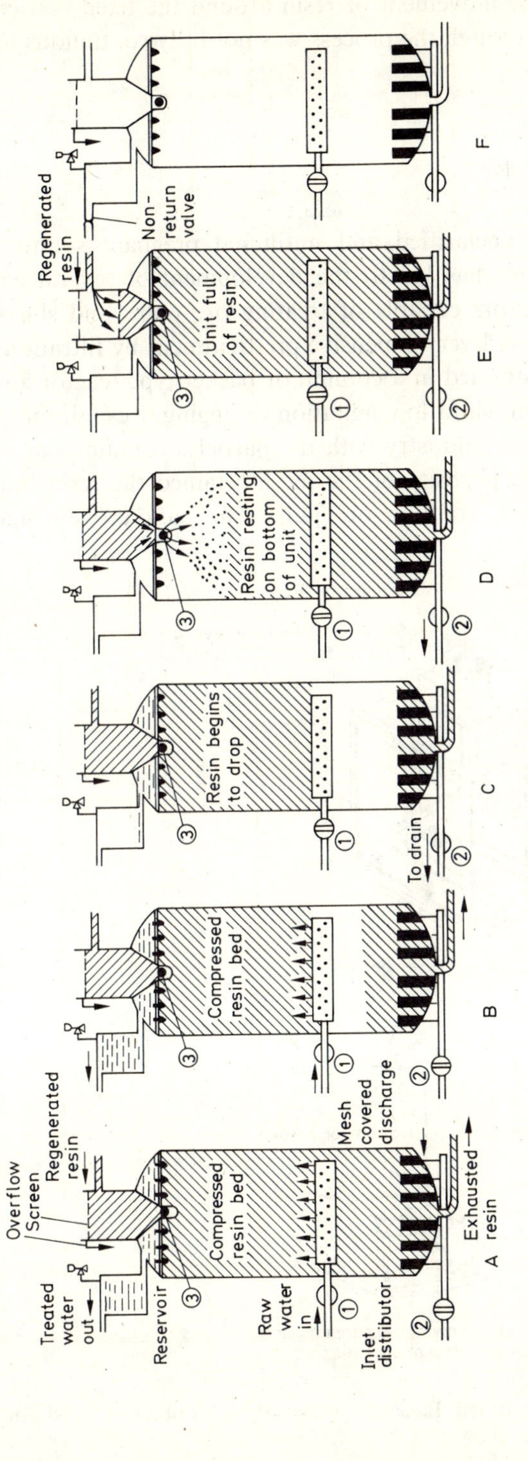

Figure 2.7 Operating sequence of the Asahi Type Countercurrent Moving Bed Contactor.

column in the cascade. The movement of resin around the fixed bed circuit was essentially countercurrent, though the process was not fully continuous by modern standards.

2.1.1.5 Agitated/jigged beds

There is a need to treat unclarified and unfiltered pregnant solutions in the metallurgical industries. This has led to the development of several agitated or jigged ion exchange contactors capable of treating ore pulps and slurries which would otherwise have required very expensive pre-treatment by filtration. The ion exchange particles were restrained in a column or basket-type reactor and periodically separated by pulsation, vibration, agitation or jigging. Resin-in-pulp reactors were operated in the uranium industry with the particles contained in baskets of stainless steel mesh [7]. The porosity of the mesh retained the resin but allowed ingress of a flowing leached uranium ore pulp. The baskets were agitated by

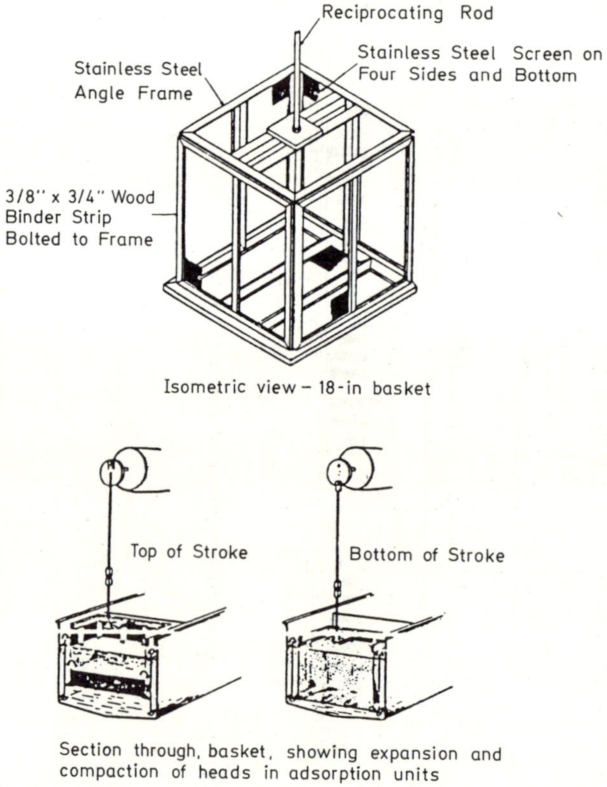

Figure 2.8 Concept of the Agitated Basket Resin-in-Pulp Contactors used for uranium recovery from ore pulps.

reciprocating action to avoid clogging of the bed (see Figure 2.8). The process operates by alternating feed, wash and eluant solutions. Though there is no countercurrent movement of the particles, this process demonstrated that an 'expanded' bed contactor could be used for metals recovery. A similar concept was developed by Swinton, Weiss and co-workers in Australia [8]. They devised an agitated or jigged bed for treatment of uranium ore pulps. Continuous pulsation of the bed (about 200 cycles/min) caused momentary separation of the particles and this allowed the slurry to pass through the bed in upflow. A screen prevented carry over of the ion exchange particles. This screen was a great disadvantage since it would blind with wood pulp and other gangue so that the pressure drop became excessive. This idea did not proceed to full-scale development. Swinton and Weiss [9] also developed a multistage pulsed or jigged ion exchange contactor for uranium recovery. The concept was feasible in principle but operation of a reasonable size reactor was difficult due to hydraulic instability. This idea was also dropped in the late 1950's. Although some resin-in-pulp contactors are still in use today, notably in the USA, there has been little further development of agitated and jigged bed systems. Modern process development requires more sophisticated techniques based on the use of fluidised bed or stirred tank contactors.

2.1.1.6 Fluidised beds

Many of the potential problems indicated in Sections 2.1.1.1–2.1.1.5 can be overcome in a well-designed fluidised bed contactor. The great advantage of a fluidised bed reactor is the ability to operate with continuous or semi-continuous countercurrent flow of resin and process solution. Control of the flow of both phases is relatively simple, requiring a minimum of process timers or controllers. The pressure drop in reactors is low since the vessels are normally open to the atmosphere and thus the mechanical design is straightforward and materials of construction can be relatively cheap. The flexibility of fluidised beds minimises the number of columns in service, often only three in an array, i. e., extraction, washing and regeneration. This also minimises resin inventory and reduces resin loss. One of the important advantages of fluidised bed design is ease of scale-up. Distributor design is not complex and hence it is possible to confidently scale-up from laboratory experiments to pilot-plant and full-scale.

By-passing and channeling of liquid, a serious problem in large diameter packed bed systems, is not a serious consideration in a multistage fluidised bed contactor. It is for this reason that the metallurgical industries, especially the uranium industry, have vigorously developed fluidised bed continuous ion exchange contactors in recent years. The most notable developments have occurred in the USA, Canada, South Africa and in the UK.

The earliest designs of fluidised beds were for water softening, since the ion exchange selectivity is extremely favourable in this case. The Dorr softener [10] comprises a series of trays on which resin is fluidised in an upward flow of water. The resin was drawn from each bed and fed to the one below using a liquid ejector driven by the product water. Transfers of resin between the extraction and regeneration columns were also carried out with liquid ejectors. The Fluicon or MAN process is also a continuous countercurrent stagewise reactor designed particularly for water softening (see Figure 2.9). Here the softener comprises a series of perforated distributor plates with downcomers for countercurrent resin flow. This concept is difficult to operate in large columns and with a reasonable number of stages, due to hydraulic instability associated with the tray and downcomer design. Fluicon installations are relatively small and water treatment plants with flow-rates of 10 – 100 m^3/h are commercially available.

Large-scale multistage fluidised bed contactors have been built based on the concept of a perforated distributor plate containing holes significantly larger than the largest resin particles and *no* downcomer. This technique was pioneered by Cloete and Streat at Imperial College in England [11] and George, Ross and Rosenbaum in the United States Bureau of Mines Laboratory, Salt Lake City, USA [12]. This early initiative forms the basis of most of the subsequent commercial development in fluidised bed technology applied to ion exchange.

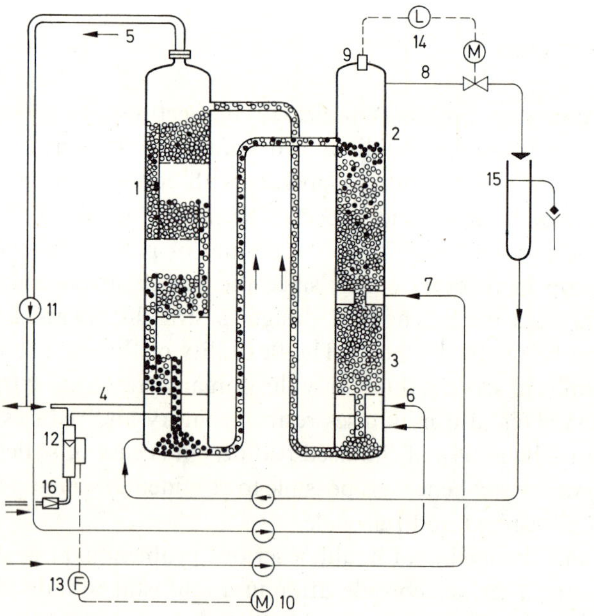

1 loading column
2 regenerating column
3 wash column
4 raw water
5 softened water
6 washwater
7 regenerating water
8 brine
9 level gauge
10 metering pumps
11 circulation pumps
12 flow meter
13 regulator
14 regulator
15 hold tank
16 pressure reducer

Figure 2.9 The Fluicon process. (Courtesy of K. Dorfner)

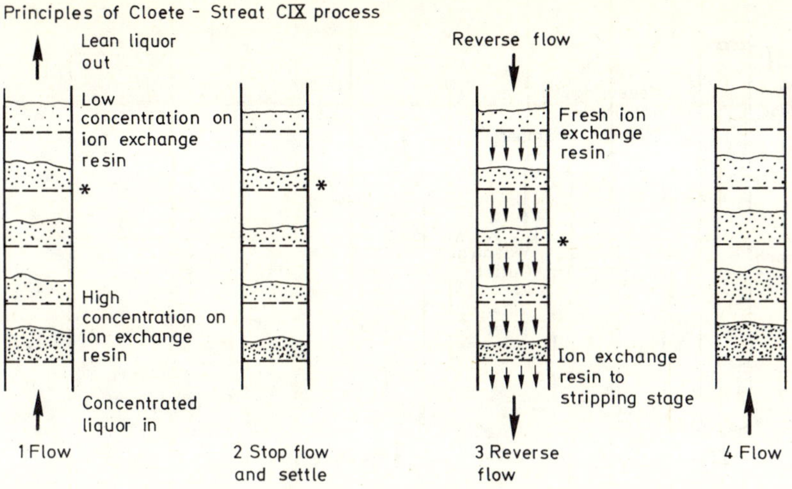

Figure 2.10 Sequence of operations of the Cloete-Streat CIX Process.

The Cloete-Streat contactor is a periodic reactor consisting of a series of perforated distributor plates containing fluidised beds of resin with feed solution normally in upflow (see Figure 2.10). Periodically, the feed solution flow is stopped for a short period (usually several minutes) to allow the fluidised beds of resin to settle onto the trays. The entire contents of the column are then allowed to flow by gravity or are pumped downwards in a reverse direction to the feed flow in order to transfer an increment of resin from plate to plate. Normal service flow is then resumed and the series of fluidised beds is re-established. In this way, the resin is enabled to move in counterflow to the net solution flow; furthermore, the operating cycle is infinitely flexible and can be controlled with simple process timers or controllers. The USBM concept is essentially similar in many respects and differs only in plate design, mode of solids transfer and resin withdrawal.

Equipment based on the Cloete-Streat principle has been widely applied in the uranium industry. The largest number of plants is to be found in South Africa, following extensive development of the concept at the National Institute for Metallurgy (NIM) now known as the Council for Mineral Technology (MINTEK). These plants are usually referred to as NIMCIX contactors and are in operation for by-product uranium recovery at several gold mines [13]. The largest plant is presently in operation at Chemwes gold mine in Stilfontein, South Africa; it consists of four columns made of glass reinforced fibre, the extraction columns are 5 m in diameter and the plant can treat 14 000 m^3/d of uranium-bearing feed solution.

A modified form of multistage fluidised bed column has been developed by Himsley in Canada [14]. The Himsley column consists of vertical stages containing resin and a single fluid inlet (see Figure 2.11). Pregnant feed solution passes in upflow from stage to stage continuously, fluidising the resin in the form of a spouted

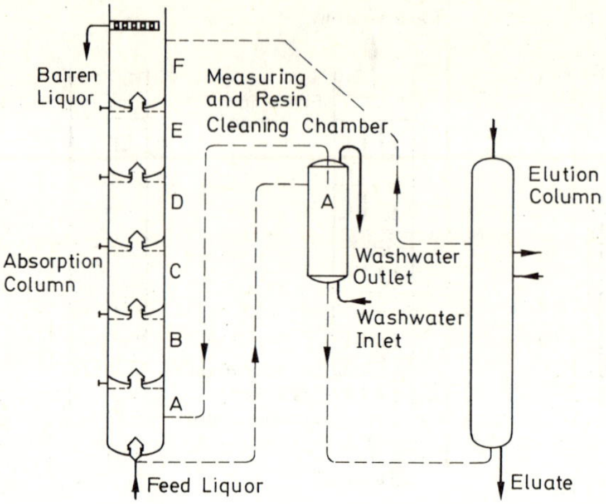

Typical Arrangement of Himsley continuous ion exchange system

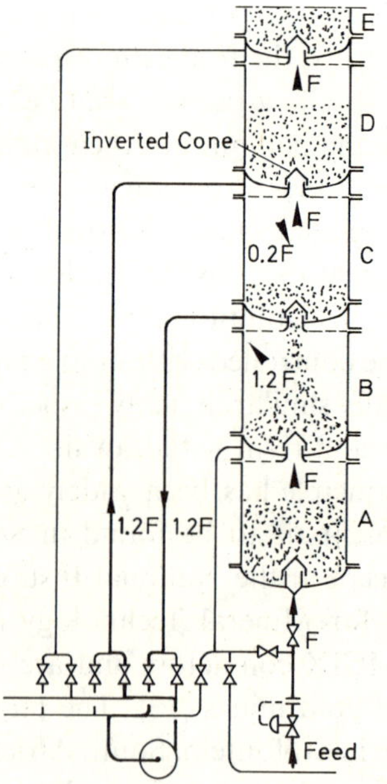

Typical resin transfer from one chamber to another

Figure 2.11 Operating sequence of the Himsley Multistage Fluidised Bed Contactor.

bed, and the feed flow is not stopped during countercurrent resin transfer. Resin is transferred sequentially from stage to stage down the column using an auxiliary pump. An empty stage at the base of the column allows this process to start and fresh resin is transferred hydraulically into the top stage. The contactor requires a series of control valves and timers to sequence the operations. Also, a screen in each stage prevents the carry-over of resin during transfer. Several large installations have been built for uranium recovery at Agnew Lake, Ontario, Canada and at Twin Buttes in Arizona, USA.

The largest fluidised bed system was designed by Porter for uranium recovery at the Rossing Uranium Mine in Namibia [15]. This comprises a cascade of open-topped circular rubber-lined steel tanks on a slight incline so that gravity flow can occur. Liquid flow is continuous, entering the tanks at low level through a pipe distributor manifold. Solution overflows weirs at the top of each tank and passes over screens to collect entrained resin. Resin is transferred in countercurrent flow by air-lift into a vacant upstream tank. The largest version of this design used rectangular rubber-lined concrete tanks about 6 m × 6 m section and 3.8 m deep. There are four parallel streams handling a total of 3600 m^3/h of pregnant feed solution.

2.1.1.7 Stirred tank reactors

Stirred tank reactors have been used to contact ion exchange resins with dense mineral pulps in uranium recovery. Stirring or agitation minimises the physical problems associated with differences in resin and pulp density. Gentle agitation is normally used to keep the resin and ore particles in suspension. This can be achieved either by air injection or by mechanical stirring both with or without draft tubes [16].

One of the early designs of this equipment was developed by Infilco in the USA and it was operated successfully on a 30% w/w uranium ore pulp [7]. The cone-shaped contactors in these units were air-agitated using a central draft tube mounted above diffusion cloths through which the air was admitted. Resin and pulp rose within the draft tube and settled back round the periphery of the tank. External vibrating screens were used for separating air-lifted resin and slurry. Resin inventory control in the contactors was difficult as resin flow was not independent of the pulp flow. More recently, a novel concept involving the use of inverted fluidised beds of floating resin in actual leach reactors was developed in South Africa. The resin-in-leach contactor or RELIX process was operated with 60% w/w ore pulps, though full-scale plant development has not yet resulted [17].

The energy for agitation can be reduced by the use of mechanical agitation or stirring, which is generally more efficient though increasing capital cost. Oldshue [18] has designed draft tube agitators with provision for re-starting after shutdown. The agitator impellor is located at the top of the draft tube above the level of settled pulp. Design of stirred tank contactors requires extensive pilot-plant work since scale-up of agitators is unreliable. Also, the effect of the agitator on attrition and fracture of resin beads must be carefully assessed.

2.1.2 Mathematical treatment

2.1.2.1 Equilibrium data

Process design requires a rational procedure for the representation of the equilibrium data of the exchanging species. The most widely employed method of expressing ion exchange equilibria has been derived from the law of mass action or the Donnan membrane theory. Consider the exchange of cations A and B between a cation exchange resin and a solution containing no other cations. Assume that the ion exchanger is initially in the B form and that solution contains ions A. The mass action expression can be written,

$$z_A \overline{B}^{z_B^+} + z_B A^{z_A^+} \rightleftharpoons z_B \overline{A}^{z_A^+} + z_A B^{z_B^+} .$$

Here the overbars denote the ionic species in the resin phase and z_A^+ and z_B^+ denote the valency and charge of counter ions A and B.

If we consider an exchange of univalent ions, i. e., $z_A = z_B = 1$, then the selectivity coefficient is given by the following expression,

$$K_{c_B}^{A} = \frac{y_A x_B}{y_B x_A} = \frac{y_A(1 - x_A)}{x_A(1 - y_A)} = \alpha_B^A .$$

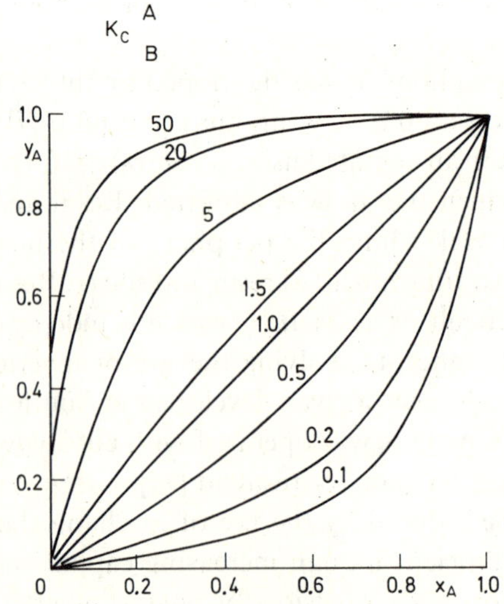

Figure 2.12 Equilibrium plot: univalent – univalent exchange.

Table 2.8 Ion exchange selectivity

Selectivity coefficient of cations in the hydrogen cycle of a sulphonic acid cation exchange resin

Cation (A)	$K_{c\,\mathrm{H}}^{\mathrm{A}}$
Li^+	0.8
Na^+	2.0
K^+	3.0
NH_4^+	3.0
Mg^{2+}	26.0
Ca^{2+}	42.0

Selectivity coefficient of anions in the hydroxide cycle of a quaternary anion exchange resin

Anion (A)	$K_{c\,\mathrm{OH}}^{\mathrm{A}}$
HCO_3^-	4.0
Cl^-	11.0
NO_3^-	40.0
SO_4^{2-}	50.0

A typical plot of equilibrium data is given in Figure 2.12 and an approximate selectivity scale for some cations and anions is given in Table 2.8.

In the industrially important case of divalent-univalent exchange i. e., $z_A = 2$, $z_B = 1$, the expression for the selectivity coefficient becomes

$$K_{c_B}^{A}\left(\frac{\bar{C}}{C}\right) = \frac{y_A(1 - x_A)^2}{x_A(1 - y_A)^2}\,.$$

Here, the relationship between y_A and x_A becomes greatly dependent on the solution concentration, C. A typical equilibrium plot is shown in Figure 2.13 and it is clear that the preference for A is greatly enhanced as the solution concentration is decreased. Extensive values of the selectivity coefficient for strongly basic anion exchange resins have been tabulated in the literature [19, 20, 21]. The theoretical treatment of multicomponent ion exchange is quite complex, especially if it involves species of different valency. The treatment of ion exchange equilibria involving only ions of the same valency is quite straightforward, since it can be assumed that the selectivity coefficient or separation factor is constant over the complete range of ionic concentration. Consider a ternary system involving three components, A, B and C. Assuming univalent ions, then

$$K_{c_B}^{A} = \alpha_B^A;\ K_{c_C}^{B} = \alpha_C^B;\ K_{c_C}^{A} = \alpha_C^A\,.$$

It follows therefore, that

$$\alpha_B^A \cdot \alpha_C^B \cdot \alpha_A^C = 1\,.$$

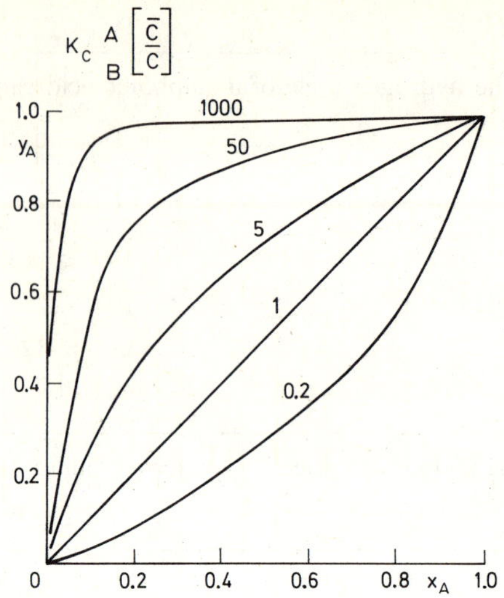

Figure 2.13 Equilibrium plot: divalent – univalent exchange.

This is known as the triangular rule and it is possible to represent the three-component data on a triangular diagram [22, 23]. The extension of this theory into multicomponent systems has been presented by Helfferich and Klein for chromatographic purposes [24]. Multicomponent systems are complex and do not always lend themselves to a rational procedure for representation. Many process designers tend to simplify the relevant equilibrium data into a more basic form. Often a Langmuir or Freundlich type adsorption isotherm can be made to fit ion exchange equilibrium data within strict limits of application. The Langmuir type isotherm is generally of the form

$$y = f(x)$$

for one species, ignoring the effect of other competing species. It is surprising that this simple type of equilibrium relationship appears to fit quite well the case of uranium sorption from aqueous sulphuric acid onto a strong base anion exchange resin [25]. A typical curve is given by the expression:

$$y = \frac{ax}{bx + 1}$$

where a and b are constants fitted from the experimental equilibrium data.

It is, of course, possible to use a polynomial fit of the data for the purpose of mathematical modelling.

$$y = p + qx + rx^2 + sx^3 + \ldots$$

This method was used by Streat and Takel [26] to facilitate the mathematical simulation of a continuous countercurrent multistage contactor.

It is invariably necessary to carry out appropriate equilibrium experiments in the concentration range of interest at the outset of process design. The data should be conveniently fitted to one or another simplified equilibrium model to facilitate handling of the data in process calculations.

2.1.2.2 Rate equations

The rate of exchange of ions with an ion exchange resin is governed by diffusion within the stagnant mass transfer boundary layer surrounding the particles or by diffusion within the polymeric matrix of the resin. If strong ion-ion or ion-matrix interactions occur, as for example in a chelating ion exchanger, then the reaction rate is slow and may be governed by chemical reaction control. However, there is little experimental evidence for this effect. Exact solutions of the diffusion equations exist in the literature but they are complex and ill-suited to process design calculations [27].

There are many simplified rate equations which approximate the more rigorous expressions found in the literature [28]. Film diffusion can be approximated by the linear driving force equation, which assumes that equilibrium exists at all times at the fluid-particle interface

$$\frac{dy}{dt} = k_1(x - x^*) = k_2 (y^* - y).$$

This equation can be solved to give the fractional conversion of the resin, X, as a function of time:

$$X = 1 - \exp\left(-\frac{3\,DCt}{r_0 \delta \bar{C}}\right)$$

$$\text{or } \ln (l - X) = - k_2 t$$

$$k_2 = \frac{3\,DC}{r_0 \delta \bar{C}}.$$

The parameter k_2, is a liquid film mass transfer parameter (s^{-1}) and is related to the liquid film mass transfer coefficient k_L($cm.s^{-1}$). The liquid film mass transfer coefficient can be independently predicted using established correlations available in the literature. Snowdon and Turner [29] and Rahman and Streat [30] have obtained a generalised correlation for k_L in fixed and fluidised beds.

Particle diffusion has been approximated by Vermeulen using a quadratic driving force model.

$$\frac{dy}{dt} = \frac{k((y^*)^2 - y^2)}{2y}.$$

This expression is a simplification of the classical diffusion equation in spherical co-ordinates solved for the case of infinite solution volume. The fractional conversion of the resin is given by the expression

$$X = \left[1 - \exp\left(-\frac{\bar{D}\pi^2 t}{r_0^2}\right)\right]^{\frac{1}{2}}$$

$$\ln(1 - X^2) = -kt$$

$$k = \frac{\bar{D}\pi^2}{r_0^2}.$$

The parameter k, is a pseudo-rate parameter (s^{-1}) within the resin phase and is constant only if the value of $\bar{D}$ is constant. This is unlikely, since the diffusion coefficient of exchanging ions in a binary system will vary with concentration and time. Order of magnitude estimates of the value of k_2 and k can be obtained using published diffusivities (see for example Table 2.9), if one bears in mind that the value of δ is indeterminate, though a value of $5 \cdot 10^{-5}$ m is a reasonable guess.

Table 2.9 Ionic diffusivities

Solution phase diffusivities at infinite dilution, 25 °C	
Ion	D ($m^2/s \times 10^9$)
H^+	9.32
K^+	1.96
Na^+	1.33
NH_4^+	1.95
$\frac{1}{2}\,Ca^{2+}$	0.79
$\frac{1}{2}\,Mg^{2+}$	0.71
OH^-	5.26
Cl^-	2.03
NO_3^-	1.90
HCO_3^-	1.18
$\frac{1}{2}\,SO_4^{2-}$	1.06

Resin phase self-diffusion coefficient, 25 °C	
Ion	$\bar{D}$ ($m^2/s \times 10^{10}$)
H^+	5.50
K^+	1.34
Na^+	2.00
Cs^+	3.00
Ca^{2+}	0.60
Cu^{2+}	0.27

2.1.2.3 Performance of fixed beds

Fixed bed columns are rarely designed from first principles, especially in the field of water treatment. Very extensive practical experience exists in the field of water softening, dealkalisation and demineralisation and this is normally made available by the ion exchange resin manufacturers. Most suppliers of ion exchange materials supply copious data books giving design information in the form of charts, graphs, nomograms and tables, which enable column design to be carried out on a strictly empirical basis. This methodology is wholly adequate where prototype plant experience is in existence, though it is possibly less acceptable for the design of new or less common processes. The manufacturers design methods invariably permit calculation of resin volume required for a specific ion exchange treatment process and this leads naturally to a selection of column diameter and bed height. Ratios of bed height/diameter from 1/1 to 2/1 are quite common, all other hydrodynamic factors considered. The mechanical design of the top and bottom distributors, valve arrangements, timers and controllers is proprietary information and is usually protected by patents.

The literature abounds with attempts to simulate and predict the performance of packed ion exchange columns. The most comprehensive review may be found in Perry and Green [31], which is a unified treatment of all the modern thinking in adsorption and ion exchange. This presentation is extremely useful to practitioners in this field, yet the treatment is sufficiently complex to make it only limited in value for process design. Nevertheless, an understanding of the performance of a packed column reactor is important and this can easily be explained with the aid of a simplified approach published much earlier [32] and still equally valid today.

Consider the hypothetical liquid phase concentration gradients in a fixed bed of resin at the various times shown in Figure 2.14. If we assume very favourable equilibrium in the sorption reaction, then the shape of the concentration profile

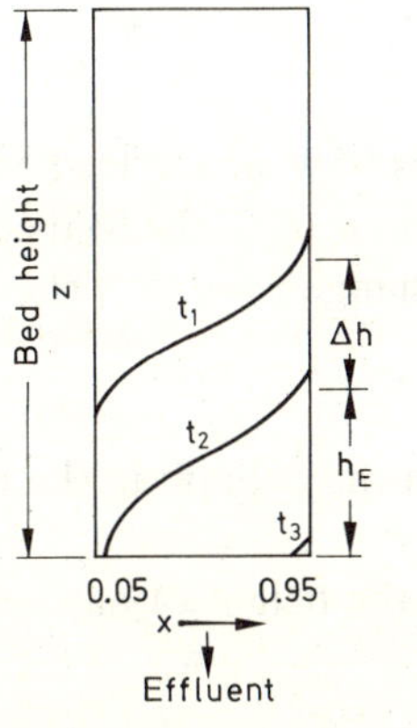

Figure 2.14 Hypothetical concentration gradient in a fixed bed: Favourable equilibria.

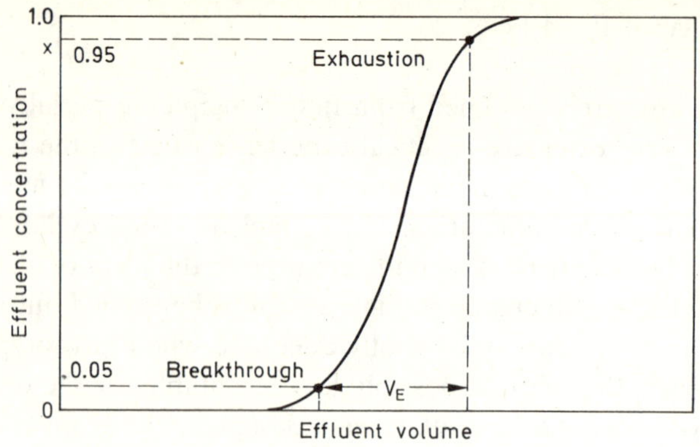

Figure 2.15 Breakthrough curve.

will be sharply defined and will remain constant as it moves down the length of the column. Michaels [32], a pioneer in this field, defined the concept of an *exchange zone*. The exchange zone is assumed to be that region in the column where the sorbing ion changes in concentration from a fixed low value in solution, say 5% of the feed to a fixed upper value which may be 95% of the concentration in the feed (see Figure 2.15). It is assumed that the conditions within the exchange zone remain unchanged with time, except during the initial period of operation when the zone is being established. This concept cannot be generalised but is completely rigorous for liquid-film diffusion controlled reactions, i. e., for processes operating at dilute solution concentrations.

As shown in Figure 2.14, the exchange zone descends a distance equal to Δh during the time interval t_1 to t_2. The position at time t_2 corresponds to breakthrough as indicated in Figure 2.15. Since the resin above the exchange zone is essentially saturated with the ion present in the feed solution, a volume of resin equal to $A\Delta h$ is exhausted in the time interval. If ΔV is the volume of solution fed to the column during the time interval, then

$$\Delta VC = A\,\Delta h\,(\bar{C} + \varepsilon C).$$

At time t_3, the concentration in the effluent reaches 95% of the feed concentration (see Figures 2.14 and 2.15). In the time interval from t_2 to t_3, the feed concentration descends a distance equal to the height of the exchange zone, h_E. Then,

$$V_E C = h_E A(\bar{C} + \varepsilon C),$$

where V_E is the volume of solution passed through the column in the time interval from t_2 to t_3.

Rearranging gives the following expression for the height of an exchange zone

$$h_E = \frac{V_E C}{A(\bar{C} + \varepsilon C)}$$

If the exchange zone concept is to be used in column appraisal and design, then it is helpful to know the rate at which the exchange zone passes down the column, u_R:

$$u_R = \frac{\Delta h}{t_2 - t_1} = \frac{\Delta VC}{A(t_2 - t_1)(\bar{C} + \varepsilon C)}$$

$$= \frac{u_L C}{\bar{C} + \varepsilon C}$$

where

$$u_L = \frac{\Delta V}{A(t_2 - t_1)}.$$

Michaels developed a relationship for the number of transfer units (N_{OL}) within the exchange zone by assuming that a fixed bed process in which the exchange zone descends through the bed at a fixed rate, u_R, is equivalent to a countercurrent ion exchange process in which the bed moves upwards at an absolute linear rate u_R, countercurrent to a liquid stream moving downwards at a superficial velocity $(u_L - u_R)$. The treatment requires an appropriate rate equation and for liquid film diffusion it is appropriate to use a linear driving force equation. The derived relationship for the number of transfer units within the exchange zone is given by the equation:

$$N_{OL} = \frac{K_c}{K_c - 1} \ln\left[\frac{(1 - x_b)}{(1 - x_e)}\left(\frac{x_e}{x_b}\right)\right] - \ln\left[\frac{(1 - x_b)}{(1 - x_e)}\right]$$

K_c = equilibrium selectivity coefficient
x_b = equivalent ionic fraction at breakthrough
x_e = equivalent ionic fraction at exhaustion
N_{OL} = number of transfer units $= \dfrac{k_L a h A}{S}$
k_L = liquid phase mass transfer coefficient
a = interfacial area for contact $= 6(1 - \varepsilon)/d$
A = cross sectional area of column
S = liquid phase volumetric flow rate
ε = porosity of bed.

If $x_b = 0.05$ and $x_e = 0.95$, then we can calculate the height of the exchange zone as follows:

$$\frac{k_L a h_E A}{S} = \frac{K_c}{K_c - 1} \ln(19)^2 - \ln(19).$$

For any value of h less than h_E, then

$$\frac{k_L a h A}{S} = \frac{K_c}{K_c - 1} \ln(19)\left(\frac{1 - x}{x}\right) - \ln\left(\frac{1 - x}{0.05x}\right).$$

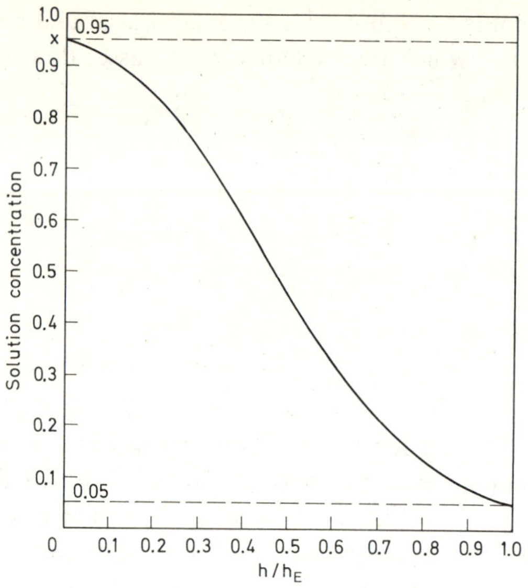

Figure 2.16 Relative change in solute concentration in solution with distance through exchange zone (idealised case for $K_c = 1.2$ and univalent ions).

Hence

$$\frac{h}{h_E} = \frac{K_c \ln\left(\frac{0.95}{x}\right) + \ln\left(\frac{1-x}{0.05}\right)}{(K_c + 1)\ln(19)}.$$

This equation relates the ionic concentration in the liquid phase to the fraction of the distance down through the bed defined as the exchange zone. This relationship yields the typical S-shaped breakthrough curve for appropriate values of K_c (see Figure 2.16). The curve is symmetrical and thus not representative of actual performance, which may be dominated by hydrodynamic factors such as by-passing, channeling, backmixing, etc. Other rate equations have been used in this treatment and have provided further useful insight into the operating characteristics of a fixed bed [33].

2.1.2.4 Continuous countercurrent ion exchange

The design of continuous countercurrent ion exchange columns is based on proprietary know-how and accumulated experience. Most of this information is available to the process developers, e. g., Chem-Seps, Himsley, MINTEK, Davy McKee, etc. and little quantitative detail has been published in the open literature. There is, however, considerable academic interest in column design and several studies have been published in recent years [34–37, 25, 26]. Simulation procedures have been published which are applicable to either moving packed bed contactors or to countercurrent multistage fluidised beds.

The performance of a moving packed bed reactor can be deduced from the theoretical development given in Section 2.1.2.3. Here, Michaels derived the operating characteristics of a fixed bed column on the basis of a countercurrent moving bed concept. Hence, the derived values of N_{OL} and the height of the exchange zone, h_E, can be regarded as analogous to the performance of a moving bed, on the premise that the exchange zone is retained stationary within the reaction vessel due to relative motion. Thus, by selection of the relative velocity in a moving fixed bed column so that the exchange zone is retained in the bed under steady-state conditions, the bed height can be predicted using the expression

$$\frac{k_L a h_E A}{S} = \frac{K_c}{K_c - 1} \ln\left[\frac{(1 - x_b)}{(1 - x_e)}\left(\frac{x_e}{x_b}\right)\right] - \ln\left[\frac{(1 - x_b)}{(1 - x_e)}\right].$$

This expression assumes favourable equilibrium data and was obtained with the simplified linear driving force expression, e. g., the sorption of a trace metal from aqueous solution. If we assume the following data:

$$x_b = 0.05$$
$$x_e = 0.95$$

Then:

$$\frac{k_L a h_E A}{S} = \frac{K_c}{K_c - 1} \ln (19)^2 - \ln (19).$$

Let $K_c = 1.2$
$d = 0.8 \times 10^{-3}$ m
$\varepsilon = 0.55$
$S/A = 60$ m^3 m^{-2} h^{-1} = 0.0167 m/s
$k_L = 7 \times 10^{-5}$ m/s (arbitrary)

then $h_E = 2.3$ m.

Increasing K_c from 1.2 to 5, thereby greatly enhancing selectivity, and assuming all other parameters remain constant, yields a revised bed height, i. e.,

$h_E = 0.31$ m.

The depth of bed in the sorption vessel is given by the value of h_E and it is quite clear that the bed depth is greatly reduced as the selectivity of the preferred ion increases. This is a simplistic approach, since it ignores the hydrodynamic effects of channeling, by-passing and back-mixing of solutions, as well as end-effects associated with the resin and solution inlet and outlet arrangements. Also, this approach is inadequate if $K_c \leqslant 1$, i. e., regeneration and a quite different procedure would then be required.

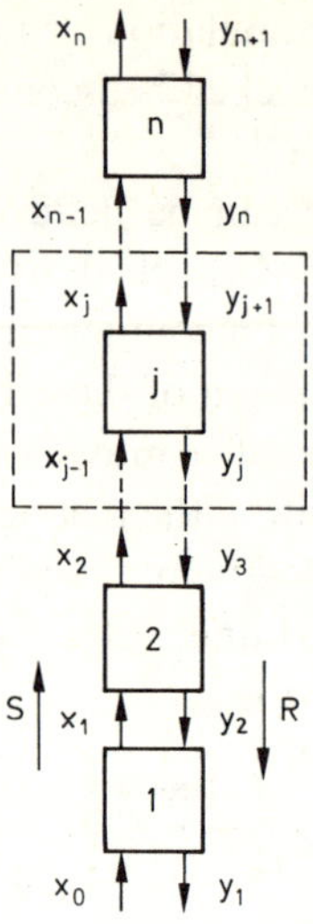

Figure 2.17 Schematic representation of a Continuous Countercurrent Multistage Ion Exchange Contactor.

Modelling a stagewise countercurrent ion exchange column is rather more difficult and frequently requires simplifying assumptions in order to obtain a realistic mathematical treatment. It is tempting to try and use the theories developed elsewhere in mass transfer operations for pseudo steady-rate processes. Treybal [37] presents an excellent approach for distillation and liquid-liquid extraction and it is relatively straightforward to adapt this procedure to ion exchange. For example, Wesselingh et al. [38] have simulated a multi-stage fluidised bed reactor using a pseudo steady-state approach and the McCabe-Thiele method of calculation. The concept of stage efficiency has been used to compensate for deviations from equilibrium stage behaviour, though this procedure will become progressively more inaccurate as the system deviates from steady-state.

The steady-state approach permits compositions to be calculated using overall and local material balances (see Figure 2.17)

$$Sx_0 - Sx_n = Ry_1 - Ry_{n+1}$$

$$Sx_0 - Sx_1 = Ry_1 - Ry_2 .$$

Assuming that the residence time of the two phases in any one stage is insufficient to equilibrate the phases, then a stage efficiency can be defined as the fractional approach to equilibrium

$$E_{MR} = \frac{y_{n+1} - y_n}{y_{n+1} - y_n^*} .$$

The liquid and the resin are assumed to be well-mixed in each stage so that the concentrations are the same everywhere. In this case the Murphree resin phase efficiency can be written in terms of the overall number of transfer units based on the resin phase, N_{OR}.

$$E_{MR} = \frac{N_{OR}}{1 + N_{OR}}$$

where

$$N_{OR} = \frac{k_{OR}a}{R};$$

k_{OR} is the overall mass transfer coefficient based on the resin phase. This approach requires a definition of the overall mass transfer coefficient. The simplest method is to assume the validity of the 'two film theory' by attributing mass transfer resistance to a liquid film mass transfer coefficient (k_L) and to some apparent 'resin phase' film mass transfer coefficient (k_R). Treybal [37] gives information for the prediction of the overall mass transfer coefficients, provided the equilibrium isotherm is linear.

The simulation of periodic fluidised bed contactors on a laboratory scale has been published by Streat et al. [26, 34, 35] and more recently by Ford [24] who was able to simulate the performance of a 2.5 m diameter pilot-plant column for the recovery of uranium from an aqueous sulphuric acid solution. A periodic reactor must be simulated with an array of non-steady-state material balance equations, since the resin phase is not in continuous flow and therefore a conversion distribution function based on a residence – time distribution of particles from stage-to-stage is required. The simulation routine requires as input the appropriate equilibrium relationship and a suitable rate equation. Uranium recovery is complex and therefore both the equilibrium data and kinetics must be simplified. The former can be done by using a Langmuir type isotherm or an empirical fit of the data. The latter uses a quadratic driving force model which seems to be adequate for a predominantly resin-phase controlled process.

It is also necessary to make hydrodynamic assumptions about the resin and solution at each stage. The most common first approach is to assume perfect mixing of both the resin and solution at each stage. This is pessimistic, since it is entirely likely that the solution is in plug flow if the distributor plates are well designed. Since it is difficult to predict the hydrodynamic behaviour in a large column, say 5 m in diameter, the assumption of perfect mixing in both phases is a useful ground rule.

Consider the notation for a multistage countercurrent fluidised bed column given in Figure 2.17. Ford [25] has adopted a momentary material balance calculation from stage-to-stage assuming perfect mixing in each stage. The resin mass balance is given by

$$y_j(t + \Delta t) = y_j(t) + r_j(t) \cdot \Delta t$$

where r_j is the reaction rate in stage j and Δt is the time increment for reaction. The solution material balance is given by the following expression

$$x_j(t + \Delta t) = x_j(t) + \frac{1}{V_s}\{S(x_{j-1}(t) - x_j(t)) \cdot \Delta t - V_R r_j(t)\}$$

where V_R and V_S are the quantities of resin and solution in each stage respectively and S is the solution flow-rate.

The procedure assumes that when the column is put into reverse flow, an increment of loaded resin is transferred countercurrent to solution and that the proportion of resin and solution transferred is the same from stage-to-stage. A computer programme solves the array of material balance calculations, if the input data is supplied, i. e., x_0, y_{n+1}, R, S, and the barrens concentration x_n. The output will give the interstage compositions and a prediction of the number of stages n, together with resin loading, y_1. The fitted data presented by Ford is in excellent agreement with actual column trials.

The model presented by Streat and Takel [25] is rather more complex since this gives the rate of change of solute concentration in the liquid phase in the form of a differential equation and the change of solute concentration in the resin phase with a partial differential equation based on the conversion distribution of the resin associated with a precise residence time distribution function. These equations can be related to Figure 2.17 as follows:

$$\frac{dx_j}{dt} = \frac{1}{\tau_j}(x_{j-1} - x_j) - \frac{(1 - \varepsilon_j)}{\varepsilon_j}\int_0^1 f_j(\bar{y}, t)\, r(\bar{y}, x_j)\, d\bar{y}$$

$$\frac{\partial f_j(\bar{y}, t)}{\partial t} + \frac{1}{\bar{C}} r(\bar{y}, x_j)\frac{\partial f_j(\bar{y}, t)}{\partial \bar{y}} = \frac{1}{\bar{C}} f_j(\bar{y}, t)\, \delta = \frac{\partial}{\partial \bar{y}} r(\bar{y}, x_j).$$

No analytical solution has been presented for this set of equations but a number of solution techniques have been tested [34]. Discretisation of the resin conversion distribution function allows the partial differential equation to be solved over discrete levels or bands. Discretisation of the above equations gives

$$\frac{dx_j}{dt} = \frac{1}{\tau_j}(x_{j-1} - x_j) - \frac{(1 - \varepsilon_j)}{\varepsilon_j}\sum_{i=1}^{N} p_j(\bar{y})\, r(\bar{y}, x_j)$$

$$\frac{d\bar{y}}{dt} = \frac{1}{\bar{C}} r(\bar{y}, x_j) \quad \text{for all} \quad i = 1 \ldots (n + 1).$$

These differential equations are solved simultaneously over the contactor from stage-to-stage and the initial conditions of each cycle are dependent on the final conditions of the previous cycle [26]. This model has been used to test small-scale pilot plant experiments performed by Davy McKee on the recovery uranium from aqueous sulphuric acid.

The model was used to simulate a small-scale pilot plant comprising eight stages each of 7 cm inside diameter and 30 cm height. A comparison between the computer prediction for cycle 24 after steady-state resin hold-up had been achieved is given in Figure 2.18. Backmixing of solution during resin transfer was assumed to involve

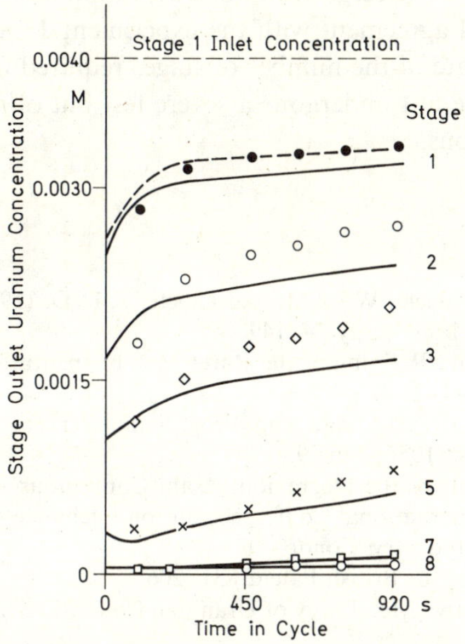

Figure 2.18 Solution concentration profiles in small-scale Cloete-Streat Contactor [26].

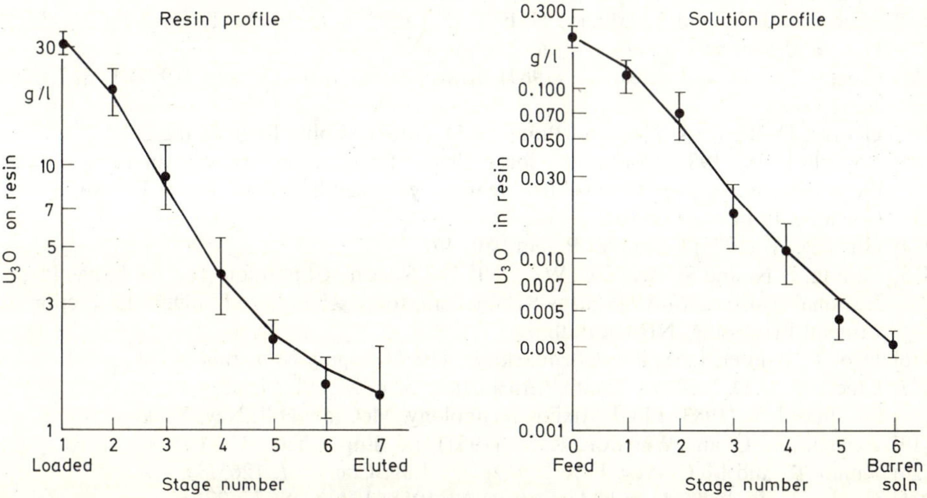

Figure 2.19 Column profiles in NIMIX Contactor [25]. (Reproduced by permission from "Ion Exchange Technology" – Fig. 3, p. 676 by M. A. Ford – published for the SCI by Ellis Horward 1984 [© SCI/Ellis Horward].)

the movement of solution from a stage to the stage below. This means that solution concentration in stage j at the beginning of a cycle is equal to the concentration in stage $j + 1$ at the end of the previous cycle. The quadratic driving force expression gives a good simulation of uranium recovery in pseudo steady-state operation.

Ford [25] has also given observed and predicted column profiles for uranium sorption (see Figure 2.19). Despite the large variations in uranium concentration, the column simulation is in good agreement with the experimental measurements and does give a reasonable estimate of the number of stages required for recovery.

These process simulations have not undergone a severe test but offer a realistic basis for process design calculations.

References

[1] Haines, A. K., Tunley, T. H., Te Riele, W. A. M. and Cloete, F. L. D. (1973), J. South African Institute of Mining and Metallurgy *74*, 149.

[2] Arden, T. V. (1957) Extraction and Refining of the Rarer Metals, Institution of Mining and Metallurgy, No. 8, pp. 119–142.

[3] Higgins, I. R. and Roberts, J. T. (1954) Chem. Eng. Prog. Symp. Ser. *50*, 87.

[4] Editorial Article, Chemical Week (1956) June 9.

[5] Bouchard, J. (1969) Development of the Degremont-Asahi Continuous Ion Exchange Process. In: Proceedings of an International Conference on Ion Exchange in the Process Industries, Society of Chemical Industry, London 91.

[6] Porter, R. R. and Arden, T. V. (1956) British Patent 831, 206.

[7] Merritt, R. C. (1971) The Extractive Metallurgy of Uranium, Colorado School of Mines Research Institute.

[8] Arden, T. V., Davies, J. B., Herwig, G. L., Stewart, R. M., Swinton, E. A. and Weiss, D. E. (1958) Proc. 2nd UN Conf. on Peaceful Uses of Atomic Energy *3*, 396.

[9] Swinton, E. A. and Weiss, D. E. (1953) Australian J. Appl. Sci. *4*, 316.

[10] Michenor, J. W. and Lundberg, H. E. (1956) in: Ion Exchange Technology (Nachod, F. C. and Schubert, J., eds.), Academic Press.

[11] Cloete, F. L. D. and Streat M. (1967) British Patent 1,070,251 and (1970) U.S. Patent 3,551,118.

[12] George, D. R., Ross, J. R. and Prater, J. D. (1968) Mining Engineering *20*, 1.

[13] Boydell, D. W. (1981) Continuous ion exchange for uranium recovery; an assessment of the achievements over the last five years. Hydrometallurgy *81*, paper E1, Society of Chemical Industry, London.

[14] Himsley, A. (1973) Canadian Patent 980, 467.

[15] Vernon, P. N. and Sylvester, C. W. (1980) The Rossing Continuous Ion Exchange Plant, National Institute for Metallurgy Symposium, Ion Exchange and Solvent Extraction in Mineral Processing, NIM Randburg.

[16] Izzo, T., Painter, L. A. and Cheminski, R. (1957) Engng. Min. Jl. *158*, 90.

[17] Cloete, F. L. D. (1981) J. South African Inst. Min. Metall. *81*, 66.

[18] Oldshue, J. Y. (1983) Fluid Mixing Technology, McGraw Hill, New York.

[19] Bauman, W. C. and Wheaton, R. M. (1951) Ind. Eng. Chem. *43*, 1088.

[20] Kunin, R. and McGarvey, F. X. (1949) Ind. Eng. Chem. *41*, 1265.

[21] Gregor, H. P., Belle, J. and Martins, R. A. (1955) J. A. C. S. *77*, 2713.

[22] Streat, M. and Brignal, W. J. (1970), Trans. Instn. Chem. Engrs. *48*, T151.

[23] Streat, M., Brignal, W. J. and Gupta, A. K. (1976) The Theory and Practice of Ion Exchange, Society of Chemical Industry, London.

[24] Helfferich, F. G. and Klein, G. (1970) Multicomponent Chromatography, Marcel Dekker, New York.
[25] Ford, M. A. (1984) in: Ion Exchange Technology (Naden, D. and Streat, M., eds) p. 668, Horwood, Chichester, UK.
[26] Streat, M. and Takel, G. N. J. (1982) Hydrometallurgy *8*, 101.
[27] Helfferich, F. (1962) Ion Exchange, McGraw Hill, New York.
[28] Vermeulen, T. (1953) Ind. Eng. Chem. *45*, 1664.
[29] Snowdon, C. B. and Turner, J. C. R. (1967), in: Proc. Internat. Symp. Fluidisation (Drinkenburg, A. A. H., ed.), p. 559, Netherlands University Press, Amsterdam.
[30] Rahman, K. and Streat, M. (1981) Chem. Eng. Sci. *36*, 301.
[31] Perry, R. H. and Green, D. (1985) The Chemical Engineers Handbook, Section 17, McGraw Hill, New York.
[32] Michaels, A. S. (1952) Ind. Eng. Chem. *44*, 1922.
[33] Baddour, R. F. and Gilliland, E. R. (1953) Ind. Eng. Chem. *45*, 330.
[34] Dodds, R., Hudson, P. I., Kershenbaum, L. S. and Streat M. (1973) Chem. Eng. Sci. *28*, 1233.
[35] Gomez-Vaillard, R., Kershenbaum, L. S. and Streat, M. (1981) Chem. Eng. Sci. *36*, 306.
[36] Gomez-Vaillard, R. and Kershenbaum, L. S. (1981), Chem. Eng. Sci. *36*, 319.
[37] Treybal, R. E. (1980) Mass Transfer Operations. Third Edition, Chapters 5 and 10. McGraw Hill, New York.
[38] Van der Meer, A. P., Woerde, H. M. and Wesselingh, J. A. (1984) Ind. Eng. Chem. Process Res. Dev. *23*, 660.

2.2 Raw Water Treatment by Ion Exchange

Thomas V. Arden
Consultant, Cobham, Surrey, Great Britain

Introduction

Since the invention of ion exchange resins in 1934, their industrial use has been closely associated with water treatment, which, in spite of the wide range of other important applications, remains the biggest user of exchangers, consuming about 90% of all materials produced. While the exchange process was first used as a more convenient or efficient replacement for existing treatment methods, it has developed far beyond them, and is now the only way in which there can be produced the highly pure water which is essential for the electrical power generation and electronic component industries.

Before that date, water treatment had followed two separate paths for potable and industrial purposes. Drinking water treatment was practised by the ancient Greeks, Romans and Egyptians, who used filtration through sand beds to remove turbidity, and give a more attractive product for drinking. It is probable that without knowing it they also achieved a considerable degree of bacterial purification. Filtration remained the only treatment process for potable water until the end of the nineteenth century, when disinfection by hypochlorite was introduced in Europe and the U.S.A., to combat the pollution caused by industry and rapidly increasing population. As the early ion exchange materials, the zeolites, resembled sand, the design of early ion exchange plants was based closely on that of existing sand filters. While this assisted the start of a new technology, it led to some features of filter design inappropriate to ion exchange equipment being retained for a very long period, and features that can still be found in many plants today.

The second historical trend was treatment of water for industry. Until 1840, industries requiring large quantities of water with particular characteristics were established only in those areas where the natural supplies were suitable. Thus, in Great Britain, a major cotton-processing industry, requiring large supplies of soft water with low dissolved solids became located in the north-west of England, where the water is of this type. During the nineteenth century, there was a massive growth of industries which, while not using water as a raw material, nevertheless required it as a heat transfer medium. It was heated in boilers to be used for space heating, process steam, the power source for steam engines, and in due course electric power generation. Industry could no longer be sited only where suitable water supplies were available, and there were severe problems of scale formation on boilers and piping in hard water areas. In 1841 the first system to improve the chemical constitution of water supplies came into action. This was the lime softening process, developed by Clark in Scotland [1]. A suspension of calcium hydroxide was added to the water in sufficient quantity to convert bicarbonate ions to carbonate, thus causing precipitation of calcium and magnesium carbonates, and giving partial softening of the water. Later, the process was modified to become the lime-soda process, which caused substantially complete removal of hardness from the water supply, and rendered it suitable for many industrial purposes.

Lime softening became widespread in the second half of the nineteenth century, and is still used to-day. Although in 1848 Thompson [2] discovered the ion exchange phenomenon, and his collaborator Way [3] developed the key laws of the subject between 1850 and 1854, it was not used for water softening until 1905, when Gans [4] in Germany first softened water on an industrial scale by passing it through beds of natural or synthetic sodium aluminium silicates, which replaced calcium and magnesium ions in the water with sodium. The mineral was subsequently regenerated by the passage of sodium chloride solution. The "base exchange" process, as it was originally known, became universally used. It did not completely replace lime softening, but by 1935, that process was confined to a small number of very large installations, while the more convenient ion exchange units were increasingly being installed by water users in many industries.

While softened water was valuable for many industrial uses, its total ionic content was the same as that of the original raw water. For many purposes, particularly in the rapidly growing chemical industry, there was a requirement for distilled water, whose production cost was high. From 1930 onwards, many workers realised that if the cation exchange process could use hydrogen instead of sodium, then all metallic salts would be converted to corresponding free acids. If there were available an anion exchanger containing hydroxide ions, then the acids formed by cation exchange could be absorbed, and the equivalent of distilled water would be produced at a potentially much lower cost. The zeolites, aluminium silicates, were unstable to acid, and hydrogen exchange was therefore impossible. No anion exchange materials existed. The demineralising or deionising process remained a theoretical concept until 1934, when two developments occurred simultaneously. First, Smit [5] and Liebknecht [6] separately produced an excellent cation exchange material, by sulphonating granular coal. The product was physically and chemically stable, and was capable of carrying out hydrogen ion exchange. In the same year, Adams and Holmes [7], in England, developed the first synthetic ion exchange resins, both cation and anion exchangers. The cation materials were phenol formaldehyde condensation products, into which sulfonic groups were introduced, the anion resins being polyamine formaldehyde condensation compounds. Both were being produced in industrial quantities within the next two years, and the first demineralising plant went into operation in 1937.

The early condensation resins were reasonably chemically stable and, in spite of their granular form, were sufficiently sturdy mechanically to serve in early plants, operating at fairly low flow rates. Users rapidly started to demand materials which would withstand much more severe conditions, and in 1945 and 1949, D'Alelio [8] and McBurney [9] produced the first polymeric materials, crosslinked polystyrene sulphonic acid, and quaternary ammonium resins. Produced by suspension polymerisation in the form of perfect spheres, these materials combined mechanical strength with chemical stability and high capacity. Together with the parallel range of polyacrylic exchangers produced later, these materials have permitted the production, from any source, of all classes of treated water up to the ultra-pure product required for power generation and silicon chip manufacture.

Water supplies. By general convention, a chemical analysis of a raw water sample is shown in a form similar to that of Table 2.10. For many purposes a separate sheet is added giving figures for bacteria and algae. For the design and operation of ion exchange plants, the analysis sheet can be greatly simplified, with only the following items being significant.

Interfering items. Suspended solids and iron figures do not influence the design of the ion exchange plant itself, but if they are high (over 0.1 mg/l Fe, or 1.0 mg/l suspended solids) they may require a clarification and filtration step ahead of the ion exchange units. O. A., i. e. oxygen absorbed in 4 h at 27 °C, which may also be recorded as "permanganate value P. V." gives a measure of the natural humic and

Table 2.10 Water analysis sheet. Items needed for plant calculations indicated as X

Received from

Lab no
File no
Date

Date of sampling
Source
Use

Appearance
Smell
Turbidity (silica scale)
Colour-hazen
Conductivity-microsiemens per cm at 20 °C
pH

	ppm					
Equivalent mineral acidity (EMA)	$CaCO_3$			C	D	E
Total alkalinity (alk. m)	$CaCO_3$		B	C		E
Phenolphthlein alkalinity (alk. p)	$CaCO_3$					
Free mineral acidity (FMA)	$CaCO_3$					
Free carbon dioxide	$CaCO_3$					E
Bicarbonate	$CaCO_3$					
Carbonate	$CaCO_3$					
Hydroxide	$CaCO_3$					
Sulphate	$CaCO_3$				D	E
Chloride	$CaCO_3$				D	E
Nitrate	$CaCO_3$				D	E
Silica	$CaCO_3$					E
Phosphate	PO_4					
Calcium	$CaCO_3$			C		
Magnesium	$CaCO_3$			C		
Sodium	$CaCO_3$			C		
Barium	Ba					
Iron total	Fe	A				
Iron in solution	Fe	A				
Manganese	Mn					
Aluminium	Al					
Free ammonia	NH_3					
Suspended solids		A	B	C	D	E
Total dissolved solids						
Oxygen absorbed 4 hours at 27 °C	O_2				D	E
Total organic carbon (TOC)	C					
Hardness total	$CaCO_3$	A	B	C		
Hardness alkaline	$CaCO_3$					
Hardness non alkaline	$CaCO_3$					

Items needed for design of plants.
A. Softening B. H^+ exchange (weak) D. Anion exchange (weak)
C. H^+ exchange (strong) E. Anion exchange (strong).

fulvic acids contained in the raw water samples. These materials can cause poisoning of anion exchange resins, and values over 2.0 mg/l may require the use of anion exchange resins designed to resist poisoning. Higher levels, over 10 mg/l, which frequently occur in surface waters, require the removal of organic matter by alum coagulation ahead of the ion exchange process.

Sodium exchange softeners. The only figure required is the total hardness, except in the cases of waters whose salinity is over 1000 mg/l, in which case total cations and sodium are also needed, as they influence the resin capacity.

Hydrogen ion exchangers. For the calculation of these units, the first stage of the deionising system, the separate figures for calcium, magnesium and sodium are required, as they each have different affinities for the resin. The total alkalinity is needed since cations are absorbed more readily from alkaline than from neutral solutions. A valuable cross-check is given by the equivalent mineral acidity (EMA), measured by passing the sample through a small cation exchange unit in the hydrogen form, which converts all metal salts to the corresponding free mineral acids $+CO_2$. The solution is then titrated with sodium hydroxide solution to pH 3.5, measuring the mineral acids only. The total of EMA and alkalinity gives the total cations, and must therefore be equal to calcium, magnesium plus sodium. In the uncommon cases in which it is higher, the difference is usually due to potassium, which is classified with sodium for plant calculation purposes.

Anion exchange units. Weakly basic resins remove from solution only mineral acids, and therefore require a knowledge only of the mineral anions, together with EMA, serving as a cross-check. When strongly basic resins are used immediately after the cation unit, they also take out the total carbon dioxide now present, which is equal to the total of the original free carbon dioxide plus the alkalinity of the raw water (which is equal to its bicarbonate content). Carbonate and hydroxide alkalinity do not normally occur in raw water supplies, but only under artificial conditions. However, for very many water supplies, the bicarbonate content forms a high proportion of the total, and the resulting CO_2 is removed in an atmospheric degassing tower which is placed ahead of the strongly basic ion exchange unit. Under these circumstances the CO_2 and alkalinity figures are not required for anion exchange plant design, and it is normal to allow a standard figure of 15 mg/l corresponding with the CO_2 left in solution after passage through the degassing tower. One of the main purposes of the strongly basic unit is to remove silicate and silica, which are not absorbed by weakly basic resins. There is in practice the problem of deciding the value to allocate to silica in terms of $CaCO_3$ equivalent. It is absorbed by the resin partly as bisilicate, and partly as silicate, the proportions varying according to circumstances. It is prudent to assume for calculation purposes that it will all be absorbed as silicate, so that each silica atom requires two active groups on the resin, and gives the highest calculated resin requirement.

2.2.1 Typical water types

From the viewpoint of ion exchange treatment for industrial purposes, water sources may be classified as four main types. A fifth source, sea water, is too concentrated to be purified by ion exchange techniques, and when needed for industrial or potable purposes must be distilled or submitted to reverse osmosis.

Table 2.11 Typical Water Types

	A. Mountain	B Wooded Hill	C Limestone	D. Brackish Well
Ca	20	60	250	40
Mg	5	20	50	10
Na	5	40	50	650
Total	30	120	350	700
Cl	10	20	50	600
SO_4	10	20	50	30
HCO_3	10	80	250	70
Total	30	120	350	700
Organics	1	10	Surface 3–5 Well 0	0

Water A (mountain). This is typical of the spring and high level surface waters found in mountainous areas, where rain and snow fall on hard insoluble rock. In inland areas, the water contains very little dissolved matter but in coastal regions it may have traces of sea salts carried by the wind. It is soft and non-scaling, but normally low in pH, from the pick-up of CO_2, and therefore very corrosive. In Scandinavia, this water has been further modified in recent years by the acid rain resulting from the burning of fossil fuels in power stations and in motor vehicles in the region, and in surrounding countries. The sulphate content is higher and the water may have pH values as low as 4.0.

Water B (wooded hills). In many areas, such as Scotland, the same water moving down from the mountain sides through rich peaty ground, picks up more inorganic salts, and above all very high levels of organic matter, in the form of humic and fulvic acids. Water B, although not very hard, would tend to form scale in heating circuits. Its sodium content, which is high as a proportion of total cations, can cause difficulties in deionising, while the high ratio of organic matter to inorganic anions can cause severe resin poisoning. This is one of the most difficult waters to treat when the objective is the production of ultra-pure water.

Water C (limestone). This is typical of the hard water found both underground and in rivers throughout the limestone region in the South-East of England, stretching eastwards across Europe. Rainfall containing free CO_2 falls upon the ground, where bacterial degradation of vegetable matter gives rise to further dissolved CO_2. As the water passes down, it attacks limestone and dolomite to give calcium and magnesium carbonates in solution. The water is very hard, and as the bicarbonates are thermally unstable, severe scaling arises in heating circuits, with deposition of calcium carbonate and liberation of carbon dioxide. As a result, scaling occurs in the hot parts of the circuit, while corrosion takes place in other parts of the system, due to acid attack from CO_2. From the viewpoint of organic content, water C falls into two sub-types. Underground water direct from artesian wells is virtually free from organic matter, while the same water found on the surface in rivers and lakes fed from springs has moderately high organic levels. These tend to vary through the year, being highest in late autumn and early winter as leaves fall from trees and decay.

Water D (brackish well). Ground waters of this quality are found all round the Mediterranean basin, and also in parts of East Anglia and in the Netherlands, due to sea water ingress into aquifers. They are unsuitable for any industrial use except crude cooling, and must be deionised for all purposes. At the level of 700 mg/l shown in the table, ion exchange methods are suitable for this purpose, but the same general type of water with higher dissolved solids up to 5000 mg/l is found in many Middle Eastern areas, and also in the interior of Algeria and Morocco. At ionic levels higher than 700 mg/l, the cost of deionising increases rapidly; and it becomes more economical to treat the waters by reverse osmosis. This yields a product resembling water A, which is subsequently treated as such.

2.2.2 Objectives of water treatment

Scum-forming conditions. In spite of the widespread use of synthetic detergents, fatty acid soaps are still widely used in industrial laundering, and in large-scale automatic dishwashing in large hotel and restaurant complexes. Even when soap is not used in these processes, it is formed by reaction of caustic alkalis in the detergent with grease and fats in the articles being washed. To avoid precipitation of calcium stearate scum, softening is required for waters B and C. In the case of A this is not always necessary, but as the cost of softening this water is very low, it is sound policy to use the process to ensure freedom from problems. Water D represents a difficult case. With its high chloride content, this water under typical laundry and other washing conditions of varying temperatures, coupled with agitation and aeration, is distinctly corrosive to unprotected mild steel. Corrosion is increased by softening. Consequently, although the most correct solution would be to reduce or eliminate the chloride content by partial or complete deionising, it is often expedient to protect the steel work as far as possible leaving the water untreated so that some scale formation will take place and give protection to the metal surfaces.

The first stage in the processing of natural wool, wool scouring, resembles an extreme case of laundering. The wool contains considerable quantities of natural fat which are broken down to fatty acids, which react with calcium salts to form insoluble calcium soaps. Complete softening of the four waters is essential, to avoid uneven deposits of scum which would interfere seriously with subsequent dyeing. Even after softening water C and D are unattractive for wool processing because of their high content of dissolved salts. For this reason the wool industry was traditionally established in regions where the water supplies were of types A and B only. Where waters C and D are the only ones available, partial demineralising to reduce the solid contents gives satisfactory results.

Automatic glass bottle washing machines in the milk and soft drinks industries normally use hot alkaline solutions containing sodium carbonate and phosphate in the form of sprays followed by a period of draining and final drying by hot air. It is important to avoid unattractive film formation on the glass surface, and the phosphate in the detergent does reduce this action to a considerable extent, since it forms flocculent and non-adherent calcium phosphate precipitates, provided the total hardness is not too high. Waters A and B can be used without ion exchange treatment, while waters C and D must be softened. A further difficulty arises with these waters. Automatic washers dry the bottles rapidly by means of hot air, and if this is applied without adequate drainage time evaporation can form films of soluble salts. If this does occur, partial deionising of waters C and D becomes necessary.

Use of water as a coolant – scale-forming conditions. The water cooling of hot metal, as in jacketed chemical reactors or injection moulding machines for plastics, gives rise to scale formation at the high temperature point of the circuit, coupled with corrosion elsewhere due to liberated CO_2. The problem does not occur in totally closed circuit systems, but very often an open circuit is used, with new water being used for cooling, and then sent to drain. As scale formation is related only to alkaline hardness, waters A and B can be used without treatment. Water C does not require complete softening, but treatment with a carboxylic resin, giving a product containing about 100 mg/l total cations, of which 50 are calcium and magnesium, present as chlorides and sulphates, and therefore not causing scale. However, chemical equipment often provides conditions for bimetallic corrosion. Stainless-steel reactors with mild steel jackets are common, and it is advisable to de-ionise water D before use to minimise corrosion.

Process water. In the chemical industry, photographic processing, metal finishing by electrophoretic deposition, mirror silvering, haemodialysis in hospitals and many other applications, water of approximately distilled quality, containing less than 5 mg/l dissolved salts, is required. In the case of waters A to C, these results are readily obtained by simple two-bed demineralising, using a weakly basic anion resin, followed by degassing to remove free carbon dioxide. The same system can be used for water D, but the process becomes less efficient at higher levels of dissolved solids, and more complex systems are used.

Boiler feedwater. The quality of water to be used in boilers depends on the type of unit, its operating pressure, and the purpose for which it is installed.

Closed circuit heating. Any water can be used without treatment, since a light calcium carbonate scale is advantageous in minimising corrosion, and under closed conditions it does not increase.

Steam-generating boilers. Make-up water must be treated to prevent four separate problems:

a) Scale formation giving localised insulation, overheating and tube failures.
b) Corrosion of boiler tubes, leading to direct perforation, or local insulation by corrosion products.
c) Concentration of salts in the boiler, causing priming, that is carry-over of contaminated water with the steam.
d) Decomposition of bicarbonates, liberating CO_2 which causes corrosion in the condensor system and return pipes.

Classical factory boilers, many of which are still in use, overcame these problems simply. They were large in relation to their steam output, and had low steam velocities. Scale formation, CO_2 liberation and corrosion were minimised by the addition of sodium phosphate, sodium hydroxide, sodium sulphite and tannin, which together gave a voluminous sludge, which settled in the bottom of the boiler and was regularly removed by opening a valve on the underside, thus releasing some water to drain, a process known as "blowing-down". The dissolved solids in the boiler were allowed to reach high levels of over 10,000 mg/l, priming being prevented by the large steam space, which allowed droplets to fall back. Blowdown and replacement with new water then prevented further rise in the dissolved solid level. No ion exchange treatment was necessary.

Modern packaged boilers are efficient, prefabricated units available for steam generation at over 10,000 kg/h, and pressures of $2 \cdot 10^6$ Pa. They are reduced to the minimum possible size, having a low total water content, small water surface area and steam space, with high steam disengagement velocity, and extremely high heat transfer rate across the metal/water interface of the boiler tubes.

The packaged boilers are extremely efficient, but are sensitive to water conditions. To minimise priming, the total dissolved solids in the boiler water must be maintained below 2500 mg/l, and if there is a 25 × concentration factor in the boiler, the feed must be under 100 mg/l. Sludge-producing anti-scalants cannot be used, as they would block the small gaps between the fire-tubes. For waters C and D, partial or complete deionising, with the addition of anticorrosive chemicals, is

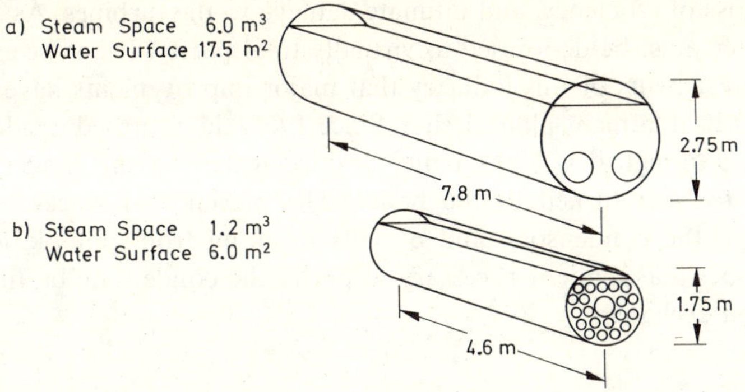

Figure 2.20 Comparison of classical (a) and packaged (b) boilers.

essential. The same may be true of waters A and B, if the proportion of returned condensate is low, giving a high concentration factor in the boiler.

High pressure boilers for power generation. The efficiency of electricity generation is in direct relation, and the cost in inverse relation, to the steam pressure, and boiler designs have constantly tended towards greater temperatures and pressures. This trend has required corresponding improvements in make-up water quality, as shown in Table 2.12. By 1950, the standards could no longer be achieved by using distilled water, and de-ionising became necessary. This was equally true of all waters, the water type influencing only the nature and the cost of the process.

In *modern very high pressure boilers*, the input water is totally converted to steam. All dissolved matter must either deposit in the tubes, causing insulation and failure, or must dissolve in the steam, to be carried over to condense on the turbine blades,

Table 2.12 Quality requirements for power generating boilers

Year	Boiler pressure (10^6 Pa)	Treatment	Water quality	
			TDS (mg/l)	Conductivity (μS/cm)
1925	1.2	Internal conditioning	350	700
1935	2.5	Softening	350	800
1945	3.5	Evaporation	5	15
1950	6.5	2-bed de-ionising	1.0	5.0
1955	10	4-bed or mixed bed de-ionising	0.3	1.0
1965	13	2-bed plus mixed bed de-ionising	0.1	0.3
1970	16	2-bed plus mixed bed de-ionising	0.03	0.1
1975	18	Counterflow 2 bed plus mixed bed	0.02	0.08
1985	22	Necessary process improvements	0.01	0.06

resulting in loss of efficiency, and ultimate damage to the turbines. As a result, the make-up water must be de-ionised to virtually total purity. It has been due to the stringent requirements of this industry that major improvements have been made in resins, and in treatment-plant design. Since 1965, the returned condensate from the turbines, although it is high-quality distilled water, has no longer been sufficiently pure to be returned to the boiler. The presence of traces of corrosion products from the condensors, and of salts resulting from pinhole leaks in the condensor tubes, has made it necessary to purify the condensate by filtration and mixed-bed ion exchange.

2.2.3 Water softening

This is the classical ion exchange process, which has led to the foundation of a complete industry. It is unique in combining high absorption efficiency with great ease of regeneration, because of the particular nature of the uni-bivalent equilibrium. The uptake of calcium and magnesium ions is strongly favoured in dilute solutions such as natural waters, while the removal of these ions is facilitated by an increase in concentration during regeneration.

Figure 2.21 shows the sodium/calcium equilibria on an 8% crosslinked sulphonic acid resin, as a function of the concentration in solution. When this is 5.0 equ/l, that is approximately 30% w/v as NaCl, the proportion of sodium ions in the resin ($\bar{X}_{Na}$) is almost identical with the proportion in solution (X_{Na}). The slight sigmoid shape of the curve is due to lack of uniformity of the polymer structure of the exchanger, which causes each ion to seek the parts of the resin most favourable to it, and therefore to have a decreasingly favourable affinity coefficient, as its occupation of the resin approaches totality.

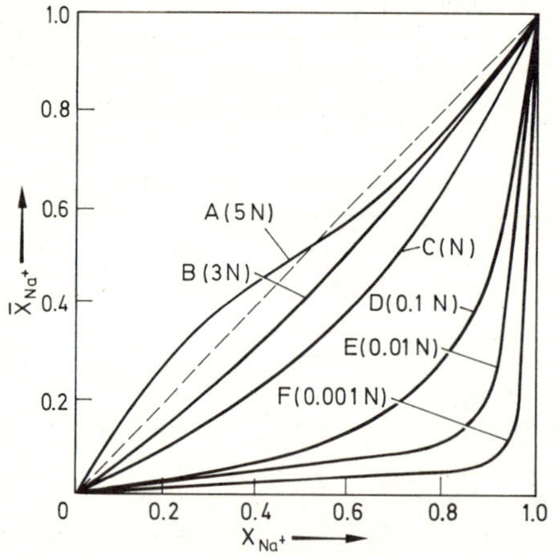

Figure 2.21 Sodium/calcium equilibrium.

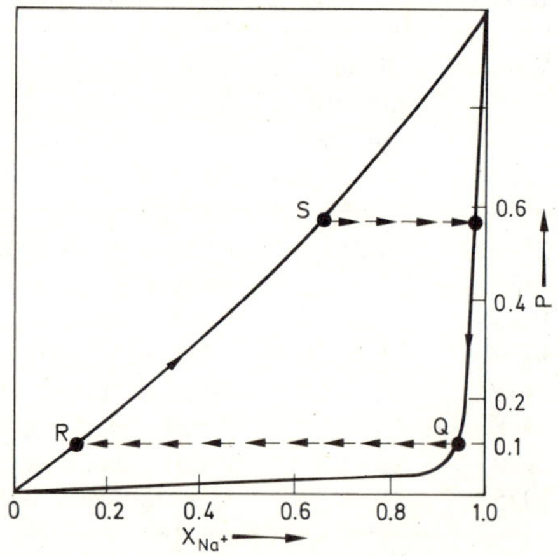

Figure 2.22 The softening cycle.

As the solution concentration decreases, the uptake of calcium is increasingly favoured, until at 0.001 N, a resin almost completely loaded with calcium will still soften water with almost total efficiency. This is illustrated in Figure 2.22.

A new resin, manufactured in the sodium form, starts its operating life at point P. As hard water is passed through it, it absorbs calcium until point Q when the resin is 90% in the calcium form, and 95% of the calcium is still being removed from solution. At this stage, the resin is taken to point R by the passage of brine at about 3 equ/l. Regeneration is carried out only sufficiently to take the resin to point S, about 55% sodium form. After rinsing out the excess brine with water, the system is at point T and all subsequent cycles follow the route TQRS. Under these circumstances, about 50% of the total capacity of the resin is used in each cycle, and the efficiency of utilisation of sodium chloride is about 70%. This figure can be improved by lowering the position of line ST, that is, by using less regenerant, with only a minor difference in treated water quality. If the water to be treated is at a higher ionic concentration, (e. g., curve E in Figure 2.21), then the shape of the operating cycle gives a lower operating capacity for the same salt consumption, together with an inferior treated water quality.

The practical effects of these equilibrium curves are shown in Figures 2.23 to 2.25. Figure 2.23 gives the operating capacity and the treated water quality as a function of sodium chloride usage, when treating a water containing 250 mg/l calcium carbonate, that is 0.005 N. The capacity curve A shows the calculated results at 100% efficiency. The operating results at the slow flow rate of 0.1 bed

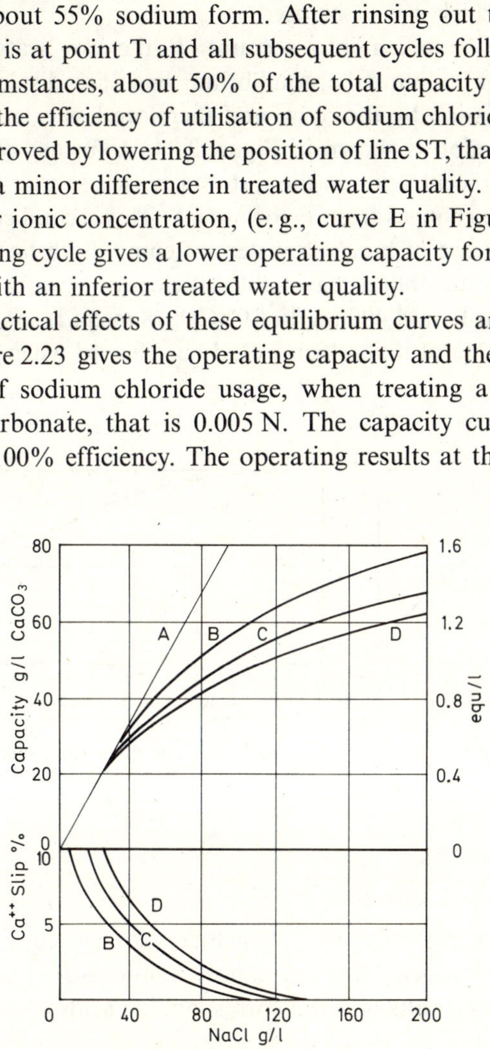

Figure 2.23 Softening. Capacity and quality.

volumes per minute (curve B) show almost theoretical salt usage up to an input of 40 g/l, with a hardness leakage of only 3%, which is acceptable in a high proportion of cases. However, this low flow rate implies a large and therefore expensive plant, and operating under the conditions of curve D reduces the plant size by a factor of 8 times. Under these circumstances it may be necessary to increase the regeneration level, in order to keep the leakage to an acceptable figure, and in practice a plant designer must choose between a range of possible solutions, in order to find the one which gives the most economical overall results for the particular water and the purpose for which it is to be used.

Two other factors must be taken into account. If the raw water contains significant concentrations of sodium, the equilibrium lines are displaced in the direction of less sodium emission by the resin, and therefore less calcium uptake. The resultant capacity reduction is shown in Figure 2.24. A further correction is required as the total ionic concentration in the water increases, so that the equilibrium line for calcium uptake changes from F towards A. This second correction is given in Figure 2.25. In the case of brackish water softening, both factors act together, but it is still possible to obtain reasonable efficiency. A water containing 1870 mg/l of NaCl, (32 mequ/l), and 400 mg/l of $CaCO_3$ (8 mequ/l) shows, for any given salt level, a capacity 72% of that which would be achieved when softening a simple water.

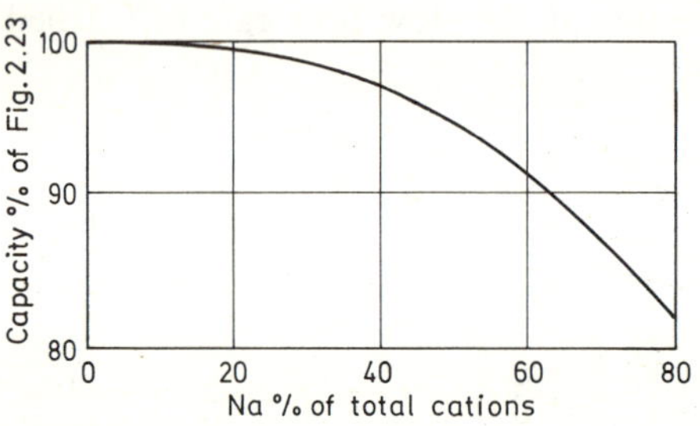

Figure 2.24 The effect of sodium.

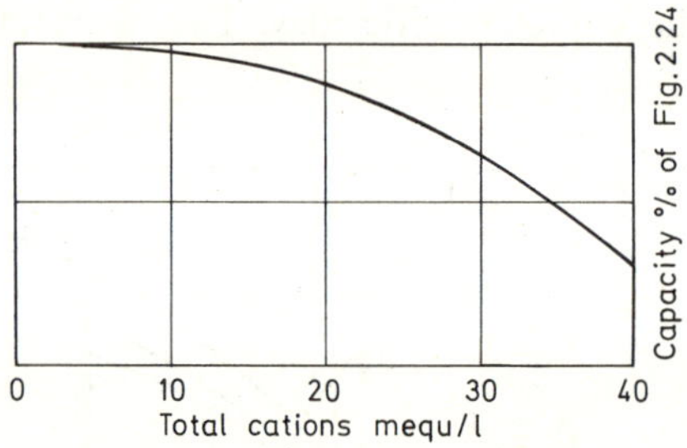

Figure 2.25 The effect of concentration.

2.2.4 Dealkalisation

The carboxylic cation exchangers, acrylic and methacrylic acid copolymers, are weakly acidic, and in the hydrogen form are almost non-ionised. They are thus incapable of converting mineral salts into the corresponding acids

$$R^-H^+ + NaCl \rightarrow \text{no action}$$
$$2R^-H^+ + CaCl_2 \rightarrow \text{no action.}$$

In the case of bicarbonate solutions, there is a marked difference between the reactions with sodium and calcium ions. The CO_2 liberated by the action of hydrogen-form resin reduces the pH sufficiently to prevent further sodium absorption, after a minor proportion of the resin capacity has been used. Calcium and magnesium, however, have such strong affinities for these resins that there is a more favourable equilibrium, and efficient uptake occurs:

$$R^-H^+ + NaHCO_3 \rightleftharpoons R^-Na^+ + H_2O + CO_2$$

$$2R^-H^+ + Ca(HCO_3)_2 \rightleftharpoons R_2^-Ca^{++} + 2H_2O + 2CO_2.$$

As a result, in a water containing both hardness and bicarbonate, such as water C, the use of a carboxylic resin removes calcium and magnesium equivalent to the bicarbonate present, thus giving in one stage a partially softened, non-scale-forming water, with reduced total dissolved solids. The liberated CO_2 is removed from the solution in an atmospheric degasser tower, and the product is suitable for such purposes as feed to packaged boilers. A further improvement in quality is obtained by softening the dealkalised water by sodium exchange. The results are shown in Table 2.13.

Table 2.13 Results of dealkalisation and softening of Water C

	Raw water	After dealkalising	After softening
Ca	250	15	0
Mg	50	50	0
Na	50	50	115
Total	350	115	115
Cl	50	50	50
SO_4	50	50	50
HCO_3	200	15	15
Total	350	115	115

Those waters (which are uncommon) in which the bicarbonate exceeds the total hardness, are almost fully softened by the dealkalisation process, yielding a product containing sodium salts only, with the excess bicarbonate remaining in solution. There is a small difference, in this connection, between the methacrylic and acrylic resins. The former are such weak acids that virtually no sodium is removed from bicarbonate solution, while the acrylic materials will remove some sodium under these conditions, provided that the whole of the calcium and magnesium, which are preferentially absorbed, have been removed by the resin higher in the column. Both types will absorb sodium from carbonate or hydroxide solutions, which do not occur in water treatment practice, but are found in special industrial processes.

The weakness of the carboxylic groups is such that sodium-form resin can be converted to hydrogen-form, albeit slowly, by the passage of deionised water only, a solution of caustic soda emerging. Calcium-form resin does not hydrolyse in this way, but conversion to hydrogen-form by means of dilute mineral acids takes place at virtually 100% efficiency, with the spent regenerant emerging as a neutral solution. As a result, the total quantity of ions removed in each cycle is exactly equivalent to the quantity of acid used for regeneration. This simplifies the calculation of dealkalisation plants, but there is an opposing extra problem, as compared with sulphonic resins. The hydrogen form of the weakly acidic resins is almost completely undissociated, the hydrogen being joined by coordinate links, whose rupture is a slower process than ionic reactions. These resins are therefore slow in action, and the operating capacity has an inverse relationship with the rate of presentation of ions, and therefore a direct relationship with the total time of the service cycle.

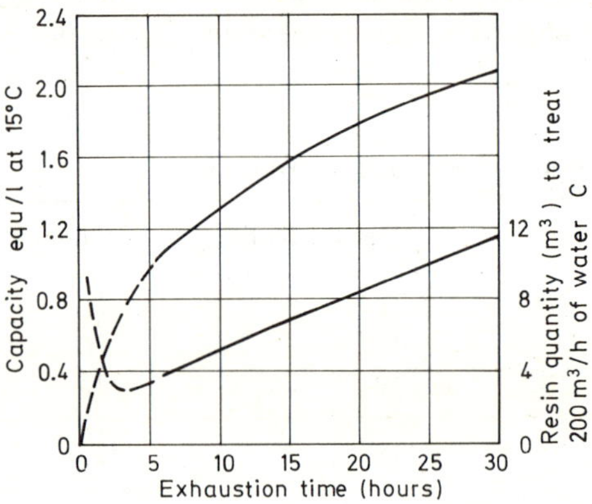

Figure 2.26 Carboxylic resins. Variation of capacity with cycle time.

This is shown in the upper curve of Figure 2.26, which gives the capacity, and therefore the regenerant quantity (which are identical), as a function of cycle duration. The lower curve gives the calculated quantity of resin to treat 200 m^3/h of water C (alkalinity 200 mg/l). The dominating rate effect gives the anomalous result that a higher capacity requires the use of more resin. As the acid consumption per unit volume treated is constant, irrespective of flow rate and cycle time, it is advantageous to use the minimum practical value of about 5 h, below which results become erratic.

2.2.5 Deionising. General considerations

The basic principles of the de-ionising process are simple. Passage of a solution through a hydrogen-form cation exchanger causes conversion of all salts to the corresponding acids, after which the solution enters a hydroxide-form anion resin, where the acids are absorbed, leaving pure water. The exchange equilibria, and the mass-action equations related to them, are straightforward, and the understanding of the process kinetics is well-advanced. Nevertheless, it is not possible to predict breakthrough curves and calculate plant dimensions with any degree of accuracy, from basic theory, for reasons imposed by practical conditions, as described in the following paragraphs.

Non-uniformity of resin structure [10–12]. During the polymerisation stage of resin manufacture, regions of tight and loose crosslinking are formed. A single resin bead thus acts as a mixture of materials, with different characteristics. As a result, "equilibrium constants" are not constant, but vary according to the ionic content of the resin. As any ion is being absorbed, the process becomes increasingly difficult as it approaches completion.

Solution complexities. Ion exchange theory is largely limited to binary and ternary systems, and therefore applies reasonably well to water softening and dealkalising, both of which can be simplified to treatment as binary equilibria of bivalent "hardness" ions with univalent sodium or hydrogen. De-ionising is more complex. The capacity and treated-water quality figures for the cation resins are governed by the calcium, magnesium, sodium and alkalinity figures in the raw water, each in separate relation to the regeneration level. It is impossible for resin suppliers to provide all the required information in one table or diagram, and the information sheets for each resin contain many pages of graphs, for different sets of conditions. The data given are produced from laboratory scale tests, and must be used with understanding for plant design. In particular, many information sheets refer to coflow conditions, and not all manufacturers provide counterflow figures.

The conditions occurring in the anion exchange unit are even more complex. A natural water supply contains silicate, bicarbonate, chloride, and sulphate, often with some nitrate and phosphate as well. Of these, only chloride and nitrate enter the resin in a simple form. Typically, dilute sulphuric acid emerging from the cation exchanger is taken up at the top of a hydroxide-form resin as sulphate. Each sulphur atom thus occupies two active sites. As the process continues, sulphate-form resin at the top of the column is converted to bisulphate, with each sulphur taking up only one site. At the point of breakthrough, the column top is in this form, while the bottom is still in the sulphate condition. The apparent capacity is thus greater than is calculated on the assumption of sulphate absorption. The same principles apply to the carbonate-bicarbonate, the silicate-bisilicate, and the triple phosphate systems. Existing theoretical studies cannot resolve these complexities, and the operating information provided by resin manufacturers results from empirical tests.

Incomplete conversion. To convert resins fully into the hydrogen and hydroxide forms is virtually impossible, and even to attempt 90% regeneration would require extremely wasteful excesses of regenerant. In practice, input of regenerant is restricted to quantities which convert less than half the resin to the required form.

While giving a reasonable regenerant usage, this procedure can give a poor quality treated water, with severe leakage of ions which should be removed by the resin. In early de-ionising practice, incomplete removal was overcome by using several cation/anion pairs in series, each taking out 80–90% of the ions presented to it. Plants of this type are still in use, but the combination of counterflow regeneration and mixed bed units has now made it possible to produce water of the highest quality with only three units.

Failure to reach equilibrium. In contrast with the slow reaction rate of the carboxylic resins, the highly ionised sulphonic and quaternary ammonium resins have very favourable kinetics. At flow rates of less than 10 bed volumes per hour, and an average water supply containing less than 50 mequ/l total salts, these resins show little rate sensitivity. The necessity in recent years of reducing capital costs has led to smaller plants, with low contact times and short operating cycles. Under these conditions, both capacity and water quality may be inferior to those quoted in the data books of the resin suppliers, not all of whom take flow rate problems into account.

2.2.6 Deionising. Hydrogen exchange with sulphonic resins

In comparison with water softening by calcium/sodium exchange, the hydrogen exchange process has two additional difficulties.

a) The remarkable efficiency of the softening process is due to the uni-bivalent equilibrium favouring calcium absorption in dilute solution. The Ca^{2+}/H^+ relationship is even more favourable to calcium absorption in dilute solution, and the de-ionising of a water which contained only calcium salts would be very efficient (Figure 2.27). Unfortunately, no such water exists and even the very hard Water C contains 50 mg/l of sodium.

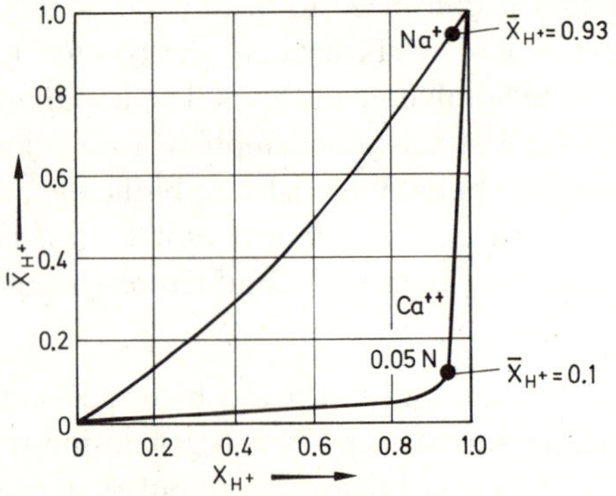

Figure 2.27 Uni-univalent and uni-bivalent equilibria.

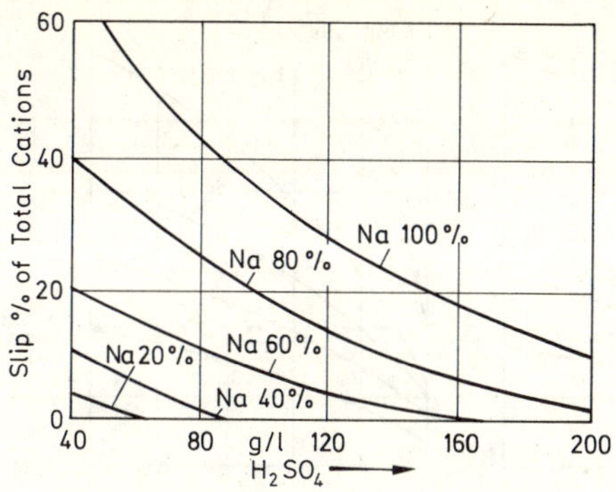

Figure 2.28 Hydrogen ion exchange. Sodium leakage with coflow regeneration.

The Na^+/H^+ relationship is totally different. K_H^{Na} is about 1.5, almost independent of concentration, and the equilibrium line, also shown in Figure 2.27, is unfavourable to sodium absorption. In order to remove 95% of the sodium from the solution, the resin must be regenerated to 93% hydrogen form, requiring an enormous excess of acid. In mixed calcium/magnesium/sodium solutions, such as normally occur, the leakage has a complex relationship with the water analysis and the regeneration level. There is considerable disagreement between resin suppliers, whose information sheets quote widely differing figures for sodium slip under identical conditions, but all show it to be high at low regeneration levels. (Figure 2.28)

b) The affinity of hydrogen ion for the resin is much less than that of any other ion, so that removal of metals from the resin is more difficult than in the case of softening, and the capacity is correspondingly lower. Figure 2.29 (a) is the first in a family of diagrams, each relating to a different magnesium proportion, but all similar in principle. With high sodium water, the maximum capacity, about 1.5 equ/l, is only slightly less than that found in softening (1.6 equ/l) but it is achieved at the expense of a very high acid usage, and a serious sodium slip. With hard waters containing only 20% sodium, which do give good leakage figures, the highest capacity is only 0.8 equ/l. This is partly due to the firmer bonding of calcium to the resin, and additionally, in the case of sulphuric acid, because it is necessary to use a dilute solution (1.5% w/v, or 0.3 equ/l) to avoid precipitation of calcium sulphate. Figures 2.29 (b) to 2.29 (d) show the increasing capacities as the Mg/Ca ratio in the raw water rises, resulting both from the lower affinity of magnesium and the use of more concentrated acid.

These problems are reduced by using hydrochloric acid, which can be employed at concentrations of up to 1.5 equ/l without damage to the resin. The overall system is less complex, (Figure 2.30), results with hard waters being better than with sulphuric acid, and not significantly affected by the Mg/Ca ratios. The figures for soft waters are only marginally better than those of Figures 2.29.

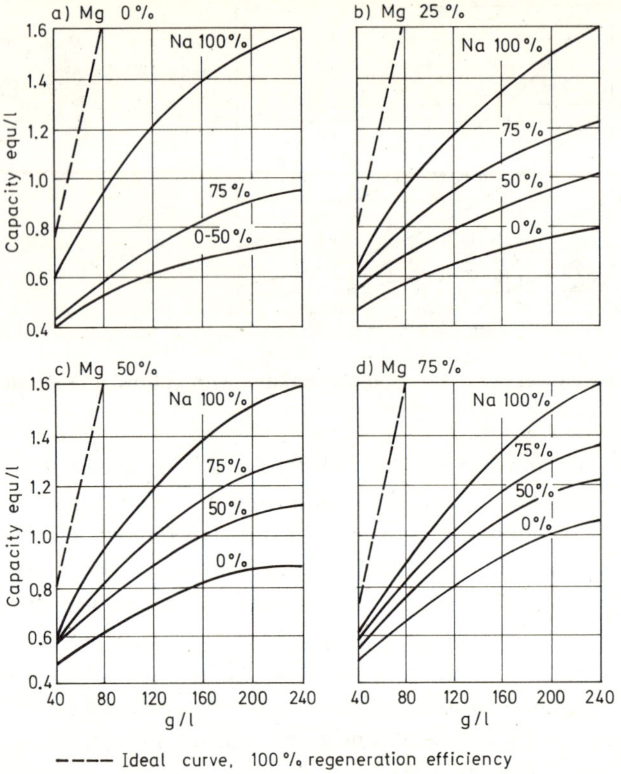

Figure 2.29 Operating capacities with sulphuric acid regeneration.

There are thus two opposing problems. Hard, low sodium waters give good treated water quality, but poor capacities and correspondingly inefficient usage of regenerant. Soft, high sodium waters yield higher capacities, and regeneration efficiencies, at low acid levels, but the sodium slip is severe. To overcome this problem, regenerant quantities must be increased to uneconomic levels, showing even lower usage efficiencies than in the case of hard waters.

The terms "slip" or "leakage", although universally used, are misleading. The sodium ions which emerge from the resin during any cycle are not those which have failed to be absorbed, but are those which were taken up during the previous cycle and left in the lower part of the column, to be displaced by the hydrogen ions leaving the resin at the top as the new cycle commences. Figure 2.31 shows the sodium slip, throughout the operating cycle, with two water types, at various regeneration levels. The treated water quality improves throughout the cycle, as the "old" sodium is removed from the resin, and then deteriorates as the resin becomes exhausted. These wide variations in apparent slip mean that the values recorded in Figures 2.28 and 2.30 have little significance in cases where they exceed 10%.

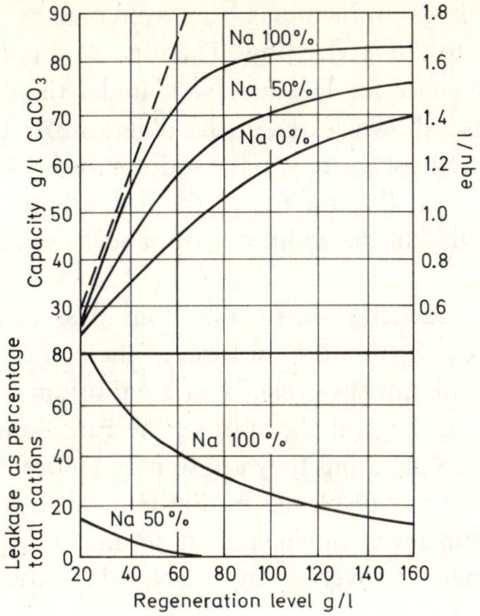

Figure 2.30 Operating capacities and quality with hydrochloric acid.

Regenerant level (g/l)		Sodium, as percentage of neutral salts
A	64	100
B	64	70
C	144	100
D	224	100
E	144	70
F	224	70

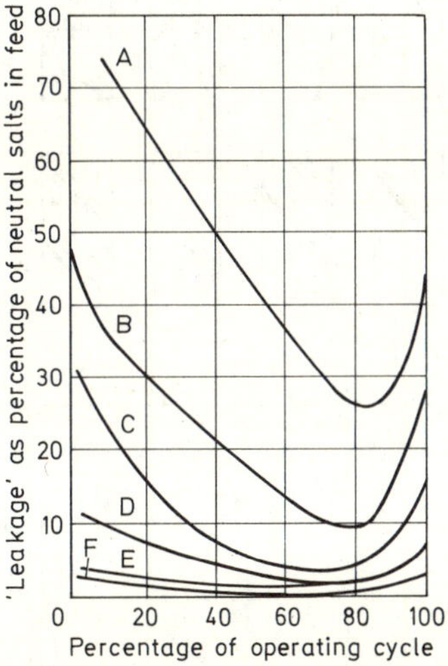

Figure 2.31 Variation in sodium leakage, throughout the cycle.

A technique which can be of some value in these cases is to mix the resin bed with air after regeneration and rinsing. This does not reduce the average or the total leakage, but makes it fairly uniform throughout the operation, thus reducing problems of control, and presenting a reasonably constant water to the subsequent unit. The method does not, however, reduce the real problem of sodium slip.

Counterflow regeneration. The leakage problem has been overcome by regenerating the resin in the opposite direction to the service flow. The top section of Figure 2.32 shows the shape of the exchange curve for H^+ and Na^+, under the conditions of curve B, Figure 2.31, which is also shown lower in the diagram. At the end of a service cycle, the bed as a whole is almost entirely in the sodium form. For simplicity, Ca^{2+} and Mg^{2+}, which are concentrated at the top of the bed, are not shown. The lowest layers of resin are about 30% in the sodium form, and the emergent water shows 30% slip.

With downflow, i. e., coflow, regeneration, metallic ions pass down the bed, leaving the lowest layers about 50% in the sodium form at the end of the process. The emergent water, during the subsequent cycle, is in equilibrium with this last layer, and there is therefore 50% leakage at the start of the run. Some Ca^{2+} and Mg^{2+} ions are also present in the column, but they are so much more strongly held than sodium that they are only slightly displaced by the H^+ ions at the top, and they then re-exchange with sodium lower down, remaining in the bed. They can therefore be ignored, and the system is governed by the Na^+/H^+ equilibria.

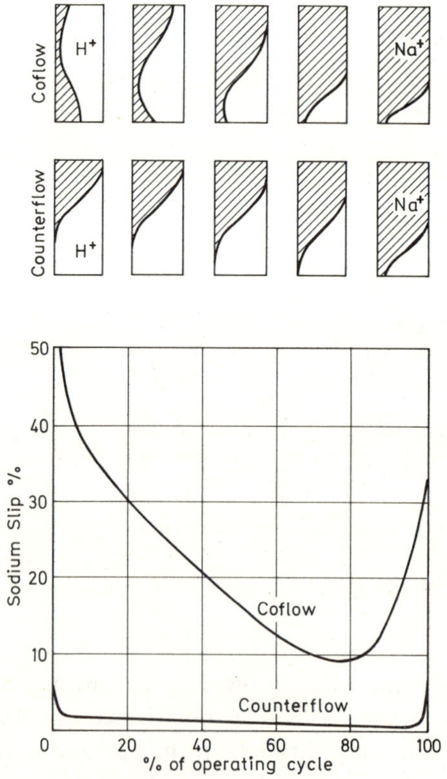

Figure 2.32 Coflow and counterflow regeneration.

Upflow, (i. e., counterflow) regeneration of the bed, with the same acid quantity, converts the same overall quantity of resin into the hydrogen form, but the residual sodium is now all at the entry into the system, while the last layers of resin, with which the emergent water is in equilibrium, are virtually 100% H^+ form. The treated water quality is good, irrespective of the raw water analysis. At any given regeneration level, the capacity and acid efficiency are somewhat improved by counterflow operation, but the difference is not great. However, the virtual elimination of sodium slip means that regeneration levels can be reduced, resulting in substantial increases in regeneration efficiency. The capacity is, of course, lower when less acid is used, balancing out the improvement given by the counterflow principle, and the total resin requirement is largely unchanged. The capital cost of counterflow units is slightly higher than corresponding coflow columns, but the reductions in raw material and effluent treatment costs resulting from improved acid usage, together with simplification of the subsequent plant permitted by the major reduction in cation load from the first unit, mean that counterflow systems are firmly established, and coflow methods should no longer be used, except in the case of very hard waters with low sodium levels, for which the disadvantages of coflow are not serious.

2.2.7 Deionising. CO_2 removal by degassing

Although not in itself an ion exchange process, a degassing tower is an essential part of the train of units, and its operation influences the results obtained in the anion exchange section. In waters such as C, with considerable proportions of bicarbonate, the CO_2 produced by the hydrogen ion stage, which remains in solution under pumping pressure, forms the majority of the anion load. By allowing the solution to pass down through a tower packed with rings or saddles of about 25 mm size, with upflow air, the CO_2 content is reduced to about 7 mg/l, equal to 15 mg/l as $CaCO_3$.

The degasser may follow a carboxylic resin column, for low pressure boiler feed, in which case there is no anion exchange stage, and the last traces of CO_2 are neutralised with sodium hydroxide. If it is situated before a strongly basic resin column, it reduces the chemical load on the unit, as well as the corresponding capital and running costs. When a weakly basic resin, which does not itself absorb CO_2, is in the circuit, the degasser is often placed after, rather than before, the unit. This is because many weakly basic resins are in practice sufficiently strong to take up some CO_2 from solution, to form the resin bicarbonate. The bicarbonate ion is then displaced by mineral anions, so that the final result is as if it had not been absorbed. Nevertheless, the presence in solution of some absorbable CO_2 increases the effective ionic concentration. As the capacity of weakly basic resins is partly dependent on concentration, the overall effect of allowing the dissolved CO_2 to pass through the column is to increase the capacity for mineral anions.

2.2.8 Deionising. Weakly basic resins

By elementary principles, the solution entering the anion resin column contains mineral acids equivalent to the salts originally present, together with CO_2 and silica. The mineral acids are absorbed completely, and the other two not at all. In practice, the situation is more complex. If the cation unit is coflow-regenerated, then there will be considerable leakage of sodium, which must, of course, be associated with an equivalent quantity of chloride or sulphate. As the solution passes through the weakly basic resin, only the free mineral acid can be taken up, since the basic strength of the resin is insufficient to convert NaCl to NaOH, and permit absorption of chloride. The quality of water emerging from the anion unit is therefore almost completely dependent on the regeneration conditions of the cation resin.

However, even when countercurrent regeneration is employed for the cation material, and sodium slip is negligible, chloride slip occurs, since the hydrochloride form of a weakly basic resin hydrolyses with the passage of water, to emit traces of HCl:

$$R\,H^{+}Cl^{-} \rightleftharpoons R + HCl.$$

The chloride leakage figures are variable, being influenced by temperature and flow rate, as well as by water analysis and regenerant usage. No practical purpose is served by attempting to calculate them, and the only important design figure is the operating capacity. If chloride values lower than 0.5 mg/l are necessary, then a strongly basic resin is required, either instead of, or in addition to, the weak base material.

Early weakly basic resins were polyamine-formaldehyde condensation products, which gave such good results that they remained in production long after the introduction of the polymeric resins. During the 1960's several manufacturers introduced polystyrene polyamine materials, made by aminating the chloromethylated polymer with tetraethylene pentamine, or another aliphatic polyamine. These resins had exceptionally high capacities and regeneration efficiencies, particularly when used with fairly rich waters; and they were widely used for deionising brackish waters such as D. They were, however, slow in action, needing cycle times of 24 h to achieve full capacity. In 1968, it could be stated [13] that the simple tertiary amine resins, then available from most suppliers, were more useful for automatic plants with short cycles, since their operating capacity was higher under these conditions, although their total capacity was lower. Polyamine resins remained preferable for cases where long cycles were acceptable. Within a few years, the trend towards small plants, operating at high flow rates with frequent automatic regeneration, had resulted in the dominance of the tertiary amine resins for all purposes in which weakly basic materials were required. The single application of deionising brackish waters similar to, but richer than, water D, for which the polyamine resins remained suitable, has disappeared, this area now being covered by reverse osmosis.

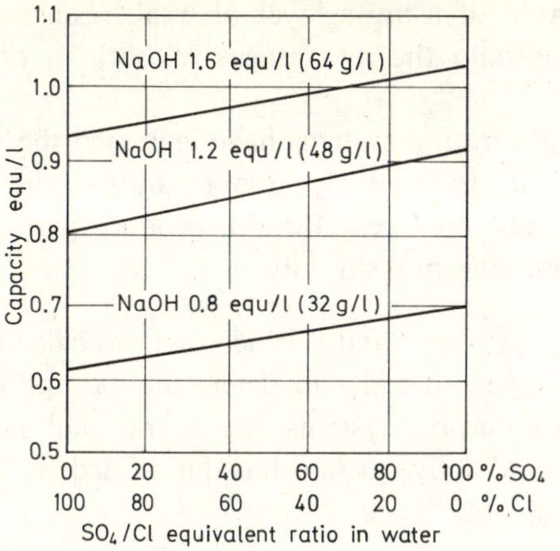

Figure 2.33 Tertiary amine resins. Capacity.

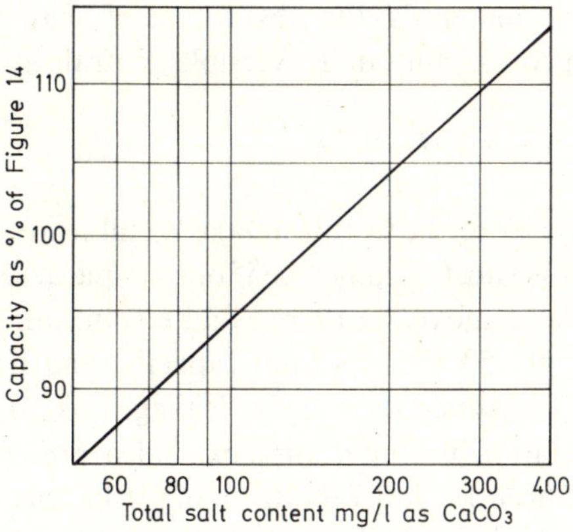

Figure 2.34 Tertiary amine resins. Effect of solution concentration.

Nevertheless, the weakly basic resins are more rate-dependent than the fully ionised strong base resins, and if capacity figures are taken from manufacturers' literature, without consideration of the cycle time, underdesign of the plant can occur. The capacity figures of Figure 2.33 for a typical polystyrene tertiary amine resin, and water with 3 mequ/l mineral anions, apply to cycles of over 8 h duration. They must be reduced by 5%, for each reduction of 1 h in cycle time. Figure 2.34 gives the corrections for total ionic strength of the solution.

The overall usage pattern of weakly basic resins has changed in recent years for a number of reasons.

a) The introduction of polyacrylic materials, with slightly higher basic strength than the polystyrene resins, has tended to extend the use of weakly basic resins.

b) The procedure, described later, of using a layer of weak resin over a bed of strongly basic material, to obtain the advantages of both in one unit, has reinforced this trend.
c) Against this, counterflow regeneration systems have increased the regeneration efficiency of strongly basic resins to be nearly equal to those of the weak resins; and many designers, particularly in Great Britain, produce plants containing sulphonic and quaternary ammonium resin only.

There is no single answer to this problem, and it is necessary to calculate sizes and costs, assuming different resin types, in order to determine the optimum system. Since the widespread use of counterflow systems, the operational significance of differences in anion resin types and active groups has diminished.

2.2.9 Deionising. Strongly basic resins

In contrast with the weakly basic materials, the absorption of acids by a strong base resin is not an equilibrium process, but an irreversible neutralisation reaction:

$$R^+OH^- + HCl \rightarrow R^+Cl^- + H_2O.$$

It might, therefore, be considered to result in full uptake of all acids, irrespective of the regeneration level, and this is substantially true if one simple acid is involved. If a strongly basic resin is used to remove HCl from dilute solution, then a resin which has been regenerated to only 50% of its total capacity, and has been well rinsed, will remove all chloride. The concept of slip or leakage has no meaning.

In practive, however, the situation is more complex. Silica presents a special problem. It is present in low concentrations (about 10 mg/l) in true or colloidal solution, in many waters, and can be much higher in individual cases. The chemistry of silica in solution is unclear. The silicate ion SiO_3^{2-} can exist only under very alkaline conditions, and silica is likely to be present as simple or polymeric bisilicate $HSiO_3^-$. Colloidal aluminium silicates, clays, and colloidal silica also occur, and can be broken down by ion exchangers to cause absorption of separate cations and anions.

Silica is taken up from pure solutions, by hydroxide-form resin, as resin silicate, further passage probably converting the top of the column to the bisilicate form. In a mixed solution resulting from cation exchange of raw water, the bed top becomes neutralised by mineral acids, and silica can be taken up only as bisilicate, which is very weakly held. It can show leakage throughout a service cycle, even when other ions are being fully absorbed, and it is always the first to break through.

The resin capacities to the silica breakthrough point are governed by the separate ratios of sulphate, chloride, nitrate and bicarbonate to silica, each as a function of regeneration level and temperature. A further factor has a major effect on results.

Any leakage of sodium salts from the cation exchanger is converted to sodium hydroxide by the strong base resin:

$$R^+OH^- + NaCl \rightleftharpoons R^+Cl^- + NaOH.$$

The effect is interfere in the uptake of ions, the most frequent result being silica slip, although in severe cases chloride leakage is also found.

A final problem with silica is due to the fact that as the top of the anion bed becomes exhausted, the conditions become acid, and siliceous ions which are being displaced by more strongly held ions, are caused to precipitate in and around the resin beads as a form of silica gel. As this material is not completely insoluble, it can leave the resin bed in an undefined form during the next cycle, reporting as silica slip. This problem is minimised by using warm NaOH, at up to 50 °C, for regeneration.

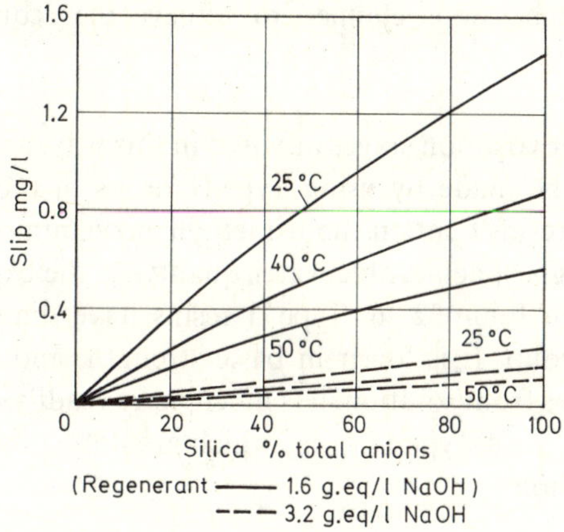

Figure 2.35 Quaternary ammonium resins. Silica slip.

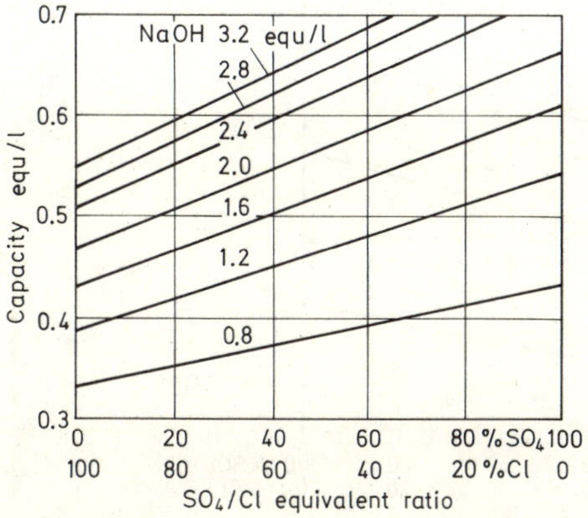

Figure 2.36 Type I quaternary ammonium resins. Capacity.

The effects of temperature and regeneration level in a coflow system are shown in Figure 2.35, while a simplified capacity/regenerant relationship is presented in Figure 2.36. Together, these show that to achieve the low silica values of 0.015 mg/l needed for boiler feedwater, extremely wasteful regenerant levels are required. These results assume the absence of all sodium, and would be less favourable in its presence. It follows that for maximum silica removal from water, the following conditions must be satisfied:

a) The cation exchange unit must be regenerated in counterflow to give minimum sodium slip.
b) Counterflow must also be used for the anion unit, so that with an economic level of 50 g NaOH per litre of bed, the last 20% of the bed is treated with 250 g/l NaOH, sufficient to strip all possible silica from it.
c) One stage of silica removal cannot in general be considered adequate, and the anion exchange unit must be followed by a mixed bed. However, the counterflow "Hipol" process referred to below is claimed to achieve the required results without a final mixed bed unit.

Type I and Type II resins. When two anion stages are used in this way, a considerable improvement in economy can be made by using Type II resins in the first stage. Type I quaternary ammonium resins contain the trimethylbenzylammonium active group, and they produce the highest achievable water quality, at the expense of the high regenerant usage shown in Figure 2.36. Type II resins have dimethylbenzylethanolammonium as active group. It is lower in basic strength, and while it has the same ability to remove anions from solution, it is much more readily regenerated. Figure 2.37 gives the capacities for a Type II resin, which are 45% higher than for a Type I under the same conditions.

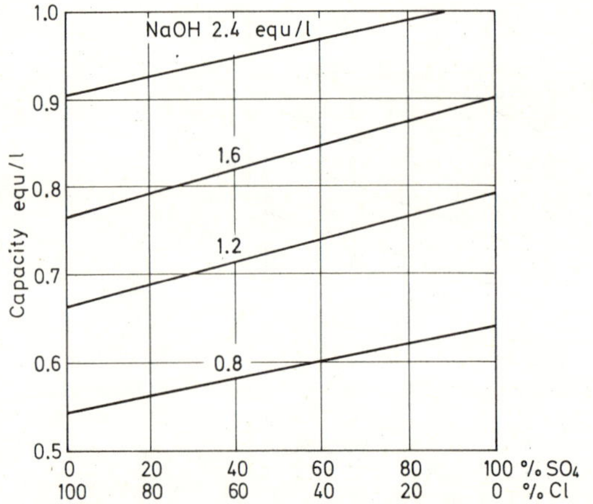

Figure 2.37 Type II quaternary ammonium resins. Capacity.

Unfortunately, the Type II materials are less chemically stable than the Type I resins, and if they are used in single units, the capacity to silica breakthrough falls off steadily, losing as much as 40% in two years. The method of degradation, however, is not mainly the loss of active groups, but a change from strong to weak base groups. The capacity for mineral anions such as Cl^- and HSO_4^- is not reduced, and may even increase slightly in the first few years of the resin life. The silica figures gradually deteriorate, but the resin continues to remove the majority of the silica, and the following mixed bed provides the final polishing.

2.2.10 Deionising. Mixed beds

If a cation-anion pair remove 90% of the salts in one pass, the subsequent pair will remove a further 9%, and so on until virtual perfection has been achieved. A mixture of cation and anion resins acts as an infinite series of de-ionising pairs, and a well-designed mixed bed will remove all ionised matter from solution down to a level of better than 20 μg/l, almost irrespective of the input level. However, it is normal to precede a mixed bed by a de-ionising pair, so that it operates under the best conditions. Regeneration is accomplished by utilising the specific gravity difference between the heavy cation resins and the lighter anion materials. Upflow water separates the resins into two layers, which are treated separately by acid and caustic soda. The successful operation of the process on a large scale requires the use of special design features which are considered in Section 2.2.15 Fixed bed plant design.

2.2.11 Deionising. Combination processes

The unit processes described above are combined in a number of ways, which are dictated by the type of water, the quality required, the scale of operations, and the balance between capital and running costs. The combinations considered below are those in most frequent use. Some of them have been rendered obsolete for large-scale use by advances in mixed bed and counterflow design. Nevertheless, as water treatment plants are very durable, many which were built 25 years ago are still in satisfactory operation. In addition, some of the simpler combinations are still used for small-scale equipment, using high regenerant levels to compensate for inefficiencies in the systems. The processes considered are summarised in Figure 2.38. The results, which are tabulated below, are indicative of trends, rather than being absolute values. In particular, the term 'NIL' represents ideal conditions, and in practice traces are observed. All four water types have been included in the tables, although in practice some processes are unnecessary for the "thin" water A, while others are impractical for the rich water D.

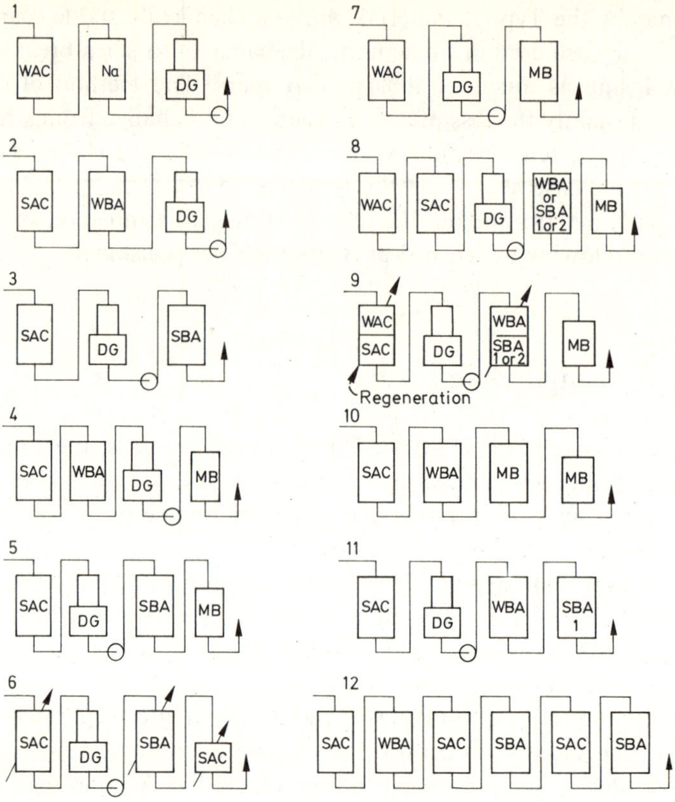

Figure 2.38 Combination processes for deionising.

Table 2.14 Results of treatment by Combination 1. (WAC-DG-Na)

	A	B	C	D
Ca	NIL	NIL	NIL	NIL
Mg	NIL	NIL	NIL	NIL
Na	30	45	115	650
Cl	10	20	50	600
SO_4	10	20	50	30
HCO_3	10	15	15	20
Total	30	45	115	650
regeneration efficiency				
Carboxylic (H^+)	Not used	100%	100%	100%
Sulphonic (Na^+)	75%	75%	75%	Not used

2.2.11.1 Weak acid cation – sodium exchange – degassing

Coflow. This is the most economical method for reducing total dissolved solids, with simultaneous softening, to produce waters suitable for low pressure boilers and a range of similar purposes.

The carboxylic resins leave about 15 mg/l bicarbonate in solution, and the first column therefore has no action for water A, while in the case of water D, in which bicarbonate exceeds hardness, virtually complete softening is achieved by the first column, and the second serves no purpose.

Counterflow. There is little purpose in using counterflow in this case, as the carboxylic resin results cannot be improved, and the sodium exchange can show only marginal change.

2.2.11.2 Strong acid cation – weak base anion – degassing

Coflow. The results of this classical method vary depending chiefly on the regeneration of the cation unit. In Table 2.15, columns H refer to a high level (160 g/l) of H_2SO_4, while columns L give results at 64 g/l. The cations and anions no longer balance, since these waters are always slightly acid, and the H^+ ion has not been included.

Water A gives good results under both conditions, with a little slip at low acid level. The increase in % Na in water B causes a higher leakage; and this figure is reduced in water C, in spite of the higher input concentration, because the Ca^{2+}/Na^+ ratio is much more favourable. During regeneration, calcium from the top of the column displaces sodium lower down, and leaves less sodium on the resin than from water B, whose total sodium is less, but with a less favourable ratio. The capacity with water C is correspondingly lower. Water D cannot be treated by this method.

Table 2.15 Results of treatment by Combination 2. (SAC-WBA-DG)

	A		B		C		D	
	H	L	H	L	H	L	H	L
Ca^{2+}	NIL	NIL	NIL	NIL	NIL	NIL	NIL	NIL
Mg^{2+}	NIL	NIL	NIL	NIL	NIL	NIL	NIL	NIL
Na^+	0.1	1.0	0.5	2.0	0.2	0.5	100	300
Cl^+	0.3	1.5	0.8	2.5	0.5	0.8	90	290
SO_4^{2+}	NIL	NIL	NIL	NIL	NIL	NIL	10	10
CO_2	10	10	15	15	15	15	15	15
Approx. pH	5	5	5	5	5	5	5	5
Conductivity	10	12	10	12	14	12	200	600
Cation-exch. capacity g/l	38	25	44	29	37	25	55	37

Counterflow. Counterflow regeneration would improve the results greatly, the Na^+ and Cl^- leakage being reduced to under 10% of the values in Table 2.15, so that even water D can be treated with moderate success. However, a water-user requiring high quality would normally need silica removal, and counterflow regeneration, in this simple form, is unlikely to be used.

2.2.11.3 Strong acid cation – degassing – strong base anion

Coflow. As compared with System 2.2.11.2, this process has unchanged cation results, but sodium leakage is converted to NaOH by the anion exchanger, and, consequently, anion leakage from the second column is eliminated, except for water D, with which the high NaOH level causes serious interference with anion exchange. Free CO_2 is removed from all waters, and silica, which is not of course affected by Systems 2.2.11.1 and 2.2.11.2, is substantially reduced in each case. The values vary so greatly (Figure 2.36) that they are not tabulated. In general, waters A, B and C, if originally containing less than 10 mg/l SiO_2, which is usual, will yield treated waters containing less than 0.5 mg/l, with water D being even less favourable. If good silica quality is required, this system is inadequate.

Table 2.16 Results of treatment by Combination 3(SAC-DG-SBA)

	A		B		C		D	
	H	L	H	L	H	L	H	L
Ca^{2+}	NIL	NIL	NIl	NIL	NIL	NIL	NIL	NIL
Mg^{2+}	NIL	NIL	NIL	NIL	NIL	NIL	NIL	NIL
Na^+	0.1	1.0	0.5	2.0	0.2	0.5	100	300
Cl^-	NIL	NIL	NIL	NIL	NIL	NIL	15	40
SO_4	NIL	NIL	NIL	NIL	NIL	NIL	2	5
CO_2	NIL	NIL	NIL	NIL	NIL	NIL	NIL	NIl
OH^-	0.1	1.0	0.5	2.0	0.2	0.5	83	255
Approx. pH	8	9.5	9	10	8	8.5	11	12
Approx. conductivity	0.4	4	2	8	0.8	2	350	1150

Counterflow. This method, even in its simple form without a following mixed bed, gives a high quality water from sources A, B and C, and it is surprising that the system has not found universal acceptance, except in Great Britain, where counterflow systems have been widely used since 1965.

With counterflow for the cation unit only, the Na^+ and OH^- figures are under 10% of those in Table 2.16, the conductivities are correspondingly reduced, and the pH values are around 7.5, the departure from neutral being due almost entirely to last traces of regenerant not rinsed from the column. Silica results in general improve, and if counterflow regeneration is used on the anion column also, silica figures are likely to be well under 0.1 mg/l.

Water D would provide a reasonable product, containing about 25 mg/l NaOH, and 1 – 2 mg/l Cl^-, but even with counterflow, it is difficult to produce fully pure water, and a following mixed bed is needed.

2.2.11.4 Strong acid cation – weak base anion – degassing – mixed bed

Coflow. This was a widely-used process during the 1950's, when system 2.2.11.2 was in common use and there was a need for improved quality, including silica removal, at low cost. A mixed bed was added to the train of units, with "thoroughfare" regeneration of the anion exchange resins. That is, the partly-used NaOH from the strong base resin in the mixed bed was pumped through the weakly basic resin, giving an overall usage of about 75%.

Since the feed water to the mixed bed was as in Table 2.15, the final treated water quality was good, even when the source water was D. However, the method has a disadvantage which prevents its use in cases where the lowest silica figures are needed. The removal of all mineral acids, and most CO_2, before the mixed bed stage, results in the mixed bed anion material being loaded almost entirely with silica. It is difficult to rinse the resin sufficiently to eliminate all traces of the sodium silicate formed during regeneration, and there is a tendency for SiO_2 slip during the next cycle.

Counterflow. As counterflow operation would not influence the silica problem, there is no object in considering it in this case.

2.2.11.5 Strong acid cation – degassing – strong base anion – mixed bed

Coflow. In a well designed plant this system gives almost the highest achievable treated water quality, as shown in Table 2.16 a, irrespective of the input water or the scale of operations.

Table 2.16 a All waters

	Coflow	Counterflow
Ca + Mg	<0.010	<0.005
Na	<0.005	<0.002
$Cl + SO_4 + OH$	<0.015	<0.015
CO_2	<0.005	<0.002
Silica	<0.010	<0.005
pH	7.0	7.0
Conductivity	<0.10	<0.07

If a Type II strongly basic resin is used in the strong base anion unit, and a Type I resin in the mixed bed, with thoroughfare regeneration, the overall efficiency of NaOH use is 70%, giving a chemical cost almost as low as in System 2.1.11.4, together with full silica removal. However, the problems of instability of the Type II resins lead many users to prefer Type I, even at a higher operating cost.

When a Type I resin is used, thoroughfare regeneration does not lead to very high chemical efficiencies, the figures being 50 – 60%. It does, however, greatly increase the quantity of caustic soda which passes through the mixed bed anion material, thus giving the best elimination of silica.

Counterflow. When the strong acid cation and strong base anion units are regenerated in counterflow, the efficiency of acid and caustic soda usage becomes over 80%, with an improvement in treated water quality. Little improvement in the final treated water is possible, since coflow gives virtual perfection. However, the chemical load on the final mixed bed is so low, that its operation becomes even more reliable, and minor operational faults, which might show a temporary deterioration in System 2.2.11.5 with coflow, make no detectable difference in System 2.2.11.5 with counterflow.

2.2.11.6 The Hipol system. Counterflow strong acid cation – degassing – strong base anion – strong acid cation

In 1977, a paper by Jackson and Smith [14] announced a new process named "Hipol", which has since progressed considerably, and is now claimed to make mixed beds unnecessary [15]. By using deep beds of strongly acidic and basic resins, at high flow rates and low regeneration levels, it is possible to obtain a treated water with a conductivity averaging just over 1.0 µS/cm, due entirely to final traces of regenerant rinsed from the anion unit, and not related to slip from the cation column. The water at this stage contains only 0.2 mg/l NaOH, which is totally removed by passage through a small, highly regenerated counterflow cation unit, giving a final conductivity of less than 0.1 µS/cm, equal to best mixed bed performance. The overall regeneration efficiency is claimed to exceed 85%.

The "Hipol" process uses the "buried collector" system described in Section 2.2.11.6, but, in principle, any satisfactory counterflow equipment would be suitable. This type of process may well supplant existing deionising methods in the next ten years.

Combinations with leading weakly acidic columns. For alkaline waters of type C, a reduction in operating cost is achieved by placing a carboxylic acid column ahead of the other units, and regenerating the cation resin of later units in thoroughfare with the first, to obtain high acid efficiency.

2.2.11.7 Weak acid cation – degassing – mixed bed

In effect, this is a simple mixed bed process in which the chemical load is reduced by a dealkalisation stage. It is a convenient means of producing mixed bed quality water at lower cost, and is particularly suitable for small installations, in which two separate, standard small units can be connected together. With a weak acid cation unit, the question of counterflow does not arise.

2.2.11.8 Weak acid cation – strong acid cation – degassing – weak base or strong base anion – mixed bed

According to the anion exchanger used, this is the equivalent of systems 2.2.11.4 or 2.2.11.5, the first unit serving only for cost reduction. With coflow regeneration of the strong acid cation unit, there is the disadvantage that as most of the Ca and Mg have been removed by the first column, the strong acid cation resin is loaded mainly with sodium, and has a high leakage, which reduces the quality from the anion unit. Counterflow regeneration of the strong acid cation unit overcomes this problem, but a better solution is in System 2.2.11.9.

2.2.11.9 Layer beds

The problem exhibited in Section 2.2.11.8 is conveniently overcome by placing a layer of carboxylic resin, on top of sulphonic, in the same unit, and regenerating in counterflow. The overall consumption of acid is 100%, and the sodium slip is minimised. However, counterflow regeneration of the sulphonic resin alone is sufficiently efficient to render the slightly more complex layer system unnecessary in most cases.

A corresponding layer system can be used for anion exchange units, with weak base anion resin over strong base anion resin, regenerated in counterflow. This process has been successful in a number of variations, but requires careful use, as the strong base anion resin becomes loaded mainly with silica. During counterflow regeneration, the waste solution containing sodium silicate and hydroxide can become acidified by the bisulphate ions on the weak base anion resin, causing precipitation of silica gel.

Many more complex systems have been developed in the past, and are still in use. Among the most successful are the following.

2.2.11.10 Strong acid cation – weak base anion – mixed bed – mixed bed

This was used to produce good quality water from brackish waters of type D, at up to 1500 mg/l ions in the raw water. This is no longer used, and the results can be obtained more economically by using reverse osmosis to convert water D to the equivalent of B, and then proceeding with a normal deionising method.

2.2.11.11 Strong acid cation – degassing – weak base anion – strong base anion

In many parts of Europe there were, for a long time, doubts as to the reliability of mixed beds, and this method is typical of a system used to remove all types of ion at reasonable cost. It gave a water of fairly high pH, due to NaOH, but this was advantageous for many purposes, such as medium-pressure boilers, in which NaOH was added in any case to reduce corrosion. The method can be considered obsolete in relation to countercurrent processes, which render two anion stages largely unnecessary.

2.2.11.12 Strong acid cation – weak base anion – strong acid cation – strong base anion – strong acid cation – strong base anion (possible mixed bed)

Six or seven bed systems, with various unit combinations, have been widely used in continental Europe, but much less in Great Britain and the USA. They are no longer used.

2.2.11.13 Organic trap columns

In any deionising process in which organic poisoning of the anion exchange resin is a problem, there may be advantages in using an "organic trap" of highly porous anion exchanger in the chloride form at the head of the train. This is a unit containing a highly porous or macroporous strongly basic anion exchanger, used in the chloride form, and regenerated with alkaline brine (see 2.2.14).

2.2.12 Partial desalination of brackish waters

Conversion of lightly brackish waters, containing up to 3000 mg/l salts, into potable water, by removing 90% of the soluble matter, has historically been an important ion exchange application giving rise to a number of interesting processes designed to reduce costs. The rapid development of membrane processes, in particular reverse osmosis, has greatly reduced the use of ion exchange for desalination, but the main processes which have been used remain valid.

2.2.12.1 Four-bed partial deionising

This, the earliest and most obvious method, was widely used in places such as Malta, where the availability of potable water was more important than cost. Many early plants used two pairs, each of sulphonated coal, followed by a high capacity condensation-type weakly basic anion exchanger. At conductivity rise after the second pair, the first was totally exhausted, and flow to service was then stopped, while the leading columns were regenerated, and replaced in the second position. Reasonable regenerant usage, up to 60%, was achieved. The objective of the following methods was to reduce costs by improving this figure.

2.2.12.2 The Sul-Bisul process

In 1965, Odland and Pabich [16] announced this process, which is concerned only with the anion exchange step. Cation exchange was exactly as above. The process depends on the relationship between the sulphate and bisulphate ions. Resin in the sulphate form is able to absorb free acids, by conversion of sulphate to bisulphate, which is not itself an ion exchange reaction:

$$R^{2+}SO_4^{2-} + HCl \rightleftharpoons R^{2+}HSO_4^{-}Cl^{-}$$

$$R^{2+}SO_4^{2-} + H_2SO_4 \rightleftharpoons R^{2+}(HSO_4^{-})_2 .$$

Thus the acids emerging from the cation exchanger are almost fully absorbed by the sulphate-form resin, yielding de-ionised water. The resin can be regenerated back to the sulphate form, simply by pumping raw water that has not been submitted to hydrogen ion exchange through the column thus reversing the equilibria above. The quantity of water required is large, and if the raw water is in short supply, the auto-regeneration process can be improved by making the regeneration water slightly alkaline, by means of lime or sodium hydroxide. Reagent efficiency is 100%, since the action is not ion exchange, but simple acid neutralisation.

The Sul-Bisul process has limitations, as follows:

a) In order to ensure that auto-regeneration returns the resin to the sulphate, not the chloride form, the ratio $SO_4^{2-} : Cl^-$ in the raw water must be over 9 : 1, which is unusual, as brackish waters are normally chloride-rich.

b) The total dissolved solids must be less than 2000 mg/l.

c) The bicarbonate content of the water must be less than 10% of the total anions, as alkalinity in the raw water causes reversal of the absorption reactions.

d) Sodium must be less than 10% of total cations.

2.2.12.3 The Desal process

Developed by Kunin and Vassiliou in 1964 [17], this process depends on the properties of an acrylic anion exchange resin, Amberlite IRA 68, which was then unique. Its basic strength was such that it could be regenerated at almost 100% efficiency with weak bases such as ammonia, and yet it was strong enough to absorb CO_2 to give a resin bicarbonate.

Three columns are used in series (Figure 2.39), containing respectively IRA 68 bicarbonate form, a carboxylic resin IRC 84 in the hydrogen form, and IRA 68, in the free base form. In principle, the reactions in the three stages are as follows:

$$R_A^+HCO_3^- + NaCl \rightarrow R_A^+Cl^- + NaHCO_3$$

$$R_C^-H^+ + NaHCO_3 \rightarrow R_C^-Na^+ + H_2O + CO_2$$

$$R_A^+OH^- + CO_2 \rightarrow R_A^+HCO_3^-.$$

The third column is converted to the bicarbonate form by this process, and is thus ready to be used as the first unit on the next cycle. The first column, on exhaustion, is regenerated at almost 100% efficiency, with ammonia or lime, and is then used as the third, with the whole train now in reverse order. The cation unit is, of course, regenerated at high efficiency with acid.

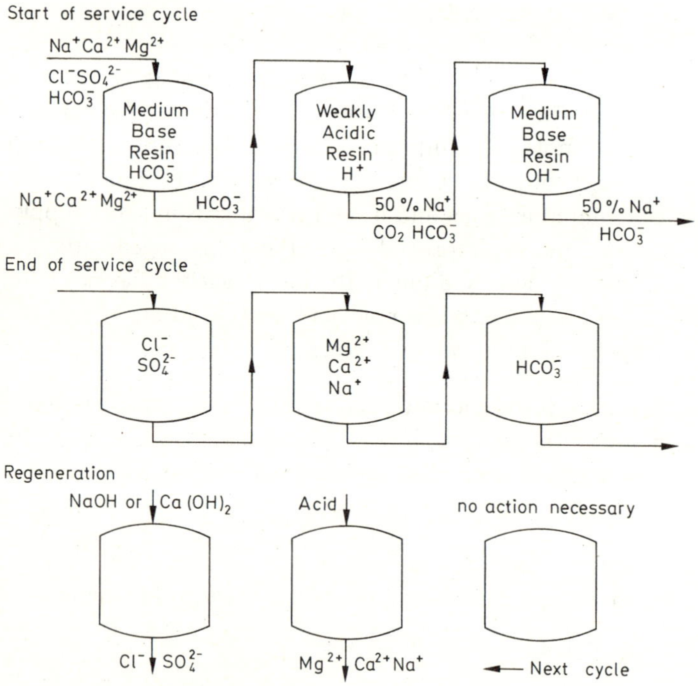

Figure 2.39 The Desal Process.

The process works well with hard, high bicarbonate waters, but there are problems in other cases. For effective operation, the whole of the CO_2 must be kept in the system, either as free CO_2 or as HCO_3^-. Any bicarbonate left on the first column at the end of the cycle is lost on regeneration and, if not replaced by bicarbonate from the raw water, requires extra CO_2 to be pumped into the system.

The cation exchanger does not remove all sodium, so that except in the rare cases of totally hard water, sodium passes the third, anion, unit and is found in the treated water, as $NaHCO_3$. Nevertheless, the method gives a considerable reduction in the total dissolved solids, at low cost, and it has been successfully used on an industrial scale.

2.2.12.4 The SIRA process

This system was developed in Italy in 1975 [18]. It bears a relationship to the Desal method, in using three columns, and depending on the manipulation of bicarbonate and carbon dioxide. It introduces a new concept of recycling CO_2, to cause an artificial increase in the bicarbonate content of the raw water.

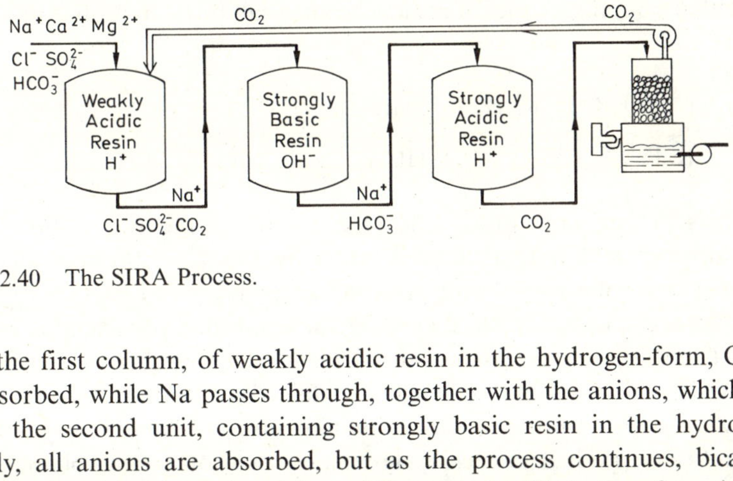

Figure 2.40 The SIRA Process.

In the first column, of weakly acidic resin in the hydrogen-form, Ca and Mg are absorbed, while Na passes through, together with the anions, which are taken up by the second unit, containing strongly basic resin in the hydroxide-form. Initially, all anions are absorbed, but as the process continues, bicarbonate is displaced by chloride and sulphate. The overall effect, therefore, is an initial emergence of NaOH, followed by Na_2CO_3, and then $NaHCO_3$.

The third column contains a strongly acidic resin in the hydrogen-form, which removes the remaining metals from solution, leaving only CO_2. This is stripped by means of a degassing tower, compressed, and returned to the head of the circuit. Theoretically, the whole of the CO_2 is kept in the system, until it has built up to the level giving the highest operational efficiency, after which the recycle is reduced to keep the system constant.

The SIRA process uses counterflow regeneration on each column, and is claimed to give 90% regeneration efficiency, in taking the total dissolved solids down from

2500 mg/l to 400 mg/l, on the industrial scale. At this input concentration, the volume of treated water produced is about 10 times the total resin volume, which is substantially better than could be achieved by any form of conventional de-ionising.

2.2.12.5 The Sirotherm process

Whereas all other partial desalination methods use ingenious chemical manipulations of simple ion exchange, and give relatively small reductions in total chemical costs, the Sirotherm process, invented by Weiss et al. in Australia in 1966 [19], has a totally new approach, regenerating the resins without the use of any chemical procedures at all.

It depends on the observation that when weakly acidic and weakly basic resins, in hydrogen- and free base-form, respectively, are mixed, each reinforces the action of the other. Acid produced by the cation exchanger, which would arrest its action if remaining in solution, is removed by the anion resin, so that the combination is able to de-ionise not only calcium bicarbonate solutions, but also sodium chloride. The key step, which led to the success of the process, was made possible by the observation that both these types of resin, when in a salt form, hydrolyse at elevated temperatures:

$$R_C^- Na^+ + H_2O \rightleftharpoons R_C^- H^+ + NaOH$$

$$R_A H^+ Cl^- \rightleftharpoons R_A \quad + HCl\,.$$

The hydrolysis process constitutes auto-regeneration of the two resins. It thus became possible to desalinate a brackish water by passing it through the mixed resin unit, and then following it with the same water, heated to 60 °C.

The early process, using standard commercial resins, had considerable capacity limitations. The Australian researchers then developed a special resin, consisting of intimately-mixed, finely-ground exchangers contained in the same bead. With this material it was possible to treat waters containing up to 1500 mg/l dissolved salts, but the quantity of product in each cycle was less than one resin volume. It was thus necessary to pass though the bed alternate small fractions of hot and cold water, and to collect alternate quantities of desalinated water and waste effluent, without mixing and without transferring heat from one fraction to another. The design problems were severe, but the Australian researchers overcame them by using a system of multiple countercurrent fraction collection. A number of successful industrial scale plants have been operated.

The Sirotherm process is probably the most efficient ion exchange system which has been developed for brackish water treatment, but scope for ion exchange systems in general has been greatly reduced by the success of reverse osmosis, which is capable of treating a much wider range of concentrations.

2.2.13 Nitrate removal

A problem which has become serious in recent years is the presence in natural water supplies, both surface and subterranean, of increasing concentrations of nitrate. Nitrate is known to cause methaemoglobinaemia ("blue babies") in small infants. It has also been suspected of possible carcinogenic effects, by reaction with body proteins to form nitrosamines, after reduction to nitrate by gut bacteria, but the evidence for such a relationship is poor [20]. Nevertheless the European Commission has set a limit of 50 mg/l in drinking water [21]. In many parts of Europe, and elsewhere in the world, this figure is seriously exceeded.

In surface waters, the effect is due mainly to run-off from farms using intensive fertilisation, and is therefore seasonal. Values may exceed 100 mg/l at the peaks. The problem with aquifers is more serious, as in many regions, nitrate is moving down through the upper layers to contaminate deep water which will be used by the end of the century.

Nitrate can be removed from water by means of a strongly basic resin, used in the chloride cycle [22, 23]. Many plants, both fixed bed and continuous, have been installed in Europe and the USA, and are operating successfully. The process presented some difficult early problems. The standard strongly basic resins were non-selective, and they thus wasted capacity, increasing regenerant costs, by taking up bicarbonate and sulphate, in addition to nitrate. The chloride emitted, equivalent to the total ionic uptake, was sufficient to make some waters undesirably corrosive. A further problem was that standard resins exhibited preference for sulphate against nitrate, which resulted in very poor nitrate uptake from the high sulphate waters (over 150 mg/l SO_4) found in limestone regions. These problems have now been largely overcome, by the use of nitrate-selective resins, Rohm and Haas IMAC HP555, and Purolite A 520-E, to treat high sulphate waters, and also by modified regeneration systems which reduce chloride emission. Nitrate removal by ion exchange is now a fully established process. A report by the Water Research Centre, Great Britain [24] has shown that it is technically and economically preferable to biological or membrane methods in virtually all cases.

2.2.14 Resin durability in water treatment

The operating life of an ion exchange resin is required to be many years, and include thousands of cycles, in constantly flowing water. They must be totally insoluble, mechanically strong, and chemically stable, while also being hydrophilic and water-permeable, having high reactivity as measured by capacity, and rapid rate of reaction. Few other polymers are required to have such conflicting properties, and the manufacturers of ion exchange resins are forced to make different compromises, according to the potential use of the materials. For this reason, their lists include many different resins of each type. For example, the standard polystyrene sulphonic acid resins, which in principle are the simplest, are supplied with two structure types, gel and macroporous, different levels of crosslinking from 4% to 20%, and a range of particle sizes.

Solubility. Linear polystyrene sulphonic acid is highly soluble, whereas the standard 8% crosslinked resin is so insoluble that no detectable loss of weight occurs after years of service. Nevertheless, new resin does contain traces of linear polymer which can be leached out during the first few weeks of service. In water softening, and in deionising with a quality target of 1 $\mu S\ cm^{-1}$, this effect is insignificant. When the ultimate in treated water quality is required, it is preferable to use a resin with 10–12% crosslinking, or a macroporous material, which is always crosslinked to at least this level. The corresponding trace leaching from anion exchange materials is less of a problem, being absorbed by the final mixed bed, or by the final cation unit in the Hipol process. Again, macroporous resins show lower leaching, and some users prefer them, in spite of their slightly lower operating capacity as compared with gel resins.

Chemical and thermal stability. In the absence of oxidising agents, sulphonic cation exchange resins are stable up to 120 °C. They resist 1 mg/1 Cl_2 at ambient temperatures, but at 80 °C this concentration causes considerable attack, with breakage of crosslinks and softening of the resin, the life being reduced to under two years in the case of standard 8% material. Under these conditions, it is preferable to use 12% crosslinked or macroporous resins, or to dechlorinate the water, by adding sodium sulphite, just before the ion exchange column.

Anion resins suffer the same oxidative attack on crosslinks, and, in addition, quaternary compounds are converted to tertiary by the action of heat. This has the anomalous effect of increasing the operating capacity slightly during the early part of the resin life, since the tertiary groups give a higher regeneration efficiency than the quaternary. Nevertheless, the useful life of a Type I resin, at ambient temperatures, is rarely more than 7 years, while Type II materials may be limited to 5 years. Tertiary amine resins are more stable, and lives of 9 years have been found in industrial practice.

Mechanical strength. During service ion exchange resins suffer two types of mechanical stress. In packed bed plants of all types, there is considerable pressure on the up-stream side of the bed, that is, the top during normal down-flow operation. If the solution being treated is not completely free from turbidity, the resin bed acts as a filter, and the top few inches rapidly become partly blocked, with a further increase in pressure drop. Any debris from broken resin beads adds to the effect. The total force on the top of the bed can amount to many tons, and can cause severe breakdown of resin particles, particularly under high flow rate conditions as in condensate polishing. In the case of the standard polystyrene cation exchange resins, the resistance to crushing forces is related to the degree of crosslinking.

When this is low, the beads are fairly soft and deformable, so that under the flow of water they pack down and give excessive pressure drop. As the crosslinking increases, the beads become firmer, but they have an increasing tendency to brittleness. There is no ideal point at which both faults disappear, and the standard 8% crosslinked material represents the best compromise for normal purposes. Even at this level, under high flow rate conditions the top of the bed can be observed to fall slightly as the resin compresses, and under these conditions a material with 10 to 11% crosslinking is preferable.

The gel anion exchange materials are in general considerably more fragile than the cation resins, due to mechanical stresses which inevitably are introduced into the beads during the manufacturing process. It is particularly important to avoid pressure on the top of the bed, and in particular any debris from breakage of the resin itself must be removed at every cycle by backwashing.

The macroporous resins are stronger than the gel resins in both cases, the difference being very noticeable for the anion material. However, the macroporous resins have the disadvantage that if they break under compressive forces they shatter totally to dust, whereas the gel resins tend to break into two or three pieces, which remain usable. In recent years, there have been general improvements in the durability of the polystyrene resins, and one manufacturer has announced a new range of tough gel materials, claimed to be stronger than either classical gel or macroporous products. The acrylic resins, both cation and anion, have a slightly elastic structure, and are very resistant to crushing forces.

The second type of stress is surface abrasion, which can occur in continuous ion exchange plants, or in fixed bed plants from which the resins are regularly removed for regeneration. With modern spherical bead resins, this effect is negligible, but some ion exchange materials are manufactured in granular form, and these are unsuitable for moving bed systems.

It should, however, be noticed that continuous ion exchange plants can put very severe compressive forces on resins. These can occur if the design is such that valves close on to resin, but even when this is not the case, the frequent opening and closing of valves under high flow rate conditions causes compressive shock waves to pass through the resin bed, resulting in considerable damage. Fluidised bed continuous systems do not suffer from either of these problems.

Osmotic shock. When gel type ion exchange resins are dried and re-wetted, they shrink and swell, resulting in shattering of the beads. Macroporous resins resist the same action, since they can accomodate the volume changes inside the resin beads.

A lesser effect can result from osmosis. When the beads are in contact with the relatively concentrated regenerant solutions, they loose water and shrink slightly. On returning to water treatment service they swell and crack. For satisfactory operation, resins must be capable of withstanding a minimum of 1000 cycles of alternate 5% HCl and 5% NaOH, with intermediate water washing. Macroporous resins resist this test satisfactorily, as do most, but not all, modern gel resins.

Resin poisoning. This phenomenon occurs only with anion exchange materials, and is due to the presence in surface waters of humic and fulvic acids, high molecular weight polycarboxylic compounds. These enter the resins by normal ion exchange, and during regeneration are partly removed with other ions. However, all resins suffer from the problem of non-uniform crosslinking, resulting from the nature of the copolymerisation process by which their structure is formed. A resin of nominal 8% crosslinking can have regions with as much as 25%. The organic acids become trapped in these parts of the structure, where they have two effects. First, they tend to slow the resin action; secondly, and more importantly, they convert part of the anion exchanger into a carboxylic acid cation exchange material, which is firmly locked into the anion exchanger (Figure 2.41).

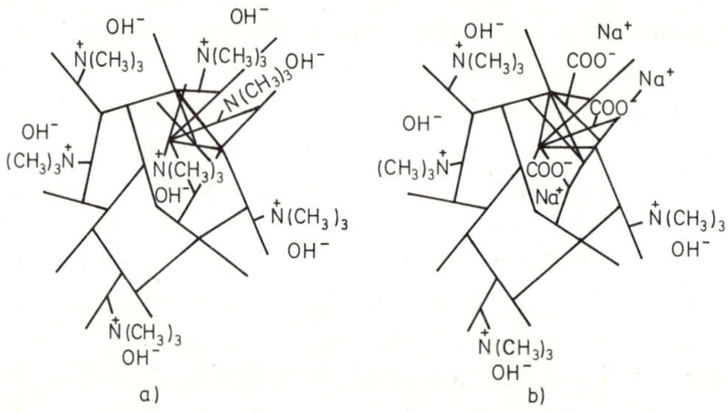

Figure 2.41 Resin poisoning. Conversion of basic to weakly acidic groups.

The carboxylic groups take up sodium from the regenerant, and lose it by hydrolysis during the following cycle. The combined effect of the two poisoning actions is to increase the conductivity of the treated water, with the results shown in Figure 2.42. This shows typical mixed bed results for a water containing 1 – 2 mg/l of organic matter. Curve A shows ideal results with new resin, while B gives readily-achievable operating results, the water quality being sufficient for the most stringent conditions. As the bed becomes poisoned, treated water quality falls to curve C, and then to D. The total quantities of ions taken up in the four cases, that is the achievable capacities, are given by the integrals between each curve and the raw water analysis. The difference between the best and worst case is, however, only 0.2%, showing that the effect of poisoning on the total available capacity of the resin is extremely small. The operating capacity is entirely dependent on the required water quality. If a reasonably good water of conductivity 1.0 μS/cm is adequate, then the ideal curve A gives a capacity of 21.8 g/l as $CaCO_3$, while D gives 17.2 g/l, a loss of only 21%. In 1955, therefore, this unit would have been in serviceable condition for power-station practice. By 1965, when the standard had become 0.3 μS/cm, it would have been giving no usable water at all, while the lightly

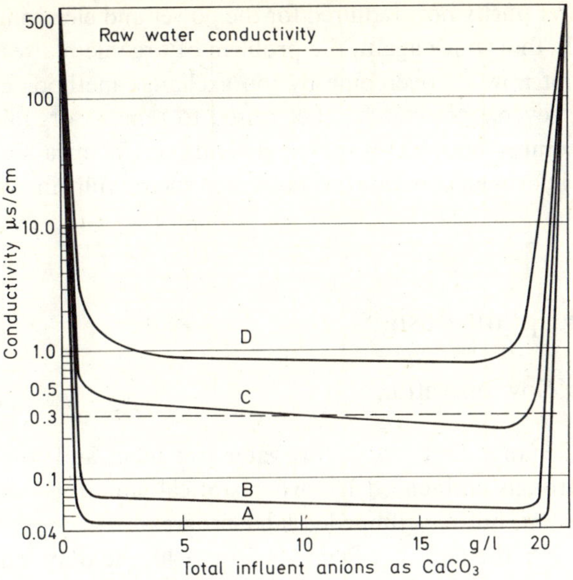

Figure 2.42 Resin poisoning. Practical effects.

poisoned unit C would have had a usable output less than 50% of the clean unit, and the slightest further deterioration would have resulted in all the treated water being rejected. Finally, a modern standard of 0.08 µS/cm can be achieved only by completely unpoisoned resin, (curves A and B).

The problem of poisoning has been countered in a number of ways, each in its turn successful by the water quality standards of its age, and each becoming inadequate as the standards became more strict. With the development of the first macroporous resins in 1959 [25–27], followed by the announcement of the isoporous exchangers [25, 26], the principle of the organic trap columns became widely used for several years. Each of these resin types was capable of absorbing organic matter and releasing it again, when regenerated with a solution containing 10% NaCl, with 1% NaOH. Columns of these resins, placed as the first units of deionising trains, removed selectively those organic acids which caused the greatest fouling problems, and treated water quality improved accordingly. By 1965, both these resin types had been developed to the stage at which they could be used as the anion exchangers in the de-ionising process itself, removing and releasing the inorganic and organic anions together.

This method applies only to the resins used in the anion units. Unprotected mixed beds remain subject to another type of organic poisoning. Humate contained in the anion exchanger can come into contact with acid used to regenerate the cation layer, causing precipitation of humic acids inside the beads, and subsequent slow leaching. For this reason, mixed beds should always be protected by an organic trap, or by a deionising pair, in which there is an organic-removing anion exchanger.

The standards of water purity now required for the power and electronics industries have become so high, that, once again, the problem of organic matter has become serious [30]. It cannot now be overcome by ion exchange methods alone. In the electronics industry, reverse osmosis has been used to remove organic molecules, and the method is coming into use in power generation. Chemical destruction by means of ozone and hydrogen peroxide has been used successfully in both industries [30, 31].

2.2.15 Fixed bed plant design

2.2.15.1 Classical coflow operation

The design of ion exchange units, both for water treatment and for other applications, has been strongly influenced by two historical situations, which greatly aided early development, but have impeded later progress.

The first of these was the fact that from 1905 to 1940, the only industrial scale exchange process was water softening, which continued to be dominant for a further ten years. The great ease, and high efficiency of calcium/sodium exchange, due to the favourable equilibria of the system, led to considerable misconceptions about the whole process and many unnecessary difficulties in de-ionising plants.

The second factor was the use of established pressure sand filter principles for ion exchange plant design. This had the advantage of immediate success, but it led to fallacies that are still accepted.

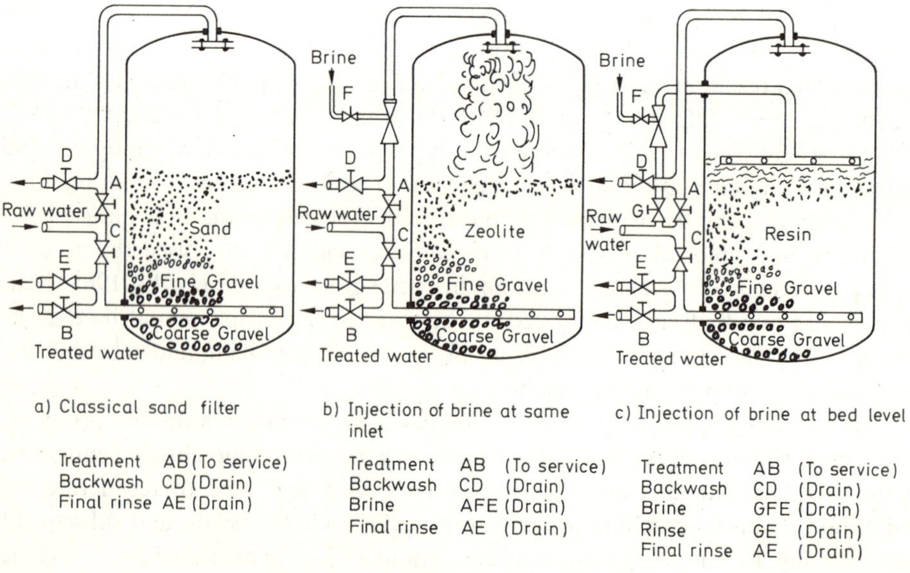

Figure 2.43 The development of downflow ion exchange units.

In a pressure sand filter, flow is downwards, suspended solids being removed mainly on the surface of the bed, with some penetration. As the bed becomes clogged, the flow rate falls off and the bed is then cleaned by using a reverse flow ("backwash"). The sand bed becomes partly fluidised and the filtered dirt is released.

The earliest softener units, Figure 2.43 b, were virtually identical, the sand being replaced by an inorganic zeolite of similar density. When the softening capacity of the unit was exhausted, flow to service was stopped, and backwash was instituted, simply because this had been filter practice. It was largely unnecessary for a softener operating on clean water, but did no harm. Regeneration was carried out using raw water to power an ejector which drew brine from a stock tank, diluting it to about 15%, and passing it into the unit through the same entry point as the raw water. Usage of the exchanger was incomplete, as the dense brine solution tended to fall directly from the entry point, and in addition was dispersed and diluted in the water above the bed. This fundamental inefficiency was overcome by using large excesses of salt, a cheap and convenient raw material.

Nevertheless, a considerable improvement was made by bringing in the brine in a new distributor just above the bed surface (Figure 2.43 c) where its high specific gravity caused it to form a pool below the static water, thus ensuring a uniform passage through the bed. This basic design, with variations in pipework and valves, has remained in use not only for softeners, but also for deionisers, in which the more dilute regenerant (1.5% w/v in the case of H_2SO_4) has too low a density to remain as in Figure 2.43 c, and mixes with the above-bed water by diffusion and turbulence.

2.2.15.2 Uniformity of flow

The considerations of the previous sections 2.2.3 – 2.2.11 become invalid if the water flow is not completely uniform across the area of the bed. Even in a laboratory column, 15 mm in diameter and 250 mm tall, severe distortions of the wave front can be observed as a result of differences in flow rate between the light and shade sides of the column (Figure 2.44 a). In an industrial column, of diameter 1.5 m, and a bed depth which must be limited to 2.5 m for practical reasons, a similar distortion could cause intolerable reductions in capacity and treated water quality (Figure 2.44 b). The achievement of uniform flow through the bed is the foremost consideration, before all other design factors. The basic principle is to ensure completely even collection from the bottom of the resin bed. Uniform distribution at the top is no problem, and flow through the bed itself does not cause distortions, since wall effects and temperature variations are neglible in a large unit.

Uniform collection is achieved by installing a large number of small, identical orifices, usually in the form of resin-proof strainer buttons. They are calculated to ensure that the headloss through the bottom collector is considerably higher than through the resin bed. A simple gravel underbed has low headloss, and does not adequately control the system.

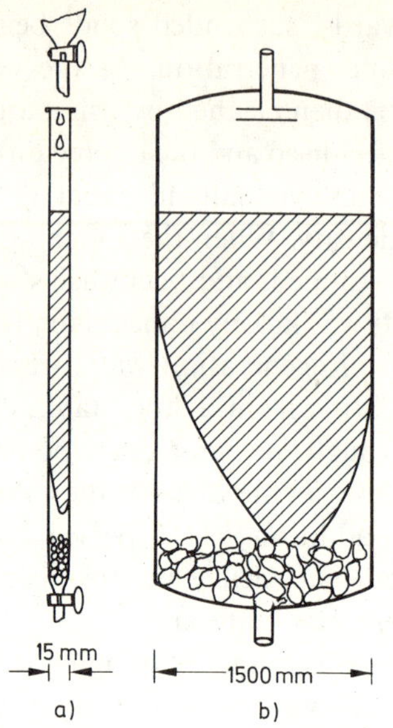

Figure 2.44 Effect of non-uniform flow.

The design of bottom collecting system must also take into account the following points:

1) The take-off system must be completely flat.
2) The take-off strainers must be as close as possible to the bottom of the unit, so that there will not be resin below them which could be incompletely rinsed.
3) There should be no obstructions to flow above the exit strainers.
4) The space downstream of the exit strainers must be as small as possible, to ensure rapid rinsing.
5) However, as the units must be rubber-lined, in contrast with filters and softeners, which are constructed of painted mild steel, there must be full access to all metal to permit rubber lining.
6) These results must be achieved at low cost.

2.2.15.3 Bottom collecting systems for downflow units

Figure 2.45 shows a number of industrial solutions to this problem:

a) The bottom dish of the mild steel shell is filled with concrete which is trowelled flat, and the whole of the inside of the unit is covered with sheet rubber over steel and concrete. A perforated pipe network is buried in silex (crushed flint) which has a higher headloss and gives a better distribution than gravel. This is an excellent system for cation units, and has been successfully used for anion units, as the attack on flint by caustic soda is very small.

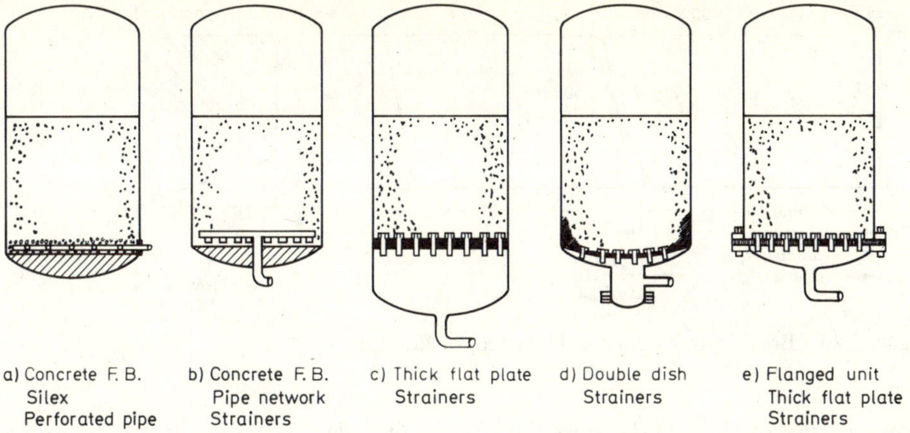

Figure 2.45 Bottom collecting systems.

b) A concrete-filled shell as before, with a pipe network having on its underside fine strainers through which resin cannot pass, and which give an appreciable headloss to ensure uniform collection. This is a good system, but care must be taken to eliminate any dead pockets of resin between the pipework and strainers.

c) A flat plate with strainers. This system gives ideal collection results, but has two significant disadvantages. First, the flat plate must be made very thick to avoid distortion, and the cost of drilling is high. Secondly, the space below the plate must be large enough to permit access to the rubber lining, which gives inefficient rinsing. The same rubber-lining problem makes it difficult to use a division plate of normal thickness, with supporting pillars. However, a solution as in d) can also be used for c).

d) A double-dished unit with strainers. Successful results have been claimed by one company for a unit with two bottom dishes welded together, the upper one being perforated for strainers. A large manhole in the outer dish permits access of an operator's head and shoulders for rubber lining. This is a fairly difficult process, and the diameter is limited to about 2.5 m. In principle, there should be a loss of efficiency due to the curvature of the division plate, and also because strainers cannot be placed near the edge, where a dead zone is formed. Nevertheless, good results have been obtained with this system.

e) A technically excellent but expensive solution uses a unit in three parts, an upper cylinder, a flat thick plate, and a lower dish, each separately rubber-covered and bolted together through flanges.

2.2.15.4 Flow rate and bed geometry

The whole of the information needed for the calculation of water treatment plant sizes is available, in principle, in the data sheets issued by the resin manufacturers.

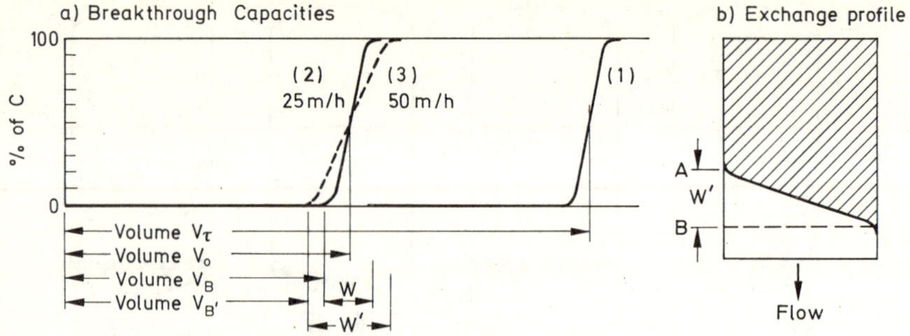

Figure 2.46 Breakthrough curves. The effect of flowrate.

However, use of these tables does not lead to one plant only, for each case, and a range of possibilities always exists. Figure 2.46 illustrates a typical case, in which it has been assumed that the regeneration level chosen gives a capacity, to the breakthrough point, exactly half the total capacity of the resin.

Curve (1), Figure 2.46 a, is the breakthrough capacity which would have been obtained if the resin capacity could have been totally used. In practice this never occurs, except in the first cycle of new resin supplied in the sodium form and used for calcium removal. Curves (2) and (3) are results observed at 25 and 50 m/h linear flowrate, the length of the breakthrough curves being *W* and *W'*. In practice, these would not be observed, as the process would be stopped at the breakthrough point.

Then, if solution concentration $= C$,

Total possible loading $L_T = C\,V_T$

Achievable loading at the chosen regeneration level $L_O = C\,V_O = 0.55\,L_T$

Breakthrough-point loading at 25 m/h $L_B = C\,V_B = 0.50\,L_T$

Breakthrough-point loading at 50 m/h $L_B = C\,V_{B'} = 0.46\,L_T$

It follows that, while the doubling of the flowrate has increased the breakthrough curve by a factor of only 1.7, the reduction in operational capacity is lower still, as $V_{B'} = 92\%$ of V_B.

This occurs because, as shown in Figure 2.46 b, the "exchange profile", in which useful action occurs, is only a small fraction of the total bed. All resin above level A is exhausted, and that below level B is not yet needed.

There are thus two potential routes to reducing plant cost. The first is to reduce the bed depth, at constant linear flowrate, to just over *W'*. The resin quantity, and therefore the duration of the service cycle, would decrease correspondingly. The practical difficulties of this approach are obvious. The smallest imperfections in uniformity of flow would cause premature breakthrough. Nevertheless, a Canadian company, EKO-TEK, has recently announced their "RECOFLO" deioniser system, using bed depths of under 15 cm, with diameters over 1.0 m.

The opposite approach is to increase linear flowrate (i. e., reduce diameter). This has two practical limitations:

a) Capacity. Typically, in the hydrogen ion exchange of London water, (7.0 mequ/l cations), a reasonable regeneration level gives an operating breakthrough loading of 700 mequ/l, and the service cycle is thus 100 bed volumes. With a normal bed depth of 1.8 m, a linear flowrate of 50 m/h gives a volume throughput of 27.8 bed volumes per hour, and a cycle time of 3.6 hours. This is the minimum practical figure, in an automatic plant, to allow for regeneration and maintenance. Manual plants normally require at least 8 hour cycles, to permit regeneration once per shift, and the flowrate would be correspondingly reduced.
b) Headloss. If the ionic content of the raw water is less, the capacity limitation disappears. With a Welsh water, at 1.0 mequ/l cations, the treated water quantity would be 700 bed volumes, and the cycle 24 hours.
The flowrate could be raised to 350 m/h without reaching capacity limits, but this figure would be impossible for hydraulic reasons. Headloss through the bed increases with flowrate, and at 50 m/h, there is a loss of 1 bar through a 1.8 m bed depth. For good bottom collection, there must be a further similar figure through the strainers, and if three units (cation, anion, mixed bed) are in hydraulic series, the pressure required at the head of the system gives high pumping costs, and excessive plate thickness in the exchanger units.

In practice, average plants have flowrates of up to 30 m/h, with bed depths of up to 2.0 m. For low TDS waters, rates of up to 60 m/h are used; conversely, for high TDS waters, where capacity is the limiting factor, bed depths of up to 3.5 m have been used with success. "Polishing" units, considered below, have special design features permitting rates up to 120 m/h.

2.2.15.5 Mixed beds

Figure 2.47 shows the regeneration sequence for a mixed bed unit. The two resins are separated by upflow water, after which the upper layer of anion exchanger is regenerated with caustic soda and then the lower layer by means of acid. The order of these two steps can be reversed, and some manufacturers use simultaneous flow, with downflow caustic and upflow acid, both leaving at the centre point. After rinsing out the majority of the excess regenerant, the resins are re-mixed with air, given a final rinse and are ready for service. The principles are therefore simple, and in small units excellent results are readily obtained. Larger units result in design complications. Figure 2.48 shows a typical design.

The water inlet, which is also the backwash outlet, is provided with a fine strainer system to prevent loss of the anion exchange material during bed separation, while permitting any general dirt or broken resin debris to leave the system. Either of these, if collecting on top of the bed, would act as a piston and impose very high stresses on the resin, leading to further breakage.

The interface distribution/collection system imposes a severe barrier to water flow, the total free flow area being as little as 40% of the total area. For this

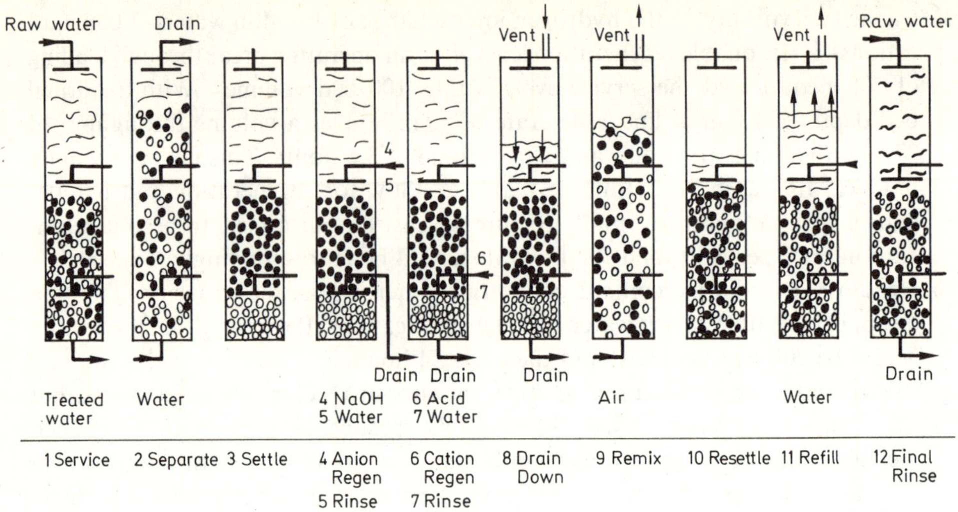

Figure 2.47 Mixed-bed regeneration sequence.
(○ Cation resin, ● anion resin).

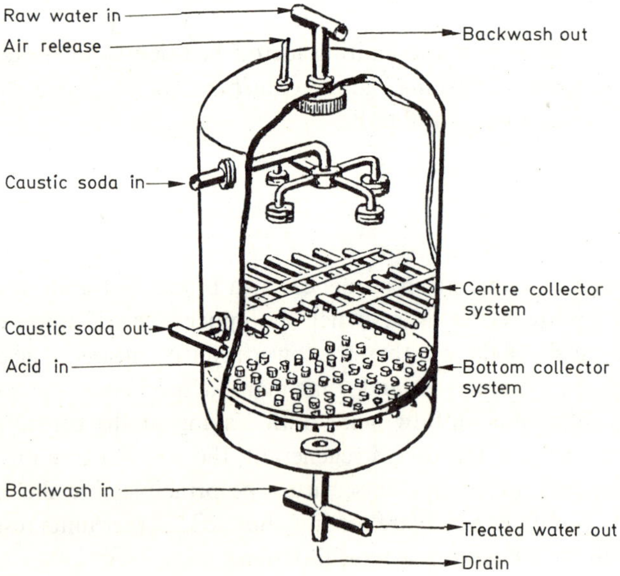

Figure 2.48 Typical mixed-bed unit.

reason, internally regenerated mixed bed units normally have a flow rate limitation of 25 m/h.

The burying of the pipe system and its strainer buttons in the resin gives potential blocking problems. With the sequence shown in Figure 2.47 the caustic flow is inwards through the strainers, and the following acid flow is outwards, thus tending to clear any blockage. When simultaneous acid and caustic treatment is practised, the flow is always inwards, so self-cleaning does not occur. It is then preferable not to use strainers, but a perforated pipe system wrapped with successive layers of plastic mesh, the outermost one being sufficiently fine to resist all penetration by broken resin particles.

The bottom-plate, as is shown in Figure 2.45 c, has the problems already mentioned, and further ones imposed by the necessity for bed separation and re-mixing. If the bed separation is not perfect, some anion resin is left at the bottom of the unit and is soaked with acid during cation regeneration. The resin is converted to the bisulphate form, which hydrolyses during the subsequent run, putting traces of acid into the treated water. If re-mixing is inefficient then both capacity and water quality suffer. A bottom-collecting system which gives uniform take-off of treated water may nevertheless be inefficient for bed separation and remixing. A difference of only 5 mm in the level of the bottom plate, due to distortion or tilting, can cause almost the whole of the air to emerge at the highest point, leaving most of the bed undisturbed. During separation, the upflow water tends to travel at an angle which leaves pyramids of unseparated resin between the buttons. A number of systems have been used to overcome this problem, a typical one being a strainer button with a tailpipe containing a plastic ball. During flow to service, the ball drops, exposing a large number of vertical slots in each strainer. During backwash it rises, blocking these slots, and exposing a single horizontal slot, which gives a high horizontal velocity to the water, effectively sweeping the bottom of the unit. The ball remains in the upper position during air input, giving a restriction on air flow such that the water level is forced down below the tailpipes, and air emerges equally, at high velocity, from each button.

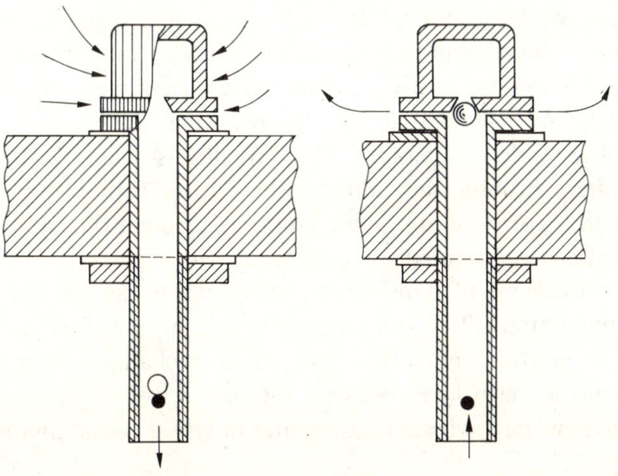

Figure 2.49 Dual-action strainer for mixed beds.

2.2.15.6 Interface problems

Difficulties are caused by regenerant entering the wrong resin layer. Sodium hydroxide diffusing into the cation resin can cause precipitation of heavy metals, while sulphuric acid entering the anion layer converts the resin into the unwanted bisulphate form, and also causes precipitation of organic matter, if it is present. If the interface effects were limited to diffusion, they might be small, but in practice serious flow malfunctions can occur.

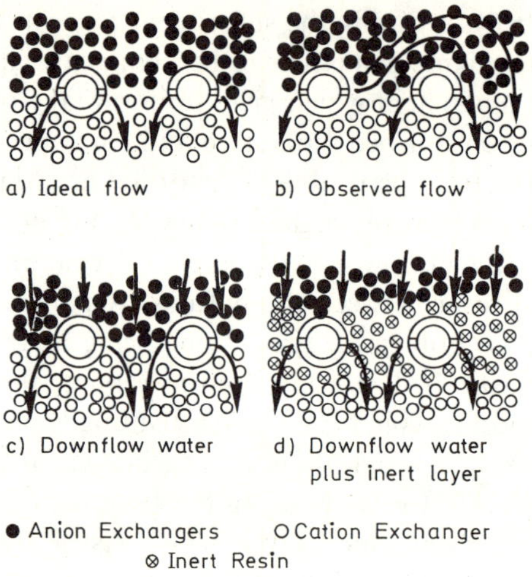

Figure 2.50 Interface flow patterns.

Figure 2.50 a shows the required flow pattern of acid regenerating the cation resin. In practice, cinematography of coloured solutions in transparent units has shown, that a chance loosening of the anion bed can cause an upward flow at one point, as in Figure 2.50 b. Repeated over the bed area, this factor causes serious contamination of the anion exchanger. The difficulty is overcome by means of a downflow of water through the anion resin during cation regeneration (see Figure 2.50 c), with a corresponding buffer upflow through the cation resin while the anion material is being treated.

A further improvement of considerable significance has been introduced by several resin manufacturers. This is an inert polymer, in bead form, with a density intermediate between those of cation and anion exchange resins. During bed separation, it forms a sharp layer between the two active resins, and when used together with buffer water it virtually eliminates interface contamination problems (Figure 2.50 d).

2.2.16 Remote regeneration of mixed beds

In large industrial installations it is now rare for mixed beds to be used alone, and combinations including preliminary deionisers are common. Under these circumstances, the chemical load on the mixed bed unit is small, and even with flowrates of 60 m/h, the maximum which can be used with a unit such as that in Figure 2.48, the operating cycles are long. In order to use even higher rates, up to 120 m/h, for "polishing" mixed beds (that is, those at the end of a train) the units are made free from all internal obstructions, and the exhausted resins are removed to separate units, designed specifically for the regeneration procedure. Similar procedures are used in the polishing of condensates from the turbines, in power generation. An advantage of remote regeneration, in large plants, is that one regenerator vessel only is needed, even when there are several parallel trains of operating units. Many of the available systems have the further advantage that remixing takes place in the external unit, thus making it unnecessary to provide above-bed space in the operating vessel.

Simple regenerator (Figure 2.51). The unit is a standard mixed bed, made tall and narrow for maximum efficiency, and with a bottom-collecting and air-distribution system designed only for regeneration, with no requirement to receive the much higher service flow rate. The system may be used as shown, or with an inert layer. In either case, remixing by air can take place either in the regenerator, or after return to the service unit.

Cation resin layer (Figure 2.52). This method is used in cases where the water entering the mixed bed contains hardness or heavy metals. In an internally regenerated unit, it is impossible to achieve perfect remixing, and the uppermost resin layers always have an excess of anion exchanger, which can cause precipitation. In the cation layer system the exhausted resins are pumped to a regenerator unit, from which, before remixing, a small amount of cation resin is pumped into a separate vessel. The mixed resin is then pumped back to the service unit, and the cation material is placed on top. Metals in solution and colloidal hydroxides are taken up in this layer.

Elimination of interface (Figure 2.53). An Australian system uses the principle of a holding tank in a different way. After remote regeneration and without re-mixing, most of the cation layer is pumped back to the service vessel, leaving the top few centimetres in the regenerator. The lowest resin in the regenerator, consisting of the remaining cation together with some anion material, is then pumped into the holding vessel while the anion exchange resin is sent to the operating unit for re-mixing. All interface problems are eliminated. This method is further improved by using an inert layer which never enters the service vessel, but takes part in the regeneration sequence only.

Fig. 2.51

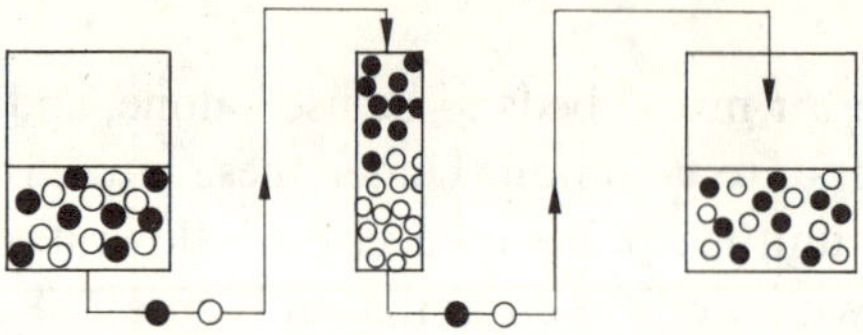

Fig. 2.52

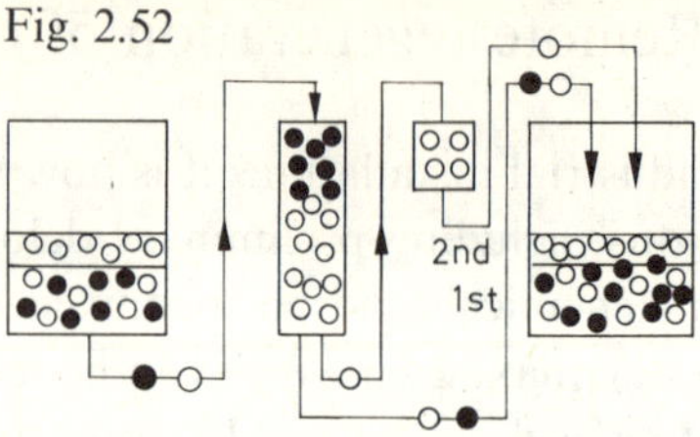

Fig. 2.53

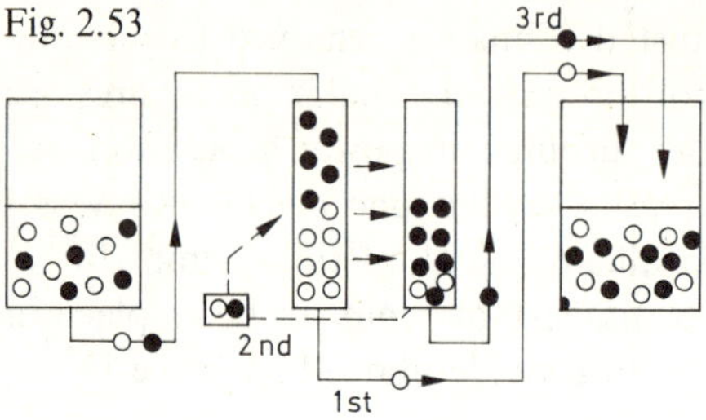

Fig. 2.54

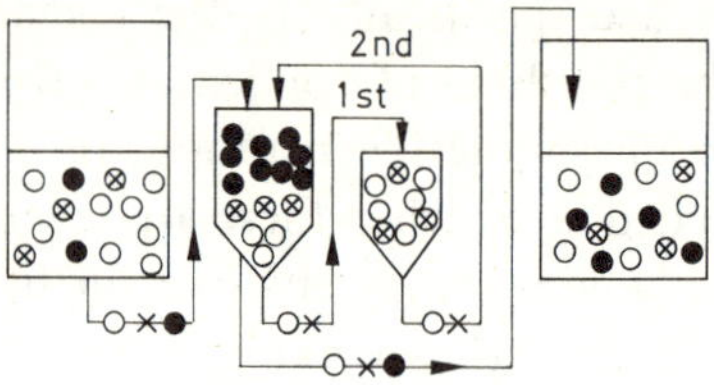

Figure 2.51 Remote regeneration of mixed beds. Simple system. (○ Cation resin, ● anion resin).

Figure 2.52 Remote regeneration of mixed beds. Cation resin layer. (○ Cation resin, ● anion resin).

Figure 2.53 Remote regeneration of mixed bed. Interface elimination. (○ Cation resin, ● anion resin).

Figure 2.54 Remote regeneration of mixed beds. Coneseps system. (○ Cation resin, ● anion resin, ⊗ inert material).

Coneseps (Figure 2.54). A most successful system is the procedure described by Emmett and Grainger [32]. The exhausted resins are transferred to a cone bottom unit, which acts both as resin separator and as regenerator for the anion exchange resin. The process normally uses an inert layer, but the separation achieved by the cones is claimed to be so good that the method can be used without it. After upflow wash, the lower layer of cation resin is pumped, free from anion material, but with most of the inert layer, into a separate unit for acid treatment. The anion resin is regenerated separately, and the cation resin is then returned to the separation vessel to be remixed and returned to the operating unit.

2.2.17 Counterflow regeneration equipment

The fundamental arguments for counterflow operation were generally accepted, and different types of successful equipment were in use by 1955. It is surprising, therefore, that the process has entered into general usage only slowly. The reason is almost certainly associated with the historical development of fixed bed units from pressure sand filters. All early designs of counterflow equipment were based on the requirement of backwashing the bed completely once every cycle. Not only did this complicate the equipment, but it also reduced the efficiency of the counterflow process, since some mixing always occurs during backwash, and the most essential feature of counterflow operation is that the bed should be undisturbed. The principle of complete backwashing has been abandoned in almost all modern designs, in some of which the resin remains totally packed, while in others, only the top 10 cm of the bed is regularly washed.

2.2.17.1 The Pressbed system [33]

Invented in 1967, this was an ingenious method of changing the bed configuration, so that it was packed for treatment and regeneration while having the necessary free space for backwash. In the first version, a long rubber bag, hanging in the column, caused little obstruction (Figure 2.55). Towards the end of backwash it was inflated under water pressure, giving a packed bed for upflow regeneration. The system was effective, but had two problems. First, the change in shape permitted disturbances in the bed which reduced the true counterflow effect; in addition, there were at times problems with bag breakage. In a second version, the vertical bag was replaced by a cylindrical one, in bellows formation, above the bed.

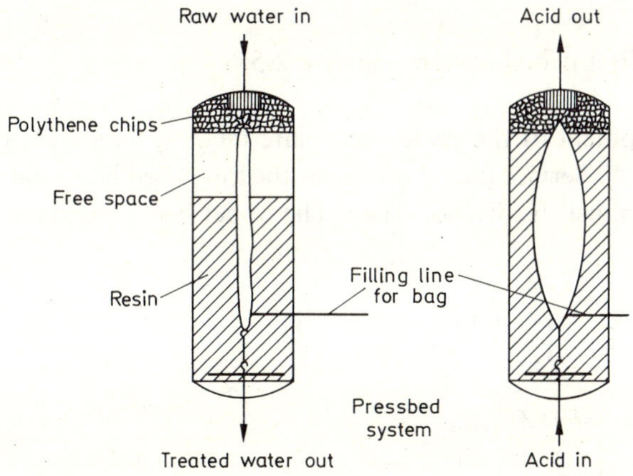

Figure 2.55 Pressbed system.

2.2.17.2 The perforated plate system

This was a successful early method, originating in Germany. The unit contained a coarsely perforated plate about 1/3 from the bottom. Treatment was carried out at an upflow rate sufficient to lift the resin to fill the upper compartment, and to be

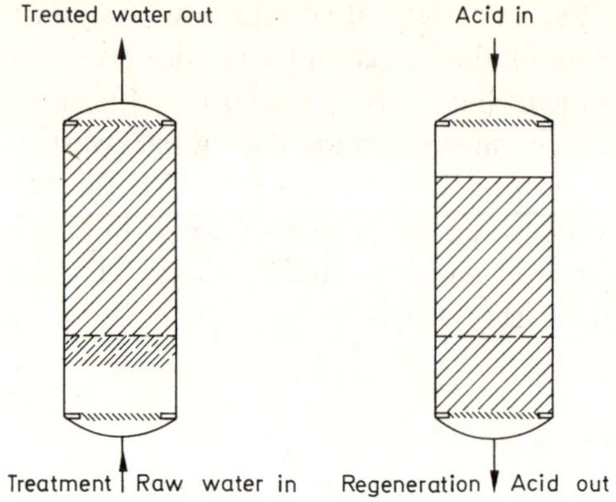

Figure 2.56 Early perforated plate system.

packed upwards against the plate in the lower section. The flow rate through the perforations was high enough to ensure that resin would not fall back, even if the service flow, once established, was reduced. For regeneration, the bed was packed downwards, by a flow of water, followed by regenerant. Some backwash was then possible, at a low flow rate, but was not normally used. The system worked well, but has been superceded by later modifications.

2.2.17.3 The Lewatit Liftbed system (Figure 2.57)

This modern development of the perforated plate method uses a plate fitted with resin-proof nozzles [34]. Service flow is upwards, the lower bed being partly fluidised, and partly packed against the division plate. The upper bed is permanently packed,

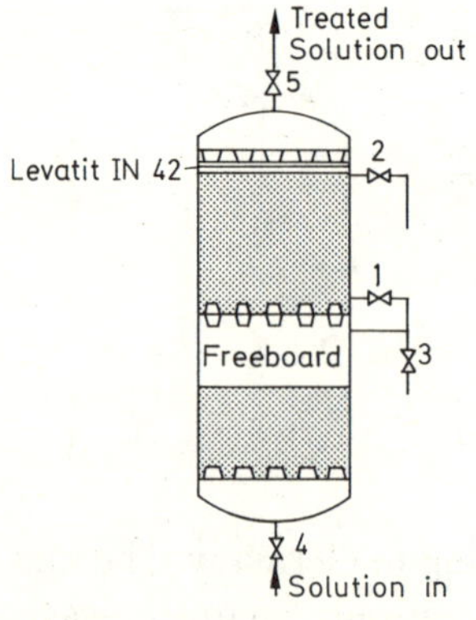

Figure 2.57 Lewatit Liftbed System (Reproduced with permission of Bayer, Leverkusen).

with flow being out at the top through a strainer plate, which may also have floating inert resin above the exchanger. Backwash is not a regular feature; when necessary, it is carried out on the lower resin only, the washings leaving through a pipe below the centre plate. Regeneration is downflow through the whole unit, being completely counterflow for the top bed, and partly so for the lower. If, for any reason, the upper resin requires cleaning, part is transferred into the lower chamber, giving freeboard in the upper one.

2.2.17.4 The Lewatit Rinsebed system (Figure 2.57 a)

In appearence this resembles the early perforated plate method, with the addition of a network of pipes and strainers just below the plate [34]. During treatment, water rises through a packed bed, leaving through this network. Regeneration is downflow through the whole unit, while backwash is upflow through the perforations and out at the top. An advantage claimed for this system is that any broken resin will collect above the level of the take-off strainers, so that no blocking can occur.

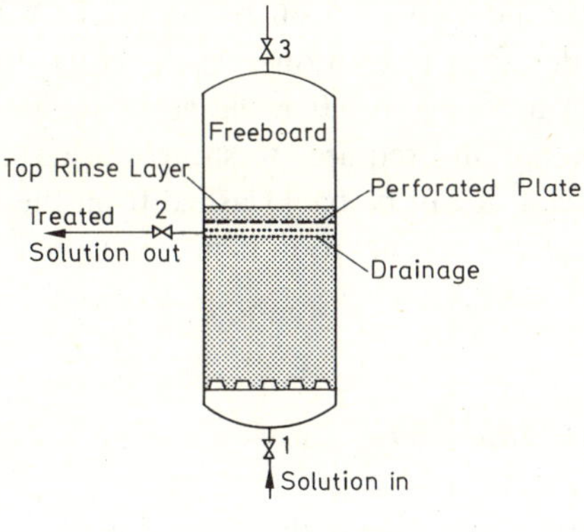

Figure 2.57a Lewatit Rinsebed System (Reproduced with permission of Bayer, Leverkusen).

2.2.17.5 The buried top collector system (Figure 2.58)

Although invented in 1955, this method was little used for about 10 years, but has since become a standard practice [35]. A pipe and strainer network is situated just below the top of the resin bed. Downflow treatment and backwash are normal, while regeneration takes place upflow, leaving through the pipe network, the resin bed being held in a stable condition by a downflow of water leaving with the waste regenerant. In some plants compressed air, entering at the top and leaving with the waste regenerant, is used to hold down the bed. The results are equally satisfactory.

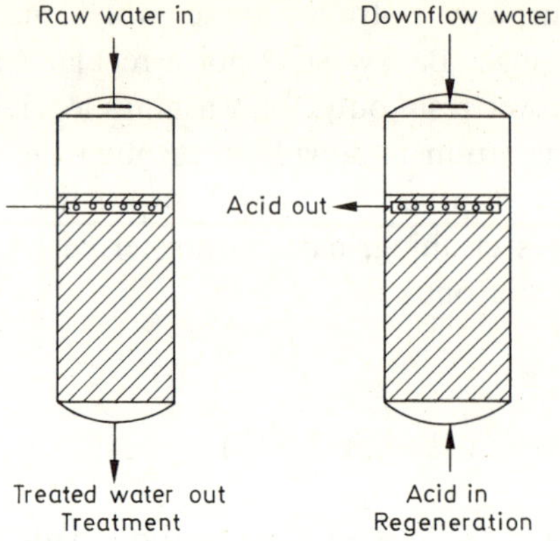

Figure 2.58 The buried top collector system.

The wide experience with this system showed that backwash was disadvantageous. Far from keeping the resin in a constantly graded condition, it tended to mix it, and therefore to destroy partially the counter-current effect. It has become standard to place the buried collector about 25 cm below the top of the bed, and to backwash only this small upper portion of resin by means of water entering through the buried collector. Under these conditions, results equal to ideal laboratory practice can be achieved in units up to 5.0 m diameter. If, for any reason, the whole bed becomes contaminated and requires washing, an event which normally occurs only after 25 – 30 cycles, it can be backwashed from the bottom inlet to the top. After such a wash, a double regeneration is necessary to overcome the effect of mixing caused by the full wash.

2.2.17.6 The split-flow system (Figure 2.59)

A disadvantage of the simple buried collector method is the waste of water used for holding down the bed. An ingenious method for overcoming this problem is the split-flow technique. The buried collector is placed about one-third the distance from the bed top. Service is downflow as before, and backwash upwards through the top one-third of the bed only. The regenerant flow is split so as to flow downwards from the top and upwards from the bottom, all leaving at the buried collector point. Two-thirds of the bed therefore has true counterflow regeneration, with all its advantages. Although the top third is coflow regenerated, a high level of efficiency can be obtained by using a low regeneration level, without affecting treated water quality, which is controlled by the bottom section of the bed. A further advantage, when using sulphuric acid, is that as calcium is absorbed largely at the top of the bed, the low acid concentration needed to avoid precipitation may be applied to the top section only, while concentrations of up to 5% w/v are used

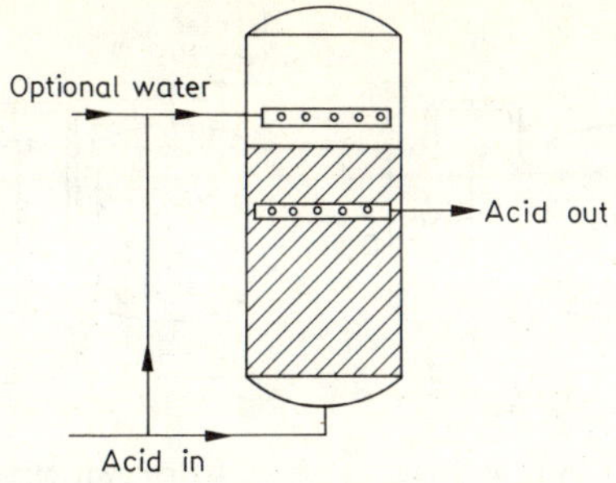

Figure 2.59 The split-flow system.

for the lower section. This system, which increases capacity and regenerant usage efficiency, is conveniently carried out by pumping 5% acid through a common pipe feeding both inlets, adding dilution water just before the upper point.

2.2.17.7 Completely filled units

If a column is provided with a resin-proof strainer system at the top and bottom, and is almost completely filled with resin, then a simple counterflow process is available by operating upflow, and regenerating downflow. If the water is not completely clean, then the bottom of the resin bed acts as a filter, and the suspended matter so collected is discharged downwards by the regenerant flow. This system has been used by several equipment suppliers for small water softeners, up to 50 cm diameter, with beds up to 1.4 m deep. Units have operated for over 15 years on waters of varying turbidity, at high efficiency and without blockage. Nevertheless, the dedication to the backwash principle has meant that few manufacturers have attempted to use large, fully-packed beds. At least two systems have now been introduced.

2.2.17.8 The Amberpack process (Figure 2.60)

Essentially, this equipment consists of two units [36]. One is a completely packed bed, in which the resin is held between upper and lower strainer systems. Treatment is upflow, and regeneration down, with no backwash or disturbance, to give the best achievable counterflow effect. The second is a simple, cone-bottomed washing unit, which can be operated in different ways according to requirements. If, from time to time, fines or broken resin beads are found at the unit top, a small quantity of resin is moved into the washer, cleaned and returned. Since, during the return process, the newly settled bed is not fully compacted, it is not possible to return all the resin to the top space. Accordingly, after upflow service has been in operation

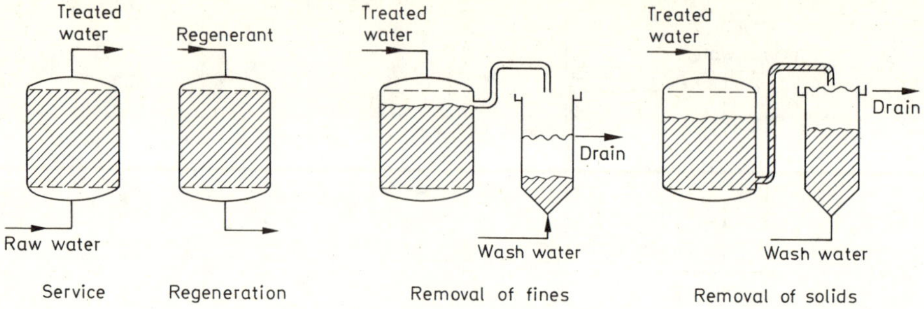

Figure 2.60 Fully packed beds. The Amberpack Process.

for some time, and the bed is fully compacted upwards, the remaining resin in the washer is pumped into the bottom space, without interrupting the service flow. If, on the other hand, the lower end of the bed, by acting as a filter, has accumulated solids which have not been discharged during regeneration and rinse, it is cleaned by transferring the bottom 20% of the bed to the washer, and returning, again in two stages, after cleaning. Both these processes are regarded as occasional, the unit operating normally as an unwashed, full-packed system.

Clearly, the success of this operation will depend greatly on the resin transfer piping. On those occasions when resin cleaning is required, it will be important to remove it as an even horizontal layer, which must be difficult in a large unit. The developers of the process, Rohm and Haas, supply resins in special particle sizes specifically for this process.

2.2.17.9 The Lewatit WS system [34]

Closely similar in principle to Amberpack, with upflow service and downflow regeneration, this system uses a unit with a small amount of free space and a floating inert layer. In appearance the column is thus almost identical with Figure 2.61, but operates in the opposite direction to UFD, etc.

2.2.17.10 Shallow packed beds. The Recoflo method

A unique system has been announced by ECO-TEC of Canada [37]. For both cation and anion units, beds of fine resin beads are used, having depths of less than 150 mm, and diameters over 1.0 m. The resin depth is only slightly greater than the exchange profile, (shown as W'' in Figure 2.46), and the profile is itself reduced in height by the use of fine mesh resins. The time cycle is also greatly reduced, as compared with normal practice, by having in the system a much smaller resin quantity. Thus the downflow service cycle can be from 5 to 60 min duration, according to the water analysis, while regeneration takes only 1 min, counterflow, followed by a recirculatory rinse for 5 minutes, resulting in water of 1.0 μS/cm conductivity. In one case, the rinse water from silicon wafer production, with a conductivity of 25 μS/cm, was treated for re-use in a plant with resin beds only 75 mm deep, resulting in water having 0.06 μS/cm quality.

Design details have not been disclosed, but the manufacturers must obviously have overcome two severe problems. First, the achievement of absolutely uniform flow patterns, which must obviously be accurate to better than 10% of bed depth; secondly, the design of resin-retaining end plates, which are not protected by inert resin layers, and which must not suffer any blocking by the very fine particles used in the bed. The method, about which little has so far been published, is clearly one of considerable interest.

2.2.17.11 The upflow Degremont (UFD), Esmil packed bed (EPB) and Dow UPCORE processes

These three processes, which are almost identical, operate in the opposite direction to Amberpack. Treatment is downflow, in a unit which is almost fully packed, but which has about 150 mm free space to accomodate changes in resin swelling resulting from differences in ionic form. There is a floating inert layer, which remains permanently at the top of the unit. When the bed is exhausted, it is lifted by a compacting upflow, followed by regenerant, with the whole bed now being held against the inert layer, and the free space underneath it. In normal operation, there is no backwash step, with any dirt that enters being filtered out by the inert layer and discharged to drain during upward compaction and regeneration. It is also

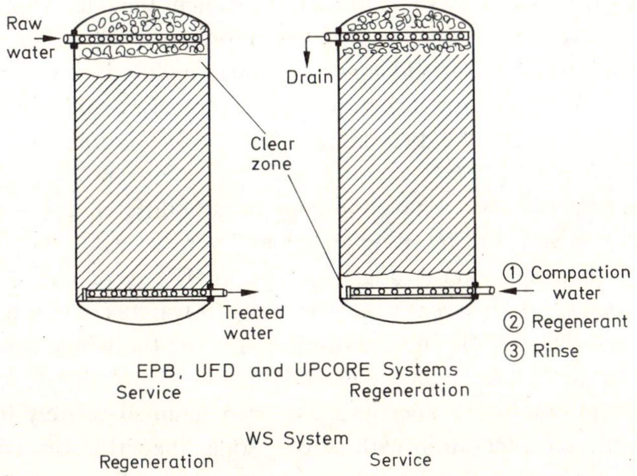

Figure 2.61 Packed beds with freeboard and floating inert resin.

claimed that any broken resin beads work their way to the top of the unit, because of the small alternation in bed position, to become discharged during regeneration. The EPB [40] and UPCORE [41] processes rely totally on this principle, whereas the UFD [38, 39] system provides for a small external unit into which the top of the resin bed is transferred and washed every 200–400 cycles.

The free space might be considered to be a disadvantage, for allowing some bed movement to the detriment of the full counterflow effect. This is, however, extremely minor, with the majority of the bed rising and falling as an undisturbed plug, and only the bottom few centimetres becoming momentarily fluidised. The method has the advantage that it can be fitted into existing units with only minor modifications; further, since the service flow is downwards, there are no problems from varying flowrates, or even stoppages.

2.2.18 Layer beds

Many of the counterflow techniques lend themselves to the use of two resins in the same unit. Normally, this requires downflow operation to service, so that the buried collector system, the split-flow technique, and the UFD, EPB and UPCORE processes are particularly suitable. In the case of cation resins, the carboxylic materials are less dense than the sulphonic resins, and during backwash (when used) form a clear layer on top. This effect has been reinforced by the resin manufacturers, who supply resins in layer form; the sulphonic material have a slightly larger particle size than standard, while the carboxylic material is finer. Counterflow regeneration gives the normal high quality treated water, and the waste acid from the sulphonic resin is consumed with almost 100% efficiency by the carboxylic material. The technique applies only to waters containing bicarbonate alkalinity; and in these cases there is, further, a problem that the upper resin layer becomes highly loaded with calcium. To avoid precipitation in this resin, it is necessary to use a low sulphuric acid concentration, which reduces the regeneration of the lower layer. The split-flow technique overcomes this problem by permitting different concentrations in the two layers. The problem does not arise when hydrochloric acid is used.

Anion exchange resins do not have the same density difference, but an adequate level of separation is achieved by using an upper layer of weakly basic material of finer particle size than the lower strongly basic resin. The system operates satisfactorily, but must be used with care for the deionising of waters containing significant concentrations of silica. At the breakthrough point for the whole bed, the upper layer of weakly basic material is in the bisulphate and hydrochloride forms, and is therefore in an acid condition. The silica has been retained entirely by the lower layer. During upflow regeneration with caustic soda, the earliest portion of regenerant is used at high efficiency, largely in removing silica from the resin. As this solution passes upwards through the weakly basic material, the residual caustic soda is neutralised, and the effluent can become acid, thus precipitating silica gel in the weakly basic resin layer. In principle, the caustic soda which follows reverses this process and removes the precipitated silica. In practice this is true with a well-designed plant, but any imperfections in flow pattern can cause problems. Since counterflow regeneration of strongly basic resins, particularly in the case of Type II materials, gives efficiencies almost identical with those achieved with weakly basic resins, it is very possible that anionic layer beds, which have been of great value in the 1960's and 1970's, may no longer be needed.

2.2.19 Combination counterflow units

In the interests of saving floor space plants have been installed with two separate units, for example cation and anion, in the same vertical cylinder, thus giving the appearence of one column, while they are actually hydraulically independent. A further stage of development is to carry out different processes in different parts of the same unit, which operate as one entity.

2.2.19.1 The Multistep system [42, 47]

This has been developed from the Lewatit WS system. Figure 2.62 shows it in use as a two-stage deioniser, and other combinations are possible. Service flow is up through the whole unit. Regeneration of the upper resin is downflow, with a buffer upflow of water through the lower vessel to prevent regenerant entering the wrong unit. Lower layer regeneration takes place downwards from the interlayer pipework, followed by a buffer water flow.

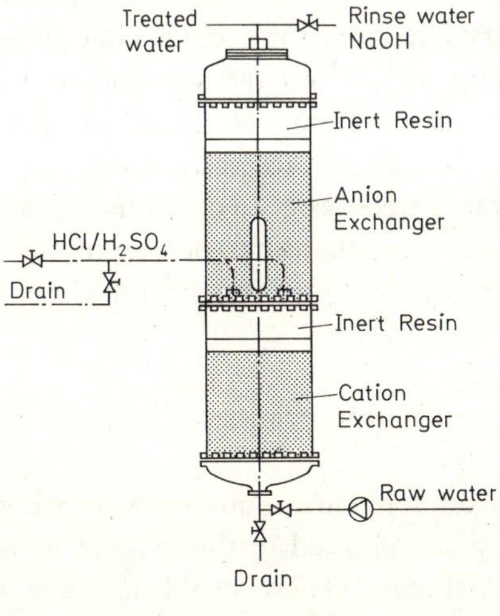

Figure 2.62 The Multistep System (Reproduced with permission of Bayer, Leverkusen).

2.2.19.2 Counterflow operation. The future position

There are no fundamental or theoretical arguments for coflow operation, which has persisted from the early days of modified sand filters only because of the simplicity of coflow plant design. With the availability of a range of different counterflow designs of proven reliability, it is probable that coflow design will disappear from future deionising plants, although it may continue for small industrial water softeners, whose operation is simple and efficient irrespective of the mode of operation.

2.2.20 Continuous countercurrent ion exchange

In most chemical engineering processes, there is an advantage to be gained by continuous operation as compared with laboratory batch procedures. Classical ion exchange technology was a compromise. It was more efficient than a simple batch contact, in that a large volume of water passed through a relatively small plant, and the interruptions to flow required for regeneration were eliminated in practice by using two or more parallel lines. Nevertheless, two of the disadvantages of batch operation remained. The first of these, inefficiency due to coflow operation, has been overcome by counterflow working; but the second, the need for much more resin than is required to contain the exchange zone, remains a fundamental problem of fixed bed practice, with the exception of the unique "Recoflo" short bed design. In a continuous countercurrent plant the absorber unit, in principle, need only be large enough to contain the exchange zone, and the resin quantity can be greatly reduced (Figure 2.63 a). In practice, however, the reduction is less than the ideal conditions of Figure 2.63 b. The height of the exchange zone varies, due to changes in water analysis and flow rate. It is necessary to design the plant for the most adverse conditions, so that it must be oversized for average conditions (see Figure 2.63 c). Moreover, a continuous plant carries out the separate operations of absorption, washing and regeneration in different vessels at the same time, and therefore carries a larger resin stock than is used for the absorption process alone. In practice, the total resin inventory in a well-designed continuous plant is less than that of a corresponding fixed bed unit, but the difference is considerably smaller than would appear from a simple consideration of zone length (Figure 2.63 d).

2.2.20.1 Mechanical problems of continuous plants

There are a number of difficulties encountered in continuous working, which caused considerable problems in the early years, and led to this type of plant falling out of favour, after widespread initial success. These problems have largely been overcome by modern designers, and continuous plants have taken their place in

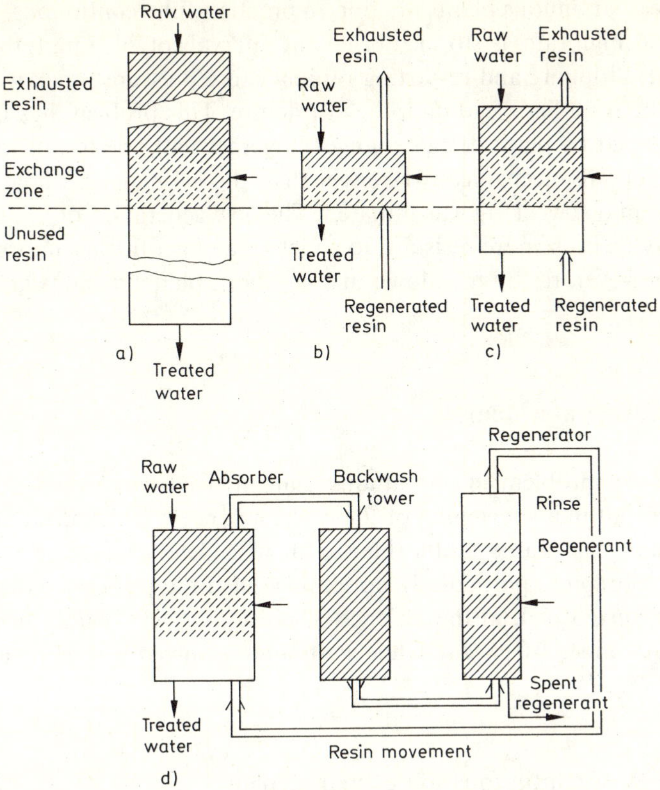

Figure 2.63 Development of continuous ion exchanger. (a) Fixed bed unit. (b) Ideal continuous absorber. (c) Practical continuous absorber. (d) Practical complete unit.

the range of industrial equipment. They have not, however, achieved the dominating position originally expected, chiefly because counterflow fixed bed plants have similar operating advantages, and greater simplicity of construction.

Resin distribution. The problems of ensuring completely even water distribution in fixed bed plants by means of multiple orifices have been discussed in Section 2.2.15. In continuous units there is the corresponding problem of achieving even input and removal of resin across the area of the bed. It is not possible to overcome this completely, and any resulting distortion of the wave front must be accomodated by an increase in plant size, as in Figure 2.63 c. Fluidised beds do not suffer from this defect, but have not so far been used in water treatment.

Resin breakdown. This was a severe problem in early plants, the replacement rate for broken resin exceeding 100% per year. There were two main causes:

a) Valves closing on resin, causing crushing. This has been eliminated in many modern designs.

b) Packed bed continuous plants are not, in practice, fully continuous. Water flow, normally at high rate, is stopped briefly at intervals of 2 – 5 min, to allow resin movement. Stopping and restarting rapidly cause pressure shock waves to pass through the resin bed, leading to bead fracture. This problem has been considerably reduced by plant design changes, together with the use of macroporous resins, which are mechanically stronger than the gel materials used at the time when continuous working was initiated. The new generation of tough gel resins, which may be even more resistant to shock, may be a further improvement. At present, however, resin breakdown in continuous plants remains greater than is desirable.

2.2.20.2 Treated water quality

This is a greater problem in continuous than in fixed bed plants. In the latter, many faults result in a shortening of the cycle, a concept which does not exist in a continuous unit. The same faults thus result in a deterioration of treated water quality, which becomes particularly noticeable when the highest quality is needed. Thus, a fixed mixed bed, in the absence of poisoning, will readily produce water of conductivity under 1.0 μS/cm, while a continuous mixed bed rarely achieves this figure.

2.2.20.3 Types of continuous ion exchange plant

The wide range of continuous equipment which has been described in the literature has been used in hydrometallurgy, and for other special purposes, more than for water treatment, but two main types have been successfully adapted for this purpose.

The Higgins contactor [43]. First announced in 1955, in a USAEC report, for metal separations in nuclear fuel processing, this equipment works in a closed loop, one version of which is shown in Figure 2.64. During normal operation all valves ABCD are closed. Raw water passes upwards through the absorber section, while regenerant and rinse are transferred down through a narrower vessel on the opposite side of the loop. At intervals of 3 – 5 minutes, valves ABC open, and a hydraulic impulse is used, for a few seconds, to force a small slug of resin round the loop in a clockwise direction, to collect in the open-topped backwash vessel just above valve D. Valves ABC then close, and after a brief interval for backwashing, valve D opens to allow resin to refill the hydraulic pulse chamber, and then recloses.

As resin movement is clockwise, while both service and regenerant flow are counterclockwise, the system is in true countercurrent, with the most highly regenerated resin making contact with the final treated water. There have been many variations in design, but all use the basic principle of countercurrent loop, with pulse transfer of resin in small slugs.

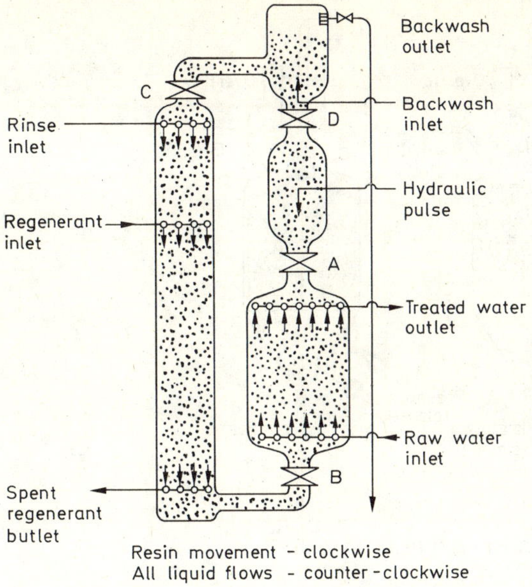

Figure 2.64 Higgins contactor.

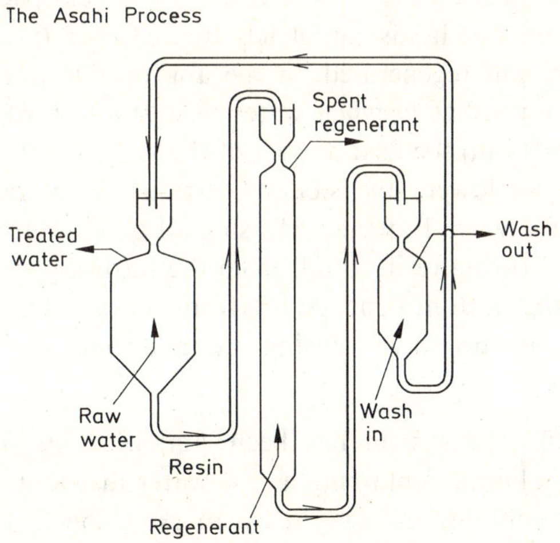

Figure 2.65 Asahi Process. Principles.

The Asahi process [44, 45]. This equipment was developed originally for copper recovery, but has been extensively used for water treatment. Again, there are many design variations, but all use the basic principle of separate vessels, for absorption, backwash and regeneration, as in Figure 2.65.

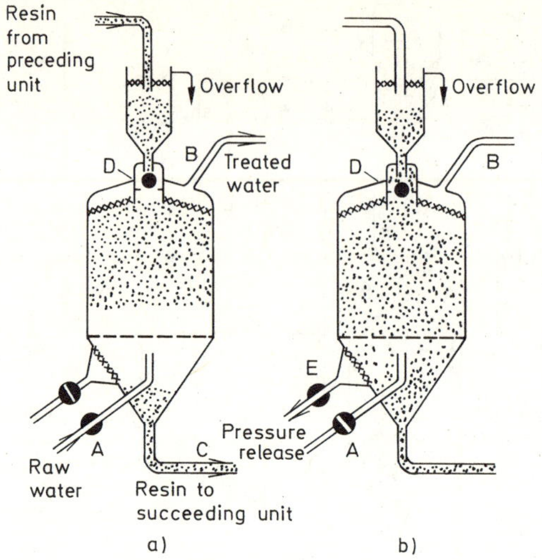

Figure 2.66 Asahi Process. Resin transfer system.

They differ in shape, according to their function, but each operates on the same broad principle (Figure 2.66). Raw water, flowing to service, or fresh regenerant, passing to drain, travel upflow though inlet valve A, a perforated plate system, an upper resin-proof screen and outlet valve B. The resin bed is held packed against the upper screen. The bottom cone holds completely treated resin (i. e., exhausted in the case of the absorber, and regenerated, in the appropriate unit) ready for transfer, which takes place immediately when water flow starts, through pipe C, which has no valves, and leads into the feed hopper of the subsequent unit. During most of the cycle, therefore, the lower cone is empty of resin. At intervals, valve A closes and the water discharge valve E opens, releasing all pressure from the unit. The floating ball D drops, allowing resin to fall from the feed hopper to fill both the operating column and the bottom cone. Apart from the ball D, which has a gentle closing action, there are no valves closing on resin, and attrition is thus minimised.

Degrémont modification [46]. The system has been simplified by changing the absorber vessel to a single column, containing a raw water distribution network, and omitting the backwash column, replacing it by an extension to the absorber feed hopper (Figure 2.67). The regenerator became a simple filled unit, operating downflow. This process was further developed to provide for continuous mixed bed deionising. The mixed resins from the absorber, Figure 2.67, are transferred in small slugs to a separation unit, and then to two separate regenerators. The rinsed resins are then sent back to the absorber feed hopper, which also acts as mixing chamber. Because the resins are dynamically mixed in this system, each absorbs the last traces of regenerant from the other, and the mixed resins are ready for use, with no final rinse stage, as in fixed bed practice.

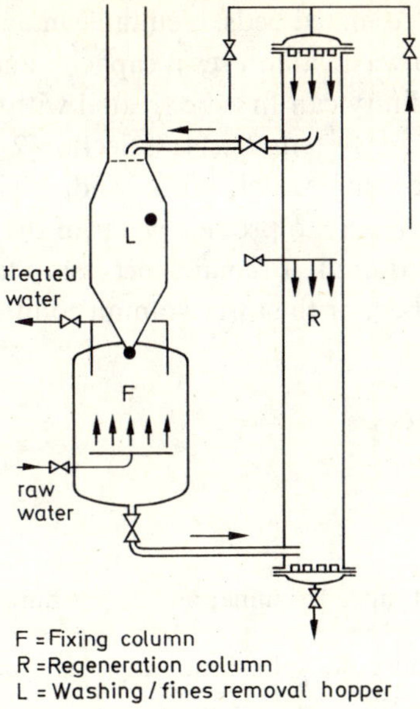

Figure 2.67 Degrémont Process (Reproduced with permission from the Degrémont Water Treatment Handbook, 1979, pages 335 and 336).

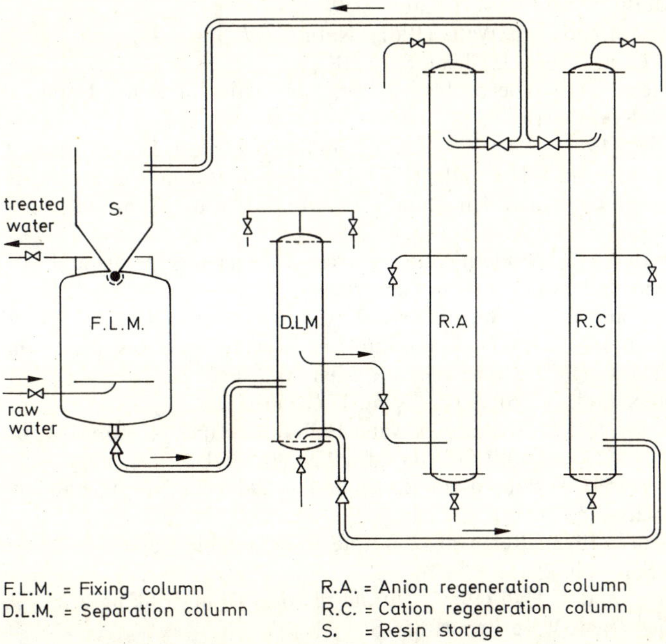

Figure 2.68 Continuous Mixed-bed Process (Reproduced with permission from the Degrémont Water Treatment Handbook, 1979, pages 335 and 336).

The continuous mixed system, like fixed mixed beds, is suitable mainly for waters containing low bicarbonate, so as not to waste anion resin capacity and regenerant on CO_2 removal. Waters with high alkalinity can first be treated with a carboxylic resin, in fixed or continuous process, as in System 2.2.11.7, Section 2.2.11.

Continuous mixed beds give good quality water, with conductivity less than 2.0 μS/cm. They are not claimed by the makers to provide water of the very highest quality, 0.05 – 0.1 μS/cm. When this is required, it remains necessary to use a three column system, with a final fixed mixed bed, or the three column counterflow Hipol method, with the final cation unit.

References

[1] W. S. Holden (1970). Water Treatment and Examination. J. A. Churchill, London, p. 383.
[2] H. S. Thompson (1850). On the absorbent power of soils. J. Roy. Agr. Soc. *11*, 68.
[3] J. T. Way (1850). J. Roy. Agr. Soc. *11*, 313.
[4] R. Gans (1905). Jahrb. Preuss. Geol. Landesanstalt *26*, 179.
[5] P. Smit (1939). US Patent 2 171 408.
[6] O. Liebknecht and United Water Softeners (1936). Brit. Patent 450 574.
[7] B. A. Adams and E. L. Holmes (1935). J. Soc. Chem. Ind. *54*, 1.
[8] G. F. D'Alelio (1945). US Patent 2 336 007.
[9] C. H. McBurney (1952). US Patent 2 591 573.
[10] E. Glueckauf and R. E. Watts (1961). Nature *191*, 904.
[11] J. Millar et al. (1963). J. Chem. Soc. 218.
[12] T. V. Arden (1965). The problem of organic matter in water demineralising. AIChE-IChem.E. Symp. Series *9*, 18.
[13] T. V. Arden (1968). Water Purification by Ion Exchange, Butterworths, London.
[14] E. W. Jackson and J. H. Smith (1977). Make-up treatment. Countercurrent regeneration experience in the United Kingdom. 38th Ann. Int. Water Conf., Pittsburgh, Penn. IWC-77-20.
[15] J. H. Smith (1980). Modern countercurrent ion exchange plants and the Hipol process. Soc. Chem. Ind. Conf., Bristol University.
[16] K. Odland and H. L. Pabich (1965). Proc. Int. Water Conf., Eng. Soc. West Penn. 143.
[17] R. Kunin and B. Vassiliou (1964). Ind. Eng. Chem. Prod. Res. Dev. *3*, 404.
[18] G. Boari et al. (1975). Ion Exch. and Membr. *2*, 127.
[19] D. E. Weiss et al. (1966). Aust. Chem. J. *19*, 561.
[20] Dept. of Env. (1986). Nitrate in Water Pollution. Paper No 26. MM 50.
[21] EC Directive (1980). Official Journal L 229, 80/778/EEC.
[22] I. Delius (1959). Verfahren zur Beseitigung schädlicher Nitratgehalte aus Trinkwasser. Gesundheits-Ing. *80*, 15.
[23] E. Korngold (1973). Removal of nitrate from potable water by ion exchange. Water, Air and Soil Poll. *2*, 15.
[24] T. V. Arden and T. Hall (1989). Nitrate removal from drinking water. A technical and economic review. WRe Report 856-S.
[25] The Permutit Co., Ltd. (1959). Brit. Patents 849 112, 860 659, 889 304.
[26] Farbenfabriken Bayer AG (1959). Brit. Patents 885 719, 885 720, 894 391.
[27] Rohm and Haas Co. (1960). Brit. Patents 932 125, 932 129.

[28] T. R. E. Kressman (1966). Eff. Water Treatm. J. *6*, 119.
[29] R. Couderc (1966). Eff. Water Treatm. J. *6*, 562.
[30] J. Brown and N. J. Ray (1984). Ion exchange in water purification. Where do we stand? In: Ion Exchange Technology. D. Naden and M. Streat, eds. Ellis Horwood Ltd., Chichester, UK; pp. 14–24.
[31] T. Mottershead (1984). Production and use of high purity water in the micro-electronics industry. In: Ion Echange Technology. D. Naden and M. Streat, eds. Ellis Horwood Ltd., Chichester, UK; pp. 25–36.
[32] J. R. Emmatt and P. M. Grainger (1979). Int. Wat. Conf. Pittsburgh IWe-79-11.
[33] International Analyser Co. (1967). Brit. Patent 1 013 069.
[34] F. Martinola and G. Siegers (1984). Experience data with Liftbed and Rinsebed processes. In: Ion Exchange Technology. D. Naden and M. Streat, eds. Ellis Horwood Ltd., Chichester, UK; p. 127.
[35] Illinois Water Treatment Co. (1955). Brit. Patent 806 107.
[36] Rohm and Haas Co. (1984). Information Sheet Amberpack.
[37] Eco-Tec Ltd., Pickering, Ontario (1984). Recoflo Bulletin; No. ET-10-84-5M.
[38] B. Causse (1985). L'eau, l'industrie, les nuisances *89*, 37.
[39] Degremont (1978). French Patent 2 443 283.
[40] Esmil Water Systems Ltd., High Wycombe, UK (1985). EPB System of counterflow regeneration. Information Sheet No. 100:85 MI.
[41] Dow Chemical GmbH, Stade, FR Germany (1984). Wasseraufbereitung. Der UPCORE Prozess im Werk Stade. Information Sheet No.6:84.
[42] Bayer AG, Leverkusen, FR Germany (1984). Lewatit^{R}. The Multistep system. Information Sheet No. OC/I 20.385e.
[43] I. R. Higgins (1955). USAEC ORNL 1907.
[44] Asahi (Japan) (1960). Brit. Patent 987 021.
[45] J. Newman (1967). Water demineralisation benefits from continuous ion exchange process. Chem. Eng. *74*, 72.
[46] Degremont (1979). Water Treatment Handbook; pp. 334-7.
[47] F. Martinola and G. Wutte (1985). The Multistep system, a flexible arrangement for water treatment with ion exchange resins. Proc. Intern. Water Conf., Pittsburgh, Pa, IWC-85-7.

2.3 Condensate Polishing

Albert Bursik
Grosskraftwerk Mannheim AG, Mannheim, Federal Republic of Germany

2.3.1 Introduction

2.3.1.1 Water in power plant cycles

The operating medium water or steam in power plant cycles comes in contact with various materials of the many components of the system, over a wide temperature range (room temperature up to 650 °C) and a wide pressure range (almost zero to more than 300 bar). The major materials used in most of the fossil and nuclear systems in contact with water or steam belong to one of the material classes listed in Table 2.17. Examples of plant cycles are given in Figure 2.69 a highly simplified flow diagram of the fossil plant cycle of a once-through boiler, and in Figure 2.70 –, the nuclear plant cycle of a boiling water reactor.

The main objective of water chemistry is to determine optimal conditions for a fail-safe, economic operation of all plant cycle components. The operation environments in power plant cycles require the use of high-purity feedwater combined with careful chemical control in order to minimize deposition and corrosion problems. This is the reason for specific quality requirements being made on the water used in large utility steam systems. As the water (or steam) is circulated within the plant cycle these requirements must be assessed not only with respect to the individual cycle components, but with regard to the total plant cycle.

Table 2.17 Materials in contact with water or steam in fossil and nuclear power plant cycles

- ▶ Carbon and alloyed steels
- ▶ Austenitic stainless steels
- ▶ Nickel alloys
- ▶ Copper and copper alloys
- ▶ Titanium and titanium alloys
- ▶ Zirkonium and zirkonium alloys

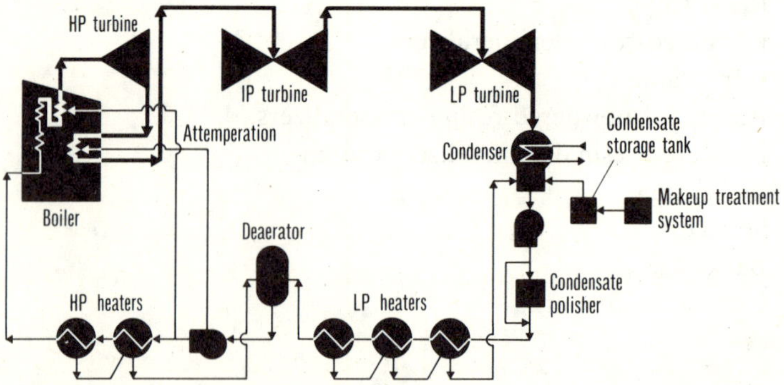

Figure 2.69 Simplified flow diagram of the fossil plant cycle of a once-through boiler.

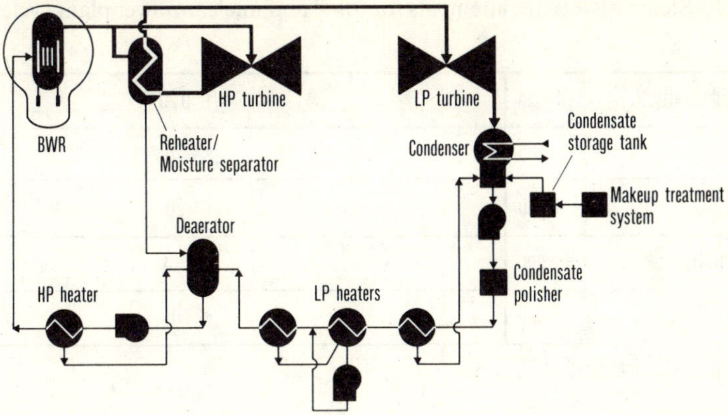

Figure 2.70 Simplified flow diagram of the nuclear plant cycle of a boiling water reactor.

2.3.1.2 Steam quality requirements

An investigation of the sensitivity of cycle components to the level of water and steam purity has shown that limiting the concentrations of ionic contaminants throughout the cycle to levels consistent with those in the steam tolerated by the turbine adequately protects the boiler and other cycle components. This means overall cycle chemistry must satisfy the turbine requirements for steam purity [1].

Table 2.18 Steam normal target values for fossil cycles of drum and once-through boilers – U.S.A. [1]

		drum boilers				once-through boiler
		without reheat		with reheat		with reheat
boiler treatment		phosphate	AVT	phosphate	AVT	
sample point		superheated steam		reheat steam		
sodium	ppb	≤ 10	≤ 6	≤ 5	≤ 3	
degased cation conductivity [a]	μS/cm	≤ 0.35	≤ 0.25	≤ 0.3	≤ 0.15	
silica	ppb	≤ 20		≤ 10		
chloride	ppb	≤ 6		≤ 3		
sulfate	ppb	≤ 6		≤ 3		
total organic carbon	ppb	≤ 100				

a — conductivity measured at 25 °C

Table 2.19 Steam quality requirements for fossil and nuclear power plant cycles – Germany [2, 3]

cation conductivity [a]	μS/cm	< 0.20
silica	ppb	< 20
iron (total)	ppb	< 20
copper (total)	ppb	< 3
sodium	ppb	< 10

a — conductivity measured at 25 °C

Normal target values for steam in fossil cycles with drum and once-through boilers, as have been recently published in the United States, are compiled in Table 2.18. Table 2.19 lists the steam quality requirements according to the respective VGB guidelines for fossil and nuclear plant cycles [2, 3]. The values indicated in Table 2.19 apply to all plants; no classification as to the type of steam generator or cycle design (fossil or nuclear, with or without reheating) is made here. These values have remained unchanged for the past 25 years. Although most power plants today clearly satisfy these requirements, it must be anticipated that more stringent regulations will be introduced after completion of the current revision of the VGB guidelines.

2.3.1.3 Fossil steam supply systems

The feedwater specifications, in addition to considering requirements for the turbines, must also take the design of the steam generating systems into account. In fossil cycles drum-type and once-through boilers are used. These types of generator differ from each other with respect to their sensitivity towards cycle contaminants.

Due to its design, the once-through boiler is considerably more sensitive. If more dissolved contaminants are present in the feedwater than corresponds to the solubility of these substances in the steam generated, the result will be serious tube failures due to the formation of deposits in the boiler tubes. This applies similarly to suspended solids in the feedwater, such as corrosion products of the structural metals.

The operation of a fossil plant cycle of a once-through boiler free of any upset conditions requires a very high feedwater purity. As a rule, this is possible only if the cycle purity can be maintained – after steam condensation in the condenser and after feeding of makeup and low pressure heater condensate into the condenser as well – by means of suitable purification methods and by adequate treatment of the purified condensate (use of chemicals to reduce the preboiler cycle corrosion).

In the drum-type boiler the steam-water mixture from the water walls is separated in the drum by means of the internal steam separation equipment. Saturated steam with the lowest possible moisture level is discharged from the drum into the superheater. The greater part of the dissolved contaminants in the feedwater remains in the boiler water and can be blown from the cycle. In this way the level of soluble contaminants present in the cycle may be controlled.

In the plant cycle of drum boilers solids suspended in the feedwater, mostly corrosion products from the preboiler cycle, are also undesirable. In the high heat flux regions of the water walls they lead to a deposit buildup. Deposits restrict the transfer of heat and cause the temperature of the tube wall and the deposit layers to increase as the concentration of contaminants and – if used – of solid boiler water-treating chemicals increase. This results in water-side corrosion failures.

2.3.1.4 Nuclear steam supply systems [4]

The nuclear plant cycles are to some extent comparable to those of a fossil plant. The plant cycle of a boiling water reactor (BWR) is very similar to that of a fossil drum boiler, even if the boiling water reactor does not – as opposed to the drum-type boiler – generate superheated, but saturated steam. The cycle impurities concentrating in the reactor water may be eliminated by means of reactor water cleanup filters and demineralizers. The effect of this approach is comparable to the blowdown in the fossil cycle of a drum boiler.

The secondary system of a nuclear plant cycle with a pressurized-water reactor (PWR) is separated from the primary system, the separating component being the steam generator. The steam generator can be either of a recirculating or a once-through design.

The once-through steam generator of the nuclear pressurized-water reactor cycle produces superheated steam. For reasons already listed for the fossil once-through boiler, this steam generator type implicates high-quality requirements to the feed-water. Adequate in-cycle purification methods must be employed. Another possibility of eliminating contaminants from the cycle is moisture separation after the high-pressure turbine. Feedwater contaminants existing in the steam are concentrated in the liquid phase due to the lower solubility in the low-pressure steam.

The recirculating steam generators produce saturated steam. Similar to the fossil cycle of drum-type boilers the contaminants entering the steam generator may be removed by means of blowdown.

2.3.1.5 Cycle chemistry control

The requirement of chemically almost pure water for power plant cycles calls for the application of appropriate chemical measures (conditioning) in order to maintain a very low corrosion product level at the inlet of the steam supply system. The only exception is in the cycle of a boiling water reactor. In such a system the low corrosion product level at the reactor inlet is accomplished not by water treatment, but by means of selecting prereactor cycle materials resistant to corrosion and corrosion erosion.

In the cycle of a pressurized-water reactor the all-volatile treatment (AVT) is used for the secondary cycle. The pH range required depends on the materials used in the cycle. Table 2.20 is a compilation of all-volatile treatment data for the secondary system.

Table 2.20 Feedwater AVT – secondary systems of cycles with PWR

		U S A [5]		Germany [3, 6]	
system		ferrous/copper	ferrous	ferrous/copper	ferrous
pH[a]		8.8 — 9.2	9.3 — 9.6	> 9	> 9.8
hydrazine	ppm	> 0.02	> 0.02	—	> 0.02
oxygen	ppm	< 0.005	< 0.005	< 0.020	< 0.005

a — pH measured at 25 °C

In fossil plant cycles several modifications of the all-volatile treatment are used to reduce preboiler cycle corrosion. In the United States a strict limitation of the oxygen concentration in the feedwater at the economizer inlet is mandated (Table 2.21) regardless of the boiler type (once-through or drum).

Table 2.21 Feedwater AVT – fossil power plant cycles [U.S.A. [1]

metallurgy		all ferrous	mixed ferrous/copper
pH[a]		9.0 — 9.6	8.8 — 9.3
oxygen	ppb	< 5	

a — pH measured at 25 °C

Table 2.22 Feedwater AVT – fossil power plant cycles of once-through boilers [7]

AVT		neutral	combined	modified alkaline	classic alkaline*	highly alkaline
pH[a]		7.0 — 7.5	8.0 — 8.5	> 9	> 9	> 9.5
oxygen	ppb	50 — 250	30 — 250	< 200	< 20	< 20

a — pH measured at 25 °C

Note: Only the classic AVT (marked *) is also used in cycles of drum boilers

In addition to volatile alkalizing substances, oxygen and hydrogen peroxide are used in Germany and in other countries for the all-volatile treatment of once-through boiler cycles (Table 2.22). Application of the individual treatment modifications is, of course, limited by the materials in the respective cycle. In the cycle of a drum boiler no oxidants are used.

The volatile alkalizing agent in power plant cycles is, in most cases, ammonia. In addition, organic neutralizing amines, such as morpholine and cyclohexylamine, are used either in combination with ammonia or alone. Hydrazine is typically used for the dissolved oxygen control. When using neutralizing amines their thermal stability must be taken into account (limited application range). For all volatile alkalizing agents performance in the in-cycle water purification must be taken into account.

The type and concentration of the solid alkalizing and/or buffering substances used in both nuclear and fossil recirculating steam generators for the steam generator or boiler water treatment do not have any substantial effect on in-cycle water purification. Therefore, neither fossil boiler water treatment nor nuclear steam generator water treatment will be discussed here.

2.3.1.6 In-cycle water purification

In modern power plant cycles the degree of cycle water purity strongly affects the system's availability and economy. Cycle water conditioning alone for corrosion control is generally not sufficient for satisfaction of the strict requirements on feedwater and steam quality. In the course of operation or during a component failure or an upset impurities escape into the plant cycle. Upset conditions resulting in accumulation of corrosion products or ingress of undesirable chemical constituents can never be excluded during the design life of power plant equipment.

Therefore, in-cycle purification of steam condensate – condensate polishing – is used for utility cycles. All once-through cycles utilize some form of condensate polishing, as do some recirculating fossil and nuclear systems [4]. There is always a full-flow condensate polishing plant in the nuclear plant cycle of a boiling water reactor in order to retain any dissolved or solid radioactive contaminants [6]. When designing condensate polishing systems chemical, mechanical, and economic considerations, together with previous plant operating experience, play an important role [4]. To produce high-quality water under all possible operating conditions, the condensate polishing system must be flexible: it must be able to function with various types of water chemical treatment and to remove all types of cycle contaminants [8].

2.3.2 Condensate polishing

2.3.2.1 Plant cycle contaminants

Contamination of the high-purity condensate and feedwater used in utility cycles can be performed either continuously during normal operation or intermittently during component failure or upset. Eliminating the sources of contaminant intrusion is obviously the most effective control, even if complete elimination is not always feasible [4].

The most important cycle contaminants are [4, 9]:

- cooling water,
- air,
- makeup water impurities, such as raw water constituents, water treatment chemicals (regenerants), and resins or resin fines,
- corrosion products,
- materials for temporary protection, such as protective coatings and vapor-phase inhibitors, and
- radioactive products in nuclear power plant cycles.

Cooling Water. Cooling water inleakage into the plant cycle is due to condenser leaks. The most frequent causes of condenser leakage are incomplete tube/tube sheet sealing, corrosion-related tube damage, and mechanical tube damage.

Tube/tube sheet weepage was the most frequent cause of condenser leakage in the United States in the seventies [10]. Today this leakage type appears to be unknown in Germany [11].

The corrosion-related tube damage includes cooling water side pitting (almost exclusively sub-crud), corrosion erosion, stress corrosion cracking at the transition of milled expansions flaring out to the tube run, and alloy depletion of admiralty and copper-nickel. The most important causes of damage emanating from the steam and condensate sides are ammonia corrosion of admiralty in the air removal section and stagnation (dead) regions, and water droplet erosion.

Today mechanical tube damage occurs only in exceptional cases. It is mostly caused by tube vibrations due to steam flow.

Cooling water inleakage has a strong impact on the contaminant levels in the plant condensate, especially in installations with cooling towers or with brackish or sea water cooling. Even if it were possible almost to preclude condenser leaks by proper material selection, accurate system fabrication, and – during operation – continuous tube cleaning, the main objective of the condensate polishing equipment would still be protection of the steam generator and turbine systems from the effects of cooling water inleakage.

Air. The most frequent sources of air inleakage into the plant cycle are turbine seals, valves (valve packings), steam generator pump seals (expansion joints), and low pressure heater expansion joints [12]. As a result of the high complexity of the systems it is extremely difficult in most cases to localize an air inleakage.

An air inleakage may sometimes cause the feedwater oxygen level to rise, since only inleakages upstream of or in the condenser may be eliminated by air ejectors. The removal of the carbon dioxide ingressed with the air by physical means is impossible in systems operated on a higher plant cycle pH (all-volatile treatment using ammonia). The only chance of removing carbon dioxide from the plant cycle then is by condensate polishers. With a strong air inleakage the removal of the carbon dioxide may lead to a reduction of the anion exchanger capacity in the condensate polishing equipment.

Makeup water impurities. In many cases makeup water may be a source of plant contaminants. The most hazardous substances are, of course, the ion exchanger resin regenerants (caustic and acids) which may escape into the plant cycle during failures in the makeup treatment system. Therefore, protection against such failures is mandatory.

The makeup water may also introduce resins and resin fines into the plant cycle. In specific plant cycle regions thermal decomposition products of the resins may bring about the formation of hazardous environments.

Operational difficulties or improper monitoring of the makeup water system can cause the introduction of suspended and dissolved inorganic impurities and organic matter from the raw water into the plant cycle. Problems with colloidal silica can be particularly troublesome with surface water supplies at certain times of the year [4].

Makeup water storage equipment can sometimes be another source of contaminants. On the one hand, makeup may be contaminated by air (oxygen, carbon dioxide) during storage, on the other, by corrosion products as well (of the storage tank materials). As a rule, the ingress of such contaminants can be precluded by modifications of the storage equipment and of makeup feeding to the plant cycle (e. g., by deaeration in the condenser).

Corrosion products. The consequences of corrosion product ingress into fossil steam generators were described in Section 2.3.1.3. Further information can be found in the literature [11, 13].

Comparable negative effects of the corrosion product transfer are found in the steam generators of the nuclear cycles of pressurized-water reactors. In cycles of boiling water reactors a high corrosion product input into the reactor is associated with an excessive fuel deposit buildup. Decreasing corrosion product transport to the reactor minimizes fuel deposits and corrosion and assists in controlling radiation buildup [14].

Of special importance is the corrosion product transport control during the startups, cycling and peaking operation of all steam supply systems. During startup the removal of suspended solids may supplement the effective layup protection. The prevention of their deposition on the heat transfer surfaces may be achieved by bypass systems and water recirculation through the condensate polishing equipment [1].

Materials of the temporary protection. Prior to the initial startup of a new power plant cycle all substances used for temporary corrosion protection, even their residues, must be removed, as these substances can contaminate the cycle water. Such a contamination may, on the one hand, irreversibly foul the ion resins used in the condensate polishing equipment [4]; on the other hand, thermal decomposition of the contaminants results in the formation of corrosive environments in specific cycle sections.

The same negative effects are caused by the intrusion of oils and of oil-based materials during operation, as described in the literature [4].

Radioactivity. Cycles of boiling water reactor systems operate with radioactive steam and condensate, since with the steam both fission and activated corrosion products are discharged from the reactor. The cycles of boiling water reactors employ full-flow condensate purification. The resins used here are contaminated.

In cycles of pressurized-water reactors the secondary system may be radioactively contaminated only by a primary-to-secondary tube leak. The degree of contamination depends on the leakage quantity. From the primary coolant, besides other substances, also boric acid can escape into the secondary system. This has to be taken into account when purifying the condensate by way of ion exchange.

2.3.2.2 Performance of condensate polishing

Condensate polishing systems have to be able to perform the following functions [15, 16]:

For normal operation
- remove dissolved and suspended solids including colloids introduced into the cycle from the makeup water supply and from condensate storage equipment,
- remove corrosion products that are generated in the cycle,
- provide a process for coping with any unusual feedwater chemistry problems due to unanticipated operating conditions, such as air inleakage and trace condenser leakage,
- maintain the cycle at optimal cleanliness, reduce dissolved solids to as close to zero as practical with filtration and ion exchange technology;

for cooling water ingress
- remove dissolved and suspended solids introduced during condenser leakage, allow proper isolation of the failed condenser section on the cooling water side or a systematic plant unit shutdown;

for startups, restarts, cycling operation
– remove crud, dissolved solids, and silica;
for commissioning of new systems
– enhance fast cleaning of the plant cycle.

A large number of steam supply systems provide process and heating steam for industry. Heating steam is also supplied to community heating systems. The condensates from industry in particular may contain a great variety of impurities that cause damage in power plant cycles. If the condensates from the supplied steam are returned to the power plant cycle, the function of condensate polishing is to treat so that they satisfy the quality requirements for steam generator feedwater.

2.3.2.3 Correlation of condensate polishing and power plant operating mode

When designing condensate polishers the intended operating mode of the power plant or plant unit must be taken into account. The operating mode has a strong bearing on the type and quantity of cycle impurities which must be eliminated by the condensate polishing system.

Base load operation. In base load plants the anticipated condensate polisher loading with suspended corrosion products is low provided unit design and in-cycle water treatment are adequate. When considering the long operating periods between boiler, and especially turbine, overhauls the quality of the polished condensate with respect to the dissolved solids level must satisfy the highest quality standards. Trace impurities and also increased cooling water inleakage must be safely trapped, even over a prolonged period of time.

Peaking and cycling operation. The cycling operation is a load-following operation. The unit load fluctuates with the system demand; a typical load variation for cycling units might range from 30% to 100% of the design capacity. Peaking operation is a form of cycling in which the unit is operated only during peak power demand periods. At offpeak hours the unit is in hot or cold standby. The loading to be anticipated for the condensate polishers in cycling or peaking operating systems clearly differs from that of base load plants.

During both cycling and peaking operation a substantial amount of time is spent at low load, startup, shutdown, and short-time layup modes, when cycle chemistry and corrosion are difficult to control. The most important function of condensate polishing is to eliminate the suspended corrosion products from the cycle.

Trapping of the dissolved impurities due to cooling water inleakage over a prolonged period of time is in most cases not required, since – for repair of the leaks – plants in the cycling and peaking operation can normally be shut down much more quickly than the base load units. Any turbine deposits are partially or entirely removed by wet steam during frequent startups.

2.3.2.4 In-cycle position, design, integration

In-cycle-positioning. The removal of dissolved ionic impurities via condensate purification in the plant cycle is achieved by ion exchange. The thermal stability of the commercial ion exchange resins must be taken into account when establishing the position of the condensate purification in the cycle.

Strongly acid cation exchangers (polystyrene-type) may be used up to a temperature of 120 °C; temperature limitation on the anion exchanger resin used in condensate polishing systems restrict the operation to a maximum of 60 °C [4] to 65 °C for a regenerable deep-bed demineralizer. It should be kept in mind that silica removal, very important for the plant cycle, is achieved optimally only at a much lower temperature (near 30 °C).

Powdered resin demineralizers may be operated at higher temperatures (up to 105 °C), since a reuse of the ion exchange resin, requiring regeneration, is not intended. However, what has been said for bead ion exchange resins about silica removal efficiency is also applicable here.

The optimal location for the removal of the suspended corrosion products would be directly before the actual steam generator; for cycles with boiling water reactors, upstream of the reactor inlet. The feedwater temperature in this location is, however, much higher than is tolerable for the ion exchange resins so that in these locations only corrosion products can be eliminated by filtration.

The positioning of the ion exchange condensate purification unit is, therefore, always a compromise between cycle requirements and the thermal stability of the resin. This is the reason why most of the condensate polishing systems are located between the condensate pump discharge and the first low-pressure feedwater heater. Depending on the thermodynamic design of the plant cycle, powdered resin demineralizers may be located also behind the first or second low-pressure feedwater heater. Recently condensate purification by means of high-temperature filtration has been reported (e. g. [17]) as being used more and more in fossil and nuclear systems. Ion exchangers are, of course, not employed in such filtration methods.

Design. For an optimal cycle chemistry it is reasonable for both fossil and nuclear once-through cycles to install a full-flow condensate polishing system. This approach is also standard for boiling water reactor cycles.

In certain recirculating fossil and nuclear systems some form of condensate polishing is utilized. With increasing pressure in the cycles of drum boilers the blowdown becomes less effective in eliminating dissolved impurities from the plant cycle. The reason for this is that the vaporous carryover increases as a function of pressure and temperature.

Some utilities try to save on the costs of a full-flow condensate polishing system by installing one system for several plant units, often provided with extensive cross connections and a large quantity of measuring and control instrumentation; they accept that such a system only polish a part of the total condensate to be treated. It is doubtful at best whether such an approach is worthwhile, considering the design life of the power plant equipment and the cost of steam generator waterside tube failures and corrosion fatigue and stress corrosion failures of steam turbine components (for failure costs see [9, 13]) due to improper water chemistry conditions.

When designing condensate polishing systems the type of cooling water used for condenser cooling must be taken into account, apart from the plant cycle design and the operating mode. An identical cooling water leakage has, depending on the total dissolved solid content of the cooling water (sea water, brackish water, cooling water in cooling tower installations, river water) diverse effects on the plant cycle chemistry or on the condensate polishing unit.

Integration into the cycle. For integrating condensate polishing into a power plant cycle various approaches are available. The most common will be briefly mentioned here.

Integration into the main condensate cycle

- low-pressure (2.5 – 16 bar) application (condensate polisher between two stages of the extraction pump) with or without booster pump,
- high-pressure application (condensate polisher behind the second stage of the extraction pump).

Installation of a full-flow bypass loop in the condensate stream

- with common booster pumps (all polishers have one common booster pump set),
- with individual booster pumps (each polisher has its own booster pump).

Each of the above possibilities has its advantages and disadvantages. For details reference is made to the literature [4, 6, 12, 15].

2.3.3 Condensate purification methods

2.3.3.1 Overview

The methods used for condensate purification may be classified by the process employed:

Filtration processes

- precoat filtration,
- element-type filtration,
- magnetic filtration.

Ion exchange processes

- deep-bed demineralization,
- powdered-resin demineralization

Combination of processes.

2.3.3.2 Filtration processes

Filtration processes are applied exclusively for the removal of suspended solids, mostly corrosion products from the plant cycle. The precoat filtration mentioned above is a surface filtration method. The element-type filtration with cartridge-type elements may be either a surface filtration or a depth filtration depending on the cartridge design. The magnetic filtration is a depth filtration.

Precoat filtration. Precoat materials for condensate filtration are inert materials (e. g. Solka Floc®, a material produced from fiber, inert fibrous acrylpolymer Lewasorb® AF 2, and others). The actual precoat filters and their filtration function correspond to those used for powdered resin demineralization. They will be described below. In the literature precoat filters using inert precoat material are also refered to as particulate filters.

Element-type filtration. The most popular element-type filters are cartridge filters. They are vertical cylindrical vessels made of steel; the filter elements comprise a great number of cartridges in a standing or suspended arrangement. The cartridges function with a flow from the outside to the inside, with the element flow rate being 8 to 15 m/h maximum.

The most common cartridges consist of a supporting core of stainless steel or synthetic material (e. g., polypropylene) wound about with cotton, nylon or polypropylene yarn. Cotton-wound cartridges can be used up to 60 °C, those of nylon up to 100 °C, of polypropylene up to 130 °C [17]. For condensate filtration cartridges in the 3 to 10 micron range are used.

Other element types, such as synthetic or metal fiber fabric, cartridges with wire fabric or made of sintered metal, and metallic screens with a pleated paper filter attached are rarely used in condensate polishing.

Magnetic filtration. In the past 10 years magnetic filters have been increasingly adopted for condensate filtration in fossil cycles, but even more in nuclear power plant cycles. The reasons are:

- possible utilization at high condensate or feedwater temperature (up to 300 °C),
- compact design due to the high filtration rate (2000 – 2500 m/h), and
- high filtration efficiency with respect to the retention of crud consisting mainly of ferromagnetic constituents.

An electromagnetic filter is a cylindrical vessel made of non-magnetic steel enclosed in a magnetic coil. The vessel is filled with steel spheres; in high-grade magnetic filters there are stacks of steel threads (steel wool) or steel nets (wire fabric) or a combination of spheres and fabric stacks.

The design, function, and cleaning of electromagnetic and high-gradient electromagnetic filters are described in the literature (e. g. [4, 6, 15, 18]).

2.3.3.3 Ion exchange processes

Only by using ion exchangers for condensate polishing it is possible to remove ionic impurities from the condensate to be polished. Both ion exchange resin beads and powdered ion exchange resins may be used in ion exchange systems. The first are applied in deep-bed demineralizers, the latter in precoat filters designed to be coated with finely ground resins in powdered-resin demineralizers.

Both types of demineralizers are capable of retaining, in addition to ionic impurities, suspended solids, such as crud, as well.

Deep-bed demineralization. Most of the deep-bed condensate polishing demineralizer installations in the United States consist of a single bed of a mixed cation and anion exchange resin combination [4]. The resins are regenerated either in the service vessel (in-place regeneration) or outside of the service vessel (external regeneration).

The utilization of two-bed demineralizers was already suggested 15 years ago [19]. However, the first two-bed and three-bed demineralizers as condensate polishers in separate vessels (see, e. g. [20, 21]) and in a common vessel (see, e. g. [22]) were built much later. In many plants the mixed-bed demineralizer is preceded in the condensate flow stream by a deep bed of cation exchange resin [4, 15].

The deep-bed demineralizer cross-sectional flow rate is around 110 to 135 m/h in most US systems [12], a flow rate of about 120 m/h is the accepted norm for high flow rate demineralizers. At several locations, however, a lower flow rate of 85 to 98 m/h has been reported to show improved performance characteristics [4, 12]. In Germany most deep-bed polishers are operated at a cross-sectional flow rate up to 80 m/h [15].

The ion exchange resins used in condensate polishing applications are generally divided into two classes – gelular and macroporous (macroreticular). A distinct trend towards application of one of the two resin types could be found neither in the United States [12] nor in Europe.

Powdered-resin demineralization. Powdered-resin demineralization in condensate purification is a much more recent procedure than the deep-bed one. This concept was first presented in 1962 under the name Powdex process [23]. Powdered-resin demineralizer systems utilize very small particles of cation and anion exchange resins, with 90% finer than 45 μm (325 mesh) which, agitated together in suspension in water, agglomerate to form large floc particles. These voluminous porous flocs can then be precoated on a filter septum.

The layers of pulverized ion exchange resins, up to 6 mm thick, are capable of removing suspended solids while simultaneously performing an ion exchange operation. Powdered-resin demineralizers are very flexible both with respect to the cation/anion ratio and also to the form of the cation resin portion (hydrogen or ammonium).

Two types of filter septa are normally used for powdered-resin demineralizers: a fiber-wound type constructed from polypropylene or nylon, and a stainless steel type with 150 × 38 μm (100 × 400 mesh) and nominal 32 μm screen openings as the most common screen size [12].

The powdered-resin demineralizer element flow rates range from 5 to 12 m/h [12]; the typical flow rate is considered to be 10 m/h [4].

2.3.3.4 Combination of processes

The first systems for condensate purification employed a combination of several processes. For example, the early US once-through boiler applications utilized full-flow precoat filtration for removing corrosion products with a downstream partial-flow deep-bed (in-place regenerated mixed bed) demineralization (Figure 2.71). The use of post-filtration for protecting the cycle from leakage of resin fragments was usual [4]. Figure 2.72 explains the resin symbol types used in Fig. 2.71 and all figures of Section 2.3.

The first plants in Germany, also in cycles of once-through boilers, were combined systems, too. Typical for that time was the combination of gravel filter, lead cation bed and mixed beds. Figure 2.73 and 2.74 illustrate the two variations of filter arrangements [24].

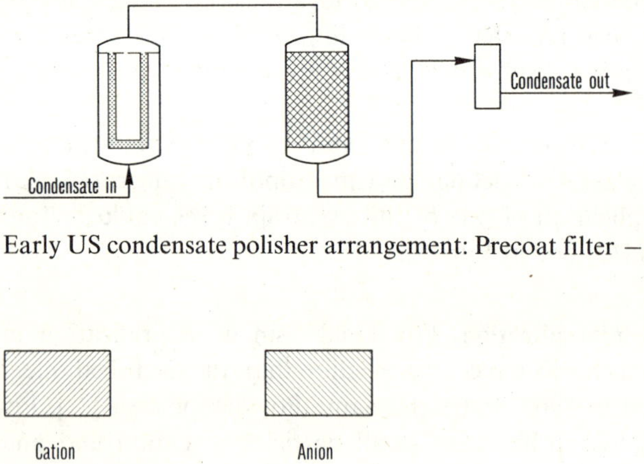

Figure 2.71 Early US condensate polisher arrangement: Precoat filter – mixed bed – post-filter.

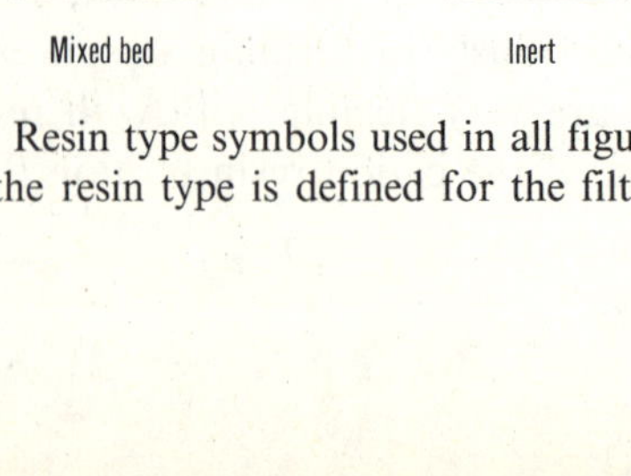

Figure 2.72 Resin type symbols used in all figures of Section 2.3.
Note: Only the resin type is defined for the filter connection variations, not the real resin depth.

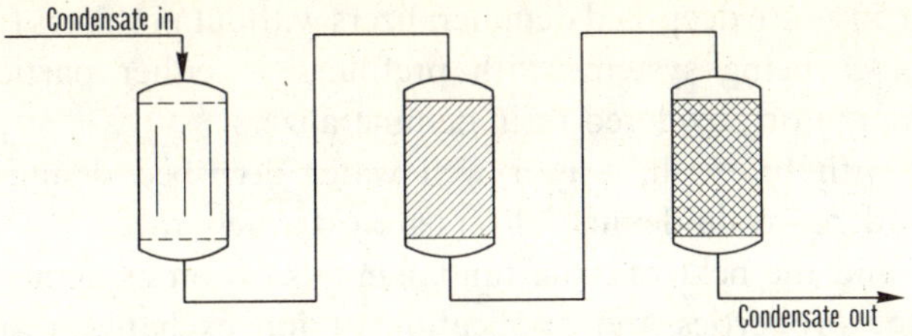

Figure 2.73 Early German condensate polisher arrangement: Gravel filter – lead cation – mixed bed.

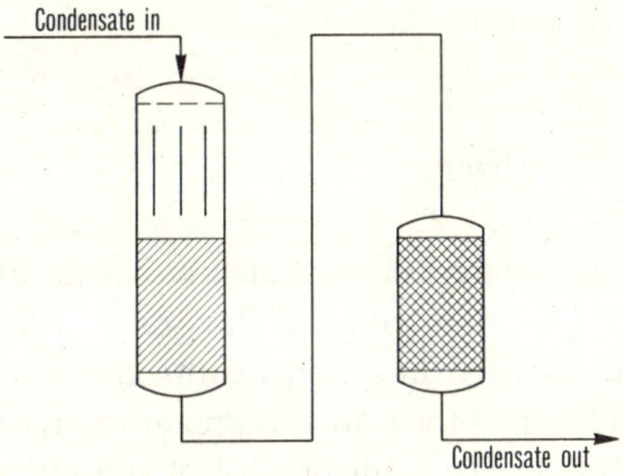

Figure 2.74 Early German condensate polisher arrangement: Double-chamber prefilter (gravel filter/lead cation) – mixed bed.

Developments in the United States resulted more or less in the elimination of prefiltration from utility polisher designs, if deep-bed demineralizers were applied. The present deep-bed demineralizers, mostly with external regeneration, are installed as 'naked' mixed beds.

In Germany, up to today combinations of filtration and ion exchange are normal for condensate purification. The most common arrangement is the combination of both deep-bed cation bed and mixed bed. The cation bed located upstream is, on the one hand, a mechanical filter for separating the corrosion products and, on the other hand, an ion exchanger for absorbing ammonia. Occasionally a separate mechanical filtration step is connected upstream of the ion exchangers [6].

Another combination of the two processes – filtration and ion exchange – was already mentioned in Section 2.3.3.3 Ion exchange processes.

2.3.4 Deep-bed demineralizers

2.3.4.1 Introduction [12]

The choice as to the type of condensate polisher depends on power plant design, economic considerations, previous utility experience and, most important, the cooling water source. The majority of US power stations have deep-bed demineralizers. More than 50% are deep-bed demineralizers without prefilters (naked mixed beds), the remainder being systems with prefilters – either particulate filters (typically Solka Floc®) or powdered-resin demineralizers.

When cooling with brackish, sea or well water deep-bed demineralizers are standard, and powdered-resin demineralizers are extremely rare.

In this section and the next one the fundamentals of ion exchange technology with respect to the properties and application of ion exchange resins in water treatment and the type of equipment required are assumed to be known. Only those questions will be dealt with which arise when ion exchange resins are used for a special type of water treatment, for condensate polishing and which are of general interest and importance.

2.3.4.2 Resins for deep-bed demineralizers

The deep-bed condensate polishers require ion exchange resins capable of maintaining a high-quality effluent at high cross-sectional flow rates. The resins must be able to withstand physically the transfer (in systems with external regeneration) and regeneration procedures, in addition to the flow and pressure surges experienced during operation in the service vessel. The high quality of the effluent must be guaranteed even in those cases when the deep-bed polisher, as a naked mixed bed, must take over the additional function of crud separation [4].

As already mentioned in Section 2.3.3.3 in deep bed demineralization both gelular and macroporous resins are used. The evaluation of the references in the literature does not allow one to draw a reasonable conclusion with regard to preference of the one or other resin type [25]. On the other hand, the selection as to the type of anion exchanger resin is undisputed. For condensate polishing applications today strongly basic anion resins of Type 1 are used exclusively because of their higher basicity. It should, however, be kept in mind that resins with the quaternary amine functional group are inherently sensitive to elevated temperatures.

Table 2.23 Ion exchange resins used in deep-bed condensate polishers

	cation exchanger		anion exchanger	
	gel-type	macroporous	gel-type	macroporous
Amberlite®	IR 132	200 252	IRA 420	IRA 900
Duolite®	C 255	C 26	A 109 A 113	A 161 A 165
Lewatit®	S 100	SP 112	M 500	MP 500 MP 600

Table 2.24 Ion exchange resin combinations for mixed-bed condensate polishers

	cation exchanger	anion exchanger
Amberlite ®	200 C	IRA 900 C
	252 C	IRA 900 C
	IR 132 C	IRA 900 C
	IR 132 C	IRA 420 C
Duolite ®	C 26 C	A 161 C
	C 255 C	A 165
	C 255	A 109
Lewatit ®	SP 112 MB	MP 500 MB
	SP 112 MB	M 500 MB
	SP 100 MB	M 500 MB

Under elevated temperature they slowly lose their strong base functionality, a loss that, over time, becomes significant [26]. The cause for this deterioration in anion exchange performance may also be the absorption of organics.

Table 2.23 lists the ion exchange resins used in condensate deep-bed polishers. This list presents, of course, only a selection. When considering the relatively small differences between the hydrated bead densities of cation and anion resins [27] for the mixed bed applications resins with a possibly uniform bead size distribution (with a possibly low uniformity coefficient) are used. Examples of the resin combinations, as applied in mixed-bed condensate polishers, are listed in Table 2.24.

For specific applications, especially in nuclear steam supply systems, pre-regenerated resins are used. The use of pre-regenerated resins and subsequent disposal after one service cycle becomes practical when the processing and disposal of radioactive regenerant chemicals is more costly than resin disposal and replacement [4]. When using pre-regenerated resins the fact that the storage or shelf life of pre-regenerated resins is limited should be considered.

2.3.4.3 Equilibria and kinetic considerations

For the application of ion exchange resins in deep-bed condensate polishing two theoretical questions are of particular importance:

- The two-phase (solid-liquid) equilibrium between the resin and the condensate, almost pure water, outside the resin phase;
- the kinetics of mixed-bed ion exchange.

The equilibria of the heterogeneous reactions between the resin phase and the outside condensate depend on the equilibrium constants of the ion exchange reactions (often referred to as selectivity coefficients), the concentrations of the components in the condensate or on the resin, and on the level of resin regeneration and loading [19, 28–31]. The resin's selectivity coefficients are affected by the degree of crosslinkage; in the case of strong base anion exchange resins, also by the type of the functional group. Early investigations of mixed bed column performance at high flow rates and low solution concentrations concluded that the ion exchange kinetics are controlled by a liquid-film mass transfer mechanism, i. e., by film diffusion through the liquid layer surrounding each resin bead (e. g. [32]). Today, more than 25 years later, condensate polisher effluent concentrations of less than 1 ppb are mandatory and many of the assumptions made then must be revised now because of this requirement. The modern model of mixed bed ion exchange kinetics for solution concentrations below 1.0×10^{-4} M takes into account the dissociation of water, the ratio of cation to anion resins, differing resin exchange rates and capacities, and reversible exchange [33, 34].

According to [29] there are two types of mixed bed demineralizer leakage, kinetic leakage and elution leakage. Kinetic leakage depends on the mixed bed design (cross-sectional flow rate, bed depth, cation to anion resin ratio), the level of resin exhaustion, and on contaminant concentration in the condensate to be treated. Given a proper design of the mixed bed demineralizer this type of leakage (which cannot be affected by modification of the regeneration) is negligible for condensate polishing.

Of greater importance for condensate polishing is elution leakage. It is a direct function of the equilibrium between the resin and the condensate phases. This is the reason why this leakage type depends on the level of resin regeneration and loading. These equilibrium considerations provide, e. g., an explanation of the so-called "leaching" of sodium from the resin even if sodium as absent from the condensate influent. A consideration of these equilibria is especially important for mixed bed on-stream ammonization and for an operation beyond ammonium breakthrough into the ammonia cycle. The elution leakage of a mixed bed may be reduced by systematic measures for upgrading the separation and regeneration of the resins. Adequate references will be given later.

2.3.4.4 Resin regenerant chemicals

Cation exchange resins are regenerated mostly with sulfuric acid solution in deep-bed condensate polishing applications. Apart from that hydrochloric (muriatic) acid solution is also used, mainly in the case of in-place regeneration. The regenerant strength is normally 4 to 8% for sulfuric acid (H_2SO_4) and 2 to 5% for hydrochloric acid (HCl).

The regenerant dose ranges from 120 to 240 g/l resin; the mean regenerant dose for US systems is 176 g/l [12]. The contact time usually varies between 30 and 75 minutes for cation resins.

Anion exchange resins are regenerated with caustic solution. The regenerant strength is normally 2 to 6% for sodium hydroxide (NaOH).

The regenerant dose range is in the same region as given for cation exchangers and the mean regenerant dose corresponds to the above [12]. The contact time varies between 60 and 75 minutes.

The regenerant doses used in the United States are substantially higher than those applied in Europe, as is stated in [25].

2.3.4.5 Mixed bed demineralizers

Service vessels [4, 25]. Service vessels for mixed bed demineralizers in condensate polishing applications do not differ substantially from the ion exchanger vessels otherwise used in water purification. Mostly they are vertical-cylindrical tanks with welded shells, made of steel and rubber-lined. They have flanged connections, structural steel legs, and an access with a bolted gasketed cover.

In systems in which the operating pressure conditions allow it sight glasses are installed in suitable locations for inspection of the tank bottom laterals or for controlling the resin transport or separation.

All vessel internals must be corrosion-resistant, made either of stainless steel, lined carbon steel, or of adequately resistant synthetic materials. It is important that they are given adequate mechanical strength to withstand the maximum differential pressure expected.

The inlet flow distributor has the function of distributing the condensate flow uniformly across the whole resin bed. The design of the underdrain system must ensure retention of the ion exchange bed.

Depending on the design of the mixed bed demineralizer with respect to the type of regeneration (external or in-place) or the flow direction through the resin bed (dual flow mixed bed) appropriate vessel internals are necessary.

Externally regenerated mixed bed (Figure 2.75 a). Vessel internals must guarantee a complete discharge of the exhausted resins from the service vessel, since residues of exhausted resins may have a negative impact on the operation of the mixed bed

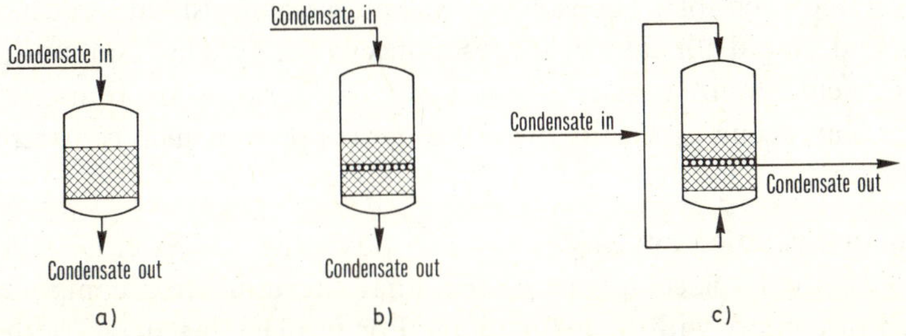

Figure 2.75 Mixed bed demineralizer
a) externally regenerated b) in-place regenerated c) dual-flow.

after charging with regenerated resins. Apart from that, the internal assembly must guarantee the uniform filling of the vessel with a mixture of regenerated resins.

In-place regenerated mixed bed (Figure 2.75 b). Vessel internals for resin transport are not required. A stable drainage system in the resin separation zone, mostly laterally designed, must be installed. The vessel internal assembly must guarantee an optimal discharge of the regenerant chemicals and rinse water without creating more than negligible turbulence during the separation of cation and anion resins.

Dual-flow mixed bed (Figure 2.75 c). The drainage located in the center of the resin mass must be designed for the full filter throughput. Both the inlet flow distributor and the underdrain system must guarantee that the condensate flow across the resin bed is uniform. Adequate measures must guarantee that the two condensate batch flows are about equal.

Resin depth, ratio and form. For all types of mixed bed polishers a resin depth of 1000 to 1200 mm may be called typical, although there is divergence from these data in both directions. The free board or expansion space is normally lower in externally regenerated mixed bed polishers (60 to 75% of the resin depth) than in those with in-place regeneration (approximately 100%).

The resin ratio in deep-bed polishers (cation to anion ratio by volume) varies from 3 :1 to 1 :2. In the United States, the practice commonly followed for naked mixed beds is a 2 : 1 ratio [12]; systems with a cation lead bed are operated at ratios of 1 : 2 to 1 : 3 in the mixed bed [15].

A comparison of the operating resin form is interesting. While in Germany operation with the H/OH-form is standard, in the United States about $\frac{1}{4}$ of the deep-bed polishers are operated with the NH_4/OH-form [12].

Resin transfer system [4]. With deep-bed polishers with external regeneration the exhausted ion exchange resin must be discharged from the service vessel into the receiving tank of the regeneration system. The necessity of completely removing the exhausted resin from the polisher has already been indicated. Resin transport may also be required within the regeneration system (three vessel installations), as is the return of the regenerated resins into the service vessel.

The transfer line piping must have a smooth internal surface; bends with a long radius are indispensible. The isolation valve assemblies should not have any dead regions, and should not damage the resins mechanically. The resin transport medium – air or demineralized water – must not fail during resin transport, since then plugging may occur, especially in vertical (riser) pipes, which is extremely difficult to eliminate.

External regeneration systems. The main advantage of external regeneration as compared with in-place regeneration is that the individual components may be designed to comply with their functions. For in-place regeneration the operation, separation and regeneration are performed in one universal service vessel. With external regeneration the service vessel may be designed taking only the operating requirements into account, while the regeneration components can be optimized for resin separation and regeneration.

Today two types of regeneration system design are common: the less familiar two-vessel type and the more common three-vessel installation [4].

Two-vessel installation. Two-vessel external regeneration installation consists of two tanks. The first one, the receiving tank, is for cleaning, separation, regeneration, and rinsing of both cation and anion resins. By design the receiving tank is similar to the service vessel. It is, however, substantially higher (3 to 4 or more meters) and distinctly smaller than the service vessel, and it has an additional drainage system (as have the in-place regenerated mixed bed polishers).

The receiving vessel is – similar to the service vessel – provided with sight glasses. Of special importance is the possibility of monitoring the separation process in the interface region. As for the structure of the receiving vessel and its internals what has been said for service vessels is applicable.

The second tank of the two-vessel installation is the resin storage tank. Regenerated resins are transferred from the receiving tank into this vessel and stored until use (i. e., transfer to the service vessel).

Three-vessel installation. External regeneration in three-vessel installation differs very much from the two-vessel type. While in the two-vessel installation both cation and anion resins are regenerated in the receiving tank, regeneration of cation and anion resin in the three-vessl assembly is performed in two separate tanks. This reduces substantially the hazard of resin cross-contamination with the regenerant chemicals.

The first tank has the function of receiving the resin from the service vessel. The resin is cleaned in this vessel and then separated. In most installations the anion resin is then transferred into a second tank, the anion regeneration tank. The cation resin remains in the receiving tank and is regenerated there. In less common regeneration systems with a bottom resin transfer technique (e. g., Conesep® regeneration process) the cation resin is transferred into the second tank, and the anion resin remains in the receiving vessel. After the regeneration and rinse of the resins in the respective tanks they are transferred into a third vessel, the resin storage tank.

In-place regeneration. Quite often hydrochloric acid solution is used as the regenerant chemical for cation exchange resin in-place regeneration. Regeneration is effected usually by one of the following regenerating techniques (very short description of only the regenerant solution flow without the rinse step):

- Caustic solution is passed from the vessel top through the anion and cation resins, then acid solution via the center drainage system through the cation exchanger;
- caustic solution from the vessel top through the anion resin and extraction through the center drainage system, then acid solution via the same drainage through the cation resin;

– caustic solution from the vessel top through the anion resin and extraction via the center drainage system, simultaneously or a little later acid solution from below via the underdrain through the cation resin and extraction also via the center drainage.

The first two regeneration methods are older than the latter, which has been more and more adopted due to the improved regeneration results.

Resin cleaning. The deep-bed polisher retains not only condensate ionic impurities, but also corrosion products. An especially heavy crud loading occurs in resin from naked deep-bed polishers. Therefore, the resin should be thoroughly cleaned prior to each regeneration cycle. With external regeneration additional cleaning stages after regeneration of the individual resins are also possible and quite frequently called for. With the cleaning procedures resin fines should also be removed. For this the following methods are used:

– Air scrubbing followed by water backwash,
– ABRO procedure (air-bump-rinse-operation),
– ultrasonic cleaning.

Air scrubbing is the most common procedure. According to statistics this procedure is used in about 70% of the stations in the United States [12].

In several US plants the ABRO procedure is used. This is a cyclic application of a one-minute upward air scrub followed by a two-minute downward rinse and drain. The air loosens crud or fluidizes the bed, and the downflow rinse forces the metal oxides out through the bottom screens [35]. With this technique in most cases more cleaning stages are required than with classic air scrubbing (mostly more than 10 scrubbing operations with ABRO as compared to less than 5 with air scrubbing [12]). However, water consumption is reduced.

Another cleaning technique is ultrasonic resin cleaning [35]. It has been used only rarely up to now [12].

Resin separation and mixing. Two operations are extremely important for the satisfactory operation of mixed bed polishers: resin separation prior to and resin mixing after regeneration. In systems with in-place regeneration und in externally regenerated polishers with two-vessel installations resin separation is performed prior to the actual regeneration (injection of regenerant chemical solutions). With externally regenerated polishers in a three-vessel installation the resins are separated prior to the transfer of one resin component from the receiving tank.

The separation of the resins is achieved in a backwash step. The lighter anion resin settles on top of the heavier cation resin. The backwash is generally accomplished in two steps: a high flow rate to separate the bulk of the resin, followed by a lower rate [4]. The flow rates adequate for the respective resins are recommended by the resin manufacturers. A satisfactory separation of the resins is the principal precondition for an operation at low ionic leakage.

Of importance is also the homogeneous mixing of the two resin components in the service vessel after completion of the regeneration. It is mandatory with the in-place regeneration or with external regeneration systems without resin mixing. In the case of external regeneration systems and external mixing the subsequent resin transfer causes a certain stratification of the resins. This has a negative effect on the mixed bed operation. Thus additional in the service vessel mixing is recommended.

For resin mixing the following operating conditions have proven reliable [36]:

air flow	$1.0\ m^3/m^2 \cdot min$
mixing time	10 min
water level above regenerated resins (related to settled regenerated resins depth)	20 – 50 mm/m.

Too high a water level in the vessel may cause the resins to separate again after mixing has been completed. Too low a water level results in an improperly mixed bed which has, in addition, gaps from residual air plugs, which may lead to piping in the mixed resin bed.

Ionic leakage. If the deep-bed polisher is to function with less than 1 ppb ionic leakage (sodium, sulfate, chloride) then the resins must contain only a limited amount of these ions. As an example, Figure 2.76 shows the dependency of sodium leakage on the regeneration level of a cation resin operating in the H/OH cycle (on the left hand side) and in the NH_4/OH cycle (on the right hand side) according to [35]. The data in Figure 2.76 are based on the equilibrium calculations mentioned in Section 2.3.4.3.

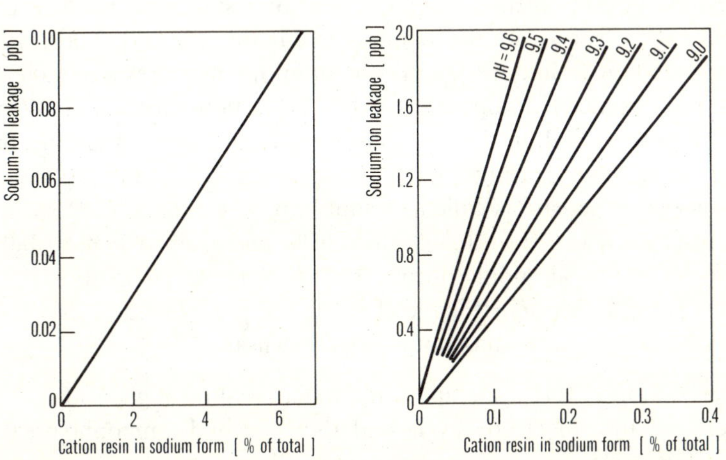

Figure 2.76 Sodium leakage as a function of the cation resin regeneration level.
right: NH_4/OH cycle
left: H/OH cycle.

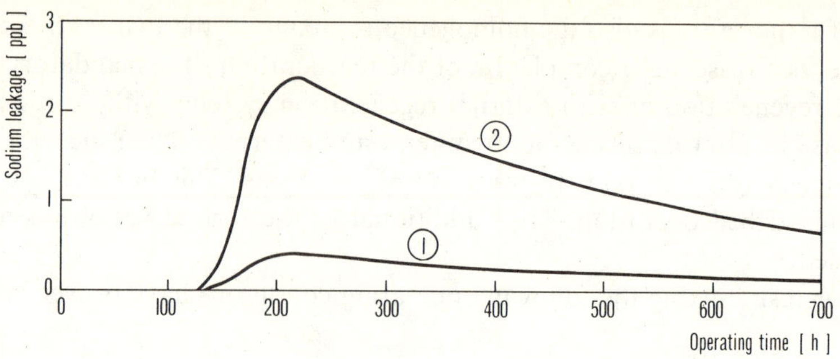

	% Na on resin		average Na leakage from NH_3 break [ppb]
	after regeneration	final	
1	0.06	0.014	0.23
2	0.35	0.08	1.31

Figure 2.77 Impact of the cation exchanger regeneration level on the sodium leakage after ammonia breakthrough.

Calculations of the equilibria for anion exchangers are similar. If a chloride leakage of 1 ppb is not to be exceeded, the ion exchanger resin may only be loaded with chloride ions at pH = 7 to 83%, at pH = 9 to only 5.2%, and at pH 9. 6 to not more than 1.3% [29].

For both cation and anion resins equilibrium calculations confirm that the regeneration level requirements for deep-bed demineralizers are much more stringent if the resin is operated in the NH_4/OH cycle than in the H/OH cycle [4].

The impact of the regeneration level of cation resin in operation beyond ammonia breakthrough is shown by the example in Figure 2.77 [37]. The curves in Figure 2.77 are based on an initial sodium content in the cation resin of 0.06% and 0.35% of total resin capacity, an operating pH of 9.4, a flow rate of 146 BV/h (BV = bed volumes), and no sodium in the polisher influent. After ammonia breakthrough (approx. after 130 operating hours) sodium is displaced from the cation resin. The sodium leakage increase is called sodium blip. The sodium content present in the cation resin decreases from 0.35% to 0.08% and from 0.06% to 0.014% due to sodium removal by ammonia during the NH_4/OH cycle operation.

With regard to the ionic leakage of mixed bed polishers there are two alternatives for operation with ammonia-containing condensate:

- Operation in H/OH cycle up to ammonia breakthrough; and
- operation first in H/OH cycle and then, beyond ammonia breakthrough, in NH_4/OH cycle.

In the first case ionic leakages below 1 ppb may be reached, as a rule, over the entire service cycle, provided resin separation and regeneration are reasonable. In the latter case, to reach the required ionic leakage below 1 ppb usually special regeneration techniques must be applied. The capacity for limiting condenser inleakage is markedly reduced. Operation in the NH_4/OH cycle is possible only if the treated condensate does not contain any ionic impurities, i. e. if the mixed bed polisher is operating exclusively as a crud separator.

Special regeneration techniques. In order to achieve a low ionic leakage, especially in cases of an intended operation beyond ammonia breakthrough, many special regeneration methods have been developed and adopted. Many have never proven practical or acceptable, but some are absolutely indispensible in deep-bed condensate polishing via mixed beds.

Improvement of regeneration with respect to the regeneration level (and, therefore, the ionic leakage level) may be achieved principally in two ways. On the one hand, the resin separation important for successful regeneration may be improved by means of specific measures; on the other, an additional resin treatment may reverse the negative consequences of incomplete separation. The most important or widely-adopted regeneration techniques will be briefly described below.

The regeneration methods for ionic (sodium) leakage reduction by a special resin treatment include, e. g.:

- Ammonex® process, and
- Calex® process.

In order to upgrade the resin separation and/or to preclude the consequences of the incomplete separation by isolation of the incompletely separated resin interface zone ("trouble resin") the following techniques among others are used:

- Seprex® process,
- three-component mixed bed,
- Conesep® regeneration process, and
- Seprex/Conesep process combination.

Ammonex® procedure. Ammonex® is the oldest of the procedures developed and adopted for the reduction of sodium leakage from the demineralizers in condensate polishing during the past 20 years. For this process an ammonia solution is passed through the freshly regenerated anion resin. The purpose here is to exchange ammonium ions for sodium ions on the small amount of residual cation resin remaining within the anion resin after separation, which was converted to the sodium form during anion regeneration, and to rinse residual caustic from the anion resin [38].

The Ammonex® process can be performed both as a once-through and a recycle procedure. In the once-through Ammonex® procedure the ammonia solution is passed only through the anion resin part of the bed and then to waste. The procedure requires the highest purity ammonia, and results in an elevated ammonia consumption; the wastewater is highly ammoniated.

In the Ammonex® recycle procedure an ammonia solution is first passed through the regenerated anion resin and then through the exhausted cation resin. The cation resin, still exhausted to the ammonium form from the service cycle, purifies the ammonia solution and allows its reuse. Of course, this assumes that the mixed bed demineralizer previously has not been exposed to any gross condenser leak or other sodium contamination.

The ammonia content of the ammonia solution is normally around 0.25 – 0.5%. The Ammonex® process may be applied for both in-place regeneration and external regeneration (with two-vessel and three-vessel systems) and also when using inert resin (e. g. [39, 40, 41, 42]).

Calex® process. The Calex® or lime wash procedure introduces a filtered lime solution wash to the anion resin following external anion resin regeneration with caustic. Here the sodium-charged residual cation resin in the anion bed is converted to the calcium form. The anion is then thoroughly backwashed before being remixed with the regenerated cation resin in the corresponding vessel of the regenerating system [43, 44].

The amount of lime required will depend on the amount of entrained cation resin. It is clear that the entrained resin should be kept to a minimum by effective separation techniques.

When using the Calex® process the fact that the solubility of lime decreases with increasing temperature must be taken into account. The lime solution concentration should be kept below the solubility limit of the highest temperature which may occur during lime washing.

Other resin treatment techniques. Out of the many other methods here only one procedure will be mentioned, which has been developed in the Netherlands. This procedure was developed for in-place regenerated mixed beds that are operated in the NH_4/OH cycle. For details of the process see [29].

Seprex® process. With the Seprex® regeneration technique complete separation of the anion resin from the cation resin is achieved by floatation separation [45]. This requires a three-vessel external regeneration system. The process steps important for resin separation are illustrated in Figure 2.78; the resin storage tank is not

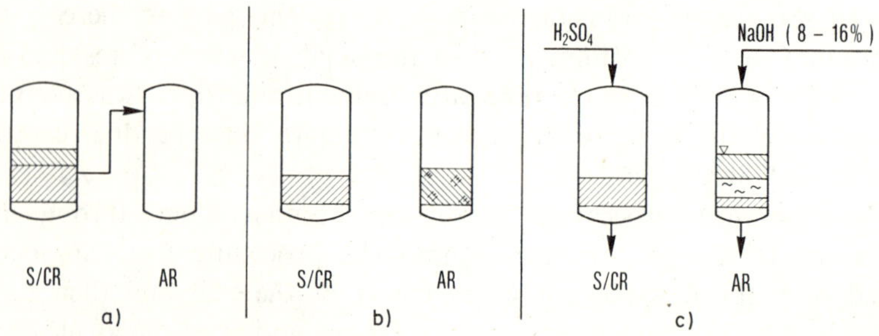

Figure 2.78 Resins separation in Seprex Process.

shown. After the normal separation of cation and anion resins in the separation/cation regeneration vessel (S/CR) the anion resin is transferred into the anion regeneration vessel (AR). The resin take-off interface is located in the cation bed portion, a small amount of the cation resin is deliberately carried over to the anion regeneration vessel to ensure that no significant quantities of anion resin remain in the separation/cation regeneration vessel. This transfer step is illustrated in Figure 2.78 a (condition at the beginning of the transfer step) and in Figure 2.78 b (post-transfer condition).

In the course of the next regeneration step in the separation/cation regeneration vessel the cation resin is regenerated in the normal way; the anion resin, however, is floated and regenerated in the anion regeneration vessel in caustic concentrations of 8 – 16% sodium hydroxide by weight (Figure 2.78 c). Due to its specific gravity, this solution causes the anion resin to float and allows the entrained cation resin to sink. The floated anion resin can then be transferred to the resin storage tank for rinsing; the rinsed regenerated cation resin from the separation/cation regeneration vessel is then also transferred into this tank. The cation resin portion from the anion regeneration vessel (cation resin in sodium form) is returned to the empty separation/cation regeneration vessel. During the next regeneration cycle it will be mixed with the emerging resin.

Three-component mixed beds. The first industrial-scale system using inert resin to improve the hydraulic separation of the cation and anion resins was installed in 1976 [46]. Today some resin manufacturers offer resin combinations with a suitable inert resin (e. g., Triobed®, Ambersep®) which may be used for systems with in-place or external regeneration. Table 2.25 shows some examples of the resin combinations used.

Table 2.25 Ion exchange resins and inert resin combinations for three-component mixed beds

	cation exchanger	inert resin	anion exchanger
Ambersep®	200	359	900
	252	359	900
	132	359	420
Duolite Triobed®	C 26 TR	S 3 TR	A 161 TR
	C 255 TR	S 3 TR	A 113 TR

The inert resin system is designed to prevent problems due to cross-contamination resulting from improper separation in mixed beds. The resin combination consists of three components – a cation resin, an anion resin, and the inert resin, which has no exchange capability. The specific gravity and particle size of the three components are designed to separate virtually completely the resins into three

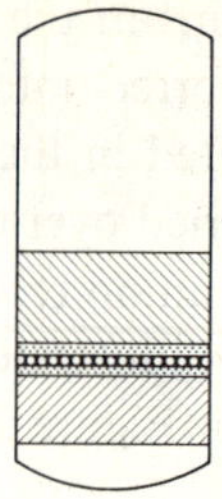

Figure 2.79 In-place regeneration three-component mixed bed – condition after resin separation.

distinct layers after backwash and settling (Figure 2.79). The layer of inert resin can be up to 20 cm in depth; a higher layer would have too great a resin capacity dilution effect. In the in-place regenerated mixed bed the inert resin layer provides a buffer zone between the cation and anion resins. Given proper resin ratios, the inert resin prevents contact between the acid regenerant solution and the anion resin and between the caustic regenerant solution and the cation resin.

With external regeneration the resin removal internals in the separation vessel will be positioned within the layer of inert resin [4] to prevent carryover of the cation with the anion resin. The inert resin depth may be optimally adjusted to the separation vessel diameter. A significantly greater depth than with the in-plance regenerated mixed beds is possible.

Conesep® regeneration process [47, 48, 49, 50].Figure 2.80 is a schematic representation of the functional principle of this regeneration process, with a bottom resin tank transfer technique (without service units). The resin requiring regeneration is transferred from the service unit into a separation/anion regeneration vessel (S/

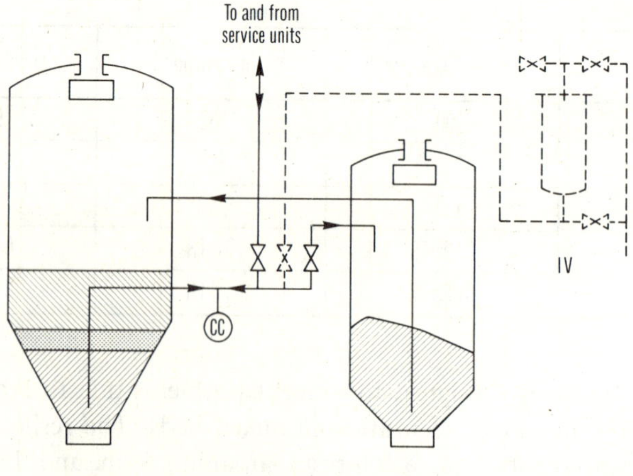

Figure 2.80 Conesep regeneration process.

AR). This vessel is a vertical cylindrical tank with a cone-shaped base. After preliminary resin cleaning and backwash separation cation resin is then transferred hydraulically to the cation regeneration vessel (CR). A conductivity cell (CC) is fitted into the transfer pipe. The completion of the cation resin transfer is detected by means of a change in the conductivity between the cation transfer water and the anion transfer water. The incompletely separated resin interface zone is isolated by an extended transfer line. This resin is not included in the regeneration process and is remixed with the regenerated main charge of resin before being transferred to the service vessel. The regeneration or further treatment of the resins after separation is performed with the standard techniques.

For the application of the NH_4/OH cycle a modified technique – double resin movement – is used. In this technique only anion resin is regenerated and rinsed after separation and transfer.

The unregenerated cation resin charge is recombined with the anion resin, remixed and again separated and transferred. The cation resin is then regenerated in the normal way, rinsed and finally recombined with the anion resin [48].

The Conesep® process system was expanded later by an isolation vessel (IV). The resin of the incompletely separated interface zone is removed from the base of the separation/anion regeneration vessel where it remains until the whole of the regeneration cycle is completed and the mixed resin returned to a service unit. The isolated resin is then flushed into the separation/anion regeneration vessel to await the next regeneration sequence. This expansion of the Conesep® system by the isolation vessel has proven successful for the ammonium cycle operation [50].

Seprex/Conesep regeneration process [51]. This process improves upon the techniques employed in the Seprex® and the Conesep® processes. The key steps in this combined regeneration process are shown in Figure 2.81.

After the separation of the resins in the separation/anion regeneration tank (S/AR) the cation resin is transferred to the cation regeneration tank (CR) – Figure 2.81 a. A conductivity cell (CC) is fitted into the transfer pipe. The transfer is terminated in response to a considerable conductivity drop when the cation resin transfer is complete and inert resin is being conveyed along the transfer line.

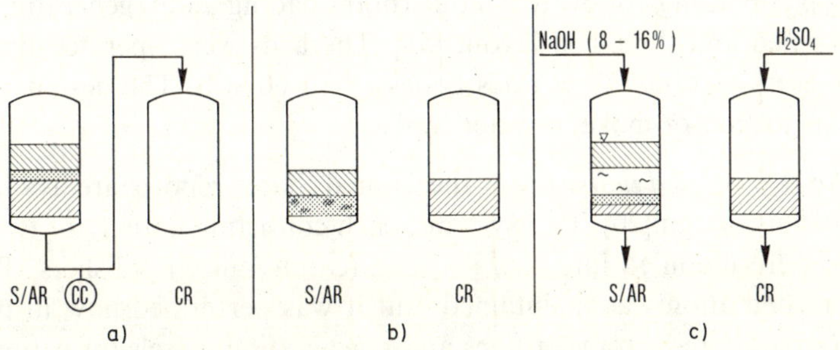

Figure 2.81 Resins separation in Seprex/Conesep regeneration process.

After this transfer step (Figure 2.81 b) the cation resin is regenerated in the cation regeneration tank and the anion resin is floated and regenerated in the separation/anion regeneration tank with caustic solution of a known concentration. Any trace of cation resin and inert resin will settle in the caustic in the separation/anion regeneration tank and remain there until the next regeneration cycle, while the floated anion resin will be discharged for rinsing.

Other resin separation techniques. Among other separation techniques only the Belco® process with a resin fines removal feature shall be mentioned. In this process in one of the steps inert resin contaminated with a small amount of cation resin (incompletely separated resin interface zone) is transferred into an additional trouble resin hold vessel. There the trouble resin is backwashed in order to remove the resin fines. For details of this Belco® separation process see the literature [52].

2.3.4.6 Multi-bed demineralizers

Apart from the mixed bed polishers with or without a lead cation bed there are also several types of demineralizer systems with two or three separate resin beds which have been in use for more than 20 years. They are far fewer in number than mixed beds, but they have advantages, especially with regard to easily attaining minimum ionic leakage.

The installation of a lead cation resin bed upstream of a mixed bed – an arrangement widely used in Germany – creates the conditions for long service runs of the mixed bed polisher. Such a mixed bed can be operated in the H/OH cycle, resulting in a very low ionic leakage when polishing high pH condensate, with a high capacity for retaining cycle ionic impurities.

The multi-bed polishers may consist of separate vessels or one common vessel. Regeneration may be performed either externally or in-place. For the latter, coflow or countercurrent techniques are available.

Mono-beds

Coflow units. Systems with an arrangement as illustrated in Figure 2.82 were already in operation at the beginning of the sixties; they were integrated into condensate polishing plants using coflow technique (both loading and regeneration downflow) in the two-bed and three-bed layout [53]. The beds were operated predominantly with low cross-sectional flow rates of less than 60 m/h. The design was more or less similar to that of makeup water trails.

Externally regenerated units. From the United States reports are available on two systems of this design [20]. Two systems, each consisting initially of two mixed bed polishers, were modified into two two-bed (cation/anion) polishers. The previous external regeneration was maintained; but it was performed now in two separate systems (separate resin transfer lines and regeneration vessels for cation and anion resin) in order to preclude any cross-contamination.

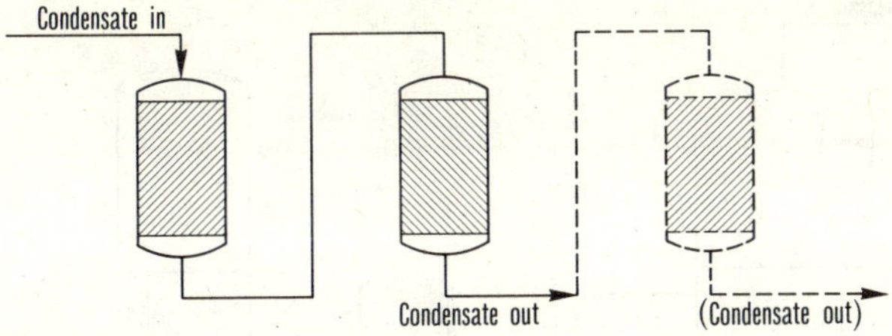

Figure 2.82 Mono-bed condensate polisher (coflow technique).

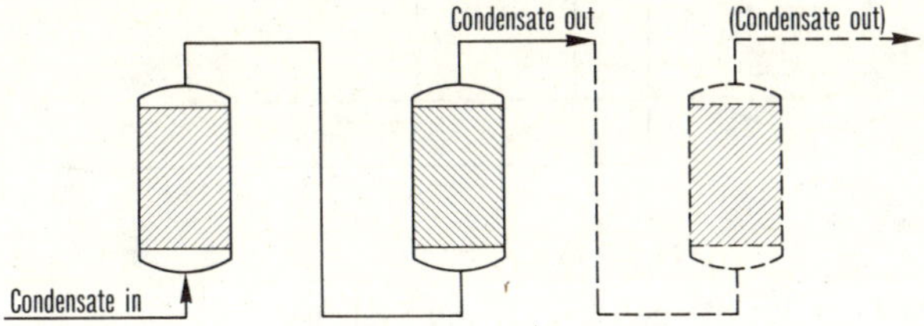

Figure 2.83 Mono-bed condensate polisher (countercurrent technique).

Countercurrent units. Two-bed and three-bed polishers may also be designed as countercurrent systems. Reference [53] reports on the first large-scale unit (commissioned in 1982). The three-bed polisher system with two parallel trains of cation and anion beds has a common trail cation downstream. The system has a capacity of 1400 m^3/h. It may be operated as a two-bed system, i. e. without the trail cation bed. All individual beds are operated in countercurrent by way of upward loading and downstream regeneration (Figure 2.83).

Use of the countercurrent technique allows a clear reduction of the regenerant chemical doses, with an ionic leakage into the polished condensate in the sub-ppb range. Both lead and trail cation beds are regenerated with a hydrochloric acid solution; the anion bed is regenerated with a caustic solution. The regenerant dose amounts to 48 g/l resin for the lead cation and to 35 g/l resin for the anion; 53 g/l resin are required for the trail cation bed.

In recent years condensate polishing in multi-bed systems has been given more and more attention in Germany [54, 55].

Triple beds

In 1981 a report was published on the first triple-bed unit, a Tripol® condensate polisher with a capacity of 180 m^3/h [56]. A simplified diagram of the Tripol® polisher is given in Figure 2.84.

All three individual beds – lead cation, anion, and trail cation – are installed in a common service vessel. The regeneration is performed externally, the cation resin in the cation regenerator (CR), and the anion resin in the anion regenerator (AR). A normal lead cation charge is temporarily stored in the cation resin storage

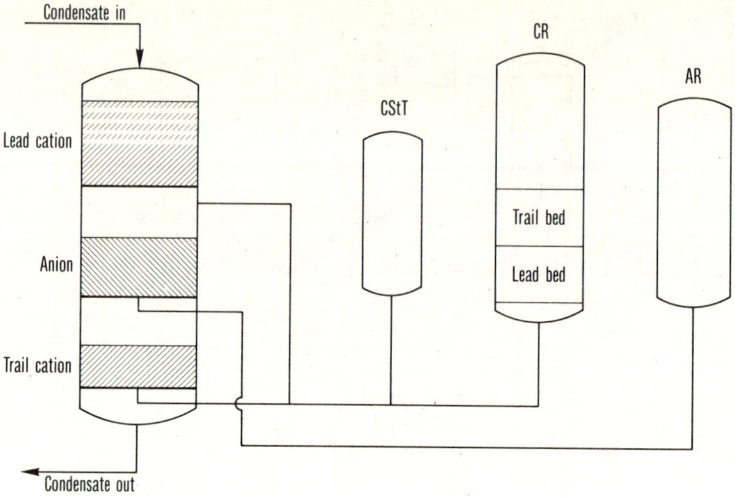

Figure 2.84 Triple-bed condensate polisher (Tripol System).

tank (CStT). The lead cation bed may be loaded and operated also with double the normal resin charge.

One report on another Tripol® condensate polisher (a system with a capacity of 185 m^3/h) states that the resin from the lead cation bed is transferred to the cation regeneration vessel first and the trail bed is then added on top, so that a pseudo-countercurrent regeneration process is achieved.

In the Tripol® process high cross-sectional flow rates (around 200 m/h) are used at a relatively low resin depth (350 to 500 mm). Operation of a Tripol® condensate polisher in both H/OH-form and NH_4/OH-form is possible [56–59].

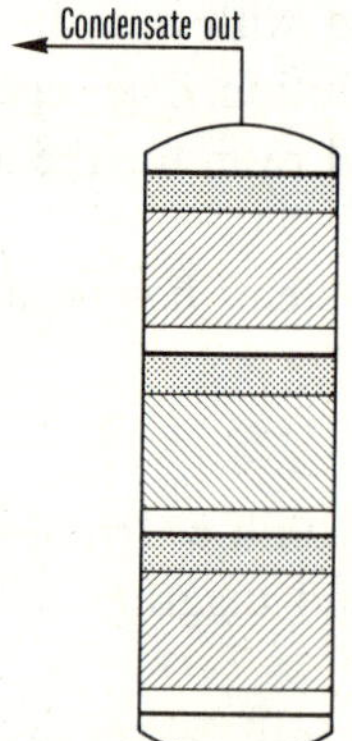

Figure 2.85 Triple-bed condensate polisher (Lewatit Multistep System).

Recently a report was published on another system of the triple-bed type, the Lewatit® multistep system [25]. While for the Tripol process the loading is performed downflow and the regeneration externally, with the Lewatit® multistep technique loading is done upflow and in-place regeneration downflow in countercurrent, Figure 2.85. As a result, the regenerant chemical doses are quite low. The system described in [25] consists of a service vessel with a diameter of 800 mm. The maximum cross-sectional flow during testing was 120 m/h, resulting in a polisher output of 60 m^3/h.

Single lead cation

Both deep-bed and powdered-resin demineralizers may have an up-stream single cation bed. This lead bed has a dual function: on the one hand, it absorbs the ammonia from the condensate to be polished, on the other hand, it retains suspended corrosion products.

As a result of the ammonia absorption in the cation bed, the actual polisher serves only to remove cycle ionic impurities from the condensate. When operated in the H/OH-form the usual problems with mixed bed ionic leakage are circumvented; the exchange capacity for trapping any cooling water inleakage is sufficient.

The lead cation bed is loaded with crud, similar to a naked mixed bed polisher. (The capacities of mixed beds for metal oxides have been estimated to be up to 5 kg/m^3 resin at flow rates of 120 m/h [4]). Adequate measures, therefore, for cleaning the cation exchange resins must be provided.

It is interesting to note that the arrangement cation bed – mixed bed in the United States is the exception [12]. In other countries, especially in Germany, this is a familiar layout [15]. The lead cation bed in both three-bed systems and triple beds, such as in the Tripol® and Lewatit® multistep systems, has the same function as when it is upstream of a mixed bed.

Multi-bed polisher equipment

The design and corrosion protection of service vessels and regenerating systems correspond largely to those of the makeup water system and mixed bed polishers so that a repetition is unnecessary.

2.3.4.7 Resin traps

Another important component of all deep-bed demineralizer systems are resin traps or resin strainers. They are located behind the condensate polishers to prevent large-scale ingress of resin beads into the feedwater.

Resin traps are typically coarse, Johnson wedge-wire grids with windings spaced at approximately 0.25 mm, or strainer baskets fitted with 600 to 300 μm (30 to 50 mesh) stainless steel screens [4].

Even though the resin strainers are specifically designed to retain catastrophic releases of resin beads, in several plants resin strainers appeared effective in retaining a portion of the resin fragments leaking from the polishers. However, in other plants the resin strainsers were not effective in retaining these resin fragments. For most of the polishers the resin fragment concentration leaving the strainer was higher than the resin fragment concentration entering it [60].

Resin bursts caused by resin traps occur predominantly in improperly maintained installations. Resin traps should be cleaned and inspected on both a time and differential pressure basis. The design and purchase specifications should be such that maintenance is facilitated [12].

The trapping of resin beads and fragments is important not only behind a deep-bed demineralizer system. Resin traps are also used between individual monobeds of multi-bed polishers in order to protect the downstream beds from any ingress of resins from the preceding beds.

2.3.5 Powdered-resin demineralizers

2.3.5.1 Introduction

Although the majority of US power plants are equipped with deep-bed demineralizers, the powdered-resin demineralizers alone or in combination with a deep bed (powdered followed by deep bed or deep bed followed by powdered) have approximately a 40% share of the total condensate polishing installations. They are used predominantly in power plants with fresh water once-through cooling systems [12].

Powdered-resin demineralizers, precoat filters designed to be coated with finely-ground ion exchange resin, were introduced into condensate polishing at the beginning of the sixties as the Graver Powdex® process. The rapid exchange rate characteristics of these fine-particle resins permit the use of thin layers that still produce the desired effluent quality. In addition to the rapid exchange rate and greater utilization of the ultimate exchange capacity of the resins, powdered resins effectively remove substantial amounts of colloidal-size materials (such as corrosion products) and non-reactive silica [23].

2.3.5.2 Resins for powdered-resins demineralizers

The average particle size of the powdered resin used in powdered-resin demineralization is in the 40 to 50 μm (approx. 325 mesh) range. The resins are highly regenerated (e. g., site composition of hydrogen form cation exchange resins 99% hydrogen minimum; site composition of hydroxide form anion exchange resins 95% hydroxide minimum).

There are several powdered resins available:

- mono-resins, such as strongly acidic cation and strongly basic anion exchange resins,

- mixtures containing both cation and anion exchange resins in varying ratios,
- mixtures containing, in addition to the exchange resin combination, an inert fibrous material.

The first group above includes Powdex® PCH and Lewasorb® A10 (hydrogen form of strongly acidic cation resin), Powdex® PCN (ammonium form of strongly acidic cation resins), and Powdex® PAO and Lewasorb® A50 (hydroxide form of strongly basic anion resin).

The second group above includes, e. g., various products of the Microionex® MB type. The products contain cation and anion resins in differing ratios depending on the application, the cation exchanger also having a variable form (hydrogen or ammonium).

Homogeneous mixtures of ion exchange resins and an organic fibrous material are products of the Ecodex® program; here, too, a selection between the form of the cation exchanger – hydrogen and ammonium – is possible.

Being directly connected with the powdered-resin precoat mixtures precoat overlay materials must also be mentioned. Normally they are organic fibrous materials without any ion exchange function. Like mixtures containing powdered resins, precoat overlay materials also form floc particles. These flocced fibers are then overlayed on filter septa which have been precoated with powdered-resin formulations.

The function of the overlay material is to inhibit fissuring of the underlying powdered-resin precoat, and to reduce the amount of crud penetrating into the precoat layer [61].

2.3.5.3 Powdered-resin demineralizer equipment

The powdered-resin demineralizer system consists of a service vessel in which the filter elements are installed, hold pump, slurry tank with mixer and appropriate baffling, precoat pump, and pertinent valve assemblies. The most frequently used filter septa are a fiber-wound and stainless steel type (see Section 2.3.3.2). The service vessel is connected to a backwash air and water supply.

The system illustrated in Figure 2.86 with all operating phases (service, backwash, precoat) has a single-tank precoat mix and feed system. With the double-tank design the powdered-resin demineralizer system has two separate tanks for precoating, a recirculating tank, and a slurry storage tank.

Filter elements are located vertically in the service vessel and attached to a tube sheet. They may be installed either above the tube sheet (tube sheet at the service vessel bottom) or below it (tube sheet at the service vessel top). The typical powdered-resin demineralizer flow rate is 10 $m^3\,h^{-1}\,m^{-2}$ of the filter elements' bare surface area [4, 12].

With a hydraulic service vessel care must be taken that all filter elements are subject to a uniform flow. The precoat film must not be washed off by the flow.

Normal filter element spacing is 25 to 40 mm. A smaller spacing of the filter septa might lead to precoat bridging, impeding the precoat backwash.

2.3.5.4 Precoating

One advantage of the powdered-resin demineralizers is the possibility of varying the precoat composition and resin ratio depending on the properties of the condensate to be treated. For normal operation, operation with condenser leaks, for startups and shutdowns precoating may be used having various resin ratios or forms.

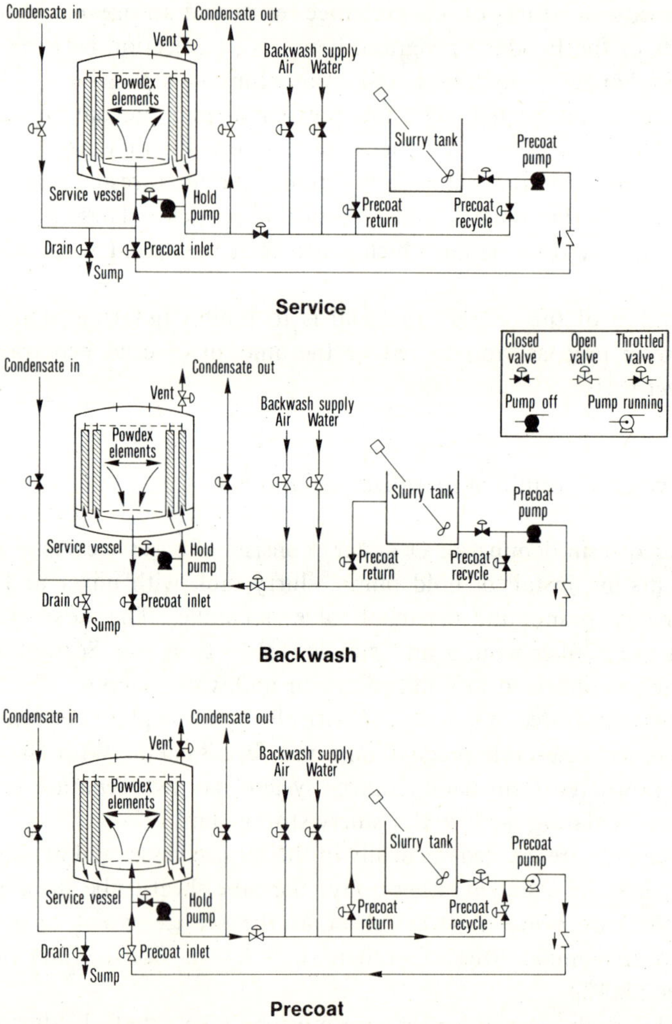

Figure 2.86 Powdered-resin demineralizer (operating phases [62]).

When preparing the slurry from powdered resins and pure water in the slurry tank, agglomerates in the form of flocculated particles form in the water due to agitation. The formation of agglomerates is enhanced by opposite electric charges of the powdered-resin surfaces. The floc size is affected by several factors, such as resin form, resin ratio, particle-size spectrum, water purity and temperature, and agitation intensity [4].

Most US plants add a slurry-conditioning polyelectrolyte in order to control the floc formation. The conditioning chemical is a polyacrylic acid (Rohm & Haas' Acrosol A-1). For the determination of the optimal amount of this additive the slurry consistency is monitored by using the "*V* over *V*" measurement technique [12]. This test compares the volume of settled slurry after a given time, usually 15 minutes, to the total (initial) volume of the sampled slurry. The desired V/V is usually specified by the manufacturer [4].

According to [62] the V/V ratios below 50% were too compact and cause a noticeable increase in differential pressure and reduced service runs. Ratios exceeding 60% seemed to be too loose; i. e., silica and iron removal were less than desired. The best ratio is approximately 55%, giving excellent control and longer service runs.

In most cases precoats are applied up to a layer thickness of 6 mm; the resin quantity required is $1-1.5$ kg/m^2 of the filter elements' surface area. Figure 2.86 (bottom) shows schematically the precoat phase.

2.3.5.5 Service

Like a bead demineralizer, a powdered-resin demineralizer operated on H/OH-form resins is more effective than one operated on a cation resin in NH_4-form in producing effluent with lowest sodium values. The separation capacity with respect to crud varies strongly with the nature of the corrosion product to be retained. The maximum crud separation capacity may be estimated to be $15-20\%$ of dry precoat weight.

The endpoint of the service cycle is reached when the differential pressure across the demineralizer rises to a given maximum or when ionic leakage (sodium, silica) or conductivity increases. When using cation resin in the NH_4-form the cation conductivity instead of the specific conductivity should be used as a criterion for the determination of the service cycle endpoint.

Figure 2.86 – the top section – is a schematic representation of the service cycle pump and valve positions.

2.3.5.6 Backwash

The backwash cycle may include either air scour, air surge or air bump procedures [4]. The selection of the procedure depends on the design of the demineralizer (position of tube sheet and filter elements), but also on the nature of the precoat loaded with crud.

The backwash cycle may present some difficulties if the precoat cracks in the presence of hydrated iron oxides and filter element plugging occurs. It may prove to be necessary to clean chemically, (e. g., by citric acid solution) the filter elements that can no longer be completely cleaned by backwashing [12].

Figure 2.86 – center section – is a schematic representation of the backwash cycle pump and valve positions.

2.3.6 Concluding remarks

Condensate polishing represents an important technical procedure for maintaining required water purity in power plant cycles of both fossil and nuclear systems. Condensate polishers of inherent good performance are – mainly in high-pressure utility units – the prerequisite for an economic and fail-safe operation of steam supply systems and turbines.

When evaluating today's condensate polishing technology one should, however, not ignore certain negative trends. Frequently, attempts are made to optimize condensate polisher economics without considering overall power plant economics. The particular polisher may well be optimized with respect to costs; however, the consequences for the power plant cycle may prove to be everything but economical.

Using completely the ion exchange capacity of a condensate polisher will naturally reduce regenerant chemical costs. However, the almost exhausted polisher will rarely be capable of coping with a major cooling water leakage. If this occurs the economic consequences may by far exceed the saving on regenerant chemicals. Such consequences would be, e. g. an unscheduled system shutdown or operation with an upset cycle chemistry. In the latter case deterioration of the power cycle efficiency due to turbine blade deposits, however light they may be, will be the result, not to mention possible boiler tube failures or turbine blade damage.

An erroneous understanding of economics has also had an influence on condensate polisher design. During the past 20 years in the United States naked mixed beds were adopted for condensate polishing [63]. The combination of two functions in one system (crud separation and ion exchange) results in a treated condensate of a quality which, in most cases, cannot match European standards, which are obtained with a different condensate polishing approach.

Here it would be worthwile to investigate the cause for the high frequency of turbine blade failures in the United States, which are almost non-existent in Europe today [9]. The impact of the operational environment, which is essentially dictated by the polished condensate quality, cannot be overlooked. At the same time the ammoniation on stream and the operation beyond ammonia breakthrough must also be mentioned here, with the known problems concerning polisher ionic leakage.

In many papers the performance of mixed beds containing ammonium-form cation resin when operating in high pH environment has already been described. From among the innumerable reports, an older one by Salem [64] should be mentioned here the results of which, despite many efforts during the past twenty years for upgrading resin separation and regeneration techniques, are still valid.

When evaluating the polishing of condensate two essential facts must be considered: the demineralization with resins in the H/OH-form presents far less problems and yields a better condensate polisher effluent quality than a demineralization with resins in the NH_4/OH-form; second, the regenerant chemicals of a deep-bed demineralizer are the most important cycle contaminants.

Taking these facts into account, a condensate polishing plant should be designed in such a way that the number of regenerations is kept to a minimum and that the demineralization is performed with resins in H/OH-form.

In order to achieve this, two major alternatives for a condensate polisher configuration are available:

- A single lead cation bed followed either by a powdered-resin or a deep-bed demineralizer both operated on resins in the H/OH-form. Whether to select a powdered-resin or a deep-bed polisher depends mostly on the cooling water type alone.
- A multi-bed polisher consisting of either mono-beds or triple beds. If possible a countercurrent design should be envisioned.

The only problem still remaining – but it is solvable – is the removal of ammonia from the regeneration effluent of the single lead cation or from the lead cation stage of the multi-bed or triple-bed polishers, possibly with recycling.

References

[1] Interim Consensus Guidelines on Fossil Plant Cycle Chemistry (1986). EPRI CS-4629.

[2] VGB-Richtlinien für Kesselspeisewasser, Kesselwasser und Dampf von Wasserrohrkesseln der Druckstufen ab 64 bar (1980). VGB Kraftwerkstechnik *60,* 793–800.

[3] VGB-Richtlinien für das Wasser in Kernkraftwerken mit Leichtwasserreaktoren (1973). VGB Kraftwerkstechnik *53,* 207–209.

[4] Paul Cohen, (ed.) The ASME Water Technology Handbook for Steam Power Systems, (1987). ASME, New York, N.Y.

[5] C. S. Welty Jr. and S. J. Green (1984) PWR Secondary Water Chemistry Guidelines. Rev. 1, Proc. Int. Water Conf. *45,* 325–331.

[6] H.-G. Heitmann (1986) Praxis der Kraftwerk-Chemie. Vulkan Verlag, Essen.

[7] A. Bursik (1986) 8 Years of Modified AVT with Elevated Oxygen Level for Once-Through Steam Generators. Proc. Int. Water Conf. *47*, 226–230.

[8] J. A. Levandusky and L. Olejar (1967) Condensate Purification. Power Engineering *71,* December, 58–61.

[9] Characterization of Operational Environment for Steam Turbine-Blading Alloys (1984). EPRI CS-2931.

[10] Questionary on Steam Surface Condenser Tube Leakage (1979). Heat Exchange Institute, New York, N. Y.
[11] Allianz-Handbuch der Schadenverhütung (1984), Allianz, Berlin, München und VDI-Verlag, Düsseldorf.
[12] State-of-the-Art Evaluation of Condensate Polisher Performance (1983). EPRI NP-2978.
[13] Manual for Investigation and Correction of Boiler Tube Failures (1985). EPRI CS-3945.
[14] W. J. Bilanin, R. L. Jones and C. S. Welty Jr. (1984) Water Chemistry Guidelines for BWRs. Proc. Int. Water Conf. *45*, 302–310.
[15] VGB-Merkblatt Nr. 18 (1971) Aufgaben und Methoden der Kondensataufbereitung. VGB, Essen (presently revised, will be issued as VG-M 412 L).
[16] J. A. Levandusky (1976) A Review of the Operating and Design Characteristics of Condensate Polishing Systems, NACE Panel Discussion on Condensate Polishing, Houston.
[17] M. Apperloo (1974) Kondensataufbereitung durch Filtration. VGB Kraftwerkstechnik *54*, 25–33.
[18] H.-G. Heitmann, V. Schneider and E. Redmann (1985) Hochtemperaturfiltration von Speisewasser zur Minimierung des Korrosionsprodukteintrages in Dampferzeugern. VGB Kraftwerkstechnik *65*, 693–699.
[19] H. W. Venderbosch, A. Snell and L. J. Overman (1971) Zur Kapazität von Ionenaustauschern bei der Entfernung von Spurenverunreinigungen aus Wasser. VGB-Speisewassertagung, 25–31.
[20] R. E. Mickel, G. R. Holmes, and P. C. Canary (1981) Different Approach to Condensate Polishing. Proc. Int. Water Conf. *42*, 209–215.
[21] A. Bursik (1985) A New GKM Approach to Makeup Water Treatment and Condensate Polishing. Proc. Int. Water Conf. *46*, 126–144.
[22] J. H. Smith and T. A. Peploe (1980) The Tripol Process – A new approach to ammonia cycle condensate polishing. BNES – Int. Conf. on the Water Chemistry of Nuclear Reactor Systems *2*, 139–144.
[23] J. H. Duff and J. A. Levandusky (1962) Powdex – A New Approach in Condensate Purification. Proc. Amer. Power Conf. *24*, 739–750.
[24] A. Bursik (1985) Historischer Abriss über den Einsatz der Ionenaustauschertechnologie im Grosskraftwerk Mannheim, VGB Kraftwerkstechnik *65*, 426–436.
[25] F. Martinola and H. R. Brost (1986) Verfahrenstechnik der Feinreinigung von Wasser durch Ionenaustausch. VGB Kraftwerkstechnik *66*, 159–170.
[26] J. T. McNulty and C. A. Bevan (1982) Anion Resin Kinetics Testing for Condensate Polishing. Proc. Int. Water Conf. *43*, 511–518.
[27] S. Fischer and G. Otten (1983) The Effect of Bead Size and Hydrated Bead Density on the Separability of Mixed Beds. Proc. Int. Water Conf. *44*, 409–416.
[28] I. M. Abrams (1976) New Requirements for Ion Exchange in Condensate Polishing. Proc. Int. Water Conf. *37*, 165–175.
[29] H. W. Venderbosch, L. J. Overmann, and A. Snel (1978) Ionenschlupf von Mischbettaustauschern in Kondensataufbereitungsanlagen – Theorie und Versuchsergebnisse. VGB Kraftwerkstechnik *58*, 228–232.
[30] H. R. Brost and F. Martinola (1989) Austauschvorgänge in Mischbettfiltern. VGB Kraftwerkstechnik *60*, 53–62.
[31] J. D. Darji and A. F. McGilbra (1980) Ion Exchange Equilibrium – a Key to Condensate Polisher Performance. Proc. Amer. Power Conf. *42*, 1101–1108.
[32] N. W. Frisch and R. Kunin (1960) Kinetics of Mixed-Bed Deionization. A.I.Ch. E. J. *6*, 640–647.
[33] C. E. Haub and G. L. Foutch (1986) Mixed-Bed Ion Exchange at Concentrations Approaching the Dissociation of Water, 1. Model Development. Ind. Eng. Chem. Fundam. *25*, 373–381.

[34] C. E. Haub and G. L. Foutch (1986) Mixed-Bed Ion Exchange at Concentrations Approaching the Dissociation of Water, 2. Column Model Applications. Ind. Eng. Chem. Fundam. *25,* 381–385.
[35] G. S. Crits (1974) High Rate Deep Bed Demineralization is Today's Best Solution to Nuclear Power. Combustion, June, 33–38.
[36] R. R. Harries and N. J. Ray (1980) Der Mischvorgang in Ionenaustausch-Mischbetten. VGB Kraftwerkstechnik *60,* 718–722. A. F. McGilbra (1984) Extend the operation of your condensate polishing system. Power *128,* 81–83.
[37] J. R. Emmet (1983) Condensate polishing ammonia cycle operation. Effluent and Water Treatment J. *33,* 507–510.
[38] T. McLaughlin, F. J. Pocock and J. F. Stewart (1969) Four Years' Operating Experience with a High Flow Rate Condensate Polishing System at the Middletown Station of the Hartford Electric Light Company. Proc. Amer. Power Conf. *31,* 647–667.
[39] R. E. Dwyer, I. E. Zoppoli, G. J. Crits and D.T. Tamaki (1976) Producing less than 1 ppb Sodium Leakage from Condensate Polishers. Proc. Amer. Power Conf. *38,* 940–948.
[40] K. L. Harner and K. H. Friederich (1979) Condensate Polishing Experience at Three Mile Island, Unit Two. Proc. Int. Water Conf. *40,* 1–5.
[41] G. J. Crits (1979) Prepared Discussion to 40. Proc. Int. Water Conf. *40,* 5–7.
[42] G. J. Crits (1982) Tri Ammonex – New Dimensions in Condensate Polishing with Inert Resin Use. Proc. Int. Water Conf. *43,* 519–522.
[43] R. Mickel, A. W. Kingsbury and C. Calmon (1970) Selective Cation Treatment in Separated Ion Exchangers for Condensate Polishing. Proc. Amer. Power Conf. *32,* 742–748.
[44] A. F. McGilbra (1984) Extend the Operation of your Condensate Polishing System. Power *128,* April, 81–83.
[45] L. L. Olejar and E. Salem (1970) Seprex – A New Technique in Ammoniated Cycle Condensate Treatment. Proc. Int. Water Conf. *31,* 195–200.
[46] S. S. Wolf (1979) Operating Experience with Triobed in the Production of Ultra High Purity Water. Proc. Inter. Water Conf. *40,* 263–267.
[47] J. R. Emmet and P. M. Grainger (1979) Ion Exchange Mechanism in Condensate Polishing. Proc. Int. Water Conf. *40,* 81–89.
[48] M. A. Sadler, D. J. O'Sullivan, J. C. Bates, and M. E. Costello (1983) Ammonium-Form Operation of Condensate Polishing Plant at Aghada Generating Station. Proc. Amer. Power Conf. *45,* 1058–1063.
[49] J. R. Emmet (1983) Condensate Polishing. Effluent and Water Treatm. J. *23,* 241–243.
[50] J. R. Emmet and A. Hebbs (1983) Factors Affecting the Performance of Condensate Polishing Plants. Proc. Int. Water Conf. *44,* 380–387.
[51] E. Salem and C. C. Scheerer (1983) A Unique Advance in Condensate Polishing at Central Illinois Public Service Company. Proc. Int. Water Conf. *44,* 151–162.
[52] J. Y. Chen and J. Nichols (1982) New Approach of Resin Separation in Condensate Polisher Application. Proc. Int. Water Conf. *43,* 21–31.
[53] A. Bursik, K. Spindler and H. Blöchl (1984) Kondensatreinigung in Monobett- und Gegenstromtechnik. VGB Kraftwerkstechnik *64,* 55–58.
[54] W. Oschmann, A. Schiffers, and V. Trümer (1982) Neue Aspekte zur Kondensatreinigung. VGB Kraftwerkstechnik *62,* 49–53.
[55] H. Vogel (1986) Ergebnisse aus einer Kondensatreinigungsanlage mit variabler Anlagenkonzeption und Prozessführung. VGB Kraftwerkstechnik *66,* 274–279.
[56] P. W. Renouf and J. H. Smith (1981) Operating Experience with a Tripol Ammonia Cycle Condensate Polisher. Proc. Int. Water Conf. *42,* 195–206.
[57] M. Ball and R. J. Burrows (1985) The Tripol Condensate Purification Process. Proc. Int. Water Conf. *46,* 112–118.

[58] M. Ball, M. A. Jenkins and R. J. Burrows (1984) Ammonium-Form Cation Resin Operation of Condensate Purification Plant. Ion Exchange Technology (D. Naden and M. Streat, eds), pp. 106–118, Ellis Horwood Ltd., Chichester.

[59] J. H. Smith, P. W. Renouf, and M. Crossen (1984) 50 Years in Separate Beds. Proc. Int. Water Conf. *45*, 105–112.

[60] Resin and Ionic Leakage from Condensate Polishers with and without Inert Resin (1986). EPRI NP-4521.

[61] L. L. Nolan (1979) Evaluation of Filter/Demineralizer Precoat Materials on Monticello Nuclear Station Condensate System. Proc. Int. Water Conf. *40*, 251–254.

[62] M. Wafdlington, J. Longo, and P. Gross (1981) 150 Million Megawatt Hours of Condensate Polishing. Proc. Int. Water Conf. *42*, 171–182.

[63] G. J. Crits (1984) Condensate Polishing with the Ammonex Procedure. Ion Exchange Technology (D. Naden and M. Streat, eds), pp. 119–126, Ellis Horwood, Ltd., Chichester.

[64] E. Salem (1969) A Study of the Chemical and Physical Characteristics of Ion Exchange Media Used in Trace Contaminant Removal – Ammonium-Form Cation Exchange Resin Mixed Bed. Proc. Amer. Power Conf. *31*, 669–684.

2.4 Treatment of Drinking Water with Ion Exchange Resins

Wolfgang Höll and Hans-Curt Flemming***

* *Kernforschungszentrum Karlsruhe GmbH, Institut für Radiochemie, Abteilung Wassertechnologie, Karlsruhe, Federal Republic of Germany*

** *Institut für Siedlungswasserbau, Wassergüte- und Abfallwirtschaft der Universität Stuttgart, Federal Republic of Germany*

2.4.1 General remarks

Drinking water is man's most important nutrient. It is always taken from the natural circulation of water in which it participates in a great variety of physical, chemical, and biological reactions. As a consequence, natural water contains a large number of either dissolved or suspended organic and inorganic substances. The most important requirement for drinking water is that it must not contain components injurious to health. Thus, in the treatment steps such substances have to

Table 2.26 Some recommended and limit concentrations in drinking water [1]

	Recommended concentration	Limit
Chloride	25 mg/l	–
Sulfate	25 mg/l	250 mg/l
Nitrate	25 mg/l	50 mg/l
Calcium	100 mg/l	–
Magnesium	30 mg/l	50 mg/l
Arsenic	–	50 µg/l
Chromium	–	50 µg/l
Cadmium	–	5 µg/l
Selenium	–	10 µg/l
C. O. D. ($KMnO_4$)	2 mg/l	5 mg/l

be eliminated or diminished below limiting concentrations prior to use. Some standards and recommended values from the Directives of the European Community are summarized in Table 2.26 [1].

Water acceptable from the perspective of health may still contain components in sufficient concentrations to cause troublesome effects. Therefore, both for economic reasons and convenience a reduction in the concentrations of such species can become necessary.

Ion exchange resins are mainly capable of removing ionic species. Thus they can be applied to eliminate components of salts or to reduce the total salinity of drinking water. Ion exchange is economic up to intermediate salt concentrations, where the costs of other processes are relatively high. As a consequence it is mainly applied to the treatment of water with less than 1200 mg/l TDS.

2.4.2 Cation exchange processes

Troublesome impurities in most water supplies not associated with health problems are those designated as water hardness. These impurities mainly include calcium and magnesium, which react with soap to form unsoluble curds, and also form scales of $CaCO_3$, $Mg(OH)_2$, and $CaSO_4$ on heat transfer surfaces. Bivalent cations can be removed by a cation exchange for sodium ions:

$$R-Na^+ + Ca^{2+},Mg^{2+} \rightarrow R-(Ca^{2+},Mg^{2+}) + Na^+.$$

All types of cation exchange resins are effective. However, for most applications strong acid resins of the styrene-divinylbenzene type are preferred. The process is efficient at most Na/Ca concentration ratios in raw water up to total concentrations of 5,000 mg/l. Regeneration is carried out with NaCl solutions – brine concentrations of 6 to 12% are recommended. In some cases sea water has also been used for regeneration [2].

With respect to corrosive effects in the distribution system, calcium and magnesium should not be eliminated completely; further, due to health risks sodium concentration should not exeed 200 mg/l. For these reasons, and also in order to reduce costs for regeneration, only part of the raw water has to be treated and blended with a bypass. Product water quality can be controlled by adjusting the flow ratio (Figure 2.87).

Softening has been introduced in many municipal water works in the U. S. A., in the U. K. and also in Belgium and in Sweden [3]. Nevertheless its main application is in homes. Home softening has become a highly developed technique that is, nonetheless, easy to handle. Regeneration may be carried out manually or automatically.

Water hardness can also be removed by applying a cation exchange resin in the hydrogen form according to:

$$R-H^+ + Ca^{2+},Mg^{2+} \rightarrow R-(Ca^{2+},Mg^{2+}) + H^+.$$

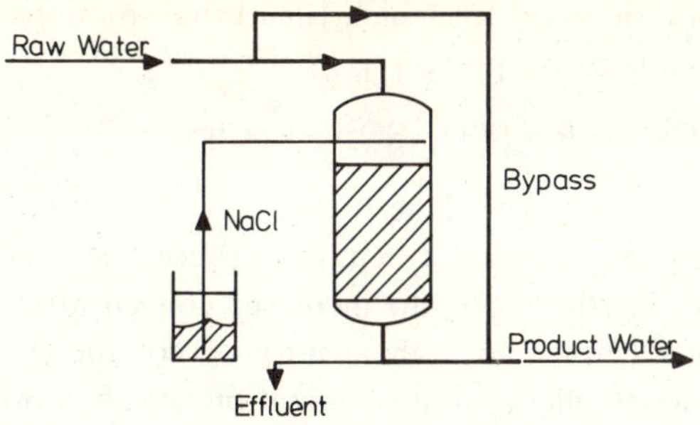

Figure 2.87 Flow scheme for partial softening.

In a secondary reaction the protons are neutralized by bicarbonate ions forming carbonic acid, which is split into CO_2 and water. Since both hardness and alkalinity are reduced at the same time, the process represents a kind of partial demineralization.

Strong acid as well as weak acid resins can be used, which are regenerated by hydrochloric or sulfuric acid. If strong acid resins are used, the degree of regeneration has to be adapted to the bicarbonate concentration of the raw water in order to avoid too low alkalinities or even strong acids in the product water. In case of weak acid resins, softening is always equivalent to raw water alkalinity [4].

Since strong acids are required for regeneration and a post-treatment is necessary for adjusting the carbonate balance, application of this kind of softening is restricted to municipal water works. It has been applied in a few water works in the U. K., in Belgium [3], and in the F. R. Germany, where some big continously operating FLUICON plants with throughputs of up to 400 m^3/h are in service. Strong acid resins are used, which are regenerated with 105% of the theoretical amount of HCl. Half of the throughput is treated and blended in a degasifier [4].

2.4.3 Anion exchange

In many surface and ground waters, nitrate levels have been increasing due to the extensive use of fertilizers in agriculture. Nitrate presents a health risk to infants since, after its reduction to nitrite, methemoglobinaemia can result. As a result, the World Health Organization recommends levels of less than 50 mg/l. Sulfate concentrations usually stem from gypsum deposits in the aquifer and are related to water hardness. Concentrations of SO_4 in excess of 250 mg/l render water unpalatable and cause purgative effects, particularly in the presence of high magnesium concentrations [5, 6].

Nitrate and sulfate ions can be removed by anion exchange for chloride ions. Usually, Type II strong base resins are recommended for this purpose, providing a good efficiency in the regeneration step for which 10% brine solutions are applied. The use of sea water has also been proposed [7].

$$R-Cl^- + NO_3^-, SO_4^{2-} \rightarrow R-(NO_3^-, SO_4^{2-}) + Cl^-.$$

The method shows some disadvantages. Due to the uptake and replacement of bicarbonate ions pH undergoes changes which may cause the precipitation of $CaCO_3$ even in the filter bed. Furthermore, the increased concentration of chloride ions produces corrosion problems. The high affinity of commercially available anion exchange resins provides excellent sulfate elimination, which, however, hinders the removal of nitrate ions. Since all natural waters contain sulfate, the exchange capacity for nitrate is reduced and the amount of NaCl required for regeneration increases. In order to reduce costs, part of the water may be treated and blended with untreated raw water as shown for softening in Figure 2.88, or all the raw water may be passed through an incompletely regenerated column [7].

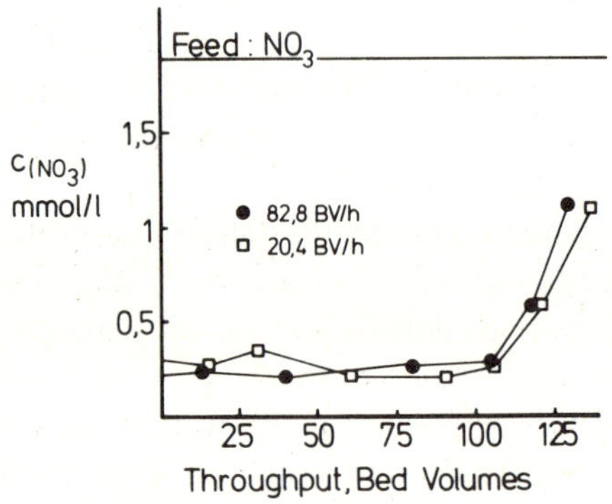

Figure 2.88 Nitrate effluent concentration after incompletely regenerated resin bed [7].

Nitrate and sulfate ions can also be replaced by bicarbonate ions, a procedure proposed in a modification of the NITREX process. In this case either $NaHCO_3$ solutions or NaOH and CO_2 are used for regeneration. As a consequence of the increase of pH, precipitation of $CaCO_3$ can occur, requiring an adjustment of the carbonate balance and filtration [8].

In various regions the groundwater contains toxic metals like arsenic, chromium, or selenium as oxide anions in concentrations that exeed the standards. Anion exchange against chloride ions is one of the possibilities of removing such species, if the affinity for the metal oxide anion is high enough compared with the anions of the background water composition. In laboratory and pilot-scale tests different types of strong base anion exchange resins have been tested for this purpose with acrylic and macroporous polystyrene resins showing the most favourable results. For example, due to the extremely high affinity of chromate for the resin, throughputs of up to 25000 bed volumes were obtained [8].

Chromate ions also may be eliminated using weak base resins in the sulfate form. The successful application of such a process in a municipal water work in Italy has been reported in the literature [3].

Removal of selenium has been tested in small pilot scale with a Type II resin in the mixed Cl^-- and HCO_3^--forms, which was obtained by regeneration using mixed $NaCl/NaHCO_3$ solutions. The particular advantage of this method was found to be the negligible variation of chloride and bicarbonate concentrations in the product water [10].

2.4.4 Partial demineralization

The human body is capable of tolerating drinking water only in a limited salinity range. The maximum permissible salt concentration depends on the type of salt, the daily water consumption and invidual factors. However, it is recommended that the salinity of drinking water should be less than 500 mg/l and under no conditions should exceed 1000 mg/l. Raw water with salt contents in excess of these values can be partially demineralized by combined cation and anion exchange. Such a process has also been proposed for nitrate and sulfate reduction. Usually a strong acid resin is used in combination with a weak base exchanger. In order to increase regenerant efficiency, a dealkalization can be carried out prior to the strong acid exchanger. Hydrochloric or sulfuric acid are used for regeneration of the cation exchangers whereas caustic, lime or even ammonia are proposed for the anion exchanger.

Since drinking water should not be completely demineralized, only part of the raw water has to be treated. The desired drinking water quality is controlled by adjusting the flow ratio (Figure 2.89).

Since complete demineralization is not necessary, salt leakages much greater than in industrial water treatment can be tolerated. This leads to the use of resin beds which are only partially regenerated, e. g. by applying weakly effective chem-

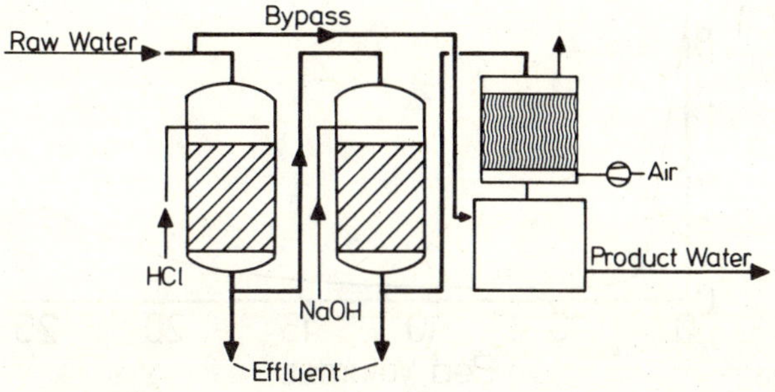

Figure 2.89 Flow scheme for partial demineralization.

icals. The Sirotherm process uses ion exchange resins of mixed weak-acid and weak-base functionality which show a temperature-dependent equilibrium with salt water [11]:

$$R_cCOOH + R_aNR_2 + NaCl \rightleftharpoons R_cCOO^-Na^+ + R_aNR_2H^+Cl^-.$$

The resins adsorb salt at ambient temperature and release it into hot water. For practical salt adsorption capacity a temperature difference of about 60 °C is required. To apply the process to the partial demineralization of water containing 1000 mg/l TDS, the water is passed through the resin bed at about 20 °C to produce a product water of about 100 – 500 mg/l (Figres 2.90 and 2.91). The resin bed is then regenerated with a smaller quantity of hot water at 80 to 90 °C to produce a regenerant effluent of 3000 – 5000 mg/l. Thermally regenerable ion exchange is applicable to the desalination of waters up to 3000 mg/l TDS. Energy required to

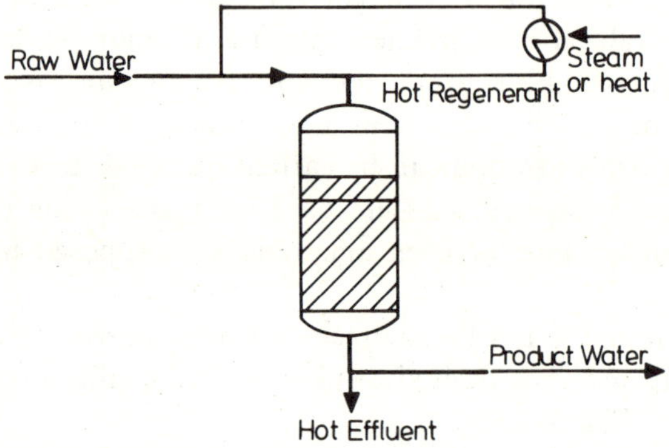

Figure 2.90 Sirotherm process principle.

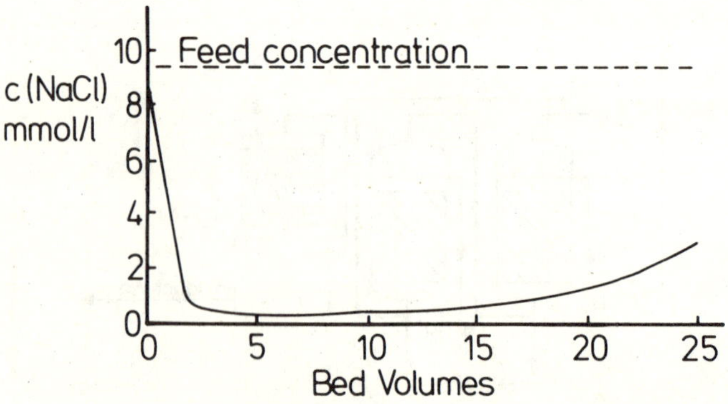

Figure 2.91 Typical breakthrough curves from Sirotherm desalination [11].

perform the work of separating salt from water can be used from low-grade or waste heat.

The CARIX® process represents another possibility of partial demineralization using weakly effective regeneration chemicals. It uses a weak acid resin in the free acid form and an anion exchange resin in the bicarbonate form which are combined in a mixed bed. Thus, dissolved salts are replaced by carbonic acid. The exhausted resins are regenerated simultaneously by CO_2 and water [12]:

$$\left.\begin{matrix} R=(COOH)_2 \\ R=(HCO_3^-)_2 \end{matrix}\right\} + CaSO_4 \rightleftharpoons \left\{\begin{matrix} R=(COO^-)Ca^{2+} \\ R=SO_4^{2-} \end{matrix}\right\} + 2\,H_2CO_3$$

Anion and cation exchanger volumes in the mixed bed can vary over a wide range of ratios. Thus the effective capacities of the filter need not be equivalent. In contrast with the Sirotherm process, elimination of neutral salts can therefore be accompanied by either a dealkalization or an elimination of anions of strong acids. As a consequence the process can be adapted to individual salinity problems by suitable choice of the resin volumes. Methacrylic weak acid resins and acrylic anion exchangers or resins of Type II showed the best results. In the pilot scale the

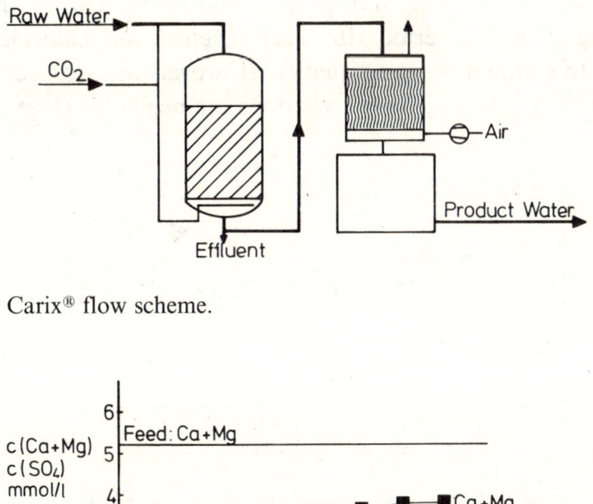

Figure 2.92 Carix® flow scheme.

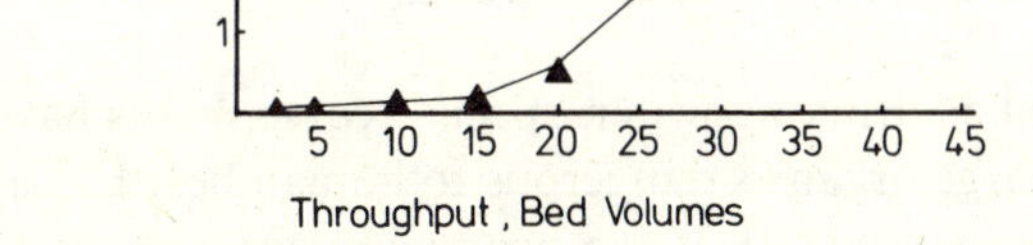

Figure 2.93 Typical Carix® breakthrough curves [11].

CARIX® process has successfully been applied to the combined removal of hardness, sulfate, and nitrate up to raw water salinities of 1000 mg/l TDS with different main objectives. It allows a considerable reduction of hardness, nitrate, and sulfate concentrations (Figures 2.92 and 2.93). The process has been employed in a 170 m^3/h installation in the F. R. Germany for the simultaneous reduction of hardness from 540 ppm (as $CaCO_3$) to 240 ppm, of nitrate form 40 mg/l to 25 mg/l, and of sulfate from 160 mg/l to 50 mg/l [12].

2.4.5 Removal of high-molecular-weight organic substances

High-molecular-weight organic substances like humic acids, in excess of 5 mg/l dissolved organic carbon (DOC), are undesirable for various reasons. In the presence of oxygen they favor the growth of bacteria in the distribution system. Furthermore, they increase the consumption of chlorine in the oxidation step which may lead to the formation of highly troublesome chlorinated hydrocarbons. Humic acids can be eliminated using macroporous anion exchange resins as adsorbents. Due to the pore structure of such resins, large organic molecules can be reversibly adsorbed. Regeneration is carried out with alkaline brine solutions [13].

The method has been utilized in a large (2500 m^3/h) German water supply to reduce the DOC from 6.5 to 3 mg/l. Typical breakthrough curves are shown in Figure 2.94. During a starting period after each regeneration, chloride is replaced by sulfate, leading to changes in water quality. However, with respect to the total throughput of 5000 bed volumes, these variations are negligible (Figure 2.94).

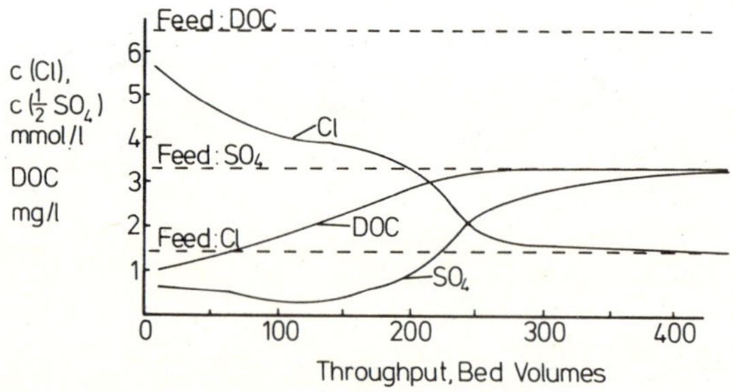

Figure 2.94 Effluent concentrations after removal of humic acids [13].

2.4.6 Hygienic and environmental aspects

If the filter effluent is used as drinking water, special requirements have to be met. The resins must not discharge organics dangerous to human health. They must not increase the chemical oxygen demand of the product water and the solid residue from evaporation tests must not exceed that of the raw water by more than 10%.

The resins used must show sufficient chemical stability, which has to be confirmed by appropriate tests. In the case of anion exchangers, the possible discharge of amines may cause problems [14].

Many cases of substantial bacterial growth in ion-exchange filters have been reported – especially during stagnation periods. The bacteria grow on the resin surfaces, since the resin pores are smaller by far than the bacteria. They are inhomogeneously distributed in the resin bed in the form of "nests". Although bacteria are not attached securely to the surface, even strong flushing and backwashing cannot rinse the filter free from bacteria. Regeneration reduces the colony number, but is usually inadequate for sanitizing an infected filter. With respect to bacterial growth it has been required that, after stagnation periods, several bed volumes of product water be drained off. Nevertheless, the method turned out to be insufficient due to the slow rinsing of bacteria [15].

In order to suppress bacterial growth, it is helpful to maintain continuous operation and to regenerate frequently. If possible, it is recommended to keep the resins bathed in the regeneration agent during inactive periods. The bactericidal properties of silver ions can be used to suppress bacterial growth. However, the microflorae are able to develop an increased tolerance towards silver ions, so that the original bactericidal effectivity may disappear [16]. Once the filter is contaminated, it is normally not possible to avoid disinfecting it in order to sanitize it. Disinfection can be carried out using formaldehyde, chlorine compounds, or peracetic acid in low concentrations. With regard to the washing-out process, it has to be taken into consideration that the efficiency is primarily dependent on the rinsing time and not on the amount of rinsing water used, since the diffusion of the disinfectant out of the resins is rate-determining [17].

One of the undesirable properties of many ion-exchange processes is the fact that more than stochiometric quantities of chemicals are required for regeneration, particularly if high-affinity counterions have to be displaced from the resin [18]. The treatment of large quantities of raw water, therefore, may lead to a considerable increase in the salinity of natural bodies of water, a condition that cannot be tolerated everywhere. From the environmental perspective, both the Sirotherm and the CARIX® process have particular advantages since the effluent only contains the amount of salt that has been eliminated in the service cycle. Precautions also have to be taken when using sulfuric acid for regeneration or when regenerating sulfate-bearing anion exchangers, since in these cases concrete corrosion may occur.

References

[1] Directives of the European Community *23* (1960), L229. DIN 2000, Nov. 1973.

[2] Kunin, R., Amber-hi-lites *95* (1966), *96* (1966), *97* (1967). Meuli, K., (1980) Gas, Wasser, Abwasser *60,* 8.

[3] Kühne, G. (1973), LEWATIT Information OC/I 20510.

[4] Baur, A. (1975) Mitt. des württ. Wasserwirtschaftsverbandes.

[5] Clifford, D. A., Weber jr., W. (1977), EPA-600/8-77-015. Johnson, M. S. and Mustermann, J. L. (1979) J. AWWA *71,* 343.
[6] Sabadell, E. (1975), AIChE Symp. Ser. *151,* 127.
[7] Guter, G. A. (1981), EPA-600/2-81-029. Clifford, D. A. (1983), Proc. ASCE, Boulder, CO, p. 14.
[8] Schmitt, S. (1984), Lebensmittel-Technologie *17.*
[9] Clifford, D. A. (1982), Water Management, May 1982; Clifford, D. A., Personal communication.
[10] Wagner, I. (DVGW), Personal communication.
[11] Bolto, B. A. (1977), Chemtech, May 1975, 303. Japan Sirotherm Ltd., Bull. No. 523.
[12] Höll, W. and Kiehling, B. (1984), in: Ion Exchange Technology (D. Naden and M. Streat, eds.), SCI, Horwood Chichester. Höll, W. and Feuerstein, W. (1986), Reactive Polymers *4,* 147. Höll, W., Kretzschmar, W. and Hagen, K. (1986), GIT Fachz. Lab. *30,* 307.
[13] Stadtwerke Hannover AG (1982), Research Project 02WT606, Final Report. Kunin, R. (1979), Amber-hi-lites *161.*
[14] Wagner, R. et al. (1984), Z. Wasser- und Abwasserforsch. *17,* 240.
[15] Slejko, F. L. and Costin, C. R. (1978), Proc. 4th Int. Symp. Contam. Control, p. 118. Flemming, H.-C. (1981), Z. Wasser- und Abwasserforsch. *14,* 132; Vom Wasser *56,* 215. Kunststoffkommission (1963), 16. Mitt., Bundesgesundheitsblatt *6,* 413.
[16] Flemming, H.-C., Z. Wasser- und Abwasserforsch. *15* (1982), 259, and *17* (1984), 229.
[17] Wagner, R. and Flemming, H.-C. (1984), Z. Wasser- und Abwasserforsch. *17,* 235.
[18] Martinola, F. (1977), Z. Wasser- und Abwasserforsch. *10*, 198. LEWATIT-LEWASORB Product informations, BAYER AG, Leverkusen.

2.5 Waste Water Treatment and Pollution Control by Ion Exchange

Friedrich Martinola
Bayer AG, Leverkusen, Federal Republic of Germany

Ion exchange resins were recognized at an early stage as being useful aids in the removal of both harmful and valuable substances from effluents. The inorganic ion exchange resins at first available, however, had limited application because they frequently showed only a very specific ion adsorption behavior and, above all, because it was hardly possible to regenerate them. In the age of nuclear reactors, use was nevertheless made of their high radiation resistance for the adsorption and subsequent storage of radioactive ions [1].

Synthetic organic ion exchange resins are chemically stable in acids and alkalies. They can be manufactured in a large number of varieties as far as their matrix and active groups are concerned, so that a specifically acting ion exchange resin can be produced for virtually any ion that has to be adsorbed. In practice, however, the variations are limited mainly to metal-selective resins. Use is generally made of cation exchangers, although anion exchangers are very effective collectors of anionic metal complexes such as cyanides, etc.

In addition to the ion exchange resins, mention should also be made of the adsorber resins, which adsorb not only ionic but also non-ionic organic products from water and effluent, thus making a contribution to water treatment. Domestic waste water and, above all, industrial effluent contains wetting agents, chlorinated compounds and colorants which can be removed using adsorbent resins.

If ion exchange resins or adsorber resins are used for waste water treatment, it must not be forgotten that the first step of ion adsorption consists merely in collecting the harmful substance on the resin. In the next step, the ion exchange resin has to be regenerated, a process in which the adsorbed substances have to be discharged from the resin. The end-product is a concentrated solution of the undesired substance, which then has to be disposed of. It is therefore not enough to integrate an ion exchange resin into the waste water treatment system; consideration also has to be given to disposing of the desorbate and recycling the rinse water.

There are, therefore, limits to the general application of ion exchange resins and adsorbents that are not directly connected with the reactions of the resins.

Application for solutions that are as dilute as possible, thus making a higher concentrating effect possible

As a rule, ion exchange resins have a capacity of approx. 1 exchange equivalent per litre of resin. Desorption is carried out with 0.5–2 bed volumes of a suitable solution. On top of this, there is 0.5–1 bed volume of rinse water which has to be recycled. This makes about 1–3 bed volumes of regenerant solution, containing, all in all, 1 equivalent of the unwanted substance – or the substance being recovered. Its concentration is then approx. 0.3 to max. 1 equivalent per litre of solution. An increase in the concentration of this substance is thus only achieved if its concentration in the effluents is less than 0.3 equivalent per litre. Normally, the limit for economical operation is, in fact, about 10 times lower than this.

One exception in this connection is where it is possible and desirable to use ion exchange for a cleaning process, e. g., through the specific adsorption of a single metal from a mixture of different cations. Copper, for example, can be separated very selectively from alkaline earth and alkali metals which are also present, and can then be recycled in high concentrations [2].

In practice, a number of processes are employed in which the regenerant solution is used several times for desorption without the desorbed substance being separated off beforehand from the regenerant solution. This is possible in cases where the solution, on passing once through the ion exchange column, is not yet in equilibrium with the ions on the resin. Such cases occur, for example, if the acid solution used for regeneration is of high concentration. It is then worthwhile to use it again for the desorption of another exhausted ion exchange column [3].

A further improvement in the concentration ratio of contaminant in the spent regenerant to contaminant in the effluent can be achieved in cases in which weakly acidic or weakly basic ion exchange resins can be used. The equilibrium between ion on the resin to ion in the solution can be shifted by changing the pH in favour of the concentration in the spent regenerant solution. In such cases, it is possible to reach the saturation limit of solubility. One example of this is the adsorption of ammonia with the aid of a weakly acidic resin and its desorption with sulphuric acid [4].

Regeneration with reagents which can be recovered

For economic reasons it is desirable to recover and recycle the eluant. This is done successfully, for example, in the desorption of metals which have been separated off by electrolysis from the regenerant acid, e. g., sulphuric acid [5]. The distillation of volatile eluants such as hydrochloric acid, ammonia and alcohol [6] is also possible and is generally used in practice. This process not only enables the regenerant chemical to be recycled, it also significantly increases the concentration of the contaminant in the residue. If a way can, therefore, be found to recover the regenerant, it is evident from what was said above that the situation for using ion exchange resins is improved.

A further limitation in the use of ion exchange resins and adsorbers lies in their selectivity. The more strongly an ion is fixed, the more difficult it is to release it again. Not only must a large excess of regenerant chemical be used, in many cases desorption does not even succeed. In such cases, if the contaminant is worthless, e. g., radioactive waste [7], the ion exchange resin has to be dumped after being used only once. It is, however, possible to incinerate organically-based ion exchange resins and to recover the adsorbed metal from the residue. This method is used, for example, to recover the gold or plating bath residues adsorbed by strongly basic ion exchange resins [8] or to distill mercury that has been collected on the resin from the incineration residues [9].

To ensure a long service-life and thus encourage economical use of ion exchange resins, attention must be paid to their stability in various media. Two factors are important for the treatment of effluent:

1. aggressiveness of the effluent and the regenerant
2. irreversible fouling of the ion exchange resins.

Aggressive effluents and the effects of regenerants

Oxidizing substances

Peroxide sulphate ($Na_2S_2O_8$) is found in effluents from the electronics industry. Cation exchange resins have greater stability than anion exchange resins. The latter take up the highly aggressive anion S_2O_8 during the exhaustion cycle, which means that the exposure time and the concentration are greatly increased. Chlorine compounds with a higher oxidation number, such as hypochlorite (NaOCl) and chlorine dioxide (ClO_2), are sometimes used for cleansing adsorbent resins. These processes must, however, be carried out under accurately controlled conditions in order to avoid irreversible damage.

Chromic acid (CrO_3) is found in effluents and concentrates from the electroplating industry. It is not only an oxidizing agent; it is also, at this stage of oxidation, a strong poison and must therefore not be allowed under any circumstances to get into the sewage system. With cation exchange resins, concentrated chromic acid leads to decrosslinkage, which means that only highly crosslinked resins are able to withstand such treatment to a reasonable extent. As far as anion exchange resins are concerned, certain medium-base types have quite good stability, but here, too, the service-life is very much shorter than it is in normal water treatment [10] (Tabl 2.27).

Table 2.27 The effect of chromic acid. Polystyrene anion resins loaded with 150 g CrO_3/l resin, after 28 days at room temperature

Resin type	Weakly basic	Medium basic	Strongly basic Type I	Strongly basic Type II
Separated chromium, g/l resin	0.52	0.21	0.52	12.3
Water content % decrease	0	3	0	20
Rinse water bed volumes	4.5	2.8	3.3	55

Reducing agents

The polystyrene and polyacrylate based resins that are in common use today generally have good resistance to reducing agents. Reducing substances such as hydrazine, sulphite and formaldehyde do not produce any change.

Regenerant chemicals

Chlorate ($NaClO_3$) may be present in the caustic regenerant solution as a direct result of the production process. It accumulates on the anion exchange resin and results in oxidative damage [11]. For this reason, caustic soda solution used for the regeneration of ion exchange resins should not contain more than 0.001% oxidizing substances, calculated on NaOH as O_2.

The effect of nitric acid on cation exchange resins

Nitric acid is a very effective regenerant, which can be used in the same way as hydrochloric acid, i. e., in equivalent quantities and corresponding concentrations.

For the desorption of many ions, e. g., silver or lead, this acid is in fact necessary to prevent precipitation. It is also very useful for recovering ammonium nitrate from a resin loaded with ammonia. It should nevertheless be borne in mind that concentrated nitric acid is above all a very strong oxidant, especially at elevated temperatures. In the presence of organic substances, particularly when catalyzed by heavy metals such as copper, a violent reaction may result which, under certain circumstances, can even take the form of an explosion.

With cation exchange resins of polystyrene sulphonic acid, 20% nitric acid has hardly any effect at all at normal temperatures, even over a longer period of time. There are appreciable changes, however, when 40% acid is used (Table 2.28).

Table 2.28 The effect of nitric acid at normal temperatures

		Polystyrene sulphonic acid resins					
HNO_3	Testing time	Gel-type resin		Gel-type resin		MP resin	
		8%	DVB	15%	DVB	18%	DVB
	days	TC%	H_2O%	TC%	H_2O%	TC%	H_20%
	new	100	100	100	100	100	100
20%	14	100	100	99	98	98	99
	28	104	104	112	104	106	100
40%	14	101	101	120	102	115	99
	28	104	106	128	110	122	102

The effect of nitric acid on anion exchange resins

As a rule, anion exchange resins contain amino groups of various levels of substitution. It is very much easier for nitric acid to act on these groups than on the sulphonic acid group of the strongly acidic resins. Consequently, the risk of a spontaneous, explosive reaction is much greater.

Where anion resins are converted completely into the nitrate form, they do not need to be heated or dried. They should never come into contact with nitric acid having a concentration of more than 1%.

Irreversible fouling of the ion exchange resins

Above all with anion exchange resins it happens that the active groups are prone to fouling by very strongly held ions or to their inner surface's becoming occupied by hydrophobic substances.

Fouling by ions

Inorganic anions are, for example, complex metal ions such as cyanides of the heavy metals, polysilicic acid or chromic acid. These can only be removed from strongly basic resins by a special treatment.

Weakly basic resins are less affected because caustic soda solution or alkaline cyanide solutions generally have a good desorption effect. If, therefore, fouling ions are present, a combination of weakly and strongly basic columns is recommended. The weakly basic ion exchange resin predominantly adsorbs cyanide complexes or chromic acid and protects the subsequent strongly basic resin, which only adsorbs any free cyanides and other weak anions that may be present. Regeneration can be carried out effectively and very economically using the compound arrangement: strongly basic → weakly basic.
Any precipitations of metal cyanides or metal hydroxides in the weakly basic resin bed are dissolved during the course of regeneration when, for example, the alkali solution coming from the strongly basic resin carries desorbed cyanide with it [12].

Fouling by organic substances

Here, too, the strongly basic resins are most affected, although cation resins and weakly basic ion exchange resins may also become fouled.

Fats and oils must be removed as far as possible from the effluent ahead of the ion exchangers. Otherwise they hydrophobize the surface of the resin beads and considerably slow down the diffusion process.

Ionic surfactants are adsorbed by both cation and anion resins. Their desorption during regeneration is, however, usually delayed, as their diffusion rate is low. As an example, for a weakly basic resin, even with a macroporous matrix, a maximum anionic surfactant content of 0.5 mg/l is specified [13].

Brighteners from electroplating baths also contribute to the fouling of anion exchange resins. When buying ready-to-use bath formulations, it is therefore important to ask for the recipe. On such occasions, information should also be obtained regarding the compatibility of the ingredients with the ion exchange resins.

Large organic molecules such as humic matter, lignin sulphonic acids or synthetic dyestuffs are known for blocking ion exchange resins. Processes have nevertheless been developed for decolourizing, for example, the effluent from paper mills or from cellulose manufacturers using special resins [14].

If the ion exchange resins are fouled by organic substances, an attempt can be made to use suitable solvents to remove them from the resin. Initial tests should be carried out with an alkaline or acid common salt solution, possibly also with alkaline cyanide solutions. In the case of degradable substances, such as humic or lignin compounds, use can also be made of alkaline solutions of free chlorine. Sugar colorants, for example, can be removed from decolourizing resins by performing an oxidation process with a calcium hypochlorite solution [15]. It is, nevertheless, important to ensure when doing this that the ion exchange resin itself is treated as carefully as possible.

Special fields of application in effluent treatment

This section will deal with a few selected applications which have not been discussed in previous chapters.

Inorganic solid waste

As has been described above, it is necessary to produce solutions of the waste substances that are as concentrated as possible. These can then be converted into a suitable form for final dumping or recycling. One process with universal application is the Intensive fraction process (IF-P), which permits the production of multimolar solutions [16]. The basic principle of the IF-P is that an excess of high-concentration regenerant is reused several times. In the eluate effluent, only the most concentrated fraction is passed on for further processing. The remainder is collected in exactly defined fractions and circulated. In the same way, only as much fresh regenerant solution is added as corresponds to the volume of discharged eluate. If the same principle is used as with the exhaustion of the ion exchange column – fractionated influent – then concentrated solutions can also be treated with ion exchange resins, which would not usually be possible. Figure 2.95 shows the principle of the process.

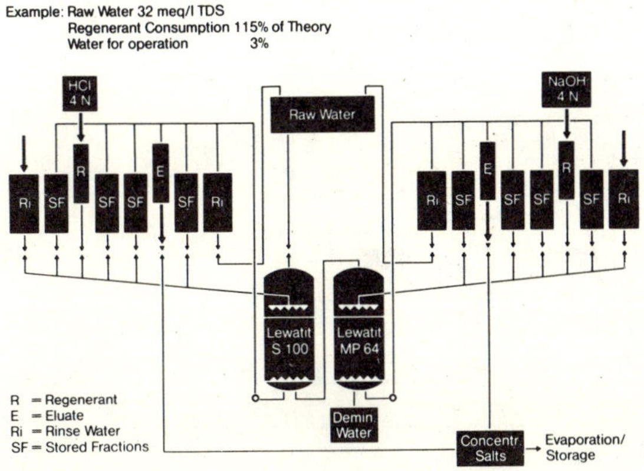

Figure 2.95 Intensive fraction process. Demineralization of water; zero-discharge-system.

A few interesting applications for the IF-P in effluent treatment are [17]:

1. the recovery of ammonium nitrate from condensates of fertilizer factories;
2. the production of hydrochloric acid from waste sulphuric acid;
3. the treatment of spent acid in titanium dioxide factories [18];
4. the demineralization of brackish water [19].

The cleansing of spent acid is a particularly interesting process because the regenerant – hydrochloric acid – is recovered by thermal treatment and recycled.

To recover the free ammonium from alkaline effluents, V. Kadlec [20] uses a weakly acidic exchanger. Here, use can be made of the high affinity of the weakly acidic ion exchange resin for H-ions in the production of highly concentrated ammonium sulphate or nitrate solutions. The ion exchange resin loaded with NH_3 is regenerated with sulphuric or nitric acid. The resultant solution is mixed with free acid and is then used for regenerating a resin saturated with NH_3. This can be performed several times, resulting in virtually saturated ammonium salt solutions.

R. Kunin et al. [21] invented an economical method of demineralizing salt-containing pit water using the Desal process. For this, use was made of weakly acidic and weakly basic ion exchange resins. Regeneration is carried out with sulphuric acid and ammonia. Carbonic acid, which is either already present or is added, serves as an important aid in generating the alkalinity, without which the weakly acidic ion exchange resin would not function.

A further example of the use of ion exchange resins for effluent reduction is the softening of oil-pumping water before it is reused [22]. The oil-pumping water is pressed into the well holes in the form of hot water or steam, collects salts below ground and then comes up to the surface again carrying the oil. Once the crude oil has been separated off, the water is softened before being heated up again in boilers. Since the water contains very high concentrations of common salt, softening with the usual strongly acidic cation exchange resins is not possible. The method therefore uses weakly acidic resins in the sodium form as the selective hardness-adsorbing exchangers. Since the Ca^{2+} and Mg^{2+} ions are held very firmly by this type of resin, a roundabout route has to be used for regeneration. This involves

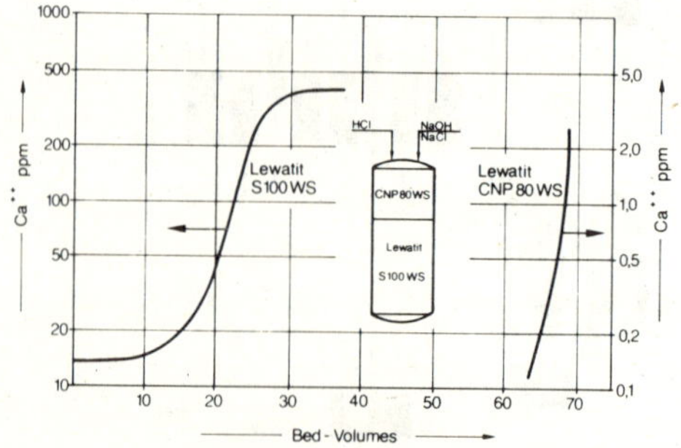

Figure 2.96 Multistep-System. Softening of brackish water.

using hydrochloric acid to release the alkaline earth salts from the resin, producing a high yield. The next step is to convert the resin, which is now present in the H-form, into the Na-form with the aid of caustic soda solution. This also brings a virtually 100% yield. Figure 2.96 shows a residual hardness diagram for "produced water" [23].

Gas flows can also be treated using ion exchange resins. Two processes which are in use are the direct adsorption of sulphur dioxide from the waste gases of sulphuric acid production [24] and the regeneration of weakly acidic resins with SO_2 from flue gases [25]. With the first process, SO_2 and SO_3 are adsorbed by weakly basic resins. Water is then used for desorption and, as a result of hydrolysis, free acid is formed in the effluent and is neutralized with lime.

In the second process, the weakly acidic ion exchange resin is used for the dealkalization of water containing bicarbonates. Following exhaustion, it is regenerated again by the SO_2 in the flue gas. The SO_2 discharged through the stack of a coal-fired power station is nevertheless produced in far greater quantities than is required for the dealkalization of boiler feedwater and cooling-tower water. Desulphuration of flue gases by the direct use of dealkalizing resins is therefore not worthwhile.

Despite this, ion exchange resins can make a valuable contribution to the desulphuration processes. The alkali scrubbing of flue gases from waste incinerators produces effluents which contain high concentrations of calcium in addition to residues of heavy metals such as Cu, Cd, Ni, Zn, Pb and Hg. Special selective resins of the iminodiacetic acid type or the aminophosphonic acid type are capable, despite the high excess of calcium, of removing the heavy metals down to almost the detection limit in the ppb range [26]. Figures 2.97 and 2.98.

Ion exchange units can, however, also make an active contribution to environmental protection. All cyclic ion exchange processes produce spent regenerants,

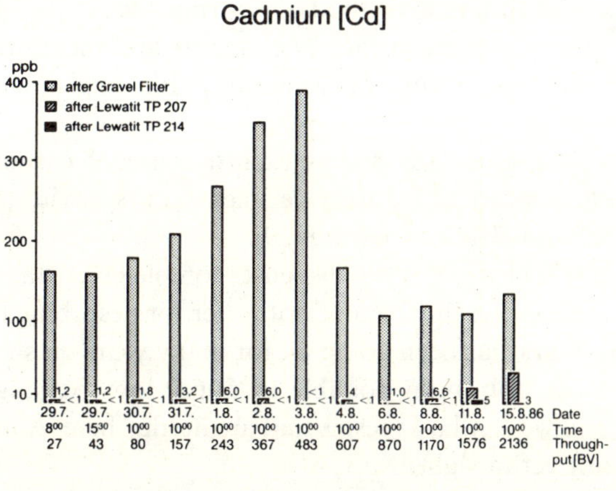

Figure 2.97 Heavy metals in waste water after treatment with selective resins. Actual incineration plant.

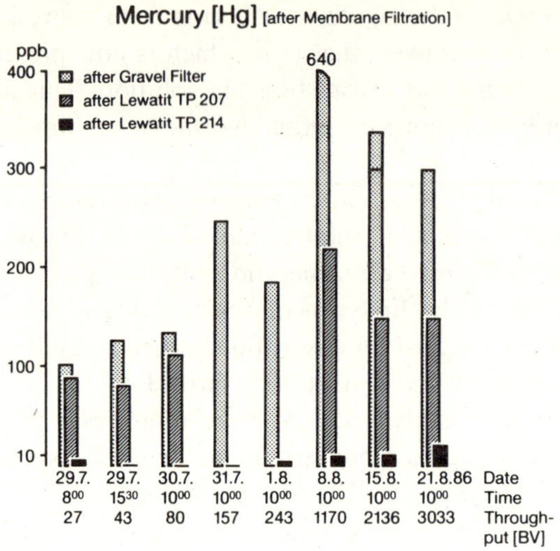

Figure 2.98 Heavy metals in waste water after treatment with selective resins. Actual incineration plant.

which contain not only the ions removed from water or from solutions, but also those of the regenerant. Thus more salts are desorbed than have been removed from the treated solution.

A whole range of proposals have been made to keep the excess of salts as low as possible or only to desorb salts which have no negative influence on the environment. Should a raw water contain sufficient bicarbonates, it will be possible to use weakly acidic resins which reduce the regenerant level [27].

One solution that can be used generally is to change over from cocurrent to countercurrent processes during regeneration. The countercurrent systems – of which there are a variety of different methods – permit the use of the minimum necessary volume of regeneration chemicals. New plants are, therefore, nowadays nearly always built to use countercurrent technology. Existing plants should be converted to these systems [28].

A further demineralization process makes exclusive use of carbonic acid for demineralization. The lime produced during regeneration is environmentally acceptable and has virtually no effect on nature [29].

A particularly elegant variation is to dispense completely with regeneration chemicals. Instead, heat is used in the form of hot water for desorbing the adsorbed cations and anions [30]. Work has been going on for many years on such processes. Until now, however, it has only been possible to utilize low exchange capacities with the weakly acidic and weakly basic resins that are needed here. A breakthrough with this technique is not yet in sight.

Organic substances

Organically-based ion exchange resins can be given a larger inner surface area during their production. The exchangeable groups make this surface hydrophilic, so that it becomes similar to activated carbon. Large organic molecules can diffuse rapidly through the wide macropores onto the inner surface and are fixed there. In such cases, an ion exchange process is not always involved in this fixing process; generally, attractive forces are also often responsible. The adsorbed organic compounds can, in this case, be displaced from the surface by other non-ionic substances, e. g., by solvents such as alcohols or ketones, and can be desorbed from the adsorber resin.

Certain macroporous resins also act as adsorbers for organic molecules without the adsorber's being capable of ion exchange. The driving force is then the affinity of the resin matrix for the organic substance to be adsorbed; e. g., with polystyrene as the matrix, organic compounds with an aromatic character. Substances which swell polystyrene are also easily adsorbed. Desorption can then be performed with a suitable solvent [31].

Phenols are very weak acids with an aromatic ring. They are adsorbed from aqueous solutions by styrene-based, weakly basic anion exchange resins. Due to the low interaction of the ionic groups, the adsorbed phenol can be removed from the resin using methanol [32]. Strongly basic ion exchange resins form phenolates, which means that caustic soda solution must be used for desorption.

Chlorinated hydrocarbons are typical representatives of substances having solvent properties. They are found in small quantities in effluents from textile cleaning plants and get from there into the receiving water. They can be removed from effluent using non-ionic adsorber resins [33]. Desorption is carried out either with solvents or with steam if the compounds are relatively volatile.

Dyes in the effluents of dyehouses frequently display a high affinity for ion exchange resins. Attempts have, therefore, repeatedly been made to use adsorber resins to free these effluents from the dyes [34]. Modern dyes, however, nearly always have a very high fastness, which means that they link up irreversibly with the adsorber. This makes it virtually impossible to remove the adsorbed dyes. The only way out is either to dump these coloured resins or to incinerate them, but this is not generally an economically viable solution. For this reason, only a few plants have been built for using adsorbers to reduce the dye content of effluents.

Crop-protection products get into the water and into the effluent during manufacture and also during use. Studies have shown that the group of organophosphorous compounds and the family of carbamates can be broken down with the aid of ion exchange resins to give less toxic or non-hazardous compounds [35]. The effect of the ion exchange resins is based on catalyzing the decomposition through H-ions or OH-ions. The toxic compounds are initially adsorbed on and in the resin and are subsequently destroyed by the catalytically acting ions which are present there.

For chemical processes of this kind, the term "reactive ion exchange" – RIEX – has been proposed. By this we mean an ion exchange process with a simultaneous or subsequent chemical reaction of the fixed ions or adsorbed compounds. A simple example of RIEX is the separation of calcium carbonate from a weakly acidic cation exchange resin saturated with calcium ions, when it is treated with a carbonic acid solution:

$$R-Ca + CO_2 + H_2O \rightarrow R-H_2 + CaCO_3.$$

This reaction leads to the formation of environmentally acceptable calcium carbonate in place of $CaSO_4$ or $CaCl_2$ formed when sulphuric acid or hydrochloric acid is used for releasing the calcium from the ion exchange resin bed [36].

References

[1] C. B. Amphlett (1964). Inorganic Ion Exchangers, p. 77, Elsevier, Amsterdam.
[2] F. Gerstner (1956). Recovery of Copper. Ion Exchange Technol., p. 340, Academic Press.
[3] L. Hartinger (1976). Taschenbuch der Abwasserbehandlung, p. 90, Hanser, München.
[4] V. Kadlec and P. Huebner (1981). Reinigung ammoniakhaltiger Wässer, Proc. Ion Exchange Symp., Hungary.
[5] D. R. Kampermann (1980). Electrolytic Metal Recovery. AES-Annual Meeting.
[6] L. Hartinger (1976). Taschenbuch der Abwasserbehandlung, p. 301, Hanser, München.
[7] H. Baatz (1976). Behandlung radioaktiver Abfälle für die Endlagerung. Atom + Strom *22,* 103.
[8] H. Benninghoff (1984). Rückgewinnung von Metallen. Metall/Oberfläche *38,* 5.
[9] H. Lüssem and G. Richter (1976). Entfernung von Silber und Quecksilber aus Laborabfällen. Z. Wasser- und Abwasserforschung *9,* 2.
[10] G. Kühne and F. Martinola (1972). Beständigkeit von Ionenaustauschern. Mitt. VGB *57,* 173.
[11] I. Abrams and L. Donnally (1959). Ind. Eng. Chem. *51,* 1043.
[12] S. Karrs et al. (1986). Ion exchange and metal recovery. Plating and Surface Finishing *73,* 60.
[13] F. Martinola (1969). Experience with scavenger resins. Proc. 2nd Sympos. Ion Exchange, Hungary, p. 262.
[14] C. H. Moebius (1981). Adsorption and ion exchange for treatment of waste water from paper mills. Water Science Techn. *13,* 681.
[15] Bayer AG (1976). Verfahren zur Reinigung von Entfärbungsharzen. Merkblatt OC/I 20302.
[16] Dynamit Nobel AG, Germany (1968). German Patent 18 12 769.
[17] S. Vajna (1970). Dechema Monograph. *64,* 173.
[18] S. Vajna (1982). German Patent 31 15 937 and 32 06 856.
[19] F. Martinola and S. Vajna (1980). Demineralization of Brackish Water. Proc. 7th Intern. Symp. Fresh Water from the Sea. *2,* 31.
[20] V. Kadlec and P. Huebner (1980). Cechoslov. Patent 186,023.
[21] R. Kunin and B. Vassiliou (1964). New deionization technique. Ing. Eng. Chem. *3,* 404.
[22] F. X. McGarvey (1981). Softening of secondary oil recovery water. Proc. Amer. Petrol. Inst.
[23] F. Martinola and G. Siegers (1984). Enthärtung von Brackwasser. Vom Wasser *62,* 101.

[24] A. Richter (1954). DDR Patent No. 9657.
[25] H. Tiger (1938). US Patent 2,227,520.
[26] F. Martinola and H. Hoffmann (1986). Ion exchange resins, application in flue gas scrubbing. Proc. 47th Intern. Water Conf, Pittsburgh.
[27] R. Brunner (1958). Wirtschaftlichkeit von Vollentsalzungsanlagen. Mitt. VGB *54,* 210.
[28] F. Martinola and P. Thomas (1986). Reduktion der Abwässer aus Ionenaustauscheranlagen. Vom Wasser *67,* 135.
[29] W. H. Höll (1985). Partial Demineralization of Ground Water by the Carix-Process. Proc. 46th Intern. Water Conf., Pittsburgh, p. 525.
[30] B. A. Bolto (1980). Novel water treatment processes. J. Macromol. Sci. Chem. *A 14,* 107.
[31] R. L. Gustafson (1970). US Patent 3,531,463.
[32] Ch. R. Fox (1978). Phenol recovery with resins. Hydrocarbon processing, p. 269.
[33] E. Isakoff and J. Bittner (1971). Resin Adsorbents for Chlororganics. Water and Sewage Works, p. 41.
[34] S. Fiala et al. (1980). Abwasserreinigung in der Textilindustrie. Melliand Textilberichte *8.*
[35] G. Janauer et al. (1984). Reactive ion exchange. Reactive Polymers *3,* 1.
[36] B. Kieling and W. Höll (1981). German Patent 31 02 693.

2.6 Ion Exchange Systems in Homes, Laboratories and Small Industries

Hans Träger
Wilhelm Werner GmbH, Bergisch Gladbach, Federal Republic of Germany

2.6.1 Ion exchangers in homes

The most common application of ion exchange systems in homes is for the softening of tap water. It is necessary to soften this water because the use of soap causes precipitation of fatty acid calcium and magnesium salts on the surface of bathroom hand-basins, etc.

Another general use of ion exchange systems in homes is for protection against lime-scaling in warm and hot water lines. The permanent hardness is usually well-dissolved in water, up to the boiling-point. Calcium sulfate (gypsum) is one part of the permanent hardness which becomes unstable in solution in concentrations of more than 1200 ppm. $CaCl_2$, $MgSO_4$, $MgCl_2$ are soluble enough in cold and hot water so as not to present a threat of lime-scaling.

At higher temperatures a greater concentration of dissolved carbon dioxide than in cold water must be present.

Applications.

- For warm water to protect from scaling due to the temporary hardness
- for water in general to protect from scaling in lines in cases of higher temporary hardness
- small indoor installations for dish-washing
- small indoor installations for laundry washing
- transportable filters to soften water for coffee machines.

Most softening installations are installed only in the feed-line of the warm water installation. Warm water makes up only 25% of the water consumption in homes. It is in warm water lines and boilers that the main precipitation is to be found. In most cases, cold water lines do not show this problem. Sometimes it is also necessary to soften all the water used in the house in cases where the temporary hardness is more than 280 ppm $CaCO_3$.

In any case, the most useful process for softening water by ion exchange to zero hardness is according to the following formula:

Softening: $CaCl_2$ + x-Na_2 → 2 NaCl + x-Ca
Regeneration: x-Ca + 2 NaCl → x-Na_2 + $CaCl_2$ (in waste water).

The fully softened water that results will then be mixed with tap water in a proportion reducing then the temporary hardness to 125 ppm $CaCO_3$. In this range the equilibrium between temporary hardness and dissolved CO_2 requires a very low amount of CO_2 to hold $Ca(HCO)_3$; in solution this is stable at an amount of CO_2 which cannot be removed by heating or degassing.

Hardness as $CaCO_3$(ppm)	Dissolved CO_2(ppm) necessary to hold the hardness in solution
90	1.83
180	11.67
270	44.11
360	108.15

Resin. A strong acid cationic exchanger is used. The regeneration will be done with salt. The maximum capacity is

1,4 mol/l i. e. 70 g $CaCO_3$/l at a regeneration consumption of 280 g salt/l resin;
1,07 mol/l i. e. 54 g $CaCO_3$/l at a regeneration consumption of 140 g salt/l resin;
0,89 mol/l i. e. 45 g $CaCO_3$/l at a regeneration consumption of 90 g salt/l resin.

The salt employed is mostly pressed into pill form. The smallest softening units have a resin volume of 10 liters (2,65 gallons). If the water consumption per week is 2.5 m^3 (2500 liters), 1250 l water will have to be softened. If the total hardness is about 340 ppm $CaCO_3$, the 10 l-softener can produce:

2000 l if regeneration is done with 2800 g salt = 280 g/l
1500 l if regeneration is done with 1400 g salt = 140 g/l
1250 l if regeneration is done with 900 g salt = 90 g/l.

It is highly recommended that one regenerates with a reduced amount of salt to prevent waste water pollution.

Regeneration. The standard home-softener has a week-timer which starts the regeneration at a fixed time. Regeneration should be done at a time of day when soft water is not going to be needed, e. g., 1 : 30 am. Home softeners supply unsoftened water, uncontaminated by the salt additive, etc., during the regeneration period.

Normally, regeneration is done in three steps:

- backwashing of resin 5 min (better, 10 – 15 min)
- regeneration by 10% brine solution
- rinsing.

In general this takes two hours.

Figure 2.99 Timer (W. Werner GmbH).

Figure 2.100 Water softener (W. Werner GmbH).

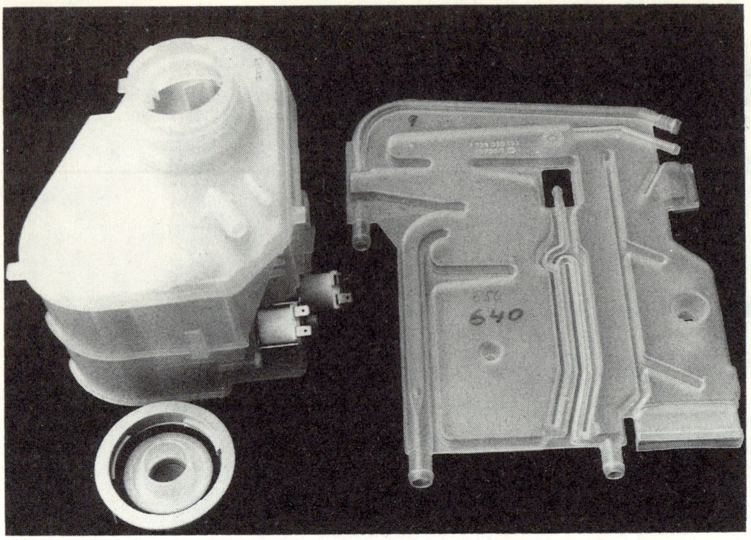

Figure 2.101 Softener of dish washer together with inlet installation (Siemens-Bosch).

In dish-washers very small softeners are installed in the bottom. Regeneration is done after each run of the machine. The typical softener in dish-washers contains 400 ml of resin. The regeneration is done using 28 g of 100% salt in a brine solution.

Disadvantage. Some authors report corrosion in zinc-lined pipes as a result of the free CO_2 not required to bind the temporary hardness in solution. Sometimes, corrosion inhibitors will be dosed after the softener, such as phosphate silicates, organics, etc.

In any case, one problem always exists: the normal growth of bacteria on the surface of the ion exchange resin [1].

The backwashing process is insufficient and does not flush out all the microorganisms, even after an extremely long rinsing time. [1]

The bacteriostatic effect of silver ions precipitated onto the resin is not effective for long because the microorganisms can adapt to higher silver concentrations in water [2].

The growth of harmless water bacteria is widespread, but the growth of pseudomonas aeruginosa in the cationic exchangers of hospital dish-washers has also been reported [3].

2.6.2 Ion exchangers in laboratory use

Ion exchange systems have many laboratory applications. Each manufacturer has a large number of purified ion exchangers for very small laboratory work. Ion exchangers are used:

- for enrichment of traces of elements on ion exchangers, such as radio isotopes, as well as enzymes, alkaloids, etc.
- as catalysts for esterification, saponification, hydrolysation, polymerisation, inversion of sugar, etc.

Ion exchangers are used in columns as well as in batch operations. In the latter case, after the reaction the exchangers are separated by filtration. The laboratory use of ion exchangers is in any case very small with regard to volume but widespread in range of application [4].

2.6.2.1 Laboratory water preparation

The most frequent application of ion exchangers in laboratories is for the deionization of water.

The older method of distilling water is no longer used, because of the high consumption of power and cooling water; only if legislation requires that the distilled water be free not only of salts but also of pyrogens and microorganisms, the older method is still used. Producing one liter of distilled water requires 700 watts of power in the form of electric energy or steam, and 15 liters of cooling water. There are large distilling apparatus which are energy-saving constructions and which do not use so much cooling water. But they need demineralized water for feeding which must be silica-free – simple demineralized water is not sufficient. Some distilling apparatus require feedwater of 1 $\mu S^{-1} \cdot cm$.

This feedwater quality is in any case good enough for most applications in the laboratory. Some examples:

- glass-washers (they have for pre-washing the same softeners as the dish-washers in homes)
- titrations
- photometry
- manual glass rinsing.

An official definition of pure water for laboratories is only specified in the USA, e. g., ASTM classification, CAP standard, ASC standard. For example, Type I water is needed for AAS, flame photometry, enzymology, blood gas analysis, pH adjustment, buffer solutions, dissolving of hydrophylized products.

Type II water is used for titrations, hematology, serology, microbiology.

Outside the USA no official standards are known and the following classification is used:

- distilled water, demineralized water
- bi-distilled water or high purity water.

The higher US standards mostly require water with a low bacteria-content. However, it is well-known that ion exchangers provide surfaces for bacteria. Membrane filtration is a good solution for lowering the level of bacteria in water. Membranes, after a period of 3 to 6 days, may show an increasing bacteria level on the downstream side.

Membrane cartridges remove well the bacteria on the upstream side, but the lifetime of such an expensive filter with regard to its removal of impurities is very short, if the cartridge has to be changed every week because bacteria grow on the downstream side. Sterilizing by chlorine, formaldehyde, etc., is possible but careful rinsing after sterilization is necessary. Some membrane cartridges can be sterilized by steam.

A safe and cheap system is the use of flat membranes in sterilizable filter holders and the use of steam-sterilized filter units of 50 mm diameter. At a pressure of 3 bar the outlet is 0,5 l/min, which is in most cases acceptable if bacteria-free demineralized water is required. Daily change of the filter is recommended.

The chemical purity of demineralized water depends on the type of ion exchange resin. In Europe, the most current procedure for demineralization uses mixed-bed cartridges, generally a pressure-stable cartridge made of plastic or stainless steel and filled with 20 l of resin. Also 14- and 42-l-cartridges are available, but a 14-l-cartridge is almost as expensive as a 20-l-cartridge, and the 42-l-cartridge causes transport problems because of its weight.

Figure 2.102 Pressure-stable stainless-steel cartridges and pressure-sensitive plastic cartridges (W. Werner GmbH).

Figure 2.103 Regeneration station (W. Werner GmbH).

A strong acid cation (40%) and a strong base anion resin (60%) are used. The flow is in general from top to bottom to prevent demixing during the run. For this reason, deaeration at the start is necessary. For an average amount of total solids of 350 ppm as $CaCO_3$, the average through-put until resistivity increases to 10 $\mu S^{-1} \cdot cm$ (100000 $\Omega \cdot cm$) is approx. 1100 l; and up to 20 $\mu S^{-1} \cdot cm$ (50000 $\Omega \cdot cm$) approx. 1250 l. The purity for more than 60% of the lifetime is for instance 0.1 $\mu S^{-1} \cdot cm$ (10 $M\Omega \cdot cm$) or better.

The exhausted resin is regenerated in big lots by central service equipment. The customer gets back regenerated resin, but not his own, receiving resin out of the big pool.

The service company, in most cases also the supplier of the cartridges, must guarantee a standard capacity for the regenerated resin. Tests have shown a capacity of 0,45 val/l until the resistance has increased to 20 µS. The removal of silica is limited because the ion exchangers can not be regenerated by warm caustic solutions.

There are also pressure-unstable low-cost cartridges. Their hydraulic form is not so favourable for ion exchange systems because their capacity is generally in the range of 0,32 val/l. See also Figure 2.102.

Figure 2.104 Customer refill station in his own laboratory (W. Werner GmbH).

Figure 2.105 Reverse Osmosis System (W. Werner GmbH).

If the demand for water is in the range of 80 – 150 l/d, customers purchase resin in big drums and refill the cartridge on their own site. Because of savings on transport costs and the purchase price of several cartridges, drums are significantly cheaper than cartridges.

If more than 200 l water, with an average total dissolved solids of 350 ppm $CaCO_3$, are to be treated the method of reverse osmosis is recommended to extend the lifetime of the cartridges.

Small reverse osmosis systems can produce 1000 l daily and the running time of a 20 l mixed bed cartridge after reverse osmosis will be between 30,000 and 50,000 l. A home softener can prepare soft water to protect membranes against scaling.

In the USA, the cartridge system has not found the same acceptance as in Europe. Bigger laboratories in the USA as well as in Europe have central D.I. plants, which are discussed in several other chapters of this book. A disadvantage of small regenerable deionizers is the need for neutralisation of the regeneration waste water.
There is a break-even point between regenerable deionizers and reverse osmosis. This break-even point varies from laboratory to laboratory. Sometimes, a neutralisation is operated that justifies demineralization by home-made regeneration. Otherwise, total solids are high and a reverse osmosis is more favourable from the perspective of running costs.

Figure 2.106 Reverse Osmosis System (Barnstead).

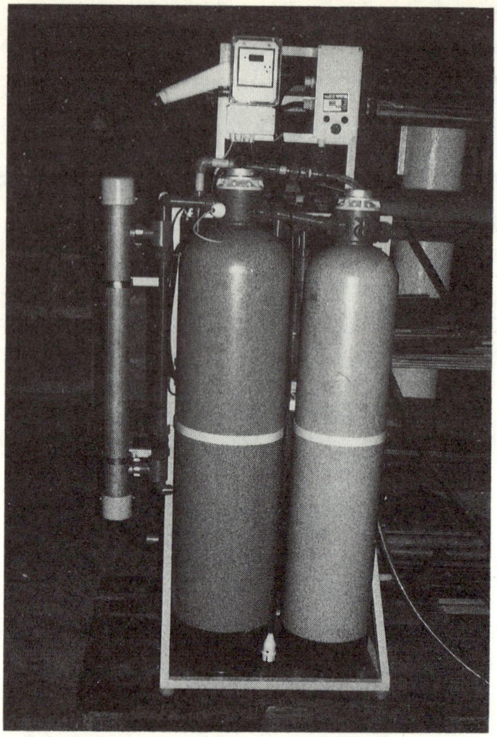

Figure 2.107 Regenerable laboratory twin-bed demineralizer together with UV light (W. Werner GmbH).

Feedwater for the reverse osmosis must fulfill certain requirements that can restrict the use of reverse osmosis:

- if the colloid index is higher than 3% per minute (K 15), sometimes special treatment is necessary; a good solution is the use of colloid removal resin
- a water softener as described in Section 2.6.1 is the best solution for protecting the membrane against scaling by temporary hardness.

The reverse osmosis permeate has to be purified because usually the 3 to 5% total residue of the feedwater does not permit the permeate to be used directly instead of distilled water. In Europe, regenerable mixed-bed cartridges will be used as a final treatment; they have a very long running time without the need for maintenance and service. In the USA, ready made cartridges are used (Barnstead), as well as regenerable cartridges.

One-way mixed-bed resin is also in use and is thrown away after use (it can be used as fertilizer) – especially in countries where no regeneration plants are in service. One-way resin also fulfills the demands made on ion exchangers but are in most cases sub-grade batches.

Figure 2.108 Ready-made cartridges (Barnstead).

2.6.2.2 High purity water

If purified water is produced by

- regenerable deionizers as twin-bed, also as mixed bed,
- reverse osmosis followed by a mixed-bed cartridge,
- mono-distillaton

it is not as pure as ASTM Type I requires.

In Europe ASTM Type I quality would be called bi-distilled or high purity water. The reason it is not pure enough is that it is not possible to regenerate cationic or anionic resin up to 100% in the H- or OH-form after a first use. The higher ion exchange technology, e. g., Triobed, is only useful in bigger installations.

To clean the water of traces of impurities, so-called "reactor-grade resin" is used. This resin fulfills the requirement of producing absolutely chloride-free water. This can be attained only if anion exchangers are nearly 100% in the OH-form and cation exchangers nearly 100% in the H-form. Every manufacturer of resins has his own highly-regenerated mixed-bed resin. This resin is generally used in the Barnstead cartridges.

In most cases, selected batches of this reactor-grade resin purify pre-purified water up to the highest purity, so-called 18 megOhm water (at 25 °C). This highly-pure water satisfies the ASTM Type I requirement. In Europe, this is used instead of bi-distilled water or high-purity water.

User-regeneration of this resin if exhausted is impossible. In Europe, the exhausted resin is sent to a service-regeneration station.

Typical applications requiring high-purity water (ASTM Type I):

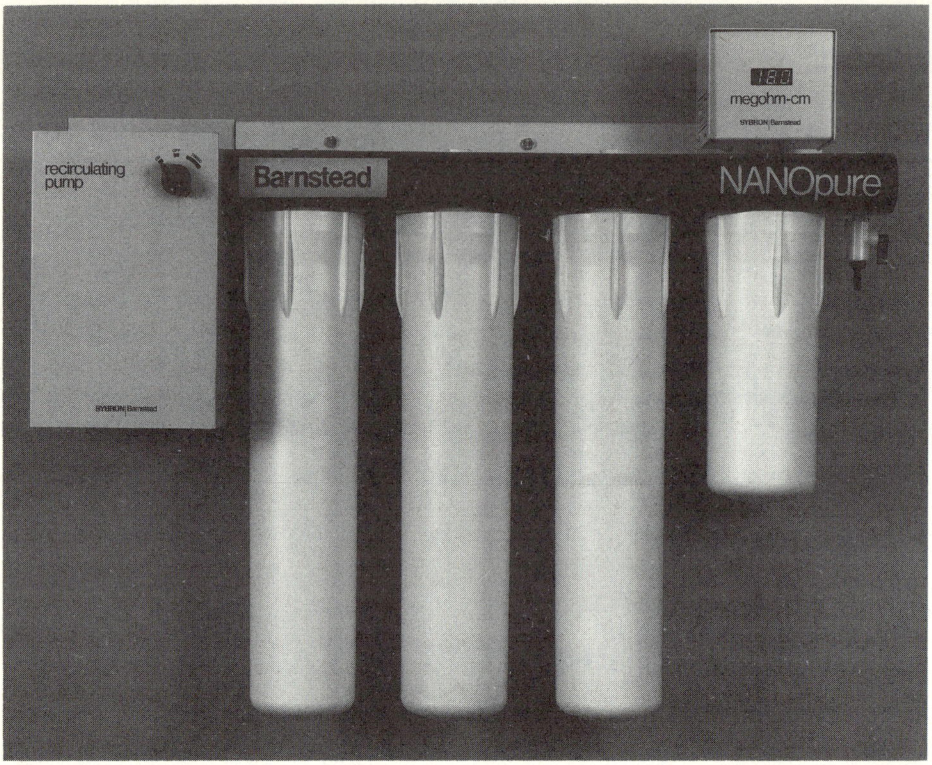

Figure 2.109 Installation to produce 3 l/min high purity water (Barnstead NANOpure).

- The growth of living animal and human cells is done in a so-called tissue-culture medium. The presence of heavy metals in this medium disturb cell growth. To produce the required medium, high-purity water is used to dissolve several nutritive salts and fetal calf serum is added. This mixture needs sterile filtration.
- Glass-washing for tissue-culture is not sufficient if the glass is rinsed with deionized water. High-purity water is much more likely to dissolve impurities from the glass walls.
- The final rinsing of semi-conductors requires highest-purity water (so-called 18 megOhm water), free of particles and with a low bacteria count. Only water of highest purity can dissolve impurities from the surface of semi-conductors. Most of the water used will be recirculated and used again after a special treatment.
- UV weathering-test machines for testing paints, plastics, etc., must be fed with high-purity water. There is no demand for 18 megOhm water, but only if the water has been purified to 18 MΩcm will it be free of silica, and silica-scaling is the most common problem in this type of test chamber.

2.6.3 Ion exchangers in small industries

In Section 2.6.2 it was said that the purification of water depends on several factors, e. g.,

- TDS in water
- possibility of using reverse osmosis
- presence of neutralisation.

These parameters have to be taken into consideration when selecting the best solution for a demineralization plant.

Another requirement is feedwater for steam boilers. This may be produced by

- softening in cases when small amounts of feedwater as additive are required
- decarbonisation by weak acid cation exchanger, regenerated by hydrochloric acid, followed by water softener.

Other applications are:

- humidification of rooms by direct adding of water by spray nozzles – demineralized water is necessary
- rinsing of surfaces after plating processes – demineralized water is used
- as an additive to cooling tower loops-softened water is used
- production of drugs in small amounts.

References

[1] Flemming, H. C. (1981) Verkeimung von Ionenaustauscher-Harz. Ergebnisse der Untersuchungen an einem stark sauren Kationen-Austauscher. Z. Wasser Abwasser Forsch. *14*, 132.

[2] Flemming, H. C. (1982) Bakterienwachstum auf Ionenaustauscher-Harz – Untersuchungen an einem stark sauren Kationen-Austauscher, Teil II. Z. Wasser Abwasser Forsch. *6*, 259–266.

[3] Ruschke, R. (1972) Pseudomonas aeruginosa – ein "Zivilisationskeim". Nachweis seiner Vermehrung in Kationenaustauschern von Geschirrspülmaschinen. Zbl. Bakt. Hyg., I. Abt. Orig. *B 156*, 391–398.

[4] Ionenaustauscher und Adsorberharze Firmenschrift 18/14596/10/1184, E. Merck, Darmstadt.

2.7 Ion Exchangers in Nuclear Technology

Günter Kühne
Bayer AG, Leverkusen, Federal Republic of Germany

Introduction

The problems connected with water treatment in nuclear power stations are entirely different from those encountered in conventional power stations. For this reason, it is useful to study the water requirements, the water consumption and the disposal of waste water, as well as the treatment of spent solutions in nuclear power stations.

Into this scheme, which includes a number of water circuits, the specific water treatment plant installations must be incorporated. If one is familiar with the conditions under which the ion exchangers have to function at the various locations, it is possible to deduce requirements based on their performance and quality. Not all ion exchange resins available on the market fulfil these demands. For this reason one will have to gather experience in actual applications and the results of different test-methods in order to select a suitable ion exchanger, as well as to define the limits of its application.

Much of what has been learned in this way in connection with nuclear power stations is again reflected in the planning and running of conventional power stations.

2.7.1 Nuclear grade ion exchange resins

2.7.1.1 Composition

From the many types of ion exchangers offered, only the resins based on polystyrene have been found suitable for use in nuclear technology. They are available in either gel-form or as macroporous resin beads. The active groups consist in the case of cation exchangers of strongly acidic sulfonic acid groups, in the case of anion exchangers of strongly basic quaternary ammonium groups, the latter as so-called Type I resins. Ion exchangers with weakly dissociated groups or other functions are in fact unimportant, if one disregards the special cases of the demineralisation of the fuel element cooling pond with a weakly basic anion exchanger [1] or selective caesium fixation with a weakly acidic condensate resin [2].

2.7.1.2 Physical form

To meet the needs of the designers of ion exchange units and of the users in nuclear power stations the bead form resins have proven to be of best stability and hydraulic advantage compared with granular resins. Therefore, the development of polymeric resins, based on crosslinked polystyrene, yielded the first possibility of success.

Special types of selective ion exchange resins produced by condensation reactions had to be used in the granular form until the manufacturers succeeded in producing these condensate resins in bead form [3].

A different kind of application is the use of ion exchange materials in precoat filtration. Fine particle size material has to be chosen because of the necessary reactivity in connection with the high flow rates. But the smaller the resin particles are, the higher the pressure loss. Polymerisation, conversion of the polymers into ion exchangers, and regeneration, etc., of particles in the size-range of microns are too difficult to consider. Therefore, the manufacturers start with standard bead production, followed by purification and regeneration, and finally grinding of the beads. Powdered resin ion exchangers with the same active groups as those of the bead resins are used, i. e., strongly acidic sulfonic acid resins and strongly basic quaternary ammonium anion exchangers. As the powdered-form has a very fine degree of particle division, a macroporous matrix beame unnecessary. Hence, in this case, gel resins form the basis. The powdered resins are then used as mixtures of cationic and anionic components which agglomerate to form large flocs of several millimeters in diameter. The resulting precoats have optimal hydraulic properties with low pressure loss but are several millimeters in thickness [4].

In addition to solid ion exchange resins, special liquid ion exchangers are of some interest. Because their functional groups are often very different in selective properties to the standard polymer resins, they must be used despite the complicated extraction procedures.

Intensive research lead to the production of impregnated resins [5] and a special development resulted in polymers with liquid ion exchangers being incorporated into the resin matrix [6, 7]. Using this development the advantages of bead materials are utilized in the exchange process with liquid ion exchangers.

2.7.1.3 Properties

Faced with the complexity of the isotopes to be removed in nuclear technology and the difficulty of fully removing them, research efforts for higher performance resins have been directed towards improving the characteristics of ion exchange resins utilized for the removal of ionic impurities.

In the theoretical calculations it should be noted that the factors controlling the efficiency with which the ions are exchanged are the coefficient of selectivity and the total capacity of the resins.

It should also be mentioned, that, in dynamic situations, the kinetics of the resins are also of particular importance, because these determine the length and the shape of the ion exchange reaction zone – hence, the exchange capacity and leakage.

The kinetics of the resins depend on a number of criteria including the internal porosity (gel structure), and the surface area of the beads (which influences the ion exchange rates at the solid/liquid boundary). It is possible to modify the composition of the ion exchangers and consequently to improve their performance.

In circuits of nuclear stations and in situations where all activity must be eliminated, it is imperative that the resins used must not release ions or materials which could result in deposits or corrosion.

All these nuclear grade resins are manufactured to very high purity specifications. They are supplied in ready-to-use form, regenerated to a very high level, and contain a minimum of residual counterions.

Table 2.29 Ion exchanger specifications for nuclear technology quality

Details of ion content and physical data		Manufacturers of ion exchangers I	II	III	IV	V	VI
H	min. %	95	95	98	95	99	95
Na	max. %	5	4	–	–	–	–
Alkali metals	max. %	–	1	–	–	1	–
Capacity	min. meq/g	4.7	–	–	4.6	–	–
Capacity	min. meq/ml	–	–	1.9	1.8	1.9	1.8
Chloride	max. mg/l	–	–	–	–	10	–
^{7}Li	min. %	99	99.98	99	95	99.9	–
Na	max. %	1	–	–	–	–	–
Capacity	min. meq/g	4.6	–	–	4.6	–	–
Capacity	min. meq/ml	–	–	–	1.8	–	–
NH_4	min. %	99	95	99	95	99	95
Na	max. %	1	4	–	–	–	–
Alkali metals	max. %	–	1	–	–	1	–
Capacity	min. meq/g	4.4	–	–	4.5	–	–
Capacity	min. meq/ml	–	–	–	1.8	2.0	1.8

Details of ion content and physical data		Manufacturers of ion exchangers I	II	III	IV	V	VI
OH	min. %	80	90	85	80	95	80
Cl	max. %	5	5	5	5	0.1	–
CO_3	max. %	15	5	15	15	5	–
Capacity	min. meq/g	3.5	–	–	3.5	4.0	–
Capacity	min. meq/ml	–	–	1.0	1.0	1.1	0.9
Fe	max.	200 mg/l	0.005%	50 ppm	0.005%	50 ppm	250 ppm
Cu	max.	100 mg/l	0.001%	25 ppm	0.0025%	–	50 ppm
Heavy metals (as Pb)	max.	100 mg/l	0.0001%	25 ppm	0.005%	20 ppm	50 ppm
Solubles	max.	–	–	0.1%	–	0.1%	1.5 mg $KMnO_4$/lH_2O
Grading	mm	0.3 – 1.2	0.3 – 0.84	0.3 – 0.84	0.3 – 1.2	0.3 – 1.5	0.3 – 1.5
Bead distribution limits		–	min. 98%	min. 95%	max. 5% > 1.2 mm	min. 98%	
Fines		max. 0.5%	–	–	max. 1%	max. 2%	max. 0.5%
Effective bead size		0.38 – 0.6	–	–	–		0.42 – 0.65
Uniformity-coefficient		1.75 – 2.0	–	–	–		
Pressure loss		–	–	–	–		max. 65 m water column
H_2O	%	55 – 60	–	–	55 – 60	50 – 65	–
Bulk density	g/l	–	670 – 800	700 – 800	720 – 850	650 – 900	–

2.7.1.4 Specifications and purity

The specifications for nuclear technology qualities are divided into four groups:

1. the degree of regeneration and the purity of the desired ion form,
2. the quantity of ions available for exchange, expressed as a percentage of the total capacity, related either to the moist or the dry resin,
3. the contamination by other substances inherent in the production,
4. the physical data.

All resins are manufactured with extreme care, which minimizes the introduction of impurities, even in trace amounts; they also undergo very thorough cleaning and regeneration processes which remove all the impurities that may have been adsorbed or fixed during manufacture.

Table 2.29, which is based on the specifications of leading manufacturers of ion exchange resins, illustrates differences in the degree of purity, which have undoubtedly come about through a change in requirements as the nuclear power stations continued to develop. If this table does not include specifications of organizations engaged in the design and construction of nuclear power stations, it is simply because manufacturers of ion exchange resins make their products to conform to the demands of these organizations. The level of regeneration and purity is therefore fully compatible with that laid down in the specifications of the various power station design engineers.

Details given in Table 2.30, showing comparison percentages and absolute values related to whether the resin is moist or dry, are important for better understanding of the values given in Table 2.29.

Table 2.30 A comparison of the different ways impurities are given for ion exchangers

Ion	% of the total capacity	mg/l moist resin	%, calculated on the dry resin	ppm, calculated on the dry resin
Na^+	5	2200	–	5400
Cl^-	1	500	–	1800
CO_3^{2-}	15	9900	–	31500
Fe^{2+}	–	18	0.005	50

The regenerants are closely connected with the purity requirements made on the ion exchangers. The manufacturers of ion exchangers are familiar with these problems and they specify the essential regenerant qualities. The situation appears to be different when the owner himself is interested in regeneration. A special example which can be quoted at this stage is the sodium-content of cation exchangers and the chloride-content of anion exchangers, together with the purity of caustic soda solution for its regeneration.

Table 2.31 shows the chloride-content of differently conditioned anion and cation exchangers. To simplify matters particulars are expressed as a percentage as well as in mg/l resin. It is remarkable that the optimum values measured on a laboratory scale at Euratom in Belgium [8] could be matched and in fact improved during large-scale production.

Table 2.31 Chloride-content of cation and anion exchangers

Strongly basic gel-type anion exchangers – Type I	Chloride content	
	as % of the total capacity	in mg/l ion exchanger
Usual delivery form (Cl form)	100	50,000 – 55,000
OH form (usual regeneration with 80 – 100 g NaOH/l)	about 50	25,000 – 28,000
General specification for nuclear technology	5	2,500
Average value for nuclear technology	about 1	500
Exchangers free from chloride	<0.1	<50
Publication by Euratom [8]	0.1 – 0.2	50 – 100
Strongly acidic cationic exchangers – gel-form		
Regeneration with HCl of the exchangers as supplied	0.15 – 0.45	100 – 300
Average values for nuclear technology	0.08 – 0.15	50 – 100
Exchangers free from chloride	about 0.015	about 10

The significance of the chloride-content of the caustic soda solution applied can be seen in Table 2.32. For examle, in the regeneration of a strongly basic anion exchanger with a large volume of caustic soda solution a considerable part of its chloride content remains on the resin. The value quoted in the DIN-standard is of importance because it was arrived at from quite a different point of view, namely, that of the conventional demineralization of water.

The degree to which the resins can be regenerated depends very much on the technique employed and the available excess of regenerant. It would go beyond the scope of this chapter to deal with these questions in detail.

The contamination introduced into ion exchangers by the manufacturing process consists mainly of metal ions from the reaction vessels and the chemicals that have been used. In the example in Table 2.33, showing the complete analysis of two ion exchangers, one can see the order of magnitude of these impurities. Checking by evaluating an arbitrarily selected batch shows that the values are at times even lower than the data given in the specifications (Table 2.29).

Table 2.32 Chloride quantities available to an anion exchanger during regeneration using varying qualities of caustic soda solution

Designation of the caustic soda solution	Chloride content as % or as mg/l	Amount of chloride available in mg/l resin on regeneration with 680 g NaOH (100%)/l resin
Purest for analytical purposes	0.0005%	3.4
Purest	0.005%	34.0
Flakes	0.015%	about 100
Chemically pure 45%	0.02%	about 270
Technically pure 50%	0.02%	about 250
Solution 44%	240 mg/l	about 370
NaOH acc. to DIN 19616	max. 0.2 weight-% of the NaOH content	up to 1360

Table 2.33 Analysis of two batches of ion exchangers for nuclear technology

Impurities in ppm, calculated on dry substance	Cation exchanger	Anion exchanger
Alkalies	210	47
Al	11	1.7
Cu	0.5	<0.5
Fe	39	49
Pb	10	7.1
Other heavy metals	3.5	15.6
P	<1	3
Cl	78	570
Soluble substances	180	620

It should be mentioned here that the problem of contamination is of practical importance in as much as these substances are eluted during the time the exchanger is in operation.

If certain elements are primarily bound to the resin matrix and therefore not present as ionic impurities, one must still observe their behaviour under radiation influence. Breaking the link to the matrix chain results in an ionic form. Thus, the ion can be eluted just as if it were an ionic impurity. Chloride, especially, is an example here. Therefore, the discussion about the chloride content of anion exchangers must include the organically bound chloride. As example of what resin manufacturers are aiming for can be found in the description of a production method described in a patent [9].

So far little is known about the rate of elution of other elements. The difficulty in answering this question lies in the lack of adequate analytical methods for determining trace substances. An answer to this question is only likely to become feasible by rendering the ions radioactive. Any data presently available will be mentioned later on.

2.7.1.5 Stability

The stability of ion exchangers is affected by a number of factors. These include mechanical, osmotic, thermal, and chemical influences, as well as radioactivity.

Resistance to chemical influences is of least importance within the scope of nuclear technology, as the affect of oxidizing agents, organic substances, detergents, metal complexes, etc., which occur in conventional water-treatment, can in fact be disregarded.

During condensate treatment the chemical influences are reduced to the presence of heavy metals and their catalytic effect in the presence of oxygen. Experience gained in nuclear power stations shows that an operating time extending over several years has had no discernible influence on the mixed bed resins in the condensate circuit.

While the temperature resistance of strongly acidic cation exchangers is claimed to be as high as 150 °C in the data sheets issued by the ion exchange resin manufacturers, the limit in the case of strongly basic anion exchangers in the OH-form remains, as it has always been, around 70 °C. It is, therefore, sufficient to study the influence exerted by temperature on the latter type of resin, as it dictates the limits for mixed beds. The loss in capacity always becomes apparent by a reduction of the strong base capacity, either by conversion into weakly basic groups or by loss of all the active groups. Autoclave tests resulted in the values that have been listed in Table 2.34. It is also known [10], that the rate of decline in capacity due to the temperature increase is higher initially than that observed when the temperature is applied continuously. It is, therefore, possible – bearing in mind this rapid loss in capacity – to operate ion exchangers above the recommended temperature limit for a short time. However, in that case one will always have to take into account a higher degree of attrition compared with that under normal conditions.

With regard to the resistance of ion exchangers to radioactivity, it has been determined that the life of an anion exchanger batch in a nuclear power station (e. g., for the purification of primary water), can be more than a year. Thereafter, the question of stability becomes important again, but it is also necessary to know what products can arise from decomposition.

Numerous studies throughout the world have been designed to produce sufficient quantities of decomposition product to carry out analyses [10, 11, 12]. The amounts of radiation chosen had to be fairly high, because up to 10^6 rad there was no detectable decomposition. The tests using cobalt radiation sources were carried out with doses of $2.7 \cdot 10^5$ rad/h up to $4.5 \cdot 10^9$ rad/h. Irradiation of dry or moist resins

Tab. 2.34 Dissociation products of a strongly basic Type I resin in OH form at 150 °C

Time in hours	0	6	12	24
Stongly basic	3.90	0.92	0.06	0.00
Weakly basic	0.10	1.43	1.81	1.84
Total mval/g	4.00	2.35	1.87	1.84
Trimethylamine mval/g	–	1.50	2.11	2.17
Methanol mval/g	–	0.14	0.20	0.20
Water content % by weight	48.7	42.8	38.3	37.5

Loss of strong base capacity for Type I resin

Type I resin in OH form. Temperature applied for about 140 hours	Loss of strong base capacity
20 °C	0%
40 °C	1.5%
60 °C	5.0%
70 °C	7.0%
80 °C	15.0%
100 °C	75.0%

in an inert gas atmosphere produces few gaseous dissociation products. In air, the dissociation is accelerated and apart form CO_2, CO, and H_2, the typical compounds from the active groups are obtained: SO_2 from the cation exchangers (Figure 2.110) and trimethylamine from the anion exchangers. In water other amounts of decomposition product occur and this is linked to the dissociation of the water under radiation and to the related formation of H_2O_2, H_2, O_2, and different radicals. This confirms that secondary reactions have an additional effect here.

The form of the resins used is also an important factor for dissociation. H- and OH-forms are more sensitive than, for example, lithium or borate-forms.

The external shape of the exchanger beads was preserved in all cases and hardly any change in colour was observed. The polymer chains partially descrosslinked and this produced an increase in volume as swelling-water was taken up, but this effect (which is due to decrosslinking) is reversed after a time as the swelling capacity of the matrix decreases again due to a loss of the active groups.

The condensation-type resins have particularly good mechanical stability, followed by the polystyrene resins. The polyacrylic resins are less stable. In all practical cases in nuclear power stations, the physical damage has remained so small that it has not caused any reduction in the life of the resin.

The active groups in the exchangers react differently to radiation. The most stable are the sulfonic acid resins and, therefore, the cation exchangers. These are followed by the resins with phosphoric acid groups. The anion exchanger groups

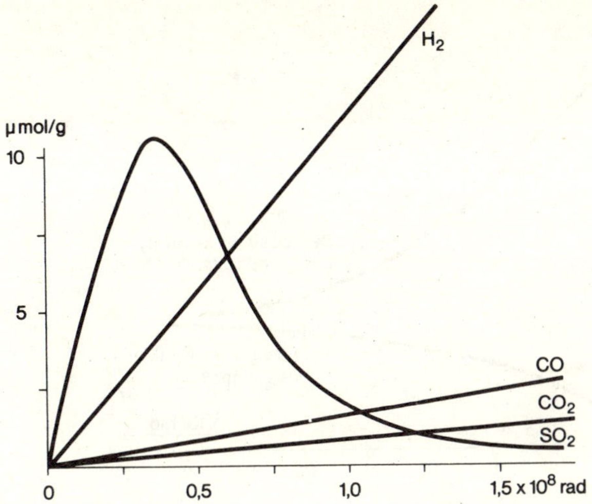

Figure 2.110 Formation of gaseous dissociation products by irradiation of a strongly acidic cation exchanger in Li-form.

are less stable, and the chemical behaviour of condensation-type resins is even less favourable.

Considering all the properties of the anion exchangers, the strongly basic resins of Type I are preferable. Under radiation the pyridine resins suffer only a slight loss of amine groups but their capacity for salt-splitting is clearly reduced. Type II resins have a very unstable functional group and are excluded from any use at elevated temperatures or in radiation fields.

With cation exchangers, where the active groups are split off, free sulphuric acid occurs and carboxyl groups or phenol hydroxyl groups are formed instead on the exchanger structure. For the most part anion exchangers lose their strongly basic group due to the loss of trimethylamine. Additionally, we find a conversion into weakly basic, mainly tertiary and secondary amine groups (see Figure 2.111). There are no reports on the types of decomposition products which are formed here.

If a comparison is made of the decomposition of ion exchangers under different dose rates it is found that if the dose rate is doubled, for example, there is no doubling of the decomposition rate [10]. This too confirms the effect of secondary reactions. There was no linear progression in the loss of SO_2, as the dose was increased for a constant dose rate [11]. This must be due to the different steric arrangement of the active groups. Of course, the main interest lies in the evaluation of the practical performance of ion exchangers. In the estimates of nuclear power stations, the ion exchangers used in the area of strongest radiation (in the units for purifying the primary reactor water) are exposed to dose-rates of about $10^7 - 10^8$ rad in a period of one year. This figure is reduced by a factor of 10 compared to the test conditions described above (where it was $10^8 - 10^9$ rad). When we consider that the majority of the tests were carried-out on resins in H- and OH-form, it

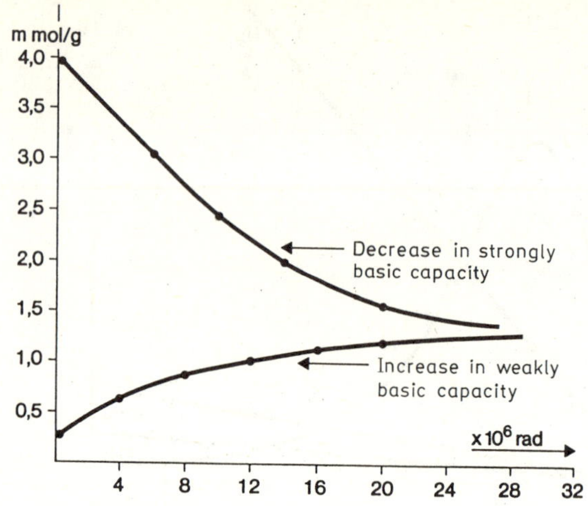

Figure 2.111 Change in the capacity of strongly basic anion exchangers (OH-form) with doses of radiation.

emerges that the mixed bed units in pressurized water reactors in the Li- and borate-form can be used over several years without decrease in performance due to radiation.

In the case of boiling-water reactors, in which resins are present to a large extent in the H- and OH-form, a shorter life would be expected. But use of up to 3 years has been shown to be possible without any risk.

2.7.1.6 Capacity

The capacity of ion exchangers varies with the type, the quantity of ions in the water to be treated, and also with the end-of-cycle criterion. In the case of radioactive circuits, nuclear grade resins have higher capacities for the ions usually encountered in water treatment than standard resins, since they are virtually completely regenerated. As an example, a strongly acidic cation exchanger can fix up to 1.7 equ/l of NH_4 or lithium ions. When two columns operate in series, the leading column can fix these ions up to the point of complete saturation, i. e., virtually up to 2 equ/l.

The capacity of ion exchangers has been the subject of several papers dealing with the adsorption of ions at conventional concentrations as well as at trace amounts [13].

In nuclear technology boric acid is of particular interest. It can only be linked to strongly basic resins in any sort of acceptable capacity. Resins with groups of lower basicity or selective properties are of little use. The capacity for boric acid is dependent on the inlet concentrations to a degree unknown with other ions. The reasons for this lie in the character of the boric acid itself, which tends to form polyborates. Table 2.35 shows the values of the quantities taken up at various influent concentrations. By analogy with silica (silicic acid) in conventional water treatment one can estimate the capacity of the exchangers in equivalents. For SiO_2 it is around 0.4 equ/l at a maximum, if regeneration is carried out economically (Table 2.36). In the case of the highly regenerated resins used in nuclear technology one can expect approx. 0.7 equ/l. If one studies the data in this context dealing with the oligomeric forms of boric acid (Table 2.37) one finds this value confirmed.

Table 2.35 Capacity of a strongly basic anion exchanger – Type I for boric acid

Boric acid influent concentrations	Exchange capacity (to breakthrough)	Quantity adsorbed	Capacity
17 ppm boron	8.5 g B/l	0.78 mol/l	
170 ppm boron	13.5 g B/l	1.25 mol/l	about 0.7 equiv/l
1700 ppm boron	23.0 g B/l	2.12 mol/l	

Table 2.36 Capacity of a strongly basic anion exchanger – Type I for SiO_2

Silica influent concentration	Exchange capacity (to breakthrough)	Quantity adsorbed	Capacity
0.3 ppm SiO_2	6 g SiO_2/l	0.1 mol/l	0.1 equiv/l
about 10 ppm SiO_2	about 24 g SiO_2/l	0.4 mol/l	0.4 equiv/l

Table 2.37 Boric acid and its oligomeric forms

Name	Undissociated form	Ionic form	Simplified form
Ortho-boric acid	$B(OH)_3$	$B(OH)_4^-$	B^-
Triborate		$B_3O_3(OH)_4^-$	B_3^-
Tetraborate		$B_4O_5(OH)_4^{2-}$	B_4^{2-}

The capacity is also influenced by increase in temperature. If the temperature rises by approx. 60 °C, the capacity reduces to 6 g/l resin. The behaviour of the boric acid should be studied when anion exchangers are used in the boric acid-form or are ordered to be delivered in that form. The conditioning of the resin must be adjusted in accordance with the boric acid content of the reactor water where it is going to be used. It can, therefore, be readily understood, that no values appear for this in the general specifications for reactor quality resins.

The capacity of ion exchange resins for other ions when being used for decontamination depends on the influent concentration of these ions, since the end-point of the cycle is based on their residual level, and on their type, both of which are variable. In this context one is concerned not with concentration but with activity, a value which it is difficult to express by means of concentration. Although the concentration of a given product can be linked with its radioactivity, a given level of activity does not present the same concentration in the case of different nuclides. Generally speaking, the activity levels normally found are low ($< 1 \cdot 10^3$ Ci/m^3), representing a total concentration of a few ppb. The quality of the ions encountered (active ions) varies from one circuit to another, from one station to another, and from one moment to the next, since it is dependent on the activity of the primary circuit, which in turn depends on a number of variables, the most important of which is fuel can breakings.

The ions most frequently found are the isotopes of caesium, iodine, cobalt, strontium, rubidium, tellurium, antimony, molybdenum, manganese, chromium and sodium, and the principal quality criteria of the resins is their specific affinity for the most dangerous ions, such as caesium 137, which has a half-life of about 30 years.

The specific affinity is translated into a decontamination factor associated with cycle length.

In practice, however, the water requiring treatment often contains a relatively large amount of conventional ions such as ammonia, morpholine, lithium, or sodium, for which the resins also have a strong affinity, and a very small quantity of dangerous ions, requiring a maximum efficiency of removal, for which the resins have no obvious affinity. This means that the actual capacity of the resins, and their decontamination factor, are essentially dependent on the processes selected. When the water to be treated contains large quantities of ammonia, morpholine or lithium which must be removed (steam generator blowdown, treatment of primary effluent) it is preferable to seperate the deionisation from the decontamination function, since the deionisation function is mainly concerned with the removal of these conditioning products. The plant is usually planned with two columns in series; the first, a cation exchanger, has the function of removing the maxiumum quantity of cations, while the second, a mixed bed, is designed to remove the active ions.

2.7.1.7 Attainable residual contents

The residue of ions following treatment with ion exchange resins is due to two causes:

displacement leakage,
equilibrium leakage.

Whereas adequate information is at hand concerning the displacement leakage, the data for which have been determined from the treatment of water, there is very little known concerning the equilibrium leakage. The reason for this is that under normal conditions it is very difficult to make a distinction between the two phenomena, in addition to which there are difficulties in the analysis of quantities in the microgram range. It will not be possible to determine and measure reliably the presence of small traces unless radioactive ions are applied, allowing the equilibrium leakage to be established when highly regenerated exchangers are being employed. In Tables 2.38 – 2.41 leakage data is recorded which were determined under various conditions.

Table 2.38 Residual contents after having passed the mixed-bed units

	Using nuclear technology quality, or optimum regeneration	Conventional regeneration
Conductivity after mixed-bed unit	0.05 μS/cm	0.1 – 0.2 μS/cm
Sodium content after mixed-bed unit	0.3 – 0.5 μg/l	1 – 50 μg/l

Table 2.39 Chloride content of a boric acid solution after passing through a strongly basic anionic exchanger of Type I

	Using an exchanger free from chloride	Using a normal nuclear technology quality
Chloride leakage	0.02 – 0.05 mg Cl/l	0.2 – 0.3 mg Cl/l

Table 2.40 Boric acid leakage after strong base anion exchanger Type I

	at 20 °C	at 60 °C
Influent > 10 ppm B	0.2 – 0.3 ppm B	0.4 – 0.6 ppm B
Influent < 1 ppm B	0.01 – 0.02 ppm B	0.02 – 0.04 ppm B

Table 2.41 Selective Caesium-137 removal by LEWATIT DN KR

	At pH 10 – 11 by NaOH and Na_2CO_3	Deionized water with 1100 ppm B as boric acid
Influent approx. 5 μCi/l	90 – 95% removal during treatment of 8000 – 11000 bed volumes	80 – 85% removal during treatment of 6000 – 8000 bed volumes

In the treatment of make-up water, one can achieve residual sodium contents of 0.3 to 0.5 µg/l, following optimum regeneration of the mixed bed units installed as a safety measure.

If regeneration has not been carried out correctly, due to incomplete separation or other difficulties, the residual sodium content in the water can rise at times from 1 to 50 µg/l (see Table 2.38). These increased sodium values cannot be discerned satisfactorily by a change in the conductivity in the treatment of make-up water. Due to interference by other substances, for example humic acid, higher values are frequently measured than those conforming to the theoretical conductivity of the most pure water of about 0.05 µS/cm (at 20 °C). The incidence of a sodium leakage of 10 µg/l can be computed if present as NaOH for approximately 0.09 µS/cm, as NaCl for about 0.05 µS/cm and as Na_2CO_3 for about 0.03 µS/cm in round figures. This demonstrates that the change in conductivity due to the sodium-leakage is in fact very small.

There are no reliable data on chloride leakage occuring with the use of strongly basic anion exchangers in demineralization plants. Using nuclear technology quality resins in a boric acid-containing reactor-water, it was possible to observe distinct differences based on the chloride content of the ion exchangers (Table 2.39). These results suggest that the chloride levels are due to displacement leakage. This occurs significantly while the exchanger unit is operative, but on starting-up following extended shut-down it is appreciably higher.

Detailed research on the dependance of chloride leakage upon boric acid concentration has led to the conclusion [14] that chloride-free resins (chloride content < 50 µg/l anion resin) are necessary, if during a cycle a boron concentration of greater than 200 ppm occurs. If there are always lower concentrations of boric acid, the standard nuclear quality is enough, with chloride-contents below 500 mg/l anion resin.

For the condition of demineralized water the chloride leakage from a strongly basic anion exchanger can be calculated [15] and is illustrated in Figure 2.112. The

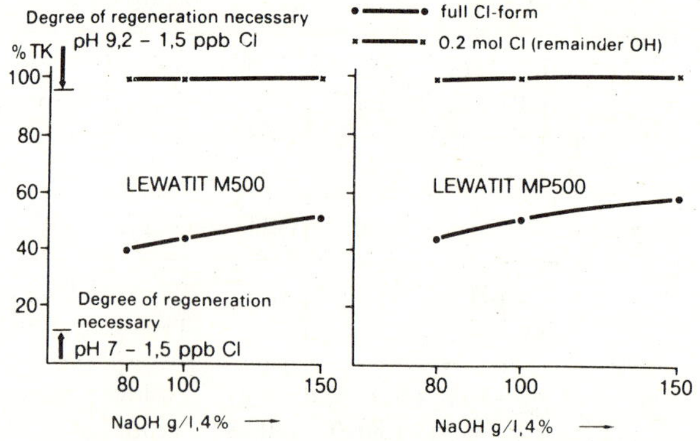

Figure 2.112 OH-ions in % TK after regeneration with NaOH.

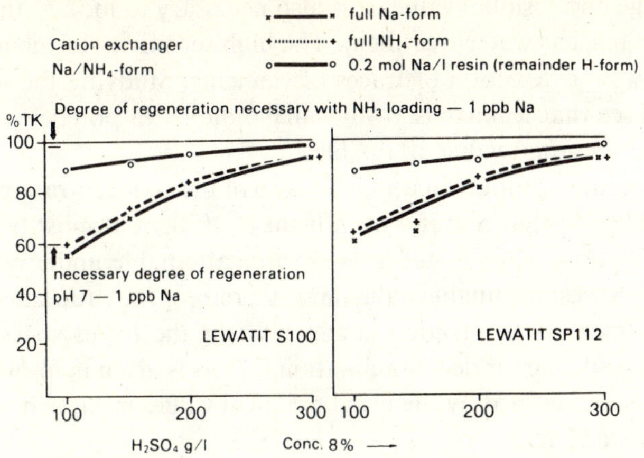

Figure 2.113 H-ions in % TK after regeneration with sulphuric acid.

importance of the residual sodium-content in strongly acidic sulfonic acid cation exchangers can also be explained (see Figure 2.113).

The necessary degree of regeneration of a cation exchanger can be derived from the example of the reaction

$$RH + Na^+ \leftrightarrows RNa + H^+ .$$

For this purpose the assumption is that the cation exchanger uses pure neutral water flow and is not intended to produce more than 1 ppb of Na-ions. In this case, the pH is 7, the hydrogen ion concentration is $1 \cdot 10^{-7}$ mol/l and the sodium ion concentration is $4.3 \cdot 10^{-8}$ mol/l (for the derivation see the following formula calculation).

$$K = \frac{[RNa]\,[H^+]}{[RH]\,[Na^+]} = 1.5$$

$$[H^+] = 1 \cdot 10^{-7}\ \text{mol/l}$$

$$[Na^+] = 0.435 \cdot 10^{-7}\ \text{mol/l} = 1\ \text{ppb}$$

$$[RNa] = [1 - RH]\ \text{mol\%}$$

$$\frac{[RH]}{1-[RH]} = \frac{1}{K} \cdot \frac{10^{-7}}{0.435 \cdot 10^{-7}} = 1.53$$

$$[RH] = 1.53\,[1 - RH]$$

$$[RH] = 0.61\ \text{or}\ 61\%\ \text{of TK}$$

Discussing leakage and residual values it is also necessary to look at the decontamination factors in nuclear water treatment. The high sensitivity of the measurement of radiation allows for detection of traces of elements. Studying the ion exchange reaction one can see that leakage is always unavoidable. In practice a limit can be seen for the lowest residual value attainable.

Because the decontamination factor is the ratio of the "concentration" of influent to effluent, indeed the radiation ratio of influent to effluent, it must be pointed out that a decontamination factor is not a fixed value attainable under all conditions. In fact, there is a lowest attainable value and, therefore, the ratio always becomes smaller, if the influent concentration comes near to the lowest possible effluent concentration. A guideline for decontamination factors is given in Table 2.42, where the influent values are defined by the usual practical values in known water circuits of nuclear power stations.

Table 2.42 Decontamination factors that can be attained with various ion exchange systems; the values given are averages over the life time of the system

	I	Cs, Rb	All other solubles
	Pressurized Water Reactors		
Mixed bed (Li-, BO_3-form)	10	2	10
Cation bed (H-form)	1	10	10^2 (cations)
Mixed bed (H-, OH-form) for radioactive waste	10^2	2	10^2
Anion bed (OH-form)	10^2	1	10^2 (anions)
	Boiling Water Reactors		
Mixed bed (H-, OH-form)			
primary coolant	10	10	10
steam condensate	10^3	10	10^3
clean waste	10^2	10	10^2
dirty waste	10^2	2	10^2
Cation bed (H-form) dirty waste	1	10	10^2 (cations)
Precoat filtration with powdered resins	10	1	10
	Other Systems		
Mixed bed (H-, OH-form) evaporator condensate	10	10	10

2.7.1.8 Treatment of used ion exchangers

If ion exchangers have fulfilled their function and have no further application consideration must be given to how to dispose of them. Here it should be remembered that these are still highly stable synthetic resins which are not inflammable and which are very stable against chemical attack. Apart from the elements they are made of – C, H, O, S, N – they also contain counterions such as Ca, Mg, Na, Cl, SO_4. However, they can also contain elements and compounds harmful to the environment, e. g., the conditioning chemicals of the nuclear circuits – lithium and boric acid – and the fission products.

These resins must be stored in underground dumps. Special provisions apply to their transport and storage and these demand leach-proof solidification. They can be encased in cement, bitumen or plastic. The weight and volume of the additives determine to a large extent the final dumping costs, while the type of solidification is what mainly determines the costs and safety of the operating plant. When they are encased in bitumen, care must be taken regarding the pyrolysis products produced by heating and the behaviour of certain forms of resin. The advantage of solidification in plastic is that there is no need for high temperatures [16]. To achieve safe solidification in plastic a special process had to be developed in which the resins could be used without prior drying. There is a special company which has portable equipment for solidification in plastic [17, 18].

Synthetic ion exchange resins can be incinerated at high temperatures. It is necessary to observe closely to the different emission products which, because of their radioactivity, are dangerous substances.

The decomposition products from the ion exchanger itself are also volatile and aggressive.

2.7.2 Powdered ion exchangers

As mentioned previously, powdered resins strongly resemble bead resins, as far as the basic types in use and the purity are concerned.

In considering performance, one must dinstinguish between ion exchange capacity, filtration performance and the residual percentages that can be attained. The operating time of a precoat of powdered resin is subject to the following criteria:

increase in conductivity,
breakthrough of cations or anions,
increase in pressure loss.

As the functions and application possibilities of the precoat filtration with powdered resins have been described fully in many papers [19–21], it will suffice here to add a few facts. At temperatures of 25 °C one can expect the values given in Table 2.43, if the concentrations at the inlet remain within the stated limits.

Table 2.43 Residual contents after passing through a powdered resin precoat filter

	Values in the raw condensate	Attainable residual contents in the treated condensate
Conductivity	0.2–0.5 μS/cm	0.1 μS/cm
Iron	unlimited	5 ppb
Copper, Nickel	unlimited	3 ppb
Silica	10–20 ppb	5 ppb

Table 2.44 Filter output of a powdered resin precoat filter at various temperatures, using 1 kg powdered resin per m^2 filtration area

Temperature of the condensate °C	Amount of suspended matter filtered out in g per m^2 precoat filter area
30	65
50	90
70	120
95	165
135	200

It was found in actual practice that the capacity to remove suspended matter showed higher values as the temperature increases. The details are shown in Table 2.44. These results, however, must be studied under the proviso that the maximum permissible pressure loss of the precoat filter is not exceeded. As water has a lower viscosity at higher temperatures, the pressure loss is correspondingly lower, which in turn extends the operating time and thus more suspended matter is filtered out.

Consideration was given in a previous section to the stability of bead resins; this is also of importance for powdered resins. In the main the same can be said as regards resistance to chemical and thermal influences, and radiation, as has been said for the bead resins. However, a few details remain to be mentioned concerning the mechanical and osmotic stability, naturally only relative to the deposited precoat layer.

It was noticed in the past that the precoats developed cracks. This observation, however, was not seriously followed up on, even when the results of filtration and demineralization were found to be unsatisfactory. Cracks can be explained by the well-known fact that ion exchange resins undergo a volume change during adsorption, and also by a shift in the electrical potential of resin particles when they come into contact with colloidal impurities carrying a potential as well. Successful countermeasures are filtration auxiliaries, such as inert fibers. Attention should also be paid to the mechanical stabilisation of precoats in connection with the breakthrough of salt-containing cooling waters inducing osmotic changes in the precoats of filters installed for condensate polishing.

2.7.3 Liquid ion exchangers

In the field of liquid ion exchangers there has been no further development since the first introduction of this type of reagent and the corresponding extraction technique.

Liquid ion exchangers are long-chain organic compounds with phosphoric acid groups or weakly basic amine groups. The nature and the number of active groups in the molecule determine the solubility of the exchanger, and the higher the dissociation of the active group, the more soluble the compound. This limits again the type of groups. On the other hand, organic compounds can be made with certain active groups which are difficult to introduce into a solid resin matrix. Therefore, liquid ion exchangers are still of interest.

From the technical point of view the decision has to be made between the ion exchange process with solid resins in filter units or with liquid exchangers and the mixer-settler process. Both materials are commercially available, as it is explained in Section 2.7.1.2.

2.7.4 Diagrammatic representation of principle types of nuclear power stations

2.7.5 Treatment of water circuits

Taking the four main reactor types one can distinguish six locations, differing in principle, where ion exchangers are used in the treatment of water, apart from

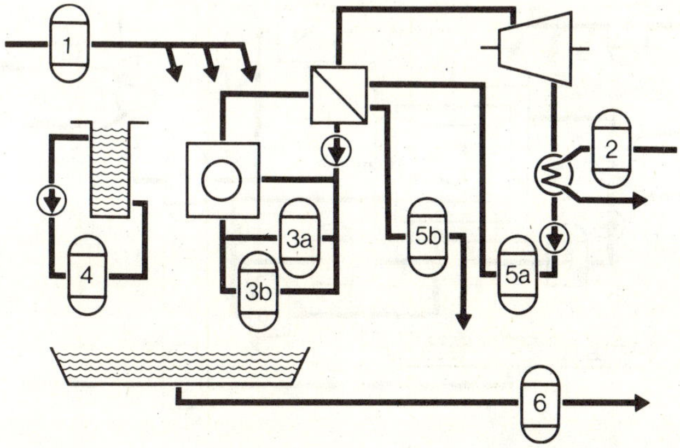

Figure 2.114 Principle of nuclear power stations with pressurized-water reactor (PWR) and steam generators, as operated, for example, at Obrigheim.
Water treatment for: 1 = make-up water, 2 = cooling water, 3a = primary circuit/purification, 3b = primary circuit/chemical shim, 4 = fuel element cooling pond, 5a = secondary-/condensate-circuit, 5b = secondary circuit blowdown treatment, 6 = waste water.

some additional ones serving to regulate reactor functions. If the flow systems shown in Figures 2.114–2.117 appear at first to be very similar, the operating conditions for the ion exchangers differ greatly in actual fact, due to existing contamination or on account of certain products fed into the water circuit.

The following paragraphs are intended to provide some explanatory details on the subject of water treatment.

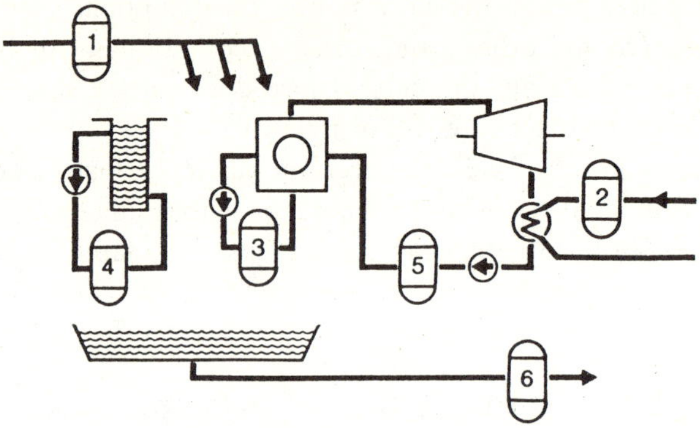

Figure 2.115 Principle of nuclear power stations with boiling-water reactor (BWR) and direct steam, as operated, for example, at Gundremmingen.
Water treatment for: 1 = make-up water, 2 = cooling water, 3 = primary circuit/coolant, 4 = fuel element cooling pond, 5 = steam-/condensate-circuit, 6 = waste water.

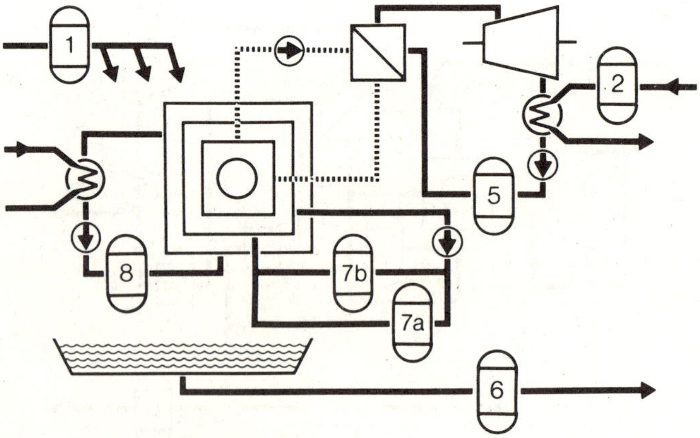

Figure 2.116 Principle of a nuclear power station with gas-cooled, heavy-water-moderated-reactor, as operated, for example, at Niederaichbach.
Water treatment for: 1 = make-up water, 2 = cooling water, 5 = secondary-/condensate circuit, 6 = waste water, 7 = D_2O-moderator/purification, 7b = D_2O-moderator/control device, 8 = shield coolant.

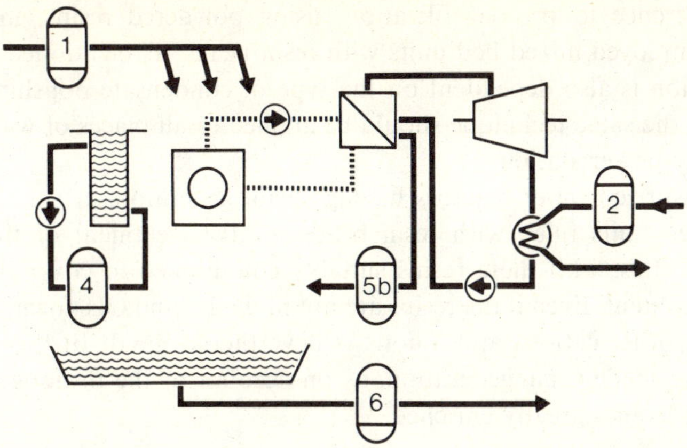

Figure 2.117 Principle of nuclear power stations with gas-cooled and graphite-moderated reactor of the Magnox type and AGR type.
Water treatment for: 1 = make-up water, 2 = cooling water, 4 = fuel element-cooling pond, 5b = steam generator-blowdown, 6 = waste water.

2.7.5.1 Make-up water

Much has been published on the possibilities for rendering underground or surface waters suitable as make-up water. Therefore, it suffices at this point to draw attention to the use of macroporous anion exchangers to ensure the essential purity and freedom from organic substances of the make-up water, and to the correct regeneration technique for mixed-bed polishing units to ensure minimum ion leakage.

2.7.5.2 Cooling water

In this area ion exchange resins play a secondary role as long as it is practical to operate a non-circulating water cooling system. However, it is important to be conversant with the cooling water circuit and the condition of the cooling water, as this has an influence on the treatment of the condensate.

2.7.5.3 Reactor coolant (primary circuit) purification

With boiling-water reactors (BWR) one distinguished originally between plants with and without steam converters. But nowadays the steam is always used directly. Special care must be taken to remove any products liable to cause corrosion and which might be introduced with the feed-water. For new projects, the tendency is to give preference to precoat filtration, using powdered resins, instead of the previously employed mixed-bed units with resin beads, as mentioned earlier.

The decision is also dependent on the type of condensate polishing units. It is obvious that the same technique should be applied to all places of water-treatment in the nuclear power station.

With pressurized water reactors having steam generators, it is conventional to use mixed-bed units filled with resin beads for the treatment of the coolant at normal flow rates, with these resins suitably conditioned to cover the boric acid and lithium content. Even if the resins are not in the H- and OH-form, the efficiency in adsorbing other cations and anions is nevertheless good. In this case also no provisions are made for regeneration, as on account of the prolonged service life the activity becomes greatly enriched.

2.7.5.4 Reactor coolant (primary circuit) treatment

This applies to pressurized-water reactors, because boiling-water reactors use demineralized water only, without any conditioning chemicals being added to the primary circuit. On account of the coolant being conditioned, two additional ion exchange units are required, apart from the purification unit: a cation exchanger with resin beads in the H-form to serve, when required, in removing lithium to regulate the pH-value, and for fixing lithium, while the fuel elements are being changed, as the reactor is flooded and has a connection with the fuel element cooling pond. Naturally, the same applies if the pH-value adjustment is carried out with potassium or ammonium ions; an anion exchanger with the resin beads in the OH-form serves to remove the boric acid from low concentrations.

2.7.5.5 Spent fuel element pond

The pond with fully demineralized water is open and in contact with the air. Hence, the water adsorbs carbon dioxide. With pressurized-water reactors, boric acid has been added. Due to these conditions, demineralization is out of the question. Filtration is the more important measure to be undertaken, because, for example, adhering corrosion particles can be dislodged during renewal of the fuel elements, which can cause cloudiness. With defective fuel elements it is furthermore important to make certain everything is efficiently decontaminated. In the first case, one would use a precoat filter, using inert precoat media, which can be replaced by powdered resin for the decontamination process as soon as it is necessary or economic. What mainly decides how the pond water will be treated is the overall design of the other purification installations. To simplify waste disposal, the aim will be a standard selection of resin beads or precoat filter media. If the choice is resin beads, the mixed bed should be filled with strongly acidic and strongly basic exchangers. The latter are converted immediately into the boric acid-form and then reacts well for removal of other strong anions. Boric acid and, of course, carbon dioxide will pass the unit without any concentration change and making no demand on resin capacity.

With gas-cooled nuclear reactors, for example the Magnox types operated in the U. K. or the types operated in France, the fuel-element cooling ponds contain as much as 200 ppm NaOH. To prevent corrosion of the fuel-element containers, which are made of magnesium alloy, the pH-value of the storage cooling water must be approx. 12. Under these conditions, however, sodium carbonate is formed by taking up carbon dioxide from the air, and the pH-value drops from 12 to 10. This calls for a demineralization by using strongly acidic cation exchangers and weakly basic anion exchangers, followed by a gas scrubber. Caustic is then re-added to the reclyced water. The high caesium activity necessitates the selective removal of this ion, which can be accomplished by using selective exchangers (see Section 2.7.1.1), and this will work effectively even if sodium is present in greater concentrations.

2.7.5.6 Condensate treatment

In boiling-water reactors with primary steam being used in the condensate circuit (Figure 2.115), it is essential to operate a 100% condensate treatment, so that, with the condensate being recycled, the specification for the reactor-water can be maintained. It has been established that precoat filtration with powdered resins is superior in filtration efficiency to treatment in a mixed bed filled with resin beads.

In later developments it was found that with systems demanding a cooling-water circuit, or where sea-water had to be used, the problem of capacity for salt in the condensate purification systems reappears in a more pronounced manner. For this reason a return to the use of mixed-bed units with bead resins can be anticipated. The addition of precoat filters here could create a longer interval before cleaning and regeneration of the mixed beds. Should salt break through, regeneration may have to be carried out at shorter intervals, but there would be no need to remove any products due to corrosion. Special regeneration techniques could be adopted, re-using the regenerants, thus helping to reduce the water consumption. Apart from the technical advantages, the decision here is also likely to be governed by economic factors.

2.7.5.7 Steam generator blowdown treatment

For a long time it was considered unnecessary to include a condensate-treatment plant in the secondary circuit of pressurized-water reactors. Some nuclear power stations, however, began to introduce the demineralization of the blowdown. This gained attention due to the fact that the water was then conditioned again with sodium phosphate or ammonia. No reliable data have become available, either on the purification effect exerted by condensate polishing mixed bed units under these conditions, or on blowdown demineralization, in particular with reference to what happens if the cycle is continued beyond the point when the sodium, ammonia and phosphate breakthrough. The idea for the demineralization of the blowdown has been a mixed bed unit filled with resin beads, preceded by an electromagnetic filter [22], the former being regenerated on exhaustion. Size and throughput vary within the limits of conventional performance.

2.7.5.8 Secondary circuit treatment

The treatment of water in a secondary circuit takes into account the principles of construction followed by the companies which make the steam generators or steam converters. Condensate treatment plants have so far only been used in exceptional circumstances.

The concepts presented in Figure 2.114 under points 3a and 3b have been transferred to a heavy-water circuit in Figure 2.116, which is the flow-sheet of nuclear power stations at many places in the world. The ion exchange resins being used at points 2.116a and b are the same – they are merely converted into the D_2O-form prior to use.

The process of purifying the circuit to cool the shields (see Figure 2.116, point 8) raises some special problems, because various products have been added to the water in an endeavour to inhibit corrosion. The temperature also rises – at times quite considerably. For these reasons the purification unit with the mixed-bed exchangers is less effective. As adequate experience gained from actual applications is still rare, it is not possible to make meaningful comments about the installation or its function.

2.7.5.9 Waste water treatment

Following collection and differentiation of the waste waters, a decision concerning further treatment is necessary. If direct disposal is not practical, it will be necessary to install precoat filters, using inert media or powdered resins, as well as mixed-bed units with resin beads, to perform the decontamination. Ion leakage is given special attention here, and the aim is to make the best possible use of the capacity of the ion exchangers.

To reduce the waste water volume during regeneration, provisions will be made for recycling of the regenerants, using well-established conventional techniques.

2.7.6 Special treatment of liquids

Besides the water circuits described in the previous sections there are many other water streams which need treatment, sometimes continuously, sometimes at intervals, either within the circuit or on disposal. The examples will be discussed because their application has proved to be effective and economic.

2.7.6.1 Removal of oxygen

Many years of practical use have shown the reliability of the catalytic removal of oxygen from boiler feedwater and stator cooling circuits. The oxygen in the water reacts with the hydrogen added on the palladium-coated surface of the catalyst resin Lewatit MC 145 [23] to form water

$$O_2 + 2\,H_2 \xrightarrow{\text{Lewatit MC 145}} 2\,H_2O\,.$$

The residual oxygen content can be reduced to < 20 µg/l during this process.

It can be used simply and economically even at elevated temperatures and with water containing salts, which may be present in heating circuits.

The system used in industrial plants for the catalytic removal of oxygen basically consists of the following:

hydrogen-metering and regulation device,
mixing section,
unit containing the catalyst resin.

This system can be incorporated into any circuit for continuous use, even at temperatures up to 70 °C. It is advisable to connect a mixed-bed unit downstream to adsorb small quantities of ions that have been split off (e. g., chloride, amines). It is also possible to simply include a cation exchange resin if only amines are to be adsorbed [24].

2.7.6.2 Removal of hydrazine

The same catalyst resin as applied for oxygen removal by dosing hydrogen can be used to destroy hydrazine by adding hydrogen peroxide. Experience in nuclear plants with a movable unit for the catalyst resin have proven suitable for overcoming problems on disposal of waters conditioned with hydrazine from special circuits [25].

Tab. 2.45 Survey of decontamination processes used by Kraftwerk Union AG, Erlangen, West Germany

Processes	Electrochemical ELPO	MOPAC	Chemical OZOX/CORD	CAPA	Mechanical
Materials	Metallic materials	Austenitic CrNi steels, Ni alloys, 17% Cr steels	Austenitic CrNi steels, Ni alloys	Unalloyed and low-alloyed steels	Metallic and non-metallic materials
Applications	PWR and BWR: system sections, removed components		PWR: primary systems, steam generator BWR: recirculation piping	BWR: main steam system, components	PWR and BWR: accessible system areas, removed components
Process principles, features and essential chemicals	Electrochemical removal: direct current, phosphoric acid as electrolyte	2-step process: oxidation with permanganate, decontamination with citric acid, oxalic acid	Multi-step process: oxidation with permanganic acid, reduction using oxalic acid; decontamination with oxalic acid	Single-step process using citric acid and formic acid	Grinding, blasting, brushing, lancing
	Short decontamination time, small amounts of waste	Short decontamination time	Very low chemical concentration (ppm region), minimum waste		

2.7.6.3 Decontamination of solutions

The increasing age of nuclear power plants together with additional inspection requirements are reasons for the importance of decontamination within the frame work of procedural development for inspection and repair of nuclear power plant components and entire systems [26].

Since 1976 numerous decontamination programs were carried out in German, other European and US nuclear power stations [27]. The processes being used are shown in Table 2.45.

A main aim of any type of process is waste treatment. If there is no sufficient capacity or equipment in the radwaste treatment systems the spent decontamination liquids can be passed through separate ion exchange units. In all these cases the aim is to obtain water which needs no further treatment before discharge or which can be used for subsequent cycles. Due to the extremely low chemical concentration of the rinse waters (in total approx. 2000 ppm) the amount of spent resin is of minor importance. Regeneration will be avoided because the resin volume is always smaller than any regenerant, even with subsequent concentration.

The removal of cations is mainly done with the highly regenerated nuclear grade resins in hydrogen-form as previously described. But there are experiments in progress to use chelating resins to remove only the polyvalent cations, e. g., the heavy metal ions. The complexing activity must of course be larger than the complex-forming tendency of the decontamination chemicals, such as citric acid.

The removal of anions from the decontamination liquids cannot be done with the OH-form but can be done with the loaded form of the main component anion of the solution. There is very often the possibility of advantageous selectivity. If, e. g., the citric acid- or oxalix acid-form is maintained other anions or anionic complexes of metals are more strongly bound. Therefore the exhaustion of anion resins is arrived at when the decontamination is lowered and not when the break-through of process anions occurs.

For the ion exchange resins it is recommended to use separate cation and anion resin units and not mixed-bed units in these cases. The individual use of resins allows for refill of one type only when necessary, while with mixed beds the exhaustion of one component makes it necessary to discharge the second one as well even before its exhaustion.

References

[1] T. Kashiwai, S. Kuroda, T. Hattori and H. Miyamoto (1983) The Behaviour of Impurity around Ion Exchanger in PWR Primary Circuit. Proc. Int. Conf., BNES (Bournemouth, 1983) *3,* Nr. 49.

[2) D. Bradbury and T. Swan (1980) Radioactive Waste Volume Reduction by Seqential Ion Exchange – Caesium Regeneration. Proc. Int. Conf., BNES (Bournemouth, 1980) *2,* Nr. 28, 209–212.

[3] Bayer AG, FR Germany (1986) Lewatit DN/KR. Product Information Nr. 5 – 3020.
[4] F. Martinola (1973) Active Filtration – Powdered Ion Exchangers and Adsorbents as Pre-coat Media. Filtration and Separation *10*, 420 – 423.
[5] A. Warshawsky (1974). Transactions of the Institution of Mining and Metallurgy *C*, *83*, June 1074, C 101 – 104.
[6] H. W. Kauczor and A. Meyer (1978) Structure and Properties of LEVEXTREL Resins. Hydrometallurgy *1*, 311 – 317.
[7] R. Kroebel and A. Meyer (1974). I. S. E. C. (Lyon) *56*, 2095 – 2107.
[8] J. Pauwels and K. F. Lauer (1968). Z. anal. Chem. *240*, 225 – 228.
[9] Strongly Basic Anion Exchange Resin and Production Method (1976) Rohm and Haas Co., Philadelphia, (1976) DOS 26 25 683, (1975) US 5866766.
[10] G. R. Hall, M. Streat and G. R. B. Creed (1968) Ion Exchange in Nuclear Chemical Processes. Trans. Instn. Chem. Engrs. *46*, 53 – 59.
[11] G. Morhocio and V. Kramer (1968). J. Polymer Sci. *C*, *16*, 4185 – 4195.
[12] G. J. Moody and J. D. R. Thomas (1972). Lab. Practice *21*, 10, 717 – 722.
[13] H. W. Venderbosch, A. Snel and L. J. Overman (1971) Zur Kapazität von Ionenaustauschern bei der Entfernung von Spurenverunreinigungen aus Wasser. (Capacity of Ion Exchangers during Removal of Trace Substances from water). VGB Conference, Essen, 1971, pp. 25 – 31.
[14] J. Rozenberg and O. Paoli (1973) Rapport DRA/SAECNI/73.187/LD/LC dated 29. 3. 73.
[15] H. R. Brost and F. Martinola (1980) Exchange Processes in Mixed Bed Units. VGB Kraftwerkstechnik *60*, 53 – 62.
[16] W. Bähr, S. H. Drobnik, W. Hild, R. Kroebel, A. Meyer and G. Naumann (1975) DOS 23 63 475, 2. Dec. 1973.
[17] R. Kroebel, A. Meyer, G. Naumann and D. Rittscher (1977). Schweizer Maschinenmarkt, *8*.
[18] H. Baatz (1976) Treatment of radioactive wastes from nuclear power stations for final disposal. Atom und Strom, *22*, 103 – 104.
[19] R. A. Vincent (1970) Experiences for application of powdered ion exchange resins. VGB Conference, Essen, 1970, pp. 35 – 40.
[20] J. Balthazar (1971) Intensive trials for continuous condensate polishing with powdered ion exchange resins. VGB Conference, Essen, 1971, pp. 47 – 56.
[21] K. Hoffmann (1971) Plant experiences with precoat-filters for condensate polishing. VGB Conference, Essen, 1971, pp. 56 – 62.
[22] H. G. Heitmann (1968) Methods and developments in treatment of condensate and make-up water. VGB Conference, Essen, 1968, pp. 43 – 53.
[23] F. Martinola, S. Oeckl and P. Thomas (1985) Catalytic reduction of oxygen in water. Vom Wasser *65*, 163 – 172.
[24] F. Martinola and P. Thomas (1980) Saving energy by catalytic reduction of oxygen in feedwater. 41st Annual Meeting Int. Water Conference, Pittsburgh, PA, Paper No. IWC-80-9.
[25] H. Hepp and G. Jacobi (1984) Entsorgung von Hydrazinlösungen höherer Konzentration durch chemische Zersetzung. VGB Conference, Essen, 1984, pp. 22 – 30.
[26] E. Schomer (1984) Verfahrensentwicklung für Prüfung, Dekontamination und Brennelement-Überwachung. Atomwirtschaft *29*, 142 – 144.
[27] W. Morell, H. O. Bertholdt, H. Operschall and K. Fröhlich (1986) Dekontamination – Stand der Technik und aktuelle Entwicklungsziele. VGB-Kraftwerkstechnik *66*, 579 – 588.

2.8 Electroplating Industry and Metal Recovery

Harold G. Fravel, jr.
The Dow Chemical Company, Separation Systems, Larkin Laboratory, Michigan, USA

Introduction

The cation and anion exchange ability of ion exchangers make them extremely important to the metal finishing industry and in numerous applications in which metal recovery is needed. The literature abounds with reports of ion exchange removal of metal ions and metal ionic complexes in a variety of solutions. Recent regulations and pollution control laws have highlighted the heavy metal arena. Heavy metals in ionic form pose serious health concerns in some cases due to acute toxicity or by their accumulation in tissue which can be linked to chronic illnesses. Non-biodegradability of these metals make conventional water treatment ineffective. Dilution of process streams is not an acceptable option. The metal ions must be removed and the use of ion exchangers offers a possible solution.

Volume reduction of the waste stream with a minimum of space requirements is a feature of ion exchange. Although the amount of metal removed from the waste stream remains the same, the concentration is increased and, in many cases, the volume for disposal is substantially reduced.

In the metal finishing industry ion exchangers are used for the purification of spent plating and processing baths for reuse by removal of metal contaminants. In some instances, the recovered metals are reused as a cost and waste-reduction step. Often ion exchangers are used as a polishing step in waste water treatment before the stream is discharged from the plant.

Purification of deionized water from rinse solutions for reuse utilizes ion exchangers. Likewise, the reconcentration of bath constituents dragged out with the treated parts and contained in used rinse solutions is an important step. Water is becoming a valued commodity and ion exchangers are playing a bigger role in close-loop systems.

Metal recovery with ion exchangers is applied in most areas in which metals are present. Many analytical applications exist in which ion exchangers are either used to concentrate dilute metal constituents or as chromatographic media for metal separation prior to detection. Generally, these methods involve multiple elution schemes or expensive eluents to remove the metals from the ion exchanger. As a result, these methods are not used in large-scale systems. The use of ion exchangers for hydrometallurgical recovery is well-known for uranium (see Chapter 2.13). In addition, systems for the recovery of gold, silver and other precious metals are being investigated.

The applications for ion exchangers in metal recovery are limitless. Table 2.46 lists a selection of examples that reflect the diversity of uses. Ion exchangers encompass several different types of materials; this chapter however will only be concerned with synthetic, polymeric ion exchange resins, which are the dominant media used in these applications.

Table 2.46 Applications for ion exchangers in metal recovery

Application	Reference
Cleanup of polluted estuary using resins to remove heavy metals from sediment	[1]
Purification of gelatin by metal removal	[2]
Purification of brine for chlor-alkali process	[3]
Wastewater treatment from pulp and paper mill	[4]
Wastewater treatment from manufacture of batteries	[5]
Wine clean-up by metal removal	[7]
Heavy metal recovery from dental clinic waste water	[8]
Clean-up of soil to allow plant growth	[9]
Treatment of locomotive cooling system water	[10]
Photographic waste solution treatment	[11]
Recovery of metals from manganese nodules	[12]

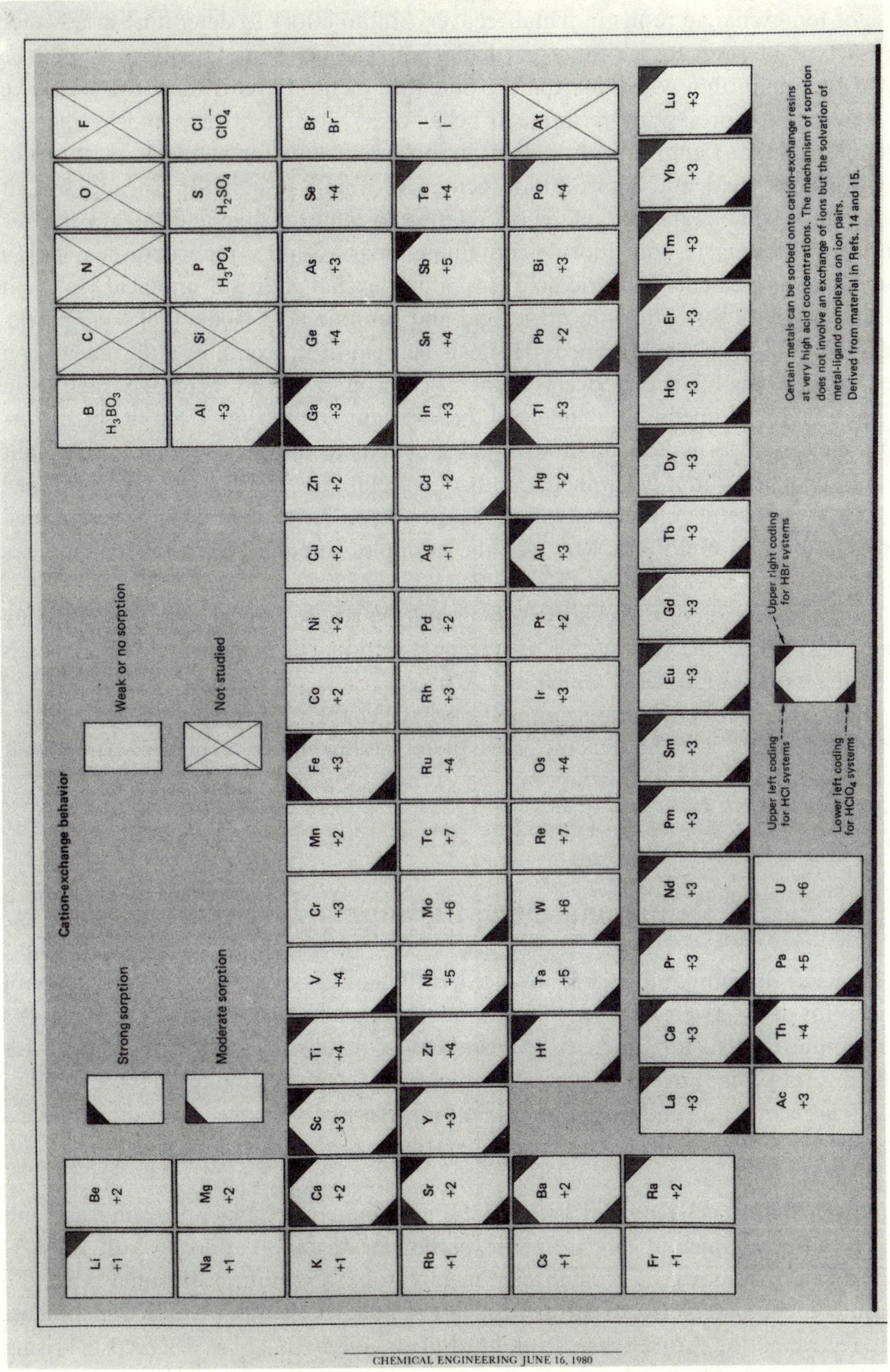

Figure 2.118 Cation-exchange behavior in strong acids (8 – 12 molar). (Excerpted by special permission from Chemical Engineering (June 16) copyright (c) (1980), by McGraw-Hill, Inc. New York, NY).

Examination of the chemical literature reveals thousands of references to the use of ion exchange resins in metal recovery. In an effort to determine in advance what type of resin to try for a particular application, some attention should be paid to equilibrium constants for the potential metal. Several excellent lists have been compiled presenting resin/metal behavior. A document prepared by the Argonne National Laboratory lists distribution coefficients for metals in mixtures of hydrochloric acid and hydrofluoric acid with anion resins [13]. In addition, it contains 404 bibliographic references relating to ion exchange and metal behavior in hydrofluoric acid and fluoride-containing media. Similar reports discuss cation exchange of elements in nitric acid [14], anion exchange in sulfuric acid [15], and resin selectivity for cations in nitric acid and sulfonic acid media [16]. One particularly useful article by Kennedy [17] can be helpful in examining the use of ion exchange resins for the recovery of metals. Through the use of charts such as in Figure 2.118, Kennedy has condensed the large amount of data from the literature. One can use it to make a qualitative judgement as to the type of ionic species that might be present in a given media and get a feeling for relative selectivity values.

Two very useful volumes on ion exchange and pollution control have been edited by Calmon and Gold [18]. Many of the examples summarized in this chapter are presented in greater detail in their work.

The first section of this chapter will discuss the electroplating and metal-finishing industry uses of ion exchange resins. The next section will examine individual metals and the variety of areas in which ion exchange resins are used.

A discussion of chelating resins will close the chapter. Through their use, greater metal selectivity is found in some cases, thereby allowing for metal separations that were not possible with conventional ion exchange resins.

2.8.1 Electroplating and metal finishing industry

The metal-finishing industry is involved with coating surfaces with thin layers of metal for decorative reasons to enhance their appearance, or for protective reasons to minimize corrosion, and in the treatment of metal surfaces in preparation for some form of coating. However, the processes are susceptible to contaminants in the treatment baths and rinse-down solutions. Basically, the metal-finishing process can be viewed as three separate unit operations, as shown in Figure 2.119.

Unit A: Plating or treatment bath. Metal contaminants build up during use and may cause detrimental effects in the plating or metal treatment. Bath reuse is improved by removal of these foreign metals from the solution. In some cases, the bath content is depleted to an ineffective level and the valuable bath constituents are recovered for future reuse. Prolonged life of the bath will save on costs arising from the purchase of new chemicals and will decrease the overall disposal needs.

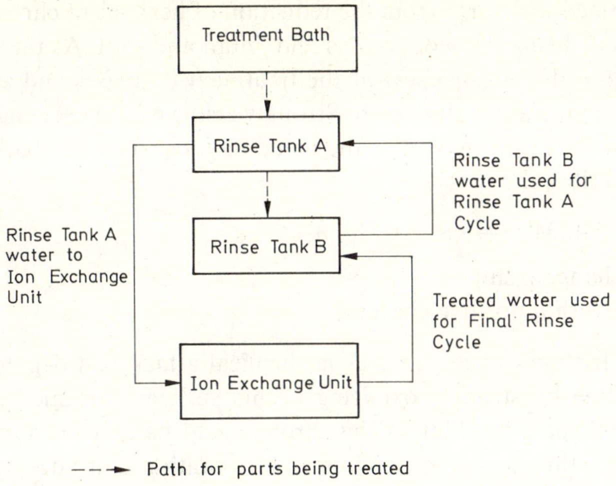

Figure 2.119 Basic metal finishing process flow sheet.

Unit B: Rinse solution. As part of the process, treated parts are moved from one operation to the next. In most cases, these parts drag-out some of the prior treatment solution and a rinse cycle is necessary before the next station. The drag-out will be composed of the treating-bath constituents and whatever metals were in it at the time of treatment. For many applications, the rinse water must be of sufficient purity to effectively clean the treated part. In actual practice, the parts go through a series of rinse stations with each effluent segregated. In this manner, the effluent from the second station can be used during the next cycle to rinse the grossly contaminated parts in the first station. Therefore, the first rinse station effluent must be treated and the purest water is used in the last station.

Unit C: Plant waste streams. Spent baths, contaminated rinse solutions, plant run-off and other streams containing ionic species must be treated down to acceptable levels before they may be discharged from the facility or disposed of by an appropriate means. In many cases, final treatment of the stream consists of using ion exchange resins to polish the ionic species down to within disposal limits.

Metal finishing and electroplating applications may encompass one or more of these units. The next section will discuss several specific examples in which ion exchange resins have been used.

2.8.1.1 Chromium plating and treatment

Plating and anodizing baths. Chromic acid solutions are used for chromium plating, electrolytic stripping of copper, anodizing of aluminum and passivation of cadmium, magnesium and zinc, as well as used in the tanning industry. Eventually, the baths become contamined with Cr^{3+} from the reduction of hexavalent chromium, as well as with the metals being treated, such as aluminum and iron. As the level of these materials increase, the effectiveness of the treatment decreases and at some point the bath must be purified or discarded. Strongly acidic cation exchange resins have been used in systems for removal of the metal cations, since the baths are at low pH. Eq. (1) describes the process:

$$nR^-H^+ + M^{n+} \rightarrow nR^-M^{n+} + nH^+ \tag{1}$$

R = cation exchange resin
M = metal ion with n positive charges.

One limitation for ion exchange resins is chemical attack and degradation of the polymer backbone by strongly oxidizing agents such as chromic acid solutions. This may be overcome by diluting the chromic acid baths from typical levels of 350 g/l to 175 g/l; this, however, requires an evaporation unit later in the process. Recent reports by Pawlowski [19, 20] describe the use of macroporous resins in undiluted systems with no ill-effect. This eliminates costly concentration operations that would otherwise be necessary. Generally, aluminum anodizing baths contain 100 g/l chromic acid and are treated directly with cation exchange resins. Removal of the cationic impurities allows reuse of the somewhat diluted bath solution in which the chromium is present as an anionic species.

The cation resin bed is usually loaded to exhaustion and regenerated with sulfuric acid. The highly concentrated first portion of the eluate is discarded from the system; the latter fractions, however, are saved for the next regeneration cycle. In this manner, the total waste stream volume is reduced for some form of final treatment prior to discharge from the facility. Solution utilization optimization is an area that can make a treating system most economical.

Rinse water. The rinse water effluents will contain various concentrations of the treating bath solution dragged out by the parts. All bath contaminants will be present, in addition to the chromate. Chromate recovery from these solutions is effected using a strongly acidic cation resin and a strongly basic anion resin. The cation resin will remove cationic impurities, whereas the anion resin is used to recover chromate for possible reuse. In this manner the rinse water is purified and available for reuse. This is becoming more important in many areas of the world where water is a valuable item.

As shown in Equations (2) and (3), the anion bed will recover mono- and dichromate. After regeneration with sodium hydroxide, the eluent goes to the cation resin where a proton is exchanged for the sodium ion. In this manner, chromic acid is recovered for further use:

$$2\overline{R}OH + CrO_4^{2-} \rightarrow \overline{R}_2CrO_4 + 2OH^- \tag{2}$$

$$\overline{R}_2CrO_4 + CrO_4^{2-} + H^+ \rightarrow \overline{R}^2Cr_2O_7 + OH^- \tag{3}$$

$\overline{R}$ = anion exchange resin.

Some recent reviews by Pawlowski [19, 21] detail several systems for the treatment of chromium-containing solutions.

2.8.1.2 Recovery of nickel

Nickel plating is used for decorative and protective coating applications, particularly those in which corrosion resistance is important. Soluble nickel is recovered from nickel-plating rinse solutions by using a cation exchange resin in the hydrogen form. Operating capacities of 19 – 28 g Ni/l of resin are found. The resin is regenerated with 96 – 160 g H_2SO_4/l of resin as a 10% H_2SO_4 solution, and nickel concentrations as high as 50 g/l have been obtained in the eluent [19]. Nickel is recovered as nickel sulfate and reused to load the plating baths. Reclaim of the water for reuse as the rinse solution results from passing the acidic cation resin effluent through a carbon bed to remove organics and then through an anion bed to remove sulfates, chlorides and borates. The carbon bed is necessary to prevent fouling of the anion bed by the organic material. Recovered nickel reuse is only practical if rinse solutions are segregated from other plating streams. Figure 2.120 shows a typical nickel recovery and water-recycle flowsheet. In many metal recovery systems stream separation is important for maintaining recovered metal purity for reuse. If the metals are being removed for eventual disposal, such efforts may be unwarranted.

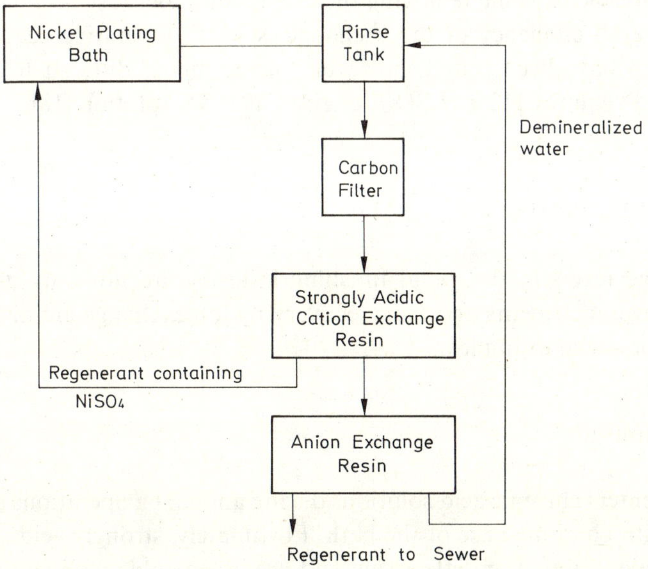

Figure 2.120 Nickel and water reuse from a nickel plating rinse system.

2.8.1.3 Phosphoric acid recovery from pickling wastes

High concentrations of phosphoric acid are used to pickle steel or to bright finish aluminum. Pickling removes metal oxides that have formed on the surface of the metal. Aluminum bright dip solutions are used to finish parts to level minute scratches and obtain a bright surface. During use, the acid becomes concentrated with metals such as aluminum and iron. Clean-up of the 10 – 15% phosphoric acid can be effected by passage of the solution through a strongly acidic cation resin in the hydrogen form to remove the metal cations. Regeneration of the resin is accomplished with sulfuric acid.

A second option is to use the acid-retardation technique in which strong acids can be separated from their salts by passage through a strongly basic anion resin [22]. The salts are eluted ahead of the acid which is "retarded by the resin." Elution of the acid is performed with water. Cycling of acid/salt solution and water can be used as a means of purifying a phosphoric acid stream for reuse.

2.8.1.4 Acetic acid-nitrate pickling of magnesium sheet

Rolled magnesium sheet is pickled using a 28% acetic acid/8% sodium nitrate solution which with use becomes contaminated with dissolved magnesium. As the oxides dissolve into the solution, magnesium acetate is generated and slows the pickling rate, as well as adversely affecting the finished surface of the treated metal. A strongly acidic cation resin in the hydrogen form is used to remove the metal from the pickling bath. Passage through the resin converts the acetate and nitrate salts to the free acids while removing the magnesium. Because acetic acid is a weak acid, the overall efficiency of the exchange is good. Operating capacities of 16 g Mg/l of resin have been found; however, magnesium is difficult to remove from the resin and requires 112 g H_2SO_4/l of resin as a 5N solution [18].

2.8.2 Metal recovery

Ion exchange resins in the metal finishing industry are not only used for metal recovery. Several elements and their recovery by ion exchange are discussed below, with some practical examples.

2.8.2.1 Aluminum

Aluminum enters chromic acid solutions during anodizing operations and can cause problems with continued use of the bath. Fortunately, strongly acidic cation resins remove trivalent aluminum effectively and are regenerable with sulfuric acid. The fact that the chromic acid concentration is lower than in plating baths allows direct treatment of the solution without dilution. This eliminates dilution tanks and reconcentrators, and saves on overall process costs.

In a metal coloring process, Ni^{2+} and Al^{3+} were contaminating the bath. A chelating resin was used to remove those cations. It was found that aluminum could be displaced by the nickel. The bed was treated with a nickel salt solution to elute off the absorbed aluminum. Finally, sulfuric acid was used to regenerate the nickel-containing resin to obtain a high purity nickel sulfate solution. [23]

Removal of aluminum from phosphoric acid solutions was discussed in Section 2.8.1.3.

2.8.2.2 Copper

There are several well-documented cases for copper removal utilizing ion exchange resins and this has been important in the Rayon fiber industries that employ the cupric ammonium process to help produce rayon from cellulose. In the process, a variety of waste solutions are generated containing copper at different levels and degrees of acidity. Weakly acidic cation resins are used to recover the copper; they are reused after regeneration with sulfuric acid. The plant effluent can be used for nutritive irrigation since it contains ammonium sulfate, but the copper must be removed to low levels.

Copper plating baths, containing sulfate or chloride, use ion exchange resins for decationization and deionization of the water, as well as carbon beds to remove organic contaminants.

Several chelating resins have been developed that show a high affinity for copper and are elutable with sulfuric acid or ammonia. These resins demonstrate high iron rejection and may find application in some hydrometallurgical operations [24]. A more detailed discussion is given in Section 2.8.3.

Modine Manufacturing reported a copper recovery system that extends the life of a copper bright dip solution and provides high grade copper for resale [25]. The system continually removes metal ions that build up in the solution and plates out the copper. They used an acid-process unit that separates the solution into two streams, one that is high in hydrogen peroxide and sulfuric acid, while the other is high in copper. The system uses small amounts of a proprietary resin and is indicative of several commercial units available that operate in short cycles to effect separations. It is claimed that these units result in a simplified operation and reduce waste treatment costs. In cases where copper plating baths contain cyanide, an anion resin can be used to recover the copper cyanide complex.

2.8.2.3 Gold

The lure of gold has generated a substantial amount of work with ion exchange resins. Gold is used in plating applications, jewelry manufacture, printed circuit board production, and a myriad of other applications. Gold from seawater has attracted interest; however, simple calculations indicate a poor economic return. Plating effluents contain gold in the cyanide complex $[Au(CN)_2]^-$. This anionic complex is collected on a strongly basic anion exchange resin. Elution of the gold is possible with various non-conventional schemes; generally, however, the resin is dried and incinerated to recover the gold. Because of the high operating capacity and high value of gold, the resin costs are insignificant.

For removal of gold from highly acidic solutions a polyisothioureanium resin can be used. Although the resin can be regenerated with thiourea or thiocyanate or perchlorate solutions, it is usually burned at 900 – 1000 °C to recover the gold [24].

In the mining industry, activated charcoal has been used to recover gold from the cyanidation process. Work is being done on the development of an economical resin to recover gold [27]. Resins offer the advantages of faster kinetics, lower energy demand regeneration and higher attrition resistance.

2.8.2.4 Iron

Because of its widespread use in equipment and its abundance in materials, iron is a very common contaminant in process streams. Removal of iron(II) (0.1 – 1%) from hydrochloric acid or phosphoric acid has been effected using strongly acid cation resins. The resin is regenerated with sulfuric acid.

In stronger chloride-containing solutions, the iron(III) chloride anion ($FeCl_4^-$) is removed from HCl or $AlCl_3$ using a strongly basic anion resin in the chloride form [22]. When the resin is eluted with water, the complex dissociates and the neutral $FeCl_3$ is removed from the resin. This method has been used as a means of obtaining hydrochloric acid as a high purity product:

$$FeCl_3 + Cl^- \rightarrow FeCl_4^- \quad [4]$$

$$\overline{R}Cl + FeCl_4^- \rightarrow \overline{R}FeCl^4 + Cl^- \quad [5]$$

$$\overline{R}FeCl_4 \xrightarrow{H_2O} \overline{R}Cl + FeCl_3 \quad [6]$$

$\overline{R}$ = strongly basic anion exchange resin.

In the appliance industry, phosphoric acid pickling before coating with porcelain enamel is preferred. Build-up of iron(II) leads to maintenance problems mainly due to metal precipitation. It was found that maintaining iron(III) at a low level minimized iron(II) formation. High-capacity, strongly acidic cation resins are used for iron removal and can be regenerated with sulfuric acid [28].

2.8.2.5 Lead

The severe toxicity of lead requires it to be reduced to ppb levels in waste water. A phosphoric acid cation resin has been used to obtain water containing 50 ppb lead. Although it was very selective for lead it was kinetically slow and required low flow rates to be efficient. In addition, nitric acid was used for regeneration due to the solubility properties of the lead salt [29].

In a separate example, a weak base anion exchange resin in the free base or sulfate form was used to remove lead from waste water with a solution pH 4–6. Lead acts like iron in a chloride solution and forms a chloride complex that can be removed with a strongly basic anion resin.

2.8.2.6 Mercury

Mercury poisoning as a result of pollution occurred in an incident in Japan. Fish bioaccumulated the mercury from water that was being contaminated from manufacturing plants. People who then ate the fish experienced the severe effects of mercury poisoning. Most of the mercury problems have come from the manufacture of chlorine and caustic that use mercury cathode cells. Recent developments in this area, such as in membrane technology, will lessen the danger of mercury contamination. Mercury levels in the effluent from chloralkali plants must be in the ppb range. The waste water is neutralized, oxidized to remove colloidal mercury and polished with anion ion exchange resins to remove the anionic mercury complex. An aqueous sulfide solution produces soluble mercury sulfide, which is removed from the resin as part of the regeneration process. Eluents from these resins are rich in mercury that can be collected with an electrolytic circuit. In this manner, the mercury is recovered for reuse rather than discharged to the environment.

Resins containing a sulfur atom seem particularly well-suited for mercury removal. Mercury is generally present as a chloride complex and the thiol group has a high affinity for Hg^{+} and $HgCl^{2+}$. Total capacities of 240 g Hg/l resin have been reported. Regeneration of the resin is accomplished with 35% hydrochloric acid [30].

2.8.2.7 Silver

Silver is mainly used in plating and photographic applications. Plating operations use the silver cyanide complex which, because of its anionic character, can be recovered from spent solutions on a strongly basic anion resin. Recovery capacity has been found in the range of 40–50 g of silver per liter of resin.

The largest area for silver recovery is the photographic industry, e. g., in photographic fixing solutions or in rinse water from washing film. Effluents containing the silver thiosulfate complex $[Ag(S_2O_3)_2]^{3-}$ are treated with strongly basic anion resins that can be regenerated with ammonium chloride [31]. In some cases, a weakly basic anion resin is used for silver recovery. Although the operating capacity is lower for weak base resins, they are easier to regenerate with alkaline solution.

2.8.2.8 Tin

In one application, waste water containing 50 ppm $SnCl_2$ was treated with a strong acid cation resin to obtain effluents with 0.1 ppm tin. The loaded resin was regenerated with 5% sulfuric acid [32].

Electrolytic coating operations may contain tin, and have used strongly acidic cation resin followed by a weakly basic anion resin to remove tin. The anionic complexes were eluted from the weak base resin with sodium hydroxide solution.

2.8.2.9 Vanadium

Pawlowski describes an ion exchange method that has been used for the recovery of vanadium from the effluent of zirconium-vanadium pigment manufacture. The effluent is passed through a strongly acidic cation resin and a weakly basic anion resin. The resins are regenerated with 6% HCl and 15–20% NH_4OH, respectively. Recovered V_2O_5 is recycled to the pigment manufacturing process after elution from the anion resin, evaporation and calcination [33].

In a separate chemical process, adipic acid is produced by the oxidation of cyclohexanol and cyclohexanone with nitric acid in the presence of copper and vanadium catalysts. The copper and vanadium are in cationic form and are removed with a strong cation exchange resin. The resin is eluted with fresh nitric acid and fed back to the reaction stage. High-strength nitric acid is used and the resin must have high oxidative stability coupled with high capacity. Nitric acid and other strong oxidizing agents can cause explosive reactions when mixed with organic materials such as ion exchange resins, and must be handled appropriately.

2.8.2.10 Zinc

The viscose process for making rayon utilizes zinc and generates a waste stream consisting of sulfuric acid and zinc sulfate with Zn^{2+} and Na^+ at levels up to 540 mg/l and 10 000 mg/l, respectively. A strong acid cation resin is used to remove the metal ions. A two-stage regeneration is used to improve resin utilization and 10% H_2SO_4 is used to elute zinc [19].

A process has been developed for the recovery of zinc from rejected galvanized parts. The "Metsep" process uses spent pickle liquors to strip zinc from those parts and absorbs the $[ZnCl_4]^{2-}$ onto a strong base anion resin. Elution with water reverses complex formation and neutral $ZnCl_2$ comes off the resin [34].

2.8.3 Chelating Resins

Inevitably, when one is faced with a metal removal problem, the question of a selective resin specific to the metal is raised. In most cases, a relative selectivity due to differences in the metals and their complexes is found; however, high selectivities are rare. The field of chelating resins lives from the dream of producing resins with such high selectivities. Chelating resins are prepared by anchoring a functional group to a polymeric backbone, either during polymerization as part of a reactive monomer, or after polymer formation by various chemical means. Generally, this functional group has been found to exhibit greater specificity and selectivity than the usual cation and anion resins have towards some metals. Their ability to remove some metals in the presence of other similar ions or in strongly acidic or alkaline media differentiate them from the more common resins. The fact that the metals are bound partly by chelation, rather than totally by ion exchange, explains some of the resin's activity. An excellent review by Sahni and Reedijk [35] lists the chelating resins that have been reported, and details the chemistry involved in the synthesis and usage of these materials.

Several chelating resins have become commercially available. The most common chelating resin is based on an iminodiacetic acid functional group; and has been prepared by several methods. These resins show a high specificity for heavy metal ions such as iron and copper. Since the functional group contains two carboxylic groups and a tertiary nitrogen, the resins are sensitive to pH and function well between pH 1.5 – 14. Regeneration consists of treatment with an acid followed by caustic to convert the resin to the sodium form.

Another type of chelating resin has been developed utilizing some weakly basic picolylamine derivatives to complex with various transition metals, especially copper [36 – 39]. The currently available resins use either bispicolylamine or *N*-(2-hydroxy-propyl)picolylamine as the functional group attached to a macroporous polystyrene-divinylbenzene copolymer, as shown in Figure 2.121. In addition to copper, metals

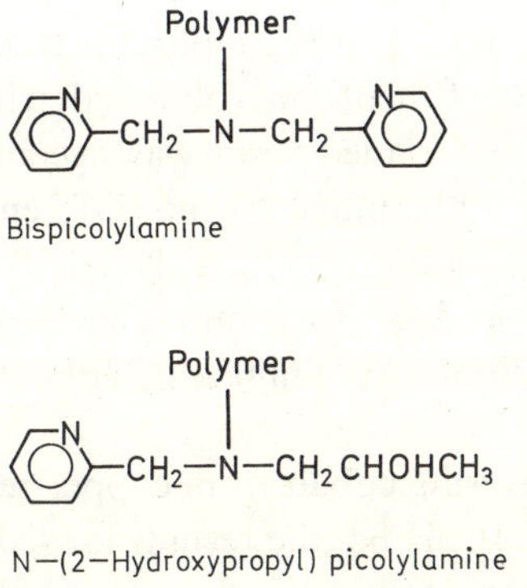

Figure 2.121 Structure of bispicolylamine and N-(2-hydroxypropyl)picolylamine resins.

Table 2.47 Conditional absorption constants (*K*) for the bispicolylamine resin [39]

Metal	pH	*K* (l/mol)
Cu(II)	2.0	1280
Ni	2.0	375
U(VI)	2.0	190
Fe(III)	2.0	181
Zn	2.0	82
	2.7	184
Co(II)	2.0	51
	3.2	280
Cd	2.0	43
	2.8	196
Fe(II)	2.3	23

such as cobalt, nickel and zinc can be removed with these resins [39, 41]. Table 2.47 shows the relative loading values of various metals for the bispicolylamine resin. As can be seen, there is a strong affinity for copper, even at low pH, whereas other metals have higher loading values at higher pH. Complexed metals can be removed with strong acid (10N H_2SO_4) or ammonium hydroxide. Some selective elution of the resin can be accomplished using different strength acid solutions for regeneration.

Although the two resins are very similar, it has been found that the *N*-(2-hydroxypropyl)picolylamine resin has a greater iron rejection while maintaining good copper loadings. This was particularly important in copper recovery from acidic leach liquors [42, 43].

Rosato et. al. [44] describe a process for the selective removal of nickel from acidic cobalt sulphate using the bispicolylamine resin. A variety of conditions were examined such as temperature, flow rate, feed concentration and acid strength for elution. Feeds of 15 – 30 g/l cobalt and 0.3 – 0.7 g/l nickel at pH 2.5 were treated. At pH 2.5 the resin has a high affinity for cobalt and nickel. The cobalt was rapidly absorbed, but with continued flow the nickel displaced the cobalt. Nickel was more strongly held than cobalt. Flow rate was found to be important since the nickel/cobalt exchange was slow. A cobalt-rich effluent was obtained with a gradual increase in nickel concentration. Elution of loaded resin was accomplished with sulfuric acid at several strengths, to take advantage of the difference in nickel-cobalt binding. Using a split elution technique, the cobalt-rich fractions were isolated to obtain a low nickel-content product. In a mini-plant operation with multiple columns and split elution, solutions containing Co/Ni > 500/l were repeatedly obtained.

Jeffers [45] describes a process for recovering cobalt from copper-recycling leach solutions using the bispicolylamine resin. At pH 3.0 the resin is loaded with several metals: cobalt, copper, nickel, iron and zinc. However, using 50 g/l sulfuric acid all but copper is eluted from the resin, which can then be removed with 2N ammonium hydroxide. Cobalt can then be extracted from the copper-free solution by solvent extraction. During the ion exchange process, cobalt is concentrated from 0.03 g/l in the feed to an eluent concentration of 0.46 g/l.

Cobalt is a strategic and critical metal that has a variety of applications. Perhaps the use of chelating resins such as those described above will make cobalt available from several new sources.

The search for new chelating resins continues with the goal of recovering valuable metals selectively. Many existing process streams contain low-level "impurities" that are currently unretrievable, but which, with the appropriate chelating resin, may some day be available for use.

References

[1] M. G. L. Van Hoek, P. J. F. Gommers and J. A. S. Overwater (1983) Removal of heavy metals from polluted sediments – ion exchange resins, a possible solution. Heavy Metal Environ., Int. Conf. 4th Vol. 2, 856–859.

[2] V. K. Nguyen and P. X. Van (1983) Refining of gelatin, Tap Chi Duoc Hoc *3*, 13–15.

[3] Y. Kataka and M. Matsuda (1982) New chelating resin for purification of brine. Sumitomo Kagaku Tokushugo *1*, 55–66.

[4] J. H. Fitch, Jr. (1983) Evaluation of ion exchange technology for toxic and nonconventional pollutant reduction in bleach plant effluents. Proc. Ind. Waste Conf. *37*, 425–433.

[5] U. M. Harder, K. D. Kaswinkel, J. W. Gould, J. K. West and M. J. Wynn (1982) Treatment of wastewater. U. S. Patent 4 341 636.

[6] G. J. Loubser (1983) Removal of impurities from beverages. S. African Patent ZA81/3988.

[7] M. H. Quick (1983) Recovery of metal values from organic reaction products. U. S. Patent 4 388 279.

[8] K. K. Nagata Denki Kogyo (1983) Heavy metal removal from wastewater from dental clinic. JP 58/180282.

[9] C. Van Assche, H. Corte, H. Heller and C. Hentschel (1976) Use of chelate forming ion exchange resins for substrate treatment. Ger. Offen. DE 2434593.

[10] J. Barcick, L. Pawlowski, D. Taranska, J. Niecko and S. Burdon (1978) Cleaning of locomotive coolant system water. Pol PL 96682.

[11] D. J. Degenkolb and F. J. Scobey (1979) Recovering heavy metals ions from dilute aqueous solutions. U. S. Patent 4 159 930.

[12] H. W. Kauczor, H. Junghanss, W. Roever and C. Kruppa (1973) Hydrometallurgy of metalliferous solutions in the processing of manganese modules. Interocean *1*, 469–473.

[13] J. P. Faris (1978) Separation of metal ion by anion exchange in mixtures of hydrochloric acid and hydrofluoric acid. Argonne National Laboratory ANL-78-78.

[14] S. F. Marsh, J. E. Alairid, C. F. Hammond, M. J. McLeod, F. R. Roensch and J. E. Rein (1978) Cation exchange of 53 elements in nitric acid. Los Alamos Scientific Laboratory LA-7083.

[15] F. W. E. Strelow and C. J. C. Bothma (1967) Anion-exchange and a selectivity scale for elements in sulfuric acid media with a strongly basic resin. Anal. Chem. *39*, 595.

[16] F. W. E. Strelow, R. Rethemeyer and C. J. Bothma (1965) Ion exchange selectivity scales for cations in nitric acid and sulfonic acid media with a sulfonated polystyrene resin. Anal. Chem. *37*, 106.

[17] D. C. Kennedy (1980) Predict sorption of metals on ion-exchange resins. Chem. Eng., 106 – 118.

[18] C. Calmon and H. Gold, eds. (1979) Ion Exchange for Pollution Control, Vol. I and II, CRC Press, Boca Raton, FL.

[19] B. A. Bolto and L. Pawlowski (1983) Reclamation of wastewater constituents by ion exchange. J. Eff. Water Treatment *23,* 157 – 167.

[20] B. A. Bolto and L. Pawlowski (1983) Reclamation of wastewaer constituents by ion exchange. J. Eff. Water Treatment *23,* 55 – 58.

[21] L. Pawlowski, M. Kotowski, B. A. Bolto and R. McNeill (1984) Reclamation of chromium from wastes. Stud. Environ. Sci. *23,* 491 – 512.

[22] M. J. Hatch and J. A. Dillon (1963) Acid retardation. Process Design and Development *2,* 253 – 263.

[23] S. Yamashita, Y. Sugihara, T. Miimura and H. Ueiima (1978) Nickel and recycling from aluminum electrolyzed wastewater. JP 78/78944.

[24] R. R. Grinstead (1984) Selective absorption of copper, nickel, cobalt and other transition metals from sulfuric acid solutions with the chelating ion exchange resin XFS-4195. Hydromet. *12,* 387 – 400.

[25] J. C. Egide (1985) Copper recovery system extends life of bright dip. Products Finishing, 72.

[26] T. E. Green and S. L. Law (1970) Properties of an ion-exchange resin with high selectivity for gold. U. S. Bur. Mines Rep. Invest., Rep. 7358.

[27] C. A. Fleming and G. Cromberge (1984) The elution of aurocyanide from strong- and weak-base resins. J. S. Afr. Inst. Min. Metall. *84,* 269 – 280.

[28] J. C. Hesler (1960) Recovery and Treatment of Metal Finishing Wastes by Ion Exchange Processes. Proc. 21st Int. Water Conf., Pittsburgh, PA, p. 89.

[29] J. Leden (1952) The use of an anion exchange resin to prove the presence of anionic complexes in some cadmium and copper salts. Svensk Kem. Tid. *64,* 145.

[30] K. Koerts (1975) Selective removal of mercury, lead, zinc, copper and silver by ion exchange resins. Proc. 2nd National Conf. Complete Water Reuse, American Institute of Chemical Engineering, NY, 260.

[31] D. J. Degenkolb and F. J. Scobey (1979) U. S. Patent 4 159 930. Process for recovering heavy metals from dilute aqueous solution.

[32] M. Tanaka (1975) Treatment of heavy metals-containing wastewater. JP 51/133957.

[33] L. Zagueski, L. Pawlowski and A. Cichocki (1980) Methods for recovery of water and vanadium compounds from wastewater. Physiochemical Methods for Water and Wastewater Treatment. Pergamon Press., pp. 229 – 236.

[34] A. K. Haines, T. H. Tunley, W. A. M. Te Riele, F. L. D. Cloete and T. D. Sampson (1976) The recovery of zinc from pickle liquors by ion exchange. National Institute for Metallurgy, Report No. 1575.

[35] S. K. Sahni and J. Reedijk (1984) Coordination chemistry of chelating resins and ion exchangers. Coord. Chem. Rvw. *59,* 1 – 139.

[36] R. R. Grinstead (1984) New developments in the chemistry of XFS-4195 and XFS-43084 chelating ion exchange resins. Ion Exchange Technology, 509 – 518; Ellis Horword Ltd.

[37] K. C. Jones and R. R. Grinstead (1977) Properties and hydrometallurgical applications of two new chelating ion exchange resins. Chem. Ind., August, 637 – 641.

[38] K. C. Jones and R. R. Grinstead (1976) New selective ion exchange resins for copper and nickel. Extractive Metallurgy of Copper, American Institute of Mining, Metallurgical and Petroleum Engineers, Inc.

[39] R. R. Grinstead (1984) Selective adsorption of copper, nickel, cobalt and other transition metal ions from sulfuric acid solutions with the chelating ion exchange resin XFS-4195. Hydrometallurgy *12,* 387 – 400.

[41] B. R. Green and R. D. Hancock (1982) Useful resins for the selective extraction of copper, nickel, and cobalt. J. S. African Institute of Mining and Metallurgy, October, 303–307.

[42] R. Grinstead (1979) Copper-selective ion exchange resin with improved iron rejection. J. Metals, *31,* 13–16.

[43] K. C. Jones and R. A. Pyper (1979) Copper recovery from acidic leach liquors by continuous ion exchange and electrowinning. J. Metals *31,* 19–25.

[44] L. Rosato, G. B. Harris and R. W. Stanley (1984) Separation of nickel from cobalt in sulphate medium by ion exchange. Hydrometallurgy *13,* 33–44.

[45] T. H. Jeffers (1985) Separation and recovery of cobalt from copper leach solutions. J. Metals, January, 47–50.

2.9 Treatment of Pickling Acids with Ion Exchange and Related Processes

George P. Herz
Newark, Notts., Great Britain

Introduction

Pickling acids are used in various areas of the metal production and treatment industries. They are acids used for the treatment of metal surfaces prior to galvanization, finishing or other surface treatment. Their specific function is the removal of scale, corrosion products and other substances formed as a result of oxidation, chemical attack, tempering, etc.

Acids commonly used for pickling include (but are not limited to) acetic, hydrochloric, hydrofluoric, nitric, phosphoric and sulfuric or mixtures of these acids.

This chapter deals specifically with the recovery and treatment of pickling acids by means of ion exchange and related processes.

Other methods, such as chemical precipitation, neutralization, chilling, incineration, evaporation, solvent extraction, etc., are not included in this discussion, nor are other facets of the metal industry such as plating baths and wastes.

2.9.1 Purpose of treatment

The purpose of the treatment roughly falls into one or more of the following categories:

recovery of the acid
recovery of metals from the acid or rinse water
reduction and/or improvement of waste water produced.

In the course of use, the pickling solution becomes contaminated with impurities, mostly consisting of the metals being pickled and their oxides, but also of organic substances having their origin in surface grease and oil. As the concentration of impurities in the acid increases, the pickling efficiency decreases. To maintain the pickling bath at a usable composition, the acid has to be replaced or rejuvenated at a rate corresponding to the rate of contamination. This can be accomplished by periodically replacing or rejuvenating the entire bath or by continuously blowing down or treating a given part of it.

The choice of the method used is a function of cost on the one hand and of environmental considerations on the other. Rejuvenation or recovery is obviously of greatest interest where the acid involved is expensive (e. g., phosphoric, hydrofluoric, etc.) but, in view of the increasingly stringent limitations upon environmental pollution, even the cost of recovery of cheaper acids, such as sulfuric and hydrochloric, is frequently justified when compared with the cost of subsequent waste treatment.

As in the case of acid recovery, the economic attractiveness of the recovery of metals is related to their value. With the increasingly stringent waste water regulations coming into force throughout the world, the recovery of less expensive metals is receiving growing attention, since their disposal may give rise to environmental pollution.

Obviously, treatment which has the improvement or reduction of waste production as its sole objective is less attractive than one which can result in other economic advantages leading to the saving of acid, improvement of operation and the production of usable or marketable by-products.

2.9.2 Problems peculiar to treatment of pickling acids

There are two problems inherent in the treatment of pickling acids by ion exchange and related processes:

stability of resins or membranes, due to the aggressiveness of the solutions to be treated; and, in the case of ion exchange resins:
unfavorable equilibrium.

2.9.2.1 Stability

The solutions to be treated are aggressive and relatively concentrated, with potential damage to ion exchange resins and/or membranes.

A few comments concerning this problem may be in order.

In the case of resins, contact with the relatively high concentrations of acid, especially in the presence of metals such as iron and copper, can result in de-crosslinking of the resin matrix [1]. This is particularly true in the case of a highly aggressive acid such as nitric.

Earlier resins, especially of the phenolic, but also of the gel styrene-divinylbenzene type, were subject to such de-crosslinking even when treating water, if the latter contained oxidants and/or metals such as iron. Physical breakdown of a purely mechanical nature also frequently occurred. Such resin deterioration could partly be reduced by increasing cross-linking of the resin matrix. The extent to which this is practical is, however, limited, since increased cross-linkage can result in a reduction of exchange and diffusion kinetics. Furthermore, increasing the gel density by higher cross-linking can result in excessive rigidity which, in turn, can lead to breakdown during the alternate swelling and shrinking process which takes place during conversion between the hydrogen (regenerated) and metal (loaded) form of the resin [2].

The advent of macroreticular or macroporous resins around the late 1950's solved such problems to an appreciable extent. Macroreticular resins are, generally speaking, resins that have been polymerized in the presence of a substance which is a good solvent for the monomers and a poor swelling agent for the polymer, resulting in the formation of a discontinuous polymeric phase [3]. From a practical standpoint, this means that the resin consists of a gel structure perforated by connected channels. It is possible for such resins to have high porosity, high surface area and, therefore, good kinetics, while, at the same time, possessing good elasticity, even if the gel matrix is reasonably highly cross-linked. Such resins are quite resistant to the chemical and osmotic attacks discussed above. They are, therefore, well suited for the treatment of pickling acids, at least from a stability standpoint. They have the further advantage of being quite resistant to organic fouling, a significant feature, since metal surfaces prior to pickling usually contain more or less appreciable quantities of grease, oil or detergents [4, 5].

Gel resins, as well, have been improved over the years, with increasing stability being attained. Many different approaches have been followed by various resin manufacturers, as can be seen from data presented in their technical literature [6].

Further discussion of this point is beyond the scope of this chapter; suffice it to say that resin stability is, presumably, not a major impediment to the use of ion exchange resins for pickling acids at present.

In the case of ion exchange membranes, there were similar difficulties in combining satisfactory membrane performance with stability. High ion exchange capacity tends to entail a high water content in the membrane, resulting in a lower fixed ion content and a physical weakening of the membrane. In recent years membranes with a homogeneous composition of a chemically stable base polymer of styrene-divinylbenzene copolymers, in some cases reinforced with thin coarse-meshed fabrics bonded to the ion exchange layer have been produced, which have combined strength and stability with high permselectivity, high transport number and low electric resistance [7]. The latter is particularly important for electrodialysis; the other characteristics are also significant for diffusion dialysis, which is discussed below.

Concerning stability, it may probably be stated that both resins and membranes currently available have improved stability over earlier ones, so that the above-mentioned impediments to the development of their application have been appreciably reduced.

2.9.2.2 Equilibrium

The second difficulty, namely, unfavorable equilibrium conditions, represents a much more serious limitation to the economical use of ion exchange resins for the treatment of pickling acids. This is due to the fact that, at the relatively high concentrations of acid from which impurities are to be removed, such acids act as regenerants during the loading cycle. Take, for example, the following equation:

$$6R-H^+ + Fe_2(SO_4)_3 = 2R_3-Fe + 3H_2SO_4 .$$

The reaction, in a dilute solution, would be expected to proceed mainly to the right, since the affinity for multivalent metallic ions is substantially higher than for the hydrogen ion [8]. A typical bath used for pickling of steel wire, however, would contain about 150 g/l of sulfuric acid and about 50–60 g/l of iron. In this case, the solution is not by any means dilute; furthermore, the solution itself contains such a high concentration of acid (H^+ ions) as to tend to drive the reaction to the left (auto-regeneration). The consequence is that the reaction proceeds relatively inefficiently, resulting in low capacity for the resin and high leakage [7].

Obviously, the discussion of ion exchange equilibrium and affinities is too complex a matter for this chapter; it has been treated elsewhere in this book. At this point, it is only important to appreciate the difficulties concerning the ion exchange treatment of pickling acid as a result of the equilibrium conditions usually prevailing in such a system, which may be an important reason why the treatment of pickling acid by ion exchange resins does not appear to have been widely used.

In some cases, however, special approaches have been employed with some success, as will be described below.

The above discussion concerning unfavorable equilibrium refers to ion exchange resins; it is not directly applicable to ion exchange membranes.

2.9.3 Types of treatment

Ion exchange and related processes applicable to the treatment of pickling acids that have been considered include:

Fixed bed columnar ion exchange
Continuous ion exchange
Liquid ion exchange
Reciprocating flow
Electrodialysis
Diffusion dialysis.

The various processes mentioned are discussed in more detail in the following sections.

2.9.4 Fixed bed columnar ion exchange

This refers to a classic system of one or more columns of ion exchange resins.

2.9.4.1 Cation exchange resin

In its simplest form, a single column of strongly acidic cation exchange resin is used with co-current regeneration. This refers to a system in which the solution to be treated, as well as the regenerant, pass through the resin in the same direction, i. e., from top to bottom.

A typical example of this is the treatment of phosphoric acid, which is widely used in the surface treatment of steel prior to painting.

Pickling baths of this type usually have a phosphoric acid concentration in the range of 10–20%. The acid is treated at a rate designed to maintain the iron concentration below 3 g/l of iron [9, 10]. Regeneration is undertaken with a 10% H_2SO_4 or HCl solution. Leakage from the ion exchange column is not critical, so that co-current regeneration has been found to be satisfactory.

Figure 2.122 is a schematic representation of this relatively simple process in which the phosphoric pickling solution is treated by a strongly acidic cation exchange resin, regenerated with HCl [4]. As mentioned, H_2SO_4 could also be used; the choice of regenerant depends upon a number of factors such as the relative cost of the regenerant acids (in some countries this difference is greater than in others) and upon the local conditions concerning disposal of the eluate.

In a Swedish steel pickling plant, the phosphoric acid was reused for over eight years employing this technique. Before such a treatment system had been installed, 4 m^3 of H_3PO_4 per month had been discarded. After installation of the system, monthly waste was limited to 10 kg of sulfuric acid containing iron; this solution, upon neutralization, caused no serious pollution problems. The treatment system thus combined recovery of the expensive phosphoric acid with improvement of the waste water, where the environmentally harmful phosphates were essentially eliminated [11].

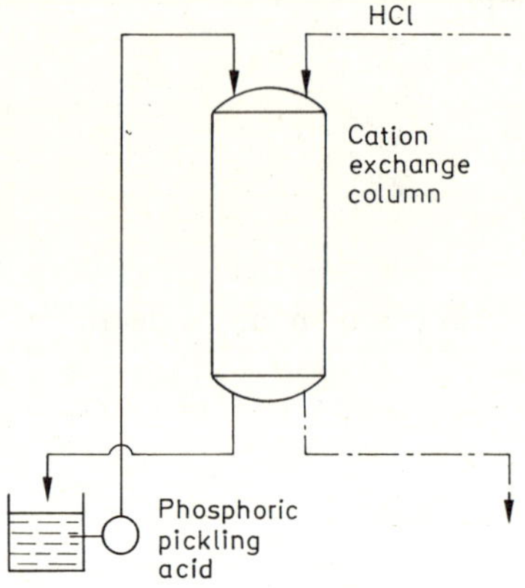

Figure 2.122 Treatment of phosphoric pickling acid by cation exchange.

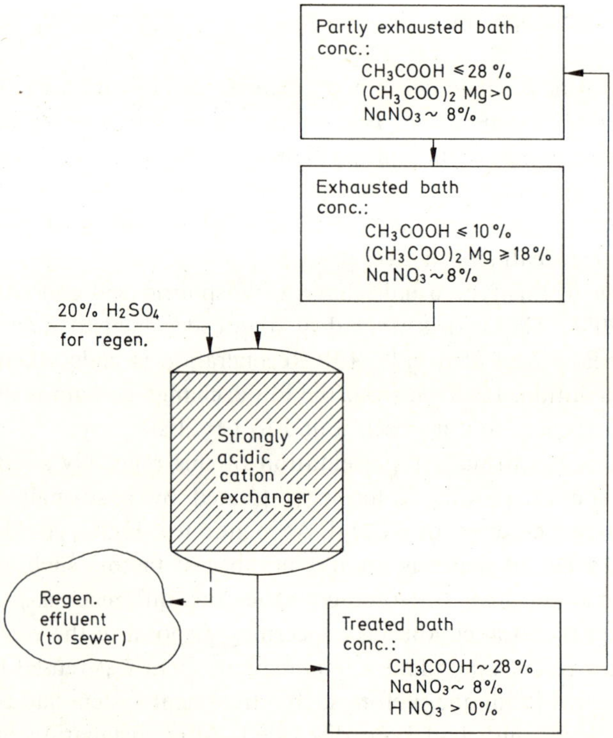

Figure 2.123 Flow sheet for the renovation of magnesium sheet pickle solution [10].

Figure 2.123 is a proposed flow-sheet for the renovation of a pickling bath consisting of acetic acid and sodium nitrate used for the pickling of magnesium sheet [10].

2.9.4.2 Anion exchange resin

One of the most fascinating fixed-bed processes is the treatment of hydrochloric acid pickling baths. This process takes advantage of the fact that iron and zinc form the complex anions $FeCl_4$ and $ZnCl_4$, respectively, in a concentrated solution of HCl. (In the case of iron, only the ferric stage forms a stable complex anion, a fact taken into consideration in the processes described below utilizing this phenomenon.) These complex anions hydrolyze easily in the presence of water. It is, therefore, possible to use an anion exchange resin to adsorb such complex ions from the concentrated acid, which is returned to the pickling section or otherwise utilized. The exhausted resin is then regenerated with water, which is not only a cheap reagent, but also does not contribute additional solids to pass into the waste water. The complex ions are recovered and utilized [10, 12, 13].

This principle has been applied by Degremont, France, for the treatment of HCl used in a steel galvanizing plant. In this particular plant, 33,000 tons of steel are processed per year, using 40 m^3 of an HCl pickling bath per 16-hour day (two shifts). The bath dissolves 250 kg of iron and 100 kg of zinc per day.

The installation adopted, which permits continuous rejuvenation of the pickling bath, comprises the following elements:

Aeration column to eliminate hydrogen gas
Reaction basin for addition of oxidizing agent
Mechanical filtration
Strongly basic ion exchange resin (two columns are used, one loading, while the other is undergoing regeneration).

Figure 2.124 is a schematic representation of the entire system [14].

The oxidation step is necessary, since only iron(III) forms the $FeCl_4$ ion which is stable in concentrated HCl.

The treated acid is returned to the pickling bath.

The operating cycle of the ion exchange system consists of a 4-hour loading period, followed by a 2-hour regeneration period. Counter-current regeneration is used, i. e., loading and elution are undertaken in opposite directions.

Regeneration has to be undertaken in a carefully controlled manner, to achieve maximum recovery of the valuable components. An essentially chromatographic separation is achieved, based upon the fact that the iron-rich portion is eluted before the zinc-rich portion. The former is stored for reconcentration, while the latter is neutralized with lime to form a precipitate of commerical zinc hydroxide.

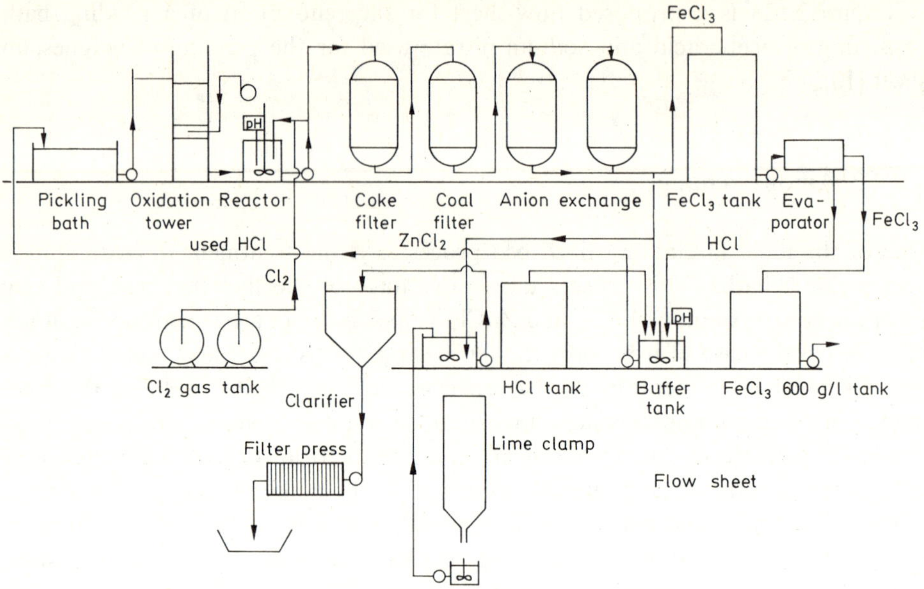

Figure 2.124 Treatment of HCl in steel galvanizing plant [14].

An economic study made when the plant was designed indicated that operating costs using the system described were about 11% lower than the classic system, while investment costs were 43% higher. The savings due to the recovery of acid and metals in the case of the IER system amounted to almost the same as the operating cost. Adding the lower operating costs to these savings resulted in a calculated pay-back period of less than two years, after which the operating costs were almost completely offset by the savings. In addition, total waste was reduced and a constant composition of the pickling bath could be attained, resulting in better performance of the latter, both qualitatively and quantitatively [14].

2.9.5 Continuous ion exchange – liquid ion exchange

Because of the high concentration of ions to be removed, resin investment costs in a conventional fixed-bed system are often excessive. To reduce the resin requirements, continuous or moving-bed ion exchange systems have been considered. In such a case, the loading and regeneration cycles are extremely short. The equipment is somewhat more complicated and resin wear and tear is higher than in a fixed-bed system, but since current technology is very well developed and modern resins of high stability are available, such processes have become feasible [15, 16].

Liquid ion exchange involves a type of solvent extraction where the organic phase contains an ion exchange material capable of adsorbing ions from an aqueous solution. The organic and aqueous phases are brought into intimate contact, whereby the ions in question pass into the organic phase as counter-ions of the ion exchange material. In theory, this is a very effective method for the treatment of relatively high concentrations, since it is, in effect, a truly continuous system, as opposed to the moving resin bed methods, in which the resins are usually moved and contacted in a series of pulses.

Liquid ion exchange has been limited, among other reasons, because it is restricted to weakly acidic or weakly basic products, the high aqueous solubility of strong electrolytes precluding their remaining in the organic phase.

Nevertheless, an interesting process for the recovery of hydrochloric pickling acid using a combination of a continuous ion exchange resin process and liquid ion exchange was developed in South Africa by Woodall-Duckham S. A. (Pty) Ltd. in cooperation with the South African National Institute for Metallurgy. The process is known as the NIM/W-D Metsep process and has been built on an industrial scale [17].

The classic method for the recovery of HCl and iron from spent pickle liquor in steel galvanization plants is pyrohydrolysis. However, the presence of zinc was found to interfere with the process as a result of the volatility of zinc chloride. Solutions containing appreciable quantities of zinc could not, therefore, be directly processed by pyrohydrolysis in an effective manner. The Metsep process combined ion exchange and pyrohydrolysis to overcome this difficulty.

Figure 2.125 is a schematic representation of the overall Metsep process, while Figure 2.126 and Figure 2.127 are flow-sheets of the continuous ion exchange resin system and liquid ion exchange plant, respectively.

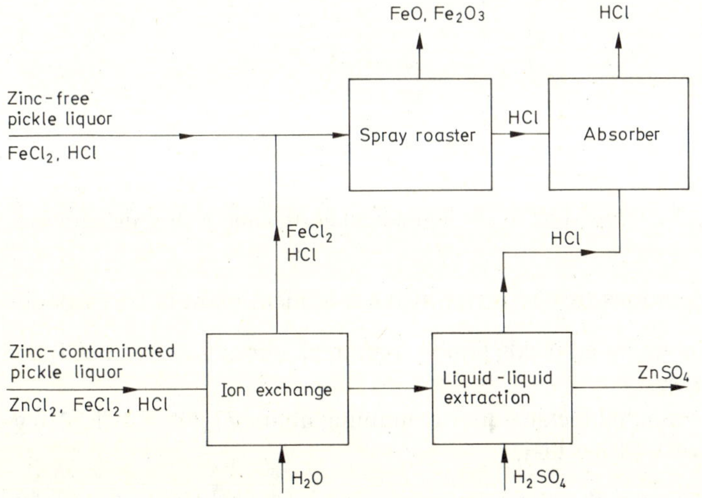

Figure 1.125 Block flow sheet of the Metsep Process.

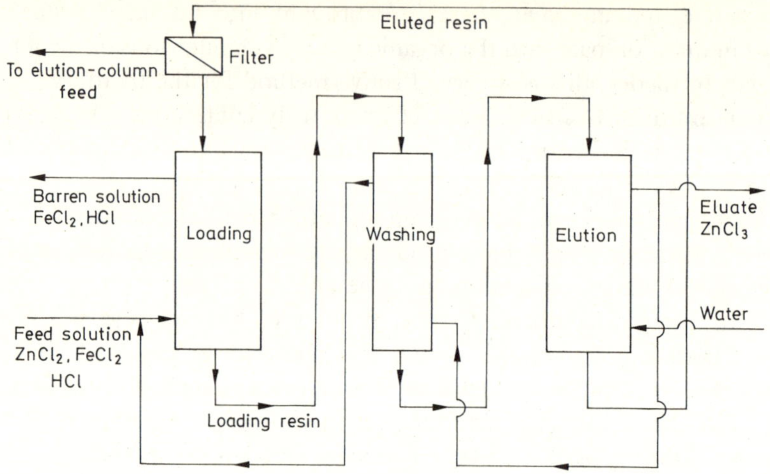

Figure 2.126 Flow sheet of the ion exchange plant (Metsep Process).

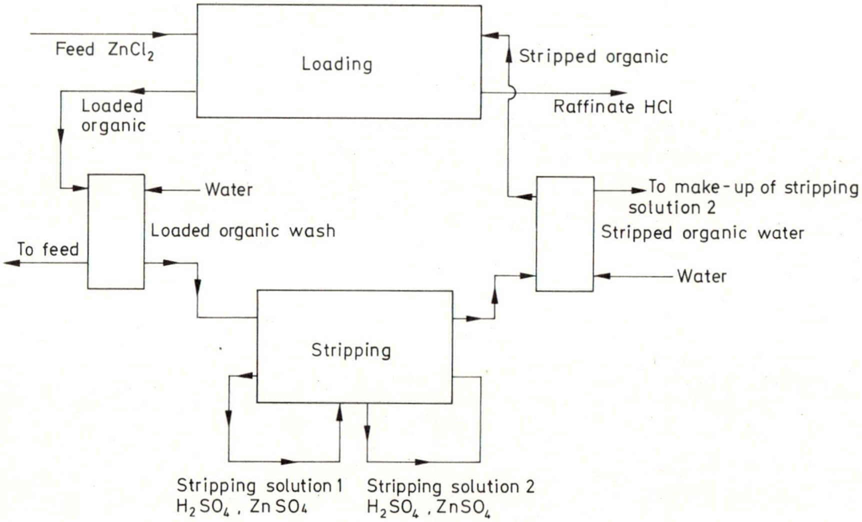

Figure 2.127 Flow sheet of the liquid-liquid extraction plant (Metsep Process).

The solutions treated arrive at the treatment plant in two separate streams:

a) uncontaminated pickle liquor containing about 220 g of $FeCl_2$ and 31 g of HCl per liter
b) contaminated pickle liquor containing about 73 g of $ZnCl_2$, 230 g of $FeCl_2$ and 30 g of HCl per liter.

Stream a) goes directly to the spray roaster for pyrohydrolysis, while stream b) is subjected to the ion exchange treatment under discussion.

As in the case of the Degremont process described in Section 2.9.4.2, the ion exchange resin portion of the Metsep process is based upon the formation of the $FeCl_4$ and $ZnCl_4$ complex anions. Where, however, the Degremont process uses an oxidation step to form an $FeCl_4$ complex to be adsorbed on the resin (separation between Fe and Zn being achieved by controlled elution), the Metsep process omits such a step, achieving instead the separation of Fe and Zn during the adsorption stage.

The continuous ion exchange resin process operates on the principle of an upward flow of solution, while the resin is periodically transported downwards to the washing and elution stages. A weakly basic resin of relatively high density is used. The choice of a weakly basic resin was based upon the fact that no strongly basic resin of sufficiently high density to avoid floating in the pickling liquor was available. From the standpoint of adsorption of the complex anions, such weakly basic resin was found to be satisfactory.

During loading, the iron chlorides, being in the unoxidized and, hence, uncomplexed state, pass through the resin and are sent directly to the spray roaster. Zinc chlorides, on the other hand, are adsorbed on the resin. The eluate goes to the liquid ion exchange plant where the zinc chloride is converted to zinc sulfate by extraction followed by elution (stripping with sulfuric acid).

The liquid ion exchange plant utilizes a cation exchanger dissolved in paraffin. A multi-stage mixer-settler system is used; the aqueous phase runs counter-current to the organic phase containing the ion exchanger. The process is continuous. About 85% of the zinc is extracted; the remaining 15%, when mixed with the raffinate from the spray roaster, represents a sufficiently low concentration to meet local market specifications for hydrochloric acid.

The plant operated in a satisfactory manner for some time, but operation was discontinued when a modification was made in the roasting process which eliminated the need for the separation achieved by the Metsep process [18].

Experiments using weakly basic liquid anion exchangers for the treatment of HCl pickling liquors have been reported in Germany, but no industrial plants have been described so far [19].

2.9.6 Reciprocating flow

With the same purpose of decreasing resin volume requirements and increasing efficiency in the treatment of relatively concentrated solutions which has given rise to continuous or moving bed ion exchange resin techniques, a special technique known as Reciprocating Flow Ion Exchange has reportedly been successfully used on an industrial scale [20]. This process, developed by Eco-Tec Limited, Canada, uses a fixed bed of resin, but the bed is extremely shallow. The particle size of the resins employed is considerably smaller than in standard columns, permitting good kinetics and, therefore, a very short contact time. Counter-current regeneration is used. Because of the extreme shallowness of the bed (2 – 3 cm), high flow rates are possible even though the particle size is very small; in a bed of conventional depth (60 cm or deeper) this would result in excessive pressure drop.

While this process falls into the category of ion exchange, it is, strictly speaking, based upon ion retardation rather than ion exchange.

Figure 2.128 shows the concept upon which the process is based, i. e., the entire loading and regeneration stages are limited to the active reaction zone of the ion exchange resin.

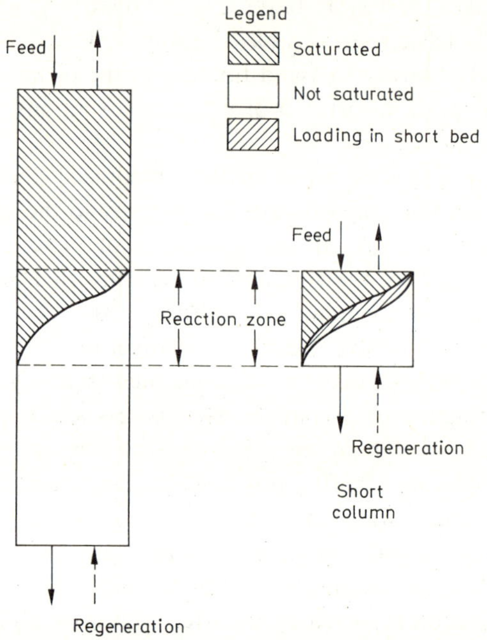

Figure 2.128 Reciprocating flow ion exchange: Concept of reaction zone.

Reciprocating Flow Ion Exchange was applied to acid retardation, a process introduced in 1963. Acid retardation involves an ion exchange bed which sorbs strong acids from solution, while excluding metallic salts of such acids. The acid can be eluted from the resin with water. A chromatographic separation of free acid from metal salts can be achieved by alternately passing contaminated acid and water through the bed.

The result of the combination of the two concepts is a two-step process in which contaminated acid is pumped up-flow through a relatively shallow resin bed (bed depth about 60 cm), followed by water being pumped down-flow through that bed. The total cycle is of about 5 min duration. The de-acidified metal salt solution is collected at the top of the bed on the "upstroke", while the purified acid is recovered from the bottom of the bed on the "down-stroke". The equipment was given the proprietory name of APU®, which stands for acid purification unit. A typical unit has a resin bed diameter of about 90 cm and can treat approximately 1700 liters of waste acid per hour. For larger flow requirements the diameter can be increased, or several units can be used in parallel.

Over 100 such units have reportedly been built, a substantial number of which are being used for the treatment of pickling acid, including the following:

sulfuric and hydrochloric acid (steel)
nitric acid/hydrofluoric acid (stainless steel)
sulfuric acid/peroxide (copper and brass)
nitric acid (copper and brass) [21].

2.9.7 Electrodialysis

The use of electrodialysis has been suggested for the treatment of sulfuric acid iron pickling liquor, which usually consists of a 15–25% solution of sulfuric acid. This acid, when exhausted and discarded, typically contains from 14–22% ferrous sulfate and 0.5–7% free acid. The most common method of treating this waste acid is by means of neutralization with lime and subsequent "lagooning" of the resulting slurry.

Since the traditional treatment method results in environmental pollution and involves the expenses of neutralization without the recovery of any useful materials, other methods have long been under investigation. Treatment by conventional ion exchange resin was not found to be economical, in view of the low value of the products recovered, although specialized methods, as discussed above, have been successfully used.

Electrodialysis appeared attractive, because it involves no addition of chemicals, a very important consideration from the standpoint of environmental protection. Horner et al., as early as 1955, conducted studies on such an approach. The basic mechanism is illustrated in Figure 2.129. Figure 2.130 shows a system they proposed for the continuous treatment of H_2SO_4 and includes a material balance [22].

A more recent article mentions work on the use of electrodialysis in the USSR for the treatment of pickling acids. The recovery of 80–90% of pickling acid for reuse was reported. The energy requirements were quoted as 600–1000 kWh/m^3 spent pickling solution [23].

Conventional electrodialysis for the treatment of pickling acid does not appear to have been adopted on a commercial basis, probably for two reasons: high energy costs and, more importantly, the fact that electrodialysis does not achieve satisfactory separation at the relatively high acid concentrations involved. In fact, while ion exchange membranes and process technology have been highly developed and are now used for a wide range of applications, most of the major manufacturers of such membranes and equipment do not advocate electrodialysis for the treatment

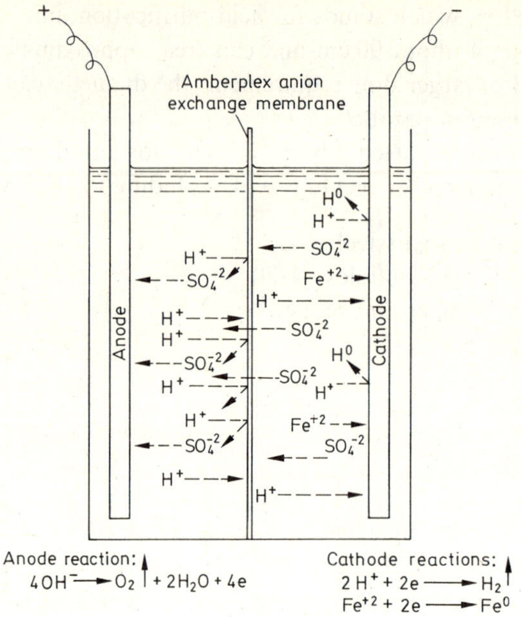

Figure 2.129 Mechanism of electrolytic treatment of waste sulfate pickle liquor [21].

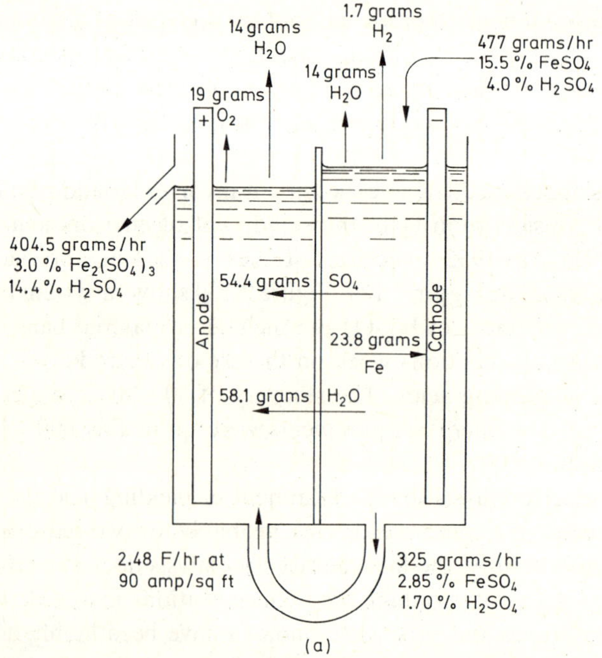

Figure 2.130 Continuous treatment of H_2SO_4 (Horner et al.).

of pickling acids [26–29]. However, an electrodialysis process combining bipolar membranes with cation and anion selective monopolar membranes has recently been reported and is discussed in the following subsection.

2.9.7.1 Bipolar membranes

The above comments regarding electrodialysis refer to techniques using "conventional" or monopolar cation- or anion-permeable membranes which are semi-permeable in the sense that they permit, across or through the entire membrane, the passage of cations or anions, respectively, while preventing or impeding the passage of the oppositely-charged ions.

Recently, however, a new electrodialysis technology based upon the combination of bipolar with monopolar membranes has been reported as suitable, among other applications, for the treatment of stainless steel pickling acids. The process has been developed in the USA under the name Aquatech® [24, 25].

The method utilizes the ability of bipolar membranes to split water into its constituent hydrogen and hydroxyl ions. Water flows over the surface of a bipolar membrane, which consists of cation- and anion-selective layers and diffuses into a central interface region (Figure 2.131). Some of the water dissociates into hydrogen and hydroxyl ions. When an electric potential is imposed across the membrane, hydrogen ions pass through the cation-permeable layer and the hydroxyl ions pass through the anion-permeable layer to adjacent chambers. The latter are bordered by conventional monopolar ion exchange membranes of opposite permeability to the side of the bipolar membrane they face, so that the hydrogen and hydroxyl ions cannot diffuse further (Figure 2.132).

Salt solution flows on the outside of these monopolar membranes. Since an electric potential is imposed across the whole system, cations from the salt solution enter the chamber containing the hydroxyl ions, forming bases, while anions enter

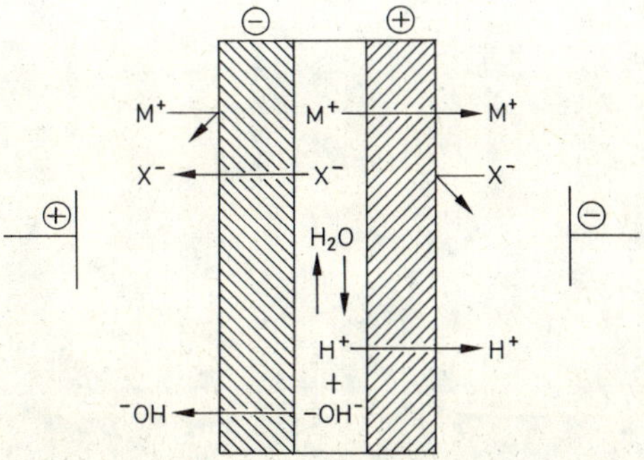

Figure 2.131 Bipolar membrane construction and operation (Courtesy of Aquatech Systems, Allied Corporation, USA).

the chamber containing the hydrogen ions, forming acids. These cations and anions are prevented from entering the bipolar membrane where they encounter a cation-impermeable and anion-impermeable layer, respectively (Figure 2.133). The chambers are arranged into cells of about 4 mm thickness.

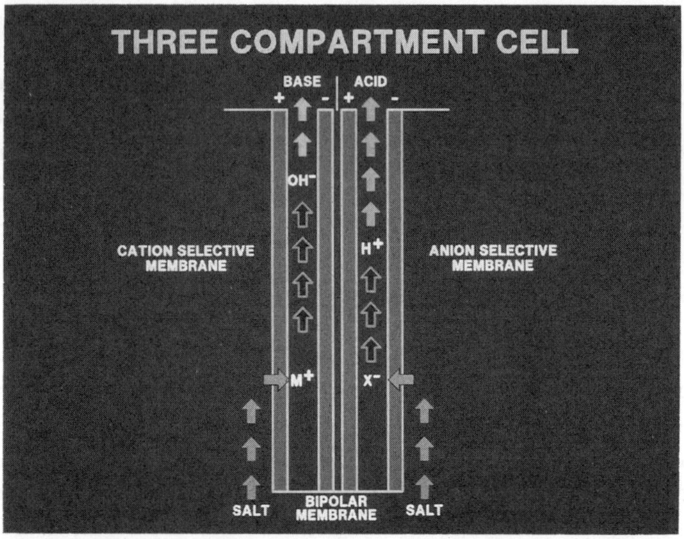

Figure 2.132 Three-compartment cell incorporating bipolar and monopolar membranes (Courtesy of Aquatech Systems, Allied Corporation, USA).

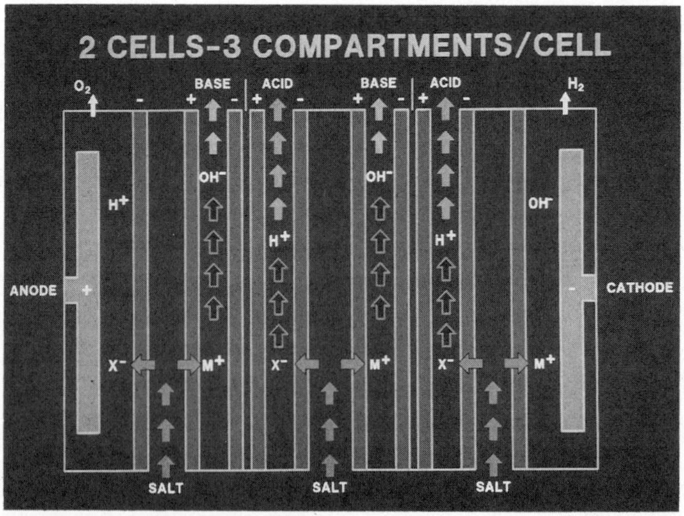

Figure 2.133 Arrangement of two 3-compartment cells into stack (Courtesy of Aquatech Systems, Allied Corporation, USA).

For a commercial application, about 200 cells are combined into a stack. The linear dimensions of the membranes are about 500 mm × 1000 mm. A single electrode pair is used for the entire stack.

The method has been adopted in an industrial plant for the regeneration of stainless steel pickle liquor (nitric/hydrofluoric acid pickle). For this purpose, the process steps are:

Neutralization/ precipitation: Potassium hydroxide is added to the spent hydrofluoric/nitric acid solution resulting in neutralization and the precipitation of metal hydroxides. Diatomaceous earth is used as a filter aid.

Filtration: Metal hydroxide solids are removed by passing the slurry through a filter press. The solids can be returned to the steel smelting operation. The water from the filter cake wash is recovered by treatment with reverse osmosis or electro-dialysis and used as make-up water for the cell stack. The concentrate from the reverse osmosis or electrodialysis system is recycled to the cell stack as salt solution feed.

Electrodialysis: The KF/KNO_3 salt solution is fed into the Auqatech® cell stack, where the HF and HNO_3 (typically 3N) and KOH (typically 2M) are regenerated (Figure 2.134). The regenerated acid is returned to the pickling operation, while the KOH is returned to the neutralization system.

The calculated economics of the process are shown in Figure 2.135.

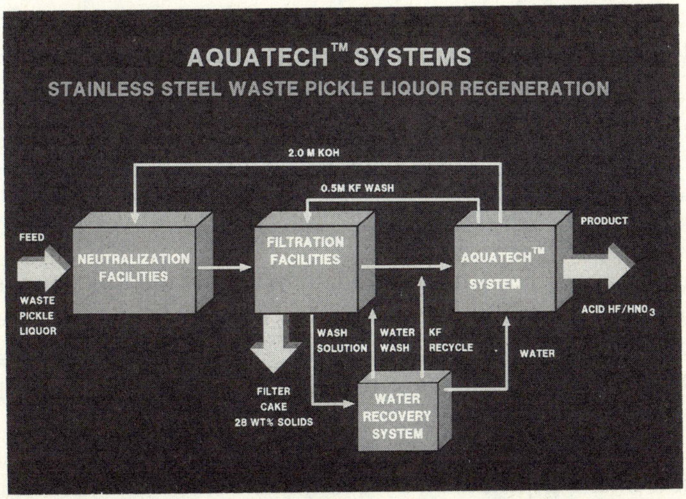

Figure 2.134 Aquatech® system for stainless steel pickle liquor regeneration (Courtesy of Aquatech Systems, Allied Corporation, USA).

COSTS

Basis: 1.5 MM US gal (5670 m3) pickle liquor/yr
(5% HF; 10% HNO_3)

Process Capital (includes 1st year's cell and initial KOH charge)	US$ 2.0 MM
Chemical Make-up 2% HF 2% HNO_3 5% KOH	$ 61,000
Cell Replacement Charge – 2nd year	$ 250,000
Electricity $\frac{5800\ \text{kwh}}{\text{ton HF produced}} \times 360$ tons HF $\times$ $ .05/kWh	$ 104,000
ANNUAL COSTS	$ 415,000

SAVINGS

Basis: 1.5 MM US gal (5670 m3) pickle liquor/yr
(5% HF; 10% HNO_3)

ACID SAVINGS	
360 tons HF (100%) US$ 1224/ton	$ 440,000/yr
719 tons HNO3 (100%) $ 210/ton	$ 151,000/yr
WASTE DISPOSAL SAVINGS	
1.5 MM US gal (5670 m3) acid waste $ 0.20/gal ($ 52.90/m3)	$ 300,000/yr
TOTAL ANNUAL SAVINGS	$ 891,000

Figure 2.135 Economics of Aquatech system for treatment of stainless steel pickle liquor (Courtesy of Aquatech Systems, Allied Corporation, USA).

2.9.8 Diffusion dialysis

Research and experience with electrodialysis led to one of the most interesting technologies to be used for the treatment of pickling acids, namely, diffusion dialysis, a technique successfully developed in Japan. This is an ion exchange membrane process based upon diffusion selectivity; it does not require the application of electricity.

Diffusion dialysis is a technique that was first developed in the early 1960's by Asahi Glass and, independently, by Tokuyama Soda. Asahi Chemical entered the field somewhat later. These three companies are all major manufacturers of electrodialysis membranes and equipment. The development of diffusion dialysis systems and the necessary membranes has been rapid and the process is now used on a standard basis in Japan; such plants have also been exported to Europe [27–29].

More than 50 diffusion dialysis plants have been constructed to date, ranging in capacity from 0.5 to 50 m^3/d of solution to be treated. Of the industrial plants in operation, more than 10 have been used for the treatment of pickling acid. This is one of the newest applications and further growth is anticipated.

Specifically, the following acids are being treated at this time:

H_2SO_4 for steel plate and wire pickling
H_2SO_4 for aluminum anodizing
HCl for aluminum etching
HNO_3 and HNO_3/HF for stainless steel pickling.

In addition, dialysis treatment for H_3PO_4 is under development.

Diffusion dialysis employs ion exchange membranes similar to those used in electrodialysis, but with very specific permselective characteristics, which permit the migration of free acids and bases while blocking salts. The principle of the process is the following:

Waste acid, containing metal ions, is passed through a cell in an up-flow direction. The cell is separated from an adjacent cell by an anion exchange membrane which is permeable to anions associated with hydrogen, (e. g., H_2SO_4) but impermeable not only to cations (as are all anion exchange membranes) but also to anions associated with salts (e. g., $Fe_2(SO_4)_3$). Water simultaneously passes through the adjacent cell in a downward direction. Because of the high concentration of ions in the waste acid cell, osmotic pressure causes them to seek their way into the dilute phase in the water cell. The acid succeeds in passing through the membrane, while the metal salts are blocked. Because of the continuous passage of fresh water through the water cell and of waste acid through the acid cell, the difference in concentration between the two cells (and, hence, the osmotic pressure) remains sufficiently great for the process to continue. Figure 2.136 depicts the principle of the process [30].

Actual equipment consists of a number of such cell pairs arranged in a "stack" of alternate dialysate chambers (carrying the waste acid or feed liquor) and diffusate chambers (carrying the water and receiving the acid molecules during the diffusion process). This is shown in Figure 2.137. The chambers are fastened by end plates such as a filter press. The membranes are separated by spacers, usually made of plastic; the cell thickness is normally in the range of 1–2 mm. In large installations several stacks comprising up to several hundred cell pairs are used, with an individual membrane area of over a square meter, resulting in a total membrane surface for the stack of several hundred square meters [31–33].

Generally speaking, diffusion dialysis membranes are plastic sheets containing ion exchange groups in the matrix of the polymer structure. The thickness is normally between 0.1 and 0.3 mm, the surface dimensions from several cm^2 to 1.0 m × 1.5 m. The membrane is usually reinforced by inert material. For acid treatment, anion exchange membranes are used: these reject cations; they are, however, very specific in also rejecting anions other than free acids [31, 32]. The maximum temperature at which the membranes can be used is limited by the anion exchange groups to about 60 °C [32].

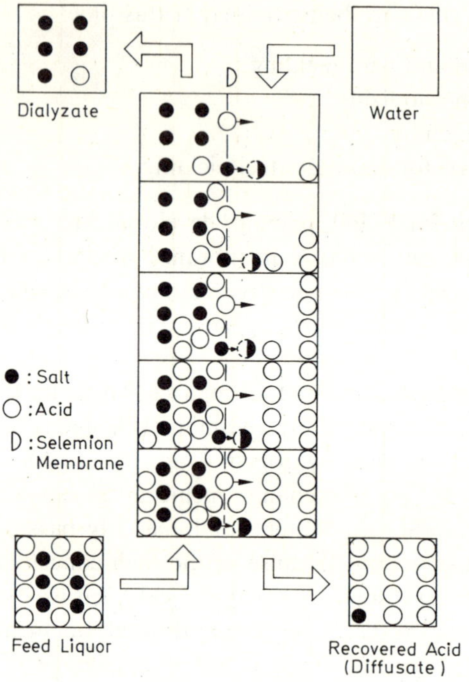

Figure 2.136 Principle of dialysis process [30].

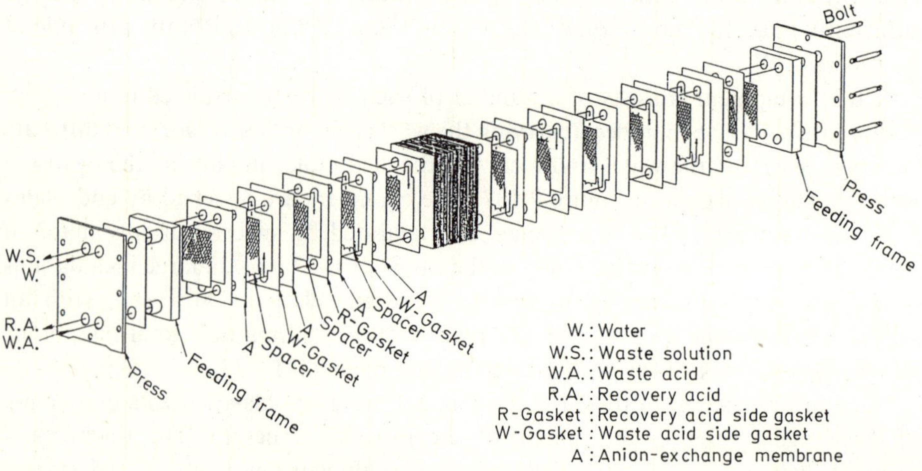

Figure 2.137 Exploded view of diffusion dialyzer stack (Courtesy of Tokuyama Soda Co.).

Figure 2.138 is a schematic representation of the mechanism of an ion exchange membrane [30].

Contrary to electrodialysis, diffusion dialysis requires no external force to cause a separation. The migration through the membrane is simply due to osmotic pressure.

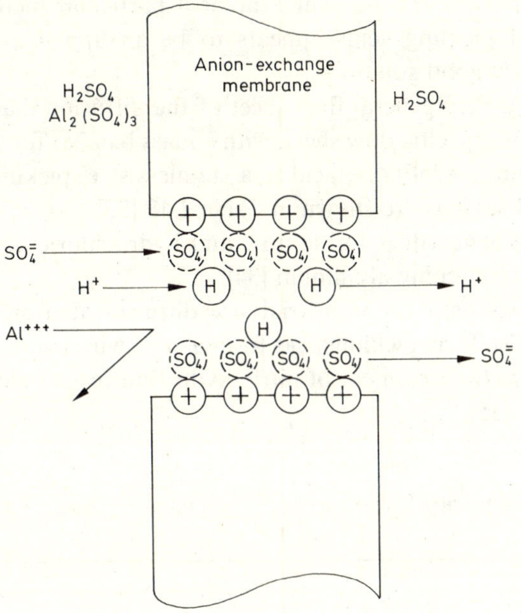

Figure 2.138 Mechanism of ion exchange membrane [30].

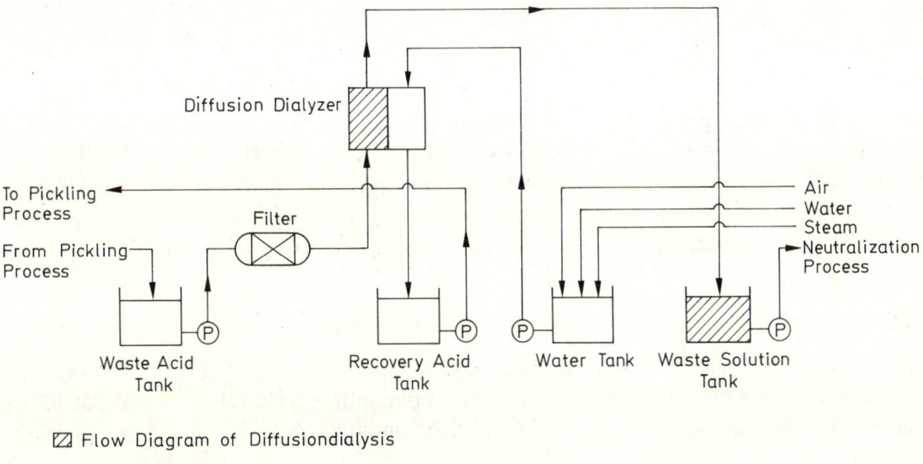

Figure 2.139 Flow diagram of diffusion dialysis [34].

Since diffusion dialysis requires no external driving force to attain separation, the only energy that must be supplied is the pumping force needed to move the liquids through the permeators. This is low compared to the pumping pressures associated with processes that oppose osmotic pressure, such as reverse osmosis and ultrafiltration. The energy costs connected with electrodialysis are also eliminated. As with many ion exchange and membrane processes, there are certain concentration and composition ranges which favor a particular method of treatment. The treatment of pickling acids appears to be an application for which diffusion dialysis is a very good solution.

Figure 2.139 depicts a very general flow sheet of the diffusion dialysis process. Figure 2.140 shows a more specific flow sheet, with a mass balance for the treatment of a mixture of nitric and hydrofluoric acid in a stainless steel pickling plant. The specifications, costs and savings are shown in Table 2.48 [33].

Table 2.49 gives treatment values for sulfuric and hydrochloric acid containing iron and for nitric acid containing aluminum [34].

Figure 2.141 [3] shows data for industrial size diffusion dialysis installations used for the recovery of sulfuric pickling acid from a steel wire plant. Figure 2.142 depicts an installation for the treatment of nitric/hydrofluoric acid pickling liquors in a stainless steel plant [35].

Table 2.48 Specification of merit of acid recovery equipment in cases of nitric acid and hydrofluoric acid [33]

		Note	
Volume of wate acid treatment	250 l/h		
Constitution of waste acid	HNO_3: 100 g/l HF: 15 g/l Fe: 25 g/l		
Rate of recovery of HNO_3	80%		
Rate of recovery of HF	70%		
Elimination of iron	5.9 kg		
Type of diffusion dialyzer	TSD-50-500	Electricity:	¥ 15/kWh
Ion exchange membrane	NEOSEPTA AFN	Water:	¥ 50/m^3
Size of equipment	3 m × 5 m × 5 m	Steam:	¥ 3,000/t
Running cost	¥ 360,000/month	Amortization:	5 years
Fixed cost (amortization and others)	¥ 440,000/month	Interest rate:	9% per annum
Total	¥ 800,000/month		
Merit of acid recovery	¥ 1,375,000/month	65% HNO_3	¥ 65,000/t
Cut-down of cost of neutralizer	¥ 345,000/month	55% HF	¥ 250,000/t
Cut-down of sludging	¥ 1,800,000/month	$Ca(OH)_2$	¥ 20,000/t
Total	¥ 3,520,000/month		

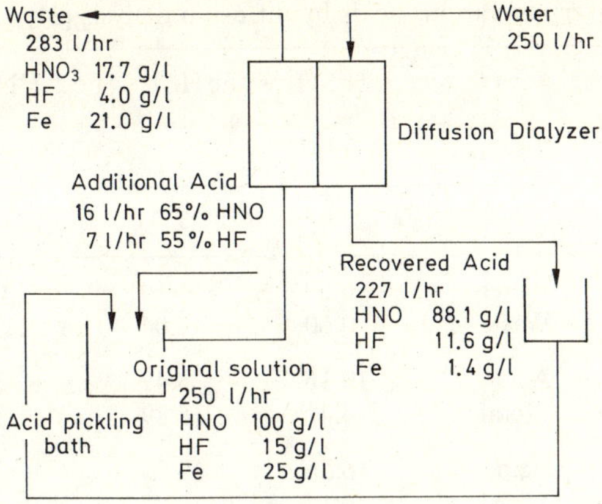

Figure 2.140 Recovery of HNO_3/HF mass balance [33].

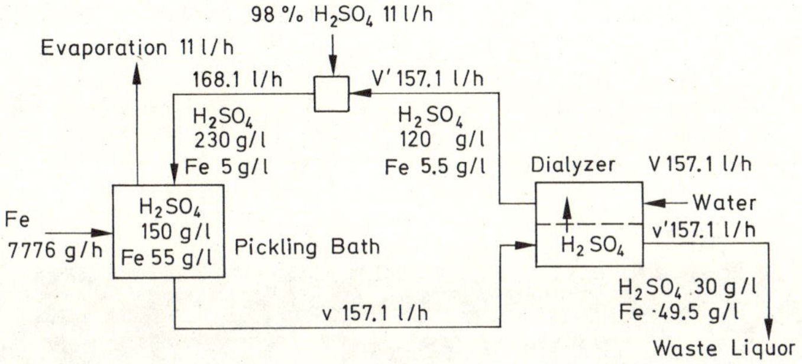

Treatment Cost of Dialysis Equipment
(Steel Wire Pickling)

A) Running Cost

1) Membrane Replacement	1,910 US$/year
2) Power	2,260 ,,
3) Others	260 ,,
Total	4,430 ,,

B) Saving of Chemicals

1) 98% Sulfuric Acid 91.5 ton/year	6,540 US$/year
2) 45% Caustic Soda	22,630 ,,
Total	29,170 ,,

Figure 2.141 Material balance and treatment cost for diffusion dialysis treatment of steel wire pickling acid [30].

Table 2.49 Example of recovery of various acids by diffusion dialysis [34]

Waste acid (Free effective acid and concomitant metal)		H_2SO_4 ~ Fe	HCl ~ Fe	HNO_3 ~ Al
Temperature (°C)		22	27	27
Capacity l/h · m²)	Water	0.59	1.10	1.05
	Waste acid	0.60	1.04	1.02
Concentration of waste acid (g/100 ml)	Acid	19.16	8.47	9.32
	Metal	4.41	5.89	1.31
Concentration of recovery acid (g/100 ml)	Acid	16.56	9.93	10.77
	Metal	0.14	0.89	0.06
Concentration of waste solution (g/100 ml)	Acid	3.92	0.37	0.38
	Metal	4.16	4.44	1.13
Recovery ratio of free acid (%)		82.1	95.0	95.0
Leak ratio of metal (%)		2.8	13.1	4.3

Figure 2.142 Industrial diffusion dialysis plant for treatment of nitric/hydrofluoric pickling acid in stainless steel plant [35].

The advantages of diffusion dialysis for the treatment of pickling acids include relatively low investment cost, which makes it suitable for smaller factories as well as large ones, low operating costs, simplicity of operation and the absence of chemicals that would contribute additional waste. The only disadvantages are that the concentration of the product is limited and the fact that maximum recovery is between 80% and 90%.

As to the concentration in the final solution, it is possible to envisage, for example, an electrodialysis unit behind the diffusion dialzyer, which should be able to increase the concentration.

2.9.9 Conclusion

Several ion exchange and related systems have been proposed, tried and, in some cases, successfully implemented over the years. Because of the nature of the solutions and the concentrations involved, success with conventional ion exchange processes has been limited. However, a number of effective specialized technologies have been developed.

At the moment, two processes that seem to be quite widely used for the treatment of pickling acids are the Reciprocating Flow Process using ion exchange resins (ion retardation) and diffusion dialysis, using ion exchange membranes. Both approaches have resulted in commercially viable unit processes and operate without the addition of minerals, such as acid, base or salt regenerants, to the waste streams. Their use has permitted the treatment of pickling acid, resulting in improved production economics while contributing to the reduction of environmental pollution. An electrodialysis method, using a combination of monopolar and bipolar ion exchange membranes, is currently being operated. This process, as well, does not seem to result in significant mineral addition to the waste stream, since the base that is added is recycled.

Further developments, such as the combination of diffusion dialysis with electrodialysis as a final concentration step, may lead to additional effectivity.

References

[1] De Jong, G. J., Brants, J. and Otten, G. (1960), Degradazione delle Resine Scambiatice., Aqua Industriale *7/8,* 68–72.

[2] Herz, G. P. (1962), The Influence of Organic Substances on Anion Exchange Resins in Water Deionization, Technische Ueberwachung, *3,* 77–85.

[3] Kun, K. A. and Kunin, R. (1974), Polymer Lett. *2,* 587–591.

[4] DIAION Manual of Ion Exchange Resins (II) Revised Edition, Mitsubishi Chemical Industries, Ltd.

[5] Herz, G. P. (1965), Field Experiences with Macroreticular Resins, Effluent and Water Journal, Sept. 1965.

[6] Cf.: New Dowex Monosphere TG, Dow Chemical Company, 1984.
[7] Seko, M., Miyauchi, H. and Omura, J. (1983), Ion Exchange Membrane Application for Electrodialysis, Electroreduction and Electrohydrodimerisation, Part 5, Chapter 12, Ion Exchange Membranes (D. S. Fleet, ed.) Society of Chemical Industry; Ellis Horwood.
[8] Kunin, R. (1972), Ion Exchange Resins, pp. 5 ff., Krieger, Melbourne, Fl.
[9] Technical Bulletin CTD-2: "Phosphoric Acid Purification", Rohm and Haas Company, USA.
[10] Bolto, B. A. and Pawlowski, L. (1983), Reclamation of Waste Water Constituents by Ion Exchange, Effluent and Water Treatment Journal, Jan. 1983.
[11] Annell, O., Akvapur, Sweden: Private communication.
[12] Dorfner, K. (1970), Ionenaustauscher, 3rd Edition, pp. 205 f, de Gruyter, Berlin, New York.
[13] Shimizu, H. (1961), Guidebook for Treatment of Sewage, Industrial Liquid Waste and Sludge, Chapt. 9, Sect 5–10, Environmental Technology Research Association (Japan).
[14] Letessant, M. and Wajsfelner, R. (1976), Purification of Hydrochloric Acid Pickling Baths with the Production of Ferric Chloride, Galvano-Organo 76/467. See also: Letessant, M. and Wajsfelder, R. (1976), Recovery of Chemicals Contained in the Waste Electroplating Baths: Purification of Hydrochloric Acid Pickling Baths with Production of Ferric Chloride, Prog. Wat. Tech. *8,* 147–152.
[15] Himsley, A. (1980), Continuous Ion Exchange for Metal Recovery and Water Treatment, Proc. 41st Internat. Water Conf. Pittsburgh, USA.
[16] Haid, E. (1980), Kontinuierlich arbeitende Ionenaustauscher- und Filtrationssysteme, Moderne Techniken zur Aufbereitung von Frischwasser und Abwasser im Kraftwerk, Seminarbericht, Hager and Elsaesser.
[17] Haines, A. K., Tunley, T. H., Te Riele, W. a. M., Cloete, F. L. D. and Sampson, T. D. (1973), The Recovery of Zinc from Pickle Liquors by Ion Exchange, J. South African Inst. Mining Metallurgy, Nov., 1973.
[18] Babcock, Woodall, Duckham: Private communication.
[19] Dorfner, K. (1970), Ionenaustauscher, 3rd Edition, pp. 195–221, de Gruyter, Berlin, New York.
[20] Spinner, I. H., Simmons, P. J., and Brown, C. J. (1979), Water and Chemical Recovery by Reciprocating Flow Ion Exchanger, presented at the 40th Internatl. Water Conf. Pittsburgh, USA.
[21] Brown, C. J. (1984), Acid/Metal Recovery by Recoflo Sorption, presented at 23rd Conf. Metallurgists of the CIM, Quebec.
[22] Kunin, R. (1972), Ion Exchange Resins, pp. 237–245, Krieger, Melbourne, Fl.
[23] Cenkin, V. E. and Belevtsev, A. N. (1985), Electrochemical Treatment of Industrial Waste Water, Effluent Water Treatment J., *25,* 987.
[24] Krieger, J. (1985), Process Converts Wastes, Pollutants Into Valuable Raw Materials, Chemical Engineering News Dec. 23.
[25] Allied Corporation (USA), Aquatech® Systems: Private communication.
[26] Kobuchi, Y., Developing Membrane Techniques and Its Applications – Ion Exchange Membranes, Technical Bulletin, Tokuyama Soda Co.
[27] General Introduction of Asahi Glass' Ion Exchange Membrane Technology, Technical Bulletin, Asahi Glass Co.
[28] Tokuyama Soda Co.: Private communication.
[29] Asahi Chemical Co: Private communication.
[30] Use of Dialysis Technique in Metal Finishing Process, Technical Bulletin, Asahi Glass Co.
[31] Nakamura, I., Kawahara, T. and Utsunomiya, T. (1983), Use of Dialysis Technique in Metal Finishing Process, New Metals and New Processes, *Vol. 2.*

[32] Kobuchi, Y., Motomura, H., Noma, Y. and Hanada, F. (1984), Application of the Ion Exchange Membranes/Acid Recovery by Diffusion Dialysis, Paper presented at Europe-Japan Congr. on Membranes and Membrane Processes, Stresa, Italy, Tokuyama Soda Co. Ltd., Tokyo, Japan.
[33] Recovery of Acid in a Process of Plating – Diffusion Dialysis Process, Technical Bulletin, Tokuyama Soda Co.
[34] Diffusion Dialyzer for Acid Recovery, Technical Bulletin, Tokuyama Soda Co.
[35] Sato, J., Onuma, M., Motomura, H., Noma, Y. (1984), Recovery of Nitric Acid and Hydrofluoric Acid from the Pickling Solution by Diffusion Dialysis, Jitsumu Hyomen Gijitsu (The Metal Finishing Society of Japan).

2.10 Ion Exchangers in the Sweetener Industry

Karlheinz W. R. Schoenrock
Consultant, Sugar and Separation Technology, 5333 Fillmore, Ogden, Utah, USA

Introduction

Ion exchange and sweetness have been closely allied since Moses sweetened the brackish waters of Marah. This affinity between ion exchange and sweetness has expanded in more modern times to include sugar. A review of the literature and activity in this field suggests that the purification of sugar-bearing liquors by ion exchange is an intensely studied subject in the application of ion exchangers.

Caloric sweeteners, or more specifically sugars, are products of photosynthesis in plants and must be separated from associated impurities to be of significant commercial value for food preparation. Conventional methods for the clarification and purification of sugar-bearing plant extracts, extract and refine only about 75 to 85% of the sugar in sugar beets or sugar cane. About 10 to 20% of the sugar can not be recovered from the molasses by crystallization because of associated impurities. Ion exchange has been recognized from the beginning of its discovery as an effective tool to recover part, or all, of the molasses sugar. Molasses represents the final residue liquor after conventional purification and clarification, and after all crystallizable sugar has been recovered. It is unlikely, however, that ion exchange can replace conventional methods of clarification.

Table 2.50 Typical Molasses Compositions

		Beet Molasses		Cane Molasses
% DS		78 – 82		75 – 83
% Water		18 – 22		17 – 25
% Sucrose		45 – 55		30 – 40
% Total Sugars		50 – 57		40 – 55
% Nonsugars		27 – 35		35 – 45
% Total Nitrogen		< 1.5 – > 2		< .5 – > 1.0
% Ash		9 – 13		7 – 15
Potassium K_2O	%/Molasses	< 4 – > 5	%/Ash	50 – 70
Sodium Na_2O	%/Molasses	< .5 – > 1	%/Ash	0 – 12
AL. and MAG. and CaO	%/Molasses	< .2 – > 2	%/Ash	30 – 90
R_2O_3	%/Molasses	< .1	%/Ash	< .5 – > 2.5
Anions and SiO_2	%/Molasses	< 1.7 – > 2	%/Ash	< 20 – > 35
Amino Acids and PCA	%/Molasses	< 3.5 – > 2		< .5
Organic Acids	%/Molasses	< 4.5 – > 5		< 2 – > 7.5
Betain	%/Molasses	< 5 – > 6		

The typical composition of sugar beet and cane molasses is given in Table 2.50 [1, 2]. The components in Table 2.50 are, therefore, those which must be addressed in any attempt to effectively utilize ion exchange in the sugar industry.

Ion exchange is especially effective for the elimination of unspecified compounds not included in Table 1, such as color. Experiments with soils and aluminum silicates (zeolites) to improve sugar recovery and sugar quality started around the middle of the nineteenth century [3 – 7]. Serious commercialization of ion exchangers in the sweetener industry was, however, not realized until after Adams and Holmes [8] discovered ion exchange properties in certain synthetic resins. With this discovery, it became possible to tailor ion exchangers for the separation of a much larger group of impurities at a higher level of efficiency.

Periodic reviews of the application of ion exchange in the sugar industry have been given by Carruthers and Oldfield [9], Schneider [10], Landi and Mantovani [11], and McGinnis [12]. Since the latest review, there have been significant advances for ion exchangers in the sweetener industry. This discourse will mainly address the later developments and refer to well-established techniques in more general terms.

In general, the use of ion exchangers in the sweetener industry may be classified into six basic divisions: Cation Exchange; Catalysis; Chromatography; Ion Exclusion; Cation/Anion Exchange; and Adsorption.

Within each of these divisions, there are several sub-divisions, each with different objectives.

The widely different physical and chemical properties of the components in sweetener-containing plant extracts have resulted in specialized applications of ion exchangers in the sweetener industry.

2.10.1 Cation exchangers

The presence of alkali salts in sugar beet molasses was suspected in 1843, by Hochstetter [13], as the cause for the inability to deplete molasses of its sugar by crystallization. The discovery of cation exchange properties in natural zeolites by Thompson [14] and Way [15] may have provided the necessary incentive to utilize clays and silicates to treat sugar beet liquors for increased sugar recovery. Claassen [7] demonstrated that the improvements from such treatment claimed by others [3–6] was a function of cation exchange. Potassium was exchanged against calcium, which lowered the solubility of sucrose in the molasses. Such was to be expected from the works of Marschall [16] and Herzfeld [17]. The interest in cation exchanger application in the sweetener industry has been maintained to the present day, and now represents, by far, the largest use of ion exchange in this field. Cation exchangers are not only commercially used in today's sweetener industry to exchange cations, but also to catalyze inversion; to separate sucrose from associated impurities by ion exclusion; and to partition fructose and glucose for the fructose enrichment of fructose-containing sweeteners by chromatography. Between 1979 and 1985, the corn sweetener industry in the United States, alone, placed about 6,000 cubic meters of cation exchanger resins on stream for the chromatographic enrichment of fructose in high-fructose corn syrups.

2.10.1.1 Decalcification (softening, deliming)

The removal of calcium hardness from clarified thin juice via cation exchanger is extensively practiced in the beet sugar industry. So-called thin juice typically contains 12 to 15% solids of which 85 to 92% is sucrose. Calcium hardness in unsoftened thin juice measures from less than 0.05% to over 0.5% CaO equivalents on dissolved solids. Normal operating temperature at that point of the sugar beet process is around 90 °C. The monovalent cation load in thin juice is about 15% of the non-sucrose components. Seventy-five per cent of the cation load is potassium, with the balance being sodium. A smaller amount of ammonia from the deamination of the amides is also present. The presence of calcium originating in the clarification process with lime can lead to significant scaling problems from lime salts during evaporative concentration. This scaling causes losses in operating capacity, poor energy utilization, and the need to de-scale with corrosive and expensive chemicals. The disposal of descalent waste leads to environmental pollution. A lime-salt scale of only 0.3 mm thickness could drop the heat transfer coefficient by as much as 50%. Sugar factories without thin juice softener frequently add soda to limit the presence of lime salts in thin juice. Such additions always lead to increased sugar loss to molasses.

Conventional decalcification. The classical softening process involving fixed bed, co-current operation of a strong cation exchanger with sulfonic acid functionality and operated over the sodium form, is most widely practiced. Readily available and inexpensive salt brine is generally used as the regenerant for this operation. The exchange kinetics for calcium are relatively poor because of the high monovalent cation load. Hence, exchanger utilization is low. Calcium leakage is usually excessive when only less than 50% of the total exchanger capacity is used. Over-all calcium elimination is typically 90 to 95% and residual lime salts are targeted around 0.02% CaO equivalent on total dissolved solids. Salt consumption for conventional fixed-bed thin juice softening is about 500% of operating capacity. The softener bed must be rinsed free of sugar before subsequent steps involving backwashing or regeneration with brine. Otherwise, general operation parallels the well-known water softening technique.

Complete de-calcification is difficult, inefficient, and usually not attempted with conventional softening because of associated operating problems downstream. Excessive foaming during evaporative concentration and accelerated corrosion in the evaporator station have been reported with totally softened juice.

Details of conventional thin juice softening have been outlined by Schneider [18] and others [19–25]. Attempts to improve the efficiency of conventional thin juice softening included counter-current operation [25] and continuous techniques [26] known as the Asahi and Degremont Cottrell Process. Although continuous processes were installed in sugar beet factories in Europe and Japan between 1960 and 1970, this technique has not succeeded, mainly because of an excessively high resin attrition attributable to the resin movement featured in this process. In some operations, a fractionated impurity stream high in monovalent cations and free of divalent cations may be used for the regeneration of the exhausted softener exchange resin. Such streams are available as an unavoidable by-product from the ion exclusion treatment of impure cane and sugar beet liquors.

The Gryllus Process. The utilization of native alkali ions in sugar syrups was proposed by Gryllus for the regeneration of thin juice softener [27–29]. Concentrated and softened juice, either before, or after, one or more crystallization steps, is used as the regenerant. Alkali concentrations in the mother liquor from second crystallization are typically 1.5 equivalents/liter at 70% total dissolved solids. These syrups were found to be more efficient in regenerating strong cation exchangers loaded with calcium than conventionally used salt brine. The calcium load in thin juice thereby by-passes the evaporator station and usually also the main crystallization to be returned to the last crystallization step. Such practice eliminates the need for salt brine and introduction of additional sodium which normally leads to increased sugar losses, avoids environmental pollution with waste chemicals, and prevents the interface dilution from sweet-on and desweetening unavoidable when salt brine is used as a regenerant. Figure 2.143 illustrates a basic flow diagram for the Gryllus Process and its integration into a sugar beet factory. Several European sugar factories utilize this process now for the softening of thin juice prior to evaporative concentration.

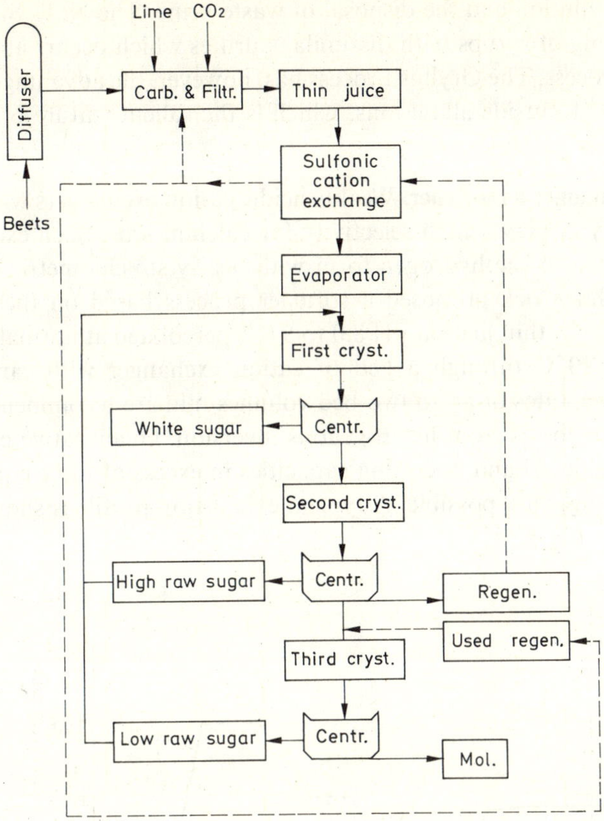

Figure 2.143 Simplified flow diagram for beet sugar factory incorporating the Gryllus Softening Process.

The N. R. S. (New Regeneration System) [30, 31]. Imacti, formerly a division of Akzo Chemie, Netherlands, proposed a variant of the Gryllus Process which has become known as the N. R. S. Process. The N. R. S. also utilizes softened sugar juices to regenerate the exhausted exchanger, but increases regeneration efficiency through caustic addition to the juice. Accordingly, one equivalent of NaOH is added per liter soft juice cooled to 40 °C. Only one bed volume of this alkalized juice is required to properly regenerate the exchanger to the alkali form with approximately 70% efficiency. The formation of soluble lime-sucrose prevents the precipitation of insoluble $Ca(OH)_2$. A three-fold regeneration efficiency is claimed over conventional salt brine usage. Exchanger utilization during exhaustion is about the same as for any of the other decalcification techniques employing strong, sulfonic-type cation exchangers. The regenerant fraction containing the calcium is returned to the carbonation step where the enhanced alkalinity promotes calcium depletion during carbonation. As with the Gryllus Process, the N. R. S. Process avoids interface dilution and the disposal of waste brine. The N. R. S. Process also eliminates blending of syrups with dissimilar purities which occurs at the interface of the Gryllus Process. The Gryllus Process has, however, the advantage of avoiding the introduction of outside alkali ions, which is the salient feature of the N. R. S. Process.

Weak cation exchanger as softener. Weakly acidic cation exchangers with carboxylic acid functionality display a high selectivity for calcium ions. Such exchangers can be easily converted to the hydrogen form with nearly stoichiometric quantities of strong acids. Schoenrock proposed a softener process based on these properties [32, 33]. Accordingly, thin juice at pH 8.0 to 9.0 is percolated at normally-prevailing temperatures of 90 °C through a bed of cation exchanger with carboxylic acid functionality. Flow rates of up to two bed volumes/min are recommended to avoid significant sucrose losses from heterogenous inversion. Quantitative calcium elimination can be achieved and operating capacities in excess of three equivalents per liter cation exchanger are possible. A typical exhaustion profile is shown in Figure 2.144.

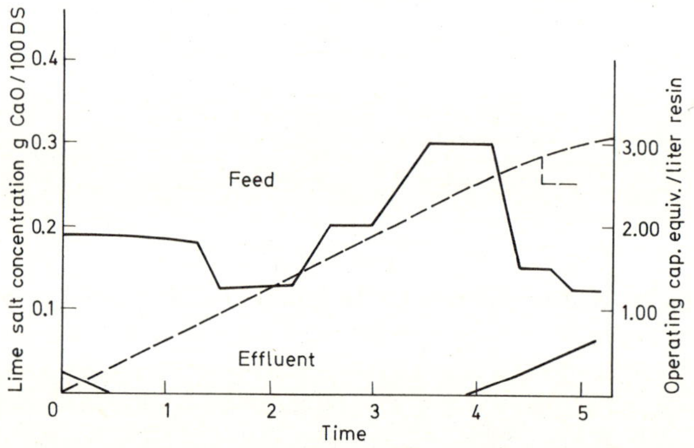

Figure 2.144 Limesalts removal from clarified sugar beet thin juice by weak cation exchanger in the hydrogen form [32]. Conditions: Variable lime salt concentrations in thin juice fed at 120 BV/h [32].

The exhausted exchanger loaded with calcium can be completely regenerated with strong mineral acids at about 110% of operating capacity. Figure 2.145 details the regeneration profile for this process in a commercial installation. Depending on the initial alkalinity, the predominance of potassium and sodium in thin juice can force initially extensive exchange of these cations against hydrogen with concurrent extreme swelling of the exchanger. Hence, systems based on this concept must be carefully designed and controlled to avoid flow restrictions from excessive bed compaction. Bed depth must be limited to about one meter. Conventional weak acid exchangers with gel matrix are unsuitable for this process.

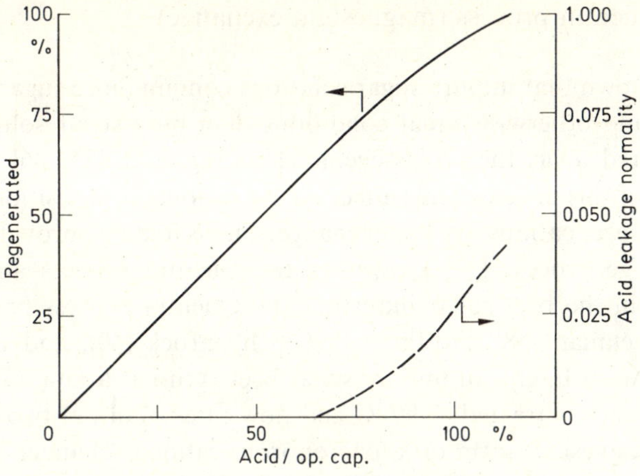

Figure 2.145 Regeneration profile for weak cation exchanger. Loaded with calcium from sugar beet thin juice [32]. Conditions: 0.1 N H_2SO_4 at 35 BV/h flow; 75 °C.

The calcium-enriched regenerant waste from this process is a suitable pressing aid in the mechanical dewatering of desugared beet pulp. This utility eliminates environmental pollution from regeneration chemicals, which is a major obstacle to the general application of ion exchangers in the sweetener industry. The direct hydrogen exchange featured by this process requires mild re-alkalization of the softened sugar juice to avoid sucrose losses from homogeneous inversion. Neutralization of the soft acidic juice with active magnesium oxide is recommended in U. S. Patent 3,887,391 [32]. U. S. Patent 3,982,956 [34] describes neutralization of the acidic juice via the OH-form of a weak anion exchanger. The anion exchanger eliminates about 5 – 10% of dissolved impurities and about 60% of the color.

This process was studied in Europe and the United States and is now commercially practiced in several factories. A typical softener installation featuring weakly acidic cation exchanger may operate with two or three columns containing 2.5 m^3 each to treat a maximum of 3000 m^3 thin juice per hour with a hardness of about 20 milliequivalents calcium salts per liter thin juice. Optimization must give due regards pertaining service and regeneration flow rates, suspended solids loads, temperatures, entrained gases such as CO_2 and air, alkalinity and acid concentration. The use of sulfuric acid requires regeneration flow rates in excess of 25 bed volumes per hour at concentrations below 0.1N to prevent fouling with gypsum. The use of hydrochloric acid would avoid this restriction. High temperatures in combination with the presence of dissolved oxygen may bring about a premature loss of operating capacity for the resin. The regeneration waste containing gypsum or calcium chloride may be used for the desweetening of the sugar beet cossettes in the diffuser.

Realkalization with caustic, soda ash or active magnesium oxide are possible option in addition to ion exchange neutralization over a weakly basic anion exchanger.

2.10.1.2 The Quentin process (magnesium exchange)

It is generally known that impure sugar solutions contain more sugar at saturation per unit water and otherwise equal conditions than pure sugar solutions. This is commonly referred to as the molassegenic effect. Quentin [35, 36] demonstrated the dependence of this molassegenic effect on the cationic composition by replacing the native sugar beet cations via ion exchange. These findings provided a basis for a cation exchange process [37] known as the Quentin Process and extensively commercialized in the beet sugar industry. The Quentin Process has been widely publicized by Neuman [38], Hoffman [39], Schoenrock [40], and Perschak [41], among others. Accordingly, an impure sugar beet syrup at about 75% purity and 70% dissolved solids is treated at 90 °C and flow rates of about two bed volumes/hour over the magnesium form of a macroporous cation exchanger with sulfonate functionality. An exchange of 30 to 50% of the native sugar beet cations is usually targeted. Relatively high operating capacities of up to two equivalent/liter resin are possible at this exchange level. Such cation exchange reduces the solubility of sucrose in final molasses sufficiently to raise sugar extraction by two to four extraction points.

In very well-controlled operations, molasses purities can be reduced to under 50%. Regeneration of the exhausted exchanger, which is then predominantly in the potassium form, proceeds as a rule with magnesium chloride brine at 5 to 6% $MgCl_2$ concentration. A regeneration efficiency of 125% on operating capacity can be achieved. Magnesium sulfate has also been proposed as a regenerant in this process [42]. Care must be exercised to prevent contact with liquids having a high pH, which could cause resin fouling from precipitated magnesium hydroxide.

The Quentin Process is dependent on the availability of low-cost magnesium salts, which is usually a by-product of the potash industry. This explains its popularity in central Europe and near the Great Salt Lake region of the United States. The extremely high osmotic shock generated by the Quentin Process required the use of special macroporous ion exchangers designed for this purpose. Tough gel resins, a new development by The Dow Chemical Company [43] have recently been successfully used in this application. Other advances in equipment design, such as those suggested by Schoenrock [44, 45], which feature multiple packed beds and counter-current operation, and by Kammerer [46] have been proposed to improve operating efficiency at reduced costs in connection with the Quentin Process but which are also applicable so to other systems. Both designs are commercially used.

2.10.1.3 The SCC process

Moebes [47] expanded on the findings of Quentin and proposed the exchange of native sugar beet cations against calcium [48 – 50]. This process became known as the SCC Process. It was operational in at least two sugar beet factories, but was soon abandoned. The salient feature of this first Moebes Process is the operation of the cation exchanger over the ammonium form and its regeneration with a suitable ammonium salt, originally 6 to 7% ammonium sulfate at about 150% of the operating capacity. Prevailing kinetics dictate a special treatment of the thin juice. The exchanged ammonium ions in the treated juice are displaced with lime addition, followed by carbonation to remove excess lime as calcium carbonate. Optimum performance for the SCC Process is difficult to maintain, which led to its disuse.

2.10.2 Catalysis

Partial or total inversion of sucrose in solution is frequently practiced to impart special properties to the sweetener:

$$C_{12}H_{22}O_{11} + H_2O \xrightarrow{H+} C_6H_{12}O_6 + C_6H_{12}O_6 .$$

Sucrose Glucose Fructose

The use of strongly acidic cation exchangers with sulfonic acid funtionality to catalyze the inversion of sucrose is practiced and has been detailed by Gilliland, Bixler, and O'Connell [51], as well as Siegers and Martinola [52], and others.

The heterogeneous hydrolysis of sucrose to the respective monosaccharides, glucose and fructose, through the use of cation exchangers, is a first-order reaction. Since this is a catalysis, using an insoluble acid, it avoids the addition of unwanted impurities associated with the conventional practice of homogeneous hydrolysis through acid addition, followed by neutralization with soda. As expected, the reaction rate increases with an increase in contact time and temperature and is inversely related to resin crosslinking and resin particle size.

Macroporous resins are usually preferred in this application because of their generally superior resistance to osmotic shock. Operating temperatures should be below 50 °C to limit the formation of undesirable hydroxymethylfurfurol, a color precursor. Viscosity considerations at temperatures below 50 °C limit the concentration of such solutions to under 60% dissolved solids. Such low solid syrups cannot be stored for extended periods and would also be unsuitable for shipment. Complete sucrose inversion in highly concentrated solution is, therefore, currently not widely practiced with the use of ion exchangers.

2.10.3 Chromatography

The chromatographic separation of fructose from glucose in the production of crystalline fructose [53] was initially based on inverted sucrose solutions as a feed stock. Low-cost production of sweeteners containing 42% fructose on solids and prepared through enzymatic isomerization from starch hydrolysates began in the late 1960's. This development moved with extraordinary speed in the United States to replace sucrose. Extensive ion exchange treatment of these starch hydrolysates made it possible to produce an ash-free, colorless, high-quality product competing with sucrose in high volume markets. Estimated production of fructose containing sweetener from starch hydrolysates reached about five million tons of solids in the United States alone in 1985. Additional fructose enrichment to about 55%/solids was, however, required to match equivalent sucrose sweetness.

Strong cation exchangers in the calcium form are known to partition glucose and fructose [54–57] to provide the means for inexpensive fructose enrichment. This separation is generally considered a function of Ligand Exchange Chromatography, in short, LEC [58]. Other possibilities, including a reaction with the calcium attached to the functional group to form a weak calcium fructosate complex must also be considered. While the non-sorbed components are pushed ahead, essentially unretarded, the adsorbed fructose is eluted with water and arrives at the exit point substantially later. Figure 2.146 illustrates the concentration profile, which may be observed for the solution leaving the column in such a system.

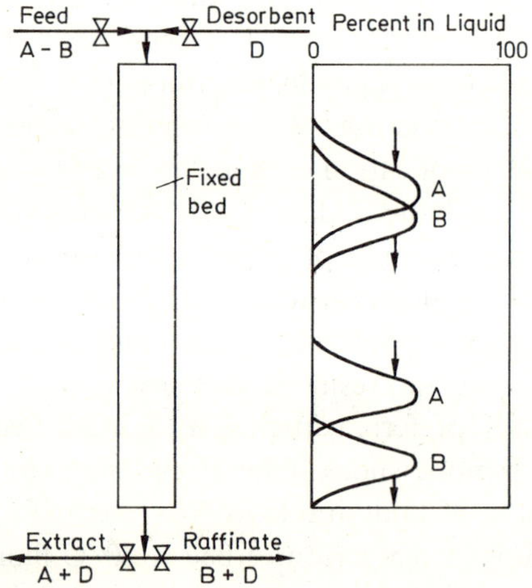

Figure 2.146 Concentration profile during chromatographic partitioning of fructose from glucose (Courtesy of UOP, Inc.).

About two-thirds of the fructose sweetener marketed in the United States in 1985 was enriched to 55% on solids via chromatography after enzymatic isomerization to 42% fructose/solids. About 6000 cubic meters cation exchange resin is required only for this fructose enrichment. A basic flow diagram for this operation is shown in Figure 2.147.

The diagram provides for a 42% fructose concentration on solids as a feedstock for the chromatographic enrichment separator. Enrichment usually exceeds 55%

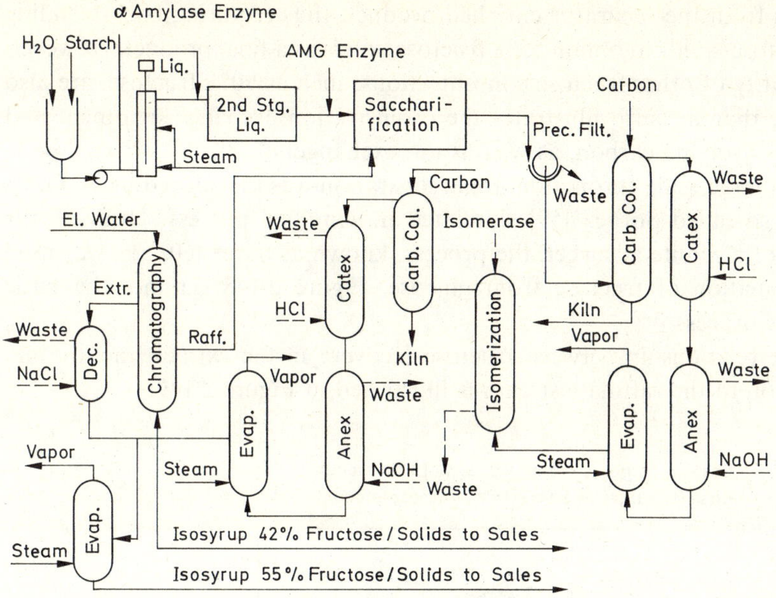

Figure 2.147 Basic flow diagram for the production of 42% and 55% isosyrup sweetener from starch.

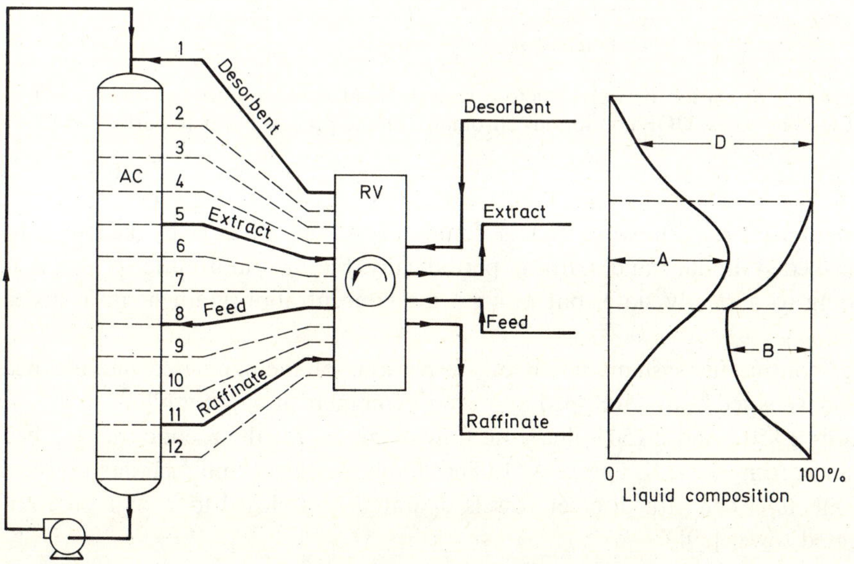

Figure 2.148 The sorbex simulated moving bed concept (Courtesy of UOP, Inc.).

fructose/solids from the separator enriched product, thus requiring back-blending with 42% fructose/solids to obtain 55% fructose/solids as a final product. However, a limited quantity of other fructose concentrations, such as 90% fructose, are also marketed. The diagram also illustrates the extensive ion exchange treatment and decolorization involving carbon, as well as ion exchangers.

Continuous large-scale chromatographic separation was introduced by U. O. P., Inc. with the so-called Sorbex [59] simulated moving-bed process. Utilizing the calcium form of a zeolite adsorber, the process, known as Sarex [60, 61], was used for the fractionation of fructose from glucose. Figure 2.148 outlines the basic concept of this process.

The inverse relationship between fructose recovery in the extract fraction and glucose rejection to the raffinate stream is illustrated in Figure 2.149.

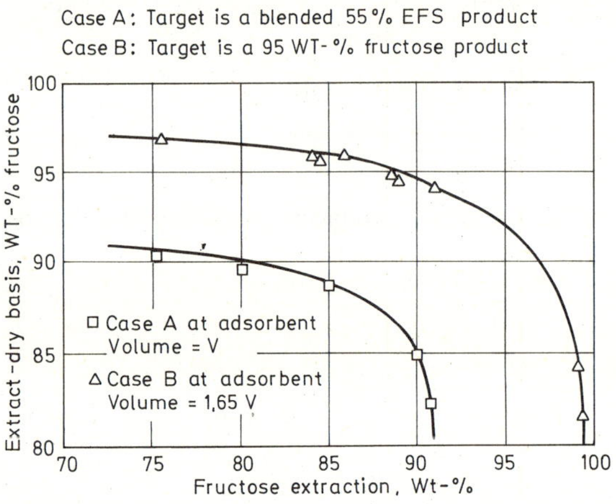

Figure 2.149 Correlation between fructose recovery and fructose purity in the extract at different loading rates. UOP continuous chromatographic process Sarex (Courtesy of UOP, Inc.).

A counter-current, moving bed is simulated in this process by keeping the adsorber-bed stationary and shifting, periodically, the inlet and outlet points corresponding to a steady state, but moving the concentration gradient in a closed loop.

Other continuous systems based on derivations of the simulated moving bed concept have since been developed and are in commercial operation [62–65].

Figures 2.150 a and 2.150 b illustrate flow diagrams for the pseudo moving bed process according to U.S. Patent 4,412,866. Other designs exhibit similar physical arrangements either with discreet vessels operated in a closed loop or as a compartmented tower [59].

The evaluation of operating efficiency for chromatographic separation must consider the following factors:

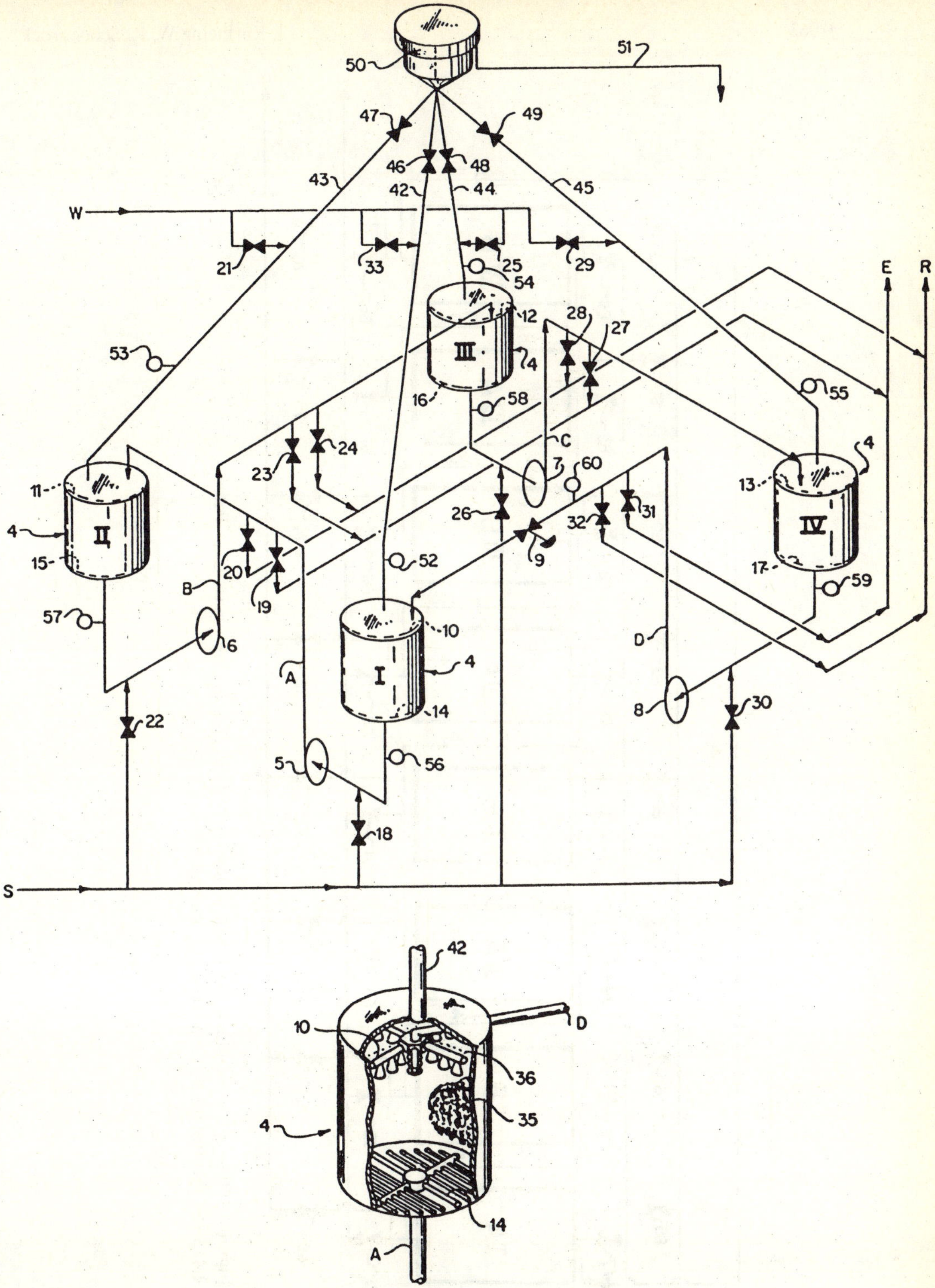

Figure 2.150 a Flow diagram for pseudo moving bed process according to U.S. Patent 4,412,866 featuring 4 columns in series operation.

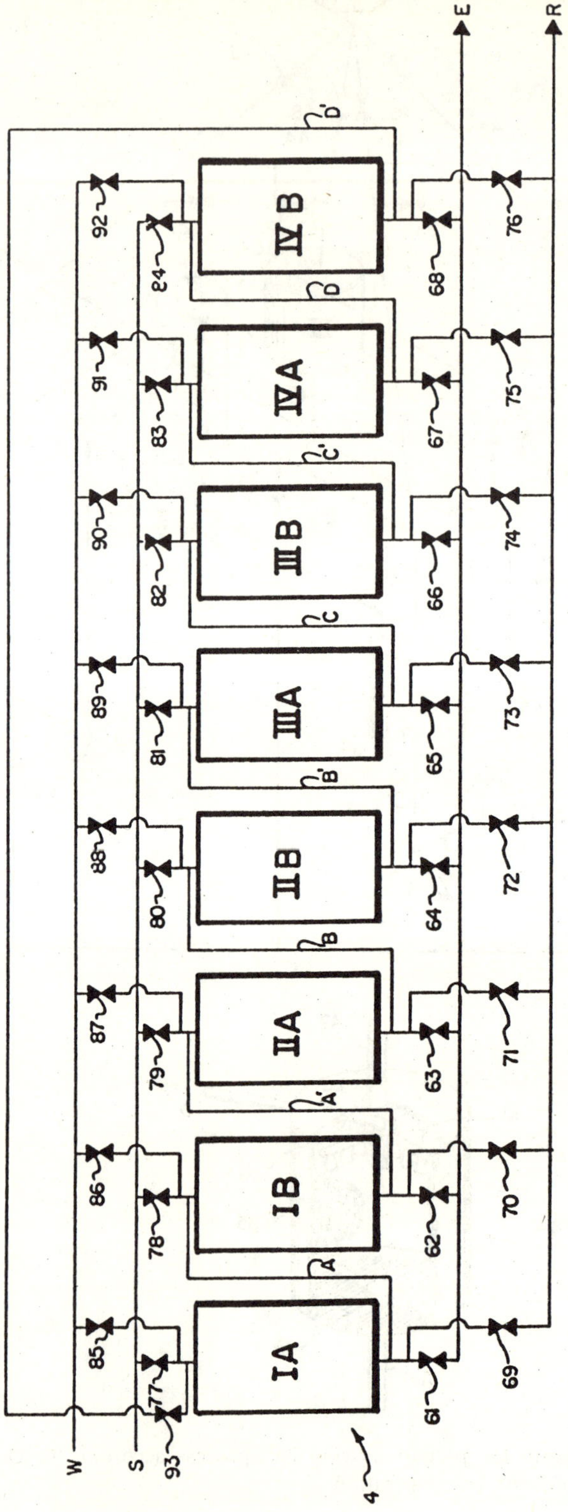

Figure 2.150 b Flow diagram for pseudo moving bed process according to U.S. Patent 4,412,866 featuring 8 columns in series operation.

a) Productivity in terms of final product per liter adsorber per day. A unit value of 1.6 kg 55% fructose/solids enriched from 42% fructose/solids feedstock is a realistic target value.
b) Product recovery in terms of feedstock fructose in the extract. A fine-tuned fructose separator system should exceed 90% of the feedstock fructose recovered in the extract.
c) Non-fructose rejection in terms of fructose purity for the extract. The extract fraction should contain about 90% fructose on solids. Fructose purities in the extract are required to be over 96% for the production of crystalline fructose. Such high purities are routinely achieved with state of the art commercial chromatographic separation.
d) Dilution. Advanced continuous systems should be able to operate with less than 1 kg water added per kg 55% fructose/solids product generated.

Although inherent kinetics are different between chromatography and ion exclusion, the basic operating techniques are very similar.

2.10.4 Ion exclusion

The largest impediment to the application of ion exchangers for separation in general and, specifically, for the purification of impure sugar solutions, is the cost of the chemicals required for regeneration and the disposal of the regenerant waste. Wheaton and Bauman [66, 67] introduced a separation process in 1953 commonly referred to as ion exclusion which promised to eliminate the need for regeneration chemicals. This discovery is based on the observation that certain ion exchangers, in particular cation exchangers, allow dissolved, but non-ionized solutes such as sugar to penetrate freely into the interior of the resin particle while rejecting electrolytes having the same charge as the exchanger's functional group. This characteristic is also known as the Donnan effect. In practical application, an impure sugar solution is applied to a cation exchanger in a selected salt form. Separation occurs as the sugar migrates to the interior of the resin particle while the electrolyte impurities are retained exterior to the resinous phase. The resin bed is then charged with water to wash and displace the sugar-depleted residue solution from the exterior of the resin particle. The aqueous phase also serves to elute the migrated sugar from the resinous phase. An interface is thus established between a front of impurities followed by a progressively enriched, but retarded and purified, sugar solution. The correct application of the various charges to the resin bed and applying the proper cut-off points for the fractions emanating from the separator bed was expected to bring about the desired purification without the use of chemical regenerants.

Earlier work studying this concept for the separation of sugar from impure sugar liquor was unsuccessful until Reents et al. [68] demonstrated a rapid decrease of resin pore size, due to multivalent cation exchange, which apparently restricted sucrose penetration, in the presence of such cations. Additionally, the salts of multivalent cations having a different charge were not repelled by the functional groups, which allowed their penetration into the resinous phase to be commingled with the sugar's entry, changing the salt form of the exchanger, thus resulting in poor separation.

Subsequent work by Prielipp and Keller [69], and Norman, Rorabaugh and Keller [70] established the feasibility of ion exclusion for the purification of impure sugar solutions.

This study also included exploration of a continuous system based on the Higgins [71] continuous contactor, featuring pulsed movement of the adsorber medium. Additional work with the Higgins contactor was reported by Keller [72].

A somewhat different approach utilizing the calcium form of a strong cation exchanger with low crosslinking was proposed by Munir [73] for the chromatographic fractionation of molasses into sugar and non-sugar fractions.

Due to the high monovalent cation load in impure sugar solutions, this process requires frequent regeneration of the adsorber to the calcium form, with associated necessary disposal of regenerant waste. An overall dilution from 85% to a composite concentration of only 4.7% has been reported. Loading capacities are also very low. No commerical application of this process has, as yet, been reported.

Commercialization of ion exclusion for the purification of impure sugar liquors in beet sugar factories was subsequently attempted by several companies using batch-type discontinuous systems. The Utah-Idaho Sugar Company, U. S. A., placed a one-column system with 22 cubic meters resin on stream at their Moses Lake Plant in 1973. The attempt failed primarily because of an inability to eliminate calcium and suspended solids from the liquors prior to the ion exclusion process and, possibly, also overloading of the adsorber. Pfeiffer and Langen of West Germany operated a prototype ion exclusion column containing 55 cubic meters at their Euskirchen plant in 1974 [74]. The plant has been in operation periodically for several years, but has been mostly idle in recent years.

A sugar recovery of about 77% in the product and a non-sugar rejection to the waste of about 80% was reported from those studies. Overall dilution was about 814% on original 80 Bx feed molasses. In actual commercial operation, completing 99 cycles over a five-day operation period, recovery of crystalline sugar was reported at 58.3% on the sugar in the feedstock. Losses to secondary molasses were 19% and about 22% to the non-sugar fraction.

The Finnish Sugar Company collaborated with Pfeiffer and Langen Company on the development of this ion exclusion process and has since placed four commercial ion exclusion plants on stream in Finland, Belgium, and Germany [75–77]. The latest and largest ion exclusion plant for the recovery of sucrose from beet molasses featuring the Finnsugar/Pfeiffer & Langen Process was erected in 1983 at Frellstedt, West Germany. It utilizes a total of about 600 cubic meters cation exchanger resin and is designed to treat 60000 tons of molasses annually [78].

Ion exclusion has also been recommended for the recovery of sugar from cane refinery molasses [79, 80]. Cane molasses is the feedstock for an ion exclusion plant at the Hokubu Factory on Okinawa. This plant came on stream in early 1985 and features Mitsubishi technology.

Besides a very high invert load, the main challenge to the de-sugarization of cane molasses via ion exclusion is the difficulty of its clarification and softening.

Hokkaido Sugar Company of Japan reported the installation and successful operation of a continuous ion exclusion process featuring simulated moving bed technology [64] for the recovery of sugar from low grade Steffen molasses at their Kitami factory [81].

State of the art ion exclusion technology should recover about 95% of introduced sugar with over 85% nonsugar elimation when loading in excess of 160 kg nonsugars per cubicmeter resin per day. Loadings in excess of 300 kg nonsugars per cubicmeter resin per day are possible under ideal conditions.

The purified fraction from a well designed and operated system may typically contain between 35–40% dissolved solids with a sugar purity between 90–96% depending on the feedstock quality. The nonsugar fraction typically averages between 5–7% dissolved solids and contains over 90% of the color with less than 5% of the sugar and essentially all cations introduced, respectively.

At least 6 ion exclusion plants featuring pseudo-moving bed technology are currently in operation or under construction in the U.S.A. and Japan with additional units on the drawing board.

This process can operate on either beet or cane molasses liquors ranging from thick juice to molasses. Feed streams to such a separator system should be free of suspended solids and entrained gases, and be nearly free of multivalent cations to prevent fouling, channelling, and plugging of the resin bed.

The adsorber beds in ion exclusion or chromatographic operations are infrequently backwashed; hence, they can suffer from bed compaction and subsequent flow restrictions unless appropriate provisions are incorporated to relieve bed compaction. Such provisions are taught by the literature [65]. The above mentioned design reverses recirculation flow in the loop momentarily only in that column which is at that time in the water phase. The continuous forward movement of the flow in the system is maintained without process interruption. Bed compaction from continuous one-directional flow is, however, avoided.

An efficient ion exclusion process requires specially designed resins. The Donnan potential, and with it the efficiency of electrolyte exclusion, increases with increasing exchange capacity and matrix crosslinking, while the capacity to retain nonelectrolytes decreases [82]. The need for high sorption capacity is usually of greater practical and economical importance. Hence, operating efficiency generally improves as crosslinking and particle size decrease. Resin stability and tolerated pressure drop through a resin column dictate a minimum crosslinking of 4% and an average resin bead size not less than 250 μm. Usually, the particle size lies between 300 and 400 μm. Particle size distribution of the resin should be as narrow as possible, preferably below a coefficient of variation of less than 10. Compromises are frequently made between these parameters. Modern installations will usually specify a crosslinking of at least 6%. Lower crosslinked resins are more susceptible to de-crosslinking through oxidation and deformation, which can lead to excess pressure drop and generally increased attrition losses. Loading capacities deteriorate if crosslinking is significantly increased above 8%. Developments in the continuous and automated operating techniques, as well as recent advances in the manufacture of ion exchange resin especially suitable for separation work by ion exclusion and commercial chromatography [43], promise to be the leading edge in the application of ion exchange resins for the efficient separation of mixed components. A major impact is already evident in the sweetener industry. Applications in the pharmaceutical and chemical industries are just now being explored.

2.10.5 Cation/anion exchange

Although the use of cation exchangers may have started in antiquity, the intentional application of anion exchangers did not commence until after Adams and Holmes [83] invented synthetic anion exchangers. The first reported work utilizing anion exchangers for the purification of sugar beet liquors began in 1940 at the beet sugar factory Oestrum, Germany [84, 85] using ion exchangers developed by IG Farbenfabrik, Wolfen, Germany. Similar studies were carried out at the Betteravia and Mount Pleasant Sugar Beet Factories in the United States during 1941 [86, 87]. This work resulted in a U. S. patent issued to Valez [88]. Work was also started to de-ionize and de-colorize starch hydrolysates [87] during this period. These studies, together with other work in the United States, led to several commercial deionization plants in the North American sweetener industry [87, 89–92].

2.10.5.1 Acid/base exchange

All of the initial work was based on the operation of the cation exchangers in the hydrogen form and the operation of the anion exchanger in the free base-form. Vast improvements have been made since that time in the quality of the ion exchange resins, and the technology of their application. The basic technique for total deionization of impure sugar liquors and starch hydrolysates, has, however, essentially remained unchanged. It involves, first, the treatment of the cooled, impure sugar solution over the hydrogen form of strong cation exchangers with sulfonic acid functionality. Cooling to below 10 °C is required to suppress inversion of the sucrose on the exchanger, referred to as heterogeneous inversion. Somewhat higher temperatures are tolerated for starch hydrolysates. During this treatment, all cationic species are adsorbed on the cation exchanger, and exchanged for hydrogen ions.

This action converts the neutral salts in the impure sugar solutions to their respective free acids. Accordingly:

$$\text{I. } R \cdot SO_3^-H^+ + Cat^+An^- \rightarrow R \cdot SO_3^-Cat^+ + H^+An^-.$$

where: R = resin matrix
Cat = Cationic species of the impurities
An = Anionic species of the impurities.

Certain amphoteric electrolytes, such as amino acids, are also extensively removed by the cation exchanger when operated over the hydrogen form:

$$\text{II. } R \cdot SO_3^-H^+ + HOOC \cdot CH\ NH_2 \cdot (CH_2)_2 \cdot COOH \rightarrow R \cdot SO_3NH_2 - CH(CH_2)_2(COOH)_2 + H^+.$$

The impure sugar solution thusly treated is then taken over the base form of an anion exchanger to remove anionic species. Although strong-base exchangers with quaternary amine functionality are more effective for removing residual impurities, weak-base exchangers with secondary or tertiary amine functionality are preferred in this application. Weak-base exchangers have inherently superior regeneration efficiency, and improved long-term stability in the functional groups. Accordingly:

$$\text{III. } R{-}NR_2' + H^+An^- \rightarrow R{-}NR_2' \cdot H^+An^-.$$

Where R′ = organic radical, or hydrogen.

Sorption of the free acid on the functional group of the weak anion exchanger is indicated. Anions must be present as free acids to be adsorbed on weak-base exchangers. Neutral salt anions can not be removed by weak-base exchangers.

Regeneration of the cation exchanger requires dilute strong mineral acids at 200 to 500% of the stoichiometric operating capacities:

$$\text{IV. } R \cdot SO_3^-Cat^+ + H^+Cl^- \rightarrow R \cdot SO_3^-H^+ + Cat^+Cl^-.$$

Dilute ammonia, soda ash, or caustic solution at 125 to 200% of the stoichiometric operating capacity are typically used for the regeneration of the weak-base exchanger:

$$\text{V. } R{-}NR_2'H^+An^- + NH_4^+OH^- \rightarrow R^-NR_2' + NH_4^+An^- + H_2O.$$

2.10.5.2 Purification of impure sugar solutions

The acid/base ion exchange process has the capacity to remove essentially all non-sugar impurities from impure sugar solutions to bring about nearly complete purification. Only non-sucrose carbohydrates and small amounts of other non-ionized components or break-down products from sugar are retained by the ion exchange-treated solution.

The average equivalent weight for the exchanged non-sugars in clarified sugar beet extracts may vary between 100 to 150. In practical application, this requires about 20 kg equivalents of strong mineral acid and between 12 to 15 kg equivalents base as regenerant chemicals to extract an additional 1.5 tons of sugar by acid/base ion exchange. Either hydrochloric or sulfuric acid is usually used for the regeneration of the cation exchanger at a concentration between 1.0 and 2.0 N. Similar regenerant concentrations prevail for the regeneration of the anion exchangers. The impurity load of sugar syrups considered for acid/base ion exchange may vary from as high as 8%/sugar for high quality thin juice to as low as 50%/sugar for low-grade final molasses. With the use of sulfuric acid and ammonia as the regenerants, the combined regenerant wastes will contain about 2.5 tons of dissolved solids at an average concentration of around 5% for the additional extraction of 1.5 tons of sucrose. Costs for regenerant chemicals and the handling of ensuing wastes are formidable.

Earlier application favored the treatment of clarified thin juice. A trend toward the treatment of more concentrated solutions including crystallization run-off may be observed for later installations. The latter represents an increased ionic load, but a substantially reduced hydraulic load.

McGarvey [93] and Bichsel et al. [94] studied the economics of ion exchange in the beet sugar industry.

The reader is referred to the extensive literature on the acid/base ion exchange process for more detailed information [9—12]. All of the acid base ion exchange installations in the U. S. A. prior to 1960 were idled after only a few years of operation. Interest in this process was rekindled between 1960 and 1980 with improved resin material and advanced technology. Many of thcsc lattcr installations in Europe, Japan, and the United States have, likewise, become inoperative in recent years as world sugar prices plummeted and environmental control requirements soared in most parts of the world.

The high ionic impurity load in clarified sugar-bearing plant extracts presents a formidable obstacle to the efforts to generate a favorable cost/benefit relation for acid/base ion exchange. In this context, it is the large amount of regenerant chemicals required and associated waste generated per unit sugar recovered which prevents the acid/base ion exchange process from expanding in the sugar industry.

Waste handling. Studies to recover spent regenerants from an acid/base ion exchange plant began as early as 1947 [95].

A similar approach was carried to a pilot plant stage by the Holly Sugar Corporation at their Hamilton City, California, plant in 1968 [96]. The process was further refined and incorporated on a large scale at the Moorhead, Minnesota, plant of The American Crystal Sugar Company [97]. The salient parts of this process involve separation of the spent regenerant waste into a so-called reconstituted non-sugar product suitable as a cattle feed and a granulated ammonium sulfate fertilizer. This is accomplished by first regenerating the exhausted anion exchanger with aqueous ammonia. The spent anion regenerant is next used to strip the exhausted cation exchanger, thereby converting it to the ammonium form. The resultant effluent, which contains essentially most of the impurities originating from the sugar liquor to be de-ionized, is concentrated and used as an additive in cattle feed. Regeneration of the stripped cation exchanger, now in the ammonium form, proceeds with sulfuric acid and generates an acidic ammonium sulfate solution. This spent cation regenerant is neutralized with additional ammonia, concentrated, crystallized, and granulated to produce a commerical-grade fertilizer. Although this approach offers a viable alternative to the waste disposal problems for the spent regenerants, it has not been sufficient to overcome associated costs and low sugar prices.

Ion exchange treatment of starch hydrolysates. With the development of isomerized starch hydrolysates, the application of acid/base ion exchange found new impetus in the caloric sweetener industry. The bulk of starch hydrolysate sweeteners is sold as syrups and can not easily be purified at this time through crystallization, as is common practice with sucrose. The well-developed acid/base ion exchange technology was ideally suited to provide the degree of purification required to compete with sucrose on certain quality factors, such as color and ash. Without crystallization, a relatively low total impurity load of around 50 equivalents per ton of solids product assures a cost-effective application of acid/base ion exchange in this industry. Figure 2.151 illustrates the common practice of deionizing before and after isomerization. A dual pass consisting of a primary and secondary acid/base exchanger combination is commonplace in both applications to maximize the degree of purification. Deionization before isomerization removes an ionic load of about 40 equivalents per ton of solid products. Small amounts of activation chemicals required for isomerization by immobilized enzymes are removed in the deionization after isomerization. The ionic load in this treatment can be restricted to under 10 equivalents per ton of solid product.

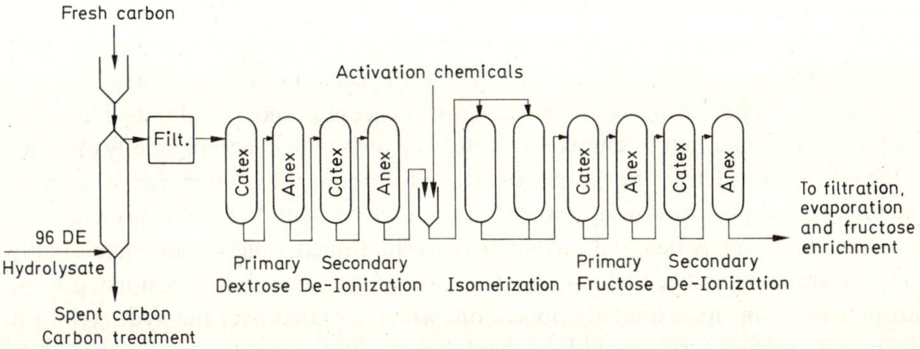

Figure 2.151 Dual pass de-ionization of starch hydrolysates in the production of high fructose sweetener.

A third ion exchange treatment primarily for decolorization is frequently exercised after fructose enrichment by chromatography. This treatment has taken the form of mixed-bed acid/base exchange, or simple decolorization via a strong base exchanger with quaternary ammonium functionality in the chloride form.

Carbon treatment before deionization is extensively practiced in the purification of starch hydrolyzates. This practice is beneficial in removing resin foulants; hence, it extends useful resin life. Even though hydrochloric acid is more expensive than sulfuric acid, it is generally preferred for cation exchange regeneration.

The weak anion exchanger is usually regenerated with soda ash solution. Ammonia is generally avoided because of associated problems in primary and secondary waste treatment. The continuous use of caustic to regenerate weak anion exchanger is also avoided because of increased cost and the tendency to accelerate the lengthening of the rinse step after regeneration. This progressive deterioration in rinse efficiency is caused by the creation of cation exchange sites in the anion exchanger. During regeneration, these exchange sites are converted to the salt form. The cation is eluted during the rinse step, leading to the formation of free base in the effluent rinse water. Rinsing of the cation exchanger is, therefore, frequently combined with the rinsing of the anion exchanger to provide de-cationized water for rinsing the anion exchanger. Such practice can reduce the rinse requirements somewhat, but is always associated with a loss in operating capacity.

2.10.5.3 The Vajna process

The high cost for regeneration chemicals and associated disposal of wastes has led to many attempts to develop schemes which generate required chemicals from the wastes. Vajna [98] proposed the operation of a cation exchanger over the ammonium form and operation of a strong base anion exchanger over the hydroxyl form. Figure 2.152 illustrates the chemical interaction between the various process streams in Vajna's proposal.

The system disposes only of the unavoidable impurity fraction originating with the untreated juice and solid, but essentially pure, calcium carbonate. The lime and the carbon dioxide required to sustain the process could be obtained through calcination of generated calcium carbonate. The process was commercially operated at the beet sugar factory, Forlimpopoli, Italy, for about three campaigns. The basic chemistry of the process was initially proven to be sound. Failure became unavoidable, however, when the strong base resin fouled quickly and irreversibly, coupled with progressive hydrolysis of its quaternary ammonium functional groups. Strong base resins have limited applications when operated over the hydroxyl form. Operating temperatures should be held below 30 °C and treatment should be restricted for syrups of relatively high purity with low concentration of irreversibly sorbed impurities when involving the hydroxyl form.

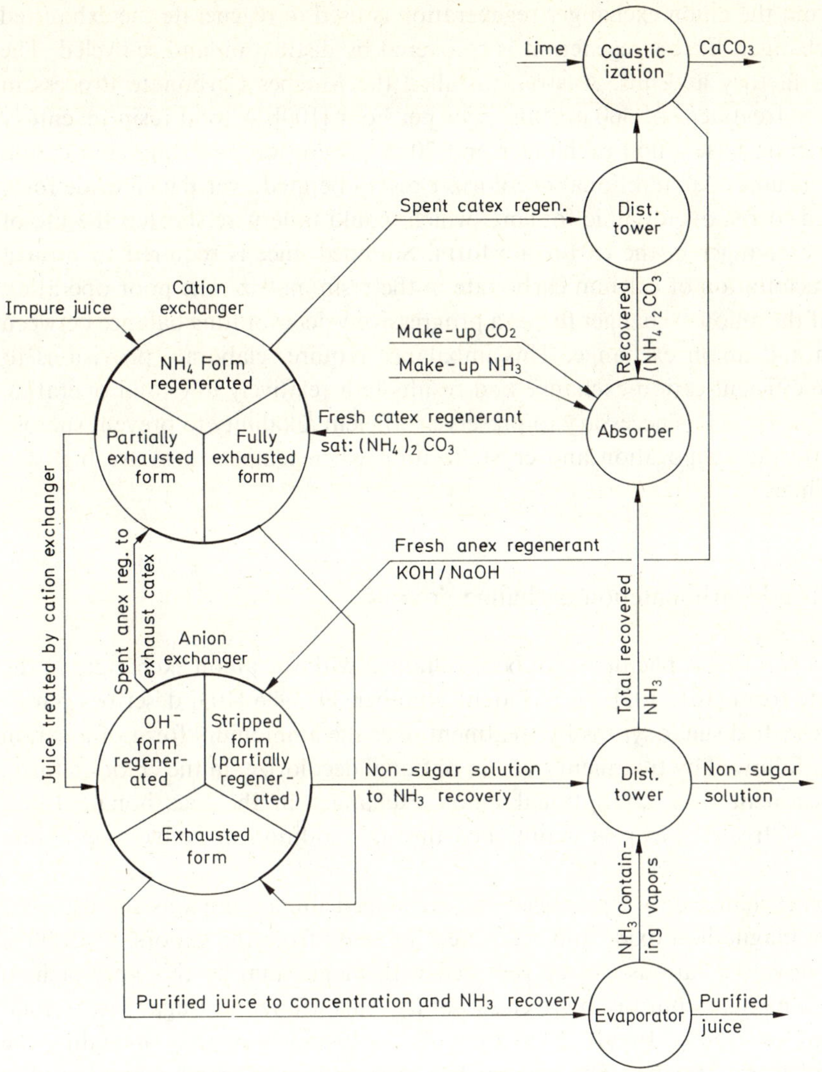

Figure 2.152 Chemical inter-action in the Vajna Process for the purification of impure sugar liquors by ion exchange.

2.10.5.4 The Moebes carbonate process

Moebes suggested avoiding the hydroxyl form of the strong base exchanger and used instead its respective salt form [99]. His carbonate process differs from the Vajna process by operating the strong anion exchanger over the carbonate form.

In the Moebes Carbonate Process, the anion exchanger is first regenerated with 20% ammonium carbonate solution at 600% of the capacity of the exchanger. The effluent from the anion exchanger regeneration is used to regenerate the exhausted cation exchanger. Excess regenerant is recovered by distillation and re-cycled. The beet sugar factory at Enns, Austria, installed the Moebes Carbonate Process in 1968 for the treatment of 300 m^3 thin juice per hour [100]. A total resin inventory of 36 m^3 strong base anion exchanger and 20 m^3 macroporous strong acid cation exchanger is used. An additional decolorizer resin operated over the chloride form is also used to remove organic foulants which would otherwise shorten the life of the anion exchanger in the carbonate form. Softened juice is required to protect against precipitation of calcium carbonate in the resin matrix. The poor operation capacity of the anion exchanger forces a progressively deteriorating balance between the cation and anion exchange. This imbalance requires elaborate provisions to bypass the efficient cation exchange and results in a relatively low total operating capacity. The bypass is necessary to preserve sufficient alkalinity to prevent sucrose inversion during evaporation and crystallization when temperatures are high for extended times.

2.10.5.5 The bicarbonate ion exchange process

Increased operating capacities have been obtained with the anion exchanger in the bicarbonate form [101, 102]. U. S. Patent Number 3973986 [103] describes purification of clarified sugar syrups by treatment over the ammonium form of a cation exchanger, followed by treatment over an optional decolorizer in the chloride form; and/or treatment over an optional anion exchanger in the bicarbonate form. Alkalinity of treated syrup is maintained through addition of active magnesium oxide.

If cation exchange only is practiced, the exchanged ammonium ions are displaced with active magnesium oxide and recovered for reuse from the vapors. Over 90% of the monovalent cations can be replaced with magnesium by this approach to reduce final molasses purities by over 20 points. This will reduce sugar lost to final molasses by over 50%. Figure 2.153 a details the complete process, including the handling of waste streams. The process has been employed in pilot plant studies and yielded increased sugar recoveries of over 10% [33].

A more recent development simplifies the bicarbonate process by featuring a mixed bed of strong anion exchangers and cation exchangers at a ratio of 2.0 to 2.5 [104]. Exchange efficiency is substantially raised in mixed beds with reduced cost for capitalization and operation. Exchanged ammonium bicarbonate is expelled with the vapors from the treated syrup during evaporation. These vapors, in turn, are used to drive a distillation column which recovers excess ammonium bicarbonate from the spent regenerant. The bottoms from the distillation contain the separated impurities which could be used as an additive in animal feed. The overhead vapors from the distillation are condensed and carbonated in an absorber to prepare the required ammonium bicarbonate regenerant. Only a small amount of ammonia

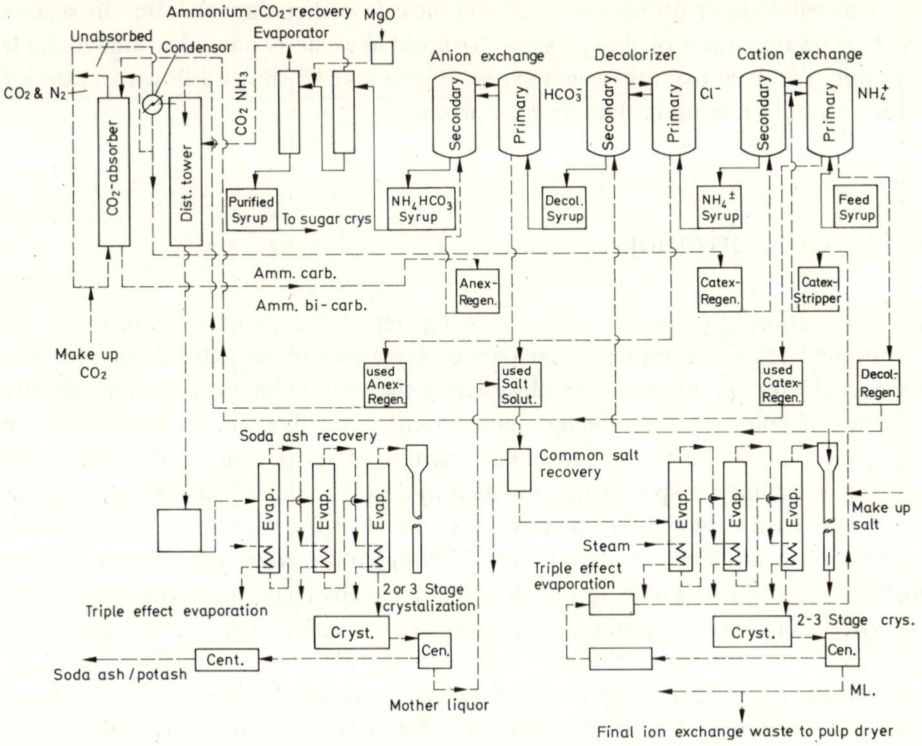

Figure 2.153 a Flow diagram for the ammonium bicarbonate ion exchange process (U.S. Patent 3973986).

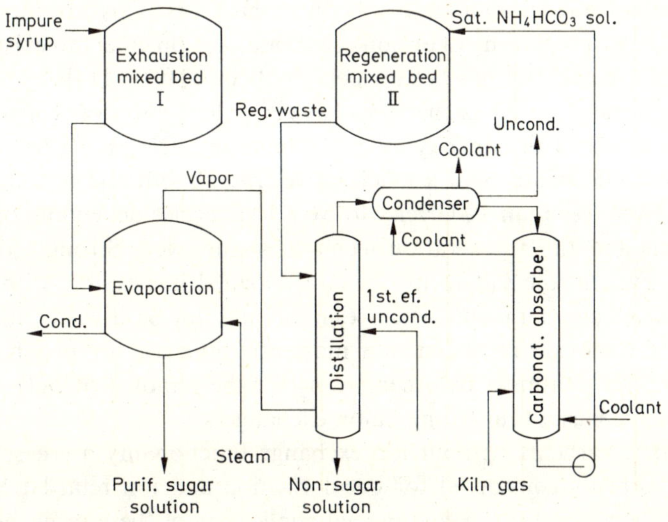

Figure 2.153 b Flow diagram for the mixed-bed ammonium bicarbonate process [104].

and somewhat larger quantities of carbon dioxide makeup are required to sustain steady state operation of this process. Disposable waste is limited to unavoidable impurities removed from the sugar syrup. Figure 2.153 b shows a flow diagram for the mixed bed ammonium bicarbonate process.

2.10.6 Decolorization

The elimination of color is a key processing step in the refining of sugar syrups. Traditionally, the industry was dependent on bone char or activated carbon to meet decolorization requirements. The observation that ion exchangers had the capacity of removing color [105] has been a major incentive for their continued exploration in sugar refining. This interest for decolorization of sugar syrups led to the development of techniques specifically designed to decolorize using ion exchange resins [106]. The main objective in this context is the need for high adsorptive capacity with complete and low cost reversibility. Although it is recognized that this ultimate goal is unachievable, efforts in this direction have advanced the application of ion exchange resins for the decolorization of sugar syrups. The conventional procedure involves highly porous strong-base anion exchangers in the chloride form. Low cost sodium chloride at a 10% concentration is universally used as a regenerant. The salt solution is frequently alkalized with about 1% sodium hydroxide to improve regeneration efficiency and maintain high adsorptive capacity. Progressive resin fouling is, however, unavoidable, leading to eventual loss of the adsorptive capacity.

Especially cane sugar refining has incorporated decolorization of refining liquors by ion exchange resin on a large scale in recent years. This development has come about either as a replacement of bone char or as a supplement to existing traditional techniques. The subject has been widely publicized [107–114].

Strongly basic anion exchangers have excellent sorption capacity when first placed in service. This capacity quickly diminishes after only a few cycles to reach a level which deteriorates at a much slower rate. Useful life of a decolorizer resin may last from less than 100 cycles to over 1000 cycles depending on the exposure to foulants and the regeneration techniques employed. Strongly basic anion exchangers, even in the salt form, are easily loaded irreversibly with organic chromophores. As a general rule, such resins should not be used for massive primary bulk color removal. They function most economically for polish decolorization behind carbon treatment or other means for the removal of bulk color, or compounds which easily foul strong anion exchangers.

Polymeric sorbents without ion exchange functionality are a special group of resinous sorbents considered for decolorization and ion retardation. There have been no reports yet for the commercial application of these resins in the sweetener industry.

Several installations in the beet sugar industry operate a mixed bed containing strong cation exchangers in the sodium form and a strong base anion exchanger in the chloride form. This operation combines decalcification and decolorization into a single step. Regeneration is, likewise, a single-step operation using sodium chloride solution [115].

The utilization of the spent regenerant from the Quentin Process as a regenerant for a decolorizer resin has also been commercially practiced.

References

[1] F. Schneider (1968). Technologie des Zuckers, pp. 977–983, M. & H. Schaper, Hannover, Germany.

[2] G. P. Mead and J. C. P. Chen (1977). Cane Sugar Handbook, pp. 361–363, Wiley, New York.

[3] F. Harms (1896). Verfahren der Reinigung von Zuckersäften mittels eisenschüssigen quarzreichen Thons. DRP No. 95447. See also: Verein Rübenzuckerind. *47,* (1897), 1089.

[4] A. Rumpler (1903). Über die Reinigung von Rübensäften durch Silikate. Z. Verein der Deutsch. Zuckerind. *53,* 798.

[5] A. Feldhoff (1906/1907). Cbl. Zuckerind. *15,* 1307.

[6] R. Gans (1907). Reinigung der Zuckersäfte von Kali und Natron vermittelst Aluminatsilikate. Z. Verein der Deutsch. Zuckerind. *57,* 206. See also: DRP No. 174097 (1905).

[7] H. Claassen (1907). Ueber die Behandlung von Zuckersäften und Melassen mit Calcium-Aluminatsilikaten und die Beschaffenheit und Eigenschaften der dadurch gewonnenen Sirupe, insbesondere die Löslichkeit und Kristallisationsfähigkeit des Zuckers in ihnen. Z. Verein Deutsch. Zuckerind. *57,* 931.

[8] B. A. Adams and E. J. Holmes (1934). Brit. Patents 450308 and 450309.

[9] A. Carruthers and J. F. T. Oldfield. Ion Exchange in the Sugar Industry, 13th An. Tech. Conf. B. S. C., Ltd.; Nottingham, England, May 1960.

[10] F. Schneider (1968). Technologie des Zuckers, pp. 589–641, M. & H. Schaper, Hannover, Germany.

[11] S. Landi and G. Mantovani (1975). Ion Exchange in the Beet Sugar Industry. Sugar Technology Reviews. *3,* Nr. 1; Elsevier, Amsterdam, The Netherlands.

[12] R. A. McGinnis, ed. (1982). Beet Sugar Technology, *3rd ed.,* pp. 300–332, Beet Sugar Development Foundation, Fort Collins, Colorado, USA.

[13] C. Hochstetter (1843). J. Pract. Chemie *29,* 1.

[14] H. S. Thompson (1850). J. Royal Agricult. Soc. *11,* 68.

[15] J. T. Way (1850). J. Royal Agricult. Soc. *11,* 313. ld. (1852). Ibid. *13,* 123.

[16] A. Marschall (1870). Z. Verein Rübenzuckerind. *20,* 339.

[17] A. Herzfeld (1892). Z. Verein Rübenzuckerind. *42,* 186.

[18] F. Schneider (1968). Technologie des Zuckers, pp. 613–615, H. & H. Schaper, Hannover, Germany.

[19] L. P. Orleans, W. A. Harris, L. W. Norman and H. W. Keller (1965). Operations of juice softener in a beet sugar factory. J. Amer. Soc. Sugar Beet Technologists *13,* 296.

[20] D. A. Muller (1966). Ion exchange deliming at Eaton Factory. Pres. 14th General Meeting American Society Sugar Beet Technologists; Minneapolis, Minn., Feb. 1966.

[21] A. Carruthers and J. F. T. Oldfield (1961). Ionenaustauschverfahren in der Zuckerindustrie. Z. Zuckerind. *11,* 23. Id. (1961), Ibid. *11,* 85.
[22] D. Ramondt (1963). Listy Cukrovarnicke, 228.
[23] J. Vlasak (1965). Listy Cukrovarnicke, 165.
[24] I. M. Litvake and D. D. Bobrownik (1961). Trudu KTIPP, 22.
[25] A. Zsigmond, E. Gryllus and E. Magyar (1966). Recent experiences on the application of ion exchangers in some branches of the food industry. Zucker, *19,* 621.
[26] J. W. Walter (1968). Deliming of beet sugar juice by Degremont-Cottrell continuous ion exchange. Pres. 15th Gen. Mtg. A. S. S. B. T., Phoenix, Ariz., Feb. 1968.
[27] V. Gryllus (1967). Verfahren zur Regenerierung des Ionenaustauschers und zur Verringerung des Alkali-Ionengehaltes von Zuckerfabriksabläufen bei der Enthärtung von Dünnsaft mittels Ionenaustausches. Austrian Pat. No. 258230.
[28] E. Felber (1970). New process for deliming thin juice without regenerating agents and waste water. Pres. 16th Gen. Mtg. A. S. S. B. T., Denver, Colorado, Feb. 1970.
[29] V. Gryllus and J. J. Delavier (1975). Das BMA-Zsigmond-Gryllus-Verfahren, ein neues Verfahren zur Dünnsaftentkalkung. Z. Zuckerind. *25,* 493, 554.
[30] W. Pannekeet (1960). Nouveau dévelopment dans les systèmes de régénération des échangeurs d'ions en sucrerie. Industrie Alimentaires et Agricoles, 757.
[31] P. L. Mottard (1983). The Imacti process for juice decalcification. Int. Sugar J. *85,* 233.
[32] K. W. R. Schoenrock, P. Richey and H. G. Rounds (1975). Process for the decalcification of sugar beet juice. U. S. Pat. 3887391.
[33] K. W. R. Schoenrock and A. C. Gupta (1977). New ion exchange systems for the beet sugar industry. I. Softener. Zucker *30,* 541. See also: La Sucrerie Belge *95,* 287 (1976).
[34] K. W. R. Schoenrock, A. Gupta and H. G. Rounds (1976). Process for the purification of impure sugar juice. U. S. Pat. 3982956.
[35] G. Quentin (1954). Über den Einfluss anorganischer Kationen auf die Löslichkeit der Saccharose in Melassen. Zucker *7,* 407.
[36] G. Quentin (1957). Der Einfluss der Kationen auf die Saccharoselöslichkeit in Melassen und die Möglichkeiten einer technologischen Auswertung der unterschiedlichen Löslichkeitsbeeinflussung. Zucker *10,* 408.
[37] G. Quentin (1961). Verfahren zur Erhöhung der Kristallisationsfähigkeit des Zuckers in Lösungen der Zuckerfabrikation. Ger. Pat. 974408.
[38] H. Neuman (1953). Das Quentinverfahren in der Zuckerfabrik Franken. Zucker *12,* 374.
[39] R. Hoffman and H. Rother (1960). Observations during operation of a Quentin plant. Zucker *13,* 380.
[40] K. W. R. Schoenrock (1971). The application and utility of magnesium exchange in the beet sugar process. J. A. S. S. B. T. *16,* 299.
[41] F. Perschak (1968). Application of Quentin process in the Hohenau Sugar Factory. Zucker *21,* 364.
[42] J. F. T. Oldfield et al. (1979). The Quentin Process. Intern. Sugar J. *81,* 103. Id. (1979). Ibid. *81,* 138.
[43] Dow Chemical Company, U. S. A. New Dowex Monosphere T. G. Pamphlet CC No. 77417; Form No. 177–13–84.
[44] K. W. R. Schoenrock (1974). Recent advances in ion exchange techniques and apparatus design. Proc. 18th General Meeting A. S. S. B. T., San Diego, California, Feb., 1974.
[45] K. W. R. Schoenrock and H. G. Rounds (1977). Ion Exchange method. U. S. Pat. 4001113.
[46] F. X. Kammerer, K. Tesch and N. Zollner (1977). Quentin-Doppelstock-Ionenaustauschanlage in der Zuckerfabrik Ochsenfurt. Z. Zuckerind. *27,* 271.

[47] E. Moebes (1960). Der Carbonat-Prozess, ein Verfahren zur Reinigung des Dünnsaftes. Zucker *13*, 254.
[48] E. Moebes and L. Wienininger (1955). Zucker *8*, 129.
[49] French Pat. 1109784 (1955).
[50] W. Wisfeld (1958). Über ein Verfahren zur Erhöhung der Zuckerausbeute unter Verwendung ionenaustauschender Stoffe. Zucker *11*, 425.
[51] E. R. Gilliland, H. J. Bixler and J. E. O'Connell (1971). Catalysis of sucrose inversion in ion exchange resins. Ind. Eng. Chem. Fund. *10*, 185.
[52] G. Siegers and F. Martinola (1985). Extensive inversion of sucrose solutions with strongly acidic cation exchange resins. Intern. Sugar J. *87*, 23.
[53] K. Lauer (1980). Technische Herstellung von Fructose. Starch/Stärke *32*, 11.
[54] G. R. Serbia (1962). U. S. Pat. 3044904.
[55] J. Lefevre (1962). U. S. Pat. 3044905.
[56] J. Lefevre (1962). U. S. Pat. 3044906.
[57] H. W. Keller, A. C. Reents and J. W. Laraway (1981). Process for fructose enrichment from fructose bearing solutions. Starch/Stärke *33*, 55.
[58] R. W. Goulding (1976). J. Chromatogr. *103*, 232.
[59] D. B. Broughton (1961). Continuous sorption process employing fixed bed of sorbent and moving inlets and outlets. U. S. Pat. 2985589.
[60] D. B. Broughton et. al. (1977). High purity fructose via continuous adsorptive separation. La Sucrerie Belge *96*, 155.
[61] H. J. Bieser and A. J. de Rosset (1977). Continuous counter-current separation of saccharides with inorganic adsorbents. Starch/Stärke *29*, 392.
[62] Anon. (1983). Food Engineering, May, 1983, p. 154.
[63] T. Hirota (1980). Continuous chromatographic separation of fructose/glucose. Sugar y Azucar *75*, 245.
[64] K. W. R. Schoenrock, M. Kearney and E. Rearick (1983). Method and apparatus for the sorption and separation of dissolved constituents. U. S. Pat. 4412866.
[65] K. W. R. Schoenrock (1985). Method for preventing compaction in sorbent beds. U. S. Pat. 4511476.
[66] R. M. Wheaton and W. S. Bauman (1953). U. S. Pat. 2684331.
[67] R. M. Wheaton and W. S. Bauman (1953). Ion exclusion, a unit operation utilizing ion exchange materials. Ind. Eng. Chem. *45*, 228.
[68] A. C. Reents and H. W. Keller (1960). U. S. Pat. 2937959.
[69] G. E. Prielipp and H. W. Keller (1956). J. Amer. Oil Chem. Soc. *33*, 103.
[70] L. Norman, G. Rorabaugh and H. W. Keller (1963). Ion exclusion purification of sugar juices. J. A. S. S. B. T. *12*, 363.
[71] Higgins (1953). U. S. Pat. 2815332.
[72] H. W. Keller (1967). Continuous ion exclusion for sugar purification. Proc. Am. Chem. Soc. An. Mtg., Miami Beach, Florida.
[73] M. Munir (1975). Desugarization of molasses by means of liquid distribution chromatography. Zucker *28*, 287.
[74] H. G. Schneider and J. Mikule (1975). Recovery of sugar from beet molasses by the P & L exclusion process. Int. Sugar J. *77*, 259 and *77*, 294.
[75] O. Adriaensen (1979). Experience pratique avec un desucrage de melasse par separation chromatographique. La Sucrerie Belge *98*, 377 and La Sucrerie Belge *98*, 377.
[76] H. Heikkila (1983). Chromatographie – ein technisches Trennverfahren. Chem. Ind. XXXV, *15*, Nr. 8, 456.
[77] H. Heikkila (1983). Separating sugars and amino acids with chromatography. Chem. Eng., Jan. 24, 1983, 50.
[78] Anon. (1983). Amino GmbH, Frellstedt, mit neuer Melasseentzuckerungsanlage. Zuckerind. *108*, 636.

[79] H. G. Schneider (1978). Ion exclusion in cane sugar refining. 37th An. Mtg. S. I. T., London, May, 1978.
[80] H. Hongistro, H. Heikkila and H. Paananen (1978). Desugarization of refinery molasses. 37th An. Mtg. S. I. T., London, May, 1978.
[81] Mitsui Trading Company (Japan). Yearbook Japan Sugar Industry 1987.
[82] F. Helfferich (1962). Ion exchange. pp. 134–146, 431–433. McGraw Hill Book Co. Inc., New York.
[83] B. A. Adams and E. J. Holmes (1935). J. Soc. Chem. Ind., *54*.
[84] Personal communication in 1962 with Dr. E. Hemmecke, Director, Oestrum Beet Sugar Factory.
[85] H. J. von der Linde (1962). Demineralization of sugar juices by means of Lewatit in the Sugar Factory Oestrum. Zucker *15*, 340.
[86] F. W. Weitz (1943). Sugar *38*, 26.
[87) Diamond Shamrock Chem. Co., (1972). Publication: Ion Exchange Resins in the Treatment of Sugar Solutions.
[88] H. A. Vallez (1945). U. S. Pat. 2388194.
[89] B. N. Dickinson (1948). Commercial ion exchange purification of sugar bearing solutions. Chem. Eng. *55*, 114.
[90] F. N. Rawlings and R. W. Shafor (1942). Sugar *37*, 1, 26.
[91] R. A. McGinnis, ed. (1982). Beet Sugar Technology. 3rd ed., p. 300. Beet Sugar Development Foundation, Fort Collins, Colorado.
[92] B. N. Dickinson (1950). Sugar *45*, 20.
[93] F. X. McGarvey (1964). An evaluation of a multiple bed deionization process for beet sugar recovery. Journal A. S. S. B. T. *13*, 252.
[94] S. E. Bichsel, H. A. Davis and A. Sandre (1979). A synopsis of the Moorhead deionization plant operating and economic characteristic. La Sucrerie Belge *78*, 11, 67.
[95] J. R. Johnson (1952). Recovery of granular fertilizer from ion exchange spent regenerant. Proc. 7th Gen. Mtg. A. S. S. B. T., 729.
[96] S. E. Bichsel and T. D. Carpenter (1972). Ion exchange treatment of sugar containing solutions and production of a liquid fertilizer. U. S. Pat. 3700460.
[97] R. F. Olson, A. M. Sandre and S. E. Bichsel (1979). Utilization of a deionization plant waste regenerant stream to produce a valuable granular fertilizer by-product. La Sucrerie Belge *78*, 11, 67.
[98] S. Vajna (1959). Saftreinigung durch Entsalzung über Ionenaustausch. Zucker *12*, 186.
[99] E. Moebes (1960). The carbonate process for purifying thin juice. Zucker *13*, 254.
[100] J. Elmer, H. Hitzel and E. Moebes (1969). Development and procedure of the carbonate process in the Enns Sugar Factory. Zucker *19*, 545, 566.
[101] K. Schoenrock, A. Gupta and D. Costesso (1976). New ion exchange systems for the beet sugar industry. La Sucrerie Belge *95*, 287, 331, 371.
[102] K. Schoenrock, A. Gupta and D. Costesso (1980). New ion exchange systems for the beet sugar industry. Zucker *30*, 541, 607, 673.
[103] K. Schoenrock and H. Rounds (1975). Process for the purification of sugar beet juice and increasing the extraction of sugar therefrom. U. S. Pat. 3973986.
[104] K. Schoenrock (1985). Work we must, but the regenerant is free in the purification of sugar liquors by ion exchange. 23rd Ann. Meeting, A. S. S. B. T., San Diego, California. February 26, 1985.
[105] I. Abrams and B. Dickinson (1949). Color removal in sugar liquors by synthetic resins. Ind. Eng. Chem. *41*, 2521.
[106] F. Harding and F. Bruder (1952). Process for decolorizing sugar solutions with a porous quaternary ammonium ion exchanger. U. S. Pat. 2785998.
[107] C. D. Conglin and A. Congelosi (1959). Use of highly basic resins in sugar refining. Proc. Sixth Tech. Sess. Bone Char., 35.

[108] F. X. McGarvey and G. M. Andrus (1965). The evaluation of ion exchange resins for sugar liquor decolorization. Meeting of Sugar Industry Technicians, New York. May 2–4, 1965.

[109] I. M. Abrams (1971). The removal of color by adsorbent resins. Sugar y Azucar, May, 31.

[110] W. W. Cooper (1976). An investigation into the use of ion exchange resins for decolorizing carbonated brown liquor in a sugar refinery. Proc. South African Sugar Tech. Ass., June, 1976.

[111] B. S. Joshi (1976). Applications of ion exchange resins in the decolorization of sugar juice. Indian Sugar *26,* 187.

[112] C. Loker (1983). Factory and pilot plant experiences with ion exchange resin employed in gross decolorization in a carbonation refinery. Proc. Ann. Meeting Sugar Ind. Technologists *42,* 211.

[113] D. S. Martin (1984). Adsorbent decolorizing system, the pros and cons. Proc. 43rd Ann. Meeting Sugar Ind. Technologists, Vol. XLIII, Houston, Texas. May, 1984.

[114] L. Ramm-Schmidt and G. Hyoky (1984). New resin decolorization station at Porkkala Refinery. 43rd Ann. Meeting Sugar Industry Technologists, Vol. XLIII, Houston, Texas. May, 1984. Also see: The Sugar Journal, February 1985, 14.

[115] G. A. Akeson and K. A. Lilja (1984). The sugar refining process at SSA, Arlov Division, Sweden (Rebuilt 1981). 43rd Ann. Meeting Sugar Industry Technologists, Vol. XLIII, Houston, Texas. May, 1984.

2.11 Ion Exchangers as Catalysts

Wilhelm Neier
Deutsche Texaco-Chemie, Moers 1, Federal Republic of Germany

2.11.1 General

2.11.1.1 Historical survey

As early as World War I, inorganic materials with ion exchanging properties were used as catalysts for gaseous phase reactions [1]. Their catalytic effectiveness was based principally on the controlled introduction of heavy metal ions into a defined space lattice. This field will not be further elaborated on in this chapter.

Catalytic reactions with proton-active zeolites have to be mentioned, e.g., cracking of hydrocarbons, condensation of methane and benzene to toluene, dehydrogenation of ethyl benzene to styrene, and dehydration of 1-propanol to propene.

The synthesis of synthetic resin-based ion exchangers, first reported in 1935 [2], made possible the use of ion-exchange resins as catalysts in a liquid phase, thus opening the vast field of acid/base catalysis. In 1911, Tacke and Süchting [3] had already reported that cane sugar can be split into its monosaccharides (glucose and fructose) by soils. The catalysis was assumed to be initiated by the humic acids contained in the soil, but in 1918 Rice and Osugi [4] proved that purely mineral soils, too, have a catalytic effect in the inversion of cane sugar.

Not until during World War II was the use of ion-exchange resins as catalysts in a liquid phase systematically examined. Most of the research work in this field was done by I. G. Farbenindustrie at Wolfen and Ludwigshafen, but the patent applications filed at that time were not published until around 1952. For the first time the hydrolysis of cane sugar and acetic acid ethyl esters, the esterification of carboxylic acids with a great number of alcohols, the formation of acetal, and the condensation of aldol were described [5]. Already at that time hydration of olefins [6] and etherification of isobutene with methanol [7] to methyl tert-butyl ether (MTBE) were performed, using so-called solid acids based on sulfonated phenol formaldehyde condensates.

In the years thereafter a number of reactions were studied in many countries around the world. At the beginning, only condensation resins were available, but after a short time these resins were replaced by copolymerizates based on styrene-divinylbenzene and acrylates or methacrylates. Basic ion exchange resins were also introduced for use as catalysts. Systematic research led to the development of gel-type exchange resins with surfaces of less than 1 m^2/g in the dry condition and pore sizes of less than 2 nm (BET) as well as of macroporous resins with surfaces greater than 250 m^2/g and pores sizes of up to 1000 nm (BET). It would go beyond the scope of this article to give a comprehensive literature survey of this subject, but references [1] and [8 – 21] are illustrative.

2.11.1.2 Advantages

As compared to the conventional homogeneous catalysis with acids or bases, the heterogeneous catalysis with ion-exchange resins offers certain significant advantages:

1. The catalyst is easy to eliminate. Hence, it is suitable for continuously-operated reactors.
2. It is not necessary to neutralize or reconcentrate the catalyst.
3. The reactions are more selective, so that higher yields are attained and fewer byproducts are formed.
4. Heterogeneous catalysis makes it easier in most instances to control the reactions better and more selectively.
5. Pollution of waste water and air is minimized.
6. Corrosion is prevented.

2.11.1.3 Disadvantages

However, certain disadvantages occurring in commercial plants also have to be mentioned. A panacea cannot be offered; each problem has to be solved individually by selecting suitable catalysts and optimizing the reaction sequence.

Particle size. To keep the beads of ion-exchange resins with grain sizes between 0.3 and 1.1 mm in the reactors it is necessary to provide special support trays with fine-mesh screens or narrow-slotted jets [23]. Experiments with larger grain sizes failed [24], because the catalytic activity decreased due to prolonged diffusion, and also the mechanical stability was lower. Incorporation in a thermoplastic resin [25–29], mixing pulverized exchange resins with inorganic material [30, 31], or coating metal pellets (e. g., aluminum) with polystyrene followed by sulfonation [32] have so far also not had the desired effect.

Grain-size distribution. Normally, ion-exchange resins have grain sizes between 0.3 and 1.1 mm. No significant differences between the individual grain sizes have been found. However, the grain-size distribution should be as uniform as possible so that smaller grains cannot obstruct the space between the big ones, thus impeding regular distribution of the reactants and leading to differential pressure build-up in tube reactors [33].

Thermal conductivity. Organic ion-exchange resins have insufficient thermal conductivity. Therefore, the method of heat removal has to be carefully selected. It is possible either to cool the reactant stream externally or to inject a cooled reaction component into the catalyst bed [34]. Operation in a fluidized bed is also recommended [35] and the use of loop reactors [36, 8] has proved particularly efficient. In some cases it may be necessary to use multitube reactors with differently heated or cooled sections [37].

Swelling capacity. The swelling capacity is strongly dependent on the degree of crosslinking [38] (Figure 2.154).

Gel-type resins are particularly sensitive to osmotic shock [14]. This problem could be solved by using macroporous resins, but these resins, too, have different swelling capacities depending on the polarities of the reactants [39] (Figure 2.155).

For characterization of the swelling agent the so-called solubility parameter can be referred to. It is calculated as follows [38, 39] (Figure 2.156).

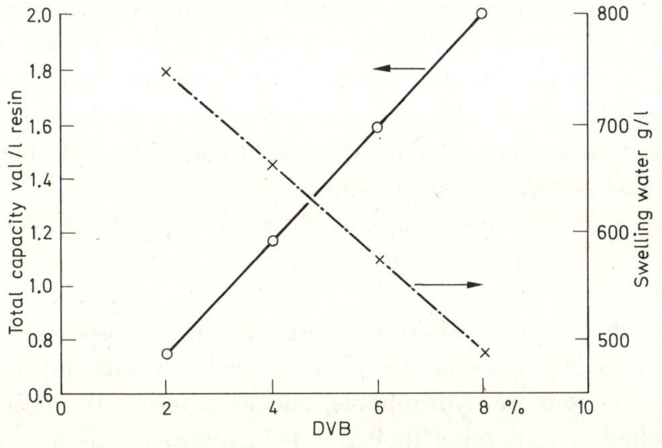

Figure 2.154 Total capacity of and swelling water content in gel-type cation exchange resins with different degrees of crosslinking.

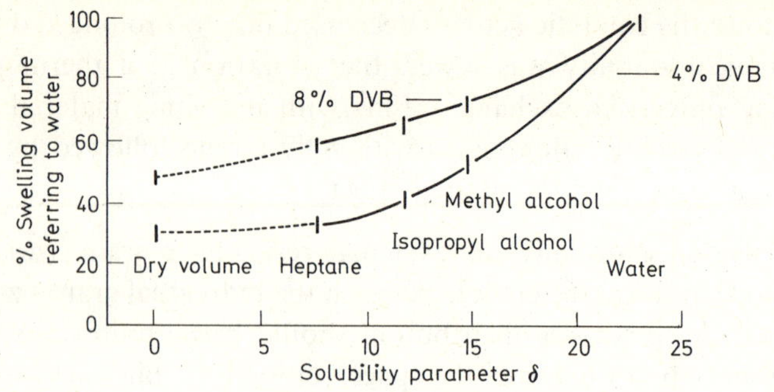

Figure 2.155 Swelling volumes of polystyrene sulfonic acid resins in different liquids.

$$\delta = \sqrt{\frac{\Delta E}{V}}$$

V = Molar volume
ΔE = Evaporation energy

Water	δ = 23.2
Methyl alcohol	δ = 14.5
Isopropyl alcohol	δ = 11.5
Heptane	δ = 7.5

Figure 2.156 Solubility parameters for polymers.

Therefore, swelling tests prior to use of ion-exchange resins as catalysts in continuously operated reactors are indispensible.

Temperature stability. Strongly basic ion-exchange resins based on polystyrene can be used at temperatures of up to about 60 °C. They are particularly affected by Hofmann degradation and are first converted into weakly basic exchange resins. But at about 90 °C even the catalytic activity of weakly basic resins decreases rapidly. Similarly, anion exchange resins based on methacrylate/acrylate are hydrolyzed and converted into weakly acidic resins containing carboxyl groups. Aminoamide-based resins have higher temperature stabilities than resins based on aminoalcohols. Polystyrene sulfonic acid-based cation exchange resins in the H-form show a drop in activity at temperatures of about 120 °C and higher. In polar media sulfuric acid is released [40] (Figure 2.157).

In nonpolar media and in the gaseous phase, sulfur and SO_2 are split off [10, 39]. Hydrolysis in polar media is dependent on the degree of crosslinking of the resins [20, 40] (Figure 2.158).

Hydrolysis of the sulfonic acid groups is an autocatalytic process [41]. Therefore, the sulfuric acid loss decreases as the operating time proceeds [40] (Figure 2.159).

The resin becomes more hydrophobic, and as a result, the selectivity of the reaction may change. According to Petras [42], hydrolysis as a function of the crosslinking degree differs in dependence on the electron-donating ethylene group of the divinylbenzene introduced by polymerization. Based on these experimental

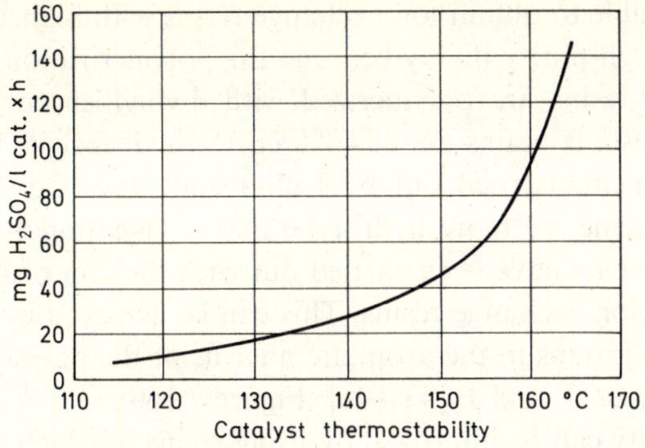

Figure 2.157 Catalyst thermostability.

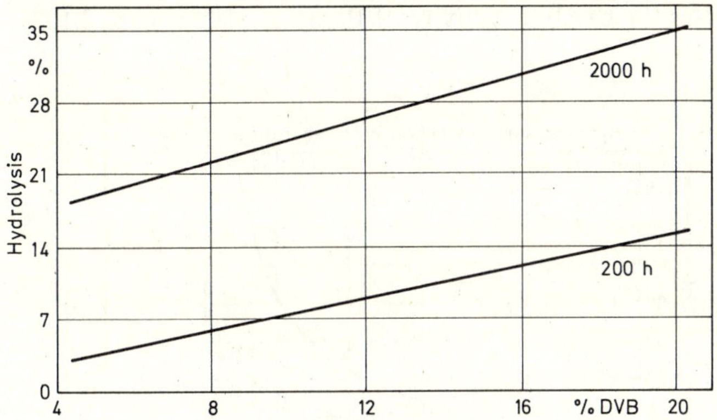

Figure 2.158 Hydrostability at 135 °C as a function of the degree of crosslinking.

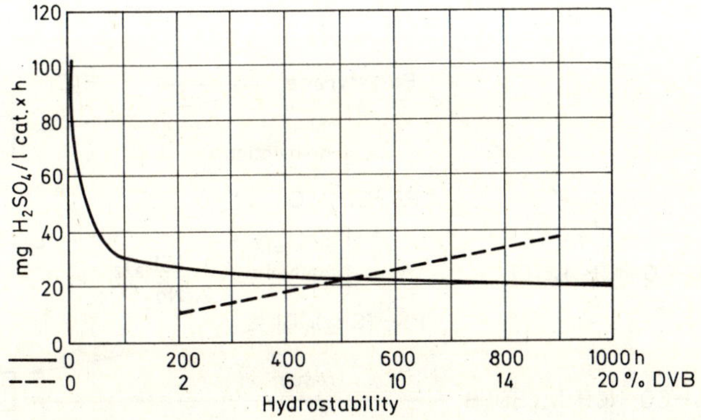

Figure 2.159 Hydrostability at 135 °C as a function of catalyst lifetime (DVB content 8%).

results it should be possible to obtain ion-exchange resins with higher temperature stabilities when using high-purity divinylbenzene for polymerization. It is known that all technical-grade resins are polymerized with divinylbenzene that is only about 60% pure. The resin contains about 40% ethyl vinylbenzene. This has also been proved by experiments carried out by Rohm and Haas Co. showing that resins based on vinyltoluene are more hydrolytic than conventional resins [43]. A great number of experiments have been carried out with the aim of increasing the temperature stability of ion-exchange resins. This can be achieved by introduction of electron-withdrawing groups in the aromatic nucleus of the polystyrene matrix, such as NO_2, C≡N, Cl, Br, I, and F [43–46] (Figure 2.160).

The hydrolysis stability can be improved by using resins in which sulfonic acids are not bound to aromatic carbon atoms [47, 48] (Figure 2.161).

An interesting alternative, providing better thermal stability than polystyrene resins, are resins with a fluor hydrocarbon matrix. Moreover, the sulfonic acid group of such resins has a high activity as a result of a neighboring CF_2 group, which increases the catalytic activity of the resin [41] (Figure 2.162).

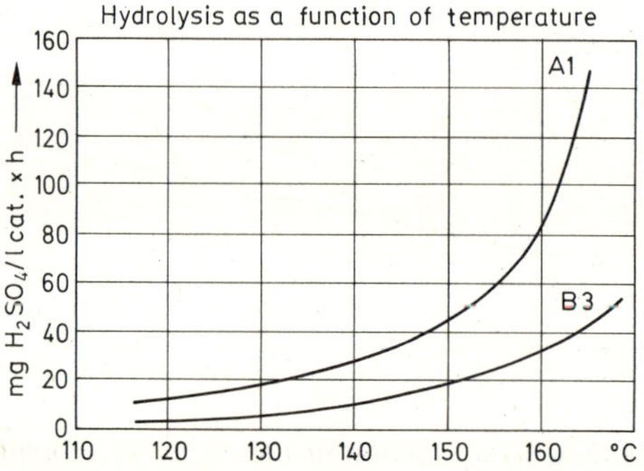

Figure 2.160 Hydrolysis stability of a conventional cation exchange resin as compared to a chlorinated resin.

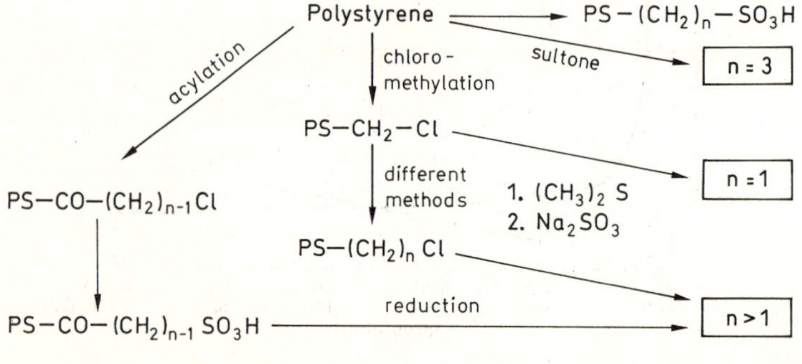

Figure 2.161 Synthesis of sulfoalkylated polystyrene resins.

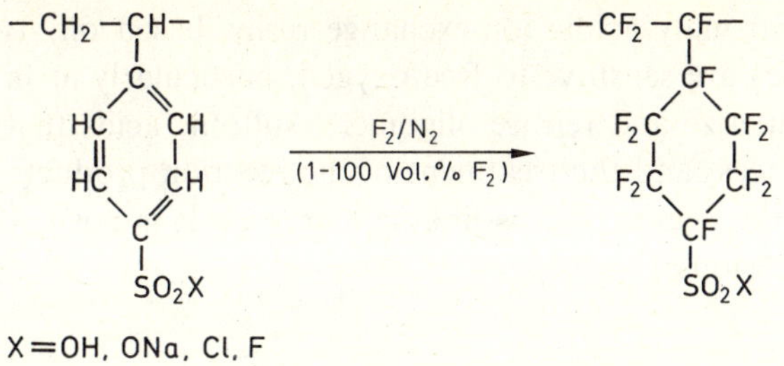

Figure 2.162 Perfluorinated polymeric acids.

For this type of matrix macroporous resins have to be used so that fluorination does take place only on the surface. Because of the pore structure these resins are superior to the known Nafion resins. In addition, they are less hydrophobic [49].

Deactivation and regeneration. Ion-exchange resins are still much more expensive than the acids and bases used in homogeneous catalysis. When used as catalysts in commercial processes at elevated temperatures, their lifetimes normally range between 6,000 and 16,000 operating hours. Only for high-margin products is a shorter lifetime justifiable from the economic point of view.

Deactivation may have the following causes:

1. Absorption of metal ions as a result of the H-form.
2. Hydrolysis of the functional groups.
3. Blocking of the active centers due to polymerization or by polycondensation products.

Reactivation is possible by the following methods:

1. Removal of the metal ions by acid treatment.
2. Reintroduction of the functional groups.
3. Treatment with solvents.

Methods 1 and 2 require removal of the catalyst from the reactor, which makes transfer from a nonaqueous to an aqueous medium necessary. Reactivation can be done batchwise or in a column.

As to method 2, reintroduction of the functional groups usually can be done solely by the resin manufacturer. Thereafter, it has to be tested whether the catalyst structure and, thus the properties, have been affected by polymer or polycondensate plugs in the pores. Moreover, experiments have shown that the orginal activity cannot be attained. Reintroduction of the functional groups has not been practised so far for economic reasons, but in future this method will draw more attention.

It has to be decided according to the circumstances whether polymers can be eliminated from the catalyst by solvent treatment.

Oxygen sensitivity. Strongly acidic ion-exchange resins based on styrene-divinylbenzene polymerizates are sensitive to free oxygen, particularly at higher temperatures. They depolymerize and release oligomeric sulfonic acids that may significantly impair the process and the treatment of the reaction products. Depolymerization decreases as the degree of crosslinking increases. It is positively influenced by catalysis with Fe^{3+} ions.

2.11.1.4 Selection and testing of ion-exchange resin catalysts

Preselection of catalysts. For the selection of optimum catalysts within the shortest possible time, the so-called micropulse reactor as described by Schleppinghoff [48] has proved effective. Actually, this apparatus is a gas chromatograph, the sample inlet tube of which is operated like a microreactor. The reaction product is condensed by deep freezing in the first section of the column and is collected for a short while in a cooling trap. After spontaneous vaporization has taken place, the product is led together with the carrier gas through the gas-chromatographic column, where it is analyzed. A catalyst sample of 1 – 20 mg is required for this test. The final result is obtained after about 50 minutes.

Fine screening of catalysts. Severe testing of a catalyst is carried out in bench-scale reactors having a capacity of about 200 ml. In discontinuous experiments the following parameters are examined: pressure, temperature, mole ratios of the reactants, residence time. The catalytic activity with respect to conversion, selectivity, and specific catalyst load is thus tested. Such extensive testing of a catalyst takes several weeks.

Testing under commercial operating conditions. The efficiency of a catalyst is tested in a continuously-operated pilot reactor. For this test the optimum operating conditions must have been determined and the effects of different operating procedures on the catalyst properties must have been studied. Product reflux lines and recycle circuits have to be installed when necessary. In the operating test catalyst lifetime, activity, and mechanical stability must be recorded. In the operating test an early analysis of the byproducts is important in order to learn as soon as possible how to purify the reaction product. It may be necessary to carry out these experiments in parallel.

Upscaling. The decisive question, namely whether to make test runs in pilot facilities prior to starting up commercial operation, has to be answered according to the circumstances. If a new product to be introduced in the market can be produced by test runs in a pilot plant, such a question is easy to decide. If ion-exchange resins are used as catalysts, test runs in pilot facilities may be indispensable in order to examine corrosion problems, technical questions of filling and dumping, reactor dimensioning, and product distribution.

2.11.1.5 Influence of the catalyst characteristics on the reaction sequence

As already mentioned in Section 2.11.1.3, the mechanical, chemical, and thermal stabilities of ion-exchange resins are determined by the physical structure of the exchanger matrix as well as by the type and number of the functional groups. These factors are also of vital importance with respect to activity and selectivity. References here are rarely found in the literature [49–51, 18].

Pore structure and pore surface. In contrast to homogeneous catalysis, the catalysis overlaps with diffusion, adsorption, and desorption processes [10]. It has to be taken into consideration that the concentration of the starting materials in the exchanger is different from that in the solution. The distribution coefficient can be changed by exchanging the solvent [9].

Since the reactive centers must be easily accessible for the reactants, the use of gel-type exchange resins is limited to those reactions in which the reactants or the solvent ensure sufficient swelling of the resin, i. e., where polar media can be used. Reduction of the crosslinking agent [14] which, as already mentioned (cf. Figure 2.157), improves the swelling ability, is limited by the reduction in mechanical stability. It is possible, for instance, by grinding or suitable polymerization techniques to diminish the grain size and, thus, to enlarge the surface area [50–55, 36], but at the same time this makes operation in commerical units more difficult.

The present state of development of polymerization techniques allows production of ion exchangers with different mesh sizes and porosities. In short-time test runs, gel-type and macroporous resins having a crosslinking degree of up to 8% divinylbenzene have not been found to differ much from each other [38] (Figure 2.163). At a higher degree of crosslinking, particularly the activity of gel-type resins decreases significantly [57]. In long-time test runs using gel-type resins noticeable activity decreases have been observed, probably due to plugging of the pores with high-molecular byproducts [57, 58]. In nonpolar media, such as in the reaction of

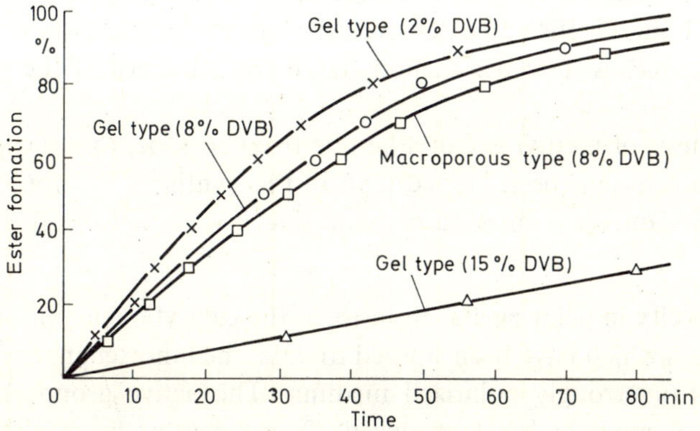

Figure 2.163 Esterification of carboxylic acids with alcohols at 110 °C.

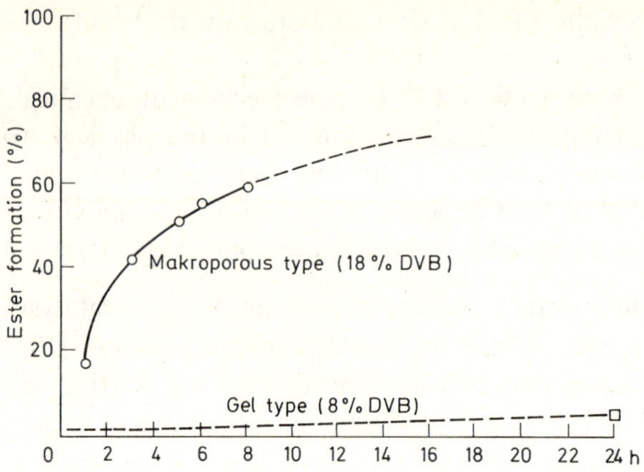

Figure 2.164 Esterification of methacrylic acid with isobutene.

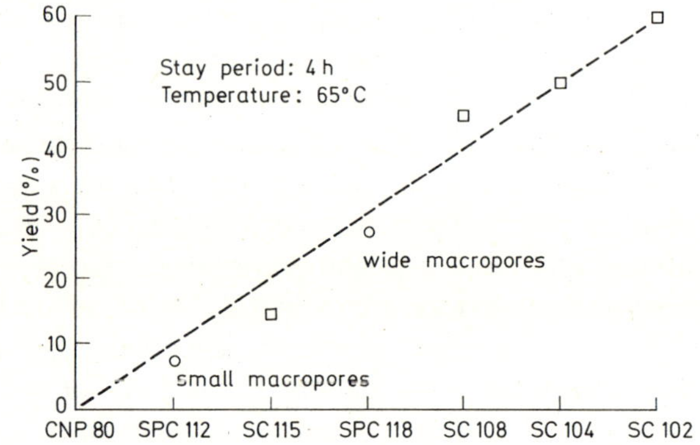

Figure 2.165 Hydrolysis of a cyclic ketoxime at 65 °C.

carboxylic acids with olefins [60, 38], macroporous ion-exchange resins are successfully used (Figure 2.164). In extensive experiments for the hydrolysis of a cyclic ketoxime to ketone, Arnold has demonstrated the influence of the pore size [38] (Figure 2.165).

In the synthesis of methyl isobutyl ketone from acetone, too, the superiority of a macroporous resin has been proved [38]. In this synthesis the pore structure has a great influence on the formation of byproducts, such as isopropyl alcohol and hexene.

Acidity and basicity in polar media. In most of the catalytic reactions performed in commercial units which have been studied to date, the ion-exchange resin is present in an aqueous or strongly polarized medium. The active group and, thus, the catalytic agent is more or less hydrolyzed. Consequently, the catalytic activity in polar media, which is comparable to that of the low-molecular Brønstedt acids, is proportional to the number of protons or of hydroxyl ions solvated per liter of catalyst. Therefore, weakly basic or weakly acidic ion exchange resins are practically of no importance. Even phosphonic acid resins are noticeably less active than sulfonic acid resins [9, 15, 18, 19, 60]. In a few cases, the reaction may have different influences on the conversion of identical starting materials. For instance, in the aldol condensation of acetaldehyde the crotonization reaction, unlike the aldolization, is favored by strongly basic resins. Weakly basic resins favor the formation of aldol. Strongly acidic resins produce para-aldehyde in high yields [10, 61].

Acidity in nonpolar media. Swelling of the resins is dependent on the polarity of the media. As a result, for reactions in nonpolar media, e. g., the alkylation of benzene, only macroporous (macroreticular) ion-exchange resins are suitable. Only those resin types are used which have a sulfonic acid group as a catalytically active species. This group is undissociated and, therefore, constitutes part of the polymeric matrix. Consequently, reactivity and selectivity may be affected. For such reactions it is particularly important that the acidity of the active groups be increased. This can be achieved by incorporating electronegative substituents (NO_2, Cl, F, Br) in the aromatic ring [43–46]. By nitration of the resin, however, the matrix is strongly attacked so that the thermal stability is affected. Rohm and Haas Company has succeeded in producing commercial batches of such resins. This may open up further interesting fields of application. It has to be pointed out, however, that the splitting-off of Cl groups as HCl is problematic and may cause noticeable corrosion unless the resin is carefully pretreated prior to use.

Another interesting alternative to polystyrene resins are resins with a stable fluorohydrocarbon matrix, e. g., Nafion from DuPont. Due to the neighboring CF_2 group, the sulfonic acid group has a higher acidity, but the porosity of such resins is very low. For commercial use they are currently unimportant because of their price. Resins with noticeably higher porosity and considerable acidity are obtained by perfluorination of conventional macroporous ion-exchange resins. Acidity and catalytic activity of such resins have been described by Klein and Widdecke [41, 62] using the cumene synthesis as an example (Figure 2.166).

Polymeric catalyst		Moles cumene/h per kg	per eq
Polystyrene sulfonic acid resin	4.8 meq/g	14.1	2.8
Fluorinated polystyrene sulfonic acid resin	1.1 meq/g	24.1	22.9
NAFION	0.8 meq/g	6.3	7.9

Figure 2.166 Alkylation of benzene with propylene.

Unfortunately, such ion-exchange resins can currently be produced only on a laboratory scale. Their commercial production would be a great step forward for ion-exchanger catalysis.

Another important step that deserves mention is the addition of metal halogenides to sulfonic acid. By this means Brønstedt acids can be combined with Lewis acids. This can be achieved by treating the dry ion-exchange resin with gaseous BF_3, or by sublimating a Lewis acid such as $AlCl_3$ on the resin [63, 64]. In this connection it has to be mentioned that efficient catalysts have been developed by binding Lewis acids to resins other than sulfonic acid ion exchangers [66].

2.11.2 Syntheses and processes

Many syntheses and reactions catalysed by ion exchangers have been carried out, but relatively few of them have been technically perfected. Some examples of experimental processes are given below, but the main emphasis will be on those processes that are industrially useful.

2.11.2.1 Functionalization

Hydration of olefins

Addition of water to an olefinic double bond is a reversible, acid-catalyzed reaction:

$$C = C + H_2O \overset{H^+}{\rightleftharpoons} H - C - C - OH$$

This addition can be treated as an acid-base interaction in which the olefin acts as base and the H_3O^+ as the acid. According to this definition, the more basic (and nucleophilic) the olefin is, the more successful the electrophilic additions will be. Therefore, +I substituents increase the reactivity of an olefin, because they increase the electron density of the double bond.

This can be confirmed by consideration of the potential energy of the intermediate carbenium ion:

$$CH_3^+ \quad 1394 \text{ kJ/mol}$$

$$CH_3CH_2^+ \quad 1254 \text{ kJ/mol}$$

$$(CH_3)_2CH^+ \quad 1159 \text{ kJ/mol}$$

$$(CH_3)_3C^+ \quad 1097 \text{ kJ/mol.}$$

As the number of alkyl groups increases, so does the stability. The more stable a carbenium ion is, the more readily it is formed, because the inductive effect also stabilizes the transition state, that is, it reduces that activation energy. The reactivity of alkenes decreases in the following series from left to right:

iso-butene > 2-butene > 1-butene > propene > ethene [66].

The relative rate constants (k_{rel}) of acid-catalysed hydration at 25 °C have been determined as follows [67]:

iso-Butene	1
Cyclohexene	1.2×10^{-4}
1-Hexene	1.2×10^{-5}
Propene	6.4×10^{-6}
Ethene	3.9×10^{-12}.

Since the addition of water to the double bond of tertiary olefins is several orders of magnitude more rapid than to unbranched olefins, the double bonds of branched olefins have a special significance which is independent of the total number of carbons in the molecule.

Tert-Butanol (TBA). TBA is is one of the major oxygenates used in gasoline. It is an octane-enhancing component with lower sensitivity than the lower molecular-weight alcohols, and is an excellent cosolvent for incorporating methanol into gasoline. TBA is used in the production of very pure iso-butene [68] and methacrylic acid [69].

TBA is still produced chiefly as a coproduct of propylene oxide synthesis by the Arco process [70], but it can be expected that, in the future, an increasing amount of TBA will be made by direct hydration processes. At present, C_4-raffinate streams from steam or catalytic crackers are used as feedstock. In future, however, iso-butene cuts from deyhdrogenation and isomerization of natural gas (NGL) will be available. In these feedstocks, the concentrations of iso-butene range from 10 to 60%. Reaction temperatures in the 60–90 °C range are needed. The reaction is inhibited by the extreme difference in polarity between the hydrocarbon and aqueous phases. The use of very acidic ion exchangers as catalysts was first patented in 1944 [6].

R. C. Odiso et al. [71] obtained a yield of only 33.6% using a 22.5% iso-butene-containing C_4 cut and a reaction temperature of 92 °C with Amberlite® IR 120. In the patent literature summarized in Table 2.51, processes are described that are said to improve the low yields. There are a number of articles in Russian journals which should also be mentioned [72–78]. For example, some methods make use of finely-suspended resins in stirred reactors. Various emulsifiers and solubilizers are used; ethylene glycol monobutyl and ethyl ethers, glycols, dioxane, ethanol, n-propanol, acetone, acetic acid and sulfolane have been recommended as solubilizers. Conversion rates better than 90% have been reported.

When such emulsifiers and solubilizers are used, it must always be determined whether they react with iso-butene, thus leading to undesired by products which reduce the yields of TBA or cause difficulties during separation of the TBA. The process developed by VEBA-Chemie, Germany, deserves special mention.

Table 2.51 Patent literature on the production of tert-butanol from iso-butene using acidic ion exchange resins

Priority		Applicant	Patents
09.02.53	DE	BASF	DE 866 191
24.01.62	DE	Bayer	DE 11 76 114 US 3 328 471
21.03.62	GB	Gulf	GB 973 832
21.04.62	DE	Bayer	DE 11 79 920
25.07.73	JPN	Mitsubishi	DE 24 30 470 US 4 011 272
22.04.74	JPN	Mitsubishi	JPN 75 137 906
20.11.74	DE	USSR	DE 24 54 998 US 4 012 456
27.08.75	DE	BASF	DE 25 38 036
02.08.76	US	Cities	US 4 096 194
20.05.77	US	Petrotex	US 4 087 471
29.07.77	JPN	Nippon Oil	DE 28 33 294 US 4 180 688
10.04.78	JPN	Nippon Oil	DE 29 13 796 US 4 208 540
06.11.78	JPN	Misubishi	DE 29 64 069 US 4 360 406 EP 10 993
15.12.78	JPN	Mitsui	JPN 80 081 825
27.04.79	JPN	Toa Nenryo	DE 30 29 739 US 4 284 831 US 4 327 231
05.07.79	JPN	Sumitomo	DE 30 25 262 US 4 011 272
31.08.79	JPN	Toa Nenryo	DE 30 31 702 US 4 270 011
18.12.79	JPN	Maruzen	JPN 81 087 526
05.05.80	US	Texaco Inc.	US 4 316 724
25.12.80	JPN	Toa Nenryo	JPN 82 108 028
03.02.81	US	Halcon	US 4 334 890

Isopropanol (IPA) and sec-butyl alcohol (SBA). Hydration of double bonds in unbranched olefins occurs much more slowly than in tertiary alkenes, for the reasons discussed above. It is not possible to produce ethanol from ethylene in the temperature range where ion-exchange resins can be used. On the other hand, propylene and butylene can react satisfactorily at 120 °C to 160 °C. The main by-products are diisopropyl and di-sec-butyl ethers. Small amounts of n-propanol or n-butanol and hexanols or octanols are also formed. If the reactant gases still contain small amounts of iso-butene, this is partially oligomerized or hydrated to TBA. Butadiene groups are hydrated to crotyl alcohol and methyl vinyl carbinol. Methyl vinyl carbinol partially rearranges to methyl ethyl ketone [20, 33, 42, 79, 86].

Table 2.52 Patent literature on the production of isopropyl alcohol and sec-butyl alcohol from propylene and butylene using acidic ion exchange resins

Priority		Applicant	Patents
09.02.53	DE	BASF	DE 866 191
15.11.54	US	ESSO	DE 11 05 403
27.09.56	US	ESSO	DE 12 10 768
20.08.57	US	Sinclair	US 2 803 667
19.11.57	US	ESSO	US 2 813 908
18.11.58	US	ESSO	US 2 861 045
08.01.60	SU	USSR	SU 125 240
26.09.61	US	Mobil	US 3 256 250
14.01.67	DE	Veba	DE 16 18 999
14.08.67	DE	Rheinpreussen	DE 12 91 729
01.08.68	US	Celanese	DE 19 39 288
27.08.69	US	Universal Oil	DE 20 41 954
27.12.69	De	Veba	DE 19 65 186
23.04.70	US	Chevron	US 4 046 520
24.09.71	DE	Deutsche Texaco	DE 21 47 737
24.09.71	DE	Deutsche Texaco	DE 21 47 738
24.09.71	De	Deutsche Texaco	DE 21 47 739
24.09.71	DE	Deutsche Texaco	DE 21 47 740
11.07.72	DE	Deutsche Texaco	DE 22 33 967 US 4 340 769
15.08.72	GB	BP	DE 23 40 816 GB 1 374 368
	US	Universal Oil	US 3 810 849
30.01.74	DE	Deutsche Texaco	DE 24 04 329 US 4 231 966
21.06.74	DE	Deutsche Texaco	DE 24 29 770 US 3 994 983
13.05.76	GB	Shell	DE 27 21 206 GB 1 518 461
08.06.76	JPN	Maruzen	JPN 77 151 106
27.06.80	DE	Deutsche Texaco	DE 30 24 146 EP 43 049
18.08.80	US	K. Kuhlmann	DE 31 31 975
31.10.80	DE	Deutsche Texaco	DE 30 40 997 EP 51 164
25.06.81	US	Chevron	DE 32 23 261

The patent literature on synthesis of IPA and SBA is summarized in Table 2.52.

The individual commercial direct hydration processes for synthesis of IPA or SBA with IER were developed by Deutsche Texaco AG. The three-phase process for IPA synthesis in a trickle reactor became operational in 1972 (Figure 2.167). At constant rates of reaction and heat removal, the temperature regime in this process can be described as quenching. The volume-specific reactor output increases with reactor size, at constant reaction rate and equivalent dosage (relative to the

Figure 2.167 Isopropyl alcohol plant reactor section.

volume) of the reactants, starting at a length of 3 m. The ion exchanger is specified in terms of porosity and internal surface. When the reactor is filled, certain conditions must be maintained in order to prevent formation of “hot spots” during operation (Figure 2.168).

Hydration can also be carried out in a sump reactor, and the alcohol removed via the supercritical olefin phase. Because the position of the phase equilibrium is favorable, this is carried out commercially in SBA production (Figure 2.169). It is advantageous to presaturate the feedstock gas with water. The direct hydration of n-butenes was first technically achieved in 1983, by application of these insights (Figure 2.170).

Other alcohols. The difference in reactivity between olefins with and without branching at the double bond is illustrated by the hydration of 2-methylbutene-2 and 1-pentene in the presence of ethylene glycol monobutyl ether as solubilizer after 3 hours (Table 2.53).

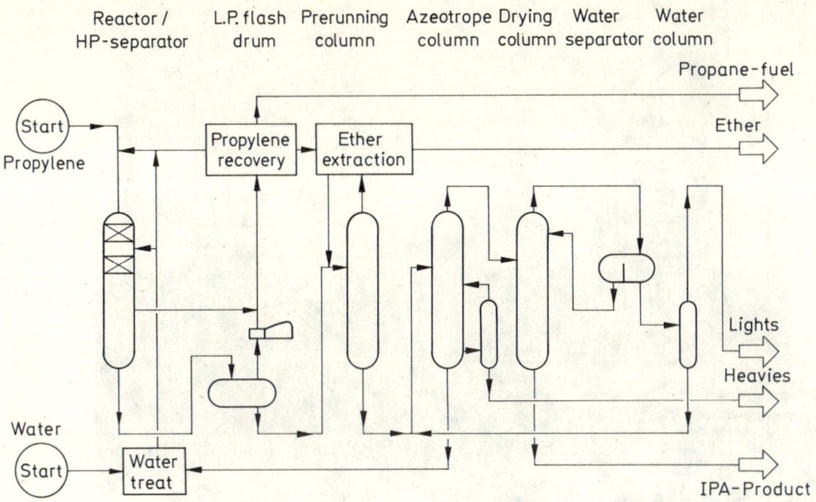

Figure 2.168 Isopropyl alcohol synthesis flow sheet.

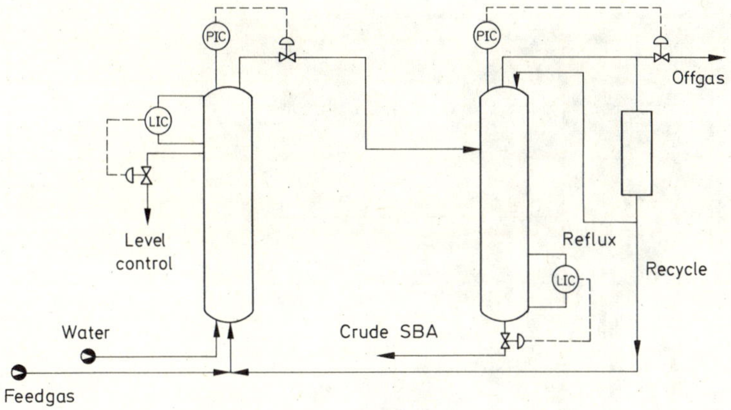

Figure 2.169 Sec-butyl alcohol synthesis flow sheet.

The selectivity of the reaction for tert-amyl alcohol decreases, as would be expected, from 96% at 40 °C to 73% at 120 °C. By contrast, only traces of pentanol-2 and pentanol-3 are formed from 1-pentene. Side reactions of the solubilizer dominate.

It was not possible to make 1-hexene react with steam in the gas phase. At 140 – 180 °C, a 1-hexene conversion of at most 0.8% was achieved.

In a heterogeneous system consisting of 2 mol 1-decene, 10 mol water and 50 g strongly acidic ion exchanger, after 3 h at 160 °C in an autoclave, neither decanols nor decene dimers were formed. After addition of a solubilizer (THF, acetic acid, n-butylglycol, dioxane, DMSO) and doubling the reaction time, only traces of decanols could be detected.

Figure 2.170
Sec-butyl alcohol plant.

Table 2.53 Reactivity of 2-methylbutene-2 and 1-pentene as a function of temperature

Temperature (°C)	2-Methylbutene-2 % Conversion	1-Pentene % Conversion
40	14	–
80	63	2
120	50	4
160	–	15

These results indicate that direct hydration of olefins with acidic IER is feasible only with tertiary olefins. For example, Universal Oil [87] performed the hydration in diethyleneglycol/dimethyl ether at 150 °C. Shell [88] succeeded in increasing the conversion of cyclohexene at 140 °C from 4% to 20% in the presence of sulfolane.

BASF hydrated isoprene (40 °C) and butadiene (100 °C) in aqueous sulfolane solution, or with added carboxylic acids [89]. The addition of water to isoprene has also been reported by Katagiri et al. [90].

Shell produced triols from *α, β*-unsaturated aldehydes via 2-hydroxydialdehydes [91]:

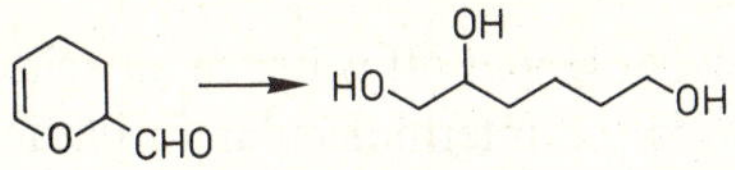

Hoechst converts 2-cyclohexenone to 3-hydroxycyclohexanone at 45 to 60 °C [92]:

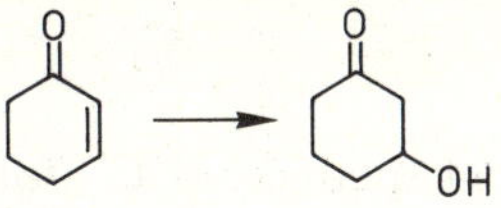

The hydration of myrcene

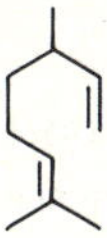

to diverse terpene alcohols has been described in a Japanese paper [93].

International Flavor and Fragrances Inc. hydrate the double bond in the side chain of myracaldehyde [94]:

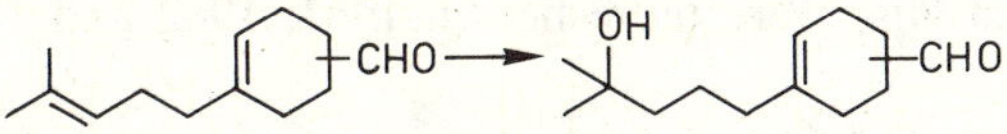

BASF hydrates citronellol to hydroxycitronellol [95]:

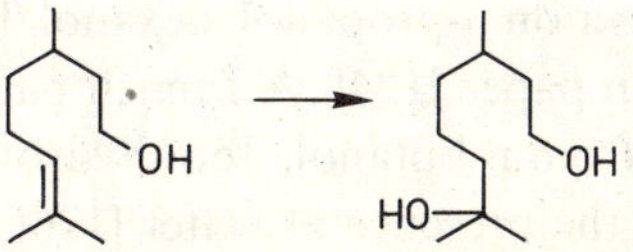

In a Soviet patent 1-methylcyclobutanol is obtained from methylenecyclobutane [96]:

The synthesis of (RS)-4-amino-3-hydroxybutyric acid was described by Pinza and Pifferi [97]:

There is a Russian-language report on the production of tert-hexyl alcohol from isohexenes [98]:

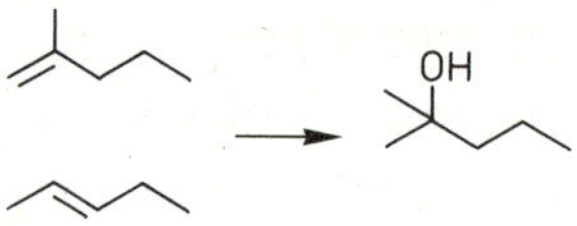

Dehydration of alcohols

In a reversal of their formation reaction, alcohols split off water in an acidic milieu, forming the olefins. A competing reaction is the intermolecular formation of the corresponding symmetric ether:

$$ROH \rightarrow H_2 + \text{olefin}$$
$$2\,ROH \rightarrow H_2O + R_2O\,.$$

Methanol, alkyl alcohols and alcohols with adjacent quaternary C atoms are a special case, because they form only ethers. Conversely, with phenyl carbinols and tert-alcohols, olefin formation predominates [60, 99–101].

Ethers and olefins from primary alcohols. Two patent publications report the synthesis of dimethyl ether from methanol on acidic IER [102, 103]. Chinese, Russian, Czechoslovakian and Spanish papers are cited [104–107]. The kinetics of dimethyl ether formation have been extensively studied and described [108–114]. In 1968, J. C. Herliny submitted his dissertation on the dehydration of ethanol to the University of Washington [115]. Other publications have appeared in Russian and Spanish [116–118]. Kinetic studies were published in 1961, 1962 and 1974 (and other years) [119–124].

The product distribution from cleavage of water out of n-propanol depends on the degree of crosslinking and the cation charge of the ion-exchange resin [122, 123]. The dehydrating effect of the ion exchanger on n-propanol, hexanol, heptanol, nonanol and decanol is described in a Russian paper [124]. A French patent [125] was granted for the synthesis of dibutyl ether from n-butanol. Tetrahydrofuran can be made from 1,4-butanediol monoacetate in the presence of water [126].

The synthesis of dioxane from diethyleneglycol

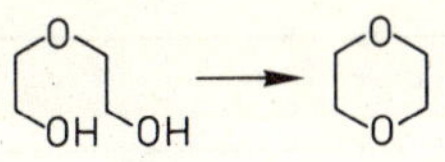

was described in a patent of BASF [127] and in a Hungarian paper [128].

Scott and Naples [129] reported the synthesis of tetrahydropyran from 1,5-pentanediol:

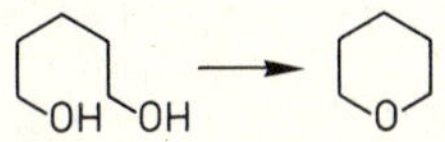

The dehydration of allyl alcohol to diallyl ether is reported in Russian [130]. The condensation of pentaerythritol was patented in 1958 [134].

Ethers and olefins from secondary alcohols. The deyhdration of IPA to propylene and diisopropyl ether is used as a test reaction for catalytic activity of ion-exchange resins with various structures; it is usually carried out in the gas phase [132–137].

The reaction of cyclohexanol to cyclohexene and dicyclohexyl ether is discussed in a Russian paper [141]. Another Russian publication treats the cleavage of water out of methylphenylcarbinol [139]:

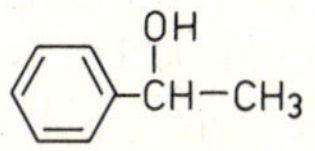

Dehydration of ricinoleic acid or castor oil can also be catalyzed by ion-exchange resins [140–143]. Carbohydrates are another natural source of raw materials. The acid dehydration of fructose yields 5-hydroxymethylfurfural [144–147]:

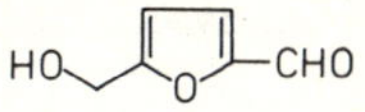

Olefins from tertiary alcohols. To produce pure iso-butene for later use in alkylation reactions, tert-butyl alcohol is dehydrated. The patent publications, in which acid ion-exchange resins are used as catalysts, are summarized in Table 2.54. The dehydration of TBA is used as a standard reaction for the study of reaction kinetics and the catalytic properties of new or modified ion-exchange resins.

Mesityl oxide is produced from diacetone alcohol

using acid IER [148–150].

Only Russian-language publications treat the use of acid IER for dehydration of tert-amyl alcohol, dimethylphenylcarbinol, 3-phenyl-1,1-butanediol and tertiary vinylacetylene alcohols [151–155].

The dehydration of terpin hydrate to α-terpineol is also possible with IER.

Table 2.54 Patent literature on the production of isobutene from tert-butyl alcohol using acidic ion exchange resins

Priority		Applicant	Patents
15.12.67	US	Arco	US 3 510 538
06.11.74	IT	USSR	DE 24 54 998
			US 4 012 456
16.07.79	JPN	Mitsubishi	DE 30 63 490
			US 4 331 824
			EP 23 119
30.01.80	JPN	Misubishi	JPN 81, 108 539
30.01.80	JPN	Mitsubishi	JPN 81, 101 540
24.12.81	DE	CWH	DE 31 51 446
			EP 82 937
			PCT 75507

Unsymmetric ethers. Intermolecular dehydration of different alcohols or phenols can be used for directed synthesis of unsymmetric ethers, if the formation of the symmetric products and (if applicable) olefins can be suppressed.

Synthesis of alkyl ethers has been reported by Bayer AG [155]. Hydroquinone monomethyl ether was synthesized by Anic from hydroquinone and methanol. The ion-exchange resins were reportedly doped with transition metals [157]. A Chinese paper describes reaction of ethylene glycol with methanol to form ethylene glycol monomethyl ether [158]. Anhydrosugars can be made from D-glucitol and D-mannitol [159].

Addition of alcohols to olefins

Like hydration, addition of alcohols to olefins is an acid-catalysed, electrophilic reaction:

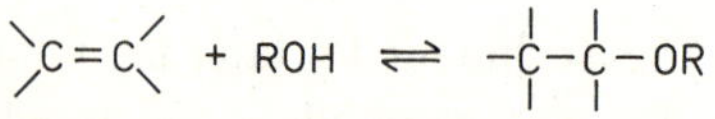

The essential difference between addition and hydration is that the anhydrous, liquid mixture of olefin and alcohol is generally homogeneous, and emulsifiers and solubilizers are not involved. The synthesis of ethers from alcohols and olefins was of little importance for a long period of time. The possibility of separating iso-butene from C_4-fractions via methyl ether, and its use as a gasoline additive, initiated a flood of patents around 1974.

Tertiary alkyl ethers. The patent literature on the synthesis of tertiary alkyl ethers from iso-olefins and alcohols is summarized in Table 2.55. By far the largest number treat the selective reaction of iso-butene in C_4 hydrocarbon with methanol. At moderate temperatures (50 – 100 °C), up to 96% of the iso-butene reacts on an acid IER. Often the formation of MTBE is part of an entire concept for C_4-refining.

Table 2.55 Patent literature on the production of tert-alkyl ether from iso-olefins and alcohols using acidic ion exchange resins

Priortiy		Applicant	DE	US	Others	Remark
06.07.44	DE	BASF	8 68 147			MTBE
09.01.61	DE	Bayer	12 24 294			MTBE
23.03.64	FR	Umel	12 68 128			MTBE
			—			
09.07.68	GB	Shell	19 34 422			MTBE and (5)
			—			
02.10.69	NL	Chevron	19 49 818	—		MTBE and (5)
23.04.70	US	Chevron	—	4 046 520		IPTBE
26.06.70	US	Chevron	—	3 849 082	—	MTBE
22.11.71	US	Sun Oil	22 46 003	3 726 942		MTBE
22.11.71	US	Sun Oil	22 46 004	—		MTBE
11.09.72	US	Sun Oil	—	—	CAN 10 10 066	MTBE
26.09.73	US	Mobil	24 45 774			MTBE
21.05.74	IT	Snamprogetti	25 21 673	4 039 590	PTC 63 798	MTBE (1)
21.05.74	IT	Snamprogetti	25 21 963	4 071 567	PTC 63 800	MTBE
21.05.74	IT	Snamprogetti	25 21 964	3 979 461	PTC 63 797	MTBE
21.05.75	IT	Snamprogetti	25 21 965	4 020 114	PTC 63 796	MTBE (1)
30.08.74	US	TDC	25 35 471	—		TAME
02.12.74	US	TDC	25 47 380	—		MTBE
16.12.74	US	Texaco Inc.	—	3 940 450		MTBE (2)
27.03.75	IT	Anic	26 12 749 26 59 942	4 150 037 4 199 516 und 4 284 567	PTC 64 931	cyclic ethers
06.05.75	US	TDC	26 20 011	4 252 541		MTBE, TAME
17.11.75	DE	DTA	—	—	J 77/062 206	MTBE
17.02.76	US	Suntech	27 06 879	4 080 180 and 4 090 885		Diisobutylene + MeOH
10.05.76	CS	Macho	—	—	CS 190 716	C_4, C_8 + Alkylalc.
20.05.76	CS	Macho	—	—	CS 190 755	GlycolTBE
28.06.76	SU	Chaplits	—	—	SU 918 290	MTBE (3)
02.07.76	DE	CWH	26 29 769	4 219 678	PTC 66 753	MTBE (2)
14.10.76	DE	BASF	26 46 333	—		MTBE
22.11.76	JPN	Nippon Oil	27 52 111	4 182 913 und 4 256 465		MTBE
16.02.77	DE	DTA	27 06 465	4 161 496		MTBE (1)
29.04.77	US	Texaco Inc.	—	4 118 425	—	MTBE
30.06.77	IT	Snamprogetti	28 28 664	—		MTBE (2)
18.01.78	US	Texaco Inc.	28 54 642	4 144 138		MTBE (2)
19.01.78	DE	BASF	28 02 198	4 287 379	EP 3305	NPTBE, NBTBE, IBTBE
19.01.78	DE	BASF	28 02 199	—		IBTBE
14.03.78	US	Gulf	29 38 222	4 218 569		
						MTBE (2)
23.03.78	NL	Shell	29 11 077	—		MTBE

Table 2.55 continued

Priortiy		Applicant	DE	US	Others	Remark
24.03.78	CS	Vybihal	–	–	CS 211 705	MTBE
27.07.78	US	Chem. Res. Lic.	–	4 215 011	EP 8860	MTBE (3)
27.07.78	US	Texaco Inc.	–	4 148 695		MTBE (2)
27.07.78	PL	"Blachownia"	–	–	PL 117 633	MTBE (1)
16.08.78	JPN	Sumitomo	–	–	J 80/027 159	MTBE
28.09.78	CAN	Gulf	29 38 738	–		MTBE (2)
02.19.78	US	Gulf	–	4 204 077	–	MTBE (2)
08.11.78	FR	IFP	29 44 914	4 267 393		(5)
13.12.78	DE	CWH	28 53 769	4 282 389	PTC 79 318	MTBE
14.02.79	FR	IFP	30 05 013	4 310 710		MTBE (2)
21.02.79	US	Chem. Res. Lic.	–	4 336 407	–	C_1-C_6-TBE (5)
05.03.79	DE	BASF	29 08 426	–	EP 15 513	NP-, NB-, IBTBE
05.03.79	JPN	Nippon Oil	–	4 404 409		MTBE
24.04.79	FR	IFP	30 15 346	4 324 924		MTBE
27.04.79	CS	Macho	30 15 882	–		MTBE (2)
09.05.79	FR	IFP	30 17 413	4 299 999		MTBE (2)
11.05.79	JPN	Mitsui	–	–	J 80/149 219	MTBE
28.05.79	DE	Davy McKee	29 21 576	4 329 516		MTBE
12.07.79	DE	BASF	29 28 098	–		Cyclododecyl TBE
14.07.79	DE	BASF	29 28 509	4 320 232	EP 22 509	MTBE + NP-, NB-, IBTBE
14.07.79	DE	BASF	29 28 510	–	EP 22 510	NP-, NB-, IBTBE
26.10.79	US	Phillips	–	4 272 823	EP 28 023	MTBE
08.11.79	US	Phillips	–	4 290 110	EP 28 824	MTBE
15.01.80	US	Phillips	–	4 262 146	–	MTBE
23.01.80	FR	IFP	31 01 703	–		C_1-C_4-TBE
29.02.80	JPN	Maruzen	–	4 299 997 4 345 102	EP 35 075	GlycolTBE
07.03.80	GB	BP	–	–	EP 36 260	MTBE (5)
10.03.80	FR	IFP	–	4 361 422	EP 35 935	(5) C_1-C_4-oles$_4$
21.03.80	US	Phillips	–	4 299 996		ArylTBE
31.03.80	IT	Snamprogetti	31 12 277	–		C_2-C_4TBE (4)
08.04.80	JPN	Sumitomo	–	–	J 81/142 218	MTBE
28.04.80	FR	IFP	31 15 496	4 366 327		MTBE
05.05.80	US	Texaco Inc.	–	4 316 724		ETBE (4) (5)
27.06.80	DE	Edeleanu	30 24 147	4 334 964	EP 43 478	MTBE (2)
12.07.80	DE	Bayer/EC	30 26 504	4 330 679	EP 43 986	MTBE
24.07.80	DE	Davy McKee	30 27 965	–		MTBE
18.08.80	US	U. Kuhlmann	31 31 974	–		Nafion
12.09.80	US	Phillips	–	–	EP 47 906	C_1-C_{12}-TBE (5)
27.09.80	DE	Bayer/EC	30 36 481	–	EP 48 893	MTBE
27.12.80	DE	BASF	30 49 213		EP 55 361	MTBE (1)

Table 2.55 continued

Priortiy		Applicant	DE	US	Others	Remark
03.02.81	US	Halcon	–	4 334 890	–	ETBE (4)
09.02.81	US	Phillips	–	4 320 233	–	MTBE
10.04.81	IT	Snamprogetti	32 13 048	–	PTC 74 729	MTBE (3)
28.04.81	DE	Veba	31 16 780	4 408 085	EP 63 814	SBTBE
28.04.81	DE	Veba	31 16 779	–	EP 63 815	IPTBE
28.04.81	DE	Veba	31 52 586	4 393 250	EP 63 813	SB-, IPTBE
19.06.81	DE	Bayer/EC	31 24 293	–	EP 68 218	MTBE (2)
19.06.81	DE	Bayer/EC	31 24 294	4 409 421	EP 68 194	MTBE (2)
01.07.81	US	Phillips	–	4 371 718	EP 68 514	MTBE (2)
27.07.81	US	Phillips	–	–	EP 71 238	MTBE
28.07.81	IT	Petroflex	–	–	EP 71 032	MTBE, ETBE
28.08.81	US	Phillips	–	–	EP 75 136	MTBE (2)
24.09.81	US	Phillips	–	–	EP 75 838	MTBE (2)
20.10.81	IT	Petroflex	–	–	EP 78 422	MTBE (2)
04.12.81	DE	Bayer/EC	31 48 109	–	EP 82 316	MTBE (2) (3)
06.01.82	JPN	Sumitomo	–	–	J 83/118 531	MTBE
26.02.82	DE	Bayer/EC	32 97 030	–	EP 87 658	MTBE (1)
12.03.82	IT	Snamprogetti	33 08 736	–		MTBE (1)
20.05.82	IT	Snamprogetti	33 18 300	–		MTBE (2)
20.05.82	IT	Samprogetti	33 18 301	–		MTBE

Remarks: ETBE = ethyl-tert-butylether; IPTBE = isopropyl-tert-butylether; SBTBE = sec-butyl-tert-butylether; NBTBE = n-butyl-tert-butylether; NPTBE = n-propyl-tert-butylether; IBTBE = iso-butyl-tert-butylether; TAME = tert-amylmethylether
(1) Feedgas contains butadiene; (2) separation process; (3) simultaneous reaction and distillation; (4) in the presence of water; (5) also C_5 and higher iso-olefins

Processes in which a feedstock-gas containing butadiene is used are indicated by a (1). Many of the publications deal with the extraction, distillation or adsorption processes used in the workup of the reaction mixture. These are indicated by (2) in Table 2.55. In some cases, the reaction is carried out simultaneously with a distillation (3) (Figures 2.171, 2.172, 2.173).

The separation of iso-butene is not limited to the reaction with methanol as the alcohol component, and some of the patents therefore are more inclusive.

On the olefin side, the reaction is not limited to 4 carbon atoms. The separation of 2-methylbutene-2 from hydrocarbons via methyl ether (TAME) is used to purify the C_5 iso-olefins (Figures 2.174, 2.175).

Aside from the improvement of octane ratings with MTBE and TAME, publications in the chemical literature deal chiefly with the kinetics of MTBE formation and the comparison of various acidic IERs [160–172]. The addition of methanol or ethanol to mesityl oxide to form diacetone alcohol ether has also been reported [173].

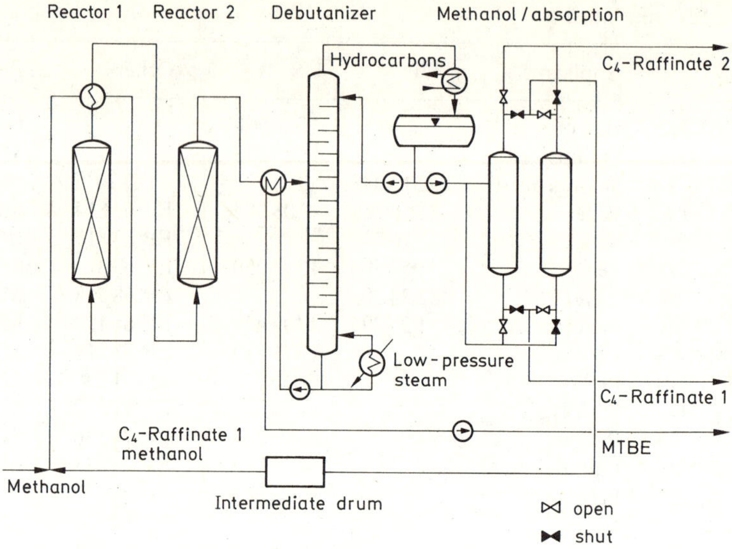

Figure 2.171 Erdölchemie's MTBE process flow sheet.

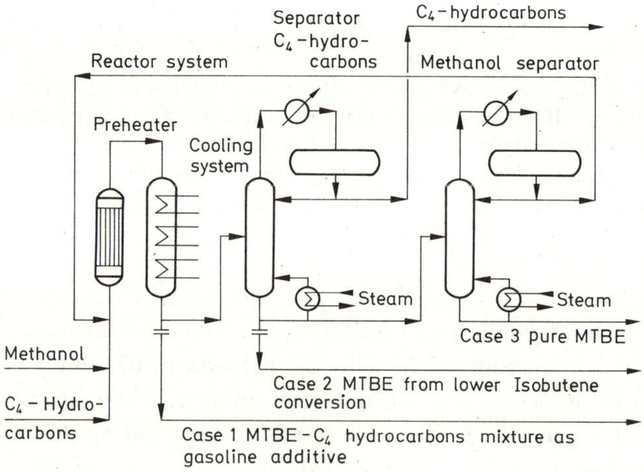

Figure 2.172 CWH's MTBE process flow sheet.

Secondary alkyl ethers. Nippon Oil has reported a process for synthesis of diisopropyl ether from propylene and IPA in the liquid phase at 107 °C/40 bar [174, 175]. The same reaction is also reported in a Polish paper [176]. Nippon Oil has also described the addition of propylene and 1-butene to glycols to make glycol monoethers [183].

Shell reacts glycols and polyglycols with internal $C_{11}-C_{12}$ olefins at 130 °C in liquid sulfur dioxide, in some cases on acidic IER. The conversion rates are less

Figure 2.173 MTBE plant Deutsche Texaco's process.

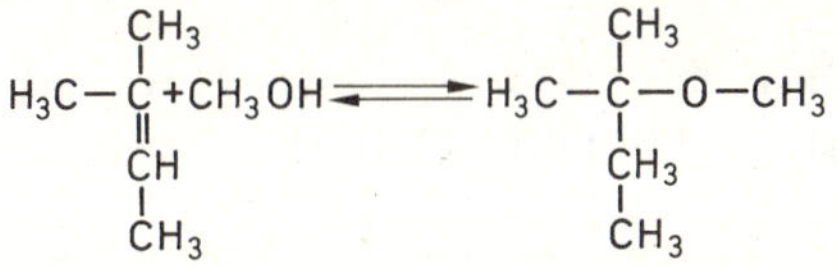

Figure 2.174 Reaction of 2-methylbutene with methanol to TAME.

than 10% [177]. Higher rates (90 – 96.5%) are achieved by BASF by reaction of polyethylene glycol monomethyl ether with propylene at 80 – 120 °C in the liquid phase [179, 180].

According to a patent granted to Rohm and Haas, the reaction of dicyclopentadiene with glycol forms exclusively the monoether products of the six-membered ring double bond [181].

Shell uses a method for synthesis of olefinic unsaturated ethers. Butadiene reacts with methanol in the presence of sulfolane to form the monomethyl ether [182].

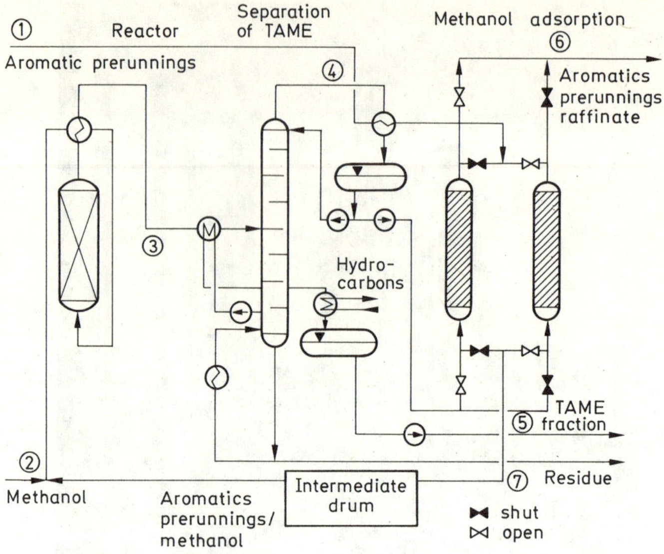

Figure 2.175 TAME process.

Conversion of ethers to olefins and alcohols

The acid-catalysed addition of alcohols to olefins is a reversible reaction. The ethers can thus be cleaved:

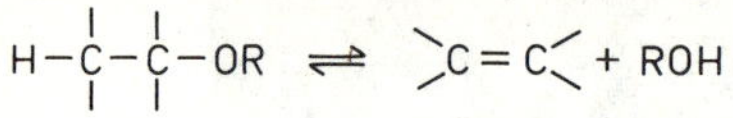

The reaction is endothermal (MTBE H_{298} = 65.3 kJ/mol). In practice, cleavage of MTBE or TAME to obtain the pure iso-olefins has been found to require temperatures above 170 °C. Organic ion exchangers for these temperatures are not available.

Davy McKee describes the cleavage of 4-tert-butoxybutyric acid to iso-butene and γ-butyrolactone with Amberlyst® 15 at 160 °C, in the course of a several-step reaction [183]:

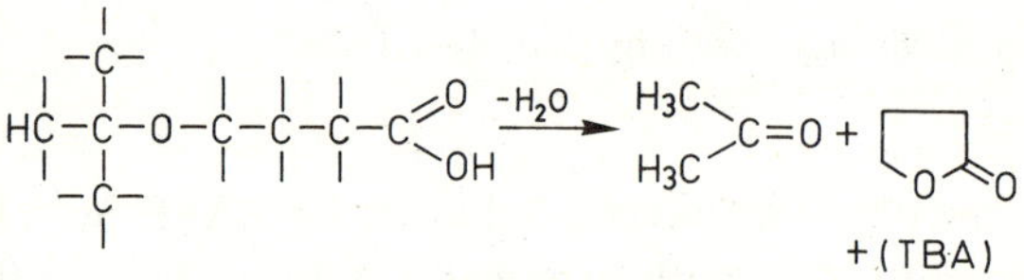

In another application, 4-tert-butoxybutanol-1 is cleaved at 115 °C at moderate yield [184]:

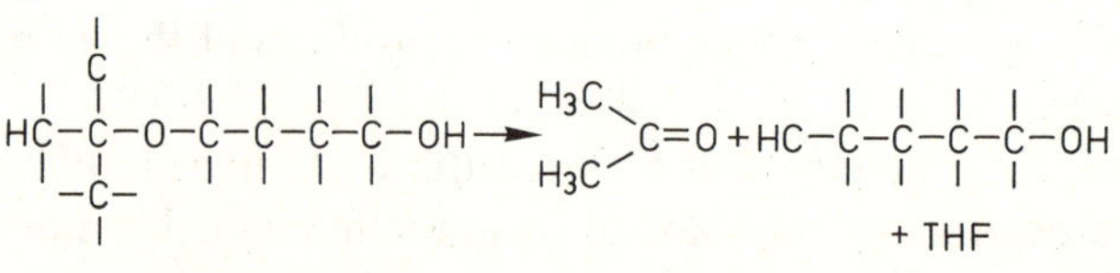

The back cleavage of diisopropyl ether (IPE) has considerable economic significance in IPA processes. Chevron cleaves IPE at 165 °C (330 °F) almost quantitatively (95%) in propylene by simultaneous dehydration in water [185]. Amberlite® 372, a chlorinated exchanger, is used as catalyst.

In this context, the possibility of hydrating ethers is even more significant. Chevron reports that IPE is hydrated in an IPA reactor [186]. Hydration of IPE is also reported by Deutsche Texaco AG [187]. Hydrolysis of glycol dialkyl ethers to the monoalkyl ether and alcohol or olefin is catalyzed by Amberlyst® 15, according to Maruzen [188].

Another way to synthesize alcohols from ethers is transetherization. Maruzen "conproportionates" ethylene glycol di-tert-butyl ether with ethylene glycol to form the monoether [189]. Texaco Inc. reacts glycol monoalkyl ethers with dialkyl ethers in the presence of Nafion and phosphonic acid exchangers to form mixed glycol diethers [190]. Bayer etherizes phenols with dialkyl ethers [191]. The reaction of vinyl isobutyl ether with 1-octanol is described by Grace [192].

Addition of carboxylic acids to olefins

Water (R = H) and alcohols (R = alkyl) are not the only hydroxy compounds (ROH) suitable for addition reactions with olefins; carboxylic acids (R = acyl) are also suitable:

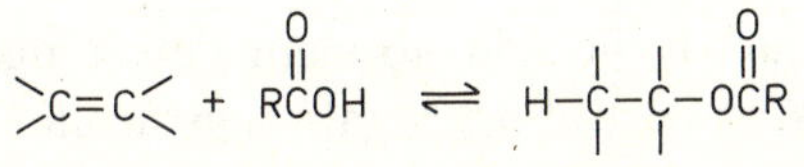

In many cases, especially with aliphatic lower carboxylic acids, the reactions can be carried out in homogeneous liquid phases. The patent literature on esterification of carboxylic acids with olefins is summarized in Table 2.56.

As early as 1954, Standard Oil described the reaction of diverse olefins with C_2-C_{12} mono- and polycarboxylic acids. Rohm and Haas make extensive use of syntheses of secondary alkyl esters from olefins and carboxylic acids. Chevron esterifies iso-butene with formic acid and glacial acetic acid. Toray describes the reaction of dicarboxylic acids (e. g., adipic acid) with cyclohexene. Hoechst reacts propylene with carboxylic acids up to C_{20} in a continuous process. A publication of Nippon Oil concerns the synthesis of isopropyl acetate from glacial acetic acid and propylene. BASF adds glacial acetic acid, propionic acid and acrylic acid to iso-butene, propylene or cyclohexene, and glacial acetic acid to allyl acetate to give the 1,2-diacetate. The synthesis of ethyl acetate with fluorinated ion-exchange resins is described by Rhône-Poulenc. A process for production of SBA from butylene via sec-butyl ester is used by Anic. Erdölchemie esterifies amylene, butadiene, isoprene, cyclopentadiene, styrene and butylene with glacial acetic acid and propionic acid. The reaction of butadiene and isoprene with glacial acetic acid is described by BASF. A Russian paper treats the kinetics of isopropyl acetate formation from propylene [193]. Monoacetoxystearic acid can be made from oleic acid [194], or monoacetoxystearic acid methyl ester from methyl oleate [195].

Table 2.56 Patent literature on the addition of carboxylic acids to olefins in the presence of acidic ion exchange resins

Priority		Applicant	Patents
11.05.54	US	Standard Oil	US 2 678 332
29.04.59	US	Rohm and Haas	US 3 037 052
26.06.70	US	Chevron	US 3 678 099
25.12.72	JPN	Toray	J 74/086 344
10.02.73	DE	Hoechst	DE 23 06 586
			US 3 922 294
22.08.75	JPN	Nippon Oil	J 77/025 710
27.01.76	DE	BASF	DE 26 02 856
02.06.76	DE	BASF	DE 26 24 628
17.05.78	FR	Rhône-Poulenc	DE 29 62 228
			US 4 275 228
			EP 5 680
30.01.79	IT	Anic	DE 30 03 126
14.02.81	DE	Erdölchemie	DE 31 05 399
15.01.82	DE	BASF	DE 32 00 990
			EP 84 133

Production of alcohols from carboxylic acid esters

The synthesis of esters by addition of carboxylic acids to olefins must logically be considered, as is the synthesis of ethers, as part of a several-stage alcohol process:

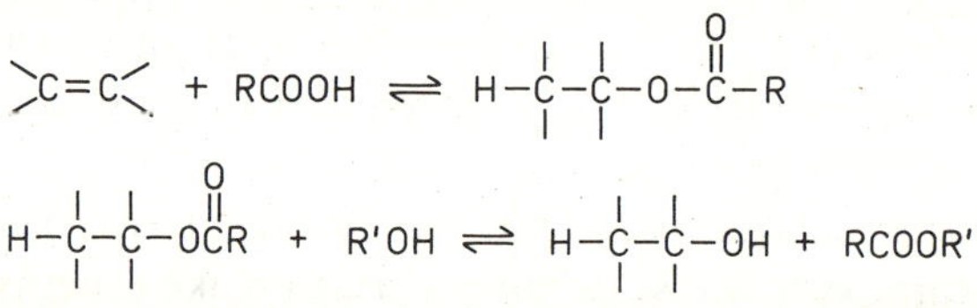

R′ = H, alkyl

If the extensive literature on saponification and alcoholysis (transesterification) of carboxylic acid esters with IER is reduced to those papers that have the goal of obtaining the alcohol component of the ester, there are only two publications left. The deacetylation of sugars succeeds well in methanolic solution in the presence of basic ion exchangers [196].

The Anic process for production of SBA via sec-butyl ester is mentioned in the section above. The crude sec-butyl acetate is saponified in a distillation apparatus with water on Amberlyst® 15 at 90 °C. The SBA is obtained off the top as a ternary mixture of water and ester.

Synthesis of epoxides from olefins

Although the IERs are made relatively sensitive to oxidation by the tertiary carbon next to the aromatic ring, there are publications on the acid-catalyzed epoxidation of olefins with hydrogen peroxide in the presence of IERs.

The epoxidation of unsaturated fatty acid derivatives is even carried out industrially [197–204].

The reaction of polybutadiene with H_2O_2 in the presence of Amberlite® IR 120 is described by Phillips [205] and bei Japan Soda [206].

Polyesters of tetrahydrophthalic acid are epoxidized with acid IER [207]. The reaction of diverse olefins has been described [208, 209]. Sugar epoxides can be synthesized in the presence of the strongly basic exchanger Amberlite® IRA 400 [210].

Oxovanadium(IV) on sulfonic acid IER is recommended for expoxidation with tert-butylhydroperoxide [217].

Cyclohexene and cyclohexene derivatives react to form the epoxides, either with coumol hydroperoxide with base catalysis [212], or directly with oxygen by acid catalysis [213].

The synthesis of propylene oxide from propylene has been described using molybdenum complexes on weakly basic ion-exchange resins [214], with molybdenum oxide on IER [215], with modified, weakly acidic IER [216] or with MoO_2Cl_2 on Amberlite® IRC 84 [217].

Solvolysis of epoxides

The epoxides synthesized from olefins can be used as intermediates in the synthesis of alcohols and ethers. There are many publications dealing with the hydration of ethylene oxide using acidic ion-exchange resin [218–228].

Yamanis and Garland describe the formation of 1.4-dioxane from ethylene oxide on dry Amberlyst® 15 [229]. The reaction of epoxides with alcohols or phenols to form hydroxyethers on acidic IER has also been reported [230–234]. Hydroxyalkyl esters of epoxides and carboxylic acids are synthesized on basic IER [235–237].

2.11.2.2 Hydrolyses and transesterifications

Saccharides. As early as 1942, the application of acidic resins for the hydrolysis of cane sugar was patented in Germany [5]. Systematic studies on diverse saccharides in the subsequent years indicated that the rate of hydrolysis depends greatly on the molecular size of the saccharide, and can be affected by the degree of crosslinking of the ion-exchange resin [238–245]. American Cyanamid obtained a patent in 1952 for a technical process for cane sugar inversion [264]. In 1965, Boehringer reported the inversion of cane sugar with simultaneous separation into fructose and glucose on a cation exchanger which was partially loaded with calcium ions [247]. This process was further developed in the following years, converted to a continuous process and optimized [248]. By careful selection of the catalyst, the reaction temperature could be reduced so far that only insignificant amounts of the fructose are converted to hydroxymethylfurfural. It was shown on gel and macroporous resins that both heteregeneous and homogeneous catalyses of these reactions are first order.

Esters. At the end of the 1930s, R. Griessbach et al. succeeded in hydrolysing esters using cation exchange resins [5]. After preliminary experiments with ethyl acetate, this process rapidly became important. After the residual acids from adipic acid synthesis have been esterified with methanol, the individual esters are separated by distillation, and the acids are recovered by continuous saponification. The resulting methanol is removed from the reaction mixture by steam [24].

Both anion and cation exchange resins are suitable catalysts. However, because the anion exchangers undergo a stoichiometric reaction, they are rapidly consumed [243, 249]. Esters of lower-molecular weight acids are hydrolysed more or less rapidly by anion exchangers. For example, the acetate ion released by hydrolysis of polyvinyl acetate migrates into the interior of the exchanger core and becomes fixed there. Polygalacturonic acid methyl ester is barely hydrolysed, because the small amounts of macromolecular anion formed block the outer surface of the catalyst.

Acid-catalyzed hydrolysis of low-molecular weight compounds occurs rapidly, because the anions formed are repelled by the like-charged cation exchanger. Attempts to hydrolyse fatty acid esters were less successful. Addition of solubilizers increases the rate of hydrolysis [238]. Low crosslinked types of gel and macroporous resins are more active than highly crosslinked resins [250, 251].

Aromatic esters are less readily saponified than aliphatic esters [252]. Many esters were hydrolysed and the effectiveness of various resins, which in some cases were also partially loaded with magnesium or amines, etc., was studied. The kinetics of the reactions were also examined [238, 239, 243, 250, 253–279].

Alcoholysis of esters is analogous to hydrolysis. Here again, it is advisable to remove one of the products, e. g., by distillation, from the reaction mixture [238, 1, 280–289]. Finally, the catalytic deacylation of sugars with strongly basic ion-exchange resins should be mentiond [196].

Peptides. Proteins are hydrolyzed by cation exchangers, but the hydrolysis is rarely complete. The type of ion exchanger and the protein have major effects on the reaction. Coffee-bean protein, casein, gelatins and milk, egg and serum albumins were studied. It was shown that the peptide bonds of different amino acids have different stabilities to hydrolysis. The peptide bond of cystine is particularly stable [290–293].

2.11.2.3 Condensation and addition reactions

Esterifications. Esterification of carboxylic acids with alcohols using ion exchangers is at present the most widely used type of reaction, and it has therefore been thoroughly studied. It is most conveniently run continuously or with the use of carrier reagents for the removal of water [1, 5, 14, 238, 294–320]. The esterification of unsaturated carboxylic acids, such as acrylic or methacrylic acids, is particularly noteworthy [311, 312, 317]. On the synthesis of esters, see also Section 2.11.2.1.

Etherifications (see also Sect. 2.11.2.1). If primary alcohols are dehydrated in the presence of cation exchangers, the main products are ethers, and only small amounts of olefins are formed. Olefin formation is much more prominent in secondary alcohols, and with tertiary alcohols it is the main reaction. The corresponding aromatic alcohols tend to form ethers [321, 322]. Cyclic ethers, such as tetrahydrofuran, can also be made in this way [323].

Formation of acetals (ketals). Continuous processes have also proven useful for the synthesis of acetals. These reactions are catalyzed by cation-exchange resins. Cyclic and mixed acetals can also be made in nearly quantitative yields [5, 9, 10, 12, 238, 322–332].

Aldol condensation and crotonization. Both aldehydes and ketones can undergo aldol condensation. Anion and cation-exchange resins are used as catalysts, with different reaction sequences in the two cases. In the presence of acid ion-exchangers, water is split off and crotonization occurs [5, 8, 9, 20, 33, 40, 302, 323, 333–352].

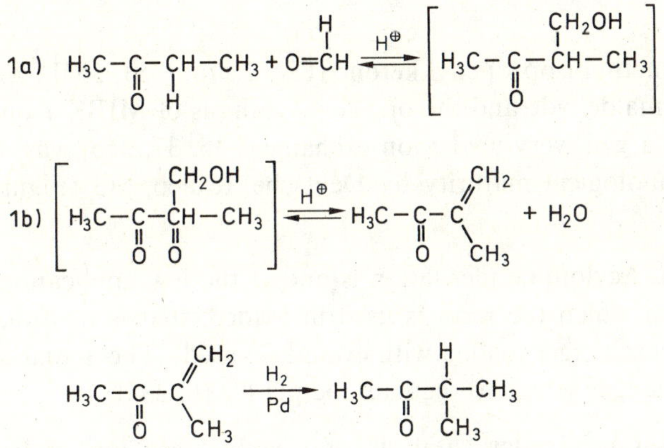

Figure 2.176 Formation of methylisopropenylketone and methylisopropylketone.

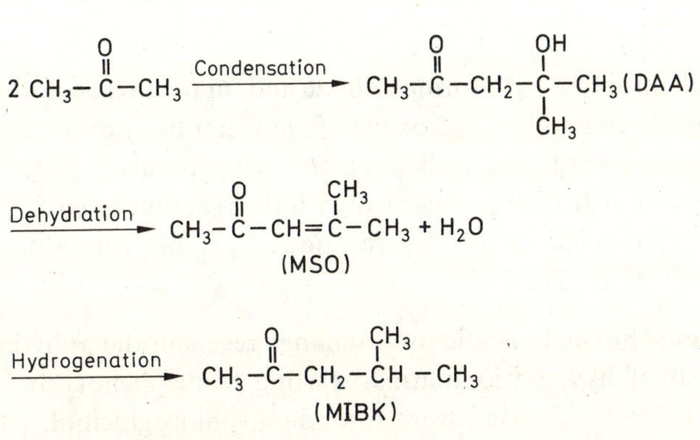

Figure 2.177 Formation of MIBK.

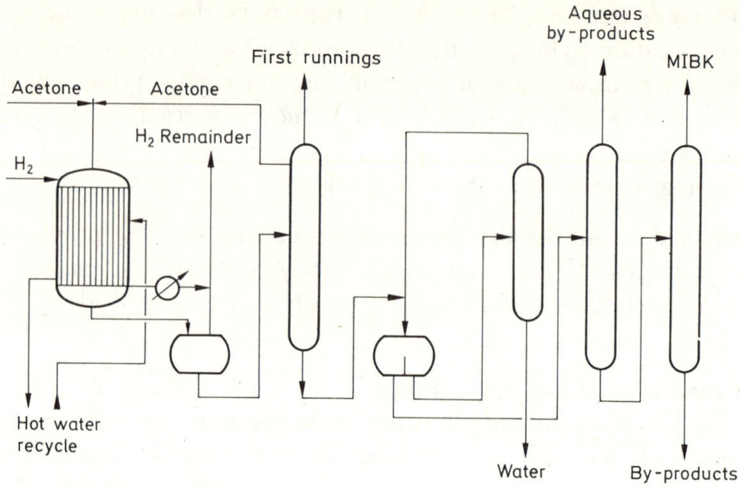

Figure 2.178 MIBK process (Deutsche Texaco AG), flow sheet.

The synthesis of methyl isopropenyl ketone (Fig. 2.176) [353, 354] from methyl ethyl ketone and formaldehyde and the one-step synthesis of MIBK from acetone using a palladium-charged, very acidic ion exchanger [35, 37, 355] was developed to the point of technological maturity by Deutsche Texaco AG (Figures 2.177, 2.178).

Acyloin condensation. Acyloin condensation is one of the few applications of ion-exchanger catalysis in which the resin is used in loaded, that is neutral, form (a strongly basic anion exchanger loaded with cyanide is used). The actual catalyst is the cyanide; the exchanger serves to regulate the pH [9, 10, 335].

Knoevenagel condensation. Under catalysis of weakly basic ion exchangers or mixtures of exchangers, aldehydes and ketones react with compounds which contain active hydrogen atoms, for example, acetoacetate, cyanoacetate or malonates [9, 10, 336].

Prins reaction. Olefins react with formaldehyde and, in the presence of ion-exchangers, the products are generally 1.3-dioxanes. These can be converted by hydrolysis to glycols or, after dehydration, to dienes [356–370] (Figure 2.179).

Work at a French institute is concerned with the reaction of enolizable carbonyl compounds with formaldehyde and the reaction of alkenes with aldehydes (Figure 2.180) [8, 358, 369, 370, 371].

Pechmann synthesis. Strongly acidic ion-exchange resins in the anhydrous state are used for synthesis of hydroxycoumarins, starting from phenols and β-ketoesters (Figure 2.181). The phenols used were resorcinol, phloroglucinol, pyrogallol and orcinol, and the β-ketoesters were ethyl acetoacetate, ethyl α-methylacetoacetate

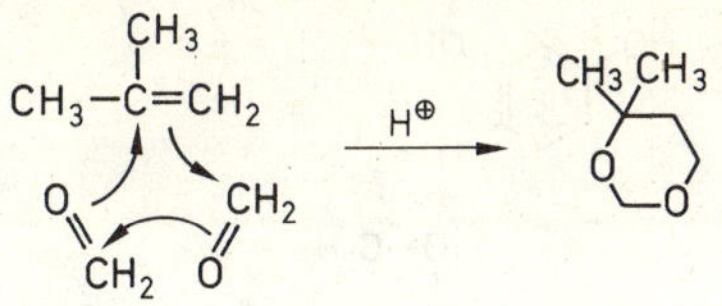

Formation of Dioxane from Isobutene and Formaldehyde

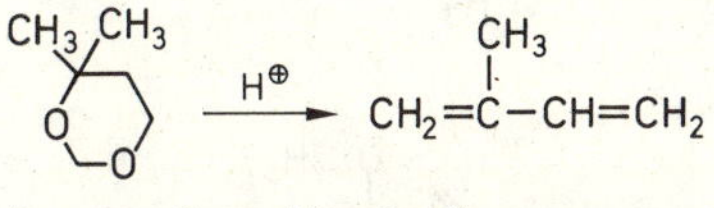

Isoprene from Dimethyldioxane

Figure 2.179 Formation of dioxane from isobutene and formaldehyde and of isopropene from dimethyldioxane.

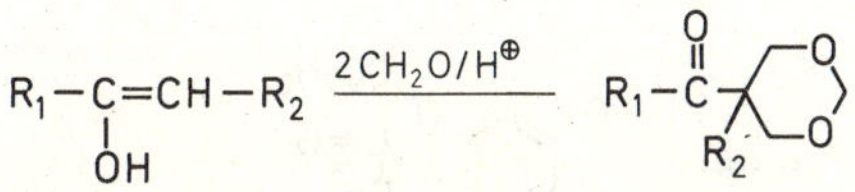

Reaction of Formaldehyde with Enols

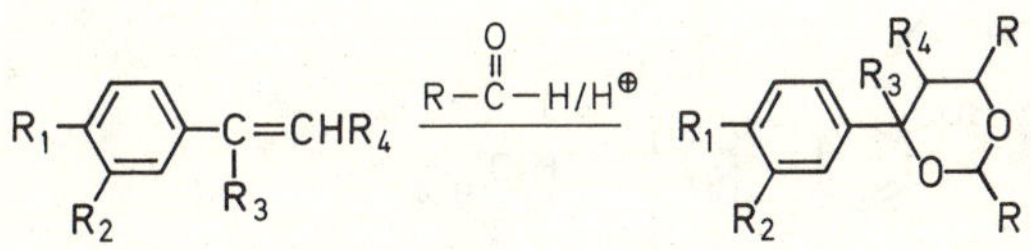

Reaction of Aldehydes with Alkenes

Figure 2.180 Reaction of formaldehyde with enols and of aldehydes with alkenes.

and ethyl α-ethyl acetoacetate. The yields varied, depending on the reactants, between 8.3 and 79.5% [373–375].

Cyan ethylation, cyanohydrin synthesis. Alcohols, ketones and water can add to the double bonds of unsaturated nitriles with the aid of anion exchangers. Aldehydes and ketones react with hydrogen cyanide or cyanide-loaded exchangers in the same way [9, 335, 376–378].

2.11.2.4 Alkylations

Alkylation reactions usually occur in nonpolar media. The catalytically active group is essentially undissociated. Only strong acid catalysts are effective; macroporous resins with large pore diameters are beneficial.

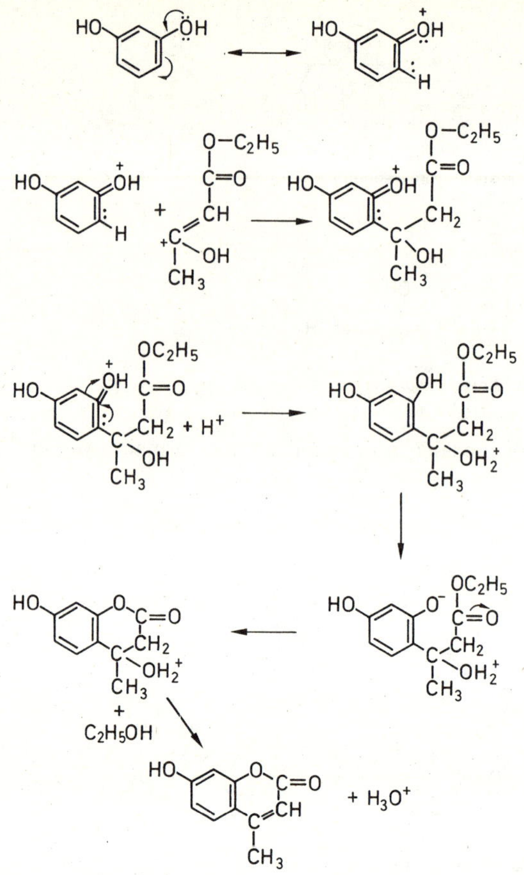

Figure 2.181 Synthesis of hydroxycoumarin from phenols and β-keto-esters.

Bisphenols/polyphenols. Bisphenol syntheses are carried out on a large scale in industry. Partial neutralization of the sulfonic acid groups with mercaptoamines is supposed to increase the reaction rate. An important problem is the isomerization of the byproducts from the first step into bisphenol. The reported increase in reactivity effected by mercapto compounds could not be confirmed in all publications [18, 41, 379–424).

Alkylphenols. Strongly acidic cation exchangers are especially suitable for alkylation of phenols. Olefins are usually used as alkylating reagents. In principle, alcohols, ethers, alkyl esters or epoxides could also be used. However, the water formed in the reaction reduces the activity of the catalyst. Since conventional gel resins swell in the presence of phenol, these can be used as catalysts, but the macroporous resins have advantages here as well.

Disubstitution products are formed as byproducts, and these can react in the next step with phenol, forming monoalkyl benzene. At higher temperatures, the formation of byproducts increases. Olefins were oligomerized [8, 14, 16, 18, 425–450]. An extensive collection of literature is given by N. P. Haran et al. [446].

Alkylbenzenes. A large number of publications are devoted to the alkylation of aromatics [18, 41, 446, 451–465) with strongly acidic ion exchangers. Haran et al. [446] give an extensive bibliography. The results with conventional resins are very unsatisfactory, because their catalytic activity is not high enough. Relatively high temperatures must be used, which causes a rapid decline in activity. Higher-molecular weight byproducts are also formed. Addition of Lewis acids can double the activity, and rates almost as high as those obtained with $AlCl_3$ are obtained.

The catalytic activity can be considerably increased by use of a sulfonated perfluoropolymerizate (Fig. 2.166). Nafion® (Du Pont), a perfluorosulfonic acid, was tested. However, this material is at present not obtainable, and it also has only a very slight porosity. Perfluorination of a conventional strongly acidic ion exchanger gave a much more active catalyst. At present, however, this catalyst can only be prepared in a very laborious process on a laboratory scale [62].

Alkylation of aliphatics. The alkylation of paraffins with olefins is also industrially significant. So far, it has been observed with ion exchangers only when Lewis acids such as BF_3, $TiCl_4$ or H_2SO_4 were also present, or when perfluorinated ion exchangers were used [64]. In the microreactor, the catalyst lost activity very rapidly, and high-molecular weight byproducts were deposited on it. Dimerization and oligomerization increased rapidly after a short time.

References

[1] R. Griessbach and G. Naumann (1953) Chem. Techn. *5,* 187.
[2] B. A. Adams and E. L. Homes (1935) J. Soc. Chem. Ind. *54,* 1 (T).
[3] B. Tacke and A. Süchting (1911) Landw. Jahrb. *41,* 717.
[4] F. E. Rice and S. Osugi (1918) Soil Sci. *5,* 333.
[5] DE 88 20 91, IG Farbenindustrie (1942).
[6] DE 86 61 91, IG Farbenindustrie, BASF (1944).
[7] DE 86 81 47, IG Farbenindustrie, BASF (1944).
[8] Kationenaustauscher als Katalysatoren in der chemischen Industrie, Techn. Inform. Bayer AG, Order no. OC/I 20392 (1984).
[9] F. Helfferich (1954) Angew. Chem. *66,* 241.
[10] G. Naumann (1959) Chem. Techn. *11,* 18.
[11] N. G. Polyanski (1962) Russ. Chem. Rev. *31,* 496.
[12] H. Spes (1966) Chemiker Ztg., *90,* 443.
[13] N. G. Polyanski (1970) Russ. Chem. Rev. *39,* 244.
[14] R. Hellmig (1971) Chem. Technik *23,* 28.
[15] G. Kühne (1972) Chemiker Ztg., *96,* 239.
[16] R. Kunin (1973) Amber-hi-lites No. 135, Rohm and Haas Company, Philadelphia, USA.
[17] F. Ancillotti, M. M. Mauri and E. Pescarello (1977) J. Cat. *46,* 49.

[18] J. Klein and H. Widdecke (1979) Chem. Ing. Techn. *51,* 560.
[19] H. Widdecke (1980) Chem. Ing. Techn. *52,* 825.
[20] W. Neier (1981) Chem. Ind. *33,* 632.
[21] S. S. Bhagade and G. D. Nageshwar (1981) Chem. and Petrochem. J. *12,* 21.
[22] J. Klein and H. Widdecke (1982) Chem. Ing. Techn. *54,* 595.
[23] DE 1 075 613, BASF (1957).
[24] F. Andreas (1959) Chem. Techn. *11,* 24.
[25] H. Spes (Wacker-Chemie) (1962) DP-Anm. 32 748.
[26] H. Spes (1963) DP-Anm. W 34 291.
[27] Th. Altenschöpfer, G. Künstel and H. Spes (Wacker Chemie) (1963) DP-Anm. W 33 960.
[28] D. N. Chaplits et al. (1975) DE 2 351 120.
[29] D. N. Chaplits et al. DE 2 549 999 (1976).
[30] P. S. Belov and V. I. Isagulyants (1962) USSR 144 854.
[31] P. S. Belov and V. I. Isagulyants (1964) Int. Chem. Eng. *4,* 618.
[32] Z. Zitny and M. Kraus (1975) Coll. Czech. Chem. Comm. *40,* 3851.
[33] W. Neier (1984) Direct Hydration of Propylene, Ion Exchange Technology, Society of Chem. Ind., p. 560, Horwood, Chichester.
[34] DE 2 147 739, Deutsche Texaco AG (1973).
[35] DE 12 38 435, Deutsche Texaco AG (1965).
[36] B. Schleppinghoff (1982) Technische Anwendung von Ionentauschern in der Katalyse, Seminarvortrag Technische Akademie Wuppertal, 29./30. 11. 1982.
[37] DE 12 60 454, Deutsche Texaco AG (1966).
[38] F. Martinola and A. Meyer (1975) Ion Exchange and Membranes *2,* 111.
[39] K. H. Arnold (1979) Der Einfluß von Vernetzung und Porosität stark saurer Kationenaustauscher auf die Katalyse chemischer Reaktionen, Diskussionsvortrag GDCh-Tagung.
[40] W. Neier (1982) Seminarvortrag Technische Anwendung von Ionentauschern in der Katalyse, Technische Akademie Wuppertal, Nov. 1982.
[41] J. Klein and H. Widdecke (1983) Erdöl und Kohle *36,* 307.
[42] L. Petras (1972) Hydration of Lower Alkenes Catalyzed by Strong Acid Ion Exchange Resins, Thesis, Universität Groningen.
[43] GB 1 393 594, Rohm and Haas (1972).
[44] G. F. D'Alelio (1949) Koppers Co. Inc., Delaware, US 2 645 621.
[45] V. J. Frilette and N. J. Erlton (1966) Socony Mobil Oil Co. Inc., US 3 256 250.
[46] C. R. Costin (1981) Rohm and Haas, Philadelphia, US 4 269 943.
[47] DE 1 291 729, Deutsche Texaco AG, (1967)
[48] H. Widdecke (1978) Thesis, TU Braunschweig.
[49] J. Klein, H. Fehrecke and H. Widdecke (1976) Chem. Ing. Techn. *48,* 173.
[50] J. Klein and H. Widdecke (1977) Proc. Int. Symp., Macromol., Dublin, p. 568.
[51] G. Scharfe (1973) Hydrocarbon Processing, *52,* 171.
[52] DE 1 194 398, Bayer AG (1960).
[53] DE 1 194 399, Bayer AG (1967).
[54] DE 1 213 837, Bayer AG (1966).
[55] F. Merger et al. (1975) BASF, DE 2 526 644.
[56] R. Hellmig (1971) Chem. Technik *23,* 28.
[57] DE 2 147 740, Deutsche Texaco AG (1971).
[58] DE 2 233 967, Deutsche Texaco AG (1972).
[59] H. J. Meyer-Stoll (1965) Thesis, TU Berlin.
[60] S. S. Bhagade and G. D. Nageshwar (1980) Chem. and Petro. Chem. J. *11,* 23.
[61] G. Durr (1956) Ann. Chim., Paris *13,* 84.

[62] F. Dröscher, J. Klein, F. Pohl and H. Widdecke (1982) DE 3 023 455.
[63] V. L. Magnotta and B. C. Gates (1974) J. Cat. *46,* 266.
[64] J. Weitkamp (1980) Compendium 80/81, Ergänzungsband Erdöl and Kohle, Erdgas-Petrochemie.
[65] J. Klein and H. Widdecke (1982) Chem. Ing. Techn. *54,* 595.
[66] Morrison/Boyd (1974) Lehrbuch der org. Chemie, Verlag Chemie, Weinheim, pp. 183, 189–190, 217.
[67] W. K. Chwang et al. (1977) J. Am. Chem. Soc. *99,* 7233.
[68] F. Obenaus, B. Greving, H. Balke and B. Scholz (1981) DE 3 151 446.
[69] T. Hasuike and H. Matszawa (1979) Hydrocarbon Processing, *58,* 105.
[70] W. Wilson (1982) Developments in Petroleum Refining, Symp. London, II 2/1.
[71] R. C. Odiso et al. (1961) Ind. Eng. Chem. *53,* 209.
[72] CA *55,* 14 285.
[73] CA *57,* 11 450.
[74] CA *58,* 56.
[75] CA *59,* 5837.
[76] CA *62,* 16 054.
[77] CA *87,* 119010.
[78] CA *99,* 21 655.
[79] J. R. Kaiser, H. Benther, C. D. Moore and R. C. Odioso (1962) Ind. Eng. Chem., Prod. Res. Dev. *1,* 296.
[80] H. J. Hiestand (1961) Hydration of propylene with a cation exchange resin, University of Michigan.
[81] J. M. Tibbit, B. C. Gates and J. R. Katzer (1975) J. Cat. *38,* 505.
[82] CA *92,* 58 160.
[83] CA *86,* 73 883.
[84] W. Neier and J. Wöllner (1972) Hydrocarb. Proc. *51,* 113.
[85] W. Neier and J. Wöllner (1973) Chem. Tech. *3,* 95.
[86] W. Neier and J. Wöllner (1975) Erdöl, Kohle, Erdgas, Petrochem. *28,* 19.
[87] DE 20 41 954, UOP.
[88] DE 27 21 206, Shell.
[89] DE 32 00 990, BASF.
[90] T. Katagiri et al. (1972) Bull. Inst. Chem. Res., Kyoto Univ. *50,* 563.
[91] BE 609 097 (14. 10. 60) Shell.
[92] DE 22 05 225 (4. 2. 72) Hoechst.
[93] CA *77,* 75 341.
[94] DE 26 43 062 (7. 10. 75), US 4 031 161 Int. Flavor and Fragrances Inc.
[95] DE 27 55 945 (15. 12. 77) BASF.
[96] USSR 666 161 (30. 12. 77).
[97] M. Pinza and G. Pifferi (1978) J. Pharm. Sci. *67,* 120.
[98] CA *92,* 110 469.
[99] Z. Prokop and K. Setinek (1977) Coll. Czech. Chem. Com. *42,* 3123.
[100] US 3 267 156 (7. 8. 61) Mobil.
[101] E. Swistak and P. Mastagli (1954) Compt. rend. *239,* 709.
[102] US 3 928 483 (23. 9. 74) Mobil.
[103] J 81 040 630 (12. 9. 79) Sumitomo.
[104] CA *93,* 113 451.
[105] CA *90,* 103 322.
[106] CA *82,* 35 310.
[107] CA *77,* 102 831.
[108] R. B. Diemer jr., K. M. Dooley, B. C. Gates and R. L. Albright (1982) J. Cat. *74,* 373.

[109] K. Klusacek and P. Schneider (1981) Chem. Eng. Sci. *36*, 517 and Sci. *36*, 523.
[110] K. Setinek and Z. Prokop (1976) Coll. Czech. Chem. Com. *41*, 1286.
[111] R. Thornton and B. C. Gates (1973) Catal., Proc. Int. Congr. 5th, 1972 *1*, 357.
[112] Le Nhu Thanh, K. Setinek and L. Beranek (1972) Coll. Czech. Chem. Com. *37*, 3878.
[113] B. C. Gates and L. N. Johanson (1971) AIChE Journal *17*, 981.
[114] B. C. Gates (1969) J. Cat. *14*, 69.
[115] J. C. Herliny (1968) Thesis, University of Washington.
[116] CA *91*, 91 101.
[117] CA *86*, 15 988.
[118] CA *82*, 47 987.
[119] R. L. Kabel and L. N. Johanson (1961) J. Chem. Eng. Data *6*, 496.
[120] R. L. Kabel and L. N. Johanson (1962) AIChE Journal *8*, 621.
[121] S. M. Hsu and R. L. Kabel (1974) J. Cat. *33*, 74.
[122] S. Kmostak and K. Setinek (1981) Coll. Czech. Chem. Com. *46*, 2354.
[123] A. Martinek, K. Setinek and L. Beranek (1978) J. Cat. *51*, 86.
[124] CA *61*, 15 965.
[125] FR 15 37 826 (15. 7. 67).
[126] CAN 10 37 485 (16. 9. 74).
[127] DE 23 00 990 (10. 1. 73).
[128] CA *86*, 120 739.
[129] L. T. Scott and J. O. Naples (1973) Synthesis 209.
[130] CA *72*, 136 892.
[131] US 3 126 356 (19. 8. 58).
[132] J. C. Gottifredi, A. A. Yeramian and R. E. Cunningham (1968) J. Cat. *12*, 245.
[133] C. A. Cooper (1979) Thesis, Univ. of Delaware.
[134] C. A. Cooper, R. L. McCullough, B. C. Gates and J. C. Seferis (1980) J. Cat. *63*, 372.
[135] S. P. Sivanand, B. V. Kamath, R. S. Singh and D. K. Chakrabary (1981) J. Cat. *69*, 502.
[136] CA *74*, 3282.
[137] CA *74*, 54 470.
[138] CA *69*, 96 031.
[139] CA *93*, 185 852.
[140] P. Mastagli, Z. Zafiriadis and E. Swistak (1935) Comp. rend. *236*, 2325.
[141] O. Grummitt and D. Marsh (1953) J. Am. Oil Chem. Soc. *30*, 21.
[142] N. A. Ghanem (1968) Paint Technol. *32*, 10.
[143] NL 67 09 386; Unilever (7. 7. 1966).
[144] Y. Nakamura and S. Morikawa (1980) Bull. Chem. Soc. Jpn. *53*, 3705.
[145] L. Rigal, A. Gaset and J. P. Gorrichon (1981) Ind. Eng. Chem. Prod. Res. Dev. *20*, 710.
[146] D. Mercadier, L. Rigal, A. Gaset and J. P. Gorrichon (1981) J. Chem. Technol. Biotechnol. *31*, 489.
[147] CA *94*, 156 646.
[148] CA *97*, 6078.
[149] N. O. Lemcoff and R. E. Cunningham (1971) J. Cat. *23*, 81.
[150] CA *68*, 29 204.
[151] CA *57*, 8418.
[152] CA *58*, 8938.
[153] CA *65*, 20 046.
[154] CA *62*, 11 672.
[155] CA *63*, 13 319.
[156] DE 26 55 826 (9. 12. 76) Bayer AG.

[157] DE 31 45 212 (13. 11. 80) Anic.
[158] CA *93,* 10 461.
[159] J. C. Goodwin, J. E. Hodge and D. Weisleder (1980) Carbohydrate Research *79,* 133.
[160] M. Voloch, M. R. Ladisch and G. T. Tsao (1986) React. Polym. Ion Exch. Sorbents *4,* 91.
[161] F. Colombo, L. Corl, L. Dalloro and P. Delogu (1983) Ind. Eng. Chem. Funndam. *22,* 219.
[162] A. Gircquel and B. Torck (1983) J. Cat. *83,* 9.
[163] W. G. Rothschild, Duolite Information (31. 8. 82).
[164] A. Convers, B. Torck, J. P. Euzen and P. Amigues (1982) Chem. Industry, 520.
[165] J. D. Chase and B. B. Galvez (1981) Hydrocarbon Processing *60,* 89.
[166] CA *93,* 70 928.
[167] CA *93,* 113 856.
[168] CA *95,* 60 932.
[169] CA *96,* 180 669.
[170] CA *97,* 215 465.
[171] CA *98,* 56 745.
[172] CA *98,* 71 181.
[173] N. B. Lorette (1958) J. Org. Chem. *23,* 937.
[174] DE 27 19 024 (28. 4. 76) Nippon Oil.
[175] DE 24 03 196 (24. 1. 73) Nippon Oil.
[176] CA *88,* 104 611.
[177] DE 24 50 667 (26. 10. 73) Nippon Oil.
[178] EP 1651 (3. 10. 77) Shell.
[179] DE 28 01 793 (17. 1. 78) BASF.
[180] DE 25 44 569 (4. 10. 75) BASF.
[181] DE 28 33 597 (11. 8. 77) Rohm and Haas.
[182] EP 25 240 (4. 9. 79) Shell.
[183] EP 18 162 (11. 4. 74) D. McKee.
[184] EP 18 163 (11. 4. 79) D. McKee.
[185] US 4 352 945 (30. 10. 81) Chevron.
[186] US 4 405 822 (30. 10. 81) Chevron.
[187] DE 33 36 644 (8. 10. 83) Deutsche Texaco AG.
[188] EP 38 129 (28. 3. 80) Maruzen.
[189] EP 35 075 (9. 9. 81) Maruzen.
[190] US 4 321 413 (30. 1. 81) Texaco Inc.
[191] DE 29 03 020 (26. 1. 79) Bayer.
[192] US 3 786 102 (9. 12. 69) Grace.
[193] CA *91,* 174 441.
[194] CA *86,* 157 396.
[195] L. T. Black and R. E. Beal (1967) J. Am. Oil Chem. Soc. *44,* 310.
[196] L. A. Reed, P. Q. Risbood and L. Goodmann (1981) J. Chem. Soc. Chem. Com., 760.
[197] R. J. Gall and F. P. Greenspan (1955) Ind. Eng. Chem. *47,* 147.
[198] R. J. Gall and F. P. Greenspan (1957) J. Am. Oil Chem. Soc. *34,* 161.
[199] US 2 919 283 (29. 12. 59) Food Machinery + Chem. Co.
[200] V. Nagiah, H. Dakshinamurthy and J. S. Aggarwal (1966) Indian J. Technol. *4,* 280.
[201] G. Wallace, W. R. Peterson, A. F. Chadwick and D. O. Barlow (1958) J. Am. Oil Chem. Soc. *35,* 205.
[202] M. Hassan El-Mallah and S. M. El-Shami (1975) SÖFW *101,* 245.
[203] M. Hassan El-Mallah and S. M. El-Shami (1975) SÖFW *101,* 573.
[204] B. M. Badran, F. M. El-Mehelmy and N. A. Ghanem (1976) J. Oil Colour Chem. Assoc. *59,* 291.

[205] US 3 022 322 (28. 9. 56) Phillips.
[206] J 71 099 477 (30. 12. 66) Japan Soda.
[207] J. W. Pearce and J. Kawa (1957) J. Am. Oil Chem. Soc. *34,* 57.
[208] GB 802 127 (1. 10. 58) S. C. Johnson + Son, Inc.
[209] W. Wood and J. Termini (1958) J. Am. Oil Chem. Soc. *35,* 331.
[210] J. Stanek Jr. adn M. Cerny (1972) Synthesis, 698.
[211] G. L. Linden and M. F. Farona (1977) Inorg. Chem. *16,* 3170.
[212] DE 24 46 216 (27. 9. 74) BASF.
[213] US 4 021 369 (29. 6. 73) Sun Ventures.
[214] CA *86,* 30 096.
[215] CA *89,* 179 752.
[216] S. Ivanov, R. Boeva and S. Tanielyan (1979) J. Cat. *56,* 150.
[217] CA *97,* 198 046.
[218] L. M. Reed, L. A. Wenzel and J. B. O'Hara (1956) Ind. Eng. Chem. *48,* 205.
[219] G. E. Hamilton and A. B. Metzner (1957) Ind. Eng. Chem. *49,* 838.
[220] DDR 13 468 (2. 7. 57).
[221] CA *57,* 2 898.
[222] D. F. Othmer and M. S. Thakar (1958) Ind. Eng. Chem. *50,* 1235.
[223] A. B. Metzner and J. E. Ehrreich (1959) AIChE Journal *5,* 496.
[224] A. B. Metzner and J. E. Ehrreich, US 3 091 647 (19. 2. 60).
[225] A. B. Metzner and J. E. Ehrreich, CA *87,* 6 397.
[226] M. Ochesenkühn, J. Starnick and R. Kerber (1975) Chem. Ing. Techn. *47,* 1031.
[227] Hydratation von Butylenoxid, Pol. 79 662 (8. 2. 73).
[228] CA *80,* 3 026.
[229] J. Yamanis and R. W. Garland (1981) Can. J. Chem. Eng. *59,* 310.
[230] DE 11 20 140 (13. 5. 59) Henkel.
[231] DE 20 36 278 (22. 7. 70) BASF.
[232] US 4 276 406 (10. 4. 80) Dow.
[233] G. A. Olah, A. P. Fung and D. Meidar (1981) Synthesis 280.
[234] J. Yamanis and B. D. Patton (1979) Can. J. Chem. Eng. *57,* 297
[235] DE 19 19 782 (19. 4. 68) Celanese.
[236] CA *92,* 76 057.
[237] CA *95,* 96 512.
[238] J. Sussmann (1946) Ind. Eng. Chem. *38,* 1228.
[239] E. Mariani (1950) Ann. Chim. (Rome) *39,* 1.
[240] E. Mariani (1950) Ann. Chim. (Rome) *40,* 1.
[241] R. Kunin (1951) Ind. Eng. Chem. *43,* 102.
[242] G. Bodamer and R. Kunin (1951) Ind. Eng. Chem. *43,* 1082.
[243] H. Deuel, J. Solms, L. Anayas and G. Huber (1951) Helv. Chim. Acta *34,* 1849.
[244] W. H. Waldman (1952) J. Chem. Soc. [London] 3051.
[245] H. Jenny (1946) J. Colloid Sci, *1,* 35.
[246] US. 2534694 (1952) Am. Cyanamid.
[247] DE 1 567 325 (1965) Boehringer, Mannheim.
[248] G. Siegers and F. Martinola (1984) Techn. Info. No. 20394 e, Bayer AG.
[249] S. S. Bhagade and G. D. Nageshwar (1977) Chem. Petro-Chem. J. *8,* 9.
[250] F. Jiracek, J. Horak and Z. Petuchova (1974) Chem. Prum. *24,* 385.
[251] S. A. Bernhard and L. P. Hammet (1953) J. Am. Chem. Soc. *75,* 6834.
[252] S. B. Desai and N. Krishnaswamy (1973) Indian J. Techn. *11,* 255.
[253] G. G. Thomas and C. W. Davies (1947) Nature [London], *159,* 372.
[254] V. C. Haskell and L. P. Hammet (1949) J. Am. Chem. Soc. *71,* 1284.
[255] N. L. Smith and N. R. Amundson (1951) Ind. Eng. Chem. *43,* 2156.

[256] C.W. Davies and G. Ġ. Thomas (1952) J. Chem. Soc. (London) 1607.
[257] S. A. Bernhard and L. P. Hammet (1953) J. Am. Chem. Soc. *75*, 1798.
[258] W. J. L. Sutton and H. E. Moore (1953) J. Am. Oil Chem. Soc. *30*, 449.
[259] S. A. Bernhard, E. Garfield and L. P. Hammet (1954) J. Am. Chem. Soc. *76*, 991.
[260] P. Riesz and L. P. Hammet (1954) J. Am. Chem. Soc. *76*, 992.
[261] H. Samelson and L. P. Hammet (1956) J. Am. Chem. Soc. *78*, 524.
[262] H. Noller and A. Hassler (1957) Z. Phys. Chem. *11*, 267.
[263] K. Klamer, C.V. Heerden and D.W. van Krevelen (1958) Chem. Eng. Sci. *9*, 1.
[264] C. C. Chen and L. P. Hammet (1958) J. Am. Chem. Soc. *80*, 1329.
[265] A. B. Metzner, P. K. Lachmet and J. E. Ehrreich (1960) Actes Intern. Congr. Catalyse 2, Paris 1, p. 735.
[266] M. J. Astle and J. A. Oscar (1961) J. Org. Chem. *26*, 1713.
[267] P. D. Bolton and T. Henshall (1962) J. Chem. Soc. (London) 1226.
[268] L. M. Goldenshtein and N. G. Freidilin (1964) Zh. Prikl. Khim *37*, 2540.
[269] L. M. Goldenshtein and N. G. Freidilin (1965) Zh. Prikl. Khim *38*, 1345.
[270] L. M. Goldenshtein and N. G. Freidilin (1965) Zh. Prikl. Khim *38*, 2538.
[271] M. P. Lee, Y. C. Yang and C. J. Huang (1966) Hua Hsueh *2*, 47.
[272] R. Tartelli (1966) Ann. chim. (Rome) *56*, 156.
[273] M. P. Vysotzkii et al. (1967) Zh. Prikl. Khim *40*, 2552.
[274] H. Noller and A. Hassler (1968) Z. Phys. Chem. *55*, 255.
[275] DE 1 933 538 (1969).
[276] DE 2 203 712 (1972.
[277] DE 3 003 126 (1980) Anic.
[278] DE 3 101 716 (1981) BASF.
[279] S. Affrossman and J. P. Murray (1966) J. Chem. Soc. (London) 1015.
[280] DE 1 518 712 (1965).
[281] K. Setinek and L. Berànek (1970) J. Cat. *17*, 306.
[282] O. Rodriguez and K. Setinek (1975) J. Cat. *39*, 449.
[283] J. Yamanis and M. Adelman (1975) Can. J. Chem. Eng. *53*, 536.
[284] DE 2 611 423 (1976).
[285] K. Setinek (1977) Coll. Czech. Chem. Comm. *42*, 979.
[286] DE 2 709 440 (1977) BASF.
[287] K. Setinek (1979) Coll. Czech. Chem. Comm. *44*, 503.
[288] K. Jerabek (1981) Coll. Czech. Chem. Comm. *46*, 1577.
[289] K. Setinek, V. Blazek, J. Hradil, F. Svek and J. Kalal (1983) J. Cat. *80*, 123.
[290] L. Lawrence and W. J. Moore (1951) J. Am. Chem. Soc. *73*, 3973.
[291] C. E. Underwood and F. E. Deatherage (1952) Science *115*, 95.
[292] J. C. Paulson, F. E. Deatherage and E. F. Almy (1953) J. Am. Chem. Soc. *75*, 2039.
[293] J. R. Whitaker and F. E. Deatherage (1955) J. Am. Chem. Soc. *77*, 3360; *77*, 5298.
[294] DE 878 348 (1942).
[295] C. L. Levesque and A. M. Craig (1948) Ind. Eng. Chem. *40*, 96.
[296] DDR 10808 (1952) V. Vailesu.
[297] DE 10 13 650 (1952) BASF.
[298] V. Rudzicka and V. Medonos (1957) Chem. Prumysl. *6*, 268.
[299] D. J. Saletan and R. R. White (1952) Chem. Eng. Prog., Symp. Ser. *48*, 59.
[300] DDR 8560, (1953) VEB Farbenfabrik Wolfen.
[301] M. J. Astle, B. Schaeffer and C. O. Obenland (1955) J. Am. Chem. Soc. *77*, 3643.
[302] R. Glenat (1956) Chim. Ind. *75*, 292.
[303] Z. Csuros, J. Fodor and Z. Hajos (1956) Act. Chim. Acad. Sci. Hun. *2*, 459.
[304] DE 1 115 240 (1958) BASF.
[305] F. Andreas (1959) Chem. Tech. *11*, 24.

[306] V. Vasilascu (1959) Chem. Techn. *11*, 29.
[307] J. Lindemann and W. Trochimcruk (1959) Chem. Techn. *11*, 32.
[308] T. Andrianova and B. Bruns (1960) Kinetika i. Kataliz, *1*, 440.
[309] F. Gracardi and E. A. Virasoro (1961) Rev. Fac. Ing. Quim. *30*, 61.
[310] G. M. Christensen (1962) J. Org. Chem. *27*, 1442.
[311] DE 2 226 829 (1972) Knappsack AG.
[312] DE 2 449 811 (1974) Hoechst AG.
[313] US 402 9675 (1975) Mobil Oil Corp.
[314] DE 2 446 753 (1976) Deutsche Texaco AG.
[315] N. P. Haran, S. S. Bhagade and G. D. Nageshwar (1979) Chem. Petrochem. J. *10*, 15.
[316] DE 2 924 186 (1979) Mallinckrodt Inc.
[317] EP 10 953 (1979) Mitsubishi Rayon Co. Ltd.
[318] DE 3 006 983 (1980) Takasago Perfumery Co. Ltd.
[319] DE 3 121 383 (1981) Chem. Werke Hüls.
[320] DE 3 221 609 (1982) Texaco Dev. Corp.
[321] DE 1 013 650 (1952) Bayer AG.
[322] F. Runge (1950) Angew. Chem. *62*, 451.
[323] P. Mastiagli, Z. Zafiriadis, G. Durr, A. Floc'h and G. Lagrange (1953) Bull. Soc. Chim. France, 693.
[324] G. Lagrange, P. Mastagli and Z. Zafiriadis (1953) C. R. hebd. Séances Acad. Sci. *236*, 616.
[325] M. J. Astle, Zalowski and Lafyatis (1954) Ind. Eng. Chem., *46*, 787.
[326] E. Swistals and P. Mastagli (1954) C. R. hebd. Séances Acad. Sci. *239*, 709.
[327] E. Swistals (1955) C. R. hebd. Séances Acad. Sci. *240*, 1544.
[328] P. Mastagli, P. Lambert and G. Francois (1957) Bull. Soc. Chim. France, 764.
[329] N. B. Lorette, W. L. Howard and J. H. Brown jr. (1959) J. Org. Chem. *24*, 1731.
[330] T. R. E. Kressmann (1960) The Industrial Chemist, p. 3.
[331] DE 1 155 780 (1961) BASF.
[332] DE 2 929 827 (1979) BASF.
[333] FP 1 023 805 (1949) Comp. de Prod. Chim. et Electromet., Alais.
[334] G. Durry (1951) C. R. hebd. Séances Acad. Sci., *234*, 1314.
[335] C. J. Schmidle and R. C. Mansfield (1952) Ind. Eng. Chem., *44*, 1388.
[336] M. J. Astle and J. A. Zaslowsky (1952) Ind. Eng. Chem., *44*, 2869.
[337] G. Durr and P. Mastagli (1952) C. R. hebd. Séances Acad. Sci. *235*, 1038.
[338] P. Mastagli, A. Floc'h and G. Durr (1952) C. R. hebd. Séances Acad. Sci. *235*, 1402.
[339] G. Durr (1953) C. R. hebd. Séances Acad. Sci. *236*, 1571.
[340] G. Durr (1953) C. R. hebd. Séances Acad. Sci. *237*, 1012.
[341] F. G. Klein (1955) Ph. D. Thesis, Univ. of Michigan.
[342] US 2 818 443 (1955) Celanese.
[343] GB 783 458 (1955) Celanese.
[344] R. Palland and V. Austerweil (1955) C. R. hebd. Séances Acad. Sci. *240*, 1218.
[345] F. G. Klein and I. T. Banchero (1956) Ind. Eng. Chem., *48*, 1278.
[346] N. B. Lorette (1958) J. Org. Chem. *22*, 346 (1958).
[347] US 3 037 052 (1959).
[348] Z. N. Verkrovskaya, M. Ya. Klimenko, E. M. Zalesskaya and I. N. Bychkova (1967) Khim. Prom., *43*, 503.
[349] H. Matyscholz and St. Ropulzynski (1968) Chem. Stosow., Ser. A 283.
[350] N. D. Lemcoff and R. E. Cunningham (1971) J. Cat. *23*, 81.
[351] DE 2 601 083 (1976) Stamicarbon B. V.
[352] DE 2 653 096 (1976) Bayer AG.
[353] DE 1 233 848 (1964) Deutsche Texaco AG.

[354] DE 1 198 814 (1963) Deutsche Texaco AG.
[355] DE 1 193 931 (1965) Deutsche Texaco AG.
[356] G. Durr (1956) C. R. hebd. Séances Acad. Sci. *242,* 1630.
[357] DE 870 271 (1943).
[358] DE 1 233 880 (1960) Bayer AG.
[359] JP 6 904 766 (1962) T. Mitsui, M. Kitahava, M. Yanayita.
[360] FR 1 364 615 (1963).
[361] J. Maurin and E. Weisang (1964) Bull. Soc. Chim. France, 3080.
[362] V. I. Isaguljanc and M. G. Safarow (1964). Ber. Akad. Wiss. Armen. SSR *39,* 235.
[363] GB. 1 026 994 (1965).
[364] V. I. Isaguljanc and M. G. Safarow (1965) [Erdölchem.] (USSR), *5,* 545.
[365] M. I. Faberov, E. P. Tepenicyna, B. N. Bobylev and A. P. Ivanovkij (1968) Chem. Ind. (USSR), *44,* 61.
[366] V. I. Isaguljanc and V. R. Melikjan (1968) Ber. Akad. Wiss. Armen. SSR *46,* 183.
[367] FR 1 554 005 (1969).
[368] J. P. Gorrichon, A. Gaset and M. Delmas (1979) Synthesis, *219.*
[369] R. El Gharbi, M. Delmas and A. Gaset (1981) Synthesis, 361.
[370] R. El Gharbi, M. Delmas and A. Gaset (1981) Chimica *35,* 478.
[371] M. Delmas and A. Gaset (1982) Inf. Chim. Nr. *232,* 151.
[372] S. S. Israelstam and E. V. O. John (1958) Chem. Ind., 1262.
[373] M. Victor, G. Austerweil and M. P. Pascal (1959) Chem. Rev., 1810.
[374] E. V. O. John and S. S. Israelstam (1961) J. Org. Chem. *26,* 240.
[375] US 3 037 052 (1962) Rohm and Haas.
[376] M. J. Astle and R. W. Etherington (1952) Ind. Eng. Chem. *44,* 2871.
[377] US 2 653 162 (1953) Rohm and Haas.
[378] M. Borrel and J. Modino (1957) Chim. Ind. *78,* 632.
[379] GB 883 391 (1958) Union Carbide.
[380] GB 842 209 (1960) Esso Res. and Eng. Co.
[381] GB 849 965 (1960) Union Carbide.
[382] US 3 049 568 (1962).
[383] US 3 153 001 (1964).
[384] US 3 172 916 (1965).
[385] FR 1 513 814 (1966) Union Carbide.
[386] US 3 394 089 (1968).
[387] US 3 634 341 (1972) Dow Chemical.
[388] US 3 760 006 (1973) Dow Chemical.
[389] FR 2 165 238 (1973) Dow Chemical.
[390] R. A. Reinicker and B. C. Gates (1974) AIChE Journal, *20,* 933.
[391] Czech 166 460 (1974) J. Pilz et al.
[392] GB 1 539 186 (1975) Shell.
[393] Czech 183 069 (1975) M. Cervinka et al.
[394] B. 836 745 (1975) Ya Gurvich et al.
[395] N. Nikolow, D. Gospodina and R. Belcheva (1976) "Neft Kim" 1st (Sekts Org. Sint.) 33.
[396] DE 2 726 762 (1976) Nautschno-issledovatelskij inst. resinovych i lateksnych izdelij.
[397] Czech 183 503 (1976) J. Kustka.
[398] DE 2 530 122 (1977) Bayer AG.
[399] DE 2 537 027 (1977) Bayer AG.
[400] DE 2 634 435 (1971) Shell BV.
[401] DE 2 722 683 (1977).
[402] DE 2 726 762 (1977) E. L. Styskin et al. (USSR).

[403] Czech 200 351 (1977) K. Jerabek et al.
[404] M. Kiedik et al. (1977) Przem. Chem. *56*, 582.
[405] DE 2 811 182 (1979) Bayer AG.
[406] K. Jerabek (1979) Coll. Czech. Chem. Comm. *44*, 2612.
[407] USSR 701 986 (1979) Z. N. Verkhovskaya et al.
[408] S. W. Marawar et al. (1979) Chem. Age 441.
[409] S. T. Meshcheryakov et al. (1979) Neftpererab Neftekhim, Moscow, 34.
[410] DE 2 811 182 (1979) Bayer AG.
[411] DE 2 931 036 (1979) Bayer AG.
[412] E. J. Grzywa et al. (1979) Chem. Stosow. 421.
[413] Czech 184 988 (1980) F. Juracka et al.
[414] EP 49411 (1980) General Electric.
[415] USSR 732 233 (1980) Z. N. Verkhovskaya et al.
[416] JP 81 131 534 (1980) Mitsubishi Chem. Ind. Co.
[417] Z. Yang (1981) Tuliao Gongye *65*, 53.
[418] DE 2 928 443 (1981) Bayer AG.
[419] EP 23325 (1981) Bayer AG.
[420] US 4 391 997 (1981) General Electric.
[421] DDR 153 680 (1982) Instytut Ciekiej, Synt. Org. Blachowina.
[422] US 4 365 099 (1982) General Electric.
[423] Czech 191 785 (1983) K. Jerabek et al.
[424] US 4 375 567 (1983) General Electric.
[425] B. Loew and J. T. Massengale (1957) J. Org. Chem. *22*.
[426] L. Boisselet, G. Paré (1958) Bull. Soc. Chim. France, 856.
[427] USSR 125 802 (1958) G. A. Ivanov et al.
[428] T. R. Edward, W. E. Parker and D. Swern (1959) J. Am. Oil Chem. Soc. *36*, 656.
[429] V. I. Isaguljanc and J. V. Panidi (1961) J. Angew. Chem. *34*, 1849.
[430] Ind 78056 (1961).
[431] GB 1 006 947 (1962), R. Y. Mixer and J. W. Wagner.
[432] P. S. Below et al. (1962) Chem. Ind. *38*, 480.
[433] US 3 037 052 (1962) Rohm and Haas.
[434] J. Kams and J. Nosek (1964) Chem. Prumsyl. *14*, 245.
[435] V. I. Isaguljanc et al. (1964) J. Angew. Chem. *37*, 1197.
[436] M. K. Julin et al. (1964) Erdölchemie *4*, 717.
[437] V. I. Isaguljanc et al. (1965) Chem. Ind. *41*, 739.
[438] A. A. Grinberg et al. (1967) Chem. Ind. *43*, 302.
[439] I. G. Gakh, (1967) C. A. *67*, 53814 h.
[440] V. Stan, A. Serbanascu, A. Nica and T. Grisan (1967) Mater. plast (Bucuresti) *4*, 179.
[441] O. N. Cvetkov et al. (1968) Ber. Akad. Wiss. (USSR) *167*, 115.
[442] F. Wolf and D. Kleinhempel (1968) Symposiums-Vortrag: 30 Jahre Ionenaustauscher, Leipzig.
[443] M. Fedtke (1968) Symposiums-Vortrag: 30 Jahre Ionenaustauscher, Leipzig.
[444] DE 2 650 888 (1976) Bayer AG.
[445] Y. Tachibana (1977) Bull. Chem. Soc. Jap. *50*, 2477.
[446] N. P. Haran, S. S. Bhagade and G. D. Nageshwar (1978) Chem. Petrochem. J. *9*, 21.
[447] H. Widdecke (1980) Chem. Ing. Tech. *52*, 825.
[448] DE 3 151 693 (1981) Chem. Werke Hüls.
[449] H. Widdecke and J. Klein (1981) Chem. Ing. Tech. *53*, 954.
[450] DE 3 130 428 (1981) Anic.
[451] US 2 915 653 (1958) Shell Dev. Corp.
[452] US 3 239 575 (1961) V. J. Frilette and W. O. Haag.

[453] US 3 037 052 (1962) Rohm and Haas.
[454] US 3 238 266 (1962) M. Skripek.
[455] US 3 326 866 (1963) Mobil.
[456] J. M. Kapura and B. C. Gates (1963) Ind. Eng. Chem., Prod. Res. Dev. *12,* 62.
[457] I. G. Gach et al. (1964) J. Allg. Chem. *34,* 2807.
[458] USSR 195 436 (1965) E. P. Babin.
[459] US 3 239 575 (1966) Socony Mobil Oil.
[460] US 3 238 266 (1966) Richfield Oil.
[461] R. B. Wesley and B. C. Gates (1974) J. Cat. *34,* 288.
[462] J. Klein, H. Fehreke and H. Widdecke (1976) Chem. Tech. *48,* 173.
[463] G. A. Olah, J. Kaspi and J. Bukala (1977) J. Org. Chem. *42,* 4187.
[464] US 4 041 090 (1977) Shell Oil Comp.
[465] DE 3 028 132 (1980) Nippon Petrochemicals.

2.12 Industrial Ion Exchange Chromatography

Frederick J. Dechow
BioCryst, Birmingham, Alabama, USA

2.12.1 Introduction

In most ion exchange operations, anion in solution is replaced by an ion from the resin and the former solution ion remains with the resin. In contrast to this, ion exchange chromatography uses the ion exchange resin as an adsorption or separation media, providing an ionic environment that allows two or more solutes in the feed stream to be separated. The feed solution is added to the chromatographic column filled with ion exchange beads and eluted with solvent, usually water. The resin beads selectively slow certain solutes, while others are eluted down the column (Figure 2.182). As the solutes move down the column, they separate and their individual purity increases. Eventually, the solutes appear at different times at the column outlet where each can be drawn off separately.

Tswett [1, 2] was the first to identify correctly the nature of the separation of colored vegetable pigments in petroleum ether when the solution was passed through a column of fine particle calcium carbonate. The process was called a chromatographic separation because of the separation of the pigments into bands of different colors. Following Tswett there have been numerous scientists who have developed ion exchange chromatography into a sophisticated analytical technique used in many scientific areas.

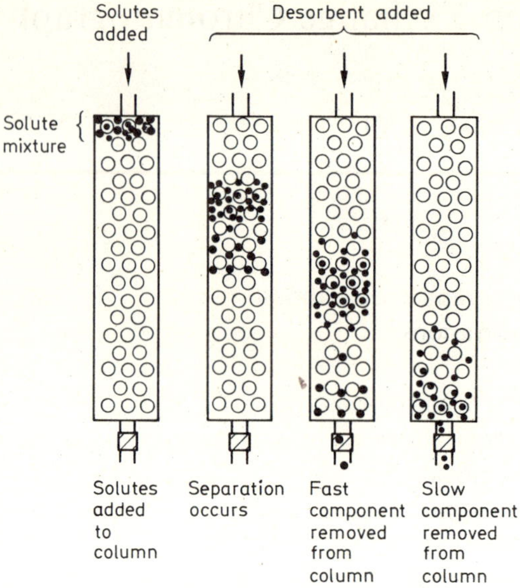

Figure 2.182 The steps of chromatographic separation are addition of the mixed solutes to the column, elution to effect separations, and removal of the separated solutes.

The first commercial use of ion exchange chromatography occurred during World War II at Iowa State University where Spedding and his co-workers [3, 4] isolated transuranium elements as their contribution to the Manhattan Project. Later, the commercial preparation of individual rare earths was performed using ion exchange chromatography [5, 6].

Despite these initial successes with chromatography in preparing pure materials in large quantities, other commercial applications were not developed until the mid- to late-1960s. Only in the 1970s were industrial separation problems identified that could not be solved by conventional, well-tried methods such as distillation or crystallization. At present, there are industrial ion exchange chromatographic units which separate amino acids, hydrocarbon isomers, glucose from fructose, sucrose from molasses, monosaccharides from di- and polysaccharides and ion exchange regenerants from salts.

There are two advances that have contributed strongly to the recent development of industrial ion exchange chromatography: (1) improvements in chromatographic equipment with recycle control systems and pseudo-moving bed control systems; and (2) improvements by resin manufacturers in developing resins with narrow, controlled-size distributions and improved osmotic resistance. General resin considerations and commercial chromatographic resins are described in Section 2.12.3. Descriptions of industrial systems will be covered in Section 2.12.5.

It is important to distinguish industrial ion exchange chromatography from gel filtration chromatography and gel permeation chromatography. Gel filtration chromatography originated in Sweden in 1939 [7] when columns packed with crosslinked polydextran gels, swollen in an aqueous solution, were used to separate water-soluble macromolecules on the basis of difference in size. These gels are still used extensively for separating water-soluble biological compounds for further characterization studies [8]. Gel permeation chromatography was developed at The Dow Chemical Company in 1964 [9] using crosslinked polystyrene gels swollen in organic solvents to separate synthetic polymers on the basis of size. Industrial chromatography also can utlize differences in size for separation, but more frequently relies upon adsorption or affinity differences to effect separations.

The low-crosslinked polymer beads used in gel filtration and gel permeation chromatography have mean diameters of 70 to 150 μm and can only be used at low flow rates and at pressures less than 17 bars (250 psi). These beads collapse at higher pressures, which restricts the flow rate, making separations impossible. Industrial chromatography uses larger mean diameter beads (200 to 450 μm) with a higher degree of crosslinking. The necessary degree of separation is achieved by removing a specific cut of the chromatogram.

2.12.2 Types of chromatographic separations

Chromatographic separations can be divided up into four groups according to the type of materials being separated: affinity difference, ion exclusion, size exclusion and ion retardation chromatography. These types of separations may be described in terms of the distribution of the materials to be separated between the phases involved.

Figure 2.183 shows a representation of the resin-solvent-solute components of a chromatographic ion exchange system. The column is filled with ion exchange beads packed together with the voids between the beads filled with solvent solution.

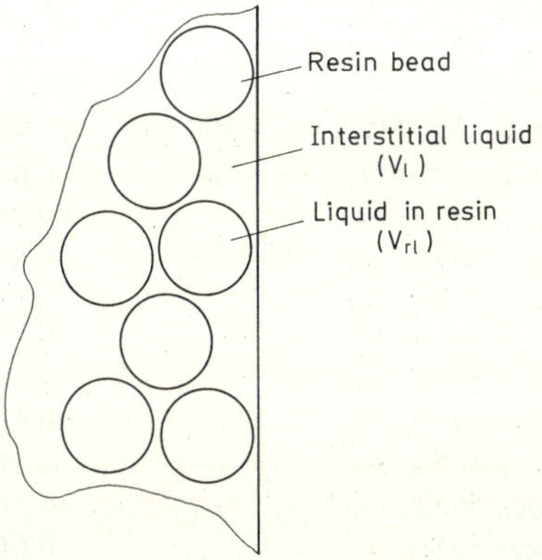

Figure 2.183 Representation of the three phases involved in chromatographic separation.

The phases of interest are: (1) the liquid phase between the resin beads; (2) the liquid phase held within the resin beads; and (3) the solid phase of the polymeric matrix of the resin beads. When the feed solution is placed in contact with the hydrated resin in the chromatographic column, the solutes distribute themselves between the liquid inside the resin beads and that between the resin beads. The distribution for component i is defined by the distribution coefficient, K_{di}:

$$K_{di} = \frac{C_{ri}}{C_{li}} \tag{1}$$

where C_{ri} is component i's concentration in the liquid within the resin bead and C_{li} is the component i's concentration in the interstitial liquid. The distribution coefficient for a given ion or molecule will depend upon the component's structure and concentration, the type and ionic form of the resin and the other components in the feed solution. The distribution coefficients for several organic compounds in aqueous solutions with ion exchange resins are given in Table 2.57 [10].

Table 2.57 Distribution constants (from Ref. 10)

Solute	Resin	K_d
Ethylene glycol	Dowex 50X8, H^+	0.67
Sucrose	Dowex 50X8, H^+	0.24
d-Glucose	Dowex 50X8, H^+	0.22
Glycerine	Dowex 50X8, H^+	0.49
Triethylene glycol	Dowex 50X8, H^+	0.74
Phenol	Dowex 50X8, H^+	3.08
Acetic acid	Dowex 50X8, H^+	0.71
Acetone	Dowex 50X8, H^+	1.20
Formaldehyde	Dowex 50X8, H^+	0.59
Methanol	Dowex 50X8, H^+	0.61
Formaldehyde	Dowex 1X7.5, Cl^-	1.06
Acetone	Dowex 1X7.5, Cl^-	1.08
Glycerine	Dowex 1X7.5, Cl^-	1.12
Methanol	Dowex 1X7.5, Cl^-	0.61
Phenol	Dowex 1X7.5, Cl^-	17.7
Formaldehyde	Dowex 1X8, SO_4^{2-}, 50–100	1.02
Acetone	Dowex 1X8, SO_4^{2-}, 50–100	0.66
Xylose	Dowex 50X8, Na^+	0.45
Glycerine	Dowex 50X8, Na^+	0.56
Pentaerythitol	Dowex 50X8, Na^+	0.39
Ethylene glycol	Dowex 50X8, Na^+	0.63
Diethylene glycol	Dowex 50X8, Na^+	0.67
Triethylene glycol	Dowex 50X8, Na^+	0.61
Ethylene diamine	Dowex 50X8, Na^+	0.57
Diethylene triamine	Dowex 50X8, Na^+	0.57
Triethylene tetramine	Dowex 50X8, Na^+	0.64
Tetraethylene pentamine	Dowex 50X8, Na^+	0.66

The ratio of individual distribution coefficients is often used as a measure of the possibility of separating two solutes and is called the separation factor α

$$\alpha = \frac{K_{d1}}{K_{d2}}. \tag{2}$$

Thus, from Table 2.57, the separation factors for acetone-formaldehyde separability would be 0.49, 0.98 and 1.54 for Dowex 50WX8(H^+), Dowex 1X8(Cl^-) and Dowex 1X8(SO_4^{2-}) resins, respectively. For comparison purposes, it may be necessary to use the inverse of α, so that the values would be 2.03 and 1.02 for Dowex 50WX8(H^+) and Dowex 1X8(Cl^-), respectively. When α is less than 1, the solute in the numerator will exit the column first. When α is greater than 1, the solute in the denominator will exit the column first. From this limited amount of data, the resin of choice would be Dowex 50WX8(H^+) and acetone would exit the column first. The separation factor is sometimes called the relative retention ratio.

The acetone-formaldehyde separation would be an example of affinity difference chromatography in which molecules of similar molecular weight or isomers of compounds are separated on the basis of differing attractions or distribution coefficients for the resin. The largest industrial chromatography application of this type is the separation of fructose from glucose to produce 55% or 90% fructose corn sweetener.

Ion exclusion chromatography involves the separation of an ionic component from a non-ionic component. The ionic component is excluded from the resin beads by ionic repulsion, while the non-ionic component is distributed into the liquid phase inside the resin beads. Since the ionic solute travels only through the interstitial volume, it reaches the end of the column before the non-ionic solute,

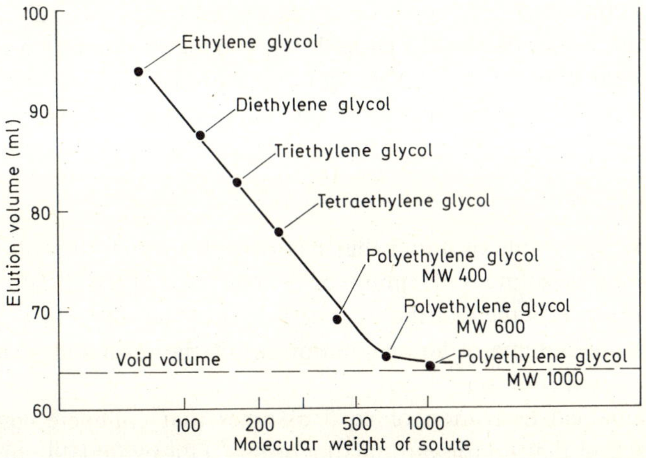

Figure 2.184 Effect of molecular weight on the elution volume required for glycol compounds (Reprinted with permission from Ref. 11).

which must travel a more tortuous path through the ion exchange beads. A major industrial chromatography application of this type is the recovery of sucrose from the ionic components of molasses.

In size exclusion chromatography, the resin beads act as molecular sieves, allowing the smaller molecules to enter the beads while the larger molecules are excluded. Figure 2.184 [11] shows the effect of molecular size on the elution time required for a given resin. An industrial chromatographic application using size exclusion is the separation of dextrose from di- and poly-saccharides in the corn wet milling industry.

Ion retardation chromatography involves the separation of two ionic solutes with a common counterion. Unless a specific complexing resin is used, the resin must be used in the form of the common counterion. The other solute ions are separated on the basis of differing affinities for the resin. Ion retardation chromatography is starting to see use in the recovery of acids from waste salts following the regeneration of ion exchange columns.

2.12.3 Theoretical considerations

Mathematical theories for ion exchange chromatography were developed in the 1940s by Wilson [12], DeVault [13], and Glueckauf [14, 15]. These theoretical developments were based on adsorption considerations and are useful in calculating adsorption isotherms from column elution data. Of more interest for understanding industrial chromatography is the theory of column processes, originally proposed by Martin and Synge [16] and augmented by Mayer and Thompkins [17], which was developed analogously to fractional distillation so that plate theory could be applied.

One of the equations developed merely expressed mathematically that the least adsorbed solute would be eluted first and that if data on the resin and the column dimensions were known, the solvent volume required to elute the peak solute concentration could be calculated. Simpson and Wheaton [18] expressed this equation as:

$$V_{\max} = K_d V_{rl} + V_l, \tag{3}$$

where $V_{\max}$ is the volume of liquid that has passed through the column when the concentration of the solute is maximum or the midpoint of the elution of the solute. K_d, defined in Eq. 1, is the distribution coefficient of the solute in a "plate" of the column, V_{rl} is the volume of liquid solution inside the resin and V_l is the volume of interstitial liquid solution.

The mathematical derivation of eq. 3 assumes that complete equilibrium has been achieved and that no forward mixing occurs. Glueckauf [19] pointed out that equilibrium is practically obtained only with resin beads of very small diameter and low flow rates. Such restricting conditions may be acceptable for analytical

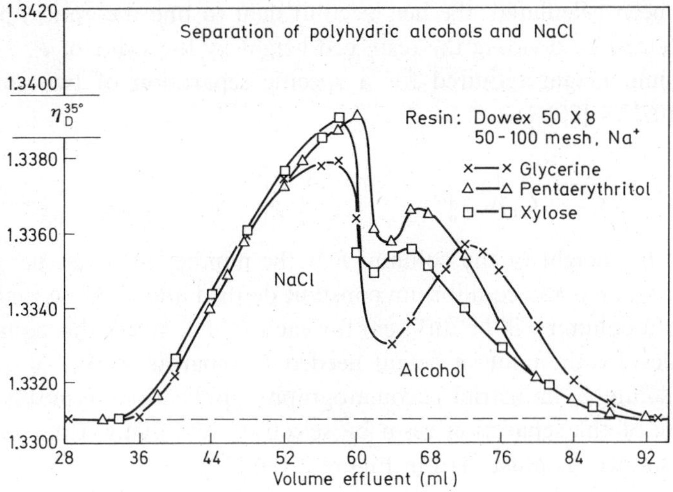

Figure 2.185 Elution chromatogram for the separation of polyhydric alcohols and NaCl using the sodium form of a cation resin (Reprinted with permission from Ref. 10).

applications but would severely limit industrial applications. For industrial chromatography, column processing conditions and solute purity requirements are often such that any deviations from these assumptions are slight enough that the equation can still be used as an approximation for the solute elution profile. An example of such an elution profile is given in Figure 2.185.

A second important equation for column processes is that used for the calculation of the number of theoretical plates, i. e., the length of column required for equilibration between the solute in the resin liquid and the solute in the interstitial liquid. If the elution curve approximates a Gaussian distribution curve, the equation may be written as:

$$P = \frac{2c(c+1)}{W^2}, \qquad (4)$$

where P is the number of theoretical plates, c $(= K_d V_{rl}/V_l)$ is the equilibrium constant, W is the half-width of the elution curve at an ordinate value of 1/e of the maximum solute concentration (with e the base of the natural logarithms). For a Gaussian distribution, $W = 4\sigma$, with σ the standard deviation of the Gaussian distribution. The equilibrium constant is sometimes called "the partition ratio".

An alternative form of this equation is:

$$P = \frac{2V_{max}(V_{max} - V_l)}{W^2}. \qquad (5)$$

Here W is measured in the same units as V_{max}. This form of the equation is probably the easiest to calculate from experimental data. Once the number of theoretical plates has been calculated, the height equivalent to one theoretical plate (HETP) can be obtained by dividing the resin bed height by the value of P.

The column height required for a specific separation of two solutes can be approximated by [20]:

$$\sqrt{H} = \frac{3.29}{C_2 - C_1}\left(\frac{C_2 + 0.5}{\sqrt{P_2}} + \frac{C_1 + 0.5}{\sqrt{P_1}}\right), \tag{6}$$

where H is the height of the column, P is the number of plates per unit of resin bed height, and c is the equilibrium constant defined above. Note that the number of plates in a column will be different for each solute. While this equation may be used to calculate the column height needed to separate 99.9% of solute 1 from 99.9% of solute 2, industrial chromatography applications typically make more efficient use of the separation resin by selectively removing a narrow portion of the eluted solutes, as illustrated in Figure 2.186 [21].

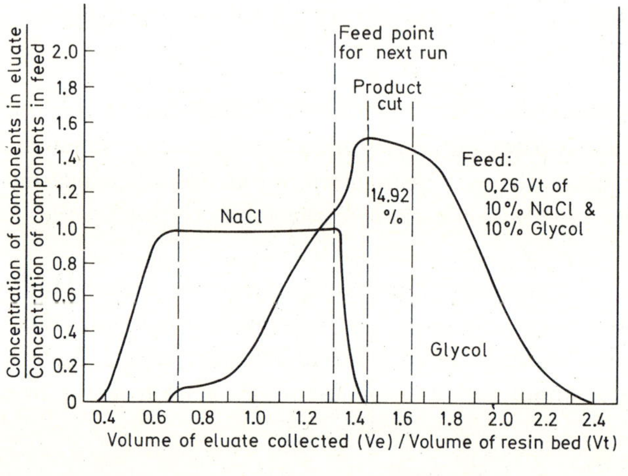

Figure 2.186 Distribution of eluate into fractions for product, recycle and waste for NaCl and glycol separation (Reprinted with permission from Ref. 21).

A variation on calculating the required column height is to calculate the resolution or degree of separation of two components. Resolution is the ratio of peak separation to average peak width:

$$R = \frac{V_{\max,2} - V_{\max,1}}{0.5\,(W_1 + W_2)}. \tag{7}$$

The numerator of eq. 7 is the separation of the two solutes' peak concentrations and the denominator is the average band width of the two peaks. This form of the equation evaluates the resolution when the peaks are separated by four standard deviations, σ. If $R = 1$ and the two solutes have the same peak concentration, it means that the adjacent tail of each peak beyond 2σ from the V_{max} would overlap with the other solute peak. In this instance, there would be a 2% contamination of each solute by the other.

Resolution can also be represented [22] by:

$$R = \frac{\sqrt{P_2}}{4}\left(\frac{\alpha - 1}{\alpha}\right)\left(\frac{C_2}{1 + C_2}\right). \tag{8}$$

Resolution can be seen to depend on the number of plates for solute 2, the separation factor for the two solutes and the equilibrium constant for solute 2.

In general, the larger the number of plates, the better the resolution. There are practical limits to the column lengths that are economically feasible in industrial chromatography. It is possible to change P also by altering the flow rate, the mean resin bead size or the bead size distribution, since P is determined by the rate processes occurring during separation. As the separation factor increases, resolution becomes greater, since the peak-to-peak separation becomes larger. Increases in the equilibrium constant will usually improve the resolution, since the ratio $C_2/(1 + C_1)$ increases. It should be noted that this is actually true only when C is small, since the ratio approaches unity asymptotically as C_2 gets larger. The separation factor and the equilibrium factor can be adjusted by temperature changes or other changes which would alter the equilibrium properties of the column operations.

Eq. 8 is only applicable when the two solutes are of equal concentration. When the solutes are of unequal concentration, a correction factor must be used:

$$(A_1^2 + A_2^2)/2A_1 A_2,$$

where A_1 and A_2 are the areas under the elution curve for solutes 1 and 2, respectively. Figure 2.187 shows the relationship between product purity (η), the separation ratio and the number of theoretical plates. This graph can be used to estimate the number of theoretical plates required to attain the desired purity of the products.

Figure 2.188 shows the effect of relative concentration of two solutes on product purity and product recovery for a few resolution values. When $R = 1.5$, the solutes are effectively completely separated, with cross-contamination less than 1%. Rarely would this level of purity be required in current industrial chromatography applications. For $R = 1.0$, the cross-contamination is less than 3%; however, the recovery of the more dilute component is significantly reduced when the solute concentration ratio is less than 1 : 4. When $R = 0.8$, the recovery and the purity are still acceptable in industrial chromatography for solutes of approximately the same concentration.

As was mentioned above both resin properties and process parameters affect the number of theoretical plates and, therefore, the separation resolution. The mean size, the size uniformity and the crosslinking of the ion exchange resin are the critical resin parameters for rate processes.

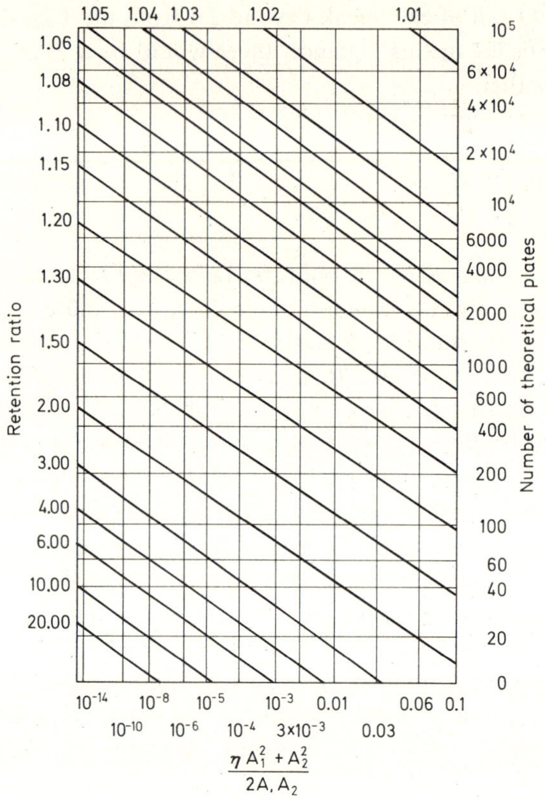

Figure 2.187 Relationship between relative retention ratio, number of theoretical plates and product purity (Reprinted with permission from Ref. 19).

The mean bead size is chosen so that the diffusion into and out of the resin beads does not become the rate-controlling step and so that the pressure drop in the column does not become excessive. Figure 2.189 [18] shows the effect of particle size on the shape of the elution profile of ethylene glycol. Small beads adsorb and release the solute quickly for sharp, distinct elution profiles. Larger beads take longer to adsorb and release the solute so that the elution profiles are wider and less distinct. The resin with the smallest mean bead size has the greatest number of theoretical plates. If the separation ratio for the solutes is large enough, a resin with coarser beads may be used, while one with fine beads must be used for solutes with a separation ratio near 1.0. The practical lower limit on bead size is determined by the pressure drop of the column.

When there is a large distribution of bead sizes for a resin of a given mean bead size, it becomes difficult to obtain proper elution profiles. The beads smaller than the mean will elute the solute more rapidly and those larger less rapidly, to cause additional spreading or tailing of the elution profile. Ideally, the resin beads should all be of the same size to amplify the attraction differences of the solutes uniformly

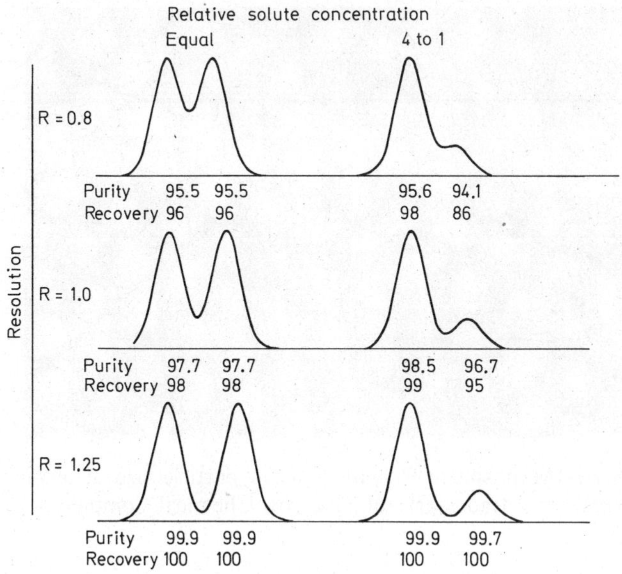

Figure 2.188 Effect of solute concentration on peak resolution.

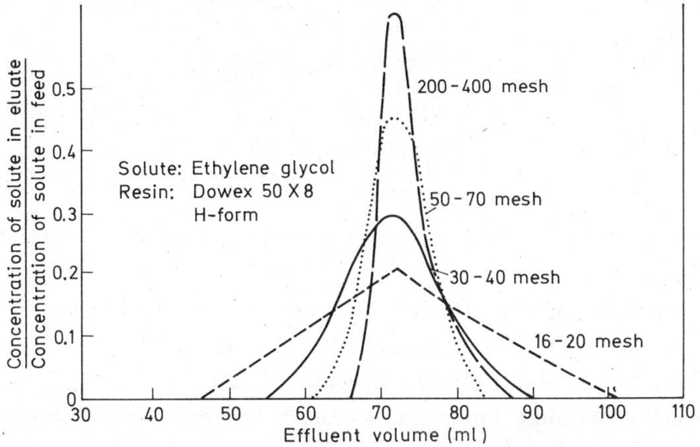

Figure 2.189 Effect of particle size on elution profile for ethylene glycol (Reprinted with permission from Ref. 18).

as they flow down the column. Separation resins are now available for industrial chromatography that have 90% of the beads within + or − 20% of the mean bead size. One resin manufacturer can even supply resins that have 90% of the beads within + or − 10% of the mean bead size (Figure 2.190). Table 2.58 shows the effect of size distribution on glucose-fructose separation.

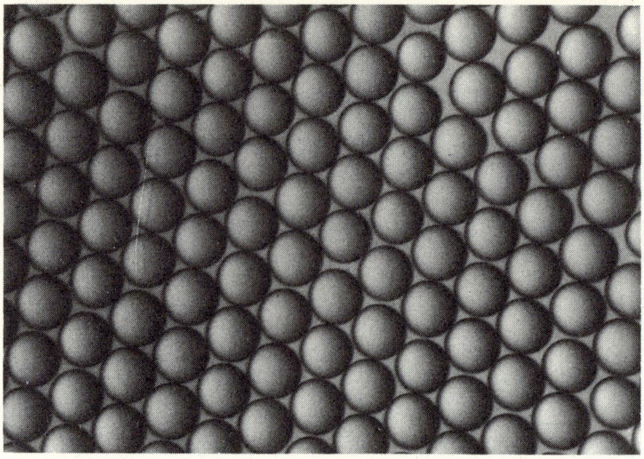

Figure 2.190 Dowex Monosphere 99 with a mean particle size of 390 μm. (Dowex and Monosphere are registered trademarks of The Dow Chemical Company).

Table 2.58 Effect of bead size distribution on glucose-fructose separation

Bead size distribution (mesh)	Volume (%)	Glucose-fructose peak separation (bed volumes)	Development length (elution volume in bed volumes from initial glucose to final fructose)
1. 35–40	55	0.9	0.96
40–50	26	0.10–0.12	0.81–0.87
2. 40–50	80	0.11–0.13	0.79–0.81
3. 40–50	90	0.12–0.15	0.76
4. 40–50	95	0.14–0.16	0.74
5. 45–50	95		

The crosslinkage of the resin affects the position of the elution profile, as is shown by Figure 2.191. This shift occurs because an increase in the crosslinking decreases the liquid volume inside the resin beads. As crosslinking is increased, the exclusion factor for ionic solutes is increased. Likewise, crosslinking decreases the rate of diffusion within the resin for non-ionic solutes. The mechanical strength and shrink/swell of organic polymeric resins is increased by increasing the resin crosslinking. The weight of the resin in the column combined with the pressure effect of flow may cause deformation of the resin beads, leading to even higher pressure drops if the resin crosslinkage is not sufficient. Industrial chromatographic separations of sugars have been achieved with resins of 3 to 8% crosslinking for resin bed depths of 3 to 6 meters [23].

The effect of porosity and resin crosslinking on glucose-salt separation was studied by Martinola and Siegers [24]. Figure 2.192 shows that at the low crosslinkage normally used in separations, the gel resins have a lower glucose loss and

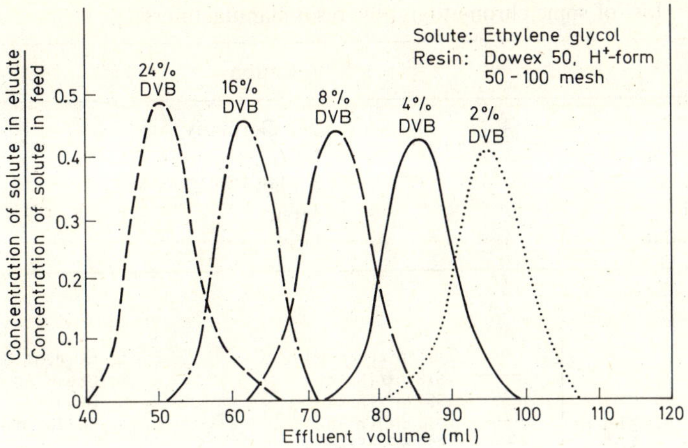

Figure 2.191 Effect of the resin crosslinkage on the position of the elution profile for ethylene glycol (Reprinted with permission from Ref. 18).

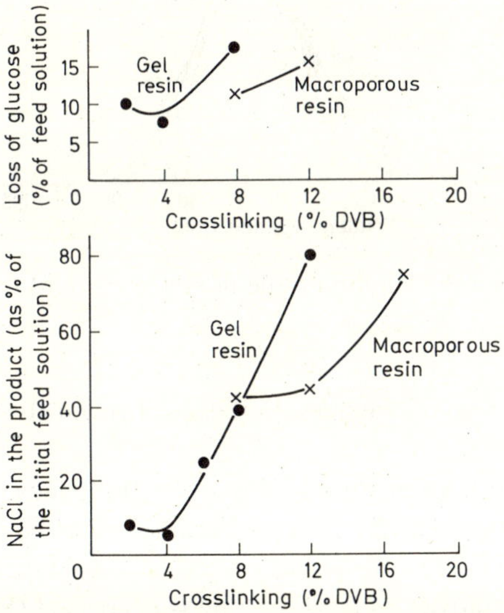

Figure 2.192 Chromatographic separation as a function of crosslinking and porosity (Reprinted with permission from Ref. 24).

have less NaCl in the product. Macroporous resins are not manufactured at the low crosslink levels (< 6%) needed for the glucose-salt separation.

Resin manufacturers have made significant advances in recent years in optimizing and customizing resin properties for specific industrial applications. Often these resins are so tailored for the separation that data sheets and resin samples are only available to the customer with the application. Table 2.59 lists the resin manufac-

Table 2.59 List of some chromatographic resin manufacturers

Company	Location
Bayer	FR Germany
Dow Chemical	USA
Finn Sugar	Finland
Mitsubishi	Japan
Rohm & Haas	USA

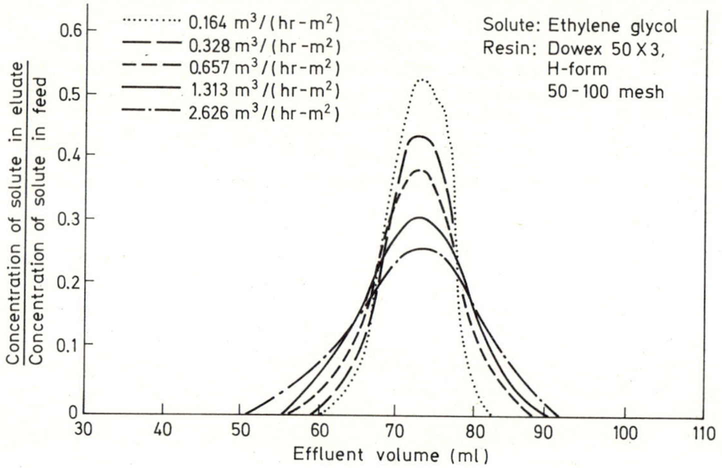

Figure 2.193 Effect of flow rate on the elution profile of ethylene glycol (Reprinted with permission from Ref. 18).

turers who are participating, to varying degrees, in these industrial chromatography advances.

The flow velocity, the solute concentrations, the cycle frequency, the pressure drop and the column temperature are the process parameters to be controlled to achieve the desired separation.

The elution profile becomes sharper as the flow rate is decreased (Figure 2.193). The flow rate should be between 0.5 and 2.0 times the critical velocity (v_c) to insure that profile "tailing", due to density and viscosity differences in the solution moving through the column, is avoided or at least minimized. The critical velocity is defined by [23]:

$$v_c = \frac{g(\varrho_2 - \varrho_1)}{k(\eta_2 - \eta_1)}, \tag{9}$$

where g is the gravity constant, ϱ is the density, η is the viscosity, $k (= \Delta P/\eta v L)$ is the permeability coefficient of the bed, Δ P is the pressure drop in the resin bed, v is the linear flow rate of the solution and L is the height of the resin bed. The

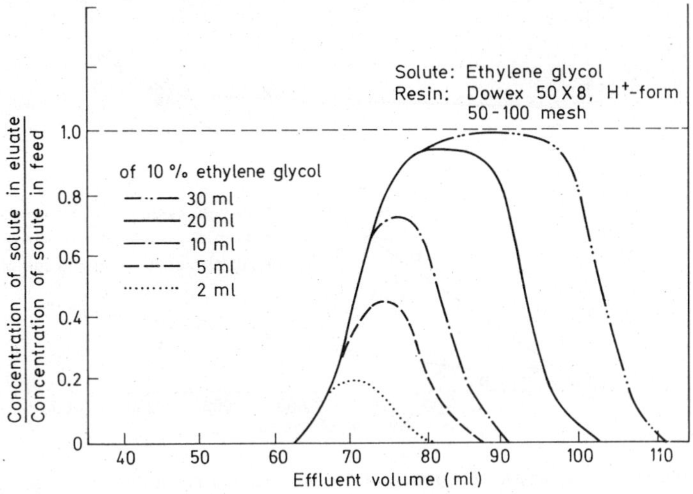

Figure 2.194 Effect of solute concentration on the elution profile of ethylene glycol (Reprinted with permission from Ref. 18).

subscripts 1 and 2 in this equation refer to the low and high extremes of these values for the solution as it cycles from feed to solvent. The permeability coefficient for industrial chromatography of sugar and polyols has been found to be 1 to $4 \cdot 10^{10}\ m^{-2}$.

The effect of solute concentration on the elution profile is shown in Figure 2.194. The concentration changes shift the point at which V_{max} occurs, but does not influence the point where the solute initially emerges from the column. Therefore, the appearance of the solute in the lowest concentration may be the limiting factor for the total dissolved solute level. For industrial chromatography to be effective and economically justifiable, the solute concentration of the most dilute component should be greater than 3% of the dissolved solids. With a total dissolved concentration of 30 to 50%, this means the relative concentration for two solutes would range from 15 to 1 to 9 to 1. Solutions with as much as 60% total dissolved solids can be separated by industrial chromatography. Beyond that concentration level the solution viscosity becomes excessive, even at temperatures of 55 °C.

As the resin beads are cycled between the feed solution and the eluting solvent, the beads will expand and contract in response to the relative moisture level of the solution. If the feed/eluent cycle occurs too rapidly, the rapid expansion and contraction can cause the resin beads to “de-crosslink”. As crosslinkage of the beads decreases, they lose strength and become more susceptible to compression under flow. A good indicator of the loss of crosslinkage is an increase in the bead water retention capacity. Figure 2.195 shows how cycle frequency affects bead crosslinkage with an increasing number of cycles. Typical cycle times for industrial chromatography are between 1 and 1.5 hours, which result in expected resin lifetimes of several years for resins of appropriate initial crosslinkage.

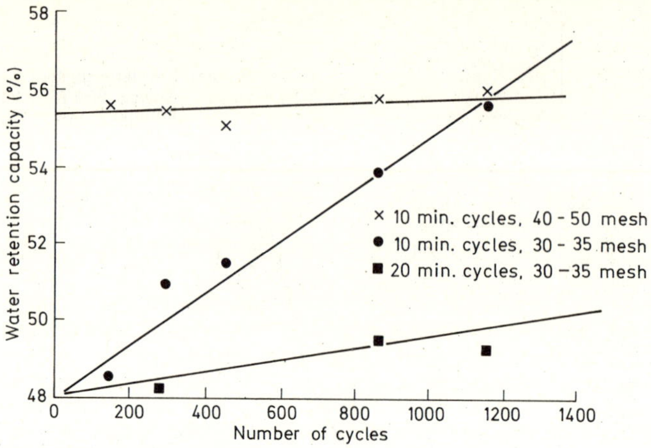

Figure 2.195 Effect of cycle frequency and bead size on resin crosslinkage changes.

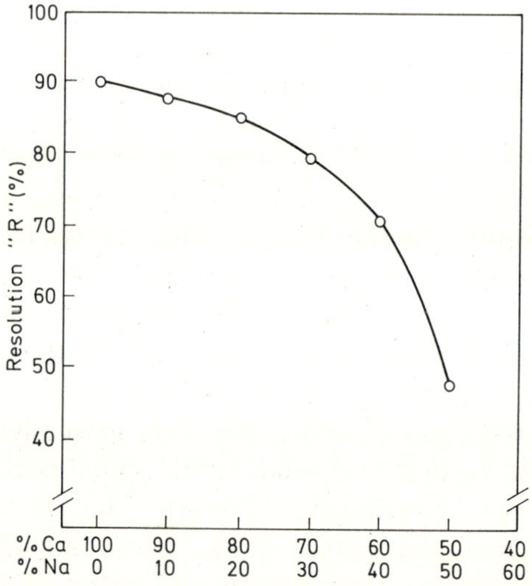

Figure 2.196 Effect of partial conversion of resin from Ca- to Na-form on the resolution of sugar and salt in molasses (Reprinted with permission from Ref. 25).

Although separation efficiency normally decreases with increases in column temperature, there are two reasons for elevating the temperature. First, elevating the temperature may decrease the viscosity of the mixture, making it possible to separate a mixture without excessive dilution. Secondly, elevated fluid temperature is often necessary to prevent microbial growth on the ion exchange beads or in the column. Generally, fouling due to microbial growth can be prevented with a column temperature greater than 50 °C.

It is necessary to backwash the resin periodically to remove small particles and to relieve compression from ion exchanger beads. For columns with organic polymer beads, some method of allowing 50% freeboard volume must be provided for bed expansion during backwashing. Inorganic chromatographic materials have the advantage that they do not compress under flow and do not have to be backwashed. However, one must monitor the effluent to insure that silicates are not being sloughed from the inorganic beads. For columns with organic polymeric beads, some method of allowing 50% freeboard volume must be provided for bead expansion during backwashing.

While chromatographic processes are not based on the exchange of ions on the resin, traces of ionic material in the feed or eluent may eventually cause enough exchange of ions to necessitate the regeneration of the resin. Figure 2.196 [25] shows the effect even partial conversion of a chromatographic resin would have on the resolution of molasses constituents.

2.12.4 Applications

2.12.4.1 Extraction of sugar from molasses

Molasses is a by-product in normal extraction and crystallization processing of cane sugar and beet sugar. The sugar content in molasses is usually between 30 and 60% as a percentage of the total dissolved solids. While it is possible to decrease this sucrose loss to molasses by typical ion exchange treatment to remove the salts prior to crystallization, this process leads to substantial waste loads [26].

When the molasses solution is added to the chromatographic column, the highly ionized ash is excluded from the resin beads while the non-ionic sugars are selectively adsorbed onto the active sites of the resin beads. The distribution constants are given in Table 2.60 for K- and Ca-forms of the resin for various salts and sugars of molasses. In addition to the non-sugar inorganics, there are several non-sugar organics in the molasses. These non-sugar organics are different, depending upon whether the source of the molasses is cane sugar or beet sugar. The non-sugar organics from cane molasses have a smaller amount of amino acids and a higher

Table 2.60 Distribution coefficients for Dowex 50WX4, 50 – 100 mesh [31]

	K^+-form	Ca^{2+}-form
KCl	0.454	0.482
$CaCl_2$	0.466	0.530
Sucrose	0.623	0.623
Glucose	0.955	0.690
Fructose	1.07	1.26

content of colored components resulting from the degradation of reducing sugars (glucose, fructose), as compared to beet molasses.

Sargent [27] studied the variables of column shape, density gradients and feed concentrations for ion exclusion purification of molasses. Stark [28] showed that both beet and cane molasses could be purified using Dowex 50WX4 in the K-ion-form. Fifty percent of the sucrose in beet molasses was recovered into a fraction with about 80% purity. Over 55% of the sucrose in cane refinery molasses was recovered with over 68% purity. At these levels of purity, the sucrose fractions could be returned to intermediate pans for sucrose recovery by crystallization. Takahashi and Takikawa [29] showed that ion exclusion combined with a recycling technique and under optimal conditions would yield an apparent purity of 90% and a sucrose recovery of 90% for beet molasses. Houssiau [30], using an ion exchange resin in the K-form or a second-strike molasses stream, was able to upgrade a sucrose stream at 72% purity to 95% purity at minimum loading, but only up to 80% purity at maximum column loading. The dissolved solids level of the recovered sucrose fraction was reduced to 6 – 18 ° Brix when the molasses feed stream was at 40 ° Brix. An engineering analysis [31] of the ion exclusion process for recovering sucrose from beet molasses showed that, under certain cost projections for molasses and sucrose, this process would be economically justified. Gross [32] reviewed many of these studies along with early patents on chromatographic separations of molasses.

Ito [33] and Hongisto [34] both have examined the two alternatives of separating molasses into a non-sugar fraction and a total sugar fraction (mixture of sucrose, glucose and fructose) or of inverting the sucrose of the molasses into glucose and fructose and then separating the inverted molasses into a non-sugar fraction and a glucose-fructose fraction.

In Ito's studies, the K-form of Dowex 50WX4 was used to separate molasses at 44 ° Brix with a 42% purity sugar into a non-sugar fraction, a sucrose fraction, and a glucose-fructose fraction. The sucrose fraction and the glucose-fructose fraction together recovered 77% of the sucrose and 96% of the glucose-fructose, with a purity of 82 to 89%. When the inverted molasses was separated by a K-form resin, 76.7% of the glucose-fructose was recovered at a 93.2% purity. When the Ca-form of the same resin was used, the separations were not as good.

Hongisto's studies showed that inverting the molasses prior to separation increases the capacity of a chromatographic column without decreasing the product purity because of the sharp separation of non-sugar from monosaccharides on an Na-form resin. The purity obtained by Hongisto's process was between 92 and 96% with a recovery of between 85 and 92%.

Many studies have been undertaken recently to examine the potential of isolating other non-sucrose organics from molasses by chromatography. For an Na-form resin, the raffinose peak occurs before, while glucose and fructose occur after, the sucrose peak. Of the main non-sugar components, betaine is eluted last.

Sayama and his coworkers [35] used a Ca-form of Dowex 50WX4 to study the chromatographic separation of betaine from beet molasses. The effects of solid level, flow rate, column temperature and ionic composition of the resin on peak resolution were examined. For this type of separation, resolution increased with increasing temperature. The effects of other variables on resolution were as expected: decreased concentration, decreased flow rate and increased percentage Ca-form resin increased betaine resolution.

To separate sucrose from molasses, a study [35] that examined resin crosslinkage, ionic form, flow rate and feed load determined that optimum conditions were the Na-form of Dowex 50WX4, 50 – 100 mesh, with a flow rate of 1.3 space velocity and a feed load of 0.125 liters of 30 ° Brix molasses per liter of resin per cycle. The chromatographic system could treat 2 tons of molasses per cubic meter of resin per day with a recovery of 85% of the sucrose at a purity of 87.5%.

Sayama and his cowokers also worked out procedures for recovering inositol [36], raffinose [37], and adenosine [38] from molasses. Over 70% of the inositol was recovered using a Ca-form resin. For raffinose recovery, the Na-form of the resin was superior to the Ca-form. Depending on the sucrose-to-raffinose ratio, from 66% to 86% of the raffinose could be recovered at a purity of 91 to 93%. The adenosine recovered on a pilot-scale chromatographic unit was 30% when betaine was recovered in a separate fraction. The Ca-form resin could be loaded with 0.250 m^3 of 40 ° Brix molasses per m^3 of resin per cycle during the recovery of betaine and adenosine.

One of the patented developments in the area of chromatographic treatment of molasses is the work of Neuzil and Fergin [39] who have applied the UOP continuous chromatography technique to recover sucrose from molasses. In this patent a non-functionalized adsorbent and an alcohol-containing eluent are utilized. From a model feed stream with 30% sucrose, 10% KCl and 10% betaine, a sucrose recovery of 90% with a purity of 99% was obtained.

2.12.4.2 Glucose-fructose separation

Glucose-fructose syrups may be obtained by the inversion of sucrose or by the enzymatic conversion of starch. While rice and wheat starch have been used to make glucose-fructose syrups, the most common starch source is corn. The resulting syrup is called high fructose corn syrup (HFCS). The enzymes convert the starch to a syrup that contains 55% glucose, 42% fructose and 3% oligosaccharides. While this level of fructose has sufficient sweetness for some applications, it is necessary to increase the fructose concentration to at least 55% of the dissolved solids in syrup to have a sweetness comparable to sucrose. The major industrial chromatographic application in the United States is this separation of HFCS into an enriched fructose stream and a glucose stream.

Table 2.61 Effect of counterion on the distribution ratio. (Dowex 1X8 was used for anion resin and Dowex 50WX8 for the cation exchanger. Resin size was 200–400 mesh) [41]

Ionic form	Glucose	Fructose
SO_3^{2-}	0.45	0.27
CO_3^{2-}	1.43	1.00
$H_2PO_4^-$	0.16	0.08
Ca^{2+}	0.30	0.80
Sr^{2+}	0.30	0.80
Zn^{2+}	0.20	0.30

The separation of fructose from glucose was first reported by LeFevre [40, 41] and Serbia [42] in 1962, using the Ba-, Sr-, Ag- and Ca-form of 4% crosslinked cation resin. The glucose and the higher saccharides do not form as strong a complex with these ions as the fructose does, so the glucose and higher saccharides appear first in the effluent, followed by the glucose-fructose mixture and then the relatively pure fructose. The Ca-form of the resin is preferred in industrial chromatography.

Ghim and Chang [43] have reported the effect of different ionic forms of anions and cations in separating fructose from glucose, using the moments of the elution curve. Table 2.61 shows the distribution ratios determined by them. It should be noted from the distribution coefficients that the order of elution would be fructose and then glucose for the anion resins and glucose followed by fructose for the cation resins. While it is possible to separate fructose from glucose using anion resins, the industrial chromatography applications only use cation resins. This is probably because the cation resins generally are more durable and less expensive than anion resins.

In spite of the great practical interest in the separation of glucose and fructose, there are only a few reports [44] covering their separation in industrial processes. Most available references deal with one or another specific chromatographic process [45, 46, 47] for carrying out the separation. These processes will be covered in the next section.

2.12.4.3 Oligosaccharide removal

The presence of 3 to 5% oligosaccharides in the high fructose syrup necessitates their removal during industrial chromatography of glucose and fructose, since the oligosaccharides would decrease the sweetness of fructose or would hinder the enzymatic isomerization of the glucose. A process for removing oligosaccharides has been described by Hirota and Shioda [48]. They used a strong cation exchange resin in the calcium-form to obtain a fructose stream that had 94 to 97% of the dissolved solids as fructose and less than 2% oligosaccharides. The glucose stream had a purity of 79 to 89% with an oligosaccharide concentration of 9 to 20%. The oligosaccharide stream had a dissolved solids oligosaccharide content of 64 to 76%. While this may not seem acceptable from an analytical chromatography standpoint, the low level of oligosaccharides in the feed solution (7 to 14%) and the high purity of the fructose product stream make the process acceptable for industrial chromatographic removal of a waste component. Such a process would also utilize the same resin and equipment that is used to separate the fructose from the glucose.

2.12.4.4 Polyhydric alcohol separation

There have also been industrial chromatographic developments in the separation and purification of monosaccharides and polyhydric alcohols from lignocellulosic materials such as wood [49, 50]. These authors were able to separate xylitol and sorbitol from galactitol, mannitol and arabinitol using a resin in the Ca-form. The Al-form of the resin could then be used to separate xylitol from sorbitol. As a further refinement of this double fractionation process, the saccharide fraction was removed to allow the purification of xylose from mannose, glucose and galactose. Other variations on these process schemes may be utilized to recover other polyhydric alcohols or monosaccharides.

2.12.4.5 Glycerol purification

One of the first industrial chromatographic separations proposed was for the purification of glycerol [51]. In this case, elevated temperatures (80 °C) were found to aid the separation of glycerol from salt, since aqueous glycerol solutions are quite viscous at the concentrations necessary for practical operation. Figure 2.197 shows the separation profile for a solution of 10% NaCl and 36.3% glycerol.

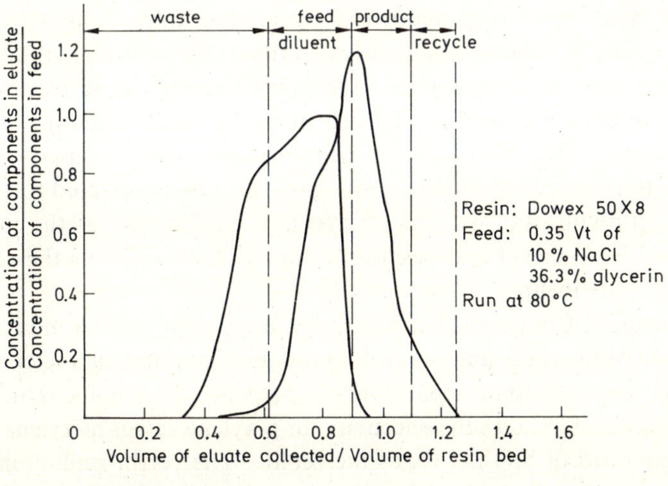

Figure 2.197 Separation profile for a solution of 10% NaCl and 36.3% glycerol (Reprinted with permission from Ref. 51).

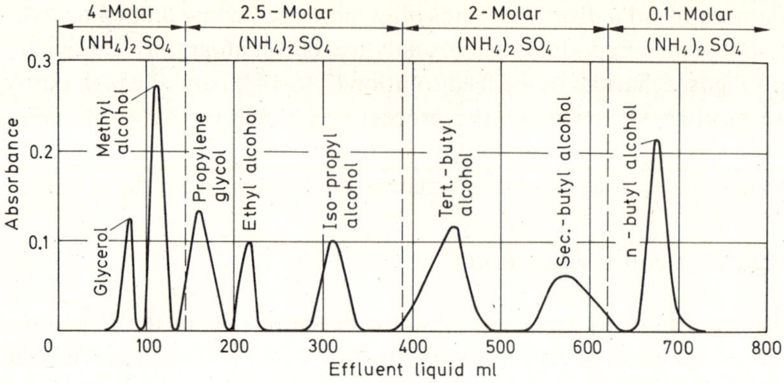

Figure 2.198 Elution profile and desorbent changes for separation of alcohols (Reprinted from Ref. 52).

Sargent and Rieman [52] proposed a chromatographic method for separating non-ionic organic solutes from one another by using aqueous electrolyte solutions as the eluent. As Figure 2.198 shows, this leads to several changes per cycle in the composition of the eluent. This approach has only been feasible for laboratory or small-scale preparatory separations. Recently, studies with ammonium carbonate solutions as the eluent for separating fermentation products containing amino acids showed that various strengths of the eluent were effective in amino acid separations and that $(NH_4)_2CO_3$ could be separated economically from the individual amino acid streams at the gaseous NH_3 and CO_2 and recycled for the next elution cycle.

2.12.4.6 Xylene isomer separation

Since 1971 there have been industrial chromatography units separating mixed isomers of xylene. Unlike the previously described systems which utilized an organic ion exchange resin and an aqueous solution as eluent, these separations use an inorganic zeolite as the insoluble adsorbent in the column and an organic eluent, such as 1-hexanol. There are now 26 such separation units licensed with a capacity of 2740000 tons/year [53]. Such systems have been demonstrated for separation of n-paraffins, n-olefins and isomers of cresol. Table 2.62 shows the product purity and recovery for p-cresol and m-cresol using Ca-X laujasite as the adsorbent and 1-hexanol as the eluent.

Asahi Chemical Company claims [54] that it has developed a modified zeolite of specific SiO_2/Al_2O_3 ratio and specific shape and granular diameter for optimum industrial chromatographic separation of p-xylene and ethylbenzene from mixed xylenes. Figure 2.199 shows the selectivity for p-xylene versus m-xylene as a function of the molar ratio of SiO_2/Al_2O_3 of the zeolite. The Asahi zeolite, in comparison with conventional zeolite materials, has a narrower particle-size distribution, a finer mean size and is only spherical (Table 2.63).

Table 2.62 Extraction of p-cresol or m-cresol at standard temperature desorbed with 1-hexanol [51]

A. Optimized for p-Cresol	Feed	Extract	Raffinate	Recovery
p-Cresol (%)	31.0	99.1	8.9	72%
m-Cresol (%)	69.8	0.8	91.0	
o-Cresol (%)	0.1	0	0.1	

B. Optimized for m-Cresol	Feed	Extract	Raffinate	Recovery
p-Cresol (%)	31.0	91.4	1.4	
m-Cresol (%)	68.9	8.6	98.5	88%
o-Cresol (%)	0.1	0	—	

Table 2.63 Comparison of zeolite materials [52]

	Asahi zeolite	Conventional zeolite
Function		
Void fraction	0.42 – 0.45 ml/g	0.21 – 0.22 ml/g
Water adsorbency	30 – 35 wt %	20 – 25 wt %
Configuration		
Size	100 – 400 μm	350 μm – several mm
Shape	Spherical	Chip or spherical

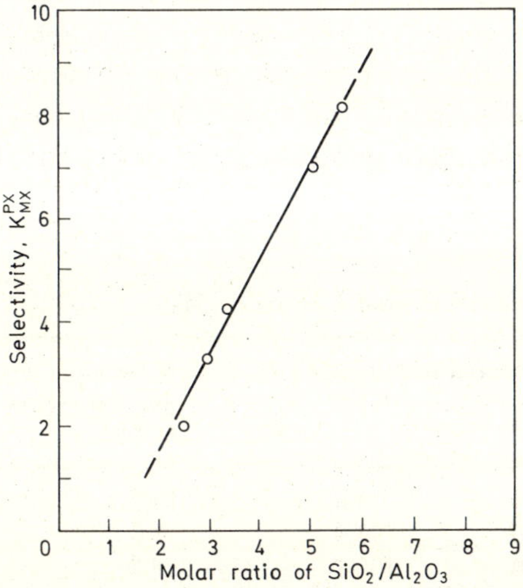

Figure 2.199 Effect of molar ratio of Si_1O_2/Al_2O_3 on the zeolite selectivity for p-xylene versus m-xylene (Ref. 54).

It is probably only in these completely organic component separations with organic desorbants that zeolites will find application. The strongly acidic or basic feed solution or desorbant that is used in many of the other industrial applications cause rapid degradation of the inorganic zeolite. Even in the pH 4 to pH 6 associated with high fructose corn syrup solutions, there is significant silica sloughage when zeolites are used as the adsorption media.

2.12.4.7 Amino acid separation

Preparatory-scale separation of complex amino acid mixtures was reported by Moore and Stein [55]. Today amino acid mixtures obtained from fermentation or from protein hydrolysis are separated by specific elution regimes after the amino acids are adsorbed onto the ion exchange resin. Although many of the commercial operations use standard ion exchange resins, such as Dowex HGR, product dilution and yield could be substantially improved by using the improved separation resins.

The separation of D- and L-isomers of amino acids may also be accomplished by using a derivatized copper form of the iminodiacetate ion exchange resin [56]. Contrary to earlier assumptions, it is not necessary to have an optically active functionality on the ion exchange resin to separate optical isomers. With the increased importance of L-phenylalanine and L-tryptophan, it is expected that industrial chromatographic separations will soon include the separation of these amino acids from their D-isomers. Table 2.64 shows the results at the preparatory-scale.

The use of non-functionalized resins [57] such as XAD-4 has been reported for the separation of amino acids and polypeptides. The authors were able to identify specific acid and base eluents that allowed di- and tri-peptide separation at high purity, with a good yield. The separation of a mixture of enkephalin peptides is shown in Figure 2.200. While this work is still on the preparatory-scale, it is anticipated that pharmaceutical companies will be utilizing this technique and other laboratory chromatographic techniques to obtain pure polypeptides for medicinal purposes.

Table 2.64 Separation and recovery of D- and L-isomers [54]

Amino acid	Recovered in pure streams		Residual solids in overlap region
	%D	%L	%DL
D,L-Tyrosine	82.4	79.2	19.2
D,L-Phenylalanine	100.0	93.0	3.6
D,L-Histidine	97.0	100.8	1.1
D,L-Tryptophan	98.8	93.6	3.8
D,L-Cystine	97.0	100.0	1.5

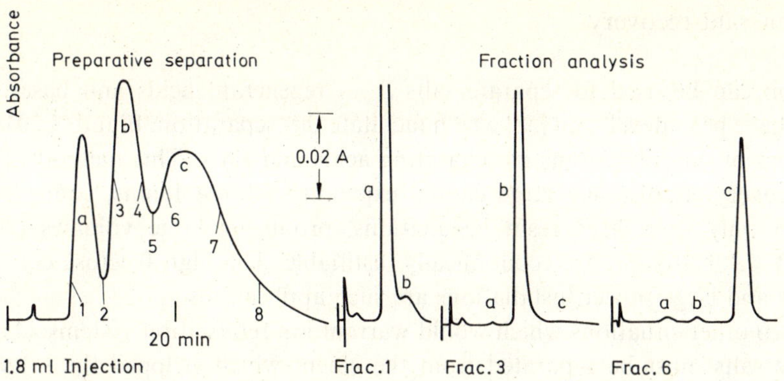

Figure 2.200 Analysis of feed and enkephalin peptide fractions separated on XAD-4 (Reprinted with permission from Ref. 57).

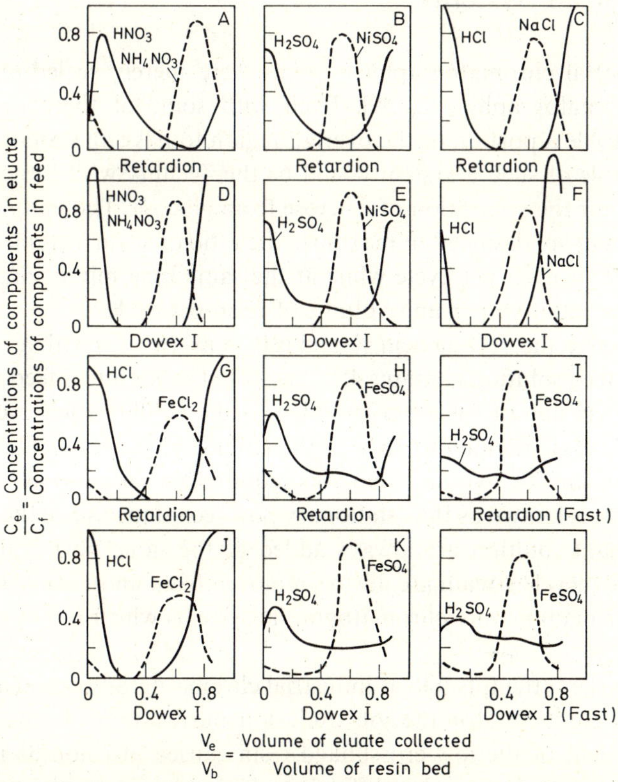

Figure 2.201 Effect of different resins on the elution profiles for various acids and salts (Reprinted with permission from Ref. 58).

2.12.4.8 Regenerant recovery

Ion retardation can be used to separate salts from regenerant acids and bases. Special resins have been developed [58] which facilitate this separation. Figure 2.201 shows the effect of different resins in separating acids and salts. This method of regenerant recovery entails such large capital expenditure for columns, controls and resin that only very large resin installations, producing large volumes of regenerant-salt waste stream, are economically justifiable. The high fructose corn syrup industry and large power installations are such applications.

There are two other situations which would warrant ion retardation systems: (1) when valuable salts must be separated from the eluent which stripped the salts from a concentrating ion exchange column or from a solvent extraction solution; and (2) when ecological considerations limit the acid, base or salt waste streams that may be discharged into the environment.

2.12.5 Industrial Systems

The first industrial chromatographic systems were merely scaled-up versions of laboratory chromatography [32, 59]. Even with some of these systems it was necessary to recycle a portion of the overlap region to have an economical process. A typical example of such a system would be the Techni-Sweet System of Technichem [45] used for the separation of fructose from glucose. The unique distributors and recycle system are designed to maximize the ratio of sugar volume feed solution per unit volume of resin per cycle while at the same time minimizing the ratio of volume of water required per unit volume of resin per cycle.

The flow through the Technichem system is 0.35 $m^3/(h \cdot m^3)$ with a column height of 3.05 m. The feed solution contains 45% dissolved solids and a feed volume equal to 22% of the volume of the resin is added to the column each cycle. The rinse water added per cycle is equal to 36% of the volume of the resin. This is much less rinse water than the 50% volume that was required by earlier systems.

This technique is known as the "stationary port technique" since the feed solution and the desorbant solution are always added at the same port and the product streams and the recycle stream are always removed from another port. Technichem and Finn Sugar manufacture chromatography systems which utilize the stationary port technique.

One of the earlier attempts [60] at industrial chromatography used an adaptation of the Higgins contactor for the ion exclusion purification of sugar juices. The physical movement of the low-crosslinked resin caused attrition as it was moved around the contactor. It was also difficult to maintain the precise control needed on flow rates because of the pressure drop changes and volume changes of the resin as it cycled from the mostly water zone to the mostly sugar solution zone.

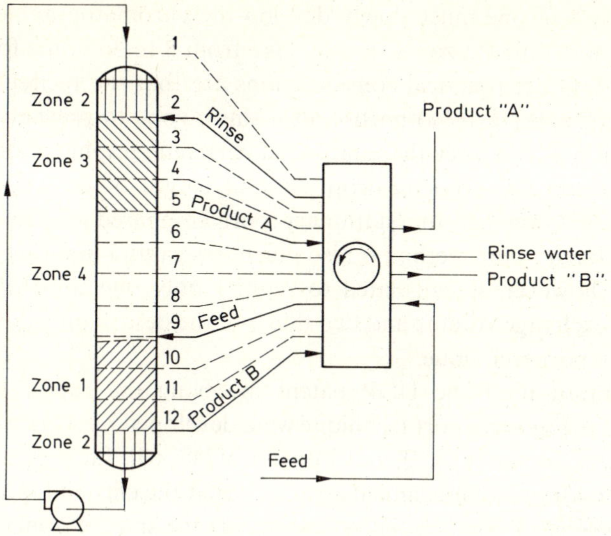

Figure 2.202 Moving port chromatographic column with four zones for continuous chromatographic separation.

An alternate approach [61] utilizes moving port or pseudo-moving bed techniques. With these techniques, the positions on the column where the feed solution is added and where the product streams are removed are periodically moved to simulate the countercurrent movement of the adsorbent material. At any given time the resin column can be segmented into four zones (Figure 2.202). Zone 1 is called the adsorption zone and is located between the point where the feed solution is added and the point where the fast or less strongly adsorbed component is removed. In this zone the slow or more strongly adsorbed component is completely adsorbed onto the ion exchange resin. The fast component may also be adsorbed, but to a much smaller extent. The second zone, Zone 2, is the purification zone and is located between the point where the fast component is removed and the point where the desorbant solution is added. Zone 3 is called the desorption zone and is between the point where the desorbant is added and the point where the slow component is removed. In this zone the slow component is removed from the resin and exits the column. The final zone, Zone 4, is called the buffer zone and is located between the point where the slow component is removed and the point where the feed solution is added. There is a circulating pump which unites the different zones in a continuous cycle.

Different sections of the column serve as a specific zone during the cycle operation. Unlike the stationary port technique, the liquid flow is not uniform throughout the column. Because of the variations in the additions and withdrawals of the different fluid streams, the liquid flow rate in each of the zones is different.

With such a system one must slowly develop the chromatographic distribution pattern through the different zones. It may take from 8 to 36 hours for the pattern to be established. Other practical considerations are that the recirculation system must represent a small (< 10%) portion of a single zone to prevent unacceptable back-mixing which would alter the established chromatographic pattern.

The flow rate and the pressure drop per unit length of the chromatographic column are much lower for the stationary port compared to the moving port system. The stationary port method is also much less capital-intensive. The moving port technique, however, is calculated to require only one third of the column volume and ion exchange volume and two thirds of the desorbant volume compared to the stationary port technique.

After the expiration of the UOP patent involving the rotary valve, several modifications of the moving port technique were developed by Amalgamated Sugar [62], Illinois Water Treatment [63], and Mitsubishi [47, 64]. Each manufacturer has its own approach for the establishment and control of the chromatographic pattern. These are the subject of confidentiality agreements for specific applications.

Mathematical models for these semi-continuous chromatographic columns have been reported by Barker and Thawait [65]. The sequential nature of the switching valves is modelled by stepping the concentration profile backwards by one zone at the end of the switch period. The typical switch time is 5 to 15 minutes for commercial units. The one difficulty with using the model is the selection of a distribution coefficient since the measured value changes with concentration.

The key items to identify when considering an industrial chromatographic project are the capital for the equipment, yield and purity of the product, the amount of dilution of the product and waste stream, the degree of flexibility the computer controls allow, the expected life of the ion exchange material and whether the equipment allows for periodic expansion of the resin.

New techniques are continually being developed which can be expected to be used in future specialized industrial applications. Multi-segmented columns have been demonstrated for the preparative purification of urokinase [66]. Other techniques [67, 68] have been developed for continuous spiral cylinder purifications which allow separation on the basis of electropotential, in addition to the selective affinity of the adsorbent resin for the components in solution.

References

[1] M. Tswett (1906) Ber. deut. botan. Ges. *24*, 316.

[2] M. Tswett (1907) The Chemistry of Chlorophyll, Phylloxanthin, Phyllocyanin and Chlorophyllane, Biochem. Z. *5*, 6.

[3] G. T. Seaborg (1946) The Transuranium Elements, Science *104* (2704), 379.

[4] F. H. Spedding, E. I. Fulmer, T. A. Butler, E. M. Gladrow, M. Gobush, P. E. Porter, J. E. Powell and J. M. Wright (1947) Separation of Rare Earth. (III) Pilot-plant Scale Separations, J. Am. Chem. Soc. *69*, 2812.

[5] F. H. Spedding and J. Powell (1954) The Separation of Rare Earths by Ion Exchange. VII. Quantitative Data for the Elution of Neodymium, J. Am. Chem. Soc. *76*, 2545.

[6] F. H. Spedding, J. Powell and E. Wheelwright (1954) The Use of Copper as the Retaining Ion in the Elution of Rare Earths with Ammonium Ethylenediamine Tetraacetate Solution, J. Am. Chem. Soc. *76*, 2557.

[7] J. Porath and P. Flodin (1959) Gel Filtration – Desalting and Group Separation, Nature *183*, 1657.

[8] J. Porath (1967) Development of Chromatography on Molecular Sieves, Lab. Pract. *16*, 838.

[9] J. C. Moore (1964) Gel Permeation Chromatography. I. Molecular Weight Distribution of High Polymers, J. Polym. Sci., Part A *2*, 835.

[10] R. M. Wheaton and W. C. Bauman (1953) Ion Exclusion, Ann. N. Y. Acad. Sci. *57*, 159.

[11] J. A. Dean (1969) Chemical Separation Methods, Van Nostrand, New York, p. 295.

[12] J. N. Wilson (1940) A Theory of Chromatography, J. Am. Chem. Soc. *62*, 1583.

[13] D. DeVault (1943) The Theory of Chromatography, J. Am. Chem. Soc. *65*, 532.

[14] E. Glueckauf (1947) Theory of Chromatography. Part V. Separation of Two Solutes Following a Freundlich Isotherm, J. Chem. Soc. 1321.

[15] E. Glueckauf (1949) Theory of Chromatography. Part VII. Theory of Two Solutes Following Nonlinear Isotherms, Disc. Faraday Soc. *7*, 42.

[16] A. J. P. Martin and R. L. M. Synge (1941) A New Form of Chromatogram Employing Two Liquid Phases. I. A Theory of Chromatography, Biochem. J. *35*, 1358.

[17] S. W. Mayer and E. R. Thompkins (1947) Ion Exchange as a Separation Method. IV. A Theoretical Analysis of the Column Separation Process, J. Am. Chem. Soc. *69*, 2866.

[18] D. W. Simpson and R. M. Wheaton (1954) Ion Exclusion – Column Analysis, Chem. Eng. Prog. *50*, 45.

[19] E. Glueckauf (1955) The Theory of Chromatography. IX. The Plate Concept in Column Separations, Trans. Faraday Soc. *51*, 34.

[20] J. Beukenkamp, W. Rieman III and S. Lindenbaum (1954) Behavior of the Condensed Phosphates in Anion-Exchange Chromatography, Anal. Chem. *26*, 505.

[21] D. W. Simpson and W. C. Bauman (1954) Concentration Effects of Recycling in Ion Exclusion, Ind. Eng. Chem. *46*, 1958.

[22] B. L. Karger (1966) Resolution in Linear Elution Chromatography, J. Chem. Educ. *43*, 47.

[23] H. J. Hongisto (1977) Chromatographic Separation of Sugar Solutions, Int. Sugar J. *79*, 100.

[24] F. Martinola and G. Siegers (1979) Chromatography on Ion Exchangers, a Large-Scale Engineering Separation Process, Verfahrenstechnik (Mainz) *13*, 32.

[25] K. Sayama, Y. Senba and T. Kawamoto (1980) Chromatographic Separation of Molasses Constituents. Part 2. Separation of Betaine from Molasses, Proc. Res. Soc. Japan Sugar Refin. Tech. *29*, 10.

[26] G. Rousseau and X. Lancrenon (1984) A More Logical Approach to Ion Exchange Sugar Technology in the Sugar Industry, Sugar y Azucar 27.

[27] R. N. Sargent (1963) Ion Exclusion – Effects of Gaseous Phase, Column Shape, Density Gradients and Feed Concentrations, Ind. Eng. Chem. Proc. Res. Dev. *2*, 89.

[28] J. B. Stark (1965) Ion Exclusion Purification of Molasses, J. Am. Soc. Sugar Beet Technol. *13*, 492.

[29] K. Takahashi and T. Takikawa (1965) Refining of Beet Molasses by Ion Exclusion, Proc. Res. Soc. Japan Sugar Refin. Tech. *16*, 51.

[30] J. Houssiau (1968) Sucrerie Belge *87*, 423.

[31] W. G. Schultz, J. B. Stark and E. Lowe (1967) Engineering Analysis of Ion Exclusion for Sucrose Recovery from Beet Molasses. I. Experimental Procedures and Data Reduction Techniques, Int. Sugar J. *69*, 35.

[32] D. Gross (1971) Purification of Sugar Products by the Ion Exclusion Process, Proc. 14th Gen. Assembly CITS, p. 445.

[33] Y. Ito (1970) Ion Exclusion Purification of Refinery Molasses, Proc. Res. Soc. Japan Sugar Refin. Tech. *22*, p. 1.

[34] H. J. Hongisto (1978) Molasses-Based Liquid Sugar, S.I.T. paper *78*, 624.

[35] K. Sayama, Y. Senba and T. Kawamoto (1980) Chromatographic Separation of Molasses Constituents. Part 1. Recovery of Sucrose from Molasses, Proc. Res. Japan Sugar Refin. Tech. *29*, 1.

[36] K. Sayama, Y. Senba and T. Kawamoto (1980) Chromatographic Separation of Molasses Constituents. Part 3. Recovering of Inositol from Molasses, Proc. Res. Japan Sugar Refin. Tech. *29*, 20.

[37] K. Sayama, Y. Senba, T. Kawamoto and T. Muratsubaki (1981) Chromatographic Separation of Molasses Constituents. Part 5. On the Manufacturing Conditions of Raffinose, Proc. Res. Japan Sugar Refin. Tech. *30*, 72.

[38] S. Oikawa, K. Sayama, Y. Senba, T. Kawamoto and T. Muratsubaki (1982) Chromatographic Separation of Molasses Constituents. Part 7. Pilot Plant Production of Raffinose, Betaine and Adenosine, Proc. Res. Japan Sugar Refin. Tech. *33*, 55.

[39] R. W. Neuzil and R. L. Fergin, Extraction of Sucrose, U.S. Patent 4426232 (Jan. 17, 1984).

[40] L. J. LeFevre, Separation of Fructose from Glucose Using Cation Exchange Resin Salts, U.S. Patent 3044905 (July 17, 1962).

[41] L. J. LeFevre, Separation of Fructose from Glucose Using a Cation Exchange Resin Salts, U.S. Patent 3,044,906 (July 17, 1962).

[42] G. R. Serbia, Separation of Dextrose and Levulose, U.S. Patent 3044904 (July 17, 1962).

[43] Y. S. Ghim and H. N. Chang (1982) Adsorption Characteristics of Glucose and Fructose in Ion Exchange Resin Columns, Ind. Eng. Chem. Fundam. *21*, 369.

[44] H. J. Bieser and A. J. deRosset (1977) Continuous Countercurrent Separation of Saccharides with Inorganic Adsorbents, Starch/Stärke *29*, 392.

[45] H. W. Keller, A. C. Reents and J. W. Laraway (1981) Process for Fructose Enrichment from Fructose Bearing Solutions, Starch/Stärke *33*, 55.

[46] S. Miyahara, S. Sakai, F. Matsuda, S. Ushikubo and K. Kono, Process for Separating Glucose from Fructose, Japan. Patent 118400 (Sept. 11, 1980).

[47] E. Katz, H. S. Davis and B. L. Scallet, High Fructose Syrup and Process for Making Same, U.S. Patent 4395292 (July 26, 1983).

[48] T. Hirota and K. Shioda, Process for Removing Oligosaccharides, Japan. Patent 48400 (April 7, 1980).

[49] A. J. Melaja and L. Hamalainen, Process for Making Xylitol, U.S. Patent 4008285 (Feb. 15, 1977).

[50] A. J. Melaja and L. Hamalainen, Process for Making Xylose, U.S. Patent 4075406 (Feb. 21, 1978).

[51] D. R. Asher and D. W. Simpson (1956) Glycerol Purification by Ion Exclusion, J. Phy. Chem. *60*, 518.

[52] R. N. Sargent and C. W. Rieman III, Method of Separating Non-Ionized Organic Solutes from One Another in Aqueous Solution, U.S. Patent 3134814 (May 26, 1964).

[53] R. W. Neuzil, D. H. Rosbach, R. H. Jensen, J. R. Teague and A. J. deRosset (1980) An Energy-Saving Separation Scheme, Chemtech, 498.

[54] M. Seko, H. Takeuchi and K. Inada (1981) Molecular Sieves for p-Xylene and Ethylbenzene Separation, Presented at Seminar "Molecular Sieves and Their Applications", New Delhi, India (Nov. 13–14).

[55] S. Moore and W. H. Stein (1954) Procedure for the Chromatographic Determination of Amino Acids on Four Percent Crosslinked Sulfonated Polystyrene Ion Exchange Resin, J. Biol. Chem. *211*, 893.

[56] W. Szczepaniak and W. Ciszewsha (1982) Resolution of Some Racemic D,L Aminoacids on an Ion Exchanger Containing Iminodi (methane phosphonic) Groups in Amino-Copper Form, Chromatographia *15*, 38.

[57] D. J. Pietrzyk, W. J. Cahill, Jr. and J. D. Stodola (1982) Preparative Liquid Chromatographic Separation of Amino Acids and Peptides on Amberlite XAD-4, J. Liquid Chrom. *5*, 443.

[58] M. J. Hatch and J. A. Dillon (1963) Acid Retardation, IEC Proc. Res. Dev. *2*, 253.

[59] R. F. Sutthoff and W. J. Nelson, Method for Separation of Water Soluble Carbohydrates, U.S. Patent 4022637 (May 10, 1977).

[60] L. Norman, G. Rorabaugh and H. Keller (1963) Ion Exclusion Purification of Sugar Juices, J. Am. Soc. Sugar Beet Tech. *12*, 363.

[61] D. B. Broughton and C. G. Gerhold, Continuous Sorption Process Employing Fixed Bed of Sorbent and Moving Inlets and Outlets, U.S. Patent 2985589 (May 23, 1961).

[62] For more information contact K. Schoenrock, Chapter 2.10 of this Volume.

[63] D. J. Burke (1983) Adsorption Separation Technologies Via Ion Exchange Resins, Presented at A.I.Ch.E. 23rd Ann. Symp. (May 12).

[64] H. Ishikawa, H. Tanabe and K. Usui, Control of Pseudo-Moving Beds, Japan. Patent 102288 (August 11, 1979).

[65] P. Barker and S. Thawait (1983) Separation of Fructose from Carbohydrate Mixtures by Semi-continuous Chromatography, Chem. Ind., 817.

[66] L. J. Novak and P. H. Bowdle, Multi-Segmented Adsorption Ion Exchange or Gel Filtration Column Apparatus and Process, U.S. Patent 4155846 (May 22, 1979).

[67] J. M. Begovich, C. H. Byers and W. G. Sisson (1983) A High-Capacity Pressurized Continuous Chromatograph, Sep. Sci. Technol. *18*, 1167.

[68] J. M. Begovich and W. G. Sisson (1984) A Rotary Annular Chromatograph for Continuous Separations, A.I.Ch.E. J. *30*, 705.

2.13 Ion Exchange Processes in Hydrometallurgy

Michael Streat
University of Technology, Department of Chemical Engineering, Loughborough, Leicestershire, Great Britain

Introduction

A list of metals that have been recovered and purified commercially by ion exchange is given in Table 2.65. In some cases, the scale of operation is relatively small, e. g., the rare earth elements, the precious and platinum group metals, though the value of the metal recovered is extremely high. Ion exchange is particularly suited to high value, low throughput metal purification processes. Alternatively, the recovery of trace amounts of metals from waste streams is carried out on a large scale, e. g., cadmium and mercury from industrial effluents, chromium from spent metal plating solutions and copper and zinc from the waste arising in the synthetic fibre industry. The application of ion exchange processes in hydrometallurgy is extensive and increasing.

The largest single application of ion exchange in hydrometallurgy is for the recovery of uranium from naturally occurring ore-bodies and as a by-product in the production of gold. It has also been shown that uranium can be recovered as a by-product during the production of wet process phosphoric acid.

Table 2.65 Metals recovered and purified commercially by ion exchange

Uranium	Chromium
Thorium	Copper
Rare earths	Zinc
Transuranic elements	Nickel
Gold	Cobalt
Silver	Tungsten
Platinum group metals	

Recent advances in the application of ion exchange in hydrometallurgy depend on the synthesis and development of new synthetic organic ion exchange resins possessing complexing ligands and selective for metals in aqueous solutions. Warshawsky [1] has given a comprehensive review of the status of chelating ion exchangers and their potential applications.

2.13.1 Ion exchange processing of uranium

Uranium is recovered from the host mineral by leaching; subsequently, the pregnant solution is clarified and purified using either solid ion exchange, liquid-liquid extraction or both processes in series. Pregnant solutions usually contain between 100 – 1000 ppm uranium as U_3O_8; as a result ion exchange treatment at the head end of the process is preferred.

Acid leaching of uranium-bearing ore-bodies tends to dissolve a wide range of metal impurities, in particular iron and vanadium. Other impurities, such as silica, are liberated, which tend to poison anion exchange resins. Likewise, the cyanide complexes of cobalt will poison anion exchange materials in by-product recovery of uranium from gold cyanide liquors. This, however, is largely overcome by adopting the reverse-leach technique whereby uranium is recovered prior to cyanide treatment of the ore to leach gold. If the host mineral contains acid-consuming material, e. g., dolomite, then leaching with an alkaline reagent is preferred. In fact, most in-situ and solution mining is performed with sodium carbonate/bicarbonate solutions.

Uranium will dissolve in sulphuric acid or in sodium carbonate in the hexavalent state and it is customary to provide an appropriate oxidant. The following reactions are typical in acid and alkaline leaching:

$$UO_2^{2+} + SO_4^{2-} \rightleftharpoons UO_2SO_4 \tag{1}$$

$$UO_2SO_4 + SO_4^{2-} \rightleftharpoons UO_2(SO_4)_2^{2-} \tag{2}$$

$$UO_2(SO_4)_2^{2-} + SO_4^{2-} \rightleftharpoons UO_2(SO_4)_3^{4-} \tag{3}$$

$$UO_2 + \frac{1}{2}O_2 \rightleftharpoons UO_3 \tag{4}$$

$$UO_3 + 3Na_2CO_3 + H_2O \rightleftharpoons Na_4UO_2(CO_3)_3 + 2NaOH \tag{5}$$

Reactions 1 – 3 relate to the acidic complexes and show that both di- and quadrivalent anionic sulphate complexes are formed. The quadrivalent complex predominates at pH values of 0.5 – 1.5 and about 0.2M sulphate concentration.

The alkaline reactions 4 and 5 show the formation of a quadrivalent anionic complex provided the pH is adjusted by the presence of sodium bicarbonate to avoid the precipitation of uranium. Sorption of these complexes onto conventional weak and strong base anion exchange resins is highly selective, since most impurities (except iron) do not form anionic species.

$$2\,(\overline{R}^+)_2\,SO_4^{2-} + UO_2(SO_4)_3^{4-} \rightleftharpoons (\overline{R}^+)_4\,UO_2\,(SO_4)_3^{4-} + 2\,SO_4^{2-} \quad (6)$$

$$2\,(\overline{R}^+)_2\,CO_3^{2-} + UO_2(CO_3)^{4-} \rightleftharpoons (\overline{R}^+)_4\,UO_2\,(CO_3)_3^{4-} + 2\,CO_3^{2-} \quad (7)$$

($\overline{R}^+$) denotes the resin matrix of a typical anion exchange resin of macroreticular or polyelectrolyte gel type.

Reactions 6 and 7 are readily reversed using either hot or cold mineral acids. Sulphuric acid is the preferred eluant for the sulphate-loaded resin and sodium nitrate is used to elute the carbonate-loaded resin because acid would cause the spontaneous evolution of CO_2 gas.

In recent years there has been a drive towards a reduction in uranium plant and operating costs, as well as a need to simplify the process flow sheet. This has encouraged the development of new extraction processes which can be operated in the presence of suspended solids, hence eliminating the costly feed-clarification stage. Fluidized bed ion exchange technology was the first development to succeed in recovering uranium from solutions containing 1 to 2 kg m^{-3} solids in suspension. This technique has been widely adopted in Southern Africa and to a lesser extent in North America [2]. The Cloete-Streat or NIMCIX concept is the most widely adopted vertical column design, whilst the Porter concept is the most widely used open horizontal tank system.

Continuous counter-current multistage fluidized bed reactors have been operated for several years on a number of uranium mills on the Witwatersrand. Twin columns each having 12 stages and a diameter of 4.85 m are in operation at Chemwes treating 320 $m^3\,h^{-1}$ of feed solution [3]. A schematic flow diagram of a typical Cloete-Streat fluid bed plant of this type is given in Figure 2.203. At Rossing Uranium Ltd., the process operates in a series of rectangular horizontal tanks each 6.2 m $\times$ 6.2 m treating 850 $m^3\,h^{-1}$ of feed solution. In both cases there is no theoretical limit to scale-up of the ion exchange contactor.

Stirred vessel resin-in-pulp (RIP) contactors in horizontal arrangement with resin separation from pulp on shaking screens have been successfully used in the uranium industry for many years. Recent technological improvements in carbon-in-pulp (CIP) processing for gold recovery have been achieved using a cascade of horizontal reactors separated by air-swept screens. This technology is now well-established in South Africa [4] and may now be considered for RIP recovery of uranium. The concept has been further developed in a novel engineering approach involving a submerged air-swept screen design which will operate in the presence of much higher concentrations of both carbon and resin than in the present generation of CIP and RIP contactors [5]. The result is a significant reduction in plant size and greater ease of operation.

Digestion of phosphate ores with sulphuric acid results in the production of phosphoric acid containing traces of cationic uranium species. The U_3O_8 content

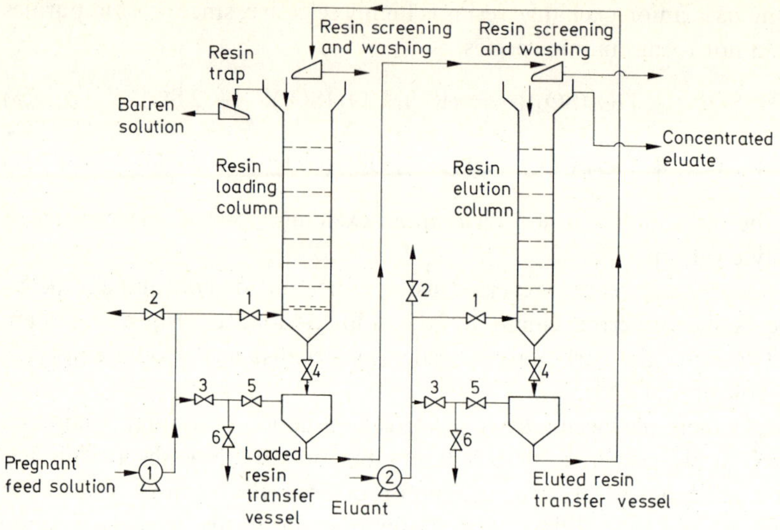

Figure 2.203 Continuous counter-current fluid bed IX in both resin loading and resin elution. (1) Loading and elution column solution forward flow to fluidize the resin in each column, valve 1 open, valves 2, 3, 4, 5 and 6 closed. (2) No solution flow, resin settling, valve 2 open to return feed solution to storage, valve 1 closed. (3) Solution back flow, resin from bottom stage transferred to resin transfer vessels, valves 2, 3, 4, 5 and 6 open, valves 1 and 3 closed. (4) Valve flush and resin transfer from resin transfer vessels to top of fluid bed columns, no forward solution flow, valves 1, 3, 4 and 5 open, valves 2 and 6 closed. (5) Resumption of solution forward flow to the fluid bed columns. (Reproduced by permission from "Ion Exchange and Sorption Processes in Hydrometallurgy", published for SCI by John Wiley, 1987 [© SCI].)

of phosphate rock from a number of locations in the world ranges from 0.005 to 0.62 wt%, which can lead to 0.01 to 0.20 kg m^{-3} U_3O_8 in solution. This compares favourably with that of typical in-situ leach streams in the USA. It is estimated that 2500 tonnes per annum is available from this source alone.

Recently Gonzalez-Luque and Streat [6] have successfully investigated the separation of uranium from wet process phosphoric acid (WPA) with polymeric ion exchange resins. A macroporous polystyrene-divinylbenzene copolymer containing amino-phosphonate functional groups (Duolite ES 467) selectively sorbs uranium from other trace metal impurities provided uranium is in the correct valency state. The chelating reaction is shown below:

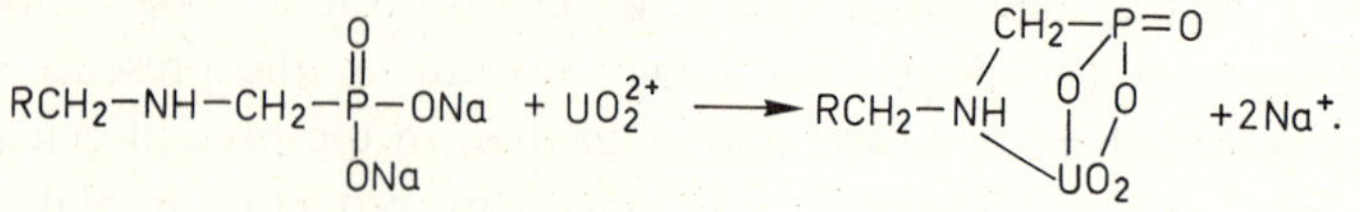

(8)

The sorption of uranium is readily reversed using either concentrated phosphoric acid or, alternatively, an ammonium carbonate solution. The product solutions can be further refined and concentrated prior to precipitation. A fixed-bed pilot plant development of this process in Israel has confirmed the feasibility of the process and it is claimed here that uranium was produced at approximately half the cost of solvent extraction [7].

2.13.2 Ion exchange processing of gold

The operation of a conventional gold or gold and uranium plant gives rise to several process streams in which the gold concentration is too low to be treated economically by the Merrill-Crowe process. Such streams are the dewatering filtrates between the gold and uranium plants, and the recycled water from the tailings ponds or slimes dams. Also, regenerated aqueous streams containing gold (30 to 100 g per tonne), which arise from the solvent-extraction circuits of certain uranium plants are not amenable to the zinc-cementation process.

Commercial plants in South Africa, which employ activated carbon in CIP plants, are successfully recovering gold from such streams. The ability of carbon to yield barren solutions with a gold concentration of 0.005 g per tonne or less also makes it a potential scavenger to recover additional gold in the 'barren' solutions resulting from the Merrill-Crowe process.

However, activated carbons have several disadvantages as compared with ion exchange resins, which can be expected to be superior in the following areas:

1) Resins are far less susceptible to poisoning by calcium or organic materials that may be present in the process streams.
2) Activated carbons require much higher temperatures for elution; they also need thermal reactivation, which is unnecessary for resins.
3) Some resins with a higher absorption capacity can ensure efficient absorption of the base-metal cyanides, thus leading to improved pollution control as well as the recovery of these metals and the cyanide from the effluent streams.

It is possible, therefore, that ion exchange resins might offer a suitable alternative to carbon.

Gold exists as an anionic aurocyanide complex $[Au(CN)_2]^-$ in cyanide leach liquors and can be recovered from solution by conventional anion exchange resins. The sorption of gold by a protonated weak base resin is given by the following equation:

$$(R_3N{:}H^+)_2SO_4^{2-} + 2\,[Au(CN)_2]^- \rightleftharpoons 2\,R_3N{:}H^+[Au(CN)_2]^- + SO_4^{2-}. \quad (9)$$

Resin selectivity for gold is adequate, though the cyanide complexes of Ag, Co, Cu, Fe, Ni, and Zn are also sorbed and will therefore affect the purity of the eluted product. Elution of gold is usually performed with sodium hydroxide solution according to the following equation:

$$R_3N{:}H^+[Au(CN)_2]^- + OH^- \rightarrow R_3N{:} + [Au(CN)_2]^- + H_2O\,. \quad (10)$$

The free base form of the eluted resin is treated with dilute sulphuric acid to protonate the resin functional groups prior to the extraction cycle:

$$2\,R_3N: + H_2SO_4 \rightarrow (R_3N{:}H^+)_2SO_4^{2-}\,. \tag{11}$$

A strong base resin contains quaternary ammonium groups and can absorb gold over the entire pH range

$$(R_4N^+)_2SO_4^{2-} + 2\,[Au(CN)_2]^- \rightleftharpoons 2\,R_4N^+[Au(CN)_2]^- + SO_4^{2-}\,. \tag{12}$$

The elution of gold is slightly more complex, since it is necessary to break the strong ligand-complex interaction by using acidified thiourea

$$R_4N^+[Au(CN)_2]^- + 2CS(NH_2)_2 + 2\,HCl \rightarrow R_4N^+Cl^- + [AuCS(NH_2)_2]^+\,Cl^- + 2\,HCN\,. \tag{13}$$

It is also possible to elute the aurocyanide complex by treatment of the resin with a strongly preferred counterion such as $Zn(CN)_4^{2-}$, though this will require further elution steps in order to recycle fresh resin to the extraction cycle (equation 12). The cost of the reagents consumed during various operations suggests that weak base resins are more suitable because the elution procedure is significantly cheaper. Also, the elution of weak base resin is easier, requiring less labour and lower temperatures [8]. Commercial weak base resins possess some process disadvantages and there is considerable independent research in progress to synthesise custom-designed weak base resins. Green et al. [9] have found very promising results with weak base resins containing imidazole functional groups. These resins load uranium at high pH values and are eluted with aqueous sodium hydroxide. However, the presence of strong base groups in the structure strongly affects the elution step, which is still a problem. No doubt, novel selective ion exchangers for gold will be available in the near future. Separation of gold from acid leach liquors is possible using ion exchange resins containing weak ester groups, e.g., Amberlite XAD-7 [10]. The mechanism of extraction is either by solvation

$$R-CO_2 + [AuCl_4]^- \rightarrow R-CO_2[AuCl_3] + Cl^- \tag{14}$$

or by ion exchange

$$R-CO_2 + H_2O \rightarrow R-COOH^+ + OH^- \tag{15}$$

$$R-COOH^+ + [AuCl_4]^- \rightarrow R-COOH^+[AuCl_4]^-\,. \tag{16}$$

Elution of the gold is performed using a mixture of hydrochloric acid and acetone; it is necessary to provide a distillation step if the eluant is to be recycled.

2.13.3 Ion exchange processing of platinum group metals

The extraction and separation of platinum group metals (PGM) by ion exchange has recently been reviewed by Warshawsky [11]. Ion exchange and solvent extraction are both used on PGM flow sheets, especially the three new process routes at Inco

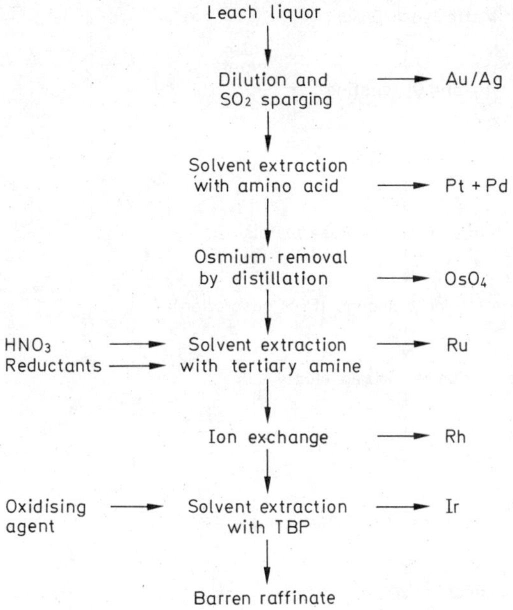

Figure 2.204 Schematic outline of the Lonhro Precious Metal Process in South Africa. (Reprinted with permission, Figure 7 p. 60 of "Hydrometallurgy-Research Development and Plant Practice", by K. Osseo-Assare and J. D. Miller, 1982.)

(UK), Matthey Rustenburg Refineres (UK) and at Lonrho (South Africa). The Inco flow sheet is based largely on solvent extraction, whereas at Matthey Rustenburg Refiners and at Lonrho an ion exchange process for the separation and recovery on rhodium from chloride solution is achieved, using a strong base anion exchange resin. A schematic outline of the Lonrho precious metal process is given in Figure 2.204.

An integrated ion exchange and solvent extraction process has been studied at MINTEK in South Africa. A typical conceptual flow sheet for the recovery of platinum from a PGM-rich Matte leach residue is shown in Figure 2.205. The concentrate is preconditioned by air roasting and coal gas roasting, followed by almost complete dissolution of Pt, Pd, Au and Ir in an $HCl-Cl_2$ leach. The PGM chloro-complexes are then concentrated on a selective resin containing polyisothiourea (PITU) groups. Both gel (Srafion NMRR) and macroreticular (Monivex) PITU resins have been developed and are found to be most effective in HCl solution (Refs. 96–109 cited by Warshawsky [11]). The explanation for the behaviour of PITU resins at different acidities is attributed to the weak base properties of the functional isothiourea groups, which are subject to the following acid-base equilibrium:

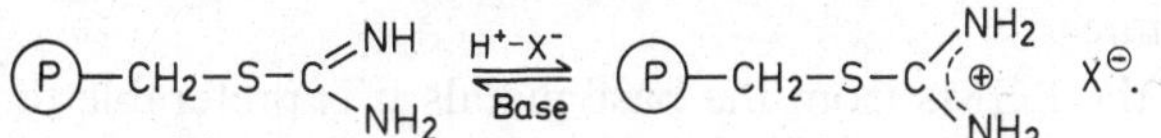

(1) (2)

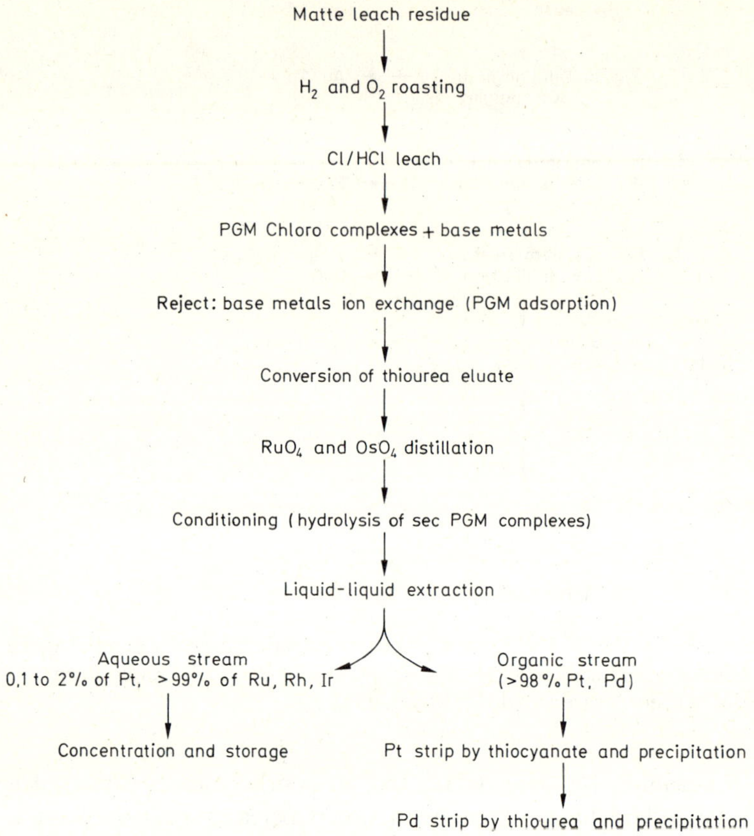

Figure 2.205 Typical conceptual flowsheet for recovery of PGMs from base metals in a PGM-rich matte leach residue. (Reprinted by permission of Kluwer Academic Publishers.)

The PITU resins in the isothiourea form (1) bind metal cations by coordination, but when converted to the isothiouronium form (2) they bind anions by an anion exchange mechanism, e. g., $[PtCl_4]^{2-}$ and $[PdCl_4]^{2-}$

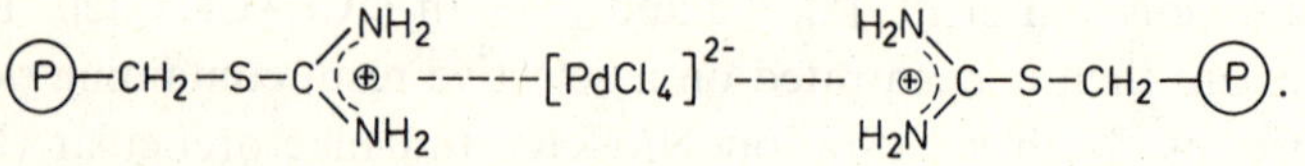

Typical pilot plant results for sorption from a solution of 2 M HCl containing about 13 g/l of PGM ions indicate that the recovery of 'primary' PGMs (i. e., Pt, Pd) is always better than 99.9%, whereas the recovery of the 'secondary' PGMs (i. e., Rh, Ru, Ir, Os) is dependent on the column operating conditions, in particular the column retention times.

After separation of the PGMs from the base metals it is preferable to separate the PGMs in chloride media, where they form soluble, relatively stable and reactive complexes. Thus, the PGM thiourea eluate has to be converted into a chloride form by oxidative hydrolysis. A procedure consisting of alkaline hydrolysis of the thiourea, precipitation of the sulphide complexes of the PGM, followed by filtration and oxidation in chlorine-water yields quantitative and consistent conversion to the chloro-complexes. The conversion procedure also includes a conditioning step, i.e., hydrolysis to give fresh PGM oxides that readily solubilize in hydrochloric acid. When conditioning is included, the Pt(IV), Pd(II), and Au(III) exist as readily extractable anionic complexes, whereas Rh(III) is cationic and Ir(III) is anionic, though the latter is difficult to extract.

Ruthenium and osmium can be distilled off as tetroxides after neutralization of the acid formed during conversion by sodium hydroxide. Very accurate pH control is required, with vigorous chlorine sparging.

The PGMs are finally separated using solvent extraction with tertiary amines. The protocol for the separation process in chloride solution is shown in Figure 2.205.

The results obtained in liquid-liquid extraction are pertinent to the anion exchange resin system. Work on Amberlite 299 showed that thiourea can be used as an eluant for PGM since it forms cationic complexes. However, the elution with thiocyanate is rather more complicated. The formation of $[Pd(SCN)_4]^{2-}$ is favoured in both the solution and polymer phases. The $[Pd(SCN_4]^{2-}$ readsorption into the ion exchange resin is the basis of a separation of platinum and palladium. However, despite considerable laboratory work, an integrated ion exchange – solvent extraction separation process for the secondary PGMs has not yet been elucidated.

2.13.4 Ion exchange processing of base metals

The separation of base metals from chloride solution is particularly attractive. A diagrammatic representation based on the periodic table of the elements has been prepared by Kraus and Nelson [12], which clearly establishes the behaviour of metal chloride complexes at trace ionic concentration in the presence of a strong base anion exchange resin (see Figure 2.206). This indicates that the alkali, alkali earth and rare earth elements do not interact, whereas the transition metals and noble metals can form anionic chloride complexes with varying affinity for an anion exchange resin. The concept of separating metals from chloride solution by continuous ion exchange has been described by Streat and Gupta [13].

Separation and recovery of the transition metals by anion exchange is possible by careful control of the ambient hydrochloric acid concentration in solution. For example, the 'METSEP' process is highly effective at low acid concentration (less than 4M) since Zn^{2+} forms a strong anionic chloride complex, whereas Fe^{2+} does not [14]. Hence, separation is possible in the extraction cycle of an ion exchange process. Alternatively, separations can be achieved by fractional elution of the anionic chloride complexes. Zinc, copper and iron(III) are strongly sorbed in strong

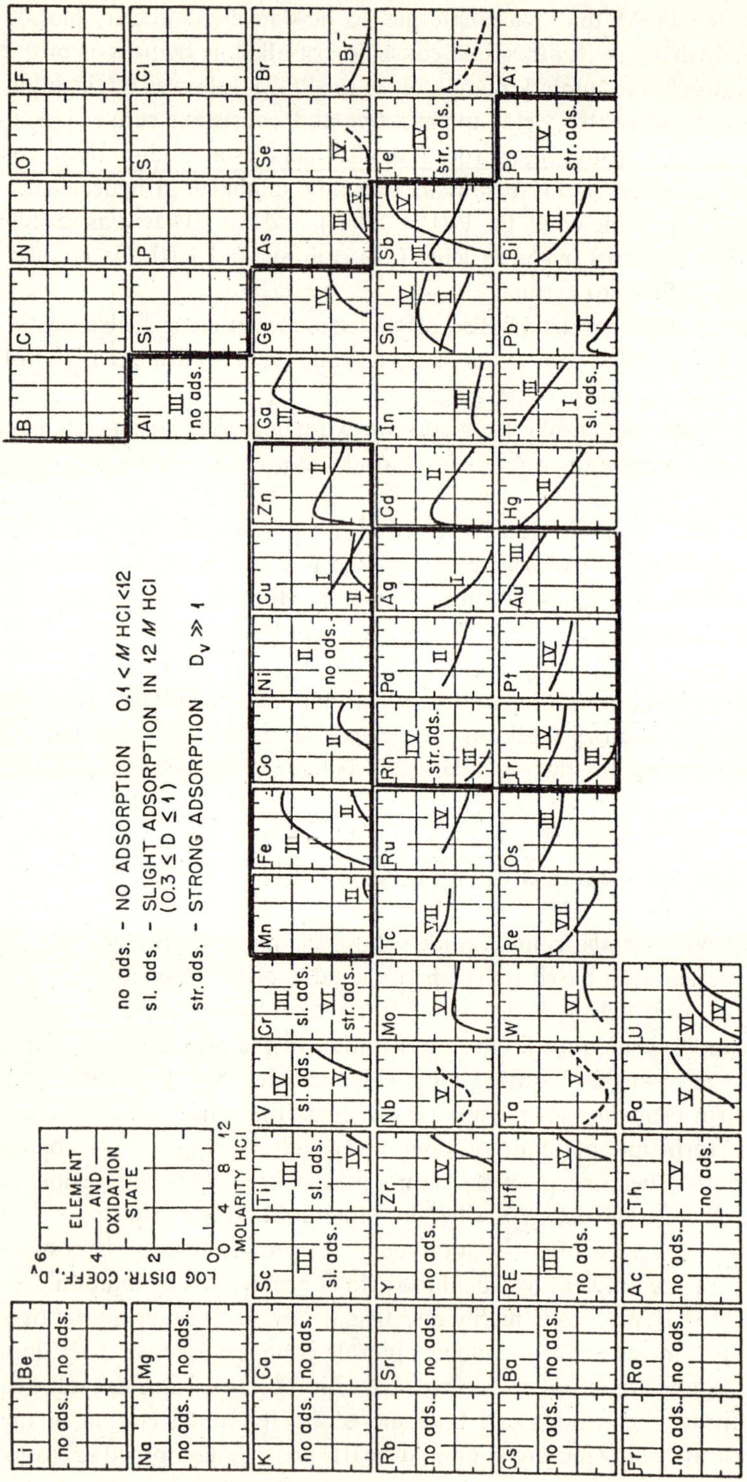

Figure 2.206 Adsorption of the elements from hydrochloric acid.

hydrochloric acid ($\geq$ 6M) and can be separated by fractional elution using 4M HCl to strip Zn^{2+}, 2M HCl to strip Cu^{2+}, and 0.05M to strip Fe^{3+}. The full potential for separating base metals from chloride solution using standard commercial anion exchange resins has not yet been realised.

The recovery of copper from sulphuric acid solutions has been widely researched and it has been found that commercially available chelating resins containing imino-diacetic acid groups show good selectivity for copper over iron(II) at pH 2:

$$RCH_2-N\begin{matrix}CH_2-COOH\\CH_2-COOH\end{matrix} + Cu^{2+} \longrightarrow RCH_2-N\begin{matrix}CH_2-C\overset{O}{\underset{}{}}\\CH_2-C\underset{O}{}\end{matrix}Cu^{2+}.$$

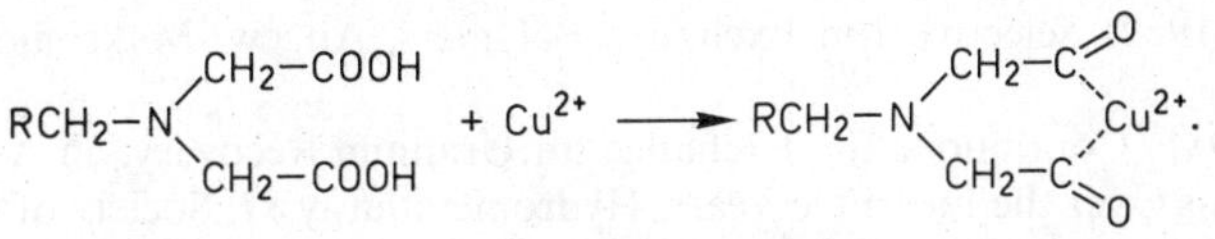

Elution of the resin is possible using sulphuric acid solutions (0.5 – 2N).

The chelating ion exchange resins Dowex XFS 4195 and XFS 43084 are also capable of separating base metals, e.g., Cu, Ni, Co, in sulphate media. The functional groups are based on picolylamine, a weak base structure, which behaves both as a weak base and as a chelating exchanger. Grinstead [15] has presented some promising results.

An alternative technique for copper recovery has been pioneered by Warshawsky [16]. He advocates the use of solvent-impregnated ion exchangers, whereby liquid extractants such as the commercially available hydroxyoxime reagents, LIX63, LIX65N, etc., are supported within the pores of a macroreticular hydrocarbon bead. Vernon and Eccles [17] have also shown that macroreticular beads containing hydroxyoxime, hydroxamic acid and hydroxyquinoline can sorb copper selectively from aqueous sulphuric acid. Solvent-supported systems do have a role to play in hydrometallurgy, but the physical properties of the material will require much improvement to overcome the gradual leaching of the reagent.

Tungsten can be recovered and purified from acid solution by ion exchange. Strong base resins have been used to recover tungstate anions from chloride solution in Portugal; process details are given by Rodrigues et al. [18].

2.13.5 Conclusions

The role of ion exchange in hydrometallurgy has been reviewed. It is seen that the major areas of application are uranium, precious metals, platinum group metals and the base metals. However, it is quite clear that significant further advances will only occur with the design and development of speciality ion exchange resins capable of selective sorption of these and other important metal values. Sufficient work has not yet been done to create selective ligands attached to a polymer backbone capable of sorbing a metal species and then readily releasing the species during elution. It is found, not surprisingly, that the more effective a ligand is for sorbing a metal ion, the more difficult is the elution step. This may be overcome by better design of the stereochemistry of the ligand group or of the polymer backbone structure. Work on the design and synthesis of selective ion exchange materials is very encouraging, based on the number of patents and publications to be found in the open literature.

References

[1] Warshawsky, A. (1982) Selective Ion Exchange Polymers, Angew. Makromol. Chem. *109–110*, 171.

[2] Boydell, D. W. (1981) Continuous Ion Exchange for Uranium Recovery: an Assessment of the Achievements Over the Last Five Years, Hydrometallurgy *81*, Society of Chemical Industry, London, paper E1.

[3] McIntosh, A. M., Viljoen, E., Craig, W. M. and Tyler, J. L. (1982) The Design, Commissioning and Performance of the NIMCIX Section of the Chemwes Uranium Plant, J. South African Inst. Min. Metall. *82*, 177–185.

[4] Laxen, P. A., Becker, G. M. and Rubin, I. (1979) The Carbon-in-pulp Gold Recovery Process – a Major Breakthrough, South African Min. Eng. J. (4152), 43–59.

[5] Naden, D., Bicker, E., Parkin, K. and Whittaker, C. (1987) Development of New Engineering and Process Design for in-pulp Recovery of Metals, in: Separation Processes in Hydrometallurgy (Davies, G. A., ed.) Ellis Horwood, Chichester, pp. 489–500.

[6] Gonzalez-Luque, S. and Streat, M. (1984) in: Ion Exchange Technology (Naden, D. and Streat, M., eds.) Ellis Horwood, Chichester, pp. 679–689.

[7] Ketzniel, Z. and Volkman, Y. (1983) Recovery of Uranium from Wet Process Phosphoric Acid by Ion Exchange, IAEA Symposium, Vienna.

[8] Mehmet, A. and Te Riele, W. A. M. (1984) in: Ion Exchange Technology (Naden, D. and Streat, M., eds.) Ellis Horwood, Chichester, pp. 637–652.

[9] Green, B. R. and Potgieter, A. H., ibid, pp. 626–636.

[10] Edwards, R. I., Haines, A. K. and Te Riele, W. A. M. (1976) The Separation of Gold from Acidic Leach Liquors with Amberlite XAD-7, in: The Theory and Practice of Ion Exchange, Society of Chemical Industry, London.

[11] Warshawsky, A. (1987) Extraction of Platinum group metals by Ion Exchange Resins, in: Ion Exchange and Sorption Process in Hydrometallurgy (Streat, M. and Naden, D., eds.) CRAC Vol. 19, Wiley, New York, pp. 127–165.

[12] Kraus, K. and Nelson, F. (1956) Proc. Int. Conf. on Peaceful Uses of Atomic Energy 7, 113.

[13] Gupta, A. K. and Streat., M. (1975) I. Chem. E. Symp. Series, No. 42.

[14] Haines, A. K., Turnley, T. H., Te Riele, W. A. M., Cloete, F. L. D. and Sampson, T. D. (1973) The Recovery of Zinc from Pickle Liquors by Ion Exchange, J. South African Inst. Min. Metall. *74*, 149.

[15] Grinstead, R. R. (1984) New Developments in the Chemistry of XFS 4195 and XFS 43084 Chelating Ion Exchange Resins, in: Ion Exchange Technology (Naden, D. and Streat, M., eds.) Ellis Horwood, Chichester, pp. 509–518.

[16] Warshawsky, A. (1974) Inst. Min. Metal, *83*, C101.

[17] Vernon, F. and Eccles, H. A. (1976) Some Hydrometallurgical Applications of Hydroxy-Oxime, Hydroxyquinoline and Hydroxamic Acid Solvent Impregnated Resins, in: The Theory and Practice of Ion Exchange, Society of Chemical Industry, London.

[18] Martins, J., Costa, C., Loureiro, J. and Rodrigues, A. (1984) Recovery of Tungsten from Hydrometallurgical Liquors by Ion Exchange, in: Ion Exchange Technology (Naden, D. and Streat, M., eds.) Ellis Horwood, Chichester, pp. 715–723.

3 Ion Exchangers in Pharmacy, Medicine and Biochemistry

3.1 Ion Exchange Resins and Polymeric Adsorbents in Pharmacy and Medicine

Marico Pirotta
Rohm and Haas Compamy, Milano, Italy

Introduction

The use of ion exchange resins and polymeric adsorbents in the pharmaceutical and medical industries is very diverse and technologically well-advanced. Many pharmaceutical companies have been pioneers in developing industrial uses of synthetic polymers such as ion exchange resins and polymeric adsorbents. The pharmaceutical and medical applications of these products utilise many – and sometimes unusual – aspects of ion exchange and adsorption technology.

Synthetic ion exchange resins were first produced in the 1940s but, for many years prior to this the medical and pharmaceutical industries had been using materials with ion exchange and adsorbent properties, for example, charcoals, alumina, kaolin clays, pectin, bentonite, alginates, homogenised milks and extracts of organs, leaves, roots, and herbs. All of these materials have been used for many centuries as pharmaceuticals, drug carriers or pharmaceutical processing aids.

In general, the applications of ion exchange resins and polymeric adsorbents can be classified as follows:

- processing aids for pharmaceutical products
- galenic formulation ingredients
- agents for protein chromatography
- biologically active substances
- blood detoxification agents
- analytical and diagnostic agents
- cosmetic agents
- enzyme immobilising agents
- pre- or post-treatments to ultrafiltration, reverse osmosis and electrodialysis
- bioprocessing agents.

3.1.1 Processing aids for pharmaceutical products

This is perhaps the largest industrial field of application of ion exchange resins and polymeric adsorbents. In the following sections the major techniques will be illustrated by reference to specific antibiotics that are processed by these methods.

The processing of antibiotics normally involves extraction, deashing, decolourisation, concentration, chromatography and salt conversion. Ion exchange resins and polymeric adsorbents are frequently employed in these steps for processing antibiotics produced from fermentation media. The choice of one technology or another is always based on experimental data. Important parameters to be considered before deciding priority in the experimental programme are:

- Antibiotic solubility in water
- Nature and concentration of the major impurities present in the extraction medium
- Possibility of adjusting the pH of the medium without de-naturing the desired product
- Possible pre-treatments such as membrane filtration, reverse osmosis and electrodialysis
- Molecular weight of the antibiotic
- Polarity of the antibiotic
- Chemical groups attached to the antibiotic molecule which could be used for specific reactions
- Viscosity and density of the extraction medium
- Suspended solids
- Nature of the product in the fermentation medium, i. e., endogenous or exogenous
- Type of fermentation organism, i. e., bacteria, fungi, yeast, etc.

3.1.1.1 Examples of specific antibiotics

Streptomycin

$C_{21}H_{39}O_{12}$ water-soluble organic strong base from streptomyces (exogenous).

Extraction

"Dog-leg"-shaped columns operated in a merry-go-round system, containing a methacrylic ion exchange resin such as Amberlite IRC-50, are normally used to extract streptomycin from its centrifuged broth.

These special columns, which are operated upflow, are equipped with some means of agitation, such as stirrers, propellers, etc., to prevent the settling of the ion exchange resin and to allow particulate matter to pass through the column without clogging.

Each ion exchange resin unit is equipped with a syphon, which returns the ion exchange resin particles that might be carried over during both upflow loading and washing. The retained syphoned resin can be hydraulically transferred back into the column at any desired moment.

During the loading, the resin bed is allowed to expand up to 25% of its original volume.

The ion exchange occurs between the two salt-form guanidine groups of streptomycin and the Na-form of the weak acid resin.

The elution is achieved by means of either strong acids – classically, sulphuric acid – or, more recently, weak acids such as oxalic or carbonic acids are used prior to H_2SO_4 elution, to eliminate the impurities of divalent cations present on the resin. For cost and yield reasons, the carbonic acid treatment is preferred.

Such a treatment is only possible on a polymethacrylic acid resin (Amberlite IRC-50), due to the particularly high pK value of such a resin. A carbonic acid solution of pH 3.5 to 4.5 is passed under moderate pressure downflow through the IRC 50 column. Divalent cations are eluted with very limited amounts of streptomycin.

Depressurisation of the solution leaving the column precipitates $CaCO_3$, which can be filtered off. The necessary quantity of CO_2 is added to the filtered solution under pressure so as to reach the desired pH (3.5 to 4.5); this can be recycled for the next treatment. The streptomycin is subsequently eluted with H_2SO_4.

After the first elution, the CO_2 solution contains a low percentage of streptomycin which is eluted at the same time as the inorganic cations. However, after the bicarbonate solution has been recycled its salinity increases continuously with the recycle, which causes a salting out effect on the streptomycin, reducing its solubility. The final consequence is that the streptomycin eluted during the first cycle is reabsorbed onto the Amberlite IRC-50 during the second and subsequent cycles. Although, after a certain number of recycles, the H_2CO_3 solution has to be discarded, no significant losses of streptomycin occur in the overall extraction process.

Gel chromatography

The main eluate from the Amberlite IRC-50 contains streptomycin sulphate along with undesirable oligomer sulphates with similar basicity but different molecular weight. By using a highly-crosslinked strong acid gel-type ion exchange resin (Amberlite IR-122 or IR-124) in the H^+-form in a cocurrent operation, a molecular exclusion phenomenon is obtained. The lower molecular weight oligomer sulphates enter the ion exchange resin matrix and are retained, whilst the higher molecular weight streptomycin sulphate is large enough to be excluded from the resin matrix and consequently passes through the resin bed unretained. The streptomycin sulphate collected from the column outlet, although purified from its oligomers, is still contaminated by free sulphuric acid and some colour bodies.

Gel chromatography is obtained by using demineralised water as a mobile phase. The strong acid ion exchange resin is regenerated with 2% H_2SO_4 solution.

Decolourisation

The colour bodies can be removed using a strongly basic gel-type resin with high porosity (Amberlite IRA-401S, Duolite A147, Imac A26). The resin is operated in its sulphate form and regenerated with sodium sulphate. From time to time it is necessary to carry out double regeneration using caustic soda and sulphuric acid.

Deacidification

The above-mentioned excess of sulphuric acid is removed from the streptomycin sulphate solution by passing it through a weak base resin in its free base form. The basicity of the resin chosen should be less than that of streptomycin to ensure that only the free acid is removed. Resins typically used are Amberlite IRA-45 or Amberlite IRA-68. The resin is regenerated with caustic soda.

Final purification and concentration

A preconcentration of the deacidified solution by reverse osmosis might be economically interesting prior to the evaporation step, necessary to produce bulk powdered streptomycin sulphate.

Sterilisation and haze removal can then be achieved by passing this RO concentrate through an ultrafiltration plant. Ultrafiltration treatment is effective, also when applied directly on the deacidified solution as such, whenever RO pretreatment is not used [1].

Gentamycin

Gentamycin and other weakly basic amino glycosidic antibiotics (AGAs) are commercially purified according to the following scheme:

- whole broth
- pH adjustment
- flocculation
- filtration

- extraction using a carboxylic ion exchange resin
- elution
- decolourisation of eluate by means of a strong base ion exchange resin
- concentration by either reverse osmosis or vacuum stripper
- colloids removal by ultrafiltration
- ion exchange chromatography using chromatographic-grade ion exchange resin (weakly acidic or strongly anionic)
- concentration by reverse osmosis of the active fractions
- crystallization or sterile filtration

Extraction

AGAs are normally extracted by the sodium or more usually the ammonium form of a carboxylic ion exchange resin such as Amberlite IRC-50 or Duolite C 464. Unlike streptomycin, the AGAs can also be eluted from the resin using diluted ammonia solution, this allowing a more selective elution and hence higher purity of the AGA in the eluate. The disadvantage of this, however, is that the volume of the eluate peak is greater than of sulphuric acid-produced eluant. After decolourisation the ammonia eluate volume can, however, be easily and economically reduced by reverse osmosis and so the total economy of the process remains in favour of the ammonia elution route. Generally, in cases where the choice lies between greater purity or higher eluate concentration the choice is always in favour of the higher purity. Although in the case of AGAs the elution routinely coincides with the regeneration step of the resin, appropriate regular cleaning treatments are also employed to maintain resin performances.

Decolourisation

For the decolourisation of AGAs it is generally preferred to use a strong base macroreticular and/or a gel-type strong base resin in the hydroxide form. The high pH of the eluate from the cation due to excess ammonia does not effect the decolourisation performance of such a resin.

The regeneration of the decolourising resin is normally achieved using a 2% sodium hydroxide solution. This lower concentration is chosen to minimise the salting out effect during the elution of the organics from the resin. Frequently it has been found that the best decolourisation performance can be achieved by using a macroreticular resin (Amberlite IRA-900) followed by a high porosity gel-type resin (Duolite A147) at a flow rate in the range of $5-10\ BV \cdot h^{-1}$.

Ion exchange chromatography

The development of the industrial production for the purification of AGAs has obliged pharmaceutical companies to scale up from laboratory chromatographic equipment to industrial columns containing thousands of litres of chromatographic-grade resins. These resins have a very narrow particle size distribution, typically between 100 and 200 US mesh, or even between 200 and 400 US mesh. The resins are high-purity chemicals matching typical analytical grade quality.

These industrial chromatographic separations are usually performed at space velocities ranging from 0.1 to 0.01 h^{-1}. At these flows the pressure drop is maintained within very reasonable limits, typically 1 to 2 bar per metre of resin-bed depth. In order to accomplish this type of separation with the optimum yield of extraction careful engineering design must be used [2].

Cephalosporin C

In 1975 Ciba-Geigy constructed and put into operation an industrial plant using a polymeric adsorbent for the extraction of cephalosporin C from its fermentation broth. This plant, located in Torre Annuziata, Italy (near Naples), was the first in the world using such large quantities of a polymeric adsorbent. The resin used was Amberlite XAD-2. The application of this resin was not a simple extraction; in fact, in one operation it was possible to eliminate the organic salts and free sulphuric acid present in the filtered broth and at the same time obtain a chromatographic separation of the cephalosporin C from desacetyl cephalosporin C (DCE) and, to a lesser extent, also from desacetoxy cephalosporin C (DCX).

Both antibiotic producer and resin manufacturer learned at lot from this experience. As a result, the close co-operation between Ciba-Geigy and Rohm and Haas in Europe with the aim of improving the process resulted in the development of two new resins between 1975 and 1986. These were, respectively, Amberlite XAD-1180 and Amberlite XAD-16. These two products are used in other chemical and pharmaceutical processes, as well as in improved industrial processes for the production of cephalosporin C. In the time between XAD-2 and XAD-1180 several cephalosporin C producers successfully used Amberlite XAD-4, which has almost twice the capacity of Amberlite XAD-2 but different hydraulic properties. Amberlite XAD-4 is still widely used in this application.

Amberlite XAD-1180 was successfully developed to improve the cephalosporin C selectivity and hence the chromatographic separation performance of XAD-2 and XAD-4.

Amberlite XAD-16, first used industrially in 1986, combines in a single resin most of the good properties of its predecessors.

It should be noted, however, that the resin is only one variable in the extraction process; therefore, in spite of the availability of such an advanced resin, individual producers sometimes find it more economic to use one of the earlier polymers.

The hydraulic design of the equipment, which will be discussed later, and the availability of certain chemicals in the factory, can also affect the choice of the process.

The original process of cephalosporin C extraction can be schematically summarised as follows [3]:

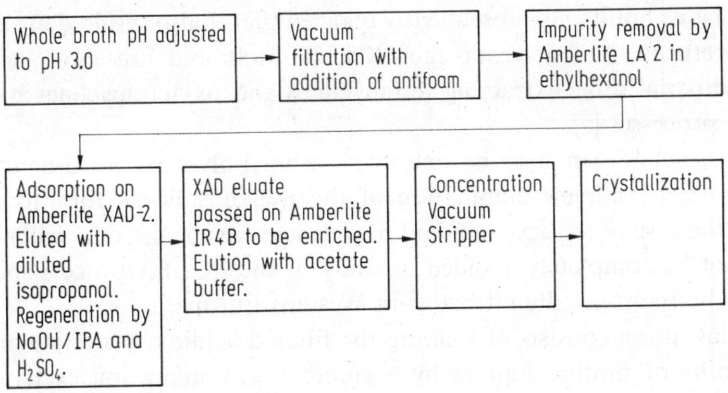

Today, a possible scheme of the extraction could be:

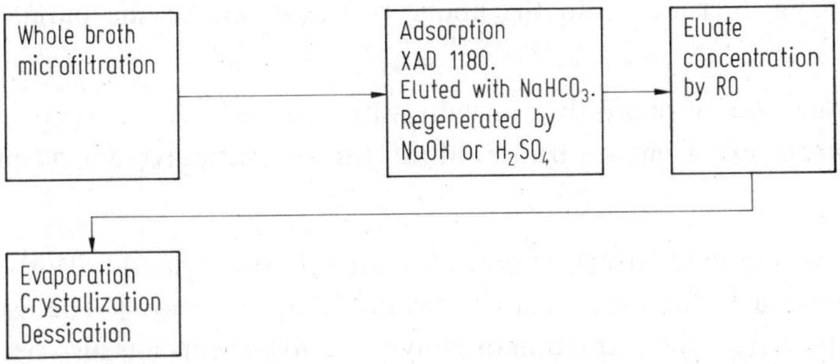

One can see that considerable process simplification has occurred. This is due to several factors acting at the same time: compositional changes of the cultural medium, increased broth potency, development of new resins, development of new engineering, field experience and, last but not least, the availability of special membranes.

In addition to the simplification of the extraction process, the quality of the final cephalosporin C salts has been significantly improved.

One goal common to all cephalosporin C producers in recent years has been to improve the ecological aspect of antibiotic extraction. Following the success of the processes described above some pharmaceutical companies have started basic investigations into the possibility of substituting more ecologically preferable processes based on polymeric adsorbents and membrane technologies for the existing solvent extraction processes based on fermentation products [4].

Tetracyclines

Although the use of polymeric adsorbents in the extraction of tetracyclines has been known for many years, these materials were never used industrially for such an extraction until very recently, due to the very low cost of tetracyclines and the simplicity of the other existing extraction processes.

A few years ago Heropolitanski and his colleagues at the Industrial Chemistry Research Institute in Poland invented a resin-based process for the extraction of oxytetracycline from its mother liquors. This was then studied by POLFA for ecological reasons but it was subsequently realized that it also resulted in significant economical returns. A few years ago POLFA built and are now successfully operating industrial units extracting tetracyclines and oxytetracyclines by ion exchange resin processes [5].

Chlorotetracycline can also be extracted using polymeric adsorbents, which allows the almost complete elimination of the use of solvents for this product. Although in the case of tetracyclines and oxytetracycline the use of a water-miscible solvent cannot be completely avoided, a study of the POLFA process shows that it might also be economically attractive in Western Europe.

The new invention consists of treating the filtered acidic tetracycline or oxytetracycline broths or mother liquors by a suitable scavenging ion exchange resin prior to feeding it to a suitable polymeric adsorbent. The recycling of the optimum cuts makes the whole process economically attractive.

The following alternatives to the above processes are at the moment under investigation:

- the combined use of microfiltration/ultrafiltration and reverse osmosis
- the combined use of microfiltration/ultrafiltration and polymeric adsorbents.

Tylosine

The previously-described extraction procedures based on polymeric adsorbents were all conducted on acid-filtered media (broths and mother liquors). The use of acid pH to affect positively the extraction on polymeric adsorbents has also been widely used in chemical processes. In these cases, it is frequently possible to elute the desired products by changing the pH to neutral or alkaline values. One exception to the above is the tylosine extraction process developed by a European pharmaceutical company, for which a patent application has been filed: in this process the antibiotic is adsorbed at a highly alkaline pH and eluted by decreasing the pH.

Clavulanic acid

Beecham Pharmaceuticals (UK) have filed several patents on the extraction of this product. Here we shall only consider part of one of these patents, in which the effect of "salting out" on the performance of the polymeric adsorbent is described.

In this process a clavulanic acid solution, obtained during the extraction process from whole broth, containing a relatively high concentration of sodium chloride, is passed onto a column of Amberlite XAD-4. Clavulanic acid is adsorbed onto the resin due to the influence of the high salinity. The elution is then performed by simply passing demineralised water through the bed: the resulting change in salt concentration causes the elution of the product [6].

Erythromycin

This antibiotic is traditionally extracted from its filtered broth using a water-insoluble solvent. Very recent unpublished studies have demonstrated that a combination of microfiltration/ultrafiltration and polymeric adsorbents is a valid alternative to this classical solvent-extraction technology.

The major advantage seems to be the possibility of avoiding the classical filtration of the whole broth, which is the most expensive step in the process due to losses of activity during the long filtration period. The use of a suitable polymeric adsorbent would significantly reduce the quantity of solvent used and the erythromycin losses.

It should be noted that without pretreatment by micro- or ultrafiltration an extraction process based on ion exchange resins and/or polymeric adsorbents becomes complicated and unattractive compared to the existing technology [7].

Ampicillin

Ion exchange resins are also used in the preparation of antibiotics such as ampicillin, cephaloridin, etc.

The process of purifying ampicillin by ion exchange resins has been described in the literature by Bristol-Myers. This process – using Amberlite LA 1, a weakly basic liquid ion exchanger – has become a classical one in the field of semi-synthetic antibiotics. The process consists of the following steps:

- Ampicillin produced by the phenylglycine chloride/hydrochloride route, is extracted with water from the chlorinated solvent used for the synthesis.
- Ampicillin is then precipitated from the aqueous phase as a sulphonic salt and filtered off.
- This precipitate contains up to a few % of dimethyl aniline (DMA) hydrochloride which has to be reduced to less than 5 ppm.
- The precipitate of ampicillin is added to a small quantity of demineralised water to obtain a slurry.
- The ampicillin slurry is reacted with a solution of MIBK (methyl isobutyl ketone) containing Amberlite LA 1 in the hydrochloride form.
- Due to the presence of the added water, ion exchange takes place and the sulphonic acid is exchanged with the hydrochloric acid giving ampicillin hydrochloride and LA 1 sulphonate. (The affinity of the sulphonic acid for the organic phase is very important for the yield of the reaction).
- DMA/HCl is extracted into the organic phase and eliminated from the LA 1/MIBK solution by contacting this solution with Amberlyst 15 (a strong acid resin) in the sodium form.
- Due to its high molecular weight Amberlite LA 1 does not react with Amberlyst 15 but DMA does. The LA 1 solution is then regenerated with caustic soda and hydrochloric acid in that order. The exchange reaction between the LA 1/MIBK and the ampicillin sulphonate water slurry, can be followed visually since the ampicillin crystals change their configuration [8].

This application of ion exchange resins was almost completely abandoned when the mixed anhydride method of ampicillin synthesis was developed; however, the basic idea has been taken and applied in other semi-synthetic antibiotic syntheses and extractions.

3.1.1.2 Vitamins

Vitamin B_{12} (Hydroxy and cyanocobalamin)

Vitamin B_{12} is recovered by a chemical extraction process applicable in principle to all endogenous materials. The major problem with such an extraction is loss of yield. Activity can be lost during cell disruption, extraction, or separation, etc.

The cell debris obtained by mechanical disruption of the cells tends to flocculate with most resins. A methacrylic carboxylic resin, however, can be used on the slurry of cell debris to perform a batch extraction. The debris and solids can then be removed from the resin by, for example, spraying pressurized water onto the resin on a vibrating screen.

Elution can be accomplished using inorganic buffers or diluted solvents. An ion exchange resin or polymeric adsorbent decolourisation step is also normally required. Very recently, some newly developed polymers have been successfully tested with the aim of improving the extraction yield of this expensive product by selectively complexing and precipitating undesired proteinaceous materials.

Whilst the same original extraction process was applied by most of the vitamin B_{12} producers, nowadays each company has developed a more specific process to fit the requirements of their own specific strain [9].

Vitamin B_1 (Thiamine)

The hydrochloride salt of this vitamin is obtained by treating (after carbon decolourisation) the sulphate salt solution with a suitable ion exchange resin in the hydrochloride form. The conversion of the vitamin sulphate to the hydrochloride form is complete. Studies have been initiated on the possible influence of ultra/microfiltration pretreatment to increase the activated carbon efficiency [10].

Vitamin C from saccharose

Ion exchange resins are mostly used in the deashing and chromatographic separation steps necessary to synthesise sorbitol, the precursor of vitamin C, in its synthesis from saccharose. In addition, chelating resins are also used to remove heavy metals leaking from the hydrogenation step.

3.1.1.3 Alkaloids

Opium alkaloids

The opium alkaloids are extracted from poppy syrup by either ion exchange resins/polymeric adsorbents or solvents. The ion exchange resin technique is also applicable to other types of natural alkaloids. There are only a few ion exchange processes for the extraction of opium alkaloids and all are based on strong cationic sulphonic resins.

- The classical process consists in extracting poppy alkaloids by means of a highly porous gel resin. Due to the fast exchange kinetics of this polymer and its high affinity for the opium alkaloids, this extraction is performed with a relatively low rate of organic fouling of ion exchange groups on the matrix. The elution is achieved using diluted solutions of sulphuric acid in sequential steps. The most diluted solution is used to remove the weak cationic impurities which have also reacted with the sulphonic groups, but at low enthalpy. A higher concentration of sulphuric acid elutes the alkaloids with a reasonable degree of selectivity and finally the solution with the highest sulphuric acid concentration elutes the strongly bound cationic impurities. Caustic soda cleaning is often necessary.
- Macroreticular sulphonic resins are also used instead of the above-mentioned gel-type. The choice of the resin is determined by the pretreatment of the poppy syrup.
- Very recently, microfiltration has been successfully tested as a preconditioning to the ion exchange extraction. Ultrafiltration is also under test as a possible way to further purify the eluates from the strong acid resins. The use of reverse osmosis for the concentration of relatively pure diluted alkaloid solutions is likely to develop.
- At a later stage in the purification process of alkaloids, polymeric adsorbents are used to separate products – for example codeine and morphine – from the impurities of their synthesis.

Ergot alkaloids
Lysergic acid and other similar alkaloids can be extracted by polymeric adsorbents, both from fermentation broths and natural product aqueous extracts.

Chromatographic separations can be achieved by passing buffered solutions of alkaloids, at a pH lower than 5, over a carboxylic chromatographic exchanger such as Amberlite CG 50 Type 1, in the hydrogen form. The resin, which is undissociated under these conditions, separates the organic acid alkaloids by a hydrogen-bonding mechanism [11].

3.1.1.4 Nucleotides

These strong acid phosphoric esters of nucleosides are also known as adenylic, guanylic, thymidylic, cytidylic and uridylic acids. In general, these acids are obtained by either the basic, or more frequently, enzymatic hydrolysis of RNA (ribonucleic acid). Alkaline hydrolysis gives isomeric nucleotides esterified at C2 or C3. The guanylic, cytidylic and uridylic acids derived from alkaline hydrolysis of RNA are also mixtures of nucleoside 2 and 3 phosphates. Pancreatic ribonuclease degrades RNA to the nucleoside 3 phosphates.

Example of nucleotide:

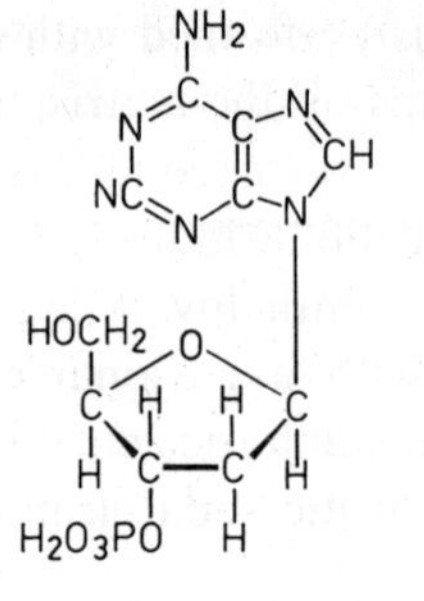

Deoxy-3-adenylic acid
(deoxyadenosine-3-phosphate)

Partial hydrolysis of nucleic acids yields compounds in which ribose, or deoxyribose, is conjugated to purine or pyrimidine base (the nucleosides).

Example of nucleoside:

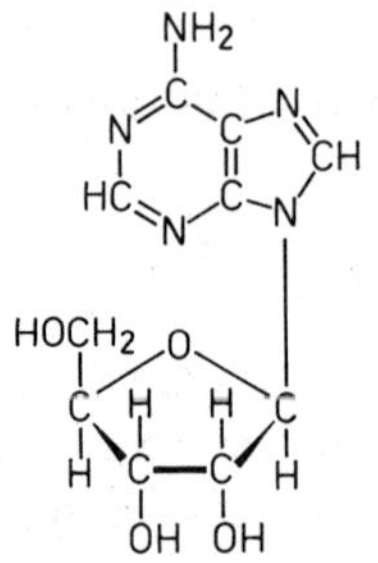

Adenosine

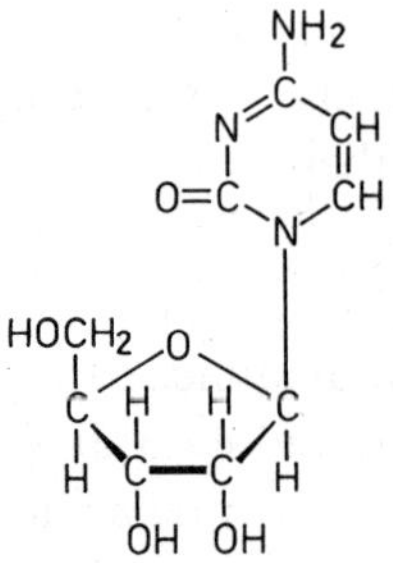

Cytidine

Nucleotides are normally extracted from an enzymatically digested, filtered liquor by means of a strong base anion exchange resin. This strong base ion exchange resin can either be of a low crosslinked gel-type, which is often physically fragile and easy to foul with organics, or, as used more recently, of a highly crosslinked highly porous type, such as Amberlite IRA-904. This resin is physically strong, has a very high resistance to organic fouling and elutes easily. The normal eluant is dilute NaCl. The eluate from this resin must be treated to eliminate the excess sodium chloride. This can be achieved by using conventional gel sulphonic cation resin and a weakly basic acrylic resin; however, a cheaper route is by sodium chloride molecular exclusion. In this method the solution containing a few percent of free sodium chloride is passed over a low crosslinked strong acid resin in the sodium form. The nucleotides are excluded during the water washing and, in the process, purified from NaCl.

The nucleotide fraction which contains only few ppm of NaCl can then be polished using a strong cation and weak anion mixed bed. Since only ppm levels of sodium chloride have to be removed now, significantly smaller columns have to be used. The strong base resin used in the initial part of the process often acts as a scavenger reducing the organic fouling of these final de-ashing units.

Final decolourisation of the demineralised solution can be achieved using either phenolic or styrenic polymeric adsorbents. Very recently, investigation of the possible use of hollow-fiber ultrafiltration has been started.

A microfiltration treatment for the filtered, digested RNA solution is also under study, with the aim of increasing the life and performance of the ion exchange resins.

The demineralised nucleotides can be further treated to obtain, by chromatographic separation, each desired nucleotide component. A high-porosity, fast kinetic, strong base gel resin is used for this purpose. It is sometimes possible to carry out this chromatographic separation on the crude microfiltered substrate, depending on the type of hydrolysis system and on the purity of the RNA used.

3.1.1.5 Amino acids and amino acid hydrolysates

For many years products such as lysine, glutamic acid and tryptophan have all been successfully extracted from their fermentation broths using ion exchange resins. More recently, other amino acids such as methionine, hydroxytryptophan, phenylglycine, phenylalanine, etc., have also been industrially extracted using similar techniques.

Phenylalanine industrial extraction has been developed only very recently. This amino acid is used for the synthesis of aspartame. Polymeric adsorbents are frequently used in the extraction and recovery processes for phenylalanine.

Ajinomoto has very recently invented and patented the use of reverse osmosis for the selective concentration of lysine from aqueous ammonia solution. The use of a low ammonia rejection factor reverse osmosis membrane, reduces the amount of ammonia to be recovered by distillation from the lysine eluate, thus improving the ecology and the economics of this extraction process.

Amino acid hydrolysates

At least three processes can be used to produce the pool of salt-free amino acids:

1. Direct hydrolysis of a protein, often casein, using an ion exchange resin with a subsequent selective elution of desired amino acid.
2. H_2SO_4 hydrolysis of the proteins, followed by a strong acid cation exchanger in the ammonium form to extract the desired amino acid.
3. Chemical or enzymatic hydrolysis of the proteins, followed by classic separation of the amino acid by precipitation of the acidic and neutral components.

Route 1
A protein or mixture of proteins is incubated in a pressurized, stirred reactor for a few hours at 120 °C with a macroporous strongly acidic sulphonic resin in the H^+-form. The protein is hydrolyzed to peptides and amino acids.

At the end of the hydrolysis, the resin is hydraulically transferred into a column and eluted with a gradient, or more frequently a stepwise sequence, of ammonia solutions. The ammonia is then stripped under vacuum and the amino acid fraction ultrafiltered to remove traces of peptides.

Route 2
A protein or mixture of proteins is hydrolyzed using a mineral acid, frequently H_2SO_4, to the corresponding amino acids. The solution of amino acids obtained is passed over a strong acid resin in the NH_4^+-form. The amino acids are fixed on the resin and the ammonium sulphate so produced can be collected at the column outlet and used as a fertilizer. The amino acid-loaded column is then eluted stepwise with diluted ammonia solutions to obtain free amino acids without any inorganic salt contamination. The solution so obtained is processed as described above in Route 1.

Route 3
When solid material proteins, such as keratines, are hydrolyzed by concentrated HCl, precipitation of acidic and neutral amino acids can be achieved by using the stoichiometric amount of a liquid ion exchanger such as Amberlite LA 2. The hydrochloride form of the basic amino acids cannot be completely transformed into the free base form using a strongly basic ion exchange resin in the OH-form. It is, therefore, necessary to use a strong acid resin eluted with ammonia, as described above in Route 2.

Enzymatic hydrolysis of proteins into amino acids can also be used. Preferentially, the protease is immobilised and therefore does not contaminate the final solution, so that it can also be reused. Ion exchange resins or chemical precipitation can then be used to extract the salt-free amino acids from the ultrafiltered hydrolyzed solution. Enzymatic hydrolyses are, in general, economically and qualitatively less attractive than chemical/ion exchange resin hydrolysis [12].

3.1.1.6 Peptide and protein chromatography

It is a well known fact that the extraction, purification and chromatography of proteinaceous materials is a very difficult task. The selection of an ion exchange resin or adsorbent polymer for such a duty is therefore a difficult problem. The major difficulty arises because biologically active proteins and peptides can be denatured by many different things, for example, changes in their three-dimensional structure, ionic charge, chemisorption, etc. The adsorption of a proteinaceous material onto a polymer occurs either by electrostatic forces or by ion exchange;

in both cases the ionic configuration and charge of the proteinaceous material can be changed. Sometimes this change is reversible, often it is not. The job of the polymer chemist is therefore to select a matrix which is compatible with the polarity of the protein in a way that the biological activity of the protein is not lost by denaturation during adsorption and desorption. Once this matrix is synthesised then it can potentially be used for protein immobilization, purification and chromatography, etc. Protein immobilization is frequently obtained by a covalent bonding reaction between the bioprotein and the functionalized polymer.
The following rules are often applicable:

1. The higher the polarity of the adsorbent, the less likely is the denaturation of the protein.
2. Non-polar adsorbents have the highest adsorption capacity.
3. The adsorption of proteinaceous materials is favourably influenced by:
   - Dissolved salt background (salting out effect)
   - pH of the solution being close to the isoelectric point
   - appropriate pore-size distribution, pore volume and pore shape of the adsorbent (very difficult to assess)
   - sufficient contact time with the adsorbent to allow good diffusion
   - presence of protein activity stabilizers, e. g., heavy metals, SH groups, Ca, etc.
   - absence of oxygen either in the resin pores or in the protein solution
   - sterility of the equipment and of the polymer.
4. The desorption of proteinaceous materials is favourably influenced by:
   - Sufficient contact time to allow rediffusion
   - low ionic strength of the eluant solution
   - pH of the eluant solution being as different as possible from the isoelectric point of the protein
   - choice of solvents, salts and/or buffer which best interact with the desired product
   - presence of organic species, e. g., detergents, amines, carboxylic acids, solvents, etc., which do not significantly denature the biologically active product.

Proteins carry a large number of ionic groups and to assess their pK values is often a difficult task. However, the nature of the ionic groups is a fundamental parameter for the choice of the ion exchange resin to be tested for a specific extraction. The pK values of the acidic and basic groups in the proteins (see Table 3.1) are different from those of the corresponding amino acids.

The isoelectric point is also an important factor in the choice of the extracting medium for a given protein.

Combinations of micro- and ultrafiltration, reverse osmosis and electro-dialysis with ion exchange resins and adsorbents are common in protein extraction and chromatography procedures. Continuous electrophoresis has also been commer-

Table 3.1 Characteristic pK values for acid and basic groups in proteins. (E. J. Cohn and J. Tedsall, Proteins, amino acids and peptides as ions and dipolar ions, Reinhold Publishing Co. New York, 1942)

	pK_a
Carboxyl (terminal)	3.0
Carboxyl (aspartic)	3.0
Carboxyl (glutamic)	−4
Imidazolium (histidine)	5.6
Amino (terminal)	7.6
Amino (lysine)	9.4
Guanidinium (arginine)	11.6
Phenolic hydroxyl (tyrosine)	9.8
Sulfhydryl (cysteine)	−8

Table 3.2 Adsorption of viruses by ion exchange resins. (See Chapter 13 of "Ion Exchangers in Organic and Biochemistry" by Calmon and Kressman, Interscience Publishers, N. Y., 1957)

Virus	Medium	Ion exchange resin	Comments
1. Lansing poliomyelitis	tissue extract	Amberlite IRP 67 C1	Virus adsorbed and eluted in viral state
2. Lansing poliomyelitis	faeces	Amberlite IRP 67 C1	Virus adsorbed and eluted in viral state
3. PRA influenza	allantoic chick embryo fluid	Amberlite IRP 64 (Na/H) pH 5.6	Virus adsorbed and eluted in viral state
4. Lee influenza	allantoic chick embryo fluid	Amberlite XE-117 (C1)	Virus adsorbed and eluted in viral state
5. Mumps	allantoic chick embryo fluid	Amberlite XE-117 (C1)	Virus adsorbed and eluted in viral state
6. Herpes simplex	allantoic chick embryo fluid	Amberlite XE-117 (C1)	Virus adsorbed and eluted in viral state
7. Coxsackie	allantoic chick embryo fluid	Amberlite XE-117 (C1)	Virus adsorbed and eluted in viral state
8. Bacteriophage	extract	Amberlite IRA-400 (Phosphate)	Virus adsorbed and eluted in viral state

cially available for a number of years. The use of selective complexing polymers is also an instrument in this new, complicated, sophisticated field of science.

Some biologically active proteins, enzymes and coenzymes currently purified by ion exchange resins are:

Biological species:	*Resin:*
Transferrin	Amberlite XAD 1180
Lysozyme	Amberlite IRC-50/XAD 2
Insulin	Amberlite CG 50
Pepsinogen	Amberlite IRA-938
Pepsin	Amberlite IRA-938
Urease	Amberlite IRC-50
Asparaginase	Amberlite IRA-938
Amylase	Amberlite XAD 1180
S. A. M.	Amberlite IRC-50/CG50/XAD 4
Glucose isomerase	Amberlite IRA-904
Cytochrome C	Amberlite IRC-50
Heparin	Amberlite XE 268/IRA/904
Lipase	Various
Serum albumin	Amberlite IRC-50
Monoclonal antibodies	Amberlite CG 50
Entherotoxines	Amberlite IRP-276

Viruses can also be adsorbed by ion exchange resins (see Table 3.2).

3.1.1.7 Other uses of resins in protein chemistry

Merrifield synthesis
Some peptides can be synthesized using the Merrifield synthesis. In this method, a highly porous chloromethylated styrene-divinylbenzene copolymer is reacted with the NH_2 group of the first amino acid of the peptide, obtaining an insoluble derivative of the amino acid. This is then used as a base for constructing the peptide structure. Various modifications of this technique are described in the literature.

Monoclonal antibodies (MCA)
MCA can be immobilised on polymers by covalent bonding and these can subsequently be used to selectively extract biologically active species. MCA can also be trapped in hollow fiber membranes which can then be used to selectively extract biologically active species.

The future of protein purification
The development of biotechnology has launched the field of protein separation, but the status of the current art in polymer chemistry and hydraulic engineering has not yet developed sufficiently to satisfy current needs. A lot of effort is still required to optimize both the chemistry and utilization of polymers for protein separations. Up until now, people have tried to develop large-scale analytical equipment and chemicals with varying degrees of success. High-performance liquid chromatography (HPLC) seems to be one of the most promising technologies to

have been developed. However, the available mechanical engineering expertise is currently insufficient to build economically attractive very large HPLC installations that can produce tons of material per day.

Functionalised silicas and celluloses are still the most widely-used chromatographic media for protein and peptides separations; however, the cost of these materials is very high for industrial application. It is expected, however, that in the future HPLC will be developed for industrial application using new chromatographic media to accomplish fast and effective separations. Liquid membrane technology also seems to be a very promising technology for the industrial future of protein purification [13].

3.1.2 Ion exchange resins in medicine and galenic applications

Ion exchange resins have been used for many years in galenic applications, e. g., as tablet disintergrants, for taste and odour masking, for drug stabilization, for sustained release, etc. Although they have been used in medical applications for treating, e. g., hyperkaliemia, hypercholesterolemia, as well as oedema reduction therapy, etc., in the following sections we shall limit our presentation to the most representative applications only: sustained release and hypercholesterolemia.

3.1.2.1 Sustained release

There are many techniques based on ion exchange resins which can be used for obtaining controlled release of a drug. In order to illustrate how ion exchange resins can be used in this area let us list how the intrinsic parameters of the resin affect the drug release pathway.

The kinetics of drug adsorption and desorption are influenced by physical parameters, e. g.:

Resin particle size
Resin crosslinking
Resin moisture
Resin porosity
Drug hinderance
Temperature
Contact time
Resin particle shape

and chemical parameters, e. g.:

Resin type (strong acid, strong base, etc.)
Resin ionic form
pH and salinity of solutions in contact with the resin
Drug ionic form and its stability.

The deeper the level of penetration of a drug into a resin particle the longer will be its rediffusion time. Therefore, different sizes of particle of the same resin can be mixed to obtain the desired drug elution profile.

The lower the crosslinking level of a particular resin type the higher will be its kinetics and vice versa. Therefore, different crosslinking levels of a particular resin type can be blended to obtain the desired release profile.

Further, combinations of different crosslinking and different particle sizes can be used.

3.1.2.2 Adsorption of adrenolytic substances

The influence of different parameters on the adsorption of adrenolytic substances (A. I.) on strong acid resins is shown as an example in Table 3.3.

Table 3.3 CORETAL hydrochloride and PRACTOLOL hydrochloride, absorbed on Na-form strong cationic resins at different crosslinking levels, different particle sizes and at different ratios of active substance to ion exchanger

Crosslinking % DVB	g of active substance / g of IER	Particle size in μm	mmoles of A. I. adsorbed by 1 g of IER (10% moisture) in 2 h contact time	
			CORETAL	PRACTOLOL
2	1	40– 100	3.10	2.20
4	1	40– 100	3.00	2.10
4	1	40– 100	2.80	2.05
8	1	150– 300	3.05	2.10
8	1	150– 300	1.60	1.20
8	1	300–1200	1.00	0.90

These changes in resin physical parameters only indirectly affect drug desorption. However, the drug bioavailability is directly influenced by the desorbing solution composition. Elution can be obtained either in the stomach or the intestine. Gastro-resistant formulations of ion exchange resins are often used when intestinal release of the drug is required.

Often one has to face the fact that the ionic strength in the intestine or stomach is not sufficient to obtain the desired release profile. When this occurs one has to consider the possibility of adding salts to the resinate formulation to obtain the desired release curves. These salts can be neutral, acidic, basic or a mixture of these depending on the final goal. Since resinate formulations have mostly been developed in the USA and only recently started to be developed in Europe, the resin used for galenic applications are normally manufactured according to U. S. Good Manufacturing Practice (GMP) and Good Laboratory Practice (GLP). However, high-purity chemical and special "clean" synthesis methods are used for the manufacturing of the resins, which are then marketed according to very tight specifications. Resins manufactured to these pharmaceutical grades are normally sold on a dry basis and within the sterility limits prescribed by US Pharmacopoeia (USP) for powders.

These pharmaceutical-grade ion exchange resins are currently ground from their original bead form to obtain the desired particle size distribution, the original beads having sizes ranging from 300 to 1200 μm. During the grinding process the beads are fragmented into irregularly shaped particles which cannot be effectively coated by standard thin film techniques.

Very recently L. P. Amsel et al. described "The Wurster Coating Chamber" as an effective device for coating ground ion exchange resins. An impregnating agent is applied to the resin particle to impart plasticity and the ethyl cellulose is applied to the same particle via an air suspension technique to form the desired coating layer around it. This technology has been developed to obtain good controlled release of drugs fixed on the ion exchange resin.

The ideal solution to the coating problem is, however, the commercial availability of uniform, perfectly spherical, ion exchanger beads of different particle sizes. These spherical particles could be coated with any of the existing techniques, including the simple thin layer procedure.

3.1.2.3 Cholestyramine

Cholestyramine as an ion exchange resin developed and used for the reduction of hypercholesterolemia. Although the treatment of hypercholesterolemia by cholestyramine had been known for some years it was only in 1983 that the National Heart, Lung and Blood Institute (NHLBI), in the USA, proved the link between high cholesterol blood level and coronary heart disease, which had long been familiar to many physicians and scientists.

A statistical study published by the World Health Organisation (WHO) predating the NHLBI study demonstrated that decreasing cholesterol levels using formulations based on chlorofibrate did, in fact, lower the risk of cardio-vascular failure, but did not substantially decrease the mortality level. This was essentially due to an increase in mortality caused by cancer and gall bladder operations.

In the case of the NHLBI study the researcher was successful by combining the treatment of the patients with cholestyramine and by controlling their diet. The results are clear and conclusive:

- a 1% reduction in cholesterol decreases by 2% the risk of cardio-vascular disease. Unlike chlorofibrate, no toxic side-effects were noticed; in particular, these was no increase in the risk of cancer and gall bladder stones.
- an important reason for this success is believed to lie in the fact that cholestyramine is not adsorbed or metabolised by the body, but goes through the intestine taking up bile acids and stimulating the liver to destroy more cholesterol.

NHLBI selected 18,000 men aged between 35 and 59 years from the 400,000 candidates examined. A further selection of statistically representative cases reduced the number of candidates from 18,000 to 3,806 volunteers. These volunteers were checked every two months for a period of time ranging from 7 to 10 years. In the cases of patients on a simple diet, without cholestyramine treatment, there was a 3.5% decrease in total cholesterol and 4% in LDL (low density lipoproteins) cholesterol. In the case of patients treated with diet and cholestyramine the obtained reductions were 13.4% and 20.3%, respectively. Thus the full drug dose could achieve close to a 50% risk reduction.

Mechanism of action of cholestyramine (Figure 3.1). Cholestyramine is a synthetic anion exchange resin which contains functional quaternary ammonium groups bound to copolymers of styrene-divinylbenzene. The gross formula is $(C_{22}H_{29}NCl)n$.

Cholestyramine is a non-absorbable, non-metabolizable anion exchange resin which, by binding the biliar acids, prevents their re-adsorption, eliminating them then with the feces. The elimination of biliar acids causes a depletion of hepatic cholesterol, since the latter represents the principal precursor of the biliar acids.

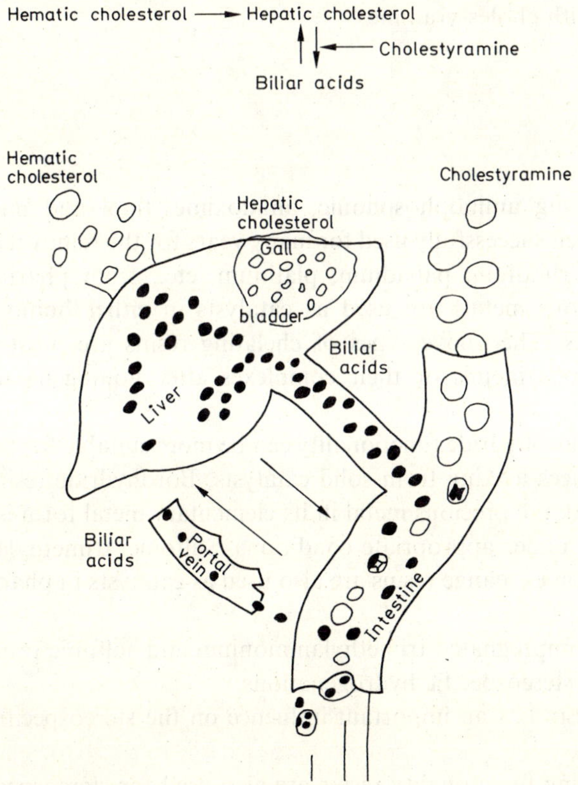

Figure 3.1 Mechanism of action of cholestyramine.

Depletion of hepatic cholesterol stimulates the transformation of hematic into hepatic cholesterol, thus lowering the total cholesterol level in the blood through low density lipoprotein (LDL) cholesterol reduction. LDL cholesterol is one of the most important risk factors in cardio-vascular disease.

Cholestyramine resin is a USP product and has other applications besides the treatment of hypercholesterolemia, e. g.:

1. Treatment of vitamin D_3 intoxication, hypercalcemia (Clinical Research *24*, 583A (1976)). Cholestyramine interrupts the enterohepatic circulation of calciferol, which is greatly increased in vitamin D intoxication, while serum 1,2,5-dihydrocholecalciferol is only minimally increased.
2. Cholestyramine improves diarrheal states by significantly reducing the toxicity of endotoxins (J. P. Nolan and M. Vilayat Ali, Digestive Diseases *17* (2) (1987)).
3. Cholestyramine has a therapeutic effect in reducing diarrhoea after ileal resection.
4. Reduction of pruritis associated with intrahepatic cholestasi.
5. The Medical College of Virginia, USA, reported that cholestyramine is very effective in treating people suffering from the toxic effects of the pesticide KEPONE (European Chemical News, February 10, 1978).
6. Very recent unpublished studies indicate a regression in arteriosclerosis in aged patients treated with cholestyramine.

3.1.3 Catalysis

Chelating resins having aminophosphonic, amidoxime, thiol and iminodiacetic functionality have been successfully used for many years for the removal of precious metals like iridium, rhodium, palladium, platinum, etc., from pharmaceuticals. Most of these precious metals are used as catalysts in either homogeneous or heterogeneous phases. The above types of chelating resins are mostly used for recovering the precious metals or their complexes after homogeneous catalytic reactions.

Polymers having borohydride functionality can be more suitable for the recovery of precious metal traces leaking from solid catalysts. Borohydride resins are normally not regenerated; the precious metal in its elementary metal form is recovered by burning the resin under appropriate conditions. Sulfonic, trimethylammonium and dimethylamine ion exchange resins are also used as catalysts in pharmaceutical reactions.

Pd, Pt, Ni, etc., - impregnated trimethylammonium and sulfonic resins are also used as supports for stereospecific hydrogenations.

Porosity of the resin has an important influence on the stereospecificity of the reaction.

Borohydride-bearing functionality resins are also used for stereospecific reductions.

Resins in Cu- and chromate-forms have been described in some organic oxidations.

Resins in K^+- and H^+-forms are commonly used for dessication of solvents or during condensation reactions.

References

[1] Hoffman, J., Hermansky, M., Parizkova, H. and Doskocil, J. (1957) Isolation of streptomycin using a carboxylic ion exchanger (in German). Chem. Techn. *9*, 151–155. Bartels, Ch. R., Kleimann, G., Korzun, J. N. and Irish, D. B. (1958) A novel ion exchange method for the isolation of streptomycin. Chem. Eng. Progr. *54*, 49–51. Kunin, R. (1975) Ion Exchange and Polymeric Adsorption Technology in Medicine, Nutrition, and the Pharmaceutical Industry, pp. 5–6. US Patent 2,528,188. Two tower process for recovery of streptomycin employing cation exchange resins.

[2] Pirotta, M. (1980) Abstr. 4th Symp. Ion Exchange Resins, Balaton Lake, Hungary, 27–30 May, Hungarian Chemical Society, pp. 241–242.

[3] US Patent 3,725,400 (Ciba-Geigy Corporation).

[4] Hood, J. D., Box, S. J. and Verrall, M. S. (1979) J. Antibiotics *32*, 295–304. Pirotta, M. (1982) Die Ang. Makr. Chem. *109/110*, 197–214. Discovery and Isolation of Microbial Products (M. S. Verrall) pp. 98–115. Ellis Horwood, Chichester, UK.

[5] Polish Patent 0.092193. EC Patent Application 83103663.7.

[6] German Patents 2559410, 2559411, 2517316.

[7] US Patent 3,629,233.

[8] US Patent 3,157,644; 17. 11. 74.

[9] Vogelmann, H. and Wagner, F. (1974) Isolation and chromatographic separation of vitamin B_{12} and other corrinoids from biological sources. Biotechnol. Bioengng. Symp., No. 4, pp. 959–975. Bi-references: Brit. Patent 1 208 693 and 1 129 125, and US Patent 3 313 693.

[10] US Patent 2,991,284.

[11] US Patent 3,932,417 and Brit. Patent 1 129 125.

[12] US Patent 2,457,117. Process for Manufacture of Amino Acids. Dominic, J. Bernardi Interchemical Corp. US Patent 2,462,597 (Feb. 22, 1949) Amino Acid Separation. R. J. Block to A. M. Armstrong, Inc.

[13] Baum, R. (1984) Biotechnology scale-up: How big a problem? Chem. Eng. News *62*, 13. Burgoyne, R. F., Bowles, D. K. and Heckendorf, A. (1984) Rapid scale up of peptide purification by preparative HPLC. Am. Biotechnol. Lab., March. (1968) Liquid chromatography looms as full scale separation method for production uses. Chem. Eng. News *46*, 62–63. Chem. Proc. (1983) November, System tackles separations at 99% purity, 0.2–15 kg/h. US Patent 2,989,438 (Heparin). Swedish Patent 334334. US Patent 4266030. French Patent 882603. Jap. Patent 40170/79.

Suggestions for further reading (Editor)

R. Kunin (1974–1975). Ion exchange technology in medicine and the pharmaceutical industry. Amber-hi-lites (Rohm and Haas Co.) Part I–IV. Issue No. 142–145.

R. K. Khar and J. S. Qudry (1979). Ion exchange resins in pharmacy. East. Pharm. *22*, 57.

Anon. (1984). New Systems for drug delivery. Targeted, sustained, cost-effective. Chem. Week (Sept. 26), 42.

G. V. Samsonov (1986). Ion exchange sorption and preparative chromatography of biologically active molecules. Booklet: Consultant Bureau Inc., New York, N. Y. 1986; 163 pp.

Yu. A. Leikin et al. (1976). Ion exchange resins in medicine. Itogi Nauki Tekh.; Khim. Tekhnol. Vysokomol. Soedin. *8*, 121. A review with 233 references on the use of ion exchange resins for separating ions from blood and blood products (CA 86: 67680).

H. J. Schneider and G. Stein (1979). Possibilities of using synthetic resin ion exchangers for therapeutic purposes. Dtsch. Gesundheitswes. *34*, 2385 (in German).

J. Cipoletti et al. (1980). Resin technology in medicine. Sorbents: Their Clin. Appl. *241*.

A. Gordon and M. Roberts (1980). Ion exchangers and their clinical applications. Sorbents: Their Clin. Appl. *249*. A review with 119 references (CA 93: 173547).

V. D. Gorchakov and Yu. A. Leikin (1981). Use of ion exchangers for hemosorption. Itogi Nauki Tekh.; Khim. Tekhnol. Vysokomol. Soedin. *16*, 213. A review with 236 references covering the use of ion exchangers in artificial kidneys for removal of toxic blood components (CA 95: 209423).

B. He and T. Qian (1983). Use of ion exchange resins in clinical therapy. Tianjin Yiyao *11*, 251 (Ch.). A review with 59 references on internal and external uses of ion exchange resins in clinical therapy, including dialysis (CA 99: 151485).

3.2 Ion Exchange Resins in Biochemistry and Biotechnology

Frederick J. Dechow
BioCryst, Birmingham, Alabama, USA

Introduction

Ion exchange resins have been used in biochemical applications for decades. Reviews covering these early studies are available [1, 2]. The biochemical applications differ from water, chemical or hydrometallurgical applications by the fact that microorganisms, cells or cell fragments are involved. The involvement of the biochemical species could have been either as the materials being separated, as the medium for production of the desired compound, or as the catalyst for microbial reaction.

In these early studies, biochemicals such as adenosine triphosphate [3], alcohols [4], alkaloids [5], amino acids [6], growth regulators [7], hormones [8], nicotine [9], penicillin [10], and vitamin B_{12} [11] were purified using ion exchange resins.

The work of Moore and Stein [12] showed how very complex mixtures of biochemicals, in this case amino acids and amino acid residues, could be isolated from each other using the ion exchange resin as a chromatographic separator. This new classic procedure has been applied to the analytical separation of thousands of biochemical preparations and to the isolation of preparative quantities of biological materials (Table 3.4). Additional reviews of these activities have been prepared by Weaver [43], by Morris and Morris [44], and by Gordon [45]. This chapter will concentrate on the unique features of biochemical solutions, on the properties of resins for these applications and on the utilization of the laboratory techniques for obtaining biochemicals for biotechnology in industry.

Table 3.4 Proteins purified by high performance ion exchange chromatography

Protein	Column	Resin type	References
Lactate dehydrogenase	DEAE-glycophase	weak base	14, 15, 16, 17, 18
isoenzymes	SynChropak AX300	weak base	16
Creatine kinase	DEAE-glycophase	weak base	17, 18, 19, 20
isoenzymes	SynChropak AX300	weak base	16, 21
Alkaline phosphatases	DEAE-glycophase	weak base	22
Hexokinase isoenzymes	SynChropak AX300	weak base	23
Arylsulfatase isoenzymes	DEAE-glycophase	weak base	24
Hemoglobins	SynChropak AX300	weak base	25, 26, 27, 28
	DEAE-glycophase	weak base	14, 15
	IEX 545 DEAE	weak base	29
	IEX 535 CM	weak cation	29
	Bio-Rex 70	weak cation	30
Cytochrome C	CM-polyamide	weak cation	31
Lysozyme	CM-polyamide	weak cation	31
Myoglobin	CM-polyamide	weak cation	31
	IEX 535 CM	weak cation	32
Soybean trypsin			
inhibitor	CM-glycophase	weak cation	15
Interferon	Partisil SCX	strong cation	33
Lipoxygenase	SynChropak AX300	weak base	34
Trypsin	DEAE-glycophase	weak base	15
Chymotrypsinogen	SP-glycophase	strong cation	15
	IEX 535 CM	weak cation	32
Immunoglobulin G	SynChropak AX300	weak base	35
Ovalbumin	SynChropak AX300	weak base	36
	IEX 545 DEAE		37
Albumin	SynChropak AX300	weak base	36
	DEAE-glycophase	weak base	15
	IEX 545 DEAE	weak base	32
Apolipoproteins	SynChropak AX300	weak base	38, 39
Adenylsuccinate			
synthetase	SynChropak AX300	weak base	40
Insulin	Partisil SCX	strong cation	41
	IEX 535 CM	weak base	32
β-Lactoglobulin	Partisil SCX	strong cation	41
Carbonic anhydrase	Partisil SCX	strong cation	41
Monoamine oxidase	SynChropak AX300	weak base	42

DEAE- and CM-glycophase are products of Pierce Chemical Company, Rockford, Illinois.
SynChropak AX300 is produced by SynChrom, Linden, Indiana.
IEX 535 CM and 545 DEAE are the products of Toyo Soda Company, Yamaguchi, Japan.
Bio-Rex 70 is supplied by Bio-Rad, San Francisco, California.
Partisil SCX is manufactured by Whatman, Clifton, New Jersey.

3.2.1 Biochemical solutions

Some of the unique features of biochemical solutions are [46]:

1. When biological tissues and cells are ruptured to make the biochemical solution, the resulting mixture will contain components, sometimes hundreds of different proteins, of widely differing sizes and properties.
2. Biochemical solutions may contain macromolecules, such as nucleic acids, polysaccharides or proteins, which form very viscous solutions at relatively low concentrations. A gel layer may be formed by these macromolecules on the ion exchange resins and associated equipment.
3. Biochemical solutions are frequently very sensitive to a wide variety of variables such as temperature, pH and oxygen.
4. It may be necessary to maintain the resin and equipment under aseptic conditions to prevent the occurance of unwanted microbial reactions.

The production preparation of biochemical compounds usually requires the processing of extremely large quantities of starting material. The typical concentrations involved are shown in Table 3.5. One of the consequences is that any method of purification and isolation of a specific biochemical compound will require a combination of separation techniques. In many instances it has been necessary to use ion exchange resins at more than one step in the biochemical product recovery.

Table 3.5 Typical product concentrations leaving fermenters

Product	Concentration (g/l)
Antibiotics (penicillin)	10– 30
Enzymes, proteins (serum protease)	2– 5
Organic acids (citric, lactic)	40–100
Riboflavin	10– 15
Vitamin B_{12}	0.02

Figure 3.2 shows the several places in the production and recovery of biochemicals where ion exchange resins might be used. It is possible not only to use ion exchange resins in the purification portion of the process but also as an enzyme support through which a feed solution is passed or in the analytical characterization or monitoring of the process fluids and products.

The first step in the purification is to separate the mycelia from the fermentation broth. If the desired compound is contained inside the mycelia, it will be necessary to lyse the cells to put the desired species into solution. The filtrate or solution must then be treated to form a crude separation of the desired family of compounds. This is called a group separation. Typical properties according to which a general "family" or group may be isolated are: molecular size, ionic properties, functionality

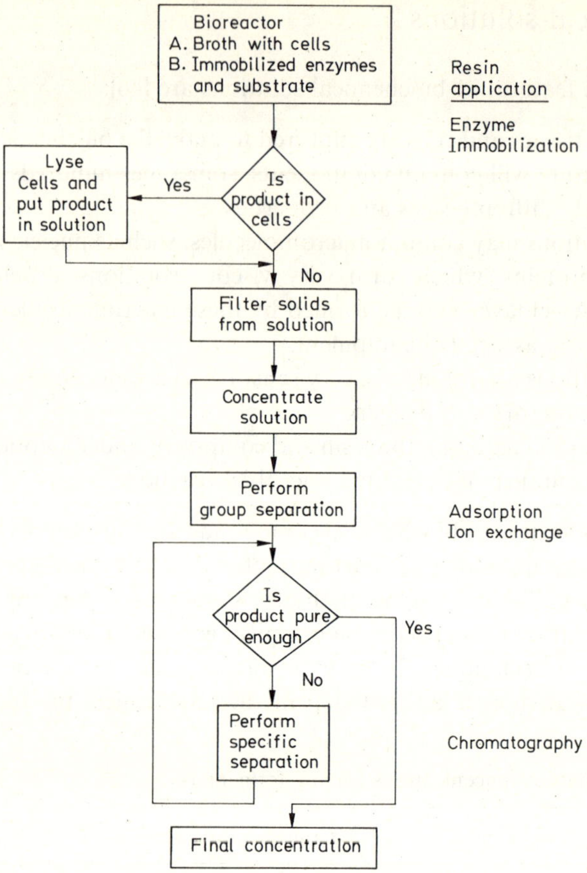

Figure 3.2 Preparation and purification of biochemical solutions with ion exchange resins.

or polarity. In addition to ion exchange resins, the group separation can be carried out using ultrafiltration membrane techniques, gel filtration, adsorption or crystallization.

Following the crude separation, it is usually necessary to fractionate the isolated group to obtain a purified single component. A chromatographic technique is frequently used to accomplish this, although solvent extraction, electrophoresis or dialysis may also be used.

The presence of color bodies and color body precursors in the final solution can have deleterious effects on the quality of the final product. The usual approach in solving this problem is to remove them by adsorption on activated carbon, either powdered or granular [47]. For some biochemicals, the use of ion exchange resins is required in place of activated carbon to provide more specific removal of the color bodies without adversely affecting the desired product. The removal of color bodies from phenylalanine, for instance, with charcoal will result in a loss of 20% of the product, whereas decolorization with an anion gel resin causes less than 2% loss of product.

3.2.2 Resin properties

The chemical and physical properties of a resin play a more important role in determining its suitability for a biochemical application than for other types of applications. The chemical properties to be considered are the matrix and the functionality of the resin. The important physical properties are the pore size, the pore volume, the surface area, the density and the particle size.

The common ion exchange matrices (styrene-divinylbenzene, methacrylic- or acrylic-divinylbenzene, phenol-formaldehyde, and epoxy amine resins) have been used in biochemical applications and even in protein purifications and enzyme immobilizations. The hydrophobic matrices have certain disadvantages: they might denature the desired biological material, or the high charge density might produce such strong binding that only a fraction of the bound material might be recovered. Until the advent of macroporous resins, these matrices were tightly crosslinked, with low porosity, so that large proteins and other macromolecules could be adsorbed or interact only with the exterior exchange sites on the matrix.

Resins with cellulosic matrices are much more hydrophilic and thus do not tend to denature proteins. Cellulosic resins have been used extensively in the laboratory analyses of biological material, enzyme immobilization and small-scale preparations. The low capacity and poor flow characteristics limited the usefulness of these matrices for larger applications.

Recently, diethylaminoethyl(DEAE)-silica gel was shown [48] to be an improvement over typical cellulosic matrix resins for the separation of acidic and neutral lipids from complex ganglioside mixtures. The specific advantages claimed were:

1. An increase in flow rate was possible through the DEAE-silica gel.
2. The DEAE-silica gel was able to be equilibrated much more rapidly with the starter buffer.
3. The DEAE-silica gel was more easily regenerated.
4. The DEAE-silica gel was less susceptible to microbial attack.
5. The preparation of DEAE-silica gel from inexpensive silica gel was described as a simple method that could be carried out in any laboratory.

The desired functionality on the selected matrix is determined by the nature of the biochemical entity to be adsorbed, its isoelectric point, the pH restrictions on the separation and the ease of eventually eluting the adsorbed species.

The mechanisms by which biochemicals adsorb to and desorb from surfaces may involve many phenomena which may be competitive or additive depending on the specific biochemical and ion exchange resin involved. The initial movement of the biochemical through the liquid phase to the solid adsorbent surface can involve non-sorptive charge behavior and diffusional, gravitational and convective transport. The subsequent attachment of biochemicals when they are near the ion exchange resin's surface may involve a balance between London-van der Waals forces and electrostatic forces of electrical double layers [49].

Table 3.6 shows the types of biochemical interactions that are possible with ion exchange resins and insoluble polymer matrices. The degree of difficulty in forming the specific interaction — without denaturing a protein or biopolymer — is related to the strength of the bond formed. The stronger the interaction, the more difficult it is to form the bond without adversely affecting the biopolymer. With ionic bonds and physical adsorption, simply bringing the resin and the biopolymer into close proximity will cause the interaction. Likewise, changing the pH or ionic strength of the solution will cause the interaction's reversal or regeneration of the resin.

Table 3.6 Biopolymer interactions with ion exchange resins and polymer matrices

	Covalent bond	Ionic bond	Physical adsorption	Entrapment
Formation	complicated	simple	simple	complicated
Possibility of biopolymer denaturing	high	low	low to high	high
Bonding strength	strong	weak	weak	strong
Regeneration	no	possible	possible	no

Specifically for interactions with ion exchange resins, the typical biopolymer can be thought of as a macroscopic ion having a number of electrostatically charged surface sites. The ionic interactions can be represented as:

$$R-N^{\oplus}(CH_3)_3 Cl^{\ominus} + {}^{\ominus}OOC-\underset{H}{\overset{R}{C}}-NH_2 \rightleftharpoons R-N^{\oplus}(CH_3)_3{}^{\ominus}OOC-\underset{H}{\overset{R}{C}}-NH_2 + Cl^{\ominus}$$

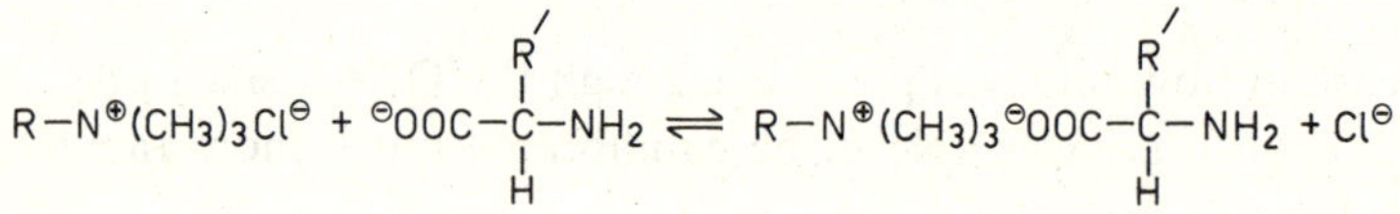

for a positively charged (anion) resin and a negatively charged biopolymer. Similarly, for a negatively charged (cation) resin and a positively charged biopolymer the interaction is represented as:

$$R-SO_3^{\ominus}H^{\oplus} + H_3N^{\oplus}-\underset{COOH}{\overset{R}{C}}-H \rightleftharpoons R-SO_3^{\ominus}H_3N^{\oplus}-\underset{COOH}{\overset{R}{C}}-H + H^{\oplus}.$$

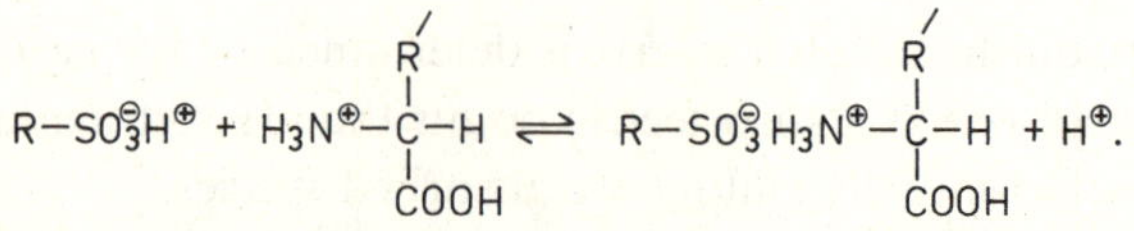

There are two other ionic interactions that are possible with the assistance of multivalent ions or polyelectrolytes in the feed stream. Thus, the presence of a multivalent cation in the solution would allow a cation resin to adsorb a negatively charged biopolymer, represented by:

$$R-SO_3^{\ominus} H^{\oplus} + P^{2\oplus} + {}^{\ominus}OOC-\underset{H}{\overset{R'}{C}}-NH_2 \rightleftharpoons R-SO_3^{\ominus\ominus} P^{\oplus\oplus}OOC-\underset{H}{\overset{R'}{C}}-NH_2 + H^{\oplus}.$$

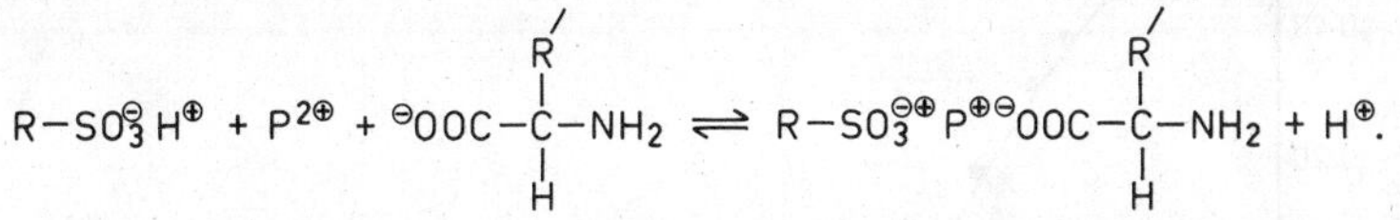

Likewise, multivalent anions allow an anion resin to adsorb a positively charged biopolymer according to:

$$R-N^{\oplus}(CH_3)_3 Cl^{\ominus} + N^{2\ominus} + H_3N^{\oplus}-\underset{COOH}{\overset{R'}{C}}-H \rightleftharpoons R-N^{\oplus}(CH_3)_3^{\ominus} N^{\ominus} H_3N^{\oplus}-\underset{COOH}{\overset{R'}{C}}-H + Cl^{\ominus}.$$

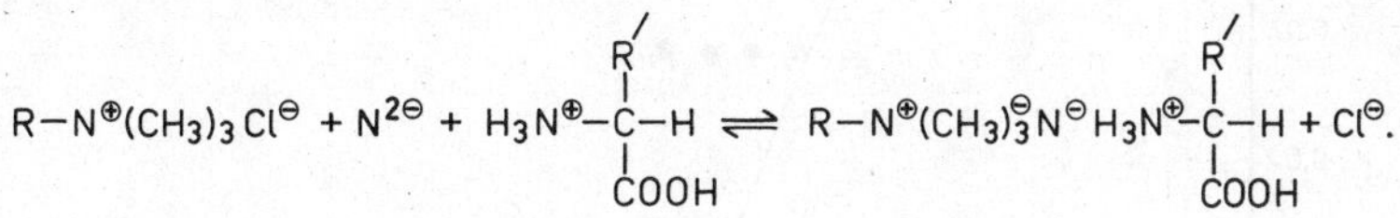

A single protein can, at the same time, participate in specific ion exchange reactions with the resin, form bonds with water-soluble polyelectrolytes and attach to other adsorbent surfaces. The mathematical representation for these complex interactions combine the kinetics of diffusional processes, ion exchange reactions and adsorption isotherms. The adsorption isotherms of Langmuir, Freundlich and others have been used in these model developments. Figure 3.3 shows the adsorption of bacterial cells on ion exchange resins and Figure 3.4 shows the desorption. It should be noted that there is usually some irreversible adsorption of biopolymers on ion exchange resins, as Figure 3.4 shows. The extent of this irreversible bonding will define the limitations of using a particular resin for purifying a biochemical solution.

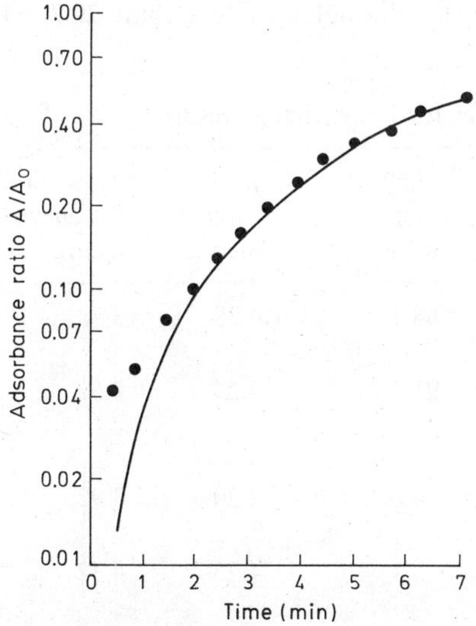

Figure 3.3 Adsorption of bacterial cells onto anion exchange resin (Dowex 1X8, Cl-form), A = 0.598, pH = 3.53 [49].

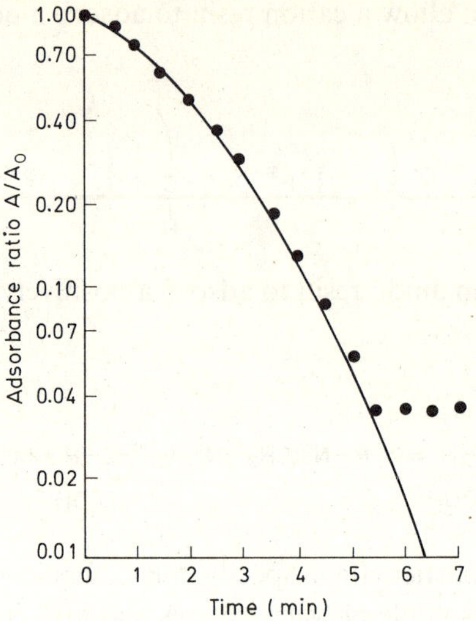

Figure 3.4 Desorption of bacterial cells from anion exchange resin (Dowex 1X8, Cl-form), A = 0.598, pH = 4.12, 1 M KCl [49].

This irreversible adsorption property can be utilized when an enzyme is to be immobilized on an ion exchange resin.

Typical macroporous ion exchange resins may have mean pore diameters ranging from 100 Å to 4000 Å. The gel resins, such as the Dowex 50 W series, rely upon hydration to provide the macroporous structure. Even at low crosslinking and full hydration, these resins have a mean pore diameter of less than 20 Å. Table 3.7

Table 3.7 Physical properties and capacities for ion exchange resins

Resin matrix	Functionality	Pore size Å	Surface area m^2/g	Resin capacity meq/g	Adorption capacity for enzyme meq/g
Phenolic	3° Polyethylene Polyamine	250	68.1	4.38	3.78
Phenolic	Partially 3° Polyethylene Polyamine	290	95.3	4.24	3.57
Polystyrene	Polyethylene Polyamine	330	4.6	4.20	3.92
Polystyrene	Polyethylene Polyamine	560	5.1	4.75	4.32
Polyvinyl chloride	Polyethylene Polyamine	1400	15.1	4.12	3.72

shows the pore sizes of several resins of different matrices that have been used in enzyme immobilization [50]. Pore volumes for macroporous resins may range from 0.1 to 2.0 cm^3/g.

Normally, as the mean pore diameter increases, the surface area of the resin decreases. These surface areas can be as low as 2 m^2/g and go as high as 300 m^2/g. Table 3.7 also shows that the total exchange capacity is not utilized in these biochemical fluid processes. Whereas in water treatment applications one can expect to utilize 95% of the available capacity, in biotechnology applications it was often possible to use only 20 to 30% of the total exchange capacity with gel resins. Only the exchange sites in the outer shell of the gel resin could be utilized because of the resin's small pore size. Macroporous resins have increased the utilization to close to 90% for the immobilization of enzymes, but even with macroporous resins biochemical fluid processing, where the fluid flows through an ion exchange resin bed, still is limited to about 60% utilization.

Table 3.8 shows the molecular size of some biopolymers for comparison to the mean pore size of the resins. When selecting the pore size of a resin for the recovery or immobilization of a specific protein, a general rule is that the optimum resin pore diameter is about 4 to 5 times the length of the major axis of the protein.

Table 3.8 Molecular size of biopolymers

Biopolymer	Molecular weight	Maximum length of biopolymer
Catalase	250000	183 Å
Glucose isomerase	25000 – 100000	75 – 100 Å
Glucose oxidase	15000	84 Å
Lysozyme	14000	40 Å
Papain	21000	42 Å

Increasing the pore size of the resin beyond that point will result in decreases in the amount of protein adsorbed because the surface area available for adsorption decreases as the pore size increases. As an example, the optimum adsorption of glucose oxidase, as defined by enzyme activity, is shown in Figure 3.5 [51]. Enzyme activity is a measure of the amount of enzyme adsorbed and accessible to substrate.

Typical resin densities may range from 0.6 g/cm^3 to 1.3 g/cm^3 for organic polymers. Silicate materials may be more dense, up to 6 g/cm^3. Since the fermentation broth or other biochemical fluid may be more dense than water, the slow flow rates that are usually involved may require resins that have a density greater than water. A minimum flow rate may be necessary to maintain a packed bed when a fluid denser than water is being processed by a medium-density resin. If this is not possible, an upflow operation or batch process may be tried. The lower density resins are usually associated with a highly porous structure which has less mechanical strength than the typical gel or macroporous resin. When the mean pore diameter

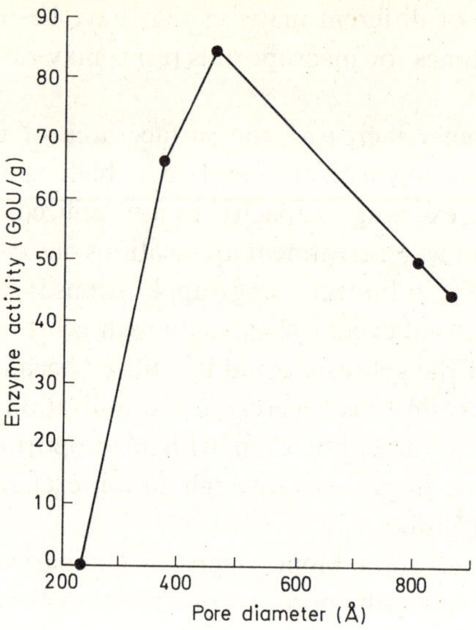

Figure 3.5 Effect of a resin's pore diameter on the enzyme activity of glucose oxidase [51].

of a resin is greater than 2000 Å, the resin may have a crush strength of less than 10 g per bead. Such resins would be subject to attrition in a stirred tank or to collapse in a tall column.

Many of the resins used in early biochemical separations were much smaller (75–300 μm) than the resins typically used in water treatment (400–1000 μm). With the development of macroporous resins, enzyme immobilization and protein purification were performed with resins of 400–1000 μm size, since the macroporous structure allowed sufficient surface area for adsorption almost independent of the bead size. For industrial chromatography of amino acids, resins in the 175–420 μm size may be used.

3.2.3 Biotechnology applications

It used to be that biotechnology was synonymous with fermentation processes. More recently, the term has been associated primarily with recombinant DNA technology. Actually, biotechnology applications range from fermentation, to cell harvesting, to enzyme immobilization, to biological sensors, to microbial manipulation. In essence, they include all applications involving microorganisms, enzymes and biological raw materials. Ion exchange resins have been used, at least in the laboratory, in all of these applications. The outgrowth of the laboratory usage of ion exchange resins has been their use in the commercial recovery and purification of biotechnology products. In fact, most purification schemes for biotechnology operations are merely changes in the scale or size for techniques developed in laboratory biochemical processes.

The industrial purification of biochemicals will be illustrated by examples for amino acids and proteins and the other biotechnology applications of ion exchange resins by enzyme immobilization.

3.2.3.1 Amino acid purification

Whereas Moore and Stein [12] used a single strong cation resin (Dowex 50 W) with changing elution conditions to separate amino acids, Winters and Kunin [2] used three resins in five columns and different elution conditions (Figure 3.6) to separate amino acids from protein hydrolyzate into their three charge groups (acid, neutral and base) and three individual amino acids (histidine, arginine and lysine). The carboxylic acid resin may offer an advantage for some amino acid purifications because of its higher buffering capacity, compared to strong cation resins, in the critical pH region.

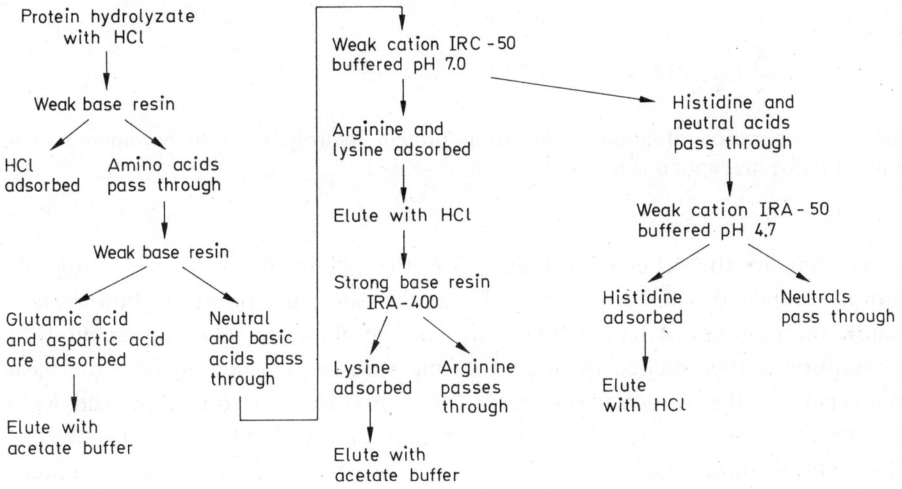

Figure 3.6 Separation of amino acids from protein hydrolyzate [2].

The use of weak base anion exchange resins for recovering glutamic acid and aspartic acid from protein hydrolyzate has been reported [52] for commercial preparations of amino acid solutions of neutral and basic amino acids for intravenous feeding.

Another scheme (Figure 3.7) has been used commercially for separating amino acids from protein hydrolyzate. In these schemes it can be seen that the cystine and cysteine have been removed prior to the ion exchange treatment of the protein hydrolyzate.

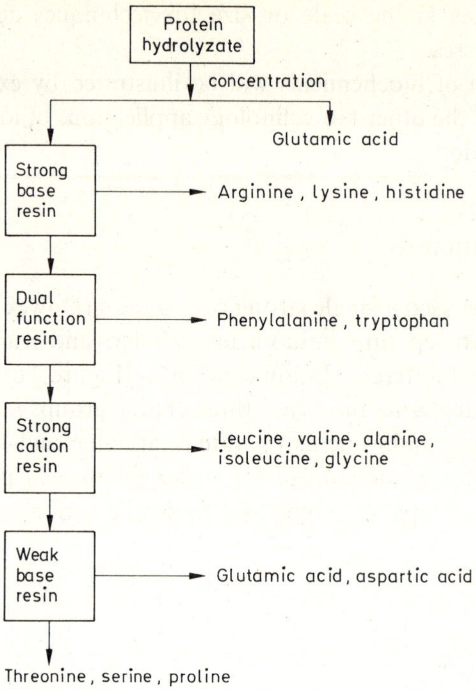

Figure 3.7 Separation of amino acids from protein hydrolyzate with chromatographic elution of individual amino acids.

According to the scheme in Figure 3.7, the pH of the protein hydrolyzate solution is adjusted with HCl to a level near the isoelectric point of glutamic acid to allow the removal of half to two-thirds of the glutamic acid by precipitation. This solution is then passed through a strong base resin which adsorbs the basic amino acids. A dual-functionality resin (or a mixture of strong base and weak cation resins) is used next to adsorb the phenylalanine and tryptophan. The majority of the neutral amino acids are adsorbed on a strong cation resin, while a weak base resin adsorbs the acidic amino acids, including any residual glutamic acid. The pH changes of the solution as it passes through the different resins and the pH of the eluting solutions which allow the chromatographic separation of the adsorbed species vary for different manufacturers.

The recovery of amino acids from these protein hydrolyzates has the limitation that the yield of each amino acid is proportional to its presence in the original proteins. While protein hydrolyzates will continue to be used as the raw material for some commercial amino acid production, the trend is more towards use of fermentations which produce a specific amino acid.

Amino acid fermentation broths have characteristics which are different from those of protein hydrolyzates. The differences are [53]:

1. In most cases, a significant amount of a single amino acid can be accumulated in the fermentation broth in its free and salt forms; other contaminant amino acids are usually small in number and quantity.
2. Amino acids produced by fermentation are usually the optically active L-form.
3. Fermentation broths are aqueous solutions of amino acids whose concentration vary from 0.1 to 10% (w/v) depending on the individual fermentation process.
4. Fairly large amounts of microbial cells and soluble protein-like biopolymers are usually contained in fermentation broths.
5. Some inorganic salts derived from microbial nutrients always exist in fermentation broths.
6. Other organic impurities derived from microbial nutrients and metabolites, such as sugars and pigments, are also present to some extent.

As an example, L-tryptophan [54] is fermented using *Arthrobacter paraffineus* which requires histidine as a nutrient. The fermentation filtrate is passed through a strong cation resin which selectively adsorbs the L-tryptophan. This is eluted from the resin with 0.5 N ammonium hydroxide solution from which it is concentrated and crystallized. The resin used and the elution conditions can be specified with more precision for purification of an amino acid from a fermentation broth since the broad spectrum of amino acids present in protein hydrolyzates are not in the fermentation broth.

The adsorption of amino acids by ion exchange resins is strongly affected by the pH of the solution. Figure 3.8 shows the decrease in the adsorption of lysine

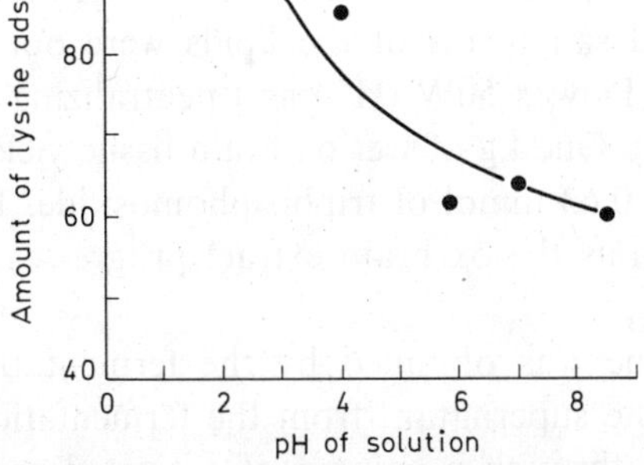

Figure 3.8 The change in the adsorption of lysine on a strong cation resin (Diaion SK1, NH_4-form) as a function of pH of fermentation broth [55].

by a strong cation resin in the ammonium form as the pH is changed. Other inorganic ions may also compete with the amino acids for exchange sites on the resin, similar to the ion exchange treatment of other mixtures of ions in solution.

3.2.3.2 Protein purification

Keay [56] described the general conditions for the purification of a protein. His specific example was for the purification of neutral and alkaline proteases from *B. subtilis* by adsorption on a cation exchange resin.

The sulfonated phenol-formaldehyde resin was adjusted to neutral pH by washing it with the solution to be used as the enzyme solvent. This solution must have a low ionic strength (below 0.5 M) to enable neutral as well as alkaline proteases to be adsorbed. The pH should be between pH 6 – 7.5.

The maximum concentration of protein in the solution should be 2.5%, with the best results obtained when it is less than 1%. This corresponds to an enzyme level of about 50000 enzyme units per milliliter.

After adsorption, the resin column is washed at neutral pH or neutral ionic strength so that the adsorbed protein is not removed while unadsorbed material is removed for subsequent treatment in the next cycle. Then the adsorbed enzymes are eluted with a 0.05 – 0.2 M solution to remove the neutral proteases, followed by elution with a 0.2 to 1.0 M solution to remove the alkaline proteases. These solutions are usually 0.1 M phosphate buffer followed by 1.0 M sodium chloride. The elution pH is usually slightly higher than the pH of the feed stream to avoid contamination of the product with any adsorbed color body impurities.

Kiselev [57] has described a preparative method for the isolation of pure fractions of di- and triphosphoinositides from ox brain. The petroleum ether lipid extract was passed through a Dowex 50W (H^+) resin column to remove the divalent metal ions. Then sodium hydroxide in methanol solution was added to the effluent to convert the lipids to the sodium salt. The resulting solution was added to a DEAE-cellulose column. Gradient elution with 0 – 0.6 M ammonium acetate in chloroform/methanol/water (20 : 9 : 1) allowed the separation of the lipids into fractions of di- and triphosphoinositides. The desired salt forms of the lipids were obtained by passing the ammonium salts through Dowex 50W (H^+) and neutralizing with the appropriate base in methanol solution. One kg of wet ox brain tissue yields about 0.35 mmol of diphosphoinositide and 0.63 mmol of triphosphoinositide. Table 3.9 shows the phospholipid concentration as the ox brain extract progresses through the preparative purification.

In another example [58], adenosine was obtained by the fermentation of a Bascillus strain, ATCC No. 21616. The supernatant from the fermentation broth was adjusted to pH 4.0, then passed through a column of activated carbon for adsorption of adenosine. The column was washed with water, then the adenosine eluted with a solution of methanol/isooctanol/ammonia/water (50 : 1 : 2 : 47). The

Table 3.9 Phospholipid concentration (mmol/l) during purification

Process step	Phosphoinositides Mono	Di	Tri
In petroleum ether extract	1.68	0.43	0.84
After first Dowex 50W treatment	1.32	0.37	0.67
After DEAE-cellulose chromatography	1.29	0.35	0.63

eluate is concentrated and then separated by chromatography with an anion exchange resin (Dowex 1 (Cl^--form)) to get a pure adenosine fraction. That fraction is concentrated and then crystallized from ethanol.

Marshall [59] used a column of DEAE-cellulose, followed by one of Ultragel AcA34, one of Sephadex G-100 and finally one of DEAE-Sepharose in his industrial-scale separation of a specific glucoamylase-type of enzyme to be used in the production of low alcohol beverages. These few examples point out the many steps involved and combinations of unit processes needed to purify a single protein from a complex mixture.

3.2.3.3 Enzyme immobilization

It was in the 1950s that ion exchange resins began to be used seriously as carriers for immobilization of enzymes [60, 61]. These initial efforts resulted in immobilized enzymes that had very low activities of no commercial use. The results were somewhat better [62] for enzymes immobilized on cellulose derivatives. The macroporous resins of the 1970s allowed them to be used in commercial immobilized enzyme processes [63]. The bonding specificity of the affinity chromatography type of resin with an enzyme has been employed to immobilize enzymes [64]. A review of the use of these immobilized enzymes in the food industry has been prepared by Kilara and Shahani [65].

The advantages of immobilized enzymes over soluble enzymes are [66]:

1. Continuous processes become practical.
2. The stability of the enzyme may be improved.
3. A product of a higher purity may be possible.
4. The sensitivity of the enzyme to changes in temperature or pH may be decreased.
5. The enzyme activity may be increased.
6. Effluent problems and material handling problems (separations) for feed and effluent are reduced.

Table 3.10 Commercial process parameters for the enzymatic production of L-amino acids

Column capacity	1000 Liters
Carrier	DEAE-Sephadex
Amount of acylase bound	333 I.U./ml
Activity of bound enzyme	157 I.U./ml
Yield of enzyme activity	47%
Operating temperature	50 °C
Operating pH	7.0
Activity loss	40% in 35 days
Column regeneration	once in 35 days
L-Methionine yield	715 kg/day at SV = 2.0

Table 3.11 Enzymes immobilized on various supports

Support	Enzyme	Reference
Dowex MSA-1	Amino acid acylase	67
DEAE-Sephadex	Glucose isomerase	68
	Glucoamylase	69
CM-Cellulose	L-Asparaginase	70
Silical gel	Acid phosphatase	71
	Phosphoglucomutase	71
Kaolinite	Chymotrypsin	72
Bentonite	Penicillin amidase	73

Table 3.10 shows the features of a commercial process, mentioned in the same article, for the production of L-amino acids using an immobilized aminoacylase enzyme. More recently [67], the aminoacylase enzyme has been immobilized on a porous, strong base resin of trimethylammonium-introduced porous silica with a pore size of about 1000 Å and a surface area of about 25 m^2/g. Additional examples of enzymes that have been immobilized on different carriers are given in Table 3.11.

Delin and coworkers [64] described an illustrative procedure for purifying and immobilizing penicillin acylase, which was then used to produce hypoallergenic penicillins from other penicillins. The penicillin acylase was purified by adsorption on a porous cation exchange resin from a solution adjusted to a pH of 4.0 – 5.0. The purified enzyme is eluted with 0.2 M ammonium acetate buffer solution (pH 6.0 – 8.0). Sephadex G 200 was treated with 5 N NaOH and then reacted with cyanogen chloride at a cold temperature (0 – 3 °C). After the reaction was complete, the resin was washed with ice water and a 0.1% borax solution. The wet polymer was added to the solution of purified penicillin acylase, more borax was added and the mixture stirred slowly for 24 hours to complete the enzyme immobilization.

Immobilizations such as this, which are due to the formation of a covalent bond, are more complex than the simple adsorption or ionic bonds which may be formed by simply passing the enzyme solution through the prepared ion exchange column.

3.2.4 Outlook for future development

The development of ion exchange resin applications in biotechnology may be expected to be in the areas of new types of resins, of adaptations of existing laboratory techniques and of new processing techniques.

Resins of very uniform size have been developed in several mean size ranges; they should be very useful in the chromatographic separation of small biochemical products such as amino acids. The use of porous glass which is silanized and functionalized [74] allows more protein-compatible resins to be developed. As resin manufacturers and specialty chemical companies seek new niches for high-value products, the development of new resins for biotechnology applications would seem a logical direction.

The stepwise change from high-pressure liquid chromatography to medium-pressure liquid chromatography, such has been described for the preparation of pectic enzyme [75], reveals the trend of progress toward industrial application of the techniques developed in analytical laboratories. The pressure for these medium-pressure chromatography applications is only 6 bar instead of the 100 to 150 bar associated with high-pressure liquid chromatography.

Studies – such as the one by Frolik and coworkers [76] – which examine the effect and optimization of variables in high-pressure liquid chromatography of proteins, can be expected to contribute to the implementation of this type of protein resolution technique into future commercial biotechnology processes.

Another new technique that offers promise for commercial biotechnology purifications is the use of parametric pumping with cyclic variations of pH and electric field. This has been described by Hollein and coworkers [77]. They worked with human hemoglobin and human serum albumin protein mixtures on a CM-Sepharose cation exchanger. The extensive equations for parametric separations reported allow analysis of other systems of two or more proteins which may be candidates for this type of separation.

The initial glamor associated with recombinant DNA and cloning biotechnology has faded with the realization that the biochemicals produced must be recovered from solutions. Such purifications, as this chapter has tried to show, are not an easy endeavor. Innovation and creativity is called for from those who are working with ion exchange resins. This is an exciting area for those who wish to turn laboratory curiosities into commercial products.

References

[1] Bersin, T. (1946) Exchange adsorption for biochemical processes. Naturwissenschaften *33*, 108.

[2] Winters, J. C. and Kunin, R. (1949) Ion exchange in the pharmaceutical field. Ind. Eng. Chem. *41*, 460.

[3] Polis, B. D. and Meyerhoff, O. (1947) Studies on the adenosinetriphosphatase in muscle. I. Concentration of the enzyme on myosin. J. Biol. Chem. *169*, 389.

[4] Carson, J. F. and Maclay, W. D. (1945) 1,4-Anhydro-D,L-xylitol. J. Am. Chem. Soc. *67*, 1808.

[5] Nagai, S. and Murakami, K. (1941) Water softening materals of Zeolite Type III. J. Soc. Chem. Ind. Japan *44*, 709.

[6] Wieland, T. (1944) Separation of basic amino acids by adsorption on "Wofatit C". Berichte *77*, 539.

[7] Bergdoll, M. S. and Doty, D. M. (1946) Chromatography in the separation and determination of the basic amino acids. Ind. Eng. Chem. Anal. Ed. *18*, 600.

[8] Lejwa, A. (1939) Biochem. Z. *256*, 236.

[9] Kingsbury, A. D., Mindler, A. B. and Gilwood, M. B. (1948) Recovery of nicotine by ion exchange. Chem. Eng. Prog. *44*, 497.

[10] Cruz-Coke, E., Gònzales, F. and Hulsen, W. (1945) Ionic exchange resins for the purification of penicillin and hypertensin. Science *101*, 340.

[11] Jackson, W. G., Whitefield, G., DeVries, W., Nelson, H. and Evans, J. (1951) The isolation of vitamin B_{12b} from neomycin fermentations. J. Am. Chem. Soc. *73*, 337.

[12] Moore, S. and Stein, W. H. (1954) Procedure for the chromatographic determination of amino acids on four per cent crosslinked sulfonated polystyrene ion exchange resin. J. Biol. Chem. *211*, 893.

[13] Regnier, F. E. (1983) High-performance ion-exchange chromatography of proteins: The current status. In: High-Perform. Liq. Chromatogr. Proteins Pept., Proc. Int. Symp., 1st 1981 (M. T. W. Hearn, F. E. Regnier and C. T. Wehr, eds.) Academic Publications, New York, N. Y.

[14] Chang, S. H., Gooding, K. M. and Regnier, F. E. (1976) HPLC of proteins. J. Chromatogr. *125*, 103.

[15] Chang, S. H., Noel, R. N. and Regnier, F. E. (1976) High speed ion exchange chromatography of peptides. Anal. Chem. *48*, 1839.

[16] Schlabach, T. D., Alpert, A. J. and Regnier, F. E. (1978) Rapid assessment of isoenzymes by High-Performance Liquid Chromatography. Clin. Chem. *24*, 1351.

[17] Schlabach, T. D., Fulton, J. A., Mockridge, P. B. and Toren, E. C., Jr. (1979) New developments in analysis of isoenzymes separated by "High Performance" Liquid Chromatography. Clin. Chem. *25*, 1600.

[18] Schlabach, T. D., Fulton, J. A., Mockridge, P. B. and Toren, E. C., Jr. (1980) Determination of serum isoenzyme activity profiles by High Performance Liquid Chromatography. Anal. Chem. *52*, 729.

[19] Fulton, J. A., Schlabach, T. D., Kerl, J. E. and Toren, E. C., Jr. (1979) Dual-detector-post-column reactor system for the detection of isoenzymes separated by High Performance Liquid Chromatography. I. Description and theory. J. Chromatogr. *175*, 269.

[20] Denton, M. S., Bostick, W. D., Dinsmore, S. R. and Mrochek, J. E. (1978) Chromatographic separation and continuously referenced, on-line monitoring of creatine kinase isoenzymes by use of an immobilized-enzyme reactor. Clin. Chem. *24*, 1408.

[21] Bostick, W. D., Denton, M. S. and Dinsmore, S. R. (1980) Liquid chromatographic separation and on-line bioluminescence detection of creatine kinase isoenzymes. Clin. Chem. *26*, 712.

[22] Schlabach, T. D., Chang, S. H., Gooding, K. M. and Regnier, F. E. (1977) Continuous-flow enzyme detector for LC. J. Chromatogr. *134*, 91.

[23] Alpert, A. J. (1979) New Materials and Techniques for HPLC of Proteins. Ph. D. Thesis, Purdue University.

[24] Bostick, W. D., Dinsmore, S. R., Mrochek, J. R. and Waalkes, T. P. (1978) Separation and analysis of arylsulfatase isoenzymes in body fluids of man. Clin. Chem. *24*, 1305.

[25] Gooding, K. M., Lu, K. C. and Regnier, F. E. (1979) High Performance Liquid Chromatography of hemoglobins. I. Determination of hemoglobin A. J. Chromatogr. *164*, 506.

[26] Hanash, S. M. and Shapiro, D. N. (1980) Separation of human hemoglobin by Ion Exchange High Performance Liquid Chromatography. Hemoglobin *5*, 165.

[27] Hanash, S. M., Kavadella, K., Amanulla, A., Scheller, K. and Bunnell, K. (1981) High Performance Liquid Chromatography of hemoglobins: Factors affecting resolutions. In: Advances in Hemoglobin Analysis (S. M. Hanash and G. J. Brewer, eds.) p. 53.

[28] Gardiner, M. B., Wilson, J. B., Carver, J., Abraham, B. L. and Huisman, T. H. J. (1981) International Symposium on HPLC of Proteins and Peptides, Paper No. 203, Washington, D. C.

[29] Umino, M., Watanabe, H., Komiya, K. and Mori, N., Ref. 28, Paper No. 208.

[30] Abraham, E. C., Cope, N. D., Braziel, N. N. and Huisman, T. H. J. (1979) On the chromatographic heterogeneity of human fetal hemoglobin. Biochim. Biophys. Acta *577*, 159.

[31] Gupta, S. P., Pfannkoch, E. and Regnier, F. E. (1983) High Performance Cation Exchange Chromatography of proteins. Anal. Biochem. *128*, 196.

[32] Umino, M., Watanabe, H. and Komiya, K., Ref. 28, Paper No. 204.

[33] Radhakrishnan, A. N., Stein, S., Licht, A., Gruber, K. A. and Udenfriend, S. (1977) High-Efficiency Cation-Exchange Chromatography of polypeptides and polyamines in the nanomole range. J. Chromatogr. *132*, 552.

[34] Vanecek, G. and Regnier, F. E. (1982) Macroporous high-performance anion-exchange supports for proteins. Anal. Biochem. *121*, 156.

[35] Lu, K. C., Gooding, K. M. and Regnier, F. E. (1979) Rapid analysis of bilirubin in neonatal serum. I. The binding of bilirubin to albumin. Clin. Chem. *25*, 1608.

[36] Vanecek, G. and Regnier, F. E. (1980) Variables in the High Performance Anion Exchange Chromatography of proteins. Anal. Biochem. *109*, 345.

[37] Kato, Y., Komiya, K. and Hashimoto, T., Ref. 28, Paper No. 214.

[38] Alpert, A. J. and Beaudet, A. L., Ref. 28, Paper No. 210.

[39] Ott, G. S. and Shore, V. G., Ref. 28, Paper No. 201.

[40] Rudolph, F. B. and Clark, S. W., Ref. 28, Paper No. 202.

[41] Frolick, C. A., Dart, L. L. and Sporn, M. B., Ref. 28, Paper No. 205.

[42] Ansari, G. A. S., Patel, N. T., Fritz, R. R. and Abell, C. W., Ref. 28, Paper No. 211.

[43] Weaver, V. C. (1969) Ion exchange in biochemistry. Chromagraphia *2*, 555.

[44] Morris, C. J. O. R. and Morris, P. (1963) Separation Methods in Biochemistry, Interscience Publ., New York, N. Y.

[45] Gordon, A. H. (1969) Electrophoresis of proteins in polyacrylamide and starch gels. In: Laboratory Techniques in Biochemistry and Molecular Biology, Vol. 1, Part 1 (T. S. Work and E. Work, eds.) North Holland Publ. Co., Amsterdam, London.

[46] Hawtin, P. (1982) Chemical Engineer *376*, 11.

[47] Kundu, S. K. (1981) DEAE-silica gel and DEAE-controlled pore glass as ion exchangers for the isolation of glycolipids. Enzymol. *72*, 174.

[48] Daniels, S. L. (1980) Mechanisms involved in sorption of microorganisms to solid surfaces. In: Adsorption of Microorganisms to Surfaces (G. Britton and K. C. Marshall, eds.) Wiley, New York, N. Y.

[49] Nagase, T., Hirohara, H. and Nabeshima, S. (1981) Enzyme-immobilization Carriers and Preparation Thereof. Brit. Pat. 1597436.

[50] Messing, R. A. (1975) Immobilized Enzymes for Industrial Reactors, Academic Press, New York.

[51] Okuda, T. and Awataguchi, S. (1973) Macrolide Antibiotics YL-704C and W from Streptomyces. U.S. Pat. 3718742.

[52] Kunin, R. (1972) Ion Exchange Resins, p. 290. Krieger, Huntington, N.Y.

[53] Yamada, K., Kinoshita, S., Tsunda, T. and Aida, K. (1972) The Microbial Production of Amino Acids, p. 228. Wiley, New York, N.Y.

[54] Nakayama, K. and Hagino, H. (1971) Process for Producing L-Tryptophan. U.S. Pat. 3594279.

[55] Hino, T., Ito, K. and Hayashi, K. (1961) Behavior of amino acids on ion exchange resins. Nippon Nogei-kagaku Kaishi *35*, 773.

[56] Keay, L. (1971) Purification and Recovery of Neutral and Alkaline Protease Using Cationic Sulfonated Phenol-Formaldehyde Resin. U.S. Pat. 3592738.

[57] Kiselev, G. V. (1982) Preparative isolation of phosphoinositide fractions from ox brain. Biochim. Biophys. Acta *712*, 719.

[58] Komatsu, K., Saijo, A., Haneda, K., Kodaira, R. and Ohsawa, H. (1973) Process for Producing Adenosine by Fermentation. U.S. Pat. 3730836.

[59] Marshall, J. J. (1982) Method of Producing Low Calorie Alcoholic Beverage with Starch-degrading Enzymes Derived from Cladosporium resinae. U.S. Pat. 4318927.

[60] Brandenberger, H. (1956) Methods for linking enzymes to insoluble carriers. Rev. Fermentations Indus. Aliment. *11*, 237.

[61] Barnett, L. B. and Bull, H. B. (1969) Optimum pH of adsorbed ribonuclease. Biochim. Biophys. Acta *36*, 244.

[62] Tosa, T., Mori, T., Fuse, N. and Chibata, I. (1966) Continuous enzyme reactions. I. Screening of carriers for preparation of water insoluble aminoacylase. Enzymologia *31*, 214.

[63] Fujita, Y., Matsumoto, A., Nishikaji, T. and Maeda, Y. (1975) Immobilized Glucose Isomerase. Japan Pat. 50-94187.

[64] Delin, P. S., Ekstrom, B. A., Nathorst-Westfield, L. S., Sjoberg, B. O. H. and Thelin, K. H. (1973) Penicillinacylase from Escherichia coli for the Manufacture of 6-Aminopenicillanic Acid. U.S. Pat. 3736230.

[65] Kilara, A. and Shahami, K. (1979) The use of immobilized enzymes in the food industry: A review. CRC Critical Reviews in Food Science and Nutrition *12*, 161.

[66] Vieth, W. R. and Venkatasubramanian, K. (1973) Enzyme engineering. I. Utility of supported enzyme systems. ChemTech, Nov., 677.

[67] Chibata, I., Tosa, T., Mori, T. and Fujimura, M. (1982) Product Containing Immobilized Aminoacylase. Brit. Pat. 2082188.

[68] Bachler, M. J., Strandberg, G. W. and Smiley, K. L. (1970) Starch conversion by immobilized glucoamylase. Biotechnol. Bioeng. *12*, 85.

[69] Nikolaev, A. Y. and Mardashev, S. R. (1961) Formation of an asparaginase-carboxymethyl cellulose complex. Biochemistry (USSR) *26*, 641.

[70] Messing, R. A. (1970) Insoluble papain prepared by adsorption on porous glass. Enzymologia *38*, 39.

[71] Goldfeld, M. G., Vorobeva, E. S. and Poltorak, O. M. (1966) Comparative study of catalase activity in adsorption layers of various type. Zh. Fiz. Khim. *40*, 2594.

[72] Ryu, D. Y., Bruno, C. F., Lee, B. K. and Venkatasubramanian, K. (1972) In: Fermentation Technology Today (G. Temi, ed.) p. 307. Society of Fermentation Technology, Japan.

[73] Broun, G., Thomas, D., Gellf, G., Domurado, D., Berjonneau, A. M. and Guillon, C. (1973) New methods for binding enzyme molecules into a water-insoluble matrix: Properties after insolubilization. Biotechnol. Bioeng. *15*, 359.
[74] Roy, A. K. and Roy, S. (1983) Synthesis of novel types of protein-compatible weak anion-exchange silica packing. Liquid Chromatography *1*, 181.
[75] Rexova-Benkova, L., Omelkova, J., Mikes, O. and Sedlackova, J. (1982) Medium Pressure Liquid Chromatography of leozym, a pectic enzyme preparation on ion exchange derivatives of spheron. J. Chromatography *238*, 183.
[76] Frolik, C. A., Dart, L. L. and Sporn, M. B. (1982) Variables in the High-Pressure Cation-exchange Chromatography of proteins. Anal. Biochem. *125*, 203.
[77] Hollein, H. C., Ma, H., Huang, C. and Chen, H. T. (1982) Parametric pumping with pH and electric field: Protein separation. Ind. Eng. Chem. Fundam. *21*, 205.

[illegible] [illegible] D. [illegible] D. [illegible] M. and [illegible] (1973). New [illegible] [illegible] insoluble [illegible]. Properties [illegible] Biotechnol. Bioeng. 15, 359.

[illegible] R[illegible] and K[illegible] (1969) Synthesis [illegible] [illegible] [illegible]

[illegible]

[illegible]

[illegible]

4 Ion Exchangers as Preparative Agents

Konrad Dorfner
Mannheim, Federal Republic of Germany

Introduction

Ion exchangers have developed into indispensible aids in laboratory and pilot plant scale development alike. Leaving aside the analytical applications and industrial chromatographic separations, which are comprehensively treated in separate chapters in this volume, ion exchangers can be used for a wide variety of simpler separations, as well as for numerous preparation processes. Some of these processes may never leave the laboratory, but others will grow into industrial processes, as has already often happened. The laboratory chemist normally gets his laboratory chemicals from reagents' and fine chemicals' suppliers and it is these supply houses that usually also offer an ion exchanger and polymeric adsorbent program covering at least the various types of ion exchangers and adsorber resins applied in water treatment and industrial chemical processes. What is of importance for the application of ion exchangers in the laboratory will be discussed first in the following and then the employment of ion exchangers in preparative chemistry will be the main part of this chapter.

4.1 Ion exchangers for the laboratory

The synthetic ion exchange resins referred to elsewhere in this volume are commercial ion exchangers and are produced by chemical industrial companies for technical purposes. The use of ion exchangers in the laboratory first emerged in the late 1940's in the field of analytical chemistry and resulted in a number of special requirements for ion exchanger materials. These are satisfied only partly by commercial products which are, besides special resins offered for instance for pharmaceutical and medicinal applications, mainly intended for industrial application. To respond to these needs laboratory ion exchangers have usually been developed in a special, closed ion exchanger production program for those working in analysis as well as in preparative chemistry to save time-consuming and costly preliminary procedures. For the laboratory the ion exchange resins must sometimes be carefully purified of very obstinate contaminations and they must have smaller and narrower particle size ranges. Normally, the specifications are: for the heavy metal content $<0.0005\%$ and for the particle sizes $>90\%$ in the range concerned. Such products can in principle be manufactured by any laboratory or fine chemicals company but there are laboratories and chemical companies that have become specialized, usually in cooperation with producers of commercial ion exchange resins. As a result, smaller quantities of resins are processed with special ionic forms for both research and pilot plant manufacturing use. The most frequently requested forms are hydroxide, acetate, formate, and halides for anion exchangers, and sodium, potassium, ammonium, tetramethyl ammonium, and lithium for cation exchangers. With regard to the specifications of ion exchange resins for the laboratory two more things should be mentioned, i. e., the specification of aliphatic amines and the so-called color throw. In anion exchangers aliphatic amines will increase slowly with time due to slight resin decomposition. Cation exchange resins in the hydrogen form normally produce a trace of red organic substance, which is a higher molecular weight sulfonate, on long standing. This is very little compared with the quantity of resin over a period, say, of six months, and may be removed with one to two bed volumes of deionized water.

Several entire production programs of ion exchangers (including tradenames used by the commercial ion exchanger producers) for the laboratory are listed in Appendix 1. They usually cover strong and weak acid cation exchangers as well as strong and weak base anion exchangers with the characteristic ionogenic groups and in standard forms. With the exception of certain types, crosslinking amounts to 8% in terms of divinylbenzene used during preparation, which is usually an optimum. The available particle sizes are suitable for general analytical purposes and can be recommended for ion exchange chromatography, since these granulometries are frequently used to determine distribution coefficients. The incorporation of color indicators permits a purely visual distinction between the nearly colorless H-form and, e. g., the red metal form, as well as between the blue OH-form and the purple-red form which is, e. g., loaded with a different anion. Also strong base anion exchangers of Type I can be found. The requirements for greater porosity are satisfied by macroporous (macroreticular) types of which strong and weak acid and strong and weak base exchangers are also available for analytical and laboratory preparative purposes.

As a result of special production techniques, the laboratory ion exchangers have a purity and heavy metal content which permit their immediate use in analysis. The fraction of intrinsic resin components which can be eluted is also strictly controlled by the manufacturing process. Therefore, for immediate laboratory use it is necessary only to provide for preswelling, which has been described elsewhere; with the use of columnar processes, transformation into the desired loading form is the only step required after packing the column.

Since the laboratory ion exchanger programs also cover products of the gel type and those of truly porous structure, the choice of a suitable type to investigate a given problem can also take this aspect into consideration, even for routine analyses. This is a particularly welcome addition for operations in nonaqueous and mixed solvent systems, since the exchanger properties of cations on cation exchangers as well after complexing on anion exchangers can be notably influenced by the use of such systems. It is not necessary to go beyond these general comments. Special applications of ion exchangers in preparative chemistry and elsewhere, which go beyond ordinary laboratory uses – in which particular exchangers offered in a laboratory ion exchanger production program can be used simply on the basis of their special properties – will be described in many examples in the appropriate sections of this monograph.

The preparation of reagents and fine chemicals by purifying technical products by ion exchange has shown that the ion exchange resin itself can be a source of impurities. For purifying aqueous solutions of technical grade copper, nickel, aluminium and manganese sulfates and various other inorganic salts the resin impurities had to be removed in fifteen cycles. The organic impurities were removed efficiently by a final treatment with charcoal [1]. But even the so-called analytical grade resins may have to be further purified, despite having been exhaustively washed and cycled in the proprietary processes of the manufacturers. Frequently, in spectrochemical analyses which involve analyzing ion exchange resins that have been ashed, resins that are purer than commercially available laboratory materials are required. A method, taken from the literature, for purifying commercial cation exchange resins follows:

One kilogram of AG 50W X8, 100–200 mesh hydrogen-form cation-exchange resin was placed in a 1-liter Teflon beaker and washed several times with distilled water. The finest particles were decanted off at each washing. The resin was then converted to the sodium form by adding 100–200 ml volumes of a 50 g/l sodium chloride solution to the resin, with stirring. The resin was allowed to settle after each addition of the sodium chloride solution and the excess liquid decanted off. This procedure avoided lowering the pH too greatly and removed some of the impurities by ion substitution. After the resin was converted to the sodium form, as evidenced by no additional lowering of the pH of the sodium chloride solution, the resin was washed several times with triply distilled water.

The resin was then washed with 1 liter of 1 : 1 HCl. This reconverted the resin to the hydrogen form and eluted more of the impurities from the resin. Because the resin shrinks when converted to the sodium form and expands when treated with the 1 : 1 HCl, these conversions were performed in a beaker rather than a column, so that the resin would not pack tightly and require a backwashing step. The backwashing step contaminates the resin with silica from the quartz frit.

The resin in 1 : 1 HCl was transferred to the quartz column and washed with an additional liter of 1 : 1 HCl. The resin was then washed with triply distilled water until the $AgNO_3$ test indicated no Cl^- in the effluent.

Any impurities added by the wash solutions would be collected by the top portion of the ion-exchange bed. For this reason, the top 1 in. of resin was removed and discarded with the apparatus shown in Figure 4.1, which is constructed of Vycor. The removed resin and water were deposited in the 500 ml filtering flask when the flask was connected to a vacuum line. Additional triply distilled water must be added during this process to facilitate removal of the resin. The resin remaining in the column, except for a 1-in. layer at the bottom of the column, was collected and saved as the purified resin.

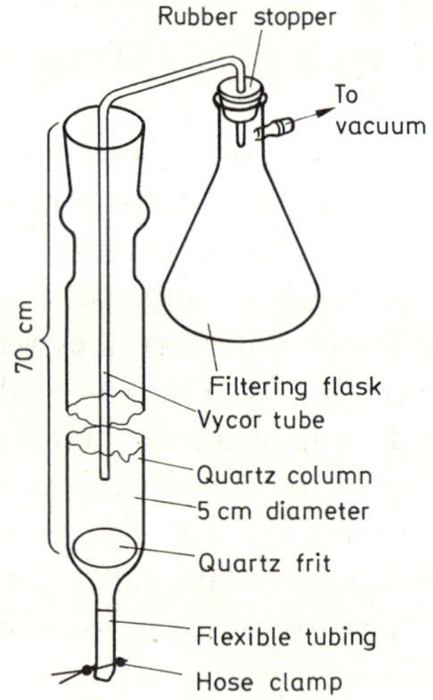

Figure 4.1 Column and apparatus for removing resin according to Wanner and Conrad.

To demonstrate the effectiveness of this purification procedure, samples were taken from both ends of the purified resin. These samples, as well as an equal weight of the starting material, were ashed and analyzed by semiquantitative spectrochemical techniques. The results are listed in Table 4.1 [2].

Table 4.1 Comparison of impurity levels of original resin versus purified resin

Element	Original resin (ppm)	Purified resin (top) (ppm)	Purified resin (bottom) (ppm)
Na	40–20	0.5–0.1	0.5–0.1
K	≈ 4	0.1–0.05	≈ 0.05
Ba	≈ 4	ND < 0.01	1–0.5
Sr	4–2	< 0.005	< 0.005
Al	≈0.2	0.1–0.05	0.1–0.05
B	ND < 0.2	0.05–0.01	< 0.005
Mg	4–2	≈ 0.05	0.05–0.01
Si	≈ 0.2	≈ 0.05	≈ 0.1
Cu	2–0.4	0.1–0.05	0.1–0.05
Ca	40–20	0.1–0.05	≈ 0.05
Ash	350	10	10

Special ion exchange polymers in the form of powders, fibers, films or foams are ready for use in the laboratory, but have not yet become commercially available. Further, ion exchangers have been used in thin layers [3]. The term flat-bed ion exchange material has been introduced for ion exchange papers containing various ion exchange resins and for ready-made thin layers that also contain ion exchange resins. For the manufacture of these layers in plates or sheets, as usual silica gel and a binder must be used. In cases where the resin is a strongly basic anion exchanger problems arise in the interpretation of research results, as silica gel is a very efficient cation exchanger (for example, for $Co(en)_3^{3+}$ complexes). For this reason in the literature a request was made to manufacturers to label completely all ion exchangers used for flat-bed ion exchange materials [4].

Extracting resins. These special ion exchange materials are a combination of a polymeric matrix with liquid ion exchangers and are obtained either by the impregnation of various polymeric particles with liquid ion exchangers (Small, 1961; Clingman and Parrish, 1963), macroreticular ion exchangers containing alkylated amidines (Dow Chemical Co., 1971), and macroporous polystyrene adsorbents containing various commercial liquid ion exchangers in the adsorbents (Warshawsky and Patchornik, 1976), or by the incorporation of liquid ion exchangers or extractants into the macroporous styrene-divinylbenzene-based copolymers during polymerization (Kroebel and Meyer, 1974; Levextrel®). Some of the extractants that may be used are aliphatic amines, phosphoric esters, and both aliphatic and aromatic amines. Because of the special structure these resins attain a high porosity, which both improves the kinetics, and increases the quantity of extractant that can be incorporated into the resins. Although the diffusion of the extractant into the aqueous medium is reduced, such a discharge also represents an economic loss. To make up for the loss of extractant in the resin, reincorporation of the extractant is practiced.

With respect to the chemical properties of these extracting resins it should be noted that their selectivity depends on the ions involved. Thus, the variety of products that can be made available is limited only by the number of liquid ion exchangers or liquid extractants available. In their physical properties the beads have a low bulk density of about 630 g/l and very low density of about 1.0 g/cm^3. The extractant can be 25 to 50% of the bulk weight. Since the extractant is hydrophobic, the resins have a very low water retention of $< 0.3\%$.

The extracting resins are an improvement over the liquid-liquid extraction process in that they eliminate the problem of phase separation and have very good capacities and good reaction rates. The resins were initially developed for the reprocessing of solutions of radioactive heavy metals like uranium, plutonium, americium, thorium, etc., especially in the reprocessing of irradiated nuclear fuels, particularly for the removal of radioactive substances from effluents. But they can be employed as well for the enrichment of heavy metals from solutions of very low concentration (in the ppm range). As further applications there are the removal of ions in heavy concentrations but of limited volumes, removal of unwanted traces of ions in concentrated baths, conversion of metal ions from one salt form to another, e. g., $ZnCl_2$ to $ZnSO_4$ through the use of sulfuric acid as the eluting agent, and separation of metal ions, thus obtaining salts from wastes in purer form so that they can be more readily recycled. There is also the possibility of utilizing such resins with extracts having an affinity for specific organic entities.

4.2 Ion exchangers in preparative chemistry

A review of the main preparative processes employing ion exchangers would include more than will be described in the following sections. More exhaustive reviews can be found in review articles and textbooks on ion exchange processes or methods. In the question of how to define ion exchange we will take a pragmatic stance: taking water as the most common solvent, we know that electrolytes dissolved in water can undergo certain changes when in contact with various solids. Under certain conditions, some of the ions present in the water exchange for ions that are present on the surface, or even in the interior, of the solid. When this exchange occurs, the total number of charges leaving the solid must equal the total number entering in order that the law of electroneutrality be satisfied. This process, in essence, is the phenomenon of ion exchange. Ion exchange is the change of ions with respect to the phases involved and nothing but this. All processes that might follow are a consequence of the application of the ion exchange principle. Depending upon need numerous procedures employing the exchange principle have been developed, as well as various methods. Who can say where the imagination will lead us?

One and the same application may be found in different fields of chemistry, and in other sciences and technologies. Preparative chemistry, being the opposite of analytical chemistry, is in many cases, including those involving ion exchange, the basis of later industrial processes. One also has, on the contrary, processes practised in large scale industrial units that may as well be used in the laboratory for preparative chemistry.

4.2.1 Ion interchange

Ion interchange (also known as conversion or metathesis) is the simplest preparative application of ion exchangers. Benefits are derived from the general ion exchanger property of exchanging counterions for other ions. Cation exchangers in the H-form and anion exchangers in the OH-form exchange the hydrogen ion or hydroxyl group for cations or anions, respectively:

$$\text{Cation exchanger} - SO_3H + MeX \rightarrow \text{Cation exchanger} - SO_3Me + HX$$

$$\text{Anion exchanger} - N(R_3)OH + MeX \rightarrow \text{Anion exchanger} - N(R_3)X + MeOH$$

Thus, a preparative possibility results of obtaining free acids and bases by a simple method. The procedure of recovering inorganic and organic acids by converting their barium salts with sulfuric acid or their lead, silver, or other heavy metal salts by conversion with H_2S is considerably simplified by the use of ion exchangers. Salts that can be obtained only through the free acids or from the alkali salts of the corresponding acids are also easily obtainable in this manner.

Generally, strong acid and strong base ion exchangers are used for this purpose. In some cases, weak base exchanger types are also applied.

Ion exchange processes are usually characterized by their simplicity and offer four additional advantages for preparative work: high rapidity, high purity of the products, suppression of insoluble precipitates, and low consumption of reagents.

4.2.1.1 Preparation of acids

The preparation of acids by ion interchange was first described by Klement [5]. He prepared thiocyanic acid, hypophosphorous acid, phosphoric monamide, triphosphoric acid, and polyphosphoric acid [6]. As a practical example, the procedure for the preparation of thiocyanic acid is described below.

Preparation of thiocyanic acid. Sulfonic acid exchange resin (70 g) in an exchange column is transformed into the H-form with 2 N HCl and washed with distilled water until the eluate remains clear after the addition of silver nitrate solution.

Subsequently, a solution of 10 g ammonium thiocyanate in 50 ml water is charged on the column, allowed to stand for 10 to 15 min, then allowed to drain. After draining, the column is washed with 50 ml water. An approximately 8% solution of the free thiocyanic acid is obtained in the eluate.

The method described by this example is generally applicable.

In this connection the research of Hein and Lilie [7], who developed the preparation of different complex acids on the basis of the mild conditions afforded by ion exchange, is of interest: when 1,5 g $K_3[Co(CN)_6]$ in 50 ml water is reacted with Wofatit P and subsequently dried in a vacuum desiccator, 0.69 g of the free acid $H_3[Co(CN)_6] \cdot aq$ is recovered. Further examples for the preparation of acids by ion interchange can be cited: there is the preparation of nitrous acid, or of diluted hydrochloric acid from common salt and sulfuric acid in New Zealand in 2 to 5% concentration containing 2 to 3% salt but still suitable for the manufacture of casein in dairies. The ion exchange preparation of hydrochloric acid on a scaled-up laboratory apparatus by using a strong acid cation exchanger in the H-form by displacement with 4.92 M NaCl solution has also been described. One can also prepare hydrazoic acid by an ion exchange technique. Sodium azide is passed through a column containing the hydrogen form of a sulfonic acid cation exchanger, the hydrazoic acid being eluted at the bottom. The method removes cationic impurities and also produces an acid of about 10–15% concentration, which has an effective margin of safety for explosivity. The ion exchange resin was regenerated with dilute sulfuric acid after washing it free of unreacted sodium azide. Preparation time for the hydrazoic acid is less than for the classical distillation technique. For the manufacture of carboxylic acids from the corresponding metallic salts in a patent example 1.44 parts sodium benzoate and 10 parts water are passed into a column containing a weak acid cation exchange resin in the acid form and eluted with metal hydroxide. Evaporation of the eluate yields crystals of benzoic acid. Terephthalic acid has also been prepared from the sodium salt. The use of ion exchange resins in chemical preparation has also been demonstrated in the manufacturing process for boric acid. In a first process alkali metal borates were passed through columns of a weak acid carboxylic cation exchange resin at elevated temperatures. Boric acid in the effluent was crystallized at ambient temperature. It was further possible to prepare extrapure boric acid of electronic grade using borax of technical grade and a strong acid cation exchange resin in the hydrogen form at 60 °C obtaining a solution of boric acid free of chloride and sulfate [8].

4.2.1.2 Preparation of bases and salts

Just as acids can be prepared from salts on cation exchangers, free bases can also be recovered with the use of anion exchangers. Thus, d'Ans et al. [9] were able to obtain the bases forming various complex salts. As an example, the conversion of hexamminocobalt(III)-chloride is described below; this can serve as a general working method.

Preparation of hexamminocobalt(III)-hydroxide. Permutit ES, Dowex 2 or Amberlite IRA-400 (50 g) is transformed into the OH-form with 1 liter 1N NaOH in a column and washed until the eluate is neutral. The yellow-brown solution of the cobalt complex in the form of its chloride is then charged on the column and allowed to percolate through the column at a flow rate of 10 ml/min. Subsequently, it is washed with about 300 ml water. The filtrate contains the free base $[Co(NH_3)_6](OH)_3$, identified by titration with 0.1 N HCl.

If the solution of free complex base obtained by this method is treated with ammonium oxalate, the slighthy soluble oxalate of the hexamminocobalt(III) ion precipitates. This method has been used to convert the chloride into the oxalate on ion exchangers.

That the preparation of bases is not limited to inorganic chemistry is demonstrated by a report of Rebek and Semlitsch [10] on the preparation of a dye base with ion exchangers. The preparation of non-aqueous solutions of hydroxides of quaternary ammonium bases by the ion exchange method has been optimized to yield solutions of these bases suitable for analytical purposes. One can proceed as follows. A hundred grams of air-dry strong base anion exchanger Type I in the Cl-form is mixed with absolute methanol. After the initial swelling it is washed with 5 l NaOH in methanol and then washed with methanol until free of sodium hydroxide in a column 30 mm diameter and 600 mm high. Then a 0.1 N solution of R_4NCl is passed through the column at 1 ml/min. After washing with methanol, R_4NOH is eluted completely with 1 N NaOH in methanol. Tetraalkylammonium chlorides, hydrogen sulfates, and formates are phase-transfer catalysts. They were prepared by using the corresponding anionic form of the anion exchanger Dowex 1X2 and a solution of tetraalkylammonium bromide [11].

A special method which must be mentioned in connection with the preparation of salts is represented by so-called ligand exchange developed by Helfferich [12]. In this procedure the complex water molecules of a metallic ion bound to a cation exchanger of the carboxylic acid type are exchanged for ammonia. The method can be extended, the ammonia being then exchanged for diamine. In contrast to ion exchange in general, this method in accordance with complex chemistry leads to an exchange of ligands on the counterion while the latter is retained in the resin phase. Carboxyl resins are used for this purpose because of their higher selectivity compared to sulfonic acid exchangers.

Additional possibilities in the preparative use of ion exchangers which go beyond their common application for purification and separation can be expected in reactions leading to the formation of salts from acids, analogous to the Solvay process. In this application of weak base anion exchangers, the exchange process depends on the relative selectivity of the anion exchanger for chlorides and bicarbonates [13].

It has been shown that anion exchange materials can be used to convert any acid to its corresponding alkali salt, when the selectivity is sufficiently favorable and the anion exchanger is sufficiently basic. Starting from potassium chloride two important chemicals – potassium nitrate and potassium silicate – can be obtained by the use of an ion exchange method. The larger scale ion exchange technique to prepare potassium nitrate uses 15 l strong base anion exchange resin in the nitrate form placed in a 14 cm inner diameter PVC column up to a height of 90 cm. In the exhaustion cycle a 2.5 N solution of potassium chloride is passed through the resin. To obtain maximum utilization of resin capacity two PVC columns were used in series. The results showed that over 85–90% resin capacity is utilized for the conversion of potassium nitrate. Regeneration of the resin was done by using 1.0 N ammonium nitrate solution, so that ammonium chloride was obtained as a by-product. The particle size of the resin and its capacity were determined at the end of twenty days of twelve hours daily operation with nearly hundred cycles. No deterioration was found in either capacity or size of the resin particles. The potassium nitrate solution was evaporated and fractionally crystallized to obtain a product with ninety-nine percent purity [8]. Potassium nitrate as well as sodium nitrate were also prepared from calcium nitrate by using cation exchangers in the corresponding form. The utility of several strong acid cation exchangers for the preparation of lithium tetraphenylborate was investigated and they were found suitable for the synthesis of $LiB(C_6H_5)_4 \cdot 2H_2O$ in acetone yielding 85–90% pure compound content, 99.5–99.8%, potassium content 0.002% and density 1.128 g/cm^3. The preparation of alkylsalicylate additives proceeded over alkaline earth metal alkylsalicylates that were obtained by an exchange reaction between aqueous solutions of alkaline earth metal chlorides and sodium alkylsalicylates. The additives were prepared making use of different excesses of alkalinity $\leq$ 100%, by adsorption of sodium carbonate on sodium alkylsalicylate before the ion exchange process. The use of ion exchangers simplified the technology of production of alkylsalicylate additives; furthermore, the products were similar in properties to those obtained by neutralization of alkylsalicylic acid with alkaline earth metal hydroxides [14]. A process has been elaborated for the production of monosodium glutamate or glutamic acid by using ion exchange resins, which is suitable for application on a factory scale. With this method, wheat gluten was hydrolyzed with sulfuric acid and neutralized with calcium carbonate. The gypsum formed was filtered off and the filtrate fed into the ion exchanger. The inorganic cations of the solution are accumulated on a cation exchanger in the hydrogen cycle, whereupon the liquid is decolorized on an adsorption resin. The cation-free, light yellow solution of the amino acid mixture is led onto a slightly basic anion exchange resin to which the glutamic acid is bound, whereas the remaining amino acids drain off in an unchanged condition. The bound glutamic acid is removed from the resin by sodium hydroxide, and monosodium glutamate obtained directly. Acidification with hydrochloric acid then gives glutamic acid.

The production of inorganic brines such as SiO_2,$Fe(OH)_3$,$Al(OH)_3$, thorium, and zirconium brines and the formation of hydrosols [15] are also examples for the preparation of bases and salts. Stable sols of silicon dioxide yielding products containing up to 40% SiO_2 and 500 : 1 $SiO_2 : Na_2O$ ratio are obtained by deionization of an aqueous solution of an alkali metal silicate containing 2.5 – 3.5 SiO_2 by passage through alternating beds of cation and anion exchange resins and further working up after stabilization by continuous or batchwise evaporation. The sol solidifies at low temperatures and liquifies above 0 ° centigrade.

4.2.1.3 Preparation of standard solutions

Another example of the use of ion exchangers as a preparative agent is for the preparation of normal solutions containing practically no carbonate or silicate.

A strong base ion exchanger such as Amberlite IRA-400 is treated with dilute hydrochloric acid in a column, washed, and subsequently transformed into the OH-form with carbonate-free 2 N NaOH. The exchanger is then washed free from alkali with boiled distilled water. When such an exchanger is loaded with a carbonate-free common salt solution, a quantitative ion interchange leads to a carbonate-sodium hydroxide solution with the equivalent quantitiy of NaOH. In practice, the method of Steinbach and Freiser [16] can be used. A solution of 2.922 g NaCl (analytical grade) in 50 – 100 ml boiled distilled water is loaded on 40 g Amberlite IRA-400 at a flow rate of 4 ml/min. The column is subsequently washed with freshly boiled distilled water. The eluate is collected in a 500 ml graduated cylinder in the absence of air and finally is brought to the mark. The 0.1 N NaOH obtained is practically free from carbonate. For the determination of silicate concentrations of 0.1 – 0.01 ppm in water by colorimetry, the blank value of an ordinary 1 N NaOH solution (analytical grade) is too high. However, with the above method an ultrapure silicate-free potassium hydroxide solution can be prepared by loading an equivalent solution of potassium sulfate on a strong base anion exchanger in the OH-form as described by Fisher and Kunin [17].

These examples offer an insight into the broad field of application of ion exchangers for the preparation of bases and salts. A few other examples will only be enumerated: silicic acid can be prepared from sodium silicate on a strong acid cation exchanger; a strong base cation exchanger in the K-form is used for the recovery of the potassium salt of penicillin; sodium citrate is obtained on a strong base anion exchanger in the citrate form; metacycline is purified or converted to other salts, and 11a-chlormetacycline-HI converted to the sulfosalicylate by adsorption onto a macroporous ion exchanger and elution with an appropriate acid.

4.2.2 Purification of solutions and substrates

Ion exchangers have found an important and rapidly growing application in the purification of solutions and substrates; thus, ionic components of solutions can be completely removed by cation or anion exchange resins. These processes, which have become known by the terms desalination, deionization, or demineralization, play an important role in the application of ion exchangers as preparative media, as well as in technology.

Laboratory applications of this type include: (1) extraction of acids and bases from neutral salts, organic materials and solvents; (2) extraction of metal salts from acids and bases; and (3) desalting of organic or biological substrates and nonaqueous solvents.

In the first two methods, cation exchangers in the H-form or anion exchangers in the OH-form are used. A mixed bed process can also be applied for desalting. This refers to a bed containing a cation as well as an anion exchanger which simultaneously removes cations and anions from the solution and replaces the salt by an equivalent quantity of water.

The practical application of such purification methods will be demonstrated by a few examples.

Example 1: Extraction of copper ions from a weakly sulfurous cupric sulfate solution on a strong acid cation exchanger. By this procedure, a copper-free solution can easily be obtained. The purification effect is easy to see because of the inherent color of copper ions.

A preswollen, strong acid cation exchanger (5 g) is slurried into a column of 15 × 1,5 cm and washed once with distilled water. Subsequently, 200 ml 2 N sulfuric acid is percolated through the column to transform the exchanger into the H-form. The column is now washed with distilled water until the eluate is neutral.

After this preparation of the column, the weakly sulfurous, blue copper(II) sulfate solution is loaded onto the column and percolated through it at a flow rate of 2 ml/min. The eluate, which is collected in a beaker, is colorless. The exchanger bed assumes a greenish opalescence as soon as it is partially transformed into the copper form. Loading of the copper(II) sulfate solution can be continued until breakthrough for copper has been attained; this can be easily recognized by the appearance of the deep-blue copper complex in an ammonia-treated receiver.

For the regeneration of the exchanger column, 300 ml 2 N sulfuric acid is charged on it and percolated through the column at 2 ml/min. After brief washing with distilled water, the column is ready for a further purification cycle.

Example 2: Extraction of small amounts of iron from technical-grade hydrochloric acid on a strong base anion exchanger. This example demonstrates that metals can be bound not only on cation exchangers but also on anion exchangers. The mechanism of action involves the conversion of the metal into an anionic complex so that it can be sorbed by an anion exchanger.

In concentrated hydrochloric acid solution, iron is present in the form of the anion $FeCl_4^-$. Therefore, if a concentrated hydrochloric acid solution containing iron(III)-chloride as an impurity is loaded onto a column of strong base anion exchanger, the eluate is free from iron.

Since the iron complex decomposes with decreasing acid concentration,

$$FeCl_4^- \rightleftharpoons FeCl_3 + Cl^-$$

the iron can be removed from the column in the form of $FeCl_3$ by simple washing with water and the column can thus be easily and inexpensively regenerated.

In the practical procedure, a column of 110 × 2,5 cm packed with anion exchangers in the OH- or Cl-form is used. To test the activity of the exchanger, the column is first loaded with about 25% hydrochloric acid containing 50 – 500 ppm iron(III)-chloride. Subsequently, 2 l water is loaded onto the column to elute the iron and regenerate the exchanger. The ion exchanger is then ready for further acid purification cycles.

Since technical grade hydrochloric acid, which is produced by various processes, always contains traces of iron, this method is of industrial importance.

Another example for the removal of metals is the purification of quaternary ammonium bases from alkali metals, which leads to final impurity concentrations of less than 1 ppm Na^+ with a yield of 80% [18].

Example 3: Extraction of small quantities of ammonia from water in a mixed bed process. Ammonia-free water can be produced from distilled water by ion exchange in a mixed bed process. For this purpose, it is best to use an exchanger mixture of 2 volume parts of Amberlite IRA-400 and 1 volume part of Amberlite IR-120.

The mixed exchange resins are used in the column process. When loaded with distilled water, for example, the nitrogen content will be reduced from 0.860 mg N_2/50 ml to 0.006 mg N_2/50 ml. The ammonia concentration in the eluate is determined by photometry with Nessler. In such cases it is more convenient in laboratory practice to discard the exhausted exchanger than to regenerate it.

The degree to which the removal of ammonia from substrates can be brought is demonstrated by the example of blood purification according to Moretti et al. [19].

Blood contains unstable ammonia substances which easily decompose, with the formation of NH_4-ions. When the strong acid cation exchanger, Dowes 50X8, was used in the Na^+-form in this procedure, the cation equilibrium in the blood shifted in favor of sodium at the expense of potassium and calcium until the exchanger was used in such a ratio of loading with sodium, potassium, calcium and magnesium that no coagulation or its consequences occurred. This demonstrates another interesting possibility of ion exchange. If such a column is installed into a heart-lung machine for extracorporeal circulation and blood is pumped through it for the removal of NH_4^+-ions, a clinical treatment of hepatic coma might well be feasible. From the purely preparative standpoint, blood – a valuable substrate – has been freed from the undesirable impurity of NH_4^+ by ion exchange.

Example 4: Extraction of pyrosulfite from noradrenaline solutions. Noradrenaline solutions for medical purposes are generally stabilized with pyrosulfite. For a quantitative determination of noradrenaline in dilute solutions (up to 0.001%), the fluorescence method of Erne and Canbäck [20] can be applied; however, pyrosulfite represents an interference substance and must be removed from the solutions. An anion exchanger can be used advantageously for this purpose.

The anion exchange resin (e. g., Amberlite IRA-400 or Dowex 2, particle size 0.4–0.8 mm) is washed with 20% sodium chloride solutions in a 100 × 10 mm column and the excess of chloride is removed with water. Thus prepared, the column is loaded with the noradrenaline solution and washed with water. The noradrenaline determination can be made in the filtrate.

The value of such an ion exchange process can be seen from the test data of Table 4.2 (taken from the cited authors) showing the fluorescence values before and after ion exchange.

Table 4.2 Fluorescence values of noradrenaline analysis with and without ion exchange

Noradrenaline	Fluorescence values	
	without ion exchange	after ion exchange
10 mg	78, 79, 80, 78	77, 79, 78, 80
10 mg with 10 mg pyrosulfite	35, 40	79, 79, 78, 79

Example 5: Extraction of formic acid from formaldehyde solutions. In the production of formaldehyde from methanol by air oxidation according to the equation:

$$2\,CH_3OH + O_2 \rightarrow 2\,HCHO + 2\,H_2O$$

a small quantity of formic acid also forms according to:

$$2\,CH_3OH + 2\,O_2 \rightarrow HCOOH\,.$$

Formic acid interferes with the production of plastics. Although it is present in aqueous formaldehyde solutions in concentrations of only about 0.1%, it can nevertheless lead to complications. An ion exchanger can be used for its extraction.

Amberlite IR-45 is transformed into the OH-form with sodium hydroxide solution and is carefully washed to neutrality. The formaldehyde solution, in which the formic acid content had been previously determined, is loaded on the column and percolated at a low flow rate. By varying the operating conditions, elevating the column temperature if necessary, the process can be controlled so that formic acid is extracted from 35–50% formaldehyde solutions to residual concentrations of 0.001%.

Example 6: Purification of a nonaqueous solvent; extraction of mercaptans from hydrocarbons. Genuinely porous or macroreticular ion exchangers can be very advantageously used in nonaqueous systems. The procedure is first carried out in water in order to transform the exchanger into the desired loading form and for regeneration. If the nonaqueous solvent is immiscible with water, an intermediate treatment is necessary with a solvent which is immiscible with water but miscible with the solvent to be treated subsequently. Methanol and ethanol are particularly suited for this purpose when used at a flow rate of 0.05 – 0.1 ml/min. After this step, the column is treated with one to two column volumes of the pure solvent and is thus ready for the purification step. When the latter has finally been completed and the exchanger is exhausted, two column volumes of alcohol are again charged through the column and displaced with water in the transition to regeneration. A practical example of a purification process which is frequently used in a similar form is the extraction of mercaptans from hydrocarbons.

Amberlyst XN-1002 (25 ml) is regenerated in a suitable column with three column volumes of aqueous N sodium hydroxide solution at a flow rate of 0.1 ml/ml/min, washed with seven column volumes of water and treated with four column volumes of alcohol. A solution of 500 ppm octylmercaptan in isooctane is passed through the column at the same flow rate. The eluate is analyzed by iodometry. Under these conditions, 150 column volumes can be percolated through the column until breakthrough is observed at 5 ppm.

There are numerous examples in the literature of ways of increasing the efficiency of ion exchange substance purification. It is usually the aim of the general investigations to elucidate the principles of ion exchange processes for the removal of impurities from dissolved substances in relation to the acidity or basicity of the exchange material and to find methods for calculating the main technological parameters such as the operating time, the degree of utilization of the ion exchanger, the total volume of purified solution before breakthrough, the length of the ion exchanger bed, and the width of the profile of an elution curve. These investigations can be based on several different models and give, in some cases, solutions as well for problems in the determination of kinetic parameters, such as diffusion coefficients, etc. There are procedures involving both complex separations and simpler purifications with the working parameters for the latter still being elaborated empirically in most cases.

The ion exchange flowsheet for final purification of B plant cesium has been made public. The primary objective of the final cesium purification process is to produce a cesium carbonate product suitable for the encapsulation process. The encapsulation specifications are designed to minimize the volume of material to be encapsulated and control corrosion rates by assuring that the salt remains in the solid form. The feed is diluted with demineralized water and is loaded onto the ion exchange bed. The impurity cations sodium, potassium and rubidium are removed from the bed with a diluted $(NH_4)_2CO_3 - NH_4OH$ solution passed through the bed in the same flow direction as the loading step. The cesium is removed from the ion exchange bed with a concentrated solution passed counter-flow through the bed and the resulting product is concentrated [21]. How important the resin itself can be for an envisaged purification process was demonstrated for the case of the ion exchange purification of sodium bromide for pharmaceutical use. Among six ion exchangers tested only two gave a product sufficiently pure to satisfy the purity requirements of the drug industry. For the purification of potassium bromide, strontium and barium fluoride for laser components the use of selective ion exchange filters was considered. The purification of aluminum chloride solutions by a weak base resin decreased the iron chloride content from 16 to 0.001 g/l in solutions containing 250 g $AlCl_3$/l. The ultra-purification of titanium trichloride that can be obtained by passing a solution through an ion exchange column containing Dowex 1X10 is chromatographic in nature. The solution thus obtained is suitable for the electrochemical preparation of high-purity titanium [22]. One can use a mixed bed of ion exchangers for the purification of boric acid solutions containing 4 – 5% impurities to obtain solutions with $4 \cdot 10^{-4}$% total impurities, which may be used directly in industry instead of solutions prepared form extra-purified acid. Ion exchange extraction and purification may be combined as it has been for the removal of indium from zinc solutions on macroporous phosphonic groups containing resins of different crosslinking and pore structure. An amphoteric ion exchange resin was found suitable for the purification of ammonium tungstate solutions. A method for the preparation of extrapure zinc sulfate for phosphors for color television has been elaborated, yielding a salt whose properties are superior to those of zinc sulfate prepared by the precipitation method. A semi-industrial ion exchange process for removing zinc from a nickel sulfate-chloride electrolyte has been developed, lowering the zinc content from 150 – 200 to <0.3 mg/l, making it possible to produce nickel containing $\leq$ 0.01 % zinc. This was achieved by using highly basic ion exchangers. One in particular containing active quaternary pyridine groups was found more suitable for industrial applications because of high capacity and mechanical strength. For nickel sulfate purification chelating and carboxylic ion exchangers have been employed. The various technologies for the purification of the individual rare earths use ion exchange as a separation process performed on column beds of organic and inorganic exchange materials [23]. Much experimental work has been done on the use of inorganic exchangers for the isolation, separation and purification of plutonium and transplutonium elements. Relevant data have been presented on the chemical stability and radiation-resistance of the sorbents, with respect to the dependence of the sorption of a whole series of elements on the concentration of acid in the solution. Methods for separating transplutonium elements and the isolation of individual elements from complex mixtures have always been in demand.

In organic chemistry the application of ion exchangers for purification processes has not come to a stillstand either. For the purification of acrylonitrile a strong acid cation exchange resin in the hydrogen form was used. The column is rinsed with a low molecular weight alcohol to remove small amounts of methyl vinyl ketone, a treatment that effectively removes the methyl vinyl ketone from the acrylonitrile. For the preparation of formalin of pure and analytical purity grade a two-column system, the first containing a strong acid cation exchanger in the H-form and the second a strong base anion exchanger Type I in the OH-form, was recommended, for the ion exchange purification of technical 36 – 38% formalin at 40 – 60 °C to give a reagent-grade product. The production of glycerol of a high degree of purity from soapy alkalies containing approx. 30% glycerol can be achieved by a three-stage ion exchange purification with six filters after electrodialysis, using the clarifying resin IA-1R, the ion exchangers KU-2 and EDE-10P in the first stage, KU-23 and AV-17P in the second and KU-2 and AV-17P in the third. For reducing the biuret content of crude urea the latter is contacted in solution with a strong base anion exchanger. Thus, on a pilot plant scale an aqueous solution containing 62.5% urea and 1.88% biuret was treated at 57 °C to give an effluent completely free of biuret. The removal of inorganic impurities from ethyl acetate and methyl ethyl ketone and their aqueous solutions by ion exchangers and activated carbon has been studied showing that copper compounds and traces of manganese, lead, chromium, iron, nickel, aluminium, and cobalt can be removed with the cation exchanger KU-2, the anion exchanger AV-17 and the optionally oxydized activated carbon BAU. The impurity concentration in the ethyl acetate was decreased by the cation exchanger by a factor of 40 – 300 to a range of $10^{-7}-10^{-6}$%. The oxidized activated carbon was suitable for removing cobalt, copper and nickel. Iron sorption was somewhat better on normal activated carbon. Ion exchange purification of the intermediate products of xylitol manufacture, i. e., of pentosan hydrolysates, is feasible with AN-1, KU-1, and EDE-10P, respectively, as ion exchangers in order to remove sulfate, acetate, cationic elements, and organic acids. If cetyldimethylbenzylammonium chloride, which has shown potential as a disinfectant in water and waste water treatment, is used for a water disinfection process it should, as only a small amount is consumend during the employment, be recovered and possibly be reused. A combined process of disinfection of water with cetyldimethylbenzylammonium chloride and its subsequent removal by an ion exchange resin showed encouraging potential as a full-scale water purification process or as a purification process for a portable water treatment unit. Aqueous solutions of cetyldimethylbenzylammonium chloride in concentrations of 4 and 8 mg/l are effective for the disinfection of water considered to be a poor source, i. e., coliform count $\geq$ 5,000/100 mg.

Purification with ion exchangers in natural and biological substances have been treated in numerous investigations. For earlier work we refer to the liteature, for later work to various chapters of this volume. The adsorptive purification of alcohols with ion exchange resins has been investigated and extractions prepared from grain and potato spirits by treatment with cation and anion exchange resins. The phenolic cation exchange resin partly removed acids, esters and aldehydes, and the treatment with styrene-divinylbenzenesulfonate KU-2 increased slightly the concentration of acids and esters and removed a part of the aldehydes. The treatment with the anion exchange resin EDE-10P in bisulfite form removed aldehydes and a part of the acids and esters. Thus, a successive treatment of spirits with KU-1 H-form, EDE-10P HSO_3-form, EDE-10P OH-form, has been recommended. Trace elements in edible fats and oils and their removal have been thoroughly investigated. The effects of different solvents, degree of dilution, type of resin, and temperature were studied to determine the optimum conditions for decationization (iron, manganese, copper, and zinc) of soy-bean oils with cation exchange resins. Prooxidant metals in crude soy-bean oils in acetone or hexane solutions can be effectively removed when passed through columns of a strong cation exchange resin. These oils were refined, bleached, deodorized, and compared with untreated oils for oxidative and flavor stabilities. Treatment of oils with resin lowered metal content and increased stability as measured by oxidative and sensory tests. Treatment of olive oil for sixteen hours with a macroreticular cation exchanger in the H-form removed practically all the iron and most of the traces of manganese, copper, and zinc present, giving a product of light color without change in any of the constants, whereas gel type cation exchangers were ineffective. Virgin olive oils for edible purposes were also decationized by means of macroreticular cation exchangers in the H-form. The treated oils were more resistant to rancidity than the original oils but their characteristic flavor was lost The treatment induced a complete destruction of the hydrogen peroxides, with a good bleaching of oils. The Na-form of the resins was less effective. Essentially, decolorization and simultaneous decationization with cation exchange resins is advantageous. This was shown by the content of tocopherols which are in conventionally neutralized and bleached oil 72 μg/g, in conventionally neutralized and decationized oil 138 μg/g, in resin deacidified and bleached oil 43 μg/g, and in the same oil decationized 140 μg/g. The different peroxide indexes for the same cases are: 60.5, 11.6, 97.0, and 8.5, respectively. The resistance to rancidity of both virgin olive oil and crude soybean oils was proved because during storage for twelve months the same oil showed a much lower peroxide value than a control sample [24].

Ion exchangers and adsorbent resins for the purification of proteins have been examined repeatedly in detail, but they will now be viewed from the perspective of large scale processes. Nevertheless are the basic principles of protein purification by ion exchangers of the same interest in preparative chemistry. It follows from these considerations that the determination of adsorption isotherms is the first step necessary in the analysis of ion exchange systems in order to be able to predict the performance of batch or column adsorption protocols. Another important aspect of adsorption performance that needs to be considered is the kinetics of the adsorption/desorption reactions. The rates of the adsorption and desorption reactions dictate the length of time that has to be allowed for batch systems to attain equilibrium and the maximum flow rates that can be used in columns. It has been found that, in general, the rates of protein adsorption are rapid, with batch systems attaining equilibrium within a few minutes. Computer methods have been used to predict the shape of adsorption breakthrough curves that would be expected for columns from the data obtained from the small-scale batch experiments. These curves have been compared with breakthrough curves obtained from experiments with small fixed beds. The effects of varying such operational parameters as flow rate, inlet protein concentration and bed dimensions have been assessed both theoretically and experimentally. The chemistry of the multivalent interactions occurring between ion exchangers and polyionic molecules such as proteins is not well understood at present. Thus it should not be surprising, perhaps, that an almost non-ionic phenol-formaldehyde resin substituted with phenolic ionogenic groups was effective for the adsorption of urokinase directly from human urine without the need of any pre-processing or concentration steps The concentration of urokinase in the urine was on the order of $2 \cdot 10^{-6}$ g/l and it and other urinary proteins were tightly bound under these conditions. Urokinase could be eluted in a purified and enzymically active form with strong solutions of ethanolamine followed by rapid neutralization. The adsorption isotherms obtained for the binding of urokinase to this resin showed marked deviation from the favorable Langmuir type. This resin (Duolite ES 762) also proved to have far superior properties as an adsorbent for urokinase than a number of other uncharged adsorbents (such as the XAD series) [25]. Soy protein extractions were subjected to activated carbon and ion exchange treatments to remove phenolic compounds. Minimum equilibrium contact time, adsorption rate, and adsorption isotherm data were established for syringic, ferulic and *p*-coumaric acids. Both activated carbon and ion exchangers removed phenolic compounds from soy protein isolates and improved their flavor and color. Activated carbon treatment resulted in soy protein isolates with superior overall odor and flavor, but with no improvement in bitter and astringent flavor properties. Ion exchange processing was in general found to be more effective than activated carbon for improving the color.

The purification of antibiotics, vitamins and alkaloids should additionally be mentioned. The use of ion exchangers for the purification of antibiotics started quite early but it was not until the development of the carboxylic acid type exchange resins that large-scale use was made of such materials. The major uses of ion exchangers have been in the commercial production of streptomycin and neomycin. Carboxylic acid type ion exchangers are comparatively selective for streptomycin, allowing all nonbasic and almost all other more weakly basic constituents of the broth to pass through the column during the loading stage. The decationization of streptomycin sulfate solutions with a carboxylic cation exchanger is carried out on its H-form as the second purification stage. The ionic reactions involved in the adsorption of streptomycin are so fast that high flow rates can be applied. The eluate obtained may have a pH of 2.3 and is neutralized with an anion exchanger in the OH-form. The solution obtained, which has a pH of 6.3, is concentrated, developing more color, which can be removed in a third purification stage by ion exchangers suitable for it, for instance styrene-divinylbenzene copolymers with quaternary trimethylammonium groups in the Cl-form or adsorbent resins without ionogenic groups. According to a patent specification penicillin G sodium was prepared in high yield and purity from penicillin G potassium salt by passing the latter in solution at 5 °C through columns of carboxylic or sulfonic type cation exchangers pretreated with sodium hydroxide at pH 10, preferably at 80 – 100 °C, incubated $\geq$ 24 h with 50 – 90% aqueous acetone and washed with pH 6.8 phosphate buffer. The pretreated resin completely removes pyrogenic impurities and largely avoids degradation and sorption of penicillin G sodium on the resin. The treatment of vitamins with ion exchangers may be summarized as follows. Ion exchangers have in many cases served as means of elimination of those substances which accompany the vitamin in its source and which interfere with an otherwise reliable method of assay for the vitamin. The application of the ion exchange process is mostly based upon the ionic nature either of the vitamin or of the undesirable contaminants. Then the purification may involve either cation or anion exchangers, or both. If the vitamin is adsorbed by the ion exchange material, while the impurities are not taken up or only partly taken up, then an easy purification of the vitamin can be expected. Corresponding data on the ion exchange behavior of a vitamin can be found in the literature. Even when the impurities accompany the vitamin onto the exchange material, the two may still be separated if they are present in different strata of an ion exchange column. In this case is has been practice to separate the layers mechanically and to elute the vitamin-loaded resin. When both vitamin and contaminants have been taken up by the exchange material, a differential elution whereby the vitamin and the contaminants are eluted at different rates may be effective in purifying the vitamin. A selective eluant and control of conditions, such as temperature, flow rate, concentration of eluant, and pH, might prove decisive. For selecting a suitable ion exchanger it should be kept in mind that among the vitamins, in addition to thiamine, which presumably could undergo cation exchange, are choline, pyridoxine, pyridoxamine, and pyridoxal. The vitamins sufficiently acidic to undergo anion exchange are ascorbic acid, nicotinic acid, biotin, *p*-aminobenzoic acid, folic and related acids, and the phosphoric or pyrophosphoric acid esters of several of the vitamins of this class. This latter group would also include the esters of riboflavin, pyridoxal, pyridoxamine, inositol, thiamine, and the disulfide form of thiamine. One also has that some of the vitamins can exist in their natural sources in more than one chemical form. The use of ion exchangers in the purification of the fat-soluble vitamins depends on adsorption phenomena rather than on ion exchange. In contrast to the rather complex situation with vitamins and their ionic state, the alkaloids all have a certain basic character. Therefore, ion exchange materials can be used in their extraction from natural sources and their purification. Disregarding the very early trials in which inorganic ion exchange materials were employed in the field of alkaloids, a considerable amount of early work had been dedicated to the use of ion exchange resins. Thus, the crude amorphous totaquine containing 23% alkaloid, prepared by alkaline precipitation of the acid extract of cinchona bark, was refined on a polycondensation cation exchanger to a white crystalline powder containing 94% alkaloids. Belladonna alkaloids, it was claimed, were purified by using the same cation exchanger, with 80 – 85% recovery. Veratrum alkaloids were first successfully treated on a Dowex 50 sulfonic acid cation exchanger and purified by elution from the column with various buffers. Further, the use of an anion exchange resin for purification of the mother liquors of d-tubocurarine chloride in order to remove excess hydrochloric acid and to crystallize the alkaloid was described. More generally, the uptake of alkaloids by the weak acid cation exchanger Amberlite IRC-50 was studied, and this exchanger has since then been recommended for the purification of alkaloids and antibiotics. With respect to alkaloids it was found that if such a compound is present as a salt in solution, it is absorbed more readily by the resin in the sodium form. If the alkaloid is present as a free base, the resin should be used in the hydrogen form. Winters and Kunin, who investigated this subject, also noted that since this resin swells considerably in polar organic solvents, it should have an appreciable exchange capacity in these solvents.

The purification of catalysts by ion exchange is an interesting example of the cleaning up of a technical product. Since it had been recognized that metals that accumulate on cracking catalysts increase the production of light gases and hydrocarbonaceous deposits at the expense of desired products a procedure was developed for removing these contaminants by ion exchange. A strong acid cation exchange resin in the H-form was slurried with the contaminated catalysts and deionized water. After the catalyst was filtered from the slurry and dried, it was determined through cracking activity that rejuvenation had been accomplished. One critical variable was found to be the effective pH of the aqueous slurry.

Purification by separation is generally possible with ion exchangers for preparative purposes and has been investigated in numerous works, especially, e. g., for amino acids and rare earths. Here the ionic constituents are removed from the ion exchanger bed in a given order through variation of the regenerant or elutant pH, concentration or composition. To separate preparatively ionic constituents from nonionic constituents the ion exclusion process can be applied. This is feasible when the exchanger tends to have less affinity for the ions in a solution than for the nonionic constituents. As a simple example one has that when a solution containing salt and glycerol is passed through a cation exchanger column, the ions will pass while the glycerol is retained by the resin particles. The purified glycerol is released by elution with water. In the ion retardation process, special snake cage ion exchangers in which both cationic and anionic groups are on the same polymeric matrices are used. The salt diffuses into the exchanger, being held there, while the purified nonionic constituents pass through the column. The salt is then released from the resin by elution with water. Similarly, acids can be separated from salts by acid retardation. For example, if a solution containing aluminum phosphate and phosphoric acid is passed through a cation exchanger bed the salt aluminum phosphate passes through the column, while the phosphoric acid is retained by the resin particles. When the capacity of separation is reached, the phosphoric acid is released by elution with water. Both constituents are purified of each other. Finally, the sieving of ions is feasible if the pores of the exchanger material are smaller than the ions in question so that the ion of interest will pass through the columnar bed while smaller ions are taken up.

4.2.3 Concentration of dilute materials

Ion exchange has also proven useful for the concentration, isolation and recovery of ionic components from highly dilute solutions. Usually, the H^+-ions of a cation exchanger or the OH^--ions of an anion exchanger are exchanged for the ions contained in a solution. In this manner, traces of ions can be concentrated on an ion exchanger. Its high capacity permits the ions contained in a relatively large volume to be stacked on the small volume of the ion exchanger and to be recovered in concentrated solution by elution with a small quantity of concentrated eluant.

The attainable effective enrichment depends on the capacity of the ion exchanger utilized and the equilibrium. It is always possible to find conditions under which the total sorbed ionic fraction can be recovered with a minimum volume of eluant.

For example, if a cation exchanger has a capacity of 1 mequ/ml, 1 ml can sorb the cations from 1000 ml 0.001 N solution. Under favorable equilibrium conditions, it may be possible to perform the elution with 5–10 ml of a 1 N solution.

Some practical examples will illustrate the resulting possibilities.

Example 1: Recovery of copper from fiber spinning dope. During the spinning of cuprammonium rayon, "blue liquor" or cuprammonium solution containing $^1/_3$ of the charged copper and a "spinning acid" containing $^2/_3$ of the charged copper are usually obtained. Copper recovery is of technological importance and is carried out with the aid of ion exchangers [26]. This can be performed in the laboratory for demonstration purposes as described below.

Blue liquor (2 l) is prepared with 0.08 g Cu/l and 0.75 g NH_3/l, while 500 ml spinning acid is prepared from 18 g Cu/l and 65 g H_2SO_4/l. A simple ion exchange column is prepared with Lewatit S 100 cation exchanger in the H-form. The copper content in the cuprammonium solution and in the spinning acid is determined by iodometry.

For the concentration of copper, the cuprammonium solution is percolated through the column, leading to a copper enrichment according to the equation:

$$2\,IE^- \cdot H^+ + [Cu(NH_3)_4]^{2+} \rightleftharpoons IE^{2-} \cdot [Cu(NH_3)_4]^{2+} + 2\,H^+ .$$

Cuprammonium solution is charged up to breakthrough. The column is then briefly washed with water. The regeneration and recovery of copper might now be carried out with acid. However, it is of industrial importance that one can also use spinning acid which already contains a part of the initially charged copper. By eluting copper with the spinning acid, the copper from the cuprammonium solution and from the acid is practically recombined in the eluate. Another copper analysis of the eluate can give information on the attained enrichment.

Example 2: Recovery of molybdenum. Metals present only in small quantities in ores or other products should be concentrated prior to their recovery. The advantages offered by ion exchange will be demonstrated with the example of molybdenum [27].

This example as well as the following are described as preparative laboratory experiments. It is clear that these methods are similarly applicable to large-scale processes.

Molybdenum is extracted from an ore (0.015% MoS_2), a copper concentrate (1.05% MoS_2), or a molybdenum concentrate (63% MoS_2) with 3% NaClO solution at 45 °C. As a rule, 30 min is sufficient to convert the molybdenum into molybdate according to the equation $7\ NaClO + MoS_2 + 4e \rightarrow MoO_4^{2-} + S_2O_3^{2-} + 7\ NaCl$; for its enrichment, the solution obtained is loaded on a strong base anion exchanger in the Cl-form. Elution should be carried out with 8% NaOH solution. After washing of the column, the exchanger is again brought into the Cl-form.

In the eluate, molybdenum is precipitated as MoO_3 with HNO_3 or as $CaMoO_4$ with $CaCl_2$.

Example 3: Separation and recovery of phenol. Organic ion exchangers sometimes adsorb organic substances with a capacity which is higher than the exchange capacity. This is also true for phenols which, as will be shown in this example, can be concentrated on ion exchangers and recovered [28]. This example is of industrial interest, since phenols and phenolic compounds present a difficult problem in wastewater treatment.

An 0.1 N phenol solution (500 ml) is prepared and loaded on a suitable exchange column with 75 ml Dowex 2 in the Cl-form. Breakthrough is controlled in the filtrate and the adsorbed quantity of phenol is calculated from this as a comparison with the exchange capacity. Pure methanol or methanol with 0.8 N sodium hydroxide is used for elution.

Example 4: Isolation of alkaloids. Ion exchangers are being used increasingly for the isolation of alkaloids from plant extracts. An example is the recovery of quinine from bark extracts, which can be carried out as in the following model experiment [29].

A 0.003 M quinine solution is passed through 100 ml of a cation exchanger in the NH_4-form until breakthrough occurs. The quinine solution contains 1% sulfuric acid. To elute the quinine and simultaneously regenerate the exchanger, ammoniacal alcohol is percolated through the column. After the exchange column has been used for two or three runs, the recovery of quinine should be quantitative with a total column capacity for quinine of 3.5–4 g.

Atropine, morphine, and strychnine can be similarly bound on a strong acid cation exchanger and simultaneously purified during elution [30].

Example 5: Enrichment of metal traces from seawater. This possibility has been repeatedly discussed in the literature. Brooks [31] offered an example to demonstrate that some trace elements in seawater can be separated by an ion exchange process. The basis of the method is that the principal elements of seawater do not form anionic chloro-complexes, while many of the trace elements, such as bismuth and thallium, form complex salts in 0.1 N HCl.

To this end, seawater is concentrated to 0.1 N HCl with distilled hydrochloric acid and is slowly percolated through a column with strong base anion exchanger (Amberlite IR-400). After loading with a sufficient quantity, the column is drained and the exchanger is subjected to combustion. The trace elements can be detected in the ash by spectroscopy.

The latter example already extends to the application of ion exchangers in analytical chemistry, where the concentration of sample components is often of special interest [32]. Trace elements can also be enriched on ion exchangers in other cases. For example, such a concentration has become important for the detection and identification of trace elements in plant materials. With the use of colorless or lightly colored exchangers, an enrichment can often be detected with the naked eye. Consequently, the analyses are simplified. The enrichment of metals from seawater has, viewed from ion exchange technologies, remained an interesting subject. In a review of the early work in this area the authors evaluated the many methods suggested for the recovery of mineral values from the large quantities of concentrated brines and salt mixtures produced through desalinization of ocean waters. It was pointed out that only a few of the many current suggestions are economically promising, but that the subject has assumed new importance as a possible way of reducing the cost of saline water conversion. A comprehensive review of all literature then available was given including works dealing with the recovery of calcium, halogens, potassium, magnesium, sodium, trace elements and mixed salts [33]. As a specific example a chelating ion exchanger was examined as to the retention of zinc, cadmium, lead and copper from seawater on a column of Chelex-100. The original seawater and the column effluent were analyzed for these metals. In agreement with preceding results complete retention was found of ionic spikes of all four metals added to seawater. However, retention of the metals naturally present in seawater was considerably less. This made it apparent that Chelex-100 chelating resin cannot be used in the manner described for the quantitative concentration of zinc, cadmium, lead and copper from seawater. Use of the resin in the sodium or ammonium, rather than the hydrogen form, improved the retention, but adsorption was still not complete [34]. The concentration of the trace elements iron, nickel, copper, zinc and uranium from seawater in a fluidized bed of 2-hydroxyphenyl-(2)-azonaphthol on bead cellulose as well as on polystyrene was studied by Lieser and Gleitsmann in parallel experiments, over a period of eight months. Samples from the incoming water and from the exchangers were taken from time to time to determine the trace element content. At the beginning the trace elements were taken up quantitatively by the specific reactive group on bead cellulose whereas about half or less was fixed on polystyrene. This is explained as due to the fast exchange on bead cellulose and the slow exchange on polystyrene. It was shown that the loading curves of the individual elements can be calculated in the range where loading is proportional to the concentration. The loading is restricted by the exchange equilibria. At the end of the experiments loading with the abovementioned trace elements corresponded to about 40% of the capacity.

The specific active group incorporated thus proved to be well-suited for the separation of uranium from seawater, whereas the network of the polystyrene beads shows pronounced exclusion of the voluminous tris-carbonato complex of uranium [35].

As far as the above already mentioned halogens are concerned their concentration on ion exchangers has also been an interesting subject for ion exchange applications. A comprehensive review of the use of ion exchangers for the extraction of bromine and iodine from natural brines has been published [36]. For the removal of dissolved fluorine ion a strong acid cation exchange resin is first converted to the H-form by hydrochloric acid, washed with water and then converted to the Th-, Zr- or Ti-form. The resin then uptakes fluorine at ten times the efficiency of an anion exchange resin. A quite interesting method for the recovery of bromine from brines using a continuous countercurrent fluid bed process was the subject of a US patent assigned to the DuPont Company. A strong base anion exchange resin is applied in the chloride or bromide form. As the resin becomes enriched with bromine during adsorption, it becomes denser, setting faster than the less dense, bromine-lean resin and is taken off at the bottom of the column. Further research work was later done on the sorption of bromine from chloride brines by an anion exchanger. A method for recovering iodine from iodide-containing salt brines, etc., was discussed in a US patent assigned to a worker at Dow Chemical Company. Aqueous iodide solutions are contacted with anion exchange resins in polyhalide forms which are used to oxidize iodide to free iodine, and which then react with such liberated iodine to form polyiodide resins from which iodine can be recovered. A process based on the sorption of elemental iodine for obtaining it in high purity from mineral waters has also been described. Elemental iodine is produced by oxidation with chlorine at pH 3–3.5, and adsorbed on a strong base anion exchange resin. The adsorbed iodine is then reduced by sulfur dioxide and eluted with twenty volumes of 8–10% sulfuric acid per volume of resin at 70–80 °C. The product is precipitated from the eluate by oxidation, filtered, and washed. The capacity of the ion exchange resin was found to be lowered by the presence of sodium chloride, but the sorption of elemental iodine is practically independent of the sodium chloride concentration [37]. For the mechanism of the uptake it was argued in another paper that for sufficiently large amounts of iodide ions in solution, the sorption of I_2 occurs in the form of I_3^- ions which displace Cl^- ions from the resin, although the concentration of Cl^- in solution exceeds that of I_3^- by more than three orders of magnitude. If the content of iodide in the original solution is less than needed to form I_3^-, this lack is made up in part by dissociation of I_2 to provide I^- ions. The sorption of iodine from aqueous solutions by a fibrous ion exchanger, prepared from polyacrylonitrile and anion exchanger AV-17, has also been studied. An empirical equation was derived for the kinetics and the mechanism of this process [38]. The use of anion exchangers for the concentration of ^{131}I and its determination in water and biological materials has been described in quite a number of papers.

For the concentration of dilute or trace metal ions the term preconcentration is often found in the ion exchange literature. That the terms concentration and preconcentration are essentially synonymous becomes obvious if it is born in mind that the main purpose of ion exchange is to collect dilute or trace ions, either for the purposes of preparation chemistry, or for analytical purposes detecting them, and then measuring and calculating their originally low concentration after they have been concentrated. A separation of the various ions collected may follow. If the main purpose here was analytical determination, the use of the term preconcentration may then be justified. With respect to preparative concentration, ion exchange is a competitive process to others and is effective only if the exchanger is selective for the ions of interest. From the selectivity behavior of metal ions it is known that the so-called heavy metals are strongly held by most resins. They are especially strongly held by carboxylic cation exchangers and by chelating polymers. In columns it is best to use the carboxylic and chelating resins in the Na-form, which is ionized and swollen, rather than the H-form, which is nonionized, nonswollen, and reacts slowly. If batch operation is applied it has the advantage that less resin is required than in columns. To optimize a preparative method using ion exchange resins much experimental work may be needed. This will be demonstrated for the preparation of ammonium perrhenate. The rhenium was sorbed from sulfate solutions on three anion exchangers that were slightly basic, one exchanger of intermediate basicity, and eight highly basic exchangers of which two had different divinylbenzene contents. The rhenium concentration in the solution was 0.1 – 0.3 g/l, and the sulfuric acid concentration was 150 – 200 g/l. The AV-17 resin with 16% DVB had the highest capacity, 8.33%; that of AM was 5.53% and of AV-17 with 6% DVB, 5.50%. For AV-17, as the percent of DVB increases up to 12%, the capacity of the resin for rhenium increases; it remains almost constant, however, from 12 to 30% DVB. As the acid concentration is increased from 30 to 100 g/l the capacity decreases somewhat (from 10 to 6%) and for 100 – 150 g/l it remains constant. The capacity of the anion exchanger decreases markedly with a decrease in the rhenium concentration in the solution. Expanded experiments were performed with the AV-17 resin with 16% DVB, resulting in an extraction of rhenium into the eluate of 98 – 100%. The extraction of rhenium in the first two fractions was 94 – 98%. These eluates were evaporated and precipitated as ammonium perrhenate [39]. With a preparative ion exchange method high-grade europium concentrates of purity above 99.9% Eu can be obtained from mixtures containing Eu_2O_3 2.35%, Gd_2O_3 60%, and Sm_2O_3 35% and small amounts of Dy, Tb, Nd, Y, Ho, Tm, Yb, Lu, Pr, and cerium. The separation can be carried out on a strong acid gel type cation exchanger with the use of α-HIBA, ammonium lactate, and EDTA and ammonium acetate solutions as eluants. The best separations are obtained with the use of 0.1 – 0.15 M solution of α-HIBA of pH 3.9 – 4.25. For a selective extraction method for tungsten from Searles Lake brines, resin beads synthesized by polymerizing 8-quinolinol, resorcinol and formaldehyde were used to extract saturated brine containing $\leq$ 80 mg WO_3/l. The tungsten oxide recovery was 92%. The solution obtained by eluting the tungsten oxide from the column contained 1 – 2 g WO_3/l. With a second ion exchange stage, a solution containing 80 – 100 g WO_3/l was obtained.

The concentration of organic compounds using ion exchange has often been carried out and is in many cases linked with the isolation of such substances. Variations in procedure can improve a process: for instance it was claimed in a patent that the isolation of vanillin was improved by shaking a mixture of the solution, the ion exchange resin, and a gas (air) in a sealed vessel at 20 – 120 Hz and 5 – 30 *g* acceleration. In another case two columns were used for the isolation of betaine from molasses, both loaded with the cation exchanger Allassion CS-NA (Romanian). Betaine was retained on the first column and eventually displaced together with some potassium and sodium, by the calcium of molasses. The second column retained betaine, potassium and sodium, and the amino acids. By eluting with 10% NH_4OH solution, a solution of betaine with 90% yield containing the amino acids was obtained. A strong base anion exchanger was then used to retain the amino acids, with the effluent containing betaine, which was concentrated and treated with hydrochloric acid. The hydrochloride was further purified by crystallization. Working with one column the isolation of an organic compound can be quite easy. For example melamine was recovered from solutions by sorption on a strong acid sulfonic cation exchanger in its H-form, followed by treatment with aqueous 2.5% ammonia solution and with hot, preferably boiling, water. Atropine on the other hand is also retained almost quantitatively from a pure solution by Amberlite IR-120, but only 82% could be recovered from a liquid extraction solution. For Amberlite IRA-410 and morphine, the recoveries were approx. 96% and 80%, respectively. By converting morphine into its anionic form, about 98% of the alkaloid was recovered. The concentrations of alkaloids by ion exchange resins are higher than by the classical methods. Concentration together with separation and purification processes have been relatively easily elaborated with single but separate cation and anion exchanger columns for tryptophan as well as for L-tryptophan, DL-tryptophan, and L-tryptophan. Purification of *Clostridium perfringens* anatoxin, for instance, on an anion exchange resin in the OH-form with a capaticy of 2.0 g/10 ml anatoxin concentrate, a contact time of 15 min, and rate of elution with 0.2 M phosphate buffer of pH 7.6 of 150 ml/h results in a concentration of the purified anatoxin 15 – 25% greater than in the original mixture. Studies on the isolation and purification of steviosides from Stevia by ion exchange led to a method yielding a concentrated product containing 95.2 – 96.4% stevioside. For this purpose Stevia leaves are extracted with water at 75 °C, the extract decolorized by precipitation with lime and iron sulfate and the filtrate passed through a column of strong acid cation exchange resin for further decolorization. Stevioside in the eluant was adsorbed by a weak acid resin in the H-form and eluted with 60 – 65% ethanol to give after evaporation the above product concentrate [40].

4.2.4 Substitution reactions with ion exchangers

In organic chemistry substitution reactions are classified as a reaction type by which a group in the starting material is replaced by a new group. Some work has also been carried out to perform substitution reactions by ion exchange resins. This reaction was found to be unique in that it involves the breaking and formation of covalent bonds on the surface of the resin with the anion from the resin becoming covalently bound in the product.

No reference to such a substitution reaction is to be found in the literature before the early nineteen-sixties, the closest example perhaps having been the conversion of methyl mercuric halides to the corresponding hydroxides. Out of the need to avoid anomalous reactions in the formation of benzyl cyanides by the reaction of the corresponding halides with alkali metal cyanides it was desirable to develop an alternative method of synthesis which would be applicable in non-polar anhydrous solvents and which would avoid any appreciable concentration of cyanide ion in solution. To achieve this, the reaction of benzyl bromides in ethanol and other solvents with the cyanide form of a strong base anion exchange resin was examined. It was known that the bromide ion is more strongly held on the resin than the cyanide ion.

Preparation of phenylacetonitrile from benzyl bromide with a strong base anion exchanger. A solution of benzyl bromide in 95% ethanol is stirred at 65 °C with an equivalent of a strong base anion exchanger in CN-form until a positive bromide test is no longer obtained (1.5 h). The exchange occurs according to the equation:

After filtration, phenylacetonitrile, b. p. 110–111 °C (15 mm Hg), identical in all respects with an authentic sample, can be isolated in 53% yield.

In a similar way, *p*-bromo- and *p*-methylbenzyl bromides and alkyl bromide were converted to the corresponding nitriles with 98%, 43% and 23% yields, respectively. The reaction of *p*-bromobenzyl bromide with the resin in ether (35 °C), tetrahydrofuran (65 °C), benzene (65 °C), and dimethylformamide gave the nitrile with an approximately 20% yield in all cases. It appears that this method is generally applicable to the synthesis of benzyl cyanides in solvents of a wide polarity range, thus offering possible advantages over conventional synthetic routes. It was assumed that yields could probably be raised, since during the first work no effort had been made to establish optimum conditions for the reaction. In subsequent investigations it was then shown that the formation of a number of phenylacetonitriles by reaction of the appropriate benzyl halide with the cyanide form of an anion exchange resin can be achieved with a good yield. The reaction can be carried out in a variety of polar and nonpolar solvents. The only anomalous result was obtained with *p*-nitrobenzyl bromide, which underwent condensation to yield 1,2,3-tris(*p*-nitrophenyl)-2-cyanopropane [41].

In extensive investigations similar substitution reactions by ion exchange resins were carried out. A strong base anion exchanger was, for instance, washed with 2 N NaOH, saturated with halogen ions by 2 N KBr, 2 N HCl, or 20% aqueous HF, and used for the exchange of halogens with alkyl iodides. Most successful were the reactions of butyl iodide at elevated temperatures (130 °C). Equilibrium was obtained after 1 h for bromide and chloride, to give nearly 100% substitution. Fluoride gave 90% substitution after 120 min. It was also possible to prepare 1-bromo-2-chloroethane from 1,2-dibromo-ethane and 1,2-dichloroethane using anion exchange resins in the Cl-form or Br-form, respectively. The formation of 1-bromo-2-chloroethane was greatest when the molar fraction of di-chloroethane was 50 – 70% for the Cl-form resin and 70 – 80% for the Br-form resin. The reaction proceeds at more than 100% of the exchange capacity, indicating that the resin acts as a catalyst. In subsequent experiments, using the resin in the Cl-form in order to exchange chloride for bromide in organic compounds, phenylbromide failed to give any phenylchloride but benzylbromide reacted easily, with 50% of the exchanger changing into the Br-form at 50 °C, yielding benzylchloride. Phenylethylbromide was similarly treated with almost the same results, but at 140 °C. Amyliodide and isopentyliodide gave comparative substitutions, but only at temperatures above 110 °C. The resin was also used in the F-form for substitution reactions; the reactivity was Br > Cl > F at the lower temperatures. The preparation of organic fluorine compounds and of ethylene from 1,2-dibromoethane and 1,2-dichloroethane by means of anion exchange resins was also investigated extensively. The reactions of anion exchange resins were extended to 1-bromo-2-chloroethane. From the latter and the resin in the Br-form, 1,2-dibromoethane is formed; with the resin in the Cl-form, 1,2-dichloroethane; the resin in the F-form yielded 1,2-difluoro-, 1-chloro-2-fluoro- and 1-bromo-2-fluoroethane with 1,2-dibromo- and 1,2-dichloroethane as byproducts. The formation of alkyl cyanides using anion exchange resins was also studied. With the resin in the CN-form methyl-, ethyl-, propyl- and butyl iodide reacted easily to give the corresponding cyanides. Increasing the chain length lowers the reactivity, but the exchange ratio is approx. 90% when the temperature is above 100 °C. Alkyl bromides react more slowly than iodides, and alkyl chlorides react the slowest. Thus butyl chloride had a 46.8% exchange ratio at 110 °C after one hour, while butyl bromide and butyl iodide had 90.0 and 76.4% exchange ratio, respectively. The reactions were then carried out using ethanol, acetone, and ethylene glycol, and it was found that using the solvents gave little better results [42].

The attempt was made to generalize the substitution reaction as a selective transformation of benzylchloride into benzylthiocyanate in the presence of ion exchange resins with it being extended to several other halogen-containing compounds. The resin as a support for thiocyanate ions changed, e. g., benzylchloride and the other substances into thiocyanates with a high yield and selectivity. A study of the effects of various physicochemical parameters governing the behavior of the ion exchanger determined the most favorable reaction conditions. The best transformation rates were obtained using a macroporous resin, appropriately hydrated, in a non-water-miscible solvent [43].

4.2.5 Dissolution of solids by ion exchangers

Although preparative ion exchange applications are mostly carried out in solution, reactions of ion exchangers with solids are also known. The example usually demonstrated to the viewers' surprise is the dissolution of barium sulfate when it is shaken with a large excess of cation exchange resin in the H-form, in the presence of water. In general, then, the total hydrogen ion exchange with a cation exchange resin can be used for the dissolution and, under analytical aspects, for the determination of weakly soluble salts. The phosphates of calcium, strontium, barium, manganese, nickel, zinc and cobalt and, at higher temperatures, even calcium fluoride and the sulfates of barium and strontium, can be dissolved. With dissolution using cation exchange resin in the H-form the concentration of the exchanging metal ions is always very low in the solution because of the very slight solubility of the solid substance to be dissolved. A film diffusion mechanism as a rate-determining factor is the consequence. The duration of the dissolution depends on the solubility of the salt, on the selectivity coefficient of the metal cation towards the hydrogen counterion of the exchanger, on the strength of the resulting acid, on the rate of stirring and on the temperature. The dissolution can be accelerated by the addition of some salt or acid, as well as by the addition of an anion exchange resin to the reaction mixture, because the latter will absorb all of the dissolved anions.

Metals may be recovered from insoluble salts and metal oxides by two different procedures. Either the sludge is treated with acid and the released or solubilized metallic ions are removed from the mixture by a resin-in-pulp process such as used in uranium recovery, or by filtering the mixture and passing the clear liquid through a column of ion exchangers; or the sludge is contacted with the exchanger for a period of time and then the mixture is backwashed at a rate sufficient to wash over the fine sludge but not the ion exchanger particles. The metal ions are recovered from the exchanger by elution with an appropriate regenerant. Calmon has subdivided the dissolution reactions in the following way, with relevant examples.

Dissolution reactions involving gas formation:

Zinc metal plus cation exchanger H-form

$$(2)RH + Zn \rightarrow RZn + H_2$$

Calcium carbonate plus cation exchanger H-form

$$(2)RH + CaCO_3 \rightarrow RCa + H_2CO_3\ (\rightarrow H_2O + CO_2)$$

Zinc sulfide plus cation exchanger H-form

$$(2)RH + ZnS \rightarrow RZn + H_2S$$

Dissolution reactions involving water formation:

Iron oxide plus cation exchanger H-form

(6)RH + $Fe_2O_3 \rightarrow$ RFe + 3 H_2O

Copper oxide plus cation exchanger H-form

(2)RH + CuO $\rightarrow$ RCu + H_2O

Silver chloride plus cation exchanger H-form and anion exchanger OH-form

RH + R′OH + AgCl $\rightarrow$ RAg + R′Cl + H_2O

Dissolution reactions involving high selectivity:

Lead chloride plus cation exchanger H-form

(2)RH + $PbCl_2 \rightarrow$ RPb + 2 HCl (can be removed by heating).

For the mechanism of the effect of free metals on ion exchange it was found that the solubility of finely powdered metals in water in the presence of ion exchangers increases with the increase of the Nernst constant of the corresponding metal ion. Metal atoms are ionized, hydrated, and then transferred to the surface of an exchanger. This shifts the equilibrium towards the continuous ionization-hydration-surface complex formation process which enhances the solubility of the powdered metal. The dissolution rate of gold and its alloys in cyanide solutions in the presence of ion exchangers has been studied in detail. Even more details have been obtained for the kinetics of the dissolution of a scheelite (69% WO_3) by means of the cation exchange resin Amberlite IR120 in an acid cycle in aqueous solution, (studied at 20, 30, 40 and 50 °C). The process appears to be of a partial order $n = 0.6$. The rate expression is $d\alpha/dt = K(1-\alpha)^n$, where α is the dissolved fraction. The apparent specific rate, K, is of the order of $10^{-7} \cdot s^{-1}$ and it increases with temperature and decreases with an increase in the initial amount of tungsten oxide (m_0,mol). The equilibrium constant, K_α, is of the order of 10^{-2}. The apparent enthalpy and entropy of the overall process are 3.1 – 3.9 kcal/mol and 2.8 – 6 cal/(K · mol), respectively. The dependence of k and K_α on the m_0 of WO_3 is of the form $1/k = A + Bm_0$ and $1/K_\alpha = A' + B'm_0$ [44]. In the above example the selectivity for lead is very high in comparison with hydrogen. A mixture of a cation exchanger and an anion exchanger could also be used, but this would involve separation of the exchangers if the lead is to be recovered later in a soluble salt form.

A quite special case of the dissolution of solids by ion exchangers is scale and deposit removal from various pieces of equipment, with their being envisaged for radioactive nuclides. This ion exchange method has also been used for the removal of mineral deposits from bone for histological investigations.

In the field of natural products and their processing, the dissolution of accompanying salts by conversion to soluble acids in the manufacture of pectin in a unique process for utilizing waste grapefruit peel may be described. Washed grapefruit peel is mixed with a strong acid cation exchanger in the H-form in a ratio of five pounds of peel to one pound of resin. Water is added to this mixture and extraction of the pectin is accomplished by agitation for one hour near the boiling point. The cation exchanger converts residual salts present in the peel to the corresponding acids, lowering the pH to about 2.7 and taking up metal cations from solution. As a result, the pectin is released into the solution without the breaking down of the pectin due to excessively low pH conditions. The exchanger is removed from the peel by washing through a screen; it is regenerated with 2% sulfuric acid for reuse. This possibility of applying ion exchangers was in operation for the manufacture of pectin on a commercial scale for nearly ten years [45].

4.2.6 Ion exchanger catalysis

Ion exchangers are used as catalysts. Industrial interest first turned to inorganic zeolites, whose catalytic activity usually depends on the presence of a metal such as manganese, iron, chromium, or vanadium. Numerous publications exist on the suitability of inorganic ion exchangers for the catalysis of oxidation reactions, hydrogenations, cracking processes, or alkylations. In this monograph, the interest is in the organic synthetic ion exchange resins which Griessbach et al. [46] introduced into catalysis. This research resulted in new organic and preparative applications for ion exchange resins. It is the counterions that are responsible for the catalytic activity. Cation exchangers in the H^+-form and anion exchangers in the OH^--form catalyze processes which are accelerated by acids and alkalis, respectively. Exchangers loaded with Hg^{2+}, cyanide or acetate exhibit the catalytic properties of these ions. The exchanger network serves solely as a catalyst support.

Ion exchange resins show a particular affinity for acid and base catalysis in organic reactions, since they can be easily transformed into the H^+- and OH^--form and their porosity and swelling offer a sufficiently large active surface, so that even large organic molecules can penetrate the interior of the resin.

In this connection, special attention should be given to the macroreticular ion exchangers, which have a structure that makes them particularly suited for heterogeneous catalysis in nonaqueous systems. This unique structure provides them with a sufficiently genuine porosity and a strong matrix together with a small volume change in different solutions and solvents. The large surface leads to high catalytic activity. The ionogenic groups on the entire exchange particle are readily accessible for liquid and gaseous reactants, a condition which is not available to the same degree in conventional gel-type ion exchangers.

Ion exchangers offer all the advantages of solid catalysts:

1. The catalyst can be easily separated from the reaction products.
2. The product remains free from impurities.
3. High yields can be obtained, since secondary reactions can be suppressed as a result of the short contact time.

4. The catalysis can be brought to proceed in a given direction as a result of the variable properties of ion exchangers.
5. The ion exchangers can be used repeatedly and in continuous processes.

Only their limited thermal and mechanical stability sets certain limits for the applicability of ion exchange resins in catalysis.

As an example, a description of the esterification of ethylene glycol with acetic acid into glycol diacetate follows. In a three-neck flask with reflux condenser, stirrer, and thermometer, 62 g of ethylene glycol and 120 g of acetic acid are treated with 2 g of a strong acid ion exchanger and heated on a boiling water bath. Every 30 min, 1 ml of the reaction mixture is removed and the conversion is determined by titration with 0.5 N NaOH. As shown by Figure 4.2, the esterification of ethylene glycol with acetic acid into glycol diacetate takes place just as rapidly with a strong acid cation exchanger as with sulfuric acid.

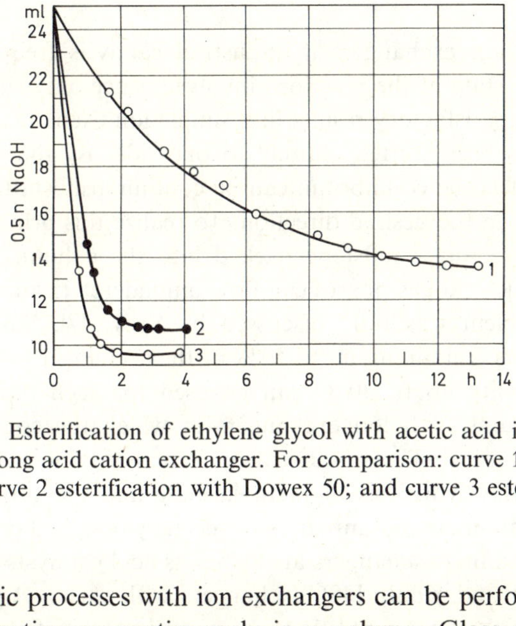

Figure 4.2 Esterification of ethylene glycol with acetic acid into glycol diacetate with the use of a strong acid cation exchanger. For comparison: curve 1 shows esterification without catalyst; curve 2 esterification with Dowex 50; and curve 3 esterification with H_2SO_4.

Catalytic processes with ion exchangers can be performed discontinuously in a batch operation or continuously in a column. Glass equipment, semiworks, or industrial metal equipment is used in batch operation; conventional glass columns or special columns can be employed (Figure 4.3). With the use of columns the raw materials are loaded onto the exchanger and the end products appear in the eluate. In batch operation the starting materials are mixed with a sufficient quantity of ion exchanger and, if necessary, are heated with refluxing. After completion of the reaction the mixture is decanted from the ion exchanger or filtered, and the solution is purified. In both types of operation the ion exchanger can be immediately used for another catalysis.

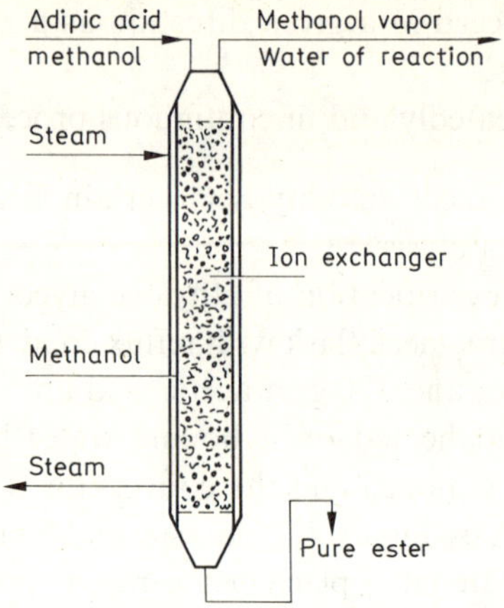

Figure 4.3 Column assembly for the continuous esterification of adipic acid with methanol catalyzed by ion exchange resin.

The full utilization of ion exchangers in industrial catalysis frequently fails because of the equilibrium state of the reactions involved, since only such conversions lead to economically satisfactory results in a simple ion exchange process in which the equilibrium has been shifted entirely to one side. By the continuous removal of reaction products the equilibrium can be continuously shifted and the reaction can be completed in the desired direction. To realize this principle in ion exchanger catalysis, the exchanger must be packed into the columns in such a configuration that the liquid and gas phases can flow unhindered through the bed. The solution of this problem was fully discussed by Spes [47]. Since molded polyethylene ion exchangers containing up to 50% of the exchanger and having a sufficient mechanical stability up to 30 °C can be used for such biphasic flow conditions, catalytic reactions with ion exchangers in which the equilibrium is shifted have become economical. Their practical performance is demonstrated as shown in Figure 4.4 by the hydrolysis of methylacetate and the production of methylal from methylacetate-methanol mixtures in specially designed columns.

Only sulfonic-acid-containing exchangers are suited as acid catalysts; according to the studies of Bodamer and Kunin [48], exchangers with carboxyl groups are too weak. For catalysts with a basic reaction, anion exchangers of various base strengths can be used. An interesting variant of this is the simultaneous use of a strong acid cation exchanger in the H^+-form as well as of hydrogen chloride gas as a dehydration catalyst; this improves the catalytic activity in terms of reaction rates and degree of conversion [49].

The reaction rate increases with the quantity of catalyst used, although a limiting value is attained as a rule beyond which the reaction can not be further accelerated. Its progress is influenced by particle size and degree of crosslinking, with the highest reaction rates being attained when these values are low [50]. The findings of Deuel

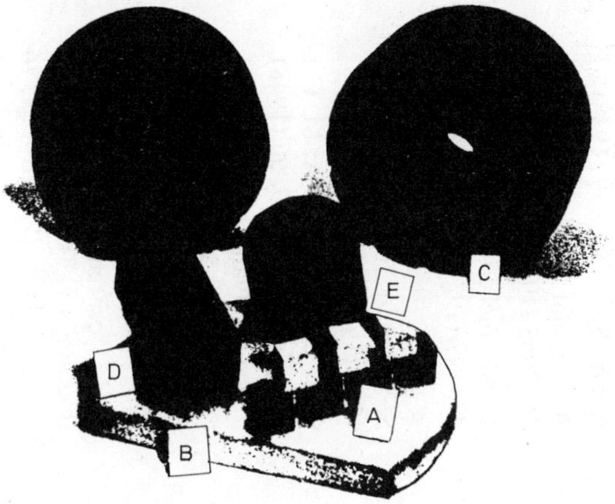

Figure 4.4 Examples of the shape of polyethylene ion exchange packings according to Spes (47). A: cubes of 3 cm sides suitable as packings; B: sheet, 3 cm thickness, post-finished; C: column base; D: right parallelipiped, 6 cm · 6 cm · 20 cm; E: cylinder, 10 cm · 10 cm.

et al. [51] are also connected with this rule; according to these authors, low molecular weight esters and saccharides can split completely, while their high molecular forms are split incompletely or not at all by the same resin. Helfferich offered an explanation of the reaction kinetics of this phenomenon. According to him, a direct relation exists between the catalytic activity of a strong acid cation exchanger and the distribution coefficient of the molecules to be converted. Thus, if the reaction takes place in the interior of the exchanger particle, the following relation must be valid:

$$K_{KA} = k_{AT} \cdot \gamma,$$

where K_{KA} is the overall conversion of the catalytic process in the presence of the resin, k_{AT} is the conversion of the reaction in the resin, and γ is the distribution coefficient of the substrate between resin and solution. For a comparison of the exchanger-catalyzed reaction with the same reaction catalyzed by a free inorganic acid, one has the relation:

$$q = \frac{K_{KA}}{k_{LS}} = \frac{k_{AT}}{k_{LS}} \cdot \gamma,$$

where q is the catalytic activity which permits a comparison of the catalytic reactions conducted with the resin and an equivalent quantity of inorganic acid [52]. Wolf [53] confirmed these relationships. The ester hydrolyses and sugar inversion carried out take place under the influence of molecular adsorption, which can be expressed by corresponding distribution coefficients in the resin and in the solution.

During operation with nonaqueous solvents, the stability of the ion exchanger with respect to the medium must be kept in mind. Before an ion exchanger is used

Table 4.3 Examples of the use of synthetic ion exchange resins as catalysts

Reaction	Exchanger; loading	Procedure	Temperature °C	Reference
Hydrolysis of various acetates	Amberlite IR-100; H^+	Batch	25	G. G. Thomas and C. W. Davies. Nature *159*, 372 (1947)
Saponification of ethyl-acetate etc.	Dowex 2; OH^- Amberlite IR-120; H^+	Batch	20 65	H. Deuel et al., Helv. Chim. Acta *34*, 1849 (1951)
Hydrolysis of proteins	Dowex 50; H^+	Batch	100	J. C. Paulson et al. J. Am. Chem. Soc. *75*, 2039 (1948)
Hydrolysis of nitrile	Amberlite IRA-400; OH^-	Reflux		A. J. Galat. J. Am. Chem. Soc. *70*, 3945 (1948)
Hydrolysis of isoamyl alcohol	Amberlite IRA-400; OH^-	Continuous tubular flow reactor		H. Manuel Llaneza. Afinidad *43*, 147 (1986)
Esterification of glycerol with ethanol	Zeo-Karb; H^+	Reflux	115	S. Sussman. Ind. Eng. Chem. *38*, 1228 (1946)
Esterification of n-butyl alcohol with acetic acid	Amberlite IR-100; H^+	Reflux	115	C. L. Levesque and A. M. Craig. Ind. Eng. Chem. *40*, 96 (1948)
Esterification of meth-acrylic acid	Cationite KU-2; H^+	Reflux and continuous plant	125	V. J. Isagulyants. Chem. Ind. (Moscow) *4*, 258 (1967)
Esterification of various carbonic acids	Polycondensation resins; H^+ and OH^-	Continuous process		V. Vasilescu. Chem. Techn. *11*, 29 (1959)
Transesterification of methyl methacrylate with nonyl alcohol	Styrene-DVB micro-spheres		110	T. D. Kozarenko and V. M. Bogolyubov. Tr. Irkutsk. Politekh. Inst. *1971, No. 69, 166*
Hydration of propylene	Sulfonic acid resin	Pressure	150	J. R. Kaiser et al. Ind. Eng. Chem. Prod. Res. Dev. *1*, 296 (1962)

Table 4.3 continued

Reaction	Exchanger; loading	Procedure	Temperature °C	Reference
Hydration of tertiary amylenes	Cationite KU-2			N. G. Polyanskii and T. I. Kozlova. Zh. Prikl. Khim. *39*, 1788 (1966)
Hydration of ethylene oxide	Lewatit SPC 118; H^+	Stirred reactor	<80	M. Ochsenkuehn et al. Chem. Ing. Techn. *47*, 1061 (1975)
Dehydration of hydroxamic acids	Wofatit KPS; H^+			F. Wolf and H. Schaaf. Z. Chem. *7*, 391 (1967)
Dehydratin of tert. butyl alcohol		Aqueous solution		J. Yamanis et al. J. Catal. *26*, 490 (1972)
Dehydration of tert. butyl alcohol	Macroreticular styrene-DVB copolymer with $PO(OH)_2$, $P(OH)_2$ and SO_3H groups	Vapor phase	95	K. Setinek. Acta Cient. Venez. Supl. *24*, 86 (1973)
Alkylation of phenols	Macroreticular styrene-DVB-SO_3H			D. Kleinhempel und F. Wolf. CA 74: 75810a.
Alkylation of xylenes	Amberlyst 15; H^+		90 – 130	Y. Morita et al. Sekiyu Gakkai Shi *15* (1972)
Alkylation of benzene	Amberlyst 15; H^+			J. Klein et al. Chem. Ing. Techn. *48*, 173 (1976)
Formation of ethers	Phenol-formaldehyde sulfonic resin; H^+		150	P. Mastagli et al. Compt. rend. *232*, 1848 (1951)
Formation of acetal and enol ethers	Polycondensation resins, sulfonated; H^+	Continuous distillation column		F. Runge, Angew. Chem. *62*, 451 (1950)
Formation of acetals and ketals	Sulfonated PS copolymer containing calcium sulfate	Catalytic dehydrator	20	V. I. Stenberg and D. A. Kubik. J. Org. Chem. *39*, 2815 (1974)
Aldol condensation and Knoevenagel condensation	Various resins; Dowex 3 plus acetic acid	Reflux	80	M. J. Astle et al. Ind Ling. Chem. *44*, 2867 (1952), J. Org. Chem. *26*, 4874 (1961)

Table 4.3 continued

Reaction	Exchanger; loading	Procedure	Temperature °C	Reference
Acyloin condensation of benzoin from benzaldehyde	Amberlite IRA · 400; CN^-	Cycle	80	C. J. Schmidle and R. C. Mansfield. Ind. Eng. Chem. *44*, 1388 (1952)
Cyanohydrin synthesis	Various anion resins as strong bases; OH^-	Reflux agitation	15 – 20	M. J. Astle and R. W. Etherington. Ind. Eng. Chem. *44*, 2871 (1952)
Hofmann decomposition	Amberlite IRA-400; OH^-	Column	180	J. Weinstock et al. J. Am. Chem. Soc. *75*, 2546 (1953)
Decomposition of diazoacetate	Sulfonic acid resin; H^+	Column	130	P. E. Gruber und N. Noller. Z. physikal. Chem. NF *38*, 184 (1963)
Preparation of polyesters	C. N. S. L.-resin; H^+	Reflux		N. D. Chatge. J. Appl. Polymer Sci. *8*, 1305 (1964)
Condensation of methylstyrene with formaldehyde	Cationite KU-2; H^2	Stirred reaction		V. J. Isagulyants. CA 67: 108 609c.
Isomerization of n-butenes	Amberlyst 15 sulfonated; H^+	Gas phase reaction		T. Uematsu. Bull. Chem. Soc. Japan *45*, 3329 (1972)
Interconversion of reducing sugars	Amberlite IRA-400; OH^- Amberlite IR-48; OH^-		20	L. Rebenfald and E. Pascu. J. Am. Chem. Soc. *75*, 4370 (1953)
Conversion of D-fructose to hydroxymethylfurfural	Sulfonic acid resin	Continuous conditions		Y. Nakamura. Noguchi Kenkyusho Jiho (24), 42 (1981)
Polymerization of unsaturated hydrocarbons	Sulfonic acid resin			N. G. Polyanski et al. Kinetics Catalysis *3*, 136 (1962)
Polymerization of α-olefins	Amberlite IRA-904			S. Nakano and M. Murayama. US Patent 3 595 849; 22 Jul 1971
Preparation of peracetic acid	Perfluorinated PS-resin; H^+	Batch	50	P. Hoffmann et al. Ger. Offen. 2 018 713; 4 Nov 1971
Aromatic nitrogen bases	Dowex 11; OH^-	Batch	100	A. Gauvreau and A. Lattes. Compt. rend. C *266*, 1126 (1968)

as a catalyst, it must have been completely transformed into the desired loading state; a regeneration is necessary only when secondary ionic components of the reaction mixture cause a charge reversal of the exchanger. The reaction solution must therefore be low in ionic impurities. If necessary, these must be extracted prior to the reaction.

Table 4.3 lists some reactions catalyzed by ion exchangers. Reported first by the former IG Farbenindustrie AG, such processes were later applied for numerous other catalytic reactions. It must be mentioned briefly that numerous publications have been published dealing with ion exchange resins as catalysts for alkylation, esterification, transesterification, hydrolysis, dehydration, hydration, oxidation, reduction, hydrogenation, condensation, oligomerization, polymerization, isomerization, and rearrangement and as polymer-bound catalysts for hydrogenation, hydroformylation, oligomerization, polymerization, organosilicon synthesis, oxidation, reduction, hydration, dehydration, alkylation, condensation, esterification and ether synthesis [54]. Recently perfluorinated ion exchange polymer resin has – because of its enhanced chemically inert backbone – been used as a strong acid catalyst for a wide variety of reactions in synthetic and thus preparative organic chemistry. In many instances there are improved yields or an increase in selectivity over other catalysts, even including the poly(styrenesulfonic acid) resins. The perfluorinated polymer has a higher thermal stability [55].

Ion exchangers as catalysts in commercial processes are of great interest. Over the years, especially in the petrochemical industry, some ion exchanger - catalyzed reactions of commercial significance have been installed. These include olefin hydrations, epoxidations, esterifications, etherifications, aldol condensations and nitrile hydrolyses. Hopefully one day alkylations will also be included. For the latter super-acid catalysis is of future interest. Other specific areas of ion exchanger catalysis include triphase catalysis [56] and metalloaded catalysis. For upscaling a preparative ion exchanger catalysis to a commercial process the decision whether to go through a pilot plant operation or not may depend on very special circumstances. The disposal of deactivated resins is becoming an increasingly important factor. As ion exchanger catalysis is now considered part of catalysis in general, it is theoretically and technically dealt with in the literature on catalysis as well.

References

[1] D. Turtoi (1973). Preparation of reagents and fine chemicals by purifying technical products by ion exchange. Rev. Chim. (Bucharest) *24*, 523.

[2] D. E. Wanner and F. J. Conrad (1967). Purification of analytical-grade cation exchange resin for spectrochemical applications. Appl. Spectrosc. *21*, 177.

[3] K. Dorfner (1969). Ion exchangers in TLC. In: Thin-Layer Chromatography (Egon Stahl, ed.) Springer-Verlag, Berlin, Heidelberg, New York, p. 44.

[4] M. Lederer (1977). Flat-bed ion-exchange materials J. Chromatogr. *144*, 275.

[5] R. Klement (1949). Anwendung von Harzaustauschern zur präparativen Darstellung freier Säuren und ihrer Salze. Z. anorgan. allgem. Chem. *260,* 267.
[6] R. Klement and R. Popp (1960). Darstellung und Eigenschaften einiger Polyphosphate. Chem. Ber. *93,* 156.
[7] F. Hein and H. Lilie (1952). Zur Darstellung von Komplexsäuren nach der Austauscher-Methode. Z. anorgan. allgem. Chem. *270,* 45.
[8] B.T. Mandalia et al. (1974). Use of ion exchange resins in chemical conversions and preparations. Proc. 3rd Symp. Ion-Exchange, Balatonfüred, 28 – 31 May.
[9] J. d'Ans et al. (1952). Die Anwendung von Anionenaustauschern in der analytischen und präparativen Chemie. Chem.-Ztg. *76,* 811.
[10] M. Rebek and M. K. Semlisch (1961). Preparation of ultrapure crystal violet base with the aid of ion exchanger. Monatsh. Chem. *92,* 214.
[11] R. Bar et al. (1983). Preparation of phase transfer catalysts by column ion exchange. Reactive Polymers *1,* 315.
[12] F. (G.) Helfferich (1962). Ligand exchange. II. Separation of ligands having different coordination valences. J. Amer. Chem. Soc. *84,* 3242.
[13] R. Kunin (1964). Chemical synthesis through ion exchange. Ind. Eng. Chem. *56,* 35.
[14] A. Ya. Levin et al. (1976). Synthesis of alkylsalicylate additives using an exchange reaction. Tr. Vses. Nauchno-Issled. Inst. Pererab. Nefti *14,* 104.
[15] A. Basinski and M. Sierocka (1955). Course of brine formation from iron(III), aluminum and chromium hydroxide with ion exchangers. Ann. Soc. chim. Polonorum *29,* 656.
[16] J. Steinbach and H. Freiser (1952). Preparation of standard sodium hydroxide solutions by use of a strong anion exchange resin. Anal. Chem. *24,* 1027.
[17] S. Fischer and R. Kunin (1956). Ion exchange preparation of low silica hydroxide solutions for colorimetric determinations of total silica. Nature *177,* 1125.
[18] E. D. Olsen and R. L. Poole (1965). Quick removal of alkali metals from quaternary ammonium bases during ion exchange. Anal. Chem. *37,* 1375.
[19] G. Moretti et al. (1966). Orientation of a cation exchange resin for trapping ammonium from blood in extracorporeal circulation. Rev. Franc. d'Etudes Clin. Biol. *11,* 938.
[20] K. Erne and T. Canbäck (1953). The fluorimetric determination of noradrenaline. J. Pharm. Pharmacol. *7,* 248.
[21] R. F. Carlstrom (1980). Ion exchange flowsheet for final purification of B plant cesium. Report RHO-F-8, 42 pp. Avail. NTIS.
[22] O. G. S. Comerzan (1978). Ulta-purification of titanium chloride by ion exchange chromatography. C. R. Hebd. Seances Acad. Sci., Ser. C. *286,* 147.
[23] E. J. Wheelwright (1974). Review of ion-exchange technology used for the purification of the individual rare earths. Report BNWL-SA-5067. Avail. Dep. NTIS.
[24] A. Vioque et al. (1966). Spurenelemente in Speisefetten. XI. Entmetallisierung von Olivenöl mit Kationenaustauschern. Fette, Seifen, Anstrichmittel *68,* 303. Id. (1968). Trace elements in edible oils. XIV. Application of the demetallization to refineable olive oils. Grasa Aceites (Seville, Spain) *19,* 81.
[25] H. A. Chase (1984). Ion exchangers and adsorbent resins for the purification of proteins. In: Ion Exchange Technology (D. Naden and M. Streat, eds.) Ellis Horwood, Chichester, UK.
[26] F. Gerstner (1953). Die Wiedergewinnung von Kupfer in der Kupferfaser-Erzeugung. Z. Elektrochem. *57,* 221.
[27] S. Sussman et al. (1945). Metal recovery by anion exchange. Ind. Eng. Chem. *37,* 618.
[28] R. E. Anderson and R. D. Hansen (1955). Phenol sorption on ion exchange resins. Ind. Eng. Chem. *47,* 71.
[29] N. Applezweig (1944). Cinchona alkaloids prepared by ion exchange. J. Amer. Chem. Soc. *66,* 1990.

[30] G. Bors et al. (1964). The isolation of atropine, morphine and strychnine by means of ion exchange resin in forensic chemistry. Farmacia *12,* 479.

[31] R. R. Brooks (1960). The use of ion exchange enrichment in the determination of trace elements in sea water. Analyst *85,* 745.

[32] J. F. Pankow and G. E. Janauer (1974). Analysis for chromium traces in natural waters. I. Preconcentration of chromate from ppb levels in aqueous solutions by ion exchange. Analyt. Chim. Acta *69,* 97.

[33] J. A. Tallmagde et al. (1964). Mineral recovery from sea water. I and EC *56*(July), 44.

[34] T. M. Florence and G. E. Batley (1975). Removal of trace metals from seawater by chelating resin. Talanta *22,* 201.

[35] K. H. Lieser and B. Gleitsmann (1982). Separation of heavy metals, in particular uranium, from sea water by use of anchor groups of high selectivity. II. Continuous flow in a fluidized bed. Fres. Z. Anal. Chem. *313,* 289.

[36] V. I. Ksenzenko et al. (1981). Use of ion-exchangers for the extraction of bromine and iodine from natural brines. Ionnyi Obmen. Edited by M. M. Senyavin, Izd. Nauk: Moscow, USSR.

[37] J. Pelikan and F. Nekvasil (1968). Recovery of iodine from mineral waters in ion exchangers. Kunstharz-Ionenaustauscher; Plenar-Diskussionsvortr. Symp. (Publ. 1970). Akad.-Verlag: Berlin, GDR.

[38] N. I. Tarchigina et al. (1986). Static sorption of iodine by an ion exchange column based on polyacrylonitrile and anionite AV-17. Zh. Fiz. Khim. *60,* 1259.

[39] V. I. Bibikova et al. (1972). Preparation of ammonium perrhenate using an ion exchange method. Nauch. Tr., Nauch.-Issled. Proekt. Inst. Redkometal. Prom. No. 38, 29.

[40] X. Zi et al. (1986). Studies on isolation and purification of steviosides from Stevia by ion exchange method. Zhongguo Tiaoweipin (1), 12 (Chin.).

[41] M. Gordon et al. (1962). The use of cyanide-form ion exchange resins in the preparation of nitriles. Chem. Ind. 1019. Id. (1963). Anion exchange resins in the synthesis of nitriles. J. Org. Chem. *28,* 698.

[42] Y. Urata et al. (1960). Substitution reaction of alkyl halides by anion exchange resin. Nippon Kagaku Zasshi *81,* 1121, Id. (1962). Substitution reactions of organic halogen compounds with anion exchange resins. Ibid. *83,* 932. Id. (1962). Preparation of organic fluorine compounds by means of anion exchange resins. Ibid. *83,* 1936. Id. (1962). Formation of alkyl cyanides using anion exchange resins. Ibid. *83,* 1105.

[43] C. Catusse et al. (1985). Selective transformation of benzyl chloride into benzyl thiocyanate in the presence of ion exchange resins. Generalization of the reaction. J. Chem. Technol. Biotechnol., Chem. Technol. *35A,* 248.

[44] C. Valenzuela Calahorro et al. (1985). The kinetics of scheelite dissolution by an acid resin. An. Quim., Ser. B. *81,* 186.

[45] H. L. Boehner and A. B. Mindler (1949). Ion exchange in waste treatment. Ind. Eng. Chem. *41,* 450.

[46] R. Grießbach et al. (1953). Ionenaustauscher und Katalyse. Chem. Techn. *5,* 187. G. Naumann (1959). Katalyse durch Ionenaustauscher. Ibid. *11,* 18. J. Remond (1960). Ion exchange resin: its use in catalysis. Rev. Prod. Chim. *63,* 417, 421. N. G. Polyanski et al. (1962). Ionenaustauscherharze als Katalysatoren für die Polymerisation ungesättigter Kohlenwasserstoffe. Kin. Kat. *3,* 136. C. N. Satterfield and T. K. Sherwood (1963). The Role of Diffusion in Catalysis. Addison-Wesley, Reading, Mass.

[47] H. Spes (1966). Katalytische Reaktionen in Ionenaustauscherkolonnen unter Verschiebung des chemischen Gleichgewichts. Chem.-Ztg. *90,* 443.

[48] G. W. Bodamer and R. Kunin (1953). Behavior of ion exchange resins in solvents other than water. Ind. Eng. Chem. *45,* 2577.

[49] F. Wolf and H. Schaaf (1967). Zur Eignung von stark sauren Kationenaustauscherharzen in Gegenwart von gasförmigem Chlorwasserstoff als Dehydratisierungskatalysatoren. Z. Chem. *7*, 391.

[50] P. E. Gruber and H. Noller (1963). Zur Dynamik der Austauscherkatalyse. Z. physikal. Chem. NF *38*, 184, 203. E. Knoezinger and H. Noller (1972). Dehydration of alcohols on poly(styrenesulfonic acid) ion exchanger. Infrared spectroscopic and microcatalytic studies. Z. phys. Chem. (Frankfurt) *79*, 130. Id. (1972). Catalytic decomposition of formic acid on the ion exchanger poly(styrenesulfonic acid). Ibid. *79*, 66.

[51] H. Deuel et al. (1951). Über selektive Reaktionen an Ionenaustauschern. Helv. Chim. Acta *34*, 1849.

[52] F. (G.) Helfferich (1954). Katalyse durch Ionenaustauscher. Angew. Chem. *66*, 241.

[53] F. Wolf (1969). Theory and application of ion exchange catalysis. I. Kinetics of ester hydrolysis and sugar inversion in the presence of strongly acidic cation-exchange resins. II. Kinetics of ester hydrolysis in the presence of strongly acidic cation-exchange resins with macroreticular structure. Conference on Ion Exchange, London.

[54] W. Dawydoff (1976). The application of polymers in the heterogeneous and homogeneous catalysis. Part 4. Faserforsch. Textiltechn. *27*, 189. N. G. Polyanskii and V. K. Sopozhnikov (1977). New advances in catalysis by ion exchangers. Usp. Khim. *46*, 445. G. Wulff (1971). Chemische Reaktionen unter Verwendung synthetischer Polymerer. Chem. unserer Zeit *5*, 170.

[55] F. J. Walter. Catalysis with a perfluorinated ion-exchange polymer. ACS Symp. Ser. 308; Polymeric reagents and catalysts. W. T. Ford, ed. American Chemical Society, Washington, DC, 1986.

[56] W. Ford (1984). Chemtech *14*, 436.

5 Ion Exchangers in Analytical Chemistry

An Introduction to Analytical Applications of Ion Exchangers

J. Inczedy
University of Chemical Engineering, Dep. of Analytical Chemistry, 8201 POB 158, Veszprém, Hungary

The history of the use of ion exchangers in analytical chemistry dates from that point in time, when ion exchange as a process on its own was recognized and the first synthetic ion exchangers were made. The first documented analytical application of synthetic inorganic ion exchangers was carried out by Folin and Bell [1] who used them for the preliminary separation and collection of ammonium ions from urine prior to its determination. Since that paper the use of ion exchangers – in particular after the development of the polymer-based synthetic organic ion exchangers – has become an indispensable tool of the modern analytical chemist.

The separation and quantitative determination of rare earth metals, the determination of amino acids in protein hydrolysates – both of which were milestones in the development of the technical and biochemical sciences – were and are today carried out by the use of ion exchange chromatography.

Separation itself was always the focus of analytical procedures, and it has remained so even up to today, because the analysis of a complicated mixture of components of similar chemical properties, or the analysis of traces in the presence of matrix compounds or elements usually cannot be accomplished without separation. Of course, the separation procedures have been developed to a great extent over the last decades, offering rapid, clear separations with acceptable performance and accuracy. The use of ion exchangers offers quite a few – in many cases very effective – ways of doing separations, especially for ionic or ionisable components.

The first monograph on the use of ion exchangers in analytical chemistry was written by Samuelson [2] and was published in 1952. Since that time many books on the subject have appeared. In the references 3–15 only monographs are listed, but not all those books in which one or more chapters are devoted to the analytical applications of ion exchangers. A complete bibliography and survey on the published books and selected papers can be found in the reviews published every two years in Analytical Chemistry [16].

Classification of ion exchange methods used in analytical chemistry

Ion exchangers are not soluble in water but, in connection with aqueous solutions, are able to exchange their ions for ions originally in the solution. They are therefore extremely useful in various procedures for the removal of interfering ions and substances, for purification, for enrichment of trace constituents, for salt-splitting procedures, for decreasing the amount of the matrix components, for separation of multicomponent mixtures, for improvement of sensitivity or selectivity in a detection step, etc., without an appreciable contamination of the analyte.

The various ion exchange methods used for analytical purposes can be classified into the following three groups [17].

1. *Ion exchange procedures based on total exchange.* To this group belong all those simple procedures, where ions are exchanged quantitatively for the enrichment of traces of metal or other ions or non-ionic organic substances; for separation (or replacement) of interfering ions or elements; or simply to change the ionic composition of an electrolyte, or to convert a salt into an acid or vice versa.

For this purpose column techniques are mainly used, but batch methods may also be employed.

2. *Separation of components of similar chemical behaviour by ion exchange chromatography.* Although in some cases it is difficult to distinguish between the simple ion exchange procedures mentioned above and chromatography, the following considerations may apply as a rule. In the procedures classified into group one, the exchange of certain ions is complete (or at least is expected to be complete) and if separations are made, the separation factors are high (usually greater than 100). Ion exchange chromatography serves, however, for the separation of ions or molecules of similar behaviour, so that the separation factors are usually low.

The technique used is mainly column chromatography, but chromatography on thin layers or on paper impregnated with ion exchangers is also employed.

3. *Ion exchangers used as carriers*. Ion exchanger beads (or powder) loaded with certain reagent ions, enzymic etc. can be used as stable reagents. There are also other methods in which the ion exchanger acts as a carrier on which a desired reaction (e. g., oxidation-reduction reaction) takes place.

The range of applications of the ion exchangers is rather large, but the most frequently and widely used methods belong to the three groups above.

Characteristic features of ion exchange methods

Ion exchange methods – belonging to the first two classes mentioned above – provide a means of separation, and for application to determinations always need preliminary sampling, sample preparation and subsequent detection or measurement steps. It is very important that all preliminary and subsequent steps should be in full accordance with the conditions and the quality (precision, accuracy and rate) of the ion exchange operation. If there are divergences in the sampling, sample preparation, separation, detection, evaluation or control steps, the complete procedure will give less reliable results. This need for uniformity is especially important if the procedure is automatised.

Ion exchange methods – as their name suggests – are used for the adsorption and separation of ionic or ionisable species from the solution. The exchange process usually can be described as a chemical reaction taking place between the ion exchanger and the solution phase, a process that can be influenced by other chemical reactions taking place in the solution phase.

One of the main advantages of ion exchange is that the process can be described by chemical reactions. Since the rate of the ionic reactions is usually high, by using the equilibrium constants of the reactions taking part in the solution as well as in the ion exchanger phase, the extent of adsorption, and the value of the distribution ratio of the distributed components can be easily calculated and, in most cases, the most adequate conditions for an analytical procedure can be predicted.

The other very important advantage of ion exchange procedures is strongly connected with the above-mentioned facts. The extent of adsorption of a component, or more appropriately the distribution ratio of the component (i. e., between the ion exchanger and solution phase), usually can be varied at will over a very large range by using proper reagents and concentrations in the solution phase. The strong adsorption of a component can be changed quickly for easy elution (desorption) of it, simply by means of changing the composition of the fluid medium properly. Binding and releasing are precisely controllable processes, usually without any hysteresis or irreversibly adsorbed remnants. Thus, in many cases, a high selectivity can be achieved in separations.

Calculation of terms used for planning and optimization of analytical methods

The most frequently used strong acid cation and strong base anion exchangers can be regarded as strong electrolytes. Thus, for a cation exchange process, where simple monovalent ions (e. g., sodium) are exchanged for divalent, (e. g., copper) ions, the following reaction takes place:

$$2\,\overline{RNa} + Cu^{2+} \overset{K^x}{\rightleftharpoons} \overline{R_2Cu} + 2\,Na^+ . \tag{1}$$

R denotes the equivalent amount of the ion exchanger, and the bar refers to the species in the solid (gel) phase. The process can be characterized by the so-called selectivity coefficient, K^x, expressed using concentrations:

$$K^x = \frac{[Cu^{2+}]_r\,[Na^+]_w^2}{[Na^+]_r^2\,[Cu^{2+}]_w}, \tag{2}$$

where r refers to the resin, and w to the aqueous solution phase.

The adsorption of the copper ions can be controlled by the concentration of the competing sodium ions. If bivalent, negatively charged complex-forming ligands (such as oxalate, tartrate ions) are present, which form neutral or negatively-charged complexes with the copper ions, the ion exchange reaction (1) will be shifted to the left. Since the negatively-charged ligands usually tend to protonate, a competition between the copper and hydrogen ions will arise

$$Cu^{2+} + L^{2-} \rightleftharpoons CuL \tag{3}$$

$$Cu^{2+} + 2\,L^{2-} \rightleftharpoons CuL_2^{2-} \tag{4}$$

$$L^{2-} + H^+ \rightleftharpoons LH^- \tag{5}$$

$$LH^- + H^+ \rightleftharpoons LH_2 . \tag{6}$$

The corresponding constants are: β_1, β_2 and K_1, K_2.

Using conditional equilibrium constants and side-reaction functions introduced by Ringbom [18], the distribution ratio of the copper, characterizing the extent of adsorption at given conditions, can be easily calculated. Using the conditional equilibrium concepts oxidation-reduction reactions in the solution phase can also be taken into account. Unfortunately, the value of K^x depends on the concentrations, but it is fairly constant in certain cases. This occurs, for example, when the amount of copper is low compared with that of the competing sodium ion and the latter is constant (the ionic strength in the solution is constant). In chromatographic separations, the amount of the eluted (separated) ions is always low compared with that of the competing (eluent) ion, and the latter ion dominates in the ion exchanger phase. In such cases the distribution ratio of the eluted ion can be calculated by a simple equation. In the case of the copper-sodium system, if trace amounts of copper are present, the following equation can be derived from eq. (2):

$$D_{Cu} = \frac{[Cu^{2+}]_r}{[Cu^{2+}]_w} = K^x \left(\frac{[Na^+]_r}{[Na^+]_w} \right)^2 \tag{7}$$
$$\log D_{Cu} = \log K^x + 2 \log Q - 2 \log [Na^+]_w \, .$$

This equation can be used for exploratory calculations in the planning of adsorption or elution of the copper ion, using a sodium-containing electrolyte as eluent, with $[Na^+]_r, \approx Q$ (with Q being the capacity of the ion-exchanger in equ/dm^3). In the presence of complex-forming ligands as mentioned above, the conditional constant K'^x is to be used instead of K^x in the above equation,

$$K'^x = \frac{K^x}{\alpha_{Cu(L)}}\,, \tag{8}$$

where $\alpha_{Cu(L)}$ is the side-reaction function

$$\alpha_{Cu(L)} = 1 + \frac{C_L}{\alpha_{L(H)}} \beta_1 + \left[\frac{C_L}{\alpha_{L(H)}} \right]^2 \beta_2 \, . \tag{9}$$

C_L is the total concentration of the complex-forming agent in the solution; $\alpha_{L(H)}$ is the side-reaction function of the ligand originating from its protonation

$$\alpha_{L(H)} = 1 + K_1 [H^+] + K_1 K_2 [H^+]^2 \, . \tag{10}$$

It is interesting to note that eq. (7) proved to be valid also in those cases, where while the eluent ion concentration remained constant, the capacity of the ion exchanger was changed.

Using similar equations the retention data of the components, the necessary size of the chromatographic column, the suitable concentration and the pH values for the required separation efficiency can be quite easily calculated, even for very complicated systems or with gradually changing pH of the eluent, etc. [19]. Equations derived for the separation of anions are similar. For more details see [20].

How to write a paper on a new analytical method based on ion exchange or ion exchange chromatography

For those who would like to publish a paper on a new analytical method using ion exchange or ion exchange chromatography detailed guidelines are given in [17]. The main points concerning the title, introduction, experimental and discussion sections of any publication on a new method are listed. It is, for example, explained that it is very important that the title should give the essence of the paper unambiguously, and that while short, should still include all the necessary words for adequate computer classification. Besides the fact that in the introduction to the paper, as usual, a brief survey of the literature must be given, and an explanation of the problem to be investigated and solved, it is at the same time very important to state clearly whether the ion exchange method presented serves to improve an existing analytical method, or is of direct use for analysis of certain substances, or is used as a model for a new separation principle. In the experimental part of the paper one should not only, as usual, give first a brief description of the reagents and instruments used but, since the reproducibility of simple ion exchange methods and especially of chromatographic separations depends highly on the properties of the ion exchanger material, full characterization of the latter is very important. The same applies to a full description of the column used, as well as of the auxiliary equipment, compositions of the sample solutions, and exact description of the pretreatment of the ion exchanger and of the preparation of the ion exchange column, etc. If a batch operation is used, it is very important to give data on the resin to liquid ratio, the mode and duration of mixing, the temperature used and the removal of the ion exchanger, rinsing, etc. In the discussions and conclusions the results obtained by the complete ion exchange analytical method should be compared with those of a reference method or other methods used for similar purposes, and the advantages of the new method emphasized, etc. In choosing the terms used in the experimental and discussion sections of the paper, close attention should be paid to the IUPAC recommendations for presentation of results, and on the nomenclature for ion exchange and chromatography.

Type of ion exchangers

At the selection of a suitable ion exchanger for a given purpose the following points should be considered.

For simple laboratory procedures (removal of interfering ions, salt-splitting, conversion, etc.) the conventional gel type strong acid and strong base ion exchangers of medium crosslinking can be applied. For ions of larger size, resins of low crosslinking, for adsorption of large organic ions (bases or acids) dextran or cellulose-based exchangers are preferred. For enrichment of metal traces conventional type ion exchangers can be used, but in order to obtain higher selectivity chelate-type adsorbents can be applied.

In the procedures concerned with very low concentrations the contamination of the used ion exchanger must be taken into consideration. In certain cases a preliminary purification step is also necessary. To remove metal impurities, EDTA-containing solution or dilute mineral acid solution can be employed. From certain technical grade resins it is almost impossible to remove the metal impurities: for the removal of iron traces the mineral acid treatment in the presence of a reducing agent may help. Usually not only metal impurities, but soluble organic constitutents must be removed. For that purpose extraction with alcohol in a Soxhlet apparatus may be suggested.

To carry out column chromatographic separations with proper efficiency, special column-packing materials are to be used. To obtain good mass transfer properties ion exchange resin beads of low particle size, surface-coated silica, or resin beads with an ion exchanger surface layer, etc., are to be applied.

Solvents other than aqueous ones

Due to the high polarity and dielectric constant of water, aqueous solutions are preferred for ion exchange procedures. In non-aqueous solutions, the degree of ionization, the swelling of the resin and, correspondingly, the exchange rate, are much lower. However, in certain cases water-organic solvent mixtures are suggested to increase separation efficiency or the degree of selectivity.

References

[1] O. Folin and R. Bell (1917) J. Biol. Chem. *29,* 329.
[2] O. Samuelson, Ion Exchangers in Analytical Chemistry, Almquist and Wiksell, Stockholm 1952; Wiley, New York 1953.
[3] E. Blasius (1958) Chromatographie und Ionenaustausch in der analytischen und präparativen Chemie, Enke, Stuttgart.
[4] K. Dorfner (1963) Ionenaustausch-Chromatographie, Akademie-Verlag, Berlin.
[5] K. Dorfner (1969) Ion Exchangers in TLC. In: Thin-Layer Chromatography (E. Stahl, ed.) Springer-Verlag, Berlin, Heidelberg, New York.
[6] B. Tremillon (1965) Les Séparations par les Resines Exchangeuses d'Ions, Gauthier-Villars, Paris.
[7] J. Inczédy (1966) Analytical Applications of Ion Exchangers, Pergamon Press, Oxford.
[8] J. Korkisch (1966) Ion exchange in mixed and nonaqueous media. Progress in Nuclear Energy Series IX. Analytical Chemistry, Vol. 6, Pergamon Press, Oxford.
[9] W. Rieman and H. F. Walton (1970) Ion Exchange in Analytical Chemistry, Pergamon Press, Oxford.
[10] R. T. Allsorp and J. A. D. Healey (1974) Chemical Analysis and Ion Exchange, Heinemann, London.
[11] J. X. Khym (1974) Analytical Ion Exchange Procedures in Chemistry and Biology, Prentice Hall, Engelwood Cliffs, N.Y.
[12] Ion-Chromatographic Analysis of Environmental Pollutants (1978) (E. Sawicki, J. D. Mulik and E. Wittgenstein, eds.) Vol. I – II. Ann Arbor.
[13] M. Marhol (1982) Ion Exchangers in Analytical Chemistry. Their properties and use in inorganic chemistry. Comprehensive Analytical Chemistry. Vol. XIV. Elsevier, Amsterdam.
[14] J. Fritz, D. T. Gjerde and C. Pohlandt (1982) Ion-Chromatography, Hüttig, Heidelberg.
[15] F. C. Smith and R. C. Chang (1983) The Practice of Ion Chromatography, Wiley, New York.
[16] R. Kunin et al. Anal. Chem. *32,* 67 R (1960); *34,* 48R (1962); *36,* 142 R (1964); *38,* 176 R (1966); *40,* 136 R (1968); H. F. Walton, Anal. Chem. *36,* 51 R (1964); *38,* 79 R (1966);

40, 51 R (1968); *42,* 85 R (1970); *44,* 256 R (1972); *46,* 398 R (1974); *48,* 52 R (1976); *50,* 36 R (1978); *52,* 15 R (1980); A. Majors, Anal. Chem. *54,* 343 R (1982); *56,* 321 R (1984).

[17] J. Inczédy (1980) Talanta *27,* 143.

[18] A. Ringbom (1963) Complexation in Analytical Chemistry, Wiley, New York.

[19] J. Inczédy (1978) J. Chromatog. *15,* 175.

[20] J. Inczédy (1976) Analytical Applications of Complex Equilibria, Section 3.9, Horwood, Chichester.

5.1 Analytical Methods Based on Ion Exchange

Günther Bonn and Ortwin Bobleter
Universität Innsbruck, Institut für Radiochemie, Innrain 52 a, 6020 Innsbruck, Austria

Symbols and definitions

$a, \overline{a}$	activity of ions in solution and in exchanger phase
An	anionic species
$C, \overline{C}$	concentration in solution and in ion exchange phase (mol/l)
C_R	exchange capacity (mol/l)
Cat	cationic species
D	diffusion coefficient for ions in solution (cm^2s^{-1})
F	cross-section of exchanger column (cm^2)
k_1, k_2	constants
k_a, k_b	constants
k_{CD}	concentration distribution coefficient
k^*_{CD}	experimental concentration distribution coefficient
k_D	weight distribution coefficient
K_a	dissociation constant

K_E	equilibrium constant of the distribution
l	litre
L	length of column (cm)
L^-	anionic ligand
m	valency state
n	valency state and exponential factor
N	number of exchanger particles per unit column volume (cm^{-3})
q	tangent of breakthrough curve (semilogarithmic plot)
Q_P	specific exchange capacity (equ/l)
r_O	radius of exchanger particles (mm)
R	ion exchanger with functional group
t_O	retention time of a non-retained solute (s)
t_R	retention time of a solute (s)
v_P	linear flow rate in liquid phase of column ($cm\ s^{-1}$)
V_O	retention volume of a non-retained solute (cm^3)
V_R	retention volume of solute (cm^3)
α	angle of the breakthrough curve (semilogarithmic plot)
β	fraction of liquid phase in exchange column
$\gamma, \bar{\gamma}$	activity coefficient of ions in solution and in exchanger phase
δ_0	thickness of the diffusion zone on the exchanger particle's surface (mm)

5.1.1 Introduction

Certain characteristics of ion exchangers, such as ion exclusion, ion adsorption and ion retardation, are widely employed in analytical chemistry. These interactions of the solutes with the ion exchange substrate not only govern chromatographic separations, as will be described in Chapter 5.2, but also facilitate other analytical processes, for example preconcentration, preseparation and purification of solvents and samples.

A necessary procedure is the cleansing of solvents, since the high sensitivity of modern detection systems requires extremely purified liquids. Ionic contaminations can be removed effectively by ion exchangers.

The low permissible concentrations of pollutants, drugs and toxic substances in human body fluids and natural waters require, in many cases, preconcentration steps in order to increase the solute concentrations to detectable limits. Ionic compounds can be easily concentrated by ion exchangers.

If ionic solutes interfere with the determination of non-ionic species, the separation is normally carried out by ion exchangers. Preseparation can also be applied in a wider sense: ion exchange separation steps, where certain solutes are isolated before the actual determination occurs, belong to this category. Ion exclusion, ion retardation, selective elution, membrane and liquid exchanger separations are the methods applied for this purpose. A special case is ion exchange sorption analysis, where the exchange behaviour allows the determination of the concentration of a solute. Molecular parameters, such as dissociation constants, lipophilicity and the characterization of the exchanger phase, can be analyzed successfully with ion exchange methods.

Books in which special aspects of this section are discussed are listed unter A 1 to A 21.

First the main ion exchange properties applied in non-chromatographic analysis will be discussed.

5.1.1.1 Ion exclusion

According to Donnan [1], the ion exclusion phenomenon can be explained as the expulsion of ions of low concentration from the exchanger phase due to the high ionic concentration of this phase. If, for example, an acid (H^+An^-) is brought into contact with a H^+-loaded cation exchanger (H^+R^-), the activities of the anions and cations in the solution are equal to those in the exchanger phase:

$$a_H \cdot a_{An} = \bar{a}_H \cdot \bar{a}_{An} \tag{1}$$

By introducing the concentrations (C) and the activity coefficients (γ), this equation becomes [2]

$$C^2_{HAn} \cdot \gamma^2_{HAn} = \bar{\gamma}^2_{HAn} \cdot \bar{C}_{HR}(\bar{C}_{HR} - n\bar{C}_{R^-}), \tag{2}$$

which can be solved as

$$\bar{C}_{HR} = \frac{\bar{C}_{R^-} \pm \sqrt{\bar{C}^2_{R^-} + 4(\gamma_{HAn}/\bar{\gamma}_{HAn})^2 C^2_{HAn}}}{2} \tag{3}$$

($\bar{C}_R$ is the exchange capacity and $\bar{C}_{HR}$ the total cation concentration of HR and HAn in the exchanger phase). In this case a monovalent ion exchange resin ($n = 1$) has been assumed for the sake of simplicity.

From eq. 3 it can be seen that, in the case of low acidic concentrations (C_{HAn} values below 1 mol/l), the expression inside the radical sign does not differ very much from the capacity of the exchanger, $\bar{C}_{R^-}$, *and therefore practically no invasion of acid into the ion exchange phase occurs. This is valid especially when the* $\gamma_{HAn}/\bar{\gamma}_{HAn}$ expression is not far from 1. The same result is obtained in the evaluation of relevant experiments. In Figure 5.1 the HCl-invasion measurements by Kraus and Moore [3] are compared with eq. 3. The theoretical curve coincides very well with the experimental points over a large part of the HCl concentration region. Only the experimental point at 0.1 molar HCl lies too high. This, however, can be explained by the experimental method applied, where the HCl is rinsed out of the exchanger column giving rise to hydrolysis of the RCl at low HCl concentrations.

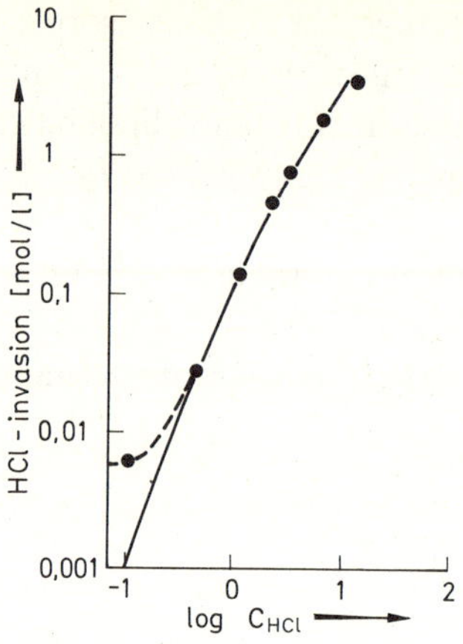

Figure 5.1 Donnan exclusion and invasion. The HCl concentration in the anion exchanger phase is plotted relative to the HCl concentration in the solution (mol/l). Measured points ● [3]; solid line according to equation 3 [2].

Figure 5.1 shows clearly that ion exclusion occurs at low concentrations. The theoretical curve indicates that, at a 0.1 molar outer HCl concentration, only a 0.001 molar HCl invasion into the exchanger phase takes place. For these calculations only the liquid part of the exchanger phase was taken into account; the organic resin part was neglected, which is considered to be a reasonable premise.

This ion exclusion is a general phenomenon. A regenerated cation exchanger (H^+-form) will therefore exchange cations, but the anions (and also the exchanged H^+ ions) remain excluded as long as dilute solutions are applied. The excluded ions can therefore easily be determined in the eluent.

At high electrolyte concentrations, strong invasion into the ion exchanger phase occurs. In this case, however, the solute is distributed in both the liquid and the ion exchange phases, and therefore this effect is of little use for analytical purposes.

5.1.1.2 Ion retardation

At relatively low ionic concentrations the adsorption and distribution isotherms show a linear concentration dependence. If a sample is injected at the top of an ion exchange column and elution is begun, it can be assumed that the low ion concentration is reached very soon and therefore the concept of the linear isotherm region is applicable.

Under these conditions the distribution coefficient (k_D) is proportional to the retention time (t_R) and the retention volume (V_R):

$$k_D = k_1(t_R - t_o) = k_2(V_R - V_o) \tag{4}$$

A substance which is not adsorbed or distributed on the ion exchanger appears non-retarded in the eluent after the time t_O in the volume V_O. It is fairly easy to choose a k_D value, so that the ionic substances are retarded and the non-retarded solutes can be collected and analyzed without interference of the ionic components. In a second step the latter can be eluted and analyzed afterwards.

The k_D values can be chosen so that fast separations are possible. For this purpose, exchanger materials (with weak or strong acidic or basic functional groups) with different exchange capacities and selected solvents are applied. At the same time, the column dimensions (inherent in k_1 and k_2 of eq. 4) can be adjusted to solve the problem concerned. If cations and anions have to be retarded, cation and anion exchanger columns, connected in series, or mixed-bed exchangers have to be applied.

Mixed-bed ion exchangers consisting of a strong anion and a weak cation (e. g., acrylic acid) copolymerisate, are also called ion retardation resins. In addition to ionic retardation, such exchangers can also be used for the desalting of protein and amino acid solutions.

5.1.1.3 Ion sorption

A special case of ion retardation is ion sorption. A relatively high distribution coefficient or a fairly large amount of exchanger is chosen so that the cations, anions or both are retained (sorbed) in the exchanger phase and only the ions with an opposite charge, or salt-free solutions, are eluted for further analysis.

In modern HPLC analysis, ion-exchanger precolumns are often applied for this purpose. Due to the smallness of the injection sample volumes (e. g. 10 μl), one-way precolumns frequently have sufficient exchanger capacity to last for several hundred injections.

5.1.1.4 Selective ion sorption

Special ion exchange materials demonstrate high selectivity for certain ions. They sometimes differentiate effectively not only between ions of varying valencies, but also between ions of the same valency.

Chelating ion exchangers. Chelating ion exchangers are a typical example. A representative model is the styrene-divinylbenzene copolymer (R), which contains paired iminodiacetate ions (e. g., Chelex 100®):

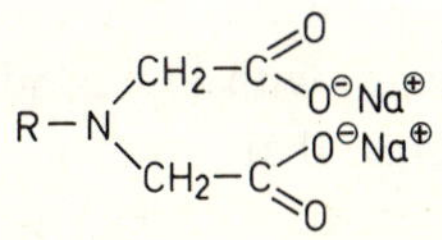

Divalent ions in contact with such exchangers may have distribution coefficients 5000 times higher than those of monovalent ions. Chelating ion exchangers can be used for the purification of buffers, ionic reagents and e. g. the separation of amines from peptides.

For certain experiments the high bond-strength of metal ions (60 to 100 kJ/mol) compared with that of normal exchangers (8 to 12 kJ/mol), has to be considered. The exchange kinetics are determined by a second-order mechanism, and to a lesser extent by diffusion.

Selective inorganic exchangers. Certain inorganic exchangers also show special selectivity. Microcrystalline ammonium molybdophosphate, with a capacity of approximately 1 mequ/g, is, for example, highly selective for heavier alkali ions. Cesium can therefore be concentrated on such exchangers from solutions with high sodium ion concentrations.

Chiral separations. The separation of enantiomers (which are mirror reflexions of the same compound) is growing in importance.

This is mainly due to the fact that enantiomers can react quite differently in certain biological systems. A mixture containing equal amounts of both enantiomers is called a "racemate". The isolation of an optically pure enantiomer and its analytical determination is a practice already widely applied. The mechanisms which govern these separations are, however, not understood very well [4].

During recent years chiral chromatographic methods have resulted in many successful separations. Since the role of the ion exchanger in this field has not yet been clearly defined, a short discussion of the subject is included in this section: in certain cases the chiral component is bound in the stationary phase, for example in cellulose triacetate. But also non-chiral columns using ion-pairing processes give good separation results [5]. The experiments of Hare and Gil-Av [6] are of special interest: an L-proline Cu(II) complex was added to the mobile phase and a non-chiral cation exchanger was used as stationary phase, to achieve chiral separations.

Many enantio-selective systems, using RP phases and chiral metal complexing agents, have been reported in the literature [7].

5.1.2 Clean-up of analytical solvents and samples

5.1.2.1 Clean-up and deionisation of solvents

In analytical procedures the quality and purity of the applied solvents (e. g., water) can be of decisive importance.

Water. For the purification of water several methods are in use. The application of cation and anion exchangers yields well deionized water with low conductivity. Mixed-bed ion exchange systems are usually more effective than two ion exchange columns filled with anion and cation exchangers connected in series. A properly packed column of mixed-bed resin produces water with a higher resistance than 10 MΩ/cm.

After ion exchange clean-up of the water the ion content is lower than in the case of bi-distilled water. The ion exchange treatment can, however, not eliminate neutral organic compounds and these residues can interfere in the analysis – for instance, through complex formation. If such complications have to be considered, it is advisable to use water which has been distilled at least three times or ion exchange deionized water purified by a further distillation step.

Mixed organic-water solutions. Other solvents, especially mixed organic-water solutions, can also be deionized by ion exchangers. This procedure is applied, for example, for the clean-up of formamid, using a mixed-bed resin [8]. In another case, sulphate-free gelatine solutions were obtained by introducing anion exchangers [9].

Reagent solutions. In many cases the chemical purity of the reagent can be very important. If, for example, hydrogen peroxide has a function as analytical reagent, its phosphate content may interfere in the analytical determination. The elemination of the phosphates from such solutions can very easily be carried out by ion exchangers [10].

5.1.2.2 Purification of analytical samples

Inorganic ions. In certain analytical procedures the inorganic ions can strongly influence the analytical results. In most cases these inorganic ions can be removed from the sample solutions by ion exchangers. The removal of phosphates from various biological samples can be of importance and is usually accomplished in this way [11].

Cations frequently interfere in the analytical determination of organic compounds. In the analysis of protein solutions the separation of metal ions with cation exchangers is a useful and often necessary method. In this way, metal-free enzyme solutions [12], cell suspensions [13], blood and urine samples [14, 15] can be obtained. For urea determination deionization steps can be introduced. The separation of ammonia and ammonium isocyanate from such solutions is described in the literature [16–18].

In modern ion exchange high-performance liquid chromatography (HPLC), ion exchange precolumns are frequently applied. They are normally in the OH^-- or H^+-form and adsorb anions and cations before the analytical separation of the remaining cations and anions on the main column is performed [19–22]. In the following Section 5.2 special applications will be described in more detail.

Organic ions. Biological samples can contain organic acids or bases that have to be eliminated before analysis. For example, the anions of oxalic and tartaric acids are adsorbed on ion exchangers before further analytical steps are taken [23, A18].

5.1.3 Preseparation processes

5.1.3.1 Preconcentration of analytical samples

The increasing necessity for the determination of ions in extremely dilute solutions requires good preconcentration methods. In many cases ion exchangers fulfil this requirement well, since they contribute very little to the contamination of the analytical sample. Cation and anion enrichment in water solutions has been carried out [A2] especially successfully. The preconcentration of many nearly carrier-free radionuclides occurring in both surface and underground waters helps to shorten the analysis time. Radioactive nuclides, such as Sr-90, I-131, Cs-137 [24–26], have been concentrated on ion exchangers and eluted as complexes.

The trace metal analyses of air, industrial waste water and biological extract samples depend to a large extent on a preconcentration step. With the higher concentrated samples thus obtained, further analytical procedures such as neutron activation and atomic adsorption are carried out without difficulty. Trace metal ions can be concentrated through adsorption by chelating resins, cation exchangers or as anion complexes on anion exchangers. Multi-element determination (Ba, Be, Cd, Co, Cu, Mn, Ni, Pb) was carried out with precision at the 10 ppb level after preconcentration utilizing a miniature ion exchange column (Chelex 100®) [27]. Preconcentration on ion exchangers is described in several papers: Uranium in natural water [28], Cd, Co, Cu, Fe, Mn, Ni, Pb, Zn in sea water [29] or Cd, Cu, Pb, Zn in environmental samples [30]. A special application of this method is the determination of Cr(VI) traces in natural waters. Cr(VI) is a toxic substance even at low concentrations and its determination in the environment is of increasing importance [31, 32]. Preconcentrated chromium samples can be determined by HPLC, atomic-absorption or plasma-emission spectroscopy.

5.1.3.2 Selective elution

In the selective elution method, the ionic substances are sorbed more strongly on to the exchanger than in the ion retardation process. A second elution medium is therefore needed to desorb these substances. In many cases, complex formation is applied to obtain the large reduction of the k_D values needed for a fast elution.

Since many results are reported in the literature, choice of the best media and conditions for selective elution has been facilitated. Nelson, Murase and Kraus [33] showed that with acids such as HCl and $HClO_4$, several elements form complexes which cause large k_D alterations (see also Section 5.2). For the same purpose, several authors (e. g. Kim [34], Peters and Del Fiore [35] and Korkisch [36]) used mixed acidic-organic-aqueous solutions. The formation of complexes can be explained by the fact that certain compounds take up ionic ligands (L^-, L^+), thus changing their valency state [37]. A cationic species (Cat) with r ligands forming an ionic compound, Cat L_r^{n+}, is given as an example. By the addition of further monovalent anionic ligands (L^-), the complex Cat $L_{r+m}^{(m-n)-}$ is built up. This can be written as

$$\text{Cat } L_r^{n+} + mL^- \rightleftharpoons \text{Cat } L_{r+m}^{(m-n)-}\,. \tag{5}$$

In this case only the complex-building step, which leads to species adsorbable by an anion exchanger, is considered. The equilibrium constant k_3 is therefore

$$k_3 = \frac{[\text{Cat } L_{r+m}^{(m-n)-}]}{[\text{Cat } L_r^{n+}]\,[L^-]^m}\,. \tag{6}$$

The negative complex can exchange with the counter-ions of an anion exchanger. A monovalent L^--loaded exchanger (RL) is assumed:

$$\text{Cat } L_{r+m}^{(m-n)-} + (m-n)\text{RL} \rightleftharpoons R_{(m-n)}\,\text{Cat } L_{r+m} + (m-n)L^-\,. \tag{7}$$

The distribution is therefore

$$K_E = \frac{[R_{(m-n)}\,\text{Cat } L_{r+m}]\,[L^-]^{m-n}}{[\text{Cat } L_{r+m}^{(m-n)-}]\,[\text{RL}]^{m-n}}\,. \tag{8}$$

However, the distribution coefficient measured, k^*_{CD}, takes both species in the solution, Cat L_r^{n+} and Cat $L_{r+m}^{(m-n)-}$, into account:

$$k^*_{CD} = \frac{[R_{(m-n)}\,\text{Cat } L_{r+m}]}{[\text{Cat } L_r^{n+}] + [\text{Cat } L_{r+m}^{(m-n)-}]}\,. \tag{9}$$

The introduction of eq. 8 and 6 into eq. 9 leads to the simple form

$$k^*_{CD} = \frac{k_3\, K_E[\text{RL}]^{m-n}\,[L^-]^n}{1 + k_3\,[L^-]^m}\,. \tag{10}$$

This equation explains the exponential increase of k^*_{CD} at lower ligand concentrations, and also the decrease at high ligand concentrations, which is frequently observed.

At the same time it can be seen from these equations that metallic complexes adsorbed onto anion exchangers can easily be removed from such columns through changes in the ligand concentration. The same is valid for cationic complexes adsorbed onto cation exchangers.

Further analytical information can be obtained from the above mathematical derivation: the evaluation of relevant experiments indicates the main complex-forming steps (the consumption of m ligand molecules) leading to the adsorbable ionic form; they show the valency of the complex $(m-n)$, the magnitude of the complex formation (k_3) and the distribution equilibria (K_E) constants involved (see also Section 5.2).

5.1.3.3 Separation by ion exchange membranes

The most frequently used detection system for ion separation methods is based on conductivity measurements. Through the removal of unwanted ions an increased sensitivity can be obtained. In the case of an anion determination, the elimination of cations and a complete transformation of the anions into their acidic form can be advantageous because considerably higher conductivity is achieved. For this purpose either ion exchange suppressor columns or ion exchange membranes can be applied.

Stevens and Davis [38] proposed a hollow-fiber ion exchange membrane device to suppress part of the conductivity phenomenon. According to Figure 5.2, where a cation hollow fiber is shown, the sample is conducted inside the fiber. The more strongly bound cations (e. g., Na^+, K^+) enter into the fiber and the exchanged hydrogen ion is released into the inner hollow section to neutralize the anion and to form the acids with their high conductivity. Outside the fiber a stream of dilute acid is led in the opposite direction. By this means a continuous regeneration of the ion exchange fiber loaded with heavier cations is obtained.

Through the use of this hollow fiber certain advantages are achieved: the continuous regeneration of the ion exchange fiber eliminates the need for the

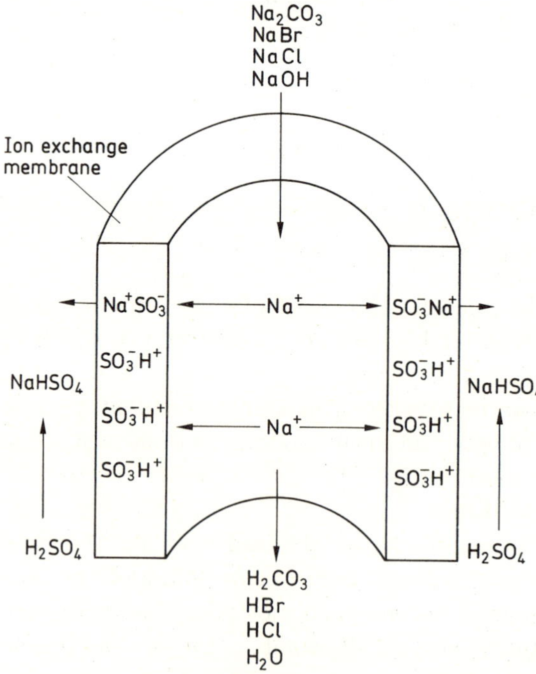

Figure 5.2 Schematic drawing of a hollow-fiber ion exchanger membrane.

periodic regeneration necessary with a suppressor column. At the same time, hollow fibers show slightly less spreading of the bands than suppressor columns.

As fiber material sulfonated polyethylene tubing, cation and anion exchangers with polytetrafluoroethylene and polyethylenevinylacetate matrices are used. A small hollow fiber (NAFION 811−X®) has, in the wet state, about 0.8 mm i. d. and 1.0 mm o. d.. Exchange capacities of 1 mequ/g are reported [39] for sulfonated polyethylene hollow fibers.

Such fibers are applied in several types of ion-chromatographic equipment [40].

A special application of ion exchange membranes is the collection of cations and anions from streams with very low ion concentrations. A typical example [41] is the adsorption of cations (e. g., Zn, Co, Na, Mn, Cu, Cs, Ba, La, W, etc.) by cation exchanger membranes, anions (e. g. I^-) by anion exchanger membranes, and certain elements (e. g. Fe and Cr) by anion-cation exchanger membranes from nuclear reactor pool waters. By the use of three ion exchange membranes in series, adsorption efficiencies between 95.6−99.8% were achieved. The radionuclides of these elements were then determined by a scintillation counting device. Through this method time-consuming evaporation as well as the loss of volatile nuclides can be avoided.

5.1.3.4 Separation by liquid ion exchangers

Solid ion exchangers are decidedly more important in separation processes than liquid ion exchangers. The main reason for this is that the latter require extraction processes, which are more difficult to apply than column procedures.

Liquid ion exchangers are high molecular bases or acids with low solubility in water and high solubility in water-immiscible solvents. Analogous to solid ion exchangers, liquid anion exchangers consist mainly of primary, secondary, tertiary and quaternary aliphatic amines. Liquid cation exchangers are usually alkyl or dialkyl compounds of phosphoric, sulfonic or carboxylic acids. Extended lists of liquid ion exchangers can be found in an article by Coleman et al. [42] and in Chapter 1.9 of this book.

In order to reduce the danger of emulsion formation, liquid ion exchangers are diluted with inert organic solvents, such as kerosene, benzene, toluene, chloroform, cyclohexane, petroleum ether, octane, etc. The ion exchanger concentration is usually kept in the range of 2 to 12% v/v.

The ion-exchange extraction behaviour of many elements is described in [A10, A11] and for uranium in [36]. Nelson et al. [33] and Peters et al. [35] plotted the distribution coefficients for solid ion exchangers in relation to different acidic concentrations, as shown in Section 5.2. These results indicate the type of exchangers and the conditions which should be chosen in order to obtain good separations.

A typical example of an analytical use of liquid ion exchangers is described in [A18]: an anionic uranium(VI) sulfato-complex is extracted by a chloroform solution of a liquid anion exchanger (Amberlite LA 1) from a sulfuric acid solution with a high iron concentration. After back-extraction of uranium with a sodium carbonate solution and an oxidation step, the uranium is determined spectrophotometrically as peruranate.

5.1.4 Special analytical methods

5.1.4.1 Ion exchange sorption analysis

If a regenerated ion exchange column is loaded with counter-ions, breakthrough curves of the type shown in Figure 5.3 a are obtained [43]. In this case the strontium ions of a 0.01 M $Sr(NO_3)_2$ acidic solution (0.243 N HNO_3) were sorbed by the cation exchanger in the H^+-form. The k_D values are kept sufficiently high to ensure that the re-exchange of sorbed strontium ions against H^+ ions is small. The corresponding equation can therefore be written

$$2\,RH + Sr^{2+} \rightarrow R_2Sr + 2\,H^+\,. \tag{11}$$

The analytical determination of the breakthrough curves was achieved by adding Sr-90 activity and counting small effluent fractions.

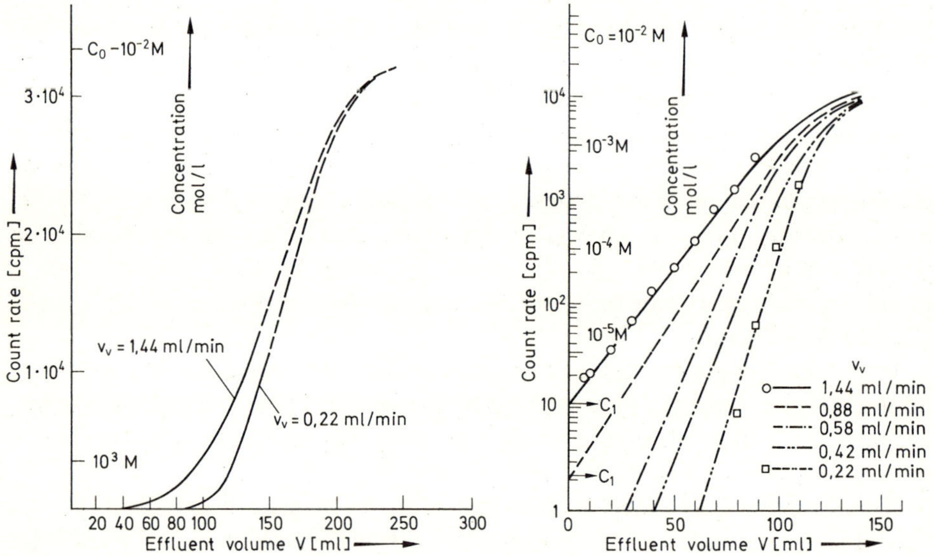

Figure 5.3 Breakthrough curves (43). A Dowex 50 WX12 cation exchanger in the H^+ form is loaded with a 0.01 M $Sr(NO_3)_2$ acidic solution (0.243 N HNO_3) containing Sr-90 tracers. Flow rate between 0.22 and 1.44 ml/min; column dimensions 6.6 × 0.5 cm.
a) linear and b) logarithmic effluent concentration (count rate) in dependence on the effluent volume.

A characteristic feature of such breakthrough curves is that in the semi-logarithmic plot (ion concentration or activity against effluent volume or time) a straight linear increase occurs until the region of ion exchange saturation is approached (Figure 5.3b). As expected, a higher flow rate gives a less steep breakthrough curve and a flatter slope (Figure 5.3a and b).

At a constant flow rate the individual slopes ($\operatorname{tg} \alpha = q$) depend directly on the ion concentration (Figure 5.4) [44]. The specific activities of the effluent fractions show this behaviour even more clearly (Figure 5.4).

The linear increase in the semi-logarithmic plot was evaluated [43] and can be given by

$$\log C = qV + \log C_1 \,, \tag{12}$$

where C is the concentration of the ions in the effluent volume V, and C_1 the intersection of the slope with the ordinate. In this equation the tangent α of the slope (q) is given by

$$q = \frac{8\,\pi\, r_0^2\, N\, D\, k_a}{2.3\, \beta \delta_0\, Q_P\, F \sqrt{L V_P}}\, C_0 \,. \tag{13}$$

If the column parameters (r_O = radius of the exchanger particles, N = number of exchanger particles per cm^3, D = diffusion coefficient of the ions in solution, k_a = desorption factor, β = fraction of liquid phase in the column, δ_o = thickness of the diffusion zone around the exchanger particle, Q_P = specific exchange capacity, F = column cross section, L = column length and v_p = flow rate in the liquid part of the column) are kept constant, the dependence of the q-values on the ion concentration of the original solution (C_O) is

$$q = k_s\, C_0 \,. \tag{14}$$

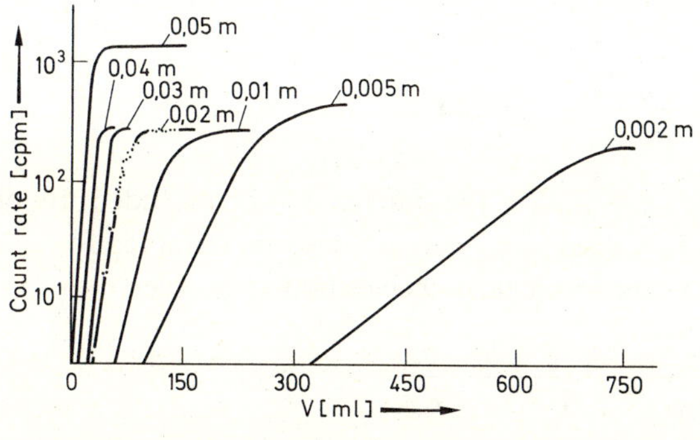

Figure 5.4 Breakthrough curves at constant flow rate (1.45 ml/min) (44). A Dowex 50X12 cation exchanger (20–50 mesh) in the H$^+$ form is loaded with strontium solutions, containing Sr-90 tracers, with concentrations between 0.002 and 0.005 mol/l. Exchange capacity 2.95 mequ; column diameter 0.5 cm.

Figure 5.4 gives the breakthrough curves which were obtained by loading a regenerated Dowex 50 exchanger in the H^+ form with Sr^{2+} ions. As can be seen from this figure, the added activity (Sr-90) does not change the q-value of the slope. This value depends on the concentration C_O of the ions, but the specific activity is sufficiently low so that the influence on C_O is practically zero. Summarizing the above, it may be said that this method can be applied to determine ion concentrations, and it can help in the analysis of exchange parameters such as diffusion constants (D), desorption factors, (k_a) and the thicknesses of the diffusion layer (δ_O).

5.1.4.2 Determination of molecular parameters

Liquid chromatography, and especially HPLC, is already widely used for the estimation of molecular parameters. In Carney's review [45] many examples for the determination of the lipophilicity, partition coefficients and even biological activities [46] of compounds, mainly using reversed phase chromatography, are given (see also [51] and [52]).

Ion exchange columns are especially suited to study ionization constants of substances in different solutions. The ionization of the compounds usually influences the aqueous solubility and can therefore be responsible for certain biological effects, such as membrane transport.

The dissociation constants, K_a, or the corresponding pK_a values, were determined for several weak acids and bases and the results compared with those obtained by a spectrophotometric method [47]. For these experiments an XAD-2 column was used. The capacity factor, k', depends as follows on the capacity factors k'_O, k'_{-1} and k'_{+1} for neutral, anionic and cationic species respectively:

$$k' = \frac{k'_0[H^+] + k'_{-1}(K_a)}{K_a + [H^+]} \quad \text{for acids} \tag{15}$$

$$k' = \frac{k'_{+1}[H^+] + k'_0(K_a)}{K_a + [H^+]} \quad \text{for bases.} \tag{16}$$

Equations 15 and 16 are only valid if ion pairing can be excluded. By determining the capacity factor, the K_a values can, however, also be obtained for solutes where ion pairing occurs. For monoprotic acid species [48] k' is given by

$$k' = \frac{k'_0}{1 + K_a/a_{H^+}} + \frac{k'_{-1} + k'_{MX}\, a_{M^+}}{1 + a_{H^+}/K_a}, \tag{17}$$

and for weak bases [49] k' is

$$k' = \frac{k'_0}{1 + a_{H^+}/K_a} + \frac{k'_{+1} + k'_{BHX}\, a_{X^-}}{1 + K_a/a_{H^+}}. \tag{18}$$

The activities of the hydrogen ions and the counter ions are a_{H^+}, a_{M^+} and a_{X^-}. The capacity factors of the ion pairs are given by k'_{MX} and k'_{BHX}.

Through this method not only the ionization constants can be determined, but also the capacity factors of ion pairs can be evaluated and the stationary phases of ion exchangers can be characterized.

With different reversed phase columns the retention behaviour of 68 purins was analyzed [50]. The substituent groups gave reproducible contributions to the capacity factor.

By further development the concept of such studies allows the correlation of capacity factors and chemical structures, and therefore may lead to interesting future applications.

References

[1] Donnan, F. G., Z. Elektrochem. Angew. Physik. Chem. *17* (1911) 572 and Z. Physik. Chem. *168* (1934) 369.
[2] Bobleter, O., G. Dincler & C. Sabau (1970) B. Bunsen-G. Physik. Chem. *74* 1050.
[3] Kraus, K. A. & G. E. Moore (1953) J. Am. Chem. Soc. *75*, 1457.
[4] Shibata, T., I. Okamoto & K. Ishii (1986) J. Liquid Chromatogr., *9*, 313.
[5] Baker, J. K., A. M. Clark & C. D. Hufford (1986) J. Liquid Chromatogr. *9*, 493.
[6] Hare, P. E. & E. Gil-Av (1979) Science *204* 1226.
[7] Lindner, F. W. & I. Hirschböck (1986) J. Liquid Chromatogr. *9*, 551.
[8] Ofengand, J. & R. Liou (1980) Biochemistry *19*, 4814.
[9] Honda, M. (1949) J. Chem. Soc. Japan *70*, 55.
[10] Lajtha, K. (1988) Plant and Soil *105*, 105.
[11] King, P. & J. R. Simmler (1964) Anal. Chem. *36*, 1837.
[12] Tunn, M. F., S. E. Pattison, M. C. Storm & E. Quiel (1980) Biochemistry *19*, 718.
[13] Bosron, W. F., F. S. Kennedy & B. L. Vallee (1975) Biochemistry *14*, 2275.
[14] Raymond, F. A. & R. M. Weinshilboum (1975) Clin. Chim. Acta *58*, 185.
[15] Agarwal, M., R. B. Bennett, I. G. Stump & J. M. D'Avria (1975) Anal. Chem. *47*, 924.
[16] Marglin, A. & R. B. Merrifield (1967) Arch. Biochem. Biophys. *122*, 748.
[17] Otieno, S. (1978) Biochemistry *17*, 546.
[18] Levine, B. J., P. D. Orphanos, B. S. Fischmann & S. Beychok (1980) Biochemistry *19*, 4808.
[19] Bennett, M. J. & C. E. Bradey (1984) Clin. Chem. *30*, 542.
[20] Pecina, R., G. Bonn, E. Burtscher & O. Bobleter (1984) J. Chromatogr. *287*, 245.
[21] Bonn, G., R. Pecina, E. Burtscher and O. Bobleter (1984) J. Chromatogr. *287*, 215.
[22] Richmond, M. L. (1982) J. Diary Sci. *65*, 1394.
[23] Vogel, A., see Ref. A 18, p. 171.
[24] Talvitie, N. A. & R. J. Demint (1965) Anal. Chem. *37*, 1605.
[25] Johnson, R. H. & T. C. Reavey (1965) Nature (London) *208*, 750.
[26] Osterried, O. (1964) Z. Anal. Chem. *199*, 260.
[27] Hartenstein, S. D., J. Ruzicka & G. D. Christian (1984) Anal. Chem. *57*, 21.
[28] Pakalns, P. (1980) Anal. Chim. Acta *120*, 289.
[29] Kingston, H. M., I. L. Barnes, T. J. Brady & T. C. Rains (1978) Anal. Chem. *50*, 2064.
[30] Figura, P. & B. McDuffy (1980) Anal. Chem. *52*, 1433.

[31] Pankow, J. F. & G. E. Janauer (1974) Anal. Chim. Acta *69*, 97.
[32] Krull, I. S., K. W. Panaro & L. L. Gershman (1983) J. Chrom. Sci. *21*, 460.
[33] Nelson, F., T. Murase & K. A. Kraus (1964) J. Chromatogr. *13*, 503.
[34] Kim, J. I. (1974) J. Inorg. Nucl. Chem. *36*, 1.
[35] Peters, J. M. & G. Del Fiore (1975) J. Chromatogr. *108*, 415.
[36] Korkisch, J. (1967) Mikrochim. Acta, 401.
[37] Bobleter, unpublished results.
[38] Stevens, T. S. & M. A. Langhorst (1982) Anal. Chem. *54*, 950.
[39] Stevens, T. S., G. L. Jewett & R. A. Bredeweg (1982) Anal. Chem. *54*, 1206.
[40] Stevens, T. S., J. C. Davis & H. Small (1981) Anal. Chem. *53*, 1488.
[41] Chan, C. C., P. C. Liu & P. S. Weng (1985) J. Radioanal. Nucl. Chem. Lett. *95*, 81.
[42] Coleman, C. F., C. A. Blake, Jr. & K. B. Brown (1962) Talanta *9*, 297.
[43] Dincler, G. & O. Bobleter (1973) Z. Physik. Chem. N. F. *86*, 156.
[44] Bobleter, O., G. Dincler & C. Sabau (1971) Mikrochim. Acta (Wien), 310.
[45] Carney, C. F. (1985) J. Liqu. Chromatogr. *8*, 2781.
[46] Unger, S. H. & G. H. Chiang (1981) J. Med. Chem. *244*, 262.
[47] Palalikt, D. & J. Block (1980) Anal. Chem. *52*, 624.
[48] Van de Venne, J. L. M. & J. L. H. M. Hendrikx (1978) J. Chromatogr. *167*, 1.
[49] Hafkenscheid, T. L. & E. Tomlinson (1984) J. Chromatogr. *292*, 305.
[50] Assenza, S. P. & P. R. Brown (1983) J. Chromatogr. *282*, 477.
[51] Melander, W. & Cs. Horvath (1980) in: High Pressure Liquid Chromatogr. Advances and Perspectives (Cs. Horvath, ed.) Vol. 2, pp. 273–279, Academic Press, New York.
[52] Kaliszan, R. (1981) J. Chromatogr. *220*, 71.

Books

A1. Helfferich, F. (1959) Ionenaustauscher, Verlag Chemie, Weinheim.
A2. Helfferich, F. (1962) Ion Exchange, McGraw-Hill, New York.
A3. Dorfner, K. (1962) Ionenaustausch-Chromatographie, Akademie-Verlag, Berlin.
A4. Samuelson, O. (1963) Ion Exchange Separations in Analytical Chemistry, Wiley, New York.
A5. Amphlett, C. B. (1964) Inorganic Ion Exchangers, Elsevier, Amsterdam.
A6. Inczedy, J. (1964) Über die Anwendung von Ionenaustauschern in der Wasseranalyse. Fortschritte der Wasserchemie, Akademie-Verlag, Berlin.
A7. Inczedy, J. (1966) Analytical Applications of Ion-Exchangers, Pergamon, New York.
A8. Marinsky, J. A. (1966) Ion Exchange, Marcel Dekker, New York.
A9. Heftmann, E., ed. (1967) Chromatography, 2nd edition, Reinhold, New York.
A10. Korkisch, J. (1969) Modern Methods for the Separation of Rare Metal Ions, Pergamon, New York.
A11. Marcus, Y. & A. S. Kertes (1969) Ion exchange and solvent extraction of metal complexes. Wiley, London.
A12. Samuelson, O. (1969) Ion Exchange and Solvent Extraction, Vol. 2, Dekker, New York.
A13. Dorfner, K. (1970) Ionenaustauscher, 3. ed., De Gruyter, Berlin.
A14. Rieman, W. & H. F. Walton (1970) Ion Exchange in Analytical Chemistry, Pergamon, Oxford.
A15. Khym, J. X. (1974) Analytical Ion-Exchange Procedures in Chemistry and Biology: Theory, Equipment, Techniques, Prentice-Hall, New Jersey.
A16. Walton, H. F., (Ed.) (1976) Ion-Exchange Chromatography, Dowden, Hudchinson and Ross, New York.

A17. Sawicki, E., J. D. Mulik & E. Wittgenstein (1978) Ion Chromatographic Analysis of Environmental Pollutants, Ann Arbor Sci. Publ., Ann Arbor, Mich.

A18. Vogel, A. I. (1979) Textbook of Qualitative Anorganic Analysis, Longman, London.

A19. Fritz, J. S., D. t. Gjerde & Ch. Pohlandt (1982) Ion Chromatography, Hüthig, Heidelberg.

A20. Pillay, K. K. S. (1982) Radiation Effects on Ion-Exchange Materials, AIChE Symp. Series.

A21. Allen, G. R. (1983) Ion Exchange of Mixed Radwaste, Proc. Int. Water Conf., Engineering Society of West, Pa.

5.2 Ion Exchange Chromatography

Ortwin Bobleter and Günther Bonn
Universität Innsbruck, Institut für Radiochemie, Innrain 52 a, 6020 Innsbruck, Austria

Symbols and definitions

A, A', A^*	terms of plate-height and retention-time formulae
B, B', B^*	terms of plate-height and retention-time formulae
C_m, C'_m, C_s	terms of plate-height and retention-time formulae
C', C^*	terms of plate-height and retention-time formulae
$\overline{C}$, C, C_{max}	Concentration, in stationary, in mobile phase, at maximum of peak (mol/l)
d_p	diameter of ion-exchange particles (mm)
D_m, D_s	diffusion coefficient in mobile (m) and stationary (s) phase (mm^2/s)
DP	degree of polymerisation
D_v	volume distribution coefficient
G^O	free energy (J)
H_T	height of a theoretical plate (mm)
H_{eff}	effective height of a theoretical plate (mm)
k_a, k_b,...	constants
k_1, k_2,...	constants
k_{CD}	concentration distribution coefficient

k_D	weight distribution coefficient
k_E	equilibrium constant
k'	capacity factor
L	length of column (mm)
m	index for mobile phase
M	amount of solute in the stationary phase (mol/g)
n	valency state or exponential factor
N	number of theoretical plates
N_c	number of plates in coupled systems
N_{eff}	number of effective theoretical plates
R	gas constant
R_f	relative migration distance
R_S	resolution
s	index for stationary phase
t_o	retention time of nonretained solutes (s)
t_R, t'_R	retention and reduced retention times of solute (s)
t_W	base line band-width in time units (s)
T	thermodynamic temperature (K)
u	linear flow rate (mm/s)
u_V	volume flow rate (cm^3/s)
V_C	volume of column (cm^3)
$V_o = V_m$	volume of mobile phase (cm^3)
V_s	volume of stationary phase (cm^3)
V_R, V'_R	retention volume and reduced retention volume (cm^3)
V_W	band width volume (cm^3)
α	relative retention
γ, ψ	constants
ϱ_S	density of solid phase (g/cm^3)
σ	standard deviation of the Gaussian

5.2.1 Introduction

In 1903 Tswett [1] reported on the brilliant experiment in which he separated individual components of a leaf extract using a polysaccharide column. This work already showed all the characteristics of modern chromatographic procedure:

1. A solid phase material (in the form of an inulin column) was used.
2. A small amount of the mixture to be separated was applied at the top of the column.

3. A medium (mobile phase) was led through the solid matter, whereby transport and separation of the mixture occurred.
4. A detection system was able to monitor the success of the separation in the solid phase (or in the course of the elution of the components).

Tswett used ligroine as the mobile phase and detected visually the compounds which passed through the column in the form of coloured rings.

The lack of theoretical understanding of chromatographic processes and the limited possibilities of detection were the major reasons why years went by until Tswett's great achievements were acknowledged, and chromatography reached the importance in the field of analytical chemistry it holds today.

A quarter of a century went by until Kuhn, Winterstein and Lederer [2] separated xantophylle on a calcium carbonate column using Tswett's method. In these experiments the column material was, however, removed in portions from which the substances were then extracted.

Little attention was paid to the first thin-layer chromatographic (TLC) work of Ismailov and Šrajber [3]. This method was rediscovered by Kirchner, Miller and Keller [4] and improved until it became a valuable analytical instrument by Stahl and his school. In 1941 Martin and Synge [5] introduced partition chromatography. A water-insoluble liquid (mobile phase) was percolated through water-loaded silica gel (stationary phase). In this way, good results were obtained in the separations of amino acids. The remark made by the authors that gases could also be used as the mobile phase, was taken as a first indication of the future gas chromatography. Most of the scientists of the Eucken school in Göttingen and in some other places, who were already working at the same time on gas separation problems by means of distillation and adsorption, were of the opinion that the diffusion in the gas phase was too high to allow an analytical use of this procedure [6].

Paper chromatography, which was used by Runge [7] approximately one hundred years before this time to form amazing ring patterns using different dye mixtures, was introduced in 1949 by Consden, Gordon and Martin [8] as an analytical method. The separation effect was again explained as a partition phenomenon between the water phase on the cellulose (filter paper) and the mobile organic phase (e. g., phenol or n-butanol/benzyl alcohol mixture). It was believed that cellulose constituted a completely inert phase. The possibilities of this method were demonstrated by the separation of 22 amino acids using a two-dimensional technique. With a simple ninhydrin reaction the acids were made visible on the paper. In the year 1951 the first gas chromatogram was published by Cremer and Prior [9] using a solid material column (gas solid chromatography = GSC). Only one year later, it was again Martin [10] who introduced gas liquid chromatography (GLC).

In the meantime considerable progress was also being achieved in the theoretical treatment of chromatography. The oft-cited works of van Deempter [11], Glueckauf [12], Snyder [13], Giddings [14], Kirkland [15] and Knox [16] deepened the understanding of essential parameters and thus were able to indicate targets where further development might bring useful results. For example, it became clear that liquid chromatography could be markedly improved by reducing both the particle size and the deviation in size of the solid phase material. This was the beginning of the high performance (pressure) liquid chromatography (HPLC) (Hamilton et al. [17]; Huber [18]).

The improvement of detection systems was a necessary step towards the enormous and world-wide application of chromatographic methods. In paper and thin-layer chromatography relatively simple methods were developed to characterise the substance spots. Gas chromatography uses a great number of universal or selective detectors such as thermistors, flame and β-ray ionisation devices, halogen sensors and mass spectrometry. During recent years decisive improvements in liquid chromatography detection have been made: e. g., high-sensitive refractive index, low-volume ultraviolet detection cells, mass sensitive devices and mass spectrometry detection.

Ion exchangers were used as chromatographic materials as soon as samples with high exchange capacities were available. Freudenberg et al. [19] succeeded in separating amino sugars on Wofatit M following an suggestion of Griessbach [20] in 1938. Wieland [21] was able to separate basic amino acids on Wofatit C and Block [22] analysed amino acids with an Amberlite column.

G. E. Boyd and coworkers demonstrated the wide analytical applicability of ion exchangers in the field of nuclear chemistry. Their work, already begun in 1942, could only be published five years later [23]. It was shown that the uranium fission products were adsorbed on Amberlite IR-1 and IR-4, and by changing the conditions they could be eluted individually. Tompkins et al. [24, 25] were able to obtain most fission products in a relatively pure form using the same exchange material. These results were also very important because they demonstrated that the larger part of these ions could be eluted selectively when complexing agents (e. g., ammonium tartrate or citrate) were applied. The exceptional possibilities of chromatographic separations with ion exchangers were shown by the experiments of Spedding and coworkers [26–30], where rare-earth elements in the kilogram range, of 99.9% purity, were isolated. Some of these elements were more difficult to separate, but even in such cases up to 100 gram amounts of great purity were obtained.

It was at this time quite clear that ions of atoms and molecules could be separated by ion exchangers, but now Samuelson et al. [31, 32] demonstrated that also non-electrolytes (e. g., aldehydes and ketones) can be separated very well by adding complexing agents (e. g., bisulfite).

Since then great progress has been made in the production of new stationary phases and detector systems. With these, much improved separations of metals, metal-complexes, peptides, and nonionic compounds have been obtained. Newer methods, such as ion chromatography and reversed phase ion-pair chromatography, have widened the important field of ion exchange chromatography considerably.

5.2.2 Theory of ion exchange chromatography

The success of a chromatographic separation depends on the sorption behaviour of a certain substance in the stationary phase, and on the difference of the sorption strength of the substances which are to be separated. In practical application, the time-span needed for a complete separation of the components to be analysed can be of utmost importance. The sorption of ions on ion exchangers is determined by the distribution coefficient (k_D).

5.2.2.1 Distribution of the solute

Distribution of ions

Separation of the two ions, (1) and (2), is very simple when the difference between the distribution coefficients k_{D1} and k_{D2} is large enough. The smaller the difference between these coefficients, the more difficult the separation operation becomes. When these difficulties cannot be solved by altering the chemical conditions to enlarge the difference of the distribution coefficients, a good separation can only be achieved by increasing the column length, and therefore usually also increasing the analysis time.

The determination of the distribution coefficients is given in more detail in Chapter 1.2 of this book. Due to the importance of this parameter its definition will be recalled:

$$k_{CD} = \frac{\text{concentration of solute X in the resin phase}}{\text{concentration of solute X in the mobile phase}} \tag{1a}$$

$$k_{CD} = \frac{\overline{C_X}}{C_X}\left[\frac{\text{mol/l}}{\text{mol/l}}\right]. \tag{1b}$$

This definition of the "concentration distribution coefficient", k_{CD}, is, however, relatively difficult to apply in batch experiments due to the swelling and shrinking properties of the resin phase. More frequently the "weight distribution coefficient" is used, given by

$$k_D = \frac{\text{amount of solute X in a unit of resin phase}}{\text{concentration of solute X in mobile phase}}, \tag{2a}$$

which can be transformed by introducing eq. 1b and taking the constants ϱ_s (density of the resin phase) and n (the valency state of the solute):

$$k_D = \frac{M_X}{C_X} = \frac{n\,\overline{C_X}}{\varrho_s\,C_X} = \frac{n}{\varrho_s}k_{CD}\cdot\left[\frac{\text{equ/kg}}{\text{mol/l}}\right] \tag{2b}$$

The "capacity factor (coefficient)", k', gives the distribution of the solute in the exchanger column:

$$k' = \frac{\text{amount of solute X in stationary phase}}{\text{amount of solute X in mobile phase}}. \tag{3a}$$

If V_s is the volume of the stationary and V_m that of the mobile phase ($V_m = V_0$), eq. 3a becomes

$$k' = \frac{V_s \overline{C_X}}{V_0 C_X} = \frac{V_s}{V_0} k_{CD} \left[\frac{\text{mol in stationary phase}}{\text{mol in mobile phase}}\right]. \tag{3b}$$

Large distribution coefficients can easily be obtained by batch procedures, but low values can be determined better by elution experiments.

From kinetic analyses it was deduced that the capacity factor, k', reflects the time spent by the solute in the solid phase in relation to that spent in the mobile phase:

$$k' = \frac{t_R - t_0}{t_0} = \frac{t_R'}{t_0}. \tag{4a}$$

The total retention time (t_R) minus the retention time of the mobile phase, t_0, (or the time a non-retained solute needs to transverse the column), gives the reduced retention time t'_R. By introducing the flow rate, u_V, k' can be expressed using the relevant retention volumes (V_R = total retention volume; $V_0 = V_m$ = retention volume of a non-retained solute; V'_R = reduced retention volume):

$$k' = \frac{V_R - V_0}{V_0} = \frac{V_R'}{V_0}. \tag{4b}$$

According to eq. 3, the distribution coefficient becomes

$$k_{CD} = \frac{V_R'}{V_s}. \tag{5}$$

In this way the coefficients k_{CD} and k' can easily be determined in elution experiments. However, it must not be forgotten that the values are only valid for the conditions applied to the ion exchanger during the experiment.

Fixed-site ion exchange.

One of the advantages of ion exchange chromatography is that, in many cases, the distribution coefficient can easily be changed due to the close relation with the equilibrium coefficient, (K_E):

$$k_{D(X)} = \frac{M_X}{C_X} = K_{E(X)} \left(\frac{M_Y}{C_Y}\right)^n \tag{6}$$

The valency state of the ions (X) is given by n. In Figure 5.5 the dependence of the k_D-values of 1-, 2- and 3-valent ions on the Li^+ concentration is given as an

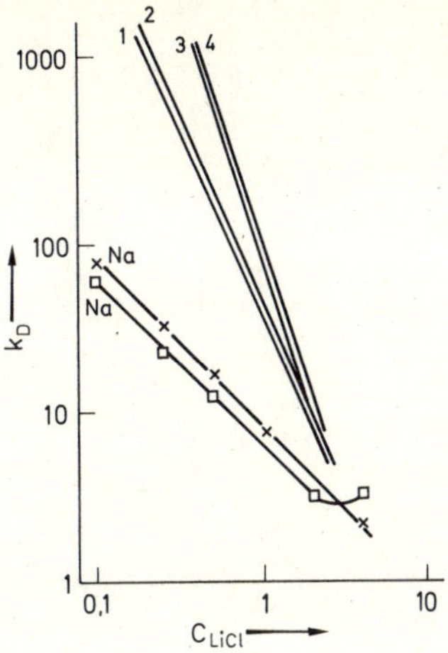

Figure 5.5 Distribution coefficients of Na^+ trace ions as dependent on the Li^+ ion concentration (measured values + on Dowex 50 WX8 and □ on Dowex 50 WX2) [33]. The plotted lines for Ca^{2+} (1), Sr^{2+} (2), Fe^{3+} (3) and La^{3+} (4) are depicted schematically.

example. As can be seen from this figure, the k_D-values are changed by orders of magnitude when the LiCl concentration is altered. At the same time, the differences of the distribution coefficients between the ions with different valencies can also be changed. A very similar picture is obtained if H^+ ions instead of Li^+ ions are applied. The target discussed earlier, of choosing certain k_D-value differences in order to obtain good separations, can very well be reached for ions with different valencies by working with the right pH. Ions with the same valency, however, differ only very little in their distribution coefficient when the pH is changed. The differences in k_D of metal ions are mainly determined by the radius of the hydrated ions, but they are generally high enough for good chromatographic separations.

The problem of applying "plate theory" to ion-exchange chromatography must be discussed. As will be described later in more detail, the number of plates *(N)* of a column is proportional to the distribution coefficient (k_D). However, Figure 5.5 shows that k_D can be altered by a factor of 100 and more, if the macroelectrolyte concentration, or the pH, is changed. In the same way the number of plates changes without any alteration of the column or the exchange material. Such a deliberate increase in the number of plates is independent of the ion-exchange material and the column length, and does not necessarily have an influence on the separation efficiency.

Ion complexes.

Compared with the fixed-site ion exchange described above, the ions formed as complexes have considerably more configuration possibilities. Ionic complexes are therefore frequently used for solving separation problems. Characteristic for many

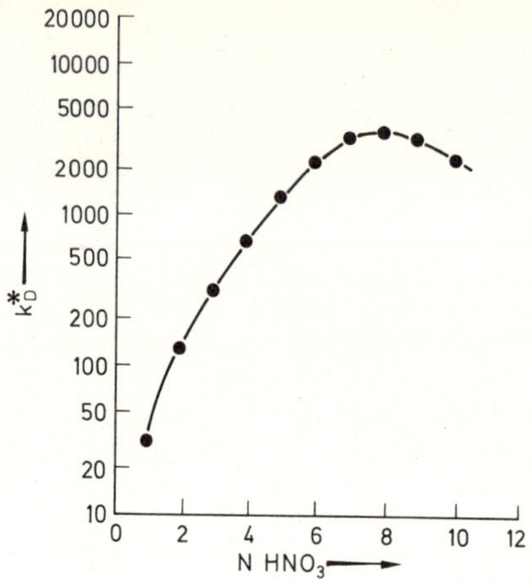

Figure 5.6 Distribution coefficients of a Np-IV complex as dependent on the HNO_3-concentration (Dowex 1X8; 100–200 mesh) [34]. In accordance with eq. 11, the increasing HNO_3-concentration causes first a rise and then a decline in the k_D-values, both determined by the expression $[NO_3^-]^2$.

metals is the formation of anion complexes. As an example the formation of the nitrate complex of neptunium-IV nitrate will be discussed.

In Figure 5.6 the distribution coefficients of neptunium(IV) on an anion exchanger are given as a function of the HNO_3 concentration. The evaluation of this figure indicates that the distribution coefficients are determined not only by two main ionic species, as described in Section 5.1.3.2, but that at least three ligand compounds are responsible for the distribution behaviour of neptunium(IV).

It can be assumed that the ligands are taken up by the metal ion in the following way:

$$Np^{4+} + 2\,NO_3^- \rightleftharpoons Np(NO_3)_2^{2+}\,. \qquad (7a)$$

The equilibrium constant k_3 is therefore

$$k_3 = \frac{[Np(NO_3)_2^{2+}]}{[Np^{4+}]\,[NO_3^-]^2}\,. \qquad (7)$$

In a further step the negative complex is formed:

$$Np(NO_3)_2^{2+} + 4\,NO_3^- \rightleftharpoons Np(NO_3)_6^{2-}\,, \qquad (8a)$$

leading to

$$k_4 = \frac{[Np(NO_3)_6^{2-}]}{[Np(NO_3)_2^{2+}]\,[NO_3^-]^4}\,. \qquad (8)$$

This negatively charged complex is now able to exchange with the nitrate ions of an anion exchanger (RNO_3):

$$Np(NO_3)_6^{2-} + 2\,RNO_3 \rightleftharpoons R_2Np(NO_3)_6 + 2\,NO_3^- , \tag{9a}$$

which gives the equilibrium constant

$$K_E = \frac{[R_2Np(NO_3)_6]\,[NO_3^-]^2}{[Np(NO_3)_6^{2-}]\,[RNO_3]^2} . \tag{9b}$$

For the experimental determination of the distribution coefficient (k_D^*) the three neptunium species in solution must be taken into account.

$$k_D^* = \frac{M_{R_2Np(NO_3)_6}}{[Np^{4+}] + [Np(NO_3)_2^{2+}] + [Np(NO_3)_6^{2-}]} . \tag{10}$$

As an approximation, $M_{R_2Np(NO_3)_6}$ can be written as $k_2\,[R_2Np(NO_3)_6]$. The introduction of eq. 7, 8 and 9 into eq. 10 leads to eq. 11; the capacity term of the exchanger $[RNO_3]^2$ is included in the new constants.

$$k_D^* = \frac{k_5\,[R_2Np(NO_3)_6]\,[NO_3]_4}{k_6\,[Np(NO_3)_2^{2+}]\,[NO_3^-]^2 + k_7\,[Np(NO_3)_6^{2-}] + k_8\,[R_2Np(NO_3)_6]\,[NO_3]^6} . \tag{11}$$

This equation fits the measured points fairly well.

Ionic complexes, such as those of neptunium(IV), can be very easily separated from ions which do not form complexes of the same charge sign or with different distribution coefficients. The formation of uranyl, neptunium and plutonium nitrate complexes, therefore, facilitates the analytical determination of these elements because most fission products do not build similar complexes.

Nelson, Murase and Kraus [35] analyzed the exchange behaviour of many elements on cation exchangers. In Figures 5.7 and 5.8 their results, obtained by applying HCl and $HClO_4$ solutions, are given. In many cases the formation of cationic species at higher acid concentrations can be observed.

Similar experiments have also been carried out using mixed acidic-organic-aqueous solutions [36, 37]. In Figure 5.9 the distribution coefficients on a cation exchanger in 1.2 M hydrochloric acid and acetone concentrations up to 90% are depicted. In these cases the distribution of the cations is strongly influenced by the growing acidity in the solvent when the organic fraction is increased.

Ligand exchange-complex formation on ion exchanger.

Khym and Zill [38] first described the chromatographic separation of sugars as borate complexes on ion-exchange resins. Sugars with vicinal hydroxyl groups build stronger complexes than those with other hydroxyl configurations. In some cases this difference in behaviour can be advantageous for certain sugar separation problems.

In recent years the complex formation of sugars on cation exchangers loaded with Li, Ca, Pb and Ag ions was investigated [39], using only water as an elution medium. Very good separation results were obtained with partially Ag-loaded sulphonic acid exchangers [40].

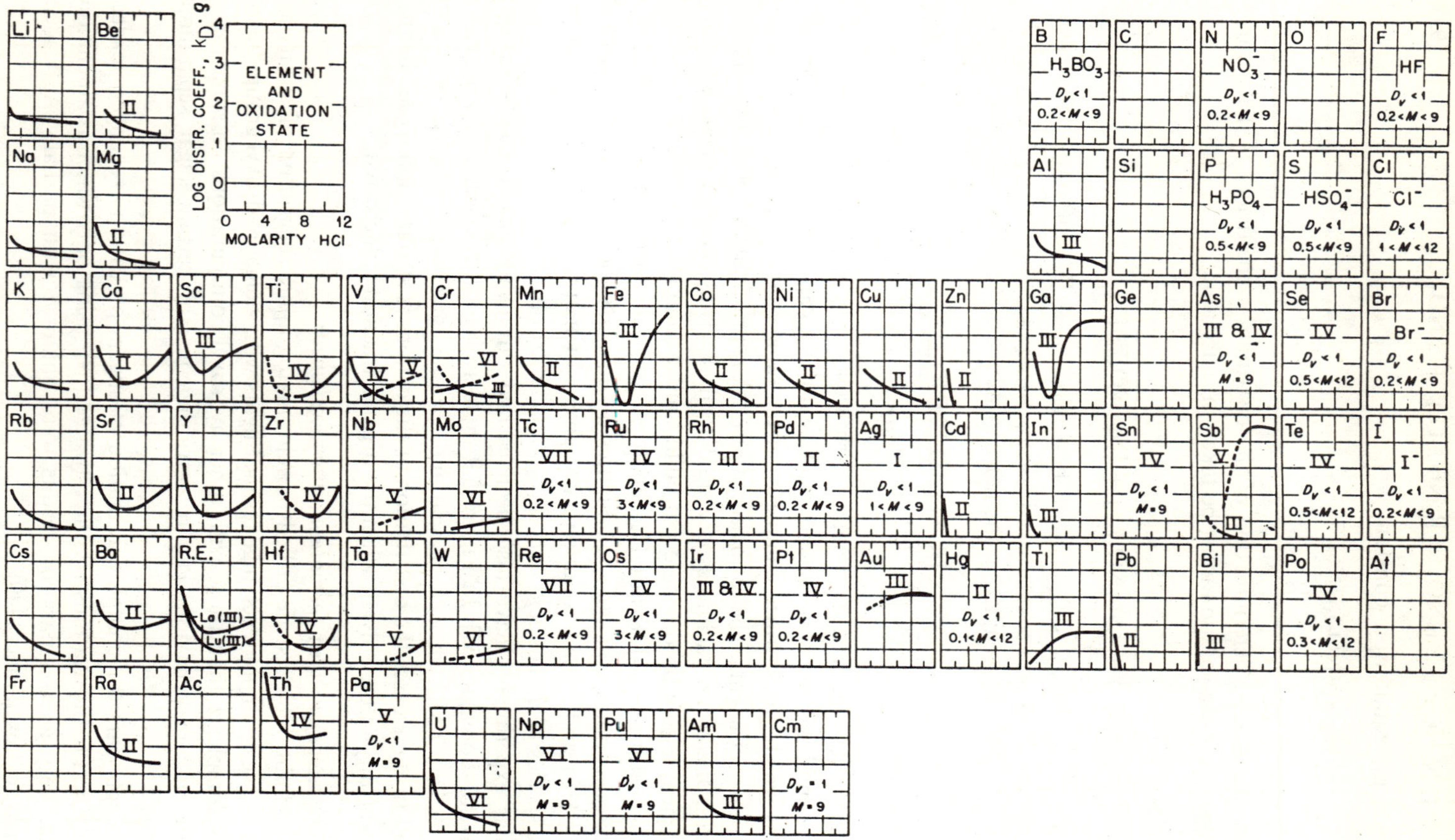

Figure 5.7 Dependence of the volume distribution coefficients (D_v) on the HCl concentration [35]. Exchanger: Dowex 50X4. $k_D = D_v/\varrho$.

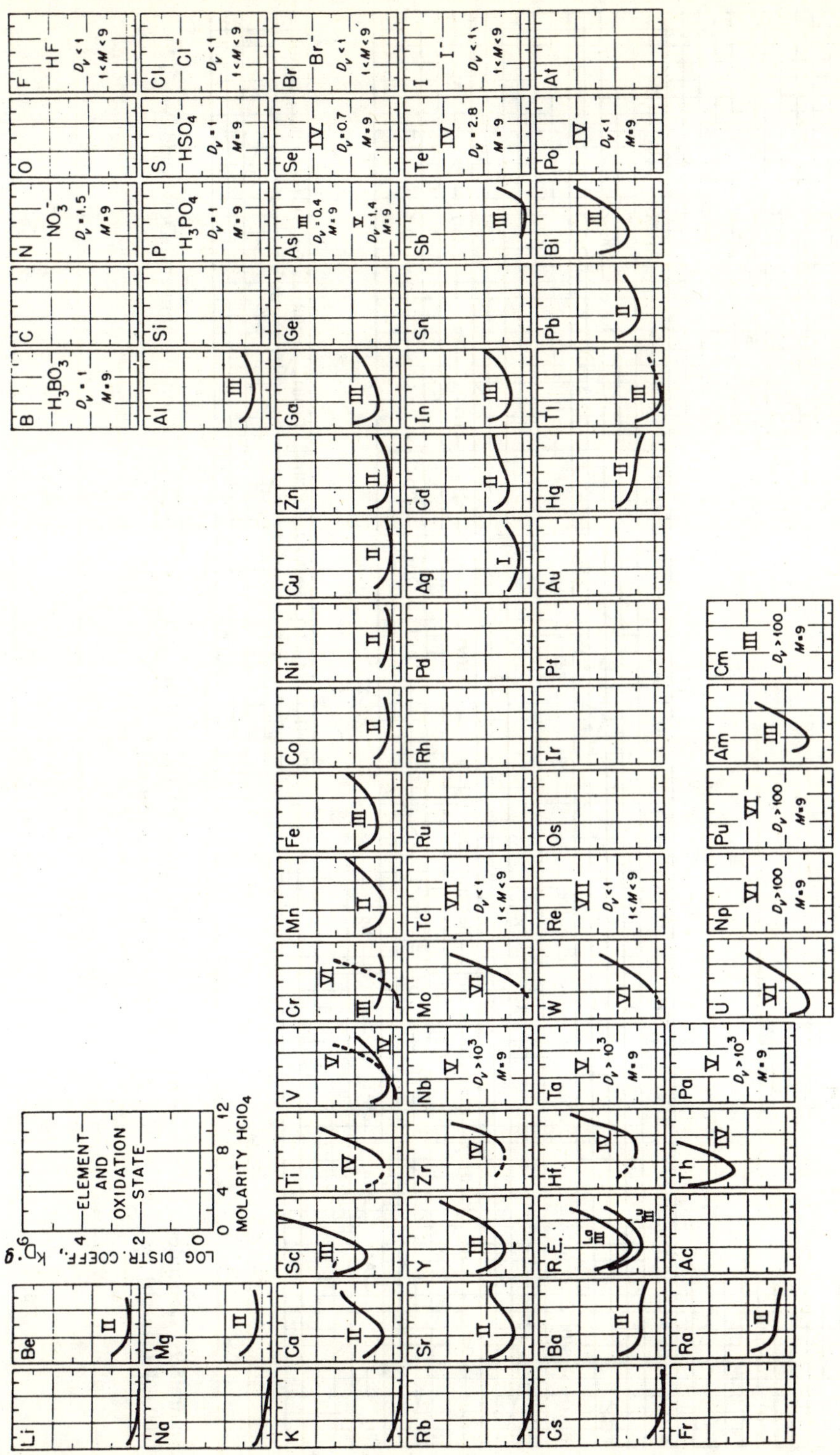

Figure 5.8 Dependence of the volume distribution coefficients (D_v) on the $HClO_4$ concentration [35]. Exchanger: Dowex 50X4. $k_D = D_v/\varrho$

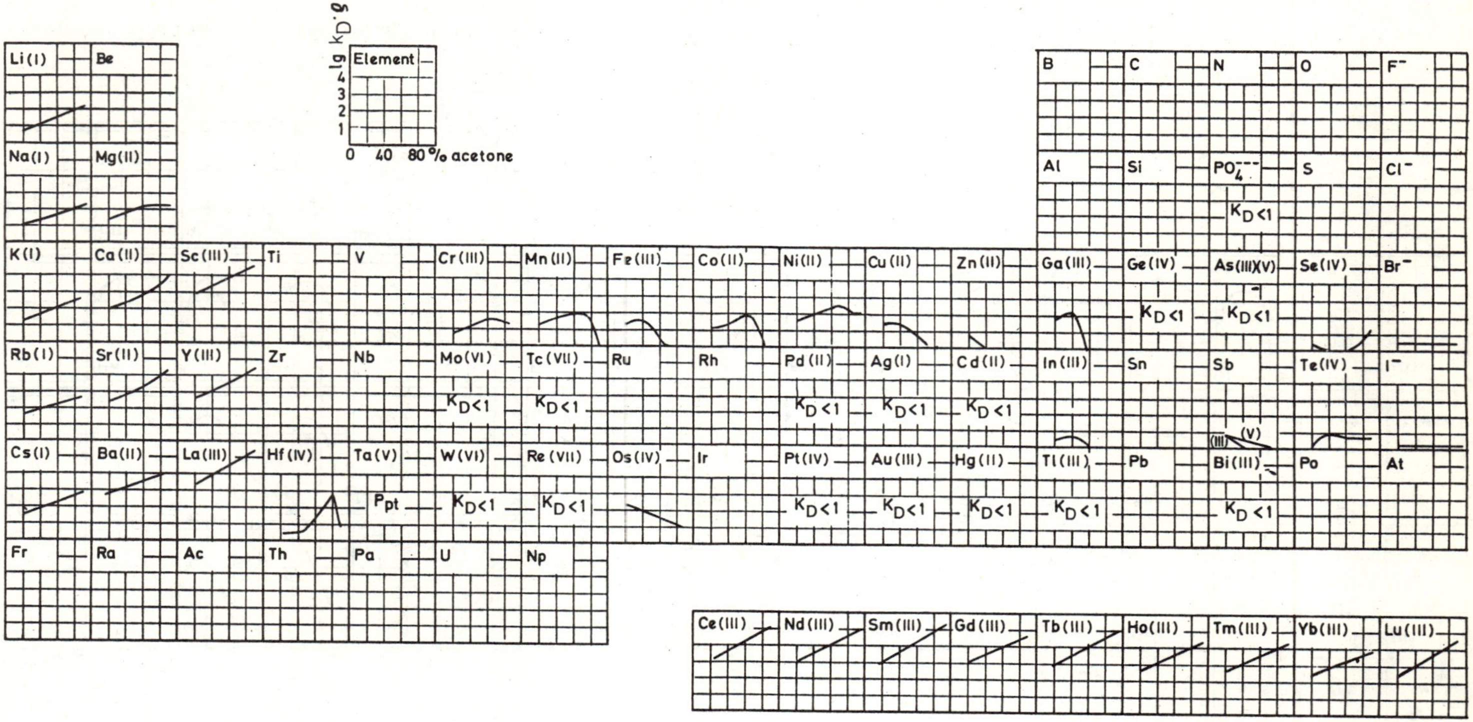

Figure 5.9 Dependence of the distribution coefficients (k_D) on the acetone concentration [37]. Cation exchangers: Dowex 50 W × 12 (200 – 400 mesh); solvent: hydrochloric acid (1.2 M)-water-acetone.

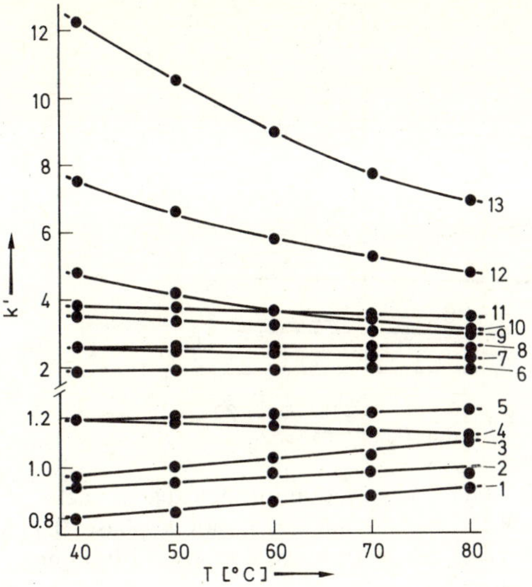

Figure 5.10 Influence of column temperature (T) on the capacity factor (k') of aldehydes and ketones. Exchanger: HPX-87H (300 × 7.8 mm I. D.); mobile phase: 0.01 N H_2SO_4; flow rate: 0.5 ml/min; 1...glyceraldehyde, 2...methylglyoxal, 3...glycolaldehyde, 4...dihydroxyacetone, 5...formaldehyde, 6...acetaldehyde, 7...acetone, 8...propanal, 9...methylethylketone, 10...5-hydroxymethylfurfural (HMF), 11...butanal, 12...furfural, 13...5-methylfurfural [41].

Alcohols, aldehydes, ketones, organic acids and carbohydrates can also be separated on H-loaded cation exchangers. In Figure 5.10 the temperature dependence of k' values for aldehydes and ketones are given. The greater part of the compounds are only little influenced by changes in temperature, which can be explained by the presence of temperature-insensitive ligand complexes. Certain other compounds, mainly the furfurals, however, show relatively strong k' decrease at higher temperatures. This effect might be attributable to an additional substrate matrix adsorption (reversed phase mechanism). In the case of organic acids, part of the sorption behaviour is assumed to be the result of ion exclusion (Donnan equilibrium [282, 283]).

Newer works show the great possibilities of ligand exchange in separating nucleosides, nucleotides, amino acids, phenolic compounds and amines with cation exchangers in the ionic form of Cu^{2+}, Ni^{2+} and Fe^{3+} [42, 43].

Reversed-phase ion pair chromatography.

This chromatography technique is based on the fact that, in this case, the fixed bed matrix has hydrophobic functional groups, which adsorb from the mobile phase the hydrophobic part of a compound containing an ionic functional group. This method is therefore also called "dynamic ion-exchange chromatography" [44], and sometimes even given the misleading name "soap-chromatography". Inorganic

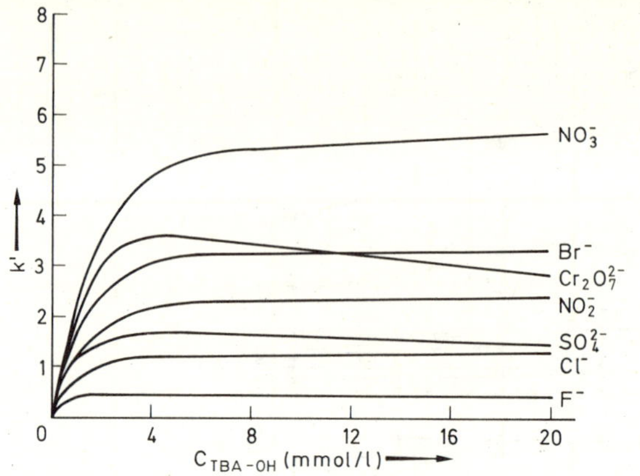

Figure 5.11 Influence of the TBA-OH concentration on the capacity factor (k') of anions. Column: 10 µm LiChrosorb RP-18 (250 × 4.6 mm I. D.); mobile phase: 0.05 M phosphate buffer, pH: 6.7 with different TBA-OH concentrations; flow rate: 2.0 ml/min, 80 bar; temperature: 25 °C; conductivity detector [46].

cations and anions, as well as many organic compounds, can be separated in this way. An indication of the ion-pair formation mechanism can be obtained from Figure 5.11. Approximately 0.006 molar $N(C_4H_9)_4^+$ OH^- (abbreviated TBA-OH) solutions are needed for a complete ion-pair formation. The "soaps" are retarded on the reversed phase, but the elution sequence of the anions is much the same as in fixed-site ion exchanges.

For a fairly long time different opinions prevailed on the actual mechanism of ion-pair chromatography. Through the work of Iskandarani and Pietrzyk [45] it became clear that the most probable explanation for the distribution of the solute is a real ion exchange, where the ion interaction may be based on a double layer formation on the stationary phase [47, 48].

Distribution of nonionic compounds

Ion exchange partition chromatography.

Samuelson [49] was successful in separating sugars and their oligomers on cation and anion exchangers. A relatively high alcohol concentration in the mobile phase was necessary to obtain sufficient variation in the k_D values to determine the oligomer sugars (Figure 5.12). The solubility of oligomer sugars in an alcohol/water mixture decreases with increasing DP. Due to the ionic functional groups the water content in the resin phase is higher than in the alcohol/water mobile phase. It is, therefore, clear that the distribution coefficient for the higher oligomers depends greatly on the alcohol concentration of the mobile phase.

Mixed mechanisms.

In modern ion-exchange chromatography of sugars excellent results are obtained by separating the carbohydrates directly on cation exchangers using HPLC tech-

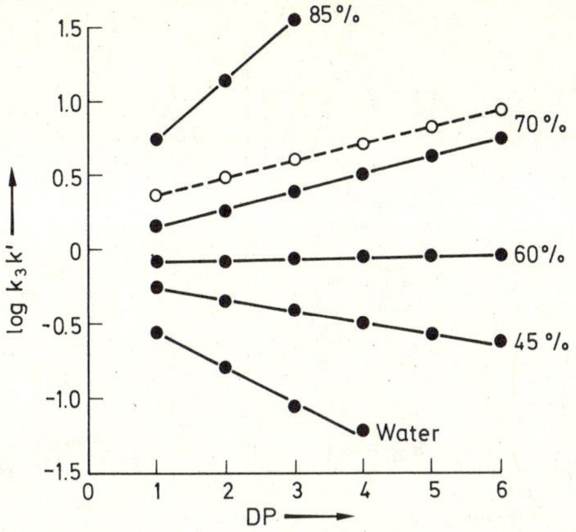

Figure 5.12 Relation between the capacity factor and the DP of oligomers, as dependent on the ethanol concentration (50). Sulphate resin (Technicon T5C, 10–17 μm); ○, xylose and oligosaccharides of β-(1→4)-linked D-xylose series; •, xylitol (at DP = 1) and oligomeric sugar alcohols of O-β-D-xylp-(1→4)-[O-β-D-xylp-(1→4)]$_{DP-2}$-D-xylitol series.

nique. It is thought that the influence on the distribution of the compounds is due to several effects: adsorption, partition, ligand exchange and molecular sieving [51, 52]. Depending on the molecular structure of the carbohydrate, the distribution is governed more by the one or the other mechanism. In the case, for example, of oligomer sugars, the molecular sieving effect dominates, whereas in the case of the low molecular sugars ligand exchange and the partition mechanism are the main influences on the distribution [53].

5.2.2.2 Kinetics of ion exchange chromatography

Martin and Synge [5] compared, in a theoretical approach, the chromatographic method with the distillation process in a column. This theory was very successful, especially at the beginning, as well as in the further development by Mayer and Thompkins [54], Rosen [55] and Glueckauf [56] (see also [57]). The introduction of the mass-transfer equation deepened the understanding of important parameters. Simplified forms of the resulting solutions are now in common use, as will be shown below.

Elution chromatography
To characterize an individual chromatographic band the symbols of Figure 5.13 are used. Equation 4a gives the reduced retention time ($t_R' = t_R - t_0$) as

$$t_R' = t_0 k' . \qquad (12)$$

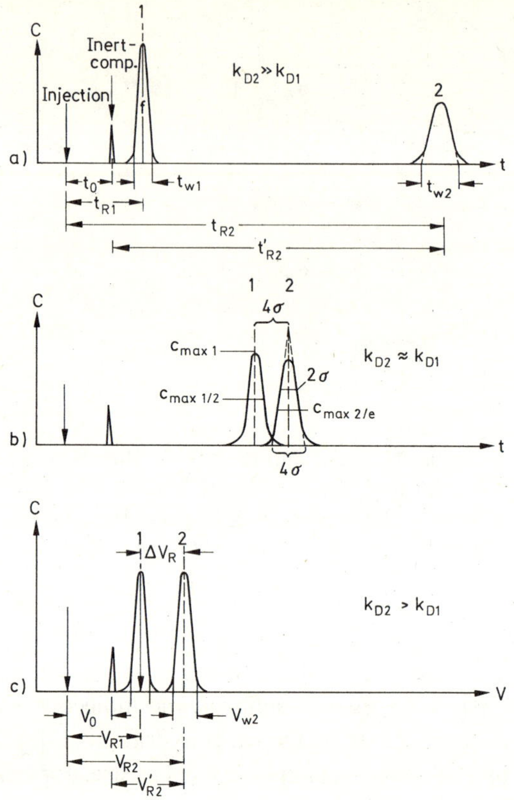

Figure 5.13 Schematic elution curves. The influence of the distribution coefficient (k_D) on the retention time (t_R, t'_R) and retention volume (V_R, V'_R) is shown.

If the distribution coefficients from eq. 2 and 3 are introduced, t'_R becomes

$$t'_R = \frac{V_s t_0}{V_0} k_{CD} = \frac{\varrho_s V_s t_0}{n V_0} k_D . \tag{13}$$

The relation between the reduced retention volume ($V'_R = V_R - V_0$) and the distribution coefficients results from eq. 4 and 5:

$$V'_R = V_s k_{CD} = V_0 k' = \frac{\varrho_s V_s}{n} k_D . \tag{14}$$

For all chromatographic work the expression for the resolution, R_S, is very important. R_S is the ratio of the difference between the retention times of two bands and the sum of their base line width (t_W):

$$R_S = 2 \frac{t'_{R2} - t'_{R1}}{t_{W1} + t_{W2}} = 2 \frac{V'_{R2} - V'_{R1}}{V_{W1} + V_{W2}} . \tag{15}$$

These band-widths are determined by the points of intersection of the tangents through the inflection points of the Gaussian peak with the base line.

Under the assumption that both peaks have the same height and are really Gaussian bands, the band-width of the curve becomes 2σ (1σ is the standard deviation of the Gaussian). The sum of both band-widths is therefore 4σ and a so-called "4σ separation" becomes

$$R_{S(4\sigma)} = 2\frac{4\sigma}{(4\sigma + 4\sigma)} = 1 \tag{16}$$

In this case, only a 2% band overlap occurs. A "6σ separation" ($R_S = 1.5$) gives a significantly increased resolution. R_S-values below 0.8 do not usually yield satisfactory separations.

The relative retention, a, can be written as

$$\alpha = \frac{t'_{R2}}{t'_{R1}} = \frac{k'_2}{k'_1} = \frac{k_{D2}}{k_{D1}}. \tag{17}$$

Therefore a characterizes the peak separation of two components and gives at the same time the thermodynamic difference of their distribution:

$$\Delta(\Delta G^0) = -RT\ln\alpha. \tag{18}$$

$\Delta(\Delta G^\circ)$ is the difference in free energy of the distribution of both components.

In the field of isotope separation a-values in the neighbourhood of 1.001 are frequently found. In certain cases good resolutions can be obtained with a-values of 1.01. Higher a-values normally cause no separation difficulties.

It is a most important feature of chromatography that the band-width increases with the square root of the retention time (or the retention volume):

$$t_W = k_b\sqrt{t'_R} = k_c\sqrt{V'_R}. \tag{19}$$

The resolution (eq. 15) can therefore be transformed

$$R_S = \frac{2(t'_{R2} - t'_{R1})}{k_b(\sqrt{t'_{R1}} + \sqrt{t'_{R2}})}. \tag{20}$$

Under the assumption that both peaks have approximately the same height and are relatively close together, the square root expressions become equal, $\sqrt{t'_{R1}} = \sqrt{t'_{R2}}$. The above equation can then, by introducing eq. 12 and a from eq. 17, be written:

$$R_S = \frac{(t'_{R2} - t'_{R1})}{k_b\sqrt{t'_{R2}}} = \frac{(k'_2 - k'_1)}{k_b k'_2}\sqrt{t_0 k'_2} = \frac{k_e(\alpha - 1)}{\alpha}\sqrt{t_0 k'_2}. \tag{21}$$

Purnell [58] obtained an analogous equation:

$$R_S = \frac{1}{4}\frac{(\alpha - 1)}{\alpha}\frac{k'_2}{(1 + k'_2)}\sqrt{N}. \tag{22}$$

It will be recalled that the plate number, N, is a direct function of t'_R, k' and k_D.

The number of plates [58] is usually determined by

$$N = 16\left(\frac{t_R}{t_W}\right)^2, \tag{23a}$$

and the number of effective plates by

$$N_{eff} = 16\left(\frac{t'_R}{t_W}\right)^2 = 16\left(\frac{V'_R}{V_W}\right)^2. \tag{23b}$$

This can be transformed, by introducing t_W from eq. 19, to

$$N_{eff} = \left(\frac{16}{k_b^2}\right) t'_R = \left(\frac{16}{k_c^2}\right) V'_R. \tag{24}$$

Great importance has always been attached to the relation of the plate number, N, or the plate height, H_T, to the flow rate, u. It was, however, frequently overlooked that the quality of column packing has a great influence on the N- or t'_R-values. Especially in high pressure chromatography an equal distribution of channel width in the mobile phase, and therefore also of the particle size, becomes very important. Otherwise the larger channels carry an over-proportional part of the mobile phase, thus reducing the retention time and frequently also introducing tailing. The plate height, H_T, in its relation to the flow rate, can be written (Huber [59]):

$$H_T = \frac{L}{N} = \left(\frac{1}{A} + \frac{1}{C_m}\right)^{-1} + \frac{B}{u} + C'_m\sqrt{u} + C_s u. \tag{25}$$

The term A depends on the Eddy-diffusion, which occurs when the mobile phase flows around the particles, and C_m takes care of the convective part of the diffusion in the flowing mobile phase. Gidding [60] showed that both these processes are coupled and lead to the form (D_m is the diffusion coefficient in the mobile phase):

$$\left(\frac{1}{A} + \frac{1}{C_m}\right)^{-1} = \left(\frac{1}{2d_p} + \frac{D_m}{d_p u}\right)^{-1}. \tag{26}$$

The term B can be expressed as

$$B = 2\gamma D_m. \tag{27}$$

For C_s the following relation is proposed [61, 62, 63]:

$$C_s = \psi \frac{k_D}{(1 + k_D)^2} \cdot \frac{d_p^2}{D_s}. \tag{28}$$

In Figure 5.14a an evaluation of the above equations is given. A nitrobenzene sample was chromatographed on a diatomaceous earth column and 2,2,4-trimethyl pentane was used as eluent. Figure 5.14b shows the plot of H_T/u versus u for the same experiments. H_T/u has the dimension of time. The lower right part of the

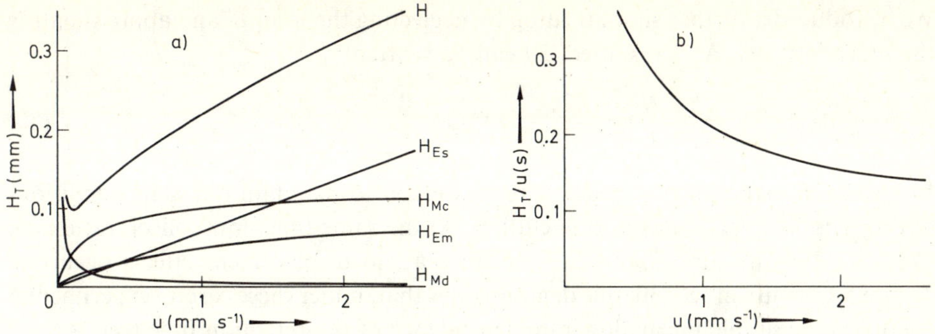

Figure 5.14 Schematic plate height curves according to [59].
a) The theoretical plate height H_T is the sum of four terms: H_{Md}...mixing by molecular diffusion; H_{Mc}...mixing by convection; H_{Em}...mass transfer in the moving phase; H_{Es}...mass transfer in the stationary phase.
b) Calculated H_T/u-values in relation to u from the H_T-curve in a).

curve indicates that in this region a saving in time can be achieved, but the plate height increases, and with it the resolution is diminished. At the upper part, on the left side of the curve, the resolution can be increased to a certain extent, but only at the expense of analysis time.

Several authors have introduced abbreviations for the complicated form of eq. 25 [64, 65, 66]. Here the results of Knox [67] are given.

$$H_T = \frac{A'}{d_p}\left(\frac{u\,d_p}{D_m}\right)^{0.33} + B'\frac{D_m}{u\,d_p} + C'\frac{u\,d_p}{D_m}. \tag{29}$$

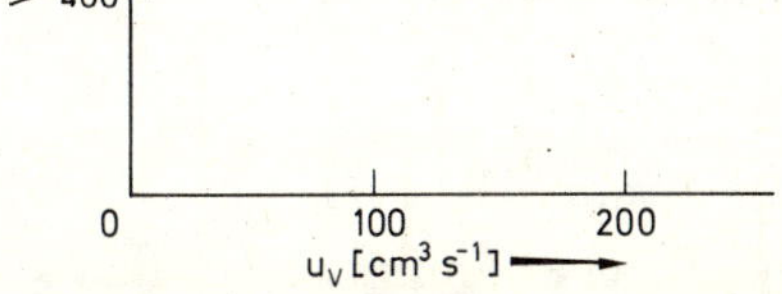

Figure 5.15 Dependence of the retention volume (V_R) on the flow rate (u_V) according to equation 30 [68]. Exchanger: Dowex 50 W × 8 (20 – 50 mesh); column: 100 × 0.3 cm I. D.; $k_{D(Co)}$ = 68.2 and $k_{D(Mn)}$ = 71.8.

We introduced a further simplification by neglecting the term B' and approximating the first term [68]. As a result eq. 29 can be written:

$$t_R' = \frac{k_1}{H_T} = k_1 \frac{N}{L} = \frac{A^* k_D}{d_p^n} - C^* u\,. \tag{30}$$

In a series of experiments special care was taken to maintain the same conditions during sample injection as during elution. At the same time, low concentrations of Co and Mn ions were applied to ensure that no displacement effects occurred. Figure 5.15 confirms eq. 30; this demonstrates that, under these careful experimental conditions, a simple linear flow-rate dependence of the retention time results.

Important relations and formulae are listed in Table 5.1.

Table 5.1 Nomenclature, units and formulae

Nomenclature	Units	Formulae	Equation
Retention time of nonretained solute	s	$t_0 = t_m$	
Retention time from start	s	$t_R = t_R' + t_0$	(12)
Reduced retention time	s	$t_R' = \frac{V_s t_0}{V_0} k_{CD} = \frac{\varrho_s V_s t_0}{n V_0} k_D = t_0 k'$	(12, 13)
		$t_R' = \frac{k_b^2}{16} N_{eff} = \frac{k_b^2}{16} \frac{L}{H_T}$	(24)
		$t_{R(Ibk)}' = \frac{k_1}{H_T} = \frac{A^* k_D}{d_p^n} - C^* u$	(30)
Retention volume of nonretained solute	cm^3	$V_o = V_m$	
Retention volume from start	cm^3	$V_R = V_R' + V_0 = (t_R' + t_0)\, u_V$	
Reduced retention volume	cm^3	$V_R' = \frac{\varrho_s V_s}{n} k_D = V_s k_{CD} = V_0 k'$	(14)
		$V_R' = \frac{k_c^2}{16} N_{eff} = \frac{k_c^2}{16} \frac{L}{H_T}$	(23)
Capacity factor	–	$k' = t_R'/t_0 = V_R'/V_0$	(12, 14)
Relative retention	–	$\alpha = \frac{t_{R2}'}{t_{R1}'} = \frac{k_{D2}}{k_{D1}} = \frac{k_2'}{k_1'} = \frac{V_{R2}'}{V_{R1}'}$	(12, 17)
Resolution	–	$R_S = 2\left(\frac{t_{R2}' - t_{R1}'}{t_{W1} + t_{W2}}\right) = 2\left(\frac{V_{R2}' - V_{R1}'}{V_{W1} + V_{W2}}\right)$	(15)

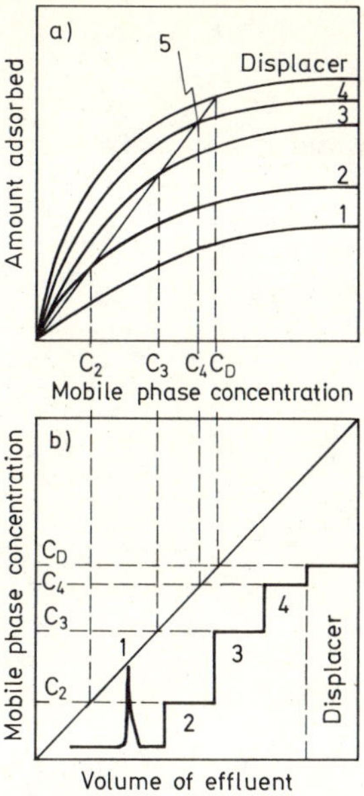

Figure 5.16 Displacement chromatography [73].
(a) Isotherms in displacement chromatography of four substances and a displacer. The operating line is indicated (5). (b) The fully developed displacement train is shown.

Displacement chromatography
Tiselius [69] made a clear distinction between elution, displacement and frontal analysis. Whereas in elution chromatography the first linear part of the adsorption isotherm is usually used, displacement chromatography operates in the upper saturation region. Figure 5.16a shows the adsorption isotherms of four different substances and the 'displacer'. The displacer demonstrates the strongest sorption behaviour of all the substances to be separated. In Figure 5.16b the corresponding displacement profile is given. As soon as the displacement train has arranged its position (isotachic conditions), all components pass through the column with the same velocity. In this situation the molar distribution coefficients are constant:

$$k_5 = \frac{\overline{C_1}}{C_1} = \frac{\overline{C_2}}{C_2} = \frac{\overline{C_3}}{C_3} = \frac{\overline{C_4}}{C_4} = \frac{\overline{C}_{\text{displacer}}}{C_{\text{displacer}}} \tag{31}$$

Spedding et al. [70] successfully applied displacement chromatography in separating rare-earth complexes on ion exchanger columns. In Figure 5.17 the separation of samarium, neodymium and praseodymium is shown. This figure also indicates that the application of a greater amount of an individual component can increase its yield, because the fraction of the overlapping region is decreased.

A considerable amount of work was also invested in separating proteins [71] and protein hydrolyzates [72]. However, at that time complete separations of

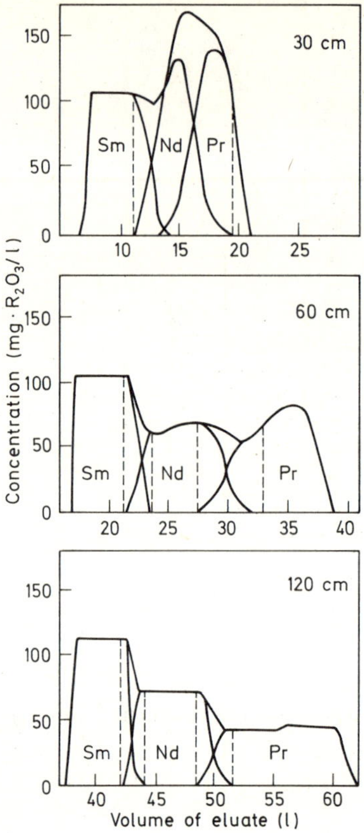

Figure 5.17 Separation by displacement chromatography of samarium, neodymium and praseodymium on 30, 60 and 100 cm ion exchange columns (22 mm I. D.). 0.1% citrate solution is used as carrier in this case. Flow rate: 0.083 mm/s [70].

complex mixtures were not possible, owing to the inefficiency of the column materials available (Figure 5.18). In newer experiments displacement chromatography was given a fresh impetus, mainly due to the fact that column materials had been much improved in the meantime. Some of these results will be discussed later.

Frontal ion exchange chromatography

Frontal chromatography is closely related to displacement chromatography. The main difference is that, in the former case, the displacing solute is applied continuously to the column. As an example the frontal analysis of Na^+ will be discussed. A regenerated cation exchanger in the H^+ form is continuously loaded with a Na^+/K^+-mixture. First of all the Na^+-ions displace the H^+-counterions completely; then the Na^+-ions are also displaced by the K^+-ions, and can be obtained in a relatively pure form. Later in the course of the experiment, the K^+-concentration increases until the concentration of the added solution is reached (Figure 5.19). Therefore

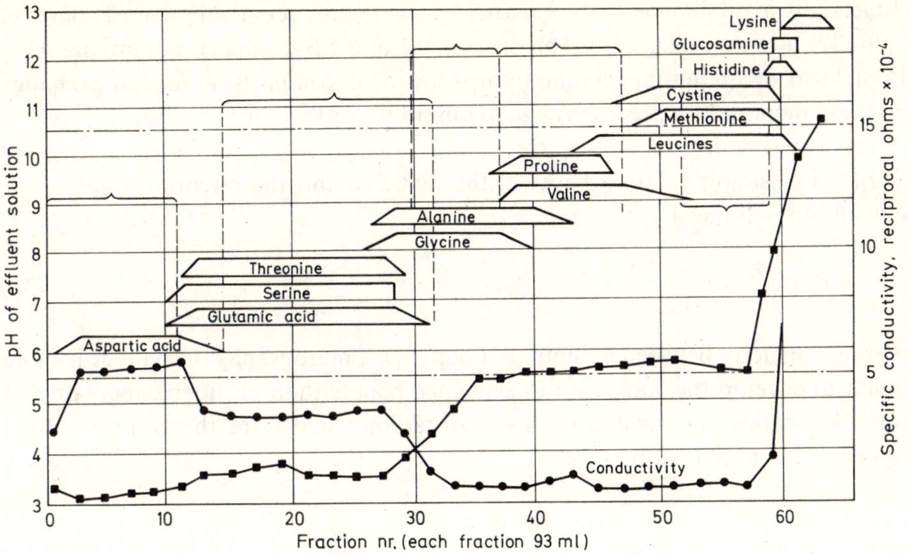

Figure 5.18 Separation of egg albumin hydrolysate by displacement chromatography on a cation exchanger (74).

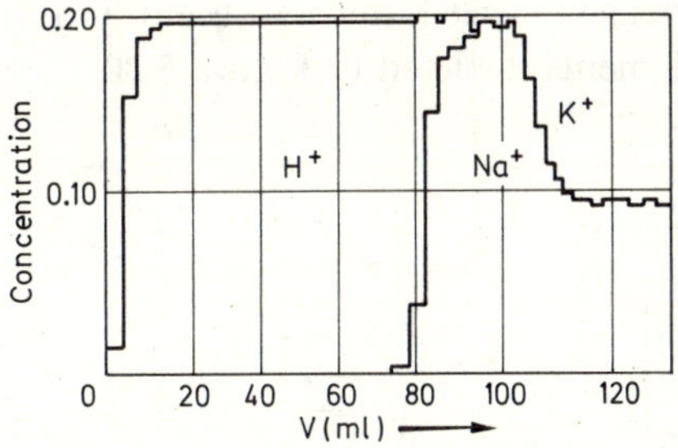

Figure 5.19 Separation of Na and K ions by frontal analysis with an ion exchanger in the H^+-form [75]. Molar concentration as dependent on the effluent volume.

only a certain amount of the Na^+-ions can be obtained in a form free from K^+ contamination, whereas the K^+-ions always contain a certain amount of Na^+.

The field of application of frontal analysis is similar to that of displacement chromatography and mainly serves for the preparative separation of chemical compounds.

Ion exchange paper chromatography

In the development of column chromatographic methods, an interruption occurred during which paper chromatography was mainly used.

Similarly, in the field of ion exchange chromatography, paper ion exchangers were examined intensively during the sixties. The ion exchange papers were either commercially available, or were prepared by the researchers themselves. Lederer et al. [76, 77], for example, used Amberlite SA-2 and SB-2 papers, containing 45% sulfonic acid or quaternary ammonium resins. For comparison, an ion exchange paper was prepared by precipitating zirconium phosphate inside a filter paper.

Such experiments show how the solutes are carried through the exchanger material. The simple relation between the R_f value and the retention time, t_R, or the retention volume, V_R,

$$R_f = \frac{t_0}{t_R} = \frac{V_0}{V_R}, \tag{32}$$

proves the analogy between column and paper chromatography. Usually it is more difficult to develop the spots on ion exchange papers than on filter papers or thin layers. Self-indicating substances, such as radionuclides, are therefore favourite solutes for this application.

5.2.3 Chromatographic equipment and procedures

Scientific and technical progress from Tswett's relatively large liquid chromatography columns and visual detection, to small columns and very sensitive detection systems, has developed with increasing speed during recent years. A modern chromatography system has the elements depicted in Figure 5.20.

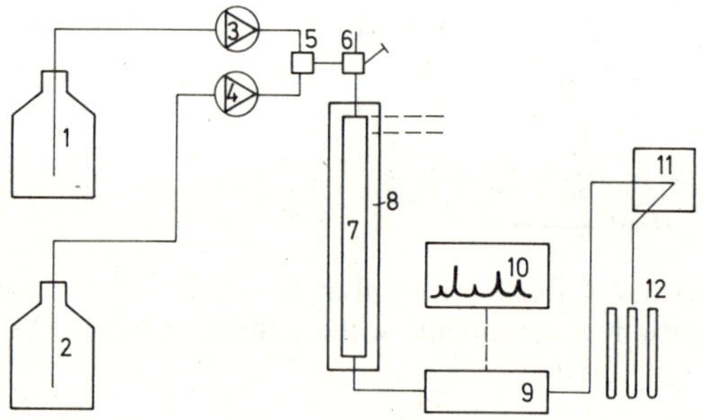

Figure 5.20 Schematic ion exchange chromatography system for gradient elution. 1, 2...reservoirs for elution media; 3, 4...pumps; 5...mixing vessel; 6...injection valve; 7...column; 8...heating unit; 9...detector; 10...recorder; 11...fraction collector; 12...collected samples.

5.2.3.1 Column and accessories

Reservoir. Depending on the elution medium, a series of different reservoir materials are available: glass, plastic (e. g., polyethylene, polytetrafluoroethylene) and stainless steel (type 304 or 316) are frequently used. It should be taken into consideration that most mobile phases have to be degassed before application. In certain cases the vessel is directly evacuated or purged with an inert gas (e. g., nitrogen).

It is a good practice to extract the mobile phase from the reservoir through a small filter unit.

Pumping systems. The pressure drop in fine grain columns can be very high. Often 50 to 100 bar are needed to guarantee a continuous flow through the column. For the purpose of filling columns, pumps should be able to deliver a pressure of up to 700 bar.

Earlier pump designs were of the screw-driven syringe type, which had the advantage of a pulse-free solvent delivery at a constant flow rate. The limited volume capacity, however, was such a disadvantage that this type was not able to withstand the competition of other makes, especially the reciprocating piston pump.

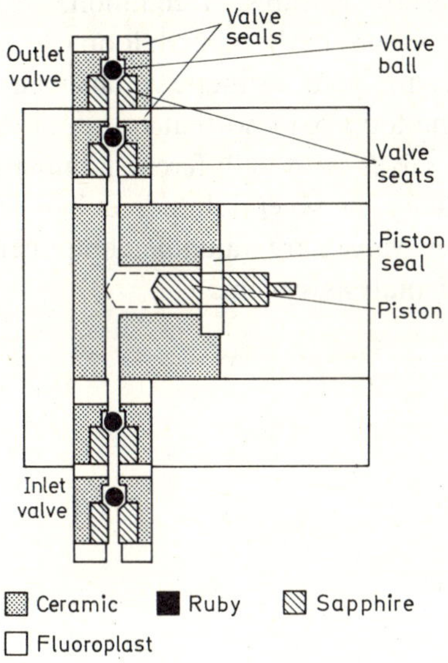

Figure 5.21 Pump head of a reciprocating piston pump. The construction consists of inert, acid-resistant materials (courtesy of LKB Inc.).

Figure 5.21 shows schematically the reciprocating piston pump head. This pump head consists to a large extent of inert materials (ceramic, ruby, sapphire and fluoroplast) to withstand highly concentrated acids. For higher pressures stainless steel heads with sapphire plungers and ball valve seats are usually applied. They have a very constant delivery rate, but the flow is not pulse-free. When low pulse amplitudes are required, pumps with more than one plunger or separate pulse dampeners are needed. Such devices, with very little pulsation, are available on the market.

Sample introduction devices. The sample can be deposited by a syringe through a septum directly in the packing of the column ("On-column injection port"). This method, however, has several disadvantages, such as plugging of the syringe needle, depositing of septum material in the column and disturbing of the column bed. Using a "swept injection port", the sample is deposited in front of the column. From there it is swept by the mobile phase into the packing. Injection ports for small (a few microlitres) and large (up to 100 microlitres) samples have been successfully designed and constructed. It is, however, almost impossible to guarantee a complete and fast elution of samples of different sizes from only one injection port.

With high-pressure syringes on-stream injection can be carried out; stop-flow injection, however, is also possible. The diffusion of the sample in the liquid mobile phase is so slow that this causes no additional problems.

In recent years, "sample valves" have been more frequently used than injection ports. The manifold designs have certain features in common: very little dead volume is involved during sample introduction; the sample loop can be changed if samples of different volumes have to be analysed; a sample volume of approximately half a millilitre is sufficient to rinse the loop part, so that no contamination from the previous sample occurs. Sample valves with a different number of ports are commercially available for pressures up to several hundred bar (Figure 5.22). Electrically- or pneumatically-operated valves are valuable components for automatic or high-speed chromatographic analysis systems.

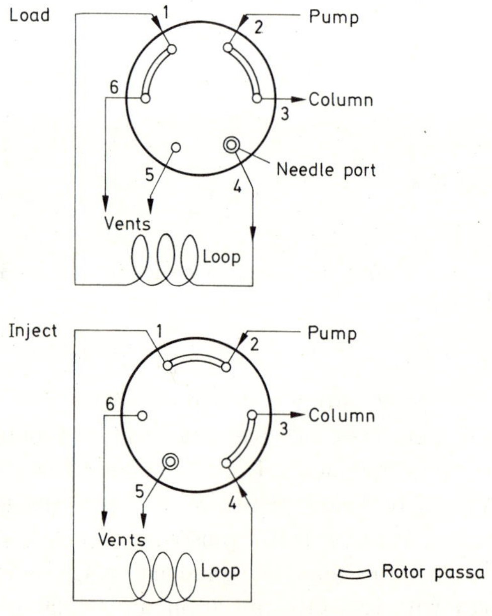

Figure 5.22 Sample valve at load and injection position (courtesy of Beckman Inc.).

Column systems. The tube material of the column can have an influence on the retention time of the solute. Under the assumption that all surfaces in use are inert, the reason for retention time deviations can only be explained by differences in the packing density of the column material. This opinion was verified [78] by experiments demonstrating that poor stainless steel tubing showed retention values similar to those of glass or special steel tubing, as soon as it was coated with Teflon® (polytetrafluoroethylene).

A porous metal frit is fastened at the outlet of the column. Porous Teflon can be used up to pressures of approximately 50 bar. Silanized quartz or glass wool are rarely used nowadays for column plugs.

The fittings have to be selected so as to ensure low dead volumes. Additional dead volumes always decrease the retention time of a solute, and therefore increase the plate height of the column. The filling of the column can be effected either by the "dry-packing" or the "slurry packing" method. Dry-packing is sometimes supported by devices that shake the column during the filling process [79]. Ion exchangers are frequently slurry-packed, because they can swell considerably during the addition of the solvent. Also high-crosslinked ion exchangers, which have relatively low swelling properties, are usually delivered in slurry form and are therefore also slurry-packed [80].

Nowadays, most column ovens have a circulating air bath. The low heat capacity of air is of no real disadvantage, because the very rapid circulation can outweigh the higher heat capacity of liquid bath media. Temperature controllers can easily keep the temperature deviation within $\pm$ 1% of the selected value in the region of the column. Within the column tubing, the deviation is still much less than 1%.

Sometimes the influence of the temperature of the mobile phase is underestimated, if there are great differences between the oven and the reservoir temperature.

Pre-column. In many cases it is advisable to use a pre-column in front of the main column, thus avoiding undesirable loading, ion -exchange or filling of the exchanger material in the main column.

Suppressor. For ion chromatography, suppressors are frequently used. They consist of a secondary ion exchanger column (sometimes containing ion exchange fibres), which is located between the separator column and the detector (usually a conductivity detector). By this means, the cations can be removed when the anions are to be determined, or the anions when the cations are to be analysed. Through this procedure, the conductivity of the solute-free eluent (background conductivity) is reduced and the ions of the solute are determined with greatly increased sensitivity.

5.2.3.2 Detectors

Owing to the characteristic differences between gaseous and liquid mobile phases, liquid chromatography detectors lagged behind those for gas chromatography in performance for many years. New developments have brought the performance of both these types of instruments surprisingly close together, with them frequently demonstrating a similar sensitivity.

Detectors can be divided into two groups: bulk-property and solute-property detector devices. The former determine the change in the overall physical properties of the mobile phase. Three main types belong to this group: refractive index, conductivity and dielectric constant detectors. The second group consists of ultraviolet (UV) absorption, radioactive, polarographic, fluorimetric, mass and mass-spectrometric detectors. Flame ionisation requires the separation of the bulk mobile phase solvent.

Sensitivity, linearity and noise are important instrument specifications to be considered. The "relative sample sensitivity" of a detector is the lowest concentration of a solute which the detector can determine. This value is usually taken as being twice the noise level. It should, however, be taken into account that this sensitivity may depend on the type of solvent used (e. g., refractive index), the dead volume in the system, or the flow rate (e. g., polarography).

Table 5.2 HPLC detector system data (2 to 10 μl cell volume)

Detector	Approx. linear range*)	Units	Sensitivity (favorable sample)	Remarks
UV absorption	$10^{-4}-2.56$	AU	0.1 ng/ml	selective; low temp. and flow sensitiv.
UV fluorescence	approx. one order of magnitude	AU	0.2 ng/ml	selective; low temp. and flow sensitiv.
Refractive index	$10^{-8}-10^{-3}$	RIU	50 ng/ml	universal: 10^{-4} RIU/°C
Conductivity	$10^{-3}-10^{3}$	μmho	10 ng/ml	selective; 2%/°C
Capacity		DC	100 ng/ml	universal; 10^{-3}/°C
Nuclear radiation		cps	(H-3 = 55% C-14 = 90% eff.)	labelled compounds
Polarography	$10^{-9}-2.10^{-5}$	A	1 ng/ml	selective; 1,5%/°C
Adsorption	$10^{-4}-10^{-1}$	°C	1 ng/s	universal
Transport		A	100 ng/ml	universal
Mass-spectrometer			extreme	universal

*) The lower value corresponds to approx. twice the noise level.

The "absolute detector sensitivity" relates the full-scale deflection of the recorder to the change of the measured physical parameter at a given noise level. The absolute sensitivity can be determined by applying a standard solution directly to the detector cell.

The detector system should give a linear response ("linearity") when the solute concentration is changed. The linearity, however, is never fulfilled perfectly. Equation 33 determines the deviation of the detector output from linearity [81]:

$$I = k_i C^n . \quad (33)$$

The solute concentration *(C)* has the exponential factor n and is related through the proportionality constant ("response factor" k_i) to the output signal *(I)*. Real linearity is achieved only when $n = 1$.

Noise is the variation in the output signal inherent in the electronic equipment and not caused by the solute.

Temperature and line voltage fluctuations, flow changes and electric circuits themselves are responsible for detector noise. Short-term noise can be seen as vibrations ("fuzz") on the recorder base line; long-term noise as "peaks and valleys". Drift is the up or down movement of the base line when no solute (e. g., only pure solvent) passes through the detector. Table 5.2 gives the approximate linear range and the sensitivity of several detector systems.

Optical detectors

Ultraviolet absorption detector.
The UV absorption detector is very sensitive for all solutes which have a reasonable absorption band in the UV region. For several substances the detection limit is below one nanogram. The sensitivity of such detectors can be in the range of 0.005 AUFS (absorbance units full scale at ± 1% noise).

In Figure 5.23 a the schematic lay-out of an UV absorption detector is given. The light source, e. g. a low or medium-pressure mercury lamp (254 to 280 nm) or an additional phosphor (reradiating at 280 nm), emits the light through lenses, a beam-splitter and the sample cell into a phototube. Part of the original beam is led by means of a mirror, lenses, and a reference cell into a second phototube.

Fluorimeter.
As in UV absorption detection, the sample is irradiated with UV light. However, the detector responds only to visible fluorescent light. The emitted fluorescent light leaves the sample cell at right-angles to the incident UV beam, or in a straight line (Figure 5.23 b). Strongly fluorescent compounds (e. g., quinine sulphate) can demonstrate high sensitivies (e. g., below 1 ng/ml). In many cases the sensitivity is similar to that of UV absorption.

Differential refractometer.
The differential refractometer measures the difference between the refractive indexes of the reference solution and the mobile phase, at the column exit. The temperature

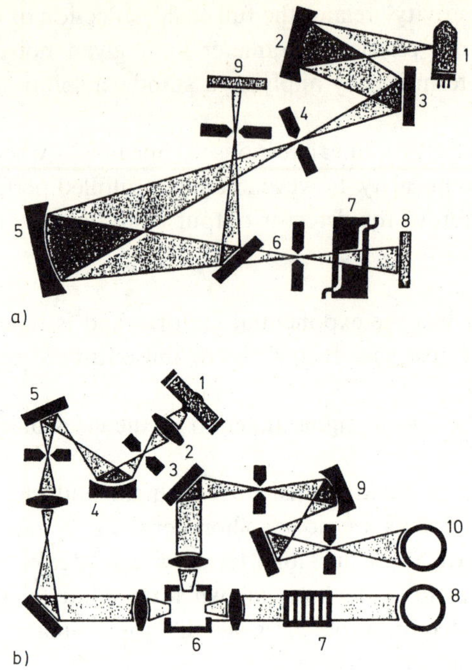

Figure 5.23 UV detectors (Courtesy of Merck-Hitachi).

a) UV absorption detection: 1...deuterium lamp, 2...mirror (toroidal), 3...mirror (plane), 4...slit, 5...grid (concave), 6...beam divider, 7...flow cell, 8...photodiode (sample), 9...photodiode (reference).

b) UV fluorimeter: 1...xenon lamp, 2...lens, 3...slit, 4...grid (concave), 5...mirror, 6...flow cell, 7...dampener, 8...photomultiplier, 9...grid (concave), 10...photomultiplier (emission).

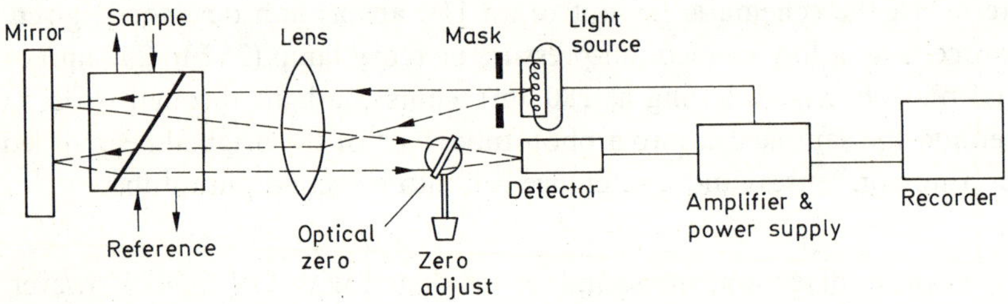

Figure 5.24 Schematic diagram of a refractive index detector system (courtesy of Waters).

must be kept within very narrow limits. Until recently it was enough to ensure a variation in cell temperature of not more than $+0.001\,°C$ to obtain sensitivities of 10^{-5} RIUFS (refractive index units full scale at + 1% noise). In the meantime, high increases in sensitivity have been achieved, due to an extremely stabilised cell temperature (see Table 5.2).

Two types of refractometers are in use: the Fresnel and the deflection refractometers. In the case of the Fresnel refractometer, the percentage of light reflected from a solid-liquid interface depends on the reflection angle and the refractive index of the liquid. The deflection refractometer makes use of the fact that a light beam shows different deflection behaviour in two compartments of the sample cell, due to the difference in the refractive index of the sample and the reference (Figure 5.24). The position-sensitive photodetector determines the relevant deflection.

Nuclear radiation detector.
Various detectors are commercially available that can be used for normal chromatographic work and HPLC separations. Such detectors are of special interest, when quantitative determinations of all degradation or transformation products of certain compounds (e. g., in chemical research, biology or pharmacy) are necessary. Low-energy pure beta-emitters, such as C-14, S-35 and H-3, are frequently used for this purpose. A favourite method of detection for these radioactive nuclei is scintillation counting. The scintillation detector is often a high-polymer plastic material (e. g. polystyrol), which contains the scintillator substance (e. g., terphenyl, POPOP). The scintillator transforms the nuclear radiation into visual light, and the adjoining photomultiplier transforms this light into electrical pulses. The scintillator cell can have canals or spirals to lead the effluent liquid. The scintillator can also be packed into a cell tube in the form of small particles. Efficiencies of 55% for C-14 and 2% for H-3 have been reported [82]. Plastic scintillation tubes are less efficient (approximately 6% for C-14).

In many cases radioactive compounds with low specific activities are used. Thus relatively low pulse rates are generated in the detection cell. In order to increase the efficiency, different methods are employed:

- If the flow is stopped at the maximum of the peak, the counting time, and with it the sensitivity, can be considerably increased.

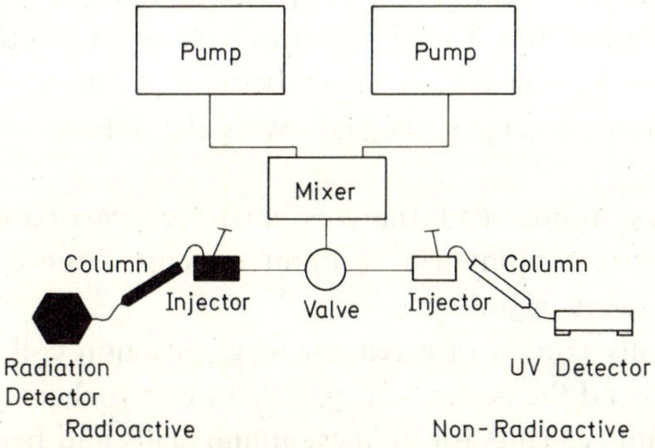

Figure 5.25 Schematic diagram of a dual detection system: pumps, mixer and valve serve either the column with the nuclear radiation detector, or the inactive line with e. g. UV detector (courtesy of Beckman Inc.).

– The effluent stream can be mixed with a scintillator liquid between the column and the detector cell. High efficiencies are obtainable in this way [83]; new instruments guarantee 90% for C-14 and 55% for H-3.

In Figure 5.25 a system is shown where either the radioactive detector or the inactive system can be used.

Electrochemical detectors

Polarography.
The current between a polarisable and a non-polarisable electrode is measured in the column effluent as dependent on time. In this case the voltage is usually kept constant, in contrast to normal polarographic analyses.

Different electrode types are described in the literature, such as dropping mercury [84], carbon-impregnated silicon rubber [85], and platinum electrodes. Many inorganic ions, amino acids, alkaloids, aldehydes and ketones have been determined by these methods.

Conductivity detector.
In order to avoid polarisation effects, which occur with dc-applied current, the greater part of the instruments are furnished with ac or high-frequency ac-detection devices. The dielectric constant of the solvent is usually the reason for an out-of-phase component of the conductivity measurement. This has to be compensated with a phase-sensitive electronic circuit.

Capacity detector.
Capacity detectors determine the difference in capacitance of the solvent and the solute-containing solvent. The dielectric constant is measured at high frequencies between two electrodes.

Reaction detectors
During recent years chemical reaction detectors have gained in importance. Several surveys (Lawrence and Frei [86], Beau and King [87], Schwedt [88]) give detailed descriptions of the advantages of this detection method. By using reaction detectors in liquid chromatography two goals are envisaged: better separation behaviour can be obtained by the reaction of certain reagents with the solute, and improved detection can be achieved.

For more than 20 years, amino acid analyses have been carried out with ion exchange columns and on-line ninhydrin reaction systems. Figure 5.26 shows schematically a new instrument of this type.

Reaction detectors usually consist of a reactor (e. g., reaction coil, bed reactor or segmented flow-reactor) and the detector (e. g., UV or electrochemical detector). Some of the more important reagents for amines, amino acids and peptides are, in addition to ninhydrin, *o*-phthalodialdehyde, potassium hexacyanoferrate(III) ethylenediamine, fluorescamin and iodine. Carbonic acids and phenols are frequently determined with oxidyzing reagents, (e. g., chromic acid, periodate, Ce(IV)). Oxy-

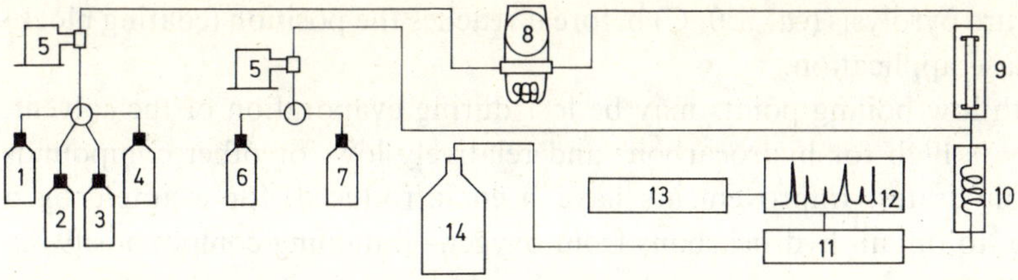

Figure 5.26 Schematic diagram of a high-performance amino acid analyzer (Courtesy of Beckman Inc.).
1,2,3,4...buffers, 5...pumps, 6...reagent, 7...solvent, 8...injection valve, 9...column, 10...post column reaction system, 11...detector, 12...recorder, 13...data system, 14...waste.

dizing media are also used for the reaction with carbohydrates, including phenol or orcin-sulphuric acid and dye reagents.

Chemical derivatization can also be carried out with steroids, pesticides and fungicides, and in certain cases enzymatic reactions (e. g., with organo phosphates and carbamate insecticides) are used.

Through these methods the detection limits of UV adsorption or fluorescence are reached in many cases, thus eliminating the disadvantages of the low sensitivity of certain compounds in liquid chromatography.

Miscellaneous detectors
The following detector systems have been developed to fulfil special demands.

Mass spectrometer [89].
Interfacing LC or HPLC with a mass spectrometer did not progress as fast as in the case of gas chromatography. The main reason for this is that mass spectrometry needs vaporized compounds, which are inherent in GC, but not in LC.

In earlier experiments part of the effluent was directly vaporized [90] or moved with a belt into the spectrometer [91]. Since 1981 it has become possible to introduce large and polar molecules into the spectrometer and excite these compounds by fast atom bombardment (FAB) [92]. The sample can be dissolved in a liquid matrix (e. g., glycerol) and transported by a belt to the FAB ion source. Vestal [93] introduced the "thermospray" technique, in which evaporation of the solute, self-ionization and protonation of the molecules occur simultaneously. With this method compounds with molecular weights beyond 1000 daltons can be analyzed.

It can be expected that HPLC-MS coupling will greatly increase in importance in the near future.

Solute transport detector.
This detector transports the effluent by chains, belts, coils or wires into an oven, in order to evaporate the solvent. The solute then reaches a further oven, where pyrolysis in a nitrogen atmosphere occurs. The gas thus formed is led into a flame ionization detector. Usually the transport system (e. g., wire) is first cleaned by

high temperature pyrolysis (ca. 750 °C) before it reaches the position (coating block) for mobile phase application.

Solutes with low boiling points may be lost during evaporation of the solvent. The sensitivity is high for hydrocarbons and relatively low for other compounds such as carbohydrates. Improvements have been introduced, for example by a reduction step, to obtain hydrocarbons from oxygen-containing compounds [94].

Heat of adsorption detectors.
These have a very general applicability, but also certain disadvantages which have so far prevented them from being used more widely.

Thermocouples or thermistors determine the adsorption and desorption heat of the solute directly in the main column or in a small secondary column. The S-shape of the response reduces the resolution, and the flow sensitivity of this device can influence the minimum detection limit.

5.2.3.3 Procedures

Gradient elution
Gradient elution or, in more general terms, solvent programming, is frequently used in ion-exchange chromatography. However, the tendency to solve separation problems by using only one mobile phase persists. One of the reasons for this is that greater accuracy is achieved in repeated analyses if isochratic conditions in the mobile phases are applied.

When gradient elution is chosen, one has to decide between low and high-pressure gradient mixers. In the case of low-pressure mixers, the two solvents are mixed at atmospheric pressure. Several methods have been reported in the literature [95].

"Exponential gradient elution" is a simple method, whereby solution A is pumped from a mixing vessel to the column and, at the same time, the solvent B is added to the mixing vessel. Often two low-pressure pumps deliver both solvents into a mixing vessel, from which a high-pressure pump supplies the column. High-pressure gradient elution is more versatile: here two high-pressure pumps are responsible for the flow of the solvents to the mixing chamber, and from there through the column. By means of such devices it is also fairly simple to program the velocity of the flow and to make step changes in the composition of the mobile phases.

Series-connected columns
Two columns connected in series have a lower number of theoretical plates than the sum of the plates of both columns. Snyder [96] describes for GC the non-additive behaviour of the coupled columns by the following equation:

$$N_c = \frac{(V_{R1} + V_{R2})}{\left(\frac{V_{R1}^2}{N_1} + \frac{V_{R2}^2}{N_2}\right)} \tag{34}$$

where N_c represents the plate number for the coupled system. With the separation of sugars on differently loaded ion-exchangers, it was shown that eq. 34 can also be applied to ion exchange chromatography [97].

Multidimensional column systems
These are also called column switching systems, multi-phase, multi-column, or coupled column chromatography. The main difference from the series-connected column systems lies in the fact that, in the case of multidimensional column procedure, only fractions of the eluent of a first column are separated on one or more other columns. In contrast to series-connected columns, different mobile phases can be applied in these procedures.

In addition to the use of multidimensional chromatography for sample clean-up and preconcentration prior to analysis, already discussed in Section 5.1, the main purpose of this method is the separation of complex samples. This technique was used, for example, for the separation of amino acid enantiomers [98] by a two step procedure.

Very elaborate equipment was designed to direct the flow of sample fractions to pre-columns, analytical columns and waste lines by automatically actuated valves [99, 100].

5.2.4 Application of ion exchange chromatography

5.2.4.1 Ion chromatography – low pressure

Separation of metal cations
Frequently, separation of cations is carried out on cation exchangers by the direct application of the uncomplexed cations. In several cases, however, the separation is facilitated when complex formation with reagents or buffer solutions is introduced. Cation, anion and mixed-bed exchangers as stationary phases are employed, depending on the ionic form of the solute.

Two varieties of cation separation reveal new dimensions for solving the analytical problems concerned: one uses highly concentrated electrolytes as mobile phase; the other mixed organic-water solutions.

Separation on cation exchangers
Over 1000 publications on the separation of rare-earth and transuranium elements document the great applicability of ion exchangers in this field.

Independently Lindner (1946) in Germany and Boyd, Tompkins, Cohn, Coryell, Marinsky and Glendenin (1947, [101, 102]) in the USA were able to separate the rare-earth elements on a resin-based ion exchanger using citric and tartaric acid as complex-forming agents. For this separation a Dowex 50 column (97 cm, 0.26 cm^2) with a relatively fine particle diameter (270 – 325 mesh) was used as a stationary phase. In the case where the mobile phase consisted of citric acid, the lower lanthanides were separated at pH 3.28 and the higher at pH 3.33 at a flow rate of 2 ml cm^{-2} min^{-1} and a temperature of 100 °C.

The above separation was improved by Nervik [103] by using the gradient elution technique. In this experiment rare-earth metals were separated with 1 M lactic acid and increasing pH at a column temperature of 90 °C. As Figure 5.27 shows, good separation was obtained within 9 hours [104–109].

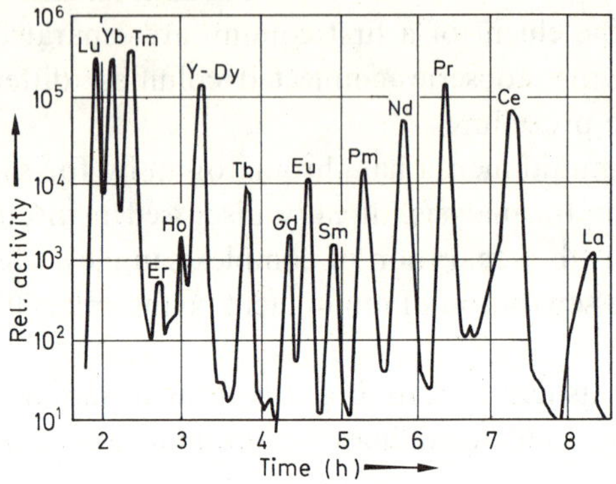

Figure 5.27 Earlier separation of rare earth metals. Column: sulphonated cation exchanger; mobile phase: gradient elution – 1 molar lactic acid from pH 3.19 to pH 7.0; column temperature: 90 °C.

The separation of transplutonium elements from each other and from iron was carried out on a cation-exchanger column with a complex-forming mobile phase [110]. The alkali metals, e. g. Li, Na, K, can also be separated by cation exchangers [111, 112]. Similar separations are described for Ca-Ti [113], Cd-Co-Ni [114], Cd-Ag [115], Fe-Co-Ni [116], Fe(II)-Fe(III) [117], Au(III)-In-Cd [118, 119], Ba from other metal ions [120], Ca and Sr from K and Rb [121], Mg from Al [122], and trace amounts of Ga from other metals [123].

Separation on anion exchangers.
For this purpose mainly anion exchangers with strong base functional groups (e. g., quaternary ammonium ions) and a styrene-divinylbenzene resin matrix (Dowex 1 and 2, Dow Chemicals Co.; Amberlite, IRA 400, 401, 410, 411, Rohm and Haas Co.; AG 1, BIO-RAD, Labs.) are used.

The HCl-formed metal ion complexes show great differences in their k_D values, depending on the HCl concentration (Kraus and Nelson [124–126]). Similar effects were also obtained by Strelow and Bothma [127–129] with sulphuric acid and by Faris and Buchanan [130] using hydrofluoric and nitric acid [131–133]. A typical example is the formation of uranium chloride complexes for the separation of uranium from many other ions. In this way uranium was determined in neutron-irradiated water [134–136]. Anionic plutonium complexes in strong nitric and hydrochloric acid solutions were separated on ion exchangers [137–139].

Special separation procedures for cations on anion exchangers have been reported: the separation of NH_4 from Na, Mg, Ca [140], of Cd from In, Ga, Zn [141], of Cu from Co by gradient elution [142], of Cr, Mo and U [143], of Fe from phosphates [144], Pd from Pt [145, 146] and of U, Am and Cm [147, 148].

In several cases the metal anion complex-formation was improved by using mixtures of organic compounds such as acetone, dioxane, methanol and ethanol with water-diluted acids [149–150]. The separation of cations transformed to complexes by organic compounds, e. g. EDTA [151, 152], citrate [153], tartrate [154], succinate [114], and malonate [155] was also successfully carried out.

Separation on mixed-bed exchangers.
The expression "mixed-bed" ion exchanger is used here in a more general sense. Firstly, the classical mixture of anion and cation exchanger material and, secondly, also the amphoteric ion exchangers fall under this heading.

Depending on the pH or the ion concentration in the mobile phase, the cations in the sample can form different complexes. Using gradient elution, various complex states can be achieved and, therefore, analytical separations can be carried out.

As an example of an amphoteric ion exchanger thorium tellurite may be taken, which can be useful for the separation of Cu from Hg and Ni, Pb from Cd and U from Hg [156]. In this case the separation effect is due to the fact that this exchanger works as a cation exchanger in an alkaline medium and as an anion exchanger in acidic solutions.

Separation of inorganic anions
For the separation of anions strong base anion exchangers are the preferred column materials. In most cases gradient elution with acids or buffer systems as mobile phase are used [157]. The distribution coefficient can be adjusted so that many anions are separable. In several cases cation exchangers are also suitable for anion separations [158–160]. Different species of phosphorous oxyanions have been chromatographed on a Dowex 50 column. This separation was possible due to the sorption behaviour of the phosphorous compound towards the counter-ion of the resin.

5.2.4.2 Ion chromatography – high pressure

The growing importance of environmental analyses has caused an enormous increase in the application of ion chromatography [161–163]. This technique is especially suitable for the determination of anions and cations at low concentrations. Several review articles have been published in recent years [164–171]. Many analytical separation procedures have been worked out for water and waste water [172–174], industrial processes [175] and plant materials used in food and beverage production [176–179]. On the one hand, as the stationary phase, ion exchange columns are directly applied; on the other, materials other than ion exchangers, such as reversed phase columns, are frequently used, whereby ion pairing is introduced through compounds in the solvent.

Separation of cations and anions

By employing newly developed exchanger materials, the separation time for rare-earth metals could be shortened from 9 h (Figure 5.27) to approx. 25 min, as shown in Figure 5.28 [180]. This analysis was carried out on bonded-phase strong acid ion exchangers (5 – 10 μm) and by high-performance liquid chromatographic technique. The metal ions in the gradient-eluted effluent were also determined after a post-column complexing reaction with a variable wavelength detector [181].

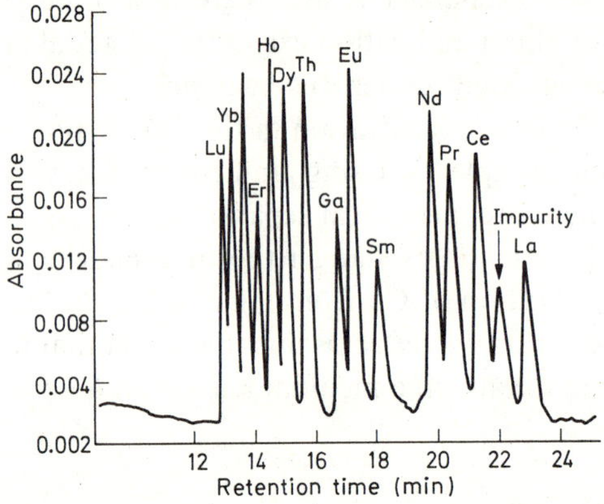

Figure 5.28 Separation of the lanthanides on Partisil-10 SCX [180]. Experimental conditions: 25 cm × 4 mm column; sample: 10 μL of a solution containing 10 $\mu g/ml^{-1}$ of each lanthanide; linear program from 0.018 mol · L^{-1} HIBA over a 20 min period at 0.8 ml · min^{-1} and pH 4.6; detection at 600 nm after post-column reaction with Arsenaza I.

Table 5.3 Selection of ion exchange hard polymer gels (182)

Support	Porous silica				Porous polymer gel	
Functional group	$-N(C_2H_5)_2$	$-COOH$	$N(C_2H_5)_2$	Polyamine	$-SO_3-Na^+$	$-N(C_2H_5)_2$
d_p (μm)	5	10	10	10	10	10
Pore size (Å)	130	240	240	300	1000	1000
pH range	2 – 7		2 – 7		2 – 12	2 – 12
Trade name	TSK gel DEAE 2SW	TSK CM3SW	DEAE 3SW	Shodex Axpak U424	TSK gel SP5PW	DEAE 5PW
Supplier	Toya Soda	Toya Soda	Toya Soda	Showa Denko	Toya Soda	Toya Soda

Different types of column materials are employed (Table 5.3). Porous polyacrylate polymer gels in the form of anion and cation exchangers, as well as porous silica gel containing bonded quaternary amino ethyl groups, are commercially available [182].

Separation of cations.

Cation exchangers were used for a fast separation of alkali earth metal and ammonia (Figure 5.29). Several procedures for higher valent metal ions have also been elaborated for fast and efficient separations [183–188].

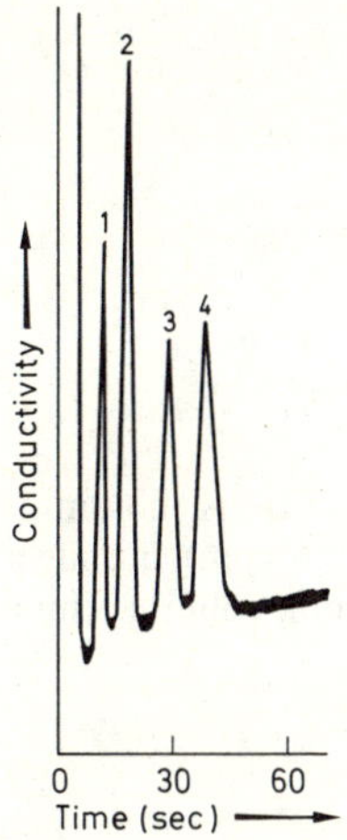

Figure 5.29 High-speed separation of cation. Column: Wescan cation/HS; mobile phase: HNO_3, pH 2.0; flow rate: 8 ml/min; detection: conductivity detector, Wescan Model 213A; 1...Li^+ 0.5 ppm, 2...Na^+ 3 ppm, 3...NH_4^+ 2 ppm, 4...K^+ 6 ppm.

Separation of anions.

Through the possibility of altering the ionic strength and the pH in the mobile phase, the interaction of the ions with the stationary phase can be changed in a relatively wide range. At the same time, the retention time can be altered in the desired direction. For the determination of the ion concentration conductivity, UV and RI detection systems are in use [189–195].

In certain cases buffer systems of such high concentration are needed as the mobile phase that only RI detection yields a high enough response [196].

Figure 5.30 shows the separation of the anions F^-, CO_3^{2-}, Cl^-, NO_2^-, Br^-, NO_3^-, PO_4^{3-} and SO_4^{2-}, with concentrations of 1 to 6 ppm [197].

Simultaneous separation of cations and anions.

In 1984 Yamamoto and coworkers [198] introduced the simultaneous determination of anions and cations. In this work the authors used a complexing agent, ethylenedinitrolotetraacetic acid, to form complexes of the divalent metals. These metals

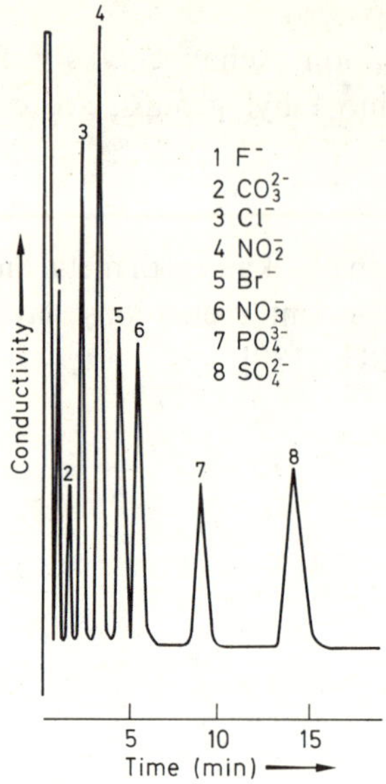

Figure 5.30 Separation of anions [197]. Experimental conditions, column: TSK-Gel IC Anion PW (4.6 × 50 mm i. D.); sample: standard solution (1 – 6 ppm); mobile phase: 1.3 mM gluconic acid, 1,3 mM boric acid, pH 8.5; flow rate: 1.2 ml/min; column temperature: 30 °C; detection: conductivity, Bio-Rad, CM-8.

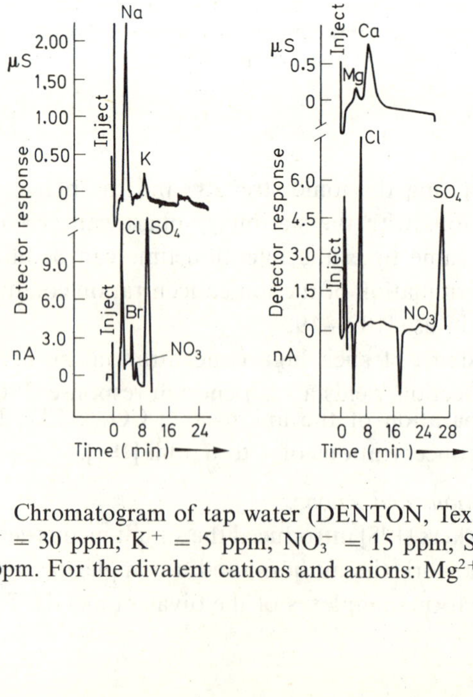

Figure 5.31 Chromatogram of tap water (DENTON, Tex.) [200]. Monovalent cations and anions: Na^+ = 30 ppm; K^+ = 5 ppm; NO_3^- = 15 ppm; SO_4^{2-} = 44 ppm; Cl^- = 23 ppm; Br^- = 0.7 ppm. For the divalent cations and anions: Mg^{2+} = 4 ppm; Ca^{2+} = 38 ppm.

were detected as anions together with the uncomplexed inorganic anions. In such devices usually a cation separator column, an anion separator column and an anion suppressor column are used. Conductivity or electrochemical detectors are operated in series [199]. An example of such separation is given in Figure 5.31, where the ions of tap water are analyzed [200].

In many cases, when the ion concentrations are very low, concentration steps as described in Section 5.1 are necessary.

Paired-ion chromatography

Hydrophobic pairs.
In practical application it is fairly easy to introduce the ion-paired substrate into reversed phase column material. However, in the case of silica columns, it is often difficult to load the column with the necessary counter-ions when water is used as the mobile phase.

Many cations and anions have been separated with columns containing hydrophobic functional groups. The ion-pairing substances were coated dynamically; the chromatographic results have been described by several authors [201–216]. For the separation of cations, C_{12} and C_{20} sulphates or C_6 and C_8 sulphonates were applied as dynamically-coating substrates, whereas for anions, C_8, C_{16}, C_{25}, C_{32} and C_{39} quaternary ammonium salts were used [217]. Molnar [46] separated cations with a RP C-18 stationary phase using n-heptylsulphonate solution as eluent. By applying tetrabutylammonium hydroxide with a phosphate buffer as the mobile phase, he was able to obtain well-resolved peaks in separating anions (Figures 5.32 and 5.33).

La-139 has been used as a fission monitor in "burn up" experiments of $(ThU)O_2$ fuels. The dissolved La-isotope was separated on a RP column which was dynamically modified with 1-octanesulphonate [218–220].

A microprocessor-controlled system was developed for the radiochemical separation of rare earths from fission product mixtures. In the effluent of a first column, the rare-earth elements were obtained using dihexaldiethylcarbamylmethylphosphate as the mobile phase and Vydac C-8 RP resin as the column material. In a

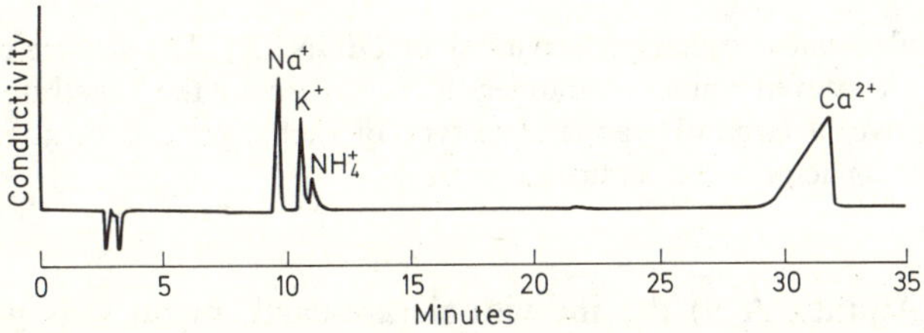

Figure 5.32 Cation separation on reversed phase with ion interactions. Column: 10 μm LiChrosorb RP-18 (3 × 250 × 4.6 mm) (103.07.23.010, Knauer). Mobile phase: 0.005 M n-heptylsulphonate solution, pH 2 (K_2 solution, 102.28, Knauer); flow rate: 2.0 mL/min; pressure: 220 bar; temperature: 25 °C [46].

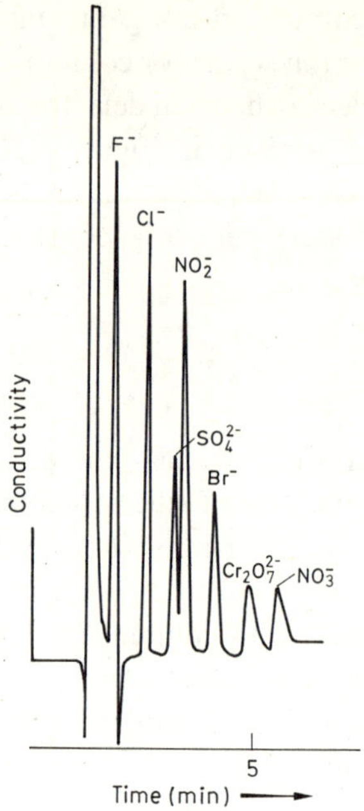

Figure 5.33 Separation of some anions with reversed-phase chromatography coupled with conductivity detection [46]. Column: 10 μm LiChrosorb RP-18 (103.07.23.010, Knauer). Mobile phase: 0.002 mol/L tetratbutylammonium hydroxide and 0.05 mol/L phosphate buffer, pH 6.7; flow-rate: 2 mL/min; pressure: 80 bar; temperature: 25 °C; conductivity detector (74.00, Knauer). Injected samples were dissolved in 0.05 mol/L phosphate buffer, pH 6.7, to give a concentration of 1 mg/mL; injection volume: 20 μL.

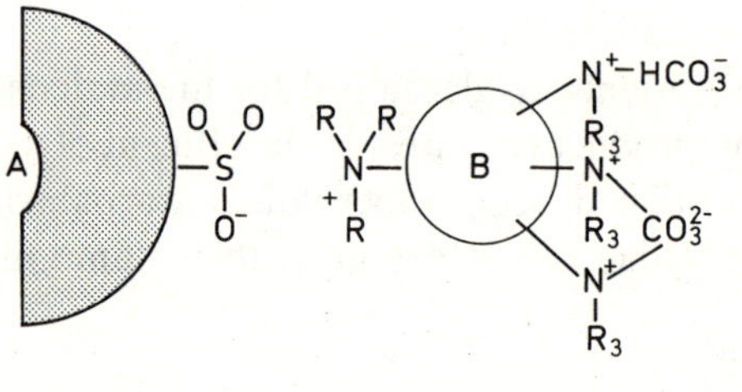

Figure 5.34 Pellicular anion exchanger (courtesy of DIONEX). The surface of a substrate, A, is sulphonated. A latexed anion exchanger, B, is sorbed to the substrate surface. The exchanger is stable over a large pH-range. The type of exchanger can be altered in several directions (e. g., low capacity, weak, cationic species).

second column (Aminex A 9) the individual rare-earth metal ions were isolated using α-hydroxyisobutyric acid as the eluent [221 – 223]. A further interesting application is the separation of NO_3^-/NO_2^- with a polystyrene-divinylbenzene stationary phase and alkylammonium salts dissolved in acetonitrile-water as eluent [224].

Ionic pairs.
High pH and pressure resistant material was introduced by attaching ionic functional groups to the surface of inert small size spheres. Latexed ion exchangers with counter-ions were then adsorbed to the spheres. In Figure 5.34 an anion exchange particle bound to the sulphonated surface is shown. Fast separations of cations and anions are obtained using such column materials [225]. The application of suppressor columns increased the sensitivity considerably.

5.2.4.3 Separation of organic compounds – low pressure chromatography

Ion exchangers as stationary phases are also very useful for the separation of organic ions and even for organic nonionic compounds.

Acids
Low molecular compounds, e. g. oxalic, tartaric, malic acid etc., are separated in two steps on anion exchangers [226, 227]; for fatty acids, ion exclusion columns are used [228, 229]. The determination of aromatic acids, for instance in the Krebs cycle, is frequently carried out by cation exchangers [230].

Amino acids
Earlier works of Moore and Stein [231] showed the excellent suitability of ion exchangers for amino acid analysis, making history in the field of analytical chemistry. The authors applied Dowex 50, 4% crosslinked, with column dimensions of 165 x 0.9 cm. Through the variation of temperature and buffer mixtures, it was possible to separate the most important amino acids in one analytical process. The detection of the amino acids was carried out in each eluent fraction photometrically after a ninhydrin reaction [232].

This method was soon automated, especially in the field of medical analysis [233]. In 1963 Hamilton described the quantitative analysis of 10^{-8} mol of amino acids with a Dowex 50 resin [234–236].

Automatic amino acid analyzers are frequently used for medical applications [237–247], especially for physiological fluids [248]. In addition to the ninhydrin reaction, the fluorimetric detection [249] is of increasing importance due to its higher sensitivity.

Proteins, peptides and nucleotides
These compounds are successfully separated on ion exchange materials. It was also possible to analyze the components of nucleic acid, such as nucleotides, nucleosides and organic bases, by the application of cation and anion exchange resins [250–252]. The cyclic nucleotides, cAMP and cGMP were separated and isolated using an anion exchanger [253–256].

Macroporous ion exchange resins are also satisfactory column materials for the separation of proteins and peptides. In several cases the exchangers showed an exclusion limit in excess of about 75 000 daltons [257–263].

Special organic nitrogen compounds

Compounds of several biochemical liquids such as polyamines [264], histamine [265], hydroxytrypramine [266], catecholamine [267–271], 5-fluoro-uracil [272], xanthines (e. g., caffeine and theobromine) [273–275] were chromatographed. The importance of these analyses for medical applications is obvious.

Phenolics

These weak acidic compounds are separated with good resolution on anion columns [276].

Carbohydrates

For the determination of carbohydrates the borate complex formation has been widely used in analytical work. Anion exchangers were applied for their separation [277]. In certain cases uronic acids and sugar degradation products were determined with the carbohydrates [278, 279].

5.2.4.4 Separation of organic compounds – high pressure chromatography

The introduction of pressure-stable stationary phase materials brought about a much increased selectivity and reduction of analysis time. Therefore, since 1978, an ever-increasing number of publications concerned with the separation of organic compounds appears every year. Most ion exchangers used for this purpose have low particle diameters (3 to 25 μm) and are in a narrow-range sieve fraction. The ion exchangers commonly used are sulphonated polystyrene crosslinked with 4–8% divinylbenzene and loaded with various counter-ions: hydroxylated polyether-based support, with covalent bound ion exchange functional groups; and silica-based weak cation and anion exchangers (Table 5.3).

Organic acids

A strong acidic ion exchange material is used for the determination of acids contained in many foodstuffs, beverages and other biological liquids [280, 281]. Figure 5.35 demonstrates such a separation on an H-loaded cation exchanger with diluted sulphuric acid as the mobile phase. The chromatographic separation mechanism is explained in part by ion-exclusion, due to the Donnan equilibrium [282, 283], as already mentioned.

Amino acids

Silica- and polyether-based exchanger materials constitute the stationary phase for amino acid separation. In recent years small particle ion exchangers have been introduced. This new material allowed the analysis time for the amino acid separations to be speeded up considerably. Twenty-one amino acids in a protein hydrolyzate were separated in less than 30 min [284–286].

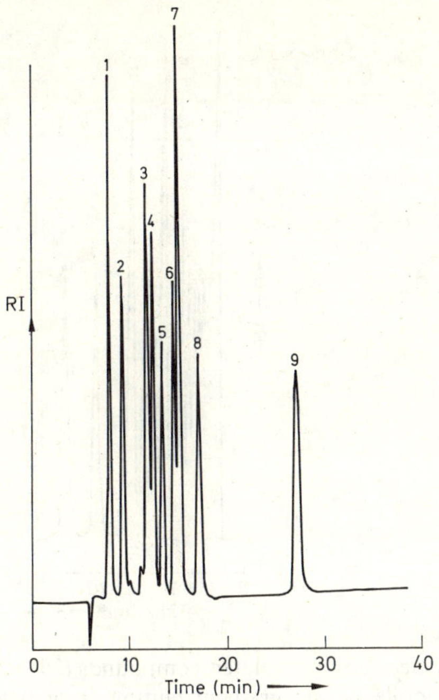

Figure 5.35 Optimized separation of acids (41). Column: HPX-87-H (300 × 7.8 mm I. D.); column temperature: 60 °C; mobile phase: 0.01 N sulphuric acid; flow-rate: 0.7 mL/min; refractive index detection.

1...2-oxoglutaric acid, 2...glyoxylic acid, 3...glycolic acid; 4...lactic acid; 5...formic acid; 6...acetic acid; 7...levulinic acid; 8...propanoic acid; 9...2-furancarbocylic acid.

Proteins, peptides and nucleotides

For these macromolecular compounds, silica- and polyether-based exchangers, as described in Table 5.3, were successfully applied for analytical purposes [287–295]. DNA restriction fragment analysis was carried out on a weak anion exchanger. This method is a very good alternative to gel filtration. Regnier et al. [296] published elegant separations of many polypeptides using silica-based anion exchangers. Perrett [297] describes a separation of 22 nucleotides in physiological extracts. He obtained a complete chromatogram within 13 min using gradient elution on an anion exchange column, as shown in Figure 5.36.

Carbohydrates

For the separation of carbohydrates sulphonated polystyrene crosslinked with divinylbenzene is usually applied (see Table 5.4) [298]. Exchangers with different counter-ions (H, Na, Ag, Ca, Pb) are commercially available and show very special separation characteristics due to their ion-moderated partition (IMP) effects. Oligomer sugars are selectively separated by a partially silver-loaded cation exchanger

Fig. 5.36

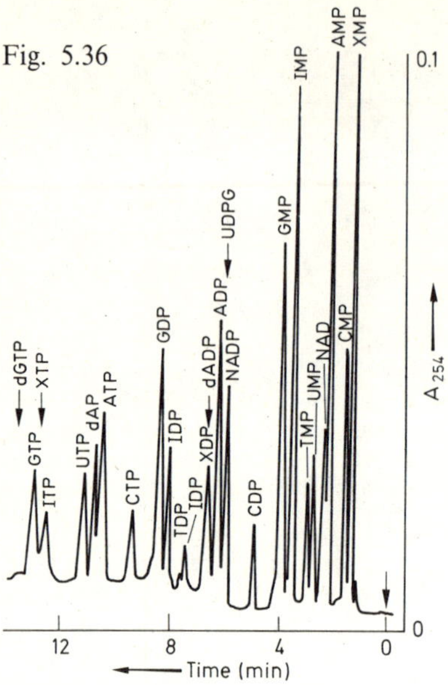

Fig. 5.37

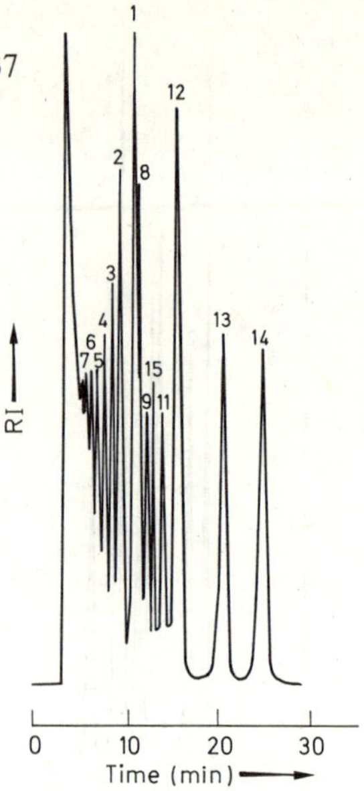

Figure 5.36 Gradient elution of 22 nucleotides and related compounds (297). Experimental conditions. Column: APS-Hypersil; mobile phase: gradient elution from 0.04M KH_2PO_4 (pH 2.9) to 0.5M KH_2PO_4 plus 0.8 M KCl (pH 2.9) in 13 minutes; flow rate: 1 mL/min; detection: UV monitor.

Figure 5.37 Optimized separation of a standard mixture (97). Chromatographic conditions: coupled column system Pb-loaded ion exchange stationary phase and Ag-loaded ion exchange stationary phase; temperature: 95 °C; mobile phase: water; flow-rate: 1.2 mL/min; refractive index detection.

1...glucose, 2–7...DP (= degree of polymerization, gluco-oligomers), 8...xylose, 9...fructose, 11...dihydroxyacetone, 12...1,6-anhydro-D-glucose, 13...hydroxymethylfurfural, 14...furfural, 15...ethanol.

Table 5.4 Selection of sulfonated polystyrene ion exchange column materials

Support	sulfonated styrene-divinylbenzene co-polymer						
Resin ionic form	Ca^{2+}	Ca^{2+}	Pb^{2+}	Ag^+	Ag^+	H^+	Na^+
Cross-linking	8%	4%	8%	4%	6%	8%	
Particle size (µm)	9	25	9	25	11	9	5–17
pH range	5–9	5–9	5–9	6–8	6–8	1–3	
Column dimension e. g. 300 × 7.8 mm							

column, whereas the low molecular sugars are better separated on calcium- and lead-loaded exchangers [40, 299–302].

In the newly revived field of biotechnology and plant biomass hydrolysis, the fast determination of sugars, sugar degradation products and alcohols is of essential importance. Cation exchange columns in the IMP technique proved to be very useful, especially when water serves as the mobile phase. Figure 5.37 shows a chromatogram, where the oligomer and monomer sugars, sugar degradation products and certain fermentation products (e. g., ethanol) are separated in less than half an hour. In this case two columns were coupled and the advantages of the silver- and lead-loaded exchangers were fully exploited [303, 304].

Miscellaneous organic compounds
In addition to the above-mentioned wide field of amino acid, polypeptide and carbohydrate separation, many other organic compounds are analyzed with ion exchanger columns under high pressure conditions. Excellent results have been obtained, for example in separating alcohols, ketones, aldehydes, acids, anhydrosugars, etc. Figure 5.38 shows a chromatogram in which 22 such compounds were determined within one hour and detected by refractive index.

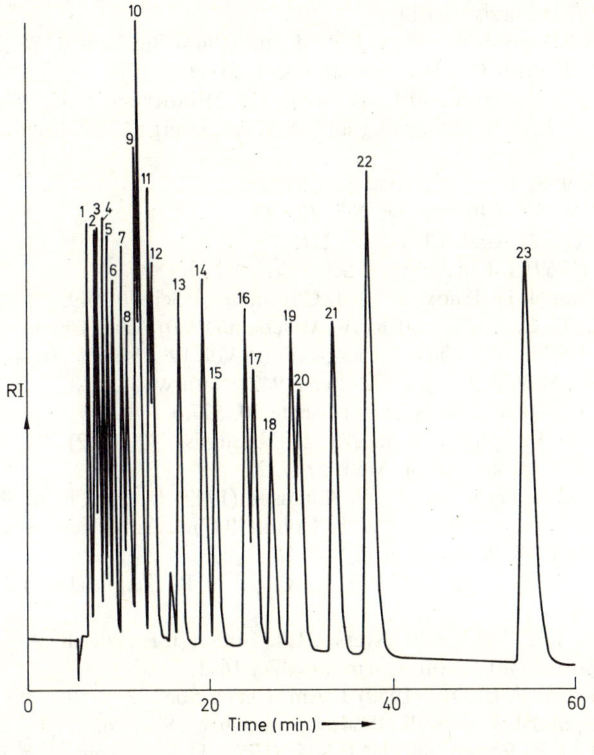

Figure 5.38 Optimized separation of a sample mixture of alcohols, aldehydes, ketones, acids and carbohydrates [41]. Column: HPX-87-H (300 × 7.8 mm I. D.); column temperature: 70 °C; mobile phase: 0.01 N sulphuric acid; flow-rate: 0.7 mL/min; refractive index detection.

1...cellobiose, 2...oxoglutaric acid, 3...D-galacturonic acid, 3...D-glucose, 5...D-galactose, 6....dulcitol, 7...glyceraldehyde, 8...glycolic acid, 9...dihydroxyacetone, 10...formic acid, 11...acetic acid, 12...levulinic acid, 13...acetaldehyde, 14...acetone, 15...propanal, 16...tert.-butanol, 17...1-propanol, 18...butanal, 19...2butanol, 20...isobutanol, 21...1-butanol, 22...furfural, 23...5-methylfurfural.

With a weak acidic ion exchanger catecholamines were separated from human urine. Using a fluorimetric detector system, sensitivity limits of 0.2 ng for epinephrine and norepinephrine, and 0.6 ng for dopamine were obtained [305].

References

[1] Tswett, M. (1906) Trans. Soc. Naturalistes Varsovie, Biol. Sec. 14, No. 6 (1903), Ber. dtsch. bot. Ges. *24,* 316.
[2] Kuhn, R., A. Winterstein and E. Lederer, Hoppe-Seylers Z. Physiol. Chem. *197* (1931) 141, Naturwiss. *19* (1931) 306, Ber. *64* (1931) 1349.
[3] Ismailov, N. A. and M. S. Šrajber (1938) Pharmazie (russ.) *3,* 1.
[4] Kirchner, J. G., J. M. Miller and G. J. Keller (1951) Anal. Chem. *23,* 420.
[5] Martin, A. J. P., and R. L. M. Synge (1941) Biochem. J. *35,* 1358.
[6] Cremer, E. (1979) J. Chromatogr. Lib. Vol. *17,* p. 21, Elsevier, New York.
[7] Runge, F. F. (1822) Thesis, Berlin.
[8] Consden, R., A. H. Gordon, and A. J. P. Martin (1944) Biochem. J. *38,* 224.
[9] Cremer, E., and F. Prior (1951) Z. Elektrochem. *55,* 66.
[10] James, A. T., A. J. P. Martin and G. H. Smith (1952) Biochem. J. *52,* 238.
[11] Van Deempter, J. J., F. J. Zuiderweg and A. Klinkenberg (1956) Chem. Eng. Sci. *5,* 271.
[12] Glueckauf, E. (1955) Trans. Faraday Soc. *51,* 34.
[13] Snyder, L. R. (1972) J. Chromatogr. Sci. *10,* 369.
[14] Giddings, J. C. (1963) Anal. Chem. *35,* 2215.
[15] Kirkland, J. J. (1969) J. Chromatogr. Sci. *7,* 7.
[16] Kennedy, G. J. and J. H. Knox (1972) J. Chromatogr. Sci. *10,* 549.
[17] Hamilton, P. B., D. C. Bogue and R. A. Anderson (1960) Anal. Chem. *32,* 1782.
[18] Huber, J. F. K. (1979) in: J. Chromatogr., Library Vol. 17, Elsevier, Amsterdam, p. 159.
[19] Freudenberg, K., H. Walch and H. Molter (1942) Naturwiss. *30,* 87.
[20] Griessbach, R. (1939) Melliands Textilberichte 20, 577.
[21] Wieland, Th. (1942) Z. phys. Chem. *273,* 24; Naturwiss. *30* (1942) 374.
[22] Block, R. J. (1942) J. Exper. Biol. Med. *51,* 252.
[23] Boyd, G. E., L. S. Myers jr. and A. W. Adamson (1947) J. Am. Chem. Soc. *69,* 2849.
[24] Tompkins, E. R., J. X. Khym, and W. E. Cohn (1947) J. Am. Chem. Soc. *69,* 2769.
[25] Tompkins, E. R. (1948) J. Am. Chem. Soc. *70,* 3520.
[26] Spedding, F. H., A. F. Voigt, E. M. Gladrow, and N. R. Sleight (1947) J. Am. Chem. Soc. *69,* 2777.
[27] Spedding, F. H., E. I. Fulmer, B. Ayers, T. A. Butler, J. E. Powell, A. D. Tevebauch and R. Thompson (1948) J. Am. Chem. Soc. *70,* 1671.
[28] Spedding, F. H., and J. L. Dye (1950) J. Am. Chem. Soc. *72,* 5350.
[29] Spedding, F. H., and J. E. Powell (1954) J. Am. Chem. Soc. *76,* 2550.
[30] Spedding, F. H., J. E. Powell, and H. J. Svec (1955) J. Am. Chem. Soc. *77,* 1393.
[31] Samuelson, O., and A. Westlin (1947) Svensk Kem. Tid. *59,* 244.
[32] Gabrielson, G., and O. Samuelson (1950) Svensk Kem. Tid. *62,* 214.
[33] Bobleter, O., G. Dincler and C. Sabau (1970) Ber. Bunsen-Ges. Physik. Chem. *74,* 1050.
[34] Steinwandter, J., O. Bobleter and F. Hecht (1971) Monatsh. Chem. *102,* 1351.
[35] Nelson, F., T. Murase and K. A. Kraus (1964) J. Chromatogr. *13,* 503.
[36] Kim, J. I. (1974) J. Inorg. Nucl. Chem. *36,* 1.
[37] Peters, J. M., and G. Del Fiore (1975) J. Chromatogr. *108,* 415.

[38] Khym, J. X. and L. P. Zill (1952) J. Am. Chem. Soc. *74,* 2090.
[39] Ladisch, M. R. and G. T. Tsao (1978) J. Chromatogr. *166,* 85.
[40] Scobell, H. D. and K. M. Brobst (1981) J. Chromatogr. *212,* 51.
[41] Pecina, R., G. Bonn, E. Burtscher and O. Bobleter (1984) J. Chromatogr. *287,* 245.
[42] Goldstein, G. (1967) Anal. Biochem. *20,* 477.
[43] Petronio, B. M., A. Lagana, M. V. Russo (1981) Talanta *28,* 215.
[44] Schmuckler, G. (1984) J. Chromatogr. *313,* 47.
[45] Iskandarani, Z. and D. J. Pyetrzyk (1982) Anal. Chem. *54,* 1065.
[46] Molnar, I., H. Knauer, and D. Wilk (1980) J. Chromatogr. *201,* 225.
[47] Bidlingmayer, B. A., J. K. Del Rios and J. Korpi (1982) Anal. Chem. *54,* 442.
[48] Scott, R. P. W. and P. Kucera (1979) J. Chromatogr. *175,* 51.
[49] Samuelson, O. and E. Sjoström (1952) Sv. Kem. Tidskr. *64,* 305.
[50] Havlicek, J. and O. Samuelson (1974) Chromatographia, *7,* 361.
[51] Jupille, Th., M. Gray, B. Black, M. Gould (1981) Int. Lab. 84.
[52] Cummings, L. J. (1978) paper No. 288, 30th Pittsburgh Conf. Analyt. Chem. Appl. Spectr..
[53] Goulding, R. W. (1975) J. Chromatogr. *103,* 229.
[54] Mayer, S. W. and E. R. Thomkins (1947) J. Am. Chem. Soc. *69,* 2866.
[55] Rosen, J. B. (1954) Ind. Eng. Chem. *46,* 1590.
[56] Glueckauf, E. (1955) in: Ion Exchange and its Applications, Soc. Chem. Ind. (London), p. 34.
[57] Helfferich, F. (1959) Ionenaustauscher, pp. 403, Verlag Chemie, Weinheim.
[58] Purnell, J. H. (1960) J. Chem. Soc. 1268.
[59] Huber, J. F. K. and J. A. R. J. Hulsman (1967) Anal. Chim. Acta *38,* 305.
[60] Giddings, J. C. (1965) Dynamics of Chromatography, Dekker, New York, p. 52.
[61] Huber, J. F. K. (1973) Ber. Bunsenges. Physik. Chem. *77,* 179.
[62] Huber, J. F. K. (1975) Z. Analyt. Chem. *277,* 341.
[63] Horvath C. and H.-J. Lin (1976) J. Chromatogr. *126,* 401.
[64] Horvath C. and H.-J. Lin (1978) J. Chromatogr. *149,* 43.
[65] Arnold, F. H., H. W. Blanch and C. R. Wilke (1985) J. Chromatogr. *330,* 159.
[66] Giddings, J. C. (1964) J. Chromatogr. *13,* 301.
[67] Kennedy, G. J. and J. H. Knox (1972) J. Chromatogr. Sci. *10,* 549.
[68] Kim, D. W., O. Bobleter and P. Brunner (1980) J. Korean Chem. Soc. *24,* 426.
[69] Tiselius, A. (1943) Arkiv Kemi Mineral. Geol. *16A(18),* 1.
[70] Spedding, F. H., J. E. Powell, E. I. Fulmer and T. A. Butler (1950) J. Am. Chem. Soc. *72,* 2354.
[71] Shepard, C. C. and A. Tiselius (1949) Disc. Faraday Soc. *7,* 275.
[72] Li, C. H., A. Tiselius, K. O. Pederson, L. Hagdahl and H. Carstensen (1951) J. Biol. Chem. *190,* 317.
[73] Horvath, C. (1985) in: J. Chromatogr. Lib., Vol. 32 (F. Bruner, ed.) p. 184, Elsevier, Amsterdam.
[74] Partridge, S. M. (1950) Chem. Ind. 383, Biochem. J. *49,* (1951) 153.
[75] Reichenberg, D. and D. J. McCanley, unpublished; D. Reichenberg (1957) in: Ion-Exchangers in Organic and Biochemistry, (C. Calmon and T. R. E. Kressman, eds.) New York.
[76] Lederer, M. and F. Rallo (1962) J. Chromatogr. *7,* 552.
[77] Lederer, M., V. Moscatelli and C. Padiglone (1963) J. Chromatogr. *10,* 82.
[78] Majors, R. E. (1970) Symp. Adv. Chromatogr. Miami Beach, Fla., June 2–5.
[79] Henry, R. A. (1971) in: Modern Practice of Liquid Chromatography, (J. J. Kirkland, ed.), p. 75, Wiley, New York.
[80] Scott, C. D. (1969) J. Chromatogr. *42,* 263.

[81] Scott, R. P. W. (1970) Course on Modern Practices of Liquid Chromatography, Wilmington, Del., April 6.
[82] Schram, E. (1970) in: Current Status of Liquid Scintillation (E. D. Bransome, Jr., ed.) Grune and Stratton, New York.
[83] Hunt, J. A. (1968) Anal. Biochem. *23,* 289.
[84] Koen, J. G., J. F. K. Huber, H. Poppe and G. DenBoef (1970) J. Chromatogr. Sci. *8,* 192.
[85] Joynes, P. L. and R. S. Maggs (1970) J. Chromatogr. Sci. 8, 427.
[86] Lawrence, J. F. and R. W. Frei (1976) Clinical Derivatization in Liquid Chromatography, Elsevier, Amsterdam.
[87] Beau, C. and G. S. King (eds.) (1977) Handbook of Derivatives for Chromatography, Heyden, London – Rheine.
[88] Schwedt, G. (1980) Chemische Reaktionsdetektoren für die schnelle Flüssigkeitschromatographie, Hüthig, Heidelberg.
[89] Biemann, K. (1985) in: J. Chromatogr. Lib., Vol. 32 (F. Bruner, ed.) p. 43, Elsevier, Amsterdam.
[90] McLafferty, F. W. and M. A. Balderin (1973) Org. Mass Spectrom. *8,* 1111.
[91] Scott, R. P. W., C. G. Scott, M. Munroe and J. Hess (1974) J. Chromatogr. *99,* 395.
[92] Barber, M., R. S. Bordoli, R. D. Sedgwick and A. N. Tyler (1981) J. Chem. Soc. Chem. Commun. *7,* 325.
[93] Blakely, C. R. and M. L. Vestal (1983) Anal. Chem. *55,* 750.
[94] Scott, R. P. W. and J. Lawrence (1970) J. Chromatogr. Sci *8,* 65.
[95] Snyder, L. R. (1965) in: Chromatographic Reviews, Vol. 7., (M. Lederer, ed.) Elsevier, New York, p. 419.
[96] Kwok, L., L. R. Snyder, J. C. Sternberg, L. R. (1968) Anal. Chem. *1,* 11.
[97] Bonn, G. (1985) J. Chromatogr. *322,* 411.
[98] Weinstein, S., M. H. Engel and P. E. Hare (1982) Anal. Biochem. *121,* 370.
[99] Beschke, K., R. Jauch, W. Roth, A. Zimmer and F. W. Koss (1982) GIT Labor-Med. *5,* 356.
[100] Riggenmann, H. J. and U. Juergens (1983) LaborPraxis *7,* 190.
[101] Cohn, W. E. and C. E. Carter (1950) J. Am. Chem. Soc. *72,* 4273.
[102] Marinsky, J. A., L. E. Glendenin and C. D. Coryell (1947) J. Am. Chem. Soc. *70,* 2781.
[103] K. Dorfner (1970) Ionenaustauscher, p. 243, de Gruyter, Berlin.
[104] Ishii, D., A. Hirose, Y. Iwasaki (1978) J. Radioanal. Chem. *46,* 41.
[105] Schaedel, M., N. Trautmann and G. Hermann (1977) Radiochim. Acta *24,* 27.
[106] Schoebrechts, F., E. Merciny and S. G. Duyckaerts (1979) J. Chromatogr. *179,* 63.
[107] Savoyant, L., F. Persin and C. Dupuy (1984) Geostand. Newslett. *8,* 159.
[108] Yang, J. T. (1950) Anal. Chim. Acta *4,* 59.
[109] Yang, J. T. (1950) J. Chim. Phys. *47,* 805.
[110] Elesin, A. A. and V. A. Karaseva (1977) Radiokhimiya *19,* 678.
[111] Delphin, W. H. and E. P. Horwitz (1978) Anal. Chem. *50,* 843.
[112] Yeager, H. L. and A. Steck (1979) Anal. Chem. *51,* 862.
[113] Zhukova, M. P. and V. T. Solomatin (1978) Zavod. Lab. *44,* 150.
[114] Dadone, A., F. Baffi, R. Frache and A. Mazzucotelli (1979) Chromatographia *12,* 38.
[115] Strelow, F. W. E. (1978) Anal. Chim. Acta *100,* 577.
[116] Dadone, A., F. Baffi and R. Frache (1976) Talanta *23,* 593.
[117] Varshney, K. G., S. Argawal and Y. K. Varshney (1984) J. Liq. Chromatogr. *7,* 201.
[118] Strelow, F. W. E. (1984) Anal. Chim. Acta *16,* 31.
[119] Strelow, F. W. E. (1984) Anal. Chem. *56,* 27.
[120] Srivastava, U., L. C. T. Eusebius and A. K. Ghose (1984) J. Indian Chem. Soc. *61,* 243.

[121] Tsuji, M. and M. Abe (1984) Radioisotope *33,* 218.
[122] Strelow, F. W. E. and van der Walt, T. N. (1984) S. Afr. J. Chem. *37,* 149.
[123] van der Walt, T. N. and F. W. E. Strelow (1983) Anal. Chem. *55,* 212.
[124] Kraus, K. A. and G. E. Moore (1949) J. Am. Chem. Soc. *71,* 3263.
[125] Kraus, K. A. and F. Nelson (1958) ASTM Special Publication *195,* 27.
[126] Kraus, K. A. and F. Nelson (1955) Proc. Intern. Conf. Peaceful Uses of Atomic Energy, Geneva *7,* 113.
[127] Strelow, F. W. E. and C. J. C. Bothma (1967) Anal. Chem. *39,* 595.
[128] Strelow, F. W. E., R. Rethemeyer and C. J. C. Bothma (1965) Anal. Chem. *37,* 106.
[129] Khopkar, S. M. and A. K. De (1960) Anal. Chim. Acta *23,* 147.
[130] Faris, J. P. and R. F. Buchanan (1964) Anal. Chem. *36,* 1158.
[131] Faris, J. P. (1960) Anal. Chem. *32,* 52.
[132] Kressin, I. K. and G. R. Waterbury (1962) Anal. Chem. *34,* 1598.
[133] James, D. B. (1963) J. Inorg. Nucl. Chem. *25,* 711.
[134] Cha Ki Won and Kim Jong Hun (1984) Taehan Hwahakhoe Chi *28,* 309.
[135] Hirose, A. and D. Ishii (1978) J. Radioanal. Chem. *46,* 211.
[136] Gladney, E. S., J. W. Owens and J. W. Starner (1976) Anal. Chem. *48,* 973.
[137] Talvitie, N. A. (1971) Anal. Chem. *43,* 1827.
[138] Lederer, M. (1983) J. Chromatogr. Libr. Vol. *22B*, p. 459.
[139] Coleman, G. H. (1965) Radiochemistry of Plutonium, National Academy of Sciences, Nuclear Science Series NAS-NS 3058.
[140] Tanaka, K., T. Ishizuka and H. Sunalara (1979) J. Chromatogr. *174,* 153.
[141] Strelow, F. W. E. (1978) Anal. Chim. Acta *97,* 87.
[142] Hicks, H. G., P. C. Stevenson and J. S. Schweiger (1978) J. Chromatogr. Sci. *16,* 527.
[143] Chakravorty, M. and S. M. Khopkar (1979) Chromatographia *12,* 459.
[144] Mitchell, J. W. and V. Gibbs (1977) Talanta *24,* 741.
[145] Petrie, R. K. and J. W. Morgan (1982) J. Radioanal. Chem. *74,* 15.
[146] Brown, R. J. and W. R. Biggs (1984) Anal. Chem. *56,* 646.
[147] Holm, E., S. Ballestra and R. Fukai (1979) Talanta *26,* 791.
[148] Sweify, F. H., R. Shabana, N. Abdel-Rahman and H. F. Aly (1985) J. Radioanal. Nucl. Chem. Art. *91,* 91.
[149] Fritz, J. S. and D. J. Pietrzyk (1961) Talanta *8,* 143.
[150] Rieman, W. and H. F. Walton (1970) Ion Exchange In Analytical Chemistry, Pergamon, New York.
[151] Dybezynski, R. (1964) J. Chromatogr. *14,* 79.
[152] Strelow, F. W. E. (1978) Anal. Chem. *50,* 1359.
[153] Korkisch, J. and H. Krivanec (1976) Anal. Chim. Acta *83,* 111.
[154] Morie, G. P. and T. R. Sweet (1964) J. Chromatogr. *16,* 201.
[155] Chakrovorty, M. and S. M. Khopkar (1979) Chromatographia *12,* 459.
[156] Rawat, J. P., M. Iqbal and H. M. A. Aziz (1984) J. Liq. Chromatogr. *7,* 1691.
[157] Grande, J. A. and J. Beukenkamp (1958) Anal. Chem. *28,* 1497.
[158] Tanaka, K. and H. Sunahara (1978) Bunseki Kagaku *27,* 95.
[159] Small, M., T. S. Stevens and W. C. Baumann (1975) Anal. Chem. *47,* 1801.
[160] Jakob, F., K. C. Park, J. Ciric and W. Riemann (1961) Talanta *8,* 431.
[161] Small, M. (1983) Anal. Chem. *55,* 235A.
[162] Gjerde, D. T., G. Schmuckler and J. S. Fritz (1980) J. Chromatogr. *187,* 35.
[163] Gjerde, D. T. and J. S. Fritz (1980) J. Chromatogr. *188,* 391.
[164] Walton, H. F. (1980) Anal. Chem. *52,* 15R. in: J. Chromatogr. Libr., Vol. 22A, Elsevier, Amsterdam 1983. p. 225.
[165] Schmuckler, G. (1984) J. Chromatogr. *313,* 47.
[166] Majors, R. E., H. G. Barth and C. H. Lochmueller (1984) Anal. Chem. *56,* 300R.

[167] Smith, F. C. Jr. and R. C. Chang (1983) Practice of Ion Chromatography, p. 218, Wiley, New York.
[168] Fritz, J. S., D. T. Gjerde and C. Pohlandt (1982) Ion Chromatography, Alfred Huethig, Heidelberg.
[169] Fritz, J. S. (1984) Ion Chromatography, LC Magazine *2,* 446.
[170] Weiss, J. (1983) CLB. Chem. Lab. Betr. *34,* 293, *34,* 342.
[171] Nickless, G. (1985) J. Chromatogr. *313,* 129.
[172] Girard, J. E. and J. A. Glatz (1981) ASTM Spec. Techn. Publ. *742,* 105.
[173] Rawa, J. A. (1981) ASTM Spec. Techn. Publ. *742,* 92.
[174] Roberts, K. M., D. T. Gjerde, and J. S. Fritz (1981) Anal. Chem. *53,* 1691.
[175] Miller, T. E., Jr. (1982) in: Autom. Stream Anal. Process Control *1* (Manka, D. P., ed.), Academic Press, New York, p. 1.
[176] Edwards, P. (1983) Food Technol. (Chicago) *37,* 53.
[177] Hertz, J. and V. Baltonsberger (1984) Fresenius Z. Anal. Chem. *318,* 121.
[178] Knudson, E. J. and K. J. Siebert (1984) J. Am. Soc. Brew. Chem. *42,* 65.
[179] Luckas, B. (1985) Fresenius Z. Anal. Chem. *320,* 519.
[180] Elchuk, S. and R. M. Cassidy (1979) Anal. Chem. *51,* 1434.
[181] Fritz, J. S. (1977) Pure Appl. Chem. *49,* 1547.
[182] Hatano, H. (1985) J. Chromatogr. *332,* 227.
[183] Suyuki, K., H. Aruga and T. Shirai (1983) Anal. Chem. *55,* 2011.
[184] Yenki, M. (1981) Anal. Chem. *53,* 968.
[185] Buechele, R. C. and D. J. Reutter (1982) J. Chromatogr. *240,* 502.
[186] Sevenich, G. J. and J. S. Fritz (1983) Anal. Chem. *55,* 12.
[187] Suzuki, K., H. Aruga and T. Shirai (1983) Anal. Chem. *55,* 2011.
[188] Fritz, J. S., D. T. Gjerde and R. M. Becker (1980) Anal. Chem. *52,* 1519.
[189] Haddad, P. R. and A. L. Heckenberg (1982) J. Chromatogr. *252,* 177.
[190] Hershcovity, H. and C. Yarnitzky, G. Schmuckler (1982) J. Chromatogr. *252,* 113.
[191] Cortes, H. J. (1982) J. Chromatogr. *234,* 517.
[192] Chang, C. A. and K. L. Fong (1984) J. Chromatogr. *312,* 99.
[193] Caude, M., J. F. Lefevre and R. Rosset (1975) Chromatographia *8,* 217.
[194] Wang, C. Y., S. D. Bunday and J. G. Tartai (1983) Anal. Chem. *55,* 1617.
[195] Williams, J. R. (1983) Anal. Chem. *6,* 851.
[196] Buytenhuys, F. A. (1981) J. Chromatogr. *218,* 57.
[197] Bio-Rad Labs. (1983) Techn. Note 1.
[198] Yamamoto, M., H. Yamamoto and Y. Yamamoto (1984) Anal. Chem. *56,* 832.
[199] Jones, V. K. and J. G. Tarter (1984) J. Chromatogr. *312,* 456.
[200] Jones, V. K. and J. G. Tarter (1985) Int. Lab. 36.
[201] Skelly, N. E. (1982) Anal. Chem. *54,* 712.
[202] Cassidy, R. M. and S. Elchuk (1983) J. Chromatogr. Sci. *21,* 454.
[203] Schmuckler, G., B. Roessner and G. Schwedt (1984) J. Chromatogr. *302,* 15.
[204] Dreux, M., M. Lafosse and M. Pequignot (1982) Chromatographia *15,* 653.
[205] Denkert, M., L. Hackyell, G. Schill and E. E. Sjoegreh (1981) J. Chromatogr. *218,* 31.
[206] Drerex, M., M. Lafosse and M. Pequignot (1982) Chromatographia *15,* 653.
[207] Barber, W. E. and P. W. Carr (1983) J. Chromatogr. *260,* 89.
[208] Cassidy, R. M. and S. Elchuk (1983) J. Chromatogr. *262,* 311.
[209] Lillig, B. (1982) GIT Fachz. Lab. 27.
[210] Giebelmann, R. (1982) Krim. Forensische Wiss. *47,* 51.
[211] Knox, J. H. and J. Jurand (1981) J. Chromatogr. *218,* 341.
[212] Knox, J. H. and J. Jurand (1981) J. Chromatogr. *218,* 355.
[213] Bidlingmeyer, B. A. and F. V. Warren (1982) Anal. Chem. *54,* 2351.
[214] Sachok, B., S. N. Deming and B. A. Bidlingmeyer (1982) J. Liq. Chromatogr. *5,* 389.

[215] Jandera, P., J. Churacek and B. Taraba (1983) J. Chromatogr. *262,* 121.
[216] Motomizu, S., I. Sawatani, M. Oshima and K. Toli (1983) Anal. Chem. *55,* 1629.
[217] Larsen, N. R. and W. B. Pedersen (1978) J. Radioanal. Chem. *45,* 135.
[218] Larsen, N. R. (1979) J. Radioanal. Chem. *52,* 85.
[219] Knight, C. H., R. M. Cassidy, B. M. Recoskie and L. W. Green (1984) Anal. Chem. *56,* 474.
[220] Baker, J. P., R. J. Gehrke, R. C. Greenwood and D. H. Meikrantz (1982) J. Radioanal. Chem. *74,* 117.
[221] Baranyi, R. (1982) Radiochem. Radioanal. Lett. *54,* 231.
[222] Martella, L. L., J. D. Navratil and M. I. Saba (1982) Radioact. Waste Management *6,* 27.
[223] Biondi, L. and J. G. Nairn (1985) J. Liq. Chromatogr. *8,* 1881.
[224] Osterloh, J. and D. Goldfield (1984) J. Liquid Chromatogr. *7,* 753.
[225] Bauer, G. M. (1985) Lebensmittel- u. Biotechnologie *1,* 18.
[226] Kyllingsbaek, A. (1984) Tidsskr. Plantearl *88,* 1984.
[227] Early, R. J., J. R. Thompson, T. Mc Allister, T. W. Fenton and R. J. Christopherson (1984) J. Chromatogr. Biomed. Appl. *310,* 1.
[228] Tanaka, K., T. Ishizuka and H. Sunalara (1979) J. Chromatogr. *174,* 153.
[229] Naikwadi, K. P., S. Rokushika and H. Hatano (1984) Chromatographia 633.
[230] Turkelson, V. T. and M. Richards (1978) Anal. Chem. *50,* 1420.
[231] Moore, S. and W. H. Stein (1948) J. Biol. Chem. *176,* 367; *178* (1949) 53; *192* (1951) 663; *211* (1954) 893.
[232] Spadaro, A. C. C., W. Draghetta, S. N. Del Lama, A. C. M. Camargo and L. J. Greene (1979) Anal. Biochem. *96,* 317.
[233] Moore, S., D. H. Spackman and W. H. Stein (1958) Anal. Chem. *30,* 1185.
[234] Hamilton, B. P. (1963) Anal. Chem. *35,* 2055.
[235] Hamilton, B. P. (1960) Anal. Chem. *32,* 1779.
[236] Hamilton, B. P., D. C. Bogue and R. A. Anderson (1960) Anal. Chem. *32,* 1782.
[237] Beecker, G. R. (1978) Adv. Exp. Biol. *105,* 827.
[238] Takeuchi, S., K. Fuzita, F. Nakaziwa and Y. Arikawa (1978) Nippon Kagaku Kaishi 64.
[239] Villanueva, V. R. and R. C. Adlakha (1978) Anal. Biochem. *91,* 264.
[240] Wright, J. C. and R. F. Evilia (1979) J. Liq. Chromatogr. *2,* 719.
[241] Kasiske, D., K. D. Klinkmueller and M. Sonneborn (1978) J. Chromatogr. *149,* 703.
[242] Doury-Berthod, M., C. Poitrenaud and B. Tremillon (1979) J. Chromatogr. *179,* 37.
[243] Schmidt, E., A. Foucault, M. Caude and R. Rosset (1979) Analysis *7,* 366.
[244] Park, Man Ki and Han, Dal Soo (1984) Yakhak Hoechi *28,* 21.
[245] Chin, Christopher C. O. (1984) Methods Enzymol. *106,* 17.
[246] Elkin, G. (1984) J. Agric. Food Chem. *32,* 53.
[247] Elkin, G. (1984) J. Assoc. Off. Anal. Chem. *67,* 1024.
[248] Vaughn, J. G. (1984) Clin. Liq. Chromatogr. *2,* 1.
[249] Ferraro, Th. N. and A. Th. Hare (1984) Anal. Biochem. *143,* 82.
[250] Plunkett, W. and S. S. Cohen (1977) J. Cell Physiology *91,* 261.
[251] Asteriadis, G. T., M. A. Armbruster and P. T. Gilham (1976) Anal. Biochem. *70,* 64.
[252] Tommel, D. K. J., J. F. G. Vliegenthart, T. J. Peders and J. F. Arens (1966) Biochem. J. *50,* 48.
[253] Krishnan, N. and G. Krishna (1976) Anal. Biochem. *70,* 18.
[254] Runeckles, V. C. and G. Das (1975) J. Chromatogr. *115,* 240.
[255] Watson, J., M. Nielsen-Hamilton and R. T. Hamilton (1976) Biochemistry *15,* 1527.
[256] Hsu, D. S. and S. S. Chen (1984) J. Chromatogr. *311,* 396.
[257] Habbal, Z. M. (1979) Clin. Chim. Acta *95,* 301.

[258] D'Anna, J. A., G. F. Strniste and L. R. Gurley (1979) Biochemistry *18,* 943.
[259] Scott, J. F. and A. Kornberg (1978) J. Biol. Chem. *253,* 3292.
[260] Stringer, E. A., A. Chaudhuri and R. Maitra (1979) J. Biol. Chem. *254,* 6845.
[261] Morin, L. G. (1976) Clin. Chem. *22,* 92.
[262] Savage, C. R. and R. Harper (1981) Anal. Biochem. *111,* 195.
[263] Lewis, S., M. R. Fennessy, F. J. Laska and D. A. Taylor (1980) Agents and Actions *10,* 197.
[264] Villanueva, V. R. and R. C. Adlakha (1978) Anal. Biochem. *91,* 264.
[265] Chambers, T. L. and W. F. Stakuszkiewicz (1978) J. Assoc. Off. Anal. Chem. *61,* 1092.
[266] Garnier, J. P., B. Bousquet and C. Dreux (1979) Analysis 7, 355.
[267] Yui, Y., M. Kimura, Y. Itokawa and C. Kawai (1979) J. Chromatogr. *177,* 376.
[268] Allenmark, S. and L. Hedman (1979) J. Liq. Chromatogr. *2,* 227.
[269] Dieker, J. W., Van der Linden and H. Poppe (1979) Talanta *26,* 511.
[270] Moyer, T. P., N. S. Jiang, G. M. Tycle and S. G. Sheps (1979) Clin. Chem. *25,* 256.
[271] Ponchow, J. L., P. Hutter, J. F. Pujol and P. Blond (1979) Analysis *7,* 376.
[272] Cohen, J. L. and R. E. Brown (1978) J. Chromatogr. *151,* 237.
[273] Walton, H. F., A. Eiceman and J. L. Otto (1979) J. Chromatogr. *180,* 145.
[274] Hsu, T. J., C. C. Liao and M. Y. Chen (1978) J. Chinese Chem. Soc. *25,* 153.
[275] Van Duijn, J. and G. H. D. Van der Stegen (1979) J. Chromatogr. *179,* 199.
[276] Bjerg, B., O. Olsen, K. W. Rasmusen and H. Sorensen (1984) J. Liq. Chromatogr. *7,* 691.
[277] Simatupang, M. H. (1979) J. Chromatogr. *178,* 588.
[278] Verhaar, L. A. T. and J. M. H. Dirkx (1977) Carbohydr. Res. *59,* 1.
[279] Simatupang, M. H., M. Sinner and H. H. Dietrichs (1978) J. Chromatogr. *155,* 446.
[280] Binder, St. R. and G. Sivorinovsky (1984) J. Chromatogr. *336,* 173.
[281] Baer, G. R., C. W. Lawley and R. M. Riddle (1984) J. Chromatogr. *302,* 65.
[282] Rajakyla, E. (1981) J. Chromatogr. *218,* 695.
[283] Woo, D. J. and J. R. Benson (1983) Liq. Chromatogr., HPLC Mag. *1,* 238.
[284] Walton, H. F. (1983) J. Chromatogr., Libr. *22A,* 225.
[285] Kuster, Th. and A. Niederwieser (1983) in: J. Chromatogr. Libr. Vol. 22B, (E. Heftmann, ed.) p. B1, Elsevier, Amsterdam.
[286] Okazaki, H., H. Wada, T. Takeuchi, K. Makino, T. Fukui and Y. Kato (1985) J. Chromatogr. *322,* 243.
[287] Cachia, P. J., J. van Eyk, P. C. S. Chong, Ashok Taneja and R. S. Hodges (1983) J. Chromatogr. *266,* 651.
[288] Jost, W., K. K. Unger, R. Lipecky and H. G. Gassen (1979) J. Chromatogr. *185,* 403.
[289] Yoshio, Kato, Koji Nakamura and Tsutomu Hashimoto (1983) J. Chromatogr. *266,* 385.
[290] Gooding, K. M. and M. N. Schmuck (1984) J. Chromatogr. *296,* 321.
[291] Fontane, J. C. (1984) Tec. Lab. *9(115),* 154.
[292] Peterson, E. A. (1970) Cellulosic Ion Exchangers, North Holland, Amsterdam.
[293] Peterson, E. A., H. A. Sober (1961) in: Biochemical Preparations (A. Meister, ed.), p. 8, 39, 43, Wiley, New York.
[294] Peterson, E. A. and H. A. Sober (1962) Methods Enzymol. *5,* 3.
[295] Cowling, G. J. (1983) in: J. Chromatogr. Libr. Vol. 22B (E. Heftmann, ed.) p. B345, Elsevier, Amsterdam.
[296] Alpert, J. and F. E. Regnier (1979) J. Chromatogr. *185,* 375.
[297] Perrett, D. (1982) Chromatographia *16,* 211.
[298] Baust, J. G., R. E. Lee, R. Rojas, D. L. Hendrix, D. Friday and H. James (1983) J. Chromatogr. *261,* 65.
[299] Smith, M., P. Monchamp and F. B. Jungalwala (1981) J. Lipid. Res. *22,* 714.

[300] Dieter, D. S. and H. F. Walton (1983) Anal. Chem. *55,* 2109.
[301] Walton, H. F. (1985) J. Chromatogr. *332,* 203.
[302] Churms, S. C. (1983) in: J. Chromatogr. Libr. Vol. 22B (E. Heftmann, ed.) p. B223, Elsevier, Amsterdam.
[303] Bonn, G. and O. Bobleter (1984) Chromatographia *18,* 445.
[304] Bonn, G., R. Pecina, E. Burtscher and O. Bobleter (1984) J. Chromatogr. *287,* 215.
[305] Seki, T., Y. Yamaguchi, K. Noguchi, Y. Yanagihara, (1985) J. Chromatogr. *332,* 9.

6 Theory of Ion Exchange

6.1 Thermodynamics

Vladimir S. Soldatov
The BSSR Academy of Sciences, Institute of Physico-Organic Chemistry, Minsk, USSR

List of symbols

a	thermodynamic activity
C	concentration
f	resinate activity coefficient
G	Gibbs' free energy
H	enthalpy
I	symbol for ion
k	equilibrium constant of hypothetical equilibrium
$\tilde{K}$	apparent equilibrium constant
K	equilibrium constant
n	number of moles
P	pressure
Q	heat effect
R	gas constant
$R_{Z_i}I^{Z_i}$	symbol for resinate
S	entropy
T	absolute temperature
V	volume
v	partial molal volume

w	symbol for solvent
W	number of moles of solvent per equivalent of ion exchanger
x	equivalent fraction
Y	property of binary system
y	specific property Y related to a certain state of ion
Z	ion charge
α	fraction of a sorbed ion
α_O	limiting degree of dissociation of polyelectrolyte
γ	activity coefficient of electrolyte
φ	formal activity coefficient
ø	osmotic coefficient
μ	chemical potential
ν	number of particles forming at dissociation of resinate
π	swelling pressure
σ	sum in least square method

Barred symbols refer to the ion exchanger phase.

Introduction

There is no generally accepted opinion on many questions associated with the thermodynamics of ion exchange. The limited length of this chapter does not allow a detailed and impartial consideration of various viewpoints. The material is therefore presented as the author sees it, based on a generalization of the material from monographs [1–10] and reviews, as well as from original papers. Despite the fact that at first the simplest concepts used in the ion exchange thermodynamics are analyzed, the reader is assumed to be familiar with chemical thermodynamics and the main points of ion exchange theory.

In this chapter, an attempt has been made to present ion exchange thermodynamics as a branch of physical chemistry that permits setting and solving problems in practice. In this connection, the chapter covers material ranging from a formal thermodynamic description of ion exchange systems to a quantitative description of equilibria in non-ideal exchange systems.

Due to the limitations, it was impossible to deal with many problems associated with ion exchange thermodynamics, such as the modelling of ion exchange processes and the prediction of equilibria; the peculiarities of exchange on mineral, liquid, and other types of ion exchangers; the exchange of organic ions; the exchange from non-aqueous media; selective ion exchangers and the thermodynamics of other processes that could be considered separately.

6.1.1 General characteristics of an ion exchange system[1]

An ion exchange system consists of an ion exchanger and a solution. The solution contains at least two kinds of exchangeable ions (counterions), $I_1^{Z_1}$ and $I_2^{Z_2}$, one kind of co-ion, $I_i^{Z_i}$, and a solvent W.

The ion exchanger contains all the above-named particles and a carrier of electric charge incapable of crossing the ion exchanger/solvent boundary.

A real ion exchange system may be more complicated: the number of counterions may be greater than 2 and that of the co-ions greater than 1, the solvent may be a mixture, the system may contain substances not involved in the ion exchange process, etc. The consideration of these possibilities raises difficulties and will not concern us here. To understand the main regularities of ion exchange, however, it is sufficient to consider the simple system described above.

Before we begin, the following question must be clarified. The expression „ion exchanger" covers substances of quite different chemical nature and physical structure. Among the most important of them are:

Crosslinked polyelectrolytes. The charge-carriers holding counterions in the ion exchanger are functional groups fixed on a spatially infinite polymeric matrix. The structure has cavities of random dimensions and forms.

Crystalline ion exchangers. The crystal lattice is the charge carrier. The charge is not fixed and the structure has cavities of definite dimensions and forms.

Ion exchange adsorbents. The functional groups located on the surface of a porous substance are the charge-carriers.

Liquid ion exchangers (ion exchange extractants). These are solutions of water-insoluble ionized organic compounds in solvents immiscible with water. The hydrophobic ion is the charge carrier. An organic phase can change its structure depending on the nature of the diluent and the counterion, ranging from a molecular to micellar solution.

A general description of ion exchange has to be, on the one hand, sufficiently abstract and, on the other, to permit the possibility of describing specific processes as particular cases.

It is convenient to present ion exchange equilibria in a generalized form where the ion exchanger, irrespective of its nature, is denoted by barred symbols:

$$\frac{1}{Z_1}\bar{I}_1^{Z_1} + \frac{1}{Z_2} I_2^{Z_2} \rightleftarrows \frac{1}{Z_1} I_1^{Z_1} + \frac{1}{Z_2}\bar{I}_2^{Z_2}. \tag{1}$$

An equivalent formulation is

$$Z_2\bar{I}_1^{Z_1} + Z_1 I_2^{Z_2} \rightleftarrows Z_2 I_1^{Z_1} + Z_1\bar{I}_2^{Z_2}. \tag{2}$$

In the following we shall use eq. (1).

[1] In this chapter, monographs and proceedings [1–10] are often used without special references.

The exchange of ions is almost always accompanied by the transfer of solvent due to swelling of the ion exchanger. This may be formulated as

$$\mathrm{W} \rightleftarrows \overline{\mathrm{W}} . \tag{3}$$

In addition, a distribution of electrolytes between the ion exchanger and the solution occurs; however, the effect of this phenomenon on the ion exchange process can be always reduced to a minimum by reducing the solution concentration. Therefore, we shall consider a "pure" ion exchange process which is not complicated by the distribution of electrolytes, which will be considered separately.

As with other chemical systems, the concept of chemical potential and the equation giving its relation to the component concentration holds a central place in ion exchange thermodynamics:

$$\mu_i = \mu_i^\circ + RT \ln a_i = \mu_i^\circ + RT \ln C_i + RT \ln f_i , \tag{4}$$

where μ_i is the chemical potential of component i, μ_i° is the standard chemical potential, a_i the thermodynamic activity, C_i the concentration, and f_i the activity coefficient.

In this equation, the chemical potential is divided up into two parts: μ_i°, independent of the mixture composition and $RT \ln a_i$, dependent on the mixture composition. In accordance with eq. (4), $\mu_i = \mu_i^\circ$ for $a_i = 1$. The applicability of eq. (4) for different values of C was discussed in detail in Ref. [11] for systems of various types. The equation is rigorously substantiated for ideal mixtures

$$\mu_i = \mu_i^\circ(T) + RT \ln C_i , \tag{5}$$

where

$$\mu_i^\circ(T) = - RT \ln \frac{(2\pi M_i kT)^{3/2} kT}{h^3} , \tag{6}$$

M_i is the molar mass, and k and h are Boltzmann and Planck's constants, respectively; it should be emphasized that C is the number of moles of component i in 1 l. In ideal solutions, the molar volume as well as the total concentration C_O do not depend on the composition; therefore, the following equation is valid:

$$\mu_i = \mu_i^\circ(T, C_0) + RT \ln N_i , \tag{7}$$

where N_i is the molar fraction, $N_i = C_i/C_\mathrm{O}$.

Eq. (5) and (7) correspond to Henry's and Raoult's law, which are valid for the whole range of compositions.

In real solutions, C_O depends on N_i. Therefore, the equation

$$\mu_i = \mu_i^\circ(P, C_0) + RT \ln N_i + RT \ln f_{N_i} \tag{8}$$

becomes inaccurate ($\mu_i^\circ(T, C_\mathrm{O}) \neq$ const.). Dependence of $\mu_i^\circ(T, C_\mathrm{O})$ on the composition is formally taken into account by the activity coefficient related to N, (f_N), whose relation with f follows from eq. (4) and (8):

$$f_N = \frac{f}{C_0} . \tag{9}$$

Hence, F_N takes into account not only interactions in a real system, as the activity method postulates, but also a change in volume, which may be considered independently.

Other methods used to express the mixture composition, e. g., molality, also lead to the dependence of μ_i° on composition. However, the use of molar fraction at low dependence of a molar volume of the system on composition and molality at low concentrations of one of the components is a convenient and acceptable approximation used in solution theory.

It is characteristic of ion exchange systems that the concentration in an ion exchanger is always high, as the structure differs from that of liquids and gases. The total concentration of components strongly depends on the ion exchanger composition. Therefore, it is to be noted that other methods distinguished from molarity used to express the composition in equations of type (4) will lead to formal activity coefficients that take into account a change in the total concentration or other properties of the ion exchanger that may be considered directly. Only quantity f is a direct measure of the non-ideality of the system.

The meaning of concepts "component", "concentration", "chemical potential" depend on the ion exchanger structure. Further, only crosslinked polyelectrolytes will be considered. Nevertheless, many conclusions drawn can be applied to other types of ion exchangers with almost only minor changes.

In crosslinked polyelectrolytes, counterions compensate the charge of one and the same infinite matrix. In the presence of two counterions, the system is described as a mixture of the polyelectrolyte salts, $R_{Z_1}I_1$ and $R_{Z_2}I_2$ (referred to as resinates) and the solvent W. R denotes a part of the matrix "belonging" to the given counterion. Concentrations of $R_{Z_1}I_1$, $R_{Z_2}I_2$, and the solvent can change independently of each other and can express any change in composition of the ion exchanger phase, hence, they are components [12]. For short, the ion exchanger components will be denoted below as $\bar{I}_1$, $\bar{I}_2$, $\overline{W}$.

Considering an ion exchanger as a polyelectrolyte of infinitely large molecular weight it has been shown in [13] that the activity in equation (4) is related to the counterion activity $\bar{a}_+$:

$$a = a_+^{\nu\varphi_0}, \tag{10}$$

where ν is the number of kinetically active particles formed at dissociation of one functional group, and $ø_0$ is the osmotic coefficient of the polyelectrolyte at infinite dilution (with $ø_0$ always being less than 1) [13, 14–25].

6.1.2 Thermodynamics of a binary mixture: crosslinked polyelectrolyte/water

There are two approaches to a description of the equilibrium between an ion exchanger and a liquid or vapour. In one of them, it is assumed that pressures in the equilibrium phases are equal; in the other, the ion exchanger phase is considered to be under additional pressure produced by the expansion of the elastic matrix with increasing volume. In Ref. [4] Guggenheim's terms "non-osmotic equilibria" [26] is suggested for the former and "osmotic equilibria" for the latter.

At equilibrium, the chemical potentials of the solvent in the resin and vapour are equal, $\bar{\mu}_w = (\mu_w)_v$. Most often a non-osmotic approach is used in practice. Then

$$\begin{aligned}\mu_w &= \mu_w^\circ + RT\ln(P/P_0) = \mu_w^\circ + RT\ln\bar{a}_w \\ &= \mu_w^\circ + RT\ln C_w + RT\ln f_w\end{aligned} \tag{11}$$

$$\bar{a}_w = \frac{P}{P_0}, \tag{12}$$

where P and P_O are the partial pressures of the solvent vapour under investigation and in the standard state (usually saturated vapour). Irrespective of the type and crosslinkage of the ion exchanger, $\bar{a}_w = 1$ when the ion exchanger is in equilibrium with a liquid solvent. The difference in the solvent state of completely swollen ion exchangers of different types is expressed by the activity coefficient equal to $\bar{C}_w^{-1}$.

The free energy change when $d\bar{n}_w$ moles of the solvent are transferred from vapour under pressure P_O into the ion exchanger containing $\bar{n}_w$ moles of the solvent is

$$dG_w = (\bar{\mu}_w - \mu_w)d\bar{n}_w = -RT\ln(P/P_0)d\bar{n}_w. \tag{13}$$

The value of $dG_w/d\bar{n}_w = \Delta G$ may be defined as the differential free energy of solvent sorption.

Denote by W the number of solvent moles per ion exchanger equivalent ($W_i = \bar{n}_w/\bar{n}_i$).

An integral change in ΔG corresponding to the change in the solvent composition from W' to W'' is expressed by the equation:

$$\Delta G_{int} = -RT\int_{W'}^{W''}\ln(P/P_0)dW. \tag{14}$$

If $W' = O$ ($P = O$) and $W'' = W_{max}$ ($P = P_O$), then $\Delta G = \Delta G^\circ$ is the standard free energy of sorption.

It is more convenient to calculate integral (14) in parts. The value of ΔG corresponding to W changing from O to W is

$$\Delta G = WRT\ln(P/P_0) - RT\int_0^{P/P_0} W d\ln(P/P_0). \tag{15}$$

A change in enthalpy (ΔH) in these processes is most reliably determined from the thermal effects of sorption, $Q(Q = -\Delta H)$. Experimentally, it is convenient to determine Q by contacting an ion exchanger containing different quantities of solvent and a liquid solvent. Then

$$-Q = \Delta H_{int} = \int_W^{W_{max}} \Delta H\, dW. \tag{16}$$

The value of ΔH corresponding to ΔG from eq. (15) may be found as $\Delta H = \Delta H^\circ - \Delta H_{int}$. The sorption entropy is given by the equation

$$T\Delta S = \Delta H - \Delta G\,. \tag{17}$$

A large number of works has been dedicated to the experimental investigation of the thermodynamics of water sorption by ion exchangers, e. g. [27–40]. We can choose from them only a typical example (Figures 6.1–6.3) and summarize some important regularities.

Two main processes occur when water is sorbed by ion exchangers: the hydration of ions (counterion and fixed ion) and the separation of ions forming an ionic pair. These processes are accompanied by opposite changes in thermodynamic functions and to a great extent compensate each other. Ultimately, the thermodynamic

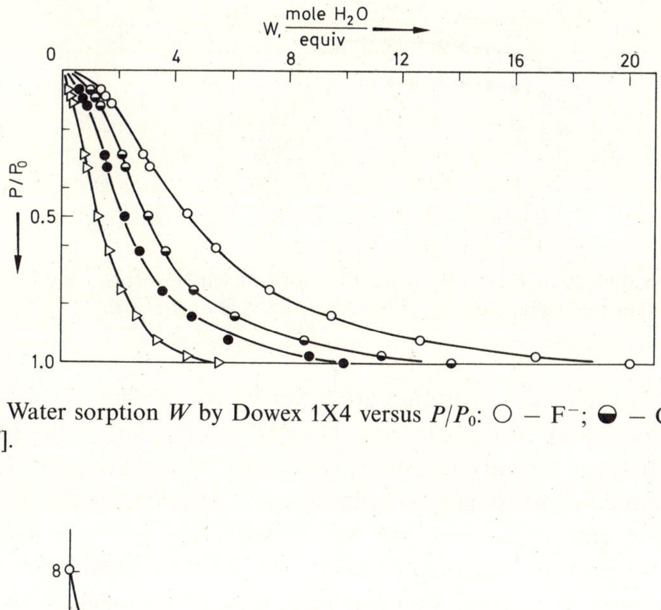

Figure 6.1 Water sorption W by Dowex 1X4 versus P/P_0: ○ – F^-; ◒ – Cl^-; ● – Br^-; △ – I^- [37].

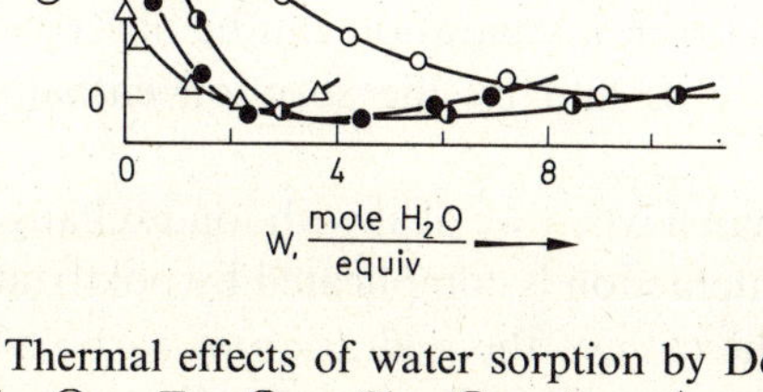

Figure 6.2 Thermal effects of water sorption by Dowex 1X4 depending on water content in the sample: ○ – F^-; ◑ – Cl^-; ● – Br^-; △ – I^- [37].

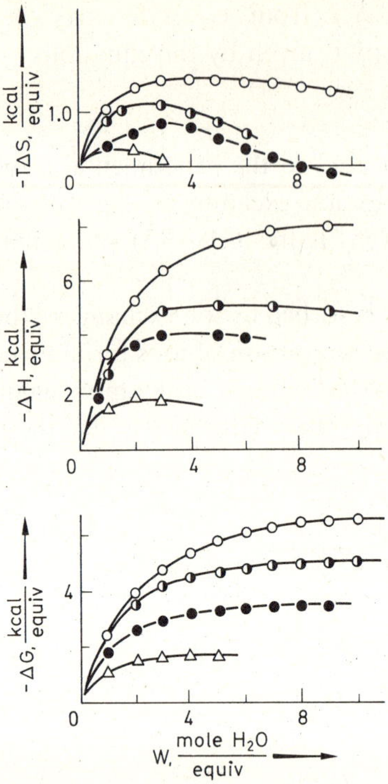

Figure 6.3 Dependence of thermodynamic functions of water sorption by Dowex 1X4 on water content *W* in the exchanger: ○ – F^-; ◑ – Cl^-; ● – Br^-; △ – I^- [37].

functions of ion exchanger hydration are lower by two orders of magnitude than those for the hydration of separate ions. The simplest regularities are observed for strong acid cation and strong base anion exchangers. At the first stage of interaction between water and an ion exchanger, hydration usually predominates. The enthalpy, free energy, and entropy are negative and they are the more negative the more strongly a counterion is hydrated. The relation among standard thermodynamic functions turns out to be quite simple (Figure 6.4), independent of the nature of the counterion and the same for cation and anion exchangers. The main changes in thermodynamic functions take place when the first 4–6 mol of water per mol of functional groups are sorbed. Therefore, the crosslinkage of the ion exchanger affects these values only minimally. In those cases where the secondary hydration of the counterions is weak, the effect of ion separation can be observed, expressed by negative thermal effects (at $W > 4$) and by the sorption enthalpy's passing through a maximum.

These simple regularities are violated when we deal with ion exchangers of other types where a simple electrostatic interaction is complicated by polarization effects. This takes place, for instance, in the case of the hydration of carboxylic resins in the alkali metals and alkaline earth metals ionic forms [33, 40].

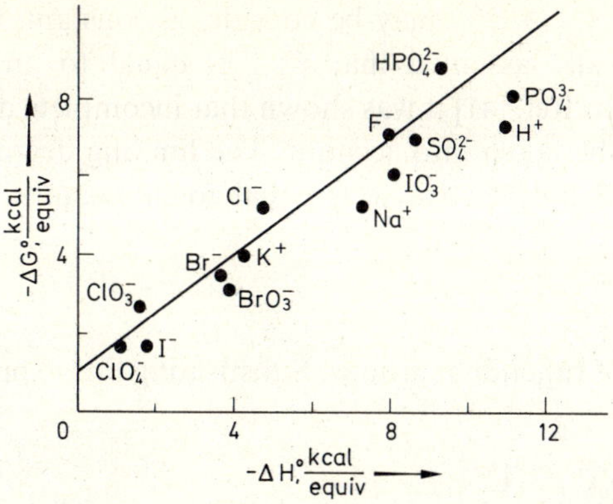

Figure 6.4 Correlation of ΔH^0 and ΔG^0 for water sorption by ion exchangers. The data on anions are obtained for Dowex 1X4 [37]; the data on cations refer to sulphostyrene ion exchangers [7, 27].

A more specific quantitative interpretation of the dependences $W = f(P/P_0)$, $\Delta G, \Delta H, T\Delta S = f(W)$ can be obtained from the model of step-wise hydration [32, 37].

6.1.3 Distribution of electrolyte in the ion exchanger/solution system

When an ion exchanger in an ionic form (for example, I^{Z_+}) and an electrolyte solution $I_{Z-}^{Z+}\, I_{Z+}^{Z-}$ are in contact, an equilibrium results that may be considered as Donnan's membrane equilibrium since a polyanion cannot cross the ion exchanger/solution boundary.

Our aim is to derive the simplest possible equation expressing the concentration of electrolyte $I_{Z-}^{Z+}\, I_{Z+}^{Z-}$ in the ion exchanger phase as a function of its concentration in the solution.

At equilibrium, the electrolyte activities in both phases are equal, if the standard states are chosen identically:

$$a_+^{Z-}\, a_-^{Z+} = \bar{a}_+^{Z-}\, \bar{a}_-^{Z+}\,. \qquad (18)$$

If $\bar{C}_{R+}$ is the concentration of counterions compensating a matrix charge of ion exchanger and γ is the activity coefficient of a distributed ion, eq. (18) may be rewritten in the form:

$$a_+^{Z-}\, a_-^{Z+} = (\bar{C}_+ + \bar{C}_{R+})^{Z-}\, \bar{C}_-^{Z+}\, \bar{\gamma}_+^{Z-}\, \bar{\gamma}_-^{Z+}\,. \qquad (19)$$

Due to a low concentration of co-ions in the ion exchanger phase, it is assumed that $\bar{\gamma}_-^{Z+} = 1$. If the concentration of sorbed electrolyte is significantly less than that of the counterions ($\overline{C}_{+} \ll \overline{C}_{R+}$), $\bar{\gamma}_{+}^{Z-}$ may be considered constant; its value is unknown, however. It is usually assumed that $\overline{C}_{R+}$ is equal to an analytical concentration of counterions. In Ref. [41] it was shown that incomplete dissociation of the polyelectrolytes should be taken into account. The limiting degree of dissociation of an infinitely diluted polyelectrolyte is equal to its osmotic coefficient ($\alpha_O = ø_O$). Then

$$\overline{C}_{R+} = \nu ø_0 \overline{C}\,, \tag{20}$$

where $\overline{C}$ is the concentration of functional groups. Substituting this expression into eq. (19), we get:

$$a_{+}^{Z-}\, a_{-}^{Z+} = (\overline{C}_{+} + \nu ø_0 \overline{C})^{Z-}\, \overline{C}^{Z+}\, \bar{\gamma}_{+}^{Z-}\, \bar{\gamma}_{-}^{Z+}\,. \tag{21}$$

The evaluation of experimental data of Ref. [42] using eq. (21) and values of $ø_0$ from [43] showed (Figure 6.5) that this equation describes the experimental data well. The experimental and computed results agree for constant values of $\bar{\gamma}_{Mg^{2+}}$ and $\bar{\gamma}_{K^+} = 1.00$ for the sorption of $MgCl_2$ and KCl onto a sulphostyrene cation exchanger and for $\bar{\gamma}_{Cl^-} = 0.48$ for the sorption of $MgCl_2$ onto a strong base anion exchanger. According to [42], the experimental and computed results obtained by use of eq. (19), supposing $\overline{C}_{R+} = \nu\overline{C}$, also agree; however, the activity coefficients satisfying eq. (19) are substantially lower, $\bar{\gamma}_{\pm(MgCl_2)} = 0.48$, $\bar{\gamma}_{\pm(KCl)} = 0.47$ (cation exchanger) $\bar{\gamma}_{\pm(MgCl_2)} = 0.29$ (anion exchanger).

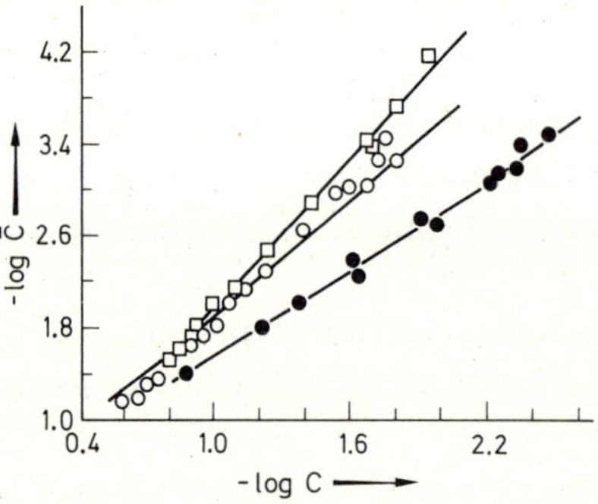

Figure 6.5 Electrolyte molality in the ion exchanger phase versus molality in the solution phase. The lines are calculated from eq. (41); the points represent experimental data of [42]. ○ – KCl; ● – $MgCl_2$; on the AG 50WX2 cation exchanger; □ – $MgCl_2$ on the AG 1X2 anion exchanger.

From this comparison, it can be concluded that introduction of $ø_0$, determined independently, decreases to a greater extent (or even excludes) an uncertainty in the constant values of $\bar{\gamma}$.

More detailed consideration of electrolyte distribution is given in Ref. [1, 4].

6.1.4 Ion exchange equilibrium equation

The equilibrium condition for an ion exchange system with ions 1 and 2 is

$$\bar{\mu}_1 d\bar{n}_1 + \bar{\mu}_2 d\bar{n}_2 + \bar{\mu}_w d\bar{n}_w + \mu_1 dn_1 + \mu_2 dn_2 + \mu_w dn_w = 0\,. \tag{22}$$

The conditions of mass balance and electrical neutrality are

$$d\bar{n}_1 = -\, dn_1\,,\; d\bar{n}_2 = -\, dn_2 \tag{23}$$

$$\frac{1}{Z_1} d\bar{n}_1 = -\frac{1}{Z_2} dn_2,\; \frac{1}{Z_1} dn_1 = -\frac{1}{Z_2} d\bar{n}_2 \tag{24}$$

$$d\bar{n}_w = -\, dn_w\,. \tag{25}$$

Substituting eq. (23)–(25) into eq. (22) we get

$$\frac{1}{Z_1}\bar{\mu}_1 - \frac{1}{Z_2}\bar{\mu}_2 - \frac{1}{Z_1}\mu_1 + \frac{1}{Z_2}\mu_2 + (\bar{\mu}_w - \mu_w)\frac{d\bar{n}_w}{d\bar{n}_1} = 0\,. \tag{26}$$

Denote by $\bar{x}_1$ an equivalent fraction of ion 1 in the ion exchanger

$$\bar{x}_1 = \frac{Z_1\bar{n}_1}{Z_1\bar{n}_1 + Z_2\bar{n}_2}\,. \tag{27}$$

Then

$$\frac{d\bar{n}_w}{Z_1 d\bar{n}_1} = \frac{dw}{d\bar{x}_1} \tag{28}$$

and eq. (26) can be rewritten in the form:

$$\frac{1}{Z_1}\bar{\mu}_1 - \frac{1}{Z_2}\bar{\mu}_2 - \frac{1}{Z_1}\mu_1 + \frac{1}{Z_2}\mu_2 + (\bar{\mu}_w - \mu_w)\frac{dw}{d\bar{x}_1} = 0\,. \tag{29}$$

Substitution of μ from eq. (4) gives an equilibrium equation:

$$K_w = \frac{\bar{a}_1^{1/Z_1}\, a_2^{1/Z_2}}{\bar{a}_2^{1/Z_2}\, a_1^{1/Z_1}}\left(\frac{\bar{a}_w}{a_w}\right)^{\frac{dw}{d\bar{x}_1}} \equiv \text{const.} \tag{30}$$

At present two approaches are most widely used in the thermodynamic description of ion exchange equilibrium. They may be referred to as "osmotic" and "non-osmotic".

The non-osmotic approach is based on the following assumptions: components 1 and 2 in the ion exchanger phase are resinates differing from the electrolytes 1 and 2, which are solution components. The solvent activities in the ion exchanger and solution phases are identical. In fact, we are dealing with a heterogeneous chemical reaction (eq. (1)). Then the equilibrium equation (30) becomes the equation of the law of mass action:

$$K = \frac{\bar{a}_1^{1/Z_1}\, a_2^{1/Z_2}}{\bar{a}_2^{1/Z_2}\, a_1^{1/Z_1}} \equiv \text{const.} \tag{31}$$

K is referred to as the thermodynamic equilibrium constant.

In the osmotic approach, the ion exchange equilibrium is considered as the distribution of ions 1 and 2 between the solution and ion exchanger phases under different pressures. Then additional conditions appear: $(\partial\mu_i/\partial P)_T = \bar{v}_i$ and the equality of the electrochemical potentials of the mobile ions

$$\mu_i + Z_i F\varepsilon_i = \bar{\mu}_i + \pi\bar{v}_i + Z_i F\bar{\varepsilon}_i\,. \tag{32}$$

The substitution of these conditions into eq. (29) for identical standard states gives

$$RT\ln K - \pi\left[\left(\frac{\bar{v}_2}{Z_2} - \frac{\bar{v}_1}{Z_1}\right) - \frac{\mathrm{d}w}{\mathrm{d}\bar{x}_1}\,\bar{v}_w\right] = 0\,, \tag{33}$$

where $\bar{v}_1$, $\bar{v}_2$ and $\bar{v}_w$ are the partial mole volumes of the corresponding components in the ion exchanger. Approximately, these values can be considered constant. Dependence of W on $\bar{x}_1$ can often be approximated by a straight line, i. e., $\mathrm{d}w/\mathrm{d}\bar{x}_1 = W_1 - W_2$, so that the following equation is valid

$$RT\ln K = \pi\left[\left(\frac{\bar{v}_2}{Z_2} + W_2\bar{v}_w\right) - \left(\frac{\bar{v}_1}{Z_1} + W_1\bar{v}_w\right)\right] = \pi(V_2 - V_1)\,, \tag{34}$$

where V_i is the equivalent volume of the swollen ion exchanger. The above equation coincides with Gregor's model [44]. Instead of the value on the right Refs. [45–48] give $\pi\ (\bar{v}_2/Z_2 - \bar{v}_1/Z_1)$. Therefore, the osmotic term in calculations in Refs. [45, 46] is small and makes an opposite contribution to the value of K as compared with that predicted by Gregor's model. The value of $\pi\ (V_2 - V_1)$ is considerably higher than $\pi\ (\bar{v}_2 - \bar{v}_1)$ and has an opposite sign. Further, it decreases rather slowly with decreasing crosslinkage, as in this case $V_2 - V_1$ increases and partially compensates a decrease in π. In this respect, it does not seem obvious that the osmotic term for the weakly crosslinked ion exchangers will be negligible in comparison with the value of ln K.

When comparing the above approaches the following points are to be considered.

Generality is clearly the main characteristic of the non-osmotic version, since it satisfies ion exchangers of all types. It is the only version if $(\bar{\mu}_w - \mu_w)\mathrm{d}w/\mathrm{d}\bar{x}_1 = 0$ in eq. (29), which is possible in case swelling of the ion exchanger is constant ($\mathrm{d}w/\mathrm{d}\bar{x}_1 = 0$) or $\bar{\mu}_w = \mu_w$. The former is valid for many mineral sorbents to a good approximation, the latter for liquid ion exchangers and ion exchangers with a surface distribution of functional groups, i. e., for the cases when the osmotic model is evidently inapplicable. However, using this method it is impossible to predict the value of K or of certain other characteristics of an ion exchange system.

The osmotic version seems to correspond well to the properties of polyelectrolytes with low and moderate crosslinkage. This approach and its variations were applied by Boyd et al. [45, 46] and later by other authors [7, 48 – 51] to predict ion exchange equilibria. The values of activity coefficients for the components and π for strongly crosslinked ion exchangers were computed from the data on weakly crosslinked ion exchangers. The ratio of resinate activity coefficients in a highly diluted ion exchanger system had to be determined empirically. In many cases the contribution of this term to the free energy appeared to be large. This is due to specific binding of the counterion in the resin.

Thus, in practice, an intermediate situation between a classical chemical reaction and Donnan's distribution is realized. In Refs. [7, 52] eq. (30) is immediately suggested to describe ion exchange equilibrium, i. e., to find $\bar{a}_w$ from the osmotic properties of polyelectrolyte solutions or weakly crosslinked ion exchangers without converting data into the value of π[2]. The value of $dw/d\bar{x}_1$ is available from independent experiments. The chemical potential of the resinates is expressed by eq. (4), for the non-osmotic version, and the standard states $\bar{\mu}_i^o$ and μ_i^o are selected differently. This means that from the osmotic viewpoint, the activity coefficients of the resinates will include a correction for a change in pressure. Only the main part of the osmotic term $-(dw/d\bar{x}_1)RT \ln \bar{a}_w$ is expressed as a separate contribution to the free energy. Thus, ion exchange is treated as a heterogenous chemical reaction accompanied and complicated by osmotic phenomena. The role of different terms in the free energy of ion exchange will be evaluated below.

6.1.5 Activity coefficients of resinates and the equilibrium constant

At present, there is no experimental means of determining the activity coefficients of the ion exchanger phase components. Therefore, the equilibrium constant K_w (eq. (30)) or K (eq. (31)) cannot be calculated from the direct experimental data and so proved to be constant. The value $\tilde{K}$ below, which does not take into account the activity coefficients of the resinates, is referred to as an "apparent equilibrium constant" and, as a rule, depends on the extent of exchange on crosslinked polyelectrolytes

$$\tilde{K} = \frac{\bar{C}_1^{1/Z_1}\, a_2^{1/Z_2}}{\bar{C}_2^{1/Z_2}\, a_1^{1/Z_1}}, \tag{35}$$

whence

$$K_w = \tilde{K}\, \frac{f_1^{1/Z_1}}{f_2^{1/Z_2}} \left(\frac{\bar{a}_w}{a_w}\right)^{\frac{dw}{d\bar{x}_1}}. \tag{36}$$

---

[2] In Refs. [4, 50] it is further pointed out that besides pressure, other forces, such as deformation stress, electrical field tension, and surface tension can be responsible for the difference between the chemical potentials inside and outside the ion exchanger.

The values of the activity coefficients f_1 and f_2 can be found by solving a set of equations consisting of eq. (36) and the Gibbs-Duhem equation for the ion exchanger phase. The procedure was used in Refs. [53–55] but, the authors described the ion exchanger with two counterions as a two-component system without allowance for a solvent. It is more correct to take three components, $\bar{I}_1$, $\bar{I}_2$, and W into account in the system. Then the Gibbs-Duhem equation is to be written as follows:

$$\bar{C}_1 \mathrm{d} \ln \bar{a}_1 + \bar{C}_2 \mathrm{d} \ln \bar{a}_2 + \bar{C}_\mathrm{w} \mathrm{d} \ln \bar{a}_\mathrm{w} = 0\,. \tag{37}$$

When one separates the terms d ln f_1, then

$$\bar{C}_1 \mathrm{d} \ln f_1 + \bar{C}_2 \mathrm{d} \ln f_2 + \bar{C}_\mathrm{w} \mathrm{d} \ln \bar{a}_\mathrm{w} + \mathrm{d}\bar{C}_0 = 0\,, \tag{38}$$

where

$$\bar{C}_0 = \bar{C}_1 + \bar{C}_2\,. \tag{39}$$

Eq. (36), after taking the logarithm and differentiating assumes the form:

$$\mathrm{d} \ln \tilde{K} + \frac{1}{Z_1} \mathrm{d} \ln f_1 + \frac{1}{Z_2} \mathrm{d} \ln f_2 + \mathrm{d}\left(\frac{\mathrm{d}w}{\mathrm{d}\bar{x}_1} \ln \frac{\bar{a}_\mathrm{w}}{a_\mathrm{w}}\right) = 0\,. \tag{40}$$

The combined solution of eq. (38) and (40) gives

$$\begin{aligned} \ln (f_1/f_{01}) = & - Z_1 \bar{x}_2 \ln \tilde{K} + Z_1 \int_{\bar{x}_1}^{1} \ln \tilde{K} \mathrm{d}\bar{x}_1 + Z_1 \int_{\bar{C}_0}^{\bar{C}_{01}} \frac{\mathrm{d}\bar{C}_0}{Z_1 \bar{C}_1 + Z_2 \bar{C}_2} \\ & - Z_1 \bar{x}_2 (\mathrm{d}w/\mathrm{d}\bar{x}_1) \ln (\bar{a}_\mathrm{w}/a_\mathrm{w}) + Z_1 W_1 \ln (\bar{a}_{\mathrm{w}1}/a_{\mathrm{w}1}) - Z_1 W \ln (\bar{a}_\mathrm{w}/a_\mathrm{w})\,. \end{aligned} \tag{41}$$

$$\begin{aligned} \ln (f_2/f_{02}) = & \; Z_2 \bar{x}_1 \ln \tilde{K} - Z_2 \int_{0}^{\bar{x}_1} \ln \tilde{K} \mathrm{d}\bar{x}_1 - Z_2 \int_{\bar{C}_{02}}^{\bar{C}_0} \frac{\mathrm{d}\bar{C}_0}{Z_1 \bar{C}_1 + Z_2 \bar{C}_2} \\ & + Z_2 \bar{x}_1 (\mathrm{d}w/\mathrm{d}\bar{x}_1) \ln (\bar{a}_\mathrm{w}/a_\mathrm{w}) + Z_2 W_2 \ln (\bar{a}_{\mathrm{w}2}/a_{\mathrm{w}2}) - Z_2 W \ln (\bar{a}_\mathrm{w}/a_\mathrm{w})\,, \end{aligned} \tag{42}$$

where $\bar{C}_{0i}$, $\bar{a}_{\mathrm{w}i}$, $a_{\mathrm{w}i}$, f_{0i} correspond to a pure i-ionic form.

The integration limits in eq. (41) and (42) are selected in such a way that $f_i = 1$ at $\bar{x}_i = 1$, $\bar{C}_0 = \bar{C}_{0i}$, $\bar{a}_\mathrm{w} = \bar{a}_{\mathrm{w}i}$, $a_\mathrm{w} = a_{\mathrm{w}i}$.

Substituting eq. (41) and (42) into (40) we get

$$\ln K_\mathrm{w} = \int_0^1 \ln \tilde{K}\, \mathrm{d}\bar{x}_1 + \int_{\bar{C}_{02}}^{\bar{C}_{01}} \frac{\mathrm{d}\bar{C}_0}{Z_1 \bar{C}_1 + Z_2 \bar{C}_2} + \ln \frac{(\bar{a}_{\mathrm{w}1}/a_{\mathrm{w}1})^{W_1}}{(\bar{a}_{\mathrm{w}2}/a_{\mathrm{w}2})^{W_2}} + \ln \frac{f_{02}^{1/Z_2}}{f_{01}^{1/Z_1}}\,. \tag{43}$$

The activity coefficients f as well as the values of K_W and $\tilde{K}$ refer to concentration scale C. It is natural that by applying other values in eq. (4) one can obtain coefficients and constants conjugated with molality f_m, $\tilde{K}_m$, $K_{m\mathrm{w}}$ or a molar fraction $N(f_N, \tilde{K}_N, K_{N\mathrm{W}})$. In contrast with this, the values obtained according to equations given in Refs. [53–55] do not refer to a particular composition scale.

The standard free energy of an ion exchange process is determined by the equation

$$\Delta G^\circ = \frac{1}{Z_1}\bar{\mu}_1^\circ - \frac{1}{Z_2}\bar{\mu}_2^\circ - \frac{1}{Z_1}\mu_1^\circ + \frac{1}{Z_2}\mu_2^\circ + (\bar{\mu}_w^\circ - \mu_w^\circ)\frac{dw}{d\bar{x}_1} = -RT\ln K_w \quad (44)$$

and it depends on the selection of standard states. The standard state of an electrolyte solution is selected so that $\mu_i = \mu_i^\circ$ for a hypothetical solution in which $a_i = 1$ at $C_i = 1$; the reference state ($f_i = 1$) is the state of a real solution for $C_i \rightarrow 0$.

If the standard state of an ion exchanger is determined in the same way, then the characteristic function $\Delta G^\circ_{\text{char}}$ is obtained, which will be equal for ion exchangers of any crosslinkage of the same chemical type, provided that the difference in their behaviour is expressed as a difference in degree of swelling. This is possible only for the osmotic version when $\bar{\mu}_w \neq \mu_w$. The choice of standard potentials depends on the last term of eq. (43). The values of f_{0i} can be found from the data of the binary resinate/solvent system [56, 57].

The free energy of ion exchange between an ion exchanger of a given cross-linkage and a standard solution (ΔG°) can be calculated if

$$\bar{\mu}_i^\circ = \bar{\mu}_i(\bar{C}_{0i}) \text{ at } \bar{a}_{0i} = \bar{C}_{0i} f_{0i} = 1 \quad \text{i.e., } f_{0i} = \frac{1}{\bar{C}_{0i}}. \quad (45)$$

Then the last term of eq. (43) can be rewritten as follows:

$$\ln(f_{01}^{1/Z_1}/f_{02}^{1/Z_2}) = \ln(\bar{C}_{02}^{1/Z_2}/\bar{C}_{01}^{1/Z_1}). \quad (46)$$

In this case either $\bar{\mu}_w \neq \mu_w$ or $\bar{\mu}_w = \mu_w$ may be assumed. In the latter case, the terms containing $\bar{a}_w$ in eq. (41)–(43) will be equal to 0. It is also to be noted that neither $\Delta G^\circ_{\text{char}}$, nor ΔG° depend on the manner of expressing the ion exchanger composition (c, m or N).

Let us consider in more detail an equation for the calculation of ΔG°.

$$\Delta G^\circ = -RT\int_0^1 \ln \tilde{K} d\bar{x}_1 - RT\int_{\bar{C}_{02}}^{\bar{C}_{01}} \frac{d\bar{C}_0}{Z_1\bar{C}_1 + Z_2\bar{C}_2} - RT\ln\frac{\bar{a}_{1w}^{W_1}}{\bar{a}_{2w}^{W_2}} + RT\ln\frac{\bar{C}_{01}^{1/Z_1}}{\bar{C}_{02}^{1/Z_2}}. \quad (47)$$

Here a_w is assumed to be equal to unity. The equation is considerably simplified when $Z_1 = Z_2$. Then the second and fourth terms (eq. (47)) cancel out, and ΔG° can be represented as two independent terms. ΔG_0° is the free energy of the redistribution of the ions

$$\Delta G^\circ = \Delta G_0^\circ + \Delta G_w^\circ = -RT\int_0^1 \ln\tilde{K} d\bar{x}_1 - RT\ln(\bar{a}_{w_1}^{w_1}/\bar{a}_{w_2}^{w_2}). \quad (48)$$

ΔG_w^o is the free energy of interphase solvent transfer accompaning the substitution of one equivalent of ion 2 by ion 1. For the non-osmotic theory, the second term is missing and we get a frequently applied expression:

$$\Delta G^o = -RT\ln K = -RT\int_0^1 \ln \tilde{K} d\bar{x}_1 . \tag{49}$$

In the case of a heterovalent exchange, a further term appears, ΔG_c^o. Eq. (47) can be represented as a sum of terms invariant under concentration for $Z_1 \neq Z_2$.

$$\begin{aligned}\Delta G^o &= \Delta G_0^o + \Delta G_c^o + \Delta G_w^o \\ &= -RT\int_0^1 \ln \tilde{K}\frac{\bar{C}_{02}^{1/Z_2}}{\bar{C}_{01}^{1/Z_1}} d\bar{x}_1 - RT\int_{\bar{C}_{02}}^{\bar{C}_{01}} \frac{d\bar{C}_0}{Z_1\bar{C}_1 + Z_2\bar{C}_2} \\ &\quad - RT\ln(\bar{a}_{w_1}^{w_1}/\bar{a}_{w_2}^{w_2}) .\end{aligned} \tag{50}$$

When $\bar{a}_{w1} = \bar{a}_{w2}$, two terms remain

$$\Delta G^o = \Delta G_0^o + \Delta G_c^o = -RT\int_0^1 \ln \tilde{K}\frac{\bar{C}_{02}^{1/Z_2}}{\bar{C}_{01}^{1/Z_1}} d\bar{x}_1 - RT\int_{\bar{C}_{02}}^{\bar{C}_{01}} \frac{d\bar{C}_0}{Z_1\bar{C}_1 + Z_2\bar{C}_2} . \tag{51}$$

The above changes in free energy do not give any information about its dependence on the degree of exchange and relate to a complete substitution of ion 2 by ion 1. More detailed information about the regularities of exchange can be obtained from the differential free energy ΔG, corresponding to the exchange of an ion equivalent on the infinite amount of ion exchanger with composition $\bar{x}_1$. The differential free energy ΔG is determined by the equation:

$$\Delta G^o = \int_0^1 \Delta G d\bar{x}_1 . \tag{52}$$

Representing eq. (47) in the form of eq. (52), we get

$$\Delta G = -RT\ln \tilde{K}\frac{\bar{C}_{02}^{1/Z_2}}{\bar{C}_{01}^{1/Z_1}} - \frac{RT}{Z_1\bar{C}_1 + Z_2\bar{C}_2}\frac{d\bar{C}_0}{d\bar{x}_1} - RT\frac{d(W\ln \bar{a}_w)}{d\bar{x}_1} . \tag{53}$$

For the special case with $\bar{a}_w = 1$, $Z_1 = Z_2$ and linear dependence of $\ln \bar{C}_0$ on $\bar{x}_1$, the second term of eq. (53) is:

$$RT\frac{1}{\bar{C}_0}\frac{d\bar{C}_0}{d\bar{x}_1} = RT\ln\frac{\bar{C}_{01}}{\bar{C}_{02}} . \tag{54}$$

Then

$$\Delta G = RT\ln \tilde{K} . \tag{55}$$

In many cases, dependence of $\ln \bar{C}_0$ on $\bar{x}_1$ is quite well approximated by a straight line and eq. (55) is valid (Figure 6.8).

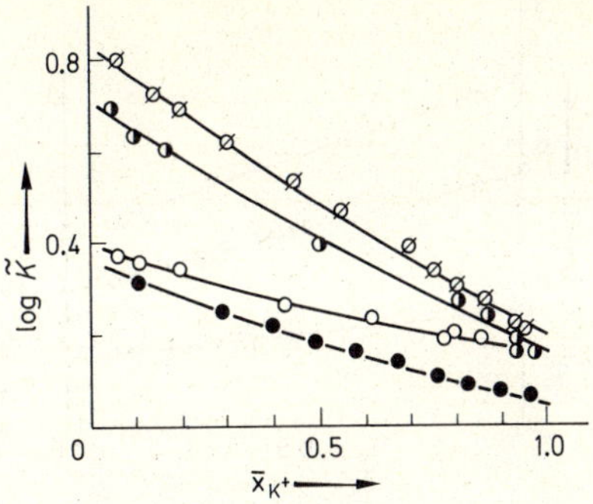

Figure 6.6 Apparent constants $\tilde{K}$ of exchange equilibrium K^+-H^+ on Dowex 50W ion exchangers versus $\bar{x}_{K^+}$.
● – 1%; ◑ – 4%; ○ 8%; ∅ – 12% DVB; t = 25 °C. Ionic strength of equilibrium solutions is 0.1 mol/l [57].

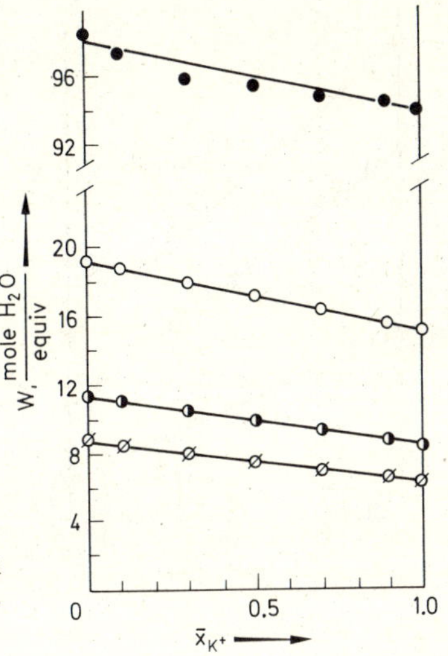

Figure 6.7 Dependence $W = f(\bar{x}_{K^+})$ at equilibrium of Dowex 50W ion exchangers with water. See caption to Figure 6.6.

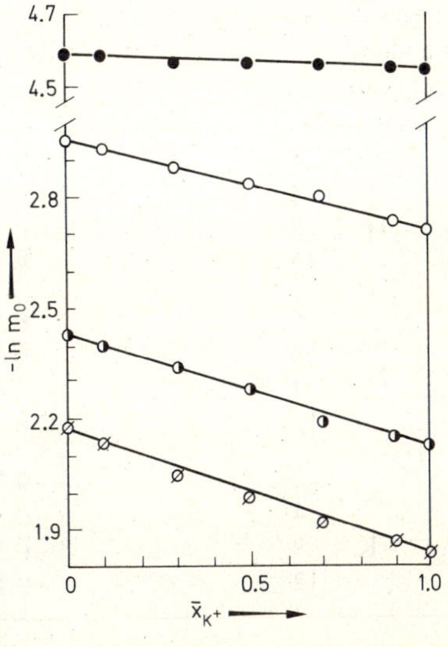

Figure 6.8 Logarithm of total molality m_0 in Dowex 50W ion exchangers versus $\bar{x}_{K^+}$ equilibrium with water. See Figure 6.6

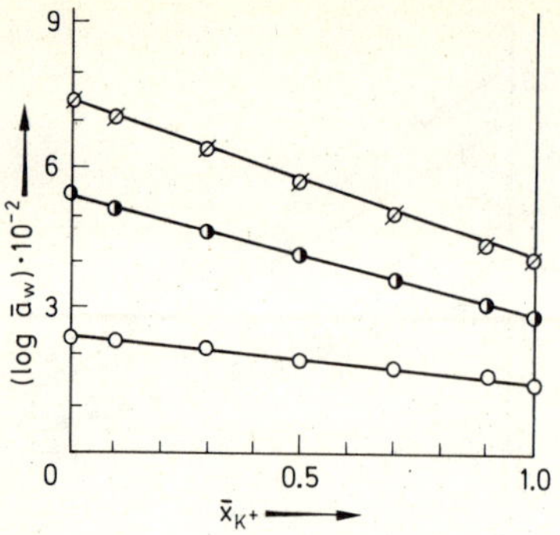

Figure 6.9 Logarithm of water activity a_w versus $\bar{x}_{K^+}$ at equilibrium of Dowex 50W ion exchangers with water. See Figure 6.6

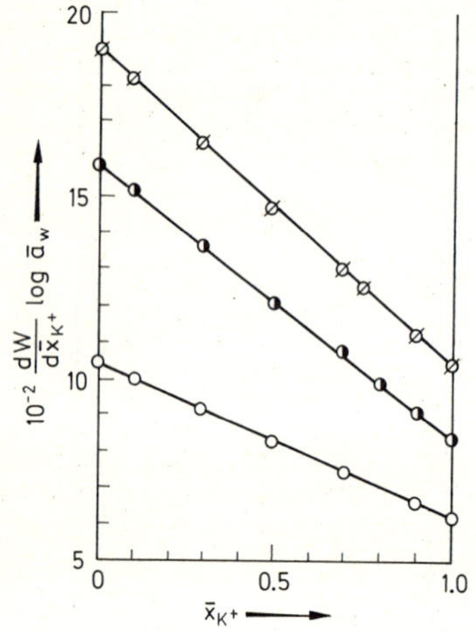

Figure 6.10 Dependence of $\left(\frac{dw}{d\bar{x}_{K^+}} \log \bar{a}_w\right) = f(\bar{x}_{K^+})$ at equilibrium of ion exchangers with water. See Figure 6.6.

Table 6.1 Standard Gibb's free energy of ion exchange processes on sulphostyrene resins (kcal/equ)

Exchange (%DVB)		ΔG°_{char} eq. (44)	ΔG° eq. (48)	ΔG° eq. (49)
	1	−0.35	−0.25	−0.25
	4	−0.29	−0.72	−0.37
K^+-H^+	8	−0.32	−1.09	−0.57
	12	−0.19	−1.20	−0.65
	1	−0.63	−0.63	−0.42
	4	−0.64	−1.24	−0.69
Ag^+-H^+	8	−0.66	−1.76	−1.09
	12	−0.58	−2.02	−1.33
	1	−0.31	−0.20	−0.20
	4	−0.36	−0.53	−0.32
Ag^+-K^+	8	−0.35	−0.68	−0.53
	12	−0.41	−0.83	−0.69

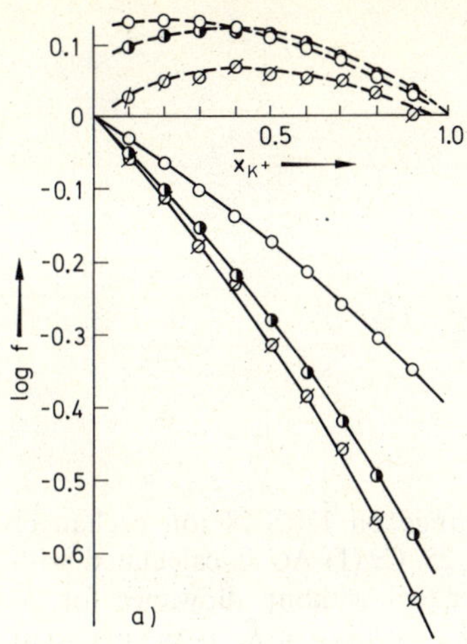

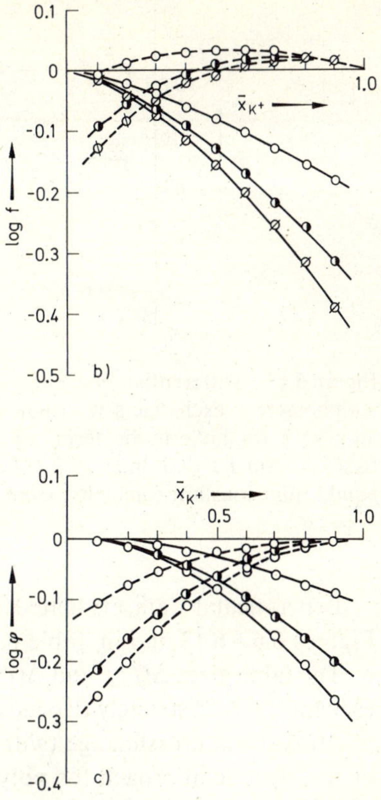

Figure 6.11 Logarithm of activity coefficients of RH (solid lines) and RK (dashed lines) resinates versus $\bar{x}_{K^+}$ calculated from eq. (41) and (42) for Dowex 50W ion exchangers with different content of DVB: ○ – 4%; ◑ – 8%; ⌀ 12%.

a) eq. (41) and (42) at $a_w = 1$, $f_{oi} = 1$; b) terms containing $\bar{a}_w$ are omitted; c) terms containing $\bar{a}_w$ and $\bar{C}_0$ are omitted.

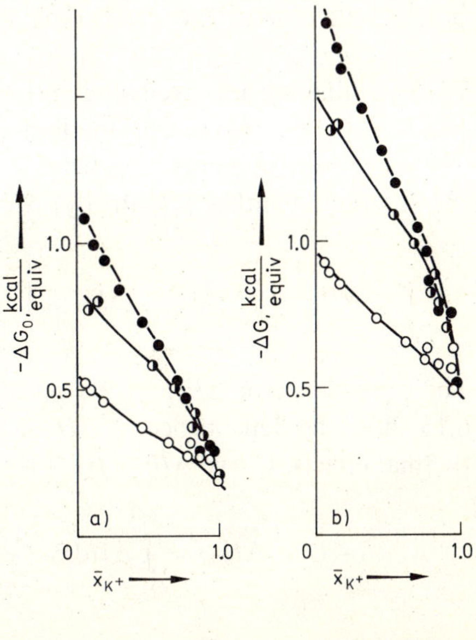

Figure 6.12 Free energy of $K^+ - H^+$ exchange on Dowex 50W ion exchangers versus $\bar{x}_{K^+}$. ○ – 4, ◑ – 8, ● – 12% DVB.

a) ΔG is calculated from eq. (55) ("non-osmotic" version)

b) ΔG is calculated from eq. (53) ("osmotic" version).

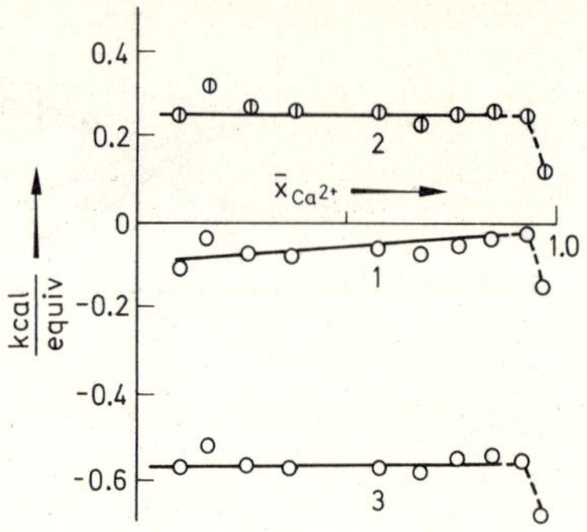

Figure 6.13 Differential free energy of $Ca^{2+}-H^+$ exchange on KRSX8 ion exchanger (sulphostyrene exchangers with 8% of para-DVB) at $t = 25\,°C$; (1) ΔG is calculated with allowance for an osmotic term, eq. (53); (2) ΔG is calculated without allowance for an osmotic term $RT\,\mathrm{d}(W \ln \bar{a}_w)\mathrm{d}\bar{x}_1$; (3) dependence $-RT \ln \tilde{K}_m = f(\bar{x}_{Ca^{2+}})$. $\tilde{K}_m$ is the apparent equilibrium constant calculated from molalities [82].

Experimental data evaluated by use of the above equations are presented in Figures 6.6–6.13 and in Table 6.1.

The table gives ΔG^o_{char} and ΔG^o for a number of exchanges on sulphostyrene ion exchangers. ΔG^o strongly depends on the degree of crosslinkage. The independence of ΔG^o_{char} from crosslinkage (with scattering $\pm$ 0.1 kcal/equ) confirms the validity of the osmotic approach for polyelectrolytes of this type up to rather high values of crosslinkage ($\approx$ 12% DVB).

6.1.6 Ion exchange enthalpy and entropy

To determine the ion exchange enthalpy, ΔH, it is necessary to measure thermal effects as $Q = -\Delta H_{int}$. This method always gives values corresponding to a change of ion exchanger composition from $\bar{x}_1'$ to $\bar{x}_1''$. Similar to eq. (52) one can determine the differential enthalpy (entropy) ΔH and ΔS:

$$\Delta H^\circ = \int_0^1 \Delta H \mathrm{d}\bar{x}_1, \quad \Delta S^\circ = \int_0^1 \Delta S \mathrm{d}\bar{x}_1 . \tag{56}$$

The dependence of ΔH and ΔS on $\bar{x}_1$ are of great importance. Figures 6.14 and 6.15 show the dependence of ΔH on $\bar{x}_1$ obtained from the experimental data on thermal effects. Curve $\Delta H = f(\bar{x}_1)$ has to satisfy the condition:

$$-Q = \Delta H_{int} = \int_{\bar{x}_1'}^{\bar{x}_1''} \Delta H \mathrm{d}\bar{x}_1 . \tag{57}$$

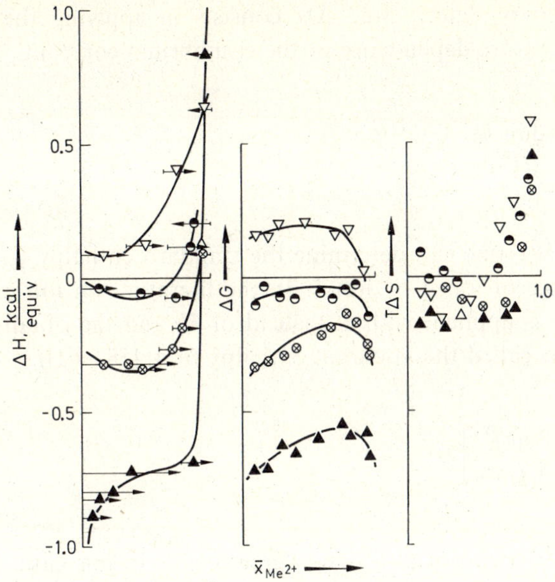

Figure 6.14 Differential thermodynamic functions of ion exchange of alkaline-earth metal ions on H^+ and KRS-p × 8 ion exchanger at $t = 25\,°C$. ΔG is caldulated from eq. (53). ▽ – Mg^{2+}; ○ – Ca^{2+}; ⊗ – Sr^{2+}; ▲ – Ba^{2+}. ΔH are found from thermal effects of exchange. The beginning of arrows corresponds to the starting ion exchanger composition; their end – to the final composition.

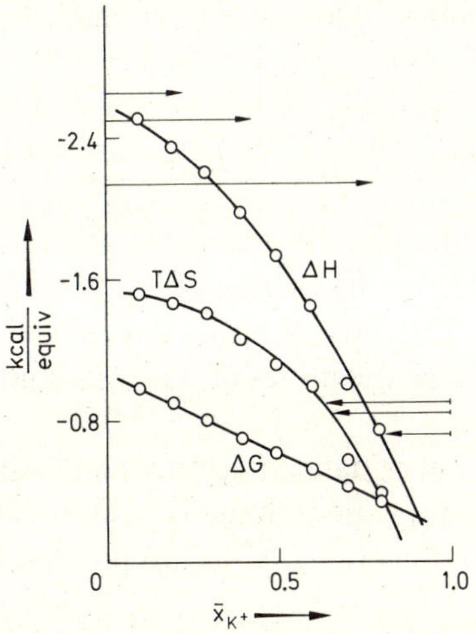

Figure 6.15 Thermodynamic functions of $K^+ - H^+$ exchange on Dowex 50X12 ion exchanger at $t = 25\,°C$. ΔH is calculated from the temperature shift of an apparent equilibrium constant, eq. (59), according to the data of [62]; equilibria were studied at 0, 25, 40, and 60 °C. The arrows show the calorimetric results. ΔG was calculated in the non-osmotic version (eq. 55).

The second method of determining ΔH consists in applying the van't Hoff equation to the temperature dependence of the equilibrium constant

$$\Delta H^\circ = -R\frac{\mathrm{d}\ln H}{\mathrm{d}(1/T)}. \tag{58}$$

According to eq. (58), one can determine the standard enthalpy corresponding to a complete exchange of ion 2 by ion 1. In the literature on ion exchange, an equation such as (58) is applied, with $\tilde{K}$ instead of K, and the obtained value of ΔH depending on $\bar{x}_1$ is called the apparent enthalpy (e. g. [58–61]).

$$\Delta H = -R\left[\frac{\partial \ln \tilde{K}}{\partial(1/T)}\right]_{\bar{x}_1}. \tag{59}$$

The differentiation is carried out at a fixed value for $\bar{x}_1$. In this case, the solution phase composition (x_1) appears to be different at different temperatures. This is justified since ΔH is the sum of the enthalpy changes of ion exchanger and solution ($\Delta H = \Delta H_{R+} + \Delta H_S$). ΔH_S is the mixing heat of diluted solutions, which is far less than ΔH_R, i. e., $\Delta H_S \cong 0$. It is, therefore, necessary to fix the composition of the ion exchanger phase rather than that of the solution [64]. In interpretations of the experimental data in the literature the value of ΔH is in fact referred to as the differential enthalpy, i. e., the enthalpy of exchange of an equivalent of ions 1 and 2 on the infinite amount of ion exchanger, with the ion exchanger composition being constant and equal to $\bar{x}_1$.

A comparison of the values of ΔH, calculated from eq. (59) and ΔH determined according to calorimetric data, shows that these values are quite close [58, 62–64]. Figures 6.14 and 6.15.

Applicability of eq. (59) for the calculation of ΔH is apparently associated with the low temperature dependence of the two last terms of eq. (53), including the values of swelling and solvent activity. The most accurate data available seems to be in Ref. [60], where the exchange of trace quantities of ions was studied over the temperature range 0–150 °C.

Up to the present, a rather large amount of data on the thermodynamic functions of ion exchange processes has been accumulated. Some typical examples for dependence of differential thermodynamic functions on $\bar{x}_i$ are presented in Figures 6.13–6.15.

The simplest regularities seem to be observed in the exchange of ions of alkali metals on sulphostyrene ion exchangers and halogenides on strong base anion exchangers. In these cases, the driving force behind the selective sorption of an ion is interaction between a counterion and a functional group. The values of ΔH, ΔG, and ΔS are negative.

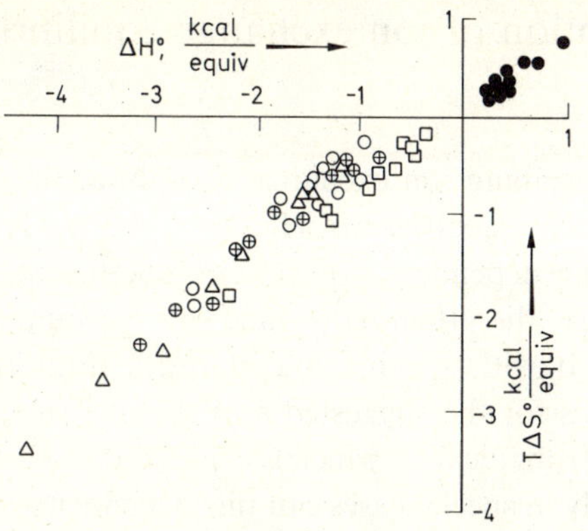

Figure 6.16 Correlation between ΔH^0 and $T\Delta S^0$ for $I^+ - H^+$ exchange on sulphostyrene exchangers. ● – Li^+; □ – Na^+; ○ – K^+; ⊗ – Rb^+; △ – Cs^+ [66]. The points are related to the resins containing 6.5, 10.5, and 25% DVB and to the temperatures of 0, 25, 60, and 90 °C. ΔG is computed according to eq. (55).

The free energy appears to be comparatively small due to a substantial decrease in entropy that accompanies the binding of a selectively absorbed ion. The relation between standard functions turns out to be very simple and independent of the ion type, temperature, and ion exchanger crosslinkage [65, 66] (Figure 6.16).

The straight line of dependence $\Delta H^\circ = f(T\Delta S^\circ)$ lies in the vicinity of the origin, i. e., at $T\Delta S^\circ = 0$, $\Delta H^\circ = 0$, which confirms that practically only the interaction is responsible for a decrease in entropy. ΔG° amounted to 35% ΔH°, hence the exchange selectivity of cations of alkali metals is comparatively low even at high ΔH°. This dependence seems to satisfy the simplest case of the electrostatic interaction between counterions and a functional group. For the exchange of weakly hydrated anions the value of ΔG° (correspondingly, the absorption selectivity) appears to be higher, as could be expected given the value of ΔH° [67, 68]. This agrees with the concept of additional "hydrophobic interaction" [69] and is caused, as shown in [68], by a decrease in the interface between "hydrophobic particles" and water when the ionic pair $PhNMe_3^+An^-$ is formed.

In sorption of counterions with a higher charge, ΔG° and selectivity also become anomalously high due to the advantageous contribution from entropy [70]. In many cases, the anomalously strong interaction of a counterion with functional groups (e. g., polarization or local hydrolysis) takes place, which changes the balance between the thermodynamic functions as compared to Figure 6.16 (e. g., exchange on sulphostyrene, Ag^+, Tl^+ [59, 67, 71, 72], carboxylic and phosphoric acid type ion exchangers [7]). We have given here very simple examples. Real changes in thermodynamic functions in the exchange of complex-forming ions or organic ions are far more complicated and at the present time cannot be predicted.

6.1.7 Quantitative description of ion exchange equilibria in non-ideal systems

In thermodynamic equations describing ion exchange equilibria, the non-ideality of the systems is expressed by activity coefficients. However, thermodynamics does not have means of predicting the dependence of activity coefficients on the ion exchanger composition. Therefore, the attempts to take into account non-ideality in practice are always associated with specific uncertainties resulting from the simplifications assumed as a basis for the suggested model. As a rule, crosslinked polyelectrolytes show deviations from ideality, which is evidenced by the dependence of $\tilde{K}$ on the exchange degree. Only in special cases can this dependence be explained by a change in the osmotic properties of the ion exchanger. In general, its interpretation is of a qualitative nature. Recently, a simple, approximate theoretical model permitting consideration of the non-ideality of a liquid ion exchanger has been suggested [73, 74]. The dependence of various properties of ion exchange systems on their composition is adequately expressed by the main equation of this model.

It is the aim of this section to derive the equation $\ln \tilde{K} = f(\bar{x}_1)$ for non-ideal systems. The dependence of the apparent constant $\tilde{K}$ on the polymer ion exchanger composition is exemplified by systems with singly-charged counterions [75, 76].

Assume that every counterion has two nearest neighbours in the swollen polyelectrolyte, interaction with which is essential when describing the system as a whole (e. g., two counterions located on both sides from it on the same polymer chain). Therefore, in the presence of two counterions (1 and 2) three states can be realized for every counterion: 111, 211 (= 112), 212, 222, 221 (= 122), 121.

Denote chemical potentials corresponding to these states as $\bar{\mu}_{1(11)}$, $\bar{\mu}_{1(12)}$, $\bar{\mu}_{1(22)}$, $\bar{\mu}_{2(22)}$, $\bar{\mu}_{2(12)}$, $\bar{\mu}_{2(11)}$. Thus, every ion in the ion exchanger phase is distributed among three energetic levels corresponding to the above states. The exchange of ions in every level is assumed to be ideal. Nonideality of the total ion exchange process is caused by unequal loading of the levels at different $\bar{x}_1$. A real ion exchange equilibrium results from superposition of the following ideal hypothetic equilibria (their constants are given on the right):

$$\overline{121} + 1 \rightleftharpoons \overline{111} + 2, \qquad k^1_{2(11)} \tag{60a}$$

$$\overline{122}(\overline{221}) + 1 \rightleftharpoons \overline{112}(\overline{211}) + 2, \quad k^1_{2(12)} \tag{60b}$$

$$\overline{222} + 1 \rightleftharpoons \overline{212} + 2, \qquad k^1_{2(22)}\,. \tag{60c}$$

In the accepted nomenclature numerals 1 and 2 denote exchangeable ions.

Let us find the free energy of exchange of an equivalent of ions 2 for ions 1 at composition $\bar{x}_1$, as a sum of the free energies at every level.

$$\Delta G = \alpha_{1(22)}\Delta G^{\circ}_{1(22)} + \alpha_{1(12)}\Delta G^{\circ}_{1(12)} + \alpha_{1(11)}\Delta G^{\circ}_{1(11)}\,, \tag{61}$$

where α_1 is a fraction of an equivalent of ions 1 distributed between levels and

$$\alpha_{1(22)} + \alpha_{1(12)} + \alpha_{1(11)} = 1\,. \tag{62}$$

The free energy on exchange of an equivalent of singly charged ions 2 for ions 1 occurring so that the nearest neighbours of the ions 1 are only ions 2 can be written as

$$dG_{1(22)} = \bar{\mu}_{1(22)}d\bar{n}_{1(22)} + \mu_1 dn_1 + \bar{\mu}_{2(22)}d\bar{n}_{2(22)} + \mu_2 dn_2\,. \tag{63}$$

If

$$d\bar{n}_{1(22)} = -\,dn_1;\; d\bar{n}_{2(22)} = -\,dn_2;\; d\bar{n}_{1(22)} = -\,d\bar{n}_{2(22)} \tag{64}$$

then

$$dG_{1(22)} = (\bar{\mu}_{1(22)} - \mu_1)d\bar{n}_{1(22)} - (\bar{\mu}_{2(22)} - \mu_2)d\bar{n}_{1(22)}\,. \tag{65}$$

Such a process occurs when an infinitesimal of ions 1 is distributed between the ion exchanger, saturated with ions 2, and the electrolyte solution containing these ions. If standard states are selected so that $\bar{\mu}_i = \bar{\mu}_i^\circ$ at $\bar{x}_i = x_i = 1$, then

$$\Delta G^\circ_{1(22)} = \bar{\mu}^\circ_{1(22)} - \mu^\circ_1 = -\,RT\ln(\bar{x}_1/x_1)_{x_1\to 0} = -\,RT\ln k^1_{2(22)}\,. \tag{66}$$

Similarly,

$$\Delta G^\circ_{1(11)} = -\,\Delta G^\circ_{2(11)} = -\,RT\ln(x_2/\bar{x}_2)_{x_2\to 0} = -\,RT\ln k^1_{2(11)} \tag{67}$$

and

$$\Delta G^\circ_{1(12)} = \bar{\mu}^\circ_{1(12)} - \mu^\circ_1 - \bar{\mu}^\circ_{2(12)} + \mu^\circ_2 = -\,RT\ln(\bar{x}_{1(12)}x_2/\bar{x}_{2(12)}x_1) = -\,RT\ln k^1_{2(12)}\,. \tag{68}$$

Assume that the counterions are statistically distributed between levels. This is acceptable since the difference in energy of interionic interactions (several kJ/mol) is far less than their absolute values (dozens of kJ/mol). Therefore, the value of $\alpha_{1(22)}$ for every composition of $\bar{x}_1$ is equal to a relative mole fraction of ions 1, surrounded by ions 2. Denote this value as $\bar{x}_{1(22)}$. The same is valid also for other ionic combinations, i. e.

$$\alpha_{1(22)} = \bar{x}_{1(22)};\; \alpha_{1(12)} = \bar{x}_{1(12)};\; \alpha_{1(11)} = \bar{x}_{1(11)}\,. \tag{69}$$

The value of $\bar{x}_{1(22)}$ is proportional to the probability that in its neighbourhood some ion 2 has ions of the same type, and this probability is equal to $\bar{x}_2^2$. Since the total sum of both probabilities and mole fractions is equal to unity, we have

$$\alpha_{1(22)} = \bar{x}_{1(22)} = \bar{x}_2^2\,. \tag{70}$$

By a similar reasoning, we find

$$\alpha_{1(12)} = 2\bar{x}_1\bar{x}_2;\; \alpha_{1(11)} = \bar{x}_1^2\,. \tag{71}$$

It follows from eq. (61), (66), (68) and (71), that

$$-\frac{\Delta G}{RT} = \ln K = \ln (k^1_{2(11)})\bar{x}_1^2 + 2\ln (k^1_{2(12)})\bar{x}_1\bar{x}_2 + \ln (k^1_{2(22)})\bar{x}_2^2 . \quad (72)$$

It has been shown that, in general, the equation

$$Y = y_{11}\bar{x}_1^2 + 2y_{12}\bar{x}_1\bar{x}_2 + y_{22}\bar{x}_2^2 \quad (73)$$

is valid for some additive property Y, where y_{11}, y_{22}, and y_{12} characterize properties Y at very level. In our case,

$$y_{11} = \ln k^1_{2(11)},\ y_{22} = \ln k^1_{2(22)},\ y_{12} = \ln k^1_{2(12)} . \quad (74)$$

In other cases, Y may be equal to ΔG, ΔH, ΔS, and W.

The values of $k^1_{2(11)}$, $k^1_{2(22)}$, and $k^1_{2(12)}$ can be found by fitting the constants of eq. (73) to experimental data. The values of y_{11} and y_{12} may be determined by extrapolation of the function $\ln \tilde{K} = f(\bar{x}_1)$, correspondingly for $\bar{x}_1 \to 0$ and $\bar{x}_1 \to 1$, in case there are no reasons for expecting a sharp change in the form of these functions in the range $\bar{x}_1 \to 0$ and $\bar{x}_1 \to 1$ due to side processes (e. g., for liquid ion exchangers [73]). In the case of highly crosslinked ion exchangers, this procedure may be unacceptable, as at low content for one of the components, impurities or irregular exchange sites can affect the form of $\ln \tilde{K} = f(\bar{x}_1)$. Therefore, examples are given below for highly crosslinked ion exchangers showing especially large deviations from ideality where coefficients y_{11}, y_{12}, and y_{22} were calculated for the main concentration range ($\bar{x}_1 = 0.1-0.9$) by the least squares fit, provided that the total sum σ is a minimum[3].

$$\begin{aligned}\sigma &= \sum_{i=1}^{n} (\ln \tilde{K}_i - Y)^2 \\ &= \sum_{i=1}^{n} [\ln \tilde{K}_i - y_{22} + 2(y_{22} - y_{12})\bar{x}_1 - (y_{11} + y_{22} - 2y_{12})\bar{x}_1^2]^2 .\end{aligned} \quad (75)$$

Figures 6.17–6.19 give experimental data evaluated by means of eq. (73). In the majority of cases this equation describes the experimental data well.

As shown in [77], eq. (73) can be applied to predict the dependence of ΔG or $\ln \tilde{K}$ on the composition of liquid ion exchangers in exchange involving organic ions. It turned out that constants y_{11}, y_{12}, and y_{22} can be split up into increments corresponding to atomic groups (CH_3, CH_2, NH_2^+, NH^+, N^+), and that the dependence of $\ln \tilde{K}$ on $\bar{x}_i$ in a real system can be obtained or predicted by the summation of the introduction of the groups in the organic ion followed by substitution into eq. (73).

[3] $\tilde{K}$ was practically used without taking activity coefficients in the solution into account since the equilibria were studied in diluted solutions at constant ionic strength.

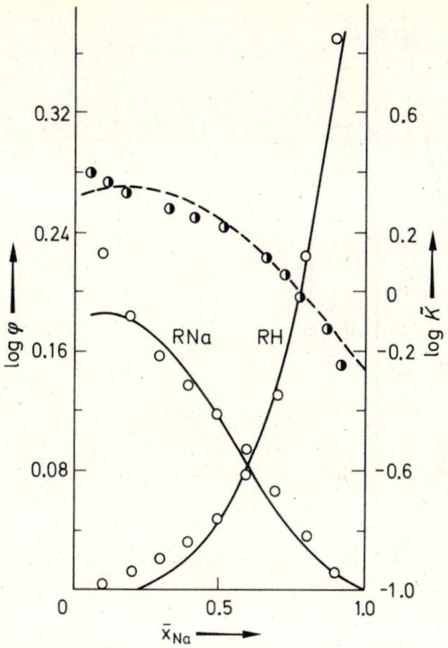

Figure 6.17 Logarithms of apparent equilibrium constant (right scale, dotted line), log $\tilde{K} = f(\bar{x}_{Na})$ and activity coefficients $\log \varphi = f(\bar{x}_{Na})$ as functions of equivalent fraction of Na^+ in resin phase of the $Na^+ - H^+$ system. Sulphostyrene exchanger with 25% DVB (KU − 2X25) [76]: ◐ – points on lines $\log K = f(\bar{x}_{Na})$ computed from experimental data; ○ – points on lines $\log \varphi = f(\bar{x}_i)$ computed by graphical integration of eq. (76). Curves computed from eq. (73) and (77) with the following constants: $y_{NaNa} = -0.276$; $y_{HH} = 0.350$; $y_{NaH} = 0.459$; ($y = \log k_{ij}$).

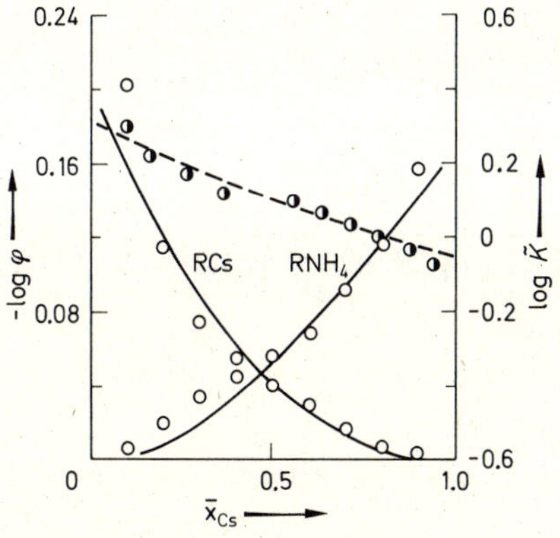

Figure 6.18 $Cs^+ - NH_4^+$ system, KRS-25; $y_{CsCs} = -0.046$; $y_{NH_4NH_4} = 0.324$; $y_{CsNH_4} = 0.067$. See Figure 6.17.

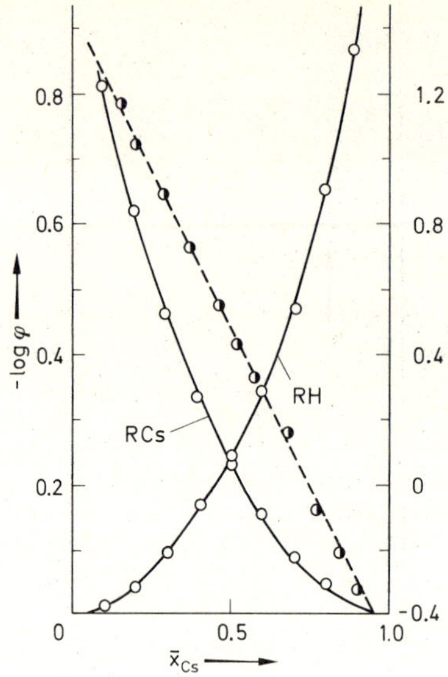

Figure 6.19 Cs^+-H^+ system, KRS-25; $y_{CsCs} = -0.503$; $y_{HH} = 1.475$; $y_{CsH} = 0.505$. See Figure 6.17.

By this manner the nonideality of an ion exchange system can be explicitly taken into account by use of the model described.

Eq. (72) makes it possible to find the dependence of resinate activity coefficients on $\bar{x}_i$ in an analytical form. Let us find these dependences for the simplest case, corresponding to the non-osmotic approach to ion exchange equilibrium, $Z_1 = Z_2$, and for the case when the ion exchanger swelling is independent of $\bar{x}_1$. Here we have well-known equations derived in [53, 54].

$$\ln \varphi_i = -(1 - \bar{x}_i) \ln \tilde{K} + \int_{\bar{x}_i}^{1} \ln \tilde{K} \, d\bar{x}_i . \tag{76}$$

Substitution of $\ln \tilde{K}$ from eq. (72) into eq. (76) gives

$$\ln \varphi_1 = -\left(\ln \frac{k^1_{2(12)}}{k^1_{2(11)}}\right)\bar{x}_2^2 - \frac{2}{3}\left(\ln \frac{k^1_{2(11)}k^1_{2(22)}}{(k^1_{2(12)})^2}\right)\bar{x}_2^3 , \tag{77a}$$

and

$$\ln \varphi_2 = \left(\ln \frac{k^1_{2(12)}}{k^1_{2(22)}}\right)\bar{x}_1^2 + \frac{2}{3}\left(\ln \frac{k^1_{2(11)}k^1_{2(22)}}{(k^1_{2(12)})^2}\right)\bar{x}_1^3 . \tag{77b}$$

Taking into account the above assumptions

$$\ln K = \int_0^1 \ln \tilde{K} \, d\bar{x}_1 = \frac{1}{3} \ln \left(k^1_{2(11)}k^1_{2(22)}k^1_{2(12)}\right) . \tag{78}$$

The particular case where

$$\ln k^1_{2(12)} = \frac{1}{2} \ln (k^1_{2(11)} k^1_{2(22)}) \tag{79}$$

deserves special attention. It corresponds to a linear dependence of $\ln \tilde{K}$ on $\bar{x}_1$. Under this condition $\ln (k^1_{2(11)} k^1_{2(22)}/(k^1_{2(12)})^2) = 0$ and eq. (77a) and (77b) become analogous to Kielland's equation:

$$\ln \varphi_1 = \frac{1}{2} \ln (k^1_{2(11)}/k^1_{2(22)})\, \bar{x}_2^2 = \text{const.}\ \bar{x}_2^2 . \tag{80a}$$

$$\ln \varphi_2 = \frac{1}{2} \ln (k^1_{2(11)}/k^1_{2(22)})\, \bar{x}_1^2 = \text{const.}\ \bar{x}_1^2 . \tag{80b}$$

Application of the stricter eq. (41) and (42) instead of eq. (76) at $\bar{a}_w = 1$ will not change the form of eq. (80), provided the dependence of $\ln \bar{C}_0$ on $\bar{x}_1$ is expressed by a polynomial not higher than the second degree, which is the case (Figure 6.8). The numerical values of φ_1 and φ_2 will change and depend on the composition scale.

Figures 6.17–6.19 illustrate the application of eq. (77) to experimental data. The values of $\ln \varphi_i$ calculated from these equations are in good agreement with the $\ln \varphi_i$ found by graphical integration of dependencies $\ln \tilde{K} = f(\bar{x}_i)$.

It is of interest to note that some systems with strong deviations from ideality exhibit regular behaviour and practically obey Kielland's equation ($NH_4^+ - H^+$, $Cs^+ - H^+$). Activity coefficients for the $Li^+ - H^+$ and $Na^+ - H^+$ systems are described only by eq. (77), though their absolute values are smaller than those in the previous case.

It seems to be true that any ion exchange system involving a monofunctional ion exchanger with statistically distributed counterions can be satisfactorily described by the model suggested. It is not applicable to polyfunctional ion exchangers or to systems with a sieve effect. In these cases, there is no simple statistical distribution of counterions among exchangeable sites.

The approach suggested allows all ion exchange systems with statistical distribution of counterions to be divided into three types:

1) Ideal systems corresponding to a constant value of $\tilde{K}$ and to activity coefficients equal to 1.
2) Regular systems corresponding to a linear function $\ln \tilde{K} = f(\bar{x}_i)$ and to Kielland's equations for the activity coefficients.
3) Irregular statistical systems corresponding to a second-degree equation for $\ln \tilde{K} = f(\bar{x}_i)$ and to a third-degree equation for the activity coefficients.

The deviation from eq. (72) and (77) can be considered as an indication of the inequivalence of the exchange groups of the ion exchanger. Consideration of data on the exchange of univalent inorganic ions on sulphostyrene ion exchangers showed [76] that it can take place only at $\bar{x}_i < 0.1$.

It is to be emphasized that eq. (73) ensures a compact and convenient means for tabulation and representation of experimental data on ion exchange.

Recently the suggested approach has been extended to multiionic systems [78–81]. It has been shown that equilibria in non-ideal multiionic systems can be accurately predicted by the use of multiparametric analogues of eq. (73) and experimental data on binary systems.

References

[1] F. Helfferich (1962) Ion Exchange, Mc Graw-Hill, New York.

[2] R. Griessbach (1957) Austauschadsorption in Theorie und Praxis, Akademie-Verlag, Berlin.

[3] Y. Marcus and A. S. Kertes (1969) Ion Exchange and Solvent Extraction of Metal Complexes, Wiley-Interscience, London.

[4] Yu. A. Kokotov and V. A. Pasechnik (1970) Ion Exchange Equilibrium and Kinetics, Khimiya, Leningrad.

[5] G. V. Samsonov, E. B. Trostyanskaya and T. E. El'kin (1969) Ion Exchange. Sorption of Organic Substances, Nauka, Leningrad.

[6] M. M. Senyavin (1980) Ion Exchange in Technology and Analysis of Inorganic Substances, Khimiya, Moscow.

[7] V. S. Soldatov (1972) Simple Ion Exchange Equilibria, Nauka i Tekhnika, Minsk.

[8] V. S. Soldatov and A. F. Pestrak; L. K. Arkhangel'skij; V. S. Soldatov (1968) in: Ion Exchange Thermodynamics, pp. 25–39; 49–59; 70–83, Nauka i Tekhnika, Minsk.

[9] V. S. Soldatov (1981) in: Ion Exchange (M. M. Senyavin, ed.) pp. 111–126, Nauka, Moscow.

[10] J. A. Marinsky; D. Reichenberg; R. M. Diamond and D. K. Witney (1968) in: Ion Exchange (J. A. Marinsky, ed.) pp. 9–76; 104–173; 174–280, Mir, Moscow.

[11] A. M. Tolmachev and V. I. Gorshkov (1966) On thermodynamics of ion exchange. Zhurn. Fiz. Khimii *40,* 1924–1928.

[12] J. W. Gibbs (1982) Thermodynamics. Statistical mechanics, pp. 61–344, Nauka, Moscow.

[13] V. S. Soldatov, V. V. Matusevich and L. V. Novitskaya (1983) Activity coefficients of polystyrenesulphonic acid salts. Zhurn. Fiz. Khimii *57,* 2926–2929.

[14] A. Katchalsky (1971) Polyelectrolytes. Pure and Appl. Chem. *26,* 327–373.

[15] G. S. Manning (1974) in: Polyelectrolytes 1972 (E. Selegny, M. Mandel & U. P. Strauss, eds.) pp. 9–37, Dordrecht-Boston.

[16] H. P. Gregor and J. M. Gregor (1977) Coulombic reactions of polyelectrolytes with counterions of different sizes. J. Chem. Phys. *66,* 1934–1939.

[17] A. Takahashi, N. Kato and M. Nagasawa (1970) The osmotic pressure of polyelectrolyte in neutral salt solutions. J. Phys. Chem. *74,* 944–946.

[18] M. Reddy and J. A. Marinsky (1970) A further investigation of the osmotic properties of hydrogen and sodium polystyrenesulphonates. J. Phys. Chem. *74,* 3884–3890.

[19] M. Reddy, J. A. Marinsky and A. Sarkar (1970) Osmotic properties of divalent metal polystyrenesulphonates in aqueous solutions. *74,* 3891–3896.

[20] D. Kozak and D. Dolar (1971) Osmotic coefficient of polyelectrolyte solutions. II. Polystyrenesulphonates with divalent counterions. Z. Phys. Chem., N. F. *76,* 93–97.

[21] S. Oman (1974) Osmotic coefficients of aqueous polyelectrolyte solutions at low concentrations. I. Polystyrenesulphonates with mono- and bivalent counterions. Macromol. Chem. *175,* 2133–2140.

[22] G. E. Boyd (1974) in: Polyelectrolytes 1972 (E. Selegny, M. Mandel and U. P. Strauss, eds.) pp. 135–155, Dordrecht-Boston.

[23] P. Chu and J. A. Marinsky (1967) The osmotic properties of polystyrenesulphonates. I. The osmotic coefficients. J. Phys. Chem. *71,* 4352–4359.

[24] V. V. Matusevich, L. V. Novitskaya and V. S. Soldatov (1982) New data on osmotic properties of polyvinylbenzyltrimethylammonium chloride and bromide. Dokl. Akad. Nauk BSSR *26,* 523–525.

[25] V. V. Matusevich and V. S. Soldatov (1984) Osmotic coefficients of aqueous solutions of hydroxide, iodate, nitrate, and sulphate of polyvinylbenzyltrimethylammonium. Vestsi Akad. Navuk BSSR, Ser. Khim. Navuk *5,* 28–30.

[26] E. A. Guggenheim (1957) Modern Thermodynamics, Elsevier, Amsterdam.

[27] D. Dolar and J. Lapanje (1958) Thermodynamic function of swelling of cross-linked polystyrenesulphonic acid resins. I. Resins in the hydrogen state. Z. Phys. Chem. N. F., *18,* 11–25. D. Dolar, S. Lapanje and S. Paljk (1962) III. Resins in the potassium state. Ibid., *34,* 360–368.

[28] S. Lapanje and D. Dolar (1959) II. Resins in the sodium state Ibid., *21,* 376–387.

[29] D. Dolar, S. Lapanje and L. Čelik (1960) Thermodynamics of water sorption by polymethylstyrenesulphonic acid and its sodium salt. Makromol. Chem. *41,* 77–85.

[30] H. P. Gregor et al. (1952) Studies on ion exchange resins. V. Water vapour sorption. J. Colloid Sci. *7,* 511–534.

[31] B. R. Sundheim, M. H. Waxman and H. P. Gregor (1953) Studies on ion exchange resins. VII. Water vapour sorption by crosslinked polystyrenesulphonic acid resins. J. Phys. Chem. *57,* 974–978.

[32] E. Glueckauf and G. P. Kitt (1955) A theoretical treatment of cation exchangers. III. The hydration of cations in polystyrene sulphonates. Proc. Roy. Soc. A *228,* 322–341.

[33] L. V. Novitskaya, V. S. Soldatov and Z. I. Sosinovich (1973) Thermodynamics of water sorption by salt forms of the KB-4P2 carboxyl cation exchangers. Kolloidnyj Zhurnal *35,* 583–586.

[34] G. E. Boyd and B. A. Soldano (1953) Osmotic free energies of ion exchangers. Z. Electrochem. *57,* 162–172.

[35] T. Matsuura (1954) Studies on some physico-chemical properties of ion exchange resin. 1. Heat of wetting. Bull. Chem. Soc. Jap. *27,* 281–287.

[36] L. S. Yurkova, A. F. Kolosova and K. M. Olshanova (1974) Sorption of water vapour by the AV-17 anion exchanger in different ionic forms. Zhurn. Fiz. Khimii *48,* 1496–1499.

[37] Z. I. Sosinovich, L. V. Novitskaya, V. S. Soldatov and E. Högfeldt (1984) in: Ion Exchange and Solvent Extraction (J. A. Marinsky and J. Marcus, eds.) vol. 9, pp. 303–338, Dekker, New York.

[38] V. S. Soldatov, L. V. Novitskaja and Z. I. Sosinovich (1974) On the state of water absorbed by anion exchanger. Kolloidnyj Zhurnal *36,* 990–992.

[39] L. V. Novitskaya, Z. I. Sosinovich and V. S. Soldatov (1975) Thermodynamics of water sorption by halogenated forms of Dowex 1X4. Kolloidnyj Zhurnal *37,* 1186–1189.

[40] K. Gärtner and J. Giesemann (1965) Thermodynamics of exchange adsorption. 3. Experimental studies on swelling thermodynamics of weakly acidic cation exchangers. Z. Phys. Chem. (DDR) *228,* 129–138.

[41] V. S. Soldatov and L. V. Novitskaya (1981) New version of Donnan's equation for calculation of unexchangeable sorption of electrolytes. Dokl. Akad. Nauk BSSR *25,* 42–44.

[42] D. H. Freeman, V. C. Patel and T. M. Buchanan (1965) Electrolyte uptake equilibria with low cross-linked ion-exchange resins. J. Phys. Chem. *69,* 1477–1481.

[43] V. V. Matusevich, V. S. Soldatov and L. V. Novitskaya (1984) Osmotic pressure of polystyrenesulphonic acid solutions and its salts. Zhurn. Fiz. Khimii *58,* 353–357.

[44] H. P. Gregor (1948) A general thermodynamic theory of ion exchange processes. J. Am. Chem. Soc. *70,* 1293; (1951) Gibbs-Donnan equilibria in ion-exchange resin systems. Ibid., *73,* 642–650.

[45] G. E. Myers and G. E. Boyd (1956) A thermodynamic calculation of cation exchange selectivities. J. Phys. Chem. *60,* 521–529.

[46] G. E. Boyd, S. Lindenbaum and G. E. Myers (1961) A thermodynamic calculation of selectivity coefficients for strong base anion exchangers. J. Phys. Chem. *65,* 577–586.

[47] G. Dickel (1960) Thermodynamische Behandlung der Ionenaustauschergleichgewichte nach dem Gibbs-Donnan-Guggenheim'schen Membran-Modell. Z. Phys. Chem. N. F. *25,* 233–252.

[48] V. I. Gorshkov and L. V. Kustova (1970) Calculation of equilibrium concentration constants for exchange of different valency ions. I. Equation derivations. Zhurn. Fiz. Khimii *44,* 257–259. II. Equilibrium of exchange $Sr^{2+}-H^+$, $Ba^{2+}-H^+$, $Ba^{2+}-K^+$ on Dowex 50. Ibid., *44,* 259.

[49] G. V. Samsonov and V. A. Pasechnik (1969) Ion exchange and ion exchanger swelling. Uspekhi Khimii *38,* 1257–1293.

[50] L. K. Arkhangel'skij (1982) in: Ion Exchangers in Chemical Technology, pp. 116–158, Khimiya, Leningrad.

[51] Yu. A. Kokotov (1979) On calculation of chemical potentials for components of ternary systems exemplified by the mixed A, B-form of ion exchanger and water. Zhurn. Fiz. Khimii *53,* 1152–1157.

[52] V. S. Soldatov (1972) On thermodynamics of ion exchange equilibria. Zhurn. Fiz. Khimii *46,* 434–438. Calculation of thermodynamic values characterizing ion exchange equilibria. Ibid., *46,* 1078–1082.

[53] E. Ekedahl, E. Högfeldt and L. G. Sillen (1950) Acta Chimica Scand. *4,* 556, 928.

[54] A. W. Davidson and W. J. Argensinger (1953) Equilibrium constants of cation exchange processes. Ann. N. Y. Acad. Sci. *57,* 105–115.

[55] G. L. Gaines and H. C. Thomas (1953) Adsorption studies on clay minerals. II. A formulation of thermodynamics of exchange adsorption. J. Chem. Phys. *21,* 714–718.

[56] V. S. Soldatov and A. F. Pestrak (1974) Activity coefficients of monoionic forms of sulpho-cation exchanger. Zhurn. Fiz. Khimii *48,* 444–446.

[57] V. S. Soldatov and A. F. Pestrak (1973) On thermodynamics of ion exchange on sulphonated styrene and divinylbenzene copolymers. XIX. Free energies of exchange of H^+, K^+, Ag^+ ions on sulphostyrene ion exchangers. Zhurn. Fiz. Khimii *47,* 1498–1505.

[58] O. D. Bonner and J. R. Overton (1961) The effect of temperature on ion exchange equilibria. IV. The comparison of enthalpy changes calculated from equilibrium measurements and calorimetrically measured values. J. Phys. Chem. *65,* 1599–1602.

[59] O. D. Bonner and R. R. Pruett (1959) The effect of temperature on ion exchange equilibria. II. The ammonium-hydrogen and thallous-hydrogen exchanges. III. Exchanges involving some divalent ions. J. Phys. Chem. *63,* 1417; 1420–1423.

[60] K. A. Kraus and R. J. Raridon (1959) Temperature dependence of some cation exchange equilibria in the range 0 to 200 °C. J. Phys. Chem. *63,* 1901–1907.

[61] G. L. Starobinets and V. S. Soldatov (1963) On thermodynamics of ion exchange on sulphonated styrene and divinylbenzene copolymers Zhurn. Fiz. Khimii *37,* 294–299.

[62] L. V. Novitskaya et al. (1971) On thermodynamics of ion exchange on sulphonated styrene and divinylbenzene copolymers. XII. Exchange heat of Ca^{2+}, K^+ and H^+ ions. Zhurn. Fiz. Khimii *45,* 124–127.

[63] G. E. Boyd, F. Vaslow and S. Lindenbaum (1964) J. Phys. Chem. *68*, 590.

[64] W. R. Heumann and D. Patterson (1966) Heats of potassium-lithium ion exchange on sulphonated polystyrene resins. Canad. J. Chem. *44*, 2139–2142.

[65] G. L. Starobinets and V. S. Soldatov (1962) On thermodynamics of ion exchange. Dokl. Akad. Nauk BSSR *6*, 233–236.

[66] V. S. Soldatov and G. L. Starobinets (1964) On thermodynamics of ion exchange on sulphonated styrene and divinylbenzene copolymers. II. Effect of ion exchanger cross-linkage, ion radius, and temperature on equilibrium. Zhurn. Fiz. Khimii *38*, 681–685.

[67] F. Vaslow and G. E. Boyd (1966) Heats of exchange of halide ions in variously cross-linked strong-base anion exchangers. J. Phys. Chem. *70*, 2507–2511.

[68] V. S. Soldatov and V. I. Sokolova (1969) Exchange selectivity and thermodynamics of some anion exchange processes. Vestsi Akad. Navuk BSSR, Ser. Khim Navuk *4*, 31–37.

[69] R. M. Diamond (1963) The aqueous solution behaviour of large univalent ions. A new type of ion-pairing. J. Phys. Chem. *67*, 2513–2517.

[70] E. H. Cruickshank and P. Meares (1958) The thermodynamics of cation exchange. Part 3. Thermodynamic properties of resins containing a mixture of cations. Trans. Faraday Soc. *54*, 174–178.

[71] J. S. Redinha and J. A. Kitchener (1963) Thermodynamics of ion exchange processes. System H, Na, and Ag with polystyrene sulphonated resins. Trans. Faraday Soc. *59*, 515–529.

[72] V. S. Soldatov and G. L. Starobinets (1964) On thermodynamics of ion exchange. Studies of Ion Exchange Material Properties, pp. 36–43, Nauka, Moscow.

[73] Z. I. Kuvaeva, V. S. Soldatov, F. Fredlund and E. Högfeldt (1978) On the properties of solid and liquid ion exchangers. VI. Application of a simple model to some ion exchange reactions on dinonylnaphtalene sulphonic acid in different solvents and for different cations with heptane as solvent. J. Inorg. Nucl. Chem. *40*, 103–108.

[74] E. Högfeldt and V. S. Soldatov (1979) A simple model for the formation of mixed micelles applied to salts of dinonylnaphtalene sulphonic acid. J. Inorg. Nucl. Chem. *41*, 575–577.

[75] V. S. Soldatov and V. A. Bichkova (1982) Analytical form of dependence of resinate activity coefficients on ion exchanger composition. Dokl. Akad. Nauk BSSR *26*, 1106–1109.

[76] V. S. Soldatov and V. A. Bichkova (1983) Quantitative description of ion exchange selectivity in non-ideal systems. Reactive Polymers *1*, 251–259.

[77] V. S. Soldatov and A. V. Mikulich (1984) Free energy additivity of alkylammonium ion exchange on the liquid dinonylnaphtalenesulphonic acid cation exchanger. Zhurn. Fiz. Khimii *58*, 889–894.

[78] V. S. Soldatov and V. A. Bichkova (1984) Ion exchange selectivity and activity coefficients as functions of the ion exchange composition. Ion Exchange Technology, pp. 179–188, Horwood, Chichester.

[79] V. S. Soldatov and V. A. Bichkova (1983) Quantitative description of ion exchange selectivity in multiionic systems. Dokl. Akad. Nauk BSSR *27*, 540–543.

[80] V. S. Soldatov and V. A. Bichkova (1985) Binary ion exchange selectivity coefficients in multiionic systems. Reactive Polymers *3*, 199.

[81] V. A. Bichkova and V. S. Soldatov (1985) A method for predicting ion exchange equilibria in the ternary ion exchange systems. Reactive Polymers *3*, 207.

[82] V. S. Soldatov and Z. I. Kuvaeva (1976) Thermodynamics of ion exchange on sulphonated styrene and divinylbenzene co-polymers. XVIII. Thermodynamic exchange functions of alkaline-earth metal ions for H^+ and Li^+ ions. Zhurn. Fiz. Khimii *50*, 1140–1143.

6.2 Ion Exchange Kinetics

Friedrich G. Helfferich and Yng-Long Hwang
The Pennsylvania State University, Department of Chemical Engineering,
133 Fenske Laboratory, University Park, PA, USA

List of symbols

C	concentration of solution
$\overline{C}$	capacity of ion exchanger – concentration of fixed ionic groups
C_i	concentration of species i in solution
$\overline{C}_i$	concentration of species i in ion exchanger
$\langle\overline{C}_i\rangle$	average concentration of species i in ion exchanger bead
D	interdiffusion coefficient in solution
$\overline{D}$	interdiffusion coefficient in ion exchanger
D_i	individual diffusion coefficient of species i in solution
$\overline{D}_i$	individual diffusion coefficient of species i in ion exchanger
F	fractional approach to equilibrium (or fractional conversion)
F	Faraday constant
He	Helfferich number (see eq. (6))
J_i	flux of species i in solution
$\overline{J}_i$	flux of species i in ion exchanger
k_f	mass transfer coefficient for liquid film

k_p	mass transfer coefficient for ion exchanger particle
R	universal gas constant
Re	Reynolds number
r_o	radius of ion exchanger bead
T	temperature
t	time
$t_{1/2}$	half time for approach to equilibrium
V	volume of solution
$\overline{V}$	volume of ion exchanger
v	flow velocity of solution
z_i	valence of species i (negative for anions)
α_{ij}	separation factor of counter-ion i relative to counter-ion j
δ	thickness of fictitious Nernst film of liquid phase
ε	fractional void volume in ion exchanger
φ	electric potential

Superscripts

*	value at interface of ion exchanger and solution (where equilibrium is normally assumed)
b	value in bulk solution
o	value at zero time (initial value)
∞	value at infinite time (equilibrium value).

Introduction

Chapter 6.2 deals with what is commonly called "ion exchange kinetics." By tradition, this encompasses the rate phenomena occurring when an ion exchange material is exposed to an electrolyte solution and thereby is partially or completely converted to another ionic form (see Figure 6.20). It does not include the much more complex dynamics of ion exchange columns or other pieces of contacting

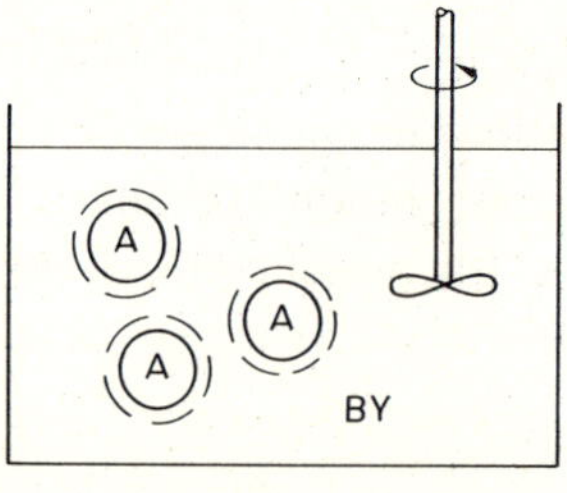

Figure 6.20 System: exchange of counter-ions A and B between exchanger beads and well-stirred liquid phase (from [1]).

equipment producing effluents whose composition is affected by a number of additional processes. Moreover, for simplicity's sake, this survey of kinetics will remain confined to solid ion exchangers in the form of spherical beads of uniform size (or materials whose behavior can be approximated within reason by assuming them to be of such a form).

For the scientist interested in mass transfer phenomena, ion exchange offers a fascinating field of study. As in no other, mass transfer by diffusion and convection is coupled with sorption equilibria, electrochemical phenomena, and, possibly, reactions such as acid dissociation, neutralization, and complex formation. The combinations are kaleidoscopic, and even in our day, new, striking facets are still being discovered. For the practical engineer, on the other hand, ion exchange kinetics is a mine field. The complexity of the dynamic problems forces him to work with greatly simplified kinetic models, but only a thorough understanding of fundamentals will enable him to avoid using what may be likened to an isothermal theory of the steam engine: an approach that dwells on minor effects while omitting the most important ones.

This section attempts to give a brief, up-to-date survey of ion exchange kinetics and its mathematical models, concentrating on phenomenological aspects rather than mathematical detail. In this way, we hope, the scientist will find stimulation and challenge; the engineer, markers to guide him through his mine field.

6.2.1 Mechanism of ion exchange

Ordinary ion exchange can be viewed as shown schematically in Figure 6.21. An ion exchanger containing mobile counter-ions of one kind (A) is exposed to a liquid containing counter-ions of a different kind (B). Some counter-ions from the liquid

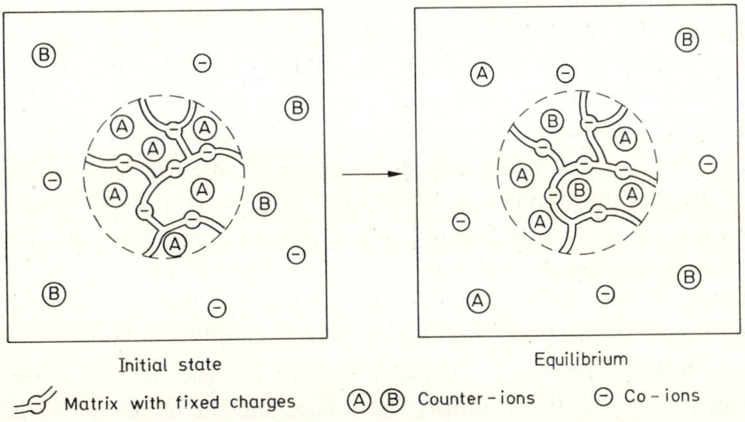

Figure 6.21 Schematic presentation of binary ion exchange as redistribution of counter-ions between exchanger and liquid phase (from [1]).

will enter the ion exchanger, in exchange for others from the bead. At equilibrium, both phases will contain both kinds of counter-ions, although usually not in the same ratio.

If we accept this simple view as a working hypothesis, ion exchange appears as a redistribution of mobile counter-ions, a mass transfer phenomenon rather than a chemical reaction. Four distinguishable mass transfer steps are involved: (1) of ion A from some initial position in the interior of the bead to its surface, (2) of that ion from the surface into the bulk liquid, (3) of ion B from the bulk liquid to the surface, and (4) of that ion from the surface into the interior of the bead. However, the fluxes of the two ions are coupled: Electroneutrality must be conserved, and this requires charge transfer by one ion to be compensated by an equivalent charge transfer by the other ion in the opposite direction. This leaves us with just two steps to consider: mass transfer of the exchanging ions within either the bead or the liquid [2].

Almost all ion exchange materials of practical interest have a microscopic pore structure that poses an enormous resistance to convection. Accordingly, mass transfer within the bead is commonly assumed to occur exclusively by diffusion. Thence the term "intraparticle diffusion" for this step.

In the liquid, on the other hand, both diffusion and convection contribute, so "liquid-phase mass transfer" seems appropriate and will be used here. However, ion exchange kinetics was first studied by physical chemists rather than engineers, and their term "film diffusion" has found wide acceptance. The "film" alludes to the fictitious Nernst film surrounding a solid particle suspended in a well-stirred liquid [3]. Actual mass transfer to and from the surface by diffusion and convection is modeled as transfer exclusively by diffusion across an adherent liquid "film." The thickness of that film is an adjustable parameter which can be related theoretically or empirically to known or measurable fluid-dynamic properties. Alternatively, a mass transfer coefficient can be used, avoiding the introduction of the film concept. Details will be shown later.

The situation is more complicated in macroporous ion exchangers, best described as having a "sand pile" structure. They consist of roughly spherical microparticles (the sand grains) fused together in the last stages of polymerization (see Figure 6.22). Here, three mass transfer steps are involved: liquid-phase mass transfer, diffusion in the macropores, and diffusion within the microspheres [5–8].

Pelletized zeolites, consisting of micron-sized crystals fused together or embedded in an inert binder [9, 10], resemble macroporous ion exchange resins with respect to the mechanim of ion exchange.

Ultimately, we shall also have to admit that our picture of ion exchange as an exclusive mass transfer phenomenon is an oversimplification. Ion exchange can be accompanied by reactions. Typical examples are the neutralization of a cation exchanger in H^+ or free-acid form by a base and the uptake of a transition-metal ion by a chelating resin. As will become apparent, such reactions can affect kinetics as profoundly as they affect equilibria.

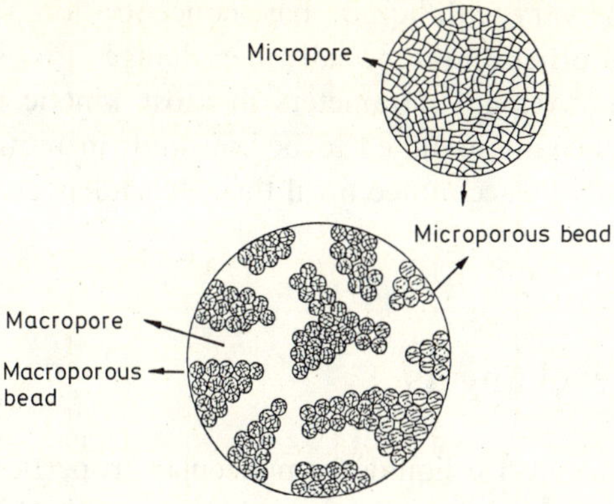

Figure 6.22 Structure of macroporous ion exchange resin (from [4]).

One further relevant facet of ion exchange is apparent from the simple picture in Figure 6.21. To the extent that Donnan exclusion is effective in keeping the concentration of co-ions in the ion exchanger negligible compared with that of the counter-ions, conservation of electroneutrality demands that the exchange between bead and liquid be stoichiometric: For each equivalent of counter-ions entering the bead, one equivalent must leave. The result is that the total concentration (in equivalents) remains constant in both bead and liquid. This constancy is assumed in most models.

6.2.2 Condition at liquid/solid interface

Ion exchangers can be viewed as "quasi-liquids" in that they are electrolytes with mobile ions in the same solvent as that of the equilibrating liquid phase. Only the crosslinks (in resins) or three-dimensional crystalline nature (in inorganic materials) of the matrix prevents them from dissolving. Accordingly, what appears macroscopically as the interface between bead and liquid is no more than the limit beyond which the matrix does not extend. No significant resistance to mass transfer can be expected per se from such an interface – unless it becomes contaminated [11]. Indeed, no mass transfer resistance of "clean" bead surfaces has ever been demonstrated, and theories commonly assume equilibrium between bead and liquid to be maintained at the interface.

Ion exchangers do, however, prefer certain counter-ions to others. This preference, or selectivity, can be characterized by binary separation factors defined as

$$\alpha_{ij} \equiv \frac{\bar{C}_i C_j}{\bar{C}_j C_i}. \tag{1}$$

Although these factors usually vary with liquid-phase concentration, ionic concentration ratios, and presence of other ions, especially in exchange of ions of different valences, they are used as convenient parameters in most kinetic models. For analytical solutions they are usually assumed to be constant; in numerical calculations their variation can easily be accounted for if their dependence on conditions is known.

6.2.3 Diffusion in ion exchangers

While data on diffusion coefficients for liquids (or molecular properties from which such diffusion coefficients can be calculated) are readily accessible [12], this is not so for ion exchangers. Diffusion in the latter is impaired by the presence of the matrix. Not only is part of the cross-sectional area blocked, but diffusion around the matrix strands has to follow a tortuous and thus longer path. There is no lack of models accounting for this effect, all using different assumptions and producing different results. At least for gel type ion exchangers the most successful has been that proposed by Mackie and Meares [13], which gives

$$\bar{D}_i = D_i[\varepsilon/(2 - \varepsilon)]^2 . \qquad (2)$$

Here, ε is, strictly speaking, the fractional intraparticle void volume, but is satisfactorily approximated by the more easily determined weight fraction of imbibed solvent (the densities of solvent and ion exchanger usually being reasonably similar).

Equation (2) has proved to be a rather good approximation for univalent counter-ions as well as for co-ions and nonelectrolytes, provided they are small. It usually overestimates the diffusion coefficients of counter-ions of higher valence and of large ions or molecules.

Apart from obstruction by the matrix, the most important facet of diffusion in ion exchangers is the action of the induced electric field. Conservation of electroneutrality imposes a constraint on the ionic fluxes: Charge transfer by the flux of one ion species must be compensated by an equivalent charge transfer of one or several others. For a true understanding of ion exchange kinetics, the physical mechanism of this coupling must be appreciated. In the course of ion exchange, if the counter-ions diffusing in opposite directions carried no charges, a greater flux of the faster ion would result. For charged ions this would amount to a net transfer of electric charge and violate the requirement of electroneutrality. The slightest deviation from electroneutrality gives rise to an electric field, which, in turn, produces electric transference of the ions. The field acts on all ions present. The direction of electric transference of counter-ions is that of diffusion of the slower counter-ion. Thus, electric transference increases the flux of the slower counter-ion and decreases that of the faster one, equalizing the net fluxes and so preventing any further build-up of net charge (see Figure 6.23).

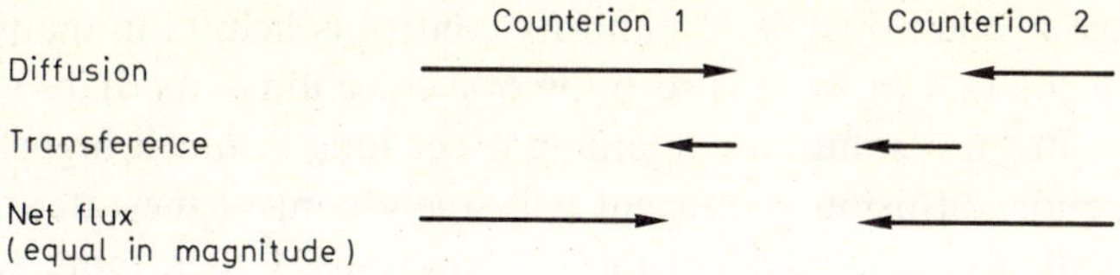

Figure 6.23 Superposition of electric transference on ordinary diffusion in Nernst-Planck equations (schematic, from [14]).

In terms of mathematics, electric transference has to be accounted for in addition to ordinary diffusion. Accordingly, instead of Fick's law

$$\bar{J}_i = -\bar{D}_i \operatorname{grad} \bar{C}_i \tag{3}$$

the Nernst-Planck equations [Refs. 15 – 17] should be used:

$$\underset{\text{net flux}}{\bar{J}_i} = \underset{\text{ordinary diffusion}}{-\bar{D}_i \operatorname{grad} \bar{C}_i} \; \underset{\text{electric transference}}{- D_i z_i \bar{C}_i (F/RT) \operatorname{grad} \varphi} \tag{4}$$

A rigorous treatment would have to be based on the Poisson-Boltzmann equation to account for the deviation from electroneutrality. However, as Planck's original work [17] has shown, this deviation is so small that electroneutrality and no net transfer of charge can be assumed without error, except for ultrathin layers or membranes, or phenomena on a micro- or millisecond time scale.

This mechanism has interesting and important implications [1]. Diffusion is a purely statistical phenomenon involving no physical force on the molecular level; as a result, its flux depends only on the concentration gradient, not on the concentration itself (first term in eq. (4)). In contrast, the corrective force of the electric field is a true physical force that acts on every ion present and so produces a flux that is proportional to the concentration of the respective ion (second term in eq. (4)). The result is that the corrective action of the electric field mainly affects the ions present as a majority; for those in the minority the electric transference term is small, so that they migrate essentially at the rate of their ordinary diffusion. For binary interdiffusion of two counter-ions A and B, in the absence of any significant amount of co-ions, the two fluxes are rigorously coupled. Not much affected by the corrective action of the electric field, the minority ion predominantly controls the rate. One might say that coupled binary diffusion is not a democratic process but, rather, is governed by the participating minority (to borrow from the vocabulary of the activist '60s) while the majority, the taxpayers, shoulder the burden. Multi-ion systems come closer to the democratic ideal: Here, the major parties haggle it out among themselves, leaving any minorities free to do their own thing[1].

[1] This principle is used in polarography, where a "supporting electrolyte" at a concentration far exceeding that of the ion of interest absorbs the transference effect of the electric field. This is

A knowledge of the principle of minority control is helpful in many situations. For instance, if the task at hand is sorption of a trace ion – perhaps a radioactive contaminant – from a solution containing other ions in much higher concentrations, the trace-ion diffusion coefficient will solely control the rate, even though the exchange calls for an equivalent flux of bulk ions. Other, more subtle consequences of minority control will become apparent later in this chapter.

6.2.4 Rate-controlling steps

Forgetting about macroporous materials and reactions for the time being, two steps appear to qualify for rate control: mass transfer in either the bead or the liquid, whichever is slower.

At first glance, we might be tempted to rule out mass transfer in the liquid. Diffusion in the bead is obstructed by the matrix, in the solution it is not. Moreover, the distance to be covered from the surface to the center of the bead is much greater than that across the liquid film or the boundary layer between the (stirred) liquid and the bead surface. Indeed, given the same concentration difference as a driving force, intraparticle diffusion would always be slower than liquid-phase mass transfer.

However, the driving force is *not* the same. In the liquid, the concentration difference is that between bulk liquid and bead surface, and so can never be larger than the bulk concentration itself. In the bead, the concentration difference is that between surface and center and can be as high as the concentration of fixed ionic groups, usually of the order of several moles per liter. The question which step controls the rate thus remains open, but the role of most of the relevant factors becomes clear. Thus, control by liquid-phase mass transfer is favored by

- low liquid-phase concentration (small driving force in the liquid),
- high ion exchange capacity (large driving force in the bead),
- small particle size (short mass transfer distance in the bead),
- open structure of ion exchanger, e. g., low degree of crosslinking (little obstruction to diffusion in the bead), and
- ineffective agitation of the liquid (low contribution of convection to liquid-phase mass transfer).

In addition, selectivity exerts a more subtle effect. If the ion exchanger prefers the entering ion, the liquid-side concentration of the latter at the bead surface remains low until substantial conversion is achieved, ions from the liquid being eagerly taken up as they arrive. As a result, the concentration difference between bulk liquid and bead surface, and thus the liquid-phase driving force, remains high until

necessary to ensure that this ion will diffuse to the mercury surface with a flux proportional to its bulk-liquid concentration.

conversion approaches completion. On the other hand, if the ion leaving the exchanger is preferred, the liquid-side concentration of the entering ion must build up for that ion to be accepted, so the concentration difference quickly diminishes and a loss in driving force results. A preference of the ion exchange for the ion it initially contains is thus an additional factor favoring rate control by liquid-phase mass transfer.

All these effects are accounted for in the following criterion [1].

$$\begin{array}{ll} \text{intraparticle diffusion control} & \text{if } He \ll 1 \\ \text{liquid-phase mass transfer control} & \text{if } He \gg 1 \end{array} \tag{5}$$

Here, He is a dimensionless quantity already named "Helfferich number" and defined by

$$He \equiv \frac{\bar{C}\,\bar{D}\,\delta}{C\,D\,\overset{\circ}{r}}(5 + 2\alpha_{AB}) \tag{6}$$

where δ is the thickness of the ficticious liquid Nernst film. This criterion, however, must be taken with a grain of salt. As most dimensionless numbers in engineering, He compares the rates of two processes; it is so defined that its value is unity when the theoretical half times of intraparticle diffusion-controlled and liquid-phase mass transfer-controlled ion exchange are equal, with the following important provisions: no reactions are involved, the ion exchanger is completely converted from the A form to the B form, and the composition of the liquid is kept constant and free of the released ion A. If these assumptions are not valid, the user interested in such a criterion as above will have to construct his own dimensionless number by equating the half times of intraparticle diffusion and liquid-phase mass transfer-controlled exchange under his conditions.

In the range $He \approx 1$, both mass transfer steps will affect the rate. The liquid-phase resistance to mass transfer is more significant at low conversion. In fact, because the rate of liquid-phase mass transfer is limited by the maximum concentration difference across the boundary layer while the rate of intraparticle diffusion theoretically is infinite at the start, the rate of any ion exchange is liquid-phase-controlled at first, though for only an insignificantly short period of time if $He \ll 1$.

In macroporous materials, the macropores essentially contain liquid at the concentration of the external liquid. Accordingly, no greater driving force compensates for obstruction of diffusion and greater distance to be covered, and rate control by mass transfer in the external liquid thus is highly unlikely. Whether diffusion in the macropores or in the microspheres controls the rate can in principle be estimated by a comparison of the half times of the two processes. Unfortunately, the required structural information (especially on size and solvent content of the microspheres) is rarely available.

Obviously, reactions can play a major role in ion exchange kinetics. One case is clear: If the reaction is very slow compared with mass transfer, the latter has time to level out any concentration gradients in the liquid and bead, and the rate will be solely controlled by the reaction – and, incidentally, be independent of particle size. However, with few exceptions, ionic reactions in solution are known to be very fast, and there is no reason to assume that this should be different in ion exchangers. Indeed, rate control by a slow reaction has never been demonstrated. Nevertheless, reactions can profoundly affect the mass transfer rate, as will become apparent later in this section, and may thus also be a significant factor to be considered in any attempt to predict the rate-controlling step.

Many criteria have been suggested to identify the rate-controlling step experimentally. Most depend on the time-honored procedure of seeking straight lines in appropriate plots and so must presuppose the validity of the mathematical models that lead to the respective straight-line relationships. More reliable are the two tests described below, best used in conjunction.

The first is the so-called "interruption test", introduced by Kressman [18] and consisting of separating the ion exchanger temporarily from the liquid while the rate is being monitored. If controlled by intraparticle diffusion, the rate will be faster upon reimmersion than it was at removal because the intraparticle concentration gradients will have had time to relax. If controlled by liquid-phase mass transfer, no such effect is observed, as the quasi-stationary gradients in the "film" are re-established in a fraction of a second.

The second test entails observation of the dependence of the rate on particle size. For reasons of geometry the rate, regardless of complicating effects that might invalidate specific models, is inversely proportional to the particle radius if controlled by liquid-phase mass transfer (being proportional to surface area per unit volume), and is inversely proportional to the square of the particle radius if controlled by intraparticle diffusion (the distance to be covered by diffusion being an additional factor).

For macroporous materials, the interruption test will be positive for both macropore and microsphere diffusion rate control. On the other hand, at constant size of the microspheres, the rate will be independent of (macro) particle size if controlled by microsphere diffusion, and inversely proportional to the square of the particle size if controlled by diffusion in the macropores.

6.2.5 Models and rate laws

In almost any field of science and technology, a model taking all conceivable effects into account would be far too complex and unwieldy for practical use. Ion exchange kinetics is no exception. Here as elsewhere, a practitioner's skill is demonstrated by his ability to choose – or construct – the simplest model that will serve his purpose. The following brief survey and discussions of models and rate laws is intended to provide background for such selections.

6.2.5.1 Driving-force models and mass transfer coefficients

In preliminary engineering estimates, e. g., for column operation, it is often desirable to have simple equations, even at the expense of accuracy. The simplest possible rate laws postulate a linear relation between flux and "driving force", taken to be the concentration difference between bead surface and bulk phase [19]:

$$J_i = k_{\mathrm{f}}(C_i^{\mathrm{b}} - C_i^*) \quad \text{for liquid phase} \tag{7}$$

$$\bar{J}_i = k_{\mathrm{p}}(\bar{C}_i^* - \langle \bar{C}_i \rangle) \text{ for particle} \tag{8}$$

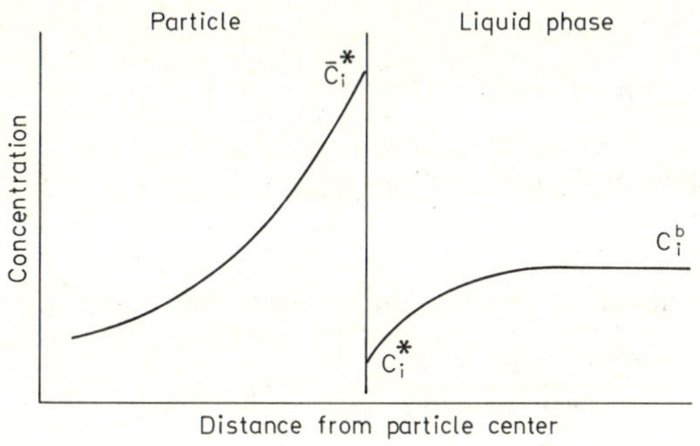

Figure 6.24 Concentration profiles of exchanging ion in particle and liquid phase (schematic).

where $\langle \bar{C}_i \rangle$ is the average concentration of ion i in the bead (see Figure 6.24). The result of interest, fractional conversion (more accurately, fractional approach to equilibrium) as a function of time, is obtained by integration over the flux. For a species i of interest:

$$F(t) = \frac{3}{r^\circ(\langle \bar{C}_i^\circ \rangle - \langle \bar{C}_i^\infty \rangle)} \int_0^t \bar{J}_i \mathrm{d}t \tag{9}$$

where $3/r^\circ$ is the surface area per unit volume of bead and the concentration difference in the denominator reflects the change from initial to equilibrium state.

Correlations suggested for the mass transfer coefficients include, among others [19–21].

$$k_{\mathrm{f}} = (10\, D/r^\circ)\, Re^{0.84} \tag{10}$$

$$k_{\mathrm{f}} = D/\delta \text{ with } \delta = 0.2r^\circ/(1 + 70r^\circ v) \text{ for low flow rate}; \quad \delta = 0.0029/v \text{ for high flow rate} \tag{11}$$

$$k_{\mathrm{p}} = 5\, \bar{D}/r^\circ \tag{12}$$

For the liquid phase, the driving-force approach is appealing, as more complex models give only marginally more reliable answers (see Section 6.2.5.3). For a constant diffusion coefficient and no selectivity ($\alpha_{AB} = 1$) the linear driving-force approach is strictly equivalent to the Fick's law model (see Section 6.2.5.2 and Table 6.2). For a constant diffusion coefficient and constant separation factor $\alpha_{AB} \neq 1$, conversion is given by Ref. [1]:

Table 6.2 Solutions of Fick's law for ion exchange with spherical beads

---

Intraparticle diffusion control:

a) fractional conversion as function of time, for constant liquid-phase composition [Ref. (25)]

$$F(t) = 1 - (6/\pi^2) \sum_{n=1}^{\infty} [(1/n^2) \exp(-\bar{D}t\pi^2 n^2/r^{\circ 2})] \tag{18}$$

b) half time, for constant liquid-phase composition

$$t_{1/2} = r^{\circ 2}/\bar{D} \tag{19}$$

c) fractional conversion as function of time, for finite volume batch [Ref. [26]]

$$F(t) = 1 - \frac{3}{3w} \sum_{n=1}^{\infty} \frac{\exp(-S_n^2 Dt/r^{\circ 2})}{1 + S_n^2/9w(w+1)} \tag{20}$$

with $w \equiv \bar{C}\bar{V}/CV$ and $S_n \cot S_n = 1 + S_n^2/3w$

Film diffusion control:

a) fractional conversion as function of time, for constant liquid-phase composition

$$F(t) = 1 - \exp(-3DCt/r^{\circ}\delta\bar{C}) \tag{21}$$

b) half time, for constant liquid-phase composition

$$t_{1/2} = 0.23\, r^{\circ}\delta\bar{C}/DC \tag{22}$$

c) fractional conversion as function of time, for finite volume batch

$$F(t) = 1 - \exp\frac{3Dt(w+1)}{r^{\circ}\delta w} \tag{23}$$

Combined intraparticle and film diffusion control:

fractional conversion as function of time, for constant liquid-phase composition [Ref. [27]]

$$F(t) = \frac{6\theta^2}{r^{\circ 2}} \sum_{n=1}^{\infty} \frac{A_n \sin^2(m_n r^{\circ})}{m_n^4} \exp(-Dm_n^2 t) \tag{24}$$

with $\theta \equiv CD/\bar{C}\bar{D}\delta, \quad A_n = \dfrac{m_n^2 r^{\circ 2} + (\theta r^{\circ} - 1)^2}{m_n^2 r^{\circ 2} + (\theta r^{\circ} - 1)\theta r^{\circ}}$

and $m_n r^{\circ} = (1 - \theta r^{\circ}) \tan m_n r^{\circ}$

$$\ln[(1 - F(t)] + (1 - 1/\alpha_{AB})F(t) = -3k_f Ct/\alpha_{AB} r^\circ \bar{C} \qquad (13)$$

(Note that $k_f = D/\delta$).

Of special interest are the limiting cases of very high preference of the ion exchanger for the entering ion ($\alpha_{AB} \ll 1$) or the leaving ion ($\alpha_{AB} \gg 1$) [22]. Except at high conversion, where $F(t) \to 1$,

$$F(t) \cong 3k_f Ct/r^\circ \bar{C} \quad \text{if} \quad \alpha_{AB} \ll 1\,. \qquad (14)$$

That is, conversion is a *linear* function of time until almost complete (see Figure 6.25). The linearity, implying a constant flux, results from the already discussed fact that high selectivity for the entering ion keeps the concentration of that ion at the bead surface very low and so essentially maintains the initial driving force. In the opposite limiting case,

$$\ln[1 - F(t)] + F(t) = -3k_f Ct/\alpha_{AB} r^\circ \bar{C} \quad \text{if} \quad \alpha_{AB} \gg 1\,. \qquad (15)$$

Here, conversion is a nonlinear function of time and, except at very low conversion, where $\ln(1 - F) \to -F$, is very slow because of the large value of α_{AB} in the denominator on the right-hand side (see Figure 6.25). Both the linear dependence of conversion on time for $\alpha_{AB} \ll 1$ and the drastic difference in rates between

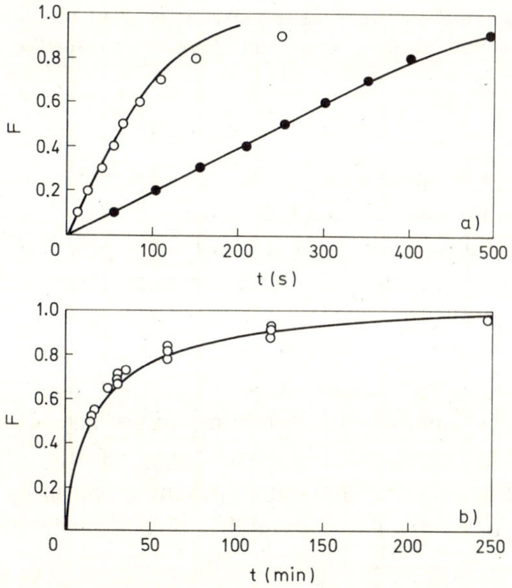

Figure 6.25 Fractional conversion versus time for forward and reverse chloride-sulfate ion exchange, illustrating asymmetry in rates and history shapes. Amerlite IRA 458 with 0.006 N solution at pH 3; (a) resin initially in chloride form, ○ 30–35 mesh, ● 18–20 mesh; (b) resin initially in sulfate form, ○ 20–30 mesh. Curves are calculated with eq. (13) (data from [22]).

forward and reverse exchange of two given ions are typical for liquid-phase mass transfer control in systems with high selectivity [22].

While the effect of selectivity is relatively easy to account for, the effect of different counter-ion mobilities is not. Interdiffusion of counter-ions of different mobilities in the boundary layer generates an electric field, which, in turn, acts on the co-ions. At the bead surface there is co-ion accumulation if the ion entering the ion exchanger is the faster one, and co-ion depletion if the opposite is true (see Figure 6.26). This makes the driving force somewhat larger in the first case, somewhat smaller in the second, even if the ion exchanger has no selectivity [23].

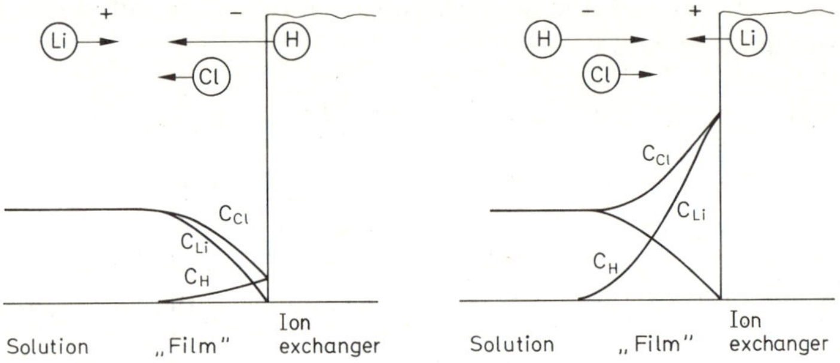

Figure 6.26 Concentration profiles in liquid phase at surface of ion exchange during forward and reverse lithium-hydrogen exchange between cation exchanger and chloride solution (schematic), illustrating co-ion accumulation or depletion at exchanger surface caused by electric field (from [1]).

A more elaborate theory would be needed to provide the dependence of the mass transfer coefficient on the mobility ratio before the linear driving-force equation could reflect this effect. In the absence of such a theory, the practical user may wish to be on the safe side by working with the diffusion coefficient of the slower of the two exchanging ions.

While providing a rather good approximation for the liquid phase, the linear driving-force approach applied to intraparticle diffusion has a fundamental flaw. In reality, diffusion in the bead is a non-steady-state process: For any given average concentration in the bead, the rate depends on the shape of the intraparticle concentration gradient and thus on prior history, as the interruption test demonstrates. Equation (8), originating from a two-film model [24] and presuming steady-state diffusion from the surface across a shell into a uniform "bulk particle", ignores this dependence. Thus, even the most elaborate nonlinear driving-force law with variable coefficient is in principle not able to reflect behavior under all possible initial and boundary conditions. Since no driving-force law can be accurate, the engineer may as well use the simplest, linear one if he cannot afford the much greater effort a transient-state calculation requires.

6.2.5.2 Fick's law models – isotopic and trace ion exchange

Another relatively simple class of kinetic models for ion exchange postulates the validity of Fick' law (eq. (3)) with constant diffusion coefficient and, where needed, proportionality of concentrations in the bead to those in the liquid. In practical terms, this amounts to demanding the diffusion coefficients of the exchanging ions to be equal and the ion exchanger to have no selectivity. These demands are rarely even approximated except in exchange of isotopic ions at otherwise constant composition.

There is, however, one exception. If the exchange is that of a trace ion and the bulk composition remains practically constant, the (constant) diffusion coefficient of the trace ion alone dictates the rate, and the separation factor usually remains essentially constant over the small concentration interval of interest.

For isotopic and trace ion exchange, a Fick's law model is thus the best choice. Otherwise, however, it can provide at best a crude approximation.

For liquid-phase mass transfer, two additional idealizations are commonly introduced. First, the curvature of the bead surface is disregarded; this simplification is justified if the film thickness is small compared with the particle radius, a condition almost always met in cases of practical interest. Second, mass transfer between bulk liquid and bead surface is assumed to be quasi-stationary; this simplification is practically always justified, as the time needed to establish quasi-stationary conditions, given by $0.5(\delta^2/D)$, is of the order of seconds or less if the liquid is agitated (δ of order of 10^{-3} cm). Granted these premises, Fick's law gives

$$J_i = (D/\delta)(C_i^{\mathrm{b}} - C_i^*), \qquad (16)$$

in strict analogy with eq. (7), and fractional conversion can be calculated with eq. (9). For the bead, a differential material balance is needed:

$$(\partial \bar{C}_i/\partial t)_{\mathrm{r}} = -\operatorname{div} \bar{J}_i \qquad (17)$$

and must be integrated. Analytical solutions for different cases are shown in Table 6.2. The condition of constant liquid-phase composition applies if either the liquid is continuosly renewed or, in batch, the amount of ions in the liquid is much larger than in the ion exchanger (i. e., $CV \gg \bar{C}\bar{V}$).

The most widely-used solution is eq. (18) for intraparticle diffusion control and constant liquid-phase composition. For high conversion, $F(t) > 0.8$, only the first term of the series need be carried out. For low to moderate conversion, $F(t) < 0.5$, the following approximation is useful:

$$F(t) \cong (6/r^\circ)(\bar{D}t/\pi)^{1/2} - 3\bar{D}t/r^{\circ 2}. \qquad (25)$$

Moreover, tables and graphs of this function are available [1, 25].

The half time for intraparticle diffusion control, given by eq. (19), is typically on the order of a few minutes. For example, with 50 mesh US screen particle size ($r^\circ = 0.015$ cm) and $\bar{D} = 10^{-6}$ cm^{2}s^{-1}, $t_{1/2} = 225$ s. In contrast, the half time for

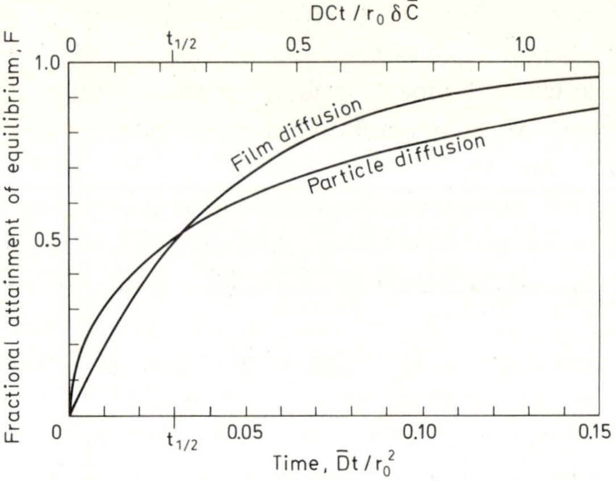

Figure 6.27 Fractional conversion versus time for isotopic exchange controlled by liquid-phase mass transfer (film diffusion) and intraparticle diffusion, calculated from eqs. (21) and (18), respectively (from [1]).

liquid-phase mass transfer control, being inversely proportional to the liquid-phase concentration, can be quite long if the latter is low.

Criterion [5] for prediction of the rate-controlling step is based on a comparison of eqs. (18) and (13) for control by liquid-phase mass transfer with arbitrary selectivity and by Fick's law intraparticle diffusion; for equal half times of these processes, $He = 1$.

Figure 6.27 shows, for comparison, conversion as a function of time for control by interparticle diffusion and liquid-phase mass transfer and illustrates the trend of intraparticle diffusion to become relatively slower as conversion progresses.

6.2.5.3 Nernst-Planck models – ion exchange without reactions

For exchange of ions of different properties, a difference in mobilities calls for the use of the Nernst-Planck equations (4) rather than Fick's law (eq. (3)). Models based on this approach are the most commonly used in ion exchange kinetics and are the standard against which new theories are measured.

The Nernst-Planck equations have been applied to the Nernst film (e. g., [23, 28–38]). However, the calculation of nonlinear concentration gradients in the fictitious film, whose thickness is derived from a linear extrapolation of gradients at the solid surface [3], greatly overburdens the model, and the modest gain in accuracy over the linear driving-force law seems to be in no proportion to the greater effort required. For more reliable results, the Nernst-Planck equations have been combined with penetration and surface-renewal models [34, 36], but its complexity has discouraged the practical use of this approach. The principal application of the Nernst-Planck equations has therefore been to intraparticle diffusion-controlled exchange.

The classic and most widely used Nernst-Planck model is for binary ion exchange, absence of co-ions, ideal intraparticle diffusion control, and constant liquid-phase composition [39, 40]. The absence of co-ions makes for rigid coupling of the two counter-ion fluxes, so that only a single differential equation need be solved. Intraparticle diffusion control and constant liquid-phase composition ensure a constant boundary condition at the bead surface. No analytical solution has been found for conversion as a function of time, but empirical approximations [1] and tabulated numerical results [41] have been given for various mobility and valence ratios and a computer program is available [42].

Mathematically, combination of the two Nernst-Planck equations for counter-ions A and B under the conditions of electroneutrality and no net charge transfer leads to [1, 39, 40]:

$$\bar{J}_A = -\bar{D}_{AB}\,\mathrm{grad}\,\bar{C}_A \tag{28}$$

where the interdiffusion coefficient $\bar{D}_{AB}$ is given by[2]

$$\bar{D}_{AB} = \frac{\bar{D}_A\bar{D}_B(z_A^2\bar{C}_A + z_B^2\bar{C}_B)}{z_A^2\bar{C}_A\bar{D}_A + z_B^2\bar{C}_B\bar{D}_B}. \tag{29}$$

This reduces to $\bar{D}_A$ for $\bar{C}_A \rightarrow O$ and to $\bar{D}_B$ for $\bar{C}_B \rightarrow O$, demonstrating the principle of minority control. The material balance (17) must now be integrated with the nonlinear eq. (28). This requires numerical methods.

The electric field, accounted for in the Nernst-Planck equations, introduces an interesting asymmetry. The exchange of two given counter-ions under otherwise identical conditions is faster if the ion entering the ion exchanger is the slower of the two (e. g., see Figure 6.28). Moreover, the transient concentration profiles in the bead are quite different in the two cases (see Figure 6.29). For extreme mobility ratios, a rather sharp concentration front moves from the surface to the center if the entering ion is the slower one. Both effects are related to minority rate control [1], discussed earlier, and have been confirmed experimentally.

Under the conditions of constant liquid-phase composition and ideal intraparticle diffusion control, the boundary condition at the bead surface is constant. Moreover, if the liquid is free of the ion leaving the bead, the surface in binary exchange is, at all times, completely converted to the form of the entering ion. Accordingly, under these conditions the rate is independent of both the selectivity of the ion exchanger and the liquid-phase concentration. If either premise is relaxed, this is no longer true, as the ion leaving the bead will then be present in finite liquid-

---

[2] Nernst [3] derived his equation for the diffusion coefficient of an electrolyte in solution (i. e., in the absence of fixed charges) in the same way. In that case, the mobile ions are cation and anion, whose concentrations in equivalents, $|z_i|C_i$, are necessarily equal, so that the equation reduces to $D_{AY} = D_A D_Y(|z_A| + |z_Y|)/(|z_A|D_A + |z_Y|D_Y)$ and yields a concentration-independent coefficient.

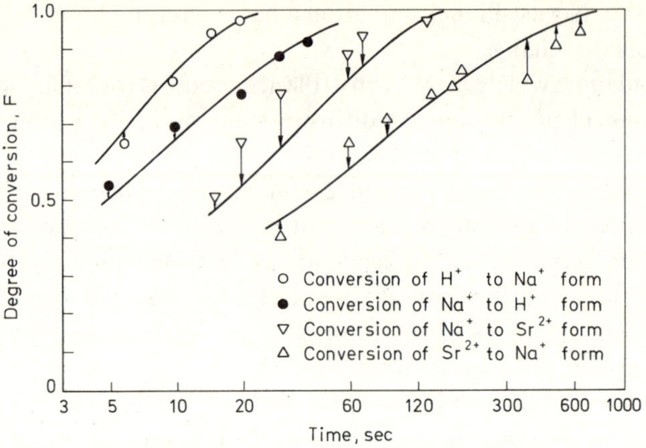

Figure 6.28 Forward and reverse sodium-hydrogen and sodium-strontium cation exchanges. Experimental data by Fedoseeva et al. on KU-2 cation exchanger; curves are calculated with Nernst-Planck equations (from [1]).

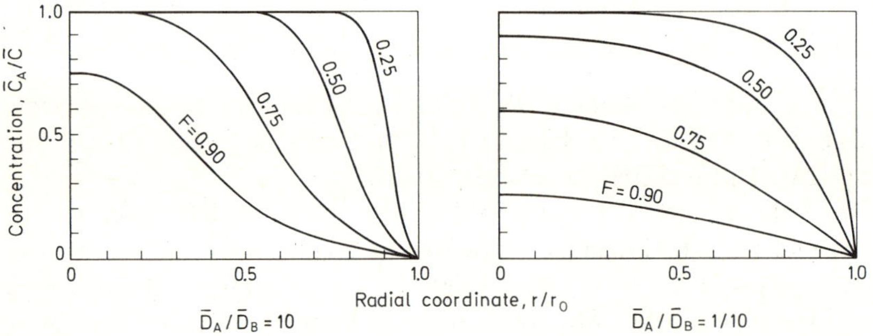

Figure 6.29 Counter-ion concentration profiles at different degrees of conversion, calculated with Nernst-Planck equations for intraparticle diffusion-controlled forward and reverse exchange of two counter-ions with mobility ratio 10 (from [1]).

phase concentration at the surface and compete with the entering ion. The additional dependence on two more variables poses no serious problem in numerical calculations, but has discouraged systematic tabulation of results. Computer programs are available that combine Nernst-Planck equations in the particle with liquid-phase mass transfer relations to cover the range in which both steps contribute [43, 44].

Each counter-ion species present contributes a Nernst-Planck equation, and only one such equation can be eliminated by use of the electric constraints (electroneutrality, no net charge transfer). Accordingly, simultaneous differential equations must be solved for ternary and higher exchange systems. Moreover, in such systems, two or more counter-ion species necessarily are present at the bead surface at all times, and the selectivity thus affects the boundary condition at that surface and thereby the rate [45–47]. In such systems it is not unusual for one or several counter-ions to overshoot their equilibrium conversion, producing transient concentration maxima or minima [47, 48]. A computer program for multispecies systems is available [49].

The absence of co-ions from the ion exchanger, postulated in the classic Nernst-Planck approach, is of course an idealization. In reality, at least a small amount of co-ions will be present. They are subject to transference by the electric field, with the result that they are drawn into the bead if the entering counter-ion is the faster one, and pushed out if the opposite is true [50]. At the same time, the electric potential gradient is reduced, slightly reducing the disparity in rate between forward and reverse exchange of two given ions [50]. Quantitative calculation requires the use of an additional Nernst-Plank equation for the co-ion, so that once again simultaneous differential equations must be solved [51]. The multispecies computer program [49] mentioned earlier can also accomodate co-ions.

6.2.5.4 Refined Nernst-Planck models

The classic Nernst-Planck equations [10] only account for diffusion and electric transference; they do not include activity coefficients, and, moreover, are based on the idea of constant "individual diffusion coefficients" of the species involved, which are expected to be similar to, if not identical with, the tracer diffusion coefficients under the same conditions. All these premises are valid only as approximations.

Activity coefficients in the ion exchanger are directly related to the separation factor, and their effect on the boundary condition at the bead surface is accounted for by this factor (or factors, in multi-ion systems). Within the bead, only gradients of activity coefficients have an effect, not the absolute values – if adjacent volume elements at different concentration have the same affinity for any given ion, no additional incentive for that ion to migrate from one to the other exists. In contrast, if the separation factor varies with ionic composition, reflecting activity coefficient gradients, the Nernst-Planck equation requires an additional term to account for this effect [1, 52, 53]. Especially for zeolites with their extreme and strongly composition-dependent selectivities, this correction term can make a major contribution to the overall flux [54, 55].

The premise of no convection in the ion exchanger is also incorrect. Since the volume elements of pore fluid carry a net charge (compensating that of the fixed groups), the electric field acts on them to produce an electro-osmotic flow [56]. The result is depletion of solvent in the ion exchanger if the entering counter-ion is the faster one, and accumulation if the opposite is true [50]. A quantitative treatment calls for addition of a convection term to the Nernst-Planck equation [57, 58]. So far, however, this approach has been used only for transport across membranes.

Electro-osmotic solvent transport opens a Pandora's Box. At least in gel type materials, solvent accumulation or depletion makes the bead swell or shrink. The least serious consequence is that the distance to be covered by intraparticle diffusion is altered. More drastic is the effect of change in solvent content on the intraparticle diffusion coefficients (see eq. (2)). What makes this situation worse is that the temporary local solvent content of a volume element of the bead will *not* be in local equilibrium but, rather, will depend on prior history – diffusion will be "non-Fickian". No theory so far has attempted to account quantitatively for this effect.

6.2.5.5 Nonequilibrium thermodynamics – Stefan-Maxwell equations

The Nernst-Planck approach is based on the idea of identifiable "individual" diffusion coefficients of species and assumes them to be constant and measurable by tracer experiments. Unfortunately, as nonequilibrium thermodynamics shows, this is at best an approximation.

There has been no lack of publications on mass transport in ion exchangers based on nonequilibrium thermodynamics (e. g. [58–69]), not even counting those who soon neglect cross-coefficients and so reduce the approach to the Nernst-Planck equations. However, none of this has so far been applied to derive rate laws for ion exchange on beads.

Nonequilibrium thermodynamics for diffusion in electrolytes can be formulated in different ways. Especially convenient for the purpose at hand appear to be the Stefan-Maxwell equations, for two reasons: Their coefficients have been found to remain reasonably constant, and the approach leads to definite conclusions about the relationship between these coefficients and the diffusion coefficients in the Nernst-Planck equations (Ref. [69]). Because of ionic interactions, the tracer diffusion coefficients vary with composition. For binary exchange of counter-ions A and B, the Nernst-Planck equations are found to give the best approximation if the coefficients D_A and D_B are taken as the tracer diffusion coefficients of A in the B form and of B in the A form, respectively, of the ion exchanger [69].

6.2.5.6 Models for mass transfer-controlled ion exchange with reactions

In many practical applications, ion exchange is accompanied by reactions, which profoundly affect equilibria and kinetics. A survey of possible types of situations is shown in Table 6.3 (from [48]) and Figure 6.30.

In the great majority of cases, the ionic reactions involved are known to be very fast in the liquid phase, and there is no reason to suspect that this may not be so in ion exchangers. Accordingly, mathematical models commonly assume such reactions to be in local equilibrium (for models with slow reaction, see next section). Mass transfer remains the slow step, but the reactions complicate the mechanism and so produce a different rate behavior.

Table 6.3 Typical processes in which ion exchange is accompanied by chemical reactions

Process type	Reactants: Resin		Reactants: Solution		Products: Resin		Products: Solution
I a	$- SO_3^- + H^+$	+	$Na^+ + OH^-$	→	$- SO_3^- + Na^+$	+	H_2O
b	$- N^+(CH_3)_3 + OH^-$	+	$H^+ + Cl^-$	→	$- N^+(CH_3)_3 + Cl^-$	+	H^2O
c	$- SO_3^- + H^+$	+	$Na^+ + AcO^-$	→	$- SO_3^- + Na^+$	+	AcOH
d	$4 - SO_3^+ + 2Ni^{2+}$	+	$4Na^+ + EDTA^{4-}$	→	$4 - SO_3^- + 4Na^+$	+	Ni_2EDTA
II a	$- COO^- + Na^+$	+	$H^+ + Cl^-$	→	$- COOH$	+	$Na^+ + Cl^-$
b	$- NH_3^- + Cl^-$	+	$Na^+ + OH^-$	→	$- NH_2$	+	$Na^+ + Cl^- + H_2O$
c	$- N(CH_2COO)_2^{2-} + 2Na^+$	+	$Ni^{2+} + 2Cl^-$	→	$- N(CH_2COO)_2Ni$	+	$2Na^+ + 2Cl^-$
III a	$- COOH$	+	$Na^+ + OH^-$	→	$- COO^- + Na^+$	+	H_2O
b	$- NH^2$	+	$H^+ + Cl^-$	→	$- NH_3^+ + Cl^-$		
c	$2 - N(CH_2COO)_2Ni$	+	$4Na^+ + EDTA^{4-}$	→	$2 - N(CH_2COO)_2^{2-} + 4Na^+$	+	Ni_2EDTA
IV	$- N(CH_2COO)_2Ni$	+	$2H^+ + 2Cl^-$	→	$- N(CH_2COOH)_2$	+	$Ni^{2+} + 2Cl^-$

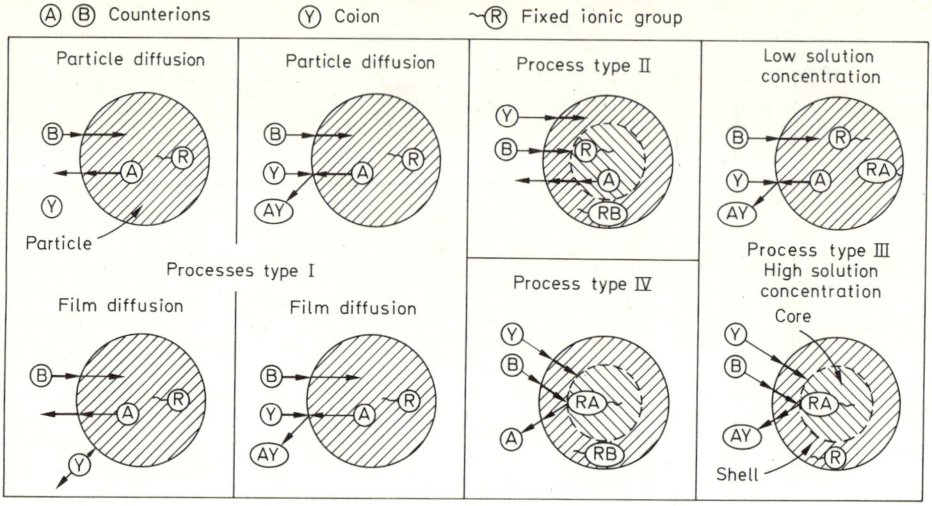

Figure 6.30 Types of combinations of ion exchange with reaction (from [48]).

The simplest case, not covered in Table 6.3 and Figure 6.30 and of only limited practical interest, is isotopic exchange of an ion that is complexed by the fixed groups, say, of a transition metal ion in a chelating resin. Here, a dissociation equilibrium is involved and leaves only a small fraction of the metal ion population in the bead free to move. Since the system is uniformly in equilibrium except for the isotope distribution, this fraction is constant, independent of conversion. Accordingly, Fick's law solutions, such as equations (18)–(20), are applicable with a constant "effective" diffusion coefficient, which is the product of the diffusion coefficient of the free ion and the degree of dissociation [70]. These two quantities can be estimated with the Mackie-Meares approximation (2) and the dissociation constant of an analogous complex in the liquid phase. Provided the complex is strong, few ions are free to move, making intraparticle diffusion so slow that it will almost invariably be the rate-controlling step.

In the simplest cases of reactions in actual ion exchange, labeled Type I in Table 6.3, the counter-ion released by the ion exchanger is consumed by a reaction with a co-ion in the liquid. Since the co-ion is barred from the ion exchanger by the Donnan effect, the reaction takes place at the bead surface as the counter-ion is released. The rate-controlling step is either mass transfer of the co-ion (Y) and entering counter-ion (B) to the bead surface, or interdiffusion of the two counter-ions (A and B) in the bead (see Figure 6.30). In the first case, the linear driving-force or Fick's law model (eq. (23) or (25)) with Nernst diffusion coefficient of the electrolyte AY is adequate [71–73] (see Figure 6.31). In the second case, the respective model for intraparticle diffusion control applies with the boundary condition for a liquid phase free of the released counter-ion even if the exchange takes place in a finite-volume batch [71, 73, 74] (see Figure 6.32).

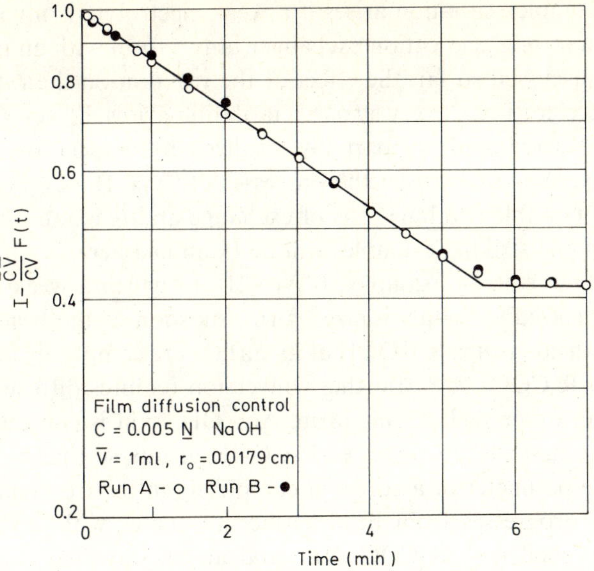

Figure 6.31 Liquid-phase mass transfer-controlled ion exchange with neutralization reaction. Dowex 50W–X8 in hydrogen form in 0.005 N aqueous sodium hydroxide; curve is theoretical prediction (from [72]).

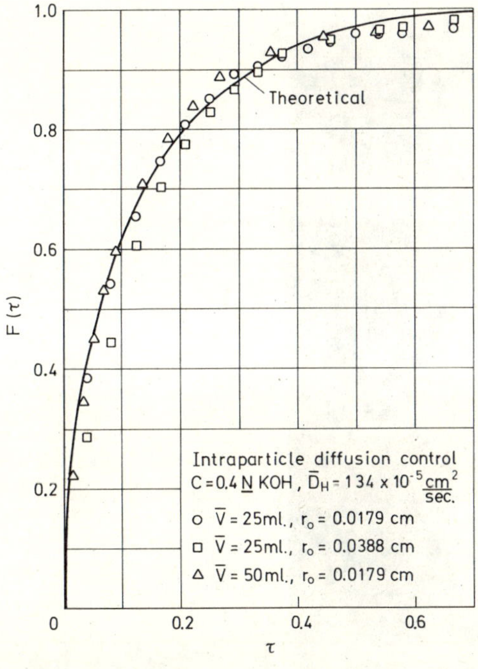

Figure 6.32 Intraparticle diffusion-controlled ion exchange with neutralization reaction. Dowex 50W–X8 in hydrogen form in 0.4 N aqueous potassium hydroxide; curve is theoretical prediction with Nernst-Planck model (from [74]).

A slightly more complex situation arises if a weak electrolyte is involved, e. g., in neutralization of a strong acid cation exchanger in H^+ form with ammonia [75].

In the examples discussed so far, the effect of the reaction on the rate is nil or a relatively minor acceleration. A very strong retardation arises, however, if species whose mass transfer is required are partially localized by association with fixed groups or the matrix, as is the case in the processes of Type II – IV in Table 6.3. There are too many possible combinations of exchange and reaction for complete coverage, and only some striking examples will be examined here.

Process IIIa may serve as an example of Type III, in which a weak-electrolyte ion exchanger is converted to a fully ionized form. As soon as the bead is placed in the liquid, the free-acid groups R-COOH at its surface are completely neutralized, that is, converted to $R\text{-}COO^-Na^+$. Further conversion requires diffusion of Na^+ and OH^- through an outer shell in this form. Any OH^- ion is consumed immediately on finding the first still unconverted R-COOH group, so that no OH^- will penetrate beyond the perimeter of a core still completely in the free-acid form. As a result, conversion progresses from bead surface to center with a sharp front between a converted shell with $R\text{-}COO^-Na^+$ and an unconverted core with R-COOH [71] (see Figure 6.30). This sharp boundary can be directly observed, confirming the validity of the shell-core model [76 – 78] (see Figure 6.33).

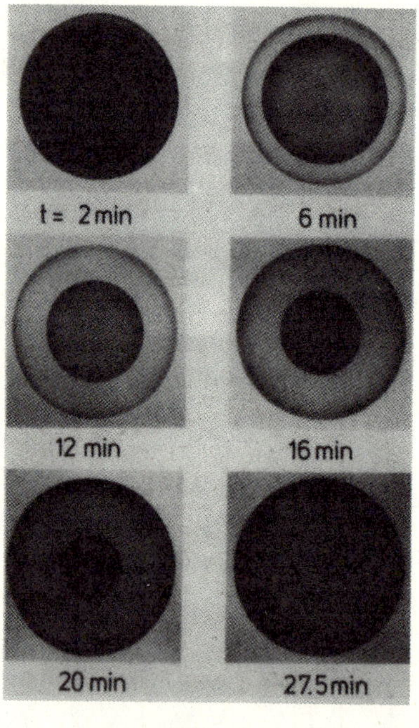

Figure 6.33 Microphotographs taken at different reaction times in cation exchange between Amberlite IRC 84 in Cu^{2+} form and 1 M HCl, clearly showing shell-core behavior (from [78]).

An interesting facet of this process is that ion exchange of H^+ for Na^+ is controlled by diffusion of the co-ion, OH^-, which is the minority ion in the converted shell. Moreover, although diffusion in the particle is the slow step and although the rate of ordinary intraparticle diffusion-controlled exchange is independent of the liquid-phase concentration, the rate here depends strongly on that concentration (proportional to its square, granted univalent ions and ideal Donnan equilibrium). The process is very slow because it relies on transfer of OH^-, largely excluded from the shell by the Donnan effect. In fact, in contact with a very dilute solution, OH^- transfer can become so slow that another potential mechanism will take over, as indicated in Figure 6.30 [71].

Under most conditions, diffusion across the shell can be considered as quasi-stationary. In this case, an analytical solution is obtained [71]:[3]

$$F(t) = 1 - \{1/2 + \sin[(1/3)\sin^{-1}(1 - 12 D_x C_x^* t / r^{\circ 2} \overline{C}]\}, \qquad (30)$$

where index x refers to the rate-controlling ion, OH^- in the example above. The more general case without the premises of a single rate-controlling ion and of quasi-stationary behavior requires numerical calculations [81–83].

A special type of shell-core behavior may arise in uptake of acids by weak-base anion exchangers in free-base form (Type IIIb in Table 6.3). For a strong, monobasic acid such as hydrochloric acid, the process is strictly analogous to that of Type IIIa described above. However, for polybasic acids such as sulfuric or phosphoric acids, an alternative and faster mechanism is possible, as illustrated in Figure 6.34. For example, interdiffusion of two HSO_4^- ions (total charge −2) and one SO_4^{2-} (same total charge), amounting to net transfer of one H_2SO_4 and no net charge transfer, provides a mechanism for proton transport across the shell that relies exclusively on counter-ions and so is unhindered by Donnan exclusion [82]. In

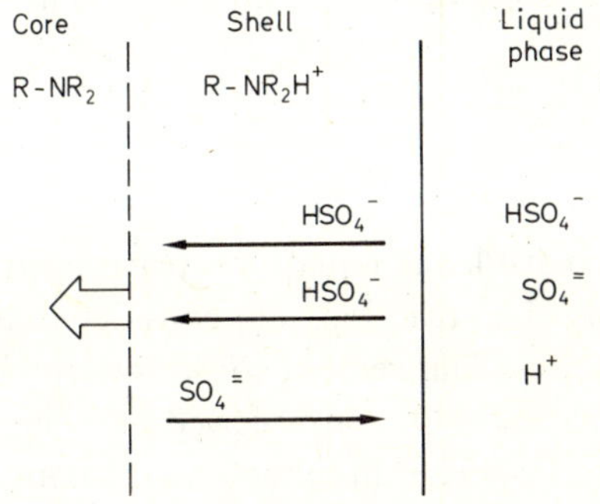

Figure 6.34 Possible mechanism of sulfuric acid uptake by weak-base anion exchanger, relying exclusively on anion diffusion and so not impeded by Donnan exclusion of co-ion.

---

[3] The mathematics is the same as for burn-off of deposits from catalyst particles, first studied by Weisz and Goodwin [79], who derived a relation for time as a function of conversion. See also "ash layer diffusion" in [80].

effect, the SO_4 moiety acts as a proton carrier. Indeed, uptake of sulfuric acid is much faster than that of hydrochloric acid [82], as had been observed previously but attributed to other causes [84–86]. Although both HSO_4^- and SO_4^{2-} may affect the rate, their diffusion coefficients are so similar that eq. (30) is applicable, provided diffusion is quasi-stationary [82].

An interesting corollary is that in aqueous systems the presence of CO_2, providing a CO_3 moiety with much the same abilitiy as SO_4, may facilitate proton transfer and so accelerate the rate of uptake. The alleged "catalytic" effect of CO_2 on acid uptake [85] may arise from this mechanism [82].[4]

Another interesting situation of practical significance is that of Type IV, in which both participating counter-ions are complexed by the fixed groups. Here, as in cases of Type II, the fixed groups in the bead or its shell are largely undissociated, so that Donnan exclusion will be weak and the absence of co-ions cannot be assumed. A shell-core behavior may or may not arise, depending on both the mobility ratio of the exchanging counter-ions and the ratio of the complex stability constants. A much higher mobility of the leaving counter-ion tends to sharpen the transient intraparticle profiles, as it does in ordinary intraparticle diffusion-controlled exchange; a much stronger complexing of the entering counter-ions has the same effect, as in the previously discussed shell-core cases, in which the other counter-ion was not complexed at all. Moreover, even if liquid-phase mass transfer is ideally fast and the solution is kept free of the counter-ion leaving the ion exchanger, the rate depends on both the liquid-phase concentration and the selectivity. These conclusions are based on extensive numerical calculations with the Nernst-Planck equations for the counter-ions and the co-ion [88–91].

A general multispecies model for systems with intraparticle diffusion control and reactions at local equilibrium has been developed [47]. A computer program for numerical calculations is available [49]. The model is based on differential material balances with source-or-sink terms $\bar{R}_i$ to account for formation or decomposition by reaction:

$$\frac{\partial \bar{C}_i}{\partial t} = -\operatorname{div} \bar{J}_i + \bar{R}_i . \tag{31}$$

The fluxes $\bar{J}_i$ are expressed by the Nernst-Planck equations, rearranged for convenient handling in multi-species systems. The reactions are considered as rearrangements of fundamental segments (moieties) that remain conserved; for example, the species H^+, H_3PO_4, $H_2PO_4^-$, HPO_4^{2-}, and PO_4^{3-} are viewed as composed of the moieties H and PO_4 in different ratios. The simultaneous algebraic equations contributed by the reaction equilibria are handled in matrix form. Allowing for reactions of counter-ions, co-ions, fixed groups, and neutral molecules in any combinations, the model is the most general developed so far and contains most

[4] The mechanism also provides facilitated proton transfer across anion exchange membranes [82, 87] and may well have implications for biological systems.

of the previously mentioned intraparticle models as special cases. It can also be used, for example, for ligand exchange, redox ion exchange, and catalysis by ion exchangers. Its essential limitations are the premises of validity of the Nernst-Planck equations and of local reaction equilibria according to mass action laws.

6.2.5.7 Reaction control models

If a chemical reaction is a much slower step than mass transfer, it will exclusively control the rate, as all concentration differences then are quickly equalized. The characteristic features of such systems are a rate independent of particle size, complete absence of intraparticle concentration gradients, and a positive interruption test.

Rate control by chemical reaction was taken for granted in the early days of ion exchange [92–96], a previous study correctly identifying diffusion [97–99] having long been forgotten. Even after the idea of mass transfer control in ion exchange, pioneered by Boyd [2], had become generally accepted, the rate equation of a reversible reaction – with the particle size-dependent rate coefficient adjusted to give best fit to solutions calculated with diffusion control – was often used by engineers for its mathematical convenience (e. g., [20, 100–102]). With the advent of high-speed computers able to handle more complex mathematics with ease, this practice has waned.

It is, of course, possible in principle that ion exchange may involve a reaction much slower than mass transfer. For instance, some complexing reactions involving Cr(III) are known to be relatively slow. Rate control by a chemical reaction has occasionally been invoked as a mechanism (e. g., [103–105]), but all such claims were later withdrawn [22, 106]. Thus, to this day, reaction-controlled ion exchange has remained a model in search of a system.

The possibility of a shell-core mechanism with rate control by a slow reaction at the core perimeter has also been considered [77]. The mathematics of such a situation have been developed for solid-fluid reactions in general [80] and lead to a rate that is inversely proportional to the particle radius. However, for this model to be applicable to ion exchange, the reaction must be slow compared with diffusion across the shell in order to be rate-controlling, yet it must be fast compared with diffusion into the core in order to keep the front sharp. In other words, the shell should pose no significant resistance to mass transfer while the core should remain impenetrable. In ion exchange, such a possibility seems remote.

6.2.6 State of the art

The practical user can draw on a large selection of theories of ion exchange kinetics, ranging from simple approximations to highly complex computer models. Even the most detailed models, however, do not provide an entirely rigorous description.

In the design of ion exchange equipment – e. g., columns – there are usually so many other factors introducing uncertainties or calling for approximations that the simplest rate law seems the best choice. Linear driving-force approximations are most frequently used. For ordinary ion exchange the mass transfer coefficients can be estimated from known properties [19]. If reactions are or could be involved, a safer course is to fit the coefficients to results of simple batch-kinetic experiments. No high accuracy can be expected, but the errors will usually not be greater than those introduced by other approximations.

In research, with greater demands on accuracy, the Nernst-Planck equations with proper constraints are the most common tool. At the cost of relatively little added complexity they achieve a significant improvement over Fick's law by accounting for the important electric field effect arising from coupled diffusion of ions of different mobilities. It is important, however, to realize that the conventional Nernst-Planck models and tabulated solutions are for binary ion exchange with a simple boundary condition (liquid-phase concentration of the counter-ion leaving the exchanger remains negligible) and absence of reactions and liquid-phase mass transfer effects. If these conditions are not met, the Nernst-Planck equations can still be used, but must be solved in a more complex setting, a task usually requiring numerical integration on a computer.

A researcher interested in ionic mass transfer as such is likely to find the Nernst-Planck equations unacceptable and be drawn to one or another form of nonequilibrium thermodynamics. This approach is more satisfying but calls for greater mathematical effort and experimental determination of more parameters and has so far not established itself as a standard method.

From the practical point of view, the key question will always be that of the point of diminishing returns. It is virtually impossible to include all potentially relevant effects in a model of ion exchange kinetics, so an entirely rigorous prediction or description is out of reach. Any refinement introduced to a model must therefore be judged by the provided gain in accuracy relative to the incurred cost in complexity.

References

[1] F. Helfferich (1962) Ion Exchange, chapters 1 and 6, McGraw-Hill, New York; reprinted 1970 by University Microfilms International, Ann Arbor, MI, #2003414.

[2] G. E. Boyd, A. W. Adamson and L. S. Myers, Jr. (1947) Exchange adsorption of ions by organic zeolites. II. Kinetics, J. Am. Chem. Soc. *69,* 2836–2848.

[3] W. Nernst (1904) Theorie der Reaktionsgeschwindigkeit in heterogenen Systemen, Z. Physik. Chem. *47,* 52.

[4] C. Calmon (1986) Recent developments in water treatment by ion exchange, Reactive Polymers *4,* 131–146.

[5] L. R. Weatherly and J. C. R. Turner (1976) Ion exchange kinetics: comparison between a macroporous and a gel resin, Trans. Inst. Chem. Eng. *54,* 89–94.

[6] S. Patell and J. C. R. Turner (1980) The kinetics of ion-exchange using porous exchangers, J. Separ. Proc. Technol. *1, (2)* 31 – 39.

[7] H. Yoshida, T. Kataoka, and S. Ikeda (1985) Intraparticle mass transfer in bidispersed porous ion exchanger: part I. isotopic ion exchange, Can. J. Chem. Eng. *63,* 422 – 429.

[8] H. Yoshida and T. Kataoka (1985) Intraparticle mass transfer in bidispersed porous ion exchanger: part II. mutual ion exchange, Can. J. Chem. Eng. *63,* 430 – 435.

[9] D. W. Breck (1974) Zeolite Molecular Sieves: Structure, Chemistry, and Use, chapter 2, Wiley, New York.

[10] R. M. Barrer (1978) Zeolites and Clay Minerals as Sorbents and Molecular Sieves, chapter 2, Academic Press, London.

[11] D. B. Spalding (1961) The prediction of mass transfer rates when equilibrium does not prevail at the phase interface, Internat. J. Heat Mass Transfer *2,* 283 – 313.

[12] P. E. Liley and W. R. Gambill (1984) Physical and chemical data, in: Perry's Chemical Engineers' Handbook, 6th ed. (R. H. Perry, D. W. Green and J. O. Maloney, eds.) section 3, McGraw-Hill, New York.

[13] J. S. Mackie and P. Meares (1955) The diffusion of electrolytes in a cation exchange membrane, Proc. Roy. Soc. (London) *A232,* 498 – 509.

[14] F. G. Helfferich (1983) Ion exchange kinetics – evolution of a theory, in: Mass Transfer and Kinetics of Ion Exchange (L. Liberti and F. G. Helfferich, eds.) pp. 157 – 179, Martinus Nijhoff, The Hague.

[15] W. Nernst (1888) Zur Kinetik der in Lösung befindlichen Körper, Z. Phys. Chem. *2,* 613.

[16] W. Nernst (1889) Die elektromotorische Wirksamkeit der Ionen, Z. Phys. Chem. *4,* 129.

[17] M. Planck (1890) Über die Erregung von Elektrizität und Wärme in Elektrolyten, Ann. Phys. Chem. *39,* 161.

[18] T. R. E. Kressman and J. A. Kitchener (1949) Cation exchange with a synthetic phenolsulphonate resin. Disc. Faraday Soc. *7,* 90 – 104.

[19] T. Vermeulen, G. Klein and N. K. Hiester (1984) Adsorption and ion exchange, in: Perry's Chemical Engineers' Handbook (see Ref. 12) section 16.

[20] E. R. Gilliland and R. F. Baddour (1953) The rate of ion exchange, Ind. Eng. Chem. *45,* 330 – 337.

[21] E. Glueckauf (1955) Principles of operation of ion-exchange columns, in: Ion Exchange and Its Applications, pp. 34 – 46, Soc. Chem. Ind., London.

[22] F. G. Helfferich, L. Liberti, D. Petruzzelli and R. Passino (1985) Anion exchange kinetics in resins of high selectivity, Israel J. Chem. *26,* 3 – 7 and 8 – 16.

[23] R. Schlögl and F. Helfferich (1957) Comment on the significance of diffusion potentials in ion exchange, J. Chem. Phys. *26,* 5 – 7.

[24] W. K. Lewis and W. G. Whitman (1924) Principles of gas absorption, Ind. Eng. Chem. *16,* 1215 – 1220.

[25] J. Crank (1975) The Mathematics of Diffusion, 2nd ed., chapter 6, Clarendon Press, Oxford.

[26] S. Paterson (1947) The heating or cooling of a solid sphere in a well-stirred fluid, Proc. Phys. Soc. (London) *59,* 50 – 58.

[27] L. R. Ingersoll, O. J. Zobel and A. C. Ingersoll (1948) Heat Conduction, pp. 169 – 174, McGraw-Hill, New York.

[28] F. A. Glaski and J. S. Dranoff (1963) Ion exchange kinetics – a comparison of models, AIChE J. *9,* 426 – 431.

[29] T. G. Smith and J. S. Dranoff (1964) Film-diffusion-controlled kinetics in binary ion exchange, Ind. Eng. Chem. Fundam. *3,* 195 – 200.

[30] J. P. Copeland, C. L. Henderson and J. M. Marchello (1967) Influence of resin selectivity on film-diffusion controlled ion exchange, AIChE J. *13*, 449–452.

[31] T. Kataoka, M. Sato and K. Ueyama (1968) Effective liquid-phase diffusivity in ion exchange, J. Chem. Eng. Japan *1*, 38–42.

[32] J. C. R. Turner and C. B. Snowdon (1968) Liquid-side mass transfer in ion exchange, Chem. Eng. Sci. *23*, 221–230 and 1099–1103.

[33] J. P. Copeland and J. M. Marchello (1969) Film-diffusion-controlled ion exchange with a selective resin, Chem. Eng. Sci. *24*, 1471–1474.

[34] L. P. Van Brocklin and M. M. David (1972) Coupled ionic migration and diffusion during liquid-phase controlled ion exchange, Ind. Eng. Chem. Fundam. *11*, 91–99.

[35] T. Kataoka, H. Yoshida and T. Yamada (1973) Liquid phase mass transfer in ion exchange based on the hydraulic radius model, J. Chem. Eng. Japan *6*, 172–177.

[36] L. P. van Brocklin and M. M. David (1975) Ionic migration effects during liquid phase controlled ion exchange, AIChE Symp. Ser. *71*, No. 152, 191–201.

[37] G. R. S. Wildhagen, R. Y. Qassim, K. Rajagopal and K. Rahman (1985) Effective liquid phase diffusivity in ion exchange, Ind. Eng. Chem. Fundam. *24*, 423–432.

[38] T. Kataoka, H. Yoshida and T. Uemura (1987) Liquid-side ion exchange mass transfer in a ternary system, AIChE J. *33*, 202–210.

[39] F. Helfferich and M. S. Plesset (1958) Ion exchange kinetics – a nonlinear diffusion problem, J. Chem. Phys. *28*, 418–424.

[40] M. S. Plesset, F. Helfferich and J. N. Franklin (1958) Ion exchange kinetics – a nonlinear diffusion problem: II. particle diffusion controlled exchange of univalent and bivalent ions, J. Chem. Phys. *29*, 1064–1069.

[41] F. Helfferich (1963) Revised tables for ion exchange kinetics, J. Chem. Phys. *38*, 1688–1691.

[42] F. G. Helfferich (1985) Diffusion With Variable Diffusion Coefficients, Istituto di Ricerca Sulle Acque, CNR, Bari, Italy, Report R/107 ISSN 03292-9671; also obtainable from Department of Chemical Engineering, The Pennsylvania State University, University Park, PA 16802.

[43] A. J. Lupa (1967) The kinetics of ion exchange with simultaneous film and particle diffusion, Ph. D. Thesis, Dept. Chem. Eng., Northwestern University, Evanston, IL.

[44] D. Petruzzelli, L. Liberti, R. Passino, F. G. Helfferich and Y.-L. Hwang (1987) Chloride-sulfate exchange kinetics: solution for combined film and particle diffusion control, Reactive Polymers *5*, 219–226.

[45] R. K. Bajpai, A. K. Gupta and M. Gopala Rao (1974) Single particle studies of binary and ternary cation exchange kinetics, AIChE J. *20*, 989–995.

[46] J. Plicka, J. Cabicar, K. Stamberg and M. Fabian (1984) The kinetics of ion exchange sorption in ternary systems, in: Ion Exchange Technology (D. Naden and M. Streat, eds.) pp. 331–336, Ellis Horwood, Chichester.

[47] Y.-L. Hwang and F. Helfferich (1987) Generalized model for multispecies ion-exchange kinetics including fast reversible reactions, Reactive Polymers *5*, 237–253.

[48] F. Helfferich (1966) Ion-exchange kinetics, in: Ion Exchange (L. A. Marinsky, ed.) chapter 2, Marcel Dekker, New York.

[49] Y.-L. Hwang and F. G. Helfferich (1986) Computer Program for Multispecies Ion-Exchange Kinetics Including Fast Reversible Reactions, Department of Chemical Engineering, The Pennsylvania State University, University Park, PA 16802.

[50] F. Helfferich (1962) Ion exchange kinetics: III. experimental test of the theory of particle-diffusion controlled ion exchange, J. Phys. Chem. *66*, 39–44.

[51] G. E. Spalding (1971) Predictive theory of coions transport accompanying particle-diffusion-controlled ion exchange, J. Chem. Phys. *55*, 4991–4995.

[52] G. Manecke and K. F. Bonhoeffer (1951) Elektrische Leitfähigkeit von Anionenaustauschermembranen, Z. Elektrochem. *55*, 475–481.

[53] Ref. 1, chapter 8.

[54] R. M. Barrer, R. F. Bartholomew and L. C. V. Rees (1963) Ion exchange in porous crystals – II. the relationship between self- and exchange-diffusion coefficients, J. Phys. Chem. Solids *24*, 309–317.

[55] N. M. Brooke and L. C. V. Rees (1968) Kinetics of ion exchange – part 1, Trans. Faraday Soc. *12*, 3383–3392.

[56] G. Schmid (1950) Zur Elektrochemie feinporiger Kapillarsysteme – I: Übersicht, Z. Elektrochem. *56*, 424–430.

[57] R. Schlögl and U. Schödel (1955) Über das Verhalten geladener Porenmembranen bei Stromdurchgang, Z. Phys. Chem. (Frankfurt) *5*, 372–397.

[58] R. Schlögl (1964) Stofftransport durch Membranen, Steinkopff, Darmstadt.

[59] A. J. Staverman (1952) Non-equilibrium thermodynamics of membrane process, Trans. Faraday Soc. *48*, 176–185.

[60] J. G. Kirkwood (1954) Transport of ions through biological membranes from the standpoint of irreversible thermodynamics, in: Ion Transport Across Membranes (H. T. Clarke, ed.) pp. 119–127, Academic Press, New York.

[61] J. W. Lorimer, E. I. Boterenbrood and J. J. Hermans (1956) Transport processes in ion-selective membranes, Disc. Faraday Soc. *21*, 141–149.

[62] K. S. Spiegler (1958) Transport processes in ionic membranes, Trans. Faraday Soc. *54*, 1408–1428.

[63] O. Kedem and A. Katchalsky (1958) Thermodynamic analysis of the permeability of biological membranes to nonelectrolytes, Biochim. Biophys. Acta *27*, 229–246.

[64] I. Michaeli and O. Kedem (1961) Description of the transport of solvent and ions through membranes in terms of differential coefficients – I. phenomenological characterization of flows, Trans. Faraday Soc. *57*, 1185–1190.

[65] E. N. Lightfoot and E. M. Scattergood (1965) Suitability of the Nernst-Planck equations for describing electrokinetic phenomena, AIChE J. *11*, 175.

[66] S. R. Caplan (1966) Transport processes in membranes, in: Ion Exchange (see Ref. 48) chapter 1.

[67] R. W. Schlögl (1983) Non-equilibrium thermodynamics – a general framework to describe transport and kinetics in ion exchange, in: Mass Transfer and Kinetics of Ion Exchange (see Ref. 14) pp. 207–212.

[68] R. Paterson and Lutfallah (1984) Simulation of membrane processes using network thermodynamics, in: Ion Exchange Technology (see Ref. 46) pp. 242–256.

[69] E. E. Graham and J. S. Dranoff (1982) Application of the Stefan-Maxwell equations to diffusion in ion exchangers, Ind. Eng. Chem. Fundam. *21*, 360–365 and 365–369.

[70] A. Schwarz, J. A. Marinsky and K. S. Spiegler (1964) Self-exchange measurements in a chelating ion-exchange resin, J. Phys. Chem. *68*, 918–924.

[71] F. Helfferich (1965) Ion-exchange kinetics: V. Ion exchange accompanied by reactions, J. Phys. Chem. *69*, 1178–1187.

[72] R. A. Blickenstaff, J. D. Wagner and J. S. Dranoff (1967) The kinetics of ion exchange accompanied by irreversible reaction: I. Film diffusion controlled neutralization of a strong acid exchanger by strong bases, J. Phys. Chem. *71*, 1665–1669.

[73] E. E. Graham and J. S. Dranoff (1972) Kinetics of anion exchange accompanied by fast irreversible reactions, AIChE J. *18*, 608–613.

[74] R. A. Blickenstaff, J. D. Wagner and J. S. Dranoff (1967) The kinetics of ion exchange accompanied by irreversible reaction: II. Intraparticle diffusion controlled neutralization of a strong acid exchanger by strong bases, J. Phys. Chem. *71*, 1670–1674.

[75] J. D. Wagner and J. S. Dranoff (1967) The kinetics of ion exchange accompanied by irreversible reaction: III. Film diffusion controlled neutralization of a strong acid exchanger by a weak base, J. Phys. Chem. *71,* 4551–4553.

[76] P. R. Dana and T. D. Wheelock (1974) Kinetis of a moving boundary ion-exchange process, Ind. Eng. Chem. Fundam. *13,* 20–26.

[77] M. Streat (1984) Kinetics of slow diffusing species in ion exchangers, Reactive Polymers *2,* 79–91.

[78] W. Höll (1984) Optical verification of ion exchange mechanism in weak electrolyte resins, Reactive Polymers *2,* 93–101.

[79] P. B. Weisz and R. D. Goodwin (1963) Combustion of carbonaceous deposits within porous catalyst particles: 1. Diffusion-controlled kinetics, J. Catalysis *2,* 397–404.

[80] O. Levenspiel (1972) Chemical Reaction Engineering, 2nd ed., chapter 12, Wiley, New York.

[81] T. Kataoka, H. Yoshida and Y. Ozasa (1977) Intraparticle ion exchange mass transfer accompanied by instantaneous irreversible reaction, Chem. Eng. Sci. *32,* 1237–1240.

[82] F. Helfferich and Y.-L. Hwang (1985) Kinetics of acid uptake by weak-base anion exchangers – mechanism of proton transfer, AIChE Symp. Ser. *81,* No. 242, 17–27.

[83] H. Yoshida, T. Kataoka and S. Fujikawa (1986) Kinetics in a chelating ion exchange, Chem. Eng. Sci. *41,* 2517–2524 and 2525–2530.

[84] G. Adams, P. M. Jones and J. R. Millar (1969) Kinetics of acid uptake by weak-base anion exchangers, J. Chem. Soc. (London), Ser. A, 2543–2551.

[85] R. E. Anderson (1975) On the pickup of hydrochloric and sulfuric acids on a polyamine resin, in: Polyelectrolytes and Their Applications (A. Renbaum and E. Sélégny, eds.) pp. 263–274, Reidel, Dordrecht.

[86] M. Gopala Rao and A. K. Gupta (1982) Kinetics of ion exchange in weak-base anion exchange resin, AIChE Symp. Ser. *78,* No. 219, 96–102.

[87] F. G. Helfferich, (1989) Facilitated proton transfer across anion exchange membranes, 197th ACS National Meeting, Dallas, Paper I & EC No. 109.

[88] A. I. Kalinitchev, T. D. Semenovskaya, E. V. Kolotinskaya, A. Ya. Pronin, and K. V. Chmutov (1981) Investigation into the kinetics of ion-exchange processes accompanied by complex formation, J. Inorg. Nucl. Chem. *43,* 787–789.

[89] E. V. Kolotinskaya, A. I. Kalinitchev and T. D. Semenovskaya (1981) Computer simulation of the kinetics of ion exchange accompanied by complex formation in ionites, J. Chromatog. *212,* 133–138.

[90] A. I. Kalinitchev, E. V. Kolotinskaja and T. D. Semenovskaya (1982) Computerized analysis of the diffusion processes in complexing ionites, J. Chromatog. *243,* 17–24.

[91] A. I. Kalinitchev, E. V. Kolotinskaya and T. D. Semenovskaya (1988) Kinetic mechanism limiting ion exchange rates in selective systems, Reactive Polymers, *7,* 123–131.

[92] J. Du Domaine, R. L. Swain and G. A. Hougen (1943) Cation-exchange water-softening rates, Ind. Eng. Chem. *35,* 546–553.

[93] F. C. Nachod and W. Wood (1944) Reaction velocity of ion exchange, J. Am. Chem. Soc. *66,* 1380–1384.

[94] F. C. Nachod and W. Wood (1945) Reaction velocity of ion exchange – II, J. Am. Chem. Soc. *67,* 629–631.

[95] H. C. Thomas (1944) Heterogeneous ion exchange in a flowing system, J. Am. Chem. Soc. *66,* 1664–1666.

[96] W. Juda and M. Carron (1948) Equilibria and velocity of the sodium-hydrogen exchange on carbonaceous exchangers in contact with chloride solutions, J. Am. Chem. Soc. *70,* 3295–3310.

[97] E. Warburg (1913) Über die Diffusion von Metallen in Glas, Ann. Physik *40,* 327–334.

[98] G. Schulze (1913) Versuche über die Diffusion von Ag in Glas, Ann. Physik *40,* 335–367.
[99] G. Schulze (1914) Die Ionendiffusion im Permutit und Natrolith, Z. Physik. Chem. *89,* 168–178.
[100] S. Goldstein (1953) On the mathematics of exchange processes in fixed columns: I. Mathematical solutions and asymptotic expansions, Proc. Roy. Soc. (London) *A219,* 151–170.
[101] T. Vermeulen (1958) Separation by adsorption methods, Advan. Chem. Eng. *2,* 147–208.
[102] J. S. Dranoff and L. Lapidus (1958) Multicomponent ion exchange column calculations, Ind. Eng. Chem. *50,* 1648–1653.
[103] R. Turse and W. Rieman III (1961) Kinetics of ion exchange in a chelating resin, J. Phys. Chem. *65,* 1821–1824.
[104] L. Liberti, G. Boari and R. Passino (1978) Chloride/sulfate exchange on anion resins – kinetic investigations: chemical control in selective exchangers, Desalination *26,* 181–194.
[105] L. Liberti, I. Madi, R. Passino and L. Walis (1980) Ion-exchange kinetics in selective systems, J. Chromatog. *201,* 43–50.
[106] A. Varon and W. Rieman III (1964) Kinetics of ion exchange in a chelating resin, J. Phys. Chem. *68,* 2716–2718.

6.3 The Influence of Polymer Structure on the Reactivity of Bound Ions

David C. Sherrington
University of Strathclyde, Department of Pure and Applied Chemistry, Glasgow, Great Britain

Introduction

Although ion exchangers have been developed primarily for use in the exchange of ions from a liquid phase there is now a wide range of applications in which suitable ion exchange and related materials are used as chemically reactive species [1–4]. Furthermore, there is every indication that the number and variety of such applications will continue to grow. In the case of ion exchange itself, it is known that the structure of the polymeric matrix which forms the basis of the exchanger can influence the exchange process, in particular the kinetics of exchange. Generally, structures are selected to minimize such effects, and the selection is facilitated to some extent since the solutions to be processed are invariably aqueous ones. In the case of reactive ion exchangers the polymer matrix can play an even more significant role. The reaction medium may be water or any of a wide range of organic solvents, and the composition of the reaction medium generally changes with time. Thus, the role of the matrix can change during a reaction as the polymer structure adjusts to the changing conditions. Alternatively, of course, the reactive ion exchanger may be used in the gas phase where again time-dependent changes may be induced in the support. This chapter will describe the various situations which can arise and evaluate them in terms of the structure of the polymer matrix. Considerable licence will be taken with regard to the bound groups of interest, and emphasis will be placed on producing a molecular 'picture' of the events, rather than on giving them an abstract and complex mathematical treatment.

6.3.1 Polymer structures

Polymeric ion exchangers consist essentially of an arrangement of linear macromolecules intermolecularly crosslinked to form an infinite network and render the system molecularly insoluble in the medium to which it is exposed. An individual polymer chain adopts, almost exclusively, a random coil conformation in solution (Figure 6.35 a) although, as a result of specific electronic and/or stereochemical factors, some natural polymers such as polypeptides and polysaccharides may retain (from the solid state) a fixed structural conformation (such as a helix) within the overall random coil geometry (Figure 6.35 b). When a dilute solution of individual polymer coils is gradually concentrated, a point is reached at which random coils start to contact each other, and beyond this point coil-coil interpenetration occurs (Figure 6.36). The critical concentration depends on the molecular weight of the polymer and the degree of interaction with the solvent. For molecular weights of $\approx 100{,}000$ in a good solvent, polymer coils may be considered as relatively isolated only in solutions up to $\approx 1-2\%$ by weight. For solution concentrations of $\approx 30\%$ interpenetration of chains is probably complete [5]. Polymer coils expand readily in a thermodynamically good solvent, and contract in a poor one, finally precipitating in a non-solvent. This variation of coil size with the quality of the solvent explains why the critical concentration for coil contact varies with the medium involved.

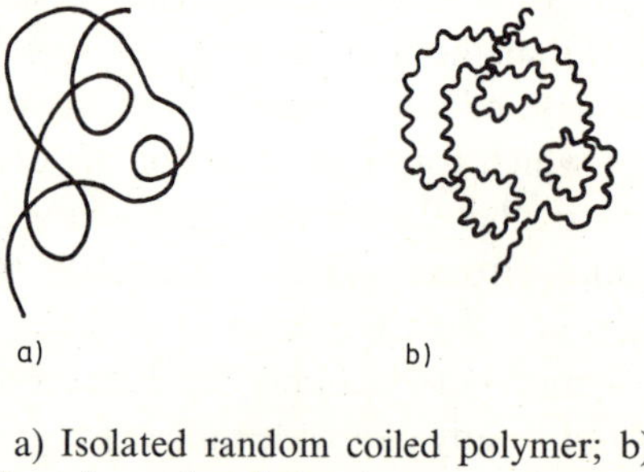

a) b)

Figure 6.35 a) Isolated random coiled polymer; b) isolated random coiled polymer with fixed internal conformational structure.

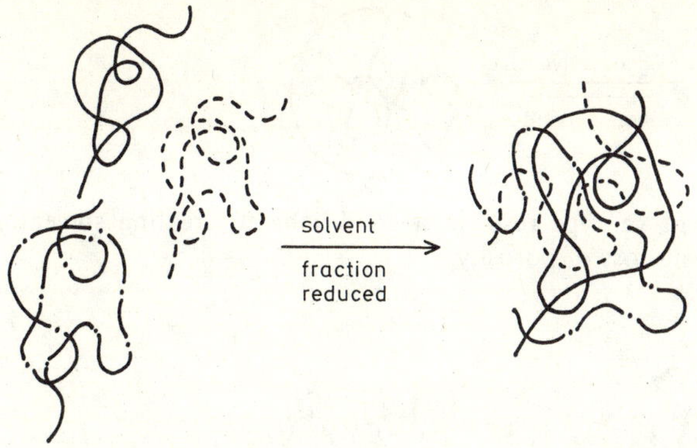

Figure 6.36 Interpenetration of polymer coils as solution is concentrated.

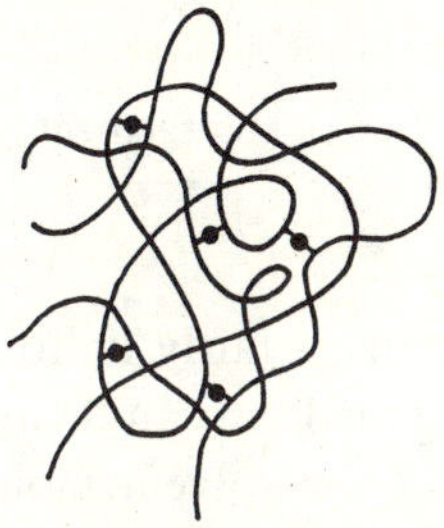

Figure 6.37 Infinite network formed by crosslinking of interpenetrating polymer coils (● = crosslink).

If a collection of interpenetrating polymer coils are formed in the presence of a crosslinking agent, or are crosslinked subsequently in a post-polymerization reaction, then an infinite network results (Figure 6.37). Such networks form the basis of ion exchange resins. Early versions were formed as a solid mass and were mechanically crushed to form irregular particles with a broad size-distribution. Various dispersion techniques have been evolved, however, yielding spherical particulate materials, both in the case of synthetic vinyl-based resins and exchangers based on natural polysaccharides (see Chapters 1.2 and 1.4).

A number of morphological variants of ion exchange matrices are also available. At one extreme there are the so-called gel type resins and at the other the macroporous species. The former are typified by polystyrene-based materials with a lower degree of crosslinking, and dextran-based species similarly crosslinked. These networks are essentially microporous in the dry state, i. e. polymer chains are separated by molecular dimensions only. Diffusion of a free molecule within these is via a molecular process involving motion of the polymer chains and is generally slow. However, in the presence of a good solvent the matrix swells reversibly to produce a gel, in the thermodynamic sense, with significant 'solvent porosity' (Figure 6.38).

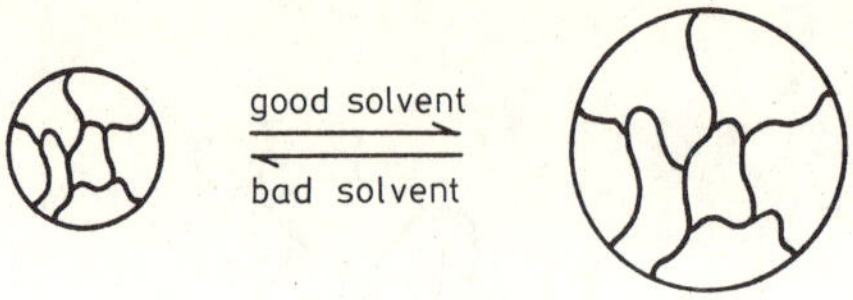

Figure 6.38 Swelling of gel-type resin in a good solvent, creating solvent porosity. Deswelling in a bad solvent, loss of porosity.

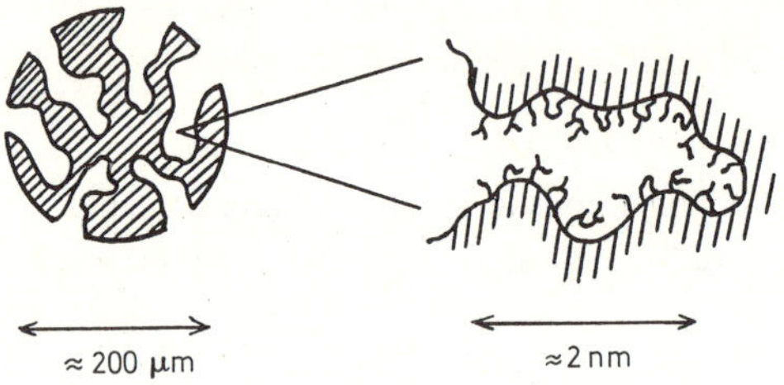

Figure 6.39 Rigid, non-compressible, macroporous resin – enlargement showing pore structure (//// = highly crosslinked regions).

In this form such matrices may be regarded as having a fairly uniform structure. Substrate molecules which are soluble in the swelling medium can then diffuse more readily through the matrix via Brownian motion of the solvent molecules rather than motion of the polymer chains. Such gel type species are soft and compressible in their swollen state, and the degree of swelling depends on the level of crosslinking and the nature of the solvent.

Macroporous exchangers are generally more rigid, less compressible and change their volume little when solvent is added. They consist of macromolecular chains which are more heavily crosslinked, but during formation of the matrix various procedures are employed to produce voids or macropores within the matrix (Figure 6.39) (see Chapter 1.2). Such structures may be regarded as having a discrete and accessible internal surface and surface area even in the dry state. In general, irrespective of the chemical structure of the primary polymer chains a wide range of solvents can penetrate the macropores following an initial wetting, if necessary. Furthermore, good solvents for the primary polymer may also swell the heavily crosslinked and entangled macromolecular chains which form the 'walls' of the macropores. Although the detailed structure of the primary polymers may differ, as indeed do the basic mechanisms for macropore formation, macroporous networks are very similar on a macroscopic scale; macroreticular polystyrene resins [6] and macroporous agarose materials [7] fall into this category. Substrate molecules which dissolve in the solvent occupying the macropores can diffuse quickly through the pore volume to reach the internal polymer surface. Where the solvent is a poor one for the polymer 'walls', diffusion into these requires motion of the polymer chains themselves and, in general, would be expected to be slow. However, where

Figure 6.40 Typical resin species. Right, clear glassy gel-type; left, opaque matt macroporous type; centre, hybrid species showing characteristics of both.

the solvent medium swells the polymer 'walls', some penetration of these may also occur fairly readily, and may contribute in subsequent reactions (see Section 6.3.5.2). Clearly the overall structure and properties of such matrices is very non-uniform and this must be considered in formulating mathematical models describing reactive exchangers based on such materials.

Between these two extreme morphological descriptions lie almost an infinite variety of structures (Figure 6.40). Relatively few of these are available commercially, and most systems conform reasonably to one description or the other. Nevertheless, potential deviation from these structural types is always possible, with concomitant changes in, for example, diffusional mechanisms. Such possibilities must always be borne in mind.

6.3.2 Chemically modified resins

The functional groups of conventional ion exchange resins are introduced by chemical modification of the pre-formed supports using organic synthetic methodologies analogous to the reactions of low molecular weight species [8, 9] (see Chapter 1.2) (Reactions 1 and 2). However, an alternative approach using a functional co-monomer can be

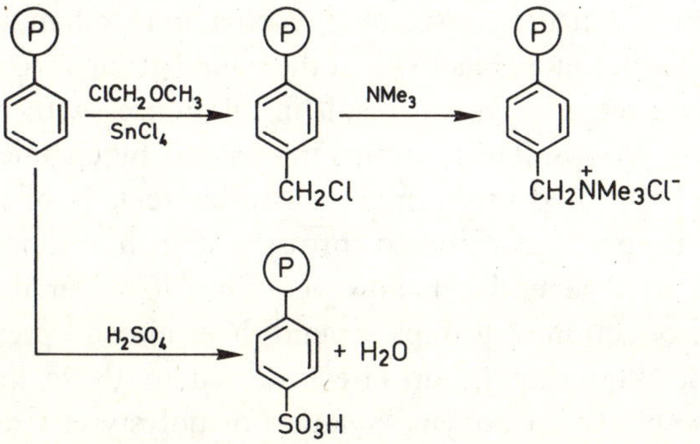

(1)

(2)

adopted. In general, the latter is likely to be a much more expensive

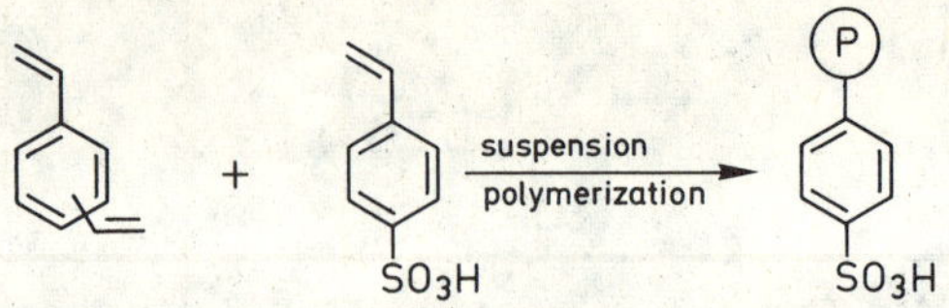

(3)

option in terms of the cost of the functional monomer, and in addition might require the development of different suspension polymerization conditions to suit each different co-monomer composition. In the case of weakly acidic carboxylic acid resins, however, acrylic and methacrylic acid monomers are cheap and readily available in large quantities, and the functional co-monomer route (Reaction 4) is preferable to one in which carboxylic groups are introduced by a chemical modification procedure (Reaction 5). However, the aromatic carboxylic

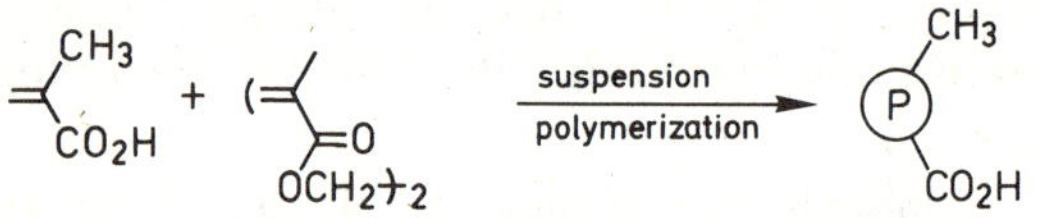

(4)

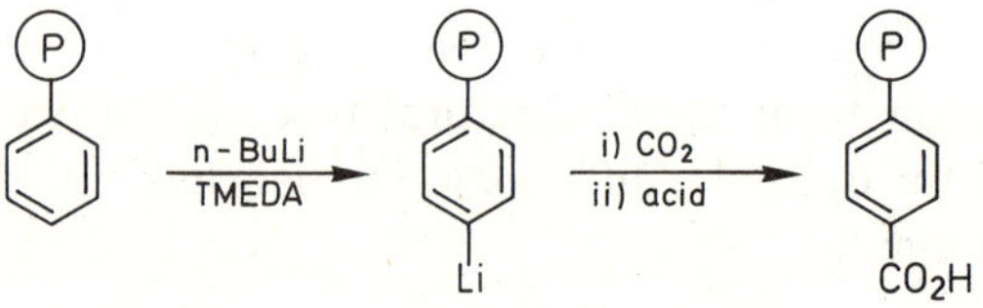

(5)

acid resin produced in the latter may well be cheaper than an analogous species made from *p*-vinyl benzoic acid (Reaction 6).

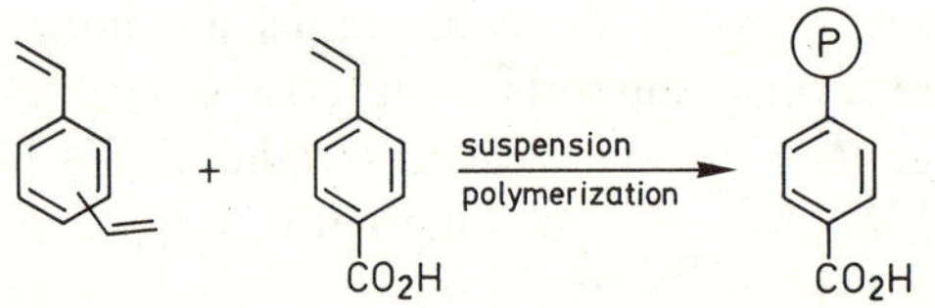

(6)

Much of this is fairly obvious, but the source of a reactive ion exchanger can be very important. As far as the subsequent reactivity of the bound groups is concerned, the history of how they were introduced can result in differences with regard to the local environment of the groups. Thus, groups introduced by chemical modification are more likely to be located in the more accessible regions of the resin, and are more likely to be uniformly distributed throughout such a support. This assumption should not be over-generalized, however, since local environmental effects could favour the introduction of groups around those already present (this might be regarded as a neighbouring group effect related to those in Section 6.3.4.5). Indeed, in the case of the carboxymethylation of polystyrene resins and the subsequent intramolecular anhydride formation, the non-uniform introduction of groups seems to be the only likely explanation for the high yield of anhydride at the very low loadings employed [10]. Nevertheless, groups introduced by chemical modification are more likely to be accessible subsequently to other molecules. Where a functional group has been built up through a sequence of chemical reactions, it is important to realise that the chemical conversions at each stage will not have been 100%, and that, indeed, some side reactions will have occurred as well. As a result the functional group purity of such species will be less than 100%, and in the case of multi-step conversions could fall rapidly to zero! This is a simple arithmetical reality brought home most vividly by those working in solid phase peptide synthesis [11], and not always recognized, or at least acknowledged, in much other published literature. Despite the fact that the coupling reactions used in peptide synthesis are some of the fastest, cleanest and complete reactions known in organic chemistry, even quite short peptide sequences can contain many omissions and truncations, resulting in a low overall purity of the cleaved peptide. As a result of this, much greater effort is now being made in polymer-supported reactions to assemble the structures required on resins (e. g., complex ligands) as pure low molecular weight species, then attach them by a single clean coupling reaction to some convenient group already on the resin. In this respect, resins made from glycidyl methacrylate monomer are more useful because the reactive epoxy group opens very efficiently with a wide variety of molecules [12].

Where the required resin-bound group is introduced via a functional monomer, the structural purity of the resultant species can be more effectively guaranteed. However, groups introduced in this way are more likely to be subject to accessibility problems, and, depending on the reactivity ratios of the co-monomers, their distribution might be very non-uniform. Thus, both approaches have their advantages and disadvantages.

As far as the materials available on a large scale from commercial sources are concerned, the criterion used in selecting which route is appropriate is solely an economic one, although in more sophisticated applications involving much smaller quantitites of materials [13] there are signs that other criteria may become more important (e. g., the functional group purity and content).

6.3.3 Classification of reactions

In constructing a model for an ion exchange process – though this may involve a number of consecutive steps – the overall process is straightforward in that one ion type enters the exchanger, undergoes exchange with the counter-ion, and the latter leaves the exchanger. In the case of chemical reactions involving ion exchangers a wide variety of reaction types can be envisaged, involving the movement in and out of resins of different numbers of species. The most common types are shown schematically in Figure 6.41.

In Figure 6.41 a, a reactant, A, enters the exchanger and undergoes some reaction with the bound ion, ⊕, or the counter-ion, ⊖, and the product remains associated with, or as a part of, the ion pair, ⊕ ⊖. Reactions which chemically modify a bound group might be regarded as examples of this type. In some instances, a second product fragment may be ejected from the resin as well. In Figure 6.41 b reactant, A, enters the exchanger and undergoes a reaction to form a product, C,

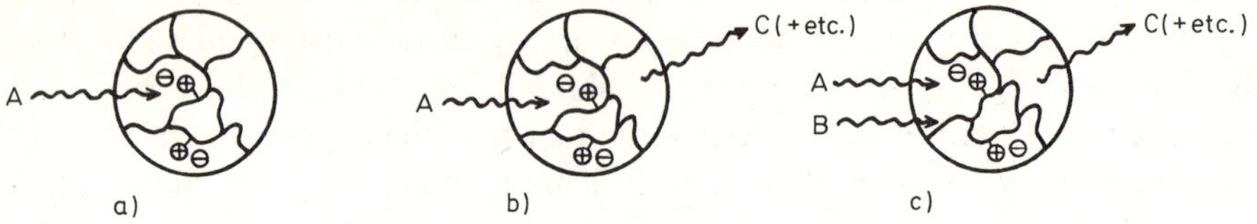

Figure 6.41 Model reactions involving polymer-bound ions.
a) Unimolecular reaction of A with bound ion, product remains attached to resin; b) unimolecular reaction of A to produce C, catalyzed or mediated by the bound ion; c) biomolecular reaction of A with B to produce C, catalyzed or mediated by the bound ion.

either mediated by or catalyzed by the bound ion, ⊕, its counter-ion, ⊖ or the ion pair, ⊕ ⊖. The product or products, C, then leave the resin. In Figure 6.41 c two reactants, A and B, enter the exchanger and undergo reaction to produce product or products, C, which subsequently leave the resin. Again ⊕, ⊖, or ⊕ ⊖ mediate in or catalyze the reaction. As with homogeneous reactions the chemical reaction itself may be complex and involve a sequence of elementary collisional reactions. Furthermore, A, B and C (+ etc.) may all be either positively or negatively charged, strongly dipolar or essentially uncharged. Hence, one can see already the great variation and complexity which can arise. As with ion exchange three fundamental processes can limit the overall reaction rate observed experimentally: (1) mass transfer of the reactant(s) from the bulk liquid (or gas) phase to the outer surface of the exchanger; (2) diffusion of the reactant(s) to the active site within the exchanger; and (3) the intrinsic reaction itself at the active site.

To evaluate the role of the matrix structure in all of these situations and in all reaction types would seem to be an impossible task. In practice, however, some of the hypothetical variants do not arise, and the various remaining situations can be categorized and simplified to some extent, because one feature, e. g., the charge on the reactant(s), can become dominant in controlling the reaction. From this point of view, in the end, each reaction must be treated independently on its own merits. Nevertheless, it is very valuable to consider separately those factors of polymer structure which might be relevant even in the case of non-crosslinked exchanger networks, and then go on to consider the additional factors which can arise with the various morphological forms of crosslinked resins.

6.3.4 Reactivity of bound ions on linear polymers

6.3.4.1 Compatibility factors

In general, the individual polymer molecules which comprise an ion exchange matrix may be regarded as random coils. In the absence of intermolecular crosslinks such coils can migrate apart in a suitable medium, when the overall polymer concentration is sufficiently low. Such species might be regarded as isolated polyelectrolyte molecules. The question then arises: can such coils always be readily penetrated by a potential reactant? Some physico-chemical evidence (e. g., fluorescence quenching measurements) suggests that indeed they can, even when the coils are present in a theta solvent (a solvent whose free energy of mixing with the polymer is zero at a given temperature). However, most of those whose experience lies predominantly in the chemical modification of, and chemical reactions involving, polymer-bound groups would suggest that there are occasions when reagents do appear to have some difficulty in penetrating polymer coils, as evidenced by rates of reaction or degrees of chemical conversion. Thus, even linear chloromethylated polystyrene reacts sluggishly with amino acids [14], whereas if the dimethylsulphonium salt is formed first by reaction with dimethyl thioether, the reaction proceeds rapidly and in good yield [15] (Reaction 7). The reaction of potassium cellulosate with chloroacetyl chloride

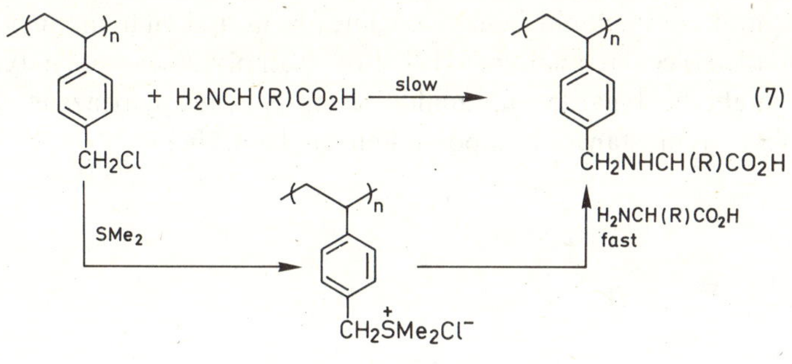

(7)

gives a good yield of chloromethyl ester, whereas the same reaction with 11-bromoundecanoyl chloride gives only a poor loading [16] (Reaction 8). In this instance the incompatibility of the

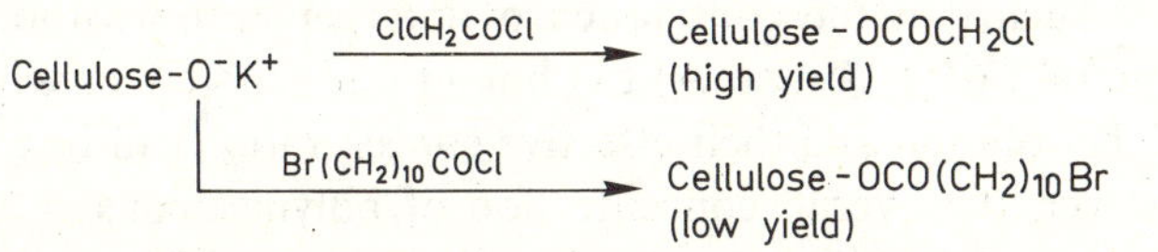

(8)

hydrophilic cellulosate and the hydrophobic long-chain bromo acid chloride seems to inhibit the reaction, despite the potentially high reactivity of the functional groups involved.

Whether such reactions are truly homogeneous is obviously an important factor in correlating the apparently different experiences of organic chemistry and physical chemistry with regard to polymer coil penetration. Certainly in the case of the cellulosate reaction a true molecular solution of the polymer is lacking, and this may be the case in the other examples of polymer chemical modification where reagent penetration seems to present a major problem. Nevertheless, the experimental observations are still very important and demonstrate the principle that in general a maximum effort must always be made to achieve optimum compatibility (with regard to polarity, hydrophobicity, hydrophilicity, etc.) between reactants and polymer-bound species, in order to maximize reaction rates and conversions.

6.3.4.2 Reactivity in freely penetrable polymer coils

Where the reactive ion exchanger is essentially a collection of freely penetrable polymer coils the question of reactivity is perhaps the simplest, and the one which is most often investigated. Under these circumstances the polymer chains play no role other than to control the volume in which the bound ion pair, ⊕ ⊖, is found and can move (Figure 6.42 b). Providing the reactant, A, is free to execute its normal solution translational motion (both inside as well as outside the coils), then the molecular circumstances are very close to those involving chemical reactions between free ion pairs and reactant, A (Figure 6.42 a). Thus, the reactive segments of linear macromolecules should behave chemically in a similar manner to some appropriate small molecular analogue [17]. For example, linear poly(styrene sulphonic acid), might be regarded as analogous to p-isopropylbenzene sulphonic acid. Under these circumstances it is possible to make a kinetic

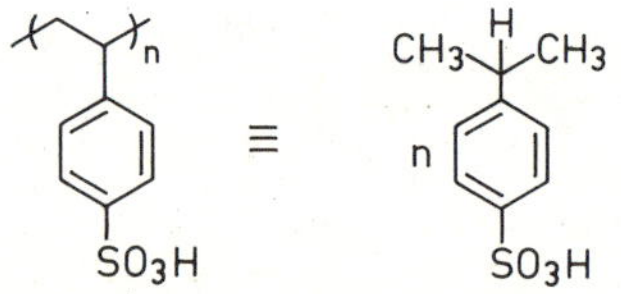

analysis of such systems analogous to that of low molecular weight counterparts, bearing in mind the constraint that the bound ion pair is restricted to the volume occupied by the expanded random polymer coils. If the overall concentration of polymer molecules is such that the coils touch or interpenetrate (without added impairment to the motion of ⊕ ⊖ and A) the bound ion pairs effectively occupy the total volume of the solution and their effective concentration can be calculated accordingly. If, however, the overall concentration of polymer coils is such that individual coils are well-separated from each other, then the bound ion pairs occupy a volume restricted to the total volume of the polymer coils (i. e., less than the total solution volume), and the *local* concentration of such bound ion pairs must be calculated upon this basis. The rate constants for reactions within polymer coils

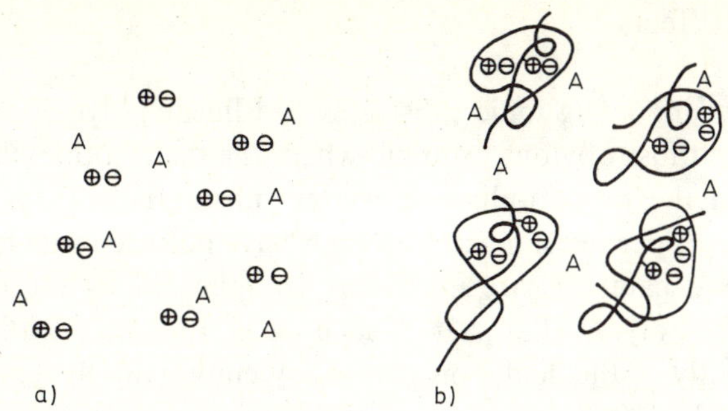

Figure 6.42 a) Homogeneous reaction of A with an ion pair, ⊕ ⊖. b) Analogous reaction in which the ion pair is attached to a linear soluble macromolecule.

(local reactions) may therefore be deduced from appropriate experimental kinetic data and, in the usual way, activation parameters may be calculated from temperature-dependence studies (accounting if necessary for the expansion/contraction of polymer coils with change in the temperature). Such kinetic parameters may then be compared with the parameters from low molecular weight model reactions, and quite valid conclusions drawn concerning similarities or differences in rate-controlling reactions, mechanisms, etc. Detailed analyses along these lines have been made for many reactions catalyzed by linear polybases [18] and reactions using polyelectrolytes [19] and polyacids [20, 21]. In general, of course, most reactions are sampled by monitoring (e. g., the disappearance of reactant A) the *bulk* solution, and some care is required in deducing the required *local* rates within polymer coils from bulk measured variables (see Section 6.3.5.1).

Many possibilities arise from deviation from the above simplistic situation, and inevitably these lead to difficulties in interpreting reactivity data and, perhaps more importantly, give rise to problems with regard to the validity of direct comparison with low molecular weight model systems. Some reports, for example, have claimed higher reactivities for matrix-bound species relative to unbound analogues as a result of an invalid comparison rather than a real molecular effect. One deviation which, in principle, is readily accommodated concerns the effective concentration of the free reactant, A. In a simple situation, A might be regarded as being distributed uniformly throughout the bulk solution and the internal volume of the polymer coils. In reality, however, A may be sorbed favourably into coils or, conversely, somewhat excluded; i. e., the thermodynamic distribution coefficient of A between the two pseudo-phases may not be unity. If this coefficient can be measured then, of course, such variation in the local concentration of A at the site of the resin reaction can be accounted for in deriving quantitative reactivity data. In the case of crosslinked polymeric networks, distribution coefficients can be determined experimentally fairly easily; in the case of molecularly soluble polymer coils, however, it is more difficult.

6.3.4.3 Electrostatic effects

The electrostatic effect involving ionic substrates and linear polymeric exchangers represents perhaps the most obvious example where the distribution coefficient of the substrate between the pseudo-phases deviates substantially from unity. The purely acidic catalytic properties of linear polyacids have not been widely examined, [20, 21] since such acid catalyses are invariably complicated by the electrostatic effects arising from the polyion character of such acids. In contrast, the latter can be studied independently of the acidic properties, by employing the corresponding alkali metal ion polysalts [19, 22, 23].

Various sulphonic acid derivatives have been examined and two distinct groups have been identified. The first possesses highly ionic character (Group 1), and the second has, in addition, varying degrees of hydrophobicity [24] (Group 2). With these species each polymer

Group 1

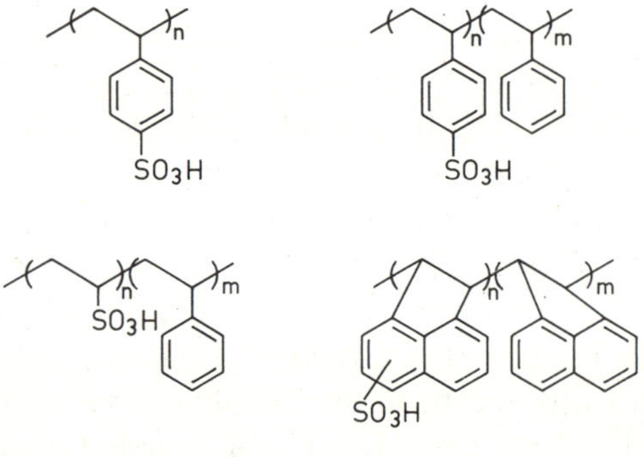

Group 2

coil represents a domain of very high electrostatic potential relative to the intervening solvent space. From a simplistic point of view it might be anticipated that ionic reactants of opposite charge to the bound ion will be relatively concentrated in the region of the polymer coil, whereas ionic reactants of the same charge will tend to be repelled into the intervening solvent. This picture predicts that the rates of bimolecular reactions between two ions of the same charge, which are *both opposite* in sign to the bound ion, would be relatively enhanced due to this local concentrating effect (Figure 6.43 b). On the other hand, reactions involving two ions of opposite charge would be adversely influenced, since the polyion would act to keep these apart (Figure 6.43 a). Such effects have been clearly demonstrated in a variety of systems [24–26].

In the case of dipolar species, rather than fully charged ones, such effects are likely to be substantially reduced, but nevertheless weaker electrostatic interactions

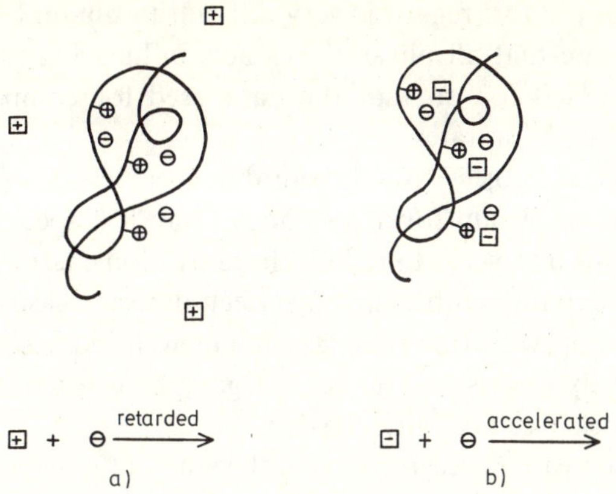

Figure 6.43 a) Reaction of charged reactants of opposite sign retarded by a cationic polyelectrolyte; b) reaction of two negatively charged reactants accelerated by a cationic polyelectrolyte.

are bound to occur and may always make some contribution in determining overall reactivity, or in favouring one reaction pathway over another. Such effects have, however, so far gone essentially undetected. In the case of aqueous bulk phases, hydrophobic interactions have been shown to play a similar role in the case of those polyacids with significant hydrophobic character [24] (i. e., Group 2 above).

6.3.4.4 Changes in activation parameters

The above are essentially the physical factors that can influence the overall reactivity of a bound ion pair, but it is important to realise that the inherent chemical reactivity of such an ion pair might also be changed simply by its immobilization on a polymeric backbone. The inherent reactivity of a chemical function with a second molecule is defined by the activation parameters for the collisional reaction, which are in turn deduced experimentally from the temperature dependence of the intrinsic rate constant for the reaction. When one of the chemical entities is attached to a macromolecule, on reaction with the second free molecule it might be expected that a greater loss of internal degrees of freedom might occur in forming the transition state for the reaction than might occur when both species were completely mobile; i. e., transition state might be more 'frozen'. Under these circumstances there would be a greater loss of entropy. Thus, the entropy of activation would be more negative, and the intrinsic rate constant for the reaction reduced. A number of classic studies have indicated that this can indeed be the case in practice [27, 28]. Furthermore, reactant molecules with many internal degrees of freedom would be expected to be most affected, and again there is some experimental evidence to confirm this [29].

In general, definitive data in this regard is very difficult to obtain because great care must be taken to ensure that all physical artifacts influencing the measured reactivity have been accounted for, so that the data used for comparison refer specifically to the inherent reaction rate.

If the reacting molecule is subject to a favourable specific adsorption in the vicinity of the polymer-bound group (over and above any less specific sorption/distribution effects), then this may serve to reduce the activation energy for reaction sufficiently to overcome the unfavourable entropy effect, in which case the inherent reactivity of the bound group with this substrate might well increase. This latter situation is much more likely to arise in the case of gas phase reactions and with crosslinked polymer supports (see Section 6.3.5.3).

Despite the enormous increase in recent years in the study of polymer-supported reactions of a wide variety of types, little progress has been made in quantifying the inherent reactivity changes described above. This has arisen partly because of the pressure to develop new systems showing interesting 'polymer effects', rather than to do detailed investigations of well-characterized systems, and partly because of the physico-chemical complexity of many of the newer systems. Certainly there is scope and need for more detailed work of this type.

6.3.4.5 Neighbouring group effects

Probably the single most important chemical factor giving rise to the modified reactivity of a matrix-bound group, and indeed to low yields and conversions and a multiplicity of side reactions, is the involvement of the neighbouring groups on a macromolecule.

In terms of reactivity the simplest example is the ionization of carboxylic acid groups, e. g., in poly(acrylic acid). Clearly the ease of ionization of a proton will decrease as the degree of ionization of neighbouring carboxylic acid groups increases, since the polymeric backbone becomes progressively more negatively charged. In fact, if K and K_0 are the ionization constants for the polymeric acid and an analogous monocarboxylic acid, respectively, K/K_0 can be as low [30, 31] as $\approx 10^{-4}$. The neighbouring groups need not, of course, be of the same type; thus, the acidity of a carboxylic acid group and a phenol residue can be mutually influenced by hydrogen bonding between the two (Reaction 9). This interaction is believed to be important in the enzyme ribonuclease [32].

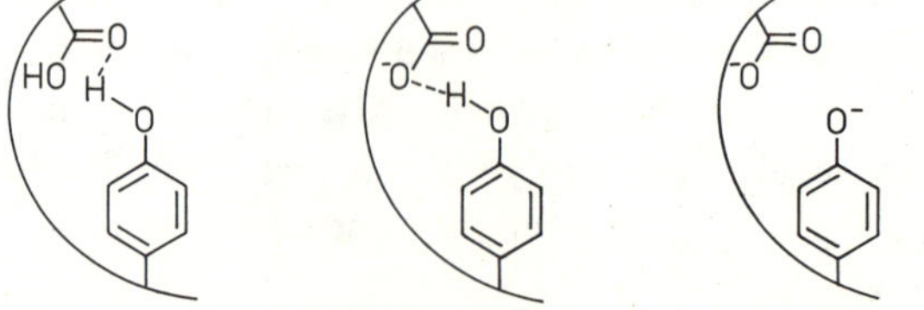

(9)

Closely related neighbouring group effects arise in the protonation and alkylation of poly(vinylpyridine)s [33]. Three rate constants are required to quantify the system: k_1, k_2 and k_3. k_1 is the rate constant for the reaction of a pyridine group flanked by two unreacted groups, while k_2 and k_3 refer to groups flanked respectively by one and two quaternized groups. Though this is a simple example it is an important one and illustrates that a great number of reactions involving polymers highly loaded with reactive groups are probably subject to such reactivity changes. Indeed it is most likely that the term 'neighbouring' should be interpreted very loosely. It may, indeed, refer to a group on an adjacent polymer segment; but it may also be applicable to a group which finds itself in close proximity to the bound group of interest simply as a result of the conformation adopted by the macromolecular chain. In this case even relatively lightly loaded polymer backbones may be subject to these effects. Again, this is an area where there has been relatively little detailed work done in recent years in the more novel systems which have been developed, and indeed it remains a difficult area to attract research resources to.

In a few important cases neighbouring groups have been shown to co-operate together to produce a pronounced increase in the reactivity of bound ions. Perhaps the most important example of this is in the dehydration of *t*-butanol catalyzed by sulphonic acid polymers [34, 35]. At high acid loadings the rate-dependence on acid concentration can reach as high as 4 or 5, and a complex concerted mechanism seems to operate. This probably involves a hydrogen-bonded network of sulphonic acid groups (Figure 6.44). Similar effects can be envisaged with linear macromolecules although, in fact, this particular example refers to a crosslinked resin species.

A somewhat different category of neighbouring group effect may also give rise to complicating side reactions involving some bound groups. For example, reaction of a primary amine with poly(chloromethyl styrene) [36] proceeds initially to form a quaternary salt (Figure 6.45). As the reaction proceeds, incoming primary amine has a choice of reacting with more $-CH_2Cl$ groups, or of deprotonating existing bound hydrochloride groups, generating polymer-bound secondary amine groups.

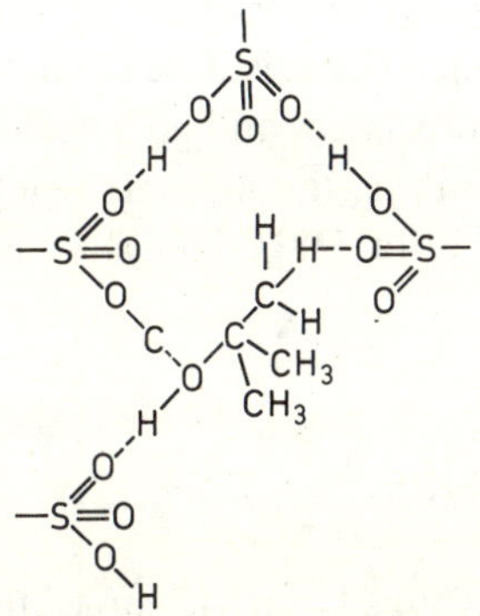

Figure 6.44 Cooperative interaction of the hydrogen-bonded network of sulphonic acid groups in an acid exchanger in the dehydration of t-butyl alcohol.

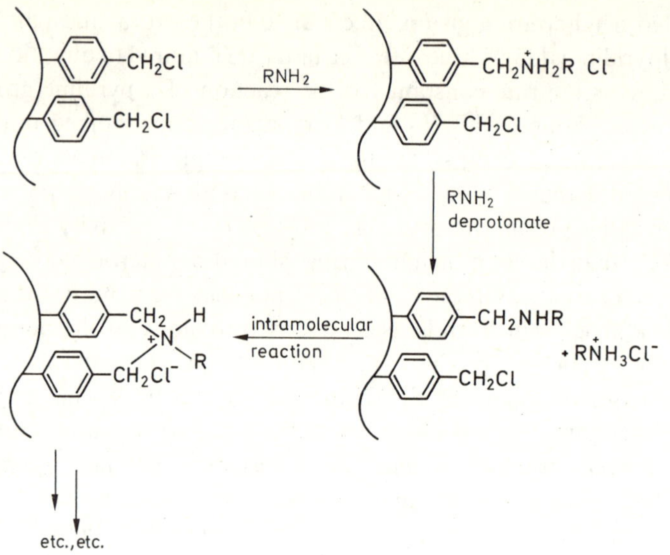

Figure 6.45 Intramolecular crosslinking reactions arising in the reaction of primary amine with poly(chloromethylstyrene).

The latter can then undergo intramolecular reaction with unreacted $-CH_2Cl$ groups to reform a quaternary nitrogen species. Clearly, deprotonation and re-alkylation can occur again, so that a complex product mixture can result. The basicities of the various amine species involved are not vastly different, and the direction of favoured reaction is thus likely to be controlled by local concentration effects and any favourable specific conformational opportunities that arise. Such reaction options arise, of course, in the corresponding homogeneous reaction, but here some external control can be exercised by, for example, choice of concentrations and the order and rate of mixing the component reactants. In the case of the polymer-bound system, that this type of neighbouring group complication occurs is shown clearly by the gelling (crosslinking) of linear polymers, and an imbalance in the elemental microanalyses of nitrogen and chlorine. There is little doubt that effects of this type are much more widespread than appreciated, and go largely undetected in the case of most resin reactions. The crosslinking (additional) reactions which arise in the chloromethylation [37] and sulphonation [38] of polystyrene resins fall into this category.

6.3.5 Reactivity of bound ions on resins

So far we have dealt with ion pairs and other groups bound to molecularly soluble polymers which, depending on the overall concentration of polymer in the bulk solution, may or may not interpenetrate one another. In the case of resins we are dealing with large collections of macromolecules permanently interpenetrated and locked together indefinitely by discrete covalent crosslinks. Right at the start it is important to realize that all of the factors already discussed in the case of linear polymers can, and very often do, arise with resin-bound species. In some cases the effects are obvious, while in others they are difficult to demonstrate, or at least quantify, because of the insoluble nature of the resin. Furthermore, because of their crosslinked nature the effects may be masked or swamped by other more dramatic changes. For example, the common occurrence of side reactions resulting from intramolecular reaction of more than one bound group, as described in Section 6.3.4.5, can go completely undetected in the case of resins, where no obvious major change in the physical properties may arise with the additional crosslinks that are formed. Where such side reactions are suspected careful chemical and physical analysis can expose them, though precise quantification can remain elusive.

6.3.5.1 Pseudo-homogenous systems

These systems represent perhaps the simplest situation which can arise with resins and the one which is most often desirable. In this case the polymeric network plays no role other than to limit the volume in which the bound group is found and can move (albeit to a restricted extent). The molecular picture is very close to the case of linear polymers with interpenetrating coils, though there is a subtle difference (Figure 6.46). Whereas in the latter case the coils can exist throughout the whole volume of the reaction medium, in the former the group of coils forming one resin bead may be well separated from the groups constituting other resin particles. From a thermodynamic point of view this represents an intriguing difference. With overlapping coils of linear polymer filling the whole solvent volume there is no net driving force for further coil expansion (i. e., polymer dilution). However, in the case of resin particles *not* filling or packing into the whole volume of the solvent medium there does remain a concentration driving force for further coil dilution, opposed by the mechanical force exerted by the permanent crosslinks. Thus it is possible for considerable osmotic forces to be generated within resins and these may well influence the reactions of bound groups. No work appears to have been carried out on this effect.

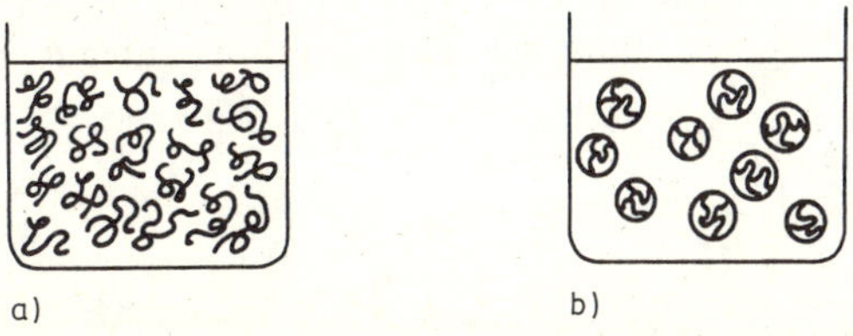

Figure 6.46 a) Distribution of soluble macromolecular species throughout an entire reaction solution volume; b) restriction of macromolecular polymer coils with the particle volume of crosslinked resins.

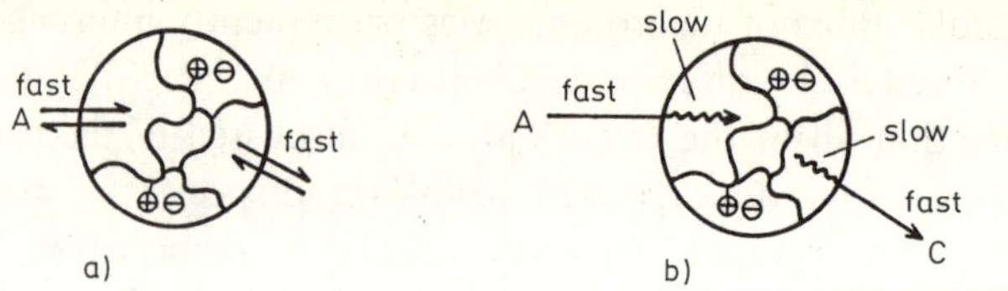

Figure 6.47 a) Resin-bound ion pair in a pseudo-homogeous reaction; b) resin-bound ion pair in a reaction limited by the diffusion of reactant (and product) through the resin matrix.

Nevertheless, a very close analogy does exist between the two systems and, in the case of the resin, the solvent within the polymer matrix may be considered as a 'homogeneous' phase within which reaction occurs and in which reactants and products are in rapid thermodynamic distribution equilibrium with the external solution. The latter acts essentially as an inert reservoir, and functions only to supply reactant and store any free product and/or by-products that are formed (Figure 6.47 a). Thus, for a bimolecular reaction between a resin-bound cation, Ⓟ—⊕, and a free reactant, A, the rate-controlling process

$$\mathrm{A} \xrightarrow[k_{\mathrm{bimol}}]{\text{Ⓟ}-\oplus} \text{Products} \qquad (10)$$

is the same as in the corresponding homogeneous model reaction involving free ⊕ and A, i. e., the bimolecular collision itself. the bimolecular rate constant, k_{bimol}, would also be the same. If the number of bound cations on the polymer matrix of volume, v, is $N_{\oplus}$, and the number of A reactant molecules is N_{A} in total solution volume, V, then the *local* rate of reaction within the resin, R_1, is given by:

$$R_1 = k_{\mathrm{bimol}} \frac{N_{\oplus}}{v} \frac{N_{\mathrm{A}}}{V}.$$

This assumes that $V \gg v$, and A is distributed uniformly throughout V and v. As the reaction proceeds A molecules will be fed rapidly into v from V, into which product molecules will emerge. Since resin reactions are generally monitored by following the appearance or disappearance of a species from the bulk volume, V, the experimentally observed rate in this volume, R_b, will be given by $R_1(v/V)$. (This is most easily appreciated by imagining the *dilution* of product as it emerges from a small volume, v, into the larger volume, V). Thus the bulk rate, R_b is given by:

$$R_b = k_{\mathrm{bimol}} \frac{\mathrm{N}_{\oplus}}{v} \frac{N_{\mathrm{A}}}{V} \frac{v}{V}$$

$$R_b = k_{\mathrm{bimol}} \frac{N_{\oplus}}{V} \frac{N_{\mathrm{A}}}{V}.$$

This, of course, is exactly the same expression applicable to two free species ⊕ and A reacting homogeneously in a volume, V. Hence, under these special circumstances the binding of ⊕ onto the polymer resin matrix has no effect on the reactivity, and the system behaves kinetically as though it were homogeneous. Though this argument has been developed for the bimolecular reaction of a free reactant, A, with a bound cation, Ⓟ—⊕, the result is a general one for other reactions, providing the pseudo-homogeneous conditions are met. These are: (1) reactant(s) and product(s) should be in rapid distribution equilibrium between the resin phase and the bulk phase, with a distribution coefficient of unity; (2) the mechanisms of reaction and respective intrinsic rate constants must be the same for the resin-supported and the non-supported reactions.

Lightly crosslinked gel type polymer resins highly swollen with a good solvent are most likely to behave in this way [39, 40], but there are also some experimental data from reactions using macroporous species which fit this model [41]. A particularly good example of a reaction which conforms to this picture is the use of resin-supported linear oligoethers as solid-liquid phase transfer catalysts [42] (Reaction 11). If the supported system is truly pseudo-homogeneous then the rate of reaction should not be dependent on the particle size

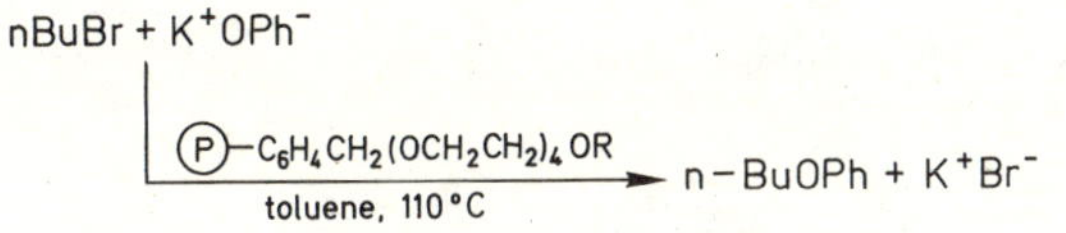

(11)

of the resin nor its structure, and for equivalent amounts of reagents, the bulk measured reaction rate should be the same as for a model homogeneous reaction, as argued above. The data in Table 6.4 shows clearly the lack of rate dependence on particle size and resin structure, and the correspondence with the rate of the homogeneous model reaction. Furthermore, for pseudo-homogeneous reactions where there are no cooperative effects involving the bound group, rates of reaction should also be independent of the loading of the bound group on the resin, provided the total number of groups remains constant in a given set of reactions. This independence of the resin loading is also apparent in the data in Table 6.4.

More often than not the factors necessary for arriving at this simple molecular picture are seldom all operative simultaneously. One obvious deviation arises when the free reactant, A, is not distributed uniformly between the support phase and the bulk solution, i. e., the distribution coefficient, λ, of A is not unity. This situation has already been described for linear macromolecules, but is more readily accounted for with resins, since λ can, in most cases, be readily established from separate sorption studies. Under these circumstances the pseudo-homogeneous rate equation for our model reaction as monitored in the bulk solution is modified to: $R_b = k_{bimol} (N_{\oplus}/V)(N_A/V)\lambda$. Clearly, where A is favourably distributed into the resin, λ is > 1 and the rate is enhanced, as in the case of some triphase catalyzed reactions [43]. On the other hand, where the resin and A are somewhat incompatible, λ falls below unity, and the rate is reduced accordingly. Where A is totally incompatible with the resin then $\lambda \rightarrow 0$ and the reaction is totally inhibited. This appears to be the

Table 6.4 Effect of the matrix parameters of the polymeric catalyst on the rate of reaction (11)

Polymer catalyst[a)]	Crosslink ratio	Loading of oligoether chains (%)	Particle size (μm)	Initial reaction[b)] rate (Ms^{-1})	Loss of *n*-BuBr (%)[c)]
I	2	61	35– 75	10.0	93
II[d)]	2	61	≪ 35	10.5	98
III	2	70	250–500	10.0	99
IV[e)]	10	71	100–200	11.0	92
V[e)]	42	29	15– 20	10.5	95
$HO(EO)_3H$[f)]	–	–	–	9.1	98

a) 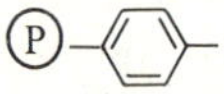 $CH_2(OCH_2CH_2)_3OCH_3$

b) $[n\text{-BuBr}]_{initial}$ = 0.24M; K^+OPh^- = 2.0 mmol; catalyst : 1.16 mmol of 'O'. Rate = Loss of n-BuBr

c) After 3 hours

d) Crushed sample of I

e) Macroporous resins

f) EO = $-CH_2CH_2O-$, homogeneous model

case, for example, with the reaction of hydroxide ion with groups on hydrophobic polystyrene resins, and is clearly the case when A is itself insoluble and the system is essentially heterogeneous. Since in many reactions the bound group itself is modified in the course of the reaction, the polarity of the resin phase itself may change significantly with time. As a result, any accompanying distribution effects may change with time, leading to a highly complex kinetic situation. Indeed, the slow shrinkage of a resin during a reaction may explain why some products or by-products can become trapped within resins [36]. Changing the reaction solvent likewise may shift any distribution factors, as can be observed with some resin-supported transition metal complex hydrogenation catalysts [44]. Indeed, the fact that some reactions proceed more effectively in one solvent than in another may often be ascribed to changes in the distribution factors.

In general, where distribution effects have been invoked they have usually involved reactant molecules. It is important to bear in mind, however, that similar factors may operate in the case of the products of a reaction involving a resin-bound species. Indeed, in the criteria which must apply for a reaction to be pseudo-homogeneous the behaviour of product(s) is included with that of reactant(s). If the product is favourably sorbed into the resin the local concentration might build up considerably. Conversely, if the product is favourably expelled, the local concentration will be maintained very low. Neither of these situations will influence significantly reactions which proceed with large and negative standard free energy changes, i. e., reactions which 'go to completion'. However, for reactions which fall into the category of 'equilibrium reactions' the distribution of the product might have a pronounced effect on the rate of the reaction and the overall position of equilibrium. Thus, build-up of product locally in the resin would inhibit the reaction, whereas expulsion of the product would tend to pull the equilibrium towards completion. There do not appear to be many examples of this behaviour reported, and indeed such an effect may not have been considered by many researchers. However, one group of reactions where it has been recognized and is likely to be important are those involving bound enzymes, since most of these operate in reaction equilibrium conditions. Indeed, it does seem that the rate and conversion of sugar to alcohol are enhanced when *Saccharomyces cerevisiae* is bound to a polysaccharide support, as a result of the expulsion of ethanol from the polymer matrix, and the displacement of the normal equilibrium [45]. Even more interesting are attempts currently being made to reverse the favoured direction in which an enzyme operates by controlling the concentration of water in the system using distribution and related effects.

In the case of alcohol dehydrations catalyzed by acidic resins [46], the inhibiting effect of the water produced might be regarded as an example of the unfavourable sorption of a product. However, this situation is further complicated by the fact that water can interact with the acid site and change the chemical nature of this from a simple proton, $H^{\oplus}$, to the hydroxonium ion, $H_3O^{\oplus}$, with a corresponding change in the catalytic activity [47].

Favourable sorption of a product into a resin catalyst may have little influence on the rate of reactions which 'got to completion' but it may give rise to complicating side reactions if the initially formed product can go on to react further to yield a second product. This seems to be the case in the cracking of methyl *tert*-butyl ether by acid resins to yield methanol and isobutene [48]. Preferential sorption of methanol allows significant acid catalysed dehydration and the formation of unwanted dimethyl ether. The problem has been ingeniously overcome by designing a resin with surface acid groups only. This has high reactivity and produces virtually no ether side product.

6.3.5.2 Diffusional effects

Although in most instances the objective of using a polymer or resin-bound species is to combine the technical convenience of a supported or heterogeneous system with the activity of a homogeneous one, this is infrequently realized. Very commonly, the reactivity of a resin-bound species is found to be altered substantially relative to the same species in homogeneous solution, and, indeed, relative to the same species attached to a soluble linear polymer. Furthermore, it is unfortunate that the change in reactivity is often downwards, although there are a number of notable exceptions. It is not unusual to find that experimental kinetic data varies with the particle size of the resin, its porosity and crosslink ratio, and with the size of the reactant molecule(s). Under these circumstances it is usual to suggest that the

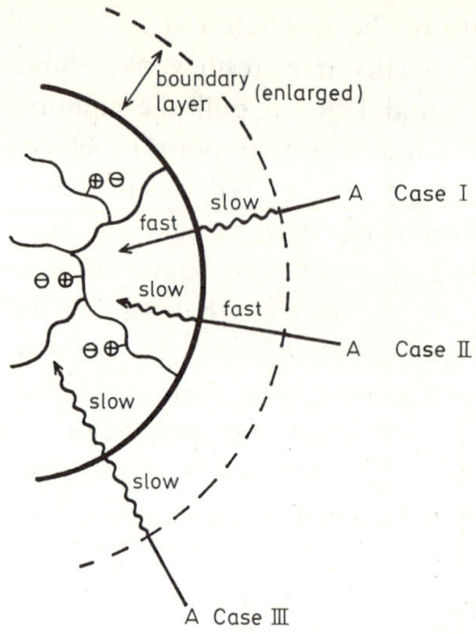

Figure 6.48 Limitation of mass transfer of a reactant A in a resin reaction; case I, diffusion through the boundary layer is slow; case II, diffusion through the resin matrix is slow; case III, diffusion through both the boundary layer and the matrix is slow.

polymer matrix provides a diffusional barrier to the reactant(s), generally in some rather ill-defined manner.

In principle such limitation of mass transfer can arise as the reactant approaches the resin through the boundary layer from the bulk reaction medium with diffusion inside the particle being relatively fast (Figure 6.48, Case I). Alternatively, the arrival of reactant from the bulk medium via the boundary layer can be rapid, corresponding to molecular collisional rates, with diffusion within the particle being slow (Case II). Finally, both diffusional processes might be slow (Case III). In each case the overall mass transfer process can be substantially slower than the reaction which follows when the reactant, A, meets the bound ion pair, ⊕ ⊖, in which case the rate of diffusion controls the overall rate of reaction. Alternatively, the rates of diffusion and of the intrinsic reaction may be comparable, in which case the overall rate will be a composite function of these. Finally, of course, if the rate of the intrinsic reaction is significantly slower than the diffusional processes, the reaction will proceed as pseudo-homogeneous, as described in Section 6.3.5.1 (provided the other criteria are also met). When diffusional limitation does occur it usually conforms to Case II. Reactant molecules are usually unchanged in most applications of reactive bound species and so boundary layer effects are rare.

An important criterion for 'diffusion control' of reactions involving resin-bound species is an inverse dependence of the reaction rate on the particle size of the resin. Where diffusion becomes limited in the boundary layer or starts immediately at the point of entry into the resin by the reactant (Figure 6.47 b), then the overall rate of reaction as monitored in the bulk solution, R_b, will be proportional to the total surface area of the particles. Thus, $R_b \alpha 4\ \pi r^2 N$, where N is the number of particles of average radius, r. The mass of a given sample of a support, m, is, of course, $(4/3)\pi r^3 \rho N$, where ρ is the density of the resin material, i. e., for a fixed value of m, $N\alpha(1/r^3)$, hence, $R_b \alpha 4\pi r^2(1/r^3)$, i. e. $R_b \alpha(1/r)$ [49]. In the case of the, 'uniform' networks characteristic of gel type resins where the dimensions between lightly solvated chains may be only of molecular size, the process of the diffusion of a reactant molecule must involve local or microenvironmental motions of the macromolecular matrix. Under these circumstances the inverse dependence, $R_b \alpha(1/r)$, should hold. Furthermore, for a fixed mass of the resin of fixed particle size, overall rates of reaction can become independent of the loading of the resin-bound group [50]. In the case of highly swollen gel type resins, 'pools' of solvent may appear within the matrix through which molecules may diffuse more quicly, without the requirement for molecular motion of the backbone. The diffusional process may thus become somewhat variable throughout the resin, and under these circumstances the $R_b \alpha(1/r)$ relationship might break down; indeed, 'diffusion control' might be lifted completely. Some resins also show evidence of their surface shell's having a different morphology than the bulk of the main particle (possibly as result of the influence of the stabilizer(s) used during preparation). With these also the porosity characteristics may be sufficiently heterogeneous for the strict $R_b \alpha(1/r)$ relationship to break down.

Processes in which reactant diffusion and the intrinsic chemical reaction proceed with comparable rates are more difficult to characterize. One generalized mathematical model has been developed [48, 51, 52] which uses the concept of an 'effectiveness factor', η. The model assumes that the diffusion of a reacting molecule can be characterized by an effective diffusivity, D, which is constant throughout a support particle, and that the intrinsic chemical reaction is represented by a true rate constant, k, again constant through the support. Thus, for reaction 10 the composite rate of reaction (i. e., the observed rate), R_{obs}, is given by:

$$R_{obs} = R_b \eta ,$$

where, R_b has the same meaning as above, and is, in effect, the hypothetical rate, if diffusion limitation plays no part, i. e., $\eta = 1$. For spherical resin particles radius, r, it can be shown that:

$$\eta = \frac{3}{\varphi}\left(\frac{1}{\tanh \varphi} - \frac{1}{\varphi}\right),$$

where $\varphi = r(k_{bimol}/D)^{\frac{1}{2}}$.

In principle, the same treatment can be used for the other reaction types described in Section 6.3.3. This model has been very successfully applied in a number of ion exchange resin catalyses [51, 53].

In the case of gel type resins, particularly those of higher crosslink ratio, this picture seems to be an appropriate one (Figure 6.47 b), and might be regarded as being associated with inefficient use of bound groups deep in the interior of the resin particles. With macroporous materials, however, where there exists a significant pore volume and a discrete interior surface which is often readily accessible to reacting molecules, poor correlation with this model might arise because of the difficulty in assigning a realistic value for r, the radius of the resin particle or, more accurately, the total distance over which diffusion can occur. Indeed, in this case, to picture diffusion as occurring through a uniform gel matrix would be misleading. Under these circumstances the inverse dependence of the reaction rate on resin size may also be expected to be invalid [50]. Some of the bound ions, Ⓟ—⊕, may be located on the interior surface of the resin and be readily accessible to a small reactant, A, (Figure 6.49). Others, Ⓟ—⊕, may lie below the interior surface in

Figure 6.49 Resin reactions involving bound ion pairs on macroporous supports; ⊕, cation on the surface of a pore, ⊕′, cation buried deep in the highly crosslinked regions; ●, small reactant molecule capable of rapidly penetrating pore volume; ⊘, large reactant molecule with dimensions comparable to the pore dimensions.

the highly crosslinked walls of the pores, and reaction of these is likely to be limited by a diffusional process similar to that already described for gel type resins. Quite clearly, to describe this situation with a simple equation applicable to all sites would be inappropriate and misleading. Recently [48], however, these circumstances have been dealt with by discriminating between the fraction of sites on the surface of macropores, γ, and the fraction located within the heavily crosslinked walls, $1-\gamma$, and defining a separate 'effectiveness factor' for each. In this case the overall 'effectiveness factor', η, is given by:

$$\eta = \eta_a[\gamma + (1 - \gamma)\eta_b],$$

where η_a and η_b are the factors for the macropore surface sites and the sites within the gel walls, respectively. η can then be used as before. Where experimentation shows no rate-dependence on particle size, and yet the rate is still depressed from its theoretical pseudo-homogeneous value, then diffusion to the macropore surface sites is not rate limiting, in which case $\eta_a \rightarrow 1$. The overall 'effectiveness factor', η, then collapses to:

$$\eta = \gamma + (1 - \gamma)\eta_b$$

and only diffusion to sites within the macropore walls reduces the overall rate below its pseudo-homogeneous value. This differentiation of sites into two clearly defined types may be somewhat naive, and in any comprehensive mathematical treatment it might well be necessary to deal with a distribution of reaction sites or even a bimodal distribution of sites with regard to accessibility and hence 'effectiveness factor'.

When diffusion does start to play an important role then, of course, the size of the reactant molecule, A, can become an important parameter. In the case of gel type supports, reaction rates (normalized to pseudo-homogeneous values) would be expected to have some inverse relationship to the size of the reactant molecule, as well as the inverse dependence on the particle size of the resin. In the case of macroporous resins some inverse dependence on the size of the reactant would also be expected though, as described above, the inverse particle size dependence might be relaxed. However, when the reactant, A, becomes very large even reaction with bound groups on the surface of macropores may be inhibited (Figure 6.49). In this case the sample diffusional model may again become sufficiently adequate to fit the experimental data. However, the movement of the reactant through the resin would be diffusion through macroscopic pores rather than through a molecular-size matrix, and might involve long-range co-operative motion of polymer support chains, and not just local motions, and be characterized by a quite different diffusion constant.

In the attachment of the alkoxides of linear oligo-ethylene oxides, $Na^{\oplus}$ $^{\ominus}O(CH_2CH_2O)_x$ R to chloromethylated polystyrene resins of various morphologies, then shorter chains, $x < 8$, appear to be attached uniformly throughout the whole volume of a resin bead. However, a longer chain species, $x \approx 30$, showed evidence of being attached only in a thin shell in the outer periphery of each bead. In this instance, where the reactant becomes bound to the resin, the initially attached species in the outer shell seem to consume enough of the available space to inhibit penetration of the remaining molecules to reactive chloromethyl groups available deeper in the resin [54].

To add further to the complexity, as the reactant, A, and the bound ion, Ⓟ—$\oplus$, react the nature of the support itself might change and therefore the macromolecular conformations too. Pore volume and pore radius may also change, and hence the assumption that any pertinent diffusion coefficient (and distribution coefficient) is constant may become invalid. One well-known and dramatic example of this is in acidic resin-catalyzed hydrations, in which water is a product. Catalyst particles may gradually swell from their initial anhydrous state. In this case a 'moving boundary' or 'shrinking core' model (Figure 6.50) has been used successfully to remove the necessity for regarding the diffusion constant and distribution coefficient of the reactant as being functions of the extent of reaction. Instead, the model assumes that as water is generated initially at the exterior surface of a bead a swollen shell of gel is formed. A distinct boundary exists between this and the unswollen glass-like core. Catalysis occurs in the gel phase and at the boundary,

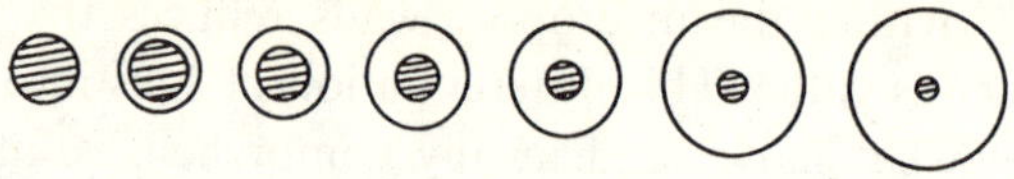

Figure 6.50 'Shrinking Core' or 'Moving Boundary Model'. ⊜ Glassy polymer; ○ water swollen gel-like polymer.

and as reaction proceeds the boundary moves inwards and the spherical core contracts or shrinks. The known inhibiting effect of water was accounted for separately, and the mathematical equations derived from this picture provided a good fit with the experimental kinetic data [55]. Though this approach has shown remarkable correlation in this case, it is doubtful whether models such as these could be more generally applicable, and indeed each system should be examined independently, and if necessary a unique analytical procedure generated.

The reactions of pyridine residues supported on gel type resins with alkyl halides of varying size display many of the features of matrix diffusion control [50] (Table 6.5). Experimental rate constants for four different alkyl halides all fall as the crosslink ratio of the resin is increased, and for a given halide and a given mass of resin, the rate of reaction is essentially independent of the loading of pyridine groups on the support. Furthermore, although all rate constants fall with the crosslink ratio of the matrix, the variation is significantly dependent on the size of the alkyl halide. Thus, the reactivity sequence for the model homogeneous reactions in solution is $C_{10}H_7CH_2Br > C_6H_5CH_2Br \gg n\text{-}C_{18}H_{37}I \approx n\text{-}C_4H_9I$. For pyridine residues on a 5% crosslinked matrix this becomes $C_{10}H_7CH_2Br > C_6H_5CH_2Br \gg n\text{-}C_4H_9I > n\text{-}C_{18}H_{37}I$; while for a 20% crosslinked matrix it further changes to $C_6H_5CH_2Br > C_{10}H_7CH_2Br > n\text{-}C_4H_9I > n\text{-}C_{18}H_{37}I$. These changes are totally consistent with the relative molecular sizes of the molecules and a diffusion-limited process. In addition, the solvent dependence of the reactivity reflects the response

Table 6.5 Relative experimental rate constants for reactions of alkyl halides with polymer-supported pyridine reagents in pentan-2-one at 102 °C

Crosslink ratio of support (%)	Relative experimental rate constants[a]			
	$n\text{-}C_4H_9I$	C_7H_7Br[b]	$n\text{-}C_{18}H_{37}I$	$C_{11}H_9Br$[b]
5	1.00	5.69	0.28	6.23
10	0.83	4.86	0.23	4.99
20	0.59	2.79	0.11	2.44
Pyridine[c]	0.42	2.42	0.07	1.51
	1.00	16.0	1.2	19.7

a) Calculated from the experimental rate law $-d[RX]_0/dt = k_{expl}[RX]_0^{0.6}$ per g of resin
b) C_7H_7Br = benzyl bromide; $C_{11}H_9Br$ = 2-bromomethyl naphthalene
c) Relative homogeneous bimolecular rate constants under same conditions

Table 6.6 Relative experimental rate constants for the reaction of 2-(bromomethyl) naphthalene with pyridine and polymer-supported pyridine species in various solvents

Crosslink ratio of support (%)	Relative experimental rate constants[a)]		
	Pentan-2-one[b)]	Toluene[c)]	Heptane[d)]
5	1.00	1.07	0.11
10	0.66	0.67	0.10
20	0.35	0.47	0.08
37	0.17	0.15	0.06
Pyridine[e)]	1.00	0.28	0.15

a) See Table 6.5 subscripts
b) 102 °C; c) 110 °C; d) 98 °C
c) homogeneous model reactions

of the polymer matrix rather than the intrinsic chemical process itself (Table 6.6). Thus, the behaviour of pentan-2-one and toluene, both good solvents for the matrix, show remarkable quantitative agreement in the resin reactions, and differ substantially from the results for heptane. In contrast, the homogeneous reaction is favoured by pentan-2-one on specific solvation grounds, while the less solvating solvents, toluene and heptane, are now comparable.

In spite of these experimental correlations the picture of the diffusional process is not a simple one. Examination under the optical microscope of sectioned samples of resins partially reacted with an alkyl iodide shows the characteristic yellow colouration of the pyridinum iodide charge transfer absorption band across the entirety of the support, with perhaps some suggestion of a higher intensity near the resin surface. In addition, the simple inverse dependence of the reaction rate on particle size is not obeyed. Thus, even with apparently uniform gel matrices the picture of diffusion occurring simply from the geometric exteriors of beads to their interiors is too naive in some cases, which shows that these models must be applied with caution.

6.3.5.3 Heterogeneous models

Under some circumstances reactions involving resin-bound ions behave essentially as though the system were truly heterogeneous, with the bound groups residing on a well-defined surface, and the reactant molecule(s) becoming adsorbed [56] prior to reaction [57]. Superficially, such a description most adequately fits gas phase reactions involving groups bound to macroporous resins [58, 59].

Unimolecular reactions can occur by fragmentation of a single reactant molecule in an adsorbed state catalyzed by the bound ion or catalyst (Figure 6.51 a). Bimolecular reactions may occur between two adsorbed molecules at the catalytic

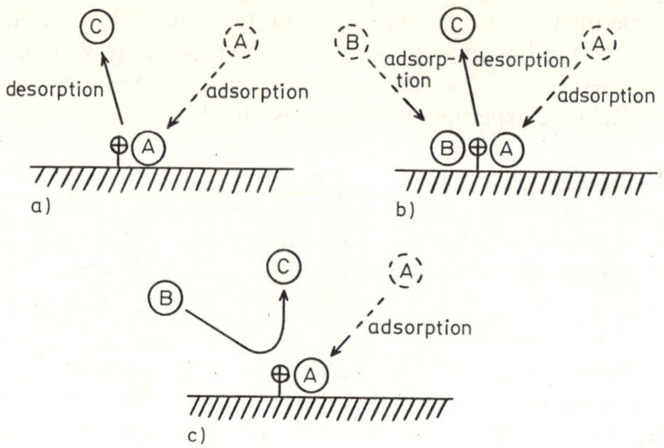

Figure 6.51 Heterogeneous model reactions. a) Unimolecular reaction of an adsorbed molecule, A, catalyzed by a bound cation; b) biomolecular reaction of two adsorbed molecules, A and B, catalyzed by a bound cation; c) biomolecular reaction between an adsorbed molecule, A, and a molecule, B, in a non-adsorbed state, catalyzed by a bound cation.

site (Figure 6.51 b), or between an adsorbed molecule and one still essentially in a non-condensed state (Figure 6.51 c), again catalyzed by the bound ion. In all cases the extent of surface coverage, and some appropriate isotherm, such as that described by Langmuir [56], is required to relate the surface coverage to the pressure of the reactant molecule which becomes adsorbed. Thus, in the reaction of a gas phase molecule, B, with an adsorbed molecule, A, (Figure 6.51 c) catalyzed by a bound cation, Ⓟ—⊕, the rate of loss of B can be expressed as:

$$-\frac{dP_B}{dt} = k_{het}\, P_B\theta_A,$$

where P_B is the pressure of B, θ_A is the fraction of surface catalytic sites occupied by A, and k_{het} is the rate constant for the heterogeneous surface reaction. If A adsorbs according to a Langmuir isotherm, then:

$$\theta_A = \frac{KP_A}{1 + KP_A},$$

where P_A is the pressure of A, and K is the ratio of the rate constants for adsorption and desorption of A. Similarly, for a reaction between two adsorbed molecules, A and B (Figure 6.51 b):

$$-\frac{dP_A}{dt} = -\frac{dP_B}{dt} = k_{het}\theta_A\theta_B,$$

where θ_A and θ_B have similar forms to the above for a Langmuir-type adsorption of each. Reactions conforming to these models are characterized by a saturation effect corresponding to the total surface coverage of sites.

With regard to the saturation effect, this model is conceptually close to the Michaelis-Menton kinetic scheme describing the binding and subsequent reaction of a substrate with its enzyme [60]. This scheme has also been widely applied to quantify the reactivity of synthetic macromolecular-bound species in liquid phase reactions [18, 19]. Its application is limited mainly to those systems where the supported species is a catalyst present in small concentrations in relation to a reactive substrate. In addition, and perhaps somewhat surprisingly, it has been applied where the system is effectively homogeneous, i. e., in the case of linear soluble macromolecular-supported species.

6.3.5.4 Site-site interaction and site isolation

When two functional groups, P and Q, are attached to an isolated linear macromolecular chain, then statistical limitations arise with regard to their ability to interact with one another. This is true for 'like' groups as well as 'unlike' ones. Providing the accessibility of all the groups is the same and remains constant, Flory [61] has shown that the maximum possible number of interactions (in the case of chemical reaction, for example) is 86.5% for a regular alternating structure (Figure 6.52 a), and 81.6% for a statistically random array (Figure 6.52 b). This applies for reactions of P and Q which are quantitative. A level of interaction of 100% is possible only for a regular vicinal structure (Figure 6.52 c). Such inherent limitations are important in studying and quantifying the neighbouring group effects described in Section 6.3.4.5.

In the case of the collection of interpenetrating polymer chains permanently crosslinked, which is typical of resins, there has been considerable experiment and debate concerning the possibility of achieving 'site isolation' of a bound group. In principle, it should be possible to introduce bound groups onto a permanent three-dimensional polymer network such that each group experiences a state analogous

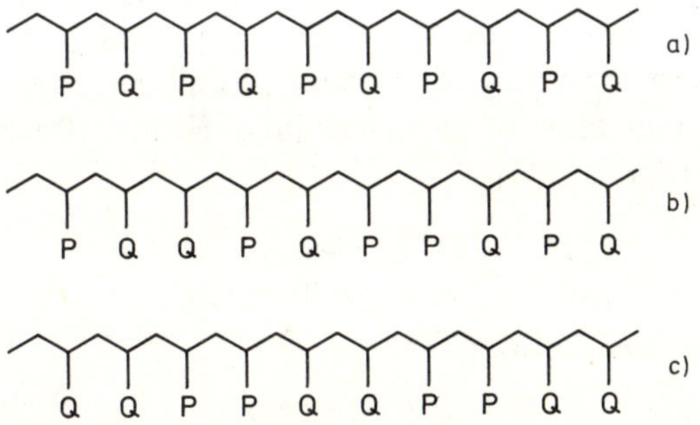

Figure 6.52 Intramolecular reactions of two groups, P and Q, on a linear macromolecule; a) groups regularly alternating along the chain; b) groups randomly arranged along the chain; c) groups in regular vicinal pairs along the chain.

to that of 'inifinite dilution' in a homogeneous solution. Such a situation could be important for producing species which, under normal homogeneous conditions, would be destroyed by rapid self-condensation reactions, e. g., coordinately unsaturated transition metal complex catalysts [62–64]. It also has considerable potential for allowing the attachment of symmetrical polyfunctional compounds by one group only, leaving the others free to be chemically transformed [65, 66]. Furthermore, such 'site isolation' of groups might inhibit bimolecular reactions of these groups, and favour some unimolecular process not normally observed in solution [67, 68]. It may also prevent unwanted side reactions, such as some of these described in Section 6.3.4.5. In view of these factors there has been a number of investigations aimed at quantifying the level of 'site-site' interaction which can occur in resins, and in particular for relating this to resin parameters such as the crosslink ratio, the pore size, the loading distribution of bound groups, and the length and conformational mobility of the linkage attaching the group to the resin polymer backbone. In addition, the effect of the reaction conditions in which the bound group is subsequently used has also been examined, e. g., the nature of the solvent, the temperature of the reaction, and the concentration of free reagent(s).

As far as the resin parameters are concerned, lightly loaded systems *favour* 'site isolation'. Typically, loadings up to 5% of pendant groups (i. e., $\approx$ 0.5 mmol bound group per gramme of resin) have been employed successfully in syntheses requiring site isolation [69]. Certainly, when loadings go beyond $\approx$ 1 mmol g^{-1}, intra-resin reactions can become dominant [70]. High loadings, however, do not necessarily preclude 'site isolation', and indeed low loadings cannot guarantee it, especially when the bound groups are non-uniformly distributed and high local concentrations occur [71]. Increasing the crosslink ratio of the support will decrease its long-range flexibility and discourage 'site-site' interaction. Gel type resins nominally 4% crosslinked are better than 2% crosslinked species [72]. However, more heavily crosslinked gels may give rise to penetration and diffusion problems. Indeed, Merrifield's early work on the 'Solid Phase' synthesis of peptides [73] showed that there are fewer exclusion complications when the crosslink ratio is restricted to 0.5%. The amination of $\approx$ 10% crosslinked chloromethylated polystyrenes to produce standard anion exchange resins can proceed very slowly.

The synthesis of dihalo-olefins using resin-bound phosphine groups and carbon tetrahalides requires the interaction of two phosphine residues (Reaction 12). As the crosslink ratio of the resin is

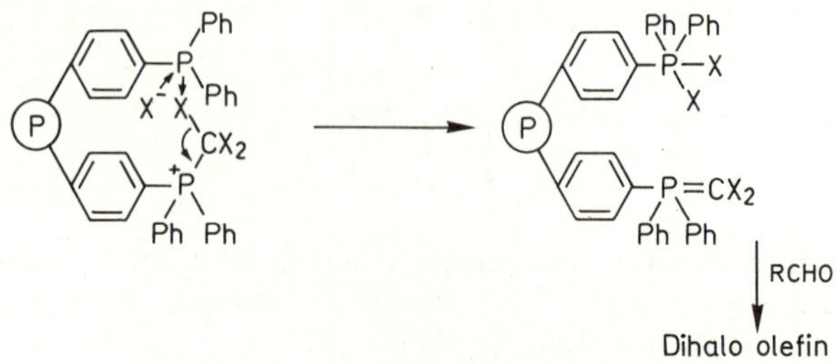

(12)

increased from 1 to 8%, and the loading simultaneously reduced from 96 to 30%, the percentage of phosphine groups which react together falls from 63 to 12% [74]. For crosslinked ratios of 15 and 35% and loadings of 12%, phosphine-phosphine interactions are virtually zero [67].

Rigid macroporous species appear at first sight to provide an optimum compromise between high crosslink ratio (and hence rigidity) and 'site isolation', along with ease of reagent penetration. However, even with these species the overall loading levels at which 'site isolation' is likely appears to be little higher than in the case of lightly crosslinked gel type resins. This is perhaps not really so surprising, bearing in mind that proportions of the macromolecular backbone and associated pendant groups are not readily accessible in these resins, but are buried in the walls of the pores. Bound groups introduced by chemical modification are likely to be found in the internal surfaces of the pores, and the local concentration of groups corresponding to an overall loading of, say, 5%, may be very much higher than for a similar overall loading on a gel type resin. In the latter case the groups will be randomly distributed along virtually the whole length of the macromolecular matrix constituting the resin. Thus, on the one hand, the prospects for 'site isolation' might be increased by the highly crosslinked and rigid structure of a macroporous resin, but, on the other, they may be reduced by the higher effective local concentration of surface groups. This may well account for the similarity in the limits of loadings, in both gel type and macroporous resins, beyond which 'site isolation' is not a good prospect.

The question of 'site isolation' and 'site-site' interaction is thus a very complex one, and although general guidelines are available to predict the situation most likely to arise with a given system, in the final analysis each case must be treated independently to achieve optimum results.

6.3.6 Conclusion

The role which the polymer structure plays in determining the reactivity of bound ions and other groups is well-understood in some respects, although remains empirical in many others. Currently the range of resin chemical structures is being expanded, as indeed are the areas in which reactive resins are currently being used [75]. Each of these situations will require its own study of the 'polymer effect', and in turn will shed further light on the behaviour of existing systems. The evolution of a general understanding will therefore continue, increasing our predictive capability. Fortunately, however, there will always be a further complication involving polymer structural effects just around the corner, and these will be guaranteed to stimulate the interests of myself, my students and my fellow polymer chemists.

References

[1] N. K. Mathur, C. K. Narang and R. E. Williams (1980) Polymers as Aids in Organic Chemistry, Academic Press, New York.

[2] P. Hodge and D. C. Sherrington (Eds.) (1980) Polymer-supported Reactions in Organic Synthesis, Wiley, London.

[3] A. Akelah and D. C. Sherrington (1981) Application of Functionalized Polymers in Organic Synthesis, Chem. Rev. *81,* 557.

[4] A. Akelah and D. C. Sherrington (1983) Recent Developments in the Application of Functionalized Polymers in Organic Synthesis, Polymer *24,* 1369.

[5] D. J. Worsfold (1974) Effect of Chain Interpenetration on Polymer-Polymer Interaction in Solution, J. Polymer Sci. Polymer Chem. *12,* 337.

[6] A. Guyot and M. Bartholin (1982) Design and Properties of Polymers as Materials for Fine Chemistry, Prog. in Polymer Sci. *8,* 277.

[7] J. K. Madden and D. Thom (1982) in: Affinity Chromatography and Related Techniques (T. C. J. Gribnau, J. Visser and R. J. F. Nivard, eds.) p. 113, Elsevier, Amsterdam.

[8] J. M. J. Frechet and M. J. Farrall (1977) in: The Chemistry and Properties of Crosslinked Polymers (S. S. Labana, ed.) p. 59, Academic Press, New York.

[9] J. R. Millar, D. G. Smith, W. E. Marr and T. R. E. Kressman (1963) The Preparation and Characterization of Expanded-network and Macroporous Styrene-Divinylbenzene Copolymers and their Sulphonates, J. Chem. Soc., 218.

[10] J. I. Crowley, T. B. Harvey and H. Rapoport (1973) Solid Phase Synthesis. Evidence for the Quantification of Intraresin Reactions, J. Macromol. Sci. Chem. *A7,* 1117.

[11] B. Merrifield (1984) The Role of the Support in Solid Phase Peptide Synthesis, Brit. Polymer J. *16,* 173.

[12] D. Lindsay and D. C. Sherrington (1985) Synthesis of Chelating Resins Based on Poly(styrene-co-divinylbenzene) and Poly(glycidyl methacrylate-co-ethylene glycol dimethacrylate), Reactive Polymers *3,* 32.

[13] Fluka AG, Chemische Fabrik, CH-9470 Buchs, Switzerland.

[14] M. A. Petit and J. Jozefonvicz (1977) Synthesis of Copper(II) Complexes of Asymmetric Resins Prepared by Attachment of α-Amino Acids to Crosslinked Polystyrene, J. Appl. Polymer Sci. *21,* 2589.

[15] D. C. Sherrington and A. Akelah (1983) Chiral Polymeric Reagents: Supported-L-proline and Polysaccharide Modified Sodium Borohydride, Polymer *24,* 147.

[16] D. C. Sherrington and A. Akelah. Unpublished results.

[17] P. J. Flory (1983) Principles of Polymer Chemistry, Chapter 3, Cornell University Press, New York.

[18] T. Kunitake, see Reference 2, Chapter 4.

[19] N. Ise (1971) The Mean Activity Coefficients of Polyelectrolytes in Aqueous Solutions and Its Related Properties, Adv. Polymer Sci. *7,* 536.

[20] K. Arai and N. Ise (1975) Studies on Hydrolysis Reaction of Model Substances of Celluloses in the Presence of Polymer Catalysts, Makromol. Chemie *176,* 37.

[21] W. Kern and B. Scherhag (1958) Hydrolysis of Peptides and Proteins with Poly(vinylsulfonic acid), Makromol. Chemie *28,* 209.

[22] H. Morawets (1975) Macromolecules in Solution, 2nd Edition, Interscience, New York.

[23] T. Okubo and N. Ise (1977) Synthetic Electrolytes as Models of Nucleic Acids and Esterases, Adv. Polymer Sci. *25,* 135.

[24] I. Sakurada, Y. Sakaguchi, T. Ono and T. Ueda (1966) Homogeneous Hydrolysis of Esters with Polymer Sulfonic Acids, Makromol. Chemie *91,* 243.

[25] C. L. Arcus, C. G. Gonzalez and D. F. C. Linnecar (1969) Hydrolysis of Esters and Ester-anions by a Polymeric Quaternary Ammonium Hydroxide, J. Chem. Soc., Chem. Commun., 1377.

[26] R. Fernandez-Prini and E. Baumgartner (1974) The Decomposition of Aspirin in Aqueous Solutions Containing Polycations, J. Amer. Chem. Soc. *96,* 4489.
[27] S. A. Bernhard, E. Garfield and L. P. Hammett (1954) Specific Effects in Acid Catalysis by Ion Exchange Resins. Some Observations on the Effect on Polyvalent Cations, J. Amer. Chem. Soc. *76,* 991.
[28] P. Riesz and L. P. Hammet (1954) Specific Effects in Acid Catalysis by Ion Exchange Resins. The Effect of Quaternary Ammonium Ions on Hydrolysis of Esters of Related Structure, J. Amer. Chem. Soc. *76,* 992.
[29] S. Affrossman and J. P. Murray (1968) Kinetics of the Hydrolysis in 1 : 3 Water-Acetone of Aliphatic Esters having Substituents in the Alkyl Group, Catalyzed by an Acid Resin or an Acidic Solution, J. Chem. Soc. B., 579.
[30] A. Oth and P. Doty (1952) Macro-Ions. Polymethacrylic Acid, J. Phys. Chem. *56,* 43.
[31] A. M. Kothiar and H. Morawetz (1955), Chelation of Copper(II) with Polyacrylic and Polymethacrylic Acid, J. Amer. Chem. Soc. *77,* 3692.
[32] H. A. Scheraga (1957) Tyrosyl-carboxylate Ion Hydrogen Bonding in Ribonuclease, Biochim. Biophys. Acta *23,* 196.
[33] J. M. Sauvage and C. Loucheux (1975) Kinetic Study of the Quaternization of Tertiary Amine as a Functional Group in the Poly(vinylpyridine) Series with Alkyl Bromide, Makromol. Chemie *176,* 315.
[34] B. C. Gates and L. N. Johanson (1969) The Dehydration of Methanol and Ethanol Catalyzed by Polystyrene Sulfonate Resins, J. Catalysis *14,* 69.
[35] B. C. Gates, J. S. Wisnouskas and H. W. Heath (1972) The Dehydration of t-Butyl Alcohol Catalyzed by Sulfonic Acid Resin, J. Catalysis *24,* 320.
[36] D. C. Sherrington. Unpublished results.
[37] B. Green and L. R. Garson (1969) Peptide Synthesis by the Soluble-polymer Technique, J. Chem. Soc. C., 401.
[38] G. D. Jones (1952) in: Styrene (R. H. Boundy and R. F. Bayer, eds.), Chapter 14, p. 674, Reinhold, New York.
[39] F. Helfferich (1954) A Quantitative Approach to Ion Exchange Catalysis, J. Amer. Chem. Soc. *76,* 5567.
[40] V. Gold and C. J. Liddiard (1977) Heterogeneous Acid-base Catalysis, J. Chem. Soc. Faraday I, *73,* 1119 and 1128.
[41] F. Ancillotti, M. M. Mauri and E. Pescarollo (1977) Ion Exchange Resin Catalyzed Addition of Alcohols to Olefins, J. Catalysis *46,* 49.
[42] W. M. MacKenzie and D. C. Sherrington (1980) Mechanism of Solid-Liquid Phase Transfer Catalysis by Polymer-supported Linear Polyethers, Polymer *21,* 791.
[43] S. L. Regen and A. Nigam (1978) Selectivity Features of Polystyrene-Based Triphase Catalysts, J. Amer. Chem. Soc. *100,* 7773.
[44] R. H. Grubbs, L. C. Kroll and E. M. Sweet (1973) The Preparation and Selectivity of a Polymer-attached Rhodium (I) Olefin Hydrogenation Catalyst, J. Macromol. Sci. *A7,* 1047.
[45] M. Wada, J. Kato and I. Chibata (1981) Continuous Production of Ethanol in High Concentration Using Immobilized Growing Yeast Cells, Eur. Appl. Microbiol. Biotechnol. *11,* 67.
[46] V. J. Frilette, E. B. Mower and M. K. Rubin (1964) Kinetics of Dehydration of t-Butyl Alcohol Catalyzed by Ion Exchange Resins, J. Catalysis *3,* 25.
[47] R. Thornton and B. C. Gates (1974) Catalysis by Matrix-Bound Sulfonic Acid Groups: Olefins and Paraffin Formation from Butyl Alcohols, J. Catalysis *34,* 275.
[48] H. Widdecke (1984) Polystyrene-supported Acid Catalysts, Brit. Polymer J. *16,* 188.
[49] F. Helfferich (1962) Ion Exchange, Chapter 11, p. 527, McGraw-Hill, New York.

[50] J. A. Greig and D. C. Sherrington (1978) Molecular Sieve Behaviour in Polymeric Reagents, Polymer *19,* 1963.
[51] E. R. Gilliland, H. J. Bixler and J. O'Connell (1971) Catalysis of Sucrose Inversion in Ion Exchange Resins, Ind. Eng. Chem. Fundam. *10,* 185.
[52] H. W. Heath and B. C. Gates (1972) Mass Transport and Reaction in Sulfonic Acid Resin Catalyst. Dehydration of t-Butyl Alcohol, AIChE J. *18,* 321.
[53] V. P. Gupta and W. J. M. Douglas (1967) Effectiveness Factors and Energies of Activation in Heterogeneous Catalysis, AIChE J. *13,* 883.
[54] W. M. MacKenzie (1979) Ph. D. Thesis. Polymer Supported Phosphines and Phase Transfer Catalysts, University of Strathclyde, U. K.
[55] H. W. Heath and B. C. Gates (1972) Mass Transport and Reaction in Sulfonic Acid Resin Catalyst, AIChE J. *18,* 321.
[56] I. Langmuir (1922) Catalysis with Special Reference to Newer Theories of Chemical Action. Heterogeneous Reactions, Trans. Faraday Soc., 607 and 621.
[57] C. N. Hinshelwood (1940) The Kinetics of Chemical Change, Oxford University Press, New York.
[58] A. Martinec, K. Setinek and L. Beranek (1978) The Effect of Crosslinking on Catalytic Properties of Macroporous Styrene-Divinylbenzene Ion Exchangers, J. Catalysis *51,* 86.
[59] J. C. Gottifredi, A. A. Yeramian and R. E. Cunningham (1968) Vapor-Phase Reactions Catalyzed by Ion Exchange Resins, J. Catalysis *12,* 245 and 257.
[60] G. W. Castellan (1970) Physical Chemistry, p. 626, Addison-Wesley, Reading, U.S.A.
[61] P. J. Flory (1939) Intramolecular Reaction Between Neighbouring Substituents of Vinyl Polymers, J. Amer. Chem. Soc. *61,* 1518.
[62] W. D. Bonds, C. H. Brubaker, E. S. Chandrasekaran, C. Gibbons, R. H. Grubbs and L. C. Kroll (1975) Polystyrene Attached Titanocene Species. Preparation and Reactions, J. Amer. Chem. Soc. *97,* 2128.
[63] F. R. W. P. Wild, G. Gubitosa and H. H. Brintzinger (1978) Supported Cyclopentadienylmetal Carbonyl Complexes, J. Organomet. Chem. *148,* 73.
[64] C. U. Pittman and Q. Ng (1978) Use of Polymer Matrices to Activate Palladium (O) Catalysts and Reduce Catalyst Agglomeration, J. Organomet. *153,* 85.
[65] J. M. J. Frechet, see Reference 2, Chapter 6.
[66] C. C. Leznoff and T. W. Hall (1982) Syntheses of a Soluble Unsymmetrical Phthalocyanine on a Polymer Support, Tetrahedron Lett. *23,* 3023.
[67] W. M. MacKenzie and D. C. Sherrington (1982) Polymer-supported Phosphines: Reactivity of Pendant and Crosslink Groups, J. Polymer Sci., Polymer Chem. *20,* 431.
[68] C. R. Harrison, P. Hodge, B. J. Hunt, E. Khoshdel and G. Richardson (1983) Preparation of Alkyl Chlorides, Acid Chlorides, and Amides Using Polymer-Supported Phosphines and Carbon Tetrachloride: Mechanism of these Reactions, J. Org. Chem. *48,* 3721.
[69] M. A. Kraus and A. Patchornik (1974) A Comparison of Various Crosslinked Polystyrenes as Diluting Media in Reactions of Ester Enolates, J. Polymer Sci., Polymer Symp. *47,* 11.
[70] J. P. Collman and C. A. Reed (1973) Synthesis of Ferrous-Porphyrin Complexes. A Hypothetical Model for Deoxymyoglobin, J. Amer. Chem. Soc. *95,* 2048.
[71] J. I. Crowley, T. B. Harvey and H. Rapoport (1973) Solid Phase Synthesis. Evidence for and Quantification of Intraresin Reactions, J. Macromol. Sci. Chem. *A7,* 1117.
[72] L. T. Scott, J. Rebek, L. Ovsyanko and C. L. Sims (1977) Organic Chemistry on the Solid Phase. Site-site Interactions on Functionalized Polystyrene, J. Amer. Chem. Soc. *99,* 625.
[73] R. B. Merrifield (1963) Solid Phase Peptide Synthesis. The Synthesis of a Tetrapeptide, J. Amer. Chem. Soc. *85,* 2149.

[74] P. Hodge and E. Khoshdel (1985) Wittig Syntheses of Dihalo-olefins by Reaction of Carbonyl Compounds with Polymer Supported Phosphines and Carbon Tetrahalides, Reactive Polymers *3,* 143.
[75] D. C. Sherrington (1984) Preparation, Modification and Characterization of Polymer-supported Species, Brit. Polymer J. *16,* 164.

7 Literature on Ion Exchangers and Ion Exchange

Konrad Dorfner
Mannheim, Federal Republic of Germany

Introduction

It might be a surprise that one chapter of this volume is dedicated to the literature on ion exchangers and ion exchange. But as in the first part the history of ion exchange and ion exchangers has been presented it may be appropriate for a relatively young part of chemistry and chemical technology, as these subjects are, also to offer an introduction to their literature. Further, the literature on ion exchangers, ion exchange and, ion exchange technology is a vast source of knowledge through which one can become acquainted both with those who broke the ground for this discipline as well as with those building its present. The chemical literature consists of books, encyclopedias and handbooks, conference reports, journals and periodicals, numeric data compilations, patent literature, government publications, market research reports and trade literature, abstract journals, and a variety of computer-based information services. Still today printed documents are more important than computerized information systems and they will be discussed first under the heading "general literature". Given the interdisciplinary character of ion exchange it is important that one has continuous access to the literature through-out one's scientific career not only for self-education and to avoid repetition but for the purpose of seeking new ideas.

Books in the field of ion exchangers and ion exchange are mainly monographs and specialized textbooks. Encyclopedias supply background information, and one can find in well-known chemical encyclopedias such as Kirk-Othmer Encyclopedia of Chemical Technology and Ullmanns Encyklopädie der technischen Chemie general introductions under the keywords "Ion Exchange" and "Ionenaustauscher", respectively. Practicing chemists and engineers will find that the information gathered there can help save time, but due to the limited space the corresponding literature references will have to be consulted for a more profound understanding of ion exchange phenomena. Handbooks hardly elaborate on ion exchangers and ion exchange, presenting at most comparison tables of commercial ion exchange resins. The trade literature distributed by ion exchanger manufacturers can be used as an introduction to ion exchange and ion exchangers.

Searching literature has occasionally been called a science and an art in its own right. The first step for anyone will always be to use the wealth of information available in the library. But the problem may still be how to find the information one needs. One should be able to run a complete literature search, both manual and on-line. Not every researcher must necessarily be a professional literature searcher, but he should know enough about the literature matrix of chemistry to ask the professional searcher for details. What makes a literature search difficult even when these problems have been solved is that much of the information needed must come from outside the library.

7.1 General literature sources

The literature of ion exchange has become so enormous that a complete bibliography is far beyond the scope of the following review. Nonetheless, the following selected literature should enable the interested reader to pursue further any branch of ion exchanger, ion exchange in general, and ion exchange technology, either to trace back the subject to the source ideas and developments or to find original contributions of individual researchers and their co-workers who paved the way in one or another direction.

After ion exchange materials had been developed and been made available for research and development, corresponding technologies and laboratory methods (the latter mainly in analytical chemistry and chromatography) emerged and were soon, usually together with the properties of ion exchangers and the principles of ion exchange, summarized in general textbooks and specialized books. Among them several have been revised in new editions; the one by F. Helfferich has become the classical monograph on the principles of ion exchangers. R. Grießbach was probably the first who planned to edit a series of monographs written by different authors, but due to unfortunate circumstances has never been completed. The following selection may be of interest to readers interested in ion exchange.

O. Samuelson
Ion Exchangers in Analytical Chemistry. Wiley, New York, 1953.

R. W. Miner, ed.
Ion Exchange Resins in Medicine and Biological Research. New York Academy of Sciences, 1953.

G. V. Austerweil
Ion Exchange and the Exchangers. Gauthiers-Villar, Paris, 1955.

G. J. Martin
Ion Exchange and Adsorption Agents in Medicine. Little, Brown and Company, Boston, 1955.

F. C. Nachod and J. Schubert, eds.
Ion Exchange Technology. Academic Press, New York, 1957.

R. Grießbach
Austauschadsorption in Theorie und Praxis. Akademie-Verlag, Berlin (East), 1957.

C. Calmon and T. R. E. Kressman, eds.
Ion Exchangers in Organic and Biochemistry. Interscience, New York, 1957.

R. Kunin
Ion Exchange Resins. Wiley, New York, 1958.

J. E. Salmon and D. K. Hale
Ion Exchange. A Laboratory Manual. Butterworths, London, 1959.

F. Helfferich
Ionenaustauscher. Band I. Grundlagen: Struktur, Herstellung, Theorie. Verlag Chemie, Weinheim, 1959.

K. M. O'Shanova
Utilization of Ion Exchangers in the Food Industry. Latvian Tekh. Inform., Riga, 1961.

B. N. Laskorin, N. M. Smirnova and M. N. Gantman
Ion Exchange Membranes and their Use. Ak. Nauk. Tekh., Moscow, 1961.

F. Helfferich
Ion Exchange, McGraw Hill, New York, 1962.

J. Stamberg and V. Radl
Ion Exchange – Principles for Laboratory and Industrial Practice. Statni. Nakl. Tech. Lit., Prague, 1962.

A. A. Gerasimenko, M. A. Abramova and P. V. Galovin
Ion Exchange Resins in the Food Industry. Izd. Akad Nauk Univ., Kiev, 1962.

K. Dorfner
Ionenaustausch-Chromatographie. Monographs on Ion Exchange. Vol. 2. R. Grießbach, ed. Akademie-Verlag, Berlin (East), 1963.

W. Rieman and F. H. Walton
Ion Exchange in Analytical Chemistry. Wiley, New York, 1963.

C. B. Amphlett
Inorganic Ion Exchangers. Elsevier, Amsterdam, 1964.

R. Schlögl
Stofftransport durch Membranen. Steinkopf-Verlag, Darmstadt, 1964.

B. Tremillon
Les Separations par les Resines Echangeuses d'Ion. Gauthiers-Villar, Paris, 1965.

J. Inczedy
Analytical Applications of Ion Exchangers. Pergamon, New York, 1966.

R. Hering
Chelatbildende Ionenaustauscherharze. Monographs on Ion Exchange. Vol. 3. R. Grießbach, ed. Akademie-Verlag, Berlin (East), 1967.

E. V. Egorov and P. D. Navikov
Action of Ionizing Radiation on Ion Exchange Materials. Davey, New York, 1967.

N. M. Korolkov and O.Vitols
Synthesis of Ion Exchangers. Text Book. Riga, Latvia, USSR, 1968.

S. Applebaum
Demineralization by Ion Exchange. Academic Press, New York, 1968.

T. V. Arden
Water Purification by Ion Exchange. Butterworths, London, 1968.

P. Smit
Ionenaustauscher und Adsorber in der Herstellung von Zuckern, Pektinen und verwandten Materialien. Monographs on Ion Exchange. Vol. 5. R. Grießbach, ed. Akademie-Verlag, Berlin (East), 1969.

J. Korkisch
Modern Methods for the Separation of Rarer Metal Ions. Pergamon, Elmsford, NY, 1969.

Y. Marcus and A. S. Kertes
Ion Exchange and Solvent Extraction of Metal Complexes. Wiley-Interscience, London, 1969.

G. V. Samsonov, E. B. Trostyanskaya and G. E. Elkin
Ion Exchange. Sorption of Organic Substances. Nauka, Leningrad, 1969.

E. V. Egorov and S. B. Makarova
Ion Exchange in Radiochemistry. Atomizdat, Moscow, 1971.

V. S. Soldatov
Simple Ion Exchange Equilibriums. Nauk Tekh., Minsk, 1972.

N. F. Chelichev
Ion Exchange Properties of Minerals. Nauk Tekh., Moscow, 1973.

J. A. Marinsky and Y. Marcus.
Ion Exchange and Solvent Extraction. Dekker, New York, 1973.

J. X. Khym
Analytical Ion Exchange Procedures in Chemistry and Biology. Prentice Hall, Englewood Cliffs, NJ, 1974.

E. Selegny, ed.
Charged Gels and Membranes. D. Reidl, Dordrecht, 1977.

M. Abe
Inorganic Ion Exchangers. Kodansha, Tokyo, 1977.

M. M. Senyavin
Ion Exchange in the Technology and Analysis of Inorganic Substances. Khimiya, Moscow, 1980.

L. K. Arkhangelskii and F. A. Belinskaya
Ion Exchangers in Chemical Technology. Khimiya, Leningrad, 1982.

N. G. Taran
Adsorbents and Ion Exchangers in the Food Industry. Legkaya i Pishchevaya Promyshlennot, Moscow, 1983.

E. F. Ergozhin and E. Zh. Menligaziev
Polyfunctional Ion Exchangers. Nauka, Alma Ata, USSR, 1986.

Conference reports have acquired more and more the dimensions of books since it has become quite difficult for one individual to write a comprehensive book on a subject comprising a chemical substance class, their properties and their applications especially when these applications are in such various fields. The more advanced ion exchangers have become, the more their applications moved into fields for which special basic and applied knowledge is a prerequisite, as for instance in medicine. Conferences as well as symposia on ion exchangers and ion exchange have been held at more or less regular intervals since about the late nineteen-fourties in various countries and their proceedings have been edited in quite large volumes. Since not all the papers presented at such conferences are later published in journals, these reports represent a very important source of first-hand information. Besides the many national and even regional symposia held over the years in different countries, organized usually by the respective Chemical Societies or Academies of Science, the following to some extent traditional international conferences deserve special mention.

Gordon Research Conference.
Since 1949. Papers not published as volumes.

Ion Exchange and its Application.
International conference on ion exchange of the Society of Chemical Industry. London, 1954 (publ. 1955).

Ion Exchange in the Process Industries.
International conference on ion exchange of the Society of Chemical Industry. London, 1969.

The Theory and Practice of Ion Exchange.
International conference and 173rd event of the European Federation of Chemical Engineering. Cambridge, 1976.

Ion Exchange Technology.
International conference on ion exchange of the Society of Chemical Industry. Cambridge, 1984.

Symposium über Ionenaustauscher.
Organized by the Chemische Gesellschaft in der Deutschen Demokratischen Republik. Merseburg, 1958.

Anomale Vorgänge an Austauschadsorbentien.
Symposium organized by Fachverband Analytische Chemie in der DDR. Weimar, 1961.

30 Jahre Kunstharz-Ionenaustauscher.
Symposium organized by the Chemische Gesellschaft in der DDR. Leipzig, 1968.

Ion Exchangers and their Application.
First Symposium of the Society of Hungarian Chemists. Balatonszeplak, 1963.

Second Symposium on Ion Exchange of the Society of Hungarian Chemists. Balatonszeplak, 1969.

Third Symposium on Ion Exchange of the Society of Hungarian Chemists. Balatonfüred, 1974.

Fourth Symposium on Ion Exchange of the Society of Hungarian Chemists. Siofok, Lake Balaton, 1980.

Fifth Symposium on Ion Exchange of the Society of Hungarian Chemists. Balatonszeplak, 1986.

Adsorption and Ion Exchange Separations.
AIChE Symposium Series. Vol. 74, No. 179, New York, 1978.

Ion Exchange Symposium
Central Salt and Marine Chemicals Research Institute. Bhavnagar, India, 1978.

Oslo Symposium on Ion Exchange and Liquid-liquid Extraction.
Chemical Institute of the University of Oslo, 1982.

Mass Transfer and Kinetics of Ion Exchange.
Nato Advanced Study Institute, Maratea, Italy, 1982. Martinus Nijhoff Publishers, The Hague, 1983.
Fundamentals and Applications of Ion Exchange.
Nato Advanced Study Institute, Bari, Italy, 1984. Martinus Nijhoff Publishers, Dordrecht, 1985.
Ion Exchange: Science and Technology.
Nato Advanced Study Institute, Troia, Portugal, 1985. Martinus Nijhoff Publishers, Dordrecht, 1986.

Numeric data represent properties of chemicals needed not only to describe but also to utilize materials. Compilations are available in abundance for all important substances and need not be dealt with in detail here. With respect to ion exchangers numeric data describing basic properties, for instance, capacity, can be found in the technical data sheets of the various manufacturers. In the selected tables of corresponding commercial ion exchanger programs in Appendix 1 of this volume basic data have been compiled for easy reference. The more special properties of ion exchangers, for instance IR, UV, or NMR spectra and thermodynamic properties, can be found so far only in papers appearing in various journals, which makes it more difficult to find them in case they may be needed. Reference is made to them in various chapters and sections of this volume. But with respect to ion exchange as a chemical process, the first attempts have been made at compiling ion exchange equilibrium constants and distribution ratios as cited below. This work has so far apparently not been continued. If it is continued it would seem advisable that it not only be revised and updated but extended as well to include diffusion coefficients and the Helfferich numbers.

Y. Marcus and D. G. Howery.
Ion exchange equilibrium constants. IUPAC, Analytical Chemistry Division, Commission on Equilibrium Data. Butterworth, London, 1975.
Eiko Akatsu.
Data on ion exchange. Nippon Genshiryoku Kenkyusho, Kakai JAERI-Memo 1977. JAERI-M-7168 (Eng.).

Specialization has resulted in periodicals and journals being in general dedicated to narrow fields of science. Since ion exchangers and ion exchange are quite varied with regard to their applications and the materials used, as well as to the explanations of the phenomena either by the employment of known natural laws or by theoretical modelling and its mathematical verification, it is understandable that the publication of such investigations are to be found throughout the literature of the natural sciences. From soil science to brewing technology, from analytical chemistry to biotechnology and the life sciences the corresponding journals contain valuable contributions with regard to ion exchangers. For the same reason many in principle simple reviews have been published in all sorts of journals, probably on request of the editors intending to give their readers an insight into the state of the art of ion exchange and ion exchangers. Review articles are of value if they are of the kind presented at regular intervals either in journals or in other periodicals, but so many have been published that only the most well-known can be cited.

R. Kunin et al.
Reviews on Ion Exchange. Unit operations and research keynotes. Industrial and Engineering Chemistry. 1948 to 1968.

R. Kunin et al.
Reviews on Ion Exchange. Analytical Chemistry. 1949 to 1966.

H. F. Walton.
Reviews on Ion Exchange. Analytical Chemistry. 1970, 1974, 1976, 1978.

T. R. E. Kressman et al.
Ion Exchange Surveys. Ion Exchange Materials. The Permutit Co., Ltd., London. For 12 years to 1969.

P. Janders and J. Churacek.
Ion-Exchange Chromatography of Carboxylic Acids. J. Chromatogr. *86,* 351 (1973).
Ion-Exchange Chromatography of Sulphur Compounds, Phenols, Phosphorous Compounds and Esters of Carboxylic Acids. J. Chromatogr. *86,* 423 (1973).
Ion-Exchange Chromatography of Nitrogen Compounds. J. Chromatogr. *98,* 1 (1974).
Ion-Exchange Chromatography of Aldehydes, Ketones, Ethers, Alcohols, Polyols and Saccharides. J. Chromatogr. *98,* 55 (1974).

For a long time ion exchangers were mainly considered as a tool for water treatment or for analytical separations, with the result that relevant research results were either published in journals more of an engineering nature, or large numbers of papers were to be found mainly in journals of analytical chemistry and chromatography. As the use of ion exchanger materials widened to include such special products as membranes and special applications as, for instance, desalination of sea water and hydrometallurgical extractions, and an understanding of ion exchangers developed that saw them as one kind of reactive polymers, journals were introduced dedicated solely to the one or the other subject. The following list contains publications on ion exchangers in these areas.

Reactive polymers, ion exchangers, sorbents.
Vol. 1, No. 1, October 1982, Quarterly. F. G. Helfferich, ed., Elsevier, Amsterdam.

Ion Exchange and Membranes.
Vol. 1, No. 1, August 1972. J. A. Mikes, ed. Gordon and Breach, New York. Only two volumes.

Solvent Extraction and Ion Exchange.
Vol. 1, No. 1, 1983. Dekker, New York.

Desalination and Membrane Science, and Desalination.
Vol. 1, No. 1, 1966 and 1976, resp. M. Balaban, ed. Elsevier, Amsterdam.

Patents are not generally considered to be a literature source for basic science; they have, however, been used as a fertile literature source for new and useful chemicals and chemical compositions, processes for the production of the same, and apparatus and equipment for manufacturing processes. Both for ion exchangers and ion exchange as a process patents are of great importance. The history of ion exchangers reveals how many new ion exchange materials have first become known over patents. Reading patents or, better, abstracts of patents first on ion exchangers is a substantial help in idea-seeking. As a sophisticated process ion exchange offers a wide field for new inventions, which are usually described in patents. How to keep up with the patent literature will be described in following sections.

Abstracting and indexing services are secondary publication operations. Without going into the history of abstracts it should nevertheless be mentioned that the first abstract journal in chemistry Pharmaceutisches Centralblatt was introduced in 1830; its name was changed to Chemisches Zentralblatt in 1856. Abstracts have always been included in chemical journals in all countries but usually a journal exclusively for abstracts was founded. As an example, the Chemical Society and the Society of Chemical Industry in England published their respective abstracts until the two societies founded British Abstracts. In Japan the Information Center for Science and Technology now publishes the Nippon Kagaku Seran, i. e., Complete Chemical Abstracts by Japan, and the American Chemical Society introduced Chemical Abstracts in 1907. Already by 1907 there were over sixty abstract journals, and by now there are well over 1,500 abstracting and indexing publications. Their number has increased with the amount of primary literature. The advantage of this development is that one has access to an abstract journal covering the literature of his specialty, but the disadvantage is that no abstract journal, not even Chemical Abstracts with its monitoring of up to 14,000 journals plus patents, books and even conference reports of most major countries, covers the whole spectrum of the literature. Within chemistry ion exchange is a new and separate discipline with interdisciplinary connections from material science to organic syntheses, as well as in many areas of engineering. Consequently, literature on ion exchangers and ion exchange is abstracted and indexed by several abstracting and indexing services. This can make it easier for the potentially interested reader to find information or result in his completely overlooking important information.

Abstracts are either indicative, informative or critical. Future developments may lean more and more to the indicative and less informative abstracts due to increasing costs of producing abstracts publications. A real informative abstract should provide enough substantial information to enable the reader to make some use of it without consulting the original itself, but with the increase in original publications the abstractors seem inclined more and more only to reprint the summaries of the papers to be abstracted, which do not necessarily contain the essential results. It would further be of value if, the more difficult it is to obtain an original paper, the more informative and elaborate the abstract should be. This applies also to certain language barriers.

Chemical Abstracts (CA) of Chemical Abstracts Service (CAS), published by the American Chemical Society, has come to have a dominant role throughout chemistry for its comprehensive abstract coverage of the chemical literature since British Chemical Abstracts in 1953 and Chemisches Zentralblatt in 1969 ceased publication.

In 1986, CA monitored 12,000 scientific and technical journals from more than 140 countries in more than 50 languages. It is not the author's purpose here to give an introduction to the various services of the Chemical Abstracts Services, as these can easily be obtained, updated yearly, from the CAS International Catalog, but it may be mentioned that with respect to ion exchange and ion exchangers the printed Chemical Abstracts can be extremely helpful. Of the above-mentioned approximately 12,000 journals monitored for CA, less than 300 are devoted entirely to chemistry and chemical engineering, but these yield more than half of the total papers abstracted in CA. About 90% of the abstracts come from slightly more than 2,000 journals. Well over 1.5 million papers, patents, and reports are currently reviewed annually for selecting material for abstracting in CA. About 30% of the documents abstracted originate inside the United States, although 54% of the total documents are published in English, with the remainder being in approximately 50 other languages. For a manual search on ion exchange and ion exchangers in CA the various indexes are the most valuable tools. These indexes are arranged by author, general subject, patent, formula, and chemical substance; every 5 years (prior to 1957, every 10 years) the volume indexes are merged and published as a Collective Index. The collective indexes now available are the 1st, 1907–1916; 2nd, 1917–1926; 3rd, 1927–1936; 4th, 1937–1946; 5th, 1947–1956; 6th, 1957–1961; 7th, 1962–1966; 8th, 1967–1971; 9th, 1972–1976; 10th, 1977–1981, and 11th, 1982–1986, which was issued in 1987. These collective indexes are indispensible for a retrospective search, in general in the subject index under the keywords ion exchange and ion exchangers. However, a problem exists in searching the older literature, as the keywords ion exchange and ion exchangers were first used only strictly in the 8th Collective Index. The 1st through 4th Collective Indexes do not cite these keywords at all, and it seems that only the appearance of the word "ionites" in the 5th Collective Index led to a cross reference to "exchanging substances" under the keyword "Ions, electrolytic". From the 6th Collective Index on the keyword "Ion exchange" is permanently used, but for the material there is still the keyword "Ion-exchange substances". This is also the case in the 7th Collective Index. In the early volumes "base exchange" instead of "ion exchange" and "zeolites" instead of "ion exchangers", were used as keywords, as well as the tradenames of commerical ion exchangers (for instance, Dowex) in the latter case. These circumstances make it more difficult to search the ion exchange and ion exchanger literature in the older CA collective indexes.

Other abstracting services which can be consulted for ion exchange will be briefly mentioned; although Chemical Abstracts is an invaluable tool for chemists and chemical engineers, it cannot serve all their information needs. Current Index to Conference Papers (CICP) is published monthly and is an alerting publication to papers which are to be delivered or which have been presented at scientific and technical meetings throughout the world in the three areas of chemistry, life sciences, and engineering. The subject index includes the title of the paper, the name(s) of the author(s), and the title of the conference. The Engineering Index publishes abstracts of research and applications relevant to engineering from journal articles, conferences, and patents. Copies and translations of most of the articles abstracted in Engineering Index are available. The index includes subject, author, and author affiliaton. The Index of Scientific Reviews publishes abstracts of papers from journals in the physical, chemical, medical, and life sciences, and in engineering, agricultural, biological, and environmental areas, which can be retrieved by keywords in the title, author, organizations, and citations. The Index to Scientific and Technical Proceedings covers papers from several thousand proceedings with indexes to those appearing in books and in journals. This indexing publication is issued monthly with semiannual cumulations. Two information centers established by governments are the Japan Information Center of Science and Technology (JICST) founded in 1957 for the publication of current bibliographies and to establish a bibliographic data base, and the National Technical Information Service (NTIS) created in 1950 by the US Congress within the US Department of Commerce as the central source for the sale and distribution of reports resulting from government-sponsored research and development. Additional abstracting and indexing services that index information on ion exchange and ion exchangers are the Analytical Abstracts issued monthly by the Chemical Society, London; Corrosion Abstracts published bimonthly by the Association of Corrosion Engineers, Houston, Texas; Dissertation Abstracts issued monthly by University Microfilms, Ann Arbor, Michigan; Index to Forthcoming Russian Books issued monthly by Scientific Information Consultants, London; and Liquid Chromatography Literature published bimonthly by Preston Publications, Inc., Niles, Illinois.

CA Selects® is the latest publication introduced by Chemical Abstracts Service. It is a series of collections of abstracts pertaining to special areas, one of which is "CA Selects Ion Exchange". This current awareness publication is published biweekly and includes those abstracts and bibliographic citations published in CA relating to the special area of ion exchange. This is a very useful service as it covers all aspects of the theory and applications of ion exchange. Included are ion exchange chromatography, ion exchange processes, techniques, and procedures, as well as ion exchanger materials, equipment and instrumentation. CA Selects is based on a concept pioneered by the Royal Society of Chemistry Information Services and is in fact a cooperative development of CAS and the RCS. This publication became feasible after printing of the abstracts from the CAS computer-readable data base was introduced. Using CA Selects Ion Exchange saves the time of searching for literature of interest in the CA printed abstracts numbers. Further, the price of CA Selects Ion Exchange is within the reach of an individual chemist and chemical engineer.

7.2 Computer-based information services

It is said that Chemical Abstracts Service Online® (CAS Online) is one's connection to the world of chemistry. And it is indeed important for the user of the printed Chemical Abstracts (CA) to know also about CAS Online, which is the chemical search system of Chemical Abstracts Service. This system provides specific access to the same substance information and bibliographic data published in CA; if the necessary equipment is available and conditions are met a CAS Online search at a terminal can accomplish in minutes what might have taken hours using the printed CA indexes. By means of CAS Online one can search the CAS Chemical Registry System, a data base with six million chemical substance records and nine million chemical names (at the and of 1983). At the end of 1983, CAS Online expanded to include the complete file of CA bibliographic information, keywords, and subject index entries from 1967 to the present. Literature accessible through CAS Online includes journal articles, patents, proceedings of meetings, symposia, new books, and other documents published in 150 nations. All references and abstracts are in English, though the original documents may be in any of fifty languages. It could perhaps be said that CAS Online is primarily substance-oriented; of the several million of substances registered, approximately 97% have a full molecular identification. Perhaps it is not so widely known that about 12% of the registry file contains substances identified by their components, namely, polymers, alloys, mixtures, addition compounds, and complex salts. CAS Online offers its information material in three different data bases, called CA File, Registry File, and CAOLD File.

CA File contains all the information in the CA SEARCH magnetic tape service, i. e., all bibliographic data since 1967, plus all abstract texts from Chemical Abstracts since 1975. This is a special feature compared to what other CAS data base brokers offer. Thus, the CA File is completely identical with the Chemical Abstracts. It is updated biweekly.

The Registry File contains the above mentioned structural formulae (totalling now almost 7 million) together with the CAS Registry Numbers, as well as nine million chemical compound names and their empirical formulae. It is updated weekly. The Registry File and the CA File are connected by the CAS Registry Number, which enables one to continue a literature search in the CA File with a CAS Registry Number found in the Registry File and thus to locate bibliographic information and the text of the abstract. The CAOLD File has been set up to record documents covered in the printed CA prior to the year 1967, which are not contained in the CA File. If, therefore, literature on a substance is to be searched going back before 1967, the Registry Number from Registry File is transferred to the CAOLD File where the CA abstract number or the CA accession number can be found. The CAOLD File is being expanded incrementally, as the pre-1965 substances are added to the CAS Registry File. In 1984, the file contained one and a half million CA Reference Numbers. One can gain entry to both the Registry File and the CA File by means of the CA Registry Number. Many ion exchanger materials have such a number. The information obtained may, because of the special chemical nature of ion exchangers and, in particular, of the ion exchange resins, come from the Registry File, and then one only gets the numbers of the latest abstracts, from the CA File, and then one gets bibliographic information for all abstracts since 1967 and the abstract texts since 1975. If one starts one's literature search with the term "ion exchange" or with bibliographic data already available, the CA File as the data base of choice, will give answers that are the same as just described. As an additional help in connecting the contemporary term "ion exchange" with terms used previously, the CA General Subject Index Headings List can be consulted, as this list contains cross references from terms of collective indexes to the valid terms of other collective indexes. When working with other data bases information material relevant for the use of the particular retrieval or command language must be employed. There are several hundred to one thousand data bases available worldwide so that online searching is fairly routine work in many organizations. An important online data base for the retrieval of information contained in patents is the Derwent patent data base, which permits the retrieval of generalized structure claims using fragment codes, which takes into account the fact that many structures are not so exactly defined.

In contrast to ion exchangers as a substance class ion exchange is often considered to be a separation technology that offers an alternative when the problem arises of how to separate a chemical species cleanly and efficiently. To find a method quickly is, therefore, often of the utmost importance to chemists and chemical engineers; it has been found, however, that retrieving information on specific chemical separations is among the most difficult problems of information management. Reasons for this are that the volume of technical literature has increased greatly and that the term ion exchange has been used for zeolite, resinous and liquid separations. These multiple meanings, that ion exchange cuts across the traditionally established scientific disciplines, as well as the inconsistencies in word usage result in inconsistent keywording which, in turn, result in the separation chemist's obtaining an unacceptably large percentage of unwanted material during such a search. Another reason for the low efficiency of data bases with respect to ion exchange separations is that a large body of information related to hydrometallurgy and analytical chemistry was generated, e. g., in the United States in government research programs originally classified secret during the mid-nineteen-forties to the mid-nineteen-sixties. This information has, at best, only a limited indexing and is not included in any computer-searchable data base. The poor availability of this reference material has certainly already resulted in duplication of research effort. Attempts to solve some of the problems of information retrieval in separations science resulted in the setting up of a Separations Science Data Base. This was established at Oak Ridge National Laboratory (ORNL) by members of the Separation Science Research Group for information relating to solvent extraction and ion exchange separation, as well as for the purpose of indexing a body of early separations science literature that was poorly and incompletely indexed. It is hoped that if indexing of this material can be accomplished while the scientists familiar with its whereabouts are still available, the repetition of a considerable amount of experimental work may be avoided.

The data base is stored on the computer system ORNL, and it can be searched and information retrieved through the Department of Energy data base management program DOE/RECON (REmote COnsole). Access is through local hardwired terminals, commercial telephone lines, TYMNET, or the Federal Telecommunications System (FTS). All Department of Energy employees engaged in energy-related work have direct access to the DOE/RECON system. Further, the intention was to make special arrangements for others to have direct access under certain circumstances. The cost of using RECON is similar to other data base search systems. In addition to direct searching, customers can have searches performed by Western Regional Information Service Center, Berkeley, California. The Separation Science Data Base has several unique features that make it especially useful to both users and abstractors in the separations science field, and these features might be valuable to data bases in other fields. The most common question asked by one being faced with a separation problem is: How do I separate A from B out of matrix M? The Separations Science Data Base has been constructed to answer this and related questions. In order to accomplish this, the subject key terms usually lumped together under "subject index" are categorized and divided into five fields: separation system, separated substance, separation agent, matrix, and type of information. Included in the separated substance field are symbols indicating the valence of the separated substance (if pertinent and available) and whether the separated substance is retained, rejected, optionally retained or rejected, or split with no separation by the system. Carefully selected descriptor terms within these fields describe the information contained in a paper or report and especially in multiple-subject reports. Dividing the subject index terms into categories in this way makes it possible, by means of the Boolean 'and-or-not' combinations, to retrieve specific information on specific separations under specific conditions. For example, iron, chloroform, or alcohol might be either a separated substance, a separation agent or a matrix, and the use of the word alone does not reveal this differentiation. The use of appropriate terms from the various categorized fields, however, allows for such distinctions and thus ensures exact indexing and retrieval. Searching of the bibliographic fields for author, title, date, and literature type can also be done, and it is also possible to search the abstracts and titles for any chosen word or words. For efficient searching, the descriptor terms in each field are presorted into an alphabetical index by RECON, and portions of this index may be viewed by an 'expand' command, thus allowing selection of the exact descriptor needed. With 'separated substance' descriptors presented in this form, the searcher can include or exclude as much material as he chooses. The process for selecting the matrix from which the separation is to be made is similar. Again, the 'expand' command permits the searcher to view the choices that are indexed. After selecting a 'matrix', combining the sets with 'and' logic produces a set of references in which a specific substance is selected or rejected from a specific matrix. Combining the selection of one element with the rejection of another (or other) element(s) and a specific matrix will yield references for the separation of A from B out of matrix M. A variety of other search combinations can be performed in a manner similar to the above. For example, separations possible for a given 'separation agent', e. g., an ion exchanger material, from a given matrix may be found by combining sets of appropriate terms from these two fields. References that list distribution coefficients for a given substance to a given organic phase from a given matrix may be found in the same way. Other useful combinations quickly become apparent. Splitting multi-subject reports into separate entries reduces the number of false returns and greatly improves the specificity of searching. In reports that describe more than one separation, each separation is indexed as a separate entry, so that cross-coupling of unrelated index terms from the different separations described does not produce an incorrect retrieval, or results in false returns that contain none of the desired information. In the Separations System Data Base each of the "splits" is a complex and separate record as far as index terms are concerned, but it does not contain the bibliographic information or abstract and simply refers back to the parent record for these, directing the searcher to the correct parent reference. Material that is located in the data base can be displayed and/or printed in several formats, ranging from a listing of accession numbers only, through brief bibliographic listings, to complete listings containing all of the index terms and the abstract.

The same features of this data base that allow advantageous searching, i. e. categorization of index terms, precise indexing, and a limited scope, make indexing a research paper for the data base remarkably simple. Indexer-abstractors are provided with a four-page input form containing all the required descriptor terms listed in the various fields, plus space for all bibliographic information and a short abstract. Since the indexer is required only to highlight or circle the appropriate word or words (descriptors) in each section, the inputting of references is very simple and rapid. A manual has also been brought out containing instructions and definitions of terms for the use of the indexer-abstractors. Individual authors may easily and accurately abstract and index their own work for placement in the data base. Thus, such a data base has the potential of becoming a means by which those involved in any of the areas of separations science can keep in closer contact with the work in their field than had been possible before. The only requirement is that each author is responsible for seeing that his work is placed in the data base.

In 1984, this data base contained approximately 4,000 entries, including progress reports from a separations research group at ORNL (1945–1967). Also covered are about twenty journals pertinent to solvent extraction and ion exchange research in the period 1978–1982. This is a small but useful fraction of the available solvent extraction and ion exchange information. Unfortunately, this data base had no further funding to expand. Work done with it, however, showed, that its indexing structure has definite advantages for those wishing to retrieve information on chemical separations and that it has features potentially applicable to data bases in other subject areas[1].

If the necessary means are available a data base on ion exchange and ion exchangers can also be built up individually. One instance has become known to the author by private communication containing approximately 25 000 entries to stored documents. To have access to such a specialized data base would be nearly ideal for all those working in the field of ion exchange and ion exchangers, despite the fact that the information it covers is only a fraction of the whole literature on the subject.

---

[1] W. J. McDowell et al. (1983). A source of solvent extraction information. Solv. Extr. Ion Exch. *1,* 1. Id. (1984). A new approach to information retrieval problems in separation science. J. Chem. Inf. Comp. Sci. *24,* 108.

Addendum

Recent developments in ion exchange. Papers presented at the International Conference on Ion Exchange Processes (ION-EX' 87). P. A. Williams and M. J. Hudson, eds. Elsevier Applied Science, London, 1987.

Ion exchange for industry. Papers presented at the Fifth International Ion Exchange Conference. Cambridge, July 1988. M. Streat, ed. Ellis Horwood, Chichester, England, 1988.

Appendix I

This appendix contains a number of tables listing commercial ion exchange materials and their sources of supply

Table I.1 Dowex™ ion exchange resins (The Dow Chemical Company, Midland, Michigan 48640, USA)

Name	Matrix	Active group	Delivery form	Shipping weight kg/m³	Particle density kg/m³	Water content %	Minimum total wet capacity eq/l	Maximum operating temperature °C	Remarks
Standard strongly acidic cation exchange resins									
HCR-S	Styrene-DVB gel	SO_3^-	Na^+ H^+	850 800	1280 1210	44–48 50–56	2.0 1.8	150	An 8% crosslinked resin, used in standard water treatment applications, catalysis and other process applications.
HCR-W2	Styrene-DVB gel	SO_3^-	Na^+ H^+	850 800	1280 1210	44–48 50–55	2.0 1.8	150	An 8% crosslinked resin with high physical stability and special size distribution. It is suited for use under stringent conditions, as e. g. high flow rate operation. It is used together with Dowex CCR/LB in layered beds and in (polishing) mixed beds.
HCR-W2	Styrene-DVB gel	SO_3^-	Na^+ H^+	870 820	1310 1230	38–43 46–49	2.2 2.0	150	A 10% crosslinked resin, with higher density and a better resistance to oxidation, also used in (polishing) mixed beds with inert interface.

MSC-1	Styrene-DVB macroporous	SO_3^-	Na^+ H^+	800 760	1220 1180	44–50 50–56	1.7 1.6	150	A highly crosslinked macroporous resin, used in water treatment and in process applications under oxidative conditions and in non-aqueous media.
MSC-1/C	Styrene-DVB macroporous	SO_3^-	Na^+ H^+	800 760	1220 1180	44–50 50–56	1.7 1.6	150	Dowex MSC-1 with a special size distribution for mixed bed and high flow rate operation.
Standard weakly acidic cation exchange resins									
CCR-2/F	Acrylic acid-DVB gel	COO^-	H^+	780	1190	42–48	4.2	120	An easily regenerable, high capacity resin, used for water softening and dealkalization.
CCR-2/LB	Acrylic acid-DVB gel	COO^-	H^+	780	1190	42–48	4.2	120	Dowex CCR-2/F with a special size distribution, suitable for use in layered beds in combination with Dowex HCR-W2.
MWC-1/F	Acrylic acid-DVB macroporous	COO^-	H^+	720	1180	44–50	3.8	120	A resin with a high physical and chemical stability, used under oxidative conditions in water treatment for the recovery of metals.
MWC-2	Acrylic acid-DVB macroporous	COO^-	H^+	750	1130	52–60	2.7	120	A resin with a good osmotic stability, used in sugar and whey treatment.

Table I.1 Dowex continued

Name	Matrix	Active group	Delivery form	Shipping weight kg/m³	Particle density kg/m³	Water content %	Minimum total wet capacity eq/l	Maximum operating temperature °C	Remarks
Standard strongly basic ion exchange resins									
SBR-P	Styrene-DVB gel	$-N^+Me_3$	Cl^- OH^-	690 640	1080 1050	53–60 60–72	1.2 1.0	100 60	Low crosslinked, porous Type 1 resin with high capacity and resistance to organic fouling. Used in standard water treatment applications.
SBR-P/C	Styrene-DVB gel	$-N^+Me_3$	Cl^- OH^-	690 640	1080 1050	50–57 60–70	1.2 1.0	100 60	Dowex SBR-P with a guaranteed high mechanical resistance used in condensate polishing.
SBR-P/LB	Styrene-DVB gel	$-N^+Me_3$	Cl^-	690	1080	50–58	1.2	100	Dowex SBR-P with a special size distribution suitable for use in layered beds in conjunction with MWA-1/LB.
SBR	Styrene-DVB gel	$-N^+Me_3$	Cl^- OH^-	705 655	1180 1070	43–48 55–65	1.4 1.0	100 60	Type 1 resin with very high capacity and very good physical and chemical stability. Used in water treatment under severe conditions and in metal separation and recovery.
SBR/C	Styrene-DVB gel	$-N^+Me_3$	Cl^- OH^-	705 655	1100 1070	43–48 60–65	1.4 1.1	100 60	Dowex SBR with a guaranteed high mechanical resistance used in condensate polishing at high flow rates.
SAR	Styrene-DVB gel	$-N^+Me_2$ $(CH_2)_2OH$	Cl^-	700	1150	38–45	1.4	75	Type 2 resin with very high capacity and good regeneration efficiency. Well-suited for general demineralisation.

11	Styrene-DVB special gel	$-N^+Me_3$	Cl^-	700	1130	50–60	1.1	100	Type 1 resin with excellent kinetics and very good resistance to organic fouling. Used for applications requiring good organic fouling resistance or as an organic screen ahead of other anion exchangers.
MSA-1	Styrene-DVB macroporous	$-N^+Me_3$	Cl^-	680	1060	56–64	1.0	100	Type 1 macroporous resin with high resistance to oxidizing agents and osmotic shock. Used in water treatment application, demineralisation of chemical solutions and in catalytic processes.
MSA-1/C	Styrene-DVB macroporous	$-N^+Me_3$	Cl^-	680	1060	56–64	1.0	100	Dowex MSA-1 with a guaranteed high mechanical resistance used in condensate polishing.
MSA-2	Styrene-DVB macroporous	$-N^+Me_2$ $(CH_2)_2OH$	Cl^-	680	1070	53–60	1.0	75	Type 2 macroporous resin with high regeneration efficiency and resistance to organic fouling used in water treatment applications.
D-1	Styrene-DVB gel + macroporous	$-N^+Me_2$ $(CH_2)_2OH$	Cl^-	690	1080	50–60	1.0	75	Specially graded mixture of Dowex 11 and Dowex MSA-2 used at critical organic matter problems.

Table I.1 Dowex continued

Name	Matrix	Active group	Delivery form	Shipping weight kg/m³	Particle density kg/m³	Water content %	Minimum total wet capacity eq/l	Maximum operating temperature °C	Remarks
Standard weakly basic ion exchange resins									
WGR	Epoxyamine	$-NR_2$, $-NHR$	FB	690	1160	50	1.6(H_2SO_4)°° 1.2(HCl)°°	60	Resin with high capacity, easily regenerable with high chemical and physical stability. Operating capacity is covered by specification.
WGR-2	Epoxyamine	$-NR_2$, $-NHR$ $-NH_2$	FB	690	1160	50	1.4(H_2SO_4)°° 1.2(HCl)°°	60	Resin with high capacity, easily regenerable with high chemical and physical stability. Both operating capacity and rinse-down are covered by specification.
MWA-1	Styrene-DVB macroporous	$-NMe_2$	FB	640	1040	50–60	1.0	60	An easily regenerable, macroporous resin with high chemical and physical stability and good organic fouling resistance used in water treatment, chemical processing and plating waste processing.
MWA-1/LB	Styrene-DVB macroporous	$-NMe_2$	FB	640	1040	50–60	1.0	60	Dowex MWA-1 with a special size distribution suitable for use in layered beds in combination with SBR-P/LB.

°° Operational capacity for loading with H_2SO_4/HCl and regenerating at 80 g NaOH/l resin.

Name and Grading	Crosslinking % DVB	Active group	Delivery form	Particle size range mm	Shipping weight kg/m^3	Water content %	Minimum tot. wet capacity eq/l
Strongly acidic resins (styrene-DVB copolymer, gel) with special grading or crosslinking							
50 WX2 (20– 50)	2	SO_3^-	H^+	0.3 –1.2	740	74–82	0.6
50 WX2 (50–100)	2	SO_3^-	H^+	0.2 –0.85	740	74–82	0.6
50 WX2 (100–200)	2	SO_3^-	H^+	0.1 –0.5	740	74–82	0.6
50 WX2 (200–400)	2	SO_3^-	H^+	0.05–0.2	740	74–82	0.6
50 WX3 (20– 50)	3	SO_3^-	Na^+	0.4 –1.2	785	67–73	1.2
50 WX4 (20– 50)	4	SO_3^-	H^+	0.3 –1.2	770	64–70	1.2
50 WX4 (50–100)	4	SO_3^-	H^+	0.2 –0.5	770	64–72	1.2
50 WX4 (100–200)	4	SO_3^-	H^+	0.1 –0.25	770	64–72	1.1
50 WX4 (200–400)	4	SO_3^-	H^+	0.05–0.15	770	64–72	1.1
50 WX8 (50–100)	8	SO_3^-	H^+	0.2 –0.5	800	50–56	1.7
50 WX8 (100–200)	8	SO_3^-	H^+	0.1 –0.25	800	50–58	1.7
21K (16– 20)	2	Type 1	Cl^-	0.7 –0.85	690	50–58	1.2
21K (16– 30)	2	Type 1	Cl^-	0.6 –1.2	690	50–58	1.2
			SO_4^{2-}	0.6 –1.2	690	–	1.2
21K (20– 40)	2	Type 1	Cl^-	0.4 –1.2	690	50–58	1.2
1X2 (16–100)	2	Type 1	Cl^-	0.15–1.2	705	70–80	0.6
1X2 (50–100)	2	Type 1	Cl^-	0.2 –0.5	705	65–75	0.7
1X2 (100–200)	2	Type 1	Cl^-	0.1 –0.5	705	70–80	0.6
1X2 (200–400)	2	Type 1	Cl^-	0.05–0.2	705	70–80	0.6
1X4 (20– 50)	4	Type 1	Cl^-	0.3 –1.2	705	50	1.0
1X4 (50–100)	4	Type 1	Cl^-	0.2 –0.5	705	50	1.0
1X4 (100–200)	4	Type 1	Cl^-	0.2 –0.25	705	55–63	1.0
1X4 (200–400)	4	Type 1	Cl^-	0.05–0.15	705	55–63	1.0
1X8 (50–100)	8	Type 1	Cl^-	0.2 –0.5	705	43–48	1.2
1X8 (100–200)	8	Type 1	Cl^-	0.1 –0.25	705	39–45	1.2
1X8 (200–400)	8	Type 1	Cl^-	0.05–0.15	705	39–45	1.2
2X8 (100–200)	8	Type 2	Cl^-	0.1 –0.25	705	34–40	1.2
2X8 (200–400)	8	Type 2	Cl^-	0.05–0.15	705	34–40	1.2

Table I.1 Dowex continued

Name	Delivery form	Minimum conversion %	Particle size range mm	Composition (1)
Ready-for-use mixed beds				
MR-3C	H^+/OH^-	99/95	0.3–1.2	Ionically balanced mixture of Dowex HCR-W2/H^+ and Dowex SBR-C/OH^- (N. G.).
MR-3N. G.	H^+/OH^-	99/95	0.4–1.2	Ionically balanced mixture of Dowex HCR-S/H^+ (N. G.) and Dowex SBR/OH^- (N. G.).
MR-4	H^+/OH^-	99/95	0.3–1.2	Ionically balanced mixture of Dowex MSC-1/H^+ and Dowex SBR/OH^- (N. G.).
MR-5/4	$^7Li^+/OH^-$	99/95	0.3–1.2	Ionically balanced mixture of Dowex HCR-S/$^7LI^+$ and Dowex SBR/OH^- (N. G.).
MR-6	Li^+/OH^-	99/95	0.3–1.2	Ionically balanced mixture of Dowex HCR-S/Li^+ and Dowex SBR/OH^- (N. G.).
MR-7	NH_4^+/OH^-	99/95	0.3–1.2	Ionically balanced mixture of Dowex HCR-S/NH_4^+ and Dowex SBR/OH^- (N. G.).
MR-12	H^+/OH^-	99/95	0.3–1.2	A mixture with volume ratio 2:1 of Dowex HCR-W2/H and Dowex SBR-P/OH^- (N. G.).
MR-12C	H^+/OH^-	99/95	0.4–1.2	A mixture with volume ratio 2:1 of Dowex HGR-W2/H^+ and Dowex SBR-C/OH^- (N. G.).
MR-13	H^+/OH^-	99/95	0.3–1.2	A mixture with volume ratio 3:1 of Dowex HCR-S/H^+ (N. G.) and Dowex SBR/OH^- (N. G.).
MR-72	H^+/OH^-	99/95	0.3–1.2	Ionically balanced mixture of Dowex C-75/H^+ (N. G.) and Dowex SBR/OH^- (N. G.)

(1) All resins are mixtures of sulphonated styrene-DVB copolymers as cation exchange resin and aminated styrene-DVB copolymers as anion exchange resins. Maximum impurity contents are specified on a dry weight basis (mg/kg dry matter) for cation exchangers (Na = 100, Fe = 200, Cu = 50, Al = 50, heavy metals = 50) and for anion exchangers (Fe = 100, Cu = 50, Al = 50, heavy metals = 50).

Name	Active group	Delivery form	Minimum conversion %	Minimum tot. wet capacity eq/l	Particle size range mm	Remarks
Dowex nuclear grade (N. G.) ion exchange resins (0)						
HCR-S/(E)	$-SO_3^-$	H^+	99	1.8	0.3–1.2	Maximum impurity levels are specified on a dry weight basis as mg/kg dry matter for Na = 100, Fe = 200, Al = 50, Cu = 50, heavy metals = 50. The standard resin for removal of cationic species.
HCR-W2	$-SO_3^-$	H^+	99	1.8	0.4–1.2	Maximum impurity levels are specified on a dry weight basis as mg/kg dry matter for Na = 100, Fe = 200, Al = 50, Cu = 50, heavy metals = 50. An 8% crosslinked resin with high physical stability: guaranteed compression resistance and high attrition resistance.
HGR-W2	$-SO_3^-$	H^+	99	2.0	0.4–1.2	Maximum impurity levels are specified on a dry weight basis as mg/kg dry matter for Na = 100, Fe = 200, Al = 50, Cu = 50, heavy metals = 50. A higher crosslinked resin with the physical stability of Dowex HCR-W2.
C-75	$-SO_3^-$	H^+	99	1.6	0.4–1.2	Maximum impurity levels are specified on a dry weight basis as mg/kg dry matter for Na = 100, Fe = 200, Al = 50, Cu = 50, heavy metals = 50. A macroporous resin with high affinity for Cs and Co.
SBR-P	$-N^+Me_3$	OH^-	95	1.0	0.3–1.2	Maximum impurity levels are specified on a dry weight basis as mg/kg dry matter for Fe = 100, Cu = 50, Al = 50, heavy metals = 50. Standard porous resin for the removal of anionic species and boron control in nuclear power plants.
SBR-P/C	$-N^+Me_3$	OH^-	95	1.0	0.4–1.2	Maximum impurity levels are specified on a dry weight basis as mg/kg dry matter for Fe = 100, Cu = 50, Al = 50, heavy metals = 50. A porous resin like Dowex SBR-P with guaranteed high compression resistance.

Table I.1 Dowex continued

Name	Active group	Delivery form	Minimum conversion %	Minimum tot. wet capacity eq/l	Particle size range mm	Remarks
SBR	$-N^+Me_3$	OH^-	95	1.1	0.3 – 1.2	Maximum impurity levels are specified on a dry weight basis as mg/kg dry matter for Fe = 100, Cu = 50, Al = 50, heavy metals = 50. A standard resin for the removal of anionic species and for boron control in nuclear power plants.
SBR/C	$-N^+Me_3$	OH^-	95	1.1	0.4 – 1.2	Maximum impurity levels are specified on a dry weight basis as mg/kg dry matter for Fe = 100, Cu = 50, Al = 50, heavy metals = 50. A resin like Dowex SBR with guaranteed high compression resistance.

(0) Refer to Table above for mixed bed resins.

Name	Particle size range mm	Particle density kg/m^3	Remarks
Dowex inert resins			
XZ 86270	1.7 – 3.4	962	Floating inert resin granules to be used in ion exchange counter-current operation, especially in self-packing systems with up-flow regeneration and down-flow production.
XZ 87423	Diamter: 1 – 1.5 Length: 1.5	970	Floating inert resin cylinders to be used in ion exchange counter-current systems, especially where inert mass blockage is required and in self-packing systems with up-flow production and down-flow regeneration.
XFS 43323	0.6(°)	1150	Inert interface resin with a uniform particle size used in high performance mixed beds as intermediate buffer bead layer.

(°) 95% of all beads between plus or minus 100 μm of the mean bead size (600 μm).

Name	Matrix	Particle size range mm	Particle density kg/m^3	Porosity	Average pore size $10^{-10}m$	Surface area m^2/g (dry)	Remarks
Dowex adsorbents							
XFS 4022	Styrene-DVB macroporous	0.3 – 1.2	1040	35	200	80 – 120	Absorbs organics from polar solvents,. e. g. benzene, toluene, aniline and many derivatives, from water. This resin resists up to temperatures of 150 °C and exhibits good regeneration characteristics, especially when the adsorbed material can be eluted with methanol.
XFS 4257	Styrene-DVB macroporous	0.3 – 1.2	1040	35	100	450	This resin resists temperatures of up to 150 °C and exhibits good regeneration characteristics, especially when the adsorbed material can be eluted with methanol.

The choice of the most appropriate adsorbing resin will depend on molecular weight and structure of the material to be adsorbed.

Dowex ion exchange resins with chelating or non-classical active groups

Name	Matrix	Active group	pH Range	Total capacity	Remarks
XFS 4195	Styrene-DVB macroporous	Weakly basic chelating agent	1 – 7 for copper adsorption	25,4 mg Cu/ml	A macroporous resin with high selectivity for copper used in copper production and purification of cobalt, nickel or zinc electrolytes by removal of trace amounts of copper.
XFS 43084	Styrene-DVB macroporous	Weakly basic chelating agent	1 – 8 for copper adsorption	35 mg Cu/ml	A macroporous resin with high selectivity for valuable metals, such as nickel, copper and cobalt with excellent Fe-rejection.
XFS 4196	Styrene-DVB macroporous	Weakly basic chelating agent	1,5 – 7 for nickel adsorption	37 mg Cu/ml	A macroporous resin with high selectivity for copper useful in nickel/copper separation.

Table I.1 Dowex continued

Dowex tough gel Monosphere™ resins

Name	Matrix	Mean particle size µm	Delivery form	Shipping weight kg/m³	Particle density kg/m³	Water content %	Minimum tot. wet capacity eq/l	Maximum operating temp. °C	Remarks
TG650C	Styrene-DVB gel	650	Na^+ H^+	800	1225	46–51	1.9	150	A resin with a specified bead-size uniformity°° combining the superior chemical performance of gel resins with the osmotic stability of macroporous resins. Used for high-performance systems like operating mixed beds for production of high-purity water, condensate polishing mixed beds, continuous ion exchange, chromatographic separation, etc.
TG550A	Styrene-DVB gel	550	Cl^- OH^- CO_3^{2-}	690	1085	46–54	1.25	100 50	
XFS43323	–	600	–	1150	–	–	–	–	An interface resin with a specific bead-size uniformity°° used in high performance mixed beds as intermediate buffer bead layer.
650-C	Styrene-DVB gel	650	Na^+ H^+	833	1310	38–45	2.1	130	Demineralization, softening, condensate polishing
550-A	Styrene-DVB gel	550	Cl^-	690	1080	44–50	1.25	100 60	Demineralization, softening condendate polishing
AMW-500	Styrene-DVB macroporous	550	Cl^- FB	640	1040	50–58	1.25	100 60	Demineralization, deacidification, removal of heavy metals

°° 95% of all beads between plus or minus 100 micron of the mean bead size.

Resins for application in the food and/or pharmaceutical industry

Resins for demineralization and deacidification

Name	Matrix	Active group	Ionic form	Water retention capacity (%)	Minimum tot. exchange cap. (meq/ml)	Maximum operating temp. °C	Remarks
DOWEX 22	Macroporous styrene-DVB	Type 2 strong base	Cl	53 – 60	1.0	40 (OH-form)	Sugar purification
DOWEX 66	Macroporous styrene-DVB	Weak base	FB	40 – 50	1.8	100	Sugar purification
DOWEX 88	Macroporous styrene-DVB	Strong acid	Na	46 – 50	1.9	150	Sugar purification
XUS-40075.01	Macroporous styrene-DVB	Weak base	FB	41 – 49	1.6	100	Low shrink/swell
XUS-40091	Epoxy-amine	Interme-diate base	FB	50	1.2	100	
XUS-40123	Epoxy-amine	Weak base	FB	50	1.2	100	Fruit juice deacification

Table I.1 Dowex continued

Name	Matrix	Active group	Ionic form	Cross linkage	Water retention capacity (%)	Minimum tot. exchange cap. (meq/ml)	Remarks
Dowex ion exchange purification resins microporous or macroporous							
XUS-40090	Microporous styrene-DVB	Strong acid	H/Na	High	50 – 55 (H)	1.8 (H)	
XUS-40111	Microporous styrene-DVB	Type 2 strong base	Cl	High	38 – 45	1.4	
XUS-40189	Microporous styrene-DVB	Type 2 strong base	Cl	Medium	40 – 46	1.3	
XUS-40196	Microporous styrene-DVB	Type 1 strong base	Cl/OH	High	46 – 54 (Cl)	1.2 (Cl)	Tough gel Monosphere
XUS-40197	Microporous styrene-DVB	Strong acid	H/Na	High	46 – 51 (H)	1.9 (H)	Tough gel Monosphere
XFS-43071.00	Microporous styrene-DVB	Strong acid	H	Medium	65 – 69	1.2	
XFS-40032	Macroporous styrene-DVB	Strong acid	Na	–	46 – 50	1.9	
XUS-40062	Macroporous styrene-DVB	Weak acid	H	–	44 – 50	3.8	
XUS-40170	Macroporous styrene-DVB	Type 2 strong base	Cl	–	53 – 60	1.0	
XUS-40184	Macroporous styrene-DVB	Strong acid	H	20	50 – 56	1.6	
XUS-40189	Macroporous styrene-DVB	Type 1 strong base	Cl	–	56 – 64	1.0	
XFS-43359	Macroporous styrene-DVB	Weak base	FB	–	40 – 50	1.8	

Dowex large pore macroporous ion exchange resins

XY-40007.00	Macroporous styrene-DVB	Weak base	FB		66–73	0.8	1800 Å copolymer pore size
XY-40007.01	Macroporous styrene-DVB	Weak base	FB		58–65	1.1	800 Å copolymer pore size
XY-40008.00	Macroporous styrene-DVB	Type 1 strong base	Cl		68–74	0.7	1800 Å copolymer pore size
XY-40008.01	Macroporous styrene-DVB	Type 1 strong base	Cl		64–70	0.8	800 Å copolymer pore size
XU-40133	Macroporous styrene-DVB	Strong acid	H		71–77	1.1	800 Å copolymer pore size

Dowex Monosphere separation resins: ligand exchange, size exclusion, ion exclusion chromatography

Dowex 99	Microporous styrene-DVB	Strong acid	Ca	Medium	57–61 (H)	1.5 (H)	Ca-form for sugar separation
XUS-40099	Microporous styrene-DVB	Type 1 strong base	Cl	Low	65–75	0.6	
XFS-43254	Microporous styrene-DVB	Type 1 strong base	Cl	Medium	42–47	1.4	
XFS-43278	Microporous styrene-DVB	Strong acid	H/Na	Low	65–75 (Na)	0.5 (Na)	
XFS-43279	Microporous styrene-DVB	Strong acid	H/Na	Medium	58–65 (Na)	1.3 (Na)	
XFS-43280	Microporous styrene-DVB	Strong acid	H/Na	Medium	57–61 (H)	1.5 (H)	
XFS-43281	Microporous styrene-DVB	Strong acid	H/Na	High	44–48 (Na)	1.9 (Na)	

Table I.1 Dowex continued

Name	Matrix	Active group	Ionic form	Cross-linkage	Water retention capacity (%)	Minimum tot. exchange capacity (meq/ml)
Dowex ion exchange fine mesh resins						
Whole bead (50 – 150 μm)						
XY-40009	Microporous styrene-DVB	Strong acid	Na	Low	74 – 82 (H)	0.6 (H)
XYS-40010	Microporous styrene-DVB	Strong acid	Na	High	44 – 52	1.9
XY-40011	Microporous styrene-DVB	Type 2 strong base	Cl	Low	65 – 70	0.7
XY-40012	Microporous styrene-DVB	Type 2 strong base	Cl	High	34 – 40	1.2
XYS-40013	Microporous styrene-DVB	Type 1 strong base	Cl	High	39 – 45	1.3
XF-43311.01	Microporous styrene-DVB	Type 1 strong base	Cl	Low	70 – 80	0.6
XFS-43361	Microporous styrene-DVB	Strong acid	Na	Very high	32 – 38	2.5

Dried and ground resins

XY-40010.01	Microporous styrene-DVB	Strong acid	Na	High	<12	4.9
XY-40013.01	Microporous styrene-DVB	Type 1 strong base	Cl	High	<12	3.5
XU-40043	Macroporous styrene-DVB	Weak base	FB	–	<12	4.3
XF-43311	Microporous styrene-DVB	Type 1 strong base	Cl	Low	<12	3.6
XF-43362	Microporous styrene-DVB	Type 2 strong base	Cl	Low	<12	3.6

DOWEX ADSORBENT RESINS

Dow produces a range of adsorbent resins which are used for the purification and isolation of pharmaceutical products and removal of contaminants from waste water and gas streams. The new type of adsorbent has a similar pore size distribution to that of activated carbon but is more easily regenerated. Among the SORBATHENE™ adsorption resins Dowex S 111 is targeted for color removal from high fructose corn syrup and Dowex S 112 is very successfully applied for the removal of 1,1,1-trichloroethane from degreasing bath vents.

Table I.2 DIAION ion exchange materials (Mitsubishi Chemical Industries, Tokyo, Japan)

Cation exchangers

		Strong acid exchangers				Weak acid exchangers	
		Gel		Macroporous		Macroporous	
		DIAION SK 1B	DIAION SK 110	DIAION PK 216	DIAION PK 228	DIAION WK 10	DIAION WK 20
Ionic form as shipped		Na	Na	Na	Na	H	H
Specific gravity (Approx.)		1.25	1.30	1.27	1.32	1.15	1.17
Shipping density (Approx.)	g/l lb/cu. ft.	825 52	845 53	780 49	805 50	615 38	690 43
Moisture content	%	45–50	40–45	46–52	37–43	50–56	40–46
Total capacity (min)	meq./ml $gCaCO_3/l$	1.9 95	2.1 110	1.8 90	2.2 110	2.5 125	3.5 175
Particle size	Micron US Std. Mesh	1190–297 16– 50	1190–297 16– 50	1190–297 16– 50	1190–297 16– 50	1190–297 16– 50	1190–297 16– 50
Max. operating temperature	°C (max)	140	140	140	140	150	150
Effective pH range		0–14	0–14	0–14	0–14	5–14	4–14

Anion exchangers

		Strong base exchangers							Weak base exchangers	
		Gel			Macroporous				Macroporous	
		DIAION SA 10A	DIAION SA 12A	DIAION SA 20A	DIAION PA 308	DIAION PA 312	DIAION PA 316	DIAION PA 416	DIAION WA 20	DIAION WA 30
Ionic form as shipped		Cl	Cl	Cl	Cl	Cl	Cl	Cl	Free base	Free base
Specific gravity (Approx.)		1.12	1.10	1.12	1.09	1.11	1.13	1.13	1.07	1.05
Functional group type		I	I	II	I	I	I	II	Sec	Tert
Shipping density (Approx.)	g/l lb/cu. ft.	685 43	675 42	700 44	655 41	670 42	680 43	690 43	650 41	615 38
Moisture content	%	43–48	50–500	40–45	57–67	50–55	44–49	40–45	39–45	49–55
Total capacity (min)	meq./ml $gCaCO_3/l$	1.4 70	1.3 65	1.4 70	1.0 50	1.2 60	1.3 65	1.3 65	2.5 125	1.5 75
Particle size	Micron US Std. Mesh	1190–297 16–50	1190–297 16–50	1190–297 16–50	1190–297 16–50	1190–297 16–50	1190–297 16–50	1190–297 16–50	1190–297 16–50	1190–297 16–50
Max. operating temperature	°C (max)	70 (OH) 90 (Cl)	70 (OH) 90 (Cl)	50 (OH) 70 (Cl)	70 (OH) 90 (Cl)	70 (OH) 90 (Cl)	70 (OH) 90 (Cl)	50 (OH) 70 (Cl)	100 (free base)	100 (free base)
Effective pH range		0–14	0–14	0–14	0–14	0–14	0–14	0–14	0–9	0–9

Variations of DIAION resins with different particle size, degrees of crosslinkage, ionic forms and degrees of purity can be supplied upon request.
Other types of DIAION resins for specific applications are also available: Chelating resins. Catalyst grade resins. Ultrapure grade resins (Reactor coolant grade, Semiconductor grade).

Table I.3 Lewatit® ion exchange resins. (Bayer AG, Geschäftsbereich OC, 5090 Leverkusen, FR Germany)

Product group		Cation exchangers								
		Strong acid							Weak acid	
Lewatit		S 100		S 100 LF	S 109	SP 112		SP 120	CNP LF	CNP 80
Structure		Gel				Macroporous				
Type										
Ionic form as supplied		Na	H	Na			H	Na	H	
Shape										
Colour		light brown, translucent				greyish light brown, opaque				
Matrix		Polystyrene							Polyacrylic	
Functional group		Sulfonic acid							Carbonic acid	
Bead size distribution	mm	0.3–1.25								
Effective size	± 0.03 mm	0.48				0.48			0.50	
Uniformity coefficient	max.	1.7				1.7				
Shipping weight	g/l	800–900				750–850			750–850	
Density	g/ml	1.28	1.22	1.28	1.24	1.27	1.21	1.20	1.18	
Moisture content	% wt	45–50	45–50	42–45	50–54	45–50	45–50	50–55	45–50	
Total capacity	min. eq/l	2.1	1.9	2.1	1.8	1.7	1.6	1.4	4.5	
Stable	at temperatures up to °C	120				120				
	at pH values between	1 – 14								
Storage (original packing)	at temperatures up to °C	− 10 to + 40				− 20 to + 40				
	min. years	2								

Anion exchangers												
Strong base									Weak base			
M 500		M 504	M 600	AP 246	AP 247 A	MP 500	MP 500 A	MP 600	AP 49	MP 35 A	MP 62	MP 64
Gel				Macroporous								
1			2	1				2				
Cl	OH	Cl		Cl/OH	Cl	Cl	Cl		OH/Cl		OH	OH/Cl
Beads												
brown, translucent				light brown, opaque								
Polystyrene				Acrylic		Polystyrene			Acrylic	Polystyrene		
Quaternary amine									Tertiary amine			
0.3–1.25									0.3–1.25			
0.48		0.48		0.48		0.48			0.50	0.50		
1.8				1.9		1.8			1.8			
670–750				700–800		660–720			700–800	600–700		
1.09	1.08	1.07	1.12	1.10		1.06		1.09	1.07	1.04		
40–45	55–60	50–55	35–40	60–65		55–60		50–55	55–60	55–65		
1.5	1.3	1.25	1.35	1.25	1.0	1.2	1.0	1.2	1.4	1.15	1.7	1.3
100			70	40	75	100		70	40	70	100	
1–14												
+ 1 to + 40				− 20 to + 40								
2			2	2				2	2			

Table I.3 Lewatit continued
Selection of Lewatit resin types: Extraction/recovery of metal ions from solutions

Metal		Cation exchange resins		Anion exchange resins			Special products				
		Lewatit®									Trial products* Levextrel
		S 100 SP 112	CNP 80	M 500 M 504 MP 500	M 600 MP 600	MP 62 MP 64	OC 1060	TP 207	TP 208	TP 214	OC 1026*
Aluminium		•						•	•		
Cadmium	chloride			•			•	•	•		
	cyanide			•			•	•	•		
	sulfate	•					•	•	•		
Cesium							•				
Chromium III		•						•**	•**		
Chromate VI						•					
Cobalt	chloride	•					•	•	•		
	cyanide					•					
	sulfate	•					•	•	•		
Copper	weak acid	•	•				•	•	•		
	ammoniacal		•				•	•	•		
	cyanide					•					
Gold	chloride			•	•	•		•[1])		•	
	cyanide			•	•	•					
Indium	chloride					•					
	sulfate	•						•	•		•
Iron III		•		•							

Lanthanides		•						•	•		•
Lead		•	•				•	•	•		
Manganese		•	•				•	•	•		
Mercury	chloride					•				•	
	cyanide			•	•	•					
	miscellaneous									•	
Molybdenum				•		•		•	•		
Nickel	weak acid	•	•				•	•	•		
	cyanide					•					
Platinum metals				•				•[1])		•	
Rhenium				•							
Silver	chloride									•	
	cyanide			•	•	•					
	sulfate	•						•	•	•	
	thiosulfate			•	•	•				•	
Tungsten						•					
Uran	sulfate			•		•		•	•		
	carbonate			•							
	nitric acid										•
Vanadium	cationic **IV**							•	•		
	anionic					•					
Zinc	chloride	•		•	•		•	•	•		
	sulfate	•					•	•	•		•
Recommended operating pH range		0–8	5–12	1–10	1–9	1–8	1,5–9	1,5–9	2–11	0–10	1–5

* The designation OC ______ refers to Trial Products of Bayer AG, therefore, all data are provisional and may be changed at any time.

** Preferably at temperatures 50–80 °C

[1]) When used in hydrochloric acid 0.1–2.0 eq/l.

Table I.3 Lewatit continued
Selection of Lewatit ion exchange resins for the sugar and foodstuffs industry

Application	Medium/solutions	Lewatit®											
			S 1428 LF	S 2528	S 3428	S 4328	S 5428	S 5428 A	S 6328	S 6328 A	S 8528 LF	VP OC 1052	S 4428
		MDS 1368								MDS 6368 A		1)	
Adsorption Decolourization	Glyerol Sugar							●		●			
Chromatography Ion exclusion	Saccharides Electrolyte/ non Electrolyte	●											
Decationization	Thin Juice		●	○						●	●		
	Liquid sugar			●									
	Gelatine		●	○									
	Glucose/HFCS			●									
	Glycerol			●									
	Whey		●	○									
Deacidification	Thin Juice					○	●						
	Liquid sugar				○	●							
	Gelatine				●								
	Glucose/HFCS				○	●	○		●*	○			●**
	Glycerol				○	●	○		●*				
	Whey				●								

Decalcification	Thin juice		•	○							•		
	Pectin		•	○									
Inversion	Liquid sugar											•	
Quentin process	B-Juice			•									

• = Best selection ○ = Well-suited

* Deacidification of sorbitol (removal of gluconic acid)

** low isomerization (change from glucose to fructose)

[1]) The designation VP OC refers to Trial Products of Bayer AG, therefore, all data are provisional and may be changed at any time.

Table I.3 Lewatit continued
Selection of strongly acidic Bayer catalyst resins

Process	Lewatit®							Lewasorb®
	SC			SPC			OC* 1038 (Pd)	AC 10 dry AC 10 FT
	102 102 BG	104 104 WS	108	108 108 BG 108 BG dry	112 112 BG	118 118 BG 118 BG dry		
	Gel			Macroporous				Powder
Acetalization						●		
Alkylation		○		●	●	○		●
Chromatography		●						
Condensation	●	●		●			●	
Conversion of esters			●			●		
Epoxidation			●					
Esterification	●	○	●	○	●	●		
Etherification	●	○	○		●	●	●	
Hydration		○	○	●		○		
Hydrogenation							●	
Hydrolysis			●					
Isomerization			●					
Oligomerization								●
Polymerization								●
Prins reaction	○	●				●		○
Purific. of phenols				●				
Saponification	○	●	●	○		●		
H_2O elimination			●					

● = Recommended ○ = Suitable
* The designation OC ________ refes to Trial Products of Bayer AG, therefore, all data are provisional and without obligation.

Table I.4 Purolite ion exchange resins. (The PUROLITE Company, 150 Monument road, Bala Cynwyd, PA 19004, USA)
Cation exchangers

PURO-LITE	Type	Ionic form	Total volume capacity min (eq/l)	Shipping weight ap-proximate (g/l)	Water retention (%)	Specific gravity moist beads	Maximum swelling	Remarks
C 100	Strong acid polystyrene	Na^+	2.0	850	44 – 48	1.29	Na → H 5%	Gel type standard. Resin with high capacity for softening and demineralisation.
C 100 E	Strong acid polystyrene	Na^+	1.9	850	46 – 50	1.27	Ca → Na 8% Na → H 5%	Gel type. Resin specially treated for domestic use and for industrial use where potable water is required.
C 100 X 10	Strong acid polystyrene	Na^+	2.2	865	40 – 44	1.30	Na → H 5%	Gel type. with excellent resistance to oxidation, gives particularly good separation from anion resin in mixed beds.
C 120 E	Strong acid polystyrene	Na^+	1.6	820	56 – 60	1.20	Ca → Na 12%	Gel type. Resin designed for softening at high flow rates. Excellent kinetics.
C 150	Strong acid polystyrene	Na^+	1.8	800	48 – 53	1.25	Na → H 4%	Macroporous type with excellent resistance to attrition and osmotic shock. For treatment of condensates, continuous processes and special application (catalysis, galvanic plating, sugar, ...).
C 155	Strong acid polystyrene	Na^+	2.2	830	39 – 44	1.28	Na → H 4%	Macroporous type. More highly crosslinked. For special applications (catalysis, ...).

Table I.4 Purolite continued

PURO-LITE	Type	Ionic form	Total volume capacity min (eq/l)	Shipping weight approximate (g/l)	Water retention (%)	Specific gravity moist beads	Maximum swelling	Remarks
C 160	Strong acid polystyrene	Na^+	2.4	840	35–40	1.30	Na → H 4%	Macroporous type. Very highly crosslinked. High capacity. For special applications (Quentin process, treatment of industrial waste, excellent resistance to oxidation, ...).
C 105	Weak acid polyacrylic	H^+	4.2	760	45–53	1.14	H → Ca 25%	Gel type. High capacity. Alkalinity removal.
C 106	Weak acid polyacrylic	H^+	3.0	750	54–59	1.15	H → Ca 15% H → Na 50%	Macroporous type with excellent resistance to osmotic shock, for special applications. Treatment of ammoniacal condensates, fixation of antibiotics.

Particle size: Cation exchangers

PURO-LITE grade	Uniformity coefficient maximum	Effective size (mm) typical	Size range limitation								Principal applications
			< 0.3 mm	< 0.42 mm	< 0.63 mm	< 0.71 mm	> 0.63 mm	> 0.85 mm	> 1.0 mm	> 1.20 mm	
STD	1.70	0.50	1%							5%	Standard quality.
MB	1.70	0.50	1%							5%	Mixed beds (MIXLITE).
PL	1.45	0.55		2%					5%		Condensate polishing (POLILITE).
TL	1.25	0.80				1%				5%	Three component mixed bed (TRILITE).
CL	1.35	0.55		1%				5%			Continuous processes (CONTILITE).
FL	1.55	0.55		2%					10%		Floating beds (FLUIDLITE).
DL strong	1.35	0.60			5%					15%	Layered beds, lower layer (DOUBLITE).
DL weak	1.35	0.40	1%				5%				Layered beds, upper layer (DOUBLITE).
S	1.50	0.55		2%						2%	Special applications (demineralisation of sugar, ...).
C	1.50	0.55		2%						2%	High flow rate applications.

Table I.4 Purolite continued

Ready-to-use Purolite mixed beds

PUROLITE	Appearence	Purolite component resins	Component type and percentage	Ionic forms	Shipping weight approximate (g/l)	Particle size		Useful capacity (eq/l)	Applications
						mm	%		
MB 450	Without indicator	**C 100** (H)	40% Strong acid cation	99% H^+	735	>1.2	<5	minimum	Production of demineralised water of high purity, silica free. Conductivity attainable less than 0.1 µS/cm.
MB 450 IND	Blue (regenerated) Beige (exhausted)	**A 450** (OH)	60% Strong base anion opaque Gel Type I	90% OH^-		<0.3	<1	0.54	
MB 450 VC	Green (regenerated) Blue (exhausted)								
MB 400	Without indicator	**C 100** (H)	40% Strong acid cation	99% H^+	750	>1.2	<5	minimum	Similar uses to those of MB 450.
MB 400 IND	Blue (regenerated) Amber (exhausted)	**A 400** (OH)	60% Strong base anion clear Gel Type I	90% OH^-		<0.3	<1	0.54	
MB 250	Without indicator	**C 100** (H)	40% Strong acid cation	99% H^+	735	>1.2	<5	minimum	Production of demineralised water of high purity, silica free. Conductivity attainable less than 0.1 µS/cm.
MB 250 VC	Green (regenerated) Blue (exhausted)	**A 250** (OH)	60% Strong base anion opaque Gel Type II	90% OH^-		<0.3	<1	0.64	

MB 37 CF	**Without indicator**	**C 100** (H) **A 400** (OH)	40% Strong acid cation 60% Strong base anion clear Gel Type I	99% H^+ 90% OH^-	735	>1.2 <0.1	<5 <0.5	minimum 0.54	Production of demineralised water, silica free of high purity. Conductivity attainable <0.1 μS/cm. Good kinetics, disposable.
MB 59 VC	Green (regenerated) Blue (exhausted)	**C 100** (H) **A 100** (FB)	60% Strong acid cation 40% Weak base anion	99% H^+ 95% OH^-	750	>1.2 <0.3	<5 <1	minimum 0.80	Production of partially demineralised water (O_2 and SiO_2 not eliminated). Capacity dependent upon % alkalinity in feed solution.

Table I.4 Purolite continued

Strong base anion exchangers

PURO-LITE	Type	Ionic form	Total volume capacity min (eq/l)	Shipping weight ap-proximate (g/l)	Water retention (%)	Specific gravity moist beads	Maximum swelling	Remarks
A 400	Polystyrene Type I	Cl^-	1.3	690	48–54	1.08	$Cl \rightarrow OH$ 20%	Clear gel type. High operating capacity. Good kinetics giving demineralisation to high purity.
A 450	Polystyrene type I	Cl^-	1.2	690	49–56	1.07	$Cl \rightarrow OH$ 20%	Opaque gel type. High porosity. Good osmotic shock resistance.
A 600	Polystyrene Type I	Cl^-	1.4	700	43–48	1.10	$Cl \rightarrow OH$ 20%	Clear gel type. Good mechanical strength. Production of ultra pure water.
A 200	Polystyrene Type II	Cl^-	1.3	690	45–51	1.08	$Cl \rightarrow OH$ 15%	Clear gel type. Good kinetics and operating capacity. Good mechanical strength. For demineralisation, giving good silica removal in countercurrent operation.
A 250	Polystyrene Type II	Cl^-	1.3	690	45–52	1.08	$Cl \rightarrow OH$ 15%	Opaque gel type. Good osmotic shock resistance. For demineralisation of water.
A 300	Polystyrene Type II	Cl^-	1.4	700	40–45	1.10	$Cl \rightarrow OH$ 10%	Clear gel type. High capacity for demineralisation of water. For floating beds (FL grade).
A 500	Polystyrene Type I	Cl^-	1.15	690	53–58	1.08	$Cl \rightarrow OH$ 15%	Macroporous type. Very good mechanical and osmotic resistance. For condensate treatment and continuous systems.

A 500 P	Polystyrene Type I	Cl^-	0.8	650	63–70	1.07	$Cl \rightarrow OH$ 20%	Macroporous type. Highly porous. For removal of organic matter. Decolorisation of sugar solutions.
A 505	Polystyrene Type I	Cl^-	1.2	700	53–58	1.08	$Cl \rightarrow OH$ 20%	Macroporous type. High operating capacity. Excellent mechanical resistance. Suitable for countercurrent operation.
A 510	Polystyrene Type II	Cl^-	1.2	700	44–51	1.08	$Cl \rightarrow OH$ 10%	Macroporous type. High operating capacity. Excellent mechanical and osmotic resistance. For demineralisation, fluidised beds and continuous systems.
A 850	Polyacrylic	Cl^-	1.25	710	57–62	1.09	$Cl \rightarrow OH$ 15%	Clear gel type. Good mechanical resistance. Reversibly removes organics with good resistance to fouling. For demineralisation of water and sugar decolorisation.

Table I.4 Purolite continued

Particle size: Anion exchangers

PUROLITE grade	Uniformity coefficient maximum	Effective size (mm) typical	Size range limitation							Principal applications
			< 0.3 mm	< 0.42 mm	< 0.63 mm	< 0.63 mm	> 0.85 mm	> 1.0 mm	> 1.2 mm	
STD	1.70	0.5	1%						5%	Standard quality.
MB	1.70	0.5	1%						2%	Mixed beds (MIXLITE).
PL	1.45	0.55		2%				5%		Condensate polishing (POLILITE).
TL	1.35	0.55		1%			5%			Three component mixed bed (TRILITE).
CL	1.35	0.55		1%			5%			Continuous processes (CONTILITE).
FL	1.55	0.55		2%				10%		Floating beds (FLUIDLITE).
DL strong	1.35	0.60			5%			25%		Layered beds, lower layer (DOUBLITE).
DL weak	1.35	0.40	3%			5%				Layered beds, upper layer (DOUBLITE).
S	1.50	0.55		2%					2%	Special applications (demineralisation of sugar, ...).
C	1.50	0.55		2%					2%	High flow rate applications.

Weak base anion exchangers

PURO-LITE	Type	Ionic form	Total capacity (eq/l)	Shipping weight approximate (g/l)	Moisture retention (%)	Specific gravity	Swelling maximum	Remarks and principal applications
A 100	Polystyrene	Free base	1.3	660	53 – 60	1.04	Free base → Cl 20%	Macroporous type. Good osmotic resistance. Resin optimised for demineralisation of water and saccharose.
A 103	Polystyrene	Free base	1.6	650	48 – 55	1.04	Free base → Cl 25%	Macroporous type. High capacity for demineralisation/decolorisation of glucose and other organic solutions.
A 104	Polystyrene	Free base	1.9	660	47 – 54	1.04	Free base → Cl 35%	Macroporous type. Particularly high capacity for demineralisation and decolorisation of organic solutions including glucose, fructose ...
A 105	Polystyrene	Free base	1.1	630	58 – 65	1.02	Free base → Cl 20%	Macroporous type. Exceptional resistance to osmotic shock and organic fouling. For use especially in continuous systems.
A 840	Polystyrene	Free base	1.6	710	52 – 60	1.10	Free base → Cl 25%	Gel type. High capacity. Very good resistance to organic fouling. Demineralisation of water and saccharose.
A 830	Polystyrene	Free base	2.7	750	47 – 53	1.10	Free base → Cl 15%	Special applications. Sulphate removal from sea water. Effluent neutralisation.

Special products

The PUROLITE range includes numerous other special products:

1. Chromatographic resins (PCR range) with specially tailored particle size and crosslinking for the separation of sugars as well as various other biochemical compounds.
2. Chelating resins for the selective removal of metals and the decalcification of brine used in the production of chlorine and caustic soda (PUROLITE S 930, S 940, S 950, ...).
3. Zeolites such as aluminosilicates (PURALCO Y, W, S) for various applications, e. g., in desalting kits; manganese zeolite (PUROLITE MZ 10) for removal of iron and manganese from natural waters.
4. Resins in bead and powder form for pharmaceutical applications (notably for dry supports and also as active medicinal agents).
5. Nitrate specific resin for the treatment of potable water supplies (PUROLITE A 520).
6. Resin for demineralisation of whey (PUROLITE A 106).
7. Highly porous adsorbent resin for removal of colloidal silica (PUROLITE A 501 P).
8. Other ranges include activated carbons, ion exchange membranes, inert polymers and adsorbents.

Table I.4 Purolite continued
Nuclear grade cation and anion exchangers
Exchangers in bead form

PUROLITE	Type	Purolite standard equivalent	Ionic form	Total capacity minimum (eq/l)	Shipping weight approximate (g/l)	Maximum operating temperature (°C)	Principal applications
NRW 100 (Cation resin)	Strong acid	**C 100**	H^+	1.8	800	120	Removal of cations including radioactive isotopes from aqueous solutions.
NRW 100 Li (Cation resin)	Strong acid	**C 100**	Li^+	1.8	800	120	Decontamination of circuits conditioned with natural lithium.
NRW 100 Li7 (Cation resin)	Strong acid	**C 100**	$^7Li^+$	1.8	800	120	Decontamination of circuits conditioned with lithium 7.
NRW 160 (Cation resin)	Strong acid	**C 160**	H^+	2.2	800	120	Removal of cations including radioactive isotopes. Highly selective for caesium 137.
NRW 160 Li7 (Cation resin)	Strong acid	**C 160**	$^7Li^+$	2.2	800	120	Decontamination of circuits conditioned with lithium 7. Selective for caesium contamination.
NRW 400 (Anion resin)	Strong base	**A 400**	OH^-	1.0	660	60	Production of ultra pure water, for the semiconductor industry.
NRW 450 (Anion resin)	Strong base	**A 450**	OH^-	1.0	660	60	Removal of anions including boric acid and radioactive isotopes.
NRW 600 (Anion resin)	Strong base	**A 600**	OH^-	1.15	690	60	High capacity for anions removal from radioactive circuits.
NRW 500 (Anion resin)	Strong base	**A 500**	OH^-	1.0	700	60	Removal of anions including boric acid, radioactive isotopes and colloids.

Particle size grading is closely controlled to give less than 5% over 1.2 mm and less than 2% under 0.42 mm.

Exchangers in powder form

MICROLITE	Type	Matrix	Ionic form	Total capacity minimum (eq/kg)	Dry matter (g/kg)	Maximum operating temperature (°C)	Principal applications
Pr CH	Strong acid	Polystyrene	H^+	4.3	450	120	For use mixed with Pr AOH on pre-coat filters for demineralisation and filtration.
Pr CN	Strong acid	Polystyrene	NH_4^+	4.3	530	120	For use mixed with Pr AOH in condensate circuits conditioned with ammonia.
Pr AOH	Strong base	Polystyrene	OH^-	3.3	400	60	For use mixed with Pr CH or Pr CN.
FC +	Inert	Fibre			420	70	Filtration in the surface coating of powdered resin to avoid crevice formation.

Purity of nuclear grade ion exchangers

Cation

Ionic form	Percentage
H^+	minimum 99.9%
Li or ^7Li	minimum 99%
NH_4^+	minimum 98%
Impurities	**ppm**
Fe^{+++}	maximum 50
Heavy metals	maximum 40
Alkali metals	maximum 40

Anion

Ionic form	Percentage
OH^-	minimum 95%
CO_3^{--}	maximum 5%
Cl^-	maximum 0.5%*
Impurities	**ppm**
Fe^{+++}	maximum 100
Alkali metals	maximum 40

* Nuclear grade anion resins can be supplied “chloride free” on demand (chloride $\leqslant$ 0.1% of ion exchange sites).

Table I.4 Purolite continued

Nuclear grade mixed beds
Exchangers in bead form

PUROLITE	PUROLITE standard equivalent	Ionic form	Shipping weight approximate (g/l)	Maximum operating temperature (°C)	Principal applications
NRW 39	**C 100/A 450**	H^+/OH^-	735	60	Demineralisation and decontamination (cooling ponds-steam generators). Effluent treatments.
NRW 39 Li	**C 100/A 450**	Li^+/OH^-	735	60	Demineralisation and decontamination of circuits conditioned with natural lithium.
NRW 39 Li7	**C 100/A 450**	$^7Li^+/OH^-$	735	60	Decontamination of pressurised water reactor (PWR) circuits.
NRW 37	**C 100/A 400**	H^+/OH^-	750	60	Demineralisation and decontamination of secondary cooling circuits/effluent. Excellent resistance to attrition.
NRW 41	**C 160/A 450**	H^+/OH^-	750	60	Treatment of radioactive waste. Highly selective for caesium 137. Purging of generators. Radioactive waste.
NRW 41 Li7	**C 160/A 450**	$^7Li^+/OH^-$	750	60	Decontamination of pressurised water circuits. Selective for caesium 137.
NRW 43	**C 160/A 500**	H^+/OH^-	750	60	Highly selective for caesium 137 and radioactive colloids. (Cooling ponds, waste waters ...).
NRW 43 Li7	**C 160/A 500**	$^7Li^+/OH^-$	750	60	Decontamination of pressurised water reactor circuits. Selective for caesium 137 and colloids.

Mixed exchangers in powder form

MICROLITE	Ionic form	Fibre/resin ratio	Cation/anion resin ratio	Dry matter (g/kg)	Maximum operating temperature (°C)	Principal applications
CG 4 H	H^+/OH^-	1/1	1/1	400–450	70	Improved filtration capacity for colloids and suspended matter.
CG 4 N	NH_4^+/OH^-	1/1	1/1	400–450	70	High filtration capacity for condensates conditioned with ammonia.
CG 6 N	NH_4^+/OH^-	2/1	2/1	400–450	70	Good filtration capacity.
CG 7 H	H^+/OH^-	1/2	3/1	400–450	70	Good filtration capacity.
CG 12 H	H^+/OH^-	1/2	4/5	400–450	70	Good capacity for demineralisation.
CG 12 N	NH_4^+/OH^-	1/2	4/5	400–450	70	Good capacity for demineralisation. For condensates conditioned with ammonia.
MB 1/1 H	H^+/OH^-	0	1/1	400	80	Good capacity for both filtration and demineralisation.
MB 3/1 H	H^+/OH^-	0	3/1	420	80	Good capacity for filtration of colloids.
MB 1/1 N	NH_4^+/OH^-	0	1/1	400	80	Combination of a good capacity for filtration and demineralisation (ammoniacal condensates).
MB 3/1 N	NH_4^+/OH^-	0	3/1	420	80	Very good capacity for filtration of colloids. For condensates conditioned with ammonia.

Note
Nuclear grade mixed beds are normally supplied with near stoichiometric equivalents of anion and cation resin sites. Other ratios can be supplied on demand.

Table I.5 Russian ion exchangers also denominated as cationites and anionites, etc. [1, 2, 3, 4]

Type	Ionogenic group(s)	Matrix (crosslinking %)	Particle size (mm)	Exchange capacity (meq/g) dry	Exchange capacity (meq/ml) wet	Bulk density (g/dm^3)	Thermal stability up to °C
Strong Acid Cation Exchangers							
Polymerization resins							
KU-2-8	$-SO_3H$	Styrene, DVB; gel-type; 8	0.5 –1	5.0	1.3	800	120
KU-2-8-cS	$-SO_3H$	Styrene, DVB; gel-type; 8	0.5 –1	5.0	1.3	800	120
KU-2-20	$-SO_3H$	Styrene, DVB; gel-type; 20					120
KU-23	$-SO_3H$	Styrene, DVB; macroporous		4.2			
KU-3	$-SO_3H$	Vinyl naphthalene, DVB; gel-type	0.3 –1.5	5.5		650	
KU-4	$-SO_3H$	Vinyl naphthalene, DVB; gel-type	0.3 –1.5	5.6		650	110
KBU-1	$-SO_3H$, $-COOH$	Styrene, methacrylic acid	0.3 –1.5	7.0			
Polycondensation resins							
KU-1	$-SO_3H$, $-OH$	Phenolsulphonic acid, formaldehyde	0.3 –2.0	4.5		740	
KU-5	$-SO_3H$, $-OH$	Phenolsulphonic acid, formaldehyde	0.3 –1.0	5.0		600	110
KU-6	$-SO_3H$, $-COOH$	Vinyl naphthalene, formaldehyde		5.5	2.0	750	
KU-6 F	$-SO_3H$, $-COOH$	Vinyl naphthylene, phenol, formaldehyde		5.5	1.9	800	
KU-7	$-SO_3H$, $-OH$	Phenol, benzaldehyde-2,4-disulphonic acid, formaldehyde		5.5	1.8	750	
KU-8	$-SO_3H$, $-OH$	Phenol, sulphonic acids of aliphatic aldehydes, formaldehyde		6.0	1.2	650	
KU-9	$-SO_3H$, $-OH$	Phenol, sulphinic acids of ketones, formaldehyde		6.0	1.2	600	
KU-21	$-SO_3H$	Naphtholsulphonic acids, formaldehyde		5.5	1.4	550	
SN, SNF	$-SO_3H$, $-OH$	Phenolic novolaks	0.3 –2.0	5.2			
Weak Acid Cation Exchangers							
Polymerization resins							
KB-1	$-COOH$	Methacrylic acid, DVB; gel-type	0.3 –1.5	10.0	3.8	600	
KB-2	$-COOH$	Methacrylic acid, DVB; gel-type; 2–3	0.3 –1.5	10–11	2.5–3.0	700	
KB-2-7P	$-COOH$	Methacrylic acid, DVB; porous; 7					

KB-2-10P	$-COOH$	Methacrylic acid, DVB; porous; 10					
KB-3	$-COOH$		0.3 −1.5	6.7	2.8	600	
KB-4	$-COOH$	Methyl methacrylic acid, DVB; gel-type; 6	0.3 −2.0	8.5	4.2	600	150
KB-4P-2	$-COOH$	Methyl methacrylic acid, DVB; 2.5					
KB-4-10P	$-COOH$	Methyl methacrylic acid, DVB; porous; 10					
KM	$-COOH$	Methacrylic acid	0.25−1.5	7.5		300	
KMD	$-COOH$	Methacrylic acid	0.25−1.5	8.5		350	
KMG	$-COOH$	Methacrylic acid		7.7		350	
KMT	$-COOH$	Methacrylic acid	0.25−0.8	10.1	2.9	550	
KMTA	$-COOH$	Methacrylic acid					
KMTB	$-COOH$	Metharcylic acid					
KN	$-COOH$	Acrylonitrile, DVB	0.25−1.5	6.0		600	
KR	$-COOH$	Methacrylic acid	0.25−1.5	7.0			
KS-1	$-COOH$	Maleic acid anhydride, methyl acrylate, DVB	0.3 −1.5	10.0	3.6	700	
Polycondensation resins							
KB-5	$-CH_2COOH$	Resorcinol, monochloracetic acid, formaldehyde	0.25−1.5	7.5		600	
KFU	$-COOH$		0.2 −0.8	6−7			
KFFU	$-OCH_2COOH$, $-OH$	Phenol, resorcinol, monochloracetic acid		2.5−4.0			
KRFU	$-OCH_2COOH$, $-OH$	Phenol, resorcinol, monochloracetic acid	0.25−1.5	4.0			
KRFFU	$-OCH_2COOH$, $-OH$	Phenol, resorcinol, monochloracetic acid		2.5−4.0			
Special Acidic Cation Exchangers							
KF-1	$-PO_3H_2$	Styrene, DVB		5.0		700	
KF-2	$-CH_2PO_3H_2$			7.0		700	
KF-3	$-PO_3H_2$			3.5		650	
KF-4	$-CH_2PO_3H_2$			5.5		650	
SF	$-PO_3H_2$	Styrene, DVB					100
FV	$-PO_3H_2$			4.5		400	
RF	$-OPO_2H_2$, $-OH$			4.3		500	60
AR	$-AsO_3H_2$, $-OH$						

Table I.5 Russian continued

Type	Ionogenic group(s)	Matrix (crosslinking %)	Particle size (mm)	Exchange capacity (meq/g) dry	Exchange capacity (meq/ml) wet	Bulk density (g/dm^3)	Thermal stability up to °C
Strong Base Anion Exchangers							
Polymerization resins							
AV-17-8	$-N(CH_3)_3OH$	Styrene, DVB; gel-type; 8	0.3 −1.0	4.3			
AV-17-8 cS	$-N(CH_3)_3OH$	Styrene, DVB; gel-type; 8	0.3 −1.0	4.3			
AV-17P	$-N(CH_3)_3OH$	Styrene, DVB; macroporous					
AV-19	$-N(CH_3)_3OH$	Vinyl naphthalene, trimethylamine		3.0	0.9	600	33
AV-21	$-N(CH_3)_3OH$	Copolymer		2.3			
AV-27	$-N(CH_3)_2C_2H_5OH$	Copolymer		4.0		600	
AV-29-12P	$-N(CH_3)_3OH$	Styrene, DVB; macroporous					
Polycondensation resins							
AV-16 GS	$=NH$, $\equiv N$, Pyridine	Pyridine, epichlorohydrin, polyethylene-polyamines		8.2		700	
PEK	$-N(CH_3)_3OH$		0.3 −2.0	6.0			
Weak Base Anion Exchangers							
Polymerization resins							
AN-4 K	$=NH$, $\equiv N$	Polyvinylchloride, ammonia	0.3 −1.5	6.5		350	
AN-7 K	$=NH$, $\equiv N$	Polyvinylchloride, polyethylenepolyamine	0.3 −1.5	7.5		420	
AN-15	$-NH_2$	Styrene, DVB, monomethylamine	0.3 −1.5	5.5		600	
AN-17	$=NH$	Styrene, DVB, monomethylamine	0.3 −1.5	4.5		600	
AN-18-8	$\equiv N$	Styrene, DVB, dimethylamine; gel-type; 8	0.3 −1.5	4.0		700	60−70
AN-18P	$\equiv N$	Styrene, DVB, dimethylamine; macroporous; 10 and 12					
AN-19	$-NH_2$, $=NH$, $\equiv N$	Copolymer	0.3 −1.5	6.0		600	
AN-20	$-NH_2$, $=NH$	Copolymer	0.3 −1.5	3.0		600	
AN-21	$-NH_2$, $=NH$	Styrene, DVB, hexamethylenediamine; gel-type; 6 and 14	0.3 −1.5	6.0		600	
AN-22-8	$-NH_2$, $=NH$	Styrene, DVB, ethylenediamine; gel-type; 8	0.3 −1.5	7.0		600	

AN-221	$-NH_2$, $=NH$	Styrene, DVB, ethylenediamine; macroporous; 12	0.3 –1.2	5.5			
AN-23	$\equiv N$	Vinylpyridine, DVB	0.3 –1.5	6.5		600	
AN-25	$-NH_2$, $=NH$	2-Methyl-5-vinylpyridine, DVB; gel-type	0.3 –1.5	6.2		600	
AN-251	$-NH_2$, $=NH$	2-Methyl-5-vinylpyridine, DVB; macroporous		5.6			120
Polycondensation resins							
AN-1	$-NH_2$, $=NH$	Melamine, formaldehyde	0.3 –2.0			800	30
AN-2 F	$-NH_2$, $\equiv N$	Phenol, formaldehyde, polyethylenepolyamine					
AN-2 FN	$=NH$, $\equiv N$	Phenol, formaldehyde, polyethylenepolyamine	0.3 –2.0	10.6		800	60
AN-9	$=NH$, $\equiv N$	Phenol, alkylamines	0.3 –2.0	4.5		450	
AN-10	$-NH_2$, $\equiv N$	Phenol, alkylamines	0.3 –1.5	14.0		600	
EDE-10 P	$-NR_2$, $-NHR$, $-NR_3$ (20%)	Epichlorohydrin, polyethylenepolyamines		8.1		600–700	45
Medium Base Anion Exchangers							
AV-16	$=NH$, $\equiv N$, $-NR_3$	Epoxypolyamine	0.3 –1.8	10.0	2.3	750	
AV-18	$\equiv N$, $-NR_3$	Styrene, pyridine	0.3 –1.6	3.0	1.0		
AV-20	$\equiv N$, $-NR_3$	Epichlorohydrin, vinylpyridine		3.5	1.1		

Remarks.
Ion exchangers manufactured in the Soviet Union have no trademarks but symbols, that indicate the places of production or manufacture.
For mainly industrial applications the following symbols are used:
KU strongly acidic cation exchangers containing the $-SO_3H$ group,
KF, RF ion exchangers containing phosphorus in the ionogenic group,
Nuclear-grade Soviet-made ion exchangers are marked cS, i.e. special purity.

References

1. A. B. Pashkov and V. S. Titov (1958). Principle characteristics of some Soviet ion exchange resins. Khim. Prom. *5*, 10.
2. Ion exchange membranes. Granules. Powders. Catalogue NIITE Chem. NPO Plastic Materials M. 1977, pp. 32.
3. M. M. Senyavin (1980). Ion exchange in industry and the analysis of inorganic materials. Moscow. Khimia. pp. 54–71.
4. E. E. Ergozhin, L. N. Prodins and Z. A. Nurkhodzhaeva (1978). Phosphorus-containing ion exchange resins. Tr. Inst. Khim. Nauk. Akad. Nauk. Kaz. SSR *47*, 23.

Table I.6 WOFATIT® Ion Exchangers Program

Producer: VEB CHEMIEKOMBINAT BITTERFELD
Bitterfeld, O-4400
Germany

Strong acid Wofatit Cation Exchangers

Wofatit	Characterization	Application
KPS	8% DVB, gel structure	Universal application: preferably for water conditioning (softening, demineralization), recovery of valuable metals, treatment of sugar juices, purification of solutions, separation of rare earths, etc.
KP 2 KP 4 KP 6 KP 16	2% DVB 4% DVB 6% DVB 16% DVB gel structure, with graded network density	For the separation and isolation of natural substances (e. g., alkaloids), the selective cation exchange of large-volume cations, sugar inversion, the separation of cations of differing size and structure.
KS 10	Macroporous	For softening and demineralization highly oxygen-containing alkaline hot waters, conditioning of condensates, treatment of rinse water in electroplating, the removal of iron from phosphoric pickling baths, the treatment and the purification of organic solutions.
KS 11	Macroporous	For softening and demineralization of warm oxygen-containing waters, the treatment of sugar juices (deliming, Quentin process).

Weak acid Wofatit Cation Exchangers

Wofatit	Characterization	Application
CA 20	H^+-form, macroporous	For the dealkalization of waters. Also suitable as a selective exchanger for heavy metal adsorption, the removal of alkalinity out of solutions and the adsorption of detergents.
CP	Na^+-form, gel structure	For the isolation of streptomycin and selective ion exchange.

Strong base Wofatit Anion Exchangers

Wofatit	Characterization	Application
SBW	Type I, gel structure	Universally applicable for removing weakly dissociated anions, in particular of silicic acid out of waters, for iodine adsorption, purifying solutions, etc.
SBK	Type II, gel structure	For demineralization desilificating in one stage and also for anionic dealkalization.
SZ 30	Type I, macroporous	Like SBW, but particularly suitable for waters highly contaminated with organic substances for catalysis and purifying organic solutions.

Table I.6 Wofatit continued

Strong base Wofatit Anion Exchangers

Wofatit	Characterization	Application
SL 30	Type II, macroporous	For demineralization and desilificating of water in one stage and also for anionic dealkalization, esp. at difficult conditions.
SN 35 L	Macroporous, triethylammonium groups	Removal of nitrate ions from drinking water, high capacity in countercurrent processes.
SN 36 L	Macroporous, triethylammonium groups	Removal of nitrate ions from drinking water. Low nitrate leakage in co-current processes.

Weak base Wofatit Anion Exchangers

Wofatit	Characterization	Application
AD 41	Macroporous	Prefereably for the adsorption of strong acids in water treatment; it displays an excellent capacity for removing organic substances contained in water, for deacidifying organic solutions, the residual detoxication of electroplating effluents, etc.
AD 42	Macroporous	With improved exchange kinetics as compared to Wofatit AD 41.

Wofatit Anion Exchangers for Hydrometallurgy

Wofatit	Characterization	Application
SBT	Strongly basic, gel structure	For the concentration of uranium out of ore pulps; it may also be applied in the filtration process.

Wofatit Adsorber Resins

Wofatit	Characterization	Application
ES	Strongly basic, gel structure	Used in water treatment for the adsorption of organic substances and for the protection of anion exchangers.
EA 60	Strongly basic, macroporous	EA 60 in the sugar industry for decolourizing liquors and juices.

Wofatit Adsorber Polymers

Wofatit	Characterization	Application
EP 61	A copolymer without ionogene groups, macroporous	For the adsorption of organic compounds from aqueous solutions, e. g. recovery of phenol from effluents.

Table I.6 Wofatit continued

Wofatit Ion Exchangers for the Floating-compact-bed Process

Wofatit	Characterization	Application
KPS-AF	Strongly acidic, gel structure	see basic type
KS 10 AF	Strongly acidic, macroporous	
CA 20 AF	Weakly acidic, macroporous	
SBW-AF	Strongly basic, Type I, gel structure	
SZ 30 AF	Strongly basic, Type I, macroporous	
SBK-AF	Strongly basic, Type II, gel structure	
SL 30 AF	Strongly basic, Type II, macroporous	
AD 41 AF AD 42 AF	Weakly basic, macroporous	
ES-AF	Adsorber resin, gel structure	
EA 60 AF	Adsorber resin, macroporous	
UD 90	Inert material as upper layer for the floating-compact-bed process	

Wofatit Ion Exchangers for the Stratified-bed Process

Wofatit	Characterization	Application
KPS-DS	Strongly acidic, gel structure	For cation exchange and for purposes of softening.
CA 20 DS	Weakly acidic, macroporous	For dealkalization.
SW 32 DS	Strongly basic, Type I, gel structure	In combination with Wofatit AD 41 DS for low levels of silicic acid. Applicable up to 60 °C.
SBK-DS	Strongly basic, Type II	In combination with Wofatit AD 41 DS, at high volume capacity, temperatures up to 40 °C and low requirements on the residual silicic acid.
AD 41 DS AD 42 DS	Weakly basic, macroporous	

Wofatit Ion Exchangers for Special Purposes

Wofatit	Characterization	Application
UC 90	Strongly basic	For the manufacture of an agent for the diagnosis of thyroid diseases.
UM 91	Mixture of strong acid and strongly basic resins, with pH indicator	For the demineralization of minor quantities of water preferably used in the shaking method.

Table I.6 Wofatit continued

Wofatit Ion Exchangers for the Mixed-bed Process

Wofatit	Characterization	Application
KPS-MB	Strongly acidic, gel structure	
KS 10 MB	Strongly acidic, macroporous	
SBW-MB	Strongly basic, Type I, gel structure	Preferred combination for the ultra-purification of water: Wofatit KPS-MB/SBW-MB.
SZ 30 MB	Strongly basic, Type I, macroporous	Preferred combination for condensate purification and external regeneration: Wofatit KS 10 MB/SZ 30 MB.
SBK-MB	Strongly basic, Type II, gel structure	Preferred combination for Wofatit KPS-MB/SBK-MB.
AD 41 MB	Weakly basic, macroporous	Preferred combination for demineralization: without desilification: Wofatit KPS-MB/AD 41 MB.

Wofatit Nuclear Grade Ion Exchangers

Wofatit	Characterization	Application
RH	Strongly acidic, gel structure	In pressurized water reactors: for lowering the pH, and in mixture with Wofatit RO-SC for cleaning the primary circuit. In boiling water reactors in conjunction with Wofatit RO in mixed-bed service for cleaning the reactor water.
RK RN	Strongly acidic, gel structure. K^+- and NH_4^--form	For pH-level control of primary circuit water in pressurized water reactors.
RO	Strongly basic, Type I, gel structure	Imixed bed with Wofatit RH for purifying the reactor water of boiling water reactors; for the purification of drain and leak waters in pressurized water reactors.
RO-SC	Strongly basic, Type I, gel structure; extremely low content of chloride	For purifying the water in the primary circuit of pressurized water reactors, preferred for chloride removal.
RS	Strongly basic, Type I, gel structure	Stable preliminary stage of Wofatit RO-SC, unlimited storage life, regeneration to be performed immediately before use.
RO 71	Weakly basic, macroporous	For removing anic contaminations from the water of the fuel element storage pool.

Wofatit Ion Exchangers for Catalytic Purposes

Wofatit	Characterization	Application
FK 2 FK 4 FK 8	Strongly acidic, gel structure, H^+-form	For acid catalysis and cation exchange in aqueous, aqueous-organic and polar solutions, e. g. hydrolyses, esterifications, epoxidations, condensation reactions, sugar-inversion.
FK 110	Strongly acidic, macroporous, H^+-form	
OK 80	Strongly acidic, macroporous, dry; H^+-form	For acid catalysis and cation exchange in moisture-free systems, e. g. alkylation, oligomerization of olefins, refining of phenol, equilibration of silicon oil.

(concluded overleaf)

Table I.6 Wofatit continued

Pulverized Wofatit Ion Exchangers		
Wofatit	**Characterization**	**Application**
PK 202	Strongly acidic, milled, H^+-form	In combination used for the mechanical and chemical fine purification of condensate and waters, preferably by means of precoat filtration.
PS 204	Strongly basic, Type I, OH^--form	

Wofatit Ion Exchangers with pH Indicators		
Wofatit	**Characterization**	**Application**
KPS-I KS 10 I	Strongly acidic	For demineralization water in exchanger columns of transparent material. pH-dependent discolorations on the exchanger allow a visual observation of the loading process.
SBW-I SBK-I	Strongly basic	
AD 42 I	Weakly basic anion exchanger	
SBW-MI	Strong base mixed bed component	

Chelating Wofatit Ion Exchangers		
Wofatit	**Characterization**	**Application**
MC 50	Macroporous, iminodiacetic-acid groups	Selective adsorption of polyvalent metals concentration or purification of substances.
MK 51	Neighbouring aliphatic hydroxyl groups	For the selective adsorption of boric acid and borate ions out of solutions with different salt content.

Table I.7. Amberlite ion exchange resins summary chart

Anion exchangers

Resin type	AMBERLITE designation	Matrix type	Functional structure	Standard ionic form	True (wet) density (g/cm^3)	Shipping weight (g/litre)	Effective size (mm)	Moisture content (%)	Operating pH range
Strongly basic Gel type	IRA 400	Styrene-DVB	$-N^+-(CH_3)_3$ (Type 1)	Cl^-	1.11	700	0.43–0.51	43–49	0–14
	IRA 402	Styrene-DVB	$-N^+-(CH_3)_3$ (Type 1)	Cl^-	1.07	690	0.43–0.51	50–57	0–14
	IRA 410	Styrene-DVB	CH_2-CH_2OH $-N^+$ $(CH_3)_2$ (Type 2)	Cl^-	1.12	700	0.43–0.51	38–44	0–14
	IRA 458	Acrylic-DVB	$-N^+-(R)_3$	Cl^-	1.06	720	0.38–0.48	57–62	0–14
	IRA 458	Acrylic-DVB	$-N^+-(R)_3$ $-N-(R)_2$	Cl^-/FB	1.07	680	0.35–0.45	57–62	0–14

Maximum operating temperature (°C)	Total exchange capacity meq/ml	Total reversible swelling %	Remarks	Application
60 (OH) 75 (Cl)	1.40	$Cl^- \rightarrow OH^-$ 20	Standard crosslinkage; usually used for the treatment of waters which are essentially free of organic materials that might be irreversibly held by the resin.	Water conditioning (deionization including silica reduction, desilicizing, deoxygenation, etc.); uranium recovery, removal of weakly acidic contaminants from process streams and chemical processing.
60 (OH) 75 (Cl)	1.25	$Cl^- \rightarrow OH^-$ 20	Chemically the same as AMBERLITE IRA 400 but with lower crosslinkage to give a better diffusion rate with large organic molecules.	Water conditioning (deionization including silica reduction, desilicizing, deoxygenating, etc.); removal of weakly acidic contaminants from process streams.
35 (OH) 75 (Cl)	1.35	$Cl^- \rightarrow OH^-$ 15	Type 2 resin (dimethyl ethanolamine functionality) of slightly lower basicity than AMBERLITE IRA 400 or AMBERLITE IRA 402, with higher regeneration efficiency. Because of its slightly lower basicity, it is less effective in the removal of silica from water.	Water conditioning (deionization, dealkalization).
35 (OH) 75 (Cl)	1.25	$Cl^- \rightarrow OH^-$ 15	Driffers from styrene-DVB resins in that it has an acrylic structure and is more hydrophilic, more resistant to organic fouling. It is recommended for treating water having a record of producing organic fouling problems for conventional, strongly basic anion exchange resins.	Water conditioning (deionization, dealkalization), decolorization of aqueous solutions, removal or recovery of acidic components of various process streams, particularity sugar.
35 (OH)	1.35	$Cl^- \rightarrow OH^-$ 5	Mixed base anion exchange resin with an exceptionally high operating exchange capacity on waters with a high free mineral acidity content.	Water conditioning, when free mineral acidity is 80% or higher.

Table I.7 Amberlite continued

Resin type	AMBERLITE designation	Matrix type	Functional structure	Standard ionic form	True (wet) density (g/cm^3)	Shipping weight (g/litre)	Effective size (mm)	Moisture content (%)	Operating pH range
Strongly basic Macroreticular	IRA 900	Styrene-DVB	$-N^+-(CH_3)_3$ (Type 1)	Cl^-	1.06	670	0.43–0.51	58–64	0–14
	IRA 904	Styrene-DVB	$-N^+-(CH_3)_3$ (Type 1)	Cl^-	1.08	670	0.43–0.51	56–62	0–14
	IRA 910	Styrene-DVB	CH_2-CH_2OH $-N^+$ $(CH_3)_2$ (Type 2)	Cl^-	1.09	670	0.43–0.51	55–60	0–14
	IRA 938 E	Styrene-DVB	$-N^+-(CH_3)_3$ (Type 1)	Cl^-	1.04	620	0.30–0.40	60–72	0–14
	IRA 958	Acrylic-DVB	$-N^+-(R)_3$	Cl^-	1.08	720	0.47–0.57	66–72	0–14
Weakly basic Gel type	IRA 67	Acrylic-DVB	$-N-(R)_2$	Free base	1.06	700	0.36–0.46	56–62	0–14
	IRA 68	Acrylic-DVB	$-N-(R)_2$	Free base	1.06	700	0.36–0.46	156–62	0–14

Maximum operating temperature (°C)	Total exchange capacity meq/ml	Total reversible swelling %	Remarks	Application
60 (OH) 75 (Cl)	1.00	$Cl^- \rightarrow OH^-$ 20	Strongly basic anion exchange resin best suited for high flow rate deionization.	Water conditioning (deionization, desilicizing, deoxygenating) recovery high molecular weight organics from process streams, sugar decolorizing and deashing. High flowrate condensate polishing.
60 (OH) 75 (Cl)	0.70	$Cl^- \rightarrow OH^-$ 5	Regeneration efficiency is low in the hydroxide cycle but because of its high pore volume and surface area, widely used in the chloride form for organic scavenging. Resistance to physical attrition is excellent.	Water conditioning (organic scavenging, the removal of high molecular weight organic materials).
35 (OH) 75 (Cl)	1.05	$Cl^- \rightarrow OH^-$ 20	This Type 2 (dimethyl ethanolamine) resin offers slightly less silica removal capability than AMBERLITE IRA 900, but offers improved regeneration efficiency.	Water conditioning (deionization, dealkalization) and deionization of chemical process streams.
60 (OH) 75 (Cl)	0.70	$Cl^- \rightarrow OH^-$ 10	Contains very large pores. Particularly suitable for the removal of colloidal silica and hydrous metal oxides from various water streams. Also has the ability to remove various high molecular weight organics from water that would not be removed by other resins.	Ultra-high quality water systems where colloid removal and extremely high electrical resistivity is mandatory, well suited for the removal of radioactive colloids (crud).
35 (OH) 80 (Cl)	1.20	$Cl^- \rightarrow OH^-$ 10 to 15	Macroreticular analogue of AMBERLITE IRA 458.	Deionization of process liquors, sugar decolorizing and deashing. Organic scavenging.
60	1.60	$FB \rightarrow Cl^-$ 15	The resins have an unusually high capacity for large organic molecules. Also they can be operated in the bicarbonate form to exchange bicarbonate for anions of strong mineral acids.	Deacidification, deionization and desalinization of water where the removal of strong mineral acids and adsorption of organics is desired.
60	1.60	$FB \rightarrow Cl^-$ 15		Deionization of process liquors; isolation of acidic natural products.

Table I.7 Amberlite continued

Resin type	AMBERLITE designation	Matrix type	Functional structure	Standard ionic form	True (wet) density (g/cm^3)	Shipping weight (g/litre)	Effective size (mm)	Moisture content (%)	Operating pH range
Weakly basic Macroreticular	IRA 35	Acrylic-DVB	$-N-(R)_2$	Free base	–	720	0.40–0.50	65–73	0–9
	IRA 60 E	Acrylic-DVB	$-N-(R)_2$	Free base	1.05	630	0.36–0.46	50–55	0–9
	IRA 93 SP	Styrene-DVB	$-N-(R)_2$	Free base	1.03	625	0.41–0.51	57–61	0–8
	IRA 94 S	Styrene-DVB	$-N-(R)_2$	Free base	1.04	640	0.40–0.50	57–61	0–8

Maximum operating temperature (°C)	Total exchange capacity meq/ml	Total reversible swelling %	Remarks	Application
60	1.00	FB→Cl^- 15 to 20	Macroreticular analogue of AMBERLITE IRA 68.	Deionization of process liquors, removal of high molecular weight acids.
60	2.3	FB→Cl^- 20	Very high capacity. Suitable for ammonia regeneration or Cl/SO_4 exchange.	Sulphate removal from brackish or sea water.
60	1.25	FB→Cl^- 23	Has excellent resistance to oxidation. Also, has high exchange capacity and exceptional resistance to organic fouling. Often used because of the long operating life that can be expected.	Deacidification, deionization of water (where the removal of strong mineral and organic acids is desired) and deionization of process liquors. Performance of this resin is outstanding in the removal of organic materials from surface water supplies. Sugar deashing and decolorization.
60	1.15	Fb→Cl^- 15	Has all the major attributes of AMBERLITE IRA 93 SP plus improved kinetics and physical stability, and lower reversible volume changes.	Deionization of surface waters and high TDS effluents.

Table I.7 Amberlite continued

Cation exchangers

Strongly acidic	Gel	IR 120	Styrene-DVB	$-SO_3^-$	Na^+	1.26	850	0.43–0.55	44–48	0–14
		IR 122	Styrene-DVB	$-SO_3^-$	Na^+	1.31	865	0.43–0.55	40–44	0–14
		IR 132 E	Styrene-DVB	$-SO_3^-$	Na^+	1.31	865	0.52–0.62	40–44	0–14
	MR	200	Styrene-DVB	$-SO_3^-$	Na^+	1.26	800	0.43–0.51	46–51	0–14
		252	Styrene-DVB	$-SO_3^-$	Na^+	1.26	800	0.43–0.51	50–54	0–14

120	1.95	$Na^+ \rightarrow H^+$ 7	Conventional strongly acidic cation exchange resin with 8% DVB by weight. This is the most widely used cation exchange resin in water conditioning.	Water conditioning (water softening, deionization, split-stream dealkalization), separation of rare earths; separation of amino acids; wine stabilization; etc.
120	2.15	$Na^+ \rightarrow H^+$ 6.5	This resin contains 10% DVB by weight, which gives it a greater resistance to oxidation than Amberlite IR 120.	Water conditioning (water softening, deionization, split-stream dealkalization, etc.).
120	2.15	$Na^+ \rightarrow H^+$ 6.5		Water conditioning, condensate polishing.
150	1.75	$Na^+ \rightarrow H^+$ 3	Excellent physical stability and resistance to oxidation. Contains over 20% DVB by weight and can be expected to provide at least three times greater resistance to oxidation than conventional gel type cation exchange resins.	Water conditioning (water softening, deionization, split-stream dealkalization, etc.); separation of rare earths; separation of amino acids; chemical processing, etc., where the maximum in operating life is desired. High flowrate deionization such as condensate polishing.
135	1.65	$Na^+ \rightarrow H^+$ 3 to 5	Combines a high operating capacity with the exceptional physical stability of the macroreticular resin. Normally used in systems where the aggressiveness does not dictate the use of AMBERLITE 200 but where the use of a gel type resin would be questionable.	Water conditioning; separation of rare earths; separation of amino acids; sugar processing, chemical processing, condensate polishing.

Table I.7 Amberlite continued

Resin type		AMBERLITE designation	Matrix type	Functional structure	Standard ionic form	True (wet) density (g/cm³)	Shipping weight (g/litre)	Effective size (mm)	Moisture content (%)	Operating pH range
Weakly acidic	Gel	IRC 84	Cross-linked acrylic	$-COO^-$	H^+	1.19	720	0.38–0.46	44–50	4–14
	MR	IRC 50	Methacrylic acid-DVB	$-COO^-$	H^+	1.25	660	0.33–0.50	47–53	5–14

Inert materials

Inert	IF 12	Polyethylene	–	–	0.962	555	2.0–3.0	–	0–14
	RF 14	Polypropylene	–	–	0.905	540	1.30–1.70	–	0–14
	AMBERSEP 359	Polymethylmethacrylate	–	–	1.17	735	0.53–0.59	10	0–14

Ambersep system

Inert material	Designation	Type	True wet density g/cm³	Particle size (mm)
	Ambersep 359	Inert Copolymer	1.17	0.42–0.71

Compared with the conventional mixed beds, the Ambersep System provides the following advantages:

- perfect physical and visible separation.
- very low ionic leakage.
- very short rinse requirements.

Cation exchangers	Designation	Type
	Ambersep 120	Gel
	Ambersep 132 E	Gel
	Ambersep 200	Macroreticular
	Ambersep 252	Macroreticular

Maximum operating temperature (°C)	Total exchange capacity meq/ml	Total reversible swelling %	Remarks	Application
120	4.00	$H^+ \rightarrow Na^+$ 65 $H^+ \rightarrow Ca^{2+}$ 15	This resin has a pK value of 5.3* and it will split the alkaline salts of monovalent cations as well as those of multivalent cations. (*This is an approximate value and is with respect to sodium in a 1 molar solution).	Water conditioning (dealkalization, deionization etc.)
120	3.10	$H^+ \rightarrow Na^+$ 100 $H^+ \rightarrow Ca^{2+}$ 35	Has a pK value of 6.1* and will selectively adsorb organic bases such as antibiotics, alkaloids, peptides, amino acids, and metals present in an alkaline solution.	Antibiotic purification and recovery; copper and nickel recovery.

80	–	0	Coarse inert material.	Inert layer for air hold-down upflow countercurrent regenerated systems.
90	–	0	Inert material for countercurrent regeneration and packed bed units.	Inert layer for air hold down upflow countercurrent regenerated systems, reverse flow packed bed systems.
80	–	0	Polymethylmethacrylate inert material, resistant to hydrolysis; high mechanical stability.	Inert resin for the complete separation of the strong acid and strong base ion exchange resins in the AMBERSEP (3 layers) mixed bed systems.

True wet density g/cm³			Particle size (mm)		Designation	Type	True wet density g/cm³	Particle size (mm)
Na^+	NH_4^+	H^+						
1.26	1.24	1.18	0.6–1.20	**Anion exchangers**	Ambersep 400	Gel	1.09	0.35–0.84
1.31	1.29	1.22	0.5–1.20		Ambersep 402	Gel	1.07	0.35–0.84
1.26	1.24	1.18	0.6–1.20		Ambersep 900	Macroreticular	1.06	0.35–0.84
1.26	1.24	1.18	0.6–1.20					

Table I.8. Duolite principal ion-exchange and adsorbent resins

	DUOLITE	Matrix	Functional group	Physical form	Total capacity eq/l	Specific gravity	Shipping (2) weight (g/l)	Moisture retention capacity %
				DUOLITE anion exchangers				
Strong base	**A 101**	Polystyrene	Quat. Ammon. Type 1	Opaque beads	1.3	1.07 (Cl)	ab. 700	50 – 55 (Cl)
Strong base	**A 102**	Polystyrene	Quat. Ammon. Type 2	Opaque beads	1.3	1.10 (Cl)	ab. 710	45 – 50 (Cl)
Strong base	**A 109**	Polystyrene	Quat. Ammon. Type 1	Translucent beads	1.4	1.12 (Cl)	ab. 705	43 – 48 (Cl)
Strong base	**A 113**	Polystyrene	Quat. Ammon. Type 1	Translucent beads	1.3	1.10 (Cl)	ab. 700	48 – 55 (Cl)
Strong base	**A 116**	Polystyrene	Quat. Ammon. Type 2	Translucent beads	1.4	1.11 (Cl)	ab. 710	45 – 50 (Cl)
Strong base	**A 132**	Polyacrylic	Quaternary Ammonium	White beads	1.25	1.09 (Cl)	ab. 720	57 – 62 (Cl)
Strong base	**A 161**	Polystyrene	Quat. Ammon. Type 1	Opaque beads	1.1	1.07 (Cl)	ab. 700	51 – 56 (Cl)
Strong base	**A 162**	Polystyrene	Quat. Ammon. Type 2	Opaque beads	1.1	1.07 (Cl)	ab. 720	48 – 53 (Cl)
Strong base	**A 165**	Polystyrene	Quat. Ammon. Type 1	Opaque beads	1.2	1.07 (Cl)	ab. 720	55 – 60 (Cl)

Maximum temp. °C	pH range	Properties	Principal uses
60 (OH) 100 (Cl)	0–14	High porosity, highest basicity, excellent osmotic resistance.	Water deionisation, silica removal, use in single and mixed beds.
35 (OH) 75 (Cl)	0–14	High porosity, high capacity, good regeneration efficiency and reversibility to organics.	Water deionisation, dealkalisation, co- and countercurrent regeneration.
60 (OH) 100 (Cl)	0–14	Higher crosslinking degree than A 113. Excellent physical aspect and mechanical resistance.	Condensate polishing, floating and stratified beds.
60 (OH) 100 (Cl)	0–14	High basicity, excellent attrition resistance.	Deionisation, silica removal, floating, stratified and mixed beds systems.
35 (OH) 75 (Cl)	0–14	High capacity, low rinse requirements.	Countercurrent regeneration systems and floating beds.
35 (OH) 70 (Cl)	0–14	Excellent organic reversibility, high regeneration efficiency.	Two-stage deionisation, sugar decolourisation.
60 (OH) 100 (Cl)	0–14	Macroporous, high basicity, exceptional osmotic and attrition stability.	C grade for condensate polishing. Ci grade for continous systems. TR grade for Triobed.
35 (OH) 75 (Cl)	0–14	Macroporous, good regeneration efficiency, outstanding physical stability.	Continuous (moving bed) systems: Ci grade.
60 (OH) 100 (Cl)	0–14	Macroporous, high basicity, good regeneration efficiency, high capacity, excellent physical resistance.	Countercurrent, floating beds, condensate polishing.

Table I.8. Duolite continued

	DUOLITE	Matrix	Functional group	Physical form	Total capacity eq/l	Specific gravity	Shipping (2) weight (g/l)	Moisture retention capacity %
	DUOLITE anion exchangers							
Medium and weak base	**A 30 B**	Polyepoxy-amine	Tertiary amine	Translucent beads	1.9	1.10 (FB)	ab. 760	58–63 (FB)
	A 303	Polystyrene	Tertiary amine	Opaque beads	1.3	1.06 (FB)	ab. 700	45–55 (FB)
	A 366	Polystyrene	Tertiary amine	Opaque beads	1.7	1.04 (FB)	ab. 720	45–52 (FB)
	A 378	Polystyrene	Tertiary amine	Opaque beads	1.3	1.03 (FB)	ab. 670	52–58 (FB)
	A 369	Polystyrene	Tertiary amine	Opaque beads	1.1	1.02 (FB)	ab. 670	56–62 (FB)
	A 374	Polyacrylic	Polyfunctional amine	White beads	2.3	1.05 (FB)	ab. 630	50–55 (FB)
	A 375	Polyacrylic	Tertiary amine	White beads	1.6	1.06 (FB)	ab. 700	56–62 (FB)
	A 561	Phenolic	Tertiary amine	Granules	1.7	1.12 (FB)	ab. 440	50–55 (FB)
	A 7	Phenolic	Secondary amine	Granules	2.0	1.12 (FB)	ab. 580 (3)	53–60 (FB)

(1) Special particle sizes available on request

(2) Approximate figures due to possible batch-to-batch slight moisture variations

(3) Shipped partially dried.

(4) N is the ratio: $\frac{\text{Organic Matter (as } KMnO_4 \text{ consumption in mg/l)}}{\text{Equivalent Mineral Acidity (in Meq/l)}}$

Maximum temp. °C	pH range	Properties	Principal uses
25 (FB) 40 (Cl)	0–9	Very high capacity for strong acids.	Deionisation of water with moderate organic load (N < 4).
60 (FB) 10 (Cl)	0–7	Outstanding resistance to oxidation.	Deionisation of strongly oxidising solutions (e. g. chromates). Deacidification of formaldehyde.
60 (FB) 100 (Cl)	0–7	Macroporous, high capacity	Protection of strong base resins from organic fouling, purification of galvanoplating rinse effluents.
60 (FB) 100 (Cl)	0–7	Macroporous, optimum capacity/stability/organic reversibility.	Standard medium base resin for deionisation of water and sugar solutions (S grade). Duobed systems (D grade).
60 (FB) 100 (Cl)	0–7	Macroporous, outstanding stability, high porosity, good kinetics.	Continuous systems (Ci grade), high flow rate.
35 (FB) 70 (Cl)	0–7	Exceptionally high capacity, good kinetics.	Process applications, e. g. ammonium nitrate recovery, demineralisation of organic solutions.
35 (FB) 70 (Cl)	0–9	Relatively high basicity, high stability, outstanding absorption/desorption of organics.	Process applications, water and sugar juice demineralisation.
50	0–6	Macroporous, high porosity, excellent adsorption efficiency. Tasteless and colourless.	Process applications. Demineralisation and decolourisation of organic solutions: phenol, citric acid, starch hydrolysates, etc.
40	0–6	Very high porosity, excellent adsorption properties.	Glycerine purification. Demineralisation and decolourisation of organic acids.

Table I.8 Duolite continued

	DUOLITE	Matrix	Functional group	Physical form (1)	Total capacity eq/l	Specific gravity	Shipping (2) weight (g/l)	Moisture retention capacity %
DUOLITE cation exchangers								
Strong acid	**C 20**	Polystyrene	Sulfonic acid	Translucent beads	2.0	1.28 (Na)	ab. 850	45–50 (Na)
Strong acid	**C 225**	Polystyrene	Sulfonic acid	Translucent beads	2.0	1.28 (Na)	ab. 850	45–50 (Na)
Strong acid	**C 204 F**	Polystyrene	Sulfonic acid	Translucent beads	1.5	1.15 (Na)	ab. 820	50–65 (Na)
Strong acid	**C 206 A**	Polystyrene	Sulfonic acid	Translucent beads	1.6	1.19 (Na)	ab. 800	55–61 (Na)
Strong acid	**C 255**	Polystyrene	Sulfonic acid	Translucent beads	2.0	1.30 (Na)	ab. 850	40–45 (Na)
Strong acid	**C 26**	Polystyrene	Sulfonic acid	Opaque beads	1.85	1.25 (Na)	ab. 800	47–52 (Na)
Strong acid	**C 264**	Polystyrene	Sulfonic acid	Opaque beads	2.3	1.36 (Na)	ab. 850	38–43 (Na)
Strong acid	**C 265**	Polystyrene	Sulfonic acid	Opaque beads	2.0	1.36 (Na)	ab. 850	35–40 (Na)
Weak acid	**C 433**	Polyacrylic	Carboxylic acid	Translucent beads	4.2	1.19 (H)	ab. 790	43–47 (H)
Weak acid	**C 464**	Polyacrylic	Carboxylic acid	Opaque beads	2.7	1.13 (H)	ab. 800	52–58 (H)
Weak acid	**C 468**	Polyacrylic	Carboxylic acid	Opaque beads	1.9	1.11 (H)	ab. 750	62–67 (H)

Maximum temp. °C	pH range	Properties	Principal uses
120	0–14	High capacity, high stability.	Standard resin; water softening (S and A grades) and demineralisation.
120	0–14	High capacity, high stability. Formerly Zerolit 225.	Standard resin for softening and demineralisation.
90	0–14	A family of fine and closely graded resins. Several qualities available.	Chromatography and ion-exclusion applications.
100	0–14	Good kinetics and physical aspect.	Domestic water softeners, high speed softening in genéral.
140	0–14	Highly stable gel type cation exchanger. Higher crosslinking degree than C 20 and C 225. Formerly Zerolit 525.	Water deionisation, mixed beds. Process applications.
140	0–14	Optimum macroporosity, extreme osmotic and mechanical stability.	Condensate polishing (C grade), continuous systems (Ci grade), process applications (e. g. electroplating rinse effluents) TR grade for Triobed.
140	0–14	High degree of crosslinking, macroporous, maximum physical strength.	Process applications (e. g. Quentin process), treatment of oxidising solutions.
140	0–14	Macroporous, very high crosslinking, good resistance to oxidation.	Special applications. Ammonium nitrate removal. Catalysis.
120	5–10	Very high capacity. Standard weak-acid cation exchanger.	Water dealkalisation, softening and demineralisation.
120	5–14	Macroporous. Outstanding resistance to osmotic shocks.	Na exchange, ammonium bicarbonate removal from condensates, recovery of antibiotics.
120	5–14	Macroporous. High porosity.	Removal of surfactants from water.

Table I.8. Duolite continued

DUO-LITE	Matrix	Functional group	Physical form	Total capacity eq/l	Specific gravity	Shipp-ing (2) weight (g/l)	Moisture retention capacity %
			DUOLITE adsorbent resins				
A 171	Polystyrene	Quaternary ammonium	Opaque beads	–	1.07 (Cl)	ab. 700	60–65 (Cl)
A 173	Polyacrylic	Quaternary ammonium	White beads	–	1.06 (Cl)	ab. 720	66–72 (Cl)
S 587(4)	Phenolic	Amine	Granules	–	1.12	ab. 450 (3)	52–60
S 761	Phenolic	–	Granules	–	1.11	ab. 610	47–51
S 861	Polystyrene	–	White beads	–	1.02	ab. 710	65–70
			DUOLITE chelating resins				
ES 346	Polyacrylic	Amidoxime	Opaque beads	–	1.10 (FB)	ab. 760	50–55
ES 465	Polystyrene	Mercaptane	Opaque beads	–	1.03 (H)	ab. 750	45–50
ES 466	Polystyrene	Aminodiacetate	Opaque beads	–	1.05 (Na)	ab. 740	55–60
C 467	Polystyrene	Amino-phosphonate	Opaque beads	–	1.12 (Na)	ab. 740	60–65

(1) Special particle sizes available on request.
(2) Approximate figures due to possible batch-to-batch slight moisture variations.
(3) Shipped partially dried.
(4) Formerly S 37.

Maximum temp. °C	pH range	Properties	Principal uses
60 (OH) 100 (Cl)	6–14	Macroporous, high porosity, good stability.	Protection of ion exchange train from organic fouling. Colour removal from sugar syrups.
80 (Cl)	0–14	Macroporous, high moisture, large pores, high colour removal capacity.	Colour removal from sugar syrups with high colour loadings.
30	2–8	Very weak-base. Porous. Resistant to attrition.	Removal of organics, including detergents, from water and waste solutions.
40	1–6	Highly porous. Almost non-ionic (phenolic function).	Removal of proteins and high molecular weight pigments from wine, fermentation broths, etc.
120	0–14	Extremely stable macroporous non-ionic structure.	Removal of colour, surfactants, from aqueous and organic solutions.
40	variable	Weakly basic chelating polymer.	Selective removal of heavy metals from acidic solutions.
60	1–13	Selective affinity for mercury (thiol function).	Selective removal of mercury from effluents.
65	variable	Macroporous chelating resin. Affinity for polyvalent metals.	Separation of metals from various solutions. Decalcification of NaCl brine.
85	variable	Macroporous. Very stable complexes with low atomic mass cations.	Decalcification of brines. Metal separations. Recovery of zinc, lead, etc.

Table I.9. IONAC ion exchange resins.
Sybron Chemicals Inc., Birmingham Road, P.O.Box 66, Birmingham, N.J. 08011, USA

Strong Acid Cation Exchangers

Trade name	**C-240**	**C-242**	**C-249**	**C-250**	**C-253**
Type of exchanger	cation strong acid	cation strong acid	cation strong acid	cation strong acid	cation strong acid
Base matrix	polystyrene	polystyrene	polystyrene	polystyrene	polystyrene
Porosity	Gel	Gel	Gel	Gel	Gel
Functional group	$-SO_3-$	$-SO_3-$	$-SO_3-$	$-SO_3-$	$-SO_3-$
Ionic form	Na^+	H^+	Na^+	Na^+	Na^+
Crosslinking	8% DVB	8% DVB	8% DVB	10% DVB	8% DVB
Particle size (mm)	0.4–1.2	0.3–1.2	0.5–1.2	0.3–1.2	0.6–1.2
Moisture (wt. %)	45–48	45–48	45–48	38–43	45–48
Capacity meq/g (dry) meq/ml (wet)	 4.6 1.9	 4.6 1.9	 4.6 1.9	 5.0 2.1	 4.6 1.9
Operational control Max. temp. (°C) pH range	 140 0–14	 120 0–14	 140 0–14	 140 0–14	 140 0–14
Remarks	**C-240** General purpose, standard cation, Na-form for water softeners	**C-242** H-form of C-240, for demineralizing water	**C-249** Household & Food Industry grade of C-240. Meets FDA Food Additives Reg. requirements.	**C-250** Higher cross-linked version of C-240 for hot water applications (see also CFS).	**C-253** Coarse grade of C-249 for high flowrate softeners

C-255	C-257	C-258	C-267	C-298 C (H)	CF
cation strong acid	cation strong acid	cation strong acid	cation strong acid	cation strong acid	cation strong acid
polystyrene	polystyrene	polystyrene	polystyrene	polystyrene	polystyrene
Gel	Gel	Gel	Gel	Gel	Gel
$-SO_3-$	$-SO_3-$	$-SO_3-$	$-SO_3-$	$-SO_3-$	$-SO_3-$
Na^+	H^+	Na^+	H^+	H^+	Na^+
10% DVB	8% DVB	10% DVB	8% DVB	8% DVB	8% DVB
0.3–1.2	0.6–1.2	0.4–1.2	0.35–1.2	0.4–1.2	0.3–1.2
35–38	45–48	38–43	49–55	52–58	43–49
5.0 2.1	4.6 1.9	5.0 2.1	4.9 1.9	4.9 2.1	5.5 2.3
140 0–14	120 0–14	140 0–14	120 0–14	120 0–14	140 0–14
C-255 Highest cross-linked version of C-240 for mild oxidizing conditions. (see also CFS, CFZ)	**C-257** Coarse grade of C-242 for higher flowrate demin.	**C-258** FDA grade of C-250	**C-267** H-form of C-249 for food industry water demin. use, etc.	**C-298 C (H)** For high flow-rate condensate demin.	**CF** Light colored, high strength variant of C-240.

Table I.9 Ionac continued

Ultrapure Grade Ion Exchangers

Trade name	**NC-10**	**NC-11**	**NA-30**	**NA-38**
Type of exchanger	cation strong acid	cation strong acid	anion strong base	anion strong base
Base matrix	polystyrene	polystyrene	polystyrene	polystyrene
Porosity				
Functional group	$-SO_3-H$	$-SO_3-H$	$-N^+ (CH_3)_3$	$-N^+ (CH_3)_3$
Ionic form	H^+	H^+	OH	OH
Crosslinking	DVB	DVB	acrylic	DVB
Particle size (mm)	0.4–1.2	0.4–1.2	0.4–1.2	0.4–1.2
Moisture (wt. %)	55 max.	55 max.	60 max.	60 max.
Capacity meq/ml (wet)	1.9	2.0	1.0	1.4
Operational control Max. temp. (°C) pH range	120 °C 0–14	130 °C 0–14	60 °C 0–14	60 °C 0–14
Remarks	**NC-10** Standard cation	**NC-11** Higher crosslinked cation	**NA-30** Ultrapure grade of A-540	**NA-38** Ultrapure grade of ASB-1

NM-40	NM-49	NM-60	NM-65	NM-75
mixed bed	mixed bed	mixed bed	mixed bed	mixed bed
$\frac{SO_3^-H^+}{R_4N^+OH^-}$	$\frac{SO_3^-H^+}{R_4N^+OH^-}$	$\frac{SO_3^-H^+}{R_4N^+OH^-}$	$\frac{SO_3^-H^+}{R_4N^+OH^-}$	$\frac{SO_3^-H^+}{R_4N^+OH^-}$
H/OH	H/OH	H/OH	H/OH	H/OH
0.4–1.2	0.4–1.2	0.4–1.2	0.4–1.2	0.4–1.2
60 max. cation 1.9 meq/ml anion 1.0 meq/ml	60 max. cation 2.1 meq/ml anion 1.0 meq/ml	60 max. cation 1.9 meq/ml anion 1.4 meq/ml	60 max. cation 1.9 meq/ml anion 1.4 meq/ml	60 max. cation 1.9 meq/ml anion 1.4 meq/ml
non reg. bed 100 reg. bed 60 0–14	non reg. bed 100 reg. bed 60 0–14	non reg. bed 100 reg. bed 60 0–14	non reg. bed 100 reg. bed 60 0–14	non reg. bed 100 reg. bed 60 0–14
NM-40 Standard mixed bed	**NM-49** Mixed bed with higher crosslinked cation.	**NM-60** High capacity mixed bed	**NM-65** Indicator mixed bed	**NM-75** Special mixed bed for stator spec.

Table I.9 Ionac continued

Strong/Weak Acid Cation Exchangers

Trade name	**CFS**	**CFZ**	**CFP-110**	**CC**
Type of exchanger	cation strong acid	cation strong acid	cation strong acid	cation weak acid
Base matrix	polystyrene	polystyrene	polystyrene	acrylic
Porosity	macroporous	macroporous	macroporous	macroporous
Functional group	$-SO_3-$	$-SO_3-$	$-SO_3-$	COO^-H^+
Ionic form	H^+	H^+	H^+	H^+
Crosslinking	DVB	DVB	DVB	DVB
Particle size (mm)	0.3–1.2	0.3–1.2	0.3–1.2	0.3–1.2
Moisture (wt. %)	45–50	40–42	40–46	46–53
Capacity meq/g (dry) meq/ml (wet)	 5.0 2.2	 5.1 2.4	 5.0 2.2	 9.5 3.5–3.7
Operational control Max. temp. (°C) pH range	 120 0–14	 120 0–14	 120 0–14	 120 0–14
Remarks	**CFS** High capacity. 2nd gen. macroporous for milder oxidizing conditions and higher temp. applications in water treatment. Also Na-cycle condensate polishing	**CFZ** Similar to CFS, recommended for strong oxidants, use with plating wastes and other special applications.	**CFP-110** Coarser grade of CFZ. Particularly for high flowrate and condensate demin.	**CC** General purpose weak acid resin. Alkalinity removal from water. Very low swelling. Buffer filters. Pharmaceutical applications.

CNN				
cation weak acid				
acrylic				
macroporous				
COO^-H^+				
H^+				
DVB				
0.3 – 1.2				
51 – 56				
10.0 3.9 – 4.1				
120 0 – 14				
CNN High capacity. Especially suited for high Na bicarbonate wates and many special applications. Different polymer structure from IONAC CC.				

Table I.9 Ionac continued

Strong Base Anion Exchangers

Trade name	**A-540**	**A-550**	**A-641**	**A-642**
Type of exchanger	anion strong base Type I	anion strong base Type II	anion strong base Type I	anion strong base Type I
Base matrix	polystyrene	polystyrene	polystyrene	polystyrene
Porosity	Gel	Gel	second genera. macroporous	second genera. macroporous
Functional group	$N^+(CH_3)_3$	$N(CH_3)_2$ (CH_2CH_2OH)	$N^+(CH_3)_3$	$N^+(CH_3)_3$
Ionic form	Cl	Cl	Cl	Cl
Crosslinking	acrylic	acrylic	DVB	DVB
Particle size (mm)	0.4–1.2	0.4–1.2	0.3–1.2	0.4–1.2
Moisture (wt. %)	48–60	43–47	54–58	57–62
Capacity meq/g (dry) meq/ml (wet)	 3.6 1.0	 3.0–3.3 1.3	 3.9 1.15	 3.9 1.1
Operational control Max. temp. (°C) pH range	 60 0–14	 100 (Cl) 40 (OH) 0–14	 100 (Cl) 70 (OH) 0–14	 100 (Cl) 70 (OH) 0–14
Remarks	**A-540** General purpose demin. resin for very low silica residuals. More fouling resistant than DVB crosslinked resins.	**A-550** General purpose demin. resin. More fouling resistant than DVB crosslinked resins.	**A-641** Resin for lowest silica residuals, best fouling resistance and operating capacity equal to gel resins, higher than other types of macropore resins.	**A-642** "Organic scavenger." Also for color removal and colloid filtration.

A-651	ASB-1/A-430	ASB-1P A-440	ASB-1PC (OH)	ASB-2/A-450
anion strong base Type II	anion strong base Type I	anion strong base Type I	anion strong base Type I	anion strong base Type II
polystyrene	polystyrene	polystyrene	polystyrene	polystyrene
second genera. macroporous	Gel	Gel	Gel	Gel
$N(CH_3)_2$ (CH_2CH_2OH)	$N^+(CH_3)_3$	$N^+(CH_3)_3$	$N^+(CH_3)_3$	$N(CH_3)_2$ (CH_2CH_2OH)
Cl	Cl	Cl	Cl	Cl
DVB	DVB	DVB	DVB	DVB
0.3 – 1.2	0.3 – 1.2	0.3 – 1.2	0.3 – 1.2	0.3 – 1.2
49 – 52	43 – 49	51 – 57	50 – 55	38 – 42
3.5 1.15	3.7 1.4	4.3 1.3 – 1.4	4.3 1.35	3.5 – 3.7 1.45 – 1.6
100 (Cl) 60 (OH) 0 – 14	100 (Cl) 60 (OH) 0 – 14	100 (Cl) 60 (OH) 0 – 14	100 (Cl) 60 (OH) 0 – 14	77 (Cl) 40 (OH) 0 – 14
A-651 Best fouling resistance and operating capacity equal to gel resins, higher than other types of macropore resins.	**ASB-1/A-430 ASB-1P A-440** Similar to IONAC A-540, but DVB crosslinked, IONAC ASB-1P is more porous than IONAC ASB-1		**ASB-1PC (OH)** Fully regenerated, condensate polishing grade.	**ASB-2/A-450** Similar to IONAC A-550 but DVB crosslinked

Table I.9 Ionac continued

Weak/Intermediate Base Anion Exchangers

Trade name	**A-305**	**A-328 (WBS)**	**AFP-329/ AFP-329K**	**A-365**
Type of exchanger	anion weak base	anion weak base	anion weak base	anion weak base
Base matrix	epoxpolyam	polystyrene	polystyrene	acrylic
Porosity	macroporous	second gener. macroporous	second gener. macroporous	Gel
Functional group	NHR, NH_2	$N(CH_3)_2$	$N(CH_3)_2$	NHR, NH_2
Ionic form	Free base	Free base	Free base	Free base
Crosslinking	COND	DVB	DVB	DVB
Particle size (mm)	0.4 – 2.0	0.3 – 1.2	0.3 – 1.2	0.3 – 1.2
Moisture (wt. %)	59 – 67	46 – 52	44 – 52	45 – 50
Capacity meq/g (dry) meq/ml (wet)	 12.0 3.5	 4.5 1.85	 4.0 (min) 1.5 (min)	 9.5 3.8
Operational control Max. temp. (°C) pH range	 40 0 – 14	 100 0 – 14	 60 0 – 14	 60 0 – 14
Remarks	**A-305** General purpose demin. resin. Good fouling resistance. Deashing starch hydrolysates, etc. Amino acid separation	**A-328 (WBS)** **AFP-329/ AFP-329K** Controlled MacroPore resins. Best fouling resistance. Also for 2-component (layered) beds, and plating rinse recovery.		**A-365** Extra high capacity and regen. efficiency. Disposable cartridges. Whey and lactic acid deashing or deacidification. High solids well water demineralization.

A-375				
anion weak base				
acrylic				
Gel				
NHR, NH_2				
Free base				
DVB				
0.3 – 1.2				
52 – 55				
5.5 1.9				
60 0 – 14				
A-375 High capacity weak base resin with improved fouling resistance.				

Table I.10. MERCK ion exchangers and adsorber resins

(E. Merck, Frankfurterstr. 250, Postfach 4119, 6100 Darmstadt, FR Germany)

Cat. No.	Name	Copolymerisate	Degree of cross-linkage (%)	Fixed ion	Counter ion/ form as supplied	Colour	Shape
Strong acid gel-type cation exchangers							
4765	Ion exchanger I	Styrene/divinylbenzene	8	$\sim SO_3^-$	H^+	Yellowish-brown, spherical transparent	
15131	Ion exchanger Amberlite® IR-120	Styrene/divinylbenzene	8	$\sim SO_3^-$	H^+	Brownish-yellow, Spherical transparent	
5257	Ion exchanger Merck-Lewatit® S 1080	Styrene/divinylbenzene	8	$\sim SO_3^-$	H^+	Light beige, transparent	Spherical

Notes: These exchangers are highly resistant to acids, bases, organic solvents, reducing substances and oxidising agents.

Strong acid macroporous cation exchangers							
5258	Ion exchanger Merck-Lewatit® SP 1080	Styrene/divinylbenzene	8	$\sim SO_3^-$	Na^+	White, opaque	Spherical
15635	Ion exchanger Amberlyst® 15	Styrene/divinylbenzene	8	$\sim SO_3^-$	H^+	White, opaque	Spherical

Notes: These exchangers are specially suitable for use in non-aqueous media. Their unusually high porosity and special surface characteristics make them available to a larger number of fixed ions than is the case with a gel-type resin.

Weak acid gel-type cation exchangers							
4835	Ion exchanger IV	Acrylic acid/divinylbenzene	5	$\sim COO^-$	H^+	Cream, opaque	Spherical
15130	Ion exchanger Amberlite® IRC-150	Methacrylic acid/divinylbenzene	4–6	$\sim COO^-$	H^+	Creamy-white, opaque	Spherical
15627	Ion exchanger Amberlite® CG-501	Methacrylic acid/divinylbenzene	5	$-COO^-$	H^+	Whitish, opaque	Spherical

Notes: Resistant to alkalis, aliphatic and aromatic solvents, oxidising agents and reducing substances. High-capacity exchanger.

Capacity (meq./ml)	Swelling at maximum charge reversal (%)	Maximum working tempera-ture (°C)	pH range	Particle size (mm) mesh ASTM	Moisture content (%)	Bulk density (g/l)	Regenerant concentration in water (%)
1.9	7	120	0–14	0.3–0.9 20–50	45–55	850	HCl 5–10
1.9	7	120	0–14	0.3–0.9 20–50	45–50	850	H_1SO_4 2–4
2.2	7	120	0–14	01–0.25 60–150	40–45	850	NaCl 8–10

Uses: Inter alia: water conditioning (dehardening, demineralisation, partial desalination), separation of rare earths, separation of amino acids, analytical applications (concentration procedures, removal of anions), food industry.

1.8	5	120	0–14	01–0.25 60–150	50	720–820	HCl 5–10
2.9	3	120	0–14	0.3–0.9 20–50	0.5	800	H_2SO_4 2–4 or NaCl 8–10

Uses: Catalysis of organic reactions in non-aqueous solutions. Deionisation of such solutions, separation-off of high molecular weight organic compounds and dyes. Water conditioning, analysis.

3.5	75–100	120	4–14	0.3–0.9 20–50	45–50	650	HCl 2–3
3.5	75–100	120	4–14	0.3–0.9 20–50	45–55	690	
3.5	75–100	120	4–14	0.075–0.150	45–55	700	H_2SO_4 0.5–1

Uses: Purification and extraction of antibiotics, vitamins and alkaloids, determination of di- and polyvalent cations, adsorption of acid dyes from acid solutions.

Table I.10. MERCK continued

Cat. No.	Name	Copolymerisate	Degree of cross-linkage (%)	Fixed ion	Counter ion/ form as supplied	Colour	Shape
Strong base gel-type anion exchangers, Type I							
4767	Ion exchanger III	Styrene/divinylbenzene	8	$\sim N^+(CH_3)_3$	OH^-	Brown, transparent	Spherical
5255	Ion exchanger Merck-Lewatit® M 5080	Styrene/divinylbenzene	8	$\sim N^+(CH_3)_3$	Cl^-	Light brown, transparent	Spherical
15128	Ion exchanger Amberlite® IRA-400	Styrene/divinylbenzene	8	$\sim N^+(CH_3)_3$	Cl^-	Light brown, transparent	Spherical

Notes: Resistant to acids and alkalis, aliphatic and aromatic solvents, oxidising agents and reducing substances.

Strong base gel-type anion exchangers, Type II							
15262	Ion exchanger Amberlite® IRA-410	Styrene/divinylbenzene	8	CH_3 $\sim N-CH_2OH$ CH_3	Cl^-	Light brown, slightly transparent	Spherical

Notes: Basicity somewhat less than that of the strong base Type I anion exchanger, hence better regenerant utilisation. Silicic acid uptake slightly lower.

Strong base macroporous anion exchangers, Type I							
5256	Ion exchanger Merck-Lewatit® MP 5080	Styrene/divinylbenzene	8	$\sim N^+(CH_3)_3$	Cl^-	Cream, opaque	Spherical
15636	Ion exchanger Amberlyst® A-26	Styrene/divinylbenzene	8	$\sim N^+(CH_3)_3$	Cl^-	Yellowish-white, opaque	Spherical

Notes: Owing to their unusually high porosity and their special surface characteristics these exchangers are available to a larger number of fixed ions in a nonaqueous system than is the case with a gel-type resin. The capacity can thus be utilised to the full.

Capacity (meq./ml)	Swelling at maximum charge reversal (%)	Maximum working temperature (°C)	pH range	Particle size (mm) mesh ASTM	Moisture content (%)	Bulk density (g/l)	Regenerant concentration in water (%)
1.4	18–22	OH : 60 Cl : 80	0–14	0.3–0.9 20–40	50–60	700	NaCl 8–10
1.6	18–22	OH : 70 Cl : 100	0–12	0.1–0.25 60–150	40–45	670–750	
1.4	18–22	OH : 60 Cl : 80	0–14	0.3–0.9 20–50	50–60	700	NaCl 8–10

Uses: Water conditioning (complete demineralisation, desilification), acidimetric determination of aqueous salt solutions, preparation of pure complex bases, isolation and determination of alkaloids, determination of pectins in fruit juices, removal of interfering ions, catalysis.

1.35	15	OH : 40 Cl : 80	0–14	0.3–0.9 20–50	40	700	NaCl 8–12 NaOH 2–4

Uses: Demineralisation of water, purification of formaldehyde, separation of amino acids, catalysis, water conditioning (complete demineralisation, decarbonation).

1.2	11	OH^- : 60 Cl^- : 80	1–14	0.1–0.25 60–150	55–60	700	NaCl 8–10
1.0	10	OH^- : 60 Cl^- : 80	0–14	0.3–0.9 20–50	60	700	NaOH 2–4

Uses: Removal of anions from non-aqueous solutions and catalysis in non-aqueous solutions.

Table I.10 MERCK continued

Cat. No.	Name	Copoly-merisate	Degree of cross-linkage (%)	Fixed ion	Counter ion/ form as supplied	Colour	Shape
Weak base gel-type anion exchanger							
4766	Ion ex-changer II	Styrene/divi-nylbenzene	5	$\sim N(R)_2$	Free base (OH^-)	Brownish-yellow, opaque	Spherical

Notes: Resistant to acids and bases, aliphatic and aromatic solvents. Sensitive to nitric acid.

Cat. No.	Name	Copoly-merisate	Degree of cross-linkage (%)	Fixed ion	Counter ion/ form as supplied	Colour	Shape
Weak base macroporous anion exchanger							
15261	Ion ex-changer Am-berlyst® A-21	Styrene/divi-nylbenzene	Not known	$\sim N(CH_3)_2$	Free base (OH^-)	Whitish-yel-low, opaque	Spherical

Notes: Excellent resistance to oxidation for a weak base exchanger. Withstands organic contaminants very well.

Cat. No.	Name	Copoly-merisate	Degree of cross-linkage (%)	Fixed ion	Counter ion/ form as supplied	Colour	Shape
Mixed bed ion exchangers							
4836	Ion ex-changer V	Styrene/divi-nylbenzene	8	CH_3 $\sim SO_3/\sim N^{+-}CH_2OH$ CH_3	H^+ and OH^-	Yellowish-brown, trans-parent	Spherical
15127	Ion ex-changer Amberlite® MB-3	Styrene/divi-nylbenzene	8	CH_3 $\sim SO_3/\sim N^{+-}CH_2OH$ CH_3	H^+ and OH^-	Light brown + blue (in-dicator), transparent	Spherical

Notes: These mixed bed exchangers are mixtures of strong acid cation exchanger and strong base anion exchanger of Types I. Amberlite® MB-3 changes from blue to brown following charging. The indicator is bound to the strong base exchanger.

Capacity (meq./ml)	Swelling at maximum charge reversal (%)	Maximum working temperature (°C)	pH range	Particle size (mm) mesh ASTM	Moisture content (%)	Bulk density (g/l)	Regenerant concentration in water (%)
1.9	10	100	1–9	0.3–0.9 20–50	45–55	670	NaOH 4 NH_3 4 Na_2CO_3 4

Uses: Separation of strong and weak acids, adsorption of basic dyes in alkaline media, deionisation of process solutions in the chemical, pharmaceutical and food industries.

1.3	20	100	1–8	0.3–0.9 20–50	45	610–680	NaOH 8–12 or NH_3 2–4 or Na_2CO_3 2–4

Uses: Neutralisation of non-aqueous solutions. Deacidification and demineralisation of water and liquid from chemical processes, whereby only strong-acid anions are removed.

0.55	–	40(OH^-)	0–14	0.3–0.9 20–50	45–60	700	*
0.55	–	40 (OH^-)	0–14	0.3–0.9 20–50	50–60	700	

Uses: Complete demineralisation of water (Use of Cat. No. 15127 means that exhaustion of the exchanger resin can be signalled in the form of a colour change).

* Following demixing, regenerate the individual components in the manner stated for strong acid cation exchangers and strong base anion exchangers.

Table I.10. MERCK continued

Liquid ion exchanger LA-2	
Cat. No.	15263
Name	Ion exchanger Amberlite® LA-2
Structure	secondary amine
Molecular weight (g/mol)	353–395
Equivalent weight of neutralisation	360–380
Colour	Clear, amber-coloured solution
Odour	Faint, pleasant odour
Capacity (meq/ml)	2.2–2.3
Form as supplied	Free base
Solidification point (°)	10
Viscosity (25 °C) (mPa.s)	18
Density (g/l)	830
Regeneration	The exchanger is regnerated in the same way as the conventional exchanger resins, e. g., with sodium carbonate, sodium hydroxide or ammonia solutions. In some cases a slurry of lime or magnesium oxide can be used.
Notes	High molecular weight, liquid, secondary amine used as a 5–40% solution in hydrocarbons. Both the free base and the salts are of very low solubility in acid, neutral or alkaline aqueous solutions. On the other hand, solubility in aliphatic and aromatic hydrocarbons, high molecular weight alcohols and many other organic solvents is excellent.
Uses	Extraction of anions and anionogenic complexes from neutral and acid aqueous solutions in hydrometallurgy, the pharmaceutical industry and food industry. Isolation or separation of valuable acids, product clean-up.

Table I.10. MERCK continued

Adsorber resins

Cat. No.	15259	15256	15257	15258
Name	Adsorber resin Amberlite® XAD-2	Adsorber resin Amberlite® XAD-4	Adsorber resin Amberlite® XAD-7	Adsorber resin Amberlite® XAD-8
Copolymerisate	Styrene/divinyl-benzene	Styrene/divinyl-benzene	Acrylate	Acrylate
Degree of crosslinkage (%)	3	3	–	–
Polarity	Non-polar	Non-polar	Medium-polar	Medium-polar
Dipole moment according to DEBYE	0.3	0.3	1.8	1.8
Specific surface (m^2/g) (dry)	330	750	450	140
Mean pore diameter (A)	90	50	80	250
Pore volume (vol. %) (cm^3/g)	42	51	55	52
Colour	Whitish, opaque	Whitish, transparent	Whitish, opaque	Whitish, opaque
Shape	Spherical	Spherical	Spherical	Spherical
Particle size (mm) mesh ASTM	0.3 – 0.9 20 – 50	0.3 – 0.9 20 – 50	0.3 – 0.9 20 – 50	0.3 – 0.9 20 – 50
Maximum working temperature (°C)	250	250	150	150
Bulk density (g/l)	640 – 700	640 – 700	660	600 – 700
Form as supplied	containing $NaCl/Na_2CO_3$	containing $NaCl/Na_2CO_3$	containing $NaCl/Na_2CO_3$	containing $NaCl/Na_2CO_3$

The adsorber resins are synthetic, mechanically stable, insoluble polystyrene/divinylbenzene polymerisates having a macroreticular structure and non-ionic character. Unlike ion exchanger resins, they neither shrink nor swell.

Separation of surfactants, e. g., detergents, emulsifiers, wetting and dispersing agents. Removal of phenol and other organic substances from water. Isolation of vitamins and antibiotics.

These adsorber resins are synthetic, mechanically stable, insoluble polyacrylate polymerisates of medium polarity. They are suitable for adsorbing polar substances.

Recovery and concentration of water-soluble steroids, enzymes, alkaloids, polypeptides and proteins.

Regenerants:
1. Methanol or other organic solvents
2. Bases for weak acids
3. Acids for weak bases
4. Water in cases where adsorption is from ionogenic solutions
5. Hot water or steam for volatile substances.

Table I.11 Serdolit® Ion Exchange Resins for the laboratory
(SERVA Heidelberg GmbH & Co, Carl Benz-Str. 7, Postfach 105260, 6900 Heidelberg, FR Germany)

Name	Character	Functional group	Structure	Capacity (meq/ml)	Particle size (mm)	Moisture (%)	Temperature limit (°C)	Notes
Serdolit Cation Exchangers								
CS-2	strongly acidic	sulfonic	Styrene-X8/DVB	1.9	0.3 –0.8	44–48	120	1
CS-2	strongly acidic	sulfonic	Styrene-X8/DVB	1.9	0.1 –0.2	10–15	120	1
CS-2	strongly acidic	sulfonic	Styrene-X8/DVB	1.9	0.05–0.1	10–15	120	1
Red	strongly acidic	sulfonic	Styrene-X8/DVB	1.8	0.3 –1.2	44–48	120	2
Red Micro	strongly acidic	sulfonic	Styrene-X8/DVB	1.8	0.05–0.1	44–48	120	2
CW-10	weakly acidic	carboxylic	Polyacrylic acid	4.2	0.3 –0.8	44–48	120	3
CW-10	weakly acidic	carbocylic	Polyacrylic acid	4.2	0.1 –0.2	10–15	120	3
CW-10	weakly acidic	carboxylic	Polyacrylic acid	4.2	0.05–0.1	10–15	120	3
CW-18	weakly acidic	carboxylic	Polyacrylic acid	1.8	0.3 –0.8	62–67	120	4
CW-18	weakly acidic	carboxylic	Polyacrylic acid	1.8	0.1 –0.2	10–15	120	4
CW-18	weakly acidic	carboxylic	Polyacrylic acid	1.8	0.05–0.1	10–15	120	4
Serdolit Anion Exchangers								
AS-6	strongly basic	quaternary ammonium Type I	Polyacrylamide	1.3	0.3 –0.8	48–55	45	5
AS-6	strongly basic	quaternary ammonium Type I	Polyacrylamide	1.3	0.1 –0.2	10–15	45	5
AS-6	strongly basic	quaternary ammonium Type I	Polyacrylamide	1.3	0.05–0.1	10–15	45	5

Blue	strongly basic	quaternary ammonium Type I	Styrene-X8/DVB	1.2	0.3 –0.9	38–43	60	6
Blue Micro	strongly basic	quaternary ammonium Type I	Styrene-X8/DVB	1.2	0.05–0.1	38–43	60	6
AW-14	medium basic	polyfunctional amines	Polyacrylamide	2.3	0.3 –0.8	50–55	35	7
AW-14	medium basic	polyfunctional amines	Polyacrylamide	2.3	0.1 –0.2	10–15	35	7
AW-14	medium basic	polyfunctional amines	Polyacrylamide	2.3	0.05–0.1	10–15	35	7
Mixed Bed	strongly basic strongly acidic	quaternary ammonium Type I; sulfonic	Polyacrylamide/ Styrene-X8/DVB	1.2	0.3 –0.8	40–46	30	8
Mixed Bed Micro	strongly basic strongly acidic	quaternary ammonium Type I; sulfonic	Polyacrylamide/ Styrene-X8/DVB	1.2	0.05–0.1	40–46	30	8

Notes:

(1) Very stable against oxidation and irreversible adsorption
(2) With exhaustion indicator
(3) High stability and exchange capacity
(4) Very useful for proteins
(5) For adsorption of colored impurities in industrial applications
(6) With exhaustion indicator, color on exhaustion: yellow
(7) For removing nitrates and sulfates from sewage
(8) With double exhaustion indicator, for demineralizing water and other aqueous process liquids

Table I.12 SERVA Cellulose Ion Exchangers
(SERVA Heidelberg GmbH & Co, Carl Benz-Str. 7, Postfach 105260, 6900 Heidelberg, FR Germany)

Name	Character	Functional group	Structure	Capacity (meq/g)	Particle size (mm)	Moisture (%)	Notes
Cellulose Anion Exchangers							
AH	weakly basic	Aminohexyl	course fibers	0.3 –0.4	0.05–0.2	60	1
BD	weakly basic	Benzoylated DEAE	granular	0.35	0.05–0.3	75	2
BND	weakly basic	Benzoylated-Naphthoylated DEAE	granular	0.2	0.05–0.3	75	2
DEAE 23 SH	weakly basic	Diethylaminoethyl	fine fibers, high capacity	0.75–0.9	0.05–0.2	–	3
DEAE 23 SN	weakly basic	Diethylaminoethyl	fine fibers, low capacity	0.4 –0.55	0.05–0.2	–	4
DEAE 23 SS	weakly basic	Diethylaminoethyl	fine fibers, standard capacity	0.55–0.75	0.05–0.2	–	5
DEAE 28 GS	weakly basic	Diethylaminoethyl	course fibers	0.3 –0.55	0.5 –1	–	6
DEAE 32	weakly basic	Diethylaminoethyl	granular	0.9 –1.0	0.1 –0.2	–	7
DEAE 52	weakly basic	Diethylaminoethyl	granular preswollen	1.0	0.1 –0.2	70	8
DEAE 80	weakly basic	Diethylaminoethyl	bead form	1.0 ±0.1	0.05–0.15	–	9
ECTEOLA 23	weakly basic	unknown	fine fibers	0.3 –0.4	0.05–0.2	–	10
PAB 23	weakly basic	p-Aminobenzyl	fine fibers	0.15–0.2	0.05–0.2	–	11
PEI	weakly basic	Polyethylenimine	course fibers	0.1	0.05–0.1	–	12
TEAE	weakly basic	DEAE, Ethylbromide treated	fine fibers	0.55–0.75	0.05–0.2	–	13

Cellulose Cation Exchangers							
CM 23	weakly acidic	Carboxymethyl	fine fibers	0.6 ±0.1	0.05–0.2	–	14
CM 32	weakly acidic	Carboxymethyl	granular	1.0	0.1 –0.2	–	15
CM 52	weakly acidic	Carboxymethyl	granular preswollen	1.0	0.1 –0.2	70	16
CM 80	weakly acidic	Carboxymethyl	bead form	1.0 ±0.1	0.05–0.15	–	17
P 23	medium acidic	Phosphorylated	fine fibers	0.8 –0.9	0.05–0.2	–	18
P 80	medium acidic	phosphorylated	bead form	1.0 ±0.1	0.05–0.15	–	17
SE 23	strongly acidic	Sulfoethyl	fine fibers	0.2 –0.3	0.05–0.2	–	18

Notes:

(1) Affinity techniques
(2) For t-RNA
(3) High capacity for proteins, peptides, RNA
(4) Universal adsorbent
(5) High selectivity, for proteins
(6) Adsorption in industry
(7) Proteins, high flow rate
(8) Proteins, fast separations
(9) Proteins and nucleotides, fast separations
(10) Reaction product of epichlorohydrin, triethanolamine and alkaline cellulose, the chemical structure is not completely known; for nucleic acids
(11) Affinity techniques (diazotizable)
(12) Oligonucleotides
(13) Esterification of carboxylic groups, quaternarization probably negligible; for acidic proteins
(14) Basic proteins
(15) High capacity for basic proteins
(16) Basic proteins, peptides
(17) Basic proteins, fast separations
(18) Very basic proteins

Table I.13 S & S Cellulose based Ion Exchangers
(Schleicher & Schuell, P.O. Box 4, 3354 Dassel, FR Germany)

Product name	Ion-exchanging group	Type	Capacity meq/g
NA 49 (DEAE-Membrane Filter)	$-O-C_2H_4-N(C_2H_5)_2$	AA	min 0,7
DEAE-Cellulose (Powder)	$-O-C_2H_4-N(C_2H_5)_2$	AA	min 0,7
NA 49 (CM-Cellulose-Membrane Filter)	$-O-CH_2-COOH$	CA	min 0,7
CM-Cellulose (Powder)	$-O-CH_2-COOH$	CA	min 0,7

Remarks:
CA: Cation exchanger; AA: Anion exchanger.
All listed ion exchangers are weak acid or weak base by design.

Table I.14 MN Ion Exchange Products
Manufacturer: Macherey–Nagel, 5160 Düren, Neumann-Neander-Str. 6–8, Postfach 307
Distributors: US: Brinkmann Instruments, Cantiague Road, Westbury, N.Y. 11590
UK: CAMLAB (Glass) LTD., Cambridge/Great Britain
Cellulose ion exchangers

Name	Designation of the CC grade[1]	Exchange group	Type[2]	Capacity meq/g	Designation of the TLC grade[3]	Designation of the exchanger-coated POLYGRAM® ready-to-use film
Carboxymethylcellulose	MN 2100 CM	$-O-CH_2-COO-$	C	0.7	–	–
Diethylaminoethylcellulose	MN 2100 DEAE	$-O-C_2H_4-N(C_2H_5)_2$	A	0.7	MN 300 DEAE	CEL 300 DEAE
ECTEOLA cellulose	MN 2100 ECTEOLA	unknown	A	0.35	MN 300 ECTEOLA	–
Polyethylenimine cellulose	MN 2100 PEI	$(-NH-CH_2-CH_2-)n$	A	1.0	MN 300 PEI	CEL 300 PEI

[1]CC = column chromatography.
[2]A = anion exchanger; C = cation exchanger.
[3]TLC = thin-layer chromatography.

NUCLEOSIL® and POLYGOSIL® modified silica gels

Type of modification	Active group	Organic Si-C group	Preferred application	Supply program Item	Particle size	pH stability at room temp.
NH_2	Amino	$-(CH_2)_3-NH_2$	weak base anion exchanger; analysis of anions and organic acids	NUCLEOSIL 5 NH_2	5 ± 1.5	1–9
				NUCLEOSIL 10 NH_2	5 ± 1.5	1–9
				NUCLEOSIL 120-7NH_2	7 ± 1.5	1–9
				NUCLEOSIL 300-7NH_2	7 ± 1.5	1–9
				POLYGOSIL 60-5NH_2	5	1–9
				POLYGOSIL 60-10NH_2	10	1–9
$N(CH_3)_2$	Dimethyl amino	$-(CH_2)_3-N(CH_3)_2$	weak base anion exchanger; same as NH_2	NUCLEOSIL 5 $N(CH_3)_2$	5 ± 1.5	1–9
				NUCLEOSIL 10 $N(CH_3)_2$	10 ± 1.5	1–9
				POLYGOSIL 60-5 $N(CH_3)_2$	5	1–9
				POLYGOSIL 60-10$N(CH_3)_2$	10	1–9
SA	Sulfonic acid	$-(CH_2)_3-SO_3Na$	strong acid cation exchanger; ion exchange chromatography; capacity ca. 1 mequ/g	NUCLEOSIL 5 SA	5 ± 1.5	1–9
				NUCLEOSIL 10 SA	10 ± 1.5	1–9
SB	Quaternary ammonium groups	$-(CH_2)_3-N(CH_3)_3Cl$	strong base anion exchanger; ion exchange chromatography; capacity ca. 1 mequ/g	NUCLEOSIL 5 SB	5 ± 1.5	1–9
				NUCLEOSIL 10 SB	10 ± 1.5	1–9

Remarks: The NH_2 packing is a multipurpose product with excellent chromatography utility in 3 different modes: normal phase, weak anion exchange and reversed-phase of polar compounds such as carbohydrates. In normal phase (using hexane, CH_2Cl_2 and isopropanol as mobile phases) the NH_2 packing separates polar compounds such as substituted anilines, esters, chlorinated pesticides etc. Anions and organic acids are analyzed in the ion exchange mode using common buffers (e. g., acetates, phosphates) in conjunction with organic modifiers (e. g., acetonitrile).

Table I.14 MN Ion Exchange Products continued

NUCLEOGEN® columns for the separation of biopolymers

Cat. No.	Column Type[3])	Pore dia. Å	Particle size µm	Approx. binding capacity[2])	Column ID × length mm
718 596	NUCLEOGEN®-DEAE 60-7	60	7	300 A_{260}	4 × 125
718 597	NUCLEOGEN® DEAE 60-7	60	7	1875 A_{260}	10 × 125
718 598	NUCLEOGEN®-DEAE 500-7	500	7	730 A_{260}	6 × 125
718 599	NUCLEOGEN®-DEAE 500-7	500	7	1940 A_{260}	10 × 125
718 600	NUCLEOGEN®-DEAE 500-7 IWC[1])	500	7	1940 A_{260}	10 × 125
718 601	NUCLEOGEN®-DEAE 4000-7	4000	7	120 A_{260}	6 × 125
718 602	NUCLEOGEN®-DEAE 4000-7	4000	7	350 A_{260}	10 × 125
718 603	NUCLEOGEN®-DEAE 4000-7 IWC[1])	4000	7	350 A_{260}	10 × 125
718 594	NUCLEOGEN® guard column				4 × 30
721 202	Guard column holder without guard column				–

[1]) Columns with inner wall coated with PTFE

[2]) A_{260}: Absorption unit at 260 nm; it corresponds to 40 µg of nucleic acid.
Type 60-7: determined with a mixture of Oligo(rA)$_n$; optimum resolution is obtained up to 50% of the binding capacity.
Type 500-7: determined with tRNA; optimum resolution is obtained up to 40% of the binding capacity.
Type 4000-7: determined with plasmid DNA; optimum resolution is obtained up to 40% of the binding capacity.

[3]) Each column is supplied with an individual test chromatogram and an operation manual.

Table I.15 Pharmacia Sephadex, Sephacell and Sepharose Ion Exchangers
(Pharmacia Fine Chemicals AB, Box 175, S-75104 Uppsala 1, Sweden)

Name	Exchanger type	Exclusion limit (globular protein)	Total capacity mmol/g	Recommended working pH range	pH stability	Cleaning and storage
DEAE Sephadex A-25	weak anion	$3 \cdot 10^4$	3 –4	2– 9	2–13	Autoclavable (out of column), 120 °C, pH 7.0, 30 min. in salt form. Regenerate and wash in the column.

CM Sephadex C-25	weak cation	$3 \cdot 10^4$	4 –5	6–10	2–13	Regenerate with high salt concentration.
QAE Sephadex A-25	strong anion	$3 \cdot 10^4$	2.6–3.4	2–10	2–13	Remove severe contamination by washing with one column volume of 0.1 M NaOH. Then wash with buffer until free from alkali.
SP Sephadex C-25	strong cation	$3 \cdot 10^4$	2.0–2.6	2–10	2–13	Store unused (swollen) gel in salt form in the presence of a suitable antimicrobial agent, e. g., 20% ethanol or 0.002% chlorohexidine (for anion exchangers) or 0.005% thiomerosal (for cation exchangers).

Remarks: Sephadex A-25, C-25: rigid spherical particles based on the crosslinked dextran matrix of Sephadex G-25. Supplied as a dry powder – to be swollen before use. Capable of very high flow rates and good separation. Designed for purification of low molecular weight proteins, polypeptides and nucleotides.

Chemical stability: Insoluble in all common solvents unless degraded.
Stable in water and aqueous salt solutions.
Can be used with non-ionic detergents.
Can be used with denaturing solvents, e. g., 8 M urea, 6 M GuHCl.
Stable in organic solvents.
Degraded by strong oxidizing agents.
Degraded by dextranases.

DEAE Sephadex A-50	weak anion	$2 \cdot 10^5$	3 –4	2– 9	2–10	Same as above, resp.
CM Sephadex C-50	weak cation	$2 \cdot 10^5$	4 –5	6–10	2–10	Same as above, resp.
QAE Sephadex A-50	strong anion	$2 \cdot 10^5$	2.6–3.4	2–10	2–10	
SP Sephadex C-50	strong cation	$2 \cdot 10^5$	2.0–2.6	2–10	2–10	

Remarks: Sephadex A-50, C-50: spherical particles, based on the crosslinked dextran matrix of Sephadex G-50. More porous but less rigid than A-25 or C-25, allowing considerably lower flow rates. Supplied as dry powder – to be swollen before use. Ideally suited to batch separations and early steps in a purification scheme. Matrix of choice where gel life is limited, e. g., radioactive contamination, crude samples, etc.

Chemical stability: As above.

Table I.15 Sephadex continued

Name	Exchanger type	Exclusion limit (globular protein)	Total capacity mmol/g	Recommended working pH range	pH stability	Cleaning and storage
DEAE Sephacell	weak anion	$1 \cdot 10^6$	95–135 (μmol/ml)	2–9	2–9	Autoclavable (out of column); 120 °C; pH 7.0; 30 min salt form. Regenerate and wash in the column. Regenerate with high salt concentrations. Store unused gel (swollen) in salt form in the presence of a suitable antimicrobial agent, e. g., 20% ethanol.
Remarks: Sephacel is based on crosslinked cellulose. Beaded-form avoids fines commonly associated with microgranular exchangers. Supplied pre-swollen, ready-to-use (no precycling required). High capacity, particularly for large molecules.				Chemical stability: Insoluble in all common solvents unless degraded. Stable in water and aqueous salt solutions. Can be used with non-ionic detergents. Can be used with denaturing solvents, e. g., 8 M urea, 6 M GuHCl. Degraded by strong oxidizing agents. Degraded by enzymes that hydrolyze β-glucosidic linkages.		
DEAE Sepharose CL-6B	weak anion	$4 \cdot 10^6$	130–170	2– 9	2–14	Autoclavable (out of column); 120 °C; pH 7.0; 30 min salt form. Regenerate and wash in the column. Regenerate with high salt concentration. Remove severe contamination by washing with one column volume of up to 1.0 M NaOH. Then wash with buffer until free from alkali. Store unused gel (swollen) in salt form in the presence of a suitable antimicrobial agent, e. g., 20% ethanol.
CM Sepharose CL-6B	weak cation	$4 \cdot 10^6$	100–140	6–10	2–14	

Remarks: Sepharose CL-6B porous, agarose-based matrix in beaded-form; crosslinked for high stability. High capacity and low non-specific effects. Supplied preswollen, ready-to-use. High exclusion limit renders it ideal for the routine separation of large biomolecules. Particularly stable during long term applications.				Chemical stability: Insoluble in all common solvents unless degraded. Stable in water and aqueous salt solutions. Can be used with non-ionic detergents. Can be used with denaturing solvents, e. g., 8 M urea, 6M GuHCl. Stable in organic solvents. Resist microbial attack. Degraded by strong oxidizing agents.		
DEAE Sepharose Fast Flow	weak anion	$4 \cdot 10^6$	100–160	2– 9	2–14	Autoclavable (out of column); 120 °C; pH 7.0; 30 min salt form. Regenerate with high salt concentration.
CM Sepharose Fast Flow	weak cation	$4 \cdot 10^6$	90–130	6–10	2–14	Regenerate and wash in column. Remove severe contamination by washing with one column volume of 1.0 M NaOH.
Q Sepharose Fast Flow	strong anion	$4 \cdot 10^6$	180–250	3–11	2–14	Then wash with buffer until free from alkali.
S Sepharose Fast Flow	strong cation	$4 \cdot 10^6$	180–250	3–11	3–11	Store unused gel (swollen) in salt form in presence of a suitable antimicrobial agent, e. g., 20% ethanol.
Remarks: Sepharose Fast Flow beaded agarose matrix; highly crosslinked for extreme chemical and physical stability. First choice for industrial applications since rapid loading and cleaning/regeneration procedures can be used. Also ideal for preparative scale laboratory work, when high resolution can be achieved at the expense of speed. Q and S Sepharose Fast Flow are optimized for protein separations. They can be used at higher ionic strength than DEAE- or CM-substituted exchangers and frequently give higher selectivity.				Chemical stability: As above.		

Table I.16 Wessalith Zeolite Ion Exchange Material
(Degussa AG, Postfach 1345, 6450 Hanau 1, FR Germany)

1. WESSALITH P (Powder)

Average particle size		μm	3.5
Sieve residue[1]) (by Mocker, 45 μm sieve)		%	0.1
Density		g/cm^3	2.0
Tapped density[2])		g/l	500
Ignition loss[3]) (1 hour at 800 °C)		%	20
Relative brightness A		%	94
pH value (aqueous suspension of 5%)			11.6
Average chemical composition:	Na_2O	%	17
	Al_2O_3	%	28
	SiO_2	%	33
	H_2O	%	22
Impurities:	As[6])	ppm	< 3
	Ions of heavy metal as Pb[6])	ppm	< 15
	Fe total	ppm	<300
	Fe free	ppm	< 50
Solubility in:	water		none
	alkali		none
	acid		soluble under decomposition
Calcium binding capacity		mg CaO/g atro[7])	160

2. WESSALITH S (Slurry)

Average particle size	μm	3.5
Density	g/cm^3	1.5
Dry substance (1 hour at 800 °C)	%	49
Sieve residue (by Mocker, 45 μm sieve)	%	0.1
Contents of stabilizer	%	3.1 (atro) = 1.5 (telquel)
pH value[5])		11.9

The stabilizer is a oxoalcohol 6 EO.

3. WESSALITH-Compounds

		CS	CN	CD
WESSALITH (atro)[7])	%	76.0	70.4	71.0
Na_2SO_4	%	2.9	–	–
NTA	%	–	8.0	–
DEGAPAS 4104 N	%	–	–	5.4
Non-ionic surfactant	%	2.6	2.4	2.4
NaOH	%	0.5	0.5	0.5
CMC	%	2.0	1.0	–
Na-silicate	%	–	–	0.2
Water	%	16.0	17.7	20.5
Average particle size[8])	μm	320	280	260
Range of 90%	μm	130–540	105–500	110–600
Sieve residue (by Mocker, 45 μm Sieve)	%	0.1	0.1	0.2
Apparent density	g/l	490	510	540
Ignition loss (1 hour at 800 °C)	%	19.1	27	25.6
Relative brightness A[4])		94	94	94
pH value[5]) (aqueous suspension of 5%)		11.0	10.9	11.2
Calcium binding capacity	mg CaO/g atro	155	177	204
Trace impurities	see WESSALITH			

[1]) acc. to DIN ISO 787/XVII or JIS K 5101/20
[2]) acc. to DIN ISO 787/XI or JIS K 5101/18
[3]) acc. to DIN 55921, ASTM D 1208 or JIS K 5101/23
[4]) acc. to DIN 53163
[5]) acc. to DIN ISO 787/IX, ASTM D 1208 or JIS K 5101/24
[6]) acc. to the Food Chemicals Codex Determination Method
[7]) atro = absolutely dry or ignited, active substance
[8]) distribution by weight

Table I.17. NEOSEPTA® ion exchange membranes

TOKUYAMA SODA

System Equipment Department
4 – 5, 1-chome Nishi-Shimbashi, Minato-ku, Tokyo Japan.
Phone: 03-597-5120 Telex: 222-3258 FAX: 03-597-5126

Properties of NEOSEPTA® of standard grades

Name	CL-25T	CM-1	CM-2	AM-1	AM-2	AM-3
Type	Strongly acidic cation permeable	Strongly acidic cation permeable	Strongly acidic cation permeable	Strongly basic anion permeable	Strongly basic anion permeable	Strongly basic anion permeable
			Little diffusion coefficient			Little diffusion coefficient
	Na-form	Na-form	Na-form	Cl-form	Cl-form	Cl-form
Electric resistance	2.2 ~ 3.0	1.2 ~ 2.0	2.0 ~ 3.0	1.3 ~ 2.0	2.0 ~ 3.0	3.0 ~ 4.0
Transport number						
• Total cation or anion	0.98 <	0.98 <	0.98 <	0.98 <	0.98 <	0.98 <
• $Na^+ + K^+$	0.70	0.70	0.70	} 0.02 >	} 0.02 >	} 0.02 >
• $Ca^{++} + Mg^{++}$	0.28	0.28	0.28			
• Cl^-	} 0.02 >	} 0.02 >	} 0.02 >	} 0.98 <	} 0.98 <	} 0.98 <
• SO_4^{--}						
Burst strength	3 ~ 5	3 ~ 5	3 ~ 5	3 ~ 5	3 ~ 5	3 ~ 5
Water content	0.25 ~ 0.35	0.35 ~ 0.40	0.25 ~ 0.35	0.25 ~ 0.35	0.20 ~ 0.30	0.15 ~ 0.25
Exchange capacity	1.5 ~ 1.8	2.0 ~ 2.5	1.6 ~ 2.2	1.8 ~ 2.2	1.6 ~ 2.0	1.3 ~ 2.0
Thickness	0.15 ~ 0.17	0.13 ~ 0.16	0.13 ~ 0.16	0.13 ~ 0.16	0.13 ~ 0.16	0.13 ~ 0.16
Reinforcing	yes	yes	yes	yes	yes	yes
Standard size (m)	1.00 × 1.50	1.00 × 1.50	1.00 × 1.50	1.00 × 1.50	1.00 × 1.50	1.00 × 1.50

Properties of NEOSEPTA® of special grades

Name	CMS	ACS	AFN	ACLE-5P	CLE-E	C66-10F
Type	Strongly acidic cation permeable	Strongly basic anion permeable	Strongly basic anion permeable	Strongly basic anion permeable	Strongly acidic cation permeable	Strongly acidic cation permeable
	Na-form	Cl-form	Cl-form	Cl-form	Na-form	Na-form
Properties	Mono-cation permselective	Mono-anion permselective	Diffusion-dialysis	Operable in high pH solution	Cation permselective	Application suggested for electrolysis of organics

			Resistant against organic fouling	Resistant against organic fouling	Resistant against organic solvent	
					Operable in high temperature solution	
Electric resistance	1.5 ~ 2.5	2.0 ~ 2.5	0.4 ~ 1.5	15 ~ 25	15 ~ 25	3.8 ~ 5.3
Transport number						
• Total cation or anion	0.98 <	0.98 <	0.98 <	0.98 <	0.98 <	0.98 <
• $Na^+ + K^+$	0.90 <	} 0.02 >	} 0.02 >	} 0.02 >	} 0.98 <	} 0.98 <
• $Ca^{++} + Mg^{++}$	0.10 <					
• Cl^-	} 0.02 >	0.98 <	} 0.98 <	} 0.98 <	} 0.02 <	} 0.02 <
• SO_4^{--}		0.005 >				
Burst strength	3 ~ 4	4 ~ 6	4 ~ 6	8 ~ 10	8 ~ 10	6 ~ 8
Water content	0.35 ~ 0.45	0.20 ~ 0.30	0.40 ~ 0.55	0.2 ~ 0.3	0.3 ~ 0.4	0.3 ~ 0.4
Exchange capacity	2.0 ~ 2.5	1.4 ~ 2.0	2.0 ~ 3.5	1.3 ~ 2.0	1.3 ~ 1.8	1.7 ~ 2.2
Thickness	0.14 ~ 0.17	0.15 ~ 0.20	0.15 ~ 0.20	0.20 ~ 0.30	0.8 ~ 1.3	0.25 ~ 0.35
Reinforcing	yes	yes	yes	yes	yes	yes
Standard size (m)	1.00 × 1.50	1.00 × 1.50	1.00 × 1.50	1.00 × 1.50	1.00 × 1.50	1.00 × 1.50

• Electric resistance: Equilibrated with 0.5N NaCl solution, at 25 °C [Ω-cm²]
Transport number: Measured by electrophoresis with sea water
Current density: 2 [A/dm²], at 25 °C
• Burst strength: [kg/cm²]
• Water content: Equilibrated with 0.5N NaCl solution [gH_2O/gNa-form dry membrane (or Cl-form)]
• Exchange capacity: [meq/gNa-form dry membrane (or Cl-form)]
• Thickness: [m/m]

Measurement basis
• Electric resistance: Equilibrated with 0.5N NaCl solution, at 25 °C [Ω-cm²]
• Transport number: Measured by electrophoresis with sea water
Current density: 2 [A/dm²], at 25 °C
• Burst strength: [kg/cm²]
• Water content: Equilibrated with 0.5N NaCl solution
[gH_2O/gNa-form dry membrane (or Cl-form)]
• Exchange capacity: [meq/gNa-form dry membrane (or Cl-form)]
• Thickness: [m/m]

Appendix II
Computing tables for practical application

Table II.1	Constants of common chemicals used in water chemistry and ion exchange
Table II.2	Conversion of weights
Table II.3	Conversion of volumes
Table II.4	Conversion of densities and concentrations
Table II.5	Chemical equivalents
Table II.6	Conversion table for water hardness units
Table II.7	pH titration indicators and determination of hydroxides/carbonates/bicarbonates by titration of the p- and m-values
Table II.8	Consumption of potassium permanganate by various substances
Table II.9	Conductivity, resistivity, and approximate electrolyte content of deionized or distilled water at 25 °C
Table II.10	Conductivity, resistivity, and approximate electrolyte content of deionized or distilled water at 25 °C
Table II.11	Concentrations and densities of solutions used as regenerants (sodium chloride, hydrochloric acid, sulfuric acid, caustic soda, ammonia, sodium carbonate)

Table II.1 Constants of common chemicals used in water chemistry and ion exchange

Name	Formula	Molecular weight	Equivalent weight
Ions			
Aluminum	Al^{3+}	27	9
Barium	Ba^{2+}	137.4	68.7
Bicarbonate	HCO_3^-	61	61
Calcium	Ca^{2+}	40.1	20
Carbonate	CO_3^{2-}	60	30
Chlorine	Cl^-	35.5	35.5
Hydroxyl	OH^-	17	17
Iron(II)	Fe^{2+}	55.8	27.9
Iron(III)	Fe^{3+}	55.8	18.6
Magnesium	Mg^{2+}	24.3	12
Manganese	Mn^{2+}	54.9	27.5
Phosphate	PO_4^{3+}	95	31.5
Sodium	Na^+	23	23
Sulfate	SO_4^{2-}	96.1	48
Bases			
Aluminum oxide	Al_2O_3	102	17
Aluminum hydroxide	$Al(OH)_3$	78	26
Ammonia	NH_3	17	17
Ammonium hydroxide	NH_4OH	35	35
Barium oxide	BaO	153.4	76.7
Barium hydroxide	$Ba(OH)_2$	171.3	85.6
Calcium oxide	CaO	56	28
Calcium hydroxide	$Ca(OH)_2$	74.1	37
Iron(II) oxide	FeO	71.8	35.9
Iron(III) oxide	Fe_2O_3	159.6	26.6
Iron(II) hydroxide	$Fe(OH)_2$	89.9	44.9
Iron(III) hydroxide	$Fe(OH)_3$	106.8	35.5
Iron(II,III) oxide	Fe_3O_4	231.4	—
Magnesium oxide	MgO	40	20
Magnesium hydroxide	$Mg(OH)_2$	58.3	29
Manganese oxide	MnO	71	35.5
Manganese dioxide	MnO_2	87	21.7
Sodium oxide	Na_2O	62	31
Sodium hydroxide	$NaOH$	40	40
Acids			
Carbon dioxide	CO_2	44	22
Carbonic acid	H_2CO_3	62	31
Hydrochloric acid	HCl	36.5	36.5
Hydrochlorous acid	$HClO$	52.5	52.5
Hydrofluoric acid	HF	20	20
Nitric acid anhydride	N_2O_5	108	54
Nitric acid	HNO_3	63	63
Nitrous anhydride	N_2O_3	76	19
Nitrous acid	HNO_2	36.5	36.5
Phosphoric anhydride	P_2O_5	142	23.7
Phosphoric acid	H_3PO_4	98	32.7
Silicic anhydride	SiO_2	60	30
Silicic acid	H_2SiO_3	78	39
Sulfuric anhydride	SO_3	80	40
Sulfuric acid	H_2SO_4	98	49
Sulfurous anhydride	SO_2	64	32
Sulfurous acid	H_2SO_3	82	41

Table II.1 continued

Name	Formula	Molecular weight	Equivalent weight
Salts			
Aluminum chloride	$AlCl_3$	133.5	44.5
Aluminum sulfate	$Al_2(SO_4)_3$	342.4	57.4
	$Al_2(SO_4)_3 \cdot 18H_2O$	668.4	122.2
Ammonium carbonate	$(NH_4)_2CO_3$		
Ammonium phosphate	$NH_4H_2PO_4$	115	115
Ammonium sulfate	$(NH_4)_2HPO_4$	122	61
	$(NH_4)_2SO_4$	132	66
Barium bicarbonate	$Ba(HCO_3)_2$	259.5	124.8
Barium carbonate	$BaCO_3$	197.4	98.7
Barium chloride	$BaCl_2$	208.3	104.1
Barium sulfate	$BaSO_4$	233.5	116.7
Calcium bicarbonate	$Ca(HCO_3)_2$	162.1	81
Calcium carbonate	$CaCO_3$	100.4	50.2
Calcium chloride	$CaCl_2$	111	55.5
Calcium phosphate	$Ca_3(PO_4)_2$	310.2	103.4
Calcium silicate	$CaSiO_3$	116.1	58
Calcium sulfate	$CaSO_4$	136.1	68
Copper hydroxide	$Cu(OH)_2$	97.6	48.8
Copper sulfate	$CuSO_4$	159.6	79.8
Hydrazine	N_2H_4	32	–
Iron bicarbonate	$Fe(HCO_3)_2$	178	89
Iron(II) chloride	$FeCl_2$	126.9	63.4
Iron(III) chloride	$FeCl_3$	162.4	54.1
	$FeCl_3 \cdot 6H_2O$	270.4	90.1
Iron(II) sulfate	$FeSO_4$	141.8	75.9
	$FeSO_4 \cdot 7H_2O$	277.8	130.9
Iron(III) sulfate	$Fe_2(SO_4)_3$	399.7	66.6
Magnesium aluminate	$Mg_3(AlO_3)_2$	223	71.7
Magnesium bicarbonate	$Mg(HCO_3)_2$	146.3	73
Magnesium carbonate	$MgCO_3$	84.3	42.1
Magnesium chloride	$MgCl_2$	95.2	47.6
Magnesium phosphate	$Mg_3(PO_4)_2$	263	87.7
Magnesium silicate	$MgSiO_3$	100.4	50.2
Magnesium sulfate	$MgSO_4$	120.4	60.2
Manganese bicarbonate	$Mn(HCO_3)_2$	177	88.5
Manganese(II) hydroxide	$Mn(OH)_2$	89	44.5
Manganese(IV) hydroxide	$Mn(OH)_4$	123	30.7
Manganese(II,IV) oxide	Mn_2O_3	157.9	52.6
Manganese sulfate	$MnSO_4$	151	75.5
Potassium permanganate	$KMnO_4$	158	31.6
Sodium aluminate	Na_3AlO_3	144	48
	$Na_2Al_2O_4$	164	164
Sodium bicarbonate	$NaHCO_3$	84	84
Sodium bisulfate	$NaHSO_3$	104	104
Sodium carbonate	Na_2CO_3	106	53
Sodium chloride	$NaCl$	58.5	58.5
Sodium fluoride	NaF	42	42
Sodium hexametaphosphate	$(NaPO_3)_6$	612	612
Sodium hypochloride	$NaClO$	74.5	74.5
Sodium phosphate	NaH_2PO_4	120	120
	Na_2HPO_4	142	71
	Na_3PO_4	164	54.5
	$Na_3PO_4 \cdot 10H_2O$	344	114.7
Sodium silicate	Na_2SiO_4	122.1	61
Sodium sulfate	Na_2SO_4	142.1	71
Sodium sulfite	Na_2SO_3	126	63
Sodium thiosulfite	$Na_2S_2O_3$	135	67.5

Table II.2 Conversion of weights

	Grain gr	Gram g	Pound lb	Kilogram kg	Metric ton t	Long ton ton (long)
1 gr	1	0.064799	$1.4286 \cdot 10^{-4}$	$6.4799 \cdot 10^{-4}$	$6.4799 \cdot 10^{-8}$	$6.378 \cdot 10^{-8}$
1 g	15.4324	1	$2.2046 \cdot 10^{-3}$	10^{-3}	10^{-6}	$9.842 \cdot 10^{-7}$
1 lb	7,000	543.592	1	0.45359	$4.5359 \cdot 10^{-4}$	$4.464 \cdot 10^{-4}$
1 kg	$1.5432 \cdot 10^{4}$	10^{3}	2.20462	1	10^{-3}	$9.842 \cdot 10^{-4}$
1 t	$1.5432 \cdot 10^{7}$	10^{6}	2,204.62	10^{3}	1	0.9842
1 ton (long)	$1.5679 \cdot 10^{7}$	$1.01605 \cdot 10^{6}$	2,240	1,016.05	1.01605	1

Table II.3 Conversion of volumes

	Milliliter ml	Liter l	Gallon (US) gal (US)	Gallon (imp) gal (imp)	Cubic foot cu ft	Cubic meter m^3
1 ml	1	10^{-3}	$264.178 \cdot 10^{-6}$	$219.976 \cdot 10^{-6}$	$3.53157 \cdot 10^{-5}$	$1.000028 \cdot 10^{-6}$
1 l	10^{3}	1	$264.178 \cdot 10^{-3}$	$219.976 \cdot 10^{-3}$	0.0353157	$1.000028 \cdot 10^{-3}$
1 gal (US)	3,785.33	3.78533	1	0.832680	0.133681	$3.78543 \cdot 10^{-3}$
1 gal (imp)	4,545.96	4.54596	1.20094	1	0.160544	$4.54609 \cdot 10^{-3}$
1 cu ft	28,316.1	28.3161	7.48047	6.22884	1	$28.3168 \cdot 10^{-3}$
1 m^3	999,972	999.972	264.170	219.969	35.3147	1

Table II.4 Conversion of densities and concentrations

	Parts per million ppm	Grain per gal (imp) gr/gal (imp)	Grain per gal (US) gr/gal (US)	Gram per liter g/l	Pound per cubic foot lb/cu ft	Gram per cubic centimeter g/cm^3
1 ppm	1	0.070157	0.058418	0.001000	$6.2426 \cdot 10^{-5}$	10^{-6}
1 gr/gal (imp)	14.2538	1	0.832680	0.0142538	$8.8983 \cdot 10^{-4}$	$1.42538 \cdot 10^{-5}$
1 gr/gal (US)	17.118	1.20094	1	0.017118	0.0010686	$1.7118 \cdot 10^{-5}$
1 g/l ($\equiv kg/m^3$)	10^{3}	70.157	58.418	1	0.062428	10^{-3}
1 lb/cu ft	$1.60189 \cdot 10^{4}$	$1.1238 \cdot 10^{3}$	935.757	16.0189	1	0.0160189
1 g/cm^3	10^{6}	$70.157 \cdot 10^{3}$	$5.8418 \cdot 10^{4}$	10^{3}	62.426	1

Table II.5 Chemical equivalents

1 lb as $CaCO_3$	= 9.08 gequ	1 gequ	= 0.11 lb as $CaCO_3$
1 kg as $CaCO_3$	= 1.3 gequ	1 gequ	= 0.77 kg as $CaCO_3$
1 lb/ft^3 as $CaCO_3$	= 0.32 gequ/l	1 gequ/l	= 3.12 kg/ft^3 as $CaCO_3$
1 g/gal as $CaCO_3$	= 0.286 mequ/l	1 mequ/l	= 35.0 g/gal as $CaCO_3$

Table II.6 Conversion table for water hardness units

	US hardness degree	British hardness degree	French hardness degree	German hardness degree	Russian hardness degree	ppm as $CaCO_3$
1° US	1	1.201	1.716	0.961	6.864	17.16
1° British	0.8324	1	1.429	0.7999	5.714	14.29
1° French	0.5828	0.700	1	0.5599	4.000	10.00
1° German	1.041	1.250	1.786	1	7.144	17.85
1° Russian	0.1457	0.175	0.2500	0.1400	1	2.500
1 ppm as $CaCO_3$	0.05828	0.070	0.1000	0.4000	0.4000	1

Expressions of ionic concentrations are based on:
United States: parts per million or grains per US gallon
United Kingdom: Clark; equivalent to a grain per British Imperial gallon
France: parts per 100 000 parts of water
Germany: parts of CaO per 100 000 parts of water.

Table II.7 pH Titration indicators and determination of hydroxides/carbonates/bicarbonates by titration of the p- and m-values

Alkalinity and acidity used as terms in water analysis may not be in accordance with the generally accepted terminology with a neutral point at pH 7. In water analysis, a pH of about 4.5 is frequently the end point for titration of alkalinity and a pH of about 8.2 for acidity.

Indicator	Range		End point	
	pH	color	pH	color
Phenolphthalein	8.0 to 10.0	colorless-red	8.2	pink
Methyl orange	3.2 to 4.4	pink-yellow	4.2	pink-orange
Methyl red	4.2 to 6.2	pink-yellow	5.5	orange
Bromcresol green	4.0 to 5.4	yellow-blue	4.5	green
Bromphenol blue	3.0 to 4.6	yellow-blue	3.7	green

Titration of 100 cm^3 water against phenolphthalein results in the p-value (in cm^3 n/10 acid) and titration of 100 cm^3 water against methyl orange results in the m-value (in cm^3 n/10 acid). The p- and m-values permit the determination of hydroxides, carbonates and bicarbonates in water.

Values	Contents in the water					
	mg/l hydroxides		mg/l carbonates		mg/l bicarbonates	
	NaOH	$Ca(OH)_2$ as CaO	Na_2CO_3	$CaCO_3$ as CaO	$NaHCO_3$	$Ca(HCO_3)_2$ as CaO
p = m	p · 40	p · 20	0	0	0	0
2p > m	(2p-m) · 40	0	(m-p) · 106	(m-p) · 56	0	0
2p = m	0	0	p · 106	p · 56	0	0
2p < m	0	0	p · 106	p · 56	(m-2p) · 84	(m-2p) · 28
p = 0; m	0	0	0	0	m · 84	m · 28

The relations of this table are not applicable if the p- and m-values are below 1 mequ/l. It is further assumed that the water does not contain other compounds influencing the p- and m-values as determined.

Table II.8 Consumption of potassium permanganate by various substances

Sometimes in the literature a relation is given between the suitability of an anion exchange resin and the N ratio as defined by:

$$N = \frac{\text{Organic matter (as } KMnO_4 \text{ consumption in mg/l)}}{\text{equivalent mineral acidity (in mequ/l)}}.$$

The following table should assist in the evaluation of the potassium permanganate consumption for various substances.

Substance 1 g/l	$KMnO_4$ consumption mg/l	Substance 1 g/l	$KMnO_4$ consumption mg/l
Urea	6	H_2S	1 850
Pyridine	100	NO_2^-	1 400
Humic acid	220	Fe^{2+}	570
Lactose	1 550		
Sucrose	2 700		
Cresol	6 200		
Phenol	10 000		

Table II.9 Conversion of flow rates

	Gallon (imp) per hour gal (imp)/h	Gallon (US) per minute gal (US)/min	Gallon (imp) per minute gal (imp)/min	Cubic meter per hour m^3/h	Cubic foot per minute cu ft/min	Liter per second l/s	Cubic foot per second cu ft/sec
1 gal (imp)/h	1	0.020016	0.016667	0.004546	0.0026757	0.0012628	0.0000446
1 gal (US)/min	49.9610	1	0.83268	0.22712	0.13368	0.06309	0.002228
1 gal (imp/min	60.00	1.20094	1	0.272765	0.16054	0.75766	0.0026757
1 m^3/h	219.969	4.4029	3.66615	1	0.588578	0.27777	0.0098096
1 cu ft/min	373.731	7.4806	6.22883	1.69901	1	0.47193	0.016667
1 l/s	791.912	15.8509	13.1985	3.600	2.11894	1	0.0353156
1 cu ft/sec	22,423.8	448.83	373.731	101.941	60.00	28.316	1

Table II.10 Conductivity, resistivity, and approximate electrolyte content of deionized or distilled water at 25 °C

Conductivity µmhos (Ω^{-1} cm^{-1} · 10^{-6})	Resistivity ohms (Ω cm)	Approximate electrolyte content ppm			
		NaCl	HCl	NaOH	CO_2
0.1	10,000,000	0.04	0.01		
0.2	5,000,000	0.08	0.02	0.03	
1	1,000,000	0.4	0.13	0.2	0.8
2	500,000	0.8	0.26	0.4	2.5
4	500,000	1.6	0.55	0.8	9.5
6	166,000	2.5	0.9	1	20
8	125,000	3.2	1.2	1.5	40
10	100,000	4	1.5	2	70
20	50,000	8	2	4	320
30	33,333	14	3	5	730
40	25,000	19	4	6	1,400
50	20,000	24	4.5	7	2,200
60	16,666	28	5.5		
70	14,286	33	6.5		
80	12,500	38	7.5	11	
90	11,111	43	8		
100	10,000	50	8	14	
200	5,000	100	18	27	

Table II.11 Concentrations and densities of solutions used as regenerants

Sodium chloride

Concentration		Density (20 °C)	Concentration		Density (20 °C)
% by weight	g/l		% by weight	g/l	
0.5	5.0	1.002	13	142.1	1.093
1	10.1	1.005	14	154.1	1.101
1.5	15.1	1.009	15	166.3	1.108
2	20.2	1.013	16	178.6	1.116
3	30.6	1.020	17	191.1	1.124
4	41.1	1.027	18	203.7	1.132
5	51.7	1.034	19	216.6	1.140
6	62.5	1.041	20	229.6	1.148
7	73.4	1.049	21	242.7	1.156
8	84.7	1.056	22	256.1	1.164
9	95.7	1.063	23	269.6	1.172
10	107.1	1.071	24	283.3	1.180
11	118.6	1.078	25	297.2	1.189
12	130.3	1.086	26	311.3	1.197

Remarks: Sodium chloride is applied as an 8−12% by weight NaCl solution for the regeneration of strong acid cation exchangers. Rock salt or common salt with specified granulometry is used, having a minimum NaCl content of 87% by weight and a water content <2% by weight. Content of solubles in distilled water: <0.1% by weight (common salt), <1% by weight (rock salt); sulfate content (expressed as SO_4^{2-}): <1% by weight; content of calcium and magnesium compounds (expressed as Ca^{2+}): <0.4% by weight; content of soluble iron compounds: not detectable; acid consumption: none.

Table II.11 continued

Hydrochloric acid

Concentration		Density (20 °C)	Concentration		Density (20 °C)
% by weight	g/l		% by weight	g/l	
0.5	5.0	1.002	19	207.1	1.093
1	10.0	1.003	20	210.6	1.098
1.5	15.1	1.006	21		1.103
2	20.2	1.008	22	243.8	1.108
3	30.4	1.013	23		1.113
4	40.7	1.018	24	268.4	1.118
5	51.1	1.023	25		1.123
6	61.7	1.028	26	293.5	1.128
7	72.3	1.033	27		1.134
8	83.0	1.038	28	319.8	1.139
9	93.8	1.043	28		1.144
10	104.8	1.048	30	344.8	1.149
11	115.8	1.053	31		1.154
12	126.9	1.058	32	371.0	1.159
13	138.1	1.063	33		1.164
14	149.5	1.068	34	397.6	1.169
15	160.9	1.073	35		1.174
16	172.4	1.078	36	424.5	1.179
17	184.1	1.083	37		1.184
18	195.8	1.088	38	451.7	1.189

Remarks: Hydrochloric acid is applied as a 5–7% by wt. HCl solution for the regeneration of strong acid cation exchangers. Purchased acid of higher percentage is tested for its concentration by a density spindle (DIN 12791) or after dilution by titration with 1 N NaOH. The technical pure quality must be free of waste acid and arsenic. Appearance: clear, colorless, at most slightly yellowish-green. HCl content: min. 30% by wt. Iron content: 0.002% by wt. (expressed as Fe^{3+}). Sulfate content: 0.5% by wt. (expresses ad SO_4^{2-}). Chloride (Cl_2): 0.01% by wt. Content of organic chlorine compounds: 0.02 g/l (expressed as Cl).

Table II.11 continued

Sulfuric acid

Concentration % by weight	Concentration g/l	Density (20 °C)	Concentration % by weight	Concentration g/l	Density (20 °C)
0.5	5.0	1.003	60	899.2	1.499
1	10.0	1.005	70	1127.4	1.610
1.5	15.1	1.008	80	1381.8	1.727
2	20.2	1.012	82	1434.3	1.749
3	30.6	1.018	84	1486.2	1.769
4	41.0	1.025	86	1537.0	1.787
5	51.6	1.032	88	1585.9	1.802
6	62.3	1.039	90	1633.0	1.814
7	73.2	1.045	91		1.820
8	84.2	1.052	92	1678.1	1.824
9	95.3	1.059	93		1.828
10	106.6	1.066	94	1721.3	1.831
11	118.0	1.073	95		1.834
12	129.6	1.080	96	1762.0	1.835
15	165.3	1.102	97		1.836
20	227.9	1.139	98	1799.4	1.836
50	697.6	1.395	100	1830.5	1.830

Remarks: Sulfuric acid is applied as a 1–2% by wt. H_2SO_4 solution for the regeneration of strong acid cation exchangers. Purchased acid of higher percentage is tested for its concentration by a density spindle (DIN 12791) or after dilution by titration with 0.5 N NaOH. The technical pure quality must be free of waste acid. Appearance: clear to slightly turbid, colorless to brownish, without deposits. H_2SO_4 content: min. 93% by weight.

Caustic soda

Concentration % by weight	Concentration g/l	Density (20 °C)	Concentration % by weight	Concentration g/l	Density (20 °C)
1	10.1	1.010	30	398.3	1.328
2	20.4	1.021	32	431.6	1.349
3	31.0	1.032	34	465.7	1.370
4	41.7	1.043	36	500.5	1.390
5	52.7	1.054	38	535.9	1.410
6	63.9	1.065	40	571.9	1.430
7	75.3	1.076	42		1.449
8	86.9	1.087	44		1.469
9	98.8	1.098	46		1.487
10	110.9	1.109	48		1.507
20	243.8	1.219	50	762.7	1.525

Remarks: Caustic soda is applied as a 4–5% by weight NaOH solution for the regeneration of strong base anion exchangers and as a 2–4% by wt. solution for the regeneration of weak base anion exchangers. Solutions of pure caustic soda with approx. 50% by wt. NaOH as purchased are diluted to 20% by wt., tested with a density spindle or titrated with 1 N HCl for actual concentration.

Table II.11 continued

Ammonia

Concentration % by weight	Concentration g/l	Density (20 °C)	Concentration % by weight	Concentration g/l	Density (20 °C)
0.5	5.0	0.996	12	114.0	0.950
1	9.9	0.993	14	132.0	0.943
1.5	14.9	0.992	16	149.8	0.936
2	19.8	0.990	18	167.3	0.930
3	29.6	0.985	20	184.6	0.923
4	39.2	0.981	22	201.6	0.916
5	48.8	0.977	24	218.4	0.910
6	58.4	0.973	26	235.0	0.904
7	67.8	0.969	28	251.4	0.898
8	77.2	0.965	30	267.6	0.892
9	86.5	0.961	32		0.886
10	95.8	0.957	34		0.881

Sodium carbonate

Concentration % by weight	Concentration g/l	Density (20 °C)	Concentration % by weight	Concentration g/l	Density (20 °C)
1	10.09	1.009	7	74.98	1.072
2	20.38	1.019	8	86.53	1.082
3	30.88	1.030	9	98.30	1.093
4	41.49	1.040	10	110.3	1.103
5	52.51	1.050	12	134.9	1.124
6	63.64	1.061	14	160.5	1.146

Appearance: clear and colorless. Quality requirements as percent of the NaOH content: Cl^- max. 0.2, oxidants (expressed as O_2) max. 0.001, SO_4^{2-} max. 0.1, SiO_2 max. 0.01, iron (expressed as Fe^{3+}) max. 0.001, aluminum (expressed as Al^{3+}) max. 0.005, and alkaline earth ions (expressed as Ca^{2+}) max. 0.01% by weight.

List of Contributors

Thomas Arden, DSc, FRSC
Consultant
11, Icklingham Road
Cobham
Surrey, KT11 2NG
Great Britain

Dr. Michael Baacke
Degussa AG
Fachbereich Forschung Chemie
Postfach 1345
D-6450 Hanau 1
FRG

Ortwin Bobleter
Professor of Radiochemistry
Institut für Radiochemie
Leopold Franzens Universität
Innrain 52 a
A-6020 Innsbruck
Austria

Dr. Günther Bonn
Assistant Professor
of Analytical Chemistry
Institut für Radiochemie
Leopold Franzens Universität
Innrain 52 a
A-6020 Innsbruck
Austria

Prof. Dr.-Ing. A. Bursik
Großkraftwerk Mannheim AG
Aufeldstraße 23
Postfach 184
D-6800 Mannheim 24
FRG

Dr. J. Frederick Dechow
Executive Director, Development
BioCryst, Inc.
1075 13th Street South
Birmingham, Alabama 35205
USA

Dr. Konrad Dorfner
Von der Au-Straße 48
D-6100 Darmstadt
FRG

Dr. Hans-Curt-Flemming
Section Chief Biofouling
Institut für Siedlungswasserbau,
Wassergüte- und Abfallwirtschaft
der Universität Stuttgart
Bandtäle 1
D-7000 Stuttgart 80
FRG

Dr. Harold C. Fravel
The Dow Chemical Company
Larkin Laboratory
Separation and Process Systems
Midland, Michigan 48674
USA

Dr. Nikolaus Grubenhofer
Serva Feinbiochemie
Postfach 105260
D-6900 Heidelberg 1
FRG

Dr. Friedrich G. Helfferich
Professor of Chemical Engineering
Department of Chemical Engineering
158 Fenske Lab
The Pennsylvania State University
University Park, Pa 16802
USA

Dr. George P. Herz
Consultant
20 Northfield Road
Maidenhead SL6 7JP
Berkshire
Great Britain

Erik Högfeldt
Associate Professor em.
Department of Inorganic Chemistry
The Royal Institute of Technology
S-100 44 Stockholm
Sweden

Dr.-Ing. habil. Wolfgang H. Höll
Section Chief Ion Exchange
Kernforschungszentrum
Karlsruhe GmbH
Postfach 3640
D-7500 Karlsruhe
FRG

Dr. Yng-Long Hwang
Department of Chemical Engineering
158 Fenske Lab
The Pennsylvania State University
University Park, Pa 16802
USA

Dr. Janos Inczedy
Professor of Chemistry
University of Chemical Engineering
Department of Analytical Chemistry
Schönherz Z.u. 12
H-8201 Veszprem
Hungary

Hideo Kawate
Assistant Director
Asahi Chemical Industry Co., Ltd.
Ion Exchange
Membrane Administration
The Imperial Tower
1-1 Uchisaiwaicho 1-chome
Chiyoda-ku
Tokyo 100
Japan

Dr. Akos Kiss
Degussa AG
Fachbereich Forschung
Anorganische Chemie
Postfach 1345
D-6450 Hanau 1
FRG

Dr. A. Günter Kühne
Quality Control of Ion Exchangers
BAYER AG, Leverkusen
Ion Exchange Application
Morgengraben 10
D-5000 Köln 80
FRG

Dr. Rober Kunin
Consultant
1318 Moon Drive
Yardley, Pa 19067
USA

Dr. Heinrich Lieser
Professor of Chemistry
Technical University
D-6100 Darmstadt
FRG

Dr. Friedrich Martinola
Bayer AG, Leverkusen
Ion Exchange Application
Im Hörnchen 7
D-5060 Bergisch Gladbach
FRG

Dr. Wilhelm Neir
RWE-DEA AG
für Mineralöl und Chemie
Chemische Werke Meerbeck
Postfach 101420
D-4130 Moers 1
FRG

Dr. Marico Pirotta
Enichem Synthesis SpA
Via Medici del Vascello 40
I-20138 Milano
Italy

Karlheinz W. R. Schoenrock
Consultant
Sugar & Separations Technology
5333 Fillmore
Ogden, Utah 84403
USA

Dr. Elfriede Sextl
Degussa AG
Fachbereich Forschung
Anorganische Chemie
Rodenbacher Chaussee 4
D-6450 Hanau
FRG

David Colin Sherrington, PhD., FRSE.
Professor of Polymer Chemistry
University of Strathclyde
Department of Pure and
Applied Chemistry
Thomas Graham Building
295, Cathedral Street
Glasgow G1 1XL
Great Britain

Dr. Hiroshi Shimizu
Senior Corporate Consultant
Japan Organo
Tokyo 100
Japan

Vladimir S. Soldatov
Professor of Chemistry
Institute of Physico-Organic Chemistry
The BSSR Academy of Sciences
Surganova Str. 13
Minsk 220603
USSR

Michael Streat
Professor of Chemical Engineering
Department of Chemical Engineering
Loughborough University
of Technology
Loughborough,
Leicestershire LE11 3TU
Great Britain

Dr. Gert-Joachim Strobel
European Consultancy
Pharmacia LKB Biotechnology AB
Munzinger Straße 9
D-7800 Freiburg
FRG

Hans Träger
Wilhelm Werner GmbH
Postfach 201007
D-5060 Bergisch Gladbach
FRG

Kazuo Tsuzura
Asahi Chemical Industry Co., Ltd.
General Manager
Ion Exchange Membrane Plant
1-3-2, Yako, Kawasaki-ku
Kawasaki 210
Japan

Dr. Dr. h. c. Armin Weiss
Professor of Chemistry
Institut für Anorganische Chemie
der Universität München
Meiserstraße 1
D-8000 München 2
FRG

Subject Index